KU-062-748

BIOLOGY
• PRINCIPLES & EXPLORATIONS •

Here's why *Biology: Principles and Explorations* is the program for you!

▶ Helps students become effective readers
- Concise lessons are structured to make complex material more accessible.
- Reading strategies teach students the skills they need to become independent learners.

See pp. T32–T33.

▶ Makes biology accessible—and relevant
- Quick Labs provide concrete experiences to help students understand concepts.
- Math Labs and Data Labs help students see how math is used in science.
- Strong graphic instruction, in features like BioGraphics, provides a visual description of complex processes.
- Real-life examples show students how biology is meaningful to *their* world.

See pp. T34–T35 and T38.

▶ Integrates new technology into the instructional plan
- NSTA's *sci*LINKS and other online resources keep students up-to-date. See p. T36.
- Holt Biology Interactive Tutor CD-ROM provides high-quality instructional support. See p. T42.

▶ Easy to teach—on any schedule
- *Biology: Principles and Explorations* One-Stop Planner CD-ROM coordinates all the resources you need into one easy-to-use package! See p. T28.
- Editable One-Stop Planner CD-ROM Lesson Plans let you customize your course for traditional or block scheduling.

▶ Prepares students for assessment
- Study Tips and Study Zones place ongoing emphasis on practical study skills and techniques for remembering important details. See pp. T33 and T46.

▶ Reaches different learning styles and abilities
- Teaching strategies for multiple intelligences help you target ALL your students.
- Basic and advanced activities make it easy to integrate different ability levels into your teaching plan.

See pp. T50–T51.

▶ Offers you flexibility
- Unique two-part table of contents allows you to tailor the course to fit your classroom needs and teaching preferences. See pp. T30–T31.

HOLT, RINEHART AND WINSTON
A Harcourt Classroom Education Company
Austin • New York • Orlando • Atlanta
San Francisco • Boston • Dallas • Toronto • London

Acknowledgments

Authors

George B. Johnson
Professor of Biology
Washington University
St. Louis, Missouri

Peter H. Raven
Director, Missouri Botanical Gardens
Engelmann Professor of Botany
Washington University
St. Louis, Missouri

Staff Credits

Editorial
Susan Feldkamp,
Executive Editor
David F. Bowman,
Managing Editor

Project Editors
Carolyn Biegert
Bill Burnside
Debra Hendricks
Laura Juárez de Ku
Brendan McInerney
Annette Ratliff

Editorial Support Staff
Jeanne Graham
Stephanie S. Sanchez
Tanu'e White

Ancillary Staff
Debra Hendricks
Brendan McInerney

Copyeditors
Dawn Spinozza,
Copyediting Supervisor
Cindy Foreman
Brooke Fugitt
Kathryn O'Shields

Art, Design, and Photo
Design, Book
Richard Metzger,
Design Director

Design Implementation and Production
Morgan-Cain & Associates

Design Team
Bob Prestwood
Stephanie Smith
Jose Garza
Rina Ouellette
Alicia Sullivan (ATE)
ATE Designer
Ruth Limon (ATE)
Joe Melomo,
Cover Design Director
Pecos Design
Cover Design

Image Acquisitions
Joe London,
Director
Elaine Tate,
Art Buyer Supervisor
Sunday Patterson,
Art Buyer
Michael Gobbi,
Photo Research

Design, New Media
Susan Michael,
Design Director
Amy Shank,
Senior Designer

Production
Mimi Stockdell,
Sr. Production Manager
Beth Sample
Amber Martin
Sara Carroll-Downs

New Media
Lydia Doty
Armin Gutzmer
Cathy Kuhles
Nina Degollado
Jessica Bega

Permissions
Janet Harrington

Research & Curriculum
Patty Kolar
Mike Tracy

For permissions to reprint copyrighted material, grateful acknowledgment is made to the following sources:

***sci*LINKS** is owned and provided by the National Science Teachers Association. All rights reserved.

Jonathan B. Losos, Kenneth I. Warheit, Thomas W. Schoener, and Nature®, a division of Macmillan Magazines Ltd.: Graph adapted from "Adaptive differentiation following experimental island colonization in *Anolis* lizards" by Jonathan B. Losos, Kenneth I. Warheit, and Thomas W. Schoener from *Nature*, vol. 387, no. 6628, May 1, 1997. Copyright © 1997 by *Nature.*

Princeton University Press: "Out of Africa" map.

Printed in the United States of America

ISBN 0-03-053834-3
2 3 4 5 6 7 8 048 05 04 03 02 01 00

Contributing Writers and Editors

Alan Eagy
Biology Teacher
The Dalles High School
The Dalles, OR

Theresa Flynn-Nason
Science Teacher
Voorhees Public Schools
Voorhees, NJ

Gary W. Goodnight
Biology Teacher
Denver South High School
Denver, CO

Erik Hahn
Duluth, MN

Christopher Hess
Freelance Editor
Austin, TX

Deborah Hix
Science Teacher
Hays CISD
San Marcos, TX

Lynne Krenicky
Science Teacher
Clifton High School
Clifton, NJ

Matt T. Lee, Ph.D.
Science Editor
Coos Bay, OR

Mitchell Leslie
Science Writer
Stanford University
Belmont, CA

Jenifer McMurray
Biology Teacher
Seneca Valley
Harmony, PA

Alison Mack
Science Writer and Editor
Wilmington, DE

James L. Middleton
Bear Creek High School
Lakewood, CO

Paul Neakrase
Biology Instructor
Washington Community High School
Washington, IL

Thomas P. Ottinger, Ph.D.
Gilmer High School
Ellijay, GA

Karen Ross
Science Writer
Georgetown, TX

Tracy Schagen
Freelance Editor
Austin, TX

Joseph B. Schiel, Jr., Ph.D.
Biology Teacher
Artesia High School
Artesia, NM

Mark A. Stallings, Ph.D.
Science Chair
Gilmer High School
Ellijay, GA

E. David Thielk
Science Educator
Port Townsend, WA

Reviewers

Neal M. Barnett
Retired Associate Professor of Botany
University of Maryland
College Park, MD

Lois Bergquist, Ph.D.
Professor–Microbiology
Los Angeles Valley College
Van Nuys, CA

Darlene Bevelhymer
Science Teacher
Napoleon High School
Napoleon, OH

Linda K. Butler, Ph.D.
Senior Lecturer
University of Texas at Austin
Austin, TX

Vern Capelle
Biology Teacher
Upsala High School
Upsala, MN

Marc Cron
Science Teacher
William Henry Harrison High School
Harrison, OH

Michael A. Crowell
Biology Teacher
Mainland Regional High School
Lynnwood, NJ

Mary Pitt Davis
Science Teacher
Long Reach High School
Dayton, MD

Marie A. DiBerardino
Professor Emerita
Department of Biochemistry
MCP Hahnemann University
Wynnewood, PA

Alan Eagy
Biology Teacher
The Dalles High School
The Dalles, OR

Tom Eddy
Associate Professor of Biology
Emporia State University
Emporia, KS

William J. Ehmann
Director, Environmental Science & Policy Program
Drake University
Des Moines, IA

Colleen Fiegel
Biology Teacher
Benjamin Franklin High School
New Orleans, LA

Reviewers, *continued*

Joseph E. Hall
Science Department Chairperson
Douglas High School
Box Elder, SD

J. D. Hand
Biology and Life Science Teacher
Augusta High School
Augusta, KS

Marilyn Hennon
Biology Teacher
Albion High School
Albion, MI

David R. Hershey, Ph.D.
Education Consultant
Hyattsville, MD

Leslie Hickok, Ph.D.
Professor
University of Tennessee
Knoxville, TN

Arthur Hulse, Ph.D.
Professor of Biology
Indiana University of Pennsylvania
Indiana, TN

Kevin Janish
Science Teacher
Scotland High School
Scotland, SD

Carolyn Jones, Ph.D.
Professor of Life Science
Vincennes University
Washington, IN

Clifford Keller, Ph.D.
University of Oregon
Eugene, OR

Hillar Klandorf, Ph.D.
Associate Professor
West Virginia University
Morgantown, WV

Kevin McCauley
Science Department Co-Chair
North Eugene High School
Eugene, OR

Edward McSweegan, Ph.D.
Crofton, MD

David Machnicki
Cuyahoga Falls, OH

Sheila Markey-DeFilippis
Science Teacher
North Eugene High School
Eugene, OR

Dana Miller
Biology Teacher
Las Lomas High School
Walnut Creek, CA

Sue Molter
Biology Teacher
Mt. Horeb High School
Mt. Horeb, WI

Paul Neakrase
Biology Teacher
Washington Community High School
Washington, IL

Martin Nickels, Ph.D.
Professor of Anthropology
Illinois State University
Normal, IL

David O. Norris, Ph.D.
Professor of Biology
University of Colorado at Boulder
Boulder, CO

Maja Nowakowski, Ph.D.
Associate Professor
Department of Pathology
SUNY–Health Science Center at Brooklyn
Brooklyn, NY

Terence M. Phillips, Ph.D.
Professor, Microbiology and Immunology
George Washington University Medical Center
Washington, DC

John Polka
Science Department Chairman
Fenwick High School
Oak Park, IL

Rick Roberts
Allan County–Scottsville High School
Scottsville, KY

Charles Robbins
Professor
Washington State University
Pullman, WA

John Richard Schrock, Ph.D.
Professor of Biology
Emporia State University
Emporia, KS

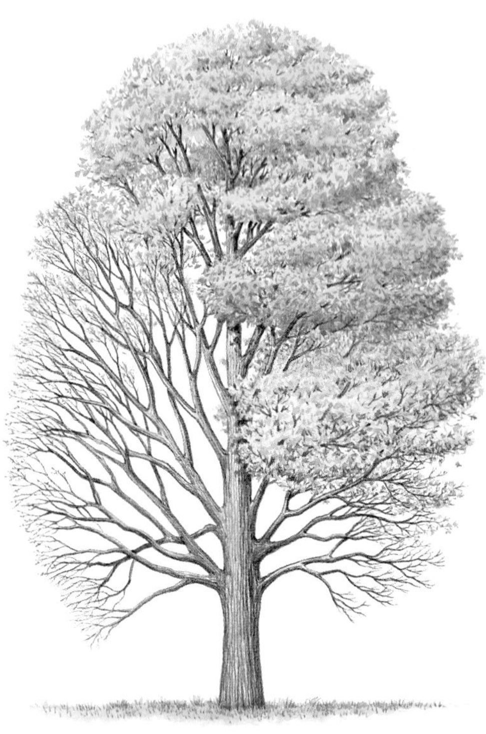

Geoffrey Smith
WARD'S Natural Science Establishment
Rochester, NY

Richard Storey, Ph.D.
Professor
Colorado College
Colorado Springs, CO

Gerald Summers, Ph.D.
University of Missouri
Columbia, MO

E. David Thielk
Science Educator
Port Townsend, WA

Judith L. Toop
Science Department Chair
Auburn Adventist Academy
Auburn, WA

E. Peter Volpe, Ph.D.
Professor of Basic Medical Sciences
Mercer University School of Medicine
Macon, GA

Linda Witulski
Science Department Co-Chair
Aurora Public Schools
Englewood, CO

Michael Zimmerman, Ph.D.
Dean of College of Letters and Sciences
University of Wisconsin
Oshkosh, WI

Contents in Brief

Turn the page to see how this program gives you flexible options for content coverage.

Part 1: Principles

Part 1 covers the fundamental content of biology.

Part 2: Explorations

Part 2 divides the world of living organisms into five units.

BIOLOGY

• PRINCIPLES & EXPLORATIONS •

Unique Two-Part Organization Offers Both Flexibility and Depth of Coverage

Part 1: Principles is devoted to *Cell Biology, Genetics, Evolution,* and *Ecology.* In these **19 core chapters,** *Biology: Principles and Explorations* covers fundamental content that lies at the heart of any biology curriculum:

Unit 1 Principles of Cell Biology teaches the basic principles of cell organization and cell energetics.

Unit 2 Principles of Genetics expands the coverage of cell biology to cover genes—what they are and how they function.

Unit 3 Principles of Evolution then explains how natural selection causes particular changes in genes to become more common, causing a species to evolve.

Unit 4 Principles of Ecology explains how coevolutionary accommodations among organisms living together create the web of relationships responsible for ecosystem stability.

Part 1: Principles

Part 2: Explorations is devoted to *Diversity. Biology: Principles and Explorations* divides the world of living organisms into **five teaching units:**

- **INTRODUCTION chapters** cover the basic principles of each unit
- **Supporting chapters** discuss particular kinds of organisms or systems in more detail.
- For example, this **INTRODUCTION chapter** describes the way in which the animal body plan has evolved, showing how evolutionary milestones have led to the animal phyla we see today.
- The four **supporting chapters** present a more detailed look at the major invertebrate phyla.

Part 2: Explorations

Unit 5
Exploring Diversity

20 **Introduction** to the Kingdoms of Life
21 Viruses and Bacteria
22 Protists
23 Fungi

Unit 6
Exploring Plants

24 **Introduction** to Plants
25 Plant Reproduction
26 Plant Structure and Function
27 Plant Growth and Development

Unit 7
Exploring Invertebrates

28 **Introduction** to Animals
29 Simple Invertebrates
30 Mollusks and Annelids
31 Arthropods
32 Echinoderms and Invertebrate Chordates

Unit 8
Exploring Vertebrates

33 **Introduction** to Vertebrates
34 Fishes and Amphibians
35 Reptiles and Birds
36 Mammals
37 Animal Behavior

Unit 9
Exploring Human Biology

38 **Introduction** to Body Structure
39 Circulatory and Respiratory Systems
40 Digestive and Excretory Systems
41 The Body's Defenses
42 Nervous System
43 Hormones and the Endocrine System
44 Reproduction and Development

UNIT 2 Principles of Genetics 138

CHAPTER 23

UNIT 6 *Exploring Plants* 516

CHAPTER 24

CHAPTER 25

CHAPTER 43

CHAPTER 44

Features

***Health Watch** features show the impact of biology on health issues.*

***Food Watch** features show you how biology relates to the foods you eat.*

***Tech Watch** features emphasize the role of technology in biology.*

Up Close

*Each **Up Close** feature provides a detailed look at an important biological organism.*

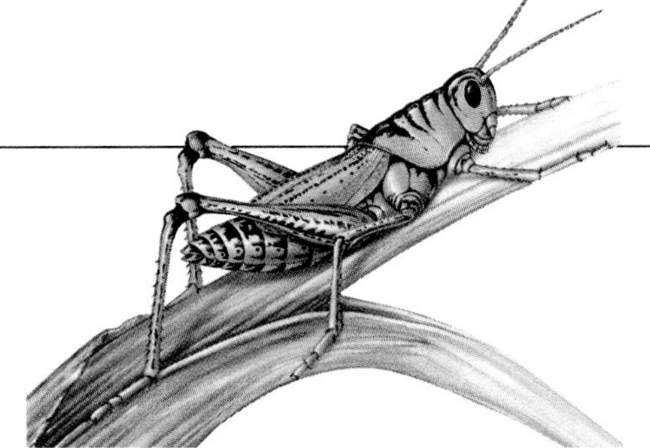

Exploring Further

***Exploring Further** features let you explore key biological topics at greater depth.*

SCIENCE • TECHNOLOGY • SOCIETY

__Science, Technology, and Society__ features examine the impact of new technologies on issues in biology.

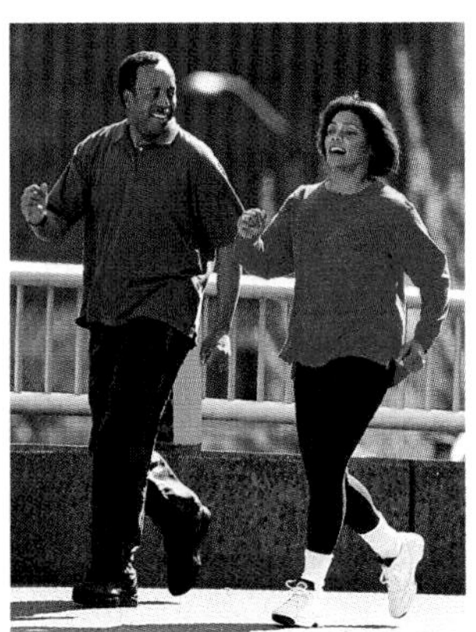

Lab Program

QUICK LAB

Quick Labs provide hands-on experience yet require few materials.

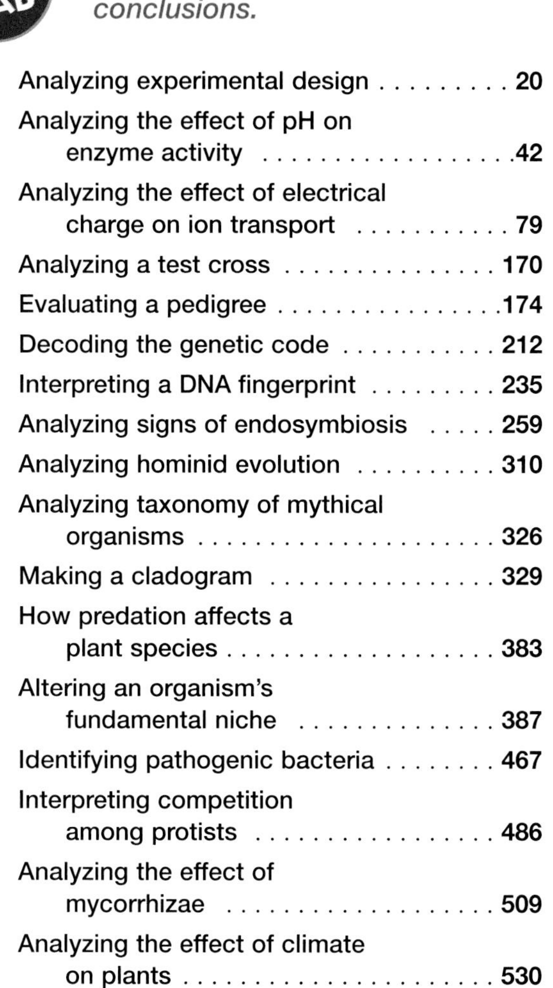

Data Labs allow you to interpret data and draw conclusions.

Math Labs let you practice real-world math skills as you analyze biological problems.

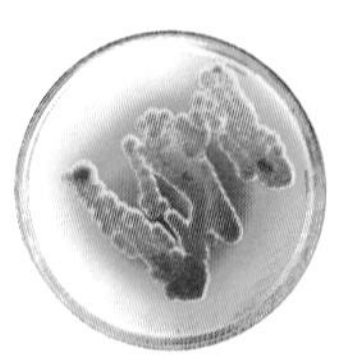

***Experimental Design** labs show you lab procedures—then let you design your own experiments.*

***Lab Technique** labs teach you lab skills used by biologists.*

***Interactive Explorations** use CD-ROM technology to enable you to see and control biological phenomena in ways never before possible.*

CHAPTER

CHAPTER

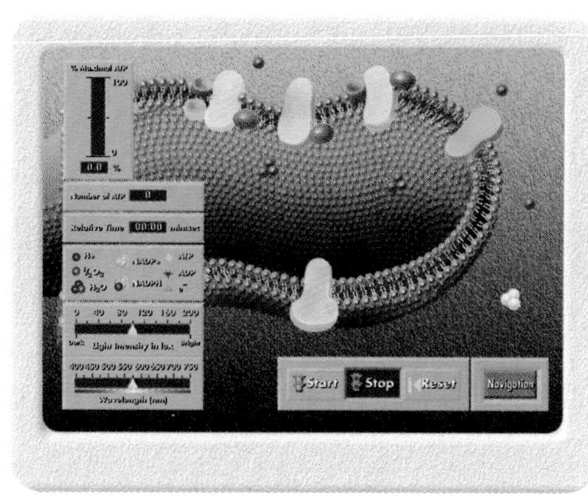
Start
Stop
Reset
Navigation

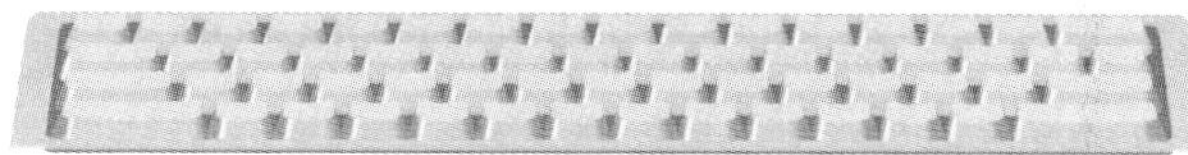

Introducing the One-Stop Planner CD-ROM

Makes Lesson Planning Quick, Easy, and Portable

Finally, a new system that saves you time in reviewing, planning, and organizing all the resources in your biology program. Starting with a suggested lesson plan for each text section, you can see how all resources are fully integrated into that lesson. You can print each resource right from the CD-ROM, so you no longer have to carry around multiple booklets when planning your lessons. Everything you need is on the disk!

Plan quickly with lesson plans that are easily changed to match your own objectives.

The suggested lesson plans are available as PDF files and as word-processing documents. You can change the word-processing lesson plans to reflect your style and schedule. These changes are easily saved as your own files for convenient storage. Lesson formats are available in Microsoft Word, Microsoft Works, Word Perfect (Windows Only), and ClarisWorks.

For Macintosh®- and Windows®-compatible computers

Sort through resources quickly with a click of a mouse.

All resources are hot-linked at appropriate points in your lesson plan. You can also view resources by category, such as labs or worksheets. All worksheet resources can be printed from the CD-ROM and photocopied for classroom use.

Now all of the resources you need are coordinated into one easy-to-use package!

Pacing suggestions consider block and traditional schedules.

Make your own custom lesson plan by selecting materials listed in the Other Resource Options section.

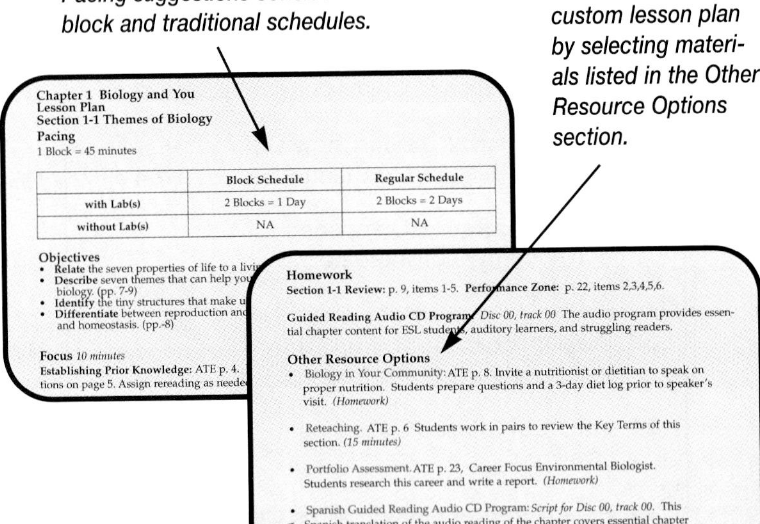

Chapter 1 Biology and You
Lesson Plan
Section 1-1 Themes of Biology
Pacing
1 Block = 45 minutes

	Block Schedule	Regular Schedule
with Lab(s)	2 Blocks = 1 Day	2 Blocks = 2 Days
without Lab(s)	NA	NA

Objectives
- **Relate** the seven properties of life to a livi
- **Describe** seven themes that can help you biology. (pp. 7-9)
- **Identify** the tiny structures that make u
- **Differentiate** between reproduction anc and homeostasis. (pp.-8)

Focus *10 minutes*
Establishing Prior Knowledge: ATE p. 4. tions on page 5. Assign rereading as neede

Homework
Section 1-1 Review: p. 9, items 1-5. **Performance Zone:** p. 22, items 2,3,4,5,6.

Guided Reading Audio CD Program: *Disc 00, track 00* The audio program provides essential chapter content for ESL students, auditory learners, and struggling readers.

Other Resource Options
- Biology in Your Community: ATE p. 8. Invite a nutritionist or dietitian to speak on proper nutrition. Students prepare questions and a 3-day diet log prior to speaker's visit. *(Homework)*
- Reteaching. ATE p. 6 Students work in pairs to review the Key Terms of this section. *(15 minutes)*
- Portfolio Assessment. ATE p. 23, Career Focus Environmental Biologist. Students research this career and write a report. *(Homework)*
- Spanish Guided Reading Audio CD Program: *Script for Disc 00, track 00.* This Spanish translation of the audio reading of the chapter covers essential chapter content for native Spanish-speaking students.
- Reading Strategies: Getting to Know Your Textbook. Students use a worksheet

Titles that appear in blue are hot-linked to the actual resource. Click to view any resource or print any worksheet.

RESOURCES INCLUDE:

- Lesson Plans
- Test Preparation Pretests
- Vocabulary Worksheets
- Science Skills Worksheets
- Directed Reading Worksheets
- Active Reading Guide
- Concept Mapping Worksheets
- Critical Thinking Review Worksheets
- Holt BioSources® Lab Program (five laboratory manuals)
- Chapter Tests with Answer Key
- Test Generator Software with Test Item Files
- CNN Presents Science in the News: Biology Connections
- Holt BioSources® Teaching Transparencies
- Audio CD Scripts (English and Spanish)
- Holt BioSources® Teaching Resources—hundreds of pages of additional worksheets!

A Comprehensive Resource Package as Diverse as Your Students

PRACTICE AND REINFORCEMENT

- **Test Preparation Pretests**
- **Vocabulary Worksheets**
- **Science Skills Worksheets**
- **Directed Reading Worksheets**
- **Active Reading Guide**
- **Concept Mapping Worksheets**
- **Critical Thinking Review Worksheets**
- **Student Review Guide**
  Contains Test Preparation Pretests, Vocabulary Worksheets, and Science Skills Worksheets

TEACHING TIPS AND STRATEGIES

- **Annotated Teacher's Edition**
- **Resource Teacher's Guides and Answer Keys**
- **Lesson Plans**
  Includes teaching tips, strategies, and activities to focus, motivate, teach, extend, and close for teaching every section in the textbook on a traditional or block schedule
- **Holt BioSources Teaching Resources**
  Customizable scoring rubrics, classroom management check lists, lab safety information, reading strategy worksheets, and more

LABORATORY RESOURCES

- **Holt BioSources Lab Program**
  Five laboratory manuals in blackline master form:
  - Quick Labs
  - Quick Labs, Teacher's Edition
  - Inquiry Skills Development
  - Inquiry Skills Development, Teacher's Edition
  - Laboratory Techniques and Experimental Design
  - Laboratory Techniques and Experimental Design, Teacher's Edition
  - Biotechnology
  - Biotechnology, Teacher's Edition
  - Interactive Explorations in Biology Laboratory Manual
  - Interactive Explorations in Biology Laboratory Manual,Teacher's Guide

ENRICHMENT AND ASSESSMENT

- **Chapter Tests with Answer Key**
- **Test Generator and Assessment Item Listing**
- **Holt BioSources Teaching Resources**
  More than 300 additional worksheets include portfolio projects, supplemental reading guides, and science research paper worksheets

MEDIA RESOURCES

- **Holt Biology Interactive Tutor**
  Seven interactive units and 42 CD-ROM topics in biology
- **Interactive Explorations in Biology: Human Biology**
- **Interactive Explorations in Biology: Cell Biology and Genetics**
  Nineteen CD-ROM interactive investigations
- **Holt Biology Videodiscs**
  Multimedia stills and video footage that are fully coordinated with *Biology: Principles and Explorations.* Includes Holt Biology Videodiscs Teacher's Correlation Guide to *Biology: Principles and Explorations*
- ***Biology: Principles and Explorations* Audio CD Program**
  Narrative summaries of the crucial concepts for all 44 chapters
- ***Biology: Principles and Explorations* Audio CD Program—Spanish Version**
- ***Biology: Principles and Explorations* Spanish Resources**
  Spanish translations of the Study Zone page from each chapter and the Glossary
- **CNN Presents Science in the News: Biology Connections**
  Video segments of newsworthy biological topics. Includes videocassette, Teacher's Guide, and a Critical Thinking Worksheet for every news story
- **Holt BioSources Teaching Transparencies and Transparency Directory**
  238 color transparencies and 71 illustrations in blackline-master form. Transparency Directory includes fullcolor views of each transparency and teaching strategies for each transparency and transparency master
- **Internet Resources**
  NSTA's sciLINKS, Smithsonian Institution Web site materials, and go.hrw.com Web resources

Alternative Courses of Study and Pacing the Course

A content sequence designed to support your curricular needs

You can cover the five Introduction chapters in any order, and then use as many supporting chapters as you wish to go into more depth. **As long as the 19 core chapters and the five Introduction chapters are covered, your students will be exposed to all the basic principles of biology and they will fulfill the requirements of the National Science Education Standards.**

The suggested pacing presumes the school year consists of approximately 180 blocks, each 45–50 minutes long. To implement one of the alternative courses of study shown, you will need to adjust the planning charts that precede each chapter to match these pacing suggestions.

If you do more lab work or add more chapters to a course sequence, you will also need to adjust the time spent on each chapter.

ALTERNATIVE COURSES	BASIC COURSE	HUMAN BIOLOGY
	24 CHAPTERS BLOCKS	30 CHAPTERS BLOCKS
Unit 1 Principles of Cell Biology		
• Chapter 1	6	6
• Chapter 2	10	5
• Chapter 3	10	6
• Chapter 4	10	4
• Chapter 5	8	5
• Chapter 6	10	4
Unit 2 Principles of Genetics		
• Chapter 7	8	6
• Chapter 8	8	7
• Chapter 9	8	6
• Chapter 10	6	4
• Chapter 11	8	7
Unit 3 Principles of Evolution		
• Chapter 12	8	6
• Chapter 13	7	5
• Chapter 14	7	5
• Chapter 15	6	4
Unit 4 Principles of Ecology		
• Chapter 16	6	4
• Chapter 17	6	6
• Chapter 18	6	6
• Chapter 19	6	6
Unit 5 Exploring Diversity		
• Chapter 20	6	6
• Chapter 21	–	–
• Chapter 22	–	–
• Chapter 23	–	–
Unit 6 Exploring Plants		
• Chapter 24	7	7
• Chapter 25	–	–
• Chapter 26	–	–
• Chapter 27	–	–
Unit 7 Exploring Invertebrates		
• Chapter 28	6	6
• Chapter 29	–	–
• Chapter 30	–	–
• Chapter 31	–	–
• Chapter 32	–	–
Unit 8 Exploring Vertebrates		
• Chapter 33	6	6
• Chapter 34	–	–
• Chapter 35	–	–
• Chapter 36		
• Chapter 37		
Unit 9 Exploring Human Biology		
• Chapter 38	9	9
• Chapter 39	–	7
• Chapter 40	–	8
• Chapter 41	–	8
• Chapter 42	–	10
• Chapter 43	–	6
• Chapter 44	–	8

BOTANY	ZOOLOGY	ZOOLOGY WITH HUMAN BIOLOGY	ECOLOGY	CELL BIOLOGY
27 CHAPTERS BLOCKS	32 CHAPTERS BLOCKS	38 CHAPTERS BLOCKS	25 CHAPTERS BLOCKS	32 CHAPTERS BLOCKS
6	6	3	6	8
6	6	6	6	10
6	6	6	4	10
6	6	6	4	10
8	7	5	6	10
6	6	6	6	10
6	6	6	6	6
8	7	5	8	6
6	8	4	8	8
4	4	3	6	8
8	8	5	6	6
6	6	4	8	6
7	5	3	7	4
7	5	3	7	4
6	6	5	8	6
6	4	3	10	4
8	6	4	10	4
6	6	4	10	4
4	4	4	10	4
6	6	5	6	4
–	–	–	4	4
–	–	–	–	4
–	–	–	–	4
8	7	5	8	4
8	–	–	–	–
8	–	–	–	4
8	–	–	–	–
4	6	5	6	4
–	4	4	–	–
–	4	4	–	–
–	6	4	–	–
–	4	4	–	–
4	6	5	6	4
–	6	5	–	–
–	4	4	–	–
–	5	5	–	–
–	5	5	–	–
6	8	8	6	6
–	–	6	–	–
–	–	6	–	–
–	–	6	–	4
–	–	6	–	4
–	–	6	–	4
–	–	6	–	4

Concise, Well-Organized Lessons Simplify Learning

Chapter openers motivate students and help them to quickly preview the upcoming chapter.

The ***chapter opener photograph*** captures students' interest, while the adjacent photo and quick introduction focus students' attention on the role of biology in everyday life.

Looking Ahead shows students what's coming up.

CHAPTER 8 Mendel and Heredity

Show jumper

Shetland ponies

Speed... strength... size...

Did you know that racehorses and show jumpers have been bred for these qualities for hundreds of years? Shetland ponies were originally bred for their strength—they were used to pull coal cars in coal mines. Today their small size makes them ideal for children to ride.

Study TIP

Ready?

Check your understanding of these topics to help you prepare for what's ahead.

Can you...

- **define** the term gamete? (Section 6-1)
- **summarize** the relationship between chromosomes and genes? (Section 6-1)
- **differentiate** between autosomes and sex chromosomes? (Section 6-1)
- **describe** how independent assortment during meiosis contributes to genetic variation? (Section 7-1)

If not, *review the sections indicated.*

Looking Ahead

- 8-1 The Origins of Genetics *It all began with pea plants.*
- 8-2 Mendel's Theory *How are traits passed from one generation to the next?*
- 8-3 Studying Heredity *You can predict how a trait will be inherited.*
- 8-4 Patterns of Heredity Can Be Complex *Can factors such as temperature influence traits?*

Labs

- Math Labs
  - Calculating Mendel's ratios (p. 163)
  - Predicting the results of crosses using probabilities (p. 172)
- Quick Lab
  - Do you have dominant or recessive traits? (p. 166)
- Data Labs
  - Analyzing a test cross (p. 170)
  - Evaluating a pedigree (p. 174)
- Interactive Exploration
  - Cystic Fibrosis (p. 184)

Features

- Exploring Further
  - Crosses That Involve Two Traits (p. 169)
- Health Watch
  - Genetic Counseling (p. 180)

BIOLOGY INTERACTIVE TUTOR Unit 5—*Heredity* Topics 1–6

internetconnect

SCILINKS. NSTA National Science Teachers Association sciLINKS Internet resources are located throughout this chapter.

Mendel and Heredity 159

Study Tip—Ready? builds self-awareness as students check their level of prior knowledge.

On pp.T32-T51, this symbol indicates a ***Key to Reading Success***

The lesson structure enables each lesson to be easily taught and readily understood.

Objectives alert students to the essential information covered in the lesson. Page numbers identify where the information can be found.

Key Terms list the lesson's boldfaced vocabulary terms in the order of their appearance.

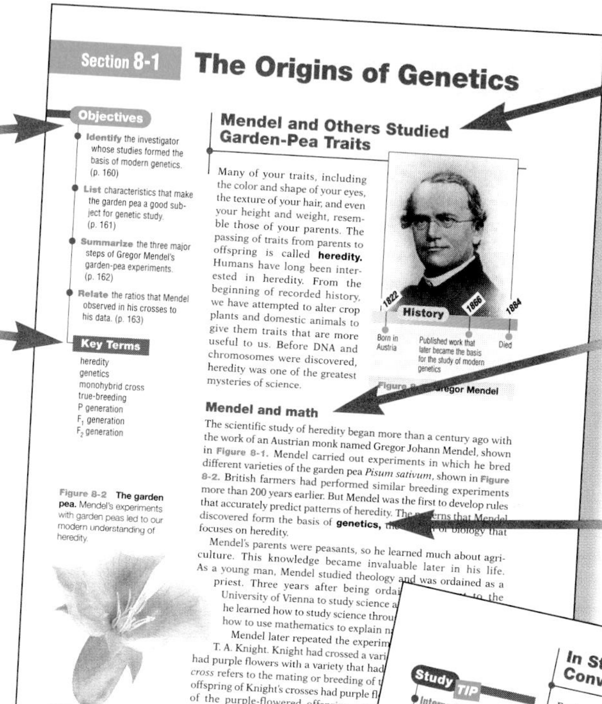

Section 8-1 **The Origins of Genetics**

Objectives

- **Identify** the investigator whose studies formed the basis of modern genetics. (p. 160)
- **List** characteristics that make the garden pea a good subject for genetic study. (p. 161)
- **Summarize** the three major steps of Gregor Mendel's garden-pea experiments. (p. 162)
- **Relate** the ratios that Mendel observed in his crosses to his data. (p. 163)

Key Terms

heredity
genetics
monohybrid cross
true-breeding
P generation
F_1 generation
F_2 generation

Mendel and Others Studied Garden-Pea Traits

Many of your traits, including the color and shape of your eyes, the texture of your hair, and even your height and weight, resemble those of your parents. The passing of traits from parents to offspring is called **heredity.** Humans have long been interested in heredity. From the beginning of recorded history, we have attempted to alter crop plants and domestic animals to give them traits that are more useful to us. Before DNA and chromosomes were discovered, heredity was one of the greatest mysteries of science.

Figure 8-1 Gregor Mendel

Mendel and math

The scientific study of heredity began more than a century ago with the work of an Austrian monk named Gregor Johann Mendel, shown in **Figure 8-1.** Mendel carried out experiments in which he bred different varieties of the garden pea *Pisum sativum*, shown in **Figure 8-2.** British farmers had performed similar breeding experiments more than 200 years earlier. But Mendel was the first to develop rules that accurately predict patterns of heredity. The patterns that Mendel discovered form the basis of **genetics,** ... of biology that focuses on heredity.

Mendel's parents were peasants, so he learned much about agriculture. This knowledge became invaluable later in his life. As a young man, Mendel studied theology and was ordained as a priest. Three years after being ordai... University of Vienna to study science a... he learned how to study science throu... how to use mathematics to explain n...

Mendel later repeated the experim... T. A. Knight. Knight had crossed a vari... had purple flowers with a variety that had... *cross* refers to the mating or breeding of t... offspring of Knight's crosses had purple fl... of the purple-flowered offspring were...

Figure 8-2 The garden pea. Mendel's experiments with garden peas led to our modern understanding of heredity.

160 Chapter 8

Located at the top of the page, ***concept heads*** make it easy to find the main concepts covered in each lesson.

Subheads in blue help students identify subtopics.

Key Terms are ***boldfaced*** within the narrative.

Reading and Study Skills are further reinforced in the Reading and Study Skills Appendix at the end of the Pupil's Edition.

- Becoming an Active Reader
- Concept Mapping
- Analyzing Word Parts

Reading and Study Skills Becoming an Active Reader

What Is an Active Reader?

Active reading means thinking and interacting with the text before, during, and after reading it. You might be wondering, "Isn't everyone *thinking* while they are reading?" Not necessarily. Many students read without consciously directing their thinking. Active readers are different—they have developed habits of thinking to use *before, during,* and *after* they read. They use these habits to successfully reach their goal: to understand what they're reading about and remember it well enough to be able to pass the test.

Before You Read

Active readers begin by getting familiar with the reading assignment. This is called previewing. The word *preview* means "to look before." Looking over the chapter before you read helps you predict what you will be learning. It can also help you recall what you already know about a subject—this can help you learn more material and remember it longer.

Scan the horizon

An active reader will turn to the first page of a chapter, ask questions, and use the information on the page to find answers. Here are some questions to ask.

1. **What's here?** If you're in an unfamiliar city, you might use a road map to find your way around. In this textbook, the section titled Looking Ahead is your map, preparing you for what's coming up. Take time to really look at the chapter title so you'll remember it. The title tells you the focus and the purpose of the chapter. Some readers skip over the title, not realizing its importance.
2. **What do I know?** Questions at the beginning of each chapter in Study TIP Ready? prompt you to remember vital information from previous chapters.
3. **What do I need to know?** Ask yourself questions, such as, "What do I expect to learn from the section I am about to read?"

Get to know the neighborhood

Now you're ready to begin the first section. Start by recognizing the elements that are there to help you:

- **Objectives** tell you what you should be able to do after reading the section. Each Objective includes the page number where the information is covered. Read these before you read the section.
- **Key Terms** are the lesson's key vocabulary words listed in order of appearance. They appear in **bold** black print within the lesson.

Find your way

As you look over the chapter, you will notice text headings in **bold** red print and text subheadings in **bold** blue print. Each red heading is a complete sentence and is a main idea for the lesson. Red headings are always located at the top of a page, making it easy for you to scan the chapter and find the main ideas. Blue subheadings are words or phrases. These subheadings identify important topics that are covered under each red heading. As you read, notice that bold blue print is used for references to nearby photos and illustrations (ex. **Figure 8-9**). Bold green print identifies tables (ex. **Table 8-1**).

1026 Reading and Study Skills: Becoming an Active Reader

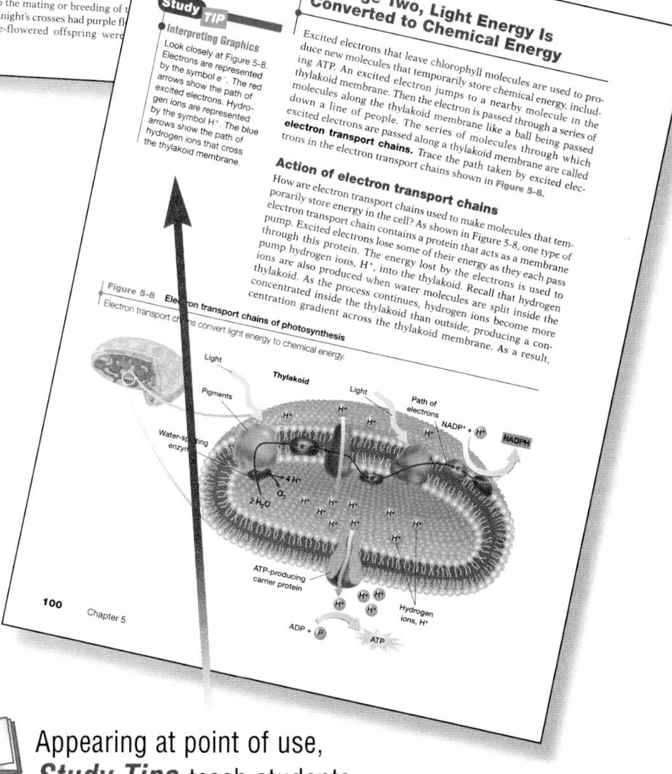

Study TIP

Interpreting Graphics

Look closely at Figure 5-8. Electrons are represented by the symbol e^-. The red arrows show the path of excited electrons. Hydrogen ions are represented by the symbol H^+. The blue arrows show the path of hydrogen ions that cross the thylakoid membrane.

In Stage Two, Light Energy Is Converted to Chemical Energy

Excited electrons that leave chlorophyll molecules are used to produce new molecules that temporarily store chemical energy, including ATP. An excited electron jumps to a nearby molecule in the thylakoid membrane. Then the electron is passed through a series of molecules along the thylakoid membrane like a ball being passed down a line of people. The series of molecules through which excited electrons are passed along a thylakoid membrane are called **electron transport chains.** Trace the path taken by excited electrons in the electron transport chains shown in **Figure 5-8.**

Action of electron transport chains

How are electron transport chains used to make molecules that temporarily store energy in the cell? As shown in Figure 5-8, one type of electron transport chain contains a protein that acts as a membrane pump. Excited electrons lose some of their energy as they each pass through this protein. The energy lost by the electrons is used to pump hydrogen ions, H^+, into the thylakoid. Recall that hydrogen ions are also produced when water molecules are split inside the thylakoid. As the process continues, hydrogen ions become more concentrated inside the thylakoid than outside, producing a concentration gradient across the thylakoid membrane. As a result,

Figure 5-8 Electron transport chains of photosynthesis

Electron transport chains convert light energy to chemical energy.

100 Chapter 5

Appearing at point of use, ***Study Tips*** teach students these six essential skills:

- Reading Effectively
- Organizing Information
- Comparing and Contrasting
- Reviewing Information
- Understanding Word Origins
- Interpreting Graphics

Outstanding illustrations and photographs deliver complex information visually.

A ***boldfaced identifier*** in each caption states the subject.

A ***concise caption*** states the point of the figure. Figures with more than one image are united by a single caption for ease of understanding.

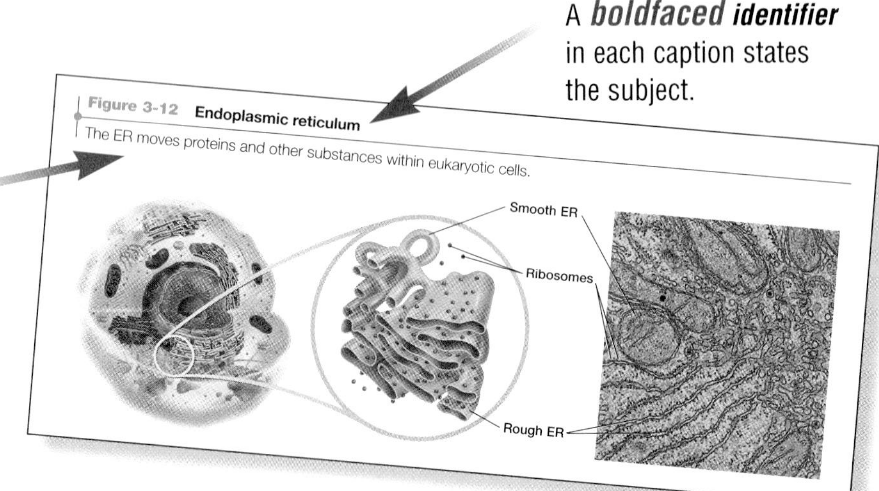

Mendel Observed that Traits Are Expressed as Simple Ratios

Mendel's initial experiments were monohybrid crosses. A **monohybrid cross** is a cross that involves *one* pair of contrasting traits. For example, crossing a plant with purple flowers and a plant with white flowers is a monohybrid cross. Mendel carried out his experiments in three steps, as summarized in **Figure 8-4**.

Step 1 Mendel allowed each variety of garden pea to self-pollinate for several generations. This method ensured that each variety was **true-breeding** for a particular trait; that is, all the offspring would display only one form of a particular trait. For example, a true-breeding purple-flowering plant should produce only plants with purple flowers in subsequent generations.

These true-breeding plants served as the parental generation in Mendel's experiments. The parental generation, or **P generation,** are the first two individuals that are crossed in a breeding experiment.

Step 2 Mendel then cross-pollinated two P generation plants that had contrasting forms of a trait, such as purple flowers [illegible] flowers. Mendel called the offspring of the [illegible] the first filial generation, or **F_1 gene**[illegible] He then examined each F_1 plant and re[illegible] number of F_1 plants expressing each trait.

Step 3 Finally, [illegible]ndel allowed the F_1 generation to self-pollinate. He called the offspring of the F_1 generation plants the second filial generation, or **F_2 generation.** Again, each F_2 plant was characterized and counted.

Study TIP

Word Origins

The word *filial* is from the Latin *filialis*, meaning "of a son or daughter." Knowing this makes it easier to remember that the F generations refer to any generation following the parental (P) generation.

BIO graphic Figure 8-4 **Three Steps of Mendel's Experiments**

Mendel studied traits in three generations of plants.

1 Producing a true-breeding P generation
2 Producing an F_1 generation
3 Producing an F_2 generation

Self-pollination
P generation
Self-pollination
P generation
Cross-pollination
F_1 generation
All purple
Self-pollination
F_2 generation
705 purple:224 white

162 Chapter 8

Each ***BioGraphic*** provides a step-by-step visual overview of an important process in biology. Each ***step*** is numbered in both the BioGraphic and the supporting text, linking together the visual and written information.

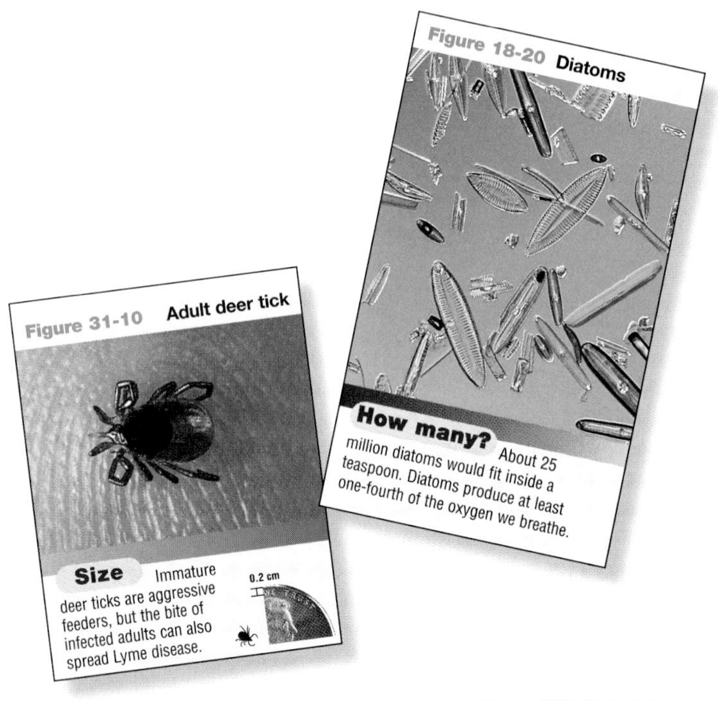

Figure 18-20 Diatoms
How many? About 25 million diatoms would fit inside a teaspoon. Diatoms produce at least one-fourth of the oxygen we breathe.
Figure 31-10 Adult deer tick
Size Immature deer ticks are aggressive feeders, but the bite of infected adults can also spread Lyme disease.
0.2 cm

Context-setting graphics help students view biological images in context of their size, number, or historical time frame.

On pp.T32-T51, this symbol indicates a ***Key to Reading Success***

Features emphasize the relevancy of biology and help you use one book to meet the needs of all of your students.

Exploring Further lets you cover complex topics in greater depth, allowing you to cover more material with your above-average students if you wish. Students can use *sci*LINKS items for additional research. Includes topics such as:

- Cancer
- Cloning
- Dihybrid crosses
- Transposons
- Endosymbiosis
- Cladistics
- Gram staining
- C_4 plants
- Cell differentiation
- Hardy-Weinberg equation
- Heart attacks
- Kidney dialysis
- . . . and more!

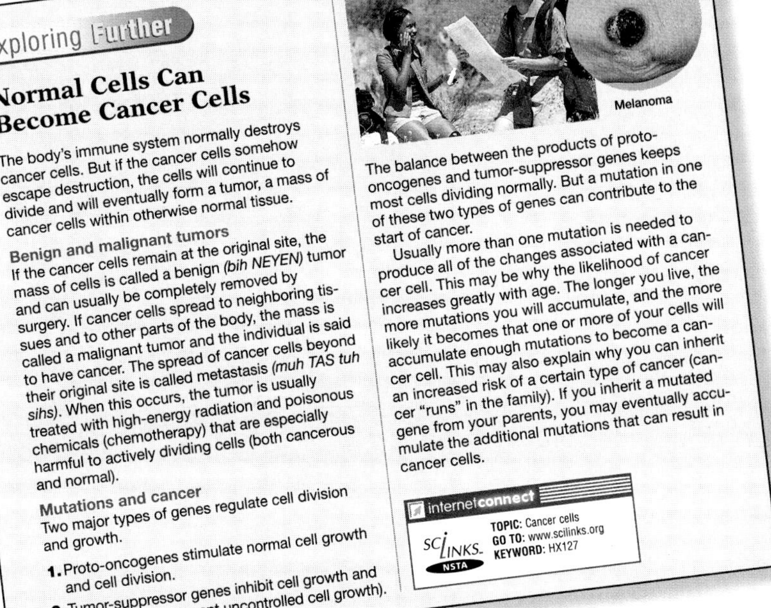
Exploring Further

Normal Cells Can Become Cancer Cells

The body's immune system normally destroys cancer cells. But if the cancer cells somehow escape destruction, the cells will continue to divide and will eventually form a tumor, a mass of cancer cells within otherwise normal tissue.

Benign and malignant tumors
If the cancer cells remain at the original site, the mass of cells is called a benign (*bih NEYEN*) tumor and can usually be completely removed by surgery. If cancer cells spread to neighboring tissues and to other parts of the body, the mass is called a malignant tumor and the individual is said to have cancer. The spread of cancer cells beyond their original site is called metastasis (*muh TAS tuh sihs*). When this occurs, the tumor is usually treated with high-energy radiation and poisonous chemicals (chemotherapy) that are especially harmful to actively dividing cells (both cancerous and normal).

Mutations and cancer
Two major types of genes regulate cell division and growth.

1. Proto-oncogenes stimulate normal cell growth and cell division.
2. Tumor-suppressor genes inhibit cell growth and cell division (to prevent uncontrolled cell growth).

Melanoma

The balance between the products of proto-oncogenes and tumor-suppressor genes keeps most cells dividing normally. But a mutation in one of these two types of genes can contribute to the start of cancer.

Usually more than one mutation is needed to produce all of the changes associated with a cancer cell. This may be why the likelihood of cancer increases greatly with age. The longer you live, the more mutations you will accumulate, and the more likely it becomes that one or more of your cells will accumulate enough mutations to become a cancer cell. This may also explain why you can inherit an increased risk of a certain type of cancer (cancer "runs" in the family). If you inherit a mutated gene from your parents, you may eventually accumulate the additional mutations that can result in cancer cells.

internetconnect
SCILINKS. NSTA
TOPIC: Cancer cells
GO TO: www.scilinks.org
KEYWORD: HX127

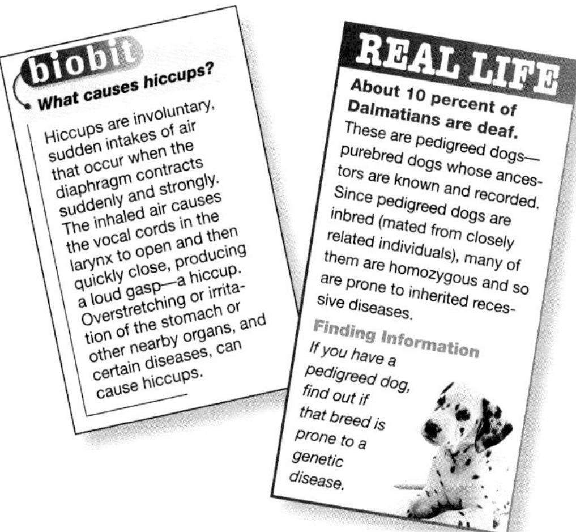
biobit
What causes hiccups?
Hiccups are involuntary, sudden intakes of air that occur when the diaphragm contracts suddenly and strongly. The inhaled air causes the vocal cords in the larynx to open and then quickly close, producing a loud gasp—a hiccup. Overstretching or irritation of the stomach or other nearby organs, and certain diseases, can cause hiccups.

REAL LIFE
About 10 percent of Dalmatians are deaf.
These are pedigreed dogs—purebred dogs whose ancestors are known and recorded. Since pedigreed dogs are inbred (mated from closely related individuals), many of them are homozygous and so are prone to inherited recessive diseases.

Finding Information
If you have a pedigreed dog, find out if that breed is prone to a genetic disease.

BioBit and ***Real Life*** show the impact of biology on daily life.

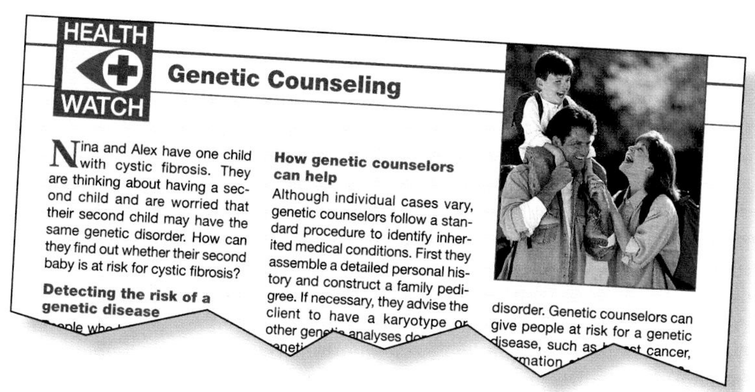
HEALTH WATCH
Genetic Counseling

Nina and Alex have one child with cystic fibrosis. They are thinking about having a second child and are worried that their second child may have the same genetic disorder. How can they find out whether their second baby is at risk for cystic fibrosis?

Detecting the risk of a genetic disease

How genetic counselors can help
Although individual cases vary, genetic counselors follow a standard procedure to identify inherited medical conditions. First they assemble a detailed personal history and construct a family pedigree. If necessary, they advise the client to have a karyotype or other genetic analyses do...

disorder. Genetic counselors can give people at risk for a genetic disease, such as ... cancer, ...mation

Health Watch, Food Watch, and ***Tech Watch*** describe real-world connections between biology and health, diet, and technology.

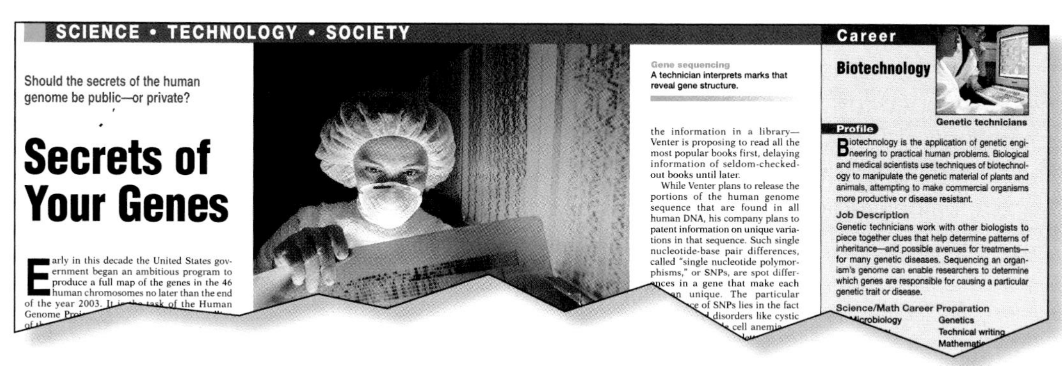
SCIENCE • TECHNOLOGY • SOCIETY

Should the secrets of the human genome be public—or private?

Secrets of Your Genes

Early in this decade the United States government began an ambitious program to produce a full map of the genes in the 46 human chromosomes no later than the end of the year 2003. It is the task of the Human Genome Pro...

Gene sequencing
A technician interprets marks that reveal gene structure.

the information in a library—Venter is proposing to read all the most popular books first, delaying information of seldom-checked-out books until later.

While Venter plans to release the portions of the human genome sequence that are found in all human DNA, his company plans to patent information on unique variations in that sequence. Such single nucleotide-base pair differences, called "single nucleotide polymorphisms," or SNPs, are spot differences in a gene that make each ...an unique. The particular ...e of SNPs lies in the fact ... disorders like cystic ... cell anemia

Career
Biotechnology

Genetic technicians

Profile
Biotechnology is the application of genetic engineering to practical human problems. Biological and medical scientists use techniques of biotechnology to manipulate the genetic material of plants and animals, attempting to make commercial organisms more productive or disease resistant.

Job Description
Genetic technicians work with other biologists to piece together clues that help determine patterns of inheritance—and possible avenues for treatments—for many genetic diseases. Sequencing an organism's genome can enable researchers to determine which genes are responsible for causing a particular genetic trait or disease.

Science/Math Career Preparation
Microbiology
Genetics
Technical writing
Mathematics

Science, Technology, and Society features consider the impact of technological developments on society.

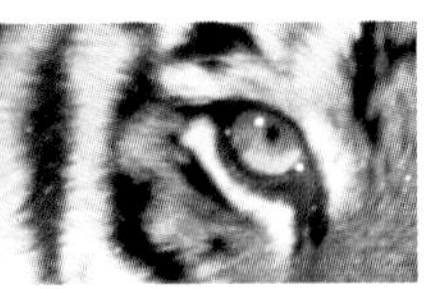

Internet Resources At Your Fingertips Keep You Up-To-Date

*sci*LINKS *does the legwork for you by screening the vast resources of the Internet.*

More than 100 NSTA *sci*LINKS are found at point of use in the margins of the pupil's edition of ***Biology: Principles and Explorations.*** *sci*LINKS websites are

- relevant.
- selected by teachers.
- up-to-date.

All *sci*LINKS topics throughout ***Biology: Principles and Explorations*** are managed and monitored by National Science Teachers Association staff. Sites are selected by teachers for appropriate content and grade level. Sites are continuously added and deleted from the system. Students don't end up at dead ends, dark sites, or sites under construction.

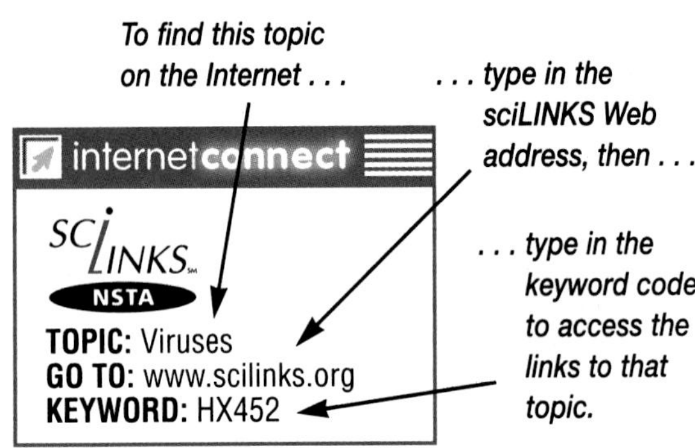

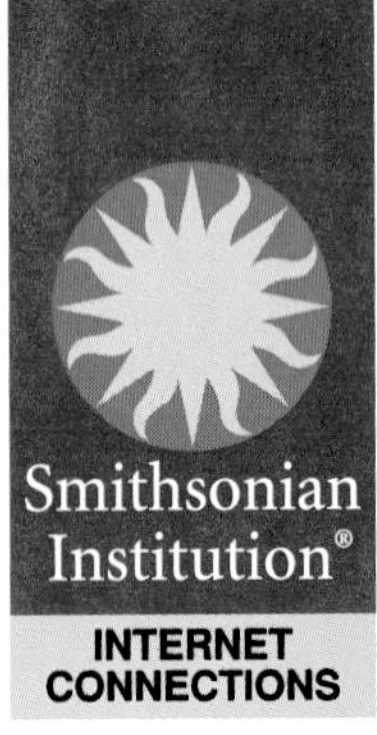

Smithsonian Institution Web resources take your students beyond the boundaries of the classroom.

The Smithsonian Institution maintains special Web sites for use with ***Biology: Principles and Explorations.*** Visit www.si.edu/hrw for a complete listing of these resources. You can find interactive exhibits, classroom activities, and interviews with scientists, along with an interesting variety of applications and extension topics.

Holt, Rinehart and Winston Web materials support and extend your resources.

Visit www.go.hrw.com to stay up to date with additional resources developed for the Biology: Principles and Explorations program. Feedback from teachers helps determine the materials you'll find on the Web site. Look for additional articles, activities, teaching suggestions, and lab tips.

The Best Biology Lab Program On The Market Just Got Better

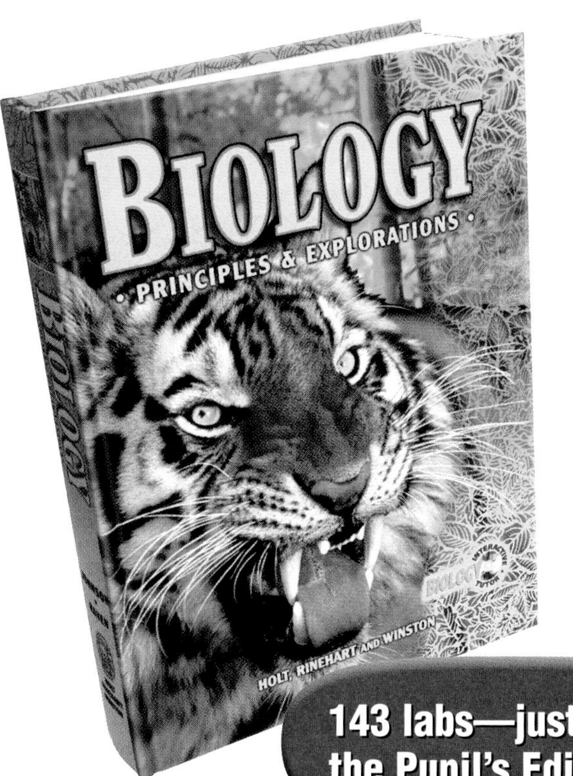

143 labs—just within the Pupil's Edition!

***From simple activities to more advanced experiments,* Biology: Principles and Explorations *meets your needs with lab options in the* Pupils Edition *and* Holt BioSources.**

Teacher Benefits

- **You will benefit from more options as Holt helps you**
  - meet state and local standards for hands-on learning.
  - save time with the labs you need right at your fingertips!
  - provide additional activities for experimental design and performance assessment.

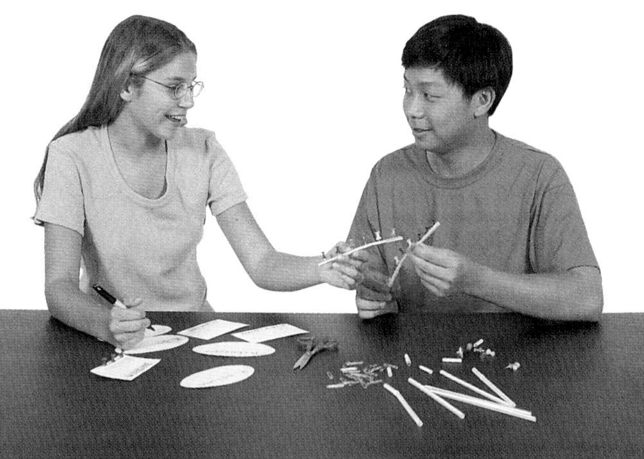

Student Benefits

Students will learn

- **skills they need to improve understanding and succeed on standardized exams.**
  - Forming hypotheses
  - Designing experiments
  - Analyzing data
  - Interpreting results
  - Evaluating conclusions
  - Interpreting tables
  - Making and interpreting graphs
  - Understanding and calculating ratios and percentages
  - Calculating and analyzing rates
  - Evaluating and defending conclusions
- **practical skills needed for achievement in the real world.**
  - Managing and allocating resources
  - Communicating effectively
  - Finding and organizing information
  - Working as a team member
  - Selecting appropriate technology

In the Pupil's Edition, three kinds of labs are found within the lessons.

Quick Labs help students form concrete models of concepts covered in the adjacent narrative. These labs provide hands-on experience yet require few materials.

When is a vegetable a fruit?

You can find out if a vegetable is a fruit by cutting it open and examining its internal structure.

Materials

apple, banana, green bean, potato, squash, tomato, plastic knife

Procedure

1. Look at several familiar fruits and vegetables. Classify each one as a commodity—either a fruit or a vegetable.
2. **CAUTION: Sharp objects can cause injury. Handle knives carefully.** Use a plastic knife to cut ... fruit and vegetable.
3. Look at the fruits and vegetables again. Classify each by its botanical function—either a fruit or a vegetative part.

Analysis

1. **Compare** the commodity and botanical classifications you gave each fruit and vegetable.
2. **SKILL Analyzing Data** Which fruits and vegetables did you classify differently?
3. **SKILL Analyzing Results** Defend the classifications you made for item 2.
4. **SKILL Drawing Conclusions** Based on your data, when is a vegetable a fruit?

Analyzing a test cross

Background

You can use a test cross to determine whether a plant with purple flowers is heterozygous (*Pp*) or homozygous dominant (*PP*). On a separate sheet of paper, copy the two Punnett squares shown below, and fill in the boxes in each square.

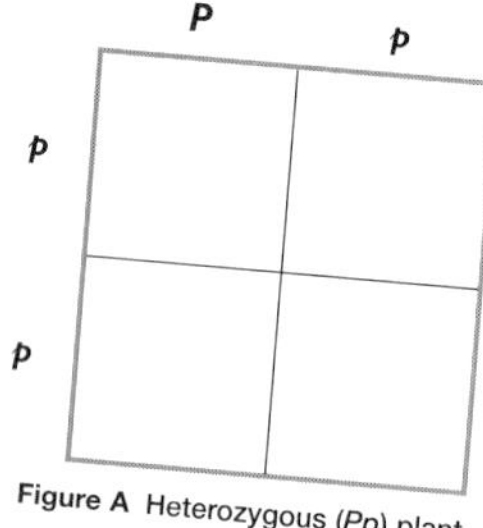

Figure A Heterozygous (*Pp*) plant

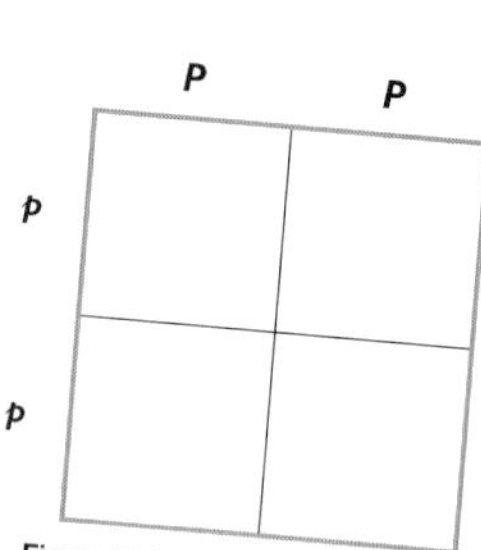

Figure B Homozygous (*PP*) plant

Is this purple flowering pea plant *Pp* or *PP*?

Analysis

1. **Determine** what the letters at the top and side of each box represent.
2. **Determine** what the letters in each box represent.
3. **Calculate** the genotypic and phenotypic ratios that would be predicted if the unknown parent of the unknown genotype were homozygous for the trait (Figure B).
4. **SKILL Predic...** Out...

Data Labs offer students opportunities to interpret data, analyze results, and evaluate conclusions, skills essential for test-taking success.

Calculating the number of cells resulting from mitosis

Background

Scientists investigating cancer might need to know the number of cells produced in a certain amount of time. In the human body the rate of mitosis is about 25 million (2.5×10^7) cells produced every second! You can calculate the number of cells produced by mitosis in a given amount of time.

1. **Calculate the number of cells produced by mitosis in the given time.** For example, to find the number of cells produced in 3 minutes, determine how many seconds there are in 3 minutes (since the rate is given in seconds).

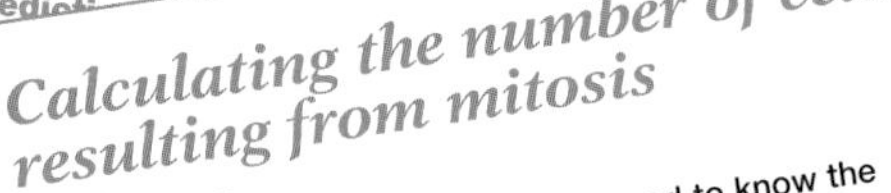

$$\frac{60 \text{ seconds}}{1 \text{ minute}} \times 3 \text{ minutes} = 180 \text{ seconds}$$

2. **Multiply the rate of mitosis by the time (in seconds) asked for in the problem (180 seconds).**

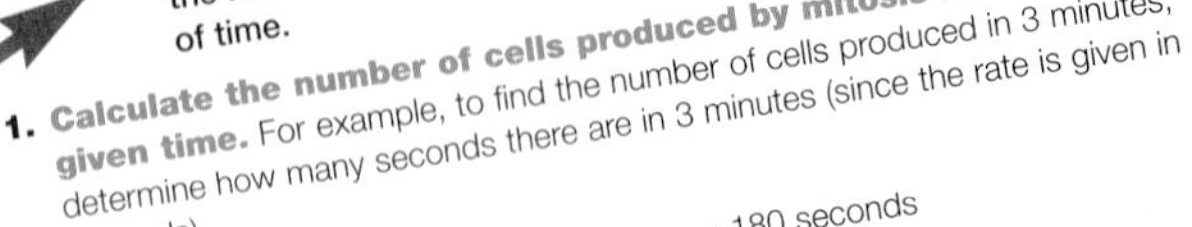

$$\frac{2.5 \times 10^7 \text{ cells}}{\text{second}} \times 180 \text{ seconds} = 4.5 \times 10^9 \text{ cells } (4{,}500{,}000{,}000 \text{ cells})$$

Analysis

1. **Calculate** the number of cells that would be produced in 1 hour.
2. **Calculate** the number of cells that would be produced in 1 day.
3. **SKILL Predicting Factors** Identify factors that might increase or decrease the rate of mitosis.

Math Labs let students use real-world math skills as they analyze biological problems. Students will gain practice in calculating ratios, estimating size, making and interpreting graphs, and other vital math skills.

Each chapter ends with one of the following kinds of labs.

Experimental Design

Chapter 40 How Can You Improve Lactose Digestion?

Interactive Exploration

Chapter 38 Life Span and Lifestyle

Lab Techniques

Chapter 26 Separating Plant Pigments

Experimental Design

labs teach students lab procedures and experimental design.

You Choose asks students to make the decisions that will enable them to design their own experiement.

Interactive Explorations

use CD-ROM technology to allow students to see biological phenomena and manipulate variables in ways never before possible.

Lab Technique

labs teach lab skills used by biologists.

On the Job describes a tool or technique used by biologists.

Holt BioSources lab manuals bring you unprecedented variety, with 135 labs that allow you to easily customize your course to meet your needs.

Selecting labs from *Holt BioSources* is easy! Appropriate labs to use with each lesson are identified in:

- The *Lesson Plans* found in the ***Biology: Principles and Explorations*** One-Stop Planner CD-ROM. See page T28.
- The Planning Guide preceding each chapter in the Annotated Teacher's Edition. See pp. T48–T49.

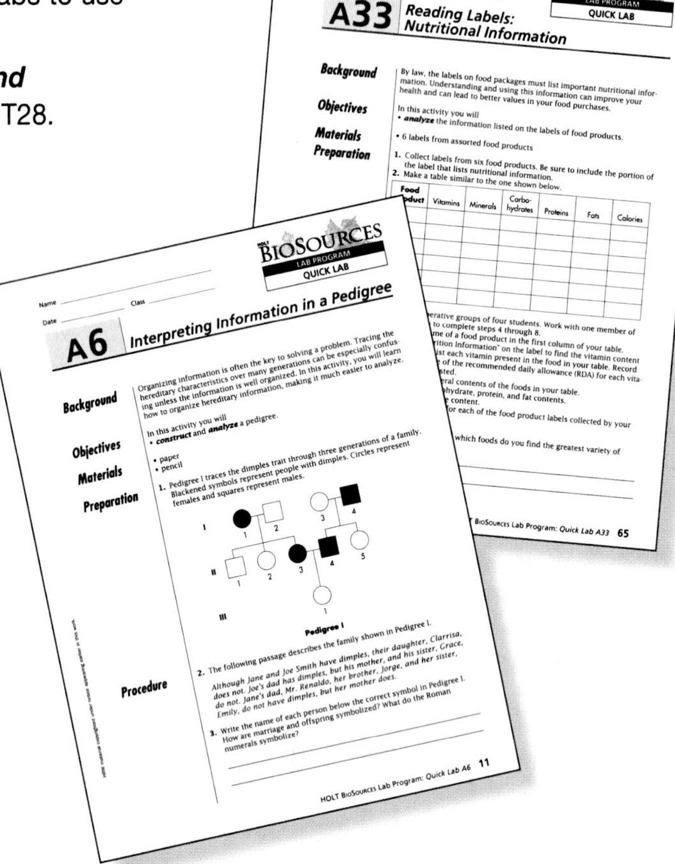

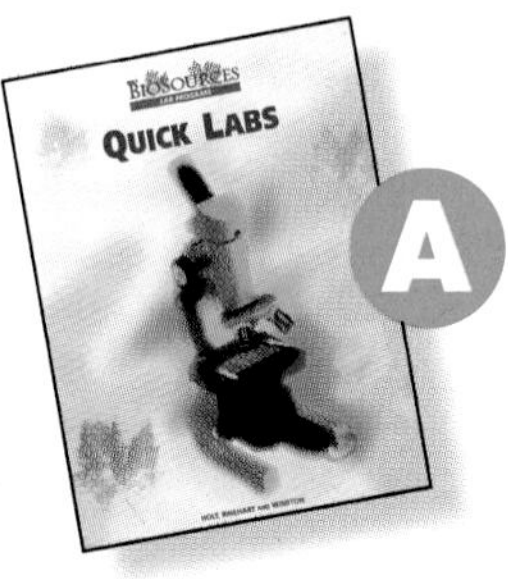

A Quick Labs

Quick Labs are just that. Inexpensive, easy-to-obtain supplies are used in each of these short lab activities. Use them as brief labs or even to introduce a new topic. 34 labs.

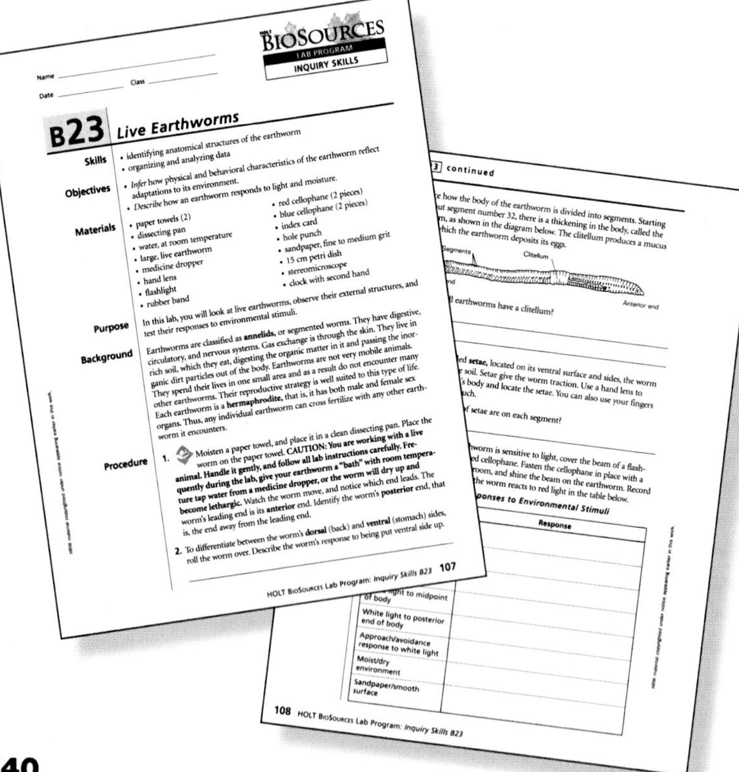

B Inquiry Skills Development

Inquiry Skills Development labs teach students practical skills and techniques. 34 labs.

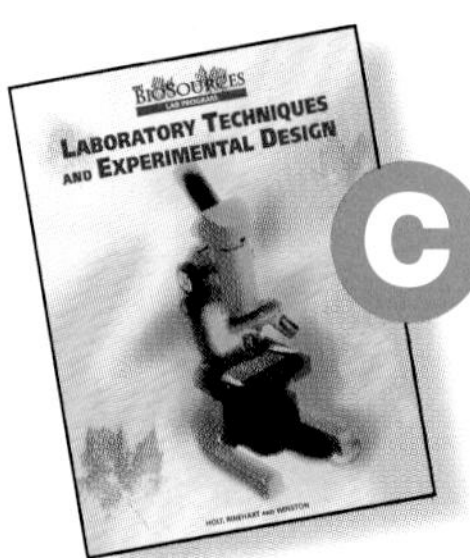

Laboratory Techniques and Experimental Design

Laboratory Techniques and Experimental Design labs require students to apply skills and techniques as they design their own experiments. 48 labs.

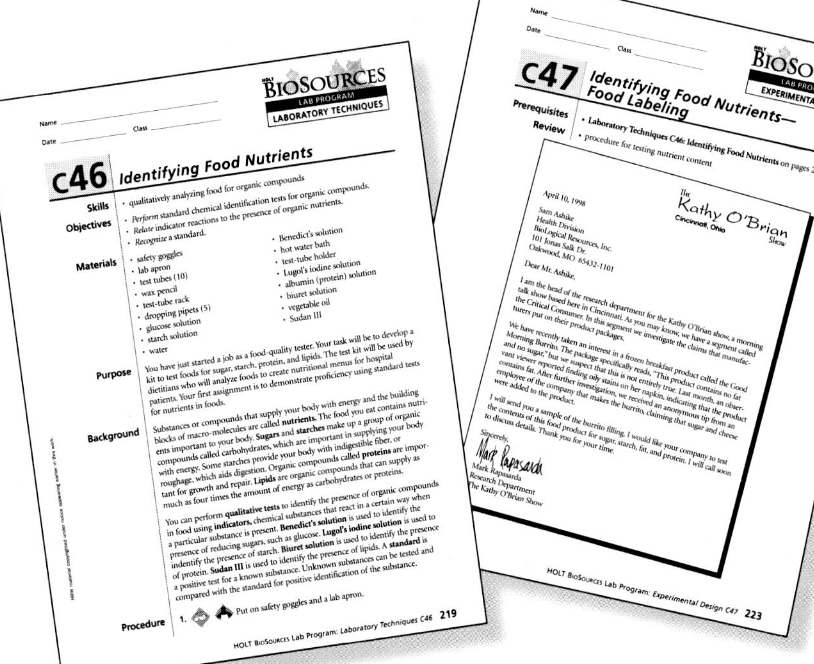

Name
Date
Class

HOLT BioSources Lab Program
LABORATORY TECHNIQUES

C46 ***Identifying Food Nutrients***

Skills
- qualitatively analyzing food for organic compounds

Objectives
- *Perform* standard chemical identification tests for organic compounds.
- *Relate* indicator reactions to the presence of organic nutrients.
- *Recognize* a standard.

Materials
- safety goggles
- lab apron
- test tubes (10)
- wax pencil
- test-tube rack
- dropping pipets (5)
- glucose solution
- starch solution
- water
- Benedict's solution
- hot water bath
- test-tube holder
- Lugol's iodine solution
- albumin (protein) solution
- biuret solution
- vegetable oil
- Sudan III

Purpose
You have just started a job as a food-quality tester. Your task will be to develop a kit to test foods for sugar, starch, protein, and lipids. The test kit will be used by dietitians who will analyze foods to create nutritional menus for hospital patients. Your first assignment is to demonstrate proficiency using standard tests for nutrients in foods.

Background
Substances or compounds that supply your body with energy and the building blocks of macro-molecules are called **nutrients.** The food you eat contains nutri- ents important to your body. **Sugars** and **starches** make up a group of organic compounds called carbohydrates, which are important in supplying your body with energy. Some starches provide your body with indigestible fiber, or roughage, which aids digestion. Organic compounds called **proteins** are impor- tant for growth and repair. **Lipids** are organic compounds that can supply as much as four times the amount of energy as carbohydrates or proteins.

You can perform **qualitative tests** to identify the presence of organic compounds in food using **indicators,** chemical substances that react in a certain way when a particular substance is present. **Benedict's solution** is used to identify the presence of reducing sugars, such as glucose. **Lugol's iodine solution** is used to indentify the presence of starch. **Biuret solution** is used to identify the presence of protein. **Sudan III** is used to identify the presence of lipids. A **standard** is a positive test for a known substance. Unknown substances can be tested and compared with the standard for positive identification of the substance.

Procedure
1. Put on safety goggles and a lab apron.

Name
Date
Class

HOLT BioSources Lab Program
EXPERIMENTAL DESIGN

C47 ***Identifying Food Nutrients—Food Labeling***

Prerequisites
- **Laboratory Techniques C46: Identifying Food Nutrients** on pages 219–222

Review
- procedure for testing nutrient content

The Kathy O'Brian Show
Cincinnati, Ohio

April 10, 1998

Sam Ashike
Health Division
BioLogical Resources, Inc.
101 Jonas Salk Dr.
Oakwood, MO 65432-1101

Dear Mr. Ashike,

I am the head of the research department for the Kathy O'Brian show, a morning talk show based here in Cincinnati. As you may know, we have a segment called the Critical Consumer. In this segment we investigate the claims that manufac- turers put on their product packages.

We have recently taken an interest in a frozen breakfast product called the Good Morning Burrito. The package specifically reads, "This product contains no fat and no sugar," but we suspect that this is not entirely true. Last month, an obser- vant viewer reported finding oily stains on her napkin, indicating that the product contains fat. After further investigation, we received an anonymous tip from an employee of the company that makes the burrito, claiming that sugar and cheese were added to the product.

I will send you a sample of the burrito filling. I would like your company to test the contents of this food product for sugar, starch, fat, and protein. I will call soon to discuss details. Thank you for your time.

Sincerely,

Mark Rapasarda
Mark Rapasarda
Research Department
The Kathy O'Brian Show

Biotechnology

Biotechnology labs bring the latest in biotechnology research techniques into your classroom. 12 labs.

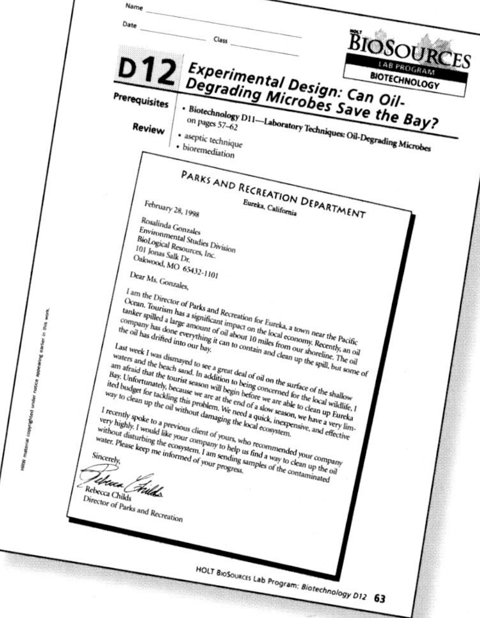

Name
Date
Class

HOLT BioSources Lab Program
BIOTECHNOLOGY

D12 ***Experimental Design: Can Oil-Degrading Microbes Save the Bay?***

Prerequisites
- **Biotechnology** D11—**Laboratory Techniques: Oil-Degrading Microbes** on pages 57–62

Review
- aseptic technique
- bioremediation

Parks and Recreation Department
Eureka, California

February 28, 1998

Rosalinda Gonzales
Environmental Studies Division
BioLogical Resources, Inc.
101 Jonas Salk Dr.
Oakwood, MO 65432-1101

Dear Ms. Gonzales,

I am the Director of Parks and Recreation for Eureka, a town near the Pacific Ocean. Tourism has a significant impact on the local economy. Recently, an oil tanker spilled a large amount of oil about 10 miles from our shoreline. The oil company has done everything it can to contain and clean up the spill, but some of the oil has drifted into our bay.

Last week I was dismayed to see a great deal of oil on the surface of the shallow waters and the beach sand. In addition to being concerned for the local wildlife, I am afraid that the tourist season will begin before we are able to clean up Eureka Bay. Unfortunately, because we are at the end of a slow season, we have a very lim- ited budget for tackling this problem. We need a quick, inexpensive, and effective way to clean up the oil without damaging the local ecosystem.

I recently spoke to a previous client of yours, who recommended your company very highly. I would like your company to help us find a way to clean up the oil without disturbing the ecosystem. I am sending samples of the contaminated water. Please keep me informed of your progress.

Sincerely,

Rebecca Childs
Rebecca Childs
Director of Parks and Recreation

Interactive Explorations

Using CD-ROM technology, ***Interactive Explorations*** enable your students to study phenomena that otherwise could not be observed in the classroom. These lab exer- cises guide students as they manipulate variables, pre- dict consequences, and study outcomes for seven of the *Interactive Explorations in Biology.* See page T43 for more information about the *Interactive Explorations in Biology* CD-ROMs. 7 labs.

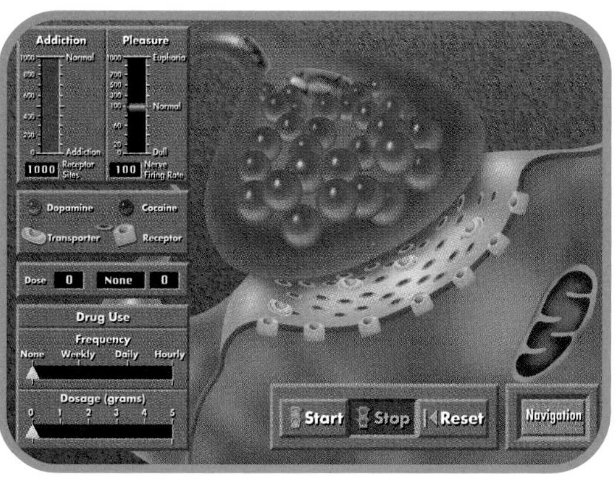

Fully-Integrated Technology Lets You Reach All of Your Students

Holt Biology Interactive Tutor

The power of CD-ROM technology lets you present key topics that are often difficult to teach and challenging for students to learn. Each of 42 different topics presents a key concept using concrete models, videos, and animations. Students can work through the topics at their own pace. Frequent evaluation lets students know if they are on the right track with diagnostic feedback.

For Macintosh®- and Windows®-compatible computers

Each Topic contains:

- **BioStory** The unit's tutor, a working professional whose job requires an understanding of the unit's biological content, introduces the topic.
- **Learn** One or more short tutorials deliver the content step-by-step.
- **Try It** and **On Your Own** Interactive activities let students show what they have learned and provide self-assessment that lets them gauge their progress.
- **Review Options** and **Assessment** Review Questions prompt students to make sure they've grasped the content before continuing on to the next topic. Summary Notes list all of the topic's on-screen content summaries.

Holt Biology Interactive Tutor also includes:

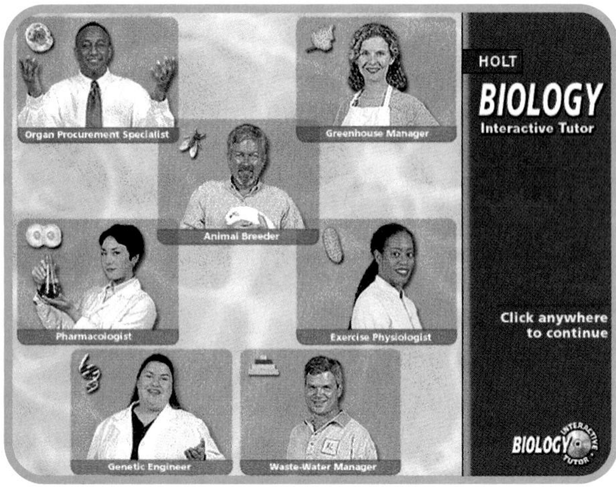

- **BioWorks**—careers in biology related to each Unit
- **Did You Know?**—interesting facts related to each unit
- **Web links**—direct acess to go.hrw.com and NSTA *sci*LINKS resources for each unit
- **BioResearch**—areas of ongoing research
- **Glossary**—audio text and definitions for more than 200 terms
- **Student and Teacher Resources**—worksheets and teaching resources in pdf format

*Each Unit begins with **Prerequisites** that state the prior knowledge required. **Objectives** identify the essential information covered in the unit.*

Each Unit includes six Topics.

UNIT

1 Cell Transport and Homeostasis
Homeostasis and Diffusion
Cell Membrane
Osmosis
Facilitted Diffusion
Active Transport
Active Uptake of Glucose

2 Photosynthesis
An Overview
Absorbing Light Energy
Electron Transport Chain
Making ATP
Calvin Cycle
Regulation

3 Cellular Respiration
An Overview
Glycolysis
Krebs Cycle
Electron Transport Chain
Fermentation
Regulation

4 Cell Reproduction
Binary Fission
Cell Cycle
Chromosomes
Mitosis
Meiosis
Gamete Production

5 Heredity
Principles of Inheritance
Determining Inheritance
Monohybrid Crosses
Dihybrid Crosses
Other Patterns of Inheritance
Pedigrees

6 Gene Expression
Genes and DNA
Replication—Copying DNA
The Genetic Code
Three Types of RNA
Transcription—Making RNA
Translation—Making Proteins

7 Ecosystem Dynamics
Ecosystems
Populations
Succession
Food Chains
Food Webs
Chemical Cycles

Interactive Explorations in Biology

These 19 CD-ROM activities enable your students to observe challenging biological phenomena that are difficult or impossible to study in the laboratory.

For Macintosh®- and Windows®-compatible computers

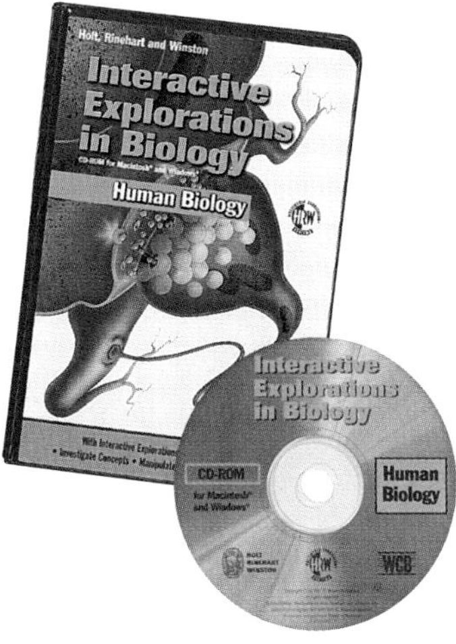

Interactive Explorations in Biology: Human Biology

1. Cystic Fibrosis
2. Active Transport
3. Life Span and Life Style
4. Evolution of the Heart
5. Diet and Weight Loss
6. Synaptic Transmission
7. Drug Addiction
8. Hormone Action
9. Immune Response
10. Heredity in Families
11. Pollution of a Freshwater Lake

Interactive Explorations in Biology: Cell Biology and Genetics

1. Hemoglobin
2. Cell Size
3. Active Transport
4. Thermodynamics
5. Oxidative Respiration
6. Photosynthesis
7. Meiosis: Down Syndrome
8. Gene Regulation

CNN Videos

CNN Presents **Science in the News: Biology Connections** includes more than 30 short video clips that show students the newsworthy nature of biology. Includes Teacher's Notes and Critical Thinking Worksheets for each segment.

Holt BioSources® Teaching Transparencies

Teaching Transparencies contain 288 full-color images of key illustrations accompanied by 71 blackline masters. Includes *Transparency Directory with Teacher's Notes.*

Holt Biology Videodiscs

Holt Biology Videodiscs offers you a comprehensive array of visuals tied directly to lessons found in ***Biology: Principles and Explorations.*** Includes:

- Four double-sided videodiscs containing thousands of images (live-action and animated videos, photographs, illustrations, and glossary terms) that reinforce material covered in the ***Biology: Principles and Explorations*** Pupil's Edition
- *Teacher's Correlation Guide to Holt* ***Biology: Principles and Explorations*** with barcodes and descriptors correlated specifically to ***Biology: Principles and Explorations***
- *Image Directory* containing a topic index, a subject matter index, and a glossary

Your Total Instructional Package Supports a Wide Range of Ability Levels and Learning Modalities

Biology: Principles and Explorations *Teaching Resources* provide the practice and reinforcement students need to improve their test scores.

Critical Thinking Skills Worksheets with Answer Key teach students how to interpret and answer analogy-based questions, skills needed for successful performance on many standardized science exams.

Vocabulary Worksheets with Answer Key provide practice in understanding and using essential terms in biology.

Science Skills Worksheets with Answer Key offer practice in key science process skills.

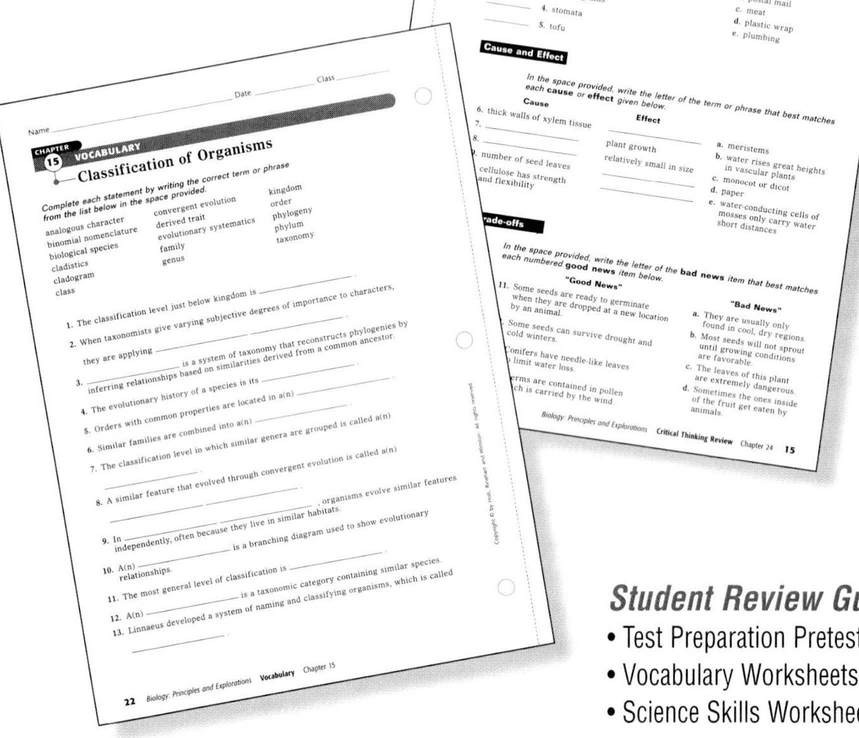

Name ______ Date ______ Class ______

CHAPTER 24 CRITICAL THINKING REVIEW

Introduction to Plants

"Work-Alikes"

In the space provided, write the letter of the term or phrase that best describes how each numbered item **functions.**

_____ 1. vascular system
_____ 2. plant cuticle
_____ 3. pollen grains
_____ 4. stomata
_____ 5. tofu

a. doorways
b. postal mail
c. meat
d. plastic wrap
e. plumbing

Cause and Effect

In the space provided, write the letter of the term or phrase that best matches each **cause** *or* **effect** *given below.*

Cause	Effect
6. thick walls of xylem tissue	______
7. ______	plant growth
8. ______	relatively small in size
9. number of seed leaves	______
cellulose has strength and flexibility	______

a. meristems
b. water rises great heights in vascular plants
c. monocot or dicot
d. paper
e. water-conducting cells of mosses only carry water short distances

Trade-offs

In the space provided, write the letter of the **bad news** *item that best matches each numbered* **good news** *item below.*

"Good News"

11. Some seeds are ready to germinate when they are dropped at a new location by an animal.
Some seeds can survive drought and cold winters.
Conifers have needle-like leaves to limit water loss.
Sperms are contained in pollen which is carried by the wind.

"Bad News"

a. They are usually only found in cool, dry regions.
b. Most seeds will not sprout until growing conditions are favorable.
c. The leaves of this plant are extremely dangerous.
d. Sometimes the ones inside of the fruit get eaten by animals.

Biology: Principles and Explorations Critical Thinking Review Chapter 24 15

Name ______ Date ______ Class ______

CHAPTER 15 VOCABULARY

Classification of Organisms

Complete each statement by writing the correct term or phrase from the list below in the space provided.

analogous character, binomial nomenclature, biological species, cladistics, cladogram, class, convergent evolution, derived trait, evolutionary systematics, family, genus, kingdom, order, phylogeny, phylum, taxonomy

1. The classification level just below kingdom is ______.
2. When taxonomists give varying subjective degrees of importance to characters, they are applying ______.
3. ______ is a system of taxonomy that reconstructs phylogenies by inferring relationships based on similarities derived from a common ancestor.
4. The evolutionary history of a species is its ______.
5. Orders with common properties are located in a(n) ______.
6. Similar families are combined into a(n) ______.
7. The classification level in which similar genera are grouped is called a(n) ______.
8. A similar feature that evolved through convergent evolution is called a(n) ______.
9. In ______, organisms evolve similar features independently, often because they live in similar habitats.
10. A(n) ______ is a branching diagram used to show evolutionary relationships.
11. The most general level of classification is ______.
12. A(n) ______ is a taxonomic category containing similar species.
13. Linnaeus developed a system of naming and classifying organisms, which is called ______.

22 Biology: Principles and Explorations Vocabulary Chapter 15

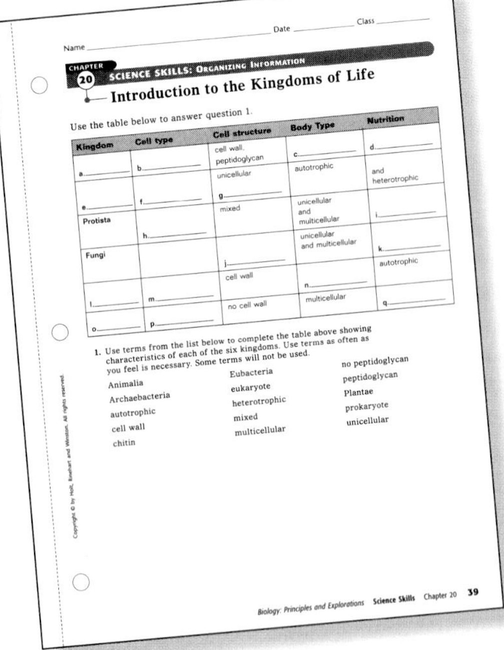

Name ______ Date ______ Class ______

CHAPTER 20 SCIENCE SKILLS: ORGANIZING INFORMATION

Introduction to the Kingdoms of Life

Use the table below to answer question 1.

Kingdom	Cell type	Cell structure	Body Type	Nutrition
a. ______	b. ______	cell wall, peptidoglycan	c. ______	d. ______
e. ______	f. ______	g. ______	autotrophic	and heterotrophic
Protista	h. ______	mixed	unicellular and multicellular	i. ______
Fungi		j. ______	unicellular and multicellular	k. ______
l. ______	m. ______	cell wall	n. ______	autotrophic
o. ______	p. ______	no cell wall	multicellular	q. ______

1. Use terms from the list below to complete the table above showing characteristics of each of the six kingdoms. Use terms as often as you feel is necessary. Some terms will not be used.

Animalia, Archaebacteria, autotrophic, cell wall, chitin, Eubacteria, eukaryote, heterotrophic, mixed, multicellular, no peptidoglycan, peptidoglycan, Plantae, prokaryote, unicellular

Biology: Principles and Explorations Science Skills Chapter 20 39

Student Review Guide

- Test Preparation Pretests
- Vocabulary Worksheets
- Science Skills Worksheets

Directed Reading Worksheets with Answer Key help students to remain focused while they read, enabling them to understand key concepts on a lesson-by-lesson basis. Contains matching, fill-in-the blank, and short answer exercises.

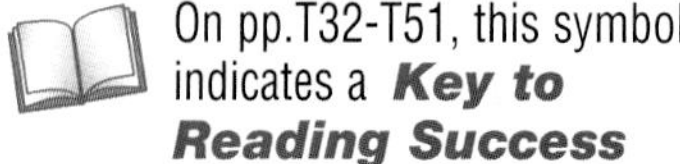

On pp.T32-T51, this symbol indicates a ***Key to Reading Success***

Active Reading Guide Worksheets with Answer Key teach students skills needed for reading success as they analyze key passages and illustrations from the ***Biology: Principles and Explorations*** Pupil's Edition. Skills emphasized:

- Recognizing Cause and Effect
- Reading Effectively
- Interpreting Graphics
- Sequencing Information
- Recognizing Similarities and Differences
- Organizing Information

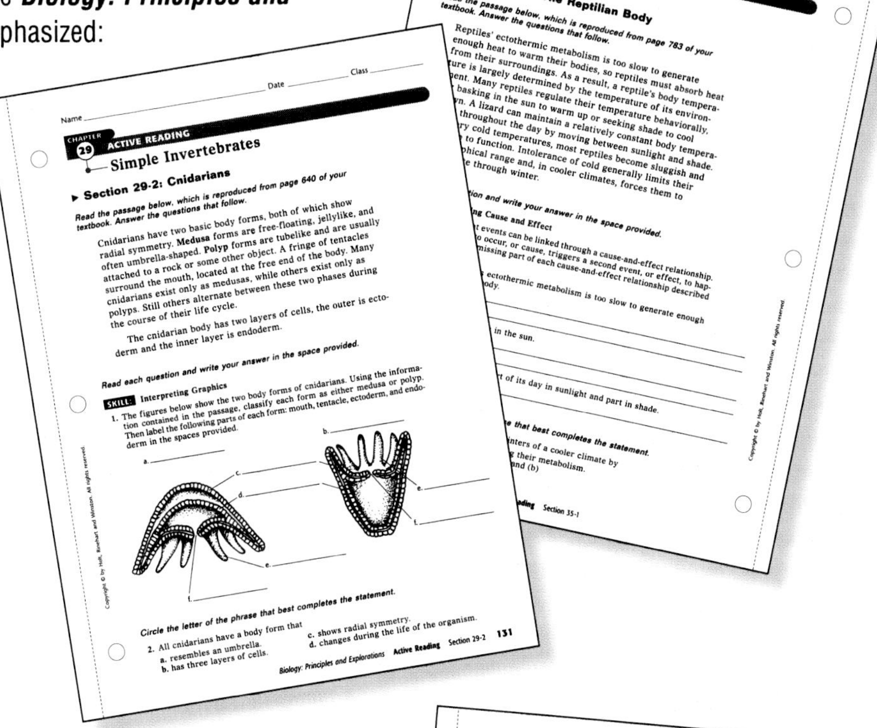

Name ________ Date ________ Class ________

CHAPTER 35 ACTIVE READING

Reptiles and Birds

▸ Section 35-1: The Reptilian Body

Read the passage below, which is reproduced from page 783 of your textbook. Answer the questions that follow.

Reptiles' ectothermic metabolism is too slow to generate enough heat to warm their bodies, so reptiles must absorb heat from their surroundings. As a result, a reptile's body temperature is largely determined by the temperature of its environ-
ent. Many reptiles regulate their temperature behaviorally,
basking in the sun to warm up or seeking shade to cool
n. A lizard can maintain a relatively constant body tempera-
throughout the day by moving between sunlight and shade.
ry cold temperatures, most reptiles become sluggish and
to function. Intolerance of cold generally limits their
phical range and, in cooler climates, forces them to
e through winter.

ion and write your answer in the space provided.

ng Cause and Effect

t events can be linked through a cause-and-effect relationship.
o occur, or cause, triggers a second event, or effect, to hap-
missing part of each cause-and-effect relationship described

ectothermic metabolism is too slow to generate enough
ody.

in the sun.

t of its day in sunlight and part in shade.

e that best completes the statement.

inters of a cooler climate by
their metabolism.
and (b)

ading Section 35-1

Copyright © by Holt, Rinehart and Winston. All rights reserved.

Name ________ Date ________ Class ________

CHAPTER 29 ACTIVE READING

Simple Invertebrates

▸ Section 29-2: Cnidarians

Read the passage below, which is reproduced from page 640 of your textbook. Answer the questions that follow.

Cnidarians have two basic body forms, both of which show radial symmetry. **Medusa** forms are free-floating, jellylike, and often umbrella-shaped. **Polyp** forms are tubelike and are usually attached to a rock or some other object. A fringe of tentacles surround the mouth, located at the free end of the body. Many cnidarians exist only as medusas, while others exist only as polyps. Still others alternate between these two phases during the course of their life cycle.

The cnidarian body has two layers of cells, the outer is ectoderm and the inner layer is endoderm.

Read each question and write your answer in the space provided.

SKILL: Interpreting Graphics

1. The figures below show the two body forms of cnidarians. Using the information contained in the passage, classify each form as either medusa or polyp. Then label the following parts of each form: mouth, tentacle, ectoderm, and endoderm in the spaces provided.

Circle the letter of the phrase that best completes the statement.

2. All cnidarians have a body form that
   a. resembles an umbrella.
   b. has three layers of cells.
   c. shows radial symmetry.
   d. changes during the life of the organism.

Copyright © by Holt, Rinehart and Winston. All rights reserved.

Biology: Principles and Explorations **Active Reading** Section 29-2 **131**

Biology: Principles and Explorations Audio CD Program provides auditory lessons that guide the listener through each lesson of the textbook.

Test Preparation Pretests with Answer Key contain multiple choice, matching, completion, and short answer questions that help students get ready for the chapter test.

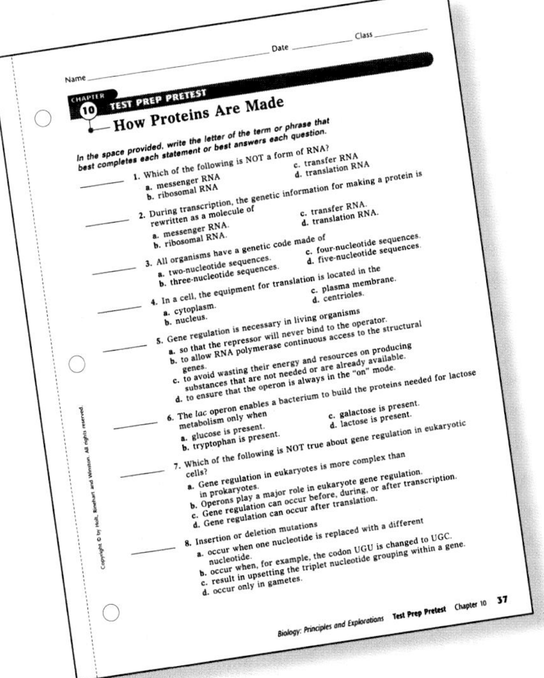

Name ________ Date ________ Class ________

CHAPTER 10 TEST PREP PRETEST

How Proteins Are Made

In the space provided, write the letter of the term or phrase that best completes each statement or best answers each question.

____ 1. Which of the following is NOT a form of RNA?
   a. messenger RNA
   b. ribosomal RNA
   c. transfer RNA
   d. translation RNA

____ 2. During transcription, the genetic information for making a protein is rewritten as a molecule of
   a. messenger RNA.
   b. ribosomal RNA.
   c. transfer RNA.
   d. translation RNA.

____ 3. All organisms have a genetic code made of
   a. two-nucleotide sequences.
   b. three-nucleotide sequences.
   c. four-nucleotide sequences.
   d. five-nucleotide sequences.

____ 4. In a cell, the equipment for translation is located in the
   a. cytoplasm.
   b. nucleus.
   c. plasma membrane.
   d. centrioles.

____ 5. Gene regulation is necessary in living organisms
   a. so that the repressor will never bind to the operator.
   b. to allow RNA polymerase continuous access to the structural genes.
   c. to avoid wasting their energy and resources on producing substances that are not needed or are already available.
   d. to ensure that the operon is always in the "on" mode.

____ 6. The *lac* operon enables a bacterium to build the proteins needed for lactose metabolism only when
   a. glucose is present.
   b. tryptophan is present.
   c. galactose is present.
   d. lactose is present.

____ 7. Which of the following is NOT true about gene regulation in eukaryotic cells?
   a. Gene regulation in eukaryotes is more complex than in prokaryotes.
   b. Operons play a major role in eukaryote gene regulation.
   c. Gene regulation can occur before, during, or after transcription.
   d. Gene regulation can occur after translation.

____ 8. Insertion or deletion mutations
   a. occur when one nucleotide is replaced with a different nucleotide.
   b. occur when, for example, the codon UGU is changed to UGC.
   c. result in upsetting the triplet nucleotide grouping within a gene.
   d. occur only in gametes.

Copyright © by Holt, Rinehart and Winston. All rights reserved.

Biology: Principles and Explorations **Test Prep Pretest** Chapter 10 **37**

Concept Mapping Worksheets with Answer Key help students to recognize the relationships among main ideas as they complete a concept map for each chapter.

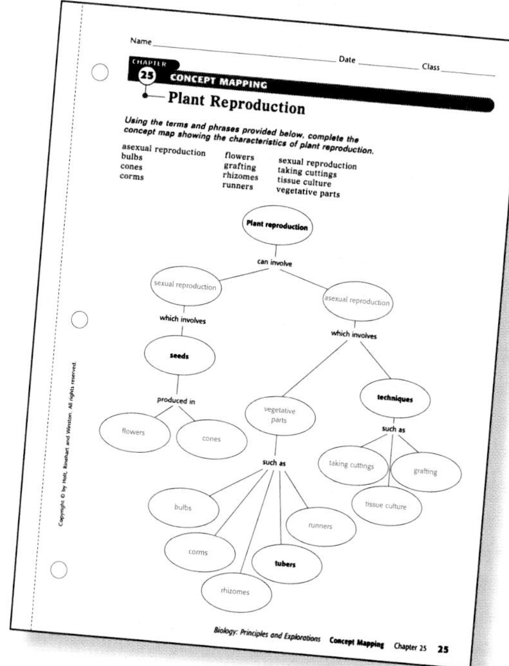

Name ________ Date ________ Class ________

CHAPTER 25 CONCEPT MAPPING

Plant Reproduction

Using the terms and phrases provided below, complete the concept map showing the characteristics of plant reproduction.

asexual reproduction, bulbs, cones, corms, flowers, grafting, rhizomes, runners, sexual reproduction, taking cuttings, tissue culture, vegetative parts

Copyright © by Holt, Rinehart and Winston. All rights reserved.

Biology: Principles and Explorations **Concept Mapping** Chapter 25 **25**

Biology: Principles and Explorations Audio CD Program—Spanish Version

Biology: Principles and Explorations Spanish Resources includes Spanish translations of the Glossary, the end-of-chapter Study Zone pages, and 160 Teaching Transparencies with accompanying worksheets.

Assessment Resources Help Students Move Easily From Practice to Performance

Section Reviews help students check their mastery of each lesson.

Study Tips throughout each lesson help students learn the lesson content.

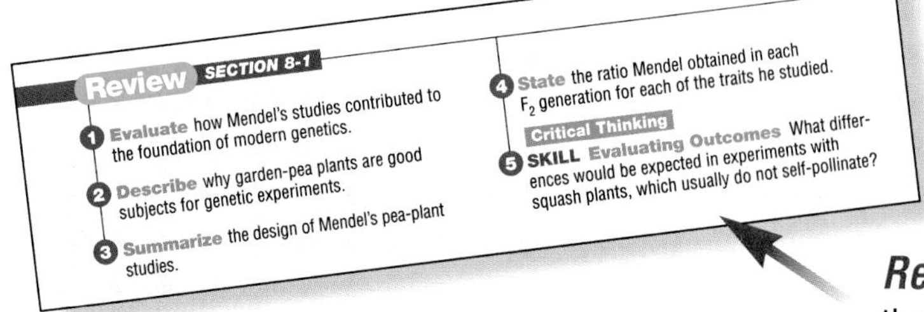

Review SECTION 8-1

1. **Evaluate** how Mendel's studies contributed to the foundation of modern genetics.
2. **Describe** why garden-pea plants are good subjects for genetic experiments.
3. **Summarize** the design of Mendel's pea-plant studies.
4. **State** the ratio Mendel obtained in each F_2 generation for each of the traits he studied.

Critical Thinking

5. **SKILL Evaluating Outcomes** What differences would be expected in experiments with squash plants, which usually do not self-pollinate?

Review questions at the end of every lesson address the lesson Objectives.

End-of-chapter materials help students practice on their own.

Study Zone helps students review the ***Key Concepts*** and ***Key Terms*** of each section.

Performance Zone measures fundamental knowledge and reinforces vocabulary development.

Skills Assessment targets proficiencies needed for successful exam performance and skills needed in everyday life.

Critical Thinking requires students to think evaluatively, using knowledge gained from the chapter to judge the accuracy, validity, or worth of the information presented.

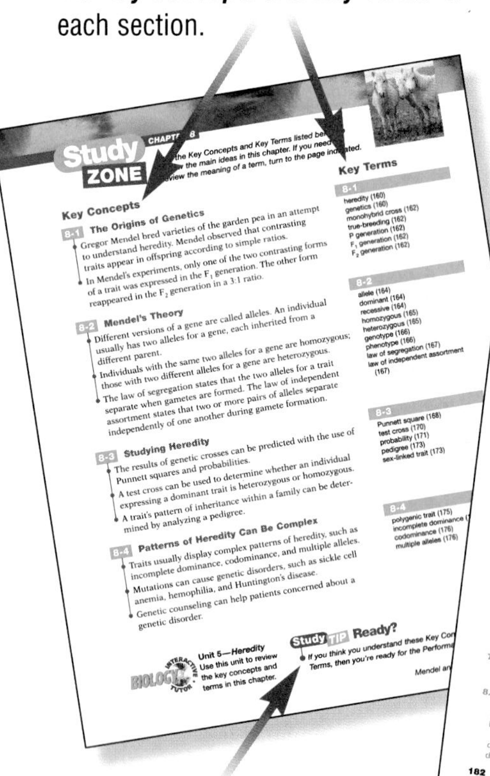

Study ZONE CHAPT[illegible] 8

[illegible] the Key Concepts and Key Terms listed be[illegible] [illegible] the main ideas in this chapter. If you need [illegible] [illegible]view the meaning of a term, turn to the page ind[illegible]ated.

Key Concepts

8-1 The Origins of Genetics
- Gregor Mendel bred varieties of the garden pea in an attempt to understand heredity. Mendel observed that contrasting traits appear in offspring according to simple ratios.
- In Mendel's experiments, only one of the two contrasting forms of a trait was expressed in the F_1 generation. The other form reappeared in the F_2 generation in a 3:1 ratio.

8-2 Mendel's Theory
- Different versions of a gene are called alleles. An individual usually has two alleles for a gene, each inherited from a different parent.
- Individuals with the same two alleles for a gene are homozygous; those with two different alleles for a gene are heterozygous.
- The law of segregation states that the two alleles for a trait separate when gametes are formed. The law of independent assortment states that two or more pairs of alleles separate independently of one another during gamete formation.

8-3 Studying Heredity
- The results of genetic crosses can be predicted with the use of Punnett squares and probabilities.
- A test cross can be used to determine whether an individual expressing a dominant trait is heterozygous or homozygous.
- A trait's pattern of inheritance within a family can be determined by analyzing a pedigree.

8-4 Patterns of Heredity Can Be Complex
- Traits usually display complex patterns of heredity, such as incomplete dominance, codominance, and multiple alleles.
- Mutations can cause genetic disorders, such as sickle cell anemia, hemophilia, and Huntington's disease.
- Genetic counseling can help patients concerned about a genetic disorder.

Key Terms

8-1
heredity (160)
genetics (160)
monohybrid cross (162)
true-breeding (162)
P generation (162)
F_1 generation (162)
F_2 generation (162)

8-2
allele (164)
dominant (164)
recessive (164)
homozygous (165)
heterozygous (165)
genotype (166)
phenotype (166)
law of segregation (167)
law of independent assortment (167)

8-3
Punnett square (168)
test cross (170)
probability (171)
pedigree (173)
sex-linked trait (173)

8-4
polygenic trait (175)
incomplete dominance ([illegible]
codominance (176)
multiple alleles (176)

BIOLOGY INTERACTIVE TUTOR **Unit 5—Heredity** Use this unit to review the key concepts and terms in this chapter.

Study TIP Ready?
- If you think you understand these Key Co[illegible] Terms, then you're ready for the Perform[illegible]

Mendel a[illegible]

Study Tip—Ready? prompts students to assess their readiness before moving on to the Performance Zone.

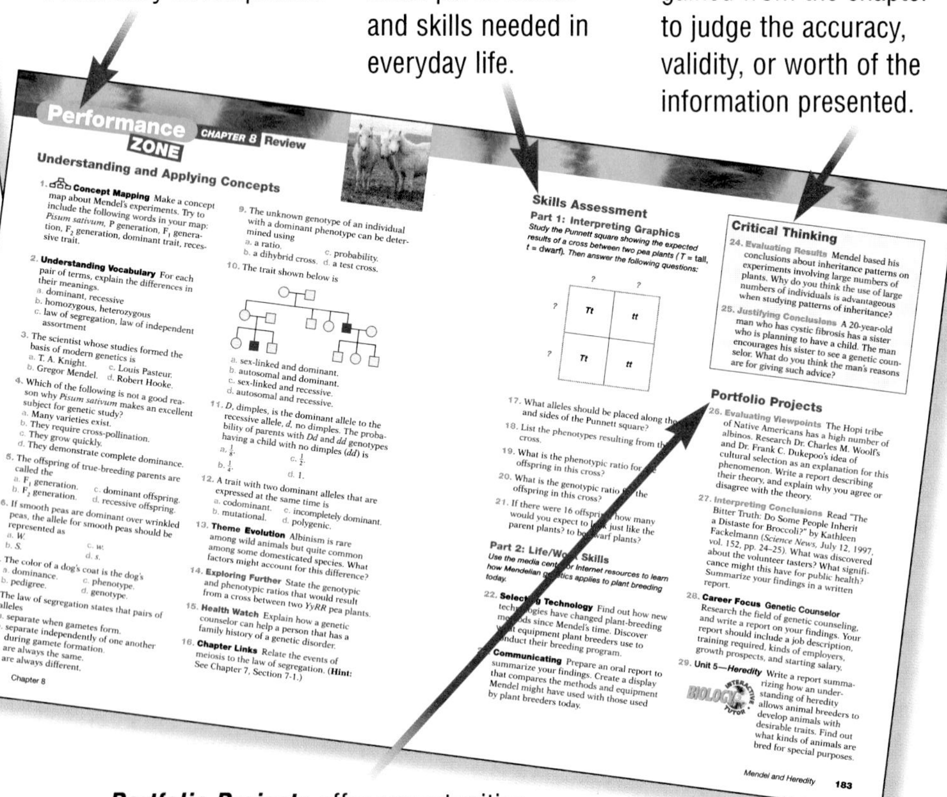

Performance ZONE CHAPTER 8 **Review**

Understanding and Applying Concepts

1. **Concept Mapping** Make a concept map about Mendel's experiments. Try to include the following words in your map: *Pisum sativum*, P generation, F_1 generation, F_2 generation, dominant trait, recessive trait.
2. **Understanding Vocabulary** For each pair of terms, explain the differences in their meanings.
   a. dominant, recessive
   b. homozygous, heterozygous
   c. law of segregation, law of independent assortment
3. The scientist whose studies formed the basis of modern genetics is
   a. T. A. Knight. c. Louis Pasteur.
   b. Gregor Mendel. d. Robert Hooke.
4. Which of the following is not a good reason why *Pisum sativum* makes an excellent subject for genetic study?
   a. Many varieties exist.
   b. They require cross-pollination.
   c. They grow quickly.
   d. They demonstrate complete dominance.
5. The offspring of true-breeding parents are called the
   a. F_1 generation. c. dominant offspring.
   b. F_2 generation. d. recessive offspring.
6. If smooth peas are dominant over wrinkled peas, the allele for smooth peas should be represented as
   a. *W*. c. *w*.
   b. *S*. d. *s*.
7. The color of a dog's coat is the dog's
   a. dominance. c. phenotype.
   b. pedigree. d. genotype.
8. The law of segregation states that pairs of alleles
   a. separate when gametes form.
   b. separate independently of one another during gamete formation.
   c. are always the same.
   d. are always different.
9. The unknown genotype of an individual with a dominant phenotype can be determined using
   a. a ratio. c. probability.
   b. a dihybrid cross. d. a test cross.
10. The trait shown below is
   a. sex-linked and dominant.
   b. autosomal and dominant.
   c. sex-linked and recessive.
   d. autosomal and recessive.
11. *D*, dimples, is the dominant allele to the recessive allele, *d*, no dimples. The probability of parents with *Dd* and *dd* genotypes having a child with no dimples (*dd*) is
   a. $\frac{1}{8}$. c. $\frac{1}{2}$.
   b. $\frac{1}{4}$. d. 1.
12. A trait with two dominant alleles that are expressed at the same time is
   a. codominant. c. incompletely dominant.
   b. mutational. d. polygenic.
13. **Theme Evolution** Albinism is rare among wild animals but quite common among some domesticated species. What factors might account for this difference?
14. **Exploring Further** State the genotypic and phenotypic ratios that would result from a cross between two *YyRR* pea plants.
15. **Health Watch** Explain how a genetic counselor can help a person that has a family history of a genetic disorder.
16. **Chapter Links** Relate the events of meiosis to the law of segregation. (**Hint:** See Chapter 7, Section 7-1.)

182 Chapter 8

Skills Assessment

Part 1: Interpreting Graphics

Study the Punnett square showing the expected results of a cross between two pea plants (T = tall, t = dwarf). Then answer the following questions:

	?	?
?	*Tt*	*tt*
?	*Tt*	*tt*

17. What alleles should be placed along th[illegible] and sides of the Punnett square?
18. List the phenotypes resulting from th[illegible] cross.
19. What is the phenotypic ratio for [illegible] offspring in this cross?
20. What is the genotypic ratio [illegible] the offspring in this cross?
21. If there were 16 offspri[illegible] how many would you expect to l[illegible] just like the parent plants? to b[illegible]warf plants?

Part 2: Life/Wo[illegible] Skills

Use the media cent[illegible]r Internet resources to learn how Mendelian g[illegible]tics applies to plant breeding today.

22. **Select[illegible] Technology** Find out how new tech[illegible]ogies have changed plant-breeding me[illegible]ds since Mendel's time. Discover [illegible]t equipment plant breeders use to [illegible]nduct their breeding program.

[illegible] **Communicating** Prepare an oral report to summarize your findings. Create a display that compares the methods and equipment Mendel might have used with those used by plant breeders today.

Critical Thinking

24. **Evaluating Results** Mendel based his conclusions about inheritance patterns on experiments involving large numbers of plants. Why do you think the use of large numbers of individuals is advantageous when studying patterns of inheritance?
25. **Justifying Conclusions** A 20-year-old man who has cystic fibrosis has a sister who is planning to have a child. The man encourages his sister to see a genetic counselor. What do you think the man's reasons are for giving such advice?

Portfolio Projects

26. **Evaluating Viewpoints** The Hopi tribe of Native Americans has a high number of albinos. Research Dr. Charles M. Woolf's and Dr. Frank C. Dukepoo's idea of cultural selection as an explanation for this phenomenon. Write a report describing their theory, and explain why you agree or disagree with the theory.
27. **Interpreting Conclusions** Read "The Bitter Truth: Do Some People Inherit a Distaste for Broccoli?" by Kathleen Fackelmann (*Science News*, July 12, 1997, vol. 152, pp. 24–25). What was discovered about the volunteer tasters? What significance might this have for public health? Summarize your findings in a written report.
28. **Career Focus Genetic Counselor** Research the field of genetic counseling, and write a report on your findings. Your report should include a job description, training required, kinds of employers, growth prospects, and starting salary.
29. **Unit 5—*Heredity*** Write a report summarizing how an understanding of heredity allows animal breeders to develop animals with desirable traits. Find out what kinds of animals are bred for special purposes.

BIOLOGY INTERACTIVE TUTOR

Mendel and Heredity 183

Portfolio Projects offer opportunities for students to extend their knowledge beyond the chapter content.

Customize your assessment with these resources.

Chapter Tests with Answer Key consist of multiple choice, matching, completion, and short-answer questions that assess content mastery for each chapter.

Test Generator with TestBuilder Software allows you to make custom tests quickly and easily.

- Contains more than 3000 multiple choice, matching, completion, and essay items.
- Sort items by topic, by chapter, or by lesson to construct your own customized tests to assess student mastery of the lesson objectives.
- *Critical Thinking* and *Analytical Thinking* items challenge students to go beyond rote recall to analyze and apply chapter content.
- Includes *Performance-Based Assessments,* laboratory-based activities that challenge students to design, perform, and analyze their own experiments to solve an assigned problem.

***Assessment Item Listing for* Biology: Principles and Explorations** is a softcover printed listing that shows you all available test items.

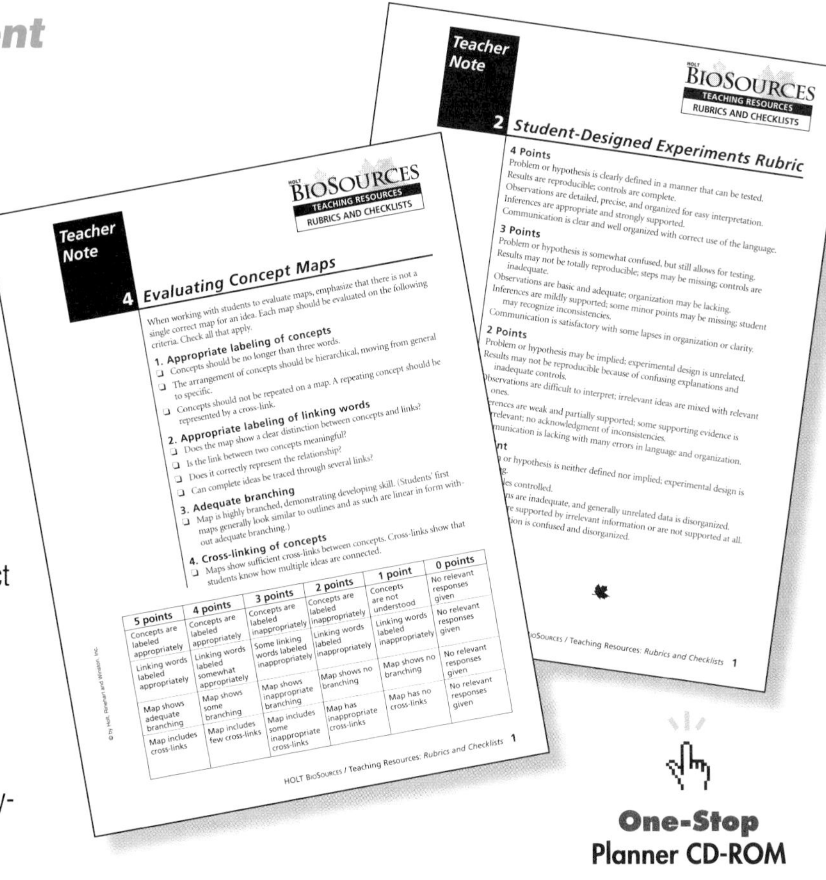

Teacher Note

HOLT BIOSOURCES TEACHING RESOURCES RUBRICS AND CHECKLISTS

4 Evaluating Concept Maps

When working with students to evaluate maps, emphasize that there is not a single correct map for an idea. Each map should be evaluated on the following criteria. Check all that apply.

1. Appropriate labeling of concepts
- ❑ Concepts should be no longer than three words.
- ❑ The arrangement of concepts should be hierarchical, moving from general to specific.
- ❑ Concepts should not be repeated on a map. A repeating concept should be represented by a cross-link.

2. Appropriate labeling of linking words
- ❑ Does the map show a clear distinction between concepts and links?
- ❑ Is the link between two concepts meaningful?
- ❑ Does it correctly represent the relationship?
- ❑ Can complete ideas be traced through several links?

3. Adequate branching
- ❑ Map is highly branched, demonstrating developing skill. (Students' first maps generally look similar to outlines and as such are linear in form without adequate branching.)

4. Cross-linking of concepts
- ❑ Maps show sufficient cross-links between concepts. Cross-links show that students know how multiple ideas are connected.

5 points	4 points	3 points	2 points	1 point	0 points
Concepts are labeled appropriately	Concepts are labeled appropriately	Concepts are labeled inappropriately	Concepts are labeled inappropriately	Concepts are not understood	No relevant responses given
Linking words labeled appropriately	Linking words labeled somewhat appropriately	Some linking words labeled inappropriately	Linking words labeled inappropriately	Linking words labeled inappropriately	No relevant responses given
Map shows adequate branching	Map shows some branching	Map shows inappropriate branching	Map shows no branching	Map shows no branching	No relevant responses given
Map includes cross-links	Map includes few cross-links	Map includes some inappropriate cross-links	Map has inappropriate cross-links	Map has no cross-links	No relevant responses given

© by Holt, Rinehart and Winston, Inc.

HOLT BIOSOURCES / Teaching Resources: *Rubrics and Checklists* 1

Teacher Note

HOLT BIOSOURCES TEACHING RESOURCES RUBRICS AND CHECKLISTS

2 Student-Designed Experiments Rubric

4 Points
Problem or hypothesis is clearly defined in a manner that can be tested.
Results are reproducible; controls are complete.
Observations are detailed, precise, and organized for easy interpretation.
Inferences are appropriate and strongly supported.
Communication is clear and well organized with correct use of the language.

3 Points
Problem or hypothesis is somewhat confused, but still allows for testing.
Results may not be totally reproducible; steps may be missing; controls are inadequate.
Observations are basic and adequate; organization may be lacking.
Inferences are mildly supported; some minor points may be missing; student may recognize inconsistencies.
Communication is satisfactory with some lapses in organization or clarity.

2 Points
Problem or hypothesis may be implied; experimental design is unrelated.
Results may not be reproducible because of confusing explanations and inadequate controls.
Observations are difficult to interpret; irrelevant ideas are mixed with relevant ones.
Inferences are weak and partially supported; some supporting evidence is irrelevant; no acknowledgment of inconsistencies.
Communication is lacking with many errors in language and organization.

[illegible]nt
… or hypothesis is neither defined nor implied; experimental design is …
…es controlled.
…ns are inadequate, and generally unrelated data is disorganized.
…e supported by irrelevant information or are not supported at all.
…ion is confused and disorganized.

…BioSources / Teaching Resources: *Rubrics and Checklists* 1

One-Stop Planner CD-ROM

Holt BioSources Teaching Resources on the One-Stop Planner includes additional assessment support.

- Scoring rubrics help you evaluate student-designed experiments, performance tasks, and concept maps.
- A Direct Observations Checklist facilitates informal assessment.
- A Portfolio Assessment Rubric and Student Evaluation Form help students and teachers evaluate portfolio assignments.
- Specialized rubrics for Holt BioSources Lab Program labs help you assess laboratory performance.

Portfolio Assessment

PORTFOLIO The "Portfolio" icon denotes items that may be appropriate for inclusion in a student portfolio. Students should select their best work and provide a self-reflective rationale *(see Portfolio Assignment Student Evaluation Form*)* for each selection. Use the *Assessment Rubric for Portfolio Items** to evaluate portfolios.

Students can make selections in the following areas:

1 **Content** One concept map from the chapter. See the rubric *Evaluating Concept Maps** for evaluation criteria.

2 **Reading Comprehension** One *Directed Reading* or *Active Reading* worksheet. Use the Answer Key to evaluate for accuracy.
Or: One *Reading Strategies Worksheet**
Or: One *Active Reading* assignment from the lesson wrap

3 **Writing** Concept map or report summarizing one of the following:
- An activity or a reading in the Performance Zone
- A newspaper or magazine article relating to the chapter topic

Or: A research paper on a subject related to the chapter topic. See *Science Research Paper Worksheets.**

4 **Performance Assessment** One lab report or *Vee Form** for an in-text lab or a **Holt BioSources Lab Program** lab. See assessment rubrics for labs.* For Vee forms, see *Assessing Student-Constructed Vee Reports** for evaluation criteria.
Or: A student-designed experiment. See *Student-Designed Experiments Rubric** for evaluation criteria.

Teachers can make selections in the following areas:

1 **Formal Assessment** Assign the appropriate chapter test from **Chapter Tests** or the **Test Generator**. The teacher-scored test should be reviewed by the student. Incorrect responses should be corrected by the student before the test becomes part of the portfolio.

2 **Informal Assessment** Use *Informal Assessment Direct Observations Checklist** during a laboratory or other cooperative learning experience.

3 **Performance Assessment** Assign the appropriate *Performance-Based Assessment Activity* from the **Test Generator.**

* Found in the **Holt BioSources® Teaching Resources** in the **One-Stop Planner CD-ROM**

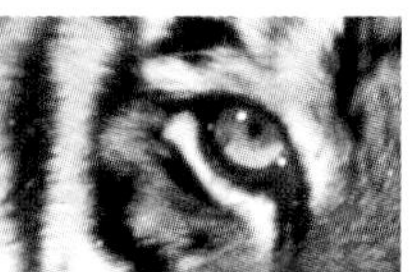

Coordinate Your Planning and Teaching With the Annotated Teacher's Edition

The Chapter Planning Guide organizes all of the program resources.

Your Annotated Teacher's Edition consists of two parts:

Weekly Planning Teacher's Interleaf pages precede each chapter in the ATE.

Daily Planning Teacher's Wrap surrounds the reduced pupil pages in the ATE.

The Planning Guide is your master plan for integrating all the materials available from ***Biology: Principles and Explorations*** and *Holt BioSources* into your teaching plan for each chapter.

Meeting Individual Needs tells you which resources are best-suited for the different learning styles and ability levels in your classroom.

Reading Skills and Strategies help you reach reluctant or struggling readers.

Plan Your Lessons

Each lesson includes options for lecture and classwork that will support various learning modalities for the longer periods required by block scheduling.

Objectives reproduced from the Pupil's Edition summarize the key points of each section.

Pacing suggestions estimate the number of 45-minute blocks required to teach each section.

Review and Assess

Help students prepare for the test by assigning any or all of the options shown here. Select from the assessment options to evaluate content mastery.

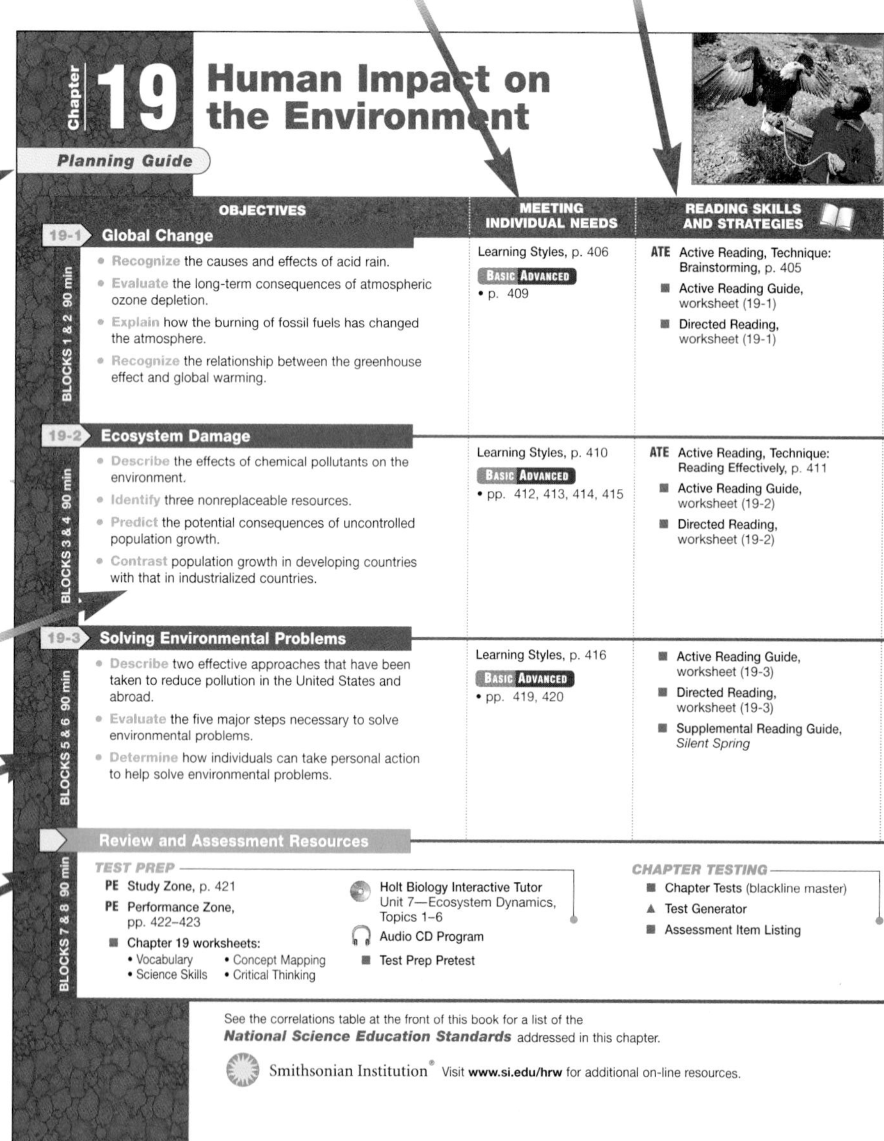

Chapter 19 Human Impact on the Environment

Planning Guide

	OBJECTIVES	MEETING INDIVIDUAL NEEDS	READING SKILLS AND STRATEGIES
BLOCKS 1 & 2 90 min	**19-1 Global Change** • Recognize the causes and effects of acid rain. • Evaluate the long-term consequences of atmospheric ozone depletion. • Explain how the burning of fossil fuels has changed the atmosphere. • Recognize the relationship between the greenhouse effect and global warming.	Learning Styles, p. 406 BASIC ADVANCED • p. 409	ATE Active Reading, Technique: Brainstorming, p. 405 ■ Active Reading Guide, worksheet (19-1) ■ Directed Reading, worksheet (19-1)
BLOCKS 3 & 4 90 min	**19-2 Ecosystem Damage** • Describe the effects of chemical pollutants on the environment. • Identify three nonreplaceable resources. • Predict the potential consequences of uncontrolled population growth. • Contrast population growth in developing countries with that in industrialized countries.	Learning Styles, p. 410 BASIC ADVANCED • pp. 412, 413, 414, 415	ATE Active Reading, Technique: Reading Effectively, p. 411 ■ Active Reading Guide, worksheet (19-2) ■ Directed Reading, worksheet (19-2)
BLOCKS 5 & 6 90 min	**19-3 Solving Environmental Problems** • Describe two effective approaches that have been taken to reduce pollution in the United States and abroad. • Evaluate the five major steps necessary to solve environmental problems. • Determine how individuals can take personal action to help solve environmental problems.	Learning Styles, p. 416 BASIC ADVANCED • pp. 419, 420	■ Active Reading Guide, worksheet (19-3) ■ Directed Reading, worksheet (19-3) ■ Supplemental Reading Guide, *Silent Spring*

Review and Assessment Resources (BLOCKS 7 & 8 90 min)

TEST PREP
- PE Study Zone, p. 421
- PE Performance Zone, pp. 422–423
- ■ Chapter 19 worksheets:
  - • Vocabulary • Concept Mapping
  - • Science Skills • Critical Thinking
- Holt Biology Interactive Tutor Unit 7—Ecosystem Dynamics, Topics 1–6
- Audio CD Program
- ■ Test Prep Pretest

CHAPTER TESTING
- ■ Chapter Tests (blackline master)
- ▲ Test Generator
- ■ Assessment Item Listing

See the correlations table at the front of this book for a list of the ***National Science Education Standards*** addressed in this chapter.

Smithsonian Institution® Visit **www.si.edu/hrw** for additional on-line resources.

403A Chapter 19

On pp.T32-T51, this symbol indicates a ***Key to Reading Success***

Visual Strategies

list the ATE *Visual Strategy* teaching tips and the Teaching Transparencies for each lesson.

Assign Classwork and Homework

Choose activities or labs for classwork and assign homework from the options shown here.

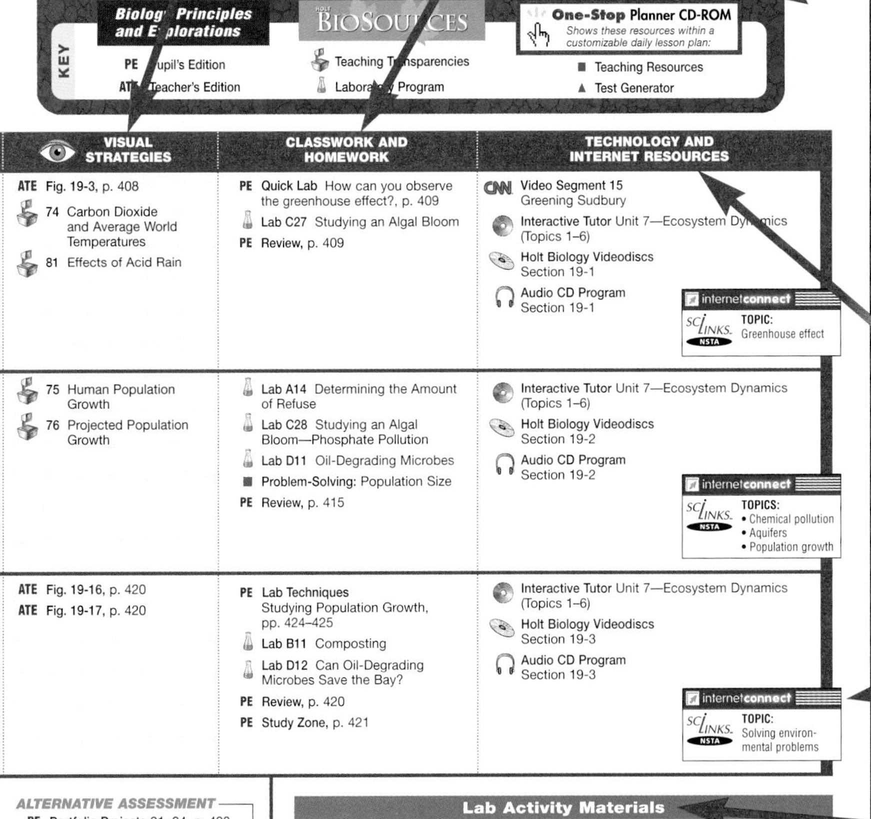

KEY

Biology: Principles and Explorations
PE Pupil's Edition
ATE Teacher's Edition

BIOSOURCES
Teaching Transparencies
Laboratory Program

One-Stop Planner CD-ROM
Shows these resources within a customizable daily lesson plan:
■ Teaching Resources
▲ Test Generator

VISUAL STRATEGIES	CLASSWORK AND HOMEWORK	TECHNOLOGY AND INTERNET RESOURCES
ATE Fig. 19-3, p. 408 74 Carbon Dioxide and Average World Temperatures 81 Effects of Acid Rain	PE Quick Lab How can you observe the greenhouse effect?, p. 409 Lab C27 Studying an Algal Bloom PE Review, p. 409	Video Segment 15 Greening Sudbury Interactive Tutor Unit 7—Ecosystem Dynamics (Topics 1–6) Holt Biology Videodiscs Section 19-1 Audio CD Program Section 19-1 internetconnect sciLINKS. NSTA TOPIC: Greenhouse effect
75 Human Population Growth 76 Projected Population Growth	Lab A14 Determining the Amount of Refuse Lab C28 Studying an Algal Bloom—Phosphate Pollution Lab D11 Oil-Degrading Microbes ■ Problem-Solving: Population Size PE Review, p. 415	Interactive Tutor Unit 7—Ecosystem Dynamics (Topics 1–6) Holt Biology Videodiscs Section 19-2 Audio CD Program Section 19-2 internetconnect sciLINKS. NSTA TOPICS: • Chemical pollution • Aquifers • Population growth
ATE Fig. 19-16, p. 420 ATE Fig. 19-17, p. 420	PE Lab Techniques Studying Population Growth, pp. 424–425 Lab B11 Composting Lab D12 Can Oil-Degrading Microbes Save the Bay? PE Review, p. 420 PE Study Zone, p. 421	Interactive Tutor Unit 7—Ecosystem Dynamics (Topics 1–6) Holt Biology Videodiscs Section 19-3 Audio CD Program Section 19-3 internetconnect sciLINKS. NSTA TOPIC: Solving environmental problems

ALTERNATIVE ASSESSMENT
PE Portfolio Projects 21–24, p. 423
ATE Alternative Assessment, p. 421
ATE Portfolio Assessment, p. 421

Lab Activity Materials

Quick Lab
p. 409 MBL or CBL system with appropriate software, temperature probes (2), 1 qt jar and lid with 0.5 cm hole in the center, tape, heat source

Lab Techniques
p. 424 safety goggles, lab apron, yeast culture, 1 mL pipets (2), test tubes (2), 1% methylene blue solution, 2 × 2 mm ruled microscope slide, coverslip, compound microscope

Lesson Warm-up
ATE photos of several forms of environmental pollution, p. 410

Demonstrations
ATE box of sand, p. 413
ATE photos of Los Angeles on a smoggy day, p. 417
ATE catalytic converter, p. 417

BIOSOURCES LAB PROGRAM
- Quick Labs: Lab A14, p. 27
- Inquiry Skills Development: Lab B11, p. 5
- Laboratory Techniques and Experimental Design: Lab C27, p. 131
- Laboratory Techniques and Experimental Design: Lab C28, p. 137
- Biotechnology: Lab D11, p. 57
- Biotechnology: Lab D12, p. 63

Human Impact on the Environment **403B**

The ***Key*** lists the components of your comprehensive resource package.

- ***Biology: Principles and Explorations*** resources are specifically correlated to each lesson and chapter of the text.
- ***Holt BioSources*** enable you to extend and customize your course to meet your needs.
- **One-Stop Planner CD-ROM** integrates the chapter resources within a customizable lesson plan format.

Select Technology and Internet Resources to Support and Extend Your Lesson

Select from these resources to enrich your lesson curriculum.

***NSTA sci*LINKS** found on the Pupil's Edition lesson pages are listed here.

Lab Activity Materials lists all the materials needed to perform the Pupil's Edition labs, the Lesson Warm-Ups and Demonstrations in the ATE Lesson Wrap, and the Holt BioSources Labs.

- These labs are found in the ***Biology: Principles and Explorations*** Pupil's Edition.
- These labs are found in the *Holt BioSources Laboratory Program.*

The lesson wrap provides daily teaching strategies.

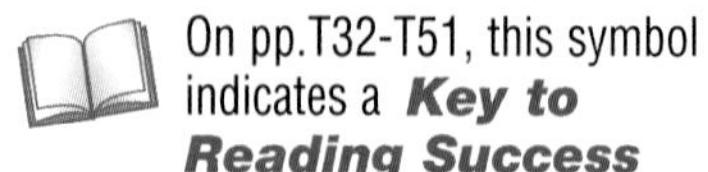

On pp.T32-T51, this symbol indicates a ***Key to Reading Success***

Chapter Theme helps you integrate one or more of the biological themes—*Cell Structure and Function, Reproduction, Metabolism, Homeostasis, Heredity, Evolution,* or *Interdependence*—important to the chapter into your lesson plan.

Establishing Prior Knowledge helps you assess how much your students know before you begin teaching.

What's Ahead—and Why lists all of the main heads in the chapter so you'll know what's coming up. Read this first.

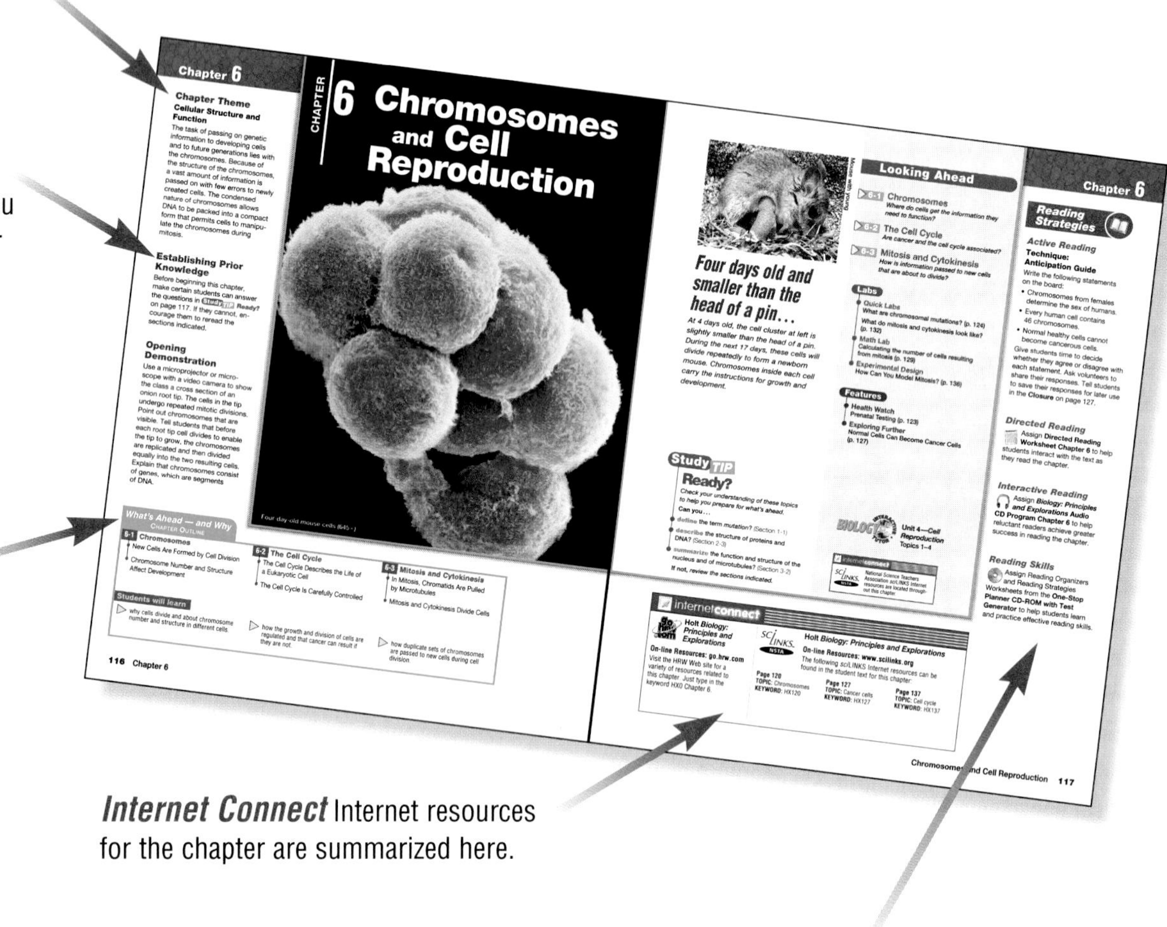

Chapter 6

Chapter Theme
Cellular Structure and Function
The task of passing on genetic information to developing cells and to future generations lies with the chromosomes. Because of the structure of the chromosomes, a vast amount of information is passed on with few errors to newly created cells. The condensed nature of chromosomes allows DNA to be packed into a compact form that permits cells to manipulate the chromosomes during mitosis.

Establishing Prior Knowledge
Before beginning this chapter, make certain students can answer the questions in Study TIP Ready? on page 117. If they cannot, encourage them to reread the sections indicated.

Opening Demonstration
Use a microprojector or microscope with a video camera to show the class a cross section of an onion root tip. The cells in the tip undergo repeated mitotic divisions. Point out chromosomes that are visible. Tell students that before each root tip cell divides to enable the tip to grow, the chromosomes are replicated and then divided equally into the two resulting cells. Explain that chromosomes consist of genes, which are segments of DNA.

CHAPTER 6 **Chromosomes and Cell Reproduction**

Four day-old mouse cells (645×)

What's Ahead — and Why
CHAPTER OUTLINE

6-1 Chromosomes
- New Cells Are Formed by Cell Division
- Chromosome Number and Structure Affect Development

6-2 The Cell Cycle
- The Cell Cycle Describes the Life of a Eukaryotic Cell
- The Cell Cycle Is Carefully Controlled

6-3 Mitosis and Cytokinesis
- In Mitosis, Chromatids Are Pulled by Microtubules
- Mitosis and Cytokinesis Divide Cells

Students will learn
- why cells divide and about chromosome number and structure in different cells.
- how the growth and division of cells are regulated and that cancer can result if they are not.
- how duplicate sets of chromosomes are passed to new cells during cell division.

116 Chapter 6

Mouse with young

Four days old and smaller than the head of a pin...
At 4 days old, the cell cluster at left is slightly smaller than the head of a pin. During the next 17 days, these cells will divide repeatedly to form a newborn mouse. Chromosomes inside each cell carry the instructions for growth and development.

Looking Ahead
- 6-1 Chromosomes *Where do cells get the information they need to function?*
- 6-2 The Cell Cycle *Are cancer and the cell cycle associated?*
- 6-3 Mitosis and Cytokinesis *How is information passed to new cells that are about to divide?*

Labs
- Quick Labs
  What are chromosomal mutations? (p. 124)
  What do mitosis and cytokinesis look like? (p. 132)
- Math Lab
  Calculating the number of cells resulting from mitosis (p. 129)
- Experimental Design
  How Can You Model Mitosis? (p. 136)

Features
- Health Watch
  Prenatal Testing (p. 123)
- Exploring Further
  Normal Cells Can Become Cancer Cells (p. 127)

Study TIP
Ready?
Check your understanding of these topics to help you prepare for what's ahead.
Can you...
- define the term *mutation*? (Section 1-1)
- describe the structure of proteins and DNA? (Section 2-3)
- summarize the function and structure of the nucleus and of microtubules? (Section 3-2)

If not, review the sections indicated.

BIOLOGY INTERACTIVE TUTOR Unit 4—*Cell Reproduction* Topics 1–4

internetconnect
SciLINKS NSTA National Science Teachers Association sciLINKS Internet resources are located throughout this chapter.

internetconnect
Holt *Biology: Principles and Explorations*
On-line Resources: go.hrw.com
Visit the HRW Web site for a variety of resources related to this chapter. Just type in the keyword HX0 Chapter 6.

SciLINKS NSTA Holt *Biology: Principles and Explorations*
On-line Resources: www.scilinks.org
The following sciLINKS Internet resources can be found in the student text for this chapter.

Page 120 **TOPIC:** Chromosomes **KEYWORD:** HX120
Page 127 **TOPIC:** Cancer cells **KEYWORD:** HX127
Page 137 **TOPIC:** Cell cycle **KEYWORD:** HX137

Chapter 6

Reading Strategies

Active Reading
Technique: Anticipation Guide
Write the following statements on the board:
- Chromosomes from females determine the sex of humans.
- Every human cell contains 46 chromosomes.
- Normal healthy cells cannot become cancerous cells.

Give students time to decide whether they agree or disagree with each statement. Ask volunteers to share their responses. Tell students to save their responses for later use in the **Closure** on page 127.

Directed Reading
Assign **Directed Reading Worksheet Chapter 6** to help students interact with the text as they read the chapter.

Interactive Reading
Assign ***Biology: Principles and Explorations* Audio CD Program Chapter 6** to help reluctant readers achieve greater success in reading the chapter.

Reading Skills
Assign Reading Organizers and Reading Strategies Worksheets from the **One-Stop Planner CD-ROM with Test Generator** to help students learn and practice effective reading skills.

Chromosomes and Cell Reproduction 117

Internet Connect Internet resources for the chapter are summarized here.

Reading Strategies help your students read more effectively and become more responsible for their own learning:

Active Reading exercises help your students become truly active readers by using the following techniques:

- Brainstorming
- K-W-L
- L.I.N.K.
- Paired Reading
- Reading Organizer
- Paired Summarizing
- Discussion
- Anticipation Guide
- Reader Response Logs
- Reading Effectively

Directed Reading prompts you to assign the Directed Reading worksheet for students to complete while they read the section.

Interactive Reading becomes easy for reluctant readers or auditory learners when they use the ***Biology: Principles and Explorations*** Audio CD Program for the chapter.

Reading Skills become more developed when you assign one or more of the Reader Organizer or Reader Strategy worksheets found on the One-Stop Planner.

Each lesson contains the following elements:

1 Prepare

Scheduling provides pacing suggestions for block and traditional scheduling.

Resource Links point out program resources.

2 Teach

Lesson Warm-Up serves as a lesson "icebreaker."

Teaching Tips appear at point of use. **BASIC** and **ADVANCED** identify tips that are especially appropriate for students of these ability levels. Tips resulting in student work that could be considered for inclusion in a student portfolio are marked **PORTFOLIO**.

Integrating Different Learning Styles helps you ensure that your lesson plans reach students of different learning styles:

- Logical/Mathematical
- Visual/Spatial
- Body/Kinesthetic
- Musical/Rhythmic
- Interpersonal
- Intrapersonal
- Verbal/Linguistic

Visual Strategies are ideas to make effective use of the outstanding instructional photographs and illustrations throughout the Pupil's Edition. A ***Visual Strategy*** for every ***BioGraphic*** helps you integrate these key visuals into your teaching program.

Making Biology Relevant shows the application of biology to health, technology, sports, medicine, and food.

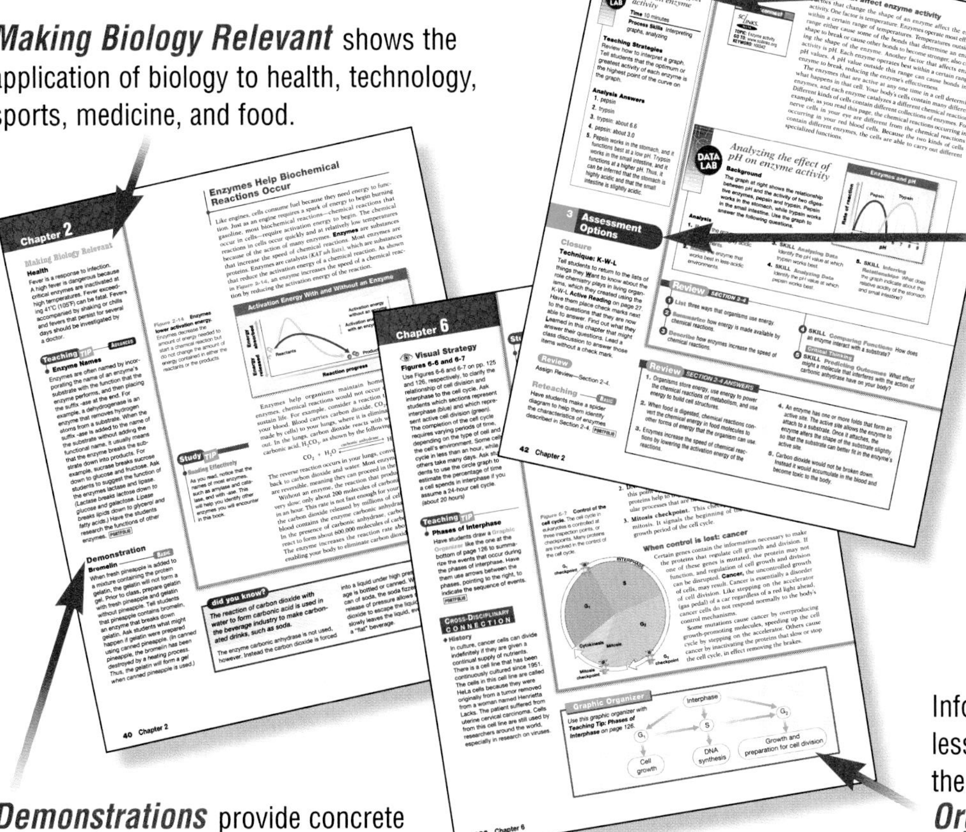

Teacher support for Quick Labs, Data Labs, and Math Labs provides Teaching Strategies and Answers right where you need them!

3 Assessment Options

Closure is a strategy for helping students summarize key concepts from the lesson.

Review assesses mastery of the section objectives.

Reteaching provides a strategy for additional instruction in a major concept of the section.

Demonstrations provide concrete modeling for abstract ideas.

Informational items along the bottom of the lesson wrap help you support and extend the lesson content. For example, ***Graphic Organizers*** are visual tools that help you describe particularly complex information.

National Science Education Standards Correlation

The following table shows the chapter correlation of Biology: Principles and Explorations with the National Science Education Standards (grades 9–12) for Life Science content.

LIFE SCIENCE CONTENT STANDARDS	PART 1: PRINCIPLES — UNIT 1 PRINCIPLES OF CELL BIOLOGY Chapters 1–6	UNIT 2 PRINCIPLES OF GENETICS Chapters 7–11	UNIT 3 PRINCIPLES OF EVOLUTION Chapters 12–15
The cell			
• Cells have particular structures that underlie their functions.	1, 2, 3, 4	7, 8, 9, 10	12, 14
• Most cell functions involve chemical reaction.	1, 2, 3, 4	7, 8, 9, 10	12
• Cells store and use information to guide their functions.	1, 2, 3, 4	7, 9, 10, 11	12
• Cell functions are regulated.	1, 2, 3, 4, 6	7, 9, 10, 11	12
• Plant cells contain chloroplasts, the site of photosynthesis.	1, 3, 5		12
• Cells can differentiate and form complete multicellular organisms.	1, 6	7, 11	12
Molecular basis of heredity			
• In all organisms, the instructions for specifying the characteristics of the organisms are carried in DNA.	2, 3, 6	7, 8, 9, 10, 11	3, 12
• Most of the cells in a human contain two copies of each of 22 different chromosomes. In addition there is a pair of chromosomes that determine sex.	6	7, 8, 9, 10, 11	
• Changes in DNA (mutations) occur spontaneously at low rates.		7, 9, 10, 11	13, 14
Biological evolution			
• Species evolve over time.	1	7	12, 13, 14, 15
• The great diversity of organisms is the result of more than 3.5 billion years of evolution.	1		12, 13, 14, 15
• Natural selection and its evolutionary consequences provide a scientific explanation for the fossil record of ancient life forms as well as for the striking molecular similarities observed among the diverse species of living organisms.	1	7	12, 13, 14, 15
• The millions of different species of plants, animals, and microorganisms that live on earth today are related by descent from common ancestors.	1		12, 13, 14, 15
• Biological classifications are based on how organisms are related.	1, 3		13, 15

***Biology: Principles and Explorations* authors George B. Johnson and Peter H. Raven were involved in the actual drafting of the content portion of the National Science Education Standards. *Biology: Principles and Explorations* was designed by them from the outset to match the standards.**

UNIT 4 PRINCIPLES OF ECOLOGY Chapters 16–19	PART 2: EXPLORATIONS UNIT 5 EXPLORING DIVERSITY Chapters 20–23	UNIT 6 EXPLORING PLANTS Chapters 24–27	UNIT 7 EXPLORING INVERTEBRATES Chapters 28–32	UNIT 8 EXPLORING VERTEBRATES Chapters 33–37	UNIT 9 EXPLORING HUMAN BIOLOGY Chapters 38–44
19	20, 21, 22, 23	24, 25, 26, 27	28, 29, 30, 31, 32	33	39, 40, 41, 42, 43, 44
	20, 21, 22, 23	24, 27	28, 30, 31		39, 40, 41, 42, 43, 44
19	21, 22	24	28	33	41, 44
19	20, 21, 22	24, 25, 26, 27	28, 29, 30, 31, 32	33	39, 40, 41, 42, 43, 44
	20	24, 25, 26, 27			
	20	25, 26, 27	28, 29		44
	20, 21, 22, 23		28		44
					44
16	20	24, 26	28, 29, 30, 31, 32	33, 34, 35, 36, 37	
				33, 34	
	20	24	32	33, 36, 37	
	20	24		33	
	20, 21, 22, 23	24	28, 31	33, 34, 36	

continued on the next page

National Science Education Standards Correlation, continued

LIFE SCIENCE CONTENT STANDARDS	PART 1: PRINCIPLES		
	UNIT 1 PRINCIPLES OF CELL BIOLOGY	**UNIT 2** PRINCIPLES OF GENETICS	**UNIT 3** PRINCIPLES OF EVOLUTION
	Chapters 1–6	Chapters 7–11	Chapters 12–15
Interdependence of organisms			
• The atoms and molecules on earth cycle among the living and nonliving components of the biosphere.		2	
• Energy flows through ecosystems in one direction, from photosynthetic organisms to herbivores to carnivores and decomposers.	2		
• Organisms both cooperate and compete in ecosystems.			12
• Living organisms have the capacity to produce populations of infinite size, but environments and resources are finite.	1		12
• Human beings live within the world's ecosystems.	1		12
Matter, energy, and organization in living systems			
• All matter tends toward more disorganized states.			
• The energy for life primarily derives from the sun.	1, 3, 5	10	12
• The chemical bonds of food molecules contain energy.	1, 2, 5		
• The complexity and organization of organisms accommodates the need for obtaining, transforming, transporting, releasing, and eliminating the matter and energy used to sustain the organism.	1, 2		12, 13
• The distribution and abundance of organisms and populations in ecosystems are limited by the availability of matter and energy and the ability of the ecosystem to recycle materials.			12, 13
• As matter and energy flows through different levels of organization of living systems—cells, organs, communities—and between living systems and the physical environment, chemical elements are recombined in different ways.	1, 2		12, 13
Behavior of organisms			
• Multicellular animals have nervous systems that generate behavior.			
• Organisms have behavioral responses to internal changes and to external stimuli.	1		12
• Like other aspects of an organism's biology, behaviors have evolved through natural selection.			12
• Behavioral biology has implications for humans, as it provides links to psychology, sociology and anthropology.			12

	PART 2: EXPLORATIONS				
UNIT 4 PRINCIPLES OF ECOLOGY	**UNIT 5** EXPLORING DIVERSITY	**UNIT 6** EXPLORING PLANTS	**UNIT 7** EXPLORING INVERTEBRATES	**UNIT 8** EXPLORING VERTEBRATES	**UNIT 9** EXPLORING HUMAN BIOLOGY
Chapters 16–19	Chapters 20–23	Chapters 24–27	Chapters 28–32	Chapters 33–37	Chapters 38–44
17	21				
17	20	24, 26	28		
17, 18	21, 22, 23	24, 25	29	33, 37	
16, 17, 18, 19					
16, 17, 19	20, 21, 22, 23	24	29		
17					
17		24, 26			
17		26			40
16, 17, 18	20, 21		28, 29, 30, 31, 32	33, 34, 35, 36, 37	38, 39, 40, 41, 42, 43
16, 17, 18					
	21	26			38
			28	33, 34, 36, 37	42
			28, 29, 30, 31	33, 34, 35, 36, 37	42, 43
				33, 36, 37	
				37	

continued on the next page

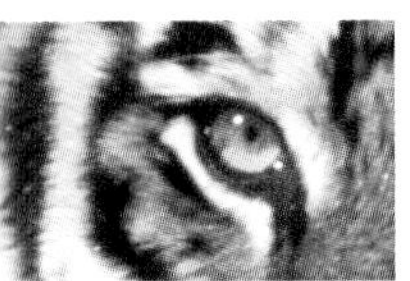

National Science Education Standards Correlation, continued

The following table shows the chapter correlation of* Biology: Principles and Explorations *with additional science content standards (grades 9–12).

ADDITIONAL SCIENCE CONTENT STANDARDS	CHAPTERS 1–44
Unifying Concepts and Processes	
Systems, order, and organization	1, 7–11, 12–15, 17, 18, 20, 21, 24, 25, 28–31, 33–44
Evidence, models, and explanation	1, 3, 7–13, 17, 18, 20, 21, 24, 25, 28, 29, 33–37
Change, constancy, and measurement	3, 7–12, 33
Evolution and equilibrium	3, 7–13, 17, 18, 20, 24, 28
Form and function	1–13, 18, 20–31, 33–44
Science as Inquiry	
Abilities necessary to do scientific inquiry	1–44
Understandings about scientific inquiry	1–44
Physical Science	
Structure of atoms	2–4
Structure and properties of matter	2–4
Chemical reactions	1–4, 12
Motions and forces	4, 21, 28
Conservation of energy and increase in disorder	2, 5, 17
Interactions of energy and matter	2–7, 12, 17, 19, 38
Earth and Space Science	
Energy in the earth system	1, 12, 17
Geochemical cycles	17, 18, 20, 30
Origin and evolution of the earth system	12
Origin and evolution of the universe	12

continued on the next page

ADDITIONAL SCIENCE CONTENT STANDARDS	CHAPTERS 1–44
Science and Technology	
Abilities of technological design	1, 11, 19
Understandings about science and technology	1–6, 11, 12
Science in Personal and Social Perspectives	
Personal and community health	1, 19–21, 24, 29, 30, 44
Population growth	1, 16–19
Natural resources	1, 16–19
Natural and human-induced hazards	1, 12, 19, 21, 29. 44
Science and technology in local, national, and global challenges	1, 11, 12, 19, 44
History and Nature of Science	
Science as a human endeavor	1–3, 8, 9, 11–13, 21, 23, 24
Nature of scientific knowledge	1–4, 11, 20
Historical perspectives	2, 8, 9, 11–14, 20, 21, 33

Running a Safe and Efficient Lab Program

General Safety Guidelines

1. Post laboratory rules in a conspicuous place in the laboratory.
2. Before the class begins an experiment, review safety rules and demonstrate proper procedures.
3. Never permit students to work in your laboratory without your supervision. No unauthorized investigations should ever be conducted, nor should unauthorized materials be in the laboratory.
4. Lock your laboratory (and storeroom) when you are not present.
5. Clearly mark locations of eyewash stations, safety shower, fire extinguishers (ABC tri-class), chemical spill kit, first-aid kit, and fire blanket in the laboratory and storeroom. Check this safety equipment prior to conducting each investigation.
6. Post an evacuation diagram and procedures by every entrance to the laboratory.
7. Provide labeled disposal containers for glass, sharps, and waste chemical reagents.
8. Allow no food or beverages in the laboratory. Caution students to keep their hands away from their face and to wash hands with soap and water before leaving the laboratory.
9. Know the location for the master shut-off for laboratory circuits. Be sure that all outlets have correct polarity and have ground-fault interception. Polarity can be tested with an inexpensive (about $5) continuity tester available from most electronic hobby shops.
   All electrical equipment should have 3-prong plugs and 3-wire cords.
10. Follow prescribed procedures for any safety incident, including full documentation. Remind students that any safety incident, no matter how trivial, must be reported directly to you.

Personal Protective Equipment

Chemical goggles: (Meeting ANSI Standard Z87.1) These should be worn with any chemical or chemical solution other than water, when heating substances, using any mechanical device, or observing physical processes that could eject an object.

Face shield: (Meeting ANSI Standard Z87.1) Use in combination with eye goggles when working with corrosives.

Contact lenses: The wearing of contact lenses for cosmetic reasons should be prohibited in the laboratory. If a student must wear contact lenses prescribed by a physician, that student should be instructed to wear eye-cup safety goggles meeting ANSI Standard Z87.1 (similar to swimmer's cup goggles).

Eye-wash station: The device must be capable of delivering a copious, gentle flow of water to both eyes for at least 15 minutes. Portable liquid supply devices are not satisfactory and should not be used. A plumbed-in fixture or a perforated spray head on the end of a hose attached to a plumbed-in outlet is suitable if it is designed for use as an eye-wash fountain and meets ANSI Standard Z358.1 It must be within a 30-second walking distance from any spot in the room.

Safety shower: (Meeting ANSI Standard Z358.1) Location should be within a 30-second walking distance from any spot in the room. Students should be instructed in the use of the safety shower in the event of a fire or chemical splash on their body that cannot simply be washed off.

Gloves: Polyethylene, neoprene rubber, or disposable plastic may be used. Nitrile or butyl rubber gloves are recommended when handling corrosives.

Apron: Rubber-coated cloth or vinyl (nylon-coated) halter is recommended.

Emergency Preparedness

What would you do if a student dropped a liter bottle of concentrated sulfuric acid? RIGHT NOW? Plan how to effectively react BEFORE you need to.

1. Post the phone number of your regional poison control center, fire department, police, ambulance, and hospital on your telephone.
2. Practice fire and evacuation drills. Also have drills on what students MUST do if they are on fire or have chemical contact.
3. Ensure that all personal and other safety equipment is available and tested frequently.
4. Compile an MSDS file for all chemicals. This reference resource should be readily available in case of a spill or other accident.
5. Provide for spill control procedures. Handle only those incidents that you

feel comfortable in handling. Situations of greater severity should be handled by trained hazardous materials professionals.

6. Students should never fight fires or handle spills.
7. Be trained in first aid and basic life support (CPR) procedures. Have first-aid kits and spill kits readily available.
8. Fully document ANY INCIDENT that occurs.

Safety With Animals

It is recommended that teachers follow the *Guidelines for the Use of Live Animals* established by the National Association of Biology Teachers, which is reproduced in the Teacher's Editions of *Holt BioSources® Lab Program Quick Labs and Inquiry Skills Development* lab manuals.

Safety in Handling Preserved Materials

The following practices are recommended when handling or dissecting any preserved specimen:

1. NEVER dissect road-kills or nonpreserved slaughterhouse materials.
2. Wear protective gloves and splash-proof safety goggles at all times when handling preserving fluids and preserved specimens and during dissection.
3. Wear lab aprons. Use of an old shirt or smock under the lab apron is recommended.
4. Conduct dissection activities in a well-ventilated area.
5. Do not allow preservation or body-cavity fluids to contact skin. Fixatives do not distinguish between living or dead tissues.

Biological supply firms use formalin-based fixatives of varying concentrations to initially fix zoological and botanical specimens. WARD'S Natural Science Establishment provides specimens that are freeze-dried and rehydrated in a 10% isopropyl alcohol solution. In these specimens, no other hazardous chemical is present.

Many suppliers provide fixed botanical materials in 50% glycerin.

Reduction of Free Formaldehyde

Currently, federal regulations mandate a permissible exposure level of 0.75 ppm for formaldehyde. Contact your supplier for an MSDS that details the amount of formaldehyde present as well as gas-emitting characteristics for individual specimens.

Prewash specimens (in a loosely covered container) in running tap water for 1–4 hours to dilute the fixative. Formaldehyde may also be chemically bound (thereby reducing danger) by immersing washed specimens in a 0.5–1.0% potassium bisulfate solution overnight or by placing them in 1% phenoxyethanol holding solutions.

Safety With Microbes

Pathogenic (disease-causing) microorganisms are not appropriate investigational tools in the high school laboratory and should never be used.

Safety With Chemicals

Label student reagent containers with the substance's name and hazard class(es) (flammable, reactive, etc.).

Dispose of hazardous waste chemicals according to federal, state, and local regulations. Refer to the Material Safety Data Sheet for recommended disposal procedures.

Remove all sources of flames, sparks, and heat from the laboratory when any flammable material is being used.

Material Safety Data Sheets

The purpose of a Material Safety Data Sheet (MSDS) is to provide readily accessible information on chemical substances commonly used in the science laboratory or in industry.

The MSDS should be kept on file and referred to BEFORE handling ANY chemical. The MSDS can also be used to instruct students on chemical hazards, to evaluate spill and disposal procedures, and to warn of incompatibility with other chemicals or mixtures.

Resources

American Chemical Society Health and Safety Service

This service will refer inquiries to appropriate resources for finding answers to questions about health and safety.

American Chemical Society (ACS)
1155 Sixteenth Street, N.W.
Washington, D.C. 20036

1-800-227-5558

Safety Information References

Gessner, G. H., ed. *Hawley's Condensed Chemical Dictionary.* 11th Ed. Van Nostrand Reinhold, 1987. (Revised by N. Irving Sax).

A Guide to Information Sources Related to the Safety and Management of Laboratory Wastes from Secondary Schools. New York State Environmental Facilities Corp., 1985.

Lefevre, J.J. *The First Aid Manual for Chemical Accidents.* Dowdwen, 1989. (Revised by Shirley A. Conibeau).

Pipitone, D., ed., *Safe Storage of Laboratory Chemicals.* John Wiley, 1984.

Prudent Practices for Disposal of Chemicals from Laboratories. Committee on Hazardous Substances in the Laboratory, National Research Council, National Academy Press, 1983.

Prudent Practices for Handling Hazardous Chemicals from Laboratories. Committee on Hazardous Substances in the Laboratory, National Research Council, National Academy Press, 1981.

Strauss, H. and M. Kaufman, ed. *Handbook for Chemical Technicians.* McGraw-Hill, 1981.

WARD'S MSDS *Users Guide.* WARD'S, 1989.

Storing Chemicals

Never store chemicals alphabetically, as this greatly increases the risk of promoting a violent reaction.

Storage Suggestions:

1. Always lock the storeroom and all its cabinets when not in use.
2. Students should not be allowed in the storeroom and preparation area.
3. Avoid storing chemicals on the floor of the storeroom.
4. Do not store chemicals above eye level or on the top shelf in the storeroom.
5. Be sure shelf assemblies are firmly secured to the walls.
6. Provide anti-roll lips on all shelves.
7. Shelving should be constructed out of wood. Metal cabinets and shelves are easily corroded.
8. Avoid metal, adjustable shelf supports and clips. They can corrode, causing shelves to collapse.
9. Acids, flammables, poisons, and oxidizers should each be stored in their own locking storage cabinet.

Master Materials List

This list indicates the quantities for a class of 32 to perform the in-text chapter labs working in eight groups of four. If you have a different number of lab groups to plan for, scale the quantities accordingly.

A number alone indicates an end-of-chapter Experimental Design, Lab Technique, or Interactive Exploration lab. **QL** indicates a Quick Lab, **DL** indicates a Data Lab, and **ML** indicates a Math Lab. These two-letter codes are followed by the page number on which the lab appears.

This materials list was prepared by **WARD'S** Natural Science Establishment, Inc., the exclusive science supplier for *Biology: Principles and Explorations* and *Holt BioSources® Lab Program.* Catalog numbers for WARD'S™ are provided. *Local* means that the item should be available locally. All investigations and lab manual activities in *Biology: Principles and Explorations* and *Holt BioSources® Lab Program* are bench tested by WARD'S technical staff to guarantee successful lab results and experiences.

WARD'S Natural Science Establishment, Inc.
5100 West Henrietta Road
Rochester, New York 14692-9012
1-800-962-2660
www.wardsci.com

Biological Supplies

Description	Item no.	Qty per class	Lab
Alfalfa seeds, pkt.	86R8002	8	**17**
Algae, mixed green (wm) slide	91R0560	4	**20**
Amphioxus—posterior to pharynx	92R3014	8	QL720
Animal kingdom survey set, general	67R9911	1	**20**
Apple and potato slices	Local	8	QL844
Apple, ripe	Local	8	QL599
Archaebacteria, mixture (sm) g(-)	90R0526	4	**20**
Bacillus megaterium culture	85R1154	1	**21**
Bacillus subtilis culture	85R0228	1	**21**
Bacillis thuringiensis culture	85R0203	1	**21**
Bean seeds, pinto, 1 lb pkg.	86R8014	400 seeds	**1**
Bean seeds, French	86R8004	48 seeds	**27**
Beta (beet) root (cs) slide	91R7980	8	QL575
Brine shrimp (*Artemia*) eggs, 6 g	87R5102	400 mL	**18**
Broth, concentrated, beef	Local	8 mL	**29**
Chameleons, American, 3/pkg.	87R8135	16	**35**
Clamshell	Local	8	**30**
Clostridium tetani (sm) g(+) slide	90R2032	4	**20**
Coleus	86R6800	16	QL599
Coleus stem (cs) slide	91R7831	8	QL575
Corn kernels, dry	Local	8	QL253
Crickets (*Acheta domestica*), 50/pkg.	87R6102	8	**17**, **34**, QL844
Daphnia magna culture, mature	87R5210	1	29, 43
Dianthus—entire leaf (cs) slide	91R7882	8	QL576
Diatomaceous earth, jar 30 oz	63R0240	8 oz	QL483
DNA extraction biolab	36R6026	1	QL193
Earthworms (*Lumbricus*), 10/pkg.	87R4660	8	**17**
Elodea, 12/pkg.	86R7500	8 sprigs	**3**, QL101
Fern antheridia and archegonia	91R4865	8	**24**, QL55
Fruits and vegetables	Local	8	QL535
Frogs, small unsexed, 12/pkg.	87R8217	8 frogs	**34**
Fungi demo plate, set of 6	85R3912	1 set	**23**
Grapes	Local	8	QL76, QL784
Grass seed, rye, 1 oz pkg.	86R8130	8 pinches	**17**
Green beans or snow peas	Local	8	**16**
Hydra culture, brown	87R2020	1	**29**
Hydra, plain (wm) slide	92R0630	8	**28**
Isopod (sowbug) culture, 45/pkg.	87R5520	32	**17**, **31**, QL838
Lemna minor (duckweed), 4 oz jar	86R7650	1	**25**
Leucosolenia (sponge) (wm) slide	92R0568	8	**28**
Lilium root (cs) slide	91R7200	8	QL575
Lilium stem (cs) slide	91R7204	8	QL575
Mealworm larvae, 100/pkg.	87R6250	8	**17**
Micrococcus luteus (sm) g(+) slide	90R0554	4	**20**
Mitosis, plant (sect) slide	93R2145	8	QL132
Monocot flower and dicot flower	Local	8 of ea.	QL556
Mung bean seed, Oriental, 1 lb pkg.	86R8013	48 seeds	**17**
Nematode, typical (wm) slide	92R1375	8 slides	**28**
Obelia medusae (wm) slide	92R0731	8	ML641
Pine ovulate cone (ls) qs slide	91R6534	8	QL554
Pine ovule—young archegonium (ls)	91R6540	8	QL554
Pine staminate cone (ls) qs slide	91R6528	8	QL554
Planaria culture, brown	87R2500	1	QL649
Planaria, plain (wm) slide	92R0820	8	**28**
Plants kept in light and dark	Local	2	QL521
Protists, mixed (wm) slide	92R0005	4	**20**
Protozoa survey mixture culture	87R1610	1	**22**
Raisins, dry	Local	8	QL766
Skin of hairy mammal (sect)	93R5023	8 slides	**36**
Skull, beaver, natural bone	65R6450	1	**36**
Skull, bullfrog, natural bone	65R5122	1	**36**
Skull, canine, adult, natural bone	65R5310	1	**36**
Skull, cat, adult, natural bone	65R5220	1	**36**

Description	Item no.	Qty per class	Lab
Skull, freshwater dogfish (*Amia*)	65R1410	8	**36**
Skull, bat, flying fox replica	80R1715	1	**36**
Skull, snake, natural bone	65R5124	1 skull	**36**
Skull, turtle, natural bone, w/key	65R5123	1 skull	**36**
Squamous epithelium (sm) slide	93R6003	8 slides	**3**
Starfish set, embryology, set of 13	95R1551	8 sets	**32**
Survey set, general, jar mount	67R9904	1	**20**, **24**
Survey set, invertebrate	62R0044	1	**28**
Sweet corn seed, untreated 200/pkt.	86R8080	48 seeds	**27**

Chemicals and Media

Description	Item no.	Qty per class	Lab
Adrenalin, epinephrine solution	38R2052	40 ml	**43**
Agar (red), 6 bottles/pkg.	88R8902	1 bottle	QL149
Ascorbic acid, L(+) vitamin C	39R2005	10 mL	**41**
Baking soda, 1 lb pkg.	37R5467	16 oz	**1**, QL899
DETAIN™ protist-slowing agent	37R7950	1	**18**, **22**
Detergent, laundry, 5 different brands	Local	1	**2**
Detergent, liquid household, 14 oz	37R2268	80 drops	QL919
Distilled water, 1 gal	88R7005	4,000 mL	**1**, **2**, **25**, QL101,QL17, QL899, QL76
Ethyl alcohol (ethanol), anhydrous	39R0273	40 mL	QL193
Gelatin, instant	Local	8 pkg.	**2**
Gelatin solution, 1%, 250 mL	32R8501	40 mL	QL257
Glucose solution, standard, 100 mL	39R1458	80 mL	**40**
Grape juice, white	Local	1	QL76
Gum arabic solution, 1%, 250 mL	32R8502	24 mL	QL257
Hydrochloric acid, 0.1 M sol.	37R9561	80 mL	**30**, QL257
Indophenol solution, 0.1%	37R9542	1	**41**
Knops solution, 500 mL	38R3520	1	**25**
Lemon juice, 8 oz	37R8542	16 oz	**1**
Liquid soap, antibacterial, 8 oz	18R1533	240 mL	QL17
Lugol's iodine solution, 500 mL	39R1685	120 mL	**3**
Methylene blue chloride, 1% aq. sol.	38R9522	80 mL	**19**, **21**
Milk treatment product liquid	Local	64 drops	**40**
Nail polish, clear, 15 cc	37R1862	1 bottle	QL521
Nutrient agar, 6 bottles /pkg.	88R1500	1 bottle	**23**
Paints, hobby, 5 colors	Local	8 sets	QL844
Petroleum jelly, 1 oz	15R9832	1 oz	**43**
Phenolphthalein agar, 800 mL	88R0928	1	**4**
Plant food, Miracle-Gro, 1.5 lb	20R6020	1 pkg.	**25**
Pond water, 1 gal	88R7010	1	**25**, QL649
Preserved specimen conditioner, 500 mL	39W1697	80 mL	**1**
Propionic acid, 500 mL	39R2761	1	**23**
Rubbing alcohol, 16 oz	39R4915	1 bottle	**21**, **23**, QL17
Silicone culture gum, 25 mL bottle	37R9810	1	**29**
Sodium carbonate, anhydrous	37R5477	0.7 g	**2**
Sodium hydroxide 0.1 % sol.	37R9562	16 oz	**1**
Spring water, 1 gal	88R7000	5 gal	**34**
Sugar, granular	39R3180	32 oz	QL76
Vegetable coloring, red	37R1597	8 drops	**30**, QL663, QL670
Vegetable oil	37R9540	160 mL	QL919
Vinegar, white	39R0138	1,200 mL	**1**, **4**, QL17
Yeast, viable baker's, 10 g packet	88R0929	10 g	**19**, QL149

Laboratory Equipment

Description	Item no.	Qty per class	Lab
Apron, polyethylene, disposable, 100/box	15R1050	32	**1**, **2**, **3**, **4**, **19**, **20**, **21**, **23**, **25**, **26**, **28**, **30**, **40**, **41**, **43**,QL17, QL257, QL940, QL988
Aquarium, 10 gal, 20 × 10 × 12 in.	1R5241	8	**34**
Aquarium/terrarium, medium, covered	1R2101	8	**34**
Aquarium/terrarium, large, covered	21R2102	8	QL844
Aquarium/terrarium, small, covered	21R2100	8	**17**
Beaker, student grade, 50 mL (Griffin)	17R4010	48	**2**, QL17, QL502
Beaker, student grade, 100 mL (Griffin)	17R4020	8	**43**, QL766
Beaker, student grade, 150 mL (Griffin)	17R4030	16	**2**, **21**, **27**
Beaker, student grade, 250 mL (Griffin)	17R4040	16	**1**, **4**, **25**, QL811, QL899, QL919
Beaker, student grade, 400 mL (Griffin)	17R4050	8	**3**,**18**, QL395
Beaker, student grade, 600 mL (Griffin)	17R4060		**8**, **34**, QL230, QL988
Blood typing kit, ABO simulated	36R0022	1	QL940

Description	Item no.	Qty per class	Lab
Bunsen burner, standard, natural gas	15R0612	8	**21**
Calculator	27R3055	8	**16**, **18**
CBL system	Local	8	QL101, QL395, QL409, QL811
CBL probe, temperature	Local	8	QL395, QL409, QL811
CBL probe, dissolved oxygen	Local	8	QL101
Chromatography of simulated plant pigments demo kit	36R6019	1	**26**
Clamp, Hoffman, screw compressor, open jaw	15R3910	24	**18**
Corks, size 2, 100/pkg.	15R8362	16	**18**
Coverslips, 22 mm plastic, 100/box	14R3555	80	**19**, **22**, **3**, **43**, QL149,QL257, QL483, QL502
Dropping bottle, 15 mL, amber glass	17R6020	8	**3**, QL257
Dropping bottle, 15 mL, flint glass	17R6018	16	**21**, **41**
Field guide, Peterson, *First Guide to Wildflowers*	32R0381	8	QL324
Filter paper, medium grade, 9.0 cm	15R2815	8	**29**
Forceps, dissecting, straight	14R1001	8	**21**, **22**, **3**, QL784
Forceps, fine point w/guide pin	14R0541	8	**29**, QL649
Funnel, utility, 60 mm, 2 3/4 in. top	18R1421	8	**18**
Gauze, wire 4 × 4 in., 12/pkg.	15R0670	8	QL988
Gloves, heat resistant, one size	15R1095	8	**2**, QL988
Gloves, latex, dispos. med., 100/pkg.	15R1071	16	**1**, **2**, **3**, **4**, **17**, **19**, **20**, **21**, **23**, **25**, **26**, **28**, **29**, **30**, **40**,**41**, **43**, QL17, QL193, QL257, QL502, QL556, QL940
Grad. cylinder, polypropylene, 10 mL	18R1705	16	**43**
Grad. cylinder, polypropylene, 100 mL	18R1730	8	**2**
Grad. cylinder, polypropylene, 25 mL	18R1710	8	QL257, QL919
Hot plate	15R7999	8	QL395, QL988
Hot water bottle, 2 qt	14R8306	8	**18**
Ice bag, 6 in. diameter	14R8305	8	**18**
Magnifier, dual 3X and 6X	24R1112	32	**20**, **24**, **25**, **28**, **36**, QL502, QL554, QL649
Meterstick, wood	15R4065	8	**12**, **13**, QL685

Description	Item no.	Qty per class	Lab
Microscope slide, single concavity, 12/pkg.	14R3510	40	**40**, **43**
Microscope slides, 72/pkg.	14R3500	16	**3**, **21**, **22**, **29**, QL149, QL257, QL483, QL502, QL521
Microscope, compund light	24R2310	8	**19**, **20**, **21**, **22**, **24**, **28**, **32**, **36**, **43**, ML641, QL132, QL149, QL257, QL483, QL502, QL521, QL550, QL554, QL575, QL576, QL720
Petri dish, disposable, 25/pkg.	18R7101	40	**18**, **23**, **25**, **30**, **31**, QL784
pH paper strips 1.0–14.0, 50/pkg.	15R3157	56 strips	**2**, QL17, QL364, QL899
Pipet, graduated polyethylene, 6 in., 500/pkg.	18R2971	40	**22**, QL483, QL502
Pipet, graduated, dropping, 2 mL	17R1706	16	**18**, **19**, **2**
Pipets, graduated, polyethylene, 100/pkg.	18R2977	8	QL193
Plant light	20R7500	8	**18**
Potting soil, growth media, 8 lb bag	20R8306	8 lb	**17**
Probe, Huber, stainless, 6 in.	14R0958	8	QL838
Probe, stainless steel	14R0950	8	**28**, **31**
Reaction chamber, 15/pkg.	15R9861	8	QL663
Rubber stopper, size 0, black, solid	15R8460	32	**18**
Rubber stopper, size 0, black, 1 hole	15R8480	16	QL395
Safety goggles, general purpose	15R3046	32	**1**, **2**, **3**, **4**, **19**, **20**, **21**, **23**, **25**, **26**, **28**, **30**, **40**, **41**, **43**, QL17, QL193, QL257, QL940, QL988
Sand, coarse, 32 oz container	45R1984	8	**17**
Sand, fine white Ottawa	20R7423	32 oz	QL395
Scalpel, sterile, No. 22 blade	14R0900	8	**27**, **30**
Scissors, economy, dissection	14R0525	8	**6**, **7**, **9**, **11**, **13**, **22**, **29**, **31**, ML737, QL124, QL145, QL208, QL216, QL663
Specimen dish, 4 1/2 × 2 3/4 in.	17R0550	8	QL649
Specimen dish, 8 1/4 × 3 1/4 in.	17R0560	8	**30**
Spirometer	14R5070	8	**39**
Spirometer mouthpieces, 250/pkg.	14R5072	32	**39**
Stage micrometer, ruled, 2 mm scale	94R9910	8	**19**

Description	Item no.	Qty per class	Lab
Stereomicroscope, beginner	24R4601	8	**23**, **27**, **29**, **30**, **31**
Stir rods, 6 in., glass, 150 × 5 mm, 10 pack	17R6005	8	**2**, **30**, QL193, QL899, QL919, QL988
Stopwatch, digital	15R0512	8	**31**, **37**, **43**, QL253, QL282, QL857
Support, rectangular with 4 × 6 in. base	15R0719	8	QL988
Support ring , 4 in. OD, zinc plated	15R0709	8	QL988
Swab applicator, 6 in., 100/pkg.	14R5502	24	**21**
Test paper, glucose, 50 strips/pkg.	14R4107	40 strips	**40**
Test-tube clamp, Stoddard, 6/pkg.	15R0841	8	QL395
Test-tube rack, Nalge, 16 mm	18R4213	8	**18**, **2**, **21**, QL193, QL395
Test-tube with rim, Pyrex, 15 × 125 mm	17R0620	32	**18**, **2**, **41**, QL101, QL193, QL257, QL395
Thermometer, lab, red alcohol	15R1462	8	**2**, QL988
Tongs, utility, 8 in.	14R0960	8	QL988
Tray w/cover, polyethylene, 1 gal	18R3650	8	**4**
Triple beam balance, Ohaus	15R6057	8	**16**, **2**, QL76, QL784, QL899
Tubing, amber latex 3/16 × 1/16 in., 10 ft	15R1133	48	QL663, QL670
Tubing, vinyl, 1/2 × 1/16 in., 10 ft	18R5084	16 ft	**18**, QL838
Tubing, vinyl, 3/16 × 1/16 in., 10 ft	18R5081	8 ft	QL663, QL670

Miscellaneous

Description	Item no.	Qty per class	Lab
Aluminum foil roll, 12 in. × 25 ft.	15R1009	16 ft.	**18**
Beads, 12 mm, approx. 100/pkg.	36R4197	8	QL230
Beads, 4 mm, approx. 100/pkg.	36R4195	8	QL230
Beads, 8 mm, approx. 100/pkg.	36R4196	8	QL230
Beads, wooden, set of 100	15R0504	200	**6**, **7**
Cardboard	Local	8	QL844
CD-ROM, *Interactive Explorations in Biology: Cell Biology and Genetics*	Local	8	**5**, **44**
CD-ROM, *Interactive Explorations in Biology: Human Biology*	Local	8	**8**, **33**, **38**, **42**
Computer with CD-ROM	Local	8	**5**, **8**, **33**, **38**, **42**, **44**
Connector, twistcock LDPE	18R1571	16	QL663, QL670
Construction paper, assorted 50/pkg.	15R9841	48	**13**, **35**
Construction paper, black, 50/pkg.	15R9825	8	**22**
Corrugated box, 10/pkg., 11 1/8 in.	18R1395	8	QL253
Drill for making holes	Local	1	**35**, QL409
Fabrics, different textures	Local	8 sets	**31**
Flashlight, 2-D cell	15R3264	8	QL101, QL483
Gas lighter, flat file, 10/pkg.	15R0683	8	**21**
Graph paper, 5 squares/inch, 100 sheets	15R3835	40 sheets	**1**, **16**, **44**, **5**, ML351, QL811
Ice	Local	8 bags	QL811, QL988
Index cards, unlined, 3 × 5 in., 100/pkg	15R9819	32	**10**, QL124, QL349, QL966
Jar, clear polystyrene, 2 oz, 53 mm	18R1632	8	QL844
Jar, clear polystyrene, 4 oz, 58 mm	18R1633	24	QL76
Jar, 1/2 gal wide-mouth, glass	17R2070	16	**35**, QL230
Jar, 1 gal wide-mouth, glass	17R2080	16	QL599
Jar, wide-mouth, glass, no cap, 950 mL	17R2060	8	QL364, QL409
Jar cap, white metal, 58 mm	17R2137	24	QL76
Jar cap, white metal, 89 mm	17R2153	8	QL364 , QL409
Jar cap, white metal, 83 mm w/liner	17R2149	16	**35**
Jar cap, 120 mm, white metal w/liner	17R2157	16	QL599
Knife, plastic	25R8128	8	**4**, QL535
Labels, lab 7/8 × 1 7/8 in.	15R1829	160	**15**
Lamp clamp with reflector	36R4168	8	**22**, QL409
Light bulb, 150 W, 120 V, clear	36R4173	8	**22**, QL409
Liver, raw, small pieces	Local	8	QL649
Marker, black lab	15R3083	8	**13**, **18**
Marking pen, black wax, for glass	15R1155 15R1157	8 8	**1**, **2**, **21**, **23**, **25**, **27**, **35**, QL521, QL76, QL899, QL919
Masking tape, 3/4 in. × 60 yds, 3/pkg.	15R9828	16 ft	**10**, **23**, QL216, QL303, QL409
Milk, whole	Local	4 L	**40**, QL364
Mirror, 3 × 4 in.	15R9860	8	**36**
Moldy bread	Local	8	QL502
Pan, aluminum foil, square, 10/pkg.	15R9891	8	ML737, QL670
Paper	Local	8	QL556

Description	Item no.	Qty per class	Lab
Paper bag, 5 in. × 10 1/2 in. × 3 in., 50/pkg.	18R6900	8	QL349
Paper clips, #1, 10 boxes/pkg.	15R9815	448 clips	**10**, **11**, **9**
Paper punch, one hole	15R9810	8	**22**
Paper towel roll, 2 ply, 90 sheets	15R9844	40 sheets	**1**, **4**, **21**, **23**, **27**, **43**, QL502, QL76, QL838
Paper, white	Local	8	**22**
Pencils, colored, set of 12	15R4690	8 pkg.	**10**, **12**, QL145, QL208, QL216
Pencils, 12/box	15R9816	16	QL303
Penny or other coin	Local	8	**13**
Photographs of wildflowers	Local	8	QL324
Pictures of various organisms	Local	8	**12**
Pipe cleaners, colored, 6 in. 12/pkg.	82R1116	8 pkg.	**6**, **7**
Pipet, dropping, 3 in., glass, 12/pkg.	17R0230	8	**3**, **29**, **30**, **40**, **43**, QL149, QL649, QL663, QL670
Plastic wrap, 12 in., 50 ft roll	15R9858	24 ft	**2**, ML737
Protractor, student, semicircle	15R4067	8	**14**
Pushpins, assorted colors, 100/box	15R0507	200	**10**, **11**, **9**
Rope, clothesline, 5/32 in. × 96 ft	15R3991	30 ft	QL437
Rubber bands, assorted, 1/4 lb	15R9824	16	**27**
Ruler, clear vinyl, 15/16 × 6 1/8 in.	14R0811	8	ML641
Ruler, standard 12 in.	15R4650	8	**1**, **2**, **4**, **9**, **14**, **16**, **18**, **22**, **27**, ML737, QL216, QL409
Screening, fiberglass, 36 × 36 in.	15R0002	8	**18**
Six-sided die	Local	8	**13**
Socks, wool and cotton	Local	8	QL811
Soft drink, clear, cold, carbonated	Local	8	QL766
Spoons, plastic, 100/pkg.	15R9800	8	**4**
Straws, plastic wrapped, 500/pkg. straws	15R9869	160	**10**, **11**, **13**, **9**, QL899
Straws, plastic, colored	Local	8	**10**
Tags, 1 3/4 × 1 1/8in., 1,000/box	15R1873	80	**6**, **7**
Tape measure, metric/English, 5 ft	15R2541	8	ML351
Tape, adding machine, 2 1/4 in., 5/pkg.	27R3076	40	**12**
Tape, transparent w/self-dispenser	15R1959	8	**2**, **10**, **13**, **18**, **31**, QL124, QL145, QL208, QL521, QL556,
Toothpicks, flat, 750/box	15R9864	40 toothpicks	**22**, **23**, **40**, ML737, QL483, QL502
Tracing paper, 40 sheets	15R9814	16 sheets	QL208
Water, boiling	Local	8	**2**
Yarn, red, 10 yd skein	15R2544	8	**6**, **7**
Zipper bags, resealable, 4 × 6 in., 10/pkg.	18R6921	40 bags	**1**

UNIT 1 Principles of Cell Biology

Chapters

As the runner sprints toward the finish line, 25 trillion red blood cells carry oxygen throughout her body.

in perspective

Blood: A River of Cells

Yesterday...

In the winter of 1667, a French physician named Jean-Baptiste Denis tried a daring new experiment. After examining a man who exhibited fits of rage, Denis transfused the patient with the blood of a gentle calf. His actions reflected the beliefs of his time—that blood carried the characteristics of the creatures in which it flowed.

Illustration from a 1692 medical textbook

Today...

Biologists know that red blood cells contain a protein called hemoglobin. **Find out on page 36 how proteins like hemoglobin form their unique shapes.**

In persons with sickle cell anemia, an incorrect form of hemoglobin is made. When this happens, red blood cells become sickle-shaped and cannot adequately perform their job. The shape of the hemoglobin molecule is the key to its role—carrying oxygen throughout the body. **Discover on page 104 why the cells of your body need oxygen.**

Hemoglobin

Tomorrow...

Researchers are discovering new treatments for sickle cell anemia. New drugs that "turn on" the body's production of normal hemoglobin have shown promise in easing the symptoms of sickle cell anemia.

Sickled red blood cell

internet**connect**

TOPIC: Sickle cell anemia
GO TO: www.scilinks.org
KEYWORD: HX003

Chapter 1

Biology and You

Planning Guide

	OBJECTIVES	MEETING INDIVIDUAL NEEDS	READING SKILLS AND STRATEGIES
BLOCKS 1 & 2 90 min	**1-1 Themes of Biology** • Relate the seven properties of life to a living organism. • Describe seven themes that can help you organize what you learn about biology. • Identify the tiny structures that make up all living organisms. • Differentiate between reproduction and heredity and between metabolism and homeostasis.	Learning Styles, p. 6 BASIC ADVANCED • pp. 8, 9	ATE Active Reading, Technique: Brainstorming, p. 5 ■ Active Reading Guide, worksheet (1-1) ■ Directed Reading, worksheet (1-1) ■ Reading Strategies, Getting to Know Your Textbook
BLOCKS 3 & 4 90 min	**1-2 Biology and Your World** • Evaluate the role biologists play in saving our tropical forests. • Describe the role biologists play in trying to increase and improve our food supply. • Describe efforts to combat three diseases that are the subject of current scientific research.	Learning Styles, p. 10 BASIC ADVANCED • pp. 12, 13	ATE Active Reading, Technique: Reading Organizer, p. 12 ■ Active Reading Guide, worksheet (1-2) ■ Directed Reading, worksheet (1-2) ■ Reading Strategies, Spider Map
BLOCKS 5 & 6 90 min	**1-3 The Scientific Process** • Describe the stages common to scientific investigation. • Distinguish between forming a hypothesis and making a prediction. • Differentiate a control group from an experimental group and an independent variable from a dependent variable. • Define the word *theory* as used by a scientist.	Learning Styles, p. 14 BASIC ADVANCED • pp. 18, 19, 20	■ Active Reading Guide, worksheet (1-3) ■ Directed Reading, worksheet (1-3) ■ Supplemental Reading Guide, *The Lives of a Cell: Notes of a Biology Watcher* and *Silent Spring*

Review and Assessment Resources

BLOCKS 7 & 8 90 min

TEST PREP

- PE Study Zone, p. 21
- PE Performance Zone, pp. 22–23
- ■ Chapter 1 worksheets:
  - Vocabulary
  - Science Skills
  - Concept Mapping
  - Critical Thinking
- Audio CD Program
- ■ Test Prep Pretest

CHAPTER TESTING

- ■ Chapter Tests (blackline master)
- ▲ Test Generator
- ■ Assessment Item Listing
- ■ Performance-Based Assessment Activity: Effect of Acid Rain on Trout Populations

See the correlations table at the front of this book for a list of the ***National Science Education Standards*** addressed in this chapter.

Smithsonian Institution® Visit **www.si.edu/hrw** for additional on-line resources.

KEY

Biology Principles and Explorations

PE Pupil's Edition

ATE Teacher's Edition

BioSources

Teaching Transparencies

Laboratory Program

One-Stop Planner CD-ROM
Shows these resources within a customizable daily lesson plan:

■ Teaching Resources

▲ Test Generator

VISUAL STRATEGIES	CLASSWORK AND HOMEWORK	TECHNOLOGY AND INTERNET RESOURCES
ATE Fig. 1-3, p. 7	Lab A2 Comparing Living and Nonliving Things Lab B1 Introduction to Experimental Design and Data Presentation **PE** Review, p. 9	Holt Biology Videodiscs Section 1-1 Audio CD Program Section 1-1
	PE Math Lab Evaluating the effectiveness of insecticides, p. 11 Lab A1 Imagining Solutions: Problem Solving **PE** Review, p.13	Holt Biology Videodiscs Section 1-2 Audio CD Program Section 1-2 internetconnect SCILINKS NSTA **TOPIC:** Population growth
ATE Fig. 1-14, p. 16 1 Revision of a Theory: Continental Drift and Plate Tectonics	**PE** Quick Lab What is the pH of some common substances?, p. 17 **PE** Data Lab Analyzing experimental design, p. 20 **PE** Experimental Design How Does Acid Rain Affect Seeds?, pp. 24–25 ■ Occupational Applications: Biology Teacher **PE** Review, p. 20 **PE** Study Zone, p. 21	Holt Biology Videodiscs Section 1-3 Audio CD Program Section 1-3 CNN Video Segment 1 Frog Pollution internetconnect SCILINKS NSTA **TOPICS:** • Biology careers • Acid rain

ALTERNATIVE ASSESSMENT

PE Portfolio Projects 24–26, p. 23

ATE Alternative Assessment, p. 21

ATE Portfolio Assessment, p. 21

Lab Activity Materials

Math Lab
p. 11 paper, pencil

Quick Lab
p. 17 paper, pencil, wide-range pH paper, 3 different solutions, beaker or small jar, water

Experimental Design
p. 24 safety goggles, lab apron, seeds (50), 250 mL beakers, 20 mL mold inhibitor, distilled water, paper towels, solutions of different pH, wax pencil or marker, resealable plastic bags, metric ruler, graph paper

Demonstrations

ATE culture of *Plasmodium polycephalum,* p. 9

ATE slide of paramecium, microprojector, p. 12

ATE collection of equipment used in biology, such as a computer, standard lab equipment, binoculars, animal traps, journal notebooks, p. 19

BIOSOURCES LAB PROGRAM
- Quick Labs: Lab A1, p. 1
- Quick Labs: Lab A2, p. 3
- Inquiry Skills Development: Lab B1, p. 1

Chapter 1

Chapter Theme

Students will encounter seven themes that occur throughout the book: Cellular Structure and Function, Reproduction, Metabolism, Homeostasis, Heredity, Evolution, and Interdependence. As students progress through their study of biology, they should become proficient at recognizing the underlying themes of each chapter.

Establishing Prior Knowledge

At the beginning of each chapter is a study tip titled **Study TIP Ready?** Except for Chapter 1, this study tip contains topics and terms students should be familiar with before reading the chapter. Make students aware of the location of this study tip. Have students locate the study tips found in the margins in Chapter 1 (*pp. 9 and 17*).

Opening Demonstration

Write the following on the board: Biology is the study of life. Under this write the following list: Taking care of your pet. Recycling your soda can. Watering your lawn. Eating breakfast. Taking medicine. Visiting a zoo. Ask students: How do all of these activities involve biology? *(Answers may vary. Encourage students to realize how biology applies to their everyday activities.)*

CHAPTER 1 Biology and You

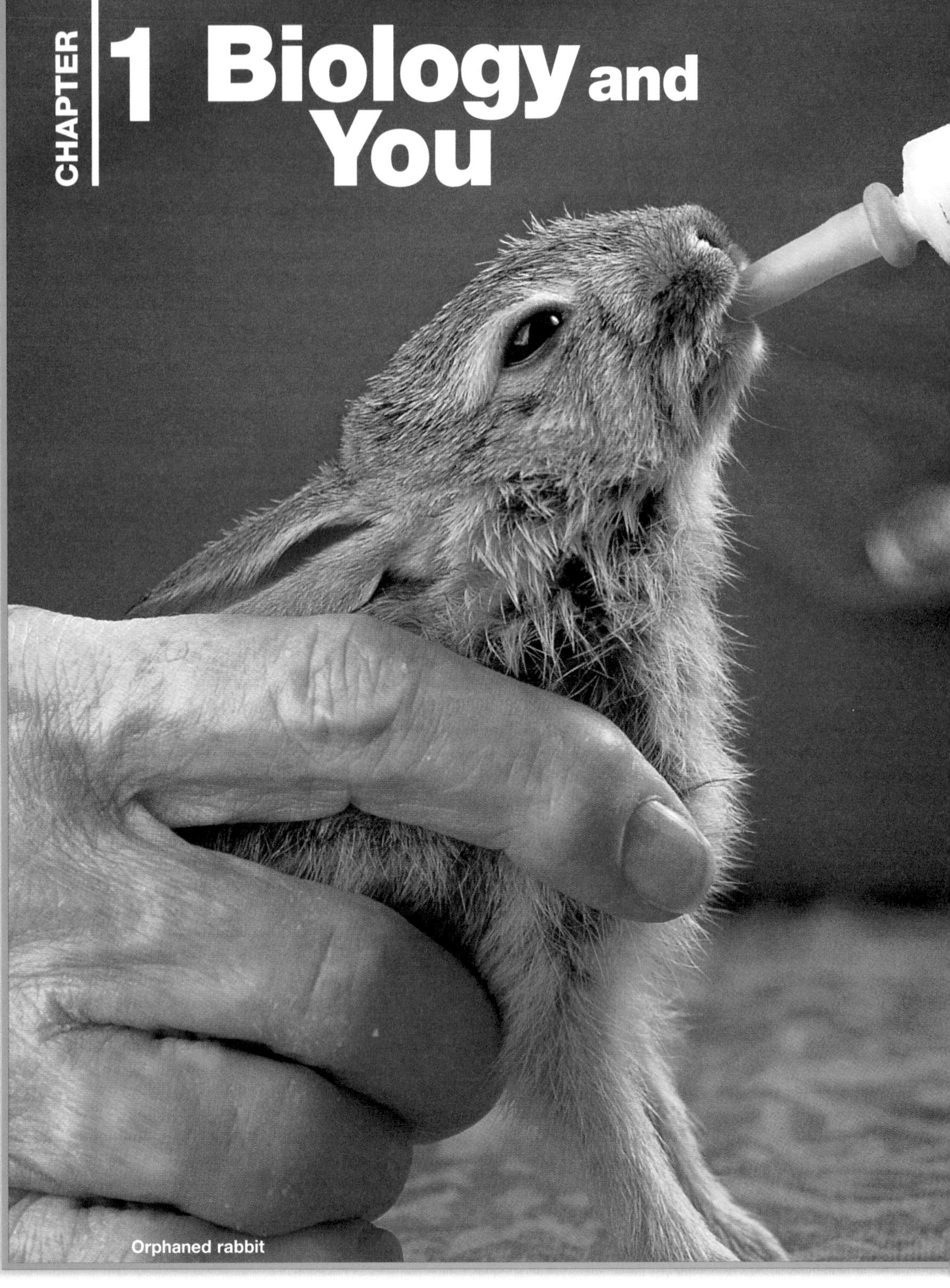

Orphaned rabbit

What's Ahead — and Why
CHAPTER OUTLINE

1-1 Themes of Biology
- Living Organisms Have Certain Characteristics in Common
- Seven Themes Unify the Science of Biology

1-2 Biology in Your World
- Biology Can Help Solve Today's Real-World Problems
- Biology Can Help Fight Diseases

1-3 The Scientific Process
- Science Is Based on Careful Observation
- Several Stages Are Common to Scientific Investigations
- Theories Are Ideas Supported by a Great Deal of Evidence

Students will learn

- how to determine if something is living or nonliving.
- how biology plays an important role in everyday life.
- how scientists plan investigations.

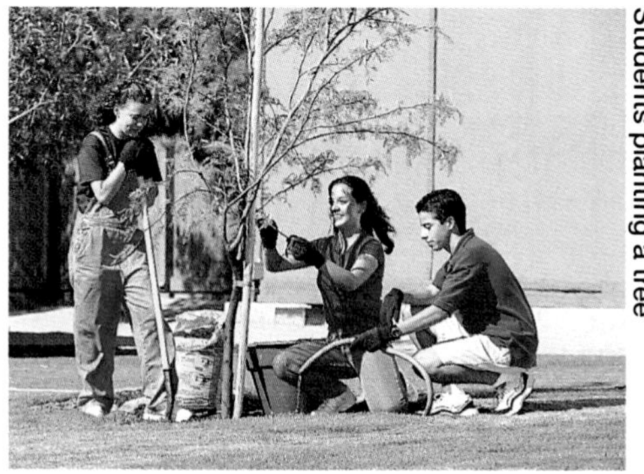
Students planting a tree

Food... Water... Oxygen...

Just as this orphaned rabbit needs food and water to live and grow, it also depends on oxygen from plants to survive. Our knowledge of biology helps us understand how all life on Earth is connected. This knowledge can be used for many purposes, from wiping out deadly diseases to making sure there is enough clean air to breathe.

Looking Ahead

Study TIP

Ready?

- *Study Tips in each chapter will help you read effectively, review and organize information, understand word origins, and interpret graphics. See the Study Skills appendix for more information.*

internet**connect**

sci*LINKS* NSTA — National Science Teachers Association *sci*LINKS Internet resources are located throughout this chapter.

Reading Strategies

Active Reading

Technique: Brainstorming

Before students begin Chapter 1, write the word *biology* on the board. Using a dictionary, ask students to define the prefix *bio-* (*life*) and the suffix *-logy* (*study of*). Next ask the students to generate a list of branches of biology that include the suffix *-logy*. (*Examples might include microbiology and histology.*) Have students use the **Appendix** on pp. 1031–1033 to define the prefixes (*small, tissue*). Finally, ask students to describe an area of biology for which they do not know the name. (*Example might include the study of the kidneys.*) Have them use the Appendix to find the specific name (*nephrology*).

Directed Reading

Assign **Directed Reading Worksheet Chapter 1** to help students interact with the text as they read the chapter.

Interactive Reading

Assign ***Biology: Principles and Explorations* Audio CD Program Chapter 1** to help reluctant readers achieve greater success in reading the chapter.

Reading Skills

Assign Reading Organizers and Reading Strategies Worksheets from the **One-Stop Planner CD-ROM with Test Generator** to help students learn and practice effective reading skills

internet**connect**

Holt *Biology: Principles and Explorations*

On-line Resources: go.hrw.com

Visit the HRW Web site for a variety of resources related to this chapter. Just type in the keyword HX0 Chapter 1.

Holt *Biology: Principles and Explorations*

On-line Resources: www.scilinks.org

The following *sci*LINKS Internet resources can be found in the student text for this chapter:

Page 10
TOPIC: Population growth
KEYWORD: HX010

Page 15
TOPIC: Biology careers
KEYWORD: HX015

Page 25
TOPIC: Acid rain
KEYWORD: HX025

Chapter 1

1 Prepare

Scheduling

- **Block:** About 90 minutes will be needed to complete this section.
- **Traditional:** About two class periods will be needed to complete this section.
- **Planning:** See the Planning Guide on pp. 3A and 3B for lecture, classwork, and assignment options for Section 1-1. Lesson plans for Section 1-1 can be found in a lesson cycle format in ***Biology: Principles and Explorations*** **One-Stop Planner CD-ROM with Test Generator.**

Resource Link

Assign **Active Reading Worksheet Section 1-1** to help students comprehend the key concepts and visuals in this lesson.

2 Teach

Lesson Warm-up

Have each student list characteristics that he or she believes all living things share. An example must be given for each item on the list. Next have the student go back over the list and determine if any nonliving object might have the same characteristic. Again, a specific example must be given. Poll the students and write the remaining items on the board. Discuss each one, removing those that are not valid. Emphasize that living organisms may not have all of the seven characteristics at one time (such as the ability to reproduce), but rather they are *capable* of having all characteristics at some point in their life. PORTFOLIO

Section 1-1 Themes of Biology

Objectives

- **Relate** the seven properties of life to a living organism. (p. 6)
- **Describe** seven themes that can help you organize what you learn about biology. (pp. 7–9)
- **Identify** the tiny structures that make up all living organisms. (p. 7)
- **Differentiate** between reproduction and heredity and between metabolism and homeostasis. (pp. 7–8)

Key Terms

biology
cell
reproduction
metabolism
homeostasis
gene
heredity
mutation
evolution
species
natural selection
ecology

Living Organisms Have Certain Characteristics in Common

You are surrounded by living things. Many of these, such as people, plants, and animals, are obvious. Other living things are so small that you cannot even see them with the naked eye. How do we know if something is alive? What does it mean to be alive?

While most people are capable of distinguishing between living and nonliving, actually defining life can be quite difficult. Perhaps you consider movement, sensitivity, development, and even death as characteristics of living organisms. While present in all living things, these properties are not enough to describe life.

Clouds, for example, move when stimulated by the wind and develop from moisture that is suspended in the atmosphere. Clouds grow and change their shapes. Some might view the breakup of clouds as being similar to death. However, disorder is not the same as death. Clouds may break up and vanish, but they do not die.

Biology is the study of life. Biologists recognize that all living organisms, such as the cheetahs shown in **Figure 1-1,** share certain general properties that separate them from nonliving things. As summarized in Figure 1-1, every living organism is composed of one or more cells, is able to reproduce, and obtains and uses energy to run the processes of life. Living organisms also maintain a constant internal environment and pass on traits to offspring. Responding and adjusting to the environment as well as growing and developing are other characteristics shared by all living organisms.

As you read further, you will have an opportunity to think more about the properties that help define life. Life is characterized by the presence of all of these properties at some stage in an organism's life, not just by one or two. Remember this fact as you attempt to determine what is living and what is not.

Figure 1-1 What does it mean to be alive? Life is characterized by the presence of all seven of these properties at some stage in an organism's life.

Properties of Life

- Cellular organization
- Reproduction
- Metabolism
- Homeostasis
- Heredity
- Responsiveness
- Growth and development

Integrating Different Learning Styles

Visual/Spatial	**ATE**	Closure, p. 9
Interpersonal	**ATE**	Biology in Your Community, p. 8
Intrapersonal	**PE**	Study Tip, p. 9
Verbal/Linguistic	**ATE**	Resource Link, p. 8; Reteaching, p. 9

Seven Themes Unify the Science of Biology

From the study of biology, themes emerge that serve to both unify and explain biology as a science. As you study biology in this text, you will repeatedly encounter these themes.

Theme 1 Cellular structure and function

All living things are made of one or more cells. **Cells** are highly organized, tiny structures with thin coverings called membranes. A cell is the smallest unit capable of all life functions. The basic structure of cells is the same in all organisms, although some cells are more complex than others. Some organisms have only a single cell, while others are multicellular (composed of many cells). Your body contains more than 100 trillion cells. **Figure 1-2** shows a single-celled organism called a paramecium.

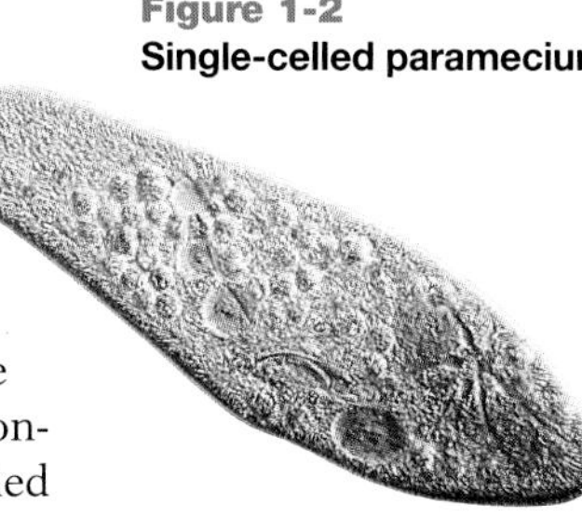

Figure 1-2
Single-celled paramecium

Theme 2 Reproduction

All living things can reproduce. **Reproduction** is the process by which organisms make more of their own kind from one generation to the next. Rapidly growing bacteria divide into offspring cells approximately every 15 minutes, and bristlecone pine trees that are 5,000 years old still produce seedlings. Since no organism lives forever, reproduction, as represented in **Figure 1-3,** is an essential part of living.

Figure 1-3
Hatchling snakes

Theme 3 Metabolism

Living organisms perform many different chemical reactions in order to obtain and use energy to run the processes of life. All living things use energy to grow, to move, and to process information. Without energy, life soon stops. **Metabolism** is the sum of all of the chemical reactions carried out in an organism.

Almost all the energy used by living organisms is captured from sunlight. Plants, algae, and bacteria capture this solar energy and use it to make complex molecules in a process called photosynthesis. These molecules then serve as the source of energy, or food, for other organisms. For example, paramecia, such as the one in Figure 1-2, eat bacteria. Humans eat plants or animals that eat plants. Energy flows from the sun to plants, from these plants to plant-eating organisms, and from plant-eating organisms to meat-eating organisms. The students in **Figure 1-4** are extracting energy from the food they eat.

Figure 1-4
Extracting energy from food

Demonstration
Is It Alive?

Obtain a culture of *Plasmodium polycephalum,* a slime mold. Do not identify it for the students. Have students use their unaided eyes to attempt to determine if it is living or nonliving. Do not let them touch or smell it. Next ask students to concentrate on how they could determine if it is a living organism. (*Answers may include reproduction, growth, and movement.*) Afterwards, explain to the students that this organism is a protist and show them the protoplasmic movement with a binocular microscope.

Visual Strategy
Figure 1-3

Explain that reproduction is not necessary to the life of an individual, but it is necessary for the survival of the species. Ask students to compare the amount of parental care provided among organisms that have large numbers of offspring to organisms that have small numbers of offspring. (*Organisms that have large numbers of offspring usually provide little parental care. Usually, many do not survive. Organisms that have fewer offspring usually provide better parental care. Usually, many of the offspring survive.*) Point out that in either case the number of offspring that survive allow the continuation of the species but extinction can still occur.

did you know?

Hummingbirds have the highest metabolic rate of all birds.

A hummingbird's metabolic rate is about 1,400 calories per kilogram. Compare that rate to the rate of a mourning dove—about 127 calories per kilogram. Hummingbirds use enormous amounts of energy in part because of their racing wings and hearts. Hummingbirds can move their wings up to 80 times per second, and their heart can beat up to 1,200 beats per minute. Because of their high metabolic rate, hummingbirds feed almost continuously throughout the day.

Resource Link BASIC

- Assign **Reading Strategies: Getting to Know Your Textbook** from the Holt Biosources Teaching Resources CD-ROM. PORTFOLIO

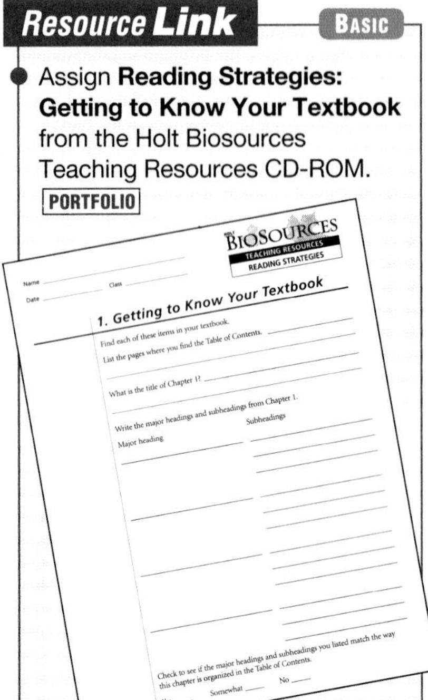
BIOSOURCES
TEACHING RESOURCES
READING STRATEGIES

Name ______ Class ______
Date ______

1. Getting to Know Your Textbook

Find each of these items in your textbook.
List the pages where you find the Table of Contents. ______

What is the title of Chapter 1? ______

Write the major headings and subheadings from Chapter 1.

Major heading	Subheadings
______	______

Check to see if the major headings and subheadings you listed match the way this chapter is organized in the Table of Contents.
Yes ____ Somewhat ____ No ____

Teaching TIP

- **Maintaining Homeostasis**
Point out to students that many animals, such as Arctic seals, that live in a cold environment are able to maintain a constant body temperature because of their thick layer of fat. Ask students for ways in which humans maintain a constant body temperature. *(Answers may include sweating, shivering, and fever.)* To illustrate the maintenance of human body temperature, ask a student to volunteer to have his or her body temperature monitored before and immediately after vigorous exercise. Human body temperature is not always 37°C (98.7°F). It fluctuates around this point.

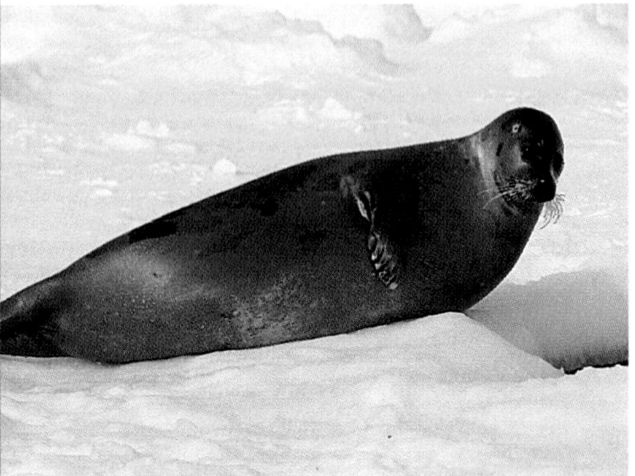
Figure 1-5 Harp seal

Theme 4 Homeostasis

All living organisms must maintain a stable internal environment in order to function properly. Organisms respond to changes in their external environment, adjusting their processes accordingly. The maintenance of stable internal conditions in spite of changes in the external environment is called **homeostasis** *(hoh mee oh STAY sihs)*. An organism unable to balance its internal conditions with its environmental conditions could become ill and die. Arctic seals, such as the one in **Figure 1-5,** are able to maintain a constant body temperature in spite of their cold environment because of their thick layer of body fat.

Theme 5 Heredity

All living things are able to pass on their characteristics (traits) to their offspring through genes that are passed from parent to offspring each generation. **Genes** are sets of inherited instructions for making proteins. Genes control when proteins are made and what proteins are made. These instructions, which are coded in a molecule called deoxyribonucleic *(dee AHKS ee reye boh nu klay ik)* acid (DNA), determine an organism's traits. The passing of traits from parent to offspring is called **heredity.** Heredity is the reason children tend to resemble their parents, as shown in **Figure 1-6.**

Sometimes damage causes genes to change. A change in the DNA of a gene is called a **mutation.** Most mutations are harmful, but sometimes mutations can help an organism survive. For example, in humans a genetic mutation for the blood protein hemoglobin, which carries oxygen to the body's cells, has both a harmful and a positive effect. The harmful effect is that the mutated form of the gene results in sickle cell anemia. Sickle cell anemia is a condition in which the defective form of hemoglobin causes many red blood cells to bend into a sickle shape that impairs the oxygen-carrying capability of the cell. The positive effect is that the mutation produces resistance to malaria, an infectious disease.

Mutations that occur in sex cells (egg and sperm) are passed on to other generations. Mutations that occur in body cells are not passed on, but they can disrupt the control of cell reproduction and result in cancer.

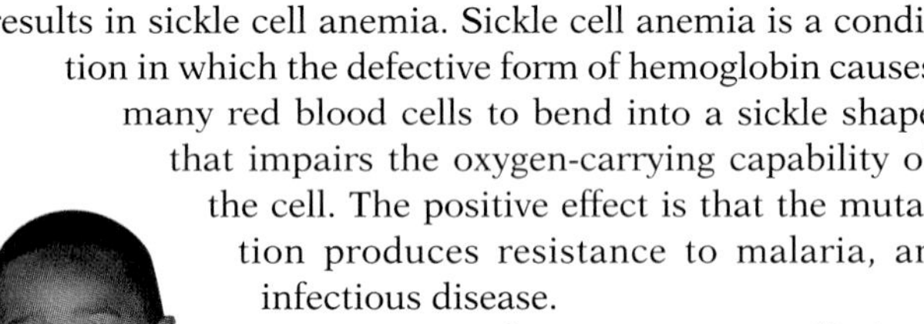
Figure 1-6 Passing on traits

Biology in Your Community

Proper Health Care
Proper nutrition is important to the well being of every individual. Invite a local clinical nutritionist or dietitian to speak on proper nutrition, dieting, and exercise for teenagers. Prior to the talk, have students prepare a list of the foods they consumed for the last three days and a list of questions for the nutritionist to answer. Ask students to compare the list of their diet to the information provided by the speaker. PORTFOLIO

Theme 6 Evolution

The great diversity of life on Earth is the result of a long history of change. Change in the inherited traits of species over time is called **evolution.** A **species** is a group of genetically similar organisms that can produce fertile offspring. Because of mutations, individuals in a species are different. Those individuals with genetic differences that better enable them to meet nature's challenges are the ones that survive, reproduce, and become more common. Darwin, the great nineteenth-century British naturalist, called this process in which organisms with favorable genes are more likely to survive and reproduce **natural selection.**

Darwin's theory of evolution by natural selection is the essence of biology, providing a consistent explanation for life's diversity. The many different species of animals, plants, and other organisms on Earth today, including humans, are the result of a long process of evolution. **Figure 1-7** shows an example of a plant that has adapted its flowers for attracting insects.

Figure 1-7
Bee pollinating flowers

Study TIP

Reviewing Information

After you read about the seven themes in this section, think about your favorite living organism. List ways in which each of the seven themes of biology plays a part in that organism's life.

Figure 1-8
Owl interacting with rat

Theme 7 Interdependence

The organisms in a biological community have evolved to live and interact with other organisms, as shown in **Figure 1-8.** A biological community is a group of interacting organisms. **Ecology** is the science that studies the interactions of living organisms with one another and with the nonliving part of their environment. Organisms are dependent on one another and their environment—they are interdependent. Interdependence within biological communities is the result of a long history of evolutionary adjustments. The complex web of interactions in a biological community is disrupted when the community is polluted and when individual species become extinct, as is happening in much of the world today.

Review SECTION 1-1

1. **Name** the seven properties all living organisms share.
2. **Relate** three of the seven major themes of biology to the life of a harp seal.
3. **Name** the tiny, organized structure with a membrane that is the basic unit of structure and function in all living organisms.
4. **Compare** metabolism with homeostasis.

Critical Thinking

5. **SKILL Recognizing Verifiable Facts** Suppose you find an object that looks like an organism. How might you determine if your discovery is indeed alive?

Section 1-1

Teaching TIP

Fossils and Evolution

Show the class a fossil specimen as a way of highlighting the theme of evolution. Explain to students that most people think of fossils as shells or old bones. Actually, fossils are any traces of dead organisms. Tracks of dinosaurs, footprints of human ancestors, insects trapped in sticky tree sap, impressions of leaves or skin, and animals buried in tar are all fossils. The transformation of one species into another by natural selection requires thousands of years. Since scientists can date fossils by using radioactive decay, fossils can contribute to our understanding of evolution.

3 Assessment Options

Closure

Seven Themes in Biology

Have students cut out magazine photos that depict one or more of the seven themes of biology. Students should glue or tape the photos on a sheet of paper and then label each photo with the theme(s). All seven themes should be represented. PORTFOLIO

Review

Assign Review—Section 1-1.

Reteaching — BASIC

Have students play a timed game in which one student tries to get his or her team to guess a particular theme of biology. The student can give clues but cannot use the actual word and certain designated words. For example, if the theme is heredity, you might designate *genetics, traits,* and *mutation* as words that cannot be given as clues.

Review SECTION 1-1 ANSWERS

1. Cellular organization, reproduction, metabolisim, homeostasis, heredity, responsiveness, growth and development.
2. Answers will vary. For example, homeostasis: fat insulates their body allowing them to survive cold temperatures; cellular structure and function: multicellular organism; reproduction: one pup
3. cell
4. Metabolism is the sum total of all the chemical reactions carried out in an organism. Homeostasis is the ability to maintain a stable internal environment in spite of changes in the external environment.
5. Answers will vary. Students should suggest testing for several of the characteristics of living organisms.

Chapter 1

Section 1-2 Biology in Your World

1 Prepare

Scheduling

- **Block:** About 90 minutes will be needed to complete this section.
- **Traditional:** About two class periods will be needed to complete this section.
- **Planning:** See the Planning Guide on pp. 3A and 3B for lecture, classwork, and assignment options for Section 1-2. Lesson plans for Section 1-2 can be found in a lesson cycle format in ***Biology: Principles and Explorations*** **One-Stop Planner CD-ROM with Test Generator.**

Resource Link

Assign **Active Reading Worksheet Section 1-2** to help students comprehend the key concepts and visuals in this lesson.

2 Teach

Lesson Warm-up

Ask students to list five issues in the area of biology that they believe are important in the world today. Tally the responses on the board to determine what the class as a whole believes are the most important issues. Next assess what students know about each problem and how scientists are working to solve it.

Objectives

- **Evaluate** the role biologists play in saving our tropical forests. (p. 10)
- **Describe** the role biologists play in trying to increase and improve our food supply. (p. 11)
- **Describe** efforts to combat three diseases that are the subject of current scientific research. (pp. 12–13)

Key Terms

HIV
cancer
cystic fibrosis

internetconnect

SCILINKS NSTA
TOPIC: Population growth
GO TO: www.scilinks.org
KEYWORD: HX010

Biology Can Help Solve Today's Real-World Problems

You are unlikely to read a newspaper or magazine today without noticing issues involving biology. We learn that biologists are working to save our tropical rain forests, to find a cure for AIDS, and to develop genetic engineering techniques that can help cure certain diseases. In this textbook you will encounter many areas in which biologists are actively working to solve some of today's problems. Like the students in **Figure 1-9,** you too can contribute to this effort.

Living in harmony with our environment

In 1999, the world's human population passed 6 billion people. The growing human population has begun to seriously harm other creatures that share the planet. For example, some people have had to clear rain-forest land in order to have space to live and to grow crops. Since the world's tropical forests are home to one-half of the world's species of plants and animals, the impact of this destruction on the diversity of plants and animals has been catastrophic.

The world's tropical forests are being destroyed at the rate of more than an acre per second. At this rate, all of the rain forests will be gone in less than 30 years! With them will be lost more than a million species, the greatest extinction event since the disappearance of the dinosaurs 65 million years ago. Extinction is forever—future generations will never see any of these animals or plants alive again. Who knows what potential medicines and foods we are discarding? Like burning a library without reading the books, extinction caused by humans is a tragedy beyond measure. Biologists are studying the plants and animals in the rain forests in order to better understand how we can maintain a balance between people's growing need for land and the need to protect these plants and animals.

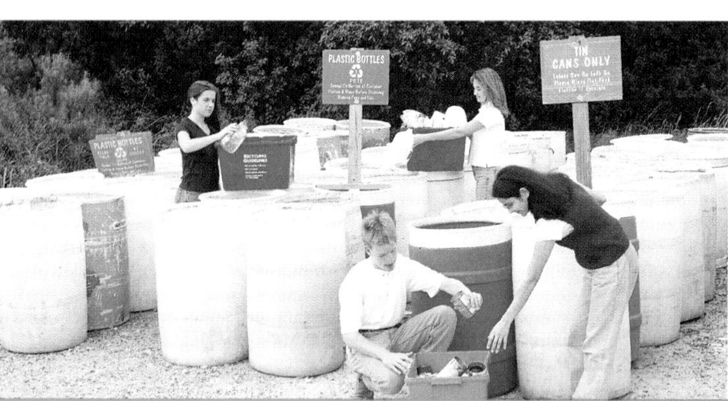

Figure 1-9 Students helping to solve today's real-world problems. Students involved in a local recycling program are practicing biology.

Integrating Different Learning Styles

Logical/Mathematical	**PE**	Math Lab, p. 11; Performance Zone items 14, 22, and 24, p. 23; Experimental Design, pp. 24–25
Visual/Spatial	**PE**	Performance Zone items 16 and 17, p. 23
	ATE	Demonstration, p. 12; Active Reading, p. 12
Interpersonal	**PE**	Performance Zone item 20, p. 23
	ATE	Biology in Your Community, p. 11
Intrapersonal	**ATE**	Closure, p. 13
Verbal/Linguistic	**PE**	Performance Zone item 19, p. 23
	ATE	Resource Link, p.11; Reteaching, p. 13

Feeding a growing population

As the population continues to grow, the demand for food is going to strain our ability to feed all the people. Biologists are vigorously seeking new crops that grow more efficiently in tropical soils and crops that grow without intensive use of fertilizers and insecticides (chemicals used to kill insects).

Genetic engineers are transplanting beneficial plant genes into other plants to create crops that are more resistant to insects and microorganisms. Although there is some apprehension about this technology, it is hoped that genetically engineered crops, such as the crops shown in **Figure 1-10,** will both reduce the need for insecticides and increase crop yields. For example, some genetically engineered plants are now resistant to frost damage because of new genes inserted into the plants. Other genetically engineered crops made resistant to insects allow farmers to decrease or avoid the use of chemical pesticides. Clearly the efforts of biologists will have an enormous impact on the world's future.

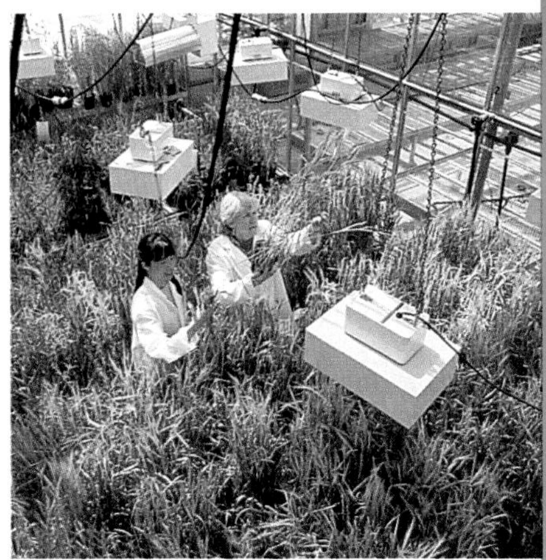

Figure 1-10 Genetically engineered crops. Crops such as barley have had their genetic material altered in an effort to help the crops resist attack by viruses.

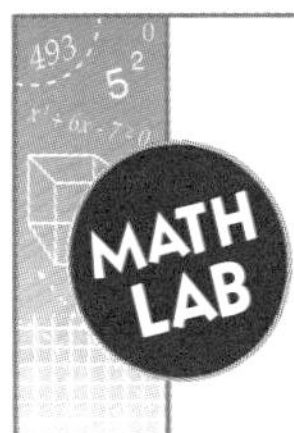

Evaluating the effectiveness of insecticides

Background

Insecticides are chemicals used to kill insects, some of which may harm valuable crops. Sometimes insects can become resistant to the insecticides—that is, the insects are not affected by the chemicals and instead survive.

Analysis

1. **Describe** how the number of insects resistant to insecticides has changed over time.
2. **Calculate** the number of years that passed between data points *A* and *B*.
3. **Calculate** the number of years that passed between data points *B* and *C*.
4. **Determine** approximately how many insect species were resistant to insecticides at points *A, B,* and *C.*

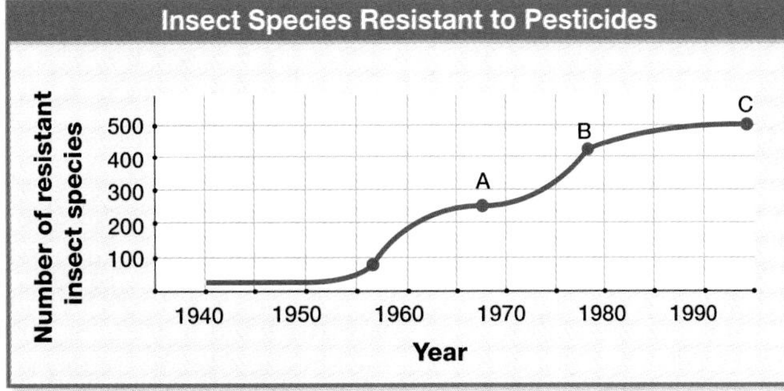

5. **Determine** whether the increase in the number of insecticide-resistant species was greater between points *A–B* or points *B–C*.
6. **Predict** the effect the numbers in the table might have on farmers who use insecticides.

Resource Link ADVANCED

Assign **Supplemental Reading: Silent Spring** from the Holt BioSources Teaching Resources CD-ROM to help students understand the impact of humans, and their products, on the environment. PORTFOLIO

Evaluating the effectiveness of insecticides

Time 10–15 minutes

Process Skills Comparing, calculating, analyzing data, predicting, interpreting graphs

Teaching Strategies

- Before they begin, show students how to use metric rulers to draw imaginary lines up to the point from the *x*-axis and over to the *y*-axis.
- Tell the students to mentally subdivide the intervals on the *x*-axis to determine the years not specifically marked on the *y*-axis.

Analysis Answers

1. As the years progress, the number of resistant insect species increases.
2. approximately 11 years
3. approximately 16 years
4. A = 250 insect species; B = 440; C = 500
5. between points A and B
6. Answers will vary. Since insects are becoming more resistant to the pesticides now in use, the farmers might expect more crop damage due to the greater numbers of insects.

Biology in Your Community

Recycling Programs

Have students research recycling programs in their community. Invite a speaker from the recycling program to discuss the effectiveness of the program, how the program helps the community financially, and how the citizens can get more involved to make the program more successful. Ask the students to write two ways your school could contribute to the recycling effort. PORTFOLIO

Demonstration

Lung Cancer

Point out, with the help of a microprojector or a videomicroscope, the cilia of a paramecium or other ciliate. Explain to students that our lungs are protected by ciliated cells that line the trachea and bronchi. These cilia trap particles and microorganisms that would otherwise reach the lungs. Chemicals in cigarette smoke paralyze the cilia. As a consequence, the lungs become susceptible to infections, and the cells sometimes become cancerous.

Active Reading

Technique: Reading Organizer — BASIC

After students read Section 1-2, have them design a spider map reading organizer to show the main ideas and supporting details of the section. Maps will vary, but students should begin by writing the main idea in the center of the diagram and the details of the section arranged around the center. For more information about reading organizers, see Reading Strategies Worksheets, in Holt BioSources Teaching Resources CD-ROM. PORTFOLIO

Teaching TIP

Politicians and Biology

Ask students to suggest what role politicians play in biological research. *(Politicians must understand the biology of diseases such as AIDS if they are to make informed decisions about providing care facilities, funding research, and protecting the rights of people with the disease.)* Emphasize that most research is supported through grants given by government agencies such as the National Institutes of Health, the National Science Foundation, and the Department of Energy.

Biology Can Help Fight Diseases

One of the most direct ways in which progress in biology affects our lives is in medicine, where scientific advances are improving health and health care every day. New technologies have enabled biologists to combat disease in ways scarcely imagined only a few years ago. Among the many diseases that you will study in this text, consider the following.

Figure 1-11 AIDS counseling

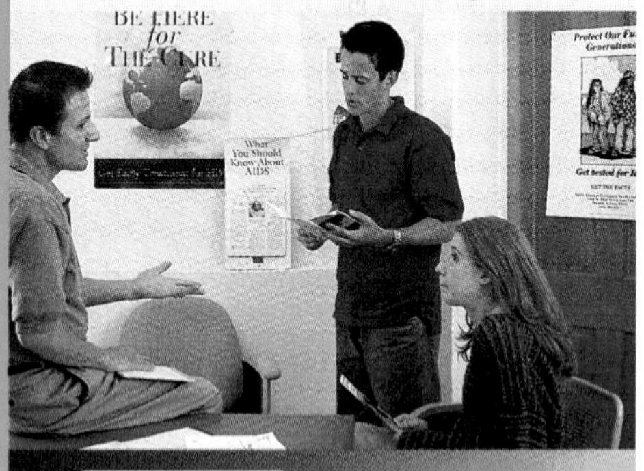

How many? AIDS is now the leading cause of death among American males between the ages of 24 and 44.

AIDS

Since 1981, when this disease was first recognized, over 50 million people worldwide have been infected with HIV (human immunodeficiency virus). **HIV** is a virus that destroys the immune system, causing acquired immune deficiency syndrome, or AIDS.

People with AIDS are not able to defend themselves from infection, so they can die of diseases that are not normally fatal. Most of the people infected with HIV are expected to develop AIDS. Fourteen million people have already died of AIDS, and about 2 million more die each year. As the number of people with HIV infection and AIDS grows, there is intense research to find a way to help those already suffering from AIDS and to halt the spread of the disease, as shown in **Figure 1-11.**

Cancer

Cancer is a growth disorder of cells that occurs when cells divide uncontrollably within the body. Although the number of people with cancer is increasing, scientists studying this disease have found that many cancer deaths are preventable. Your diet and the chemicals that you are exposed to can affect whether or not you get a particular form of cancer. For example, research indicates that most lung cancers are caused by the use of tobacco. Eliminating cigarette smoking would prevent many lung cancers.

biobit

Can trees cure cancer?

In the early 1990s, the bark of the Pacific yew tree was found to contain an anticancer chemical called taxol. Today taxol is obtained from different sources and is used to treat different types of cancer, including ovarian and breast cancers.

Cystic fibrosis

Cystic fibrosis is a fatal disorder in which abnormally thick mucus builds up in many organs, including the lungs. The accumulation of mucus causes difficulty in breathing. Cystic fibrosis is caused by a defective gene.

In cystic fibrosis patients, the gene for making a protein that helps pump chloride ions into and out of cells has the wrong instructions. Because the protein is not made correctly, the cells cannot pump chloride ions into and out of cells. This eventually results in chloride ions building up within cells. Like sponges, these cells soak up the surrounding water, turning the fluid around the outside of the cells into thick mucus.

MULTICULTURAL PERSPECTIVE

AIDS Researcher

The 1980s saw the emergence of AIDS. Many scientists have researched this virus to discover its structure and how it attacks the body. One such person involved in this research is Flossie Wong-Staal. Born in China in 1946, she came to the United States in 1965. She earned her Ph.D. and in 1973 worked for the National Institutes of Health, where she joined the team of Robert Gallo. In the 1980s the team isolated the AIDS virus. Wong-Staal helped identify the virus's structure, was the first to clone the virus, discovered how the virus regulated its growth and that of the cells it attacks, and discovered the parts of the virus that frequently mutate.

Since 1990, biologists have been attempting to transfer a healthy version of the gene into cystic fibrosis patients. They are hopeful that the healthy gene will enable the body's cells to efficiently pump chloride ions.

The replacement of a defective gene with a normal gene is called gene therapy. Other serious genetic disorders, such as muscular dystrophy, are also good candidates for gene therapy. Transferring normally functioning genes into affected individuals offers the first real hope of curing genetic disorders. The difficulty lies in introducing the normal cells into the correct cells and inserting them into the cells' DNA. These efforts continue, and progress is being reported in solving both of these problems. Although still experimental, the future of gene therapy looks very bright.

Biologists actively work to solve world problems

It is clear that biology must play a major role if we are to improve the quality of life for ourselves and for future generations. Addressing problems like the growing human population, increasing damage to the environment, and life-threatening diseases requires a thorough understanding of biological systems.

As you proceed through this text, you will encounter biologists in many settings, as laboratory and field researchers, as technicians performing analyses, as animal keepers in zoos, and as doctors in hospitals. Many careers are open to biologists, who can make important contributions to improving the future. Even if you do not become a biologist, a knowledge of biology will help you to make informed decisions as a citizen in shaping our future world.

Exploring Further

Biologists at Work

The word *biologist* often calls to mind an image of a person wearing a white coat in a windowless laboratory filled with strange chemicals. While many biologists do work in laboratories, some biologists work in other environments, such as outer space, swamps, forests, deserts, jungles, oceans, volcanoes, and computer labs.

Biologists also differ in the tools they use. For example, a biologist involved in deep-sea research might use a submarine with sophisticated cameras, and a biologist who tracks animals to study their behavior might use radio transmitters.

What do biologists do?

Regardless of the environment in which they work or the tools they use, all biologists study the origin, development, structure, function, geographic distribution, and other basic aspects of living organisms. For example, botanists study the growth, structure, and classification of plants. Ecologists study the relationships between organisms and their environment, including their relationships with other organisms. Marine biologists specialize in the study of organisms that inhabit the sea.

Biologists apply their knowledge in many ways: by drawing scientific illustrations, by guiding visitors through national parks, by working in greenhouses, by treating people who are ill. Yet others do research to advance the fields of medicine, agriculture, and industry. Biologists are at work all around us.

Ecologist

Review *SECTION 1-2*

1. **Relate** two ways that the growing human population is disturbing the Earth's resources.
2. **Evaluate** why saving tropical forests is such an important issue.
3. **Apply** genetic engineering to the development of new crops.
4. **Describe** why mucus builds up in cystic fibrosis patients.

Critical Thinking

5. **SKILL Evaluating Viewpoints** Your teacher believes that a knowledge of biology is essential in the battle against diseases. Do you agree or disagree? Explain.

Review *SECTION 1-2 ANSWERS*

1. An increasing population is responsible for pollution and for clearing land that contains plant and animal resources.
2. The rain forest contains many species of animals and plants not found in other environments. The rain forest also has many potential medicines and foods.
3. Genetic engineering can be used to develop crops that are more tolerant to certain conditions, are insect resistant, and produce a higher crop yield.
4. The thick mucus builds up because the protein responsible for pumping the chloride ions into and out of the cell is not functional. The chloride ion concentration eventually builds up inside of cells. The cells take in the surrounding water, turning the fluid around the outside of the cells into thick mucus.
5. Students should agree. In order to create treatments or cures for diseases, the researchers must understand the biology of the symptoms and the cause of each disease.

Exploring Further

Biologists at Work

Teaching Strategies

- Bring science magazines to class. Have students mark the pages that have photos of scientists. On the board, make a list of the environment where the scientist is shown working; the type of equipment he or she is using; and, if shown, the organism the scientist is working with. Describe the various work environments. Point out how biologists work with many different types of organisms.
- Have students use the *sci*LINKS Web site on page 15 to explore more about careers in biology.

Discussion

- What would be your favorite environment if you were a biologist? *(Answers will vary.)*
- Summarize the gender and ethnicity of the biologists in the photos. *(Answers will vary.)*
- What organism would you like to work with? *(Answers will vary.)*

3 Assessment Options

Closure

Protecting the Earth

Have each student make a list of 10 ways that they can contribute to protecting the Earth. *(Student suggestions might include bringing their lunches in reusable containers, carpooling, and riding a bicycle.)* Ask students to discuss their answers and ways their school can help protect the Earth. PORTFOLIO

Review

Assign Review—Section 1-2.

Reteaching — BASIC

Have students work with a partner to review all the headings, subheadings, and boldface words in this section. Then have each pair list problems that may someday be solved by scientists.

1 Prepare

Scheduling

- **Block:** About 90 minutes will be needed to complete this section.
- **Traditional:** About two class periods will be needed to complete this section.
- **Planning:** See the Planning Guide on pp. 3A and 3B for lecture, classwork, and assignment options for Section 1-3. Lesson plans for Section 1-3 can be found in a lesson cycle format in ***Biology: Principles and Explorations* One-Stop Planner CD-ROM with Test Generator.**

Resource Link

Assign **Active Reading Worksheet Section 1-3** to help students comprehend the key concepts and visuals in this lesson.

2 Teach

Lesson Warm-up

Ask students to think of some current biological questions or problems. Then have students describe the steps they think might be involved in a scientific investigation of those questions or problems. Explain that the scientific process outlines a general method of investigating questions.

Section 1-3 The Scientific Process

Objectives

- **Describe** the stages common to scientific investigations. (pp. 14–20)
- **Distinguish** between forming a hypothesis and making a prediction. (p. 16)
- **Differentiate** a control group from an experimental group and an independent variable from a dependent variable. (p. 17)
- **Define** the word *theory* as used by a scientist. (p. 19)

Key Terms

observation
hypothesis
prediction
pH
experiment
control group
independent variable
dependent variable
theory

Science Is Based on Careful Observation

Recognizing the properties of living organisms and knowing why biology is important in your world are good first steps in your exploration of biology. All scientists, including biologists, have a certain way of investigating the world. Studying an actual scientific investigation is an exciting way to learn how science is done. Our story begins many years ago with two biologists, David Bradford and John Harte, but the story continues to develop even today.

Shared observations help to solve scientific puzzles

In the summer of 1988, Bradford reflected on the silence that surrounded him. He had spent the summer looking for a small frog in the many lakes of the Sequoia & Kings Canyon National Parks. The frog had lived in the parks' lakes for as long as anyone had kept records. In the last count of the frog's populations, the frogs had been everywhere. Now, for some reason, they had disappeared from 98 percent of the lakes.

Observation is the act of noting or perceiving objects or events using the senses. As Bradford reported his observations to other biologists, he found that local populations of amphibians (frogs, toads, and salamanders) elsewhere were also disappearing. Amphibians have been around for 370 million years. The disappearance of amphibians from their natural homes sounded an alarm among biologists that something was seriously damaging the environment. Amphibians are particularly sensitive to their environment; their moist skin absorbs chemicals from water.

Figure 1-12 Biologist and the disappearing salamander. Biologist John Harte first observed the decline in number of tiger salamanders in the early 1980s.

Tiger salamander, *Ambystoma tigrinum*

Between the years 1984 and 1988, John Harte, a biology professor at the University of California, Berkeley, was also studying amphibians. He was studying the tiger salamander, *Ambystoma tigrinum*, shown in **Figure 1-12.** Tiger salamanders live in ponds high on the western slopes of the Rocky Mountains of Colorado. Harte had seen their numbers fall by 65 percent as he and his students had collected and analyzed water samples from the ponds in the area over the years.

Integrating Different Learning Styles

Logical/Mathematical	**PE**	Real Life, p. 16; Quick Lab, p. 17; Data Lab, p. 20; Performance Zone items 13, 21, and 23, pp. 22, 23; Experimental Design, pp. 24–25
	ATE	Closure, p. 20; Reteaching, p. 20
Visual/Spatial	**PE**	Performance Zone item 1, p. 22
	ATE	Cross-Disciplinary Connection, p. 15; Teaching Tip, p. 19; Demonstration, p. 19
Intrapersonal	**ATE**	Demonstration, p. 16
Verbal/Linguistic	**PE**	Study Tip, p. 17; Performance Zone items 25 and 26, p. 23
	ATE	Resource Link, p. 18

Several Stages Are Common to Scientific Investigations

Harte wanted to discover the facts surrounding the disappearance of the salamanders. Like other scientists, Harte began a scientific investigation that combined knowledge, imagination, and intuition to get a sense of what might be true. Even though scientists might expect certain results, they do not form conclusions until they have enough evidence to support them.

Although there is no single "scientific method," all scientific investigations can be said to have common stages: collecting observations, asking questions, forming hypotheses and making predictions, confirming predictions (with controlled experiments when appropriate), and drawing conclusions. These stages are summarized in **Figure 1-13.**

Collecting observations

The heart of scientific investigation is careful observation. Harte had studied the Colorado salamander population for years. He had learned what they eat, how they behave, when they reproduce, and the conditions they thrive in. His students had helped him collect water samples from the ponds, as shown in Figure 1-13. Frequent visits to the ponds helped him realize the salamander population was decreasing in number. Keeping careful records of the lakes' conditions helped him find an explanation.

Asking questions

Observations of the natural world often raise questions. Harte questioned why the number of salamanders was dropping. He talked to other scientists, carefully observed the organisms and environment in the Rocky Mountains of Colorado, and read scientific reports. He answered many of his questions through his observations, but some key questions remained unanswered.

In the natural world, the moisture that falls as rain and snow are neither acidic nor basic. In the Rocky Mountains of Colorado, however, the moisture is high in sulfuric acid from power plants that burn high-sulfur coal. This acidic moisture, called acid rain, is released into mountain ponds each spring when the snow melts,

internetconnect

SCILINKS. NSTA

TOPIC: Biology careers
GO TO: www.scilinks.org
KEYWORD: HX015

Scientific Process

- Collecting observations
- Asking questions
- Forming hypotheses and making predictions
- Confirming predictions (with experiments when needed)
- Drawing conclusions

Figure 1-13 Testing the acidity of water. Asa Bradman, a student of John Harte, helps in Harte's scientific investigation by collecting water samples from a Colorado pond.

Teaching TIP

Identifying Factors

Students should recognize that an observable phenomenon, such as the salamander population's decline, may be the result of any one of a number of factors. A scientist's task in such a situation is to identify the factors responsible. Have each student list as many factors as possible that may have been responsible for the population's decline. *(Possibilities include increased predation, drastic temperature changes, habitat pollution, and decreased food availability.)*

CROSS-DISCIPLINARY CONNECTION

◆ Geography

Have students locate Arizona, California, and the Rocky Mountains of Colorado on a map of the United States. Inform them that acidic compounds from Arizona smelters and the California smog may be responsible for the decline of the tiger salamander in the Colorado Rockies. Now have them locate Wheeling, West Virginia. Inform them that the pH of precipitation in Wheeling has been recorded at 1.5. Ask what might be the source of the acidity. (*Acidic compounds probably originated from industries in the upper Midwest and coal mines in the more immediate areas.*)

did you know?

The Office of Orphan Products Development (OOPD) is responsible for the development of products that demonstrate promise for the diagnosis and treatment of rare diseases and conditions.

The office was created in 1982 and is located in the Office of the Commissioner, Food and Drug Administration. The OOPD works in conjunction with the medical profession, research and pharmaceutical facilities, and rare disease groups. Since 1983 over 170 drugs and biological products have been developed and brought to market. In the decade prior to that, there were only 10.

Chapter 1

REAL LIFE

Answer

A 5K race is about 16,393 ft, or 5,464 yd, or 3.11 mi. Students should find the metric system easier to use because of the ease with which conversions between metric units are accomplished.

Demonstration

Using the Scientific Process

Have students perform the following exercise and identify each stage of a scientific investigation. Ask students to record the number of times they breathe during a 1-minute period (*stage: collecting observations*). Next have them suggest how exercise will affect this number (*stage: forming a hypothesis*). Then ask them to note the number of breaths they think they will take in the same period of time after they have jogged in place for one minute (*stage: making predictions*). Instruct them to carry out the exercise and immediately record the number of breaths they take (*stage: verifying predictions*). Ask the class how they know that exercise was a factor that affected the number of breaths they took (*stage: performing controlled experiments*). Finally, have them suggest a connection between exercise and the rate of breathing (*stage: drawing conclusions*). PORTFOLIO

Making Biology Relevant

Medicine

Point out to students that in clinical trials the control group receives a placebo so that the subjects do not know whether or not they are receiving the drug being tested. In a single blind study only the subjects are unaware of whether they are taking the drug or the placebo. Ask students to describe what constitutes a double blind study. (*Both the experimenters and the subjects are unaware of which subjects are receiving the drug.*)

REAL LIFE

Scientists and many people use the metric system. In the metric system of measurement (SI system), the units of measurement are based on powers of 10. Conversion between units is easily done by moving the decimal point and changing the prefix of the unit. For example, a runner in a 5K race runs 5 kilometers, 5,000 meters, or 5,000,000 millimeters.

Calculating *Calculate the length of a 5K race using the English system of measurement in feet, yards, and miles. Which system is easier to use?*

causing the water in the ponds to become more acidic in late May. Most of the mountains' annual moisture falls as snow. Harte thought acid rain was important in the puzzle of the declining salamander population, but he needed evidence.

Forming hypotheses and making predictions

A **hypothesis** *(heye PAHTH uh sihs)* is an explanation that might be true—a statement that can be tested by additional observations or experimentation. In that respect, a hypothesis (plural form, hypotheses) is not just a guess, it is an educated guess based on what is already known. Harte formed two hypotheses that together he believed explained the disappearance of the amphibians:

1. Acids that were formed in the upper atmosphere by pollutants were falling onto the mountains in the winter snows.
2. Melting snow was making the ponds acidic and harming the salamander embryos.

If Harte's hypotheses were correct, he could expect several possible outcomes. A **prediction** is the expected outcome of a test, assuming the hypothesis is correct. For his first hypothesis, Harte predicted he would find acid in the ponds after the snow melted. For his second hypothesis, he predicted that there would be enough acid in the ponds to harm salamander embryos. Using his predictions as a starting point, Harte set out to test his hypotheses.

Confirming predictions

Harte gathered data from many years of observations, including measurements of the acidity of the ponds before, during, and after snowmelt. Harte and his students had taken water samples at frequent intervals from several ponds before and after snowmelt. Data for part of one year, after snowmelt, are shown in **Figure 1-14.**

To describe how acidic a solution is, scientists use a number between 0 and 14 to represent **pH,** which is a relative measure of the hydrogen ion concentration within a solution. Solutions with a low pH (below 7) are acidic, solutions above 7 are basic, and solutions at pH 7 are neutral. Acid rain usually has a pH of between 2 and 6. A solution with a pH of 2 is far more acidic than one with a pH of 6.

Figure 1-14 Pond pH after snowmelt. The pH levels in a pond in the Rocky Mountains are acidic (low pH) at the same time of the year that salamander eggs are developing.

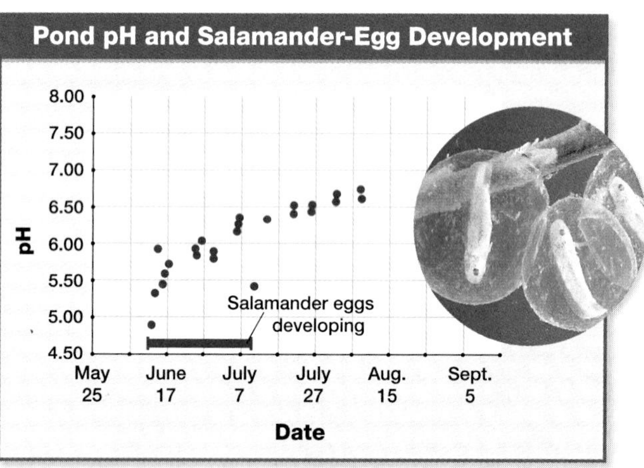

Harte's data indicated that the ponds became more acidic when the snow melted. After a few weeks, the pH rose and then leveled off. The data confirmed Harte's first prediction and supported his first hypothesis—melting snow caused acid to be released into the ponds at snowmelt, as shown in Figure 1-14. After snowmelt, the acid was

Overcoming Misconceptions

Hypotheses

Students often equate a hypothesis with a guess. Emphasize that a hypothesis must have some basis in fact, and often stems both from observations the scientist has noted and from prior knowledge. A hypothesis is an explanation that can be tested through specific and sometimes controlled observation. Hypotheses are an initial part of formulating a problem and setting down strategies for studying it. Scientists continually modify their hypotheses to accommodate new data.

neutralized, probably by minerals that dissolved from the rocks in the ponds, and pond acidity returned to normal for the rest of the year.

To confirm his second hypothesis (melting snow was making the ponds acidic and harming the salamander embryos), Harte did an **experiment**—a planned procedure to test a hypothesis. Salamanders lay eggs in the ponds once a year, as soon as pond ice melts. Harte wanted to test whether exposure to the pH levels he had recorded at that time of year would harm the salamanders that hatched from the eggs.

Harte performed a controlled experiment. In a controlled experiment, an experimental group (a group that receives some type of experimental treatment) is compared with a control group. A **control group** is a group in an experiment that receives no experimental treatment. The control and experimental groups are designed to be identical except for one factor, or variable. The factor that is varied in an experiment is called the **independent variable.** In Harte's experiment, the independent variable was the acid (pH) level. The variable that is measured in an experiment is called the **dependent variable.** Harte's dependent variable was the number of salamanders that hatched from the eggs.

Harte allowed captive salamanders to lay eggs in regular pond water. He collected and then divided the eggs into five groups. One group, the control group, contained eggs placed in pond water with a neutral pH. Each of the other four groups, the experimental groups, contained eggs placed in pond water with different amounts of acid added, similar to the acid levels found in the ponds after snowmelt.

Study TIP

Reviewing Information

On a separate sheet of paper, make a table with two columns. List the stages common to scientific investigations in the left-hand column. In the right-hand column, describe in your own words what is actually done at that stage and why that is important to scientific inquiry.

What is the pH of some common substances?

QUICK LAB

You can use pH indicator paper to determine the pH of various solutions. The pH indicator paper changes color when it is exposed to a solution. The change in color indicates how acidic or basic the solution is.

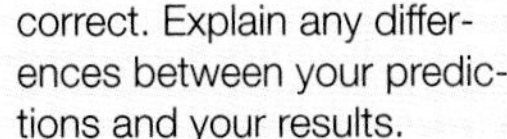

Materials

paper, pencil, wide-range pH paper, three different solutions, beaker or small jar, water

Procedure

1. Make a data table with three columns and the following headings: *Solution, Predicted pH,* and *Measured pH.* Make a row for each solution to be tested.
2. Predict the pH (acid or base) of each solution, and record your predictions in your data table.
3. Test each solution with pH paper, and record the results in the appropriate row in your data table.

Analysis

1. **Summarize** your findings in two sentences.
2. **Determine** whether the predictions that you made were correct. Explain any differences between your predictions and your results.
3. **Compare** your results with those of the rest of the class. Explain any differences.
4. **List** the steps of the scientific method that you followed in doing this activity.

What is the pH of some common substances?

QUICK LAB

Time 15–20 minutes

Process Skills Observing, predicting, analyzing, comparing, measuring

Teaching Strategies

- Warn the students not to get the liquids on their hands. Hands should be washed after the lab is completed.
- Containers (jars or beakers) with "mystery" solutions should be labeled "1," "2," or "3" to help the students form consistent charts. This will allow for easier class comparisons later.
- Suggested solutions include apple juice (acid), orange juice (acid), and ammonia (base, diluted 4:1). Be careful to use only dilute solutions.
- All solutions mentioned above can be safely put into school drainage systems.

Analysis Answers

1. Answers will depend on the solutions used. Students should state which solutions were acidic, which were basic, and which were neutral.
2. Answers will vary. Explanation of the differences might include the student's perception of the substance.
3. Answers will vary. Reasons for differences might include human error, contamination, and incorrect color matching.
4. prediction, experimentation, drawing conclusions

MULTICULTURAL PERSPECTIVE

Importance of Frogs to Various Cultures

The ancient Egyptians worshipped a frog goddess named Heket who was considered to be an ancestor of the other gods. Heket symbolized the human embryo and the germination of grain. Prehistoric Germans also had a frog goddess, named Holla. This goddess brought the red apple of life back to Earth from the cave into which it fell each year at harvest time. A frog is an important character in a story told by the Haida Indians of the American Northwest coast about the first Kwakuitl totem pole. This special frog led the Haida chieftain to a spectacular totem pole. The frog called it the sky-supporting pole.

Chapter 1

Teaching TIP

Biology Journals

Point out to students that biology magazines such as *Scientific American, Discover,* and *Science News* are available in most libraries and newsstands; they are written at a level that most high school students can understand.

CAREER

Biology Teacher

Biology teachers help students discover the world of living things. Through lectures, laboratory investigations, field work, and field trips, biology teachers help the students to gain a better appreciation and understanding of the world in which they live. Although certification requirements vary from state to state, most biology teachers will need at least a bachelor of arts or bachelor of science degree. Have students contact a college or university that offers a biology teaching certification program in your state to obtain information about course and certification requirements.

Resource Link

Assign **Occupational Applications: Biology Teacher** from the Holt BioSources Teaching Resources CD-ROM PORTFOLIO

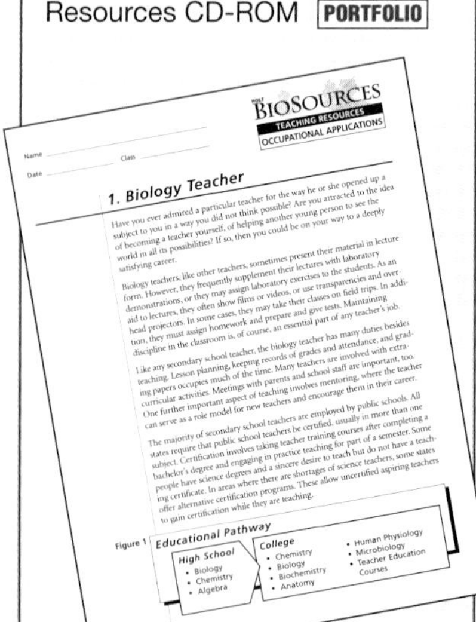

HOLT BIOSOURCES TEACHING RESOURCES OCCUPATIONAL APPLICATIONS

Name ______ Class ______

Date ______

1. Biology Teacher

Have you ever admired a particular teacher for the way he or she opened up a subject to you in a way you did not think possible? Are you attracted to the idea of becoming a teacher yourself, of helping another young person to see the world in all its possibilities? If so, then you could be on your way to a deeply satisfying career.

Biology teachers, like other teachers, sometimes present their material in lecture form. However, they frequently supplement their lectures with laboratory demonstrations, or they may assign laboratory exercises to the students. As an aid to lectures, they often show films or videos, or use transparencies and overhead projectors. In some cases, they may take their classes on field trips. In addition, they must assign homework and prepare and give tests. Maintaining discipline in the classroom is, of course, an essential part of any teacher's job.

Like any secondary school teacher, the biology teacher has many duties besides teaching. Lesson planning, keeping records of grades and attendance, and grading papers occupies much of the time. Many teachers are involved with extracurricular activities. Meetings with parents and school staff are important, too. One further important aspect of teaching involves mentoring, where the teacher can serve as a role model for new teachers and encourage them in their career.

The majority of secondary school teachers are employed by public schools. All states require that public school teachers be certified, usually in more than one subject. Certification involves taking teacher training courses after completing a bachelor's degree and engaging in practice teaching for part of a semester. Some people have science degrees and a sincere desire to teach but do not have a teaching certificate. In areas where there are shortages of science teachers, some states offer alternative certification programs. These allow uncertified aspiring teachers to gain certification while they are teaching.

Figure 1 Educational Pathway

High School
- Biology
- Chemistry
- Algebra

College
- Chemistry
- Biology
- Biochemistry
- Anatomy
- Human Physiology
- Microbiology
- Teacher Education Courses

Figure 1-15 Deformities in amphibians. Scientists are investigating factors that may play a role in the development of deformities found in amphibians throughout the United States and other parts of the world.

Harte found that acid did indeed affect development. Many of the salamanders never hatched from the eggs placed in acidic water. Some of the salamanders that did hatch were born with developmental abnormalities. Other scientists have found abnormalities in amphibians, as shown in **Figure 1-15.**

Drawing conclusions

Once data are collected and analyzed, a conclusion is made as to whether the data support the hypothesis. The hypothesis is either supported or rejected. A hypothesis can be supported but never proven because another experiment with new data and new information may alter the conclusion.

Harte's data supported both of his hypotheses. The pH levels in the ponds before and after snowmelt indicated that the ponds became more acidic after the snow melted. This supported his first hypothesis—acids that were formed in the upper atmosphere by pollutants were falling onto the mountains in the winter snows.

Harte's controlled experiment showed that acidic water reduces the number of salamanders that hatch from eggs. This supported his second hypothesis—melting snow could make the ponds acidic and harm the salamander embryos. Harte concluded that melting snow in the Rocky Mountains of Colorado could cause acid absorbed from atmospheric pollution to be released into the ponds at snowmelt, harming salamander embryos.

Publishing results in scientific journals

Once a scientist completes an experiment, he or she often writes a report for publication in a scientific journal. Before publication, the research report is reviewed by other scientists, who ensure that the experiment was carried out with the appropriate controls, methods, and data analysis. The reviewers also check that the conclusions reached by the author are justified by the data obtained. Publishing the experiment in one of the hundreds of scientific journals allows other scientists to use the information for hypotheses they are forming. They can also repeat the experiments and confirm the validity of the report. Reading scientific papers allows scientists to keep up with new information and provides a firm basis of experimental knowledge upon which others can build.

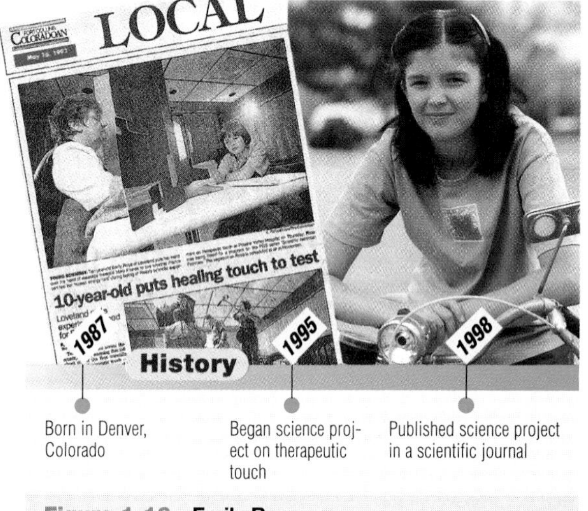

Figure 1-16 Emily Rosa

The young girl in **Figure 1-16,** Emily Rosa, conducted a controlled experiment for her school's science fair. Her work was later published in a scientific journal. At 11 years old, she became the youngest person to have work published in a scientific journal!

Theories Are Ideas Supported by a Great Deal of Evidence

Scientific progress is made the same way a marble statue is, by chipping away the unwanted bits. If a hypothesis does not provide a reasonable explanation for what has been observed, the hypothesis is rejected. Harte was able to show that enough acid was being introduced into the ponds to kill the salamander embryos. Therefore, he could keep the hypothesis that acid from melting snow was killing the salamanders. Scientists routinely make predictions and attempt to confirm them by testing one or more hypotheses.

A **theory** is a set of related hypotheses that have been tested and confirmed many times by many scientists. A theory unites and explains a broad range of observations. The hypothesis that acid rain is contributing to the loss of amphibian populations will require much more evidence before becoming accepted as a theory. Many other environmental factors may play important roles, such as increased ultraviolet (UV) radiation due to ozone depletion or a fatal skin fungus that has killed and deformed frogs in parts of Australia, Central America, and the United States.

It is important in science not to be misled by an isolated observation. Only after many studies like Harte's will scientists be able to assemble a picture that accurately reveals what is harming the amphibians. **Figure 1-17** summarizes the steps in the development of a theory.

Constructing a theory often involves contrasting ideas and conflicting hypotheses. For example, Harte's conclusions have been questioned by scientists who suggest that his observations may be only coincidence. Argument, disagreement, and unresolved questions are a healthy part of science, a true reflection of how science is done.

As you study biology, it is important to remember that the word *theory* is used very differently by scientists than by the general public. To scientists, a theory represents that of which they are most certain. To the general public, *theory* may imply a lack of knowledge, a guess. How often have you heard someone say, "It's only a theory" to imply lack of certainty? As you can imagine, confusion often results. In this textbook, the word *theory* will always be used in its scientific sense, as a generally accepted scientific principle.

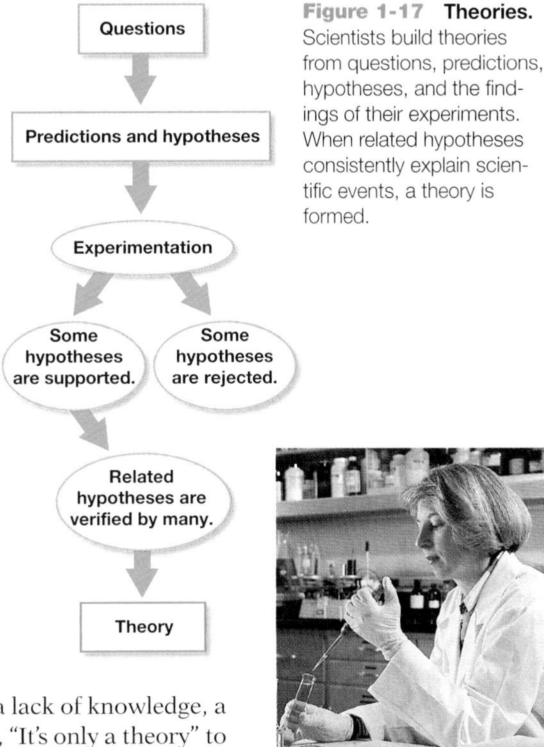

Figure 1-17 Theories. Scientists build theories from questions, predictions, hypotheses, and the findings of their experiments. When related hypotheses consistently explain scientific events, a theory is formed.

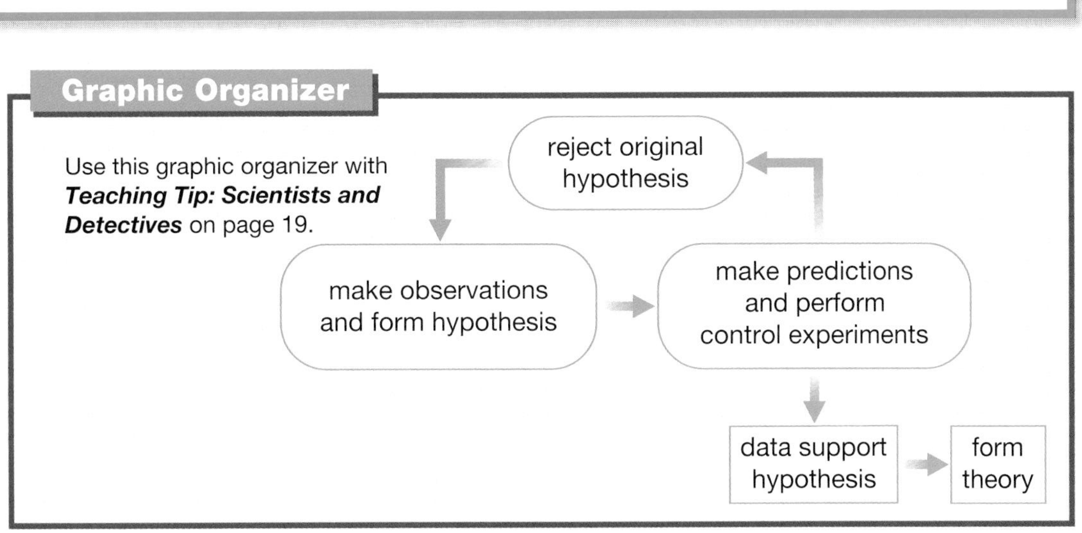

Teaching TIP — ADVANCED

Scientists and Detectives

Scientists use many of the same skills to solve a problem that detectives use to solve a crime. For example, scientists reject hypotheses that are not supported by all the observations; detectives reject suspects that are not implicated by all the evidence. Both investigators will eventually arrive at a conclusion to explain what happened. Students will clearly see the common thread between scientists' and detectives' work by reading one of the case studies found in *Medical Detectives* by Berton Rouche or in Sir Arthur Conan Doyle's *Chronicles of Sherlock Holmes.* Have students select one such case study and write a summary documenting the similarities between the investigations carried out by both scientists and detectives. Have students construct a **Graphic Organizer** similar to the one at the bottom of page 19, but filling in specific information from the case study. PORTFOLIO

Demonstration

Scientific Equipment

Organize a collection of scientific equipment commonly used in biology. Display each piece, with the name written on an index card, around the room. In addition to standard laboratory equipment, be sure to include items such as a computer, binoculars, small animal traps, and journal notebooks. Have students walk around the room to observe these pieces. Afterwards, discuss what type of biologist might use each piece of equipment.

Analyzing experimental design

Time 20 minutes

Process Skills Comparing, predicting outcomes, analyzing, proposing

Teaching Strategies
Set up a well slide containing *Daphnia* and use a videomicroscope to show how you can count the number of heartbeats.

Analysis Answers

1. dependent variable: heart rate; independent variable: substance tested
2. groups that received tea, coffee, or alcohol
3. water
4. Answers may include that the number of drops added should be the same for each substance; there should be more than one trial per substance; and a control group is needed.
5. Answers will vary. Students need to compare the heart rate of *Daphnia* in water (control) to the heart rate in coffee. There should be more than one trial for the entire experiment.

3 Assessment Options

Closure

Designing a Research Project

Tell students that an insurance company is claiming that white cars are safer than other cars. Have students work in groups to form a model for a research project that would determine the validity of the "white-car claim." PORTFOLIO

Review

Assign Review—Section 1-2.

Reteaching — BASIC

Have students design an experiment, using five apples, that would support the hypothesis that the skin of an apple functions to protect the apple from decay. PORTFOLIO

There is, however, no absolute certainty in a scientific theory. The possibility always remains that future evidence will cause a scientific theory to be revised or rejected. A scientist's acceptance of a theory is always provisional.

Some theories are so strongly supported that the likelihood of their being rejected in the future is small. The theory of evolution, for example, is backed by such a wealth and diversity of evidence that all but a few scientists accept it with as much confidence as they do the theory of gravity.

DATA LAB *Analyzing experimental design*

Background

To study the effects of common substances on the heart rate of a tiny aquatic organism known as *Daphnia,* students placed a *Daphnia* in a drop of water on a glass slide. The students then added 1 or more drops of a test substance dissolved in water to the slide, waited 10 seconds, then counted heart beats for 10 seconds. The students used a clean slide and a new *Daphnia* each time. Their data table is shown below.

Magnification: 16×

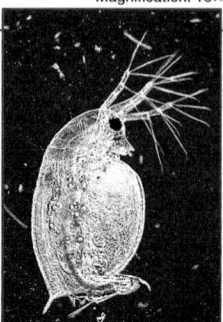

Daphnia

Heart Rate of *Daphnia* in Different Solutions

Substance tested	Heart rate (beats per minute)
Tea	62
Coffee	65
Ethanol	50

Analysis

1. **Identify** the dependent and independent variables in the experiment.
2. **Identify** the experimental groups in the experiment.
3. **Propose** a liquid that could be used for a control group.
4. **Evaluate** how the instructions could be changed to improve the design of the experiment.
5. **Design** an experiment that students can perform to verify the prediction that coffee will increase heart rate in *Daphnia*.

Review SECTION 1-3

1. **Relate** why observations and communication were so important among the biologists who studied amphibians.
2. **Summarize** how scientists use hypotheses, predictions, and experiments in scientific investigations.
3. **Differentiate** independent variables from dependent variables.
4. **Define** the word *theory* in a scientific sense and then in a more general sense.

Critical Thinking

5. **SKILL Evaluating Results** Ducks and geese in Harte's study area were not affected by acid rain. Explain why.

Review SECTION 1-3 ANSWERS

1. Observations allowed the biologists to note that the population of amphibians they were studying was declining. Communication allowed the biologists to realize that other types of amphibians were experiencing a similar population decline.
2. Hypotheses are used to explain what might be true. Predictions are used to state the expected outcome of the hypotheses. Experiments are used to test the hypotheses.
3. Independent variables are those factors that are varied in an experiment. A dependent variable is the factor that is measured in an experiment.
4. Scientific sense: A theory is a set of related hypotheses that has been tested and confirmed many times by scientists. General sense: A theory is a guess.
5. Duck and geese eggs are not laid in water and are enclosed by a thick shell. Unlike amphibians, duck and geese do not have moist skin that absorbs chemicals—their skin is protected by feathers.

Study ZONE CHAPTER 1

Use the Key Concepts and Key Terms listed below to review the main ideas in this chapter. If you need to review the meaning of a term, turn to the page indicated.

Key Concepts

1-1 Themes of Biology

- Living organisms are diverse but share certain characteristics.
- All living organisms are composed of cells, grow and develop, and are able to maintain homeostasis.
- Living organisms reproduce, producing offspring similar to themselves.
- Living organisms obtain and use energy to stay alive, and they respond to their environment.
- Seven themes unify the science of biology: cellular structure and function, reproduction, metabolism, homeostasis, heredity, evolution, and interdependence.

1-2 Biology in Your World

- Pollution of the atmosphere, extinction of plants and animals, and a growing demand for food are current environmental problems caused by the growing human population.
- Biologists are using genetic engineering to develop crops that require fewer fertilizers and pesticides and to develop new crops.
- Biological research and new technologies will help scientists battle diseases such as AIDS, cancer, and cystic fibrosis.

1-3 The Scientific Process

- Scientists add to scientific knowledge by sharing observations and posing questions about those observations.
- Although there is no single method, observing, asking questions, and forming and testing hypotheses are important in planning a scientific experiment.
- In a controlled experiment, the independent variable is varied between the experimental and control groups. The measured variable is the dependent variable.
- A collection of hypotheses that have been repeatedly tested and are supported by a great deal of evidence form a theory.

Key Terms

1-1

biology (6)
cell (7)
reproduction (7)
metabolism (7)
homeostasis (8)
gene (8)
heredity (8)
mutation (8)
evolution (9)
species (9)
natural selection (9)
ecology (9)

1-2

HIV (12)
cancer (12)
cystic fibrosis (12)

1-3

observation (14)
hypothesis (16)
prediction (16)
pH (16)
experiment (17)
control group (17)
independent variable (17)
dependent variable (17)
theory (19)

Study TIP Ready?

If you think you understand these Key Concepts and Key Terms, then you're ready for the Performance Zone.

Answer to Concept Map

The following is one possible answer to Performance Zone item 1 on page 22.

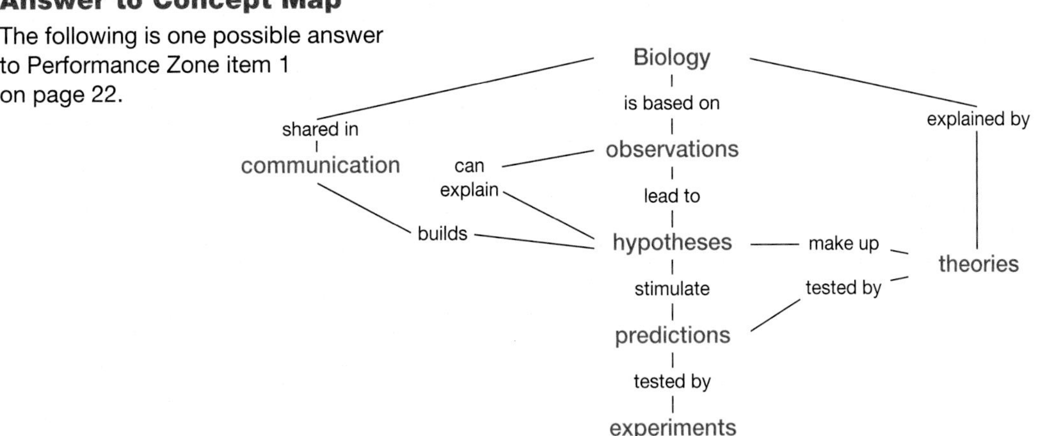

Study ZONE CHAPTER 1

Review and Practice

Assign the following worksheets for Chapter 1:

- **Student Review Guide**
- **Concept Mapping Worksheet**
- **Test Preparation Pretest**
- **Vocabulary Worksheet**
- **Critical Thinking Review Worksheet**
- **Science Skills Worksheet**

Assessment Options

The following assessment options are available for this chapter:

Resource Link

- Assign **Chapter Test 1** to assess students' understanding of the chapter.

Alternative Assessment

Have students find a current news report about the work of a scientist or look up a chronicle about the discoveries of a famous scientist in history. Ask students to draw a flowchart that shows the events leading up to the scientist's hypothesis or explanation of his or her scientific work. Then have students write a brief paragraph that compares the events in their flowchart with the elements of scientific investigation discussed in this chapter. PORTFOLIO

Portfolio Assessment

- *Portfolio Assessment* at the front of this book provides options for building and evaluating student portfolios. Students and teachers can select items from the areas listed for inclusion in a portfolio.
- See items labeled PORTFOLIO on pp. 6, 8, 9, 11, 12, 13, 16, 18, 19, 20, and 21.

Performance ZONE
CHAPTER 1 REVIEW ANSWERS

Assignment Guide

Section	Review
1-1	2, 3, 4, 5, 6
1-2	7, 8, 9, 14, 16, 17, 18, 19, 20, 22, 24
1-3	1, 10, 11, 12, 13, 15, 21, 23, 25, 26

Understanding and Applying Concepts

1. The answer to the concept map is found at the bottom of page 21.
2. **a.** Evolution is the change in the inherited traits of a species over time. These changes occur because of mutations in certain genes. The individuals possessing more favorable genes are more likely to survive and reproduce in a process called natural selection.

   b. Metabolism is the sum of all chemical reactions that are carried out in an organism. Homeostasis is the ability of the organism to maintain a stable internal environment in spite of changes in the external environment.

   c. Control and experimental groups are identical except for one factor, which is varied in the experimental group.
3. d
4. a
5. b
6. a
7. d
8. b
9. c
10. a
11. c
12. c

Performance ZONE CHAPTER 1 Review

Understanding and Applying Concepts

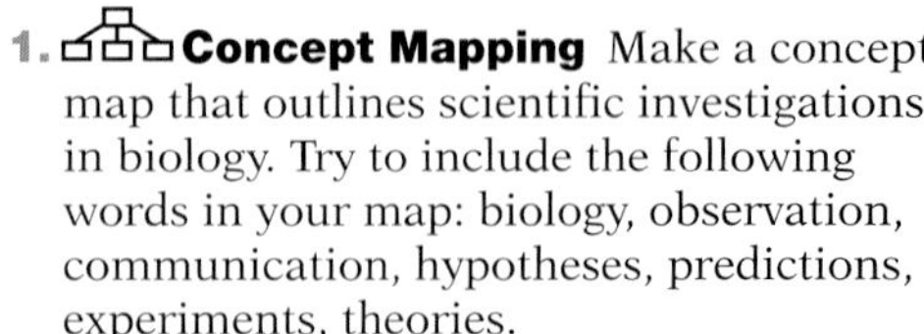

1. **Concept Mapping** Make a concept map that outlines scientific investigations in biology. Try to include the following words in your map: biology, observation, communication, hypotheses, predictions, experiments, theories.
2. **Understanding Vocabulary** For each set of terms, write one or more sentences summarizing what you learned in this chapter.
   a. mutation, evolution, natural selection
   b. metabolism, homeostasis
   c. control group, experimental group
3. The sets of instructions for making proteins that are coded in DNA and passed from parent to offspring each generation are called
   a. cells. c. hypotheses.
   b. species. d. genes.
4. The sum of all the chemical reactions carried out in an organism is called
   a. metabolism. c. reproduction.
   b. homeostasis. d. sensitivity.
5. Toads that live in hot, dry regions bury themselves in the soil during the day. What theme of biology does this describe?
   a. interdependence c. evolution
   b. homeostasis d. heredity
6. Change in the inherited traits of species over time is called
   a. evolution. c. homeostasis.
   b. reproduction. d. responsiveness.
7. Which of the following are real-world problems that biologists can help solve?
   a. destruction of rain forests
   b. extinction of plants and animals
   c. AIDS
   d. all of the above
8. The demand for more food is due to
   a. the spread of disease.
   b. the growth of the human population.
   c. uncontrolled scientific experimentation.
   d. extinction.
9. The disorder characterized by cells dividing uncontrollably within the body is called
   a. AIDS. c. cancer.
   b. cystic fibrosis. d. homeostasis.
10. A proposed explanation for a scientific observation that can be tested by additional observations or experimentation is
    a. a hypothesis. c. a theory.
    b. a prediction. d. a variable.
11. The factor that is varied in a controlled experiment is called the ______ variable.
    a. control c. independent
    b. dependent d. hypothesis
12. Which statement is false?
    a. Observations are an important part of the scientific process.
    b. A solution with a low pH is more acidic than a solution with a high pH.
    c. A hypothesis can be proven with a well-designed experiment.
    d. A prediction is the expected outcome of a test.
13. **Applying Information** Is the word *theory* in the newspaper headline shown below used in a scientific sense or in a more general sense? Explain.

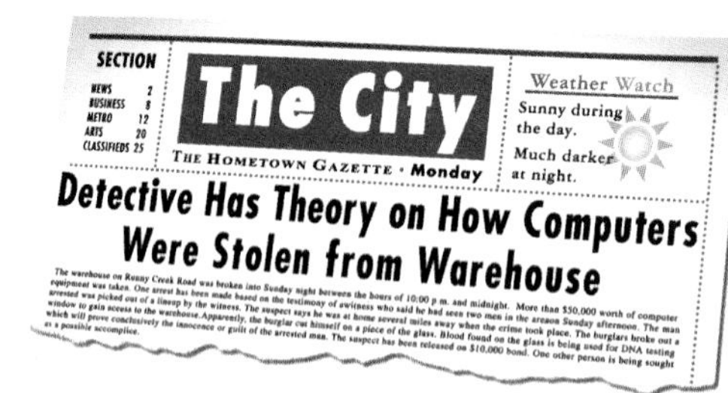

SECTION
NEWS 7
BUSINESS 9
METRO 12
ARTS 20
CLASSIFIEDS 25

The City

THE HOMETOWN GAZETTE · Monday

Weather Watch
Sunny during the day.
Much darker at night.

Detective Has Theory on How Computers Were Stolen from Warehouse

14. **Theme Interdependence** What arguments could you offer against forest clearing for the development of neighborhoods?
15. **Exploring Further** What types of biologists have you seen portrayed on television shows or in films? Were they realistic?

13. It is used in a general sense since the headline is referring to a theory as a guess. In science, a theory is a hypothesis that has been tested and confirmed many times.
14. Answers will vary but might include destruction of plant and wildlife habitat, destruction of lumber resources, and loss of potential medicines and foods.
15. Answers will vary. Students should defend their answers.

Performance ZONE
CHAPTER 1

Skills Assessment

Part 1: Interpreting Graphics

Every year Americans generate more than 4 billion tons of solid waste (any material that is thrown away that is not a liquid or a gas). The pie chart below provides a breakdown of the types of municipal solid waste in the United States. Use the pie chart to answer the following questions:

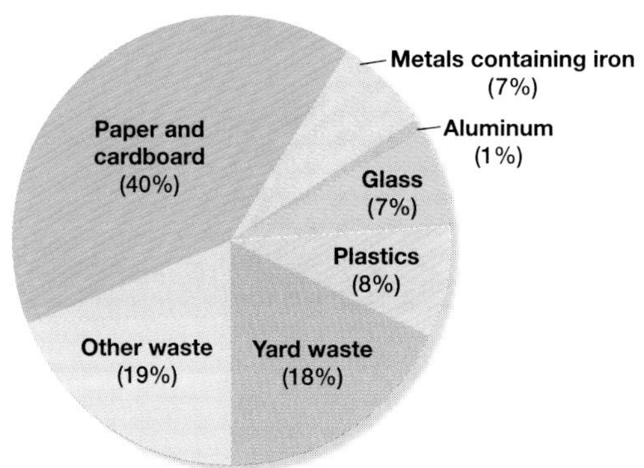

16. What type of waste makes up the major portion of waste in the United States?
17. Even though more people are recycling, most of our waste (80 percent) ends up in landfills. How much of the waste shown in this chart could be recycled?

Part 2: Life/Work Skills

Use the media center or Internet resources to explore community gardens and community garden associations.

18. **Finding Information** What is the mission of the American Community Gardening Association?
19. **Communicating** Prepare an oral report that describes how community gardening works to ease the stress on the environment.
20. **Being a Team Member** Try to find a recycling center in your community that uses volunteers. Find out how the volunteers work as a team, and learn what kinds of tasks they perform for the center. If you can, volunteer to work with the team and write a two-page typewritten report about your experience.

Critical Thinking

21. **Evaluating an Argument** A scientist on television states that it is inappropriate to say that a hypothesis has been proven. Do you agree or disagree? Explain.
22. **Forming Reasoned Opinions** Some people believe that scientists should not tamper with a person's genes. Do you think biologists should use gene therapy to try to cure diseases? Explain.
23. **Recommending Methods** A student wants to determine which of two foods produces the greatest weight gain in gerbils. Design an experiment that the student could perform. Specify a hypothesis, independent and dependent variables, controlled and uncontrolled variables, and the control and experimental groups.

Portfolio Projects

24. **Evaluating Viewpoints** Tropical rain forests are located mainly in developing nations in Central America, South America, and Africa. They are home to at least half of the world's animal and plant species. It is predicted that in fewer than 100 years, the rain forests will be gone as a result of lumbering and clearing for farms. Many people in wealthy nations argue that helping the economic growth of the nations with rain forests is the best way to save these forests. Write a report that explores both sides of this argument.
25. **Interpreting Information** Read "Frogs in Peril" by Robert Gannon (*Popular Science*, Dec. 1997, vol. 251, pp. 84–88). Describe five factors that scientists think could be responsible for the disappearance of frog populations in various locations around the world.
26. **Career Focus** Environmental Biologist Research the field of environmental biology, and write a report on your findings. Your report should include a job description, training required, kinds of employers, growth prospects, and starting salary.

Skills Assessment

16. paper and cardboard
17. Answers may vary. A little less than 81 percent of the waste shown in the chart could be recycled since "other waste" is undefined and may not be recyclable. In addition, not all plastics are recyclable.
18. The mission is to promote and support all aspects of community food and ornamental gardening, urban forestry, preservation and management of open space, and integrated planning and management of developing urban and rural lands.
19. Answers will vary. The stress on the environment is eased by forming and expanding state and regional community gardening networks, developing resources in support of community gardening, encouraging research, and conducting educational programs.
20. Answers will vary.

Critical Thinking

21. Students should agree. It is inappropriate to say that a hypothesis has been proven. A hypothesis can never be proven, only supported or rejected.
22. Answers will vary. Students should defend their answer.
23. Answers will vary. Students should include adequate sample size; identical living conditions; using gerbils with the same age, gender, and initial weight, and at least three groups (control and two experimental groups). Hypothesis might state: Food A will cause an increase in weight when fed to gerbils; independent variable: type of food; dependent variable: weight; control group: gerbils receiving a regular diet; experimental groups: two groups of gerbils, each receiving one of the two foods.

Portfolio Projects

24. Reports will vary.
25. The students should describe each of the following factors: habitat destruction, acid rain, endocrine disrupters, thinning of the ozone layer, and an infestation of parasitic flukes.
26. Environmental biologists conserve biological diversity, manage wildlife and natural resources, control pollution, and research the impact of human activities on the environment. Employers may include national and state parks, environmental organizations, government agencies, private industries, and universities. Training varies, but may include a four-year college degree and post-graduate training. Growth prospects are good. Starting salary will vary by region.

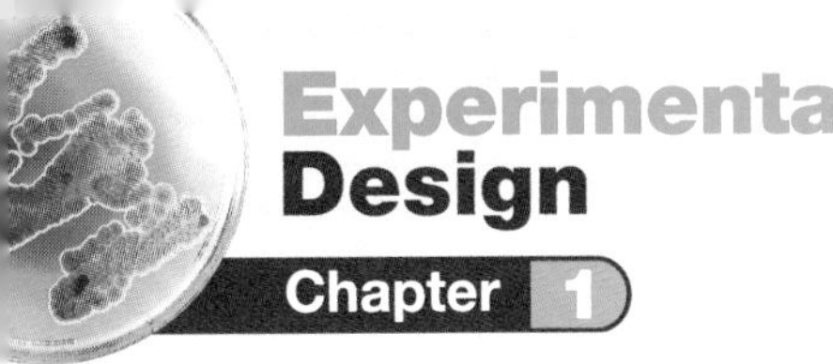

Experimental Design

Chapter 1

How Does Acid Rain Affect Seeds?

Purpose

Students will design and conduct an experiment on the effect of acidic solutions on seeds and seedlings.

Preparation Notes

Time Required: Allow 7–10 days: Day 1: 45–55 minutes; Days 3–10: 15 minutes, every other day.

Materials

Materials for this lab can be ordered from WARD'S. See *Master Materials List* at the front of this book for catalog numbers.

Preparation Tips

- Place enough materials for a class on a supply table. Beans, peas, or corn germinate quickly and are usually mold resistant.
- Set up labeled containers for the disposal of solutions and broken glass.
- Prepare solutions under a ventilated hood. Wear goggles, a face shield, impermeable gloves, and a lab apron.
- Prepare 600 mL of mold inhibitor for 25 students (one part concentrated bleach to four parts water).
- Prepare solutions of different pH as follows. Use distilled water to dilute 5 mL of 1.0 M sulfuric acid (H_2SO_4) to 1 L to prepare a 0.01 M H_2SO_4 solution with a pH of 2. Dilute 50 mL of the 0.01 M H_2SO_4 solution to 1 L to make a solution with a pH of 3. Repeat this procedure using 5 mL and 0.5 mL of the 0.01 M H_2SO_4 solution to make pH 4 and pH 5 solutions, respectively. Verify the pH of each solution.

Safety Precautions

- Have students read *Safe Laboratory Practices* in the **Appendix** before beginning the lab.
- Remind students to read the **ChemSafety** section before beginning the lab. Discuss all safety symbols and caution statements with students.

Procedural Tips

Allow two days between the start of the experiment and the first observation.

Experimental Design

Chapter 1

How Does Acid Rain Affect Seeds?

SKILLS

- Using scientific methods
- Collecting, organizing, and graphing data

OBJECTIVES

- **Use** a scientific method to investigate a problem.
- **Predict** how acid rain might affect seed germination and seedling growth.

MATERIALS

- safety goggles
- lab apron
- 50 seeds
- 250 mL beakers
- 20 mL mold inhibitor
- distilled water
- paper towels
- solutions of different pH
- wax pencil or marker
- zip-lock plastic bags
- metric ruler
- graph paper

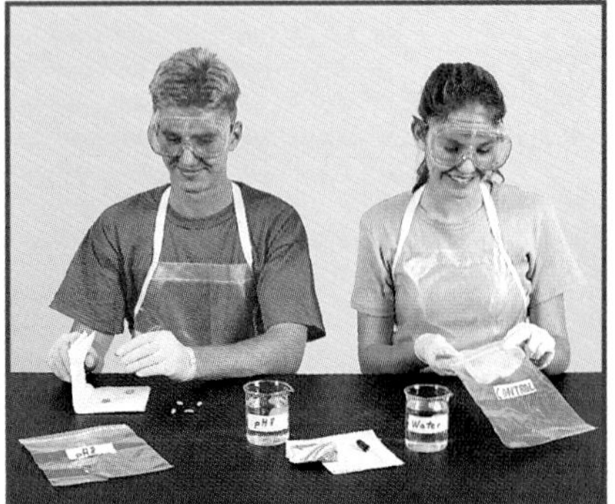

ChemSafety

CAUTION: Always wear safety goggles and a lab apron to protect your eyes and clothing.

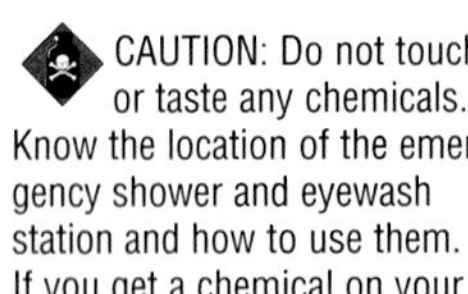

CAUTION: Do not touch or taste any chemicals. Know the location of the emergency shower and eyewash station and how to use them. If you get a chemical on your skin or clothing, wash it off at the sink while calling to the teacher. Notify the teacher of a spill. Spills should be cleaned up promptly, according to your teacher's directions.

CAUTION: Glassware is fragile. Notify the teacher of broken glass or cuts. Do not clean up broken glass or spills with broken glass unless the teacher tells you to do so.

Before You Begin

Living things, such as salamander embryos, can be damaged by **acid rain** at certain times during their lives. In this lab, you will investigate the effect of acidic solutions on seeds. One way to investigate a problem is to design and conduct an **experiment.** We begin a scientific investigation by making **observations** and asking questions.

1. Write a definition for each boldface term in the paragraph above and for each of the following terms: pH, hypothesis, prediction, variable, control group.
2. Based on the objectives for this lab, write a question you would like to explore about the effect of acid rain, for example, When is a plant most susceptible to acid rain?

Procedure

PART A: Design an Experiment

1. Work with members of your lab group to explore one of the questions written for step 2 of **Before You Begin.** To explore the question, design an experiment that uses the materials listed for this lab.

You Choose

As you design your experiment, decide the following:

a. what question you will explore
b. what hypothesis you will test
c. how to simulate growing seeds in soil moistened by acid rain
d. how to keep seeds moist during the experiment
e. what your test solutions and control will be
f. how to measure seedling growth
g. what to record in your data table

Disposal

- Collect solutions in a single container and neutralize to pH 7 with 1.0 M base before discarding in sink. Dilute acid spills with water; mop up with wet cloths designed for spill cleanup.

Before You Begin

Answers

1. *acid rain*—rain with a pH lower than 7 due to the pollutants it contains

   experiment—a planned procedure to test a hypothesis

   observation—the act of noting or perceiving objects or events using the senses

   pH—a relative measure of the hydrogen ion concentration within a solution

   hypothesis—an explanation that might be true and that can be tested by additional observations or experimentation

   prediction—the expected outcome of a test, assuming the hypothesis is correct

   variable—factor that can change

   control group—a group in an experiment that receives no experimental treatment and serves as a standard with which experimental groups can be compared

2. Write a procedure for your experiment. Make a list of all the safety precautions you will take. Have your teacher approve your procedure and safety precautions before you begin the experiment.

PART B: Conduct Your Experiment

3. Put on safety goggles and a lab apron.
4. Place your seeds in a 250 mL beaker, and slowly add enough mold inhibitor to cover the seeds. **CAUTION: The mold inhibitor contains household bleach, which is a base.** Soak the seeds for 10 minutes, and then pour the mold inhibitor into the proper waste container. Gently rinse the seeds with distilled water, and place them on clean paper towels.
5. Set up your group's experiment. **CAUTION: Solutions with a pH below 7.0 are acids.** Conduct your experiment for 7–10 days. Make observations every 1–2 days, and note any changes. Record each day's observations in a data table, similar to the one below.

DATA TABLE

Solution	Date	Observations

PART C: Cleanup and Disposal

6. Dispose of solutions, broken glass, and seeds in the designated waste containers. Do not pour chemicals down the drain or put lab materials in the trash unless your teacher tells you to do so.
7. Clean up your work area and all lab equipment. Return lab equipment to its proper place. Wash your hands thoroughly before you leave the lab and after you finish all work.

Analyze and Conclude

1. **Summarizing Results** Describe any changes in the look of your seeds during the experiment. Discuss seed type, average seed size, number of germinated seeds, and changes in seedling length.
2. **Analyzing Results** Were there any differences between the solutions? Explain.
3. **Analyzing Methods** What was the control group in your experiment?
4. **Analyzing Data** Make graphs of your group's data. Plot seedling growth (in millimeters) on the *y*-axis. Plot number of days on the *x*-axis.
5. **Relating Concepts** What scientific methods did you use to design and conduct your experiment?
6. **Evaluating Methods** How could your experiment be improved?
7. **Inferring Conclusions** How do acidic conditions appear to affect seeds?
8. **Predicting Outcomes** How might acid rain affect the plants in an ecosystem?
9. **Further Inquiry** Write a new question about the effect of acid rain that could be explored with another investigation.

Do You Know?

Do research in the library or media center to answer these questions:

1. Which parts of the United States are most affected by acid rain, and why?
2. How have factories been changed to reduce the amount of acid rain?

Use the following Internet resources to explore your own questions about acid rain.

2. Answers will vary. For example: Are bean plants more susceptible to acid rain than corn plants?

Procedure

Sample Procedure

1. Label plastic bags "pH 3," "pH 4," "pH 5," and "control."
2. Moisten three layers of paper towels with each solution.
3. Arrange 10 seeds that have been treated with mold inhibitor on one half of each set of treated paper towels. Fold the other half over the seeds. Place paper towels and seeds in the proper bag and seal the bag.
4. Record the number of seeds germinated and the length of each seedling. Note any other changes in the seedlings.
5. After each observation, re-wet the paper towels with the same solution as noted on each bag. Return the seeds to the bag.

Sample Data: Seedling Growth (mm)

DAY	PH 3	PH 4	PH 5	CONTROL
0	0*	0	0	0
3	12	15	21	25
5	24	27	37	43
7	32	35	50	58

*Lengths recorded are to the nearest millimeter.

Analyze and Conclude

Answers

1. Answers will vary.
2. Answers will vary. Plants usually grow best in a pH of 4.5 to 6.5.
3. seeds germinated in distilled water
4. Answers will vary. For example:

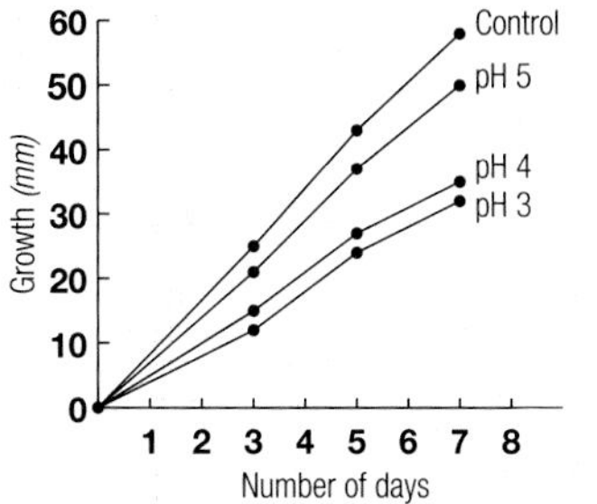

5. Answers should include collecting observations, asking questions, forming hypotheses and making predictions, confirming predictions with experiments, and drawing conclusions.
6. Answers will vary. For example, students may suggest increasing the sample size in each group.
7. Answers will vary. In general, very acidic conditions inhibit seedling growth.
8. Answers will vary. Acidic rain might inhibit plant growth or kill plants.
9. Answers will vary. For example: What are the effects of acidic solutions on mature plants?

Do You Know?

Answers

1. upstate New York and other eastern states, the Pacific Northwest, and several midwestern states
2. Industries have installed devices that remove sulfur and nitrogen compounds from industrial emissions before they reach the atmosphere. Also, lime is added to lakes and rivers to temporarily neutralize their acidity.

Chapter 2

Chemistry of Life

Planning Guide

	OBJECTIVES	MEETING INDIVIDUAL NEEDS	READING SKILLS AND STRATEGIES
2-1 Nature of Matter — BLOCK 1 45 min	• **Differentiate** between atoms and elements. • **Analyze** how compounds are formed. • **Distinguish** between covalent bonds, hydrogen bonds, and ionic bonds.	Learning Styles, p. 28 BASIC ADVANCED • p. 30	ATE Active Reading, Technique: K-W-L, p. 27 ■ Active Reading Guide, worksheet (2-1) ■ Directed Reading, worksheet (2-1) ■ Reading Strategies, K-W-L worksheet
2-2 Water and Solutions — BLOCK 2 45 min	• **Analyze** the properties of water. • **Describe** how water dissolves substances. • **Distinguish** between acids and bases.	Learning Styles, p. 31 BASIC ADVANCED • pp. 32, 33	■ Active Reading Guide, worksheet (2-2) ■ Directed Reading, worksheet (2-2)
2-3 Chemistry of Cells — BLOCK 3 45 min	• **Summarize** the characteristics of organic compounds. • **Distinguish** between carbohydrates, lipids, and proteins. • **Describe** the structure and function of nucleic acids and ATP.	Learning Styles, p. 34 BASIC ADVANCED • pp. 36, 37	ATE Active Reading, Technique: Paired Summarizing, p. 35 ■ Active Reading Guide, worksheet (2-3) ■ Directed Reading, worksheet (2-3)
2-4 Energy and Chemical Reactions — BLOCKS 4, 5, & 6 135 min	• **Evaluate** the importance of energy to living things. • **Relate** energy and chemical reactions. • **Describe** the role of enzymes in chemical reactions.	Learning Styles, p. 38 BASIC ADVANCED • pp. 40, 42	■ Active Reading Guide, worksheet (2-4) ■ Directed Reading, worksheet (2-4) ■ Supplemental Reading Guide, *The Lives of a Cell: Notes of a Biology Watcher*

Review and Assessment Resources

BLOCKS 7 & 8 90 min

TEST PREP

- PE Study Zone, p. 43
- PE Performance Zone, pp. 44–45
- ■ Chapter 2 worksheets:
  - Vocabulary
  - Concept Mapping
  - Science Skills
  - Critical Thinking
- Audio CD Program
- ■ Test Prep Pretest

CHAPTER TESTING

- ■ Chapter Tests (blackline master)
- ▲ Test Generator
- ■ Assessment Item Listing

ALTERNATIVE ASSESSMENT

- PE Portfolio Projects 27–29, p. 45
- ATE Alternative Assessment, p. 43
- ATE Portfolio Assessment, p. 43

See the correlations table at the front of this book for a list of the ***National Science Education Standards*** addressed in this chapter.

Smithsonian Institution® Visit **www.si.edu/hrw** for additional on-line resources.

KEY

Biology Principles and Explorations

PE Pupil's Edition
ATE Teacher's Edition

Teaching Transparencies
Laboratory Program

One-Stop Planner CD-ROM
Shows these resources within a customizable daily lesson plan:

■ Teaching Resources
▲ Test Generator

VISUAL STRATEGIES	CLASSWORK AND HOMEWORK	TECHNOLOGY AND INTERNET RESOURCES
2 Ionic Bonds 3 Hydrogen Bonds in Water 4 Structure of an Atom	■ **Problem-Solving:** Operations with Small and Large Numbers **PE** **Review,** p. 30	Holt Biology Videodiscs Section 2-1 Audio CD Program Section 2-1
ATE **Fig. 2-7,** p. 33	**PE** **Review,** p. 33	Holt Biology Videodiscs Section 2-2 Audio CD Program Section 2-2 internet**connect** SCILINKS NSTA **TOPIC:** Properties of water
ATE **Fig. 2-9,** p. 35	**PE** **Review,** p. 37	Holt Biology Videodiscs Section 2-3 Audio CD Program Section 2-3 internet**connect** SCILINKS NSTA **TOPIC:** Foods as fuel
ATE **Fig. 2-13,** p. 39 **ATE** **Fig. 2-15,** p. 41 4A Molecular Structure of ATP	**PE** **Data Lab** Analyzing the effect of pH on enzyme activity, p. 42 **PE** **Experimental Design** Do Enzyme Detergents Work?, pp. 46–47 Lab E3 Thermodynamics **PE** **Review,** p. 42 **PE** **Study Zone,** p. 43	Holt Biology Videodiscs Section 2-4 Audio CD Program Section 2-4 internet**connect** SCILINKS NSTA **TOPICS:** • Chemical reactions • Enzyme activity • Enzymes

Lab Activity Materials

Data Lab
p. 42 paper, pencil

Experimental Design
p. 46 safety goggles, lab apron, balance, graduated cylinder, glass stirring rod, 150 mL beaker, 18 g regular instant gelatin or 1.8 g sugar-free instant gelatin, 0.7 g Na_2CO_3, tongs or a hot mitt, 50 mL boiling water, thermometer, pH paper, test tubes (6), test-tube rack, pipet with bulb, plastic wrap, tape, 50 mL beakers (6), 50 mL distilled water, 1 g each of 5 brands of laundry detergent, wax pencil, metric ruler

Lesson Warm-up
ATE substances such as table salt, sugar, copper wire, aluminum, water, p. 28

ATE sewing needle, petroleum jelly, water, beaker, p. 31
ATE foods such as oil, sugar, meat, p. 34
ATE tuning fork, flashlight, vinegar, baking soda, p. 38

Demonstrations
ATE inflated balloon, p. 30
ATE stalk of celery, water, red food coloring, p. 32
ATE fresh pineapple, gelatin, p. 40

BIOSOURCES LAB PROGRAM
- Lab Techniques and Experimental Design: Lab C8, p. 35

Chapter 2

Chapter Theme

Metabolism

Metabolism refers to all the chemical reactions that occur in an organism. Inherent to those chemical reactions are atoms, the fundamental units of matter. Atoms will join together to form compounds. Much of the body is composed of organic compounds, such as proteins, lipids, and carbohydrates. These compounds are involved in many chemical reactions that help organisms maintain homeostasis.

Establishing Prior Knowledge

Before beginning this chapter, make certain students can answer the questions in **Study TIP Ready?** on page 27. If they cannot, encourage them to reread the sections indicated.

Opening Question

Ask students what the science of chemistry is, and have them suggest how chemistry relates to biology. *(Chemistry is the study of the composition and properties of matter. Chemistry is related to biology because all organisms are composed of chemical substances.)* Tell students that the properties and reactions of chemical substances are essential to all living things.

CHAPTER 2 Chemistry of Life

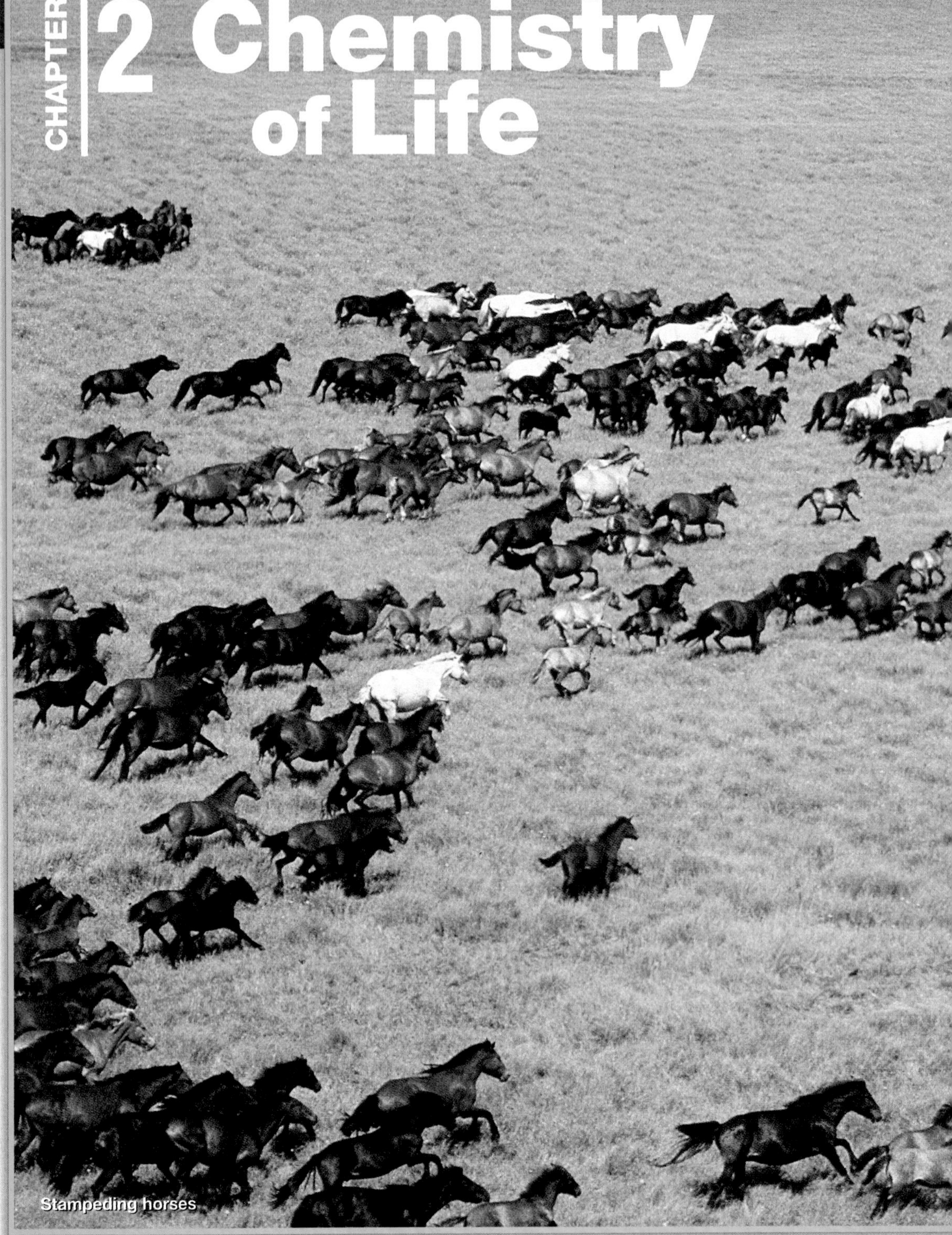

Stampeding horses

What's Ahead — and Why
CHAPTER OUTLINE

Students will learn

- ▷ about the structure of atoms and how atoms form ions and compounds.
- ▷ how the properties of water make it an important substance in living things.
- ▷ about the roles of carbohydrates, lipids, proteins, and nucleic acids in cell functions.
- ▷ how energy can be stored or released by chemical reactions.

Sea turtle

Matter... Energy... Life...

What do this sea turtle and these wild horses have in common? They and all other organisms are made of chemical substances, including water, carbohydrates, proteins, and fats. Organisms also get the energy they need to live from these substances. Water is in and around every cell of your body. Without water, cells cannot function or survive.

Study TIP

Ready?

Check your understanding of these topics to help you prepare for what's ahead.

Can you...

- **identify** seven properties of life? (Section 1-1)
- **list** seven themes of biology? (Section 1-1)
- **distinguish** between metabolism and homeostasis? (Section 1-1)

If not, *review the section indicated.*

Looking Ahead

Labs

- **Data Lab**
  Analyzing the effect of pH on enzyme activity (p. 42)
- **Experimental Design**
  Do Enzyme Detergents Work? (p. 46)

Features

- **Food Watch**
  Foods as Fuel (p. 36)

internet connect

National Science Teachers Association *sci*LINKS Internet resources are located throughout this chapter.

Chapter 2

Reading Strategies

Active Reading

Technique: K-W-L

Before students read this chapter, have each write a short list of all the things they already **K**now (or think they know) about the chemistry of organisms. Ask students to contribute their entries to a group list on a large poster board. Then have students list things they **W**ant to know about chemistry in living things. Have students save their lists as well as the poster boards for later use in the **Closure** activity on page 42.

Directed Reading

Assign **Directed Reading Worksheet Chapter 2** to help students interact with the text as they read the chapter.

Interactive Reading

Assign ***Biology: Principles and Explorations*** **Audio CD Program Chapter 2** to help reluctant readers achieve greater success in reading the chapter.

Reading Skills

Assign Reading Organizers and Reading Strategies Worksheets from the **One-Stop Planner CD-ROM with Test Generator** to help students learn and practice effective reading skills.

internet connect

Holt *Biology: Principles and Explorations*

On-line Resources: go.hrw.com

Visit the HRW Web site for a variety of resources related to this chapter. Just type in the keyword HX0 Chapter 2.

Holt *Biology: Principles and Explorations*

On-line Resources: www.scilinks.org

The following *sci*LINKS Internet resources can be found in the student text for this chapter:

Page 32
TOPIC: Properties of water
KEYWORD: HX032

Page 36
TOPIC: Foods as fuel
KEYWORD: HX036

Page 39
TOPIC: Chemical reactions
KEYWORD: HX039

Page 42
TOPIC: Enzyme activity
KEYWORD: HX042

Page 47
TOPIC: Enzymes
KEYWORD: HX047

Chapter 2

1 Prepare

Scheduling

- **Block:** About 45 minutes will be needed to complete this section.
- **Traditional:** About one class period will be needed to complete this section.
- **Planning:** See the Planning Guide on pp. 25A and 25B for lecture, classwork, and assignment options for Section 2-1. Lesson plans for Section 2-1 can be found in a lesson cycle format in ***Biology: Principles and Explorations*** **One-Stop Planner CD-ROM with Test Generator.**

Resource Link

Assign **Active Reading Worksheet Section 2-1** to help students comprehend the key concepts and visuals in this lesson.

2 Teach

Lesson Warm-up

Show students various substances, such as table salt, sugar, water, copper wire, and a piece of aluminum. Ask students to identify each substance as a compound or element. *(Salt is a compound; sugar is a compound; water is a compound; copper is an element; and aluminum is an element.)* At the end of this section, come back to the question. Emphasize that ions and molecules are important to the functioning of living things.

Section 2-1 Nature of Matter

Objectives

- **Differentiate** between atoms and elements. (p. 28)
- **Analyze** how compounds are formed. (p. 29)
- **Distinguish** between covalent bonds, hydrogen bonds, and ionic bonds. (pp. 29–30)

Key Terms

atom
element
compound
molecule
ion

Matter Is Made of Atoms

Cooking requires an understanding of how the ingredients in foods interact. A cook's application of chemical principles while preparing recipes affects the flavor and texture of foods. Just as a cook can benefit from a knowledge of basic chemistry, you can better understand principles of biology if you also understand the fundamentals of chemistry. Chemistry will help you learn about biology because organisms, including yourself, are chemical machines.

What does all matter have in common? Matter consists of atoms. An **atom** is the smallest unit of matter that cannot be broken down by chemical means. Atoms are so small and dynamic that their exact structure is difficult to determine. Scientists have developed models, such as the one shown in **Figure 2-1,** to explain the structure and properties of atoms.

As shown in Figure 2-1, atoms consist of three kinds of particles: electrons, protons, and neutrons. Protons, shown in red, and neutrons, shown in blue, make up the core, or nucleus, of an atom. The space around the nucleus that electrons may occupy at any time is called the electron cloud, shown as a blue haze around the nucleus. Electrons are negatively charged, so the electron cloud has a negative charge. Protons are positively charged and neutrons have no charge, so the nucleus has a positive charge. Because protons and electrons are oppositely charged, they attract one another. Atoms typically have one electron for each proton, so they have no electrical charge.

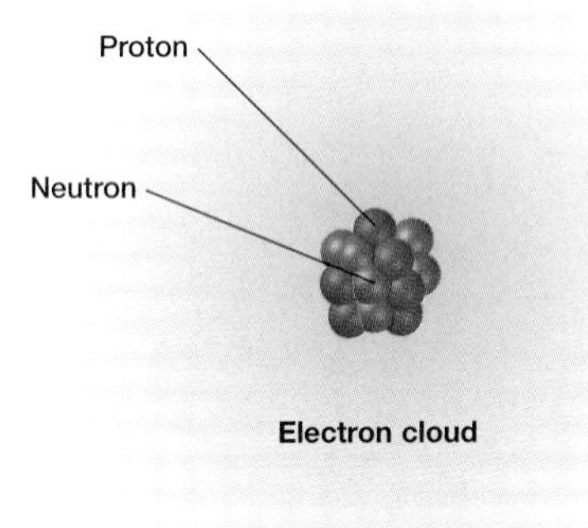

Figure 2-1 **Atom.** The electron cloud is the region of an atom where electrons are most likely to be found. The nucleus of this atom contains six protons and six neutrons.

Elements are the simplest pure substances

An **element** is a substance made of only one kind of atom, and it is therefore a pure substance. There are more than 100 known elements, such as gold and helium. Each element is represented by a one-, two-, or three-letter symbol. For example, the elements hydrogen, oxygen, and carbon are represented by the symbols H, O, and C, respectively. Elements differ in the number of protons their atoms contain. Atoms of the simplest element, hydrogen, each contain one proton and one electron. In contrast, oxygen atoms contain eight protons and eight electrons. The number of neutrons in an atom is often but not always equal to the number of protons and electrons in the atom. Atoms of an element that contain different numbers of neutrons are called isotopes of that element. For example, carbon, C, has three isotopes—carbon-12, carbon-13, and carbon-14—each containing six protons.

Integrating Different Learning Styles

Logical/Mathematical	**ATE**	Teaching Tip, p. 29
Visual/Spatial	**PE**	Performance Zone item 1, p. 44
	ATE	Lesson Warm-up, p. 28; Closure, p. 30; Reteaching, p. 30
Body/Kinesthetic	**ATE**	Demonstration, p. 30

Atoms Can Bond Together

Atoms can join with other atoms to form stable substances. A force that joins atoms is called a chemical bond. A **compound** is a substance made of the joined atoms of two or more different elements. For example, when sodium atoms, Na, bond with chlorine atoms, Cl, the compound sodium chloride (table salt) forms. Every compound is represented by a chemical formula that identifies the elements in the compound and their proportions. The formula for sodium chloride, NaCl, shows that there is one sodium atom for every chlorine atom in the compound.

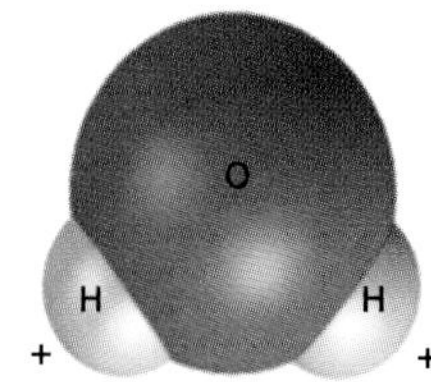

Figure 2-2 Water molecule. Each water molecule is held together by covalent bonds between two hydrogen atoms and one oxygen atom.

Covalent bonds

Covalent bonds form when two or more atoms share electrons to form a molecule. A **molecule,** such as the water molecule shown in **Figure 2-2,** is a group of atoms held together by covalent bonds. Like the rivets and welds that connect steel girders in a skyscraper, covalent bonds join the atoms in molecules. Because the number of protons is equal to the number of electrons in a molecule, the molecule has no net electrical charge. Other examples of molecules include carbon dioxide, CO_2, and oxygen gas, O_2.

In molecules, the arrangement of electrons determines how the atoms bond together. Electrons are grouped into different levels. The inner levels (closest to the nucleus) have less energy than the outer levels (farther from the nucleus). Electron levels can hold a limited number of electrons. With the exception of hydrogen and helium, which have outer electron levels that can hold up to two electrons, all atoms have outer electron levels that can hold up to eight electrons. An atom becomes stable when its outer electron level is full. If the outer electron level is not full, an atom will react readily with atoms that can provide electrons to fill its outer level. As Figure 2-2 shows, water, H_2O, forms when an oxygen atom, which has six outer electrons, combines with two hydrogen atoms, which have one outer electron each.

Hydrogen bonds

The electrons in a water molecule are shared by oxygen and hydrogen atoms. However, the shared electrons are attracted more strongly by the oxygen nucleus than by the hydrogen nuclei. The water molecule therefore has partially positive and negative ends, or poles. As shown in **Figure 2-3,** the partially positive end of one water molecule is attracted to the negative end of another water molecule. Molecules with an unequal distribution of electrical charge, such as water molecules, are

Figure 2-3 Hydrogen bonds in water

Water molecules are attracted by hydrogen bonds.

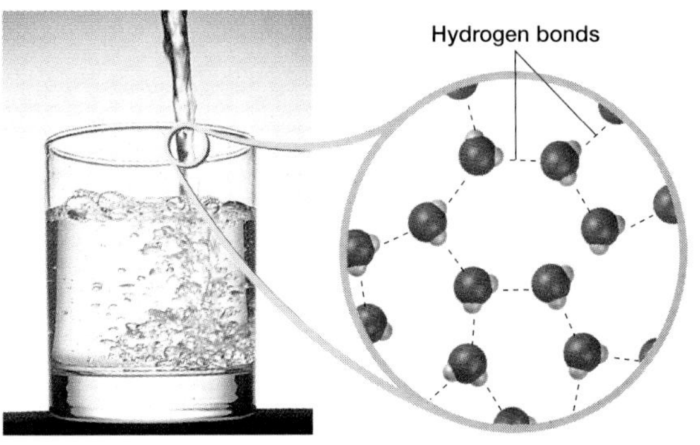

Teaching TIP

Periodic Table of the Elements

Give a brief introduction to the periodic table of the elements. Explain how to determine the atomic number, the atomic mass, and the number of protons, electrons, and neutrons in an atom. For example, the element fluorine, F, has an atomic number of 9, which is equal to both the number of protons and the number of electrons in each atom of fluorine. The atomic mass of fluorine is 18.99, making the number of neutrons about 10 (19–9). After a few examples, randomly pick some elements and have the students determine the answers. Elements found in living things include carbon, C, hydrogen, H, oxygen, O, and nitrogen, N. Have students find these elements on the periodic table.

Making Biology Relevant

Health

Nuclear medicine uses radioisotopes to treat certain diseases. Isotopes are atoms of the same element that have different numbers of neutrons. A radioisotope is an isotope that has an unstable nucleus and undergoes radioactive decay. Have students research radioisotopes and their uses.

PORTFOLIO

Overcoming Misconceptions

Atomic Structure

Students often do not understand that atoms consist mostly of empty space. If the nucleus of an atom were the size of a typical marble, the first electron level would be about 0.8 km (0.5 mi) away from the nucleus.

Chapter 2

REAL LIFE

Answer

Humans are endothermic, and we therefore maintain a fairly constant internal temperature. In response to cold, the body decreases blood flow near the skin's surface. Muscle contractions known as shivering may also occur to help generate heat.

Demonstration

Opposite Charges

To demonstrate the attraction between opposite charges, have students gently rub an inflated balloon through their hair. Then have students bring the balloon close to their hair without the two touching. Ask students to describe their observations. Explain that their hair is attracted to the balloon because the negative charges on the balloon are attracted to the positive charges on their hair.

3 Assessment Options

Closure

Atomic Structure

Have students imagine that you have magnified an atom to the size of the classroom. Hang a marble or a lump of modeling clay in the center of the room to represent the nucleus. Tell students that the farther away from the nucleus an electron is, the more energy it has.

Review

Assign Review—Section 2-1.

Reteaching — BASIC

Assign students to groups of four. Have each group make flash cards listing the key terms from this chapter. Some of the cards should have the definition on one side and the term on the other. Other flash cards should have questions appropriate to this chapter on one side and the answer on the other. Upon completion, have the groups quiz each other.

REAL LIFE

Because cells contain so much water, they are vulnerable to freezing. Some species of marine fish can swim in icy water because their body fluids contain chemicals similar to automotive antifreeze.

Analyzing Information *Humans lack these chemicals. Why then can we tolerate short-term exposure to subzero temperatures?*

called polar molecules. Nonpolar molecules have an equal distribution of electrical charge. This attraction between two water molecules is an example of a hydrogen bond. A hydrogen bond is a weak chemical attraction between polar molecules.

Ionic bonds

Sometimes atoms or molecules gain or lose electrons. An atom or molecule that has gained or lost one or more electrons is called an **ion** *(EYE ahn)*. Ions have an electrical charge because they contain an unequal number of electrons and protons. An atom that has lost electrons is positively charged, whereas an atom that has gained electrons is negatively charged.

Ions of opposite charge may interact to form an ionic bond. For example, an atom of sodium is unstable because it has only one electron in its outer level. Sodium readily gives up this electron to become a stable, positively charged sodium ion, Na^+. An atom of chlorine is also unstable because it has seven electrons in its outer level. Chlorine readily accepts an electron to become a stable, negatively charged chloride ion, Cl^-. The negative charge of a chloride ion is attracted to the positive charge of a sodium ion. Thus, sodium atoms and chlorine atoms readily form an ionic bond to become sodium chloride, as shown in **Figure 2-4.**

Figure 2-4 Ionic bonds in sodium chloride

Ionic bonds in sodium chloride, NaCl, are formed by the interaction between sodium ions, Na^+, and chloride ions, Cl^-.

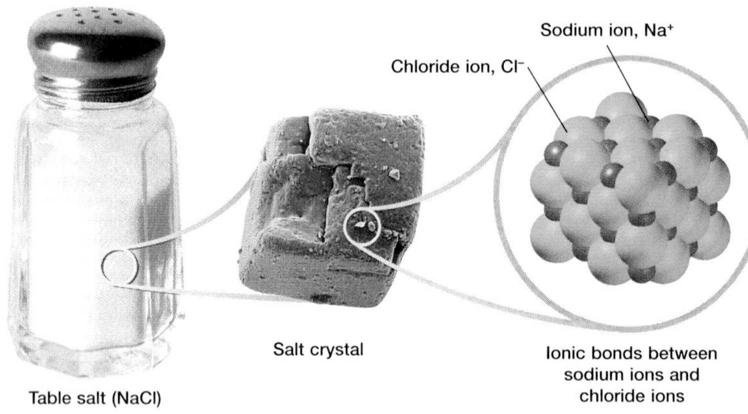

Review SECTION 2-1

1. **Differentiate** between atoms and elements.
2. **Describe** how an atom differs from a molecule.
3. **Distinguish** between covalent bonds and ionic bonds.
4. **SKILL Recognizing Differences** Explain the difference between polar molecules and nonpolar molecules. Give an example of a polar molecule.
5. **SKILL Relating Concepts** Explain the relationship between hydrogen bonding and polarity.

Review SECTION 2-1 ANSWERS

1. An atom is the smallest unit of matter that cannot be broken down by chemical means. An element is a substance made of only one kind of atom.
2. An atom is a single unit, whereas a molecule is a group of atoms held together by covalent bonds.
3. Covalent bonds form when two or more atoms share electrons. Ionic bonds form when ions of opposite charge interact with one another.
4. Polar molecules have an unequal distribution of electrical charge. Nonpolar molecules have an equal distribution of electrical charge. An example of a polar molecule is water.
5. Because polar molecules have an unequal distribution of electrical charge, the more positive end of one polar molecule will attract the more negative end of another polar molecule. This attraction is called a hydrogen bond.

Water and Solutions

Section 2-2

Water Is a Major Component of Cells

You may not realize it, but nearly 70 percent of your body is made of water. About two-thirds of the molecules in your body are water molecules. Your body's cells are filled with water, and water is the medium in which most cellular events take place. Your cells are also surrounded by water, and water helps move nutrients and other substances into and out of your cells. What are some of the properties of water that make it such an important substance for life?

Water stores heat efficiently

Water heats more slowly and retains heat longer than many other substances. For example, a pot of boiling water removed from a stove takes a long time to cool down to room temperature. Many organisms release excess heat through water evaporation. For example, humans cool themselves by sweating. The water vapor lost through the evaporation of sweat carries heat away from the body. In organisms, this ability to control temperature enables cells to maintain a constant internal temperature when the external temperature changes drastically. Water thus helps cells maintain homeostasis.

Water bonds to itself and other substances

The hydrogen bonds between water molecules cause the cohesion of liquid water. **Cohesion** *(koh HEE zhuhn)* is an attraction between substances of the same kind. Because of cohesion, water and other liquids form drops and thin films, as shown in **Figure 2-5.** Molecules at the surface of water are linked together by hydrogen bonds like a crowd of people linked by holding hands. This attraction between water molecules causes a condition known as surface tension. Surface tension prevents the surface of water from stretching or breaking easily.

Water molecules are also attracted to many other similarly polar substances. **Adhesion** *(ad HEE zhuhn)* is an attraction between different substances. Because of adhesion, some substances get wet. Adhesion powers a process, called capillary action, in which water molecules move upward through a narrow tube, such as the stem of a plant. The attraction of water to the walls of the tube sucks the water up more strongly than gravity pulls it down. Water moves upward through a plant from roots to leaves through a combination of capillary action, cohesion, and other factors.

Objectives

- **Analyze** the properties of water. (p. 31)
- **Describe** how water dissolves substances. (p. 32)
- **Distinguish** between acids and bases. (p. 33)

Key Terms

cohesion
adhesion
solution
acid
base

Figure 2-5 Cohesion. Because of cohesion, water forms drops like those on this plant.

Integrating Different Learning Styles

Logical/Mathematical	**PE**	Performance Zone items 18–21, p. 45
Visual/Spatial	**ATE**	Lesson Warm-up, p. 31; Demonstration, p. 32; Teaching Tip, p. 32; Closure, p. 33; Reteaching, p. 33
Verbal/Linguistic	**PE**	Study Tip, p. 33

1 Prepare

Scheduling

- **Block:** About 45 minutes will be needed to complete this section.
- **Traditional:** About one class period will be needed to complete this section.
- **Planning:** See the Planning Guide on pp. 25A and 25B for lecture, classwork, and assignment options for Section 2-2. Lesson plans for Section 2-2 can be found in a lesson cycle format in ***Biology: Principles and Explorations*** **One-Stop Planner CD-ROM with Test Generator.**

Resource Link

- Assign **Active Reading Worksheet Section 2-2** to help students comprehend the key concepts and visuals in this lesson.

2 Teach

Lesson Warm-up

Prior to class, coat a sewing needle with petroleum jelly. Ask students to predict whether you can float the needle on the surface of water in a beaker. The grease keeps the water from wetting the needle; that is, the grease keeps the water molecules from being more attracted to the needle than they are to each other. Using a pair of forceps, carefully lay the needle flat on the surface of the water. The needle will float. Have students observe the results without touching the beaker. Then push the needle slightly beneath the surface of the water. The needle will sink. Ask students why the needle floated on the water at first, but then sank when the water's surface was disturbed. *(At first, the surface tension of the water supported the needle. When the surface tension was broken, the needle sank.)*

internet connect

SciLinks NSTA Have students research the properties of water using the Web site in the Internet Connect box on page 32. Have students write a report summarizing their findings. PORTFOLIO

Demonstration

Capillary Action — BASIC

Place one end of a freshly cut stalk of celery in a beaker of water with a high concentration of red food coloring. Let the celery sit in the water overnight. Have students observe the results. Have each student pull one of the "strings" out of the celery stalk and examine it with a hand lens. Ask students to describe the structure of the "strings." Ask students how the water was able to move up through the celery stalk? *(Capillary action and other factors enable water to move upward through the celery stalk.)*

Teaching TIP — ADVANCED

Water Density

Hydrogen bonds in ice, unlike those in liquid water, are stable. In ice, the water molecules form a crystalline structure. The molecules in an ice crystal are spaced apart, while in liquid water they are very close together. A volume of ice contains fewer water molecules than the same volume of liquid water. Thus, ice is less dense than water. Fill a drinking glass with water, and put ice cubes in the glass. Ask students why the ice cubes float in the water. *(The ice cubes float because they are less dense than the liquid water.)*

Water Dissolves Many Substances

Many substances dissolve in water. For example, when you add salt to water, the resulting mixture is a saltwater solution. A **solution** is a mixture in which one or more substances are evenly distributed in another substance. Many important substances in the body have been dissolved in blood or other aqueous fluids. Because these substances can dissolve in water, they can more easily move within and between cells. For example, sugar could not be delivered to your cells if it were not dissolved in water.

Polarity

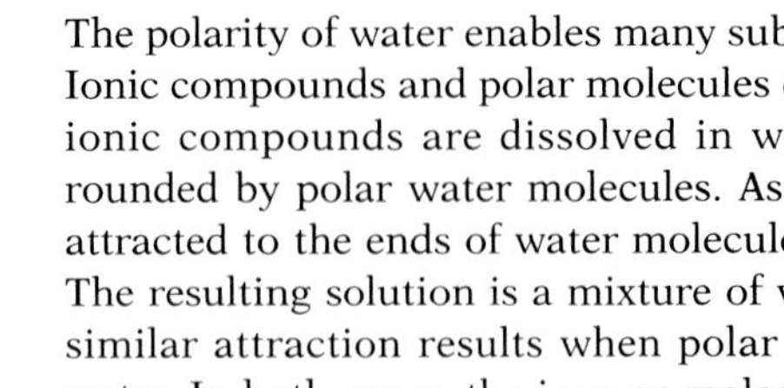

The polarity of water enables many substances to dissolve in water. Ionic compounds and polar molecules dissolve best in water. When ionic compounds are dissolved in water, the ions become surrounded by polar water molecules. As **Figure 2-6** shows, ions are attracted to the ends of water molecules with the opposite charge. The resulting solution is a mixture of water molecules and ions. A similar attraction results when polar molecules are dissolved in water. In both cases, the ions or molecules become evenly distributed in the water.

Nonpolar molecules do not dissolve well in water. When nonpolar substances, such as oil, are placed in water, the water molecules are more atttracted to each other than to the nonpolar molecules. As a result, the nonpolar molecules are shoved together. This explains why oil forms clumps or beads in water. The inability of nonpolar molecules to dissolve in polar molecules is important to organisms. For example, the shape and function of cell membranes depend on the interaction of polar water with nonpolar membrane molecules.

Figure 2-6 Water dissolves ionic compounds

When sodium chloride, NaCl, is dissolved in water, sodium ions, Na^+, and chloride ions, Cl^-, become surrounded by water molecules, H_2O.

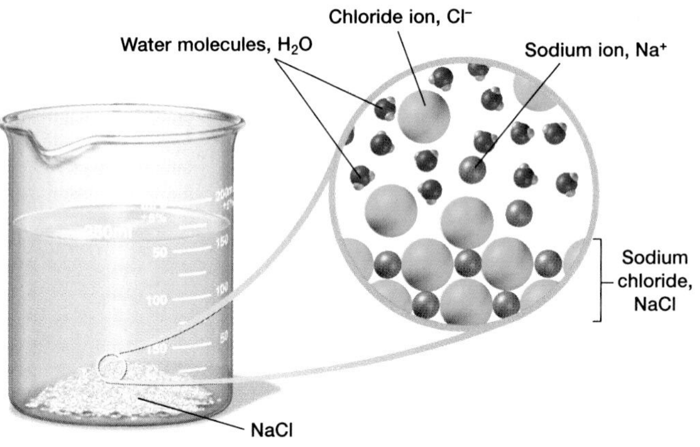

did you know?

Tardigrades, more commonly called "water bears," are small animals usually less than 1 mm long.

Tardigrades live in moist habitats, such as thin films of water on mosses. They are classified in their own phylum, Tardigrada. When subjected to dry conditions, they assume a barrel-shaped form called a tun. Tardigrades may remain in that state for as long as 100 years. When water becomes plentiful, they assume their normal appearance and activities.

Acids and bases

While the bonds in water molecules are strong, at any given time a tiny fraction of those bonds might break, forming a hydrogen ion, H^+, and a hydroxide ion, OH^-:

$$H_2O \longrightarrow H^+ + OH^-$$

As a result, pure water always has a low concentration of hydrogen ions and hydroxide ions. Notice from the equation above that in pure water the number of hydrogen ions and hydroxide ions would have to be equal. Compounds that form hydrogen ions when dissolved in water are called **acids.** When an acid is added to water, the concentration of hydrogen ions in the solution is increased above that of pure water. In contrast, compounds that reduce the concentration of hydrogen ions in a solution are called **bases.** Many bases form hydroxide ions when dissolved in water. Such bases lower the concentration of hydrogen ions because hydroxide ions react with hydrogen ions to form water molecules.

The pH scale shown in **Figure 2-7** measures the concentration of hydrogen ions in solutions. Most solutions have a pH value between 0 and 14. Pure water has a pH value of 7. Acidic solutions have pH values below 7, and basic solutions have pH values above 7. Each whole number represents a factor of 10 on the scale. A solution with a pH value of 5, for example, has 10 times as many hydrogen ions as one with a pH value of 6. Stomach acid has a pH value of about 2 (very acidic). Blood has a pH value of about 7.5 (slightly basic). Your body constantly monitors the pH of these and other fluids.

Study TIP

Reading Effectively

Solutions with pH values below 7 are usually referred to as acidic solutions. Solutions with pH values above 7 are often called basic solutions but are also called alkaline solutions. As you read, you may encounter the terms *alkaline* or *alkalinity.* Both of these terms may indicate the presence of a base or a basic solution.

Figure 2-7 **The pH scale.** The pH scale measures the concentration of hydrogen ions in a solution.

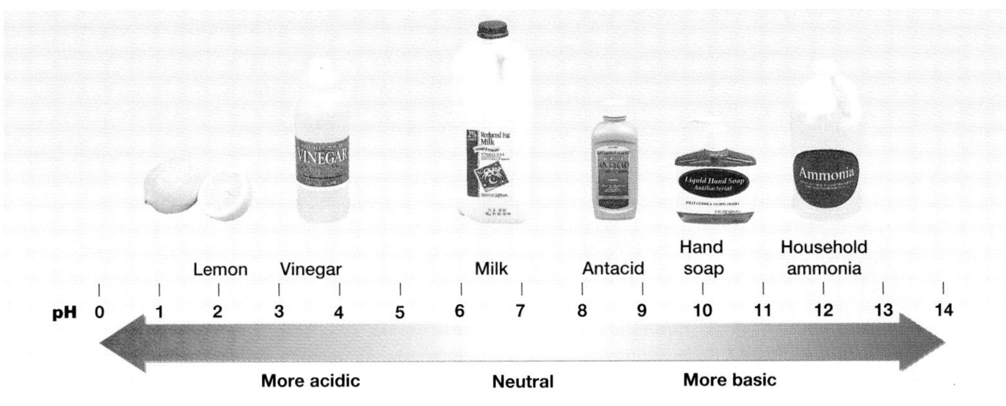

Review SECTION 2-2

1. **Distinguish** between adhesion and cohesion.
2. **Name** a substance that would not dissolve well in water. Explain why.
3. **Differentiate** between acids and bases.
4. **SKILL Relating Concepts** Describe how water helps cells maintain homeostasis.

Critical Thinking

5. **SKILL Inferring Relationships** When salt is added to water, the freezing point of the water decreases. Explain why this occurs.

Review SECTION 2-2 ANSWERS

1. Adhesion is an attraction between different substances. Cohesion is an attraction between substances of the same kind.
2. Oil, for example, would not dissolve well in water because oil is nonpolar. Water molecules are polar and are more attracted to each other than to the nonpolar oil molecules.
3. An acid forms hydrogen ions when dissolved in water. A base forms hydroxide ions when dissolved in water.
4. The properties of water help cells and organisms maintain homeostasis. Refer to pp. 31–32.
5. When salt is added to water, the salt dissolves in the water. The dissolved ions interfere with the hydrogen bonds between water molecules, which normally become stable when ice is formed. When the temperature of the solution is lowered, it is therefore more difficult for ice to form after salt has been added to the water. Thus, salt lowers the freezing point of water.

Teaching TIP ADVANCED

Buffers

A buffer is a substance that resists changes in pH when an acid or base is added. Human blood has a pH of about 7.5, and it must be maintained as such. When carbon dioxide enters the blood, which is mostly water, it reacts with the water to form a buffer called carbonic acid. Carbonic acid dissociates into a hydrogen ion, H^+, and a bicarbonate ion, HCO_3^-. If the hydrogen ion concentration of the blood begins to decrease (the pH rises), more carbonic acid dissociates, lowering the pH. If the hydrogen ion concentration rises (the pH drops), the bicarbonate ion removes excess hydrogen ions, raising the pH.

3 Assessment Options

Closure

Immiscible Substances

Bring a bottle of vinaigrette salad dressing to class. Ask students to identify the major ingredients in the dressing *(oil and vinegar).* Shake the bottle vigorously and then let the bottle rest during the class period. Ask students to explain why the oil and vinegar eventually separate. *(The oil, which is nonpolar, is repelled by the water in the vinegar, which is polar.)*

Review

Assign Review—Section 2-2.

Reteaching BASIC

Show students photographs in which the properties of water are illustrated. Use pictures from various magazines. Ask students to identify the property being illustrated in each picture. *(Answers will vary. For example, a picture of a tree could be used as an example of capillary action, in that water moves through the tree. A picture of a glass filled with water indicates adhesion between water and the glass, as well as cohesion of the water molecules.)*

Section 2-3 Chemistry of Cells

1 Prepare

Scheduling

- **Block:** About 45 minutes will be needed to complete this section.
- **Traditional:** About one class period will be needed to complete this section.
- **Planning:** See the Planning Guide on pp. 25A and 25B for lecture, classwork, and assignment options for Section 2-3. Lesson plans for Section 2-3 can be found in a lesson cycle format in ***Biology: Principles and Explorations*** **One-Stop Planner CD-ROM with Test Generator.**

Resource Link

Assign **Active Reading Worksheet Section 2-3** to help students comprehend the key concepts and visuals in this lesson.

2 Teach

Lesson Warm-up

Show students several foods, such as oil, sugar, and a piece of meat. Ask students what all these substances have in common. *(Answers may include that each food contains macromolecules, including lipids, carbohydrates, and proteins, or a combination of molecules.)* Emphasize that all contain the element carbon.

Objectives

- **Summarize** the characteristics of organic compounds. (p. 34)
- **Distinguish** between carbohydrates, lipids, and proteins. (pp. 34–36)
- **Describe** the structure and function of nucleic acids and ATP. (p. 37)

Key Terms

carbohydrate
monosaccharide
lipid
protein
amino acid
nucleic acid
nucleotide
DNA
RNA
ATP

Carbon Compounds Are Found in Living Things

Most matter in your body that is not water is made of organic compounds. Organic compounds contain carbon atoms that are covalently bonded to other elements—typically hydrogen, oxygen, and other carbon atoms. Four principal classes of organic compounds are found in living things: carbohydrates, lipids, proteins, and nucleic acids. Without these compounds, cells could not function.

Carbohydrates

Carbohydrates are organic compounds made of carbon, hydrogen, and oxygen atoms in the proportion of 1:2:1. Carbohydrates are a key source of energy, and they are found in most foods—especially fruits, vegetables, and grains. The building blocks of carbohydrates are single sugars, called **monosaccharides** *(mahn oh SAK uh reyedz),* such as glucose, $C_6H_{12}O_6$, and fructose. Glucose is a major source of energy in cells. Disaccharides are double sugars formed when two monosaccharides are joined. For example, sucrose, or common table sugar, consists of both glucose and fructose. Polysaccharides such as starch, shown in **Figure 2-8,** are chains of three or more monosaccharides. A polysaccharide is an example of a macromolecule, a large molecule made of many smaller molecules.

In organisms, some polysaccharides function as storehouses of the energy contained in sugars. Two polysaccharides that store energy in this way are starch, which is made by plants, and glycogen, which is made by animals. Both starch and glycogen are made of hundreds of linked glucose molecules. Cellulose is a polysaccharide that provides structural support for plants. Humans cannot digest cellulose. Thus, you cannot digest wood, which is mostly cellulose.

Figure 2-8 Structure of polysaccharides. Starch is a long chain of many linked glucose molecules.

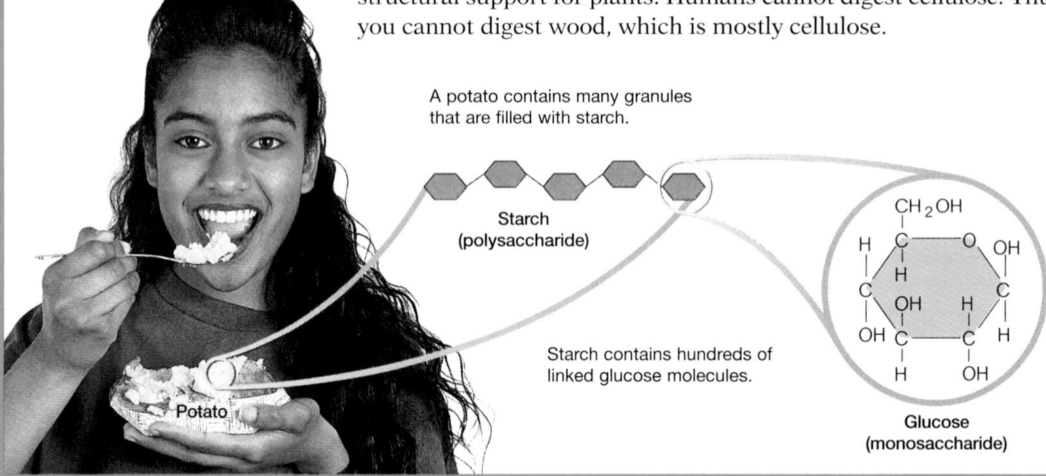
A potato contains many granules that are filled with starch.
Starch (polysaccharide)
Starch contains hundreds of linked glucose molecules.
Potato
CH_2OH
Glucose (monosaccharide)

Integrating Different Learning Styles

Logical/Mathematical	**PE**	Study Tip, p. 37
	ATE	Teaching Tip, p. 35
Visual/Spatial	**ATE**	Lesson Warm-up, p. 34; Visual Strategy, p. 35; Reteaching, p. 37
Interpersonal	**ATE**	Active Reading, p. 35; Closure, p. 37
Intrapersonal	**PE**	Real Life, p. 35; Food Watch, p. 36
Verbal/Linguistic	**PE**	Performance Zone item 2, p. 44

Figure 2-9 Structure of fats

Fatty acids can be saturated or unsaturated.

Saturated fatty acid

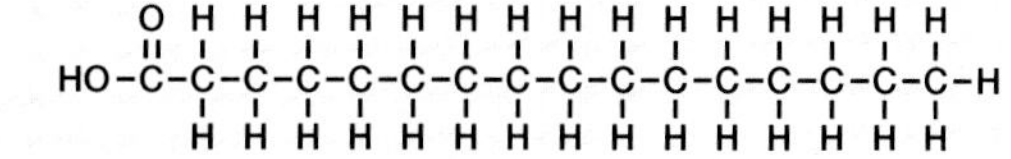

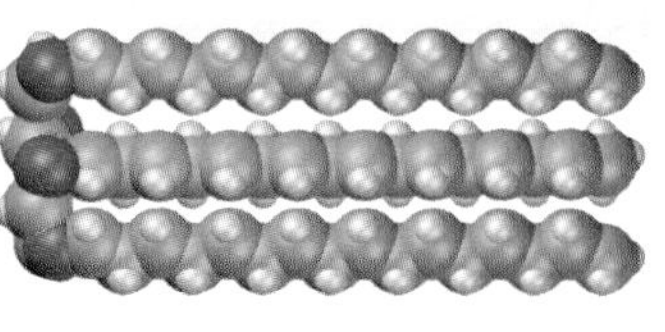

Saturated fats, such as butter, are solid at room temperature.

Unsaturated fatty acid

O H H H H H H H H H H H H H H H H
HO-C-C-C-C-C-C-C-C-C=C-C-C=C-C-C=C-C-C-H
H H H H H H H H H H H H H H

Unsaturated fats, such as olive oil, are liquid at room temperature.

Lipids

Lipids *(LIHP ihdz)* are nonpolar molecules that are not soluble in water. They include fats, phospholipids, steroids, and waxes. Lipids are an important part of the structure and functioning of cell membranes. Phospholipids make up the lipid bilayer of cell membranes. Steroids include cholesterol, which is found in animal cell membranes. Other lipids include some light-absorbing compounds called pigments, such as the plant pigment chlorophyll.

Fats Fats are lipids that store energy. As **Figure 2-9** shows, a typical fat contains three fatty acids bonded to a glycerol molecule. Glycerol is an alcohol with three carbon atoms. A fatty acid is a long chain of carbon atoms, shown in green, with hydrogen atoms bonded to them. Most carbon atoms in a fatty acid are bonded to either one or two hydrogen atoms, shown in blue. Because bonds between carbon and hydrogen are rich in energy, fats can store a lot of energy.

In a saturated fatty acid, all of the carbon atoms in the chain are bonded to two hydrogen atoms (except the one on the end, which is bonded to three hydrogen atoms). Most animal fats—such as those in butter, lard, and grease from cooked meats—contain primarily saturated fatty acids. Saturated fatty acids are straight molecules and are generally solid at room temperature.

In an unsaturated fatty acid, some of the carbon atoms are linked by a "double" covalent bond, each with only one hydrogen atom, producing kinks in the molecule, as shown in Figure 2-9. Most plant oils, such as olive oil, and some fish oils contain mainly unsaturated fatty acids and are generally liquid at room temperature. Hydrogenated vegetable oils contain naturally unsaturated fatty acids that have been saturated artificially by the addition of hydrogen atoms. Thus, hydrogenated vegetable oils, such as those in margarine and vegetable shortening, are generally solid at room temperature.

REAL LIFE

Fat-free potato chips fried in artificial fats contain fewer calories than those fried in natural fats. Unfortunately, some artificial fats may reduce vitamin absorption and cause indigestion in some people.

Finding Information *Research the benefits and potential shortcomings of artificial fats.*

Overcoming Misconceptions

Cholesterol

Not all cholesterol is bad. LDL, or "bad," cholesterol contributes to atherosclerosis, a form of arteriosclerosis characterized by the buildup of fatty substances on the interior surfaces of artery walls. There is also "good" cholesterol that is essential to the body's functions. For example, the body needs cholesterol to produce lipid-based molecules, such as steroid hormones, which are synthesized from cholesterol.

Visual Strategy

Figure 2-9

Point out that the saturated fatty acid contains no double bonds in its hydrocarbon chain, whereas the unsaturated fatty acid does contain double bonds. Ask students what causes the kinks in an unsaturated fatty acid. *(The double bonds between carbon atoms in the hydrocarbon chain cause the kinks in the chain.)* Ask students why unsaturated fats, such as olive oil, are liquid at room temperature. *(In unsaturated fats, the kinks caused by the double bonds in fatty acids prevent the fat molecules from packing closely together. Thus, unsaturated fats have a low melting point; they cannot solidify at room temperature.)*

REAL LIFE

Answer

Artificial fats are used in certain foods. In some people, artificial fats cause stomach cramps, diarrhea, and the loss of fat-soluble vitamins. Ask students to weigh the benefits of eating foods with artificial fats against the possible side effects. *(Artificial fats would seem to provide a healthy alternative to a high-calorie, high fat-food diet, but the associated loss of essential nutrients in growing young people may outweigh these advantages.)*

Active Reading

Technique: Paired Summarizing

Have students read Section 2-3 silently. After a designated time, assign students to cooperative pairs. Have one student summarize what he or she read while the other student listens. The listener should point out inaccuracies and add concepts that the other left out. Both students should work together during the final clarification process, referring to the text as needed.

CAREER

Organic Chemist

An organic chemist identifies and analyzes the chemical processes related to organic compounds. Organic chemists may perform either pure or applied research. Job opportunities vary. They may, for example, work in universities, dividing their time between teaching and research, or in pharmaceutical companies, where their expertise is needed in the discovery, development, and evaluation of drugs. A bachelor's degree is required for entry-level positions. PORTFOLIO

Foods As Fuel

Teaching Strategies

- Use the Food Watch on page 36 to stimulate discussion about nutrition and diet.
- Have students make a table listing foods that are high in carbohydrates, fats, or proteins.

Discussion

What is the difference between essential amino acids and nonessential amino acids? *(Essential amino acids can only be acquired from foods. The body makes the nonessential amino acids.)*

Resource Link — ADVANCED

- Assign **Problem Solving: Using Food Labels to Calculate Percentage of Nutrients and Calories** from the Holt BioSources Teaching Resources CD-ROM. PORTFOLIO

Proteins

A **protein** *(PROH teen)* is a chain of molecules called amino acids linked together like pearls on a necklace. **Amino acids** are the building blocks of proteins. Twenty different amino acids are found in proteins. Each amino acid has a different chemical structure. Some amino acids are polar, and others are nonpolar. Some amino acids are electrically charged, and others are not charged. As **Figure 2-10** shows, proteins tend to fold into compact shapes determined by how the amino acids in the "arms" of the protein interact with water and one another.

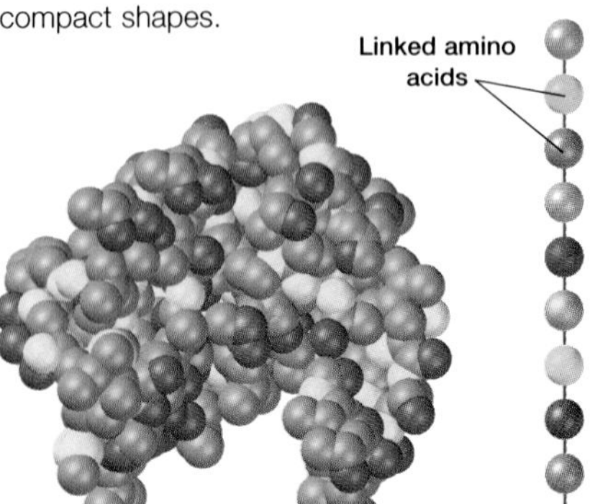

Figure 2-10 Structure of proteins. Proteins are chains of amino acids folded into compact shapes.

Some proteins called enzymes promote chemical reactions. Other proteins have important structural functions. For example, the most abundant protein in your body is collagen *(KAHL uh juhn)*, which is found in skin, ligaments, tendons, and bones. Your hair and muscles contain structural proteins and so do the fibers of a blood clot. Other proteins called antibodies help your body defend against infection. Specialized proteins in muscles enable your muscles to contract, making body movement possible. In your blood, a protein called hemoglobin carries oxygen from your lungs to body tissues.

FOOD WATCH: Foods as Fuel

Most foods contain a mixture of carbohydrates, proteins, and fats. The body can use these molecules to build new tissues, but it uses them mostly as an energy source. Your body's cells harvest the energy in food molecules for metabolism. The energy value of food molecules is measured in kilocalories (kcal).

The minimal rate of energy use per hour (h), called the basal metabolic rate, is about 70 kcal/h for men and 60 kcal/h for women. Typically, walking uses about 200 kcal/h and jogging uses about 600 kcal/h. If more kilocalories are consumed than are used, the body will store the excess kilocalories as fat, regardless of whether the consumed kilocalories are contained in carbohydrates, proteins, or fats.

Carbohydrates

Most carbohydrates in foods come from plant products, such as fruits, grains, and vegetables. Other sources are milk, which contains the sugar lactose, and various meats, which contain some glycogen. Candy and soft drinks also contain sugars. About 4 kcal of energy is supplied by 1 gram (g) of carbohydrates.

Proteins

Primary sources of dietary protein include legumes, eggs, milk, fish, poultry, and meat. As with carbohydrates, proteins supply about 4 kcal/g. Dietary protein is an important source of amino acids. Proteins also provide raw materials for other compounds, such as nucleic acids.

Fats

Fats are found mainly in vegetable oils, such as olive oil; dairy products, such as milk and butter; and meat, such as beef and pork. Fats contain more energy per gram than do carbohydrates and proteins; fats supply about 9.5 kcal/g of energy.

MULTICULTURAL PERSPECTIVE

A Cultural Dependence on Milk?

Lactose, the sugar found in milk, is a disaccharide made of the monosaccharides glucose and galactose. Many adults lack lactase, the enzyme needed to digest lactose. This enzyme deficiency is commonly known as lactose intolerance. For example, it is estimated that 80–90 percent of African-Americans and Asian Americans lack lactase and therefore cannot metabolize lactose. Biologists think that this difference may reflect a cultural dependency on milk products. The diets of Asian and African cultures often do not include as many dairy products as do Western diets.

Figure 2-11 Structure of nucleic acids

DNA is made of two strands of many linked nucleotides.

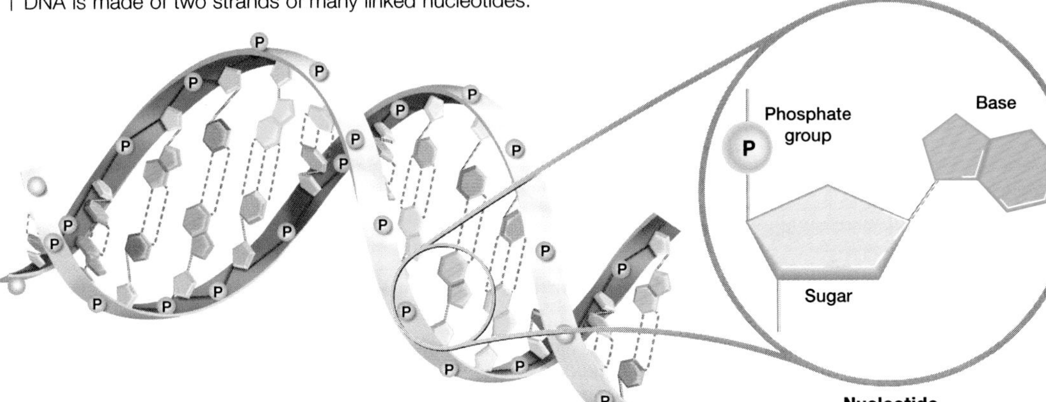

Nucleic acids

All of your cells contain nucleic acids. A **nucleic acid** is a long chain of smaller molecules called nucleotides. A **nucleotide** has three parts: a sugar, a base, and a phosphate group, which contains phosphorus and oxygen atoms. There are two types of nucleic acids—DNA and RNA—and each type contains four kinds of nucleotides.

DNA, or deoxyribonucleic acid, consists of two strands of nucleotides that spiral around each other, as shown in **Figure 2-11.** The two strands of a DNA molecule are held together by hydrogen bonds between bases across from one another. Your chromosomes consist of very long strands of DNA, which stores hereditary information that can be used to make proteins. **RNA,** or ribonucleic acid, consists of a single strand of nucleotides. RNA plays several roles in cell function, including the manufacture of proteins.

ATP carries energy in cells

Another important biological molecule is ATP. **ATP,** or adenosine *(uh DEHN uh seen)* triphosphate, is a single nucleotide with two extra energy-storing phosphate groups. ATP is the main energy currency of cells. When food molecules, such as carbohydrates and fats, are broken down inside cells, some of the energy in the molecules is stored temporarily in ATP. Some of that energy is used by the cell. Cells need a steady supply of ATP to function.

Study TIP

Reading Effectively

Make a table that lists information about carbon compounds. Along the top, write *Carbohydrates, Lipids, Proteins,* and *Nucleic acids.* Along the sides, write *Characteristics* and *Examples.* Add information to the table as you review this section.

Review SECTION 2-3

1. **List** two polysaccharides used to store energy.
2. **Distinguish** between unsaturated fatty acids and saturated fatty acids.
3. **Compare** the structure of DNA with the structure of RNA.
4. **SKILL Organizing Information** Describe two different functions of proteins.

Critical Thinking

5. **SKILL Inferring Relationships** Why do some athletes eat high-carbohydrate foods the day before a competition?

Review SECTION 2-3 ANSWERS

1. Both starch and glycogen are polysaccharides used to store energy.
2. In an unsaturated fatty acid, some of the carbon atoms in the hydrocarbon chain are bonded to only one hydrogen atom, and the chain contains one or more double bonds. In saturated fatty acids, there are no double bonds in the hydrocarbon chain.
3. Both DNA and RNA are nucleic acids, which are made of nucleotides. DNA consists of a double strand of nucleotides held together by hydrogen bonds. RNA is a single-stranded nucleic acid.
4. Some proteins help chemical reactions occur, while others are important in structural functions.
5. Carbohydrates are a rich energy source that meets the high-energy needs of athletes during a competition.

Teaching TIP

ATP

Explain to students that energy is stored in the phosphate bonds of ATP molecules. When these bonds are broken, energy is released and can be used by cells. Ask students if cells have an unlimited supply of ATP. *(Explain to students that when the outer phosphate bond of an ATP molecule breaks, ATP becomes ADP, or adenosine diphosphate. ADP can once again become ATP with the addition of a phosphate group.)*

3 Assessment Options

Closure

Food Labels

Bring some food labels to class to show students how they can check for carbohydrate, fat, and protein content in foods. Assign students to groups of four. Have each group examine the labels. Review the four principal classes of macromolecules. Explain that because the body manufactures nucleic acids, they are not considered a nutrient.

Review

Assign Review—Section 2-3.

Reteaching — BASIC

Have students make a concept map that includes the following terms: carbon, carbohydrate, monosaccharide, disaccharide, polysaccharide, lipids, fats, fatty acid, proteins, amino acid, nucleic acid, nucleotide, glucose, sucrose, starch. *(Maps will vary.)* PORTFOLIO

Chapter 2

1 Prepare

Scheduling

- **Block:** About 135 minutes will be needed to complete this section.
- **Traditional:** About three class periods will be needed to complete this section.
- **Planning:** See the Planning Guide on pp. 25A and 25B for lecture, classwork, and assignment options for Section 2-4. Lesson plans for Section 2-4 can be found in a lesson cycle format in ***Biology: Principles and Explorations*** **One-Stop Planner CD-ROM with Test Generator.**

Resource Link

Assign **Active Reading Worksheet Section 2-4** to help students comprehend the key concepts and visuals in this lesson.

2 Teach

Lesson Warm-up

Demonstrate different examples of energy forms. For example, use a tuning fork to demonstrate sound energy, a flashlight to demonstrate light energy, and a mixture of vinegar and baking soda to demonstrate chemical energy. Pictures can also be used to illustrate some forms of energy, such as one of an explosion to demonstrate heat energy.

Section 2-4

Energy and Chemical Reactions

Objectives

- **Evaluate** the importance of energy to living things. (p. 38)
- **Relate** energy and chemical reactions. (p. 39)
- **Describe** the role of enzymes in chemical reactions. (pp. 40–42)

Key Terms

energy
activation energy
enzyme
substrate
active site

Organisms Need Energy for Life Processes

You are surrounded by energy. Energy is in food, in the motion of a speeding car, in the sound of a guitar, and in the warmth of a blazing fire. **Energy** is the ability to move or change matter. Energy exists in many forms—including light, heat, chemical energy, mechanical energy, and electrical energy—and it can be converted from one form to another. If you kick a ball, for example, the energy of your kick makes the ball move. If you cook an egg in a hot skillet, heat causes the egg to change color and solidify, as shown in **Figure 2-12.** The energy transferred to the egg by heat rearranges the atoms and molecules in the egg. The bioluminescent click beetle, also shown in Figure 2-12, uses energy to produce light.

Energy can be stored or released by chemical reactions. A chemical reaction is a process during which chemical bonds between atoms are broken and new ones are formed, producing one or more different substances. At any moment, thousands of chemical reactions are occurring in every cell of your body. The starting materials for chemical reactions are called reactants. The newly formed substances are called products. Chemical reactions are summarized by chemical equations, which are written in the following form:

$$\text{Reactants} \longrightarrow \text{Products}$$

The arrow is read as "changes to" or "forms." For example, dissolving sodium chloride in water causes the following reaction:

$$NaCl \longrightarrow Na^+ + Cl^-$$

Figure 2-12 Evidence of chemical reactions. An egg becomes solid when it is heated. A chemical reaction causes the bioluminescent click beetle, *Pyrophorus noctilucus,* to give off light energy.

Integrating Different Learning Styles

Logical/Mathematical	**PE**	Data Lab, p. 42
	ATE	Closure, p. 42
Visual/Spatial	**ATE**	Teaching Tip, p. 39; Visual Strategies, pp. 39 and 41; Demonstration, p. 40, Reteaching, p. 42
Musical/Rhythmic	**ATE**	Lesson Warm-up, p. 38
Verbal/Linguistic	**ATE**	Study Tip, p. 40
	PE	Teaching Tip, p. 40

Figure 2-13 Energy and chemical reactions

Chemical reactions absorb or release energy.

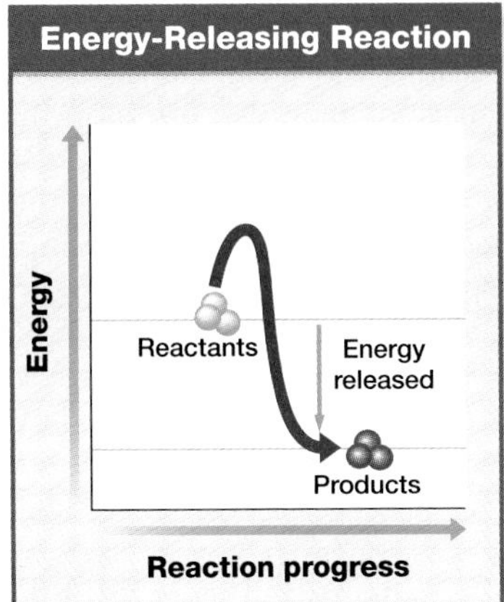

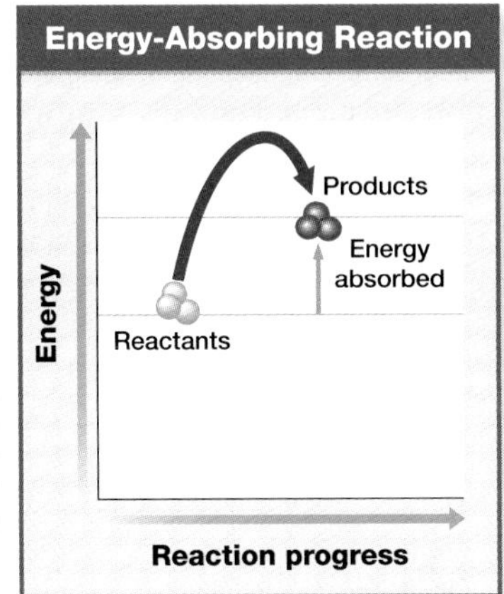

Chemical reactions absorb or release energy

In chemical reactions, energy is absorbed or released when chemical bonds are broken or formed. The graphs shown in **Figure 2-13** compare a chemical reaction that releases energy with a chemical reaction that absorbs energy. The freezing and melting of water are good examples of how energy is released or absorbed during chemical reactions. When water freezes, the process that leads to the formation of ice crystals causes heat energy to be released. When you fill an ice-cube tray with water and place it in the freezer to make ice, heat is released from the water as the water freezes. When you remove ice cubes from the freezer, the ice begins to melt. When ice melts, it absorbs heat from the environment. When you hold a piece of ice, your hand gets cold and heat is transferred from your hand to the ice as the ice begins to melt.

Metabolism *(muh TAB uh lihz uhm)* is the term used to describe all of the chemical reactions that occur within an organism. Your cells get most of the energy needed for metabolism from the food you eat. As food is digested, chemical reactions convert the chemical energy in food molecules to forms of energy that can be used by cells.

TOPIC: Chemical reactions
GO TO: www.scilinks.org
KEYWORD: HX039

Energy is needed to start a chemical reaction

The heat from a flame transfers enough energy to ignite the logs in a campfire. The spark from a spark plug causes the gasoline in an automobile engine to ignite. In both cases, energy is needed to start a chemical reaction. The energy needed to start a chemical reaction is called **activation energy.** To better understand activation energy, think of rolling a boulder down a hill. To get the boulder rolling downhill, you must first push it. Activation energy is simply a chemical "push" that starts a chemical reaction. Even in a chemical reaction that releases energy, activation energy must be supplied before the reaction can occur.

Teaching TIP

Energy Transformation

Have students draw a **Graphic Organizer** similar to the one at the bottom of page 39 showing an example of energy transformation. They should list at least two transformations in their diagrams. Ask students to include a paragraph tracing the flow of energy they have charted. Emphasize that energy transformations occur in living things. PORTFOLIO

Visual Strategy

Figure 2-13

Energy-releasing reactions, represented by the graph on the left, are called exergonic reactions. Energy-absorbing reactions, represented by the graph on the right, are called endergonic reactions. Ask students to give examples of energy-releasing reactions and energy-absorbing reactions *(For example, the burning of wood gives off heat and light energy.)*

SCILINKS NSTA Have students research chemical reactions using the Web site in the Internet Connect box on page 39. Have students write a report summarizing their findings. PORTFOLIO

Graphic Organizer

Use this graphic organizer with ***Teaching Tip: Energy Transformation*** *on page 39.*

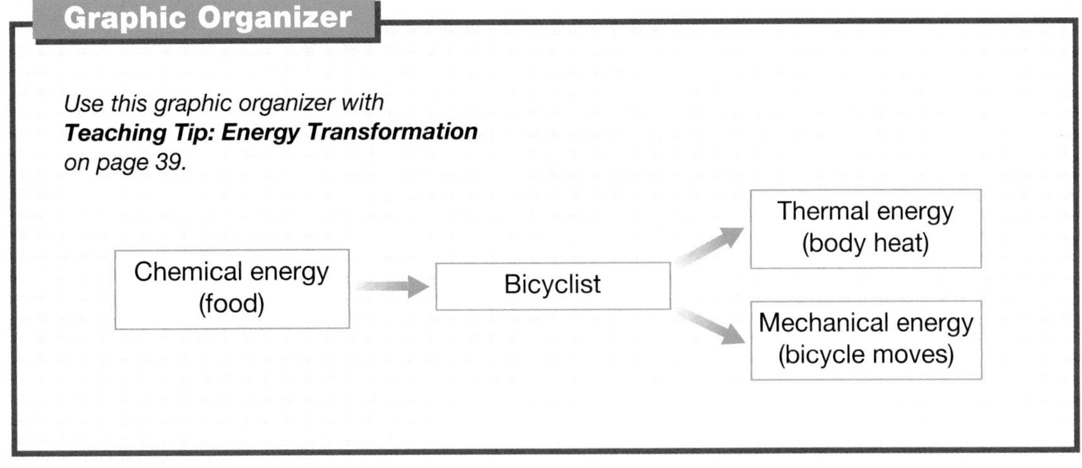

Making Biology Relevant

Health

Fever is a response to infection. A high fever is dangerous because critical enzymes are inactivated at high temperatures. Fever exceeding 41°C (105°F) can be fatal. Fevers accompanied by shaking or chills and fevers that persist for several days should be investigated by a doctor.

Teaching TIP — ADVANCED

Enzyme Names

Enzymes are often named by incorporating the name of an enzyme's substrate with the function that the enzyme performs, and then placing the suffix *-ase* at the end. For example, a dehydrogenase is an enzyme that removes hydrogen atoms from a substrate. When the suffix *-ase* is added to the name of the substrate without adding the functional name, it usually means that the enzyme breaks the substrate down into products. For example, sucrase breaks sucrose down to glucose and fructose. Ask students to suggest the function of the enzymes lactase and lipase. *(Lactase breaks lactose down to glucose and galactose. Lipase breaks lipids down to glycerol and fatty acids.)* Have the students research the functions of other enzymes. PORTFOLIO

Demonstration

Bromelin — BASIC

When fresh pineapple is added to a mixture containing the protein gelatin, the gelatin will not form a gel. Prior to class, prepare gelatin with fresh pineapple and gelatin without pineapple. Tell students that pineapple contains bromelin, an enzyme that breaks down gelatin. Ask students what might happen if gelatin were prepared using canned pineapple. *(In canned pineapple, the bromelin has been destroyed by a heating process. Thus, the gelatin will form a gel when canned pineapple is used.)*

Enzymes Help Biochemical Reactions Occur

Like engines, cells consume fuel because they need energy to function. Just as an engine requires a spark of energy to begin burning gasoline, most biochemical reactions—chemical reactions that occur in cells—require activation energy to begin. The chemical reactions in cells occur quickly and at relatively low temperatures because of the action of many enzymes. **Enzymes** are substances that increase the speed of chemical reactions. Most enzymes are proteins. Enzymes are catalysts *(KAT uh lists),* which are substances that reduce the activation energy of a chemical reaction. As shown in **Figure 2-14,** an enzyme increases the speed of a chemical reaction by reducing the activation energy of the reaction.

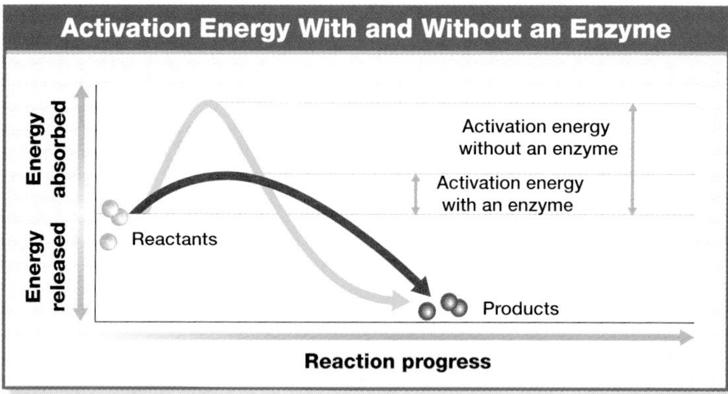

Figure 2-14 Enzymes lower activation energy. Enzymes decrease the amount of energy needed to start a chemical reaction but do not change the amount of energy contained in either the reactants or the products.

Study TIP

Reading Effectively

As you read, notice that the names of most enzymes, such as amylase and catalase, end with *-ase.* This will help you identify other enzymes you will encounter in this book.

Enzymes help organisms maintain homeostasis. Without enzymes, chemical reactions would not occur quickly enough to sustain life. For example, consider a reaction that takes place in your blood. Blood carries carbon dioxide, CO_2, (a waste product made by cells) to your lungs, where it is eliminated as you breathe out. In the lungs, carbon dioxide reacts with water, H_2O, to form carbonic acid, H_2CO_3, as shown by the following equation:

$$CO_2 + H_2O \underset{}{\overset{\text{carbonic anhydrase}}{\rightleftharpoons}} H_2CO_3$$

The reverse reaction occurs in your lungs, converting carbonic acid back to carbon dioxide and water. Most enzyme-assisted reactions are reversible, meaning they can proceed in the opposite direction.

Without an enzyme, the reaction that produces carbonic acid is very slow; only about 200 molecules of carbonic acid are produced in an hour. This rate is not fast enough for your blood to carry away the carbon dioxide released by millions of cells. Fortunately, your blood contains the enzyme carbonic anhydrase *(an HEYED rays).* In the presence of carbonic anhydrase, carbon dioxide and water react to form about 600,000 molecules of carbonic acid per second! The enzyme increases the reaction rate about 10 million times, enabling your body to eliminate carbon dioxide efficiently.

did you know?

The reaction of carbon dioxide with water to form carbonic acid is used in the beverage industry to make carbonated drinks, such as soda.

The enzyme carbonic anhydrase is not used, however. Instead the carbon dioxide is forced into a liquid under high pressure as the beverage is bottled or canned. When you open a can of soda, the soda fizzes because the release of pressure allows some of the carbon dioxide to escape the liquid. Carbon dioxide slowly leaves the liquid, eventually resulting in a "flat" beverage.

Enzymes affect specific substances

A substance on which an enzyme acts during a chemical reaction is called a **substrate** *(SUHB strayt)*. Enzymes act only on specific substrates. For example, the enzyme amylase *(AM uh lays)* assists in the breakdown of starch to glucose in the following chemical reaction. In this reaction, starch is amylase's substrate.

$$\text{starch} \underset{}{\overset{\text{amylase}}{\rightleftharpoons}} \text{glucose}$$

The enzyme catalase *(KAT uh lays)* assists in the breakdown of hydrogen peroxide, H_2O_2, a toxin formed in cells. In this case, hydrogen peroxide is broken down to water, H_2O, and oxygen gas, O_2. In this reaction, hydrogen peroxide is catalase's substrate.

$$2H_2O_2 \overset{\text{catalase}}{\rightleftharpoons} 2H_2O + O_2$$

An enzyme's shape determines its activity. Typically, an enzyme is a large protein with one or more deep folds on its surface. These folds form pockets called **active sites.** As shown in **Figure 2-15,** an enzyme's substrate fits into the active site. An enzyme acts only on a specific substrate because only that substrate fits into its active site.

Step 1 When an enzyme first attaches to a substrate during a chemical reaction, the enzyme's shape changes slightly so that the substrate fits more tightly in the enzyme's active site.

Step 2 At an active site, an enzyme and a substrate interact in a way that reduces the activation energy of the reaction, making the substrate more likely to react.

Step 3 The reaction is complete when products have formed. The enzyme is now free to catalyze further reactions.

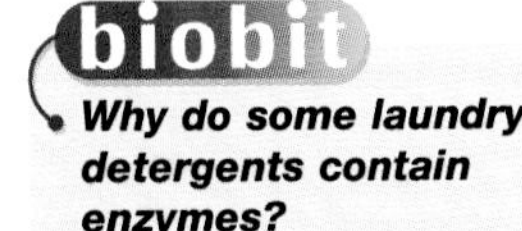

Why do some laundry detergents contain enzymes?

Enzymes break down carbohydrates or proteins in common stains such as food and blood, making the stains easier to remove. The enzymes are purified from mutant bacteria that can tolerate the high temperatures and alkaline conditions required for cleaning fabrics.

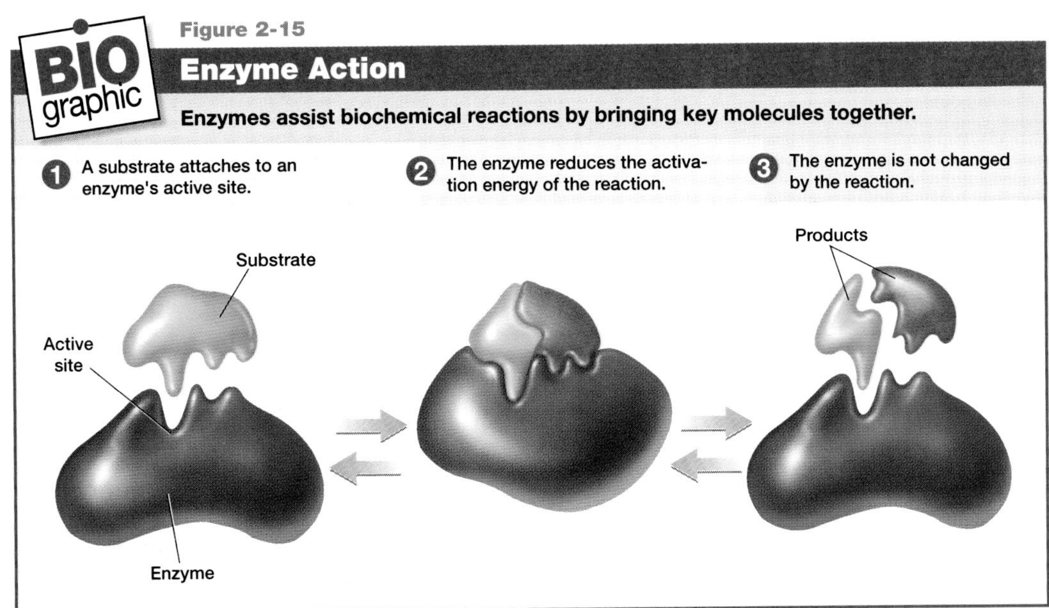
BIO graphic
1 A substrate attaches to an enzyme's active site.
2 The enzyme reduces the activation energy of the reaction.
3 The enzyme is not changed by the reaction.
Substrate
Active site
Enzyme
Products

Figure 2-15

Enzyme Action

Enzymes assist biochemical reactions by bringing key molecules together.

did you know?

Onions can cause some people to "cry."

Onions contain sulfur compounds called amino acid sulfoxides. When an onion is cut, peeled, or crushed, enzymes called allinases are released from the tissue of the onion. These enzymes convert the amino acid sulfoxides into sulfenic acids, which form a chemical that triggers tear production in the eyes.

CROSS-DISCIPLINARY CONNECTION

◆ Physics

X rays, discovered in 1895, were the first means of producing a photographic image of internal organs, such as bone. However, there are major drawbacks in using X rays. For example, X rays do not make soft tissues observable; they only produce a two-dimensional image. Exposure to large doses of X rays can cause cancer.

Since the discovery of X rays, other imaging techniques have been developed. One is magnetic resonance imaging, or MRI, which uses neither X rays nor high-energy radiation. MRI actually uses water molecules in the body. The nuclei of the hydrogen atoms in water molecules are randomly oriented. MRI uses powerful magnets to align these nuclei, then uses a brief pulse of radio waves to knock them out of alignment. The nuclei immediately realign and in the process give out a faint signal, which is picked up by the MRI scanner and translated by a computer into an image. MRI produces images of soft tissues because soft tissues consist mainly of water. Bone, on the other hand, contains very little water and is not picked up as clearly by MRI.

Visual Strategy

BIO graphic Figure 2-15 Tell students that an enzyme cannot completely eliminate the activation energy of a chemical reaction the enzyme catalyzes. As you guide students through **Step 1**, **Step 2**, and **Step 3** of this figure, emphasize that enzyme-assisted reactions are reversible. This is indicated by the double arrows between each step.

Analyzing the effect of pH on enzyme activity

Time 10 minutes

Process Skills Interpreting graphs, analyzing

Teaching Strategies
Review how to interpret a graph. Tell students that the optimum or greatest activity of each enzyme is the highest point of the curve on the graph.

Analysis Answers
1. pepsin
2. trypsin
3. trypsin: about 6.6
4. pepsin: about 3.0
5. Pepsin works in the stomach, and it functions best at a low pH. Trypsin works in the small intestine, and it functions at a higher pH. Thus, it can be inferred that the stomach is highly acidic and that the small intestine is slightly acidic.

3 Assessment Options

Closure

Technique: K-W-L
Tell students to return to the lists of things they Want to know about the role chemistry plays in living organisms, which they created using the K-W-L **Active Reading** on page 27. Have them place check marks next to the questions that they are now able to answer. Find out what they Learned in this chapter that might answer their questions. Lead a class discussion to answer those items without a check mark.

Review

Assign Review—Section 2-4.

Reteaching — BASIC

Have students make a spider diagram to help them identify the characteristics of enzymes described in Section 2-4. PORTFOLIO

internetconnect

SCILINKS. NSTA

TOPIC: Enzyme activity
GO TO: www.scilinks.org
KEYWORD: HX042

Factors that affect enzyme activity

Factors that change the shape of an enzyme affect the enzyme's activity. One factor is temperature. Enzymes operate most efficiently within a certain range of temperatures. Temperatures outside this range either cause some of the bonds that determine an enzyme's shape to break or cause other bonds to become stronger, also changing the shape of the enzyme. Another factor that affects enzyme activity is pH. Each enzyme operates best within a certain range of pH values. A pH value outside this range can cause bonds in an enzyme to break, reducing the enzyme's effectiveness.

The enzymes that are active at any one time in a cell determine what happens in that cell. Your body's cells contain many different enzymes, and each enzyme catalyzes a different chemical reaction. Different kinds of cells contain different collections of enzymes. For example, as you read this page, the chemical reactions occurring in nerve cells in your eye are different from the chemical reactions occurring in your red blood cells. Because the two kinds of cells contain different enzymes, the cells are able to carry out different specialized functions.

DATA LAB: Analyzing the effect of pH on enzyme activity

Background

The graph at right shows the relationship between pH and the activity of two digestive enzymes, pepsin and trypsin. Pepsin works in the stomach, while trypsin works in the small intestine. Use the graph to answer the following questions.

Enzymes and pH

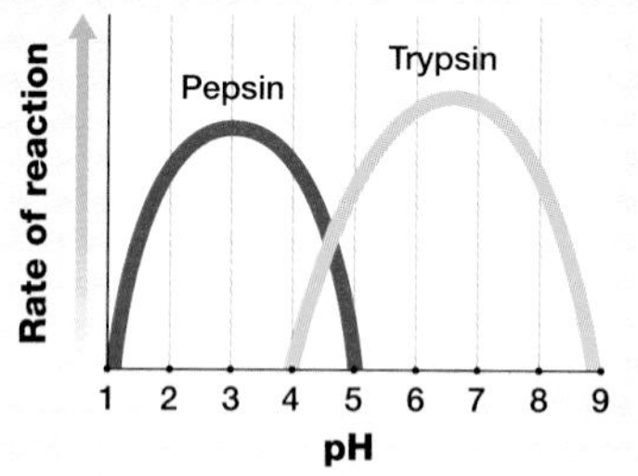

Analysis

1. **Name** the enzyme that works best in highly acidic environments.
2. **Name** the enzyme that works best in less-acidic environments.
3. **SKILL Analyzing Data** Identify the pH value at which trypsin works best.
4. **SKILL Analyzing Data** Identify the pH value at which pepsin works best.
5. **SKILL Inferring Relationships** What does the graph indicate about the relative acidity of the stomach and small intestine?

Review SECTION 2-4

1. **List** three ways that organisms use energy.
2. **Summarize** how energy is made available by chemical reactions.
3. **Describe** how enzymes increase the speed of chemical reactions.
4. **SKILL Comparing Functions** How does an enzyme interact with a substrate?

Critical Thinking

5. **SKILL Predicting Outcomes** What effect might a molecule that interferes with the action of carbonic anhydrase have on your body?

Review SECTION 2-4 ANSWERS

1. Organisms store energy, use energy to power the chemical reactions of metabolism, and use energy to build cell structures.
2. When food is digested, chemical reactions convert the chemical energy in food molecules to other forms of energy that the organism can use.
3. Enzymes increase the speed of chemical reactions by lowering the activation energy of the reactions.
4. An enzyme has one or more folds that form an active site. The active site allows the enzyme to attach to a substrate. Once it attaches, the enzyme alters the shape of the substrate slightly so that the substrate can better fit in the enzyme's active site.
5. Carbon dioxide would not be broken down. Instead it would accumulate in the blood and become toxic to the body.

Study ZONE

CHAPTER 2

Use the Key Concepts and Key Terms listed below to review the main ideas in this chapter. If you need to review the meaning of a term, turn to the page indicated.

Key Concepts

2-1 Nature of Matter

- All matter is made of atoms. Atoms consist of electrons, protons, and neutrons.
- Molecules are groups of atoms linked by covalent bonds.
- Hydrogen bonding occurs between polar molecules.
- An ion is a charged atom or molecule. Ions of opposite charge may form an ionic bond.

2-2 Water and Solutions

- Water, which is essential for life, stores heat efficiently and binds to itself and other substances.
- Water dissolves polar molecules and ionic compounds.
- Acids increase the hydrogen ion concentration of a solution.
- Bases decrease the hydrogen ion concentration of a solution.
- The pH scale measures the strength of acids and bases.

2-3 Chemistry of Cells

- Organic compounds are found in living things.
- Carbohydrates, such as glucose, are a source of energy and are used as structural materials in organisms.
- Lipids are nonpolar molecules that store energy and are an important part of cell membranes.
- Proteins are chains of amino acids. The sequence of amino acids determines a protein's shape and specific function.
- Nucleic acids store and transmit hereditary information.
- ATP is the main energy currency of cells.

2-4 Energy and Chemical Reactions

- Chemical reactions absorb or release energy.
- Starting a chemical reaction requires activation energy.
- Enzymes speed up chemical reactions by decreasing the activation energy of the reactions.
- Enzymes bind only certain substrates.
- Factors such as temperature and pH affect enzyme activity.

Key Terms

2-1

atom (28)
element (28)
compound (29)
molecule (29)
ion (30)

2-2

cohesion (31)
adhesion (31)
solution (32)
acid (33)
base (33)

2-3

carbohydrate (34)
monosaccharide (34)
lipid (35)
protein (36)
amino acid (36)
nucleic acid (37)
nucleotide (37)
DNA (37)
RNA (37)
ATP (37)

2-4

energy (38)
activation energy (39)
enzyme (40)
substrate (41)
active site (41)

Study TIP Ready?

- *If you think you understand these Key Concepts and Key Terms, then you're ready for the Performance Zone.*

Answer to Concept Map

The following is one possible answer to Performance Zone item 1 on page 44.

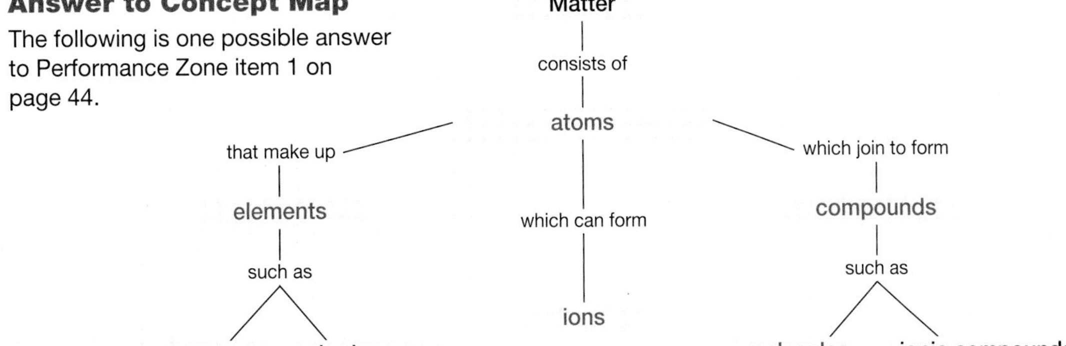

Study ZONE

CHAPTER 2

Review and Practice

Assign the following worksheets for Chapter 2:

- **Student Review Guide**
- **Concept Mapping Worksheet**
- **Test Preparation Pretest**
- **Vocabulary Worksheet**
- **Critical Thinking Review Worksheet**
- **Science Skills Worksheet**

Assessment Options

The following assessment options are available for this chapter:

Resource Link

- Assign **Chapter Test 2** to assess students' understanding of the chapter.

Alternative Assessment

Assign students to groups of four. Have each group make a puzzle that emphasizes the key terms and concepts presented in this chapter. Examples of puzzles include crossword puzzles, word scrambles, and word searches. An answer key must also be included. When the puzzles are complete, have the groups exchange the puzzles. Each group should then solve the puzzle given to them. PORTFOLIO

Portfolio Assessment

- *Portfolio Assessment* in the front of this book provides options for building and evaluating student portfolios. Students and teachers can select items from the areas listed for inclusion in a portfolio.
- See items labeled PORTFOLIO on pp. 29, 32, 36, 37, 39, 40, 42, and 43.

Performance ZONE
CHAPTER 2
REVIEW ANSWERS

Assignment Guide

Section	Review
2-1	1, 3, 4, 5, 14
2-2	6, 7, 13, 18, 19, 20, 21, 27
2-3	2, 8, 10, 16, 17, 22, 23, 25, 26
2-4	9, 11, 12, 15, 24, 28

Understanding and Applying Concepts

1. The answer to the concept map is found at the bottom of page 43.
2. Answers will vary. (a): Carbohydrates are organic compounds that include monosaccharides, disaccharides, and polysaccharides. (b): A nucleic acid is a long chain of nucleotides. (c): A lipid is a nonpolar molecule that is not soluble in water. (d): A protein is a chain of amino acids.
3. d
4. a
5. d
6. c
7. b
8. c
9. c
10. d
11. c
12. a
13. Surface tension forms across the surface of the water because of the attraction between individual water molecules. Because of surface tension, the water strider can stand on the water's surface.
14. In a covalent bond, atoms share electrons. Atoms held together by covalent bonds form molecules. An ionic bond occurs between atoms or molecules that have lost or gained one or more electrons. Ions of opposite charge held together by ionic bonds form ionic compounds.

Performance ZONE
CHAPTER 2 **Review**

Understanding and Applying Concepts

1. **Concept Mapping** Make a concept map that illustrates the structure of matter. Include the following key terms in your map: atom, element, compound, molecule, ion. Include additional key terms in your map as needed.

2. **Understanding Vocabulary** Write a sentence that shows your understanding of each of the following terms:
   a. carbohydrate
   b. nucleic acid
   c. lipid
   d. protein

3. Atoms contain one or more
   a. neutrons.
   b. electrons.
   c. protons.
   d. All of the above

4. Atoms share electrons in a(n)
   a. covalent bond.
   b. hydrogen bond.
   c. ionic bond.
   d. nuclear bond.

5. A weak attraction between polar molecules is called a(n)
   a. nuclear bond.
   b. covalent bond.
   c. ionic bond.
   d. hydrogen bond.

6. Water dissolves ionic compounds because water molecules
   a. are nonpolar.
   b. have a pH value of 14.
   c. have partially charged ends that interact with ions.
   d. do not contain atoms.

7. A substance that forms hydrogen ions when dissolved in water is called a(n)
   a. atom.
   b. acid.
   c. base.
   d. carbohydrate.

8. In cells, ATP temporarily stores
   a. amino acids.
   b. DNA.
   c. energy.
   d. lipids.

9. The energy needed to start a chemical reaction is called
   a. electrical energy.
   b. mechanical energy.
   c. activation energy.
   d. chemical energy.

10. Energy needed for metabolism comes from
    a. food.
    b. lipids.
    c. carbohydrates.
    d. All of the above

11. Most enzymes are
    a. lipids.
    b. carbohydrates.
    c. proteins.
    d. nucleic acids.

12. Enzymes speed up chemical reactions by
    a. decreasing the activation energy.
    b. storing glycogen.
    c. repelling substrates.
    d. increasing the activation energy.

13. Look at the water strider in the photograph below. Using what you have learned about the properties of water, explain how the insect can stand on the water's surface.

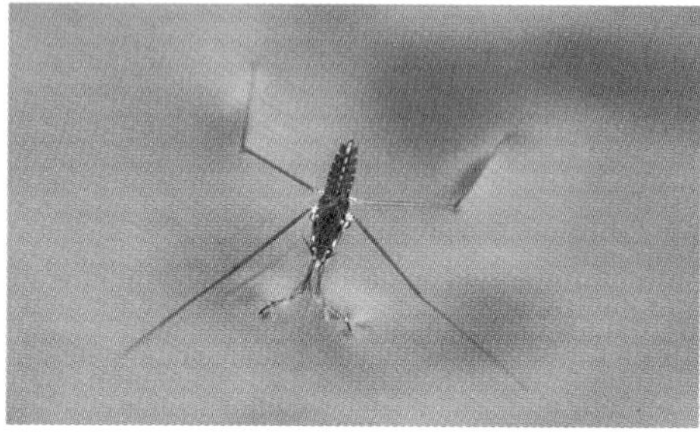

14. **Recognizing Differences** What are two differences between ionic bonds and covalent bonds?

15. **Theme Metabolism** Explain how ATP, carbohydrates, lipids, and enzymes are involved in metabolism.

16. **Food Watch** Ruby-throated hummingbirds migrate 2,000 km every fall. Before migrating, they eat nectar and convert much of the sugar in the nectar to fat. Why is it advantageous for these birds to store energy as fat rather than as glycogen?

17. **Chapter Links** Describe how molecules such as carbohydrates and lipids are important in homeostasis. (**Hint:** See Chapter 1, Section 1-1.)

15. Lipids and carbohydrates are a source of energy used for metabolism. They are broken down inside the cell by enzymes. Some of the energy released is temporarily stored in molecules of ATP.
16. Fats store more energy per gram than do carbohydrates, such as glycogen. Thus, it is more advantageous for the birds to store energy as fat rather than as glycogen.
17. Lipids are good insulators and help maintain a constant internal temperature. Lipids also store energy efficiently. Carbohydrates, such as glucose, can be stored by the body and released when energy is needed.

Skills Assessment

Part 1: Interpreting Tables

Use the table below to answer the questions that follow:

pH Values of Solutions	
Solution	**pH value**
A	3.6
B	7.2
C	1.4
D	13.8
E	9.0

18. Which solution is the most acidic?
19. Which solution is the most basic?
20. Rank the solutions from highest to lowest in hydrogen ion concentration.
21. Which solution has a pH closest to that of pure water?

Part 2: Life/Work Skills

Use the media center or Internet resources to find out how organic compounds are used in biotechnology and in food processing.

22. **Finding Information** Investigate the laboratory techniques of cell fractionation, centrifugation, and electrophoresis. Find out how each technique enables biologists to experiment with cells and to analyze the substances that cells produce. Describe the equipment used for each technique and what can be accomplished when the technique is employed. Prepare an oral report, using graphics to interpret and summarize your findings.
23. **Analyzing Information** Study the nutrition labels on various packaged food products in your home. Record the percentage of carbohydrates, fats, and proteins in each product. Also analyze the ingredients of each food product. List any additives that the products contain. Research whether the additives are natural or artificial, and find out why they are added to particular foods.

Critical Thinking

24. **Evaluating Results** In an experiment you conducted, the rate of an enzyme-catalyzed reaction increased as you increased the substrate concentration. But the reaction rate increased by only a small amount. Explain why this happened.
25. **Evaluating Conclusions** On a recent Arctic expedition, a biochemist discovered an unknown substance. After performing several experiments, the biochemist determined that the substance had the following characteristics: it contained carbon, hydrogen, and oxygen and was soluble in oil but not in water. What kind of substance did the biochemist discover? Explain your answer.
26. **Justifying Conclusions** Animal fats are usually solid, and plant fats are usually (liquid) oils. However, in many animals of the Arctic and Antarctic, animal fats are mostly oils. What adaptive advantage would the storage of body fat as oil instead of as a solid be to animals that live in freezing climates?

Portfolio Projects

27. **Summarizing Information** Write and perform a series of skits for your class to demonstrate the unique properties of water that make water essential for life.
28. **Interpreting Information** Read the article titled "The Suicide Seeds" (*Time*, February 1, 1999, pp. 44–45). How is the biotechnology company mentioned in the article using enzymes to genetically engineer crop seeds that can be used only once? What are the benefits and potential shortcomings of this technology?
29. **Career Focus Biochemist** Research the field of biochemistry, and write a report on your findings. Your report should include a job description, the training required, names of employers, growth prospects, and an average starting salary. If possible, interview a person who works in this field.

Performance ZONE

CHAPTER 2

Critical Thinking

24. There is a limited amount of enzyme molecules available. When the amount of substrate exceeds the amount of the enzyme, there are no additional enzyme molecules available to bind the substrate molecules. The enzymes will become available when the other reactions have been completed.
25. Answers will vary. For example, the substance could be a lipid, which is fat-soluble, or a nonpolar amino acid.
26. Answers will vary. Oil, which is mostly unsaturated fat, provides better protection against freezing than do solid fats, which are mostly saturated fats, because oil has a lower freezing point than a solid fat.

Portfolio Projects

27. Skits will vary. For example, the skits could demonstrate capillary action, cohesion, adhesion, solubility, and heat storage.
28. Enzymes are used to remove DNA sequences from genes of a seed, causing the production of a toxin that kills the seed. Thus, the seed can only be used once.
29. Answers will vary. Biochemists study chemical reactions that occur in cells. Biochemists may perform either pure or applied research. They may, for example, work in universities teaching and doing research. Many biochemists work for pharmaceutical or biotechnological companies. A bachelor's degree is required for entry-level positions. Starting salary will vary by region.

Skills Assessment

18. C
19. D
20. C, A, B, E, D
21. B
22. Cell fractionation is used to release the components of a cell. Cells can be fractionated in a blender. Centrifugation separates cell parts by spinning fractionated cells at high speeds inside a centrifuge. The process of electrophoresis separates fragments of large molecules, such as proteins and DNA, based on the electrical charge and size of the fragments. The molecules move through a porous material in an electrical field.
23. Answers will vary. Have the students bring in their food labels, and check their answers for accuracy.

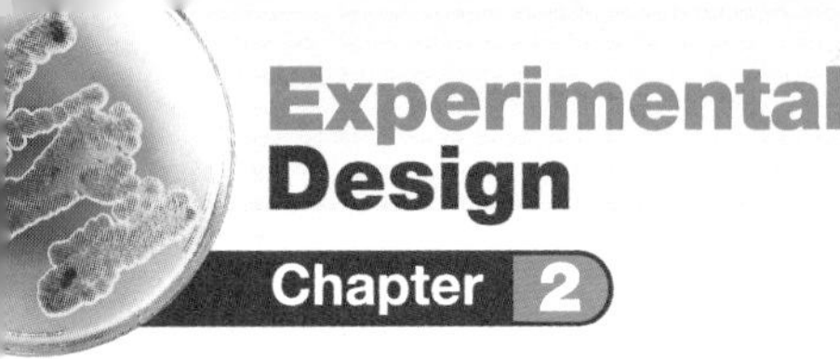

Do Enzyme Detergents Work?

Purpose

Students will investigate the effectiveness of laundry detergents that contain enzymes.

Preparation Notes

Time Required: 45 minutes for Day 1; 30 minutes for Day 2; and 15 minutes for Day 3.

Materials

Materials for this lab can be ordered from WARD'S. See *Master Materials List* at the front of this book for catalog numbers.

Preparation Tips

- Before starting the lab, ask students why enzymes are added to detergents *(to help break down proteins and other substances from foods that may stain clothing)*. Ask students why detergent enzymes are stable during the hot water cycle. *(The enzyme found in commercial laundry soap has been genetically engineered.)*
- The rate of gelatin hydrolysis is slower in instant gelatin that contains sugar than in sugar-free gelatin.
- Students should bring in labeled samples (3–5 tablespoons) of each of five different laundry detergents. Labels should include active ingredients listed on the container.
- Students should have a control test tube with 15 drops (1 mL) of water (no detergent) added to the gelatin surface.

Safety Precautions

- Be sure that students have read and understand all safety rules for working in the lab.
- Remind students to read the ChemSafety section before beginning the lab. Discuss all safety symbols and caution statements with students.
- Make sure students wear safety goggles and a lab apron.
- Caution students to avoid burns by working carefully when heating and pouring boiling water.

Experimental Design

Chapter 2 *Do Enzyme Detergents Work?*

SKILLS

- Using scientific methods
- Measuring volume, mass, and pH

OBJECTIVES

- **Recognize** the function of enzymes in laundry detergents.
- **Relate** temperature and pH to the activity of enzymes.

MATERIALS

- safety goggles and lab apron
- balance
- graduated cylinder
- glass stirring rod
- 150 mL beaker
- 18 g regular instant gelatin or 1.8 g sugar-free instant gelatin
- 0.7 g Na_2CO_3
- tongs or a hot mitt
- 50 mL boiling water
- thermometer
- pH paper
- 6 test tubes
- test-tube rack
- pipet with bulb
- plastic wrap
- tape
- 50 mL beakers (6)
- 50 mL distilled water
- 1 g each of 5 brands of laundry detergent
- wax pencil
- metric ruler

ChemSafety

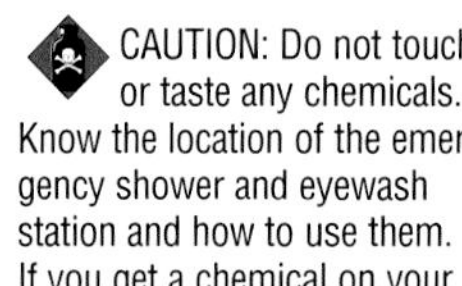

CAUTION: Always wear safety goggles and a lab apron to protect your eyes and clothing.

CAUTION: Do not touch or taste any chemicals. Know the location of the emergency shower and eyewash station and how to use them. If you get a chemical on your skin or clothing, wash it off at the sink while calling to the teacher. Notify the teacher of a spill. Spills should be cleaned up promptly, according to your teacher's directions.

CAUTION: Glassware is fragile. Notify the teacher of broken glass or cuts. Do not clean up broken glass or spills with broken glass unless the teacher tells you to do so.

Before You Begin

Enzymes are substances that speed up chemical reactions. Each enzyme operates best at a particular **pH** and temperature. Substances on which enzymes act are called **substrates.** Many enzymes are named for their substrates. For example, a **protease** is an enzyme that helps break down proteins. In this lab, you will investigate the effectiveness of laundry detergents that contain enzymes.

1. Write a definition for each boldface term in the paragraph above.
2. Bring a sample of laundry detergent from home in a plastic bag. Note any use of the word *enzyme* on the label.
3. Based on the objectives for this lab, write a question you would like to explore about enzyme detergents.

Procedure

PART A: Make a Protein Substrate

1.   Put on safety goggles and a lab apron.
2.    **CAUTION: Use tongs or a hot mitt to handle heated glassware.** Put 18 g of regular (1.8 g of sugar-free) instant gelatin in a 150 mL beaker. Slowly add 50 mL of boiling water to the beaker, and stir the mixture with a stirring rod. Test and record the pH of this solution.

Procedural Tips

- Require students to present a written procedure for their experiment and a list of all safety precautions before allowing them to gather materials for the lab.
- When students add the detergent solutions to the gelatin, the mixture will foam. During this reaction, carbon dioxide gas is released. The addition of washing soda (Na_2CO_3) to the gelatin raises the pH of the gelatin from 4 to 8, which is the optimum pH for protease activation in the detergent samples.
- Have students use a wax pencil to mark the test tubes at the uppermost level of the cooled gelatin in each tube. They will use this mark to measure the hydrolysis of the gelatin each day. Label the test tubes 1–6.
- To prepare a 10 percent solution of laundry detergent, students should dissolve 1 g of detergent in 9 mL of distilled water. Have students record the pH for each numbered detergent sample.
- Students can measure protein hydrolysis after 24 hours by using a wax pencil to draw a second line at the top of the gelatin layer, and then measuring the distance (in mm) between the first line and the second line. This indicates the amount of hydrolysis of the protein in the gelatin by the enzymes in the detergent.

3. Very slowly add 0.7 g of Na_2CO_3 to the hot gelatin while stirring. Note any reaction. Test and record the pH of this solution.
4. Place 6 test tubes in a test-tube rack. Pour 5 mL of the gelatin-Na_2CO_3 mixture into each tube. Use a pipet to remove any bubbles from the surface of the mixture in each tube. Cover the tubes tightly with plastic wrap and tape. Cool the tubes, and store them at room temperature until you begin **Part C.** Complete step 12.

PART B: Design an Experiment

5. Work with members of your lab group to explore one of the questions written for step 3 of **Before You Begin.** To explore the question, design an experiment that uses the materials listed for this lab.

> **You Choose**
> **As you design your experiment, decide the following:**
> a. what question you will explore
> b. what hypothesis you will test
> c. what detergent samples you will test
> d. what your control will be
> e. how much of each solution to use for each test
> f. how to determine if protein is breaking down
> g. what data to record in your data table

6. Write a procedure for your experiment. Make a list of all the safety precautions you will take. Have your teacher approve your procedure and safety precautions before you begin the experiment.

PART C: Conduct Your Experiment

7.  Put on safety goggles and a lab apron.
8. 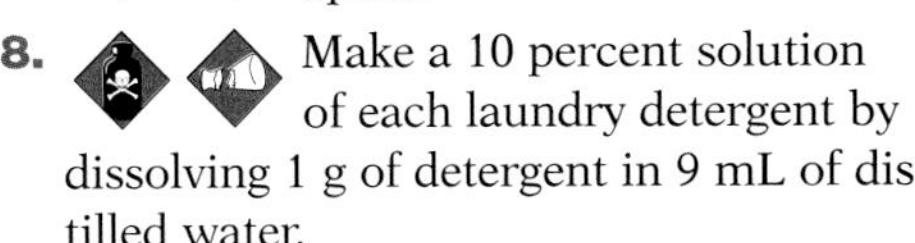 Make a 10 percent solution of each laundry detergent by dissolving 1 g of detergent in 9 mL of distilled water.
9. Set up your experiment. Repeat step 12.
10. Record your data after 24 hours.

PART D: Cleanup and Disposal

11. Dispose of solutions, broken glass, and gelatin in the designated waste containers. Do not pour chemicals down the drain or put lab materials in the trash unless your teacher tells you to do so.
12. Clean up your work area and all lab equipment. Return lab equipment to its proper place. Wash your hands thoroughly before leaving the lab and after finishing all work.

Analyze and Conclude

1. **Analyzing Methods** Suggest a reason for adding Na_2CO_3 to the gelatin solution.
2. **Analyzing Results** Make a bar graph of your data. Plot the amount of gelatin broken down (change in the depth of the gelatin) on the *y*-axis and detergent on the *x*-axis.
3. **Inferring Conclusions** What conclusions did your group infer from the results? Explain.
4. **Further Inquiry** Write a new question about enzyme detergents that could be explored with another investigation.

Do You Know?

Do research in the library or media center to answer these questions:

1. What other household products contain enzymes, and what types of enzymes do they contain?
2. What type of organic compound is broken down by each enzyme that you identified?

Use the following Internet resources to explore your own questions about products that contain enzymes.

SciLINKS NSTA
TOPIC: Enzymes
GO TO: www.scilinks.org
KEYWORD: HX047

Procedure

Sample Procedure

1. Prepare a 10 percent solution of each of five different detergents.
2. Test the pH of each detergent solution with pH paper. Record the pH for each sample in a data table similar to the one below.
3. Add about 15 drops (1 mL) of the first detergent solution to the gelatin surface of the first test tube. Repeat for each of the other samples. Reseal the test tubes and place in a test-tube rack for observation.
4. Add 15 drops of water to the gelatin surface of the sixth test tube.
5. Put the tubes aside for 24 hours at room temperature.
6. Draw another wax pencil line at the top of the gelatin layer. Measure the distance in mm between the first line and the second line. Record the data in a data table.

Sample Data

DETERGENT	PH	AMOUNT OF PROTEIN BROKEN DOWN
1	8	7 mm
2	8	0 mm
3	8	9 mm
4	8	0 mm
5	8	6 mm

Analyze and Conclude

Answers

1. The washing soda increases the pH of the gelatin from 4 to 8—the optimum pH for enzyme activity in this reaction.
2. Graphs will vary.
3. Answers will vary. For example, enzymes operate best at a certain temperature and pH.
4. Answers will vary. For example: Are enzymes in detergent stable in the presence of bleach?

Do You Know?

Answers

1. Answers will vary. For example, dairy digestive supplements contain the enzyme lactase.
2. Lactase breaks down lactose, the sugar in milk.

Disposal

- Pour all laundry detergents down the drain, and then flush the drain with water.
- Place gelatin in the trash.

Before You Begin

Answers

1. *enzymes*—molecules (usually proteins) that speed up chemical reactions

pH—indicates the hydrogen ion concentration of a solution

substrate—a reactant in a reaction that binds to an enzyme's active site

protease—an enzyme that aids in the breakdown of proteins

3. Answers will vary. For example: How can you tell if a detergent contains enzymes?

Chapter 3 Cell Structure

Planning Guide

	OBJECTIVES	MEETING INDIVIDUAL NEEDS	READING SKILLS AND STRATEGIES
BLOCK 1 45 min	**3-1 Looking at Cells** • Describe how scientists measure the length of objects. • Relate magnification and resolution in the use of microscopes. • Analyze how light microscopes function. • Compare light microscopes with electron microscopes. • Describe the scanning tunnelling microscope.	Learning Styles, p. 50 BASIC ADVANCED • pp. 53, 54	**ATE** Active Reading, Technique: Anticipation Guide, p. 49 ■ Active Reading Guide, worksheet (3-1) ■ Directed Reading, worksheet (3-1)
BLOCKS 2, 3, & 4 135 min	**3-2 Cell Features** • List the three parts of the cell theory. • Determine why cells must be relatively small. • Compare the structure of prokaryotic cells with that of eukaryotic cells. • Describe the structure of cell membranes.	Learning Styles, p. 55 BASIC ADVANCED • pp. 56, 57, 59, 60	**ATE** Active Reading, Technique: Reading Organizer, p. 58 ■ Active Reading Guide, worksheet (3-2) ■ Directed Reading, worksheet (3-2)
BLOCKS 5 & 6 90 min	**3-3 Cell Organelles** • Describe the role of the nucleus in cell activities. • Analyze the role of internal membranes in protein production. • Summarize the importance of mitochondria in eukaryotic cells. • Identify three structures in plant cells that are absent from animal cells.	Learning Styles, p. 61 BASIC ADVANCED • pp. 62, 64, 66	■ Active Reading Guide, worksheet (3-3) ■ Directed Reading, worksheet (3-3) ■ Supplemental Reading Guide, *The Lives of a Cell: Notes of a Biology Watcher*

Review and Assessment Resources

BLOCKS 7 & 8 90 min

TEST PREP

- **PE** Study Zone, p. 67
- **PE** Performance Zone, pp. 68–69
- ■ Chapter 3 worksheets:
  - Vocabulary
  - Concept Mapping
  - Science Skills
  - Critical Thinking
- Holt Biology Interactive Tutor Unit 1—Cell Transport and Homeostasis, Topics 1–2
- Audio CD Program
- ■ Test Prep Pretest

CHAPTER TESTING

- ■ Chapter Tests (blackline master)
- ▲ Test Generator
- ■ Assessment Item Listing

See the correlations table at the front of this book for a list of the ***National Science Education Standards*** addressed in this chapter.

Smithsonian Institution® Visit **www.si.edu/hrw** for additional on-line resources.

KEY

Biology Principles and Explorations

PE Pupil's Edition

ATE Teacher's Edition

HOLT **BIOSOURCES**

 Teaching Transparencies

Laboratory Program

One-Stop Planner CD-ROM

Shows these resources within a customizable daily lesson plan:

■ Teaching Resources

▲ Test Generator

VISUAL STRATEGIES	CLASSWORK AND HOMEWORK	TECHNOLOGY AND INTERNET RESOURCES
ATE **Fig. 3-2,** p. 51	■ **Basic Skills:** Microscope Magnification **PE** **Review,** p. 54	Interactive Tutor Unit 1—Cell Transport and Homeostasis (Topics 1 and 2) Holt Biology Videodiscs Section 3-1 Audio CD Program Section 3-1 internet**connect** SCiLINKS NSTA **TOPIC:** Microscopes
ATE **Fig. 3-8,** p. 58 **ATE** **Fig. 3-9,** p. 59 8 Eukaryotic Cell Structure: Chloroplast, Mitochondrion, and Cytoskeleton 9 Structure of Lipid Bilayer 5A Surface-Area-to-Volume Ratio	**PE** **Math Lab** Calculating surface area and volume, p. 56 **Lab A3** Modeling Cells: Surface Area to Volume **Lab C1** Using a Microscope **Lab E1** Cell Size **PE** **Review,** p. 60	Interactive Tutor Unit 1—Cell Transport and Homeostasis (Topics 1 and 2) Holt Biology Videodiscs Section 3-2 Audio CD Program Section 3-2 internet**connect** SCiLINKS NSTA **TOPIC:** Cell features
ATE **Fig. 3-13,** p. 63 **ATE** **Fig. 3-14,** p. 64 **ATE** **Fig. 3-15,** p. 65 7 Eukaryotic Cell 120 Plant Cell Structure	**PE** **Lab Techniques** Studying Animal Cells and Plant Cells, pp. 70–71 **Lab B2** Cell Structures **PE** **Review,** p. 66 **PE** **Study Zone,** p. 67	Interactive Tutor Unit 1—Cell Transport and Homeostasis (Topics 1 and 2) Holt Biology Videodiscs Section 3-3 Audio CD Program Section 3-3 internet**connect** SCiLINKS NSTA **TOPICS:** • Proteins • Laser surgery

ALTERNATIVE ASSESSMENT

PE **Portfolio Projects** 26–29, p. 69

ATE **Alternative Assessment,** p. 67

ATE **Portfolio Assessment,** p. 67

Lab Activity Materials

Math Lab
p. 56 pencil, paper

Lab Techniques
p. 70 compound light microscope, prepared slide of human epithelial cells from the lining of the mouth, safety goggles, lab apron, sprig of *Elodea,* forceps, microscope slides and coverslips, dropper bottle of Lugol's iodine solution

Lesson Warm-up

ATE specimens of cork, p. 50

ATE plain yogurt, microscope slide with coverslip, water, p. 55

ATE prepared slide of cheek cell, micrographs of cells and organelles, p. 61

Demonstrations

ATE micrographs of different kinds of cells, p. 48

ATE sheet of clear plastic wrap, sheet of wax paper, p. 54

ATE balloon, small cardboard box, p. 65

BIOSOURCES LAB PROGRAM

- Quick Labs: Lab A3, p. 5
- Inquiry Skills Development: Lab B2, p. 5
- Laboratory Techniques and Experimental Design: Lab C1, p. 1
- Interactive Explorations in Biology Laboratory Manual: Lab E1, p. 1

Chapter 3

Chapter Theme

Cellular Structure and Function

The relationship between structure and function is a unifying theme in biology. In this chapter, the relationship is evident at the cellular and subcellular levels. Students will learn the common features of cells and the major differences between prokaryotic and eukaryotic cells. Students will also learn that proper functioning of all cell structures is needed to maintain homeostasis.

Establishing Prior Knowledge

Before beginning this chapter, make certain students can answer the questions in **Study TIP Ready?** on page 49. If they cannot, encourage them to reread the sections indicated.

Opening Demonstration

Using micrographs, show students various cells from a multicellular organism, such as neurons, erythrocytes, and sperm cells. First have students attempt to identify the cells. After the cells have been identified, have students suggest the function of each cell and how the structure of the cell might be involved in that function.

CHAPTER 3 Cell Structure

Paramecium (790×)

What's Ahead — and Why
CHAPTER OUTLINE

3-1 Looking at Cells
- Microscopes Reveal Cell Structure
- Microscopes Have Different Uses and Limitations

3-2 Cell Features
- The Cell Theory Has Three Parts
- Prokaryotes Do Not Contain Internal Compartments
- Eukaryotic Cells Are Organized
- The Structure and Function of Cell Membranes Are Closely Related

3-3 Cell Organelles
- The Nucleus Directs Cell Activities and Stores DNA
- An Internal Membrane System Processes Proteins
- Mitochondria Produce ATP
- Plant Cells Contain Structures That Animal Cells Lack

Students will learn

- how microscopes are used to study cellular structure and function.
- features that are common to all cells.
- the names and functions of cell organelles.

Automobile factory

Production... Packaging... Distribution...

Cells have all the equipment necessary to perform the essential functions of life. The structures of a cell serve specific functions, much as the different stations in this automobile factory have different roles in the production of cars. The parts of a cell also function together as a unit, just as the stations in the factory work together in production.

Study TIP

Ready?

Check your understanding of these topics to help you prepare for what's ahead.

Can you...

- **distinguish** between polar and nonpolar molecules? (Section 2-1)
- **differentiate** between carbohydrates, lipids, proteins, and nucleic acids? (Section 2-3)
- **describe** the function of ATP? (Section 2-3)

If not, *review the sections indicated.*

Looking Ahead

Labs

- **Math Lab**
  Calculating surface area and volume (p. 56)
- **Lab Techniques**
  Studying Animal Cells and Plant Cells (p. 70)

Features

- **Tech Watch**
  Laser Microsurgery (p. 66)

Unit 1—*Cell Transport and Homeostasis*
Topics 1–2

internetconnect

National Science Teachers Association *sci*LINKS Internet resources are located throughout this chapter.

Chapter 3

Reading Strategies

Active Reading

Technique: Anticipation Guide

Write the titles of this chapter and the three sections on the board. Have students copy these titles into their notebooks, leaving a few lines after each section title. Next ask students to write what they think they will learn in each section. Check their lists for completeness and tell them to save their lists for later use in the **Closure** on page 66.

Directed Reading

Assign **Directed Reading Worksheet Chapter 3** to help students interact with the text as they read the chapter.

Interactive Reading

Assign ***Biology: Principles and Explorations* Audio CD Program Chapter 3** to help reluctant readers achieve greater success in reading the chapter.

Reading Skills

Assign Reading Organizers and Reading Strategies Worksheets from the **One-Stop Planner CD-ROM with Test Generator** to help students learn and practice effective reading skills.

internetconnect

Holt *Biology: Principles and Explorations*

On-line Resources: go.hrw.com

Visit the HRW Web site for a variety of resources related to this chapter. Just type in the keyword HX0 Chapter 3.

Holt *Biology: Principles and Explorations*

On-line Resources: www.scilinks.org

The following *sci*LINKS Internet resources can be found in the student text for this chapter:

Page 52
TOPIC: Microscopes
KEYWORD: HX052

Page 56
TOPIC: Cell features
KEYWORD: HX056

Page 62
TOPIC: Proteins
KEYWORD: HX062

Page 66
TOPIC: Laser surgery
KEYWORD: HX066

Section 3-1 Looking at Cells

1 Prepare

Scheduling

- **Block:** About 45 minutes will be needed to complete this section.
- **Traditional:** About one class period will be needed to complete this section.
- **Planning:** See the Planning Guide on pp. 47A and 47B for lecture, classwork, and assignment options for Section 3-1. Lesson plans for Section 3-1 can be found in a lesson cycle format in ***Biology: Principles and Explorations*** **One-Stop Planner CD-ROM with Test Generator.**

Resource Link

Assign **Active Reading Worksheet Section 3-1** to help students comprehend the key concepts and visuals in this lesson.

2 Teach

Lesson Warm-up

Pass some specimens of cork around the classroom for students to look at. Ask students to identify the source of the cork. *(It is the bark of a particular type of oak tree,* Quercus suber, *which grows in Spain and Portugal.)* Show students a magnified section of cork and point out the cell walls. Lead students to the idea that bark is dead and that the cell walls enclose air.

Cross-Disciplinary Connection

◆ **Language Arts**

When Hooke observed cork using a microscope, he observed tiny compartments. He gave them the Latin name *cellulae,* meaning "small rooms." That is the origin of the term *cell.*

Objectives

- **Describe** how scientists measure the length of objects. (p. 50)
- **Relate** magnification and resolution in the use of microscopes. (p. 51)
- **Analyze** how light microscopes function. (p. 52)
- **Compare** light microscopes with electron microscopes. (pp. 52–53)
- **Describe** the scanning tunneling microscope. (p. 54)

Key Terms

light microscope
electron microscope
magnification
resolution
scanning tunneling microscope

Microscopes Reveal Cell Structure

Most cells are too small to see with the naked eye; a typical human body cell is many times smaller than a grain of sand. Scientists became aware of cells only after microscopes were invented, in the 1600s. When the English scientist Robert Hooke used a crude microscope to observe a thin slice of cork in 1665, he saw "a lot of little boxes." The boxes reminded him of the small rooms in which monks lived, so he called them cells. Hooke later observed cells in the stems and roots of plants. Ten years later, the Dutch scientist Anton van Leeuwenhoek used a microscope to view water from a pond, and he discovered many living creatures. He named them "animalcules," or tiny animals. Today we know that they were not animals but single-celled organisms.

Measuring the sizes of cell structures

Measurements taken by scientists are expressed in metric units. Scientists throughout the world use the metric system. The official name of the metric system is the International System of Measurements, abbreviated as SI. SI is a decimal system, so all relationships between SI units are based on powers of 10. For example, scientists measure the sizes of objects viewed under a microscope using the SI base unit for length, which is the meter. A meter, which is about 3.28 ft (a little more than a yard), equals 100 centimeters (cm), or 1,000 millimeters (mm). A meter also equals 0.001 kilometer (km). Most SI units have a prefix that indicates the relationship of that unit to a base unit. For example, the symbol "μ" stands for the metric prefix *micro.* A micrometer (μm) is a unit of linear measurement equal to one-millionth of a meter, or one-thousandth of a millimeter. **Table 3-1** summarizes the SI units used to measure length.

Table 3-1 Metric Units of Length and Equivalents

Unit	Prefix	Metric equivalent	Real-life equivalent
Kilometer (km)	*Kilo-*	1,000 m	About two-thirds of a mile
Meter (m)		1 m (SI base unit)	A little more than a yard
Centimeter (cm)	*Centi-*	0.01 m	About half the diameter of a Lincoln penny
Millimeter (mm)	*Milli-*	0.001 m	About the width of a pencil tip
Micrometer (μm)	*Micro-*	0.000001 m	About the length of an average bacterial cell
Nanometer (nm)	*Nano-*	0.000000001 m	About the length of a water molecule

Integrating Different Learning Styles

Logical/Mathematical	**ATE**	Did You Know?, p. 51; Closure, p. 54
Visual/Spatial	**PE**	Performance Zone item 13, p. 68
	ATE	Lesson Warm-up, p. 50; Visual Strategy, p. 51; Biology in Your Community, p. 52; Teaching Tip, p. 53; Demonstration, p. 54
Verbal/Linguistic	**PE**	Performance Zone items 2 and 22, pp. 68–69
	ATE	Cross-Disciplinary Connections, pp. 50, 53; Internet Connect, p. 52

Characteristics of microscopes

Since Robert Hooke first observed cork cells, microscopes have unveiled the details of cell structure. These powerful instruments provide biologists with insight into how cells work—and ultimately how organisms function. Biologists use different microscopes depending on the organisms they wish to study and the questions they want to answer. Two common kinds of microscopes are light microscopes and electron microscopes. In a **light microscope,** light passes through one or more lenses to produce an enlarged image of a specimen. An **electron microscope** forms an image of a specimen using a beam of electrons rather than light.

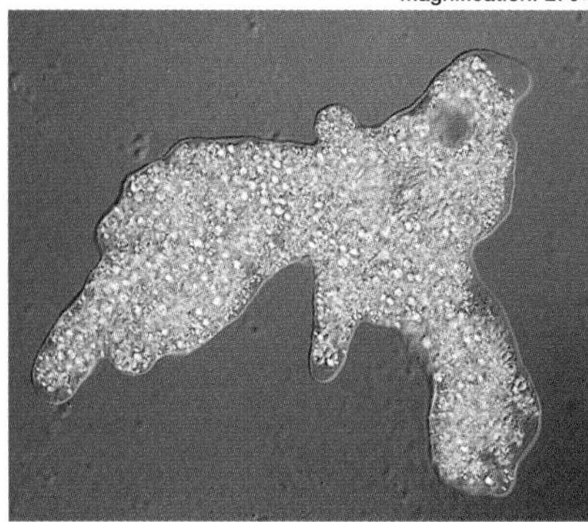

Figure 3-1 Micrograph. Micrographs often include the name of the object, the magnification of the image, and the type of microscope used. This light micrograph (LM) shows an amoeba.

An image produced by a microscope, such as the one shown in **Figure 3-1,** is called a micrograph. Many micrographs specify which kind of microscope produced the image—such as a light micrograph (LM), a transmission electron micrograph (TEM), or a scanning electron micrograph (SEM). Micrographs will often include the magnification value of the image. **Magnification** is the ability to make an image appear larger than its actual size. For example, a magnification value of 200× indicates that the object in the image appears 200 times larger than the object's actual size. **Resolution** is a measure of the clarity of an image. Both high magnification and good resolution are needed to view the details of extremely small objects clearly. In a microscope with poor resolution, images viewed under high magnification will appear blurred or fuzzy. As shown in **Figure 3-2,** electron microscopes have much higher magnifying and resolving powers than light microscopes.

Figure 3-2 Magnifying power of microscopes. The scale below shows the size range of objects that can be viewed with electron microscopes, light microscopes, and the unaided eye.

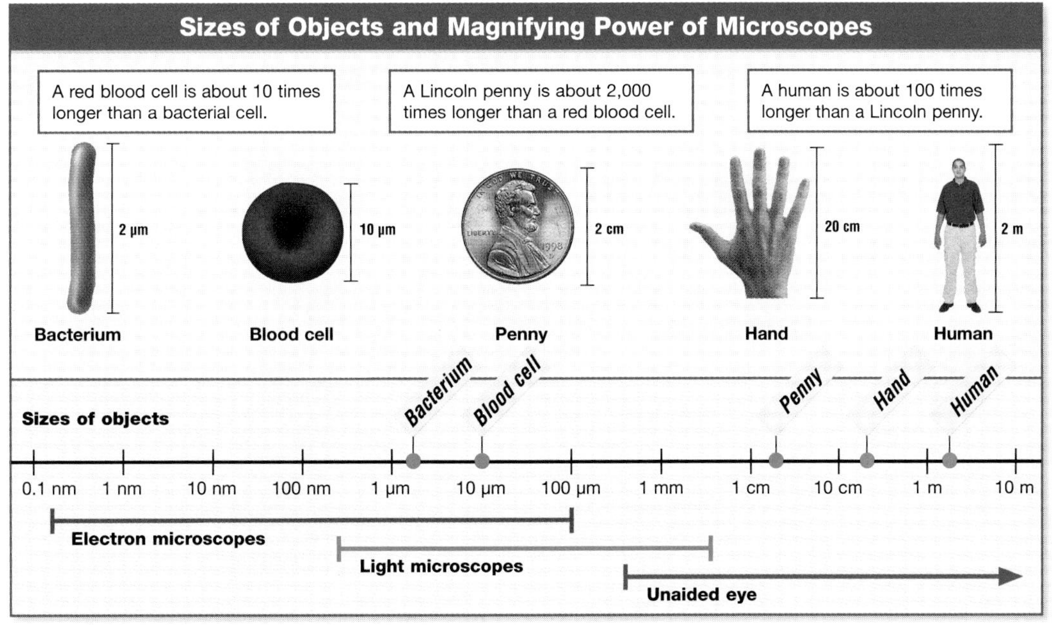

did you know?

The International Bureau of Weights and Measures in France provides the standards of SI measurement.

For example, it has a piece of metal that is the standard for the kilogram and a metal bar that is the standard for the meter. A nickel has a mass of about 5 g, and a dime is about 1 mm thick. Have students determine the mass, thickness, and volume of some of their own coins using SI measurements.

Career

Cytotechnologist

A cytotechnologist stains, mounts, and studies cells to detect malignant cells and other abnormal conditions. The cytotechnologist then reports all findings to a pathologist. A cytotechnologist must be able to obtain precise results and work under stress during emergency situations. An undergraduate degree in cytotechnology or medical technology is usually needed for entry-level positions. Invite a cytotechnologist from a local hospital to talk to your class. Discuss ahead of time what should be included in the talk, such as recommended high school courses and other requirements, working conditions, employment opportunities, and salary ranges.

Visual Strategy

Figure 3-2

This would be a good time to review the metric system with students. Be certain they understand how to convert from one unit to another. Tell students that 1 micrometer (µm) is equal to 1 millionth, or 10^{-6}, of a meter (0.000001 m). Ask: If a cell is 10 µm in diameter, what is its diameter in meters? in centimeters? *(0.00001 m, 0.001 cm)*

Be sure students understand the connection between the objects in the figure—the bacterium, the blood cell, the penny, the hand, and the human—and the metric scale at the bottom of the figure. Note that the scale is logarithmic because it is based on factors of 10. Have students compare the size ranges of objects that can be viewed with electron microscopes, light microscopes, and the unaided eye.

internet connect

SCILINKS. NSTA Have students research microscopes using the Web site in the Internet Connect box on page 52. Have students write a report summarizing their findings. PORTFOLIO

CROSS-DISCIPLINARY CONNECTION

◆ **History**

The first microscope was built in the Netherlands between 1590 and 1608 by Hans Janssen, Zacharias Janssen, and Hans Lippershey. It could magnify objects from 3× to 10×. Significant improvements to the microscope were made by Anton van Leeuwenhoek, who developed lenses with a magnification of up to 300×. Van Leeuwenhoek produced about 400 microscopes. He used them to study many things, including yeast, muscles, plants, and insects.

Resource Link

Assign **Basic Skills: Microscope Magnification** from the Holt BioSources Teaching Resources CD-ROM.

Microscopes Have Different Uses and Limitations

Different types of microscopes have different qualities and uses. Microscopes vary in magnification and resolution capabilities, which affect the overall quality of the images they produce. Microscopes also have different limitations. For example, electron microscopes have high magnifying power, but they cannot be used to view living cells. Light microscopes have lower magnifying power, but they can be used to view living cells.

Compound light microscope

Light microscopes that use two lenses are called compound light microscopes. In a typical compound light microscope, such as the one shown in **Figure 3-3,** a light bulb in the base shines light up through the specimen. The objective lens, closest to the specimen, collects the light, which then travels to the ocular *(AHK yoo luhr)* lens, closest to the viewer's eye. Both lenses magnify the image. Thus, a microscope with a 40× objective lens and a 10× ocular lens produces a total magnification of 400×.

Why not add a third lens and magnify even more? This approach does not work because you cannot distinguish between two objects, or "resolve" them, when they are closer together than a few hundred nm. When the objects are this close, the light beams from the two objects start to overlap!

Figure 3-3 Compound light microscope

In a compound light microscope, a specimen is mounted on a glass slide and is illuminated with a beam of light from below.

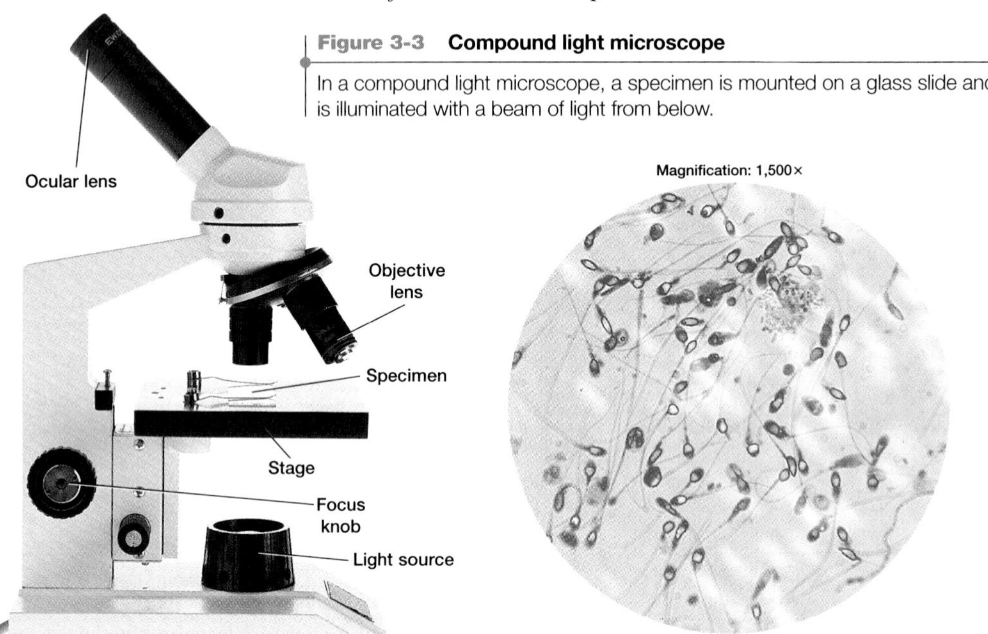

Biology in Your Community

Electron Microscopy

Contact a local college or university and arrange for a field trip that will introduce your students to electron microscopy. Specifically request that the parts of the electron microscope be shown and explained to students and that the students also be shown how a specimen is prepared for viewing and how it is viewed. Also check to see if the institution has any other microscopes normally not available to high schools so that these too can be shown to the students.

The most powerful compound light microscopes have a total magnification of up to 2,000×, which is sufficient for viewing objects as small as 0.5 µm in diameter. For you to see smaller objects, the wavelength of the light beam must be shorter than the wavelength of visible light. Electron beams have a much shorter wavelength than that of visible light, so electron microscopes are much more powerful than light microscopes.

Electron microscopes

Electron microscopes can magnify an image up to 200,000×, and they can be used to study even the smallest structures inside cells or on cell surfaces. In electron microscopes, both the electron beam and the specimen must be placed in a vacuum chamber so that the electrons in the beam will not bounce off gas molecules in the air. Because living cells cannot survive in a vacuum, they cannot be viewed using electron microscopes.

Transmission electron microscope In a transmission electron microscope, shown in **Figure 3-4,** the electron beam is directed at a very thin slice of a specimen stained with metal ions. Some structures in the specimen become more heavily stained than others. The heavily stained parts of the specimen absorb electrons, while those that are lightly stained allow electrons to pass through. The electrons that pass through the specimen strike a fluorescent screen, forming an image on the screen. A TEM, such as the micrograph of sperm cells shown in Figure 3-4, can reveal a cell's internal structure in fine detail. TEM images are always in black and white. However, with the help of computers, scientists often add artificial colors to make certain structures more visible.

biobit

What are the smallest objects visible with a microscope?

Using the Scanning Interferometric Apertureless Microscope (SIAM), researchers in 1997 viewed features that were about four atoms (1 nm) in diameter. This technology could be used to code information. For example, as many as 30 movies could be stored on a disk the size of a penny.

Figure 3-4 Transmission electron microscope

In a transmission electron microscope, electrons pass through a specimen, forming an image of the specimen on a fluorescent screen.

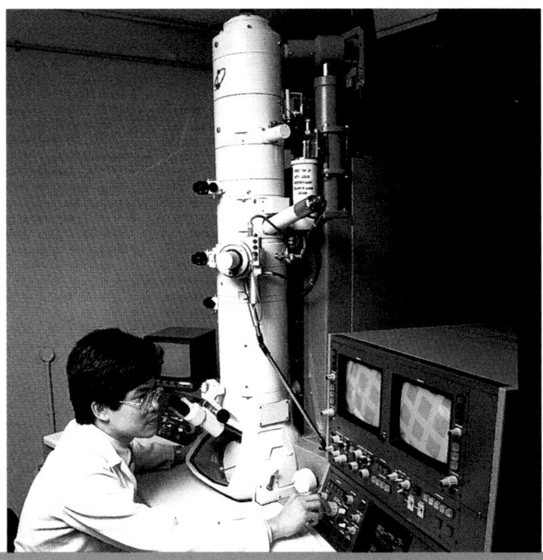

Magnification: 7,730×

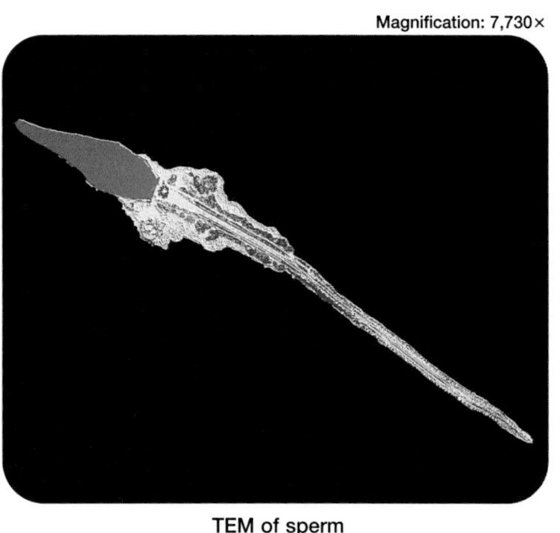

TEM of sperm

Section 3-1

Teaching TIP BASIC

Making a Wet Mount

Have students practice preparing wet mounts. **CAUTION: Slides and coverslips break easily and have sharp edges.** Provide students with glass slides, coverslips, small beakers filled with water, and eyedroppers. Students can begin by viewing cut-out letters from newspapers. Challenge students to make a bubble-free slide. Then have students prepare wet mounts using drops of pond water. Emphasize that it will be much easier to find and watch organisms if students make their slides bubble-free.

CROSS-DISCIPLINARY CONNECTION

◆ Earth Science

Electron microscopes are used in areas of research other than biology. For example, they are an important tool in Earth science. In this field, electron microscopes have been used to determine the effects of weathering on the microstructural arrangement of the mineral components of rocks, and the effects of chemicals on rock formation and stability. Have students prepare a report that describes uses of electron microscopes in fields other than biology. PORTFOLIO

MULTICULTURAL PERSPECTIVE

Development of the Electron Microscope

The transmission electron microscope was invented in 1931 by two Germans, Ernst Ruska and Rheinhold Ruedenberg. In 1933, Ruska built the first electron microscope that was more powerful than the light microscope. In 1986, Ruska was awarded the Nobel prize in physics for this achievement. The electron microscope did not become available until 1935 in England. In 1945, Albert Claude, a Belgian-American cytologist, began electron microscope studies of the cell and discovered the endoplasmic reticulum and the details of mitochondria, for which he was awarded the Nobel prize in physiology or medicine in 1974.

Demonstration

Resolution

Hold a sheet of clear plastic wrap over your hand. Then replace the plastic with a sheet of wax paper. Ask students to identify the material through which it was easier to distinguish individual hand features, such as fingers, hair, and veins. *(the plastic wrap)* Tell students that the ability to distinguish the individual components of an object clearly is called resolution.

3 Assessment Options

Closure

Working with SI Units

Assign students values of SI units and ask them to convert the units into other SI units using Table 3-1. For example, ask students how many centimeters are in 10 mm *(1 cm).*

Review

Assign Review—Section 3-1.

Reteaching — BASIC

Ask students which kind of microscope—the compound light microscope, the transmission electron microscope (TEM), or the scanning electron microscope (SEM)—would be most useful to high school students studying the general structure of a leaf. *(The light microscope shows good detail for general observations.)* Ask students which kind of microscope would be most useful for studying three-dimensional details of a small organism, such as a unicellular protist. *(The SEM shows the surfaces of objects in great detail.)* Ask them which microscope would be most useful in research on the internal structures of a bacterial cell. *(The TEM shows internal detail with great magnification.)*

Figure 3-5 Scanning electron microscope

In a scanning electron microscope, electrons bounce off a specimen, forming a three-dimensional image of the specimen on a fluorescent screen.

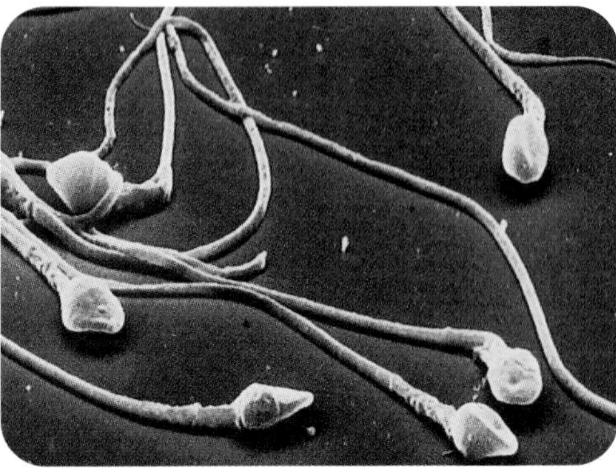

SEM of sperm

Scanning electron microscope In a scanning electron microscope, shown in **Figure 3-5,** the electron beam is focused on a specimen coated with a very thin layer of metal. The electrons that bounce off the specimen form an image on a fluorescent screen. An SEM shows three-dimensional images of cell surfaces, such as the image of sperm cells shown in Figure 3-5. As with the transmission electron microscope, images produced by the scanning electron microscope are black and white, but often they are artificially colored.

Scanning tunneling microscope

New video and computer techniques are increasing the resolution and magnification of microscopes. The **scanning tunneling microscope** uses a needle-like probe to measure differences in voltage caused by electrons that leak, or tunnel, from the surface of the object being viewed. A computer tracks the movement of the probe across the object, enabling objects as small as individual atoms to be viewed. The computer generates a three-dimensional image of the specimen's surface. The scanning tunneling microscope can be used to study living organisms.

Review SECTION 3-1

1. **Describe** the relationship between a meter, a millimeter, and a micrometer.
2. **Describe** how magnification and resolution affect the appearance of objects viewed under a microscope.
3. **Compare** the magnifying power of a light microscope with that of an electron microscope.
4. **SKILL Recognizing Differences** Explain why electron microscopes cannot be used to view the structure of living cells.
5. **SKILL Comparing Functions** A scientist wants detailed images of the internal structure of a bacterium. What type of microscope would be best for that task? Explain your answer.

Review SECTION 3-1 ANSWERS

1. A millimeter is equal to 0.001 m, or one-thousandth of a meter. A micrometer is equal to 0.000001 m, or one-millionth of a meter.
2. As the magnification increases, the resolution decreases; therefore the clarity of the image is reduced.
3. Light microscopes can magnify an object up to 2,000×. Electron microscopes can magnify an object up to 200,000×.
4. Specimens must be placed in a vacuum as part of the preparation for viewing with electron microscopes; living things cannot survive in a vacuum.
5. The scientist would use a transmission electron microscope because the electron beam passes through the bacterium, revealing the cell's internal structure in great detail.

Cell Features

Section 3-2

The Cell Theory Has Three Parts

It took scientists more than 150 years to fully appreciate the discoveries of Hooke and Leeuwenhoek. In 1838, the German botanist Mattias Schleiden concluded that cells make up not only the stems and roots but every part of a plant. A year later, the German zoologist Theodor Schwann claimed that animals are also made of cells. In 1858, Rudolph Virchow, a German physician, determined that cells come only from other cells. The observations of Schleiden, Schwann, and Virchow form the **cell theory,** which has three parts:

1. All living things are made of one or more cells.
2. Cells are the basic units of structure and function in organisms.
3. All cells arise from existing cells.

Cells must be small

Small cells function more efficiently than large cells. There are about 100 trillion cells in the human body, most ranging from 5 μm to 20 μm in diameter. What is the advantage of having so many tiny cells instead of fewer large ones? All substances that enter or leave a cell must cross that cell's surface. If the cell's surface-area-to-volume ratio is too low, substances cannot move through the cell quickly enough to meet the cell's needs. Small cells can exchange substances more rapidly than large cells because small objects have a higher surface-area-to-volume ratio than larger objects, as shown in **Table 3-2.** As a result, substances do not need to travel as far to reach the center of a smaller cell.

Table 3-2 Relationship Between Surface Area and Volume

Side length	Surface area	Volume	Surface area/ volume ratio
1 mm	6 mm^2	1 mm^3	6:1
2 mm	24 mm^2	8 mm^3	3:1
4 mm	96 mm^2	64 mm^3	3:2

1 mm
2 mm
4 mm

Objectives

- **List** the three parts of the cell theory. (p. 55)
- **Determine** why cells must be relatively small. (p. 55)
- **Compare** the structure of prokaryotic cells with that of eukaryotic cells. (p. 57)
- **Describe** the structure of cell membranes. (p. 59)

Key Terms

cell theory
cell membrane
cytoplasm
cytoskeleton
ribosome
prokaryote
cell wall
flagellum
eukaryote
nucleus
organelle
cilium
phospholipid
lipid bilayer

Integrating Different Learning Styles

Logical/Mathematical	**PE**	Math Lab, p. 56
Visual/Spatial	**ATE**	Lesson Warm-up, p. 55; Teaching Tips, pp. 56, 58; Visual Strategies, pp. 58, 59
Body/Kinesthetic	**ATE**	Closure, p. 60
Interpersonal	**ATE**	Reteaching, p. 60
Verbal/Linguistic	**PE**	Study Tip, p. 57; Performance Zone items 2, 28 and 29 pp. 68–69
	ATE	Active Reading, p. 57; Teaching Tip, p. 60

Section 3-2

1 Prepare

Scheduling

- **Block:** About 135 minutes will be needed to complete this section.
- **Traditional:** About three class periods will be needed to complete this section.
- **Planning:** See the Planning Guide on pp. 47A and 47B for lecture, classwork, and assignment options for Section 3-2. Lesson plans for Section 3-2 can be found in a lesson cycle format in ***Biology: Principles and Explorations*** **One-Stop Planner CD-ROM with Test Generator.**

Resource Link

Assign **Active Reading Worksheet Section 3-2** to help students comprehend the key concepts and visuals in this lesson.

2 Teach

Lesson Warm-up

Smear some yogurt with live cultures on a microscope slide. Mix a drop of water into the smeared yogurt and add a coverslip. Allow students to examine the slide under high power using a compound light microscope. They should see bacteria of the genus *Lactobacillus.* Ask them to compare the size of these cells with the sizes of cells on prepared slides of human and plant tissues.

Making Biology Relevant

Food

Mechanical digestion of food is just as important as chemical digestion. Chewing food into small pieces increases the surface-area-to-volume ratio of the pieces, making digestion more efficient.

Teaching TIP BASIC

Effect of Cell Size

Cut three cubes of varying sizes from a potato. Let each cube represent a cell. Pour food coloring into three beakers, and place a cube in each beaker. After about 5 min., remove each cube and cut it in half. Have students observe the results. Ask them to determine how the surface-area-to-volume ratio of each cube affected the penetration of the food coloring into the cube. *(Students should see that the food coloring traveled the same distance in each cube. However, the smallest cube will be more completely colored than the others because it has the greatest surface-area-to-volume ratio.)*

MATH LAB: Calculating surface area and volume

Time 15–20 minutes

Process Skills Calculating, analyzing, applying

Teaching Strategies

- Guide students through the calculations to be sure they understand the mathematics involved.
- Bring in cube-shaped game dice of three different sizes. Have students measure the sides of the dice. Then have them calculate the surface-area-to-volume ratios of the dice and compare their results.

Analysis Answers

1. total surface area = 24 mm^2
   volume = 8 mm^3
   surface-area-to-volume ratio is 3:1
2. total surface area = 6 mm^2
   volume = 1 mm^3
   surface-area-to-volume ratio is 6:1
3. By being flat, a *paramecium* spreads its volume over a large area. The surface-area-to-volume ratio is increased because there is more surface area per unit volume.

internetconnect
SciLINKS NSTA
TOPIC: Cell features
GO TO: www.scilinks.org
KEYWORD: HX056

Common features of cells

Cells share common structural features, including an outer boundary called the **cell membrane** (technically referred to as the plasma membrane). The cell membrane encloses the cell and separates the cell interior, called the **cytoplasm** *(SEYET oh plaz uhm),* from its surroundings. The cell membrane also regulates what enters and leaves a cell—including gases, nutrients, and wastes. Inside the cytoplasm are many structures, often suspended in a system of microscopic fibers called the **cytoskeleton.** All cells have ribosomes. **Ribosomes** *(REYE buh sohmz)* are the cellular structures on which proteins are made. All cells also have DNA, which provides instructions for making proteins, regulates cellular activities, and enables cells to reproduce.

MATH LAB: Calculating surface area and volume

Background

You can improve your understanding of the relationship between a cell's surface area and its volume by practicing with the large cube in Table 3-2.

1. **Find the total surface area of the cube.**
   - *side length* $(l) = 4$ mm
   - *surface area of one side* $= l \times l = l^2$
   - *surface area of one side* $(l^2) = 4 \text{ mm} \times 4 \text{ mm} = 16 \text{ mm}^2$
   - *total surface area* $= 6 \times l^2 = 6 \times 16 \text{ mm}^2 = 96 \text{ mm}^2$

2. **Calculate the volume of the cube.**
   - *height* $(h) = l = 4$ mm
   - *volume* $= l^2 \times h = 16 \text{ mm}^2 \times 4 \text{ mm} = 64 \text{ mm}^3$
3. **Determine the surface-area-to-volume ratio.** A ratio compares two numbers by dividing one number by the other. A ratio can be expressed in three ways:

in words	as a fraction	with a colon
x to y	$\frac{x}{y}$	$x{:}y$

For the surface-area-to-volume ratio, divide total surface area by volume.

$$\frac{\textit{total surface area}}{\textit{volume}} = \frac{96}{64}$$

Divide both numbers by their greatest common factor:

$$\frac{(96 \div 32)}{(64 \div 32)} = \frac{3}{2}$$

Magnification: 230×

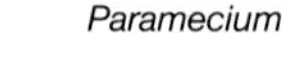

Paramecium (SEM)

Analysis

1. **Calculate** the surface-area-to-volume ratio of the cube with a side length of 2 mm in Table 3-2.
2. **Calculate** the surface-area-to-volume ratio of the cube with a side length of 1 mm in Table 3-2.
3. **SKILL Relating Concepts** How does the flatness of the single-celled *Paramecium* shown above affect the cell's surface-area-to-volume ratio?

Overcoming Misconceptions

Shape vs. Size

Large cells do not necessarily have small surface-area-to-volume ratios. Many large cells have shapes that allow them to maintain a large surface-area-to-volume ratio. A cell could grow large in one or two dimensions but remain small in others. For example, parts of neurons are very long cylinders with small diameters, whereas epithelial cells are broad and flat. In both cases, the surface-area-to-volume ratio is quite large.

Prokaryotes Do Not Contain Internal Compartments

The smallest and simplest cells are prokaryotes. A **prokaryote** *(proh KAIR ee oht)* is a single-celled organism that lacks a nucleus and other internal compartments. Without separate compartments to isolate materials, prokaryotic cells cannot carry out many specialized functions. Early prokaryotes lived at least 3.5 billion years ago. For nearly 2 billion years, prokaryotes were the only organisms on Earth. They were very simple and small (1–2 μm in diameter). Like their ancestors, modern prokaryotes—bacteria—are also very small (1–15 μm). In prokaryotes, the genetic material is a single, circular molecule of DNA.

Characteristics of bacteria

Modern prokaryotes are informally called bacteria. Bacteria can exist in a broad range of environmental conditions. Many bacteria, including those that cause infection in humans, grow and divide very rapidly. Some bacteria do not need oxygen to survive. Other bacteria cannot survive in the presence of oxygen. Some bacteria can even make their own food.

The cytoplasm of a bacterial cell includes everything inside the cell membrane. As **Figure 3-6** shows, a bacterium's enzymes and ribosomes are free to move around in the cytoplasm because there are no internal structures that divide the cell into compartments. The DNA of a bacterium is circular, and it is often located near the center of the cell. Bacterial DNA is suspended within the cytoplasm.

Bacterial cells have a **cell wall** surrounding the cell membrane that provides structure and support. The cells of fungi and plants also have cell walls; only animal cells lack cell walls. Bacteria lack an internal supporting skeleton, so they must depend on a strong cell wall to give the cell shape. A bacterial cell wall is made of strands of polysaccharides connected by short chains of amino acids. Some bacterial cell walls are surrounded by a structure called the capsule, which is also made of polysaccharides. The capsule enables bacteria to cling to almost anything, including teeth, skin, and food.

Many bacteria have **flagella** *(fluh JEL uh),* which are long, threadlike structures that protrude from the cell's surface and enable movement. Bacterial flagella rotate, propelling a bacterium through its environment at speeds of up to 20 cell lengths per second. Figure 3-6 shows a bacterium with several flagella.

Study TIP

Reading Effectively

For many words ending in *-um,* the plural is formed by changing the *-um* to *-a.* For example, the plural of *bacterium* is *bacteria,* and the plural of *flagellum* is *flagella.*

Figure 3-6 Bacteria. Bacterial cells have little internal structure. Many also have a capsule and flagella.

Magnification: 61,850×

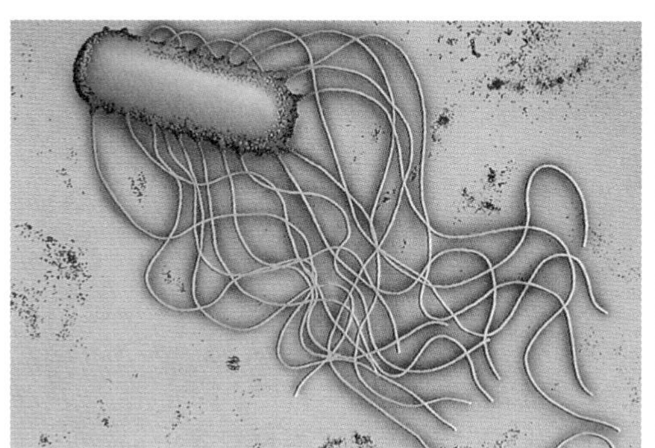

Teaching TIP ADVANCED

Cell Walls

Tell students that the cell wall varies among different types of organisms. In plants, the cell wall is composed of cellulose fibers embedded in a polysaccharide and protein matrix. In eubacteria, the cell wall is composed of peptidoglycan, a polymer of sugars cross-linked by a short polypeptide. In fungi, the cell wall is composed of chitin, a polysaccharide.

Resource Link

Assign **Basic Skills: Length, Area, and Volume** from the Holt BioSources Teaching Resources CD-ROM.

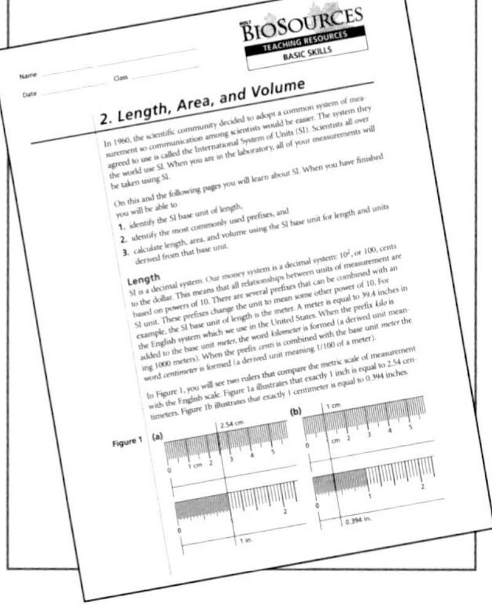

BioSources Teaching Resources Basic Skills

2. Length, Area, and Volume

did you know?

Prokaryotes evolved about 3.5 billion years ago, whereas eukaryotic cells did not appear until about 1.5 billion years ago.

Modern prokaryotes include archaebacteria and eubacteria. They can be found in nearly any environment on Earth, including volcanic vents at the bottom of the ocean and ice in Arctic and Antarctic regions.

Chapter 3

Teaching TIP

Prokaryotic Cells vs. Eukaryotic Cells

Have students make a **Graphic Organizer** similar to the one on page 58 that compares prokaryotic cells with eukaryotic cells. PORTFOLIO

Visual Strategy

Figure 3-8

This figure shows most of the organelles that are commonly found in animal cells. Explain that for the sake of clarity only a few of the microtubules and microfilaments have been drawn. In reality, these structures form a dense network (the cytoskeleton) inside the cell. Point out that the nuclear envelope actually consists of two lipid bilayers.

Active Reading

Technique: Reading Organizer

As students read pp. 57–58, have them make a descriptive reading organizer. Students should design the organizer to identify and describe the differences between prokaryotic and eukaryotic cells. PORTFOLIO

Eukaryotic Cells Are Organized

The first cells with internal compartments were primitive eukaryotic cells, which evolved about 1.5 billion years ago. A **eukaryote** *(yoo KAIR ee oht)* is an organism whose cells have a nucleus. The **nucleus** *(NOO klee uhs)* is an internal compartment that houses the cell's DNA. Other internal compartments, or organelles, enable eukaryotic cells to function in ways different from bacteria. An **organelle** is a structure that carries out specific activities in the cell.

Many single-celled eukaryotes use flagella for movement. Short hairlike structures called **cilia** *(SIL ee uh)* protrude from the surface of some eukaryotic cells. Flagella or cilia propel some cells through their environment. In other cells, cilia and flagella move substances across the cell's surface. For example, cilia on cells of the human respiratory system, shown in **Figure 3-7,** sweep mucus and other debris out of the lungs.

The major organelles in an animal cell are shown in **Figure 3-8.** The cytoplasm includes everything inside the cell membrane but outside the nucleus. A web of protein fibers makes up the cytoskeleton. The cytoskeleton holds the cell together and keeps the cell membrane from collapsing or folding. Protein fibers of the cytoskeleton include microtubules and microfilaments. Microtubules are long, hollow tubes that extend throughout the cytoplasm, supporting the cell. Microtubules may also help move organelles around the cell. Microfilaments are extremely thin protein strands that also support the cell's shape and structure.

Figure 3-7 Cilia. Cilia on cells lining the respiratory system remove debris from air passages.

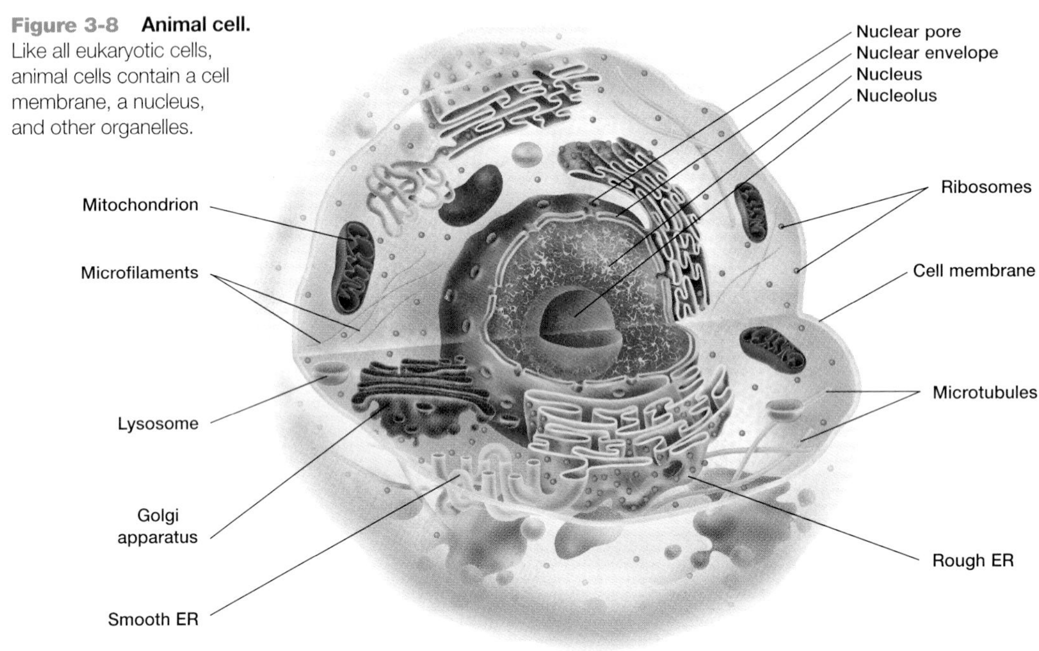

Figure 3-8 Animal cell. Like all eukaryotic cells, animal cells contain a cell membrane, a nucleus, and other organelles.

Graphic Organizer

Use this graphic organizer with ***Teaching Tip: Prokaryotic Cells vs. Eukaryotic Cells*** *on page 58.*

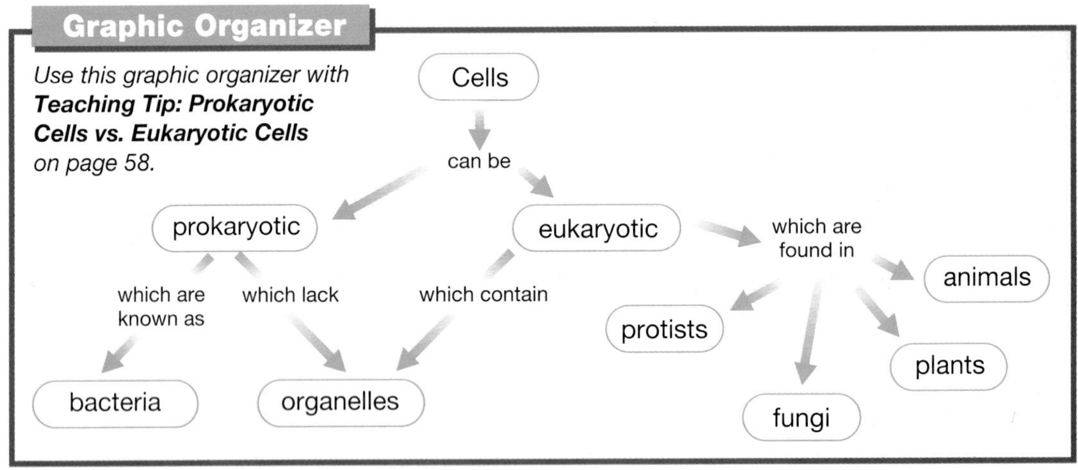

The Structure and Function of Cell Membranes Are Closely Related

The cytoplasm and organelles of a cell are contained by its membrane. Cell membranes are not rigid like an eggshell. Rather, they are fluid like a soap bubble. The fluidity of cell membranes is caused by lipids, which form the foundation of membranes. The lipids form a barrier that separates the inside of the cell from the outside of the cell. This barrier allows only certain substances in the cell's environment to pass through. This selective permeability of the cell membrane determines which substances enter and leave the cell.

The cell membrane as a barrier

The selective permeability of the cell membrane is caused mainly by the way phospholipids interact with water. A **phospholipid** is a lipid made of a phosphate group and two fatty acids. As shown in **Figure 3-9,** a phospholipid has both a polar "head" and two nonpolar "tails." Recall from Chapter 2 that the polar ends of water molecules will form weak bonds with other polar substances. The head of a phospholipid, which contains a phosphate group, is polar and is attracted to water. In contrast, the two fatty acids, or tails, are nonpolar and therefore are repelled by water.

In a cell membrane, the phospholipids are arranged in a double layer called a **lipid bilayer,** as shown in Figure 3-9. The nonpolar tails of the phospholipids make up the interior of the lipid bilayer. Because water both inside and outside the cell repels the nonpolar tails, they are forced to the inside of the lipid bilayer. Ions and most polar molecules, including sugars and some proteins, are repelled by the nonpolar interior of the lipid bilayer.

REAL LIFE

Donated blood is frozen in a special process called cryopreservation. Similar methods are used to preserve human eggs, embryos, and blood from the umbilical cord, a rich source of immune-system cells.

Finding Information
Research how cryopreservation methods enable cells to withstand freezing.

Figure 3-9 Lipid bilayer

Cell membranes are made of a double layer of phospholipids, called a lipid bilayer.

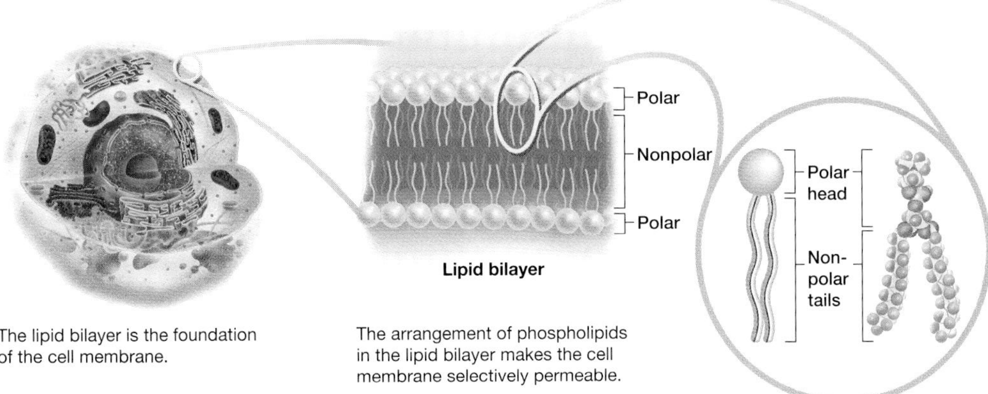

The lipid bilayer is the foundation of the cell membrane.

The arrangement of phospholipids in the lipid bilayer makes the cell membrane selectively permeable.

A phospholipid's "head" is polar, and its two fatty acid "tails" are nonpolar.

did you know?

Phospholipids in a lipid bilayer can move about 2 μm in one second.

Proteins also move, but because of their greater size, their movement is slower. The membrane remains fluid unless the temperature decreases to the point at which the membrane solidifies. If the phospholipids are rich in unsaturated fatty acids, the membrane will remain fluid at lower temperatures. If the phospholipids are rich in saturated fatty acids, the membrane will be more viscous at any given temperature.

REAL LIFE

Answer

The cells are frozen very quickly, which prevents the formation of large ice crystals which would damage the cells.

Teaching TIP BASIC

Composition of Cell Membranes

Ask students to identify two major classes of organic compounds that are found in cell membranes. *(lipids and proteins)* Ask students to describe the functions of these compounds in cell membranes. *(Lipids make up the bilayer, which is the foundation of cell membranes. Proteins serve as markers, receptors, enzymes, and transporters.)*

Visual Strategy

Figure 3-9

Have students review the terms *polar* and *nonpolar.* Be certain students understand that the nonpolar interior of the lipid bilayer repels ions and polar substances. Remind them that even though water molecules are polar, they are small enough to move through the lipid bilayer. Make sure students understand that membranes are found in many places in cells, such as around the entire cell, mitochondria, the ER, the Golgi apparatus, lysosomes, and the nucleus. Ask students to list cell structures that lack membranes. *(ribosomes, microfilaments, microtubules, and cell walls)*

Teaching TIP — ADVANCED

Cell-Membrane Proteins

Ask students which types of cell-membrane proteins are involved in the rejection of transplanted organs *(marker proteins),* the movement of glucose into cells *(transport proteins),* and the triggering of allergic *(receptor proteins)* reactions.Have students research additional examples of the activity of cell-membrane proteins. Have them prepare a newsletter that includes a title, a summary of the function of each cell-membrane protein, illustrations, and references. PORTFOLIO

3 Assessment Options

Closure

Cell Membrane

Have students make a model of the cell membrane by cutting off half of one "leg" of several wooden clothespins. Different colors can be used to represent the hydrophobic (nonpolar) and hydrophilic (polar) ends. Proteins can be represented by buttons, bottle caps, and cardboard tubes.

Review

Assign Review—Section 3-2.

Reteaching — BASIC

Assign students to two cooperative groups. Have group A represent a prokaryotic cell, and have group B represent a eukaryotic cell. Have each group take turns stating a characteristic that its cell has or something that its cell is able to do. Write these characteristics on the board. Give the other group an opportunity to challenge the answers. Award one point for each correct answer, and deduct one point for each incorrect answer. Award one point to a group that challenges and is correct, and deduct one point if that challenge is incorrect. Add up each group's total number of points at the end.

Figure 3-10 Membrane proteins

The cell membrane contains various proteins with specialized functions.

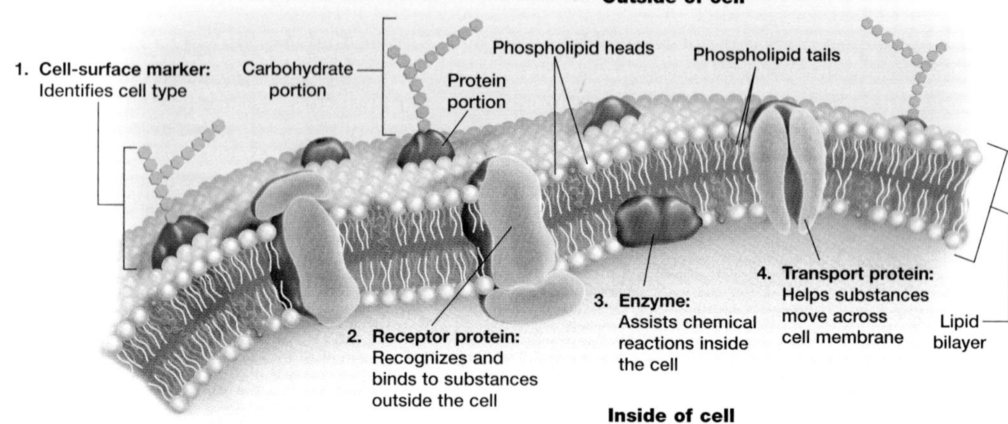

Proteins are embedded in the lipid bilayer

Various proteins are located in the lipid bilayer of a cell membrane. What keeps these proteins within the lipid bilayer? Recall from Chapter 2 that proteins are made of amino acids and that some amino acids are polar, while others are nonpolar. The nonpolar part of a membrane protein is attracted to the interior of the lipid bilayer but is repelled by the water on either side of the lipid bilayer. In contrast, the polar parts of the protein are attracted to the water on either side of the lipid bilayer. This dual attraction holds the protein in the lipid bilayer. However, the motion and fluidity of phospholipids enable membrane proteins to move around within the lipid bilayer.

As shown in **Figure 3-10,** cell membranes contain different types of proteins. Each type plays a vital role in the life of a cell. Marker proteins, which are attached to a carbohydrate on the cell's surface, help other cells recognize their cell type—liver cell or heart cell, for example. Receptor proteins recognize and bind to specific substances, such as signal molecules, outside the cell. Enzymes embedded in the cell membrane are involved in important biochemical reactions in the cell. Transport proteins aid the movement of substances into and out of the cell.

Review SECTION 3-2

1. **Summarize** observations that formed the cell theory.
2. **Describe** the importance of the surface-area-to-volume ratio of a cell.
3. **Compare** the structure of a eukaryotic cell with that of a prokaryotic cell.
4. **SKILL Recognizing Relationships** How does the arrangement of phospholipids influence the permeability of the lipid bilayer?
5. **SKILL Comparing Functions** Describe the functions of two types of cell-membrane proteins.

Review SECTION 3-2 ANSWERS

1. Schleiden, Schwann, and Virchow observed that plants as well as animals are composed of cells, which are the basic units of structure and function, and that cells arise only from other cells.
2. If the surface-area-to-volume ratio of a cell is too small, substances cannot penetrate the cell quickly enough to meet the needs of the cell.
3. Prokaryotic cells lack organelles, which enable eukaryotic cells to carry out unique, specialized functions.
4. Phospholipids are arranged in a lipid bilayer such that the polar heads face the outside and inside of the cell, and the nonpolar tails face each other. Most polar molecules cannot cross the lipid bilayer without the aid of membrane proteins.
5. Answers will vary, but may include marker proteins, which help cells recognize cells of their own type; receptor proteins, which bind to signal molecules; transport proteins, which move substances into and out of the cell; and enzymes, which aid in biochemical reactions.

Cell Organelles

Section 3-3

The Nucleus Directs Cell Activities and Stores DNA

Most functions of a eukaryotic cell are controlled by the cell's nucleus. As shown in **Figure 3-11,** the nucleus is surrounded by a double membrane called the nuclear envelope, also called the nuclear membrane. The nuclear envelope is made of two lipid bilayers that separate the nucleus from the cytoplasm. Scattered over the surface of the nuclear envelope are many small channels through the envelope called nuclear pores. Substances that are made in the nucleus, including ribosomal proteins and RNA, move into the cytoplasm by passing through the nuclear pores. Ribosomes are partially assembled in a region of the nucleus called the nucleolus, which is also shown in Figure 3-11. Recall from Section 3-2 that ribosomes are the structures on which proteins are made.

The hereditary information of a eukaryotic cell is coded in the cell's DNA, which is stored in the nucleus. Eukaryotic DNA is wound tightly around proteins. Most of the time, DNA exists as elongated and thin strands. When a cell is about to divide, however, the DNA strands wind up into a more compact form and appear as dense, rod-shaped structures called chromosomes. You will learn more about DNA and chromosomes in Chapter 6.

Objectives

- **Describe** the role of the nucleus in cell activities. (p. 61)
- **Analyze** the role of internal membranes in protein production. (pp. 62–63)
- **Summarize** the importance of mitochondria in eukaryotic cells. (p. 64)
- **Identify** three structures in plant cells that are absent from animal cells. (p. 65)

Key Terms

endoplasmic reticulum
vesicle
Golgi apparatus
lysosome
mitochondrion
chloroplast
central vacuole

Figure 3-11 **Nucleus**

The nucleus is surrounded by a double membrane called the nuclear envelope.

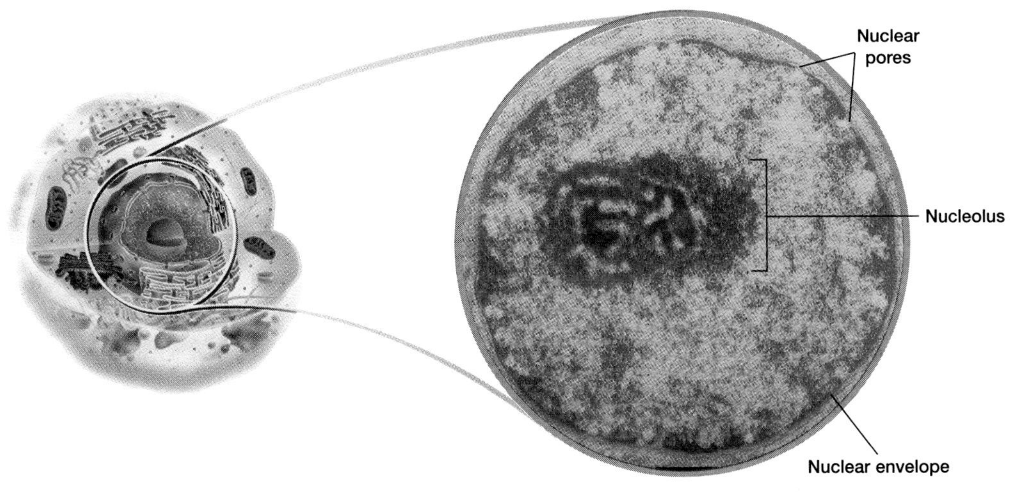

Integrating Different Learning Styles

Logical/Mathematical	**PE**	Performance Zone items 24 and 25, p. 69
Visual/Spatial	**PE**	Performance Zone items 1, 17, 18, 19, and 20, pp. 68–69; Lab Techniques, pp. 70–71
	ATE	Lesson Warm-up, p. 61; Visual Strategies, pp. 63, 64, 65; Demonstration, p. 65
Body/Kinesthetic	**PE**	Performance Zone item 26, p. 69
	ATE	Reteaching, p. 66
Verbal/Linguistic	**PE**	Study Tip, p. 63; Performance Zone items 2 and 21, pp. 68–69
	ATE	Internet Connect, p. 62; Teaching Tip, p. 63

1 Prepare

Scheduling

- **Block:** About 90 minutes will be needed to complete this section.
- **Traditional:** About two class periods will be needed to complete this section.
- **Planning:** See the Planning Guide on pp. 47A and 47B for lecture, classwork, and assignment options for Section 3-3. Lesson plans for Section 3-3 can be found in a lesson cycle format in ***Biology: Principles and Explorations* One-Stop Planner CD-ROM with Test Generator.**

Resource Link

Assign **Active Reading Worksheet Section 3-3** to help students comprehend the key concepts and visuals in this lesson.

2 Teach

Lesson Warm-up

Using a light microscope, show students a prepared slide of a cheek cell stained with methylene blue. Then show students electron micrographs of various cell parts and organelles. Have students compare these micrographs with the illustrations in the textbook. Ask them to attempt to find these structures in the cheek cell. Only the cytoplasm, nucleus, and cell membrane will be observed easily. Remind students that the other structures either are too small to be seen with the light microscope or may need to be stained in a different way to be seen.

internetconnect

SciLinks NSTA Have students research protein production using the Web site in the Internet Connect box on page 62. Have students write a report summarizing their findings. PORTFOLIO

Teaching TIP

Protein Production

Tell students that proteins made in the ER are packaged into vesicles before they are taken to other parts of the cell. Vesicles are small, membrane-bound sacs that are formed by pinching off from another membrane. Proteins on the surface of a vesicle determine the vesicle's contents and play a role in directing the vesicle toward its destination in the cell.

Teaching TIP ADVANCED

Endoplasmic Reticulum

Enzymes found in the smooth ER help detoxify drugs. Detoxification usually involves the addition of a hydroxyl group, which makes the drug more soluble and allows it to leave the body more easily. Many drugs, including barbiturates and alcohol, cause the smooth ER and its enzymes to increase in number. This effect increases a person's tolerance to the drug, meaning that higher doses of the drug are needed to achieve the desired response.

An Internal Membrane System Processes Proteins

Unlike prokaryotic cells, eukaryotic cells have a system of internal membranes that play an essential role in the processing of proteins. Cells make proteins on ribosomes. Each ribosome is made of dozens of different proteins as well as RNA. Some of the ribosomes in a eukaryotic cell are suspended in the cytoplasm, as they are in prokaryotic cells. These "free" ribosomes make proteins that remain inside the cell, such as proteins used to build new organelles.

Production of proteins

Proteins that are exported from the cell, such as some signal molecules, are made by ribosomes on the surface of the endoplasmic reticulum, shown in Figure 3-12. The **endoplasmic reticulum** *(ehn doh PLAZ mihk rih TIHK yuh luhm),* or ER, is an extensive system of internal membranes that move proteins and other substances through the cell. Like the cell membrane, the membranes of the ER are made of a lipid bilayer with embedded proteins.

The part of the ER with attached ribosomes is called rough ER because it has a rough appearance when viewed in the electron microscope. The rough ER helps transport the proteins that are made by its attached ribosomes. As each protein is made, it crosses the ER membrane and enters the ER. The portion of the ER that contains the completed protein then pinches off to form a vesicle. A **vesicle** is a small, membrane-bound sac that transports substances in cells. By enclosing certain proteins inside vesicles, the eukaryotic cell keeps these proteins separate from proteins that are produced by free ribosomes in the cytoplasm.

The rest of the ER is called smooth ER because it lacks ribosomes and thus appears smooth when viewed in the electron microscope. The smooth ER performs various functions, such as making lipids and breaking down toxic substances.

Figure 3-12 Endoplasmic reticulum

The ER moves proteins and other substances within eukaryotic cells.

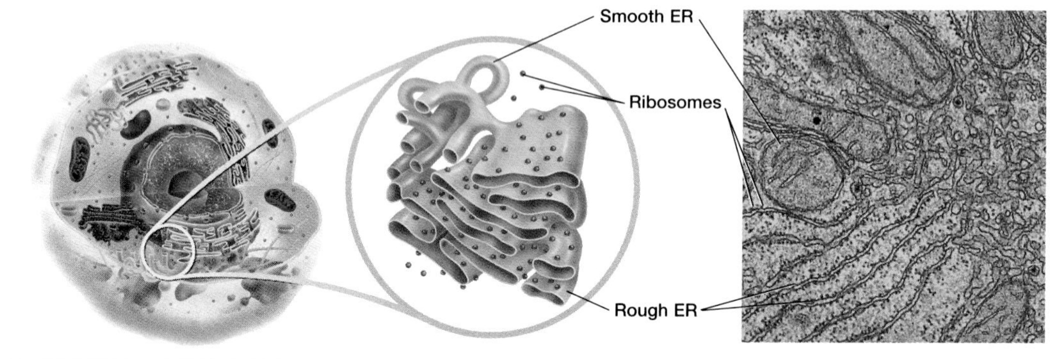

MULTICULTURAL PERSPECTIVE

Camillo Golgi

Camillo Golgi was born in 1844 in Corteno, Italy. After he received his medical degree in 1865, he began his work as a psychiatrist. His real interest, however, was in the microscopic anatomy of the nervous system. Golgi developed a technique for staining nervous tissue using silver salts. He also described the morphological features of glial cells. Golgi discovered in nerve cells the organelle that is now known as the Golgi apparatus. He was awarded the Nobel prize in physiology or medicine in 1906, along with Santiago Ramón y Cajal, for work on the structure of the nervous system. Golgi died at the age of 83.

Packaging and distribution of proteins

Vesicles that contain newly made proteins move through the cytoplasm from the ER to an organelle called the Golgi apparatus. The **Golgi** *(GOHL jee)* **apparatus** is a set of flattened, membrane-bound sacs that serves as the packaging and distribution center of the cell. Enzymes inside the Golgi apparatus modify the proteins that are received in vesicles from the ER. The modified proteins are then enclosed in new vesicles that bud from the surface of the Golgi apparatus. Some of these vesicles include **lysosomes** *(LEYE seh sohms),* which are small, spherical organelles that contain the cell's digestive enzymes. The ER, the Golgi apparatus, and lysosomes work together in the production, packaging, and distribution of proteins, as summarized in **Figure 3-13.**

Step 1 Ribosomes make proteins on the rough ER. The proteins are packaged into vesicles.

Step 2 The vesicles transport the newly made proteins from the rough ER to the Golgi apparatus.

Step 3 In the Golgi apparatus, proteins are processed and then packaged into new vesicles.

Step 4 Many of these vesicles move to the cell membrane and release their contents outside the cell.

Step 5 Other vesicles, including lysosomes, remain within the cytoplasm. Lysosomes digest and recycle the cell's used components by breaking down proteins, nucleic acids, lipids, and carbohydrates.

Study TIP

Interpreting Graphics

As you read, use Steps 1–5 in the text, shown in red, to help you follow the same numbered steps shown in Figure 3-13.

BIO graphic

Figure 3-13

Processing of Proteins

Proteins are processed by an internal system of membranes.

Nucleus
1 Proteins are made by ribosomes on the rough ER.
2 Vesicles carry proteins from the rough ER to the Golgi apparatus.
3 Proteins are modified in the Golgi apparatus and enter new vesicles.
4 Some vesicles release their proteins from the cell.
5 Other vesicles remain in the cell and become lysosomes.

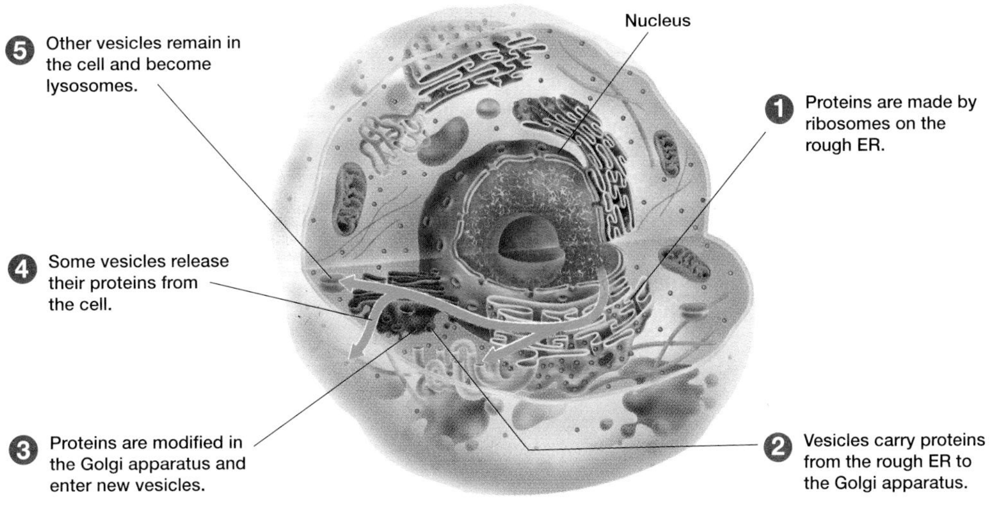

Teaching TIP

What Am I?

Have each student choose one cell part or organelle and write a "What Am I?" essay in which the student pretends to be that structure. The essay should be written in the first person and include the function and structure of that cell part. Encourage students to be creative in their writing styles. Have each student read his or her essay aloud in class. Other students should try to guess the structure or organelle presented in the essay. PORTFOLIO

Visual Strategy

BIO graphic

Figure 3-13

To help students understand the processing of proteins, compare the process with a classroom situation:

1. In the library (ER), students (ribosomes) produce homework (proteins).
2. Backpacks (vesicles) carry the homework (proteins) from the library (ER) to the teacher (Golgi apparatus).
3. The teacher (Golgi apparatus) grades (modifies) the homework, which enters folders (new vesicles).
4. Some folders (vesicles) release their homework (proteins) from the school (cell).
5. Other folders (vesicles) remain in the school (cell).

did you know?

Several diseases have been attributed to improperly functioning lysosomes.

Tay-Sachs disease is caused by the deficiency of a lysosomal enzyme that digests lipids. As a result, cells in the brain become filled with lipids, causing impairment. Pompe's disease results when a lysosomal enzyme that breaks down glycogen is absent, causing liver damage.

Cross-Disciplinary Connection

◆ **History**

Dr. Lynn Margulis was born in 1938 and received her PhD from the University of California, Berkeley, in 1963. Her theory of endosymbiosis is a widely accepted explanation of the origin of mitochondrial DNA. The theory states that chloroplasts evolved from photosynthetic bacteria and that mitochondria evolved from aerobic bacteria. Scientists think that the organisms that evolved into chloroplasts provided eukaryotic cells with the ability to use carbon dioxide to produce organic molecules. This theory is supported by the fact that both chloroplasts and mitochondria have their own DNA and can be killed with antibiotics that affect prokaryotes but not eukaryotes.

Visual Strategy

Figure 3-14

Point out the folds in the inner membrane of the mitochondrion. Ask students to speculate on how effective a mitochondrion without the folds would be. *(With less surface area to carry out its chemical reactions, it would be much less efficient.)* Next have students predict what would happen to an organism if the mitochondria in its cells suddenly became only half as efficient. *(Life processes that require energy would slow down.)*

biobit

Do organelles get sick?

Several diseases have been linked to malfunctioning organelles. The movie *Lorenzo's Oil* describes how parents struggled to find a cure for their son's deadly disease of the peroxisome, an organelle that regulates fat metabolism.

Mitochondria Produce ATP

Nearly all eukaryotic cells contain many **mitochondria** *(meyet uh KAHN dree uh)*, like the one shown in **Figure 3-14.** A mitochondrion is an organelle that harvests energy from organic compounds to make ATP, the main energy currency of cells. Although some ATP is made in the cytoplasm, most of a cell's ATP is made inside mitochondria. Cells that have a high energy requirement, such as a muscle cell, may contain hundreds or thousands of mitochondria. Figure 3-14 shows that a mitochondrion has two membranes. The outer membrane is smooth, while the inner membrane is greatly folded. The two membranes form two compartments in which the chemical reactions that produce ATP during cell metabolism take place.

Mitochondrial DNA

The nucleus is not the only organelle in the cell that contains nucleic acids. Mitochondria also have DNA and ribosomes, and mitochondria make some of their own proteins. However, most mitochondrial proteins are made by free ribosomes in the cytoplasm. The fact that mitochondria contain their own DNA—independent of nuclear DNA and similar to the circular DNA of prokaryotic cells—reflects the widely accepted theory that primitive prokaryotes are the ancestors of mitochondria. You will learn more about the origin of mitochondrial DNA in Chapter 12.

Figure 3-14 Mitochondrion

In a eukaryotic cell, mitochondria make most of the ATP.

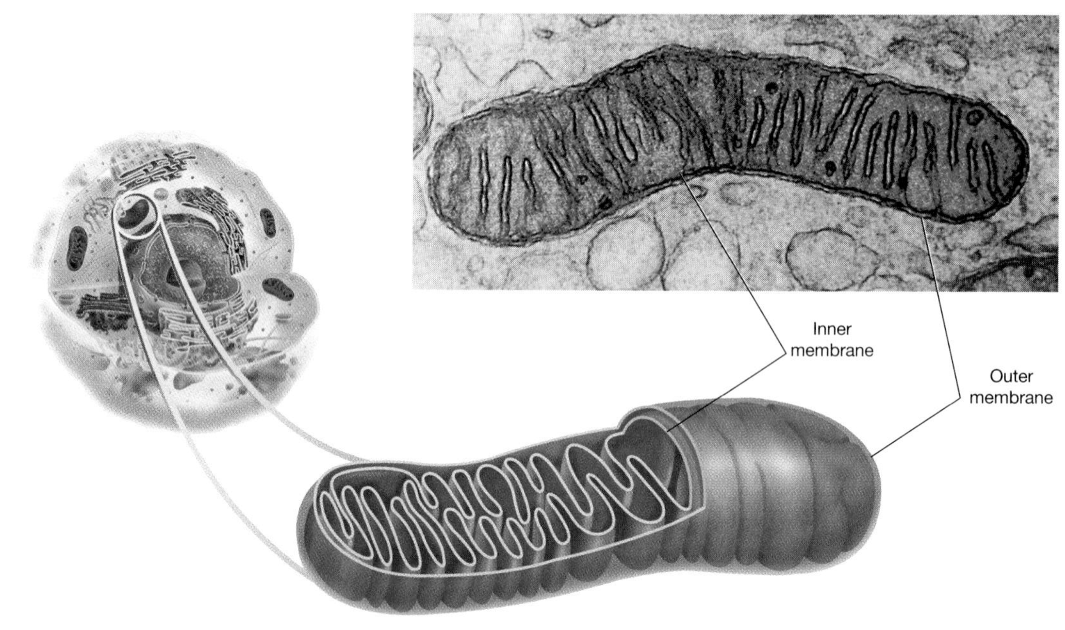

Overcoming Misconceptions

Mitochondria in Plant Cells

Make sure students understand that plant cells contain mitochondria. Many students think that because plants perform photosynthesis, plant cells do not contain mitochondria. It is important for students to understand that plants also produce ATP through aerobic respiration, which requires mitochondria.

Plant Cells Contain Structures That Animal Cells Lack

All of the organelles described so far in this section are found in both animal cells and plant cells. However, plant cells have three additional structures that are not found in animal cells: a cell wall, several chloroplasts, and a central vacuole. These three structures are shown in the plant cell in **Figure 3-15.**

Unique features of plant cells

Cell wall The cell membrane of a plant cell is surrounded by a thick cell wall, which supports and protects the cell. Although the cell wall surrounds the cell membrane, the cell wall does not prevent the movement of substances across the cell membrane. The cell wall consists of a mixture of proteins and carbohydrates, including the polysaccharide cellulose. The cell wall helps support and maintain the cell's shape, protects the cell from damage, and connects the cell with adjacent cells.

Chloroplasts Plant cells contain one or more **chloroplasts.** Chloroplasts are organelles that use light energy to make carbohydrates from carbon dioxide and water. Plants use these carbohydrates to make other organic compounds, which in turn provide the energy they need for metabolism. You will learn more about this process, called photosynthesis, in Chapter 5. Chloroplasts are found not only in plants but also in a wide variety of eukaryotic algae, such as seaweed. Some scientists think that plants have evolved from a

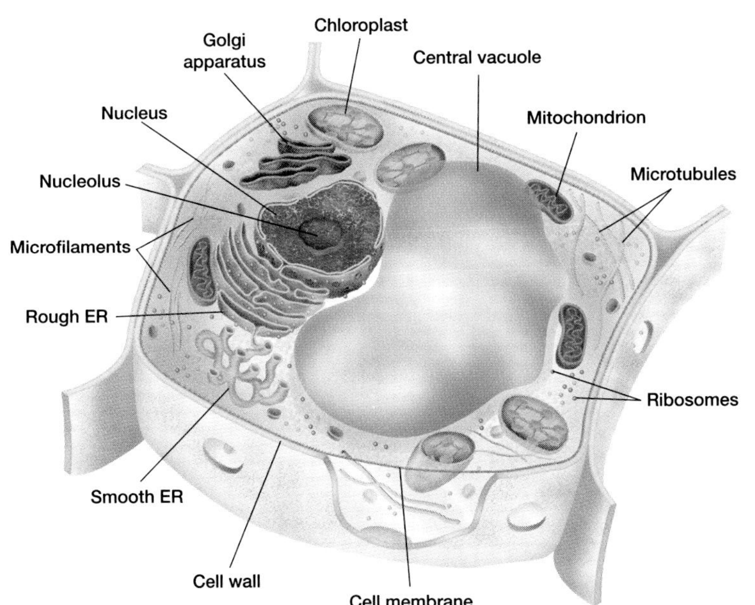

Figure 3-15 Plant cell. Plant cells have a cell wall, chloroplasts, and a large central vacuole.

REAL LIFE

Termites can turn wood into a meal. Protists in their gut enable them to digest cellulose. The Formosan termite threatens to destroy older buildings in many cities.

Predicting Outcomes
What might be the consequences of eradicating termites from Earth?

Section 3-3

REAL LIFE

Answer

The breakdown of dead wood and the release of its nutrients in forests might be slowed.

Demonstration

Role of Cell Wall

To show how a plant cell wall provides strength and support, inflate a balloon inside a small cardboard box. The balloon represents the cell membrane, and the box represents the cell wall. Filling the balloon with air is analogous to filling the central vacuole with water.

Visual Strategy

Figure 3-15

Have students compare this plant cell with the animal cell shown in Figure 3-8 on page 58. Have students make a list of the similarities and differences between the two cells. Then ask students to volunteer their answers and write them on the board.

did you know?

Plants differ fundamentally from animals in that plants continue to grow throughout their lives.

In contrast, animals grow only until they reach maturity. Plants also differ from animals in that plants do not have reproductive structures during their entire mature lifetime.

Laser Microsurgery

Teaching Strategies

- Explain that the tail of the sperm cell shown in the photo has been cut by an "optical scalpel" in a procedure involving laser microsurgery.
- Have students research other applications of lasers in biology.

Discussion

What is the significance of optical scalpels and optical tweezers in biology? *(Answers will vary. For example, optical tweezers can be used to transfer genetic material between cells.)*

3 Assessment Options

Closure

Technique: Anticipation Guide

Have students return to the lists that they made for the **Active Reading** activity on page 49. Ask them to check everything they learned that was on their list. Review any unchecked items with the students. Perhaps the material will be covered in a future chapter. If not, spend some time discussing these items.

Review

Assign Review—Section 3-3.

Reteaching — BASIC

Have students use food to make a model of a eukaryotic cell. Gelatin (cytoplasm) could be poured into a large, clear cellophane bag (cell membrane). Organelles could be represented by kidney beans (mitochondria), poppy seeds (ribosomes), sections of lasagna noodles (ER and Golgi apparatus), peppercorns (lysosomes), spaghetti noodles (cytoskeleton), and an orange (nucleus).

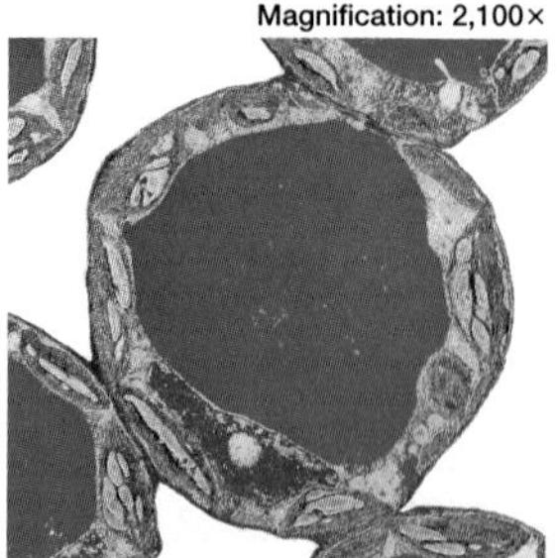

Figure 3-16 Central vacuole. The central vacuole stores substances and gives a plant rigidity.

green algae. Chloroplasts, along with mitochondria, supply much of the energy needed to power the activities of plant cells. Like mitochondria, chloroplasts are surrounded by two membranes, contain their own DNA, and are thought to be the descendents of ancient prokaryotic cells.

Central vacuole As shown in **Figure 3-16,** much of a plant cell's volume is taken up by a large, membrane-bound space called the **central vacuole** *(VAK yoo ohl)*. Vacuoles are found in both plant and animal cells, but only plant cells contain a central vacuole. The central vacuole stores water and may contain many substances, including ions, nutrients, and wastes. When the central vacuole is full, it presses the cytoplasm against the cell wall, making the cell rigid. This rigidity enables a plant to stand upright.

TECH WATCH: Laser Microsurgery

Most people know that surgeons use tweezers and scalpels to operate on the body. But did you know that cell biologists are now using lasers to operate on individual cells?

Optical tweezers

When directed through a light microscope, a very narrow laser beam exerts a small push against the specimen that is being viewed. Thus, the laser can move tiny objects, such as living cells and their organelles. Scientists can use these "optical tweezers" to position two cells near each other, to move individual chromosomes, and to transfer DNA between cells. Because the laser beam is not absorbed by living tissue, cells and organelles are not damaged by the procedure.

Optical scalpel

In contrast, the "optical scalpel" uses a laser beam that is absorbed by living tissue. When a cell or organelle absorbs the laser's energy, it heats up and is destroyed. Because the laser beam is a very brief, highly focused pulse, the damage is limited to an area less than 1 mm in diameter. The optical scalpel can be used to make small holes in cell membranes in order to remove specific parts of cells and to cut chromosomes into segments.

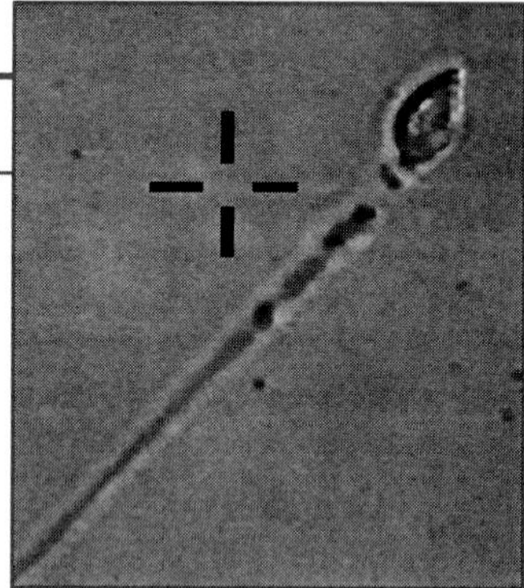

Sperm cell severed by an optical scalpel

internetconnect
SCiLINKS
NSTA
TOPIC: Laser surgery
GO TO: www.scilinks.org
KEYWORD: HX066

Review SECTION 3-3

1. **Describe** the role of the nucleus in cell activities.
2. **Sequence** the course of newly made proteins from the rough ER to the outside of the cell.
3. **Describe** the role of mitochondria in the metabolism of eukaryotic cells.
4. **SKILL Comparing Structures** How do a plant cell's central vacuole and cell wall help make the cell rigid?

Critical Thinking

5. **SKILL Inferring Relationships** What is the importance of a cell enclosing its digestive enzymes inside lysosomes?

Review SECTION 3-3 ANSWERS

1. The nucleus contains DNA, directs cell activities, and assembles ribosomal proteins and RNA.
2. The proteins, which are made by ribosomes, pass through the ER. The proteins are packaged into vesicles. The vesicles transport the proteins to the Golgi apparatus, where the proteins are modified. The Golgi apparatus packages the modified proteins into new vesicles. Some of these vesicles release their contents from the cell, while other vesicles remain in the cell.
3. Mitochondria produce most of the ATP made by eukaryotic cells.
4. The cell wall contains cellulose and is thick and supportive. The central vacuole is filled with water. Both provide rigidity in plant cells.
5. The digestive enzymes must be enclosed by lysosomes because if the enzymes were free inside the cell, the cell would digest itself.

Study ZONE CHAPTER 3

Use the Key Concepts and Key Terms listed below to review the main ideas in this chapter. If you need to review the meaning of a term, turn to the page indicated.

Key Concepts

3-1 Looking at Cells

- Microscopes enable biologists to examine the details of cell structure and to understand how organisms function.
- Scientists use the metric system to measure the size of objects.
- Light microscopes have a low magnification and can be used to examine living cells.
- Electron microscopes have a high magnification but cannot be used to examine living cells.
- The scanning tunneling microscope uses a computer to generate a three-dimensional image of an object.

3-2 Cell Features

- The cell theory has three parts.
- Small cells function more efficiently than large cells because small cells have a higher surface-area-to-volume ratio than large cells.
- All cells have a cell membrane, cytoplasm, ribosomes, and DNA.
- Prokaryotic cells lack internal compartments.
- Eukaryotic cells have a nucleus and other organelles, as well as a cytoskeleton of microscopic protein fibers.
- The lipid bilayer of a cell membrane is made of a double layer of phospholipid molecules.
- Proteins in cell membranes include enzymes, receptor proteins, transport proteins, and cell-surface markers.

3-3 Cell Organelles

- The nucleus of a eukaryotic cell directs the cell's activities and stores DNA.
- In eukaryotic cells, an internal membrane system produces, packages, and distributes proteins.
- Mitochondria harvest energy from organic compounds to ATP.
- Lysosomes digest and recycle a cell's used components.
- Plant cells have three structures that animal cells lack: a cell wall, chloroplasts, and a central vacuole.

Key Terms

3-1

light microscope (51)
electron microscope (51)
magnification (51)
resolution (51)
scanning tunneling microscope (54)

3-2

cell theory (55)
cell membrane (56)
cytoplasm (56)
cytoskeleton (56)
ribosome (56)
prokaryote (57)
cell wall (57)
flagellum (57)
eukaryote (58)
nucleus (58)
organelle (58)
cilium (58)
phospholipid (59)
lipid bilayer (59)

3-3

endoplasmic reticulum (62)
vesicle (62)
Golgi apparatus (63)
lysosome (63)
mitochondrion (64)
chloroplast (65)
central vacuole (66)

Unit 1—Use this unit to review the key concepts and terms in this chapter.

Study TIP Ready?

If you think you understand these Key Concepts and Key Terms, then you're ready for the Performance Zone.

Study ZONE CHAPTER 3

Review and Practice

Assign the following worksheets for Chapter 3:

- **Student Review Guide**
- **Concept Mapping Worksheet**
- **Test Preparation Pretest**
- **Vocabulary Worksheet**
- **Critical Thinking Review Worksheet**
- **Science Skills Worksheet**

Assessment Options

The following assessment options are available for this chapter:

Resource Link

- Assign **Chapter Test 3** to assess students' understanding of the chapter.

Alternative Assessment

Assign students to cooperative pairs. Have each pair devise a test that includes the material covered in this chapter. Each pair must also provide an answer key. Set the parameters for the test. For example, designate the number and type of questions allowed. After the tests are collected, exchange them with the other pairs, and have the pairs take the tests. Let the pairs check their own answers when they have completed the test. PORTFOLIO

Portfolio Assessment

- *Portfolio Assessment* at the front of this book provides options for building and evaluating student portfolios. Students and teachers can select items from the areas listed for inclusion in a portfolio.
- See items labeled PORTFOLIO on pp. 52, 53, 57, 58, 60, 62, 63 and 67.

Answer to Concept Map

The following is one possible answer to Performance Zone item 1 on page 68.

Eukaryotic cells
include
plant cells
animal cells
which contain
which contain
chloroplasts
cell wall
central vacuole
cell membrane
mitochondria

Performance ZONE
CHAPTER 3
REVIEW ANSWERS

Assignment Guide

SECTION	REVIEW
3-1	2, 3, 13, 22, 27
3-2	2, 4, 5, 6, 7, 14, 23, 28, 29
3-3	1, 2, 8, 9, 10, 11, 12, 15, 16, 17, 18, 19, 20, 21, 24, 25, 26

Understanding and Applying Concepts

1. The answer to the concept map is found at the bottom of page 67.
2. **a.** Light microscopes use light that passes through lenses to produce an enlarged image of a specimen. Electron microscopes use beams of electrons to form an enlarged image of a specimen.

   b. A flagellum is a long, threadlike structure that protrudes from the cell's surface. A cilium is a much shorter, hairlike structure.

   c. The cytoplasm is the cell interior. The cytoskeleton is a system of microscopic fibers in which all structures are suspended.

   d. A chloroplast is a structure found in the cells of photosynthetic eukaryotes that uses light energy to make organic compounds. A mitochondrion is a structure found in eukaryotic cells that makes ATP.
3. c
4. a
5. d
6. d
7. d
8. b
9. b
10. c
11. d
12. a

Performance ZONE CHAPTER 3 Review

Understanding and Applying Concepts

1. **Concept Mapping** Make a concept map that compares plant cells with animal cells. Include the following terms in your map: cell membrane, cell wall, central vacuole, chloroplasts, mitochondria.

2. **Understanding Vocabulary** For each pair of terms, explain the differences in their meanings.
   a. light microscope, electron microscope
   b. flagellum, cilium
   c. cytoplasm, cytoskeleton
   d. chloroplast, mitochondria

3. The main advantage of the transmission electron microscope is that it shows
   a. three-dimensional images of cell surfaces.
   b. the organelles of living cells.
   c. a cell's internal structure in fine detail.
   d. the actual colors of a cell's components.

4. The maximum size of a cell is determined by the ratio between the cell's
   a. surface area and volume.
   b. volume and organelles.
   c. organelles and cytoplasm.
   d. cytoplasm and nucleus.

5. Eukaryotic cells differ from prokaryotic cells in that eukaryotic cells
   a. lack organelles.
   b. have DNA but not ribosomes.
   c. are smaller than prokaryotic cells.
   d. have a nucleus.

6. In the cell membrane, the fatty acids of phospholipid molecules
   a. face the cytoplasm.
   b. face the outside of the cell.
   c. are on both sides of the membrane.
   d. are in the interior of the membrane.

7. Cell-membrane proteins include
   a. enzymes. c. markers.
   b. transporters. d. All of the above

8. The nucleus houses a eukaryotic cell's
   a. organelles. c. ribosomes.
   b. DNA. d. ER.

9. Proteins are produced by
   a. vesicles.
   b. ribosomes.
   c. lysosomes.
   d. smooth ER.

10. One function of the Golgi apparatus is to
    a. store DNA.
    b. make carbohydrates.
    c. modify proteins.
    d. digest and recycle the cell's wastes.

11. Mitochondria produce most of a cell's
    a. proteins. c. RNA.
    b. lipids. d. ATP.

12. Structures present in plant cells but absent from animal cells include
    a. chloroplasts and the central vacuole.
    b. mitochondria and the cell wall.
    c. ribosomes and ER.
    d. lysosomes and the Golgi apparatus.

13. What kind of microscope produced the image of cilia shown below?

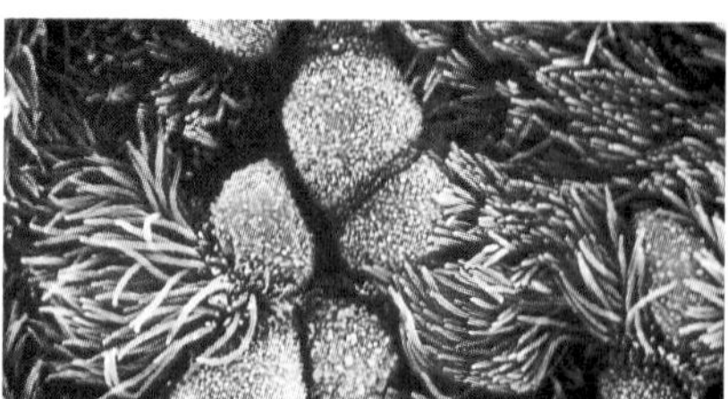

14. **Theme Homeostasis** Explain how the cell membrane contributes to a cell's ability to maintain homeostasis.

15. **Tech Watch** Lasers can be used to manipulate cell organelles without disrupting the cell membrane. Why is that an advantage?

16. **Chapter Links** Transport proteins in the membrane of a lysosome move hydrogen ions into the lysosome. Use this information to predict whether digestive enzymes in a lysosome work best in a neutral, a basic, or an acidic environment. (**Hint:** See Chapter 2, Section 2-2.)

---

13. A scanning electron microscope produced the image.
14. The cell membrane regulates what substances enter and leave the cell, thus maintaining the integrity of the cell.
15. Because the cell membrane is not disrupted, the cell will stay intact and alive.
16. Because of the presence of hydrogen ions in lysosomes, one can infer that the inside of the lysosome is acidic. Digestive enzymes in a lysosome therefore work best in an acidic environment.

Skills Assessment

Part 1: Interpreting Graphics

Use the illustration of the animal cell below to answer the questions that follow:

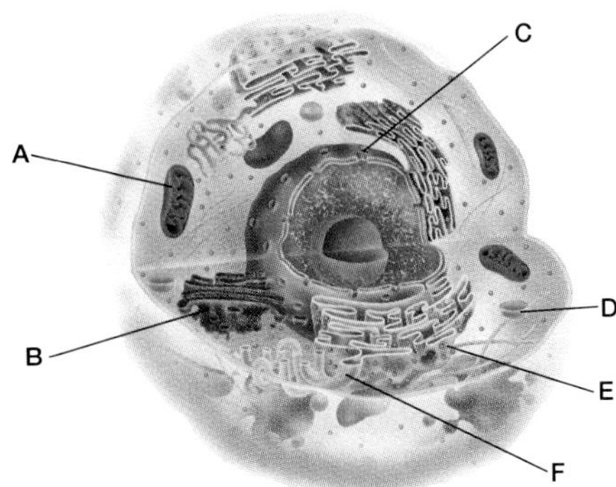

17. Identify the structures labeled *A–F*.
18. Name two substances that pass through the structure labeled *C*.
19. What is the function of the structure labeled *B*?
20. What critical role does the structure labeled *A* play in cell metabolism?

Part 2: Life/Work Skills

Use the media center or Internet resources to find out how different cell structures are involved in various diseases.

21. **Finding Information** Research the causes of a human disease that involves lysosomes. Explain how the disease affects lysosomes, and describe the effects of the disease on health. Prepare an oral report using graphics to interpret and summarize your findings.
22. **Using Technology** Electron microscopes can be used to view viruses, bacteria, and other small organisms in amazing detail. Find references that show details of microorganism structures discovered during the last 70 years. Make a table listing the different kinds of microorganisms and the structures revealed.

Critical Thinking

23. **Applying Information** Using your knowledge of the relationship between surface area and volume, explain why small pieces of a food cook faster than larger pieces of the same food.
24. **Evaluating Results** The chemical erythromycin effectively inhibits protein synthesis in bacterial cells, but it does not inhibit protein synthesis in human cells. Suggest a possible use for erythromycin.
25. **Inferring Relationships** Explain why some kinds of eukaryotic cells may contain more mitochondria than other kinds of eukaryotic cells.

Portfolio Projects

26. **Summarizing Information** Make a model of a cell organelle using modeling clay, and present your model to the class. In your presentation, describe the organelle's structure, function, and relationship to other organelles.
27. **Analyzing Methods** Read the article titled "Adult Human Brains Add New Cells" (*Science News*, October 31, 1998, p. 276). What techniques are researchers using to study the possibility that adult human brains develop new nerve cells?
28. **Career Focus Microbiologist** Use the media center or the Internet to research the field of microbiology. Your report should include a job description, training required, kinds of employers, growth prospects, and starting salary. If possible, interview a person who works in this field. Write a report to share with your class.
29. **Unit 1—*Cell Transport and Homeostasis*** Write a report summarizing the role of the cell membrane in the preservation of body organs donated for transplant.

Skills Assessment

17. A—mitochondrion; B—Golgi apparatus; C—nuclear pore; D—lysosome; E—rough ER; F—smooth ER
18. ribosomal proteins, RNA
19. The Golgi apparatus serves as the packaging and distribution center of the cell. It also modifies proteins.
20. Mitochondria produce most of a eukaryotic cell's ATP, which supplies much of the energy needed to power the cell's functions.
21. Answers will vary. Examples include Tay-Sachs disease and Pompe's disease.
22. Answers will vary. Have students access current microbiology books and other resources for this information.

Critical Thinking

23. Small pieces of food have less volume and therefore a larger surface-area-to-volume ratio than larger pieces. Thus, small pieces heat up more quickly than larger pieces.
24. Erythromycin, which is used as an antibiotic, kills bacterial cells by inhibiting protein production, while it does not affect human cells.
25. Cells with more mitochondria, such as muscle and liver cells, might have a greater need for energy.

Portfolio Projects

26. Models and presentations will vary. Check both for accuracy and clarity against the descriptions and illustrations in the chapter.
27. The researchers have found that the drug used in chemotherapy is present in the hippocampus, a structure of the brain involved in learning and memory. These cells were labeled to differentiate them from other kinds of brain cells. This finding indicates that the adult human brain can form new neurons, a phenomenon previously thought not to occur.
28. Answers will vary. Microbiologists study the growth, structure, development, and characteristics of bacteria and other microorganisms. They also study the action of microorganisms on living and dead tissue. The career requires at least a two-year technical training degree from a community college or technical institution. Many microbiologists have a four-year bachelor's degree plus a master's or doctoral degree. Employers include universities and various biomedical companies. Growth potential is very good. Starting salary will vary by region.
29. Answers will vary. Students may focus on the role of various types of transport proteins in maintaining the organs.

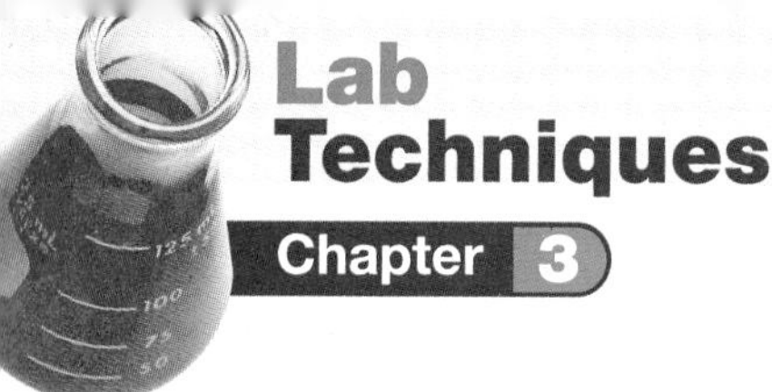

Lab Techniques

Chapter 3

Studying Animal Cells and Plant Cells

Purpose

Students will use a light microscope to examine animal and plant cells.

Preparation Notes

Time Required: Allow 45 minutes for this experiment.

Materials

Materials for this lab can be ordered from WARD'S. See *Master Materials List* at the front of this book for catalog numbers.

Preparation Tips

- Set up labeled containers for the disposal of broken glass and materials stained with Lugol's solution.
- Place enough materials for one class on a supply table.

Safety Precautions

- Be sure that students read and understand all safety rules for working in the lab.
- Remind students to read the ChemSafety section before beginning the lab. Discuss all safety symbols and caution statements with students.
- Make sure students wear safety goggles and a lab apron. **CAUTION: Tell students that they must wear safety goggles when working with Lugol's iodine solution because iodine can cause severe eye damage.**
- If students are using microscopes with mirrors, caution them not to look directly at the reflected light from the mirror.

Procedural Tips

- Allow students to take off their safety goggles when looking through the microscope.
- Have students use a scale of 1 µm = 1 cm when drawing cells. Help students take measurements of cell size.

Lab Techniques

Chapter 3

Studying Animal Cells and Plant Cells

SKILLS

- Using a compound microscope
- Drawing

OBJECTIVES

- **Identify** the structures you can see in animal cells and plant cells.
- **Compare and Contrast** the structure of animal cells and plant cells.

MATERIALS

- compound light microscope
- prepared slide of human epithelial cells from the lining of the mouth
- safety goggles
- lab apron
- sprig of *Elodea*
- forceps
- microscope slides and coverslips
- dropper bottle of Lugol's iodine solution

Magnification: 530×

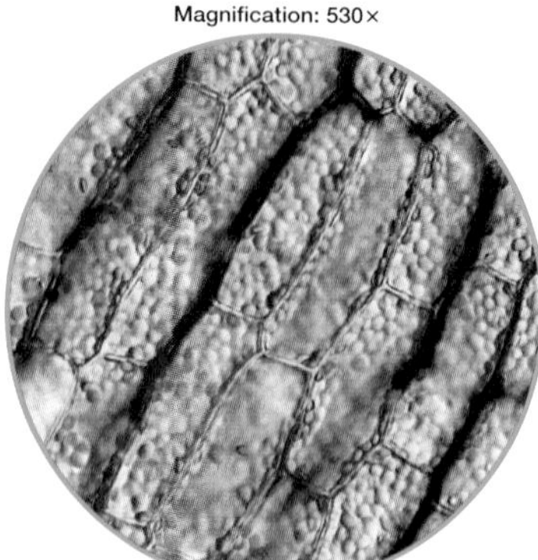

Plant cells

ChemSafety

CAUTION: Always wear safety goggles and a lab apron to protect your eyes and clothing.

CAUTION: Do not touch or taste any chemicals. Know the location of the emergency shower and eyewash station and how to use them. If you get a chemical on your skin or clothing, wash it off at the sink while calling to the teacher. Notify the teacher of a spill. Spills should be cleaned up promptly, according to your teacher's directions.

CAUTION: Glassware is fragile. Notify the teacher of broken glass or cuts. Do not clean up broken glass or spills with broken glass unless the teacher tells you to do so.

Before You Begin

You can see many cell parts with a **light microscope.** In animal cells, the **cytoplasm, cell membrane, nucleus, nucleolus,** and **vacuoles** can be seen. In plant cells, the **cell wall** and **chloroplasts** can also be seen. **Stains** add color to cell parts and make them more visible with a light microscope. A stain can even make the **endoplasmic reticulum** visible. In this lab, you will use a light microscope to examine animal and plant cells.

1. Write a definition for each boldface term in the paragraph above.
2. Why might a stain be needed to see cell parts under a microscope?
3. Based on the objectives for this lab, write a question you would like to explore about cell structure.

Procedure

PART A: Animal Cells

1. Examine a prepared slide of human epithelial cells under low power with a compound light microscope. Find cells that are separate from each other, and place them in the center of the field of view. Switch to high power, and adjust the diaphragm until you can see the cells more clearly. Identify as many cell parts as you can. *Note: Remember to use only the fine adjustment to focus at high power.*
2. Draw two or three epithelial cells as they look under high power. Label the cell membrane, the cytoplasm, the nuclear envelope, and the nucleus of at least one of the cells. Make a second drawing of these cells as you imagine they might look in the lining of your mouth.

Disposal

Combine all wastes containing Lugol's solution. To this mixture add a few drops of a strong acid, such as 1.0 M sulfuric acid (H_2SO_4), to make the mixture slightly acidic. Then slowly add 0.1 M sodium thiosulfate ($Na_2S_2O_3$) while stirring until the mixture loses its yellowish-orange color. Pour the resulting mixture down the drain.

Before You Begin

Answers

1. *light microscope*—a microscope that uses a beam of light passing through one or more lenses

   cytoplasm—the material inside a cell between the cell membrane and the nuclear membrane

   cell membrane—a lipid bilayer with embedded proteins that encloses a cell

   nucleus—an organelle found only in eukaryotic cells that houses the DNA

Part B: Plant Cells

3. Using forceps, carefully remove a small leaf from near the top of an *Elodea* sprig. Place the whole leaf in a drop of water on a slide, and add a cover slip.
4. Observe the leaf under low power. Look for an area of the leaf in which you can see the cells clearly, and move the slide so that this area is in the center of the field of view. Switch to high power, and, if necessary, adjust the diaphragm. Identify as many cell parts as you can.
5. Find an *Elodea* cell in which you can see the chloroplasts clearly. Draw this cell. Label the cell wall, a chloroplast, and any other cell parts that you can see.
6. Notice if the chloroplasts are moving in any of the cells. If you do not see movement, warm the slide in your hand or under a bright lamp for a minute or two. Look for movement of the cell contents again under high power. Such movement is called cytoplasmic streaming.
7. Put on safety goggles and a lab apron. Make a wet mount of another *Elodea* leaf, using Lugol's iodine solution instead of water. **CAUTION: Lugol's solution will stain your skin and clothing. Promptly wash off spills to minimize staining.** Observe these cells under low and high power.
8. Draw a stained *Elodea* cell. Label the cell wall and a chloroplast, as well as the central vacuole, the nucleus, and the cell membrane if they are visible.

Part C: Cleanup and Disposal

9. Dispose of solutions, broken glass, and *Elodea* leaves in the waste containers designated by your teacher. Do not pour chemicals down the drain or put lab materials in the trash unless your teacher tells you to do so.
10. Clean up your work area and all lab equipment. Return lab equipment to its proper place. Wash your hands thoroughly before you leave the lab and after you finish all work.

Analyze and Conclude

1. **Recognizing Patterns** In what observable ways are animal and plant cells similar in structure, and in what observable ways are they different?
2. **Comparing Structures** Compare and contrast the cytoplasm of epithelial cells and *Elodea* cells.
3. **Analyzing Methods** What is the reason for staining *Elodea* cells with iodine?
4. **Inferring Conclusions** Lugol's iodine solution causes the movement of chloroplasts to stop. Explain why.
5. **Inferring Conclusions** If some of the epithelial cells were folded over on themselves but were still transparent, what could you conclude about their thickness?
6. **Further Inquiry** Write a new question about cell structure that could be explored with another investigation.

On the Job

Microscopy is an important tool for biologists who study cell structure. Do research in the media center or on the Internet to learn more about how biologists use specialized microscopes to study cell structure. Summarize your findings in a written or oral report.

nucleolus—a structure in the nucleus of eukaryotic cells in which ribosomal proteins form

vacuole—a membrane-enclosed sac that contains substances, such as nutrients, needed by the cell

cell wall—a protective layer outside the cell membrane found in cells of bacteria, plants, fungi, and some protists

chloroplast—an organelle found only in the cells of plants and photosynthetic protists that uses light energy to drive the synthesis of organic compounds from carbon dioxide and water

stain—a substance that adds color to cell parts and makes them more visible with a light microscope

endoplasmic reticulum—an extensive system of internal membranes that move proteins and other substances through the cell

2. Because many cell structures are transparent, a stain may be needed to distinguish certain cell parts.
3. Answers will vary. For example, How does the size of animal cells compare with the size of plant cells?

Lab Techniques

Chapter 3

Sample Data

Drawings will vary. For the epithelial (animal) cells, make sure students label the cell membrane, cytoplasm, nucleus, nuclear envelope, and any other visible cell parts. For the *Elodea* (plant) cells, make sure students label the cell wall, cell membrane, chloroplasts, central vacuole, nucleus, and any other visible cell parts.

Analyze and Conclude

Answers

1. Animal and plant cells both have cytoplasm and a nucleus. Both types of cells are bound by a cell membrane. Unlike animal cells, plant cells have chloroplasts, a cell wall, and a central vacuole. Plant cells are also more consistent in shape than animal cells.
2. The cytoplasm of epithelial cells appears grainy. The cytoplasm of *Elodea* has observable chloroplasts present, is usually located near the edge of the cell, and typically flows in one direction.
3. The stain makes the nucleus, cell wall, and nucleolus more visible.
4. Lugol's solution is a poison that kills the cell and therefore stops the chloroplasts from moving.
5. A reasonable conclusion is that epithelial cells are extremely thin.
6. Answers will vary. For example: How do other animal and plant cells compare?

On the Job

Answers

Answers will vary. For example, the atomic force microscope is used to study biological surfaces with a resolution of single atoms.

Chapter 4

Cells and Their Environment

Planning Guide

OBJECTIVES	MEETING INDIVIDUAL NEEDS	READING SKILLS AND STRATEGIES
4-1 Passive Transport (BLOCKS 1 & 2 90 min) • Relate concentration gradients, diffusion, and equilibrium. • Predict the direction of water movement into and out of cells. • Describe the importance of ion channels in passive transport. • Identify the role of carrier proteins in facilitated diffusion.	Learning Styles, p. 74 BASIC ADVANCED • pp. 77, 80	ATE Active Reading, Technique: Reading Effectively, p. 73 ATE Active Reading, Technique: Paired Reading, p. 77 ■ Active Reading Guide, worksheet (4-1) ■ Directed Reading, worksheet (4-1)
4-2 Active Transport (BLOCKS 3 & 4 90 min) • Compare active transport with passive transport. • Describe the importance of the sodium-potassium pump. • Distinguish between endocytosis and exocytosis. • Identify three ways that receptor proteins can change the activity of a cell.	Learning Styles, p. 81 BASIC ADVANCED • pp. 82, 83, 86	■ Active Reading Guide, worksheet (4-2) ■ Directed Reading, worksheet (4-2) ■ Supplemental Reading Guide, *The Lives of a Cell: Notes of a Biology Watcher*

Review and Assessment Resources

BLOCKS 5 & 6 90 min

TEST PREP

- PE **Study Zone**, p. 87
- PE **Performance Zone**, pp. 88–89
- ■ **Chapter 4 worksheets:**
  - Vocabulary
  - Concept Mapping
  - Science Skills
  - Critical Thinking

 **Holt Biology Interactive Tutor** Unit 1—Cell Transport and Homeostasis, Topics 1–6

 **Audio CD Program**

■ **Test Prep Pretest**

CHAPTER TESTING

- ■ **Chapter Tests** (blackline master)
- ▲ **Test Generator**
- ■ **Assessment Item Listing**
- ■ **Performance-Based Assessment Activity:** Diffusion Through a Cell Membrane

See the correlations table at the front of this book for a list of the ***National Science Education Standards*** addressed in this chapter.

 Smithsonian Institution® Visit **www.si.edu/hrw** for additional on-line resources.

KEY

Biology Principles and Explorations

PE Pupil's Edition

ATE Teacher's Edition

HOLT BIOSOURCES

Teaching Transparencies

Laboratory Program

One-Stop Planner CD-ROM
Shows these resources within a customizable daily lesson plan:

■ Teaching Resources

▲ Test Generator

VISUAL STRATEGIES	CLASSWORK AND HOMEWORK	TECHNOLOGY AND INTERNET RESOURCES
ATE Fig. 4-1, p. 75 **ATE** Fig. 4-3, p. 78 **ATE** Fig. 4-4, p. 80 5 Diffusion 6 Osmosis 7A Gated Channels 10A Three Types of Solutions 10 Active and Passive Transport	**PE** Quick Lab Under what conditions does osmosis occur?, p. 76 **PE** Data Lab Analyzing the effect of electrical charge on ion transport, p. 79 Lab A4 Demonstrating Diffusion Lab B3 Diffusion and Cell Membranes **PE** Review, p. 80	Interactive Tutor Unit 1—Cell Transport and Homeostasis (Topics 1, 2, and 3) Holt Biology Videodiscs Section 4-1 Audio CD Program Section 4-1 internetconnect SCILINKS NSTA TOPICS: • Water movement in cells • Ion channels
ATE Fig. 4-5, p. 82 **ATE** Fig. 4-8, p. 84 6A Endocytosis and Exocytosis 11 Active Transport: Sodium-Potassium Pump	**PE** Experimental Design How Does Cell Size Affect Diffusion?, pp. 90–91 ■ Occupational Applications: Pharmacist **PE** Review, p. 86 **PE** Study Zone, p. 87	CNN Video Segment 2 Cystic Fibrosis Interactive Tutor Unit 1—Cell Transport and Homeostasis (Topics 4, 5, and 6) Holt Biology Videodiscs Section 4-2 Audio CD Program Section 4-2 internetconnect SCILINKS NSTA TOPICS: • Autoimmune diseases • Receptor proteins • Cell membrane

ALTERNATIVE ASSESSMENT

PE Portfolio Projects 24–27, p. 89

ATE Alternative Assessment, p. 87

ATE Portfolio Assessment, p. 87

Lab Activity Materials

Quick Lab
p. 76 grapes (3), small jars with lids (3), saturated sugar solution, grape juice, tap water, marking pen, paper towel, balance

Data Lab
p. 79 pencil, paper

Experimental Design
p. 90 safety goggles, lab apron, disposable gloves, block of phenolphthalein agar (3 × 3 × 6 cm), plastic knife, metric ruler, 250 mL beaker, 150 mL of vinegar, plastic spoon, paper towel

Lesson Warm-up
ATE beaker, warm water, red or blue food coloring, p. 74
ATE air pump, inflatable ball, p. 81

Demonstration
ATE potato, distilled water, tap water, salt water, beakers (3), p. 77

BIOSOURCES LAB PROGRAM
- Quick Labs: Lab A4, p. 7
- Inquiry Skills Development: Lab B3, p. 9

Chapter 4

Chapter Theme

Homeostasis

Homeostasis, one of the dominant themes of biology, is the focus of this chapter. Emphasize that homeostasis is the ability of an organism to maintain a constant internal condition even though its external environment changes. Remind students that the cell membrane is selectively permeable to different substances. Ask students what the cell membrane is composed of. *(mainly proteins and phospholipids)* Tell students that the transport of substances into and out of cells is essential to maintaining homeostasis.

Establishing Prior Knowledge

Before beginning this chapter, make certain students can answer the questions in **Study TIP Ready?** on page 73. If they cannot, encourage them to reread the sections indicated.

Opening Questions

Ask students to name four kinds of membrane proteins. *(cell-surface markers, receptor proteins, enzymes, and transport proteins)* Ask students why transport proteins are important to cell function. *(They enable ions and polar molecules to pass through the cell membrane.)*

CHAPTER 4 Cells and Their Environment

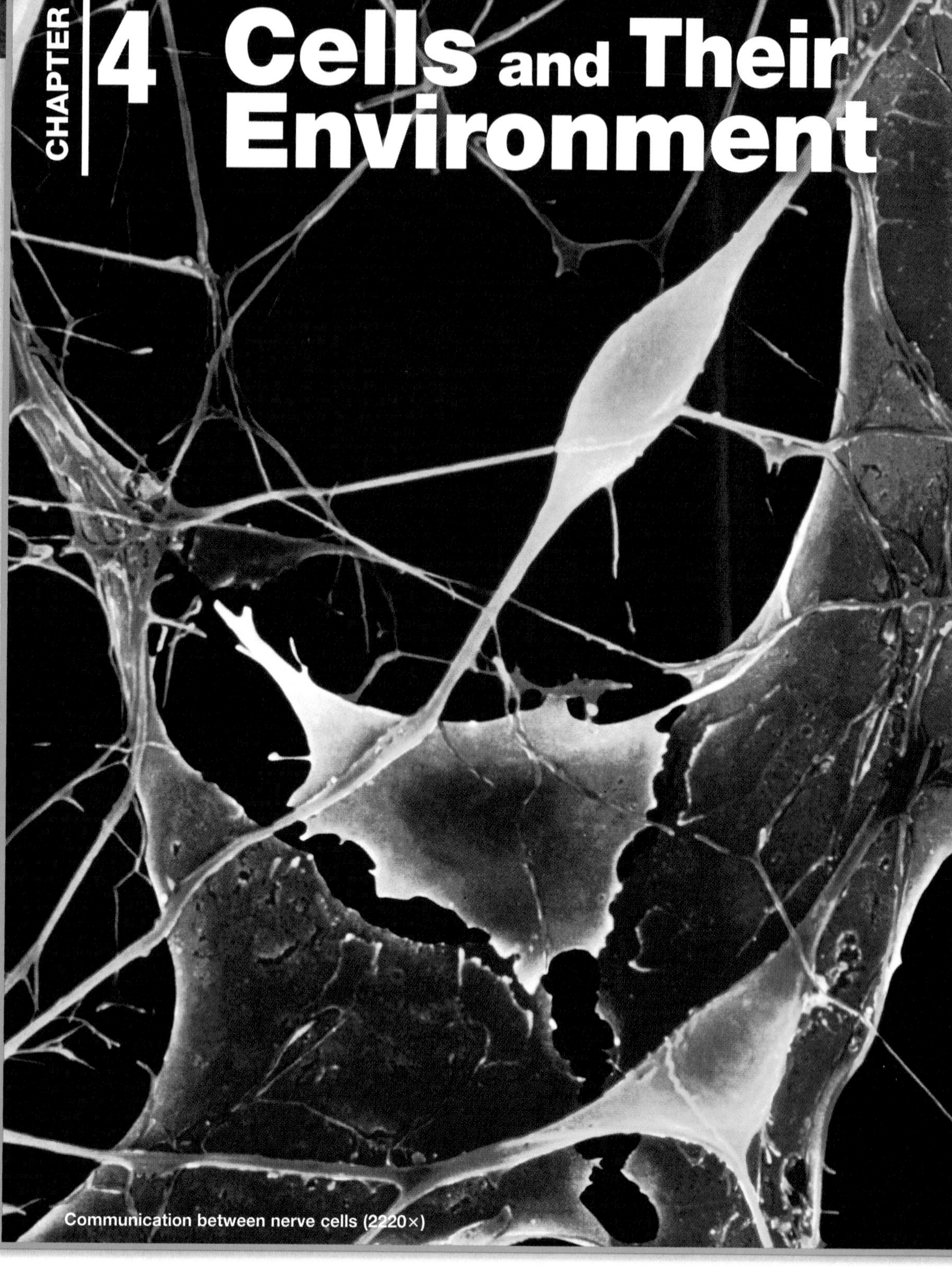

Communication between nerve cells (2220×)

What's Ahead — and Why

CHAPTER OUTLINE

4-1 Passive Transport

- Diffusion Is Caused by the Random Movement of Particles
- Water Diffuses into and out of Cells by Osmosis
- Proteins Help Some Substances Cross the Cell Membrane

4-2 Active Transport

- Some Substances Are Transported Against a Concentration Gradient
- Vesicles Move Substances Across Membranes
- Membrane Receptor Proteins Receive Information

Students will learn

- how cell membranes use diffusion, osmosis, and facilitated diffusion in passive transport.
- how substances that cannot be moved through cell membranes by passive transport can be moved by active transport.

You are made of billions

of eukaryotic cells. Each cell is encased in its own cell membrane. Just as a ticket taker controls who comes and goes from this ski area, the cell membrane regulates what goes into and out of a cell. Only people with tickets can cross the boundary, and only certain substances in your body can get into or out of your cells.

Looking Ahead

Labs

- **Quick Lab**
  Under what conditions does osmosis occur? (p. 76)
- **Data Lab**
  Analyzing the effect of electrical charge on ion transport (p. 79)
- **Experimental Design**
  How Does the Size of a Cell Affect Diffusion? (p. 90)

Features

- **Health Watch**
  Myasthenia Gravis (p. 86)

Study TIP

Ready?

Check your understanding of these topics to help you prepare for what's ahead.

Can you . . .

- **distinguish** between polar and nonpolar substances? (Section 2-1)
- **describe** the role of ATP in cells? (Section 2-3)
- **identify** different kinds of cell-membrane proteins? (Section 3-2)
- **summarize** the functions of vesicles and the Golgi apparatus? (Section 3-3)

If not, *review the sections indicated.*

Unit 1—*Cell Transport and Homeostasis*
Topics 1–6

internetconnect

National Science Teachers Association *sci*LINKS Internet resources are located throughout this chapter.

Reading Strategies

Active Reading

Technique: Reading Effectively

Strongly suggest that students take a break after reading each section to study the figures in the section. Discuss the importance of reading figure captions. Prepare questions that correspond to each figure. For example, in reference to Figure 4-1 on page 75, you might ask students to summarize the series of events in their own words, using molecules instead of balls. Refer to the figures when you talk about the concepts presented in each section.

Directed Reading

Assign **Directed Reading Worksheet Chapter 4** to help students interact with the text as they read the chapter.

Interactive Reading

Assign ***Biology: Principles and Explorations* Audio CD Program Chapter 4** to help reluctant readers achieve greater success in reading the chapter.

Reading Skills

Assign Reading Organizers and Reading Strategies Worksheets from the **One-Stop Planner CD-ROM with Test Generator** to help students learn and practice effective reading skills.

internetconnect

Holt *Biology: Principles and Explorations*

On-line Resources: go.hrw.com
Visit the HRW Web site for a variety of resources related to this chapter. Just type in the keyword HX0 Chapter 4.

Holt *Biology: Principles and Explorations*

On-line Resources: www.scilinks.org
The following *sci*LINKS Internet resources can be found in the student text for this chapter:

Page 75
TOPIC: Water movement in cells
KEYWORD: HX075

Page 79
TOPIC: Ion channels
KEYWORD: HX079

Page 86
TOPIC: Receptor proteins
KEYWORD: HX085

Page 86
TOPIC: Autoimmune diseases
KEYWORD: HX086

Page 91
TOPIC: Cell membrane
KEYWORD: HX091

Chapter 4

Section 4-1 Passive Transport

1 Prepare

Scheduling

- **Block:** About 90 minutes will be needed to complete this section.
- **Traditional:** About two class periods will be needed to complete this section.
- **Planning:** See the Planning Guide on pp. 71A and 71B for lecture, classwork, and assignment options for Section 4-1. Lesson plans for Section 4-1 can be found in a lesson cycle format in ***Biology: Principles and Explorations*** **One-Stop Planner CD-ROM with Test Generator.**

Resource Link

Assign **Active Reading Worksheet Section 4-1** to help students comprehend the key concepts and visuals in this lesson.

2 Teach

Lesson Warm-up

Fill a beaker with lukewarm water. Place two drops of red or blue food dye in the beaker. As the dye disperses in the water, ask students what they are observing. *(Diffusion is one of the processes they are observing.)* Explain that diffusion works very quickly over short distances but slowly over long distances. Most of the mixing in this demonstration is caused by convection and bulk flow of the water caused by slight air currents and temperature differences in the water. Therefore, emphasize that this demonstration is a *model* of diffusion rather than *actual* diffusion.

Objectives

- **Relate** concentration gradients, diffusion, and equilibrium. (pp. 74–75)
- **Predict** the direction of water movement into and out of cells. (pp. 76–77)
- **Describe** the importance of ion channels in passive transport. (pp. 78–79)
- **Identify** the role of carrier proteins in facilitated diffusion. (p. 80)

Key Terms

passive transport
concentration gradient
equilibrium
diffusion
osmosis
hypertonic solution
hypotonic solution
isotonic solution
ion channel
carrier protein
facilitated diffusion

Study TIP

Reading Effectively

As you read this chapter, write the objectives for each section on a sheet of paper. Rewrite each objective as a question, and answer these questions as you read the section.

Diffusion Is Caused by the Random Movement of Particles

You constantly interact with your environment, whether you are eating or putting on a raincoat to help keep you dry. Your body also responds to external conditions to maintain a stable internal condition. Just as you must respond to your environment to maintain stability, all other organisms and their cells must respond to external conditions to maintain a constant internal condition. Recall that when organisms adjust internally to changing external conditions, they are maintaining homeostasis. One way cells maintain homeostasis is by controlling the movement of substances across the cell membrane. Cells must use energy to transport some substances across the cell membrane. Other substances move across the cell membrane without any use of energy by the cell.

Random motion and concentration

Movement across the cell membrane that does not require energy from the cell is called **passive transport.** To understand passive transport, imagine two rooms of equal size separated by a wall with a closed door, as shown in **Figure 4-1.** Suppose you release several rubber balls into the first room. The balls move randomly, bouncing off the walls, the floor, the ceiling, and each other. Also suppose the balls can bounce forever without slowing down. The balls become evenly distributed throughout the room. What happens when you open the door between the rooms? Some of the balls in the first room bounce through the doorway and into the second room, as shown in Figure 4-1. You do not have to use energy to make the balls move into the second room. They enter the second room because of their own random motion. Occasionally, a ball will bounce back into the first room. However, most of the balls that pass through the doorway move from the first room, where their concentration is high, to the second room, where their concentration is low. A difference in the concentration of a substance, such as the balls, across a space is called a **concentration gradient.**

As more balls enter the second room, the concentration of balls in the second room increases, while the concentration of balls in the first room decreases. Eventually the concentration of balls in the two rooms will be equal. The balls will still bounce around the rooms, but they will move from the second room to the first room just as often as they move from the first room to the second room. At this point, the system is said to be in equilibrium, as shown in Figure 4-1. **Equilibrium** *(ee kwih LIHB ree uhm)* is a condition in which the concentration of a substance is equal throughout a space.

Integrating Different Learning Styles

Logical/Mathematical	**PE**	Data Lab, p. 79; Experimental Design, pp. 90–91
Visual/Spatial	**PE**	Quick Lab, p. 76; Performance Zone items 1 and 11, p. 88
	ATE	Lesson Warm-up, p. 74; Visual Strategies, pp. 75, 78, 80; Demonstration, p. 77
Interpersonal	**ATE**	Active Reading, p. 77; Reteaching, p. 80
Intrapersonal	**ATE**	Closure, p. 80
Verbal/Linguistic	**PE**	Study Tips, pp. 74, 77; Performance Zone items 2, 20, and 27 pp. 88–89
	ATE	Internet Connect, p. 75; Cross-Disciplinary Connection, p. 78

Figure 4-1 Models of diffusion

Because of diffusion, food coloring (blue) will gradually move through gelatin (yellow), as shown in the beakers below.

1. Randomly bouncing balls are distributed evenly throughout a closed room.

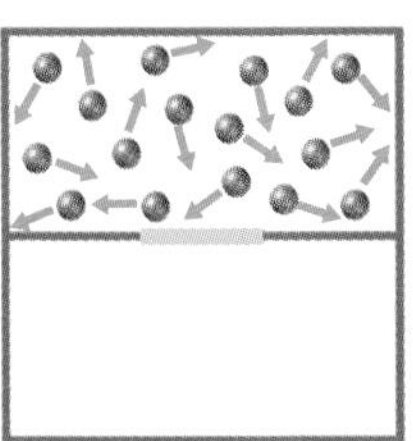

2. If the door to an adjoining room is opened, the balls begin to enter, or diffuse into, that room.

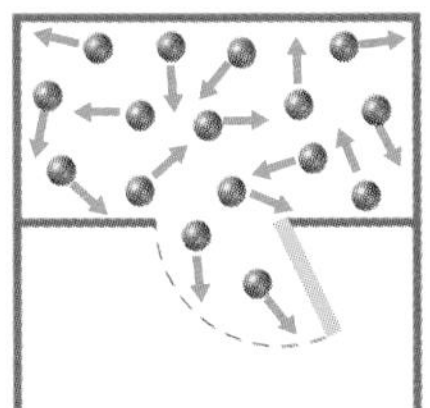

3. At equilibrium, the concentration of balls inside the two rooms will be equal.

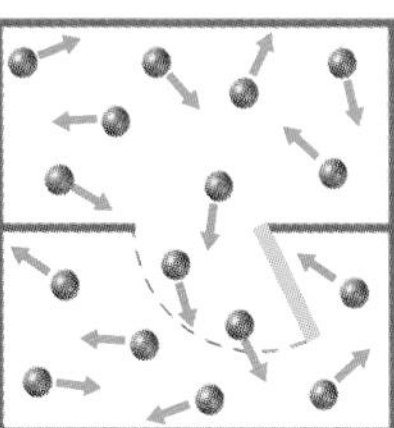

Movement of substances

Like these imaginary rubber balls, particles of a substance in a solution also move around randomly. If there is a concentration gradient in the solution, the substance will move from an area of high concentration to an area of lower concentration. The movement of a substance from an area of high concentration to an area of lower concentration caused by the random motion of particles of the substance is called **diffusion** *(dih FYOO zhuhn)*. If diffusion is allowed to continue, equilibrium eventually results.

Many substances, such as molecules and ions dissolved in the cytoplasm and in the fluid outside cells, enter or leave cells by diffusing across the cell membrane. Inside the cell, the concentrations of most of these substances are different from their concentrations outside the cell. Thus, for each of these substances a concentration gradient exists across the cell membrane. To diffuse "down" its concentration gradient—from an area of high concentration to an area of lower concentration—a substance must be able to pass through the cell membrane.

The cell membrane is selectively permeable to substances. The nonpolar interior of the lipid bilayer repels ions and most polar molecules, preventing these substances from diffusing across the cell membrane. In contrast, molecules that are either very small or nonpolar can diffuse across the cell membrane down their concentration gradient. The diffusion of such molecules across the cell membrane is the simplest type of passive transport.

internetconnect

SCILINKS NSTA

TOPIC: Water movement in cells
GO TO: www.scilinks.org
KEYWORD: HX075

Section 4-1

Visual Strategy

Figure 4-1

Correlate the model of balls moving between the two rooms with what is happening in the beakers filled with gelatin. Explain that food coloring moves through the gelatin because of diffusion caused by the random motion of particles of the food coloring. Emphasize that the individual particles of food coloring move randomly and independently and that the net movement of the food coloring is also caused by this random motion. Because of the concentration gradient of the food coloring, the food coloring moves through the gelatin from an area of high concentration to an area of lower concentration.

Making Biology Relevant

Medicine

The function of the kidneys is to remove wastes from the blood. However, during renal failure this cannot be done, and wastes remain in the blood. In such cases, hemodialysis is used. Hemodialysis is a process that uses a semipermeable membrane to separate large substances, such as blood cells, from smaller ones, such as urea. Hemodialysis is often referred to as the artificial kidney. Blood from an artery of a patient passes through a semipermeable cellophane tube that is inside a container. Here the tube is surrounded by a solution that contains varying concentrations of electrolytes and other chemicals. Urea passes through the pores of the membrane and into a wash solution for removal. Blood cells cannot pass through and are returned to the patient's body through a vein.

Overcoming Misconceptions

Diffusion

Some students may think that particles of a substance in solution stop moving when equilibrium is reached. Although at equilibrium there is no concentration gradient because the concentration of the substance is the same throughout the solution, the particles of the substance will continue to move randomly and independently. Once equilibrium is reached, it must be maintained.

internetconnect

Have students research osmosis in cells using the Web site in the Internet Connect box on this page. Have students write a report summarizing their findings. PORTFOLIO

Teaching TIP

Cells and Osmosis

A common remedy for a sore throat is to gargle with salt water. Use the concept of osmosis,to explain how this remedy might work. Some of the pain associated with a sore throat is caused by swelling of the throat tissues, which contain water. When a person gargles with salt water, which contains a lower concentration of water than do the throat tissues, water moves by osmosis from the throat tissues into the salt water. This decreases the swelling and relieves some of the pain.

Under what conditions does osmosis occur?

Time about 15 minutes to set up; about 24 hours for completion

Process Skills Observing, analyzing, evaluating

Teaching Strategies

Review how to use a balance to find the mass of objects.

Analysis Answers

1. Answers will vary. The tap water should cause water to move into the grape. The sugar solution should cause water to move out of the grape. Results with the grape juice may vary depending on the sugar content of the juice.
2. Whether osmosis occurs can be determined by noting an increase or a decrease in the mass of the grapes.
3. Answers will vary. Students who had a clear understanding of osmosis before the experiment probably will not change their thinking.

Water Diffuses into and out of Cells by Osmosis

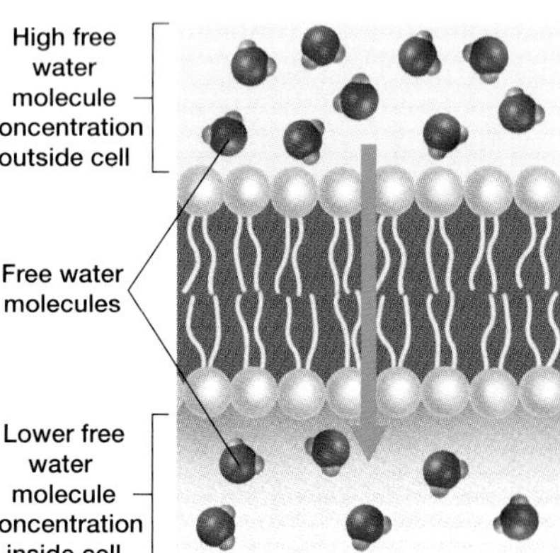

Figure 4-2 **Osmosis.** Water diffuses across the cell membrane by osmosis.

Water molecules are small and can diffuse through the cell membrane, as shown in **Figure 4-2.** The diffusion of water through a selectively permeable membrane is called **osmosis** *(ahz MOH sihs).* Like other forms of diffusion, osmosis involves the movement of a substance—water—down its concentration gradient. Osmosis is a type of passive transport.

What causes osmosis? Recall that a solution is a substance dissolved in another substance. In the solutions on either side of the cell membrane, many ions and polar molecules are dissolved in water. When these substances dissolve in water, some water molecules are attracted to them and so are no longer free to move around. If the solutions on either side of the cell membrane have different concentrations of dissolved particles, they will also have different concentrations of "free" water molecules. Then osmosis will occur as free water molecules move into the solution with the lower concentration of free water molecules.

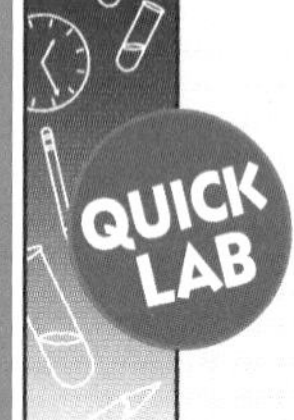

Under what conditions does osmosis occur?

You can observe the movement of water into or out of a grape under different conditions.

Materials

3 grapes, 3 small jars with lids, saturated sugar solution, grape juice, tap water, marking pen, paper towel, balance

Procedure

1. Make a data table with four columns (Solution, Original mass, Predicted mass, and Actual mass) and a row for each solution (Grape juice, Sugar solution, and Water).
2.  Fill one jar with grape juice. (The grape will be more visible inside the jar if you fill the jar with white grape juice, as shown in the middle jar in the photo at right.) Fill a second jar with the sugar solution. Fill the third jar with tap water. Label each jar according to the solution it contains.
3. Using the balance, find the mass of each grape. Place one grape in each jar, and record the mass of each jar in your data table. Put a lid on each jar.
4. Predict whether the mass of each grape will increase or decrease over time. Explain your predictions.
5. After 24 hours, remove each grape from its jar, and dry it gently with a paper towel. Using the balance, find its mass again. Record your results.
6. Clean up your materials before leaving the lab.

Analysis

1. **Identify** the solutions in which osmosis occurred.
2. **SKILL Evaluating Conclusions** How did you determine whether osmosis occurred in each of the three solutions?
3. **SKILL Evaluating Hypotheses** Has your thinking changed after having seen the data? Explain how.

Multicultural Perspective

Homeostasis

The process of osmosis was discovered in 1748 by the French physicist Abbé Jean Antoine Nollet, who discovered that water spontaneously diffused through parchment paper into a glass tube. In 1865, the French physiologist Claude Bernard published his theories on the stability of the internal environment of organisms. His concept, now called homeostasis, has given rise to modern theories that explain the body's control mechanisms. The American physiologist Walter B. Cannon, building on Bernard's work, published postulates governing homeostasis.

Table 4-1 Hypertonic, Hypotonic, and Isotonic Solutions			
If the fluid outside the cell has...	**Then outside fluid is...**	**Water diffuses...**	**Effect on cell**
...lower free water molecule concentration than cytoplasm	...hypertonic.	...out of cell. H_2O	Cell shrinks.
...higher free water molecule concentration than cytoplasm	...hypotonic.	...into cell. H_2O	Cell swells.
...same free water molecule concentration as cytoplasm	...isotonic.	...into and out of cell at equal rates. H_2O	Cell stays same size.

The direction of water movement across the cell membrane depends on the relative concentrations of free water molecules in the cytoplasm and in the fluid outside the cell. There are three possibilities for the direction of water movement:

1. **Water moves out.** When water diffuses out of the cell, the cell shrinks. A solution that causes a cell to shrink because of osmosis is called a **hypertonic** *(heye puhr TAHN ihk)* **solution.** If the fluid outside the cell has a higher concentration of dissolved particles than the cytoplasm has, then the outside fluid also has a lower concentration of free water molecules than the cytoplasm.
2. **Water moves in.** When water diffuses into the cell, the cell swells. A solution that causes a cell to swell because of osmosis is called a **hypotonic** *(heye poh TAHN ihk)* **solution.** If the fluid outside the cell has a lower concentration of dissolved particles than the cytoplasm has, then the outside fluid also has a higher concentration of free water molecules than the cytoplasm.
3. **No net water movement.** If the cytoplasm and the fluid outside the cell have the same concentration of free water molecules, water diffuses into and out of the cell at equal rates. This results in no net movement of water across the cell membrane, and the cell stays the same size—a state of equilibrium. A solution that produces no change in cell volume because of osmosis is called an **isotonic** *(eye soh TAHN ihk)* **solution.** Table 4-1 summarizes the effects of hypertonic, hypotonic, and isotonic solutions on cells.

If left unchecked, the swelling caused by a hypotonic solution could cause a cell to burst. Different kinds of cells have evolved different ways of dealing with this problem. The cells of plants and fungi have rigid cell walls that keep the cells from expanding too much. Some unicellular eukaryotes have contractile vacuoles *(kuhn TRAK tihl VAK yoo ohlz),* which are organelles that collect excess water inside the cell and force the water out of the cell. Animal cells have neither cell walls nor contractile vacuoles. However, many animal cells can avoid swelling caused by osmosis by removing dissolved particles from the cytoplasm. The removal of dissolved particles from a cell increases the concentration of free water molecules inside the cell. You will learn how this happens in Section 4-2.

Study TIP

Word Origins

The words *hypertonic, hypotonic,* and *isotonic* have the same ending, *-tonic,* which is from the Greek *tonos,* meaning "tension." The prefix *hyper-* is from the Greek *hyper,* meaning "over." The prefix *hypo-* is from the Greek *hypo,* meaning "lower." The prefix *iso-* is from the Greek *isos,* meaning "same."

did you know?

On humid days, wooden drawers in dressers absorb water from the air because of osmosis.

This absorption warps the wood, making opening and closing the drawers more difficult. In dry weather, water is lost from the wood, and the drawers return to their original size.

Demonstration

Osmosis in Potato Slices

Prepare three potato slices, each approximately 1 cm thick. Place one slice in distilled water, one in tap water, and one in salt water. Ask students to classify the solutions using the terms *hypotonic, hypertonic,* and *isotonic. (Distilled water represents a hypotonic solution; tap water represents an isotonic solution; and salt water represents a hypertonic solution.)* Ask students to predict what will happen to each slice. *(The slice in the distilled water should swell and become crisp. The slice in the tap water should be unchanged. The slice in the salt water should shrink.)*

Teaching TIP — ADVANCED

Salting Roads

During the winter, salt is often spread on icy roads. The plants along the sides of these roads often do not survive. Ask students to explain the relationship between the plants and the salt on the roads. *(The higher-than-normal salt concentration in the external environment of these plants causes their cells to lose water because of osmosis. Without sufficient water, the plants die.)*

Active Reading

Technique: Paired Reading

Assign students to cooperative pairs, and have each student read pp. 76–77 silently. As they read, have students indicate passages they understand with a check mark and passages they do not understand with a question mark. Then have the partners discuss what they did or did not understand.

REAL LIFE

Answer

The molecules move more rapidly as they heat up, causing the scent to spread more quickly.

CROSS-DISCIPLINARY CONNECTION

◆ **Earth Science**

If fresh water and salt water were separated by a semipermeable membrane, the fresh water would flow through the membrane and combine with the salt water because of osmosis. In a process called reverse osmosis, pressure is applied to the salt water, forcing pure water through the membrane and leaving salt and other impurities behind. This process involves several membranous filters. Reverse osmosis is gaining more recognition as a means of desalination. It is also used in many home filtration systems to purify tap water. A major drawback of reverse osmosis is that the system removes fluoride from the water. Ask students why this is important. *(Fluoride helps reduce the incidence of tooth decay.)* Have students research and write a report on the mechanisms of water purification. PORTFOLIO

Visual Strategy

Figure 4-3

Remind students that the cell membrane is selectively permeable to different substances. Emphasize that only certain ions can pass through each type of ion channel. For example, only sodium ions can pass through the ion channel shown in this figure. Tell students that ion channels span the thickness of the membrane so ions, which are charged particles, can pass through the membrane. When an ion passes through the pore of an ion channel, the ion is shielded from the nonpolar interior of the lipid bilayer.

REAL LIFE

Does temperature affect how odors travel?

Odor-causing molecules travel across a room by diffusing through the air. If you cook a pizza, its aroma will fill the kitchen.

Predicting Outcomes

Describe the motion of odor-causing molecules as they heat up.

Proteins Help Some Substances Cross the Cell Membrane

In Chapter 3, you learned that ions and most polar molecules cannot diffuse across the cell membrane because they cannot pass through the nonpolar interior of the lipid bilayer. However, such substances can cross the cell membrane when they are aided by transport proteins. Transport proteins called channels provide polar passageways through which ions and polar molecules can move across the cell membrane. Each channel allows only a specific substance to pass through the cell membrane. For example, some channels allow only one type of ion to cross the cell membrane, while others transport a particular kind of sugar or amino acid. This selectivity is one of the most important properties of the cell membrane because it enables a cell to control what enters and leaves.

Diffusion through ion channels

Ions such as sodium, Na^+, potassium, K^+, calcium, Ca^{2+}, and chloride, Cl^-, are involved in many important cell functions. For example, ions are essential to the ability of nerve cells to send electrical signals throughout your body. Muscle cells in your heart could not make your heart beat without the movement of ions between the cells. Although ions cannot diffuse through the nonpolar interior of the lipid bilayer, they can cross the cell membrane by diffusing through ion channels. An **ion channel** is a doughnut-shaped transport protein with a polar pore through which ions can pass. As **Figure 4-3** shows, the pore of an ion channel spans the thickness of the cell membrane. Thus, an ion that enters the pore can cross the cell membrane without contacting the nonpolar interior of the lipid bilayer.

The pores of some ion channels are always open. In other ion channels, the pores can be closed by ion channel gates. A model of an ion channel with a gate is shown in Figure 4-3. Ion channel gates may open or close in response to several stimuli, such as stretching of the cell membrane, a change in electrical charge, or the binding of

Figure 4-3 Ion channels

Ion channels allow certain ions to pass through the cell membrane.

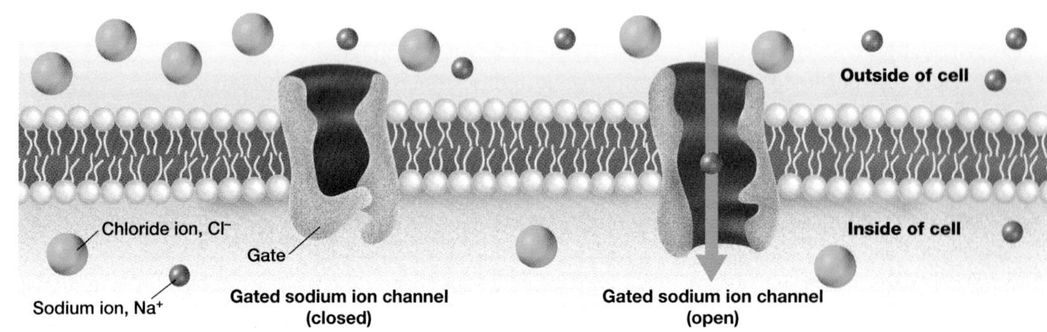

did you know?

Calcium channel blockers are drugs that inhibit the flow of calcium ions through ion channels in cardiac muscle cells.

Calcium channel blockers decrease the heart rate and the force of heartbeats. These drugs are used to decrease blood pressure, to relieve chest pain, and to stabilize abnormal heart rhythms.

specific molecules to the ion channel. In this way, the stimuli are able to control the ability of particular ions to cross the cell membrane. Like the diffusion of small molecules and nonpolar molecules through the lipid bilayer, the diffusion of ions through ion channels is a form of passive transport. No use of energy by the cell is required because the ions move down their concentration gradients.

TOPIC: Ion channels
GO TO: www.scilinks.org
KEYWORD: HX079

Effect of electrical charge on ion transport The rate of movement of a substance across the cell membrane is generally determined by the concentration gradient of the substance. The movement of a charged particle, such as an ion, across the cell membrane is also influenced by the particle's positive or negative electrical charge. The inside of a typical cell is negatively charged with respect to the outside of the cell. Opposite charges attract, and like charges repel. Thus, a more positively charged ion located outside the cell is more likely to diffuse into the cell, where the charge is negative. Conversely, a more negatively charged ion located inside the cell is more likely to diffuse out of the cell. The direction of movement caused by an ion's concentration gradient may oppose the direction of movement caused by the ion's electrical charge. Thus, an ion's electrical charge often affects the diffusion of the ion across the cell membrane. This is extremely important to the functioning of nerve cells in animals, as you will see in Chapter 42.

Analyzing the effect of electrical charge on ion transport

Background

The electrical charge of an ion affects the diffusion of the ion across the cell membrane. Some ions are more concentrated inside cells, and some ions are more concentrated outside cells. Use the table below to answer the following questions:

Magnification: 13,000×

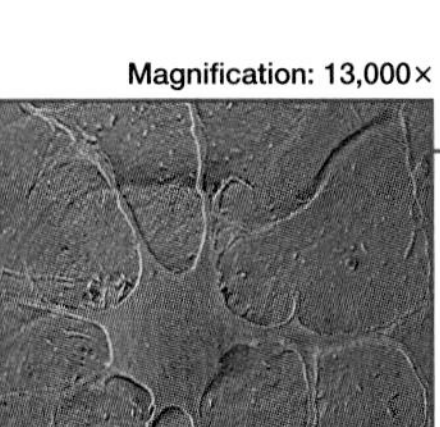

Nerve cell

Ion Charges and Concentration Inside and Outside Cell

Ion	Charge of ion	Concentration of ion outside cell : inside cell
Sodium (Na^+)	Positive	10:1
Potassium (K^+)	Positive	1:20
Calcium (Ca^{2+})	Positive	10,000:1
Chloride (Cl^-)	Negative	12:1

Analysis

1. **State** which ion is more concentrated inside the cell than outside the cell.
2. **State** which ions are more concentrated outside the cell than inside the cell.
3. **SKILL Recognizing Relationships** Do the positive charges of calcium ions and sodium ions make these ions more likely to move into or out of the cell?
4. **SKILL Inferring Relationships** Which ions' electrical charges oppose the direction of movement that is caused by their concentration gradient?

Section 4-1

Making Biology Relevant

Food

Ask students why some athletes prefer sports drinks during and after a practice or a game. Explain that these thirst quenchers replace both water and ions lost through perspiration. Have students check the labels on such drinks and make a list of the substances they contain. PORTFOLIO

Analyzing the effect of electrical charge on ion transport

Time about 10 minutes

Process Skills Analyzing data, inferring relationships

Teaching Strategies

Review the principles of ions and ionic bonding. Remind students that ions of opposite charge attract one another, and that ions of similar charge repel one another.

Analysis Answers

1. potassium ion
2. sodium ion, calcium ion, and chloride ion
3. Calcium ions and sodium ions are more likely to move into the cell because of their positive charges.
4. Potassium ions move down their concentration gradient out of the cell, but their concentration gradient is opposed by their positive charge, which is attracted to the negatively charged interior of the cell. Chloride ions move down their concentration gradient into the cell, but their concentration gradient is opposed by their negative charge, which is repelled by the negatively charged interior of the cell.

MULTICULTURAL PERSPECTIVE

Sweat Lodges

Different cultures throughout the world use water and mineral salts to purify their bodies. Some Native American tribes use sweat lodges or community steam baths for physical and spiritual purposes. The dome-shaped sweat lodge contains hot rocks placed in the center of the earthen floor. To purify the air, small pieces of sage, sweet grass, or juniper are burned on the rocks as incense. Bathers drink tea made with medicinal plants and sprinkle the tea on the rocks. The tea increases perspiration, causing the bathers to lose more than 2 liters (half a gallon) of water.

Visual Strategy

Figure 4-4 Review the process of facilitated diffusion with students. Compare the carrier proteins in this figure with the ion channels in Figure 4-3 on page 78. Tell students that when the concentration of glucose in the cell is low, glucose quickly enters the cell through carrier proteins by facilitated diffusion. Glucose is too large to pass through the cell membrane.

3 Assessment Options

Closure

Diffusion

Spray air freshener at the front of the room. Ask students seated in the back of the room why they can smell the fragrance. *(The fragrance is spread mainly by air currents. Remind students that diffusion works very slowly over such long distances.)*

Review

Assign Review—Section 4-1.

Reteaching — BASIC

Assign students to cooperative groups of three. Write the following questions on the board. Have each group discuss each question and attempt to answer correctly.

1. Why are green leafy vegetables sprinkled with water at the supermarket? *(This prevents them from wilting due to water loss.)*
2. Why is salt sometimes used to preserve foods? *(Because a salt solution is hypertonic, the microorganisms present on the foods will shrink and die.)*
3. Why should you not drink sea water? *(Sea water contains high concentrations of solutes, so drinking sea water would increase the concentration of solutes outside body cells. The cells would lose water because of osmosis and possibly die.)*

Facilitated diffusion

Most cells also have a different kind of transport protein that can bind to a specific substance on one side of the cell membrane, carry the substance across the cell membrane, and release it on the other side. Such proteins are called **carrier proteins.** When carrier proteins are used to transport specific substances—such as amino acids and sugars—down their concentration gradient, that transport is called facilitated diffusion. **Facilitated** *(fah SIHL uh tayt ehd)* **diffusion** is a type of passive transport because it moves substances down their concentration gradient without using the cell's energy. **Figure 4-4** shows one model of how facilitated diffusion works.

Step 1 The carrier protein changes shape when it binds to a specific molecule on one side of the cell membrane.

Step 2 This change in the shape of the carrier protein exposes the molecule to the other side of the cell membrane. The molecule is transported across the cell membrane by the carrier protein.

Step 3 The carrier protein shields the molecule from the interior of the lipid bilayer. The molecule is then released from the carrier protein, which returns to its original shape.

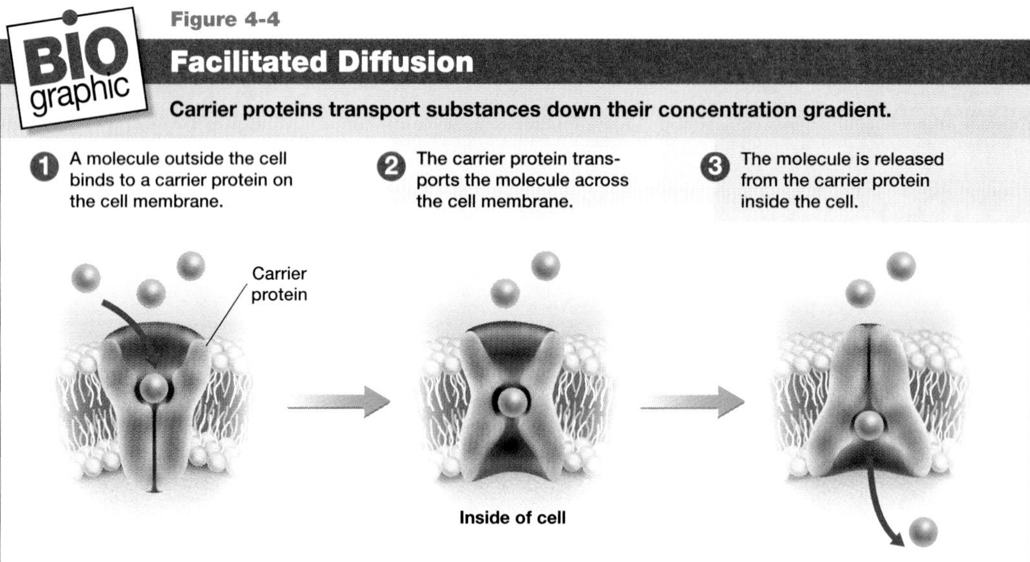

BIO graphic
1 A molecule outside the cell binds to a carrier protein on the cell membrane.
2 The carrier protein transports the molecule across the cell membrane.
3 The molecule is released from the carrier protein inside the cell.
Carrier protein
Inside of cell

Figure 4-4 **Facilitated Diffusion**

Carrier proteins transport substances down their concentration gradient.

Review SECTION 4-1

1. **Distinguish** between diffusion and equilibrium.
2. **Predict** what would happen to a cell that is placed in a hypertonic solution.
3. **Describe** how the diffusion of ions across a cell membrane differs from the diffusion of nonpolar molecules across the cell membrane.
4. **SKILL Predicting Outcomes** If a substance is less concentrated outside a cell than inside the cell, in which direction is that substance likely to diffuse across the cell membrane?
5. **SKILL Summarizing Information** Explain how some substances cross the cell membrane by facilitated diffusion.

Review SECTION 4-1 ANSWERS

1. Diffusion is the movement of a substance from an area of high concentration to an area of lower concentration caused by the random motion of particles of the substance. Equilibrium is a condition in which the concentration of a substance is equal throughout a space.
2. Water will diffuse out of the cell, causing it to shrink.
3. Because ions are charged, they cannot pass through the nonpolar interior of the lipid bilayer. Instead they must pass through ion channels embedded in the lipid bilayer. Nonpolar molecules, however, can pass directly through the lipid bilayer.
4. The substance will diffuse out of the cell.
5. Facilitated diffusion involves carrier proteins that transport substances across the cell membrane down their concentration gradient. A carrier protein exposes a substance to the other side of the membrane, where the substance is released.

Active Transport

Section 4-2

Section 4-2

Some Substances Are Transported Against a Concentration Gradient

Although facilitated diffusion can help move amino acids and sugars across the cell membrane, it can only transport these substances down their concentration gradient. Cells must transport certain amino acids, sugars, and other substances into their cytoplasm from the surrounding fluid. But many of these substances have a low concentration outside cells and a higher concentration inside cells. This concentration gradient would cause these important substances to move out of the cell rather than into the cell. So in addition to passive transport, cells must also have a way to move some substances against their concentration gradient—from an area of low concentration to an area of higher concentration.

The transport of a substance across the cell membrane against its concentration gradient is called **active transport.** Unlike passive transport, active transport requires the cell to use energy because the substance is being moved against its concentration gradient. Most often, the energy needed for active transport is supplied directly or indirectly by ATP.

Some active-transport processes involve carrier proteins. Like the carrier proteins used in facilitated diffusion, the carrier proteins used in active transport bind to specific substances on one side of the cell membrane and release them on the other side of the cell membrane. But in active transport, the substances bind to carrier proteins where they are low in concentration and are released where they are higher in concentration. Thus, carrier proteins in active transport function as "pumps" that move substances against their concentration gradient. For this reason, these carrier proteins are often called membrane pumps.

Sodium-potassium pump

One of the most important membrane pumps in animal cells is a carrier protein called the sodium-potassium pump. In a complete cycle, the **sodium-potassium pump** transports three sodium ions, Na^+, out of a cell and two potassium ions, K^+, into the cell. Sodium ions are usually more concentrated outside the cell than inside the cell, and potassium ions are typically more concentrated inside the cell than outside the cell. Thus, the sodium-potassium pump actively transports both sodium ions and potassium ions against their concentration gradient. The energy needed to power sodium-potassium pumps is supplied by ATP. In some cells, sodium-potassium pumps are so active that they use much of the ATP produced by the cells.

Objectives

- **Compare** active transport with passive transport. (p. 81)
- **Describe** the importance of the sodium-potassium pump. (pp. 81–82)
- **Distinguish** between endocytosis and exocytosis. (p. 83)
- **Identify** three ways that receptor proteins can change the activity of a cell. (pp. 84–85)

Key Terms

active transport
sodium-potassium pump
endocytosis
exocytosis
receptor protein
second messenger

biobit

Why don't saltwater frogs get pickled?

Certain types of frogs keep urea—a salty product of metabolism that is usually excreted as urine—in their blood, making them nearly as salty as the surrounding water. Therefore, the frogs become dehydrated less quickly in salt water.

1 Prepare

Scheduling

- **Block:** About 90 minutes will be needed to complete this section.
- **Traditional:** About two class periods will be needed to complete this section.
- **Planning:** See the Planning Guide on pp. 71A and 71B for lecture, classwork, and assignment options for Section 4-2. Lesson plans for Section 4-2 can be found in a lesson cycle format in ***Biology: Principles and Explorations*** **One-Stop Planner CD-ROM with Test Generator.**

Resource Link

Assign **Active Reading Worksheet Section 4-2** to help students comprehend the key concepts and visuals in this lesson.

2 Teach

Lesson Warm-up

Bring an air pump and an inflatable ball to class. Ask a student to inflate the ball with the pump. Relate this activity to a cell membrane pump. *(The air pump uses energy supplied by the student to move air against a pressure gradient.)* Tell students that membrane pumps use energy supplied by the cell to move substances against their concentration gradient.

Integrating Different Learning Styles

Visual/Spatial	**PE**	Study Tip, p. 83; Performance Zone items 1, 15, 16, 17, 18, and 19, pp. 88–89
	ATE	Visual Strategies, pp. 82, 84; Teaching Tip, p. 82
Body/Kinesthetic	**ATE**	Lesson Warm-up, p. 81
Interpersonal	**ATE**	Closure, p. 86; Reteaching, p. 86
Verbal/Linguistic	**PE**	Study Tip, p. 82; Performance Zone items 2, 21, and 26 pp. 88–89
	ATE	Making Biology Relevant, p. 83

Teaching TIP

Sodium-Potassium Pump

Have students construct a **Graphic Organizer** similar to the one at the bottom of page 82 that illustrates the function of the sodium-potassium pump.

PORTFOLIO

Visual Strategy

Figure 4-5

Review ATP with students. Notice that the sodium-potassium pump goes through four stages to move three sodium ions out of a cell and two potassium ions into the cell. Remind students that these ions are being transported against their concentration gradients. Emphasize that cells must use energy to power the sodium-potassium pump. Point out that ATP provides that energy.

Teaching TIP — ADVANCED

Hydrogen Ion Pump

During cellular respiration (covered in detail in Chapter 5), in which cells harvest energy from food molecules, the production of ATP depends on the active transport of hydrogen ions. In the electron transport chain of cellular respiration, the hydrogen ion pump is used to generate a hydrogen ion concentration gradient that drives the production of ATP. Ask students to suggest where these hydrogen ion pumps are located in animal cells. *(Many are located in mitochondria.)*

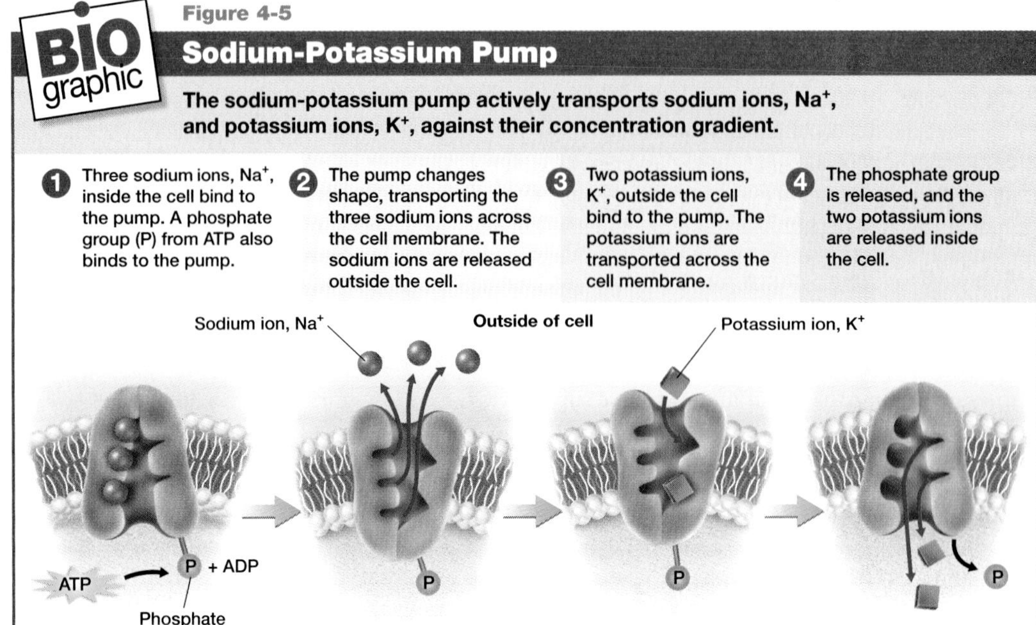

BIO graphic
Figure 4-5
Sodium-Potassium Pump
The sodium-potassium pump actively transports sodium ions, Na⁺, and potassium ions, K⁺, against their concentration gradient.
1 Three sodium ions, Na⁺, inside the cell bind to the pump. A phosphate group (P) from ATP also binds to the pump.
2 The pump changes shape, transporting the three sodium ions across the cell membrane. The sodium ions are released outside the cell.
3 Two potassium ions, K⁺, outside the cell bind to the pump. The potassium ions are transported across the cell membrane.
4 The phosphate group is released, and the two potassium ions are released inside the cell.
Sodium ion, Na⁺
Outside of cell
Potassium ion, K⁺
ATP
P + ADP
Phosphate group
P
P
P
Inside of cell

A model of the sodium-potassium pump is shown in **Figure 4-5.**

Step 1 Three sodium ions inside the cell bind to the sodium-potassium pump. Because energy is needed to move the sodium ions against their concentration gradient, a phosphate group is removed from ATP and also binds to the pump.

Step 2 The pump changes shape, transporting the three sodium ions across the cell membrane and releasing them outside the cell.

Step 3 The pump is now exposed on the surface of the cell. Two potassium ions outside the cell bind to the pump. The phosphate group is released, changing the shape of the pump.

Step 4 The pump is again exposed to the inside of the cell. The two potassium ions are transported across the cell membrane and are released inside the cell. The pump is ready to bind more sodium ions.

The sodium-potassium pump is important for two main reasons. First, the pump prevents sodium ions from accumulating in the cell; this would be toxic to the cell. Sodium ions continuously diffuse into the cell through ion channels embedded in the lipid bilayer of the cell membrane. If not removed from the cell, sodium ions would accumulate. The increased concentration of sodium ions would then cause water to enter the cell by osmosis, causing the cell to swell or even burst. Second, the sodium-potassium pump helps maintain the concentration gradients of sodium ions and potassium ions across the cell membrane. Many cells use the sodium-ion concentration gradient to help transport other substances, such as glucose, across the cell membrane.

Study TIP

Interpreting Graphics

As you follow the steps in Figure 4-5, describe in writing how the pump changes shape twice. Notice that both the binding of ions and the binding of a phosphate group to the pump change the shape of the pump.

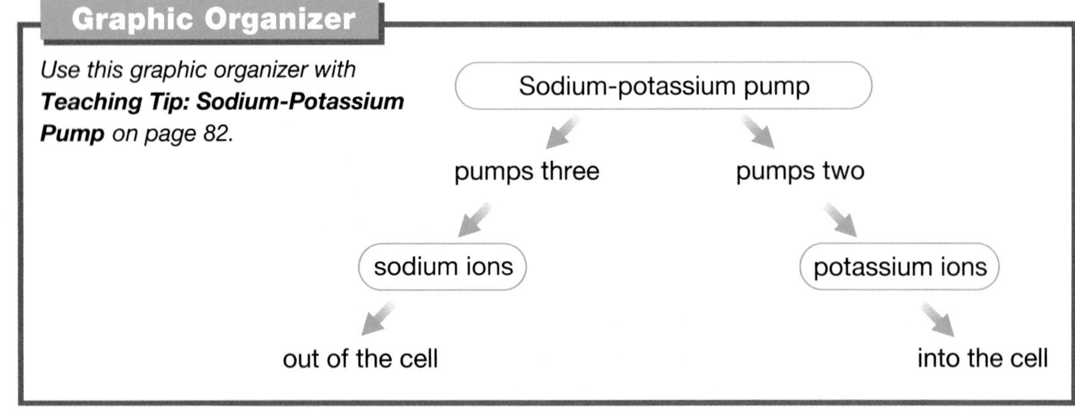

Vesicles Move Substances Across Membranes

Many substances that are too large to be transported by carrier proteins, such as proteins and polysaccharides, are moved across the cell membrane by vesicles. The movement of a substance into a cell by a vesicle is called **endocytosis** *(ehn doh seye TOH sihs)*. During endocytosis, the cell membrane forms a pouch around a substance, as shown in **Figure 4-6.** The pouch then closes up and pinches off from the membrane to form a vesicle. Vesicles formed by endocytosis may fuse with lysosomes or other organelles.

The movement of a substance by a vesicle to the outside of a cell is called **exocytosis** *(ek soh seye TOH sihs),* also shown in Figure 4-6. During exocytosis, vesicles in the cell fuse with the cell membrane, releasing their contents. Cells use exocytosis to export proteins that are modified by the Golgi apparatus. Nerve cells and cells of various glands, for example, release proteins by exocytosis.

Figure 4-6 Endocytosis and exocytosis

Vesicles transport substances into and out of cells.

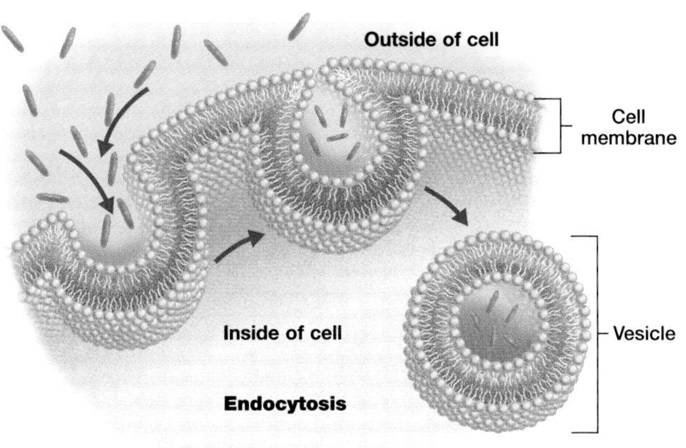

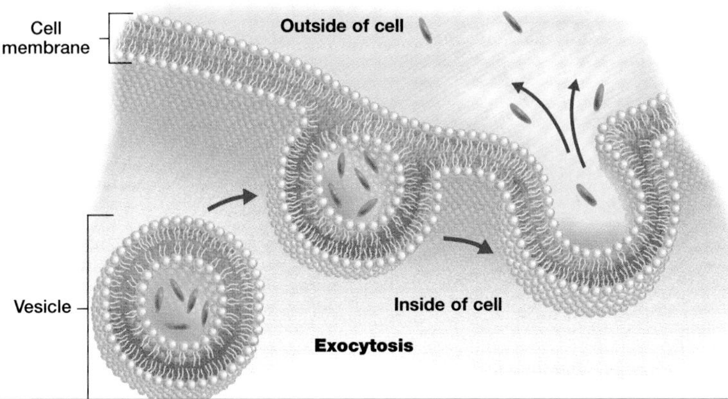

Study TIP

Interpreting Graphics

As you look at Figure 4-6, notice that during endocytosis, the cell membrane pinches off to become the vesicle membrane. Conversely, during exocytosis, the vesicle membrane becomes part of the cell membrane. Vesicle membranes often contain transport proteins that help move substances into and out of the vesicle.

Teaching TIP ADVANCED

Endocytosis

Tell students that there are actually two types of endocytosis. One is phagocytosis, in which the material brought into the cell is another cell or fragments of organic matter. Phagocytosis is common among unicellular eukaryotes, such as the amoeba, which devours its prey by engulfing it. Ask students if they can think of another example of phagocytosis. *(Macrophages engulf invading pathogens.)* Another type of endocytosis is pinocytosis, in which the material brought into the cell is a liquid that contains dissolved particles. For example, maturing human egg cells use pinocytosis to take in dissolved nutrients secreted by surrounding cells.

Making Biology Relevant

Health

Tell students that cholesterol is transported into cells by endocytosis. Cholesterol travels in the blood partly as low-density lipoprotein (LDL). LDL binds to receptor sites on the cell membrane, triggering endocytosis. LDL is composed of an outer lipid-protein membrane that surrounds cholesterol molecules. LDL, or "bad" cholesterol, contributes to atherosclerosis, a form of arteriosclerosis characterized by the buildup of fatty substances on the interior surfaces of artery walls. There is also "good" cholesterol, which is essential to the body's functions. For example, the body needs cholesterol to produce lipid-based molecules, such as steroid hormones. Have students research and write a report on the differences between good cholesterol and bad cholesterol.

PORTFOLIO

did you know?

Cells transport glucose using a sodium ion concentration gradient.

This gradient is established by sodium-potassium pumps. Eventually the sodium ions that moved outside the cell during active transport will reenter the cell through another transport protein down their concentration gradient. When this happens, glucose enters the cell coupled with sodium ions.

Making Biology Relevant

Technology

A type of reverse osmosis technology is being used by the United States Army to develop lightweight chemical and biological protective clothing. This will eliminate the need for carbon, which is currently used in such protective clothing. The new clothing will not only be lightweight, but it will also take up less space, wash easily, be waterproof, and allow moisture to pass through to help cooling. It will provide protection from toxic chemicals and biological agents, including viruses and bacteria. Eventually this clothing may be used by emergency hospital personnel and industries that deal with chemicals and pesticides.

Visual Strategy

Figure 4-8

The receptor protein in this figure is coupled with an ion channel. Be sure students understand that the binding of a specific signal molecule to the receptor protein causes the ion channel to open. When the signal molecule binds to the receptor protein, it causes a conformational change in the structure of the receptor protein, which opens the ion channel. When the ion channel is open, sodium ions move through the ion channel down their concentration gradient.

Resource Link

Assign **Occupational Applications: Pharmacist** from the Holt BioSources Teaching Resources CD-ROM to help students learn more about careers in pharmacy. PORTFOLIO

Membrane Receptor Proteins Receive Information

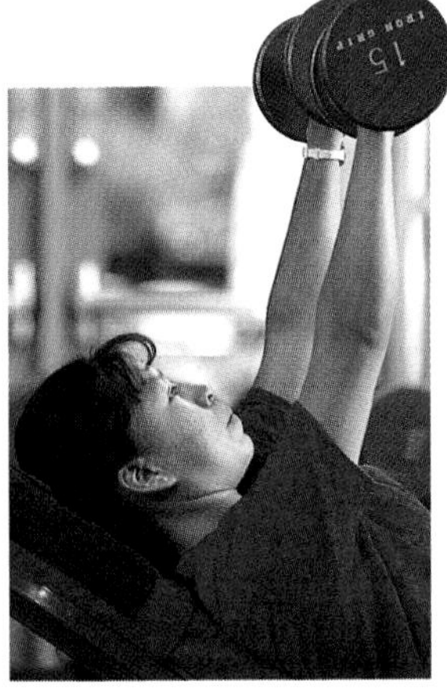

Figure 4-7 Action of signal molecules. When you exercise, signal molecules bind to receptor proteins on your muscle cells, signaling your muscles to contract.

We are constantly bombarded with information from other people and through television, the Internet, and many other media. To interpret information, we must be able to communicate and to distinguish between important and unimportant information. Similarly, your body's cells must communicate with each other to coordinate your growth, metabolism, and other activities. Cells that do not lie next to each other cannot communicate directly. Instead, some cells release signal molecules that carry information to nearby cells and throughout the body. Hormones are one familiar example of signal molecules. Hormones are made in one part of the body and carried in the bloodstream to other parts, where they have their effects.

Cells must also respond to important information and filter out unimportant information. To receive the messages carried by signal molecules, the cell membrane contains specialized proteins that are able to bind to these signal molecules. Such proteins are called receptor proteins. A **receptor protein** is a protein that binds to a specific signal molecule, enabling the cell to respond to the signal molecule. For example, the muscles of the person exercising in **Figure 4-7** could not contract without receptor proteins and signal molecules that tell the muscles when to contract and when to relax.

Functions of receptor proteins

When a signal molecule encounters a cell, the signal molecule binds to the receptor protein that is specialized to fit that molecule, as shown in **Figure 4-8.** Most receptor proteins are embedded in the lipid bilayer of the cell membrane, and the part of the protein that fits the signal molecule faces the outside of the cell.

Figure 4-8 Changes in permeability

Some receptor proteins are coupled with ion channels.

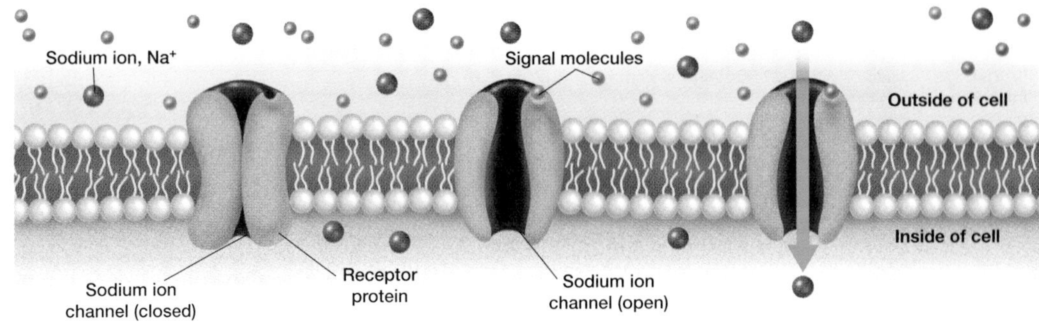
Sodium ion, Na$^+$
Signal molecules
Outside of cell
Inside of cell
Sodium ion channel (closed)
Receptor protein
Sodium ion channel (open)
1. The ion channel is closed, so no ions can move through the channel.
2. When a signal molecule binds to the receptor protein, the ion channel opens.
3. Sodium ions diffuse into the cell through the open ion channel.

Multicultural Perspective

Curare

Curare is a poison obtained from a woody vine, *Struchnos toxifer,* found in South America. Certain tribes of that continent use curare as an arrow poison. Curare blocks acetylcholine receptors in muscle cells. This results in paralysis and usually death because muscles involved in respiration, such as the diaphragm, stop functioning. Curare was used as a drug especially after World War II. Have students hypothesize its medicinal use. *(Curare was used as a muscle relaxant during surgery; the patient was monitored and the lungs were kept working by artificial means. Curare is not an anesthetic.)*

Figure 4-9 Second messengers

Some receptor proteins trigger the production of second messengers.

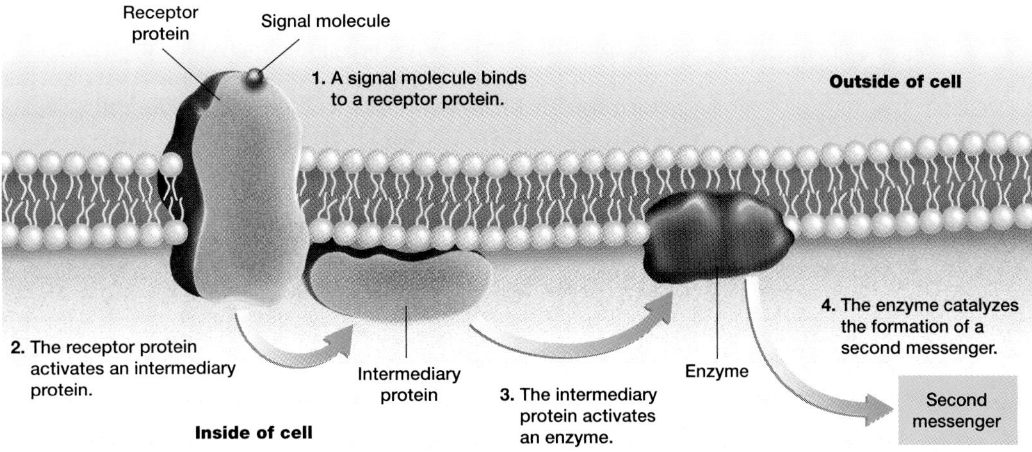
Receptor protein
Signal molecule
1. A signal molecule binds to a receptor protein.
Outside of cell
2. The receptor protein activates an intermediary protein.
Intermediary protein
3. The intermediary protein activates an enzyme.
Enzyme
4. The enzyme catalyzes the formation of a second messenger.
Second messenger
Inside of cell

The binding of a signal molecule to its matching receptor protein causes a change in the activity of the receiving cell. This change can occur in the following three ways: by causing changes in the permeability of the receiving cell, as shown in Figure 4-8; by triggering the formation of second messengers inside the cell; and by activating enzymes inside the cell.

Changes in permeability The receptor protein may be coupled with an ion channel, as shown in Figure 4-8. The binding of a signal molecule to the receptor protein causes the ion channel to open, allowing specific ions to cross the cell membrane. As you will learn in Chapter 42, this type of receptor protein is especially important in the nervous system.

Second messengers The receptor protein may cause the formation of a second messenger inside the cell, as shown in **Figure 4-9.** When it is activated, a **second messenger** acts as a signal molecule in the cytoplasm. The second messenger amplifies the signal of the first (messenger) signal molecule. Second messengers can change the functioning of a cell in several ways. For example, some second messengers activate enzymes, triggering a series of biochemical reactions in the cell. Other second messengers change the permeability of the cell by opening ion channels in the cell membrane. You will learn more about second messengers in Chapter 43.

Enzyme action The receptor protein may act as an enzyme. When a signal molecule binds to the receptor protein, the receptor protein speeds up chemical reactions inside the cell. Receptor proteins may also activate other enzymes located inside the cell or in the cell membrane, triggering chemical reactions in the cell. In this way, the signal molecule can cause many changes in the functioning of the receiving cell.

REAL LIFE

Many medicines are drugs that bind to receptor proteins. Some of these drugs may interfere with the receptor's ability to bind to signal molecules.

Finding Information

Research some medicines that bind to receptor proteins.

CAREER

Pharmacist

A pharmacist not only dispenses medications prescribed by physicians, but also provides information or counseling about medications to patients. Pharmacists inform patients about such things as drug interactions, side effects, dosages, and how to take the medication. They also provide information about over-the-counter drugs. The employment conditions of pharmacists vary. A pharmacist may work, for example, in a hospital or in a pharmacy. Most employers prefer a master's degree and sometimes a Ph.D. in pharmacology. To become licensed, a prospective pharmacist must pass a state exam and then serve as an intern under a licensed pharmacist. The demand for pharmacists is expected to increase over the next few years. Because the population is aging, the need for medication is expected to increase. Earnings can be very high, but the hours may be very long.

REAL LIFE

Answer

Examples of medicines that bind to receptor proteins include heart medications such as beta blockers and calcium channel blockers, and narcotics such as morphine and codeine.

Biology in Your Community

Receptor-Binding Drugs

Contact a local pharmaceutical company or the biology department of a local university. Arrange for a speaker to give a presentation to your students on medicines that work by affecting the functions of receptor proteins. Before the presentation, consult with the speaker on the topics that will be discussed. Be sure the speaker includes the name of each drug, the receptor it affects, its mode of action, the conditions or diseases it alleviates, and the symptoms of the conditions or diseases.

HEALTH WATCH

Myasthenia Gravis

Teaching Strategies
Use this feature to stimulate discussion about the functions of receptor proteins.

Discussion
What kinds of receptor proteins does myasthenia gravis affect? *(The disease affects acetylcholine receptors on muscle cells. These receptor proteins are coupled with ion channels. When acetylcholine binds to a receptor protein, ions move through the channel, simulating muscle contraction.)*

3 Assessment Options

Closure

Crossword Puzzles
Assign students to cooperative groups of four. Have each group make a crossword puzzle with clues using the terms introduced in this section. Make copies of all the puzzles, and have each group try to complete the puzzles.

Review

Assign Review—Section 4-2.

Reteaching — BASIC

Assign students to cooperative groups of three. Write the following terms on the board: active transport, sodium-potassium pump, endocytosis, exocytosis, and receptor protein. Have the students in each group discuss and write down what they know or think they know about each term. After they have completed the assignment, let them check their answers using the textbook.

internetconnect
SCILINKS NSTA
TOPIC: Receptor proteins
GO TO: www.scilinks.org
KEYWORD: HX085

Many drugs affect the binding of signal molecules to receptor proteins. Some drugs, such as the illegal drug heroin, imitate signal molecules by binding to receptor proteins on a receiving cell, altering the function of the cell. Other drugs block or interfere with receptor proteins, preventing signal molecules from binding to the receptor proteins. For example, signal molecules that bind to receptor proteins on heart-muscle cells stimulate the cells, causing the heart rate to increase. Beta blockers, which are drugs prescribed to patients with a rapid heartbeat, bind to some of these receptor proteins. Beta blockers therefore interfere with the binding of signal molecules to the receptor proteins, preventing the heart rate from increasing too rapidly. In Chapter 42, you will learn much more about the effects of drugs on the body.

HEALTH WATCH Myasthenia Gravis

After her 15th birthday, Karen began to have difficulty speaking, swallowing, and keeping her eyes open. At first her symptoms appeared only at bedtime, but they soon began to appear earlier in the day. After performing several tests, including a blood test, Karen's doctor diagnosed her with the muscle disorder myasthenia gravis.

Myasthenia gravis, or MG, is a disease that affects about three of every 100,000 people. The disease is characterized by abnormal fatigue and severe muscle weakness after only mild physical exertion, such as walking a short distance.

In people with MG, muscles of the eyes, the face, and the mouth become weak and tend to fatigue quickly. Other muscles may also be affected. Strangely, the weakness comes and goes.

An attack from within

In a person with MG, the immune system attacks receptor proteins on muscle cells. As a result, muscle cells have fewer receptor proteins to bind to signal molecules released by nerve cells. Thus, fewer muscle cells respond when the nervous system commands them to contract.

People with MG are usually given drugs that prolong the effects of signal molecules. This increases the chance that the signal molecules will bind to the remaining receptor proteins on muscle cells.

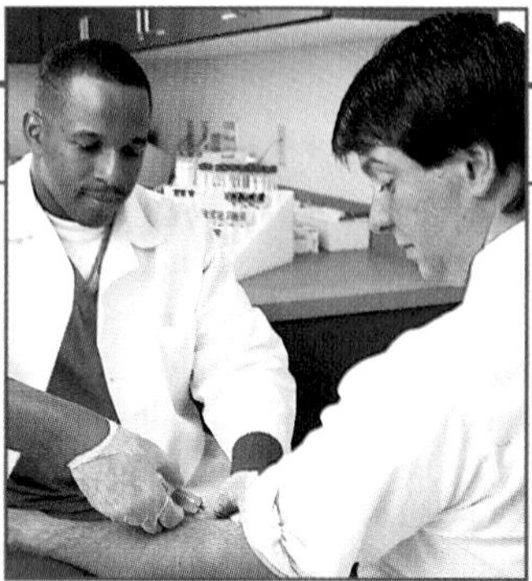

Blood test

internetconnect
SCILINKS NSTA
TOPIC: Autoimmune diseases
GO TO: www.scilinks.org
KEYWORD: HX086

Review SECTION 4-2

1. **Distinguish** between passive transport and active transport.
2. **Describe** how the sodium-potassium pump helps prevent animal cells from bursting.
3. **Compare** two ways that the binding of a signal molecule to a receptor protein causes a change in the activity of the receiving cell.
4. **SKILL Comparing Functions** Distinguish between endocytosis and exocytosis.

Critical Thinking

5. **SKILL Applying Information** During exercise, potassium ions accumulate in the fluid surrounding muscle cells. Which cell-membrane protein helps muscle cells counteract this tendency? Explain your answer.

Review SECTION 4-2 ANSWERS

1. In passive transport, substances cross cell membranes by moving down their concentration gradients. In active transport, cells use energy to move substances against their concentration gradients.
2. If sodium ions were not removed from the cell, their increased concentration would cause water to enter the cell because of osmosis. The cell would swell and could burst.
3. A signal molecule outside the cell binds to a receptor protein, which in turn causes an ion channel to open. A signal molecule may bind to a receptor protein, which in turn activates a second messenger inside the cell. This second messenger may activate enzymes or open ion channels.
4. Endocytosis is the movement of substances into a cell by a vesicle. Exocytosis is the movement of substances out of a cell by a vesicle.
5. The sodium-potassium pump helps counteract this tendency by transporting potassium ions into muscle cells.

CHAPTER 4

Use the Key Concepts and Key Terms listed below to review the main ideas in this chapter. If you need to review the meaning of a term, turn to the page indicated.

Key Concepts

4-1 Passive Transport

- Passive transport is the movement of substances across the cell membrane without the use of energy by the cell.
- Diffusion is the movement of a substance from an area of high concentration to an area of lower concentration.
- Osmosis is the diffusion of free water molecules across a selectively permeable membrane.
- Ion channels are proteins that have a pore through which ions can cross the cell membrane.
- In facilitated diffusion, a carrier protein transports a substance across the cell membrane down the concentration gradient of the substance.

Key Terms

4-1

passive transport (74)
concentration gradient (74)
equilibrium (74)
diffusion (75)
osmosis (76)
hypertonic solution (77)
hypotonic solution (77)
isotonic solution (77)
ion channel (78)
carrier protein (80)
facilitated diffusion (80)

4-2 Active Transport

- Active transport is the movement of a substance against the concentration gradient of the substance. Active transport requires cells to use energy.
- In animal cells, the sodium-potassium pump uses energy supplied by ATP to transport sodium ions out of the cell and potassium ions into the cell.
- During endocytosis, substances are moved into a cell by a vesicle that pinches off from the cell membrane.
- During exocytosis, substances inside a vesicle are released from a cell as the vesicle fuses with the cell membrane.
- Communication between cells often involves signal molecules that bind to receptor proteins on cells.
- A signal molecule that binds to a receptor protein on a cell can change the activity of the cell in three ways: by enabling specific ions to cross the cell membrane, by causing the formation of a second messenger, or by speeding up chemical reactions inside the cell.

4-2

active transport (81)
sodium-potassium pump (81)
endocytosis (83)
exocytosis (83)
receptor protein (84)
second messenger (85)

Unit 1—*Cell Transport and Homeostasis* Use Topics 1–6 in this unit to review the key concepts and terms in this chapter.

Ready?

- *If you think you understand these Key Concepts and Key Terms, then you're ready for the Performance Zone.*

Answer to Concept Map

The following is one possible answer to Performance Zone item 1 on page 88.

Homeostasis
is aided by
cell transport
which includes
passive transport — active transport
passive transport: such as — osmosis, diffusion, facilitated diffusion; of substances down their — concentration gradient
active transport: of substances against their — concentration gradient; such as — endocytosis, exocytosis, sodium-potassium pump
facilitated diffusion, sodium-potassium pump: which use a — carrier protein

Study ZONE CHAPTER 4

Review and Practice

Assign the following worksheets for Chapter 4:

- **Student Review Guide**
- **Concept Mapping Worksheet**
- **Test Preparation Pretest**
- **Vocabulary Worksheet**
- **Critical Thinking Review Worksheet**
- **Science Skills Worksheet**

Assessment Options

The following assessment options are available for this chapter:

Resource Link

- Assign **Chapter Test 4** to assess students' understanding of the chapter.

Alternative Assessment

Have each student list the types of passive transport mechanisms —diffusion, osmosis, ion channels, and facilitated diffusion—and types of active transport mechanisms—sodium-potassium pump, exocytosis, and endocytosis—discussed in this chapter. Ask students to define each one, to state the method of movement (down or against the concentration gradient), and to give an example of each. Have them write their answers on paper and make drawings to illustrate the mechanisms. After the assignment is completed, assign students to cooperative groups of three. Have the students in each group use the textbook to check their answers and to critique their drawings. PORTFOLIO

Portfolio Assessment

- *Portfolio Assessment* at the front of this book provides options for building and evaluating student portfolios. Students and teachers can select items from the areas listed for inclusion in a portfolio.
- See items labeled PORTFOLIO on pp. 75, 78, 79, 82, 83, 84, and 87.

Performance ZONE

CHAPTER 4

REVIEW ANSWERS

Assignment Guide

Section	Review
4-1	1, 2, 3, 4, 5, 6, 7, 11, 20, 22, 23, 24, 25, 27
4-2	1, 2, 8, 9, 10, 12, 13, 14, 15, 16, 17, 18, 19, 21, 26

Understanding and Applying Concepts

1. The answer to the concept map is found at the bottom of page 87.
2. **a.** Diffusion is the movement of a substance from an area of high concentration to an area of lower concentration. Equilibrium is a condition in which the concentration of a substance is equal throughout a space.

   b. A receptor protein binds to a signal molecule outside a cell, causing changes in the activity of the cell. A carrier protein binds to a substance on one side of the cell membrane and then releases that substance on the other side of the membrane.

   c. Endocytosis is the movement of a substance into a cell by a vesicle. Exocytosis is the movement of a substance out of a cell by a vesicle.

   d. Active transport is the movement of a substance against its concentration gradient. Active transport requires cells to use energy. Passive transport is the movement of a substance down its concentration gradient. Passive transport does not require cells to use energy.

3. d 7. d
4. a 8. b
5. b 9. c
6. d 10. d

11. The solution is hypertonic. The concentration of free water molecules outside the cell was less than the concentration of free water molecules in the cytoplasm, causing water to move out of the cell because of osmosis.

Performance ZONE

CHAPTER 4 Review

Understanding and Applying Concepts

1. **Concept Mapping** Make a concept map that shows how cells maintain homeostasis. Include the following terms in your map: concentration gradient, diffusion, osmosis, carrier protein.

2. **Understanding Vocabulary** For each pair of terms, explain the difference in their meanings.
   a. diffusion, equilibrium
   b. receptor protein, carrier protein
   c. endocytosis, exocytosis
   d. active transport, passive transport

3. In diffusion, a substance tends to move
   a. against its concentration gradient.
   b. from an area of low concentration to an area of higher concentration.
   c. independent of its concentration.
   d. down its concentration gradient.

4. When the concentration of a substance is the same throughout a space, the substance is
   a. in equilibrium.
   b. in homeostasis.
   c. not moving.
   d. in a gradient.

5. If a cell swells when placed in a solution, the solution is
   a. hypertonic.
   b. hypotonic.
   c. isotonic.
   d. None of the above

6. Substances enter or leave a cell through the
   a. cytoplasm.
   b. Golgi apparatus.
   c. nucleus.
   d. cell membrane.

7. Facilitated diffusion
   a. is driven by energy from ATP.
   b. is a type of active transport.
   c. involves receptor proteins.
   d. involves carrier proteins.

8. The sodium-potassium pump moves
   a. sodium ions into the cell and potassium ions out of the cell.
   b. sodium ions out of the cell and potassium ions into the cell.
   c. both sodium and potassium into the cell.
   d. both sodium and potassium out of the cell.

9. Which of the following is not an example of active transport?
   a. endocytosis
   b. exocytosis
   c. facilitated diffusion
   d. sodium-potassium pump

10. The binding of a signal molecule to a receptor protein can
    a. activate a second messenger inside the receiving cell.
    b. trigger enzyme activity in the cell.
    c. change the permeability of the cell.
    d. All of the above

11. The drawing below shows a plant cell that has become shriveled after having been placed in a solution. Is the solution most likely hypertonic, hypotonic, or isotonic? Explain your reasoning.

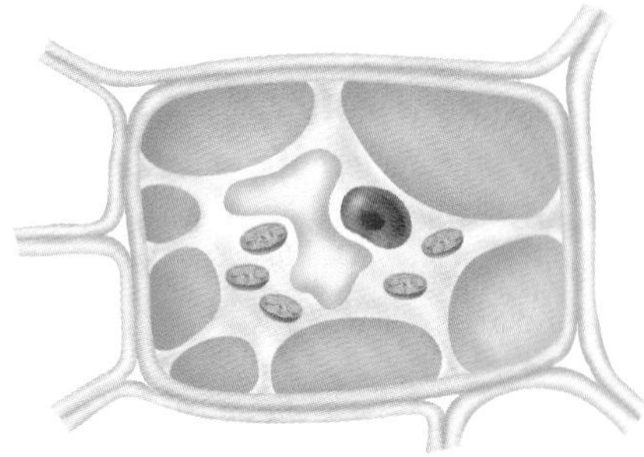

12. **Theme Homeostasis** Explain how the sodium-potassium pump contributes to homeostasis in an animal.

13. **Health Watch** When you breathe, you draw air into your lungs by contracting a muscle called the diaphragm. Why do some patients with severe myasthenia gravis need artificial respiration to help them breathe?

14. **Chapter Links** When a cell takes in a food particle by endocytosis, the vesicle that is formed may fuse with a lysosome. How would that help the cell digest the food particle? (**Hint:** See Chapter 3, Section 3-3.)

---

12. The sodium-potassium pump prevents sodium ions from accumulating in cells. Without the pump, the cells would die because water would begin to diffuse into the cells. The pump also maintains the sodium concentration gradient, which helps transport important substances, such as glucose, into cells.
13. In a person with myasthenia gravis, muscle cells have fewer receptor proteins. Therefore, when the nervous system signals muscles to contract, the muscles do not respond fully. Because contractions of muscles that aid in breathing are impaired, artificial respiration is needed.
14. Lysosomes contain digestive enzymes. When the vesicle fuses with a lysosome, the enzymes can digest the food.

Skills Assessment

Part 1: Interpreting Graphics

Use the illustration of the sodium-potassium pump below to answer the following questions:

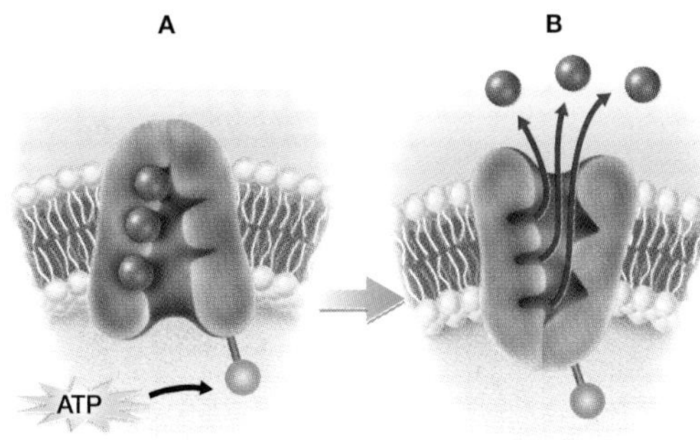

15. What ions bind to the pump in step *A*?
16. Where are the three ions released in step *B*?
17. Why does the sodium-potassium pump require energy to transport these ions?
18. What happens to a cell if its sodium-potassium pumps do not function properly?
19. What supplies the energy needed to power the pump?

Part 2: Life/Work Skills

Use the media center or Internet resources to make a list of several human diseases that involve various membrane proteins.

20. **Finding Information** Explain how these diseases affect the function of specific transport proteins. Describe the symptoms of each disease. Also describe any treatments for the diseases. Summarize your findings in a written report.
21. **Summarizing Information** List several different medicinal drugs that interact with receptor proteins. Describe the effects of each drug and list the diseases or illnesses that the drug treats. Prepare an oral report, using graphics to interpret and summarize your findings.

Critical Thinking

22. **Evaluating Results** A gelatin block is prepared with a chemical indicator that turns pink in the presence of a base. The block is enclosed in a membrane and placed in a beaker of ammonium hydroxide solution (a base). After half an hour, the block begins to turn pink. Explain why the gelatin turns pink.
23. **Analyzing Information** Use your understanding of osmosis and diffusion to explain why cooking dried pasta in boiling water makes the pasta soft.

Portfolio Projects

24. **Finding Information** Research the attitudes of Japanese people toward eating the puffer fish. Find out how tetrodotoxin, a toxin sometimes present in the fish, can cause paralysis and even death.
25. **Interpreting Information** Read the article titled "Stopgap measure could limit stroke damage" (*Science News,* October 10, 1998, p. 230). Describe the possible role of gap junctions in the damage of nerve cells during a stroke.
26. **Career Focus Molecular Biologist** Research the field of molecular biology, and write a report on your findings. Your report should include a job description, the training required, kinds of employers, growth prospects in the field, and an average starting salary. If possible, interview a person who works in this field.
27. **Unit 1—*Cell Transport and Homeostasis***

Write a report summarizing the roles of osmosis and diffusion in the preservation of body organs donated for transplants. Why must the organs be preserved in special solutions prior to a transplant? Find out what kinds of substances these solutions contain.

Skills Assessment

15. sodium ions
16. outside the cell
17. The sodium and potassium ions are moved against their concentration gradients.
18. Sodium ions accumulate in the cell, causing water to diffuse into the cell. As a result, the cell might burst.
19. ATP
20. Answers will vary. Some examples of diseases that involve membrane proteins include cystic fibrosis, hypercholesterolemia, and myasthenia gravis.
21. Answers will vary. Examples include beta blockers and calcium channel blockers. Check oral reports and graphics for accuracy. Encourage students to ask the presenters questions.

Critical Thinking

22. The ammonium hydroxide solution diffuses through the membrane down its concentration gradient into the gelatin block. The gelatin block then turns pink because the indicator reacts with the solution.
23. The water diffuses into the dried pasta, in which the concentration of water is low.

Portfolio Projects

24. Answers will vary. Tetrodotoxin is a poison that blocks sodium channels in nervous tissue, preventing the flow of sodium ions into nerve cells. This prevents nerve impulses from occurring in some nerve cells.
25. In animals, some cells are linked together by intercellular channels called gap junctions. Dying cells that are linked to other cells by gap junctions may cause their neighboring cells to die. The researchers mentioned in the article are studying ways that gap junctions can be inhibited to limit the amount of cell death in stroke victims.
26. Answers will vary. Molecular biologists study the functions of molecules found in cells. Most positions require at least a bachelor's degree with courses in biology and chemistry. Many positions require a master's or doctoral degree as well. Employers include colleges, universities, and various biomedical companies. Prospects for career growth are very good. Starting salary will vary by region.
27. Answers will vary. The solutions in which body organs are preserved must have specific concentrations of solutes to keep the organ tissues functioning prior to transplant.

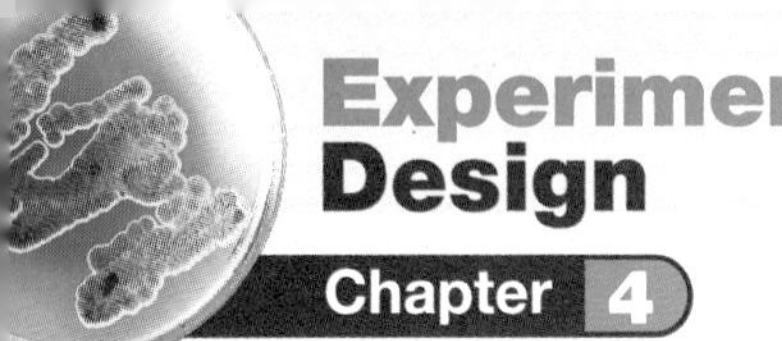

Experimental Design

Chapter 4

How Does the Size of a Cell Affect Diffusion?

Purpose

Students will investigate how cell size affects the diffusion of substances into a cell.

Preparation Notes

Time Required: Allow 45 minutes for this experiment.

Materials

Materials for this lab can be ordered from WARD'S. See *Master Materials List* at the front of this book for catalog numbers.

Preparation Tips

- Phenolphthalein agar can be prepared in the laboratory. Add drops of 0.1 M sodium hydroxide solution (prepared by using 4 g NaOH diluted to 1 L) to turn the agar red. Pour the mixture into a flat pan to a depth of slightly more than 3 cm. After the agar hardens, cut it into 3 × 3 × 6 cm rectangles.
- Place enough materials for a class on a supply table.
- Set up labeled containers for the disposal of solutions, broken glass, and phenolphthalein agar.

Safety Precautions

Phenolphthalein solutions are flammable and the vapors can explode when mixed with air. Make sure that there are no flames or sources of ignition, such as sparks, present when you are using the phenolphthalein solution.

- Be sure that students read and understand all safety rules for working in the lab.
- Remind students to read the ChemSafety section before beginning the lab.
- Make sure students wear safety goggles and a lab apron.

Procedural Tips

1. Require students to present a written procedure for their experiment and a list of all safety precautions before allowing them to gather materials for the lab.
2. Make sure students rinse off their knives between each cutting of different cubes to prevent vinegar from the previous cube from contaminating the next cube.

Disposal

- Wrap phenolphthalein agar in newspaper, and place the newspaper in the garbage.
- Dilute vinegar with water, and then pour the vinegar down the sink.

Experimental Design

Chapter 4 ***How Does the Size of a Cell Affect Diffusion?***

SKILLS

- Using scientific methods
- Collecting, organizing, and graphing data

OBJECTIVES

- **Relate** the size of a cell to its surface-area-to-volume ratio.
- **Predict** how the surface-area-to-volume ratio of a cell will affect the diffusion of substances into the cell.

MATERIALS

- safety goggles
- lab apron
- disposable gloves
- block of phenolphthalein agar (3 × 3 × 6 cm)
- plastic knife
- metric ruler
- 250 mL beaker
- 150 mL of vinegar
- plastic spoon
- paper towel

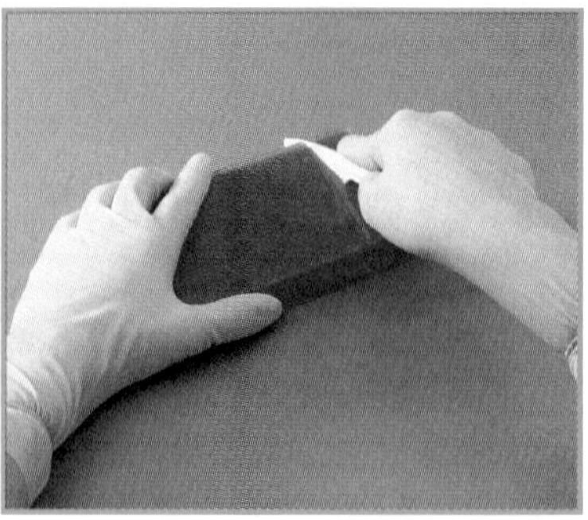

ChemSafety

CAUTION: Always wear safety goggles and a lab apron to protect your eyes and clothing.

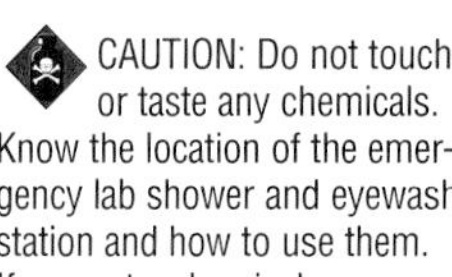

CAUTION: Do not touch or taste any chemicals. Know the location of the emergency lab shower and eyewash station and how to use them. If you get a chemical on your skin or clothing, wash it off at the sink while calling to the teacher. Notify the teacher of a spill. Spills should be cleaned up promptly, according to your teacher's directions.

CAUTION: Glassware is fragile. Notify the teacher of broken glass or cuts. Do not clean up broken glass or spills with broken glass unless the teacher tells you to do so.

Before You Begin

Substances enter and leave a cell in several ways, including by **diffusion.** How efficiently a cell can exchange substances depends on the **surface-area-to-volume ratio** (surface area ÷ volume) of the cell. **Surface area** (length × width × number of sides) is the size of the outside of an object. **Volume** (length × width × height) is the amount of space an object takes up. In this lab, you will investigate how cell size affects the diffusion of substances into a cell. To do this, you will make cell models using agar that contains an indicator. This indicator will change color when an acidic solution diffuses into it.

1. Write a definition for each boldface term in the paragraph above.
2. Based on the objectives for this lab, write a question you would like to explore about cell size and diffusion.

Procedure

PART A: Design an Experiment

1. Work with members of your lab group to explore one of the questions written for step 2 of **Before You Begin.** To explore the question, design an experiment that uses the materials listed for this lab.

You Choose

As you design your experiment, decide the following:

a. what question you will explore
b. what hypothesis you will test
c. how many "cells" (agar cubes) you will have and what sizes they will be
d. how long to leave the "cells" in the vinegar
e. how to determine how far the vinegar diffused into a "cell"
f. how to prevent contamination of agar cubes as you handle them
g. what data to record in your data table

Before You Begin

Answers

1. *diffusion*—the random movement of particles of a substance from an area of high concentration to an area of lower concentration

   surface-area-to-volume ratio—the ratio of the surface area of an object to its volume

   surface area—the area of the exterior surface of an object

   volume—the amount of space an object takes up
2. Answers will vary. For example: How does surface-area-to-volume ratio affect the diffusion of a substance into a cell?

2. Write a procedure for your experiment. Make a list of all the safety precautions you will take. Have your teacher approve your procedure and safety precautions before you begin the experiment.

PART B: Conduct Your Experiment

3. Put on safety goggles, a lab apron, and disposable gloves.

4. Carry out the experiment you designed. Record your observations in your data table.

PART C: Cleanup and Disposal

5. Dispose of solutions, broken glass, and agar in the designated waste containers. Do not pour chemicals down the drain or put lab materials in the trash unless your teacher tells you to do so.

6. Clean up your work area and all lab equipment. Return lab equipment to its proper place. Wash your hands thoroughly before you leave the lab and after you finish all work.

Analyze and Conclude

1. **Summarizing Results** Describe any changes in the appearance of the cubes.
2. **Summarizing Results** Make a graph using your group's data. Plot "Diffusion Distance (mm)" on the vertical axis. Plot "Surface-Area-to-Volume Ratio" on the horizontal axis.
3. **Analyzing Results** Using the graph you made in item 2, make a statement about the relationship between the surface-area-to-volume ratio and the distance a substance diffuses.
4. **Summarizing Results** Make a graph using your group's data. Plot "Rate of Diffusion (mm/min)" (distance vinegar moved ÷ time) on the vertical axis. Plot "Surface-Area-to-Volume Ratio" on the horizontal axis.
5. **Analyzing Results** Using the graph you made in item 4, make a statement about the relationship between the surface-area-to-volume ratio and the rate of diffusion of a substance.
6. **Evaluating Methods** In what ways do your agar models simplify or ignore the features of real cells?
7. **Calculating** Calculate the surface area and volume of a cube with a side length of 5 cm. Calculate the surface area and volume of a cube with a side length of 10 cm. Determine the surface-area-to-volume ratio of each of these cubes. Which cube has the greater surface-area-to-volume ratio? If you need help with these calculations, refer to the Math Lab titled "Calculating surface area and volume," in Chapter 3, page 56.
8. **Evaluating Conclusions** How does the size of a cell affect the diffusion of substances into the cell?
9. **Further Inquiry** Write a new question about cell size and diffusion that could be explored with another investigation.

Do You Know?

Do research in the library or media center to answer these questions:

1. How does cell transport in prokaryotic cells differ from cell transport in eukaryotic cells?
2. Which of the following molecules can diffuse across the cell membrane without the help of a transport protein: water, carbohydrates, lipids, proteins?

Use the following Internet resources to explore your own questions about cell size and cell transport.

internet**connect**

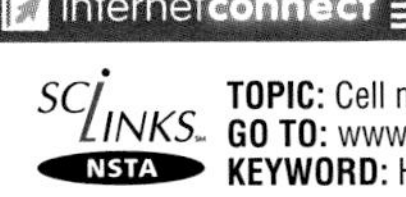

TOPIC: Cell membrane
GO TO: www.scilinks.org
KEYWORD: HX091

Procedure

Sample Procedure

1. Trim the agar block with a knife to make three cubes with side lengths of 1 cm, 2 cm, and 3 cm. Calculate the surface area, volume, and surface-area-to-volume ratio of each cube.
2. Place the three cubes in the beaker, and cover them with vinegar. Using the plastic spoon, turn the cubes frequently.
3. After 10 minutes, use the spoon to remove the agar cubes. Blot the cubes dry using a paper towel.
4. Cut each cube in half. Measure the distance (in mm) the vinegar diffused.
5. Calculate the rate of diffusion (mm/min).

Sample Data

Table 1 Agar Cube Comparisons

Side Length	Surface Area	Volume	Ratio
1 cm	6 cm^2	1 cm^3	6:1
2 cm	24 cm^2	8 cm^3	3:1
3 cm	54 cm^2	27 cm^3	2:1

Table 2 Rate of Diffusion

Side Length	Diffusion Distance	Time	Rate
1 cm	5 mm	10 min	0.5 mm/min
2 cm	5 mm	10 min	0.5 mm/min
3 cm	5 mm	10 min	0.5 mm/min

Analyze and Conclude

Answers

1. Cubes should appear light pink near their surfaces.
2. The line plotted on the graph should be horizontal.
3. The diffusion distance is the same regardless of the surface-area-to-volume ratio.
4. The line plotted on the graph should be horizontal.
5. The rate of diffusion is constant regardless of the surface-area-to-volume ratio.
6. The agar models ignore the selective permeability of cell membranes, the role of membrane proteins in facilitated diffusion and active transport, and other mechanisms of cell transport.
7. For cube with a side length of 5 cm: surface area = 150 cm^2; volume = 125 cm^3; surface-area-to-volume ratio = 6:5. For cube with side length of 10 cm: surface area = 600 cm^2; volume = 1,000 cm^3; surface-area-to-volume ratio = 3:5. The small cube has the greater surface-area-to-volume ratio.
8. In a small cell, substances do not have to diffuse as far to reach the center of the cell, even though th the same.
9. Answers will vary. For example, How can the cell membrane be modified to take in materials more efficiently?

Do You Know?

Answers

1. Eukaryotic cells have an internal system of membranes that transports materials, while prokaryotic cells do not.
2. water and lipids

Chapter 5

Photosynthesis and Cellular Respiration

Planning Guide

OBJECTIVES	MEETING INDIVIDUAL NEEDS	READING SKILLS AND STRATEGIES
5-1 Energy and Living Things (BLOCKS 1 & 2 90 min) • **Analyze** the flow of energy through living systems. • **Compare** the metabolism of autotrophs with that of heterotrophs. • **Describe** the role of ATP in metabolism. • **Describe** how energy is released from ATP.	Learning Styles, p. 94 BASIC ADVANCED • pp. 95, 96	ATE Active Reading, Technique: K-W-L, p. 93 ■ Active Reading Guide, worksheet (5-1) ■ Directed Reading, worksheet (5-1)
5-2 Photosynthesis (BLOCK 3 45 min) • **Summarize** how energy is captured from sunlight in the first stage of photosynthesis. • **Analyze** the function of electron transport chains in the second stage of photosynthesis. • **Relate** the Calvin cycle to carbon dioxide fixation in the third stage of photosynthesis. • **Identify** three environmental factors that affect the rate of photosynthesis.	Learning Styles, p. 97 BASIC ADVANCED • pp. 101, 103	■ Active Reading Guide, worksheet (5-2) ■ Directed Reading, worksheet (5-2) ■ Reading Strategies, K-W-L worksheet
5-3 Cellular Respiration (BLOCKS 4 & 5 90 min) • **Summarize** how glucose is broken down in the first stage of cellular respiration. • **Describe** how ATP is made in the second stage of cellular respiration. • **Identify** the role of fermentation in the second stage of cellular respiration. • **Evaluate** the importance of oxygen in aerobic respiration.	Learning Styles, p. 104 BASIC ADVANCED • pp. 106, 110	ATE Active Reading, Technique: Anticipation Guide, p. 105 ■ Active Reading Guide, worksheet (5-3) ■ Directed Reading, worksheet (5-3) ■ Supplemental Reading Guide, *The Lives of a Cell: Notes of a Biology Watcher*

Review and Assessment Resources

BLOCKS 6 & 7 90 min

TEST PREP

- **PE** Study Zone, p. 111
- **PE** Performance Zone, pp. 112–113
- ■ Chapter 5 worksheets:
  - Vocabulary
  - Concept Mapping
  - Science Skills
  - Critical Thinking
- Holt Biology Interactive Tutor
  Unit 2—Photosynthesis, Topics 1–6
  Unit 3—Cellular Respiration, Topics 1–6
- Audio CD Program
- ■ Test Prep Pretest

CHAPTER TESTING

- ■ Chapter Tests (blackline master)
- ▲ Test Generator
- ■ Assessment Item Listing
- ■ Performance-Based Assessment Activity: Fermentation by Yeast

See the correlations table at the front of this book for a list of the ***National Science Education Standards*** addressed in this chapter.

Smithsonian Institution® Visit **www.si.edu/hrw** for additional on-line resources.

KEY

Biology Principles and Explorations

PE Pupil's Edition

ATE Teacher's Edition

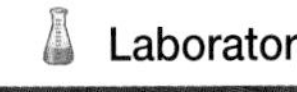

Teaching Transparencies

Laboratory Program

One-Stop Planner CD-ROM
Shows these resources within a customizable daily lesson plan:

■ Teaching Resources

▲ Test Generator

VISUAL STRATEGIES	CLASSWORK AND HOMEWORK	TECHNOLOGY AND INTERNET RESOURCES
ATE Fig. 5-1, p. 95 13 Energy Flow Through an Ecosystem	Lab C7 Measuring the Release of Energy from Sucrose **PE** Review, p. 96	Interactive Tutor Unit 2—Photosynthesis (Topic 1) Interactive Tutor Unit 3—Cellular Respiration (Topic 1) Holt Biology Videodiscs Section 5-1 Audio CD Program Section 5-1
ATE Fig. 5-7, p. 99 **ATE** Fig. 5-8, p. 100 **ATE** Fig. 5-9, p. 102 11A Converting Light Energy to Chemical Energy 12 Overview of Photosynthesis 17 Electron Transport Chain in Mitochondria 18 Photosystems and Electron Transport 19 Calvin Cycle	**PE** Quick Lab Do photosynthetic organisms give off oxygen?, p. 101 **PE** Review, p. 103 **PE** Interactive Exploration Photosynthesis, pp. 114–115	Interactive Tutor Unit 2—Photosynthesis (Topics 1–6) Holt Biology Videodiscs Section 5-2 Audio CD Program Section 5-2 internetconnect SCILINKS NSTA TOPICS: • Light absorption • Factors affecting photosynthesis
ATE Fig. 5-11, p. 105 **ATE** Fig. 5-12, p. 106 **ATE** Fig. 5-13, p. 107 16 Overview of Cellular Respiration 20 Photosynthesis—Cellular Respiration Cycle 21 Glycolysis 22 Krebs Cycle 23 Two Pathways of Respiration	Lab A5 Interpreting Labels: Stored Food Energy Lab D9 Introduction to Fermentation ■ Occupational Applications, Physical Therapist **PE** Review, p. 110 **PE** Study Zone, p. 111	Interactive Exploration: Cell Biology and Genetics, Photosynthesis Interactive Tutor Unit 3—Cellular Respiration (Topics 1–6) Holt Biology Videodiscs Section 5-3 Audio CD Program Section 5-3 internetconnect SCILINKS NSTA TOPICS: • Aerobic respiration • Fermentation • Carbon monoxide • Photosynthesis

ALTERNATIVE ASSESSMENT

PE Portfolio Projects 22–25, p. 113

ATE Alternative Assessment, p. 111

ATE Portfolio Assessment, p. 111

Lab Activity Materials

Quick Lab
p. 101 MBL or CBL system with appropriate software, test tube or small glass jar, sprig of *Elodea*, distilled water, cool light source, dissolved oxygen (DO) probe

Interactive Exploration
p. 114 CD-ROM, computer, graph paper

Demonstrations
ATE 5 cm chocolate mints, p. 99

ATE one-half packet of yeast, 400 mL of warm water, flask or beaker, p. 109

BIOSOURCES LAB PROGRAM
- Quick Labs: Lab A5, p. 9
- Laboratory Techniques and Experimental Design: Lab C7, p. 31
- Biotechnology: Lab D9, p. 45

Chapter Theme

Metabolism

Energy is stored and released in many different forms. Many autotrophs use light energy during photosynthesis. They pass that energy on to heterotrophs as chemical energy in the form of food. In this chapter, students will learn how energy is harvested and used by cells to power metabolism.

Establishing Prior Knowledge

Before beginning this chapter, make certain students can answer the questions in **Study TIP Ready?** on page 93. If they cannot, encourage them to reread the sections indicated.

Opening Question

Ask students why organisms need a constant supply of energy. *(Answers will vary, but may include that energy is needed for movement, growth, and metabolism. Students should understand that organisms ultimately use energy to maintain homeostasis.)*

CHAPTER 5 Photosynthesis and Cellular Respiration

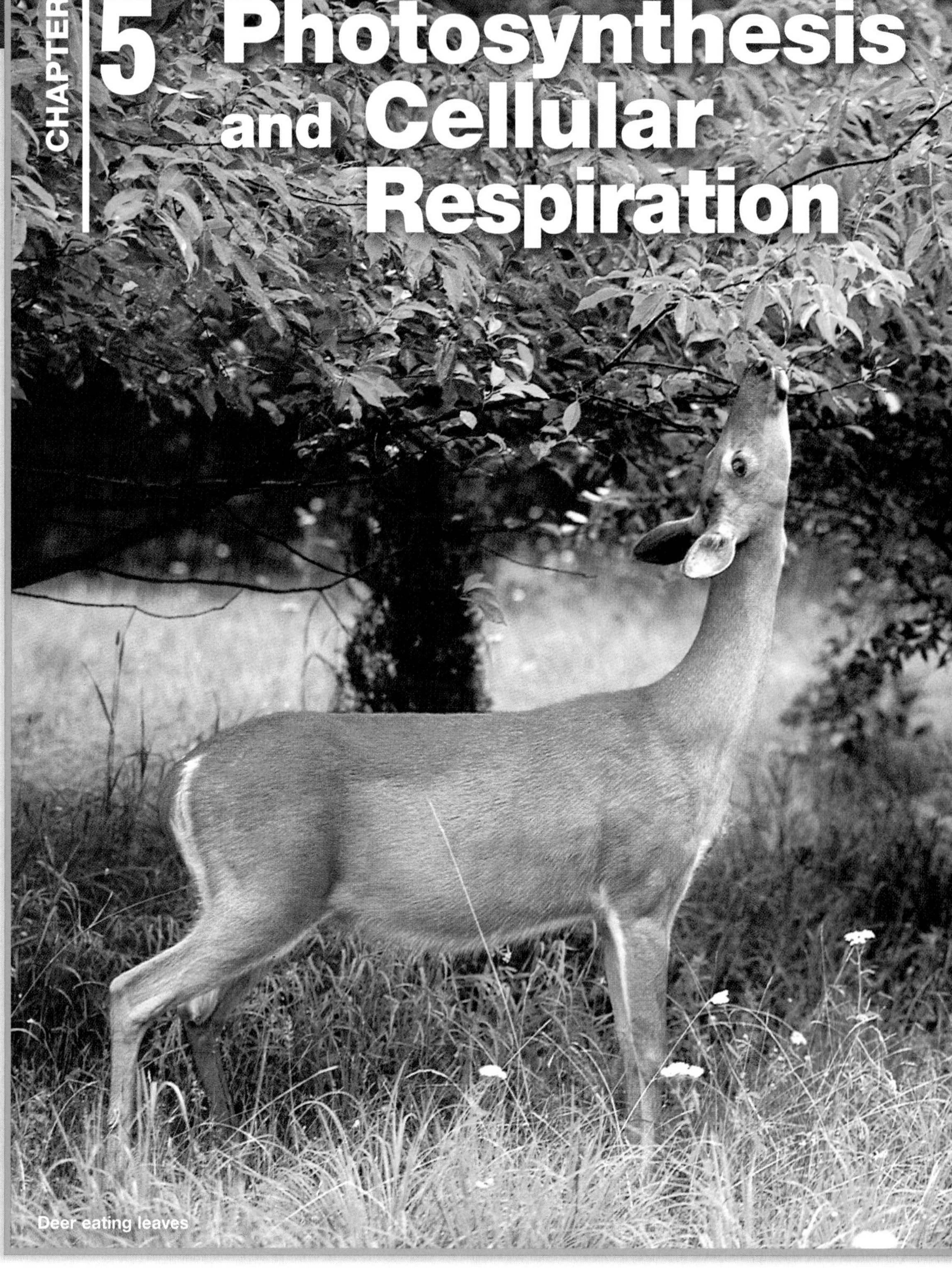

Deer eating leaves

What's Ahead — and Why

CHAPTER OUTLINE

Students will learn

- 5-1: how autotrophs and heterotrophs are involved in the flow of energy.
- 5-2: the role of sunlight, electrons, water, and carbon dioxide in photosynthesis.
- 5-3: the role of glucose, oxygen, and electrons in cellular respiration.

Shopping for food

A few leaves from a tree

may not sound as appetizing to you as pasta or a sandwich or whatever you bought at a grocery store, but leaves provide much of this deer's diet. All living things need energy to survive. Food supplies the energy needed for cellular activities.

Study TIP

Ready?

Check your understanding of these topics to help you prepare for what's ahead.

Can you . . .

- **describe** different kinds of chemical bonds? (Section 2-1)
- **list** the properties of organic compounds? (Section 2-3)
- **distinguish** between mitochondria and chloroplasts? (Section 3-3)
- **differentiate** between passive transport and active transport? (Section 4-2)

If not, *review the sections indicated.*

Looking Ahead

Labs

Quick Lab
Do photosynthetic organisms give off oxygen? (p. 101)

Interactive Exploration
Photosynthesis (p. 114)

Features

Health Watch
How Carbon Monoxide Kills (p. 109)

Unit 2—*Photosynthesis*
Topics 1–6
Unit 3—*Cellular Respiration*
Topics 1–6

National Science Teachers Association *sci*LINKS Internet resources are located throughout this chapter.

Reading Strategies

Active Reading

Technique: K-W-L

Before students read this chapter, have them write short individual lists of all the things they already **K**now (or think they know) about photosynthesis and cellular respiration. Ask students to contribute their entries to a group list on a large poster board. Then have students list things they **W**ant to know about these processes. Have students save their lists as well as the poster boards for use later in the **Closure** on page 110.

Directed Reading

Assign **Directed Reading Worksheet Chapter 5** to help students interact with the text as they read the chapter.

Interactive Reading

Assign ***Biology: Principles and Explorations* Audio CD Program Chapter 5** to help reluctant readers achieve greater success in reading the chapter.

Reading Skills

Assign Reading Organizers and Reading Strategies Worksheets from the **One-Stop Planner CD-ROM with Test Generator** to help students learn and practice effective reading skills.

internetconnect

Holt *Biology: Principles and Explorations*

On-line Resources: go.hrw.com

Visit the HRW Web site for a variety of resources related to this chapter. Just type in the keyword HX0 Chapter 5.

Holt *Biology: Principles and Explorations*

On-line Resources: www.scilinks.org

The following *sci*LINKS Internet resources can be found in the student text for this chapter:

Page 99
TOPIC: Light absorption
KEYWORD: HX099

Page 103
TOPIC: Factors affecting photosynthesis
KEYWORD: HX103

Page 106
TOPIC: Aerobic respiration
KEYWORD: HX106

Page 109
TOPIC: Fermentation
KEYWORD: HX108

Page 109
TOPIC: Carbon monoxide
KEYWORD: HX109

Page 115
TOPIC: Photosynthesis
KEYWORD: HX115

Chapter 5

1 Prepare

Scheduling

- **Block:** About 90 minutes will be needed to complete this section.
- **Traditional:** About two class periods will be needed to complete this section.
- **Planning:** See the Planning Guide on pp. 91A and 91B for lecture, classwork, and assignment options for Section 5-1. Lesson plans for Section 5-1 can be found in a lesson cycle format in ***Biology: Principles and Explorations* One-Stop Planner CD-ROM with Test Generator.**

Resource Link

Assign **Active Reading Worksheet Section 5-1** to help students comprehend the key concepts and visuals in this lesson.

2 Teach

Lesson Warm-up

Ask students to list as many different forms of energy as they can. *(Answers will vary but should include heat, light, chemical energy, mechanical energy, and electrical energy.)* Ask students to give an example of each kind of energy. *(Answers will vary. For example, heat and light from the sun, chemical energy in food, mechanical energy of a moving car, and electrical energy from a battery.)*

Section 5-1 Energy and Living Things

Objectives

- **Analyze** the flow of energy through living systems. (p. 94)
- **Compare** the metabolism of autotrophs with that of heterotrophs. (pp. 94–95)
- **Describe** the role of ATP in metabolism. (p. 95)
- **Describe** how energy is released from ATP. (p. 96)

Key Terms

photosynthesis
autotroph
heterotroph
cellular respiration

Energy Flows Between Organisms in Living Systems

You get energy from the food you eat. Where does the energy in food come from? Directly or indirectly, almost all of the energy in living systems needed for metabolism comes from the sun. **Figure 5-1** shows how energy flows through living systems. Energy from the sun enters living systems when plants, algae, and certain bacteria absorb sunlight. Some of the energy in sunlight is captured and used to make organic compounds. These organic compounds store chemical energy and can serve as food for organisms.

Building molecules that store energy

Metabolism involves either using energy to build molecules or breaking down molecules in which energy is stored. **Photosynthesis** is the process by which light energy is converted to chemical energy. Organisms that use energy from sunlight or inorganic substances to make organic compounds are called **autotrophs** *(AWT oh trohfs)*. Most autotrophs, especially plants, are photosynthetic organisms. Some autotrophs, including certain bacteria, use chemical energy from inorganic substances to make organic compounds. Bacteria found near deep-sea volcanic vents live in perpetual darkness. Because sunlight does not reach the bottom of the ocean, these bacteria get energy instead from chemicals flowing out of the vents.

Figure 5-1 Flow of energy

Energy flows from sunlight or inorganic substances to autotrophs, such as grasses, and then to heterotrophs, such as rabbits and foxes.

Light energy
1. Plants convert light energy to chemical energy. Plants use this chemical energy for cellular respiration.
2. Rabbits get energy by eating plants. Rabbits use this energy for cellular respiration.
3. Foxes get energy by eating rabbits. Foxes use this energy for cellular respiration.

Integrating Different Learning Styles

Visual/Spatial	**ATE**	Visual Strategy, p. 95; Reteaching, p. 96
Interpersonal	**ATE**	Closure, p. 96
Verbal/Linguistic	**PE**	Study Tip, p. 95; Performance Zone items 2 and 22, pp. 112–113
	ATE	Making Biology Relevant, p. 96

Breaking down food for energy

The chemical energy in organic compounds can be transferred to other organic compounds or to organisms that consume food. Organisms that must get energy from food instead of directly from sunlight or inorganic substances are called **heterotrophs** *(HEHT uhr oh trohfs)*. Heterotrophs, including humans and other animals, get energy from food through the process of cellular respiration. **Cellular respiration** is a metabolic process similar to burning fuel. While burning converts almost all of the energy in a fuel to heat, cellular respiration releases much of the energy in food to make ATP. This ATP provides cells with the energy they need to carry out the activities of life.

Study TIP

Word Origins

The words *autotroph* and *heterotroph* have the same suffix, *-troph*, which is from the Greek word *trophikos*, meaning "to feed." The prefix *auto-* is from the Greek word *autos*, meaning "self," and the prefix *hetero-* is from the Greek word *heteros*, meaning "other."

Cells transfer energy from food to ATP

The word *burn* is often used to describe how cells get energy from food. Although the overall processes are similar, the "burning" of food in living cells clearly differs from the burning of a log in a campfire. When a log burns, the energy stored in wood is released quickly as heat and light. But in cells, chemical energy stored in food molecules is released gradually in a series of enzyme-assisted chemical reactions. As shown in **Figure 5-2**, the product of one chemical reaction is a reactant in the next reaction. In the breakdown of starch, for example, each reaction releases energy.

When cells break down food molecules, some of the energy in the molecules is released as heat. Much of the remaining energy is stored temporarily in molecules of ATP. Like money, ATP is a portable form of energy "currency" inside cells. ATP delivers energy wherever energy is needed in a cell. The energy released from ATP can be used to power other chemical reactions, such as those that build molecules. In cells, most chemical reactions require less energy than is released from ATP. Therefore, enough energy is released from ATP to drive most of a cell's activities.

Figure 5-2 Breakdown of starch

Energy is released from starch in a series of enzyme-assisted chemical reactions.

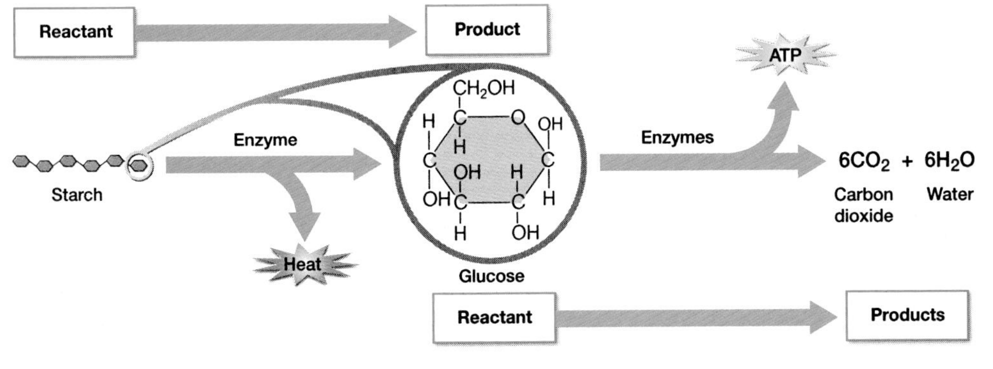

Section 5-1

Visual Strategy

Figure 5-1

Ask students to identify the autotrophs *(grasses)* and heterotrophs *(rabbits, fox)* in Figure 5-1. Next ask students how the fox is indirectly using the sun to get energy. *(The plants convert the energy in sunlight to chemical energy. Rabbits eat the plants, getting this chemical energy in the form of food. The fox eats the rabbit to get the energy it needs to survive. Thus, the fox indirectly gets energy from the sun.)*

Teaching TIP — BASIC

Autotrophs Versus Heterotrophs

Make two columns on the board. In one column write the word *Autotroph,* and in the other column write the word *Heterotroph.* Ask each student to name one specific autotroph and one specific heterotroph. Write the students' responses in the columns. If you notice a pattern—such as students naming mostly mammals—bring this to their attention. Tell them other kinds of heterotrophs exist, such as some protists and fungi.

Overcoming Misconceptions

Transfer of Energy

Because ATP supplies most of the energy that drives metabolism, ATP is sometimes called an energy-rich compound, and the bonds between its phosphate groups are sometimes called "high-energy" bonds. These terms are misleading because they imply that ATP contains an unusually large amount of energy. ATP serves as the cell's energy currency. The bonds between phosphate groups are unstable and therefore break easily.

Making Biology Relevant

Health

Have students research the "alternative" forms of medicine known as acupressure, acupuncture, and zero balancing. Ask students how these techniques are thought to relate to the flow of energy in the human body. *(Acupressure: pressure releases blocked "energy" centers in the body; acupuncture: needles release the "energy" at blocked channels; and zero balancing: aligns the body's "energy" with the body's physical structure.)* Have students present their findings in a written report. PORTFOLIO

3 Assessment Options

Closure

Flow of Energy

Assign students to cooperative groups of three or four. Have each group create a display that shows the transfer of energy through living systems. For example, students may decide to assemble a collage of photographs or drawings, make a computer-generated display, or produce a video. PORTFOLIO

Review

Assign Review—Section 5-1.

Reteaching — BASIC

Set up a demonstration that includes several organisms, such as a sponge, a worm, a Venus flytrap, and a mushroom. Have students examine the display. Do not let them touch any of the items. Ask students if each item is or once was an autotroph or a heterotroph.

biobit

Why do some foods cause indigestion?

Undigested carbohydrates often cause indigestion. Humans cannot digest carbohydrates called oligosaccharides, which are found in beans and other legumes. Bacteria in the digestive system break down oligosaccharides, producing large amounts of gases.

ATP Stores and Releases Energy

Recall that ATP (adenosine triphosphate) is a nucleotide with two extra energy-storing phosphate groups. As shown in **Figure 5-3,** the three phosphate groups in ATP form a chain that branches from a five-carbon sugar called ribose *(REYE bohs)*. This phosphate "tail" is unstable because the phosphate groups are negatively charged and therefore repel each other. The phosphate groups store energy like a compressed spring does. This energy is released when the bonds that hold the phosphate groups together are broken.

As shown in Figure 5-3, breaking the outer phosphate bond requires an input of energy, but much more energy is released during the reaction. The removal of a phosphate group from ATP produces adenosine diphosphate, or ADP. This reaction releases energy in a way that enables cells to use the energy. The following equation summarizes this reaction:

$$\text{ATP} \rightarrow \text{ADP} + \text{P} + \text{energy}$$

Cells use the energy released by this reaction to power metabolism. In some chemical reactions, two phosphate groups are cleaved off of ATP instead of just one. This tends to make the reaction irreversible because the pair of phosphate groups that is cleaved off is not available for the reverse reaction; the pair is quickly split into two single phosphate groups. In the following sections of this chapter, you will see how ATP is produced in both heterotrophs and autotrophs.

Figure 5-3 ATP releases energy

When the outer phosphate group detaches from ATP, energy is released.

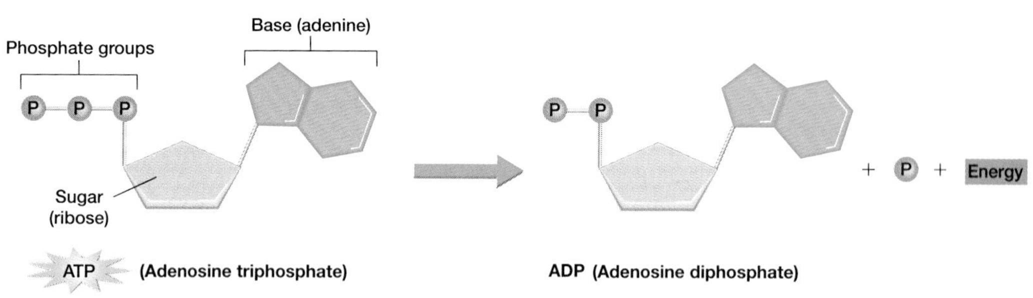

Review SECTION 5-1

1. **Identify** the primary source of energy that flows through living systems.
2. **Compare** the metabolism of autotrophs with that of heterotrophs.
3. **Describe** how energy is released from ATP.
4. **SKILL Inferring Relationships** How can the energy in the food that a fox eats be traced back to the sun?

Critical Thinking

5. **SKILL Recognizing Patterns** Explain how life involves a continuous flow of energy.

Review SECTION 5-1 ANSWERS

1. the sun
2. Autotrophs use the energy in sunlight or inorganic substances to make organic compounds. Heterotrophs must consume food sources to get energy needed to power their metabolism.
3. When the outer phosphate bond in a molecule of ATP is broken, energy is released.
4. Foxes eat other animals to get the energy needed to power metabolism. Those animals eat plants to get the energy they need. Plants convert the energy in sunlight to chemical energy, which is used to power metabolism. Thus, foxes get energy from the sun indirectly.
5. The flow of energy between organisms is continuous because energy passes from the sun to autotrophs, then to heterotrophs, and then to other heterotrophs.

Photosynthesis

Photosynthetic Organisms Use the Energy in Sunlight

When you eat a hamburger, you get energy from the sun indirectly. Plants, such as grass, capture the energy in sunlight. The beef in a hamburger comes from a cow that ate grass. The bun, lettuce, and tomato come directly from plants. With few exceptions, you end up with plants whenever you trace your food back to its origin. Plants, algae, and some bacteria capture about 1 percent of the energy in the sunlight that reaches Earth and convert it to chemical energy through the process of photosynthesis.

OVERVIEW Photosynthesis

Photosynthesis is the process that provides energy for almost all life. As **Figure 5-4** shows, photosynthesis has three stages:

Stage 1 Energy is captured from sunlight.

Stage 2 Light energy is converted to chemical energy, which is temporarily stored in ATP and the energy carrier molecule NADPH.

Stage 3 The chemical energy stored in ATP and NADPH powers the formation of organic compounds, using carbon dioxide, CO_2.

Photosynthesis occurs in the chloroplasts of plant cells and algae and in the cell membrane of certain bacteria. Photosynthesis can be summarized by the following equation:

$$\underset{\text{carbon dioxide}}{3CO_2} + \underset{\text{water}}{3H_2O} \xrightarrow{\text{light}} \underset{\text{3-carbon sugar}}{C_3H_6O_3} + \underset{\text{oxygen gas}}{3O_2}$$

However, this equation does not show how photosynthesis occurs. It merely says that three carbon dioxide molecules, three water molecules, and light are needed to form one three-carbon organic compound (a sugar) and three molecules of oxygen. Plants use the organic compounds they make during photosynthesis to carry out their life processes. For example, some of these sugars are used to form starch, which can be stored in stems or roots. The plant may later break down the starch to make ATP used to power metabolism. All of the proteins, nucleic acids, and other molecules of the cell are assembled from fragments of these sugars.

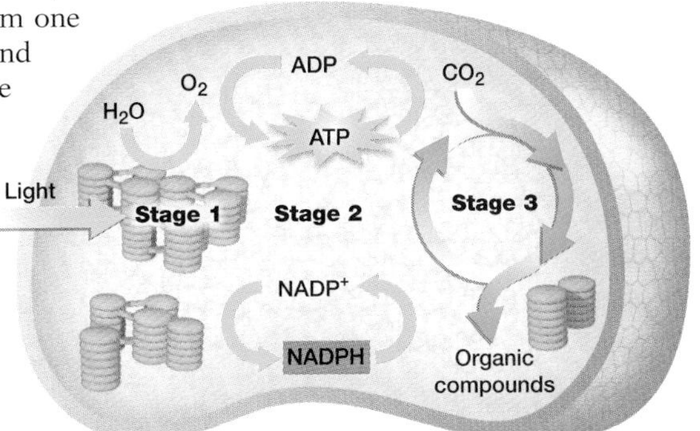

Figure 5-4
Photosynthesis. The process of photosynthesis occurs in three stages.

Objectives

- **Summarize** how energy is captured from sunlight in the first stage of photosynthesis. (pp. 97–99)
- **Analyze** the function of electron transport chains in the second stage of photosynthesis. (pp. 100–101)
- **Relate** the Calvin cycle to carbon dioxide fixation in the third stage of photosynthesis. (pp. 102–103)
- **Identify** three environmental factors that affect the rate of photosynthesis. (p. 103)

Key Terms

pigment
chlorophyll
carotenoid
thylakoid
electron transport chain
NADPH
carbon dioxide fixation
Calvin cycle

Section 5-2

1 Prepare

Scheduling

- **Block:** About 45 minutes will be needed to complete this section.
- **Traditional:** About one class period will be needed to complete this section.
- **Planning:** See the Planning Guide on pp. 91A and 91B for lecture, classwork, and assignment options for Section 5-2. Lesson plans for Section 5-2 can be found in a lesson cycle format in ***Biology: Principles and Explorations*** **One-Stop Planner CD-ROM with Test Generator.**

Resource Link

Assign **Active Reading Worksheet Section 5-2** to help students comprehend the key concepts and visuals in this lesson.

2 Teach

Lesson Warm-up

Ask students how sunlight is important to living systems. *(Sunlight is the main source of energy in living systems.)* Ask students to define photosynthesis. *(Answers will vary but should indicate that photosynthesis is the process by which plants and some other autotrophs use light energy to make their own food.)*

Integrating Different Learning Styles

Logical/Mathematical	**PE**	Interactive Exploration, pp. 114–115; Quick Lab, p. 101
Visual/Spatial	**PE**	Performance Zone items 1, 12, 16, and 17, pp. 112–113
	ATE	Demonstration p. 99; Visual Strategies, pp. 99, 100, 102; Biology in Your Community, p. 100; Teaching Tips, pp. 101, 102
Interpersonal	**ATE**	Reteaching, p. 103
Verbal/Linguistic	**PE**	Performance Zone items 2, 18, and 24, pp. 112–113
	ATE	Making Biology Relevant, p. 98; Internet Connect, p. 99; Teaching Tip, p. 101; Cross-Disciplinary Connection, p. 102

Chapter 5

Making Biology Relevant
Food

Yellow and orange vegetables are rich sources of carotenoids. A carotenoid called beta carotene is an important dietary source of vitamin A, which is necessary for proper eyesight, for maintaining the health of membranes, and for tooth and bone development. Have students research and write a report on the effectiveness of beta carotene as an antioxidant. Also have them compare the effectiveness of dietary sources of beta carotene with that of other sources, such as dietary supplements. PORTFOLIO

Teaching TIP
Shade-Grown Plants Versus Sun-Grown Plants

Plants grown in the shade often produce larger leaves than plants grown in full sunlight. Have students propose a hypothesis to explain this. *(The larger leaves of shade-grown plants gather more sunlight because they have a larger surface area available to absorb light. This is advantageous because less light reaches the leaves of shade-grown plants.)*

Resource Link
Assign **Reading Strategies: K-W-L** from the Holt BioSources Teaching Resources CD-ROM. PORTFOLIO

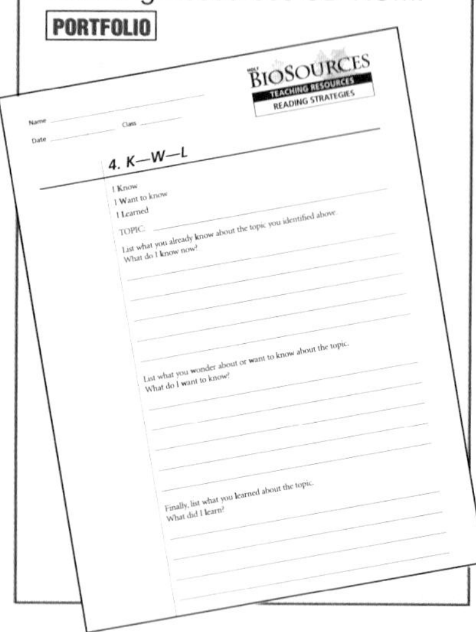

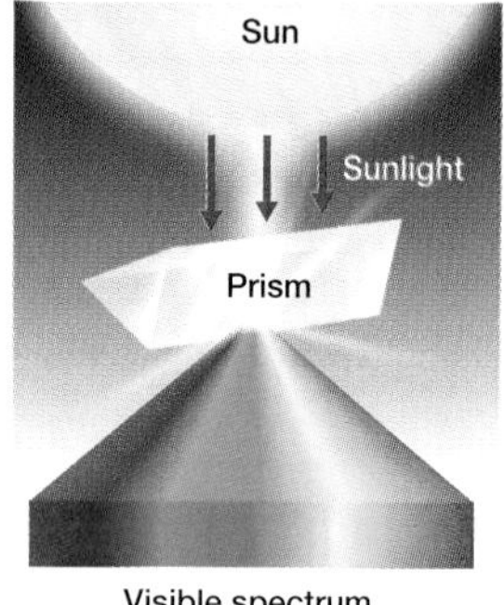

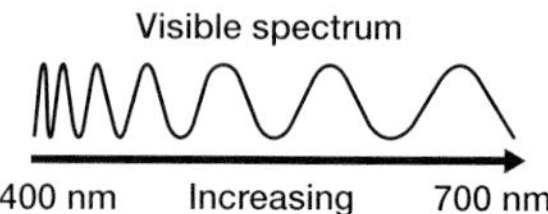

Figure 5-5 Visible spectrum. Sunlight contains a mixture of all the wavelengths (colors) of visible light. When sunlight passes through a prism, the prism separates the light into different colors.

In Stage One, Light Energy Is Absorbed

The chemical reactions that occur in the first and second stages of photosynthesis are sometimes called "light reactions," or light-dependent reactions. Without the absorption of light, these reactions could not occur. Light energy is used to make energy-storing compounds. Light is a form of radiation, energy in the form of waves that travel through space. Different types of radiation, such as light and heat, have different wavelengths (the distance between two consecutive waves). When the sun shines on you, your body is bombarded by many kinds of radiation from the sun. However, you can see only radiation known as visible light. You see wavelengths of visible light as different colors. As shown in **Figure 5-5,** sunlight contains all the wavelengths of visible light.

Pigments absorb different wavelengths of light

How does a human eye or a leaf absorb light? These structures contain light-absorbing substances called **pigments.** Pigments absorb only certain wavelengths and reflect all the others. **Chlorophyll** *(KLOR uh fihl)*, the primary pigment involved in photosynthesis, absorbs mostly blue and red light and reflects green and yellow light. This reflection of green and yellow light makes many plants, especially their leaves, look green. Plants contain two types of chlorophyll, chlorophyll *a* and chlorophyll *b*. Both types of chlorophyll play an important role in plant photosynthesis. The pigments that produce yellow and orange fall leaf colors, as well as the colors of many fruits, vegetables, and flowers, are called **carotenoids** *(kuh RAH tuh noydz)*. Carotenoids absorb wavelengths of light different from those absorbed by chlorophyll, so using both pigments enables plants to absorb more light energy during photosynthesis. The graph in **Figure 5-6** shows the wavelengths of light absorbed by chlorophyll *a*, chlorophyll *b*, and carotenoids.

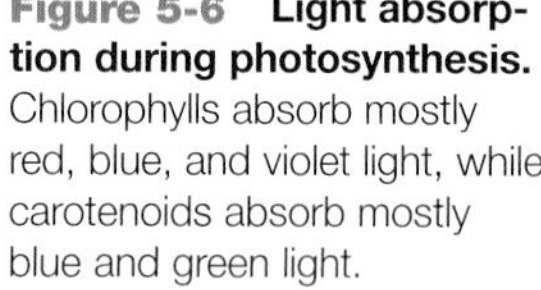

Figure 5-6 Light absorption during photosynthesis. Chlorophylls absorb mostly red, blue, and violet light, while carotenoids absorb mostly blue and green light.

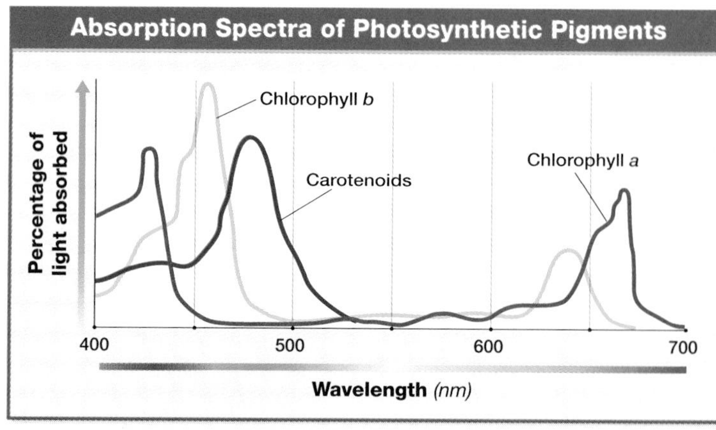

did you know?

Spectroscopy is the study and measurement of specific interactions of light with matter.

The spectrophotometer is the instrument used to accomplish this. Many areas of biology use spectroscopy. The ability of a pigment to absorb various wavelengths of light can be measured using the spectrophotometer. A graph plotting the pigment's light absorption versus wavelength is called an absorption spectrum.

Production of oxygen

As shown in **Figure 5-7**, pigments involved in plant photosynthesis are located in the chloroplasts of leaf cells. Clusters of pigments are embedded in the membranes of disk-shaped structures called **thylakoids** *(THEYE luh koydz)*. When light strikes a thylakoid in a chloroplast, energy is transferred to electrons in chlorophyll and other pigments. This energy transfer causes the electrons to jump to a higher energy level. Electrons with extra energy are said to be "excited." This is how plants first capture energy from sunlight.

Excited electrons jump from chlorophyll molecules to other nearby molecules in the thylakoid membrane, where the electrons are used to power the second stage of photosynthesis. The excited electrons that leave chlorophyll molecules must be replaced by other electrons. Plants get these replacement electrons from water molecules, H_2O. Water molecules are split by an enzyme inside the thylakoid. When water molecules are split, chlorophyll molecules take the electrons from the hydrogen atoms, H, leaving hydrogen ions, H^+. The remaining oxygen atoms, O, from the disassembled water molecules combine to form oxygen gas, O_2.

Figure 5-7 Chloroplast

Pigment molecules are embedded in thylakoid membranes, as are other molecules that participate in plant photosynthesis.

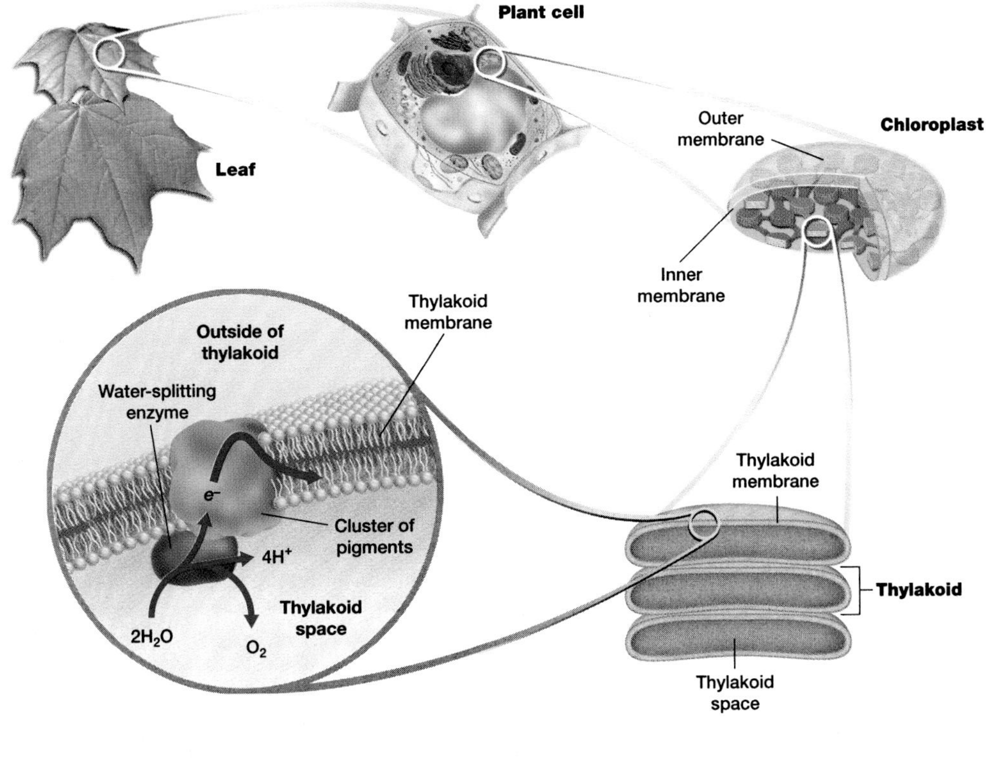

internetconnect

SCILINKS NSTA Have students research light absorption using the Web site in the Internet Connect box on page 99. Have students write a report summarizing their findings. PORTFOLIO

Demonstration

Chocolate Mints as Thylakoids

Use chocolate mints approximately 5 cm in diameter with a center of contrasting color. Make stacks of four or five mints. On one or two of the stacks, cut the top mint in half to expose the center before putting it on the stack. Point out that each stack of mints represents a column of thylakoids. To make the model more realistic, connect one stack to another using strips of paper to represent the membranous connections between thylakoids. Point out that a fluid surrounds the thylakoids. Ask students why it is more advantageous for the thylakoids to be in a stack than in a single unit. *(Stacks increase the surface area available for light absorption by pigment molecules.)*

Visual Strategy

Figure 5-7

This figure breaks down the structure of a chloroplast. Point out that leaves are a primary site of photosynthesis in plants because leaf cells contain many chloroplasts. Leaves are generally thin, allowing sunlight to penetrate the cells. Openings in the leaf surface enable carbon dioxide to enter and oxygen to leave.

Overcoming Misconceptions

Why Plants Look Green

Many people think that plants are green because plants use green light during photosynthesis. In fact, plants absorb and use mainly red and blue light for photosynthesis.

Tell students that plants look green because they contain large amounts of chlorophyll, which reflects green and yellow light and absorbs blue and red light.

CROSS-DISCIPLINARY CONNECTION

◆ **Chemistry**

When an electron in an atom is boosted to a higher energy level, it gains energy. When an electron drops back to a lower energy level, it emits energy. These emissions often take the form of electromagnetic waves that are always specific to the magnitude of the drop in energy. Emissions may be above the visible range of the electromagnetic spectrum, within the visible range, or below the visible range. Scientists use these specific emissions as a "fingerprint" to identify the atoms emitting the waves.

Visual Strategy

Figure 5-8

This figure follows the path of electron transport during the light-dependent reactions of photosynthesis. Guide students through the electron transport chains of photosynthesis by asking the following questions: What is the source of the excited electrons? *(pigment molecules)* What is the source of some of the replacement electrons? *(split water molecules)* What type of transport occurs when hydrogen ions are pumped into the thylakoid? *(active transport)* What type of transport occurs when hydrogen ions move out of the thylakoid? *(passive transport)* What kind of membrane protein is involved? *(carrier protein)* Have students record the questions and answers in their notebooks.

Study TIP

Interpreting Graphics

Look closely at Figure 5-8. Electrons are represented by the symbol e^-. The red arrows show the path of excited electrons. Hydrogen ions are represented by the symbol H^+. The blue arrows show the path of hydrogen ions that cross the thylakoid membrane.

In Stage Two, Light Energy Is Converted to Chemical Energy

Excited electrons that leave chlorophyll molecules are used to produce new molecules that temporarily store chemical energy, including ATP. An excited electron jumps to a nearby molecule in the thylakoid membrane. Then the electron is passed through a series of molecules along the thylakoid membrane like a ball being passed down a line of people. The series of molecules through which excited electrons are passed along a thylakoid membrane are called **electron transport chains.** Trace the path taken by excited electrons in the electron transport chains shown in **Figure 5-8.**

Action of electron transport chains

How are electron transport chains used to make molecules that temporarily store energy in the cell? As shown in Figure 5-8, one type of electron transport chain contains a protein that acts as a membrane pump. Excited electrons lose some of their energy as they each pass through this protein. The energy lost by the electrons is used to pump hydrogen ions, H^+, into the thylakoid. Recall that hydrogen ions are also produced when water molecules are split inside the thylakoid. As the process continues, hydrogen ions become more concentrated inside the thylakoid than outside, producing a concentration gradient across the thylakoid membrane. As a result,

Figure 5-8 **Electron transport chains of photosynthesis**

Electron transport chains convert light energy to chemical energy.

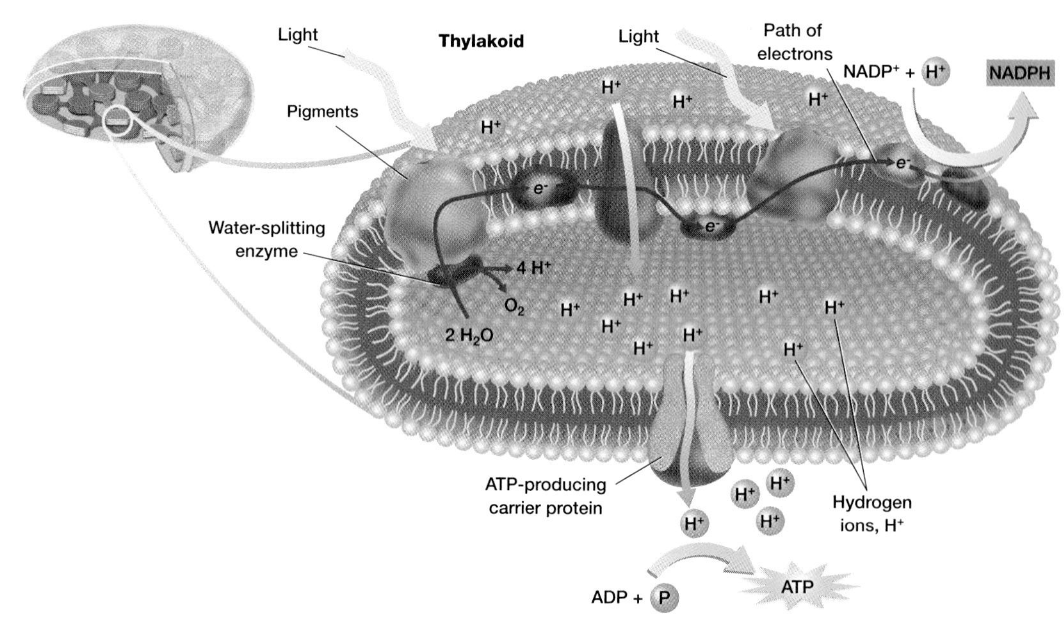

Biology in Your Community

Botanical Gardens

Arrange for a guided tour to a nearby botanical garden. Here students can see various plants that have different light, temperature, and mineral requirements. If possible, make arrangements ahead of time with the guide to discuss any specific topics you want mentioned to your students.

hydrogen ions have a tendency to diffuse back out of the thylakoid down their concentration gradient through specialized carrier proteins. These carrier proteins are unique because they function both as an ion channel and as an enzyme. As hydrogen ions pass through the channel portion of the protein, the protein catalyzes a reaction in which a phosphate group is added to a molecule of ADP, making ATP. Thus, the movement of hydrogen ions across the thylakoid membrane through these carrier proteins provides the energy needed to make ATP used to power the third stage of photosynthesis.

While one electron transport chain provides energy used to make ATP, a second electron transport chain provides energy used to make NADPH. **NADPH** is an electron carrier that provides the high-energy electrons needed to make carbon-hydrogen bonds in the third stage of photosynthesis. In this electron transport chain, also shown in Figure 5-8, excited electrons combine with hydrogen ions as well as an electron acceptor called $NADP^+$, forming NADPH.

The light-dependent reactions of photosynthesis can be summarized as follows. Pigment molecules in the thylakoids of chloroplasts absorb light energy. Electrons in the pigments are excited by light and move through electron transport chains in thylakoid membranes. These electrons are replaced by electrons from water molecules, which are split by an enzyme. Oxygen atoms from water molecules combine to form oxygen gas. Hydrogen ions accumulate inside thylakoids, setting up a concentration gradient that provides the energy to make ATP and NADPH.

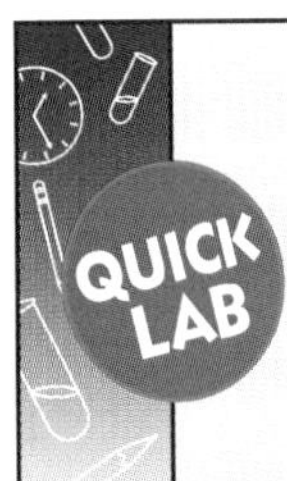

Do photosynthetic organisms give off oxygen?

You can use the following procedure to identify the gas given off by a photosynthetic organism.

Materials

MBL or CBL system with appropriate software, test tube or small glass jar, sprig of *Elodea*, distilled water, cool light source, dissolved oxygen (DO) probe

Procedure

1. Set up an MBL/CBL system to collect and graph data from a dissolved oxygen probe at 30-second intervals for 60 data points. Calibrate the DO probe.
2. Place a sprig of *Elodea* in a test tube or glass jar, and fill the test tube or jar with distilled water.
3. Place the test tube or glass jar under a cool light source, and lower a DO probe into the water. Collect data for 30 minutes.
4. When data collection is complete, view the graph of your data. If possible, print the graph. Otherwise, sketch the graph on paper.

Analysis

1. **Infer** the cause of any change you observed.
2. **Propose** a control for this experiment.
3. **SKILL Evaluating Hypotheses** Explain how your data support or do not support the hypotheses that photosynthetic organisms do or do not give off oxygen.

Graphic Organizer

*Use this graphic organizer with **Teaching Tip: Stages of Photosynthesis** on page 101.*

Stages of Photosynthesis		
	Used	Produced
Stage 1	Light, water	Oxygen, hydrogen ions
Stage 2	Electrons, hydrogen ions	ATP, NADPH
Stage 3	ATP, NADPH, carbon dioxide	Organic compounds

Teaching TIP BASIC

Photosynthesis

Help students reinforce the concepts in the three stages of photosynthesis by having them make flash cards with a structure or process on one side and a description of its role on the other side. Topics should include pigment molecules, thylakoids, electron transport chains, ATP, and NADPH.

Do photosynthetic organisms give off oxygen?

Time 30 minutes

Process Skills Inferring, analyzing, evaluating

Teaching Strategies
The *Elodea* should be placed in the test tube cut-side up; the cut should be fresh. Be careful working with glass.

Analysis Answers

1. Answers will vary. For example, oxygen is produced as water molecules are split in the electron transport chains of *Elodea.*
2. A control would be the same setup without a light source.
3. If the probe detects an increase in the oxygen content of the water, that would support the hypothesis that photosynthetic organisms give off oxygen.

Teaching TIP

Stages of Photosynthesis

Have students make a **Graphic Organizer** similar to the one at the bottom of page 101 to illustrate the substances used and produced during each stage of photosynthesis. PORTFOLIO

Chapter 5

CROSS-DISCIPLINARY CONNECTION

◆ History

Melvin Calvin was born in St. Paul, Minnesota, in 1911. He received his B.S. degree in chemistry in 1931 and his Ph.D. in chemistry in 1935. Calvin's studies of organic molecular structure eventually led to an interest in photosynthesis, especially the role of carbon dioxide. In 1949 Calvin began using carbon dioxide labeled with carbon-14 to trace the path of carbon dioxide during photosynthesis. In 1961 Calvin won a Nobel prize in chemistry for his work. He died in 1997. Have students research and write a report on his work. PORTFOLIO

Visual Strategy

BIO graphic **Figure 5-9**

Have students count the total number of carbon atoms present at each step in this summary of the Calvin cycle. Ask students the following questions: How are ATP and NADPH important to the Calvin cycle? *(They supply energy used to form new compounds.)* How many carbon dioxide molecules are needed? *(3)* How many 3-carbon sugars are made? *(6)* How many of these sugars are actually used to make organic compounds that the plant uses for energy? *(1)* Emphasize that the main product of the Calvin cycle is not glucose but rather a 3-carbon sugar that is used to produce other organic compounds, such as starch and sucrose.

Teaching TIP

Storing Energy

Use a scalpel to make thin cross sections of a potato. Have students examine the potato under a compound microscope. Ask them to look for starch granules, which should appear as large, translucent structures inside the potato cells. Explain that the organic compounds plants make during photosynthesis are stored mainly as starch.

In Stage Three, Energy Is Stored in Organic Compounds

In the first and second stages of photosynthesis, light energy is used to make ATP and NADPH, which temporarily store chemical energy. These stages are therefore considered light-dependent. In the third (final) stage of photosynthesis, however, carbon atoms from carbon dioxide in the atmosphere are used to make organic compounds, in which chemical energy is stored. The transfer of carbon dioxide to organic compounds is called **carbon dioxide fixation.** The reactions that "fix" carbon dioxide are sometimes called "dark reactions," or light-independent reactions. Among photosynthetic organisms, there are several ways in which carbon dioxide is fixed.

Calvin cycle

The most common method of carbon dioxide fixation is the Calvin cycle. The **Calvin cycle** is a series of enzyme-assisted chemical reactions that produces a three-carbon sugar. The Calvin cycle is summarized in **Figure 5-9.**

BIO graphic Figure 5-9

Calvin Cycle

The Calvin cycle is a common method of carbon dioxide fixation.

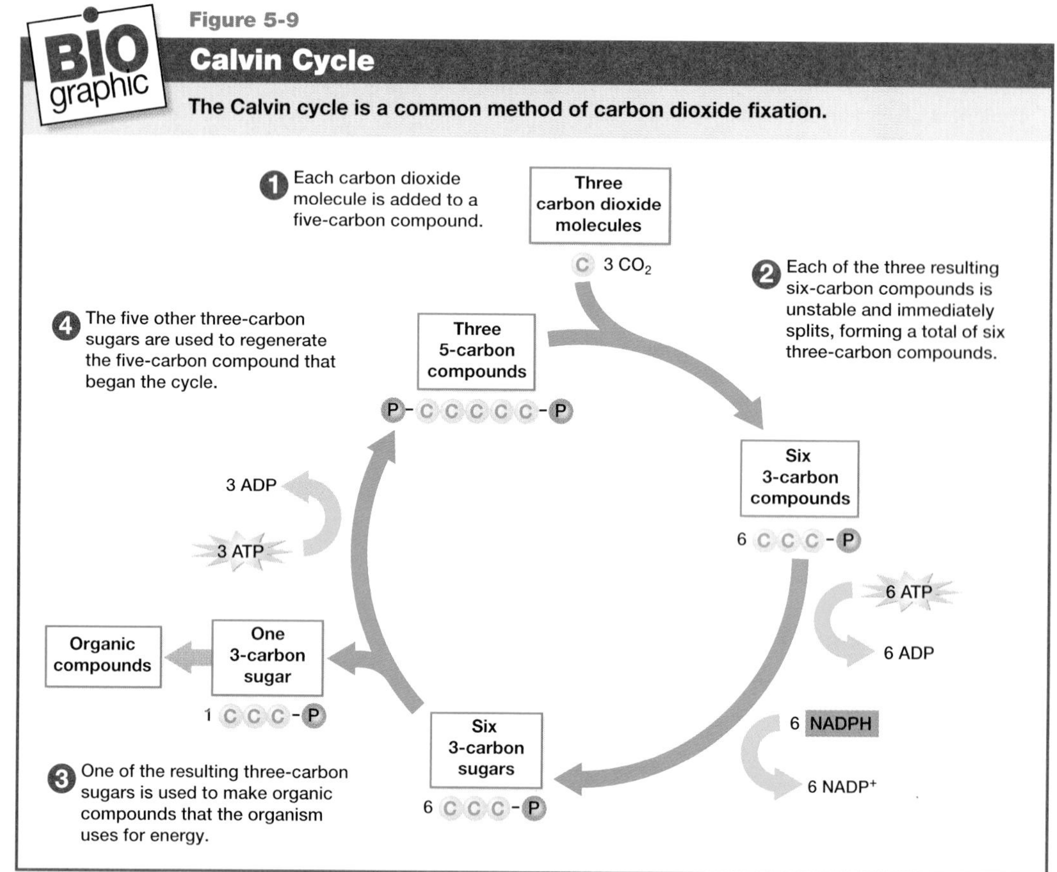

Overcoming Misconceptions

Plant Respiration

Many students think plants undergo photosynthesis but not cellular respiration. Point out that plants use the organic compounds they produce during photosynthesis for their metabolism. Plants store carbohydrates, such as starch and sucrose, which can be broken down and used in cellular respiration. Remind students that plants contain both chloroplasts and mitochondria. Mitochondria produce most of the ATP made during cellular respiration.

Step 1 In carbon dioxide fixation, each molecule of carbon dioxide, CO_2, is added to a five-carbon compound by an enzyme.

Step 2 The resulting six-carbon compound splits into two three-carbon compounds. Phosphate groups from ATP and electrons from NADPH are added to the three-carbon compounds, forming three-carbon sugars.

Step 3 One of the resulting three-carbon sugars is used to make organic compounds—including carbohydrates such as starch and sucrose—in which energy is stored for later use by the organism.

Step 4 The other three-carbon sugars are used to regenerate the initial five-carbon compound, thereby completing the cycle.

The Calvin cycle is named for Melvin Calvin, the American biochemist who worked out the chemical reactions in the cycle. The reactions are cyclic because they recycle the five-carbon compound needed to begin the cycle again. A total of three carbon dioxide molecules must enter the Calvin cycle to produce each three-carbon sugar that will be used to make other organic compounds. These organic compounds provide the organism with energy for growth and metabolism. The energy used in the Calvin cycle is supplied by ATP and NADPH made during the second stage of photosynthesis.

Factors that affect photosynthesis

Photosynthesis is directly affected by various environmental factors. The most obvious of these factors is light. In general, the rate of photosynthesis increases as light intensity increases until all the pigments are being used. At this saturation point, the rate of photosynthesis levels off because pigments cannot absorb any more light. The carbon dioxide concentration affects the rate of photosynthesis in a similar manner. Once a certain concentration of carbon dioxide is present, photosynthesis cannot proceed any faster.

Photosynthesis is most efficient within a certain range of temperatures. Like all metabolic processes, photosynthesis involves many enzyme-assisted chemical reactions. Recall that enzymes operate properly only within certain temperature ranges. Unfavorable temperatures may inactivate certain enzymes. Light intensity, carbon dioxide concentration, and temperature determine the optimum level of photosynthesis for a particular plant.

REAL LIFE

Some houseplants thrive in dim light, but others require bright light.

A plant in a living room may receive 100 times less bright light than it would if it were grown outdoors.

Recognizing Patterns

Examine several species of houseplants in a store or nursery. Which features are common among houseplants?

internetconnect

TOPIC: Factors affecting photosynthesis
GO TO: www.scilinks.org
KEYWORD: HX103

Review SECTION 5-2

1. **Summarize** how photosynthetic organisms capture the energy in sunlight.
2. **Compare** the roles of water molecules and hydrogen ions in electron transport chains.
3. **Describe** the role of the Calvin cycle in the third stage of photosynthesis.
4. **SKILL Organizing Information** Make a table in which you identify the role of each of the following in photosynthesis: light, water, pigments, ATP, NADPH, and carbon dioxide.
5. **SKILL Inferring Relationships** What combination of environmental factors affects the rate of photosynthesis?

Review SECTION 5-2 ANSWERS

1. Pigment molecules absorb specific wavelengths of light.
2. As excited electrons pass along the thylakoid membrane, their energy is used to pump hydrogen ions into the thylakoid. These electrons are replaced by electrons from water molecules, which are split by an enzyme. The hydrogen ion gradient is used to produce ATP.
3. Carbon dioxide is used in the Calvin cycle to produce a 3-carbon sugar that will be used to produce organic compounds.
4. Answers will vary but may be similar to the table below.

Light	Excites electrons
Water	Provides hydrogen ions and replacement electrons
Pigments	Absorb light
ATP and NADPH	Store chemical energy
Carbon dioxide	Used to produce organic compounds

5. Factors include light intensity, carbon dioxide concentration, and temperature.

REAL LIFE

Answers

Most houseplants are small and are able to grow in a small amount of soil.

3 Assessment Options

Closure

Identifying Variables

Have students list three factors that would increase the rate of photosynthesis. Have them identify what stage of photosynthesis each factor would affect. *(Examples might include increasing light intensity, which would affect Stage 1; providing more water, which would affect Stage 2; and increasing the carbon dioxide concentration, which would affect Stage 3.)*

Review

Assign Review—Section 5-2.

Reteaching — BASIC

Form six cooperative groups of students. Assign one of the following factors to each group: water, chlorophyll, light, carbon dioxide, NADPH, and ATP. Have each group decide on the role their factor plays in photosynthesis. After a designated time, select a person in each group to read the decision of the group to the rest of the class. Students should use this time to ask any relevant questions.

Chapter 5

Section 5-3 Cellular Respiration

1 Prepare

Scheduling

- **Block:** About 90 minutes will be needed to complete this section.
- **Traditional:** About two class periods will be needed to complete this section.
- **Planning:** See the Planning Guide on pp. 91A and 91B for lecture, classwork, and assignment options for Section 5-3. Lesson plans for Section 5-3 can be found in a lesson cycle format in ***Biology: Principles and Explorations*** **One-Stop Planner CD-ROM with Test Generator.**

Resource Link

Assign **Active Reading Worksheet Section 5-3** to help students comprehend the key concepts and visuals in this lesson.

2 Teach

Lesson Warm-up

Ask students how the products of photosynthesis are related to cellular respiration. *(The products of photosynthesis are the starting materials for cellular respiration.)* Ask students what kinds of organisms undergo cellular respiration. *(All organisms, including photosynthetic organisims, undergo cellular respiration.)*

Objectives

- **Summarize** how glucose is broken down in the first stage of cellular respiration. (pp. 104–105)
- **Describe** how ATP is made in the second stage of cellular respiration. (pp. 106–107)
- **Identify** the role of fermentation in the second stage of cellular respiration. (pp. 108–109)
- **Evaluate** the importance of oxygen in aerobic respiration. (p. 110)

Key Terms

aerobic
anaerobic
glycolysis
NADH
Krebs cycle
$FADH_2$
fermentation

Cellular Respiration Produces ATP

Most of the foods we eat contain usable energy. Much of the energy in a hamburger, for example, is stored in proteins, carbohydrates, and fats. But before you can use that energy, it is transferred to ATP. Like in most organisms, your cells transfer the energy in organic compounds, especially glucose, to ATP through a process called cellular respiration. Oxygen in the air you breathe makes the production of ATP more efficient, although some ATP is made without oxygen. Metabolic processes that require oxygen are called **aerobic** *(ehr OH bihk)*. Metabolic processes that do not require oxygen are called **anaerobic** *(AN ehr oh bihk)*, meaning "without air."

OVERVIEW Cellular Respiration

Cellular respiration harnesses the energy in organic compounds, particularly glucose. The breakdown of glucose during cellular respiration can be summarized by the following equation:

$$\underset{\text{glucose}}{C_6H_{12}O_6} + \underset{\text{oxygen gas}}{6O_2} \xrightarrow{\text{enzymes}} \underset{\text{carbon dioxide}}{6CO_2} + \underset{\text{water}}{6H_2O} + \underset{\text{ATP}}{\text{energy}}$$

As **Figure 5-10** shows, cellular respiration occurs in two stages:

Stage 1 Glucose is converted to pyruvate *(PEYE roo vayt)*, producing a small amount of ATP and NADH.

Stage 2 When oxygen is present, pyruvate and NADH are used to make a large amount of ATP. This is called aerobic respiration. Aerobic respiration occurs in the mitochondria of eukaryotic cells and in the cell membrane of prokaryotic cells. When oxygen is not present, pyruvate is converted to either lactate *(LAK tayt)* or ethanol (ethyl alcohol) and carbon dioxide.

The equation above does not show how cellular respiration occurs. It simply shows that the complete enzyme-assisted breakdown of a glucose molecule uses six oxygen molecules and forms six carbon dioxide molecules, six water molecules, and ATP. Aerobic respiration produces most of the ATP made by cells. Many intermediate products of aerobic respiration are used to make organic compounds, which in turn help build and maintain cells.

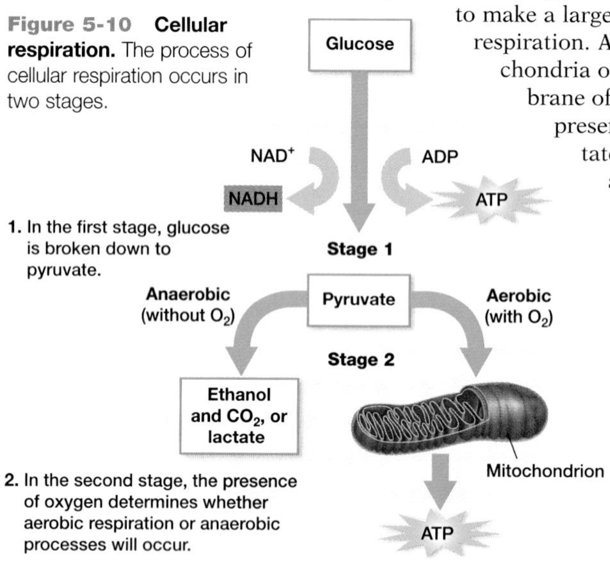

Figure 5-10 Cellular respiration. The process of cellular respiration occurs in two stages.

Integrating Different Learning Styles

Logical/Mathematical	**ATE**	Teaching Tip, p. 110
Visual/Spatial	**PE**	Performance Zone item 1, p. 112
	ATE	Visual Strategies, pp. 105, 106, 107; Demonstration, p. 109; Biology in Your Community, p. 109
Body/Kinesthetic	**PE**	Real Life, p. 107
Interpersonal	**ATE**	Reteaching, p. 110
Verbal/Linguistic	**PE**	Study Tip, p. 105; Performance Zone items 2, 19, 24, and 25, pp. 112–113
	ATE	Internet Connect, p. 106; Multicultural Perspective, p. 106; Career, p. 108

In Stage One, Glucose Is Broken Down During Glycolysis

The primary fuel for cellular respiration is glucose, which is formed when carbohydrates such as starch and sucrose are broken down. If too few carbohydrates are available to meet an organism's glucose needs, other molecules, such as fats, can be broken down to make ATP. In fact, one gram of fat contains more energy than two grams of carbohydrate. Proteins and nucleic acids can also be used to make ATP, but they are normally used for building important cell parts.

Study TIP

Word Origins

The word *glycolysis,* meaning "the splitting of glucose," is from the Greek words *glykys,* meaning "sweet," and *lysis,* meaning "to dissolve."

Glycolysis

In the first stage of cellular respiration, glucose is broken down in the cytoplasm during a process called **glycolysis** *(gleye KAHL uh sihs)*. Glycolysis is an enzyme-assisted anaerobic process that breaks down one six-carbon molecule of glucose to two three-carbon pyruvates. Recall that a molecule that has lost or gained one or more electrons is called an ion. Pyruvate is the ion of a three-carbon organic acid called pyruvic acid. The pyruvate produced during glycolysis still contains some of the energy that was stored in the glucose molecule.

As glucose is broken down, some of its hydrogen atoms are transferred to an electron acceptor called NAD^+. This forms an electron carrier called **NADH.** For cellular respiration to continue, the electrons carried by NADH are eventually donated to other organic compounds. This recycles NAD^+, making it available to accept more electrons. Glycolysis is summarized in **Figure 5-11.**

Step 1 Phosphate groups from two ATP molecules are transferred to a glucose molecule.

Step 2 The resulting six-carbon compound is broken down to two three-carbon compounds, each with a phosphate group.

Step 3 Two NADH molecules are produced, and one more phosphate group is transferred to each three-carbon compound.

Step 4 Each three-carbon compound is converted to a three-carbon pyruvate, producing four ATP molecules in the process.

Glycolysis uses two ATP molecules but produces four ATP molecules, yielding a net gain of two ATP molecules. Glycolysis is followed by another set of reactions that use the energy temporarily stored in NADH to make more ATP.

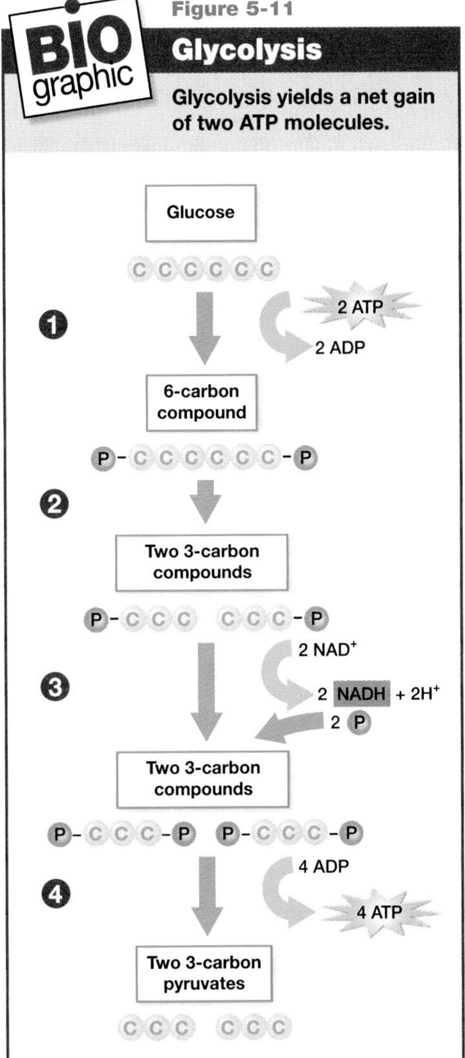

Figure 5-11
Glycolysis
Glycolysis yields a net gain of two ATP molecules.

did you know?

The first prokaryotes probably used glycolysis to make ATP long before oxygen was present in Earth's atmosphere.

According to fossil records, prokaryotes were present on Earth 3.5 billion years ago, but oxygen was not abundant in the atmosphere until around 2.5 billion years ago. Because glycolysis is an anaerobic metabolic pathway that occurs in all cells and not in organelles, glycolysis most likely occurred in early cells.

Teaching TIP

Metabolic Pathways

Review the term *metabolism* with students. (*Metabolism is the term used to describe the sum of all chemical reactions in an organism.*) Tell students that glycolysis is an example of a metabolic pathway. In glycolysis, the energy stored in glucose is gradually released in a series of enzyme-assisted chemical reactions.

Visual Strategy

BIO graphic Figure 5-11
Have students examine this figure, and ask them to identify the starting material in glycolysis. *(glucose)* Work with students to summarize the events that take place in each step. Explain that as glucose is broken down during glycolysis, some of the energy contained in glucose is transferred to the products of glycolysis. Ask students to identify those products. *(pyruvate, ATP, and NADH)* Explain that pyruvate is the ion of the organic molecule pyruvic acid.

Active Reading

Technique: Anticipation Guide

Write the following statements on the board:

1. Organisms do not need oxygen to get energy from organic molecules.
2. It is unlikely that humans will ever run 1500 m (1 mi) in under 2 minutes.

Ask students to think about each statement and decide whether they agree or disagree. Students should point to specific passages in the text that support their reasoning. Then have them see if their opinions have changed.

internetconnect

SCLINKS. NSTA Have students research aerobic respiration using the Web site in the Internet Connect box on page 106. Have students write a report summarizing their findings. PORTFOLIO

Teaching TIP ADVANCED

Coenzymes

A coenzyme is an organic chemical that is necessary for the action of many enzymes. Ask students why it is important for glucose to be partially broken down to pyruvate. *(Pyruvate is small enough to diffuse across the mitochondrial membranes, while glucose is too large to do so.)* Tell students that when pyruvate enters a mitochondrion and is broken down to a 2-carbon acetyl group, coenzyme A attaches to the acetyl group, forming acetyl-CoA. Coenzyme A enables the acetyl group to enter the Krebs cycle.

Visual Strategy

BIO graphic **Figure 5-12**

Guide students through the steps of this illustration. Have students count the number of carbon atoms present at each step during the Krebs cycle. Ask them where the Krebs cycle occurs. *(in mitochondria)* Remind students that for every molecule of glucose that is broken down, two pyruvates are produced. Thus, the Krebs cycle occurs for each pyruvate. Also tell them that a specific enzyme is involved in each step. Emphasize the role of the Krebs cycle as a precursor of the electron transport chain.

internetconnect

SCLINKS. NSTA

TOPIC: Aerobic respiration
GO TO: www.scilinks.org
KEYWORD: HX106

In Stage Two, More ATP Is Made by Aerobic Respiration

When oxygen is present, pyruvate produced during glycolysis enters a mitochondrion and is converted to a two-carbon compound. This reaction produces a carbon dioxide molecule, one NADH molecule, and a two-carbon acetyl *(uh SEET uhl)* group. The acetyl group is attached to a molecule called coenzyme A (CoA), forming a compound called acetyl-CoA *(uh SEET uhl-koh ay).*

Krebs cycle

Acetyl-CoA enters a series of enzyme-assisted reactions called the **Krebs cycle,** summarized in **Figure 5-12.** The cycle is named for the biochemist Hans Krebs, who first proposed the cycle in 1937.

Step 1 Acetyl-CoA combines with a four-carbon compound, forming a six-carbon compound and releasing coenzyme A.

Step 2 Carbon dioxide, CO_2, is released from the six-carbon compound, forming a five-carbon compound. Electrons are transferred to NAD^+, making a molecule of NADH.

Step 3 Carbon dioxide is released from the five-carbon compound,

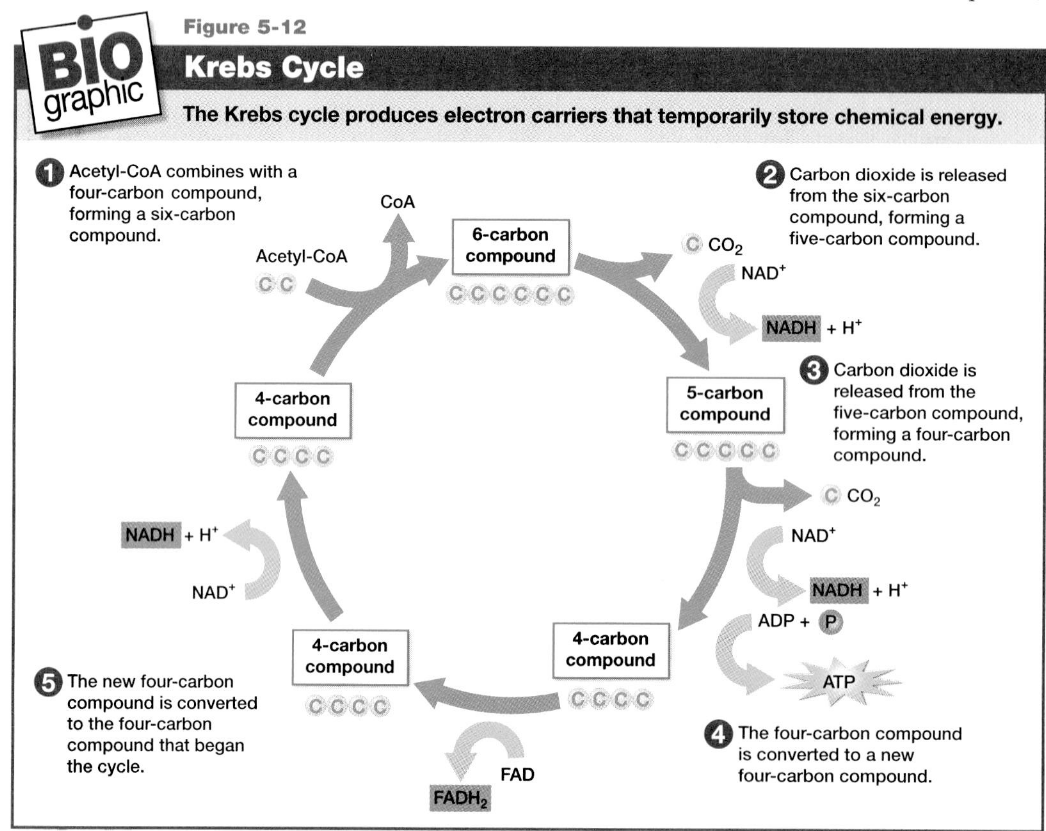

1 Acetyl-CoA combines with a four-carbon compound, forming a six-carbon compound.
CoA
Acetyl-CoA
C C
6-carbon compound
C C C C C C
2 Carbon dioxide is released from the six-carbon compound, forming a five-carbon compound.
C CO2
NAD+
NADH + H+
3 Carbon dioxide is released from the five-carbon compound, forming a four-carbon compound.
5-carbon compound
C C C C C
C CO2
NAD+
NADH + H+
ADP + P
ATP
4 The four-carbon compound is converted to a new four-carbon compound.
4-carbon compound
C C C C
FAD
FADH2
4-carbon compound
C C C C
5 The new four-carbon compound is converted to the four-carbon compound that began the cycle.
NAD+
NADH + H+
4-carbon compound
C C C C

Figure 5-12

Krebs Cycle

The Krebs cycle produces electron carriers that temporarily store chemical energy.

MULTICULTURAL PERSPECTIVE

Hans Adolph Krebs

In 1937 Hans Adolph Krebs discovered the details of the Krebs cycle. In 1953 Krebs was awarded the Nobel prize in physiology or medicine for his discovery. Have students research and prepare a report on the life of Krebs, who had to interrupt his work and leave Nazi Germany prior to World War II because he was Jewish. PORTFOLIO

forming a four-carbon compound. A molecule of ATP is made, and a molecule of NADH is also produced.

Step 4 The four-carbon compound is converted to a new four-carbon compound. Electrons are transferred to an electron acceptor called FAD, making a molecule of $FADH_2$. **$FADH_2$** is another type of electron carrier.

Step 5 The new four-carbon compound is then converted to the four-carbon compound that began the cycle. Another molecule of NADH is produced.

After the Krebs cycle, NADH and $FADH_2$ now contain much of the energy that was previously stored in glucose and pyruvate. When the Krebs cycle is completed, the four-carbon compound that began the cycle has been recycled, and acetyl-CoA can enter the cycle again.

Electron transport chain

In aerobic respiration, electrons donated by NADH and $FADH_2$ pass through an electron transport chain, as shown in **Figure 5-13.** In eukaryotic cells, the electron transport chain occurs in the inner membranes of mitochondria. The energy of these electrons is used to pump hydrogen ions out of the inner mitochondrial compartment. Hydrogen ions accumulate in the outer compartment, producing a concentration gradient across the inner membrane. Hydrogen ions diffuse back into the inner compartment through a carrier protein that adds a phosphate group to ADP, making ATP. At the end of the electron transport chain, hydrogen ions and spent electrons combine with oxygen molecules, O_2, forming water molecules, H_2O. Oxygen, therefore, serves as the final electron acceptor.

Figure 5-13 **Electron transport chain of aerobic respiration**

In the inner membranes of mitochondria, electron transport chains make ATP.

Outer compartment

3. ATP is produced as hydrogen ions diffuse into the inner compartment through a channel protein.

ATP-producing carrier protein

Inner mitochondrial membrane

$NADH + H^+$ → NAD^+

$4H^+ + O_2$ → $2H_2O$

$ADP + P$ → ATP

Inner compartment

1. The electron transport chain pumps hydrogen ions, H^+, out of the inner compartment.

2. At the end of the chain, electrons and hydrogen ions combine with oxygen, forming water.

did you know?

The human body uses about 1 million molecules of ATP per cell per second.

Much of the body's energy is released as heat. *Symplocarpus foetidus,* commonly known as the skunk cabbage, also generates significant amounts of heat. The skunk cabbage is a plant that lives in North American swamplands. It flowers in February and March when the ground may still be covered by snow. The heat it produces allows it to melt the snow and to attract pollinators when it flowers.

REAL LIFE

Regular aerobic exercise increases the efficiency of oxygen use. A couch potato's heart would beat faster than an athlete's heart during physical exertion.

Designing an Experiment
Devise a test to determine the relative fitness of your classmates based on their heart rate.

Teaching TIP

Electron Transport

Ask students how the electron transport chains of photosynthesis and cellular respiration are similar. *(In both processes, electrons are passed along an electron transport chain and are picked up by an electron acceptor. The energy of these electrons is used to produce a hydrogen ion concentration gradient, which provides the energy needed to make ATP.)*

Visual Strategy

Figure 5-13

Have students follow the path of electrons, shown by the red arrows, through the electron transport chain. Point out that the energy of these electrons is used to pump hydrogen ions out of the inner compartment. Ask students to identify this type of transport. *(active transport)* These ions then diffuse back into the inner compartment through the specialized carrier protein (ATP synthase), providing enough energy to make ATP. Ask students to identify this type of transport. *(passive transport)* Ask students why the folds of the mitochondrion are important. *(They increase the surface area of the membranes, allowing more ATP to be produced.)* Ask students to identify the role of oxygen in the electron transport chain. *(Oxygen is the final electron acceptor, and water is produced when the spent electrons, hydrogen ions, and oxygen combine.)*

REAL LIFE

Answer

For example, students could devise a test in which heart rate is measured before, during, and after 10 minutes of aerobic exercise, such as pedaling a stationary bicycle.

Chapter

Making Biology Relevant

Health

Cyanide is a fast-acting poison that blocks the action of the electron transport chain. It exists as hydrogen cyanide gas or cyanide salts used in gold and other metal extractions, electroplating, and metal cleaning. Cyanide enters the body by absorption through the lungs, skin, or gastrointestinal tract. It is highly toxic, and symptoms appear soon after exposure. Ingesting as little as 3 g of cyanide can result in death in a short period of time.

CAREER

Fitness Trainer

Many people use health clubs to exercise or work out. These clubs employ fitness trainers to direct exercise programs for groups of people or custom-tailored programs to suit an individual's needs. Have students investigate and write a report on the requirements needed to become a fitness trainer and the responsibility of the trainer to clients. PORTFOLIO

Resource Link

- Assign **Occupational Applications: Physical Therapist** from the Holt BioSources Teaching Resources CD-ROM.

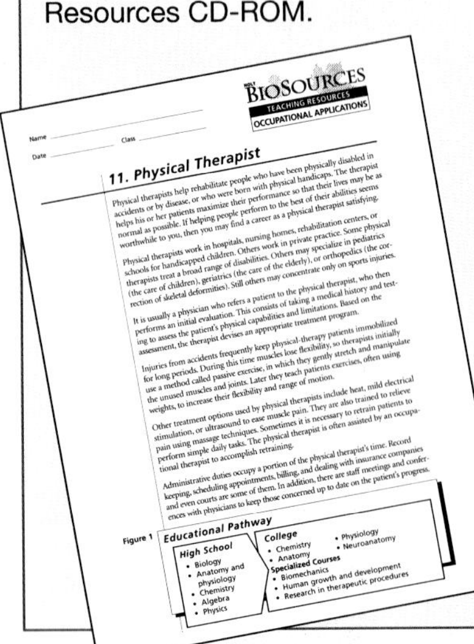

HOLT BIOSOURCES TEACHING RESOURCES OCCUPATIONAL APPLICATIONS

Name ____ Class ____
Date ____

11. Physical Therapist

Physical therapists help rehabilitate people who have been physically disabled in accidents or by disease, or who were born with physical handicaps. The therapist helps his or her patients maximize their performance so that their lives may be as normal as possible. If helping people perform to the best of their abilities seems worthwhile to you, then you may find a career as a physical therapist satisfying.

Physical therapists work in hospitals, nursing homes, rehabilitation centers, or schools for handicapped children. Others work in private practice. Some physical therapists treat a broad range of disabilities. Others may specialize in pediatrics (the care of children), geriatrics (the care of the elderly), or orthopedics (the correction of skeletal deformities). Still others may concentrate only on sports injuries.

It is usually a physician who refers a patient to the physical therapist, who then performs an initial evaluation. This consists of taking a medical history and testing to assess the patient's physical capabilities and limitations. Based on the assessment, the therapist devises an appropriate treatment program.

Injuries from accidents frequently keep physical-therapy patients immobilized for long periods. During this time muscles lose flexibility, so therapists initially use a method called passive exercise, in which they gently stretch and manipulate the unused muscles and joints. Later they teach patients exercises, often using weights, to increase their flexibility and range of motion.

Other treatment options used by physical therapists include heat, mild electrical stimulation, or ultrasound to ease muscle pain. They are also trained to relieve pain using massage techniques. Sometimes it is necessary to retrain patients to perform simple daily tasks. The physical therapist is often assisted by an occupational therapist to accomplish retraining.

Administrative duties occupy a portion of the physical therapist's time. Record keeping, scheduling appointments, billing, and dealing with insurance companies and even courts are some of them. In addition, there are staff meetings and conferences with physicians to keep those concerned up to date on the patient's progress.

Figure 1 **Educational Pathway**

High School
- Biology
- Anatomy and physiology
- Chemistry
- Algebra
- Physics

College
- Chemistry
- Anatomy
- Physiology
- Neuroanatomy

Specialized Courses
- Biomechanics
- Human growth and development
- Research in therapeutic procedures

Fermentation Follows Glycolysis in the Absence of Oxygen

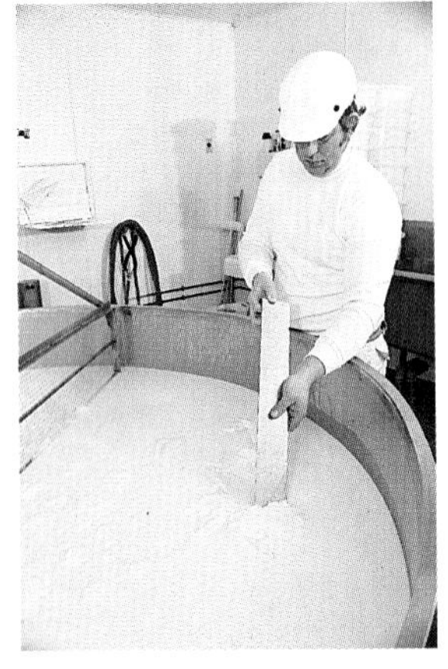

Figure 5-14
Fermentation. In cheese making, fungi or bacteria added to milk carry out lactic acid fermentation on some of the sugar in the milk.

What happens when there is not enough oxygen for aerobic respiration to occur? The electron transport chain does not function because oxygen is not available to serve as the final electron acceptor. Electrons are not transferred from NADH, and NAD^+ therefore cannot be recycled. When oxygen is not present, NAD^+ is recycled in another way. Under anaerobic conditions, electrons carried by NADH are transferred to pyruvate produced during glycolysis. This process recycles NAD^+ needed to continue making ATP through glycolysis. The recycling of NAD^+ using an organic hydrogen acceptor is called **fermentation.** Bacteria carry out more than a dozen kinds of fermentation, all using some form of organic hydrogen acceptor to recycle NAD^+. Two important types of fermentation are lactic acid fermentation and alcoholic fermentation. Lactic acid fermentation by some bacteria and fungi is used in the production of foods such as yogurt and some cheeses, as shown in **Figure 5-14.**

Lactic acid fermentation

In some organisms, a three-carbon pyruvate is converted to a three-carbon lactate through lactic acid fermentation, as shown in **Figure 5-15.** Lactate is the ion of an organic acid called lactic acid. For example, during vigorous exercise pyruvate in muscles is converted to lactate when muscle cells must operate without enough oxygen. Fermentation enables glycolysis to continue producing ATP in muscles as long as the glucose supply lasts. Blood removes excess lactate from muscles. Lactate can build up in muscle cells if it is not removed quickly enough, sometimes causing muscle soreness.

Figure 5-15 **Two types of fermentation**

When oxygen is not present, cells recycle NAD^+ through fermentation.

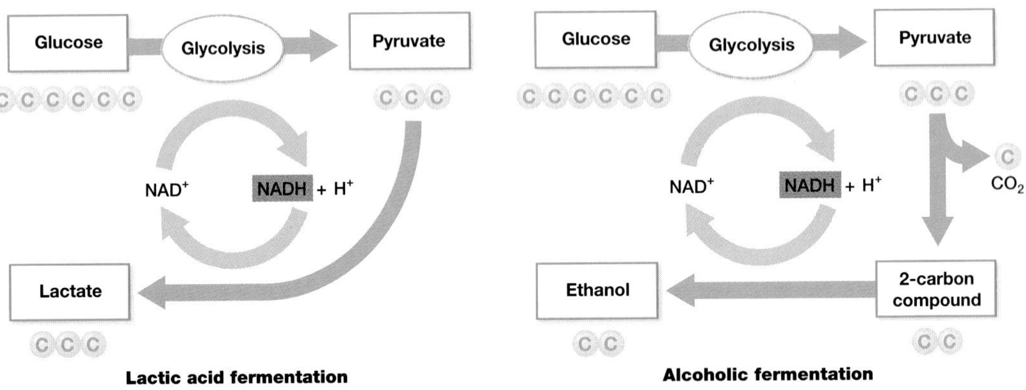

Biology in Your Community

Exercise

Invite one of the physical education teachers from your school to discuss the physiological effects of exercise on the body. Be certain that the speaker discusses oxygen debt, muscle fatigue, the role of myoglobin in muscles, and the role of lactate in muscle soreness.

Alcoholic fermentation

In other organisms, the three-carbon pyruvate is broken down to ethanol (ethyl alcohol), a two-carbon compound, through alcoholic fermentation. Carbon dioxide is released during the process. As shown in Figure 5-15, alcoholic fermentation is a two-step process. First, pyruvate is converted to a two-carbon compound, releasing carbon dioxide. Second, electrons are transferred from a molecule of NADH to the two-carbon compound, producing ethanol. As in lactic acid fermentation, NAD^+ is recycled so glycolysis can continue to produce ATP.

For centuries, alcoholic fermentation by yeast, a fungus, has been used in the preparation of many foods and beverages. Wine and beer contain ethanol made during alcoholic fermentation by yeast. Carbon dioxide released by the yeast causes the rising of bread dough and the carbonation of some alcoholic beverages, such as beer. Ethanol is actually toxic to yeast. Ethanol kills yeast at a concentration of about 12 percent. Thus, naturally fermented wine contains about 12 percent ethanol.

internetconnect

SCiLINKS. NSTA

TOPIC: Fermentation
GO TO: www.scilinks.org
KEYWORD: HX108

HEALTH WATCH

How Carbon Monoxide Kills

One function of the blood is to carry oxygen from your lungs to your body's cells. However, only about 1.5 percent of the oxygen in your blood is dissolved in the liquid portion of blood, called plasma. The remaining 98.5 percent of oxygen travels with red blood cells, bound to molecules of hemoglobin.

Carbon monoxide

Carbon monoxide, CO, is an odorless, colorless gas produced when a fuel, such as gasoline or wood, is burned without enough oxygen. If carbon monoxide is inhaled, it competes with oxygen for binding sites on hemoglobin. In fact, hemoglobin binds to carbon monoxide more than 200 times more strongly than it binds to oxygen. Thus, even at very low concentrations, carbon monoxide can displace oxygen from hemoglobin, severely reducing the amount of oxygen the blood can carry. A condition known as carbon monoxide poisoning results. Each year, nearly 5,000 people in the United States are treated for carbon monoxide poisoning, and about 300 people die from the condition. Carbon monoxide poisoning is the leading cause of death from fire.

Symptoms of carbon monoxide poisoning

People with low levels of carbon monoxide in their blood typically experience headaches, nausea, drowsiness, and vomiting. Higher levels of carbon monoxide may cause disorientation, unconsciousness, and even death. Most symptoms of carbon monoxide poisoning result from depriving the brain of oxygen; oxygen deprivation may cause cells in the brain to die.

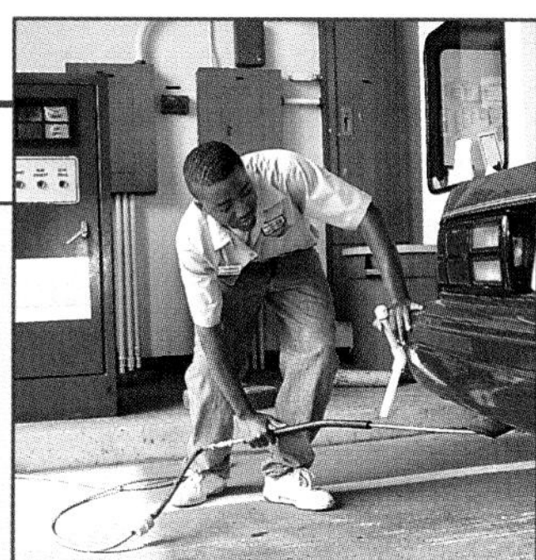

Automobile emissions test

Effect on cellular respiration

Without oxygen to accept electrons from the mitochondrial electron transport chains, aerobic respiration cannot occur. As a result, brain cells must rely on glycolysis alone to produce ATP. Glycolysis is much less efficient than aerobic respiration, and the cells cannot make enough ATP to meet their needs.

internetconnect

SCiLINKS. NSTA

TOPIC: Carbon monoxide
GO TO: www.scilinks.org
KEYWORD: HX109

Demonstration

Alcoholic Fermentation

Thoroughly mix half a packet of yeast with approximately 400 mL of warm water. Pour the mixture into an Erlenmeyer flask or a beaker. Add a few drops of bromothymol blue indicator. Set the flask aside. Have students note the formation of bubbles. Ask students what the bubbles indicate. *(Carbon dioxide is being released.)* By the end of the class period, the blue yeast mixture should turn yellow. Ask students why this happens. *(The indicator turns yellow in the presence of an acid; when carbon dioxide is produced by the yeast, carbonic acid is formed in the water.)*

HEALTH WATCH

How Carbon Monoxide Kills

Teaching Strategies

Emphasize the role of oxygen in the electron transport chain. Ask students why hemoglobin is so important to cells. *(It is the molecule on which oxygen is transported to body tissues by red blood cells.)*

Discussion

- How does carbon monoxide affect the level of oxygen in the blood? *(The oxygen level is severly reduced because oxygen cannot bind to hemoglobin on red blood cells.)*
- How does carbon monoxide poisoning cause oxygen deprivation? *(Oxygen deprivation results because oxygen cannot be delivered to cells.)*

Biology in Your Community

Fermentation

Arrange a field trip to visit a winery, brewery, bakery, or cheese-making facility to show students commercial applications of fermentation. Have students prepare a list of questions such as the following: What are the starting materials for the commercial processes? How do changes in conditions, such as temperature and pH, affect the finished product?

Teaching TIP

ATP Production

Tell students that up to 36 molecules of ATP can be produced during aerobic respiration. Have students calculate how much more efficient aerobic respiration is (at its maximum ATP production capacity) than glycolysis, which yields two ATP molecules. *(36 ATP÷2 ATP = 18 times more efficient)*

3 Assessment Options

Closure

Technique: K-W-L

Refer to the **Active Reading** on page 93. Have students review their lists and determine if they Learned everything on the lists. If they did not, open a class discussion to answer the questions.

Review

Assign Review—Section 5-3.

Reteaching — BASIC

Assign students to cooperative pairs. Have each pair evaluate the following scenario: Suppose you are an organism that can carry out either aerobic respiration or anaerobic processes. Which one would be more beneficial to you and why? *(If oxygen is present, aerobic respiration is more beneficial because more ATP can be produced.)*

Figure 5-16 Effect of oxygen on ATP production

Most ATP is produced during aerobic respiration.

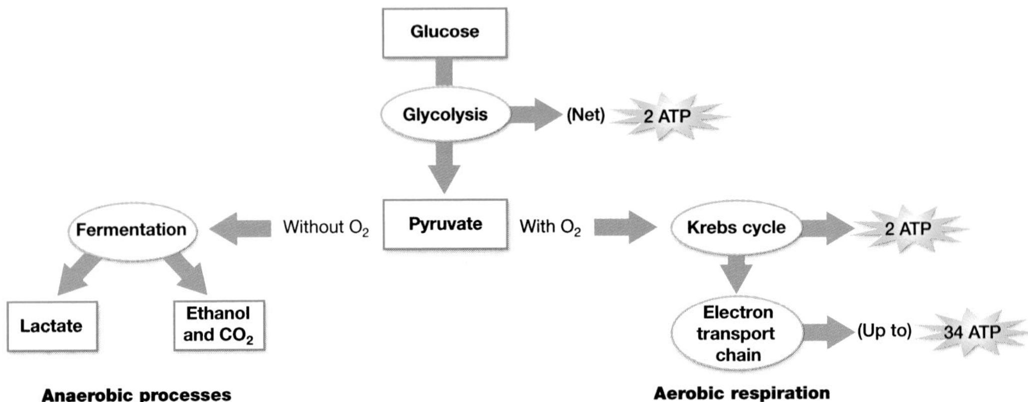

Comparing anaerobic processes with aerobic respiration

The total amount of ATP that a cell is able to harvest from each glucose molecule that enters glycolysis depends on the presence or absence of oxygen. As shown in **Figure 5-16,** cells use energy most efficiently when oxygen is present. In the first stage of cellular respiration, glucose is broken down to pyruvate during glycolysis. Glycolysis is an anaerobic process. There is a net gain of two ATP molecules during glycolysis. In the second stage of cellular respiration, the pyruvate passes through either aerobic respiration or (anaerobic) fermentation. When oxygen is present, it is aerobic respiration that occurs. When oxygen is not present, fermentation occurs instead. The NAD^+ that gets recycled during fermentation enables glycolysis to continue producing ATP. Thus, a small amount of ATP is produced even during fermentation. However, most of a cell's ATP is made during aerobic respiration. For each molecule of glucose that is broken down, as many as two ATP molecules are made directly during the Krebs cycle, and up to 34 ATP molecules are produced later by the electron transport chain.

Review SECTION 5-3

1. **List** the products of glycolysis. What is the role of each of these products in cellular respiration?
2. **Summarize** the roles of the Krebs cycle and the electron transport chain during aerobic respiration.
3. **Describe** the role of fermentation in the second stage of cellular respiration.
4. **Distinguish** between lactic acid fermentation and alcoholic fermentation.
5. **SKILL Comparing Functions** Explain why cellular respiration is more efficient when oxygen is present in cells.

Critical Thinking

6. **SKILL Inferring Conclusions** Excess glucose in your blood is stored in your liver as glycogen. How do you suppose your body "knows" when to convert glucose to glycogen and glycogen back to glucose?

Review SECTION 5-3 ANSWERS

1. Pyruvate: if oxygen is present, pyruvate will enter the Krebs cycle; if oxygen is absent, pyruvate will undergo fermentation. NADH: if oxygen is present, NADH will enter the electron transport chain. ATP: temporarily stores energy.
2. The Krebs cycle produces electron carriers that donate electrons to the electron transport chain. The electron transport chain produces most of the ATP that is produced in cellular respiration.
3. Fermentation recycles NAD^+, which is needed to continue ATP production in the absence of oxygen.
4. In lactic acid fermentation, pyruvate is converted to lactate. In alcoholic fermentation, pyruvate is broken down to ethanol, releasing carbon dioxide.
5. If oxygen is present, aerobic respiration can occur. Much more ATP can be produced in aerobic respiration than in anaerobic processes.
6. Sensors in the body monitor the level of glucose in the blood. When the blood glucose level is high, the storage of glycogen is stimulated. When the blood glucose level is low, glucose is released into the blood.

Study ZONE CHAPTER 5

Use the Key Concepts and Key Terms listed below to review the main ideas in this chapter. If you need to review the meaning of a term, turn to the page indicated.

Key Concepts

5-1 Energy and Living Things

- Energy from sunlight flows through living systems, from autotrophs to heterotrophs.
- Photosynthesis and cellular respiration form a cycle because one process uses the products of the other.
- ATP supplies cells with energy needed for metabolism.

5-2 Photosynthesis

- Photosynthesis has three stages. First, energy is captured from sunlight. Second, energy is temporarily stored in ATP and NADPH. Third, organic compounds are made using ATP, NADPH, and carbon dioxide.
- Pigments absorb light energy during photosynthesis.
- Electrons excited by light travel through electron transport chains, in which ATP and NADPH are produced.
- Through carbon dioxide fixation, often by the Calvin cycle, carbon dioxide in the atmosphere is used to make organic compounds, which store energy.
- Photosynthesis is directly affected by environmental factors such as the intensity of light, the concentration of carbon dioxide, and temperature.

5-3 Cellular Respiration

- Cellular respiration has two stages. First, glucose is broken down to pyruvate during glycolysis, making some ATP. Second, a large amount of ATP is made during aerobic respiration. When oxygen is not present, NAD^+ is recycled during the anaerobic process of fermentation.
- The Krebs cycle is a series of reactions that produce energy-storing molecules during aerobic respiration.
- During aerobic respiration, large amounts of ATP are made in an electron transport chain.
- When oxygen is not present, fermentation follows glycolysis, regenerating NAD^+ needed for glycolysis to continue.

Key Terms

5-1

photosynthesis (94)
autotroph (94)
heterotroph (95)
cellular respiration (95)

5-2

pigment (98)
chlorophyll (98)
carotenoid (98)
thylakoid (99)
electron transport chain (100)
NADPH (101)
carbon dioxide fixation (102)
Calvin cycle (102)

5-3

aerobic (104)
anaerobic (104)
glycolysis (105)
NADH (105)
Krebs cycle (106)
$FADH_2$ (107)
fermentation (108)

Unit 2, Unit 3—Use Topics 1–6 in these units to review the key concepts and terms in this chapter.

Study TIP Ready?

If you think you understand these Key Concepts and Key Terms, then you're ready for the Performance Zone.

Answer to Concept Map

The following is one possible answer to Performance Zone item 1 on page 112.

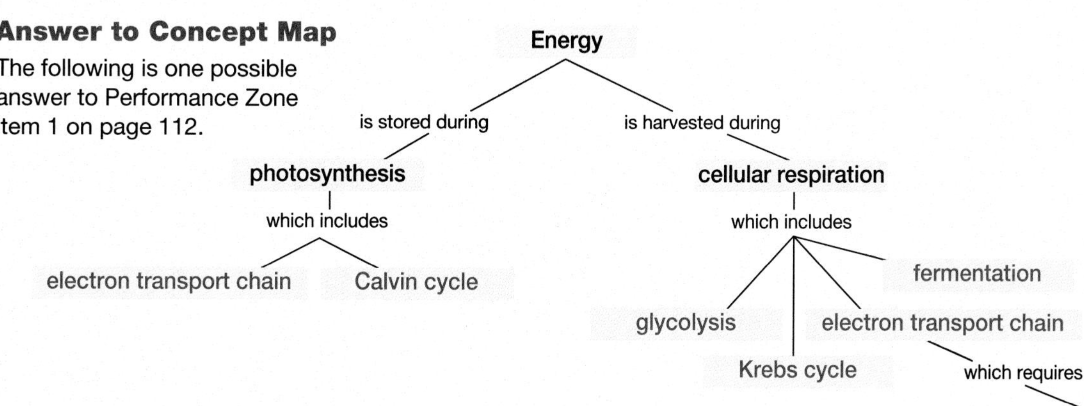

Study ZONE CHAPTER 5

Review and Practice

Assign the following worksheets for Chapter 5:

- **Student Review Guide**
- **Concept Mapping Worksheet**
- **Test Preparation Pretest**
- **Vocabulary Worksheet**
- **Critical Thinking Review Worksheet**
- **Science Skills Worksheet**

Assessment Options

The following assessment options are available for this chapter:

Resource Link

Assign **Chapter Test 5** to assess students' understanding of the chapter.

Alternative Assessment

Have each student write a question on an index card based on the information presented in this chapter. Each student should also record an answer to the question on a second index card. Have students trade question cards with a partner and try to answer their partner's question. Then have them confer with each other about their answers. Encourage students to refer to the textbook to settle any disagreements.

Portfolio Assessment

- *Portfolio Assessment* at the front of this book provides options for building and evaluating student portfolios. Students and teachers can select items from the areas listed for inclusion in a portfolio.
- See items labeled PORTFOLIO on pp. 96, 98, 99, 101, 102, 106, and 108.

Performance ZONE
CHAPTER 5
REVIEW ANSWERS

Assignment Guide

SECTION	REVIEW
5-1	2, 3, 13, 22
5-2	1, 2, 4, 5, 6, 7, 8, 12, 16, 17, 18, 21, 24
5-3	1, 2, 9, 10, 11, 14, 15, 19, 20, 23, 24, 25

Understanding and Applying Concepts

1. The answer to the concept map is found at the bottom of page 111.
2. **a.** An autotroph uses the energy in sunlight or inorganic substances to make organic compounds. A heterotroph must eat food to get energy.

   b. Glycolysis is the first stage of cellular respiration. When oxygen is not present, fermentation recycles NAD^+, allowing glycolysis to continue.

   c. Chlorophyll is the primary pigment of photosynthesis. Chlorophyll reflects green light and absorbs blue and red light. Carotenoids are pigments that reflect yellow and orange light and absorb mostly blue and green light.

   d. The term *aerobic* refers to a metabolic process that requires oxygen. The term *anaerobic* refers to a process that does not require oxygen.
3. c
4. a
5. c
6. d
7. d
8. b
9. a
10. a
11. c
12. The structures are thylakoids. During photosynthesis, hydrogen ions are more concentrated inside thylakoids.

Performance ZONE CHAPTER 5 Review

Understanding and Applying Concepts

1. **Concept Mapping** Make a concept map that shows how photosynthesis and cellular respiration are related. Try to include the following terms in your map: glycolysis, Krebs cycle, electron transport chain, Calvin cycle, fermentation, NADH.

2. **Understanding Vocabulary** For each pair of terms, explain the difference in their meanings.
   a. autotroph, heterotroph
   b. glycolysis, fermentation
   c. chlorophyll, carotenoid
   d. aerobic, anaerobic

3. Energy flows through living systems from
   a. the sun to heterotrophs and then to autotrophs.
   b. autotrophs to the environment and then to heterotrophs.
   c. the sun to autotrophs and then to heterotrophs.
   d. the environment to heterotrophs and then to autotrophs.

4. The products of photosynthesis that begin cellular respiration are
   a. organic compounds and oxygen.
   b. carbon dioxide and water.
   c. $NADP^+$ and hydrogen.
   d. ATP and water.

5. Carotenoids are pigments that
   a. cause plants to look green.
   b. are carried by NADPH.
   c. trap light that chlorophyll cannot absorb.
   d. do not play a role in photosynthesis.

6. The thylakoid membranes of a chloroplast are the sites where
   a. electron transport chains operate.
   b. NADPH and ATP are produced.
   c. pigments are located.
   d. All of the above

7. The electron transport chains of photosynthesis produce
   a. pyruvate. c. glucose.
   b. water. d. ATP and NADPH.

8. The oxygen produced during photosynthesis comes directly from the
   a. splitting of carbon dioxide molecules.
   b. splitting of water molecules.
   c. mitochondrial membranes.
   d. absorption of light.

9. Which of the following is the correct pairing of a process and its need for oxygen?
   a. glycolysis: no oxygen required
   b. fermentation: oxygen required
   c. Krebs cycle: no oxygen required
   d. none of the above

10. Aerobic respiration involves all of the following *except*
    a. glycolysis. c. mitochondria.
    b. the Krebs cycle. d. ATP.

11. Most of the ATP made during cellular respiration is produced in
    a. glycolysis. c. mitochondria.
    b. the Krebs cycle. d. fermentation.

12. Study the micrograph of a chloroplast shown below. Identify the structures labeled *X* in the micrograph. During photosynthesis, are hydrogen ions more concentrated in these structures or in the spaces surrounding them?

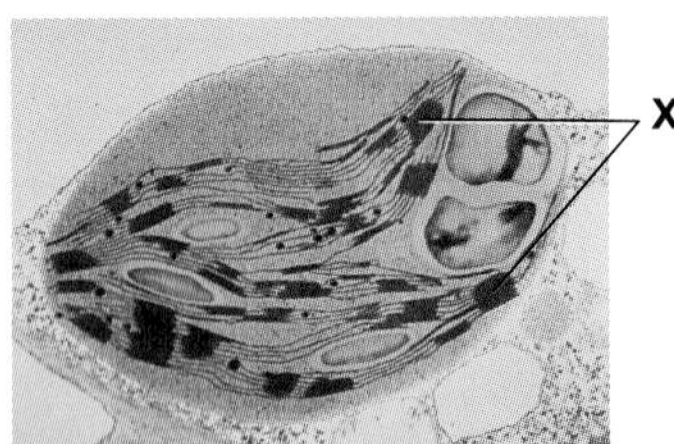

13. **Theme Metabolism** Describe the flow of energy that enables you to get energy from the food you eat.

14. **Health Watch** How does carbon monoxide affect the level of oxygen in the blood?

15. **Chapter Links** How is starch broken down to glucose prior to glycolysis? (**Hint:** See Chapter 2, Section 2-4.)

13. Autotrophs, such as plants, get energy from the sun. Heterotrophs, such as humans, eat the plants, getting some of that energy. Humans can also eat other heterotrophs that eat plants, such as cows.
14. Carbon monoxide causes the level of oxygen in the blood to decrease.
15. Starch is composed of hundreds of glucose molecules. It is broken down by enzymes, such as amylase, yielding maltose. Maltose, a disaccharide that consists of two glucose molecules, is then broken down by the enzyme maltase, yielding glucose.

Skills Assessment

Part 1: Interpreting Graphs

The graph below shows the effect of temperature on the rate of photosynthesis. Study the graph, and then answer the following questions:

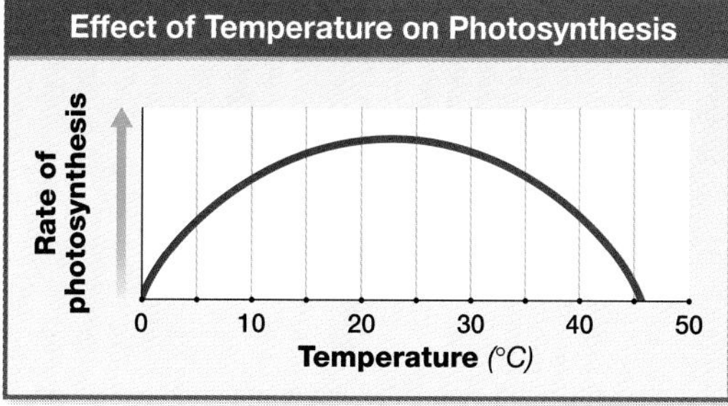

16. What is the effect of high or low temperatures on the rate of photosynthesis?

17. What is the optimum temperature range for photosynthesis?

Part 2: Life/Work Skills

Use the media center or Internet resources to learn more about cell metabolism.

18. **Finding Information** Scientists know a lot about the chemical processes of photosynthesis. Prior to the seventeenth century, however, scientists knew very little about how plants make organic compounds. Research the discoveries that led to the understanding of photosynthesis. Look up the contributions made by the English chemist Joseph Priestley, the Dutch physician Jan Ingenhousz, the Swiss chemist Nicolas de Saussure, the German physicist Julius Robert von Mayer, and the American biochemist Melvin Calvin. Summarize your findings in a written report.

19. **Analyzing Methods** Research several ways that fermentation is used in food preparation. Find out what kinds of microorganisms are used in cultured dairy products, such as yogurt, sour cream, and some cheeses. Research the role of alcoholic fermentation by yeast in bread making. Prepare an oral report to summarize your findings.

Critical Thinking

20. **Distinguishing Relevant Information** The enzyme that aids in the conversion of pyruvate to acetyl-CoA requires vitamin B_1, also called thiamine. Thiamine is not made in the human body. From this information, what can you infer about the nutritional requirements of humans? How would a deficiency of thiamine in cells affect cellular respiration?

21. **Evaluating Viewpoints** State whether you think the following viewpoint can be supported, and justify your answer. "If Earth's early atmosphere had been rich in oxygen, photosynthetic organisms would not have been able to evolve."

Portfolio Projects

22. **Recognizing Relationships** Research the following martial arts: tai chi, tae kwan do, and aikido. What do they have in common? How do they relate to the flow of energy? Where and how did each originate? Report your findings to your class.

23. **Interpreting Information** Read the article titled "Making Calories Count" (*Newsweek* Special Edition, Spring/Summer 1999, pp. 88–89). What are the potential benefits and shortcomings of the special diets discussed in the article?

24. **Career Focus Enzymologist** Research the educational background necessary to become an enzymologist. List the courses required and describe additional degrees or training that are recommended for this career. Write a report on your findings.

25. **Unit 2—*Photosynthesis***
Unit 3—*Cellular Respiration*

Write a report summarizing how exercise physiologists regulate the diet and training of athletes. Find out how diet varies according to the needs of each athlete. Research the relationship between exercise and metabolism.

Skills Assessment

16. The rate of photosynthesis decreases at both high and low temperatures.
17. The optimum temperature range for photosynthesis is 22–23°C.
18. Answers will vary. Priestley discovered oxygen and showed that plants "restore" air from which oxygen has been removed. Ingenhousz demonstrated that plants must have light to restore air. De Saussure developed the first balanced equation for photosynthesis. Von Mayer helped formulate the law of conservation of energy. Calvin traced the path of carbon dioxide in photosynthesis.
19. Answers will vary. Bacteria are used to make yogurt, sour cream, and some cheeses. Other cheeses are made with the help of fungi. During bread making, alcoholic fermentation by yeast produces alcohol, which evaporates, and carbon dioxide, which makes the bread rise.

Performance ZONE

CHAPTER 5

Critical Thinking

20. One can infer that humans must obtain thiamine from the food they eat. Insufficient thiamine in cells may decrease the efficiency of the conversion of pyruvate to acetyl-CoA in the Krebs cycle, possibly impairing aerobic respiration.
21. Answers will vary. However, students should acknowledge that photosynthetic organisms can carry out photosynthesis in the presence of oxygen.

Portfolio Projects

22. Answers will vary. Tai-chi is a Chinese martial art that may have originated around the beginning of the fifteenth century. Its practitioners believe it can enhance the flow of *qi*, or "life force." Tae kwon do is a Korean martial art that was founded in 1955. Aikido is a Japanese martial art that was developed in 1925. It stresses the harmony between *ki* (energy) and *tai* (the body).
23. Shortcomings may include malnutrition and side effects from prescription or over-the-counter weight-loss drugs.
24. Answers will vary. Enzymologists study the structure and function of enzymes and the effects of enzyme deficiencies. The career requires bachelor's and advanced degrees in chemistry or biology. Employers include universities and companies such as chemical and pharmaceutical manufacturers. Growth prospects are good. Starting salary will vary by region.
25. Answers will vary. The diet of an athlete depends on the energy requirements of the athlete's sport. Some sports, such as weight lifting, involve mainly anaerobic metabolism. Others, such as jogging and swimming, involve aerobic respiration.

Photosynthesis

Purpose

Students will explore how the intensity and wavelength of light affect the rate of ATP production during photosynthesis.

Preparation Notes

Time Required: 45 minutes

Preparation Tips

- Depending on how many computers are available, set up work stations that enable students to work individually or in teams of two to four.
- Encourage students to read the lab and complete the **Before You Begin** section before they come to class on the day of the lab.

Before students begin the lab, ask them the following questions:

- What is chlorophyll? *(Chlorophyll is a pigment found in plants and other photosynthetic organisms that absorbs some wavelengths of light but reflects green light.)*
- What is a chloroplast, and what function does it perform in a plant cell? *(A chloroplast is an organelle with a complex series of internal membranes. Chloroplasts are the site of photosynthesis.)*
- Describe the main steps of photosynthesis. *(Light energy is captured by pigments in the chloroplasts. This energy is used to make ATP and NADPH, which are then used along with carbon dioxide to make organic compounds.)*
- Write the equation used to summarize photosynthesis. *($3CO_2 + 3H_2O \rightarrow C_3H_6O_3 + 3O_2$)*

Procedural Tips

The exploration can be run at two speeds. Selecting *Slow Animation* from the *Speed* menu on the toolbar will allow students to follow the events occurring within the chloroplast.

Interactive Exploration

Chapter 5 *Photosynthesis*

CD-ROM Lab

SKILL

- Computer modeling

OBJECTIVES

- **Simulate** how the rate of photosynthesis is affected by variations in the intensity and wavelength of light.
- **Relate** the rate of photosynthesis to the production of ATP in a chloroplast.

MATERIALS

- computer with CD-ROM drive
- CD-ROM *Interactive Explorations in Biology: Cell Biology and Genetics*
- graph paper

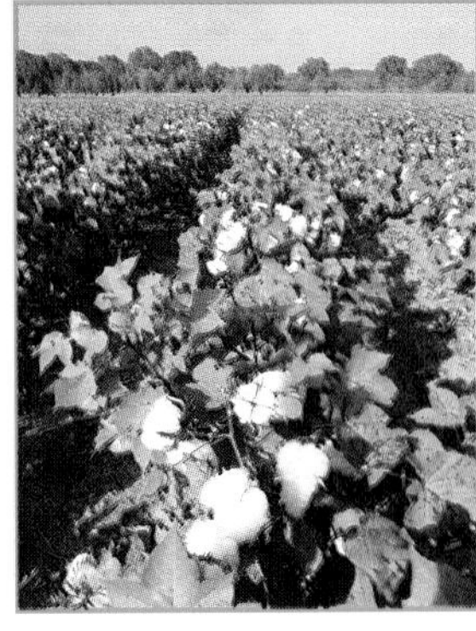

Before You Begin

Photosynthesis is the process by which plants, some bacteria, and some protists use energy from sunlight to make organic compounds. First, light energy is captured. Second, light energy is used to make ATP and NADPH. Third, ATP and NADPH are then used to make organic compounds. This Interactive Exploration enables you to explore how the intensity and wavelength of light affect the rate of ATP production during photosynthesis.

1. Select *Photosynthesis* from the menu. You will see a diagram like the one below. Click *Navigation,* and then click the key on the navigation palette. Read the focus questions, and review the following concepts: pigments, photons, and the action spectrum of photosynthesis.
2. Click *Help* at the top left of the screen, and select *How to Use This Exploration*. Listen to the instructions. Click the magnifying glass on the navigation palette to begin the exploration.

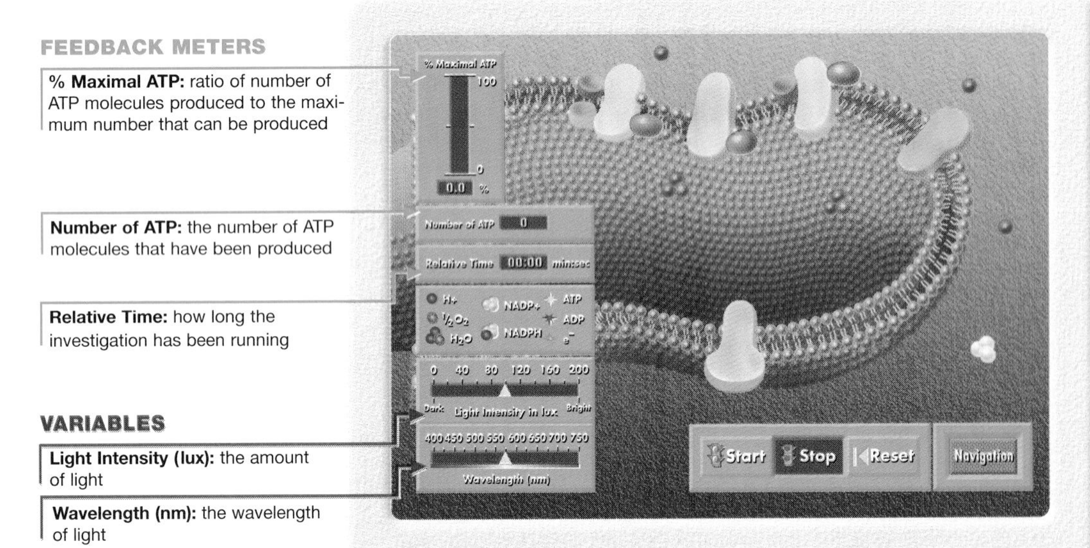
FEEDBACK METERS
% Maximal ATP: ratio of number of ATP molecules produced to the maximum number that can be produced
Number of ATP: the number of ATP molecules that have been produced
Relative Time: how long the investigation has been running
VARIABLES
Light Intensity (lux): the amount of light
Wavelength (nm): the wavelength of light
Start
Stop
Reset
Navigation

Additional Background

In the light-dependent reactions of photosynthesis, chlorophyll and other photosynthetic pigments absorb light, and electrons in these pigment molecules are excited to higher energy levels. These excited electrons then pass through electron transport chains and are ultimately used to make NADPH. ATP is also produced during the light-dependent reactions. ATP and NADPH are transferred to the Calvin cycle, in which carbon dioxide is incorporated into organic compounds. This exploration presents a cross section of a thylakoid membrane—the site of the light-dependent reactions—within the chloroplast of a cell. Students will manipulate the wavelength and intensity of light that falls on this chloroplast and observe the effects on the rate of ATP production.

Procedure

Answers

3. ATP is made at the carrier protein.

3. Make two data tables like the ones below.

DATA TABLE A

Light intensity (lux)	% Maximal ATP	Number of ATP
0		
40		
80		

DATA TABLE B

Wavelength (nm)	% Maximal ATP	Number of ATP
400		
450		
500		

Procedure

PART A: Effect of Light Intensity

1. Click *Speed* at the top right of the screen, and then select *Slow*.
2. Click and slide the wavelength indicator to 650 nm. Click and slide the light intensity indicator to 40 lux.
3. Click the *Start* button. Observe the carrier protein at the bottom of the screen. Which molecule is made at this carrier protein?
4. Allow the simulation to run for 30 seconds, as indicated on the *Relative Time* meter, and then click the *Stop* button. In Data Table A, record the percent maximal ATP and the number of ATP molecules that were produced. Note below your table that the wavelength of light was 650 nm.
5. Click the *Reset* button. Then click and slide the light intensity indicator to another setting. Repeat step 4. Continue testing until you have recorded data for 0, 40, 80, 120, 160, and 200 lux.

PART B: Effect of Wavelength

6. Click the *Reset* button. Then click and slide the light intensity indicator to 200 lux. Slide the wavelength indicator to 400 nm.
7. Click the *Start* button. Allow the simulation to run for 30 seconds, as indicated on the *Relative Time* meter, and then click the *Stop* button. In Data Table B, record the percent maximal ATP and number of ATP molecules produced. Note below your table that the light intensity was 200 lux.
8. Click the *Reset* button. Then click and slide the wavelength indicator to another setting. Repeat step 4. Continue testing until you have recorded data for 400, 450, 500, 550, 600, 650, 700, and 750 nm.

Analyze and Conclude

1. **Analyzing Methods** Why was one of the variables constant in Part A and Part B?
2. **Organizing Data** Make a bar graph of the data in Data Table A. Plot *% Maximal ATP* on the *y*-axis and *Light intensity (lux)* on the *x*-axis. Above each bar, indicate the number of ATP molecules produced.
3. **Inferring Conclusions** How does light intensity affect the rate of photosynthesis?
4. **Organizing Data** Make a bar graph of the data in Data Table B. Plot *% Maximal ATP* on the *y*-axis and *Wavelength of light (nm)* on the *x*-axis. Above each bar, indicate the number of ATP molecules produced.
5. **Analyzing Data** Which wavelengths of light are most effective for photosynthesis, and which wavelengths are least effective?
6. **Predicting Outcomes** Which combination of light intensity and wavelength would cause the highest possible rate of photosynthesis?
7. **Relating Concepts** What other factors affect the rate of photosynthesis?

Use the following Internet resources to explore your own questions about the process of photosynthesis.

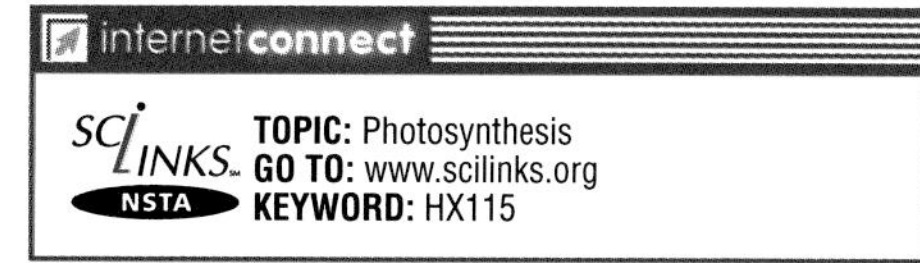

Interactive Exploration

Analyze and Conclude

Answers

1. In Part A, the wavelength was kept constant so that the effect of varying the light intensity could be evaluated. In Part B, the intensity of light was kept constant so the effect of varying the wavelength could be evaluated.
2. **Sample Graph**

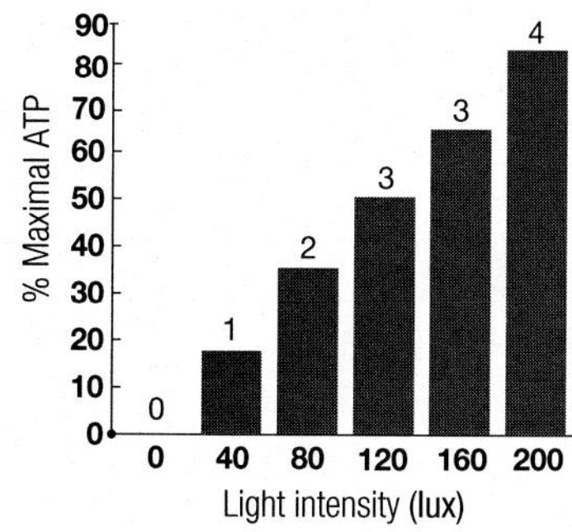

3. The rate of photosynthesis increases linearly with light intensity, assuming that the rate of photosynthesis is proportional to the rate of ATP production.
4. **Sample Graph**

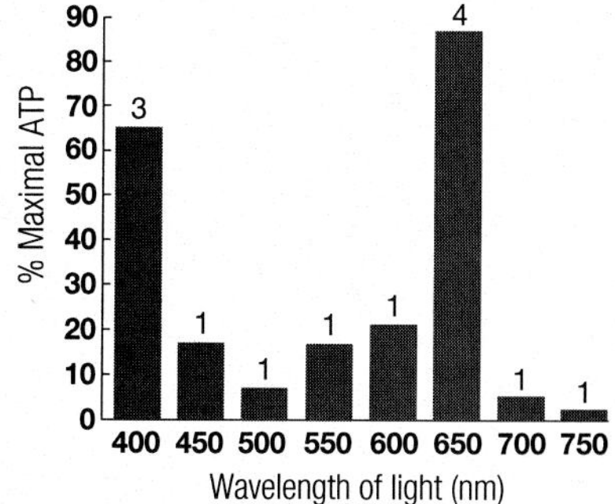

5. The most effective wavelengths are 400 nm and 650 nm. The least effective wavelength is 750 nm.
6. 650 nm at 200 lux
7. Temperature and carbon dioxide concentration also affect the rate of photosynthesis.

4. Sample Data Table

LIGHT INTENSITY (LUX)	% MAXIMAL ATP	NUMBER OF ATP
0	0	0
40	17	1
80	34	2
120	51	3
160	68	3
200	85	4

8. Sample Data Table

WAVELENGTH (NM)	% MAXIMAL ATP	NUMBER OF ATP
400	65	3
450	15	1
500	7.5	1
550	15	1
600	20	1
650	85	4
700	5	1
750	3	1

Chapter 6

Chromosomes and Cell Reproduction

Planning Guide

OBJECTIVES	MEETING INDIVIDUAL NEEDS	READING SKILLS AND STRATEGIES
6-1 Chromosomes (BLOCK 1 45 min) • Relate binary fission to cell division in prokaryotes. • Differentiate between a gene, a DNA molecule, a chromosome, and a chromatid. • Compare haploid and diploid cells. • Differentiate between homologous chromosomes, autosomes, and sex chromosomes. • Predict how changes in chromosome number or structure can affect development.	Learning Styles, p. 118 BASIC ADVANCED • pp. 119, 122, 124	ATE Active Reading, Technique: Anticipation Guide, p. 117 ATE Active Reading, Technique: Reading Organizer, p. 119 ■ Active Reading Guide, worksheet (6-1) ■ Directed Reading, worksheet (6-1) ■ Reading Strategies, Compare/Contrast Matrix
6-2 The Cell Cycle (BLOCK 2 45 min) • Identify the major events that characterize each of the five phases of the cell cycle. • Describe how the cell cycle is controlled in eukaryotic cells. • Differentiate between normal cells and cancer cells.	Learning Styles, p. 125 BASIC ADVANCED • p. 127	■ Active Reading Guide, worksheet (6-2) ■ Directed Reading, worksheet (6-2) ■ Reading Strategies, Recognizing Cycles
6-3 Mitosis and Cytokinesis (BLOCKS 3 & 4 90 min) • Describe the structure and function of the spindle during mitosis. • Summarize the events of the four stages of mitosis. • Differentiate cytokinesis in animal and plant cells.	Learning Styles, p. 128 BASIC ADVANCED • pp. 128, 131, 132	■ Active Reading Guide, worksheet (6-3) ■ Directed Reading, worksheet (6-3) ■ Reading Strategies, Series-of-Events Chain ■ Supplemental Reading Guide, *The Lives of a Cell: Notes of a Biology Watcher*

Review and Assessment Resources

BLOCKS 5 & 6 90 min

TEST PREP

- PE Study Zone, p. 133
- PE Performance Zone, pp. 134–135
- ■ Chapter 6 worksheets:
  - Vocabulary
  - Concept Mapping
  - Science Skills
  - Critical Thinking

 Holt Biology Interactive Tutor Unit 4—Cell Reproduction, Topics 1–4

 Audio CD Program

■ Test Prep Pretest

CHAPTER TESTING

- ■ Chapter Tests (blackline master)
- ▲ Test Generator
- ■ Assessment Item Listing

See the correlations table at the front of this book for a list of the ***National Science Education Standards*** addressed in this chapter.

Smithsonian Institution® Visit **www.si.edu/hrw** for additional on-line resources.

logy Principles
Explorations

Pupil's Edition
Teacher's Edition

HOLT BIOSOURCES

 Teaching Transparencies
  Laboratory Program

 **One-Stop Planner CD-ROM**
Shows these resources within a customizable daily lesson plan:

■ Teaching Resources
▲ Test Generator

VISUAL STRATEGIES	CLASSWORK AND HOMEWORK	TECHNOLOGY AND INTERNET RESOURCES
ATE Fig. 6-3, p. 120 14 Chromosome and DNA Structure	**PE** Quick Lab What are chromosomal mutations?, p. 124 **PE** Review, p. 124	Interactive Tutor Unit 4—Cell Reproduction (Topic 3) Holt Biology Videodiscs Section 6-1 Audio CD Program Section 6-1 internetconnect SCILINKS NSTA TOPIC: Chromosomes
ATE Fig. 6-7, p. 126	**PE** Review, p. 127	Interactive Tutor Unit 4—Cell Reproduction (Topic 2) Holt Biology Videodiscs Section 4-2 Audio CD Program Section 4-2 internetconnect SCILINKS NSTA TOPICS: • Cancer cells • Cell cycle
ATE Fig. 6-9, p. 130 **ATE** Fig. 6-10, p. 131 24 Mitosis	**PE** Math Lab Calculating the number of cells resulting from mitosis, p. 129 **PE** Quick Lab What do mitosis and cytokinesis look like?, p. 132 **PE** Experimental Design How Can You Model Mitosis?, pp. 136–137 Lab B5 Mitosis **PE** Review, p.132 **PE** Study Zone, p. 133	CNN Video Segment 3 Split Liver Donor Interactive Tutor Unit 4—Cell Reproduction (Topics 1 and 4) Holt Biology Videodiscs Section 6-3 Audio CD Program Section 6-3

ALTERNATIVE ASSESSMENT

PE Portfolio Projects 24–27, p. 135
ATE Alternative Assessment, p. 133
ATE Portfolio Assessment, p. 133
■ Portfolio Projects: Genetics Project

Lab Activity Materials

Quick Lab
p. 124 pencil, paper, notecard pieces (14), tape

Math Lab
p. 129 pencil, paper

Experimental Design
p. 136 pipe cleaners of at least two different colors, yarn, wooden beads, white labels, scissors

Lesson Warm-up
ATE photo of person with a visible chromosomal disorder, photo of a normal person, p. 118

ATE paper, pencil, p. 125

Demonstrations
ATE microprojector, cross section slide of onion root tip, p. 116
ATE long piece of yarn, scissors, paper clip, p. 119
ATE models or photos of a human brain and a human bone, p. 126
ATE football, p. 129

BIOSOURCES LAB PROGRAM
• Inquiry Skills Development: Lab B5, p. 17

Chapter 6

Chapter Theme

Cellular Structure and Function

The task of passing on genetic information to developing cells and to future generations lies with the chromosomes. Because of the structure of the chromosomes, a vast amount of information is passed on with few errors to newly created cells. The condensed nature of chromosomes allows DNA to be packed into a compact form that permits cells to manipulate the chromosomes during mitosis.

Establishing Prior Knowledge

Before beginning this chapter, make certain students can answer the questions in Study TIP **Ready?** on page 117. If they cannot, encourage them to reread the sections indicated.

Opening Demonstration

Use a microprojector or microscope with a video camera to show the class a cross section of an onion root tip. The cells in the tip undergo repeated mitotic divisions. Point out chromosomes that are visible. Tell students that before each root tip cell divides to enable the tip to grow, the chromosomes are replicated and then divided equally into the two resulting cells. Explain that chromosomes consist of genes, which are segments of DNA.

CHAPTER 6 Chromosomes and Cell Reproduction

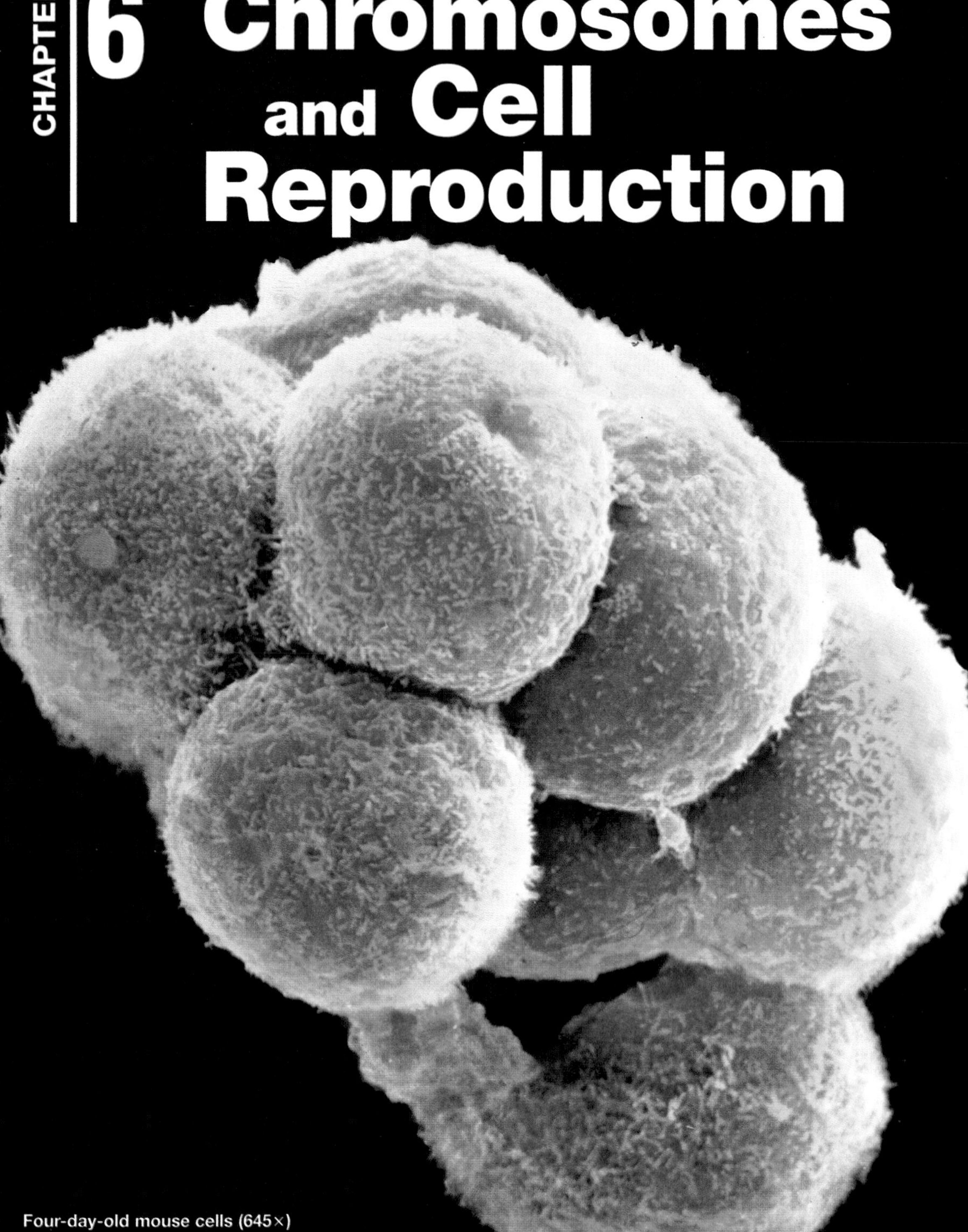

Four-day-old mouse cells (645×)

What's Ahead — and Why

CHAPTER OUTLINE

6-1 Chromosomes
- New Cells Are Formed by Cell Division
- Chromosome Number and Structure Affect Development

6-2 The Cell Cycle
- The Cell Cycle Describes the Life of a Eukaryotic Cell
- The Cell Cycle Is Carefully Controlled

6-3 Mitosis and Cytokinesis
- In Mitosis, Chromatids Are Pulled by Microtubules
- Mitosis and Cytokinesis Divide Cells

Students will learn

- ▷ why cells divide and about chromosome number and structure in different cells.
- ▷ how the growth and division of cells are regulated and that cancer can result if they are not.
- ▷ how duplicate sets of chromosomes are passed to new cells during cell division.

Mouse with young

Four days old and smaller than the head of a pin...

At 4 days old, the cell cluster at left is slightly smaller than the head of a pin. During the next 17 days, these cells will divide repeatedly to form a newborn mouse. Chromosomes inside each cell carry the instructions for growth and development.

Looking Ahead

Study TIP

Ready?

Check your understanding of these topics to help you prepare for what's ahead.

Can you...

- **define** the term *mutation*? (Section 1-1)
- **describe** the structure of proteins and DNA? (Section 2-3)
- **summarize** the function and structure of the nucleus and of microtubules? (Section 3-2)

If not, *review the sections indicated.*

Unit 4—*Cell Reproduction*
Topics 1–4

internetconnect

National Science Teachers Association *sci*LINKS Internet resources are located throughout this chapter.

Reading Strategies

Active Reading

Technique: Anticipation Guide

Write the following statements on the board:

- Chromosomes from females determine the sex of humans.
- Every human cell contains 46 chromosomes.
- Normal healthy cells cannot become cancerous cells.

Give students time to decide whether they agree or disagree with each statement. Ask volunteers to share their responses. Tell students to save their responses for later use in the **Closure** on page 127.

Directed Reading

Assign **Directed Reading Worksheet Chapter 6** to help students interact with the text as they read the chapter.

Interactive Reading

Assign ***Biology: Principles and Explorations* Audio CD Program Chapter 6** to help reluctant readers achieve greater success in reading the chapter.

Reading Skills

Assign Reading Organizers and Reading Strategies Worksheets from the **One-Stop Planner CD-ROM with Test Generator** to help students learn and practice effective reading skills.

internetconnect

Holt *Biology: Principles and Explorations*

On-line Resources: go.hrw.com

Visit the HRW Web site for a variety of resources related to this chapter. Just type in the keyword HXO Chapter 6.

Holt *Biology: Principles and Explorations*

On-line Resources: www.scilinks.org

The following *sci*LINKS Internet resources can be found in the student text for this chapter:

Page 120	Page 127	Page 137
TOPIC: Chromosomes	**TOPIC:** Cancer cells	**TOPIC:** Cell cycle
KEYWORD: HX120	**KEYWORD:** HX127	**KEYWORD:** HX137

Chapter 6

1 Prepare

Scheduling

- **Block:** About 45 minutes will be needed to complete this section.
- **Traditional:** About one class period will be needed to complete this section.
- **Planning:** See the Planning Guide on pp. 115A and 115B for lecture, classwork, and assignment options for Section 6-1. Lesson plans for Section 6-1 can be found in a lesson cycle format in ***Biology: Principles and Explorations*** **One-Stop Planner CD-ROM with Test Generator.**

Resource Link

Assign **Active Reading Worksheet Section 6-1** to help students comprehend the key concepts and visuals in this lesson.

2 Teach

Lesson Warm-up

Show the class two pictures or photos. One picture should show a person without a genetic disorder. The second picture should show a person with recognizable physical abnormalities due to a genetic disorder caused by the presence or absence of a chromosome (such as Klinefelter's or Turner's syndrome). Discuss with students how a single chromosome contains thousands of genes that code for proteins involved in determining how a person's body develops and functions.

Section 6-1 Chromosomes

Objectives

- **Identify** four examples of cell division in eukaryotes and one example in prokaryotes. (pp. 118–119)
- **Differentiate** between a gene, a DNA molecule, a chromosome, and a chromatid. (p. 119)
- **Differentiate** between homologous chromosomes, autosomes, and sex chromosomes. (pp. 120 and 123)
- **Compare** haploid and diploid cells. (p. 121)
- **Predict** how changes in chromosome number or structure can affect development. (pp. 122–124)

Key Terms

gamete
binary fission
gene
chromosome
chromatid
centromere
homologous chromosome
diploid
haploid
zygote
autosome
sex chromosome
karyotype

New Cells Are Formed by Cell Division

About 2 trillion cells are produced by an adult human body every day. This is about 25 million new cells per second! These new cells are formed when older cells divide. Cell division, also called cell reproduction, occurs in humans and other organisms at different times in their life. In **Figure 6-1,** the cells of the fawn that is growing and developing and the cells in the wound that is healing are undergoing cell division. The type of cell division differs depending on the organism and why the cell is dividing. For example, bacterial cells undergoing reproduction divide by one type of cell division. Eukaryotic organisms undergoing growth, development, repair, or asexual reproduction divide by a different type of cell division. And the formation of gametes involves yet a third type of cell division. **Gametes** are an organism's reproductive cells, such as sperm or egg cells.

Regardless of the type of cell division that occurs, all of the information stored in the molecule DNA (deoxyribonucleic acid) must be present in each of the resulting cells. Recall from Chapter 3 that DNA stores the information that tells cells which proteins to make and when to make them. This information directs a cell's activities and determines its characteristics. Thus, when a cell divides, the DNA is first copied and then distributed. Each cell ends up with a complete set (copy) of the DNA.

Figure 6-1 **Cell division**

The cells of these organisms are undergoing some type of cell division.

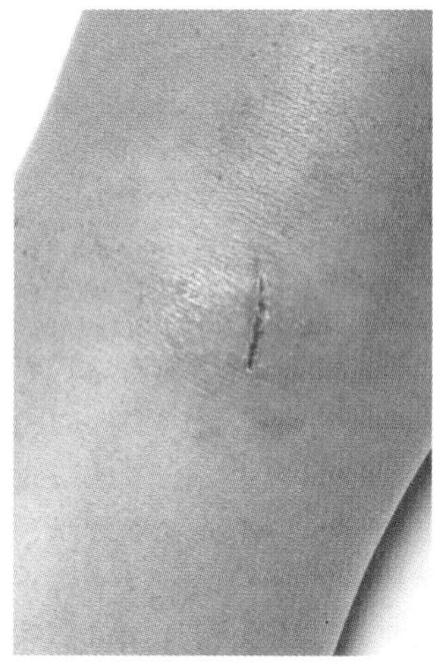

Repair

Growth and development

Integrating Different Learning Styles

Logical/Mathematical	**ATE**	Did You Know?, p. 119
Visual/Spatial	**PE**	Quick Lab, p. 124; Performance Zone item 6, p. 134
	ATE	Opening Demonstration, p. 116; Lesson Warm-up, p. 118; Demonstration, p. 119; Visual Strategy, p. 120; Teaching Tip, p. 122; Closure, p. 124
Verbal/Linguistic	**PE**	Performance Zone item 25, p. 135
	ATE	Active Reading, p. 119; Internet Connect, p. 120; Biology in Your Community, p. 121; Career, p. 122

Bacterial cells divide to reproduce

A bacterium's single DNA molecule is circular and is attached to the inner cell membrane. Bacteria reproduce by a type of cell division called binary fission. **Binary fission** is a form of asexual reproduction that produces identical offspring. In asexual reproduction, a single parent passes exact copies of all of its DNA to its offspring.

Binary fission occurs in two stages: first, the DNA is copied (so that each new cell will have a copy of the genetic information), and then the cell divides. The bacterium divides by adding a new cell membrane to a point on the membrane between the two DNA copies. As new material is added, the growing cell membrane pushes inward and the cell is constricted in the middle, like a long balloon being squeezed near the center. A new cell wall forms around the new membrane. Eventually the dividing bacterium is pinched into two independent cells. Each cell contains one of the circles of DNA and is a complete functioning bacterium.

REAL LIFE

***E. coli* cells can produce 1 million new cells in less than 7 hours.** A variety of *E. coli* known as O157:H7 is sometimes found in raw or undercooked meat. When such meat is eaten, the bacteria can cause life-threatening intestinal bleeding and kidney failure. Thorough cooking is necessary to destroy the bacteria in meat.

Finding Information
Research outbreaks of E. coli *O157:H7 in your community or state.*

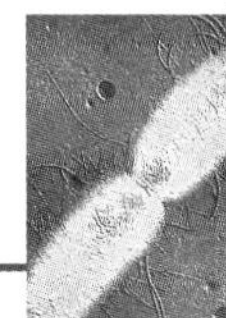

Eukaryotic cells form chromosomes before cell division

The vast amount of information encoded in DNA is organized into units called genes. A **gene** is a segment of DNA that codes for a protein or RNA molecule. A single molecule of DNA has thousands of genes lined up like train cars. Genes play an important role in determining how a person's body develops and functions. When genes are being used, the DNA is stretched out so that the information it contains can be used to direct the synthesis of proteins.

As a eukaryotic cell prepares to divide, the DNA and the proteins associated with the DNA coil into a structure called a **chromosome,** as shown in Figure 6-2. Before the DNA coils up, however, the DNA is copied. The two exact copies of DNA that make up each chromosome are called **chromatids** *(KROH muh tihdz)*. The two chromatids of a chromosome are attached at a point called a **centromere.** The chromatids, which become separated during cell division and placed into each new cell, ensure that each new cell will have the same genetic information as the original cell.

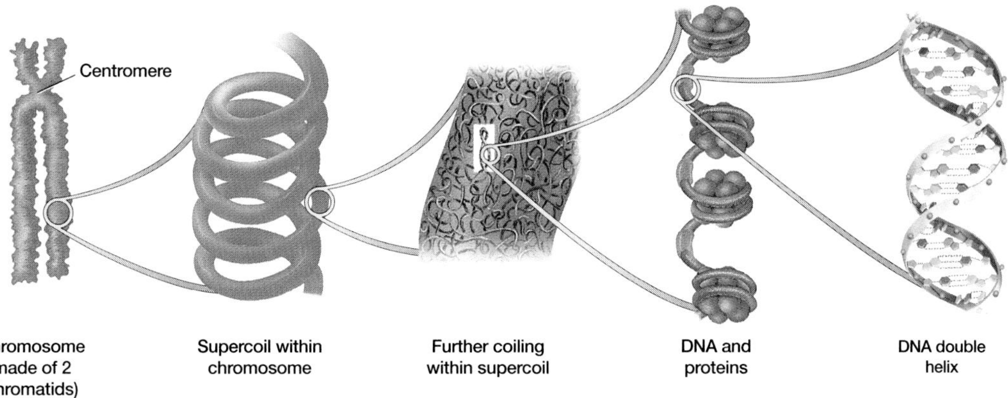

Figure 6-2 Chromosome structure. A chromosome consists of DNA tightly coiled around proteins. The chromosomes are formed as a cell prepares to divide.

did you know?

Prokaryotic chromosomes are hundreds of times longer than the cell that contains them.

For example, if a chromosome in an *E. coli* bacterium were fully extended, it would measure about 1 mm in length. The cell itself is only about 0.002 mm in length. Have students calculate how much longer *E. coli*'s chromosome is than the cell itself. *(1 mm/0.002 mm = 500 times longer than the cell)* PORTFOLIO

Section 6-1

Active Reading

Technique: Reading Organizer

Have students create a reading organizer to compare reproduction in bacterial cells and eukaryotic cells. A description of reproduction in eukaryotic cells is given in Sections 6-2 and 6-3. PORTFOLIO

REAL LIFE

Answer

Answers will vary by community. The strain was first identified in 1982. Between 1992 and 1998 there were about 30 outbreaks each year in various communities including areas in the Pacific Northwest; Alpine, Wyoming; Milwaukee, Wisconsin; and Indianapolis, Indiana.

Demonstration

DNA Coiling BASIC

Use a long piece of yarn to represent chromatin (DNA with proteins). Point out to students how the yarn is long and stringy and not very visible to those in the back of the class. Then cut a second piece of yarn and tell students that the chromatin has now been copied. Have four students twist the ends of each strand in opposite directions. Tell students that this represents the chromosomes becoming condensed and visible. Match up the yarn pairs and join them with a paper clip. Tell students the paper clip represents a centromere.

Chapter 6

internet connect

SciLINKS NSTA Have students research chromosomes using the Web site in the Internet Connect box on page 120. Have students write a report summarizing their findings. PORTFOLIO

Teaching TIP

Missing Homologue

Ask students to hypothesize what might happen if a human sperm or egg cell did not contain one member of each homologous pair. *(The resulting zygote would not have all of its chromosomes. The zygote might fail to develop. If the zygote did develop, the individual would not be normal because its cells would lack the important information contained in the missing genes.)*

Visual Strategy

Figure 6-3

Use this figure to point out how fertilization restores the diploid number. Ask students why sexually reproducing organisms possess an even number of chromosomes in the diploid state. *(If students do not realize that 2n must be an even number, then ask them to calculate the haploid number in each gamete as if the diploid number in humans were 47 and not 46.)*

Chromosome Number and Structure Affect Development

Each human somatic cell (any cell other than a sperm or egg cell) normally has two copies of 23 different chromosomes, for a total of 46 chromosomes. The 23 chromosomes differ in size, shape, and set of genes. Each chromosome contains thousands of genes that play important roles in determining how a person's body develops and functions. For this reason, possession of all chromosomes is essential to survival.

Sets of chromosomes

Each of the 23 pairs of chromosomes consists of two homologous *(hoh MAHL uh gus)* chromosomes, or homologues *(HOH muh logs)*. **Homologous chromosomes** are chromosomes that are similar in size, shape, and genetic content. Each homologue in a pair of homologous chromosomes comes from one of the two parents, as shown in **Figure 6-3**. Thus, the 46 chromosomes in human somatic cells are actually two sets of 23 chromosomes—one set from the mother and one set from the father. A human chromosome is shown in **Figure 6-4**.

Figure 6-3 Fertilization

When haploid gametes fuse, they produce a diploid zygote.

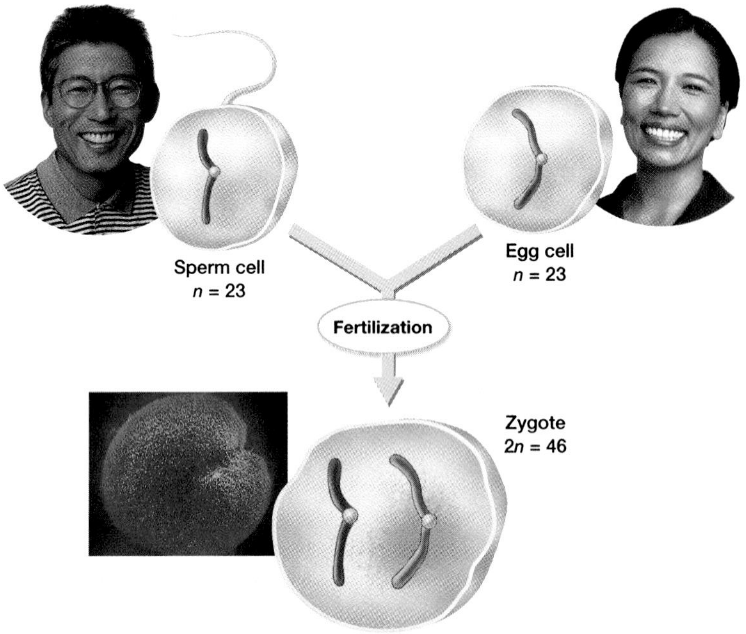

MULTICULTURAL PERSPECTIVE

Mormons and Mapping Genes

The Human Genome Project is an effort to map the over 100,000 genes in human cells to each chromosome and to sequence each gene. Scientists throughout the world are now engaged in this effort. One group has traced genetic markers through three generations of 60 Mormon families living in Utah. Their work with the Mormon families has resulted in the mapping of almost 500 genes.

All of the cells in the body, other than gametes, are somatic cells. When a cell, such as a somatic cell, contains two sets of chromosomes, it is said to be **diploid** *(DIHP loyd).* Unlike somatic cells, human gametes contain only one set of chromosomes (23 total). When a cell, such as a gamete, contains one set of chromosomes, it is said to be **haploid** *(HAP loyd).* Biologists use the symbol n to represent one set of chromosomes. The haploid number in a human gamete can be written as $n = 23$. The diploid number in a somatic cell can be written as $2n = 46$. The fusion of two haploid gametes—a process called fertilization—forms a diploid zygote, as shown in Figure 6-3. A **zygote** *(ZY goht)* is a fertilized egg cell, the first cell of a new individual.

As seen in **Table 6-1,** each organism has a characteristic number of chromosomes. The number of chromosomes in cells is constant within a species. Fruit flies, for example, have only eight chromosomes in each cell. Although most species have different numbers of chromosomes, some species by chance have the same number. For example, potatoes, plums, and chimpanzees all have 48 chromosomes in each cell. Many plants have far more chromosomes. Some ferns have more than 500. A few kinds of organisms—such as the Australian ant *Myrmecia,* the plant *Haplopappus* (a desert relative of the sunflower), and the fungus *Penicillium* (from which the antibiotic penicillin is obtained)—have only one pair of chromosomes.

Figure 6-4
Chromatids and centromeres

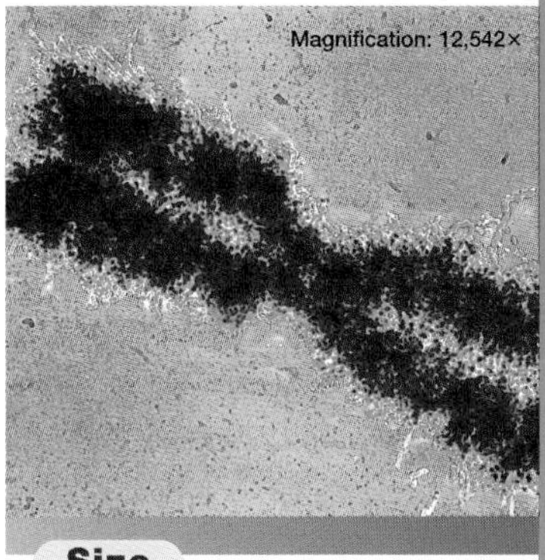

Magnification: 12,542×

Size As many as 500 chromosomes lined up end to end would fit in a 0.2 cm space. The chromosome above has replicated and consists of two identical chromatids.

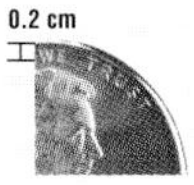

0.2 cm

Table 6-1 Chromosome Number of Various Organisms

Organism	Number of chromosomes
Penicillium	1–4
Saccharomyces (yeast)	18
Mosquito	6
Housefly	12
Garden pea	14
Corn	20
Adder's tongue fern	1,262
Frog	26
Human	46
Orangutan	48
Dog	78

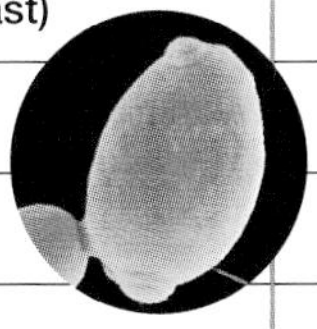

Teaching TIP

Prefixes

Have students use the prefixes in the words *haploid* and *diploid* to help them remember the meanings of these words. Have them relate the words to the algebraic terms n and $2n$. Haploid, or n, represents one set of chromosomes, as in a gamete. Diploid, or $2n$, represents two sets of chromosomes, as in somatic cells. Tell students that the prefix *hapl-* means "single" and that the prefix *dipl-* means "double." Add that the prefix *poly-* means "many." Then have students predict what would be true of a polyploid cell. *(It would have many sets of chromosomes.)*

Cross-Disciplinary Connection

◆ History

Approximately 100 years ago, biologists first began counting the number of chromosomes in human cells. In the 1920s, a biologist reported the diploid number in human cells to be 48. For many years following this report, textbooks stated that 48 was the normal diploid number of human chromosomes. It was not until 1956 that J. H. Tjio and A. Levan showed 46 to be the correct number. These two scientists worked out a procedure to culture white blood cells. Their methods increased the number of cells undergoing mitosis and improved the spreading of the chromosomes, which allowed the scientists to accurately count the number of human chromosomes.

Biology in Your Community

Genetic Counselor

Invite a genetic counselor to speak to the class. Before the speaker attends, have the students research chromosomal abnormalities that result from too few or too many chromosomes. Have the students prepare a list of questions on their findings. Ask the speaker to bring sample karyotypes of the abnormalities the students researched, if possible. PORTFOLIO

Chapter 6

CAREER

Cytogenetic Technologist

Cytogenetic technologists prepare specimens, such as karyotypes, for cell genetic analysis to aid physicians in patient diagnosis of genetic disorders. Have students use library references or on-line databases to find information about the requirements necessary to become a cytogenetic technologist and the duties a person with this career might perform. *(prepare cells, prepare slides in the correct mitotic stage, stain slides, and prepare accurate karyotypes from photographic prints or computer images)* Have students present their findings to the class. PORTFOLIO

Teaching TIP — ADVANCED

Karyotypes

Obtain copies of normal human karyotypes from the library or on-line resources. Enlarge them so that each chromosome is no smaller than 2.5 cm (1 in.) in length. Paste the copies of the karyotype on a poster board. Cut out each chromosome and mix up the pieces. Allow the students to arrange the chromosomes using size, length, position of centromere, and banding patterns. Have them compare the karyotypes with each other and with the karyotype shown in Figure 6-5. PORTFOLIO

Study TIP

Word Origins

The word *chromosome* is from the Greek *chroma,* meaning "color," and *soma,* meaning "body." Knowing this makes it easier to remember that chromosomes were so named because they absorbed a colored dye that made them more visible under a microscope.

Chromosomes determine your sex

Of the 23 pairs of chromosomes in human somatic cells, 22 pairs are called autosomes. **Autosomes** are chromosomes that are not directly involved in determining the sex (gender) of an individual. The **sex chromosomes,** one of the 23 pairs of chromosomes in humans, contain genes that will determine the sex of the individual.

In humans and many other organisms, the two sex chromosomes are referred to as the X and Y chromosomes. The genes that cause a fertilized egg to develop into a male are located on the Y chromosome. Thus, any individual with a Y chromosome is male, and any individual without a Y chromosome is female. For example, in human males, the sex chromosomes are made up of one X chromosome and one Y chromosome (XY). The sex chromosomes in human females consist of two X chromosomes (XX). Because a female can donate only an X chromosome to her offspring, the sex of an offspring is determined by the male, who can donate either an X or a Y.

The structure and number of sex chromosomes vary in different organisms. In some insects, such as grasshoppers, there is no Y chromosome—the females are characterized as XX and the males are characterized as XO (the O indicates the absence of a chromosome). In birds, moths, and butterflies, the male has two X chromosomes and the female has only one.

Change in chromosome number

Each of an individual's 46 chromosomes has thousands of genes. Since genes play an important role in determining how a person's body develops and functions, the presence of all 46 chromosomes is essential for normal development and function. A person must have the characteristic number of chromosomes in his or her cells. Humans who are missing even one of the 46 chromosomes do not survive. Humans with more than two copies of a chromosome, a condition called trisomy *(try SOH mee),* will not develop properly. Abnormalities in chromosome number can be detected by analyzing a **karyotype** *(KAR ee uh teyep),* a photo of the chromosomes in a dividing cell that shows the chromosomes arranged by size. **Figure 6-5** shows a typical karyotype. A portion of a karyotype from an individual with an extra copy of chromosome 21 is also shown in Figure 6-5. This condition is called Down syndrome, or trisomy 21. Short stature, a round

Figure 6-5 A human karyotype

Karyotypes are used to examine an individual's chromosomes.

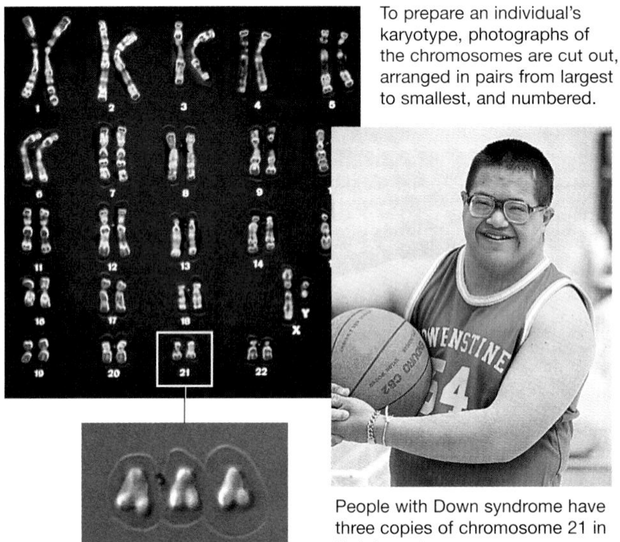

To prepare an individual's karyotype, photographs of the chromosomes are cut out, arranged in pairs from largest to smallest, and numbered.

People with Down syndrome have three copies of chromosome 21 in their karyotype.

Overcoming Misconceptions

Sex Chromosomes

Students sometimes have the misconception that all traits associated with sex are controlled by genes on the sex chromosomes, while genes controlling traits that have nothing to do with sex are located on the autosomes. Point out that genes for red-green colorblindness and hemophilia are located on the sex chromosomes, while those controlling the production of sex hormones are found on the autosomes.

face with upper eyelids that cover the inner corners of the eyes, and varying degrees of mental retardation are characteristics of people with Down syndrome.

In mothers younger than 30, Down syndrome occurs in about 1 in 1,500 births. In mothers 30 to 35 years old, the incidence doubles to 1 in 750 births. In mothers over 45, the risk is as high as 1 in 16 births. Older mothers are more likely to have a baby with Down syndrome because all the eggs a female will ever produce are present in her ovaries when she is born, unlike males who produce new sperm throughout adult life. As a female ages, her eggs can accumulate an increasing amount of damage. Because of this risk, a pregnant woman over the age of 35 may be advised to undergo prenatal testing that includes fetal karyotyping.

What events can cause an individual to have an extra copy of a chromosome? When sperm and egg cells form, each chromosome and its homologue separate, an event called disjunction *(dihs JUHNK shuhn)*. If one or more chromosomes fail to separate properly—a process called nondisjunction—one new gamete ends up receiving both chromosomes and the other gamete receives none. Trisomy occurs when the gamete with both chromosomes fuses with a normal gamete during fertilization, resulting in offspring with three copies of that chromosome instead of two. In Down syndrome, nondisjunction involves chromosome 21.

Prenatal Testing

Will our baby be normal? For many expectant parents, prenatal testing can help answer this question. In prenatal testing, the cells of a fetus are tested for normal chromosome number and cell structure by a procedure called fetal karyotyping. Fetal karyotyping allows parents and doctors to view the chromosomes found in the cells of the fetus. The doctor can then check for any abnormalities, such as Down syndrome. There are two ways to obtain fetal cells.

Amniocentesis

In amniocentesis *(am nee oh sehn TEE sihs)*, a needle and syringe are used to remove a small amount of the amniotic fluid that surrounds the fetus. The fluid contains fetal cells. The fetal cells are grown in a laboratory for 1–4 weeks to obtain enough actively dividing fetal cells to make a karyotype, which is then analyzed.

Chorionic villi sampling (CVS)

In chorionic villi *(kawr ee AHN ihk VIHL eye)* sampling, a tissue sample is collected from the chorionic villi, fingerlike extensions of the placenta that grow into the mother's uterus. Enough actively dividing cells are obtained to produce a karyotype without having to culture cells. Since the villi have the same genetic makeup as the fetus, the doctor is able to detect abnormalities in the fetal chromosome number.

Multicultural Perspective

Cultural Attitudes Toward Genetic Defects

From the Middle Ages through the nineteenth century, most Europeans believed that genetic defects reflected inner corruption. However, many other cultures, including the Celtic people of Europe and Native Americans, thought individuals with such defects had a special insight and a closer connection with nature. These unique members of the community were given responsibility as tribal leaders or healers. Many physically disabled leaders made careful astronomical observations of the sun and moon that helped them advise their communities on the optimum times for planting, harvesting, or migrating to another area.

Making Biology Relevant

Medicine

Most nondisjunctions involving autosomes are lethal, but those involving sex chromosomes usually are not. However, a YO combination in humans is lethal, leading biologists to suspect that at least one X chromosome is necessary for development and survival.

Prenatal Testing

Teaching Strategies

- Use the figure to differentiate the two procedures.
- Point out that genetic counselors are available before, during, and after the procedures to explain the procedures and results and to offer emotional support.

Discussion

- What risks might be involved in each type of procedure? *(infection or injury to the fetus)* Tell students that ultrasound techniques are used to help locate the fetus during testing to avoid injury to the fetus.
- Which procedure takes longer to produce results. *(amniocentesis)* Why? *(Amniocentesis requires the cells to be cultured, whereas with CVS the cells come directly from the fetus, which provides a quicker way to examine the cells.)*
- What are some of the difficult decisions that must be made after a genetic disorder is diagnosed? *(Answers will vary.)*

What are chromosomal mutations?

QUICK LAB

Time 15–20 minutes

Process Skills Modeling, sequencing, interpreting

Teaching Strategies
Cut note card pieces prior to the lab.

Analysis Answers
deletion: cell would be missing a gene, which could prove fatal; duplication: cell would have an extra gene, which could prove fatal or result in malfunctioning of the cell; inversion: cell may not be able to use gene because it is located in a different area on the chromosome, which could prove fatal; translocation: cell may not be able to use the gene because it is located on a different chromosome, which could prove fatal

3 Assessment Options

Closure

Predicting Outcomes
Ask students to differentiate each of the following terms by defining and sketching them: *chromosomes, chromatids, DNA,* and *genes.* Students should label their sketches. PORTFOLIO

Review

Assign Review—Section 6-1.

Reteaching — BASIC

Write the following table on the board. Then have students determine the haploid number, *n,* for each organism listed.

Organism	Diploid number, 2*n*
Yeast	18
Mosquito	6
Orangutan	48
Dog	78

(haploid number, n, *= 9, 3, 24, 39)*

Change in chromosome structure

Changes in an organism's chromosome structure are called mutations. Breakage of a chromosome can lead to four types of mutations. In a deletion mutation, a piece of a chromosome breaks off completely. After cell division, the new cell will lack a certain set of genes. In many cases this proves fatal to the zygote. In a duplication mutation, a chromosome fragment attaches to its homologous chromosome, which will then carry two copies of a certain set of genes. A third type of mutation is an inversion mutation, in which the chromosome piece reattaches to the original chromosome but in a reverse orientation. If the piece reattaches to a nonhomologous chromosome, a translocation mutation results.

QUICK LAB

What are chromosomal mutations?

You can use paper and a pencil to model the ways in which chromosome structure can change.

Materials

14 note-card pieces, pencils, tape

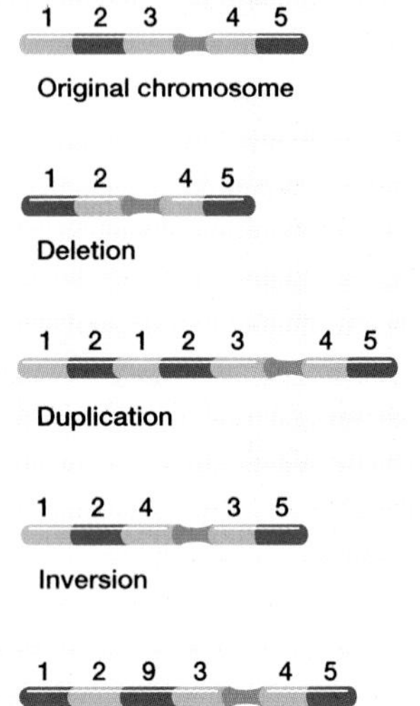

Procedure

1. Write the numbers 1–8 on note-card pieces (one number per piece). Tape the pieces together in numerical order to model a chromosome with eight genes.
2. Use the "chromosome" you made to model the four alterations in chromosome structure discussed on this page and illustrated at right. For example, remove the number 3 and reconnect the remaining chromosome pieces to represent a deletion.
3. Reconstruct the original chromosome before modeling a duplication, an inversion, and a translocation. Use the extra note-card pieces to make the additional numbers you need.

Analysis

Describe how a cell might be affected by each mutation if the cell were to receive a chromosome with that mutation.

Review SECTION 6-1

1. **Summarize** how bacterial cells divide by binary fission.
2. **Identify** the point in a eukaryotic cell's life at which DNA coils up to form chromosomes.
3. **Summarize** the difference between a haploid cell and a diploid cell.
4. **Compare** the sex chromosomes of human females with those of human males.
5. **Compare** the karyotype of an individual with Down syndrome with a normal karyotype.

Critical Thinking

6. **SKILL Evaluating Conclusions** A student concludes that homologous chromosomes are found in gametes. Do you agree or disagree? Explain.

Review SECTION 6-1 ANSWERS

1. DNA is first copied. Then the bacterium divides into equal halves by adding new cell membrane between the two DNA copies. The growing cell membrane pushes inward, and the cell constricts to form two new, identical cells. A new cell wall forms around each new membrane.
2. Chromosomes become visible in a eukaryotic cell when the cell prepares to divide.
3. A haploid cell, *n,* contains one set of chromosomes. A diploid cell, 2*n,* contains two sets of chromosomes.
4. Human females have XX sex chromosomes and males have XY sex chromosomes.
5. normal human karyotype: 23 pairs of chromosomes (46 total); karyotype of an individual with Down syndrome: 47 chromosomes, with three chromosomes for chromosome number 21
6. Students should disagree. Homologous chromosomes are pairs of similar chromosomes. Because gametes are haploid, *n,* they contain only one set of chromosomes. Thus, homologous chromosomes are not normally found in gametes.

The Cell Cycle

Section 6-2

The Cell Cycle Describes the Life of a Eukaryotic Cell

Cell division in eukaryotic cells is more complex than cell division in bacteria because it involves dividing both the cytoplasm and the chromosomes inside the nucleus. Many internal organelles must be correctly rearranged before the eukaryotic cell can properly divide and form two fully functioning cells.

The cell cycle

The life of a eukaryotic cell is traditionally shown as a cycle, as illustrated in Figure 6-6. The **cell cycle** is a repeating sequence of cellular growth and division during the life of an organism. A cell spends 90 percent of its time in the first three phases of the cycle, which are collectively called **interphase.** A cell will enter the last two phases of the cell cycle only if it is about to divide. The five phases of the cell cycle are summarized below:

1. **First growth (G_1) phase.** During the G_1 phase, a cell grows rapidly and carries out its routine functions. For most organisms, this phase occupies the major portion of the cell's life. Cells that are not dividing remain in the G_1 phase. Some somatic cells, such as most muscle and nerve cells, never divide. Therefore, if these cells die, the body cannot replace them.
2. **Synthesis (S) phase.** A cell's DNA is copied during this phase. At the end of this phase, each chromosome consists of two chromatids attached at the centromere.
3. **Second growth (G_2) phase.** In the G_2 phase, preparations are made for the nucleus to divide. Hollow protein fibers called microtubules are assembled. The microtubules are used to move the chromosomes during mitosis.
4. **Mitosis.** The process during cell division in which the nucleus of a cell is divided into two nuclei is called **mitosis** *(meye TOH sihs)*. Each nucleus ends up with the same number and kinds of chromosomes as the original cell.
5. **Cytokinesis.** The process during cell division in which the cytoplasm divides is called **cytokinesis** *(SEYET oh kih nee sihs).*

Mitosis and cytokinesis produce new cells that are identical to the original cells and allow organisms to grow, replace damaged tissues, and, in some organisms, reproduce asexually.

Objectives

- **Identify** the major events that characterize each of the five phases of the cell cycle. (p. 125)
- **Describe** how the cell cycle is controlled in eukaryotic cells. (p. 126)
- **Relate** the role of the cell cycle to the onset of cancer. (pp. 126–127)

Key Terms

cell cycle
interphase
mitosis
cytokinesis
cancer

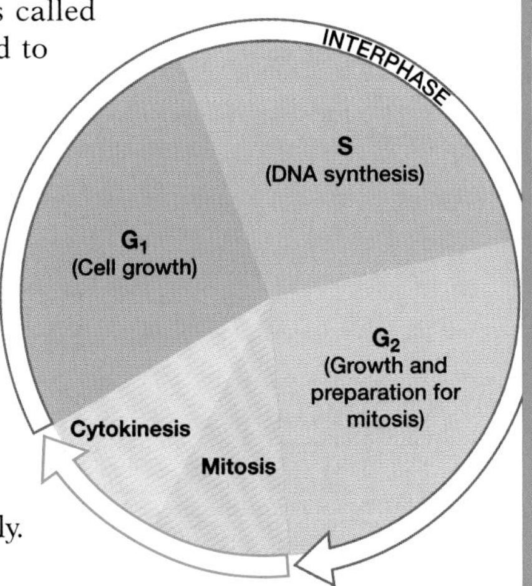

Figure 6-6 The eukaryotic cell cycle. The cell cycle consists of phases of growth, DNA replication, preparation for cell division, and division of the nucleus and cytoplasm.

1 Prepare

Scheduling

- **Block:** About 45 minutes will be needed to complete this section.
- **Traditional:** About one class period will be needed to complete this section.
- **Planning:** See the Planning Guide on pp. 115A and 115B for lecture, classwork, and assignment options for Section 6-2. Lesson plans for Section 6-2 can be found in a lesson cycle format in ***Biology: Principles and Explorations*** **One-Stop Planner CD-ROM with Test Generator.**

Resource Link

- Assign **Active Reading Worksheet Section 6-2** to help students comprehend the key concepts and visuals in this lesson.

2 Teach

Lesson Warm-up

Display either a model or a photograph of a human brain. Point out that once the brain is fully formed, most of the nerve cells do not divide again. These cells are arrested in a certain stage in the life of a cell (the G_1 phase). Then display either a model or a photograph of a human bone. Point out that red blood cells are produced from cells in the marrow of long bones. An average red blood cell lives for about 120 days. Each second, about 2 million red blood cells are produced by cell division in the bone marrow. Cells in the marrow, unlike those in the brain, continue going through the life of a cell, or the cell cycle, as long as a person lives.

Integrating Different Learning Styles

Logical/Mathematical	**ATE**	Visual Strategy, p. 126
Visual/Spatial	**PE**	Performance Zone items 1, 17, 18, and 19, pp. 134–135
	ATE	Lesson Warm-up, p. 125; Teaching Tip, p. 126; Reteaching, p. 127
Verbal/Linguistic	**PE**	Performance Zone items 20, 21, 24, 26, and 27, p. 135

Chapter 6

Visual Strategy

Figures 6-6 and 6-7

Use Figures 6-6 and 6-7 on pp. 125 and 126, respectively, to clarify the relationship of cell division and interphase to the cell cycle. Ask students which sections represent interphase *(blue)* and which represent active cell division *(green).* The completion of the cell cycle requires varying periods of time, depending on the type of cell and the cell's environment. Some cells cycle in less than an hour, while others take many days. Ask students to use the circle graph to estimate the percentage of time a cell spends in interphase if you assume a 24-hour cell cycle. *(about 20 hours)*

Teaching TIP

Phases of Interphase

Have students draw a Graphic Organizer like the one at the bottom of page 126 to summarize the events that occur during the phases of interphase. Have them use arrows between the phases, pointing to the right, to indicate the sequence of events. PORTFOLIO

Cross-Disciplinary Connection

◆ History

In culture, cancer cells can divide indefinitely if they are given a continual supply of nutrients. There is a cell line that has been continuously cultured since 1951. The cells in this cell line are called HeLa cells because they were originally from a tumor removed from a woman named Henrietta Lacks. The patient suffered from uterine cervical carcinoma. Cells from this cell line are still used by researchers around the world, especially in research on viruses.

Study TIP

Reviewing Information

Learn the stages of interphase by reviewing the steps numbered 1–5 on page 125. You can see in Figures 6-6 and 6-7 that the cell cycle is a repeating series of three steps followed by mitosis and cytokinesis.

The Cell Cycle Is Carefully Controlled

If a cell spends 90 percent of its time in interphase, how do cells "know" when to divide? How is the cycle controlled? Just as traffic lights control the flow of traffic, cells have a system that controls the phases of the cell cycle. Cells have a set of "red light–green light" switches that are regulated by feedback information from the cell. The cell cycle has key checkpoints (inspection points) at which feedback signals from the cell can trigger the next phase of the cell cycle (green light). Other feedback signals can delay the next phase to allow for completion of the current phase (yellow or red light).

The cell cycle in eukaryotes is controlled by many proteins. Control occurs at three principal checkpoints, as shown in **Figure 6-7.**

1. **Cell growth (G_1) checkpoint.** This checkpoint makes the key decision of whether the cell will divide. If conditions are favorable for division and the cell is healthy and large enough, certain proteins will stimulate the cell to begin the synthesis (S) phase. During the S phase, the cell will copy its DNA. If conditions are not favorable, cells can typically stop the cell cycle at this checkpoint. The cell cycle will also stop at this checkpoint if the cell needs to pass into a resting period. Certain cells, such as some nerve and muscle cells, remain in this resting period permanently and never divide.
2. **DNA synthesis (G_2) checkpoint.** DNA replication is checked at this point by DNA repair enzymes. If this checkpoint is passed, proteins help to trigger mitosis. The cell begins the many molecular processes that are needed to proceed into mitosis.
3. **Mitosis checkpoint.** This checkpoint triggers the exit from mitosis. It signals the beginning of the G_1 phase, the major growth period of the cell cycle.

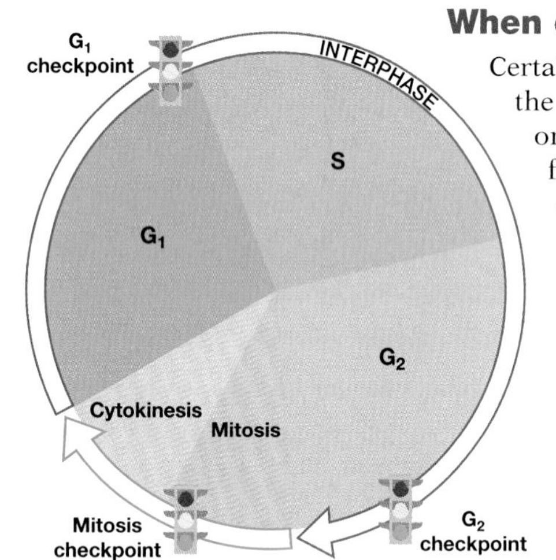

Figure 6-7 Control of the cell cycle. The cell cycle in eukaryotes is controlled at three inspection points, or checkpoints. Many proteins are involved in the control of the cell cycle.

When control is lost: cancer

Certain genes contain the information necessary to make the proteins that regulate cell growth and division. If one of these genes is mutated, the protein may not function, and regulation of cell growth and division can be disrupted. **Cancer,** the uncontrolled growth of cells, may result. Cancer is essentially a disorder of cell division. Like stepping on the accelerator (gas pedal) of a car regardless of a red light ahead, cancer cells do not respond normally to the body's control mechanisms.

Some mutations cause cancer by overproducing growth-promoting molecules, speeding up the cell cycle by stepping on the accelerator. Others cause cancer by inactivating the proteins that slow or stop the cell cycle, in effect removing the brakes.

Graphic Organizer

Use this graphic organizer with ***Teaching Tip: Phases of Interphase*** *on page 126.*

Interphase → G_1, S, G_2

G_1 → S → G_2

G_1 → Cell growth

S → DNA synthesis

G_2 → Growth and preparation for cell division

Although mutations can occur spontaneously, many occur as a result of environmental influences. This is why some of the more significant cancer risk factors are linked to lifestyle. For example, the use of tobacco products and exposure to ultraviolet radiation are known risk factors. Certain viruses can also induce cancer. Recently there has been a great deal of interest in the effects of diet on cancer.

Exploring Further

Normal Cells Can Become Cancer Cells

The body's immune system normally destroys cancer cells. But if the cancer cells somehow escape destruction, the cells will continue to divide and will eventually form a tumor, a mass of cancer cells within otherwise normal tissue.

Benign and malignant tumors

If the cancer cells remain at the original site, the mass of cells is called a benign *(bih NEYEN)* tumor and can usually be completely removed by surgery. If cancer cells spread to neighboring tissues and to other parts of the body, the mass is called a malignant tumor and the individual is said to have cancer. The spread of cancer cells beyond their original site is called metastasis *(muh TAS tuh sihs)*. When this occurs, the tumor is usually treated with high-energy radiation and poisonous chemicals (chemotherapy) that are especially harmful to actively dividing cells (both cancerous and normal).

Mutations and cancer

Two major types of genes regulate cell division and growth.

1. Proto-oncogenes stimulate normal cell growth and cell division.
2. Tumor-suppressor genes inhibit cell growth and cell division (to prevent uncontrolled cell growth).

Melanoma

The balance between the products of proto-oncogenes and tumor-suppressor genes keeps most cells dividing normally. But a mutation in one of these two types of genes can contribute to the start of cancer.

Usually more than one mutation is needed to produce all of the changes associated with a cancer cell. This may be why the likelihood of cancer increases greatly with age. The longer you live, the more mutations you will accumulate, and the more likely it becomes that one or more of your cells will accumulate enough mutations to become a cancer cell. This may also explain why you can inherit an increased risk of a certain type of cancer (cancer "runs" in the family). If you inherit a mutated gene from your parents, you may eventually accumulate the additional mutations that can result in cancer cells.

Review SECTION 6-2

1. **Differentiate** between the G_1, G_2, and S phases of the eukaryotic cell cycle.
2. **Relate** the onset of cancer to three major checkpoints at which the cell cycle is controlled.
3. **Identify** the type of environmental factors that can cause mutations that may lead to cancer.

Critical Thinking

4. **SKILL Evaluating Information** Why are individual chromosomes more difficult to see during interphase than during mitosis?

Review SECTION 6-2 ANSWERS

1. G_1 phase: cell grows rapidly and carries out routine functions; G_2 phase: mitochondria and other organelles replicate, microtubules are assembled; S phase: DNA is copied
2. Each checkpoint determines whether cells should continue through the cell cycle. Cancer can occur if one of the proteins involved at one of the checkpoints is mutated.
3. Environmental factors that can cause mutations include tobacco products, ultraviolet radiation, certain viruses, and certain foods.
4. Chromosomes are formed right before a cell divides in mitosis. During most of interphase, the DNA exists as chromatin, which is more elongated and harder to see with a microscope.

Exploring Further

Normal Cells Can Become Cancer Cells

Teaching Strategies

- Ask students if they know anyone who has cancer. Ask them what type of cancer each individual has or had. Lead students to realize that there are many different types of cancer.
- Tell students that most anticancer drugs interfere with the cell cycle of cancer cells. Unfortunately, the drugs also interfere with healthy cells, which is why the patient suffers side effects.

Discussion

- Where in the cell cycle could scientists target anticancer drugs? *(Answers will vary but may include proteins involved in the three checkpoints, DNA replication, or cytokinesis.)*
- What type of environmental factors have been associated with the onset of cancer? *(Answers may include diet, UV radiation, hormones, and environmental pollution.)*

3 Assessment Options

Closure

Technique: Anticipation Guide

Tell students to return to their responses in the **Active Reading** on page 117. Have them change their responses as needed. Discuss the changes. PORTFOLIO

Review

Assign Review—Section 6-2.

Reteaching — BASIC

Assign students to groups. Have each group visually interpret a stage in the cell cycle. Have each group display their graphics in the appropriate space in a circle graph that represents the cell cycle. PORTFOLIO

Chapter 6

Section 6-3

Mitosis and Cytokinesis

1 Prepare

Scheduling

- **Block:** About 90 minutes will be needed to complete this section.
- **Traditional:** About two class periods will be needed to complete this section.
- **Planning:** See the Planning Guide on pp. 115A and 115B for lecture, classwork, and assignment options for Section 6-3. Lesson plans for Section 6-3 can be found in a lesson cycle format in ***Biology: Principles and Explorations*** **One-Stop Planner CD-ROM with Test Generator.**

Resource Link

Assign **Active Reading Worksheet Section 6-3** to help students comprehend the key concepts and visuals in this lesson.

2 Teach

Lesson Warm-up

Draw a football field horizontally on the board, with a 50 yd line and goal posts at either end. Tell students that the 50 yd line represents the area of the cell called the equator. The goal posts represent the poles of the cell. Tell students that these areas are important when discussing how chromosomes are distributed during cell division.

Objectives

- **Describe** the structure and function of the spindle during mitosis. (pp.128–129)
- **Summarize** the events of the four stages of mitosis. (pp. 130–131)
- **Differentiate** cytokinesis in animal and plant cells. (pp. 131–132)

Key Term

spindle

In Mitosis, Chromatids Are Pulled by Microtubules

Every second about 2 million new red blood cells are produced in your body by cell divisions occurring in the bone marrow. These cells have received the signal to divide. The cells continue past the G_2 phase and enter into the last two phases of the cell cycle—mitosis and cytokinesis. During mitosis the nucleus divides to form two nuclei, each containing a complete set of the cell's chromosomes. During cytokinesis the cytoplasm is divided between the two resulting cells.

During mitosis, the chromatids on each chromosome are physically moved to opposite sides of the dividing cell with the help of the spindle, shown in **Figure 6-8. Spindles** are cell structures made up of both centrioles and individual microtubule fibers that are involved in moving chromosomes during cell division.

Forming the spindle

Animal cells usually have one pair of centrioles, with the centrioles at right angles to each other. During the G_2 phase of the cell cycle, the centriole pair is replicated so that the cell has two pairs of centrioles as it enters the mitotic phase. When a cell enters the mitotic phase, the centriole pairs start to separate, moving toward opposite poles of the cell. As the centrioles move apart, the spindle begins to form.

Centrioles and spindle fibers are both made of hollow tubes of protein called microtubules. Each spindle fiber is made of an individual microtubule. Each centriole, however, is made of nine triplets

Figure 6-8 The spindle

The spindle, made up of centrioles and spindle fibers, helps move chromosomes apart during mitosis.

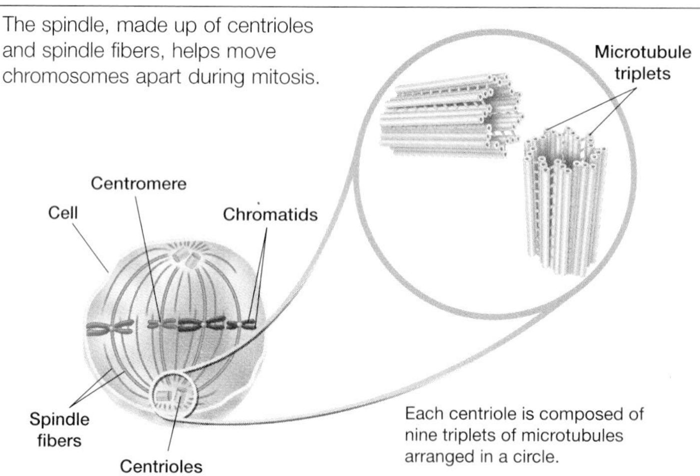
Microtubule triplets
Centromere
Cell
Chromatids
Spindle fibers
Centrioles
Each centriole is composed of nine triplets of microtubules arranged in a circle.

Integrating Different Learning Styles

Logical/Mathematical	**PE**	Math Lab, p. 129
Visual/Spatial	**PE**	Quick Lab, p. 132; Experimental Design, pp. 136–137
	ATE	Lesson Warm-up, p. 128; Demonstration, p. 129; Visual Strategy, p. 130; Teaching Tip, p. 131; Reteaching, p. 132
Verbal/Linguistic	**PE**	Study Tip, p. 130
	ATE	Teaching Tip, p. 130; Resource Link, p. 130

of microtubules arranged in a circle. Unlike animal cells, plant cells do not have centrioles, but they form a spindle that is almost identical to that of an animal cell.

Separating chromatids by attaching spindle fibers

Some of the microtubules in the spindle interact with each other; others attach to a protein structure found on each side of the centromere. The two sets of microtubules extend out toward opposite poles of the cell. Once the microtubules attach to the centromeres and poles, the two chromatids in each chromosome can be separated.

The chromatids are moved to each pole of the cell in a manner similar to bringing in a fish with a fishing pole. When the microtubule "fishing line" is "reeled in," the chromatids are dragged to opposite poles. The reeling in occurs because the ends of the spindle fibers are broken down bit by bit at each of the poles. As the fibers become shorter, the chromatids they are pulling move closer and closer to the poles.

As soon as the chromatids separate from each other they are called chromosomes. When the chromosomes finally arrive, each pole has one complete set of chromosomes.

Some drugs can damage spindle fibers. Griseofulvin *(grihs ee oh FOOL vihn)* is a chemical used to treat fungal infections of the skin, such as ringworm. Griseofulvin binds to microtubules, disrupting the formation of the mitotic spindle and preventing cell division by the fungus that causes ringworm.

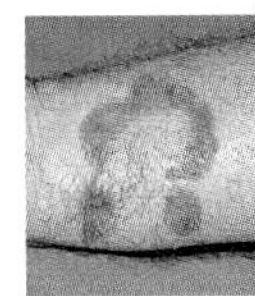

Finding Information
What other drugs function by damaging microtubules, such as those that make up the spindle?

Calculating the number of cells resulting from mitosis

Background

Scientists investigating cancer might need to know the number of cells produced in a certain amount of time. In the human body the rate of mitosis is about 25 million (2.5×10^7) cells produced every second! You can calculate the number of cells produced by mitosis in a given amount of time.

1. **Calculate the number of cells produced by mitosis in the given time.** For example, to find the number of cells produced in 3 minutes, determine how many seconds there are in 3 minutes (since the rate is given in seconds).

$$\frac{60 \text{ seconds}}{1 \text{ minute}} \times 3 \text{ minutes} = 180 \text{ seconds}$$

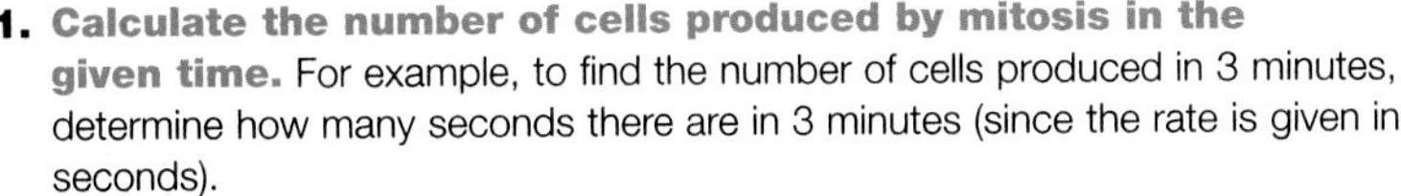

2. **Multiply the rate of mitosis by the time (in seconds) asked for in the problem (180 seconds).**

$$\frac{2.5 \times 10^7 \text{ cells}}{\text{second}} \times 180 \text{ seconds} = 4.5 \times 10^9 \text{ cells } (4,500,000,000 \text{ cells})$$

Analysis

1. **Calculate** the number of cells that would be produced in 1 hour.
2. **Calculate** the number of cells that would be produced in 1 day.
3. **SKILL Predicting Factors** Identify factors that might increase or decrease the rate of mitosis.

did you know?

The fruit fly* Drosophila melanogaster *has been used to study chromosomal mutations since 1933.

The salivary glands of these flies, as in many insects, are composed of cells that do not divide during the larval stage. However, the chromosomes continue to replicate, producing many copies. The copies of each chromosome are closely aligned, resulting in thick chromosomes that are easy to study with a microscope.

Demonstration

Spindle Shape — BASIC

To give students an idea of the three-dimensional shape of the spindle, hold up a football and point out that the ends of the football represent the centrioles and that spindle fibers span the length of the football.

REAL LIFE

Answer

Drugs that function by damaging microtubules include colchicine and taxol. Colchicine binds to free tubulin and prevents its polymerization into microtubules. Taxol binds to microtubules and prevents them from losing subunits and disassembling.

Calculating the number of cells resulting from mitosis

Time 10 minutes

Process Skills Calculating, predicting

Teaching Strategies

- Review numbers in scientific notation before starting the lab.
- Review with students how to cancel units and set up equal unit proportions. For example, 60 seconds/1 minute is an equal proportion.

Analysis Answers

1. 9.0×10^{10} cells = 90,000,000,000 cells
2. 2.16×10^{12} cells
3. Factors that might increase or decrease the rate of mitosis include mutated genes, diet, and exposure to ultraviolet light and tobacco products.

Teaching TIP

Word Origins

To help students master vocabulary, relate the word mitosis to its Greek origin, *mitos,* meaning "thread." Ask students to explain why mitosis might be named for that word. *(The hereditary material consists of long, threadlike molecules.)*

Visual Strategy

Figure 6-9

Lead students through each stage shown in Figure 6-9, focusing on the behavior of chromosomes. Point out that the various stages are not of equal duration. Using movie film as an analogy, help students avoid the misconception that mitosis "jumps" from stage to stage. Although a movie consists of individual frames of film, the images on the film appear to change continuously when the film is projected on a screen. Explain that mitosis progresses in a similar fashion.

Resource Link

Assign **Reading Strategies: Series-of-Events Chain** from the Holt BioSources Teaching Resources CD-ROM to help students summarize the stages of mitosis. PORTFOLIO

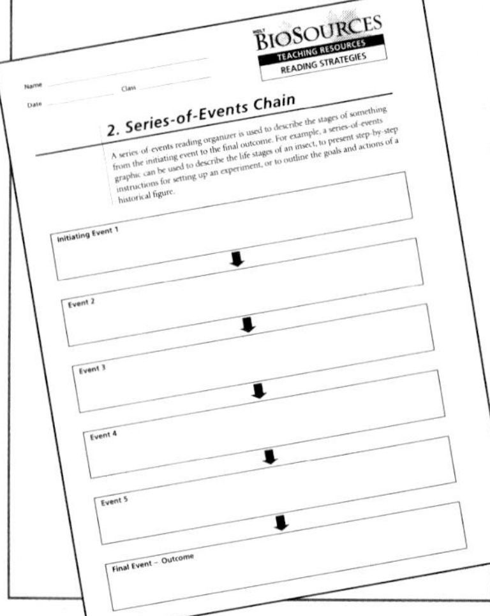
BIOSOURCES
TEACHING RESOURCES
READING STRATEGIES

2. Series-of-Events Chain

Initiating Event 1
Event 2
Event 3
Event 4
Event 5
Final Event – Outcome

Study TIP

Organizing Information

Make a table to organize information about mitosis and cytokinesis. Across the top write the headings *Main idea* and *Details*. Along the sides write *Mitosis* and *Cytokinesis*.

Mitosis and Cytokinesis Divide Cells

Although mitosis is a continuous process, biologists traditionally divide it into four stages, as shown in **Figure 6-9.**

Mitosis

Step 1 Prophase Chromosomes coil up and become visible during prophase. The nuclear envelope dissolves and a spindle forms.

Step 2 Metaphase During metaphase the chromosomes move to the center of the cell and line up along the equator. Spindle fibers link the chromatids of each chromosome to opposite poles.

Step 3 Anaphase Centromeres divide during anaphase. The two chromatids (now called chromosomes) move toward opposite poles as the spindle fibers attached to them shorten.

Step 4 Telophase A nuclear envelope forms around the chromosomes at each pole. Chromosomes, now at opposite poles, uncoil

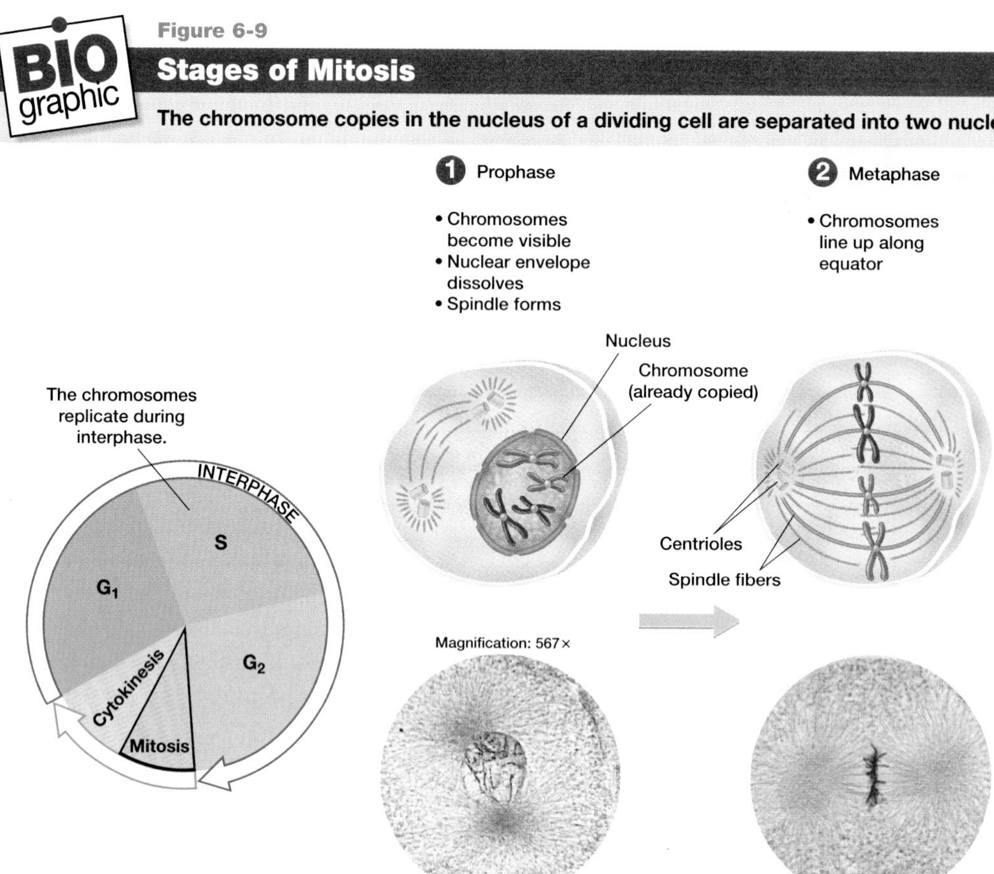

Figure 6-9
Stages of Mitosis
The chromosome copies in the nucleus of a dividing cell are separated into two nuclei.

and the spindle dissolves. The spindle fibers break down and disappear. Mitosis is complete.

Cytokinesis

As mitosis ends, cytokinesis begins. During cytokinesis, the cytoplasm of the cell is divided in half, and the cell membrane grows to enclose each cell, forming two separate cells as a result. The end result of mitosis and cytokinesis is two genetically identical cells where only one cell existed before.

During cytokinesis in animal cells and other cells that lack cell walls, the cell is pinched in half by a belt of protein threads, as shown in **Figure 6-10.**

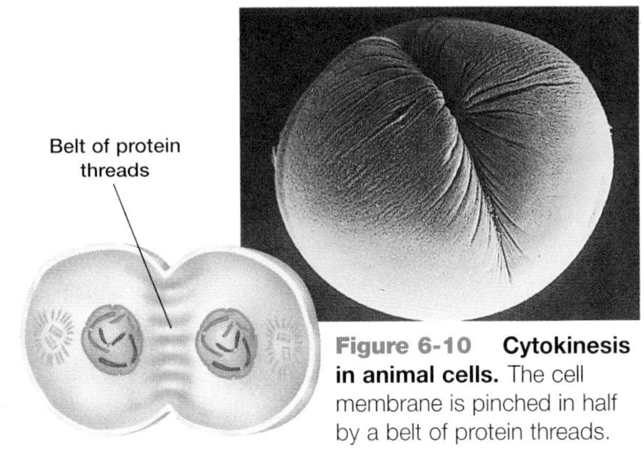

Belt of protein threads

Figure 6-10 Cytokinesis in animal cells. The cell membrane is pinched in half by a belt of protein threads.

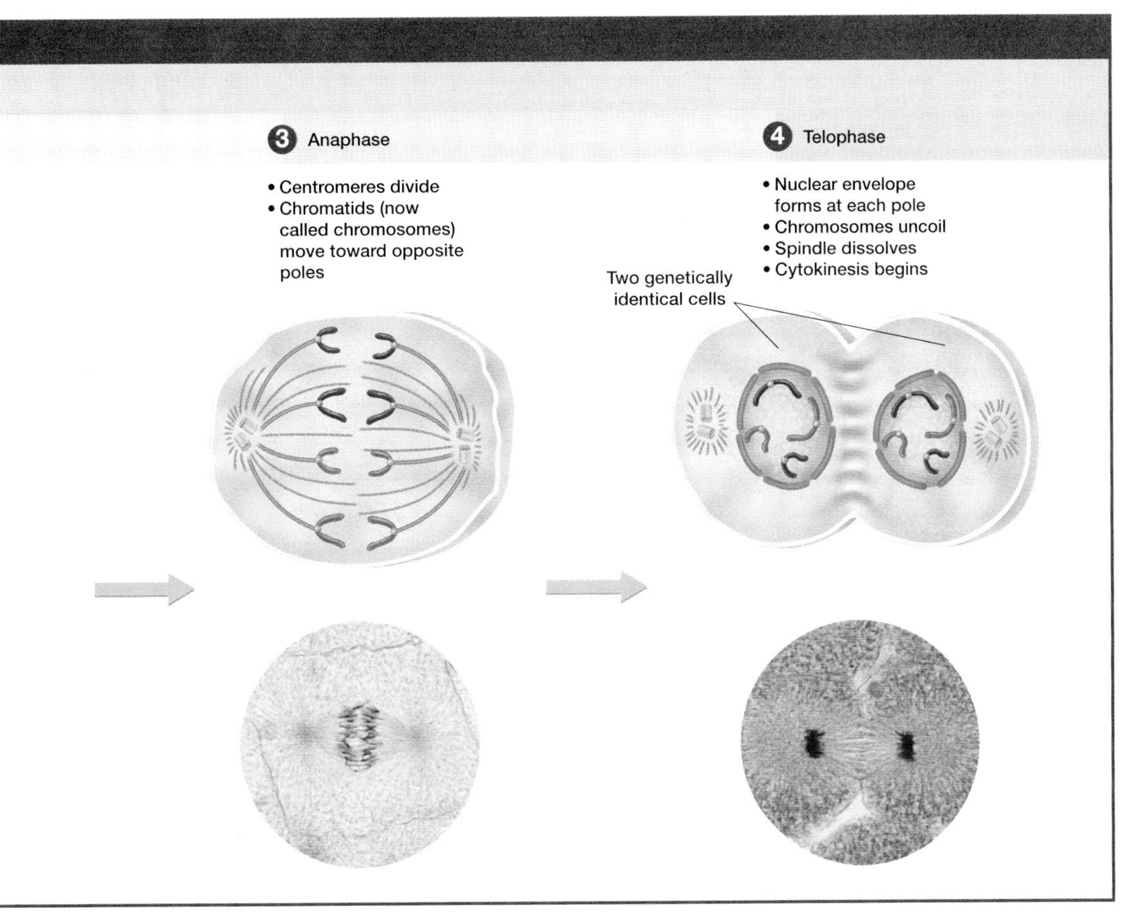

3 Anaphase
• Centromeres divide
• Chromatids (now called chromosomes) move toward opposite poles
4 Telophase
• Nuclear envelope forms at each pole
• Chromosomes uncoil
• Spindle dissolves
• Cytokinesis begins
Two genetically identical cells

Teaching TIP BASIC

Stages of Mitosis

Have students use map pencils to draw pictures of each stage of mitosis on four note cards. On one side, have students write the name of the stage. On the opposite side, have them draw an accurate sketch of the stage. They can then use these as flash cards to help each other learn the stages of mitosis. PORTFOLIO

Visual Strategy

Figure 6-10

Tell students that the belt of protein threads that are labeled in Figure 6-10 is actually a ring of the proteins actin and myosin, the same two proteins that interact to contract muscle cells. The furrowing of the cell membrane occurs perpendicular to the long axis of the spindle. Cytokinesis usually begins in anaphase but is not completed until after the two nuclei have formed.

Overcoming Misconceptions

Mitosis, Cytokinesis, and Cell Division

Students often think that mitosis is the same as cell division. Be sure they understand that mitosis refers strictly to the division of the chromosomes, while cytokinesis refers to the division of the cytoplasm. Remind students that cell division is just one of the four events that make up the cell cycle.

What do mitosis and cytokinesis look like?

Time 20 minutes

Process Skills Identifying, comparing, inferring, predicting

Teaching Strategies

- Use the information on pp. 1040–1041 of the **Appendix** to review the correct procedure for operating a microscope.
- Caution students to use care when handling prepared slides.

Analysis Answers

1. prophase: chromosomes are visible as dark threads; metaphase: chromatids line up along the equator; anaphase: chromosomes appear to pull toward opposite poles; telophase: chromosomes are at opposite poles
2. Answers will vary depending on slide preparation but should indicate more cells in interphase than in the other stages.
3. Cells spend the majority of their time in interphase.

3 Assessment Options

Closure

Relating Concepts

Have students hypothesize about how a cell's ATP use changes during mitosis. *(The events of mitosis require a lot of additional energy, which is supplied by ATP.)*

Review

Assign Review—Section 6-3.

Reteaching — Basic

Provide students with yarn to represent chromosomes, nuclear envelopes, and cell membranes; tie tabs to represent centromeres; and string to represent spindle fibers. Have them recreate on their desktops a cell undergoing mitosis. PORTFOLIO

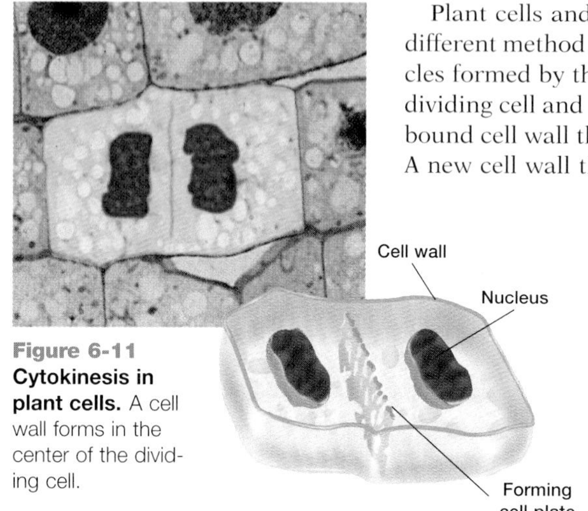

Figure 6-11
Cytokinesis in plant cells. A cell wall forms in the center of the dividing cell.

Plant cells and other cells that have rigid cell walls have a different method of dividing the cytoplasm. In plant cells, vesicles formed by the Golgi apparatus fuse at the midline of the dividing cell and form a cell plate. A cell plate is a membrane-bound cell wall that forms across the middle of the plant cell. A new cell wall then forms on both sides of the cell plate, as shown in **Figure 6-11.** When complete, the cell plate separates the plant cell into two new plant cells.

In both animal and plant cells, offspring cells are about equal in size. Each offspring cell receives an identical copy of the original cell's chromosomes. Each offspring cell also receives about one-half of the original cell's cytoplasm and organelles.

What do mitosis and cytokinesis look like?

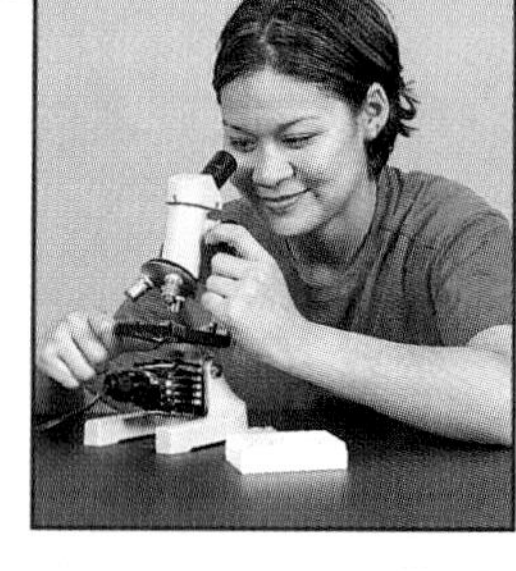

You can identify the stages of mitosis and the process of cytokinesis by observing slides of tissues undergoing mitosis under a compound microscope.

Materials

compound microscope, prepared slide of mitosis, paper, pencil

Procedure

1. View a prepared slide of cells undergoing mitosis under low power of a compound microscope.
2. Move the slide until you find a section where different stages of mitosis are visible.
3. Switch to high power. Use the photos in Figure 6-9 to help you locate and identify cells in interphase and in each stage of mitosis.
4. On a separate piece of paper, sketch an example of each stage. Label each sketch with the following terms where appropriate: chromosomes, cell membrane, cytoplasm, nucleus, spindle, cell wall.
5. Switch to low power, and estimate how many cells are clearly in interphase and how many cells are in one of the stages of mitosis.

Analysis

1. **Describe** the activity of chromosomes in each stage of mitosis.
2. **Compare** the number of cells in interphase with the number of cells in one of the stages of mitosis.
3. **SKILL Inferring Relationships** What does your answer to item 2 indicate about the relative length of interphase?

Review SECTION 6-3

1. **Describe** the function of the microtubules during anaphase.
2. **Describe** the events that occur during each of the four stages of mitosis.
3. **Compare** how cytokinesis occurs in plant cells and animal cells.
4. **SKILL Predicting Outcomes** What would happen if cytokinesis were omitted from the cell cycle?

Review SECTION 6-3 ANSWERS

1. The microtubules, attached to the centromeres, shorten and pull the chromatids to opposite poles, similar to a fishing line reeling in a fish.
2. prophase: chromosomes become visible, nuclear envelope dissolves, spindle forms; metaphase: chromosomes line up at the equator, spindle fibers attach to each chromatid; anaphase: centromeres divide, chromatids move to opposite poles due to shortening spindle fibers; telophase: nuclear envelope forms around the chromatids at each pole, chromosomes uncoil, spindle fibers break down and disappear
3. In plant cells, vesicles formed by the Golgi apparatus fuse at the equator and form the cell plate. A new cell wall forms on both sides of the cell plate. In animal cells, the cell is pinched in half by a belt of protein threads.
4. The resulting cell would have two nuclei, and twice as much genetic information. The cell would probably not survive.

Study ZONE CHAPTER 6

Use the Key Concepts and Key Terms listed below to review the main ideas in this chapter. If you need to review the meaning of a term, turn to the page indicated.

Key Concepts

6-1 Chromosomes

- Cell division allows organisms to reproduce asexually, grow, replace worn-out or damaged tissues, and form gametes.
- Bacteria reproduce by binary fission.
- Before cell division, DNA coils tightly around proteins and forms chromosomes. At cell division, each chromosome consists of two chromatids attached at the centromere.
- Each organism has a characteristic number of chromosomes.
- Human somatic cells are diploid, with 23 pairs of homologous chromosomes. Human gametes are haploid, with 23 chromosomes.
- Sex chromosomes carry information that determines an organism's sex.
- Changes in chromosome number or structure can cause abnormal development. Karyotypes are used to examine an individual's chromosomes.

6-2 The Cell Cycle

- The life of a eukaryotic cell—the cell cycle—includes interphase, mitosis, and cytokinesis.
- Interphase consists of 3 phases: growth, DNA synthesis (replication), and preparation for cell division. A cell about to divide enters the mitosis and cytokinesis phases of the cell life cycle.
- The cell cycle is carefully controlled; failure of cellular control can result in cancer.

6-3 Mitosis and Cytokinesis

- During mitosis, spindle fibers drag the chromatids to opposite poles of the cell. A nuclear envelope forms. Each resulting nucleus contains a set of the original cell's chromosomes.
- Cytokinesis in animal cells occurs when a belt of protein threads pinches the cell membrane in half. Cytokinesis in plant cells occurs when vesicles from the Golgi apparatus fuse to form a cell plate.

Key Terms

6-1

gamete (118)
binary fission (119)
gene (119)
chromosome (119)
chromatid (119)
centromere (119)
homologous chromosome (120)
diploid (121)
haploid (121)
zygote (121)
autosome (122)
sex chromosome (122)
karyotype (122)

6-2

cell cycle (125)
interphase (125)
mitosis (125)
cytokinesis (125)
cancer (126)

6-3

spindle (128)

BIOLOGY INTERACTIVE TUTOR: Unit 4—*Cell Reproduction* Use Topics 1–4 in this unit to review the key concepts and terms in this chapter.

Study TIP Ready?

If you think you understand these Key Concepts and Key Terms, then you're ready for the Performance Zone.

Study ZONE CHAPTER 6

Review and Practice

Assign the following worksheets for Chapter 6:

- **Student Review Guide**
- **Concept Mapping Worksheet**
- **Test Preparation Pretest**
- **Vocabulary Worksheet**
- **Critical Thinking Review Worksheet**
- **Science Skills Worksheet**

Assessment Options

The following assessment options are available for this chapter:

Resource Link

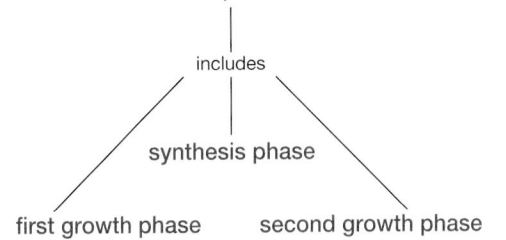

- Assign **Chapter Test 6** to assess students' understanding of the chapter.

Alternative Assessment

- Set up a lab practical with stations. Include microscopes with slides at various stages of mitosis, pictures of chromosomes, and karyotypes for interpretation. PORTFOLIO

Portfolio Assessment

- *Portfolio Assessment* at the front of this book provides options for building and evaluating student portfolios. Students and teachers can select items from the areas listed for inclusion in a portfolio.
- See items labeled PORTFOLIO on pp. 119, 120, 121, 122, 124, 126, 127, 130, 131, and 133.

Answer to Concept Map

The following is one possible answer to Performance Zone item 1 on page 134.

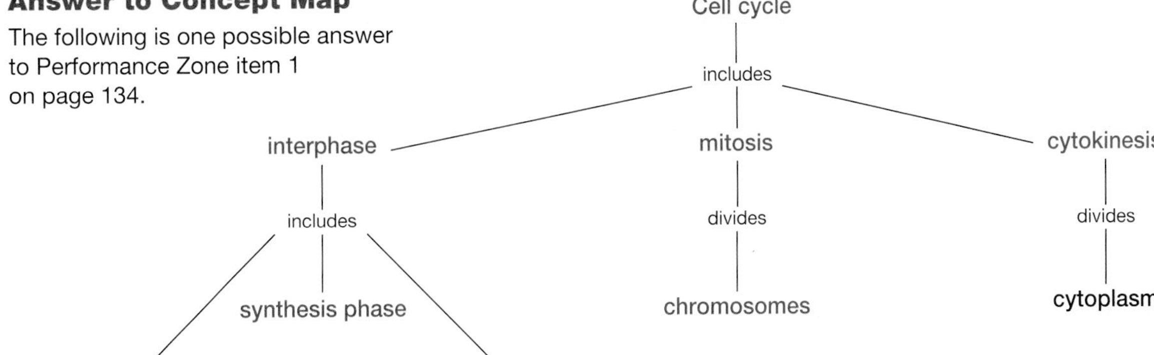

Performance ZONE
CHAPTER 6 REVIEW ANSWERS

Assignment Guide

SECTION	REVIEW
6-1	2, 3, 4, 5, 6, 14, 23, 25
6-2	1, 7, 8, 15, 17, 18, 19, 20, 21, 22, 24, 26, 27
6-3	9, 10, 11, 12, 13, 16

Understanding and Applying Concepts

1. The answer to the concept map is found at the bottom of page 133.
2. a. Gametes, or reproductive cells, are haploid, *n*, and contain one set of chromosomes. When gametes unite, they form a zygote, which is diploid, 2*n*, and contains two sets of chromosomes.

   b. Of the 23 pairs of chromosomes, 22 pairs are called autosomes. The individual's sex is determined by the 23rd pair, the sex chromosomes. A karyotype is a photograph of chromosomes arranged by size.

   c. In humans, females are represented by XX chromosomes and males are represented by XY chromosomes.
3. b
4. c
5. b
6. d
7. a
8. c
9. a
10. b
11. d
12. a
13. Mitosis increases the chance that both cells will receive complete sets of chromosomes.
14. chromosome abnormalities
15. Normal cells can become cancer cells if a gene that codes for a protein that regulates cell growth and division is mutated. The mutation may cause uncontrolled cell growth—cancer.
16. Answers will vary but should include organelles such as lysosomes, the Golgi apparatus, the endoplasmic reticulum, ribosomes, and vacuoles.

Performance ZONE CHAPTER 6 Review

Understanding and Applying Concepts

1. **Concept Mapping** Make a concept map that shows the events in the cell cycle. Try to include the following words in your map: cell cycle, interphase, synthesis phase, chromosomes, cytokinesis, mitosis, second growth phase, first growth phase.

2. **Understanding Vocabulary** For each set of terms, write one or more sentences summarizing information learned in this chapter.
   a. haploid, diploid, gametes, zygote
   b. autosome, sex chromosome, karyotype
   c. XX chromosomes, XY chromosomes

3. Bacteria reproduce asexually by
   a. disjunction.
   b. binary fission.
   c. cytokinesis.
   d. mitosis.

4. Chromatids are
   a. dense patches within the nucleus.
   b. bacterial chromosomes.
   c. two exact copies of DNA that make up each chromosome.
   d. structures that move chromosomes during mitosis.

5. In humans, females have _______ sex chromosomes.
   a. XY
   b. XX
   c. YY
   d. XO

6. The diagram below represents a(n) _______ mutation.
   a. deletion
   b. translocation
   c. inversion
   d. duplication

1 2 3 4 5
Original chromosome

1 2 1 2 3 4 5
? mutation

7. The stage of the cell cycle in which a cell's DNA is copied is called the _______ phase.
   a. S
   b. G_1
   c. G_2
   d. mitosis

8. When the cell cycle is not controlled, _______ may result.
   a. Down syndrome
   b. binary fission
   c. cancer
   d. a spindle

9. As a result of mitosis, each resulting cell
   a. receives an exact copy of all of the chromosomes present in the original cell.
   b. receives most of the chromosomes from the original cell.
   c. donates a chromosome to the original cell.
   d. receives exactly half the chromosomes from the original cell.

10. During the metaphase stage of mitosis,
    a. the cell membrane folds inward.
    b. chromosomes line up at the cell's equator.
    c. spindle fibers shorten, pulling chromosomes to the poles of the cell.
    d. chromosomes are at opposite ends of the cell.

11. The process in which the cytoplasm of a cell is divided is called
    a. disjunction.
    b. interphase.
    c. binary fission.
    d. cytokinesis.

12. How does cell division differ between animal and plant cells?
    a. Plant cells do not have centrioles.
    b. Animal cells form a cell plate.
    c. Animal cells are always haploid.
    d. Animal cells do not have centrioles.

13. **Theme Evolution** Bacteria simply split, but eukaryotic cells reproduce by mitosis. How has the evolution of mitosis in eukaryotic cells aided their reproduction?

14. **Health Watch** What information, besides chromosome number, can amniocentesis and chorionic villi sampling reveal?

15. **Exploring Further** Summarize how normal cells can become cancer cells.

16. **Chapter Links** List five organelles that must divide or fragment before the cytoplasm divides. (**Hint:** See Chapter 3, Section 3-2.)

Skills Assessment

17. cytokinesis
18. G_1
19. DNA replicates.
20. Answers will vary. For example: In skin cancer, direct or artificial sources of ultraviolet (UV) radiation are the main cause. Symptoms include a change on the skin, especially a new growth or a sore that does not heal. Treatment usually involves some type of surgery and possibly radiation therapy or chemotherapy. Skin cancer is cured in 85–95 percent of all cases. Screening includes regular checks by people for new growths or other changes in the skin or by doctors during routine exams.

Skills Assessment

Part 1: Interpreting Graphics

Use the illustration of the cell cycle below to answer the following questions:

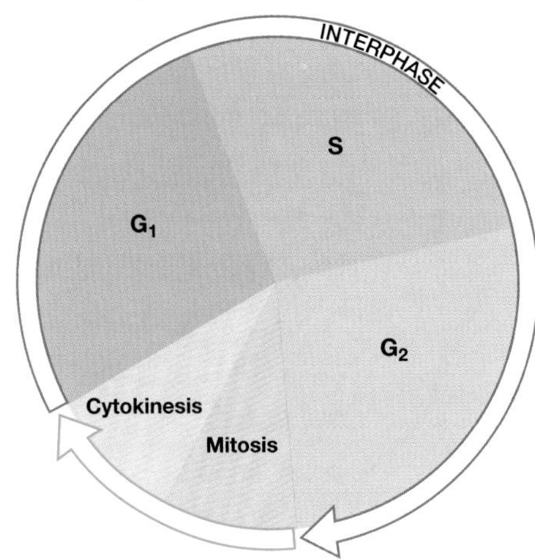

17. During which phase of the cell cycle does the cytoplasm divide?
18. During which phase(s) of the cell cycle would one expect to find a rapidly growing cell?
19. Describe what occurs during the S phase of the cell cycle.

Part 2: Life/Work Skills

Use the media center or Internet resources to investigate different types of cancer.

20. **Communicating** Write a report on one type of cancer and its causes, symptoms, treatments, and survival rate. Include information on screening methods to detect the cancer.
21. **Finding Information** Scientists have determined that telomeres (the tips of chromosomes) are shaved down slightly every time a cell divides. When the telomeres reach a certain length, the cell may lose its ability to divide. Find out what scientists have recently uncovered about telomeres and their association with cell division and cancer. Prepare a brief written report to share with your class.

Critical Thinking

22. **Evaluating Conclusions** Damage to the brain or the spinal cord is usually permanent. Use your knowledge of the cell cycle to explain why damaged cells in the brain or spinal cord are not replaced.
23. **Evaluating Viewpoints** A newspaper article describes the concern that more infants with Down syndrome will be born in the United States as more women delay having children. Do you agree or disagree with this concern? Explain.

Portfolio Projects

24. **Interpreting Information** Read "Diverse Strategies to Vanquish Cancer," by Kathleen Fackelmann (*Science News*, May 3, 1997, vol. 151, pp. 274–275). Write a report explaining the problems associated with using chemotherapy to kill tumor cells. Summarize three approaches that attempt to overcome these problems.
25. **Summarizing Information** Read "It's a Girl! Is Sex Selection the First Step to Designer Children?" by Kathleen Fackelmann (*Science News*, Nov. 28, 1998, vol. 154, pp. 350–351). Write a report summarizing the research and the ethical debate over the technique that allows researchers to sort sperm carrying a specific type of sex chromosome.
26. **Career Focus Oncologist** Research the field of oncology, and write a report on your findings. Your report should include a job description, training required, kinds of employers, growth prospects, and starting salary.
27. **Unit 4—*Cell Reproduction*** Write a report summarizing how different cancer-fighting drugs kill cancer cells by interrupting the life cycle of the cells. How is cell division related to the spread of cancer?

21. Answers will vary. Cells from large tumors often have unusually short telomeres. Telomerase, an enzyme that catalyzes the lengthening of telomeres, stabilizes telomere length, especially in cancer cells. Thus, researchers are focusing on telomerase as a target for cancer diagnosis and chemotherapy.

Critical Thinking

22. Most nerve cells remain in interphase G_1 permanently. Because they do not undergo mitosis, most damaged nerve cells are not replaced.
23. Answers will vary but should include statistics about how risk factors change as maternal age increases.

Portfolio Projects

24. Answers will vary. Because chemotherapy drugs kill healthy cells in addition to malignant cells, they have many side effects. Also, some cancer cells survive the chemotherapy. The surviving cancer cells may be resistant to further rounds of chemotherapy and can subsequently form lethal tumors. Three approaches to chemotherapy include tailoring drugs to kill tumor cells that contain certain cancer-causing genes, using viruses to destroy cells with certain genetic defects, and blocking the growth of blood vessels that supply blood to tumors.
25. Answers will vary. The researchers treated sperm with a dye that attaches to DNA and glows under laser light. When exposed to a laser beam, sperm carrying an X chromosome glow brighter than sperm carrying a Y chromosome, allowing the researchers to sort the sperm. Ethical arguments include concerns about altering sex ratios, the influence of sexism in the decision to select certain sperm, and increasing parental control over their children's characteristics.
26. Oncologists treat and manage cancer patients by using radiation therapy or drugs. Training includes an M.D. or D.O. degree, a licensing exam, an internship, and a residency program. Employers include solo, partnership, or group practices, universities, and government agencies. Growth prospects are excellent. Starting salary will vary by region.
27. Answers will vary. For example: Vincristine and taxol block the mitosis spindle microtubules from functioning. Cell division is related to the spread of cancer in that cancer cells do not respond to the normal signals that regulate the cell cycle.

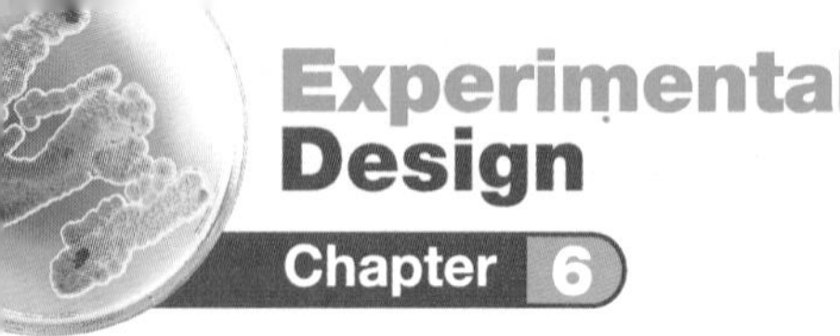

How Can You Model Mitosis?

Purpose

Students will build a model that will help them understand the events of mitosis. Students will also use the model to demonstrate the effects of nondisjunction and mutations.

Preparation Notes

Time Required: 45–55 minutes

Materials

Materials for this lab can be ordered from WARD'S. See *Master Materials List* at the front of this book for catalog numbers.

Preparation Tips

- For each model, students need four pipe cleaners of one color, four pipe cleaners of another color, four wooden beads, 90 cm of yarn, and 16 small white labels.
- You may want to cut the string before the start of the lab.

Safety Precautions

Discuss all safety symbols and caution statements with students.

Procedural Tips

- Review mitosis before starting the lab.
- If students have difficulty in beginning their models, ask the following questions:
  1. Which of these materials would make the best cell membrane? *(yarn)*
  2. Which would make the best spindle fibers? *(yarn)*
  3. Which would make the best chromosomes? *(pipe cleaners)*
  4. Which would make the best centromeres? *(beads)*
- Ask students what the cells they make were like before the chromosomes were duplicated. Emphasize that each pair of chromatids represents one chromosome.

Disposal

Most of the supplies for this lab can be saved and used again in Chapter 7.

Experimental Design

Chapter 6 How Can You Model Mitosis?

SKILLS

- Modeling
- Using scientific methods

OBJECTIVES

- **Describe** the events that occur in each stage of mitosis.
- **Relate** mitosis to genetic continuity.

MATERIALS

- pipe cleaners of at least two different colors
- yarn
- wooden beads
- white labels
- scissors

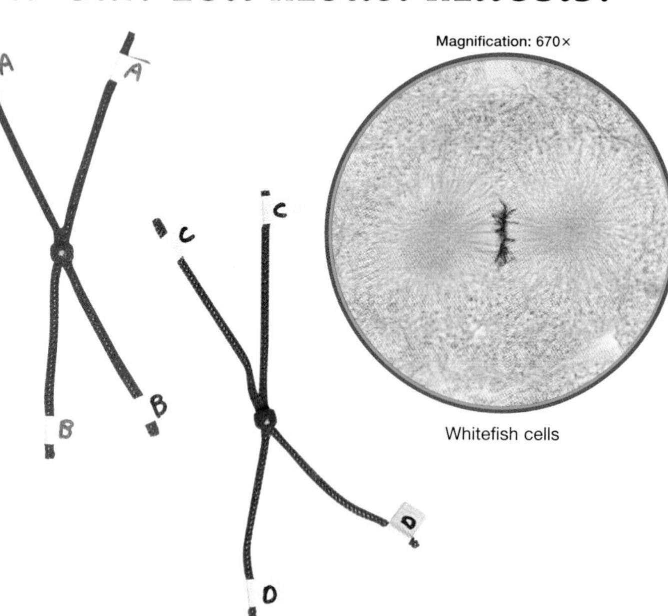

Whitefish cells

Before You Begin

The cell cycle includes all of the phases in the life of a cell. The **cell cycle** is a repeating sequence of cellular growth and division during the life of an organism. Mitosis is one of the phases in the cell cycle. **Mitosis** is the process by which the material in a cell's nucleus is divided during cell reproduction. In this lab, you will build a model that will help you understand the events of mitosis. You can also use the model to demonstrate the effects of **nondisjunction** and **mutations.**

1. Write a definition for each boldface term in the paragraph above and for the following terms: chromatid, centromere, spindle fiber, cytokinesis.
2. Where in the human body do cells undergo mitosis?
3. How does a cell prepare to divide during interphase of the cell cycle?
4. Based on the objectives for this lab, write a question you would like to explore about mitosis.

Procedure

PART A: Design a Model

1. Work with the members of your lab group to design a model of a cell that uses the materials listed for this lab. Be sure your model cell has at least two pairs of chromosomes and is about to undergo mitosis.

> **You Choose**
>
> **As you design your model, decide the following:**
>
> a. what question you will explore
> b. how to construct a cell membrane
> c. how to show that your cell is diploid
> d. how to show the locations of at least two genes on each chromosome
> e. how to show that chromosomes are duplicated before mitosis begins

2. Write out the plan for building your model. Have your teacher approve the plan before you begin building the model.

Before You Begin

Answers

1. *cell cycle*—a repeating five-phase sequence of eukaryotic cellular growth and division

   mitosis—the process during cell division in which the nucleus of a cell divides into two nuclei, each with the same number and kind of chromosomes

   nondisjunction—the failure of two chromosomes or chromatids to separate during nuclear division

   mutation—a change in the DNA of a gene or chromosome

   chromatid—one of a pair of strands of DNA that make up a chromosome

   centromere—a section of a chromosome where two chromatids are joined

   spindle fiber—a structure made of microtubules that helps pull chromatids apart during cell division

   cytokinesis—the process of dividing the cytoplasm of a cell
2. everywhere in the human body except in most nerve and skeletal muscle cells
3. During interphase, the cell grows, duplicates its chromosomes and organelles, and assembles microtubules.
4. Answers will vary. For example: How many chromosomes will each new nucleus have after mitosis has occurred?

3. Build the cell model your group designed. **CAUTION: Sharp or pointed objects can cause injury. Handle scissors carefully. Promptly notify your teacher of any injuries.** Use your model to demonstrate the phases of mitosis. Draw and label each phase you model.
4. Use your model to explore one of the questions written for step 4 of **Before You Begin.** Describe the steps you took to explore the question.

PART B: Test Hypotheses

Answer each of the following questions by writing a hypothesis. Use your model to test each hypothesis, and describe your results.

5. Cytokinesis follows mitosis. How will the size of each new cell that is formed following cytokinesis compare with that of the original cell?
6. Sometimes two chromatids fail to separate during mitosis. How might this failure affect the chromosome number of the two new cells?
7. A mutation is a permanent change in a gene or chromosome. What effect might a mutation in a parent cell have on future generations of cells that result from the parent cell?

PART C: Cleanup and Disposal

8. Dispose of paper and yarn scraps in the designated waste container.
9. Clean up your work area and all lab equipment. Return lab equipment to its proper place. Wash your hands thoroughly before you leave the lab and after you finish all work.

Analyze and Conclude

1. **Analyzing Results** How do the nuclei you made by modeling mitosis compare with the nucleus of the model cell you started with? Explain your result.
2. **Evaluating Methods** How could you modify your model to better illustrate the process of mitosis?
3. **Recognizing Patterns** How does the genetic makeup of the cells that result from mitosis compare with the genetic makeup of the original cell?
4. **Inferring Conclusions** How is mitosis important?
5. **Further Inquiry** Write a new question about mitosis or the cell cycle that could be explored with your model.

Do You Know?

Do research in the library or media center to answer these questions:

1. How often do different cells of the human body undergo mitosis?
2. What are some common chemicals that disrupt the cell cycle?

Use the following Internet resources to explore your own questions about mitosis or the cell cycle.

internet**connect**

SCILINKS NSTA
TOPIC: Cell cycle
GO TO: www.scilinks.org
KEYWORD: HX137

3. The ways the students model the stages of mitosis may vary. See diagram below for an example.
4. Answers will vary.
5. Each new cell will initially be smaller than the original cell that divides.
6. Nondisjunction of one of the chromosomes will result in one of the new cells having two copies of that chromosome and the other cell having none.
7. All of the cells in the subsequent generations of cells will carry the mutation.

Analyze and Conclude

Answers

1. The nuclei students made should be the same as the nucleus of the original cell except that the chromosomes in the original cell are not replicated until right before mitosis.
2. Answers will vary. Students could mention that the pipe cleaners do not show how the shapes of the chromosomes change as they are pulled apart.
3. The genes found in the cells that result from mitosis are the same as the genes in the original cell. The chromosomes replicate before cell division. One copy of each gene goes to each new cell.
4. Answers will vary. Students should mention that mitosis preserves the chromosome number and genetic makeup of cells.
5. Answers will vary. For example: What happens if the DNA is not replicated before mitosis begins?

Do You Know?

Answers

1. The frequency at which cells in the human body undergo mitosis varies from tissue to tissue. Epidermal cells divide several times a week, while most mature skeletal muscle cells and nerve cells never undergo mitosis.
2. Chemicals that disrupt the cell cycle include nicotine and griseofulvin, an antifungal drug.

Procedure

Answers

2. The cell model the students build may vary. For example: The students may use the yarn to represent the cell membrane and spindle fibers, the pipe cleaners to represent the chromosomes, the beads to represent the centromeres, and the labels to indicate the genes on the chromosomes.

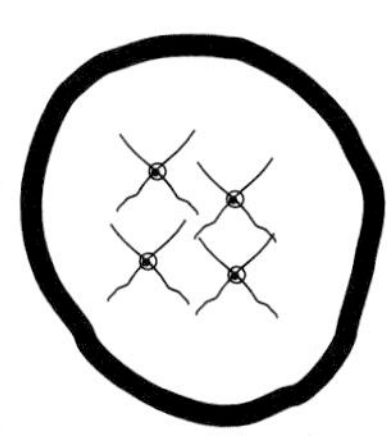

Prophase

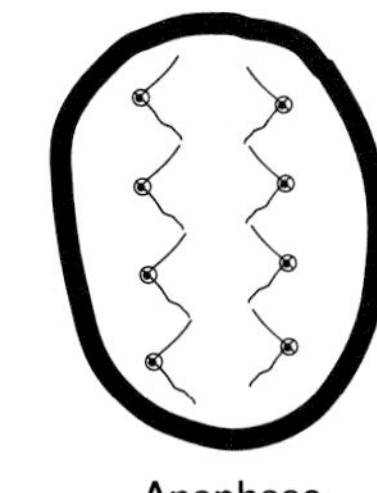

Metaphase

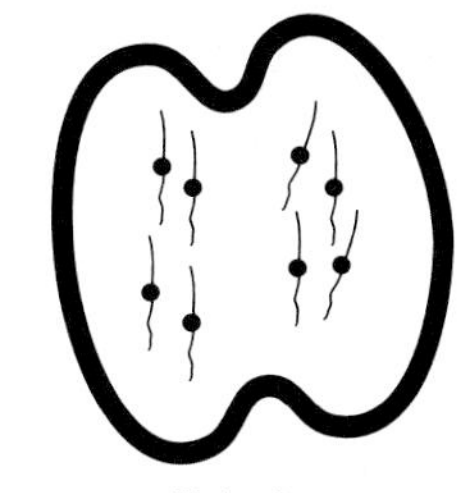

Anaphase

Telophase

UNIT 2 Principles of Genetics

Chapters

A DNA sequencing gel, shown here being examined by a technician, reveals an individual's unique chemical fingerprint.

in perspective

Hemophilia

Yesterday... Queen Victoria of England carried the gene for hemophilia. "Our poor family," she wrote in her diary, "seems persecuted by this awful disease, the worst I know." One son died of hemophilia and two daughters inherited the gene for the disease. **Why are carriers of hemophilia always female? Find out on pp. 173 and 178.**

Queen Victoria

Today... The blood of normal individuals contains a protein called factor VIII that enables the blood to clot after an injury. But in people with hemophilia, the gene that produces factor VIII is defective and bleeding continues uncontrolled. **Discover on pp. 206-212 how a gene controls the production of a protein.** Injections of factor VIII at the first sign of bleeding allow many hemophiliacs to control bleeding episodes.

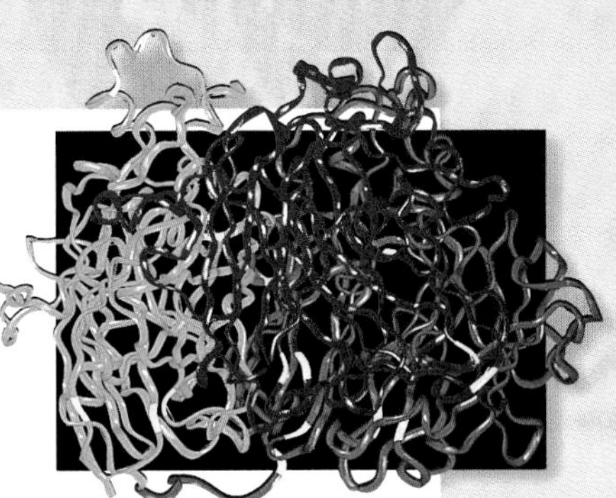

Factor VIII protein

Tomorrow... Biologists have recently learned to make a genetically engineered form of factor VIII. **What are the safety advantages of genetically-engineered factor VIII? Find out on pages 231-232.** New research in genetic engineering has yielded female pigs that produce factor VIII in their milk.

internet**connect**

SCiLINKS NSTA

TOPIC: Hemophilia
GO TO: www.scilinks.org
KEYWORD: HX139

Chapter 7

Meiosis and Sexual Reproduction

Planning Guide

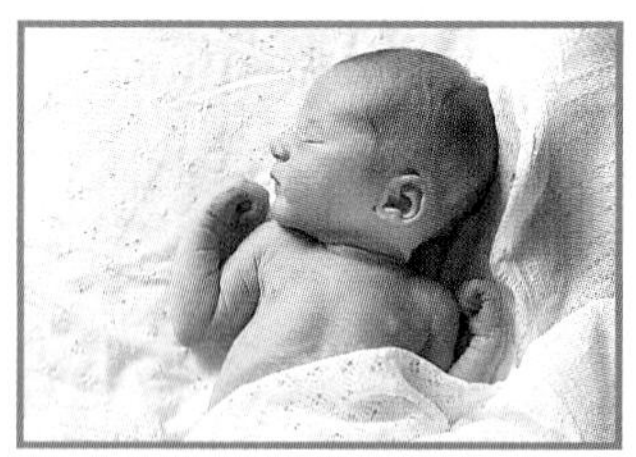

OBJECTIVES	MEETING INDIVIDUAL NEEDS	READING SKILLS AND STRATEGIES
7-1 Meiosis (BLOCKS 1 & 2 90 min) • **Summarize** the events that occur during meiosis I and meiosis II. • **Relate** how crossing-over, independent assortment, and random fertilization contribute to genetic variation. • **Compare** the formation of gametes in male and female animals.	Learning Styles, p. 142 BASIC ADVANCED • pp. 144, 147	ATE Active Reading, Technique: Brainstorming, p. 141 ■ Active Reading Guide, worksheet (7-1) ■ Directed Reading, worksheet (7-2) ■ Reading Strategies, Compare/Contrast Matrix—Alike/Different
7-2 Sexual Reproduction (BLOCKS 3 & 4 90 min) • **Differentiate** between asexual and sexual reproduction. • **Identify** three types of asexual reproduction. • **Evaluate** the relative genetic and evolutionary advantages and disadvantages of asexual and sexual reproduction. • **Differentiate** between the three major sexual life cycles found in eukaryotes.	Learning Styles, p. 148 BASIC ADVANCED • p. 152	ATE Active Reading, Technique: Reading Organizer, p. 149 ■ Active Reading Guide, worksheet (7-2) ■ Directed Reading, worksheet (7-2) ■ Reading Strategies, Recognizing Cycles ■ Supplemental Reading Guide, *The Double Helix*

Review and Assessment Resources

BLOCKS 5 & 6 90 min

TEST PREP

PE Study Zone, p. 153

PE Performance Zone, pp. 154–155

■ Chapter 7 worksheets:
- Vocabulary
- Concept Mapping
- Science Skills
- Critical Thinking

 Holt Biology Interactive Tutor Unit 4—Cell Reproduction, Topics 5–6

 Audio CD Program

■ Test Prep Pretest

CHAPTER TESTING

■ Chapter Tests (blackline master)

▲ Test Generator

■ Assessment Item Listing

See the correlations table at the front of this book for a list of the ***National Science Education Standards*** addressed in this chapter.

Smithsonian Institution® Visit **www.si.edu/hrw** for additional on-line resources.

KEY

Biology Principles and Explorations

PE Pupil's Edition

ATE Teacher's Edition

HOLT BIOSOURCES

Teaching Transparencies

Laboratory Program

One-Stop Planner CD-ROM
Shows these resources within a customizable daily lesson plan:

■ Teaching Resources

▲ Test Generator

VISUAL STRATEGIES	CLASSWORK AND HOMEWORK	TECHNOLOGY AND INTERNET RESOURCES
ATE Fig. 7-1, p. 143 27 Events of Meiosis I 28 Events of Meiosis II	**PE** **Quick Lab** How can you model crossing-over?, p. 145 **Lab C11** Modeling Meiosis **PE** **Review,** p. 147	CNN **Video Segment 4** Organ Cloning **Interactive Tutor** Unit 4—Cell Reproduction (Topics 5 and 6) **Holt Biology Videodiscs** Section 7-1 **Audio CD Program** Section 7-1 internet**connect** SCiLINKS NSTA **TOPICS:** • Genetic variation • Environmental toxins
ATE Fig. 7-6, p. 150 **ATE** Fig. 7-7, p. 151 **ATE** Fig. 7-8, p. 152	**PE** **Quick Lab** How do yeast reproduce?, p. 149 **PE** **Experimental Design** How Can You Model Meiosis?, p. 156 **PE** Review, p.152 **PE** Study Zone, p. 153	**Interactive Tutor** Unit 4—Cell Reproduction (Topics 5 and 6) **Holt Biology Videodiscs** Section 7-2 **Audio CD Program** Section 7-2 internet**connect** SCiLINKS NSTA **TOPIC:** Meiosis

ALTERNATIVE ASSESSMENT

PE Portfolio Projects 26–29, p. 155

ATE Alternative Assessment, p. 153

ATE Portfolio Assessment, p. 153

■ Portfolio Projects: Genetics Project

Lab Activity Materials

Quick Lab
p. 145 paper strips (4), pens or pencils (two colors), scissors, tape

Quick Lab
p. 149 microscope, microscope slides, dropper, culture of yeast

Experimental Design
p. 156 pipe cleaners of at least two different colors, yarn, wooden beads, white labels, scissors

Lesson Warm-up
ATE pictures of plants, animals, protists, bacteria, p. 148

Demonstrations
ATE 2 identical decks of playing cards, p. 143

BIOSOURCES LAB PROGRAM
- Lab Techniques and Experimental Design: Lab C11, p. 49

Chapter 7

Chapter Theme

Reproduction

Reproduction is the process by which organisms make more of their own kind. In sexual reproduction, two parents each form haploid reproductive cells, which join to form offspring. Meiosis halves the number of chromosomes when reproductive cells are formed. It also allows for the rapid generation of new genetic combinations.

Establishing Prior Knowledge

Before beginning this chapter, make certain students can answer the questions in **Study TIP Ready?** on page 141. If they cannot, encourage them to reread the sections indicated.

Opening Demonstration

Write the following table on the board:

Organism	Diploid chromosome number
Mosquito	6
Corn	20
Human	46
Horse	64

On the board, outline what would happen at fertilization if sperm and egg cells were diploid. Ask students to calculate the chromosome number of each organism after five generations. *(mosquito—12, 24, 48, 96, 192; corn—40, 80, 160, 320, 640; human—92, 184, 368, 736, 1472; horse—128, 256, 512, 1024, 2048)*

CHAPTER 7 Meiosis and Sexual Reproduction

Sperm on the surface of an egg (2890×)

What's Ahead — and Why
CHAPTER OUTLINE

7-1 Meiosis

- Meiosis Forms Haploid Cells
- Meiosis Contributes to Genetic Variation
- Gamete Formation in Male and Female Animals Involves Meiosis

7-2 Sexual Reproduction

- Similarity to Parents Is Determined by the Type of Reproduction
- Eukaryotes Have Three Kinds of Sexual Life Cycles

Students will learn

- how haploid cells are formed, how genetic variation occurs, and how gametes form in males and females.
- how types of reproduction produce either identical or genetically varied individuals and that eukaryotic life cycles alternate between haploid and diploid chromosome numbers.

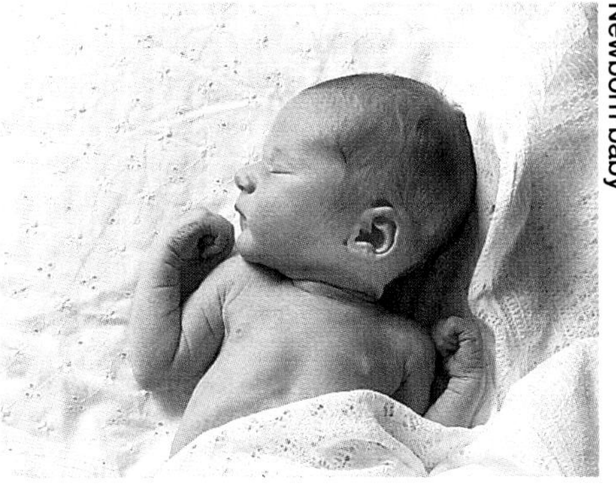
Newborn baby

Looking Ahead

7-1 Meiosis
How are 8 million different gene combinations possible from one cell that divides?

7-2 Sexual Reproduction
Why are some offspring identical to their parents, while others are not?

Labs

Quick Labs
How can you model crossing-over? (p. 145)
How do yeast reproduce? (p. 149)

Experimental Design
How Can You Model Meiosis? (p. 156)

Features

Health Watch
Egg Cells and Toxins (p. 147)

Exploring Further
Cloning by Parthenogenesis (p. 151)

Unit 4—*Cell Reproduction*
Topics 5 and 6

internet**connect**
*sci*LINKS NSTA
National Science Teachers Association *sci*LINKS Internet resources are located throughout this chapter.

***Only one out of several hundred million** sperm cells will complete its journey to the egg. A special form of cell reproduction forms sperm and egg cells. When the sperm and egg cell join, the genetic instructions in both cells are brought together and a new individual is formed.*

Study TIP

Ready?

Check your understanding of these topics to help you prepare for what's ahead.

Can you...

- **define** the terms *evolution* and *natural selection*? (Section 1-1)
- **define** the term *homologous chromosomes* and identify the chromatids? (Section 6-1)
- **differentiate** between haploid and diploid cells? (Section 6-1)
- **describe** the structure and function of the spindle? (Section 6-3)
- **summarize** the steps of mitosis? (Section 6-3)

If not, *review the sections indicated.*

Chapter 7

Reading Strategies

Active Reading

Technique: Brainstorming

Pair each student with a partner. Tell students to study pp. 142–143 (Section 7-1) and Figure 7-1 on pp. 142–143. Have student pairs read pp. 142–143 aloud, sharing the reading equally. Then ask students to answer the following questions:

- What is the subject of this section?
- What are the similarities between meiosis I and meiosis II?
- What are the differences between meiosis I and meiosis II?

Directed Reading

Assign **Directed Reading Worksheet Chapter 7** to help students interact with the text as they read the chapter.

Interactive Reading

Assign ***Biology: Principles and Explorations* Audio CD Program Chapter 7** to help reluctant readers achieve greater success in reading the chapter.

Reading Skills

Assign Reading Organizers and Reading Strategies Worksheets from the **One-Stop Planner CD-ROM with Test Generator** to help students learn and practice effective reading skills.

internet**connect**

Holt *Biology: Principles and Explorations*

On-line Resources: go.hrw.com
Visit the HRW Web site for a variety of resources related to this chapter. Just type in the keyword HX0 Chapter 7.

Holt *Biology: Principles and Explorations*

On-line Resources: www.scilinks.org
The following *sci*LINKS Internet resources can be found in the student text for this chapter:

Page 144	Page 147	Page 157
TOPIC: Genetic variation	**TOPIC:** Environmental toxins	**TOPIC:** Meiosis
KEYWORD: HX144	**KEYWORD:** HX147	**KEYWORD:** HX157

Chapter 7

1 Prepare

Scheduling

- **Block:** About 90 minutes will be needed to complete this section.
- **Traditional:** About two class periods will be needed to complete this section.
- **Planning:** See the Planning Guide on pp. 139A and 139B for lecture, classwork, and assignment options for Section 7-1. Lesson plans for Section 7-1 can be found in a lesson cycle format in ***Biology: Principles and Explorations*** **One-Stop Planner CD-ROM with Test Generator.**

Resource *Link*

Assign **Active Reading Worksheet Section 7-1** to help students comprehend the key concepts and visuals in this lesson.

2 Teach

Lesson Warm-up

On the board, draw a sperm cell on the edge of an egg cell. Each cell should have three homologous chromosome pairs. Tell students that when the sperm fertilizes the egg, the number of chromosomes in the resulting cell, the first cell of an individual, will double. Lead students to the conclusion that the number of chromosomes in sperm and egg cells must be halved, or the number of chromosomes in each subsequent generation will continue to double. Tell students that the process by which the chromosome number is halved when reproductive cells are formed is called meiosis.

Section 7-1 Meiosis

Objectives

- **Summarize** the events that occur during meiosis. (pp. 142–143)
- **Relate** crossing-over, independent assortment, and random fertilization to genetic variation. (pp. 144–145)
- **Compare** the formation of gametes in male and female animals. (pp. 146–147)

Key Terms

meiosis
crossing-over
independent assortment
spermatogenesis
sperm
oogenesis
ovum

Meiosis Forms Haploid Cells

Some organisms reproduce by joining gametes to form the first cell of a new individual. The gametes are haploid—they contain one set of chromosomes. Imagine how the chromosome number would increase with each generation if gametes were not haploid!

Meiosis *(meye OH sihs)* is a form of cell division that halves the number of chromosomes when forming specialized reproductive cells, such as gametes or spores. Meiosis involves two divisions of the nucleus—meiosis I and meiosis II—and each division is subdivided into prophase, metaphase, anaphase, and telophase stages.

Before meiosis begins, the DNA in the original cell is replicated. Thus, meiosis starts with homologous chromosomes. Recall that homologous chromosomes are similar in size, shape, and genetic content. The stages of meiosis are summarized in **Figure 7-1.**

Step 1 Prophase I The chromosomes condense, and the nuclear envelope breaks down. Homologous chromosomes pair all along their length and then cross-over. **Crossing-over** occurs when portions of a chromatid on one homologous chromosome are

Figure 7-1

Stages of Meiosis

Four cells are produced, each with half as much genetic material as the original cell.

1 Prophase I

Chromosomes become visible. The nuclear envelope breaks down. Crossing-over occurs.

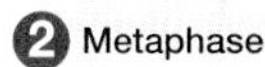

Pairs of homologous chromosomes move to the equator of the cell.

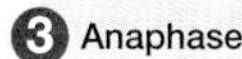

Homologous chromosomes move to opposite poles of the cell.

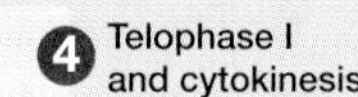

Chromosomes gather at the poles of the cell. Cytoplasm divides.

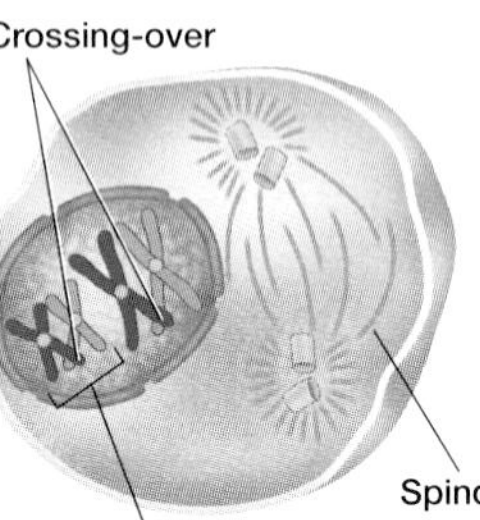

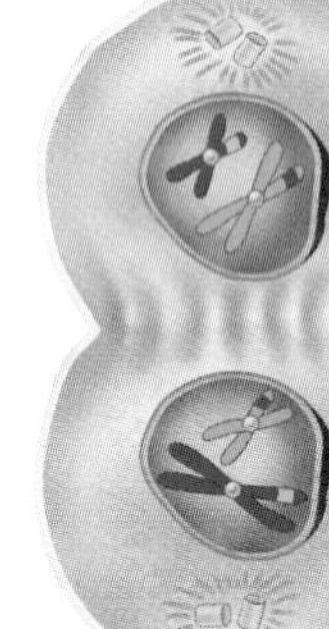

Integrating Different Learning Styles

Logical/Mathematical	**PE**	Performance Zone items 24 and 25, p. 155
	ATE	Opening Demonstration, p. 140; Cross-Disciplinary Connection, p. 144
Visual/Spatial	**PE**	Quick Lab, p. 145; Performance Zone item 13, p. 154; Experimental Design, pp. 156–157
	ATE	Lesson Warm-up, p. 142; Demonstration, p. 143; Visual Strategy, p. 143; Teaching Tips, pp. 144, 145, 146; Reteaching, p. 147
Interpersonal	**PE**	Performance Zone item 23, p. 155
Verbal/Linguistic	**PE**	Study Tip, p. 145; Performance Zone items 2, 21, 22, 26, 28, and 29, pp. 154–155
	ATE	Internet Connect, p. 144; Career, p. 146; Closure, p. 147

broken and exchanged with the corresponding portions on one of the chromatids of the other homologous chromosome.

Step ❷ Metaphase I The pairs of homologous chromosomes are moved by the spindle to the equator of the cell. The homologous chromosomes, each made up of two chromatids, remain together.

Step ❸ Anaphase I The homologous chromosomes separate. As in mitosis, the chromosomes of each pair are pulled to opposite poles of the cell by the spindle fibers. *But the chromatids do not separate at their centromeres—each chromosome is still composed of two chromatids. The genetic material, however, has recombined.*

Step ❹ Telophase I Individual chromosomes gather at each of the poles. In most organisms, the cytoplasm divides (cytokinesis), forming two new cells. Both cells or poles contain one chromosome from each pair of homologous chromosomes. *The chromosomes do not replicate between meiosis I and meiosis II.*

Step ❺ Prophase II A new spindle forms around the chromosomes.

Step ❻ Metaphase II The chromosomes line up along the equator and are attached at their centromeres to spindle fibers.

Step ❼ Anaphase II The centromeres divide, and the chromatids (now called chromosomes) move to opposite poles of the cell.

Step ❽ Telophase II A nuclear envelope forms around each set of chromosomes. The spindle breaks down, and the cell undergoes cytokinesis. The result of meiosis is four haploid cells.

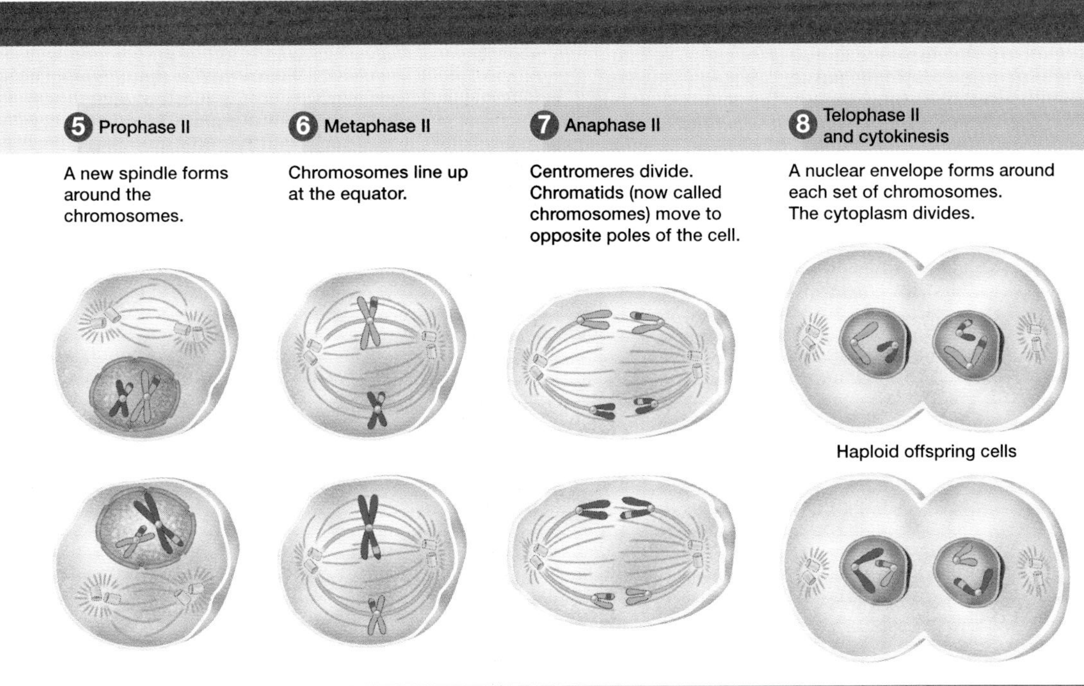

5 Prophase II
A new spindle forms around the chromosomes.
6 Metaphase II
Chromosomes line up at the equator.
7 Anaphase II
Centromeres divide. Chromatids (now called chromosomes) move to opposite poles of the cell.
8 Telophase II and cytokinesis
A nuclear envelope forms around each set of chromosomes. The cytoplasm divides.
Haploid offspring cells

Overcoming Misconceptions

Reducing the Chromosome Number

Students may be confused when they learn that the cells produced by meiosis I as well as those produced by meiosis II are haploid. Point out that both sets of cells are haploid because they contain one-half the number of chromosomes as the original cell. Ask students to determine the diploid chromosomal number for the cell that is in prophase I in Figure 7-1 on pp. 142–143. *(The diploid number is 4.)* Remind students that prior to prophase I, the original cell's chromosomes replicate. Because the cells produced at the end of meiosis I and meiosis II each contain two chromosomes, they are haploid cells.

Visual Strategy

Figure 7-1

BIO graphic

Ask students to describe what is happening to the chromosomes at each stage of meiosis. Ask students to explain the difference between the metaphase stage of mitosis as seen in Figure 6-9 on pp. 130–131 and the metaphase I stage of meiosis as seen in **Step ❷** of Figure 7-1. *(In metaphase I of meiosis, the homologous chromosomes pair up rather than line up independently along the equator, as they do in mitosis.)* Ask students to identify the stage in meiosis when the diploid number of chromosomes is reduced to the haploid number. *(telophase I)* Ask them why the haploid cells produced by meiosis I and meiosis II look different from each other. *(Prior to meiosis I, the chromosomes replicate. Thus, in telophase I, the chromosome number has been halved, but each chromosome has twice the original amount of DNA. Prior to meiosis II, the chromosomes do not replicate.)* Have students compare the stages of meiosis II as illustrated in Figure 7-1 to those of mitosis as drawn in Figure 6-9. *(The stages are identical except in meiosis II, the genetic material has recombined.)*

Demonstration

Modeling Meiosis I

Obtain two identical decks of playing cards. Tape together, side-by-side, cards of the same number and suit from both decks. Tell students that each card pair represents a chromosome made of two chromatids. (Each individual card is a chromatid.) Place pairs of cards with the same number but a different suit next to each other to represent homologous chromosomes. Have students match several of the pairs (homologous chromosomes) and then separate the cards to model the steps of meiosis.

internet**connect**

SciLINKS NSTA Have students research genetic variation using the Web site in the Internet Connect box on page 144. Have students write a report to summarize their findings. PORTFOLIO

CROSS-DISCIPLINARY CONNECTION

◆ **Mathematics**

At metaphase I, the orientation of homologous pairs is random: the paternal or maternal homologue can be on either side of the equator. Because of this random orientation, the number of possible chromosome combinations in the gametes is 2^n, where 2 represents the number of homologues in a pair and n represents the haploid chromosome number. Therefore, the number of possible chromosome combinations for 23 chromosomes is 2^{23}. Have students use their calculators to determine how many combinations are possible. *(8,388,608 combinations)* PORTFOLIO

Teaching TIP BASIC

Crossing-Over

Draw two large chromosomes on the board with colored chalk. Each chromosome should have two chromatids and a centromere. Use different colors along the length of each chromatid to represent genes. Ask students to show crossing-over in two different places by erasing and redrawing the chromosomes with the corresponding gene color. Point out that because of the random orientation of chromosomes at metaphase I and because of crossing-over, children resemble their parents but never look exactly like them.

Meiosis Contributes to Genetic Variation

internet**connect**

SciLINKS NSTA

TOPIC: Genetic variation
GO TO: www.scilinks.org
KEYWORD: HX144

Meiosis is an important process that allows for the rapid generation of new genetic combinations. Three mechanisms make key contributions to this genetic variation: independent assortment, crossing-over, and random fertilization.

Chromosome pairs separate independently

Most organisms have more than one chromosome. In human beings, for example, each gamete receives one chromosome from each of 23 pairs of homologous chromosomes. But, the chromosome that an offspring receives from each of the 23 pairs is a matter of chance. This random distribution of homologous chromosomes during meiosis is called **independent assortment.** Independent assortment is summarized in **Figure 7-2.** Each of the 23 pairs of chromosomes segregates (separates) independently. Thus, 2^{23} (about 8 million) gametes with different gene combinations can be produced from one original cell by this mechanism.

Crossing-over and random fertilization

The DNA exchange that occurs during crossing-over adds even more recombination to the independent assortment of chromosomes that occurs later in meiosis. Thus, the number of genetic combinations that can occur among gametes is practically unlimited.

Figure 7-2
Independent assortment. The same cell (with two homologous chromosome pairs) is shown twice. Since each pair of homologous chromosomes separates independently, four different gametes can result in each case.

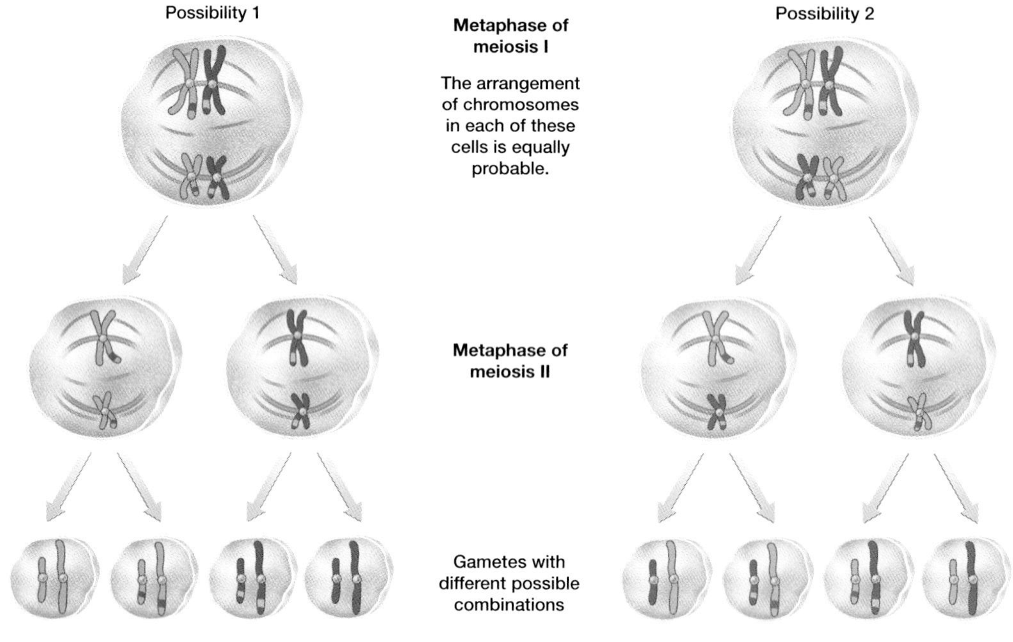

did you know?

An unfertilized ovum can survive for about 12 to 24 hours after its release, but sperm may remain viable in the female reproductive tract for up to 72 hours.

The timing is fairly crucial because fertilization must take place in the upper third of a fallopian tube to be successful. If sexual intercourse occurs between 72 hours before and 24 hours after ovulation, fertilization is likely to take place. However, it is difficult to determine when ovulation has occurred or will occur.

How can you model crossing-over?

QUICK LAB

You can use paper strips and pencils to model the process of crossing-over.

Materials

4 paper strips, pens or pencils (two colors), scissors, tape

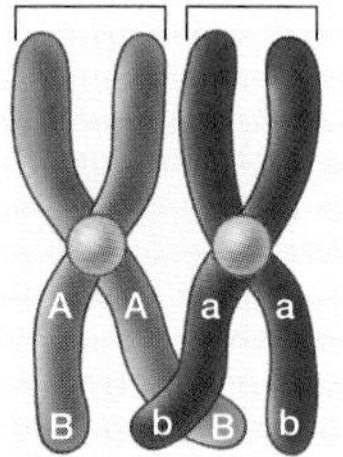
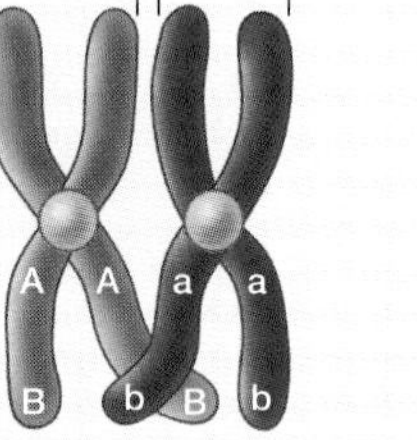

Procedure

1. Using one color, write the letters *A* and *B* on two paper strips. These two strips will represent one of the two homologous chromosomes shown above.
2. Using a second color, write the letters *a* and *b* on two paper strips. These two strips will represent the second homologous chromosome shown above.
3. Use your chromosome models, scissors, and tape to demonstrate crossing-over between the chromatids of two homologous chromosomes.

Analysis

1. **Determine** what the letters *A, B, a,* and *b* represent.
2. **Infer** why the chromosomes you made are homologous.
3. **Compare** the number of different types of chromatids (combinations of *A, B, a,* and *b*) before crossing-over with the number after crossing-over.
4. **SKILL Applying Information** How does crossing-over relate to genetic recombination?

Furthermore, because the zygote that forms a new individual is created by the random joining of two gametes (each gamete produced independently), fertilization squares the number of possible outcomes ($2^{23} \times 2^{23}$ = 64 trillion).

Importance of generating variation

Meiosis and the joining of gametes are essential to evolution. No genetic process generates variation more quickly. In many cases, the pace of evolution appears to increase as the level of genetic variation increases. For example, when genetic variation is used in agricultural breeding programs to increase the size of domesticated animals such as cattle and sheep, many large animals are produced at first. But as the existing genetic combinations become used up, the ability to obtain larger and larger animals slows down. Further progress must then wait for the formation of new gene combinations.

Racehorse breeding provides another example. Thoroughbred racehorses are all descendants of a small number of individuals, and selection for speed has accomplished all it can with this limited amount of genetic variation. The winning times in major races stopped improving decades ago.

Strangely enough, the evolutionary process is both revolutionary and conservative. It is revolutionary in that the pace of evolutionary change is quickened by genetic recombination. It is conservative in that evolutionary change is not always favored by natural selection, which may instead favor existing combinations of genes.

Study TIP

Reviewing Information

Each chapter in this book builds on concepts you explored in earlier chapters. *Study Tip: Ready?* on page 141 identifies the concepts this chapter builds on. Can you explain each concept in your own words? If not, return to the section indicated for a quick review.

How can you model crossing-over?

QUICK LAB

Time 15 minutes

Process Skills Inferring, interpreting, comparing, applying information

Teaching Strategies

Prepare the paper strips ahead of time.

Analysis Answers

1. individual genes
2. The chromosomes are of similar size, shape, and genetic content.
3. before crossing-over: AB, AB, ab, and ab; after crossing-over: answers will vary (for example, AB, Ab, aB, and ab)
4. Crossing-over causes genetic recombination.

Teaching TIP

Distinguishing Between Mitosis and Meiosis

Have students make a **Graphic Organizer** like the one at the bottom of page 145 to summarize the difference in chromosomal number between mitosis and meiosis. Ask students to include a brief paragraph that compares the two processes. PORTFOLIO

Graphic Organizer

*Use this graphic organizer with **Teaching Tip: Distinguishing Between Mitosis and Meiosis** on page 145.*

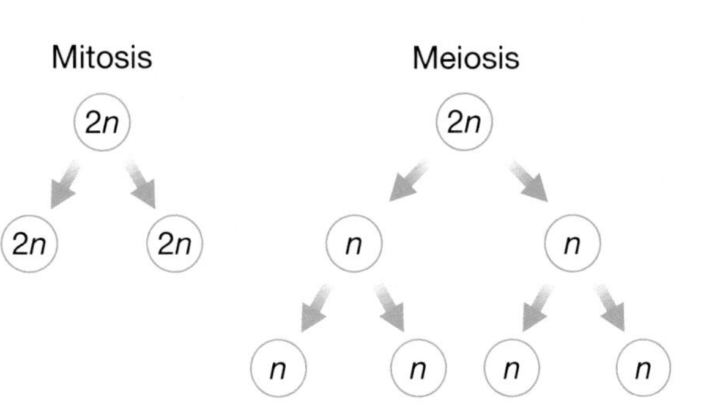

Chapter 7

Teaching TIP

Gamete Development

Have students make a Graphic Organizer similar to the one at the bottom of page 146 that explains the formation of gametes in humans. Be sure they include the terms *spermatogenesis, oogenesis, diploid germ cell, sperm, polar bodies,* and *egg cell.* PORTFOLIO

Making Biology Relevant

Sports

In human females, the ovaries usually release only one egg cell every 28 days, a process called ovulation. Under certain conditions, however, ovulation can be suppressed. For example, many women who participate in vigorous sports, such as long distance running and gymnastics, do not ovulate. This effect is part of a condition known as athletic menstrual cycle irregularity (AMI). The mechanism of AMI is unknown, although research suggests that rapid weight loss, decreased body fat content, and stress may be involved.

Career

Cytotechnologist

A cytotechnologist is a person trained to observe abnormalities in cells that might indicate cancer or a genetic deformity. Have students use library references or on-line resources to find out about becoming a cytotechnologist. Have them learn what kind of equipment cytotechnologists use and where they work. Students should research what kind of education and how much education is required for the job. Have them prepare a written report summarizing their findings. PORTFOLIO

biobit

Can animals be both male and female?

Hermaphrodites *(huhr MAF roh deyetz)* are animals with both male and female reproductive organs. Most mate with another member of the same species, but each animal donates and receives sperm. Some hermaphrodites fertilize themselves. Others can reverse their sex each breeding season. Many fish and invertebrates are hermaphroditic.

Gamete Formation in Male and Female Animals Involves Meiosis

The fundamental events of meiosis occur in all sexually reproducing organisms. However, organisms vary in timing and structures associated with gamete formation. Meiosis is the primary event in the formation of gametes—gametogenesis.

Meiosis in males

The process by which sperm are produced in male animals is called **spermatogenesis** *(spur mat uh JEHN uh sihs).* Spermatogenesis occurs in the testes (male reproductive organs). As illustrated in **Figure 7-3,** a diploid cell first increases in size and becomes a large immature cell (germ cell). The large cell then undergoes meiosis I. Two cells are produced, each of which undergoes meiosis II to form a total of four haploid cells. The four cells change in form and develop a tail to become male gametes called **sperm.**

Meiosis in females

The process by which gametes are produced in female animals is called **oogenesis** *(oh oh JEHN uh sihs).* Oogenesis, summarized in Figure 7-3, occurs in the ovaries (female reproductive organs). Notice that during cytokinesis following meiosis I, the cytoplasm divides unequally. One of the resulting cells gets nearly all of the cytoplasm. It is this cell that will ultimately give rise to an egg cell.

Figure 7-3 **Meiosis in male and female animals**

Meiosis of a male diploid cell results in four haploid sperm, while meiosis of a female diploid cell results in only one functional haploid egg cell.

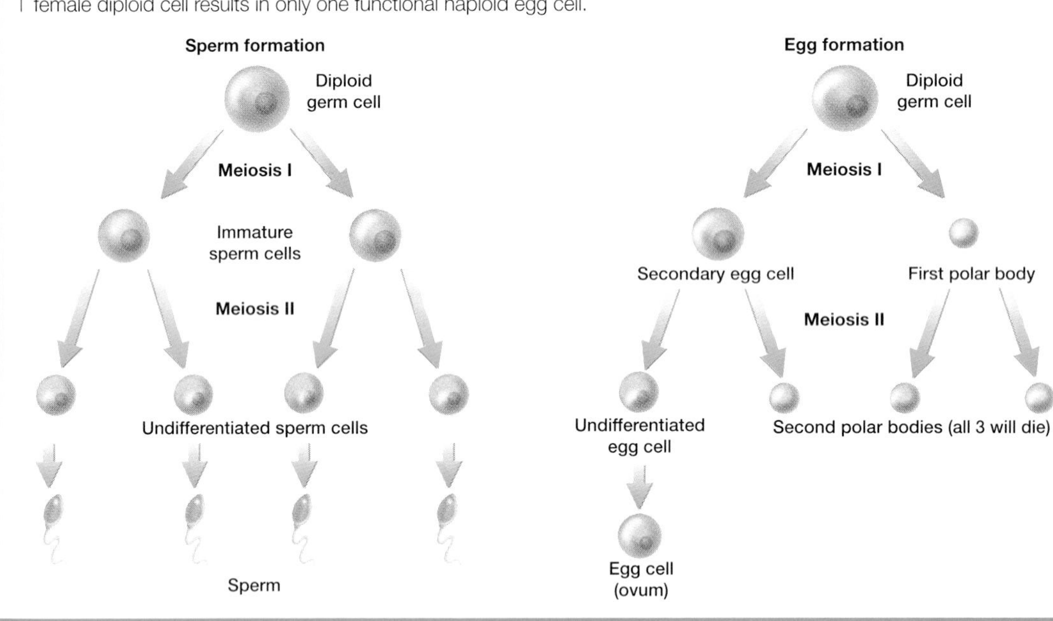

Graphic Organizer

*Use this graphic organizer with **Teaching Tip: Gamete Development** on page 146.*

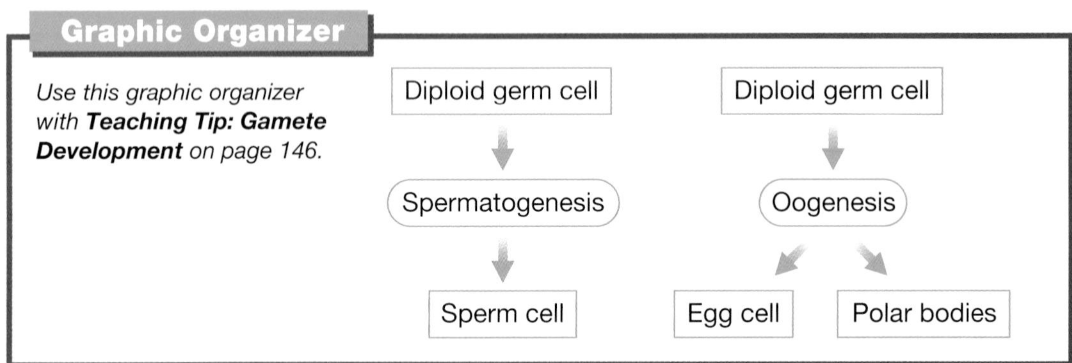

The other cell is very small and is called a polar body. The polar body may divide again, but its cells will not survive.

The larger cell undergoes meiosis II, and the division of the egg cell during cytokinesis is again unequal. The larger cell develops into a gamete called an **ovum** (plural, ova) or, more commonly, egg. The smaller cell, the second polar body, dies. Because of its larger share of cytoplasm, the mature ovum has a rich storehouse of nutrients. These nutrients nourish the young organism that develops if the ovum is fertilized.

Egg Cells and Toxins

A woman's gametes are especially vulnerable to damage by drugs, environmental toxins, radiation, and anything else that could potentially harm reproductive cells. The reason has to do with the timing of oogenesis as compared with spermatogenesis in males.

Egg formation starts early

In men, the cells in the testes, from which sperm eventually develop, are continuously formed throughout the male's reproductive years. In women, however, egg development begins before birth, and eggs are not continually produced by the ovary. During development before birth, cells divide by mitosis to produce as many as 2 million immature egg cells. The ovaries of a newborn female contain all of the immature egg cells she will ever have. When a female is born, the immature egg cells she carries have begun meiosis I but remain paused at prophase I.

Egg formation ends at fertilization

The immature egg cells do not change again until puberty, when a hormone stimulates them to complete meiosis I. Meiosis then pauses again. *It is not until an immature egg cell is fertilized by sperm that an egg begins to undergo meiosis II.* Not until fertilization is oogenesis actually complete. Thus, an immature egg may be more than 40 years old before it completes meiosis!

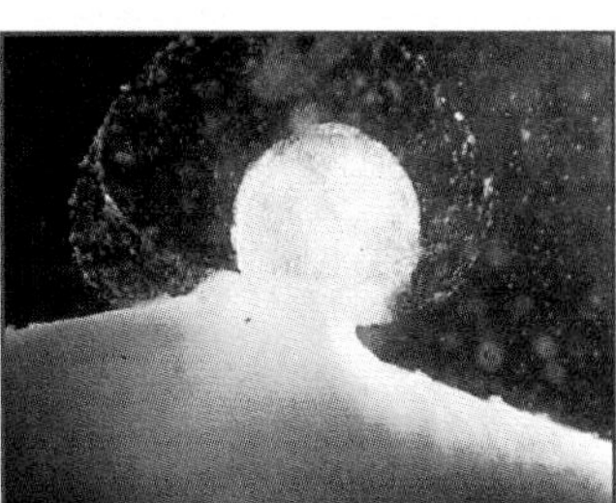

Ovary releasing an egg cell

Consequently, if a woman's immature egg cells are exposed to drugs, toxins, or radiation, which can induce mutations, she has no further chance to produce healthy egg cells.

Review SECTION 7-1

1. **Summarize** why meiosis is necessary in organisms that reproduce sexually.
2. **Name** the stage of meiosis during which chromatids are separated to opposite poles of the cell.
3. **Compare** the processes of crossing-over and independent assortment.
4. **Differentiate** gamete formation in male animals from gamete formation in female animals.
5. **SKILL Summarizing Information** How do independent assortment, crossing-over, and random fertilization affect the rate of evolution?

Critical Thinking

6. **SKILL Evaluating Information** If a cell in a dog ($2n = 78$) undergoes meiosis, how many chromosomes will each resulting cell contain?

Review SECTION 7-1 ANSWERS

1. Meiosis halves the number of chromosomes so that when two gametes unite to form a zygote, the zygote contains the diploid number of chromosomes.
2. anaphase II
3. Crossing-over occurs when part of a chromatid is broken and exchanged with the corresponding part on one of the chromatids of the homologous chromosome. Independent assortment is the random distribution of homologous chromosomes during meiosis.
4. In males, a diploid cell forms four haploid cells, which develop tails and become sperm. In females, a diploid cell forms one egg cell and three polar bodies. The polar bodies die.
5. They increase the rate of evolution by increasing the level of genetic variation.
6. Each resulting cell will contain 39 chromosomes.

HEALTH WATCH

Egg Cells and Toxins

Teaching Strategies

Use Figures 44-2 and 44-6 on pp. 1007 and 1010 to show students that oogenesis and spermatogenesis occur in the ovaries and testes, respectively.

Discussion

- Compare the time involved in the maturation of a sperm cell with that of an egg cell. *(Sperm cells mature in a couple of months, whereas egg cells may take more than 40 years to mature.)*
- Have students use the Web site on page 147 to find specific examples of toxins known to harm reproductive cells.

3 Assessment Options

Closure — BASIC

Answering Objectives

Ask students to rewrite each objective on page 142 as a question. Students should then work in pairs to answer each question. PORTFOLIO

Review

Assign Review—Section 7-1.

Reteaching — BASIC

Prior to class, make three or four homologous chromosome pairs of different lengths using different colors of construction paper. Include an X chromosome and a Y chromosome. Divide the chromosomes into "genes" with a marker. Have students use the chromosomes to model meiosis, including crossing-over. Ask students why sexual reproduction provides more diversity in organisms. *(New organisms have genetic information from two parents.)*

Section 7-2 Sexual Reproduction

1 Prepare

Scheduling

- **Block:** About 90 minutes will be needed to complete this section.
- **Traditional:** About two class periods will be needed to complete this section.
- **Planning:** See the Planning Guide on pp. 139A and 139B for lecture, classwork, and assignment options for Section 7-2. Lesson plans for Section 7-2 can be found in a lesson cycle format in ***Biology: Principles and Explorations*** **One-Stop Planner CD-ROM with Test Generator.**

Resource Link

Assign **Active Reading Worksheet Section 7-2** to help students comprehend the key concepts and visuals in this lesson.

2 Teach

Lesson Warm-up

Display pictures of different plants, animals, protists, and bacteria. Have students suggest a means of reproduction for each. Ask them to formulate a general rule about reproduction in eukaryotes and prokaryotes. *(Prokaryotes reproduce asexually; eukaryotes can reproduce either asexually, sexually, or both, depending on the organism.)*

Objectives

- **Differentiate** between asexual and sexual reproduction. (p. 148)
- **Identify** three types of asexual reproduction. (p. 148)
- **Evaluate** the relative genetic and evolutionary advantages and disadvantages of asexual and sexual reproduction. (p. 149)
- **Differentiate** between the three major sexual life cycles found in eukaryotes. (pp. 150–152)

Key Terms

asexual reproduction
clone
sexual reproduction
life cycle
fertilization
sporophyte
spore
gametophyte

Similarity to Parents Is Determined by the Type of Reproduction

Some organisms look exactly like their parents and siblings. Others share traits with family members but are not identical to them. Some organisms have two parents, while others have one. The type of reproduction that produces an organism determines how similar the organism is to its parents and siblings. Reproduction, the process of producing offspring, can be asexual or sexual.

In **asexual reproduction** a single parent passes copies of all of its genes to each of its offspring; there is no fusion of haploid cells such as gametes. An individual produced by asexual reproduction is a **clone,** an organism that is genetically identical to its parent. As discussed in Chapter 6, prokaryotes reproduce by a type of asexual reproduction called binary fission. Many eukaryotes, as shown in **Figure 7-4,** also reproduce asexually.

Figure 7-4
Asexual reproduction

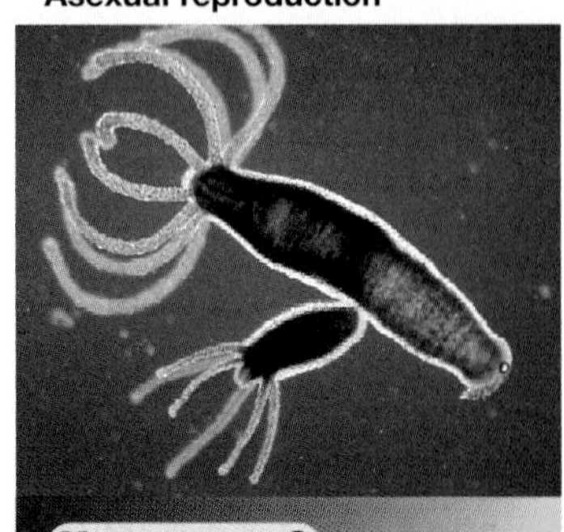

How many? A hydra colony can consist of hundreds of individuals that arise through asexual reproduction. The colony originates when one hydra becomes permanently attached to a surface.

In contrast, in **sexual reproduction** two parents each form haploid reproductive cells, which join to form offspring. Since both parents contribute genetic material, the offspring have traits of both parents but are not exactly like either parent. Sexual reproduction, with the formation of haploid cells, occurs in eukaryotic organisms, as shown in **Figure 7-5.**

Types of asexual reproduction

There are many different types of asexual reproduction. For example, amoebas reproduce by fission, the separation of a parent into two or more individuals of about equal size. Some multicellular eukaryotes undergo fragmentation, a type of reproduction in which the body breaks into several pieces. Some or all of these fragments later develop into complete adults when missing parts are regrown. Other organisms, like the hydra shown in Figure 7-4, undergo budding, in which new individuals split off from existing ones. The bud may break from the parent and become an independent organism, or it may remain attached to the parent. An attached bud can eventually give rise to a group of many individuals.

Integrating Different Learning Styles

Logical/Mathematical	**PE**	Review item 6, p. 147; Review item 6, p. 152
Visual/Spatial	**PE**	Quick Lab, p. 149; Study Tip, p. 150; Performance Zone items 1, 18, 19, and 20, pp. 154–155
	ATE	Lesson Warm-up, p. 148; Visual Strategies, pp. 150, 151, 152; Resource Link, p. 150; Teaching Tip, p. 152; Closure, p. 152; Alternative Assessment, p. 153
Verbal/Linguistic	**PE**	Performance Zone items 2 and 27, pp. 154–155
	ATE	Active Reading, p. 149

Advantages and disadvantages of each type of reproduction

Asexual reproduction is the simplest and most primitive method of reproduction. In a stable environment, asexual reproduction allows organisms to produce many offspring in a short period of time, without using energy to produce gametes or to find a mate. However, the DNA of these organisms does not vary much between individuals. This may be a disadvantage in a changing environment because they may not be able to adapt to a new environment.

On the other hand, sexual reproduction provides a powerful means of quickly making different combinations of genes among individuals. Such genetic diversity is the raw material for evolution.

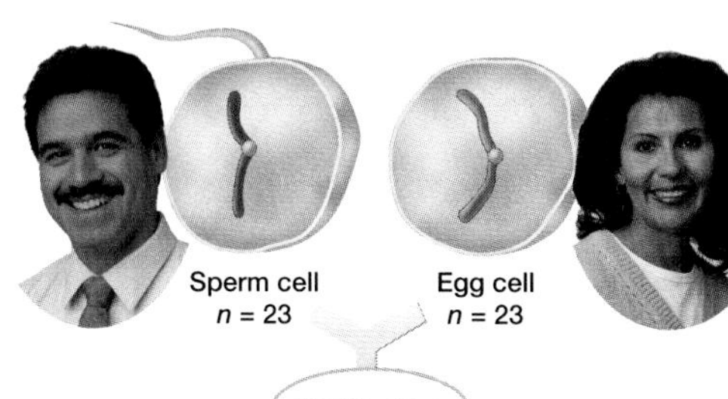

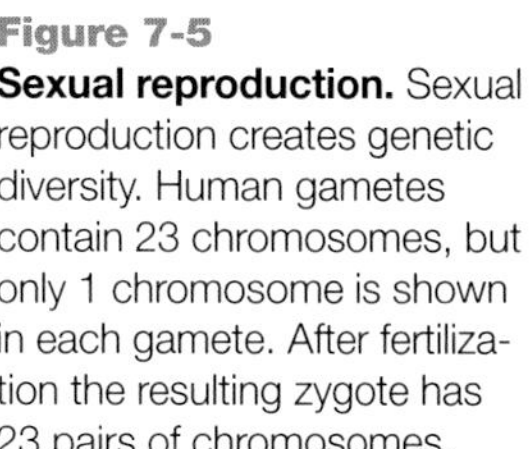

Figure 7-5
Sexual reproduction. Sexual reproduction creates genetic diversity. Human gametes contain 23 chromosomes, but only 1 chromosome is shown in each gamete. After fertilization the resulting zygote has 23 pairs of chromosomes.

How sexual reproduction evolved

Biologists have studied protists to determine how sexual reproduction evolved. Many protists are haploid most of the time, and so reproduce asexually. They form a diploid cell only in response to stress in the environment. Biologists think this occurs because only diploid cells can repair certain kinds of chromosome damage such as breaks in both strands of DNA. As organisms became larger, it was important for them to be able to repair such damage.

The pairing of homologous chromosomes during meiosis may have originally been a way to repair damaged DNA. Indeed, many enzymes that repair DNA damage are also involved during the time when chromosomes exchange genetic material during meiosis.

QUICK LAB

How do yeast reproduce?

Yeast are unicellular organisms that live in liquid or moist environments. You can examine a culture of yeast to observe one of the types of reproduction that yeast can undergo.

Materials

microscope, microscope slides, dropper, culture of yeast

Procedure

1. Make a wet mount of a drop of yeast culture.
2. Observe the yeast with a compound microscope under low power.
3. Look for yeast that appear to be in "pairs."
4. Observe the pairs under high power, and then make drawings of your observations.

Analysis

1. **Infer** the type of reproduction you observed when the yeast appeared to be in pairs.
2. **Identify** the reason for your answer.
3. **SKILL Finding Information** Use your textbook to determine the name of the type of reproduction you observed.

Section 7-2

QUICK LAB

How do yeast reproduce?

Time 15 minutes

Process Skills Observing, inferring, locating information

Teaching Strategies

- Caution students to use care when working with a microscope, glass slides, and living organisms.
- Refer students to page 1041 in the **Appendix** to review the proper procedure for preparing a wet-mount slide.
- Review with students the correct technique for lab drawings.

Analysis Answers

1. asexual reproduction
2. Only one parent was observed. The new organism will split off from the parent.
3. budding

Active Reading

Technique: Reading Organizer

Point out to students that they will encounter three reading organizers on pp. 150–152. Ask students if they can spot the reading organizers provided. *(They depict the haploid life cycle, the diploid life cycle, and alternation of generations.)* Have them explain why these organizers are appropriate graphic representations of the text.

Overcoming Misconceptions

Reproduction Versus Sex

Students often think that reproduction and sex are the same thing. Reproduction is the formation of new individuals, whether of unicellular or multicellular organisms. Sex, on the other hand, is the combining of genetic material from two cells.

Visual Strategy

Figure 7-6

Use this figure to discuss the steps involved in a haploid life cycle. Be sure students know the difference between a diploid cell and a haploid cell. Ask students to point out the diploid stage in the life cycle. *(the zygote)* Ask them what the result of meiosis is. *(haploid individuals)* Have students explain why this type of life cycle is called a haploid life cycle. *(The haploid cells occupy the major portion of the life cycle.)*

Resource Link

- Assign **Reading Strategies: Recognizing Cycles** from the Holt BioSources Teaching Resources CD-ROM. Have students summarize the three kinds of sexual life cycles. PORTFOLIO

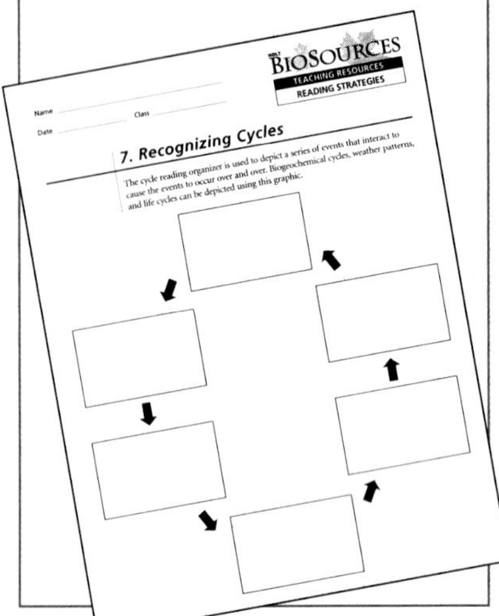

Study TIP

Interpreting Graphics

After reading this chapter, trace or make a sketch of Figures 7-6, 7-7, and 7-8 without the labels. On separate pieces of paper, write down the labels. Without referring to your book, match the labels with the correct part of your sketch.

Eukaryotes Have Three Kinds of Sexual Life Cycles

The entire span in the life of an organism from one generation to the next is called a **life cycle.** The life cycles of all sexually reproducing organisms follows a basic pattern of alternation between the diploid and haploid chromosome numbers. The type of sexual life cycle that a eukaryotic organism has depends on the type of cell that undergoes meiosis and on when meiosis occurs. Eukaryotes that undergo sexual reproduction can have one of three types of sexual life cycles: haploid, diploid, or alternation of generations.

Haploid life cycle

The haploid life cycle is the simplest of sexual life cycles. In this life cycle, shown in **Figure 7-6,** haploid cells occupy the major portion of the life cycle. The zygote is the only diploid cell, and it undergoes meiosis immediately after it is formed, creating new haploid cells. The haploid cells give rise to haploid multicellular individuals that produce gametes by mitosis (not meiosis). In a process called fusion, the gametes fuse to produce a diploid zygote, and the cycle continues.

When the diploid zygote undergoes meiosis it provides an opportunity for the cell to correct any genetic damage, as discussed earlier. The damage is repaired during meiosis, when the two homologous chromosomes are lined up side-by-side in preparation for crossing over. Special repair enzymes remove any damaged sections of double stranded DNA, and fill in any gaps. This type of life cycle is found in many protists, as well as in some fungi and algae, such as the unicellular *Chlamydomonas (KLUH mih duh moh nuhs),* shown in Figure 7-6.

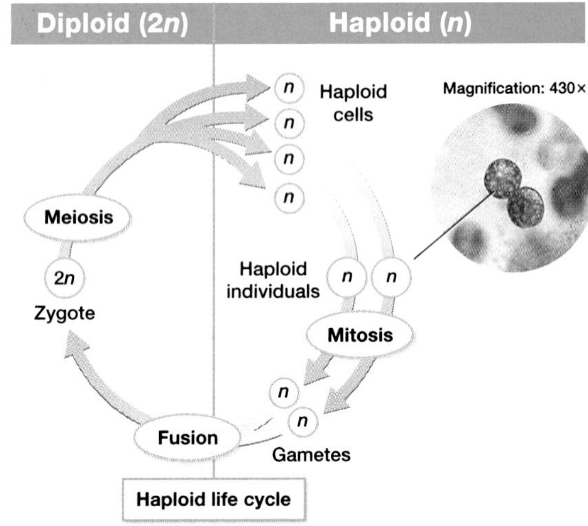

Figure 7-6 Haploid life cycle. Some organisms, such as *Chlamydomonas,* have haploid cells as a major portion of their life cycle.

did you know?

Although Darwin (1809–1882) recognized the importance of genetic variation in natural selection (the mechanism for evolution), he could not explain why offspring resembled but were not identical to their parents.

Gregor Mendel (1822–1884) later published his theory of inheritance which helped explain genetic variation. But Mendel's work was not recognized by biologists until more than 15 years after both Darwin's and Mendel's deaths.

Diploid life cycle

The outstanding characteristic of the diploid life cycle is that adult individuals are diploid, each individual inheriting chromosomes from two parents. In most animals, including humans, a diploid reproductive cell undergoes meiosis to produce gametes.

As shown in **Figure 7-7,** the gametes (sperm and egg cells) join in a process called **fertilization,** which results in a diploid zygote. After fertilization, the resulting zygote begins to divide by mitosis. This single diploid cell eventually gives rise to all of the cells of the adult. The cells of the adult are also diploid since they are produced by mitosis.

The diploid individual that develops from the zygote occupies the major portion of the diploid life cycle. The gametes are the only haploid cells in the diploid life cycle; all of the other cells are diploid.

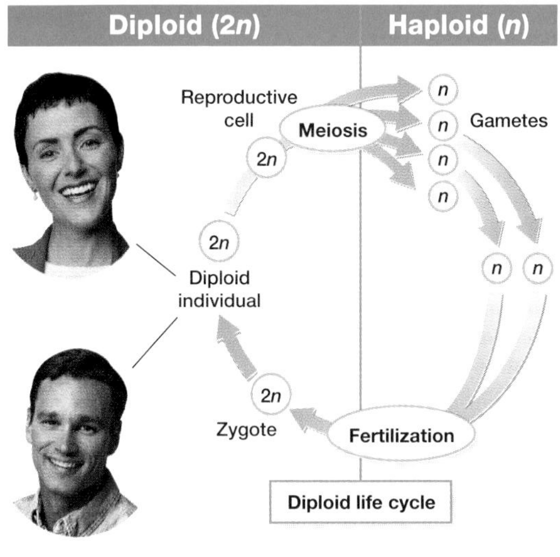

Figure 7-7 Diploid life cycle. Humans and other organisms have a life cycle dominated by a diploid individual.

Visual Strategy

Figure 7-7

Use this figure to discuss the steps involved in a diploid life cycle. Ask students to point out the diploid stage in the life cycle. *(the zygote, which grows into an individual)* Ask students what the result of meiosis is. *(haploid gametes)* Ask students why this type of life cycle is called a diploid life cycle. *(The diploid individual occupies the major portion of the life cycle.)*

Exploring Further

Cloning by Parthenogenesis

A snake is born to a mother that did not have a mate. Although this may sound impossible, or like some headline in a tabloid magazine, this can actually occur in nature. Parthenogenesis *(pahr thuh noh JEHN uh sihs)* is a type of reproduction in which a new individual develops from an unfertilized egg. Since there is no male that contributes genetic material, the offspring is a clone (genetically identical) of the mother. Clones are usually produced in nature by asexual reproduction. Parthenogenesis, however, is a special form of cloning.

Why does parthenogenesis occur?

Parthenogenesis in snakes has usually occurred in older females that have lived many years without male companionship, such as those in a zoo. It is hypothesized that the mother snake copies her own chromosomes in place of the missing father's chromosomes, thereby self-fertilizing her egg. Other scientists think that after a long absence of males, some unknown signal (such as a hormone) triggers the egg to start dividing.

Whiptail lizard

Many different organisms undergo parthenogenesis

Organisms capable of reproducing by parthenogenesis include dandelions, hawkweeds, and some fishes, lizards, and frogs. Whiptail lizards are all females that lay eggs that hatch without any male contributions. Honeybees also produce male drones by parthenogenesis.

Parthenogenesis is not thought to be possible in mammals. Embryos of mammals that do not have genes from both a female and a male parent do not develop normally. The only natural mammalian clones known are identical twins, which develop when a fertilized egg splits and two individuals develop.

Exploring Further

Cloning by Parthenogenesis

Teaching Strategies

Before students read the article, tell them that parthenogenesis is a type of *reproduction in which a new individual develops from an unfertilized egg.* Ask students what organisms they think exhibit this type of reproduction. (Answers will vary.)

Discussion

- Why do scientists think parthenogenesis is not possible in mammals? *(Mammalian embryos that do not have genes from both a female and a male do not develop normally.)*
- Hypothesize why parthenogenesis occurs in older female snakes that have lived without male companionship. *(Some unknown signal, like a hormone, may trigger the egg to start dividing.)*
- List organisms that undergo parthenogenesis. *(dandelions, hawkweeds, and some fishes, lizards, and frogs)*
- Name the only natural mammalian clones. *(identical twins)*

did you know?

Parthenogenesis is not the only variation on sexual reproduction that occurs among vertebrates.

Among many species of fish, individuals can change their sex. Some begin life as females and then change into males, a phenomenon called protogyny. Others change from males into females, which is known as protandry. Even more unusual, some deep-sea fish are hemaphrodites—both male and female at the same time.

Teaching TIP

Alternation of Generations
Show students examples of live plants, including flowering plants, conifers, and bryophytes. Point out the parts of each plant that would be a gametophyte or a sporophyte.

Visual Strategy

Figure 7-8
Use this figure to discuss the steps involved in alternation of generations. Ask students to point out the diploid stage in the life cycle. *(the sporophyte)* Ask students what the result of meiosis is. *(haploid spores that grow into the gametophyte)* Ask students why this type of life cycle is called alternation of generations. *(The organisms alternate between haploid and diploid phases.)*

3 Assessment Options

Closure

Haploid and Diploid Life Cycles
Before class, prepare flashcards of the life cycles in Figures 7-6 and 7-7. For example, for the haploid life cycle, write an "H" on the back of each card. On the front, write the words enclosed in the circles. Do the same for the diploid life cycle. After students have studied the two life cycles, have them arrange the cards in the correct order for each life cycle.

Review

Assign Review—Section 7-2.

Reteaching — BASIC

Ask students to identify the major phase in each of the different life cycles. *(haploid life cycle—haploid; diploid life cycle—diploid; alternation of generations—haploid and diploid alternate)*

Alternation of generations

Plants, algae, and some protists have a life cycle that regularly alternates between a haploid phase and a diploid phase. As shown in **Figure 7-8,** in plants, the diploid phase in the life cycle that produces spores is called a **sporophyte** *(SPOH ruh feyet)*. Spore-forming cells in the sporophyte undergo meiosis to produce spores. A **spore** is a haploid reproductive cell produced by meiosis that is capable of developing into an adult without fusing with another cell. Thus, unlike a gamete, a spore gives rise to a multicellular individual called a gametophyte *(guh MEET uh feyet)* without joining with another cell.

In the life cycle of a plant, the **gametophyte** is the haploid phase that produces gametes by mitosis. The gametophyte produces gametes that fuse and give rise to the diploid phase. Thus, the sporophyte and gametophyte generations take turns, or alternate, in the life cycle.

In moss, for example, haploid spores develop in a capsule at the tip of the sporophyte "stalk." When the lid of the capsule pops off, the spores scatter. The spores germinate by mitosis and eventually form sexually mature gametophytes. The male gametophytes release sperm which swim through a film of moisture to the eggs in the female gametophyte. The diploid zygote develops as a sporophyte within the gametophyte and the life cycle continues.

It is important not to lose sight of the basic similarity of all three types of sexual life cycles. All three involve an alternation of haploid and diploid phases. The three types of sexual life cycles differ from each other only in which phases become multicellular.

Figure 7-8
Alternation of generations. Some organisms, such as roses, have a life cycle that alternates between diploid and haploid phases.

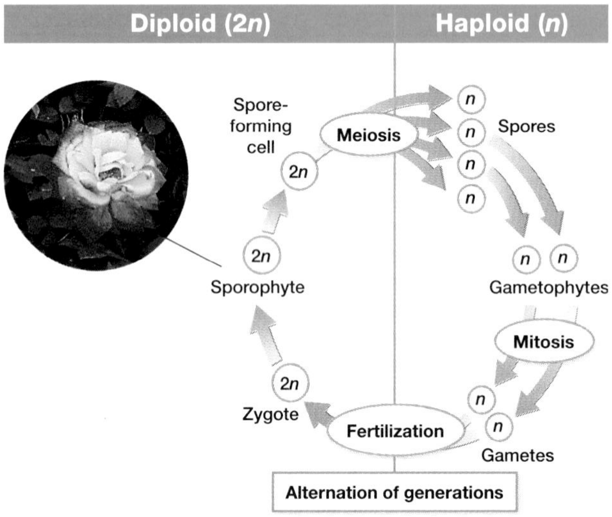

Review SECTION 7-2

1. **Identify** the type of reproduction that results in offspring that are genetically identical to their parent.
2. **Describe** two different types of eukaryotic asexual reproduction.
3. **Compare** the haploid life cycle found in *Chlamydomonas* with a diploid life cycle.
4. **Summarize** the process of alternation of generations.
5. **SKILL Summarizing Information** Summarize the role that the repair of damaged chromosomes may have played in the evolution of sexual reproduction.

Critical Thinking

6. **SKILL Evaluating Viewpoints** A student states that organisms that reproduce asexually are at a disadvantage in a stable environment. Do you agree or disagree? Explain.

Review SECTION 7-2 ANSWERS

1. asexual
2. fission—parent separates into two or more individuals of equal size; fragmentation—the body breaks into several pieces; budding—new individuals split off from existing ones
3. The haploid cells of *Chlamydomonas* give rise to haploid multicellular individuals that produce gametes by mitosis. The gametes fuse to form a diploid zygote, which undergoes meiosis, creating new haploid cells. In a diploid life cycle, haploid gametes are produced by meiosis. The gametes join by fertilization to form a diploid zygote.
4. Spore-forming cells in the sporophyte undergo meiosis to produce spores, which develop into gametophytes. The gametophyte produces gametes by mitosis. The gametes fuse and give rise to the sporophyte.
5. Sexual reproduction is believed to have evolved originally as a way for cells to repair damaged DNA.
6. Most students should disagree. In a stable environment, asexual reproduction allows organisms to produce offspring without using energy to produce gametes or to find a mate.

Study ZONE

CHAPTER 7

Use the Key Concepts and Key Terms listed below to review the main ideas in this chapter. If you need to review the meaning of a term, turn to the page indicated.

Key Concepts

7-1 Meiosis

- Meiosis reduces the number of chromosomes by half to form reproductive cells. When the reproductive cells unite in fertilization, the normal diploid number is restored.
- During meiosis I, homologous chromosomes separate. Crossing-over during prophase I results in the exchange of genetic material between homologous chromosomes.
- During meiosis II, the two chromatids of each chromosome separate. As a result of meiosis, four haploid cells are produced from one diploid cell.
- Independent assortment, crossing-over, and random fertilization contribute to produce genetic variation in sexually reproducing organisms.
- In sexually reproducing eukaryotic organisms, gametes form through the process of spermatogenesis in males and oogenesis in females.

7-2 Sexual Reproduction

- Asexual reproduction is the formation of offspring from one parent. The offspring are genetically identical to the parent.
- Sexual reproduction is the formation of offspring through the union of gametes. The offspring are genetically different from their parents.
- A disadvantage to asexual reproduction in a changing environment is the lack of genetic diversity among the offspring.
- Sexual reproduction increases variation in the population by making possible genetic recombination.
- Sexual reproduction may have evolved as a mechanism to repair damaged DNA.
- Eukaryotic organisms can have one of three kinds of sexual life cycles, depending on the type of cell that undergoes meiosis and on when meiosis occurs.

Key Terms

7-1

meiosis (142)
crossing-over (142)
independent assortment (144)
spermatogenesis (146)
sperm (146)
oogenesis (146)
ovum (147)

7-2

asexual reproduction (148)
clone (148)
sexual reproduction (148)
life cycle (150)
fertilization (151)
sporophyte (152)
spore (152)
gametophyte (152)

Unit 4—*Cell Reproduction*
Use Topics 5–6 in this unit to review the key concepts and terms in this chapter.

Study TIP Ready?

- *If you think you understand these Key Concepts and Key Terms, then you're ready for the Performance Zone.*

Study ZONE

CHAPTER 7

Review and Practice

Assign the following worksheets for Chapter 7:

- **Student Review Guide**
- **Concept Mapping Worksheet**
- **Test Preparation Pretest**
- **Vocabulary Worksheet**
- **Critical Thinking Review Worksheet**
- **Science Skills Worksheet**

Assessment Options

The following assessment options are available for this chapter:

Resource Link

- Assign **Chapter Test 7** to assess students' understanding of the chapter.

Alternative Assessment

Have students design a concept map that relates the two forms of cell division to the human life cycle. Be sure they include the terms *haploid, diploid, mitosis, meiosis,* and *gamete.* PORTFOLIO

Portfolio Assessment

- *Portfolio Assessment* at the front of this book provides options for building and evaluating student portfolios. Students and teachers can select items from the areas listed for inclusion in a portfolio.
- See items labeled PORTFOLIO on pp. 144, 145, 146, 147, 150 and 153.

Answer to Concept Map

The following is one possible answer to Performance Zone item 1 on page 154.

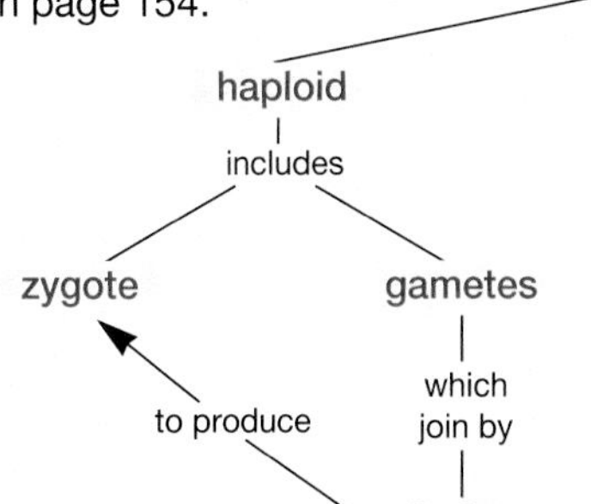

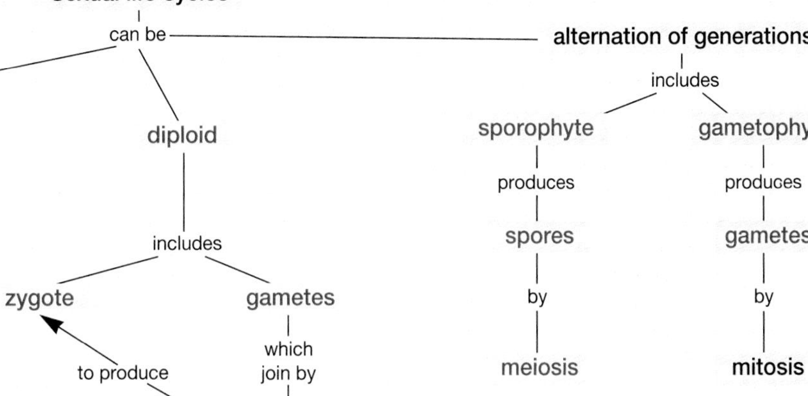

Performance ZONE
CHAPTER 7
REVIEW ANSWERS

Assignment Guide

Section	Review
7-1	3, 4, 5, 6, 7, 8, 13, 15, 17, 21, 22, 23, 24, 25, 26, 28, 29
7-2	1, 2, 9, 10, 11, 12, 14, 16, 18, 19, 20, 27

Understanding and Applying Concepts

1. The answer to the concept map is found at the bottom of page 153.
2. **a.** Spermatogenesis is the formation of male gametes, or sperm. Oogenesis is the formation of female gametes, or eggs.

   b. In asexual reproduction, a single parent passes copies of all its genes to its offspring. In sexual reproduction, two parents each form haploid cells, which join to form offspring.

   c. The sporophyte is the diploid, spore-producing phase in the life cycle of a plant. The gametophyte is the haploid, gamete-producing phase.
3. c
4. a
5. b
6. d
7. b
8. a
9. d
10. b
11. c
12. a
13. ABcdE, ABcde, aBCDE, aBCDe
14. Sexual reproduction introduces genetic variation, which is not favorable for well-adapted organisms.
15. All of a woman's gametes are formed before birth. If a woman's egg cells are exposed to certain drugs, she has no further chance to produce healthy egg cells.

Performance ZONE CHAPTER 7 Review

Understanding and Applying Concepts

1. **Concept Mapping** Make a concept map that shows the three sexual life cycles in eukaryotic organisms. Try to include the following words in your map: meiosis, fusion, gametes, spores, fertilization, zygote, gametophyte, sporophyte, haploid, diploid.

2. **Understanding Vocabulary** For each pair of terms, explain the differences in their meanings.
   a. spermatogenesis, oogenesis
   b. asexual reproduction, sexual reproduction
   c. sporophyte, gametophyte

3. Most multicellular eukaryotes form specialized reproductive cells by
   a. binary fission. c. meiosis.
   b. mitosis. d. fragmentation.

4. Which of the following events occurs during prophase I of meiosis?
   a. crossing-over
   b. chromatids are duplicated
   c. reduction in chromosome number
   d. chromatids separate to opposite poles

5. Genes are exchanged between homologous chromosomes during
   a. fertilization. c. fission.
   b. crossing-over. d. telophase II.

6. The random distribution of homologous chromosomes during meiosis is called
   a. fission.
   b. budding.
   c. random fertilization.
   d. independent assortment.

7. Homologous pairs of chromosomes move to opposite poles during
   a. prophase I. c. metaphase II.
   b. anaphase I. d. anaphase II.

8. Spermatogenesis produces
   a. four haploid cells.
   b. four diploid cells.
   c. four polar bodies.
   d. two haploid cells.

9. Which of the following does not relate to the subject of asexual reproduction?
   a. fission c. budding
   b. fragmentation d. meiosis

10. Sexual reproduction is believed to have evolved originally as a way for cells to
    a. shuffle genetic material.
    b. repair damaged DNA.
    c. produce diploid individuals.
    d. increase their population growth at a maximum rate.

11. In humans, gametes join during ______ to form a diploid zygote.
    a. mitosis c. fertilization
    b. meiosis d. crossing-over

12. In plants, the sporophyte generation produces ______ spores through meiosis.
    a. haploid c. diploid
    b. triploid d. mutated

13. **Interpreting Graphics** After crossing-over where shown below, what would the sequence of genes be for each of the chromatids?

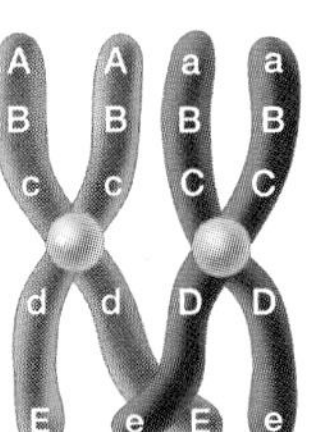

14. **Theme Reproduction** Explain why sexual reproduction does not favor well-adapted organisms.

15. **Health Watch** Explain why certain drugs are more dangerous to women's gametes than to men's.

16. **Exploring Further** What hypotheses have been proposed to explain parthenogenesis?

17. **Chapter Links** Compare and contrast mitosis and meiosis. (**Hint:** See Chapter 6, Section 6-3.)

16. Proposed hypotheses include copying of egg chromosomes (self-fertilization) by older mothers who have lived many years without male companionship, and an unknown signal (such as a hormone) triggers eggs to start dividing in the absence of males.

17. Mitosis produces two genetically identical diploid cells. Meiosis produces four haploid cells that are not genetically identical to the original cell. In metaphase I of meiosis, homologous chromosomes pair up along the equator. In metaphase of mitosis, homologous chromosomes line up independently along the equator. The stages of meiosis II are identical to mitosis.

Skills Assessment

Part 1: Interpreting Graphics

Use the illustration below to answer the following questions:

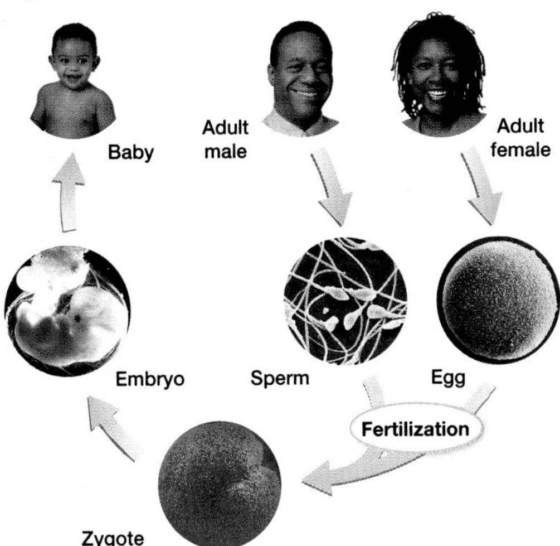

18. What type of eukaryotic sexual life cycle does the diagram above represent?

19. What is the chromosome number for each of the following stages of the life cycle?
   a. adult female
   b. adult male
   c. egg
   d. sperm
   e. zygote
   f. embryo
   g. baby

20. Will new cells that develop from the zygote result from mitosis or meiosis?

Part 2: Life/Work Skills

Use the media center or Internet resources to investigate captive-breeding programs for endangered species in zoos.

21. **Finding Information** How do captive-breeding programs affect genetic diversity? What can researchers do to help maintain genetic diversity in light of these programs?

22. **Communicating** Prepare an oral report to summarize your findings.

23. **Being a Team Member** Volunteer to help a zoo spread information to the public about their captive-breeding program.

Critical Thinking

24. **Evaluating Results** Occasionally homologous chromosomes fail to separate during meiosis I. Using the hypothetical example of an adult organism with two pairs of chromosomes, describe the chromosomal makeup of the eggs that would result from this error in meiosis.

25. **Evaluating Results** If these eggs were fertilized by normal sperm, what would the chromosomal makeup of the resulting zygote be?

Portfolio Projects

26. **Interpreting Information** Read the article "Brave New Egg," by John Travis (*Discover,* April 1, 1997, pp. 76–80). Some women are incapable of providing eggs for fertilization because of either disease, sterilizing cancer therapy, or advancing age. Write a report explaining the new methods being tested for storing and supplying human eggs for later fertilization by these women.

27. **Inferring Relationships** The bacterium *Escherichia coli* usually reproduces by binary fission, but it can also transfer genetic material through a sex pilus. Research how this transfer of genetic material takes place. Write a report that summarizes this process. In your report, speculate about how the transfer of genetic material might affect the use of antibiotics.

28. **Career Focus Cell Biologist** Research the field of cell biology, and write a report on your findings. Your report should include a job description, training required, kinds of employers, growth prospects, and starting salary.

29. **Unit 4—*Cell reproduction*** Write a report summarizing the effects of various treatments for infertility. Find out how the production of gametes may be affected in some people with infertility.

Performance ZONE

CHAPTER 7

Critical Thinking

24. If one of the pairs of chromosomes failed to separate, eggs would have either three chromosomes or one chromosome.
25. Zygotes would have either five chromosomes or three chromosomes.

Portfolio Projects

26. Answers will vary. Reports should describe the transplanting of ovarian tissue and the maturing of oocytes *in vitro.*
27. Answers will vary. Students should describe how a "male" bacterium extends a sex pilus, which attaches to a "female" bacterium. Later the mates briefly join by a cytoplasmic bridge, through which the "male" transfers DNA to the "female." Genes that confer antibiotic resistance can be transferred by this method.
28. Cell biologists study how cells are constructed and how they grow, divide, and communicate. A four-year degree in the sciences is required. Many jobs also require postgraduate training. Employers include universities, health care clinics, pharmaceutical industries, and private companies. Growth prospects are very good. Starting salary will vary by region.
29. Answers will vary but might include summaries of one or more of the following types of treatments for infertility: *in vitro* fertilization (IVF); donor egg IVF; artificial insemination; tubal embryo transfer (TET); gamete intrafallopian transfer (GIFT); and drug treatments (such as clomiphine citrate, pergonal, and metrodin).

Skills Assessment

18. diploid life cycle
19. a. 46
   b. 46
   c. 23
   d. 23
   e. 46
   f. 46
   g. 46
20. mitosis
21. Answers will vary. Captive breeding programs attempt to increase genetic diversity. Researchers can document the genetic makeup of each individual in a breeding program in order to mate individuals that are genetically varied.
22. Answers will vary.
23. Answers will vary.

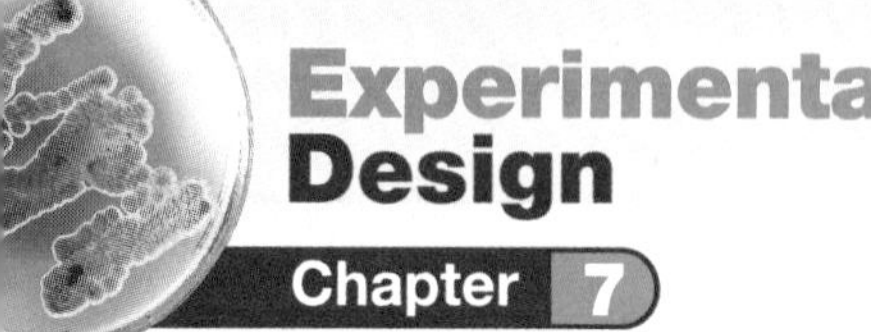

How Can You Model Meiosis?

Purpose
Students will build a model that will help them understand the events of meiosis and crossing-over.

Preparation Notes
Time Required: 45–75 minutes

Materials
Materials for this lab can be ordered from WARD'S. See *Master Materials List* at the front of this book for catalog numbers.

Preparation Tips
For each model, students need 4 pipe cleaners of one color and 4 pipe cleaners of another color, 4 wooden beads, 90 cm of yarn, and 16 small white labels.

Safety Precautions
Discuss all safety symbols and caution statements with students.

Procedural Tips
- Review the stages of meiosis before beginning the lab.
- If students have difficulty in building their models, ask the following questions:
  1. Which of these materials would make the best cell membrane? *(yarn)*
  2. Which would make the best spindle fibers? *(yarn)*
  3. Which would make the best chromosomes? *(pipe cleaners)*
  4. Which would make the best centromeres? *(beads)*
- Emphasize that each pair of chromatids represents one chromosome.
- Be sure students show that homologous chromosomes pair during metaphase I of meiosis.

Disposal
Dispose of paper and yarn scraps in the trash.

Before You Begin
Answers
1. *meiosis*—type of nuclear division that halves the chromosome number when forming reproductive cells

sexual reproduction—reproduction that involves the union of two gametes

crossing-over—the exchange of a portion of a chromatid on one homologous chromosome with the corresponding portion on one of the chromatids of the other homologous chromosome

genetic recombination—the formation of new combinations of genes

homologous chromosomes—chromosomes that are the same size and shape and carry genes for the same traits

gamete—a reproductive cell, either a sperm or an egg

2. Meiosis occurs in the ovaries and testes.
3. A cell's chromosomes duplicate, and certain organelles replicate.
4. Answers will vary. For example: How many chromosomes will each new nucleus have after meiosis has occurred?

Procedure
2. Plans will vary. For example, students may use the yarn to represent the cell membrane and spindle fibers, the pipe cleaners to represent the chromosomes, and the beads to represent the centromeres.
3. Drawings will vary. See example on page 157 (spindle fibers and labels not shown).

Experimental Design

Chapter 7 How Can You Model Meiosis?

SKILLS
- Modeling
- Using scientific methods

OBJECTIVES
- **Describe** the events that occur in each stage of the process of meiosis.
- **Relate** the process of meiosis to genetic variation.

MATERIALS
- pipe cleaners of at least two different colors
- yarn
- wooden beads
- white labels
- scissors

Glass frog with eggs

Before You Begin
Meiosis is the process that results in the production of cells with half the normal number of chromosomes. It occurs in all organisms that undergo **sexual reproduction.** In this lab, you will build a model that will help you understand the events of meiosis. You can also use the model to demonstrate the effects of events such as **crossing-over** to explain results such as **genetic recombination.**

1. Write a definition for each boldface term in the paragraph above and for the following terms: homologous chromosomes, gamete.
2. In what organs in the human body do cells undergo meiosis?
3. During interphase of the cell cycle, how does a cell prepare for dividing?
4. Based on the objectives for this lab, write a question you would like to explore about meiosis.

Procedure
Part A: Design a Model
1. Work with the members of your lab group to design a model of a cell using the materials listed for this lab. Be sure that your model cell has at least two pairs of chromosomes.
2. Write out the plan for building your model. Have your teacher approve the plan before you begin building the model.

> **You Choose**
> **As you design your experiment, decide the following:**
> a. what question you will explore
> b. how to construct a cell membrane
> c. how to show that your cell is diploid
> d. how to show the locations of at least two genes on each chromosome
> e. how to show that chromosomes are duplicated before meiosis begins

3. Build the cell model your group designed. **CAUTION: Sharp or pointed objects can cause injury. Handle scissors carefully.** Use your model to demonstrate the phases of meiosis. Draw and label each phase you model.
4. Use your model to explore one of the questions written by your group for step 4 of **Before You Begin.** Describe the steps you took to explore your question.

PART B: Test Hypotheses

Answer each of the following questions by writing a hypothesis. Use your model to test each hypothesis, and describe your results.

5. In humans, gametes (eggs and sperm) result from meiosis. Will all gametes produced by one parent be identical?
6. When an egg and a sperm fuse during sexual reproduction, the resulting cell (the first cell of a new organism) is called a zygote. How many copies of each chromosome and each gene will be found in a zygote?
7. Crossing-over frequently occurs between the chromatids of homologous chromosomes during meiosis. Under what circumstances does crossing-over result in new combinations of genes in gametes?
8. Synapsis (the pairing of homologous chromosomes) must occur before crossing-over can take place. How would the outcome of meiosis be different if synapsis did not occur?

PART C: Cleanup and Disposal

9. Dispose of paper and yarn scraps in the designated waste container.
10. Clean up your work area and all lab equipment. Return lab equipment to its proper place. Wash your hands thoroughly before you leave the lab and after finishing all work.

Analyze and Conclude

1. **Analyzing Results** How do the nuclei you made by modeling meiosis compare with the nucleus of the cell you started with? Explain your result.
2. **Recognizing Relationships** How are homologous chromosomes different from chromatids?
3. **Forming Reasoned Opinions** How is synapsis important to the outcome of meiosis? Explain.
4. **Evaluating Methods** How could you modify your model to better illustrate the process of meiosis?
5. **Drawing Conclusions** How are the processes of meiosis similar to those of mitosis? How are they different?
6. **Predicting Outcomes** What would happen to the chromosome number of an organism's offspring if the gametes for sexual reproduction were made by mitosis instead of by meiosis?
7. **Further Inquiry** Write a new question about meiosis or sexual reproduction that could be explored with your model.

Do You Know?

Do research in the library or media center to answer these questions:

1. What types of human abnormalities arise when chromosomes do not separate properly during meiosis?
2. How do chemicals such as nicotine affect meiosis?

Use the following Internet resources to explore your own questions about meiosis and gamete formation.

internetconnect

SciLINKS NSTA
TOPIC: Meiosis
GO TO: www.scilinks.org
KEYWORD: HX157

4. Answers will vary.
5. Hypotheses will vary. Students should find that the gametes will be identical only if the parent is homozygous for every one of its genetic traits.
6. Hypotheses will vary. Students should find two copies of each chromosome and two copies of each gene in a zygote.
7. Hypotheses will vary. Students should find that crossing-over can produce new combinations of genes when an organism has different versions of the genes on the parts that cross over.
8. Hypotheses will vary. Students may find the wrong chromosome number in a gamete if the homologous pairs do not separate properly during anaphase I. Also, there would be less genetic variation.

Meiosis

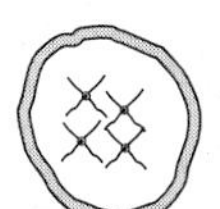

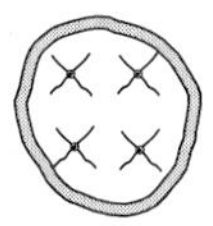
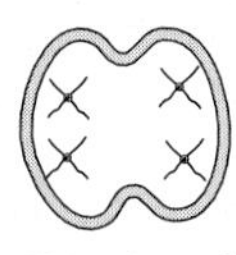
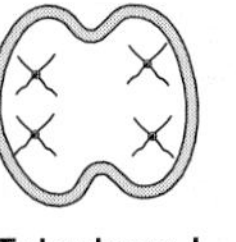
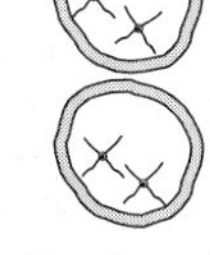
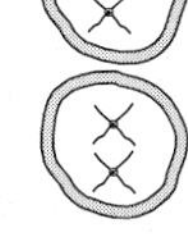
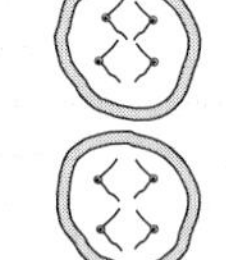
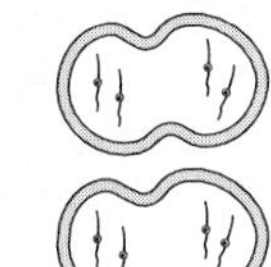

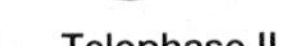

Prophase I Metaphase I Anaphase I Telophase I Prophase II Metaphase II Anaphase II Telophase II

Analyze and Conclude

Answers

1. The nuclei made by meiosis have half the original chromosome number. Two divisions of the nuclear material occur.
2. Homologous chromosomes are the same size and have genes for the same traits, but they are not identical as are chromatids.
3. Synapsis ensures that each new cell will get one member of each pair of homologous chromosomes.
4. Answers will vary. Students may suggest using different materials.
5. Mitosis and meiosis are both forms of nuclear division. Both consist of prophase, metaphase, anaphase, and telophase. Mitosis consists of one division and results in two cells that are genetically the same and have the same number of chromosomes as the original cell. Meiosis consists of two divisions and results in four cells that each have half the number of chromosomes as the original cell and are not the same genetically.
6. The chromosome number would be twice that in its parents' cells.
7. Answers will vary. For example: How do mitosis and meiosis compare?

Do You Know?

Answers

1. Answers will vary but might include Down syndrome, Klinefelter syndrome, and Turner syndrome.
2. Chemicals can affect the ability of chromosomes to separate normally during meiosis.

Chapter 8

Mendel and Heredity

Planning Guide

OBJECTIVES	MEETING INDIVIDUAL NEEDS	READING SKILLS AND STRATEGIES
8-1 The Origins of Genetics (BLOCKS 1 & 2 90 min) • **Identify** the investigator whose studies formed the basis of modern genetics. • **List** four characteristics that make the garden pea a good subject for genetic study. • **Summarize** the three major steps of Gregor Mendel's garden-pea experiments. • **Relate** the ratios that Mendel observed in his crosses.	Learning Styles, p. 160 BASIC ADVANCED • pp. 161, 162, 163	**ATE** Active Reading, Technique: K-W-L, p. 159 ■ Active Reading Guide, worksheet (8-1) ■ Directed Reading, worksheet (8-1) ■ Reading Strategies, K-W-L worksheet
8-2 Mendel's Theory (BLOCKS 3 & 4 90 min) • **Describe** the four major hypotheses Mendel developed. • **Relate** Mendel's work to the terms *homozygous, heterozygous, genotype,* and *phenotype*. • **Compare** Mendel's two laws of heredity.	Learning Styles, p. 164 BASIC ADVANCED • pp. 165, 167	■ Active Reading Guide, worksheet (8-2) ■ Directed Reading, worksheet (8-2)
8-3 Studying Heredity (BLOCKS 5 & 6 90 min) • **Predict** the results of monohybrid genetic crosses by using Punnett squares. • **Apply** a test cross to determine the genotype of an organism with a dominant phenotype. • **Predict** the results of monohybrid genetic crosses by using probabilities. • **Analyze** a simple pedigree.	Learning Styles, p. 168 BASIC ADVANCED • p. 174	■ Active Reading Guide, worksheet (8-3) ■ Directed Reading, worksheet (8-3)
8-4 Patterns of Heredity Can Be Complex (BLOCKS 7 & 8 90 min) • **Identify** five factors that influence patterns of heredity. • **Describe** how mutations can cause genetic disorders. • **List** two genetic disorders, and describe their causes and symptoms. • **Evaluate** the benefits of genetic counseling.	Learning Styles, p. 175 BASIC ADVANCED • p. 180	**ATE** Active Reading, Technique: K-W-L, p. 180 ■ Active Reading Guide, worksheet (8-4) ■ Directed Reading, worksheet (8-4) ■ Supplemental Reading Guide, *The Double Helix*

Review and Assessment Resources

BLOCKS 9 & 10 90 min

TEST PREP

- **PE** Study Zone, p. 181
- **PE** Performance Zone, pp. 182–183
- ■ Chapter 8 worksheets:
  - Vocabulary
  - Science Skills
  - Concept Mapping
  - Critical Thinking

- Holt Biology Interactive Tutor Unit 5—Heredity, Topics 1–6
- Audio CD Program
- ■ Test Prep Pretest

CHAPTER TESTING

- ■ Chapter Tests (blackline master)
- ▲ Test Generator
- ■ Assessment Item Listing
- ■ Performance-Based Assessment Activity: *Identifying Dominant and Recessive Traits*

See the correlations table at the front of this book for a list of the ***National Science Education Standards*** addressed in this chapter.

Smithsonian Institution® Visit **www.si.edu/hrw** for additional on-line resources.

KEY

Biology Principles and Explorations

PE Pupil's Edition
ATE Teacher's Edition

HOLT BIOSOURCES

 Teaching Transparencies
Laboratory Program

 **One-Stop Planner CD-ROM**
Shows these resources within a customizable daily lesson plan:

■ Teaching Resources
▲ Test Generator

 VISUAL STRATEGIES	CLASSWORK AND HOMEWORK	TECHNOLOGY AND INTERNET RESOURCES
ATE **Fig. 8-4,** p. 162 25 Mendel's Experimental Design	**PE** **Math Lab** Calculating Mendel's ratios, p. 163 **Lab E5** Heredity in Families ■ **Problem-Solving:** Ratios and Proportions Worksheet **PE** **Review,** p. 163	 **Interactive Tutor** Unit 5—Heredity (Topic 1)  **Holt Biology Videodiscs** Section 8-1  **Audio CD Program** Section 8-1
 **ATE** **Fig. 8-6,** p. 165  29 Crossing Homozygous Pea Plants  30 Crossing Heterozygous Pea Plants	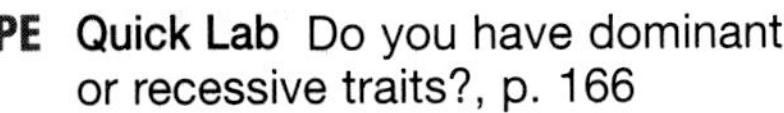 **PE** **Quick Lab** Do you have dominant or recessive traits?, p. 166  **PE** **Review,** p. 167	 **Interactive Tutor** Unit 5—Heredity (Topic 2)  **Holt Biology Videodiscs** Section 8-2  **Audio CD Program** Section 8-2
ATE **Fig. 8-9,** p. 168 **ATE** **Fig. 8-12,** p. 172 **ATE** **Fig. 8-13,** p. 173  31 Monohybrid Crosses  32 Dihybrid Crosses	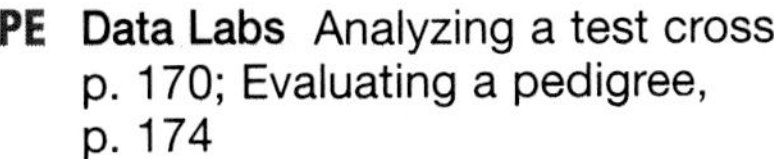 **PE** **Data Labs** Analyzing a test cross, p. 170; Evaluating a pedigree, p. 174 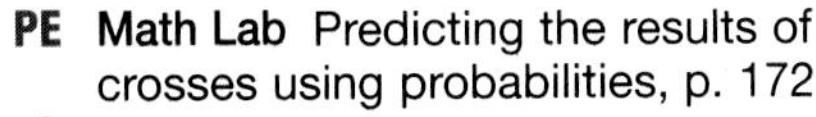 **PE** **Math Lab** Predicting the results of crosses using probabilities, p. 172  **Lab A6** Interpreting Information in a Pedigree **PE** **Review,** p. 174	 **Interactive Tutor** Unit 5—Heredity (Topics 3, 4, and 6) 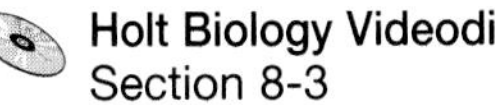 **Holt Biology Videodiscs** Section 8-3  **Audio CD Program** Section 8-3
ATE **Fig. 8-18,** p. 177  38 Tracking Inherited Traits (Family Tree)  30A Important Genetic Disorders	 **PE** **Interactive Exploration** Cystic Fibrosis, pp. 184–185 **PE** Review, p. 180 **PE** Study Zone, p. 181	CNN **Video Segment 5** Cloning Mice 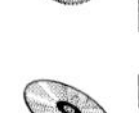 **Interactive Exploration:** Human Biology, Cystic Fibrosis  **Interactive Tutor** Unit 5—Heredity (Topic 5)  **Holt Biology Videodiscs** Section 8-4 **Audio CD Program** Section 8-4

internet**connect**

TOPICS:
- Genetic disorders
- Genetic counseling
- Cystic fibrosis

ALTERNATIVE ASSESSMENT

PE **Portfolio Projects** 26–29, p. 183
ATE **Alternative Assessment,** p. 181
ATE **Portfolio Assessment,** p. 181
■ **Portfolio Projects:** Genetics Project, "A Day in the Life of Mendel"

Lab Activity Materials

Math Labs
pp. 163, 172 paper and pencil

Quick Lab
p. 166 paper, pencil

Data Labs
pp. 170, 174 paper and pencil

 **Interactive Exploration**
p. 184 CD-ROM, computer

Lesson Warm-up
ATE photos of animals, p. 164

Demonstrations
ATE photos of flowering plants, p. 161
ATE deck of playing cards, p. 171
ATE flasks (4), food coloring, water, p. 176

 **BIOSOURCES LAB PROGRAM**
- Quick Labs: Lab A6, p. 11
- Interactive Explorations in Biology Laboratory Manual: Lab E5, p. 25

Chapter Theme

Evolution

The passing on of different genetic traits is fundamental to evolutionary change. The genetic variation provided by the possible combinations of parental genes and by crossing-over offers a means to change and an increase in the speed of evolution within a given species. Moreover, genetic mutations make evolution possible by introducing random changes into the inherited information that is passed on.

Establishing Prior Knowledge

Before beginning this chapter, make certain students can answer the questions in **Study TIP Ready?** on page 159. If they cannot, encourage them to reread the sections indicated.

Opening Demonstration

Have students bring photographs of parents or siblings to class. Ask them to match the names of fellow students with the photographs of their family members. After the students have matched the photographs, show each picture individually and ask the student whose relative is pictured to stand. Ask students to identify how the two are similar and different. Then ask them what techniques they think scientists might use to find a family connection.

CHAPTER 8 Mendel and Heredity

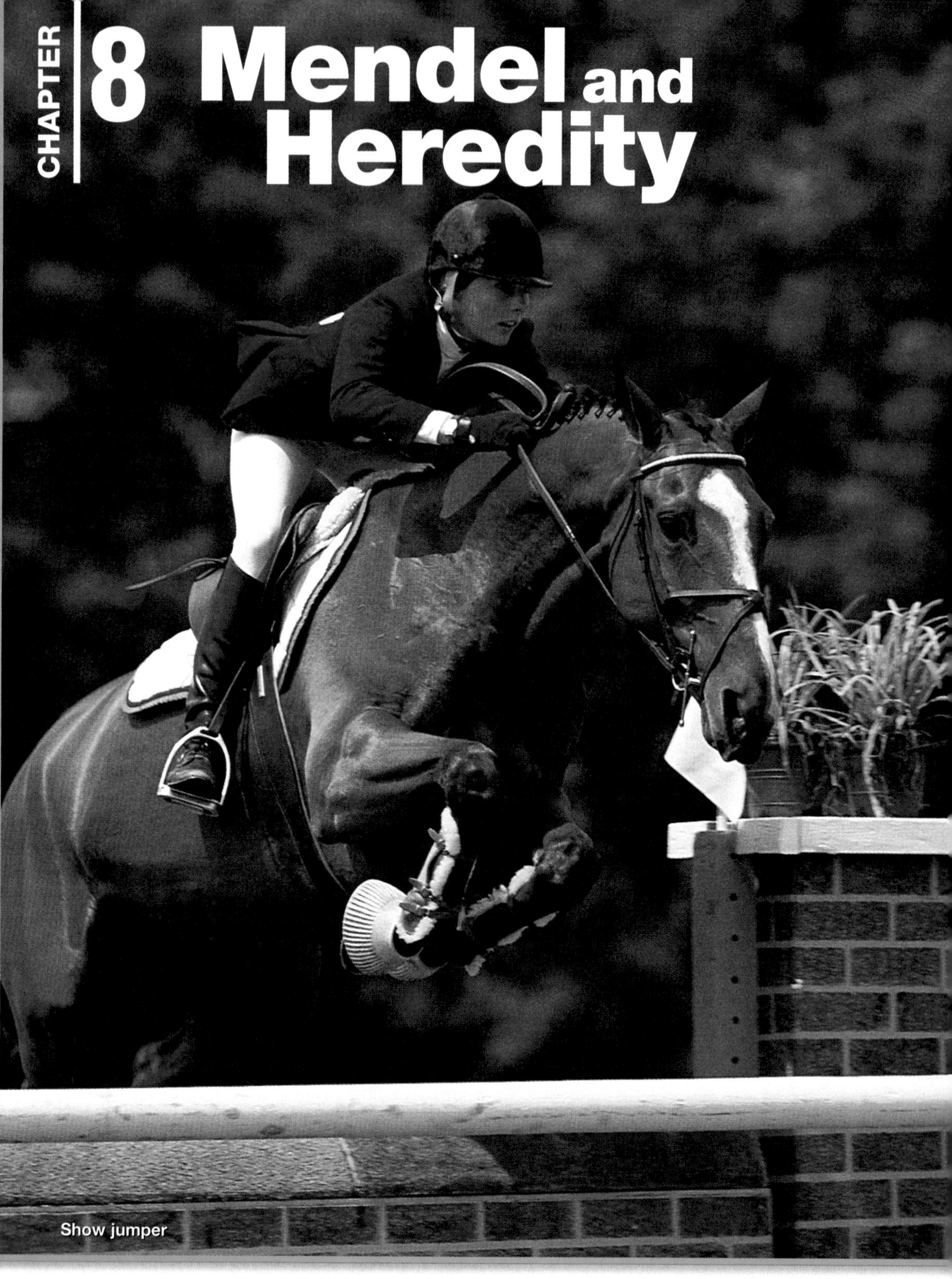

Show jumper

What's Ahead — and Why

CHAPTER OUTLINE

8-1 The Origin of Genetics
- Mendel and Others Studied Garden-Pea Traits
- Mendel Observed that Traits Are Expressed as Simple Ratios

8-2 Mendel's Theory
- Mendel's Work Became a Theory of Heredity
- Mendel's Ideas Gave Rise to the Laws of Heredity

8-3 Studying Heredity
- Punnett Squares Can Predict the Expected Results in Crosses
- Probabilities Can Also Predict the Expected Results of Crosses
- Family Pedigrees Can Be Used to Study How Traits Are Inherited

8-4 Patterns of Heredity Can Be Complex
- Most Traits Are Not Controlled by Simple Dominant-Recessive Alleles
- Some Traits Are Caused by Mutations

Students will learn

- 8-1: how heredity was first studied scientifically.
- 8-2: the terms that are used when studying heredity.
- 8-3: how different methods are used to study heredity.
- 8-4: that many factors can control the expression of traits.

Shetland ponies

Speed... strength... size...

Did you know that racehorses and show jumpers have been bred for these qualities for hundreds of years? Shetland ponies were originally bred for their strength—they were used to pull coal cars in coal mines. Today their small size makes them ideal for children to ride.

Study TIP

Ready?

Check your understanding of these topics to help you prepare for what's ahead.

Can you...

- **define** the term *gamete*? (Section 6-1)
- **summarize** the relationship between chromosomes and genes? (Section 6-1)
- **differentiate** between autosomes and sex chromosomes? (Section 6-1)
- **describe** how independent assortment during meiosis contributes to genetic variation? (Section 7-1)

If not, *review the sections indicated.*

Looking Ahead

Labs

- **Math Labs**
  Calculating Mendel's ratios (p. 163)
  Predicting the results of crosses using probabilities (p. 172)
- **Quick Lab**
  Do you have dominant or recessive traits? (p. 166)
- **Data Labs**
  Analyzing a test cross (p. 170)
  Evaluating a pedigree (p. 174)
- **Interactive Exploration**
  Cystic Fibrosis (p. 184)

Features

- **Exploring Further**
  Crosses That Involve Two Traits (p. 169)
- **Health Watch**
  Genetic Counseling (p. 180)

Unit 5—*Heredity*
Topics 1–6

internet connect

National Science Teachers Association *sci*LINKS Internet resources are located throughout this chapter.

Reading Strategies

Active Reading

Technique: K-W-L

Before they read this chapter, have each student write a short list of all the things they already Know (or think they know) about inheritance. Ask them to contribute their entries to a group list on the board or the overhead projector. Then have students list things they Want to know about inheritance. Have students save their lists for later use in the **Closure** on page 180.

Directed Reading

Assign **Directed Reading Worksheet Chapter 8** to help students interact with the text as they read the chapter.

Interactive Reading

Assign ***Biology: Principles and Explorations*** **Audio CD Program Chapter 8** to help reluctant readers achieve greater success in reading the chapter.

Reading Skills

Assign Reading Organizers and Reading Strategies Worksheets from the **One-Stop Planner CD-ROM with Test Generator** to help students learn and practice effective reading skills.

internet connect

Holt *Biology: Principles and Explorations*

On-line Resources: go.hrw.com
Visit the HRW Web site for a variety of resources related to this chapter. Just type in the keyword HX0 Chapter 8.

Holt *Biology: Principles and Explorations*

On-line Resources: www.scilinks.org
The following *sci*LINKS Internet resources can be found in the student text for this chapter:

Page 179
TOPIC: Genetic disorders
KEYWORD: HX179

Page 180
TOPIC: Genetic counseling
KEYWORD: HX180

Page 185
TOPIC: Cystic fibrosis
KEYWORD: HX185

Section 8-1 The Origins of Genetics

1 Prepare

Scheduling

- **Block:** About 90 minutes will be needed to complete this section.
- **Traditional:** About two class periods will be needed to complete this section.
- **Planning:** See the Planning Guide on pp. 157A and 157B for lecture, classwork, and assignment options for Section 8-1. Lesson plans for Section 8-1 can be found in a lesson cycle format in ***Biology: Principles and Explorations*** **One-Stop Planner CD-ROM with Test Generator.**

Resource Link

Assign **Active Reading Worksheet Section 8-1** to help students comprehend the key concepts and visuals in this lesson.

2 Teach

Lesson Warm-up

Tell students that a gardener noticed that some of the flowers on her pea plants were white. In previous years, the flowers had been purple. Ask students to suggest an explanation for this difference. Tell them Section 8-1 might give some insight. At the end of the chapter, come back to the question. *(Instead of buying the more expensive F_1 hybrid seeds as she usually did in the past, she decided to save money and plant pea seeds from the crop she harvested the previous year. Her plants were the F_2 generation, which shows a 3:1 ratio of purple to white flowers.)*

Objectives

- **Identify** the investigator whose studies formed the basis of modern genetics. (p. 160)
- **List** characteristics that make the garden pea a good subject for genetic study. (p. 161)
- **Summarize** the three major steps of Gregor Mendel's garden-pea experiments. (p. 162)
- **Relate** the ratios that Mendel observed in his crosses to his data. (p. 163)

Key Terms

heredity
genetics
monohybrid cross
true-breeding
P generation
F_1 generation
F_2 generation

Mendel and Others Studied Garden-Pea Traits

Many of your traits, including the color and shape of your eyes, the texture of your hair, and even your height and weight, resemble those of your parents. The passing of traits from parents to offspring is called **heredity.** Humans have long been interested in heredity. From the beginning of recorded history, we have attempted to alter crop plants and domestic animals to give them traits that are more useful to us. Before DNA and chromosomes were discovered, heredity was one of the greatest mysteries of science.

Figure 8-1 **Gregor Mendel**

Mendel and math

The scientific study of heredity began more than a century ago with the work of an Austrian monk named Gregor Johann Mendel, shown in **Figure 8-1.** Mendel carried out experiments in which he bred different varieties of the garden pea *Pisum sativum,* shown in **Figure 8-2.** British farmers had performed similar breeding experiments more than 200 years earlier. But Mendel was the first to develop rules that accurately predict patterns of heredity. The patterns that Mendel discovered form the basis of **genetics,** the branch of biology that focuses on heredity.

Mendel's parents were peasants, so he learned much about agriculture. This knowledge became invaluable later in his life. As a young man, Mendel studied theology and was ordained as a priest. Three years after being ordained, he went to the University of Vienna to study science and mathematics. There he learned how to study science through experimentation and how to use mathematics to explain natural phenomena.

Mendel later repeated the experiments of a British farmer, T. A. Knight. Knight had crossed a variety of the garden pea that had purple flowers with a variety that had white flowers. (The term *cross* refers to the mating or breeding of two individuals.) All of the offspring of Knight's crosses had purple flowers. However, when two of the purple-flowered offspring were crossed, their offspring

Figure 8-2 **The garden pea.** Mendel's experiments with garden peas led to our modern understanding of heredity.

Integrating Different Learning Styles

Logical/Mathematical	**PE**	Math Lab, p. 163; Performance Zone items 22 and 24, p. 183
	ATE	Resource Link, p. 162
Visual/Spatial	**PE**	Performance Zone item 1, p. 182
	ATE	Demonstration, p. 161; Visual Strategy, p. 162
Interpersonal	**ATE**	Biology in Your Community, p. 162; Resource Link, p. 162; Closure, p. 163
Verbal/Linguistic	**PE**	Performance Zone item 23, p. 183
	ATE	Teaching Tip, p. 161

showed both white and purple flowers. The white trait had reappeared in the second generation!

Mendel's experiments differed from Knight's because Mendel counted the number of each kind of offspring and analyzed the data. Quantitative approaches to science—those that include measuring and counting—were becoming popular in Europe. Mendel's method was on the cutting edge of research at the time.

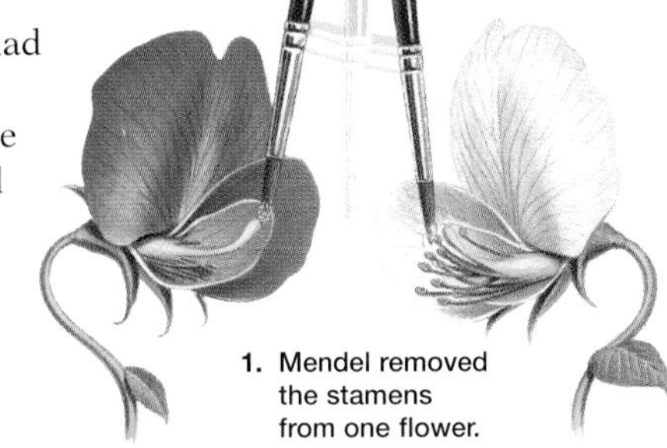

1. Mendel removed the stamens from one flower.
2. Mendel transferred pollen from a different plant to the pistil of the original flower.

Figure 8-3 Pollen transfer in Mendel's experiments. Mendel cross-pollinated flowers that had clearly different forms of a trait.

Useful features in peas

The garden pea is a good subject for studying heredity for several reasons:

1. The garden pea has many traits that have two clearly different forms that are easy to tell apart. For example, the flower color is either purple or white—there are no intermediate forms. **Table 8-1** shows the seven traits in garden peas that Mendel chose to study.
2. The mating of the garden-pea flowers can be easily controlled because the male and female reproductive parts are enclosed within the same flower. You can allow a flower to fertilize itself (self-fertilization), or you can transfer the pollen to another flower on a different plant (cross-pollination). Mendel had to control the cross-pollination of his pea plants carefully. To cross two pea plants, Mendel removed the stamens (the male reproductive organs that produce pollen) from the flower of one plant. As shown in **Figure 8-3,** he then dusted the pistil (the female reproductive organ that produces eggs) of that plant with pollen from a different pea plant. In this way, Mendel was able to control the matings.
3. The garden pea is small, grows easily, matures quickly, and produces many offspring. Thus, results can be obtained quickly, and there are plenty of subjects to count.

Table 8-1 The Seven Traits Mendel Studied and Their Contrasting Forms

Flower color	Seed color	Seed shape	Pod color	Pod shape	Flower position	Plant height

Section 8-1

Demonstration

Inherited Traits — BASIC

Display large pictures of a few flowering plants (or bring in real plants). Ask students to come up with a list of traits that could be inherited. Encourage students to think of many different traits, such as flower shape, flower color, position of flowers on stems, leaf shape, leaf color, pattern of veins, pattern of stem growth, presence of hairs on stems, and inner structure of flowers.

Teaching TIP — ADVANCED

Ad for Garden Peas

Have students design newspaper ads that would have attracted someone like Mendel to purchase peas for genetic research. The ads should mention all of the benefits of *Pisum sativum* that make it useful for genetic research. Ask students to use illustrations in their ads. PORTFOLIO

did you know?

Thoroughbred horses have been selectively bred since the 1600s.

All thoroughbreds can be traced back to three stallions (the Darley Arabian, the Godolphin Barb, and the Byerly Turk). With such a limited gene pool, inbreeding becomes a significant problem. The fertility rate is around 67 percent, which is much lower than in other animals, such as cattle (85 percent). In addition, the performance of the animals is not improving, even though they are being bred for speed.

Visual Strategy

BIO graphic

Figure 8-4

Work with students to help them summarize the events that took place during each step of Mendel's experiments. Check for understanding of the term *true-breeding* by having students identify the step in which Mendel used such plants (**Step 1**). To personalize the steps of Mendel's experiments, ask student volunteers to identify P, F_1, and F_2 generations using their own family as an example. Make sure students understand that each experiment began with a cross between two parent plants having different traits, and that the F_2 generation was produced by self-pollination of only one plant (**Step 3**).

Resource Link

- Assign **Portfolio Project: A Day in the Life of Mendel** from the Holt BioSources Teaching Resources CD-ROM. PORTFOLIO
- Assign **Problem Solving: Ratios and Proportions** from the Holt BioSources Teaching Resources CD-ROM. PORTFOLIO

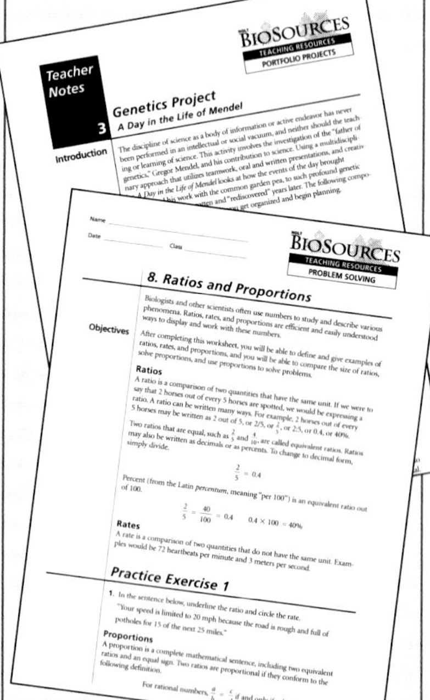

BioSources Teaching Resources Portfolio Projects

Teacher Notes

3 Genetics Project
A Day in the Life of Mendel

BioSources Teaching Resources Problem Solving

8. Ratios and Proportions

Practice Exercise 1

Study TIP

Word Origins

The word *filial* is from the Latin *filialis,* meaning "of a son or daughter." Knowing this makes it easier to remember that the F generations refer to any generation following the parental (P) generation.

Mendel Observed that Traits Are Expressed as Simple Ratios

Mendel's initial experiments were monohybrid crosses. A **monohybrid cross** is a cross that involves *one* pair of contrasting traits. For example, crossing a plant with purple flowers and a plant with white flowers is a monohybrid cross. Mendel carried out his experiments in three steps, as summarized in **Figure 8-4.**

Step 1 Mendel allowed each variety of garden pea to self-pollinate for several generations. This method ensured that each variety was **true-breeding** for a particular trait; that is, all the offspring would display only one form of a particular trait. For example, a true-breeding purple-flowering plant should produce only plants with purple flowers in subsequent generations.

These true-breeding plants served as the parental generation in Mendel's experiments. The parental generation, or **P generation,** are the first two individuals that are crossed in a breeding experiment.

Step 2 Mendel then cross-pollinated two P generation plants that had contrasting forms of a trait, such as purple flowers and white flowers. Mendel called the offspring of the P generation the first filial generation, or **F_1 generation.** He then examined each F_1 plant and recorded the number of F_1 plants expressing each trait.

Step 3 Finally, Mendel allowed the F_1 generation to self-pollinate. He called the offspring of the F_1 generation plants the second filial generation, or **F_2 generation.** Again, each F_2 plant was characterized and counted.

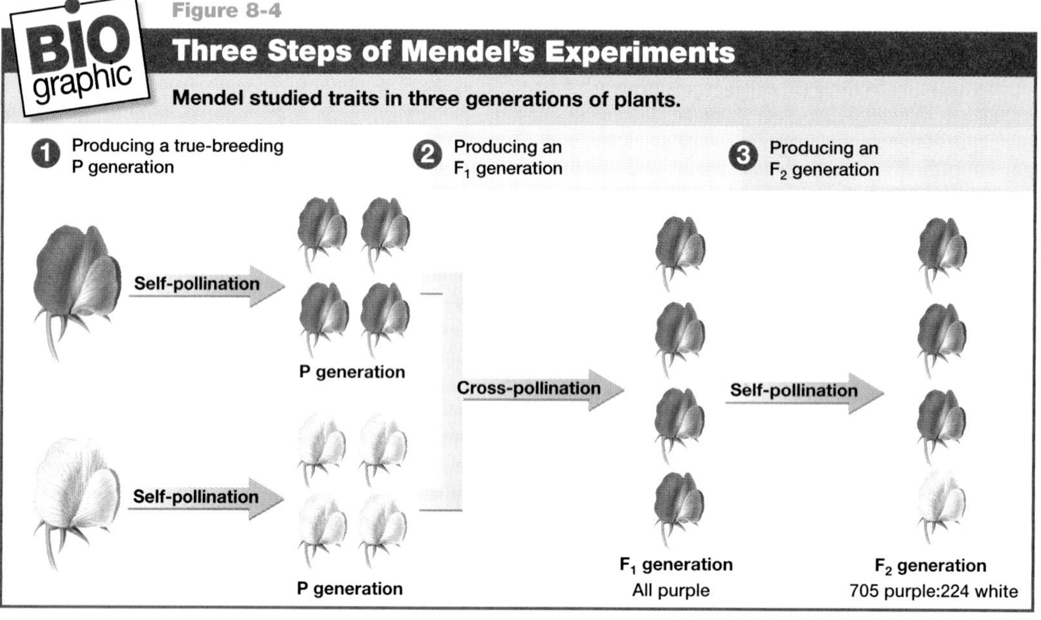

BIO graphic Figure 8-4 **Three Steps of Mendel's Experiments**

Mendel studied traits in three generations of plants.

Biology in Your Community

Animal Breeder

Invite a local animal breeder to speak to your class. Possible speakers could be breeders of dogs, cats, horses, birds, or any livestock. Have the guest discuss traits that are desirable, traits that are undesirable, and the reasons for these distinctions. Students may be interested in the market value of animals with certain traits. Ask the breeder to bring photographs of the animals, so students can see some of the traits being discussed.

Mendel's results

Each of Mendel's F_1 plants showed only one form of the trait. The contrasting form of the trait had disappeared! But when the F_1 generation was allowed to self-pollinate, the missing trait *reappeared* in some of the plants in the F_2 generation. When Mendel crossed purple flowers with white flowers, all of the offspring in the F_1 generation had purple flowers. In the F_2 generation, 705 plants had purple flowers and 224 plants had white flowers—a ratio of 705 to 224.

A ratio is a comparison of two numbers that have the same unit. A ratio can be written in words (705 to 224), as a fraction ($\frac{705}{224}$), or with a colon (705:224). When analyzing data, you can see patterns easier by reducing a ratio to its simplest form—divide each number by the smaller of the two numbers. For example, 705:224 reduced to its simplest form would be 3:1.

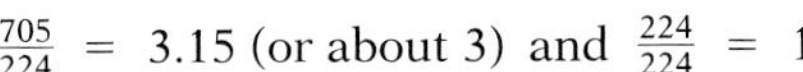

$$\frac{705}{224} = 3.15 \text{ (or about 3) and } \frac{224}{224} = 1$$

For each of the seven traits Mendel studied, he found the same 3:1 ratio of plants expressing the contrasting traits in the F_2 generation.

MATH LAB: *Calculating Mendel's ratios*

Background

You can calculate the ratios Mendel obtained in the F_2 generation for the traits he studied. First copy the partially completed table below on a separate piece of paper.

Analysis

1. **Calculate** the ratio for each contrasting trait. Use colon form.
2. **State** the ratio for each contrasting trait in words and as a fraction.
3. **SKILL Interpreting Results** Do the data confirm that Mendel obtained a 3:1 ratio in the F_2 generation for each of the traits he studied?

Contrasting traits	F_2 generation results		Ratio
Flower color	705 purple	224 white	3.15:1
Seed color	6,022 yellow	2,001 green	
Seed shape	5,474 round	1,850 wrinkled	
Pod color	428 green	152 yellow	
Pod shape	882 round	299 constricted	
Flower position	651 axial	207 top	
Plant height	787 tall	277 dwarf	

Review SECTION 8-1

1. **Evaluate** how Mendel's studies contributed to the foundation of modern genetics.
2. **Describe** why garden-pea plants are good subjects for genetic experiments.
3. **Summarize** the design of Mendel's pea-plant studies.
4. **State** the ratio Mendel obtained in each F_2 generation for each of the traits he studied.

Critical Thinking

5. **SKILL Evaluating Outcomes** What differences would be expected in experiments with squash plants, which usually do not self-pollinate?

Review SECTION 8-1 ANSWERS

1. Mendel was the first to develop rules that accurately predict patterns of heredity.
2. Garden-pea plants have many traits with two clearly different forms that are easy to see. The mating of the plants is easy to control. The plants are small and easy to grow, and they mature quickly and produce many offspring.
3. Mendel self-pollinated different varieties of pea flowers for several generations to produce true-breeding P generations. He then cross-pollinated P generation plants with contrasting forms of a trait and recorded the number of F_1 plants expressing each trait. He also self-pollinated F_1 plants and recorded the number of F_2 plants expressing each trait.
4. about 3:1
5. Without self-pollination, one cannot ensure that the P generation is true-breeding. Thus, there would not be a consistent ratio of traits in the F_1 or F_2 generations.

Calculating Mendel's ratios

Time 10–15 minutes

Process Skills Calculating, interpreting results, applying information

Teaching Strategies

Before students begin, provide an example of how to calculate ratios. Ask your class to count the number of students in the class (for example, 20 students). Then ask, "How many students in the class own a dog?" (Assume the answer is 5 students.) Tell students that the ratio of the number of students in the class to the number who own a dog is 20 to 5. Explain that the ratio is simplified by dividing each number by the smaller of the two numbers (20/5 and 5/5). The simplest ratio is 4 to 1 (stated in words), $\frac{4}{1}$ (stated as a fraction), and 4:1 (written with a colon).

Analysis Answers

1. 3.15:1, 3.01:1, 2.96:1, 2.82:1, 2.95:1, 3.14:1, 2.84:1
2. 3.15 to 1, $\frac{3.15}{1}$; 3.01 to 1, $\frac{3.01}{1}$; 2.96 to 1, $\frac{2.96}{1}$; 2.82 to 1, $\frac{2.82}{1}$; 2.95 to 1, $\frac{2.95}{1}$; 3.14 to 1, $\frac{3.14}{1}$; 2.84 to 1, $\frac{2.84}{1}$
3. yes

3 Assessment Options

Closure

Mendel's Journal

Have students write journal pages that represent comments Mendel might have written as he did his research. PORTFOLIO

Review

Assign Review—Section 8-1.

Reteaching — BASIC

Write the following terms on the board and ask students to define them: *monohybrid cross, P generation, F_1 generation, F_2 generation.*

Section 8-2 Mendel's Theory

1 Prepare

Scheduling

- **Block:** About 90 minutes will be needed to complete this section.
- **Traditional:** About two class periods will be needed to complete this section.
- **Planning:** See the Planning Guide on pp. 157A and 157B for lecture, classwork, and assignment options for Section 8-2. Lesson plans for Section 8-2 can be found in a lesson cycle format in ***Biology: Principles and Explorations* One-Stop Planner CD-ROM with Test Generator.**

Resource Link

Assign **Active Reading Worksheet Section 8-2** to help students comprehend the key concepts and visuals in this lesson.

2 Teach

Lesson Warm-up

Bring in photos from magazines or other sources of animals with different traits. Use these photos to emphasize that many genes are involved in giving an animal its overall appearance, and that the genes for most traits have two or more versions.

Objectives

- **Describe** the four major hypotheses Mendel developed. (pp. 164–165)
- **Describe** the terms *homozygous, heterozygous, genotype,* and *phenotype.* (pp. 165–166)
- **Compare** Mendel's two laws of heredity. (p. 167)

Key Terms

allele
dominant
recessive
homozygous
heterozygous
genotype
phenotype
law of segregation
law of independent assortment

Mendel's Work Became a Theory of Heredity

Before Mendel's experiments, many people thought offspring were just a *blend* of the characteristics of their parents. For example, if a tall plant were crossed with a short plant, the offspring would be medium in height.

Mendel's results did not support the blending hypothesis. Mendel correctly concluded that each pea has two separate "heritable factors" for each trait—one from each parent. As shown in **Figure 8-5,** when gametes (sperm and egg cells) form, only one of the two factors for each trait is given to a gamete. When the two gametes fuse during fertilization, the resulting offspring has two factors for each trait. Today these factors are called genes.

Mendel's hypotheses

The four hypotheses Mendel developed were based directly on the results of his experiments. The hypotheses are summarized below using modern terminology. These four hypotheses now make up the Mendelian theory of heredity, which forms the foundation of genetics.

Figure 8-5 Mendel's factors

Each parent has two separate "factors," or genes, for a particular trait.

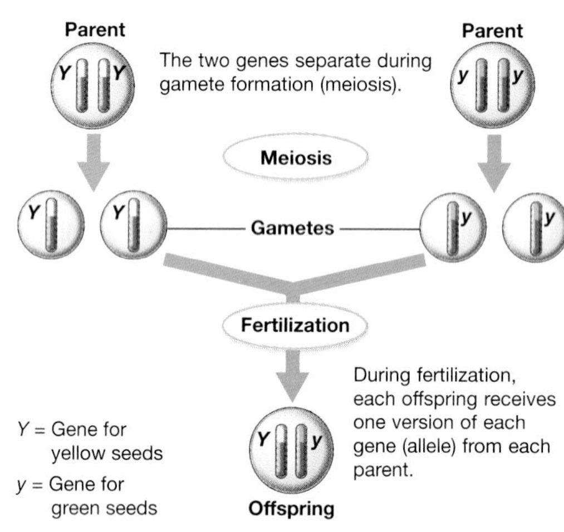

Parent
Parent
The two genes separate during gamete formation (meiosis).
Meiosis
Gametes
Fertilization
During fertilization, each offspring receives one version of each gene (allele) from each parent.
Y = Gene for yellow seeds
y = Gene for green seeds
Offspring

1. *For each inherited trait, an individual has two copies of the gene—one from each parent.*
2. *There are alternative versions of genes.* For example, the gene for flower color in peas can exist in a "purple" version or a "white" version. Today the different versions of a gene are called its **alleles.** As shown in Figure 8-5, an individual receives one allele from each parent. Each allele can be passed on when the individual reproduces.
3. *When two different alleles occur together, one of them may be completely expressed, while the other may have no observable effect on the organism's appearance.* Mendel described the expressed form of the trait as **dominant.** The trait that was not expressed when the dominant form of the trait was present was described as **recessive.** For every pair of contrasting forms of a trait that Mendel studied, the allele for one form of the trait was always dominant and the allele for the other form of the trait was always recessive. For example, if

Integrating Different Learning Styles

Logical/Mathematical	**PE**	Performance Zone item 16, p. 182
	ATE	Reteaching, p. 167
Visual/Spatial	**ATE**	Lesson Warm-up, p. 164; Teaching Tips, pp. 165, 167; Visual Strategy, p. 165
Intrapersonal	**PE**	Quick Lab, p. 166
Verbal/Linguistic	**PE**	Performance Zone item 2, p. 182

a plant has both purple and white alleles for flower color but blooms purple flowers, then purple is the dominant form of the trait; white is the recessive form. This is shown in **Figure 8-6.**

4. *When gametes are formed, the alleles for each gene in an individual separate independently of one another. Thus, gametes carry only one allele for each inherited trait. When gametes unite during fertilization, each gamete contributes one allele.* As shown in Figure 8-5 and described in Chapter 7, each parent can contribute only one of the alleles because of the way gametes are produced during the process of meiosis.

Mendel's findings in modern terms

Geneticists have developed specific terms and ways of representing an individual's genetic makeup. For example, letters are often used to represent alleles. Dominant alleles are indicated by writing the first letter of the trait as a capital letter. For instance, in pea plants, purple flower color is a dominant trait and is written as *P.* Recessive alleles are also indicated by writing the first letter of the dominant trait, but the letter is lowercase. For example, white flower color is recessive and is written as *p*.

If the two alleles of a particular gene present in an individual are the same, the individual is said to be **homozygous** *(hoh moh ZEYE guhs)* for that trait. For example, a plant with two white flower alleles is homozygous for flower color, as shown in Figure 8-6. The allele for yellow peas, *Y*, is dominant to the allele for green peas, *y*. A plant with two yellow-pea alleles, *YY,* is homozygous for seed color.

If the alleles of a particular gene present in an individual are different, the individual is **heterozygous** *(heht uhr oh ZEYE guhs)* for that trait. As shown in Figure 8-6, a plant with one "purple flower" allele and one "white flower" allele is heterozygous for flower color. A plant with one "yellow pea" allele and one "green pea" allele is heterozygous for seed color.

biobit

Who has a cleft chin?

Like the traits in pea plants, many human traits are simple dominant or recessive traits. For example, the allele for a cleft chin (*C*) is dominant to the allele for a chin without a cleft (*c*).

Figure 8-6 **Recessive alleles can be present but not expressed**

The allele for purple flowers, *P,* is dominant to the recessive allele, *p*.

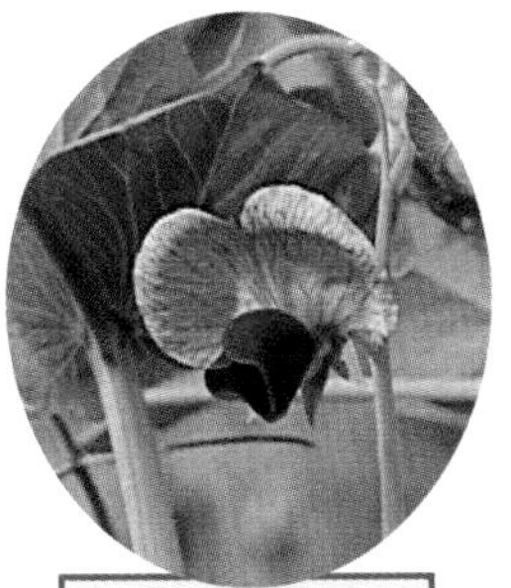

PP
Purple flowers, homozygous dominant

pp
White flowers, homozygous recessive

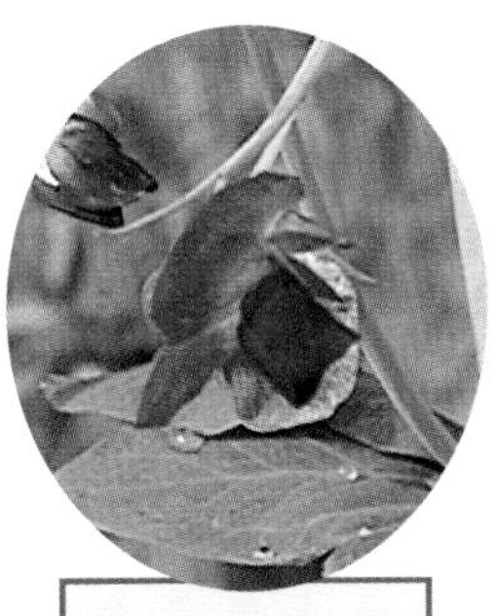

Pp
Purple flowers, heterozygous

did you know?

Two different forms of the same species of fire ant have been found to select queens based on phenotype.

One type of colony allows several queens to survive, while the other type selects a single queen. Workers in the colony with multiple queens eliminate queens with a particular homozygous recessive genotype from the colony. However, this genotype is prevalent in the single-queen colony. Why are homozygous recessive queens eliminated in one colony and not in the other one? Queens homozygous for the recessive gene reproduce at a faster rate. This trait is beneficial in a single-queen colony but detrimental in a multiple-queen colony.

Teaching TIP

Concept Mapping

Have students use the boldface vocabulary words in Section 8-2 to draw a concept map. The boldface vocabulary words are listed under **Key Terms** on page 164. Students may also include other words on the map, but the vocabulary words should be highlighted in color. Have students draw three cross-links on the map to show how the terms are related. PORTFOLIO

Visual Strategy

Figure 8-6

Point out to students that you cannot always know an organism's genotype by looking at its phenotype. For example, one purple flower in Figure 8-6 on page 165 is homozygous dominant, while the other purple flower is heterozygous.

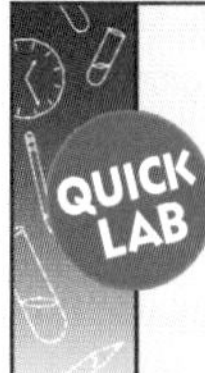

Do you have dominant or recessive traits?

Time 10–15 minutes

Process Skills Summarizing, calculating, applying information, drawing conclusions

Teaching Strategies

- Emphasize that dominant phenotypes are not more common than recessive phenotypes.
- Point out that the expression of some phenotypes (such as freckles) may be influenced by the environment.

Analysis Answers

1. Answers will vary.
2. Answers will vary.
3. The recessive traits. Recessive traits must be homozygous to be expressed.

Teaching TIP

Genotype and Phenotype

Have students practice using the boldface terms on pp. 164–166 by providing several examples. For example, tell them that the gene for plant height has two versions: T = tall and t = dwarf. Ask students to identify the two alleles for plant height. *(T and t)* Write *Tt, tt,* and *TT* on the board and ask students to identify the genotype and phenotype of each set of alleles. *(genotypes—Tt, tt, and TT; phenotypes—tall, dwarf, and tall, respectively)* Ask students to identify whether a plant with *TT* alleles is homozygous or heterozygous. *(homozygous)*

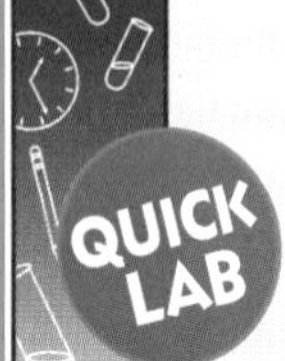

Do you have dominant or recessive traits?

You can determine some of the genotypes and all of the phenotypes for human traits that are inherited as simple dominant or recessive traits.

Materials

pencil, paper

Procedure

1. **Make** a table like the one at right. For each trait, circle the phenotype that best matches your own phenotype.
2. **Determine** how many students in your class share your phenotype by recording your results in a table on the chalkboard.

Dominant trait	Recessive trait
Cleft chin	No cleft
Dimples	No dimples
Hair above knuckles	Hairless fingers
Freckles	No freckles

Analysis

1. **Summarize** the class results for each trait.
2. **Calculate** the class dominant:recessive ratio for each trait.
3. **SKILL Applying Information** For which phenotypes in the table can you determine a person's genotype without ever having seen his or her parents? Explain.

Study TIP

Reading Effectively

Read the symbol for a genotype by saying the words for that genotype. For example, read *Pp* as "heterozygous purple," read *pp* as "homozygous white," and read *PP* as "homozygous purple."

In heterozygous individuals, only the dominant allele is expressed; the recessive allele is present but unexpressed. An example of a human trait that is expressed in a heterozygous individual is freckles. Freckles *F,* is a dominant allele. The recessive allele is *f,* no freckles. The recessive allele may be present but not expressed. As shown in **Figure 8-7,** people who are heterozygous for freckles (*Ff*) will have freckles even though they also have the allele for no freckles, *f.*

The set of alleles that an individual has is called its **genotype** *(JEE noh teyep).* The physical appearance of a trait is called a **phenotype** *(FEE noh teyep).* Phenotype is determined by which alleles are present. For example, if *Pp* is the genotype of a pea plant, its phenotype is purple flowers. If *pp* is the genotype of a pea plant, its phenotype is white flowers. When considering seed color, if *Yy* is the genotype of a pea plant, its phenotype is yellow seeds. If *yy* is the genotype of a pea plant, its phenotype is green seeds. Note that by convention, the dominant form of the trait is written first, followed by the lowercase letter for the recessive form of the trait.

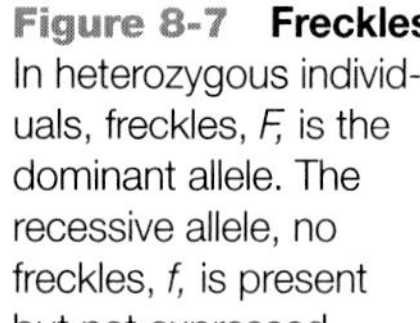

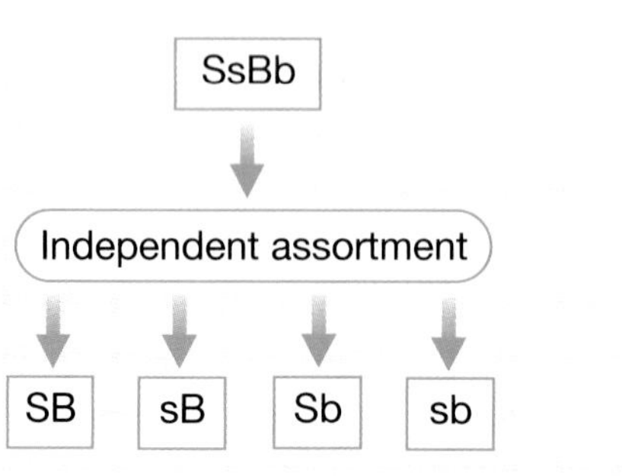

Figure 8-7 Freckles. In heterozygous individuals, freckles, *F,* is the dominant allele. The recessive allele, no freckles, *f,* is present but not expressed.

Graphic Organizer

Use this graphic organizer with ***Teaching Tip: Independent Assortment*** *on page 167.*

SsBb → Independent assortment → SB | sB | Sb | sb

Mendel's Ideas Gave Rise to the Laws of Heredity

Mendel's hypotheses brilliantly predicted the results of his crosses and also accounted for the ratios he observed. Similar patterns of heredity have since been observed in countless other organisms. Because of their importance, Mendel's ideas are often referred to as the laws of heredity.

The law of segregation

The first law of heredity describes the behavior of chromosomes during meiosis, when homologous chromosomes and then chromatids are separated. The first law, the **law of segregation,** states that the two alleles for a trait segregate (separate) when gametes are formed (as shown in Figure 8-5).

The law of independent assortment

Mendel went on to study whether the inheritance of one trait (such as plant height) influenced the inheritance of a different trait (such as flower color). To study how different pairs of genes are inherited, Mendel conducted dihybrid crosses. A dihybrid cross is a cross that considers two pairs of contrasting traits. For example, a cross that considers both plant height and flower color is a dihybrid cross. The plant in **Figure 8-8** demonstrates these two traits.

Mendel found that for the pairs of traits he studied, the inheritance of one trait did not influence the inheritance of any other trait. This observation eventually became known as the law of independent assortment. The **law of independent assortment** states that the alleles of different genes separate independently of one another during gamete formation. For example, the alleles for plant height separate independently of the alleles for flower color. We now know that this law applies only to genes that are located on different chromosomes or that are far apart on the same chromosome.

The search for the physical nature of Mendel's "factors" dominated biology for more than half a century after Mendel's work was rediscovered in 1900. We now know that the units of heredity are portions of DNA called genes, which are found on the chromosomes that an individual inherits from its parents.

Figure 8-8 The law of independent assortment. Mendel found that the inheritance of one trait, such as plant height, did not influence the inheritance of another trait, such as flower color.

biobit

When was Mendel's work recognized?

Mendel published his results in 1866. Unfortunately, the results failed to create much interest and were quickly forgotten. In 1900, sixteen years after Mendel's death, several scientists independently rediscovered the pioneering paper.

Review SECTION 8-2

1. **Differentiate** between alleles and genes.
2. **Apply** the terms *homozygous, heterozygous, dominant,* or *recessive* to describe a plant with the genotype *Pp*.
3. **Identify** the phenotype of a rabbit with the genotype *Bb,* where *B* = black coat and *b* = brown coat.
4. **Determine** whether the rabbit in item 3 is heterozygous or homozygous.

Critical Thinking

5. **SKILL Recognizing Relationships** How are Mendel's two laws explained in terms of meiosis? (**Hint:** See Figure 8-5.)

Section 8-2

Teaching TIP

Independent Assortment

Have students make a **Graphic Organizer** similar to the one at the bottom of page 166 to demonstrate the law of independent assortment. Ask students to illustrate their graphic organizers with at least one example showing the inheritance of two pairs of contrasting traits. Have them write a brief explanation. PORTFOLIO

3 Assessment Options

Closure

Identifying Variables

Tell students that they have been presented with one of Mendel's purple-flowering pea plants. Ask them if they can identify the genotype of the plant with regard to flower color. Then have them propose a method for discovering the genotype of the plant.

Review

Assign Review—Section 8-2.

Reteaching BASIC

Have students review the steps involved in scientific investigations discussed in Section 1-3 in Chapter 1. Then have them apply these methods to Mendel's discoveries about heredity. Ask students to summarize Mendel's hypotheses and predictions.

Review SECTION 8-2 ANSWERS

1. Genes are pieces of DNA that code for a particular trait. There are alternative versions, or alleles, for each gene.
2. The plant is heterozygous, with one dominant *P* allele and one recessive *p* allele.
3. The rabbit has a black coat.
4. The rabbit is heterozygous.
5. The law of segregation states that the members of each pair of alleles separate when gametes are formed. This occurs when homologous chromosomes and then chromatids separate during meiosis. The law of independent assortment states that pairs of alleles separate independently during gamete formation. This results from the random arrangement of homologous chromosomes and chromatids during metaphase of meiosis I and meiosis II.

Section 8-3 Studying Heredity

1 Prepare

Scheduling

- **Block:** About 90 minutes will be needed to complete this section.
- **Traditional:** About two class periods will be needed to complete this section.
- **Planning:** See the Planning Guide on pp. 157A and 157B for lecture, classwork, and assignment options for Section 8-3. Lesson plans for Section 8-3 can be found in a lesson cycle format in ***Biology: Principles and Explorations*** **One-Stop Planner CD-ROM with Test Generator.**

Resource Link

Assign **Active Reading Worksheet Section 8-3** to help students comprehend the key concepts and visuals in this lesson.

2 Teach

Lesson Warm-up

Have students determine whether or not they have hair above their knuckles. Tell students that the presence of hair above the knuckles is caused by a dominant allele, *H*. Then ask them to identify the genotype of a person who does not have hair above their knuckles. *(hh)* Have students determine under what circumstances a parent without hair above their knuckles can produce a child with hair above their knuckles. *(The second parent must have the hair above knuckles allele.)*

Visual Strategy

Figure 8-9

Point out to students that the genotype of a parent determines the possible alleles that can be found in their gametes. The possible gametes are written along the top and left sides of the square. Review with students how the genotype in each square was obtained. Assign several monohybrid crosses for students to practice.

Objectives

- **Predict** the results of monohybrid genetic crosses by using Punnett squares. (pp. 168–169)
- **Apply** a test cross to determine the genotype of an organism with a dominant phenotype. (p. 170)
- **Predict** the results of monohybrid genetic crosses by using probabilities. (pp. 171–172)
- **Analyze** a simple pedigree. (pp. 173–174)

Key Terms

Punnett square
test cross
probability
pedigree
sex-linked trait

Punnett Squares Can Predict the Expected Results in Crosses

Animal breeders have to breed animals with very specific characteristics. Thus, breeders must be able to predict how often a trait will appear when two animals are crossed (bred). Likewise, horticulturists (plant breeders) need to produce plants with very specific characteristics. One simple way of predicting the expected results (not necessarily the actual results) of the genotypes or phenotypes in a cross is to use a Punnett square.

A **Punnett square** is a diagram that predicts the expected outcome of a genetic cross by considering all possible combinations of gametes in the cross. Named for its inventor, Reginald Punnett, the simplest Punnett square consists of four boxes inside a square. As shown in **Figure 8-9,** the possible gametes that one parent can produce are written along the top of the square. The possible gametes that the other parent can produce are written along the left side of the square. Each box inside the square is filled with two letters obtained by combining the allele along the top of the box with the allele along the side of the box. The letters in the boxes indicate the possible genotypes of the offspring.

Crosses involving one pair of contrasting traits

Punnett squares can be used to predict the outcome of a monohybrid cross (a cross that considers one pair of contrasting traits between two individuals). For example, a Punnett square can be used to predict the outcome of a cross between a pea plant that is homozygous for yellow seed color (*YY*) and a pea plant that is homozygous for green seed color (*yy*). Figure 8-9 shows that 100 percent of the offspring in this type of cross are expected to be heterozygous (*Yy*), expressing the dominant trait of yellow seed color.

Figure 8-9 Monohybrid cross: homozygous plants

A cross between a pea plant that is homozygous for yellow seeds (*YY*) and a pea plant that is homozygous for green seeds (*yy*) will produce only yellow heterozygous offspring (*Yy*).

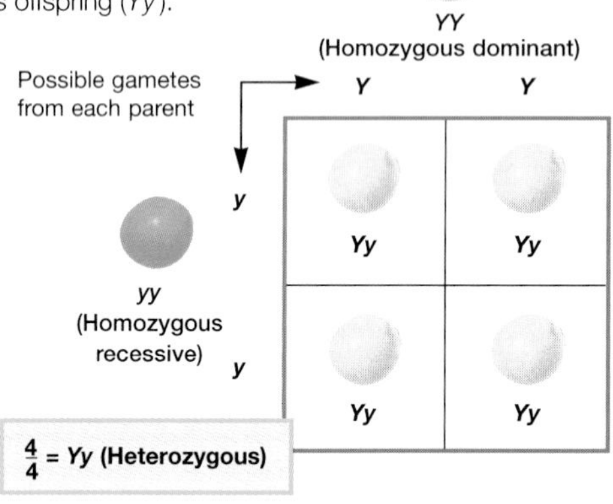

Figure 8-10 shows a Punnett square that predicts the results of a monohybrid cross between two pea plants that are both heterozygous (*Yy*) for seed color. One-fourth of the offspring would be expected to have the genotype *YY*, two-fourths (or one-half) would be expected to have the genotype *Yy*, and

Integrating Different Learning Styles

Logical/Mathematical	**PE**	Exploring Further, p. 169; Real Life, p. 170; Data Lab, p. 170; Math Lab, p. 172; Performance Zone items 14, 17, 18, 19, 20, and 21, pp. 182–183
	ATE	Teaching Tips, pp. 171, 173; Demonstration, p. 171; Resource Link, p. 171; Cross-Disciplinary Connection, p. 173; Reteaching, p. 174
Visual/Spatial	**PE**	Data Lab, p 174; Performance Zone item 10, p. 182
	ATE	Visual Strategies, pp. 168, 172, and 173
Interpersonal	**ATE**	Alternative Assessment, p. 181
Intrapersonal	**ATE**	Lesson Warm-up, p. 168
Verbal/Linguistic	**PE**	Performance Zone items 26 and 29, p. 183
	ATE	Biology in Your Community, p. 170

one-fourth would be expected to have the genotype *yy*. Another way to express this is to say that the genotypic ratio is 1 *YY*: 2 *Yy* : 1 *yy*. Since the *Y* allele is dominant over the *y* allele, three-fourths of the offspring would be yellow, and one-fourth would be green. The phenotypic ratio is 3 yellow : 1 green.

Punnett squares allow direct and simple predictions to be made about the outcomes of genetic crosses. Although animal breeders and horticulturists are not always certain what characteristics will turn up in the offspring, they can use the predictions from Punnett squares to cross individuals that they know will be most likely to produce offspring with the desired phenotypes.

Figure 8-10 Monohybrid cross : heterozygous plants

Crossing two pea plants that are heterozygous for seed color (*Yy*) will produce offspring in the ratio shown in the Punnett square.

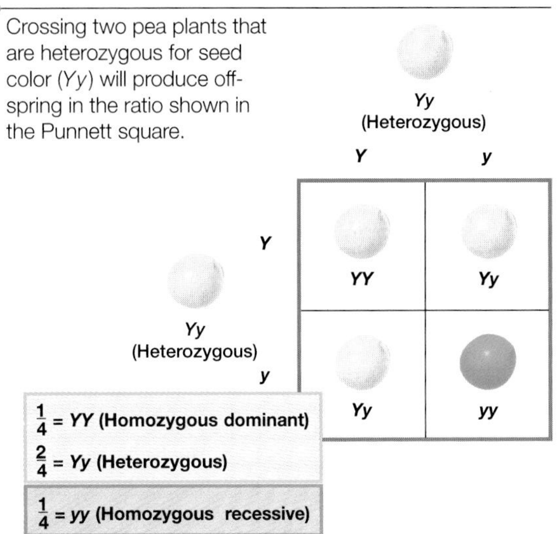

$\frac{1}{4}$ = *YY* (Homozygous dominant)

$\frac{2}{4}$ = *Yy* (Heterozygous)

$\frac{1}{4}$ = *yy* (Homozygous recessive)

Exploring Further

Crosses That Involve Two Traits

Suppose a horticulturist has two traits that she wants to consider when crossing two plants. A cross that involves two pairs of contrasting traits is called a dihybrid cross. For example, she may want to predict the results of a cross between two pea plants that are heterozygous for seed shape (*R* = round, *r* = wrinkled) and seed color (*Y* = yellow, *y* = green).

Determine possible gametes

To use a Punnett square to predict the results of this cross, first consider how the four alleles from either parent (*RrYy*) can combine to form gametes that are either *RY, Ry, rY,* or *ry* (Figure A).

Figure A Gametes

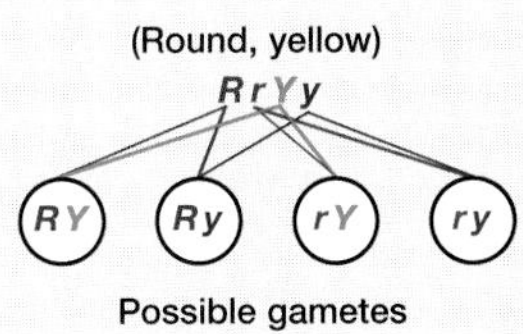

Then write these gametes on the top and left sides of a Punnett square (Figure B).

Complete the Punnett square

On a separate sheet of paper, make a copy of the Punnett square in Figure B, which has been partially filled in with the predicted genotypes. Fill in the remaining genotypes, then do the following:

- **List** all of the possible genotypes that can result.
- **Calculate** the genotypic ratio for this cross.
- **List** all of the possible phenotypes that can result.
- **Calculate** the phenotypic ratio for this cross.

Figure B Punnett square

Possible gametes from each parent — Yellow (top), Yellow (left)

	RY	*Ry*	*rY*	*ry*
RY	*RRYY*	*RRYy*	*RrYY*	*RrYy*
Ry	*RRYy*		*RrYy*	
rY	*RrYY*	*RrYy*		
ry	*RrYy*			

Multicultural Perspective

Blood Lines

The Greek philosopher Aristotle associated inheritance with blood. He thought the blood carried hereditary information from the body's various structures to the reproductive organs. We know this is not true, but the idea is ingrained in many languages. For example, "blue blood," "blood stock," and "It is in the blood" (English); "Corre en la sangre" (Spanish); "Bon sang ne peut mentir" and "celle est dans le sang" (French); and "Es liegt im Blute" and "von gutem Blut" (German) all associate inheritance with blood.

Exploring Further

Crosses That Involve Two Traits

Teaching Strategies

- Start by having students practice listing possible gametes when given the genotype of a dihybrid parent.
- Point out that the possible gametes are written along the top and left sides of a Punnett square, just as with a monohybrid cross.
- Explain how genotypes are written in dihybrid crosses: the two alleles for one gene are written together, followed by the alleles for the second gene (for example, *RrYy,* not *RYrY*).

Discussion

- Review the answers to the cross. Possible genotypes—*RRYY, RRYy, RrYY, RrYy, Rryy, RRyy, rrYY, rrYy, rryy.* Genotypic ratio—1 *RRYY* : 2 *RRYy* : 1 *RRyy* : 2 *RrYY* : 4 *RrYy* : 2 *Rryy* : 1 *rrYY* : 2 *rrYy* : 1 *rryy.* Possible phenotypes—round, yellow; round, green; wrinkled, yellow; wrinkled, green. Phenotypic ratio—9 round, yellow : 3 round, green : 3 wrinkled, yellow : 1 wrinkled, green.
- How do we know that these parents can produce four different gametes? *(Mendel's law of independent assortment states that during gamete formation, pairs of alleles separate independently.)*
- A female Labrador retriever with white fur and a pink nose is bred with a male Labrador retriever with black fur and a black nose. What type of cross does this example represent? *(dihybrid cross)*

REAL LIFE

Answer

It would take at least 39 years or more, depending on the age at which mating begins each time.

Teaching TIP

Experimental Animals

Have students read the **Real Life** on page 170. Discuss why, in addition to the reasons cited, scientists use experimental animals to study heredity. *(Answers may vary. For example, scientists can control their diet, breeding, and environmental factors.)* Ask students what problems might arise if scientists used humans to study heredity. *(In addition to moral and ethical dilemmas, humans have long generation times and it is difficult to control environmental factors.)*

Analyzing a test cross

Time 15–20 minutes

Process Skills Analyzing, interpreting, inferring, drawing conclusions, predicting outcomes

Teaching Strategies

Encourage students to recognize the importance of sample size in making conclusions about the genotype of an unknown individual.

Analysis Answers

1. possible alleles each parent can produce
2. the genotype of each possible kind of offspring
3. The genotypic ratio will be 4 *Pp* : 0 *PP* : 0 *pp*. The phenotypic ratio for the offspring will be 4 purple : 0 white.
4. heterozygous

REAL LIFE

Some people love flies and worms. Many scientists who study genetics use the fruit fly *Drosophila melanogaster* or the roundworm *Caenorhabditis elegans* in their research. These organisms show a variety of traits, are easy to obtain and breed, have short generation times (less than 2 weeks for fruit flies; less than 3 days for roundworms), and produce large numbers of offspring.

Calculating *How long would it take to study three generations of humans?*

Determining unknown genotypes

Animal breeders, horticulturists, and others involved in breeding organisms with very specific characteristics often need to know whether an organism with a dominant phenotype is heterozygous or homozygous for a trait. How do they determine this? For example, how might a horticulturist determine whether a pea plant with a dominant phenotype, such as yellow seeds, is homozygous (*YY*) or heterozygous (*Yy*)? The horticulturist could perform a **test cross,** a cross in which an individual whose phenotype is dominant, but whose genotype is not known, is crossed with a homozygous recessive individual.

For example, a plant with yellow seeds but of unknown genotype (*Y?*) is test-crossed with a plant with green seeds (*yy*). If all of the offspring produce yellow seeds, the genotype of the "unknown" plant must be *YY.* If half of the offspring produce yellow seeds and half produce green seeds, the genotype of the unknown plant must be *Yy.* In reality, if the cross produces even one plant that produces green seeds, the genotype of the unknown parent plant is likely to be heterozygous. After performing a test cross, the horticulturist can continue breeding the original plant with more certainty of its genotype.

Analyzing a test cross

Background

You can use a test cross to determine whether a plant with purple flowers is heterozygous (*Pp*) or homozygous dominant (*PP*). On a separate sheet of paper, copy the two Punnett squares shown below, and fill in the boxes in each square.

Is this purple flowering pea plant *Pp* or *PP*?

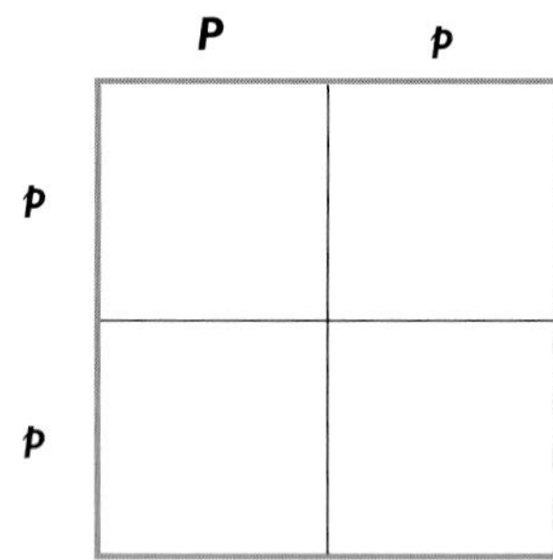

Figure A Heterozygous (*Pp*) plant

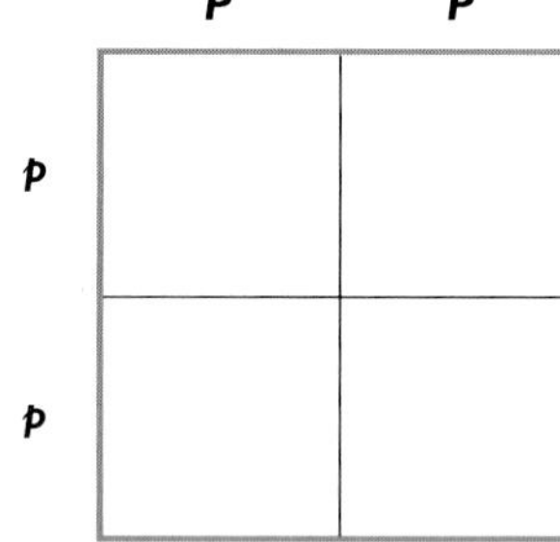

Figure B Homozygous (*PP*) plant

Analysis

1. **Determine** what the letters at the top and side of each box represent.
2. **Determine** what the letters in each box represent.
3. **Calculate** the genotypic and phenotypic ratios that would be predicted if the parent of the unknown genotype were homozygous for the trait (Figure B).
4. **SKILL Predicting Outcomes** If the test cross shows that half of the offspring have white flowers, what is the genotype of the plant with purple flowers?

Biology in Your Community

Mendel and Selective Breeding

Since the dawn of agriculture, people have used selective breeding to improve crops and domestic animals. Modern applications of Mendelian genetics and gene technology have resulted in major changes in crops and animals. Have students investigate the history of a regionally important agricultural product or domestic animal. PORTFOLIO

Probabilities Can Also Predict the Expected Results of Crosses

Like Punnett squares, probability calculations can be used to predict the results of genetic crosses. **Probability** is the likelihood that a specific event will occur. Probabilities can be expressed in words, as decimals, as percentages, or as fractions. For example, if an event will definitely occur, its probability can be expressed as either 1 out of 1 (in words), 1 (as a decimal numeral), 100 percent (as a percentage), or $\frac{1}{1}$ (as a fraction). If an event will definitely not occur, its probability can be expressed as either 0 out of 0, 0, 0 percent, or $\frac{0}{0}$.

In order to simplify our discussion of probability, we will express probabilities as fractions. Probability can be determined by the following formula:

$$\text{Probability} = \frac{\text{number of one kind of possible outcome}}{\text{total number of all possible outcomes}}$$

Consider the possibility that a coin tossed into the air will land on heads (one possible outcome). The total number of all possible outcomes is two—heads or tails. Thus, the probability that a coin will land on heads is $\frac{1}{2}$, as shown in **Figure 8-11.**

Probability of a specific allele in a gamete

The same formula can be used to predict the probability of an allele being present in a gamete. If a pea plant has two alleles for seed color, that individual can contribute either allele (yellow or green) to the gamete it produces (the law of independent assortment). For the plant with two alleles for seed color, the probability that a gamete will carry the allele for green seed color is $\frac{1}{2}$. There is one possible outcome—green—and the total number of possible outcomes is two—green or yellow. The probability that a gamete from this plant will carry the allele for yellow seed color is also $\frac{1}{2}$.

Probability of the outcome of a cross

Since two parents are involved in a genetic cross, both parents must be considered when calculating the probability of the outcome of a genetic cross. Consider the analogy of two coins being tossed at the same time. The probability of a penny landing on heads is $\frac{1}{2}$, and the probability of a nickel landing on heads is $\frac{1}{2}$. The way one coin falls does not depend on how the other coin falls. Similarly, the allele carried by the gamete from the first parent does not depend on the allele carried by the gamete from the second parent. The outcomes are independent of each other.

To find the probability that a combination of two independent events will occur, multiply the separate probabilities of the two events. Thus, the probability that a nickel and a penny will both land on heads is

$$\frac{1}{2} \times \frac{1}{2} = \frac{1}{4}$$

Study TIP

Reviewing Information

Notice that the probability problems in Figure 8-12 are set up with the use of a Punnett square. When practicing genetics probability problems, use Punnett squares to set up your problems.

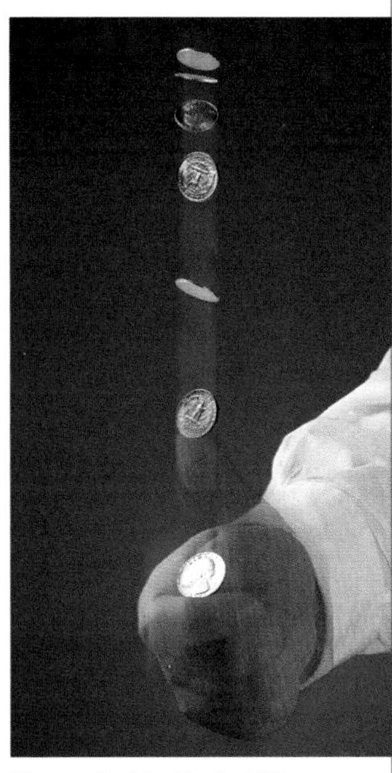

Figure 8-11 Probability of heads or tails. The probability that a tossed coin will land on heads is $\frac{1}{2}$. The probability that a tossed coin will land on tails is $\frac{1}{2}$.

Section 8-3

Demonstration

Card Games

Shuffle a deck of playing cards. Ask students to determine the probability of drawing an ace from the deck. *(Students may suggest $\frac{4}{52}$ or $\frac{1}{13}$.)* Ask how they arrived at this conclusion. Deal 13 cards from the top of the deck. Count the number of aces in those 13 cards and compare that number with the students' prediction. If the number varies from the prediction, have students speculate about the reasons for the difference.

Resource Link

Assign **Problem Solving: Genetics and Probability** from the Holt BioSources Teaching Resources CD-ROM. PORTFOLIO

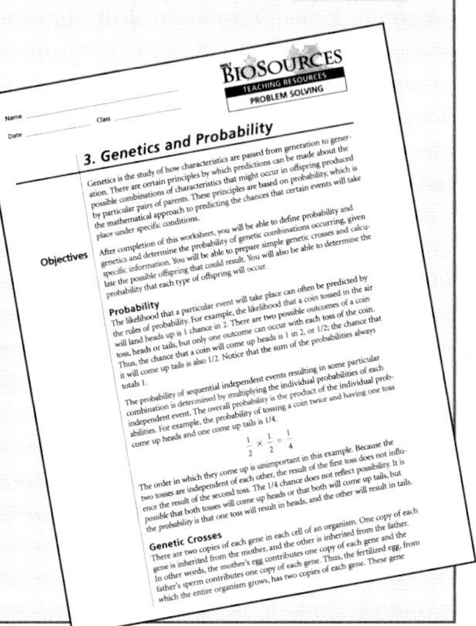

BIOSOURCES TEACHING RESOURCES PROBLEM SOLVING

Name ______ Class ______ Date ______

3. Genetics and Probability

Genetics is the study of how characteristics are passed from generation to generation. There are certain principles by which predictions can be made about the possible combinations of characteristics that might occur in offspring produced by particular pairs of parents. These principles are based on probability, which is the mathematical approach to predicting the chances that certain events will take place under specific conditions.

Objectives After completion of this worksheet, you will be able to define probability and genetics and determine the probability of genetic combinations occurring, given specific information. You will be able to prepare simple genetic crosses and calculate the possible offspring that could result. You will also be able to determine the probability that each type of offspring will occur.

Probability

The likelihood that a particular event will take place can often be predicted by the rules of probability. For example, the likelihood that a coin tossed in the air will land heads up is 1 chance in 2. There are two possible outcomes of a coin toss, heads or tails, but only one outcome can occur with each toss of the coin. Thus, the chance that a coin will come up heads is 1 in 2, or 1/2; the chance that it will come up tails is also 1/2. Notice that the sum of the probabilities always totals 1.

The probability of sequential independent events resulting in some particular combination is determined by multiplying the individual probabilities of each independent event. The overall probability is the product of the individual probabilities. For example, the probability of tossing a coin twice and having one toss come up heads and one come up tails is 1/4.

$$\frac{1}{2} \times \frac{1}{2} = \frac{1}{4}$$

The order in which they come up is unimportant in this example. Because the two tosses are independent of each other, the result of the first toss does not influence the result of the second toss. The 1/4 chance does not reflect possibility. It is *possible* that both tosses will come up heads or that both will come up tails, but the *probability* is that one toss will result in heads, and the other will result in tails.

Genetic Crosses

There are two copies of each gene in each cell of an organism. One copy of each gene is inherited from the mother, and the other is inherited from the father. In other words, the mother's egg contributes one copy of each gene and the father's sperm contributes one copy of each gene. Thus, the fertilized egg, from which the entire organism grows, has two copies of each gene. These gene

Overcoming Misconceptions

Understanding Probability

Students may think that probabilities in genetic crosses show the definite outcome of a genetic cross. Point out that probabilities are used only to predict the *possible* outcome of a genetic cross.

Visual Strategy

Figure 8-12

Make sure students understand that the probabilities in each square were obtained by multiplying the probability at the top of the box by the probability along the side of the box.

Teaching TIP

Using Probabilities in Genetic Crosses

Point out that the probability of a specific genotype occurring in a cross can be obtained by setting up a Punnett square similar to those in Figures 8-9 and 8-10 on pp. 168 and 169. The probability of finding a specific allele in a gamete is written next to the possible allele.

Predicting the results of crosses using probabilities

Time 10–15 minutes

Process Skills Calculating, applying information

Teaching Strategies
Have students set up Punnett squares similar to those in Figures 8-9 and 8-10 on pp. 168 and 169. Then ask them to write in the probabilities of finding a specific allele in a gamete.

Analysis Answers

1. $\frac{1}{4}$
2. $\frac{1}{2}$
3. 1
4. 0

Figure 8-12 Probability with two coins

Probability helps predict the likely results of flipping two coins at the same time.

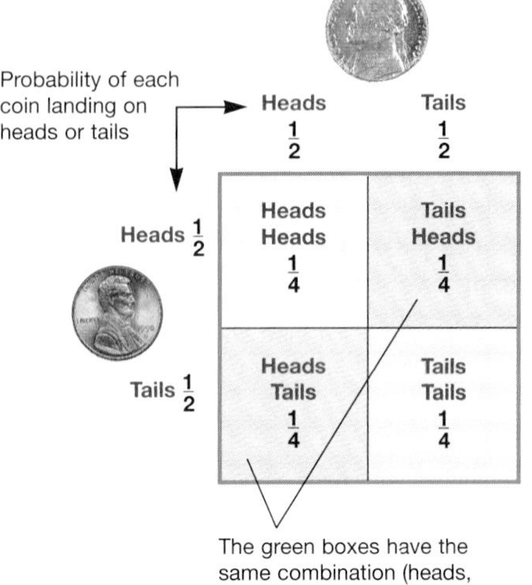

The possible results of tossing a nickel and a penny at the same time and the probability of each outcome are shown in **Figure 8-12.** Since the combination of heads and tails can occur in two possible ways, those two probabilities are added together.

$$\frac{1}{4} + \frac{1}{4} = \frac{2}{4} \text{ or } \frac{1}{2}$$

Consider the possible results that can occur in a cross between two pea plants that are heterozygous for seed shape (*Rr*). The *R* allele for round seed shape is dominant over the *r* allele for wrinkled seed shape. The probability of each parent carrying gametes with *R* or *r* alleles is $\frac{1}{2}$. The probability of offspring with *RR* alleles is

$$\frac{1}{2} \times \frac{1}{2} = \frac{1}{4}$$

Similarly, the probability of offspring with *rr* alleles is

$$\frac{1}{2} \times \frac{1}{2} = \frac{1}{4}$$

The combination of *Rr* alleles can occur in two possible ways. One parent can contribute the *R* allele, and the second parent the *r* allele, or vice versa. Thus, the probability of offspring with *Rr* alleles is

$$\frac{1}{4} + \frac{1}{4} = \frac{1}{2}$$

Predicting the results of crosses using probabilities

Background

In rabbits, the allele *B* for black hair is dominant over the allele *b* for brown hair. You can practice using probabilities to predict the outcome of genetic crosses by completing the genetic problems below. Draw Punnett squares for each problem.

Analysis

1. **Calculate** the probability of homozygous dominant (*BB*) offspring resulting from a cross between two heterozygous (*Bb*) parents.
2. **Calculate** the probability of heterozygous offspring resulting from a cross between a heterozygous parent and a homozygous recessive (*bb*) parent.
3. **Calculate** the probability of heterozygous offspring resulting from a cross between a homozygous dominant parent and a homozygous recessive parent.
4. **Calculate** the probability of homozygous dominant offspring resulting from a cross between a heterozygous parent and a homozygous recessive parent.

did you know?

The basenji is a breed of dog noted for its inability to bark.

The ability to bark is a dominant trait in dogs. All basenjis have two recessive genes for this trait, so they cannot bark. They can make a yodeling type of sound, however. Basenjis are small dogs with pointed ears that stand straight up. They have short, silky hair of various colors and rows of wrinkles on their foreheads.

Family Pedigrees Can Be Used to Study How Traits Are Inherited

Imagine that you want to learn about an inherited trait present in your family. How would you find out the chances of passing the trait to your children? Geneticists often prepare a **pedigree,** a family history that shows how a trait is inherited over several generations. Pedigrees are particularly helpful if the trait is a genetic disorder and the family members want to know if they are carriers or if their children might get the disorder. Carriers are individuals who are heterozygous for an inherited disorder but do not show symptoms of the disorder. Carriers can pass the allele for the disorder to their offspring.

Figure 8-13 shows an example of a pedigree for a family with albinism. In the genetic disorder albinism, the body is unable to produce an enzyme necessary for the production of melanin. Melanin is a pigment that gives dark color to hair, skin, scales, eyes, and feathers. Without melanin, an organism's surface coloration may be milky white and its eyes may be pink, as shown in Figure 8-13.

Scientists can determine several pieces of genetic information from a pedigree:

Autosomal or sex-linked? If a trait is autosomal, it will appear in both sexes equally. Recall that an autosome is a chromosome other than an X or Y sex chromosome. If a trait is sex-linked, it is usually seen only in males. A **sex-linked trait** is a trait whose allele is located on the X chromosome. Most sex-linked traits are recessive. Because males have only one X chromosome, a male who carries a recessive allele on the X chromosome will exhibit the sex-linked condition.

A female who carries a recessive allele on one X chromosome will not exhibit the condition if there is a dominant allele on her other X chromosome. She will express the recessive condition only if she

REAL LIFE

About 10 percent of Dalmatians are deaf. These are pedigreed dogs—purebred dogs whose ancestors are known and recorded. Since pedigreed dogs are inbred (mated from closely related individuals), many of them are homozygous and so are prone to inherited recessive diseases.

Finding Information
If you have a pedigreed dog, find out if that breed is prone to a genetic disease.

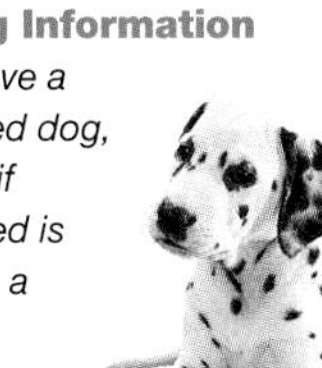

Figure 8-13 Albinism pedigree

Albinism is a genetic disorder transmitted by a recessive allele.

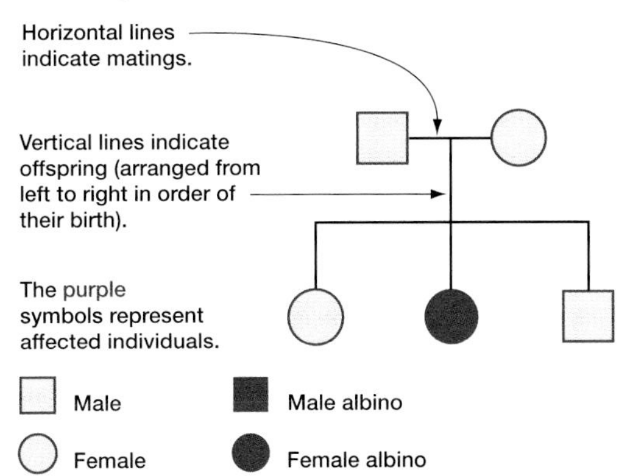

Horizontal lines indicate matings.
Vertical lines indicate offspring (arranged from left to right in order of their birth).
The purple symbols represent affected individuals.
Male
Female
Male albino
Female albino

In the wild, albino animals have little chance of survival. They lack the pigments that provide protection from the sun's ultraviolet rays.

MULTICULTURAL PERSPECTIVE

Albinism in Hopi Tribes

A survey of a Hopi tribe in Arizona found the frequency of albinism to be 1 in 277. In contrast, albinism is very rare to nonexistent in other Native American communities in Arizona and New Mexico. Why is the frequency so high among the Hopi? The Hopi people have always had a high regard for albinos, and clan leaders have taken special care to protect them from the harsh desert sun. This type of selection could explain the increase in albinism in the community.

REAL LIFE

Answer

Answers will vary. Pedigreed dogs that are prone to genetic diseases include Irish setters (blindness), German shepherds (hip dysplasia), and daschunds (dwarfism).

CROSS-DISCIPLINARY CONNECTION

◆ Mathematics

Compare the probability involved in tossing a coin with the segregation of alleles during meiosis. Ask students how many heads and tails are expected if a coin is flipped 100 times. *(50 heads, 50 tails)* Ask them what numbers they may actually obtain. *(Answers will vary.)* Emphasize that observed results often differ from predicted ones, although the differences should be minor. Relate this to the number of gametes containing a dominant allele and the number containing a recessive allele, if 100 gametes are produced by a heterozygous individual.

Visual Strategy

Figure 8-13

Before teaching students to interpret a pedigree, introduce the symbols: male (square), female (circle), trait expressed (shaded circle or square), and trait not expressed (circle or square not shaded). Once students are comfortable with the meanings of the symbols, have them interpret the pedigree in Figure 8-13 on page 173. Have students develop stories from which a pedigree can be drawn. Encourage them to be creative in thinking of characters and traits that they choose to follow through several generations. To illustrate their pedigree stories, students can add "family portraits." Read some of the stories in class, and have students draw the pedigrees from the information given in each story.

PORTFOLIO

Evaluating a pedigree

Time 10 minutes

Process Skills Analyzing, interpreting, drawing conclusions, applying information

Teaching Strategies
Encourage students to use "If-then" statements to organize their thoughts and interpret the pedigree. Example: *If* a trait is expressed by an offspring but not by either parent, *then* the trait must be recessive.

Analysis Answers

1. autosomal recessive
2. homozygous
3. $\frac{1}{2}$

3 Assessment Options

Closure

Geneticists' Tools

Ask students to differentiate between Punnett squares, probabilities and pedigrees. *(Punnett squares predict the expected outcome of a cross by considering all possible combinations of gametes in a cross. Probabilities predict the mathematical likelihood that a specific event, such as the outcome of a cross, will occur. Pedigrees provide a visual representation of how a trait is inherited over several generations.)*

Review

Assign Review—Section 8-3.

Reteaching — BASIC

Write the following genotypes on the board: (1) *PP*, (2) *Pp*, and (3) *pp*. Pair each student with a partner. Assign two of the genotypes to one partner. Assign a different pair of genotypes to the other partner. Have students construct a Punnett square showing a cross between their assigned genotypes. Ask them to express the offspring's phenotypes and genotypes as ratios. Have them share their results with their partners. PORTFOLIO

inherits two recessive alleles. Thus, her chances of inheriting and exhibiting a sex-linked condition are significantly less.

Dominant or recessive? If the trait is autosomal dominant, every individual with the trait will have a parent with the trait. If the trait is recessive, an individual with the trait can have one, two, or neither parent exhibit the trait.

Heterozygous or homozygous? If individuals with autosomal traits are homozygous dominant or heterozygous, their phenotype will show the dominant characteristic. If individuals are homozygous recessive, their phenotype will show the recessive characteristic. Two people who are heterozygous carriers of a recessive mutation will not show the mutation, but they can produce children who are homozygous for the recessive allele.

DATA LAB

Evaluating a pedigree

Background

The photo shows a family with an albino member. Pedigrees, such as the one below, can be used to track different genetic traits, including albinism. Use the pedigree below to practice interpreting a pedigree.

Analysis

1. **Interpret** the pedigree to determine whether the trait is sex-linked or autosomal and whether the trait is inherited in a dominant or recessive manner.
2. **Determine** whether Female A is homozygous or heterozygous for the trait.
3. **SKILL Applying Information** If Female B has children with a homozygous individual, what is the probability that the children will be heterozygous?

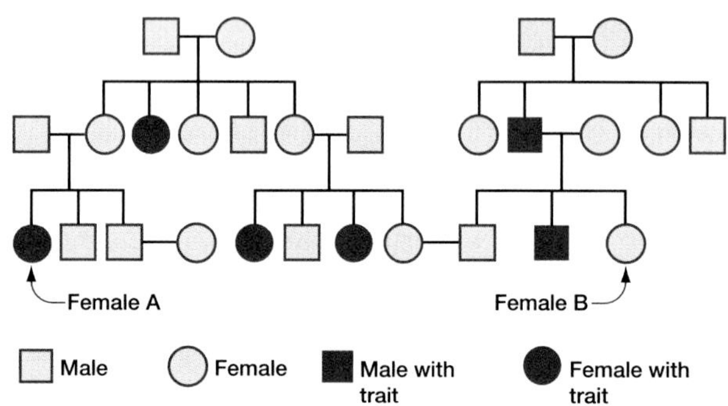

Review SECTION 8-3

1. **Predict** the expected phenotypic and genotypic ratios among the offspring of two individuals who are heterozygous for freckles (*Ff*) by using a Punnett square.
2. **Summarize** how a test cross can reveal the genotype of a pea plant with round seeds.
3. **Calculate** the probability that an individual heterozygous for a cleft chin (*Cc*) and an individual homozygous for a cleft chin (*cc*) will produce offspring that are homozygous recessive for a cleft chin (*cc*).
4. **SKILL Analyzing Graphics** When analyzing a pedigree, how can you determine if an individual is a carrier (heterozygous) for the trait being studied?

Review SECTION 8-3 ANSWERS

1. 3 freckles:1 no freckles; 1 *FF* : 2 *Ff* : 1 *ff*
2. If, after a testcross, all of the offspring have round seeds, the parent with the unknown genotype is likely to be homozygous dominant. If, after a testcross, any of the offspring have wrinkled seeds, the parent with the unknown genotype is likely to be heterozygous.
3. $\frac{1}{2}$
4. The individual does not show the trait and one of his or her parents is homozygous recessive and the other parent does not express the trait.

Patterns of Heredity Can Be Complex

Section 8-4

Most Traits Are Not Controlled by Simple Dominant-Recessive Alleles

A horse with red hair mates with a horse with white hair, and their offspring has both red and white hair. How can this be? If traits are controlled by single genes with simple dominant and recessive alleles, the colt's hair should be one color or the other. Not always! Most of the time, traits, such as hair color in horses, display more-complex patterns of heredity than the simple dominant-recessive patterns discussed so far.

Traits influenced by several genes

When several genes influence a trait, the trait is said to be a **polygenic trait.** The genes for a polygenic trait may be scattered along the same chromosome or located on different chromosomes. Determining the effect of any one of these genes is difficult. Due to independent assortment and crossing-over during meiosis, many different combinations appear in offspring. Familiar examples of polygenic traits in humans include eye color, height, weight, and hair and skin color. All of these characteristics have degrees of intermediate conditions between one extreme and the other, as shown in **Figure 8-14.**

Intermediate traits

Recall that in Mendel's pea-plant crosses, one allele was completely dominant over another. In some organisms, however, an individual displays a trait that is intermediate between the two parents, a condition known as **incomplete dominance.** For example, incomplete dominance in plants occurs when a snapdragon with red flowers is crossed with a snapdragon with white flowers to produce a snapdragon with pink flowers. Neither the red nor the white allele is completely dominant over the other allele. The flowers appear pink because they have less red pigment than the red flowers.

In Caucasians, the child of a straight-haired parent and a curly-haired parent will have wavy hair. Straight and curly hair are homozygous dominant traits. Wavy hair is heterozygous and is intermediate between straight and curly hair.

Objectives

- **Identify** five factors that influence patterns of heredity. (pp. 175–177)
- **Describe** how mutations can cause genetic disorders. (p. 178)
- **List** two genetic disorders, and describe their causes and symptoms. (pp. 178–179)
- **Evaluate** the benefits of genetic counseling. (pp. 179–180)

Key Terms

polygenic trait
incomplete dominance
codominance
multiple alleles

Figure 8-14 Polygenic traits. Many traits—height, weight, hair color, and skin color—are traits that are influenced by many genes.

Integrating Different Learning Styles

Logical/Mathematical	**PE**	Performance Zone item 25, p. 183
Visual/Spatial	**PE**	Interactive Exploration, pp. 184–185
	ATE	Demonstration, p. 176; Visual Strategy, p. 177
Interpersonal	**ATE**	Reteaching, p. 180
Verbal/Linguistic	**PE**	Performance Zone items 27 and 28, p. 183
	ATE	Career, p. 176; Teaching Tip, p. 177; Active Reading, p. 178; Teaching Tip, p. 179; Internet Connect, p. 179; Closure, p. 180

Section 8-4

1 Prepare

Scheduling

- **Block:** About 90 minutes will be needed to complete this section.
- **Traditional:** About two class periods will be needed to complete this section.
- **Planning:** See the Planning Guide on pp. 157A and 157B for lecture, classwork, and assignment options for Section 8-4. Lesson plans for Section 8-4 can be found in a lesson cycle format in ***Biology: Principles and Explorations* One-Stop Planner CD-ROM with Test Generator.**

Resource Link

Assign **Active Reading Worksheet Section 8-4** to help students comprehend the key concepts and visuals in this lesson.

2 Teach

Lesson Warm-up

Have students read the **BioBit** on page 178. Discuss the idea that genes are influenced by environmental factors, such as hormones, and that most traits are not controlled by simple dominant-recessive alleles.

Teaching TIP

Breeding Snapdragons

Ask students whether a plant breeder could produce only pink-flowering snapdragons by crossing pink-flowering snapdragons and white-flowering snapdragons. Lead students to understand that since all pink-flowering snapdragons are heterozygous, mating a pink-flowering snapdragon with a white-flowering one would produce pink-flowering and white-flowering offspring in a 1:1 ratio.

CAREER

Agricultural Science

Agricultural scientists apply their knowledge of genetics and inheritance to improve the marketability of farm animals or plant products. Have students work in teams to use library references or on-line databases to find information about the role of genetics in careers in agricultural science. Assign a specific career (such as animal breeder, botanist, dairy scientist, horticulturist, plant breeder, or poultry scientist) to each team, and have students present their findings to the class. For additional information, students can write to the American Society of Agronomy, 667 South Segoe Road, Madison, WI 53711. PORTFOLIO

Demonstration

Blood Types

To convey the concepts of universal donor and universal acceptor, set up four flasks to represent each blood type, and label them appropriately. The flasks should contain the following: "A" blood (water with red food color); "B" blood (water with blue food color); "AB" blood (water with both red and blue food color); and "O" blood (water only). Take an empty beaker and pour "O" blood into it. Show students that pouring "A" blood into the beaker containing "O" blood will contaminate the "O" blood (change its color). Demonstrate the possible mixtures, and have students derive which blood types are compatible with each other.

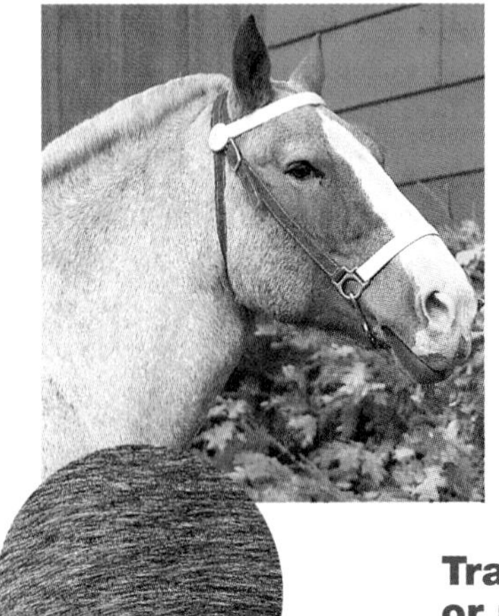

Figure 8-15
Codominance in horse-coat color. The roan coat of this horse consists of red hairs and white hairs because two dominant alleles are expressed at the same time.

Traits with two forms displayed at the same time

For some traits, two dominant alleles are expressed at the same time. When two dominant alleles are expressed at the same time, both forms of the trait are displayed, a phenomenon called **codominance.** Codominance is different from incomplete dominance because both traits are displayed. An example of codominance is the roan coat in horses, as shown in **Figure 8-15.** A cross between a homozygous red horse and a homozygous white horse results in heterozygous offspring with both red and white hairs in approximately equal numbers, producing the mixed color called roan.

Traits controlled by genes with three or more alleles

Genes with three or more alleles are said to have **multiple alleles.** For example, in the human population, the ABO blood groups (blood types) are determined by three alleles, I^A, I^B, and i. The letters A and B refer to two carbohydrates on the surface of red blood cells. The i allele means that neither carbohydrate is present. The I^A and I^B alleles are both dominant over i, which is recessive. But neither I^A nor I^B is dominant over the other. When I^A and I^B are both present in the genotype, they are codominant. When traits are controlled by genes with multiple alleles, an individual can have only two of the possible alleles for that gene. **Figure 8-16** shows how combinations of the three different alleles can produce four different blood types—A, B, AB, and O. Notice that a person who inherits two i alleles has type O blood.

Figure 8-16 **Multiple alleles control the ABO blood groups**

Different combinations of the three alleles I^A, I^B, and i result in four different blood phenotypes, A, AB, B, and O. For example, a person with the alleles I^A and i would have blood type A.

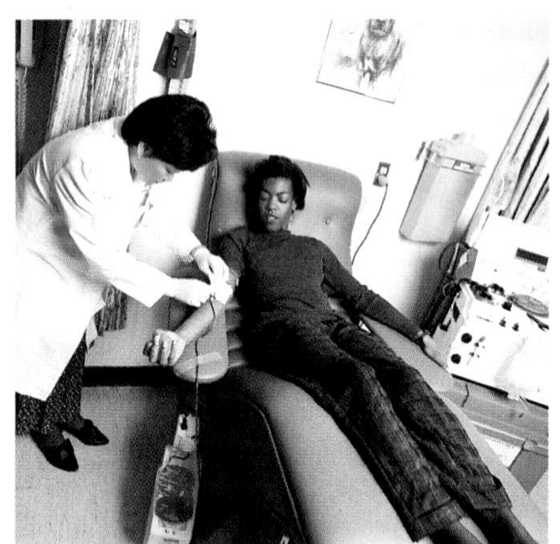

Possible alleles (columns) / Possible alleles (rows)

	I^A	I^B	i
I^A	I^AI^A	I^AI^B	I^Ai
I^B	I^AI^B	I^BI^B	I^Bi
i	I^Ai	I^Bi	ii

Blood types A AB B O

did you know?

Scientists are currently working on techniques to convert any blood type to O negative, the blood type known as the universal donor.

The blood groups A, B, AB, and O all begin with the same basic red blood cell, and they all have the same sugar chain on the cell surface. A cells have an additional sugar, B cells have a different additional sugar, and AB cells have both of the additional sugars. O cells lack the extra sugars. Researchers hope to produce O cells by using enzymes to remove these additional sugars from A, B, and AB cells.

Traits influenced by the environment

An individual's phenotype often depends on conditions in the environment. In plants, hydrangea *(heye DRAYN juh)* flowers of the same genetic variety range in color from blue to pink, depending on the acidity of the soil. As seen in **Figure 8-17,** hydrangea plants in acidic soil bloom blue flowers, while those in neutral to basic soil will bloom pink flowers.

The color of the arctic fox is affected by temperature. During the warm temperatures of summer, the fox produces enzymes that make pigments. These pigments darken the fox's coat to a reddish brown, as shown in **Figure 8-18,** enabling the fox to blend in with the summer landscape. During the winter, the pigment-producing genes of the arctic fox do not function because of the cold temperature. As a result, the coat of the fox is white, and the animal blends in with the snowy background.

Fur color in Siamese cats is also influenced by temperature. In a Siamese cat, the fur on its ears, nose, paws, and tail is darker than on the rest of its body. The Siamese cat has a genotype that results in dark fur at locations on its body that are cooler than the normal body temperature. Thus, the darkened parts have a lower body temperature than the light parts.

In humans many traits, such as height, are influenced by the environment. For example, height is influenced by nutrition, an internal environmental condition. Exposure to the sun, an external environmental condition, alters the color of the skin. Many aspects of human personality, such as aggressive behavior, are strongly influenced by the environment, although genes appear to play an even more important role. Because identical twins have identical genes, they are often used to study environmental influences. Since identical twins are genetically identical, any differences between them are attributed to environmental influences.

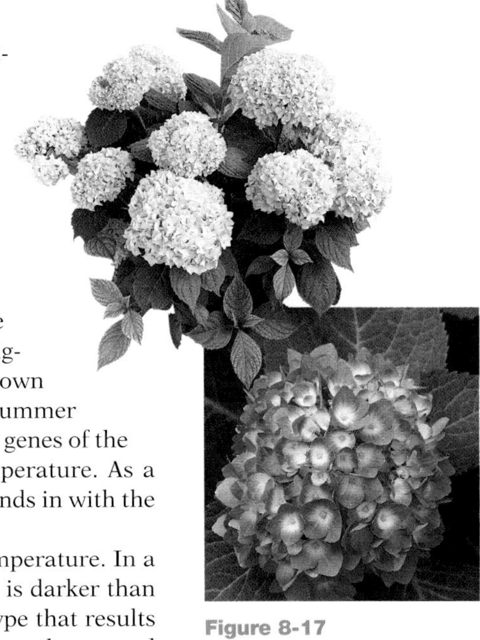

Figure 8-17
Environmental influences on flower color. Hydrangea with the same genotype for flower color express different phenotypes depending on the acidity of the soil.

Figure 8-18 Environmental influences on fur color

Can the same fox look so different? Many arctic mammals, such as the arctic fox, develop white fur during the winter and dark fur during the summer.

Section 8-4

Visual Strategy

Figure 8-18 ADVANCED

Ask students to consider how the environment and genes affect these animals so that they change colors with changing seasons. Help students understand that the temperature triggers enzymes involved in hormonal responses that influence the genes.

Teaching TIP

Traits with Complex Inheritance Patterns

Ask students to make a table to organize information about patterns of heredity that are more complex than simple dominant-recessive inheritance patterns. The students should write the following headings across the top: *Explanation, Example(s).* Along the sides the students should write the following: *Polygenic traits, Incomplete dominance, Codominance, Multiple alleles,* and *Environmentally-influenced traits.* Have students add information to the table as they read pp. 175–177. PORTFOLIO

did you know?

Mendel's work with garden-pea plants showed that the traits he studied are controlled by single genes.

In humans, single-factor inheritance has been found in about 600 recessively inherited traits and in such dominant conditions as Huntington's disease and Marfan's syndrome. However, many more conditions are determined by polygenic inheritance, which involves several genes. Such conditions include cleft lip and palate, schizophrenia, hypertension, and diabetes.

Teaching TIP

Hemoglobin

Ask students how a faulty gene can alter a hemoglobin molecule. *(Hemoglobin is a protein. A mutation in a hemoglobin gene that results in a change in the amino acid sequence of the gene can alter the structure of the protein, and ultimately the protein's function.)*

Teaching TIP

Royal Blood

Tell students that there is a high frequency of hemophilia among members of royal families throughout Europe. Explain that Queen Victoria was a carrier of sex-linked hemophilia. Because members of the European nobility usually married within their own social class, the hemophilia gene was passed via Queen Victoria's daughters to the Russian, German, and Spanish royal families, increasing the frequency of the recessive allele among European nobility.

Active Reading

Technique: Reading Organizer

Have students make a reading organizer describing the cause and effect of each of the genetic disorders discussed in this section. Students should construct a cause/effect graphic for each disease. PORTFOLIO

Cause → Effect

biobit

Why are most bald people men?

Both males and females can inherit a "baldness" allele, but the male hormone testosterone activates the allele and eventually leads to baldness. Women produce small amounts of testosterone. However, baldness does not occur in females unless they have both alleles for baldness. The presence of only one allele for baldness causes men to become bald.

Some Traits Are Caused by Mutations

In order for a person to develop and function normally, the proteins encoded by his or her genes must function precisely. Unfortunately, sometimes genes are damaged or are copied incorrectly, resulting in faulty proteins. Changes in genetic material are called mutations. Mutations are rare because cells have efficient systems for correcting errors. But mutations sometimes occur, and they may have harmful effects.

The harmful effects produced by inherited mutations are called genetic disorders. Many mutations are carried by recessive alleles in heterozygous individuals. This means that two phenotypically normal people who are heterozygous carriers of a recessive mutation can produce children who are homozygous for the recessive allele. In such cases, the effects of the mutated allele cannot be avoided. Several human genetic disorders are summarized in **Table 8-2.**

Sickle cell anemia

An example of a recessive genetic disorder is sickle cell anemia, a condition caused by a mutated allele that produces a defective form of the protein hemoglobin. Hemoglobin is found within red blood cells, where it binds oxygen and transports it through the body. In sickle cell anemia, the defective form of hemoglobin causes many red blood cells to bend into a sickle shape, as seen in **Figure 8-19.** The sickle-shaped cells rupture easily, resulting in less oxygen being carried by the blood. Sickle-shaped cells also tend to get stuck in blood vessels; this can cut off blood supply to an organ.

Magnification: 13,600×

Figure 8-19 Sickle cell

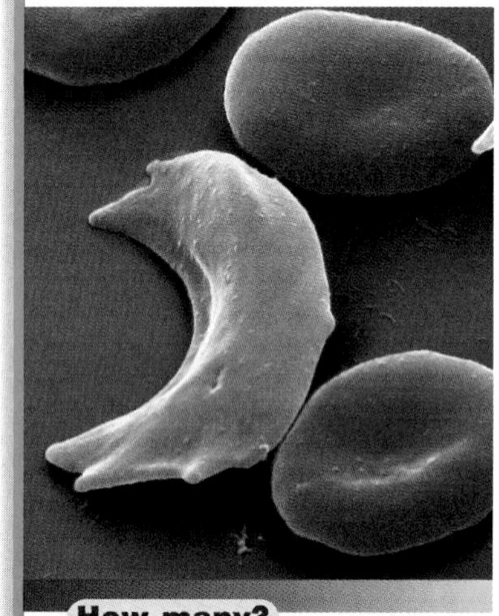

How many? One out of 500 African Americans has sickle cell anemia, which is caused by a gene mutation that produces a defective form of hemoglobin.

The recessive allele that causes sickle-shaped red blood cells also helps protect the cells of heterozygous individuals from the effects of malaria. Malaria is a disease caused by a parasitic protozoan that invades red blood cells. The sickled red blood cells of heterozygous individuals cause the death of the parasite. But the individual's normal red blood cells can still transport enough oxygen. Therefore, these people are protected from the effects of malaria that threaten individuals who are homozygous dominant for the hemoglobin gene.

Hemophilia

Another recessive genetic disorder is hemophilia *(hee moh FIHL ee uh),* a condition that impairs the blood's ability to clot. Hemophilia is a sex-linked trait. More than a dozen genes code for the proteins involved in blood clotting. A mutation on one of these genes on the X chromosome causes the form of hemophilia called hemophilia A. If the mutation appears on the X chromosome, which a male receives from his mother, he does not have a normal gene on the Y chromosome to compensate. Therefore, he will develop hemophilia.

MULTICULTURAL PERSPECTIVE

Understanding Sickle Cell Anemia

Working with limited laboratory facilities and a strong determination to fight the disease that was killing their friends and families, two African-American researchers, Dr. Angela Ferguson and Dr. Roland Scott, published a paper on sickle cell anemia in the 1940s—25 years ahead of other researchers. Dr. Scott, known as the "father of sickle cell anemia research," is the founder and former director of Howard University's Center for Sickle Cell Anemia Research. Dr. Ferguson was an associate professor of pediatrics at Howard University.

Huntington's disease (HD)

Huntington's disease is a genetic disorder caused by a dominant allele located on an autosome. The first symptoms of HD—mild forgetfulness and irritability—appear in victims in their thirties or forties. In time, HD causes loss of muscle control, uncontrollable physical spasms, severe mental illness, and eventually death. Unfortunately, most people who have the HD allele do not know they have the disease until after they have had children. Thus, the disease is unknowingly passed on from one generation to the next.

internetconnect
SCILINKS NSTA
TOPIC: Genetic disorders
GO TO: www.scilinks.org
KEYWORD: HX179

Detecting and treating genetic disorders

Most genetic disorders cannot be cured, although progress is being made. A person with a family history of genetic disorders may wish to undergo genetic counseling before becoming a parent. Genetic counseling is a form of medical guidance that informs people about genetic problems that could affect them or their offspring.

In some cases, therapy is available to treat a genetic disorder if it is diagnosed early enough. For example, an individual with the genetic disorder phenylketonuria (PKU) lacks an enzyme that converts the amino acid phenylalanine into the amino acid tyrosine. As a result,

Table 8-2 Some Human Genetic Disorders

Disorder	Dominant or recessive	Symptom	Defect	Frequency among human births
Sickle cell anemia	Recessive	Poor blood circulation	Abnormal hemoglobin molecules	1:500 (African Americans)
Tay-Sachs disease	Recessive	Deterioration of central nervous system in infancy; affected individuals die in early childhood	Defective form of a brain enzyme	1:3,500 (Ashkenazi Jews)
Cystic fibrosis	Recessive	Mucus clogs many organs, including the lungs, liver, and pancreas; affected individuals usually do not survive to adulthood	Defective chloride-ion transport protein	1:2,500 (Whites)
Hemophilia A (classical)	Sex-linked recessive	Failure of blood to clot	Defective form of a blood-clotting factor	1:10,000 (White males)
Huntington's disease	Dominant	Gradual deterioration of brain tissue in middle age; shortened life expectancy	Inhibitor of brain-cell metabolism is made	1:10,000

did you know?

In the United States, many cases of Huntington's disease can be traced back to two brothers.

The two men immigrated to North America from England in the 1600s because of accusations of witchcraft in their family. The family members of these brothers were apparently persecuted because of their strange behaviors, which are now understood to be symptoms of Huntington's disease. The disease is characterized by constant dancelike movements in its victims.

internetconnect
SCILINKS NSTA
Have students research genetic disorders using the Web site in the Internet Connect box on page 179. Have students write a report summarizing their findings. PORTFOLIO

Making Biology Relevant
Medicine

Women who have PKU often have babies with mental retardation, not because the baby has PKU, but because the mother's body chemistry is altered during pregnancy. These babies cannot be helped with a special diet. However, the mental retardation can be avoided if the mother follows a low-phenylalanine diet before and during pregnancy.

Teaching TIP — ADVANCED
Medical Brochures

Assign each student a genetic disorder and have students design an informative brochure about the disorder, similar to brochures found in a doctor's office. Set guidelines on information you expect students to include, such as symptoms, causes, prognosis, and support groups. Encourage students to use colors and illustrations that would interest teenage readers. PORTFOLIO

HEALTH WATCH

Genetic Counseling

Teaching Strategies
Use the genetic disorders summarized in Table 8-2 (page 179) to stimulate discussion about genetic counseling.

Discussion
- What personality traits would be essential in a genetic counselor? *(Answers will vary.)*
- Would you want to know if you had a genetic disorder? *(Answers will vary.)*

3 Assessment Options

Closure

Technique: K-W-L

Tell students to return to their lists of things they Want to know about inheritance, which they created using the **Active Reading** exercise on page 159. Have them place check marks next to the questions that they are now able to answer. Students should finish by making a list of what they have Learned. Conclude by asking students which questions are still unanswered. Ask if they have any new questions. PORTFOLIO

Review

Assign Review—Section 8-4.

Reteaching — BASIC

Place students into groups and assign each group one of the following topics: incomplete dominance, codominance, and multiple alleles. Ask students to think of a concrete example for teaching their assigned topics to the class. Give each group 5 to 10 minutes to explain its strategy to the class.

phenylalanine builds up in the body and causes severe mental retardation. If PKU is diagnosed soon after birth, the newborn is placed on a low-phenylalanine diet, ensuring that the baby will get enough phenylalanine to make proteins, but not enough to cause damage. Because this disorder can be easily diagnosed by inexpensive laboratory tests, many states require PKU testing of all newborns.

Gene technology may soon make it possible for scientists to correct certain genetic disorders by replacing defective genes with copies of healthy ones, a technique called gene therapy. Gene therapy is being used by scientists seeking cures for many genetic disorders, including cystic fibrosis and muscular dystrophy.

HEALTH WATCH Genetic Counseling

Nina and Alex have one child with cystic fibrosis. They are thinking about having a second child and are worried that their second child may have the same genetic disorder. How can they find out whether their second baby is at risk for cystic fibrosis?

Detecting the risk of a genetic disease

People who have a family history of a genetic disease might want to see a genetic counselor to find out if they or their children are at risk of having the genetic defect. Women over the age of 35 have an increased risk of having a baby with Down syndrome, so physicians may advise genetic counseling for them as well.

How genetic counselors can help

Although individual cases vary, genetic counselors follow a standard procedure to identify inherited medical conditions. First they assemble a detailed personal history and construct a family pedigree. If necessary, they advise the client to have a karyotype or other genetic analyses done. The genetic counselor explains the results and the probability that the disorder will affect offspring, then outlines the options for dealing with those risks.

Genetic counselors can also offer emotional support to people who are coping with disappointing news about a genetic disorder. Genetic counselors can give people at risk for a genetic disease, such as breast cancer, information about early detection, possible preventive measures, and treatment options.

Review SECTION 8-4

1. **Differentiate** between incomplete dominance and codominance.
2. **Identify** two examples of traits that are influenced by environmental conditions.
3. **Summarize** how a genetic disorder can result from a mutation.
4. **Describe** how males inherit hemophilia.
5. **SKILL Applying Information** Why might a couple undergo genetic counseling?

Critical Thinking

6. **SKILL Justifying Conclusions** A nurse states that a person cannot have the blood type ABO. Do you agree or disagree? Explain.

Review SECTION 8-4 ANSWERS

1. Incomplete dominance produces traits that are intermediate between two contrasting forms of a trait. In codominance, both dominant forms of a trait are displayed at the same time.
2. Answers will vary but may include fur color in Siamese cats, flower color in hydrangea plants, fur color in arctic mammals, and height and skin color in humans.
3. A genetic disorder results when a mutation is inherited and the mutation produces harmful effects.
4. The male receives from his mother an X chromosome with a mutated blood-clotting gene.
5. People at risk of having children with a genetic disorder go to genetic counselors to determine the probability of having a child with a genetic disorder, and to learn about treatments or support groups for the genetic disorder.
6. Students should agree. It would require that an individual have three alleles—I^A, I^B, and i.

Study ZONE CHAPTER 8

Use the Key Concepts and Key Terms listed below to review the main ideas in this chapter. If you need to review the meaning of a term, turn to the page indicated.

Key Concepts

8-1 The Origins of Genetics

- Gregor Mendel bred varieties of the garden pea in an attempt to understand heredity. Mendel observed that contrasting traits appear in offspring according to simple ratios.
- In Mendel's experiments, only one of the two contrasting forms of a trait was expressed in the F_1 generation. The other form reappeared in the F_2 generation in a 3:1 ratio.

8-2 Mendel's Theory

- Different versions of a gene are called alleles. An individual usually has two alleles for a gene, each inherited from a different parent.
- Individuals with the same two alleles for a gene are homozygous; those with two different alleles for a gene are heterozygous.
- The law of segregation states that the two alleles for a trait separate when gametes are formed. The law of independent assortment states that two or more pairs of alleles separate independently of one another during gamete formation.

8-3 Studying Heredity

- The results of genetic crosses can be predicted with the use of Punnett squares and probabilities.
- A test cross can be used to determine whether an individual expressing a dominant trait is heterozygous or homozygous.
- A trait's pattern of inheritance within a family can be determined by analyzing a pedigree.

8-4 Patterns of Heredity Can Be Complex

- Traits usually display complex patterns of heredity, such as incomplete dominance, codominance, and multiple alleles.
- Mutations can cause genetic disorders, such as sickle cell anemia, hemophilia, and Huntington's disease.
- Genetic counseling can help patients concerned about a genetic disorder.

Key Terms

8-1

heredity (160)
genetics (160)
monohybrid cross (162)
true-breeding (162)
P generation (162)
F_1 generation (162)
F_2 generation (162)

8-2

allele (164)
dominant (164)
recessive (164)
homozygous (165)
heterozygous (165)
genotype (166)
phenotype (166)
law of segregation (167)
law of independent assortment (167)

8-3

Punnett square (168)
test cross (170)
probability (171)
pedigree (173)
sex-linked trait (173)

8-4

polygenic trait (175)
incomplete dominance (175)
codominance (176)
multiple alleles (176)

Unit 5—*Heredity*
Use this unit to review the key concepts and terms in this chapter.

Study TIP Ready?

If you think you understand these Key Concepts and Key Terms, then you're ready for the Performance Zone.

Study ZONE CHAPTER 8

Review and Practice

Assign the following worksheets for Chapter 8:

- **Student Review Guide**
- **Concept Mapping Worksheet**
- **Test Preparation Pretest**
- **Vocabulary Worksheet**
- **Critical Thinking Review Worksheet**
- **Science Skills Worksheet**

Assessment Options

The following assessment options are available for this chapter:

Resource Link

- Assign **Chapter Test 8** to assess students' understanding of the chapter.

Alternative Assessment

Assign students to cooperative work groups. Have each group create a genetics problem based on the information presented in this chapter. Students can use **Review** questions 1 and 3 on page 174 as models. Collect the problems, write them on the board, and have each group work on the solutions. PORTFOLIO

Portfolio Assessment

- *Portfolio Assessment* at the front of this book provides options for building and evaluating student portfolios. Students and teachers can select items from the areas listed for inclusion in a portfolio.
- See items labeled PORTFOLIO on pp. 161, 162, 163, 165, 167, 170, 171, 173, 174, 176, 177, 178, 179, 180, and 181.

Answer to Concept Map

The following is one possible answer to Performance Zone item 1 on page 182.

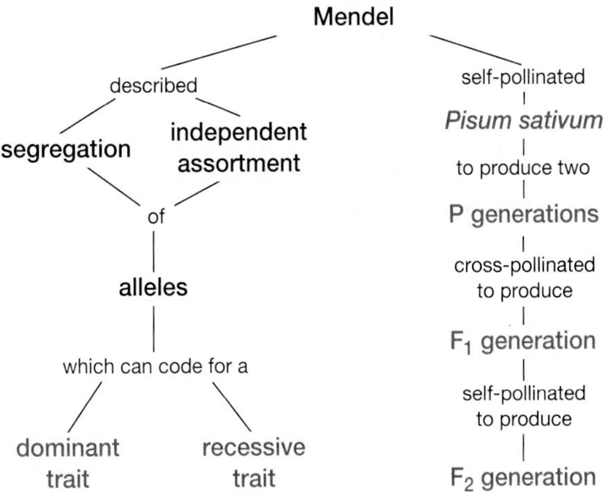

Performance ZONE
CHAPTER 8
REVIEW ANSWERS

Assignment Guide

Section	Review
8-1	1, 3, 4, 5, 22, 23, 24
8-2	2, 6, 7, 8, 16
8-3	9, 10, 11, 13, 14, 17, 18, 19, 20, 21, 26, 29
8-4	12, 15, 25, 27, 28

Understanding and Applying Concepts

1. The answer to the concept map is found at the bottom of page 181.
2. **a.** A dominant trait appears in a heterozygous individual; a recessive trait is hidden in a heterozygous individual.
   b. Homozygous refers to an individual with two identical alleles for a trait. Heterozygous refers to an individual with two different alleles for a trait.
   c. The law of segregation states that the two alleles for a trait separate when gametes are formed. The law of independent assortment states that the alleles of different genes separate independently of one another during gamete formation.
3. b
4. b
5. a
6. b
7. c
8. a
9. d
10. c
11. c
12. a
13. Since domesticated animals are more likely to be inbred, many are homozygous and thus prone to inherited recessive traits, such as albinism.
14. 1 *YYRR* : 2 *YyRR* : 1 *yyRR*; 3 yellow, round : 1 green, round

Performance ZONE CHAPTER 8 Review

Understanding and Applying Concepts

1. **Concept Mapping** Make a concept map about Mendel's experiments. Try to include the following words in your map: *Pisum sativum,* P generation, F_1 generation, F_2 generation, dominant trait, recessive trait.

2. **Understanding Vocabulary** For each pair of terms, explain the differences in their meanings.
   a. dominant, recessive
   b. homozygous, heterozygous
   c. law of segregation, law of independent assortment

3. The scientist whose studies formed the basis of modern genetics is
   a. T. A. Knight. c. Louis Pasteur.
   b. Gregor Mendel. d. Robert Hooke.

4. Which of the following is not a good reason why *Pisum sativum* makes an excellent subject for genetic study?
   a. Many varieties exist.
   b. They require cross-pollination.
   c. They grow quickly.
   d. They demonstrate complete dominance.

5. The offspring of true-breeding parents are called the
   a. F_1 generation. c. dominant offspring.
   b. F_2 generation. d. recessive offspring.

6. If smooth peas are dominant over wrinkled peas, the allele for smooth peas should be represented as
   a. *W.* c. *w.*
   b. *S.* d. *s.*

7. The color of a dog's coat is the dog's
   a. dominance. c. phenotype.
   b. pedigree. d. genotype.

8. The law of segregation states that pairs of alleles
   a. separate when gametes form.
   b. separate independently of one another during gamete formation.
   c. are always the same.
   d. are always different.

9. The unknown genotype of an individual with a dominant phenotype can be determined using
   a. a ratio. c. probability.
   b. a dihybrid cross. d. a test cross.

10. The trait shown below is

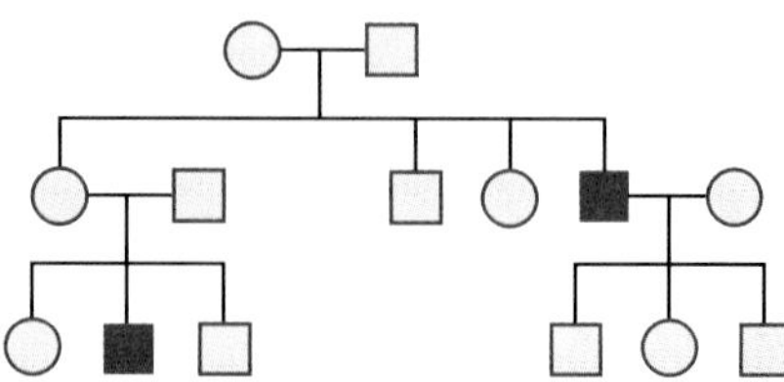

   a. sex-linked and dominant.
   b. autosomal and dominant.
   c. sex-linked and recessive.
   d. autosomal and recessive.

11. *D,* dimples, is the dominant allele to the recessive allele, *d,* no dimples. The probability of parents with *Dd* and *dd* genotypes having a child with no dimples (*dd*) is
   a. $\frac{1}{8}$. c. $\frac{1}{2}$.
   b. $\frac{1}{4}$. d. 1.

12. A trait with two dominant alleles that are expressed at the same time is
   a. codominant. c. incompletely dominant.
   b. mutational. d. polygenic.

13. **Theme Evolution** Albinism is rare among wild animals but quite common among some domesticated species. What factors might account for this difference?

14. **Exploring Further** State the genotypic and phenotypic ratios that would result from a cross between two *YyRR* pea plants.

15. **Health Watch** Explain how a genetic counselor can help a person that has a family history of a genetic disorder.

16. **Chapter Links** Relate the events of meiosis to the law of segregation. (**Hint:** See Chapter 7, Section 7-1.)

---

15. A genetic counselor explains the effects of a genetic disorder to people who have a family history of the disorder. A genetic counselor helps a couple with a family history of a genetic disorder determine if their children are at risk of having the disorder. A genetic counselor also offers emotional support and gives information about early detection, possible prevention, and treatment options for genetic disorders.
16. During meiosis II, the members of each pair of alleles separate when gametes are formed during meiosis as described in the law of segregation.

Skills Assessment

Part 1: Interpreting Graphics

Study the Punnett square showing the expected results of a cross between two pea plants (T = tall, t = dwarf). Then answer the following questions:

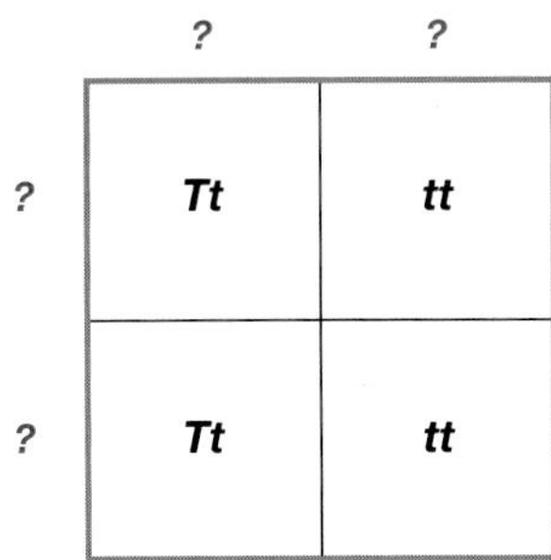

17. What alleles should be placed along the top and sides of the Punnett square?

18. List the phenotypes resulting from this cross.

19. What is the phenotypic ratio for the offspring in this cross?

20. What is the genotypic ratio for the offspring in this cross?

21. If there were 16 offspring, how many would you expect to look just like the parent plants? to be dwarf plants?

Part 2: Life/Work Skills

Use the media center or Internet resources to learn how Mendelian genetics applies to plant breeding today.

22. **Selecting Technology** Find out how new technologies have changed plant-breeding methods since Mendel's time. Discover what equipment plant breeders use to conduct their breeding program.

23. **Communicating** Prepare an oral report to summarize your findings. Create a display that compares the methods and equipment Mendel might have used with those used by plant breeders today.

Critical Thinking

24. **Evaluating Results** Mendel based his conclusions about inheritance patterns on experiments involving large numbers of plants. Why do you think the use of large numbers of individuals is advantageous when studying patterns of inheritance?

25. **Justifying Conclusions** A 20-year-old man who has cystic fibrosis has a sister who is planning to have a child. The man encourages his sister to see a genetic counselor. What do you think the man's reasons are for giving such advice?

Portfolio Projects

26. **Evaluating Viewpoints** The Hopi tribe of Native Americans has a high number of albinos. Research Dr. Charles M. Woolf's and Dr. Frank C. Dukepoo's idea of cultural selection as an explanation for this phenomenon. Write a report describing their theory, and explain why you agree or disagree with the theory.

27. **Interpreting Conclusions** Read "The Bitter Truth: Do Some People Inherit a Distaste for Broccoli?" by Kathleen Fackelmann (*Science News,* July 12, 1997, vol. 152, pp. 24–25). What was discovered about the volunteer tasters? What significance might this have for public health? Summarize your findings in a written report.

28. **Career Focus Genetic Counselor** Research the field of genetic counseling, and write a report on your findings. Your report should include a job description, training required, kinds of employers, growth prospects, and starting salary.

29. **Unit 5—*Heredity*** Write a report summarizing how an understanding of heredity allows animal breeders to develop animals with desirable traits. Find out what kinds of animals are bred for special purposes.

INTERACTIVE BIOLOGY TUTOR

Performance ZONE

CHAPTER 8

Skills Assessment

17. along top—*T, t;* along sides—*T, t*
18. *Tt* = tall; *tt* = dwarf
19. 1 tall : 1 dwarf
20. 1 *Tt* : 1 *tt*
21. 8, 8
22. Answers will vary. Gene technology is now used in plant-breeding. Many plant breeders now use gene technology equipment to conduct their breeding program.
23. Answers will vary.

Critical Thinking

24. Patterns obtained from large samples are less likely to be distorted by rare events that can occur by chance.
25. Cystic fibrosis is a recessive autosomal disorder. Thus, each parent must have the recessive allele. Chances are increased that his sister is a carrier (heterozygote) for cystic fibrosis.

Portfolio Projects

26. According to Dr. Dukepoo, a Hopi who grew up among albino Native Americans in the Southwest, the reason for the high number of albinos is cultural selection. Because Hopi people have a high regard for albinos, clan leaders have taken special care to protect them. Hopi albinos are not required to work out in the fields. Instead, they stay at home to weave, make pottery and other objects, and help prepare meals.
27. About 25 percent of the American population are supertasters—people with a genetically determined dislike of bitter compounds found in many fruits and vegetables. About 50 percent are regular tasters and 25 percent are nontasters. Supertasters shun many foods that scientists think prevent cancer and ward off heart attacks.
28. Genetic counselors use various types of information, including pedigrees, laboratory tests, and karyotypes, to determine the odds of a person or a couple's child having a genetic disorder. Genetic counselors also outline the options for dealing with those risks and offer emotional support. Genetic counseling requires a specialized graduate degree and experience in the areas of medical genetics and counseling. Employers include university medical centers, private hospital settings, health maintenance organizations, and laboratories. Growth prospects are good. Starting salary will vary by region.
29. Answers will vary. Animal breeders use genetics to predict how often a trait will appear when two animals are bred. Animals bred for special purposes include dogs, cats, horses, goats, rabbits, and cattle.

Interactive Exploration

Chapter 8

Cystic Fibrosis

Purpose

Students will explore how an individual's genotype affects the membrane proteins involved in transporting Cl^- ions through the cell membrane.

Preparation Notes

Time Required: 40–50 minutes

Preparation Tips

- Depending on how many computers are available, set up work stations that enable students to work individually or in teams of two to four.
- Encourage students to read the lab and complete the **Before You Begin** section on page 184 before coming to class on the day of the lab.
- Before students begin the lab, conduct a class discussion in which you ask the following questions:
  - What is cystic fibrosis? *(Cystic fibrosis is a fatal inherited disease in which large amounts of thick, sticky mucus are produced.)*
  - What causes cystic fibrosis? *(It is caused by an autosomal recessive allele.)*
  - What is the function of protein channels that pass through the membrane of a cell? *(They allow substances that cannot pass through the cell membrane to enter or leave the cell.)*

Procedural Tips

- Remind students that the upper part of the screen (above the cell membrane) represents the outside of the cell and the lower part of the screen represents the inside of the cell.
- After students finish the activity, encourage them to read and take notes on the article titled "Closing in on Cystic Fibrosis: Researchers Are Learning to Replace a Faulty Gene." The article can be found by clicking the Navigation button and then the Readings button, which is indicated by the "book" icon.

Interactive Exploration

Chapter 8 *Cystic Fibrosis*

CD-ROM Lab

SKILL

- Computer modeling

OBJECTIVES

- **Relate** the symptoms of cystic fibrosis to the ability of water molecules and chloride ions to cross cell membranes.
- **Simulate** the effect of genotype on the permeability of a cell membrane to chloride ions.

MATERIALS

- computer with CD-ROM drive
- CD-ROM *Interactive Explorations in Biology: Human Biology*

Magnification: 2,000×

Mucus of partially blocked airway in lung of cystic fibrosis patient

Before You Begin

Cystic fibrosis is a fatal genetic disease that results from the failure of chloride ions (Cl^-) to pass through the cell membrane. As a result, chloride ions accumulate inside the cells of the body. When this happens in the lungs, pancreas, and liver, mucus builds up on the lining of these organs. In this Interactive Exploration, you will explore how an individual's genotype affects the membrane proteins involved in transporting Cl^- ions through the cell membrane.

1. Select *Cystic Fibrosis* from the menu. Click on the key near the top of the navigation palette. Read the focus questions, and review these concepts: transport channels, osmosis, mutation, and cystic fibrosis.
2. Click *File* at the top of the screen, and select *Interactive Exploration Help*. Listen to the instructions. Click *Exploration* at the bottom right of the screen to begin the exploration. You will see a diagram like the one below.

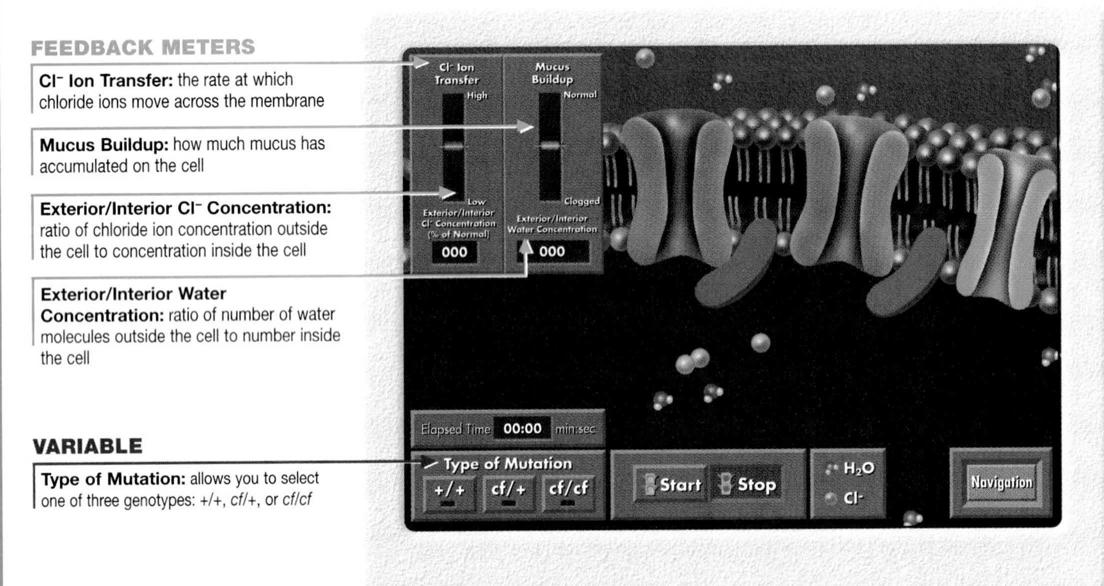

Additional Background

Cystic fibrosis is the most common genetic disorder among Caucasians in the United States. About 1 in 2,500 Caucasian children born in the United States has cystic fibrosis, and about 1 in 20 adults is a carrier of the disease.

Unusually thick and sticky mucus builds up in the liver, pancreas, lungs, and bronchial passages of people with cystic fibrosis. This mucus may block delivery of pancreatic enzymes to the small intestine or disrupt liver function, but most deaths are the result of infections or damage to the lungs or bronchial passages. Recent advances in treatment have improved the prospects for cystic fibrosis patients, but the mortality rate is still high—about 90 percent of patients will die before reaching age 30.

The protein encoded by the cystic fibrosis allele appears to function in the transport of chloride ions across cell membranes. Defects in the protein prevent chloride ions from leaving cells in the pancreas, liver, and lungs. Cells with abnormally high chloride ion concentrations will retain water, and this may account for the thick mucus that causes so many problems for cystic fibrosis patients.

3. Make a data table like the one below.

DATA TABLE

Genotype	Exterior/interior water concentration	Exterior/interior Cl^- concentration	Mucus buildup
+/+			
cf/+			
cf/cf			

Procedure

PART A: Two Normal Alleles (+/+)

1. Move the pointer to the *Type of Mutation* box, and click the +/+ button. What is the phenotype of a person with this genotype? Note that water molecules and Cl^- ions are located outside the cell (the part of the screen above the cell membrane) and inside the cell (the part of the screen below the cell membrane). Also note the position of the two CF proteins in the cell membrane.
2. Click the *Start* button, and allow the simulation to run 3–4 minutes. Describe how water molecules cross the cell membrane. How does this differ from the way Cl^- ions cross the membrane?
3. Click the *Stop* button. In your table, record the exterior/interior Cl^- concentration, the exterior/interior water concentration, and the amount of mucus buildup outside the cell membrane.

PART B: One Mutant Allele and One Normal Allele (cf/+)

4. Move the pointer to the *Type of Mutation* box, and click the cf/+ button. What is the phenotype of a person with this genotype? Describe what happened to the CF protein shown on the left. How did this affect the Cl^- ion channel?
5. Click the *Start* button, and allow the simulation to run 3–4 minutes. Describe any changes in the way water molecules and Cl^- ions cross the cell membrane. How will these changes affect the external surface of the cell?
6. Click the *Stop* button. Record the same three variables as you did in step 3. On which side of the membrane is the Cl^- ion concentration higher? On which side is the water concentration higher?

PART C: Two Mutant Alleles (cf/cf)

7. Move the pointer to the *Type of Mutation* box, and click the cf/cf button. What is the phenotype of a person with this genotype? Describe what has happened to both CF membrane proteins. How did this affect the Cl^- ion channels?
8. Click the *Start* button, and allow the simulation to run 3–4 minutes. Describe any changes in the movement of water molecules and Cl^- ions through the membrane. How will these changes affect the external surface of the cell?
9. Click the *Stop* button. Describe the difference between the external and internal environments of the cell. Record the same three variables as you did in steps 3 and 6. On which side of the membrane is the Cl^- ion concentration higher? On which side is the water concentration higher?

Analyze and Conclude

1. **Relating Concepts** Why are the Cl^- ion channels, but not the water channels, called gated channels?
2. **Analyzing Data** Compare the results you obtained in Parts A, B, and C, and explain any differences.
3. **Inferring Conclusions** Why do people with one normal allele and one mutant allele for cystic fibrosis have a normal life expectancy?

Use the following Internet resources to find out more about cystic fibrosis.

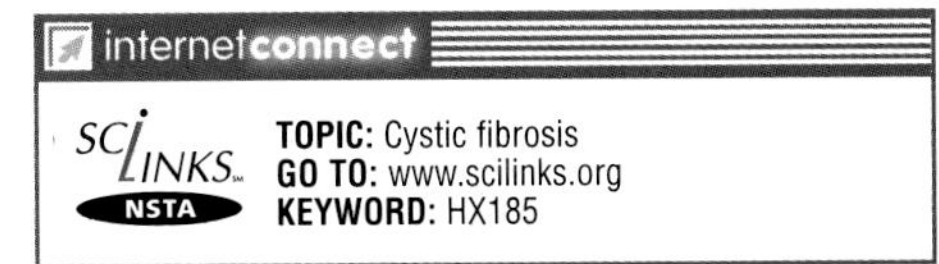

Procedure

Answers

1. The phenotype is normal.
2. Water molecules move into and out of the cell through the membrane. Chloride ions leave the cell through a protein channel.
3. **Sample Data Table**

GENOTYPE	EXTERIOR/INTERIOR WATER CONCENTRATION	EXTERIOR/INTERIOR CL^- CONCENTRATION	MUCUS BUILDUP
+/+	100%	normal	normal
cf/+	50%	normal	normal
cf/cf	20%	low	clogged

4. This phenotype is also normal. The CF protein on the left pivoted and closed off the Cl^- ion channel.
5. Fewer Cl^- ions leave the cell, because one of the Cl^- ion channels is blocked. As a result, more water molecules remain inside the cell. More mucus builds up on the cell's surface.
6. See data table at bottom of page 185. The Cl^- ion concentration is higher inside the cell. The water concentration is also higher inside the cell.
7. This person has cystic fibrosis. Both CF proteins have changed position and closed off both Cl^- ion channels.
8. Chloride ions cannot leave the cell, and twice as many water molecules remain inside the cell as normal. The external surface of the cell becomes clogged with mucus.
9. Inside the cell, there are high concentrations of Cl^- ions and water. See data table at bottom of page 185. Outside the cell, there are lower-than-normal concentrations of Cl^- ions and water and, as a result, a buildup of thick mucus.

Analyze and Conclude

Answers

1. Chloride ion channels are gated because they can be closed off by the CF protein. The water channel is not gated because it cannot be closed off.
2. In Part A, the Cl^- ion and water concentrations on both sides of the membrane were equal, because both Cl^- ion channels were open, and more Cl^- ions could move out of the cell. In Part B, one Cl^- ion channel was open, and the Cl^- ion concentration remained the same on both sides of the membrane, but more water remained inside the cell. In Part C, Cl^- ions could not leave the cell, because both channels were blocked. The cell retained water, causing the buildup of mucus outside the cell.
3. Although Cl^- ions and water do accumulate in their cells, neither reaches very high concentrations, and mucus does not build up.

Chapter 9

DNA: The Genetic Material

Planning Guide

OBJECTIVES	MEETING INDIVIDUAL NEEDS	READING SKILLS AND STRATEGIES
9-1 Identifying the Genetic Material (BLOCKS 1 & 2 90 min) • **Relate** Griffith's conclusions to observations he made during the transformation experiments. • **Summarize** the steps involved in Avery's transformation experiments, and state the results. • **Evaluate** the results of the Hershey and Chase experiment.	Learning Styles, p. 188 BASIC ADVANCED • pp. 190, 191	ATE Active Reading, Technique: K-W-L, p. 187 ATE Active Reading, Technique: Paired Reading, p. 189 ■ Active Reading Guide, worksheet (9-1) ■ Directed Reading, worksheet (9-1) ■ Reading Strategies, K-W-L worksheet
9-2 The Structure of DNA (BLOCKS 3 & 4 90 min) • **Describe** the three components of a nucleotide. • **Develop** a model of the structure of a DNA molecule. • **Evaluate** the contributions of Chargaff, Franklin, and Wilkins in helping Watson and Crick to determine the double-helical structure of DNA. • **Relate** the role of the base-pairing rules to the structure of DNA.	Learning Styles, p. 192 BASIC ADVANCED • pp. 193, 194, 195	■ Active Reading Guide, worksheet (9-2) ■ Directed Reading, worksheet (9-2)
9-3 The Replication of DNA (BLOCKS 5 & 6 90 min) • **Summarize** the process of DNA replication. • **Describe** how errors are corrected during DNA replication. • **Compare** the number of replication forks in prokaryotic and eukaryotic DNA.	Learning Styles, p. 196 BASIC ADVANCED • pp. 197, 198	ATE Active Reading, Technique: K-W-L, p. 198 ■ Active Reading Guide, worksheet (9-3) ■ Directed Reading, worksheet (9-3) ■ Supplemental Reading Guide, *The Double Helix*

Review and Assessment Resources

BLOCKS 7 & 8 90 min

TEST PREP

- PE Study Zone, p. 199
- PE Performance Zone, pp. 200–201
- ■ Chapter 9 worksheets:
  - Vocabulary
  - Concept Mapping
  - Science Skills
  - Critical Thinking
- Holt Biology Interactive Tutor Unit 6—Heredity, Topics 1–2
- Audio CD Program
- ■ Test Prep Pretest

CHAPTER TESTING

- ■ Chapter Tests (blackline master)
- ▲ Test Generator
- ■ Assessment Item Listing

See the correlations table at the front of this book for a list of the ***National Science Education Standards*** addressed in this chapter.

Smithsonian Institution® Visit **www.si.edu/hrw** for additional on-line resources.

KEY

Biology Principles and Explorations

PE Pupil's Edition

ATE Teacher's Edition

HOLT BIOSOURCES

Teaching Transparencies

Laboratory Program

One-Stop Planner CD-ROM
Shows these resources within a customizable daily lesson plan:

■ Teaching Resources

▲ Test Generator

VISUAL STRATEGIES	CLASSWORK AND HOMEWORK	TECHNOLOGY AND INTERNET RESOURCES
ATE Fig. 9-2, p. 189 **ATE** Fig. 9-3, p. 190 25A Griffith's Experiment	Lab D1 Staining DNA and RNA **PE** Review, p. 191	Interactive Tutor Unit 6—Gene Expression (Topic 1) Holt Biology Videodiscs Section 9-1 Audio CD Program Section 9-1
ATE Fig. 9-8, p. 195 2A Biologically Important Molecules	**PE** Quick Lab How can you extract DNA from onion cells?, p. 193 Lab D2 Extracting DNA **PE** Review, p. 195	Interactive Tutor Unit 6—Gene Expression (Topic 1) Holt Biology Videodiscs Section 9-2 Audio CD Program Section 9-2
ATE Fig. 9-9, p. 196	**PE** Math Lab Analyzing the rate of DNA replication, p. 197 **PE** Experimental Design How Can You Model DNA Structure?, pp. 202–203 **PE** Review, p. 198 **PE** Study Zone, p. 199	CNN Video Segment 6 Teen Discovery Interactive Tutor Unit 6—Gene Expression (Topic 2) Holt Biology Videodiscs Section 9-3 Audio CD Program Section 9-3 internet**connect** SCILINKS NSTA **TOPICS:** • DNA replication • DNA

ALTERNATIVE ASSESSMENT

PE Portfolio Projects 23–26, p. 201

ATE Alternative Assessment, p. 199

ATE Portfolio Assessment, p. 199

Lab Activity Materials

Math Lab
p. 197 pencil, paper

Experimental Design
p. 202 plastic soda straws in 3 cm sections, metric ruler, pushpins (red, blue, yellow, green), paper clips

Lesson Warm-up
ATE photos of several organisms including chimpanzees and humans, p. 188

Demonstrations
ATE stack of books, p. 186
ATE 6 colors of chalk or DNA model, p. 193

BIOSOURCES LAB PROGRAM
- Biotechnology: Lab D1, p. 1
- Biotechnology: Lab D2, p. 5

Chapter 9

Chapter Theme

Evolution

Patterns within DNA suggest that similar genes originated from ancestral genes that differentiated and became more specific with time. Humans have 12 different hemoglobin genes with enough similarity to suggest a common source of origin.

Establishing Prior Knowledge

Before beginning this chapter, make certain students can answer the questions in Study TIP **Ready?** on page 187. If they cannot, encourage them to reread the sections indicated.

Opening Demonstration

Make a stack of books totaling about 10,000 pages. Tell students that the stack of books represents only about one-fiftieth of the information contained in the DNA of every human cell. Correlate this with the amount of information required to describe a human being.

CHAPTER 9 DNA: The Genetic Material

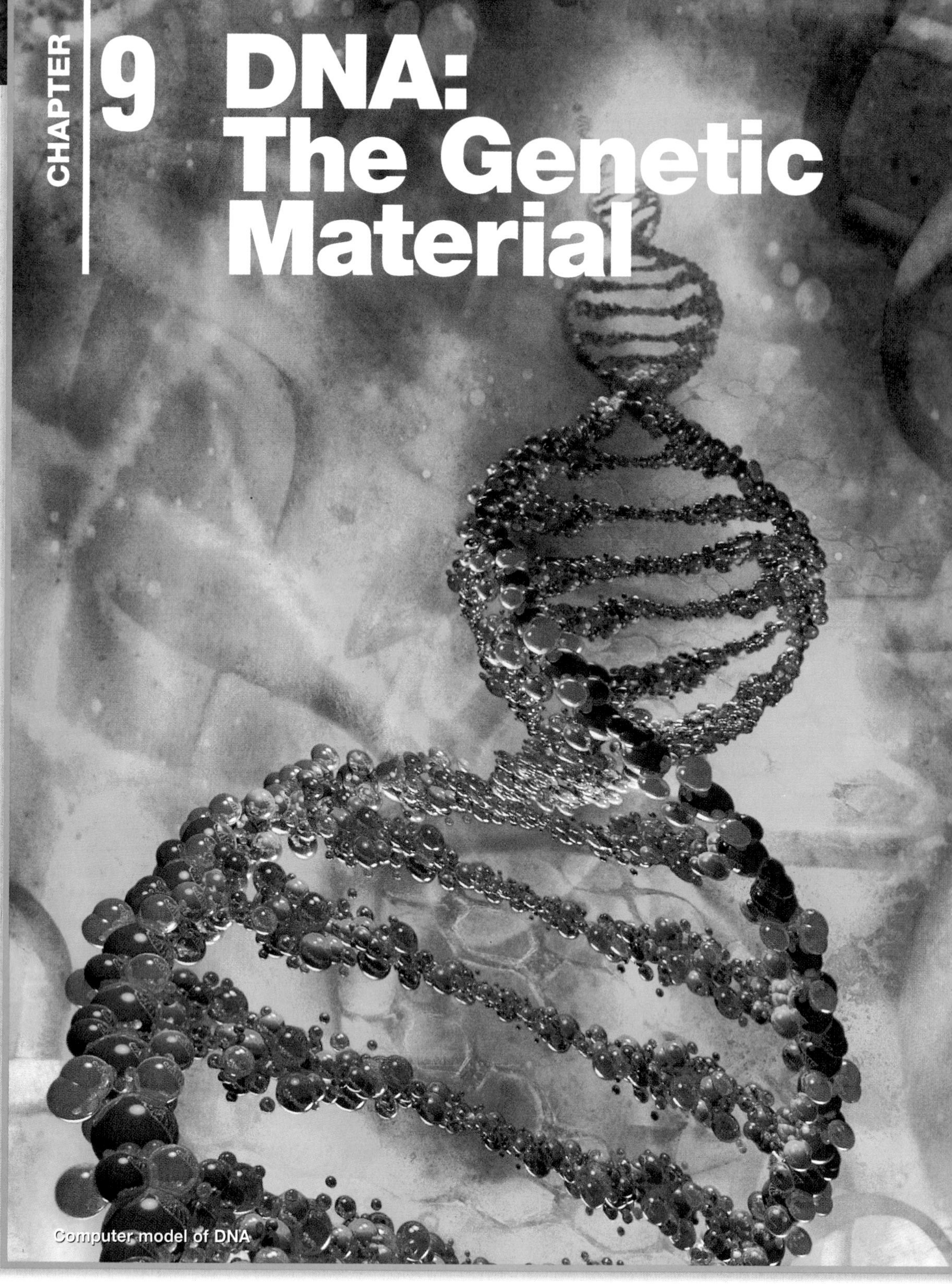

Computer model of DNA

What's Ahead — and Why

CHAPTER OUTLINE

Students will learn

- about the experiments that led to the discovery that DNA is the genetic material.
- about the interaction of structure and function in DNA.
- how DNA is copied before a cell divides.

Identical twins, identical DNA

Six billion pairs...

Each human cell contains about 6 billion pairs of nucleotides—the building blocks that make up DNA. The pairs of nucleotides are identical in these identical twins. DNA contains the genes that determine the physical traits and regulate the body functions of the twins.

Looking Ahead

Labs

- **Quick Lab**
  How can you extract DNA from onion cells? (p. 193)
- **Math Lab**
  Analyzing the rate of DNA replication (p. 197)
- **Experimental Design**
  How Can You Model DNA Structure? (p. 202)

Unit 6—*Gene Expression*
Topics 1–2

internet connect

National Science Teachers Association *sci*LINKS Internet resources are located throughout this chapter.

Study TIP

Ready?

Check your understanding of these topics to help you prepare for what's ahead.

Can you...

- **describe** the structure of chromosomes? (Section 6-1)
- **define** the term *gene*? (Section 6-1)
- **identify** the stage in the cell cycle when DNA is copied? (Section 6-2)
- **describe** the effects of mutations? (Section 6-3)
- **summarize** Mendel's theory of heredity? (Section 8-2)

If not, *review the sections indicated.*

Reading Strategies

Active Reading

Technique: K-W-L

Before students read this chapter, have them write short individual lists of all the things they already **K**now (or think they know) about DNA. Ask them to contribute their entries to a group list on the board or overhead. Then have students list things they **W**ant to know about DNA. Have students save their lists for later use in the **Closure** on page 198.

Directed Reading

Assign **Directed Reading Worksheet Chapter 9** to help students interact with the text as they read the chapter.

Interactive Reading

Assign ***Biology: Principles and Explorations*** **Audio CD Program Chapter 9** to help reluctant readers achieve greater success in reading the chapter.

Reading Skills

Assign Reading Organizers and Reading Strategies Worksheets from the **One-Stop Planner CD-ROM with Test Generator** to help students learn and practice effective reading skills.

internet connect

Holt *Biology: Principles and Explorations*

On-line Resources: go.hrw.com
Visit the HRW Web site for a variety of resources related to this chapter. Just type in the keyword HX0 Chapter 9.

Holt *Biology: Principles and Explorations*

On-line Resources: www.scilinks.org
The following *sci*LINKS Internet resources can be found in the student text for this chapter:

Page 198
TOPIC: DNA replication
KEYWORD: HX198

Page 203
TOPIC: DNA
KEYWORD: HX203

Chapter 9

1 Prepare

Scheduling

- **Block:** About 90 minutes will be needed to complete this section.
- **Traditional:** About two class periods will be needed to complete this section.
- **Planning:** See the Planning Guide on pp. 185A and 185B for lecture, classwork, and assignment options for Section 9-1. Lesson plans for Section 9-1 can be found in a lesson cycle format in ***Biology: Principles and Explorations*** **One-Stop Planner CD-ROM with Test Generator.**

Resource Link

Assign **Active Reading Worksheet Section 9-1** to help students comprehend the key concepts and visuals in this lesson.

2 Teach

Lesson Warm-up

Show photographs of various organisms, including chimpanzees and humans. Ask students if the structure of DNA is the same in all organisms, and whether the composition of DNA differs between organisms. *(The double stranded helical and subunit structure is the same in all organisms. Prokaryotic DNA is circular. Eukaryotic DNA is linear. The nucleotide sequences vary between organisms.)* Point out that human and chimpanzee DNA nucleotide sequences differ by only 1.6 percent. The differences observed between two individuals of the same species are about 1 percent.

Section 9-1

Identifying the Genetic Material

Objectives

- **Relate** Griffith's conclusions to the observations he made during the transformation experiments. (pp. 188–189)
- **Summarize** the steps involved in Avery's transformation experiments, and state the results. (p. 189)
- **Evaluate** the results of the Hershey and Chase experiment. (pp. 190–191)

Key Terms

vaccine
virulent
transformation
bacteriophage

Griffith and Avery Helped Identify DNA As the Genetic Material

Mendel's experiments and results answered the question of why you resemble your parents. You resemble your parents because you have copies of their chromosomes, which contain sets of instructions called genes. But Mendel's work created more questions, such as, What are genes made of? Scientists believed that if they could answer this question they would understand how chromosomes function in heredity.

Griffith's experiments

In 1928, an experiment completely unrelated to the field of genetics led to an astounding discovery about DNA. Frederick Griffith, a bacteriologist, was trying to prepare a vaccine *(vahk SEEN)* against the pneumonia-causing bacterium, *Streptococcus pneumoniae* (abbreviated *S. pneumoniae*), shown in **Figure 9-1.** A **vaccine** is a substance that is prepared from killed or weakened microorganisms and is introduced into the body to protect the body against future infections by the microorganisms.

Griffith worked with two types, or strains, of *S. pneumoniae,* as shown in **Figure 9-2.** The first strain is enclosed in a capsule made of polysaccharides. The capsule protects the bacterium from the body's defense systems; this helps make the microorganism **virulent** *(VIHR yoo luhnt)*, or able to cause disease. Because of the capsule, this strain of *S. pneumoniae* grows as smooth-edged (*S*) colonies when grown in a Petri dish. The second strain of *S. pneumoniae* lacks the polysaccharide capsule and does not cause disease. When grown in a Petri dish, the second strain forms rough-edged (*R*) colonies.

Griffith knew that mice infected with the *S* bacteria grew sick and died, while mice infected with the *R* bacteria were not harmed, as shown in Figure 9-2. To determine whether the capsule on the *S* bacteria was causing the mice to die, Griffith injected the mice with dead *S* bacteria. The mice remained healthy. Griffith then prepared a vaccine of weakened *S* bacteria by raising their temperature to a point at which the bacteria were "heat-killed," meaning that they could no longer reproduce (the capsule remained on the bacteria).

Figure 9-1
Streptococcus pneumoniae. Certain types of *S. pneumoniae* bacteria can cause pneumonia in animals.

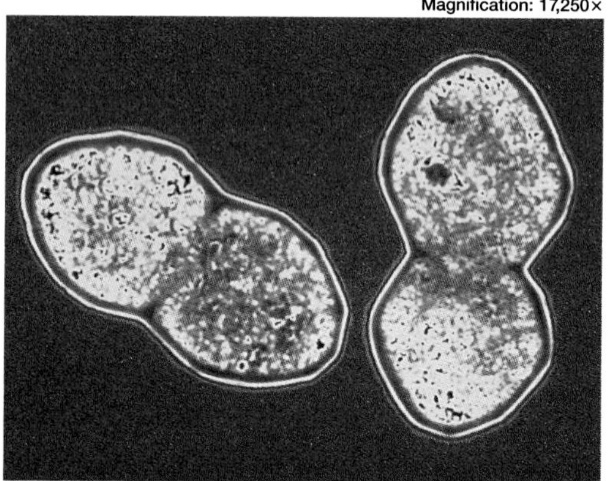

Integrating Different Learning Styles

Visual/Spatial	**PE**	Study Tip, p. 191
	ATE	Opening Demonstration, p. 186; Lesson Warm-up, p. 188; Visual Strategies, pp. 189, 190; Teaching Tip, p. 190; Reteaching, p. 191
Interpersonal	**ATE**	Biology in Your Community, p. 189; Teaching Tip, p. 190
Verbal/Linguistic	**PE**	Real Life, p. 191; Performance Zone item 25, p. 201
	ATE	Active Reading, p. 189

Figure 9-2 Griffith's discovery of transformation

Griffith discovered that harmless bacteria could turn virulent when mixed with bacteria that cause disease.

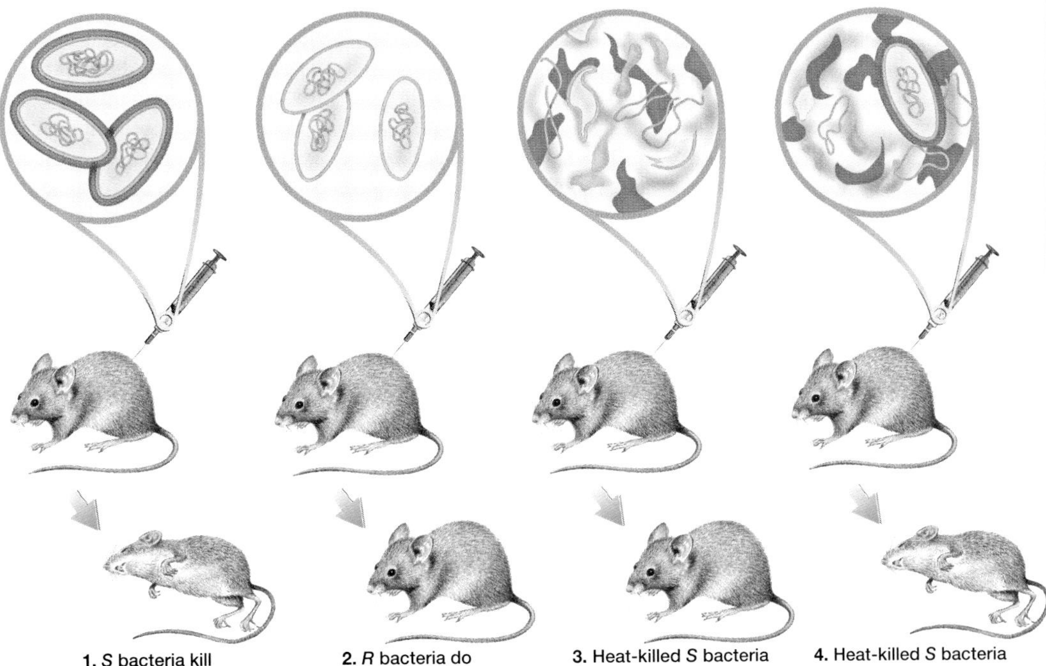

1. *S* bacteria kill mouse.
2. *R* bacteria do not kill mouse.
3. Heat-killed *S* bacteria do not kill mouse.
4. Heat-killed *S* bacteria and *R* bacteria kill mouse.

When Griffith injected the mice with heat-killed *S* bacteria, the mice still lived. Thus, Griffith knew it was not the capsule on the *S* bacteria that killed the mice. He then mixed the harmless live *R* bacteria with the harmless heat-killed *S* bacteria. Mice injected with this mixture of previously harmless preparations died. When Griffith examined the blood of the dead mice, he found that the live *R* bacteria had acquired polysaccharide capsules. Somehow, the harmless *R* bacteria had changed and became virulent *S* bacteria. Griffith had discovered what is now called **transformation,** a change in phenotype caused when bacterial cells take up foreign genetic material. But the cause of the transformation was not known at the time.

Avery's experiments

The search for the material responsible for transformation continued until 1944. Then, an elegant series of experiments showed that the activity of the material responsible for transformation was not affected by protein-destroying enzymes, but the activity was stopped by a DNA-destroying enzyme. In this way, almost 100 years after Mendel's experiments were performed, Oswald Avery and his co-workers, biologists at the Rockefeller Institute, in New York City, demonstrated that DNA is the material responsible for transformation. DNA had the instructions for the making of the capsule in the *S* strain of *S. pneumoniae.*

biobit

Do S. pneumoniae live in your throat?

S. pneumoniae live in the throats of healthy people. When a person's defenses are weakened by such factors as a bout of flu or poor nutrition, the virulent *S* bacteria can invade the lungs and cause pneumonia— inflammation of the lungs.

Biology in Your Community

Vaccines

Invite a health care worker to talk about current types of vaccines. Ask the speaker to discuss DNA vaccines and the latest findings on their effectiveness. Ask the students to prepare a list of questions they would like to ask the speaker. PORTFOLIO

Section 9-1

Visual Strategy

Figure 9-2

Use Figure 9-2 to review each step of Griffith's experiment. Do not tell the students why Griffith obtained the results he did. Have them ask questions to which you can answer "yes" or "no." Encourage students to put the explanation into their own words so that they will better understand the results and the implications of the experiment.

Teaching TIP

The Heat Factor

DNA can tolerate temperatures around 90°C without being altered, whereas proteins are denatured around 60°C. Ask students what effect temperature had on Griffith's work. *(The DNA of the heat-killed bacteria survived and transformed the DNA of the live bacteria.)*

Teaching TIP

Nuclein Versus DNA

Tell students that DNA was first isolated in 1869 by a German physician named Friedrich Miescher. Miescher reported that the substance he isolated from white blood cells was white, sugary, and slightly acidic. Because he found it only in the nuclei of these cells, he called it nuclein.

Active Reading

Technique: Paired Reading

Give the students self-stick notes, and have them mark areas of difficulty as they silently read Section 9-1. Have students use a question mark to signify an area that is unclear and a check mark to indicate understanding. Remind students to study the captions, diagrams, and illustrations as they read. After completing the section, have students work with a partner to discuss what they did or did not understand. Have students write their remaining questions on the board for later discussion.

Teaching TIP

Hershey and Chase's Experiment

To summarize Hershey and Chase's experiment, have students design a **Graphic Organizer** like the one shown at the bottom of page 190. Have them include a paragraph that explains how the experiment showed that DNA is the genetic material. PORTFOLIO

Visual Strategy

Figure 9-3

Use Figure 9-3 to reinforce the text description of Hershey and Chase's experiment. Make sure students understand how the researchers arrived at their conclusion in **Step 4**. *(Because only DNA entered the bacteria, where the new viruses were produced, DNA must carry the genetic instructions needed to make the new viruses.)* Some viruses also inject some or all of their protein into the cell. Ask students to predict the outcome if Hershey and Chase had worked with one of these types of viruses instead of with T2 phages. *(Answers will vary. Students should realize that Hershey and Chase would have found both the ^{32}P and ^{35}P labels in the lower layer of the test tube containing the* E. coli. *They would not have been able to conclude which of the two, the DNA or the protein, was the genetic material.)*

Teaching TIP ADVANCED

Role Playing

Divide the class into three groups: one group represents the scientists who did the research described on pp. 190–191, the second group represents a team of scientists who will question the first group on their results, and the third group is a committee that will decide if the work is to be published in a recognized scientific journal. The questions of the second group will help the committee make a decision. The committee writes a summary supporting its decision.

Hershey and Chase Show that Virus Genes Are Made of DNA

Even though Avery's experiments clearly indicated that the genetic material is composed of DNA, many scientists remained skeptical. Scientists knew that proteins were important to many aspects of cell structure and metabolism, so most of them suspected that proteins were the genetic material. They also knew very little about DNA, so they could not imagine how DNA could carry genetic information.

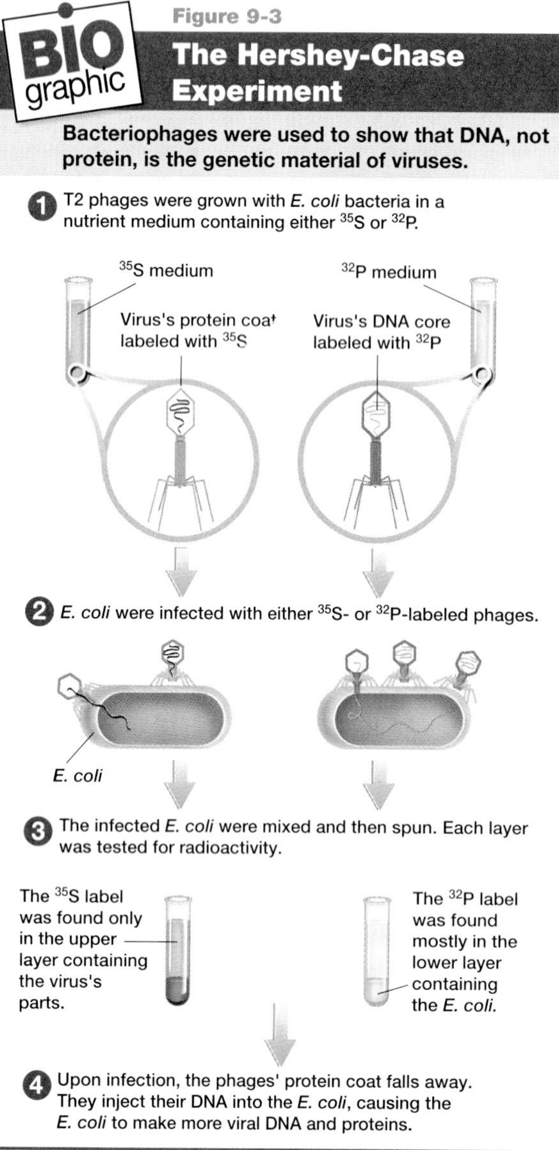
BIO graphic
Figure 9-3
The Hershey-Chase Experiment
Bacteriophages were used to show that DNA, not protein, is the genetic material of viruses.
1 T2 phages were grown with E. coli bacteria in a nutrient medium containing either 35S or 32P.
35S medium
32P medium
Virus's protein coat labeled with 35S
Virus's DNA core labeled with 32P
2 E. coli were infected with either 35S- or 32P-labeled phages.
E. coli
3 The infected E. coli were mixed and then spun. Each layer was tested for radioactivity.
The 35S label was found only in the upper layer containing the virus's parts.
The 32P label was found mostly in the lower layer containing the E. coli.
4 Upon infection, the phages' protein coat falls away. They inject their DNA into the E. coli, causing the E. coli to make more viral DNA and proteins.

Viruses reveal DNA's role

In 1952, Alfred Hershey and Martha Chase, scientists at Cold Spring Harbor Laboratory, in New York, performed an experiment that settled the controversy. It was known at that time that viruses, which are much simpler than cells, are made of DNA (or sometimes RNA) surrounded by a protective protein coat. **Bacteriophage** *(bak TIHR ee uh fayj)*, also referred to as phage *(fayj)*, are viruses that infect bacteria. It was also known that when phage infect bacterial cells, the phage are able to produce more viruses, which are released when the bacterial cells rupture.

What was not known at the time was how the bacteriophage reprogrammed the bacterial cell to make viruses. Was it the phage DNA, the protein, or both that issued instructions to the bacteria during infection?

Hershey and Chase used the bacteriophage T2 (T2 phage for short), shown in **Figure 9-3**, to answer this question. Hershey and Chase knew that the only molecule in the phage that contained phosphorus was its DNA. Likewise, the only phage molecules that contained sulfur were the proteins in its coat. Hershey and Chase used these differences in chemical composition to carry out the experiment shown in Figure 9-3.

Step 1 Hershey and Chase first grew T2 with *Escherichia coli* (abbreviated *E. coli*) bacteria in a nutrient medium that contained radioactive sulfur

Graphic Organizer

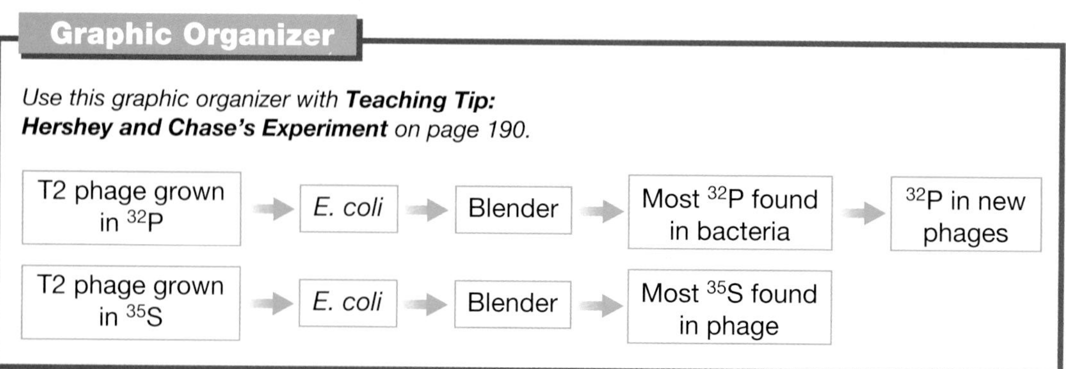
Use this graphic organizer with Teaching Tip: Hershey and Chase's Experiment on page 190.
T2 phage grown in 32P
E. coli
Blender
Most 32P found in bacteria
32P in new phages
T2 phage grown in 35S
E. coli
Blender
Most 35S found in phage

(^{35}S). The protein coat of the virus would incorporate the ^{35}S. They grew a second batch of phage with *E. coli* bacteria in a nutrient medium that contained radioactive phosphorus (^{32}P). The radioactive phosphorus would become part of the phages' DNA.

Step 2 The ^{35}S-labeled and ^{32}P-labeled phages were used to infect two separate batches of *E. coli* bacteria. Because radioactive elements release particles that can be detected with machines, they can be followed, or traced, in a biological process. Scientists could determine whether it was the DNA, the protein, or both that were being transferred into the bacterial cells to reprogram the bacteria.

Step 3 After a few minutes, the scientists tore the ^{35}S-labeled phages off the surfaces of the bacteria (with the help of a blender). The bacteria infected with the ^{32}P-labeled phage were likewise mixed in a blender. The investigators used a centrifuge to separate the bacteria and phages. The heavier, bacterial cells formed a solid layer at the bottom of the centrifuge tubes. The lighter, viral parts remained in the upper, liquid layer.

When the layers from the ^{35}S-infected bacteria were examined, Hershey and Chase found that most of the ^{35}S label was still part of the phage (the upper layer), meaning the protein was not injected into the bacteria. When the layers from the ^{32}P-infected bacteria were examined, they found that the ^{32}P label was mostly part of the layer containing the bacterial cells (the lower layer), meaning the DNA had been injected into the hosts. The new generation of phages that was produced by these bacteria also contained radioactive DNA.

Step 4 Hershey and Chase concluded that the DNA of viruses is injected into the bacterial cells, while most of the proteins remain outside. The injected DNA molecules caused the bacterial cells to produce more viral DNA and proteins. This meant that the DNA, rather than proteins, was the hereditary material, at least in viruses.

These important experiments, and many others since, have shown that DNA is the molecule that stores genetic information in living cells. As you will see in the next section, the structure of DNA makes DNA particularly well suited to this function.

REAL LIFE

Many viruses infect humans. Although the T2 viruses used by Hershey and Chase infect bacteria, there are many viruses that infect humans and cause diseases. For example, viruses are responsible for causing polio, the common cold, measles, and rabies.

Finding Information
Research other human diseases that are caused by viruses.

Study TIP

Organizing Information
Create a timeline that summarizes the people and events that led to the discovery that DNA is the molecule where genetic information is stored. Start with 1928, and end with 1952.

Review SECTION 9-1

1. **Summarize** Griffith's transformation experiments.
2. **Describe** how Oswald Avery's experiment supplied evidence that DNA, and not protein, is the genetic material.
3. **Describe** how Hershey and Chase altered the DNA and proteins in the T2 phage in their experiments.
4. **Summarize** the steps of the experiment that led Hershey and Chase to conclude that DNA is the genetic material.
5. **Describe** how Hershey and Chase concluded that DNA is the genetic material.

Critical Thinking

6. **SKILL Evaluating Methods** Why did heat kill the *S* bacteria in Griffith's experiments? (**Hint:** See Chapter 2, Section 2-4.)

Review SECTION 9-1 ANSWERS

1. Mice were injected with either *S* bacteria, *R* bacteria, heat-killed *S* bacteria, or a mixture of heat-killed *S* bacteria and *R* bacteria. The mice injected with the *S* bacteria or the mixture of heat-killed *S* bacteria and *R* bacteria died; those injected with the *R* or the heat-killed *S* bacteria lived.
2. Avery showed that DNA-destroying enzymes, but not protein-destroying enzymes, stopped the transformation.
3. Radioactive sulfur (^{35}S) and radioactive phosphorous (^{32}P) were incorporated into the bacteriophage's proteins and DNA, respectively.
4. T2 phages were grown with *E. coli* bacteria in a nutrient medium containing either ^{35}S or ^{32}P. The *E. coli* were infected with either ^{35}S- or ^{32}P-labeled phages. The infected *E. coli* were mixed and then spun. Each layer was tested for radioactivity.
5. The ^{32}P label was found mostly in the lower layer containing the *E. coli*. Conversely, most of the ^{35}S label was still part of the phage—it had remained on the outside of the bacteria.
6. The heat denatured their proteins.

REAL LIFE

Answer

Answers will vary. Other human diseases caused by viruses include rubella, mumps, AIDS, influenza, hepatitis B, and infectious mononucleosis. Different viruses have also been associated with several types of cancer.

3 Assessment Options

Closure

Heredity and DNA

Explain to students that many scientists were reluctant to accept the results of experiments that suggested DNA was the hereditary material. Ask students to explain why most scientists initially thought that the hereditary material was protein, not DNA. Have them consider the structures of both proteins and DNA. *(Students should recall that proteins are made from 20 different amino acids, which can be combined in a variety of ways. DNA, on the other hand, has a relatively simple structure that consists basically of four nucleotide bases. Scientists reasoned that the complex processes involved in heredity must be controlled by a complex molecule—protein—rather than DNA.)*

Review

Assign Review—Section 9-1.

Reteaching — BASIC

Pair students into groups. Provide each group with materials to make flashcards. Have them recreate Griffith's experiment using Figure 9-2 on page 189 as a guide. After they have finished making flashcards, tell them to shuffle the cards and place them face up on their tables in the correct order.
PORTFOLIO

Chapter 9

1 Prepare

Scheduling

- **Block:** About 90 minutes will be needed to complete this section.
- **Traditional:** About two class periods will be needed to complete this section.
- **Planning:** See the Planning Guide on pp. 185A and 185B for lecture, classwork, and assignment options for Section 9-2. Lesson plans for Section 9-2 can be found in a lesson cycle format in ***Biology: Principles and Explorations*** **One-Stop Planner CD-ROM with Test Generator.**

Resource Link

Assign **Active Reading Worksheet Section 9-2** to help students comprehend the key concepts and visuals in this lesson.

2 Teach

Lesson Warm-up

Tell students that DNA is often compared to a ladder. Ask students to use Figure 9-4 on page 192 to explain how the structure of DNA is similar to that of a ladder. *(The rails of a ladder represent the sugar-phosphate backbone of DNA, and the rungs of a ladder represent the nitrogen bases.)* Ask students how a ladder is different from DNA. *(DNA is twisted, while a ladder is flat. A "rung" in the DNA molecule is made of two bases, while a ladder's rung is a single unit.)*

Section 9-2 The Structure of DNA

Objectives

- **Describe** the three components of a nucleotide. (p. 192)
- **Develop** a model of the structure of a DNA molecule. (pp. 192–193)
- **Evaluate** the contributions of Chargaff, Franklin, and Wilkins in helping Watson and Crick determine the double-helical structure of DNA. (p. 194)
- **Relate** the role of the base-pairing rules to the structure of DNA. (p. 195)

Key Terms

double helix
nucleotide
deoxyribose
base-pairing rules
complementary

DNA Has the Structure of a Winding Staircase

By the early 1950s, most scientists were convinced that genes were made of DNA. They hoped that the mystery of heredity could finally be solved by understanding the structure of DNA. The research of many scientists led two young researchers at Cambridge University, James Watson and Francis Crick, to piece together the structure of DNA. The discovery of DNA's structure was an incredibly important finding because it clarified *how* DNA could serve as the genetic material.

Watson and Crick determined that DNA is a molecule that is a **double helix**—two strands twisted around each other, like a winding staircase. As shown in **Figure 9-4,** each strand is made of linked nucleotides *(NOO klee oh teyedz).* **Nucleotides** are the subunits that make up DNA. Each nucleotide is made of three parts: a phosphate group, a five-carbon sugar molecule, and a nitrogen-containing base. Figure 9-4 shows how these three parts are arranged to form a nucleotide. The five-carbon sugar in DNA nucleotides is called **deoxyribose** *(dee ahk see REYE bohs),* from which DNA gets its full name, deoxyribonucleic acid.

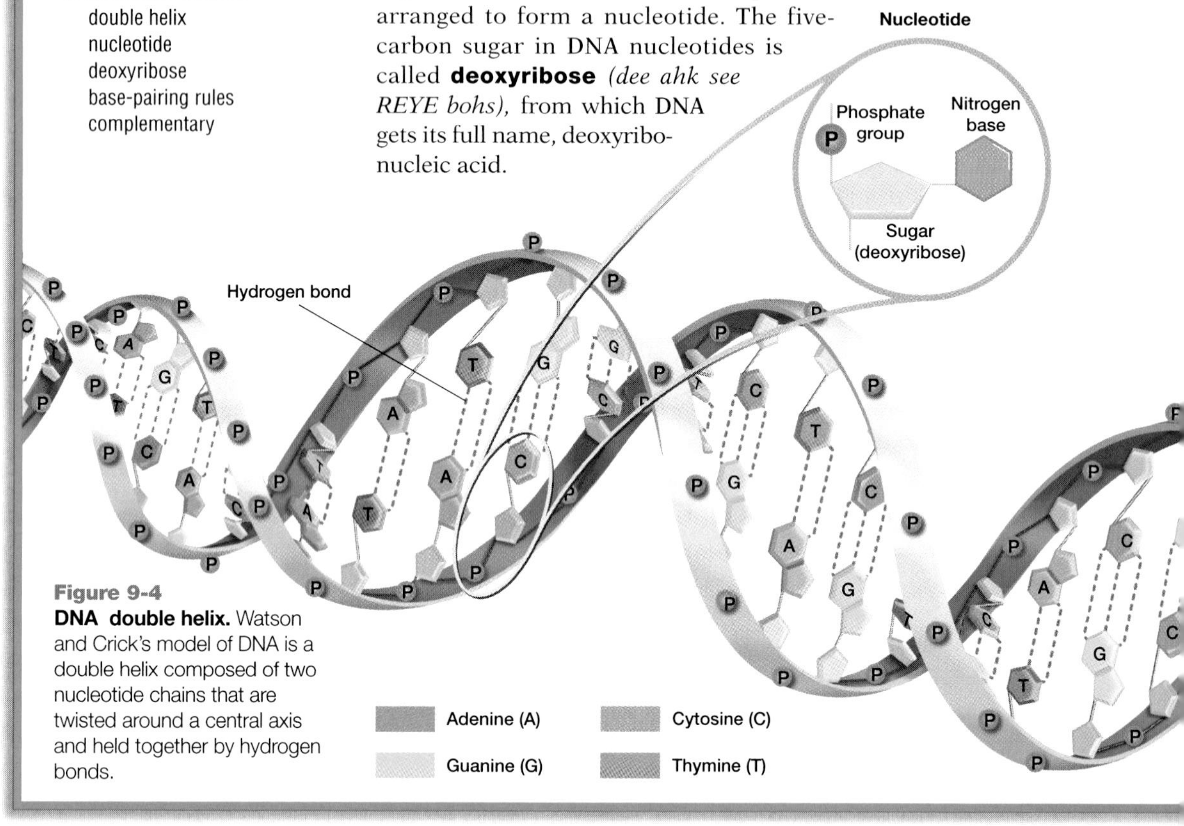

Figure 9-4
DNA double helix. Watson and Crick's model of DNA is a double helix composed of two nucleotide chains that are twisted around a central axis and held together by hydrogen bonds.

Integrating Different Learning Styles

Logical/Mathematical	**PE**	Performance Zone items 13, 21, and 22, pp. 200–201
Visual/Spatial	**PE**	Quick Lab, p. 193; Experimental Design, pp. 202–203
	ATE	Lesson Warm-up, p. 192; Demonstration, p. 193; Visual Strategy, p. 195; Closure, p. 195; Reteaching, p. 195; Performance Zone items 1 and 20, pp. 200–201
Interpersonal	**PE**	Performance Zone item 23, p. 201
Verbal/Linguistic	**PE**	Performance Zone item 19, p. 201
	ATE	Resource Link, p. 194

While the sugar molecule and the phosphate group are the same for each nucleotide in a molecule of DNA, the nitrogen base may be any one of four different kinds. **Figure 9-5** illustrates the four different nitrogen bases in DNA: adenine *(AD uh neen),* guanine *(GWAH neen),* thymine *(THEYE meen),* and cytosine *(SEYET oh seen).* Adenine (A) and guanine (G) are classified as purines *(PYUR eenz),* nitrogen bases made of two rings of carbon and nitrogen atoms. Thymine (T) and cytosine (C) are classified as pyrimidines *(pih RIHM uh deenz),* nitrogen bases made of a single ring of carbon and nitrogen atoms.

Note how the DNA shown in Figure 9-4 resembles a ladder twisted like a spiral staircase. The sugar-phosphate backbones are similar to the side rails of a ladder, and the paired nitrogen bases are similar to the rungs of the ladder. The nitrogen bases face each other. The double helix is held together by weak hydrogen bonds between the pairs of bases.

Figure 9-5 Purines and pyrimidines

The nitrogen base in a nucleotide can be either a bulky, double-ring purine, or a smaller, single-ring pyrimidine.

Purines

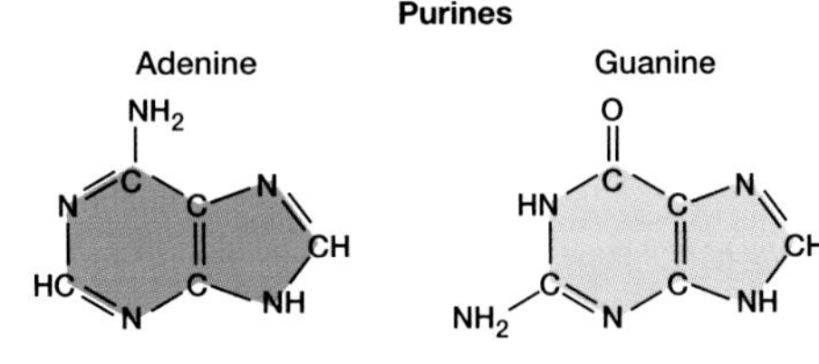

Pyrimidines

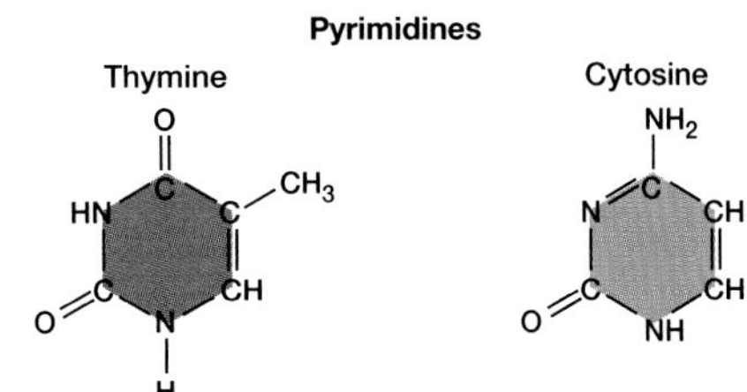

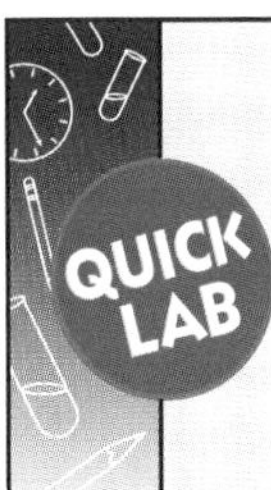

How can you extract DNA from onion cells?

You can analyze the structure of DNA by using ethanol and a stirring rod to extract DNA from onion cells.

Materials

safety goggles and plastic gloves, 5 mL of onion extract, test tube, 5 mL of ice-cold ethanol, plastic pipet, glass stirring rod, test tube rack

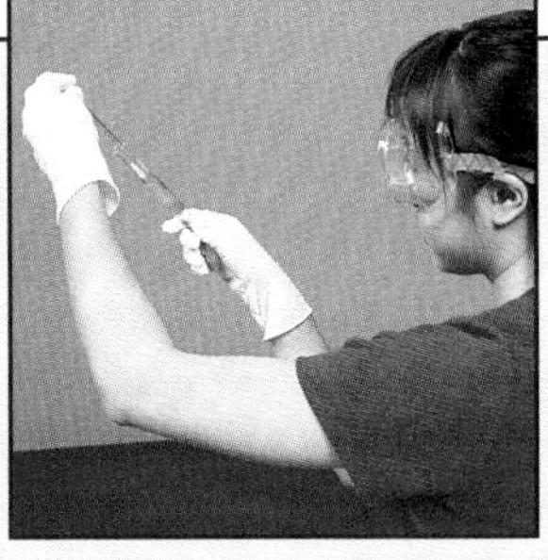

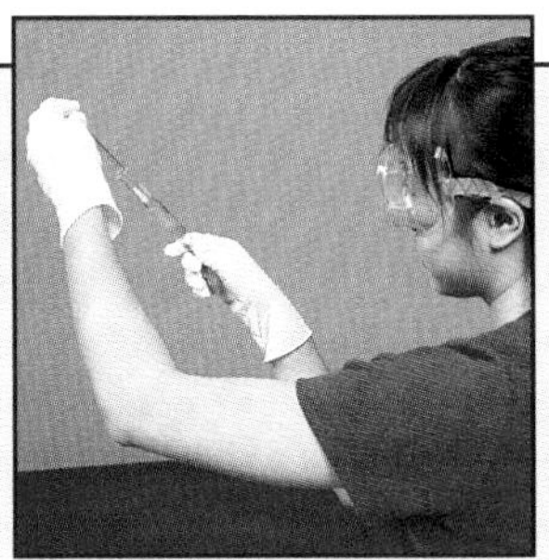

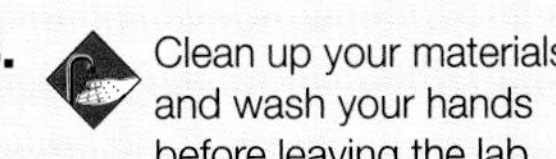

Procedure

1. Place 5 mL of onion extract in a test tube.
2. **CAUTION: *Ethanol is flammable. Do not use it near a flame.*** Hold the test tube at a 45° angle. Use a pipet to add 5 mL of ice-cold ethanol to the tube one drop at a time. *NOTE: Allow the ethanol to run slowly down the side of the tube so that it forms a distinct layer.*
3. Let the test tube stand for 2–3 minutes.
4. Insert a glass stirring rod into the boundary between the onion extract and ethanol. Gently twirl the stirring rod by rolling the handle between your thumb and finger.
5. Remove the stirring rod from the liquids, and examine any material that has stuck to it. Touch the material to the lip of the test tube, and observe how the material acts as you try to remove it.
6. Clean up your materials and wash your hands before leaving the lab.

Analysis

1. **Describe** any material that stuck to the stirring rod.
2. **Relate** the characteristics of your sample to the structural characteristics of DNA.
3. **Propose** a way to determine if the material on the stirring rod is DNA.

Teaching TIP BASIC

DNA Diagram

Draw a diagram of DNA on the board using different colors for the phosphate, deoxyribose sugar, and each nitrogen base (or use a model of DNA). Show students the structure of the molecule, and point out that certain bases always pair with each other. Ask students to determine the base-pairing rules from the diagram or model. *(An adenine on one strand always pairs with a thymine on the opposite strand. A cytosine on one strand always pairs with a guanine on the opposite strand.)* Emphasize that the pairing allows the DNA to be copied.

How can you extract DNA from onion cells?

Time 20 minutes

Process Skills Observing, measuring, relating, analyzing

Teaching Strategies

- Review all safety symbols with students.
- Order the DNA Extraction Biolab kit from WARD'S.
- Premeasure 5 mL of onion extract. Tell students not to shake or mix the onion extract–ethanol combination.
- Keep ethanol in an ice bath to maintain its temperature.
- Demonstrate the lab procedure for adding ethanol to the test tube and for using the stirring rod.

Analysis Answers

1. The material is clear, viscous, and elastic.
2. It sticks to itself and to the test tube, and it forms long strings.
3. Students cannot know for sure that this is DNA. Students should suggest that a laboratory test might confirm whether the material is DNA.

Overcoming Misconceptions

How Different Are We?

Even though people can seem very different, humans are very similar to each other biologically. The differences observed between two individuals are 2 million to 10 million nucleotide base pairs out of 3 billion—only 1 percent of the total DNA. Because these differences in base pairs are insignificant in terms of mapping the genome, the information sought by the Human Genome Project will be useful to all humans.

CROSS-DISCIPLINARY CONNECTION

◆ **History**

Ask students to name some prestigious awards. *(Answers may include the Oscar or the Heisman Trophy.)* Ask students what they think would be a prestigious award for a leading scientist. *(Answers may vary.)* Discuss the Nobel prize. Alfred Nobel (1833–1896), the Swedish inventor of dynamite, provided a $9 million fund in his will to be used to award Nobel prizes in the areas of chemistry, physics, physiology or medicine, and literature. Later, prizes in peace and economics were added. James Watson and Francis Crick received the Nobel prize in 1962 for their involvement in elucidating the structure of DNA. Encourage students to read about the careers and awards of other scientists they encounter in their study of biology.

Resource Link — **ADVANCED**

- Assign **Supplemental Reading: The Double Helix** from the Holt BioSources Teaching Resources CD-ROM. PORTFOLIO

BIOSOURCES TEACHING RESOURCES SUPPLEMENTAL READING

3 *The Double Helix* — A Personal Account of the Discovery of the Structure of DNA —*James D. Watson*

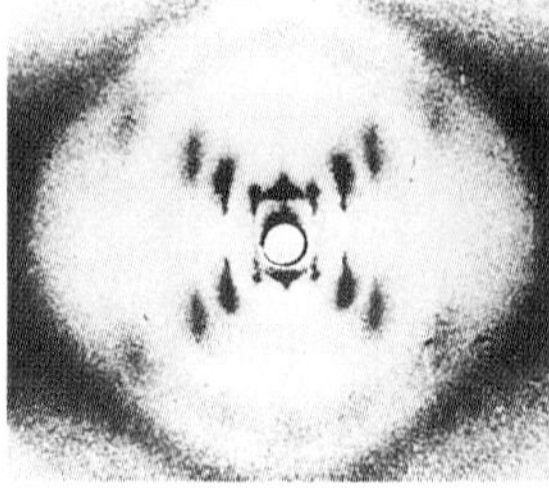

Figure 9-6 Franklin and her X-ray diffraction photo. The photographs revealed the X pattern characteristic of a helix. Franklin died of cancer when she was 37 years old.

Many Scientists Contributed to the Discovery of DNA's Structure

How were Watson and Crick able to determine the double helical structure of DNA? As with most discoveries in science, other scientists provided crucial pieces that helped them solve this puzzle.

Chargaff's observations

In 1949, Erwin Chargaff, a biochemist working at Columbia University, in New York City, made an interesting observation about DNA. Chargaff's data showed that for each organism he studied, the amount of adenine always equaled the amount of thymine (A=T). Likewise, the amount of guanine always equaled the amount of cytosine (G=C). However, the amount of adenine and thymine and of guanine and cytosine varied between different organisms.

Wilkins and Franklin's photographs

The significance of Chargaff's data became clear in the 1950s when scientists began using X-ray diffraction to study the structures of molecules. In X-ray diffraction, a beam of X rays is directed at an object. The X rays bounce off the object and are scattered in a pattern onto a piece of film. By analyzing the complex patterns on the film, scientists can determine the structure of the molecule (much like shining a light on an object and then analyzing its shadow).

In the winter of 1952, Maurice Wilkins and Rosalind Franklin, two scientists working at King's College in London, developed high-quality X-ray diffraction photographs of fibers of the DNA molecule. These photographs, such as the one in **Figure 9-6,** suggested that the DNA molecule resembled a tightly coiled helix and was composed of two or three chains of nucleotides.

1949	1952	1953	1958	1962
Chargaff's data show that A = T and G = C	Franklin and Wilkins photograph DNA	Watson and Crick propose double helix model	Franklin dies	Watson, Crick, and Wilkins receive the Nobel Prize

Figure 9-7 Watson and Crick

Watson and Crick's DNA model

The three-dimensional structure of the DNA molecule, however, was yet to be discovered. Any model had to take into account both Chargaff's findings and Franklin and Wilkins's X-ray diffraction data. In 1953, Watson and Crick used this information, along with their knowledge of chemical bonding, to come up with a solution. With tin-and-wire models of molecules, they built a model of DNA with the configuration of a double helix, a "spiral staircase" of two strands of nucleotides twisting around a central axis. **Figure 9-7** shows Watson and Crick next to their tin-and-wire model of DNA.

did you know?

Rosalind Franklin could have determined the structure of DNA.

Studying the work of their colleagues, Watson and Crick pooled knowledge acquired in their different backgrounds to determine the structure of DNA. However, Watson has since commented that Rosalind Franklin would have been able to determine the structure of DNA if she and Crick had simply talked with one another for an hour. Although Franklin's X-ray diffraction photograph was crucial to solving the puzzle of the structure of DNA, she never received the credit that Watson and Crick did. She died from cancer at the age of 37, before her accomplishments were fully recognized.

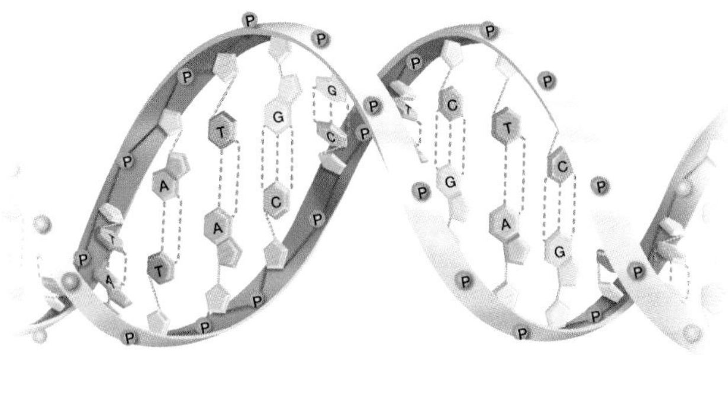

Figure 9-8 Base-pairing in DNA. The diagram of DNA on the *bottom* makes it easier to visualize the base-pairing that occurs between DNA strands.

Pairing between bases

As you can see in **Figure 9-8,** Watson and Crick determined that a purine on one strand of DNA is always paired with a pyrimidine on the opposite strand. More specifically, an adenine on one strand always pairs with a thymine on the opposite strand, and a guanine on one strand always pairs with a cytosine on the opposite strand. The structure and size of each of the nitrogen bases allows only these two pairs. This pairing arrangement of the nitrogen bases between the two strands, now known as the **base-pairing rules,** is supported by Chargaff's observations. One easy way to visualize base-pairing is by simplifying the way in which DNA structure is represented. Figure 9-8 shows one way that DNA structure can be represented in a simpler way.

Adenine forms two hydrogen bonds with thymine, and cytosine forms three hydrogen bonds with guanine. The hydrogen bonds between the nitrogen bases keep the two strands of DNA together. The strictness of base-pairing results in two strands that are **complementary** to each other; that is, the sequence of bases on one strand determines the sequence of bases on the other strand. For example, if the sequence of nitrogen bases on one strand of a DNA molecule is TCGAACT, the sequence of nitrogen bases on the other strand must be AGCTTGA.

biobit

Does the sun damage our DNA?

Repeated exposure of our skin cells to certain ultraviolet (UV) light produced by the sun can cause breaks and then unnatural links between the nucleotides in the skin cells' DNA. When the DNA is copied, errors occur. If such errors occur in genes that regulate cell division, the normal skin cells may divide uncontrollably, leading to cancer.

Review SECTION 9-2

1. **Name** the three parts of a nucleotide.
2. **Relate** the base-pairing rules to the structure of DNA.
3. **Identify** the two pieces of information that enabled James Watson and Francis Crick to discover the double-helical structure of DNA.
4. **Explain** why the two strands of the double helix are described as "complementary."
5. **SKILL Applying Information** Suppose a strand of DNA has the nucleotide sequence CCAGATTG. What is the nucleotide sequence of the complementary strand?

Review SECTION 9-2 ANSWERS

1. The three parts of a nucleotide are a phosphate group, a sugar, and a nitrogen base. The nitrogen base can be adenine, guanine, cytosine, or thymine.
2. Thymine on one strand of DNA base pairs with adenine on the opposite strand. Cytosine on one strand of DNA base pairs with guanine on the opposite strand. The hydrogen bonding that occurs between the base pairs keeps the two strands of DNA together.
3. Chargaff's data on the amounts of nitrogen bases in different organisms and Franklin and Wilkin's X-ray photos of DNA enabled Watson and Crick to propose the structure of DNA.
4. The two strands are described as complementary (not identical) because each strand provides one member of an obligatory pair.
5. GGTCTAAC

Section 9-2

Visual Strategy

Figure 9-8

Use Figure 9-8 to show students how the structure of DNA can be simplified in a drawing such as the one on the bottom. Ask students to follow one of the strands on the drawing on the top and compare the nitrogen base sequence to the sequence on the simplified diagram on the bottom. Point out that the sequence on both drawings is the same—they must follow the twist of the double helix. Have students verify that Chargaff's base-pairing rules are indeed followed in each DNA molecule. Ask students to identify the deoxyribose and the phosphate groups in each model. *(In the simplified version, the long line above or below the nitrogen base represents the sugar and phosphate groups.)*

3 Assessment Options

Closure — BASIC

Complementary Bases

Draw a sequence of eight bases on the board. Have students copy it onto a piece of paper and draw the base sequence for the complementary strand. PORTFOLIO

Review

Assign Review—Section 9-2.

Reteaching — BASIC

Provide pairs of students index cards with the components of DNA written on them: sugar on one card, phosphate on one card, and each of the four nitrogen bases on four different cards. Each set of cards should include at least eight sugars, eight phosphates, and two of each of the nitrogen bases. Ask students to put the cards in a pattern to demonstrate the structure of DNA. Students should be able to reconstruct a DNA molecule. PORTFOLIO

1 Prepare

Scheduling

- **Block:** About 90 minutes will be needed to complete this section.
- **Traditional:** About two class periods will be needed to complete this section.
- **Planning:** See the Planning Guide on pp. 185A and 185B for lecture, classwork, and assignment options for Section 9-3. Lesson plans for Section 9-3 can be found in a lesson cycle format in ***Biology: Principles and Explorations*** **One-Stop Planner CD-ROM with Test Generator.**

Resource Link

Assign **Active Reading Worksheet Section 9-3** to help students comprehend the key concepts and visuals in this lesson.

2 Teach

Lesson Warm-up

DNA is considered to be a relatively stable molecule. Ask students what gives it this stability even though the hydrogen bonds between the nitrogen bases are easily broken. *(The bonds connecting the deoxyribose and phosphate groups, which form the backbone of DNA, are covalent bonds. Covalent bonds are strong and keep the strands of nucleotides intact.)*

Visual Strategy

Figure 9-9

Use the overhead projector to help students identify the replication fork and helicase enzymes in **Step 1**. Ask students to identify the original DNA strands, the new strands, and the location of the DNA polymerases in **Step 2**. Have students compare the two DNA molecules in **Step 3** with the DNA in Step 1. *(They are identical.)* Then ask students to identify the new strands of DNA in Step 3.

Section 9-3 The Replication of DNA

Objectives

- **Summarize** the process of DNA replication. (pp. 196–197)
- **Describe** how errors are corrected during DNA replication. (p. 197)
- **Compare** the number of replication forks in prokaryotic and eukaryotic DNA. (p. 198)

Key Terms

DNA replication
DNA helicase
replication fork
DNA polymerase

DNA Is Copied with the Help of Many Enzymes

When the double helix structure of DNA was first discovered, scientists were very excited about the complementary relationship between the sequences of nucleotides. They predicted that the complementary structure was used as a basis to make exact copies of the DNA each time a cell divided. Watson and Crick proposed that one DNA strand serves as a template, or pattern, on which the other strand is built. Within five years of the discovery of DNA's structure, scientists had firm evidence that the complementary strands of the double helix do indeed serve as templates for building new DNA.

The process of making a copy of DNA is called **DNA replication.** DNA replication is summarized in **Figure 9-9.** Recall from Chapters 6 and 7 that DNA replication occurs during the synthesis (S) phase of the cell cycle, before a cell divides.

Step 1 Before DNA replication can begin, the double helix unwinds. This is accomplished by enzymes called **DNA helicases,** which open the double helix by breaking the hydrogen bonds that link the complementary nitrogen bases between the two strands. Once

Figure 9-9

DNA Replication

DNA replication results in two identical DNA strands.

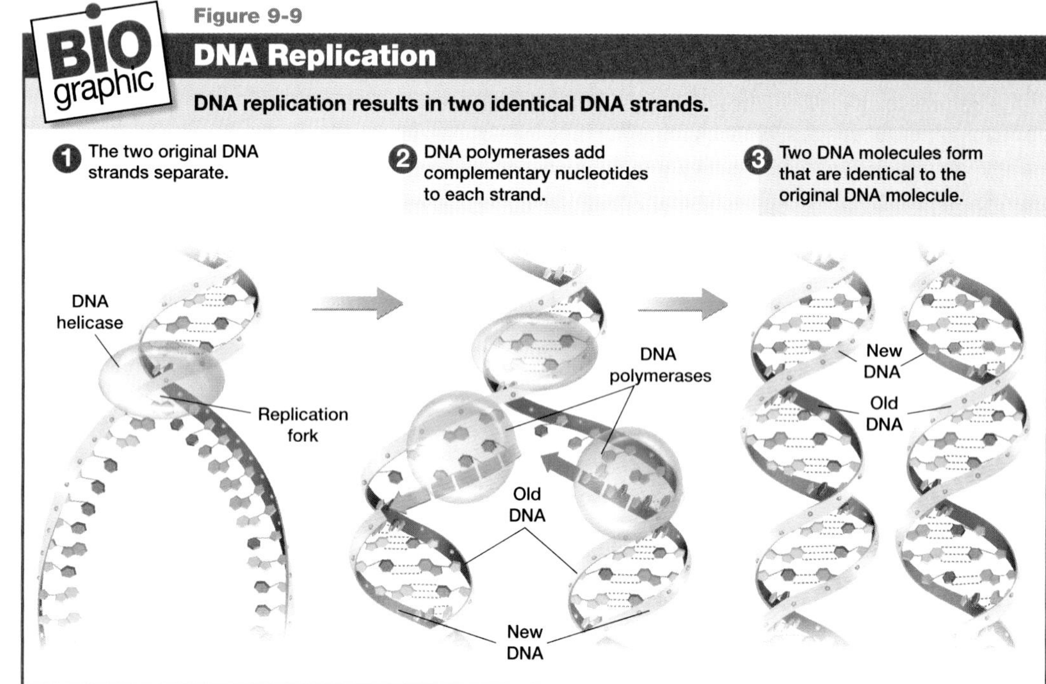
BIO graphic
1 The two original DNA strands separate.
2 DNA polymerases add complementary nucleotides to each strand.
3 Two DNA molecules form that are identical to the original DNA molecule.
DNA helicase
Replication fork
DNA polymerases
Old DNA
New DNA
New DNA
Old DNA

Integrating Different Learning Styles

Logical/Mathematical	**PE**	Math Lab, p. 197
Visual/Spatial	**PE**	Performance Zone items 16, 17, and 18, p. 201; Experimental Design, pp. 202–203
	ATE	Visual Strategy, p. 196; Teaching Tip, p. 197; Reteaching, p. 198; Alternative Assessment, p. 199
Verbal/Linguistic	**PE**	Study Tip, p. 197; Performance Zone items 2, 24, and 26, pp. 200–201
	ATE	Career, p. 198; Closure, p. 198

the two strands are separated, additional proteins attach to each strand, holding them apart and preventing them from twisting back into their double-helical shape. The two areas on either end of the DNA where the double helix separates are called **replication forks** because of their Y shape, as shown in Figure 9-9.

Step ❷ At the replication fork, enzymes known as **DNA polymerases** move along each of the DNA strands, adding nucleotides to the exposed nitrogen bases, according to the base-pairing rules. As the DNA polymerases move along, two new double helixes are formed.

Step ❸ Once a DNA polymerase has begun adding nucleotides to a growing double helix, the enzyme remains attached until all of the DNA has been copied and it is signaled to detach. This process produces two DNA molecules, each composed of a new and an original strand. The nucleotide sequences in both of these DNA molecules are identical to each other and to the original DNA molecule.

Checking for errors

In the course of DNA replication, errors sometimes occur and the wrong nucleotide is added to the new strand. An important feature of DNA replication is that DNA polymerases have a "proofreading" role; they can add nucleotides to a growing strand only if the previous nucleotide is correctly paired to its complementary base. In the event of a mismatched nucleotide, DNA polymerase is capable of backtracking, removing the incorrect nucleotide and replacing it with the correct one. This proofreading prevents most errors in DNA replication. Indeed, only one error per 1 billion nucleotides typically occurs. Recall that these errors, or changes in nucleotide sequence, are called mutations.

Study TIP

Reviewing Information

Using your own words, write two sentences to describe each of the three steps summarizing DNA replication in Figure 9-9.

MATH LAB

Analyzing the rate of DNA replication

Background

Cancer is a disease caused by cells that divide uncontrollably. Scientists studying drugs that prevent cancer often measure the effectiveness of a drug by its effect on DNA replication. During normal DNA replication, nucleotides are added at a rate of about 50 nucleotides per second in mammals and 500 nucleotides per second in bacteria.

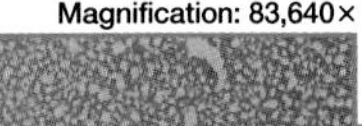
Magnification: 83,640×

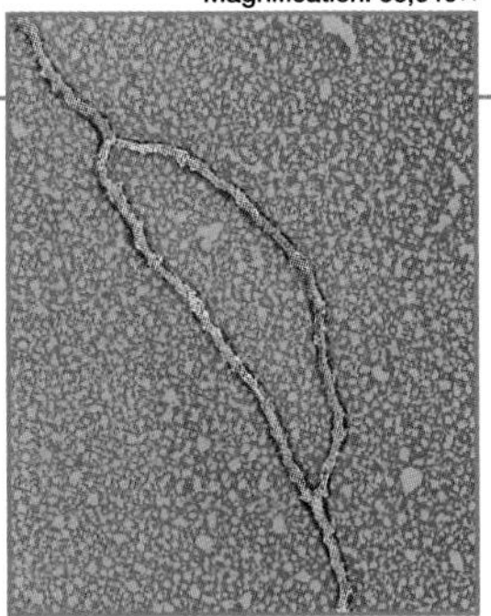
DNA replication forks

Analysis

1. **Calculate** the time it would take a bacterium to add 4,000 nucleotides to one DNA strand undergoing replication.
2. **Calculate** the time it would take a mammalian cell to add 4,000 nucleotides to one DNA strand undergoing replication.
3. **SKILL Predicting Outcomes** How would the total time needed to add the 4,000 nucleotides be affected if a drug that inhibits DNA polymerases was present?

MULTICULTURAL PERSPECTIVE

True Ancestors of Modern Japanese

The Jomon settled Japan more than 10,000 years ago. It appears that the Yayoi people inhabited Japan at a later date. There is a debate about which group of people are the true ancestors of the modern Japanese and where these people originated. By studying and comparing the genes in modern Japanese and the people in different parts of Asia, researchers are beginning to discover the lineage of the Japanese people. Data suggest that the Jomon originated in northern Asia and that both the Jomon and Yayoi peoples have contributed genes to the modern Japanese.

Making Biology Relevant

Health

Ask students to identify where DNA is located in eukaryotic cells. *(the nucleus of a cell)* Point out that mitochondria also contain DNA (the DNA is smaller, circular, and double stranded). Mitochondrial genes code for proteins the mitochondria use during cellular respiration. Researchers have discovered that mutations in mitochondrial DNA can cause certain human disorders, including young-adult blindness (Leber's hereditary optic neuropathy) and certain brain and muscle disorders (such as Alzheimer's disease, dystonia, and Leigh's syndrome). Have students research other human disorders attributed to mutations in mitochondrial DNA. PORTFOLIO

Teaching TIP BASIC

Modeling Replication

Draw a sequence of eight bases on the board. Have students copy it onto a piece of paper and draw the base sequence for the complementary strand. Then, using Figure 9-9 on page 196 as a model, have students separate the two strands by drawing each sequence of eight bases. Finally, have students "replicate" each strand and compare the two new molecules with the original sequence. PORTFOLIO

MATH LAB

Analyzing the rate of DNA replication

Time 10 minutes

Process Skills Calculating, predicting

Teaching Strategies

Remind students that rates are ratios that compare quantities of different units, such as miles per hour.

Analysis Answers

1. 8 seconds
2. 80 seconds
3. The total time would increase considerably.

Career

DNA Researcher

DNA researchers are attempting to determine the nucleotide sequence of the entire human genome and to map the location of every gene on each chromosome by the year 2003 or sooner. This large cooperative effort is called the Human Genome Project. Have students use on-line databases to access information about institutes worldwide that are involved in this project. Have them find out the requirements to become a DNA researcher and the opportunities for employment in international scientific institutes. PORTFOLIO

3 Assessment Options

Closure

Technique: K-W-L

Have students return to their lists of things they Want to know about DNA, which they created as they began reading the chapter. (See **Active Reading,** page 187.) Have them place check marks next to questions that they are able to answer. Students should finish by making a list of what they have Learned. Have students identify unanswered questions, as well as new questions they may have. PORTFOLIO

Review

Assign Review—Section 9-3.

Reteaching — BASIC

Assign students to cooperative groups. Have students build their own DNA models. Let students choose their own construction materials to represent the sugars, phosphates, and different nitrogen bases, as well as the connections between the components of the molecules. Once the molecules are finished, let students demonstrate DNA replication using their models. PORTFOLIO

internetconnect

SciLINKS NSTA
TOPIC: DNA replication
GO TO: www.scilinks.org
KEYWORD: HX198

Multiple Replication Forks Increase the Speed of Replication

Replication does not begin at one end of the DNA molecule and end at the other. The circular DNA molecules found in bacteria usually have two replication forks that begin at a single point. The replication forks move away from each other until they meet on the opposite side of the DNA circle, as shown in **Figure 9-10.**

In eukaryotic cells, each chromosome contains a single, long strand of DNA. The length presents a challenge: The replication of a typical human chromosome with one pair of replication forks spreading from a single point, as occurs in bacteria, would take 33 days! To understand how eukaryotes meet this challenge, imagine that your class has to carry 25 boxes to another building. Carrying one box over, returning, carrying the second box, and so on, would be very slow. It would be much faster if everyone in the class picked up a box so that all of the boxes could be carried in one trip. That is similar to replication in eukaryotic cells, as shown in Figure 9-10. Each human chromosome is replicated in about 100 sections that are 100,000 nucleotides long, each section with its own starting point. With multiple replication forks working in concert, an entire human chromosome can be replicated in about 8 hours.

Figure 9-10 Replication forks

Circular and linear DNA have a different number of replication forks.

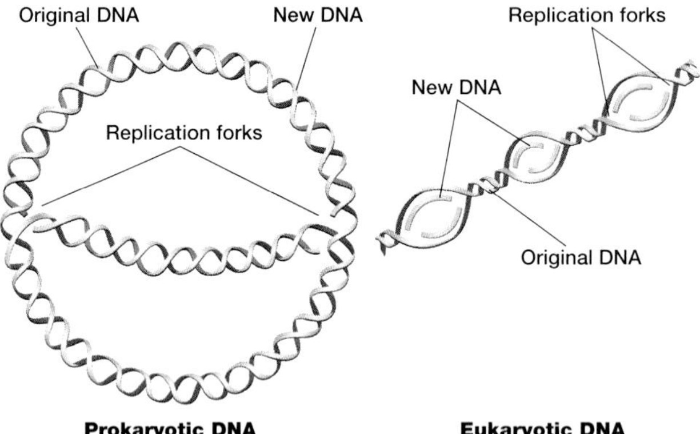

Review SECTION 9-3

1. **Describe** the role of DNA helicases during DNA replication.
2. **Summarize** the two roles that DNA polymerases play in replication.
3. **State** the effect of multiple replication forks on the speed of replication in eukaryotes.

Critical Thinking

4. **SKILL Applying Information** If a mutation occurs during DNA replication in the formation of an egg or sperm cell, why is that mutation inherited? (**Hint:** See Chapter 7, Section 7-1.)

Review SECTION 9-3 ANSWERS

1. DNA helicases are enzymes that open up or unwind the double helix by breaking the hydrogen bonds between the nitrogen bases.
2. DNA polymerases move along each DNA strand and add complementary nucleotides. They also act as a "proofreader" by checking for mismatched nucleotides. If a nucleotide is placed incorrectly, polymerases remove the incorrect nucleotide and replace it with the correct nucleotide.
3. Multiple replication forks increase the speed of replication.
4. The mutation is inherited because the egg and sperm cell unite during fertilization, and the developing zygote inherits the information contained on the chromosomes in the gametes.

Study ZONE — CHAPTER 9

Use the Key Concepts and Key Terms listed below to review the main ideas in this chapter. If you need to review the meaning of a term, turn to the page indicated.

Key Concepts

9-1 Identifying the Genetic Material

- The experiments of Griffith and of Avery yielded results that suggested DNA was the genetic material.
- Hershey and Chase used the bacteriophage T2 and radioactive labels to show that virus genes are made of DNA, not protein.
- DNA stores the information that tells cells which proteins to make and when to make them.

9-2 The Structure of DNA

- DNA is made of two strands of nucleotides twisted into the form of a double helix.
- Each nucleotide in DNA is made of the sugar deoxyribose, a phosphate group, and one of four nitrogen bases. The four nitrogen bases found in DNA nucleotides are adenine (A), thymine (T), guanine (G), and cytosine (C).
- The two strands of DNA are complementary—each A on one strand pairs with a T on the opposite strand, and each G on one strand pairs with a C on the opposite strand.
- Watson and Crick determined the structure of DNA in 1952 with the help of data gathered by Wilkins, Franklin, and Chargaff.

9-3 The Replication of DNA

- Before a cell divides, it copies its DNA by a process called DNA replication.
- In DNA replication, enzymes work to unwind and separate the double helix and add complementary nucleotides to the exposed strands.
- The result of DNA replication is two exact copies of the cell's original DNA. Each new double helix is composed of one original DNA strand and one new DNA strand.
- DNA polymerase proofreads DNA during its replication so that very few errors occur.

Key Terms

9-1

vaccine (188)
virulent (188)
transformation (189)
bacteriophage (190)

9-2

double helix (192)
nucleotide (192)
deoxyribose (192)
base-pairing rules (195)
complementary (195)

9-3

DNA replication (196)
DNA helicase (196)
replication fork (197)
DNA polymerase (197)

Unit 6—*Gene Expression*
Use Topics 1–2 in this unit to review the key concepts and terms in this chapter.

Study TIP Ready?

- *If you think you understand these Key Concepts and Key Terms, then you're ready for the Performance Zone.*

Study ZONE — CHAPTER 9

Review and Practice

Assign the following worksheets for Chapter 9:

- **Student Review Guide**
- **Concept Mapping Worksheet**
- **Test Preparation Pretest**
- **Vocabulary Worksheet**
- **Critical Thinking Review Worksheet**
- **Science Skills Worksheet**

Assessment Options

The following assessment options are available for this chapter:

Resource Link

- Assign **Chapter Test 9** to assess students' understanding of the chapter.

Alternative Assessment

Provide students with three-dimensional objects that represent sugars, phosphate groups, and eight bases. Provide a key for what each object represents. Instruct students to construct a DNA model with the objects. Have them demonstrate replication by separating the bases. Instruct them to draw the model they have made on paper and assign names to the bases. Then have them draw the finished product of replication with the sugar, phosphate, and bases labeled. PORTFOLIO

Portfolio Assessment

- *Portfolio Assessment* at the front of this book provides options for building and evaluating student portfolios. Students and teachers can select items from the areas listed for inclusion in a portfolio.
- See items labeled PORTFOLIO on pp. 190, 191, 194, 195, 197, 198, and 199.

Answer to Concept Map

The following is one possible answer to Performance Zone item 1 on page 200.

DNA
is composed of — nucleotides
is shaped like — double helix
undergoes — replication
makes up — genes
nucleotides — twist into — double helix
nucleotides — made of — five-carbon sugar, phosphate group, nitrogen base
nitrogen base — can be a — purine, pyrimidine
replication — involves — DNA polymerases
DNA polymerases — add — nucleotides

Performance ZONE

CHAPTER 9

REVIEW ANSWERS

Assignment Guide

SECTION	REVIEW
9-1	3, 4, 5, 6, 25
9-2	1, 7, 9, 10, 13, 15, 19, 20, 21, 22, 23
9-3	2, 8, 11, 12, 14, 16, 17, 18, 24, 26

Understanding and Applying Concepts

1. The answer to the concept map is found at the bottom of page 199.
2. Answers will vary. For example: Before replication can begin, DNA helicases unwind the double helix. The unwinding forms replication forks that enable DNA polymerases to attach the correct nucleotides to form two identical DNA molecules.
3. a
4. c
5. b
6. b
7. c
8. b
9. a
10. c
11. a
12. b
13. **a.** about 1:1
    b. A and T, C and G
    c. yes
14. both
15. Genes are made of DNA. Thousands of genes, which are packaged with proteins and coiled up before cell division, make up a chromosome. Chromatids are replicated chromosomes and are formed before a cell undergoes cell division.

Skills Assessment

16. **a.** E
    b. A
    c. C
    d. D
    e. F
17. thymine
18. DNA polymerases add complementary nucleotides to a DNA strand during DNA replication, and they check and correct base-pairing errors during DNA replication.
19. Answers will vary. Techniques may include the Maxam-Gilbert method or the more frequently used Sanger (dideoxy) sequencing method. Fluorescent dyes are now used in the Sanger method to tag the modified nucleotides. Sequencing technology can be used to find mutations that result in genetic disorders.
20. Answers will vary.

Critical Thinking

21. X rays could have caused DNA mutations that led to her cancer.
22. The same number of mouse kidney cells would yield 4.6 picograms of DNA; 2.3 picograms would be extracted from the same amount of mouse sperm. Sperm cells are haploid, *n*, and contain half the amount of genetic material found in body cells.

Performance ZONE — CHAPTER 9 Review

Understanding and Applying Concepts

1. **Concept Mapping** Make a concept map that shows the structure of DNA and how it is copied. Try to include the following words in your map: nucleotides, phosphate group, five-carbon sugar, nitrogen base, purine, pyrimidine, double helix, replication, DNA polymerases, gene.

2. **Understanding Vocabulary** Write a sentence that shows your understanding of each of the following terms: DNA helicase, replication fork, DNA polymerase.

3. In his experiments with *Streptococcus pneumoniae,* Griffith found that
   a. the DNA of the heat-killed *S* bacteria entered some of the *R* bacteria.
   b. the *S* bacteria were transformed.
   c. the capsule did not protect the bacterium.
   d. mice injected with the *R* bacteria died.

4. What is the name of the process that was involved in changing Griffith's *R* bacteria to *S* bacteria?
   a. DNA replication c. transformation
   b. polymerization d. crossing-over

5. Hershey and Chase showed that the genetic material of the T2 bacteriophage was
   a. protein. c. DNA helicase.
   b. DNA. d. DNA polymerase.

6. Hershey and Chase showed that
   a. bacteriophages can infect human cells.
   b. DNA controls heredity.
   c. bacteria undergo transformation.
   d. a vaccine for pneumonia could be produced.

7. A DNA nucleotide does not include a
   a. five-carbon sugar. c. double helix.
   b. phosphate group. d. nitrogen base.

8. DNA is replicated before
   a. crossing-over. c. cell death.
   b. cell division. d. the G_1 phase.

9. James Watson and Francis Crick
   a. built a structural model of DNA.
   b. discovered DNA replication.
   c. used X-ray diffraction.
   d. discovered DNA polymerases.

10. If the sequence of nucleotides on one strand of a DNA molecule is GCCATTG, the sequence on the complementary strand is
    a. GGGTAAG. c. CGGTAAC.
    b. CCCTAAC. d. GCCATTC.

11. The enzymes that add complementary nucleotides during DNA replication and proofread the new DNA strand are called
    a. DNA polymerases. c. DNA helicases.
    b. phages. d. nitrogen bases.

12. Multiple replication forks along the DNA
    a. correct replication errors.
    b. reduce DNA replication time.
    c. ensure that the new and old DNA strands are complementary.
    d. signal DNA polymerase to stop.

13. **Analyzing Data** The table below summarizes the percentage of each nitrogen base found in an organism's DNA.

Percentage of Each Nitrogen Base

	A	T	G	C
Human	30.4	30.1	19.6	19.9
Wheat	27.3	27.1	22.7	22.8
E. coli	24.7	23.6	26.0	25.7

   a. What is the ratio of purines to pyrimidines?
   b. Within each organism, which nucleotides are found in similar percentages?
   c. Do the ratios and percentages in (a) and (b) follow Chargaff's rule?

14. **Theme Reproduction** Does DNA replication occur right before asexual reproduction, sexual reproduction, or both?

15. **Chapter Links** Differentiate between DNA, genes, chromatids, and chromosomes. (**Hint:** See Chapter 6, Section 6-1.)

Skills Assessment

Part 1: Interpreting Graphics

Use the illustration below, which shows the process of DNA replication, to answer the following questions:

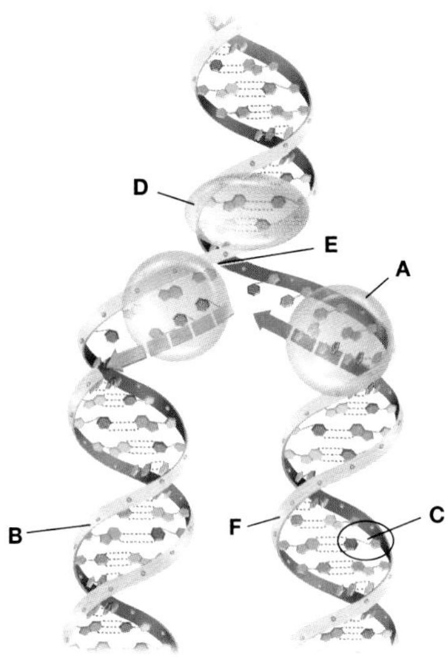

16. Match each of the following to the corresponding letter label in the picture.
a. replication fork
b. DNA polymerase
c. nucleotide
d. DNA helicase
e. newly made DNA strand

17. Which nitrogen base is complementary to adenine?

18. What are two functions of DNA polymerases during DNA replication?

Part 2: Life/Work Skills

Use the media center or Internet resources to investigate different methods used to sequence the nucleotides in a gene.

19. Selecting Technology Research two methods used to sequence the nucleotides in a gene. Compare and contrast the two methods. Give examples of how this technology might be used in a clinical setting.

20. Communicating Prepare a poster to summarize the nucleotide-sequencing methods you researched.

Critical Thinking

21. Forming Reasoned Opinions Rosalind Franklin died of cancer at an early age. How might her work with X-ray diffraction have contributed to her death?

22. Evaluating Results A scientist extracted 4.6 picograms (or 4.6×10^{-12} grams) of DNA from mouse muscle cells. She also extracted 4.6 picograms of DNA from an equal amount of mouse skin cells. How much DNA could be extracted from the same amount of mouse kidney cells? How much DNA could be extracted from the same amount of mouse sperm? Explain.

Portfolio Projects

23. Summarizing Information Research the life of Rosalind Franklin. Discuss, in a written report, the kinds of obstacles that she faced in her career. What kind of discrimination might she have had to overcome?

24. Summarizing Information Read "Crystal Clear: X-ray Snapshots Illuminate How Enzymes Stitch Together DNA" by John Travis (*Science News,* February 14, 1998, vol. 153, pp. 106–107). How is the shape of DNA polymerase analogous to a hand? How will the new images of DNA polymerase help in cancer, HIV, and antibiotic research?

25. Career Focus Molecular Biologist Research the field of molecular biology, and write a report on your findings. Your report should include a job description, training required, kinds of employers, growth prospects, and starting salary.

26. Unit 6—*Gene Expression* Write a report summarizing the use of polymerase chain reaction (PCR) in genetic engineering. Research how PCR analysis is applied in procedures such as DNA fingerprinting and genetic screening.

Performance ZONE

CHAPTER 9

Portfolio Projects

23. Answers will vary. Students may recognize that female researchers faced social obstacles during the time Rosalind Franklin lived.

24. The palm of the hand represents the part of DNA polymerase where the DNA sits. Two fingers "wrap" around the DNA. Many gene mutations arise because of copying errors by a DNA polymerase. Many of these mutations can cause cancer. The new images of DNA polymerase will help researchers to better understand how DNA is copied accurately. The new images also can be used to help create drugs targeted to interfere with HIV's DNA polymerase (reverse transcriptase). Antibiotics can be created based on their effect on DNA polymerase.

25. Molecular biologists study cells at the molecular level. Many molecular biologists participate in DNA research. Most molecular biologists have at least a 4-year degree in one of the fields of science. Many continue in a research-based graduate program. Often they progress to a postdoctoral appointment at a university. Employers include universities, pharmaceutical companies, and health care facilities. The growth potential of this field is good. Starting salary will vary by region.

26. The polymerase chain reaction (PCR) is a technique that allows a small sample of DNA to be copied many times. The DNA is treated with DNA polymerase and with primers, which are small segments of nucleic acid required for the initiation of replication. When these ingredients are incubated, the DNA is amplified (copied repeatedly), resulting in billions of copies of the target gene or genes. Analysis of patterns in the DNA (DNA fingerprinting) can then be performed. Genetic analysis, which determines the presence of mutations on specific genes, can also be performed with PCR-amplified DNA samples.

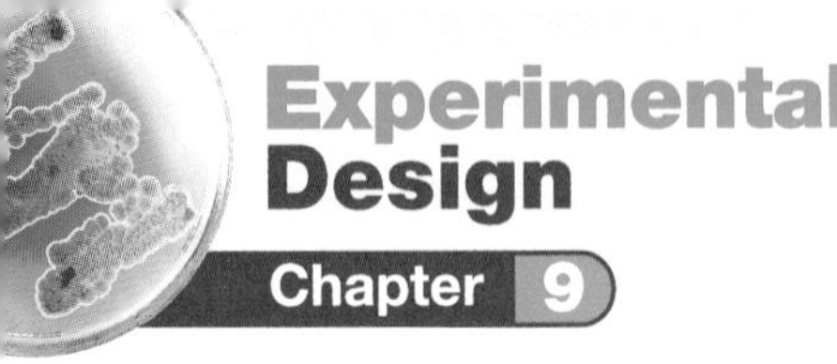

Experimental Design

Chapter 9

How Can You Model DNA Structure?

Purpose:

Students will build a model of DNA to better understand DNA structure, DNA replication, and mutations.

Preparation Notes

Time Required: 40–50 minutes

Materials

Materials for this lab can be ordered from WARD'S. See *Master Materials List* at the front of this book for catalog numbers

Preparation Tips

Obtain enough plastic soda straws to make 48 3-cm sections for each group. Students will also need 48 standard-size paper clips and 48 colored pushpins (start each group with 12 red, 12 blue, 12 yellow, and 12 green pushpins, but let them know they may need more or less of each color as they decide on the sequence of their DNA).

Safety Precautions

Review all safety symbols with students before beginning the lab. Caution students to be careful of the sharp tips of the pushpins. Account for all materials at the end of the lab.

Procedural Tips

- Make sure each group of students has enough materials to complete the activity before they begin.
- If desired, ask students to save the assembled "nucleotides" for use in the lab at the end of Chapter 10.

Disposal

The supplies for this lab can be saved and used again in Chapter 10.

Before You Begin

Answers

1. *DNA*—double-stranded, helical nucleic acid that stores hereditary information

   protein—an organic molecule composed of linked amino acids

   nucleotides—subunits of nucleic acids; consist of a nitrogenous base, a sugar, and a phosphate group

   double helix—a spiral structure characteristic of the DNA molecule

   complementary—characteristic of nucleic acids in which the sequence of bases on one strand determines the sequence of bases on the other

   replication—the process by which DNA in a cell is copied

   mutation—a change in the nucleotide sequence of a gene or chromosome

   replication fork—a Y-shaped structure that results when a double helix of DNA separates so that it can be copied

   base-pairing rules—the pairing arrangement of the nitrogen bases between two DNA strands; adenine on one strand always pairs with a thymine on the opposite strand, and a guanine on one strand always pairs with a cytosine on the opposite strand
2. a nitrogen base, a sugar, a phosphate
3. adenine, guanine, cytosine, and thymine
4. Answers will vary. For example: Will the sequence of nucleotides on the two strands of DNA be identical?

Experimental Design

Chapter 9 How Can You Model DNA Structure?

SKILLS
- Modeling
- Using scientific methods

OBJECTIVES
- **Design** and analyze a model of DNA.
- **Describe** how replication occurs.
- **Predict** the effect of errors during replication.

MATERIALS
- plastic soda straws, 3 cm sections
- metric ruler
- pushpins (red, blue, yellow, and green)
- paper clips

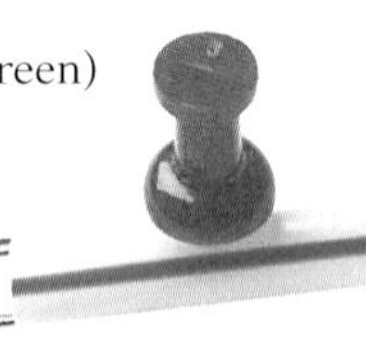

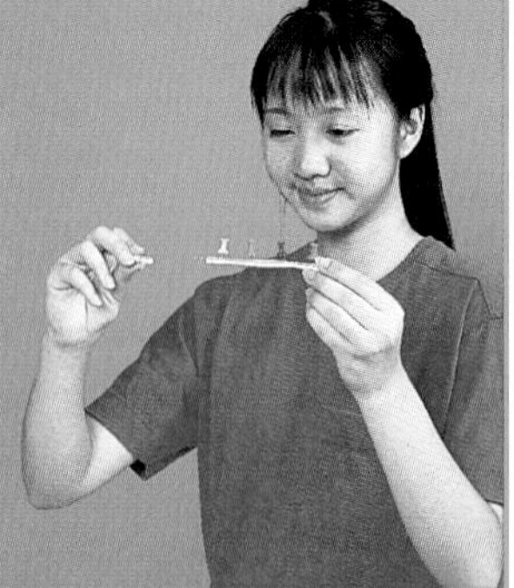

Before You Begin

DNA contains the instructions that cells need in order to make every **protein** required to carry out their activities and to survive. DNA is made of two strands of **nucleotides** twisted around each other in a **double helix.** The two strands are **complementary,** that is, the sequence of bases on one strand determines the sequence of bases on the other strand. The two strands are held together by hydrogen bonds.

In this lab, you will build a model to help you understand the structure of DNA. You can also use the DNA model to illustrate and explore processes such as **replication** and **mutation.**

1. Write a definition for each boldface term in the paragraphs above and for each of the following terms: replication fork, base-pairing rules.
2. Identify the three different components of a nucleotide.
3. Identify the four different nitrogen bases that can be found in DNA nucleotides.
4. Based on the objectives for this lab, write a question you would like to explore about DNA structure.

Procedure

PART A: Design a Model

1. Work with the members of your lab group to design a model of DNA that uses the materials listed for this lab. Be sure that your model has at least 12 nucleotides on each strand.

> **You Choose**
>
> **As you design your model, decide the following:**
>
> a. what question you will explore
>
> b. how to use the straws, pushpins, and paper clips to represent the three components of a nucleotide
>
> c. how to link (bond) the nucleotides together
>
> d. in what order you will place the nucleotides on each strand

2. Write out the plan for building your model. Have your teacher approve the plan before you begin building the model.
3. Build the DNA model your group designed. **CAUTION: Sharp or pointed objects may cause injury. Handle pushpins carefully.** Sketch and label the parts of your DNA model.
4. Use your model to explore one of the questions written for step 4 of **Before You Begin.**

PART B: DNA Replication

5. Discuss with your lab group how the model you built for Part A may be used to illustrate the process of replication.
6. Write a question you would like to explore about replication. Use your model to explore the question you wrote. Sketch and label the steps of replication.

PART C: Test Hypothesis

Answer each of the following questions by writing a hypothesis. Use your model to test each hypothesis, and describe your results.

7. Mitosis follows replication. How might the cells produced by mitosis be affected if nucleotides on one DNA strand were incorrectly paired during replication?
8. What would happen if only one strand in a DNA molecule were copied during replication?

PART D: Cleanup and Disposal

9. Dispose of damaged pushpins in the designated waste container.
10. Clean up your work area and all lab equipment. Return lab equipment to its proper place. Wash your hands thoroughly before you leave the lab and after you finish all work.

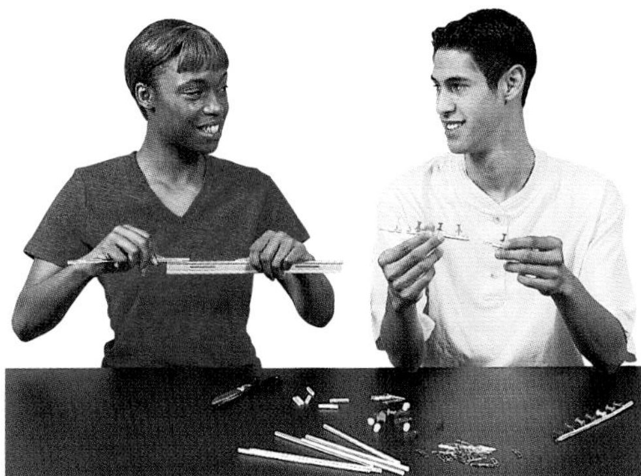

Analyze and Conclude

1. **Analyzing Results** In your original DNA model, were the two strands identical to each other?
2. **Relating Concepts** How does DNA structure ensure that the two DNA molecules made by replication are the same as the original DNA molecule?
3. **Drawing Conclusions** Did the two DNA molecules you made in step 6 have the same nitrogen-base sequence as your original model DNA molecule?
4. **Inferring Relationships** The order of nitrogen bases on a DNA strand is a code for making proteins. What does this mean has happened to the "code" in one of the DNA molecules you made in step 7?
5. **Predicting Outcomes** What would happen if the DNA in a cell that is about to divide were not replicated?
6. **Inferring Information** What are the advantages of having DNA remain in the nucleus of a cell?
7. **Further Inquiry** Write a new question about DNA that could be explored with your model.

Do You Know?

Do research in the library or media center to answer these questions:

1. Are there any pollutants in the environment that disrupt replication when an organism is exposed to the pollutant?
2. How do DNA molecules differ among various species of animals and plants? How are they similar?

Use the following Internet resources to explore your own questions about DNA.

Experimental Design

8. Hypotheses will vary. The resulting DNA molecules would not be identical.

Analyze and Conclude

Answers

1. No, they were complementary.
2. The DNA structure is such that the two strands are complementary to each other.
3. yes
4. The code has been changed.
5. One of the resulting cells would die because it would not have any protein-making instructions.
6. The nucleus provides an isolated and protected area for storing genetic information.
7. Answers will vary. For example: How do DNA molecules differ among various species of animals and plants?

Do You Know?

Answers

1. Answers will vary and may include benzo [a] pyrene and other polyaromatic compounds.
2. The number of chromosomes varies among species, but the relative chromosome number is not an indication of the kingdom to which an organism belongs. Lettuce has 18 chromosomes, while an adder's tongue fern has 1,262 chromosomes. A human has 46 chromosomes, and a sand dollar has 52 chromosomes. The DNA sequences differ between plants and animals. The DNA structure in plants and animals is similar.

Procedure

Answers

3. Answers will vary. Students might use straw segments to model the sugar-phosphate backbone of DNA, different-colored pushpins to represent the nitrogen bases, and paper clips to represent the bonds that hold the nucleotides in a chain. See sketch at right.
5. Answers will vary. See sketch at far right.
6. Answers will vary. For example: What happens to the new DNA strand if the wrong nucleotide is added during replication?
7. Hypotheses will vary. The new cells that result from cell division will not be identical.

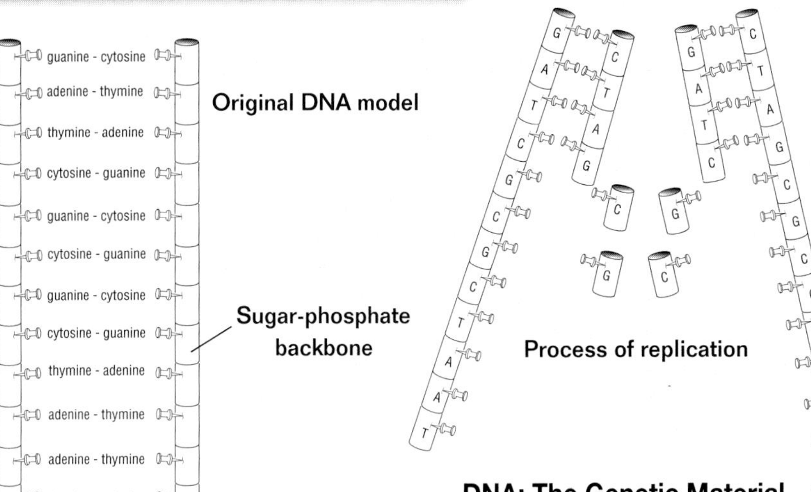

Chapter 10

How Proteins Are Made

Planning Guide

OBJECTIVES	MEETING INDIVIDUAL NEEDS	READING SKILLS AND STRATEGIES
10-1 From Genes to Proteins (BLOCKS 1 & 2 90 min) • **Compare** the structure of RNA with that of DNA. • **Summarize** the process of transcription. • **Relate** the role of codons to the sequence of amino acids that results after translation. • **Outline** the major steps of translation. • **Discuss** the evolutionary significance of the genetic code.	**Learning Styles,** p. 206 BASIC ADVANCED • pp. 208, 210, 211, 212	**ATE** **Active Reading, Technique: K-W-L,** p. 205 **ATE** **Active Reading, Technique: Brainstorming,** p. 208 ■ **Active Reading Guide,** worksheet (10-1) ■ **Directed Reading,** worksheet (10-1) ■ **Reading Strategies,** K-W-L worksheet
10-2 Gene Regulation and Structure (BLOCKS 3 & 4 90 min) • **Relate** how the *lac* operon is turned on or off. • **Summarize** the role of transcription factors in regulating eukaryotic gene expression. • **Describe** how eukaryotic genes are organized. • **Evaluate** three ways that point mutations can alter genetic material.	**Learning Styles,** p. 213 BASIC ADVANCED • pp. 214, 217, 218	■ **Active Reading Guide,** worksheet (10-2) ■ **Directed Reading,** worksheet (10-2) ■ **Supplemental Reading Guide,** *The Double Helix*

Review and Assessment Resources

BLOCKS 5 & 6 90 min

TEST PREP

PE **Study Zone,** p. 219

PE **Performance Zone,** pp. 220–221

■ **Chapter 10 worksheets:**
- Vocabulary
- Science Skills
- Concept Mapping
- Critical Thinking

 **Holt Biology Interactive Tutor** Unit 6—Gene Expression, Topics 3–6

 **Audio CD Program**

■ **Test Prep Pretest**

CHAPTER TESTING

■ **Chapter Tests** (blackline master)

▲ **Test Generator**

■ **Assessment Item Listing**

See the correlations table at the front of this book for a list of the ***National Science Education Standards*** addressed in this chapter.

 Smithsonian Institution® Visit **www.si.edu/hrw** for additional on-line resources.

KEY

Biology Principles and Explorations

PE Pupil's Edition
ATE Teacher's Edition

HOLT BIOSOURCES

Teaching Transparencies
Laboratory Program

One-Stop Planner CD-ROM
Shows these resources within a customizable daily lesson plan:

■ Teaching Resources
▲ Test Generator

VISUAL STRATEGIES	CLASSWORK AND HOMEWORK	TECHNOLOGY AND INTERNET RESOURCES
ATE Fig. 10-2, p. 207 **ATE Fig. 10-5,** p. 210 34 Transcription 35 Translation: Forming the First Peptide Bond 36 Translation: Completing the Protein 37 The Genetic Code	**PE Quick Lab** How does transcription occur?, p. 208 **PE Data Lab** Decoding the genetic code, p. 212 **Lab C12** Analyzing Corn Genetics **Lab E4** Gene Regulation **PE Review,** p. 212	**Interactive Tutor** Unit 6—Gene Expression (Topics 3, 4, 5, and 6) **Holt Biology Videodiscs** Section 10-1 **Audio CD Program** Section 10-1 internetconnect SCiLINKS NSTA **TOPIC:** Genetic code
26A Removing Introns from Eukaryotic Genes 27A Mechanism of the *Lac* Operon 28A Chromosome Mutations	**PE Quick Lab** How can you represent introns and exons?, p. 216 **PE Experimental Design** How Can You Model Protein Synthesis?, pp. 212–213 **Lab B6** Effect of Environment on Gene Expression **Lab B7** Gene Expression **PE Review,** p. 218 **PE Study Zone,** p. 219	CNN **Video Segment 7** Gene Progress **Holt Biology Videodiscs** Section 9-2 **Audio CD Program** Section 9-2 internetconnect SCiLINKS NSTA **TOPIC:** Genetic disorders

ALTERNATIVE ASSESSMENT

PE Portfolio Projects 27–30, p. 221
ATE Alternative Assessment, p. 219
ATE Portfolio Assessment, p. 219

Lab Activity Materials

Quick Lab
p. 208 paper, scissors, pens or pencils (two colors), tape

p. 216 masking tape, pens or pencils (two colors), metric ruler, scissors

Data Lab
p. 212 pencil, paper

Experimental Design
p. 222 masking tape, plastic soda-straw pieces of two different colors, plastic soda-straw pieces of a different color, paper clips, pushpins of five different colors, marking pens of the same colors as the pushpins, 3 × 5 in. note cards, oval-shaped card, transparent tape

Demonstrations
ATE colored paper, scissors, p. 214

BIOSOURCES LAB PROGRAM
- Inquiry Skills Development: Lab B6, p. 21
- Inquiry Skills Development: Lab B7, p. 25
- Laboratory Techniques and Experimental Design: Lab C12, p. 53
- Interactive Explorations in Biology Laboratory Manual: Lab E4, p. 19

Chapter 10

Chapter Themes

Homeostasis

Cells waste energy if they constantly transcribe every gene. Genes often are turned off in a cell unless a protein the genes code for is needed. Feedback systems regulate which genes are turned on and which genes are turned off in a cell. Thus, feedback systems prevent overproduction of unnecessary proteins in the cell.

Cellular Structure and Function

Control of gene expression is required for cells to become specialized. Liver cells differ from brain cells because different genes are expressed within different cells. Gene regulation allows cells to differentiate in the early developmental stages of an organism.

Establishing Prior Knowledge

Before beginning this chapter, make certain students can answer the questions in **Study TIP Ready?** on page 205. If they cannot, encourage them to reread the sections indicated.

Opening Questions

Ask students to name some proteins. Help them by giving clues. *(Their answers might include keratin, hemoglobin, insulin, and enzymes.)* Next ask students where these proteins originated. *(Expect a variety of answers; ultimately lead students to the conclusion that DNA contains the code to make proteins.)*

CHAPTER 10 How Proteins Are Made

Firefly (19×)

What's Ahead — and Why
CHAPTER OUTLINE

10-1 From Genes to Proteins

- Proteins Are Made by Decoding the Information in DNA
- Transcription Transfers Information from DNA to RNA
- The Genetic Code Is Written in Three-Nucleotide "Words"
- Many RNAs Are Used to Make a Protein

10-2 Gene Regulation and Structure

- Protein Synthesis in Prokaryotes Is Controlled by "On-Off" Switches
- The Control of Protein Synthesis in Eukaryotes Is Complex
- Genes in Eukaryotes Often Have Intervening DNA
- Mutations Can Result in a Nonfunctional Protein

Students will learn

- that DNA, RNA, and various enzymes are involved in the processes of transcription and translation.
- how prokaryotic and eukaryotic genes are controlled and that mutations may affect gene expression.

Teen with a light stick

Light sticks and fireflies both *give off light because of a chemical reaction. In a light stick the reaction occurs after the stick is bent. In a firefly the reaction is activated by a protein (enzyme) made by its cells.*

Looking Ahead

10-1 From Genes to Proteins
How is the information in DNA used to make proteins, such as those that make up hair?

10-2 Gene Regulation and Structure
How do cells know when to make a protein and when not to?

Labs

- **Quick Labs**
  How does transcription occur? (p. 208)
  How can you represent introns and exons? (p. 216)
- **Data Lab**
  Decoding the genetic code (p. 212)
- **Experimental Design**
  How Can You Model Protein Synthesis? (p. 222)

Features

- **Exploring Further**
  Jumping Genes (p. 214)
- **Tech Watch**
  Gene Sequencing (p. 218)

Study TIP

Ready?

Check your understanding of these topics to help you prepare for what's ahead.

Can you . . .

- **summarize** the structure and function of proteins? (Section 2-3)
- **describe** the function of ribosomes? (Section 3-2)
- **differentiate** between DNA and genes? (Section 6-1)
- **summarize** the structure and function of DNA? (Section 9-2)
- **state** the base-pairing rules? (Section 9-2)

If not, *review the sections indicated.*

Unit 6—*Gene Expression*
Topics 3–6

National Science Teachers Association *sci*LINKS Internet resources are located throughout this chapter.

Reading Strategies

Active Reading

Technique: K-W-L

Before students read this chapter, have them write short individual lists of all the things they already Know (or think they know) about protein synthesis. When students have finished their lists, ask them to contribute their entries to a group list on the board or overhead. Then have students list the things they Want to know about DNA. Have students save their lists for later use in **Closure** on page 218.

Directed Reading

Assign **Directed Reading Worksheet Chapter 10** to help students interact with the text as they read the chapter.

Interactive Reading

Assign ***Biology: Principles and Explorations*** **Audio CD Program Chapter 10** to help reluctant readers achieve greater success in reading the chapter.

Reading Skills

Assign Reading Organizers and Reading Strategies Worksheets from the **One-Stop Planner CD-ROM with Test Generator** to help students learn and practice effective reading skills.

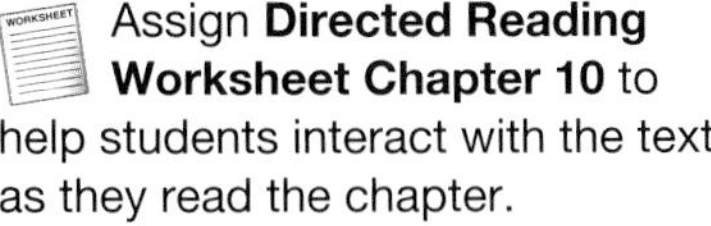

internetconnect

Holt *Biology: Principles and Explorations*

On-line Resources: go.hrw.com

Visit the HRW Web site for a variety of resources related to this chapter. Just type in the keyword HX0 Chapter 10.

Holt *Biology: Principles and Explorations*

On-line Resources: www.scilinks.org

The following *sci*LINKS Internet resources can be found in the student text for this chapter:

Page 209
TOPIC: Genetic code
KEYWORD: HX209

Page 223
TOPIC: Genetic disorders
KEYWORD: HX223

Section 10-1 From Genes to Proteins

1 Prepare

Scheduling

- **Block:** About 90 minutes will be needed to complete this section.
- **Traditional:** About two class periods will be needed to complete this section.
- **Planning:** See the Planning Guide on pp. 203A and 203B for lecture, classwork, and assignment options for Section 10-1. Lesson plans for Section 10-1 can be found in a lesson cycle format in ***Biology: Principles and Explorations*** **One-Stop Planner CD-ROM with Test Generator.**

Resource *Link*

Assign **Active Reading Worksheet Section 10-1** to help students comprehend the key concepts and visuals in this lesson.

2 Teach

Lesson Warm-up

Draw the structures of deoxyribose and ribose on the board. Ask students to compare the two sugars. *(Students should notice that the structures are almost identical except deoxyribose has one less oxygen than ribose.)* Have students correlate the structure of deoxyribose with its name.

Deoxyribose

HOH_2C, O, OH, C, H, H, C, H C—C H, OH, H

Ribose

HOH_2C, O, OH, C, H, H, C, H C—C H, OH, OH

Objectives

- **Compare** the structure of RNA with that of DNA. (p. 206)
- **Summarize** the process of transcription. (pp. 207–208)
- **Relate** the role of codons to the sequence of amino acids that results after translation. (p. 209)
- **Outline** the major steps of translation. (pp. 210–211)
- **Discuss** the evolutionary significance of the genetic code. (p. 212)

Key Terms

ribonucleic acid (RNA)
uracil
transcription
translation
gene expression
RNA polymerase
messenger RNA
codon
genetic code
transfer RNA
anticodon
ribosomal RNA

Proteins Are Made by Decoding the Information in DNA

Traits, such as eye color, are determined by proteins that are built according to instructions specified in an organism's DNA. Recall that proteins have many functions, including acting as enzymes and forming cell membrane channels. How are the instructions in DNA actually carried out? Proteins are not built directly from DNA. Ribonucleic *(reye boh noo KLAY ihk)* acid molecules are also involved.

Like DNA, **ribonucleic acid (RNA)** is a nucleic acid—a molecule made of nucleotides linked together. RNA differs from DNA in three ways. First, RNA consists of a single strand of nucleotides instead of the two strands found in DNA, as shown in **Figure 10-1.** Second, RNA nucleotides contain the five-carbon sugar ribose *(REYE bohs)* rather than the sugar deoxyribose, which is found in DNA nucleotides. Ribose contains one more hydrogen atom than deoxyribose contains. And third, in addition to the A, G, and C nitrogen bases found in DNA, RNA nucleotides can have a nitrogen base called **uracil** *(YUR uh sihl)*—abbreviated as U. No thymine (T) bases are found in RNA. Like thymine, uracil is complementary to adenine whenever RNA base-pairs with another nucleic acid.

A gene's instructions for making a protein are coded in the sequence of nucleotides in the gene. The instructions for making a protein are transferred from a gene to an RNA molecule in a process called **transcription.** Cells then use two different types of RNA to read the instructions on the RNA molecule and put together the amino acids that make up the protein in a process called **translation.** The entire process by which proteins are made based on the information encoded in DNA is called **gene expression,** or protein synthesis. This process is summarized in Figure 10-1.

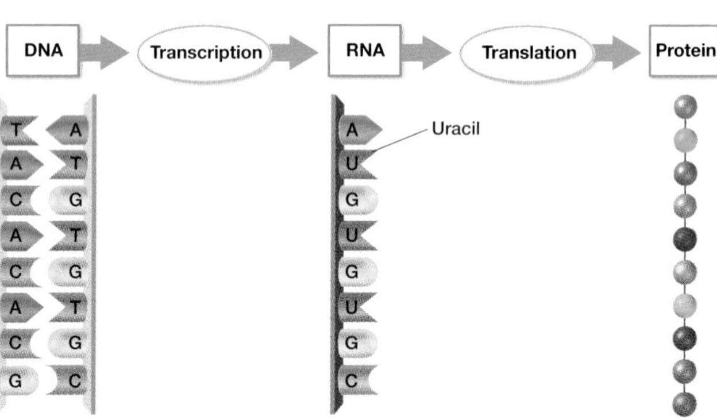

Figure 10-1 Gene expression. The instructions for building a protein are found in a gene and are "rewritten" to a molecule of RNA during transcription. The RNA is then "deciphered" during translation.

Integrating Different Learning Styles

Logical/Mathematical	**PE**	Data Lab, p. 212; Performance Zone item 25, p. 221
Visual/Spatial	**PE**	Quick Lab, p. 208; Study Tip, p. 211; Performance Zone items 1, 13, 18, 19, 20, 21 and 22, pp. 220 and 221; Experimental Design, pp. 222–223
	ATE	Visual Strategies, pp. 207 and 210; Teaching Tips, pp. 207, 208, 211
Musical/Rhythmic	**PE**	Performance Zone item 27, p. 221
Interpersonal	**ATE**	Biology in Your Community, p. 207
Verbal/Linguistic	**PE**	Performance Zone items 23, 24, and 30, p. 221
	ATE	Active Reading, p. 208; Internet Connect, p. 209; Making Biology Relevant, p. 211

Transcription Transfers Information from DNA to RNA

The first step in the making of a protein, transcription, takes the information found in a gene in the DNA and transfers it to a molecule of RNA. **RNA polymerase,** an enzyme that adds and links complementary RNA nucleotides during transcription, is required. **Figure 10-2** summarizes the steps of transcription.

Step 1 Transcription begins when RNA polymerase binds to the gene's promoter—a specific sequence of DNA that acts as a "start" signal for transcription.

Step 2 RNA polymerase then unwinds and separates the two strands of the double helix, exposing the DNA nucleotides.

Step 3 RNA polymerase adds and then links complementary RNA nucleotides as it "reads" the gene. RNA polymerase moves along the nucleotides of the DNA strand that has the gene, much like a train moves along on a track. Transcription follows the base-pairing rules for DNA replication except that uracil, rather than thymine, pairs with adenine.

Transcription proceeds until the RNA polymerase reaches a "stop" signal on the DNA—a sequence of bases that signals the end of transcription. At this point the RNA polymerase detaches from the DNA and releases the RNA molecule for the next stage of gene expression—translation.

biobit

Why are "death cup" mushrooms deadly if eaten?

One of the poisons in death cup mushrooms *(Amanita phalloides)* is taken up by liver cells, where the poison binds tightly to one of the RNA polymerases. This prevents the making of RNA by the liver cells and thus, the making of proteins. Liver failure—and death—can result.

BIO graphic **Figure 10-2**

Transcription: Making RNA

RNA polymerase adds complementary RNA nucleotides as it reads the gene.

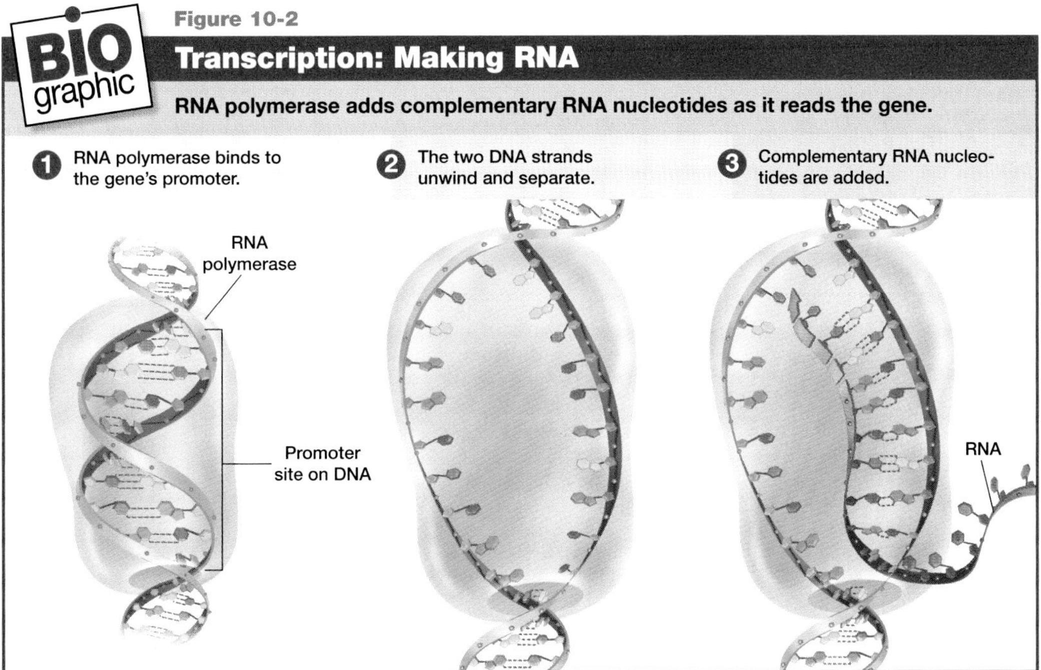

1 RNA polymerase binds to the gene's promoter.
2 The two DNA strands unwind and separate.
3 Complementary RNA nucleotides are added.
RNA polymerase
Promoter site on DNA
RNA

Biology in Your Community

Genetic Disorders

Contact a local laboratory where genetic testing is performed, or contact an organization such as the March of Dimes. Arrange for a speaker to give a presentation on genetic diseases (such as sickle cell anemia) that are the result of point mutations (mutations that change one or just a few nucleotides in a gene). Ask the speaker to discuss the specific point mutation, the gene and chromosome involved, and the consequences of the mutation for each disorder.

Section 10-1

Visual Strategy

Figure 10-2

BIO graphic Use Figure 10-2 to review the steps of transcription. In **Step 1** point out the large size of the RNA polymerase molecule. The RNA polymerase recognizes the gene's promoter site (the sequence that signals where transcription is to begin). In **Step 2** the DNA strands unwind and separate, with the help of RNA polymerase. In **Step 3** the complementary RNA nucleotides are added according to the base-pairing rules, except that a uracil nucleotide is added whenever an adenine occurs on the DNA strand.

Teaching TIP

Gene Expression I

Have students divide a sheet of notebook paper into five columns. The columns are to be labeled with the following headings: *DNA, Complementary DNA, mRNA, tRNA,* and *Amino acid.* In the first column, have students write the following sequence of nucleotides vertically on the page: T A C A A C G G T C T C A G C A C G A T T. An overhead transparency of the same information should be prepared ahead of time. In the second column, have students write the sequence complementary to the first. After a minute or two, have a student write the answer on the overhead transparency. *(A T G T T G C C A G A G T C G T G C T A A)* Next, have the students complete the third column by filling in the bases that would compose the mRNA if the first DNA strand were transcribed. Have another student write the answer on the overhead transparency. *(A U G U U G C C A G A G U C G U G C U A A)* Ask students to name the process by which mRNA is made. *(transcription)* Tell students to save this chart for later use in **Teaching Tip: Gene Expression II** on page 211. PORTFOLIO

Teaching TIP — BASIC

Comparing Transcription and Replication

Have students make a **Graphic Organizer** similar to the one at the bottom of page 208 to demonstrate the differences between transcription and DNA replication. PORTFOLIO

Active Reading

Technique: Brainstorming

Tell students that the first page of Section 10-1, page 206, provides an overview of the contents of the section and hints at the organizational pattern of the section. Pair each student with a partner. Have each pair of students read the first page of the section aloud, sharing the reading equally. Then ask students to answer the following questions:

- What is the subject of this section?
- What passages or words led you to this conclusion?
- How will this section be organized?
- What words or sentences support this conclusion?

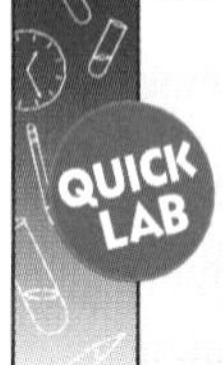

How does transcription occur?

Time 15 minutes

Process Skills Analyzing, predicting

Teaching Strategies

- To save time, prepare the squares for each group prior to the start of class.
- Review the safety symbol with students before starting the lab.

Analysis Answers

1. Different colors are used to distinguish between the sugars in DNA and RNA nucleotides.
2. The mRNA would not be the same as the one constructed in the activity.
3. Students should observe that their second mRNA is different from the first mRNA.

Figure 10-3
Multiple copies of RNA

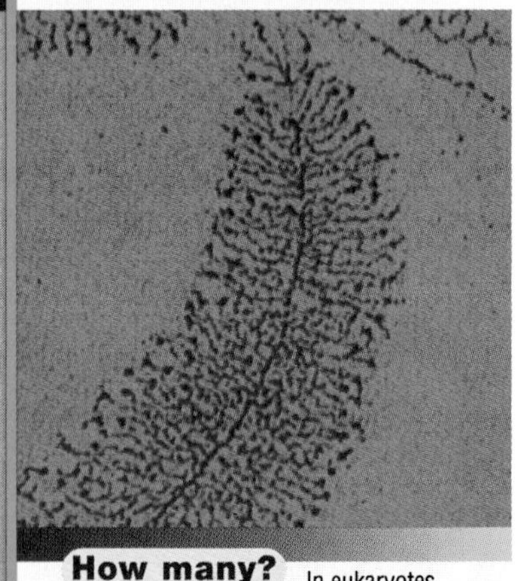

How many? In eukaryotes, RNA polymerase adds about 60 nucleotides per second. There are typically about 100 RNA polymerase molecules per gene.

When the RNA nucleotides are added during transcription, they are linked together with covalent bonds. As RNA polymerase moves down the strand, a single strand of RNA grows. Behind RNA polymerase, the two strands of DNA close up by forming hydrogen bonds between complementary bases, re-forming the DNA double helix.

Like DNA replication, transcription uses DNA nucleotides as a template for making a new molecule. However, in DNA replication, the new molecule made is DNA. In transcription, the new molecule made is RNA. In addition, in DNA replication, both strands of DNA serve as templates, whereas in transcription, only part of one of the two strands of DNA (a gene) serves as a template.

Transcription in prokaryotic cells occurs in the cytoplasm (because prokaryotic cells have no nucleus); transcription in eukaryotic cells occurs in the nucleus, where the DNA is located. During transcription, many identical RNA molecules are made simultaneously from a single gene, as shown in **Figure 10-3.** The RNA being made fans out from the gene to give a "feathery" appearance. The long line along the length of the "feather" is the DNA being transcribed. The circles along the length are the RNA polymerase molecules. The "hairs" on the feather are the RNA chains being made.

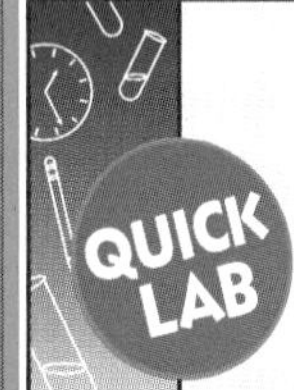

QUICK LAB: *How does transcription occur?*

You can use paper and pens to model the process of transcription.

Materials

paper, scissors, pens or pencils (two colors), tape

Procedure

1. 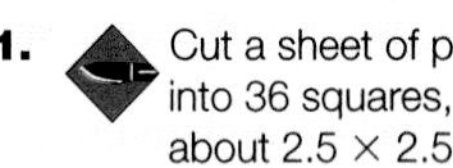 Cut a sheet of paper into 36 squares, each about 2.5 × 2.5 cm (1 × 1 in.) in size.
2. To make one side of your DNA model, line up 12 squares in a column. Using one color, randomly label each square with one of the following letters: A, C, G, or T. Each square represents a DNA nucleotide. Use tape to keep the squares in a column.
3. To make the second side of your DNA model, line up 12 squares next to the first column. Use the same color you used in step 2 to label each square with the complementary DNA nucleotide. Tape the squares together in a column.
4. Separate the two columns. The remaining 12 squares represent RNA nucleotides. Use a different color to "transcribe" one of the DNA strands.

Analysis

1. **Propose** a reason for using different colors for the DNA and RNA "nucleotides."
2. **Predict** how a change in the sequence of nucleotides in a DNA molecule would affect the mRNA transcribed from the DNA molecule.
3. **SKILL Applying Information** Use your model to test your prediction. Describe your results.

Graphic Organizer

*Use this graphic organizer with **Teaching Tip: Comparing Transcription and Replication** on page 208.*

Transcription	DNA replication
RNA polymerase is used.	DNA polymerase is used.
RNA nucleotides are linked.	DNA nucleotides are linked.
An RNA molecule is made.	A DNA molecule is made.
Only one part of one strand (a gene) is used as a template.	Both DNA strands serve as templates.

The Genetic Code Is Written in Three-Nucleotide "Words"

Different types of RNA are made during transcription, depending on the gene being expressed. When a cell needs a particular protein, it is messenger RNA that is made. **Messenger RNA** (mRNA) is a form of RNA that carries the instructions for making a protein from a gene and delivers it to the site of translation. The information is translated from the language of RNA—nucleotides—to the language of proteins—amino acids. The RNA instructions are written as a series of three-nucleotide sequences on the mRNA called **codons** *(KOH dahnz)*. Each codon along the mRNA strand corresponds to an amino acid or signifies a start or stop signal for translation.

In 1961, Marshall Nirenberg, an American biochemist, deciphered the first codon by making artificial mRNA that contained only the base uracil (U). The mRNA was translated into a protein made up entirely of phenylalanine amino-acid subunits. Nirenberg concluded that the codon UUU is the instruction for the amino acid phenylalanine. Later, scientists deciphered the other codons. **Figure 10-4** shows the **genetic code**—the amino acids and "start" and "stop" signals that are coded for by each of the possible 64 mRNA codons.

internetconnect

TOPIC: Genetic code
GO TO: www.scilinks.org
KEYWORD: HX209

Figure 10-4 **Interpreting the genetic code**

The amino acid coded for by a specific mRNA codon can be determined by following the three steps below.

1. Find the first base of the mRNA codon along the left side of the table.
2. Follow that row to the right until you are beneath the second base of the codon.
3. Move up or down in that section until you are even, on the right side of the chart, with the third base of the codon.

Codons in mRNA

First base	Second base U	Second base C	Second base A	Second base G	Third base
U	UUU, UUC – Phenylalanine UUA, UUG – Leucine	UCU, UCC, UCA, UCG – Serine	UAU, UAC – Tyrosine UAA, UAG – Stop	UGU, UGC – Cysteine UGA – Stop UGG – Tryptophan	U C A G
C	CUU, CUC, CUA, CUG – Leucine	CCU, CCC, CCA, CCG – Proline	CAU, CAC – Histidine CAA, CAG – Glutamine	CGU, CGC, CGA, CGG – Arginine	U C A G
A	AUU, AUC, AUA – Isoleucine AUG – Start	ACU, ACC, ACA, ACG – Threonine	AAU, AAC – Asparagine AAA, AAG – Lysine	AGU, AGC – Serine AGA, AGG – Arginine	U C A G
G	GUU, GUC, GUA, GUG – Valine	GCU, GCC, GCA, GCG – Alanine	GAU, GAC – Aspartic Acid GAA, GAG – Glutamic Acid	GGU, GGC, GGA, GGG – Glycine	U C A G

Overcoming Misconceptions

Translation

When given an mRNA strand to translate, many students mistakenly use the anticodons, instead of the codons, to determine the amino acid sequence. Point out to students that the genetic code is based on the codons found on the mRNA and not on the anticodons. The codon sequence is the genetic code (DNA language) rewritten or transcribed in RNA language.

Teaching TIP

Triplet Spelling

Write the four letters *A, E, R,* and *T* on the board. Ask students to write as many meaningful three letter words as they can using any three of the four letters. *(Answers may include* are, art, rat, tar, tea, eat, ate, ear.*)* Continue by asking students to form words that use the letters more than once. *(Answers may include* tat *and* tee.*)* Students should see that many three-letter combinations can be produced. Compare this exercise to the table in Figure 10-4 on page 209. Ask students how the two compare. *(In the exercise, letters are combined, but in the table, nucleotides are combined.)* PORTFOLIO

Making Biology Relevant

Medicine

The prokaryotic ribosome is somewhat smaller than the eukaryotic ribosome. Its protein and RNA content are also dissimilar. These differences between prokaryotic and eukaryotic ribosomes allow selective antibiotics, such as tetracycline and chloramphenicol, to bind to the ribosomes in prokaryotic cells and interfere with prokaryotic protein synthesis. Eukaryotic ribosomes are unaffected. Ask students why these antibiotics can be given safely to humans. *(Protein synthesis in human cells is not affected.)*

internetconnect

SCILINKS NSTA

Have students research information about the genetic code using the Web site in the Internet Connect box on page 209. Have students write a report summarizing their findings. PORTFOLIO

Chapter 10

Teaching TIP ADVANCED

Ribozymes

A substance called peptidyl transferase catalyzes the formation of a peptide bond that joins the polypeptide chain from the P site to the amino acid in the A site. There is evidence to suggest that peptidyl transferase is not a protein, as might be assumed, but an rRNA component that acts as an enzyme. RNA catalysts are known as ribozymes. Ask students what impact this evidence has on previously held beliefs about biological catalysts. *(All biological catalysts were once thought to be protein in nature; ribozymes are nucleic acids.)* Sidney Altman and Thomas R. Cech were awarded the Nobel prize in Chemistry in 1989 for their discovery of the catalytic properties of RNA.

Visual Strategy

BIO graphic Figure 10-5

Work with students to help them summarize the events that take place during translation. Point out the complementarity of the codons on the mRNA and the anticodons on the tRNA, as shown in **Step 1**. Refer students to the chart in Figure 10-4 on page 209 to make the connection between the amino acid the tRNA is carrying and the codon it binds to. Be sure students understand the function of each of the following: mRNA *(carries the instructions for making a protein)*, tRNA *(temporarily carries a specific amino acid on one end)*, and the ribosome *(coordinates the assembly of the protein)*. Some students may find this process confusing. Encourage questions. When you are satisfied that they understand the process, ask students to close their books and write a summary of translation. PORTFOLIO

Many RNAs Are Used to Make a Protein

Translation takes place in the cytoplasm. Here transfer RNA molecules and ribosomes help in the synthesis of proteins. **Transfer RNA** (tRNA) molecules are single strands of RNA that temporarily carry a specific amino acid on one end. Each tRNA is folded into a compact shape and has an anticodon *(an tee KOH dahn)*. An **anticodon** is a three-nucleotide sequence on a tRNA that is complementary to an mRNA codon. As shown in **Figure 10-5,** the amino acid that a tRNA molecule carries corresponds to a particular mRNA codon.

Ribosomes, shown in Figure 10-5, are composed of both proteins and ribosomal RNA (rRNA). **Ribosomal RNA** molecules are RNA molecules that are part of the structure of ribosomes. A cell's cytoplasm contains thousands of ribosomes. Each ribosome temporarily holds one mRNA and two tRNA molecules. Figure 10-5 summarizes the process of translation:

Step 1 Translation begins when the mRNA leaves the nucleus and enters the cytoplasm. The mRNA, the two ribosomal subunits, and a tRNA carrying a modified form of the amino acid methionine *(muh THEYE uh neen)* together form a functional ribosome. The mRNA "start" codon AUG, a codon that signals the beginning of a protein chain, is oriented in a region of the ribosome called the P site, where the tRNA molecule carrying methionine can bind to the start codon.

Figure 10-5

BIO graphic Translation: Assembling Proteins

Amino acids are assembled from information encoded in mRNA.

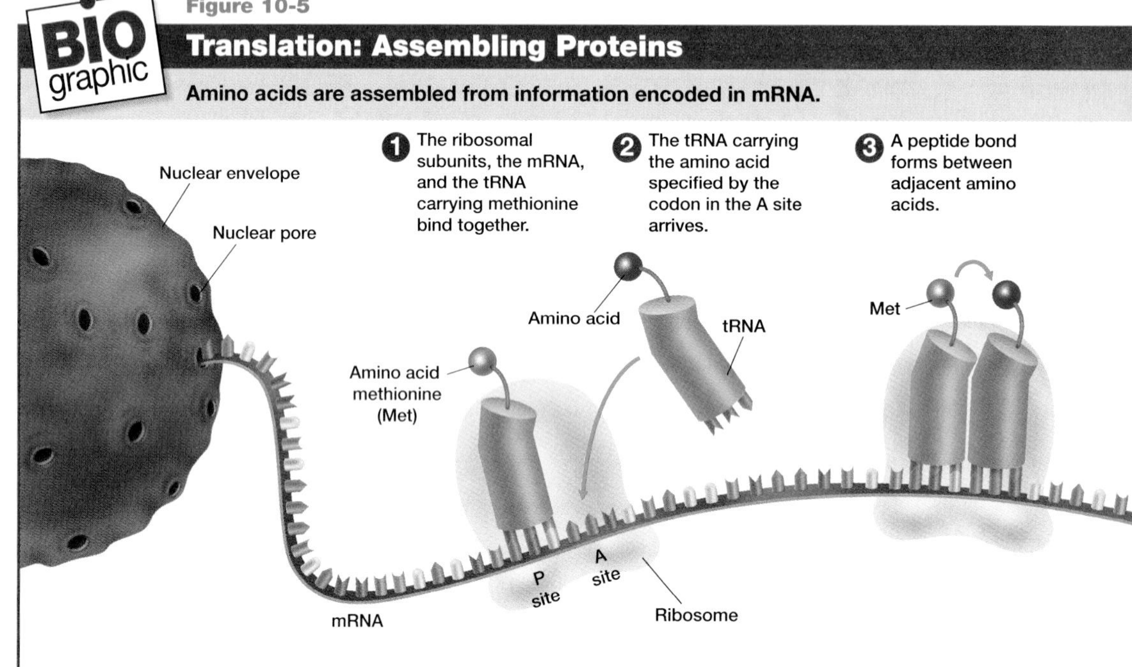

did you know?

In addition to the A and P sites, researchers have discovered a third tRNA-binding site on the ribosome—the E, or exit, site.

When a tRNA molecule detaches, as shown in Step 4 of Figure 10-5 on page 211, the tRNA first moves from the P site to the E site on the ribosome. The tRNA is then released from the ribosome.

Step 2 The codon in the area of the ribosome called the A site is ready to receive the next tRNA. A tRNA molecule with the complementary anticodon arrives and binds to the codon. The tRNA is carrying its specific amino acid.

Step 3 Now both the A site and the P site are holding tRNA molecules, each carrying a specific amino acid. Enzymes then help form a peptide bond between the adjacent amino acids carried by the adjacent tRNA molecules.

Step 4 Afterward, the tRNA in the P site detaches, leaves behind its amino acid, and moves away from the ribosome.

Step 5 The tRNA (with its protein chain) in the A site moves over to fill the empty P site. Because the anticodon remains attached to the codon, the tRNA molecule and mRNA molecule move as a unit. As a result, a new codon is present in the A site, ready to receive the next tRNA and its amino acid. An amino acid is carried to the A site by a tRNA and then bonded to the growing protein chain.

Step 6 The tRNA in the P site detaches and leaves behind its amino acid.

Step 7 Steps 2 through 6 are repeated until a stop codon is reached. A stop codon is one of three codons (UAG, UAA, or UGA) for which there is no tRNA molecule with a complementary anticodon. Since there is no tRNA to fit into the empty A site in the ribosome, no more amino acids can be added to the chain, and protein synthesis stops. The newly made protein is released into the cell.

Study TIP

Interpreting Graphics

Compare Figure 9-8 with Figure 10-5. In this text, the DNA sugar phosphate backbone strands are always colored blue. The RNA backbone is always colored red. The A, G, and C bases are the same color and shape in both DNA and RNA. The T bases (found in DNA but not RNA) are colored red, and the U bases (found in RNA but not in DNA) are colored purple. Remembering this will help you when studying figures with DNA or RNA.

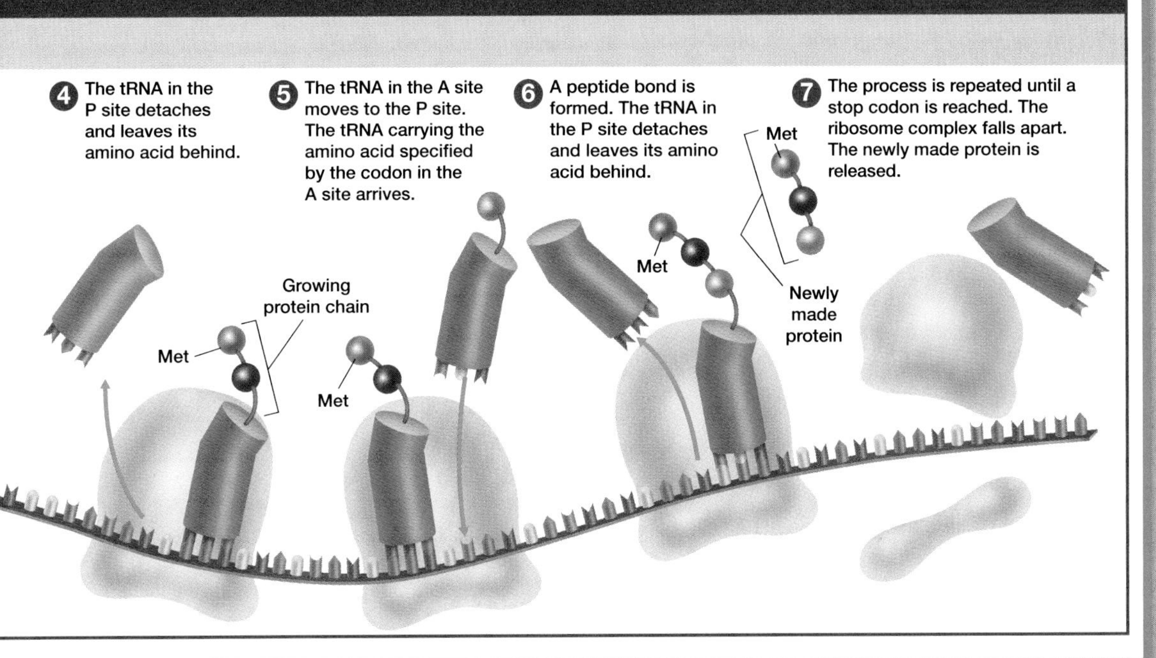

4 The tRNA in the P site detaches and leaves its amino acid behind.
5 The tRNA in the A site moves to the P site. The tRNA carrying the amino acid specified by the codon in the A site arrives.
6 A peptide bond is formed. The tRNA in the P site detaches and leaves its amino acid behind.
7 The process is repeated until a stop codon is reached. The ribosome complex falls apart. The newly made protein is released.
Growing protein chain
Met
Met
Met
Met
Newly made protein

MULTICULTURAL PERSPECTIVE

Albrecht Kossel

Albrecht Kossel was born in Germany in 1853. In 1879 he began investigating "nuclein," which was actually DNA discovered by Friedrich Miescher 10 years earlier. By 1901 he determined the bases in the DNA molecule—adenine, guanine, thymine, and cytosine—were components of nucleic acids. His discovery opened the door for further scientific research in this area, one that would have profound effects. Kossel was awarded the Nobel prize in 1910 for his work on proteins and nucleic acids. He died in 1927.

Making Biology Relevant

Health — ADVANCED

Retroviruses are RNA viruses. They contain RNA rather than DNA as their nucleic acid, and they use RNA as a template to make DNA. To do this, an enzyme called reverse transcriptase is used. HIV is an example of one such virus. Several drugs have been developed to slow the replication of the virus. Some of these drugs include lamivudine and zidovudine. Have students research how these drugs function and have them write a brief report on their findings. *(These drugs inhibit the action of reverse transcriptase. In retroviruses, DNA is made from viral RNA. This DNA is incorporated into the host DNA, which is then transcribed, resulting in the viral mRNA being translated to make specific viral proteins.)* PORTFOLIO

Teaching TIP

Gene Expression II

Refer to **Teaching Tip: Gene Expression I** on page 207, and have students return to the charts they constructed. Ask students to use the chart in Figure 10-4 on page 209 to complete the amino acid column using the mRNA sequence they formed. Explain that the first three bases represent the first triplet, the next three bases represent the second triplet, and so on. Ask them to decipher the code. *(start [methionine]–leucine–proline–glutamic acid–serine–cysteine–stop)* Ask students to fill in the fourth column, tRNA. *(U A C A A C G G U C U C A G C A C G A U U)* Have a volunteer write the answers on an overhead transparency. PORTFOLIO

Decoding the genetic code

Time 15 minutes

Process Skills Recognizing patterns, interpreting information

Teaching Strategies

- Use Figure 10-5 on pp. 210–211 to review the process of translation.
- Make sure students understand the terms *codon* and *anticodon.*
- Point out that because of space limitations, the start and stop codons are not included on the mRNA strand that is shown.

Analysis Answers

1. serine-arginine-glutamic acid-phenylalanine-serine
2. AGA, GCA, CUU, AAA, AGG
3. A G A G C A C T T A A A A G G
4. T C T C G T G A A T T T T C C

3 Assessment Options

Closure

Types of RNA

Ask students to compare the function of mRNA, tRNA, and rRNA. *(mRNA carries the instructions for making a protein; tRNA temporarily carries a specific amino acid on one end; rRNA is part of the ribosome, which coordinates the assembly of the protein)*

Review

Assign Review—Section 10-1.

Reteaching — BASIC

Write the following on eight different pieces of paper: *transcription, translation, DNA, RNA, mRNA, tRNA, codon,* and *anticodon.* Put these pieces of paper in a small container. Write the question *How are they linked?* on the board. Have a student pick two pieces of paper from the container. Show the terms to the class and give them a couple of minutes to write down the answer. Repeat at least 16 times. PORTFOLIO

As the mRNA moves across the ribosome, another ribosome can find the AUG codon on the same mRNA and begin making a second copy of the same protein. In this way many copies of the same protein are made from a single mRNA molecule.

With few exceptions, the genetic code is the same in all organisms. For example, the codon GUC codes for the amino acid valine in bacteria, in eagles, in plants, and in your own cells. For this reason, the genetic code is often described as being nearly universal. It appears that all life-forms have a common evolutionary ancestor with a single genetic code. Some exceptions include the ways cell organelles that contain DNA (such as mitochondria and chloroplasts) and a few microscopic protists read "stop" codons.

DATA LAB: Decoding the genetic code

Background

Keratin is one of the proteins in hair. The gene for keratin is transcribed and translated by certain skin cells. The series of letters below represents the sequence of nucleotides in a portion of an mRNA molecule transcribed from the gene for keratin. This mRNA strand and the genetic code in Figure 10-4 can be used to determine some of the amino acids in keratin.

U C U C G U G A A U U U U C C

Analysis

1. **Determine** the sequence of amino acids that will result from the translation of the segment of mRNA above.
2. **Determine** the anticodon of each tRNA molecule that will bind to each codon in this mRNA segment.
3. **SKILL Recognizing Patterns** Determine the sequence of nucleotides in the segment of DNA from which the mRNA strand above was transcribed.
4. **SKILL Recognizing Patterns** Determine the sequence of nucleotides in the segment of DNA that is complementary to the DNA segment described in item 3.

Review SECTION 10-1

1. **Distinguish** two differences between RNA structure and DNA structure.
2. **Describe** how RNA is made during transcription.
3. **Interpret** the genetic code to determine the amino acid coded for in the codon CCU.
4. **Compare** the roles of the three different types of RNA during translation.
5. **Infer** the evolutionary significance in describing the genetic code as "nearly universal."

Critical Thinking

6. **SKILL Justifying Conclusions** Evaluate the following statement: The term *transcription* is appropriate for describing the process of making RNA, and the term *translation* is appropriate for describing the synthesis of proteins.

Review SECTION 10-1 ANSWERS

1. RNA is single stranded; DNA is double stranded. RNA contains the sugar ribose; DNA contains the sugar deoxyribose; RNA contains uracil but not thymine; DNA contains thymine but not uracil.
2. RNA polymerase binds to the promoter, unwinds and separates the DNA strands, then adds and links complementary RNA nucleotides as it "reads" the gene.
3. proline
4. mRNA carries the instructions for making a protein to a ribosome; rRNA is part of the ribosome; and tRNA brings the specific amino acids to the mRNA on the ribosome.
5. All organisms might have had a common evolutionary ancestor.
6. *Transcription* means "the process of writing out." In transcription, the instructions on the gene are written out as mRNA. *Translation* means "to put into the words of a different language." In translation, the instructions for making a protein are written in the language of nucleic acids—nucleotides. The instructions are translated to the language of proteins—amino acids.

Gene Regulation and Structure

Section 10-2

Protein Synthesis in Prokaryotes Is Controlled by "On-Off" Switches

Although prokaryotic organisms, such as bacteria, might seem simple because of their small size, prokaryotic cells can have up to several thousand genes. Eukaryotic cells can have more than 100,000 genes. However, not all of the genes are transcribed and translated all of the time; this would be wasteful of the cell's energy and materials. Both prokaryotic and eukaryotic cells are able to regulate (control) which genes are expressed and which are not, depending on the cell's needs.

An example of gene regulation that is well understood in prokaryotes is found in the bacterium *Escherichia coli*. When you eat or drink a dairy product, the disaccharide lactose ("milk sugar") reaches the intestinal tract and becomes available to the *E. coli* living there. The bacteria can absorb the lactose and break it down for energy or for making other compounds. In *E. coli*, breaking down lactose into its two components, glucose and galactose, requires three different enzymes, each of which is coded for by a different gene.

As shown in **Figure 10-6,** the three lactose-metabolizing genes are located next to each other and are controlled by the same promoter site. There is an on-off switch that "turns on" (transcribes and

Objectives

- **Describe** how the *lac* operon is turned on or off. (pp. 213–214)
- **Summarize** the role of transcription factors in regulating eukaryotic gene expression. (p. 215)
- **Describe** how eukaryotic genes are organized. (p. 216)
- **Evaluate** three ways that point mutations can alter genetic material. (pp. 217–218)

Key Terms

operator
operon
lac operon
repressor
intron
exon
point mutation

Figure 10-6 Turning prokaryotic genes on and off

The *lac* operon is a cluster of genes that enables a bacterium to build the proteins needed for lactose metabolism only when lactose is present.

Lactose absent: The repressor protein is bound to the operator, and the *lac* operon is switched off.

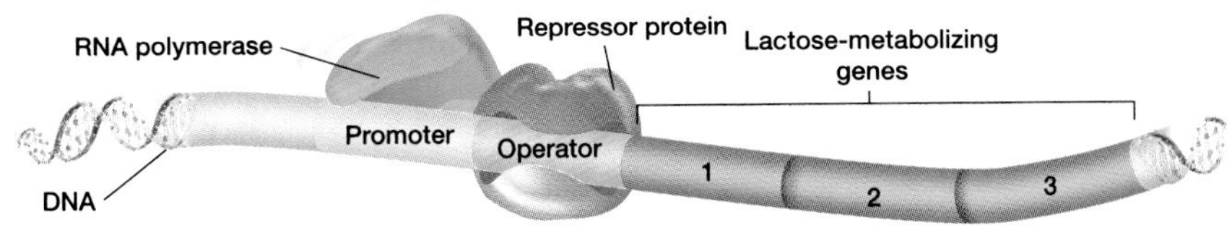

Lactose present: The repressor protein detaches from the operator, and the *lac* operon is switched on.

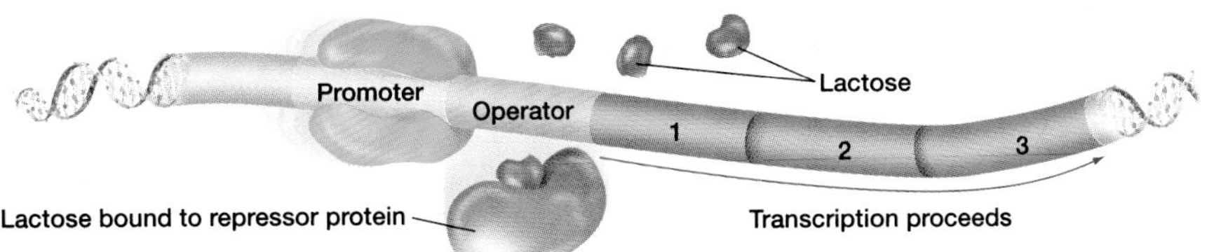

Integrating Different Learning Styles

Logical/Mathematical	**PE**	Performance Zone item 26, p.221
Visual/Spatial	**PE**	Quick Lab, p. 216
	ATE	Demonstration, p. 214; Teaching Tips, pp. 216, 217
Interpersonal	**ATE**	Exploring Further, p. 214; Reteaching, p. 218
Verbal/Linguistic	**PE**	Study Tips, pp. 215, 216; Performance Zone items 2, 28, and 29, pp. 220–221
	ATE	Career, p. 215; Teaching Tip, p. 217; Cross-Disciplinary Connection, p. 217; Closure, p. 218

1 Prepare

Scheduling

- **Block:** About 90 minutes will be needed to complete this section.
- **Traditional:** About two class periods will be needed to complete this section.
- **Planning:** See the Planning Guide on pp. 203A and 203B for lecture, classwork, and assignment options for Section 10-2. Lesson plans for Section 10-2 can be found in a lesson cycle format in ***Biology: Principles and Explorations*** **One-Stop Planner CD-ROM with Test Generator.**

Resource Link

- Assign **Active Reading Worksheet Section 10-2** to help students comprehend the key concepts and visuals in this lesson.

2 Teach

Lesson Warm-up

Ask students to recall the differences and similarities between prokaryotes and eukaryotes. *(Prokaryotes lack most organelles, such as the nucleus, but do contain ribosomes like eukaryotic cells. Both types of cells contain DNA, but prokaryotic DNA is circular. Eukaryotic cells contain much more DNA, and the DNA is linear.)* Ask students how the differences between prokaryotes and eukaryotes might relate to the study of gene expression. *(It is easier to study gene expression in prokaryotes because their structure is simpler.)*

Chapter 10

REAL LIFE

Answer

Answers will vary. The commercial products include over-the-counter lactase tablets, capsules, liquid, and caplets. Lactase-supplemented milk and milk products are also available.

Demonstration

The Operon Model

Cut shapes from colored paper to represent the components of the *lac* operon as shown in Figure 10-6 on page 213. These shapes should be large enough for students to see from their seats. Secure these shapes to the board and use them to demonstrate the process by which genes are turned on and off in the *lac* operon. Emphasize the role of the feedback system in gene regulation.

Exploring Further

Jumping Genes

Teaching Strategies

- Capture the students' interest by telling them that one type of transposon, in which a piece of chromosome 22 breaks off and binds to chromosome 9, has been identified as a cause of leukemia.
- Tell students that transposase, an enzyme encoded by transposons, is responsible for transposition. The movement of transposase to different sites of the genome is random and rare.

Discussion

- What effect do transposons have on other genes? *(Transposons can inactivate the genes they jump into.)*
- McClintock's findings were far ahead of her time and beyond the understanding of her fellow scientists. Assign **Supplemental Reading: A Feeling for the Organism** in Holt BioSources Teaching Resources CD-ROM. The biography of Barbara McClintock will help students understand her as both a scientist and a person. PORTFOLIO

REAL LIFE

Bloating, gas, and diarrhea can occur after eating dairy products.
This occurs in people with lactose intolerance, a shortage of the human digestive enzyme called lactase. Lactase breaks down the lactose in dairy products. Bacteria in the intestine use lactose-metabolizing enzymes to break down undigested lactose, releasing gas in the process.

Finding Information
Research commercial products that enable people with lactose intolerance to consume dairy products without discomfort.

then translates) the three genes when lactose is available and "turns off" the genes when lactose is not available. The piece of DNA that overlaps the promoter site and serves as the on-off switch is called an **operator.** Because of its position, the operator is able to control RNA polymerase's access to the three lactose-metabolizing genes.

In bacteria, a group of genes that code for enzymes involved in the same function, their promoter site, and the operator that controls them all function together as an **operon** *(AHP uhr ahn).* The operon that controls the metabolism of lactose is called the ***lac* operon** and is shown in Figure 10-6.

What determines whether the *lac* operon is in the "on" or "off" mode? When there is no lactose in the bacterial cell, a repressor turns the operon off. A **repressor** is a protein that binds to an operator and physically blocks RNA polymerase from binding to a promoter site. The blocking of RNA polymerase consequently stops the transcription of the genes in the operon, as shown in Figure 10-6.

When lactose is present, the lactose binds to the repressor and changes the shape of the repressor. The change in shape causes the repressor to fall off of the operator, as shown in Figure 10-6. Now RNA polymerase is not blocked and the bacterial cell can begin transcribing the genes that code for the lactose-metabolizing enzymes. By producing the enzymes only when the nutrient is available, the bacterium saves energy.

Exploring Further

Jumping Genes

The spotted and streaked patterns seen in Indian corn result from genes that have moved from one chromosomal location to another. Such genes are called transposons *(trans POH zahns)*. When a transposon jumps to a new location, it often inactivates a gene or causes mutations. In Indian corn, some pigment genes are not expressed in some cells because they have been disrupted by jumping genes.

The discovery of transposons
In the 1950s, the geneticist Barbara McClintock discovered transposons while studying corn. Most scientists rejected her ideas for more than 20 years. The idea that genes could change locations on the chromosome contradicted the prevailing view that genes and chromosomes are stable parts of the cell. Over time, additional research supported her hypothesis, and her model gradually gained acceptance. In 1983, McClintock received a Nobel Prize for her discoveries involving transposons.

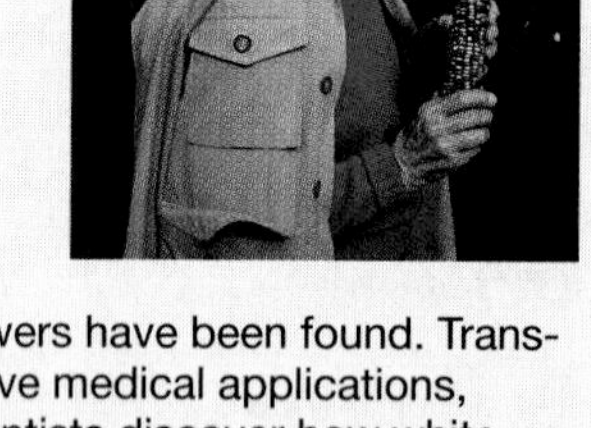

Importance of transposons
All organisms, including humans, appear to have transposons. Transposons probably play a role in spreading genes for antibiotic resistance among bacteria. Transposons that affect flower color in morning glory flowers have been found. Transposons may also have medical applications, such as helping scientists discover how white blood cells make antibodies and what causes cancer.

Although the movement of transposons is very rare, transposons are important because they can cause mutations and bring together different combinations of genes. The transfer of these mobile genes could be a powerful mechanism in evolution and could help solve certain mysteries about evolution, such as how larger organisms developed from single cells and where new species come from.

MULTICULTURAL PERSPECTIVE

The *lac* Operon
Two French scientists, Francois Jacob and Jacques Monod, were responsible for the discovery of operons. Jacob began his studies in the hopes of pursuing a career in medicine. However, he joined the Free French Forces in London during World War II and was severely injured. He received the highest French military decoration for his services and completed his medical studies after the war. But due to his injuries, he was unable to practice medicine. Instead, Jacob turned his attention to biology and in 1950, joined the staff of Institut Pasteur.

Monod developed an interest in biology very early in life, having been influenced by his father who used to read the books of Charles Darwin to him. He received his Ph.D. and eventually joined the staff of Institut Pasteur. By 1960 these two men, working with *E. coli,* were able to determine how genes influence the making or degrading of certain substances—a system referred to as an operon. They were awarded a Nobel prize in 1965.

The Control of Protein Synthesis in Eukaryotes Is Complex

Eukaryotic cells contain much more DNA than prokaryotic cells do. Like prokaryotic cells, eukaryotic cells must continually turn certain genes on and off in response to signals from their environment. Operons have not been found often in eukaryotic cells. Instead, genes with related functions are often scattered on different chromosomes.

Because a nuclear envelope physically separates transcription from translation in a eukaryotic cell, more opportunities exist for regulating gene expression. For example, gene regulation can occur before, during, and after transcription. Gene regulation can also occur after mRNA leaves the nucleus or after translation, when the protein is functional.

Controlling the onset of transcription

Most gene regulation in eukaryotes controls the onset of transcription—when RNA polymerase binds to the beginning of a gene. Like prokaryotes, eukaryotic cells use regulatory proteins. But many more proteins are involved in eukaryotes, and the interactions are more complex. The regulatory proteins in eukaryotes are called transcription factors.

As shown in **Figure 10-7,** transcription factors help arrange RNA polymerases in the correct position on the promoter. A gene can be influenced by many different transcription factors. For example, an enhancer is a sequence of DNA that influences transcription and is located thousands of nucleotide bases away from the promoter. A loop in the DNA may bring the enhancer and its attached transcription factor (called an activator) into contact with the transcription factors and RNA polymerase at the promoter. When the strand of DNA is looped in the correct manner, as shown in Figure 10-7, transcription factors bound to enhancers are able to activate transcription factors bound to the promoter.

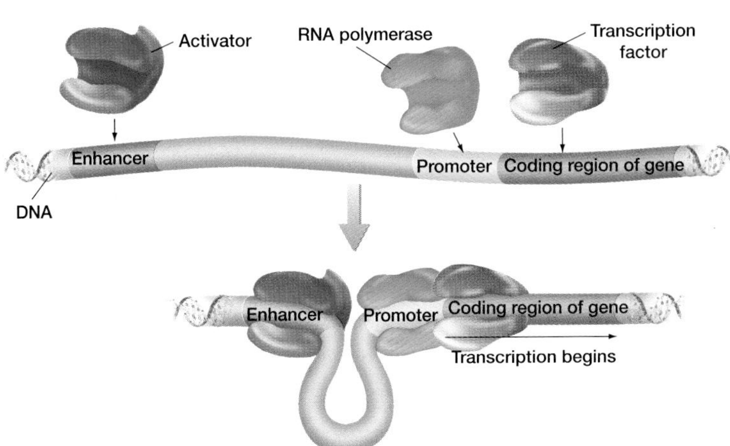

Figure 10-7 Controlling transcription in eukaryotes. Transcription factors bind to the enhancer and to the RNA polymerase. The binding activates transcription factors bound to the promoter.

Study TIP

Organizing Information

Make a table to organize information about the regulation of protein synthesis. Across the top write the headings *Prokaryotes* and *Eukaryotes*. Along the sides write *Protein(s) that regulate(s) the genes* and *Details of regulation.* Add information to the table as you read Section 10-2.

CAREER

Molecular Genetics

Molecular genetics deals with the molecular nature of genes and their role in the function and development of an organism. This information can be used in medicine, botany, and animal studies. Molecular geneticists with a bachelor of science degree can work as laboratory technicians. With advanced degrees, molecular geneticists can design and supervise research projects. Many molecular geneticists work on research projects at universities or for government agencies such as the National Institutes of Health. Agricultural, pharmaceutical, and biotechnological firms also hire molecular geneticists for their skills in genetic engineering techniques. Have students conduct an on-line search of colleges and universities that offer a degree or a concentration in the area of molecular genetics. Students should then prepare a brief report that includes a list of the science courses required for the degree. PORTFOLIO

Teaching TIP

Adult Versus Fetal Genes

Fetal hemoglobin, which differs somewhat from adult hemoglobin, has a higher affinity for oxygen. A short time after birth, fetal hemoglobin is replaced by adult hemoglobin. We do not know what mechanism stops the production of fetal hemoglobin and begins the production of adult hemoglobin. Apparently, the genes for fetal hemoglobin are turned off, and those for adult hemoglobin are turned on just after birth.

did you know?

A typical human cell only expresses 3–5 percent of its genes at any given time.

Also, not all mRNA is translated once it is made. The egg cells of many organisms synthesize and store large amounts of mRNA molecules, which are translated just after fertilization.

Teaching TIP

Introns and Exons

Tell students that two scientists independently found evidence that introns and exons existed. These two scientists, Richard Roberts and Phillip Sharp, shared a Nobel Prize in 1993 for this discovery. We now know that the cutting and pasting of introns and exons is carried out by a complex of proteins and RNA molecules together called snRNPs (pronounced "snurps"), or small ribonucleoproteins.

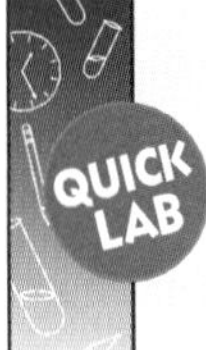

How can you represent introns and exons?

QUICK LAB

Time 10–15 minutes

Process Skills Applying information, predicting outcomes

Teaching Strategies

- Review the terms intron and exon with the students.
- Ask students to explain the graphic before beginning the lab.

Analysis Answers

1. The strip with the letters *appriae-lyjoed* represents the introns. The strip with the letters that spell *protein* represent the exons.
2. Answers will vary. Because the function of a protein is ultimately a result of its amino acid sequence, a protein with additional amino acids will most likely not function.

Study TIP

Word Origins

The "int" in the word *intron* comes from the "int" in the word *intervening*. The "ex" in the word *exon* comes from the "ex" in the word *expressed*.

Genes in Eukaryotes Often Have Intervening DNA

While it is tempting to think of a gene as an unbroken stretch of nucleotides that code for a protein, this simple arrangement is usually found only in prokaryotes. In eukaryotes, many genes are interrupted by **introns** *(IHN trahnz)*—long segments of nucleotides that have no coding information. **Exons** (*EHK sahnz*) are the portions of a gene that are translated (expressed) into proteins. After a eukaryotic gene is transcribed, the introns in the resulting mRNA are cut out by enzymes. The exons that remain are "stitched" back together to form a smaller mRNA molecule that is then translated.

Many biologists think this organization of genes adds evolutionary flexibility. Each exon encodes a different part of a protein. By having introns and exons, cells can occasionally shuffle exons between genes and make new genes. The thousands of proteins that occur in human cells appear to have arisen as combinations of only a few thousand exons. Some genes in your cells exist in multiple copies, in clusters of as few as three or as many as several hundred. For example, your cells each contain 12 different hemoglobin genes, all of which arose as duplicates of one ancestral hemoglobin gene.

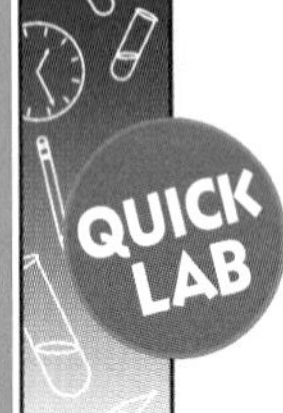

How can you represent introns and exons?

QUICK LAB

You can use masking tape to model introns and exons.

Materials

masking tape, pens or pencils (two colors), metric ruler, scissors

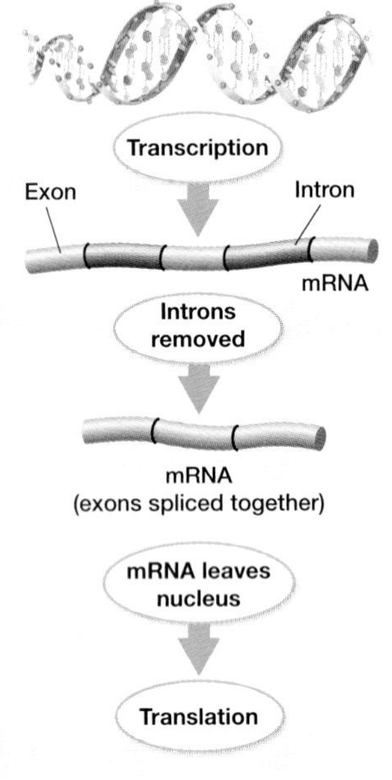

Procedure

1. Place a 15–20 cm strip of masking tape on your desk. The tape represents a gene.
2. Use two colors to write the words *appropriately joined* on the tape exactly as shown in the diagram below. Space the letters so that they take up the entire length of the strip of tape. The segments in one color represent introns; those in the other color represent exons.

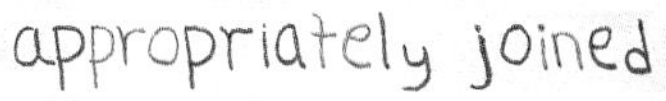

appropriately joined

3. Lift the tape. Working from left to right, cut apart the groups of letters written in the same color. Stick the pieces of tape to your desk as you cut them, making two strips according to color and joining the pieces in their original order.

Analysis

1. **Determine** from the resulting two strips which strip is made of "introns" and which is made of "exons."
2. **SKILL Predicting Outcomes** Predict what might happen to a protein if an intron were not removed.

did you know?

In eukaryotic cells, mRNA is modified as the transcription process occurs.

At one end of the mRNA molecule, a cap consisting of a nucleotide called 7-methylguanylate attaches to the molecule. At the other end of mRNA, a poly A tail, consisting of many adenines, attaches. Scientists think that the cap prevents the mRNA from degradation, making it more stable than prokaryotic DNA. In a eukaryotic cell, mRNA can persist for hours, and sometimes for days or weeks. Compare this to two minutes in a prokaryotic cell. The function of the poly A tail is unknown, although it is thought to aid in the export of mRNA from the nucleus and prevent degradation in the cytoplasm.

Mutations Can Result in a Nonfunctional Protein

Although changes in an organism's hereditary information are rare, they can occur. As you learned in Chapter 6, a change in the DNA of a gene is called a mutation. Mutations in gametes can be passed on to offspring of the affected individual, but mutations in body cells affect only the individual in which they occur.

Mutations that change one or just a few nucleotides in a gene on a chromosome are called **point mutations,** as shown in **Figure 10-8.** There are two general types of point mutations. In the first type, one nucleotide in a gene is replaced with a different nucleotide. Such a point mutation is called a substitution. For example, consider a gene containing the nucleotide sequence ACA. After transcription, the codon would be UGU, which codes for the amino acid cysteine. If a substitution mutation alters the ACA sequence to ACT, after transcription the codon would be UGA. Because the codon UGA codes for a stop codon, translation would end prematurely.

A point mutation can result in a protein not being made at all or a protein with a different amino acid so that the protein does not function. Sometimes substitutions have little or no effect. For example, if a mutation results in a change from UGU to UGC, both codons translate into the same amino acid—cysteine—and there would be no loss of function.

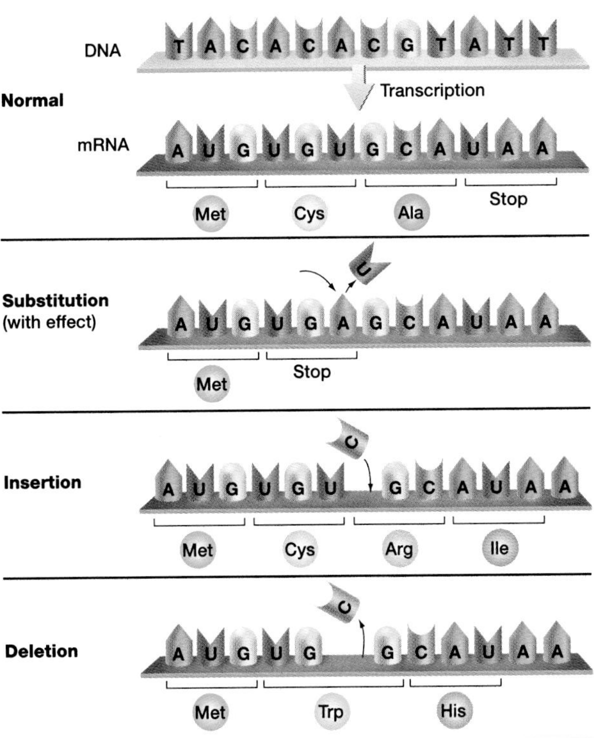

Figure 10-8 Point mutations. The substitution, addition, or removal of a nucleotide is called a point mutation. If the mutation changes the original amino acid sequence in the protein, the protein may not function normally.

did you know?

Siblings and their mother share the same mitochondrial DNA.

Mothers and their offspring have identical mitochondrial DNA because sperm mitochondria are destroyed in the developing zygote. Mutations rarely occur in mitochondrial DNA. Families can be traced by this DNA because it remains essentially unchanged from generation to generation.

Teaching TIP

Mutations

Refer to the DNA nucleotides given in the **Teaching Tip: Gene Expression II** on page 207 and have students return to their charts. Give students large pieces of construction paper. Ask them to take the DNA strand and substitute a nucleotide. Then have students complete the processes of transcription and translation, comparing the new protein with the original. Have them repeat this exercise two more times with an addition of a nucleotide and a deletion of a nucleotide. Display the papers around the room so that all students can see how the random mutations affected the original sequence of amino acids. PORTFOLIO

Cross-Disciplinary Connection

◆ History

In 1986, there was an explosion of a nuclear reactor at the Chernobyl power plant, in Ukraine. Scientists predicted that the residents of the surrounding towns would develop high rates of thyroid cancer and leukemia as a result of radiation exposure. Have students read articles about the effects of the explosion and then discuss their findings in brief reports that link the elevated incidence of cancer to radiation damage to DNA. PORTFOLIO

Teaching TIP — Advanced

Mutagens

Many environmental factors (such as X rays and gamma rays) and chemicals can cause mutations. Have pairs of students research a specific mutagen. Their report should include the type of mutation caused and the ill effects of the mutation on the human body. PORTFOLIO

TECH WATCH

Gene Sequencing

Teaching Strategies

- Tell students about the Human Genome Project—a collaboration between many scientists around the world. The project has many goals, including trying to sequence all 3 billion nucleotides in the human genome.
- Tell students that DNA sequencing is made possible with the help of enzymes (restriction enzymes), which cut DNA into pieces.

Discussion

DNA sequences are collected in computer databases. What benefit does this provide to researchers? *(Long sequences can be scanned to find particular genes or promoters.)*

3 Assessment Options

Closure

Technique: K-W-L

Have students return to their lists of things they Want to know about protein synthesis, which they created as they started to read this chapter (See **Active Reading** on page 205.) Have them place check marks next to questions that they are now able to answer. Students should finish by making a list of what they have Learned. Have students identify unanswered questions, as well as new questions, they may have. PORTFOLIO

Review

Assign Review—Section 10-2.

Reteaching — BASIC

Divide the class into groups of three. Assign each group one of the following topics: the *lac* operon, eukaryotic gene expression, and mutations. Have each group prepare a 10-minute presentation that summarizes their topic and prepare a four-question quiz. Read the quiz questions to the class, and have the students answer them in writing. PORTFOLIO

In a second type of point mutation, insertions and deletions, one or more nucleotides are added to or deleted from a gene. Because the genetic message is read as a series of triplet nucleotides, insertions and deletions of one or two nucleotides can upset the triplet groupings. Imagine deleting the letter C from the sentence "THE CAT ATE." Keeping the triplet groupings, the message would read "THE ATA TE," which is meaningless. A mutation that causes a gene to be read in the wrong three-nucleotide sequence is called a frameshift mutation because the reading pattern is displaced one or two positions.

TECH WATCH Gene Sequencing

Many genetic disorders, such as sickle cell anemia, are caused by single nucleotide mutations. Today, certain genetic disorders can be detected by comparing the sequence of nucleotides in the genes involved to the sequence in corresponding healthy genes.

Special nucleotides

One technique to find the sequence of nucleotides in a gene uses nucleotides that each have a different colored fluorescent dye "tagged" on. The tagged nucleotides are added to a test tube containing single strands of the gene of interest, "untagged" nucleotides, enzymes needed to make DNA, and small single-stranded pieces of DNA called primers.

The primers base-pair to the single strands of the gene of interest. The tagged and untagged nucleotides compete to make the primer longer by matching the nucleotides on the gene of interest, according to the base-pairing rules. The tagged nucleotides are altered so that once a tagged nucleotide is added on to the primer, the synthesis reaction stops on that primer strand.

Base-pairing rules help

The researcher separates the different-sized strands using a method called gel electrophoresis. The different fluorescent dyes help the researcher determine the sequence of the nucleotides on the gene. Using the base-pairing rules, the researcher then figures out the sequence of the gene of interest.

Sequence information is important

Today, the sequence of nucleotides in genes from many different organisms are kept on databases on the Internet. Scientists use this information to look for similarities to the gene they are studying. Comparing the sequences can help them find genes with similar functions, help them classify organisms, and determine evolutionary relationships.

Review SECTION 10-2

1. **Describe** the effect a repressor has on the *lac* operon when lactose is present.
2. **Explain** the role of transcription factors and enhancers in eukaryotic gene expression.
3. **Differentiate** between exons and introns.
4. **SKILL Predicting Outcomes** Which type of mutation would have a greater effect on the sequence of amino acids in a protein, a base-pair substitution or a frameshift mutation? Explain.

Review SECTION 10-2 ANSWERS

1. Lactose binds to the repressor, which changes the shape of the repressor and causes the repressor to fall off of the operator. RNA polymerase is able to bind, allowing transcription to occur.
2. Transcription factors are regulatory proteins. Some help to arrange RNA polymerase in the correct position on the promoter, while others, called activators, bind to an enhancer (a segment of DNA). Transcription begins when the activator bound to the enhancer comes in contact with the transcription factor and RNA polymerase at the promoter.
3. Exons are portions of a eukaryotic gene that are translated into proteins. Introns are portions of a eukaryotic gene that contain no coding information and are cut out before translation occurs.
4. A frameshift mutation would have the greater effect because the insertion or deletion of a nucleotide causes a disruption in the triplet groupings, which results in the incorrect sequence of amino acids. A base-pair substitution may or may not change the amino acid coded for in the triplet.

Study ZONE

CHAPTER 10

Use the Key Concepts and Key Terms listed below to review the main ideas in this chapter. If you need to review the meaning of a term, turn to the page indicated.

Key Concepts

10-1 From Genes to Proteins

- The instructions needed to make proteins are coded in the nucleotides that make up a gene. The instructions are transferred to an mRNA molecule during transcription. The RNA is complementary to the gene, and the RNA nucleotides are put together with the help of RNA polymerase.
- During translation, the mRNA molecule binds to a ribosome, and tRNAs carry amino acids to the ribosome according to the codons on the mRNA. Each codon specifies an amino acid. The amino acids are joined to form a protein.
- The genetic code (codons) used by most organisms to translate mRNA is nearly universal.

10-2 Gene Regulation and Structure

- Prokaryotic and eukaryotic cells are able to control which genes are expressed and which are not, depending on the cell's needs.
- In prokaryotes, gene expression is regulated by operons. Gene expression is switched off when repressor proteins block RNA polymerase from transcribing a gene.
- In eukaryotes, an enhancer must be activated for a eukaryotic gene to be expressed. Transcription factors initiate transcription by binding to enhancers and to RNA polymerases.
- Many eukaryotic genes are interrupted by segments of DNA that do not code for proteins; these segments are called introns. The segments of DNA that are expressed are called exons. After transcription, the introns are cut out, and the exons are joined. The exons are then translated.
- Mutations are changes in DNA. Point mutations alter one or just a few nucleotides in a gene.

Key Terms

10-1

ribonucleic acid (RNA) (206)
uracil (206)
transcription (206)
translation (206)
gene expression (206)
RNA polymerase (207)
messenger RNA (209)
codon (209)
genetic code (209)
transfer RNA (210)
anticodon (210)
ribosomal RNA (210)

10-2

operator (214)
operon (214)
lac operon (214)
repressor (214)
intron (216)
exon (216)
point mutation (217)

INTERACTIVE TUTOR BIOLOGY Unit 6—*Gene Expression*
Use Topics 3–6 in this unit to review the key concepts and terms in this chapter.

Study TIP Ready?

- *If you think you understand these Key Concepts and Key Terms, then you're ready for the Performance Zone.*

Answer to Concept Map

The following is one possible answer to Performance Zone item 1 on page 220.

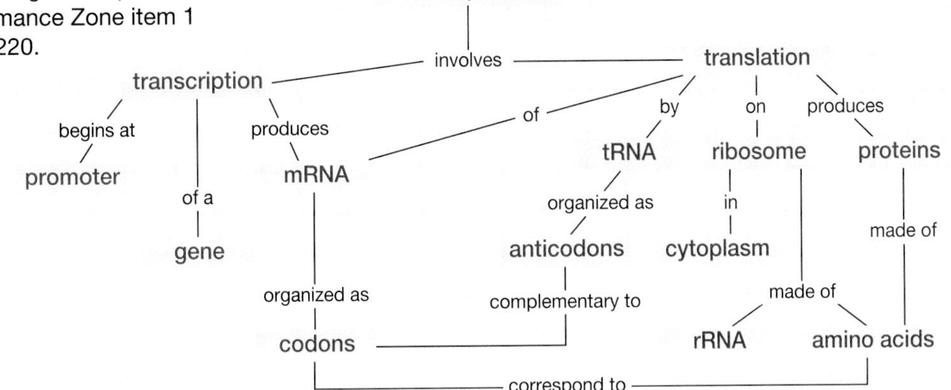

Study ZONE

CHAPTER 10

Review and Practice

Assign the following worksheets for Chapter 10:

- **Student Review Guide**
- **Concept Mapping Worksheet**
- **Test Preparation Pretest**
- **Vocabulary Worksheet**
- **Critical Thinking Review Worksheet**
- **Science Skills Worksheet**

Assessment Options

The following assessment options are available for this chapter:

Resource Link

- Assign **Chapter Test 10** to assess students' understanding of the chapter.

Alternative Assessment

Have students work in groups of four. Provide each group with a large poster board. Ask each group to create its own DNA sequence by writing the names of 24 bases on the poster. The first three bases must be T A G, and the last three bases must be either A T T, A T C, or A C T. The bases in between should be random but when reading the bases as triplets, should not be any of the four triplets indicated above. Have each group determine the mRNA sequence that would form and the protein that would be produced. Next have the group choose a point mutation—substitution, insertion, or deletion—and "mutate" its DNA, and determine how the mutation affected the protein. PORTFOLIO

Portfolio Assessment

- *Portfolio Assessment* at the front of this book provides options for building and evaluating student portfolios. Students and teachers can select items from the areas listed for inclusion in a portfolio.
- See items labeled PORTFOLIO on pp. 207, 208, 209, 210, 211, 212, 214, 215, 217, 218, and 219.

Performance ZONE
CHAPTER 10 REVIEW ANSWERS

Assignment Guide

SECTION	REVIEW
10-1	1, 2, 3, 4, 5, 6, 7, 13, 18, 19, 20, 21, 22, 23, 24, 25, 27, 30
10-2	8, 9, 10, 11, 12, 14, 15, 16, 17, 26, 28, 29

Understanding and Applying Concepts

1. The answer to the concept map is found at the bottom of page 219.
2. **a.** A codon is a three-nucleotide sequence on the mRNA that codes for a specific amino acid or a start or stop codon. An anticodon is a three-nucleotide sequence on tRNA that is complementary to a codon on the mRNA.

   b. The instructions for making a protein that are encoded on a gene are rewritten to a molecule of mRNA. The tRNA molecules carry specific amino acids to the ribosomes; rRNA is a component of ribosomes, where proteins are made.

   c. A promoter is a sequence of DNA that signals the start of transcription. In prokaryotes, the operator is a piece of DNA that overlaps the promoter and acts as an on-off switch. An operon is a group of prokaryotic genes involved in the same function, together with their promoter and operator. A repressor is a protein involved in regulating prokaryotic gene expression—it binds to an operator and blocks RNA polymerase from binding to a promoter.

   d. Exons are portions of a eukaryotic gene that are translated into a protein. Introns are the noncoding regions of a eukaryotic gene that are removed before translation.

3. b
4. c
5. d
6. b
7. d
8. c
9. a
10. b
11. a
12. d

Performance ZONE CHAPTER 10 Review

Understanding and Applying Concepts

1. **Concept Mapping** Make a concept map that shows the role of RNA in gene expression. Try to include the following words in your map: transcription, translation, mRNA, tRNA, rRNA, gene, promoter, codons, anticodons, proteins, amino acids, ribosome, cytoplasm.
2. **Understanding Vocabulary** For each set of terms, write one or more sentences summarizing information learned in this chapter.
   a. codon, anticodon
   b. mRNA, tRNA, rRNA
   c. promoter, operator, operon, repressor
   d. exon, intron
3. RNA differs from DNA in that RNA
   a. has the sugar deoxyribose.
   b. has the nitrogen base uracil.
   c. has a phosphate group.
   d. has nucleotides.
4. The making of RNA based on the sequence of nucleotides in DNA is called
   a. DNA replication. c. transcription.
   b. translation. d. gene regulation.
5. A short chain of DNA has the nucleotide sequence ATA CCG. Its complementary mRNA nucleotide sequence is
   a. TAT GCC. c. TUT GCC.
   b. UAU GCC. d. UAU GGC.
6. The making of proteins from the information carried by mRNA is called
   a. DNA replication. c. transcription.
   b. translation. d. gene regulation.
7. Anticodons are found on ______ molecules.
   a. mRNA c. rRNA
   b. DNA d. tRNA
8. The *lac* operon allows a bacterium to build the proteins needed for lactose metabolism when
   a. glucose is present. c. lactose is present.
   b. lactose is absent. d. glucose is absent.
9. Transcription of lactose-metabolizing genes is blocked when ______ is bound to the operator.
   a. the repressor c. the inducer
   b. the operon d. the enhancer
10. In eukaryotes, gene expression can be regulated by
    a. mutations. c. repressors.
    b. transcription factors. d. operons.
11. A change in the genetic code is called
    a. a mutation. c. a codon.
    b. an operon. d. an operator.
12. Mutations that change one or just a few nucleotides in a gene are called
    a. operon mutations. c. repressor proteins.
    b. codon mutations. d. point mutations.
13. **Interpreting Graphics** Does the drawing below represent a strand of RNA or a strand of DNA? Explain.

14. **Theme Homeostasis** How does gene regulation of the *lac* operon promote homeostasis in intestinal *E. coli* bacteria?
15. **Exploring Further** Compare the way transposons and exons affect genes.
16. **Tech Watch** A researcher trying to determine the sequence of nucleotides on a particular gene obtained the following sequence with the primer and tagged nucleotide: TCCGGAAG. What was the sequence of nucleotides on the gene of interest?
17. **Chapter Links** Compare and contrast chromosomal mutations with point mutations. (**Hint:** See Chapter 6, Section 6-3.)

13. RNA—DNA does not contain uracil.
14. When lactose is taken into the cell, the *lac* operon is activated and the necessary enzymes needed to metabolize lactose are produced. If lactose is not present, then the enzymes are not produced, conserving needed energy.
15. Transposons can inactivate a gene or cause a mutation. Exons do not affect genes—they are the parts of a gene that are translated.
16. A G G C C T T C
17. Chromosomal mutations are mutations that involve sections of a chromosome (thousands of nucleotides). Point mutations involve changes in a fewer number—one or more—of nucleotides.

Skills Assessment

Part 1: Interpreting Graphics

Study the illustration below showing the events of translation. Then answer the following questions:

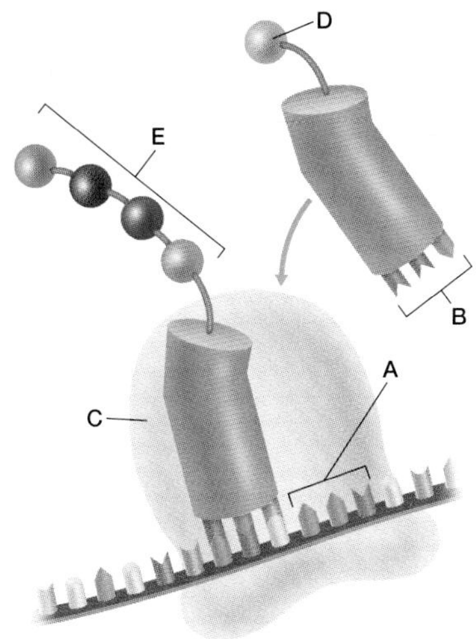

18. What does *A* represent in the model?
19. What is the name of organelle *C*?
20. What is the relationship between *A* and *B* during translation?
21. What structure is represented by *D*?
22. What structure is represented by *E*?

Part 2: Life/Work Skills

Use the media center or Internet resources to learn about drugs that interfere with protein synthesis.

23. **Finding Information** Antibiotics are substances made by or derived from microorganisms. Antibiotics destroy or inhibit the growth of certain microorganisms. Antibiotics are used to treat infections caused by organisms (such as bacteria and fungi) that are sensitive to them. Tetracyclines are antibiotics that are commonly used as antibacterial drugs. How do tetracyclines fight bacterial infection?
24. **Communicating** Prepare an oral report using graphics to interpret and summarize your findings.

Critical Thinking

25. **Evaluating Results** A molecular biologist isolates mRNA from the brain and liver of a mouse and finds that the mRNA molecules are different. Can these results be correct, or has the biologist made an error? Explain your answer.
26. **Evaluating an Argument** A classmate states that damage to exons is very likely to affect the synthesis of a protein, while damage to introns is not. Evaluate the statement.

Portfolio Projects

27. **Forming a Model** Like musical notes that are arranged for a song, bases can be arranged into a variety of codons that code for amino acids. Assign a different musical note to each of the four mRNA bases. Use the following mRNA sequence, and translate the sequence into a song using the notes that you selected: AUG, GGG, GGA, GGU, GGA, GGC, GAG, GAA, GAU, GAC, GUG, GUA, GUU, GUC, GCG, GCA, GCU, GCC, AGG, AGA, GCU, GCC, AGG, AGA, AGU, AGC, AAG. Make a recording of your song to share with the class.
28. **Interpreting Information** Read "...And a Gene That Causes Hair Loss" (*Science News,* March 7, 1998, vol. 153, p. 159). What is alopecia universalis? Describe how Christiano and her colleagues made their discoveries. What type of gene is affected?
29. **Career Focus Protein Chemist** Research the field of protein chemistry, and write a report on your findings. Your report should include a job description, training required, kinds of employers, growth prospects, and a starting salary.
30. **Unit 6—*Gene Expression*** Write a report summarizing how antibiotics inhibit protein synthesis in bacteria. How do some antibiotics interfere with translation?

BIOLOGY INTERACTIVE TUTOR

Skills Assessment

18. a codon
19. ribosome
20. Complementary base pairing will occur between the two.
21. an amino acid
22. a protein
23. Tetracyclines bind to bacterial ribosomes, interfering with bacterial protein synthesis.
24. Answers will vary.

Critical Thinking

25. The results can be correct. A cell only expresses a specific gene when it needs the protein coded by the gene. Two different cells may not need the same proteins. Thus, their mRNAs will differ.
26. Mutations of introns are not likely to affect the synthesis of a protein because introns are cut out before the mRNA is translated. A mutation in an exon is more likely to affect the synthesis of a protein.

Portfolio Projects

27. Check students' recordings for accuracy.
28. Alopecia universalis is a condition that results in no scalp or body hair growth after birth. Christiano and her collegues were studying a family in Pakistan that has many affected members. Using the DNA sequence of a mutated mouse gene from hairless rodents, the researchers found the gene region in the human DNA from the affected family members. The gene that is involved codes for a transcription factor.
29. Protein chemists use computer models and genetic engineering to design synthetic compounds. They may design drugs or other bioactive compounds. Protein chemists attend college, followed by a research-based graduate program. Often they progress to a postdoctoral appointment at a university. They are employed by research organizations, including universities, and by private companies, such as drug manufacturers. The growth potential of this field is good. Starting salary will vary by region.
30. Students' reports will vary. Many antibiotics inhibit protein synthesis by combining with ribosomal proteins. Erythromycin, chloramphenicol, clindamycin, and lincomycin combine with the 50S ribosomal subunit. The tetracyclines, spectinomycin, streptomycin, gentamicin, kanamycin, amikacin, and the nitrofurans combine with the 30S ribosomal subunit. Mupirocin and puromycin inhibit protein synthesis at the tRNA level.

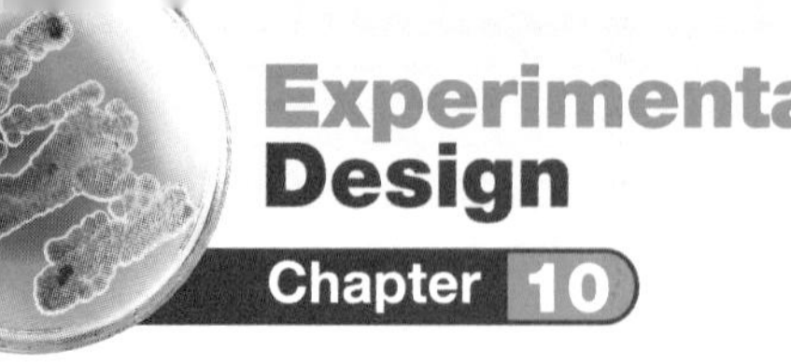

Experimental Design

Chapter 10

How Can You Model Protein Synthesis?

Purpose

Students will model the process of protein synthesis and the effect of the mutation that causes sickle cell anemia.

Preparation Notes

Time Required: 40–45 minutes

Materials

Materials for this lab can be ordered from WARD'S. See *Master Materials List* at the front of this book for catalog numbers.

Preparation Tips

Use materials from Chapter 9 Experimental Design, pp. 202–203, if they were saved. Each group will need enough plastic soda straws to make about 40 sections of one color (3 cm each) and 15 sections of a different color. Each group will also need about 40 standard-sized paper clips, 40 colored pushpins, and enough materials to make at least eight tRNA/amino acid cards.

Safety Precautions

- Discuss all safety symbols and caution statements with students.
- Caution students to be careful of the sharp tips of the pushpins.

Procedural Tips

- Make sure each group of students has enough materials to complete the activity before they begin.
- Before beginning the lab, use Figures 10-2 and 10-5 on pp. 207 and 210–211 to review the steps in gene expression.

Disposal

Account for all materials at the end of the lab, and have students store the pushpins by color. If desired, ask students to save the assembled "nucleotides" for use in Chapter 11 Lab Techniques, pp. 244–245.

Before You Begin

Answers

1. *protein*—an organic molecule made of amino acids

Experimental Design

Chapter 10

How Can You Model Protein Synthesis?

SKILLS

- Modeling
- Using scientific methods

OBJECTIVES

- **Compare and Contrast** the structure and function of DNA and RNA.
- **Model** protein synthesis.
- **Demonstrate** how a mutation can affect a protein.

MATERIALS

- masking tape
- plastic soda-straw pieces of one color
- plastic soda-straw pieces of a different color
- paper clips
- pushpins of five different colors
- marking pens of the same colors as the pushpins
- 3 × 5 in. note cards
- oval-shaped card
- transparent tape

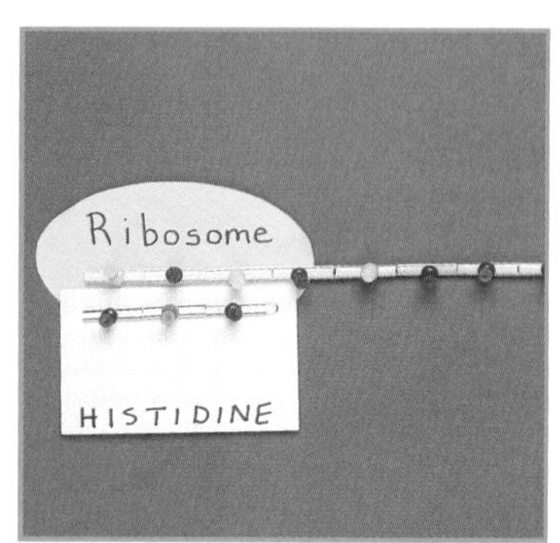

Before You Begin

The nature of a **protein** is determined by the sequence of amino acids in its structure. During **protein synthesis,** the sequence of nitrogen bases in an **mRNA** molecule is used to assemble **amino acids** into a protein chain.

A **mutation** is a change in the nitrogen-base sequence of DNA. Many mutations lead to altered or defective proteins. For example, the genetic blood disorder **sickle cell anemia** is caused by a mutation in the gene for **hemoglobin.**

In this lab, you will build models that will help you understand how protein synthesis occurs. You can also use the models to explore how a mutation affects a protein.

1. Write a definition for each boldface term in the paragraph above and for each of the following terms: transcription, translation, tRNA, ribosome, codon, anticodon.
2. Describe three differences between DNA and RNA.
3. Based on the objectives for this lab, write a question you would like to explore about protein synthesis.

Procedure

PART A: Design a Model

1. Work with the members of your lab group to design models of DNA, RNA, and a cell. Use the materials listed for this lab.

> **You Choose**
>
> **As you design your models, decide the following:**
>
> a. what question you will explore
> b. how to represent DNA nucleotides
> c. how to represent RNA nucleotides
> d. how to represent five different nitrogen bases
> e. how to link (bond) nucleotides together
> f. how to represent tRNA molecules with amino acids
> g. how to represent the locations of DNA and ribosomes

2. Write out the plan for building your models. Have your teacher approve the plan before you begin building the models.
3. Build the models your group designed. **CAUTION: Sharp or pointed objects may cause injury. Handle pushpins carefully.** Start your model of DNA with a strand of nucleotides that has the following sequence of nitrogen bases: TTTGGTCTCCTC.

protein synthesis—the process by which proteins are made based on the information encoded in a gene

mRNA—an RNA copy of a gene that carries the instructions for making a protein from a gene and delivers it to a ribosome for translation

amino acids—organic molecules that are the building blocks of proteins

mutation—a change in the DNA of a gene or chromosome

sickle cell anemia—a condition caused by a mutant allele that produces a defective form of hemoglobin

hemoglobin—a protein found in red blood cells that binds with and carries oxygen throughout the body

transcription—a stage of gene expression in which the information in DNA is transferred to an RNA molecule

translation—a stage of gene expression in which the information in mRNA is used to make a protein

tRNA—an RNA molecule that temporarily carries a specific amino acid to a ribosome during translation

ribosome—a cytoplasmic organelle on which proteins are made

Part B: Model Protein Synthesis

4. Use your models and Figure 10-5 on pp. 210–211 to demonstrate how transcription and translation occur. Draw and label the steps of each process.
5. Use your models to explore one of the questions written for step 3 of **Before You Begin.**

Part C: Test Hypothesis

Answer each of the following questions by writing a hypothesis. Use your models to test each hypothesis, and describe your results.

6. The DNA model you built for step 3 represents a portion of a gene for hemoglobin. Sickle cell anemia results from the substitution of an A for the T in the third codon of the nitrogen-base sequence given in step 3. How will this substitution affect a hemoglobin molecule?
7. The addition of a nucleotide to a strand of DNA is a type of mutation called an *insertion.* What happens when an insertion occurs in the first codon in a DNA strand, before the DNA strand is transcribed?

Part D: Cleanup and Disposal

8. Dispose of damaged pushpins in the designated waste container.
9. Clean up your work area and all lab equipment. Return lab equipment to its proper place. Wash your hands thoroughly before you leave the lab and after you finish all work.

Analyze and Conclude

1. **Comparing Structures** How did the nitrogen-base sequence of the mRNA you made compare with that of the DNA it was transcribed from?
2. **Recognizing Relationships** How is the nitrogen-base sequence of a gene related to the structure of a protein?
3. **Recognizing Patterns** What is the relationship between the anticodon of a tRNA and the amino acid the tRNA carries?
4. **Drawing Conclusions** How does a mutation in the gene for a protein affect the protein?
5. **Further Inquiry** Write a new question about protein synthesis that could be explored with your model.

Do You Know?

Do research in the library or media center to answer these questions:

1. What are mutagens, and how do they affect DNA?
2. What are two other genetic disorders that result from mutations?

Use the following Internet resources to explore your own questions about DNA.

SCiLINKS NSTA
TOPIC: Genetic disorders
GO TO: www.scilinks.org
KEYWORD: HX223

codon—a three-nucleotide sequence in mRNA that encodes an amino acid or signifies a start or stop signal

anticodon—a three-nucleotide sequence on tRNA that recognizes a complementary codon on mRNA

2. RNA is single stranded, and DNA is double stranded. RNA nucleotides contain the sugar ribose and the nitrogen bases uracil, guanine, cytosine, or adenine. DNA nucleotides contain the sugar deoxyribose and the nitrogen bases thymine, guanine, cytosine, or adenine.
3. Answers will vary. For example: How does the cell's DNA determine the proteins made by a cell?

Procedure

Answers

2. Answers may vary. Students might use straw segments to model the sugar-phosphate backbone of DNA, different-colored pushpins to represent the four different nitrogen bases, paper clips to represent the bonds that hold the nucleotides in a chain, note cards to model tRNA molecules with amino acids attached, and oval-shaped cards to represent ribosomes.
4. Answers may vary. Check for students' understanding of each step in gene expression.
5. Students' results will vary. Some students might produce a mutation and demonstrate how it affects protein synthesis. Some mutations cause the wrong amino acid to be inserted in the protein. Other mutations might insert a start or stop codon at an inappropriate place.
6. Hypotheses will vary. The substitution will change the amino acid sequence from lysine–proline–glutamic acid–glutamic acid to lysine–proline–valine–glutamic acid.
7. Hypotheses will vary. The codon triplets are shifted.

Analyze and Conclude

Answers

1. They are complementary.
2. The nitrogen-base sequence of a gene contains a code that determines the amino-acid sequence of a protein.
3. A tRNA with a particular anticodon always carries the same amino acid.
4. A mutation can change the amino acid sequence of a protein, and thus may change its activity.
5. Answers will vary. For example: What happens if a mutation occurs that changes a codon that codes for an amino acid to a codon that codes for a stop codon?

Do You Know?

Answers

1. Mutagens are agents, such as certain chemicals and radiation, that cause mutations.
2. Answers will vary but may include cystic fibrosis and Huntington's disease.

Chapter 11

Gene Technology

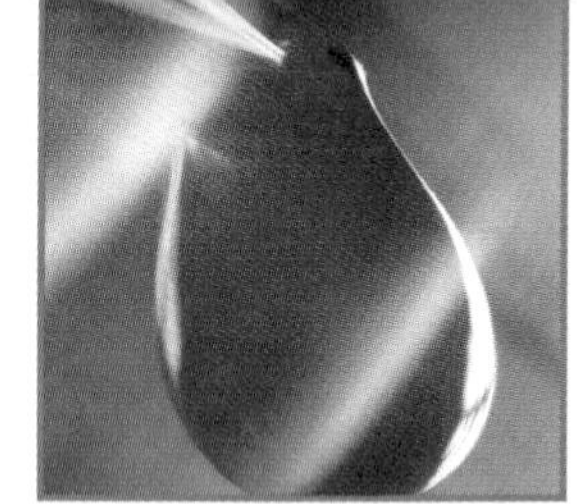

Planning Guide

	OBJECTIVES	MEETING INDIVIDUAL NEEDS	READING SKILLS AND STRATEGIES
BLOCKS 1, 2, & 3 135 min	**11-1 Genetic Engineering** • Describe four basic steps commonly used in genetic engineering experiments. • Evaluate how restriction enzymes are used in genetic engineering. • Relate the role of electrophoresis and probes in identifying a specific gene.	Learning Styles, p. 226 BASIC ADVANCED • pp. 227, 230	**ATE** Active Reading, Technique: K-W-L, p. 225 **ATE** Active Reading, Technique: Reading Effectively, p. 228 ■ Active Reading Guide, worksheet (11-1) ■ Directed Reading, worksheet (11-1)
BLOCKS 4, 5, & 6 135 min	**11-2 Genetic Engineering in Medicine and Society** • Describe how drugs produced by genetic engineering are being used. • Summarize the steps involved in making a genetically engineered vaccine. • Describe how gene therapy is being used to try to cure genetic disorders. • Identify two different uses for DNA fingerprints. • Summarize two major goals of the Human Genome Project.	Learning Styles, p. 231 BASIC ADVANCED • pp. 234, 236	■ Active Reading Guide, worksheet (11-2) ■ Directed Reading, worksheet (11-2)
BLOCKS 7 & 8 90 min	**11-3 Genetic Engineering in Agriculture** • Describe three ways in which genetic engineering has been used to improve plants. • Summarize two ways in which genetic engineering techniques have been used to modify farm animals. • Summarize the cloning of sheep through the use of differentiated cells.	Learning Styles, p. 237 BASIC ADVANCED • pp. 238, 240	■ Active Reading Guide, worksheet (11-3) ■ Directed Reading, worksheet (11-3) ■ Supplemental Reading Guide, *The Double Helix*

Review and Assessment Resources

BLOCKS 9 & 10 90 min

TEST PREP

- **PE** Study Zone, p. 241
- **PE** Performance Zone, pp. 242–243
- ■ Chapter 11 worksheets:
  - • Vocabulary
  - • Concept Mapping
  - • Science Skills
  - • Critical Thinking
- Audio CD Program
- ■ Test Prep Pretest

CHAPTER TESTING

- ■ Chapter Tests (blackline master)
- ▲ Test Generator
- ■ Assessment Item Listing

See the correlations table at the front of this book for a list of the ***National Science Education Standards*** addressed in this chapter.

Smithsonian Institution® Visit **www.si.edu/hrw** for additional on-line resources.

KEY

Biology Principles and Explorations

PE Pupil's Edition

ATE Teacher's Edition

HOLT BIOSOURCES

Teaching Transparencies

Laboratory Program

One-Stop Planner CD-ROM

Shows these resources within a customizable daily lesson plan:

■ Teaching Resources

▲ Test Generator

VISUAL STRATEGIES	CLASSWORK AND HOMEWORK	TECHNOLOGY AND INTERNET RESOURCES
ATE Fig. 11-2, p. 227 **ATE Fig. 11-3,** p. 228 **ATE Fig. 11-5,** p. 229 39 Cleaving DNA	**PE Quick Lab** How can you model gel electrophoresis?, p. 230 **Lab A7** Making Models **Lab D5** Introduction to Gel Electrophoresis **Lab D6** DNA Fragment Analysis **PE Review,** p. 230	CNN Video Segment 8 Alzheimer's Mutation Holt Biology Videodiscs Section 11-1 Audio CD Program Section 11-1 internetconnect SCINKS NSTA **TOPIC:** Genetic engineering
40 Four Steps in Genetic Engineering 31A Polymerase Chain Reaction 32A Making Genetically Engineered Drugs 33A Genetically Engineered Medicines 43A Comparing the Hemoglobin Gene Among Species	**PE Data Lab** Interpreting a DNA fingerprint, p. 235 **Lab C15** Karyotyping—Genetic Disorders **Lab D3** Genetic Transformation of Bacteria **Lab D4** Genetic Transformation—Antibiotic Resistance **Lab D7** DNA Ligation **PE Review,** p. 236	Holt Biology Videodiscs Section 11-1 Audio CD Program Section 11-1
41 Genetic Engineering and Cotton Plants	**PE Lab Techniques** Modeling Genetic Engineering, pp. 244–245 **Lab A8** Making a Genetic Engineering Model **Lab D8** Comparing DNA Samples **PE Review,** p. 240 **PE Study Zone,** p. 241	Holt Biology Videodiscs Section 11-3 Audio CD Program Section 11-3 internetconnect SCINKS NSTA **TOPIC:** Cloning

ALTERNATIVE ASSESSMENT

PE **Portfolio Projects 26–28,** p. 243

ATE **Alternative Assessment,** p. 241

ATE **Portfolio Assessment,** p. 241

■ **Portfolio Projects:** Genetics Project

Lab Activity Materials

Quick Lab
p. 230 500 mL graduated cylinder, large jar, beads (3 sets, each a different size and color)

Data Lab
p. 235 pencil, paper

Lab Techniques
p. 244 plastic soda-straw pieces (56), pushpins (15 red, 15 green, 13 blue, and 13 yellow), paper clips (56)

Lesson Warm-up

ATE bowl of fruits and vegetables, p. 226

ATE ink pad, white paper, p. 231

Demonstrations

ATE yarn, pipe cleaners, tape, scissors, p. 227

ATE photos of farm animals produced by selective breeding and by genetic engineering, p. 238

BIOSOURCES LAB PROGRAM
- Quick Labs: Lab A7, p. 13
- Quick Labs: Lab A8, p. 15
- Laboratory Techniques and Experimental Design: Lab C15, p. 67
- Biotechnology: Lab D3, p. 9
- Biotechnology: Lab D5, p. 19
- Biotechnology: Lab D6, p. 25
- Biotechnology: Lab D7, p. 33
- Biotechnology: Lab D8, p. 41

Chapter Theme

Homeostasis

Some genetic diseases are due to a failure to produce a particular substance, which in turn leads to the loss of homeostasis in the body. Diabetes leads to loss of control of blood sugar levels. PKU is a failure to metabolize an amino acid, leading to its buildup in the body. Hemophilia causes a breakdown in blood clotting reactions, which leads to blood loss. With gene therapy, a type of gene technology, scientists hope to restore homeostasis.

Establishing Prior Knowledge

Before beginning this chapter, make certain students can answer the questions in Study TIP Ready? on page 225. If they cannot, encourage them to reread the sections indicated.

Opening Question

Ask students to discuss why gene technology is controversial. *(Lead students to suggest that DNA technology allows researchers to produce new life-forms with specific characteristics or abilities.)* Students should understand that gene technology is a potentially powerful tool for fighting diseases and for understanding organisms. Explain that people are excited about the prospects of using this tool but concerned about its misuse.

CHAPTER 11 Gene Technology

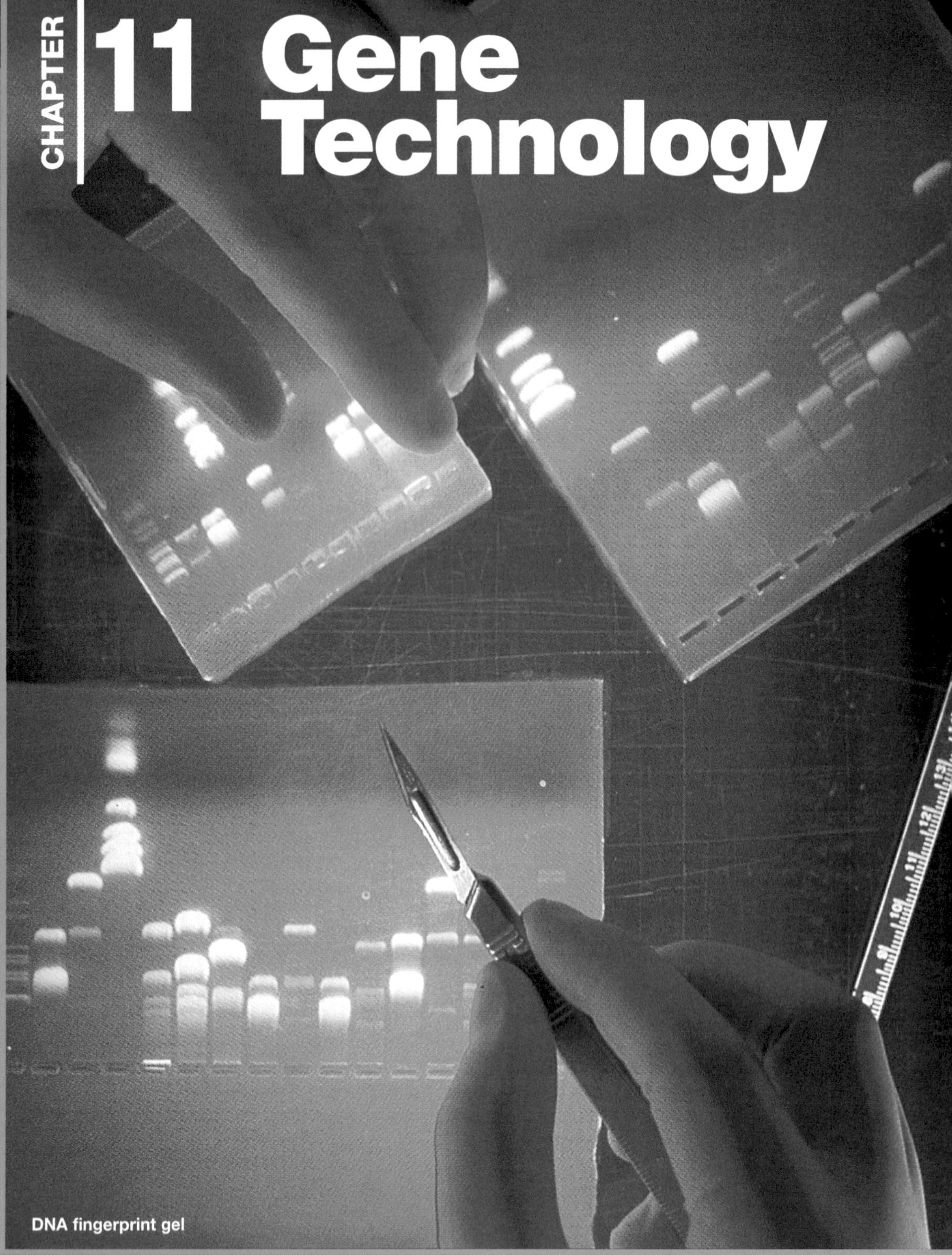

DNA fingerprint gel

What's Ahead — and Why

CHAPTER OUTLINE

Students will learn

- how a particular gene is isolated and cloned.
- how genetic engineering can be used to treat and cure certain diseases and identify organisms.
- how genetic engineering can be used to improve food crops and farm animals and to have farm animals make medically useful proteins.

Chapter 11

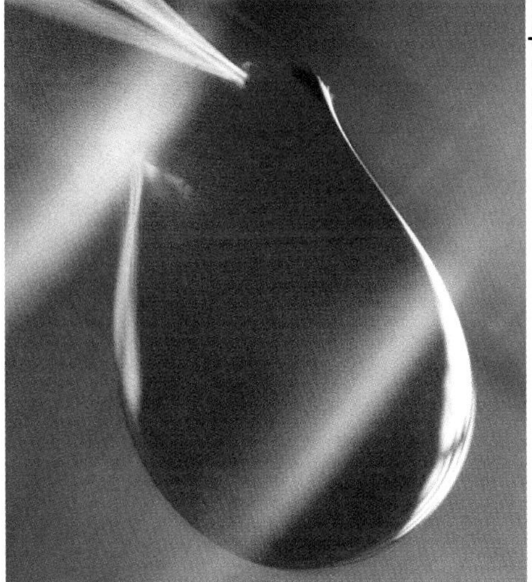
Drop of blood

One drop of blood *has enough DNA to use gene technology to determine a person's DNA profile. Thanks to gene technology, many criminal cases have been solved and many paternity disputes have been settled.*

Looking Ahead

11-1 Genetic Engineering
How are clones made?

11-2 Genetic Engineering in Medicine and Society
How is your life affected by genetic engineering?

11-3 Genetic Engineering in Agriculture
Are you eating fruits and vegetables that were cloned through gene technology?

Labs

- **Quick Lab**
  How can you model gel electrophoresis? (p. 230)
- **Data Lab**
  Interpreting a DNA fingerprint (p. 235)
- **Lab Techniques**
  Modeling Genetic Engineering (p. 244)

Features

- **Tech Watch**
  Polymerase Chain Reaction (PCR) (p. 233)
- **Exploring Further**
  A Breakthrough in Cloning (p. 240)

Study TIP

Ready?

Check your understanding of these topics to help you prepare for what's ahead.

Can you . . .

- **define** the term *gene*? (Section 6-1)
- **describe** the structure of DNA? (Section 9-2)
- **state** the base-pairing rules that determine the structure of DNA? (Section 9-2)
- **describe** how the genetic code is universal? (Section 10-1)

If not, *review the sections indicated.*

internet connect

National Science Teachers Association *sci*LINKS Internet resources are located throughout this chapter.

Reading Strategies

Active Reading

Technique: K-W-L

Before students read this chapter, have them write short individual lists of all the things they already Know (or think they know) about gene technology. Ask them to contribute their entries to a group list on the board. Then have students list things they Want to know about gene technology. Have students save their lists for use later in **Closure** on page 240.

Directed Reading

Assign **Directed Reading Worksheet Chapter 11** to help students interact with the text as they read the chapter.

Interactive Reading

Assign ***Biology: Principles and Explorations*** **Audio CD Program Chapter 11** to help reluctant readers achieve greater success in reading the chapter.

Reading Skills

Assign Reading Organizers and Reading Strategies Worksheets from the **One-Stop Planner CD-ROM with Test Generator** to help students learn and practice effective reading skills.

internet connect

Holt *Biology: Principles and Explorations*

On-line Resources: go.hrw.com

Visit the HRW Web site for a variety of resources related to this chapter. Just type in the keyword HX0 Chapter 11.

Holt *Biology: Principles and Explorations*

On-line Resources: www.scilinks.org

The following *sci*LINKS Internet resources can be found in the student text for this chapter:

Page 230
TOPIC: Genetic engineering
KEYWORD: HX230

Page 238
TOPIC: Cloning
KEYWORD: HX238

Chapter 11

1 Prepare

Scheduling

- **Block:** About 135 minutes will be needed to complete this section.
- **Traditional:** About three class periods will be needed to complete this section.
- **Planning:** See the Planning Guide on pp. 223A and 223B for lecture, classwork, and assignment options for Section 11-1. Lesson plans for Section 11-1 can be found in a lesson cycle format in ***Biology: Principles and Explorations*** **One-Stop Planner CD-ROM with Test Generator.**

Resource Link

Assign **Active Reading Worksheet Section 11-1** to help students comprehend the key concepts and visuals in this lesson.

2 Teach

Lesson Warm-up

Show students a bowl with various types of fruits and vegetables. Ask students to identify their favorite fruits and vegetables. *(Answers will vary.)* Ask students what characteristics they would change in their favorites, if they could. *(Answers will vary and may include making the fruit or vegetable sweeter, firmer, easier to bite into, or a different color.)* Discuss with students how scientists are now able to manipulate many characteristics of fruits and vegetables using genetic engineering techniques. Once the gene that controls the trait is found and isolated, scientists can try to manipulate it in various ways.

Section 11-1 Genetic Engineering

Objectives

- **Describe** four basic steps commonly used in genetic engineering experiments. (pp. 226–227)
- **Evaluate** how restriction enzymes and the antibiotic tetracycline are used in genetic engineering. (p. 228)
- **Relate** the role of electrophoresis and probes in identifying a specific gene. (pp. 229–230)

Key Terms

genetic engineering
recombinant DNA
restriction enzyme
vector
plasmid
gene cloning
electrophoresis
probe

Genetic Engineering Involves Four Basic Steps

Not too long ago, using bacteria to produce human insulin and inserting genes into tomatoes and human cells were ideas that existed in science fiction books and movies. But now, the techniques required to carry out these ideas have been developed and are used daily.

In 1973, Stanley Cohen and Herbert Boyer conducted an experiment that revolutionized genetic studies in biology. They isolated the gene that codes for ribosomal RNA from the DNA of an African clawed frog and then inserted it into the DNA of *Escherichia coli* bacteria, as summarized in **Figure 11-1.** During transcription, the bacteria produced frog rRNA, thereby becoming the first genetically altered organisms. The process of manipulating genes for practical purposes is called **genetic engineering.** Genetic engineering involves building **recombinant DNA**—DNA made from two or more different organisms.

The basic steps in genetic engineering can be explored by examining how the human gene for insulin is transferred into bacteria. Insulin is a protein hormone that controls sugar metabolism. Diabetics cannot produce enough insulin, so they must take regular injections of insulin. Before genetic engineering, insulin was extracted from the pancreases of slaughtered cows and pigs and then purified. Today, the insulin gene is transferred to bacteria through genetic engineering. Because the genetic code is universal, bacteria can transcribe and translate a human insulin gene the same way a human cell can in order to produce human insulin.

Figure 11-1 Genetic alteration of an organism

Cohen and Boyer produced the first genetically engineered organisms.

1. Cohen and Boyer used an African clawed frog as their experimental organism.

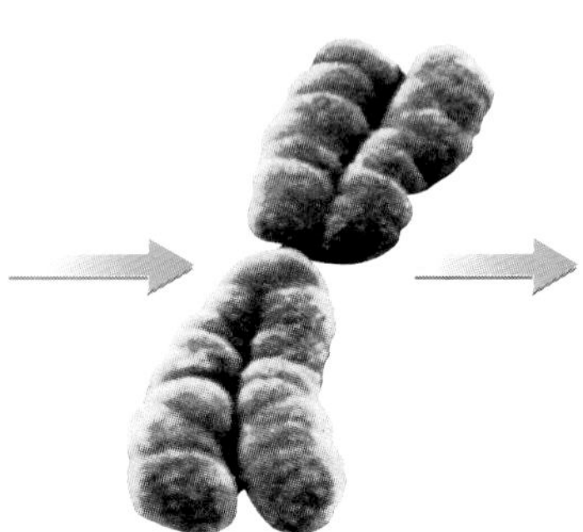

2. They isolated an rRNA gene from one of its chromosomes.

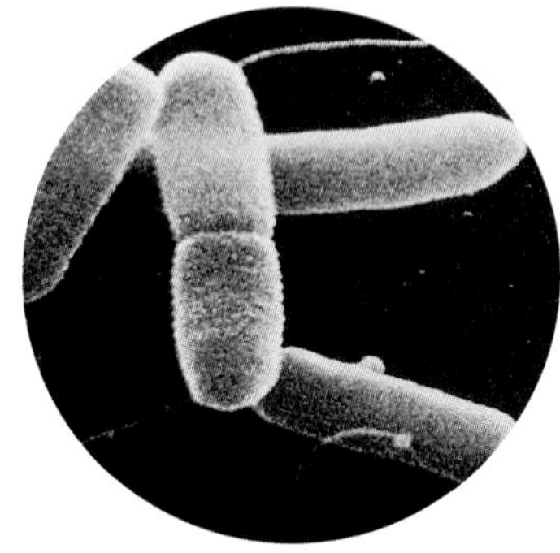

3. They inserted the gene into bacteria. The bacteria produced frog rRNA.

Integrating Different Learning Styles

Logical/Mathematical	**PE**	Performance Zone item 24, p. 243
Visual/Spatial	**PE**	Quick Lab, p. 230; Performance Zone items 1, 14, 19, 20, and 21, pp. 242–243; Lab Techniques, pp. 244–245
	ATE	Lesson Warm-up, p. 226; Demonstration, p. 227; Visual Strategies, pp. 227, 228, 229
Verbal/Linguistic	**PE**	Study Tip, p. 229; Performance Zone item 29, p. 243
	ATE	Did You Know?, p. 227; Cross-Disciplinary Connection, p. 227; Active Reading, p. 228; Reteaching, p. 230

Steps in a genetic engineering experiment

Genetic engineering experiments use different approaches, but most share four basic steps, as illustrated in **Figure 11-2.**

Step 1 Cutting DNA Both the DNA from the organism containing the gene of interest (in our example, the insulin gene) and the DNA from a vector are cut. The DNA is cut by restriction enzymes. **Restriction enzymes** are bacterial enzymes that recognize and bind to specific short sequences of DNA, and then cut the DNA between specific nucleotides within the sequences. A **vector** is an agent that is used to carry the gene of interest into another cell. Commonly used vectors include viruses, yeast, and plasmids. **Plasmids,** shown in Figure 11-2, are circular DNA molecules that can replicate independently of the main chromosomes of bacteria.

Step 2 Making recombinant DNA The DNA fragments from the organism containing the gene of interest are combined with the DNA fragments from the vector. An enzyme called DNA ligase is added to help bond the DNA fragments together. In our example, human DNA fragments are combined with plasmid DNA fragments. After the recombinant DNA is produced, the host cells are treated so that they can take up the recombinant DNA.

Step 3 Cloning In a process that is called **gene cloning,** many copies of the gene of interest are made each time the host cell reproduces. Recall from Chapter 6 that bacteria reproduce by binary fission, producing identical offspring. When a bacterial cell replicates its DNA, it also replicates its plasmid DNA.

Step 4 Screening Cells that have received the particular gene of interest are distinguished, or screened, from the cells that did not take up the vector with the gene of interest. This process is called screening. Each time the cells reproduce, they make a copy of the gene of interest (in our example, the insulin gene). The cells can transcribe and translate the gene to make the protein coded for in the gene.

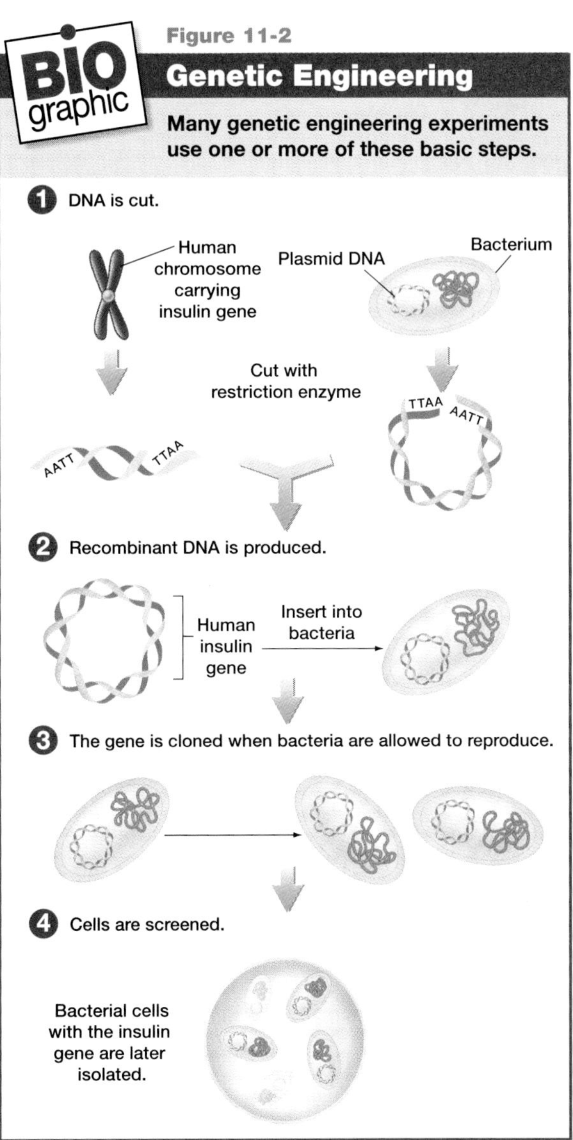
BIO graphic
Figure 11-2
Genetic Engineering
Many genetic engineering experiments use one or more of these basic steps.
1 DNA is cut.
Human chromosome carrying insulin gene
Plasmid DNA
Bacterium
Cut with restriction enzyme
AATT
TTAA
2 Recombinant DNA is produced.
Human insulin gene
Insert into bacteria
3 The gene is cloned when bacteria are allowed to reproduce.
4 Cells are screened.
Bacterial cells with the insulin gene are later isolated.

Demonstration

Recombinant DNA — BASIC

Give pairs of students a piece of yarn (to represent a human chromosome), a long pipe cleaner a different color from the yarn (to represent plasmid DNA), and some tape. Ask the students to form a circle with the pipe cleaner, and twist the ends once to secure the circle. Have students use scissors to cut a "gene" from the yarn and to untangle the twisted end of the pipe cleaner. Ask the students what the scissors represent. *(restriction enzymes)* Tell the students to insert the human gene (piece of yarn) into the opening of the plasmid and secure the ends with tape. Ask the students the following: What does the tape represent? *(DNA ligase as it helps bond the DNA pieces together)* What does the plasmid represent? *(a vector)* Why is this recombinant DNA? *(The plasmid now contains foreign DNA.)* What is the next step in genetic engineering? *(The plasmid is inserted into a bacterium.)* Why? *(The bacterium will reproduce; each time it does, all the genetic material will be copied, including the human gene.)*

CROSS-DISCIPLINARY CONNECTION

◆ **Language Arts**

The word *clone* comes from the Greek *klon* meaning "twig." Pieces of a plant can be cut off and rooted, producing a new and identical plant—a clone.

Visual Strategy

Figure 11-2

Use this figure to review the steps used in many genetic engineering experiments. Emphasize in **Step 4** that the bacterial cells with the insulin gene will later be isolated. These cells will transcribe and translate the insulin gene just as they would their own gene. Thus, scientists can isolate the insulin from the cells, purify it, and then use it for medicinal purposes.

did you know?

Restriction enzymes are named after the specific bacteria from which they are isolated.

For example, the restriction enzyme *Eco*RI is named after the bacteria *Escherichia coli*. The first letter, *E*, is the initial letter of the genus name of the organism from which the enzyme was isolated *(Escherichia)*. The second and third letters, *co*, are usually the first two letters of the species name *(coli)*. These three letters are always italicized since they are part of the scientific name. The fourth letter, *R*, if present, represents the strain of the organism (strain RY 13). The Roman numerals (I) indicate the order of discovery (first endonuclease isolated in this strain of bacteria).

Visual Strategy

Figure 11-3

Use this illustration to reinforce the importance of sticky ends and how sticky ends are complementary. Ask students why the genes for replication and tetracycline resistance must be present in the vector. *(replication gene—so that the plasmid can replicate; tetracycline resistance gene—to screen the bacterial cells that take up the recombined vector)* Tell students that there are thousands of different restriction enzymes, and that each recognizes a different DNA sequence.

Active Reading

Technique: Reading Effectively

Ask students to review pp. 227–228 and to write a summary of the basic steps in genetic engineering. Tell students not to write their names on the paper and to use the following terms: *chromosome, plasmid, bacterium, gene, clone, restriction enzyme, sticky ends, vector,* and *DNA ligase.* Collect the summaries, choose one at random, and read it to the class. Ask the class to correct any inaccuracies or to contribute additional information. Repeat by reading a few more papers until the students are confident about outlining the steps. PORTFOLIO

Teaching TIP

Natural Selection and Screening

Screening cells involves events similar to those of natural selection. Only certain organisms will survive in a given environment. In screening, only bacteria with the gene for antibiotic resistance will survive in a culture medium containing an antibiotic. As in natural selection, the bacteria with the successful trait are the ones that survive and produce future generations.

Figure 11-3 Restriction enzymes cut DNA

The restriction enzyme *Eco*RI recognizes the nucleotide sequence GAATTC and makes its cut between the G and the A.

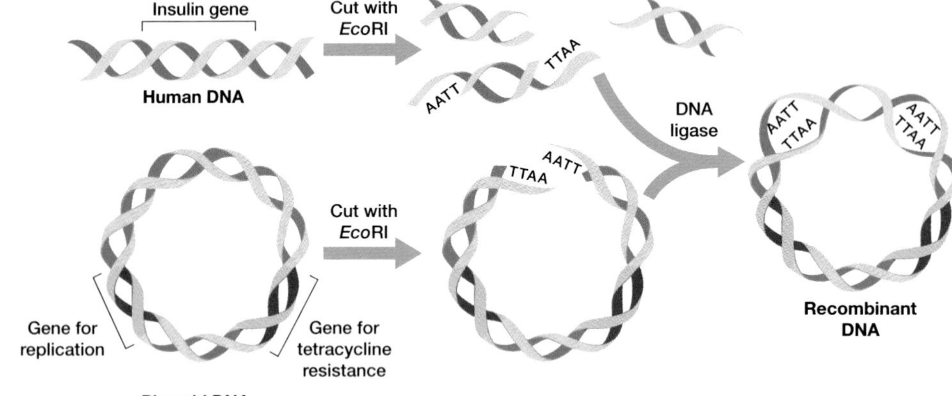

Cutting DNA and making recombinant DNA

An example of how restriction enzymes work is shown in **Figure 11-3.** The enzyme recognizes a specific sequence of DNA. The sequence the enzyme recognizes and the sequence on the complementary DNA strand are palindromes—they read the same backward as they do forward (such as the word *toot*).

The cuts of most restriction enzymes produce pieces of DNA with short single strands on each end that are complementary to each other. The ends are called sticky ends. As illustrated in Figure 11-3, the vectors that are used contain only one nucleotide sequence that the restriction enzyme recognizes. Thus, vectors such as the circular plasmids "open up" with the same sticky ends as that of the cut human DNA. With the help of an enzyme called DNA ligase, the two DNA molecules bond together by means of complementary base pairing at the sticky ends. The plasmid DNA has the gene for plasmid DNA replication and the gene that makes the cell carrying the plasmid resistant to the antibiotic tetracycline.

Figure 11-4 Screening. Only the cells that take up the vectors are resistant to tetracycline and survive when tetracycline is added.

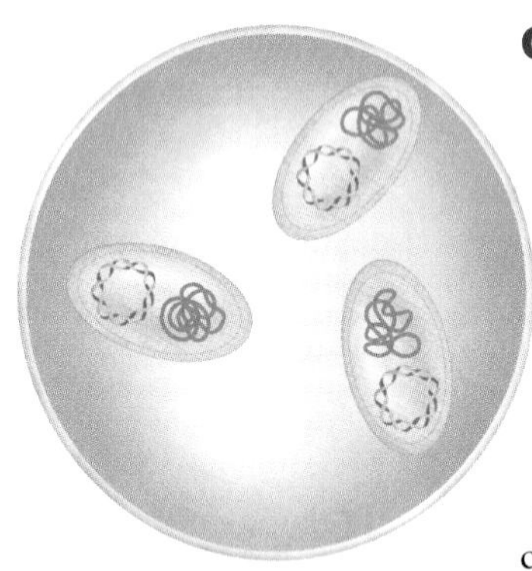

Cloning and screening cells

One difficult part in a genetic engineering experiment is finding and isolating the cells that contain the gene of interest. The cells that have taken up the recombined plasmid must first be identified. The bacterial cells that take up the recombined plasmid are identified by adding the antibiotic tetracycline to the bacterial cultures. As shown in **Figure 11-4,** only the cells that have taken up the vectors (which contain the gene for tetracycline resistance) survive when tetracycline is added to the bacterial cultures. Each surviving cell makes a copy of the gene of interest each time it reproduces. Eventually a colony of genetically identical cells, or clones, is formed for each surviving cell.

Overcoming Misconceptions

Determining Gene Function

Many people assume that knowing the location of every gene on every chromosome means that their functions are also known. In fact, the functions of many genes discovered in genomic sequencing and gene expression studies are unknown. A technique called *in vitro* mutagenesis is used to aid in the determination of the function of an unknown gene. Specific changes to the sequence of a cloned gene are made. These mutations can alter or destroy the function of the protein product. The mutated gene is then returned to the host cell. After a period of time the function of the protein is determined by examining the cell's physiology. The mutated genes can also be inserted into the early embryonic cells of multicellular organisms, such as mice, allowing researchers to study the role of the gene in the development and functioning of the organism.

Southern Blots Confirm a Cloned Gene Is Present

The surviving bacterial colonies are tested for the presence of the gene of interest. One method used to identify a specific gene is a technique called a Southern blot, as summarized in **Figure 11-5.**

Step 1 In a Southern blot, the DNA from each bacterial clone colony is isolated and cut into fragments by restriction enzymes.

Step 2 The DNA fragments are separated by gel **electrophoresis** *(eh lehk troh fuh REE sihs),* a technique that uses an electrical field within a gel to separate molecules by their size and charge. The gel looks like a rectangular slab of gelatin with a line of little rectangular "pits" scooped out. The DNA sample is placed in the pits. Because DNA is negatively charged, it migrates toward the positive pole when the electric field is applied. The DNA fragments move through the gel, with the smallest DNA fragments moving fastest. A pattern of bands is formed. The gel is soaked in a chemical solution that separates the double strands in each DNA fragment into single-stranded DNA fragments.

Step 3 The DNA bands are then transferred (blotted) directly onto a piece of filter paper. The filter paper is moistened with a probe solution. **Probes** are radioactive- or fluorescent-labeled RNA or single-stranded DNA pieces that are complementary to the gene of interest.

Step 4 Only the DNA fragments complementary to the probe will bind with the probe and form visible bands.

Study TIP

Word Origins

The word *electrophoresis* is from the Latin *electrocus,* meaning "electricity," and the Greek *phoresis,* meaning "to carry." Knowing this makes it easier to remember that electrophoresis uses electricity to separate DNA fragments.

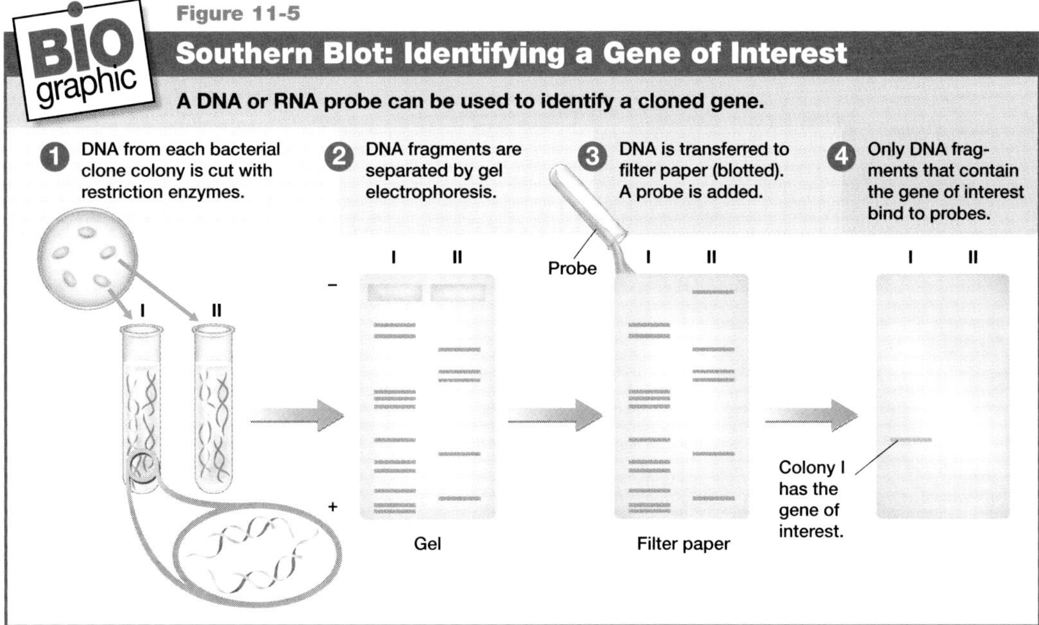

BIO graphic
1 DNA from each bacterial clone colony is cut with restriction enzymes.
2 DNA fragments are separated by gel electrophoresis.
3 DNA is transferred to filter paper (blotted). A probe is added.
4 Only DNA fragments that contain the gene of interest bind to probes.
I
II
–
+
Probe
Gel
Filter paper
Colony I has the gene of interest.

Figure 11-5

Southern Blot: Identifying a Gene of Interest

A DNA or RNA probe can be used to identify a cloned gene.

Teaching TIP

The Discovery of Restriction Enzymes

Tell students that restriction enzymes were actually discovered in the 1950s after scientists realized that bacteria have a very primitive immune system. They observed that certain strains of *E. coli* were able to resist infection by various bacteriophages. It was later discovered that the bacteria possessed enzymes that recognized and destroyed bacteriophage DNA. These enzymes were isolated and called restriction endonucleases (restriction enzymes).

Visual Strategy

BIO graphic **Figure 11-5**

Remind students that the bacterial colonies in **Step 1** were produced from the basic genetic engineering steps described in Figure 11-2. Point out that the fragments on the gel, in **Step 2** are all of different sizes, with the smallest fragments closest to the positive pole. The transfer or blotting of the DNA fragments that occurs in **Step 3** is the reason for the second part of the name of this technique—Southern blot. The first part of the name, Southern, is named after E. M. Southern, who developed the technique. In **Step 4** the colony that contains the gene of interest is actually identified. Because the original Petri dish (from Step 1) is stored while the researcher conducts the Southern blot, the researcher can then return to the original Petri dish and conduct further research on the colony containing the gene of interest.

did you know?

Two other types of "blot" techniques also exist—Northern blot and Western blot.

The Northern blot is similar to a Southern blot but is used to identify RNA, not DNA, fragments. The Western blot is used to identify proteins. These two techniques were named as a play on the name *Southern blot,* which was developed in 1975 by E. M. Southern.

How can you model gel electrophoresis?

Time 15 minutes

Process Skills Modeling, relating information, forming conclusions

Teaching Strategies
The large beads should be big enough so that when placed in the jar there are spaces between them. The smallest beads should flow through the spaces of the large beads.

Analysis Answers
1. the smallest beads
2. The smaller beads represent the smaller DNA fragments.
3. bottom; DNA is negatively charged and will flow to the pole with the opposite charge.
4. The smaller size allows them to flow through the spaces in between the large beads.

3 Assessment Options

Closure

Enzymes and Gene Technology

Have students explain the role of restriction enzymes and DNA ligases in genetic engineering. *(Restriction enzymes are used to cut genes from chromosomes, to "open" bacterial plasmids, and to cut DNA for gel electrophoresis studies; DNA ligases join the ends of the vector with the gene of interest.)*

Review

Assign Review—Section 11-1.

Reteaching — BASIC

Ask students to rewrite each **Objective** on page 226 as a question. Then ask the students to answer each question. PORTFOLIO

internet**connect**

SCILINKS NSTA
TOPIC: Genetic engineering
GO TO: www.scilinks.org
KEYWORD: HX230

Once the bacterial colonies containing the gene of interest are identified, the researcher can manipulate the genetically engineered bacteria in many different ways. For example, the gene of interest can be isolated so that the researcher has pure DNA to use in genetic studies. The researcher can then study how the gene is controlled. Pure DNA allows the researcher to determine the sequence of nucleotides that make up the gene. By comparing the nucleotide sequence of several different organisms, researchers can study the evolution of a particular gene.

The gene of interest can also be isolated and then transferred to other organisms. The bacterial colonies can be used to produce large quantities of the protein coded for by the gene so that the protein can be studied further or used to make drugs, such as insulin.

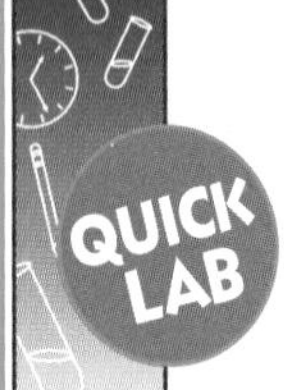

QUICK LAB

How can you model gel electrophoresis?

You can use beads to model how DNA fragments of different sizes are separated in a gel during electrophoresis.

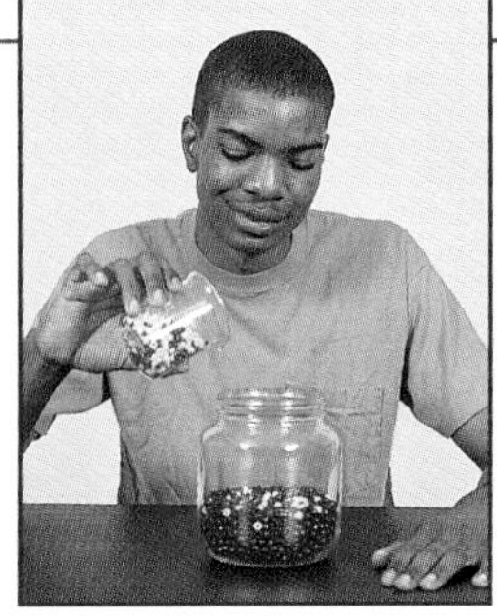

Materials

500 mL beaker, large jar, 3 sets of beads—each set a different size and different color

Procedure

1. Fill a large jar with the largest beads. The filled jar represents a gel.
2. Mix the two smaller beads in the beaker and then pour them slowly on top of the "gel." The two smaller size beads represent DNA fragments of different sizes.
3. Observe the flow of the beads through the "gel." Lightly agitate the jar if the beads do not flow easily.

Analysis

1. **Identify** which beads flowed through the "gel" the fastest.
2. **Relate** the sizes of the beads to the sizes of DNA fragments.
3. **Determine** whether the top or the bottom of the jar represents the side of the gel with the positively charged pole.
4. **SKILL Forming Conclusions** Why do the beads you identified in item 1 pass through the "gel" more quickly?

Review SECTION 11-1

1. **Apply** the four steps commonly used in genetic engineering experiments to describe the cloning of a human gene.
2. **Relate** the role of DNA "sticky ends" in the making of recombinant DNA.
3. **Summarize** how cells are screened in genetic engineering experiments.
4. **Evaluate** the role of probes in identifying a specific gene.

Critical Thinking

5. **SKILL Evaluating Conclusions** A student performing electrophoresis on a DNA sample believes that her smallest DNA fragment is the band nearest the negative pole of the gel. Do you agree with her conclusion? Explain.

Review SECTION 11-1 ANSWERS

1. Human and plasmid DNA are cut and then combined to produce recombinant DNA. The recombinant DNA is inserted into bacteria. The bacteria reproduce, cloning the human gene. The bacterial cells are screened for the presence of the human gene.
2. The bases of the sticky ends are complementary, which allows the DNA from the two different organisms to combine.
3. The plasmids contain the gene for tetracycline resistance. The bacteria with the recombined plasmid are grown in the presence of tetracycline. Only the bacteria that take up the plasmid will survive.
4. The probes are pieces of single-stranded DNA or RNA complementary to the gene of interest. The DNA fragments on the gel are made single stranded. The probe will bind to any complementary fragments. The genes are identified because the probes are made with radioactive or fluorescent labeled tags.
5. Students should disagree. DNA is negatively charged and opposite charges attract. The smallest fragments will move fastest through the gel.

Genetic Engineering in Medicine and Society

Section 11-2

Genetically Engineered Drugs and Vaccines Are Now Commonplace

Much of the excitement about genetic engineering has focused on its potential uses in our society. The possibilities for the applications of these techniques in medicine and research are endless. Many applications are already commonplace, such as the production of genetically engineered proteins used to treat illnesses. Others, such as the replacement of defective human genes with healthy ones, are still being developed.

Medicines

Since many genetic disorders and other human illnesses occur when the body fails to make critical proteins that are essential for proper functioning, the development of genetically engineered proteins is important. Most proteins are present in the body in very low amounts, making the large quantities needed for pharmaceuticals difficult and expensive to obtain. As summarized in **Figure 11-6,** today hundreds of pharmaceutical companies around the world produce medically important proteins in bacteria using genetic engineering techniques. These products include anticoagulants (proteins involved in dissolving blood clots), which are effective in treating heart attack patients, and insulin, which is used to treat diabetes.

Factor VIII, a protein that promotes blood clotting, is also made by genetic engineering and sold as a drug. A deficiency in factor VIII leads to one type of hemophilia, an inherited disorder characterized by prolonged bleeding. For a long time, hemophiliacs received blood factors that had been isolated from donated blood. Unfortunately, some of the donated blood was infected with

Objectives

- **Describe** how drugs produced by genetic engineering are being used. (pp. 231–232)
- **Summarize** the steps involved in making a genetically engineered vaccine. (pp. 232–233)
- **Describe** how gene therapy is being used to try to cure genetic disorders. (p. 234)
- **Identify** two different uses for DNA fingerprints. (p. 235)
- **Summarize** two major goals of the Human Genome Project. (p. 236)

Key Terms

vaccine
gene therapy
DNA fingerprint
Human Genome Project

Genetically Engineered Medicines

Product:	Used for treatment of:
Erythropoetin	Anemia
Growth factors	Burns, ulcers
Human growth hormone	Growth defects
Interleukins	HIV infection, cancer, and immune deficiencies
Interferons	Viral infections and cancer
Taxol	Ovarian cancer

Figure 11-6 Use of genetically engineered medicines. Many medicines, such as medicines used to treat burns, are produced by genetic engineering techniques.

Integrating Different Learning Styles

Logical/Mathematical	**PE**	Performance Zone items 16 and 25, p. 242–243
Visual/Spatial	**PE**	Data Lab, p. 235
	ATE	Teaching Tips, pp. 232, 236
Interpersonal	**ATE**	Biology in Your Community, p. 233
Intrapersonal	**PE**	Biobit, p. 232
	ATE	Lesson Warm-up, p. 231; Teaching Tip, p. 232
Verbal/Linguistic	**PE**	Study Tip, p. 234; Performance Zone items 22 and 23, p. 243
	ATE	Teaching Tips, pp. 232, 234; Reteaching, p. 236

Section 11-2

1 Prepare

Scheduling

- **Block:** About 135 minutes will be needed to complete this section.
- **Traditional:** About three class periods will be needed to complete this section.
- **Planning:** See the Planning Guide on pp. 223A and 223B for lecture, classwork, and assignment options for Section 11-2. Lesson plans for Section 11-2 can be found in a lesson cycle format in ***Biology: Principles and Explorations*** **One-Stop Planner CD-ROM with Test Generator.**

Resource Link

- Assign **Active Reading Worksheet Section 11-2** to help students comprehend the key concepts and visuals in this lesson.

2 Teach

Lesson Warm-up

Use an ink pad and white paper to take a fingerprint of each student. Display the prints. Ask the class what is special about each print. *(They are all different.)* Explain that just as everyone (except identical twins) has different fingerprints, we all have different DNA prints. Inform students that the DNA prints are called DNA fingerprints because they create a pattern similar to the way that the ridges on the fingers leave a pattern in a fingerprint. Obtain autoradiographs of DNA fingerprints from science journals or magazines and demonstrate them to the class.

Teaching TIP

Making Genetically Engineered Drugs

Have students make a **Graphic Organizer** similar to the one at the bottom of page 232 to outline how genetically engineered drugs are made. PORTFOLIO

Making Biology Relevant

Health

Ask students to read the **Biobit** on page 232. Ask them why this type of vaccine would be beneficial for people living in developing countries. *(Developing countries are economically disadvantaged. People cannot afford medical care. The genetically engineered vaccines should be less expensive to produce and might even be distributed through various agencies. These vaccines should be easy to distribute and administer, especially to children.)*

Teaching TIP

Ethical Issues in Genetic Engineering

Obtain the October 1999 issue of *National Geographic* (vol. 196, no. 4). For each of the photos in the article "Secrets of the Gene," discuss with the students the ethical and moral problems that have arisen. Emphasize that environmental factors play a role in many diseases—just because you have the gene for a disease does not mean you will get the disease. Ask the students to discuss the following in relation to their own lives: Should insurance companies have access to our DNA profiles? Would you want to know if you have a gene for a disease or a type of cancer? Should you have children if you know that you have a gene for a disease or a type of cancer?

biobit

Could I get a vaccination from a banana?

Genetic engineers are putting genes from disease-causing microbes into fruits and vegetables. They want to create oral vaccines that are inexpensive and easy to take. Such vaccines would be especially useful in developing countries. Clinical trials using different foods, including potatoes, are underway.

viruses such as HIV and hepatitis B. The viruses were sometimes unknowingly transmitted to people who received blood transfusions. Today the use of genetically engineered factor VIII eliminates these risks.

Vaccines

Many viral diseases, such as smallpox and polio, cannot be treated effectively by existing drugs. Instead they are combated by prevention—using vaccines. A **vaccine** is a solution containing all or part of a harmless version of a pathogen (disease-causing microorganism). When a vaccine is injected, the immune system recognizes the pathogen's surface proteins and responds by making defensive proteins called antibodies. In the future, if the same pathogen enters the body, the antibodies are there to combat the pathogen and stop its growth before it can cause disease.

Traditionally, vaccines have been prepared either by killing a specific pathogenic microbe or by making the microbe unable to grow. This ensures that the vaccine itself will not cause the disease. The problem with this approach is that there is a small but real danger that a failure in the process to kill or weaken a pathogen will result in the transmission of the disease to the very patients seeking protection. This danger is one of the reasons why, for example, rabies vaccines are administered only when a person has actually been bitten by an animal suspected of carrying rabies.

Vaccines made by genetic engineering techniques side step this danger. As illustrated in **Figure 11-7,** the genes that encode the pathogen's surface proteins can be inserted into the DNA of harmless

Figure 11-7 Making a genetically engineered vaccine

A person vaccinated with a genetically engineered vaccine, such as the genital herpes vaccine, will make antibodies against the virus.

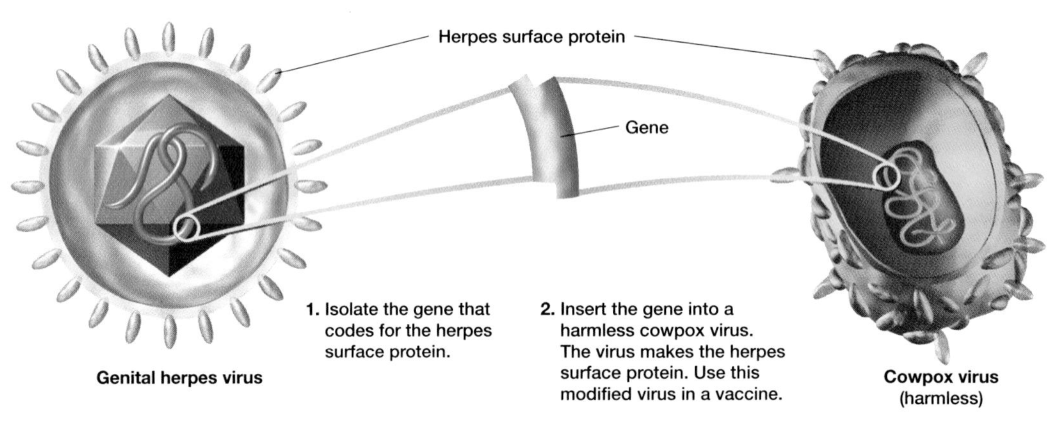

Herpes surface protein
Gene
1. Isolate the gene that codes for the herpes surface protein.
2. Insert the gene into a harmless cowpox virus. The virus makes the herpes surface protein. Use this modified virus in a vaccine.
Genital herpes virus
Cowpox virus (harmless)

Graphic Organizer

*Use this graphic organizer with **Teaching Tip: Making Genetically Engineered Drugs** on page 232.*

Desired gene identified → Gene transferred to bacterial plasmid → Plasmid containing desired gene introduced into bacteria → Bacteria grown in large numbers to produce needed drug

bacteria (or viruses). The modified but harmless bacteria become an effective and safe vaccine.

Vaccines for the herpes II virus and for the hepatitis B virus are now being made through genetic engineering. The herpes II virus produces small blisters on the genitals (the external sex organs). The hepatitis B virus causes an inflammation of the liver that can be fatal. A major effort is underway to produce a vaccine that will protect people against malaria, a protozoan-caused disease for which there is currently no effective protection.

Polymerase Chain Reaction (PCR)

A detective finds a single hair as the only evidence left behind at a crime scene. Will this hair provide enough DNA to analyze? For DNA fingerprinting and many of the genetic engineering uses discussed throughout this chapter, a certain amount of DNA is needed. Sometimes, however, only a very tiny amount of DNA is available.

Today scientists use a technique called the polymerase chain reaction (PCR) to quickly make many copies of selected segments of the available DNA. With PCR, a scientist can produce a billionfold increase in DNA material within a few hours!

Heating and replication cycles

In PCR the double-stranded DNA sample to be copied is heated, which separates the strands. The mixture is cooled, and short pieces of artificially made DNA called primers are added. The primers bind to places on the DNA where the copying can begin.

DNA polymerase and free nucleotides are added to the mixture. The DNA polymerase extends the DNA by adding on complementary free nucleotides to the primer. The result is two strands of DNA that are identical to each other and to the original strand. The heating and replication process is repeated over and over again. Every 5 minutes, the sample of DNA doubles again, resulting in many copies of the sample in a short amount of time.

PCR has many uses

PCR can duplicate DNA from as few as 50 white blood cells, which might be found in a nearly invisible speck of blood. PCR is important for diagnosing genetic disorders and for solving crimes. PCR is also used in different types of research and for studying ancient fragments of DNA found in fossils or in preserved material.

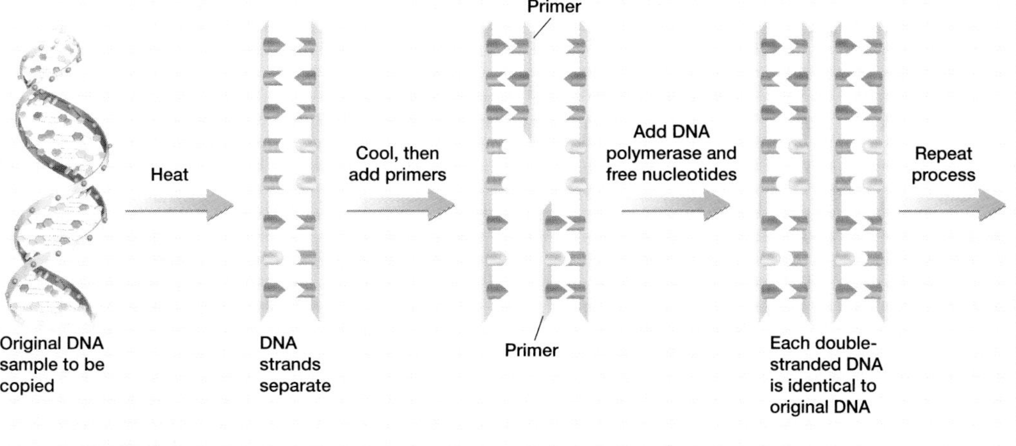

TECH WATCH

Polymerase Chain Reaction (PCR)

Teaching Strategies

- Continue the diagram at the bottom of this feature to give students an idea of how quickly copies of the original DNA strand are made.
- Let students know that PCR is now a very common technique used to obtain sufficient DNA samples not just in criminal cases but for basic research as well.
- When the source of DNA is meager or impure, PCR is a quicker and more selective method than gene cloning. Because PCR is performed completely *in vitro* (in test tubes), cells are not used as with gene cloning. Billions of copies of DNA can be made in a few hours with PCR, whereas gene cloning would take days.

Discussion

- If most proteins (such as enzymes) denature or break down with high heat, how can the DNA polymerase used in PCR withstand the heating cycles? *(The DNA polymerase is isolated from bacteria living in hot springs. The DNA polymerase from these bacteria can withstand the heat needed to separate the DNA strands at the start of each cycle.)*
- Under conditions of PCR amplification, DNA replicates about every 5 minutes. Determine how many copies of a DNA fragment will result from 85 minutes of PCR. *(85 min $\div$ 5 min/cycle $=$ 17 cycles; 2^{17} or 1.31×10^5 copies)*

Biology in Your Community

Research Lab

Contact a local university or pharmaceutical company and arrange a field trip for students to visit a lab that uses genetic engineering techniques. Have students briefly research the area in which the investigator is working. Ask students to prepare a list of questions about the research topic, requirements for a career in genetic engineering, the type of jobs available, and applications of the techniques. Ask the researcher if it is possible for the students to practice some of the techniques, such as loading DNA on an electrophoresis gel. PORTFOLIO

Teaching TIP — ADVANCED

In the Future

Have students write a newspaper article that may be written in 30 years, dealing with some aspect of genetic engineering. The article should include a photo. Students can either draw the photo or cut and paste other photos to create their final photo. Before starting the project, remind students that 30 years ago no one would have thought possible some of the genetic engineering techniques that are available today.

PORTFOLIO

Making Biology Relevant

Medicine

Xenotransplantation is the transplanting of an organ from an animal species into a human. Since human organs are not widely abundant, researchers have tried using primate organs, such as chimpanzee kidneys and baboon livers, without success. Pigs, on the other hand, are physiologically similar to humans but are too genetically different and produce rejection. The cells lining a pig's blood vessels contain a sugar molecule called alpha-galactose, which the human immune system immediately recognizes and attacks. The focus is now on developing a line of donor animals that are genetically engineered to prevent rejection by the immune system of the recipient. This is done by inserting into the pig the human gene for H-transferase, an enzyme that converts the pig alpha-galactose into a harmless human sugar that appears on human type O blood cells. Much work still needs to be done in this area since rejection may occur over a period of time.

Study TIP

Reading Effectively

Write in your notes the red subheadings at the top of the pages to form an outline of the main concepts in Section 11-2. Under each subheading, list the Objectives that address each subhead.

A Key Goal of Genetic Engineering Is to Cure Genetic Disorders

Many genetic disorders occur when an individual lacks a functioning copy of a particular gene. **Gene therapy** is a technique that involves putting a healthy copy of a gene into the cells of a person whose copy of the gene is defective. Cells are removed from the patient, healthy genes are inserted into the cells, and the cells are returned to the patient's body. The substance lacking in the person is then produced by the cells with the new genes.

One of the first gene-therapy attempts involved the young girl shown in **Figure 11-8,** who suffered from an immune-system disorder caused by a defective gene. This gene failed to produce an important immune-system enzyme. Doctors extracted bone marrow cells from the girl and replaced the defective gene with a normal gene. These cells were returned to the girl's bones and began to produce the missing enzyme. Because this kind of bone-marrow cell actively divides, researchers hope that offspring of the genetically engineered cells will continue to secrete the enzyme into the blood.

Cancer and cystic fibrosis

Gene therapy is also providing a new weapon in the battle against cancer. All humans have white blood cells that secrete a protein called tumor necrosis factor (TNF). TNF attacks and kills cancer cells. Unfortunately, this attack does not happen often. Molecular biologists recently developed a method of adding the TNF gene to a kind of white blood cell that is very effective at locating cancer cells. The modified white blood cells secrete TNF and kill cancer cells.

Cystic fibrosis is another genetic disorder that is being treated with gene therapy. Cystic fibrosis is caused by one defective gene that results in the malfunction of a single protein. Some success has been reported with nasal sprays that carry a normal cystic fibrosis gene to the cells in the nose and lungs—cells that are particularly affected by the defective protein.

Researchers still must overcome many obstacles, including how and where to insert genes safely and directly into cells. Researchers hope to use gene therapy to provide permanent cures for genetic disorders. Diseases that are currently being treated through gene therapy are listed in Figure 11-8.

Figure 11-8 Gene therapy. The diseases listed are currently being treated in clinical trials using gene therapy. Cynthia Cutshall was one of the first patients to receive gene therapy.

Diseases Being Treated with Gene Therapy

- Cancer
- SCID (Severe Combined Immunodeficiency)
- Cystic fibrosis
- Familial hypercholesterolenemia
- Hemophilia
- Fanconi's anemia
- Rheumatoid arthritis

Multicultural Perspective

Incan Ice Mummies

Found at 22,057 ft above sea level in the Andes mountains in Argentina, the ice mummies of three Incan children were sacrificial victims offered to the Incan gods. Unlike other ice mummies that have been found, the bodies have been exceptionally well preserved. Two of the bodies contain internal organs that are frozen rather than freeze-dried. Scientists hope that DNA fingerprinting of the mummies might trace genetic links between Incas and other societies and determine how South America was settled.

Genetic Engineering Techniques Are Used to Identify Organisms

Other than identical twins, no two individuals have the same genetic material. Because of random mutations and recombinations during sexual reproduction, any two individuals will differ in their DNA nucleotide sequence. Since the places a restriction enzyme can cut depend on the DNA sequence, the DNA fragments that result will differ between two individuals. Such DNA fragments of different lengths (polymorphisms) are called restriction fragment length polymorphisms, or RFLPs (pronounced "riflips").

RFLPs can be used to identify individuals and to determine how closely related members of a population are to one another. The RFLPs are treated in the same manner as a Southern blot, shown in Figure 11-5. The result is called a DNA fingerprint. A **DNA fingerprint** is a pattern of dark bands on photographic film that is made when an individual's DNA fragments (RFLPs) are separated by gel electrophoresis, probed, and then exposed to an X-ray film. Because restriction enzymes cut the DNA from different individuals into DNA fragments with different lengths, each individual (other than identical twins) has a unique pattern of banding or DNA fingerprint.

The banding patterns from two different individuals can be compared to establish whether they are related, such as in a paternity case. Because DNA fingerprinting can be performed on a sample of DNA found in blood, semen, bone, or hair, it is useful in forensics. Forensics is the scientific investigation of the causes of injury and death when criminal activity is suspected. DNA fingerprints are also valuable for identifying the genes that cause genetic disorders, such as Huntington's disease and sickle cell anemia.

REAL LIFE

A soldier buried in a memorial was identified by DNA fingerprinting.
The Tomb of the Unknown Soldier memorial honors four unidentified American soldiers, one each from World War I, World War II, the Korean War, and the Vietnam War. The soldier who was identified served in the Vietnam War.

Finding Information
Research other ways DNA fingerprinting is used to identify individuals.

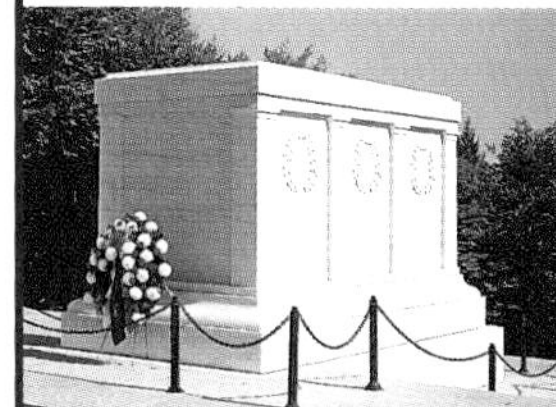

DATA LAB

Interpreting a DNA fingerprint

Background

The photograph shows seven columns, or lanes, on a photographic film from a DNA fingerprinting experiment. The experiment was conducted to determine whether one of four suspects was involved in a crime.

1 2 3 4 5 6 7

KEY

1 Control	4 Suspect 1
2 Blood at crime scene	5 Suspect 2
3 Victim	6 Suspect 3
	7 Suspect 4

Analysis

1. **Determine** which suspect's DNA fingerprint matches the DNA in the blood found at the crime scene.
2. **Infer** why the DNA from the victim in the crime was included in the experiment.
3. **Discuss** why a control was included in lane 1.
4. **SKILL Relating Conclusions** How likely is it that blood found at the crime scene belongs to one of the suspects?

MULTICULTURAL PERSPECTIVE

Genetic Diseases of Different Cultures

Scientists are currently using the knowledge of DNA technology to find a prevention or cure for four serious genetic defects that affect people from particular environments: sickle cell anemia (affecting people from Africa, the Mediterranean, India, and Asia); thalassemia (affecting people in the malarial areas of the Middle East, Africa, the Mediterranean, and India); Tay-Sachs (affecting Ashkenazi Jews); and tyrosinemia (affecting French-Canadian children in the isolated Lac St. Jean-Chicoutimi region of Quebec).

REAL LIFE

Answer

Answers will vary. For example: DNA fingerprinting is used to identify unrelated individuals (especially endangered species) in zoos or breeding programs (to increase genetic diversity when mating individuals); to determine the paternity of populations in behavioral ecology studies; for the isolation and characterization of genes for various human diseases; and for forensic investigations.

CROSS-DISCIPLINARY CONNECTION

◆History

Czar Nicholas II, emperor of Russia from 1894–1917, his wife, and five children were shot to death by the Bolsheviks in the aftermath of the Russian revolution in 1918. Their bodies were not found until 1991 when Russian anthropologists discovered the bones. To be certain, DNA samples were taken from the corpses and living members of the Romonov family. From this, it was determined that the bodies were indeed the royal family.

DATA LAB

Interpreting a DNA fingerprint

Time 10 minutes

Process Skills Analyzing, inferring, relating conclusions

Teaching Strategies

- To capture students' interest, discuss a current high-profile event that has involved DNA fingerprinting (such as a crime or a paternity suit).
- Before students begin, draw a sample DNA fingerprint on the board and show students how to match the DNA bands.

Analysis Answers

1. suspect 2 (lane 5)
2. to determine if the DNA gathered at the crime scene belonged to the victim
3. to ensure that the DNA fragments are moving as expected for their size
4. very likely

Chapter 11

Teaching TIP — ADVANCED

Timeline

Tell students that the first sequencing of an entire human chromosome occurred in 1999 (chromosome 22). Over 500 genes were found on the chromosome. Divide students into groups. Assign each group a time span such as 1980–1985, 1986–1990, 1991–1996, etc. Ask students to use the Internet and other resources to research the discovery of genes in the human genome. The chromosome location for the genes should be included whenever possible. *(For example: cystic fibrosis on chromosome 7 in 1989 and Werner's syndrome—premature aging—on chromosome 8 in 1996)* Have students plot their findings on a timeline that stretches across one wall in the classroom. Then ask them whether the discovery of genes has increased as time passes. *(yes)* PORTFOLIO

3 Assessment Options

Closure

Recognizing Differences

Ask students to explain how gene therapy differs from other methods of dealing with human genetic disorders. *(Gene therapy is an attempt to cure the disorder by replacing the defective gene; other methods involve treating only the symptoms of the disorder.)*

Review

Assign Review—Section 11-2.

Reteaching — BASIC

Have students write one sentence for each of the **Key Terms** on page 231. Each sentence should demonstrate the meaning of the term as it is defined in the text. PORTFOLIO

The Sequence and Location of All Human Genes Are Being Studied

The potential for gene technology to help in the fight against disease is great. One of the most significant efforts to increase the usefulness of gene technology is the Human Genome Project. The goal of the **Human Genome Project** is to determine the nucleotide sequence of the entire human genome and to map the location of every gene on each chromosome by the year 2003 or sooner. The word *genome* is used to refer to all of an organism's DNA. Remember that human cells have their nuclear DNA in 23 pairs of chromosomes. Because the human genome contains about 3 billion nucleotide base-pairs and about 100,000 genes, the task is challenging.

Scientists hope that the knowledge gained from the Human Genome Project will lead to improvements in diagnoses, treatments, and even cures for the approximately 4,000 human genetic disorders. Scientists have already discovered specific genes responsible for several genetic disorders, including cystic fibrosis, Duchenne muscular dystrophy, and colon cancer. All human chromosomes have been sequenced, and now the functions of the genes are being identified. **Figure 11-9** shows several genes and genetic disorders that have been mapped on the X chromosome.

Figure 11-9
Chromosome map. The Human Genome Project has revealed the location of many genes. Although more than 450 genes and 200 genetic disorders have been located on the X chromosome, only a few are shown here.

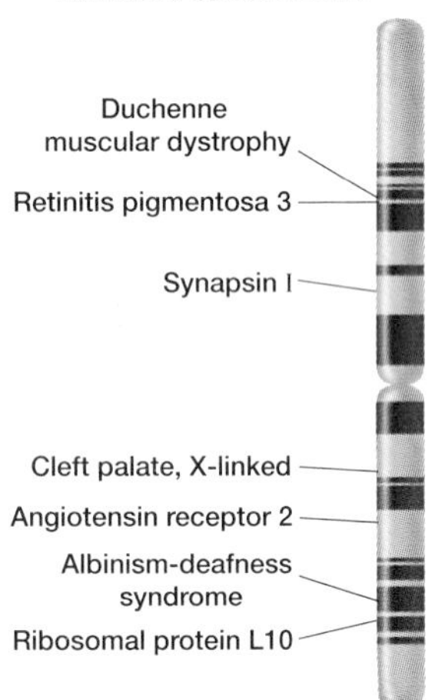

Mapping and sequencing other genomes

In addition to the work on the human genome, the genomes of many other organisms are being mapped and sequenced. The genomes of more than 15 organisms have already been completely sequenced. For example, the genome of the bacterium *Haemophilus influenzae* was sequenced in 1995, and the genome for brewer's yeast *(Saccharomyces cerevisiae)* was sequenced in 1996. In 1998, the sequencing of the genome of the first multicellular animal—the roundworm *Caenorhabditis elegans*—was completed. The genomes of several key organisms used in biological research, including the mouse, the fruit fly, *Drosophila* and the plant *Arabidopsis*, are being sequenced and mapped.

Review SECTION 11-2

1. **Relate** the use of genetic engineering to the treatment of human illnesses such as hemophilia.
2. **Relate** genetic engineering techniques to the making of vaccines.
3. **Name** the process in which a healthy copy of a gene is inserted into the cells of a person with a defective version of the gene.
4. **List** two ways in which DNA fingerprinting has been useful to society.
5. **SKILL Applying Information** Why is the Human Genome Project important to medical research?

Critical Thinking

6. **SKILL Distinguishing Relevant Information** A student states that genetic engineering is"perfectly safe and sound." What safety and ethical issues do you think might arise over the use of genetic engineering?

Review SECTION 11-2 ANSWERS

1. Many genetically engineered proteins are used to treat illnesses. For example, Factor VIII, a protein that promotes blood clotting, is now made by genetic engineering and sold as a drug to hemophiliacs. Genetic engineers are also attempting to replace defective human genes with healthy ones.
2. Instead of using a killed or weakened pathogen, the genes that code for the proteins found on the surface of the pathogen are inserted into the DNA of harmless bacteria or viruses. People are then vaccinated with the modified virus or bacteria.
3. gene therapy
4. DNA fingerprinting has been useful in forensics, in paternity suits, and in identifying the genes that cause genetic disorders.
5. By identifying and mapping every gene, it might be possible to diagnose, treat, and cure genetic disorders.
6. Answers will vary. For example: How will the information be used? Will genetically engineered crops be controlled to stay in a specific environment?

Genetic Engineering in Agriculture

Section 11-3

Transporting Genes into Plants Can Improve Crops

Farmers began primitive genetic breeding by selecting seeds from their best plants, replanting them, and gradually improving the quality of successive generations. In the twentieth century, plant breeders started using the principles of genetics to select plants. Today, genetic engineers can add favorable characteristics to a plant by manipulating the plant's genes, as shown in **Figure 11-10.**

Genetic engineers can change plants in many ways, including making plants more tolerant to drought conditions and creating plants that can adapt to different soils, climates, and environmental stresses. Genetic engineers can control how fast some fruits ripen. Genetic engineers have been able, in some instances, to improve the nutritional value of plants. For example, several strains of rice with high levels of beta-carotene (which is broken down in the body to vitamin A) and iron have been genetically engineered. This will have a major impact in Asia, where millions of people suffer from vitamin A and iron deficiency.

Makers of a biodegradable weedkiller called glyphosate have developed crop plants that are resistant to glyphosate. This has enabled farmers to apply glyphosate to kill weeds without killing their crops. Because the field does not need to be tilled to control weeds, less topsoil is lost to erosion. Scientists have developed crops that are resistant to insects by inserting a certain gene into crop plants. This gene makes a protein that injures the gut of chewing insects. Crops that are resistant to insects do not need to be sprayed with pesticides, many of which can harm the environment.

Finding a plant vector

For years, genetic engineers lacked a suitable vector to carry a gene into plant cells. The breakthrough came in the form of an unusual bacterial plasmid responsible for crown gall, a plant disease characterized by large bulbous tumors on the plant. This plasmid is called the Ti plasmid ("Ti" stands for tumor-inducing). The Ti plasmid easily infects broad-leaved crop plants, such as tomatoes, tobacco, and soybeans, by inserting itself into the plant's cells. Scientists removed the tumor-causing genes from the Ti plasmid and filled the space with specific DNA. More recently, molecular biologists successfully "shot" the Ti plasmid into cells of wheat plants using a "gene gun."

Objectives

- **Describe** three ways in which genetic engineering has been used to improve plants. (p. 237)
- **Summarize** two ways in which genetic engineering techniques have been used to modify farm animals. (pp. 238–239)
- **Summarize** the cloning of sheep through the use of differentiated cells. (p. 239–240)

Key Terms

transgenic animal

Figure 11-10
Genetically engineered plants

How many? At least 50 plants have been genetically engineered, including potatoes, soybeans, and corn. A researcher named Athanasios Theologis genetically engineered tomatoes to ripen without becoming soft.

Integrating Different Learning Styles

Logical/Mathematical	**PE**	Performance Zone items 17 and 27, pp. 242–243
Visual/Spatial	**ATE**	Demonstration, p. 238
Verbal/Linguistic	**PE**	Performance Zone items 26 and 28, p. 243
	ATE	Teaching Tip, p. 238; Closure, p. 240; Reteaching, p. 240; Alternative Assessment, p. 241

Section 11-3

1 Prepare

Scheduling

- **Block:** About 90 minutes will be needed to complete this section.
- **Traditional:** About two class periods will be needed to complete this section.
- **Planning:** See the Planning Guide on pp. 223A and 223B for lecture, classwork, and assignment options for Section 11-3. Lesson plans for Section 11-3 can be found in a lesson cycle format in ***Biology: Principles and Explorations*** **One-Stop Planner CD-ROM with Test Generator.**

Resource Link

- Assign **Active Reading Worksheet Section 11-3** to help students comprehend the key concepts and visuals in this lesson.

2 Teach

Lesson Warm-up

Tell students that the Flavr-Saver™ tomato was the first genetically altered fruit to reach supermarket shelves. It was altered to last longer and taste better than other types of tomatoes. Genetically engineered crops also present an alternative to the use of pesticides. A large percentage of the world's food supply is lost to pests. Pesticide use, though effective, has its disadvantages. Ask students what some of the disadvantages might be. *(Over time the pests become resistant to the pesticide; other plant and animal life may be harmed, causing a disruption in the food web; human life may also be affected.)* Ask students to point out some concerns about using genetically engineered crops. *(The genetically engineered plant may pass the genes on to close relatives in the wild.)*

Teaching TIP ADVANCED

Ad for Gene Products

Have groups of students choose and then research a genetically engineered product such as human insulin. Ask students to create a full-page advertisement. The ad must be creative, name the product and its use, list side effects, and convince the public that the genetically engineered product is safe. PORTFOLIO

Demonstration

Selective Breeding Versus DNA Technology

Bring in pictures of farm animals produced by selective breeding and by genetic engineering. Discuss the advantages of genetic engineering over selective breeding. *(A new variety is developed in a shorter period of time using DNA technology instead of selective breeding. You can also insert medically useful human proteins so that the animal produces the protein in the milk.)* Emphasize that the development of organisms to enhance certain features has been occurring for many years through selective breeding. Ask students to envision ways in which genetic engineering might *not* be beneficial. *(may lead to a loss of biodiversity or create unforeseen problems)*

Gene Technology Is Being Used in Animal Farming

Farmers have long tried to improve farm animals and crops through traditional breeding and selection programs. In the past, the cow that produced the most milk on a farm was mated to sons of high producers in hopes that her offspring would also produce a lot of milk. But these traditional processes were slow and inefficient.

Now many farmers use genetic-engineering techniques to improve or modify farm animals. Some farmers add growth hormone to the diet of cows to increase milk production. Previously, the growth hormone was extracted from the brains of dead cows. But now the cow growth hormone gene is introduced into bacteria. The bacteria produce the hormone so cheaply that it is practical to add it as a supplement to the cows' diet.

Extra copies of the growth-hormone gene have been introduced directly into the chromosomes of both cattle and hogs to increase their weight. Though these procedures are still new, they may lead to the creation of new breeds of very large and fast-growing cattle and hogs.

Making medically useful proteins

Another way in which gene technology is used in animal farming is in the addition of human genes to the genes of farm animals in order to get the farm animals to produce human proteins in their milk. This is used especially for complex human proteins that cannot be made by bacteria through gene technology. The human proteins are

Figure 11-11 Cloning a sheep from mammary cells

In 1997 scientists announced the first successful cloning using differentiated cells—a lamb named Dolly.

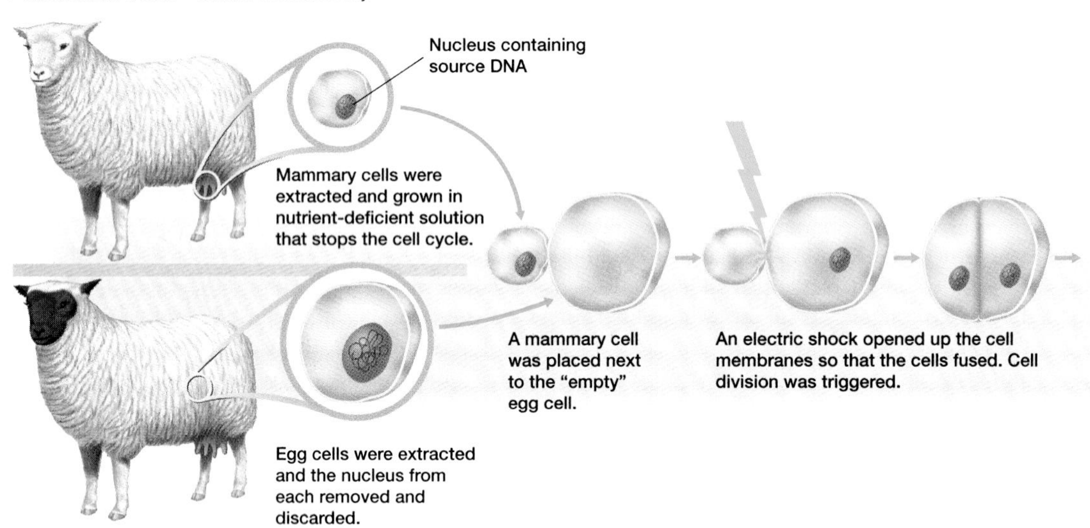

Nucleus containing source DNA
Mammary cells were extracted and grown in nutrient-deficient solution that stops the cell cycle.
A mammary cell was placed next to the "empty" egg cell.
An electric shock opened up the cell membranes so that the cells fused. Cell division was triggered.
Egg cells were extracted and the nucleus from each removed and discarded.

did you know?

Scientists have genetically engineered the common research plant,* Arabidopsis, *to resist cold weather.

Researchers have genetically engineered these plants to overproduce a protein. The protein activates several genes that help plants resist damage caused by freezing temperatures. The researchers are currently working on genetically engineering crop plants, such as corn and soybeans, so that they can withstand the sudden cold snaps that often cause large crop losses.

extracted from the animals' milk and sold for pharmaceutical purposes. The animals are called **transgenic animals** because they have foreign DNA in their cells.

Most recently, scientists have turned to cloning animals as a way of creating herds of identical animals that can make medically useful proteins. The intact nucleus of an embryonic or fetal cell (whose DNA has been recombined with a human gene) is placed into an egg whose nucleus has been removed. The egg with the new nucleus is then placed into the uterus of a surrogate, or substitute mother, and is allowed to develop.

Cloning from specialized cells

In 1997, a scientist named Ian Wilmut captured worldwide attention when he announced the first successful cloning using differentiated cells from an adult animal. A differentiated cell is a cell that has become specialized to become a specific type of cell (such as a liver or udder cell). As summarized in **Figure 11-11,** a lamb was cloned from the nucleus of a mammary cell taken from an adult sheep. Previously, scientists thought that cloning was possible only using embryonic or fetal cells that have not yet differentiated. Scientists thought that differentiated cells could not give rise to an entire organism. Wilmut's experiment proved otherwise.

An electric shock was used to fuse mammary cells from one sheep with egg cells without nuclei from a different sheep. The fused cells divided to form embryos, which were implanted into surrogate mothers. Only one embryo survived the cloning process. Dolly, born on July 5, 1996, was genetically identical to the sheep that provided the mammary cell.

REAL LIFE

Scientists are trying to clone the first dog. The project is called the Missyplicity Project in honor of Missy, the dog whose DNA will be used to produce the first cloned offspring. The researchers have many goals including cloning service dogs such as seeing-eye dogs.

Finding Information *Use Internet sources to find out the current status of this project that is being led by researchers at Texas A&M University.*

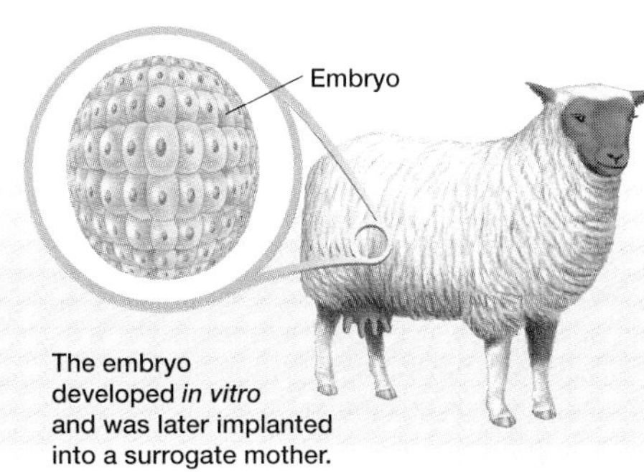

Embryo
The embryo developed *in vitro* and was later implanted into a surrogate mother.

After a 5-month pregnancy, a lamb was born that was genetically identical to the sheep from which the mammary cell was extracted.

did you know?

Scientists are trying to clone and analyze the DNA of a 23,000-year-old woolly mammoth.

Researchers found the mammoth, which was perfectly preserved as a result of being frozen soon after dying, in the frozen tundra of northern Siberia. Although it is unlikely the DNA has remained intact, the researchers hope to analyze the mammoth's genes, as well as the genes of biological material (such as pollen grains and grasses) that were frozen alongside the animal.

Teaching TIP

Dolly Aging Prematurely

Dolly, the lamb cloned from undifferentiated cells, has developed some unforeseen problems. Her chromosomes are showing signs of premature aging. Ask students why the chromosomes would appear abnormally old when Dolly is relatively young. *(Dolly was cloned from mammary cells from an adult sheep.)* How does this information affect other cloning experiments? *(It appears that when organisms are cloned from adult cells, their chromosomes age prematurely. This in turn may eventually affect the metabolism of the organism.)*

REAL LIFE

Answer

Search the Internet for the Missyplicity Web site to obtain the latest developments on this project.

CAREER

Agricultural Scientist

An agricultural scientist works to improve the quantity and quality of farm crops and animals. These scientists often use the ever-increasing knowledge of biotechnology in their field. Agricultural scientists may work in research and development, oversee the development of research programs, or act as consultants to various agencies. For more information, students can contact organizations such as Food and Agricultural Careers for Tomorrow, Purdue University, 1140 Agricultural Administration Building, West Lafayette, IN 47907-1140.

Other scientists have since used a similar approach to clone cows and mice, showing cloning in this manner is possible with other mammals. As future research continues, cloning using differentiated cells from transgenic cloned animals will likely become commonplace in agriculture.

It is impossible not to speculate on the possibility of cloning a human. There is no reason to believe that such an experiment would fail, but many question whether it should be done. The ethics of human cloning have not been fully addressed and will undoubtedly continue to generate considerable controversy.

Exploring Further

A Breakthrough in Cloning

Teaching Strategies

- Emphasize to students that Dr. Wilmut's goal was to clone animals that could make medically useful proteins in the milk.
- Have students research the latest developments that have occurred in cloning with differentiated cells. Have them include the different types of animals that have been successfully cloned using this method. *(Answers may include mice, cows, and bulls.)*

Discussion

- Do you think that scientists will soon be able to clone humans? *(Answers will vary.)*
- Under what circumstances do you think cloning of humans should be allowed? *(Answers will vary.)*

Exploring Further

A Breakthrough in Cloning

Ian Wilmut's sheep-cloning experiment generated a lot of controversy. People began to discuss the possibility that humans could be "easily" cloned from differentiated cells, such as skin or hair cells. Why did the experiments generate so much controversy? How was this experiment different from previous cloning experiments?

Wilmut reset the cell-cycle stage

Wilmut's experiment marked the first time that a clone was produced from a differentiated cell, rather than from an embryonic cell. The key was that the egg whose nucleus had been removed and the donated nucleus of the adult cell needed to be at the same stage in the cell cycle. Wilmut was able to do this by starving the adult udder cell of nutrients. He found that when cells are starved, they pause at the beginning of the cell cycle.

Wilmut removed mammary cells from the udder of a 6 year-old sheep. The cells were grown in tissue culture and then were starved for 5 days. At the same time, the nuclei were removed from eggs obtained from a ewe. The mammary cells and egg cells were surgically combined. The researchers applied a brief electrical shock, causing the plasma membrane to become leaky. The contents of the mammary cell passed into the egg cell. The shock also kick-started the cell cycle, causing the egg cell to begin dividing.

Cloned calf

Many tries were necessary

After 6 days, in 29 of 277 tries, the dividing embryos reached a stage at which they could be implanted into surrogate mother sheep. A little over 5 months later one sheep gave birth to a lamb, Dolly. Dolly later gave birth to a lamb, which confirmed that clones produced in this way are able to breed normally and produce healthy offspring.

3 Assessment Options

Closure

Technique: K-W-L

Tell students to return to the list of things they Want to know about gene technology from the **Active Reading** exercise on page 225. Have them place check marks next to the questions they can now answer. Students should finish by making a list of what they have Learned. Conduct a discussion of the remaining questions that have gone unanswered. PORTFOLIO

Review

Assign Review—Section 11-3.

Reteaching — BASIC

Ask students to create a table to summarize the ways in which genetic engineering has been used to improve food crops and farm animals and to make medically useful proteins in the milk of farm animals. PORTFOLIO

Review SECTION 11-3

1. **List** three ways in which food crops have been improved through genetic engineering.
2. **Summarize** how the Ti plasmid is used to insert genes into plant cells.
3. **Compare** the cloning of sheep through the use of differentiated cells with the cloning of sheep through the use of embryonic cells.
4. **SKILL Analyzing Methods** In the movie *Jurassic Park,* scientists used DNA to bring back extinct species. How is that different from the creation of cloned sheep using differentiated cells?

Critical Thinking

5. **SKILL Forming Reasoned Opinions** List reasons you would or would not be concerned about consuming milk from cows treated with growth hormone.

Review SECTION 11-3 ANSWERS

1. making food crops more tolerant to drought conditions, different soils, climates, or environmental stresses; enhancing their nutritional value; controlling the process by which fruit ripens; making them resistant to the weedkiller glyphosate
2. The tumor-causing genes from the Ti plasmid are removed, and the space is filled with specific DNA. The plasmid then infects the cells of plants.
3. Cloning with embryonic cells allows an organism to develop from cells that have not undergone specialization. Cloning with differentiated cells requires that the cell be manipulated first so that the cell cycle is stopped.
4. The DNA used in *Jurassic Park* was fragmented. The researchers filled it in with DNA from other avian and amphibian species. Thus, the dinosaurs had foreign DNA and were not truly clones.
5. Answers will vary. Students may not be concerned if evidence provided by the company indicates the hormone is safe. Concerns may include that the hormone may contribute to breast and prostate cancer and that cows get more infections and have to be given antibiotics, which could enter the milk.

Study ZONE CHAPTER 11

Use the Key Concepts and Key Terms listed below to review the main ideas in this chapter. If you need to review the meaning of a term, turn to the page indicated.

Key Concepts

11-1 Genetic Engineering

- Genetic engineers manipulate DNA for practical purposes.
- Restriction enzymes cleave DNA into fragments that have short sticky ends. Sticky ends allow DNA fragments from different organisms to join together to form recombinant DNA.
- Recombinant DNA is inserted into host cells. The cells are screened to identify cells that have the recombinant DNA. Each time the cells reproduce, the gene of interest is cloned.
- Electrophoresis uses an electrical field within a gel to separate DNA fragments by their size and charge.
- Specific genes can be identified with the Southern blot technique.

11-2 Genetic Engineering in Medicine and Society

- Genetic engineering is used to manufacture human proteins for use as drugs and to make safer and more effective vaccines.
- Some human genetic disorders are being treated through gene therapy.
- DNA fingerprinting is used to identify individuals and determine relationships between individuals.
- The Human Genome Project is an effort to determine the nucleotide sequence of and map the location of every gene on each human chromosome by the year 2003. The sequence of the genomes of many organisms have already been determined.

11-3 Genetic Engineering in Agriculture

- Crop plants can be genetically engineered to add favorable characteristics, including improved yields and resistance to herbicides and destructive pests.
- The Ti plasmid is used as a plant vector for certain plants.
- Genetically engineered growth hormone increases milk production in dairy cows and weight gain in cattle and hogs.
- The cloning of farm animals using differentiated cells was accomplished in 1997. In addition, transgenic animals can be cloned and used to make proteins that are useful in medicine.

Key Terms

11-1

genetic engineering (226)
recombinant DNA (226)
restriction enzyme (227)
vector (227)
plasmid (227)
gene cloning (227)
electrophoresis (229)
probe (229)

11-2

vaccine (232)
gene therapy (234)
DNA fingerprint (235)
Human Genome Project (236)

11-3

transgenic animal (239)

Study TIP Ready?

- *If you think you understand these Key Concepts and Key Terms, then you're ready for the Performance Zone.* 

Answer to Concept Map

The following is one possible answer to Performance Zone item 1 on page 242.

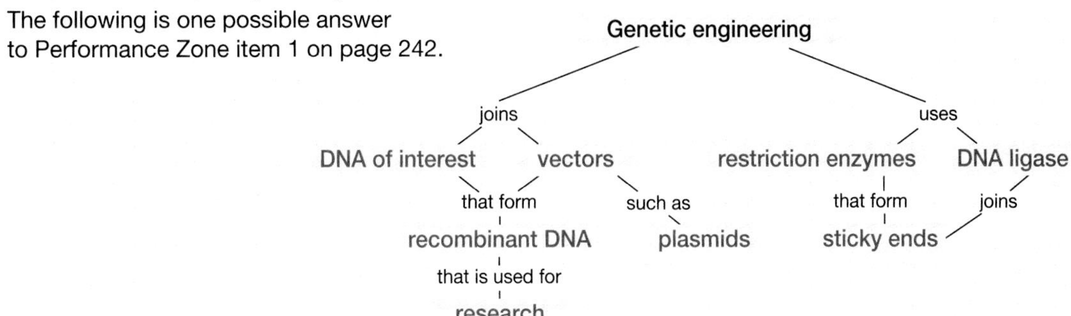

Study ZONE CHAPTER 11

Review and Practice

Assign the following worksheets for Chapter 11:

- **Student Review Guide**
- **Concept Mapping Worksheet**
- **Test Preparation Pretest**
- **Vocabulary Worksheet**
- **Critical Thinking Review Worksheet**
- **Science Skills Worksheet**

Assessment Options

The following assessment options are available for this chapter:

Resource Link

- Assign **Chapter Test 11** to assess students' understanding of the chapter.

Alternative Assessment

The use of DNA technology presents many ethical and moral issues. Divide your class into two groups. Ask one group to prepare information that supports DNA technology, including specifics about the benefits of genetic engineering and gene therapy to society. Have the other group prepare information that demonstrates the negative aspects of these technologies. Have both groups debate their stand on the issue. **PORTFOLIO**

Portfolio Assessment

- *Portfolio Assessment* at the front of this book provides options for building and evaluating student portfolios. Students and teachers can select items from the areas listed for inclusion in a portfolio.
- See items labeled **PORTFOLIO** on pp. 228, 230, 232, 233, 234, 236, 238, 240, and 241.

Performance ZONE
CHAPTER 11
REVIEW ANSWERS

Assignment Guide

SECTION	REVIEW
11-1	1, 2, 3, 4, 5, 6, 14, 15, 18, 19, 20, 21, 24, 26, 28, 29
11-2	7, 8, 9, 10, 16, 22, 23, 25
11-3	11, 12, 13, 17, 26, 27, 28

Understanding and Applying Concepts

1. The answer to the concept map is found at the bottom of page 241.
2. **a.** Recombinant DNA is DNA made from two different organisms. Restriction enzymes are enzymes used to cut DNA.
   b. A vector is an agent that carries the gene of interest into another cell. A vaccine is a solution containing a harmless version of a pathogen.
   c. Cloning is a process that results in an exact copy of a gene or an organism. Screening is a process used to separate bacterial cells that contain the gene of interest from those that do not.
3. a
4. b
5. d
6. a
7. b
8. a
9. b
10. c
11. d
12. d
13. c
14. Molecule A was produced through genetic engineering. Plasmid DNA and DNA from a different organism are cut with restriction enzymes and then combined to produce recombinant plasmid DNA. The recombinant plasmid DNA molecules are then inserted into bacterial cells.
15. In natural selection, organisms that are best suited to their environment will reproduce more successfully than others. By changing the genetic makeup of individuals, genetic engineering alters the course of natural selection. The fittest members of a naturally evolving population might not be able to compete with individuals containing newly engineered genes. In addition, the newly engineered genes are passed on to offspring.
16. DNA is extracted from the bone. A small sample of DNA is heated to separate the strands. DNA primers, DNA polymerase, and nucleotides are added. In a short time the original DNA from the fossilized bone is replicated. The process is continued until an adequate sample size is obtained.
17. Dolly's lamb is not a clone. Dolly was bred with a ram; therefore, the lamb contains genetic information from both parents.
18. By determining the structure of the DNA molecule and how the component parts are arranged, scientists were able to establish that DNA is the genetic material. DNA could now be studied and manipulated.

Performance ZONE CHAPTER 11 Review

Understanding and Applying Concepts

1. **Concept Mapping** Make a concept map about genetic engineering. Try to include the following words in your map: DNA of interest, vectors, recombinant DNA, plasmids, restriction enzymes, DNA ligase, sticky ends, research.

2. **Understanding Vocabulary** For each pair of terms, explain the difference in their meanings.
   a. recombinant DNA, restriction enzyme
   b. vector, vaccine
   c. cloning, screening

3. A molecule containing DNA from two different organisms is called
   a. recombinant DNA.
   b. RFLP DNA.
   c. a plasmid.
   d. a probe.

4. A ______ is an extra ring of DNA in bacteria.
   a. probe
   b. plasmid
   c. clone
   d. restriction enzyme

5. Genetic engineering depends on the ability of ______ to cut DNA at specific sites.
   a. the Ti plasmid
   b. DNA ligase
   c. plasmid DNA
   d. restriction enzymes

6. Gel electrophoresis is used to ______ DNA fragments.
   a. separate
   b. join
   c. cut
   d. copy

7. Which of the following human illnesses can be treated using a product of genetic engineering?
   a. malaria
   b. hemophilia
   c. flu
   d. a sinus cold

8. ______ is used to identify individuals in paternity cases and criminal cases.
   a. DNA fingerprinting
   b. Gene therapy
   c. A vaccine
   d. The Ti plasmid

9. Injecting a healthy copy of a gene into a person with a defective gene is called
   a. probing.
   b. gene therapy.
   c. PCR.
   d. DNA cloning.

10. The major effort to map and sequence all human genes by the year 2003 is called
    a. the RFLP Project.
    b. the PCR Project.
    c. the Human Genome Project.
    d. DNA fingerprinting.

11. The ______ plasmid is used as a plant vector to transfer genes into broad-leaved crop plants.
    a. factor VIII
    b. gene therapy
    c. glyphosate
    d. Ti

12. Genetic engineers can make plants
    a. resistant to insects.
    b. more tolerant to droughts.
    c. that adapt to different soils.
    d. All of the above

13. The 1997 sheep-cloning experiments differed from previous cloning experiments because ______ cells were used.
    a. fetal
    b. embryonic
    c. differentiated
    d. cow

14. **Applying Information** Describe how molecule *A* was produced.

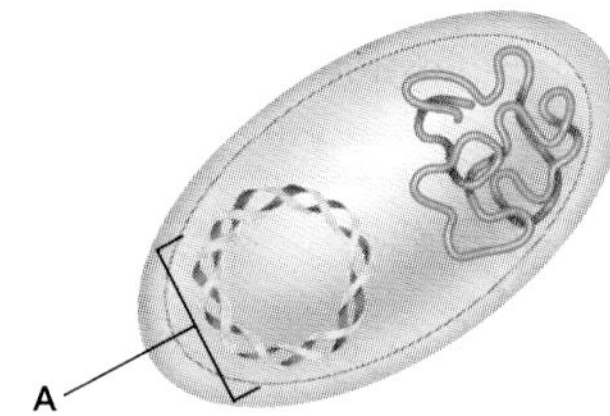

15. **Theme Evolution** How is natural selection affected by genetic engineering?

16. **Tech Watch** You have discovered a fossilized bone. How can you use PCR to obtain sufficient DNA for DNA analysis?

17. **Exploring Further** Is Dolly's lamb a clone of Dolly? Explain.

18. **Chapter Links** How did Watson and Crick's double-helix model of DNA lead to genetic engineering? (**Hint:** See Chapter 9, Section 9-2.)

Skills Assessment

Part 1: Interpreting Graphics

The illustration below shows two DNA molecules. The DNA on top represents plasmid DNA after it has been cut with a restriction enzyme. The DNA on the bottom represents a piece of human DNA resulting from a larger piece of human DNA that was cut. Use the illustration to answer the questions below:

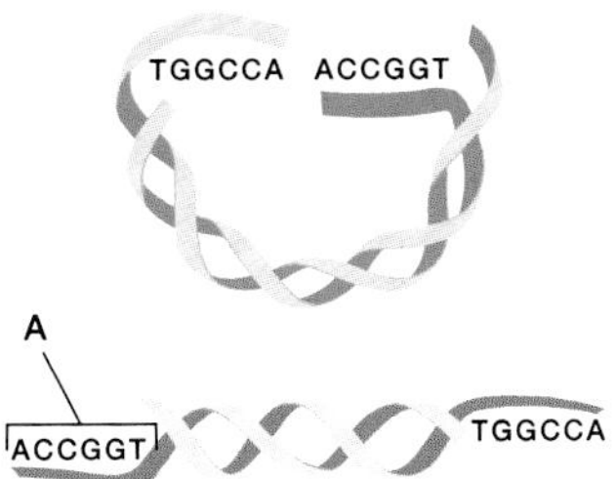

19. Identify the area on the human DNA fragment indicated by *A*.
20. Use colored pencils to draw a DNA molecule that represents how the two pieces of DNA would join together.
21. What is the name for the DNA molecule that would result if the two pieces of DNA were joined together?

Part 2: Life/Work Skills

The question of awarding patents on genetically engineered organisms arose when a microbiologist named Ananda Chakrabarty filed for a patent on a bacterium capable of digesting the components of crude oil. Use the media center or Internet resources to learn more about this.

22. **Finding Information** Find out about the bacterium that Chakrabarty genetically engineered and about the court battle he waged for the right to obtain a patent for a genetically engineered organism.
23. **Communicating** Prepare an oral report using graphics to summarize your findings about the bacterium that Chakrabarty genetically engineered and tried to patent. Present the report to your class.

Critical Thinking

24. **Forming Reasoned Opinions** In the United States, government regulations require researchers to contain experimental genetically engineered organisms inside a laboratory and to ensure that the organisms could not survive outside the laboratory. Why do you think these strict regulations are necessary?
25. **Distinguishing Fact from Opinion** A judge presiding over a highly publicized murder trial dismissed the prosecution's request to admit DNA fingerprints as evidence, calling it "unproven scientific mumbo-jumbo." Do you agree with the judge? Explain your answer.
26. **Distinguishing Relevant Information** Organize and videotape a class debate about the safety questions raised by the potential release of genetically engineered plants, bacteria, and animals into the environment. Use library references and online databases to back up your arguments.

Portfolio Projects

27. **Analyzing Data** Obtain normal tomato seeds and seeds from a genetically engineered variety of tomato. Plant the seeds in labeled containers. Grow the plants under similar conditions, and compare the number and quality of tomatoes produced by each plant. Create a poster that summarizes your findings.
28. **Interpreting Information** Read the article "Ewe Again? Cloning from Adult DNA" (*Science News,* March 1, 1997, vol. 151, p. 132). Why are researchers interested in Dolly's aging process? What are some of the research priorities for cloning technology?
29. **Career Focus Genetic Engineer** Research the field of genetic engineering, and write a report on your findings. Your report should include a job description, training required, kinds of employers, growth prospects, and a starting salary.

Performance ZONE
CHAPTER 11

Skills Assessment

19. sticky end

20.

21. recombinant DNA

22. Chakrabarty identified enzymes that degrade different components of crude oil and added the enzymes to *Pseudomonas* bacteria. His patent request was brought before the U.S. Supreme Court, which ruled in 1980 that human-engineered organisms are patentable under federal law.

23. Answers will vary.

Critical Thinking

24. Answers will vary. The regulations were prompted by concerns that genetically engineered organisms might cause disease or have harmful effects on the environment. By limiting their survival outside the laboratory, scientists are preventing the organism from possibly harming people or other organisms and interfering in the food chain.

25. Students should disagree. A match of DNA fingerprints from two different samples has only two explanations—the prints are from the same person or from identical twins.

26. Answers will vary. For example: Will genetically engineered foods contain new proteins that are allergenic or toxic to some people? Will genetically engineered crops pass their new genes to close relatives in nearby wild areas and create weeds that are very difficult to control?

Portfolio Projects

27. Answers will vary.

28. Because the DNA used in the cloning process was 6 years and 7 months old, scientists were not certain at that time how Dolly's development would be affected. Research priorities include determining if other cells can be used to clone the organism, applying this technique to other organisms, and cloning animals for the purpose of producing needed pharmaceuticals and organs for human transplants.

29. Genetic engineers are scientists that alter genes or genetic material to produce desirable new traits in organisms or to eliminate undesirable traits. Training required depends on the specific job responsibilities. Training can vary from an associate's degree to more advanced degrees. Employers include universities, private industries, and pharmaceutical companies. Growth prospects are excellent. Starting salary will vary by region.

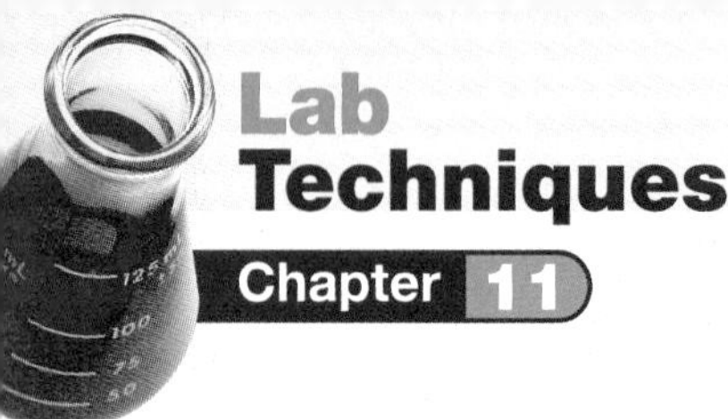

Lab Techniques

Chapter 11

Modeling Genetic Engineering

Purpose

Students will use straws, pushpins, and paper clips to model the making of recombinant DNA.

Preparation Notes

Time Required: 40–50 minutes

Materials

Materials for this lab can be ordered from WARD'S. See *Master Materials List* at the front of this book for catalog numbers.

Preparation Tips

Make sure each group has enough materials before they begin the activity. Use materials from **Experimental Design** in Chapter 10, pp. 222–223, if they were saved.

Safety Precautions

- Review all safety symbols and caution statements with students before beginning the lab.
- Caution students to be careful of the sharp tips on the pushpins.

Procedural Tips

- Have students work in cooperative groups of four students. Divide each group into pairs. One two-person team should complete steps 2–4 while the other team completes steps 5 and 6. The entire group should work together to complete steps 7–11.
- Before beginning the lab, use Figure 9-8 on page 195 to review the structure of DNA and Figure 11-2 on page 227 to review the steps of genetic engineering.

Disposal

Account for all materials at the end of the lab. Have the students store the pushpins by color.

Lab Techniques

Chapter 11 *Modeling Genetic Engineering*

SKILLS

- Modeling
- Comparing

OBJECTIVES

- **Construct** a model that can be used to explore the process of genetic engineering.
- **Describe** how recombinant DNA is made.

MATERIALS

- plastic soda straw pieces (56)
- pushpins (15 red, 15 green, 13 blue, and 13 yellow)
- paper clips (56)

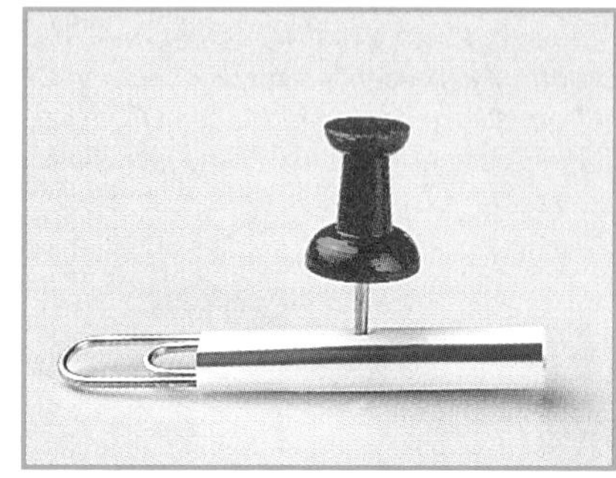

Before You Begin

Genetic engineering is the process of taking a gene from one organism and inserting it into the DNA of another organism. The gene is delivered by a **vector,** such as a virus, or a bacterial **plasmid.**

First, a fragment of a chromosome that contains the gene is isolated by using a **restriction enzyme,** which cuts DNA at a specific nucleotide-base sequence. Some restriction enzymes cut DNA unevenly, producing single-stranded **sticky ends.** The DNA of the vector is cut by the same restriction enzyme. Next the chromosome fragment is mixed with the cut DNA of the vector. Finally, an enzyme called **DNA ligase** joins the ends of the two types of cut DNA, producing **recombinant DNA.**

In this lab, you will model genetic engineering techniques. You will simulate the making of recombinant DNA by inserting a human gene into the DNA of a plasmid.

1. Write a definition for each boldface term in the paragraph above and for the term *base-pairing rules.*
2. Based on the objectives for this lab, write a question you would like to explore about the process of genetic engineering.

Procedure

PART A: Model Genetic Engineering

1. Make 56 model nucleotides. To make a nucleotide, insert a pushpin midway along the length of a 3 cm piece of a soda straw. **CAUTION: Handle pushpins carefully. Pointed objects can cause injury.** Push a paper clip into one end of the soda-straw piece until it touches the pushpin.
2. Begin a model of a bacterial plasmid by arranging nucleotides for one DNA strand in the following order: blue, red, green, yellow, red, red, blue, blue, green, red, blue, green, red, blue, blue, green, yellow, and red. Join two adjacent nucleotides by inserting the paper clip end of one into the open end of the other.
3. Using your first DNA strand and the base-pairing rules, build the complementary strand of plasmid DNA. **Note:** *Yellow is complementary to blue, and green is complementary to red.*
4. Complete your model of a circular plasmid by joining the opposite ends of each DNA strand. Make a sketch showing the sequence of bases in your model plasmid. Use the abbreviations B, Y, G, and R for the pushpin colors. Your sketch should be similar to the one at the top of the next page.

Before You Begin

Answers

1. *genetic engineering*—the process of isolating a gene from the DNA of one organism and inserting it into the DNA of another organism

   vector—agent such as a virus or plasmid used to carry a DNA fragment into a cell

   plasmid—small circular DNA molecule, usually found in bacteria, that can replicate independently from the main chromosome

   restriction enzyme—bacterial enzyme that cuts DNA at a specific sequence of nucleotides

   sticky ends—single-stranded ends of DNA that are produced when a restriction enzyme cuts DNA unevenly

   DNA ligase—enzyme that joins ends of DNA

   recombinant DNA—DNA that contains DNA segments from different organisms

   base-pairing rules—rules stating that in DNA, adenine on one strand always pairs with thymine on the opposite strand and cytosine on one strand always pairs with guanine on the opposite strand
2. Answers will vary. For example: What steps are involved in transferring a gene from one organism to another one?

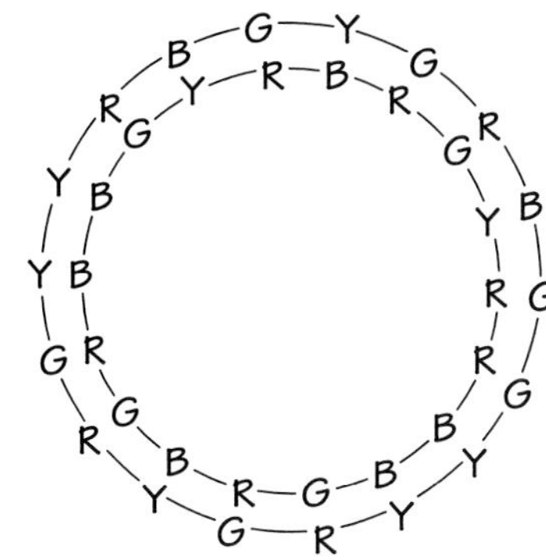

5. Begin a model of a human chromosome fragment made by a restriction enzyme. Place nucleotides for one DNA strand in the following order: BBRRYGGBRY. Build the second DNA strand by arranging the remaining nucleotides in the following order: BRRYGBYYGG.
6. Match the complementary portions of the two strands of DNA you made in step 5. Make a sketch showing the sequence of bases in your model of a human chromosome fragment.
7. Imagine that the restriction enzyme that cut the human chromosome fragment you made in steps 5 and 6 is moving around your model plasmid until it finds the sequence YRRBBG and its complementary sequence, BGGYYR. Find such a section in your sketch of your model plasmid's DNA.
8. Simulate the action of the restriction enzyme on the section you identified in step 7. Open both strands of your model plasmid's DNA by pulling apart the adjacent green and blue nucleotides in each strand. Make a sketch of the split plasmid DNA molecule.
9. Move your model human DNA fragment into the break in your model plasmid's DNA molecule. Imagine that a ligase joins the ends of the human and plasmid DNA. Make a sketch of your final model DNA molecule.

PART B: Cleanup and Disposal

10. Dispose of damaged pushpins in the designated waste container.
11. Clean up your work area and all lab equipment. Return lab equipment to its proper place. Wash your hands thoroughly before you leave the lab and after you finish all work.

Analyze and Conclude

1. **Comparing Structures** Compare your models of plasmid DNA and human DNA.
2. **Relating Concepts** What do the sections of four unpaired nucleotides in your model human DNA fragment represent?
3. **Comparing Structures** How did your original model plasmid DNA molecule differ from your final model DNA molecule?
4. **Drawing Conclusions** What does the molecule you made in step 9 represent?
5. **Further Inquiry** Write a new question that could be explored with another investigation.

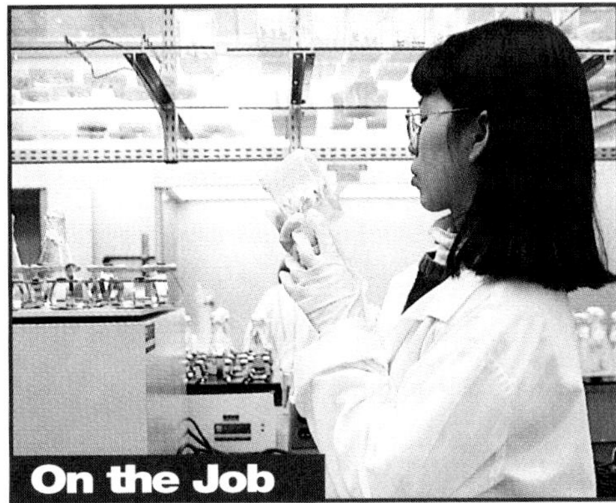

On the Job

Genetic engineering is used to produce many products that are useful to humans. Do research in the media center or on the Internet to discover how genetic engineering is used to better our lives. Summarize your findings in a written or oral report.

Lab Techniques

11 Chapter

Analyze and Conclude

Answers

1. Both the plasmid DNA model and the human DNA model are double stranded. The plasmid DNA model is circular, and the human DNA model is linear. The human DNA model represents a fragment of a DNA molecule that has been cut with a restriction enzyme and has sticky ends.
2. sticky ends
3. The original model plasmid DNA was smaller (had fewer nucleotides) and did not contain a gene from a human chromosome.
4. recombinant DNA
5. Answers will vary. For example: What happens if the plasmid and human DNA do not have complementary sticky ends?

On the Job

Answer

Answers will vary. Advances have been made in the production of proteins used to treat illnesses, in the development of new vaccines to combat diseases, in the replacement of defective genes with healthy ones, and in the improvement of food crops.

Procedure

Answers

4.

Model of plasmid DNA molecule

6.

Human donor DNA fragment

7. The sketch should show the split at the nucleotides indicated by the two arrows in item 4 under **Procedure Answers** on page 245.

9.

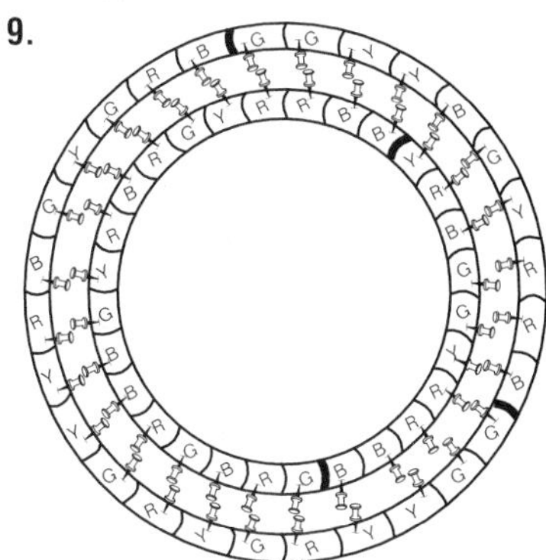

Final recombined DNA molecule

SCIENCE • TECHNOLOGY • SOCIETY

Secrets of Your Genes

Background

The race to sequence the human genome has been one of the most controversial areas of biological research today. The sequencing of the human genome is seen as a vital first step in developing possible treatments and cures for inherited diseases. Private genome initiatives are fueled by the potential earnings that await companies who develop new drugs to treat genetic diseases and disorders. However, in order to maintain exclusive rights to their discoveries (and thus ensure financial reward), private initiatives usually seek patents on the gene sequences they discover. Such exclusive "ownership" of human genomic information is controversial—many scientists feel that withholding potentially critical information from the scientific community will hamper research. Finding the right balance between publicly funded government efforts and private research will help ensure continued progress.

Teaching Strategies

- Ask students to brainstorm about what kinds of medicines or drug treatments might be designed if certain gene sequences were known. Have students debate who should control the information regarding these gene sequences and who should receive potential profits derived from the sales of new medicines or drug treatments.
- This feature focuses on one of the fastest-growing areas of biological research today. Have students use on-line resources to find the latest information on gene sequencing projects that are currently underway.
- Unlocking the secrets of the human genome raises difficult questions that frequently involve the individual's right to privacy. Have students read the article "DNA Detectives," by Jeffrey Kluger, found on pp. 62–63 of the January 11, 1999, issue of *Time*. Ask students how knowledge of an individual's genome may conflict with rights guaranteed under the Fourth Amendment to the United States Constitution.
- The presence of certain gene sequences may indicate that an individual will develop a disease later in life or that he or she has a predisposition for a certain disease or a particular behavior. Have students discuss whether genetic testing should be mandatory.

Point out to students that the Equal Employment Opportunity Commission has made genetic discrimination illegal in the United States, ruling that employers cannot terminate or refuse to hire an individual because of a genetic predisposition to an illness. This ruling is based on the Americans with Disabilities Act, which prohibits employers from discriminating against anyone who is disabled. Insurance companies, however, are exempt from the act and therefore are also exempt from the EEOC ruling.

Should the secrets of the human genome be public—or private?

Secrets of Your Genes

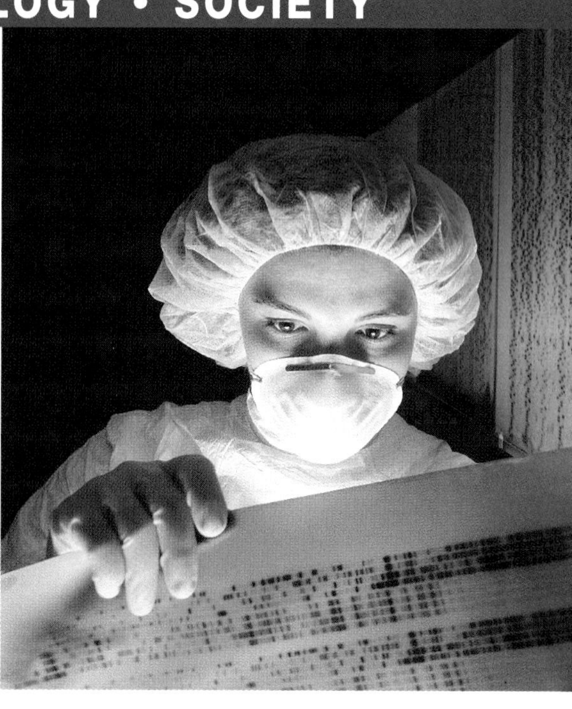

Early in the 1990s the United States government began an ambitious program to produce a full map of the genes in the 46 human chromosomes no later than the end of the year 2003. It is the task of the Human Genome Project, directed by Dr. Francis Collins of the National Institutes of Health, to read each and every gene, writing down all the nucleotide-base pairs in each gene in proper order. Because the human genome contains some 3 billion nucleotide-base pairs, this task has presented no small challenge, but work on the sequence was finished ahead of schedule.

However, in 1998, a pioneer in gene sequencing made a startling announcement. Dr. J. Craig Venter, a researcher with his own commercial company, said that his company would sequence the entire 3 billion nucleotide-base pairs of human DNA by the end of the year 2001! Had anyone else made this claim, he or she might not have been taken seriously, but Venter has proven himself a highly successful sequencer, sequencing the first bacterial genome in 1995 and more complete genomes than any other researcher.

A New Strategy

How could Venter have hoped to move so fast? Instead of just starting at one end of a chromosome and writing all the nucleotide-base pairs in linear order, Venter focused on the most interesting bits first. Imagine trying to grasp all

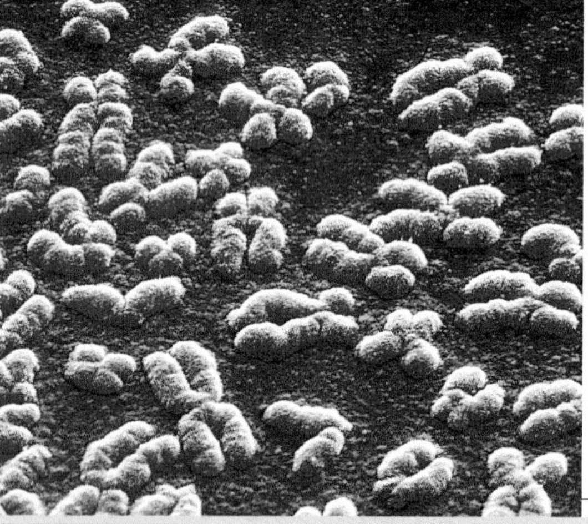

Valuable knowledge? These human chromosomes could contain information worth millions to pharmaceutical companies.

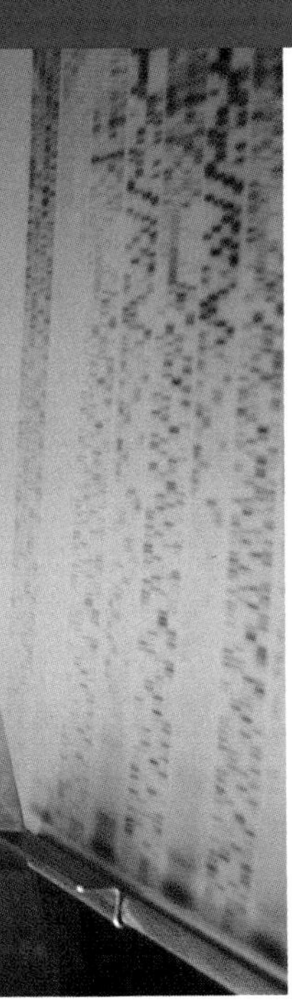

Gene sequencing
A technician interprets marks that reveal gene structure.

the information in a library—Venter proposed reading all the most popular books first, delaying information of seldom-checked-out books until later.

In 2000, Dr. Collins and Dr. Venter jointly announced completion of the human genome sequence. But, while both plan to release the portions of the human genome sequence that are found in all human DNA, Dr. Venter's company plans to patent information on unique variations in that sequence. Such single nucleotide-base pair differences, called "single nucleotide polymorphisms," or SNPs, are spot differences in a gene that make each human unique. The particular importance of SNPs lies in the fact that inherited disorders like cystic fibrosis and sickle cell anemia are caused by such single-letter typos in genes. To access this information, pharmaceutical companies and other researchers will have to pay licensing fees to Venter's company.

Public Information?

However, many research scientists feel that genetic information should remain public. Granting patents for SNPs, these individuals argue, is the equivalent of letting companies patent basic scientific information. These scientists compare paying a fee for genetic information with paying a fee to examine the periodic table of the elements. Many scientists also argue that licensing costs will slow medical research.

Responding to this challenge, major pharmaceutical companies contributed funding to the Human Genome Project to pay for the added sequencing needed to find SNPs. The research labs of the Human Genome Project abruptly changed direction, choosing to seek a "blueprint" of the genome before a private company. By publishing the interesting SNPs first, the Human Genome Project plans to place the information forever in the hands of the public. ■

Career

Biotechnology

Genetic technicians

Profile

Biotechnology is the application of genetic engineering to practical human problems. Biological and medical scientists use techniques of biotechnology to manipulate the genetic material of plants and animals, attempting to make commercial organisms more productive or disease resistant.

Job Description

Genetic technicians work with other biologists to piece together clues that help determine patterns of inheritance—and possible avenues for treatments—for many genetic diseases. Sequencing an organism's genome can enable researchers to determine which genes are responsible for causing a particular genetic trait or disease.

Science/Math Career Preparation

Microbiology	Genetics
Chemistry	Technical writing
Biochemistry	Mathematics

References

1. Human Genome Project Information. On-line. World Wide Web. Available http://www.ornl.gov/TechResources/Human_Genome/home.html.
2. Biotechnology Information Center. On-line. World Wide Web. Available http://www.nal.usda.gov/bic/.
3. National Human Genome Research Institute (NHGRI). On-line. World Wide Web. Available http://www.nhgri.nih.gov/index.html.
4. Beardsley, Tim. "An Express Route to the Genome?" *Scientific American,* August 1998, 30–32.
5. Shreeve, James. "Secrets of the Gene." *National Geographic,* October 1999, 42–75.
6. Thompson, Dick. "Gene Maverick." *Time,* January 11, 1999, 54–55.

Analyzing STS Issues

Science and Society

1. **How should the human genome be sequenced?** Read "Racing to Map Our DNA," by Michael D. Lemonick and Dick Thompson (*Time,* January 11, 1999, pp. 44–50). How do private companies differ from the Human Genome Project in their approach toward sequencing the human genome?

Technology: Gene Sequencers

2. **How do genetic researchers figure out the gene sequence of an organism?** Read "The Code Breaker," by James Shreeve (*Discover,* May 1998, pp. 44–51). How did the TIGR Assembler enable Venter and Smith to determine the genome of *Haemophilus influenzae*?

Analyzing STS Issues Answers

1. Private companies are zeroing in on particular stretches of human DNA, hoping to pick out specific genes whose malfunctions actually cause disease. The Human Genome Project, however, has sought to sequence the entire genome with high precision.
2. The TIGR assembler determined where tiny DNA fragments overlap and then patched these fragments together at the overlapping segments until the full gene sequence for *Haemophilus* was assembled.

UNIT 3 Principles of Evolution

Chapters

Tropical rain forests contain more than one-half of all the world's animal and plant species, such as this red-eyed tree frog from Central America.

in perspective

Tropical Rain Forests

Yesterday...

During their travels, early European visitors were fascinated by the variety and number of tropical rain-forest plants and animals, so different from those they knew at home. **Discover how early scientists named and classified living things on page 320.**

Exploring the Brazilian jungle in 1820

Today...

Today's explorers are faced with the challenge of classifying living things in lesser-known areas, such as tropical rain forests, that have not yet been studied by scientists. **What is the impact of human activity on tropical rain-forest species? To find out, see page 263.**

Rain-forest researcher at Madre de Dios, Peru

Tomorrow...

To preserve species, new ways of protecting tropical habitats are being explored. For example, modern coffee-growing techniques eliminate native species as fields are planted solely with coffee, a major source of revenue for many tropical countries. But shade coffee plantations grow coffee in the midst of the forest canopy, thus preserving habitat for tropical species.

Coffee beans

internet**connect**

TOPIC: Rain forests
GO TO: www.scilinks.org
KEYWORD: HX249

Chapter 12 History of Life on Earth

Planning Guide

OBJECTIVES	MEETING INDIVIDUAL NEEDS	READING SKILLS AND STRATEGIES
12-1 How Did Life Begin? (BLOCKS 1 & 2 90 min) • **Summarize** how radioisotopes can be used in determining Earth's age. • **Compare** two models that describe how the chemicals of life originated. • **Describe** how cellular organization might have begun. • **Recognize** the importance of the development of heredity to the development of life.	Learning Styles, p. 252 BASIC ADVANCED • p. 257	**ATE** Active Reading, Technique: K-W-L, p. 251 ■ Active Reading Guide, worksheet (12-1) ■ Directed Reading, worksheet (12-1) ■ Reading Strategies, K-W-L worksheet
12-2 Complex Organisms Developed (BLOCKS 3 & 4 90 min) • **Distinguish** between the two groups of prokaryotes. • **Describe** the evolution of eukaryotes. • **Recognize** an evolutionary advance first seen in protists. • **Summarize** how mass extinctions have affected the evolution of life on Earth.	Learning Styles, p. 258 BASIC ADVANCED • p. 263	■ Active Reading Guide, worksheet (12-2) ■ Directed Reading, worksheet (12-2)
12-3 Life Invaded the Land (BLOCKS 5 & 6 90 min) • **Relate** the development of ozone to the adaptation of life to the land. • **Identify** the first multicellular organisms to live on land. • **Name** the first animals to live on land. • **List** the first vertebrates to leave the oceans.	Learning Styles, p. 264 BASIC ADVANCED • p. 268	■ Active Reading Guide, worksheet (12-3) ■ Directed Reading, worksheet (12-3) ■ Supplemental Reading Guide, *The Dinosaur Heresies*

Review and Assessment Resources

BLOCKS 7 & 8 90 min

TEST PREP

- **PE** Study Zone, p. 269
- **PE** Performance Zone, pp. 270–271
- ■ Chapter 12 worksheets:
  - Vocabulary
  - Science Skills
  - Concept Mapping
  - Critical Thinking
- Audio CD Program
- ■ Test Prep Pretest

CHAPTER TESTING

- ■ Chapter Tests (blackline master)
- ▲ Test Generator
- ■ Assessment Item Listing

See the correlations table at the front of this book for a list of the ***National Science Education Standards*** addressed in this chapter.

Smithsonian Institution® Visit **www.si.edu/hrw** for additional on-line resources.

KEY

Biology Principles and Explorations

PE Pupil's Edition
ATE Teacher's Edition

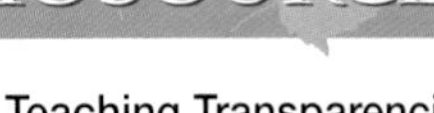

Teaching Transparencies
Laboratory Program

One-Stop Planner CD-ROM
Shows these resources within a customizable daily lesson plan:

■ Teaching Resources
▲ Test Generator

VISUAL STRATEGIES	CLASSWORK AND HOMEWORK	TECHNOLOGY AND INTERNET RESOURCES
ATE Fig. 12-1, p. 252 42A Theories of Life's Origins 43 Lerman's Bubble Model 45 Rate of Decay in Potassium-40 46 Miller-Urey Apparatus 47 Spontaneous Assembly of RNA	PE Quick Lab How can you model radioactive decay?, p. 253 PE Quick Lab How can you model coacervates?, p. 257 ■ Basic Skills: Mass and Density PE Review, p. 257	Holt Biology Videodiscs Section 12-1 Audio CD Program Section 12-1 internetconnect SCILINKS NSTA TOPIC: Radioactive decay
ATE Fig. 12-5, p. 259 61 Theory of Endosymbiosis	PE Data Lab Analyzing signs of endosymbiosis, p. 259 Lab B8 Fossil Study PE Review, p. 263	Holt Biology Videodiscs Section 12-2 Audio CD Program Section 12-2 internetconnect SCILINKS NSTA TOPICS: • Endosymbiosis • Extinction
	PE Lab Techniques Making a Timeline of Life on Earth, pp. 272–273 Lab A10 Analyzing Adaptations: Living on Land PE Review, p. 268 PE Study Zone, p. 269	CNN Video Segment 9 Dino Egg Discovery Holt Biology Videodiscs Section 12-3 Audio CD Program Section 12-3

ALTERNATIVE ASSESSMENT

PE Portfolio Projects 24–25, p. 271
ATE Alternative Assessment, p. 269
ATE Portfolio Assessment, p. 269

Lab Activity Materials

Quick Lab
p. 253 dry corn kernels (approximately 100 per group), cardboard box, clock or watch with a second hand

p. 257 safety goggles, lab apron, graduated cylinder, 1% gelatin solution, 1% gum arabic solution, test tube, 0.1 M HCl, pipet, microscope slide and coverslip, microscope

Data Lab
p. 259 pencil, paper

Lab Techniques
5 m roll of adding machine tape, meterstick, colored pens or pencils, photographs or drawings of organisms from ancient Earth to present day

Lesson Warm-up
ATE photographs of human fossils, p. 252

Demonstrations
ATE several types of sunblocks, sunscreens, suntan lotions, p. 265
ATE soft, gelatin-based candy (worm shaped); toothpicks; tape, p. 267
ATE chicken eggs (2), Petri dishes (2), p. 268

BIOSOURCES LAB PROGRAM
- Quick Labs: Lab A10, p. 19
- Inquiry Skills Development: Lab B8, p. 29

Chapter Theme

Evolution

The history of life on Earth began approximately 3.5 billion years ago. How this occurred has been and will continue to be a topic for inquiry. Scientific evidence indicates that since life's beginning, each organism that has inhabited this planet has been the product of evolution—the change of organisms over time.

Establishing Prior Knowledge

Before beginning this chapter, be certain that students can answer the questions in **Study TIP Ready?** on page 251. If they cannot, encourage them to reread the sections indicated.

Opening Demonstration

To help students visualize the enormity of 1 billion of anything, ask students if they could physically carry 1 million dollars. They typically answer that they would like to try! Until recently, $1,000 bills were obtainable. One million dollars would then be a stack of $1,000 bills approximately 1 m in height (1,000 bills). A billion dollars would be a stack of the same to a height of the Washington Monument (1 million bills). Wow! Now relate this image to the first billion years of our planet's existence without any form of life.

CHAPTER 12 History of Life on Earth

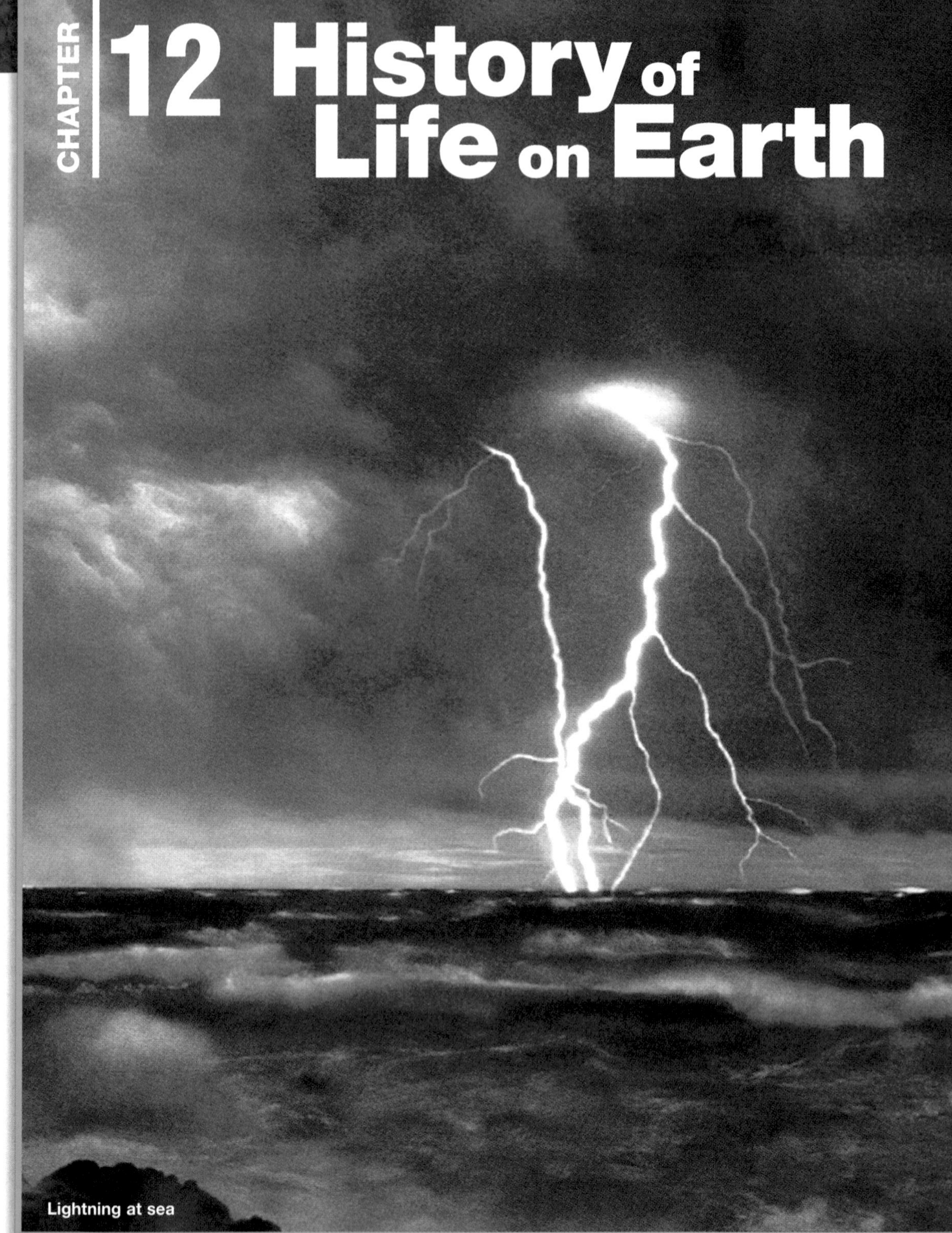

Lightning at sea

What's Ahead — and Why

CHAPTER OUTLINE

12-1 How Did Life Begin?
- The Early Earth Was Lifeless
- How the Basic Chemicals of Life Were Formed
- Organic Chemicals Became Complex

12-2 Complex Organisms Developed
- Prokaryotes Are the Oldest Group of Organisms
- The First Eukaryotes Evolved
- Multicellularity Evolved Many Times
- Mass Extinctions Have a Major Impact

12-3 Life Invaded the Land
- Ozone Enables Survival on Land
- Plants and Fungi Began Living on the Land Together
- Arthropods Crawled out of the Sea
- Vertebrates Followed onto Land

Students will learn

- how scientists estimate the age of Earth and how life may have begun.
- that life progressed in complexity from its first forms on Earth.
- about the types of environments inhabited by Earth's early organisms.

Like all life-forms, *hunting dogs, such as the one above, require energy and complex organic molecules in order to survive. Scientists think that billions of years ago the combination of energy from various sources, such as lightning, and simple molecules gave rise to the first life-forms on Earth.*

Study TIP

Ready?

Check your understanding of these topics to help you prepare for what's ahead.

Can you . . .

- **describe** the structure of proteins, lipids, and nucleic acids? (Section 2-3)
- **describe** the role of enzymes in catalyzing chemical reactions? (Section 2-4)
- **contrast** prokaryotes and eukaryotes? (Section 3-2)
- **identify** the structure and function of chloroplasts and mitochondria? (Section 3-3)
- **summarize** the role of DNA in heredity? (Section 9-1)

If not, *review the sections indicated.*

Looking Ahead

internetconnect

National Science Teachers Association *sci*LINKS Internet resources are located throughout this chapter.

Reading Strategies

Active Reading

Technique: K-W-L

Before students read this chapter, have them write short individual lists of all the things they already Know (or think they know) about how life began and then developed into what exists today. Ask them to contribute their entries to a group list placed on the board or overhead. Then have students list things they Want to know about the development of life on this planet. Have students save their lists for use later in **Closure** on page 268.

Directed Reading

Assign **Directed Reading Worksheet Chapter 12** to help students interact with the text as they read the chapter.

Interactive Reading

Assign ***Biology: Principles and Explorations* Audio CD Program Chapter 12** to help reluctant readers achieve greater success in reading the chapter.

Reading Skills

Assign Reading Organizers and Reading Strategies Worksheets from the **One-Stop Planner CD-ROM with Test Generator** to help students learn and practice effective reading skills.

internetconnect

Holt *Biology: Principles and Explorations*

On-line Resources: go.hrw.com

Visit the HRW Web site for a variety of resources related to this chapter. Just type in the keyword HX0 Chapter 12.

Holt *Biology: Principles and Explorations*

On-line Resources: www.scilinks.org

The following *sci*LINKS Internet resources can be found in the student text for this chapter:

Page 253
TOPIC: Radioactive decay
KEYWORD: HX253

Page 260
TOPIC: Endosymbiosis
KEYWORD: HX260

Page 263
TOPIC: Extinction
KEYWORD: HX263

Chapter 12

1 Prepare

Scheduling

- **Block:** About 90 minutes will be needed to complete this section.
- **Traditional:** About two class periods will be needed to complete this section.
- **Planning:** See the Planning Guide on pp. 249A and 249B for lecture, classwork, and assignment options for Section 12-1. Lesson plans for Section 12-1 can be found in a lesson cycle format in ***Biology: Principles and Explorations*** **One-Stop Planner CD-ROM with Test Generator.**

Resource Link

Assign **Active Reading Worksheet Section 12-1** to help students comprehend the key concepts and visuals in this lesson.

2 Teach

Lesson Warm-up

Different radioactive isotopes have different decay rates. Show students photographs of human fossils, and explain that carbon-14, frequently used for age determination of human tools and bones, gives accurate measurements for fossils up to 50,000 years old. Other radioactive isotopes are used to date older samples. Potassium-40 is used in the example on this page. Remember, for very old samples, measurement and computation may over- or underestimate an actual date by several million years and still have an error of 1 percent or less.

Section 12-1 How Did Life Begin?

Objectives

- **Summarize** how radioisotopes can be used in determining Earth's age. (p. 252)
- **Compare** two models that describe how the chemicals of life originated. (pp. 254–255)
- **Describe** how cellular organization might have begun. (p. 256)
- **Recognize** the importance of the development of heredity to the development of life. (p. 257)

Key Terms

radiometric dating
radioisotope
half-life
spontaneous origin
microsphere

The Early Earth Was Lifeless

When Earth formed, about 4.5 billion years ago, it was a fiery ball of molten rock. It could not have supported the development of life. Eventually, the planet's surface cooled and formed a rocky crust. Water vapor in the atmosphere condensed to form vast oceans. Many scientists think life first evolved in these oceans. Scientists think the evolution of life took hundreds of millions of years to occur. Evidence that Earth has existed long enough for this evolution to have taken place can be found by measuring the age of the Earth.

Measuring the Earth's age

Scientists are able to calculate the age of the Earth through the use of a technique called radiometric dating. **Radiometric dating** is the calculation of the age of an object by measuring the proportions of the radioactive isotopes *(EYE soh tohps)* of certain elements. Isotopes are forms of an element that differ in atomic mass (the mass of each individual atom). Radioactive isotopes, or **radioisotopes,** are unstable elements (parent) that break up and give off energy in the form of charged particles (radiation). This change, or decay, results in isotopes that are smaller and more stable (product). For example, certain rocks contain traces of potassium-40, an isotope of the element potassium, K. As Figure 12-1 shows, it takes about 1.3 billion years for one-half of the potassium-40 in a rock to change into other isotopes. The time it takes for one-half of a given amount of a radioisotope to change is called the radioisotope's **half-life.** By measuring how many half-lives have passed since a rock was formed, scientists can estimate the rock's age.

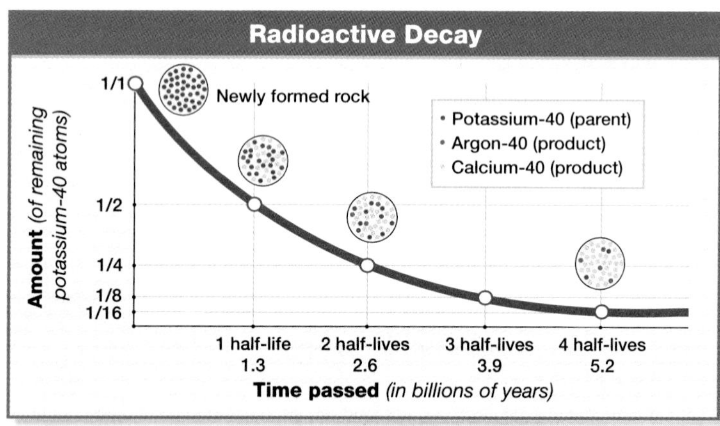

Figure 12-1 Rate of decay for potassium-40. This graph shows the rate of decay for the radioisotope potassium-40. After one *half-life* has passed, half of the original amount of the radioisotope remains.

Integrating Different Learning Styles

Logical/Mathematical	**PE**	Quick Labs, pp. 253, 257; Performance Zone items 14 and 22, pp. 270–271
	ATE	Opening Demonstration, p. 250
Visual/Spatial	**PE**	Performance Zone items 1, 17, 18, pp. 270–271
	ATE	Lesson Warm-up, p. 252; Visual Strategy, p. 253; Reteaching, p. 257
Interpersonal	**PE**	Quick Labs, pp. 253, 257; Performance Zone item 20, p. 271
Verbal/Linguistic	**PE**	Study Tip, p. 255; Performance Zone items 2, 24, pp. 270–271
	ATE	Teaching Tip, p. 254; Multicultural Perspective, p. 254

How the Basic Chemicals of Life Were Formed

Most scientists think that life on Earth developed through natural chemical and physical processes. **Spontaneous origin** is the process through which life is thought to have developed when molecules of nonliving matter reacted chemically during the first billion years of Earth's history. These molecules formed many different simple, organic molecules. Energized by the sun and volcanic heat, these simple, organic molecules formed more-complex molecules that eventually became the building blocks of the first cells. The hypothesis that many of the organic building blocks of life were made from molecules of nonliving matter has been tested and confirmed using scientific methods in laboratory experiments.

TOPIC: Radioactive decay
GO TO: www.scilinks.org
KEYWORD: HX253

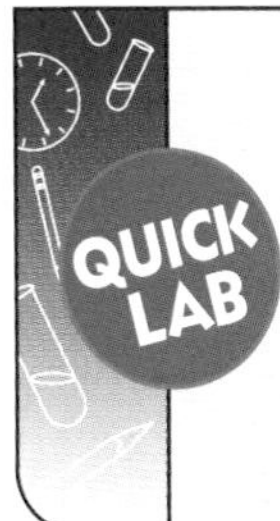

How can you model radioactive decay?

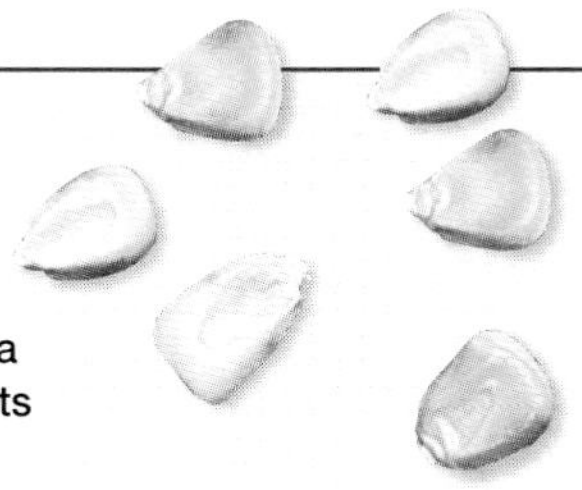

You can use some dried corn, a box, and a watch to make a model of radioactive decay that will show you how scientists measure the age of objects.

Materials

approximately 100 dry corn kernels per group, cardboard box, clock or watch with a second hand

Procedure

1. On a separate sheet of paper, make a data table like the one below.
2. Assign one member of your team to keep time.
3. Place 100 dry corn kernels into a box.
4. Shake the box gently from side to side. Stop after 10 seconds.
5. Remove and count the kernels that "point" to the left side of the box, as shown below. Record in your data table the number of kernels you removed.
6. When you have finished counting the kernels, repeat steps 4 and 5 until all kernels have been counted and removed.
7. Calculate the cumulative number of kernels removed for each time interval.
8. Make a graph using your group's data. Plot "time (seconds)" on the *x*-axis. Plot "cumulative number of kernels removed" on the *y*-axis.

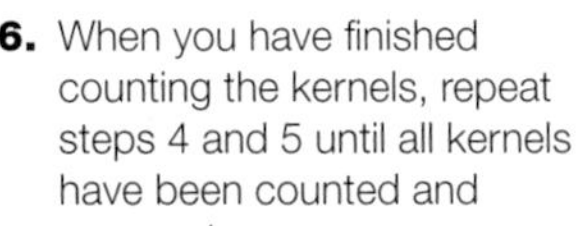

Analysis

1. **Explain** what the kernels at the beginning of the experiment represent in terms of radioactive decay.
2. **Identify** what the removed kernels represent in each step.
3. **Calculate** the half-life of your sample, in seconds, that is represented in this activity.
4. **Calculate** the age of your sample, in years, if each 10-second interval represents 5,730 years.

DATA TABLE

Time (seconds)	Number of kernels removed	Cumulative number of kernels removed
10		
20		
30		

Section 12-1

Visual Strategy

Figure 12-1

Ask students to explain why the rate of decay shown in Figure 12-1 is *not* a straight line. *(The rate of radioactive decay is based on the half-life of a radioactive material. The amount of material that decays in one half-life is one-half of the undecayed portion of any size sample and, therefore, continuously changes. The line would be straight only if the decay rate was fixed.)*

internet**connect**

Have students research radioactive decay using the Web site in the Internet Connect box on page 253. Have students write a report summarizing their findings. PORTFOLIO

How can you model radioactive decay?

Time 20 minutes

Process Skills Forming a model, inferring, calculating

Teaching Strategies

- Have kernels precounted and placed in paper or plastic cups.
- If time allows, multiple trials (two or three) would provide a better set of data.
- Make sure all kernels are accounted for; they can be slippery if stepped on.

Analysis Answers

1. The kernels represent molecules of a radioactive isotope.
2. The kernels that are removed with each step represent the molecules that decayed.
3. The half-life of the sample is the number of trials it took to remove 50 kernels multiplied by 10 sec/trial.
4. Age of the sample = number of trials × 5,730 years/trial

REAL LIFE

Answer

Scientists are working on technology to eventually examine (by remote telescopes) the planets in and beyond our solar system whose spectral "signatures" indicate the presence of oxygen, water, and carbon dioxide. NASA plans a sample recovery mission to Mars for 2008.

Teaching TIP

Analyzing Information

Focus students' attention on the wording of the text near the end of "The 'primordial soup' model" section on this page. Point out the sentence "These results indicate that some of the basic chemicals of life *could have* formed spontaneously on the early Earth..." Ask the students why this sentence does not read "...that some basic chemicals of life formed spontaneously on the early Earth..." *(Miller and Urey's experiments did not prove that organic molecules formed more than 3.5 billion years ago under the conditions that they modeled in the laboratory. The experiments showed only that they could have formed in such a manner. Only by observing firsthand could one say that organic molecules actually formed in this way.)*

Making Biology Relevant

Health

Remind students that ultraviolet (UV) radiation in sunlight can be very damaging to their health. Sunburn indicates that skin has been damaged by UV radiation. The American Cancer Society reminds us that people who have been seriously sunburned have an increased risk of skin cancer. Remind students that if they are going to be exposed to the sun for long periods of time, they should wear a high SPF sunscreen or protective clothing to reduce UV damage. Explain that the "peeling" that occurs after sunburn is the body's way of getting rid of cells too damaged to repair themselves.

REAL LIFE

Is there life on other planets? Many scientists think life could have arisen on other planets the same way it did on Earth.

Analyzing Information
Find out about research on extraterrestrial life, and compare scientists' predictions about possible life-forms on other planets.

The "primordial soup" model

In the 1920s, the Russian scientist A. I. Oparin and the British scientist J.B.S. Haldane each suggested that the early Earth's oceans once contained large amounts of organic molecules. This model became known as the primordial *(preye MAWR dee uhl)* soup model; the Earth's vast oceans were thought to be filled with many different organic molecules, like a soup is filled with many different vegetables and meats. Oparin and Haldane hypothesized that these molecules formed spontaneously in chemical reactions activated by energy from solar radiation, volcanic eruptions, and lightning.

Oparin, the American scientist Harold Urey, and other scientists also proposed that Earth's early atmosphere lacked oxygen. They hypothesized that the early atmosphere was instead rich in nitrogen, N_2, hydrogen, H_2, and hydrogen-containing gases such as water vapor, H_2O, ammonia, NH_3, and methane, CH_4. Electrons in these gases would have been frequently pushed to higher energy levels by light particles from the sun or by electrical energy in lightning. Today high-energy electrons are quickly soaked up by the oxygen in Earth's atmosphere because oxygen atoms have a great "thirst" for such electrons. But without oxygen, high-energy electrons would have been free to do other things, such as reacting with other hydrogen-rich molecules.

In 1953, the primordial soup model was tested by Stanley Miller, who was then working with Urey. Miller placed the proposed gases into a device like the one seen in **Figure 12-2.** To simulate lightning, he provided electrical sparks. After a few days, Miller found a complex "chemical zoo" in his device. These chemicals included some of life's basic building blocks: amino acids, fatty acids, and other hydrocarbons (molecules made of carbon and hydrogen). These results indicate that some basic chemicals of life could have formed spontaneously on the early Earth under conditions like those in the experiment.

Figure 12-2 Miller-Urey experiment. Miller simulated the early Earth's conditions as hypothesized by Oparin, Urey, and other scientists. His experiment produced the chemicals of life.

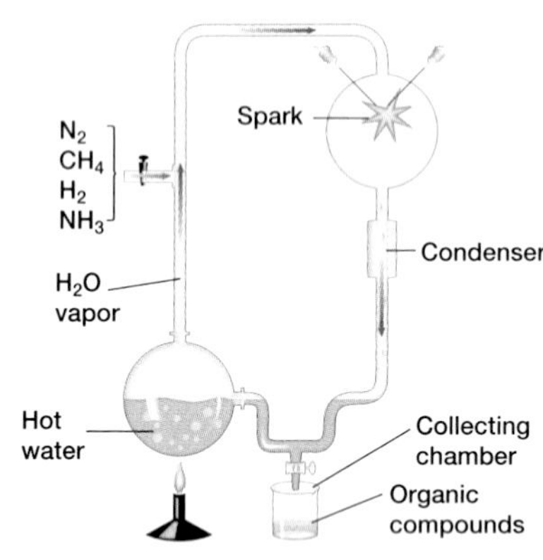

Reevaluating the primordial soup model

Recent discoveries have caused scientists to reevaluate the primordial soup model. At the time of Miller's experiment, scientists thought that life had taken more than 1 billion years to begin. However, measurements of Earth's age and discoveries of 3.5-billion-year-old fossils indicate that life began much earlier. Thus, it appears that life developed much faster than was previously assumed.

Another problem is we now know that the mixture of gases used in Miller's experiment could not have existed on early Earth. Four billion years ago, Earth did not have a protective layer of ozone gas, O_3. Today ozone protects Earth's surface from most of the sun's damaging ultraviolet radiation. Without ozone, ultraviolet radiation would have destroyed any ammonia and methane present in the atmosphere. When these gases are absent from the

MULTICULTURAL PERSPECTIVE

Hindu and Hopi Creation Stories

The theme that Earth gave rise to life appears in the legends of many cultures. For example, according to a Hindu proverb, "The Earth is our mother, and we are all her children." A Hopi proverb states, "Earth gives life and seeks the man who walks gently upon it."

Miller-Urey experiment, key biological molecules are not made. This raises a very important question: If the chemicals needed to form life were not in the atmosphere, where did they come from?

The bubble model

In 1986, the geophysicist Louis Lerman suggested that the key processes that formed the chemicals needed for life took place within bubbles on the ocean's surface, as shown in **Figure 12-3.**

Step 1 Ammonia, methane, and other gases resulting from the numerous eruptions of undersea volcanoes were trapped in underwater bubbles.

Step 2 Inside the bubbles, the methane and ammonia needed to make amino acids might have been protected from damaging ultraviolet radiation. Chemical reactions would take place much faster in bubbles (where reactants would be concentrated) than in Oparin and Haldane's primordial soup.

Step 3 Bubbles rose to the surface and burst, releasing simple organic molecules into the air.

Step 4 Carried upward by winds, the simple organic molecules were exposed to ultraviolet radiation and lightning, which provided energy for further reactions.

Step 5 More complex organic molecules that formed by further reactions fell into the ocean with rain, starting another cycle.

Thus, life could have begun much more quickly than is estimated by the primordial soup model.

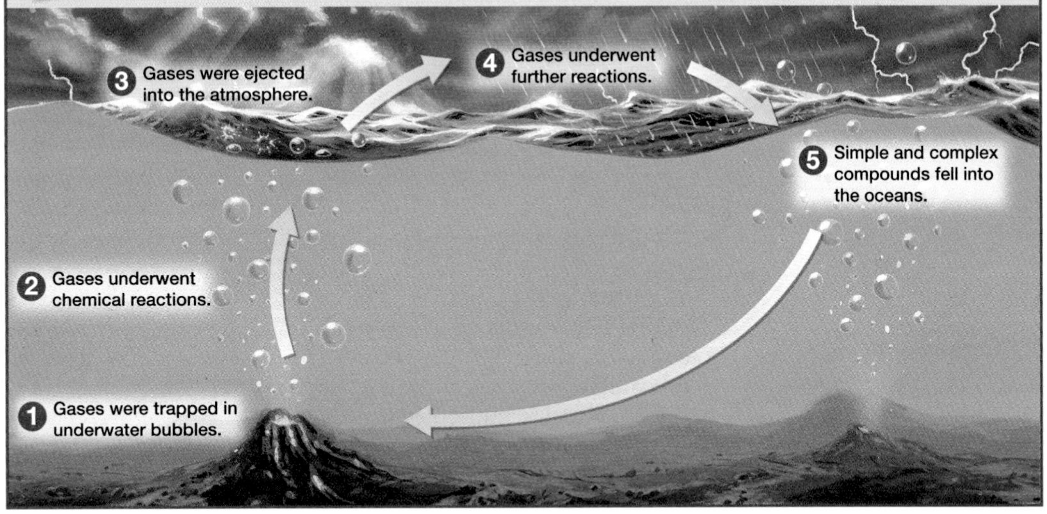

Study TIP

Reading Effectively

Before reading this chapter, write the Objectives for each section on a sheet of paper. Rewrite each Objective as a question, and answer these questions as you read the section.

BIO graphic

Figure 12-3

Lerman's Bubble Model

Lerman proposed that gases formed simple organic molecules.

3 Gases were ejected into the atmosphere.
4 Gases underwent further reactions.
5 Simple and complex compounds fell into the oceans.
2 Gases underwent chemical reactions.
1 Gases were trapped in underwater bubbles.

Teaching TIP

A Spark of Life

Experiments similar to Miller's that used UV light instead of electrical sparks as an energy source also produced amino acids.

CROSS-DISCIPLINARY CONNECTION

◆ Geology

The origin of life on this planet is a concern of many scientific fields. In the past few years, we have sent remote-controlled machines to our planetary neighbor Mars. Scientists are trying to answer questions about that planet similar to the questions asked about Earth in this chapter. Investigators in the fields of biology, chemistry, physics, geology, and astrophysics are all working together to learn more about other planets in the quest to learn more about our own. Indeed, whole new fields of study, such as planetary geology, were unheard of a few years ago. Ask students what new fields of study might arise from our exploration of the planets in our solar system. *(Students' answers might include planetary mining and refining, mass transit between planets, and so on.)*

Overcoming Misconceptions

Atmospheric Resources

Students may think that the atmosphere is an infinite resource that will always be present as it is today. Challenge this misconception by having students compare the changes in the atmosphere of early Earth with the changes that are occurring in our present atmosphere. Begin by describing how oxygen levels steadily increased in the ancient atmosphere due to the activity of photosynthetic bacteria. Then ask students what kind of gaseous products human activity is producing today. *(carbon dioxide, methane, sulfur dioxide, nitrous oxide, and others)* Ask which gas of particular importance is increasing worldwide and which is decreasing worldwide. *(Carbon dioxide is increasing, and upper-atmospheric ozone is decreasing.)*

Chapter 12

Teaching TIP

Ancient Oceans

Early oceans were probably similar in salt content to modern freshwater lakes. The salinity of today's oceans is the result of millions of years of water running over rocks, picking up mineral salts, and flowing to the sea.

Teaching TIP

Organic Compounds

Carbon atoms can combine to form an almost infinite number of compounds by combining with each other in long chains and in other types of large, complex molecules. They not only form a total of four bonds, they can also join other atoms in single, double, and triple bonds. The structure of silicon is far more rigid. Implications of the importance of these differences can be seen by comparing carbon-based plastics to silicon-based ceramics.

Organic Chemicals Became Complex

Though scientists disagree about the details of the process that led to the origin of life, most scientists accept that by adding energy, the basic molecules of life could have formed spontaneously through simple chemistry. But there are enormous differences between organic molecules and living cells. How did amino acids link to form proteins? How did nucleotides form the long chains of DNA that store the instructions for making proteins? In the laboratory, scientists have not been able to make either of these macromolecules form spontaneously in water. However, short chains of RNA, the nucleic acid that helps carry out DNA's instructions, have been made to form on their own in water.

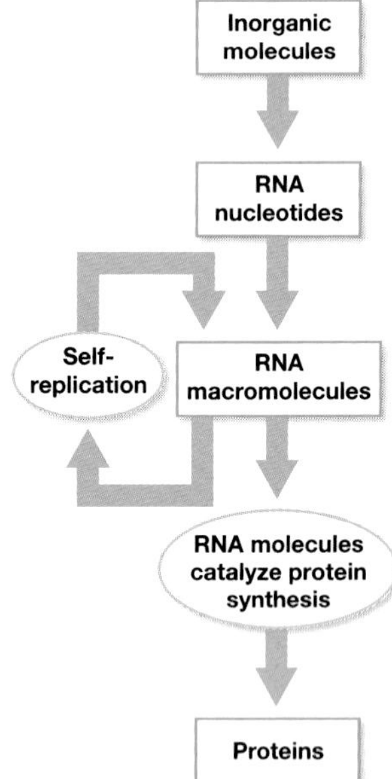

Figure 12-4 Stages leading to RNA self-replication and protein synthesis. Chemical reactions between inorganic molecules formed RNA nucleotides. The nucleotides assembled into RNA macromolecules. These molecules might have been able to self-replicate and to catalyze the formation of proteins.

A possible role for catalysts

In the 1980s, the American scientists Thomas Cech, of the University of Colorado, and Sidney Altman, of Yale University found that certain RNA molecules can act like enzymes. RNA's three-dimensional structure provides a surface on which chemical reactions can be catalyzed. Also, recall from Chapter 10 that messenger RNA acts as an information-storing molecule. As a result of Cech's and Altman's work and other experiments showing that RNA molecules can form spontaneously in water, a simple hypothesis was formed: Perhaps RNA was the first self-replicating information-storage molecule. After such a molecule had formed, it could also have catalyzed the assembly of the first proteins, as shown in **Figure 12-4.** More important, such a molecule would have been capable of changing from one generation to the next.

Microspheres may have led to cells

Observations show that lipids, which make up cell membranes, tend to gather together in water. By shaking up a bottle of oil and vinegar, you can see something similar happen—small balls of oil form. A lipid molecule looks a bit like a ball with a tail. Lipid molecules align themselves to form a hollow sphere that resembles a cell membrane. Similarly, laboratory experiments have shown that short chains of amino acids tend to gather into tiny vesicles called **microspheres.** Another type of vesicle, called a coacervate *(koh A sur vayt),* is made of molecules of different types, including linked amino acids and sugars.

Scientists think that formation of microspheres might have been the first step toward cellular organization. According to this hypothesis, microspheres formed, persisted for a while, and then dispersed. Over millions of years, those microspheres that could survive longer by incorporating molecules and energy would have become more common than here today-gone tomorrow kinds. However, microspheres could not be considered alive unless they had developed the ability to transfer their characteristics to offspring.

Overcoming Misconceptions

The First Hereditary Molecule

Armed with the knowledge of the function of RNA in protein synthesis, many students may not be able to understand immediately how RNA, which relies on a template provided by DNA, could have developed before DNA.

Explain to students that today certain viruses (called retroviruses), including HIV, contain only RNA as their genetic material. Their viral RNA, when released into a host cell, is used as a template to then make DNA.

Origin of heredity

Although scientists disagree about the exact origin of heredity, most agree that double-stranded DNA probably evolved after RNA and that RNA "enzymes" catalyzed the assembly of the earliest proteins. Scientists do not agree about whether RNA first formed inside or outside of microspheres. Most tentatively accept the hypothesis that some microspheres that contained RNA developed a means of transferring their characteristics to offspring. Once the mechanism of heredity was developed, life as we know it began.

Because researchers do not yet understand how DNA, RNA and hereditary mechanisms first developed, how life might have originated naturally and spontaneously remains a subject of intense interest, research, and discussion among scientists.

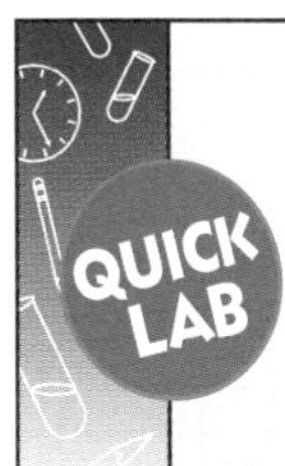

How can you model coacervates?

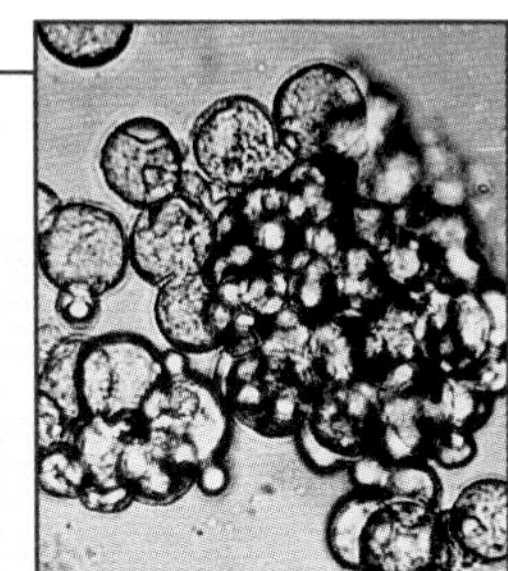

By using simple chemistry, you will see how coacervates may have led to the origin of the first cells.

Materials

safety goggles and lab apron, graduated cylinder, 1 percent gelatin solution, 1 percent gum arabic solution, test tube, 0.1 M HCl, pipet, microscope slide and coverslip, microscope

Procedure

1. **CAUTION: Hydrochloric acid is corrosive. Be certain to wear safety goggles and apron. Avoid contact with skin and eyes. Avoid breathing vapors. If any of this solution should spill on you, immediately flush the area with water, and then notify your teacher.**
2. Mix 5 mL of a 1 percent gelatin solution with 3 mL of a 1 percent gum arabic solution in a test tube.
3. Add 0.1 M HCl to the gelatin–gum arabic solution one drop at a time until the solution turns cloudy.
4. Prepare a wet mount of the cloudy solution, and examine it under a microscope at high power.
5. Prepare a drawing of the structures that you see. They should resemble the structures in the photo above.

Analysis

1. **Describe** what happened to the solutions after the acid was added.
2. **Compare** the appearance of coacervates with that of cells.
3. **Predict** what would happen to the coacervates if a base were added to the solution.
4. **SKILL Evaluating Hypotheses** Based on the evidence you obtained, defend the hypothesis that coacervates could have been the basis of life on Earth.

Review SECTION 12-1

1. **Identify** how radioisotopes are used to determine the age of a rock.
2. **Contrast** two scientific models that explain the origin of life.
3. **Describe** the first step that led toward cellular organization.
4. **Relate** how the development of heredity contributed to the successful formation of life.

Critical Thinking

5. **SKILL Evaluating Viewpoints** Must one model of the origin of life's chemicals exclude the possibility of another model?

Review SECTION 12-1 ANSWERS

1. The amount of the radioisotope in the rock is compared to the amounts of other species that resulted from decay.
2. In the "primordial soup" model, the early atmosphere contained large amounts of nitrogen and hydrogen-containing gases. Energy from the sun or lightning stimulated chemical reactions that produced amino acids and other hydrocarbons. In the bubble model, ammonia, methane, and other gases from underwater volcanoes were trapped in bubbles in the seas. Chemical reactions were stimulated by the proximity of reactants to each other, and the underwater bubbles were protected from UV radiation.
3. The formation of microspheres might have led to cellular organization.
4. Heredity is a necessary feature of living things. Microspheres that contained replicating RNA might have transferred both their structure and their RNA to offspring.
5. Scientific models are explanations based on all available data. As more data are found, models must change to incorporate them.

How can you model coacervates?

Time 20 minutes

Process Skills Forming a model, predicting outcomes

Teaching Strategies

- Solutions must be fresh and prepared accurately.
- You and your students should always wear safety goggles and lab aprons when mixing chemicals. Tell students to notify you in case of an accident. Supervise the cleanup of all spills.

Analysis Answers

1. The solution became cloudy.
2. Coacervates are spherical and membrane-bound but have no internal structures.
3. The coacervates might break up.
4. Answers will vary. Students might state that coacervates are discrete structures that resemble cells and that arise in certain chemical environments.

3 Assessment Options

Closure

Evolution Events

Ask students to write an answer to this question: Were the first cells that evolved on Earth prokaryotes or eukaryotes? Have them justify their choice. PORTFOLIO

Review

Assign Review—Section 12-1.

Reteaching — BASIC

Ask students to draw a rough diagram of the Miller-Urey apparatus. *(Students should include in their sketch an area for holding water, an area where water and gases can combine and be subjected to electrical sparks, an area where any by-products can cool, and an area where the by-products can be collected.)* PORTFOLIO

Section 12-2 Complex Organisms Developed

1 Prepare

Scheduling

- **Block:** About 90 minutes will be needed to complete this section.
- **Traditional:** About two class periods will be needed to complete this section.
- **Planning:** See the Planning Guide on pp. 249A and 249B for lecture, classwork, and assignment options for Section 12-2. Lesson plans for Section 12-2 can be found in a lesson cycle format in ***Biology: Principles and Explorations* One-Stop Planner CD-ROM with Test Generator.**

Resource Link

Assign **Active Reading Worksheet Section 12-2** to help students comprehend the key concepts and visuals in this lesson.

2 Teach

Lesson Warm-up

Some students may have difficulty understanding the timeline of Earth's history that begins at the bottom of page 258. This analogy should provide some perspective. Draw a large circle on the board to represent a clock. Place an arrow at 12:01 to indicate when Earth was formed. Place additional arrows at 1:15 (the first bacteria appeared); 8:00 (the first eukaryotes); 11:00 (the first life on land); 11:50 (dinosaurs became extinct); and 11:59 (first humans appeared).

Teaching TIP

Evolutionary Timeline

After they read each section's subheading, have students use the timeline on the bottom of pp. 258–268 to review events discussed in that section. Students should construct a table with two columns and then describe a particular event in one column and record the approximate date in the second column.

Objectives

- **Distinguish** between the two groups of prokaryotes. (p. 258)
- **Describe** the evolution of eukaryotes. (pp. 259–260)
- **Recognize** an evolutionary advance first seen in protists. (p. 261)
- **Summarize** how mass extinctions have affected the evolution of life on Earth. (p. 263)

Key Terms

fossil
cyanobacteria
eubacteria
archaebacteria
endosymbiosis
protist
mass extinction

Prokaryotes Are the Oldest Group of Organisms

When did the first organisms form? To find out, scientists study the best evidence of early life that we have, fossils. A **fossil** is the preserved or mineralized remains (bone, tooth, or shell) or imprint of an organism that lived long ago. The oldest fossils, which are microscopic fossils of prokaryotes, come from 3.5-billion-year-old rock deposits found in western Australia.

Recall that prokaryotes are single-celled organisms that lack internal membrane-bound organelles. Prokaryotes are also known as bacteria. Among the first of the bacteria to appear were cyanobacteria. **Cyanobacteria** *(SEYE an oh bak TIHR ee ah)* are photosynthetic bacteria. Before cyanobacteria appeared, oxygen gas was rare on Earth. But as ancient cyanobacteria carried out photosynthesis, they released oxygen gas into Earth's oceans. After hundreds of millions of years, the oxygen produced by cyanobacteria began to escape into the air, as shown in **Figure 12-5.** Over the billions of years that followed, more oxygen was added to the air. Today oxygen gas makes up 21 percent of the Earth's atmosphere.

Two groups of bacteria

Early in the history of life, two different groups of prokaryotes evolved—eubacteria and archaebacteria. Living examples include *Escherichia coli*, a species of eubacteria, and *Sulfolobus*, a group of archaebacteria. **Eubacteria** are prokaryotes that contain a chemical called peptidoglycan *(pep tih doh GLEYE kan)* in their cell walls and have the same type of lipids in their cell membranes as eukaryotes do. Eubacteria include many bacteria that cause disease and decay. **Archaebacteria** are prokaryotes that lack peptidoglycan in their cell walls and have unique lipids in their cell membranes. Archaebacteria are thought to be closely related to the first bacteria to have existed on Earth. Chemical evidence indicates that the first eukaryotic cells are more likely to have evolved from archaebacteria than from eubacteria.

Figure 12-5 Evolutionary timeline. This timeline shows some of the major events that occurred during the evolution of life on Earth.

Age (in millions of years ago)

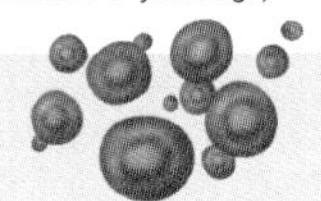

Earliest fossil bacteria

3,500

Origin of O_2 by photosynthesis

2,500

PRECAMBRIAN ERA

Integrating Different Learning Styles

Logical/Mathematical	**PE**	Data Lab, p. 259; Exploring Further, p. 260
Visual/Spatial	**ATE**	Lesson Warm-up, p. 258; Teaching Tip, p. 259
Interpersonal	**ATE**	Closure, p. 263
Verbal/Linguistic	**PE**	Internet Connect, pp. 260, 263; Performance Zone item 21, p. 271
	ATE	Teaching Tips, pp. 258 and 260; Career, p. 262

The First Eukaryotes Evolved

About 1.5 billion years ago, the first eukaryotes appeared. Recall that a eukaryotic cell has a complex system of internal membranes and has DNA that is enclosed within a nucleus. A third characteristic of almost all eukaryotes is the presence within the cells of complex bacteria-sized organelles called mitochondria and chloroplasts. These organelles contain their own DNA. Except for one rare primitive form, all eukaryotes have mitochondria. Chloroplasts, which carry out photosynthesis, are found only in protists and plants.

The evolution of mitochondria and chloroplasts

To explain the origins of mitochondria, almost all biologists now accept the once-controversial theory of **endosymbiosis** that was first presented in 1966 by the American biologist Lynn Margulis. The theory of endosymbiosis proposes that mitochondria are the descendants of symbiotic, aerobic (oxygen-requiring) eubacteria.

Analyzing signs of endosymbiosis

Background

As you will recall from Chapter 3, mitochondria have their own DNA and produce their own proteins. The data below were collected by several different scientists studying the proteins produced by mitochondrial DNA. These scientists found that the three-nucleotide sequences (codons) in the nucleus of an organism's cell can code for different amino acids than the codons of the mitochondria code for. Examine the data below, and answer the questions that follow.

Magnification: 6930×

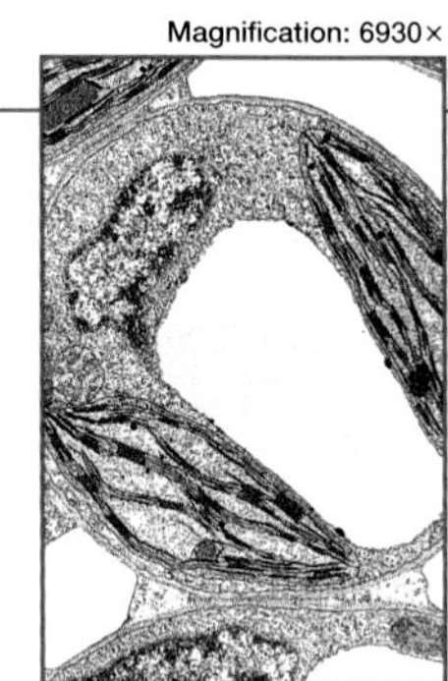

Analysis

1. **Defend** the theory of endosymbiosis using these data.
2. **Infer** what these data indicate about the evolution of plant cells.
3. **Describe** how these data can be used to support the idea that more than one type of cell evolved early in the history of life.

Amino Acids Made in the Nucleus and Mitochondria

	Amino acids coded for in the nucleus	Amino acids coded for in mitochondria	
Codon	**Plants and mammals**	**Plants**	**Mammals**
UGA	Stop	Stop	Tryptophan
AGA	Arginine	Arginine	Stop
AUA	Isoleucine	Isoleucine	Methionine
AUU	Isoleucine	Isoleucine	Methionine
CUA	Leucine	Leucine	Leucine

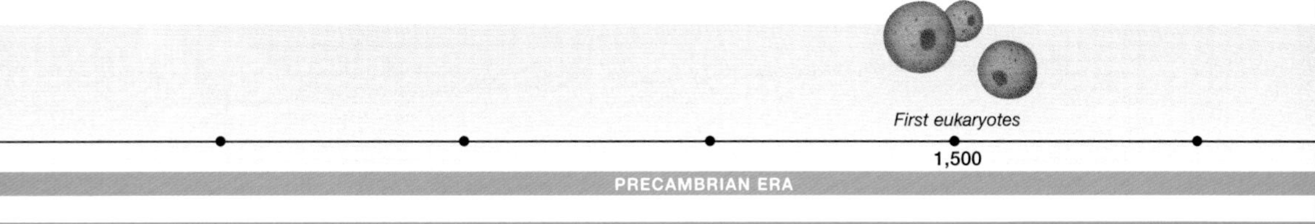

Graphic Organizer

Use this graphic organizer with ***Teaching Tip: Remembering What Came from What*** *on page 259.*

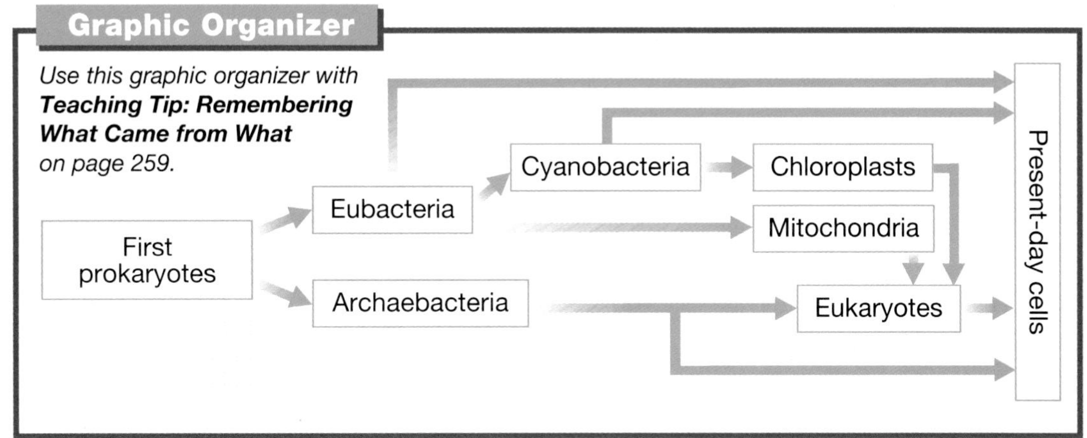

Analyzing signs of endosymbiosis

Time 20 minutes

Process Skills Analyzing data, relating conclusions

Teaching Strategies

Point out to students that they are looking at a "doubled double" set of data. That is, they are considering the nucleus with respect to mitochondria and plants with respect to mammals.

Analysis Answers

1. The fact that in mammals the genetic code of nuclear DNA differs from that of mitochondrial DNA suggests that mitochondria have evolved separately from the animal cells that contain them.
2. We can infer that plant mitochondria and plant cells together have been subjected to selective pressures different from those that acted on mammalian cells and on mammalian mitochondria.
3. We can infer that the prokaryotic ancestors of mammalian mitochondria probably diverged very early from the prokaryotic ancestors of plant mitochondria because the genetic codes are quite different in the two types.

Teaching TIP

Remembering What Came from What

Use a **Graphic Organizer** similar to the one at the bottom of page 259 to help students grasp the progression of the complexity of life. Reproduce the diagram, without the eight terms, on the board or overhead. Next, write an alphabetical list of the eight terms beside the graphic organizer. Ask students to insert the terms into their correct rectangles. To help students begin, add two terms to the diagram and then cross them off the list. *(Perhaps add* First prokaryotes *and* Present-day cells *or* Eukaryotes.*)* Conclude by asking each student to write a paragraph that summarizes this graphic organizer. PORTFOLIO

A second successful invasion of pre-eukaryotic cells by photosynthetic bacteria gave rise to chloroplasts. Chloroplasts are the organelles in plant cells where photosynthesis takes place. The invading bacteria were probably closely related to cyanobacteria. Both mitochondria and chloroplasts have characteristics that are similar to those of bacteria—for example, their size and structure, genetic material, method of reproduction, and ribosome structure.

Teaching TIP

Endosymbiosis

The term *endosymbiosis* refers to any symbiotic relationship in which one organism lives inside another. Students will better understand Margulis's hypothesis for the endosymbiotic origin of mitochondria and chloroplasts if they review other examples of endosymbiosis. *(For example, bacteria within the digestive tracts of termites digest wood, allowing the termites to use wood as food.)*

Exploring Further

Why Endosymbiosis?

Teaching Strategies

Tell students that the points noted in statements 1–4 are factors that contributed to the development of Margulis's theory. Remind them that a hypothesis (which precedes a theory) is a plausible explanation that takes into account everything that is known about a situation.

Discussion

- What is the distinction between endosymbiosis and symbiosis? *(Endosymbiosis is a type of symbiosis in which one organism lives inside another.)*
- Identify another example of endosymbiosis. *(Answers will vary. Students should cite a symbiotic relationship involving an endosymbiont, such as that of the normal bacteria that live in the human gut, which not only feed on but also help to digest food.)*

Exploring Further

Why Endosymbiosis?

Did you know that the cotton in your clothes was produced as a result of the sun's energy and endosymbiosis? Recall that plants, such as the cotton plant, collect energy through photosynthesis. The organelle that carries out photosynthesis is the chloroplast. Both the chloroplast and the mitochondrion evolved through the process of endosymbiosis. According to Lynn Margulis's theory of endosymbiosis, bacteria entered large cells either as parasites or as undigested prey. Instead of being consumed, the bacteria began to live inside the host cell, where they performed cellular respiration and photosynthesis. The following observations support the idea that mitochondria and chloroplasts descended from bacteria:

1. **Size and structure.** Mitochondria are about the same size as most eubacteria. They are surrounded by two membranes. The smooth outer membrane of mitochondria is thought to be derived from the endoplasmic reticulum of the larger host cell. The inner membrane of mitochondria is folded into many layers, so it looks like the cell membranes of aerobic eubacteria. Inside this membrane are proteins that carry out cellular respiration.
2. **Genetic material.** Mitochondria have circular DNA similar to the chromosomes found in bacteria. Both chloroplasts and mitochondria contain genes that are different from those found in the nucleus of the cell in which they are found.
3. **Ribosomes.** Mitochondrial ribosomes and bacterial ribosomes have a similar size and structure.
4. **Reproduction.** Like bacteria, chloroplasts and mitochondria reproduce by simple fission. This replication takes place independently of the cell cycle of the cell in which chloroplasts and mitochondria are located.

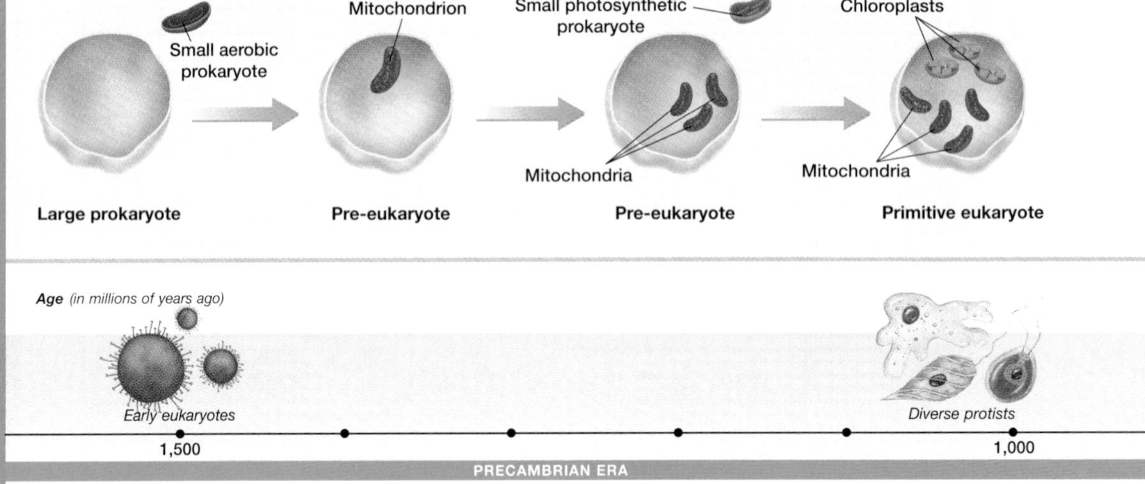

Multicellularity Evolved Many Times

Most biologists group all living things into six broad categories called kingdoms. The two oldest kingdoms, Eubacteria and Archaebacteria, are made up of single-celled prokaryotes. The first eukaryotic kingdom was the kingdom Protista. **Protists** are members of the kingdom Protista, a large, varied group that includes both multicellular and unicellular organisms. The other three kingdoms (fungi, plants, and animals) evolved later and also consist of eukaryotes.

The unicellular body plan has been tremendously successful, with unicellular organisms today constituting about half the biomass (the total weight of all living things) on Earth. But a single cell has limits. Distinct types of cells in one body allow an organism to specialize in different functions. For example, some organisms may have specific cells that help the organism protect itself from predators or disease and other cells that help it resist drying out. Other examples of specialized cells include cells that help a multicellular organism move about in order to find a mate or prey. With all these advantages, it is not surprising that multicellularity has arisen independently many times.

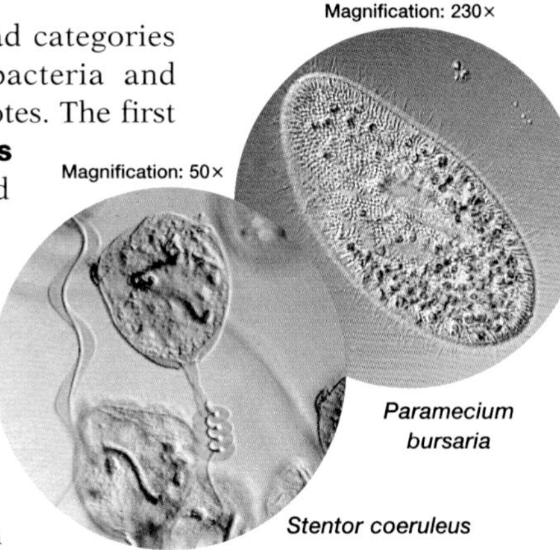

Figure 12-6 Single-celled protists. Single-celled protists occur in many shapes and can live in many different types of environments, including water and land.

Almost every organism large enough to see with the naked eye is multicellular. Most protists, such as those shown in **Figure 12-6,** are single-celled, but there are many multicellular forms. The development of multicellular organisms in the kingdom Protista marked an important step in the evolution of life on Earth. The oldest known fossils of multicellular organisms were found in 700-million-year-old rocks.

Some of the multicellular lines that resulted did not produce diverse groups of organisms. Among those groups of organisms that survived until today are the red, green, and brown algae, shown in **Figure 12-7.** You may have seen algae lying along the seashore as seaweed. Three of the multicellular groups that evolved from the protists were very successful, producing three separate kingdoms—Fungi, Plantae, and Animalia. Each of these three kingdoms evolved independently from a different kind of protist.

Figure 12-7 Brown algae. Brown algae, called kelps, are multicellular protists that form vast underwater "forests" in some coastal waters.

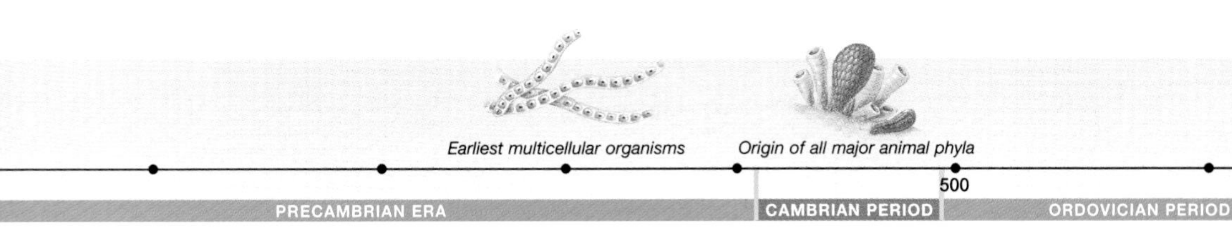

Teaching TIP
Multicellularity and Cellular Specialization

It is vital that students recognize the relationship between multicellularity and the division of labor among different types of cells. Illustrate these concepts with the following analogy. Modern industry uses individual workers' skills and abilities to perform different tasks. Use as an example a local industry (such as an automotive plant, steel mill, canning company, meat packing plant, and so on) to show how more is produced, both in quantity and complexity, by breaking the total task into smaller parts. Point out that cells exhibit the same type of departmentalizing.

Teaching TIP
Favorite Fossils

Have students pick an extinct species that interests them and write a short report about it. Ask students to address the following questions: When, where, and how did the species live? What evidence do we have that the species lived at all? Why might it have become extinct? Do you see a relationship between the species and a modern species? PORTFOLIO

Teaching TIP

Examining Fossils

Students love to examine real fossils. Petrified wood, diatomaceous earth, trilobites, casts and molds of small fish, mollusks, and coprolites (petrified feces) usually are inexpensive and easy to obtain. Fossils are available through biological supply houses or other retail dealers. Assign identification or classification activities using the specimens.

Teaching TIP

Burgess Shale

Have students use library or Internet resources to research the Burgess Shale. Fossil evidence from the shale suggests that animals with a wide variety of body plans existed very early in evolutionary history. Have students search for organisms that would be considered unusual (even bizarre) today and make simple sketches of them.

CAREER

Paleontologist

Paleontologists are scientists who study early life on Earth using fossils. The prefix *paleo-* comes from the Greek word *palai,* which means "ancient." There are several branches of paleontology, including paleoanthropology, which is the study of early humans; paleobotany, the study of early plants; paleoecology, the study of early ecosystems; and paleozoology, the study of early animals. Have students work in teams to use library or Internet resources to find information about the work of paleontologists. Assign a specific branch of paleontology, such as one of those listed above, to each team, and have students write a short report and present their findings to their classmates. PORTFOLIO

Today's organisms originated in the Cambrian period

Most groups of organisms that exist today seem to have originated sometime during the Cambrian period, which lasted from about 540 million to about 505 million years ago. The Cambrian period was a time of great evolutionary expansion, as shown in **Figure 12-8.** Many unusual marine animals also appeared at this time, animals for which there are no close living relatives. A very rich collection of Cambrian fossils was uncovered in 1909 on a rocky mountainside in Canada called the Burgess Shale. The fossils in the Burgess Shale show strange creatures that are not like anything alive today.

The Ordovician period, which followed the Cambrian period, lasted from about 505 million to 438 million years ago. During this time, many different animals continued to abound in the seas. Among them were trilobites, marine arthropods that became extinct about 250 million years ago, shown in Figure 12-8.

Figure 12-8 Possible scene from the Cambrian period. By studying fossils from the Burgess Shale, such as the trilobite fossils below, artists were able to re-create a scene from the shallow seas of the Cambrian period.

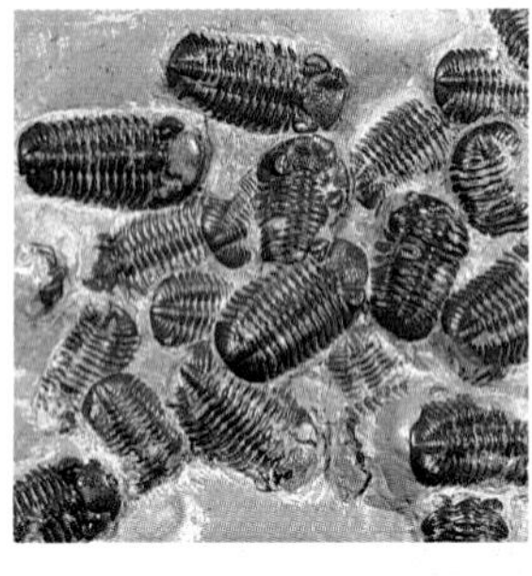

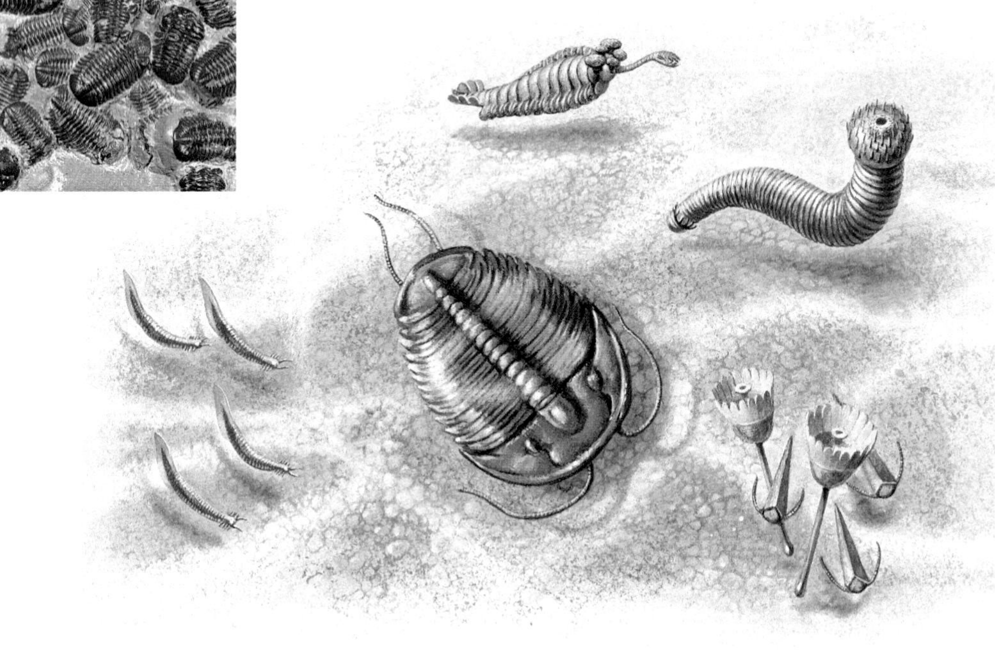

Age (in millions of years ago)

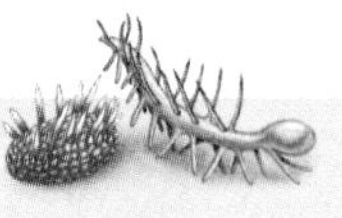

Mass Extinctions Have a Major Impact

The fossil record indicates a sudden change occurred at the end of the Ordovician period. About 440 million years ago, a large percentage of the organisms on Earth suddenly became extinct. This was the first of five major mass extinctions that have occurred on Earth. A **mass extinction** is the death of all members of many different species, usually caused by a large ecological disaster.

Another mass extinction of about the same size happened about 360 million years ago. The third and most devastating of all mass extinctions occurred at the end of the Permian period, about 245 million years ago. About 96 percent of all species of animals living at the time became extinct. About 35 million years later, a fourth, less devastating mass extinction occurred. Although the specific causes of these extinctions are unknown, evidence indicates that worldwide geological and weather changes were likely factors.

The fifth mass extinction will be discussed in more detail in Chapter 33. It occurred 65 million years ago and brought about the extinction of about two-thirds of all land species, including the dinosaurs.

Many scientists think that another mass extinction is occurring today. These scientists reason that this new extinction is taking place because the Earth's ecosystems, especially tropical rain forests, are being destroyed by human activity, as shown in **Figure 12-9.** The world has already lost half its tropical rain forests, and at the rate they are being destroyed; all will be gone in your lifetime. As many as 50,000 species of plants, one quarter of the world's total, will be lost, along with 2,000 of the world's 9,000 species of birds and countless insect species. This is an astonishing rate of loss.

internet**connect**

SCILINKS NSTA

TOPIC: Extinction
GO TO: www.scilinks.org
KEYWORD: HX263

Figure 12-9 Rain forests are being destroyed at an alarming rate. Although tropical rain forests cover only 7 percent of the Earth's land surface, they contain more than one-half of all the world's animal and plant species.

Review SECTION 12-2

1. **Contrast** the two major groups of prokaryotes.
2. **Describe** how eukaryotes evolved.
3. **Compare** bacteria with eukaryotes.
4. **Summarize** how multicellularity advanced the evolution of protists.

Critical Thinking

5. **Justify** the argument that today's organisms would not exist if mass extinctions had not occurred.

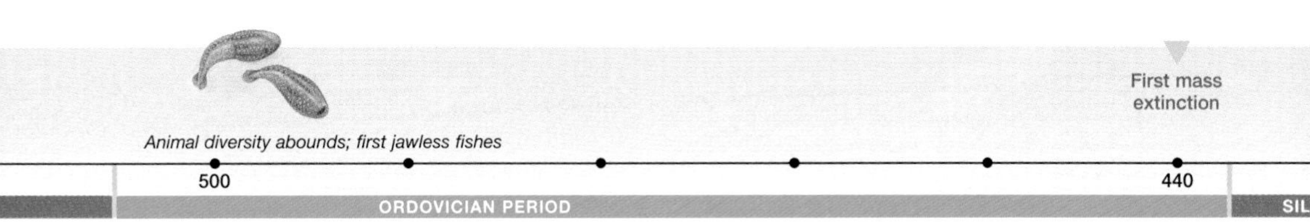

Review SECTION 12-2 ANSWERS

1. Eubacteria contain peptidoglycan in their cell walls and have the same kind of lipids in their cell membranes as do eukaryotes. Archaebacteria lack peptidoglycan and their cell-membrane lipids differ from those found in eukaryotes.
2. Eukaryotes probably evolved through endosymbiosis—the inclusion of small prokaryotes responsible for photosynthesis and cellular respiration inside of larger prokaryotes.
3. Bacteria have no membrane-bound cellular inclusions as do eukaryotes; bacterial DNA is in a single loop, unlike the chromosomal DNA of eukaryotes; and bacteria have cell walls, unlike eukaryotic protistan and animal cells.
4. Multicellularity allowed for cell specialization in multicellular protists.
5. The ancestors of many living species arose and filled niches vacated by organisms that died in mass extinctions.

internet**connect**

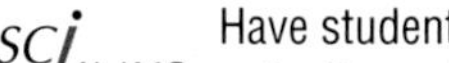

SCILINKS NSTA Have students research extinction using the Web site in the Internet Connect box on page 263. Ask students to write a report summarizing their findings. PORTFOLIO

Teaching TIP

Mass Extinctions Have Had a Major Impact

Have students identify the five mass extinctions on the timeline in this chapter. This will lead to conjectures about the next mass extinction and its cause. Perhaps students have seen movies about calamities on a global scale. Point out that this is not something to fear, but mass extinctions are part of Earth's history.

3 Assessment Options

Closure

Quiz Show

Divide the class into two teams. Have each team write 10 true-false questions about the material in this section. A spokesperson for one team should pose the questions to the other team. Give two points for each correct answer. Give students one point for each of their own questions that was answered correctly. Tally the scores, and declare a winning team.

Review

Assign Review—Section 12-2.

Reteaching — BASIC

Assign students to cooperative groups of four. Each group should choose a leader, timer, recorder, and reporter. Ask each group to design a hypothetical animal that would be well adapted for making the transition from the ocean to land. Have each group's reporter describe to the class its organism and any specialized adaptations.

1 Prepare

Scheduling

- **Block:** About 90 minutes will be needed to complete this section.
- **Traditional:** About two class periods will be needed to complete this section.
- **Planning:** See the Planning Guide on pp. 249A and 249B for lecture, classwork, and assignment options for Section 12-3. Lesson plans for Section 12-3 can be found in a lesson cycle format in ***Biology: Principles and Explorations*** **One-Stop Planner CD-ROM with Test Generator.**

Resource Link

- Assign **Active Reading Worksheet Section 12-3** to help students comprehend the key concepts and visuals in this lesson.

2 Teach

Lesson Warm-up

Ask students if ozone is a desirable part of our atmosphere. *(Most will answer yes.)* Then ask if ozone is always a desirable part of our atmosphere. *(At lower elevations, ozone contributes to pollution and smog; at higher elevations, however, ozone absorbs much of the damaging ultraviolet radiation that would otherwise reach Earth.)*

Section 12-3 Life Invaded the Land

Objectives

- **Relate** the development of ozone to the adaptation of life to the land. (p. 264)
- **Identify** the first multicellular organisms to live on land. (p.265)
- **Name** the first animals to live on land. (p. 266)
- **List** the first vertebrates to leave the oceans. (p. 267)

Key Terms

mycorrhiza
mutualism
arthropod
vertebrate
continental drift

Ozone Enables Survival on Land

The sun provides both life-giving light and dangerous ultraviolet radiation. Early in Earth's history, life formed in the seas, where early organisms were protected from ultraviolet radiation. These organisms could not leave the water because ultraviolet radiation made life on dry ground unsafe. What enabled life-forms to leave the protection of the seas and live on the land?

Formation of the ozone shield

During the Cambrian period and for millions of years afterward, organisms did not live on the dry, rocky surface of Earth. However, a slow change was taking place. About 2.5 billion years ago, photosynthesis by cyanobacteria began adding oxygen to Earth's atmosphere. As oxygen began to reach the upper atmosphere, the sun's rays caused some of the molecules of oxygen, O_2, to bond and form molecules of ozone, O_3. In the upper atmosphere, ozone blocks the ultraviolet radiation of the sun, as shown in **Figure 12-10.** After millions of years, enough ozone had accumulated to make the Earth's land a safe place to live.

Figure 12-10 Ozone shields the Earth

As ancient cyanobacteria added oxygen to the atmosphere, ozone began to form.

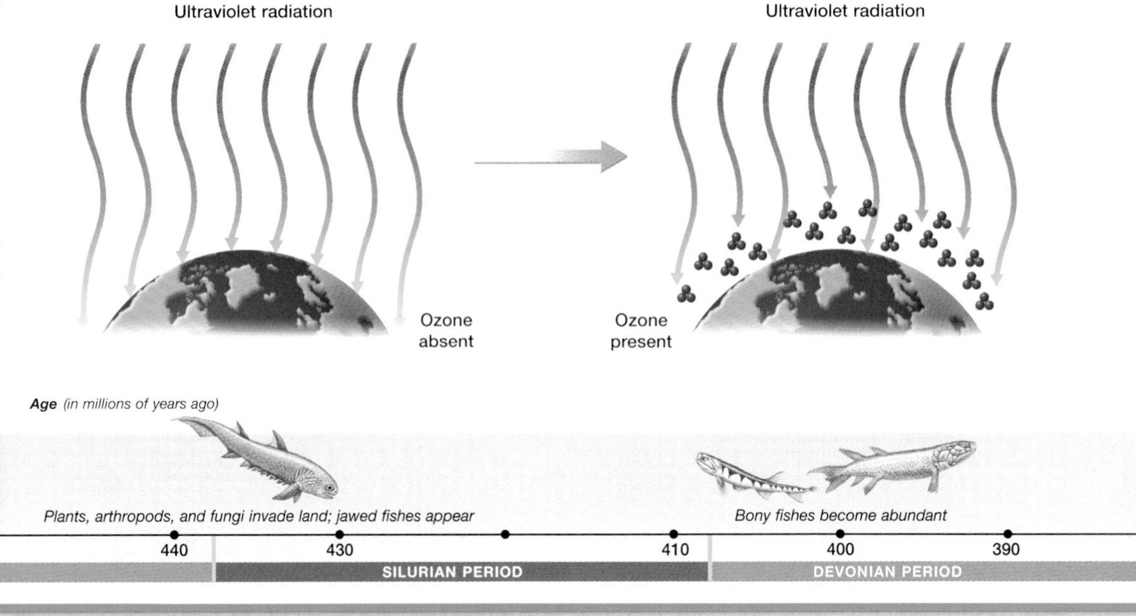

Integrating Different Learning Styles

Visual/Spatial	**ATE**	Demonstrations, pp. 265, 267, 268
Interpersonal	**ATE**	Demonstration, p. 267
Verbal/Linguistic	**ATE**	Lesson Warm-up, p. 264; Teaching Tip, p. 265; Overcoming Misconceptions, p. 266

Plants and Fungi Began Living on the Land Together

The first multicellular organisms to live on land are thought to have been plants and fungi living together. Both groups of organisms were able to live on land because each group evolved an ability that the other needed.

Plants, which evolved from photosynthetic protists, could carry out photosynthesis. Recall from Chapter 5 that in photosynthesis, plants use the energy from sunlight to make their own nutrients. They could not, however, harvest needed minerals from bare rock. Fungi could not make nutrients from sunlight but could absorb minerals—even from bare rock.

Plants and fungi evolved biological partnerships called mycorrhizae *(MEYE koh REYE zee)*, which enabled them to live on the harsh habitat of bare rock. **Mycorrhizae** are associations between fungi and the roots of plants, as shown in **Figure 12-11.** The fungi provide minerals to the plant, and the plant provides nutrients to the fungi. This kind of partnership is called mutualism. **Mutualism** is a relationship in which both organisms benefit. Plants and fungi began living on the surface of the land at the same time, about 430 million years ago.

REAL LIFE

The ozone layer is still needed. Over the past century, pollution has caused the ozone layer to become thinner. Scientists think that the increasing number of people who develop skin cancer is related to the thinning ozone layer.

Finding Information
Gather information on how you can protect yourself from ultraviolet radiation, and share your tips with your classmates.

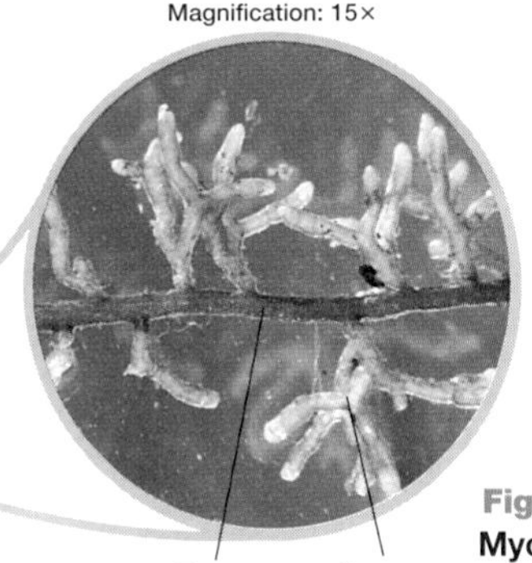

Figure 12-11
Mycorrhizae form on plant roots. The roots of this pine tree, held by a U.S. Forest Service worker in Idaho, are covered with a fungus. The plant and the fungus provide each other needed materials for survival.

did you know?

Ancestors of modern algae might have been quite diverse.

The 16th International Botanical Congress, which met in St. Louis in August 1999, concluded that there should be at least three separate kingdoms replacing the single kingdom Protistae that we have today. This conclusion is based on data that show three separate origins—all one-celled—for green, red, and brown algae.

Section 12-3

Demonstration

Personal Experience with UV Radiation

Display several types of skin protection agents including sun-blocking agents, sunscreens, and suntan lotions. Ask students what effect UV radiation has on people. *(It causes tanning and burning of the skin.)* Have they ever known anyone hurt by sunlight? *(They may have been sunburned or know someone stricken with skin cancer.)* What would happen to them if they had to spend their entire lives outside, as most other organisms do? *(They would be more likely to feel the effects of the sun such as sunburns, skin cancers, and dehydration.)* Explain to the students how sun blocking agents, sunscreens, and tanning lotions differ in their ability to protect them from the sun's UV rays. *(Sun-blocking agents are opaque and reflect ultraviolet radiation, offering the greatest protection. Sunscreens absorb ultraviolet radiation. Suntan lotions offer essentially no protection from the sun—some that contain oils even promote sunburning.)*

Teaching TIP

Hard As a Rock

Ask students what kind of organisms could survive on bare rock. What would it take to be successful? *(Students may suggest an association similar to a lichen in which a fungus and a cyanobacterium coexist in a symbiotic relationship.)* Remind students that as life emerged from the oceans, living things were confronted with this problem.

Teaching TIP

Who's on First?

Pose this question to students: As life moved from water to land, why did plants occupy the land before animals? Could animals have preceded plants? Why or why not? *(Food chains begin with producers; and plants, not animals, are producers.)*

Teaching TIP

Arthropods Invade Earth

Ask students to identify some of the characteristics that may have enabled an arthropod (the scorpion) to be the first or one of the first animals to live on land successfully. *(Their jointed appendages make it possible for them to walk. Their strong exoskeleton protects them and supports their body weight.)*

Teaching TIP

Millions and Millions

Harvard-based naturalist E. O. Wilson estimates that there are 1 million ants for every human on Earth.

Arthropods Crawled out of the Sea

Figure 12-12
An arthropod

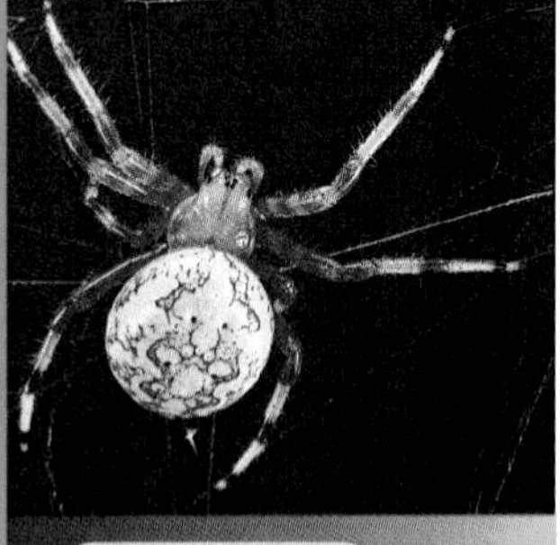

How many? This marbled spider is a member of a group called arthropods. This group includes about 1 billion billion (10^{18}) individuals in about 1.5 million described species.

By 100 million years after their first union with fungi, plants had covered the surface of the Earth, forming large forests. These land plants provided a food source for animals, enabling the evolution of land-dwelling animals. The first animals to successfully invade land from the sea were **arthropods,** a kind of animal with a hard outer skeleton and jointed limbs. Examples of arthropods include lobsters, crabs, insects, and spiders, like the one in **Figure 12-12.** Biologists think a type of scorpion was the first arthropod to live on land.

A unique kind of terrestrial arthropod—the insect—evolved from the first land dwellers. Insects have since become the most plentiful and diverse group of animals in Earth's history. The success of the insects is probably connected to their ability to fly. Insects were the first animals to evolve wings. Early insects, like the dragonfly shown in **Figure 12-13,** had two pairs of wings. Flying allowed insects to efficiently search for food, mates, and nesting sites. It also led to the partnership between insects and flowering plants. The oldest fossils of flowering plants are from about 127 million years ago, but flowering plants may be much older than that.

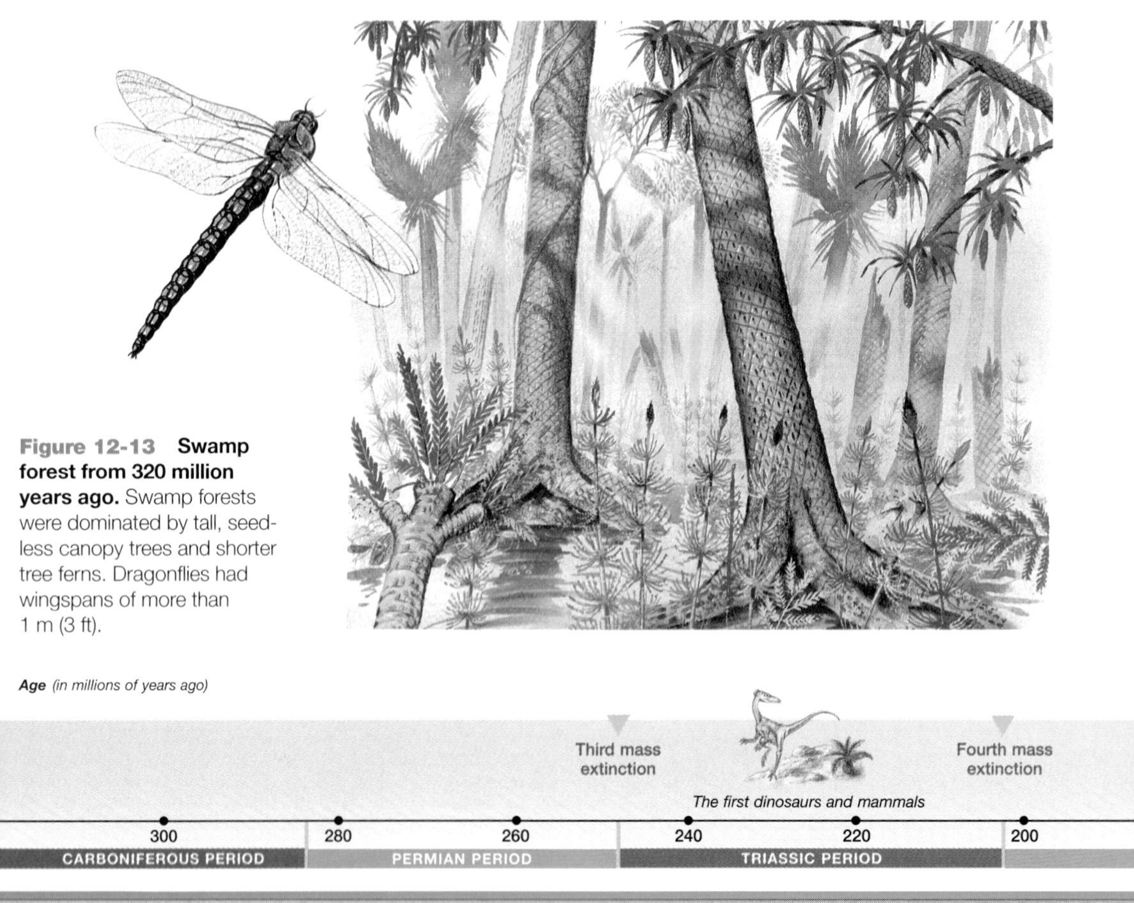

Figure 12-13 **Swamp forest from 320 million years ago.** Swamp forests were dominated by tall, seedless canopy trees and shorter tree ferns. Dragonflies had wingspans of more than 1 m (3 ft).

Age (in millions of years ago)

Third mass extinction

The first dinosaurs and mammals

Fourth mass extinction

300 | 280 | 260 | 240 | 220 | 200

CARBONIFEROUS PERIOD | PERMIAN PERIOD | TRIASSIC PERIOD

Overcoming Misconceptions

Populations Change

Some students mistakenly view evolution as individuals changing into new forms, rather than populations changing over periods of time longer than the lifetime of an individual organism. It may be necessary to review what students have already learned about how changes in DNA occur, and to relate those changes to adaptive traits present in a population.

Vertebrates Followed onto Land

Of all animals, vertebrates are the most familiar to us. Humans are vertebrates, and almost all other land animals bigger than our fist are vertebrates as well. **Vertebrates** are animals with a backbone.

Fishes evolved

The first vertebrates were small, jawless fishes that evolved in the oceans about 500 million years ago. Jawed fishes first appeared about 430 million years ago. Jaws enabled fishes to bite and chew their food instead of sucking up their food. As a result, jawed fishes became efficient predators. An example of a jawed fish is shown in **Figure 12-14.** Fishes soon came to be among the most abundant animals in the seas, and for hundreds of millions of years that is where vertebrates stayed. Fishes are the most successful living vertebrates—they make up more than half of all living vertebrate species. After nearly 200 million years of living in the sea, fishes have become uniquely adapted for success in water. Major changes had to occur in a fish's body design to evolve into an animal capable of living on land.

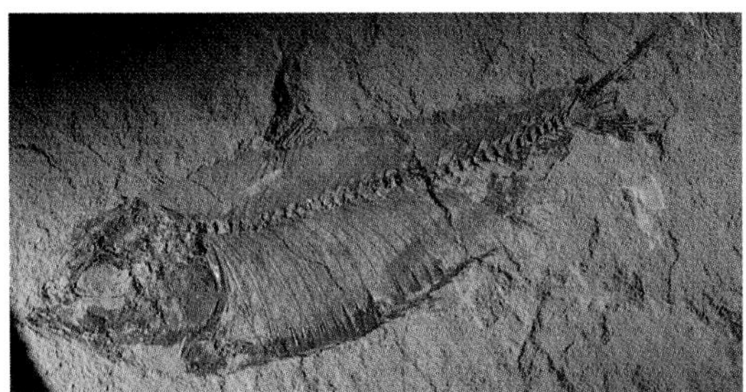

Figure 12-14 Fossilized fish skeleton. This fish skeleton clearly shows the backbone, the structure that is characteristic of all vertebrate animals.

Amphibians were the first vertebrates on land

The first vertebrates to inhabit the land did not come out of the sea until 370 million years ago. Those first land vertebrates were early amphibians. Amphibians are smooth-skinned organisms that include frogs, toads, and salamanders.

Amphibians were able to adapt to land because of the development of several structural changes in their bodies. Early amphibians had moist breathing sacs called lungs, which they used to absorb oxygen from air. The limbs of amphibians are thought to be derived from the bones of fish fins. The evolution of a strong support system of bones in the region just behind the head made walking possible. This system of bones provided a rigid base for the limbs to work against. Because of their strong, flexible internal skeleton, vertebrates can be much larger than insects. While amphibians were well adapted to their environment, a new group of animals more suited to a drier environment evolved from them.

biobit

Why do some flowers resemble insects?

Some flowering plants, including several species of orchid, have evolved to produce blooms that look so much like insects that they fool real insects. When insects try to mate with buglike blossoms, the plants get pollinated.

Appearance of flowering plants

200 | 180 | 160 | 140 | 120 | 100

JURASSIC PERIOD | CRETACEOUS PERIOD

Teaching TIP

Clamping Down Advantages

The lamprey, a jawless fish alive today, is a parasite. Help students appreciate the importance of jaws by having each name a vertebrate predator. *(Answers will vary.)* How does each predator catch and eat its prey? For every case, direct the discussion to the important role of the predator's jaws in its successful lifestyle. *(If not directly involved in capture, jaws are important in ingesting food.)*

Demonstration

Inside or Out?

Distribute soft gelatin-based candy such as "gummy worms" to the class and challenge small groups of students to create an internal skeleton for some worms and an external skeleton for others using toothpicks and tape. Caution students not to eat the candy. Have the groups share their progress and results with the class. Ask which type of skeleton requires less material and weighs less. *(the internal [endo-] skeleton)*

did you know?

There is a worldwide problem with declining amphibian populations.

Numbers of amphibians are declining even in areas that are protected from habitat destruction. A recent study indicates that increased UV radiation may be responsible for killing amphibians early in life. These data underscore the importance of the ozone shield for the survival of terrestrial life-forms.

Chapter 12

Demonstration

Egg Watch

Crack open an egg in a Petri dish. In another Petri dish, place an intact egg. Weigh each egg-and-dish combination and record the weight. At the end of 2 days, crack open the intact egg and compare it to the exposed egg. Record the similarities and differences between the two; then reweigh both Petri dishes. Have students explain which egg setup showed the greater difference in weight and why. *(The opened egg should weigh less and may show signs of decomposition.)*

3 Assessment Options

Closure

Technique: K-W-L

Tell students to return to the lists of things they want to know about how life began, which they created using the **Active Reading** exercise on page 251. Have them place check marks next to the questions that they are now able to answer. Students should finish by making a list of what they have learned. Ask students the following:

- Which questions are still unanswered?
- What new questions do you have?

You may wish to assign projects that require students to research their unanswered questions.

Review

Assign Review—Section 12-3.

Reteaching — BASIC

Draw a table on the board as follows. Across the top of the table write these four headings: *Evolutionary History, Structural Features, Special Adaptations,* and *Examples.* Down the left side of the table write these seven classes: *Jawless fishes, Sharks (cartilaginous fishes), Bony fishes, Amphibians, Reptiles, Birds,* and *Mammals.* Have students take turns filling in the information.

Figure 12-15 Reptiles. Reptiles, such as this crocodile, were the largest group of land-dwelling organisms until the end of the Cretaceous period.

Reptiles arose from amphibians

Reptiles evolved from amphibians about 350 million years ago. Reptiles include snakes, lizards, turtles, dinosaurs, and crocodiles. These animals evolved a watertight skin, which protected them from losing moisture to the atmosphere. Reptiles also developed a watertight egg, such as the one shown in **Figure 12-15.** Unlike amphibians, reptiles can lay their eggs on dry land because the eggs are surrounded by a watertight shell. Amphibians still need to lay their eggs in water or in very moist soil because their eggs are unable to retain enough water to remain alive.

The next 50 million years after the reptiles evolved was a period of widespread drought. Reptiles, better adapted for dry climate, had an advantage over amphibians. Gradually, reptiles became the more abundant vertebrates on Earth until about 65 million years ago.

Evolution of mammals and birds

Birds evolved from reptiles during the Jurassic period. Therapsids, animals with complex teeth and legs positioned beneath their body, gave rise to mammals during the Triassic period. Sixty-five million years ago, during the fifth mass extinction, most species (including all of the dinosaurs) disappeared forever. The smaller reptiles, mammals, and birds survived. Although many resources were available to the surviving animals, the world's climate was no longer dry, and the reptiles' advantages in dry climates were not so important. Birds and mammals then became the dominant vertebrates on land.

In addition to extinctions, continental drift played an important role in evolution. **Continental drift** is the movement of Earth's land masses over geologic time. Continental drift resulted in the present-day position of the continents. The movement of continents helps explain why there are a large number of marsupial (pouched) mammal species in Australia and South America, continents that were once connected.

biobit

Do the continents still move?

Yes; every year Europe and North America move apart about 2 cm.

Review SECTION 12-3

1. **Summarize** why ozone was important in enabling organisms to live on land.
2. **Name** the first multicellular organisms that colonized land.
3. **Identify** the first kinds of animals to live on land.
4. **Describe** the first kinds of vertebrates that inhabited land.

Critical Thinking

5. **Defend** the argument that invasion of land could not have happened until well after the evolution of cyanobacteria.

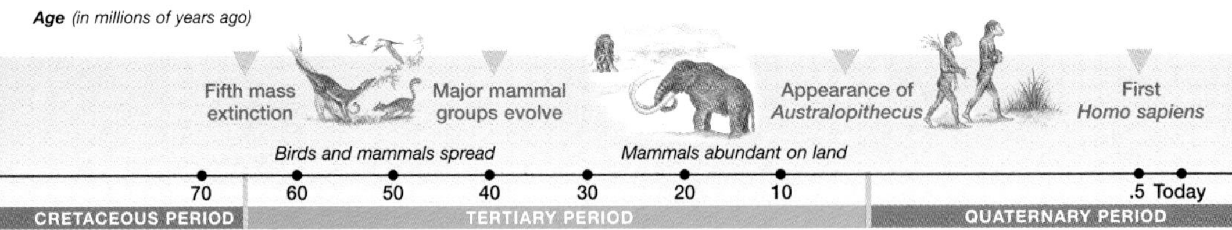

Review SECTION 12-3 ANSWERS

1. Ozone was essential to protect early land organisms from UV rays from the sun.
2. plants and fungi (or mycorrhizae)
3. arthropods
4. The first land-dwelling vertebrates were amphibians, which had lungs, an endoskeleton, and smooth, skin.
5. Cyanobacteria contributed oxygen gas to the atmosphere. Some of this oxygen gas reacted and formed the ozone layer, which was necessary to make the land hospitable to any multicellular organism.

Study ZONE CHAPTER 12

Use the Key Concepts and Key Terms listed below to review the main ideas in this chapter. If you need to review the meaning of a term, turn to the page indicated.

Key Concepts

12-1 How Did Life Begin?

- The Earth formed about 4.5 billion years ago according to evidence obtained by radiometric dating.
- The primordial soup model and the bubble model propose explanations of the origin of the chemicals of life.
- Scientists think RNA formed before DNA or proteins formed.
- Scientists think that the first cells may have developed from microspheres.
- The development of heredity made it possible for organisms to pass traits to subsequent generations.

12-2 Complex Organisms Developed

- Prokaryotes are the oldest organisms and are divided into two groups, archaebacteria and eubacteria.
- Prokaryotes gave rise to eukaryotes through the process of endosymbiosis.
- Mitochondria and chloroplasts evolved through endosymbiosis.
- Multicellularity arose many times and resulted in many different groups of multicellular organisms.
- Extinctions influenced the evolution of the species alive today.

12-3 Life Invaded the Land

- Ancient cyanobacteria produced oxygen, some of which became ozone. Ozone enabled organisms to live on land.
- Plants and fungi formed mycorrhizae and were the first multicellular organisms to live on land.
- Arthropods were the first animals to leave the ocean.
- The first vertebrates to invade dry land were amphibians.
- The extinction of many reptile species enabled birds and mammals to become the dominant vertebrates on land.
- The movement of the continents on the surface of the Earth has contributed to the geographic distribution of some species.

Key Terms

12-1

radiometric dating (252)
radioisotope (252)
half-life (252)
spontaneous origin (253)
microsphere (256)

12-2

fossil (258)
cyanobacteria (258)
eubacteria (258)
archaebacteria (258)
endosymbiosis (259)
protist (261)
mass extinction (263)

12-3

mycorrhiza (265)
mutualism (265)
arthropod (266)
vertebrate (267)
continental drift (268)

Study TIP Ready?

If you think you understand these Key Concepts and Key Terms, then you're ready for the Performance Zone.

Answer to Concept Map

The following is one possible answer to Performance Zone item 1 on page 270.

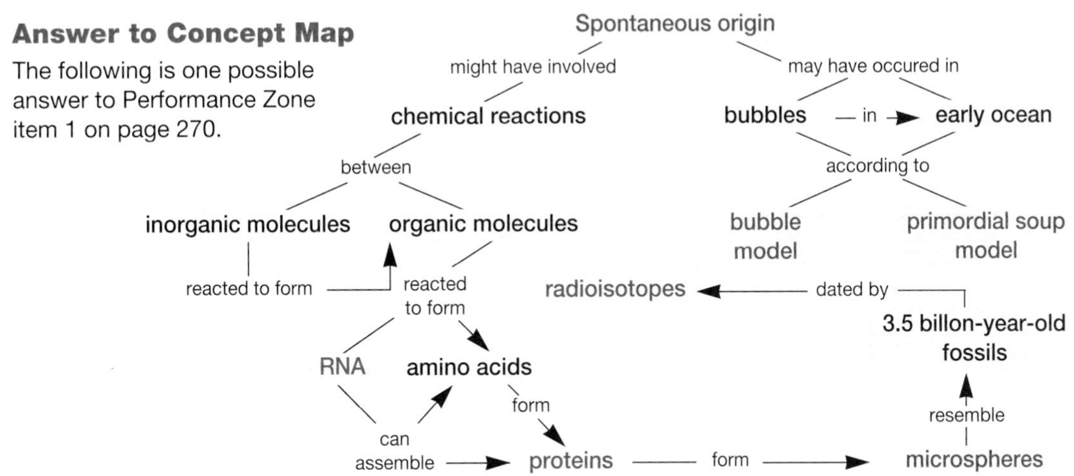

Study ZONE CHAPTER 12

Review and Practice

Assign the following worksheets for Chapter 12:

- **Study Review Guide**
- **Concept Mapping Worksheet**
- **Test Preparation Pretest**
- **Vocabulary Worksheet**
- **Critical Thinking Review Worksheet**
- **Science Skills Worksheet**

Assessment Options

The following assessment options are available for this chapter:

Resource Link

- Assign **Chapter Test 12** to assess students' understanding of the chapter.

Alternative Assessment

Have students prepare a blank bingo card with 25 squares, and label the center square *FREE*. Write 24 terms or short phrases on the board. Have students write in each term or phrase in random order, filling all of the squares on their bingo cards. Next, have students cross out the appropriate square each time a definition or clue is provided. Students call out "bingo" whenever a horizontal, vertical, or diagonal line is completed. PORTFOLIO

Portfolio Assessment

- *Portfolio Assessment* at the front of this book provides options for building and evaluating student portfolios. Students and teachers can select items from the areas listed for inclusion in a portfolio.
- See items labeled PORTFOLIO on pp. 253, 257, 259, 261, 262, 263, and 269.

Performance ZONE
CHAPTER 12 REVIEW ANSWERS

Assignment Guide

SECTION	REVIEW
12-1	1, 2, 3, 4, 5, 14, 15, 17, 18, 19, 20, 22, 24
12-2	6, 7, 8, 12, 13, 16, 21, 23
12-3	9, 10, 11, 25

Understanding and Applying Concepts

1. The answer to the concept map is found at the bottom of page 269.
2. **a.** Radiometric dating is a technique used to measure the age of objects. Radioisotopes are unstable forms of elements that give off charged particles at known rates of decay. Half-life is the time needed for one-half of a given amount of radioistope to decay.
   b. The chemical reactions between nonliving molecules might have produced the molecules of life in the process of spontaneous origin. Primordial soup refers to the large number of organic molecules in Earth's early oceans. In the Miller-Urey experiment, conditions thought to resemble those on early Earth were simulated.
   c. Microspheres are vesicles that contain chains of amino acids. A cell is the smallest unit of life that is able to exist independently. Microspheres floating in sea water might have given rise to the first cells.
   d. Symbiotic aerobic eubacteria may have given rise to mitochondria by a process called endosymbiosis. Archaebacteria are thought to be closely related to the earliest bacteria found on Earth and probably gave rise to the first eukaryotic cells. A protist is a member of the earliest group of eukaryotic cells and can be either single-celled or multicellular.

3. b **7.** b
4. d **8.** d
5. a **9.** b
6. b **10.** c

Performance ZONE CHAPTER 12 Review

Understanding and Applying Concepts

1. **Concept Mapping** Construct a concept map that shows how life might have originated by natural forces. Try to include the following items in your map: spontaneous origin, primordial soup model, bubble model, RNA, proteins, microspheres, radioisotopes. Include additional terms in your map as needed.

2. **Understanding Vocabulary** For each set of terms, write one or more sentences summarizing information learned in this chapter.
   a. radiometric dating, radioisotope, half-life
   b. spontaneous origin, primordial soup, Miller-Urey experiment
   c. microsphere, cell, sea water
   d. endosymbiosis, archaebacteria, protist

3. Unlike the primordial soup model, Lerman's bubble model takes ______ into account.
   a. ozone
   b. ultraviolet radiation
   c. lightning
   d. volcanoes

4. Cells are different from microspheres because cells
   a. contain amino acids.
   b. have a two-layer outer boundary.
   c. grow by taking in molecules from their surroundings.
   d. transfer information through heredity.

5. The development of heredity allowed organisms to store and pass on
   a. information.
   b. energy.
   c. radioisotopes.
   d. ultraviolet radiation.

6. All eukaryotic cells evolved from
   a. eubacteria.
   b. archaebacteria.
   c. cyanobacteria.
   d. endosymbiosis.

7. Eukaryotes originated as a result of
   a. mass extinctions.
   b. endosymbiosis.
   c. multicellularity.
   d. photosynthesis.

8. Cell specialization came about as a result of
   a. endosymbiosis.
   b. archaebacteria.
   c. the Cambrian period.
   d. multicellularity.

9. The oxygen that became ozone in Earth's atmosphere was originally formed through photosynthesis by
   a. mycorrhizae.
   b. cyanobacteria.
   c. arthropods.
   d. eubacteria.

10. The first multicellular organisms to invade the land were
    a. reptiles.
    b. amphibians.
    c. fungi and plants.
    d. mammals.

11. **Theme Interdependence** Explain how the relationship between plants and fungi contributed to the colonization of dry land by other organisms.

12. **Theme Homeostasis** Explain how multicellularity can be seen as an answer to the problem of a cell's surface-area-to-volume ratio limiting the size that unicellular organisms can achieve.

13. **Exploring Further** Describe the evidence that supports the theory of endosymbiosis.

14. **Calculating** The half-life of carbon-14 is 5,730 years. If a sample originally had 26 g of carbon-14, how much would it contain after 22,920 years?

15. **Recognizing Relationships** After each of the five mass extinctions of Earth, there was a large increase of evolutionary activity. Explain why these bursts of evolution occurred after the extinctions.

16. **Chapter Links** Relate the order in which different types of organisms invaded the land to the flow of energy. (**Hint:** See Chapter 5, Section 5-1.)

---

11. The symbiotic relationship produced when fungi and plants live together provides nutrients for the fungi and minerals for the plants, allowing them to live together on rock.
12. As a cell doubles in size, the volume grows by an 8:1 ratio when compared to the surface area of the cell. Materials for cellular processes must come through the cell surface, and the cell's ability to get enough materials falls as the cell grows in size. In addition, the nucleus must communicate with all parts of the cell, and this is most efficient when cells are small. Thus, the number of cells (not the size of the cell) increases as organisms become larger.
13. The hypothesis that mitochondria are descended from aerobic prokaryotes that came to live within other prokaryotes is supported by the observations that mitochondria reproduce by binary fission, and that they have their own genes that are separate and distinct from nuclear genes.
14. 1.625 g
15. There were unexploited niches to fill.
16. Producers (plants) colonized the land first, providing food for consumers of plants and in turn, consumer of the consumers, or predators.
17. A. N_2, CH_4, H_2, NH_3; B. electric spark; C. water vapor; D. organic compounds

Skills Assessment

Part 1: Interpreting Graphics

Use the drawing below to answer the following questions:

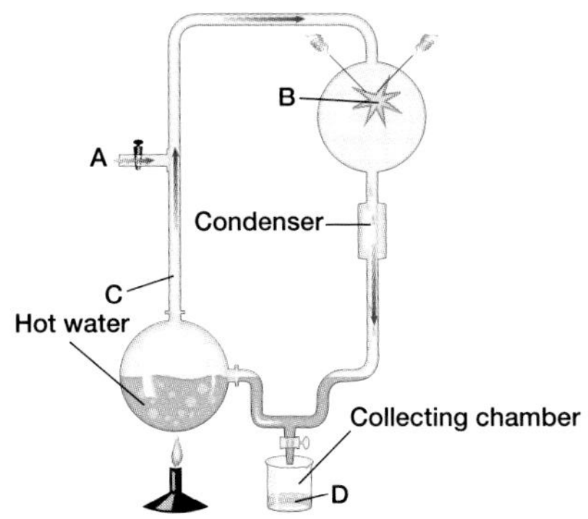

17. Identify the substances labeled *A–D*.

18. What would probably be the outcome of the experiment in the figure above if *B* were missing from the apparatus?

Part 2: Life/Work Skills

Use the media center or Internet resources to learn about the conditions on Earth that scientists think existed before life formed.

19. **Finding Information** Identify which compounds Miller and Urey formed in their experiment. Prepare a report describing which of the compounds on early Earth would have contributed to the types of compounds Miller and Urey made.

20. **Being a Team Member** Work together in groups to design a poster to illustrate the different models that describe how life's chemicals may have originated. Show how the compounds on early Earth would have participated in each of these models.

21. **Communicating** Thomas Cech and Sidney Altman shared a Nobel prize in 1989 for their work on RNA. Research their work and the rewards associated with winning a Nobel prize. Relate your findings in an oral report.

Critical Thinking

22. **Evaluating Results** A sample of a certain type of rock contains 25 g of a radioisotope. The same amount of this type of rock contains 200 g of the radioisotope when it forms. Using the graph below that shows the radioisotope's decay rate, a student infers that the radioisotope's half-life is 1.5 billion years. Is the student correct? Explain.

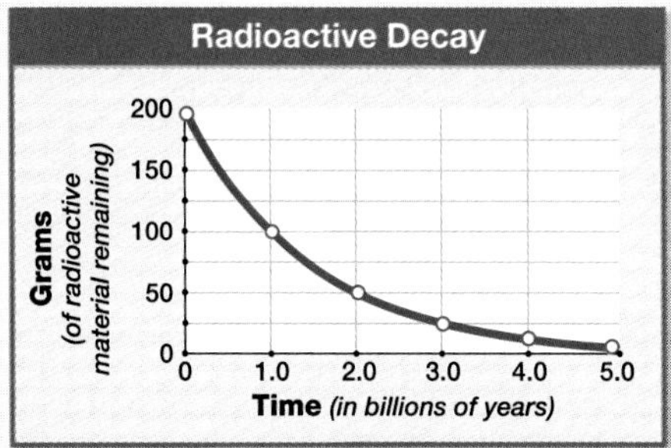

23. **Evaluating Viewpoints** In the late 1990s astronomers identified many asteroids that cross Earth's orbital path. These scientists said that if a large asteroid struck the Earth, the impact could result in a mass extinction. If an asteroid impact did not kill all organisms, would evolution continue or stop? Explain.

Portfolio Projects

24. **Summarizing Information** Read the article "Life's Crucible," by Peter Radetsky (*Earth,* February 1998, vol. 7, pp. 34–41). Write a report explaining the different hypotheses that scientists say could explain the origin of life. In your report, describe the flaws in the hypotheses and explain why scientists think there could be life on Europa or Mars. Share your report with the class.

25. **Career Focus Museum Curator** Research the career of museum curator, and write a report on your findings. Your report should include a job description, training required, kinds of employers, growth prospects, and starting salary.

Critical Thinking

22. No; one-half of the original material remained (the definition of half-life) after just 1 billion years.
23. It would continue. In the remaining organisms, there would be traits that helped ensure the organism's survival to the point of reproduction. These traits would be amplified within populations; this process is evolution.

Portfolio Projects

24. Answers will vary. Our direct evidence—one sampling from one meteor from one planet (Mars)—is too slight to justify making far-reaching conclusions. It will be interesting and exciting to watch the continued exploration of Mars. Upcoming missions will include flying a remote-controlled ship to the red planet and bringing back Martian rock and soil samples.
25. A museum curator establishes and oversees the operation of a museum. Curators in life science museums may have a college degree in a science, such as zoology or botany, as well as postgraduate academic training in the sciences and in aspects of curatorship such as restoration and preservation of materials and assembly and organization of exhibits. Science museum curators are employed by public and private museums, including those associated with universities. Job growth is somewhat limited in this field. Starting salary will vary by region.

18. No organic compounds would be generated.
19. Miller and Urey produced a variety of organic compounds, including some amino acids. These organic compounds were formed from the water, hydrogen gas, methane, and ammonia that Oparin had postulated existed on early Earth.
20. Answers will vary. Posters should illustrate the concepts developed in this chapter.
21. Thomas Cech and Sidney Altman discovered that RNA molecules could function as catalysts as well as templates. RNA catalysts called ribozymes catalyze synthesis of RNA and remove introns from sections of DNA.

    The six Nobel prizes are awarded annually in the categories of physics, chemistry, physiology or medicine, literature, economics, and peace. A Nobel prize is given to one person or shared among up to three people (except for the peace prize, which can be awarded to an institution). A Nobel prize consists of a gold medal, a written certificate, and a monetary award of varying amounts.

Lab Techniques

Chapter 12

Making a Timeline of Life on Earth

Purpose

Students will construct timelines showing geologic time and the major events in the evolution of life on Earth.

Preparation Notes

Time Required: 90 minutes

Materials

Materials for this lab can be ordered from WARD'S. See *Master Materials List* at the front of this book for catalog numbers.

Preparation Tips

- Set up stations in various parts of the classroom with specimens, fossils, photographs, or slides of different types of organisms.
- Provide compound light microscopes if needed.

Procedural Tips

- During the first hour (or lab period), have students make their timelines and observe and record data for as many organisms as possible.
- During the second period, have students complete the investigation.

Before You Begin

Answers

1. *fossil*—preserved remains or traces of a past life-form
   fossil record—the record of past life-forms found in fossils
   timeline—a series of events arranged in chronological order
3. Answers will vary. For example: How did life on Earth change after oxygen entered the atmosphere?

Lab Techniques

Chapter 12

Making a Timeline of Life on Earth

SKILLS

- Observing
- Inferring relationships
- Organizing data

OBJECTIVES

- **Compare** and contrast the distinguishing characteristics of representative organisms of the six kingdoms.
- **Organize** the appearance of life on Earth in a timeline.

MATERIALS

- 5 m roll of adding-machine tape
- meterstick
- colored pens or pencils
- photographs or drawings of organisms from ancient Earth to present day

Before You Begin

About 4.5 billion years ago, Earth was a ball of molten rock. As the surface cooled, a rocky crust formed and water vapor in the atmosphere condensed to form rain. By 3.9 billion years ago, oceans covered much of the Earth's surface. Rocks formed in these oceans contain **fossils** of bacterial cells that lived about 3.5 billion years ago. The **fossil record** shows a progression of life-forms and contains evidence of many changes in Earth's surface and atmosphere.

In this lab, you will make a **timeline** showing the major events in Earth's history and in the history of life on Earth, such as the evolution of new groups of organisms and the mass extinctions. This timeline can be used to study how living things have changed over time.

1. Write a definition for each boldface term in the paragraphs above.
2. Make a data table similar to the one at right.
3. Based on the objectives for this lab, write a question you would like to explore about the history of life on Earth.

Procedure

PART A: Making a Timeline

1. Make a mark every 20 cm along a 5 m length of adding-machine tape. Label one end of the tape "5 billion years ago" and the other end "Today." Write "20 cm = 200 million years" near the beginning of your timeline.
2. Locate and label a point representing the origin of Earth on your timeline. Use your textbook as a reference. See the timeline at the bottom of pp. 258–268. Also locate and label the 11 periods of the geologic time scale beginning with the Cambrian period.
3. Using your textbook as a reference, mark the following events on your timeline: the first cyanobacteria appear; oxygen enters the atmosphere; the five mass extinctions; the first eukaryotes appear; the first multicellular organisms appear; the first vertebrates appear; the first plants, fungi,

DATA TABLE

Organism	Kingdom	Characteristics/adaptation for life on Earth

Procedure

Answers

2. The Earth is estimated to be more than 4 billion years old. Students should label the 11 periods of the geologic time scale approximately as follows: Cambrian—550 million to 505 million years ago (mya); Ordovician—505 to 438 mya; Silurian—438 to 408 mya; Devonian—408 to 360 mya; Carboniferous—360 to 286 mya; Permian—286 to 245 mya; Triassic—245 to 208 mya; Jurassic—208 to 144 mya; Cretaceous—144 to 67 mya; Tertiary—67 to 2 mya; and Quaternary—2 mya to the present.
3. Cyanobacteria appear—3.5 billion years ago; oxygen enters the atmosphere—2.75 billion years ago; five mass extinctions—440 million years ago (mya), 360 mya, 245 mya, 210 mya, and 65 mya; eukaryotes appear—1.5 billion years ago; multicellular organisms appear—700 mya; vertebrates appear—500 mya; land plants and fungi appear—430 mya; land animals appear—370 mya; dinosaurs and mammals appear—210 mya; flowering plants appear—120 mya; humans appear—500,000 years ago
4. Answers will vary. See sample data table at the bottom of page 273.

and land animals appear; the first dinosaurs and mammals appear; the first flowering plants appear; the first humans appear.

4. Look at the photographs of organisms provided by your teacher. Identify the major characteristics of each organism. Record your observations in your data table.
5. Lay out your timeline on the floor in your classroom. Place photographs (or drawings) of the organisms you examined on your timeline to show when they appeared on Earth.
6. Fold the timeline at the mark representing 4.8 billion years ago. This leaves 24 segments, each representing 200 million years, in your timeline. Now you can think of each segment as 1 hour in a 24-hour day.
7. When you are finished, walk slowly along your timeline. Note the sequence of events in the history of life on Earth and the relative amount of time between each event.

Part B: Cleanup and Disposal

8. Dispose of paper scraps in the designated waste container.
9. Clean up your work area and all lab equipment. Return lab equipment to its proper place.

Analyze and Conclude

1. **Analyzing Information** Think of each segment of your timeline as 1 hour in a 24-hour day as you answer each of the following questions:
   a. How long has life existed on Earth?
   b. For what part of the day did only unicellular life-forms exist?
   c. At what time of day did the first plants appear on Earth?
   d. At what time of day did mammals appear on Earth?
   e. How long have humans been on Earth?
2. **Summarizing Information** Identify the major developments in life-forms that have occurred over the last 3.5 billion years.
3. **Inferring Relationships** How do mass extinctions appear to be related to the appearance of new major groups of organisms?
4. **Justifying Conclusions** Cyanobacteria are thought to be responsible for adding oxygen to Earth's atmosphere. Use your timeline to justify this conclusion.
5. **Calculating** Determine the amount of time, as a percentage of the time that life has existed on Earth, that humans (*Homo sapiens*) have existed.
6. **Further Inquiry** Write a new question about the history of life on Earth that could be explored in another investigation.

On the Job

Timelines are used to organize events in chronological order. Do research in the library or media center or on the Internet to discover how other scientists use timelines in their work. Summarize your findings in a written or oral report.

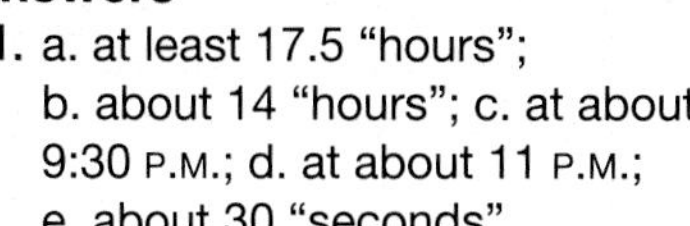

Analyze and Conclude

Answers

1. a. at least 17.5 "hours"; b. about 14 "hours"; c. at about 9:30 P.M.; d. at about 11 P.M.; e. about 30 "seconds"
2. Life has changed from all prokaryotic, unicellular forms to complex multicellular forms composed of eukaryotic cells. A vast variety of living things has lived on Earth during the last 3.5 billion years. Most life-forms that evolved have become extinct.
3. Mass extinctions appear to be followed closely by the evolution of many new types of organisms.
4. Cyanobacteria appeared before the accumulation of oxygen in the atmosphere, and cyanobacteria produce oxygen. Therefore, it is reasonable to conclude that they are responsible for adding oxygen to Earth's atmosphere.
5. 0.014 percent
6. Answers will vary. For example: How have Earth's climates changed since Precambrian times?

On the Job

Answers

Many types of scientists use timelines in their work. For example, an epidemiologist (a scientist who tracks the emergence and spread of disease) might arrange the known cases of a disease on a timeline to track the events of an outbreak.

Sample Data Table

Organism	Kingdom	Characteristics/Adaptations for Life on Earth
Insect (fossil in amber)	Animal	Segmented body, jointed legs, and wings
Fern (fossil mold in shale)	Plant	Large surface area to capture sunlight; reproductive structures on fronds

5. Students should place the photograph or specimen at the point on the timeline where that type of organism appeared. For example, a photograph of a flowering plant should be placed on the timeline at 120 million years ago.

Chapter 13

The Theory of Evolution

Planning Guide

	OBJECTIVES	MEETING INDIVIDUAL NEEDS	READING SKILLS AND STRATEGIES
BLOCK 1 45 min	**13-1 The Theory of Evolution by Natural Selection** • Identify several observations that led Darwin to conclude that species evolve. • Relate the process of natural selection to its outcome. • Summarize the main points of Darwin's theory of evolution by natural selection as it is stated today.	Learning Styles, p. 276 BASIC ADVANCED • pp. 277, 280, 282	**ATE** Active Reading, Technique: Reader Response Logs, p. 275 **ATE** Active Reading, Technique: Paired Summarizing, p. 281 ■ Active Reading Guide, worksheet (13-1) ■ Directed Reading, worksheet (13-1)
BLOCKS 2 & 3 90 min	**13-2 Evidence of Evolution** • Describe how the fossil record supports evolution. • Summarize how biological molecules such as proteins and DNA are considered evidence of evolution. • Infer how comparing the anatomy and development of living species provides evidence of evolution. • Contrast the gradualism and punctuated equilibrium models of evolution.	Learning Styles, p. 283 BASIC ADVANCED • pp. 288, 289	■ Active Reading Guide, worksheet (13-2) ■ Directed Reading, worksheet (13-2)
BLOCKS 4 & 5 90 min	**13-3 Examples of Evolution** • Identify five elements in the process of natural selection. • Describe how natural selection has affected the European peppered moth. • Relate natural selection to the beak size of finches. • Summarize the process of species formation.	Learning Styles, p. 290 BASIC ADVANCED • pp. 291, 292, 293, 294	■ Active Reading Guide, worksheet (13-3) ■ Directed Reading, worksheet (13-3) ■ Supplemental Reading Guide, *Origin of Species*

Review and Assessment Resources

BLOCKS 6 & 7 90 min

TEST PREP

PE Study Zone, p. 295

PE Performance Zone, pp. 296–297

■ Chapter 13 worksheets:
- Vocabulary
- Concept Mapping
- Science Skills
- Critical Thinking

Audio CD Program

■ Test Prep Pretest

CHAPTER TESTING

■ Chapter Tests (blackline master)

▲ Test Generator

■ Assessment Item Listing

See the correlations table at the front of this book for a list of the ***National Science Education Standards*** addressed in this chapter.

Smithsonian Institution® Visit **www.si.edu/hrw** for additional on-line resources.

KEY

Biology Principles and Explorations

PE Pupil's Edition

ATE Teacher's Edition

BIOSOURCES

Teaching Transparencies

Laboratory Program

One-Stop Planner CD-ROM *Shows these resources within a customizable daily lesson plan:*

■ Teaching Resources

▲ Test Generator

VISUAL STRATEGIES	CLASSWORK AND HOMEWORK	TECHNOLOGY AND INTERNET RESOURCES
ATE **Fig. 13-3,** p. 278 **ATE** **Fig. 13-6,** p. 281 44 Darwin's Voyage 52 Process of Natural Selection	**PE** **Quick Lab** How can you model natural selection?, p. 282 **PE** **Review,** p. 282	CNN Video Segment 10 Galápagos Holt Biology Videodiscs Section 13-1 Audio CD Program Section 13-1 internetconnect SciLINKS NSTA TOPICS: • Theory of evolution • Natural selection
ATE **Fig. 13-9,** p. 284 **ATE** **Fig. 13-10,** p. 286 **ATE** **Fig. 13-11,** p. 286 **ATE** **Fig. 13-14,** p. 286 49 Whale Evolution 50 Forelimbs of Vertebrates 44A Comparing Vertebrate Embryo Development	**Lab A9** Comparing Observations of Body Parts **Lab C17** Analyzing Blood Serum to Determine Evolutionary Relationships **PE** **Review,** p. 289	Holt Biology Videodiscs Section 13-2 Audio CD Program Section 13-2 internetconnect SciLINKS NSTA TOPICS: • Paleontology • Fossil record • Evolution
ATE **Fig. 13-16,** p. 291	**PE** **Math Lab** Analyzing change in lizard populations, p. 294 **PE** **Lab Techniques** Modeling Natural Selection, pp. 298–299 **Lab B9** Peppered Moth Survey **PE** **Review,** p. 294 **PE** **Study Zone,** p. 295	Holt Biology Videodiscs Section 13-3 Audio CD Program Section 13-3 internetconnect SciLINKS NSTA TOPIC: Species formation

ALTERNATIVE ASSESSMENT

PE **Portfolio Projects** 19–21, p. 297

ATE **Alternative Assessment,** p. 295

ATE **Portfolio Assessment,** p. 295

Lab Activity Materials

Quick Lab
p. 282 paper, pencil, watch or stopwatch

Math Lab
p. 294 pencil, paper

Lab Techniques
p. 298 scissors, construction paper, cellophane tape, soda straws, felt-tip marker, meterstick or tape measure, penny or other coin, six-sided die

Demonstrations
ATE several photos of varieties of plants and animals, p. 277

ATE samples or photos of fossils, p. 284

ATE roll of aluminum foil, scissors, white paper, p. 292

BIOSOURCES LAB PROGRAM
- Quick Labs: Lab A9, p. 17
- Inquiry Skills Development: Lab B9, p. 35
- Laboratory Techniques and Experimental Design: Lab C17, p. 79
- Laboratory Techniques and Experimental Design: Lab C20, p. 93

Chapter 13

Chapter Theme

Evolution

The theory of evolution by natural selection states that organisms change over time. This occurs when the organisms best adapted to their particular environment produce more offspring than organisms that are less well adapted. Eventually, organisms in different environments can change enough that they actually become different species. There is a great deal of scientific evidence for large changes in organisms over time as well as for the small changes in biological molecules within cells that cause the larger changes.

Establishing Prior Knowledge

Before beginning this chapter, make sure students can answer the questions in **Study TIP Ready?** on page 275. If they cannot, encourage them to reread the sections indicated.

Opening Question

Ask students what the word *evolution* means. Students should recall that *evolution* means "change over time." Ask students to list examples of types of animals or plants that have changed over time. *(Domesticated plants and animals are the most likely examples.)*

CHAPTER 13 The Theory of Evolution

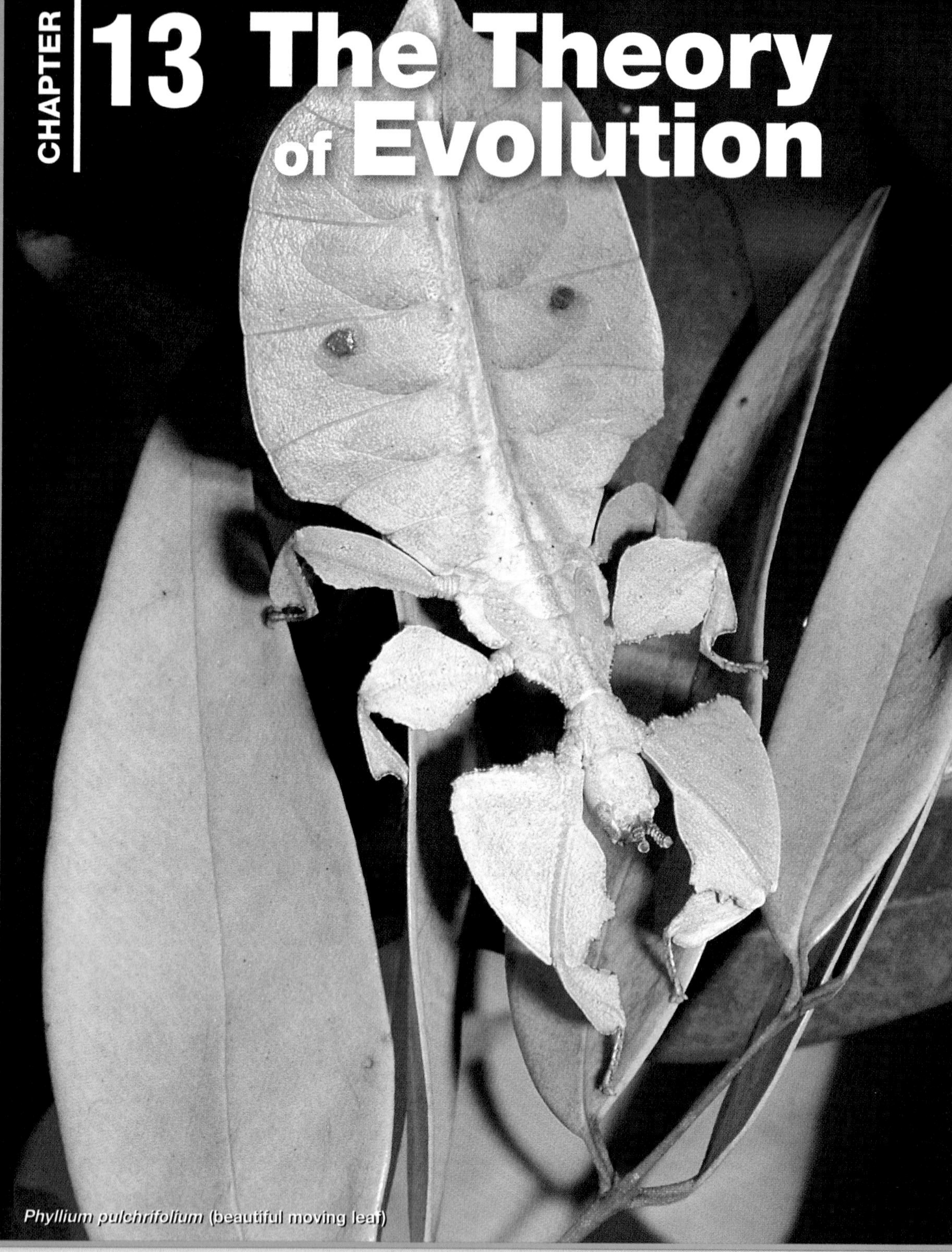

Phyllium pulchrifolium (beautiful moving leaf)

What's Ahead — and Why
CHAPTER OUTLINE

13-1 The Theory of Evolution by Natural Selection

- Darwin Proposed a Mechanism for Evolution
- Darwin Developed His Idea
- Darwin Sought a Reasonable Explanation for Evolution
- Darwin's Ideas Have Been Updated

13-2 Evidence of Evolution

- Fossils Provide an Objective Record of Evolution
- Biological Molecules Contain a Record of Evolution
- Anatomy and Development Suggest Common Ancestry

13-3 Examples of Evolution

- Natural Selection Changes the Makeup of Populations
- The Formation of New Species Begins with Small Changes

Students will learn

- how Darwin developed the theory of evolution by natural selection and what the basic elements of the theory are.
- about evidence from fossils, molecules, anatomy, and developing organisms that supports the theory of evolution.
- about examples of changes in natural populations, including the formation of new species, that illustrate the process of evolution.

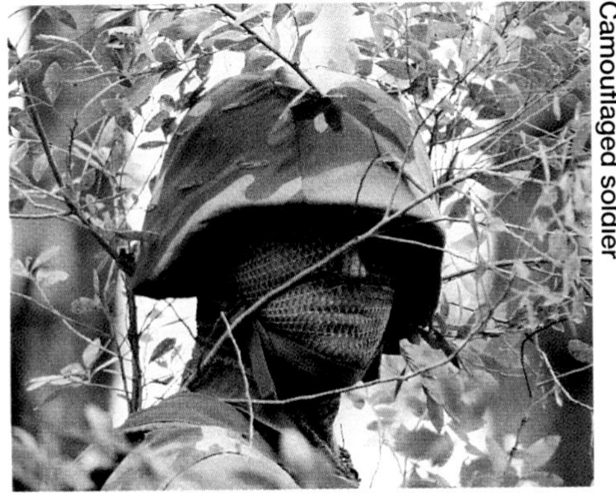
Camouflaged soldier

***Like a camouflaged** soldier hiding from the enemy, the moving leaf in the photo at left is able to hide from predators. This trait enables these insects to live long enough to have offspring. Through genetic inheritance, their offspring will then have the same kind of successful camouflage.*

Looking Ahead

Labs

- **Quick Lab**
  How can you model natural selection? (p. 282)
- **Math Lab**
  Analyzing change in lizard populations (p. 294)
- **Lab Techniques**
  Modeling Natural Selection (p. 298)

Features

- **Exploring Further**
  The Rate of Evolution (p. 289)

Study TIP

Ready?

Check your understanding of these topics to help you prepare for what's ahead.

Can you...

- **describe** the structure of proteins? (Section 2-3)
- **relate** the sequence of nucleotides to the amino acid sequence in protein? (Section 10-1)
- **define** *genetic mutations*? (Section 10-2)
- **describe** gene sequencing? (Section 11-1)
- **summarize** the concept of radiometric dating? (Section 12-1)

If not, *review the sections indicated.*

internetconnect

National Science Teachers Association *sci*LINKS Internet resources are located throughout this chapter.

Reading Strategies

Active Reading

Technique: Reader Response Logs

Students may already hold opinions about the origin of species. So that students can make personal responses to the concepts presented in this chapter, have them create a Reader Response Log. Ask students to divide their paper in half. On the left side of the paper, have them copy a word, phrase, or passage from the text. On the right side, have them write their reactions, thoughts, or questions about the entries on the left side of the paper.

Directed Reading

Assign **Directed Reading Worksheet Chapter 13** to help students interact with the text as they read the chapter.

Interactive Reading

Assign ***Biology: Principles and Explorations*** **Audio CD Program Chapter 13** to help reluctant readers achieve greater success in reading the chapter.

Reading Skills

Assign Reading Organizers and Reading Strategies Worksheets from the **One-Stop Planner CD-ROM with Test Generator** to help students learn and practice effective reading skills.

internetconnect

Holt *Biology: Principles and Explorations*

On-line Resources: go.hrw.com

Visit the HRW Web site for a variety of resources related to this chapter. Just type in the keyword HX0 Chapter 13.

Holt *Biology: Principles and Explorations*

On-line Resources: www.scilinks.org

The following *sci*LINKS Internet resources can be found in the student text for this chapter:

Page 276
TOPIC: Theory of evolution
KEYWORD: HX276

Page 279
TOPIC: Natural selection
KEYWORD: HX279

Page 285
TOPIC: Paleontology
KEYWORD: HX285

Page 289
TOPIC: Fossil record
KEYWORD: HX288

Page 289
TOPIC: Evolution
KEYWORD: HX289

Page 294
TOPIC: Species formation
KEYWORD: HX294

Chapter 13

1 Prepare

Scheduling

- **Block:** About 45 minutes will be needed to complete this section.
- **Traditional:** About one class period will be needed to complete this section.
- **Planning:** See the Planning Guide on pp. 273A and 273B for lecture, classwork, and assignment options for Section 13-1. Lesson plans for Section 13-1 can be found in a lesson cycle format in ***Biology: Principles and Explorations* One-Stop Planner CD-ROM with Test Generator.**

Resource Link

Assign **Active Reading Worksheet Section 13-1** to help students comprehend the key concepts and visuals in this lesson.

2 Teach

Lesson Warm-up

Have students study the photograph on page 274 of *Phyllium pulchrifolium,* which means "beautiful moving leaf." Ask students to think of an advantage of looking like a leaf. *(Leaf-dwelling insects that resemble leaves may avoid being eaten by predators.)* Ask students to predict what would happen to this insect's chances of survival if it found itself in an environment that had only plants with red leaves. *(It would not blend in as well, so would be more likely to be captured by predators.)*

Section 13-1 The Theory of Evolution by Natural Selection

Objectives

- **Identify** several observations that led Darwin to conclude that species evolve. (p. 278)
- **Relate** the process of natural selection to its outcome. (pp. 279–280)
- **Summarize** the main points of Darwin's theory of evolution by natural selection as it is stated today. (p. 281)

Key Terms

population
natural selection
adaptation
isolation
extinct

Darwin Proposed a Mechanism for Evolution

The idea that life evolves was first proposed by Lucretius, a Roman philosopher who lived 1,900 years before the modern theory of evolution was invented. He thought that organisms could not continue their species unless they were able to survive and produce healthy offspring. His idea was not seriously considered until it was revived by several eighteenth- and nineteenth-century scientists. Because these scientists lacked a reasonable mechanism for evolution, other scientists of the time firmly opposed the idea. Then in 1859, the English naturalist Charles Darwin published convincing evidence that species evolve, and he proposed a reasonable mechanism explaining how evolution occurs.

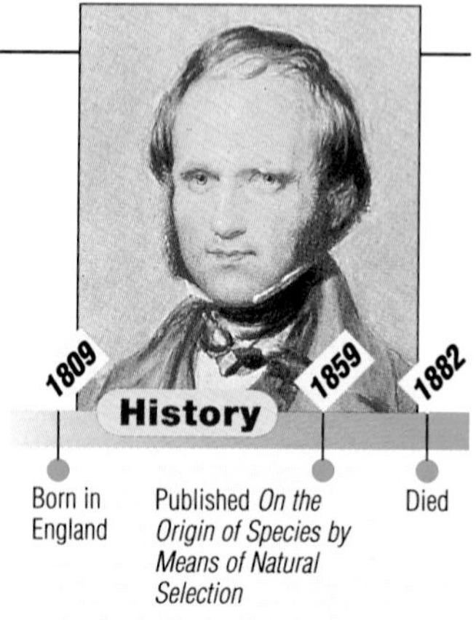

Figure 13-1 Charles Darwin

Like all scientific theories, the theory of evolution has developed through decades of scientific observation and experimentation. The modern theory of evolution began to take shape as a result of the work of Charles Darwin, shown in **Figure 13-1.** Today almost all scientists accept that evolution is the basis for the diversity of life on Earth.

OVERVIEW Theory of Evolution

The theory of evolution is supported by four major points:

1. Variation exists within the genes of every population or species (the result of random mutation).
2. In a particular environment, some individuals of a population or species are better suited to survive (as a result of variation) and have more offspring (natural selection).
3. Over time, the traits that make certain individuals of a population able to survive and reproduce tend to spread in that population.
4. There is clear proof from fossils and many other sources that living species evolved from organisms that are extinct.

internet**connect**
SCILINKS NSTA
TOPIC: Theory of evolution
GO TO: www.scilinks.org
KEYWORD: HX276

Integrating Different Learning Styles

Logical/Mathematical	**ATE**	Cross-Disciplinary Connection, p. 279
	PE	Quick Lab, p. 282
Visual/Spatial	**ATE**	Lesson Warm-up, p. 276; Visual Strategy, p. 278
	PE	Performance Zone item 1, p. 296
Interpersonal	**ATE**	Active Reading, p. 281
	PE	Performance Zone item 14, p. 297
Verbal/Linguistic	**PE**	Study Tip, p. 277; Performance Zone items 15 and 16, p. 297

Darwin Developed His Idea

As a youth, Darwin struggled in school. His father was a wealthy doctor who wanted him to become either a doctor or a minister. Not interested in the subjects his father urged him to study, Darwin frequently spent more time outdoors than in class. At the age of 16, Darwin was sent to Edinburgh, Scotland, to study medicine. Repelled by surgery, which at the time was done without anesthetics, Darwin repeatedly skipped lectures to collect biological specimens. In 1827, Darwin's father sent him to Cambridge University, in England, to become a minister. Although he completed a degree in theology, Darwin spent much of his time with friends who were also interested in natural science.

In 1831, one of Darwin's professors at Cambridge recommended him as an unofficial naturalist on a naval voyage of the HMS *Beagle*. Although the ship had an official naturalist, the *Beagle*'s captain preferred to have someone aboard who was of his own social class, as was Darwin. At the age of 22, Darwin set off on a journey that would both change his life and forever change how we think of ourselves. The ship and its route are shown in **Figure 13-2.**

Science before Darwin's voyage

When the *Beagle* sailed on December 27, 1831, most scientists and nonscientists thought that each species was a divine creation, unchanging and existing as it was originally created. This view of divine creation was thought to explain why species were uniquely suited to their environments. But scientists had begun to appreciate that traditional views of divine creation could not explain the kinds and distributions of fossils that had been found. Some scientists tried to explain their observations by changing traditional explanations of creation, while others (including Darwin's own grandfather) proposed various mechanisms to explain how evolution occurs.

Study TIP

Reading Effectively

Before reading this chapter, write the objectives for each section on a sheet of paper. Rewrite each Objective as a question, and answer these questions as you read the section.

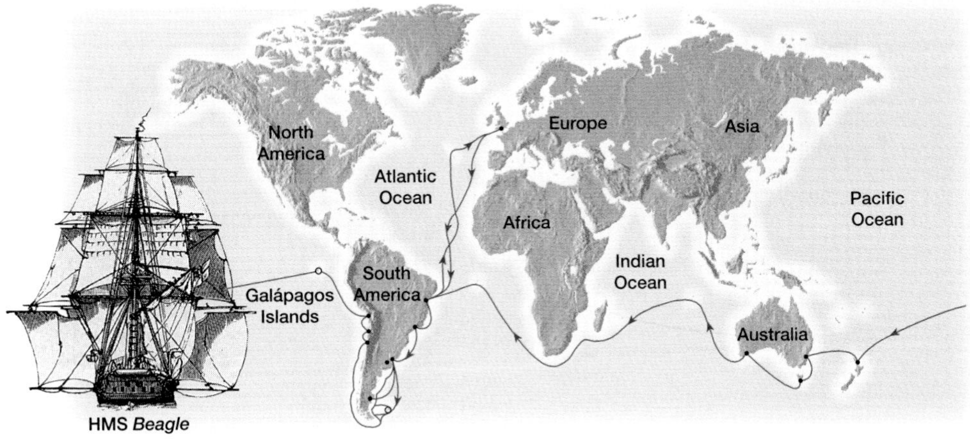

Figure 13-2 The route of the HMS *Beagle*. The HMS *Beagle* sailed around the world along the route shown on this map. The purpose of the ship's 5-year voyage was to survey the coast of South America.

Demonstration

Organisms That Have Changed

Show students pictures of several different varieties (breeds) of a commercially important plant (e.g., cotton, wheat, roses) or animal (e.g., cattle, horses, domesticated dogs and cats). Ask students how these varieties originated. *(All originated through selective breeding by humans.)* If possible, show students pictures of the ancestors of these varieties. *(For example, scientists think that dogs originated from wolves.)* Tell students that observations of change in domesticated animals and plants helped people recognize that species can change over time.

Teaching TIP — BASIC

Role of Naturalists

Tell students that while most scientists today must specialize in one field, many of the great scientists of the past were called naturalists—people who studied nature from a variety of different perspectives. Darwin's role on the *Beagle* was not just to study the native plants and animals encountered at the ship's many stops but to study the geology, climate, and people of those areas as well. In fact, his interest in geology led him to collect many fossils, some high in the Andes Mountains of South America. Such findings helped to stimulate his thinking about how environments and organisms might have changed over time.

Chapter 13

Visual Strategy

Figure 13-3

Have students examine Figure 13-3 and describe the features of the finches' bills. Have students describe how each finch's bill is adapted for the bird's diet. *(For example, the large ground finch's bill is thick and heavy, which is useful for cracking seeds.)*

Teaching TIP

Predicting Changes in Finch Populations

In 1977, the Galápagos Islands suffered a severe drought. Plants that produced small, soft seeds were devastated. In turn, a species of finch that fed on these seeds was similarly affected. As its source of food became depleted, the finch's population declined 85 percent. The species changed so that the average body and beak size of its members increased dramatically. Ask students to describe a type of finch that could have survived this drought. *(finches with beaks strong enough to crack large, hard seeds)*

Teaching TIP

Law of Use and Disuse

Tell students that in 1809 the French scientist Jean Baptist de Lamarck proposed an explanation for how evolution occurs. Lamarck recognized that the environment plays an important role in evolution. According to Lamarck, when an organism uses a part of its body, that part becomes more developed as a result of its use. He reasoned that the modified part is then passed on to the organism's offspring. For example, he might have argued that a bird that eats tough seeds will develop a thicker beak and will in turn have offspring with thicker beaks. Tell students that if Lamarck were correct, the offspring of bodybuilders would be born with enormously developed muscles.

In 1809, the French scientist Jean Baptiste Lamarck (1744–1829) proposed a novel mechanism to explain how evolution occurs. Lamarck suggested that evolution occurs through the use and disuse of physical features by individual members of a species. Lamarck believed that in the lifetime of an individual, structures increase in size because of use or reduce in size because of disuse.

According to Lamarck, these acquired traits are then passed to the offspring of that individual. This part of Lamarck's hypothesis is now known to be incorrect. However, Lamarck correctly believed that the cause of evolution is linked to the "physical conditions of life," referring to an organism's environmental conditions.

What Darwin saw during his voyage

During his voyage on the *Beagle,* Darwin found evidence that challenged the traditional belief that species are unchanging. During the voyage, Darwin read Charles Lyell's book *Principles of Geology,* which contains an explanation of Lamarck's theory of evolution. Lyell supported the hypothesis that the surface of the Earth changed slowly over many years. As Darwin visited different places, he also saw things that he thought could be explained only by a process of gradual change (evolution). For example, in South America, Darwin found fossils of extinct armadillos. These fossilized animals closely resembled, but were not identical to, the armadillos living in the area.

On the Galápagos Islands, located about 1,000 km (620 mi) off the coast of Ecuador, Darwin found more signs that species evolve. Darwin was struck by the fact that the plants and animals of the Galápagos Islands and those of the nearby coast of South America resembled each other, as shown in **Figure 13-3.** Darwin later suggested that the simplest explanation was that the ancestors of Galápagos species must have migrated to the islands from South America long ago and changed after they arrived. Darwin later called such a change "descent with modification"—evolution.

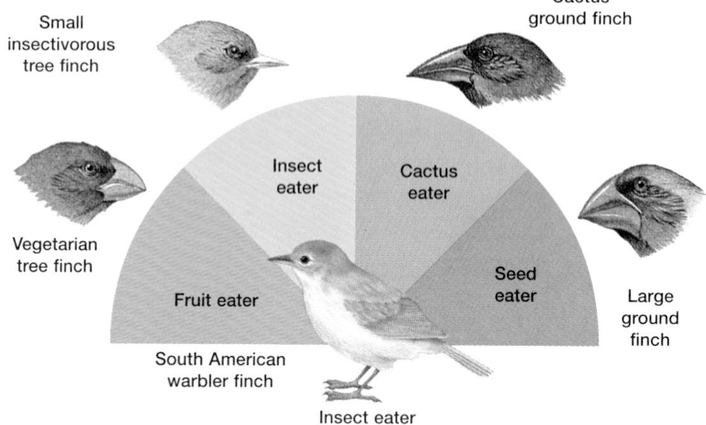

Figure 13-3 Darwin's finches. Darwin discovered that these finches closely resembled the South American finches, even though their diets differed.

did you know?

Darwin never used the finches he observed and collected on the Galápagos Islands as evidence for his theory.

Most biology books cite "Darwin's finches" as a perfect example of species formation. However, it wasn't until the 1940s that the biologist David Lack wrote the classic study that cemented the Galápagos finches' unique place in the history of evolutionary thought.

Darwin Sought a Reasonable Explanation for Evolution

When Darwin returned from his voyage at the age of 27, he continued his lifelong study of plants, animals, and geology. However, he did not report his ideas about evolution until many years later. During those years, Darwin studied the data from his voyage. As Darwin studied his data, his confidence that organisms had evolved grew ever stronger. But he was still deeply puzzled about how evolution occurs.

Malthus's contribution

The key that unlocked Darwin's thinking about how evolution takes place was an essay written in 1798 by the English economist Thomas Malthus. Malthus wrote that human populations are able to increase faster than the food supply. Malthus pointed out that population size grows by a geometric progression, as shown in **Figure 13-4.** However, according to Malthus, food supply increases by an arithmetic progression at best, also shown in Figure 13-4. He suggested that the human population would cover Earth's entire surface within a very short period of time if it could reproduce unchecked. Humans do not cover the Earth, Malthus pointed out, because death caused by disease, war, and famine slows population growth.

The term *population*, as it is used today, does not refer only to the number of people living in a particular area. In biology, a **population** consists of all the individuals of a species that live together in one place at one time. For example, the alligators in the Everglades make up a population because they live together in a particular area.

Figure 13-4 Geometric and arithmetic progressions. The blue graph line shows uncontrolled population growth, in which the numbers increase by a multiplied constant. The red graph line shows increased food supply, in which the numbers increase by an added constant.

Two Rates of Progression

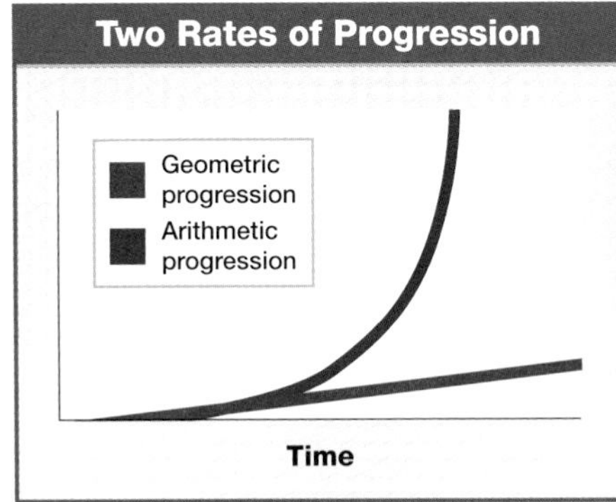

Natural selection

Darwin realized that Malthus's ideas about the human population apply to all species. Every organism has the potential to produce many offspring during its lifetime. In most cases, however, only a limited number of those offspring survive to reproduce. Adding Malthus's view to what Darwin saw on his voyage and to his own experiences in breeding domestic animals, Darwin made a key association: *Individuals that have physical or behavioral traits that better suit their environment are more likely to survive and reproduce than those that do not have such traits.* Darwin suggested that by surviving long enough to reproduce, individuals have the opportunity to pass on their favorable characteristics to offspring. In time, these favorable characteristics will increase in a population, and the nature of the population will gradually change. Darwin called this process by which populations change in response to their environment **natural selection.**

internetconnect

TOPIC: Natural selection
GO TO: www.scilinks.org
KEYWORD: HX279

Biology in Your Community

Roach Reproduction

As **Figure 13-4** illustrates, populations grow geometrically and therefore have the potential to increase dramatically within a few generations. Most species, including most insects, produce many offspring and have short generation times. If the cockroaches living in or near your home successfully reproduced at their full potential, they would soon cover the floor and walls.

Cross-Disciplinary Connection

◆ **Mathematics**

Have students draw a graph with numerical data to reinforce the concepts of geometric progression and arithmetic progression and the differences between them. Ask students to graph data for the following situation: A population of animals feeds primarily on one particular food source. Assume that at a particular point in time (time "zero" on the graph), there are 10 individuals in the population. The population doubles in size every 5 years. Have students draw a graph that depicts population size over a 30-year time period and connect the points on their graph with a line. Then tell them that at time "zero" there are 10 units of food in the animal's environment and that the food supply increases by 10 units every 5 years. Have students plot points and a connecting line for the food supply on their graph. Ask them which line depicts a geometric progression. *(the line depicting population size)* Ask them which line on the graph depicts an arithmetic progression. *(the line depicting units of food)* PORTFOLIO

internetconnect

SciLinks NSTA Have students research natural selection using the Web site in the Internet Connect box on page 279. Have students write a report summarizing their findings. PORTFOLIO

C

Teaching TIP

Darwin's Theory of Evolution

Students may wonder why the theory of evolution by natural selection is usually credited solely to Charles Darwin. Point out that Darwin worked on his ideas for more than 20 years before he published them. Darwin also published his ideas with many examples of how natural selection can cause evolution. In addition to *On the Origin of Species,* Darwin published several other books on the subject and is responsible for the theory's widespread acceptance among scientists.

Making Biology Relevant

Medicine — ADVANCED

Antibiotics are powerful drugs that have saved many lives. Penicillin, the first antibiotic discovered, was observed to inhibit bacterial growth on Petri dishes by the scientist Alexander Fleming in 1928. Many other antibiotics have been discovered or synthesized since then. In the last several decades, the effectiveness of antibiotics has become compromised by the evolution of antibiotic-resistant strains of many species. Some diseases must now be treated with a "cocktail" of several different antibiotics to cure patients. Antibiotic resistance has become a problem because of the widespread use of antibiotics, not only for treatment of human diseases, but also for prevention of disease in humans and domestic animals. With such widespread use, a mutation in a bacterium's DNA that allows it to survive treatment will be perpetuated. Soon bacteria with this mutation may constitute the majority of the species alive.

biobit

Was Darwin the only biologist influenced by Malthus?

In February 1858, Alfred Russel Wallace was sick with malaria. During the fevers that accompany malaria, he remembered Malthus's essay on population (the same one Darwin had read). The memory caused him to realize the importance of competition in evolution. Within days he sent the essay to Darwin, prompting Darwin to publish his ideas.

Darwin suggested that organisms differ from place to place because their habitats present different challenges to, and opportunities for, survival and reproduction. Each species has evolved in response to its particular environment. The changing of a species that results in its being better suited to its environment is called **adaptation** *(ad uhp TAY shun)*. Darwin also noticed that organisms more closely resemble those living in nearby geographic locations than those living in similar yet widely separated parts of the world. Darwin concluded that the species in a particular place evolved from species that previously lived there or that migrated from nearby areas.

Other books influenced Darwin

In 1844, Darwin finally wrote down his ideas about evolution and natural selection in an early outline that he showed to only a few scientists he knew and trusted. In that year, however, a book called *Vestiges of the Natural History of Creation* was published by Robert Chambers (1802–1871), a Scottish engineer. In one chapter Chambers claimed that evolution took place in Earth's past. Although a bestseller, the book was intensely criticized by both the government and the church. Lamarck's theory of evolution by the inheritance of acquired characteristics had also been severely criticized. Shrinking from such controversy, Darwin put aside his manuscript. For the next 14 years, he continued to enlarge and improve his notes on evolution, but he said nothing about his ideas in public. While Darwin published several books on other scientific subjects, his manuscript on evolution remained unpublished.

Figure 13-5 Political cartoon of Charles Darwin. This 1874 cartoon of Darwin with a monkeylike "ancestor" is an example of how some people ridiculed Darwin because of his work.

Darwin published his ideas in 1859

Darwin decided to publish after he received a letter and essay in June 1858 from the young English naturalist Alfred Russel Wallace (1823–1913), who was in Malaysia at the time. Wallace's essay described a hypothesis of evolution by natural selection! In his letter, he asked if Darwin would help him get the essay published. Darwin's scientific friends urged him to get his own work ready. They arranged for a summary of Darwin's manuscript to be presented with Wallace's paper at a public scientific meeting. Neither Darwin nor Wallace attended the meeting on July 1, 1858. At the time, few people took notice of the two papers, but Darwin had finally begun to prepare what he considered to be a "short abstract of his work."

Darwin's book *On the Origin of Species by Means of Natural Selection* appeared in November of 1859. Many people were deeply disturbed by certain suggestions of Darwin's theory, such as the implication that humans are related to apes. While recognizing that humans closely resemble apes, they found the possibility of a direct evolutionary link unacceptable, as **Figure 13-5** suggests. But Darwin's arguments and evidence were very convincing, and his view that evolution occurs gained acceptance slowly from biologists around the world.

did you know?

The most famous statement describing Darwin's theory, "survival of the fittest," did not appear in Darwin's original work.

The phrase was coined by another biologist, Herbert Spencer, upon learning about Darwin's theory. Darwin liked the phrase and used it to summarize his theory's implications. Unfortunately, others misunderstood and misinterpreted the phrase. Soon the meaning of *fittest* was distorted to mean "most powerful or worthy." Political leaders and industrialists used it to justify conquest, colonialism, and oppression under the guise of "natural law."

Darwin's Ideas Have Been Updated

Since Darwin's work was published, his hypothesis—that natural selection explains how evolution happens—has been carefully examined by biologists. New discoveries, especially in the area of genetics, have given scientists new insight into how natural selection brings about the evolution of species. Darwin's ideas, restated in modern terms, are summarized here.

Natural selection causes change *within* populations

Darwin's key inference was based on the idea that in any population, individuals that are best suited to survive and do well in their environment will produce the most offspring. So the traits of those individuals will become more common in each new generation.

Scientists now know that genes are responsible for inherited traits. Therefore, certain forms of a trait become more common in a population because more individuals in the population carry the alleles for those forms. In other words, natural selection causes the *frequency* of certain alleles in a population to increase or decrease over time. Mutations and the recombination of alleles that occurs during sexual reproduction provide endless sources of new variations for natural selection to act upon. **Figure 13-6** shows the extent to which evolution can change a species.

Figure 13-6 Evolution in crop breeding. These vegetables belong to the same species, *Brassica oleracea.* Each was developed through selective breeding.

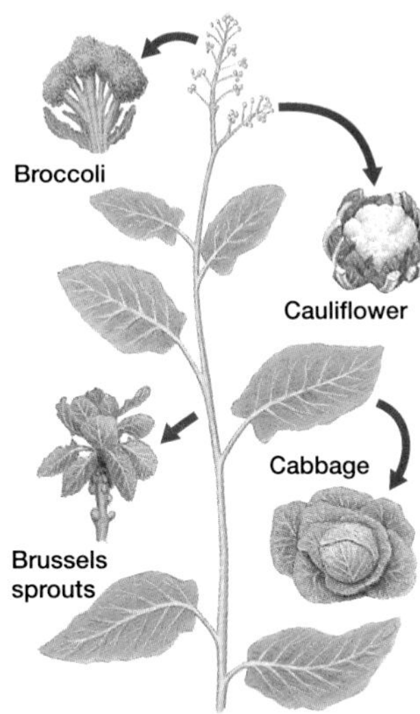

Isolation leads to species formation

The environment differs from place to place. Therefore, populations of the same species living in different locations tend to evolve in different directions. **Isolation** is the condition in which two populations of the same species cannot breed with one another, as shown in **Figure 13-7.** As two isolated populations of the same species become more different over time, they may eventually be unable to breed with one another. Generally, when the individuals of two related populations can no longer breed with one another, the two populations are considered to be different species. The Kaibab squirrel, which lives on the North Rim of the Grand Canyon in Arizona, evolved a black belly and other characteristics that distinguish it from the Abert squirrel. The Abert squirrel, which evolved a white belly, lives on the South Rim of the Grand Canyon. Because they are so isolated from one another, they have become different enough that biologists consider them separate species.

Figure 13-7 Isolation in action

These two squirrel populations became isolated from each other 10,000 years ago, thus preventing their interbreeding.

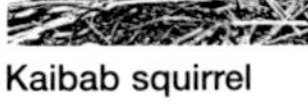
Kaibab squirrel

Abert squirrel

Visual Strategy

Figure 13-6

Point out to students that many different varieties of vegetables have been bred from the same species, *Brassica oleracea.* Ask students how these varieties of the same species could look so different. *(There is so much variation in the genes of this species.)* Is it possible for these varieties to become separate species? *(yes)* How? *(Through isolation, their genetic makeups may become so different that they can no longer interbreed.)*

Active Reading

Technique: Paired Summarizing

Group students into pairs and have them silently read about the updating of Darwin's ideas on pp. 281 and 282. Then have one student summarize the refinements in evolutionary theory. The other student should listen to the retelling and point out any inaccuracies or ideas that were left out. Have students work with their partners during this clarification process and refer to the text as needed.

did you know?

Charles Darwin's grandfather, Erasmus Darwin, wrote about evolution more than 60 years before his grandson's theory was presented.

Erasmus Darwin cited things such as the metamorphosis of insects, the new varieties produced by selective breeding, the variations among similar organisms in different climates, and the similarities of vertebrate structure as evidence that all life was "produced from a similar living filament." Unlike his famous grandson, Erasmus Darwin attributed change among organisms to the inheritance of characteristics acquired either naturally or at the will of the organism.

Chapter 13

How can you model natural selection?

Time 20 minutes

Process Skills Collecting data, interpreting, analyzing, summarizing, evaluating

Teaching Strategies
Remind students that individuals have two copies of each gene, and most mutations code for recessive traits, meaning that the genes are "expressed" only if an individual contains two copies of them.

Analysis Answers

1. the different things that can happen to an organism if it is exposed to change in its environment
2. Most mutations will be passed on because they are harmful only if an individual has two copies.
3. The survivors avoided chance death (a "die" card) and two copies of the mutation, which is "lethal" when expressed.
4. sample answer: does not distinguish between beneficial and harmful mutations; does not distinguish between living and reproducing

3 Assessment Options

Closure

Relating Concepts
Ask students what causes some individuals to produce more offspring than other individuals. *(They are better adapted to the environment in which they live, so more survive and produce more offspring.)*

Review

Assign Review—Section 13-1.

Reteaching — BASIC

Point out that Wallace, like Darwin, traveled extensively and collected many specimens. While in Indonesia, Wallace wrote a paper describing his idea about how evolution occurred and sent it to Darwin. Ask students to write a letter that Darwin could have written to Wallace in response. PORTFOLIO

Extinction leads to species replacement

Over long periods of time, events such as climate changes and natural disasters result in some species becoming **extinct,** which means that they disappear permanently. Species that are better suited to the new conditions may replace those that became extinct. For example, the extinction of the dinosaurs was followed by the evolution of modern mammals and birds over millions of years.

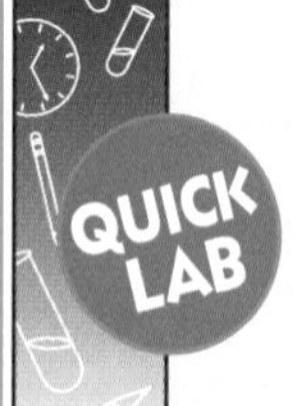

QUICK LAB

How can you model natural selection?

By making a simple model of natural selection you can begin to understand how natural selection changes a population.

Materials
paper, pencil, watch or stopwatch

Procedure

1. On a chalkboard or overhead, make a data table like the one shown below.
2. Write each of the following words on separate pieces of paper: live, die, reproduce, mutate. Fold each piece of paper in half twice so that you cannot see the words. Shuffle your folded pieces of paper.
3. Exchange two of your pieces of paper with those of a classmate. Make as many exchanges with additional classmates as you can in 30 seconds. Mix your pieces of paper between each exchange you make.
4. Look at your pieces of paper. If you have two pieces that say "die" or two pieces that say "mutate," then sit down. If you do not, then you are a "survivor." Record your results in your class table.
5. If you are a "survivor," record the words you are holding in the data table. Then refold your pieces of paper and repeat steps 2 and 3 two more times with other "survivors."

Analysis

1. **Identify** what the four slips of paper represent.
2. **Describe** what happens to most mutations in this model.
3. **Identify** what factor(s) determined who "survived." Explain.
4. **Evaluate** the shortcomings of this model of natural selection.

DATA TABLE

Student name	Trial 1	Trial 2	Trial 3

Review SECTION 13-1

1. **List** two observations made by Charles Darwin during his 5-year voyage that led him to conclude that living species evolved from extinct species.
2. **Describe** how natural selection occurs.
3. **Summarize** the modern theory of evolution by natural selection.

Critical Thinking

4. **SKILL Evaluating an Argument** Why is evolution called a theory and not a hypothesis? (**Hint:** See Section 1-3.)
5. **Defend** the theory that natural selection causes species to evolve.

Review SECTION 13-1 ANSWERS

1. Darwin found fossils of armadillos that closely resembled living armadillos. He also observed the resemblance between organisms on the Galápagos Islands and those on the nearest coast.
2. Individuals with traits well-suited to their environment are more likely to survive and reproduce than individuals without such traits. Over time, the population will change as more individuals possess the adaptive traits.
3. Organisms in a population vary genetically. These variations enable some organisms to reproduce more offspring than others. Over time, populations evolve to reflect the survival of organisms with the most advantageous heritable traits.
4. Hypothesis; because it is supported by a great deal of scientific evidence
5. In a given environment, natural selection results in improved survival and reproduction of the better-adapted members of a population. As populations change in response to these pressures, the makeup of the species as a whole will change. This change over time is evolution.

Evidence of Evolution

Section 13-2

Fossils Provide an Objective Record of Evolution

Have you ever looked at a series of maps that show how a city has grown? Buildings and streets are added, changed, or destroyed as the years pass by. In the same way, fossils of animals show a pattern of development from early ancestors to their modern descendants. Fossils offer the most direct evidence that evolution takes place. As you will recall, a fossil is the preserved or mineralized remains (bone, petrified tree, tooth, or shell) or imprint of an organism that lived long ago. Fossils, therefore, provide an actual record of Earth's past life-forms. Change over time (evolution) can be seen in the fossil record. Fossilized species found in older rocks are different from those found in newer rocks, as you can see in **Figure 13-8.**

After observing such differences, Darwin predicted that "missing links" (intermediate forms) between the great groups of organisms would eventually be found. Since Darwin, many of these links have been found. For example, fossil links have been found between fishes and amphibians, between reptiles and birds, and between reptiles and mammals, adding valuable evidence about the fossil history of the vertebrates.

Today Darwin's theory is almost universally accepted by scientists as the best available explanation for the biological diversity on Earth. Based on a large body of supporting evidence, most scientists agree on the following three major points:

1. Earth is about 4.5 billion years old.
2. Organisms have inhabited Earth for most of its history.
3. All organisms living today evolved from earlier, simpler life-forms.

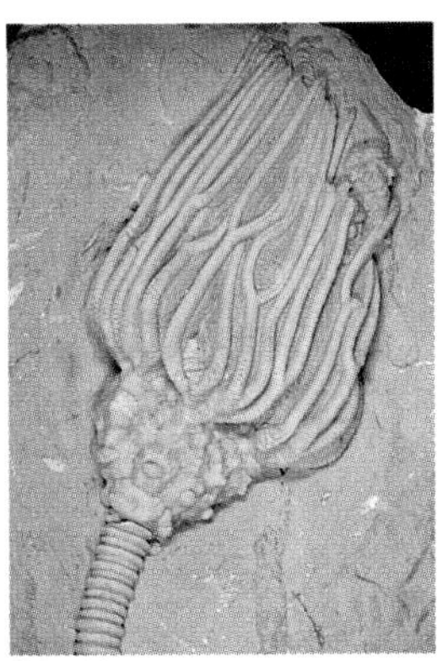

Crinoid

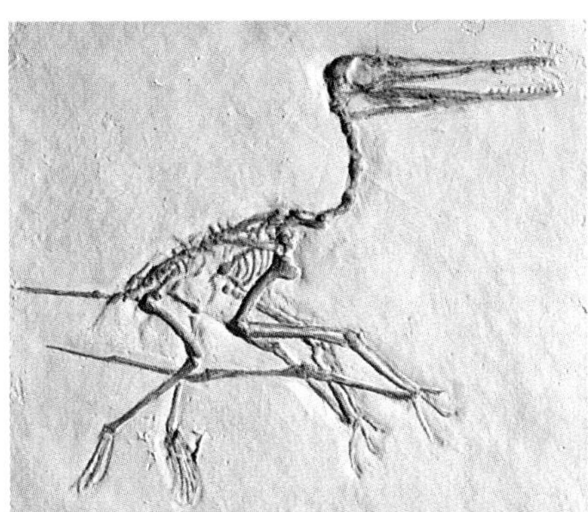

Pterodactyl

Figure 13-8 Fossils. Fossils of multicellular life-forms, such as the crinoid, occur in 800-million-year-old rocks found in Indiana. Fossils of the pterodactyl occur in 140- to 210-million-year-old rocks.

Objectives

- **Describe** how the fossil record supports evolution. (pp. 283–285)
- **Summarize** how biological molecules such as proteins and DNA are considered evidence of evolution. (p. 286)
- **Infer** how comparing the anatomy and development of living species provides evidence of evolution. (pp. 287–288)
- **Contrast** the gradualism and punctuated equilibrium models of evolution. (p. 289)

Key Terms

paleontologist
vestigial structure
homologous structure
gradualism
punctuated equilibrium

1 Prepare

Scheduling

- **Block:** About 90 minutes will be needed to complete this section.
- **Traditional:** About two class periods will be needed to complete this section.
- **Planning:** See the Planning Guide on pp. 273A and 273B for lecture, classwork, and assignment options for Section 13-2. Lesson plans for Section 13-2 can be found in a lesson cycle format in ***Biology: Principles and Explorations*** **One-Stop Planner CD-ROM with Test Generator.**

Resource Link

Assign **Active Reading Worksheet Section 13-2** to help students comprehend the key concepts and visuals in this lesson.

2 Teach

Lesson Warm-up

Read the following passage from Darwin's *On the Origin of Species:* "Of this history [of the world], we possess the last volume alone.... Of this volume, only here and there a short chapter has been preserved; and of each page, only here and there a few lines." Ask students to describe what Darwin meant by this passage. *(Darwin was referring to the incompleteness of the fossil record unearthed so far.)*

Integrating Different Learning Styles

Logical/Mathematical	**PE**	Performance Zone items 9, 10, and 18, pp. 296–297
Visual/Spatial	**ATE**	Demonstration, p. 284; Visual Strategy, p. 284; Visual Strategy, p. 288
	PE	Performance Zone items 12 and 13, p. 297
Verbal/Linguistic	**ATE**	Lesson Warm-up, p. 283
	PE	Study Tip, p. 284; Performance Zone items 17 and 21, p. 297

Chapter 13

Demonstration

Examining Fossils

Have students study samples or pictures of different types of fossils. Ask how these organisms are similar to modern organisms. *(Answers will vary. Students should note the similarities in bone and shell structure.)* Diatomaceous earth used in aquarium filters provides excellent examples of microfossils for examination.

Visual Strategy

Figure 13-9

Tell students that the first three animals in Figure 13-9 are known only from their fossil remains. The fourth animal is a living species. Note that the backbone of the *Rodhocetus kasrani* skeleton is not complete, as the tailbones shown in gray indicate. Tell students that it is rare to find a complete skeleton in any one fossil. As more skeletons of this species are found, all the species' bones may eventually be found. Using their knowledge of anatomy, paleontologists project what these bones will look like. Ask students how the backbone changed in relation to the time these animals spent in water? *(The backbone became heavier.)* What is the advantage of this change? *(Whales use up-and-down motions of their bodies to swim. A heavier backbone better supports the muscles used for this motion.)*

Study TIP

Reading Effectively

Read the heading "Fossils Provide an Objective Record of Evolution," and ask one or more Who, What, Where, When, Why, or How questions. For example, Where are fossils found? As you read, answer your questions.

In the 1990s, the discoveries of the fossilized remains of two whale ancestors, shown in **Figure 13-9,** provided new evidence of the evolution of whales from four-legged land mammals.

However, the fossil record, and thus the record of the evolution of life, is not complete. Many species have lived in environments where fossils do not form. Most fossils form when organisms and traces of organisms are rapidly buried in fine sediments deposited by water, wind, or volcanic eruptions. The environments that are most likely to cause fossil formation are wet lowlands, slow-moving streams, lakes, shallow seas, and areas near volcanoes that spew out volcanic ash. The chances that organisms living in upland forests, mountains, grasslands, or deserts will die in just the right place to be buried in sediments and fossilized are very low. Even if an organism lives in an environment where fossils can form, the chances are slim that its dead body will be buried in sediment before it decays. For

Figure 13-9 **Evidence of whale evolution**

Whales are thought to have evolved from four-legged mammals, which are represented here by their fossils and artistic reconstructions showing what scientists think they may have looked like.

Mesonychids are the hypothesized link between modern whales and hoofed mammals. They were about 2 m (6 ft) long. They are thought to have lived about 60 million years ago.

Ambulocetus natans apparently walked on land like modern sea lions and swam by flexing its backbone and paddling with its hind limbs (as do modern otters). They were about 3 m (10 ft) long. They existed about 50 million years ago.

did you know?

Soft-bodied organisms are not well represented in the fossil record.

Hard body parts, such as shells, bones, and branches, fossilize most readily. Therefore, soft-bodied organisms, such as invertebrates and nonwoody plants, are not well represented in the fossil record.

example, it may be eaten and scattered by scavengers. Furthermore, the bodies of some organisms decay faster than others. For example, an animal with a hard exoskeleton (such as a crab) would have a better chance of becoming fossilized than would a soft-bodied organism, such as an earthworm.

Although the fossil record will never be complete, it presents strong evidence that evolution has taken place. **Paleontologists,** scientists who study fossils, can determine the age of fossils fairly accurately by using radiometric dating, as you learned in Chapter 12. Radiometric dating enables paleontologists to arrange fossils in sequence from oldest to youngest. When this is done, orderly patterns of evolution can be seen. Based on existing fossils, Figure 13-9 shows an artist's idea of what a group of extinct whale ancestors looked like compared with a modern whale. They are arranged in the order that they evolved, based on their age as determined by radiometric dating.

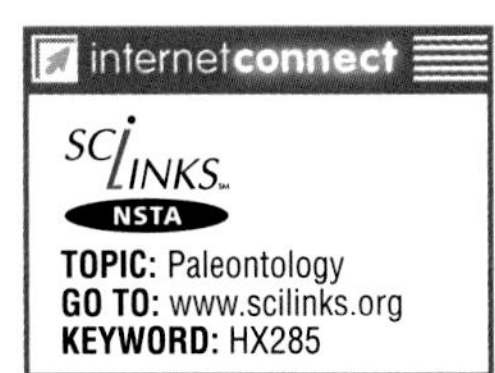

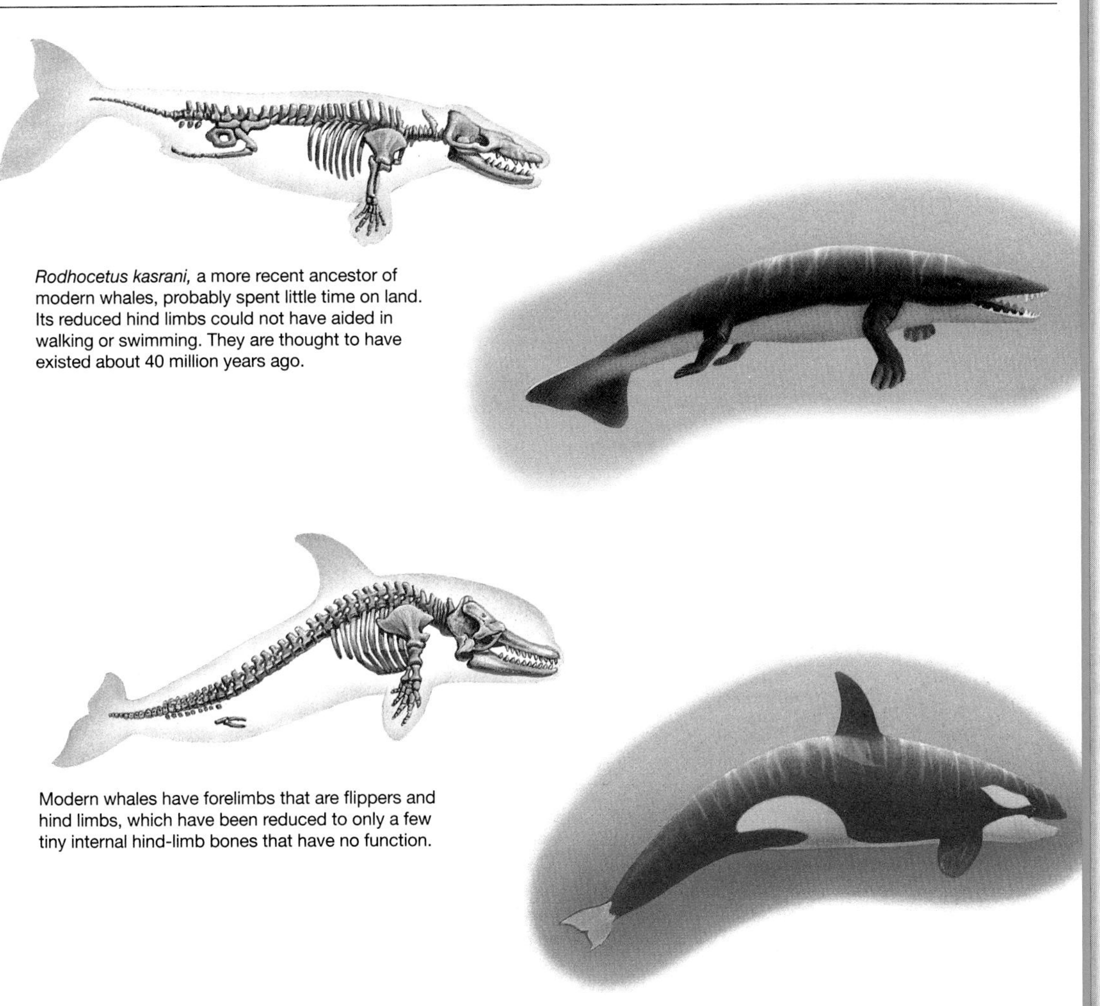

Rodhocetus kasrani, a more recent ancestor of modern whales, probably spent little time on land. Its reduced hind limbs could not have aided in walking or swimming. They are thought to have existed about 40 million years ago.

Modern whales have forelimbs that are flippers and hind limbs, which have been reduced to only a few tiny internal hind-limb bones that have no function.

Cross-Disciplinary Connection

◆ Chemistry

Scientists can determine the ages of rocks and fossils by measuring the amount of radioactive decay, or breakdown, of radioactive atoms in the rock. A radioactive atom contains an unstable combination of protons and neutrons. Since it is unstable, a radioactive atom will eventually change into a more stable atom. For example, carbon-14, a rare form of carbon found in tiny amounts in living things, decays into nitrogen. (The most common form of carbon is carbon-12.) The term *half-life* describes how long it takes for one-half of the radioactive atoms in a sample to decay. For example, the half-life of carbon-14 is 5,730 years. When an organism dies, it no longer accumulates carbon, and the carbon-14 it has steadily decreases as it decays. Because carbon-12 is stable, the ratio of carbon-14 to carbon-12 decreases over time. If a fossilized organism is found to have one-fourth the carbon-14 to carbon-12 ratio of a living organism, a scientist would conclude that the fossil was 11,460 years old. (This is two half-lives of carbon-14.) Because the half-life of carbon-14 is relatively short, this radioisotope is only used for dating fossils less than about 50,000 years old. To date older fossils, scientists use radioactive isotopes with longer half-lives. For example, uranium-235 has a half-life of 704 million years and potassium-40 has a half-life of 1.25 billion years.

did you know?

Most of Earth's organisms are now extinct.

Scientists estimate that 99 percent of all animal and plant species that ever existed are now extinct. Most people do not realize how many species have been found by paleontologists. In the past few years, many new species have been discovered. For example, in one small area in Wyoming, early Eocene rocks have yielded fossils of more than 50 species of animals, only a small portion of the animals that lived in the area at the time.

Chapter 13

Visual Strategy

Figures 13-10 and 13-11

Have students compare the data in Figure 13-10 with the family tree in Figure 13-11. Ask them if a family tree produced with the amino acid data in Figure 13-10 would show the same relationships as the family tree based on nucleotide substitutions. *(yes)* Why? *(A nucleotide sequence determines an amino acid sequence.)*

Teaching TIP

Making Mutations

To show students that the amino acid sequence of a protein is determined by the nucleotide sequence of a gene, put the following sequence of nucleotide bases on the board or overhead: *CUU, GUU, CCU, GGC, AGG.* Have students look up the amino acids encoded by these triplets in the table on page 290. *(leucine, valine, proline, glycine, and arginine)* Have students take turns substituting one of the other three nitrogen bases for the base in the first, second, or third position of a triplet. Then have students look up the name of the amino acid that would be encoded by the new triplet. Tell students that the substitutions they made represent mutations. Ask how mutations affect the proteins encoded by DNA. *(Mutations change the proteins produced by changing the blueprints for their production.)*

Hemoglobin Comparison

Species	Amino acid differences
Gorilla	1
Rhesus monkey	8
Mouse	27
Chicken	45
Frog	67
Lamprey	125

Figure 13-10 Hemoglobin differences. Species that share a more recent ancestor with humans have fewer amino acid differences with human hemoglobin.

Biological Molecules Contain a Record of Evolution

The picture of successive change seen in the fossil record allows scientists to make a prediction that can be tested. If species have changed over time, then the genes that determine their characteristics should also have changed. As species evolved, one change after another should have become part of their genetic instructions through mutation. Therefore, more and more changes in a gene's nucleotide sequence should build up over time.

Proteins

This prediction was first tested by analyzing the amino acid sequences of similar proteins found in several species. Recall that the amino acid sequence of proteins is genetically determined. If evolution has taken place, then species that descended from a common ancestor in the distant past should have more amino acid sequence differences between their proteins than do species that shared a common ancestor more recently. A common ancestor is a species from which two or more species have diverged. Comparing one chain of human hemoglobin with that of other species reveals the predicted pattern, as shown in **Figure 13-10.** Notice that species which shared a common ancestor more recently (for example, humans and gorillas) have few amino acid sequence differences. However, those species that share a common ancestor in the more distant past (such as humans and frogs) have many amino acid sequence differences.

Figure 13-11 Nucleotide changes

The length of the branches on this molecular phylogenetic tree for the hemoglobin gene indicates the number of nucleotide changes that have occurred.

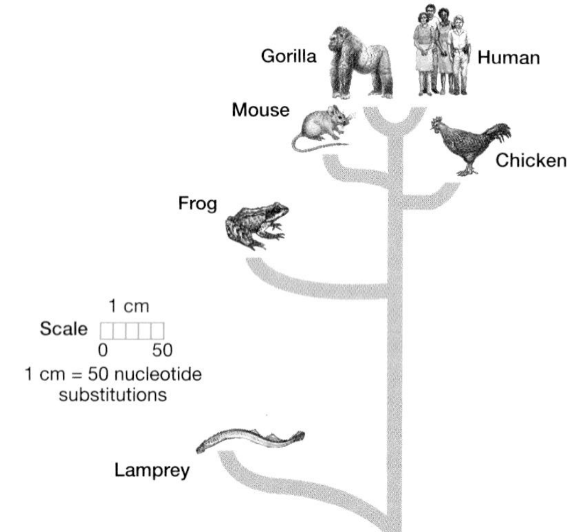

Nucleic acids

As you learned in Chapter 10, nucleotide changes (such as nucleotide substitutions) cause changes in the amino acid sequence of a protein. Scientists can directly estimate the number of nucleotide changes that have taken place in a gene since two species diverged from a common ancestor by comparing the exact nucleotide sequence of genes. Using the data obtained from proteins and nucleotides, scientists produce diagrams like the one in **Figure 13-11.** These diagrams, or phylogenetic *(fy loh juh NEHT ihk)* trees, show how organisms are related through evolution. Phylogenetic trees provide very strong evidence supporting evolution because they show the same relationships indicated by the fossil record.

Overcoming Misconceptions

Vestigial Is Not Necessarily Useless

Many people think that structures are labeled as vestigial because they are useless. Many such structures do, in fact, have functions. Structures are labeled vestigial if they are smaller, less functional, or perform a different function as homologous structures did in ancestral life-forms. For example, the yolk sac of a mammal is homologous to the yolk sac of birds and reptiles, but it is considered vestigial because it does not provide nutrients for the growing embryo. However, the mammalian yolk sac does produce blood cells.

Anatomy and Development Suggest Common Ancestry

Comparisons of the anatomy (structures) of different types of organisms often reveal basic similarities in body structures, even though their functions may be very different. For example, sometimes bones (or other structures) are present in an organism but are reduced in size and either have no use or have a less important function than they do in other, related organisms. Such structures, which are considered to be evidence of an organism's evolutionary past, are called **vestigial** *(vehs TIJ ee uhl)* **structures.** The word *vestigial* comes from the Latin word *vestigium,* meaning "footprint." An example of a vestigial structure can be seen in **Figure 13-12.**

Figure 13-12 Flightless cormorant. The vestigial wings of this flightless cormorant are too small to enable it to fly.

As different vertebrates evolved, particular sets of bones evolved differently. But similarities in bone structure can still be seen, suggesting that all vertebrates share a common ancestor. As you can see in **Figure 13-13,** the forelimbs of all vertebrates are made from the same basic groups of bones. Such structures are called homologous *(hoh MAHL uh guhs),* from the Greek word *homologos,* meaning "agreeing." **Homologous structures** are structures that share a common ancestry. That is, a similar structure in two organisms can be found in their common ancestor.

Figure 13-13 Homologous structures

The forelimbs of vertebrates contain the same kinds of bones, called homologous structures, which form in the same way during embryological development.

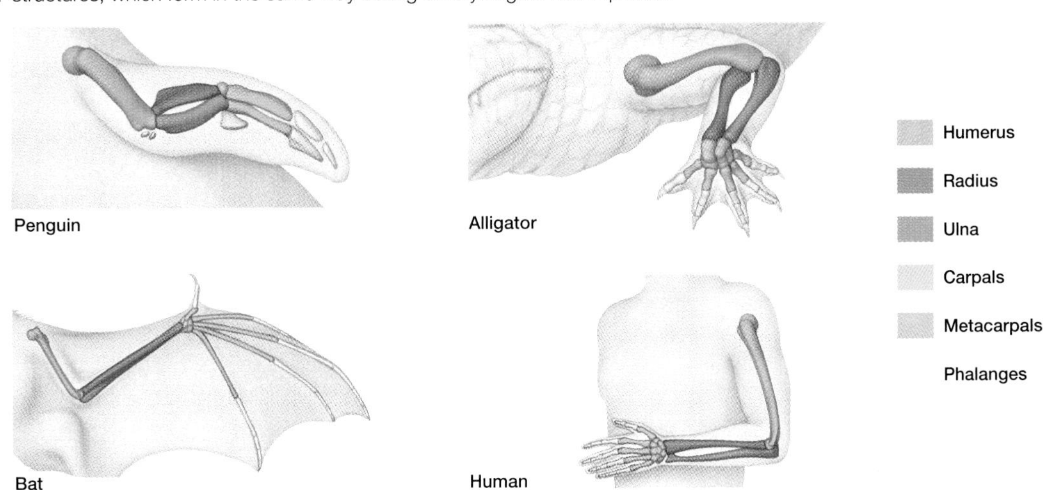

Teaching TIP

Analyzing the Selective Advantage

Many cave-dwelling animals have vestigial eyes. Ask students if it is a selective advantage for a cave dweller to invest energy in well-developed eyes. *(no)* Might it be a selective advantage for the cave dweller to invest that energy elsewhere? *(yes, because investing the energy elsewhere might provide a better "return" on its investment)* Which individuals, ones with large eyes or ones with vestigial eyes, will need to invest more energy in eye development as they grow? *(ones with large eyes)* In a cave, which individuals are likely to leave more offspring? *(ones with vestigial eyes)* How will a population of organisms change over time if it moves into a cave? *(There will be more and more individuals in the population with vestigial eyes.)*

did you know?

Penguins "fly" underwater.

Although penguins are not able to fly in the air, the movements of their wings under water resemble the motion of the wings of birds that do fly. The hefty bone structure of a penguin's wings is an advantage for moving through water, which is far denser than air. Also, heavier bones are not a disadvantage because of the buoyancy of water. The bones of birds and bats that fly in the air must be delicate to cut down the weight.

Visual Strategy

Figure 13-14

Have students locate the bony tail and pharyngeal pouches on each of the embryos. Ask: In which organism do the pharyngeal pouches persist the longest? *(fish)* In which organisms does the tail persist the longest? *(The tail persists into the adult stages of the fish and tortoises. It also persists well into the embryological development of the chicken but is not obvious in the adult stage.)* Tell students that different sets of genes are activated as an animal's embryological development proceeds, and that the most basic characteristics of an organism appear the earliest in its development. Ask: What is the most basic characteristic of all vertebrates? *(a backbone)* How might the differences among the vertebrates have been acquired? *(They might have been acquired through natural selection of traits that conferred reproductive advantages on different organisms in different environments.)*

Teaching TIP — ADVANCED

Evidence of Evolution

Have students design a **Graphic Organizer** similar to the one on page 288 that describes two kinds of physical traits that can be used to support the theory of evolution by natural selection.

PORTFOLIO

The pelvic bones found in modern whales are homologous to the pelvises that are found in land vertebrates. Although the whale pelvis bones are located near their reproductive organs, these bones do not function like a pelvis in a land vertebrate. The whale pelvis is located far from the vertebrae and has no apparent function. Thus, the whale pelvis is a vestigial structure.

The evolutionary history of organisms is also seen in the development of embryos. Compare the development of a human embryo with that of the other vertebrate embryos shown in **Figure 13-14.** Each embryo develops a tail, buds that become limbs, and pharyngeal *(fair IHN jee uhl)* pouches (which contain the gills of fish and amphibians). The tail remains in most adult vertebrates. Although the structures develop at different rates in different groups of vertebrates, they are homologous. Only adult fish and immature amphibians retain pharyngeal pouches. In humans, the tail disappears by the time of birth, and pharyngeal pouches develop into other structures.

Figure 13-14 **Vertebrate embryos**

Early in development, vertebrate embryos have similar characteristics. As development continues, various structures are modified until they take on their characteristic adult forms.

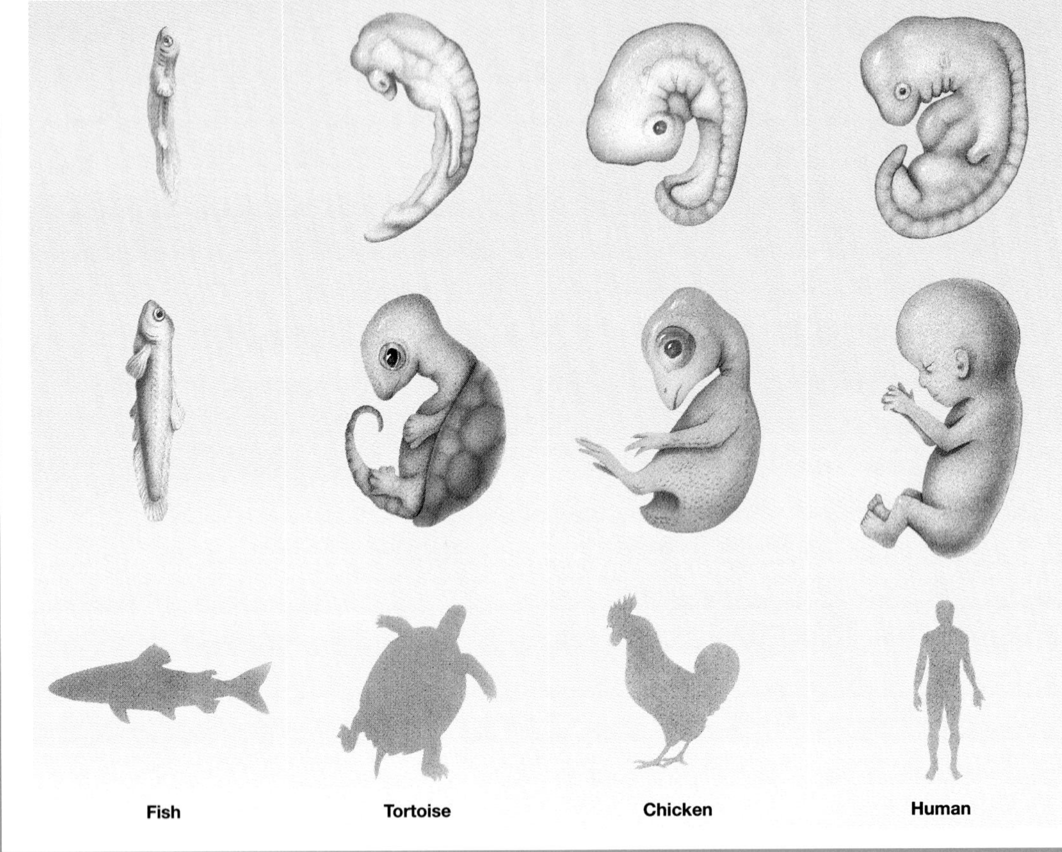

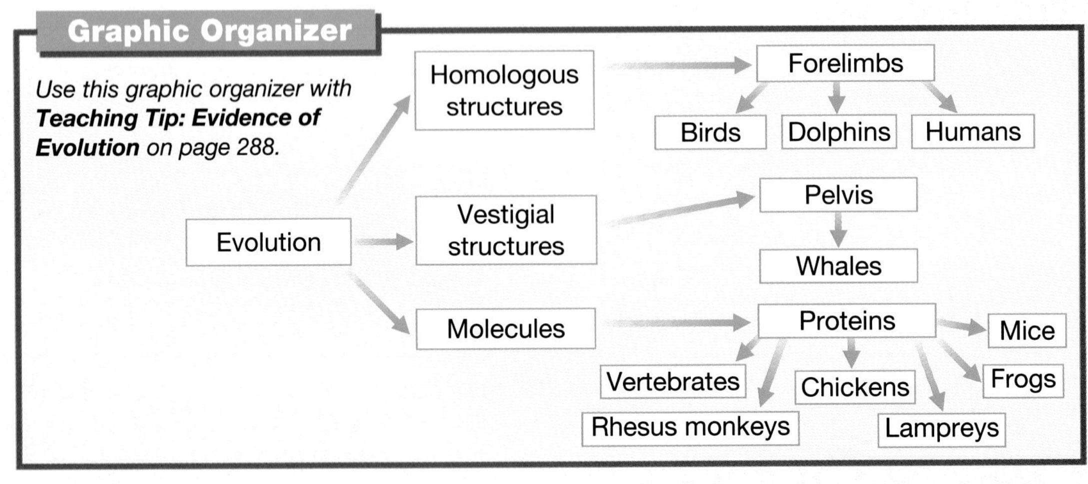

Does evolution occur in spurts?

For decades, most biologists have understood evolution as a gradual process that occurs continuously. The model of evolution in which gradual change over a long period of time leads to species formation is called **gradualism.** But American biologists Stephen Jay Gould and Niles Eldredge have suggested that successful species stay unchanged for long periods of time. Gould and Eldredge hypothesize that major environmental changes in the past have caused evolution to occur in spurts. This model of evolution, in which periods of rapid change in species are separated by periods of little or no change, is called **punctuated equilibrium.**

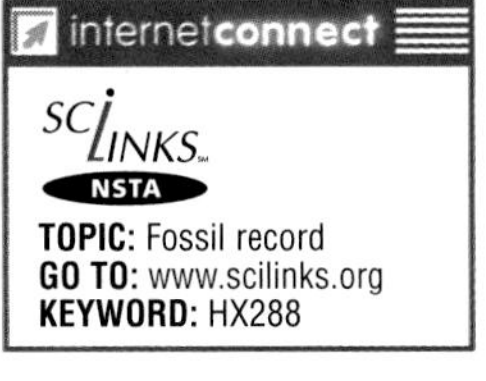

Exploring Further

The Rate of Evolution

How could major environmental changes lead to spurts in evolution? The fossil record shows that drastic environmental changes have occurred very infrequently, separated by periods of time that often last tens of millions of years. Events such as volcanic eruptions, asteroid impacts, and ice ages have been linked to sudden and drastic changes in climate. Such changes have also been linked to the extinction of many groups of organisms. As a result, environments that were once inhabited became empty. This provided opportunities for colonization by species that could quickly adapt to the new conditions through natural selection.

What fossils reveal

Although there are large gaps in the fossil record as a result of erosion and other destructive geologic processes, the fossil record seems to provide evidence of gradualism and punctuated equilibrium. Many groups of organisms appear suddenly in the fossil record. Some of these groups remain virtually unchanged for millions of years, while other groups disappear as suddenly as they appear. Still other groups of organisms appear to change slowly through time, as predicted by the gradualism model of evolution. More study of the fossil record may reveal additional examples of one or both types of evolution.

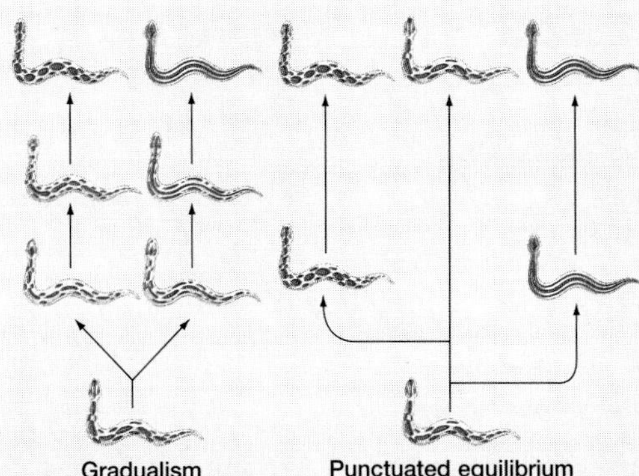

Review SECTION 13-2

1. **Relate** how the fossil record provides evidence that evolution has occurred.
2. **State** how comparing the amino acid sequence of a protein can provide evidence that evolution has taken place.
3. **Describe** how comparing the anatomy of living species provides evidence of evolution.
4. **Compare** the punctuated equilibrium model of evolution with the gradualism model.

Critical Thinking

5. **SKILL Justifying Conclusions** Explain why the wings of the flightless cormorant shown in Figure 13-12 are considered to be vestigial structures.

Review SECTION 13-2 ANSWERS

1. Answers will vary. Sample answer: Fossilized remains of organisms differ from those of organisms alive today. However, the fossil record shows changes in organismal form that can be traced forward and backward in time along different ancestral lines.
2. Identical organisms share the same genes. Since protein amino acid sequences are determined genetically, changes in these sequences indicate that two organisms differ. If the organisms share most of the same genes, however, it is reasonable to assume that they share a common ancestor and therefore that evolution has occurred.
3. Anatomical similarities between living species, such as similar bones used for similar functions, indicate that these species have evolved from a common ancestor.
4. According to the punctuated equilibrium model, evolution occurs in spurts in response to strong environmental pressures. According to gradualism, species evolve over long periods of time.
5. They are too small for the cormorant to fly.

Exploring Further

The Rate of Evolution

Teaching Strategies

Explain that these models are just that: abstractions that describe and help us understand the mechanics of evolutionary change. Emphasize that a good scientist tries to make a model fit his or her research results, not the other way around.

Discussion

Ask students to think about the similarities between these two models. *(Answers will vary. Sample answer: Both models assume that evolution precedes through the agent of natural selection. The effect of both models should produce a branching pattern of species evolution, as diagrammed over time.)*

3 Assessment Options

Closure

Looks Do Not Tell All

Tell students that the earliest phylogenetic trees were constructed using only evidence of morphological characteristics. Ask students why such trees might not reflect true evolutionary relationships. *(Similar morphological characteristics might reflect adaptation to a similar environment, rather than a common ancestry.)*

Review

Assign Review—Section 13-2.

Reteaching — BASIC

Show students a photo of the Grand Canyon, and explain that digging for fossils in the canyon is like traveling back in time. The rocks near the top of the canyon contain animal track imprints. Successively lower layers contain plant and reptile fossils, then the bones of primitive fish, then shell fossils, and finally no fossils.

1 Prepare

Scheduling

- **Block:** About 90 minutes will be needed to complete this section.
- **Traditional:** About two class periods will be needed to complete this section.
- **Planning:** See the Planning Guide on pp. 273A and 273B for lecture, classwork, and assignment options for Section 13-3. Lesson plans for Section 13-3 can be found in a lesson cycle format in ***Biology: Principles and Explorations*** **One-Stop Planner CD-ROM with Test Generator.**

Resource Link

Assign **Active Reading Worksheet Section 13-3** to help students comprehend the key concepts and visuals in this lesson.

2 Teach

Lesson Warm-up

Point out that the animal phylum Arthropoda includes more species than any other phylum. Within the phylum, the class Insecta has more species than any other class. In fact, about one-third of all animals are beetles, which constitute just one order of insects. To date, over 1 million species of insects have been classified, and many scientists estimate that there are probably several million more. Ask students how they think so many different kinds of species of insects could have evolved. *(Insects have adapted to many different kinds of environments and rely on many different kinds of food sources.)*

Section 13-3 Examples of Evolution

Objectives

- **Identify** five elements in the process of natural selection. (p. 290)
- **Describe** how natural selection has affected the European peppered moth. (pp. 290–291)
- **Relate** natural selection to the beak size of finches. (p. 292)
- **Summarize** the process of species formation. (pp. 293–294)

Key Terms

industrial melanism
divergence
speciation
ecological race
reproductive isolation

Natural Selection Changes the Makeup of Populations

Can you imagine an effect without a cause? The heart of Darwin's theory of evolution is that natural selection is the mechanism (cause) that drives evolution (effect). Darwin wrote: "Can we doubt . . . that individuals having any advantage, however slight, over others, would have the best chance of surviving and of procreating their kind? On the other hand, we may feel sure that any variation in the least degree injurious would be rigidly destroyed. This preservation of favorable variations, I call Natural Selection." In his writings, Darwin offered examples of how natural selection has shaped life on Earth. There are now many well-known examples of natural selection in action.

The key lesson scientists have learned about evolution is that the environment dictates the direction and amount of change, as shown in **Figure 13-15.** Just as success determines which plays a football coach keeps in his team's game plan, so success determines which changes in a species are "kept" through natural selection.

Industrial melanism

A well-studied example of natural selection in action is **industrial melanism,** the darkening of populations of organisms over time in response to industrial pollution. The best-known case of industrial melanism involves the European peppered moth, *Biston betularia.* Among the members of this species, there are two color variations,

Figure 13-15 Polar bear. The polar bear's white fur enables it to hunt successfully in its snowy environment.

OVERVIEW Natural Selection

The process of natural selection depends on five main elements:

1. All species have genetic variation.
2. The environment presents many different challenges to an individual's ability to reproduce.
3. Organisms tend to produce more offspring than their environment can support; thus, individuals of a species often compete with one another to survive.
4. Individuals within a population that are better able to cope with the challenges of their environment tend to leave more offspring than those less suited to the environment.
5. The traits of the individuals best suited to a particular environment tend to increase in a population over time.

Integrating Different Learning Styles

Logical/Mathematical	**PE**	Real Life, p. 293; Math Lab. p. 294
Visual/Spatial	**ATE**	Visual Strategy, p. 291
	PE	Performance Zone item 19, p. 297
Interpersonal	**ATE**	Demonstration, p. 292
Verbal/Linguistic	**PE**	Study Tip, p. 292; Performance Zone item 20, p. 297

Figure 13-16 **Kettlewell's moths**

Two color variations occur among European peppered moths, *Biston betularia.*

This graph shows the results of Kettlewell's experiments. In the polluted woods near Birmingham, two-thirds of the surviving moths were dark. In rural Dorset, two-thirds of the surviving moths were light.

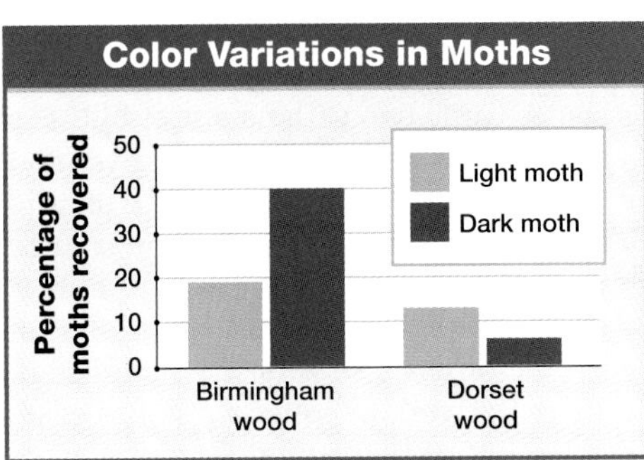

as shown in **Figure 13-16.** The darker moths have alleles for increased production of melanin (a pigment that makes them dark). Once treasured by butterfly collectors, the dark variety of *Biston betularia* was very rare until the 1850s. Starting around 1850, however, dark peppered moths began to appear more often, usually in heavily industrialized areas. After 100 years, almost all of the peppered moths near industrial centers were dark.

A hypothesis explaining the replacement of light moths by dark moths can be formed using Darwin's theory of evolution by natural selection. Dark peppered moths are common in industrial areas where tree trunks are darkened by the soot of pollution. Dark moths are camouflaged on the soot-darkened bark and so are not eaten by birds. Light moths, on the other hand, would stand out against a dark background and would be easy prey for hungry birds.

Testing natural selection of moths

To see if natural selection could have caused the color change in the peppered-moth populations, the British ecologist H.B.D. Kettlewell performed an experiment during the late 1950s. Kettlewell raised populations of light and dark peppered moths in a laboratory. He then marked the underside of their wings with a dot of paint so they could be recognized later. Next he released light and dark moths in two separate wooded areas of England. One of the wooded areas, near the city of Birmingham, was heavily polluted. The other wooded area, in the rural county of Dorset, was unpolluted. Finally, Kettlewell set traps around the woods to catch the moths and see which kind had survived. As the graph in Figure 13-16 shows, more of the moths matching the color of the tree trunks in each location survived. Many later experiments confirmed these results. Hidden observers even saw birds passing by dark moths on polluted tree trunks and attacking the more easily seen light moths. Kettlewell concluded that natural selection does cause industrial melanism in peppered moths.

Overcoming Misconceptions

Individuals Do Not Evolve

The story of the peppered moth illustrates an important though often misunderstood concept: Individuals do not evolve, populations do. The darkened trees did not cause the darker variant of moth to appear. That characteristic was already present in the natural genetic variation of the moth population. The "new" environment merely favored the survival of the darker moths.

Visual Strategy

Figure 13-16 BASIC

Instruct students to examine the photograph of the two color variations of peppered moths. Ask: For which of these moths is color an advantage in this environment? *(The light-colored moth has an advantage here because it blends in with the light-colored bark and is therefore less likely to be detected by a predator.)* Under what circumstances would the dark coloration be an advantage? *(Dark color would be an advantage if the moths were resting on a dark-colored tree trunk.)*

Teaching TIP

What If There Were No Dark Peppered Moths?

When the Industrial Revolution began, air pollution upset the homeostasis of the peppered moth population of Great Britain. Ask: What would likely have happened to the peppered moth population during the Industrial Revolution if a dark variety had not existed in the gene pool of the moths? *(They likely would have become extinct in industrial areas.)*

Making Biology Relevant

Technology

Natural selection has worked to make insect pests harder to fight. When DDT was first introduced, for example, it was a highly effective insecticide. Over time, DDT became less and less effective; individuals that were resistant to the insecticide survived and produced the next generation. In fact, many populations of insects are now resistant to DDT. Although DDT is now banned in this country because of its persistent toxicity, farmers have repeatedly had to deal with insect populations that develop resistance to insecticides.

Demonstration

Modeling Natural Selection

Divide students into small groups and provide each group with one 30 cm (1 ft) square of aluminum foil and 10 each of 2.5 cm (1 in.) squares of aluminum foil and white paper. (Crinkle the foil and flatten it out before cutting the squares.) Have one student in each group spread out the 20 small squares on the larger square of aluminum. Give another student in each group 10 seconds to pick up and remove as many squares as possible, one at a time, from the big sheet. Have each group count and report the number of aluminum squares and paper squares "captured." Keep a tally on the board or overhead. White paper squares should get "picked off" in greater numbers. Thus, the camouflaged aluminum squares have an adaptive survival advantage. Ask: If the little squares could reproduce, which type would be more numerous in the next generation? *(small aluminum squares)* Why? *(More of them were left to reproduce.)*

Teaching TIP — ADVANCED

Selection Pressure and Rates of Evolution

The term *selection pressure* refers to the impact of natural selection on a particular trait. If individuals expressing a particular trait all die before they reach reproductive age, then we say that there is very strong selection pressure against that trait. An example is the human disease cystic fibrosis, whose victims have until only recently rarely survived to reproductive age. An allele that has strong selection pressure against it is not likely to persist in a species. However, if the trait is recessive, individuals who are heterozygous will not be subjected to selection pressure but will pass the allele on to future generations. In this way, harmful alleles can persist for a long time in a species.

Study TIP

Organizing Information

Make a table to organize information about how natural selection causes evolution. Across the top write the headings *Moth* and *Finch.* Write the following headings along the sides:

1. *Types of variations*
2. *Environmental challenge*
3. *Successful trait*
4. *Effect of selection*

Add information to the table as you read Section 13-3.

The beaks of finches evolve

Darwin collected 31 specimens of finches from three islands when he visited the Galápagos Islands, in 1835. In all, he collected 9 distinct species, all very similar to one another except for their bills. Two ground finches with large bills feed on seeds that they crush in their beaks, while two with narrower bills eat insects. One finch is a fruit eater, one is a cactus eater, and yet another creeps up on sea birds and uses its sharp beak to drink their blood.

Darwin suggested that an original ancestral species of bird evolved in different ways to adapt to different food sources. This idea was first tested in 1938 by the naturalist David Lack. He watched the birds closely for five months and found little evidence to support Darwin's hypothesis. Stout-beaked finches and slender-beaked finches were feeding on the same sorts of seeds. A second, far more thorough study was carried out over 25 years beginning in 1973 by Peter and Rosemary Grant of Princeton University. The Grants' study presents a much clearer picture that supports Darwin's interpretation.

It was Lack's misfortune to study the birds during a wet year, when food was plentiful. The size of the beak of the finch is of little importance in such times. Slender and stout beaks both work well to gather the small, soft seeds which were plentiful.

During dry years, however, plants produce few seeds, large or small. During these leaner years, few small, tender seeds were available. The difference between survival and starvation is the ability to eat the larger, tougher seeds that the birds usually pass by. The Grants measured the beaks of many birds every year. They found that after several dry years, the birds that had longer, more-massive beaks had better feeding success and thus reproduced more effectively.

When wet seasons returned, birds tended to have smaller beaks again, as shown in **Figure 13-17.** The numbers of birds with different beak shapes are changed by natural selection in response to the available food supply, just as Darwin had suggested.

Figure 13-17 Natural selection in finches

By relating the environment to beak size, the Grants showed that natural selection influences evolution.

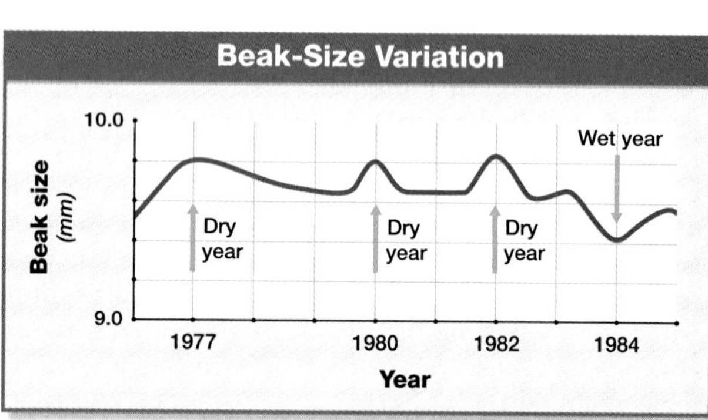

MULTICULTURAL PERSPECTIVE

Is All Variation Adaptive?

Based on his research on population genetics and molecular evolution, the Japanese geneticist Motoo Kimura proposed a neutral theory of evolution. Kimura argued that much of the variation in natural populations is selectively neutral. In other words, the different forms of genetic traits neither help nor hinder survival and reproduction. According to Kimura, changes in the frequency of these traits are due to chance, not to natural selection.

The Formation of New Species Begins with Small Changes

Species formation occurs in a series of stages. Because natural selection favors changes that increase reproductive success, evolution continuously molds and shapes a species to improve its "fit" to its environment. The accumulation of differences between groups is called **divergence** *(deye VUR jehns)*. Within populations, divergence leads to the formation of new species. Biologists call the process by which new species form **speciation** *(spee see AY shun)*.

Forming ecological races

A species often lives in several different kinds of environments. In each environment, natural selection acts on individuals of the species, resulting in offspring that are better adapted to that environment. If their environments differ enough, local populations can become very different. Over time, populations of the same species that differ genetically because of adaptations to different living conditions become what biologists call **ecological races.** Although the members of different ecological races are not yet different enough for the groups to be called different species, they have taken the first step toward speciation. Eventually, the races may become so different that they can no longer interbreed successfully. Biologists then consider them separate species.

Maintaining new species

What keeps new species separate? Why are even closely related species usually unable to interbreed? Once ecological races become different enough, a barrier to reproduction, like the one shown in **Figure 13-18,** usually prevents different groups from breeding with each other.

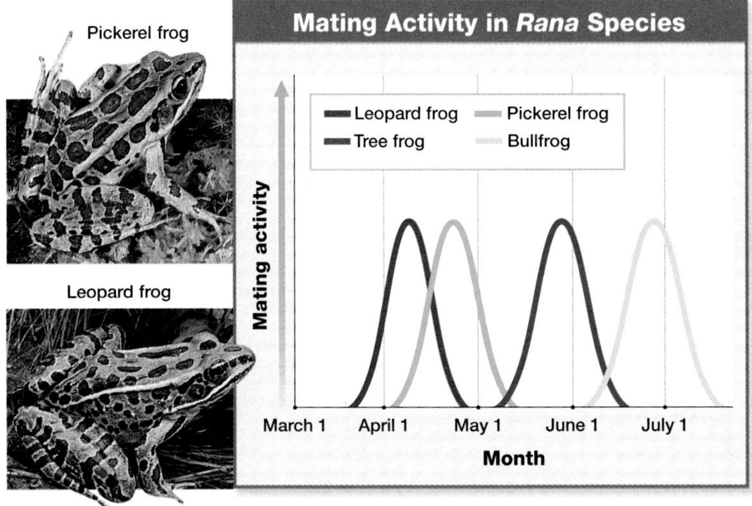

Figure 13-18 Mating activity in various *Rana* species. Though they appear to be similar, pickerel frogs (*Rana palustris*) and leopard frogs *(Rana pipiens)* are different species. The graph shows that the time of peak mating activity varies between several species of frogs.

REAL LIFE

Why do we find certain kinds of faces pretty or handsome? Evolutionary biologists think that many traits that contribute to a person's attractiveness actually reveal their fitness as a mate.

Finding Information
Research and evaluate competing theories concerning biological reasons for attractiveness.

REAL LIFE

Answer

Answers will vary. Sample answers might include the hypothesis that body form, especially the degree of left-right symmetry, correlates with genetic fitness. Studies have confirmed that people find symmetry an attractive physical quality.

CAREER — **ADVANCED**

Evolutionary Psychologist

Evolutionary psychologists study how people think and behave considering their evolutionary heritage. It is a relatively new subfield of psychology that seeks to put human cognition in an explicitly evolutionary context. Evolutionary psychologists draw from many fields in their research, including genetics and anthropology.

Teaching TIP

Sequential Diagram of Speciation

Have students make a **Graphic Organizer** similar to the one on page 293 that is a sequential diagram for speciation. They should use the following terms: *divergence, isolation, natural selection, new species,* and *variation.*

PORTFOLIO

Graphic Organizer

*Use this graphic organizer with **Teaching Tip: Sequential Diagram of Speciation** on page 293.*

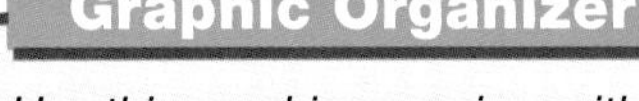

Variation → Isolation → Natural selection → Divergence → New species

Chapter 13

Analyzing change in lizard populations

Time 20 minutes

Process Skills Interpreting, analyzing, concluding, predicting

Teaching Strategies
Point out to students that the graph shown here is a scatter plot. The data points on such a graph are not connected by lines. Rather, these graphs illustrate the distribution and patterns of the data.

Analysis Answers

1. The average hind limb length of each population changed from that of the original population in response to differences in the average perch diameter of plants on the different islands.
2. The population would evolve to have longer average hind limbs.
3. The experiment illustrates that characteristics of populations can change over time in response to environmental pressures.

3 Assessment Options

Closure

Speciation

Ask students to explain why islands such as the Galápagos favor the evolution of new species. *(The islands are physically separated and have different environments. Therefore, populations on different islands face different selection pressures.)*

Review

Assign Review—Section 13-3.

Reteaching — BASIC

Have students suggest why sharks and alligators, which are are considered "living fossils," have changed little over millions of years. *(Their environments, to which they are well adapted, have remained fairly constant over this time period.)*

internetconnect

SCILINKS. NSTA

TOPIC: Species formation
GO TO: www.scilinks.org
KEYWORD: HX294

Reproductive isolation is the inability of formerly interbreeding groups to mate or produce fertile offspring. There are several types of barriers that may isolate two or more closely related groups. For example, groups may be geographically isolated or may reproduce at different times. Physical differences may also prevent mating, or they may not be attracted to one another for mating. The hybrid offspring may not be fertile or suited to the environment of either parent.

Species formation

Biologists have seen the stages of speciation in many different organisms. Thus, the way that natural selection leads to the formation of new species has been thoroughly documented. As changes continue to build up over time, living species may become very different from their ancestors and from other species that evolved from the same common ancestor, leading to the appearance of new species.

Analyzing change in lizard populations

Background

In 1991, Jonathan Losos, an American scientist, measured hind-limb length of lizards from several islands and the average perch diameter of the island plants. The lizards were descended from a common population 20 years earlier, and the islands had different kinds of plants on which the lizards perched. Examine the graph at right and answer the following questions:

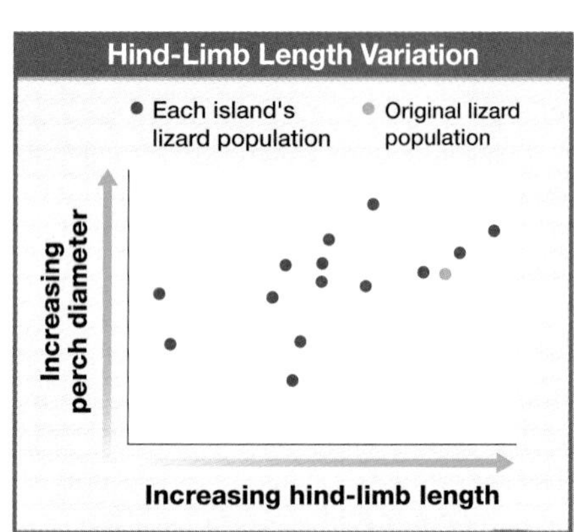

Analysis

1. **SKILL Interpreting Graphics** How did the average hind-limb length of each island's lizard population change from that of the original population?
2. **Predict** what would happen to a population of lizards with short hind limbs if they were placed on an island with a larger average perch diameter than from where they came.
3. **Justify** the argument that this experiment supports the theory of evolution by natural selection.

Review SECTION 13-3

1. **List** five elements of natural selection.
2. **Describe** the mechanism that caused the population changes in the European peppered moth.
3. **Identify** what caused the change in the finch's beaks as seen in the Grants' study.
4. **Describe** how speciation takes place.

Critical Thinking

5. **SKILL Evaluating Results** Based on the results of David Lack's study and the Grants' study of finches, what conclusion can you make about the length of time required for evolution of a new species to take place?

Review SECTION 13-3 ANSWERS

1. The five elements of natural selection are genetic variation, environmental challenges to reproduction, overproduction of offspring and a struggle for survival, survival of the fittest, and rise in the number of individuals with characteristics suited to the environment.
2. The moths had genetic variations that enabled some to be successful in an unpolluted environment and others in a polluted environment.
3. The changes in the finches' beaks were caused by climate changes that affected the food sources that the birds ate.
4. As populations of a species spread throughout an environment, they are exposed to varying conditions (environmental pressures). Over time, the separate populations become distinct and split into ecological races. Gradually, they can become so different from one another that they may become separate species.
5. It would take at least several years, but this depends on the life span of the organism and how long it takes to reach reproductive maturity.

Study ZONE

CHAPTER 13

Use the Key Concepts and Key Terms listed below to review the main ideas in this chapter. If you need to review the meaning of a term, turn to the page indicated.

Key Concepts

13-1 The Theory of Evolution by Natural Selection

- Charles Darwin concluded that animals on the coast of South America that resembled those on the nearby islands evolved differences after separating from a common ancestor.
- Darwin was influenced by Thomas Malthus, who wrote that populations tend to grow as much as the environment allows.
- Darwin proposed that natural selection favors individuals that are best able to survive and reproduce.
- Change within a species eventually leads to new species.

13-2 Evidence of Evolution

- Evidence of orderly change can be seen when fossils are arranged according to their age.
- Differences in amino acid sequences and DNA sequences are greater between species that are more distantly related than between species that are more closely related.
- The presence of homologous structures and vestigial structures suggests that all vertebrates share a common ancestor.
- Gradualism is a process of evolution in which speciation occurs gradually, and punctuated equilibrium is a process in which speciation occurs rapidly between periods of little or no change.

13-3 Examples of Evolution

- Individuals that have traits that enable them to survive in a given environment can reproduce and pass those traits to their offspring.
- Experiments show that evolution through natural selection has occurred within populations of the European peppered moth and in Darwin's finches.
- Speciation begins as a population adapts to its environment.
- Reproductive isolation keeps newly forming species from breeding with one another.

Key Terms

13-1

population (279)
natural selection (279)
adaptation (280)
isolation (281)
extinct (282)

13-2

paleontologist (285)
vestigial structure (287)
homologous structure (287)
gradualism (289)
punctuated equilibrium (289)

13-3

industrial melanism (290)
divergence (293)
speciation (293)
ecological race (293)
reproductive isolation (294)

Study TIP Ready?

- *If you think you understand these Key Concepts and Key Terms, then you're ready for the Performance Zone.*

Study ZONE

Review and Practice

Assign the following worksheets for Chapter 13:

- **Student Review Guide**
- **Concept Mapping Worksheet**
- **Test Preparation Pretest**
- **Vocabulary Worksheet**
- **Critical Thinking Review Worksheet**
- **Science Skills Worksheet**

Assessment Options

The following assessment options are available for this chapter:

Resource Link

- Assign **Chapter Test 13** to assess students' understanding of the chapter.

Alternative Assessment

Have students list the key terms used in this chapter and a brief definition of each term. Have them construct crossword puzzles using the definitions as the clues. PORTFOLIO

Portfolio Assessment

- *Portfolio Assessment* at the front of this book provides options for building and evaluating student portfolios. Students and teachers can select items from the areas listed for inclusion in a portfolio.
- See items labeled PORTFOLIO on pp. 279, 282, 288, 293, and 295.

Answer to Concept Map

The following is one possible answer to Performance Zone item 1 on page 296.

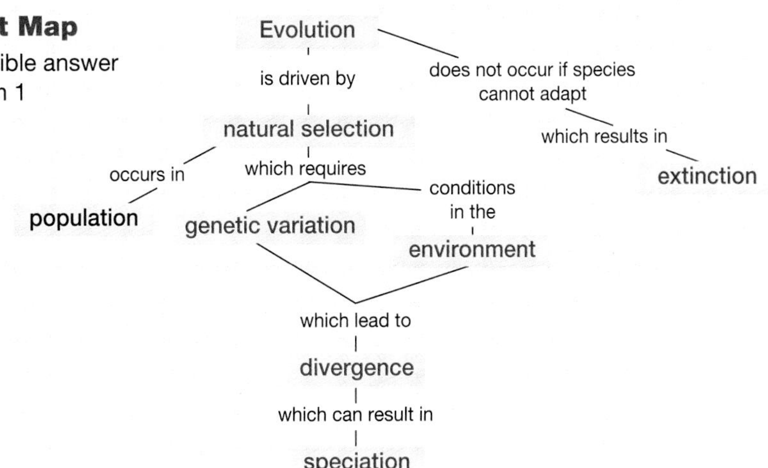

Performance ZONE
CHAPTER 13 REVIEW ANSWERS

Assignment Guide

SECTION	REVIEW
13-1	1, 4, 11, 14, 15, 16
13-2	3, 5, 9, 10, 12, 13, 17, 18, 21
13-3	2, 6, 7, 8, 19, 20

Understanding and Applying Concepts

1. The answer to the concept map is found at the bottom of page 295.
2. a. An adaptation is a change in a characteristic of an organism that enables it to better survive in its environment; natural selection is the process by which populations change in response to their environment.
   b. Extinction refers to the death of all members of a species; isolation refers to the separation of populations of the same species and their resulting inability to interbreed.
   c. Populations are groups of individuals of the same species living in the same area; ecological races are populations of the same species that differ genetically because of adaptations to different conditions.
   d. The word homologous refers to structures that share a common ancestry. The word vestigial refers to structures that are reduced in size and have no function or a less important function than they do in other, related organisms.
   e. Divergence is the accumulation of differences between groups; speciation is the process by which new species form.
3. b
4. b
5. b
6. a
7. d
8. a
9. It indicates that these animals probably share a common ancestor.

Performance ZONE CHAPTER 13 Review

Understanding and Applying Concepts

1. **Concept Mapping** Make a concept map that shows how natural selection leads to speciation. Try to include the following terms in your map: evolution, natural selection, genetic variation, environment, speciation, divergence, extinction. Include additional terms in your map as needed.

2. **Understanding Vocabulary** For each pair of terms, explain the differences in their meanings.
   a. adaptation, natural selection
   b. extinction, isolation
   c. population, ecological races
   d. homologous, vestigial
   e. divergence, speciation

3. According to the modern theory of evolution,
   a. Lamarck was completely wrong.
   b. random gene mutation is a component of evolution.
   c. punctuated equilibrium has replaced natural selection.
   d. the diversity of life-forms resulted from the inheritance of acquired characteristics.

4. The process by which a species becomes better adapted to its environment is called
   a. gradualism.
   b. adaptation.
   c. natural selection.
   d. isolation.

5. Anatomical structures that share a common ancestry are called _______ structures.
   a. vestigial
   b. homologous
   c. analogous
   d. evolutionary

6. Kettlewell's experiments proved
   a. the concealment hypothesis.
   b. that light-colored moths favor industrial areas.
   c. that birds prefer to eat dark-colored moths.
   d. that the theory of evolution by natural selection was wrong.

7. In Lack's study, the effect of weather on the size of the finch's beak is an example of
   a. industrial melanism.
   b. isolation.
   c. fossilization.
   d. natural selection.

8. The process by which isolated populations of the same species become new species is called
   a. speciation.
   b. industrial melanism.
   c. genetic variation.
   d. natural selection.

9. **Theme Evolution** Adult lobsters and barnacles look very different, as shown below. However, the larvae of barnacles and lobsters are practically identical. What does this indicate about the evolutionary history of these organisms?

10. **Exploring Further** Other than punctuated equilibrium, what naturally occurring phenomena might explain large gaps in the fossil record?

11. **Chapter Links** How is meiosis beneficial to the evolution of a species by natural selection? (**Hint:** See Chapter 7, Section 7-1.)

10. erosion or other destructive geological processes
11. Meiosis is beneficial to the evolution of a species by natural selection because it provides a source of genetic variation.

Skills Assessment

12. b
13. The birds; the length of the branch between the birds and their recent common ancestor is shorter than between armadillos and their recent common ancestor. This shorter length indicates that less time has passed, and therefore that the birds diverged more recently than the armadillos did.
14. Answers will vary.
15. Answers will vary. Students should note that Wallace collected insects on an 1848 expedition to the Amazon. He also made observations in the Malay Archipelago between 1854 and 1862. Wallace discovered that animals on the western islands of the Malay archipelago differed sharply from those on the eastern islands. This demarcation, which was vital to his understanding of evolution, is now known as Wallace's line.

Skills Assessment

Part 1: Interpreting Graphics

Study the phylogenetic tree below showing the relationships between finches and armadillos. Then answer the following questions:

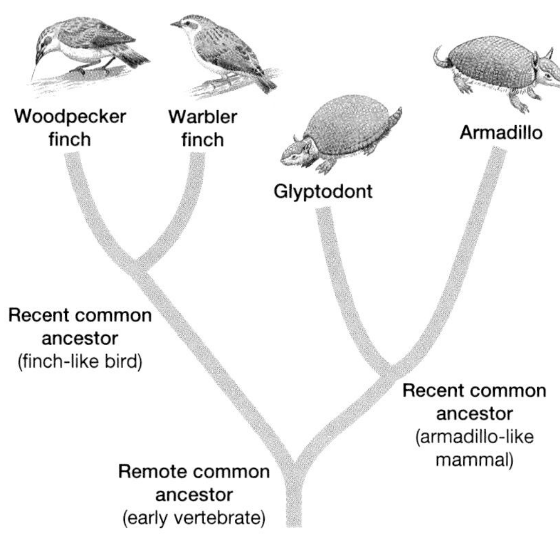

12. The phylogenetic tree implies that modern finches and armadillos
 a. are unrelated.
 b. share a remote common ancestor.
 c. share a recent common ancestor.
 d. did not evolve from older forms of life.

13. Which group is most recently diverged from its recent common ancestor, the birds or the armadillo? Explain your answer.

Part 2: Life/Work Skills

Use the media center or Internet resources to learn about the biological data that Darwin and Wallace used to draw their conclusions about natural selection and evolution.

14. **Being a Team Member** With two to three other students, locate and examine photographs and drawings of the tortoises that Darwin observed on the Galápagos Islands. Plan and produce a mural showing the tortoises in their natural environment.

15. **Communicating** Prepare an oral report on the biological research of Alfred Russel Wallace. Present your findings to your class.

Critical Thinking

16. **Evaluating** Explain why the relationship between evolution and natural selection is so important.

17. **Justifying Conclusions** About 40 years after the publication of *On the Origin of Species,* genetics was recognized as a science. At that time, support for Darwin's theory of evolution by natural selection increased among scientists. Explain how genetics supports Darwin's theory.

18. **Evaluating an Argument** Use the data in the table to evaluate the claim that Species A is more closely related to humans than is Species C.

Number of Amino Acid Differences Compared With Human Hemoglobin	
Humans	0
Species A	8
Species B	17
Species C	39

Portfolio Projects

19. **Identifying Variables** Take a walk in the woods or in a park. Prepare a report identifying various forces you see that could be influencing the evolution of species. Explain how these forces can influence evolution of species. Include weather and the influence of other organisms in the report.

20. **Interpreting Information** Read the article "Evolving Backward" by Jarrod Diamond (*Discover,* September, 1998, pp. 64–69). Write a report describing the evolutionary pressures that can cause a species to lose some characteristics.

21. **Career Focus Paleontologist** Research the field of paleontology, and write a report on your findings. Your report should include a job description, training required, kinds of employers, growth prospects, and starting salary.

Critical Thinking

16. The relationship between evolution and natural selection is so important because the pace and nature by which populations change over time, or evolve, is largely driven by natural selection. That is, the natural selection of traits within a population determines which traits are favored and the rate at which they spread.

17. Genetics reveals how traits are passed from one generation to another. This understanding supports Darwin's theory by explaining how offspring can have characteristics found in their parents, whether or not such characteristics are expressed. Genetics also supports Darwin's theory by providing tools for measuring how closely organisms are related to one another.

18. Species A has fewer amino acid differences compared with human hemoglobin than does species C. Amino acids are coded for by DNA. Therefore, differences in amino acids indicate differences in DNA. DNA differences are caused by mutations. Because mutations occur randomly over evolutionary time, organisms with few differences in their DNA are more closely related than those with many differences (that is, they diverged from a common ancestor more recently). Therefore, species A is more closely related to humans than species C is.

Portfolio Projects

19. Answers will vary. In general, students should note the local environmental forces that are likely to exert evolutionary pressure. Examples include the shade in a dense forest, the heat in a desert, and the impact of human disturbance in a park.

20. Answers will vary, but should mention that evolutionary pressures apparently result in the loss of unused characteristics because they cost the organism energy to produce and thus are not adaptive.

21. Answers will vary. Paleontologists are scientists who study fossils and other remains of past life. They are concerned with all aspects of ancient life, including the environments that existed at the time. Paleontologists usually have at least an undergraduate degree in zoology and/or geology, including training in chemistry and physics. University and museum jobs usually require a Ph.D. Most paleontologists are employed by universities, museums, or large oil and construction companies. Growth potential for this field is fair. Starting salary will vary by region.

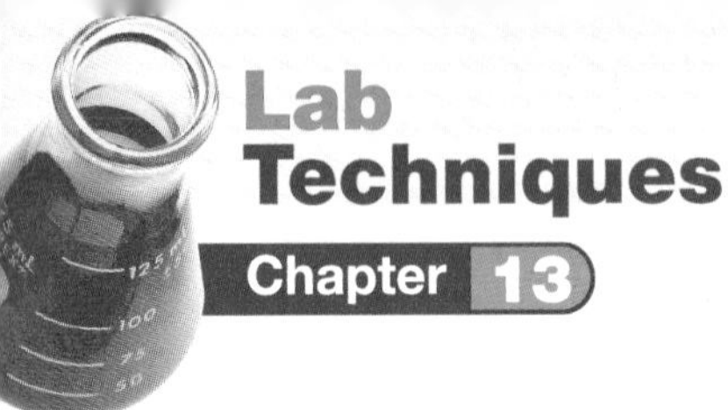

Lab Techniques
Chapter 13

Modeling Natural Selection

Purpose
Students will model the selection of favorable traits in a new generation using a paper model of a bird as their study species.

Preparation Notes
Time Required: one 50-minute class period

Materials
Materials for this lab can be ordered from WARD'S. See *Master Materials List* at the front of this book for catalog numbers.

Preparation Tips
Students will simulate the breeding of several generations of the birds and observe the effect of various phenotypes on the evolutionary success of these animals. The random nature of mutations is demonstrated by randomly changing the anterior and posterior wing position and wing circumference of the birds. Each student group will need one sheet of construction paper, a meter stick, a ruler, a pair of scissors, a straw, cellophane tape, a coin, and a die.

Safety Precautions
Remind students to be careful with scissors and always cut in a direction away from the face and body.

Procedural Tips
1. Have students work in pairs or small groups.
2. To save time, have students create their tables before beginning the lab.
3. You may wish to review and clarify the principles of selection. This investigation reinforces the concept of natural selection as differential *reproduction* rather than merely differential survival.

Disposal
Have students dispose of their paper scraps and paper birds in the trash.

Lab Techniques
Chapter 13 Modeling Natural Selection

SKILLS
- Modeling a process
- Inferring relationships

OBJECTIVES
- **Model** the process of selection.
- **Relate** favorable mutations to selection and evolution.

MATERIALS
- scissors
- construction paper
- cellophane tape
- soda straws
- felt-tip marker
- meterstick or tape measure
- penny or other coin
- six-sided die

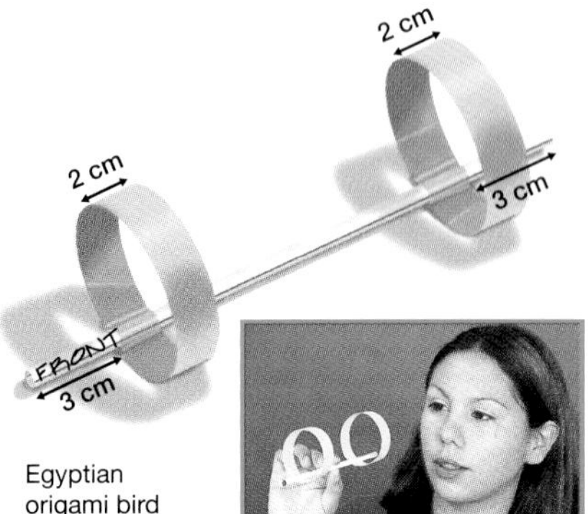

Egyptian origami bird

Before You Begin
Natural selection occurs because some organisms have **traits** that improve their ability to survive and produce offspring. A population evolves when individuals with different **genotypes** survive or reproduce at different rates. In this lab, you will model the selection of favorable traits in a new generation by using a paper model of a bird—the fictitious Egyptian origami bird *(Avis papyrus)*, which lives in dry regions of North Africa. Assume that only birds that can successfully fly the long distances between water sources will live long enough to breed successfully.

1. Write a definition for each boldface term in the preceding paragraph.
2. Make a data table similar to the one shown below.
3. Based on the objectives for this lab, write a question you would like to explore about the process of selection.

Procedure
PART A: Parental Generation
1. Cut two strips of paper, 2 × 20 cm each. Make a loop with one strip of paper, letting the paper overlap by 1 cm, and tape the loop closed. Repeat for the other strip.

DATA TABLE

Bird	Coin flip (H or T)	Die throw (1–6)	Anterior wing (cm)			Posterior wing (cm)			Average distance flown (m)
			Width	Circum.	Distance from front	Width	Circum.	Distance from back	
Parent	NA	NA	2	19	3	2	19	3	
Generation 1									
Chick 1									
Chick 2									
Chick 3									
Generation 2									
Chick 1									
Chick 2									
Chick 3									

Before You Begin
Answers
1. *natural selection*—process by which populations change in response to their environment as individuals better adapted to the environment leave more offspring
   traits—distinguishing characteristics
   genotype—the genetic constitution of an organism as indicated by its set of alleles
3. Answers will vary. For example, Are the effects of natural selection obvious in only a few generations?

2. Tape one loop 3 cm from each end of the straw, as shown above. Mark the front end of the bird with a felt-tip marker. This bird represents the parental generation.
3. Test how far your parent bird can fly by releasing it with a gentle overhand pitch. Test the bird twice. Record the bird's average flight distance in your data table.

PART B: First (F_1) Generation

4. Each origami bird lays a clutch of three eggs. Assume that one of the chicks is a clone of the parent. Use the parent to represent this chick in step 6.
5. Make two more chicks. Assume that these chicks have mutations. Follow Steps A–C below for each chick to determine the effects of its mutation.

   Step A Flip a coin to determine which end is affected by a mutation.

   Heads = anterior (front)

   Tails = posterior (back)

   Step B Throw a die to determine how the mutation affects the wing.

   - ⚀ = Wing position moves 1 cm toward the end of the straw.
   - ⚁ = Wing position moves 1 cm toward the middle of the straw.
   - ⚂ = Wing circumference increases by 2 cm.
   - ⚃ = Wing circumference decreases by 2 cm.
   - ⚄ = Wing width increases by 1 cm.
   - ⚅ = Wing width decreases by 1 cm.

   Step C A mutation is lethal if it causes a wing to fall off the straw or a wing with a circumference smaller than that of the straw. If you get a lethal mutation, disregard it and produce another chick.
6. Record the mutations and the wing dimensions of each offspring.
7. Test each bird twice by releasing it with a gentle overhand pitch. Release the birds as uniformly as possible. Record the distance each bird flies. The most successful bird is the one that flies the farthest.

PART C: Subsequent Generations

8. Assume that the most successful bird in the previous generation is the sole parent of the next generation. Repeat steps 4–7 using this bird.
9. Continue to breed, test, and record data for eight more generations.

PART D: Cleanup and Disposal

10. Dispose of paper scraps in the designated waste container.
11. Clean up your work area and all lab equipment. Return lab equipment to its proper place. Wash your hands thoroughly before you leave the lab and after you finish all work.

Analyze and Conclude

1. **Analyzing Results** Did the birds you made by modeling natural selection fly farther than the first bird you made?
2. **Inferring Conclusions** How might this lab help explain the variety of species of Galápagos finches?
3. **Further Inquiry** Write another question about natural selection that could be explored with another investigation.

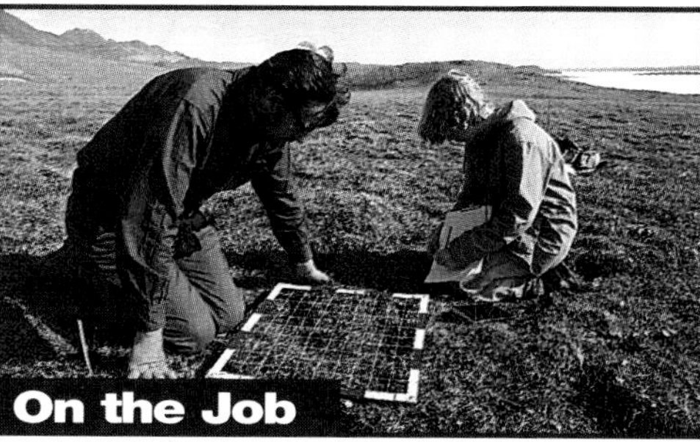

On the Job

Population biology is the study of populations. Do research in the library or media center or on the Internet to discover how population biologists study evolution. Summarize your findings in a written or oral report.

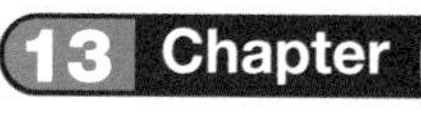

Lab Techniques

13 Chapter

Procedure

Answers

3. Answers will vary.
6. Answers will vary.
7. Answers will vary.
8. Answers will vary.
9. Answers will vary.

Analyze and Conclude

Answers

1. Most students should answer "yes" as the best-flying birds are selected as the sole parents of the next generation.
2. This lab demonstrates that organisms can change significantly over only a few generations. The lab therefore shows how isolated populations could diverge to the point that they constitute different species, as happened to finches on the Galápagos Islands.
3. Answers will vary. For example, Does natural selection act on one trait at a time, or can selection pressure affect the evolution of several traits at once?

On the Job

Answers

Population biologists study evolution by examining changes in population size over time. Techniques population biologists use to study evolution include population sampling, computer simulations, and various statistical techniques for analyzing data.

Chapter 14

Human Evolution

Planning Guide

OBJECTIVES	MEETING INDIVIDUAL NEEDS	READING SKILLS AND STRATEGIES
14-1 The Evolution of Primates (BLOCK 1 45 min) • **Describe** the mammal that gave rise to the first primates. • **List** two unique features of primates. • **Contrast** prosimians with monkeys. • **Distinguish** monkeys from apes • **Describe** the evolutionary relationship between humans and apes.	Learning Styles, p. 302 BASIC ADVANCED • pp. 304, 305	**ATE** Active Reading, Technique: Brainstorming, p. 301 **ATE** Active Reading, Technique: Reading Organizer, p. 304 ■ Active Reading Guide, worksheet (14-1) ■ Directed Reading, worksheet (14-1)
14-2 Early Hominids (BLOCKS 2 & 3 90 min) • **Contrast** apes with australopithecines. • **Identify** the evidence that indicates human ancestors walked upright before their brains enlarged. • **Describe** why Dart's discovery was so important in supporting the theory of evolution. • **Name** the earliest hominid fossil and where it was found.	Learning Styles, p. 306 BASIC ADVANCED • pp. 307, 308	■ Active Reading Guide, worksheet (14-2) ■ Directed Reading, worksheet (14-2)
14-3 The Genus Homo (BLOCKS 4 & 5 90 min) • **Compare** *Homo habilis* with australopithecines. • **Describe** the characteristics of *Homo erectus.* • **Outline** two hypotheses that explain the origin of *Homo sapiens.* • **Compare** Neanderthals with modern humans.	Learning Styles, p. 309 BASIC ADVANCED • pp. 311, 312	■ Active Reading Guide, worksheet (14-3) ■ Directed Reading, worksheet (14-3) ■ Supplemental Reading Guide, *Origin of Species*

Review and Assessment Resources

BLOCKS 6 & 7 90 min

TEST PREP

- **PE** Study Zone, p. 313
- **PE** Performance Zone, pp. 314–315
- ■ Chapter 14 worksheets:
  - • Vocabulary • Concept Mapping
  - • Science Skills • Critical Thinking
- Audio CD Program
- ■ Test Prep Pretest

CHAPTER TESTING

- ■ Chapter Tests (blackline master)
- ▲ Test Generator
- ■ Assessment Item Listing

See the correlations table at the front of this book for a list of the ***National Science Education Standards*** addressed in this chapter.

Smithsonian Institution® Visit **www.si.edu/hrw** for additional on-line resources.

KEY

Biology Principles and Explorations

PE Pupil's Edition
ATE Teacher's Edition

Teaching Transparencies
Laboratory Program

One-Stop Planner CD-ROM
Shows these resources within a customizable daily lesson plan:

■ Teaching Resources
▲ Test Generator

VISUAL STRATEGIES	CLASSWORK AND HOMEWORK	TECHNOLOGY AND INTERNET RESOURCES
ATE Fig. 14-4, p. 305 53 Evolutionary Relationships of Anthropoids 54 Comparison of Chimpanzee and Human Jaws 59 Cladogram of Seven Vertebrates	**PE** Quick Lab How are binocular vision and the ability to grasp advantages?, p. 303 Lab A11 Comparing Primate Features **PE** Review, p. 305	Holt Biology Videodiscs Section 14-1 Audio CD Program Section 14-1 internetconnect SCILINKS NSTA TOPICS: • Prosimian • Phylogenetic tree
ATE Fig. 14-7, p. 308 55 Comparison of Gorilla and Australopithecine Skeletons 62 Two Hypotheses on the Evolution of Hominids	Lab C18 Analyzing Blood Serum—Evolution of Primates **PE** Review, p. 308	Holt Biology Videodiscs Section 14-2 Audio CD Program Section 14-2
	PE Data Lab Analyzing hominid evolution, p. 310 **PE** Lab Techniques Comparing Hominid Skulls, pp. 316–317 Lab C19 Analyzing Amino Acid Sequences to Determine Evolutionary Relationships **PE** Review, p. 312 **PE** Study Zone, p. 313	Holt Biology Videodiscs Section 14-3 Audio CD Program Section 14-3 internetconnect SCILINKS NSTA TOPIC: Evolution

ALTERNATIVE ASSESSMENT

PE Portfolio Projects 23–25, p. 315
ATE Alternative Assessment, p. 313
ATE Portfolio Assessment, p. 313

Lab Activity Materials

Quick Lab
p. 303 masking tape, unsharpened pencils (2)

Data Lab
p. 310 pencil, paper

Lab Techniques
p. 316 metric ruler, protractor

Lesson Warm-up
ATE butcher paper, tape, markers, p. 309

Demonstrations
ATE large, clear bottles or jars (8), labeling marker, water, p. 310

BIOSOURCES LAB PROGRAM
- Quick Labs: Lab A11, p. 21
- Laboratory Techniques and Experimental Design: Lab C18, p. 83
- Laboratory Techniques and Experimental Design: Lab C19, p. 87

Chapter 14

Chapter Theme

Evolution

The first primates were small mammals that lived in trees and had grasping hands and binocular vision. Monkeys were the first primates to evolve opposable thumbs. The apes, which lack tails, evolved later. Next some primates evolved upright posture. We call these primates hominids. Hominids have large brains and can use tools. Modern humans have been a very successful species largely because human cultures evolve to meet our environmental needs.

Establishing Prior Knowledge

Before beginning this chapter, make certain students can answer the questions in Study TIP **Ready?** on page 301. If they cannot, encourage them to reread the sections indicated.

Opening Question

To stimulate student thinking, ask them to list the characteristics that make them different from other primates. Have them save their list of answers and revisit them at the end of the chapter to see if their ideas have changed. *(Their revised answers should include: have large brains, walk upright, use fire, make and use complex tools, speak, write symbolic language, and possess culture.)*

CHAPTER 14 Human Evolution

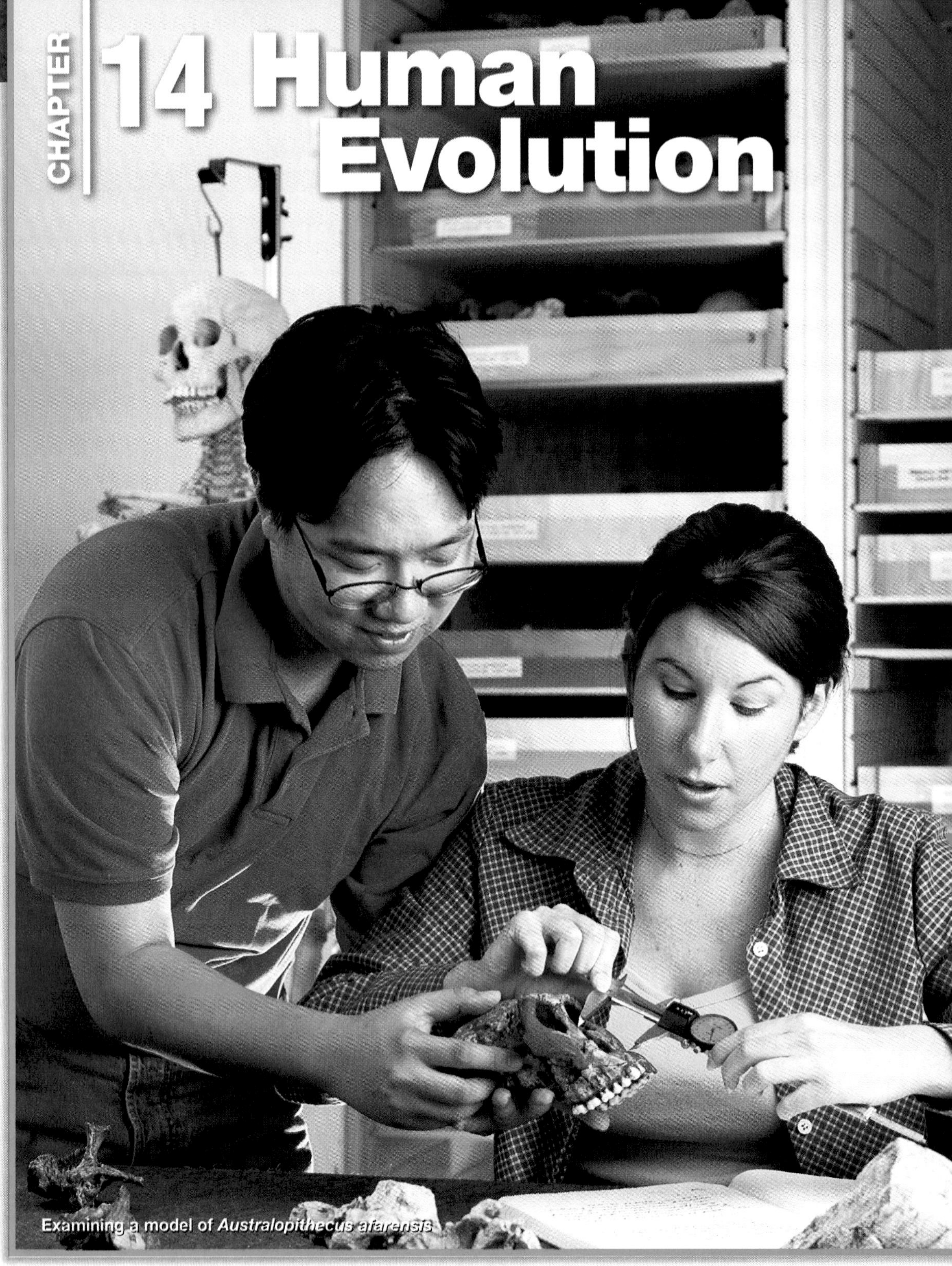

Examining a model of *Australopithecus afarensis*

What's Ahead — and Why

CHAPTER OUTLINE

14-1 The Evolution of Primates

- Primate Ancestors Evolved During the Age of Dinosaurs
- The First Primates Were Prosimians
- Primates Evolved Specialized Traits

14-2 Early Hominids

- Primates Evolve Upright Walking

14-3 The Genus *Homo*

- *Homo habilis* Is the Oldest Member of the Genus *Homo*
- The First Hominids That Left Africa
- The Species *Homo sapiens* Arose Fairly Recently

Students will learn

- 14-1: about the characteristics of early primates and about primate characteristics that evolved more recently.
- 14-2: about the differences between early hominids and nonhominid primates and about the hominid family tree.
- 14-3: about the characteristics of early humans and about the differences between early humans and modern humans.

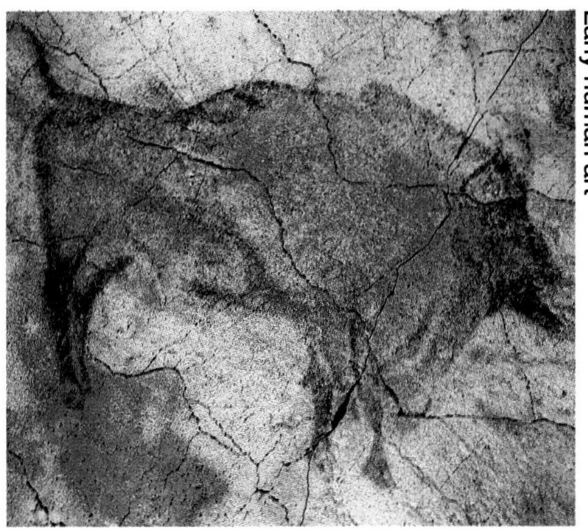
Early human art

Millions of years *of evolution can change the structure of a species considerably. Fossils, such as the models of early prehuman fossils these students are examining, indicate that over millions of years, the process of natural selection gave rise to modern humans.*

Looking Ahead

Labs

- **Quick Lab**
  How are binocular vision and the ability to grasp advantages? (p. 303)
- **Data Lab**
  Analyzing hominid evolution (p. 310)
- **Lab Techniques**
  Comparing Hominid Skulls (p. 316)

Features

- **Tech Watch**
  Human History in Our Cells (p. 311)

Study TIP
Ready?

Check your understanding of these topics to help you prepare for what's ahead.

Can you . . .

- **describe** protein structure? (Section 2-3)
- **describe** DNA structure? (Section 9-2)
- **summarize** the function of genes? (Section 10-1)
- **recognize** the importance of phylogenetic relationships? (Section 13-2)
- **define** the theory of natural selection? (Section 13-3)

If not, *review the sections indicated.*

internet**connect**

sciLINKS NSTA — National Science Teachers Association *sci*LINKS Internet resources are located throughout this chapter.

Chapter 14

Reading Strategies

Active Reading

Technique: Brainstorming

Before students begin reading, explain that the focus of this chapter is the evolution of humans. Work with the class to create a list of those characteristics that humans share with other living primates and those characteristics unique to humans. Incorporate amendments that students suggest as they read the chapter.

Directed Reading

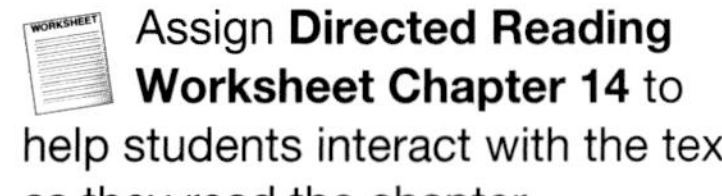

Assign **Directed Reading Worksheet Chapter 14** to help students interact with the text as they read the chapter.

Interactive Reading

Assign ***Biology: Principles and Explorations*** **Audio CD Program Chapter 14** to help reluctant readers achieve greater success in reading the chapter.

Reading Skills

Assign Reading Organizers and Reading Strategies Worksheets from the **One-Stop Planner CD-ROM with Test Generator** to help students learn and practice effective reading skills.

internet**connect**

Holt *Biology: Principles and Explorations*

On-line Resources: go.hrw.com

Visit the HRW Web site for a variety of resources related to this chapter. Just type in the keyword HX0 Chapter 14.

Holt *Biology: Principles and Explorations*

On-line Resources: www.scilinks.org

The following *sci*LINKS Internet resources can be found in the student text for this chapter:

Page 303	Page 305	Page 311
TOPIC: Prosimian	**TOPIC:** Phylogenetic tree	**TOPIC:** Evolution
KEYWORD: HX303	**KEYWORD:** HX305	**KEYWORD:** HX311

1 Prepare

Scheduling

- **Block:** About 45 minutes will be needed to complete this section.
- **Traditional:** About one class period will be needed to complete this section.
- **Planning:** See the Planning Guide on pp. 299A and 299B for lecture, classwork, and assignment options for Section 14-1. Lesson plans for Section 14-1 can be found in a lesson cycle format in ***Biology: Principles and Explorations*** **One-Stop Planner CD-ROM with Test Generator.**

Resource Link

Assign **Active Reading Worksheet Section 14-1** to help students comprehend the key concepts and visuals in this lesson.

2 Teach

Lesson Warm-up

Have students demonstrate for themselves the importance of forward-facing eyes for depth perception. Have students try to touch things at arm's length with only one eye open, then with both eyes open. Tell them that each eye sends a different view of the world to the brain, which then uses this information to judge how close things are.

Section 14-1 The Evolution of Primates

Objectives

- **Describe** the mammal that gave rise to the first primates. (p. 302)
- **List** two unique features of primates. (p. 302)
- **Contrast** prosimians with monkeys. (pp. 303–304)
- **Distinguish** monkeys from apes. (p. 305)
- **Describe** the evolutionary relationship between humans and apes. (p. 305)

Key Terms

primate
prosimian
diurnal
anthropoid
opposable thumb

Primate Ancestors Evolved During the Age of Dinosaurs

Some early science fiction movies showed "cave men" running for their lives from hungry dinosaurs. But this could not have happened because humans did not evolve until long after the dinosaurs became extinct. Very early mammals did live during the age of dinosaurs, however.

In 1871, 12 years after *The Origin of Species,* Charles Darwin published *The Descent of Man.* He proposed that humans, gorillas, and chimpanzees all evolved from a common ancestor. Although the fossil record of human origins is not complete, many fossil finds since Darwin's time have strongly confirmed his hypothesis.

Fossil evidence indicates that a small insect-eating mammal with big eyes and tiny, sharp teeth, resembling a shrew, as shown in **Figure 14-1,** lived about 80 million years ago, during the age of dinosaurs. These ancient mammals were the ancestors of the first **primates,** the mammalian group that includes prosimians, monkeys, apes, and humans.

The first primates evolved about 50 to 60 million years ago. These animals had two features that enabled them to stalk and capture insect prey in the branches of trees.

1. **Grasping hands and feet** Unlike the clawed, unbendable toes of their ancestors, primates have grasping hands and feet that enable them to cling to their mothers when they are young, grip limbs, hang from branches, and seize food.
2. **Forward placement of the eyes** Unlike the eyes of their ancestors, which were located on the sides of the head, the eyes of primates are positioned at the front of the face. This forward placement of the eyes produces overlapping "binocular vision" that enables the primate brain to judge distance more precisely (depth perception).

Depth perception is a very important ability for an animal that leaps from branch to branch high above the ground. Other mammals have binocular vision, but only primates have both binocular vision and grasping hands. The abilities to judge distances and manipulate objects are thought to have been very important in the evolution of increased intelligence in primates.

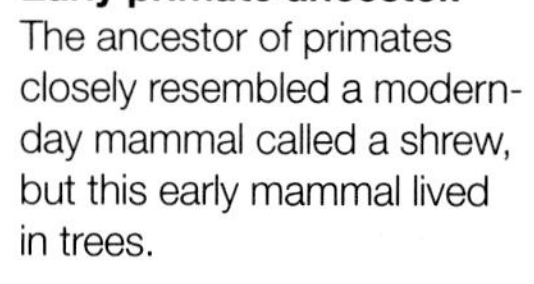

Figure 14-1
Early primate ancestor. The ancestor of primates closely resembled a modern-day mammal called a shrew, but this early mammal lived in trees.

Integrating Different Learning Styles

Visual/Spatial	**PE**	Visual Strategy. p. 305
	ATE	Lesson Warm-up, p. 302; Teaching Tip, p. 304
Body/Kinesthetic	**PE**	Quick Lab, p. 303
Verbal/Linguistic	**PE**	Study Tip, p. 304; Performance Zone item 12, p. 314
	ATE	Active Reading, p. 304

The First Primates Were Prosimians

According to the fossil record, the group of modern primates that most closely resembles early primates is the prosimians. A **prosimian** is a member of a group of mostly night-active primates that live in trees. Modern prosimians include lorises, lemurs, and tarsiers, such as the one shown in **Figure 14-2.** Fossil evidence indicates that prosimians were common about 38 million years ago in North America, Europe, Asia, and Africa.

Figure 14-2 Prosimian. The tarsier, *left*, has bendable, clawed fingers and toes and forward-facing eyes, which are key adaptations for life in the trees.

Only a few kinds of prosimians survive today, and their present range is very limited. A lemur is about the size of a house cat, and it has a long tail that it uses for balance as it climbs through the trees. But lemurs are in great danger today, and they may soon become extinct in the wild. The forest homes of these animals are being destroyed rapidly by activities associated with an expanding human population. And as the lemurs' forest homes disappear, so does perhaps the oldest living link to our ancestors.

internet**connect**

SCILINKS NSTA

TOPIC: Prosimian
GO TO: www.scilinks.org
KEYWORD: HX303

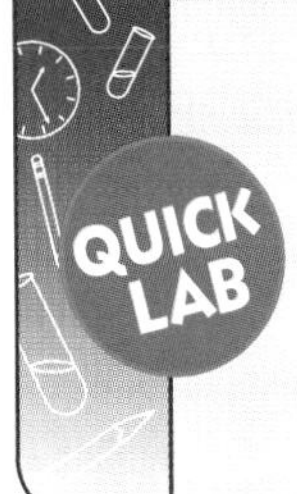

QUICK LAB

How are binocular vision and the ability to grasp advantages?

By temporarily giving up your binocular vision and ability to grasp objects, you can understand how having these abilities is an advantage to survival.

Materials

masking tape, 2 pencils (unsharpened)

Procedure

Binocular vision

1. Have your partner hold two pencils at different distances in front of you so that both pencils can be plainly seen.
2. With both eyes open, try to grab the pencil that is farthest from you.
3. Repeat steps 1 and 2 several times. Have your partner change the pencils' distance with each trial.
4. Repeat steps 1–3 with one eye closed.

Grasping ability

5. Practice picking up items such as books, paper, and pencils using your dominant hand. Your dominant hand is the one that you usually write with.
6. Have your partner tape the thumb of your dominant hand so that it cannot move, as shown in the figure above.
7. Repeat step 5.

Analysis

1. **Describe** why grabbing the pencil was more difficult with one of your eyes closed.
2. **SKILL Predicting Outcomes** How would the lack of binocular vision affect the survival of a primate?
3. **SKILL Evaluating Conclusions** Explain how having grasping ability and binocular vision are evolutionary advantages.

did you know?

Primates are unique among mammals in their ability to see so many colors so well.

Although some vertebrates, like birds and fishes, are able to distinguish colors, most mammals are colorblind. For example, bulls cannot see the red of a cape, but instead just a shade of gray.

Section 14-1

CROSS-DISCIPLINARY CONNECTION

◆ **Geography**

Have students locate Madagascar on a world map. Tell them that lemurs only occur in Madagascar and the Comoros islands, which also lie off the coast of Africa.

QUICK LAB

How are binocular vision and the ability to grasp advantages?

Time 20 minutes

Process Skills Comparing, describing, analyzing, predicting, evaluating

Teaching Strategies

Point out to students that objects are in clearest focus with the human eye if they are at approximately arm's length. This suggests that our binocular vision and manual dexterity evolved together and have been advantageous in combination.

Analysis Answers

1. Answers will vary, but students should indicate that they had difficulty judging the distance between their hand and the pencil.
2. The lack of binocular vision would make it difficult or impossible for primates to move safely through treetops and areas with uneven terrain. It would also make it more difficult for meat-eating primates to capture prey.
3. These adaptations endow primates with excellent hand-eye coordination and the ability to perform finely tuned tasks. These adaptations may also have aided the development of increased intelligence in primates.

Teaching TIP BASIC

Look at Your Own Hands

Have students look at their own hands and observe how their fingers bend and how their nails are attached to their fingers. Ask students to name some uses for hands and some uses for paws. *(hands: grasping, climbing, pinching, holding, pushing, pointing, waving, slapping, using tools; paws: running, hopping, digging, climbing, pushing)* Then ask why a dog cannot climb a tree, while a monkey can. *(A dog cannot grasp with its paws, while a monkey can use its hands to grasp. If students mention that cats climb well, point out that cats have sharp, retractable claws that can grasp.)*

Active Reading

Technique: Reading Organizer

Students may need an organizational framework to help them remember the hominids discussed in Sections 14-2 and 14-3. Have students make a chart with these column headings: *Hominid, Discoverer, Place Discovered,* and *Date Discovered.* Tell students to complete as many columns as possible for each of the hominid species mentioned in these sections.
PORTFOLIO

CAREER

Physical Anthropologist

Physical anthropologists, also called biological anthropologists, study humans and other primates as biological organisms. Most of the research on hominid fossils is performed by physical anthropologists. These scientists are so named because they concentrate on the physical aspects of humans, rather than on human culture or cognition.

Study TIP

Word Origins

The word *diurnal* is from the Latin *diurnis,* meaning "daily." Knowing this makes it easier to remember that diurnal primates are active during the day.

Primates Evolved Specialized Traits

About 36 million years ago, a revolution occurred in how primates live. Many primate species became diurnal. **Diurnal** *(deye UR nuhl)* primates are active during the day and sleep at night. The evolution of a diurnal pattern gave primates more opportunities to feed and enabled them to better detect predators. The evolution of daytime vision favored changes in the structure of the eye.

One of these changes was an increase in specialized cells called cone cells. Cone cells are found in the sensory tissue that lines the back of the eye. Cone cells make it possible for the eye to detect color, and an increase in the number of cone cells in the eye sharpens daytime vision. The rise of color vision was associated with the development of a larger, more complex brain. These day-active primates are called monkeys. Monkeys, together with apes and humans, are called **anthropoids.**

Monkeys

Feeding mainly on fruits and leaves rather than on insects, monkeys were among the first primates to have opposable thumbs. An **opposable thumb** stands out at an angle from the other fingers and can be bent inward toward them to hold an object. This gives the hand a greatly increased level of skill at manipulating objects.

Monkeys, such as those shown in **Figure 14-3,** appear to have evolved first in central Africa and spread to Asia. Modern African and Asian species are called Old World monkeys. Scientists think that before continental drift separated Africa from the Americas, some monkeys migrated to Central America and South America, where they evolved in isolation. These monkeys are called New World monkeys.

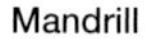
Mandrill

Woolly spider monkey

Figure 14-3 Old World and New World monkeys. Baboons, like the mandrill, are from Africa and are called Old World monkeys. Many New World monkey species like the woolly spider monkey, which is from Brazil, grasp objects with their long, flexible tails.

Overcoming Misconceptions

Ape Size

Many people incorrectly assume that apes are large by definition. Like all organisms, apes are classified according to their evolutionary relationships, not according to their size. Gibbons, the smallest apes, are actually much smaller than many species of monkeys.

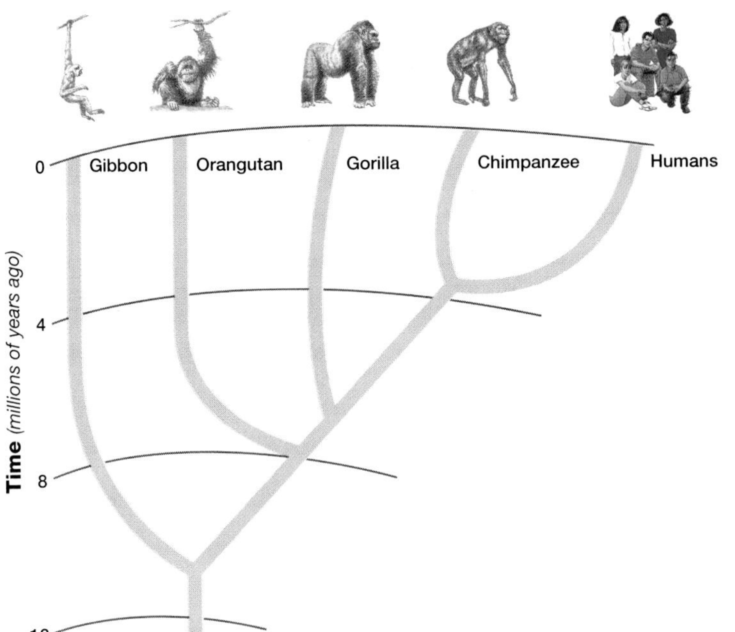

Figure 14-4
Phylogenetic tree of apes and humans. The evolutionary group leading to gibbons diverged from other apes about 10 million years ago. The key split between the gorillas and chimpanzees and the group leading to humans occurred about 5–7 million years ago.

Emergence of the apes

Fossil evidence indicates that humans evolved from the same evolutionary group that gave rise to apes. Apes, which share a common ancestor with monkeys, first appeared about 30 million years ago. Modern apes include gibbons, orangutans, gorillas, and chimpanzees. Apes have larger, more developed brains than monkeys, and none of the apes have tails.

The evolution of apes and humans is illustrated in **Figure 14-4.** The split between the group that gave rise to humans and the group that gave rise to the gorillas and chimpanzees occurred relatively recently. As a result, the genes of humans and chimpanzees have not evolved many differences. For example, human and chimpanzee DNA nucleotide sequences differ by only 1.6 percent. All 287 amino acids that make up two kinds of chains of protein in your hemoglobin are identical to the amino acids in chimpanzee hemoglobin. However, humans and gorillas have two amino acid differences in their hemoglobin.

Review SECTION 14-1

1. **Describe** the ancestors of primates.
2. **Relate** two unique primate characteristics to primate survival.
3. **Describe** one characteristic that separates monkeys from prosimians.
4. **Compare** monkeys with apes.
5. **Identify** and explain the evidence that closely links humans to chimpanzees.

Section 14-1

Visual Strategy

Figure 14-4

Some people think that according to evolutionary theory, humans descended from gorillas or chimpanzees. Use Figure 14-4 to point out that according to modern evolutionary theory, gorillas and chimpanzees share a common apelike ancestor with humans. Modern chimpanzees, gorillas, and other apes followed a different evolutionary path than humans.

3 Assessment Options

Closure

Primate Relationships

Ask students which modern primates are most distantly related to humans. *(prosimians)* Then ask them which anthropoids are most distantly related to humans. *(monkeys)* Finally, ask them which apes are most distantly related to humans. *(gibbons)*

Review

Assign Review—Section 14-1.

Reteaching — BASIC

Have students make a chart with three column headings: *Prosimians, Monkeys,* and *Apes.* Have them write down the names of the members of each category in the appropriate column. Then have them look back through the chapter to check their answers and to add any names they omitted. PORTFOLIO

Review SECTION 14-1 ANSWERS

1. The ancestors of primates were small mammals that lived in trees, ate insects, and resembled modern shrews.
2. Grasping hands enable primates to seize and manipulate food. Binocular vision allows primates to judge distances accurately.
3. Answers will vary. Sample answer: Monkeys have opposable thumbs, while prosimians do not.
4. Apes have larger, more highly developed brains than monkeys. Monkeys have tails, while apes are tailless.
5. The DNA nucleotide sequences of humans and chimpanzees differ by only 1.6 percent. Also, humans and chimps share all 287 amino acids that make up two kinds of hemoglobin protein chains.

Section 14-2 Early Hominids

1 Prepare

Scheduling

- **Block:** About 90 minutes will be needed to complete this section.
- **Traditional:** About two class periods will be needed to complete this section.
- **Planning:** See the Planning Guide on pp. 299A and 299B for lecture, classwork, and assignment options for Section 14-2. Lesson plans for Section 14-2 can be found in a lesson cycle format in ***Biology: Principles and Explorations*** **One-Stop Planner CD-ROM with Test Generator.**

Resource Link

Assign **Active Reading Worksheet Section 14-2** to help students comprehend the key concepts and visuals in this lesson.

2 Teach

Lesson Warm-up

Ask students why it is highly unlikely that the first hominids evolved in North or South America. *(Students should remember that anatomical and biochemical evidence indicates that humans have evolved from the evolutionary line that gave rise to apes. Because apes never lived in North and South America, it is highly unlikely that the hominids originated there.)*

Objectives

- **Contrast** apes with australopithecines. (pp. 306–307)
- **Identify** the evidence that indicates human ancestors walked upright before their brains enlarged. (p. 307)
- **Describe** why Dart's discovery was so important in supporting the theory of evolution. (p. 307)
- **Name** the earliest hominid fossil and where it was found. (p. 308)

Key Terms

hominid
bipedal

Primates Evolve Upright Walking

The early primates probably walked on all four limbs, but what caused our ancestors to stand up and become two-legged? Fifteen million years ago, the world's climate began to cool, and the great forests of Africa were largely replaced by savannas (treeless plains). In response to these environmental changes, human ancestors began to diverge from the group leading to chimpanzees and bonobos about 5 to 7 million years ago. The members of the group that led to the evolution of humans are called hominids. **Hominids** *(HAHM ih nihds)* are primates that can walk upright on two legs.

Australopithecines

Our earliest known direct ancestors belong to the group *Australopithecus.* Fossils indicate that australopithecines *(aw stray loh PIHTH uh seenz)* had two key hominid characteristics. First they were fully **bipedal,** meaning they were able to walk upright on two legs. As Table 14-1 shows, the ape and australopithecine skeletons are different. The structure of the ape skeleton makes walking upright for a long period of time difficult for apes. However, the structure of australopithecine skeletons enabled them to be fully bipedal.

The second hominid characteristic australopithecines exhibited was a large brain, which had a greater volume, relative to body weight, than the brain of an ape. Some australopithecines weighed

Table 14-1 Comparison of Gorilla and Australopithecine Skeletons

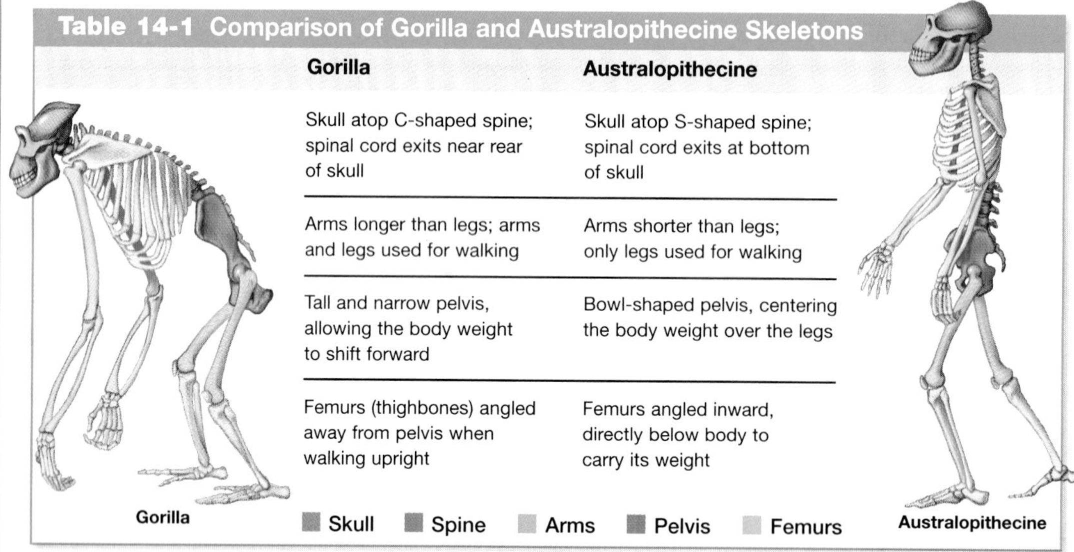

Gorilla	Australopithecine
Skull atop C-shaped spine; spinal cord exits near rear of skull	Skull atop S-shaped spine; spinal cord exits at bottom of skull
Arms longer than legs; arms and legs used for walking	Arms shorter than legs; only legs used for walking
Tall and narrow pelvis, allowing the body weight to shift forward	Bowl-shaped pelvis, centering the body weight over the legs
Femurs (thighbones) angled away from pelvis when walking upright	Femurs angled inward, directly below body to carry its weight

Integrating Different Learning Styles

Logical/Mathematical	**PE**	Review item 4, p. 308; Performance Zone item 21, p. 315
	ATE	Lesson Warm-up, p. 306; Teaching Tip, p. 307
Visual/Spatial	**PE**	Performance Zone item 10, p. 314
	ATE	Visual Strategy, p. 308
Verbal/Linguistic	**PE**	Performance Zone items 2, 15, and 20, pp. 314–315
	ATE	Closure, p. 308

Figure 14-5 **Four fossil australopithecine skulls**

Most species of *Australopithecus* had brains quite a bit larger than those of apes, and their teeth were more like those of humans than of apes.

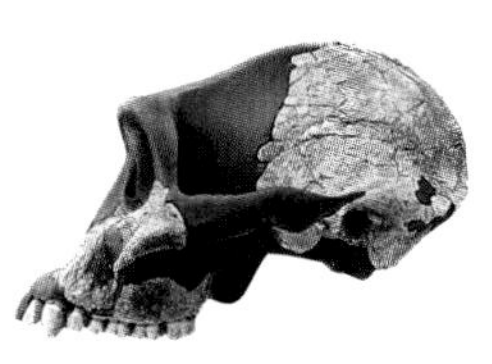
Australopithecus afarensis

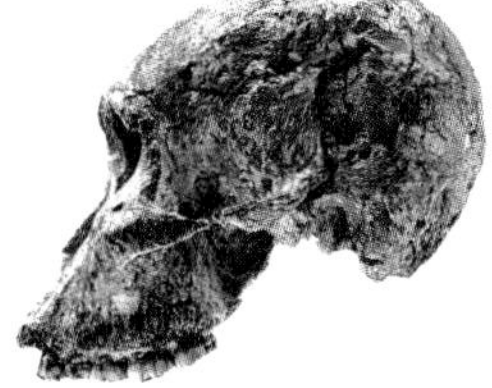
Australopithecus africanus

Australopithecus robustus

Australopithecus boisei

about 18 kg (40 lb) and were approximately 1.1 m (3.5 ft) tall, about the size of a small chimpanzee. Other australopithecine species were larger, weighing more than 45 kg (100 lb) and standing over 1.5 m (5 ft) tall. Their brains, 400 cm^3 in volume, were generally larger than a chimp's and in some species were even larger than a gorilla's. Australopithecine brains were much smaller than your brain, however, which is about 1,350 cm^3 (83 $in.^3$). **Figure 14-5** shows fossil skulls of several different australopithecines.

Discovery of *Australopithecus*

The first australopithecine skull ever found was discovered in 1924 by Raymond Dart, a South African anatomy professor. The fossil that Dart found has a rounded jaw, unlike the pointed jaw of apes, and the brain case is far larger than the brain case of an ape of similar age. Dart noticed that the spinal cord opening was located at the bottom of the skull, which indicated the ability to walk upright. The skull is about 2 million years old.

Dart named his discovery *Australopithecus africanus*. He argued that *Australopithecus africanus* was a bipedal "manlike ape," the long-sought evolutionary link between humans and apes.

Other kinds of australopithecines

In 1938, another *Australopithecus* was found in South Africa. Called *Australopithecus robustus*, it had very large teeth and jaws. Then, in 1959 in East Africa, Mary Leakey discovered *Australopithecus boisei*. It had a great bony ridge on the crest of the head that anchored immense jaw muscles. Like the other australopithecines, *Australopithecus boisei* lived long ago—almost 2 million years ago.

Lucy

In northern Ethiopia in 1974, the American anthropologist Donald Johanson found what is thought to be the most complete and best-preserved skeleton of a prehuman hominid yet discovered. Nicknamed "Lucy," the skeleton is shown in **Figure 14-6**. The skeleton, which was named *Australopithecus afarensis*, is nearly 3 million years old.

Figure 14-6 **Lucy**

Size Lucy's leg bones indicate that she must have walked upright. She stood about 1 m (3 ft) tall.

CROSS-DISCIPLINARY CONNECTION

◆ History

Tell students that most scientists of Dart's day rejected his contention that *Australopithecus africanus* was in the human lineage. One reason was that another very humanlike fossil had been unearthed about a decade before—the famous "Piltdown man," from southeastern England. The skull fragments of Piltdown man indicated a very large brain, but the jaw was similar to that of an ape. In the 1950s, Piltdown man was revealed as a hoax. The skull fragments were from a modern human skull that had been stained to look ancient, the jaw came from an orangutan, and the teeth had been filed to appear less apelike.

Teaching TIP — ADVANCED

Lucy's Place

Have students list the characteristics of Lucy that were apelike and those that were humanlike. Then have students evaluate the two lists and make an argument for or against placing Lucy on the hominid family tree. Be sure they use data from their lists to support or reject their stance.

Teaching TIP — BASIC

Comparison of Primate Adaptations

Have students prepare a **Graphic Organizer** similar to the one at the bottom of page 307. After reading about each type of primate, have students check the appropriate boxes in the table. PORTFOLIO

Graphic Organizer

*Use this graphic organizer with **Teaching Tip: Comparison of Primate Adaptations** on page 307.*

Type of primate	Grasping hands	Binocular vision	Color vision	No tail	Walks upright	Uses tools
Prosimian	✓	✓				
Monkey	✓	✓	✓			
Ape	✓	✓	✓	✓		✓
Hominid	✓	✓	✓	✓	✓	✓

Chapter 14

Visual Strategy

Figure 14-7

Homo erectus and *H. habilis* are thought to be ancestors of modern humans; *Australopithecus robustus* and *Australopithecus boisei* are thought to be offshoots of the hominid family tree. Ask students to identify two physical characteristics that distinguish *A. robustus* and *A. boisei* from humans. *(They had small braincases and massive, apelike jaws.)*

3 Assessment Options

Closure

Interpreting Darwin

Read students the following passage from Darwin's *Descent of Man*. "Thus we have given to man a pedigree of prodigious length . . . . The world . . . appears as if it had long been preparing for the advent of man: and this, in one sense, is strictly true, for he owes his birth to a long line of progenitors. If any single link in this chain had never existed, man would not have been exactly what he is now." Ask students to interpret this passage. *(The evolution of humans can be traced through a long chain to the origin of the first living things. If any ancestors in this chain had not evolved, modern humans would not be the same.)*

Review

Assign Review—Section 14-2.

Reteaching — BASIC

Use a model or picture of a human skeleton to point out the parts of a skeleton that differ between gorillas and hominids. Paraphrase descriptions from Table 14-1, on page 306, and have students determine whether each description applies to gorillas or to hominids.

Study TIP

Word Origins

The word *ardipithecus* is from the Afar (the people who live in the area where this fossil was found) word *ardi*, meaning "ground" or "floor," and from the Greek word *pithecus*, meaning "ape." Knowing this makes it easier to remember that *Ardipithecus* is the genus that represents the oldest hominid.

The oldest hominid

In 1992, another nearly complete fossil skeleton was found in Ethiopia by the American anthropologist Tim White. The fragile skeleton was found in many pieces and is still being assembled. Estimated to be 4.4 million years old, it is the most ancient hominid yet discovered. The skeleton is more apelike than australopithecine and was therefore assigned to a new group, *Ardipithecus*.

In 1995, hominid fossils about 4.2 million years old were found in Kenya by Meave Leakey. While clearly australopithecine, the fossils are intermediate in many ways between *Ardipithecus* and *Australopithecus afarensis*. Named *Australopithecus anamensis*, it represents the oldest known member of the group *Australopithecus*.

Although scientists agree that these fossils represent human ancestors and their relatives, they still disagree about their exact relationship. **Figure 14-7** shows two possible phylogenetic trees.

Figure 14-7 Differing interpretations of human evolution

Because of an incomplete fossil record, scientists differ in their interpretations of how australopithecines evolved.

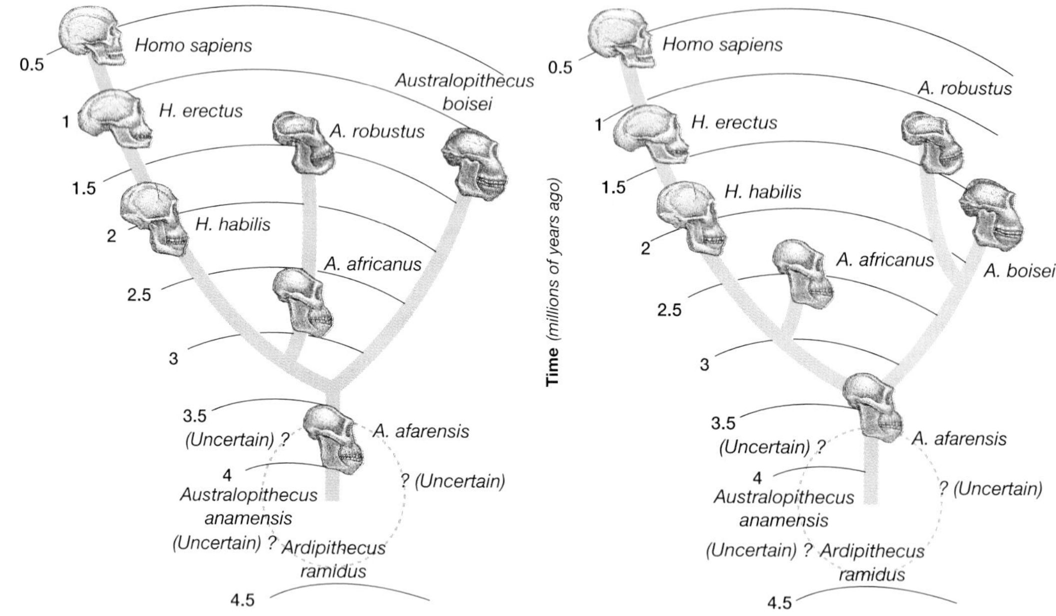

Review SECTION 14-2

1. **SKILL Comparing Structures** Summarize the differences between the skeleton of an ape and the skeleton of an australopithecine.
2. **SKILL Evaluating Information** Explain what evidence scientists look for in a skull to determine if a hominid was bipedal.
3. **Name** the oldest hominid fossils that have been found and where they were discovered.

Critical Thinking

4. **Justify** Dart's claim that his fossil find supported Darwin's theory of evolution by means of natural selection.

Review SECTION 14-2 ANSWERS

1. They differ in the way their skulls are attached to their spines, in the relative lengths of their arms and legs, by the shape and tilt of their hipbones, and by the shapes and angles of their femurs.
2. In bipedal hominids, the opening from which the spinal cord emerges from the brain is at the base of the skull.
3. A 4.4 million-year-old fossil of *Ardipithecus ramidus* was found in Ethiopia, and a 4.2 million-year-old fossil of *Australopithecus anamensis* was found in Kenya.
4. Dart's fossil had some humanlike characteristics and some apelike characteristics. As a transitional form between apes and humans, it supports the theory that organisms evolve from other organisms over time.

The Genus *Homo*

Section 14-3

Homo habilis Is the Oldest Member of the Genus *Homo*

Our genus, *Homo,* is composed of at least three species. We humans are the only surviving species. The first members of the genus *Homo* evolved from australopithecines more than 2 million years ago.

In the early 1960s, more hominid bones were discovered near the site where *Australopithecus boisei* had been unearthed. Stone tools were scattered among the bones. Although the fossils were crushed, reconstruction of the pieces indicated a skull with a volume of about 640 cm^3 (39 in.^{3}), larger than the australopithecine brain volume of 400–550 cm^3 (24–34 in.^{3}). Scientists disagreed about whether this fossil was an early human or an australopithecine.

In 1972, Richard Leakey, shown in **Figure 14-8,** discovered a similar skull that was virtually complete. Many of the critics who doubted that the first skull was an early human were silenced by Leakey's new discovery. Because of its association with tools, this early prehuman was named *Homo habilis.* The name *Homo habilis* is from the Latin *homo,* meaning "man," and *habilis,* meaning "handy." Skeletons discovered in 1987 indicate that *Homo habilis* stood about 1.2 m (4 ft) tall. *Homo habilis* lived in Africa for about 500,000 years and then became extinct.

Objectives

- **Compare** *Homo habilis* with australopithecines. (p. 309)
- **Describe** the characteristics of *Homo erectus.* (p. 310)
- **Outline** two hypotheses that explain the origin of *Homo sapiens.* (p. 311)
- **Compare** Neanderthals with modern humans. (p. 312)

Key Terms

None

Figure 14-8 **Richard Leakey and *Homo habilis* skull**

Homo habilis lived in East Africa.

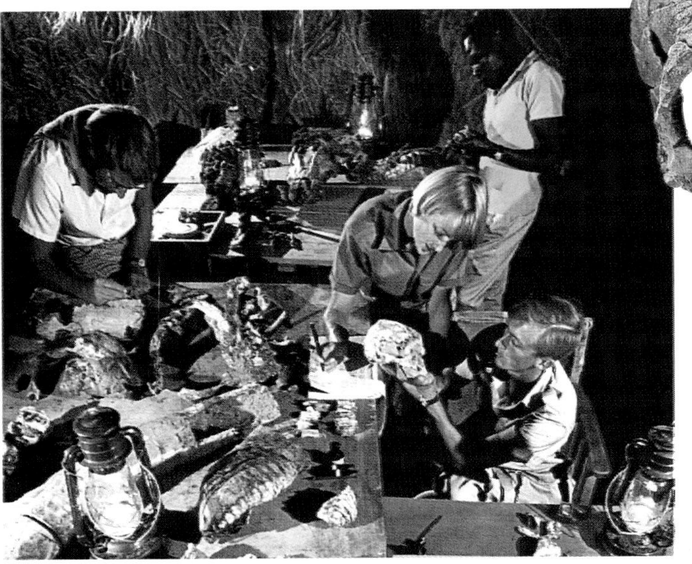

The skull of *Homo habilis* had a brain volume of about 775 cm^3 (47 in.^{3}) and many of the characteristics of modern human skulls.

Richard Leakey, far right, and his associates found one of the most complete *Homo habilis* skulls to date.

1 Prepare

Scheduling

- **Block:** About 90 minutes will be needed to complete this section.
- **Traditional:** About two class periods will be needed to complete this section.
- **Planning:** See the Planning Guide on pp. 299A and 299B for lecture, classwork, and assignment options for Section 14-3. Lesson plans for Section 14-3 can be found in a lesson cycle format in ***Biology: Principles and Explorations* One-Stop Planner CD-ROM with Test Generator.**

Resource Link

- Assign **Active Reading Worksheet Section 14-3** to help students comprehend the key concepts and visuals in this lesson.

2 Teach

Lesson Warm-up

Split the class into five groups to make life-size hominid outlines on butcher paper. Tape several large sheets of butcher paper on the wall. Have one group make a grid that is marked off every 10 cm from the floor to 200 cm up the wall. The other four groups should pick *Australopithecus afarensis* (1.1 m tall), *H. habilis* (1.2 m tall), *H. erectus* (1.5 m tall), or *H. sapiens* (1.7 m tall) and create a life-size paper outline of their chosen hominid. Have students label each species and attach it to the grid on the wall.

Integrating Different Learning Styles

Logical/Mathematical	**PE**	Data Lab, p. 310; Performance Zone item 22, p. 315
Visual/Spatial	**PE**	Performance Zone item 1, p. 314
	ATE	Demonstration, p. 310
Interpersonal	**PE**	Performance Zone item 24, p. 315
	ATE	Lesson Warm-up, p. 309; Closure, p. 312
Verbal/Linguistic	**PE**	Performance Zone items 13, 14, and 25, pp. 314–315

Demonstration

Comparing Brain Sizes

Hominid brain sizes, measured as cranial capacities, are given in the **Data Lab** table on page 310. Increasing brain size is an important aspect of human evolution. To help students appreciate how much hominid brain size has increased, get eight large, clear jars or bottles, label each container for a hominid species, and fill each with the volume of water equal to the hominid's estimated brain size.

Analyzing hominid evolution

Time 20 minutes

Process Skills Organizing, analyzing, making conclusions, evaluating

Teaching Strategies

Point out to students that brain size will have an upper limit that is dictated by several considerations. A major limitation on brain size is the size of a baby's skull at birth, when it must travel through an opening in the pelvic bones of its mother. Obviously there will be an upper limit to the size of a human female's pelvis.

Analysis Answers

1. The overall trend is that cranial capacity increased as more-modern hominids evolved.
2. The cranial capacity of *A. robustus* seems smaller than it should be for its relatively recent appearance on Earth. Also, the larger cranial capacity of Neanderthals than that of modern humans seems inconsistent with what we know about the mental capabilities of these two groups.
3. Answers will vary. Sample answer: Brain size might not be a survival advantage if larger brains limit the use of instinctual survival reactions.

Figure 14-9 ***Homo erectus.*** The skull of *Homo erectus* had prominent brow ridges. Like modern humans, *Homo erectus* had smaller teeth than *Australopithecus* and *Homo habilis.*

The First Hominids That Left Africa

Our understanding of *Homo habilis* lacks detail because it is based on only a few fossil specimens. However, scientists have found many fossils of the species that replaced *Homo habilis—Homo erectus.*

The first fossil of *Homo erectus,* called Java man, was found in 1891 on the island of Java, located in southeast Asia. The structure of the leg bones and the size of the skull cap indicated upright walking and a larger brain. The bones are thought to be as much as 500,000 years old.

Homo erectus was larger than *Homo habilis*—about 1.5 m (5 ft) tall. *Homo erectus* had a large brain of about 1,000 cm^3 (60 $in.^3$), walked upright, and used crude tools and fire. **Figure 14-9** shows an artists representation of what *Homo erectus* might have looked like.

In 1976, a complete 1.5-million-year-old *Homo erectus* cranium was discovered in East Africa. The cranium is 1 million years older than Java man. The cranium is the part of the skull that surrounds the brain. Far more numerous than *Homo habilis, Homo erectus* became widespread in Africa and migrated into Asia and Europe.

Homo erectus lived in tribes of 20 to 50 individuals. They hunted large animals using flint and bone tools, and cooked them. *Homo erectus* survived for more than 1 million years. However, it disappeared about 200,000 years ago, as early modern humans emerged. *Homo erectus* was the direct ancestor of our species, *Homo sapiens.*

Analyzing hominid evolution

Background

Hominids differ from apes in two important ways—hominids are bipedal and have much larger brains. One indicator of brain size is cranial capacity. Cranial capacity is the amount of room in the skull for the brain. The table below lists several hominids and their cranial capacities.

Analysis

1. **SKILL Organizing Data** Prepare a bar graph of hominid cranial capacity using the information in the table. Place the species next to one another in the order listed. What trend do you see in the graph?
2. **SKILL Evaluating Conclusions** Does any of the information in the chart seem contradictory to you? Explain.
3. **SKILL Evaluating Results** Suggest a reason that brain size might not be a survival advantage.

Cranial Capacity of Hominids

Species name	Cranial capacity*
Australopithecus afarensis	375–400 cm^3 (22.5–24 $in.^3$)
Australopithecus africanus	400–500 cm^3 (24–30 $in.^3$)
Australopithecus boisei	650 cm^3 (39 $in.^3$)
Australopithecus robustus	530 cm^3 (32 $in.^3$)
Homo habilis	700 cm^3 (42 $in.^3$)
Homo erectus	1000 cm^3 (60 $in.^3$)
Homo sapiens (Neanderthal)	1400 cm^3 (84 $in.^3$)
Homo sapiens (modern human)	1350 cm^3 (81 $in.^3$)

*Approximate values

MULTICULTURAL PERSPECTIVE

Fossils and Cultural Pride in Africa

African museums and governments carefully monitor and safeguard access to many hominid fossil specimens and sites. These fossils are highly valued by African people as important cultural resources that, both literally and metaphorically, help define their sense of self. Understandably, they want to preserve the fossils for current and future generations.

The Species *Homo sapiens* Arose Fairly Recently

The name *Homo sapiens* is from the Latin *homo,* meaning "man," and *sapiens,* meaning "wise." *Homo sapiens* is a newcomer to the hominid family. *Homo sapiens* has not existed as long as *Homo erectus* had. Early *Homo sapiens* left behind many fossils and artifacts, including the first known paintings by humans. Despite abundant evidence, scientists disagree about the details of how *Homo sapiens* evolved.

Two hypotheses

Based on fossil evidence, some scientists have argued that independent *Homo erectus* groups living in Africa, Europe, and Asia interbred and that *Homo sapiens* thus arose as a new species simultaneously all over the globe. Most scientists, however, argue instead that *Homo sapiens* appeared in one place (Africa), then migrated to Europe and Asia, replacing *Homo erectus* as they migrated.

Human History in Our Cells

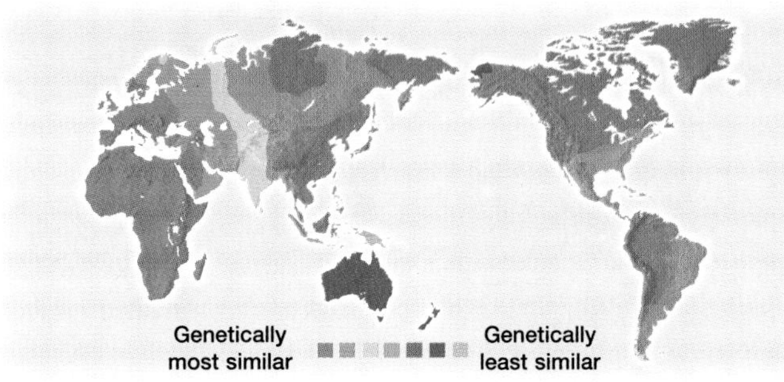

To answer the question about where *Homo sapiens* originated, scientists studied the DNA of modern human mitochondria. Because mitochondrial DNA is inherited only from the mother's egg, it does not undergo genetic recombination. As a result of this single-parent inheritance pattern, there is little genetic variation in mitochondrial DNA from one generation to the next. Any mutations that do occur would happen at a steady rate, acting as a biological clock. The mitochondrial DNA with the oldest lineage would show the most mutations.

The study showed that the greatest number of different mitochondrial DNA sequences occurs among modern Africans. This indicates that *Homo sapiens* have been living in Africa longer than on any other continent. This evidence thus supports the hypothesis that *Homo sapiens* evolved first in Africa.

Nuclear DNA

In studies of nuclear DNA, scientists demonstrated that greater variation among populations indicates a newer population and less variation indicates an older population. Unlike mitochondrial DNA, nuclear DNA is inherited from both parents and so genetic recombination occurs. Older, more similar populations will tend to maintain DNA patterns, but younger populations, composed of parents with different genetic backgrounds, tend to develop new DNA patterns. This study indicates that *Homo sapiens* migrated from Africa to the rest of the world, as shown in the map above.

internetconnect
SciLinks NSTA
TOPIC: Evolution
GO TO: www.scilinks.org
KEYWORD: HX311

Teaching TIP — BASIC

Comparing Species
Inform students that the tools made by *H. habilis* were crude by modern standards and included only a few varieties. Scientists have found about a dozen types used by *H. erectus*. Many of these tools had characteristic shapes. For example, teardrop-shaped hand axes are often found at sites where *H. erectus* lived.

TECH WATCH

Human History in Our Cells

Teaching Strategies
Remind students that scientists think mitochondria contain DNA because these organelles descended from ancient eubacteria. As eubacteria, these cells would have contained their own genetic information.

Discussion

- Why is mitochondrial DNA inherited only from the mother? *(Sperm mitochondria are destroyed in the fertilized egg.)*
- Why do people living in different regions have different physical characteristics? *(Physical characteristics represent adaptations to local conditions that have changed over time and over distance from Africa.)*
- How would the results of the nuclear DNA studies mentioned here have been different if *Homo sapiens* simultaneously arose from *H. erectus* in several different continents? *(Human nuclear DNA patterns throughout the world would be more similar.)*

Chapter 14

Making Biology Relevant

Health

How do scientists know that Neanderthals cared for their sick and injured? Neanderthal skeletons have been found with healed broken bones. The fractures probably incapacitated the individuals, so they must have been cared for by others until their bones healed. Furthermore, scientists found an adult male's skeleton with severe deformities. The individual would probably have been dependent on others his entire life.

3 Assessment Options

Closure

Summarizing the Section

Pair each student with a partner. Have one student in each group summarize orally the information presented in this section without referring to the text. The other student should listen and then ask clarifying questions to ensure that the final account is accurate and complete.

Review

Assign Review—Section 14-3.

Reteaching — Basic

Assume that a skull has just been found. Ask students what characteristics would indicate that the fossil was of a Neanderthal skull. *(The skull would need a large cranial capacity; protruding mouth bones; and heavy, bony brow ridges.)*

biobit

Did different species of Homo exist simultaneously?

Some anthropologists think that *Homo sapiens, Homo erectus,* and Neanderthals might have shared the Earth about 30,000 years ago. A few researchers even speculate that *Homo erectus* could have preyed on an ancestor of *Homo sapiens.*

Homo sapiens in Europe

Members of the genus *Homo* first appeared in Europe about 130,000 years ago. The first *Homo* fossils, called Neanderthals *(nee AN dur THALZ)* were found in 1856 in the Neander Valley of Germany. The Neanderthals were short and powerfully built. Their skulls were massive, with protruding faces and heavy, bony ridges over the brows. Their brains were larger than those of modern humans, and many scientists that study human evolution now assign them to a separate human species, *H. neanderthalensis.*

Rare at first outside of Africa, Neanderthals became more and more abundant in Europe and Asia, and by 70,000 years ago they had become fairly common. Neanderthals took care of their injured and sick and commonly buried their dead, often placing food, weapons, and even flowers with the bodies. Such attention to the dead suggests that they may have believed in a life after death. Neanderthals were the first hominids to show evidence of abstract thinking.

Modern *Homo sapiens*

About 34,000 years ago, the European Neanderthals were abruptly replaced by *Homo sapiens* of essentially modern appearance. These early humans are thought to have evolved first in Africa, then migrated to Europe and Asia. Early modern humans lived by hunting. They had complex patterns of social organization and are thought to have been capable of using sophisticated language. They eventually spread across Siberia and reached North America at least 12,000 years ago, when a land bridge still connected Siberia and Alaska. There were no more than several million people living in the entire world 10,000 years ago, compared with more than 6 billion living in 1999.

Our ability to make and use tools effectively has been important to our success in the animal kingdom. Humans use symbolic language and can shape concepts out of experience. Language has enabled us to transmit concepts from one generation to another as shown in **Figure 14-10**. Humans have what no other animal is thought to have ever had—cultural evolution. Through culture, we have found ways to change our environment to suit our needs, rather than changing ourselves in response to the environment. This is both an exciting potential and an enormous responsibility.

Figure 14-10 Use of language. While not the only animal capable of conceptual thought, human beings have refined and extended this ability until it has become a characteristic of our species

Review SECTION 14-3

1. **Describe** how *Homo habilis* differs from australopithecines.
2. **Describe** the evidence that identifies *Homo erectus* as the first human species to have left Africa.
3. **Contrast** the characteristics of *Homo erectus* with those of *Homo habilis.*
4. **Summarize** the two hypotheses of the origin of *Homo sapiens.*
5. **Contrast** Neanderthals with modern humans.
6. **SKILL Recognizing Relationships** Associate brain size with evolutionary success.

Review SECTION 14-3 ANSWERS

1. *H. habilis* had a larger brain and made and used tools.
2. *H. erectus* skeletons have been found in Asia and Europe in addition to Africa, whereas *H. habilis* fossils have only been found in Africa.
3. *H. erectus* had a larger brain and was taller than *H. habilis.*
4. One hypothesis holds that *H. sapiens* arose independently and simultaneous in many locations in Africa, Europe, and Asia from interbreeding groups of *H. erectus.* A second hypothesis holds that *H. sapiens* arose in only one place, Africa, and then migrated to Europe and Asia, replacing *H. erectus* as they migrated.
5. Neanderthals had slightly larger brains than modern humans. They were also shorter and stockier and had massively boned faces.
6. It is unreasonable to argue for a correlation between brain size and evolutionary success in the genus *Homo* because humans have only existed for a relatively short time.

Use the Key Concepts and Key Terms listed below to review the main ideas in this chapter. If you need to review the meaning of a term, turn to the page indicated.

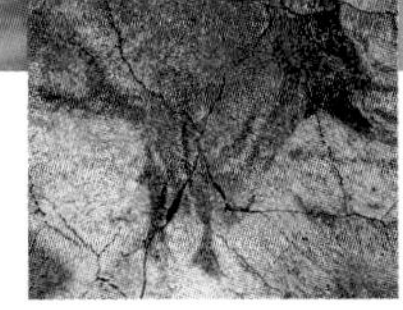

Key Concepts

14-1 The Evolution of Primates

- Primate ancestors lived about 80 million years ago.
- The first primates evolved 50–60 million years ago.
- Primates are characterized by the presence of grasping hands and binocular vision.
- Monkeys are day-active primates with opposable thumbs.
- Apes have larger, more complex brains than monkeys have, and do not have a tail.
- Human ancestors diverged from the group that led to gorillas and chimpanzees about 5 to 7 million years ago.
- DNA evidence indicates that humans are more closely related to chimpanzees than to any other primate species.

14-2 Early Hominids

- Australopithecines, unlike apes, were bipedal and had large brains.
- Australopithecine fossils represent the evolutionary link between humans and apes.
- In australopithecines bipedalism evolved before larger brains appeared.
- The oldest hominid ever discovered is *Ardipithecus ramidus*; it was found in Ethiopia.

14-3 The Genus *Homo*

- *Homo habilis* had a much larger brain than australopithecines and probably used tools.
- *Homo erectus* evolved in Africa about 1.5 million years ago, had a large brain, and became extinct about 200,000 years ago.
- *Homo sapiens* evolved from *Homo erectus* about 200,000 years ago.
- *Homo sapiens* either evolved from different groups of *Homo erectus* simultaneously or evolved in Africa and migrated to the rest of the world.
- Neanderthals were short, stout, had larger brains than modern humans, and became extinct about 34,000 years ago.

Key Terms

14-1

primate (302)
prosimian (303)
diurnal (304)
anthropoid (304)
opposable thumb (304)

14-2

hominid (306)
bipedal (306)

14-3

None

Study TIP Ready?

- *If you think you understand these Key Concepts and Key Terms, then you're ready for the Performance Zone.*

Study ZONE CHAPTER 14

Review and Practice

Assign the following worksheets for Chapter 14:

- **Student Review Guide**
- **Concept Mapping Worksheet**
- **Test Preparation Pretest**
- **Vocabulary Worksheet**
- **Critical Thinking Review Worksheet**
- **Science Skills Worksheet**

Assessment Options

The following assessment options are available for this chapter:

Resource Link

- Assign **Chapter Test 14** to assess students' understanding of the chapter.

Alternative Assessment

Ask students to draw a phylogenetic tree that shows the evolution of primates leading to modern humans. Have them include the following groups and species on the tree: first primates, prosimians, monkeys, gorillas, chimpanzees, australopithecines, *H. habilis, H. erectus*, and *H. sapiens.* PORTFOLIO

Portfolio Assessment

- *Portfolio Assessment* at the front of this book provides options for building and evaluating student portfolios. Students and teachers can select items from the areas listed for inclusion in a portfolio.
- See items labeled PORTFOLIO on pp. 304, 305, 307, and 313.

Answer to Concept Map

The following is one possible answer to Performance Zone item 1 on page 314.

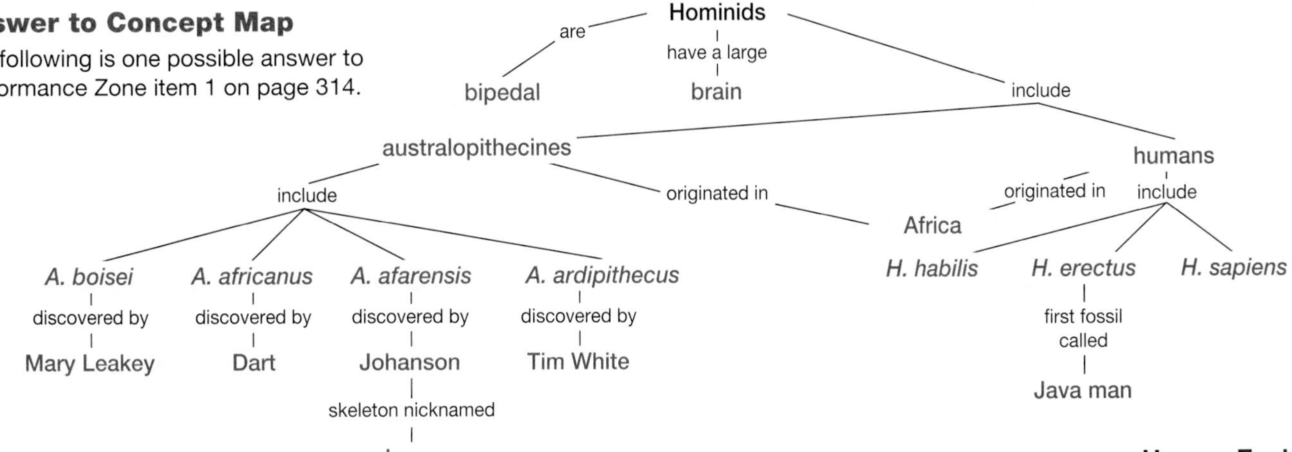

Performance ZONE
CHAPTER 14 REVIEW ANSWERS

Assignment Guide

Section	Review
14-1	3, 4, 5, 6, 12
14-2	2, 10, 15, 16, 17, 19, 20, 21, 23
14-3	1, 7, 8, 9, 11, 13, 14, 18, 22, 24, 25

Understanding and Applying Concepts

1. The answer to the concept map is found at the bottom of page 313.
2. **a.** The earliest primates were the prosimians; the more recent group of primates to evolve were the anthropoids, which have opposable thumbs and are diurnal.

   b. Hominids include all of the primates that are bipedal; the australopithecines were the first hominids.
3. b
4. c
5. c
6. d
7. c
8. d
9. d
10. You can tell that it is a primate skeleton because it has grasping hands and feet and forward-facing eyes. You can tell that it is a hominid because it is bipedal.
11. Studies of nuclear DNA indicate that the oldest human populations exist in Africa and that populations are progressively newer the further from Africa one goes. This evidence correlates with the evidence from mitochondrial DNA studies that points to Africa as being the geographical origin of *Homo sapiens.*
12. The eyes of the first primates faced forward. The resulting binocular vision gave primates excellent depth perception, which enabled them to judge distances accurately when moving through trees.

Performance ZONE CHAPTER 14 Review

Understanding and Applying Concepts

1. **Concept Mapping** Construct a concept map that shows the hominids and their relationships to each other. Use as many terms as needed from the vocabulary list. Try to include the following items in your map: australopithecines, humans, Africa, *Australopithecus boisei, Australopithecus africanus, Australopithecus afarensis, Homo habilis, Homo erectus, Homo sapiens,* bipedal, brain, Mary Leakey, Dart, Johanson, Tim White, Lucy, Java man. Include additional terms in your map as needed.
2. **Understanding Vocabulary** Write a sentence that shows your understanding of each of the following terms:
   a. primate, prosimian, opposable thumbs, diurnal
   b. australopithecines, hominids, bipedal
3. The mammal that gave rise to the first primates resembled a modern-day
   a. ape. c. baboon.
   b. shrew. d. monkey.
4. Monkeys differ from most prosimians in that monkeys
   a. are colorblind.
   b. have opposable thumbs.
   c. sleep at night.
   d. live alone.
5. Apes have larger, more developed brains than monkeys but lack
   a. opposable thumbs.
   b. bipedalism.
   c. a tail.
   d. fingernails.
6. Among the apes, the most distant relatives of humans are the
   a. chimpanzees. c. gorillas.
   b. orangutans. d. gibbons.
7. The first hominid to use tools was
   a. *Homo erectus.* c. "handy man."
   b. "Lucy." d. *Homo sapiens.*
8. Compared with australopithecines, *Homo habilis*
   a. used a more sophisticated language.
   b. developed better hand dexterity due to opposable thumbs.
   c. was more apelike.
   d. had greater brain volume.
9. Neanderthals differ most from modern humans in their
   a. tooth structure. c. foot structure.
   b. walking gait. d. forehead shape.
10. **Theme Evolution** Explain how you can tell that the skeleton shown at right is from a primate. How do you know it is from a hominid?

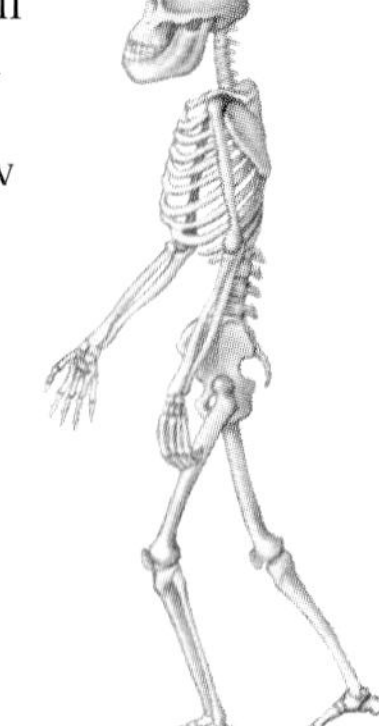

11. **Tech Watch** How do studies of nuclear DNA confirm mitochondrial DNA studies that have helped determine the geographical origin of *Homo sapiens?*
12. **Recognizing Relationships** Describe the positioning of the eyes in the first primates. How did this positioning of the eyes improve the primates' ability to survive?
13. **Describe** the evidence that scientists use to defend the two different hypotheses about the evolution of *Homo sapiens* discussed in this chapter. Which hypothesis is supported by molecular data?
14. **Summarize** the evidence that indicates Neanderthals were the first hominids to show evidence of abstract thinking.
15. **Chapter Links** Explain the importance of Dart's discovery to the theory of evolution. (**Hint:** See Chapter 13, Section 13-2.)

13. Scientists have relied on fossil evidence to argue that *H. sapiens* arose simultaneously at several places around the world. Other scientists have used fossil and molecular evidence to argue that *H. sapiens* arose in Africa and migrated to other areas. The hypothesis that *H. sapiens* originated in Africa and migrated to other continents is supported by molecular data.
14. Neanderthals often buried their dead, placing food, weapons, and even flowers with the bodies. Such acts indicate that Neanderthals believed in a life after death, which indicates a capacity for abstract thought.
15. Darwin predicted that scientists would eventually find "missing links" between groups of organisms in the form of transitional fossils. Dart's discovery of a skull with human and apelike characteristics provides objective evidence of one such link— that between humans and ancient primates.

Skills Assessment

Part 1: Interpreting Graphics

Study the phylogenetic tree below. Then answer the following questions:

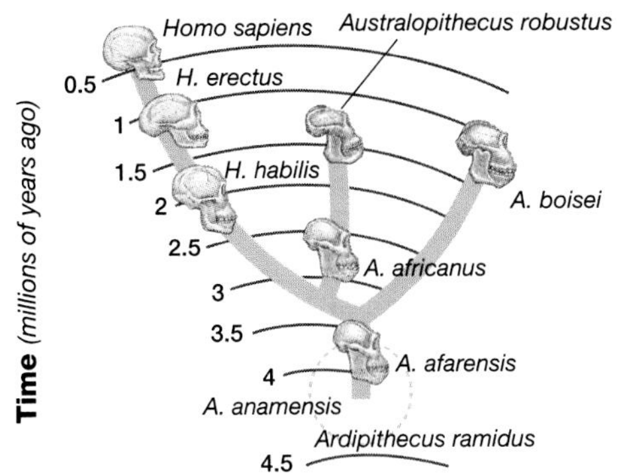

16. Which species is the common ancestor of all the other species?

17. Which species is the direct ancestor of *Australopithecus robustus*?

18. Which species exists today?

Part 2: Life/Work Skills

Use the media center or Internet resources to learn about the evidence that supports the theory of human evolution.

19. Selecting Technology On a field expedition, you have discovered a number of fossilized bones that you suspect are *Australopithecus africanus.* In order to be certain, the fossils must be scientifically dated. Use Internet sources and the media center to research dating methods and to determine which would be best for your needs. Write a report explaining your conclusions.

20. Finding Information A colleague has called to tell you she is returning to the lab with an early hominid skull she believes to be *Australopithecus boisei.* Use the media center and Internet resources to find detailed information about the structure and characteristics of an *Australopithecus boisei* skull, and prepare the information for the skull's arrival.

Critical Thinking

21. Making Predictions It is not known why our ancestors became bipedal. But it has been proposed that bipedalism evolved as an adaptation that kept the brain cool on the hot savanna. Develop a hypothesis that explains how being bipedal and being hairless except for the top of the head function as interrelated adaptations to protect the brain from overheating.

22. Making Inferences Using new dating techniques, anthropologists have determined that flint tools used by modern humans are about 92,000 years old. Flint tools used by Neanderthals were determined to be much younger—about 60,000 years old. What do these data suggest about the assumption that Neanderthals lived before modern humans?

Portfolio Projects

23. Evaluating Hypotheses Read the article "The Dawn of Humans," by Lee Berger (*National Geographic,* August 1998, pp. 90–99). Evaluate the argument that the author uses to support his hypothesis that the phylogenetic tree of humans should be redrawn.

24. Cooperative Group Project Select a hominid described in this chapter, and write a report about who found the fossils and when the fossils were found. Be sure to include a map of where the fossils were found, a description of the fossils, and a drawing of a phylogenetic tree that shows how the hominid is related to modern humans. Present your findings in the form of an oral report to the class.

25. Career Focus Molecular Anthropologist Use the media center or Internet sources to research the educational background and field experience necessary to become a molecular anthropologist. List the high school courses required, and describe additional degrees or training recommended for this career. Summarize the employment outlook for this field.

Portfolio Projects

23. Answers will vary. Sample answer: The author argues that *A. africanus* is the direct ancestor of the genus *Homo* and that *A. afarensis* forms the base of another branch that eventually died out. He cites fossil evidence and the idea that evolution rarely works backward to support his hypothesis. His argument is plausible but by no means airtight. There does not seem to be enough fossil evidence to firmly prove which drawing of the hominid family tree is correct. Also, there is no inherent reason why body forms could not repeat later in evolutionary time. In fact, there might be strong selective pressure to re-evolve certain traits if the environment changes to resemble prior conditions.

24. Answers will vary.

25. Answers will vary. Molecular anthropologists are scientists who use knowledge and techniques of molecular biology to investigate human origins and study modern human populations. Many of the biological techniques they use come from the field of genetics. Molecular anthropologists usually have a Ph.D. and training in genetics, anthropology, and population biology. Most molecular anthropologists work for universities and museums. This is a fairly new field, so the growth potential is uncertain. Starting salary will vary by region.

Skills Assessment

16. *Ardipithecus ramidus*

17. *A. africanus*

18. *H. sapiens*

19. Answers will vary but should include radiometric dating using an isotope with a relatively long half-life (much longer than that of carbon-14).

20. Answers will vary but should include that *A. boisei* had a bony ridge on the crest of its head.

Critical Thinking

21. Answers will vary but should account for the fact that hair is an excellent insulator.

22. It suggests that this assumption may be wrong and that modern humans may have evolved before Neanderthals.

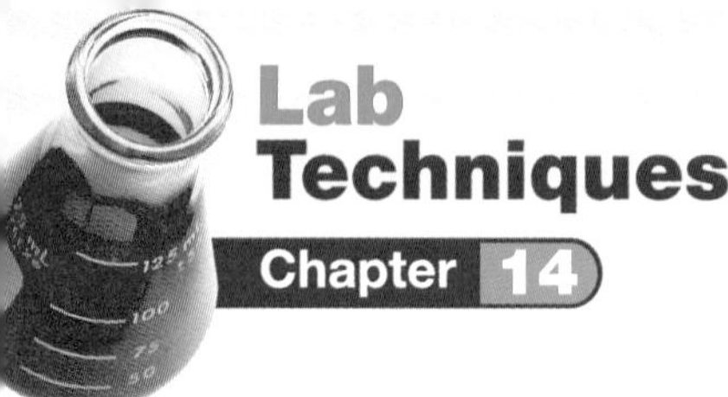

Lab Techniques

Chapter 14

Comparing Hominid Skulls

Purpose

Students will identify differences and similarities between the skulls of apes, early hominids, and humans.

Preparation Notes

Time Required: One 40-minute class period

Materials

Materials for this lab can be ordered from WARD'S. See *Master Materials List* at the front of this book for catalog numbers.

Preparation Tips

If life-size casts of skulls are available, substitute them for the diagrams provided or use them to supplement the investigation.

Procedural Tips

Since the drawings of the human, ape, and fossil hominid skulls are approximately $\frac{1}{6}$ to $\frac{1}{7}$ of life size, the checklist of features has incorporated conversion factors of 40 for calculating square area and 1000 for calculating cubic volume of life-size skulls. If life-size casts of skulls are available for students to measure, eliminate the factor of 40 in calculating square area. Substitute 4.2 or $\frac{4}{3}$ × pi for the factor of 1000 in calculating cubic volume.

Lab Techniques

Chapter 14 Comparing Hominid Skulls

SKILLS

- Measuring
- Comparing anatomical features

OBJECTIVES

- **Identify** differences and similarities between the skulls of apes and the skulls of humans.
- **Identify** differences and similarities between the fossilized skulls of hominids.
- **Classify** the features of hominid skulls as apelike, humanlike, or intermediate.

MATERIALS

- metric ruler
- protractor

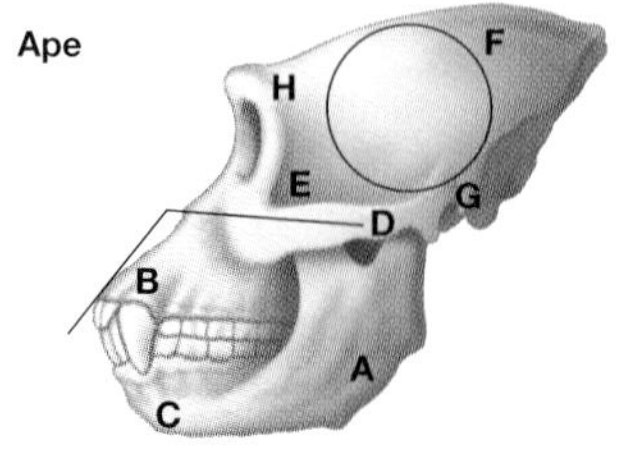

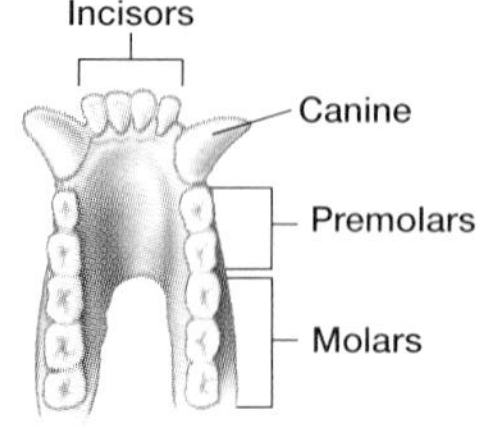

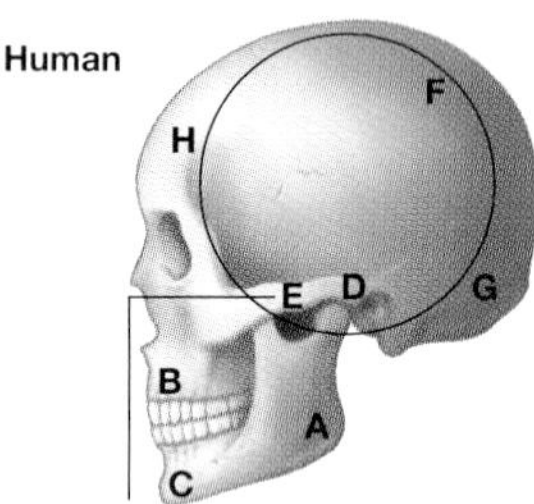

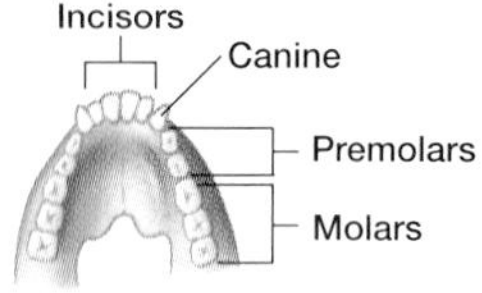

Before You Begin

Modern **apes** and humans share a **common ancestor.** Much of our understanding of human evolution is based on the study of the fossilized remains of **hominids.** By studying fossilized bones and identifying similar and dissimilar structures, scientists can infer the **anatomy,** or body structure, of a species. In this lab, you will identify differences and similarities between the skulls of apes, early hominids, and humans.

1. Write a definition for each boldface term in the paragraph above.
2. Make a data table similar to each one shown at right.
3. Based on the objectives for this lab, write a question you would like to explore about human evolution.

DATA TABLE 1

Name	Cranial capacity (cm^3)	Lower face area (cm^2)	Brain area (cm^2)	Jaw angle (degrees)	Brow ridge	Teeth
Ape						
Human						

DATA TABLE 2

Name				
Australopithecus robustus				
Australopithecus africanus				
Homo erectus				
Neanderthal				

Before You Begin

Answers

1. *apes*—tailless primates with long arms, a broad chest, and a larger, more developed brain than monkeys. Apes belong to the family Pongidae, which includes the chimpanzee, gorilla, and orangutan; or to the family Hylobatidae, which includes the gibbon and the siamang.

   common ancestor—species from which two or more species evolved

   hominids—member of the family Hominidae of the order Primates; characterized by opposable thumbs, no tail, longer lower limbs, and erect bipedalism

   anatomy—body structure of an organism
2. See Procedure on page 317.
3. Answers will vary. For example: How have facial features changed as hominids have evolved?

How to Interpret the Features of a Skull

Cranial capacity: Use the circles drawn on the skulls to estimate brain volume, or cranial capacity. Measure the radius of each circle in centimeters. Then cube this number, and multiply the result by 1,000 to calculate the approximate life-size cranial capacity in cubic centimeters.

Lower face area: Measure *A* to *B* and *C* to *D* in centimeters for each skull. Multiply these two numbers together, and multiply the product by 40 to approximate the life-size lower face area in square centimeters.

Brain area: Measure *E* to *F* and *G* to *H* in centimeters for each skull. Multiply these two numbers and multiply the product by 40 to approximate the life-size brain area in square centimeters.

Jaw angle: Note the two lines that come together near the nose of each skull. Use a protractor to measure the inside angle made by the lines and to determine how far outward the jaw projects.

Brow ridge: Note the presence or absence of a bony ridge above the eye sockets.

Teeth: Count the number of each kind of teeth in the lower jaw.

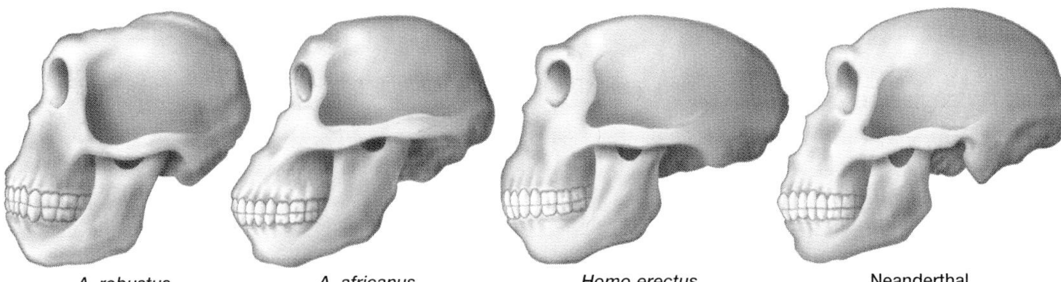

Procedure

PART A: Ape Skulls and Human Skulls

1. Examine the diagrams of the skull and jaw of an ape and a human. Look for similarities and differences between the features listed in the chart *How to Interpret the Features of a Skull.* Record your observations and measurements for each feature listed in Data Table 1.

PART B: Fossil Hominids

2. Examine the four fossil hominid skulls. On the hominid skulls, observe and measure four features that are listed in the chart *How to Interpret the Features of a Skull.* Use the human skull as a model for taking measurements. Record your observations and measurements in Data Table 2.
3. Compare your data for the hominids with your data for the modern ape and human. Classify each feature of the hominid skulls as being apelike, humanlike, or intermediate by writing an *A, H,* or *I* next to your observation or measurement for that feature.
4. Using your data, predict the order in which the hominids shown here evolved.

Analyze and Conclude

1. **Summarizing Results** How did skull structure change as hominids evolved?
2. **Drawing Conclusions** Which fossil skull is most apelike? most humanlike?
3. **Further Inquiry** Write a new question about human evolution that could be explored with another investigation.

On the Job

Anthropology is the scientific study of humans. Do research in the library or media center or on the Internet to learn about a famous anthropologist, such as Louis or Mary Leakey. Summarize your findings in a written or oral report.

Lab Techniques

14 Chapter

Analyze and Conclude

Answers

1. The brow ridge disappeared, the brain area increased, and the jaw became less prominent.
2. Students should recognize that the Neanderthal skull is most humanlike because of its large brain area and nonprotruding jaw. Students may disagree about which hominid skull is most apelike, *A. robustus* or *A. africanus.* Both skulls have a small brain area, a brow ridge, and a protruding jaw, albeit to varying degrees.
3. Answers will vary. For example: What was the probable diet of fossil hominids based on their tooth and jaw structure?

On the Job

Answers

Answers will vary. Sample answer: Louis and Mary Leakey were a husband-and-wife team. Their son Richard was also an anthropologist. Louis Leakey discovered fossil hominids in Tanzania that showed that human evolution was centered in Africa, not Asia. Mary Douglas Leakey described the sequence in which early stone tools came into use. She also discovered one of the first humanlike skulls that gave indication that humanlike beings lived in prehistoric eastern Africa. She led a team that discovered a trail of footprints preserved in volcanic ash thought to be from humanlike creatures that lived approximately 3,700,000 years ago. Richard Leakey has found many important fossils at Lake Turkana, in Kenya, including one of the oldest known *H. habilis* fossils and an almost complete *H. erectus* skeleton.

Procedure

Answers

Sample Data Table 1

	Cranial capacity (cm^3)	Lower face area (cm^3)	Brain area (cm^2)	Jaw angle (degrees)	Brow ridge	Teeth
Ape	422	420	230	130	Yes	6 molars, 4 premolars 2 canines, 4 incisors
Human	1953	161	264	80	No	Same as above

Sample Data Table 2

Hominid	Lower face area (cm^2)	Brain area (cm^2)	Jaw angle (degrees)	Brow ridge
A. robustus	300 (I)	260 (I)	115 (I)	Yes
A. africanus	252 (I)	190 (A)	125 (I)	Yes
A. erectus	312 (I)	252 (I)	111 (I)	Yes
Neanderthal	300 (I)	308 (H)	115 (I)	Yes

1. See Sample Data Table 1.
2. See Sample Data Table 2.
3. See Sample Data Table 2.
4. Students should predict that *A. africanus* evolved first, followed by *A. robustus, H. erectus,* and Neanderthal.

Chapter 15

Classification of Organisms

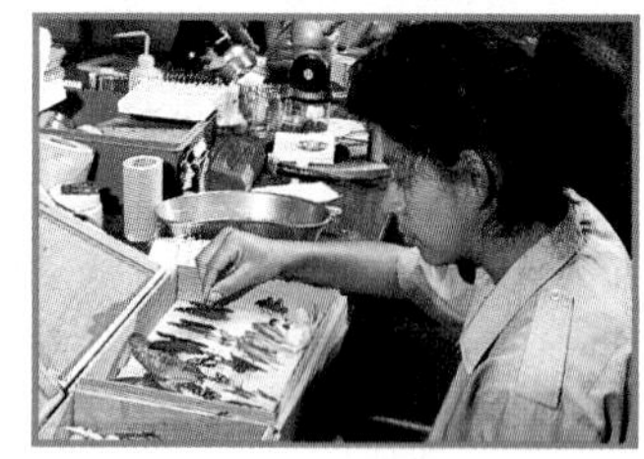

Planning Guide

OBJECTIVES	MEETING INDIVIDUAL NEEDS	READING SKILLS AND STRATEGIES
15-1 Categories of Biological Classification (BLOCKS 1 & 2 90 min) • Describe Linnaeus's role in developing the modern system of naming organisms. • Summarize the scientific system for naming a species. • List the seven levels of biological classification.	Learning Styles, p. 320 BASIC ADVANCED • pp. 321, 324	ATE Active Reading, Technique: Brainstorming, p. 319 ■ Active Reading Guide, worksheet (15-1) ■ Directed Reading, worksheet (15-1)
15-2 How Biologists Classify Organisms (BLOCKS 3, 4, & 5 135 min) • List the characteristics that biologists use to classify organisms. • Summarize the biological species concept. • Relate analogous structures to convergent evolution. • Describe how biologists use cladograms to determine evolutionary histories.	Learning Styles, p. 325 BASIC ADVANCED • pp. 327, 330	ATE Active Reading, Technique: Anticipation Guide, p. 325 ■ Active Reading Guide, worksheet (15-2) ■ Directed Reading, worksheet (15-2) ■ Supplemental Reading Guide, *The Dinosaur Heresies*

Review and Assessment Resources

BLOCKS 6 & 7 90 min

TEST PREP

- PE Study Zone, p. 331
- PE Performance Zone, pp. 332–333
- ■ Chapter 15 worksheets:
  - • Vocabulary • Concept Mapping
  - • Science Skills • Critical Thinking

- Audio CD Program
- ■ Test Prep Pretest

CHAPTER TESTING

- ■ Chapter Tests (blackline master)
- ▲ Test Generator
- ■ Assessment Item Listing

See the correlations table at the front of this book for a list of the ***National Science Education Standards*** addressed in this chapter.

Smithsonian Institution® Visit **www.si.edu/hrw** for additional on-line resources.

KEY

Biology Principles and Explorations

PE Pupil's Edition
ATE Teacher's Edition

 Teaching Transparencies

One-Stop Planner CD-ROM
Shows these resources within a customizable daily lesson plan:

■ Teaching Resources
▲ Test Generator

VISUAL STRATEGIES	CLASSWORK AND HOMEWORK	TECHNOLOGY AND INTERNET RESOURCES
ATE Fig. 15-1, p. 321  56 Biological Hierarchy of Classification	**PE Quick Lab** How do you use a field guide?, p. 324 **Lab A15** Grouping Things You Use Daily ■ **Basic Skills:** Words and Roots **PE Review,** p. 324	**Holt Biology Videodiscs** Section 15-1 **Audio CD Program** Section 15-1 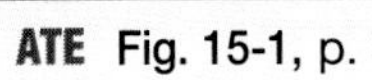 internet**connect**  SCILINKS NSTA **TOPICS:** • Naming species • Systems of classification
ATE Fig. 15-6, p. 327  57 Classification of Modern Humans	**PE Data Lab** Analyzing taxonomy of mythical organisms, p. 326 **PE Data Lab** Making a cladogram, p. 329 **PE Experimental Design** How Do You Make a Dichotomous Key?, pp. 334–335 **Lab B12** Classification **Lab C29** Classifying Mysterious Organisms **PE Review,** p. 330 **PE Study Zone,** p. 331	CNN **Video Segment 11** Something Worth Saving **Holt Biology Videodiscs** Section 15-2 **Audio CD Program** Section 15-2  internet**connect**  SCILINKS NSTA **TOPICS:** • Taxonomy • Evolution systematics • Classification

ALTERNATIVE ASSESSMENT

PE Portfolio Projects 23–25, p. 333
ATE Alternative Assessment, p. 331
ATE Portfolio Assessment, p. 331

Lab Activity Materials

Quick Lab
p. 324 paper, pencil, plant or animal field guide, specimens or photographs of specimens (3 to 6)

Data Lab
pp. 326, 329 pencil, paper

Experimental Design
p. 334 objects found in the classroom, such as shoes, books, writing instruments (6 to 10); stick-on labels; pencil

Lesson Warm-up
ATE handful of nuts, bolts, screws, washers, p. 325

Demonstrations
ATE two obviously different plants, p. 318
ATE photo of *Felis concolor,* p. 320
ATE photos of *Carnegia gigantea, Nymphaea ordata, Canis familiaris, Viola tricolor, Peromyscus californicus,* p. 321
ATE map of Belgium, p. 326

BIOSOURCES LAB PROGRAM
- Quick Labs: Lab A15, p. 29
- Inquiry Skills Development: Lab B12, p. 51
- Laboratory Techniques and Experimental Design: Lab C29, p. 141

Chapter 15

Chapter Theme

Evolution

The bewildering diversity of life on Earth reflects the ongoing process of evolution. Scientists have developed classification systems to make sense of this diversity and to illustrate evolutionary relationships between organisms.

Establishing Prior Knowledge

Before beginning this chapter, make certain students can answer the questions in **Study TIP Ready?** on page 319. If they cannot, encourage them to reread the sections indicated.

Opening Demonstration

Show students two plants that have obvious differences such as an ivy or a fern and a plant in bloom. Ask the students to devise a single-word name that could be used to describe both plants. Next have them distinguish the plants from each other by adding a second word that is unique to each plant. Have students share their names and point out similarities and differences in their nomenclature.

CHAPTER 15 Classification of Organisms

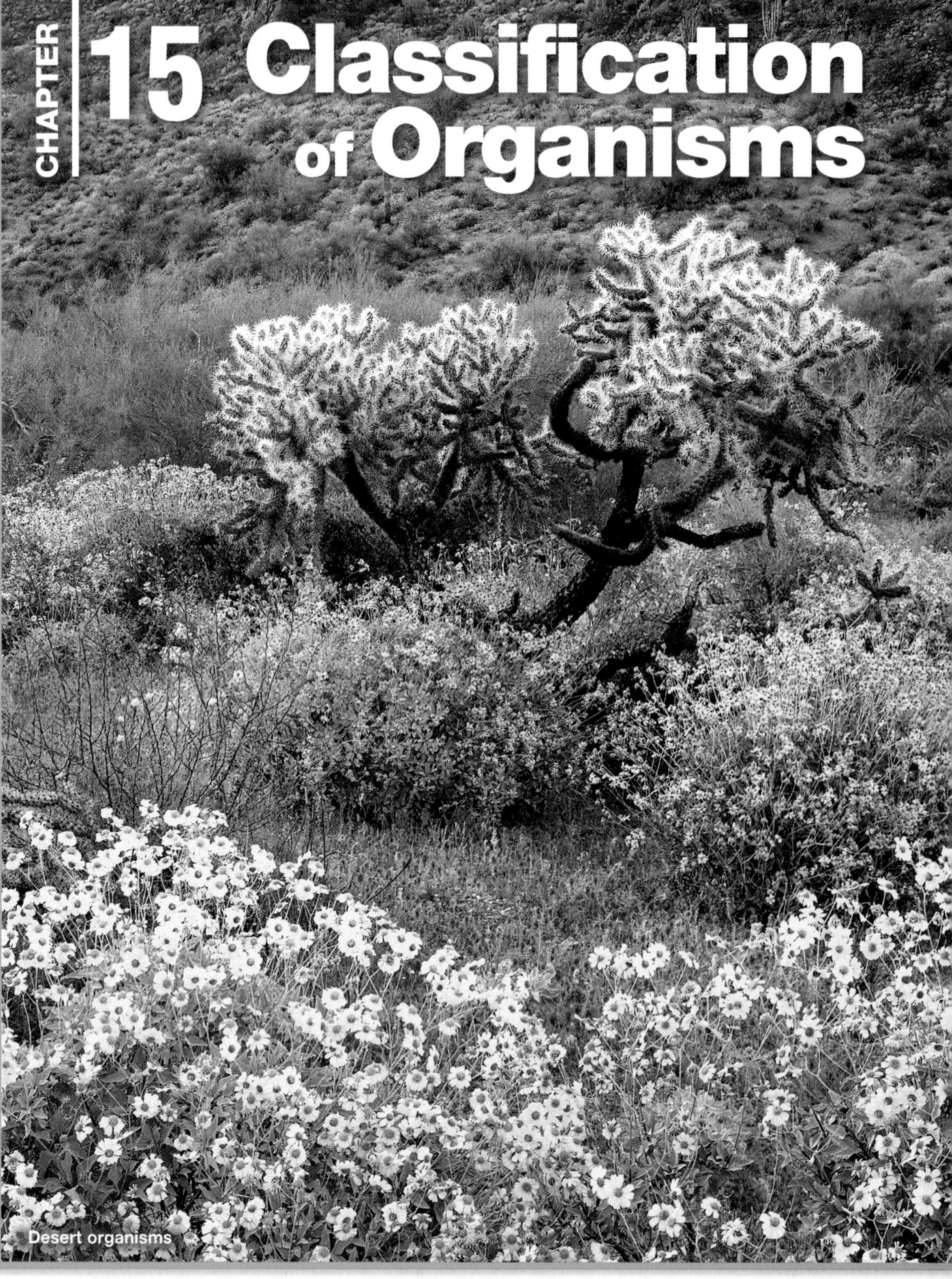

Desert organisms

What's Ahead — and Why

CHAPTER OUTLINE

15-1 Categories of Biological Classification

- Scientists Assign Organisms Two-Word Names
- Scientists Use a System to Classify Organisms

15-2 How Biologists Classify Organisms

- How Biologists Recognize Species
- Taxonomy Reveals Evolutionary History
- Phylogenies Give Extra Weight to Important Characters

Students will learn

- how each organism is described by a unique two-word name.
- how a species is defined and how physical traits can be used to categorize an organism.

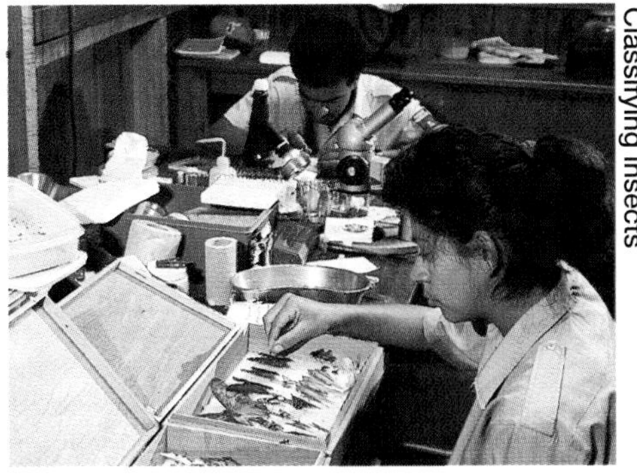
Classifying insects

Groups within groups...

Of the millions of different kinds of organisms on Earth, many groups share similarities. These similarities form the basics of a classification system that allows scientists to group organisms. For example, different butterflies share more traits with one another than do cactuses and bushes.

Looking Ahead

Labs

- **Quick Lab**
  How do you use a field guide? (p. 324)
- **Data Labs**
  Analyzing taxonomy of mythical organisms (p. 326)
  Making a cladogram (p. 329)
- **Experimental Design**
  How Do You Make a Dichotomous Key? (p. 334)

Features

- **Exploring Further**
  Cladograms (p. 328)

Study TIP

Ready?

Check your understanding of these topics to help you prepare for what's ahead.

Can you...

- **define** the term *species*? (Section 1-1)
- **relate** macromolecules to evolutionary history? (Section 13-2)
- **relate** homologous structures to evolutionary relationships? (Section 13-2)
- **summarize** speciation? (Section 13-3)

If not, *review the sections indicated.*

internetconnect
SCILINKS
NSTA
National Science Teachers Association sciLINKS Internet resources are located throughout this chapter.

Reading Strategies

Active Reading

Technique: Brainstorming
Explain that in situations with an overwhelming amount of information, a useful classification system might include both broad and specific categories. Geography provides a good example. A broad category might be North America, followed by the United States, then the Midwest, then Minnesota, then Minneapolis, and so on, until you reach a specific street address. Work with the class as a whole to develop a list of other things that are categorized in this way.

Directed Reading

Assign **Directed Reading Worksheet Chapter 15** to help students interact with the text as they read the chapter.

Interactive Reading

Assign ***Biology: Principles and Explorations* Audio CD Program Chapter 15** to help reluctant readers achieve greater success in reading the chapter.

Reading Skills

Assign Reading Organizers and Reading Strategies Worksheets from the **One-Stop Planner CD-ROM with Test Generator** to help students learn and practice effective reading skills.

internetconnect

Holt *Biology: Principles and Explorations*

On-line Resources: go.hrw.com
Visit the HRW Web site for a variety of resources related to this chapter. Just type in the keyword HX0 Chapter 15.

Holt *Biology: Principles and Explorations*

On-line Resources: www.scilinks.org
The following *sci*LINKS Internet resources can be found in the student text for this chapter:

Page 321
TOPIC: Naming species
KEYWORD: HX321

Page 322
TOPIC: Systems of classification
KEYWORD: HX322

Page 327
TOPIC: Taxonomy
KEYWORD: HX327

Page 330
TOPIC: Evolutionary systematics
KEYWORD: HX330

Page 335
TOPIC: Classification
KEYWORD: HX335

Chapter 15

1 Prepare

Scheduling

- **Block:** About 90 minutes will be needed to complete this section.
- **Traditional:** About two class periods will be needed to complete this section.
- **Planning:** See the Planning Guide on pp. 317A and 317B for lecture, classwork, and assignment options for Section 15-1. Lesson plans for Section 15-1 can be found in a lesson cycle format in ***Biology: Principles and Explorations*** **One-Stop Planner CD-ROM with Test Generator.**

Resource Link

Assign **Active Reading Worksheet Section 15-1** to help students comprehend the key concepts and visuals in this lesson.

2 Teach

Lesson Warm-up

Ask students if they have ever met someone or heard of someone who shares their name. Have them consider what kinds of problems could arise from this situation. Ask students what other information they could use to differentiate themselves from an individual with the same name. *(social security number, driver's license number, address, age, and so on)*

Demonstration

Communicating

Show students a picture of *Felis concolor,* and ask them to name the animal. *(lion, mountain lion, puma, cougar, panther, or catamount)* Emphasize that besides being ambiguous, common names can be misleading. Ask students to give examples of misleading common names. *(Some examples might include the starfish, sea horse, silverfish, horseshoe crab, and ringworm.)*

Section 15-1 Categories of Biological Classification

Objectives

- **Describe** Linnaeus's role in developing the modern system of naming organisms. (p. 320)
- **Summarize** the scientific system for naming a species. (p. 321)
- **List** the seven levels of biological classification. (p. 322)

Key Terms

taxonomy
binomial nomenclature
genus
family
order
class
phylum
kingdom

Scientists Assign Organisms Two-Word Names

Just as it is impossible for postal workers to sort mail bearing only the addressee's first name, it is impossible for biologists to memorize every name of the estimated 10–30 million organisms on Earth. To make sorting mail easier, postal workers sort first by zip code, then by street name and house number. In the same way, biologists group organisms into large categories that in turn are assigned to smaller and more specific categories.

More than 2,000 years ago, the Greek philosopher and naturalist Aristotle grouped plants and animals according to their structural similarities. Later Greeks and Romans grouped plants and animals into basic categories such as oaks, dogs, and horses. Eventually each unit of classification was called a genus *(JEE nuhs),* the Latin word for "group." Starting in the Middle Ages, genera were named in Latin. The science of naming and classifying organisms is called **taxonomy** *(tak SAH nuh mee).*

Until the mid-1700s, biologists named a particular type of organism by adding descriptive phrases to the name of the genus. These phrases sometimes consisted of 12 or more Latin words. They were called polynomials (from *poly,* meaning "many," and *nomen,* meaning "name"). As you can see in **Figure 15-1,** the polynomial for the European honeybee became very large and awkward. Polynomials were often changed by biologists, so organisms were rarely known to everyone by the same name.

Figure 15-1 European honeybee. The European honeybee once had a 12-part scientific name.

Apis pubescens, thorace subgriseo, abdomine fusco, pedibus posticis glabis, untrinque margine ciliatus

A simpler system

A simpler system for naming organisms was developed by the Swedish biologist Carolus Linnaeus. His ambition was to catalog all the known kinds of organisms. In the 1750s, he wrote several books that used the polynomial system. But Linnaeus also included a two-word Latin name for each species. Linnaeus's two-word system for naming organisms is called **binomial** *(beye NOH mee uhl)* **nomenclature** (from *bi,* meaning "two"). His shorthand name for the European honeybee was *Apis mellifera,* the genus name followed by a single descriptive word. Over the past 250 years since Linnaeus first used two-part binomial species names, his approach has been universally adopted. Most of the species he described in 1753 still have the two-part names he gave them.

Integrating Different Learning Styles

Logical/Mathematical	**PE**	Quick Lab, p. 324; Performance Zone items 19, 21, and 23, p. 333
Visual/Spatial	**PE**	Performance Zone items 1, 15, 16, and 17, pp. 332–333
	ATE	Opening Demonstration, p. 318; Teaching Tips, pp. 321, 323; Demonstrations, pp. 320, 321
Interpersonal	**PE**	Performance Zone item 18, p. 333
Verbal/Linguistic	**PE**	Study Tips, pp. 321, 322; Internet Connect, p. 321; Real Life, p. 323; Performance Zone items 24 and 25, p. 333
	ATE	Reading Strategies, p. 319; Lesson Warm-up, p. 320; Teaching Tips, pp. 322, 323; Resource Link, p. 323; Closure, p. 324; Reteaching, p. 324

Scientific names are universal

The unique two-word name for a species is its scientific name. The first word is the genus to which the organism belongs. A **genus** is a taxonomic category containing similar species. An organism is assigned to a genus based on its major characteristics. For example, oak trees are placed in the genus *Quercus.* The second word identifies one particular kind of organism within the genus, called a species. A species is the basic biological unit in the Linnaean system of classification. Table 15-1 lists and describes two species of oaks in the genus *Quercus.*

The first letter of the genus name is always capitalized, and the first letter of the second word is always lowercased. Scientific names are italicized or underlined. After the first use of the full scientific name, the genus name can be abbreviated as a single letter. For example, *Quercus rubra* can be abbreviated *Q. rubra.*

The scientific name of an organism gives biologists a common way of communicating, regardless of their native language. One species may have many common names, and one common name may be used for more than one species. For example, the bird called a robin in Great Britain is *Erithacus rubicula,* and the bird called a robin in North America is *Turdus migratorius.*

The name given to a species must conform to the rules established by an international commission of scientists. All scientific names must have two Latin words or terms created according to the rules of Latin grammar. Two different organisms cannot have the same scientific name. Since all the members of a genus will share the genus name, the second word in the name of each member of that genus must be different. For example, only one species of the genus *Homo* can be given the name *sapiens.*

Study TIP

Word Origins

The word *Quercus* is Latin for "oak." The word *rubra* is Latin for "red," and the word *phellos* is Greek for "cork."

Table 15-1 Two Species of Oak

Common name	Genus	Scientific name	Traits
Red oak	*Quercus*	*Quercus rubra*	Lobed leaves; produces acorns approximately 25 mm (1 in.) long
Willow oak	*Quercus*	*Quercus phellos*	Unlobed leaves; produces acorns approximately 15 mm (0.6 in.) long

did you know?

A bestiary is an illustrated medieval text that names and describes animals in Latin.

Many of the ancient Latin names are still used today as modern genus names. Bestiaries often made no distinction between real animals and mythical ones. For example, although the Aberdeen Bestiary (written around the year 1200) describes the eagle, wolf, hyena, and bear (among many others), it also describes the satyr, the phoenix, and the dragon.

Teaching TIP

Setting the Stage

Carolus Linnaeus arranged organisms into groups based on similarities, although he did not associate the similarities with evolution. His work, however, was used a century later as a supporting argument for evolution.

internetconnect

SCILINKS NSTA

Have students research how species are named using the Web site in the Internet Connect box on page 321. Have students write a report summarizing their findings. PORTFOLIO

Demonstration

What's in a Name? — BASIC

Display photos of the following organisms, and write their scientific names on the board: *Carnegiea gigantea* (giant saguaro cactus), *Nymphaea ordata* (fragrant water lily), *Canis familiaris* (domestic dog), *Viola tricolor* (pansy with three-colored flowers), and *Peromyscus californicus* (a mouse common in California). Before telling them the common names, have students suggest why these organisms were given these particular scientific names.

Teaching TIP

Mnemonic Device

Mnemonic devices that are highly visual are often easiest to remember. Have students devise their own mnemonic device to learn the sequence of the classification levels from kingdom to species. One classic example: **K**ing **P**hilip **c**ame **o**ver **f**rom **G**ermany, **s**wimming. Students should picture a king in a crown swimming his way across the ocean from Germany. Or they may have an easier time picturing: **K**ing **P**hilip **c**ame **o**ver **f**or **g**reat **s**paghetti.

internet connect

SCiLINKS NSTA Have students research systems of classification using the Web site in the Internet Connect box on page 322. Have students write a report summarizing their findings. PORTFOLIO

CAREER

Museum Curator

Natural history museums, zoos, gardens, and aquariums hire curators to maintain and expand exhibits and collections, create new exhibits, conduct research, and provide public programs. There is overlap with the fields of research and education. A bachelor's degree in the sciences may be required, and many curators hold advanced degrees. Encourage students to find out more about the work of curators. Visit a local museum, or invite a curator from a nearby natural history or science museum to speak to the class.

Study TIP

Reviewing Information

Create your own mnemonic device (a phrase formed from the first letters of a series of words) that will help you remember the system of classification. For information about mnemonics, see the Study Skills appendix.

Scientists Use a System to Classify Organisms

Linnaeus worked out a broad system of classification for plants and animals in which an organism's form and structure are the basis for arranging specimens in a collection. He later organized the genera and species that he described into a ranked system of groups that increase in inclusiveness. The different groups into which organisms are classified have expanded since Linnaeus's time and now consist of seven levels, as shown in **Figure 15-2.**

Similar genera are grouped into a **family.** Similar families are combined into an **order.** Orders with common properties are united in a **class.** Classes with similar characteristics are assigned to a **phylum** *(FEYE luhm).* Similar phyla are collected into a **kingdom.** There are six kingdoms of living things. The two kingdoms of bacteria—Archaebacteria and Eubacteria—are prokaryotes. The other four kingdoms are Protista, Fungi, Plantae, and Animalia. The seven-level system can be divided into more-specific categories, such as superclass, subclass, superorder, and suborder. In all, more than 30 taxonomic levels are recognized.

In order to remember the seven categories of classification in their proper order, it may prove useful to memorize a phrase, such as Kindly Pay Cash Or Furnish Good Security, to remember Kingdom Phylum Class Order Family Genus Species.

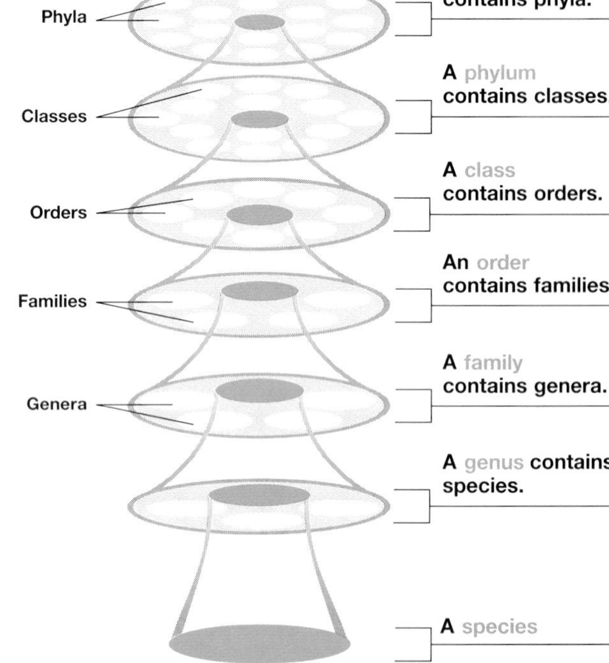

Phyla
Classes
Orders
Families
Genera
A kingdom contains phyla.
A phylum contains classes.
A class contains orders.
An order contains families.
A family contains genera.
A genus contains species.
A species

Figure 15-2 System of classification. Each living thing is assigned to a series of groups, beginning with species (least inclusive) and ending with kingdom (most inclusive).

MULTICULTURAL PERSPECTIVE

A Rose by Any Other Name . . .

Common names in different cultures can cause confusion. The flowering plant *Mertensia virginica* is known in the United States as the bluebell. In Europe and Asia, however, bluebells are plants from the genus *Endymion*. Elsewhere, certain species of the genera *Campanula, Clemitis,* and *Polemonium* also are known as bluebells.

Classification of the honeybee

Scientific names are particularly powerful because they tell you so much about the organism. Each level of classification is based on characteristics shared by all the organisms it contains. For example, consider the classification of the honeybee, shown in **Figure 15-3.** All of the world's commercial honey production is carried out by harvesting honey from hives of European varieties of this species, *Apis mellifera.* Not all varieties of the species are exactly the same, however, so a species name doesn't tell you everything. For example, Africanized "killer" bees are also *Apis mellifera.* Its scientific name, *Apis mellifera,* indicates that it belongs to the genus *Apis,* which is classified in the family Apidae. Knowing the family is Apidae tells you a great deal about members of that family. All members of the family Apidae are bees that live either alone or in hives, as does *Apis mellifera.* Knowing the order bees belong in tells you even more. The order to which the honeybee belongs, Hymenoptera, includes ants, bees, and wasps, which usually have two pairs of wings and are likely to be able to sting. At each higher level, the information becomes more general. At the next higher level of classification, *A. mellifera* belongs to the class Insecta, meaning it is an insect with three major body parts and three pairs of legs. Its phylum,

REAL LIFE

You named what kind of organism for me? Several plant and animal names honor famous people. *Shakesperia* was the original name of a genus of parasitic wasp (see *The Taming of the Shrew,* Act 2, Scene I), and cartoonist Gary Larson has both a beetle and a biting louse named for him.

Finding Information

Look up the scientific name of your favorite plant or animal, and create a binomial based on your own name.

Kingdom
Animalia

Phylum
Arthropoda

Class
Insecta

Order
Hymenoptera

Family
Apidae

Genus
Apis

Species
Apis mellifera

Figure 15-3
Classification of a bee. In the ranked system of taxonomy, the European honeybee has seven categorical names.

Graphic Organizer

Use this graphic organizer with ***Teaching Tip: Comparing Classifications*** *on page 323.*

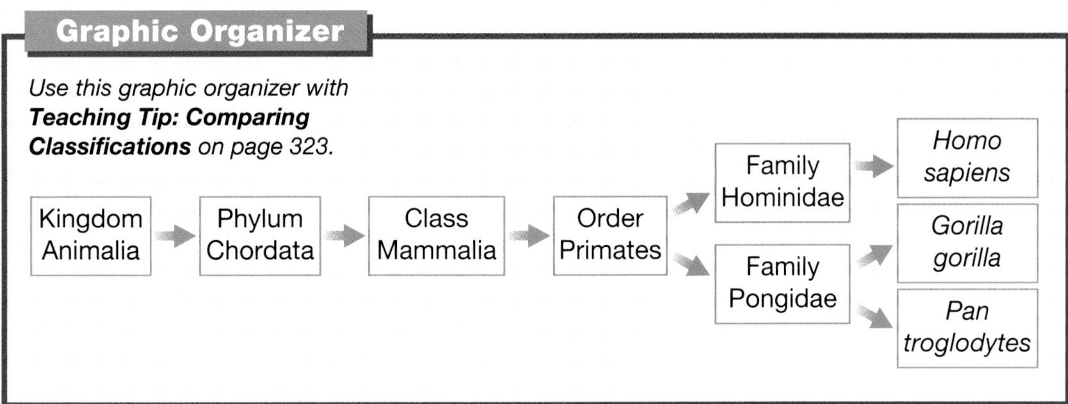

Section 15-1

REAL LIFE

Answer

Students' names should be Latinized. Point out that Carolus Linnaeus Latinized his own name; it originally was Carol von Linné.

Teaching TIP

Comparing Classifications

Ask students to devise a **Graphic Organizer** similar to the one at the bottom of page 323 that compares humans, chimpanzees, and gorillas in a hierarchy, beginning with kingdom Animalia. Provide books that help students identify the kingdom, phylum, class, order, family, genus, and species for each organism. PORTFOLIO

Teaching TIP

These Are the Names of My Favorite Things

Have students write down the common names of their favorite plants and animals. Then have them research the classification of these organisms and report to the class the scientific name of each organism, as well as its kingdom, phylum, class, order, and family. Make sure that students choose different organisms to research. PORTFOLIO

Resource Link

Assign **Basic Skills: Words and Roots** from the Holt BioSources Teaching Resources CD-ROM. PORTFOLIO

Chapter 15

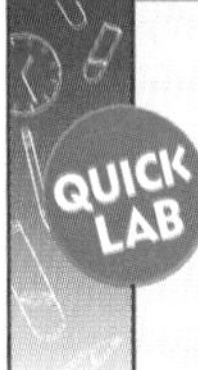

QUICK LAB: How do you use a field guide?

Time 15–20 minutes

Process Skills Observing, recognizing relationships, inferring conclusions

Teaching Strategies
- It is easiest to identify plants when they are flowering.
- Some field guides ask a series of questions to narrow down the possible choices. These will encourage students to examine specimens more carefully.

Analysis Answers
1. Answers will vary but may include branching pattern and leaf shape.
2. Species from the same genus have the same genus name.
3. Answers will vary.
4. Two organisms of the same genus share more key characteristics than two organisms of different genera.

3 Assessment Options

Closure

The Name Game

Have students determine what is incorrect about each of the following "scientific" names: spotted leopard *(common name), Tabanus b. (species name missing), Acer Rubrum (species name is capitalized),* and lumbricus terrestris *(genus name is lowercase, and names are not italic).*

Review

Assign Review—Section 15-1.

Reteaching — BASIC

Write the following scientific names on the board or overhead: *Panthera leo, Panthera tigris, Canis lupus, Canis rufus, Canis familiaris.* Have students research the meanings of each word using dictionaries and then give the common name of each organism. *(lion, tiger, timber wolf, red wolf, domestic dog)*

Arthropoda, indicates that the honeybee is an arthropod, an organism with a coelom, segmented body, jointed appendages, and a hard outer skin made of a complex carbohydrate called chitin. Arthropods have been the most successful of all animals. Two-thirds of all the named species on Earth are arthropods. Its kingdom, Animalia, tells you that *A. mellifera* is a multicellular heterotroph whose cells lack walls.

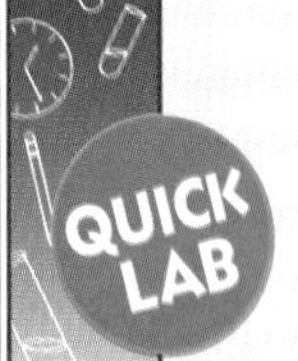

QUICK LAB: How do you use a field guide?

You can use a field guide to help you identify species of plants, animals, or other organisms.

Materials

paper and pencil, a plant or animal field guide, 3 to 6 specimens or photographs of specimens

Procedure

1. Prepare a table on your paper like the one below.
2. **CAUTION: Some plants cause ill effects when touched or eaten. Never eat any part of an unknown plant.** Observe each of the specimens or pictures provided to you. In your group, discuss the characteristics they share and the characteristics that are unique to each.
3. Using the field guides provided, determine and record the binomial name of each specimen in your table.
4. Read the description of each species in the field guide. Determine the set of characteristics that characterize each specimen. Write these characteristics in your table.

Analysis

1. **List** the characteristics shared by two specimens that are in the same genus but are different species.
2. **Describe** how the binomial name of these two species show that they are members of the same genus.
3. **Identify** the key characteristics your field guide uses to tell these two species apart.
4. **SKILL Analyzing Data** Based on your observations, are two species from the same genus more similar or less similar than two species from different genera?

Specimen	Genus name	Binomial species name	Identifying characteristics
A			
B			
C			

Review SECTION 15-1

1. **Identify** how Linnaeus helped develop the modern system of taxonomy.
2. **List** the rules that scientists use to name organisms.
3. **SKILL Inferring Relationships** Explain why *Escherichia coli* and *Entamoeba coli* are not closely related.
4. **Sequence** the seven levels of biological classification, beginning with the most specific level of organization.

Critical Thinking

5. **Justify** the preference of the binomial system of naming organisms over the polynomial system.

Review SECTION 15-1 ANSWERS

1. Linnaeus originated the practice of identifying species with two words and of grouping species into broader, more inclusive categories.
2. The first letter of the genus name is capitalized; the first letter of the species name is not. Both names are printed in italics or underlined. The names must be Latinized, and no two species of the same genus can have the same species name.
3. Although *Escherichia coli* and *Entamoeba coli* have the same descriptive species name, they belong to different genera.
4. species, genus, family, order, class, phylum, kingdom
5. The binomial system is universally accepted, and binomial names are brief. Polynomial names are long and cumbersome, and many biologists disagree about the names of certain organisms.

How Biologists Classify Organisms

Section 15-2

How Biologists Recognize Species

Have you ever wondered how scientists tell one species from another? For example, how can you tell a kind of fungus that makes antibiotics from a similar looking fungus that is poisonous? Scientists usually use differences in appearance and structure to group organisms. Sometimes, however, these differences cannot be used to determine if two organisms are of the same species. For example, the single-celled protist *Paramecium syngens* was once thought to be a single species, but it is actually composed of many different species that look the same. They now have different names.

In 1942, the biologist Ernst Mayr of Harvard University proposed a definition of species, called the biological species concept. Mayr explained that a **biological species** is a group of actually or potentially interbreeding natural populations that are reproductively isolated from other such groups. Reproductive isolation occurs when a barrier separates two or more groups of organisms and prevents them from interbreeding. However, reproductive barriers between sexually reproducing species are not always complete. Sometimes individuals of different species interbreed and produce offspring called hybrids. For example, wolves and dogs are members of separate species in the genus *Canis*. But interbreeding between wolves and dogs produces fertile offspring, such as the hybrid shown in **Figure 15-4.** Another example of a fertile hybrid is triticale, a hybrid of wheat and rye. It is important to remember that when reproductive barriers between two species are not complete, the two species are closely related.

Objectives

- **List** the characteristics that biologists use to classify organisms. (p. 325)
- **Summarize** the biological species concept. (p. 325)
- **Relate** analogous structures to convergent evolution. (p. 327)
- **Describe** how biologists use cladograms to determine evolutionary histories. (pp. 327–330)

Key Terms

biological species
convergent evolution
analogous character
phylogeny
cladistics
derived trait
cladogram
evolutionary systematics

Figure 15-4 Dog-wolf hybrid. Wolves and dogs can produce fertile offspring.

Dog *(Canis familiaris)*

Wolf *(Canis lupus)*

Dog-wolf hybrid

Integrating Different Learning Styles

Logical/Mathematical	**PE**	Data Labs, pp. 326, 329; Exploring Further, p. 328; Performance Zone items 20 and 22, p. 333
	ATE	Experimental Design, pp. 334–335
Visual/Spatial	**PE**	Study Tip, p. 329; Data Lab, p. 329; Performance Zone items 1 and 23, pp. 332–333
	ATE	Demonstration, p. 326; Visual Strategy, p. 327; Closure, p. 330
Intrapersonal	**ATE**	Lesson Warm-up, p. 325
Verbal/Linguistic	**PE**	Performance Zone item 2, p. 332
	ATE	Active Reading, p. 325

Section 15-2

1 Prepare

Scheduling

- **Block:** About 135 minutes will be needed to complete this section.
- **Traditional:** About three class periods will be needed to complete this section.
- **Planning:** See the Planning Guide on pp. 317A and 317B for lecture, classwork, and assignment options for Section 15-2. Lesson plans for Section 15-2 can be found in a lesson cycle format in ***Biology: Principles and Explorations*** **One-Stop Planner CD-ROM with Test Generator.**

Resource Link

Assign **Active Reading Worksheet Section 15-2** to help students comprehend the key concepts and visuals in this lesson.

2 Teach

Lesson Warm-up

Allow students to work with a partner. Give each pair of students a small handful of hardware—nuts, bolts, screws, washers, and so on. Have students group similar pieces into "genera," and give each piece a unique "species" name. Have students share with the class their genus and species names for their various pieces of hardware.

Active Reading

Technique: Anticipation Guide

Before reading this section, have students write the answer to the following question in their notebooks: How do scientists decide if similar organisms are the same species? Discuss their answers, and have students make any necessary corrections or additions after reading Section 15-2.

Demonstration

Deforestation of Rain Forest

Show students a map of Belgium. Tell them that an area of rain forest three times as large as Belgium is destroyed each year in the Amazon. This is roughly 50 acres per minute.

Teaching TIP

Loss of Biodiversity

Give students the following data regarding the diversity of life in the rain forest: 1,200 different species of beetles have been found on just 19 trees; 80,000 species of plants have been identified in the Amazon Basin (compared to 1,200 species in New England); Brazil has more terrestrial vertebrates than any other nation. Ask students to write an essay explaining why the South American rain forests are so rich in diversity. Have them share their ideas, then discuss the impact of deforestation on the region and the world. PORTFOLIO

Analyzing taxonomy of mythical organisms

Time 10–15 minutes

Process Skills Applying information, classifying, evaluating

Teaching Strategies

- If time allows, have students create a field guide to accompany the given information.
- Ask students to place the mythical creatures into real-life classes and orders. Ask students to list the traits they used to categorize each creature.

Analysis Answers

1. Answers will vary.
2. Answers will vary. Students may group the creatures according to size, body covering, presence of wings, or other features.
3. The biological species concept cannot be used to classify the beasts without knowledge of their breeding compatibility with other groups.

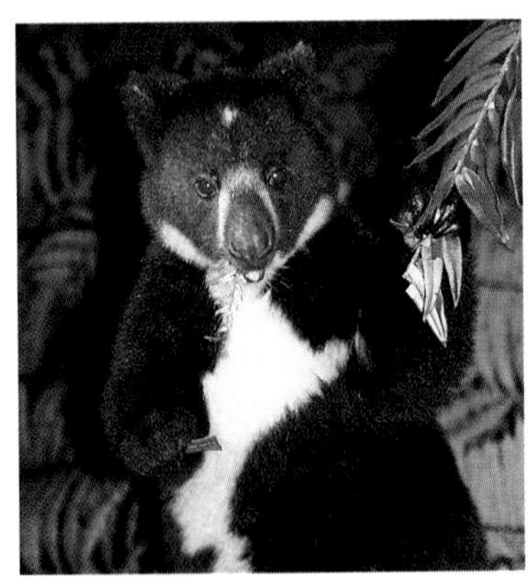

Figure 15-5 New species of tree kangaroo. *Dendrolagus mbaiso* is a black-and-white tree kangaroo that lives in the mountains of New Guinea.

Evaluating the biological species concept

The biological species concept works well for most members of the kingdom Animalia, in which strong barriers to hybridization usually exist. For example, Asian elephants and African elephants do not interbreed in nature. But the biological species concept fails to describe species that reproduce asexually, such as all species of bacteria and some species of protists, fungi, plants, and even some animals.

Within many groups of organisms, there are no barriers to interbreeding between the species. Most species of plants, some mammals, and many fishes are able to form fertile hybrids with one another. In practice, modern biologists recognize species by studying an organism's features.

How many species do we recognize?

The number of species in the world is probably greater than the number described. Large numbers of species are still being discovered, such as the kangaroo shown in **Figure 15-5.** Only about 1.5 million species have been described, and scientists estimate that many of them (more than 6 million species) live in the tropics. Since no more than 500,000 tropical species have been named, our knowledge of these organisms is limited.

Analyzing taxonomy of mythical organisms

Background

Classification of organisms often requires grouping organisms based on their characteristics. Use the following list of mythological organisms and their characteristics to complete the analysis.

- **Pegasus** stands 6 ft tall, has a horse's body, a horse's head, four legs, and two wings.
- **Centaur** stands 6 ft tall, has a horse's body with a human torso, a male human head, and four legs.
- **Griffin** stands 4–6 ft tall, has a lion's body, an eagle's head, four legs, two wings, fur on its body, and feathers on its head and wings.
- **Dragon** can grow to several hundred feet, has a snake-like body, from 1–3 reptile-like heads, four legs, scales, and breathes fire.
- **Chimera** stands 6 ft tall, has a goat's body, snake's tail, four legs, a lion's head, fur on its body and head, scales on its tail, and breathes fire.
- **Hydra** is several hundred feet long, has a long body with four legs and a spiked tail, 100 snake heads, scales, and is poisonous.

Analysis

1. **Identify** the characteristics that you think are the most useful for grouping the organisms into separate groups.
2. **Classify** the organisms into at least three groups based on the characteristics you think are most important.
3. **Evaluate** the use of the biological species concept to classify these mythical organisms.

did you know?

The Endangered Species Act of 1973 has a clear goal to protect endangered species.

A closer look at the protected animals, however, shows that some may be hybrids. Hybrids do not currently qualify for protection under the act, and some animals risk removal from the protection list. One example is the red wolf, which is thought to be a hybrid of the gray wolf and the coyote.

Taxonomy Reveals Evolutionary History

Linnaeus's classification system was based on the fact that organisms have different degrees of similarity. For instance, tigers resemble gorillas more closely than they resemble fish. According to Darwin's views, organisms that are similar descended from a common ancestor; therefore, classification provides strong evidence of evolution. However, making evolutionary connections based on similar traits can be misleading because not all traits are inherited from a common ancestor. Consider the wings of a bird and the wings of an insect. Both enable flight, but the two kinds of wings are built differently, and, based on fossil evidence, we know they evolved independently of each other. Similar traits that evolve independently, such as wings, are the result of convergent evolution. In **convergent evolution,** organisms evolve similar features independently, often because they live in similar habitats. Similar features that evolved through convergent evolution are called **analogous** *(ah NAHYL uh guhs)* **characters.** **Figure 15-6** shows an example of convergent evolution found in two different plant families, the cactus family and the spurge family.

Biologists must be able to distinguish homologous traits from analogous traits in order to reconstruct an evolutionary history. The evolutionary history of a species is called its **phylogeny** *(feye LAHJ uh nee).* Taxonomists determine the phylogeny of a species by the overall similarity between the characteristics of different kinds of organisms.

Unique characteristics help distinguish groups

One way biologists reconstruct phylogenies is through cladistics *(kluh DIHS tihks).* **Cladistics** is a system of taxonomy that reconstructs phylogenies by inferring relationships based on similarities. Cladistics is used to determine the sequence in which different groups of organisms evolved. To do this, cladistics focuses on a set of unique characteristics found in a particular group of organisms.

Figure 15-6 **Similar structures in the cactus and spurge families**

Although they evolved in different parts of the world, cactuses and spurges have similar structures.

Organ pipe cactus, *Sternocereus thurber*

Smelly euphors, *Euphorbia virosa*

biobit

Are sharks and dolphins related?

Both sharks and dolphins have body plans adapted to swimming. But these body plans and structures are analogous because sharks are fish and dolphins are mammals. Therefore, their common characters evolved along different phylogenies.

internetconnect

SCILINKS NSTA

TOPIC: Taxonomy
GO TO: www.scilinks.org
KEYWORD: HX327

Section 15-2

Teaching TIP — **ADVANCED**

Analogous Structures

Let students brainstorm and derive a list of analogous structures that are found in organisms. *(examples: insect wing/bird wing; platypus bill/duck bill)* Have them explain their hypotheses about the adaptive value of each structure.

internetconnect

Have students research taxonomy using the Web site in the Internet Connect box on page 327. Have students write a report summarizing their findings. PORTFOLIO

Visual Strategy

Figure 15-6

Point out that the organ pipe cactus is found in some of the deserts of the Western Hemisphere, and the cactuslike spurges are found in the Eastern Hemisphere. Ask students how the organisms could be so similar yet not be closely related. *(The two organisms evolved in similar environments, developing similar adaptations for survival in hot, dry deserts.)*

did you know?

Aristotle classified dolphins with land creatures rather than with fish.

Aristotle recognized that dolphins and land mammals shared many characteristics, a fact overlooked by other biologists for almost 2,000 years. It is now thought that the ancestors of modern dolphins lived on land and that their legs became flippers as they became aquatic.

Chapter 15

Teaching TIP

Cladistics and DNA

Review DNA's role as the carrier of genetic information. Emphasize that much of what taxonomists use to classify organisms—basic structures, reproductive patterns, life cycles, and development from fertilization to adulthood—is controlled by DNA. Ask students to discuss which method of grouping organisms they think is more accurate—constructing a cladogram based on anatomical features or a cladogram based on DNA. Have them justify their answer.

Exploring Further

Cladograms

Teaching Strategies

Point out that DNA analysis is changing some long-held beliefs about evolutionary relationships based on anatomical traits alone. For example, the cheetah was thought to be a unique standout in the cat family. DNA studies show, however, that it is closely related to the pumas, including the mountain lion.

Discussion

- How can DNA analysis alone be used to construct a cladogram? *(A similar chart and graph showing evolutionary relationships could be created based on the presence or absence of certain genes.)*
- How can the presence or absence of certain traits be misleading when making judgments about evolutionary relationships? *(An organism's phenotype does not necessarily reveal all of the genes that it carries. Some organisms have lost traits found in their ancestors. For example, ancestors of modern whales once had legs.)*

These unique characteristics are called **derived traits.** Using patterns of shared derived traits, a biologist using cladistics constructs a branching diagram called a **cladogram,** which shows the evolutionary relationships among groups of organisms. The key to cladistics is identifying morphological, physiological, molecular, or behavioral traits that differ among the organisms being studied and that can be attributed to a common ancestor.

Animals or plants, which share derived characters, are grouped together on the cladogram. As groups evolve, new derived characters appear on the cladogram that were not present in earlier organisms.

Exploring Further

Cladograms

How many different ways can you organize your possessions? For example, should all your clothes be grouped according to their type, or should your clothes be grouped according to color? Biologists sometimes disagree with one another about how to organize groups of organisms.

Why study cladograms?

Some biologists use cladograms to study the evolutionary relationships among certain groups of organisms, such as species within a genus or genera within a family. Cladograms show how closely two or more groups are related based on important characteristics. Cladograms convey comparative information about relationships. Organisms that are grouped more closely on a cladogram share a more recent common ancestor than those farther apart. Because the analysis is comparative, a cladogram deliberately includes an organism that is only distantly related to the other organisms. This distantly related organism is called an out-group. The out-group serves as a base line for comparisons with the other organisms being evaluated, the in-group.

Organisms	Derived traits		
	Vascular tissue	Seeds	Flowers
Mosses (out-group)	0	0	0
Pine trees	1	1	0
Flowering plants	1	1	1
Ferns	1	0	0
Total	3	2	1

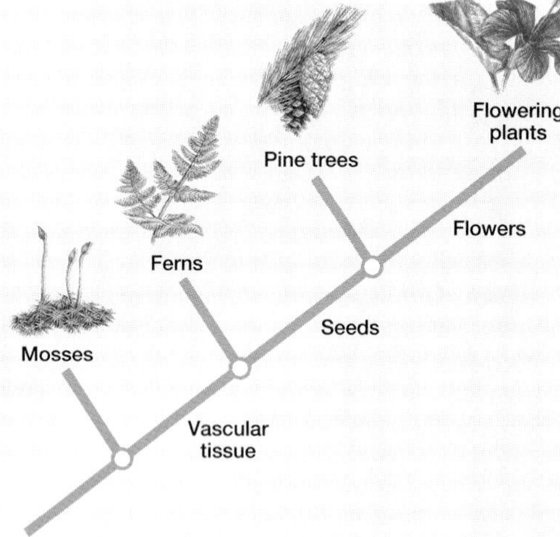

Constructing a cladogram

This example shows the evolutionary relationships among plants.

1. In the table at below left, the traits in the row for the out-group are marked with a zero. When a plant has a trait not found in the out-group, the trait is considered a derived trait and is marked with a one. Next the numbers of shared derived traits are totaled.
2. Starting with a diagonal line, as shown above, the out-group (mosses) is placed on the first branch of the cladogram. Just past this first branch, the most common derived trait is listed—vascular tissue. Vascular tissue is a series of tubes and vessels within a plant.
3. Next the second most common derived trait is determined, which in this case is seeds. The ferns lack seeds and so are placed in the second branch of the cladogram.
4. The third most common derived trait is flowers. Conifers do not have flowers and so are placed in the third branch above the second branch on the cladogram. The flowering plants are placed at the end of the cladogram.

Phylogenies Give Extra Weight to Important Characters

The great strength of a cladogram is its objectivity. If a computer is fed a certain amount of data, it will make exactly the same cladogram every time. The great disadvantage of a cladogram is that it ignores too much information. It simply indicates that a character does or does not exist.

Important characters and cladistics

A cladogram cannot take into account variations in the "strength" of a character, such as the size or location of a fin, the effectiveness of a lung, and so on. Each character is treated the same. Because evolutionary success depends so much on radical changes, such as the evolution of feathers, cladograms sometimes fail to look at information of great potential importance. So a cladogram of vertebrate evolution will group birds with reptiles, accurately reflecting their true ancestry but ignoring the immense evolutionary impact of a derived character like feathers.

Because evolutionary success depends so critically on such high-impact events, some modern cladistic studies attempt to weigh the evolutionary significance of the characters being studied. Most, however, continue to apply a "no-bias, all-characters-count-the-same" type of approach.

Study TIP

Organizing Information

The numbered list in the Exploring Further feature tells you how to make a cladogram. Use this list to prepare a cladogram of the organisms listed in the Data Lab.

DATA LAB

Making a cladogram

Background

A cladogram is one type of model used to hypothesize the order in which major lines of organisms evolved from a common ancestor. Scientists construct a cladogram by first listing the presence of derived characteristics in a data table, like the one shown on the previous page. Use the data below to construct a cladogram on a separate sheet of paper, and then complete the analysis.

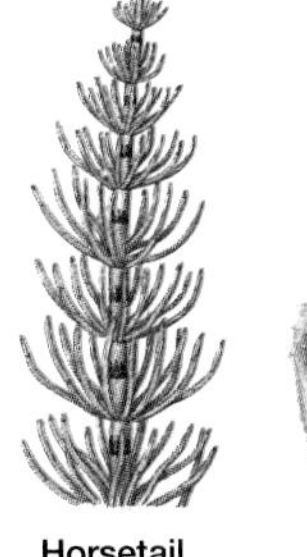

Horsetail

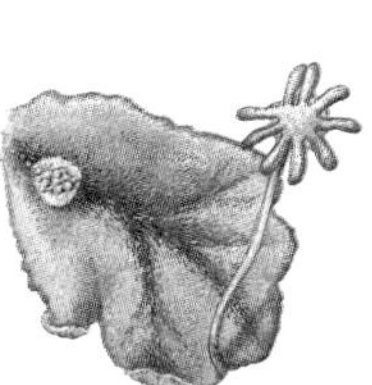

Liverwort

Analysis

1. **Identify** the out-group.
2. **Name** the least common derived trait.
3. **List** the order in which the plants in the table would be placed on a cladogram.

	Derived characteristics	
Plants	Seeds	Vascular system
Horsetails	No	Yes
Liverworts	No	No
Pine trees	Yes	Yes

Pine tree

Section 15-2

CROSS-DISCIPLINARY CONNECTION

◆ **Medicine**

Cytochrome *c* is a protein found in all aerobic organisms. The amino acid sequences of this protein have been determined for a wide variety of species and provide data about the degree of relatedness among species. For example, human and canine cytochrome *c* differ by 13 amino acids, but the 104 amino acids of human and chimpanzee cytochrome *c* match exactly.

DATA LAB

Making a cladogram

Time 5–15 minutes

Process Skills Interpreting results, applying information

Teaching Strategies

- Inform students that millions of years ago, horsetails were more treelike than are modern ones. Pines and flowering trees, which developed later, outcompeted horsetails. Today, horsetails grow primarily in wet areas and are usually less than 1 m (39 in.) tall.
- Ask students to place the horsetail and liverwort on the cladogram they constructed in the **Data Lab** on page 328. *(Students should place the horsetail with the ferns and the liverwort with the mosses. Not enough information is given to place them more precisely.)*

Analysis Answers

1. Liverworts are the out-group.
2. Seeds are the least common derived group.
3. liverwort, horsetail, pine tree

Overcoming Misconceptions

Cladistics Versus Classical Taxonomy

Taxonomy is a hotly-debated topic, and the introduction of cladistics has divided scientists into at least two camps. Students might not understand that the debates about cladistics and classical taxonomy are (or should be) less concerned with the *method* than the *result.* We can agree that there is only one true history of life on Earth. Finding a universally accepted way of representing their ideas of what occurred, however, continues to vex evolutionary biologists.

Chapter 15

internet**connect**

SCI LINKS NSTA

Have students research evolution systematics using the Web site in the Internet Connect box on page 330. Have students write a report summarizing their findings. PORTFOLIO

Making Biology Relevant

Medicine

Recent DNA analysis of cats places them surprisingly close to humans on the family tree, a relationship that taxonomists might have overlooked if they used only more easily observed physical traits to classify animals. Except for apes and monkeys, cats are more similar genetically to humans than are all other mammals. Researchers are using this knowledge to explore the cat's immune response to some of the cancers and viruses we have in common, including a virus that leads to feline AIDS.

3 Assessment Options

Closure

Categorization

Have students prepare a graphic organizer to show the relationships of the following shapes: triangle, hexagon, trapezoid, parallelogram, square, pentagon, rectangle, quadrilateral, and polygon. Ask the students to justify the rationale for their graphic organizer. Suggest that the students begin by grouping the shapes according to their general structure, and then group them according to their more specific shapes. PORTFOLIO

Review

Assign Review—Section 15-2.

Reteaching — BASIC

Let students work with a partner to devise a table and cladogram for the following animals: eagle, frog, goldfish, dog, and mouse. They should compare at least five different traits. Allow students to rank the animals as they choose. PORTFOLIO

internet**connect**

SCI LINKS NSTA

TOPIC: Evolutionary systematics
GO TO: www.scilinks.org
KEYWORD: HX330

Many taxonomists give varying degrees of importance to characters and thus produce a subjective analysis of evolutionary relationships called **evolutionary systematics.** Evolutionary systematics places birds in an entirely separate class from reptiles, as shown in **Figure 15-7,** giving more importance to characters like feathers that made powered flight possible. Evolutionary systematics involves the full observational power of the biologist, along with any biases he or she may have.

In practice, evolutionary systematics is the approach of choice when a great deal of information about a group of organisms is available. You cannot give appropriate evolutionary importance to a character unless you have enough information to make an accurate judgment. So when little information is available about how a character affects the life of an organism, cladistics is the better choice.

Figure 15-7 Evolutionary systematics and cladistics

Biologists differ in the ways that they classify organisms.

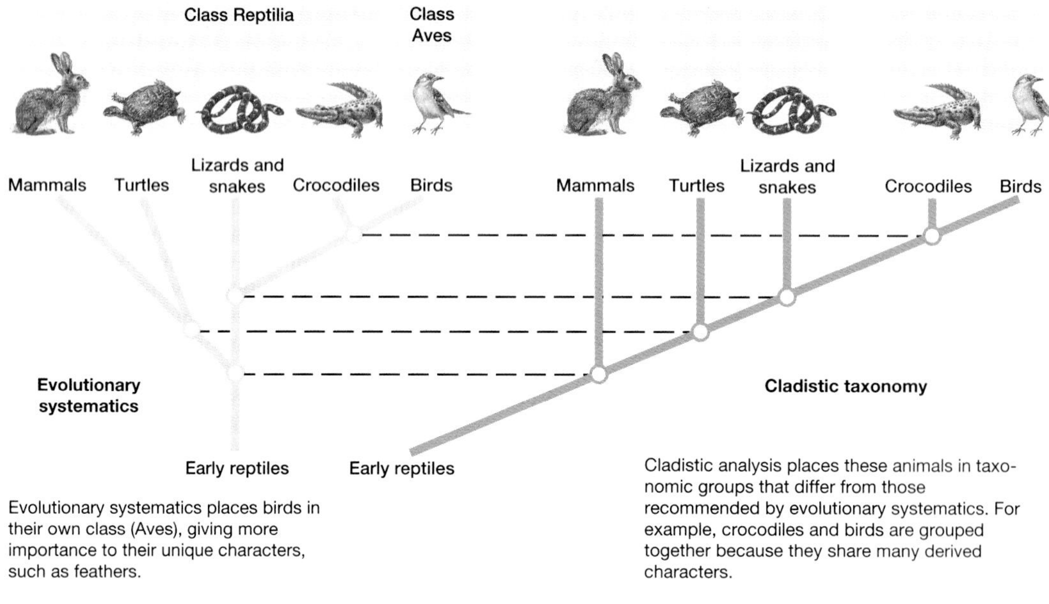

Evolutionary systematics places birds in their own class (Aves), giving more importance to their unique characters, such as feathers.

Cladistic analysis places these animals in taxonomic groups that differ from those recommended by evolutionary systematics. For example, crocodiles and birds are grouped together because they share many derived characters.

Review SECTION 15-2

1. **Identify** the kinds of information that scientists use to classify organisms.
2. **Summarize** what scientists mean by the biological species concept.
3. **Define** *cladistics,* and list the kind of information it reveals about evolutionary histories.
4. **SKILL Inferring Relationships** Explain the relationship between convergent evolution and analogous characters.

Critical Thinking

5. **SKILL Forming Reasoned Opinions** Evaluate whether a scientist should use evolutionary systematics or cladistics.

Review SECTION 15-2 ANSWERS

1. Scientists classify organisms based on appearance and structure, breeding patterns, and, more recently, similarities of DNA sequences.
2. The biological species concept defines a species as a group of naturally interbreeding populations that is reproductively isolated from other such groups.
3. Cladistics is a taxonomic system that uses similarities of organisms to infer evolutionary relationships. Scientists infer these relationships from morphological, physiological, molecular, and behavioral characteristics.
4. In convergent evolution, similar structures called analogous characters evolve in relatively unrelated species living in similar environments.
5. If enough information is available, a scientist can get a more realistic picture of evolutionary relationships using evolutionary systematics. If little information is available, then cladistics is the better choice.

Study ZONE

CHAPTER 15

Use the Key Concepts and Key Terms listed below to review the main ideas in this chapter. If you need to review the meaning of a term, turn to the page indicated.

Key Concepts

15-1 Categories of Biological Classification

- Swedish biologist Carolus Linnaeus developed binomial nomenclature, the two-word system of naming organisms.
- Scientific names are written in Latin and give biologists a common way of communicating, regardless of the language they speak.
- The scientific name of an organism consists of its genus name followed by a second name, which identifies its species.
- Each category of classification is based on characteristics that are shared by all the organisms in the category.
- Scientists use a seven-level system to classify organisms, and these levels can be further subdivided into 30 different taxonomic levels.
- The modern system of classification includes the groups kingdom, phylum, class, order, family, genus, and species.

15-2 How Biologists Classify Organisms

- Biologists usually define species according to their appearance and structure.
- The biological species concept defines species according to their sexual reproductive potential.
- The biological species concept cannot be used to classify asexually reproducing species.
- Similar organisms may have analogous structures that arose through convergent evolution.
- Cladistics focuses on sets of unique characteristics found in a particular group of organisms to reconstruct an evolutionary history.
- Evolutionary systematics is a more subjective method of classification than is cladistics, but evolutionary systematics allows greater evolutionary importance to be placed on certain characters.

Key Terms

15-1

taxonomy (320)
binomial nomenclature (320)
genus (321)
family (322)
order (322)
class (322)
phylum (322)
kingdom (322)

15-2

biological species (325)
convergent evolution (327)
analogous character (327)
phylogeny (327)
cladistics (327)
derived trait (328)
cladogram (328)
evolutionary systematics (330)

Study TIP Ready?

- *If you think you understand these Key Concepts and Key Terms, then you're ready for the Performance Zone.*

Study ZONE

CHAPTER 15

Review and Practice

Assign the following worksheets for Chapter 15:

- **Student Review Guide**
- **Concept Mapping Worksheet**
- **Test Preparation Pretest**
- **Vocabulary Worksheet**
- **Critical Thinking Review Worksheet**
- **Science Skills Worksheet**

Assessment Options

The following assessment options are available for this chapter:

Resource Link

- Assign **Chapter Test 15** to assess students' understanding of the chapter.

Alternative Assessment

Have students use a field guide to identify plants and trees in and around the schoolyard. Using weather-resistant materials, have them make identification labels that include the common and scientific names of the identified organisms. Place the labels on or near the appropriate specimens.

Portfolio Assessment

- *Portfolio Assessment* at the front of this book provides options for building and evaluating student portfolios. Students and teachers can select items from the areas listed for inclusion in a portfolio.
- See items labeled PORTFOLIO on pp. 321, 322, 323, 327, and 330.

Answer to Concept Map

The following is one possible answer to Performance Zone item 1 on page 332.

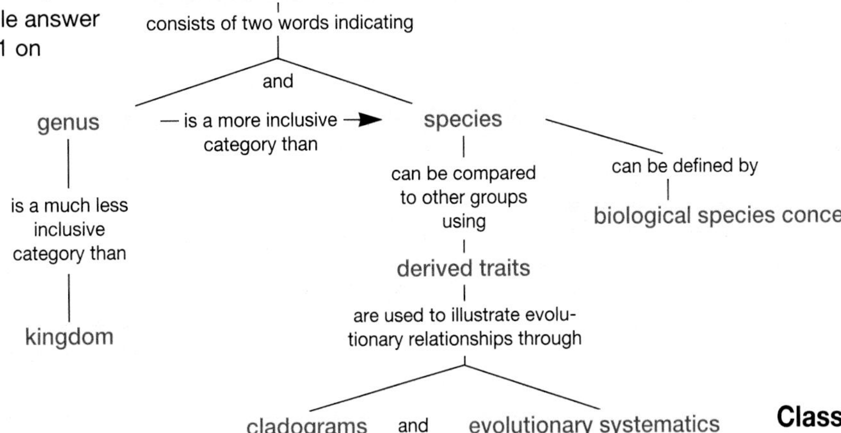

Performance ZONE
CHAPTER 15 REVIEW ANSWERS

Assignment Guide

Section	Review
15-1	3, 4, 5, 6, 7, 15, 16, 17, 18, 19, 21, 24, 25
15-2	1, 2, 8, 9, 10, 11, 12, 13, 14, 20, 22, 23

Understanding and Applying Concepts

1. The answer to the concept map is found at the bottom of page 331.
2. In convergent evolution, similar features evolve in relatively unrelated organisms as they respond to similar environments. Similar features that evolve in relatively unrelated organisms are called analogous characters. An organism's phylogeny is its evolutionary history. A graphic device that shows an organism's phylogeny and relationship to other organisms and is based on derived traits is called a cladogram. Derived traits are features of organisms that are considered when postulating the organism's phylogeny.
3. a
4. c
5. c
6. a
7. d
8. d
9. b
10. a
11. Yes, according to the biological species concept, the horse and donkey would have *fertile* offspring if they were the same species.
12. Bats and birds both have wings. Many bats and birds also migrate seasonally.
13. A cladogram can show the sequence of the development of certain traits over evolutionary time, thereby illustrating evolutionary relationships between groups of organisms.
14. Analogous structures are similar characteristics that evolved separately in two relatively unrelated organisms. Homologous structures are characteristics that may or may not appear similar in two organisms but developed from a common ancestor.

Performance ZONE CHAPTER 15 Review

Understanding and Applying Concepts

1. **Concept Mapping** Construct a concept map that shows how biologists determine the name of a new species. In constructing your map, use the following terms: genus, species, binomial nomenclature, kingdom, biological species concept, derived traits, cladogram, evolutionary systematics. Include additional concepts in your map as needed.

2. **Understanding Vocabulary** Write a sentence that shows your understanding of each of the following terms: convergent evolution, analogous characters, phylogeny, cladogram, derived traits.

3. The system of binomial nomenclature was developed by
   a. Linneaus.
   b. Aristotle.
   c. Mayr.
   d. Darwin.

4. The scientific name for humans is correctly written as
   a. Homo sapiens.
   b. Homo Sapiens.
   c. *Homo sapiens.*
   d. *Homo Sapiens.*

5. A difference between the scientific name of an organism and the classification of that organism is that
   a. the scientific name includes the family and class of the organism.
   b. the scientific name always contains three words (trinomial nomenclature).
   c. the classification includes more categories than the scientific name.
   d. classification can vary from place to place.

6. For grasshoppers and locusts to be in the same family, they must also be in the same
   a. order.
   b. group.
   c. genus.
   d. species.

7. Two members of the same _______ would be most closely related.
   a. order
   b. class
   c. family
   d. genus

8. Biologists classify organisms based on
   a. their appearance.
   b. their structure.
   c. their ability to interbreed.
   d. All of the above.

9. Convergent evolution can lead to features called
   a. cladograms.
   b. analogous characters.
   c. derived traits.
   d. wings.

10. Cladograms reveal _______ relationships.
    a. evolutionary
    b. convergent
    c. binomial
    d. analogous

11. **Theme Reproduction** The offspring of a donkey *(Equus asinus)* and a horse *(Equus caballus)* is a mule, shown below. The mule is sterile. Is the classification of donkeys and horses into different species justified according to the biological species concept? Explain.

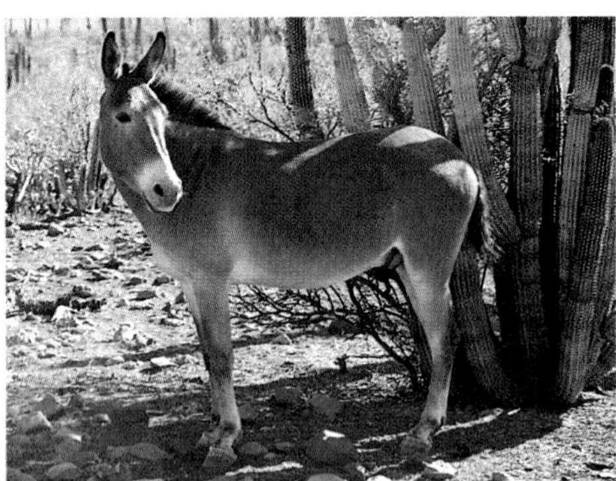

12. **Theme Evolution** What features common to both bats and birds suggest that convergent evolution has occurred?

13. **Exploring Further** List what a cladogram does to help scientists understand evolutionary relationships.

14. **Chapter Links** Contrast analogous structures with homologous structures. (**Hint:** See Chapter 13, Section 13-2.)

Skills Assessment

15. A—kingdom; B—phylum; C—class; D—order; E—family; F—genus; G—species
16. Subclass would be placed between C and D.
17. no
18. Answers will vary. Students may select fossilized or preserved specimens or photographs of organisms. They should have a representation of each of the six kingdoms of living things.
19. Answers will vary. For example, students might choose to categorize movies as action, drama, comedy, and so on. Students' choices of categories should be clear and nonoverlapping.

Skills Assessment

Part 1: Interpreting Graphics

Study this diagram, which represents the seven levels of classification. Then answer the following questions:

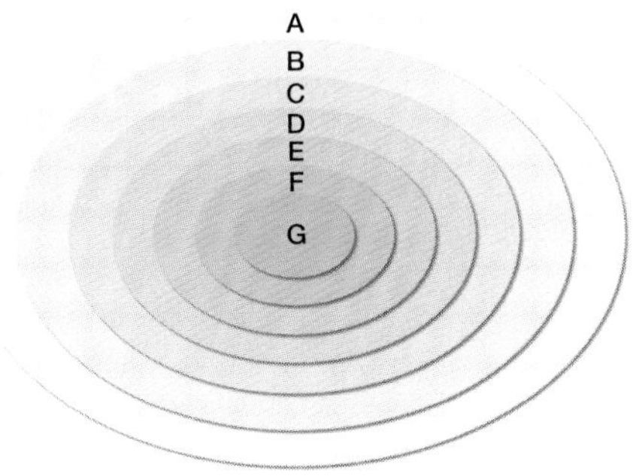

15. Identify the levels of classification labeled *A–G*.

16. Where would a subclass be placed?

17. If two different species were classified in different groups at level *E*, could they share the same name at level *G*?

Part 2: Life/Work Skills

Use the media center or Internet resources to learn how scientists apply the science of taxonomy today.

18. **Allocating Resources** You and your classmates have been allotted $500 to buy enough laboratory specimens to demonstrate the six kingdoms of life. Look at scientific supply catalogs to get ideas of what is available and how much they cost. Make a poster of your choices to "sell" your biology teacher on your ideas.

19. **Using Technology** Using computer software, work with two or three of your classmates to develop a system of classification to organize your favorite books, movies, sports teams, or musical groups. For each system of classification, include at least three levels of distinguishing characteristics. Make a poster of the classification system to share with your class.

Critical Thinking

20. **Evaluating Hypotheses** Scientists infer that groups of organisms that have homologous traits must be related, so they use cladograms to study relationships. Explain the reasoning that supports this inference.

21. **Evaluating Viewpoints** Explain this statement: Diversity is the result of evolution; classification systems are the inventions of humans.

22. **Forming Reasoned Opinions** In the laboratory, a scientist studied two identical-looking daisies that belonged to the genus *Aster.* The two plants produced fertile hybrids in the laboratory, but they never interbreed in nature because one plant flowers only in the spring and the other only in autumn. Are the plants the same species? Explain.

Portfolio Projects

23. **Organizing Information** Create a cladogram on a poster board using at least five examples of animal species common to the area where you live. Explain how you created your cladogram in a presentation to your class. Explain why you chose the derived characters that you used in preparing your cladogram.

24. **Interpreting Information** Read "A Flowering of Finds for American Botanists," by Laura Tangley (*U.S. News and World Report,* Nov. 16, 1998, vol. 125, p. 64). How do botanists explain why new plants are still being discovered in North America? What do the estimates of undiscovered plants indicate about the future for plant taxonomists?

25. **Career Focus Taxonomist** Research the field of taxonomy, and write a report on your findings. Your report should include a job description, training required, kinds of employers, growth prospects, and starting salary.

Portfolio Projects

23. Answers will vary. Students' cladograms should follow the rules for construction of cladograms presented on pp. 328 and 329.

24. Many undiscovered plant species are found in remote, isolated microhabitats. If estimates of yet-undiscovered plant species are correct, plant taxonomists have their work cut out for them for years to come.

25. Taxonomists are scientists who work on the classification of organisms. Taxonomists usually are quite specialized; that is, a scientist would focus on the taxonomy of only plants, animals, or bacteria. Many taxonomists have college degrees in a life science, such as botany or zoology, followed by postgraduate work in the same field and in taxonomy. Taxonomists often are employed by research organizations such as universities. The field is growing as the methods of classical taxonomy are being augmented by molecular biology techniques. Starting salary will vary by region. For more information, contact:

American Society of Plant Taxonomists
St. Mary's College
Department of Biology
Notre Dame, IN 46556

American Association for Zoological Nomenclature
c/o National Museum of Natural History MRC
Smithsonian Institution
Washington, DC 20560

Critical Thinking

20. Homologous traits may develop differently in different organisms, but such traits originated in a common ancestor. By arranging related organisms on the basis of the presence or absence of certain traits, scientists can gain insight into their evolutionary background.

21. The diversity that we observe is a result of ongoing evolutionary processes. Humans have developed classification systems to make sense out of the bewildering diversity.

22. No, the plants are not the same species because they do not *naturally* interbreed in the wild. They are reproductively isolated.

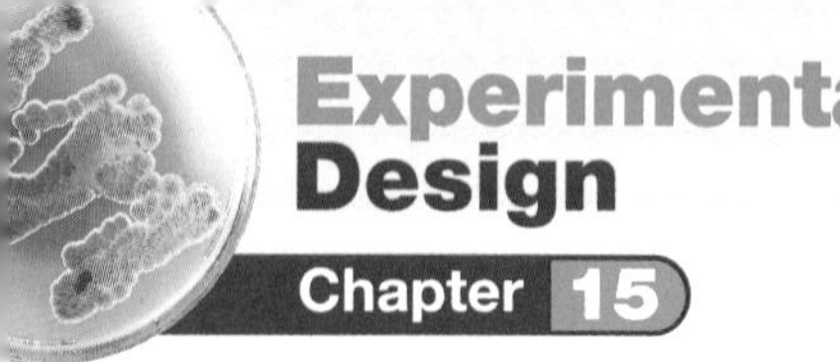

Experimental Design

Chapter 15

Dichotomous Keys

Purpose

Students will construct a dichotomous key to identify a set of objects based upon physical characteristics.

Preparation Notes

Time Required: about 50 minutes

Preparation Tips

- Having students make dichotomous keys allows them to practice observing, recording, and organizing data. Lead students through a discussion on how to use the process of elimination when working with a dichotomous key.
- You may want to have students complete parts A and B on two different days. Part A will take about 10 minutes to complete. Part B will require about 40 minutes.

Procedural Tips

- Encourage students to select objects that can be easily distinguished from one another.
- Emphasize that a dichotomous key includes pairs of descriptions that lead to the identification of an object.

Before You Begin

Answers

1. *identification key*—a tool that can be used to identify an unknown organism or object
   dichotomous key—a key that uses pairs of contrasting descriptive statements to lead to the identification of an organism or some other object
2. Answers will vary. For example: How do you use a dichotomous key?

Experimental Design

Chapter 15 *How Do You Make a Dichotomous Key?*

SKILLS

- Identifying and comparing
- Organizing data

OBJECTIVES

- **Identify** objects using dichotomous keys.
- **Design** a dichotomous key for a group of objects.

MATERIALS

- 6 to 10 objects found in the classroom (e.g., shoes, books, writing instruments)
- stick-on labels
- pencil

A

B

C

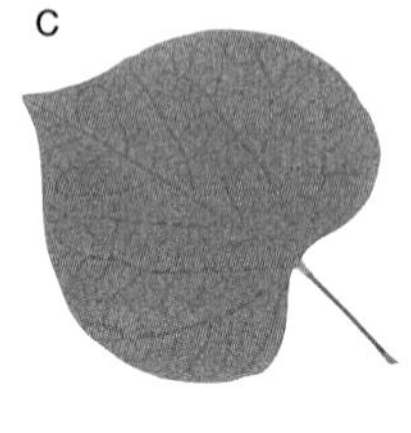

D

E

F

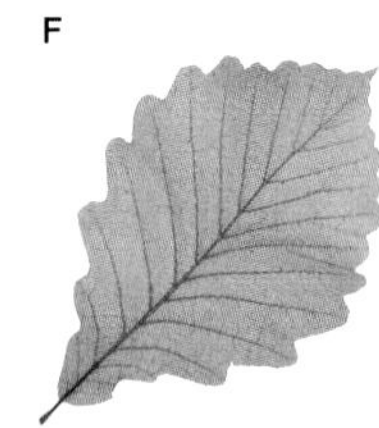

Before You Begin

One way to identify an unknown organism is to use an **identification key,** which contains the major characteristics of groups of organisms. A **dichotomous key** is an identification key that contains pairs of contrasting descriptions. After each description, a key either directs the user to another pair of descriptions or identifies an object. In this lab, you will design and use a dichotomous key. A dichotomous key can be written for any group of objects.

1. Write a definition for each boldface term in the paragraph above.
2. Based on the objectives for this lab, write a question you would like to explore about making or using a dichotomous key.

Procedure

PART A: Using a Dichotomous Key

1. Use the **Key to Forest Trees** to identify the tree that produced each of the leaves shown here. Identify one leaf at a time. Always start with the first pair of statements (**1***a* and **1***b*). Follow the direction beside the statement that describes the leaf. Proceed through the key until you get to the name of a tree.

Key to Forest Trees

1*a*	Leaf edge has no teeth, waves, or lobes	go to **2**
1*b*	Leaf edge has teeth, waves, or lobes	go to **3**
2*a*	Leaf has a bristle at its tip	**shingle oak**
2*b*	Leaf has no bristle at its tip	go to **4**
3*a*	Leaf edge is toothed	**Lombardy poplar**
3*b*	Leaf edge has waves or lobes	go to **5**
4*a*	Leaf is heart-shaped	**red bud**
4*b*	Leaf is not heart-shaped	**live oak**
5*a*	Leaf edge has lobes	**English oak**
5*b*	Leaf edge has waves	**chestnut oak**

PART B: Design a Dichotomous Key

2. Work with the members of your lab group to design a dichotomous key using the materials listed for this lab.

Procedure

Answers

3. Objects chosen for the dichotomous key should be in a single category but different from each other in appearance.
4. Students' dichotomous keys should include sufficient pairs of contrasting statements to distinguish one object from the next.
5. Answers will vary.
6. Students may find that their keys need refining.

You Choose

As you design your key, decide the following:

a. what question you will explore
b. what objects your key will identify
c. how you will label personal property
d. what distinguishing characteristics the objects have
e. which characteristics to use in your key
f. how you will organize the data you will need for writing your key

3. Before you begin writing your key, have your teacher approve the objects your group has decided to work with.
4. Using the **Key to Forest Trees** as a guide, write a key for the objects your group selected. Remember, a dichotomous key includes pairs of contrasting descriptions.
5. Use your key to explore one of the questions written for step 2 of **Before You Begin.**
6. After each group has completed step 5, exchange keys and the objects they identify with another group. Use the key you receive to identify the objects. If the key does not work, return it to the group so corrections can be made.

Part C: Cleanup

7. Clean up your work area and all lab equipment. Return lab equipment to its proper place. Wash your hands thoroughly before you leave the lab and after you finish all work.

Analyze and Conclude

1. **Drawing Conclusions** What tree produced each of the leaves shown in this lab?
2. **Forming Hypotheses** What other characteristics might be used to identify leaves with a dichotomous key?
3. **Analyzing Methods** How was the key your group designed dichotomous?
4. **Evaluating Results** Were you able to use another group's key to identify the objects for which it was written? If not, describe the problems you encountered.
5. **Analyzing Methods** Does a dichotomous key begin with general descriptions and then proceed to more specific descriptions or vice versa? Explain your answer, giving an example from your key.
6. **Further Inquiry** Write a new question about making or using keys that could be explored with another investigation.

Do You Know?

Do research in the library or media center to answer these questions:

1. Are the identification keys used in biology always dichotomous?
2. What other types of identification keys do biologists use?

Use the following Internet resources to explore your own questions about the diversity of life on Earth.

internet**connect**

SCILINKS NSTA

TOPIC: Classification
GO TO: www.scilinks.org
KEYWORD: HX335

Experimental Design

15 Chapter

Analyze and Conclude

Answers

1. A. shingle oak; B. Lombardy poplar; C. red bud; D. live oak; E. English oak; and F. chestnut oak
2. Other leaf characteristics might include whether the leaf is compound or simple, whether the leaf is needlelike, the arrangement of leaves on the branches, and the leaf's vein pattern.
3. The key was dichotomous because the descriptive statements were written as pairs of contrasting statements.
4. Answers will vary. Students might have discovered that some of the opposing statements were not contrasting, that some of the objects were not correctly described, or that the key did not include a sufficient number of descriptions to distinguish all the objects from one another.
5. Dichotomous keys proceed from general characteristics to specific characteristics. Examples from keys will vary but should reflect this gradation.
6. Answers will vary. For example: Can the members of any kind of group be identified using a dichotomous key?

Do You Know

Answers

1. No, many keys will use groups of three or more contrasting descriptions instead of only two.
2. Some keys, such as field guides, use descriptive paragraphs, diagrams, and/or photos for comparison to identify organisms.

internet**connect**

SCILINKS NSTA

Have students research classification using the Web site in the Internet Connect box on page 335. Have students write a report summarizing their findings. PORTFOLIO

UNIT 4 Principles of Ecology

Chapters

Fewer than 3,000 gray wolves exist in the lower 48 states.

in perspective

The Gray Wolf

Yesterday...

The gray wolf once roamed North America from coast to coast. In an effort to protect their livestock, early settlers hunted and killed wolves nearly to the point of extinction. By the mid-1950s, few wolves existed in the lower 48 states. **How does the relationship between predator and prey help to maintain an ecosystem? See page 368.**

Steel trap

Today...

Gray wolves play a vital role in ensuring the diversity and health of an ecosystem. Because they tend to prey upon the old, the sick, and the injured, wolves help to produce stronger populations of deer, elk, and moose. **Discover on pp. 385–390 how competition between species affects a biological community.** In each of the lower 48 states except Minnesota, the gray wolf is listed as endangered.

Biologist David Mech

Tomorrow...

Because the numbers and range of the gray wolf have been increasing, the U.S. Fish and Wildlife Service is reviewing potential changes to the Endangered Species Act protection for gray wolves. Researchers must continue to study wolf populations before a long-term plan to conserve gray wolves can be implemented. **See page 390 to learn why species diversity is important.**

Wolf researcher Diane Boyd

Chapter 16 Populations

Planning Guide

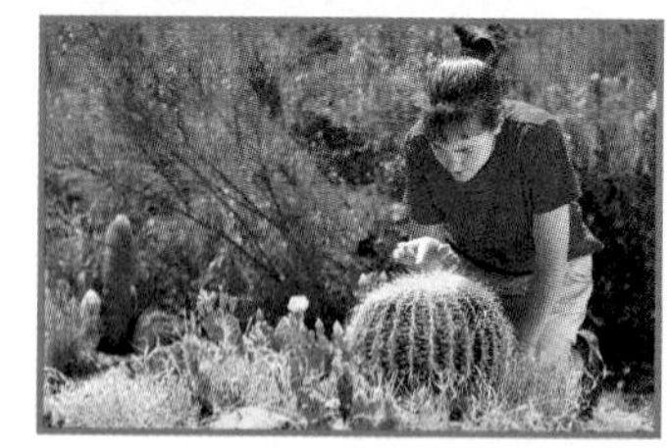

OBJECTIVES	MEETING INDIVIDUAL NEEDS	READING SKILLS AND STRATEGIES
16-1 How Populations Grow (BLOCKS 1 & 2 90 min) • **Distinguish** among the three patterns of dispersion in a population. • **Contrast** exponential growth and logistic growth. • **Differentiate** *r*-strategists from *K*-strategists.	Learning Styles, p. 340 BASIC ADVANCED • pp. 343, 345	**ATE** Active Reading, Technique: Anticipation Guide, p. 339 **ATE** Active Reading, Technique: Reading Organizer, p. 344 ■ Active Reading Guide, worksheet (16-1) ■ Directed Reading, worksheet (16-1) ■ Reading Strategies, Compare/Contrast Matrix
16-2 How Populations Evolve (BLOCKS 3 & 4 90 min) • **Summarize** the Hardy-Weinberg principle. • **Describe** the five forces that cause genetic change in a population. • **Identify** why selection against unfavorable recessive alleles is slow. • **Contrast** directional and stabilizing selection.	Learning Styles, p. 346 BASIC ADVANCED • pp. 348, 350, 352	■ Active Reading Guide, worksheet (16-2) ■ Directed Reading, worksheet (16-2) ■ Reading Strategies, Spider Map ■ Supplemental Reading Guide, *Silent Spring*

Review and Assessment Resources

BLOCKS 5 & 6 90 min

TEST PREP

- **PE** Study Zone, p. 353
- **PE** Performance Zone, pp. 354–355
- ■ Chapter 16 worksheets:
  - Vocabulary
  - Science Skills
  - Concept Mapping
  - Critical Thinking

 Holt Biology Interactive Tutor Unit 7—Ecosystem Dynamics, Topic 2

 Audio CD Program

■ Test Prep Pretest

CHAPTER TESTING

- ■ Chapter Tests (blackline master)
- ▲ Test Generator
- ■ Assessment Item Listing

See the correlations table at the front of this book for a list of the ***National Science Education Standards*** addressed in this chapter.

 Smithsonian Institution® Visit **www.si.edu/hrw** for additional on-line resources.

KEY

Biology Principles and Explorations
- **PE** Pupil's Edition
- **ATE** Teacher's Edition

BIOSOURCES
- Teaching Transparencies
- Laboratory Program

One-Stop Planner CD-ROM
Shows these resources within a customizable daily lesson plan:
- 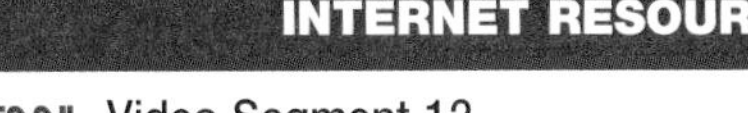■ Teaching Resources
- ▲ Test Generator

VISUAL STRATEGIES	CLASSWORK AND HOMEWORK	TECHNOLOGY AND INTERNET RESOURCES
ATE Fig. 16-4, p. 342 **ATE** Fig. 16-5, p. 343  63A Three Patterns of Population Dispersion  64A Two Types of Population Growth	Lab A13 Using Random Sampling ■ **Problem-Solving:** Population Size **PE** Review, p. 345	Video Segment 12 Britain's Living Plan Interactive Tutor Unit 7—Ecosystem Dynamics (Topic 2) Holt Biology Videodiscs Section 16-1 Audio CD Program Section 16-1 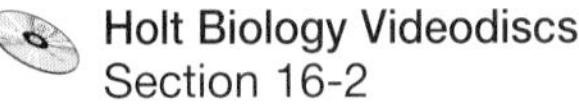internetconnect — SCILINKS NSTA TOPICS: • Population characteristics • Population growth factors • Population pyramids
ATE Fig. 16-13, p. 352 75 Human Population Growth 76 Projected Population Growth	**PE** **Quick Lab** How can you demonstrate the Hardy-Weinberg principle?, p. 349 **PE** **Math Lab** How can you build a normal distribution curve?, p. 351 **PE** **Experimental Design** How Does Natural Selection Affect a Population?, pp. 356–357 ■ **Problem-Solving:** Genetics and Probability **PE** **Review,** p. 352 **PE** **Study Zone,** p. 353	Interactive Tutor Unit 7—Ecosystem Dynamics (Topic 2) Holt Biology Videodiscs Section 16-2 Audio CD Program Section 16-2 internetconnect — SCILINKS NSTA TOPICS: • Hardy-Weinberg equation • Natural selection • Populations

ALTERNATIVE ASSESSMENT
- **PE** Portfolio Projects 22–25, p. 355
- **ATE** Alternative Assessment, p. 353
- **ATE** Portfolio Assessment, p. 353

Lab Activity Materials

Quick Lab
p. 349 equal numbers of cards marked *A* or *a* to represent the dominant and recessive alleles for a trait, paper bag

Math Labs and Data Labs
p. 351 paper, pencil, measuring tape, graph paper

Experimental Design
p. 356 metric ruler, graph paper (optional), green beans or snow peas, calculator, balance

Lesson Warm-up
ATE population statistics for your town or city and other cities, p. 340

Demonstrations
ATE photo of human population of different phenotypes, photo of a population of penguins, p. 350

BIOSOURCES LAB PROGRAM
- Quick Labs: Lab A13, p. 25

Chapter 16

Chapter Theme

Evolution

Soon after Mendel's contribution to genetics was recognized, biologists began to extend the study of heredity beyond the analysis of the genetic makeup of individuals to the genetic makeup of populations. Population genetics allows scientists to determine when evolution is occurring in populations. Scientists have identified five forces that cause populations to evolve—natural selection, mutation, gene flow, genetic drift, and nonrandom mating.

Establishing Prior Knowledge

Before beginning this chapter, make certain students can answer the questions in Study TIP Ready? on page 339. If they cannot, encourage them to reread the sections indicated.

Opening Question

Tell students some genetic abnormalities such as brachydactylism (abnormally short fingers) or achondroplasia (hereditary dwarfism) are caused by dominant alleles. Ask them why the dominant allele does not "drive out" the recessive allele in the population until most or all of the people inherit the abnormality. *(Scientists were puzzled by this question until 1908 when Hardy and Weinberg showed that the relative number of people with the normal condition should remain the same generation after generation—as long as certain conditions are satisfied.)*

CHAPTER 16 Populations

Atlantic puffins

What's Ahead — and Why

CHAPTER OUTLINE

16-1 How Populations Grow

- The Individuals of a Species Live in Populations
- Scientists Use Models to Explain How Populations Grow
- Real Populations Exhibit a Range of Growth Patterns

16-2 How Populations Evolve

- Forces That Change Population Allele Frequencies
- Natural Selection Acts Only on Phenotypes
- Natural Selection Changes Trait Distribution in a Population

Students will learn

- how populations are defined and the factors that determine population size and growth rate.
- how populations evolve.

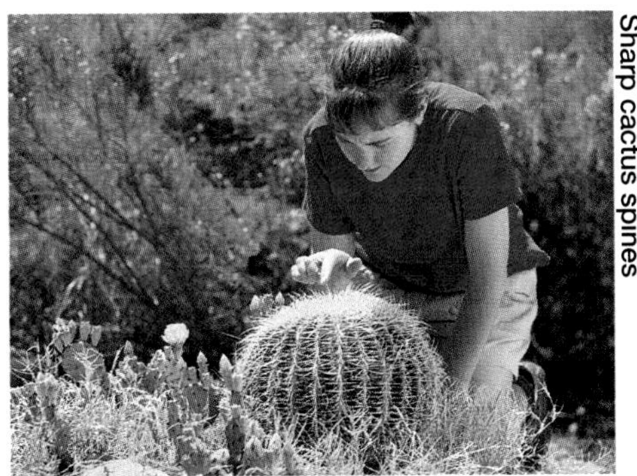
Sharp cactus spines

Feathers... thorns... survival...

Animals and plants alike must be adapted to their environment in order to survive. Webbed feet and waterproof feathers enable the puffin to thrive on cold seashores. The cactus must live without water for most of the year. Spiny armor protects its stored water from thirsty animals in the desert.

Study TIP

Ready?

Check your understanding of these topics to help you prepare for what's ahead.

Can you...

- **describe** Mendel's laws of inheritance? (Section 8-2)
- **define** *phenotype* and *genotype*? (Section 8-2)
- **define** *probability*? (Section 8-3)
- **identify** the importance of mutations? (Section 10-2)
- **define** *natural selection*? (Section 13-1)

If not, *review the sections indicated.*

Looking Ahead

Unit 7—*Ecosystem Dynamics*
Topic 2

internet**connect**

National Science Teachers Association *sci*LINKS Internet resources are located throughout this chapter.

Chapter 16

Active Reading

Technique: Anticipation Guide

Before students begin this chapter, write the following statements on the board or overhead. Have students write *true* or *false* on a sheet of paper for each statement. Students should then read the chapter to see if their predictions are correct.

1. Very small populations are more likely to become extinct than larger populations. *(true)*
2. A single bacterium dividing every 30 minutes will become a population of more than a million in only 10 hours. *(true)*
3. Natural selection acts only on genes themselves, not phenotypes. *(false)*

Directed Reading

Assign **Directed Reading Worksheet Chapter 16** to help students interact with the text as they read the chapter.

Interactive Reading

Assign ***Biology: Principles and Explorations* Audio CD Program Chapter 16** to help reluctant readers achieve greater success in reading the chapter.

Reading Skills

Assign Reading Organizers and Reading Strategies Worksheets from the **One-Stop Planner CD-ROM with Test Generator** to help students learn and practice effective reading skills.

internet**connect**

Holt *Biology: Principles and Explorations*

On-line Resources: go.hrw.com

Visit the HRW Web site for a variety of resources related to this chapter. Just type in the keyword HXO Chapter 16.

Holt *Biology: Principles and Explorations*

On-line Resources: www.scilinks.org

The following *sci*LINKS Internet resources can be found in the student text for this chapter:

Page 341
TOPIC: Population characteristics
KEYWORD: HX341

Page 344
TOPIC: Population growth factors
KEYWORD: HX344

Page 345
TOPIC: Population pyramids
KEYWORD: HX345

Page 347
TOPIC: Hardy-Weinberg equation
KEYWORD: HX347

Page 350
TOPIC: Natural selection
KEYWORD: HX350

Page 357
TOPIC: Populations
KEYWORD: HX357

Section 16-1 How Populations Grow

1 Prepare

Scheduling

- **Block:** About 90 minutes will be needed to complete this section.
- **Traditional:** About two class periods will be needed to complete this section.
- **Planning:** See the Planning Guide on pp. 337A and 337B for lecture, classwork, and assignment options for Section 16-1. Lesson plans for Section 16-1 can be found in a lesson cycle format in ***Biology: Principles and Explorations*** **One-Stop Planner CD-ROM with Test Generator.**

Resource Link

- Assign **Active Reading Worksheet Section 16-1** to help students comprehend the key concepts and visuals in this lesson.

2 Teach

Lesson Warm-up

Have the class calculate the population density of its local town or city. Ask students what information they need to get started. *(the size of the population and the size of the geographic area in question)* Supply the information, and compare their results to other nearby towns as well as to national and international cities.

Objectives

- **Distinguish** among the three patterns of dispersion in a population. (p. 341)
- **Contrast** exponential growth and logistic growth. (pp. 342–343)
- **Differentiate** *r*-strategists from *K*-strategists. (pp. 344–345)

Key Terms

population
population size
population density
dispersion
population model
exponential growth curve
carrying capacity
density-dependent factor
logistic model
density-independent factor
r-strategist
K-strategist

The Individuals of a Species Live in Populations

The commuters crowded on the New York sidewalk shown in **Figure 16-1** are members of a population. Since 1930, the world's human population has nearly tripled. What causes populations to grow? What determines how fast they grow? What factors can slow their growth?

A **population** consists of all the individuals of a species that live together in one place at one time. This definition allows scientists to use similar terms when speaking of the world's human population, the population of *Escherichia coli* bacteria that live in your intestine, or the population of Devil's Hole pupfish that swim in the tiny pool shown in **Figure 16-2.**

Figure 16-1 Rush-hour commuters

How many? In the late 1990s, more than 17 million people were living in the greater New York City area. That's more people than lived on Earth 10,000 years ago.

Every population tends to grow because pairs of individuals tend to have multiple offspring over their lifetime. But eventually, limited resources in an environment limit the growth of a population. The statistical study of all populations is called demography *(dih MAH gruh fee)*. Demographers can predict how the size of a population will change.

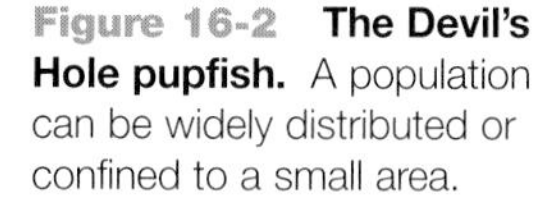

Figure 16-2 The Devil's Hole pupfish. A population can be widely distributed or confined to a small area.

Integrating Different Learning Styles

Logical/Mathematical	**PE**	Performance Zone items 8, 9, 10, 21, and 24, pp. 354–355
	ATE	Lesson Warm-up, p. 340; Visual Strategy, p. 343; Resource Link, p. 343; Health Watch, p. 345; Closure, p. 345; Reteaching, p. 345
Interpersonal	**PE**	Performance Zone item 18, p. 355
	ATE	Biology in Your Community, p. 341
Intrapersonal	**PE**	Performance Zone item 23, p. 355
	ATE	Career, p. 341
Verbal/Linguistic	**PE**	Study Tip, p. 341; Real Life, p. 342; Performance Zone items 3, 4, and 22, pp. 354–355
	ATE	Overcoming Misconceptions, p. 343; Active Reading, p. 344

Three key features of populations

Every population has features that help determine its future. One of the most important features of any population is its size. The number of individuals in a population, or **population size,** can affect the population's ability to survive. Studies have shown that very small populations are among those most likely to become extinct. Random events or natural disturbances, such as a fire or flood, endanger small populations more than they endanger larger populations. Small populations also tend to experience more inbreeding (breeding with relatives) because only relatives are available as mates. Inbreeding produces a more genetically uniform population and is therefore likely to reduce the population's fitness—more individuals will be homozygous for harmful recessive traits. For example, the worldwide cheetah population is very small, and the individuals are almost genetically identical. Biologists think that a disaster, such as a new disease, could cause their extinction.

A second important feature of a population is its density. **Population density** is the number of individuals that live in a given area. If the individuals of a population are few and are spaced widely apart, they may seldom encounter one another, making reproduction rare.

A third feature of a population is the way the individuals of the population are arranged in space. This feature is called **dispersion.** Three main patterns of dispersion are possible within a population, and each is shown in **Figure 16-3.** If individuals are randomly spaced, the location of each individual is self-determined. If individuals are evenly spaced, they are located at regular intervals. In a clumped distribution, individuals are bunched together in clusters. Each of these patterns reflects the interactions between the population and its environment.

Study TIP

Reading Effectively

Before reading this chapter, write the Objectives for each section on a sheet of paper. Rewrite each Objective as a question, and answer these questions as you read the section.

internet**connect**

TOPIC: Population characteristics
GO TO: www.scilinks.org
KEYWORD: HX341

Figure 16-3 **Patterns of dispersion**

These are the three possible patterns of dispersion in a population.

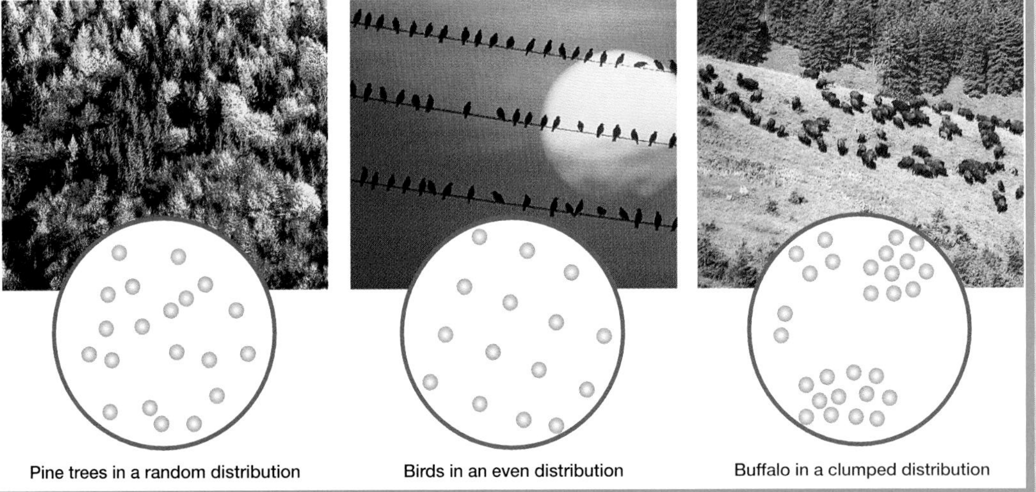

Pine trees in a random distribution | Birds in an even distribution | Buffalo in a clumped distribution

Section 16-1

CAREER

Ecologist

Ecologists are biologists who specialize in the study of how species interact within a population, with other species, and with the environment. Ecologist Dr. Carmen Cid compares her work to that of a detective: "You need to study a scene, search for clues, and test your ideas with investigations." Have students use library or internet references to find information about the work of ecologists. Have students present their findings to their classmates. PORTFOLIO

internet**connect**

SCILINKS NSTA Have students research population characteristics using the Web site in the Internet Connect box on page 341. Have students write a report summarizing their findings. PORTFOLIO

Biology in Your Community

Wildlife Biologist

Invite a wildlife biologist from your state or a federal wildlife agency to speak to your class. Have the guest discuss how population density, dispersion, and size can affect the health of animal populations and the resources they depend on. Students may be interested in how the work of wildlife biologists affects hunting seasons and bag limits. Ask the scientist to bring photographs of some of the local wildlife.

Chapter 16

REAL LIFE

Answer

Census information is used to obtain federal and state funding for many public projects. Improvements include scanning responses into computers and cooperating with state, local, and tribal governments to correct census maps and address lists.

CROSS-DISCIPLINARY CONNECTION

◆ **History**

Tell students that rapid human population growth is a relatively recent phenomenon. About 10,000 years ago, cultivation of food crops began and populations started to grow more rapidly. By the mid 1600s, advances in hygiene, health care, agriculture, commerce, and technology accelerated population growth. During the period from 1750 to 1900, the population doubled from 800 million to 1.6 billion. From 1900 to 1965, it doubled again to 3.2 billion. The population in 1999 was about 6 billion.

Teaching TIP

What Values Can *r* Take?

Explain to students that the value for *r* (growth rate) may be negative, positive, or 0. For a population in which the birthrate and death rate are the same, *r* equals 0. For those in which the birthrate exceeds the death rate, *r* is positive, and for those in which the death rate exceeds the birthrate, *r* is negative.

REAL LIFE

Uncle Sam wants to count you. The United States census, conducted every 10 years, collects detailed information on the country's population.

Finding Information

Explore Internet resources to find out more about the 2000 census. Why should every household complete a census form? What steps did the government take to improve the accuracy of the 2000 census?

Scientists Use Models to Explain How Populations Grow

When demographers try to predict how a population will grow, they make a model of the population. A **population model** is a hypothetical population that attempts to exhibit the key characteristics of a real population. By making a change in the model and observing the outcome, demographers can predict what might occur in a real population. To learn how demographers study a population, consider a simple model of population growth in three stages of complexity.

Growth rate

A population grows when more individuals are born than die in a given period. So a simple population model describes the rate of population growth as the difference between the birthrate and the death rate. This population model is called a stage I model. For human populations, birth and death rates are usually expressed as the number of births and deaths per thousand people per year.

Growth rate affects population size

When population size is plotted against time on a graph, the population growth curve resembles a J-shaped curve and is called an exponential *(ehks poh NEHN shuhl)* growth curve. An **exponential growth curve** is a curve in which the rate of population growth stays the same, as a result the population size increases steadily. This model of population growth is called a stage II model.

Figure 16-4 shows an exponential growth curve. For example, a single bacterial cell that divides every 30 minutes will produce more than 1 million bacteria after only 10 hours. To calculate the number of individuals that will be added to the population as it grows, multiply the size of the current population (N) by the rate of growth (r).

However, populations do not usually grow unchecked. Their growth is limited by predators, disease, and the availability of resources. Eventually, growth slows, and the population may stabilize. The population size that an environment can sustain is called the **carrying capacity** (K).

Figure 16-4 Exponential growth. This J-shaped curve is characteristic of exponential growth.

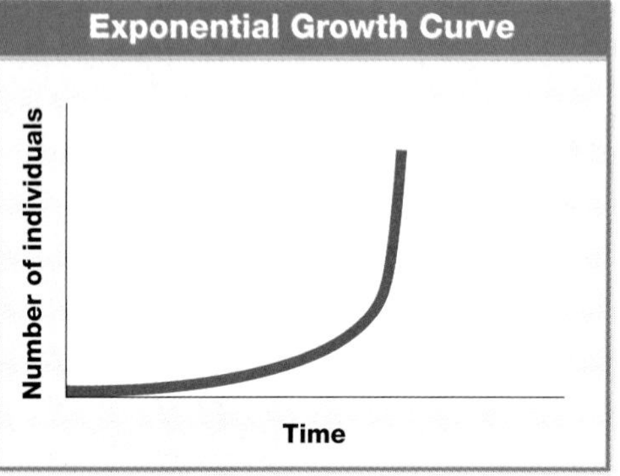

Resources affect population size

As a population grows, limited resources (that is, resources in short supply) eventually become depleted. When this happens, the growth of the population slows. The population model can be adjusted to account for the effect of limited resources, such as food and water. These resources are called **density-dependent factors** because the rate at which they become depleted depends upon the population density of the population that uses them.

did you know?

Elephants are the slowest reproducers in the animal kingdom.

Charles Darwin once calculated that a single pair of breeding elephants, which are among the slowest reproducers in the animal kingdom, would produce 19 million elephants in just 750 years. This was based on the assumption that all the offspring would live, survive a normal life span, and reproduce the normal number of offspring. Yet Darwin recognized that the average number of elephants remains fairly constant over the years. Obviously, as Darwin concluded, various forces (such as limiting factors) must be operating to keep the number constant from generation to generation.

The population model that takes into account the declining resources available to populations is called the logistic model of population growth, or stage III model. The **logistic model** is a population model in which exponential growth is limited by a density-dependent factor. The word *logistics* refers to the ability to obtain, maintain, and transport materials. In other words, logistics is about solving the day-to-day problems of living. Unlike the stage I and stage II models, the logistic model assumes that birth and death rates vary with population size. When a population is below carrying capacity, the growth rate is rapid. However, as the population approaches the carrying capacity, death rates begin to rise and birthrates begin to decline. As a result, the rate of growth slows. The population eventually stops growing when the death rate equals the birthrate. In real situations, the population may, for a short time, actually exceed the carrying capacity of its environment. If this happens, deaths will increase and outnumber births until the population falls down to the carrying capacity. Environmentalists are concerned that the Earth's human population, which passed 6 billion in 1999, may have exceeded its carrying capacity. A curve that shows logistic growth is illustrated in **Figure 16-5.**

The logistic model of population growth, though simple, provides excellent estimates of how populations grow in nature. Competition for food, shelter, mating sites, and limited resources tends to increase as a population approaches its carrying capacity. The accumulation of wastes also increases. Demographers can make logistic models based on current population sizes and predict how much a population will increase. **Figure 16-6** summarizes the three stages of a population model.

Logistic Growth

Carrying capacity

Population size

Time

Figure 16-5 Logistic growth. The curve of logistic growth looks like a stretched-out letter *S.*

Figure 16-6 Stage I, II, and III models. A population model can be described using three stages.

Population Growth Model

Stage I model: calculating the population growth rate

r (rate of growth) = birthrate − death rate

The rate of population growth equals the rate of births minus the rate of deaths.

Stage II model: exponential growth curve

ΔN (change in population, read as "delta N") = rN

Once r has been determined for a population (using the stage I model), the number of individuals that will be added to a population as it grows is equal to the rate of growth multiplied by the number of individuals in the current population (N).

Stage III model: logistic model

$$\Delta N = rN\frac{(K - N)}{K}$$

Population size calculations often need to be adjusted by the amount of resources available (K).

Visual Strategy

Figure 16-5 ADVANCED

Have students identify the three growth rates of a logistic growth curve: **1.** the initial slow growth during the establishment of a population; **2.** rapid exponential growth; and **3.** stabilization around the carrying capacity. Emphasize that fluctuations (periods of growth and decline) continue after the population reaches the carrying capacity.

Teaching TIP

Human Population Growth

Ask students in what stage is human population growth. *(Globally, the human population is still growing exponentially, but locally, the population may be stabilizing or even declining.)*

Resource Link ADVANCED

Assign **Problem Solving: Population Size** from the Holt BioSources Teaching Resources CD-ROM. PORTFOLIO

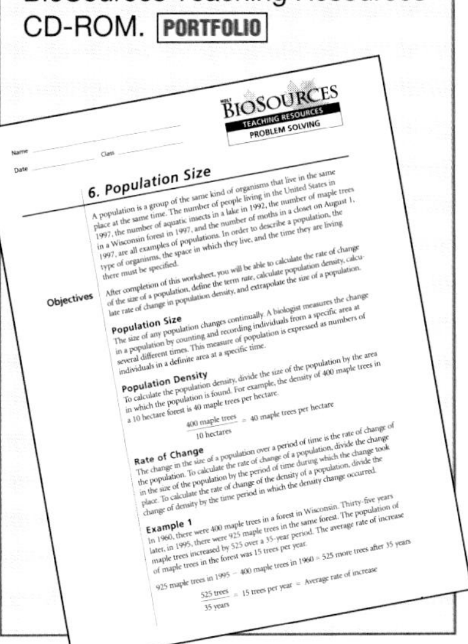
BioSources Teaching Resources Problem Solving

6. Population Size

Overcoming Misconceptions

Zero Population Growth

Many environmentalists think that we should achieve zero population growth (ZPG) to prevent widespread suffering and environmental damage as we approach the limits of our carrying capacity. Many students have the misconception that people must cease having offspring to reach ZPG. In reality, if each couple has two children, then the population would achieve ZPG.

Active Reading

Technique: Reading Organizer

Before reading this section, have students prepare a two-column table to compare the features of *r*- and *K*-strategists. Students should fill in their tables as they read. Headings for the tables are: *Growth pattern (r-: exponential,* K-: *slow); Population size (r-: temporarily large,* K-: *small); Environment (r-: unpredictable,* K-: *stable); Reproductive strategy* (r-: *early in life when conditions are favorable,* K-: *later in life under most conditions); Offspring characteristics (r-: many in number, small in size, mature rapidly, and little or no parental care;* K-: *few in number, large in size, mature slowly, and receive much parental care).* PORTFOLIO

internet connect

Have students research population growth factors using the Web site in the Internet Connect box on page 344. Have students write a report summarizing their findings. PORTFOLIO

Real Populations Exhibit a Range of Growth Patterns

biobit

How fast do cockroaches reproduce?

Blatella germanica, also called the German cockroach, produces 80 offspring every 6 months.

Many species of plants and insects reproduce rapidly. Their growth is usually limited not by density-dependent factors but by environmental conditions, also known as **density-independent factors.** Weather and climate are the most important density-independent factors. For example, mosquito populations increase in the summer, while the temperature is warm, but decrease in the winter. The growth of many plants and insects is often described by an exponential growth model. The population growth of slower growing organisms, such as bears, elephants, and humans, is better described by the logistic growth model. Most species have evolved a strategy somewhere between the two models; other species change from one strategy to the other as their environment changes.

Rapidly growing populations

internet connect

SCILINKS NSTA

TOPIC: Population growth factors
GO TO: www.scilinks.org
KEYWORD: HX344

Many species, including bacteria, some plants, and many insects like cockroaches and mosquitos, are found in rapidly changing environments. Such species, called ***r*-strategists,** grow exponentially when environmental conditions allow them to reproduce. This strategy results in temporarily large populations. When environmental conditions worsen, the population size drops quickly. In general, *r*-strategists have a short life span. In addition they reproduce early in life and have many offspring each time they reproduce. Their offspring are small, and they mature rapidly with little or no parental care. An example of a species that is an *r*-strategist is shown in **Figure 16-7.**

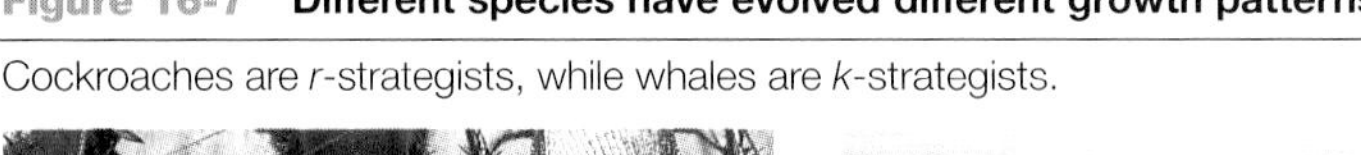

Figure 16-7 Different species have evolved different growth patterns.

Cockroaches are *r*-strategists, while whales are *k*-strategists.

Cockroaches

Humpback whale

Overcoming Misconceptions

Arctic Mosquitoes

Most people associate mosquitoes with warm weather and southern climates. However, many students may not realize that some species of mosquitoes have adapted to living in the Arctic tundra. Although the summers are short, mosquitoes living there are able to reproduce rapidly.

Slowly growing populations

Populations that grow slowly, such as whales and redwood trees, have small population sizes. These species are called ***K*-strategists** because their population density is usually near the carrying capacity (*K*) of their environment. *K*-strategists are characterized by a long life span, few young, a slow maturing process, and reproduction late in life. *K*-strategists often provide extensive care of their young. These populations tend to live in stable environments. Many endangered species, such as tigers and gorillas, are *K*-strategists. This strategy is summarized in Figure 16-7.

Population Pyramids

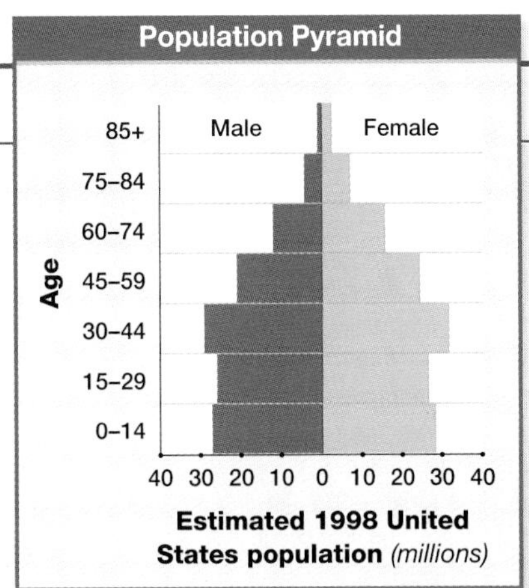

A picture is worth a thousand words, according to an old proverb. Some kinds of pictures are worth more. For example, one way of representing the structure of a large human population is a graph in which age groups are plotted on the *y*-axis and the numbers of individuals are plotted on the *x*-axis. The younger age groups appear at the bottom, and the older groups appear at the top. The resulting graphic often resembles a pyramid and thus is called a population pyramid.

Predicting future health needs

The construction of a population pyramid has many applications. From the late 1940s until 1960, for example, population pyramids for the United States were bottom-heavy with "baby boom" children, who were born up to 15 years after World War II. During this period, there was an increased demand for child-care products and pediatric care.

By 1997, the baby-boom segment of the population had moved up to the 30–54 age-group levels. Baby boomers were competing for opportunities to work, marry, and buy houses. Demands for goods and services by this age group showed increases over previous years. As the baby-boom generation ages, the need for geriatric medical care will increase.

internetconnect
SciLINKS NSTA
TOPIC: Population pyramids
GO TO: www.scilinks.org
KEYWORD: HX345

Review SECTION 16-1

1. **Identify** whether fans attending a basketball game are randomly dispersed, evenly dispersed, or clumped. Explain your answer.

2. **Differentiate** a logistic growth pattern from an exponential growth pattern.

3. **Describe** why an *r*-strategist might be better suited for an unpredictable environment than a *K*-strategist.

Critical Thinking

4. **SKILL Evaluating an Argument** According to Thomas Malthus, the human population could increase without ending if there were no wars, famines, or disease. Determine which population model best describes Malthus's views, and justify your choice.

Review SECTION 16-1 ANSWERS

1. The fans may be clumped into different sections (best viewing areas) but are evenly dispersed within each section (due to even dispersal of seats).
2. Logistic growth begins slowly, then increases exponentially, and finally stabilizes as the population nears the carrying capacity. Exponential growth shows slow growth at first then increases very rapidly.
3. If environmental conditions become favorable for even a short time, *r*-strategists are able to mature and have offspring because they reproduce and mature rapidly. For a slow-growing *K*-strategist, long, favorable periods are required for maturation and reproduction.
4. Without the increased death rate caused by famine, war, and disease, the human population would grow unchecked as characterized by the exponential growth model.

Population Pyramids

Teaching Strategies
Ask students to draw a population pyramid for a country with a very high birth rate and a high death rate for the young and those over 50 years of age. *(Their pyramid should have a very wide base and should angle sharply upwards to a narrow peak.)*

Discussion
What will the population pyramid for the United States look like if we continue to have fewer children and more adults continue to live into their 60s, 70s and 80s? *(The pyramid will become more column shaped, possibly even somewhat inverted, with a base smaller than the levels above it.)*

3 Assessment Options

Closure

Relating Concepts
Ask students to explain how dispersion and resource availability help determine the carrying capacity for a species. *(The density of a population helps determine the amount of resources available to that population. The amount of resources helps determine the carrying capacity.)*

Review

Assign Review—Section 16-1.

Reteaching — BASIC

Have students decide whether the following population is an example of an *r*-strategist or a *K*-strategist: rapid population growth under ideal conditions, no parental care, and short life span. (r-*strategist)* How would a *K*-strategist differ? *(slow population growth, parental care, and long life span)* What do the *r*- and *K*- refer to in these two types? *(The* r- *refers to the growth rate variable r because this type shows such rapid growth.* K- *refers to the carrying capacity variable K.)*

Section 16-2 How Populations Evolve

1 Prepare

Scheduling

- **Block:** About 90 minutes will be needed to complete this section.
- **Traditional:** About two class periods will be needed to complete this section.
- **Planning:** See the Planning Guide on pp. 337A and 337B for lecture, classwork, and assignment options for Section 16-2. Lesson plans for Section 16-2 can be found in a lesson cycle format in ***Biology: Principles and Explorations*** **One-Stop Planner CD-ROM with Test Generator.**

Resource Link

Assign **Active Reading Worksheet Section 16-2** to help students comprehend the key concepts and visuals in this lesson.

2 Teach

Lesson Warm-up

Ask students what the possible outcomes for two consecutive coin tosses are. *(two heads, two tails, or a head and a tail).* Point out that this is much like the possible outcomes of the combination of two alleles for a given gene. This example is identical to the Hardy-Weinberg principle they will be learning. Point out that although the probabilities of heads or tails are equal in this example, the frequencies of two alleles in a population probably will not be. Remind students that probability was discussed in Chapter 8.

Objectives

- **Summarize** the Hardy-Weinberg principle. (p. 346)
- **Describe** the five forces that cause genetic change in a population. (pp. 347–349)
- **Identify** why selection against unfavorable recessive alleles is slow. (p. 350)
- **Contrast** directional and stabilizing selection. (p. 352)

Key Terms

Hardy-Weinberg principle
gene flow
nonrandom mating
genetic drift
polygenic trait
normal distribution
directional selection
stabilizing selection

Forces That Change Population Allele Frequencies

In the 100 years after Darwin died, the science of genetics has allowed biologists to construct models of how natural selection alters the proportions of alleles within populations. But before you can understand how populations change in response to evolutionary forces, you need to learn how populations behave in the absence of these forces.

Allele frequencies

When Mendel's work was rediscovered in 1900, biologists began to study how alleles change their frequency in a population. Specifically, they wondered if dominant alleles, which are usually more common than recessive alleles, would spontaneously replace recessive alleles within populations. In 1908, the English mathematician G. H. Hardy and the German physician Wilhelm Weinberg independently demonstrated that dominant alleles do not automatically replace recessive alleles. Using algebra and a simple application of the theories of probability, they showed that the frequency of alleles in a population and the ratio of heterozygous individuals to homozygous individuals do not change from generation to generation unless the population is acted on by processes that favor particular alleles. If a dominant allele is lethal, for example, it will not become more common just because it is dominant—in fact, it will become more rare as dominant individuals die. Their discovery, called the **Hardy-Weinberg principle,** states that the frequencies of alleles in a population do not change unless evolutionary forces act on the population.

Figure 16-8 Fruit fly mutation

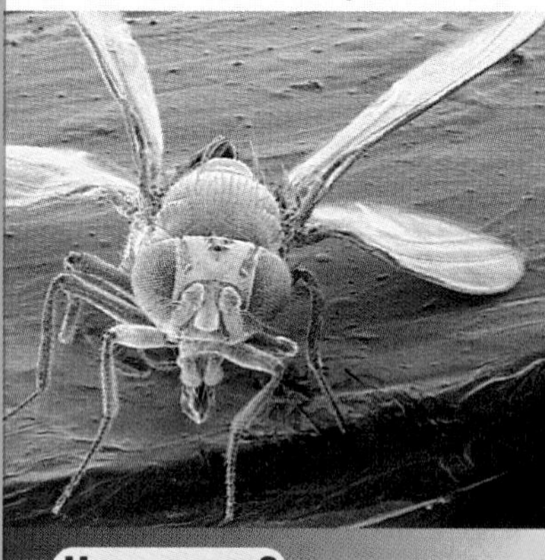

How many?

Fruit flies, such as *Drosophila melanogaster,* experience one DNA mutation in every 840,000,000,000 base pair replications.

Hardy-Weinberg

The Hardy-Weinberg principle holds true for any population as long as the population is large enough that its members are not likely to mate with relatives and as long as evolutionary forces are not acting. There are five principle evolutionary forces: mutation, shown in **Figure 16-8,** gene flow, nonrandom mating, genetic drift, and natural selection. These evolutionary forces can cause the ratios of genotypes in a population to differ significantly from those predicted by the Hardy-Weinberg principle. The Hardy-Weinberg principle can also be written as an equation that can be used to predict genotype frequencies in a population.

Integrating Different Learning Styles

Logical/Mathematical	**PE**	Exploring Further, p. 347; Quick Lab, p. 349; Math Lab, p. 351; Performance Zone items 11, 13, 14, and 15, pp. 354–355
	ATE	Resource Link, p. 348; Visual Strategy, p. 352; Closure, p. 352; Reteaching, p. 352
Visual/Spatial	**PE**	Performance Zone item 1, p. 354
	ATE	Resource Link, p. 348; Teaching Tip, p. 349; Demonstration, p. 350
Verbal/Linguistic	**PE**	Study Tips, pp. 349, 352; Performance Zone items 16 and 24, p. 355

Mutation

Although mutation from one allele to another can eventually change allele frequencies, mutation rates in nature are very slow. Most genes mutate only about 1 to 10 times per 100,000 cell divisions, so mutation does not significantly change allele frequencies, except over very long periods of time. Furthermore, not all mutations result in phenotypic changes. Recall that more than one codon can code for the same amino acid. Therefore, some mutations may result in no change in the amino acid coded for in a protein, and other changes in an amino acid that do occur may not affect how the protein works. Mutation is, however, the source of all variation and thus makes evolution possible.

TOPIC: Hardy-Weinberg equation
GO TO: www.scilinks.org
KEYWORD: HX347

Exploring Further

Using the Hardy-Weinberg Equation

You can use the Hardy-Weinberg principle to predict genotype frequencies. The Hardy-Weinberg principle is usually stated as an equation.

p^2	+	$2pq$	+	$q^2 = 1$
frequency of individuals that are homozygous for allele *A*		frequency of heterozygous individuals with alleles *A* and *a*		frequency of individuals that are homozygous for allele *a*

By convention, the frequency of the more common of the two alleles is referred to as *p*, and the frequency of the rarer allele is referred to as *q*. A frequency is the proportion of a group that is of one type. The frequency of allele *A* is the proportion of all alleles that are *A* for this gene in the population. Similarly, the frequency of allele *a* is the proportion of alleles that are *a*. The sum of the allele frequencies must always equal 1.

Individuals that are homozygous for allele *A* occur at a frequency of *p* times *p*, or p^2. Individuals that are homozygous for allele *a* occur at the frequency of *q* times *q*, or q^2. Heterozygotes have one copy of *A* and one copy of *a*, but heterozygotes can occur in two ways—*A* from the father and *a* from the mother or *a* from the father and *A* from the mother. Therefore, the frequency of heterozygotes is *2pq*.

Calculating the frequency of cystic fibrosis

How do you calculate the number of people in a crowd, like the one below, who are likely to be carriers of the cystic fibrosis gene?

1. **Calculate the frequency of the recessive allele.** Recall from Chapter 9 that cystic fibrosis is caused by the recessive allele *c*. If q^2, the frequency of recessive homozygotes, is 0.00048, then *q* is $\sqrt{0.00048}$, or 0.022.
2. **Calculate the frequency of the dominant allele *C*.**

Because
$$p + q = 1, p = 1 - q.$$
So
$$p = 1 - 0.022,$$
or 0.978.

3. **Determine the frequency of heterozygotes.**

$$2pq = 2 \times 0.978 \times 0.022 = 0.043$$

This means that 43 of every 1,000 Caucasian North Americans are predicted to carry the cystic fibrosis allele unexpressed (without disease).

Hardy-Weinberg proportions seldom, if ever, occur in nature because at least one of the five causes of evolution is always affecting populations.

Multicultural Perspective

Allele Frequencies

Allele frequencies in humans often differ among populations. The A, B, and O blood type alleles are a well-studied example. This chart shows the A-B-O frequencies for several populations.

Group	Frequency of A	Frequency of B	Frequency of O
United States (Caucasians)	0.258	0.70	0.673
United States (African-Americans)	0.168	0.131	0.701
China	0.181	0.158	0.662
Peru (Native Americans)	0.073	0.000	0.928

Section 16-2

Teaching TIP

What If Your Genes Mutate?

Tell students that when a mutation occurs in a gamete (egg or sperm cells or their predecessors), the mutation may be transmitted to the next generation. But if the mutation occurs in a nonreproductive cell, it will not be transmitted to subsequent generations. Such a mutation may disrupt the functioning of that cell, which can result in cancer. Or it may have little effect on the cell at all.

Exploring Further

Using the Hardy-Weinberg Equation

Teaching Strategies

- Ask students how do we know that $p + q = 1$? *(We are assuming that there are only two alleles for this gene, and they are found in the frequencies of* p *and* q. *Because there are no other possibilities, the sum of the two frequencies must be 1, or 100 percent of the possible alleles. Note also that if both sides of the equation* $p + q = 1$ *are squared, the equation still holds true because we treat both sides of the equation in the same manner. The result is the Hardy-Weinberg equation:* $p^2 + 2pq + q^2 = 1$.*)*
- Tell students that the Hardy-Weinberg equation applies to any trait, no matter how many alleles are involved. For example, blood type is controlled by three alleles (A, B, and O). In this case, the Hardy-Weinberg equation becomes $p + q + r = 1$, where *p*, *q*, and *r* represent the frequencies of the three alleles.

Discussion

- Which of the five forces of evolution affect the outcome of the Hardy-Weinberg proportions in human populations? *(all of them)*
- Which of the factors play the strongest role? *(Answers will vary.)*

Chapter 16

Teaching TIP

Chance Events

Tell students that in a small population, chance events, such as storms, mudslides, and fires, can change allele frequencies significantly. Have students imagine a population of 25 individuals in which two alleles, *A* and *a*, are present in frequencies of 0.8 and 0.2, respectively. Tell students there is 1 homozygous recessive individual, 8 heterozygous, and 16 homozygous dominant individuals in this population. Explain that if a mudslide kills just the one homozygous recessive individual, the frequency of this allele falls by 17 percent, to 0.167. Then tell students the loss of the same individual in a population of 100 reduces the frequency of the allele by only 4 percent. In conclusion, point out that in a population of 1,000, the frequency would decline by only 0.4 percent.

Resource Link ADVANCED

- Assign **Reading Strategies: Spider Map** from the Holt BioSources Teaching Resources CD-ROM. PORTFOLIO
- Assign **Problem-Solving: Genetics and Probability** from the Holt BioSources Teaching Resources CD-ROM. PORTFOLIO

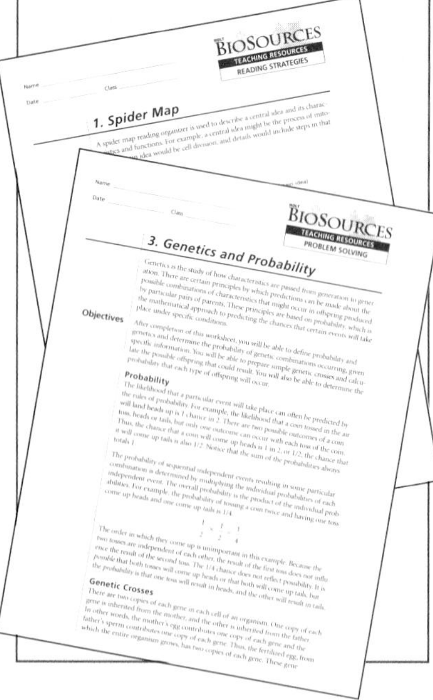
BioSources Teaching Resources Reading Strategies

1. Spider Map

BioSources Teaching Resources Problem Solving

3. Genetics and Probability

Gene flow

The movement of individuals from one population to another can cause genetic change. The movement of individuals to or from a population, called migration, creates **gene flow,** the movement of alleles into or out of a population. Gene flow occurs because new individuals (immigrants) add alleles to the population and departing individuals (emigrants) take alleles away.

Figure 16-9 Nonrandom mating. Male widowbirds, such as the one shown above, have evolved long tails because female widowbirds prefer to mate with males that have long tails than with males that have short tails.

Nonrandom mating

Sometimes individuals prefer to mate with others that live nearby or are of their own phenotype, a situation called **nonrandom mating.** Mating with relatives (inbreeding) is a type of nonrandom mating that causes a lower frequency of heterozygotes than would be predicted by the Hardy-Weinberg principle. Inbreeding does not change the frequencies of alleles, but it does increase the proportion of homozygotes in a population. For example, populations of self-fertilizing plants consist mostly of homozygous individuals. Nonrandom mating also results when organisms choose their mates based on certain traits. In animals, females often select males based on their size, color, ability to gather food, or other characteristics, as shown in **Figure 16-9.**

Genetic drift

In small populations the frequency of an allele can be greatly changed by a chance event. For example, a fire or landslide can reduce a large population to a few survivors. When an allele is found in only a few individuals, the loss of even one individual from the population can have major effects on the allele's frequency. Because this sort of change in allele frequency appears to occur randomly, as if the frequency were drifting, it is called **genetic drift.** Small populations that are isolated from one another can differ greatly as a result of genetic drift.

The cheetah, shown in **Figure 16-10,** is a species whose evolution has been seriously affected by genetic drift. Cheetahs have undergone

Figure 16-10 Cheetahs are endangered. Cheetahs have gone through at least two drastic declines in population size.

MULTICULTURAL PERSPECTIVE

Inbreeding

Tell students that inbreeding can affect a small population. Explain that the entire population of the Old Order Amish of Pennsylvania has about 17,000 people. The Old Order Amish are descended from only a few individuals whose offspring continued to marry within the group. As a result, approximately 13 percent of the Old Order Amish are homozygous recessive for a very rare allele. These individuals have a combination of dwarfism and polydactylism (extra fingers). Since the group was founded in the early 1770s, some 61 cases of this rare condition have been reported, about as many as in the rest of the world.

drastic population declines over the last 50–100 years. As a result, the cheetahs alive today are descendants of only a few individuals, and each cheetah is almost genetically uniform with other members of the population. One consequence of this genetic uniformity is reduced disease resistance—cheetah cubs are more likely to die from disease than are the cubs of lions or leopards. This reduction in genetic diversity of cheetahs may hasten their extinction.

Natural selection

Natural selection causes deviations from the Hardy-Weinberg proportions by directly changing the frequencies of alleles. The frequency of an allele will increase or decrease, depending on the allele's effects on survival and reproduction. For example, the allele for sickle cell anemia is slowly declining in frequency in the United States because individuals who are homozygous for this allele rarely have children. Heterozygotes, who are resistant to malaria, however do not have an advantage over normal homozygotes as they would have in a malaria area. As a result, homozygotes are selected against in the United States, and the frequency of the sickle cell allele decreases. Natural selection is one of the most powerful agents of genetic change.

Study TIP

Compare and Contrast

To compare and contrast the five forces that cause evolution, list each force and describe how it causes populations to evolve. Write the ways in which the five forces are alike and the ways in which they are different.

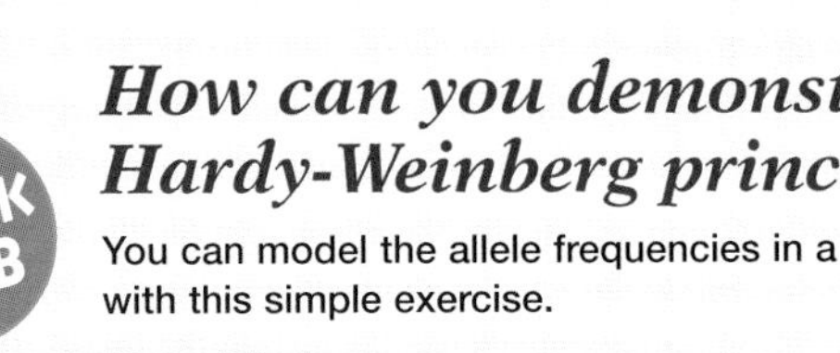

How can you demonstrate the Hardy-Weinberg principle?

You can model the allele frequencies in a population with this simple exercise.

Materials

equal numbers of cards marked *A* or *a* to represent the dominant and recessive alleles for a trait, paper bag

Procedure

1. Make a data table like the one below.
2. Work in a group, which will represent a population. Count the individuals in your group, and obtain *that* number of *both A* and *a* cards.
3. Place the cards in a paper bag, and mix them. Have each individual draw two cards, which represent a genotype. Record the genotype and phenotype in your data table.
4. Randomly exchange one "allele" with another individual in your group. Record the resulting genotypes.
5. Repeat Step 4 four more times.

Analysis

1. **Determine** the genotype and phenotype ratios in your group for each trial. Do the ratios vary among the trials?
2. **Hypothesize** what could cause a change in the "genetic makeup" of your group. Test one of your hypotheses.

DATA TABLE

	Trial 1	Trial 2	Trial 3	Trial 4	Trial 5
Genotype					
Phenotype					

Graphic Organizer

Use this graphic organizer with ***Teaching Tip: How Populations Evolve*** *on page 349.*

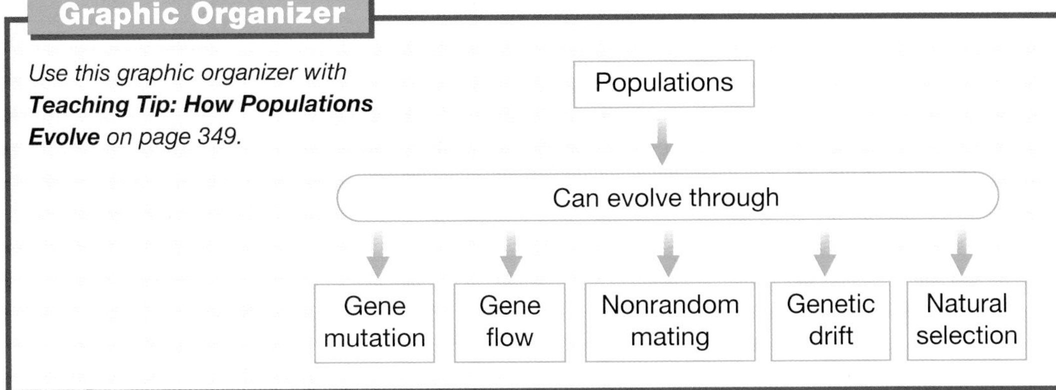

Section 16-2

Teaching TIP

How Populations Evolve

Have students create a graphic organizer that represents the five forces that cause populations to evolve: mutation, gene flow, genetic drift, nonrandom mating, and natural selection. A completed Graphic Organizer is shown at the bottom of page 349.

PORTFOLIO

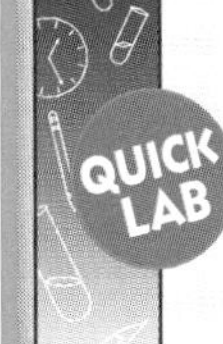

How can you demonstrate the Hardy-Weinberg principle?

Time 15–20 minutes

Process Skills Observing, applying, identifying variables, drawing conclusions

Teaching Strategies

- To increase interest, allow groups to choose the recessive and dominant phenotypes (for example, straight or curly hair).
- If groups have less than five members, then students will have to exchange alleles with the same person more than once.
- The exchange represents a different trial for both participants.

Analysis Answers

1. Unless all group members are heterozygous or all are homozygous (a rare situation, but more likely in small groups), the ratio of genotypes and phenotypes should remain constant for each trial.
2. A new individual could join the group, or a member could leave (gene flow). Homozygous recessive individuals could be prevented from passing on their alleles (natural selection). Groups may also consider genetic drift and nonrandom mating.

Chapter 16

Demonstration

Selecting Phenotypes BASIC

Reinforce the concept that natural selection operates on phenotype and not genotype. Show the class two photographs, one of a human population in which differences among phenotypes are obvious, and one in which phenotypic differences are not very apparent. The latter could be one of penguins, which, like humans, form large populations of sexually reproducing individuals. Point out that no two penguins are exactly alike, but every penguin is able to recognize its mate, its offspring, and its nesting-ground neighbors. Ask students what type of variations in the phenotypes might be favored by natural selection.

CROSS-DISCIPLINARY CONNECTION

◆ **History**

Victoria, Queen of the United Kingdom, carried the gene for hemophilia and because so many of her children married into other royal families in Europe, many royal families in Europe carry the gene for hemophilia. Have students use the Internet or other resources to research hemophilia in Queen Victoria's descendants and construct a family pedigree of her family showing how the hemophilia trait has been inherited. PORTFOLIO

Natural Selection Acts Only on Phenotypes

Natural selection constantly changes populations through actions on individuals within the population. However, natural selection does not act directly on genes. It enables individuals who express favorable traits to reproduce and pass those traits on to their offspring. This means that natural selection acts on phenotypes, not genotypes.

When selection acts

internetconnect

SciLINKS NSTA

TOPIC: Natural selection
GO TO: www.scilinks.org
KEYWORD: HX350

Think carefully about how natural selection might operate on a mutant allele. Only characteristics that are expressed can be targets of natural selection. Therefore, selection cannot operate against rare recessive alleles, even if they are unfavorable. Only when the allele becomes common enough that heterozygous individuals come together and produce homozygous offspring does natural selection have an opportunity to act. For example, Alexei Nikolayevich, the only son of the last Tsar of Russia, shown in **Figure 16-11,** suffered from hemophilia, a disease caused by a recessive gene. A cut could have led to uncontrollable bleeding and death. This kind of selection would remove a homozygous person from the gene pool. However, such an act of natural selection does not affect heterozygotes, who do not express hemophilia. Therefore, the gene is not eliminated from the population.

Why genes persist

To better understand this limitation on natural selection, consider this example. If a recessive allele (a) is homozygous in only 1 out of 100 individuals, then 18 out of 100 individuals will be heterozygous (Aa) and will therefore carry the allele unexpressed. So natural selection can act on only 1 out of every 19 individuals that carry the allele. As a result, this leaves 18 individuals that maintain the allele in the population. Many human diseases caused by recessive alleles have frequencies similar to this. For example, cystic fibrosis, the most common fatal genetic disorder among Caucasians, produces a thick mucus that clogs the lungs and organs. About 1 in 20 Caucasians has a copy of the defective gene but shows no symptoms. Homozygous recessive individuals, which include about 1 in 2,500, die from the disease. Genetic conditions are not eliminated by natural selection because very few of the individuals bearing the alleles express the recessive phenotype.

Figure 16-11
Alexei Nikolayevich and his family history

Overcoming Misconceptions

Individuals Do Not Evolve

A common misconception is that individuals evolve, changing to match changes in the environment. While the appearance of some species may change in response to the environment, the genotype of individuals does not change. Evolution is change in the genetic makeup of a population or species over time.

Natural Selection Changes Trait Distribution in a Population

Natural selection shapes populations affected by phenotypes that are controlled by one or by a large number of genes. A trait that is influenced by several genes is called a **polygenic** *(pah lee JEHN ihk)* **trait.** Human height and human skin color, for example, are influenced by dozens of genes. Natural selection can change the allele frequencies of many different genes governing a single trait, influencing most strongly those genes that make the greatest contribution to the phenotype. Like following one duck in a flock, it is difficult to keep track of a particular gene. Biologists measure changes in a polygenic trait by measuring each individual in the population and adding these measurements to calculate the average value of the trait for the population as a whole.

Because genes can have many alleles, polygenic traits tend to exhibit a range of phenotypes clustered around an average value. If you were to plot the height of everyone in your class on a graph, the values would probably form a hill-shaped curve called a **normal distribution,** with the average value at the summit, as illustrated in **Figure 16-12.**

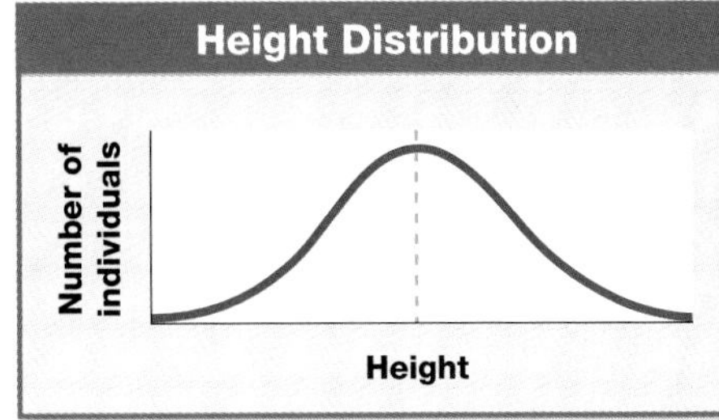

Figure 16-12 Normal distribution. This hill-shaped curve represents a normal distribution.

MATH LAB

Building a normal distribution curve

Background

You can help your class build a normal distribution curve by measuring the length of your shoes and plotting the data.

Materials

paper, pencil, measuring tape, graph paper

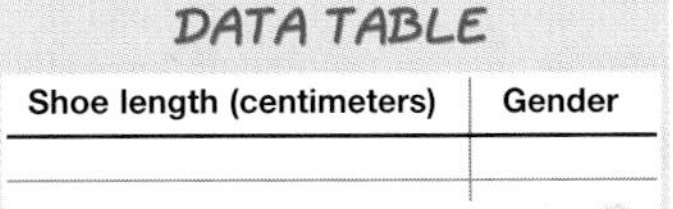

DATA TABLE

Shoe length (centimeters)	Gender

Procedure

1. Prepare a table like the one above.
2. Measure and record the length of one of your shoes to the nearest centimeter. Record your measurement and your gender.
3. Determine the number of students at each length.
4. Make a graph showing the distribution of shoe length in your class. Show the number of students on the *y*-axis and shoe length on the *x*-axis. Your graph should resemble the graph in Figure 16-12.
5. Make a second graph using data only from females.
6. Make a third graph using data only from males.

Analysis

1. **Describe** the shape of the curve that resulted from the graph you made in step 4.
2. **Distinguish** how the distribution curve for shoe length of females differs from the curve for the shoe length of males.
3. **Predict** how the distribution curve that you made in step 4 would change if the data for males were deleted.

Building a normal distribution curve

Time 20 minutes

Process Skills Measuring, applying, drawing conclusions, predicting

Teaching Strategies

- Emphasize to students that many human characteristics are polygenic, so individuals will have a range of traits. The distribution of that range usually conforms to a normal curve.
- Have students enter their data into a chart for the whole class, as well as one for their gender. This can be done on the board or an overhead to facilitate the graphing.

Analysis Answers

1. Their graphs should resemble Figure 16-12.
2. Although the curves should appear similar, the peak of the graph of female shoe lengths should be to the left of (lower than) the peak of the graph of male lengths.
3. The curve would match the curve from step 5.

Overcoming Misconceptions

Human Heredity

Most students transfer what they learn about Mendelian genetics to the inheritance pattern for all traits, including those in humans. Tell students that very few traits in humans involve only one gene with two alleles. In fact, most traits in humans are the result of polygenic inheritance. In addition to height, other polygenic traits include weight, skin color, and metabolic rate. How genes interact in most polygenic traits is not well understood.

Chapter 16

Making Biology Relevant

Food

Tell students that directional selection is often used to improve domestic plants and animals. For example, farmers will select the varieties of corn plants that produce the highest yields and cross them to produce even higher-yielding varieties.

Visual Strategy

Figure 16-13 — ADVANCED

Point out that in directional selection, the entire curve shifts in one direction, so the average becomes higher (or lower) and so do the lower and upper ends of the distribution. In stabilizing selection, individuals become more and more alike.

3 Assessment Options

Closure

Technique: Anticipation Guide

Have students refer back to their answers to the **Active Reading** exercise on page 339. Ask them if their answers to the statements were changed or confirmed.

Review

Assign Review—Section 16-2.

Reteaching — BASIC

Tell students that a certain fictional trait (fuzzy wings) is caused by a recessive allele and prevents birds from flying swiftly. Ask them if they would expect rapid or slow selection against this allele. Why? *(slow; Only recessive homozygotes would be selected against.)* Ask students if the Hardy-Weinberg principle applies to the frequency of this allele. *(no)* Ask students why or why not. *(A recessive, fuzzy wings phenotype either makes the bird unable to catch food or unable to flee from predators. So natural selection is acting on this population.)*

Study TIP

Compare and Contrast

To compare and contrast directional selection with stabilizing selection, make a two-column list. In one column write the ways the two are alike. In the other column, write the ways they are different.

Directional selection

When selection eliminates one extreme from a range of phenotypes, the alleles promoting this extreme become less common in the population. For example, when fruit flies were raised in the dark and were then allowed to fly toward light, some flew toward light and some did not. In each generation, the flies that had the strongest tendency to fly toward light were allowed to reproduce. After 20 generations, the average tendency to fly toward light increased.

This form of selection that causes the frequency of a particular trait to move in one direction is called **directional selection.** Directional selection is illustrated in the upper panel of **Figure 16-13.** This type of selection also characterizes the evolution of single-gene traits, such as pesticide resistance and antibiotic resistance in disease-causing bacteria.

Stabilizing selection

When selection eliminates extremes at both ends of a range of phenotypes, the frequencies of the intermediate phenotypes increase. As a result, the population contains fewer individuals that have alleles promoting extreme types.

As you can see in the lower panel of Figure 16-13, in **stabilizing selection,** the distribution becomes narrower, tending to "stabilize" the average by increasing the proportion of similar individuals. Stabilizing selection is very common in nature.

Figure 16-13 Two kinds of selection. This figure shows the kinds of selection for polygenic traits. Directional selection is the change of the average value of a population. Stabilizing selection is the increase of the number of average individuals in a population.

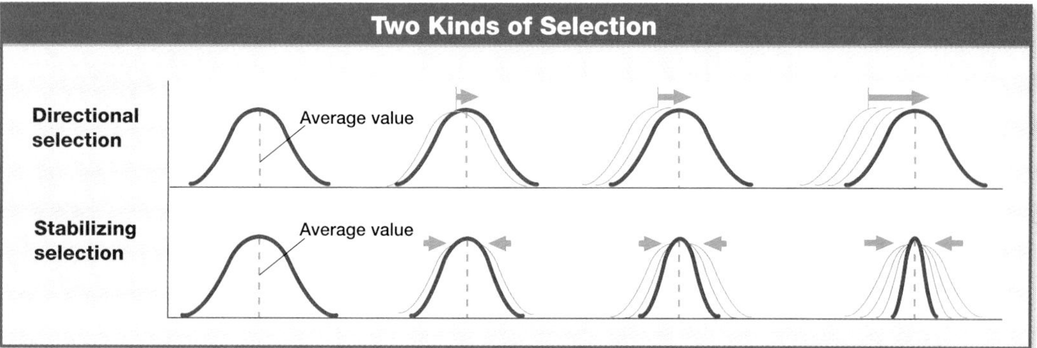

Review SECTION 16-2

1. **Describe** the Hardy-Weinberg principle.
2. **List** the five forces that cause genetic change in a population.
3. **Describe** why natural selection against an unfavorable recessive allele is a slow process.
4. **SKILL Comparing Concepts** Explain how directional selection and stabilizing selection differ.

Critical Thinking

5. **SKILL Justifying Conclusions** Recall from Chapter 8 that individuals who are heterozygous for the sickle cell allele are resistant to malaria. Does this fact mean that the frequency of the sickle cell allele will change in an area where malaria occurs? Explain why.

Review SECTION 16-2 ANSWERS

1. The Hardy-Weinberg principle states that for large populations in which evolutionary forces are not a factor, the frequencies of alleles do not change over time.
2. mutation, gene flow, nonrandom mating, genetic drift, and natural selection
3. Natural selection only operates on expressed phenotypes. Only those individuals who were homozygous recessive would be selected against, while heterozygotes would continue to maintain the allele within the population.
4. In directional selection, one extreme phenotype is favored over the other extreme, resulting in a shift of the average value of the normal distribution toward the favored trait. In stabilizing selection, the average phenotype is favored, resulting in an increase of individuals with the average value.
5. The frequency of the sickle cell allele will tend to increase in such areas because the heterozygotes are favored by natural selection. The Hardy-Weinberg principle does not apply because natural selection (in the form of malaria) acts on the population.

Study ZONE CHAPTER 16

Use the Key Concepts and Key Terms listed below to review the main ideas in this chapter. If you need to review the meaning of a term, turn to the page indicated.

Key Concepts

16-1 How Populations Grow

- A population consists of all the individuals of a species that live together in one place at one time.
- A population's future survival is determined by its size, density, and dispersion.
- Though a population's growth is limited by factors such as predation and availability of resources, a population can grow rapidly and may eventually stabilize at a size that the environment can sustain.
- Some populations grow quickly in response to density-independent factors, and other populations grow more slowly and their size is controlled by density-dependent factors.

16-2 How Populations Evolve

- The Hardy-Weinberg principle states that the frequencies of alleles and genotypes remain constant in populations in which evolutionary forces are absent.
- Allele frequencies in a population can change if evolutionary forces, such as mutation, migration, nonrandom mating, genetic drift, and natural selection, act on the population.
- Natural selection acts only on phenotype, not on genotype.
- Natural selection reduces the frequency of a harmful recessive allele slowly; very few individuals are homozygous recessive, so very few express the allele.
- The range of phenotypes that are controlled by polygenic traits result in a normal distribution when plotted on a graph.
- Directional selection results in the range of phenotypes shifting toward one extreme.
- Stabilizing selection results in the range of phenotypes narrowing.

Key Terms

16-1

population (340)
population size (341)
population density (341)
dispersion (341)
population model (342)
exponential growth curve (342)
carrying capacity (342)
density-dependent factor (342)
logistic model (343)
density-independent factor (344)
r-strategist (344)
K-strategist (345)

16-2

Hardy-Weinberg principle (346)
gene flow (348)
nonrandom mating (348)
genetic drift (348)
polygenic trait (351)
normal distribution (351)
directional selection (352)
stabilizing selection (352)

Unit 7—*Ecosystem Dynamics*

Use Topic 2 in this unit to review the key concepts and terms in this chapter.

Study TIP Ready?

- *If you think you understand these Key Concepts and Key Terms, then you're ready for the Performance Zone.*

Study ZONE CHAPTER 16

Review and Practice

Assign the following worksheets for Chapter 16:

- **Student Review Guide**
- **Concept Mapping Worksheet**
- **Test Preparation Pretest**
- **Vocabulary Worksheet**
- **Critical Thinking Review Worksheet**
- **Science Skills Worksheet**

Assessment Options

The following assessment options are available for this chapter:

Resource Link

- Assign **Chapter Test 16** to assess students' understanding of the chapter.

Alternative Assessment

- Have each student create a fictitious trait controlled by a recessive allele. If 1 in every 20,000 people express the trait, what is the frequency of heterozygotes? *(2pq = 0.014).* Have students describe any evolutionary factors at play that may affect the Hardy-Weinberg prediction. *(Answers will vary.)*

Portfolio Assessment

- *Portfolio Assessment* at the front of this book provides options for building and evaluating student portfolios. Students and teachers can select items from the areas listed for inclusion in a portfolio.
- See items labeled PORTFOLIO on pp. 341, 343, 344, 349, and 350.

Answer to Concept Map

The following is one possible answer to Performance Zone item 1 on page 354.

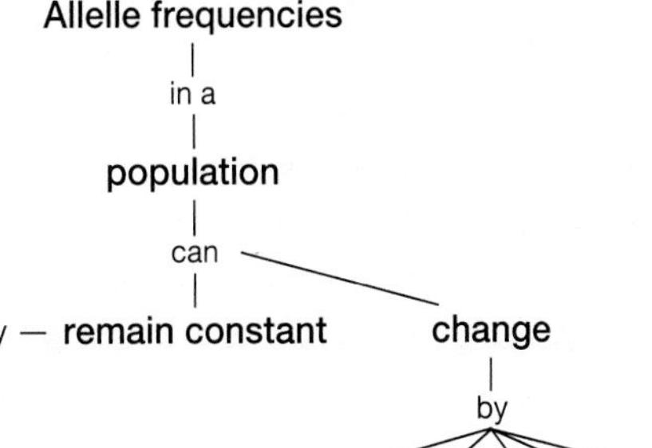

Performance ZONE
CHAPTER 16
REVIEW ANSWERS

Assignment Guide

SECTION	REVIEW
16-1	2, 3, 4, 8, 9, 10, 21, 22, 23, 24
16-2	1, 2, 5, 6, 7, 11, 12, 13, 14, 15, 16, 17, 18, 19, 20, 24

Understanding and Applying Concepts

1. The answer to the concept map is found at the bottom of page 353.
2. **a.** Demography is the statistical study of populations. Dispersion is the way individuals of a population are arranged in space.

   b. Density-dependent factors limit population size, and the abundance of these factors is affected by the density of the population that uses them. Density-independent factors limit population size, but their abundance is independent of population density.

   c. Gene flow is the movement of alleles into or out of a population as *individuals* move into or out of a population. Genetic drift is the change of allele frequency following a chance event in a small population.

   d. Directional selection favors one extreme of a given trait, shifting the normal distribution toward that extreme. In stabilizing selection, the average expression of a given trait is favored, shifting the normal distribution toward the center.
3. a
4. b
5. d
6. b
7. a
8. The carrying capacity is about 1,600 individuals, based on the way the logistic curve stabilizes then fluctuates around this value.

Performance ZONE CHAPTER 16 Review

Understanding and Applying Concepts

1. **Concept Mapping** Construct a concept map that shows how the forces of genetic change cause evolution. Use as many terms as needed from the vocabulary list. Try to include the following terms in your map: Hardy-Weinberg principle, genetic drift, nonrandom mating, natural selection, mutation, gene flow. Include additional terms in your map as needed.
2. **Understanding Vocabulary** For each pair of terms, explain the differences in their meanings.
   a. demography/dispersion
   b. density-dependent factors/density-independent factors
   c. gene flow/genetic drift
   d. directional selection/stabilizing selection
3. If spacing of individuals in a population is self-determined, then the dispersion is
   a. random.
   b. regular.
   c. clumped.
   d. None of the above
4. The growth exhibited by a colony of bacteria that has a limited food supply will most likely be
   a. exponential.
   b. logistic.
   c. natural.
   d. random.
5. According to the Hardy-Weinberg principle, allele frequencies in randomly mating populations without selection
   a. change when birthrate exceeds death rate.
   b. increase and then decrease.
   c. decrease and then increase.
   d. do not change.
6. Which of the following is *not* a cause of genetic change?
   a. genetic drift
   b. random mating
   c. natural selection
   d. mutation
7. Why is it unlikely that the frequency of hemophilia in the human population will be reduced quickly by natural selection?
   a. Natural selection acts only on recessive homozygotes.
   b. Hemophilia is not a genetic disorder.
   c. The frequency of recessive homozygotes is too great.
   d. Dominant homozygotes can have affected children.
8. **Interpreting Data** Biologists introduced pheasants onto an island in Washington State in the 1930s. Using the data shown below, determine the island's carrying capacity. How did you reach your answer?

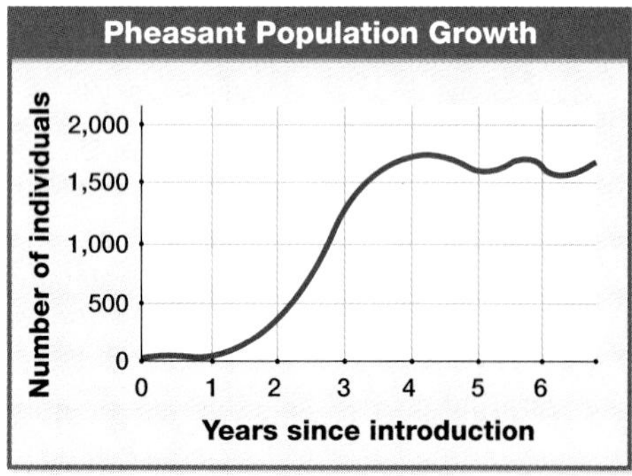

9. **Theme Evolution** After a forest fire, plants quickly recolonize the burned area. Are these plants more likely to be *r*-strategists or *K*-strategists? Explain your answer.
10. **Health Watch** By the year 2030, where will the "baby boom" age group be on the population pyramid?
11. **Exploring Further** The frequency of homozygous recessive albino rats in a population is 0.01. Calculate the expected frequency of the dominant allele in this population.
12. **Chapter Links** Is industrial melanism, as seen in the peppered moth, a case of directional or stabilizing selection? Explain your answer. (**Hint:** See Chapter 13, Section 13-3.)

9. They are more likely to be *r*-strategists, which grow and reproduce quickly, making them better suited to colonize vacant habitats.
10. Baby boomers will be in the age 60–84 portion of the pyramid (near the top).
11. The frequency of the dominant allele is 0.9.
12. This is an example of directional selection, as the moths of one extreme trait (dark) are favored over the other extreme trait (light).

Skills Assessment

13. directional selection
14. Short animals are being eliminated, as the other extreme trait (tall) is being favored.
15. Multiple genes are at work because such traits show a range of expression and are illustrated by a normal curve.
16. Answers will vary. Students should cite specific examples of evolution, such as the development of drug resistance in bacteria.
17. Answers will vary. Students' presentations should include clear visual aids that support their examples of evolution.

Skills Assessment

Part 1: Interpreting Graphs

Study the graph below, and answer the questions that follow:

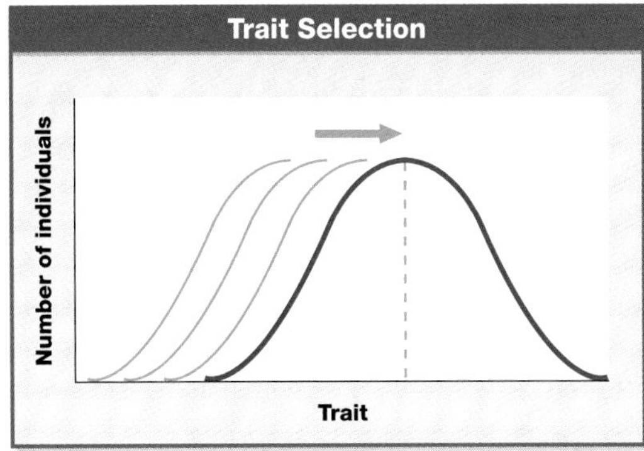

13. What type of selection is illustrated in the graph above?
14. If the trait that is measured is height, are short animals, tall animals, or both short and tall animals being eliminated from the population. Explain.
15. Is the trait that is measured more likely to be controlled by a single gene or by multiple genes? Explain your answer.

Part 2: Life/Work Skills

Use the media center or Internet resources to learn what studies scientists have done to demonstrate how natural selection causes populations to evolve.

16. **Finding Information** For each example that you find, describe what caused the population to evolve.
17. **Communicating** Prepare a presentation using multimedia software or other graphics to interpret and summarize your findings. Present your work to the class.
18. **Being a Team Member** With two or three of your classmates, visit a museum and take notes of the kinds of displays that show how natural selection causes populations to evolve.

Critical Thinking

19. **Evaluate** the argument of Thomas Malthus that famine, war, and disease keep the human population in check. Use the terminology found in this chapter to criticize or defend Malthus's ideas.
20. **Justify** the conclusion that natural selection causes populations to evolve.

Portfolio Projects

21. **Identifying Variables** Use library or Internet resources to find estimates of the current rate of human population growth. Find out if these estimates include a forecast for future growth. Present the information in the form of a report to your class.
22. **Interpreting Information** Read the article "Return of the Gray Wolf" (*National Geographic,* May 1998, pp. 72–99). What caused the disappearance of the gray wolf from many of its habitats in the United States? What characteristics of wolf populations made wolves a perceived threat to humans in those areas? What steps are being taken to reintroduce the gray wolf to these habitats? What factors stand in the way of this process? Present your findings and any other opinions you have about the reintroduction of gray wolves in a report to your class.
23. **Career Focus Population Biologist** Research the field of population biology, and write a report on your findings. Your report should include a job description, training required, kinds of employers, growth prospects, and starting salary.
24. **Unit 7—*Ecosystem Dynamics*** Write a report summarizing how the availability of resources on a ranch helps ranchers determine the carrying capacity of the ranch. What are the risks of maintaining a population of livestock that exceeds the carrying capacity of the ranch?

18. Answers will vary. Some natural history exhibits may present real data on evolution at work in short-lived organisms, such as the peppered-moth.

Critical Thinking

19. Answers will vary, but students should cite increasing death rates and falling birthrates at various times in human history as factors that limit human population.
20. If evolution is defined as the change in frequency of alleles in a population over time, and we accept that natural selection changes the frequency of alleles by selecting individuals with certain traits (alleles) over others, then it follows that natural selection causes populations to evolve.

Portfolio Projects

21. According to the U.S. Census brief *World Population Profile: 1998,* the human population reached 6 billion in 1999 and will reach 8 billion by 2026 (a growth rate of 1.22 percent per year) and 9.3 billion by 2050 (a growth rate of 0.66 percent per year). Note that the rate of growth is decreasing.
22. Wolves disappeared due to people hunting, trapping, and poisoning them. Their strength, intelligence, and pack behavior makes them competitors for meat and to some, a perceived threat. Wolves have been moved from Canada, where they are plentiful, to parts of Idaho and Yellowstone National Park. Ranchers fear these wolves will kill livestock and are fighting reintroduction efforts.
23. Population biologists study the dynamics of populations. Population biologists generally have a graduate degree in a life science such as botany or zoology. Employers include universities, government agencies, and various private industries. As concern for endangered species increases, this field may grow. Starting salary will vary by region.
24. Answers will vary. Reports should include information about resources needed to raise cattle and how too large a herd might exhaust those resources. The result would most likely be decreased profits.

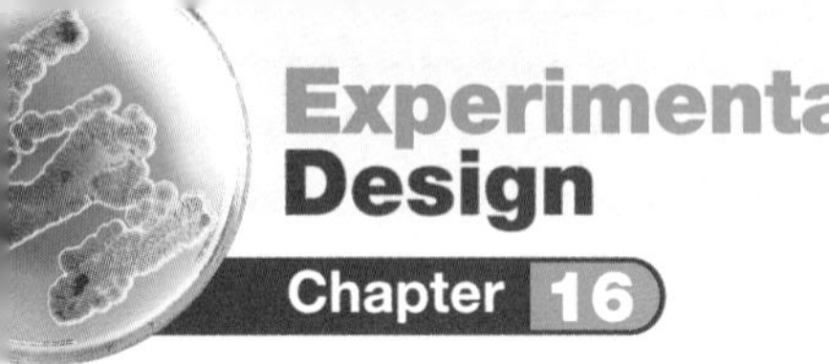

Experimental Design

Chapter 16

How Does Natural Selection Affect a Population?

Purpose

Students will examine various characteristics of an organism to analyze the variation of certain traits in a population.

Preparation Notes

Time Required: two 40-minute periods

Materials

Materials for this lab can be ordered from WARD'S. See *Master Materials List* at the front of this book for catalog numbers.

Preparation Tips

- Encourage students to read the lab and complete the **Before You Begin** section before coming to class on the day of the lab.
- Provide enough metric rulers so each lab group has one. Standard graph paper will be sufficient for the graph. The balance should be sensitive enough to measure 0.1 g.

Procedural Tips

- During the first period, have students make the measurements and record their data. During the second period, have students graph their data and answer the questions.
- Pooling the data will increase the size of the populations studied, thus making students' statistical analyses more valid. To compile class data, record each group's data on the board and have students that measured a similar trait pool the data.
- Make sure students choose a sufficient number of beans or peas to measure so that their data will be more meaningful. Students might choose to measure length of beans or peas, number of seeds in beans or peas, etc.

Experimental Design

Chapter 16

How Does Natural Selection Affect a Population?

SKILLS

- Using scientific methods
- Collecting, graphing, and analyzing data

OBJECTIVES

- **Measure** and collect data for a trait in a population.
- **Graph** a frequency distribution curve of your data.
- **Analyze** your data by determining its mean, median, mode, and range.
- **Predict** how natural selection can affect the variation in a population.

MATERIALS

- metric ruler
- graph paper (optional)
- green beans or snow peas
- calculator
- balance

Before You Begin

Natural selection can occur when there is **variation** in a **population.** You can analyze the variation in certain traits of a population by determining the mean, median, mode, and range of the data collected on several individuals. The **mean** is the sum of all data values divided by the number of values. The **median** is the midpoint in a series of values. The **mode** is the most frequently occurring value. The **range** is the difference between the largest and smallest values. The variation in a characteristic can be visualized with a **frequency distribution curve.** Two kinds of natural selection—**stabilizing selection** and **directional selection**—can influence the frequency and distribution of traits in a population. This changes the shape of a frequency distribution curve. In this lab, you will investigate variation in fruits and seeds.

1. Write a definition for each boldface term in the paragraph above.
2. Based on the objectives for this lab, write a question you would like to explore about variation in green beans or snow peas.

Procedure

PART A: Design an Experiment

1. Work with the members of your lab group to explore one of the questions written for step 2 of **Before You Begin.** To explore the question, design an experiment that uses the materials listed for this lab.

> **You Choose**
>
> **As you design your experiment, decide the following:**
>
> a. what question you will explore
> b. what hypothesis you will test
> c. which trait (length, color, weight, etc.) you will measure
> d. how you will measure the trait
> e. how many members of the population you will measure (keep in mind that the more data you gather, the more revealing your frequency distribution curve will be)
> f. what data you will record in your data table

2. Write a procedure for your experiment. Make a list of all the safety precautions you will take. Have your teacher approve your procedure and safety precautions before you begin the experiment.
3. Conduct your experiment.

Before You Begin

Answers

1. *natural selection*—the process by which populations change in response to their environment, as better adapted individuals leave more offspring

   variation—a difference in traits among individuals in a population

   population—a group of individuals that belong to the same species, live in the same area, and breed with others in the group

   mean—a mathematical average

   median—the midpoint in a series of values

   mode—a value that occurs most frequently

   range—the difference between the largest and smallest values

   frequency distribution curve—a type of graph that maps the number of occurrences of each of a set of variations of a trait

   stabilizing selection—a type of selection that increases the proportion of individuals with the average trait

   directional selection—a type of selection that increases one extreme form of a trait over another form
2. Answers will vary. For example: How does the size of a population of green beans that has limited resources change over time?

PART B: Cleanup and Disposal

4.  Dispose of seeds in the designated waste containers. Do not put lab materials in the trash unless your teacher tells you to do so.
5. Clean up your work area and all lab equipment. Return lab equipment to its proper place. Wash your hands thoroughly before you leave the lab and after you finish all work.

Analyze and Conclude

1. **Summarizing Results** Make a frequency distribution curve of your data. Plot the trait you measured on the *x*-axis (horizontal axis) and the number of times that trait occurred in your population on the *y*-axis (vertical axis).
2. **Calculating** Determine the mean, median, mode, and range of the data for the trait you studied.
3. **Analyzing Results** How does the mean differ from the mode in your population?
4. **Drawing Conclusions** What type of selection appears to have produced the type of variation observed in your experiment?
5. **Evaluating Data** The graph below shows the distribution of wing length in a population of birds on an island. Notice that the mean and the mode are quite different. Is the mean always useful in describing traits in a population? Explain.

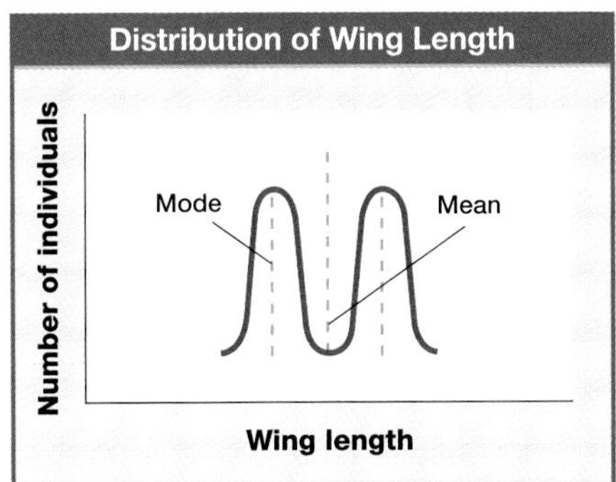

6. **Forming Hypotheses** What type of selection (stabilizing or directional) would be indicated if the mean of a trait you measured shifted, over time, to the right of a frequency distribution graph?
7. **Further Inquiry** Write a new question about variation in populations that could be explored in another investigation.

Do You Know?

Do research in the library or media center to answer these questions:

1. What factors have contributed to the rise in bacterial resistance to antibiotic drugs?
2. How do farmers use directional selection to improve domestic plants and animals?

Use the following Internet resources to explore your own questions about how populations change.

internetconnect

SciLinks NSTA
TOPIC: Populations
GO TO: www.scilinks.org
KEYWORD: HX357

Analyze and Conclude

Answers

1. Answers will vary, but most will approximate a normal distribution.
2. Answers will vary.
3. Answers will vary, but most will have similar values for the means and modes.
4. Answers will vary, but most will show stabilizing selection.
5. No, in the example given, the two modes in the population are very different from the mean. Thus, the mean is not always useful in describing traits in a population. Variation is the differences in the expressions of genetic traits among the individuals of a population.
6. directional selection
7. Answers will vary. For example: How does the size of a population of snow peas that has been grown in hot conditions compare to a population of snow peas grown in normal weather conditions?

Do You Know?

Answers

1. Answers may vary, but students should cite the use of antibiotics as a selective force in which the antibiotic eliminates the susceptible bacteria but leaves those bacteria that can resist the effects of the antibiotic.
2. Answers may vary, but students should cite desirable characteristics for which farmers may breed their animals and plants, such as growing quicker, producing more seeds (row crops), or producing more milk (in cows).

Procedure

Sample Procedure

1. Measure the lengths of 20 green beans or snow peas to the nearest millimeter. Measure the weight of the 20 green beans or snow peas.
2. Record the measurements. Organize the measurement values on a single line from the lowest to the highest value, and round to the nearest whole number. Beneath each value, write the number of green beans of that length. On the graph paper, make a graph showing the distribution curve.
3. Calculate the mean, median, and range of distribution for each type of measurement taken.
4. Clean up materials and wash hands before leaving the lab.

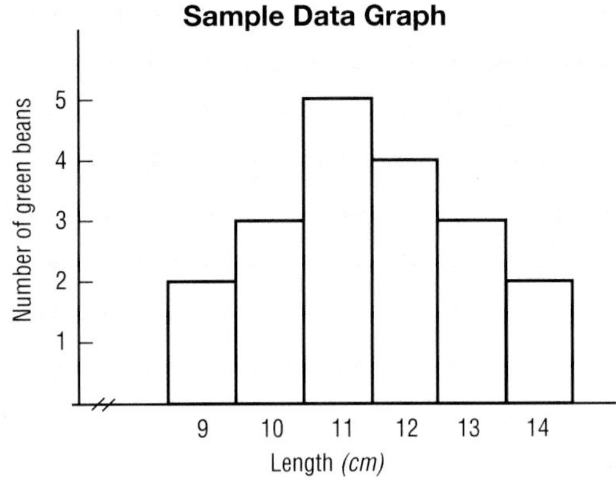

Chapter 17

Ecosystems

Planning Guide

OBJECTIVES	MEETING INDIVIDUAL NEEDS	READING SKILLS AND STRATEGIES
17-1 What Is an Ecosystem? (BLOCKS 1 & 2 90 min) • **Distinguish** an ecosystem from a community. • **Describe** the diversity of a representative ecosystem. • **Sequence** the process of succession.	Learning Styles, p. 360 BASIC ADVANCED • pp. 363, 364	**ATE** Active Reading, Technique: Anticipation Guide, p. 359 ■ Active Reading Guide, worksheet (17-1) ■ Directed Reading, worksheet (17-1) ■ Reading Strategies, One-Minute Quick Write
17-2 Energy Flow in Ecosystems (BLOCKS 3 & 4 90 min) • **Distinguish** between producers and consumers. • **Compare** food webs with food chains. • **Describe** why food chains are rarely longer than three or four links.	Learning Styles, p. 365 BASIC ADVANCED • pp. 368, 369	■ Active Reading Guide, worksheet (17-2) ■ Directed Reading, worksheet (17-2)
17-3 Ecosystems Cycle Materials (BLOCKS 5 & 6 90 min) • **Summarize** the role of plants in the water cycle. • **Relate** the carbon cycle to the flow of energy. • **Identify** the role of bacteria in the nitrogen cycle.	Learning Styles, p. 370 BASIC ADVANCED • p. 374	**ATE** Active Reading, Technique: Paired Reading, p. 372 ■ Active Reading Guide, worksheet (17-3) ■ Directed Reading, worksheet (17-3) ■ Supplemental Reading Guide, *Silent Spring*

Review and Assessment Resources

BLOCKS 7 & 8 90 min

TEST PREP

PE Study Zone, p. 375

PE Performance Zone, pp. 376–377

■ Chapter 17 worksheets:
- Vocabulary
- Science Skills
- Concept Mapping
- Critical Thinking

 Holt Biology Interactive Tutor Unit 7—Ecosystem Dynamics, Topics 1, 3–6

 Audio CD Program

■ Test Prep Pretest

CHAPTER TESTING

■ Chapter Tests (blackline master)

▲ Test Generator

■ Assessment Item Listing

See the correlations table at the front of this book for a list of the ***National Science Education Standards*** addressed in this chapter.

Smithsonian Institution® Visit **www.si.edu/hrw** for additional on-line resources.

Biology Principles and Explorations

HOLT BIOSOURCES

One-Stop Planner CD-ROM
Shows these resources within a customizable daily lesson plan:

KEY

PE Pupil's Edition
ATE Teacher's Edition

 Teaching Transparencies
 Laboratory Program

■ Teaching Resources
▲ Test Generator

VISUAL STRATEGIES	CLASSWORK AND HOMEWORK	TECHNOLOGY AND INTERNET RESOURCES
ATE Fig. 17-3, p. 361 **ATE** Fig. 17-5, p. 363 70 Ecological Succession at Glacier Bay 80 Succession in a Developing Ecosystem 82 Making an Ecosystem Model 83 Effect of Area on Ecosystem Diversity	**PE** **Quick Lab** How can you determine biodiversity around your home?, p. 362 **PE** **Quick Lab** How can you model succession?, p. 364 **Lab C24** Mapping Biotic Factors in the Environment ■ **Occupational Applications:** Wildlife Biologist **PE** Review, p. 364	**Interactive Tutor** Unit 7—Ecosystem Dynamics (Topics 1 and 3) **Holt Biology Videodiscs** Section 17-1 **Audio CD Program** Section 17-1 internetconnect — SciLinks NSTA — **TOPIC:** Biodiversity
ATE Fig. 17-7, p. 366 **ATE** Fig. 17-8, p. 367 67 Grassland Food Web 69 Food Web in an Antarctic Ecosystem 77 Four Trophic Levels in an Aquatic Ecosystem 85 Amount of Energy at Four Trophic Levels	**Lab A12** Making a Food Web **Lab C25** Assessing Abiotic Factors in the Environment **PE** **Review,** p. 369	**Interactive Tutor** Unit 7—Ecosystem Dynamics (Topics 4 and 5) **Holt Biology Videodiscs** Section 17-2 **Audio CD Program** Section 17-2 internetconnect — SciLinks NSTA — **TOPICS:** • Food chains • Energy pyramids
ATE Fig. 17-13, p. 372 **ATE** Fig. 17-14, p. 373 71 Water Cycle 72 Carbon Cycle 73 Nitrogen Cycle	**PE** **Experimental Design** How Does an Ecosystem Change over Time?, pp. 378–379 **Lab B10** Ecology Scavenger Hunt **PE** **Review,** p. 374 **PE** **Study Zone,** p. 375	CNN **Video Segment 13** Fat Wolves **Interactive Tutor** Unit 7—Ecosystem Dynamics (Topic 6) **Holt Biology Videodiscs** Section 17-3 **Audio CD Program** Section 17-3 internetconnect — SciLinks NSTA — **TOPICS:** • Water cycle • Carbon cycle • Sustainable agriculture

ALTERNATIVE ASSESSMENT

PE **Portfolio Projects** 23–26, p. 377
ATE **Alternative Assessment,** p. 375
ATE **Portfolio Assessment,** p. 375

Lab Activity Materials

Quick Labs
p. 362 notepad, pencil
p. 364 1 qt glass jar with a lid, one-half quart of pasteurized milk, pH strips

Experimental Design
p. 378 coarse sand or pea gravel, terrarium, soil, grass seeds, clover seeds, mung bean seeds, earthworms, isopods (pill bugs), mealworms (beetle larva), crickets

Lesson Warm-up
ATE log on which fungus is growing, p. 370

Demonstrations
ATE aquarium, terrarium, or photo of an ecosystem, p. 358

ATE rock covered with lichens or mosses, p. 363

BIOSOURCES LAB PROGRAM
- Quick Labs: Lab A12, p. 23
- Inquiry Skills Development: Lab B10, p. 39
- Laboratory Techniques and Experimental Design: Lab C24, p. 115
- Laboratory Techniques and Experimental Design: Lab C25, p. 121

Chapter 17

Chapter Theme

Interdependence

The various organisms in an ecosystem depend upon other organisms and the nonliving environment for the materials and energy required for life. This energy flows from the sun through producers and then through consumers. Energy is lost as heat at each transfer through the food chain, so producers must constantly capture energy, primarily through photosynthesis. Materials, however, are continually recycled from the nonliving environment to the living organisms and back to the nonliving environment.

Establishing Prior Knowledge

Before beginning this chapter, make certain students can answer the questions in **Study TIP Ready?** on page 359. If they cannot, encourage them to reread the sections indicated.

Opening Demonstration

Provide students with an ecosystem to observe. Use a functioning aquarium/terrarium if one is available. If not, display a photograph of an ecosystem. Then have students describe examples of interdependence in the system. For example, oxygen supplied by a plant is used by a fish, which in turn supplies carbon dioxide needed by the plant.

CHAPTER 17 Ecosystems

Coral-reef ecosystem

What's Ahead — and Why

CHAPTER OUTLINE

17-1 What Is an Ecosystem?

- Organisms Interact with Each Other and Their Environment
- Ecosystems Support Diverse Communities
- Ecosystems Change over Time

17-2 Energy Flow in Ecosystems

- How Energy Moves Through Ecosystems
- Energy Is Lost in a Food Chain

17-3 Ecosystems Cycle Materials

- Materials Cycle Between Living and Nonliving Things
- The Water Cycle Is Driven by the Sun
- The Carbon Cycle Is Linked to Energy
- Plants and Bacteria Take Part in the Phosphorous and Nitrogen Cycles

Students will learn

- what components make up an ecosystem and how those components may change over time through a process called succession.
- how energy flows from one trophic level of a food chain to the next, and that an ever smaller portion of that energy is available at each subsequent trophic level.
- how materials vital to life are cycled among living organisms and between living organisms and the nonliving environment.

Buffalo and lions

Autotrophs... prey... predator...

Buffaloes eat grass, lions eat the buffaloes, and vultures clean up the mess. Nutrients are cycled on a coral reef much in the same way they are cycled on the African plains. All life-forms are part of the cycle. This complex relationship of organisms and their physical environment makes up an ecological system, or ecosystem.

Study TIP

Ready?

Check your understanding of these topics to help you prepare for what's ahead.

Can you...

- **contrast** autotrophs with heterotrophs? (Section 5-1)
- **summarize** the process of photosynthesis? (Section 5-2)
- **describe** the process of cellular respiration? (Section 5-3)

If not, *review the sections indicated.*

Looking Ahead

Labs

- **Quick Labs**
  How can you determine biodiversity around your home? (p. 362)
  How can you model succession? (p. 364)
- **Experimental Design**
  How Does an Ecosystem Change over Time? (p. 378)

Features

- **Food Watch**
  Sustainable Agriculture (p. 374)

Unit 7—*Ecosystem Dynamics*
Topics 1, 3–6

internetconnect

National Science Teachers Association *sci*LINKS Internet resources are located throughout this chapter.

Chapter 17

Reading Strategies

Active Reading

Technique: Anticipation Guide

Before students read this chapter, write the following on the board or overhead:

1. In an ecosystem, there is more energy stored in plants than in consumers.
2. The extinction of one species in an ecosystem can have an impact on all other species.

Let students think a minute about the statements and decide if they agree with them. Have students read the chapter to see if their opinions are confirmed.

Directed Reading

Assign **Directed Reading Worksheet Chapter 17** to help students interact with the text as they read the chapter.

Interactive Reading

Assign ***Biology: Principles and Explorations*** **Audio CD Program Chapter 17** to help reluctant readers achieve greater success in reading the chapter.

Reading Skills

Assign Reading Organizers and Reading Strategies Worksheets from the **One-Stop Planner CD-ROM with Test Generator** to help students learn and practice effective reading skills.

internetconnect

Holt *Biology: Principles and Explorations*

On-line Resources: go.hrw.com

Visit the HRW Web site for a variety of resources related to this chapter. Just type in the keyword HX0 Chapter 17.

Holt *Biology: Principles and Explorations*

On-line Resources: www.scilinks.org

The following *sci*Links Internet resources can be found in the student text for this chapter:

Page 361
TOPIC: Biodiversity
KEYWORD: HX361

Page 366
TOPIC: Food chains
KEYWORD: HX366

Page 368
TOPIC: Energy pyramids
KEYWORD: HX368

Page 371
TOPIC: Water cycle
KEYWORD: HX371

Page 372
TOPIC: Carbon cycle
KEYWORD: HX372

Page 374
TOPIC: Sustainable agriculture
KEYWORD: HX374

Page 379
TOPIC: Ecosystems
KEYWORD: HX379

Chapter 17

1 Prepare

Scheduling

- **Block:** About 90 minutes will be needed to complete this section.
- **Traditional:** About two class periods will be needed to complete this section.
- **Planning:** See the Planning Guide on pp. 357A and 357B for lecture, classwork, and assignment options for Section 17-1. Lesson plans for Section 17-1 can be found in a lesson cycle format in ***Biology: Principles and Explorations*** **One-Stop Planner CD-ROM with Test Generator.**

Resource Link

Assign **Active Reading Worksheet Section 17-1** to help students comprehend the key concepts and visuals in this lesson.

2 Teach

Lesson Warm-up

Begin to explore the diversity within an ecosystem by asking the class to select a local ecosystem. If possible, visit the area to conduct a survey of species diversity. Otherwise, have students identify as many species as they can imagine that would be found within that ecosystem. Compare their results with the diversity of a tropical rain forest. For example, there may be nearly 100 species of trees in just 1 hectare (2.5 acres) of tropical rain forest, while a temperate forest would contain just a few species of trees per hectare.

Section 17-1 What Is an Ecosystem?

Objectives

- **Distinguish** an ecosystem from a community. (p. 360)
- **Describe** the diversity of a representative ecosystem. (pp. 361–362)
- **Sequence** the process of succession. (pp. 363–364)

Key Terms

ecology
habitat
community
ecosystem
abiotic factor
biotic factor
biodiversity
pioneer species
succession
primary succession
secondary succession

Organisms Interact with Each Other and Their Environment

It is easy to think of the environment as being around but not part of us—something we always use, sometimes enjoy, and sometimes damage. But in fact, we are part of the environment along with all of Earth's other organisms. All of Earth's inhabitants are interwoven in a complex web of relationships, as illustrated in **Figure 17-1.** To understand how the interactions of the parts can affect a whole system, think about how a computer operates. Removing one circuit from a computer can change or limit the interactions of the computer's many components in ways that influence the computer's overall operation. In a similar way, removing one species from our environment can have many consequences, not all of them easily predictable.

In 1866, the German biologist Ernst Haeckel gave a name to the study of how organisms fit into their environment. He called this study *ecology,* which comes from the Greek words *oikos,* meaning "house," or "place where one lives," and *logos,* meaning "study of." **Ecology** is the study of the interactions of living organisms with one another and with their physical environment (soil, water, climate, and so on). The place where a particular population of a species lives is its **habitat.** The many different species that live together in a habitat are called a **community.** An **ecosystem,** or ecological system, consists of a community and all the physical aspects of its habitat, such as the soil, water, and weather. The physical aspects of a habitat are called **abiotic** *(ay beye AHT ihk)* **factors,** and the living organisms in a habitat are called **biotic factors.**

Figure 17-1 Organisms interact within an ecosystem. Organisms within an ecosystem continually change and adjust. This plant species is dependent on the bat for its reproduction, and the bat uses part of the flower for food.

Integrating Different Learning Styles

Logical/Mathematical	**PE**	Quick Lab, p. 364; Performance Zone items 18 and 20, p. 377
Visual/Spatial	**PE**	Quick Lab, p. 362; Real Life, p. 362; Performance Zone item 6, p. 376
	TE	Teaching Tip, p. 363
Verbal/Linguistic	**PE**	Study Tips, pp. 361, 364; Performance Zone items 3, 18, 25, and 26, pp. 376–377

Ecosystems Support Diverse Communities

The number of species living within an ecosystem is a measure of its **biodiversity.** Consider a pine forest in the southeastern United States, such as the one shown in **Figure 17-2.** If you could fence in a square kilometer (0.4 mi^2) of this forest and then collect every organism, what would you expect to get?

Figure 17-2 Pine forest. Pine forests like this one are common in the southeastern United States.

Ecosystem inhabitants

Large animals in the forest might include a bear or a white-tailed deer. The woods also contain smaller mammals—raccoons, foxes, gray squirrels, rabbits, and chipmunks. Snakes and toads often remain hidden among the leaves. Many birds can be found, including hawks, warblers, and sparrows. If the square kilometer included a lake, you might find catfish, bass, perch, a variety of turtles, and perhaps an alligator.

There are pine trees, a variety of smaller trees, and shrubs. Beneath the trees, grasses and many kinds of flowers grow on the forest floor. You might find many kinds of fungi growing on fallen trees and spreading as fine threads through the decaying material on the forest floor, as illustrated in **Figure 17-3.** Other fungi are found on the surface of trees or rocks as lichens. Lichens are associations between fungi and algae or cyanobacteria.

Many of the life-forms in the soil and water of a pine forest are too small to see without a microscope. Protists include algae and related microscopic eukaryotes that thrive in water. There may be billions of bacteria in a handful of soil.

The soil contains an immense number of worms. Hidden under the bark of trees and beneath the leaves covering the ground are

Figure 17-3 Forest fungi

These fungi digest plants and other materials they find in the forest.

Mushrooms are often found on moist forest floors.

Shelf fungi grow on and digest trees.

Study TIP

Reading Effectively

Use the objectives and the red subheadings at the top of the pages to help you identify the main ideas presented in Section 17-1.

internet connect

SCiLINKS. NSTA

TOPIC: Biodiversity
GO TO: www.scilinks.org
KEYWORD: HX361

Section 17-1

Visual Strategy

Figure 17-3

Tell students that in a forest community, more than 90 percent of the available energy from producers (plants and algae) is consumed by bacteria or fungi and not by herbivores, such as deer. Bacteria and fungi belong to a group of organisms known as detritivores, which live on detritus (dead remains of plants and animals).

Making Biology Relevant

Food

It is often difficult for students to see themselves as part of an ecosystem. Tell them that the next time they bite into a piece of fruit, they should give thanks to the bee that pollinated the flower that gave rise to the fruit. And the late-flowering plants, such as asters and goldenrod, deserve their respect as well. Without these late-blooming flowers, the bees would not survive the winter.

internet connect

SCiLINKS. NSTA Have students research coral reef biodiversity using the Web site in the Internet Connect box on page 361. Have students write a report summarizing their findings. PORTFOLIO

Multicultural Perspective

Native Taxonomy in New Guinea

Indigenous peoples often have a more complete knowledge of their local plants and animals than Western biologists have. In the 1920s, American biologist Ernst Mayr, who would later propose the biological species concept, visited a remote area of New Guinea to collect birds. He hired hunters of the Arfak tribe to bring him specimens and found that Mayr and the local tribe usually recognized the same species of birds. Many biologists since then have relied on indigenous peoples around the world for valuable information.

CAREER

Wildlife Biologist

Wildlife biologists are hired by federal, state, and sometimes local government agencies to study and manage wildlife. Common game and nongame species, as well as rare and endangered species, are the focus of this career. Knowledge of each species' role in the ecosystem and its relation to other species is critical.

REAL LIFE

Answer

Answers will vary depending on many factors, including the size and successional state of the parcel surveyed.

QUICK LAB

How can you determine biodiversity around your home?

Time 20 minutes

Process Skills Observing, applying, identifying variables, drawing conclusions

Teaching Strategies

- Have each student use an area of equal dimensions, at least 50 × 50 m, to make the task more manageable and to allow comparisons between different ecosystems.
- Encourage students to take their time, as they will notice much more.

Analysis Answers

1. Answers will vary.
2. Answers will vary.
3. Answers will vary.
4. Answers will vary. In general, the abiotic factors in an ecosystem provide organisms (biotic factors) with a physical place to live, energy, nutrients, and water. The organisms alter and recycle some of these abiotic factors, changing the landscape in the process.

Figure 17-4 A variety of arthropods. Arthropods have different structures that help them survive and reproduce. For example, the jumping spider lies in wait and pounces on passing insects. The male stag beetle uses its large jaws to help them compete against other male stag beetles for mates.

Jumping spider

Male stag beetle

many different species of insects and spiders, such as those shown in **Figure 17-4.**

If you were to remove every organism from your square kilometer, the nonliving surroundings that remain make up the abiotic factor. This would include the minerals, organic compounds, water, wind that blows over the Earth, and rain and sunlight.

REAL LIFE

The amount of biodiversity can vary. A single acre in New Hampshire may have eight species of trees, but the same amount of land in a tropical rain forest may have 120 different species of trees.

Observing *If you live near a forest or a park, find out how many different species of trees can be seen there.*

Ecosystem boundaries

The physical boundaries of an ecosystem are not always obvious, and they depend on how the ecosystem is being studied. For example, a scientist might consider a single rotting log on the forest floor to be an ecosystem if he or she is interested only in the fungi and insects living in the log. Often individual fields, forests, or lakes are studied as isolated ecosystems. Of course, no location is ever totally isolated. Even oceanic islands get occasional migrant visitors, such as birds blown off course.

QUICK LAB

How can you determine biodiversity around your home?

By making simple observations, you can draw some conclusions about biodiversity in an ecosystem.

Materials

note pad, pencil

Procedure

1. Use your note pad and pencil to prepare a list of biotic and abiotic factors that you observe around your house or in a nearby park. Record the number of factors that you see.

Analysis

1. **Identify** the habitat and community that you observed.
2. **Calculate** the number of different species as a percentage of the total number of organisms that you saw.
3. **Rank** the importance of biotic factors within the ecosystem you observed.
4. **Infer** what the relationships are between biotic factors and abiotic factors in the observed ecosystem.

Biology in Your Community

Introduced Species

Moving a species from a distant ecosystem into another where it is not normally found can have devastating effects on the local ecosystem. One such example is the kudzu vine, which was introduced into the United States from Japan as an ornamental plant and cattle feed. Kudzu grows quickly and now competes aggressively with native vegetation through much of the South. Have students write a short report identifying what problem species have been introduced into their local ecosystems, why these species are a problem, and what is being done to control the problem. They may wish to contact a naturalist or the state department of natural resources.

PORTFOLIO

Ecosystems Change over Time

When a volcano forms a new island, a glacier recedes and exposes bare soil, or a fire burns all of the vegetation in an area, a new habitat is created. This change sets off a process of colonization and ecosystem development. The first organisms to live in a new habitat are small, fast-growing plants, called **pioneer species.** They may make the ground more hospitable for other species. Later waves of plant immigrants may then outcompete and replace the pioneer species.

Succession

A somewhat regular progression of species replacement is called **succession.** Succession that occurs where plants have not grown before is called **primary succession.** Succession that occurs in areas where there has been previous growth, such as in abandoned fields or forest clearings, is called **secondary succession.** It was once thought that the stages of succession were predictable and that succession always led to the same final community of organisms within any particular ecosystem. Ecologists now recognize that initial conditions and chance play roles in the process of succession. For example, if two species are in competition, a sudden change in the climate may favor the success of one species over the other. For this reason, no two successions are alike.

Figure 17-5 Glacier Bay

A receding glacier makes primary succession possible.

Recently exposed land has few nutrients.

Alders, grasses, and shrubs later take over from pioneer plants.

As the amount of soil increases, spruce and hemlock trees become plentiful.

Glacier Bay

A good example of primary succession is a receding glacier because land is continually being exposed as the face of the glacier moves back. The glacier that composes much of the head of Glacier Bay, Alaska, has receded some 100 km (62 mi) over the last 200 years. Figure 17-5 shows the changes that have taken place as time passed.

The most recently exposed areas are piles of rock and gravel that lack the usable nitrogen essential to plant and animal life. The seeds and spores of pioneer species are carried in by the wind. These include lichens, mosses, fireweed, willows, cottonwood, and *Dryas,* a sturdy plant with clumps about 30 cm (1 ft) across. At first all of these plants grow close to the ground, severely stunted by mineral deficiency, but *Dryas* eventually crowds out the other plants.

After about 10 years, alder seeds blown in from distant sites take root. Alder roots have nitrogen-fixing nodules, so they are able to grow more rapidly than *Dryas.* Dead leaves and fallen branches from the alder trees add more usable nitrogen to the soil. The added nitrogen allows willows and cottonwoods to invade and grow with vigor. After about 30 years, dense thickets of alder, willow, and cottonwood shade and eventually kill the *Dryas.*

Demonstration

Pioneers

Show the class a rock covered with lichens or mosses. Ask students why these organisms are known as "pioneers." *(Like our pioneer settlers, they are the first to inhabit a new area.)* Point out that a lichen is a partnership between a fungus and an alga or cyanobacterium. Each organism contributes to the survival of the other. The fungus can even absorb nutrients from bare rock. The alga or cyanobacterium produces carbohydrates through photosynthesis.

Teaching TIP — BASIC

Succession

Have each student draw a timeline representing succession that occurs after a forest fire has burned all of the vegetation in an area. Starting at time zero, the land should be barren, followed by the appearance of grasses and weeds, then small bushes, and finally trees. PORTFOLIO

Visual Strategy

Figure 17-5

Tell students that primary succession such as this is a slow process, sometimes very slow. Scientists estimate that the primary succession from sand dunes to the beech-maple forest along the shores of Lake Michigan took about 1,000 years. In contrast, secondary succession may take less than 100 years.

did you know?

When Mount St. Helens erupted on May 18, 1980, shock waves leveled all of the fir trees in an area of about 18,000 hectares (44,000 acres).

Mudflows, traveling at speeds of up to 80 km per hour (50 mph), buried almost all of the remaining vegetation. Today, small plants and some animals have returned to many of the devastated areas. Scientists estimate that it will take about 200 years for the forest to return to its pre-eruption state.

Chapter 17

How can you model succession?

QUICK LAB

Time 10–15 minutes for day 1, 5–10 minutes for each following day

Process Skills Observing, applying, inferring, evaluating

Teaching Strategies

- Make sure students do not screw the lid down too tightly because the growing culture will suffocate, or the jar may explode from anaerobically produced gas.
- Have students include *odor* as part of their daily description.

Analysis Answers

1. The pH dropped as the environment became more acidic.
2. By-products from the metabolism of the microorganisms change the pH. Then new organisms better adapted to the changed pH begin to thrive.
3. Like the Glacier Bay model, organisms colonize and slowly change a new environment such that it becomes more suitable for other organisms.

3 Assessment Options

Closure

What an Ecosystem Is Not

Have students strengthen their understanding of what an ecosystem is by defining what it is not. Have them write their ideas in short essay format and present them to the class.

Review

Assign Review—Section 17-1.

Reteaching — BASIC

Show students a picture of a forest ecosystem. Have them describe the events that would take place after a strong fire scorched the area. *(First, only fast-growing weeds and grasses would be found. After a number of years, shrubs would grow back, which in years would give way to trees.)*

Study TIP

Reviewing Information

Prepare flash cards for each of the **Key Terms** in this chapter. On each card, write the term on one side and its definition on the other side. Use the cards to review the meanings of the Key Terms.

About 80 years after the glacier first exposes the land, Sitka spruce invades the thickets. Spruce trees use the nitrogen released by the alders and eventually form a dense forest. The spruce blocks the sunlight from the alders, and the alders then die, just as the *Dryas* did before them.

After the spruce forest is established, hemlock trees begin to grow at the site. Hemlocks are very shade tolerant and have a root system that competes well against spruce for soil nitrogen. Hemlock trees soon become dominant in the forest. This community of spruce and hemlock proves to be a very stable ecosystem from the perspective of human time scales, but it is not permanent. As local climates change, this forest ecosystem may change too.

QUICK LAB

How can you model succession?

You can create a small ecosystem and measure how organisms modify their environment.

Materials

1 qt glass jar with a lid, one-half quart of pasteurized milk, pH strips

Procedure

1. Prepare a table like the one below.
2. Half fill a quart jar with pasteurized milk, and cover the jar loosely with a lid. Measure and record the pH. Place the jar in a 37°C incubator.
3. Check and record the pH of the milk with pH strips every day for seven days. As milk spoils, its pH changes. Different populations of microorganisms become established, alter substances in the milk, and then die off when conditions no longer favor their survival.
4. Record any visible changes in the milk each day.

Analysis

1. **Identify** what happened to the pH of the milk as time passed.
2. **Infer** what the change in pH means about the populations of microorganisms in the milk.
3. **SKILL Evaluating Results** How does this model confirm the model of succession in Glacier Bay?

DATA TABLE

Day	pH	Appearance
1		
2		
3		

Review SECTION 17-1

1. **Identify** what components of an ecosystem are not part of a community.
2. **Classify** the parts of the pine forest ecosystem as biotic or abiotic factors.
3. **Relate** how gardening or agriculture affects succession.
4. **Differentiate** primary succession from secondary succession.

Critical Thinking

5. **SKILL Forming Reasoned Opinions** Why do so many ecosystems remain stable for centuries, while others (such as Glacier Bay) undergo succession?

Review SECTION 17-1 ANSWERS

1. the abiotic (nonliving) components
2. Biotic factors include pine trees and other trees, animals, herbs, lichen, fungi, and bacteria. Abiotic factors include water, soil, climate, air, and rocks.
3. Because the land is kept in a constant state of disturbance, (secondary) succession in gardens and farms does not usually get beyond the first stage of fast-growing weeds and grasses.
4. In primary succession, pioneer plants colonize surfaces that have never before been populated. In secondary succession, a disturbance clears or creates a new (usually rock) surface. Secondary succession is relatively rapid because the soil has already been developed and contains seeds for new growth. In primary succession, rock or soil may not be suitable for vascular plants until the pioneer plants have been established for decades.
5. Answers may vary, but students should cite factors such as climate, geology, and human intervention.

Energy Flow in Ecosystems

Section 17-2

How Energy Moves Through Ecosystems

Everything that organisms do in ecosystems—running, breathing, burrowing, growing—requires energy. The flow of energy is the most important factor that controls what kinds of organisms live in an ecosystem and how many organisms the ecosystem can support. In this section you will learn where organisms get their energy.

The primary energy source

Most life on Earth depends on photosynthetic organisms, which capture some of the sun's light energy and store it as chemical energy in organic molecules. These organic compounds are what we call food. The rate at which organic material is produced by photosynthetic organisms in an ecosystem is called **primary productivity.** Primary productivity determines the amount of energy available in an ecosystem. Most organisms in an ecosystem can be thought of as chemical machines driven by the energy captured in photosynthesis. Organisms that first capture energy, the **producers,** include plants, some kinds of bacteria, and algae. Producers make energy-storing molecules. All other organisms in an ecosystem are consumers. **Consumers** are those organisms that consume plants or other organisms to obtain the energy necessary to build their molecules.

Trophic levels

Ecologists study how energy moves through an ecosystem by assigning organisms in that ecosystem to a specific level, called a **trophic** *(TROHF ihk)* **level,** in a graphic organizer based on the organism's source of energy. Energy moves from one trophic level to another, as illustrated in **Figure 17-6.**

Figure 17-6 **Trophic levels**

The sun is the ultimate source of energy for producers and all consumers.

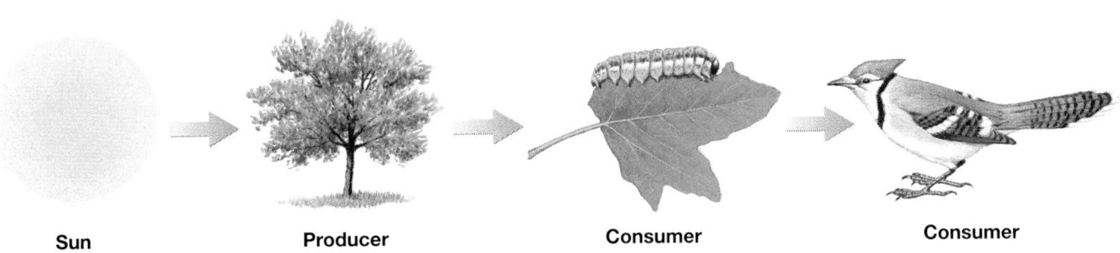

Objectives

- **Distinguish** between producers and consumers. (p. 365)
- **Compare** food webs with food chains. (pp. 366–367)
- **Describe** why food chains are rarely longer than three or four links. (pp. 368–369)

Key Terms

primary productivity
producer
consumer
trophic level
food chain
herbivore
carnivore
omnivore
detritivore
decomposer
food web
energy pyramid
biomass

Section 17-2

1 Prepare

Scheduling

- **Block:** About 90 minutes will be needed to complete this section.
- **Traditional:** About two class periods will be needed to complete this section.
- **Planning:** See the Planning Guide on pp. 357A and 357B for lecture, classwork, and assignment options for Section 17-2. Lesson plans for Section 17-2 can be found in a lesson cycle format in ***Biology: Principles and Explorations*** **One-Stop Planner CD-ROM with Test Generator.**

Resource Link

Assign **Active Reading Worksheet Section 17-2** to help students comprehend the key concepts and visuals in this lesson.

2 Teach

Lesson Warm-up

List the following organisms that can be found in an open field: robin, hawk, snake, frog, grasshopper, mouse, and rabbit. Have students draw arrows to show what eats what in this field ecosystem. Students should see the complexity of even this simple food web in which each predator can take more than one type of prey and each type of prey can be exploited by several different species of predators.

Integrating Different Learning Styles

Logical/Mathematical	**PE**	Performance Zone items 7 and 22, pp. 376–377
	ATE	Cross-Disciplinary Connection, p. 368; Teaching Tip, p. 368
Visual/Spatial	**PE**	Performance Zone items 1 and 10, p. 376
	ATE	Visual Strategy, p. 366; Visual Strategy, p. 367; Teaching Tip, p. 367; Teaching Tip, p. 369
Verbal/Linguistic	**PE**	Study Tip, p. 366; Performance Zone item 2, p. 376
	ATE	Teaching Tip, p. 366
Interpersonal	**PE**	Study Tip, p. 368

Teaching TIP

Enzymes

Remind students that chemical reactions in all organisms are carried out by enzymes. Humans do not possess the enzyme that can hydrolyze cellulose. The bacteria that live in the gut of cows and termites do, and therefore the cow can digest the cellulose in grass and the termite can digest the cellulose in wood. Ask students to imagine how much more food would be available if humans had this enzyme. Ask students what our meals would look like.

Teaching TIP — BASIC

Trophic Levels

Ask a volunteer from the class to describe his or her dinner last night. Then have students describe the trophic level of each different food item in the meal. What was the trophic level of the student as he or she ate each item?

Visual Strategy

Figure 17-7

Have students identify the producers, herbivores, and carnivores in this diagram. *(The algae are producers; krill are herbivores; and cod, leopard seals, and killer whales are carnivores.)* Then ask students which organisms, if any, are omnivores or detritivores. *(There are no obvious omnivores or detritivores in this food chain.)*

Study TIP

Reading Effectively

The **Key Terms** at the beginning of each section are in boldface in that section. As you read, make sure that you can define each Key Term (such as **consumer** and **decomposer**) before you continue reading.

internetconnect
SCiLINKS NSTA
TOPIC: Food chains
GO TO: www.scilinks.org
KEYWORD: HX366

The path of energy through the trophic levels of an ecosystem is called a **food chain.** An example is shown in **Figure 17-7.** The lowest trophic level of any ecosystem is occupied by the producers, such as plants, algae, and bacteria. Producers use the energy of the sun to build energy-rich carbohydrates. Many producers also absorb nitrogen gas and other key substances from the environment and incorporate them into their biological molecules.

At the second trophic level are **herbivores** *(HUHR beh vohrz),* animals that eat plants or other primary producers. They are the primary consumers. Cows and horses are herbivores, as are caterpillars and some ducks. A herbivore must be able to break down a plant's molecules into usable compounds. However, the ability to digest cellulose is a chemical feat that only a few organisms have evolved. As you will recall, cellulose is a complex carbohydrate found in plants. Most herbivores rely on microorganisms, such as bacteria and protists, in their gut to help digest cellulose. Humans cannot digest cellulose because we lack these particular microorganisms.

At the third trophic level are secondary consumers, animals that eat herbivores. These animals are called **carnivores.** Tigers, wolves, and snakes are carnivores. Some animals, such as bears, are both herbivores and carnivores; they are called **omnivores** *(AHM nih vohrz).* They use the simple sugars and starches stored in plants as food, but they cannot digest cellulose. Humans, for example, are omnivores.

In every ecosystem there is a special class of consumers called detritivores, which include worms and fungal and bacterial decomposers. **Detritivores** *(deh TRIH tih vohrz)* are organisms that obtain their energy from the organic wastes and dead bodies that are

Figure 17-7 Aquatic food chain

This food chain shows one path of energy flow in an Antarctic ecosystem.

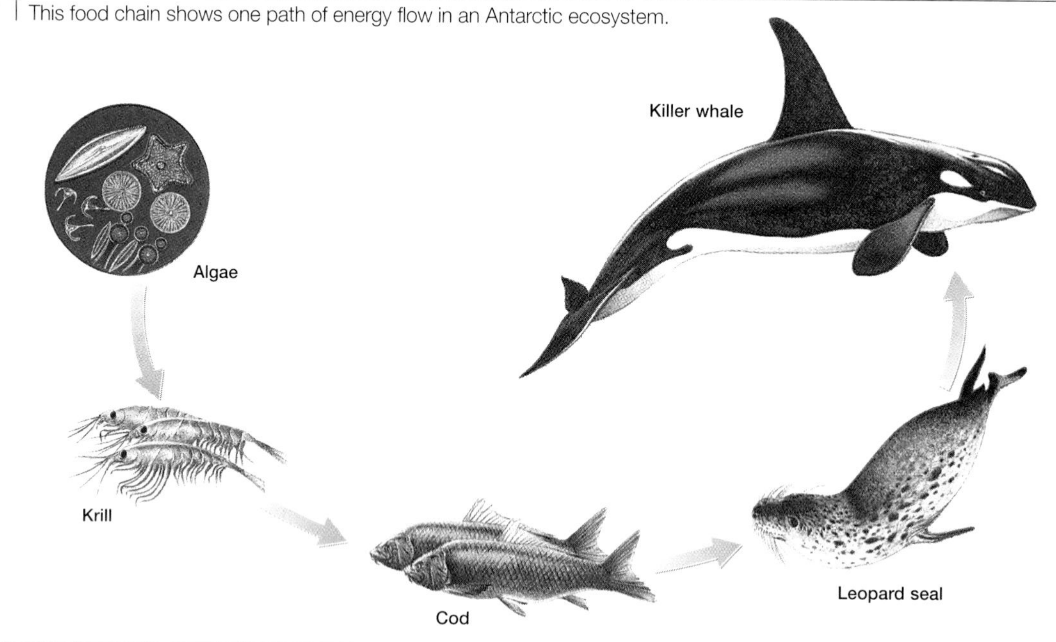

Overcoming Misconceptions

Plants Respire Too

Emphasize that plants, like animals, respire 24 hours a day but carry out photosynthesis only during the daylight. Healthy plants capture more energy than they metabolize, using the excess for building tissue or energy storage molecules, such as starch. Also point out that the carbon dioxide produced when plants respire is released into the atmosphere.

produced at all trophic levels. Bacteria and fungi are known as **decomposers** because they cause decay. Decomposition of bodies and wastes releases nutrients back into the environment to be recycled by other organisms.

Many ecosystems contain a fourth trophic level made of carnivores that consume other carnivores. They are called tertiary consumers, or top carnivores. A hawk that eats a snake is a tertiary consumer. Very rarely do ecosystems contain more than four trophic levels.

In most ecosystems, energy does not follow simple straight paths because individual animals often feed at several trophic levels. This creates a complicated, interconnected group of food chains called a **food web,** as illustrated in **Figure 17-8.**

biobit

Are all producers photosynthetic?

At the bottom of the oceans near volcanic vents live bacteria that harvest energy from the reduced sulfur compounds ejected by the volcanic vents. These chemosynthetic bacteria are at the bottom of their food chain.

Figure 17-8 Aquatic food web

This food web shows a more complete picture of the feeding relationships in an Antarctic ecosystem.

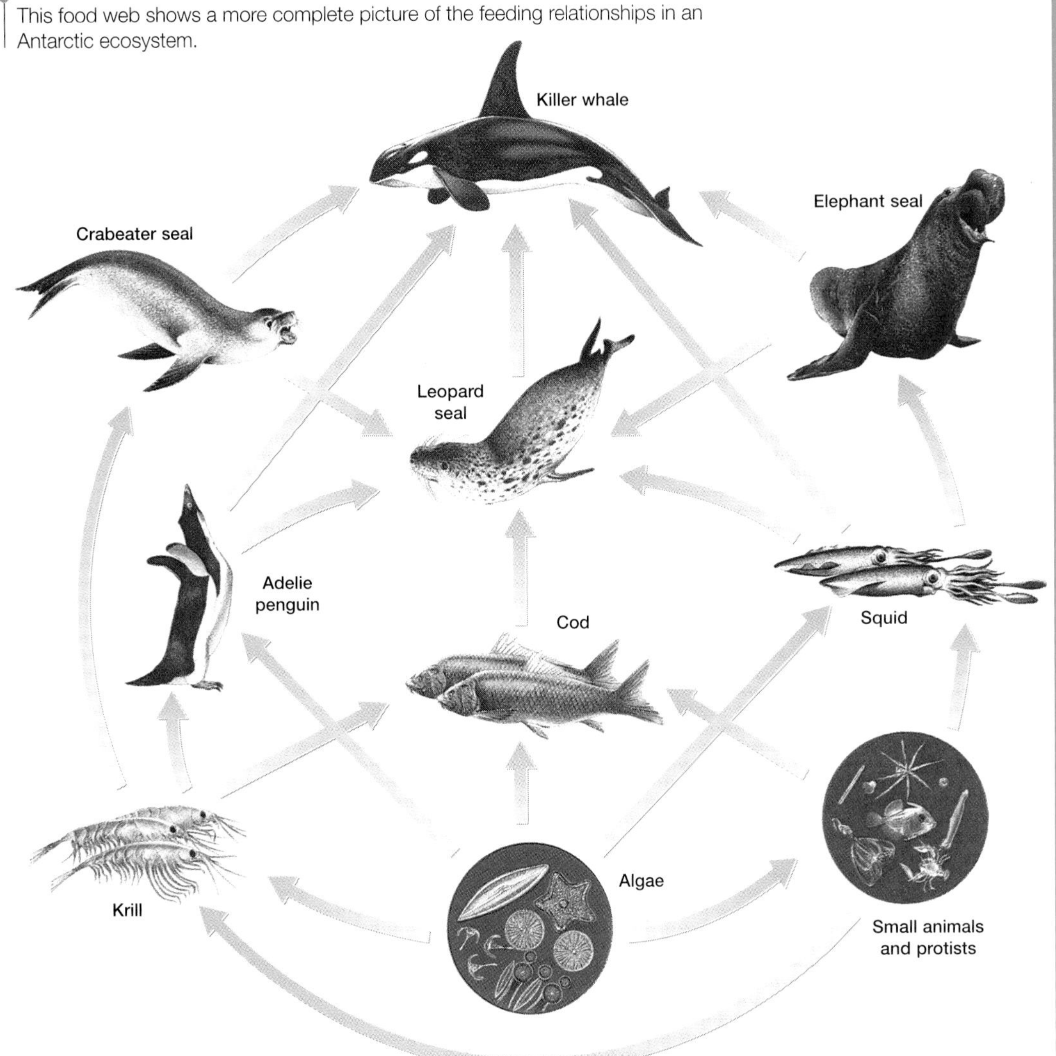

Section 17-2

Visual Strategy

Figure 17-8

Have students describe how this ecosystem would be affected if algae were eliminated. Students should recognize that eliminating algae from this ecosystem is analogous to destroying all of the plants in a terrestrial ecosystem. Like plants, algae are producers that capture the energy needed by the other members of the ecosystem. Eliminating algae would remove the ecosystem's energy source, and the other organisms would eventually starve.

Teaching TIP

Understanding the Flow of Energy

Have students draw a graphic organizer that summarizes the flow of energy from the sun to producers to herbivores, omnivores, carnivores, and detritivores. A completed **Graphic Organizer** is shown at the bottom of page 367.

Graphic Organizer

*Use this graphic organizer with **Teaching Tip: Understanding the Flow of Energy** on page 367.*

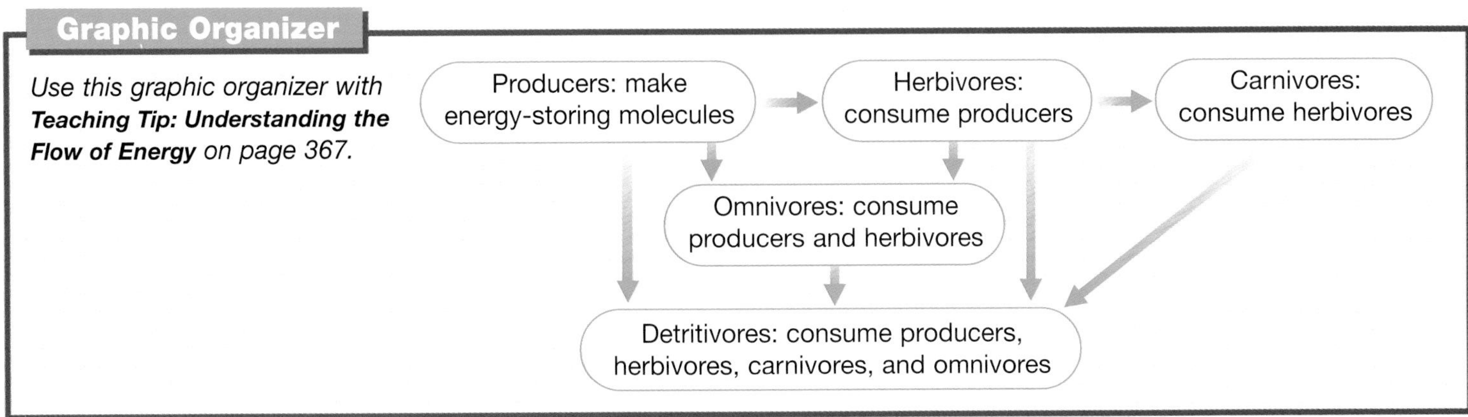

CROSS-DISCIPLINARY CONNECTION

◆ **Physics**

Remind students that the first law of thermodynamics states that energy cannot be created or destroyed but only changed from one form into another form. The second law states that energy conversion between different forms is never 100 percent efficient.

Teaching TIP **ADVANCED**

Energy Through the Trophic Levels

Tell students that if an average of 1,500 kilocalories of light energy per day falls on a square meter of land surface covered by plants, only about 15–30 kilocalories become incorporated into chemical compounds through photosynthesis. How much of this energy could end up in a person who eats these plants? *(1.5 kilocalories)* How much of this energy could end up in a person who eats a steak from a steer that ate the plants? *(0.15 kilocalories)*

Study TIP

Reviewing Information

Work with a partner to review why a food chain cannot have more than four trophic levels. Review how energy is lost when energy is transferred by quizzing each other on the trophic levels shown in Figure 17-9.

Energy Is Lost in a Food Chain

A deer browsing on leaves is acquiring energy. Potential energy is stored in the chemical bonds within the molecules of the leaves. Some of this energy is transformed to other forms of potential energy, such as fat. Some of it aids in the deer's mechanical work, such as running, breathing, and eating more leaves. But almost half of the energy is lost to the environment as heat.

Energy transfer

During every transfer of energy within an ecosystem, energy is lost as heat. Although heat can be used to do work (as in a steam engine), it is generally not a useful source of energy in biological systems. Thus, the amount of useful energy available to do work decreases as energy passes through an ecosystem. The loss of useful energy limits the number of trophic levels an ecosystem can support. When a plant harvests energy from sunlight, it stores in chemical bonds only about one-half of the energy it is able to capture. When a herbivore uses plant molecules to make its own molecules, only about 10 percent of the energy in the plant ends up in the herbivore's molecules. And when a carnivore eats the herbivore, 90 percent of the energy is lost in making carnivore molecules. At each trophic level, the energy stored by the organisms in a level is about one-tenth of that stored by the organisms in the level below.

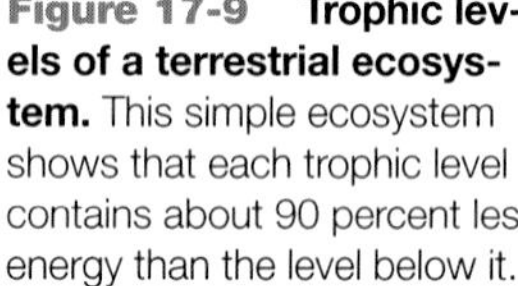

internetconnect
SCiLINKS
NSTA
TOPIC: Energy pyramids
GO TO: www.scilinks.org
KEYWORD: HX368

The pyramid of energy

Ecologists often illustrate the flow of energy through ecosystems with a block diagram called an energy pyramid. An **energy pyramid** is a diagram in which each trophic level is represented by a block, and the blocks are stacked on top of one another, with the lowest trophic level on the bottom. The width of each block is determined by the amount of energy stored in the organisms at that trophic level. Because the energy stored by the organisms at each trophic level is about one-tenth the energy stored by the organisms in the level below, the diagram takes the shape of a pyramid, as shown in **Figure 17-9.**

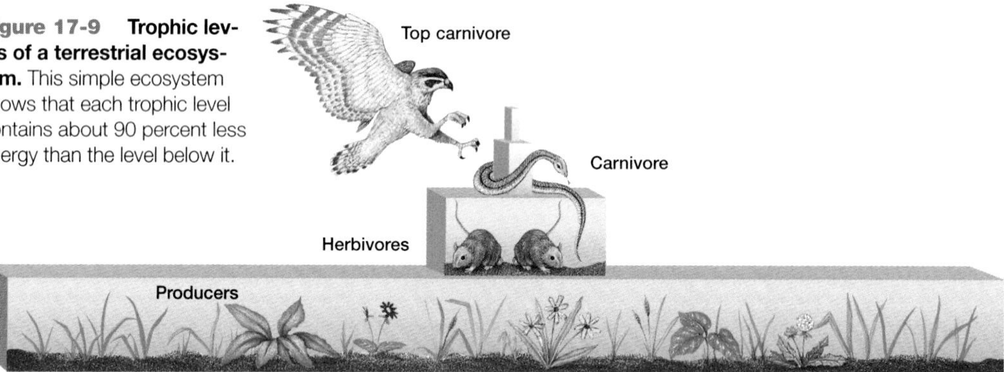

Figure 17-9 Trophic levels of a terrestrial ecosystem. This simple ecosystem shows that each trophic level contains about 90 percent less energy than the level below it.

did you know?

The transfer of energy from one trophic level to another is not a set value, but rather varies around 10 percent.

Although the *average* value for the efficiency of energy transfer from one trophic level to the next is 10 percent, actual measurements of the transfer efficiency range from less than 1 percent to more than 20 percent.

Trophic levels are limited

Most terrestrial ecosystems involve only three or, on rare instances, four levels. Too much energy is lost at each level to allow more levels. For example, a large human population could not survive by eating lions captured on the Serengeti Plain of Africa because there are too few lions to make this possible. The amount of grass in that ecosystem cannot support enough zebras to maintain a large enough population of lions to feed lion-eating humans. In other words, the number of trophic levels that can be maintained in an ecosystem is limited by the loss of potential energy.

As we seek ways to maintain ever larger human populations, we may benefit from remembering the lions' situation. Humans are omnivores, and unlike the lion, we can choose to eat either meat or plants. As illustrated in **Figure 17-10**, 10 kg of grain are needed to build 1 kg of human tissue if the grain is directly ingested by a human. If a cow eats the grain and a human eats the cow, then 100 kg of grain are needed to build 1 kg of human tissue.

Also, the number of individuals in a trophic level can be an inaccurate indicator of the amount of energy in that level. Some organisms are much bigger than others and therefore use more energy. Because of this, the number of organisms often does not form a pyramid when one compares different trophic levels. For instance, caterpillars and other insect herbivores greatly outnumber the trees they feed on. To better determine the amount of energy present in trophic levels, ecologists measure biomass. **Biomass** is the dry weight of tissue and other organic matter found in a specific ecosystem. Each higher level on the pyramid contains only 10 percent of the biomass found in the trophic level below it.

Figure 17-10 Energy efficiency in food consumption

Adding a trophic level to a food chain increases the energy demand by consumers by 10 times.

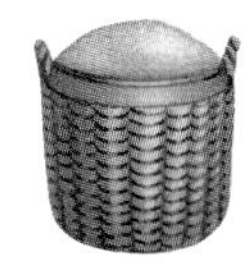
It takes a certain amount of grain

to produce enough bread

to provide one person with a certain amount of energy.

It takes 10 times more grain

to feed one cow

to make enough beef

to provide one person with the same amount of energy.

Review SECTION 17-2

1. **Describe** how producers differ from consumers.
2. **Construct** a food chain, and use it to make a new food web.
3. **Describe** why eating vegetables is more efficient, in terms of energy, than eating meat.
4. **List** the reasons why a food chain cannot exceed four links.

Critical Thinking

5. **SKILL Justifying an Argument** Explain why scientists believe that most animals would become extinct if all plants died.

Review SECTION 17-2 ANSWERS

1. Unlike consumers, which must consume organic matter for food energy, producers use energy, usually from the sun, to create their own energy-storing molecules.
2. Answers will vary. The food web should show more connections between organisms than the simple food chain.
3. A person could eat 100 Cal of vegetables or 100 Cal of meat. Because only 10 percent of the energy in one trophic level is available to the next consumer, the cow must consume 1000 Cal of vegetables to pass on 100 Cal in its meat to the person. It is far more energy-efficient to eat the 100 Cal of vegetables.
4. The enormous loss of energy as energy moves between trophic levels limits the number of trophic levels most ecosystems can support to around four.
5. Animals are consumers, so they rely on producers to create the energy-storing molecules they eat. Most producers are plants. Without plants, most animals, at all trophic levels, would eventually starve to death.

Teaching TIP — ADVANCED

Pyramid of Energy

Have students draw an energy pyramid that accurately reflects the following energy transfer: 1,000 calories (algae) → 150 calories (small aquatic animals) → 30 calories (smelt) → 6 calories (trout) → 1.2 calories (human). PORTFOLIO

3 Assessment Options

Closure

Diagramming Energy Flow

Have students draw a food chain, a food web, and an energy pyramid for an ecosystem of their choosing. Tell them to choose enough organisms to represent several trophic levels and feeding pathways. PORTFOLIO

Review

Assign Review—Section 17-2.

Reteaching — BASIC

Have students construct a food web and energy pyramid for the school yard. Ask students to identify producers, herbivores, omnivores, carnivores, and detritivores. Then ask them what species are missing that they would expect to find if the area had been a natural park.

Chapter 17

1 Prepare

Scheduling

- **Block:** About 90 minutes will be needed to complete this section.
- **Traditional:** About two class periods will be needed to complete this section.
- **Planning:** See the Planning Guide on pp. 357A and 357B for lecture, classwork, and assignment options for Section 17-3. Lesson plans for Section 17-3 can be found in a lesson cycle format in ***Biology: Principles and Explorations*** **One-Stop Planner CD-ROM with Test Generator.**

Resource Link

Assign **Active Reading Worksheet Section 17-3** to help students comprehend the key concepts and visuals in this lesson.

2 Teach

Lesson Warm-up

Show the class a small log on which fungi are growing. Remind students that the fungi, as consumers, digest the log to obtain the nutrients they need. However, as they do so, they also help to decompose the log and thus return materials to the ecosystem. Draw an analogy between the fungi and a person who recycles metal, glass, and plastic. Humans have taken a lesson from nature and are increasingly recycling the materials they use.

Section 17-3 Ecosystems Cycle Materials

Objectives

- **Summarize** the role of plants in the water cycle. (p. 371)
- **Relate** the carbon cycle to the flow of energy. (p. 372)
- **Identify** the role of bacteria in the nitrogen cycle. (p. 373)

Key Terms

biogeochemical cycle
ground water
transpiration
nitrogen fixation

Materials Cycle Between Living and Nonliving Things

Humans throw away tons of garbage every year as unwanted, unneeded and unusable. However, nature does not throw anything away. Unlike energy, which flows through the Earth's ecosystems in one direction (from the sun to producers to consumers), the physical parts of the ecosystems cycle constantly. Carbon atoms, for example, are passed from one organism to another in a great circle of use. Producers are eaten by herbivores, herbivores are eaten by predators, and predators are eaten by top predators. Eventually the top predators die and decay; their carbon atoms then become part of the soil to feed the producers in a long and complex cycle that reuses this important element. Carbon is not the only element that is constantly recycled in this way. Other recycled elements include many of the inorganic (noncarbon) substances that make up the soil, water, and air, such as nitrogen, sulfur, calcium, and phosphorus. All materials that cycle through living organisms are important in maintaining the health of ecosystems, but four substances are particularly important: water, carbon, nitrogen, and phosphorus. All organisms require carbon, hydrogen, oxygen, nitrogen, phosphorus, and sulfur in relatively large quantities. They require other elements, such as magnesium, sodium, calcium, and iron, in smaller amounts. Some elements, such as cobalt and manganese, are required in trace amounts.

The paths of water, carbon, nitrogen, and phosphorus pass from the nonliving environment to living organisms, such as the trees in **Figure 17-11,** and then back to the nonliving environment. These paths form closed circles, or cycles, called biogeochemical *(beye oh jee oh KEHM ih kuhl)* cycles. In each **biogeochemical cycle,** a pathway forms when a substance enters living organisms such as trees from the atmosphere, water, or soil; stays for a time in the living organism; then returns to the nonliving environment. Ecologists refer to such substances as cycling within an ecosystem between a living reservoir (an organism that lives in the ecosystem) and a nonliving reservoir. In almost all biogeochemical cycles, there is much less of the substance in the living reservoir than in the nonliving reservoir.

Figure 17-11 Trees and the carbon cycle

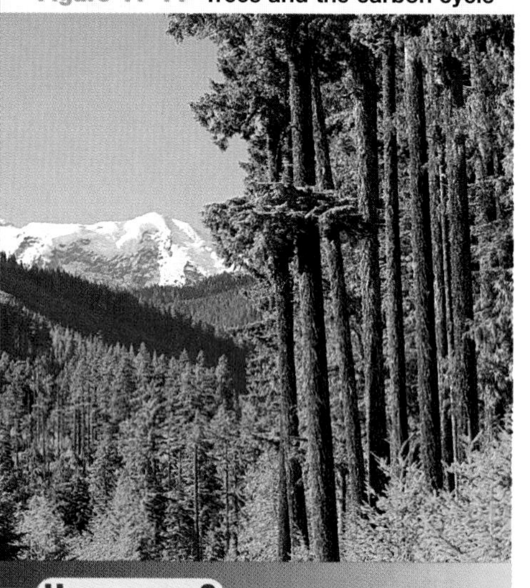

How many? Approximately 500 million tons of carbon were taken up as a result of forest regrowth in the Northern Hemisphere between 1980 and 1989.

Integrating Different Learning Styles

Logical/Mathematical	**PE**	Performance Zone items 14, 15, 16, 21, p. 377
	ATE	Teaching Tip, p. 371; Visual Strategy, p. 373
Visual/Spatial	**PE**	Biobit, p. 371; Performance Zone item 23, p. 377
	ATE	Visual Strategy, p. 372; Teaching Tip, p. 373; Visual Strategy, p. 373
Verbal/Linguistic	**PE**	Study Tip, p. 373; Food Watch, p. 374; Performance Zone items 8, 9, 13, and 24, pp. 376–377
	ATE	Active Reading, p. 371; Reteaching, p. 374
Interpersonal	**ATE**	Active Reading, p. 371
Intrapersonal	**ATE**	Teaching Tip, p. 372

The Water Cycle Is Driven by the Sun

Of all the nonliving components of an ecosystem, water has the greatest influence on the ecosystem's inhabitants. In the nonliving portion of the water cycle, water vapor in the atmosphere condenses and falls to the Earth's surface as rain or snow. Some of this water seeps into the soil and becomes part of the **ground water,** which is water retained beneath the surface of the Earth. Most of the remaining water that falls to the Earth does not remain at the surface. Instead, heated by the sun, it reenters the atmosphere by evaporation. The path of water within an ecosystem is shown in **Figure 17-12.**

In the living portion of the water cycle, much water is taken up by the roots of plants. After passing through a plant, the water moves into the atmosphere by evaporating from the leaves, a process called **transpiration.** Transpiration is also a sun-driven process. The sun heats the Earth's atmosphere, creating wind currents that draw moisture from the tiny openings in the leaves of plants.

In aquatic ecosystems (lakes, rivers, and oceans), the nonliving portion of the water cycle is the most important. In terrestrial ecosystems, the nonliving and living parts of the water cycle both play important roles. In thickly vegetated ecosystems, such as tropical rain forests, more than 90 percent of the moisture in the ecosystem passes through plants and is transpired from their leaves. In a very real sense, plants in rain forests create their own rain. Moisture travels from plants to the atmosphere and falls back to the Earth as rain.

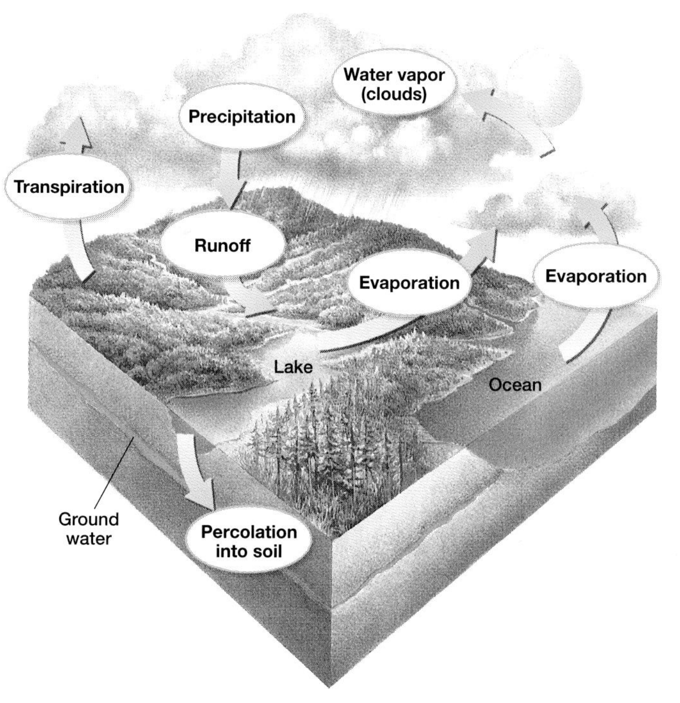

Figure 17-12 Water cycle. This diagram shows the major steps in the water cycle.

biobit

What is an aquifer?

Aquifers are natural storage tanks, large areas of buried porous rock or soil that receive, store, and release ground water. Humans are draining many major aquifers faster than they can refill. As a result, some aquifers have partially collapsed; they cannot be restored.

internetconnect

SCILINKS. NSTA

TOPIC: Water cycle
GO TO: www.scilinks.org
KEYWORD: HX371

did you know?

Ninety-eight percent of the water on Earth is in liquid form—in oceans, lakes, or other bodies of water. The remaining two percent is found as ice in the polar regions, as water vapor in the atmosphere, and as a liquid in the soil and in the bodies of living organisms.

Section 17-3

Active Reading

Technique: Paired Reading

Pair each student with a partner. Have each student read about the water cycle silently while making a question mark on a sticky note next to the passages they find confusing. After they finish reading, ask one student to summarize and the second to add anything omitted. Both readers should then help each other with any passages that either (or both) did not understand. Have them create a list of questions to ask the class. Students should repeat this process for the carbon and nitrogen cycles.

Teaching TIP

Cycles

Earlier in their study of biology, students examined various cycles, including the Krebs cycle in respiration and the Calvin cycle in photosynthesis. Ask what advantage a cycle provides for living systems. *(Students should recognize that because of cycles, materials are constantly regenerated, thus avoiding the need for a continuous supply that could, at some time, run out.)*

Teaching TIP **ADVANCED**

Transpiration

Have each student make a list of as many factors as possible that affect transpiration, and have students justify their choices. Answers should include temperature *(higher temperatures increase transpiration)*, soil *(low water content reduces transpiration)*, wind *(wind increases transpiration)*, humidity *(low humidity increases transpiration)*, and leaf structure *(for example, hairs or a waxy cuticle reduce transpiration)*. **PORTFOLIO**

Visual Strategy

Figure 17-13

Have students trace the steps in the carbon cycle. Point out that an increase of "activity" along one path or section of the cycle will affect other parts of the cycle. For example, increased combustion will increase the amount of carbon dioxide in the atmosphere, which, in turn, may affect rates of photosynthesis.

Teaching TIP

Atmospheric Carbon Dioxide

Have each student make a list of all the activities in their lives that contribute to rising levels of carbon dioxide. Answers could include burning gasoline in a car, propane in a stove, charcoal in a barbecue, and fuel oil or natural gas in a furnace. Ask them about their electricity use. How is electrical power generated in your area? If fossil fuels are burned, then their use of electricity also contributes (indirectly) to carbon dioxide levels. Have students identify ways to reduce the amount of carbon dioxide they contribute. PORTFOLIO

The Carbon Cycle Is Linked to Energy

Carbon also cycles between the nonliving environment and living organisms. You can follow the carbon cycle in **Figure 17-13.** Carbon dioxide in the air or dissolved in water is used by photosynthesizing plants, algae, and bacteria as a raw material to build organic molecules. Carbon atoms may return to the pool of carbon dioxide in the air and water in three ways.

1. **Respiration.** Nearly all living organisms, including plants, engage in cellular respiration. They use oxygen to oxidize organic molecules during cellular respiration, and carbon dioxide is a byproduct of this reaction.
2. **Combustion.** Carbon also returns to the atmosphere through combustion, or burning. The carbon contained in wood may stay there for many years, returning to the atmosphere only when the wood is burned. Sometimes carbon can be locked away beneath the Earth for thousands or even millions of years. The remains of organisms that become buried in sediments may be gradually transformed by heat and pressure into fossil fuels—coal, oil, and natural gas. The carbon is released when the fossil fuels are burned.
3. **Erosion.** Marine organisms use carbon dioxide dissolved in sea water to make calcium carbonate shells. Over millions of years, the shells of the dead organisms form sediments, which form limestone. As the limestone becomes exposed and erodes, the carbon becomes available to other organisms.

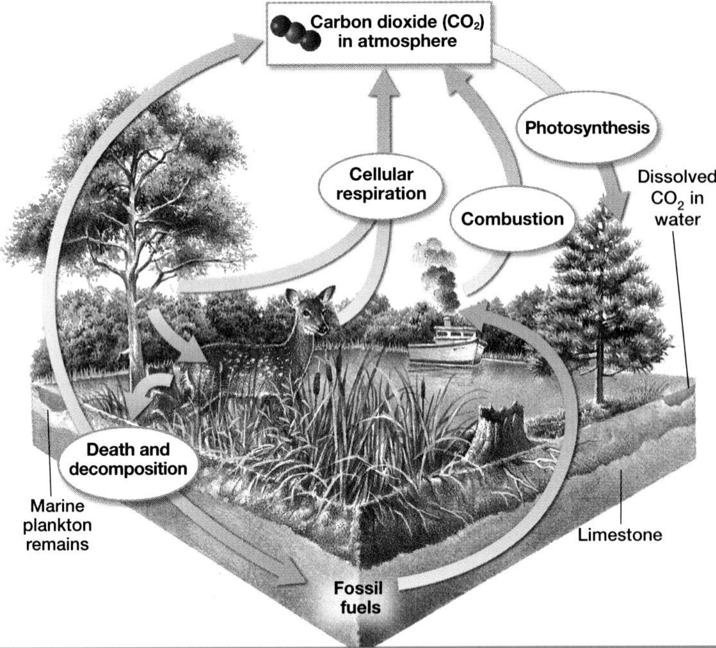

Figure 17-13 Carbon cycle. This diagram shows the major steps of the carbon cycle. In addition to the 700 billion tons of carbon dioxide in the atmosphere, approximately 1 trillion tons are dissolved in the oceans.

did you know?

The burning of fossil fuels and the destruction of forests affects the carbon cycle.

These activities add carbon dioxide to the atmosphere, where it acts as a "greenhouse gas," trapping heat energy that the Earth absorbed as sunlight. Many scientists think this added CO_2 is primarily responsible for the warming trend the Earth has experienced in recent years.

Plants and Bacteria Take Part in the Phosphorus and Nitrogen Cycles

Organisms need nitrogen and phosphorus to build proteins and nucleic acids. Phosphorus is an essential part of both ATP and DNA. Phosphorus is usually present in soil and rock as calcium phosphate, which dissolves in water to form phosphate ions, PO_4^{3-}. This phosphate is absorbed by the roots of plants and used to build organic molecules. Animals that eat the plants reuse the organic phosphorus.

The atmosphere is 79 percent nitrogen gas, N_2. However, most organisms are unable to use it in this form. The two nitrogen atoms in a molecule of nitrogen gas are connected by a strong triple covalent bond that is very difficult to break. However, a few bacteria have enzymes that can break it, and they bind nitrogen atoms to hydrogen to form ammonia, NH_3. The process of combining nitrogen with hydrogen to form ammonia is called **nitrogen fixation.** Nitrogen-fixing bacteria live in the soil and are also found within swellings, or nodules, on the roots of beans, alder trees, and a few other kinds of plants.

The nitrogen cycle, diagramed in **Figure 17-14,** is a complex process with four important stages.

1. **Assimilation** is the absorption and incorporation of nitrogen into plant and animal compounds.
2. **Ammonification** is the production of ammonia by bacteria during the decay of nitrogen-containing urea (found in urine).
3. **Nitrification** is the production of nitrate from ammonia.
4. **Denitrification** is the conversion of nitrate to nitrogen gas.

Study TIP

Reviewing Information

Using your own words, write four sentences, each one describing one of the four biogeochemical cycles.

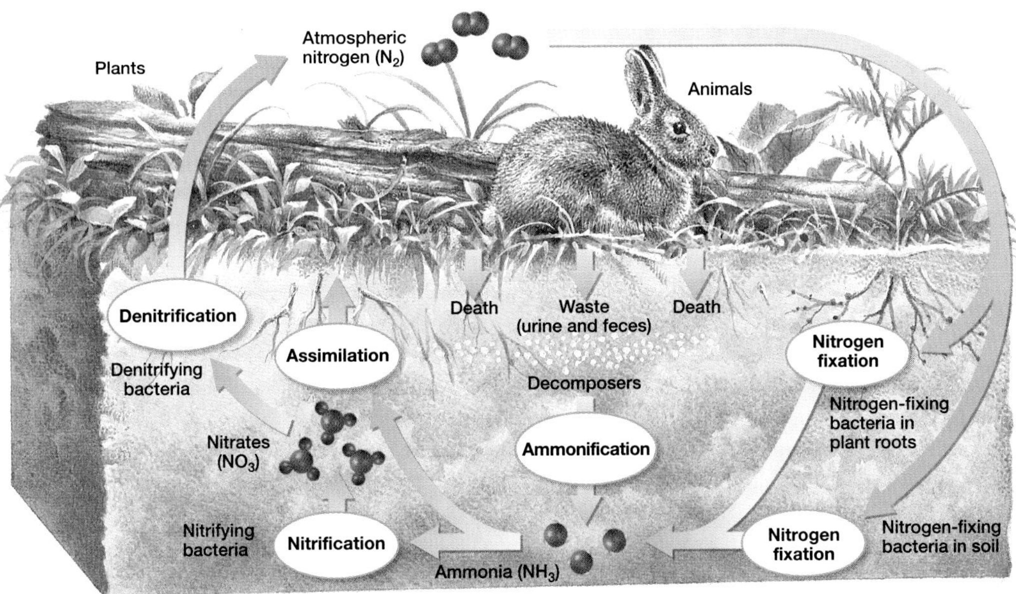

Figure 17-14 Nitrogen cycle. Bacteria carry out many of the important steps in the nitrogen cycle, including the conversion of atmospheric nitrogen into a usable form, ammonia.

Teaching TIP

Legumes

Tell students that beans are leguminous plants, the roots of which have nodules containing nitrogen-fixing bacteria. Then tell them that other such plants include clover, peas, alfalfa, lupines, and locust trees. If any of these plants are available nearby, find out if you can dig up an example to show the class any nodules that may be present.

Visual Strategy

Figure 17-14

Point out the difference between nitrogen fixation (nitrogen gas → ammonia) and nitrification (ammonia → nitrates). Tell students that lightning also changes nitrogen gas to ammonia, but such atmospheric action amounts to less than 10 percent of that carried out by organisms through nitrogen fixation. Finally, have students recognize that denitrification returns nitrates to the atmosphere as nitrogen gas.

Resource Link

Assign **Reading Strategies: Recognizing Cycles** from the Holt Biosources Teaching Resources CD-ROM.

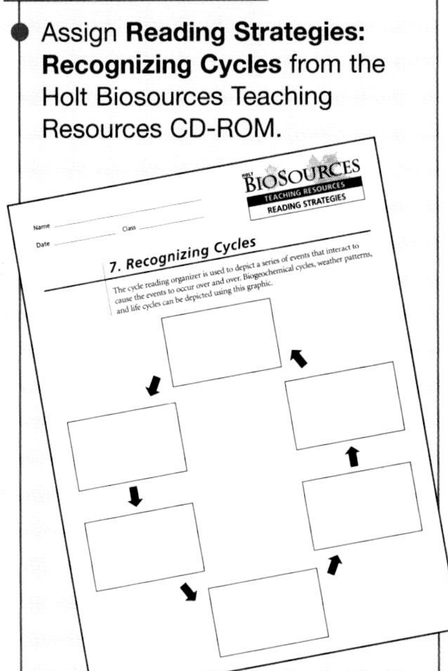

did you know?

Farmers often rotate a nonleguminous crop, such as corn, with a leguminous one, such as alfalfa.

The alfalfa will fix nitrogen and release some of it into the soil. Moreover, if a crop of alfalfa is plowed back into the soil, it may add as much as 350 kg (770 lb) of nitrogen per hectare (2.5 acres) of soil, enough to grow a crop of nonleguminous plants without the need for additional fertilizer.

Sustainable Agriculture

Teaching Strategies
Tell students that organic farming is a form of sustainable agriculture that does not use chemical fertilizers or pesticides.

Discussion
Ask students if they think practicing sustainable agriculture would lead to bigger profits for farmers. *(Sustainable agriculture is probably less profitable in the short term, as farmers invest in cover crops or limit the size of their herds to manage the pastures. But in the long run, they may come out ahead because they will still have healthy soil and will not be overusing costly fertilizers. Likewise, livestock farmers will not need as much feed for their livestock.)*

3 Assessment Options

Closure

Technique: Anticipation Guide
Have students review their answers to the **Active Reading** on page 359 and decide whether their ideas have changed after reading the chapter.

Review
Assign Review—Section 17-3.

Reteaching — BASIC

Ask students to write a short essay describing what the world would look like without fungus or bacteria. *(Dead organisms would pile up, and the materials of their bodies would not be recycled. Plants would become sickly without the nitrogen-fixing bacteria, carbon and nutrients would not be recycled, and many animals would become extinct without the bacteria in their gut to help them digest cellulose.)* PORTFOLIO

The growth of plants in ecosystems is often limited by the availability of nitrate and ammonia in the soil. Today most of the ammonia and nitrate that farmers add to soil is produced chemically in factories, rather than by bacterial nitrogen fixation. As you learned in Chapter 11, genetic engineers are trying to place nitrogen-fixing genes from bacteria into the chromosomes of crop plants. If these attempts are successful, the plants themselves will be able to fix nitrogen, thus eliminating the need for nitrogen-supplying fertilizers. Some farmers adjust their farming methods to increase natural recycling of nitrogen.

FOOD WATCH: Sustainable Agriculture

In an ecosystem, decomposers return mineral nutrients to the soil. However, when the plants are harvested and shipped away, there is a net loss of nutrients from the soil where the plants were growing. The amount of organic matter in the soil also decreases, making the soil less able to hold water and more likely to erode.

What is sustainable agriculture?
Sustainable agriculture refers to farming that remains productive and profitable through practices that help replenish the soil's nutrients, reduce erosion, and control weeds and insect pests.

Use of cover crops
After harvest, farmers can plant cover crops, such as rye, clover, or vetch, instead of letting the ground lie bare. Cover crops keep the soil from compacting and washing away, and they help the soil absorb water. They also provide a habitat for beneficial insects, slow the growth of weeds, and keep the ground from overheating. When cover crops are plowed under, as illustrated in the figure at right, they return nutrients to the soil.

Rotational grazing
Farmers who raise cattle and sheep can divide their pastures into several grazing areas. By rotating their livestock from one area to another, they can prevent the animals from overgrazing the pasture. This allows the plants on which the animals feed to live longer and be more productive. Water quality improves as the pasture vegetation becomes denser. Animals distribute manure more evenly with rotational grazing than they do in feed lots or unmanaged pastures.

There are many other methods used in sustainable agriculture. Farmers must determine which methods work best for their crops, soil conditions, and climate.

Review SECTION 17-3

1. **Identify** the process that plants use to transfer water to the atmosphere.
2. **Summarize** the carbon cycle's relationship to the flow of energy.
3. **Describe** how bacteria participate in the nitrogen cycle.

Critical Thinking

4. **Defend** the argument that nutrients can cycle but energy cannot.

Review SECTION 34-3 ANSWERS

1. transpiration
2. Carbon and energy both move through ecosystems. Energy from sunlight is captured by photosynthesizers, which use it to build organic molecules using carbon dioxide in air or water. Energy flows out of ecosystems primarily as heat during cellular respiration and combustion. These processes also liberate carbon, but the released carbon does not "leave" the ecosystem; instead it is recycled by photosynthesizers.
3. Nitrogen-fixing bacteria make ammonia, a nitrogen compound that some plants can use, from nitrogen gas. Nitrifying bacteria create nitrates, which are also used by plants, from ammonia. Denitrifying bacteria turn the nitrates into nitrogen gas.
4. Nutrients can cycle because they are always in a form usable by at least some organisms, which keeps the nutrient moving through (but still within) the ecosystem. Most energy, on the other hand, only flows in one direction. By the end of the food chain, nearly all the original energy has been "lost" as unusable heat.

Study ZONE CHAPTER 17

Use the Key Concepts and Key Terms listed below to review the main ideas in this chapter. If you need to review the meaning of a term, turn to the page indicated.

Key Concepts

17-1 What Is an Ecosystem?

- Ecology is the study of how organisms interact with each other and with their environment.
- A community of organisms and their nonliving environment constitute an ecosystem.
- Ecosystems contain diverse organisms.
- Ecosystems change through the process of succession.
- Succession on a newly formed habitat is primary succession.
- Secondary succession occurs on a habitat that has previously supported growth.

17-2 Energy Flow in Ecosystems

- Energy moves through ecosystems in food chains, passing from photosynthesizers (producers) to herbivores (consumers) to carnivores (consumers), creating a food web.
- Energy transfers between trophic levels transfer only 10 percent of the energy in a trophic level to the next level.
- Most terrestrial ecosystems have only three or four trophic levels because energy transfers between trophic levels are inefficient.

17-3 Ecosystems Cycle Materials

- Minerals and other materials cycle within ecosystems among organisms and between organisms and the physical environment.
- In the water cycle, water falls as precipitation and either evaporates from bodies of water, is stored in ground water, or cycles through plants and then evaporates.
- Carbon enters the living portion of the carbon cycle through photosynthesis. Organisms release carbon through cellular respiration. Carbon trapped in rocks and fossil fuels is released by erosion and burning.
- Bacteria fix atmospheric nitrogen, thus making ammonia available to other organisms.
- Phosphorus is cycled through plants into animals.

Key Terms

17-1

ecology (360)
habitat (360)
community (360)
ecosystem (360)
abiotic factor (360)
biotic factor (360)
biodiversity (361)
pioneer species (363)
succession (363)
primary succession (363)
secondary succession (363)

17-2

primary productivity (365)
producer (365)
consumer (365)
trophic level (365)
food chain (366)
herbivore (366)
carnivore (366)
omnivore (366)
detritivore (366)
decomposer (367)
food web (367)
energy pyramid (368)
biomass (369)

17-3

biogeochemical cycle (370)
ground water (371)
transpiration (371)
nitrogen fixation (373)

Unit 7—*Ecosystem Dynamics*
Use Topics 1, 3–6 in this unit to review the key concepts and terms in this chapter.

Ready?

- *If you think you understand these Key Concepts and Key Terms, then you're ready for the Performance Zone.*

Answer to Concept Map

The following is one possible answer to Performance Zone item 1 on page 376.

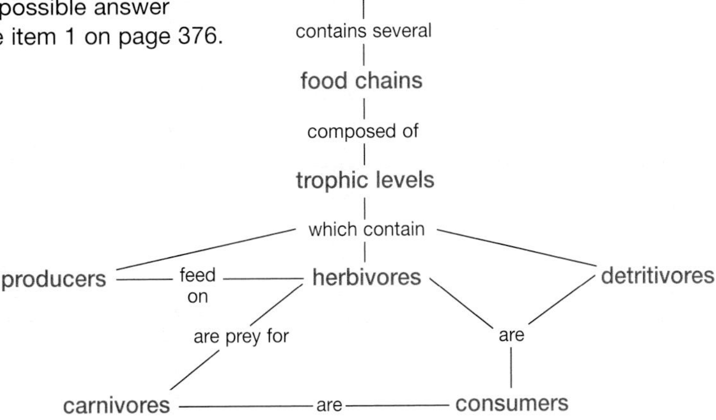

Study ZONE CHAPTER 17

Review and Practice

Assign the following worksheets for Chapter 17:

- **Student Review Guide**
- **Concept Mapping Worksheet**
- **Test Preparation Pretest**
- **Vocabulary Worksheet**
- **Critical Thinking Review Worksheet**
- **Science Skills Worksheet**

Assessment Options

The following assessment options are available for this chapter:

Resource Link

- Assign **Chapter Test 17** to assess students' understanding of the chapter.

Alternative Assessment

Have students choose an ecosystem from anywhere in the world. *(For example, students could choose the Costa Rican rain forest, Sonoran Desert, Kalahari Plain, or Florida Everglades.)* Have them research and report on the living and nonliving components. Tell them to include descriptions of four producers, three herbivores, two small carnivores, one top carnivore, and one detritivore. Have them draw a food chain, food web, and energy pyramid for this ecosystem. PORTFOLIO

Portfolio Assessment

- *Portfolio Assessment* at the front of this book provides options for building and evaluating student portfolios. Students and teachers can select items from the areas listed for inclusion in a portfolio.
- See items labeled PORTFOLIO on pp. 361, 362, 363, 369, 371, 372, 374, and 375.

Performance ZONE
CHAPTER 17
REVIEW ANSWERS

Assignment Guide

Section	Review
17-1	3, 6, 17, 18, 19, 20, 25, 26
17-2	1, 2, 4, 5, 7, 10, 11, 22
17-3	8, 9, 12, 13, 14, 15, 16, 21, 23, 24

Understanding and Applying Concepts

1. The answer to the concept map is found at the bottom of page 375.
2. **a.** Biodiversity describes the number of different species living in an area. Biomass is a measure of the total dry mass of living organisms in an area.

   b. Ecosystem is a more inclusive term, encompassing all the living and nonliving factors in a given area. A community describes the different populations living in a given area.

   c. Producers are organisms that create their own food, usually through photosynthesis. Consumers are organisms that must digest other living organisms (or their remains) for energy.

   d. A food chain is a simple concept or diagram illustrating the flow of energy from one trophic level to the next. A food web is a more complex concept or diagram that illustrates the various possible food chains/ energy flows for a given set of organisms in an ecosystem.
3. c
4. b
5. d
6. a
7. c
8. b
9. d
10. b
11. Greater flexibility in diet allows omnivores to eat whatever food is available.
12. Plowing the corn into the field would return organic matter and nutrients to the soil—a more sustainable approach.

Performance ZONE CHAPTER 17 Review

Understanding and Applying Concepts

1. **Concept Mapping** Make a concept map that shows the role of organisms in the flow of energy through an ecosystem. Use as many terms as needed from the vocabulary list. Try to include the following terms in your map: trophic level, food web, food chain, producer, consumer, carnivore, detritivore, herbivore. Include additional terms in your map as needed.
2. **Understanding Vocabulary** For each pair of terms, explain the differences in their meanings.
   a. biodiversity, biomass
   b. ecosystem, community
   c. producers, consumers
   d. food chain, food web
3. Ecosystems differ from a community in that ecosystems also contain
   a. a habitat.
   b. several communities.
   c. all the physical aspects of the habitat.
   d. succession.
4. What critical role is played by fungi and bacteria in any ecosystem?
   a. primary production
   b. decomposition
   c. boundary setting
   d. physical weathering
5. A mountain lion is a(n)
   a. omnivore. c. detritivore.
   b. herbivore. d. carnivore.
6. Which sequence shows the correct order of succession at Glacier Bay, Alaska?
   a. *Dryas,* alder, Sitka spruce
   b. mosses, hemlock, Sitka spruce
   c. *Dryas,* hemlock, alder
   d. alder, *Dryas,* hemlock
7. How much energy would be available to organisms at the third trophic level of an energy pyramid if there were 1,000 kcal available in the first trophic level?
   a. 1,000 kcal c. 10 kcal
   b. 100 kcal d. 1 kcal
8. Plants return water to the atmosphere by
   a. condensation. c. assimilation.
   b. transpiration. d. desiccation.
9. Which role is not performed by bacteria in the nitrogen cycle?
   a. fixing nitrogen
   b. changing urea to ammonia
   c. turning nitrates into nitrogen gas
   d. changing nitrates to ammonia
10. How do you think the food web below would be affected if the grass and the shrub were both eliminated?
    a. Birds and mice would starve.
    b. The food web would completely collapse.
    c. The food web would collapse partially.
    d. Nothing would happen.

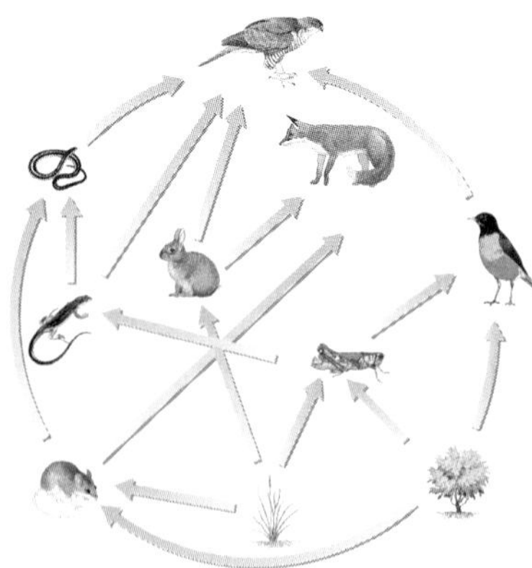

11. **Theme Evolution** Humans, raccoons, and bears are omnivores. What adaptive advantage might this feeding strategy provide these organisms?
12. **Food Watch** After harvesting a field of corn, a farmer could either plow the cornstalks into the field or haul them off the field and burn them. Which option is best for sustainable agriculture? Explain.
13. **Chapter Links** Relate the importance of photosynthesis to the nitrogen cycle. (**Hint:** See Chapter 5, Section 5-2.)

---

13. Photosynthesis enables plants to incorporate nitrogen into proteins, making it available to consumers. When consumers excrete waste or decompose, plants (with the aid of bacteria) may take up the nitrogen again, completing the cycle.

Skills Assessment

14. As trees begin photosynthesis in the spring, they take up atmospheric carbon dioxide, so levels drop. In the fall, as trees lose their leaves, photosynthesis stops and leaves decompose, putting carbon dioxide back into the atmosphere.
15. The fluctuation would be less severe because evergreens keep most of their leaves (needles) during the winter.
16. The graph would be inverted, with the highest amounts of water transpired in the summer and the lowest in the winter.
17. Answers will vary. In tree-dominated ecosystems, trees affect the water and carbon cycles through evapotranspiration, cellular respiration, and decomposition. They also provide habitat for a variety of species.
18. Answers will vary.

Skills Assessment

Part 1: Interpreting Graphs

Study the graph below showing the amount of atmospheric carbon dioxide in a Michigan deciduous forest (deciduous trees, such as elms and maples, shed their leaves in the fall); then answer the following questions:

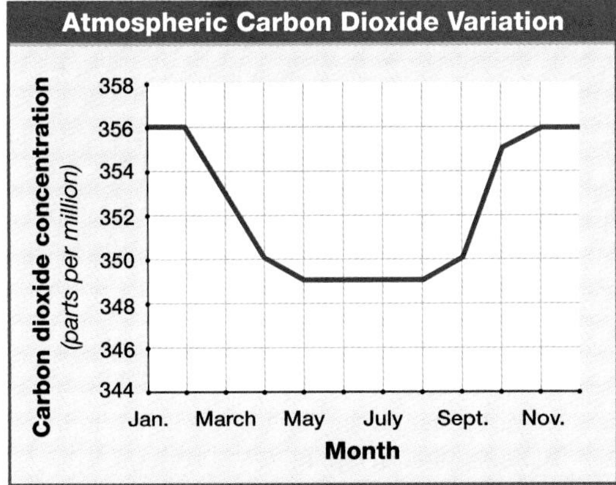

14. A scientist finds that the carbon dioxide levels fluctuate during the year, as shown in the graph above. Using your knowledge of the carbon cycle, explain the cause of these fluctuations.

15. Would you expect these results to be the same if evergreen trees in Michigan were monitored instead of deciduous trees? Explain.

16. What would you expect the graph to look like if the *y*-axis measured the amount of water transpired instead? Explain your reasoning.

Part 2: Life/Work Skills

Use the library or Internet resources to learn how scientists study the role of trees in ecosystems.

17. Finding Information For each ecosystem you find information on, identify the kinds of trees that each ecosystem has. Identify the role that trees play in each ecosystem.

18. Communicating Prepare an oral report to summarize your findings. Create a display that shows the methods and equipment scientists use today to show how the trees affect the ecosystem.

Critical Thinking

19. Evaluating Viewpoints Ecologists once referred to stable ecosystems as a final or climax community. Now most ecologists say that no ecosystem can truly have a final end point. Explain why ecologists changed their viewpoint.

20. Evaluate the role of herbivores in primary succession.

21. Choose whether nitrogen cycling or carbon cycling is more important to a pioneer species during primary succession.

22. Evaluate the importance of food chains to the water supply.

Portfolio Projects

23. Identifying Functions Obtain photocopies of nature paintings by American painters such as John James Audubon or Edward Hicks. Write a report that describes the various organisms, their places in the various cycles, and the trophic level they occupy.

24. Interpreting Information Read the article "Leaf Litter," by Arthur Fisher (*Popular Science,* December 1997, vol. 251, p. 18). Write a report explaining what ecologist Bruce Wallace believed prior to his experiment. In your report describe what he did to test the hypothesis and what his results were.

25. Career Focus Ecologist Use library and Internet resources to research the educational background necessary to become an ecologist. List the high school courses required, and describe additional degrees or training that is recommended for this career. Summarize the employment outlook for this field.

26. Unit 7—*Ecosystem Dynamics* Write a report summarizing how artificial ecosystems used in the management and treatment of waste water and pollutants, can demonstrate succession.

Critical Thinking

19. Changes in climate, large-scale disturbances, such as fire or volcanic eruption, and even changes in biotic factors, such as an insect pest outbreak, can change the conditions of the ecosystem. Ecologists realized that ecosystems are dynamic by nature.

20. The activities of herbivores play an important role in primary succession, especially during later successional stages. Before soil is established, herbivores may inadvertently deposit seeds on the ground in their droppings. Once a plant community begins to develop, grazers affect the mix of plants through their eating habits, while seed gatherers (such as squirrels) inadvertently plant new seeds. Herbivores can also introduce plant seeds caught on their fur and in their droppings.

21. Many pioneer organisms, such as lichens, have the ability to fix nitrogen, but *all* pioneer plants are able to fix carbon during photosynthesis. Therefore, nitrogen cycling is probably more important to a pioneer species during primary succession.

22. Food chains, especially the producer links, play an important role in the water supply. Although most of the Earth's water is found in lakes and oceans, plants return huge amounts of water to the atmosphere through transpiration—a vital link in the water cycle.

Portfolio Projects

23. Answers will vary depending on the organisms pictured.

24. Wallace believed that leaf litter was important to the ecology of streams, but its importance had not been tested. Wallace and his colleagues made a canopy over a stream to keep leaf litter out. The researchers then compared the covered stream to an uncovered stream. They found that the covered stream had fewer organisms than did the uncovered stream.

25. Answers will vary. Ecologists are scientists who study how organisms interact with their living and nonliving environments. Ecologists usually earn at least an undergraduate degree, and many have a master's or a Ph.D. The largest employers of ecologists are government agencies (federal, state, and local), universities, and consulting firms. Growth potential for this field is fair. Starting salary will vary by region.

26. Answers will vary. Student reports should note that artificial wastewater-treatment ecosystems tend to undergo eutrophication, just as natural wetlands do. However, the high nutrient levels in waste water promote rapid algae growth in many artificial wastewater-treatment systems. If the systems are not manipulated, they will eventually fill with algae and decaying organic matter, forming a marsh and eventually a meadow.

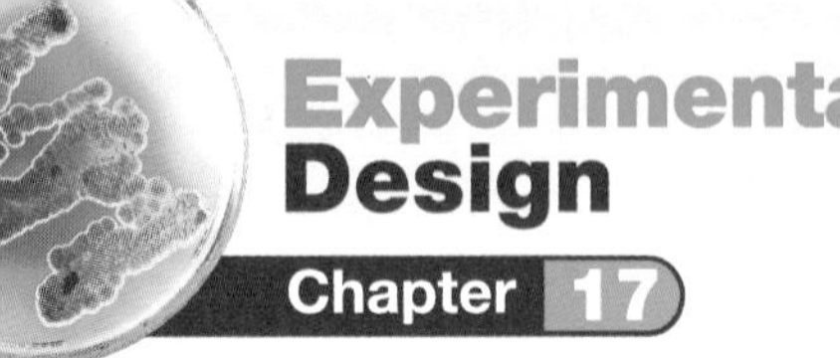

Experimental Design

Chapter 17

How Does an Ecosystem Change over Time?

Purpose

Students will build a closed ecosystem and observe how it changes over time.

Preparation Notes

Time required: About 20 minutes on day 1 and about 10 minutes each day thereafter over a period of a few weeks

Materials

Materials for this lab can be ordered from WARD'S. See *Master Materials List* at the front of this book for catalog numbers.

Preparation Tips

- Plastic soda bottles may be substituted for glass bottles. Have students bring clear plastic 2 or 3 L soda bottles from home, or obtain some from a local recycling center. Labels and bottoms can be removed by using a hair dryer or by soaking the bottles in warm water.
- Soil can be collected from around the school, brought from home, or purchased from a supermarket or garden center.
- Organisms may be obtained from a biological supply house. You may also be able to find earthworms and crickets in the local environment or at pet stores or bait shops.
- Providing small lizards or snakes will provide one more level in the food web for students to observe. These can be purchased at local pet stores.
- Allow plants and algae to grow for one week before adding animals to the ecosystem.

Safety Precautions

- Warn students to take care when handling insects and other small animals. Small animals are easily harmed, and some are capable of biting when disturbed.
- Remind students that the ecosystems are dependent on humans for care. They should not be permitted to be overheated or become too cold. If water evaporates from the ecosystem, it should be replenished.
- If glass jars are used, remind students that cut glass resulting from broken jars is dangerous and that they should notify you immediately if any jars break.

Procedural Tips

- Students will need a few minutes each day for observations over a period of a few weeks.
- Recording the number of organisms may be tricky in some cases and estimates may be required. Discuss the sources of error in estimates versus direct counts.
- The ecosystems may be kept for the rest of the school year and then given to students to take home over the vacation.

Disposal

Soil and plant material can be disposed of in the trash or in a garden. Place any animals in a garden or a nearby natural area.

Experimental Design

Chapter 17 *How Does an Ecosystem Change over Time?*

SKILLS

- Using scientific methods
- Modeling
- Observing

OBJECTIVES

- **Construct** a model ecosystem.
- **Observe** the interactions of organisms in a model ecosystem.
- **Predict** how the number of each species in a model ecosystem will change over time.
- **Compare** a model ecosystem with a natural ecosystem.

MATERIALS

- coarse sand or pea gravel
- large glass jar with a lid or terrarium
- soil
- pinch of grass seeds
- pinch of clover seeds
- mung bean seeds
- earthworms
- isopods (pill bugs)
- mealworms (beetle larva)
- crickets

Before You Begin

Organisms in an **ecosystem** interact with each other and with their environment. One of the interactions that occurs among the organisms in an ecosystem is feeding. A **food web** describes the feeding relationships among the organisms in an ecosystem. In this lab, you will model a natural ecosystem by building a **closed ecosystem** in a bottle or a jar. You will then observe the interactions of the organisms in the ecosystem and note any changes that occur over time.

1. Write a definition for each boldface term in the paragraph above and for each of the following terms: producer, decomposer, consumer, herbivore, carnivore, trophic level.
2. Based on the objectives for this lab, write a question you would like to explore about ecosystems.

Procedure

PART A: Building an Ecosystem in a Jar

1. Place 2 in. of sand or pea gravel in the bottom of a large, clean glass jar with a lid. **CAUTION: Glassware is fragile. Notify your teacher promptly of any broken glass or cuts. Do not clean up broken glass or spills with broken glass unless your teacher tells you to do so.** Cover the gravel with 2 in. of soil.
2. Sprinkle the seeds of two or three types of small plants, such as grasses and clovers, on the surface of the soil. Put a lid on the jar, and place it in indirect sunlight. Let the jar remain undisturbed for a week.
3. After one week, place the animals into the jar and replace the lid. Place the lid on the jar loosely to enable air entry for the animals. You may also put small holes in the lid.

You Choose

As you design your experiment, decide the following:

a. what question you will explore
b. what hypothesis you will test
c. how you will plant the seeds
d. where you will place the ecosystem for one week so that it remains undisturbed and in indirect sunlight
e. how often you will add water to the ecosystem after the first week
f. how many of each organism you will use
g. what data you will record in your data table

PART B: Design an Experiment

4. Work with the members of your lab group to explore one of the questions written for step 2 of **Before You Begin.** To explore the question, design an experiment that uses the materials listed for this lab.
5. Write a procedure for your experiment. Make a list of all the safety precautions you will take. Have your teacher approve your procedure and safety precautions before you begin the experiment.
6. Set up your group's experiment. Conduct your experiment for at least 14 days.

PART C: Cleanup and Disposal

7. Dispose of solutions, broken glass, and other materials in the designated waste containers. Do not put lab materials in the trash unless your teacher tells you to do so.
8. Clean up your work area and all lab equipment. Return lab equipment to its proper place. Wash your hands thoroughly before you leave the lab and after you finish all work.

Analyze and Conclude

1. **Summarizing Results** Make graphs showing how the number of individuals of each species in your ecosystem changed over time. Plot time on the x-axis and the number of organisms on the y-axis.
2. **Analyzing Results** How did your results compare with your hypothesis? Explain any differences.
3. **Inferring Conclusions** Construct a food web for the ecosystem you observed.
4. **Recognizing Relationships** Does your model ecosystem resemble a natural ecosystem? Explain.
5. **Analyzing Methods** How might you have built your model ecosystem differently to better represent a natural ecosystem?
6. **Evaluating Methods** Was your model ecosystem truly a "closed ecosystem"? List your model's strengths and weaknesses as a closed ecosystem.
7. **Further Inquiry** Write a new question about ecosystems that you could explore with another investigation.

Do You Know?

Do research in the library or media center to answer these questions:

1. What is Biosphere 2?
2. What problems were encountered by the Biosphere 2 crew during the 1991–1993 project?

Use the following Internet resources to explore your own questions about ecosystems.

TOPIC: Ecosystems
GO TO: www.scilinks.org
KEYWORD: HX379

Before You Begin

Answers

1. *ecosystem*—an ecological system encompassing a biological community and all the physical aspects of its habitat

   food web—a network of feeding relationships in an ecosystem

   closed ecosystem—an ecosystem that does not exchange materials outside of itself
2. Answers will vary. For example, What are the effects of continuous exposure to bright light on the organisms in an ecosystem?

Procedure

Sample Procedure

Place the jar to the side of a window such that it receives indirect sunlight throughout the day. After one week, add three earthworms, five isopods, three mealworms, and three crickets to the jar. Using a mister, add four squirts of water per square decimeter of soil surface every other day. Record population data every other day for two weeks.

Analyze and Conclude

Answers

1. Answers will vary. Students should make one graph for each species observed or use different colors to indicate each species.
2. Answers will vary.
3. Answers will vary. All plants are producers (primary trophic level); earthworms feed on dead plant material in the soil; crickets feed on plants; mealworms (beetle larva) feed on plants; isopods (pill bugs) eat wood.
4. Yes and no. Natural ecosystems and the model ecosystem both contain organisms at several trophic levels, have living and nonliving components, and depend on the sun for energy. However, the model ecosystem is less diverse, much younger, and has more definite boundaries than a natural ecosystem.
5. Answers will vary.
6. No, strengths are that the organisms in the model ecosystem did not leave the ecosystem and that other organisms could not enter from the outside. Weaknesses are that water and air probably had to be added to maintain a healthy ecosystem.
7. Answers will vary. For example: What are the effects of certain abiotic factors, such as temperature, light, and moisture, on the organism in an ecosystem?

Do You Know?

Answers

1. Biosphere 2 is a large artificial ecosystem near Tucson, Arizona. Now managed by Columbia University, the 204,000 m^3 glass and steel structure contains seven ecosystems. The structure is used for research on the biosphere, the sum of all of Earth's ecosystems.
2. Answers will vary. Students should list several examples of problems, including food shortages, oxygen shortages, and population explosions of microorganisms, ants, and cockroaches.

Chapter **18**

Biological Communities

Planning Guide

	OBJECTIVES	MEETING INDIVIDUAL NEEDS	READING SKILLS AND STRATEGIES
BLOCKS 1 & 2 90 min	**18-1 How Organisms Interact in Communities** • Describe coevolution. • Predict how coevolution can affect interactions between species. • Identify the distinguishing features of symbiotic relationships.	**Learning Styles,** p. 382 BASIC ADVANCED • pp. 383, 384	**ATE** **Active Reading,** **Technique: Brainstorming,** p. 381 ■ **Active Reading Guide,** worksheet (18-1) ■ **Directed Reading,** worksheet (18-1)
BLOCKS 3 & 4 90 min	**18-2 How Competition Shapes Communities** • Describe the role of competition in shaping the nature of communities. • Distinguish between fundamental and realized niches. • Contrast character displacement with competitive exclusion. • Describe how competition affects an ecosystem. • Summarize the importance of biodiversity.	**Learning Styles,** p. 385 BASIC ADVANCED • p. 390	**ATE** **Active Reading,** **Technique: Paired Summarizing,** p. 389 ■ **Active Reading Guide,** worksheet (18-2) ■ **Directed Reading,** worksheet (18-2)
BLOCKS 5 & 6 90 min	**18-3 Major Biological Communities** • Recognize the role of climate in determining the nature of a biological community. • Describe how elevation and latitude affect the distribution of biomes. • Summarize the key features of the Eath's major biomes. • Compare and contrast the major freshwater and marine habitats.	**Learning Styles,** p. 391 BASIC ADVANCED • pp. 392, 398	■ **Active Reading Guide,** worksheet (18-3) ■ **Directed Reading,** worksheet (18-3) ■ **Supplemental Reading Guide,** *Silent Spring*

Review and Assessment Resources

BLOCKS 7 & 8 90 min

TEST PREP

PE **Study Zone,** p. 399

PE **Performance Zone,** pp. 400–401

■ **Chapter 18 worksheets:**
- Vocabulary
- Science Skills
- Concept Mapping
- Critical Thinking

Holt Biology Interactive Tutor Unit 7—Ecosystem Dynamics, Topics 1–6

Audio CD Program

■ **Test Prep Pretests**

CHAPTER TESTING

■ **Chapter Tests** (blackline master)

▲ **Test Generator**

■ **Assessment Item Listing**

See the correlations table at the front of this book for a list of the ***National Science Education Standards*** addressed in this chapter.

Smithsonian Institution® Visit **www.si.edu/hrw** for additional on-line resources.

KEY

Biology Principles and Explorations

PE Pupil's Edition
ATE Teacher's Edition

HOLT BIOSOURCES

Teaching Transparencies
Laboratory Program

One-Stop Planner CD-ROM
Shows these resources within a customizable daily lesson plan:

■ Teaching Resources
▲ Test Generator

VISUAL STRATEGIES	CLASSWORK AND HOMEWORK	TECHNOLOGY AND INTERNET RESOURCES
ATE Fig. 18-3, p. 384	**PE** Data Lab How does predation affect a plant species?, p. 383 Lab C22 Examining Owl Pellets **PE** Review, p. 384	Interactive Tutor Unit 7—Ecosystem Dynamics (Topics 1–6) Holt Biology Videodiscs Section 18-1 Audio CD Program Section 18-1
ATE Fig. 18-4, p. 385 **ATE** Fig. 18-5, p. 386 **ATE** Fig. 18-7, p. 389 86 Earthworm Niche 87 Effect of Competition on an Organism's Niche	**PE** Data Lab What might alter an organism's fundamental niche?, p. 387 Lab B4 Plant and Animal Interrelationships **PE** Review, p. 390	CNN Video Segment 14 Tropical Reforestation Interactive Tutor Unit 7—Ecosystem Dynamics (Topics 1–6) Holt Biology Videodiscs Section 18-2 Audio CD Program Section 18-2 internetconnect SCILINKS NSTA TOPICS: • Symbiosis • Extinction
ATE Fig. 18-10, p. 392 **ATE** Fig. 18-11, p. 393 65 Global Distribution of Seven Biomes 65A Causes of Earth's Climate 78 Warbler Foraging Zones 79 Three Habitats in a Marine Environment	**PE** Quick Lab What factors influence the cooling of the Earth's surface?, p. 395 **PE** Experimental Design How Do Brine Shrimp Select a Habitat?, pp. 402–403 **PE** Review, p. 398 **PE** Study Zone, p. 399	Interactive Tutor Unit 7—Ecosystem Dynamics (Topics 1–6) Holt Biology Videodiscs Section 18-3 Audio CD Program Section 18-3 internetconnect SCILINKS NSTA TOPICS: • Biomes • Estuaries • Adaptation

ALTERNATIVE ASSESSMENT

PE Portfolio Projects 19–22, p. 401
ATE Alternative Assessment, p. 399
ATE Portfolio Assessment, p. 399

Lab Activity Materials

Data Labs
pp. 383, 387 pencil, paper

Quick Lab
p. 395 MBL or CBL system with appropriate software, temperature probes, test tubes, beaker, hot plate, one-holed stoppers, water, sand, test-tube tongs, test-tube rack

Experimental Design
p. 402 clear, flexible plastic tubing, metric ruler, marking pen, corks to fit tubing, brine shrimp culture, screw clamps, test tubes with stoppers and test-tube rack, pipet, Petri dish, Detain™ or methyl cellulose, aluminum foil, calculator, fluorescent lamp or grow light, funnel, graduated cylinder or beaker, hot-water bag, ice bag, pieces of screen, tape

Lesson Warm-up
ATE photograph of a tapeworm, p. 382

Demonstrations
ATE photograph of poison ivy, p. 384
ATE photographs of a starling and a bluebird, p. 388
ATE photographs of 5 major biomes, p. 392

BIOSOURCES LAB PROGRAM
- Inquiry Skills Development: Lab B4, p. 13
- Laboratory Techniques and Experimental Design: Lab C22, p. 103

Chapter 18

Chapter Theme

Evolution

The biological communities on Earth today have been shaped by millions of years of evolution. The members of these communities are well adapted to their environments for a reason: they are descended from a long line of survivors. Their bodies and behaviors have evolved, through natural selection, in response to each other and to their physical environment.

Establishing Prior Knowledge

Before beginning this chapter, make certain students can answer the questions in **Study TIP Ready?** on page 381. If they cannot, encourage them to reread the sections indicated.

Opening Question

Ask each student to choose two animals that live in the same community and that interact. Prepare a list on the board or overhead with three headings: *Both benefit, One benefits/One suffers, One benefits/One not affected.* Ask in which category students would place the interaction of each animal pair.

CHAPTER 18 Biological Communities

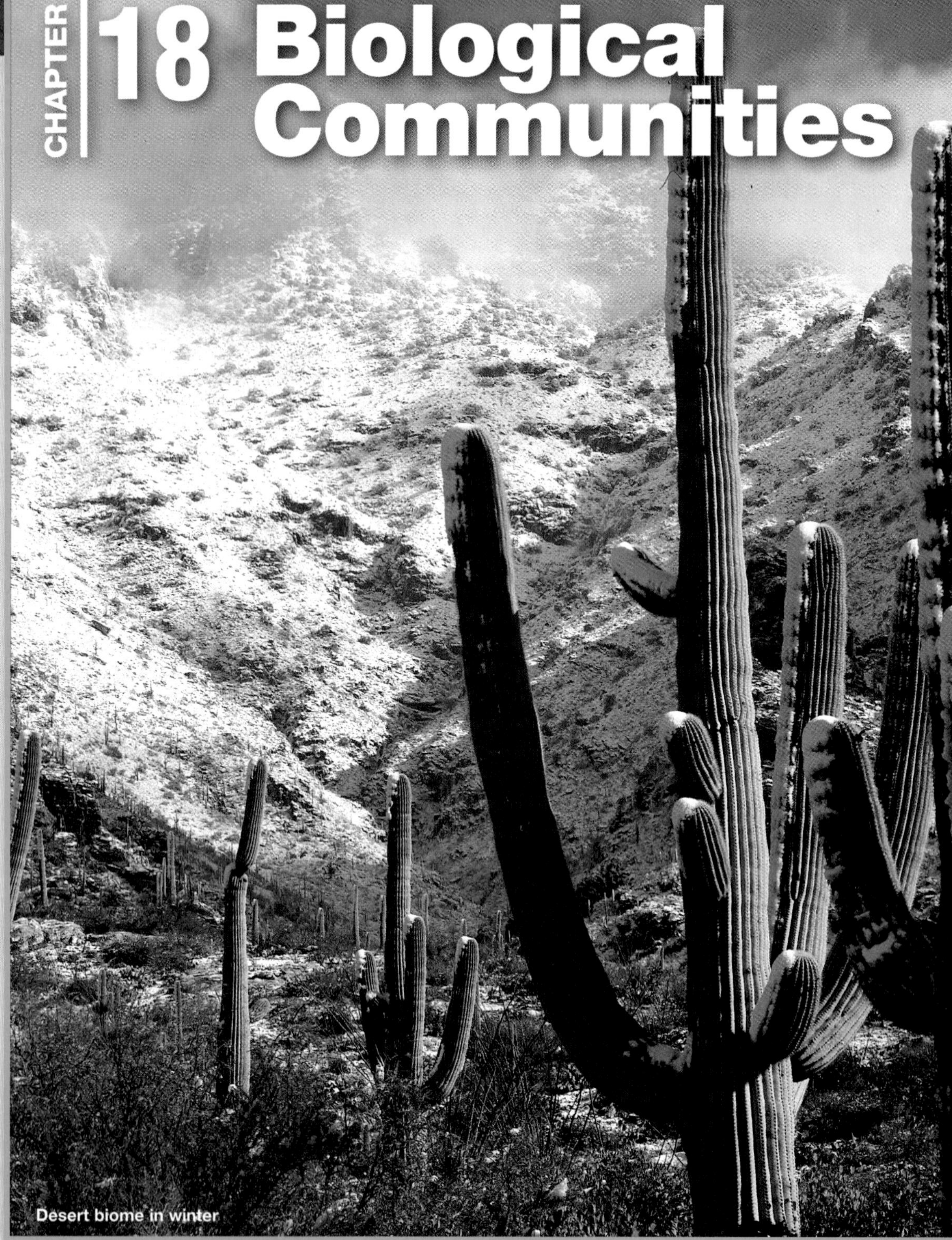

Desert biome in winter

What's Ahead — and Why
CHAPTER OUTLINE

18-1 How Organisms Interact in Communities

- Species Evolve in Response to One Another
- Symbiotic Species Are Shaped by Long-term Relationships

18-2 How Competition Shapes Communities

- Common Use of Scarce Resources Leads to Competition
- Competition Can Limit How Species Use Resources
- Competition Without Division of Resources Leads to Extinction

18-3 Major Biological Communities

- Climate Largely Determines Where Species Live
- There Are Seven Major Biomes
- Aquatic Communities Are Linked to Terrestrial Communities

Students will learn

- 18-1: that species within communities coevolve.
- 18-2: how competition shapes and limits the organisms in an ecosystem.
- 18-3: how physical conditions determine where species live and about the major terrestrial and aquatic communities.

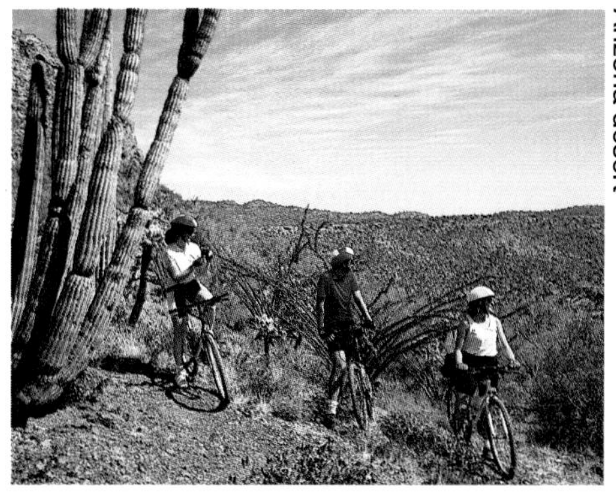
Arizona desert

Harsh and unforgiving, the *desert is home to plants and animals uniquely equipped to thrive in the face of environmental challenges. No other ecosystem displays a wider range of extreme conditions.*

Looking Ahead

18-1 How Organisms Interact in Communities
How do interactions among organisms affect ecosystems?

18-2 How Competition Shapes Communities
Winner take all? Not always.

18-3 Major Biological Communities
Which of these seven communities do you call home?

Labs

- **Data Labs**
  How predation affects a plant species (p. 383)
  Altering an organism's fundamental niche (p. 387)
- **Quick Lab**
  What factors influence the cooling of the Earth's surface? (p. 395)
- **Experimental Design**
  How Do Brine Shrimp Select a Habitat? (p. 402)

Features

- **Exploring Further**
  Estuaries (p. 397)

Study TIP

Ready?

Check your understanding of these topics to help you prepare for what's ahead.

Can you . . .

- **identify** the roles of photosynthesis and respiration in the flow of energy in the living world? (Section 5-1)
- **differentiate** between the terms *habitat, community,* and *ecosystem*? (Section 17-1)
- **recognize** the relationship between primary productivity and energy flow in ecosystems? (Section 17-2)

If not, *review the sections indicated.*

Unit 7—*Ecosystem Dynamics*
Topics 1–6

internetconnect

National Science Teachers Association *sci*LINKS Internet resources are located throughout this chapter.

Chapter 18

Reading Strategies

Active Reading

Technique: Brainstorming

Before students begin reading, work with the class to create a list of all the ways that two species in an ecosystem can interact. Then develop a list of all the different types of communities/ecosystems students can recall. Can any of the ecosystems be grouped into larger systems (biomes)? Have students read the chapter to see how scientists have defined interactions and biomes.

Directed Reading

Assign **Directed Reading Worksheet Chapter 18** to help students interact with the text as they read the chapter.

Interactive Reading

Assign ***Biology: Principles and Explorations* Audio CD Program Chapter 18** to help reluctant readers achieve greater success in reading the chapter.

Reading Skills

Assign Reading Organizers and Reading Strategies Worksheets from the **One-Stop Planner CD-ROM with Test Generator** to help students learn and practice effective reading skills.

internetconnect

Holt *Biology: Principles and Explorations*

On-line Resources: go.hrw.com

Visit the HRW Web site for a variety of resources related to this chapter. Just type in the keyword HX0 Chapter 18.

Holt *Biology: Principles and Explorations*

On-line Resources: www.scilinks.org

The following *sci*LINKS Internet resources can be found in the student text for this chapter:

Page 386
TOPIC: Symbiosis
KEYWORD: HX386

Page 389
TOPIC: Extinction
KEYWORD: HX389

Page 392
TOPIC: Biomes
KEYWORD: HX392

Page 397
TOPIC: Estuaries
KEYWORD: HX397

Page 403
TOPIC: Adaptation
KEYWORD: HX403

Chapter 18

1 Prepare

Scheduling

- **Block:** About 90 minutes will be needed to complete this section.
- **Traditional:** About two class periods will be needed to complete this section.
- **Planning:** See the Planning Guide on pp. 379A and 379B for lecture, classwork, and assignment options for Section 18-1. Lesson plans for Section 18-1 can be found in a lesson cycle format in ***Biology: Principles and Explorations*** **One-Stop Planner CD-ROM with Test Generator.**

Resource Link

- Assign **Active Reading Worksheet Section 18-1** to help students comprehend the key concepts and visuals in this lesson.

2 Teach

Lesson Warm-up

Show the class a photograph of a tapeworm. Point out the various adaptations that make it a successful parasite. *(Hooks on the head enable the tapeworm to attach to the intestinal wall of its host. The tapeworm's body wall is permeable, allowing easy absorption of nutrients from the host. Most of the body consists of segments specialized for reproduction.)* Tapeworms in humans may grow as long as 6 m (20 ft). Ask students why it is not in the tapeworm's best interest to kill its host. *(Killing its host would destroy its home and food supply.)*

Section 18-1 How Organisms Interact in Communities

Objectives

- **Describe** coevolution. (p. 382)
- **Predict** how coevolution can affect interactions between species. (pp. 382–384)
- **Identify** the distinguishing features of symbiotic relationships. (p. 384)

Key Terms

coevolution
predation
parasitism
secondary compound
symbiosis
mutualism
commensalism

Species Evolve in Response to One Another

What are the most important members of an ecosystem? When you try to answer this question, you soon realize that you cannot view an ecosystem's inhabitants as single organisms, but only as members of a web of interactions.

Interactions among species

Interactions among species are the result of a long evolutionary history in which many of the participants adjust to one another over time. Thus, flowering plants evolved that promoted efficient dispersal of their pollen by insects and other animals. In turn, traits developed in pollinators that enabled them to obtain food or other resources from the flowers they pollinate. Natural selection has often led to a close match between the characteristics of the flowers of a plant species and its pollinators, as you can see in **Figure 18-1.** Back-and-forth evolutionary adjustments between interacting members of an ecosystem are called **coevolution.**

Predators and prey coevolve

Predation is the act of one organism feeding on another. Familiar examples of predation include lions eating zebras and snakes eating mice. Less familiar, but no less important, examples occur among insects. Spiders are exclusively predators, as are centipedes.

A special case of predation is **parasitism,** in which one organism feeds on and usually lives on or in another, typically larger, organism. Parasites do not usually kill their prey (known as the "host") because they depend on it for food and a place to live. The host also serves to transmit the parasite's offspring to new hosts. Many parasites (such as lice) feed on the host's outside surface. Among the external parasites that may have fed on you at some time are ticks, mosquitoes, and fleas. More highly specialized parasites like hookworms live entirely within the body of their host.

Figure 18-1 Coevolution. With its long beak and tongue, the hummingbird is able to reach the nectar deep within this flower.

Integrating Different Learning Styles

Logical/Mathematical	**PE**	Data Lab, p. 383; Performance Zone item 3, p. 400
Visual/Spatial	**ATE**	Lesson Warm-up, p. 382; Closure, p. 384
Verbal/Linguistic	**PE**	Real Life, p. 383; Performance Zone items 12, 18, and 20, pp. 400–401

Plant defenses against herbivores

How would you expect predators and prey to coevolve? As you might expect, prey species have evolved ways to escape, avoid, or fight off predators. For example, the most obvious way that plants protect themselves from herbivores is with thorns, spines, and prickles. But the use of chemical compounds to discourage herbivores is even more common. Virtually all plants contain defensive chemicals called **secondary compounds.** For some plants, secondary compounds are the primary means of defense.

As a rule, each group of plants makes its own special kind of defensive chemical. For example, the mustard plant family produces a characteristic group of chemicals known as mustard oils. These oils give pungent aromas and tastes to such plants as mustard, cabbage, radish, and horseradish. The same tastes that we enjoy signal the presence of chemicals that are toxic to many groups of insects.

How herbivores overcome plant defenses

Surprisingly, certain herbivores are able to feed on plants that are protected by particular defensive chemicals. For example, the larvae of cabbage butterflies feed almost exclusively on plants of the mustard and caper families. Yet these plants produce mustard oils that are toxic to many groups of insects. How do the butterfly larvae manage to avoid the chemical defenses of the plants? Cabbage butterflies have evolved the ability to break down mustard oils and thus feed on mustards and capers without harm.

REAL LIFE

Leaflets three, let it be. Some of the best-known plants with defensive chemicals are members of the genus *Toxicodendron,* which includes poison ivy. These plants produce a gummy oil called urushiol *(OO roo shee awl),* which causes a severe, itchy rash in some people.

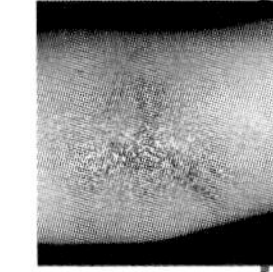

Finding Information
Do research to discover effective treatments for the rash caused by poison ivy.

How predation affects a plant species

Background

Grazing is the predation of plants by animals. Some plant species, such as *Gilia*, respond to grazing by growing new stems. Consider a field in which a large number of these plants are growing and being eaten by herbivores.

Analysis

1. Which of the plants is likely to produce more seeds? Why?
2. How would the production of more seeds provide an advantage to a plant?
3. How does grazing affect this plant species? Explain.
4. How might this plant species be affected if individual plants did not produce new stems in response to grazing?

did you know?

Coevolution takes place when species interact so closely that they exert a strong selective force on one another.

When the hummingbird inserts its beak into the deep, bell-shaped flower, it brushes against the female parts of the flower, leaving any pollen it already carries on their surface. It also contacts the male pollen-bearing structures, becoming dusted with pollen, which it will take to another flower.

REAL LIFE

Answer

Calamine lotion, Epsom salts, and bicarbonate of soda may all reduce the severity of a poison-ivy rash.

Cross-Disciplinary Connection

◆ History

When Europeans began claiming land in the New World in the late 1400s, they introduced infectious diseases. By some estimates, 80 percent of the indigenous people of the Western Hemisphere died from diseases such as smallpox, measles, and influenza brought from the Old World. Why were native peoples so susceptible? *(Native Americans had not been exposed to European microbes, so they had not developed natural resistance to the microbes' effects.)*

DATA LAB: How predation affects a plant species

Time 10–15 minutes

Process Skills Interpreting results, applying information, predicting

Teaching Strategies

Point out that a grazed *Gilia* plant has more stems than an ungrazed plant.

Analysis Answers

1. the grazed plant because it will have more stems and therefore more seeds
2. The plant would produce more offspring.
3. Grazing induces the plant to create more seeds and therefore more offspring.
4. The grazed plants would produce few, if any, seeds. Over the years, the plant might become rare or extinct in areas of heavy grazing.

Teaching TIP

Overcoming Defenses

Show the class a specimen or photograph of a monarch butterfly, which can feed on poisonous milkweed without being poisoned. (The caterpillars, rather than the adults, eat the plants.) The toxic plant compounds are incorporated into the butterfly, making it slightly toxic as well. The bitter, white sap of milkweed plants contains a compound that acts as a heart poison to vertebrates.

3 Assessment Options

Closure

Drawing a Food Web

Remind students that predation is only one of several types of interactions that bind natural communities and shape the evolution of their inhabitants. Challenge them to draw a simple food web for an ecosystem of their choice. Then have them use differently colored arrows to identify nonfeeding relationships. Have students code these arrows in a key below their diagram. PORTFOLIO

Review

Assign Review—Section 18-1.

Reteaching — BASIC

Have students make a chart to help them remember the different types of species interactions—predation, parasitism, mutualism, and commensalism. Tell them to use the four interactions as headings. Then have them place plus and minus signs in the row below to indicate whether each of the two organisms is helped or harmed by the interaction above. For example, mutualism would have a + / + because both organisms benefit.

Figure 18-2 Mutualism. The small green insects on this plant stem are aphids. They are protected by their ant guards, which feed on the sugary fluid the aphids secrete.

Symbiotic Species Are Shaped by Long-Term Relationships

In **symbiosis** *(SIM beye OH sis)*, two or more species live together in a close, long-term association. Symbiotic relationships can be beneficial to both organisms or benefit one organism and leave the other harmed or unaffected. Parasitism, mentioned earlier, is one type of symbiotic relationship that is detrimental to the host organism. While it is relatively easy to determine that an organism in a symbiotic relationship is being helped, it can be difficult to determine that an organism is neither harmed nor helped.

Mutualism

Mutualism is a symbiotic relationship in which both participating species benefit. A well-known instance of mutualism involves ants and aphids, as shown in **Figure 18-2.** Aphids are small insects that use their piercing mouthparts to suck fluids from the sugar-conducting vessels of plants. They extract a certain amount of the sucrose and other nutrients from this fluid. However, much of the fluid—so-called honeydew—runs out in an altered form through their anus. Certain ants have taken advantage of this fact and "milk" the aphids for the honeydew, which they use as food. The ants, in turn, protect the aphids against insect predators. Thus, both species benefit from the relationship.

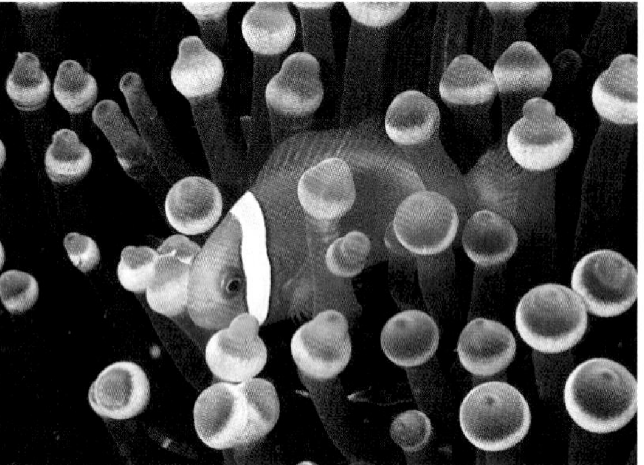

Figure 18-3 Commensalism. The clown fish can survive the stings of the sea anemone, which protects it from predators. The anemone is apparently unaffected by the fish.

Commensalism

A third form of symbiosis is **commensalism,** a symbiotic relationship in which one species benefits and the other is neither harmed nor helped. Among the best-known examples of commensalism are the relationships between certain small tropical fishes and sea anemones, marine animals that have stinging tentacles. These fishes, such as the clown fish shown in **Figure 18-3,** have evolved the ability to live among and be protected by the tentacles of the sea anemones, even though these tentacles would quickly paralyze other fishes.

Review SECTION 18-1

1. **Describe** why predator-prey coevolution can be described as an "arms race."
2. **SKILL Predicting Outcomes** In a relationship that is an example of commensalism, would the species that is neither helped nor harmed evolve in response to the other species? Defend your answer.

Critical Thinking

3. **SKILL Evaluating Results** In Japan, native honeybees have evolved an effective defense strategy against giant Japanese hornets. Imported European honeybees, however, are unable to defend themselves. Explain this in terms of coevolution.

Review SECTION 18-1 ANSWERS

1. Natural selection favors both predators and prey that have features that will give them an advantage over their "opponent."
2. No, the unaffected species neither gains nor suffers from the relationship, so there is no selective pressure for coevolution.
3. The native honeybees have been exposed to the Japanese hornets for thousands of years, enough time for a defense to have evolved. The European honeybees have only recently been exposed, so there has not been enough time for significant coevolution.

How Competition Shapes Communities

Section 18-2

Section 18-2

1 Prepare

Scheduling

- **Block:** About 90 minutes will be needed to complete this section.
- **Traditional:** About two class periods will be needed to complete this section.
- **Planning:** See the Planning Guide on pp. 379A and 379B for lecture, classwork, and assignment options for Section 18-2. Lesson plans for Section 18-2 can be found in a lesson cycle format in ***Biology: Principles and Explorations*** **One-Stop Planner CD-ROM with Test Generator.**

Resource Link

Assign **Active Reading Worksheet Section 18-2** to help students comprehend the key concepts and visuals in this lesson.

Objectives

- **Describe** the role of competition in shaping the nature of communities. (p. 385)
- **Distinguish** between fundamental and realized niches. (pp. 386–387)
- **Describe** how competition affects an ecosystem. (pp. 388–390)
- **Summarize** the importance of biodiversity. (p. 390)

Key Terms

competition
niche
fundamental niche
realized niche
competitive exclusion
biodiversity

Common Use of Scarce Resources Leads to Competition

When two species use the same resource, they participate in a biological interaction called **competition.** Resources for which species compete include food, nesting sites, living space, light, mineral nutrients, and water. For competition to occur, however, the resource must be in short supply. In Africa, for example, lions and hyenas compete for prey. Fierce rivalry between these species can lead to battles that cause injuries to both sides. But most competitive interactions do not involve fighting. In fact, some competing species never encounter one another. They interact only by means of their effects on the abundance of resources.

To understand how competition influences the makeup of communities, you must focus on the day-to-day events within the community. What do organisms eat? Where do they live? The functional role of a particular species in an ecosystem is called its **niche** *(NICH)*. A niche is how an organism lives, the "job" it performs within the ecosystem.

A niche may be described in terms of space utilization, food consumption, temperature range, requirements for moisture or mating, and other factors. A niche is not to be confused with a habitat, the place where an organism lives. A habitat is a location; a niche is a pattern of living. **Figure 18-4** summarizes some aspects of the jaguar's niche in the Central American rain forest.

A niche is often described in terms of how the organism affects energy flow within the ecosystem in which it lives. For example, the niche of a deer that eats a shrub is that of a herbivore. The niches of some organisms overlap. If the resources that these organisms share are in short supply, it is likely that there will be competition between the organisms.

A Jaguar's Niche

- **Diet** Jaguars feed on mammals, fish, and turtles.
- **Reproduction** Jaguars give birth from June to August, during the rainy season.
- **Time of activity** Jaguars hunt by day and by night.

Figure 18-4 Each organism has its own niche. All of the ways that this jaguar interacts with its environment make up its niche.

2 Teach

Lesson Warm-up

Ask students why competition is usually most intense between closely related organisms. *(Closely related organisms are likely to be very similar and therefore are likely to use resources in similar ways.)*

Visual Strategy

Figure 18-4

Point out that a niche is very complex and includes all the ways an organism affects and is affected by its environment. This figure shows only four aspects of the jaguar's niche. Have students make a list of other aspects of the jaguar's niche.

Integrating Different Learning Styles

Logical/Mathematical	**PE**	Data Lab, p. 387; Performance Zone item 17, p. 401
	ATE	Lesson Warm-up, p. 385; Teaching Tip, p. 387; Visual Strategy, p. 389
Visual/Spatial	**PE**	Performance Zone item 16, p. 401
	ATE	Demonstration, p. 388; Reteaching, p. 390
Verbal/Linguistic	**PE**	Performance Zone items 6 and 9, p. 400
	ATE	Visual Strategy, p. 386; Active Reading, p. 389

internet**connect**

SCiLINKS. NSTA Have students research symbiosis using the Web site in the Internet Connect box on page 386. Assign students to write a report summarizing their findings. PORTFOLIO

Visual Strategy

Figure 18-5 ADVANCED

Ecologists use the phrase "resource partitioning" to describe the patterns of resource use in a community. Ask students why resource partitioning is an appropriate description of the feeding behavior of the five warbler species. *(The warblers partition, or divide up, the insect populations on which they feed by foraging in different parts of the tree.)*

Teaching TIP

Warblers

Tell students that all five warbler species that MacArthur studied belong to the same genus *(Dendroica)*. Because closely related species are often competitors, MacArthur was interested in how these very similar species could coexist.

internet**connect**

TOPIC: Symbiosis
GO TO: www.scilinks.org
KEYWORD: HX386

A species' niche can vary in size

To gain a better understanding of what a niche is, you must look more closely at a particular species. Imagine a Cape May warbler (a small, insect-eating songbird) flying in a forest and landing to search for dinner in a spruce tree. The niche of this bird includes several variables. These variables include the temperature it prefers, the time of year it nests, what it likes to eat, and where on the tree it finds its food. (The Cape May warbler spends its summers almost exclusively in the northeastern United States and Canada. It nests in midsummer, eats small insects, and searches for food high on spruce trees at the tips of the branches.) The entire range of conditions an organism is potentially able to occupy within an ecosystem is its **fundamental niche.**

Dividing resources among species

Now reconsider what the Cape May warbler is doing. It feeds mainly at the very top of the spruce tree even though insects that the warbler could eat are located all over the tree. **Figure 18-5** shows the part of the spruce tree where Cape May warblers feed. In other words, Cape May warblers occupy only a portion of their fundamental niche. Why?

Closer study reveals that this surprising behavior is part of a larger pattern of niche restriction. In the late 1950s, Robert MacArthur, an ecologist from Princeton University, carried out a classic investigation of niche usage, summarized in Figure 18-5. He studied the feeding habits of five warbler species—the Cape May warbler and four of its potential competitors. MacArthur found that all five species fed on insects in the same spruce trees at the same time. As Figure 18-5 shows, each species concentrated on a different

Figure 18-5 **Niche restriction**

Each of these five warbler species feeds on insects in a unique portion of the same tree, as indicated by the five colors shown below.

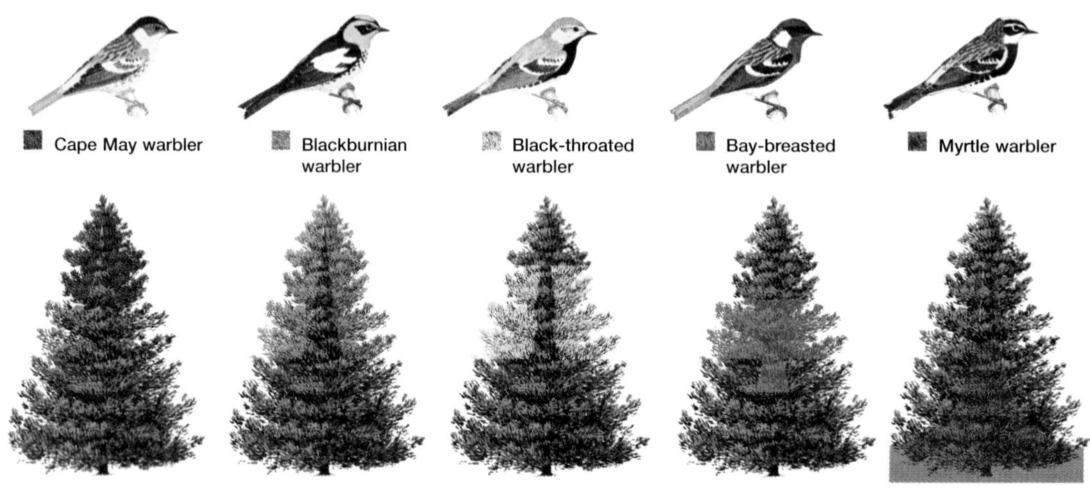

Biology in Your Community

Habitat Versus Niche

An analogy may help students understand the difference between habitat and niche. Point out that their house or apartment and the places they frequent make up their habitat. What they do in their habitat is their niche.

Ask students to write a short description of their habitat and their niche. What would be different if they lived in another habitat (such as a very rural or very urban habitat)? PORTFOLIO

part of the tree. Although all five species of warbler had very similar fundamental niches, they did not use the same resources. In effect, they divided the range of resources among them, each taking a different portion. A different color is used to represent the feeding areas of each of the five warbler species shown in Figure 18-5.

The part of its fundamental niche that a species occupies is called its **realized niche.** Stated in these terms, the realized niche of the Cape May warbler is only a small portion of its fundamental niche.

How does this species of warbler benefit from hunting for food in only a portion of the tree? MacArthur suggested that this feeding pattern reduces competition among the five species of warblers. Because each of the five warbler species uses a different set of resources by occupying a different realized niche, the species are not in competition with one another. MacArthur concluded that natural selection has favored a range of preferences and behaviors among the five species that "carve up" the available resources. Most ecologists agree with this conclusion.

Study TIP

Reading Effectively

To better understand the relationship between fundamental and realized niches for MacArthur's warblers, draw two circles, one within the other. Label the larger circle "Fundamental niche, entire tree." Label the smaller circle "Realized niche of Cape May warbler, upper branches."

DATA LAB

Altering an organism's fundamental niche

Background

Two features of a niche that can be readily measured are the prey size of a species and the location where the species feeds. The darkest shade in the center of the graph below indicates the prey size and feeding location most frequently selected by one bird species (called Species A).

Analysis

1. **State** the preferred length of Species A's prey.
2. **Identify** the maximum height at which Species A will feed.
3. **SKILL Analyzing Information** Species B is introduced into Species A's feeding range. Species B has exactly the same feeding preferences, but it hunts at a slightly different time of day. How might this affect Species A?
4. **SKILL Interpreting Graphics** Species C is now introduced into Species A's feeding range. Species C feeds at the same time of day as Species A, but it prefers prey that are between 10 and 13 mm long. How might this affect Species A?
5. **SKILL Predicting Outcomes** How would the introduction of a species with exactly the same feeding habits as Species A affect the graph?
6. **SKILL Interpreting Graphics** What does the lightest shade at the edge of the contour lines represent?

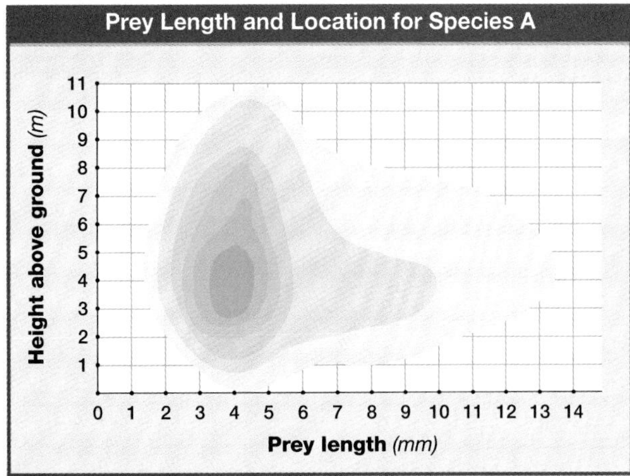

Teaching TIP

Realized Niches

Point out that an organism's realized niche is a subset of its fundamental niche. Thus, the realized niche can be smaller than (or the same size as) the fundamental niche, but never larger.

Altering an organism's fundamental niche

Time 10–15 minutes

Process Skills Interpreting results, applying information, analyzing information, predicting outcomes

Teaching Strategies

Explain to students that this type of graph is a convenient way to represent three variables: height above the ground, prey length, and relative number of predators preferring the combination of the two.

Analysis Answers

1. 4 mm
2. 11 m
3. It might reduce the prey available to Species A.
4. Species C will reduce Species A's realized niche by competing with Species A for large prey.
5. Accept any well-reasoned answer. Sample answer: It would not affect the graph. One of the two species would be forced to extinction in this area.
6. The lightest shade represents the combination of feeding height and prey length least favored but still exploited by Species A.

did you know?

The coyote's broad fundamental niche has allowed it to thrive in areas of human encroachment.

For example, coyotes can survive on a variety of animal and plant foods, including most of the food humans throw away. The coyote's brazenness is a behavioral adaptation and therefore part of its niche as well. Its boldness pays off by allowing it access to areas that competitors, such as bobcats, usually avoid.

Demonstration

Winner Takes Your Home

Competition between two species can result in one being eliminated from the community. Show the class photographs of a starling and a bluebird. Starlings were first introduced into Central Park in New York City in 1890. Today starlings are found throughout the continental United States, and in many areas they have bested bluebirds for nesting sites, causing a drastic decline in bluebirds. Make bluebird house plans available to students and select an area to post them. Emphasize the importance of monitoring the houses to keep out non-native competitors like starlings.

Career

Naturalist

Parks (local, state, and national), as well as school districts and private nature centers, employ naturalists. Passionate about the workings of living communities, naturalists educate youths and adults about natural history, ecology, and environmental science in some of the most beautiful settings in the world. Strong communications skills and knowledge of ecology and natural history are required.

Study TIP

Interpreting Graphics

In Figure 18-6, arrows of two different colors represent two species of barnacles. The first arrow for each species shows the water depth at which that species can survive. The second arrow for each species shows the water depth where each species actually lives.

Competition Can Limit How Species Use Resources

A very clear case of competition was shown by experiments carried out in the early 1960s by Joseph Connell of the University of California. Connell worked with two species of barnacles that grow on the same rocks along the coast of Scotland. Barnacles are marine animals that are related to crabs, lobsters, and shrimp. Young barnacles attach themselves to rocks and remain attached there for the rest of their lives. As you can see in **Figure 18-6,** one species, *Chthamalus stellatus,* lives in shallow water, where it is often exposed to air by receding tides. A second species, *Semibalanus balanoides,* lives lower down on the rocks, where it is rarely exposed to the atmosphere.

When Connell removed *Semibalanus* from the deeper zone, *Chthamalus* was easily able to occupy the vacant surfaces. This indicates that it was not intolerance of the deeper environment that prevented *Chthamalus* from becoming established there. *Chthamalus*'s fundamental niche clearly includes the deeper zone. However, when *Semibalanus* was reintroduced, it could always outcompete *Chthamalus* by crowding it off the rocks. In contrast, *Semibalanus* could not survive when placed in the shallow-water habitats where *Chthamalus* normally occurs. *Semibalanus* apparently lacks the adaptations that permit *Chthamalus* to survive long periods of exposure to air. Connell's experiments show that *Chthamalus* occupies only a small portion of its fundamental niche. The rest is unavailable because of competition with *Semibalanus.* As MacArthur suggested, competition can limit how species use resources.

Figure 18-6 Effects of competition on two species of barnacles

The realized niche of *Chthamalus* is smaller than its fundamental niche because of competition from the faster-growing *Semibalanus*.

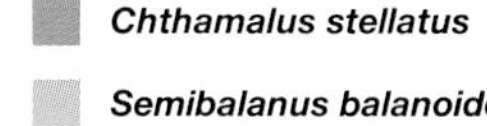

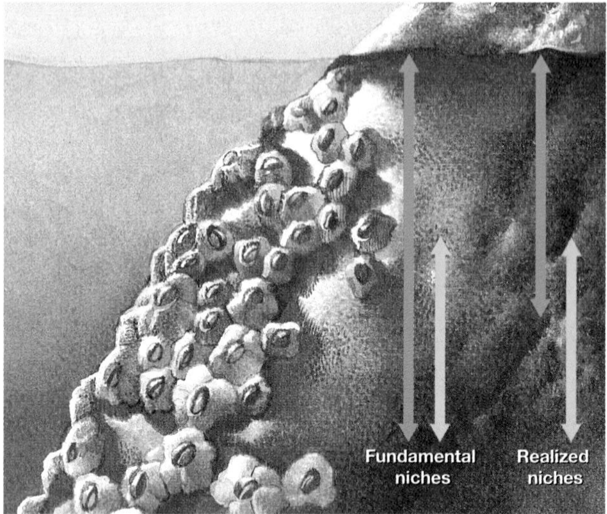

Biology in Your Community

Guarding the Cat Food

Anyone who has lived with an outdoor cat knows that other animals, such as raccoons and opossums, will eat a cat's food if given the chance. These wild animals are competing with domestic cats for resources. When we move a cat's food inside or shoo away unwanted diners, we are artificially manipulating the cat's realized niche.

Competition Without Division of Resources Leads to Extinction

In nature, shortage is the rule, and species that use the same resources are almost sure to compete with each other. Darwin noted that competition should be most acute between very similar kinds of organisms because they tend to use the same resources in the same way. Can we assume, then, that when very similar species compete, one species will always become extinct locally? In a series of carefully controlled laboratory experiments performed in the 1930s, the Russian biologist G. F. Gause looked into this question.

In his experiments, Gause grew two species of *Paramecium* in the same culture tubes, where they had to compete for the same food (bacteria). Invariably, the smaller of the two species, which was more resistant to bacterial waste products, drove the larger one to extinction, as shown in the first graph in **Figure 18-7.** Gause realized that if two species are competing, the species that uses the resource more efficiently will eventually eliminate the other. This elimination of a competing species is referred to as **competitive exclusion.**

internetconnect
SCILINKS. NSTA
TOPIC: Extinction
GO TO: www.scilinks.org
KEYWORD: HX389

When can competitors coexist?

Is competitive exclusion the inevitable outcome of competition for limited resources, as Gause suggests? No. When it is possible for two species to avoid competing, they may coexist.

In a revealing experiment, Gause challenged *Paramecium caudatum*—the defeated species in his earlier experiments—with a third species, *P. bursaria.* These two species were also expected to compete for the limited bacterial food supply. Gause thought one species would win out, as had happened in his previous experiments.

But that's not what happened. As shown in the second graph in Figure 18-7, both species survived in the culture tubes. Like MacArthur's warblers, the two species of *Paramecium* divided the food resources. How did they do it? In the upper part of the culture tubes, where oxygen concentration and bacterial density were high, *P. caudatum* was dominant. It was better able to feed on bacteria than was *P. bursaria.* But in the lower part of the tubes, the lower oxygen concentration favored the growth of a different potential food—yeast. *Paramecium bursaria* was better able to eat the yeast, so it used this resource more efficiently. The fundamental niche of each species was the whole culture tube, but the realized niche of each species was only a portion of the tube. Because the niches of the two species did not overlap too much, both species were able to survive.

Figure 18-7 Gause's experiments. Gause's studies revealed that the outcome of competition depends on the degree of similarity between the fundamental niches of the competing species.

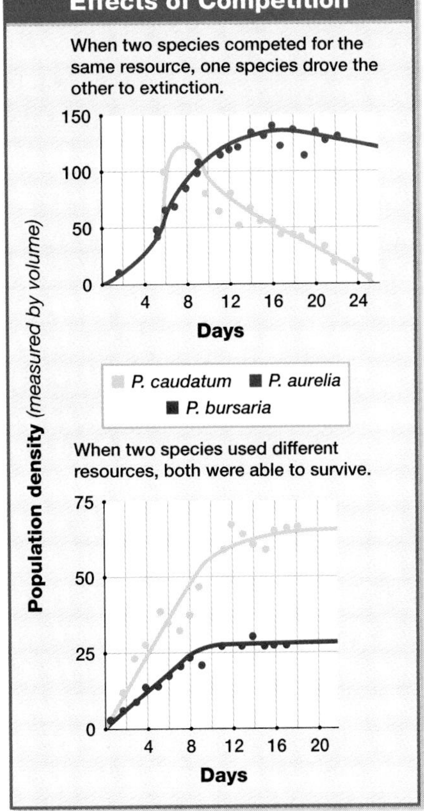

Teaching TIP

Competitive Exclusion

Inform students that the elimination of one competing species may be the result of direct aggressive interactions between the two species (interference competition) or the result of the one species using more of a given resource (exploitative competition).

Active Reading

Technique: Paired Summarizing

Pair students with a partner. Have each student read silently about the competition experiments of Connell, Gause, and Paine on pp. 388–390. Then have one student summarize aloud what has been read without referring to the text. The partner should listen without interrupting and be prepared to point out any inaccuracies or omissions in the summary. At this point students can refer to the text. Have students switch roles for each of the three experiments.

Visual Strategy

Figure 18-7

Be sure students do not develop the misconception that the same species of a competing pair will always eliminate the other species. Point out that it is possible to alter the environment so that the outcome is reversed. For example, Thomas Park and his colleagues at the University of Chicago conducted competition experiments on two species of flour beetles *(Tribolium).* The temperature and humidity at which the beetles were raised determined which species was the superior competitor.

Overcoming Misconceptions

Predation Can Increase Diversity

The notion that the removal of a predator can reduce the diversity of an ecosystem may seem counterintuitive to students. An example occurred in England when a viral epidemic wiped out the rabbit population. The grasses, once controlled by the rabbits, grew out of control. At the same time, the many species of wildflowers that once thrived in the chalky soils disappeared.

Chapter 18

Teaching TIP

Ecosystem Complexity and Fragility

Paine's experiments show that even a low-diversity ecosystem (only 15 species) is complex and highly interconnected and that disturbing such a system can have unexpected and harmful effects. To predict that the removal of the sea star would result in the disappearance of seven other species, Paine would have needed to understand the relationships between Pilaster and its prey, as well as the relationships among the prey species. Humans regularly make far greater changes to ecosystems that are less well understood. Students will see some of the consequences of these changes in the next chapter.

3 Assessment Options

Closure — ADVANCED

Ecological Versus Economic Competition

Have students write a short report contrasting competition between organisms in a natural community with competition between businesses in a human community.

PORTFOLIO

Review

Assign Review—Section 18-2.

Reteaching — BASIC

Show students pictures of the Galapagos Island finches and explain how Darwin postulated that competition led to the evolution of their realized niches.

Predation can lessen competition

Many studies of natural ecosystems have shown that predation lessens the effects of competition. A very clear example is provided by the studies of Robert Paine of the University of Washington. Paine examined how sea stars affect the numbers and types of species within marine intertidal communities. Sea stars are fierce predators of marine animals such as clams and mussels. When sea stars were kept out of experimental plots, the number of their prey species fell from 15 to 8. The 7 eliminated species were crowded out by the sea stars' chief prey, mussels, shown in **Figure 18-8.** Mussels can outcompete other species for space on the rocks. By preying on mussels, sea stars keep the mussel populations too low to drive out other species.

Because predation can reduce competition, it can also promote **biodiversity,** the variety of living organisms present in a community. Biodiversity is a measure of both the number of different species in a community (species richness) and the relative numbers of each of the species (species diversity).

Figure 18-8 Effect of removing sea stars. When the sea star *Pisaster* was removed from an ecosystem, the diversity of its prey species decreased. Mussels, the superior competitor, crowded seven other prey species out of the ecosystem.

Biodiversity and productivity

A key investigation carried out in the early 1990s by David Tilman of the University of Minnesota illustrates the relationship between biodiversity and productivity. Tilman and some co-workers and students tended 147 experimental plots in a Minnesota prairie. Each plot contained a mix of up to 24 native prairie plant species. The biologists monitored the plots, measuring how much growth was occurring. Tilman found that the more species a plot had, the greater the amount of plant material produced in that plot. Tilman's experiments clearly demonstrated that increased biodiversity leads to greater productivity.

In addition to increased productivity, Tilman also found that the plots with greater numbers of species recovered more fully from a major drought. Thus, the biologically diverse plots were also more stable than the plots with fewer species.

Review SECTION 18-2

1. **Distinguish** between niche and habitat.
2. **SKILL Applying Information** Can an organism's realized niche be larger than its fundamental niche? Justify your answer.
3. **Summarize** how competition affected the ecosystems studied by Connell and Paine.
4. **Describe** how Tilman's experiments demonstrate the effects of biodiversity on productivity and stability.

Critical Thinking

5. **SKILL Evaluating Conclusions** A scientist finds no evidence that species in a community are competing and concludes that competition never played a role in the development of this community. Is this conclusion valid? Justify your answer.

Review SECTION 18-2 ANSWERS

1. A habitat is where an organism lives. Its niche is the role it plays, or its way of life, within that habitat.
2. No, the realized niche can be smaller than, or the same size as, the fundamental niche, but never larger.
3. Connell showed that the barnacle *Chthamalus* was able to live within a larger portion of its habitat, but could not survive in the same zone as the barnacle *Semibalanus*, the superior competitor. In Paine's study, sea stars were removed from experimental plots. The sea stars' primary prey, the mussel, grew unchecked and soon outcompeted the other marine animals for space. Diversity dropped from 15 species to eight.
4. Tilman showed that the plots with the greatest species diversity displayed greater productivity and were more resilient to drought conditions than plots with lower diversity.
5. The scientist may not be aware of competition taking place. Also, competition that is no longer apparent may have played a key role in the past development of the ecosystem, even driving one or more species out of the area.

Major Biological Communities

Section 18-3

Climate Largely Determines Where Species Live

If you traveled across the country by car you would notice dramatic changes in the plants and animals outside your window. For example, the drought-tolerant cactuses in the deserts of Arizona do not live in the wetlands of Florida. Why is this? The climate of any physical environment determines what organisms live there. **Climate** refers to the prevailing weather conditions in any given area.

Temperature and moisture

The two most important elements of climate are temperature and moisture. **Figure 18-9** illustrates the different types of ecosystems that will occur under particular temperature and moisture conditions.

Temperature Most organisms are adapted to live within a particular range of temperatures and will not thrive if temperatures are colder or warmer. The growing season of plants, for example, is primarily influenced by temperature.

Moisture All organisms require water. On land, water is sometimes scarce, so patterns of rainfall often determine an area's life-forms. The moisture-holding ability of air increases when it is warmed and decreases when it is cooled.

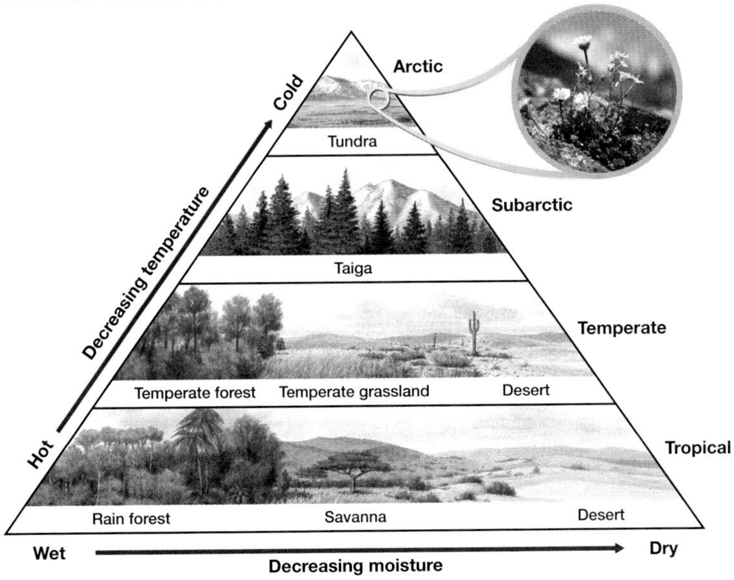

Figure 18-9 Elements of climate. Temperature and moisture help determine ecosystem distribution. For example, the asters and the saxifrage are able to produce flowers and seeds in the cold temperatures of the tundra.

Objectives

- **Recognize** the role of climate in determining the nature of a biological community. (p. 391)
- **Describe** how elevation and latitude affect the distribution of biomes. (p. 392)
- **Summarize** the key features of the Earth's major biomes. (pp. 393–395)
- **Compare and contrast** the major freshwater and marine habitats. (pp. 396–398)

Key Terms

climate
biome
littoral zone
limnetic zone
profundal zone
plankton

1 Prepare

Scheduling

- **Block:** About 90 minutes will be needed to complete this section.
- **Traditional:** About two class periods will be needed to complete this section.
- **Planning:** See the Planning Guide on pp. 379A and 379B for lecture, classwork, and assignment options for Section 18-3. Lesson plans for Section 18-3 can be found in a lesson cycle format in ***Biology: Principles and Explorations*** **One-Stop Planner CD-ROM with Test Generator.**

Resource Link

- Assign **Active Reading Worksheet Section 18-3** to help students comprehend the key concepts and visuals in this lesson.

2 Teach

Lesson Warm-up

The kangaroo rat can survive the hot, dry climate of the deserts of Mexico and the southwestern United States without drinking water. Ask students to hypothesize how the kangaroo rat obtains the water it needs to survive. *(The rat survives on the water in its food, supplemented with water it produces through cellular respiration. It also conserves water by spending the day in a cool, humid burrow and excreting very concentrated urine.)*

Integrating Different Learning Styles

Logical/Mathematical	**PE**	Quick Lab, p. 395; Performance Zone items 11 and 14, pp. 400–401
Visual/Spatial	**PE**	Study Tip, p. 396
	ATE	Teaching Tip, p. 392; Teaching Tip, p. 393
Verbal/Linguistic	**PE**	Performance Zone items 15 and 22, p. 401
	ATE	Teaching Tips, pp. 392–393, Internet Connect, p. 397
Interpersonal	**PE**	Performance Zone item 19, p. 401

Chapter 18

Demonstration

Biological Communities

Show the class photographs of the five major biomes found in the lower 48 states and Alaska—desert, temperate grassland, temperate deciduous forest, tundra, and taiga. Ask students what they can deduce about the physical and biological characteristics of these biomes from examining the photos.

Teaching TIP

Rainfall

Have students check library or Internet resources to identify areas in the world that receive the highest amounts of rainfall. Have them write a report that explains the physical factors responsible for the heavy rainfalls in these areas. PORTFOLIO

Teaching TIP

Latitude and Longitude

Using a globe, remind students that longitude indicates east-west position and that latitude indicates north-south position. Draw on the board a Graphic Organizer similar to the one on page 392. Lead students to conclude that latitude profoundly affects climate, but longitude is essentially irrelevant to climate.

Visual Strategy

Figure 18-10

Emphasize that a biome is a category and not a place. A tropical rain forest does not refer to any specific geographical location on Earth, but rather to all regions of the planet where such a biome can be found. Point out that biomes are characterized by their dominant vegetation (drought-tolerant plants in deserts, coniferous trees in the taiga, and so on). Also emphasize that the boundaries of biomes are not as well defined as they are shown in this figure.

Temperature and moisture help determine biological communities

If you were to tour the world and look at biological communities on land and in the oceans, you would soon learn a general rule of ecology: very similar communities occur in many different places that have similar climates and geographies. A major biological community that occurs over a large area of land is called a **biome.**

A biome's structure and appearance are similar throughout its geographic distribution. While there are different ways of classifying biomes, the classification system used here recognizes seven widely occurring biomes: (1) tropical rain forest, (2) desert, (3) savanna, (4) temperate deciduous forest, (5) temperate grassland, (6) taiga, and (7) tundra. These biomes differ greatly from one another because they have developed in regions with very different climates. The global distribution of these biomes is shown in **Figure 18-10.**

Many factors such as soil type and wind play important roles in determining where biomes occur. Two key factors are particularly important: temperature and precipitation. In general, temperature and available moisture decrease as latitude (distance from the equator) increases. They also decrease as elevation (height above sea level) increases. As a result, mountains often show the same sequence of change in ecosystems that is found as one goes north or south from the equator.

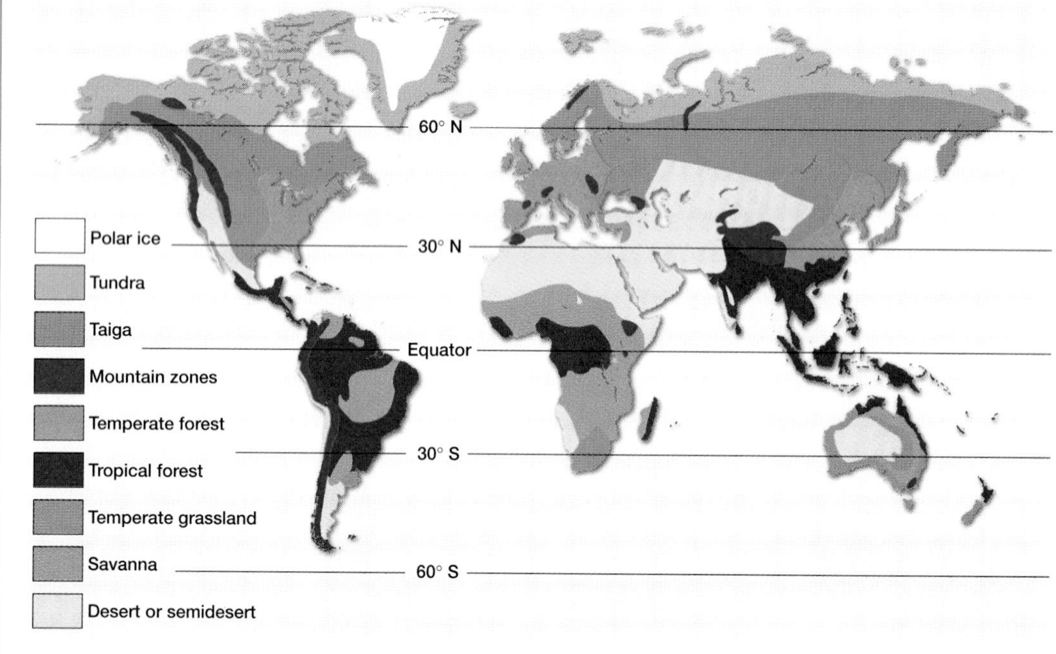

Figure 18-10 Earth's biomes. Seven major biomes cover most of the Earth's land surface. Because mountainous areas do not belong to any one biome, they are given their own designation.

Graphic Organizer

*Use this graphic organizer with **Teaching Tip: Latitude and Longitude** on page 392.*

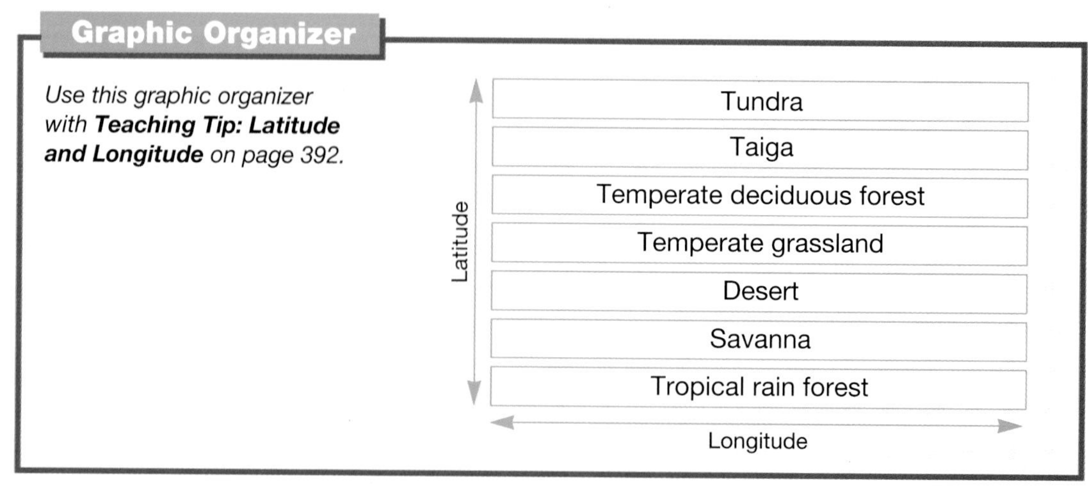

There Are Seven Major Biomes

If you were to tour the world and visit each of the seven biomes described below, you would quickly discover that each biome has unique characteristics reflecting its climate and geography. The characteristics of each biome are summarized below.

Biome 1 Tropical rain forests

The rainfall in tropical rain forests is generally 200–450 cm (80–180 in.) per year, with little difference in distribution from season to season. The tropical rain forest, shown in **Figure 18-11,** is the richest biome in terms of number of species, probably containing at least half of the Earth's species of terrestrial organisms—more than 2 million species. In a single square mile (just over 2.5 sq km) of tropical forest in Peru or Brazil, there may be 1,200 or more species of butterflies, twice the total number of species found in the United States and Canada combined.

Figure 18-11 Tropical rain forests in Puerto Rico

Tropical rain forests have a high primary productivity even though they exist mainly on quite infertile soils. Most of the nutrients are held within the plants themselves. Roots of the tall trees spread through the top 1 or 2 cm (less than 1 in.) of the soil and extract the nutrients from decomposing leaves and other plant parts. Because decomposing matter recycles so quickly into living plants, the soil itself contains few nutrients.

Biome 2 Deserts

Typically, fewer than 25 cm (10 in.) of precipitation falls annually in the world's desert areas. The scarcity of water is the overriding factor influencing most biological processes in the desert. In desert regions, the vegetation is characteristically sparse. Deserts are most extensive in the interiors of continents, especially in Africa (the Sahara), Asia, and Australia. Less than 5 percent of North America is desert. In deserts, shown in **Figure 18-12,** the amount of water that actually falls at a particular place can vary greatly, both during a given year and between years.

Figure 18-12 Desert in California

Biome 3 Savannas

Dry climates often favor the development of grassland. The world's great dry grasslands, called savannas, are found in tropical areas that have relatively low annual precipitation or prolonged annual dry seasons. On a global scale, the savanna biome is found in a transitional zone located between tropical rain forest and desert. Annual rainfall is generally 90–150 cm (35–60 in.)

Teaching TIP

Pattern of Biomes

Have students make a simplified map that illustrates the relative zones for the seven biomes of the Northern Hemisphere. Starting with a line to represent the equator, students should draw horizontal bands to represent the boundaries of the biomes and label them appropriately. Point out that this is an oversimplification of the real world biomes.

Visual Strategy

Figure 18-11

Point out that the strongest competition among trees in a rain forest is for light. In the rain forest, about 70 percent of all plant species are trees. Their trunks are usually slender and tall. Their leaves are usually large and dark green. All these are adaptations to obtain and utilize the most sunlight for photosynthesis.

Teaching TIP

Deserts in the United States

Desert covers less than 5 percent of North America. The North American deserts are the Mojave, Sonoran, Great Basin, and Chihuahuan. Have students write a report that contrasts the animals, plants, geography, and climate of these four deserts.

PORTFOLIO

MULTICULTURAL PERSPECTIVE

The Joshua Tree

Show the class a photograph of a Joshua tree, a North American desert plant. This tree was named by early Mormon colonists who thought it resembled the biblical prophet Joshua waving them toward the promised land.

Teaching TIP BASIC

Desert Temperatures
Temperatures in the desert undergo drastic changes over a 24-hour period. Daytime temperatures can be 30°C (54°F) higher than nighttime readings. These fluctuations occur due to the lack of heat-retaining moisture in the desert air.

Making Biology Relevant

Food
Many of today's domesticated plants and animals, including cattle, sheep, horses, wheat, barley, and oats, are descendants of the wild grasslands community of Eurasia.

Teaching TIP

Numbers Versus Diversity
A deciduous forest usually has only a few species of trees, but there are numerous individuals of each species. In contrast, a tropical rain forest usually has many species of trees, but each species may be represented by only one individual.

Figure 18-13 Savanna in East Africa

in savannas. There is a wider fluctuation in temperature during the year than in the tropical rain forests, and there is seasonal drought. These factors have led to an open landscape with widely spaced trees. Many of the animals are active only during the rainy season. **Figure 18-13** shows the huge herds of grazing mammals that are found on the savannas of East Africa.

Biome 4 Temperate deciduous forests

Relatively mild climates and plentiful rain promote the growth of forests. Temperate deciduous forests (deciduous trees shed their leaves all at once in the fall) grow in areas with relatively warm summers, cold winters, and sufficient precipitation. The annual precipitation generally ranges from 75 to 250 cm (30 to 100 in.) and is well distributed throughout the year. Moisture is generally unavailable to animals and plants in the winter because it is usually frozen.

Temperate deciduous forests cover very large areas, including much of the eastern United States, southeastern Canada, and extensive areas of Europe and Asia. In North America, deciduous forests are home to deer, bears, beavers, raccoons, and the other familiar animals of the temperate regions. The trees are hardwoods (oak, hickory, and beech). Shrubs and herbs grow on the forest floor.

In other temperate areas, drier weather and different soil conditions favor the growth of evergreens. For example, large portions of the southeastern and western United States have extensive areas where pine forests predominate over deciduous forests. Where conditions are even drier, temperate forests give way to areas of dry shrubs, such as in the chaparral areas of coastal California and in the Mediterranean.

Figure 18-14 Temperate deciduous forest in Pennsylvania

Figure 18-15 A prairie dog, a grassland resident

Biome 5 Temperate grasslands

Moderate climates halfway between the equator and the poles promote the growth of rich temperate grasslands. Temperate grasslands once covered much of the interior of North America and were widespread in Eurasia and South America. Such grasslands are often highly productive when converted to agriculture. Much of the rich agricultural land in the United States and southern Canada was originally occupied by prairie, another name for temperate grassland. The roots of grasses characteristically penetrate far into the soil, which tends to be deep and fertile. Temperate grasslands are often populated by herds of grazing animals. In North America, huge herds of bison once inhabited the prairies.

Overcoming Misconceptions

Rain Forests
Rain forests are not limited to regions near the equator. In fact, the Olympic National Park in western Washington is a temperate rain forest that receives about 350 cm (about 140 in.) of rain per year. Temperate rain forests are also found in Australia and New Zealand. However, the largest rain-forest areas lie in the tropics.

Biome 6 Taiga

Cold, wet climates promote the growth of coniferous forests. A great ring of northern forests of coniferous trees, primarily spruce and fir, extends across vast areas of Eurasia and North America. This biome, one of the largest on Earth, is called by its Russian name, taiga *(TEYE guh)*. Winters in the taiga are long and cold, and most of the precipitation falls in the summer. Many large mammals live in the taiga, including herbivores such as elk, moose, and deer and carnivores such as wolves, bears, lynxes, and wolverines.

Figure 18-16 Moose

QUICK LAB

What factors influence the cooling of the Earth's surface?

Time 20 minutes

Process Skills Observing, interpreting results, applying information, analyzing results, predicting outcomes

Teaching Strategies

- Preparation tip: Be sure that each lab station has the required materials before the start of class.
- Safety notes: Have students wear safety goggles. Remind them to handle the hot test tubes with tongs and to use extreme care. Remind them to be careful of the hot plate at all times.
- Moisten the ends of the probes to facilitate pushing them through the stoppers.
- When test tubes are heated, the sand and water may not reach 70°C at the same time. Students may remove each tube as it reaches 70°C. Remind them that they are interested in observing the rate of cooling, and that the starting temperature is not critical.

Analysis Answers

1. The sand should cool faster, as water has a greater capacity to store heat and therefore a slower cooling rate.
2. The desert would cool more rapidly due to the lack of water on and in the ground and in the atmosphere.

Figure 18-17 Tundra in Denali National Park, Alaska

Biome 7 Tundra

Between the taiga and the permanent ice surrounding the North Pole is the open, sometimes boggy biome known as the tundra. This enormous biome covers one-fifth of the Earth's land surface. Annual precipitation in the tundra is very low, usually less than 25 cm (10 in.), and water is unavailable for most of the year because it is frozen. During the brief Arctic summers, the surface of the tundra is often extremely boggy. The permafrost, or permanent ice, usually exists within 1 m (about 3 ft) of the surface and prevents the drainage of water from melting snow. Foxes, lemmings, owls, and caribou are among the vertebrate inhabitants.

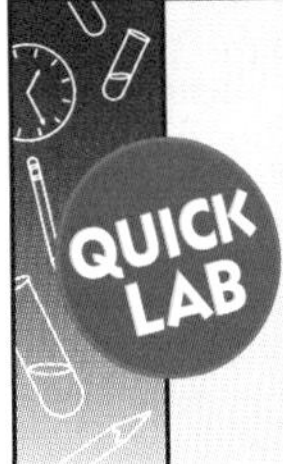

QUICK LAB

What factors influence the cooling of the Earth's surface?

You can discover how the amount of water in an environment affects the rate at which that environment cools.

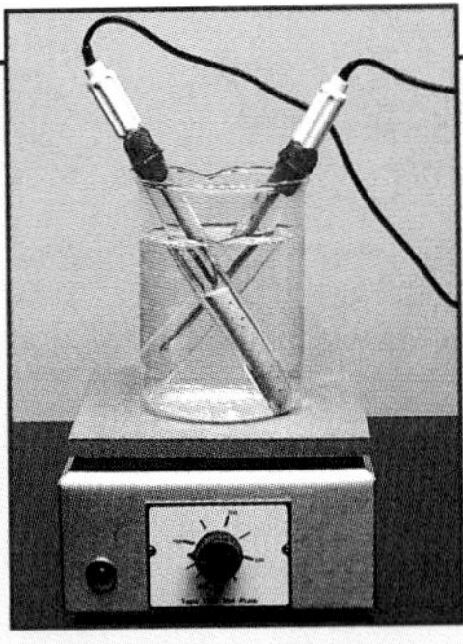

Materials

MBL or CBL system with appropriate software, temperature probes, test tubes, beaker, hot plate, one-holed stoppers, water, sand, test-tube tongs, test-tube rack

Procedure

1. Set up an MBL/CBL system to collect and graph data from each temperature probe at 5-second intervals for 240 data points. Calibrate the probe using stored data.
2. Fill one test tube with water. Fill another test tube halfway with sand.
3. Place a temperature probe in the sand, and suspend another temperature probe at the same depth in the water, using one-holed stoppers to hold each temperature probe in place.
4. Place both test tubes in a beaker of hot water. Heat them to a temperature of about 70ºC.
5. Using test-tube tongs, remove the test tubes and place them in the test-tube rack. Record the drop in temperature for 20 minutes.

Analysis

1. **SKILL Analyzing Results** Did the two test tubes cool at the same rate? Offer an explanation for your observations.
2. **SKILL Predicting Outcomes** In which biome—tropical rain forest or desert—would you expect the air temperature to drop most rapidly? Explain your answer.

did you know?

Hours of sunlight per day range widely through the year in northern biomes.

The northernmost areas of taiga receive only 6 to 8 hours of sunlight during the winter but nearly 19 hours during the summer. The extremes are even more exaggerated in the tundra, which receive less energy from the sun than does any other biome.

Teaching TIP

Deep-Water Detritivores

In freshwater communities, there are no photosynthetic organisms in the deep-water zone. Most of its inhabitants are detritivores, including scavenging fish, bacteria, and aquatic worms.

Teaching TIP

Photosynthetic Organisms

When sunlight strikes the ocean surface, red, orange, and yellow wavelengths of light are absorbed first. Only blue and green can penetrate deeply. Thus, below a depth of a few meters, only those photosynthetic organisms capable of using these shorter wavelengths of light can survive.

Resource Link

Assign **Reading Strategies: Compare/Contrast Matrix** from the Holt BioSources Teaching Resources CD-ROM. PORTFOLIO

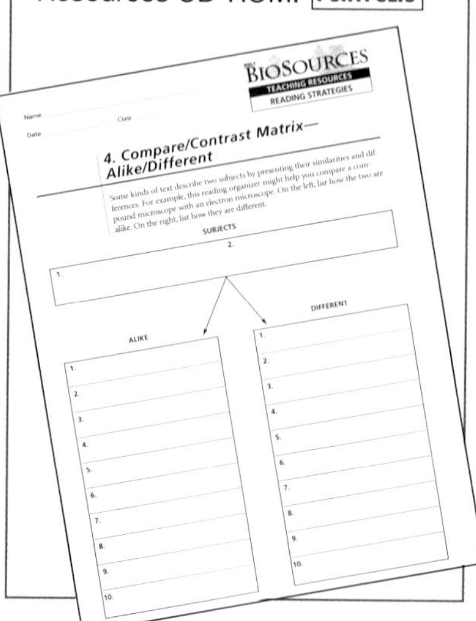

BIOSOURCES
TEACHING RESOURCES
READING STRATEGIES

Name ______ Class ______
Date ______

4. Compare/Contrast Matrix—Alike/Different

Some kinds of text describe two subjects by presenting their similarities and differences. For example, this reading organizer might help you compare a compound microscope with an electron microscope. On the left, list how the two are alike. On the right, list how they are different.

SUBJECTS

1. ______
2. ______

ALIKE	DIFFERENT
1.	1.
2.	2.
3.	3.
4.	4.
5.	5.
6.	6.
7.	7.
8.	8.
9.	9.
10.	10.

Study TIP

Organizing Information

Make a concept map that describes the zones of a pond or lake as described at right and in Figure 18-18. For each zone, include the plant and animal life found there.

Aquatic Communities Are Linked to Terrestrial Communities

At a glance, you might at first think that freshwater and marine communities are separate from terrestrial biomes. Yet large amounts of organic and inorganic material continuously enter both bodies of fresh water and ocean habitats from communities on the land.

Freshwater communities

Freshwater habitats—lakes, ponds, streams, and rivers—are very limited in area. Lakes cover only about 1.8 percent of the Earth's surface, and rivers and streams cover about 0.3 percent. All freshwater habitats are strongly connected to terrestrial ones, with freshwater marshes and wetlands constituting intermediate habitats. Many kinds of organisms are restricted to freshwater habitats, including plants, fish, and a variety of arthropods, mollusks, and other invertebrates too small to see without a microscope.

Ponds and lakes have three zones in which organisms live, as illustrated in **Figure 18-18.** The **littoral zone** is a shallow zone near the shore. Here, aquatic plants live along with various predatory insects, amphibians, and small fish. The **limnetic zone** refers to the area that is farther away from the shore but close to the surface. It is inhabited by floating algae, zooplankton, and fish. The **profundal zone** is a deep-water zone that is below the limits of effective light penetration. Numerous bacteria and wormlike organisms that eat debris on the lake's bottom live in this zone. The breakdown of this debris releases large amounts of nutrients. Not all freshwater systems are deep enough to include a profundal zone.

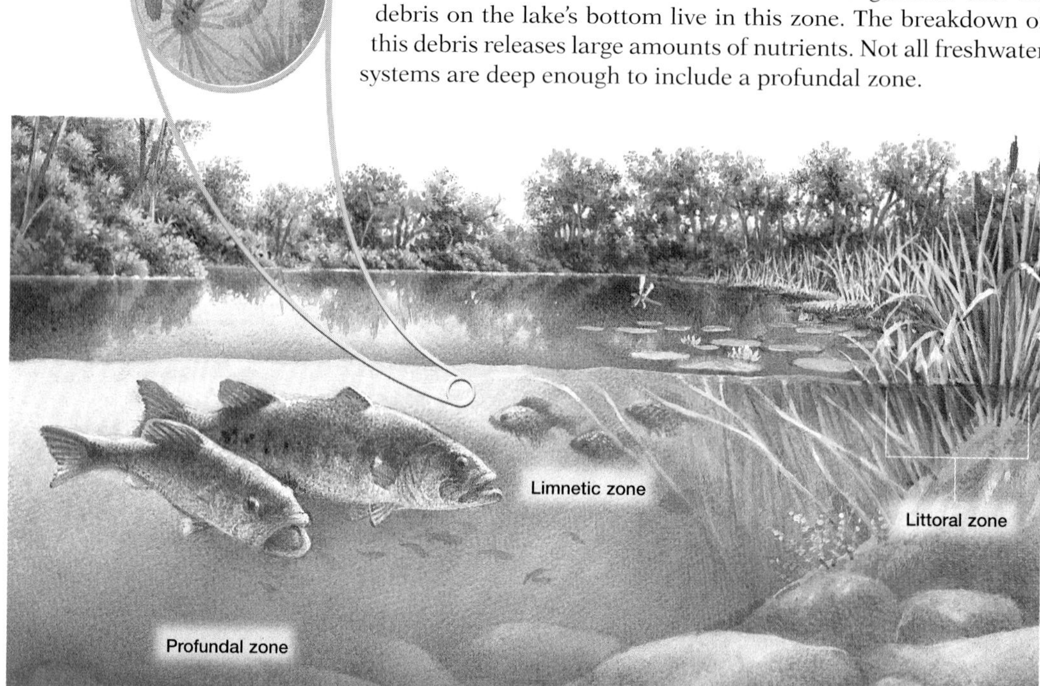

Figure 18-18 Three lake zones. Each region, or zone, of a lake contains characteristic organisms.

did you know?

The physical link between rivers or streams and their surrounding terrestrial communities is the stream bank, also called the riparian zone.

Riparian zones often appear as ribbons of vegetation around waterways. These areas provide humans with many services, including filtering pollution and absorbing flood water. Riparian zones also provide vital habitat for a broad array of plant and animal species.

Marine communities

Figure 18-19 **Shallow ocean species.** Schools of fish, such as these blue stripe snapper near Hawaii, live in shallow waters.

Nearly three-fourths of the Earth's surface is covered by ocean. The oceans have an average depth of more than 3 km (1.9 mi). They are, for the most part, cold and dark. Photosynthetic organisms are confined to the upper few hundred meters of water. The heterotrophic organisms that live below this level obtain almost all of their food from organic debris that drifts downward. Even in the lightless, icy depths of the Marianas Trench, in the western Pacific Ocean (almost 11 km, or 6 mi, deep), life exists.

The marine environment consists of three major kinds of habitats: (1) the shallow ocean waters along the coasts of the continents, (2) the surface of the open sea, and (3) the depths of the open ocean.

Shallow ocean waters The zone of shallow water is small in area, but compared with other parts of the ocean, it is inhabited by large numbers of species. Coral reef communities occur in shallow, tropical waters as shown in **Figure 18-19.** The world's great fisheries are located in the coastal zones of cooler waters. Here, nutrients washed out from the land and large numbers of plankton support huge numbers of fishes.

Exploring Further

Estuaries

If you've ever eaten seafood caught in a saltwater marsh, you've experienced one of the benefits of estuaries. Estuaries are unique transition zones between marine and freshwater environments. Nutrients washed from nearby land stimulate the growth of plants and algae. As a result, estuaries are among the most productive ecosystems on Earth. One hundred acres of healthy estuary can produce 4–10 times as much organic matter as a cultivated cornfield of the same size! The estuary's plants, invertebrates, fishes, birds, mammals, and other animals are part of a complex food web.

Where the river meets the sea

In addition to serving as wildlife habitats, estuaries filter sediment and nutrients, purifying the water that drains off the land. Porous salt-marsh soils absorb floodwaters and protect coastal communities from erosion. Estuaries also provide jobs for people in the seafood and recreation industries.

Estuaries in peril

Sadly, the public has long regarded estuaries as wastelands. People have drained estuaries to provide land for housing and agriculture. Pollutants and improperly treated sewage have

poisoned the estuaries' habitats. As a result, estuaries in the United States have greatly decreased in size over the past century. However, a variety of environmental organizations are now working to restore estuaries by cleaning up pollutants and replanting native vegetation. For example, restoration efforts in Tampa Bay, Florida, improved sewage treatment facilities, enabling sea grass meadows to return. When the sea grasses returned, so did the fishes and other animals that depend on them.

internet**connect**

SCILINKS NSTA
TOPIC: Estuaries
GO TO: www.scilinks.org
KEYWORD: HX397

Section 18-3

Exploring Further

Estuaries

Teaching Strategies

- Inform students that a healthy estuary is third only to coral reefs and tropical rain forests in terms of biological productivity.The Chesapeake Bay has received national attention as an example of an imperiled estuary. Overfishing combined with urban and agricultural pollution has nearly destroyed one of the most productive ecosystems on the planet. The problem is large scale: rivers and creeks from six different states flow into the bay.
- Have students research a major estuary, reporting on its historical significance, the seafood harvested, its current environmental health, and efforts to preserve or restore the ecosystem. PORTFOLIO

Discussion

Ask students why they think estuaries have been overharvested and polluted if they are so important to humans and wildlife. *(Answers may vary. Students should be aware that estuaries tend to be found in highly populated areas and occur downstream from agricultural areas. Also, people tend to overuse public land or water because it is "free" for the taking.)*

internet**connect**

SCILINKS NSTA Have students research estuaries using the Web site in the Internet Connect box on page 397. Assign students to write a report summarizing their findings. PORTFOLIO

Chapter 18

Teaching TIP

Deep-sea Waters

Humans first reached the floor of the Marianas Trench in 1960. Two scientists used the bathyscaphe Trieste to descend nearly 11 km (7 mi) below the ocean's surface. The craft carried gasoline for buoyancy and buckshot for ballast. It descended when water was pumped into air tanks at either end, and it ascended when water was pumped out and buckshot released. The Trieste obtained much data about the conditions at such depths and its scientists observed many never-before-seen deep-water organisms.

3 Assessment Options

Closure

Biome Inhabitants

Have students select one biome and two interacting species that can be found living there. Have them explain how each species is adapted to its biome and how the two species interact.

Review

Assign Review—Section 18-3.

Reteaching — BASIC

Show students pictures of various biomes. Ask them to identify each and to discuss its physical and biological characteristics, as well as where on the planet such a biome might be found.

Surface of the open sea Drifting freely in the upper waters of the ocean is a diverse community of bacteria, algae, fish larvae, and certain species of small animals called **plankton.** Some of the members of the plankton, including algae such as diatoms and some bacteria, are photosynthetic, as shown in **Figure 18-20.** These photosynthetic organisms account for about 40 percent of all the photosynthesis that takes place on Earth. Because these organisms are photosynthetic, they are restricted to approximately the upper 100 m (328 ft) of the ocean. These surface waters are also home to many types of commercial fishes, such as herring, anchovies, cod, tuna, and swordfish. These fishes provide the food for birds, seals, and some whales. Many invertebrates, such as jellyfish, also live here.

Figure 18-20 Diatoms

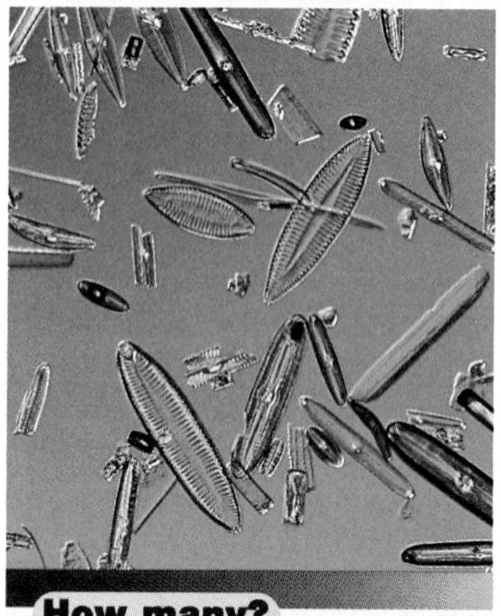

How many? About 25 million diatoms would fit inside a teaspoon. Diatoms produce at least one-fourth of the oxygen we breathe.

Figure 18-21 Unusual residents. A deep-sea inhabitant, the angler fish has a projection from its head that emits light.

Depths of the open ocean In the deepest waters of the sea, below the depth of 300 m (984 ft), live some of the most bizarre organisms found on Earth. Many of these animals, such as the angler fish shown in **Figure 18-21,** can produce their own light. They use the light to communicate with one another or to attract prey. Diverse communities flourish around deep sea vents, cracks in the ocean floor through which hot water escapes. The producers in this community are bacteria that use hydrogen sulfide gas ("rotten egg" gas) or methane leaking from the cracks as an energy source. On the cold dark ocean floor, at an average depth of more than 3 kilometers lives an unusual community of marine invertebrates such as tube worms, clams, and mussels. These invertebrates live under enormous pressure. The unexpected abundance of species, recently discovered, is so diverse it may rival that of tropical rain forests.

Review SECTION 18-3

1. **Describe** the relationship between climate and location of species.
2. **State** the common environmental hardship that confronts organisms in deserts and tundra.
3. **SKILL Analyzing Information** Why can't photosynthesis occur in the deepest parts of the ocean or a deep lake?

Critical Thinking

4. **SKILL Forming Reasoned Opinions** The equator passes across the country of Ecuador. But the climate there can range from hot and humid to cool and dry. What might explain this?

Review SECTION 18-3 ANSWERS

1. The climate of a region determines what species can live in that region. (Only those well adapted to the conditions will prevail.) Similar climates produce similar communities of species.
2. Organisms in both deserts and tundra encounter extreme temperatures, scarcity of available water, and lack of trees and other producers of food and shelter.
3. Photosynthesis cannot occur because light is completely absorbed before it reaches these depths.
4. Mountains create climate conditions different from those associated with latitude alone.

Use the Key Concepts and Key Terms listed below to review the main ideas in this chapter. If you need to review the meaning of a term, turn to the page indicated.

Key Concepts

18-1 How Organisms Interact in Communities

- Species within communities coevolve, making many adjustments to living together.
- In a predator-prey interaction, prey often evolve ways to escape being eaten. Predators evolve ways to overcome the defenses of the prey.
- In mutualism and commensalism, species evolve in ways that benefit one or both parties.

18-2 How Competition Shapes Communities

- Interactions among species help shape ecosystems.
- Competition occurs when two species use the same limited resource.
- An organism's niche is its way of life. An organism may occupy only a part of its fundamental niche, which is called its realized niche.
- Competition can limit how species use resources.
- Biodiversity tends to promote stability and productivity.

18-3 Major Biological Communities

- Climate largely determines where species live.
- Temperature and moisture are key factors in determining where biomes occur.
- The seven major biomes are tropical rain forest, desert, savanna, temperate deciduous forest, temperate grassland, taiga, and tundra.
- Freshwater communities have three zones of life—littoral, limnetic, and profundal.
- The three major marine communities are shallow ocean waters, open sea surface, and deep-sea waters.

Key Terms

18-1

coevolution (382)
predation (382)
parasitism (382)
secondary compound (383)
symbiosis (384)
mutualism (384)
commensalism (384)

18-2

competition (385)
niche (385)
fundamental niche (386)
realized niche (387)
competitive exclusion (389)
biodiversity (390)

18-3

climate (391)
biome (392)
littoral zone (396)
limnetic zone (396)
profundal zone (396)
plankton (398)

BIOLOGY INTERACTIVE TUTOR Unit 7—*Ecosystem Dynamics* Use Topics 1–6 in this unit to review the key concepts and terms in this chapter.

Study TIP Ready?

- *If you think you understand these Key Concepts and Key Terms, then you're ready for the Performance Zone.*

Review and Practice

Assign the following worksheets for Chapter 18:

- **Student Review Guide**
- **Concept Mapping Worksheet**
- **Test Preparation Pretest**
- **Vocabulary Worksheet**
- **Critical Thinking Review Worksheet**
- **Science Skills Worksheet**

Assessment Options

The following assessment options are available for this chapter:

Resource Link

- Assign **Chapter Test 18** to assess students' understanding of the chapter.

Alternative Assessment

Have students imagine that a large island has suddenly appeared on the surface of the planet. Have students create climate conditions for the island, then predict what biome(s) would be found on the island in the future. What are some of the species they could expect to find? How might those species interact?

Portfolio Assessment

- *Portfolio Assessment* at the front of this book provides options for building and evaluating student portfolios. Students and teachers can select items from the areas listed for inclusion in a portfolio.
- See items labeled PORTFOLIO on pp. 384, 386, 390, 392, 393, and 397.

Answer to Concept Map

The following is one possible answer to Performance Zone item 1 on page 400.

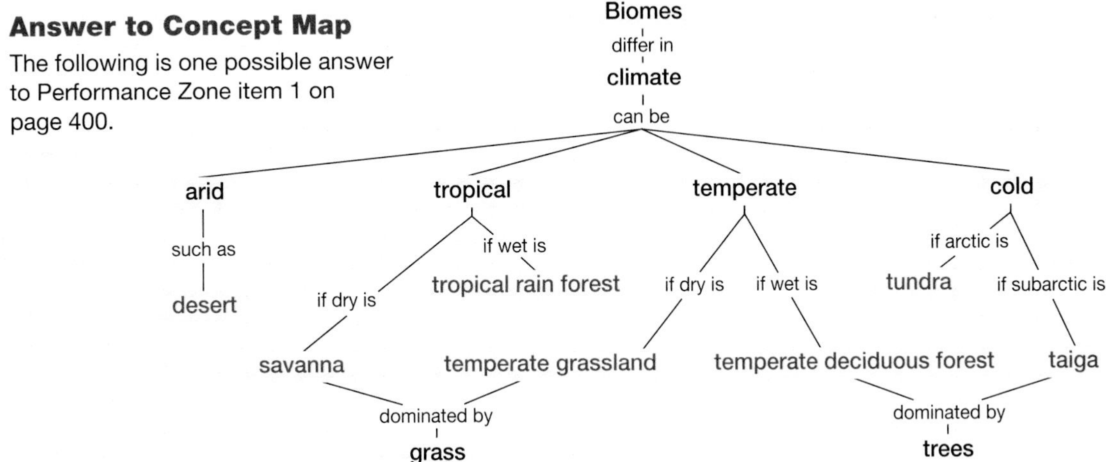

Performance ZONE
CHAPTER 18
REVIEW ANSWERS

Assignment Guide

Section	Review
18-1	3, 12, 18, 20
18-2	4, 5, 6, 9, 16, 17
18-3	1, 2, 7, 8, 10, 11, 13, 14, 15, 19, 21, 22

Understanding and Applying Concepts

1. The answer to the concept map is found at the bottom of page 399.
2. **a.** Predation is a type of interaction between organisms in which one organism, the predator, kills and eats the other organism, the prey. Parasitism is a form of predation in which the predator, or parasite, does not immediately kill the other member, or host, but feeds on it while living on or in its body.

   b. Mutualism is a symbiotic relationship between two organisms in which both organisms benefit. Commensalism is a symbiotic relationship in which one organism benefits while the other is unaffected.

   c. Competition is an interaction between two organisms that are attempting to use the same resource. Competitive exclusion occurs when one organism outcompetes the other, thereby "forcing" the other out of an area.

   d. Climate refers to the prevailing weather conditions in a given area. A biome is a major biological community that occurs over a large area of land.
3. a
4. d
5. a
6. c
7. b
8. a

Performance ZONE CHAPTER 18 Review

Understanding and Applying Concepts

1. **Concept Mapping** Make a concept map that shows how the biomes can be classified based on precipitation, temperature, and geographical location. Use as many terms as needed from the vocabulary list. Try to include the following terms in your map: tropical rain forest, savanna, desert, temperate deciduous forest, temperate grassland, taiga, tundra. Include additional terms in your map as needed.

2. **Understanding Vocabulary** For each pair of terms, explain the differences in their meanings.
   a. parasitism, predation
   b. mutualism, commensalism
   c. competition, competitive exclusion
   d. climate, biome

3. In predator-prey coevolution, if the prey evolves a defense to stop predation, then the predator likely will evolve
   a. in a way that enables it to overcome the prey's defense.
   b. so that it can parasitize the prey.
   c. countermeasures to secondary compounds.
   d. into the prey.

4. The interaction between a spruce tree and a hemlock tree, both of which require nitrogen from the soil, is an example of
   a. mutualism. c. succession.
   b. commensalism. d. competition.

5. The ways in which an organism interacts with its environment make up its
   a. niche. c. habitat.
   b. space. d. ecosystem.

6. Competitive exclusion indicates that
   a. a niche can be shared by two species if their niches are very similar.
   b. niche subdivision may occur.
   c. one species will eliminate a competing species if their niches are very similar.
   d. competition ends in worldwide elimination of a species.

7. Which of the following is a transitional zone between tropical rain forest and desert?
   a. taiga c. temperate deciduous forest
   b. savanna d. tundra

8. Cold and long winters, very few trees, and little precipitation describe the
   a. tundra. c. deciduous forest.
   b. taiga. d. grasslands.

9. **Theme Metabolism** Describe the niches of a lion, a zebra, and the grass that grows on the African plain in terms of how each species affects energy flow in the ecosystem.

10. **Analyzing Data** Describe how elevation and latitude affect the distribution of biomes in a figure similar to the one shown below.

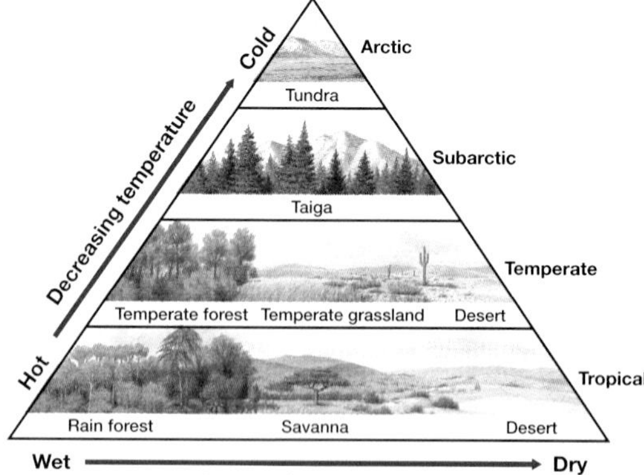

11. **Exploring Further** Data show that although waterfowl are breeding in a particular estuary, the percentage of young that survive is decreasing. Two years ago, engineers dug a deep channel in this estuary. How might these two events be related?

12. **Chapter Links** How does the flow of energy through living systems help determine the components of a biological community? (**Hint:** See Chapter 5, Section 5-1.)

9. The grass grows well in strong sunlight and recovers quickly from dry spells and grazing. A portion of the energy from the sun is stored in its tissues and passed on to feeding zebra. Zebra are herbivores that feed during the day and live in herds, which offer some protection from predators such as lions. Lions feed primarily on herbivores, capturing only a small portion of the grazed energy. Lions hunt in small groups, often at night, which enables them to capture large prey, such as zebra.

10. Temperatures generally decrease with increasing elevation (height above sea level) and latitude (distance from the equator). In other words, increasing elevation and latitude positively correlate with the transition from tropical to temperate to arctic biomes. Therefore, either variable could be interchanged with "decreasing temperature" in such a diagram.

Skills Assessment

Part 1: Interpreting Graphics

Study the diagram below, which shows the comparative productivity of several biomes and aquatic communities. Then answer the following questions.

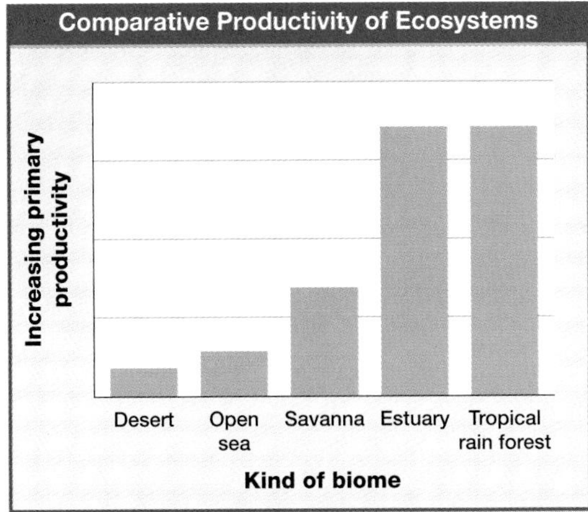

13. Which two biological communities show the highest primary productivity?

14. Identify the least productive biological community shown, and suggest a probable cause for its low ranking.

Part 2: Life/Work Skills

Use the media center or Internet sources to find information about the tropical rain-forest biome.

15. Finding Information Identify two or three factors that cause the tropical rain forest to have high productivity. List factors that are responsible for the biodiversity of this community. Describe reasons for the decline of tropical rain forests, and discuss actions that some countries have taken to protect tropical rain forests.

16. Communicating Prepare a visual presentation that summarizes your findings about the tropical rain-forest biome. After studying your presentation, viewers should understand productivity and its relationship to biodiversity.

Critical Thinking

17. Justifying Conclusions In Gause's experiments, *Paramecium caudatum* could coexist with *P. bursaria* but not with *P. aurelia.* Predict what would happen if *P. aurelia* and *P. bursaria* were grown together, and justify your conclusions.

18. Justifying Conclusions Early human settlers of Hawaii introduced predators—including cats, dogs, rats, and pigs—that the native animals had never encountered. These introduced predators proved devastating to the native animals, many of which became extinct. Explain why prey are often more vulnerable to introduced predators than to native predators.

Portfolio Projects

19. Summarizing Information Work with a small group of students to develop a map that shows the most prominent terrestrial and aquatic communities within your state. Be certain to include any large swamps or wetlands that connect terrestrial and aquatic communities.

20. Interpreting Information Read the article "El Niño Strikes Again" (*Current Science,* January 16, 1998, pp. 4–77). Explain why El Niño can be life-threatening to some animal species that live in or by the Pacific Ocean. What species have been adversely affected by El Niño?

21. Career Focus Park Ranger Use the media center or Internet resources to research the career of park ranger. Your report should describe the job duties, required educational background, growth prospects, and an average starting salary.

22. Unit 7—*Ecosystem Dynamics* Write a report summarizing how artificial ecosystems are used to eliminate pollutants from waste water. Find out what kinds of factories or plants use artificial ecosystems.

11. Changing the characteristics of the ecosystem may have created a more hostile environment for the young waterfowl. Perhaps cover from predators is now limited, the new conditions favor more predators, or the deeper estuary is colder, all of which might affect hatchling survival.
12. The flow of energy through ecosystems ensures that biological communities have the same basic structure: energy flows from sunlight or inorganic substances to autotrophs, such as plants, and then to heterotrophs, such as rabbits and foxes.

Skills Assessment

13. estuary and tropical rain forest
14. desert; relatively sparse vegetation because of the chronic lack of water
15. Answers will vary. Sample answer: Wet climates and long growing seasons help make tropical rain forests so productive. One reason these forests are so diverse is that they have so many potential ecological roles, so there are more potential niches to exploit. Tropical rain forests have declined because they lie within countries that want to use their natural resources to develop economically. Tropical nations have established nature reserves and have encouraged sustainable forest harvesting to help preserve their forests.
16. Answers will vary.

Critical Thinking

17. The two could coexist if *P. bursaria* continues to feed on yeast near the bottom of the tube and *P. aurelia* feeds on bacteria near the top.
18. Such prey have not had enough time to coevolve with the new predators and develop defenses.

Portfolio Projects

19. Answers will vary.
20. The warm surface-water temperatures caused by El Niño reduce anchovy populations, thereby threatening populations of species that depend on anchovies for food. Such species include boobies, pelicans, and southern fur seals.
21. Answers will vary. Park rangers carry out a range of duties necessary for the maintenance and functioning of national and state parks. Rangers enforce park rules, maintain roads and trails, and staff information centers. Rangers have a high school degree, and many have two or more years of college. Growth potential for this field is limited. Starting salary will vary by region.
22. Answers will vary. People create artificial wetland ecosystems to treat waste water. Wetland plants take up nitrogen and phosphorus, and both plants and soils absorb heavy metals. Cities use such artificial wetlands to treat municipal waste water.

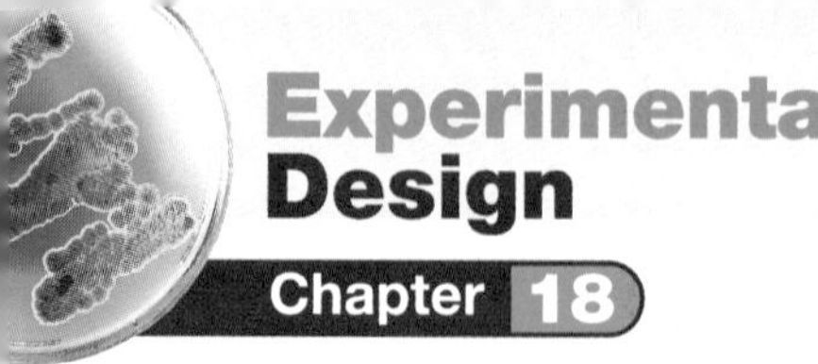

How Do Brine Shrimp Select a Habitat?

Purpose

Students will investigate habitat selection by brine shrimp and determine which environmental conditions the shrimp prefer.

Preparation Notes

Time Required: Two lab periods of 40–50 minutes each

Materials

Materials for this lab can be ordered from WARD'S. See *Master Materials List* at the front of this book for catalog numbers.

Preparation Tips

Students should begin by observing the movement of brine shrimp through water. Tell students to allow the shrimp enough time to distribute themselves in response to environmental factors.

Safety Precautions

- Make sure students do not use very hot water.
- Students should not put their hands near their face while handling the brine shrimp.
- Tell students to wash their hands thoroughly after concluding the experiment.

Procedural Tips

- This activity works well with groups of four students.
- When tubing is divided as directed, it allows for space taken up by stoppers and forms equal quarters.
- Students may count shrimp by looking in the Petri dish or in the pipet. Students might hold the pipet up to a light for better visibility.
- The **Procedure** section provides students with one method for randomly dividing brine shrimp into groups that can be experimentally manipulated in the tubing or in the Petri dishes.

Experimental Design

Chapter 18 *How Do Brine Shrimp Select a Habitat?*

SKILLS

- Using scientific methods
- Collecting, organizing, and graphing data

OBJECTIVES

- **Observe** the behavior of brine shrimp.
- **Assess** the effect of environmental variables on habitat selection by brine shrimp.

MATERIALS

- clear, flexible plastic tubing
- metric ruler
- marking pen
- corks to fit tubing
- brine shrimp culture
- screw clamps
- test tubes with stoppers and test-tube rack
- pipet
- Petri dish
- Detain™ or methyl cellulose
- aluminum foil
- calculator
- fluorescent lamp or grow light
- funnel
- graduated cylinder or beaker
- hot-water bag
- ice bag
- pieces of screen
- tape

Magnification 20×

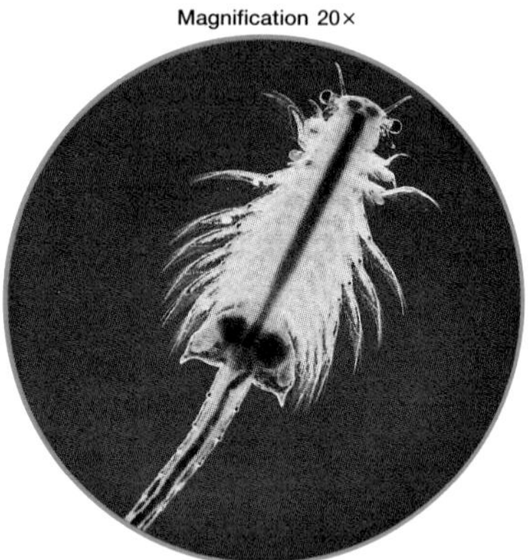

Brine shrimp

Before You Begin

Different organisms are adapted for life in different **habitats.** For example, **brine shrimp** are small crustaceans that live in salt lakes. Given a choice, organisms select habitats that provide the conditions (e.g., temperature, light, pH, salinity) to which they are adapted. In this lab, you will investigate habitat selection by brine shrimp and determine which environmental conditions they prefer.

1. Write a definition for each boldface term in the paragraph above.
2. Based on the objectives for this lab, write a question you would like to explore about habitat selection by brine shrimp.

Procedure

Part A: Making and Sampling a Test Chamber

1. Divide a piece of plastic tubing into 4 sections by making a mark at 12 cm, 22 cm, and 32 cm from one end. Label the sections *1, 2, 3,* and *4.*
2. Place a cork in one end of the tubing. Then transfer 50 mL of brine shrimp culture to the tubing. Place a cork in the open end of the tubing. The chamber can now be used to conduct an experiment.
3. When you are ready to count shrimp, divide the tubing into four sections by placing a screw clamp at each mark on the tubing. *While someone holds the corks firmly in place,* first tighten the middle clamp and then the outer clamps.
4. Starting at one end, pour the contents of each section into a test tube labeled with

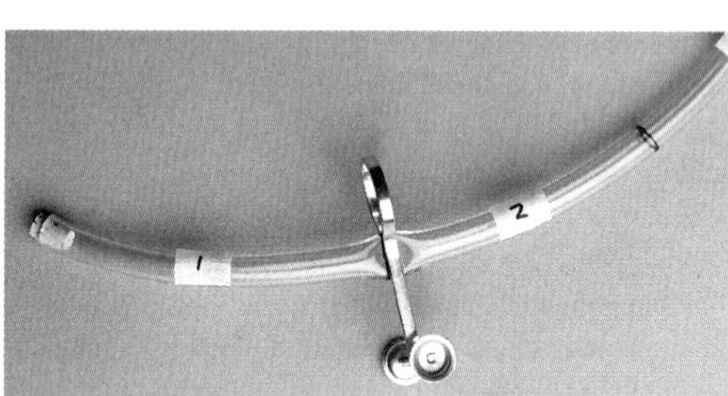

Disposal

- Return the brine shrimp to an aquarium or a pet supply store.
- Carefully dispose of solutions, and wash out lab ware in the sink.

Before You Begin

Answers

1. *habitats*—areas where organisms live and to which they are adapted

   brine shrimp—small crustaceans that live in salt lakes
2. Answers will vary. For example: Will brine shrimp prefer a warm habitat or a cool habitat?

the same number. After you empty a section, loosen the adjacent clamp and fill the next test tube.

5. Stopper one test tube, and invert it gently to distribute the shrimp. Use a pipet to transfer a 1 mL sample of shrimp culture to a Petri dish. Add a few drops of Detain™ to the sample. Count and record the number of live shrimp.
6. Repeat step 5 three more times for the same test tube. Record the average number of shrimp for this test tube.
7. Repeat steps 5 and 6 for each of the remaining test tubes.

Part B: Design an Experiment

8. Work with the members of your lab group to explore one of the questions written for step 2 of **Before You Begin.** To explore the question, design an experiment that uses the materials listed for this lab.

> **You Choose**
>
> **As you design your experiment, decide the following:**
>
> a. what question you will explore
> b. what hypothesis you will test
> c. how to set up your control
> d. how to expose the brine shrimp to the conditions you chose
> e. how long to expose the brine shrimp to the environmental conditions
> f. how you will set up your data table

9. Write a procedure for your group's experiment. Make a list of all the safety precautions you will take. Have your teacher approve your procedure and safety precautions before you begin the experiment.
10. Set up and conduct your group's experiment. Do *not* use water over 70°C, which can burn you. **CAUTION: If you are working with the hot-water bag, handle it carefully. If you are working with a lamp, do not touch the bulb. Light bulbs get very hot and can burn your skin.**

Part C: Cleanup and Disposal

11. Dispose of broken glass in the designated waste container. Put brine shrimp in the designated container. Do not pour chemicals down the drain or put lab materials in the trash unless your teacher tells you to do so.
12. Clean up your work area and all lab equipment. Return lab equipment to its proper place. Wash your hands thoroughly before you leave the lab and after you finish all work.

Analyze and Conclude

1. **Summarizing Results** Make a bar graph of your data. Plot the environmental variable on the x-axis and the number of shrimp on the y-axis.
2. **Analyzing Results** How did the shrimp react to changes in the environment?
3. **Analyzing Methods** Why was a control necessary?
4. **Analyzing Methods** Why was it necessary to take many counts in each test tube (step 6 of Part A)?
5. **Further Inquiry** Write a new question about brine shrimp that could be explored with another investigation.

Do You Know?

Do research in the library or media center to answer these questions:

1. What are some predators of brine shrimp?
2. What is the ideal habitat for one of your favorite animals?

Use the following Internet resources to explore your own questions about habitat selection.

SCILINKS NSTA
TOPIC: Adaptation
GO TO: www.scilinks.org
KEYWORD: HX403

Analyze and Conclude

Answers

1. Answers will vary.
2. Answers will vary depending on the species of *Artemia* used.
3. The control was necessary to show that the brine shrimp did not inherently prefer one part of the tube.
4. Many counts were taken to make allowances for variations in populations and to provide data for calculating an average.
5. Answers will vary. For example: How do brine shrimp react to water movement?

Do You Know?

Answers

1. Fish are the main predators of brine shrimp.
2. Answers will vary depending on the species of animal chosen.

internetconnect

SCILINKS NSTA Have students research adaptation using the Web site in the Internet Connect box on page 403. Assign students to write a report summarizing their findings. PORTFOLIO

Procedure

Sample Procedure

Perform steps 1–4 of part A. Record the number of shrimp and temperature in each test tube, then calculate the average number and temperature. This is the control sample. Then make a new test chamber using the procedure outlined in part A steps 1 and 2. Tape the tubing to the desk top and cover it with aluminum foil. Mark the foil to show the approximate positions of sections 1 and 4. Carefully place a bag of very cold water over section 1 and a bag of very warm water over section 4. After 10 minutes, quickly complete steps 4 and 5 of part A. Immediately read and record the temperature of the solution in the two test tubes. Then count the shrimp in each tube. Make a histogram plotting temperature on the x-axis versus number of shrimp on the y-axis versus. Include a bar for the control sample.

Sample Data Table

Treatment	No. of Shrimp
control	even distribution . . . about 25
warm water	about 35
room temperature	about 17

Chapter 19 Human Impact on the Environment

Planning Guide

BLOCKS	OBJECTIVES	MEETING INDIVIDUAL NEEDS	READING SKILLS AND STRATEGIES
BLOCKS 1 & 2 90 min	**19-1 Global Change** • Recognize the causes and effects of acid rain. • Evaluate the long-term consequences of atmospheric ozone depletion. • Explain how the burning of fossil fuels has changed the atmosphere. • Recognize the relationship between the greenhouse effect and global warming.	Learning Styles, p. 406 BASIC ADVANCED • p. 409	ATE Active Reading, Technique: Brainstorming, p. 405 ■ Active Reading Guide, worksheet (19-1) ■ Directed Reading, worksheet (19-1)
BLOCKS 3 & 4 90 min	**19-2 Ecosystem Damage** • Describe the effects of chemical pollutants on the environment. • Identify three nonreplaceable resources. • Predict the potential consequences of uncontrolled population growth. • Contrast population growth in developing countries with that in industrialized countries.	Learning Styles, p. 410 BASIC ADVANCED • pp. 412, 413, 414, 415	ATE Active Reading, Technique: Reading Effectively, p. 411 ■ Active Reading Guide, worksheet (19-2) ■ Directed Reading, worksheet (19-2)
BLOCKS 5 & 6 90 min	**19-3 Solving Environmental Problems** • Describe two effective approaches that have been taken to reduce pollution in the United States and abroad. • Evaluate the five major steps necessary to solve environmental problems. • Determine how individuals can take personal action to help solve environmental problems.	Learning Styles, p. 416 BASIC ADVANCED • pp. 419, 420	■ Active Reading Guide, worksheet (19-3) ■ Directed Reading, worksheet (19-3) ■ Supplemental Reading Guide, *Silent Spring*

Review and Assessment Resources

BLOCKS 7 & 8 90 min

TEST PREP

- PE Study Zone, p. 421
- PE Performance Zone, pp. 422–423
- ■ Chapter 19 worksheets:
  - • Vocabulary • Concept Mapping
  - • Science Skills • Critical Thinking
- Holt Biology Interactive Tutor Unit 7—Ecosystem Dynamics, Topics 1–6
-  Audio CD Program
- ■ Test Prep Pretest

CHAPTER TESTING

- ■ Chapter Tests (blackline master)
- ▲ Test Generator
- ■ Assessment Item Listing

See the correlations table at the front of this book for a list of the ***National Science Education Standards*** addressed in this chapter.

Smithsonian Institution® Visit **www.si.edu/hrw** for additional on-line resources.

KEY

Biology Principles and Explorations

PE Pupil's Edition

ATE Teacher's Edition

HOLT BIOSOURCES

 Teaching Transparencies

 Laboratory Program

One-Stop Planner CD-ROM
Shows these resources within a customizable daily lesson plan:

■ Teaching Resources

▲ Test Generator

VISUAL STRATEGIES	CLASSWORK AND HOMEWORK	TECHNOLOGY AND INTERNET RESOURCES
ATE **Fig. 19-3,** p. 408 74 Carbon Dioxide and Average World Temperatures 81 Effects of Acid Rain	**PE** **Quick Lab** How can you observe the greenhouse effect?, p. 409 **Lab C27** Studying an Algal Bloom **PE** **Review,** p. 409	CNN **Video Segment 15** Greening Sudbury **Interactive Tutor** Unit 7—Ecosystem Dynamics (Topics 1–6) **Holt Biology Videodiscs** Section 19-1 **Audio CD Program** Section 19-1 internet**connect** SCILINKS NSTA **TOPIC:** Greenhouse effect
75 Human Population Growth 76 Projected Population Growth	**Lab A14** Determining the Amount of Refuse **Lab C28** Studying an Algal Bloom—Phosphate Pollution **Lab D11** Oil-Degrading Microbes ■ **Problem-Solving:** Population Size **PE** **Review,** p. 415	**Interactive Tutor** Unit 7—Ecosystem Dynamics (Topics 1–6) **Holt Biology Videodiscs** Section 19-2 **Audio CD Program** Section 19-2 internet**connect** SCILINKS NSTA **TOPICS:** • Chemical pollution • Aquifers • Population growth
ATE **Fig. 19-16,** p. 420 **ATE** **Fig. 19-17,** p. 420	**PE** **Lab Techniques** Studying Population Growth, pp. 424–425 **Lab B11** Composting **Lab D12** Can Oil-Degrading Microbes Save the Bay? **PE** **Review,** p. 420 **PE** **Study Zone,** p. 421	**Interactive Tutor** Unit 7—Ecosystem Dynamics (Topics 1–6) **Holt Biology Videodiscs** Section 19-3 **Audio CD Program** Section 19-3 internet**connect** SCILINKS NSTA **TOPIC:** Solving environmental problems

ALTERNATIVE ASSESSMENT

PE **Portfolio Projects** 21–24, p. 423

ATE **Alternative Assessment,** p. 421

ATE **Portfolio Assessment,** p. 421

Lab Activity Materials

Quick Lab
p. 409 MBL or CBL system with appropriate software, temperature probes (2), 1 qt jar and lid with 0.5 cm hole in the center, tape, heat source

Lab Techniques
p. 424 safety goggles, lab apron, yeast culture, 1 mL pipets (2), test tubes (2), 1% methylene blue solution, 2 × 2 mm ruled microscope slide, coverslip, compound microscope

Lesson Warm-up
ATE photos of several forms of environmental pollution, p. 410

Demonstrations
ATE box of sand, p. 413
ATE photos of Los Angeles on a smoggy day, p. 417
ATE catalytic converter, p. 417

BIOSOURCES LAB PROGRAM
- Quick Labs: Lab A14, p. 27
- Inquiry Skills Development: Lab B11, p. 45
- Laboratory Techniques and Experimental Design: Lab C27, p. 131
- Laboratory Techniques and Experimental Design: Lab C28, p. 137
- Biotechnology: Lab D11, p. 57
- Biotechnology: Lab D12, p. 63

Chapter 19

Chapter Theme

Homeostasis

One feature of ecosystems is stability, the capacity to absorb impacts without wholesale change. In terrestrial ecosystems, for example, increased carbon dioxide levels stimulate increased photosynthesis by plants, which in turn helps stabilize carbon dioxide levels. In this chapter students will see that human activities often overwhelm the capacity of ecosystems to respond, resulting in widespread environmental instability and change.

Establishing Prior Knowledge

Before beginning this chapter, make certain students can answer the questions in **Study TIP Ready?** on page 405. If they cannot, encourage them to reread the sections indicated.

Opening Demonstration

So much has been written and said about the negative impact of humans on the environment that students may be "turned off" when the issue is raised. To stimulate their interest, show them a picture that reflects the magnitude of the problem. For example, show them a picture of the Fresh Kills landfill, in New York State, and emphasize the enormity of the site. It covers nearly 20 km^2 (8 mi^2), and it receives 34 million pounds of garbage every day.

CHAPTER 19 Human Impact on the Environment

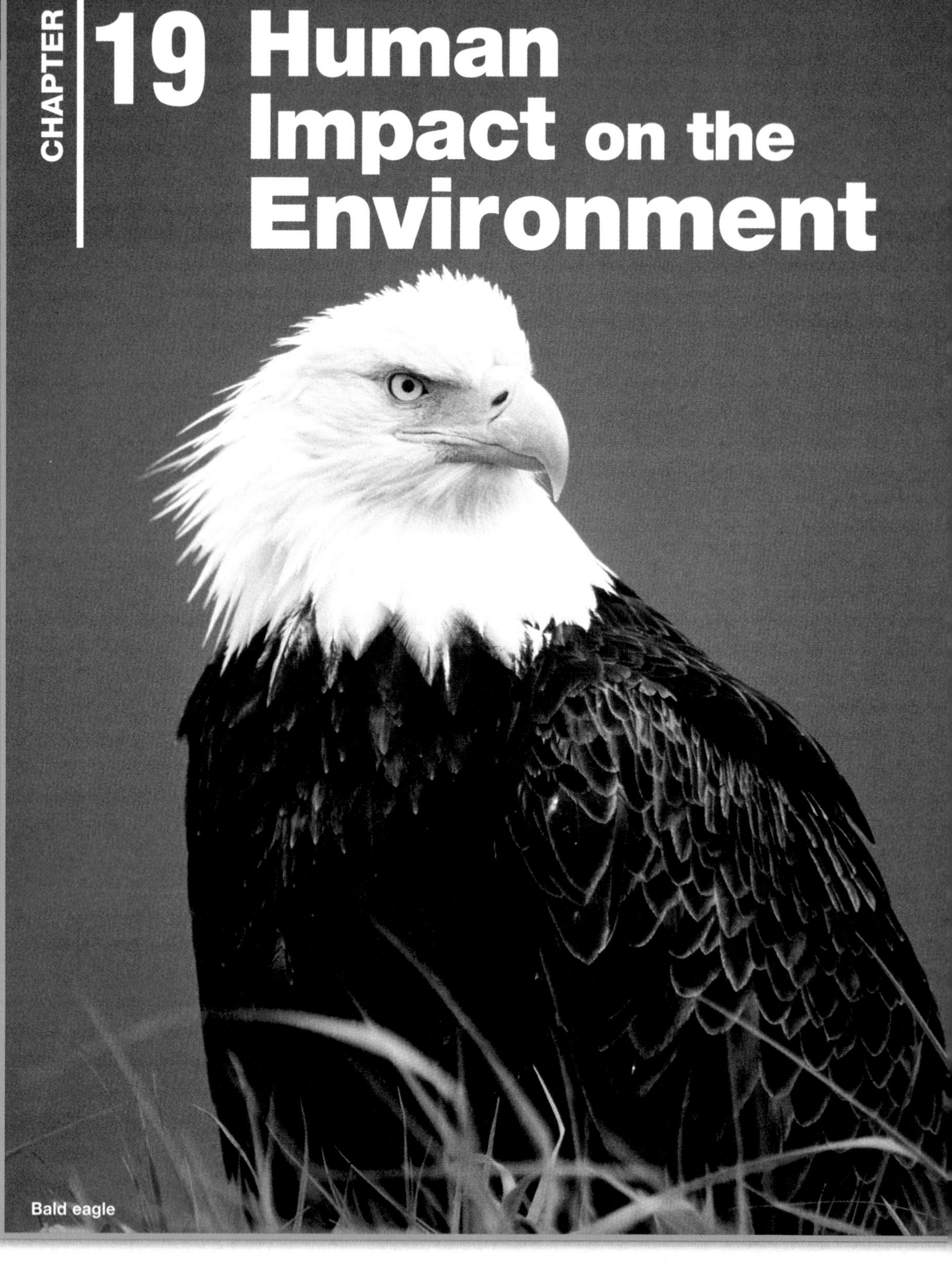

Bald eagle

What's Ahead — and Why
CHAPTER OUTLINE

19-1 Global Change
- Global Change Affects Ecosystems Worldwide
- Release of CO_2 and Other Greenhouse Gases Has Led to Global Warming

19-2 Ecosystem Damage
- Chemical Pollution Can Have Disastrous Results
- Loss of Resources Is a Serious Environmental Problem
- Human Population Is a Key Environmental Concern

19-3 Solving Environmental Problems
- Reducing Pollution Will Take a Worldwide Effort
- Environmental Problems Can Be Solved

Students will learn

- 19-1: that certain human activities have had international and even worldwide environmental impacts.
- 19-2: how human activities can damage local ecosystems and deplete non-replaceable resources.
- 19-3: what is needed to solve environmental problems from an individual and collective viewpoint.

Bald eagle with handler

A proud symbol of our natural heritage, *the bald eagle was a familiar sight in colonial America. By 1963, fewer than 1,000 bald eagles remained in the lower 48 states. Their numbers had been diminished by loss of habitat, sport hunting, and in the mid-twentieth century, the pesticide DDT. Efforts to protect and manage our natural resources have enabled the bald eagle to make a remarkable comeback.*

Study TIP

Ready?

Check your understanding of these topics to help you prepare for what's ahead.

Can you . . .

- **define** *pH* and describe acid rain? (Section 1-3)
- **identify** the role of the ozone layer in Earth's atmosphere? (Section 12-1)
- **describe** how carrying capacity limits the size of a population? (Section 16-1)
- **recognize** the importance of biodiversity? (Section 17-1)
- **summarize** the events of the water cycle and the carbon cycle? (Section 17-3)

If not, *review the sections indicated.*

Looking Ahead

19-1 Global Change
Humans are changing the Earth—perhaps forever.

19-2 Ecosystem Damage
How has human activity affected ecosystems?

19-3 Solving Environmental Problems
What can you do to help?

Labs

- **Quick Lab**
  How can you observe the greenhouse effect? (p. 409)
- **Lab Techniques**
  Studying Population Growth (p. 424)

Features

- **Exploring Further**
  Why Pollution Is Profitable (p. 417)

Unit 7—*Ecosystem Dynamics*
Topics 1–6

internet**connect**

National Science Teachers Association *sci*LINKS Internet resources are located throughout this chapter.

Chapter 19

Reading Strategies

Active Reading

Technique: Brainstorming

Most students will be familiar with or will have participated in efforts to preserve the environment. Working with the class, create a student-generated list of issues and efforts. Tell students that they will read about global environmental problems in this chapter and learn how they can help solve these problems.

Directed Reading

Assign **Directed Reading Worksheet Chapter 19** to help students interact with the text as they read the chapter.

Interactive Reading

Assign ***Biology: Principles and Explorations*** **Audio CD Program Chapter 19** to help reluctant readers achieve greater success in reading the chapter.

Reading Skills

Assign Reading Organizers and Reading Strategies Worksheets from the **One-Stop Planner CD-ROM with Test Generator** to help students learn and practice effective reading skills.

Holt *Biology: Principles and Explorations*

On-line Resources: go.hrw.com

Visit the HRW Web site for a variety of resources related to this chapter. Just type in the keyword HX0 Chapter 19.

Holt *Biology: Principles and Explorations*

On-line Resources: www.scilinks.org

The following *sci*LINKS Internet resources can be found in the student text for this chapter:

Page 408
TOPIC: Greenhouse effect
KEYWORD: HX408

Page 411
TOPIC: Chemical pollution
KEYWORD: HX411

Page 413
TOPIC: Aquifers
KEYWORD: HX413

Page 414
TOPIC: Population growth
KEYWORD: HX414

Page 417
TOPIC: Solving environmental problems
KEYWORD: HX417

Chapter 19

1 Prepare

Scheduling

- **Block:** About 90 minutes will be needed to complete this section.
- **Traditional:** About two class periods will be needed to complete this section.
- **Planning:** See the Planning Guide on pp. 403A and 403B for lecture, classwork, and assignment options for Section 19-1. Lesson plans for Section 19-1 can be found in a lesson cycle format in ***Biology: Principles and Explorations*** **One-Stop Planner CD-ROM with Test Generator.**

Resource Link

Assign **Active Reading Worksheet Section 19-1** to help students comprehend the key concepts and visuals in this lesson.

2 Teach

Lesson Warm-up

Ask students to explain the major steps of the water cycle. *(The water cycle is illustrated on page 371 in Chapter 17.)* Point out that precipitation can transport substances such as sulfuric acid from the atmosphere to the surface. Then ask the class to list the steps of the carbon cycle that remove carbon dioxide from the atmosphere *(photosynthesis and absorption by the ocean)* and those steps that add carbon dioxide to the atmosphere *(cellular respiration, combustion, and decomposition).*

Section 19-1 Global Change

Objectives

- **Recognize** the causes and effects of acid rain. (p. 406)
- **Evaluate** the long-term consequences of atmospheric ozone depletion. (p. 407)
- **Explain** how the burning of fossil fuels has changed the atmosphere. (p. 408)
- **Recognize** the relationship between the greenhouse effect and global warming. (p. 409)

Key Terms

acid rain
chlorofluorocarbon
greenhouse effect

Global Change Affects Ecosystems Worldwide

How do your actions affect others? You may be surprised to learn that things you do—and actions taken by other individuals—can ultimately influence every ecosystem on Earth. Human-induced environmental changes that affect ecosystems worldwide are referred to as global change. Three examples of human-induced global change are acid rain, ozone destruction, and global warming.

Burning high-sulfur coal creates acid rain

As you learned in Chapter 1, acid rain adversely affects many species of plants and animals. Acid rain begins when coal-burning power plants send smoke high into the atmosphere through smokestacks more than 65 m (210 ft) tall. This smoke contains high concentrations of sulfur because the coal that the plant burns is rich in sulfur. The intent of those who designed the power plant was to release the sulfur-rich smoke high into the atmosphere, where winds would disperse and dilute it. Tall smokestacks were first introduced in England in the mid-1950s. They rapidly became popular in the United States and in Europe.

Scientists have since discovered that the sulfur introduced into the atmosphere by smokestacks combines with water vapor to produce sulfuric acid. Rain and snow carry the sulfuric acid back to Earth's surface. This acidified precipitation is called **acid rain.** In North America, acid rain is most severe in the northeastern United States and in southeastern Canada, areas that are downwind from the coal-burning plants in the Midwest. Tests performed in 1989 showed that in the northeastern United States, rain and snow had a pH of about 3.8. This is almost 100 times as acidic as the typical values for precipitation in the rest of the United States.

Acid rain destroys many forms of life. In the United States and Canada, tens of thousands of lakes are "dying" as their pH levels fall below 5.0. Forests in the eastern United States and southern Canada are being seriously damaged, as shown in **Figure 19-1.** The acid pH kills symbiotic fungi in tree roots.

Figure 19-1 One effect of acid rain. The bare trees in this North Carolina forest died as a result of acid rain. Acid rain can interfere with nutrient absorption by plants.

Integrating Different Learning Styles

Logical/Mathematical	**PE**	Quick Lab, p. 409; Performance Zone items 8 and 19, pp. 422–423
	ATE	Quick Lab, p. 409
Visual/Spatial	**PE**	Performance Zone item 1, p. 420
	ATE	Cross-Disciplinary Connection, p. 407; Visual Strategy, p. 408; Graphic Organizer, p. 408
Verbal/Linguistic	**PE**	Real Life, p. 407; Performance Zone item 21, p. 423
	ATE	Lesson Warm-up, p. 406; Reteaching, p. 409

Certain manufactured chemicals promote destruction of the ozone layer

As you learned in Chapter 12, living organisms were able to leave the oceans and colonize the land only after a protective shield of ozone, O_3, had developed in the upper atmosphere. Imagine what would happen if that shield were destroyed. Alarmingly, it appears that this is just what is happening. The ozone layer is disappearing, and we are destroying it ourselves.

The ozone hole

In 1985, a researcher in Antarctica noticed that ozone levels in the atmosphere seemed to be as much as 30 percent lower than they had been 10 years earlier. Satellite images taken over the South Pole revealed that the ozone concentration was unexpectedly lower over Antarctica than elsewhere in the Earth's atmosphere, as shown in **Figure 19-2.** It was as if an "ozone eater" were causing a mysterious zone of below-normal concentration, an area that researchers called the ozone hole. Alarmed, scientists examined satellite images taken in previous years. They found that the disintegration of the Earth's ozone shield was evident as far back as 1978. Every year since then, more ozone has disappeared, and the ozone hole has grown larger. Moreover, a smaller hole has appeared over the Arctic.

Because the decrease in ozone allows more ultraviolet radiation to reach the Earth's surface, scientists expect an increased incidence of diseases caused by exposure to ultraviolet radiation. These diseases include skin cancer, cataracts (a disorder in which the lens of the eye becomes cloudy), and cancer of the retina, the light-sensitive part of the eye. In fact, in the United States, the number of cases of malignant melanoma, a potentially lethal form of skin cancer, has almost doubled since 1980.

What is destroying the ozone?

The major cause of ozone destruction is a class of chemicals called **chlorofluorocarbons** (CFCs). Invented in the 1920s, CFCs were considered extremely stable, supposedly harmless, and a nearly ideal heat exchanger. Throughout the world, CFCs were commonly used as coolants in refrigerators and air conditioners, as aerosol propellants in spray cans, and as foaming agents in the production of plastic-foam cups and containers. Though CFCs were escaping into the atmosphere, at first no one worried.

By 1985, the scientific community had learned that CFCs are the primary cause of the ozone hole. High in the atmosphere, ultraviolet radiation from the sun is able to break the usually stable bonds in CFCs. The resulting free chlorine atoms then enter into a series of reactions that destroys ozone. As a result of this discovery, CFCs have been banned as aerosol propellants in spray cans in the United States. Today many countries limit or ban the use of CFCs.

REAL LIFE

Ozone can be harmful. In many cities and towns, summer weather reports include warnings of high ozone levels. Ozone irritates people's noses, throats and lungs and may harm the immune system.

Finding Information *Find out why ground-level ozone is increasing even though the ozone layer in the upper atmosphere is decreasing.*

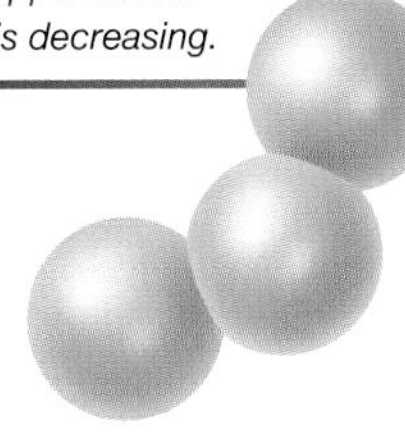

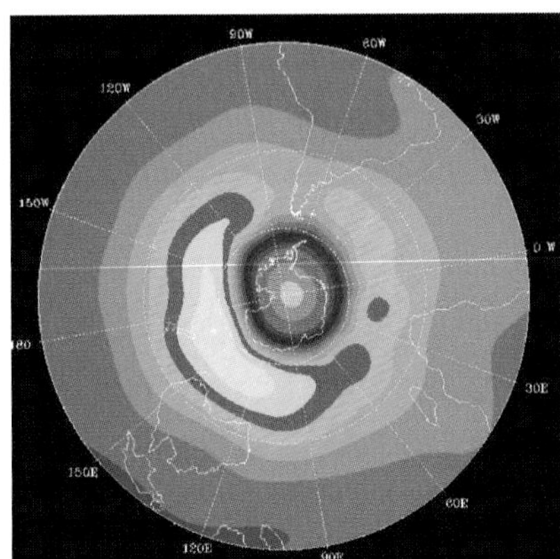

Figure 19-2 Ozone "hole" over Antarctica. In this satellite view of the South Pole, the pink area indicates the region with the least amount of ozone.

REAL LIFE

Answers

Ground-level ozone is increasing because it is produced when sunlight strikes car exhaust, and there are more cars on the road each year. Ground-level ozone rarely makes it to the upper atmosphere, where ozone levels are decreasing due to the effects of CFCs.

CROSS-DISCIPLINARY CONNECTION

◆ Chemistry

Students have had some exposure to chemical equations. Write the equation for the formation of acid rain to help them visualize how acid rain forms. Sulfur released by smokestacks reacts with oxygen in the atmosphere to form sulfur trioxide: $2S + 3O_2 \rightarrow 2SO_3$. In turn, the sulfur trioxide reacts with water vapor in the atmosphere to form sulfuric acid: $SO_3 + H_2O \rightarrow H_2SO_4$. Burning coal also produces sulfur *di*oxide, which reacts with water to form sulfurous acid: $SO_2 + H_2O \rightarrow H_2SO_3$.

Teaching TIP

Acid Rain

The United States, Canada, and many European nations have reduced sulfur emissions in recent years. However, lakes and forests damaged by acid rain have not recovered as quickly as hoped. One reason is that acid rain alters soil chemistry, an effect which may take many decades to reverse.

Overcoming Misconceptions

"Good" and "Bad" Ozone

Students might not understand the difference between "good" ozone and "bad" ozone. Explain that the ozone layer lies between 17 and 26 km (11–16 mi) above sea level, where it absorbs incoming UV radiation. Ozone also forms lower in the atmosphere when emissions from cars or factories react with sunlight. Ozone is a corrosive, reactive gas that is a serious pollutant at surface levels. Ozone irritates and damages the lungs and eyes, suppresses the immune system, and aggravates respiratory and heart diseases. It also damages many plants.

Teaching TIP
Ozone Depletion

It is estimated that every 1 percent drop in upper atmosphere ozone concentration will lead to a 6 percent rise in skin cancer. In addition, many important crops are easily damaged by UV radiation. But perhaps most seriously, certain marine algae (the producers in marine ecosystems) are very sensitive to UV levels. Reductions in algae populations would have a severe impact on marine ecosystems. Have students draw a **Graphic Organizer** similar to the one on page 408 that shows the impact on human health of releasing CFCs into the atmosphere.

PORTFOLIO

Teaching TIP
Warming Effects

If the global warming trend continues, many scientists think average global temperatures will increase at least 2°C (about 4°F) by 2050. This increase would probably be enough to trigger droughts in some inland regions as rainfall patterns change. It would also cause flooding in coastal regions as the polar icecaps melt. Powerful storms may become more frequent worldwide due to the increased energy in the lower atmosphere.

Visual Strategy
Figure 19-3

A greenhouse effect also exists on Venus but to a much greater extent. Early in their formation, Earth and Venus had similar concentrations of carbon dioxide, but Earth's oceans and plants eliminated much of the gas from the atmosphere. Venus lacks Earth's oceans and plants, so the dense carbon dioxide gas remains, causing surface temperatures to reach 460°C (865°F).

Release of CO_2 and Other Greenhouse Gases Has Led to Global Warming

When fossil fuels are burned, carbon atoms of the fuel molecules combine with oxygen atoms from air, yielding carbon dioxide, CO_2. The burning of fossil fuels for industrial purposes has released huge amounts of carbon dioxide into the atmosphere. Additional carbon dioxide has been released by the burning of vegetation to clear land for agriculture. As with CFCs, carbon dioxide was thought to be harmless. The atmosphere was thought to be able to absorb and disperse any amount of carbon dioxide. We now know this is not true.

The greenhouse effect

The chemical bonds in carbon dioxide molecules absorb solar energy, trapping heat within the atmosphere in the same way glass traps heat within a greenhouse. Because of this heat-trapping ability, gases such as carbon dioxide, methane, and nitrous oxide are known as greenhouse gases. The warming of the atmosphere that results from greenhouse gases is known as the **greenhouse effect,** as shown in **Figure 19-3.** Earth's moderate climate results, at least in part, from the insulating effect of greenhouse gases.

Global warming

Human activities are greatly increasing the concentrations of greenhouse gases in the atmosphere. Studies show that average global temperature is directly correlated with the increasing concentration of these gases in the atmosphere. Temperature records for the last 50 years seem to support these studies. The correlation of increasing

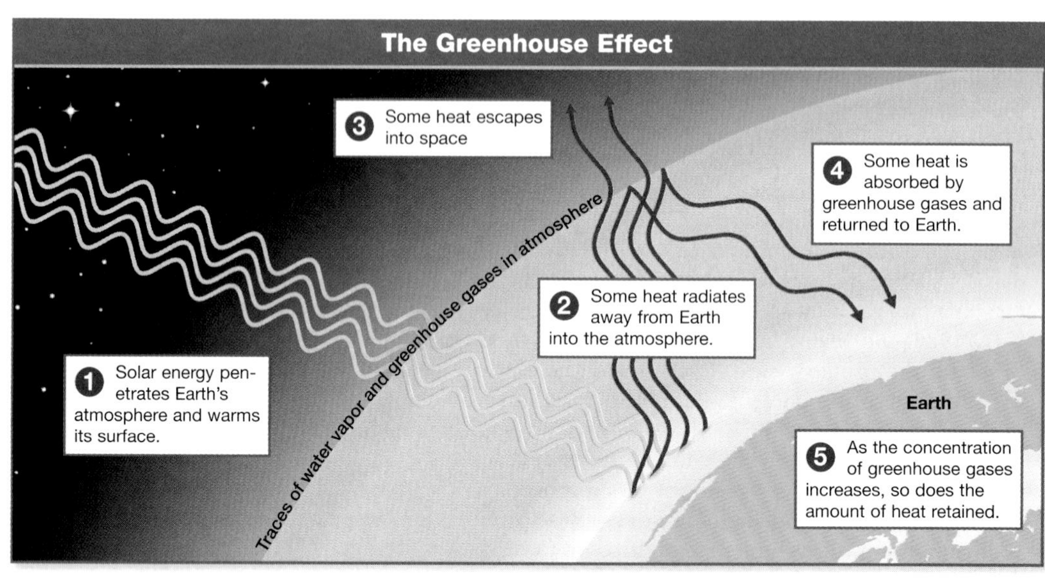

Figure 19-3 Earth's atmosphere traps heat. Just as the glass panes of a greenhouse retain heat, CO_2 and other greenhouse gases in the atmosphere capture heat radiated from Earth's surface.

Graphic Organizer

*Use this graphic organizer with **Teaching Tip: Ozone Depletion** on page 408.*

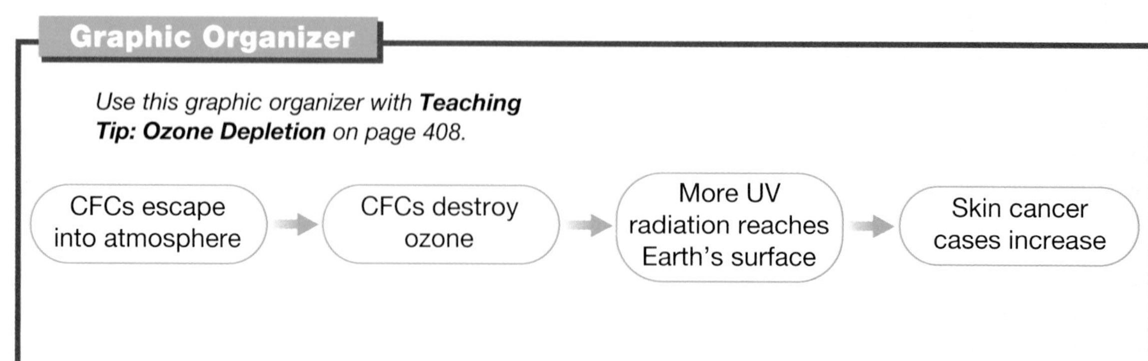

carbon dioxide levels with increasing temperatures is so close that most of the scientific community is now convinced that the two are related. **Figure 19-4** shows the concentration of atmospheric carbon dioxide and the average change in global temperature over the last 130 years. The concentration of carbon dioxide in the atmosphere has steadily increased, particularly since the 1950s. So, too, has the average global temperature. Major problems associated with global warming include rising sea levels and changes in patterns of rainfall. Sea levels may have already risen 5 centimeters from global warming. If the polar ice caps melt, sea levels would rise by more than 150 meters, flooding the entire Atlantic coast of North America inland for several hundred kilometers!

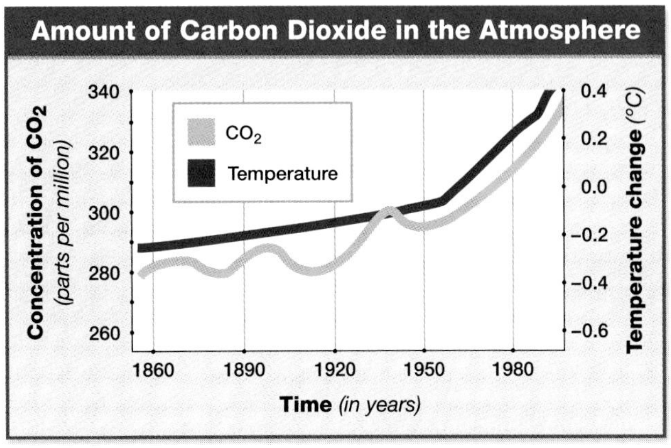

Amount of Carbon Dioxide in the Atmosphere

Figure 19-4 Global changes. Atmospheric CO_2 concentration and the average change in global temperature have risen together over the last 130 years.

How can you observe the greenhouse effect?

QUICK LAB

Time 20 minutes

Process Skills Observing, interpreting results, applying information, predicting outcomes

Teaching Strategies

- If MBL/CBL system is not available, have students record temperature manually every minute for 15 minutes.
- Use lamps as a heat source if direct sunlight is not available.

Analysis Answers

1. The jar's glass kept heat from escaping, raising the temperature inside the jar.
2. Like the glass of the jar, carbon dioxide retains heat energy from the sun that would otherwise dissipate.
3. The temperature on Earth would be much colder.

QUICK LAB

How can you observe the greenhouse effect?

You can use a quart jar to explore the greenhouse effect.

Materials

MBL or CBL system with appropriate software, 2 temperature probes, 1qt jar, lid with a 0.5 cm hole in the center, tape, heat source

Procedure

1. Set up an MBL/CBL system to collect data from two temperature probes at 6 second intervals for 150 data points.
2. Insert the end of one probe into the hole in the lid of a quart jar, and tape the probe in place. Place the other probe about 4 in. from the jar and at the same height as the first probe.
3. Place the jar about 30 cm from a heat-radiating source, and begin collecting data.
4. After 5 minutes, turn off (or remove) the heat source. Collect data for another 10 minutes.

Analysis

1. **Propose** an explanation for any differences between the two probes.
2. **SKILL Comparing Functions** How does carbon dioxide gas in the atmosphere act like the glass jar?
3. **SKILL Predicting Outcomes** How would the temperature on Earth be different if there were no carbon dioxide in the atmosphere?

Review SECTION 19-1

1. **Summarize** two harmful effects of acid rain.
2. **SKILL Predicting Outcomes** How will the depletion of the ozone layer affect human health?
3. **Describe** the greenhouse effect and its effect on the Earth's climate.

Critical Thinking

4. **SKILL Justifying Conclusions** Some of the pollution from power plants in the United States causes acid rain in Canada. Should the United States pay Canada for damage to its ecosystems?

Review SECTION 19-1 ANSWERS

1. Acid rain can weaken or kill trees and reduce the pH of lakes and streams, sometimes to levels deadly to most aquatic life.
2. Ozone depletion will result in an increased incidence of skin and retinal cancer and cataracts.
3. The greenhouse effect is a warming of Earth's climate as a result of the heat-retaining properties of certain atmospheric gases. The greenhouse effect causes Earth's climate to be much warmer, and its daily temperature fluctuations to be much smaller.
4. Answers will vary. Many feel that it is time that governments and businesses are assessed the total cost of doing business, including damage to the environment.

3 Assessment Options

Closure

Global Effects

Ask students to discuss ecosystem interdependence by hypothesizing how damage to one ecosystem can affect other ecosystems. *(For example, destruction of the rain forest in South America can destroy the wintering grounds of migratory birds, affecting North American ecosystems where the birds spend summers.)*

Review

Assign Review—Section 19-1.

Reteaching BASIC

Invite a guest speaker to address how your community might be affected by one of the following global changes: acid rain, ozone depletion, or global warming. Have students prepare questions ahead of time.

Section 19-2 Ecosystem Damage

1 Prepare

Scheduling

- **Block:** About 90 minutes will be needed to complete this section.
- **Traditional:** About two class periods will be needed to complete this section.
- **Planning:** See the Planning Guide on pp. 403A and 403B for lecture, classwork, and assignment options for Section 19-2. Lesson plans for Section 19-2 can be found in a lesson cycle format in ***Biology: Principles and Explorations* One-Stop Planner CD-ROM with Test Generator.**

Resource Link

Assign **Active Reading Worksheet Section 19-2** to help students comprehend the key concepts and visuals in this lesson.

2 Teach

Lesson Warm-up

Display pictures of different forms of environmental pollution, such as water pollution, air pollution, and solid waste. Encourage students to add their own examples to the display. Ask students to consider the causes and effects of each type of pollution. For example, sewage, agricultural runoff, and industrial waste cause water pollution. Such water may harm the plants, animals, and humans that use it.

Objectives

- **Describe** the effects of chemical pollutants on the environment. (pp. 410–411)
- **Identify** three nonreplaceable resources. (pp. 412–413)
- **Predict** the potential consequences of uncontrolled population growth. (pp. 414–415)
- **Contrast** population growth in developing countries with that in industrialized countries. (p. 415)

Key Terms

biological magnification
aquifer

Chemical Pollution Can Have Disastrous Results

Examples of global change such as ozone depletion or global warming occur on a large scale. But what about serious environmental problems that occur in our own backyard? For example, one important urban environmental problem is chemical pollution. Until recently, people assumed that the environment can absorb any amount of pollution. Lake Erie and other large lakes became polluted because of the assumption that they could absorb unlimited amounts of industrial chemicals. Because of this incorrect assumption, pollution has often risen to a serious level.

In a highly publicized example of pollution, an oil tanker named the *Exxon Valdez* ran aground off the coast of Alaska in 1989. Oil from the tanker heavily polluted 1,600 km (1,000 mi) of coastline and killed thousands of marine animals, as shown in **Figure 19-5.** If the tanker had been loaded no higher than the waterline, little oil would have been lost. But it was loaded far higher than that, and the weight of the oil above the waterline forced 42 million liters (11 million gallons) of oil out the hole in the ship's hull. Despite costly cleanup efforts, damage to local wildlife was extensive. This dramatic example is not an isolated occurrence. Smaller oil spills and leaks that receive little or no publicity account for more than 90 percent of all pollution from oil seepage.

Many of the most disastrous incidents of pollution involve industrial chemicals that are toxic or carcinogenic (cancer-causing). Until recently, there has been relatively little regulation of the manufacture, transportation, storage, and destruction of such chemicals. A particularly clear example of this problem occurred in Basel, Switzerland, in 1986. Firefighters putting out a warehouse fire accidentally washed 27,000 kg (30 tons) of mercury and pesticides that were stored in the warehouse into the Rhine River. These poisons flowed down the Rhine, through Germany and the Netherlands, and into the North Sea, killing fish and other aquatic animals and plants. Today the river is recovering, but its species diversity remains far lower than it was before the disaster.

Figure 19-5 Oil spill victim. This common scoter was one of thousands of animals injured or killed when the *Exxon Valdez* ran aground off the Alaskan coast in 1989.

Agricultural chemicals

In many countries, modern agriculture introduces large amounts of chemicals into the global ecosystem. These chemicals include pesticides, herbicides, and fertilizers. Industrialized countries, like the United

Integrating Different Learning Styles

Logical/Mathematical	**PE**	Study Tip, p. 415; Performance Zone items 9, 11, and 20, pp. 422–423
Visual/Spatial	**PE**	Performance Zone items 14, 15, and 16, p. 423
	ATE	Lesson Warm-up, p. 410, Demonstration, p. 413
Verbal/Linguistic	**PE**	Study Tip, p. 413; Performance Zone items 2, 10, and 24, pp. 422–423
	ATE	Active Reading, Overcoming Misconceptions, p. 411; Teaching Tip, p. 414; Internet Connect, p. 414

States, now attempt to carefully monitor side effects of these chemicals. Unfortunately, large quantities of many toxic chemicals that are no longer manufactured still circulate in the ecosystem.

For example, molecules of chlorinated hydrocarbons—a class of compounds that includes DDT, chlordane, lindane, and dieldrin—break down slowly in the environment. They also accumulate in the fatty tissue of animals. As these molecules pass up through the trophic levels of the food chain, they become increasingly concentrated. This process is called **biological magnification,** illustrated in **Figure 19-6.** The presence of DDT in birds causes thin, fragile eggshells, most of which break during incubation. Because of DDT use, many predatory birds in the United States and elsewhere failed to reproduce, and their numbers dwindled. In 1972, the use of DDT was severely restricted in the United States, and threatened bird populations slowly began to increase. However, chlorinated hydrocarbons are still manufactured in the United States and exported to other countries, where their use continues.

In order for us to meet the needs of an increasingly crowded world, the use of chemicals is necessary. We must learn to use them as intelligently as possible. Doing so will enable us to protect the productive capacity of the Earth. Failure is not a rational option.

internetconnect
SCI LINKS NSTA
TOPIC: Chemical pollution
GO TO: www.scilinks.org
KEYWORD: HX411

Figure 19-6 Biological magnification of DDT

Because DDT accumulates in fatty tissue, DDT concentrations (in parts per million, ppm) increase as this chemical moves up the food chain.

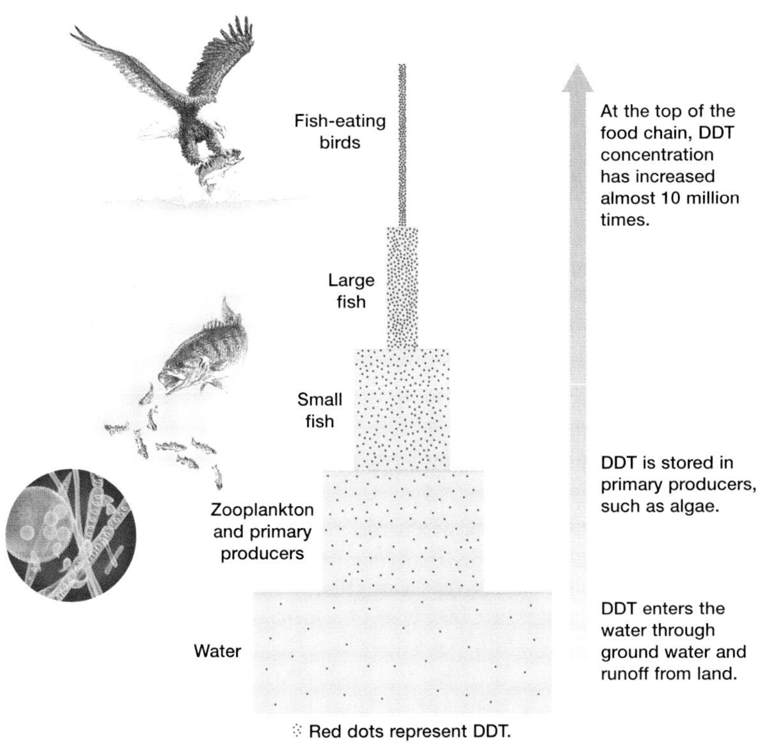

Fish-eating birds
Large fish
Small fish
Zooplankton and primary producers
Water
At the top of the food chain, DDT concentration has increased almost 10 million times.
DDT is stored in primary producers, such as algae.
DDT enters the water through ground water and runoff from land.
Red dots represent DDT.

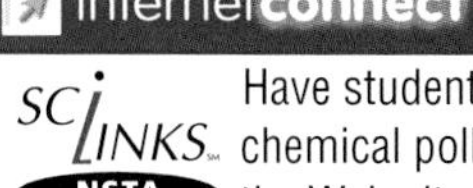

Section 19-2

internetconnect
SCI LINKS NSTA Have students research chemical pollution using the Web site in the Internet Connect box on page 411. Have students write a report summarizing their findings. PORTFOLIO

Active Reading
Technique: Reading Effectively

Have students write down the main issues of this section and briefly describe the impact or importance of each issue as they read. Tell them to look for section headings to help identify the main points.

Teaching TIP
Freshwater Ecosystems

When rainfall washes nitrogen-containing fertilizers or plant and animal wastes into a body of fresh water, the ecosystem can be seriously disrupted. Large amounts of nitrates and phosphates stimulate explosive algae growth. Only the topmost algal layer receives enough light for photosynthesis, while the lower layers die and are decomposed by bacteria. These bacteria use up large amounts of dissolved oxygen, which slowly suffocates fish and other lake organisms.

Overcoming Misconceptions
Agriculture and Water Pollution

Most people think industry is the primary source of water pollution and would be surprised to learn that agriculture has the biggest impact. The runoff of silt, pesticides, and fertilizers from farmland is difficult to control because it comes from so many sources rather than a single point, such as a pipe. That is why such pollution is known as *nonpoint-source* pollution.

Chapter 19

Teaching TIP ADVANCED

The Endangered Species Act

In the United States, the Endangered Species Act of 1973 is the most powerful legislation protecting rare and endangered species. Have students write a report that summarizes the provisions of this law, its history, the recent controversies about its application, and a discussion of what species it protects in your area. PORTFOLIO

Teaching TIP

Balancing Development and Conservation

One way to preserve a species is to set aside habitat as a park or reserve. However, "locking away" large tracts of land may not be practical, especially in developing countries with large, rapidly growing populations that depend on the land for subsistence. A better solution may be extractive reserves, which preserve habitat while allowing the sustainable harvest of products such as fruits, seeds, and rubber. Local populations gain a sustainable income that depends on keeping the habitat intact for all species.

Teaching TIP

How Many Species Are Becoming Extinct?

Inform students that the precise rate of extinction today is not known because no one knows how many species exist. About 1.6 million species have been described and named, but millions more certainly await description. Estimates for the total number of species on Earth range from 5 million to 100 million.

biobit

What is the biosphere worth?

In 1997, a group of economists, ecologists, and political scientists calculated that the Earth provides "ecosystem services"—including water supply, climate regulation, soil formation, food production and flood control—worth about $33 trillion per year.

Loss of Resources Is a Serious Environmental Problem

Among the many ways that ecosystems are being damaged, one problem stands out as potentially more serious than the rest—the consumption or destruction of resources that we cannot replace. Though a polluted stream can be cleaned up, no one can restore an extinct species. Three kinds of nonreplaceable resources are being consumed or destroyed at alarming rates: species of living things, topsoil, and ground water.

Extinction of species

Over the last 50 years, about half of the world's tropical rain forests have been burned to make pasture and farmland or cut for timber. About 150,000 km^2 (58,000 mi^2, about the area of the state of Georgia) will be destroyed this year. At this rate, all of the rain forests of the world will be gone within your lifetime. As the rain forests disappear, so do their inhabitants. Worldwide, the loss of habitat is the largest single cause of the loss of species. It is estimated that at least one-fifth of the world's species of animals and plants—about 1 million species—will become extinct during the next 50 years. An extinction of this size has not occured in at least 65 million years, since the end of the age of the dinosaurs.

As species disappear, so do our chances to learn about them and their possible benefits. **Figure 19-7** shows two species that have great potential benefits to humans, the rosy periwinkle and *Zea diploperennis*.

Figure 19-7 Beneficial species. Two potent anticancer drugs have been isolated from the leaves of the rosy periwinkle. The desirable genes of *Zea diploperennis*, a wild relative of corn, could be incorporated into domestic corn.

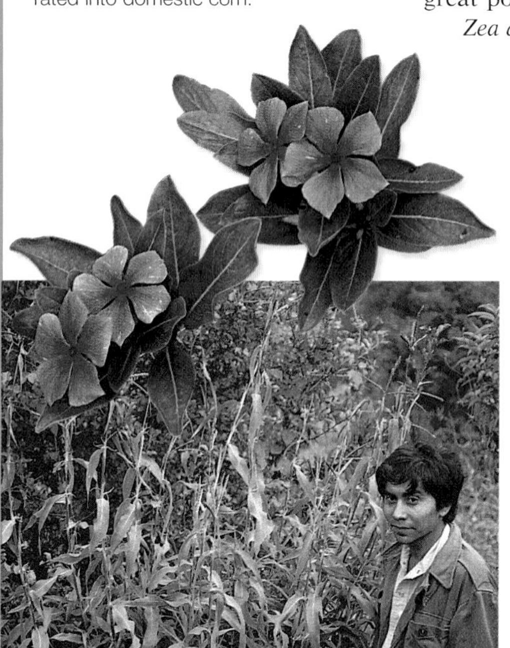

The rosy periwinkle grows in Madagascar, a country already devastated by deforestation. Two anticancer drugs have been isolated from its leaves. With these drugs, a child with leukemia has a 95 percent chance of survival. Without them, the child would have only a 20 percent chance to live.

Zea diploperennis, a perennial relative of wild corn, was discovered in 1977. Most corn *(Zea mays)* is so genetically uniform that it is difficult to produce improved strains by crossing different varieties. But *Zea diploperennis* and *Zea mays* can be crossed to produce fertile hybrids. This is important because *Zea diploperennis* is resistant to five of the seven main types of viral diseases that damage corn. Disease-resistant, fertile hybrids of the two species could revolutionize the farming industry, but this great opportunity was almost lost. *Zea diploperennis* existed in only one mountain field in Mexico and was nearly destroyed by farmers clearing land in the area. Only the chance arrival of a botanist saved it.

did you know?

The first humans to reach North America encountered an assortment of large mammalian species including woolly mammoths, sabertooth cats, camels, and large ground sloths.

Within 1,000 years nearly all such species had disappeared. Did human hunters exterminate the large North American mammals? Yes, say some scientists. According to Harvard University's E. O. Wilson, "Mankind soon disposed of the large, the slow, and the tasty." Critics of this view point out that these extinctions coincide with large-scale climatic changes associated with the retreat of the glaciers following the last ice age. This debate is hard to resolve because both sides base their arguments on the same circumstantial evidence.

Loss of topsoil

The United States is one of the most productive agricultural countries on Earth, largely because of its fertile soils. These soils have accumulated over tens of thousands of years. The Midwestern farm belt sits astride what was once a great prairie. The topsoil of that ecosystem accumulated slowly as the remains of countless animals and plants decayed. By the time humans came to plow the prairie, the topsoil was more than a meter thick.

This rich topsoil cannot be replaced, and it is being lost at a rate of several centimeters each decade. Turning over the soil to eliminate weeds, allowing animals to overgraze ranges and pastures, and practicing poor land management all permit wind and rain to remove more and more of the topsoil. Since 1950, the United States has lost one-fourth of its topsoil, primarily because of human activity.

Ground-water pollution and depletion

A third resource that we cannot replace is ground water. Much ground water is stored within porous rock reservoirs called **aquifers** *(AHK wuh furz),* as shown in **Figure 19-8.** Water seeps into aquifers too slowly to replace the large amount of water now being withdrawn. In most areas of the United States, there is relatively little effort to conserve ground water. Consequently, a very large portion of it is wasted on watering lawns, on washing cars, and through leaky and inefficient faucets and toilets. A great deal more ground water is being polluted by irresponsible disposal of chemical wastes. Once pollution enters the ground water, there is no effective way to remove it.

Figure 19-8 Aquifer

Large amounts of ground water are being removed from many aquifers far faster than natural processes can replenish it.

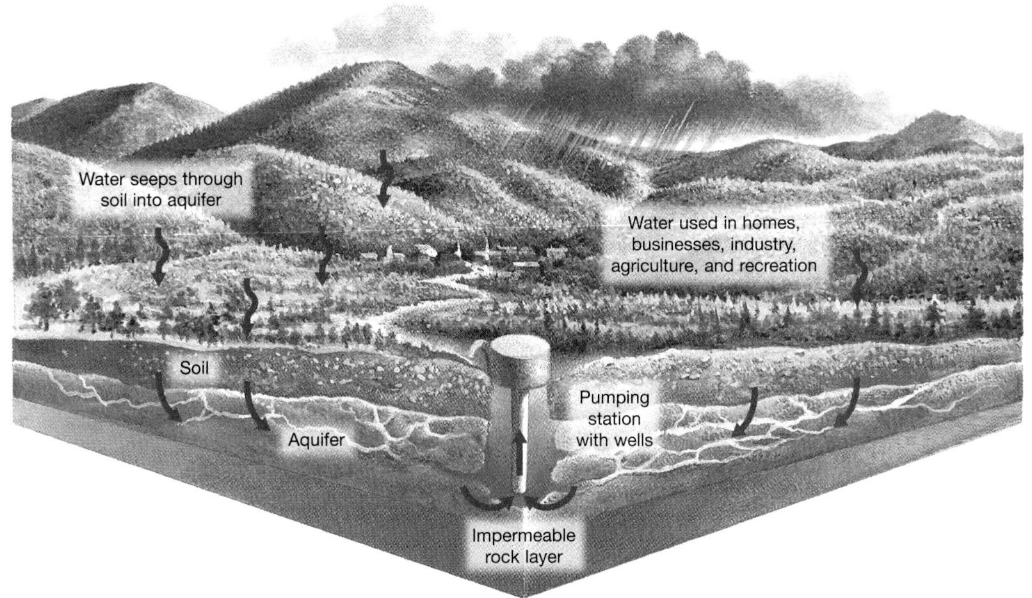

Study TIP

Organizing Information

Make a table to organize information about *loss of resources*. Across the top, write the headings *Extinction of species, Loss of topsoil,* and *Ground-water pollution and depletion.* Along the sides, write *Causes, Effects,* and *Possible solutions.* Add information to the table as you read pages 412 and 413.

TOPIC: Aquifers
GO TO: www.scilinks.org
KEYWORD: HX413

Teaching TIP

Sustainability

The idea of sustainability is key to preventing the loss of our natural resources. When a resource is used sustainably, it is consumed only as fast as it is naturally replenished, and therefore it remains abundant. Some resources, such as fossil fuels, groundwater, and topsoil, are replenished on a time scale of thousands or millions of years. The creation of these resources proceeds far too slowly to keep up with today's rapid rate of consumption. Living resources, such as trees, are easier to use sustainably because they replace themselves more quickly.

Teaching TIP BASIC

Topsoil

Point out that soil typically has three layers. The top layer (topsoil) contains the most organic matter, or humus. The second layer (subsoil) consists of inorganic particles and minerals that have washed down from the topsoil. The bottom layer consists of loose rock that extends down to the bedrock.

did you know?

Ground-water contamination in urban areas is often caused by leaking underground storage tanks.

Underground storage tanks hold liquid fuels and industrial chemicals below ground. More than one quarter million tanks at service stations in the United States are thought to be leaking. Disposing of the tanks and contaminated soil is often very costly and difficult.

Teaching TIP ADVANCED

Humans and Carrying Capacity

Ask students why the human population has been able to grow beyond what the natural environment could support. *(Humans have more control over our environment than other animals and have been able to produce more food and habitat as it was needed. In a sense, we continue to increase the carrying capacity of the environment, although it is unclear how long we can continue in this manner.)*

Teaching TIP

The Cairo Population Conference

Have students use the library to research the decisions of the Cairo Population Conference, which took place in late 1994. Students should write a report that describes the nature of the conference debate and summarizes the proposals agreed upon. PORTFOLIO

internetconnect

Have students research population growth using the Web site in the Internet Connect box on page 414. Have students write a report summarizing their findings. PORTFOLIO

Figure 19-9
Population growth

How many? The world's population reached 6 billion in late 1999.

Human Population Is a Key Environmental Concern

If all of the problems associated with pollution mentioned earlier in this chapter were solved, would the future of our environment be free from danger? Not necessarily. Scientists recognize that we would still need to address a more fundamental problem: the rapid growth of the human population, as shown in **Table 19-1.** Humans first reached North America at least 12,000 years ago, by crossing the land bridge between Siberia and Alaska. Humans then spread throughout North America and South America. Ten thousand years ago the continental ice sheets that covered northern Europe and North America receded, and agriculture soon developed. There were only about 5 million people on Earth then. As agriculture produced more dependable sources of food, the human population began to grow. By about 2,000 years ago, there were an estimated 130 million people on Earth. By 1650, the world's population had reached 500 million.

Since then, the average global birthrate has remained near 30 births per 1,000 people per year. However, with the development of technology to ensure better sanitation and improved medical care, the death rate has fallen steadily. In 1994, the estimated death rate was about 9 deaths per 1,000 people per year. The difference between the current annual birthrate (now estimated to be 26 births per 1,000 people) and death rate results in an annual worldwide increase in the human population of approximately 1.7 percent. This number may seem small, but don't be deceived. The world's population will double in just over 40 years if it continues to grow at this rate!

internetconnect

TOPIC: Population growth
GO TO: www.scilinks.org
KEYWORD: HX414

Table 19-1 Number of Years to Add 1 Billion People

Human population	Year	Years to add
1 billion	1804	All of human history
2 billion	1927	123
3 billion	1960	33
4 billion	1975	15
5 billion	1987	12
6 billion	1999	12
Projected		
7 billion	2012	13
8 billion	2026	14

MULTICULTURAL PERSPECTIVE

India's High Birthrate

In India, children traditionally provide security for their parents in old age. Statistical studies have shown that it is necessary for an Indian mother to have five children if she and her husband are to be 95 percent certain that at least one son will survive the father's 65th birthday.

Worldwide rates of growth

The world's population exceeded 6 billion in October 1999, and the annual increase is now about 94 million people. About 260,000 people are added to the world population each day, or about 180 every minute. Population growth is fastest in the developing countries of Asia, Africa, and Latin America. It is slowest in the industrialized countries of North America, Europe, Japan, New Zealand, and Australia. In industrialized countries like the United States, about one fifth of the population is under 15 years of age. In developing countries like Nigeria, the percentage is typically twice as high, leading to explosive population growth in coming decades. **Figure 19-10** shows the population growth rates of developing and developed countries. The population growth rate in the United States is only 0.8 percent, less than half of the global rate. Most European countries are growing even more slowly, and the populations of Germany and Russia are actually declining. By contrast, as of 1996, Nigeria's population was increasing by about 3.05 percent per year.

Though the global rate of population growth has been declining, experts project that the world's population will increase to 8.5 billion by the year 2025. It is estimated by the United Nations that the population may stabilize by the end of the next century at about 10 billion to 15 billion people. Population growth tends to be the highest in countries that can least afford it. Already limited resources are strained further, and natural resources—ground water, land for farming, forests—are ever more quickly depleted or polluted.

No one knows whether the Earth can support six billion people indefinitely, much less the far larger population that lies in our future. Building a sustainable world is the most important task facing humanity's future. The quality of life available to your children in the new century will depend to a large extent on our success.

Study TIP

Interpreting Graphics

To better understand the rate of population growth, use Table 19-1 and Figure 19-10 to estimate how many people have been born since the year of your birth. Compare this figure with the projected world population for the year when you will be twice your current age.

Figure 19-10 World population growth patterns. Most of the world population increase since 1950 has been in the developing countries.

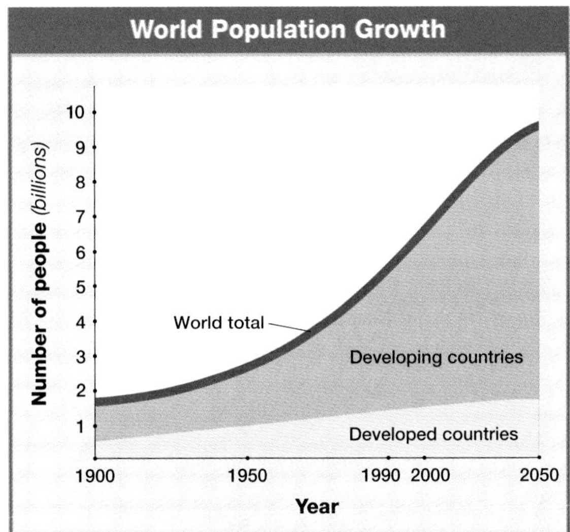

Review SECTION 19-2

1. **Describe** why chlorinated hydrocarbons are an environmental threat even though their use has been banned in the United States.

2. **Summarize** why supplies of soil and ground water are dwindling even though these resources are replenished by natural processes.

3. **SKILL Inferring Relationships** Describe two instances in which technology has caused the growth rate of the human population to increase.

Critical Thinking

4. **SKILL Evaluating Viewpoints** A classmate claims that the growth of the human population will not affect populations of other species. Evaluate your classmate's viewpoint.

Section 19-2

Resource Link

Assign **Problem Solving: Population Size** from the Holt BioSources Teaching Resources CD-ROM. PORTFOLIO

3 Assessment Options

Closure

Impacts at Home

Have students hypothesize about which environmental concern will have the greatest impact on their community in the next decade. Ask them to justify their choice. *(Answers will vary. For example, if the community is located near a beach, then the rise in skin cancer caused by the loss of ozone could result in fewer people at the beach, and the community could suffer financially.)*

Review

Assign Review—Section 19-2.

Reteaching — BASIC

Have students prepare an outline for a movie script about a real ecological disaster. Students should include environmental conditions as well as factual consequences of the disaster. *(Accept any well-reasoned scenarios that are based on fact.)* PORTFOLIO

Review SECTION 19-2 ANSWERS

1. Despite being banned in the United States, chlorinated hydrocarbons continue to be used elsewhere in the world.
2. Soil and ground-water supplies are dwindling because they are being used much faster than they are being replenished.
3. Technological advances in farming have increased the food supply dramatically, allowing an increase in population size. Medical advancements have also increased the population due to reduced death rates.
4. The growing human population will inevitably use more land, which other organisms use for habitat. The loss of habitat will necessarily affect the number of organisms (population size) that a given area can support.

Chapter 19

1 Prepare

Scheduling

- **Block:** About 90 minutes will be needed to complete this section.
- **Traditional:** About two class periods will be needed to complete this section.
- **Planning:** See the Planning Guide on pp. 403A and 403B for lecture, classwork, and assignment options for Section 19-3. Lesson plans for Section 19-3 can be found in a lesson cycle format in ***Biology: Principles and Explorations*** **One-Stop Planner CD-ROM with Test Generator.**

Resource Link

Assign **Active Reading Worksheet Section 19-3** to help students comprehend the key concepts and visuals in this lesson.

2 Teach

Lesson Warm-up

Have students discuss their answers to this question: Could an organism exist that does not have an impact on its environment in any way? *(The answer is no. All organisms require substances from their environment and change the abundance of these substances by using them. Likewise, all organisms produce wastes. Students should see that humans are not alone in exploiting the environment. However, no other organism disrupts ecosystems on the scale that humans do.)*

Section 19-3 Solving Environmental Problems

Objectives

- **Describe** two effective approaches that have been taken to reduce pollution in the United States and abroad. (pp. 416–417)
- **Evaluate** the five major steps necessary to solve environmental problems. (p. 418)
- **Determine** how individuals can take personal action to help solve environmental problems. (pp. 419–420)

Key Terms

None

Reducing Pollution Will Take a Worldwide Effort

As you have seen, environmental problems affect all inhabitants of an ecosystem without regard to state or national boundaries. As human activities continue to place severe stresses on ecosystems, worldwide attention must be focused on solving these problems.

One of the most encouraging developments of the early 1990s was the global increase in efforts to reduce pollution. International agreements to stop CFC production are one example. And the release of many dangerous industrial and agricultural chemicals—notably the insecticide DDT and the carcinogens asbestos and dioxin—has been restricted in the United States. A great deal of progress has also been made in reducing air pollution, as shown in **Figure 19-11.** Emissions of sulfur dioxide, carbon monoxide, and soot—three pollutants produced by the burning of coal—have been cut by more than 30 percent in 10 years. The number of secondary sewage treatment facilities, which remove chemicals as well as bacteria from sewage, has increased 72 percent in the last 10 years. However encouraging, this progress represents only a beginning. A serious attempt to address the overall problem of pollution requires more fundamental changes in the way the economy of our society operates.

Figure 19-11 Reducing air pollution

Devices called scrubbers remove many pollutants from factory emissions.

Smokestack without scrubber in New Mexico

Smokestack with scrubber in New Mexico

Integrating Different Learning Styles

Logical/Mathematical	**PE**	Performance Zone item 17, p. 423
Visual/Spatial	**ATE**	Demonstrations, p. 417; Visual Strategy, p. 420
Interpersonal	**ATE**	Lesson Warm-up, p. 416; Reteaching, p. 420
Intrapersonal	**ATE**	Teaching Tip, p. 419
Verbal/Linguistic	**PE**	Performance Zone items 18, 22, and 23, p. 423
	ATE	Internet Connect, p. 417; Resource Link, p. 419; Making Biology Relevant, p. 417

Two effective approaches have been taken to reduce pollution in the United States. The first approach has been to pass laws forbidding it. In the last 20 years, laws have begun to significantly slow the spread of pollution. These laws impose strict standards for what can be released into the environment. For example, all cars are required to have catalytic converters to reduce emissions. Similarly, the Clean Air Act of 1990 requires that power plants install "scrubbers" on their smokestacks to reduce sulfur emissions. Converters and scrubbers make cars and energy more expensive. The effect is that the consumer pays to avoid polluting the environment, lowering consumption to more appropriate levels.

A second effective approach to reducing pollution has been to make it more expensive by placing a tax on it. The gasoline tax is a good example of such a tax. To be fully effective, however, a tax must be high enough to reflect the actual cost of the pollution. The tax is in effect an artificial price hike imposed by the government to lower consumption. By adjusting the tax, the government attempts to balance the conflicting demands of environmental safety and economic growth. Such taxes, often imposed on industry in the form of "pollution permits," are becoming increasingly common.

Figure 19-12 Carpooling. Some cities reserve certain lanes for carpool use during peak travel times.

Exploring Further

Why Pollution Is Profitable

To find economic solutions to environmental problems, it is first necessary to understand that the economy of much of the industrialized world is based on a system of supply and demand. As something gets scarce, its price increases. This increased profit on an item in short supply acts as an incentive for the production of more of the item. If too much of the item is available, the price falls. Because it is less profitable to produce the item, less of it is made.

A price too low?
This system works very well and is responsible for the economic strength of our nation. But it has one great weakness. If demand is set by price, then it is very important that all of the production costs be included in the price of an item. If a person selling an item were able to pass off part of its production cost to a third person, the seller would be able to set a lower price and sell more of the item. Stimulated by the lower price, the buyer would purchase more of the item. This pricing system has driven industry's pollution of the environment over the last century.

The true cost of pollution
The true costs of energy and manufactured goods are composed of direct and indirect production

costs. Direct costs include materials and wages. Indirect costs include pollution and other damage to the environment. For example, the true costs of fossil fuels include the indirect costs of reduced harvests of fish and shellfish due to oil spills. But the indirect costs are *not* included in the price that the consumer pays for fossil fuels. As a result, far more is consumed than if these costs were included. The indirect costs do not disappear because we ignore them. They are passed on to future generations, who must pay the bill in terms of damage to their own health and to the ecosystems on which they depend.

did you know?

The first auction of pollution permits in the United States was held in 1993.

Some 150,000 permits were sold for over $21 million. The top buyer was California Power and Light, which spent $11.5 million for permits. Some public interest groups also bid in an effort to keep the pollution permits out of the hands of industry. Their efforts amounted to less than 1 percent of the bidding.

Section 19-3

Demonstration
Catalytic Converter

Show the class a catalytic converter or a cutaway diagram of one. Tell students that a few grams of platinum, palladium, or rhodium are embedded as pellets in the catalytic converter. The pellets act as catalysts in reactions that break down toxic emissions that would otherwise pass through the exhaust system and into the atmosphere.

Exploring Further

Why Pollution Is Profitable

Teaching Strategies
- Inform students that, if forced to pay for indirect costs, most companies would pass additional costs on to consumers to avoid losing profits.
- Indirect costs may be hard to identify. For example, how many asthma and respiratory ailments are caused by or aggravated by pollution?

Discussion
Ask students to identify some of the indirect costs associated with oil production and with agriculture. *(Oil production costs include loss of habitat from drilling, air pollution, water pollution from spills, global warming from combustion, and smog production. Agricultural costs include water pollution, soil loss, aquifer depletion, and groundwater contamination.)*
- Ask students if they would be willing to pay higher prices for food, gas, and clothes to pay for some of the indirect costs of these goods.

REAL LIFE

Answers

Pros: Recycling conserves natural resources, saves land from mining and logging, and reduces solid waste production. Cons: Recycling can be expensive and it uses energy resources.

CAREER

Toxicologist

Toxicologists protect the public health by studying and identifying toxic substances in the environment. Usually employed by state or federal agencies, toxicologists provide vital information and education to health departments, physicians, legislators, and citizens. Most toxicologists receive a degree in biology, chemistry, or environmental science and then pursue advanced degrees in their chosen field.

Making Biology Relevant

Technology

Have students use library resources to locate information concerning how an individual, group, or community solved an environmental problem using knowledge of biology. Ask each student to make an oral presentation to the class summarizing what he or she learned.

REAL LIFE

Better the second time around? Although many communities have recycling programs, some researchers argue that recycling doesn't pay off.

Evaluating Viewpoints
Find out the main arguments for and against recycling, and decide what kinds of waste-management programs would work in your community.

Environmental Problems Can Be Solved

It is easy to get discouraged when considering the world's many serious environmental problems. But do not lose track of the conclusion that emerges from our examination of these environmental problems—each of the world's many problems is solvable. If one looks at how environmental problems have been overcome, a clear pattern emerges.

Five steps to success

Viewed simply, there are five components to successfully solving any environmental problem.

1. **Assessment.** The first stage is scientific analysis of the problem, the gathering of information about what is happening. To construct a model of the ecosystem, data must be collected and experiments must be performed. A model makes it possible to describe how the ecosystem is responding to the situation. It is then used to make predictions about the future course of the ecosystem.
2. **Risk analysis.** Using the information obtained by scientific analysis, it is possible to predict the consequences of environmental intervention, that is, what can be expected to happen if a particular course of action is followed. It is also necessary to evaluate any adverse effects that a plan of action might cause.
3. **Public education.** When a clear choice can be made among alternative courses of action, the public must be informed. This involves explaining the problem in understandable terms, presenting the alternative actions available, and explaining the probable costs and results of the different choices.
4. **Political action.** The public, through its elected officials, selects and implements a course of action. Individuals can be influential at this stage by exercising their right to vote and by contacting their elected officials.
5. **Follow-through.** The results of any action should be carefully monitored to see if the environmental problem is being solved.

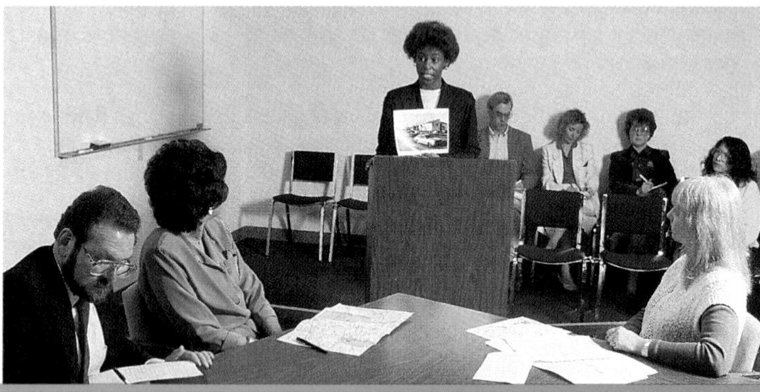

Figure 19-13 Public participation. Public discussion of environmental problems helps citizens to evaluate alternative courses of action.

did you know?

Recent studies have extended the history of large-scale pollution back several thousand years.

Sediment samples from the beds of lakes in Sweden show that lead concentrations were well above natural levels beginning about 2,600 years ago. Ice cores removed from Greenland's icecap also have higher-than-normal lead levels starting at about the same time. What caused this ancient pollution? The pollution is most likely the fallout from early metal smelting operations in what is now Spain.

Two success stories

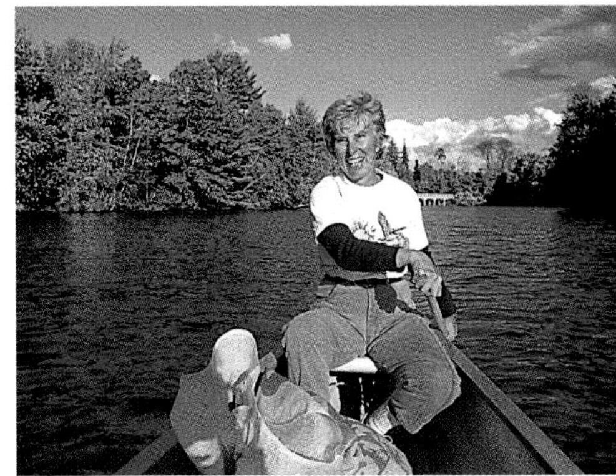

Figure 19-14 **Marion Stoddart.** The recovery of the Nashua River shows that polluted environments can be restored when committed individuals, like Marion Stoddart, work to bring about a change.

The development of appropriate solutions to the world's environmental problems often rests partly on the shoulders of politicians, economists, bankers, and engineers. However, it is important not to lose sight of the key role often played by informed individuals. Two examples serve to illustrate the point.

The Nashua River Running through the heart of New England, the Nashua River was severely polluted by mills established in Massachusetts in the early 1900s. When Marion Stoddart, shown in **Figure 19-14,** moved to a town along the river in 1962, she was appalled. Stoddart organized the Nashua River Cleanup Committee. The committee presented bottles of dirty river water to politicians, spoke at town meetings, recruited business people to help finance a waste treatment plant, and began to clean garbage from the Nashua River's banks. This citizen's campaign contributed to the passage of the Massachusetts Clean Water Act of 1966. Industrial dumping into the river is now banned, and the river has largely recovered.

Lake Washington Following World War II, this very large freshwater lake east of Seattle became surrounded by a ring of 10 suburbs, each with its own municipal sewage treatment plant. Between 1940 and 1953, these 10 municipal sewage plants discharged their treated outflow into the lake. Safe enough to drink, the outflow was believed to be harmless. Starting in the early 1940s, the combined daily discharge in the lake was 75 million liters (20 million gallons).

In 1954, an ecology professor at the University of Washington in Seattle, W. T. Edmondson, noted that his research students were reporting blue-green algae growing in the lake. Such algae require an abundance of the nutrients nitrogen and phosphorus to grow. Because deep freshwater lakes like Lake Washington usually lack these nutrients, the presence of the algae was surprising. They found that phosphates and nitrates in the sewage had been fertilizing the lake! Edmondson was alarmed and began a campaign in 1956 to educate public officials about the danger: Bacteria decomposing the dead algae would soon deplete the lake's oxygen. This would kill all life in the lake, and it would never recover. After 5 years, as a direct result of his efforts, joint municipal taxes financed the cleanup of Lake Washington with a massive trunk sewer that rings the lake and carries treated discharge far out into Puget Sound. Today, through the efforts of many people, the lake is healthy, its waters clean and blue, as shown in **Figure 19-15.**

Figure 19-15 **Lake Washington.** Once choked with algae that were nourished by the outflow from sewage treatment plants, Lake Washington is a healthy lake today.

Section 19-3

Teaching TIP — BASIC

Making a Commitment

Have students consult books and magazines to identify some ways they can personally help the environment. Then have each student carefully consider and select four actions he or she is not already doing and carry out these actions for 2 weeks. At the end of this period, the students should write a brief and honest appraisal of their efforts, addressing how consistently they performed their actions, why some actions were easier than others, and any unexpected benefits or impediments they encountered. PORTFOLIO

Resource Link

- Assign **Supplemental Reading: Silent Spring—Rachel Carson** from the Holt BioSources Teaching Resources CD-ROM. PORTFOLIO

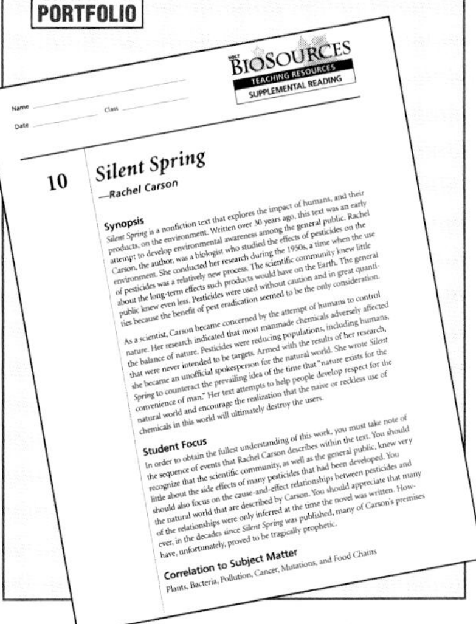

HOLT BIOSOURCES
TEACHING RESOURCES
SUPPLEMENTAL READING

Name ______ Class ______
Date ______

10 *Silent Spring*
—Rachel Carson

Synopsis
Silent Spring is a nonfiction text that explores the impact of humans, and their products, on the environment. Written over 30 years ago, this text was an early attempt to develop environmental awareness among the general public. Rachel Carson, the author, was a biologist who studied the effects of pesticides on the environment. She conducted her research during the 1950s, a time when the use of pesticides was a relatively new process. The scientific community knew little about the long-term effects such products would have on the Earth. The general public knew even less. Pesticides were used without caution and in great quantities because the benefit of pest eradication seemed to be the only consideration.

As a scientist, Carson became concerned by the attempt of humans to control nature. Her research indicated that most manmade chemicals adversely affected the balance of nature. Pesticides were reducing populations, including humans, that were never intended to be targets. Armed with the results of her research, she became an unofficial spokesperson for the natural world. She wrote *Silent Spring* to counteract the prevailing idea of the time that "nature exists for the convenience of man." Her text attempts to help people develop respect for the natural world and encourage the realization that the naive or reckless use of chemicals in this world will ultimately destroy the users.

Student Focus
In order to obtain the fullest understanding of this work, you must take note of the sequence of events that Rachel Carson describes within the text. You should recognize that the scientific community, as well as the general public, knew very little about the side effects of many pesticides that had been developed. You should also focus on the cause-and-effect relationships between pesticides and the natural world that are described by Carson. You should appreciate that many of the relationships were only inferred at the time the novel was written. However, in the decades since *Silent Spring* was published, many of Carson's premises have, unfortunately, proved to be tragically prophetic.

Correlation to Subject Matter
Plants, Bacteria, Pollution, Cancer, Mutations, and Food Chains

Chapter 19

Visual Strategy

Figures 19-16 and 19-17

Ask students if they participate in any of the activities illustrated on this page. Have them brainstorm other actions they could take that would help the environment. *(Possible suggestions include building a compost pile; using reusable plates, cups, and napkins; buying recycled products; and drying clothes on a line.)*

3 Assessment Options

Closure — ADVANCED

Pollution Permits

Have students research the practice of selling pollution permits and write a report about what they learn. In their reports, students should present the arguments for and against this practice and evaluate its effectiveness as a method of pollution control. PORTFOLIO

Review

Assign Review—Section 19-3.

Reteaching — BASIC

Have the class organize and implement a schoolwide project that addresses one of the environmental concerns discussed in this chapter. Tell students that their project must be organized according the five-step plan outlined on page 418.

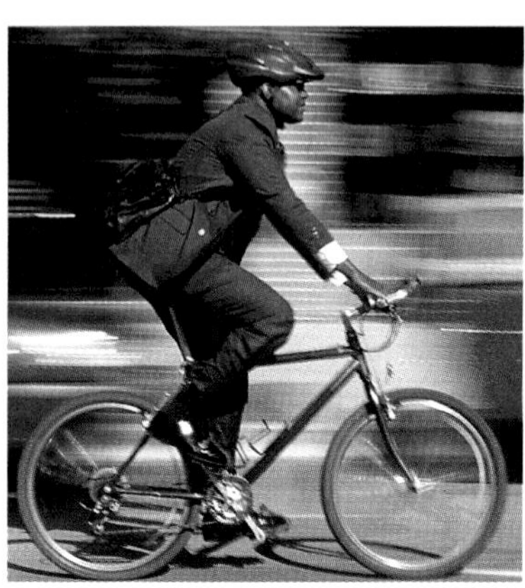

Figure 19-16 Conserving energy. By riding a bicycle instead of traveling in a car, bus, or train, this commuter helps save energy.

What you can contribute

You cannot hope to preserve what you do not understand. Humans rely on the Earth's ecosystems for food and all of the other materials our civilization depends on. It has been said that we do not inherit the Earth from our parents but borrow it from our children. Therefore, we must preserve for them a world in which they can live.

Although solving the world's environmental problems will take the efforts of many people, including politicians, economists, and engineers, the issues are largely biological. When all is said and done, your knowledge of ecology is the essential tool that you will need to contribute to the effort. If economists and politicians understood as much ecology as you have learned in these chapters, the world might not be as polluted as it is today. **Figures 19-16** and **19-17** show some simple ways you can participate in solving the problems described in this chapter by conserving energy and reducing pollution and waste. For example, you can save energy by walking, riding a bicycle, or taking public transportation to work or school. Newspapers, aluminum products, glass containers, and many plastic containers can be recycled. Performing an energy-use inventory of your home can identify additional ways to help the environment. For example, installing inexpensive, low-flow shower heads reduces shower water use by 50 percent.

Figure 19-17 Your lifestyle affects the environment. Choices that you make in your day-to-day activities can benefit the environment.

You Can Help

- Conserve energy by walking, riding a bicycle, or taking public transportation
- Do an energy-use inventory of your home
- Learn about the environment
- Write letters to elected officials to support environmental protection
- Recycle newspapers, aluminum products, glass, and plastic
- Create rich soil by making your own compost heap from leaves, grass, and fruit peelings.

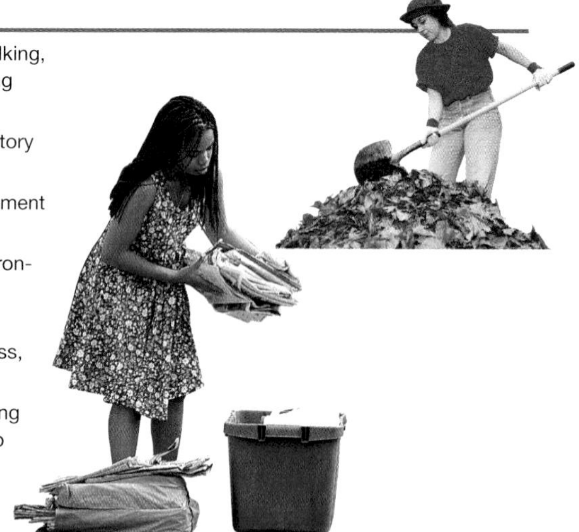

Review SECTION 19-3

1. **Describe** how a tax can reduce pollution.
2. **SKILL Analyzing Information** At which step in the solution of an environmental problem could you have the greatest influence? Explain your answer.

Critical Thinking

3. **SKILL Justifying Conclusions** Of the five steps to solving environmental problems, explain which step you think might be the most difficult to implement.

Review SECTION 19-3 ANSWERS

1. A tax will raise the price of the good, in turn reducing its consumption and production. The money from the tax can also be used to implement pollution controls.
2. Accept any reasonable answer. For example: Individuals can probably have the most influence at step 4—political action—because voting and contacting elected officials provide direct channels of influence.
3. Accept any reasonable answer. For example: step 5—follow-through—may be the most difficult to implement because it requires a clear commitment of money and resources as well as cooperation among many parties.

Study ZONE

CHAPTER 19

Use the Key Concepts and Key Terms listed below to review the main ideas in this chapter. If you need to review the meaning of a term, turn to the page indicated.

Key Concepts

19-1 Global Change

- Acid rain, which is caused by airborne pollutants that lower the pH of rain, has damaged many forests and lakes, especially in the Northeast.
- Destruction of the ozone layer is caused by chlorofluorocarbons (CFCs) and several other manufactured chemicals.
- The greenhouse effect occurs when greenhouse gases, such as carbon dioxide, trap heat within Earth's atmosphere.
- Scientists think that increased concentrations of CO_2 and other greenhouse gases in the atmosphere has led to global warming.

19-2 Ecosystem Damage

- The release of toxic chemicals into the environment can have serious effects, particularly when their concentration is magnified by food chains.
- Three nonreplaceable resources—species of living things, topsoil, and ground water—are being consumed or destroyed at a rapid rate.
- Rapid growth of the human population places serious stress on the Earth's ecosystems.

19-3 Solving Environmental Problems

- Worldwide efforts to reduce pollution are being made, but they are only part of the solution to the overall pollution problem.
- Taxing products or services that create pollution and creating laws requiring pollution-control devices are two methods that have been used to reduce pollution.
- Each of the world's many environmental problems can be solved if seriously addressed. A combination of scientific investigation and public action can solve environmental problems that seem hopeless.

Key Terms

19-1

acid rain (406)
chlorofluorocarbon (407)
greenhouse effect (408)

19-2

biological magnification (411)
aquifer (413)

19-3

None

Unit 7—*Ecosystem Dynamics*

Use this unit to review the Key Concepts and Terms in this chapter.

Study TIP Ready?

- *If you think you understand these Key Concepts and Key Terms, then you're ready for the Performance Zone.*

Study ZONE

CHAPTER 19

Review and Practice

Assign the following worksheets for Chapter 19:

- **Student Review Guide**
- **Concept Mapping Worksheet**
- **Test Preparation Pretest**
- **Vocabulary Worksheet**
- **Critical Thinking Review Worksheet**
- **Science Skills Worksheet**

Assessment Options

The following assessment options are available for this chapter:

Resource Link

- Assign **Chapter Test 19** to assess students' understanding of the chapter.

Alternative Assessment

Have the class create a display titled "Human Impact on the Environment" to inform the rest of the school about the severity of the environmental issues raised in this chapter. The display should cover a range of issues, provide specific information, and show possible solutions and how an individual can make a difference. Make arrangements to place the display where it will be viewed by the entire student body.

Portfolio Assessment

- *Portfolio Assessment* at the front of this book provides options for building and evaluating student portfolios. Students and teachers can select items from the areas listed for inclusion in a portfolio.
- See items labeled PORTFOLIO on pp. 408, 411, 412, 414, 415, 419, 420, and 425.

Answer to Concept Map

The following is one possible answer to Performance Zone item1 on page 422.

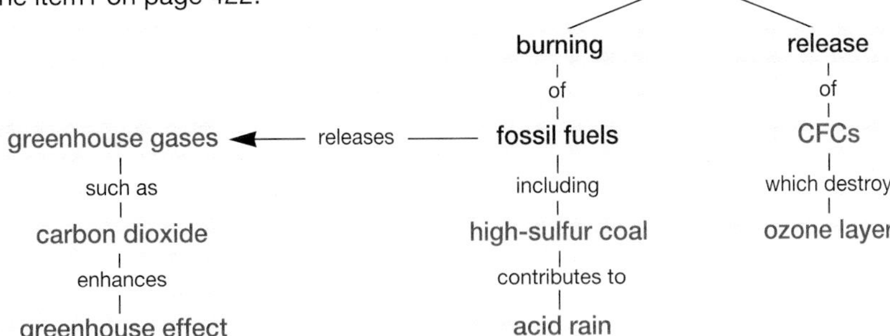

Performance ZONE
CHAPTER 19 REVIEW ANSWERS

Assignment Guide

SECTION	REVIEW
19-1	1, 3, 4, 19, 21
19-2	2, 5, 6, 8, 9, 10, 11, 13, 14, 15, 16, 20, 24
19-3	7, 12, 17, 18, 22, 23

Understanding and Applying Concepts

1. The answer to the concept map is found at the bottom of page 421.
2. Accept any reasonable answer. Sample answers follow.
   a. Sulfur from smokestacks combines with water vapor in the atmosphere to form acidified precipitation, which is commonly called acid rain.
   b. The biological magnification of DDT has been most harmful to animals at the top of the food chain, like fish-eating birds.
   c. It is very difficult to remove pollutants from water stored in an aquifer.
   d. The greenhouse effect helps provide Earth with a moderate climate.
3. b
4. d
5. c
6. c
7. b
8. Ultraviolet radiation from sunlight could result in sunburn. Repeated exposure to UV radiation can damage the DNA of skin cells, causing long-term damage and possibly cancer.
9. Loss of topsoil diminishes the productivity of agricultural land. Species extinctions may prevent the discovery of new food sources and wild strains of domestic crops. Wild crop relatives could improve food production when crossed with domestic varieties.
10. *Zea diploperennis* may impart its viral resistance to *Zea mays* through crossbreeding, resulting in a hardier plant with greater productivity.
11. Through biological magnification, DDT became increasingly concentrated as it moved up through the food chain, with the highest levels found in bald eagles. The chemical reduced the eagles' birth rate by weakening the shells of their eggs. With fewer eagles born, the population eventually declined.
12. Industries release pollutants at no cost to consumers because we as a society ignore the indirect costs associated with pollution, choosing instead to pass them on to future generations.
13. The loss of topsoil and ground water reduces the resources that plants need to grow. Less plant tissue means less food for consumers, thereby reducing the carrying capacity of Earth.

Performance ZONE — CHAPTER 19 Review

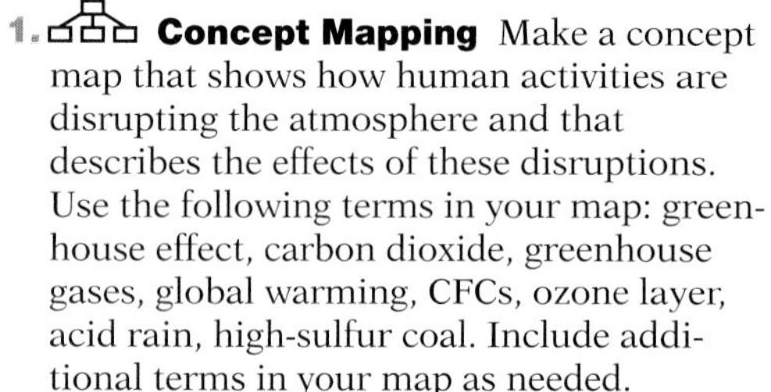

Understanding and Applying Concepts

1. **Concept Mapping** Make a concept map that shows how human activities are disrupting the atmosphere and that describes the effects of these disruptions. Use the following terms in your map: greenhouse effect, carbon dioxide, greenhouse gases, global warming, CFCs, ozone layer, acid rain, high-sulfur coal. Include additional terms in your map as needed.
2. **Understanding Vocabulary** Write a sentence that shows your understanding of each of the following terms:
   a. acid rain
   b. biological magnification
   c. aquifer
   d. greenhouse effect
3. Which of the following causes acid rain?
   a. releasing chlorofluorocarbons
   b. burning high-sulfur coal
   c. polluting ground water
   d. scrubbing smokestack emissions
4. The burning of fossil fuels has changed the atmosphere by
   a. increasing the global concentration of ozone in the upper atmosphere.
   b. reducing the amount of CFCs.
   c. producing an ozone hole.
   d. increasing the concentration of carbon dioxide.
5. Which of the following is *not* true of the drastic extinction of species that is now occurring?
   a. It is the largest extinction event since the dinosaurs disappeared.
   b. One of its causes is the destruction of tropical rain forests.
   c. It is confined to tropical countries.
   d. Potentially useful species are becoming extinct.
6. Which of these countries has the largest population growth rate?
   a. Germany c. Nigeria
   b. United States d. Russia
7. Of the five major steps to solving environmental problems, which involves determining the potential outcomes of an environmental plan before it is tried?
   a. assessment c. follow-through
   b. risk analysis d. political action
8. **Theme Cell Structure and Function** Why is it wise to wear a hat and apply sunscreen to exposed areas of your body when you plan to be outdoors for an extended period of time?
9. **Theme Interdependence** Why are the loss of topsoil and the extinction of living species a threat to increased food production?
10. **Theme Heredity** Explain the possible role of *Zea diploperennis*, a wild relative of corn, in developing an improved strain of the modern species of corn, *Zea mays*.
11. **Theme Interdependence** What is the relationship between biological magnification and the reduction of the bald eagle population by the pesticide DDT?

12. **Exploring Further** Explain how industries can release pollutants into the environment at no cost to the consumer
13. **Chapter Links** Describe how the global decrease in available ground water and the loss of topsoil could affect the carrying capacity of the Earth. (**Hint:** See Chapter 16, Section 16-1.)

Performance ZONE
CHAPTER 19

Skills Assessment

Part 1: Interpreting Graphics

Grains make up most of the world's food supply. The graph below shows the relationship of world grain production to grain production per person. Use the graph to answer the following questions:

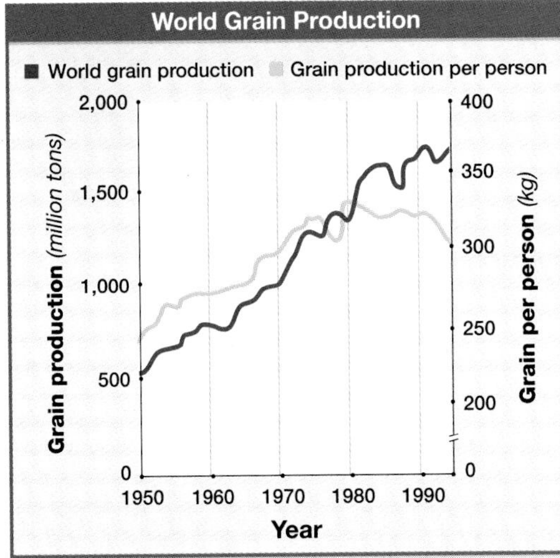

14. Approximately how many kilograms of grain were produced per person in 1960? in 1990?
15. In approximately what year did world grain production first exceed grain production per person?
16. Identify the trend in grain production since the early 1980s.

Part 2: Life/Work Skills

Use the media center or Internet resources to find out more about how trends in world grain production affect world population.

17. **Finding Information** If the trend shown in the graph above continues, what would be the consequences for the human population? Find out what worldwide efforts are being made to reverse this trend.
18. **Communicating** Prepare an oral report using graphics to interpret and summarize your answer to the previous item.

Critical Thinking

19. **Evaluating Results** Tall smokestacks are a part of many coal-burning power plants. State the purpose of these tall smokestacks, and evaluate their effectiveness.
20. **Evaluating Viewpoints** Species diversity has declined drastically several times during Earth's history. Each time, new species evolved. Some people argue that we need not worry about species loss because new species will evolve. Evaluate this argument.

Portfolio Projects

21. **Summarizing Information** Write a report that identifies several alternatives to CFCs and describes the requirements established for CFC replacements.
22. **Interpreting Information** Read the article " 'No-Take' Zones Spark Fisheries Debate" (*Science,* July 25, 1997, pp. 489–491). By banning fishing in "no-take" zones, what do officials hope to accomplish? Why is there opposition to these zones? What did James Bohnsack mean when he said, "The reserves are the controls, and everything else is the experiment"?
23. **Career Focus Environmental Scientist** Use library or Internet resources to research the educational background necessary to become an environmental scientist. List the high school courses required and describe additional degrees or training that is recommended for this career. Summarize the employment outlook for this field.
24. **Unit 7—*Ecosystem Dynamics*** Write a report summarizing ways that humans can work to reduce the depletion and pollution of ground water. How would more-efficient use and recycling of ground water benefit ecosystems?

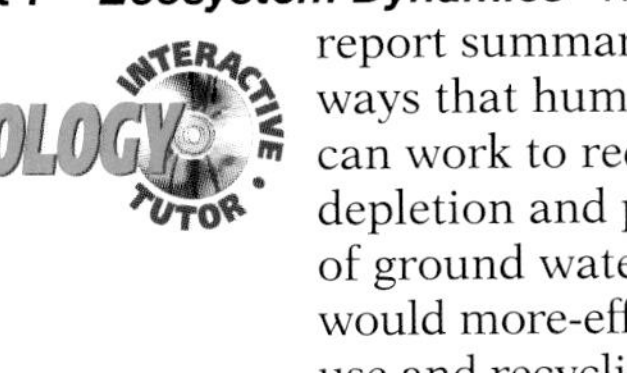

Skills Assessment

14. about 275 kg; about 340 kg
15. about 1977
16. Overall grain production continues to rise, but grain production per person is dropping.
17. Answers will vary. Sample answer: More people, especially those in poor countries, will go without enough food. Researchers are developing better crop plants and improved farming techniques to help reverse this trend.
18. Answers will vary. Students' viewpoints should be well-supported by current data.

Critical Thinking

19. The purpose of such smokestacks is to disperse smoke into the atmosphere and away from people. Much of the pollution, however, is carried back to Earth downwind of the source by precipitation.
20. New species would evolve but only over millions of years. Meanwhile, millions of unique species alive today could be lost, causing irreparable harm to the ecosystems that they help compose and on which we rely.

Portfolio Projects

21. Alternatives to CFCs include hydrochlorofluorocarbons (HCFCs) and hydrofluorocarbons (HFCs). To be used as alternative refrigerants and propellants, the replacements must have: (1) similar boiling points to CFCs, (2) nonflammability and low toxicity, (3) technical and economic viability, (4) chemical stability, and (5) low potential for causing ozone depletion and global warming.
22. Officials hope no-fish zones will serve as sources of fish, larvae, and eggs for surrounding marine areas. People oppose the zones because of their economic implications and because no one is sure the zones will provide the intended benefits. James Bohnsack meant that no-take reserves are not experimental at all. Instead, these zones closely resemble the ocean's natural state, whereas other areas, which are increasingly overfished and polluted, represent a huge experiment with our marine resources.
23. Answers will vary. Environmental scientists use scientific techniques to study the natural and human-altered environment, often with the aim of understanding what effect a certain action produced. Most positions require a bachelor's degree and often a graduate degree in science or engineering. Recommended high school courses include as much math and science as your school offers. The employment outlook for this field is fair. Starting salary will vary by region.
24. Answers will vary. Humans can conserve ground water by planting crop, yard, and garden plants suited to the local climate, among other things. Using and recycling ground water efficiently would help preserve ecosystems that depend on springs for water.

Studying Population Growth

Purpose

Students will study the concepts of population growth, decline, and carrying capacity by growing and observing yeast.

Preparation Notes

Time Required: 20–30 minutes per day for 4 consecutive days

Materials

Materials for this lab can be ordered from WARD'S. See *Master Materials List* at the front of this book for catalog numbers.

Preparation Tips

- Prepare the yeast population by dissolving a fresh cake of yeast in 80 mL of warm water. Add 20 g of sugar and stir. Fresh yeast may be prepared up to 4 hours before the investigation. If substituting dried yeast or freeze-dried yeast (available from WARD'S), prepare about a week in advance. Keep the yeast in a warm, dark area for the duration of the investigation.
- Ruled microscope slides can be purchased from WARD'S. Alternatively make a transparency copy of a piece of graph paper and then cut transparency into coverslip-size pieces. Students can use these transparency pieces as their coverslips.
- Use gloves when preparing methylene blue and avoid creating dust while working. Make a 1.0 % solution by dissolving 0.1 g of methylene blue in 100 mL of distilled water.

Safety Precautions

- Caution students to treat all microorganisms as potential pathogens. Remind students to keep their hands away from their faces as they handle the yeast cultures. Remind students to wash their hands after this lab.

Lab Techniques

Chapter 19 *Studying Population Growth*

SKILLS

- Using a microscope
- Collecting, graphing, and analyzing data
- Calculating

OBJECTIVES

- **Observe** the growth and decline of a population of yeast cells.
- **Determine** the carrying capacity of a yeast culture.

MATERIALS

- safety goggles
- lab apron
- yeast culture
- (2) 1 mL pipets
- 2 test tubes
- 1% methylene blue solution
- ruled microscope slide (2 × 2 mm)
- coverslip
- compound microscope

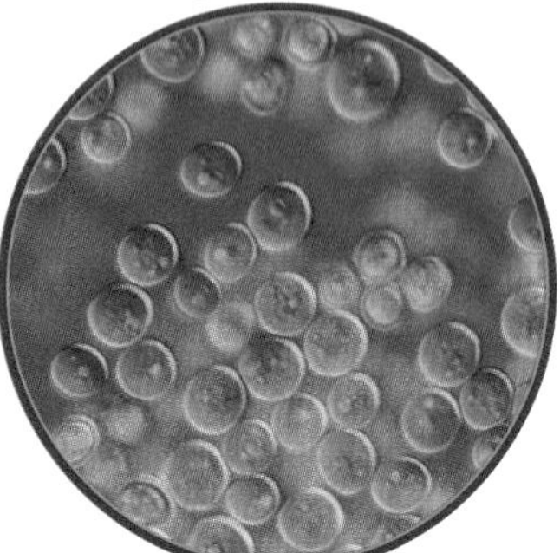

Yeast cells

ChemSafety

CAUTION: Always wear safety goggles and a lab apron to protect your eyes and clothing.

CAUTION: Do not touch or taste any chemicals. Know the location of the emergency shower and eyewash station and how to use them. If you get a chemical on your skin or clothing, wash it off at the sink while calling to the teacher. Notify the teacher of a spill. Spills should be cleaned up promptly, according to your teacher's directions.

CAUTION: Glassware is fragile. Notify the teacher of broken glass or cuts. Do not clean up broken glass or spills with broken glass unless the teacher tells you to do so.

Before You Begin

The growth of the human **population** has reduced Earth's **carrying capacity** for many kinds of organisms. Recall that population size is controlled by **limiting factors**—environmental resources such as food, water, oxygen, light, and living space. **Population growth** occurs when a population's **birthrate** is greater than its **death rate.** A **decline** in population size occurs when a population's death rate surpasses its birthrate. In this lab, you will study the concepts of population growth, decline, and carrying capacity by growing and observing yeast.

1. Write a definition for each boldface term in the previous paragraph.
2. Make a data table similar to the one below.
3. Based on the objectives for this lab, write a question about population growth that you would like to explore.

DATA TABLE

Time (hours)	Number of cells per square: Squares 1–6						Average	Population size (cells/0.1 mL)
0								
24								
48								
72								
96								

Procedure

PART A: Counting Yeast Cells

1. Put on safety goggles and a lab apron.
2.  Transfer 1 mL of a yeast culture to a test tube. Add 2 drops of methylene blue to the tube. **Caution: Methylene blue**

Procedural Tips

Have students work in teams of two.

Disposal

Solutions of yeast and of methylene blue may be rinsed down the drain. Wash thoroughly and air dry all glassware.

Before You Begin

Answers

1. *population*—a group of individuals that belong to the same species, lives in the same area, and breeds with others in the group

   carrying capacity—population size that an environment can sustain

   limiting factors—resources needed for survival obtained from or provided by the environment

   population growth—increase in population size

will stain your skin and clothing. The methylene blue will remain blue in dead cells but will turn colorless in living cells.

3. Make a wet mount by placing 0.1 mL (one drop) of the yeast and methylene blue mixture on a ruled microscope slide. Cover the slide with a coverslip.
4. Observe the wet mount under the low power of a compound microscope. Notice the squares on the slide. Then switch to the high power. *Note: Adjust the light so that you can clearly see both stained and unstained cells.* Move the slide so that the top left-hand corner of one square is in the center of your field of view. This will be area 1, as shown in the diagram below.

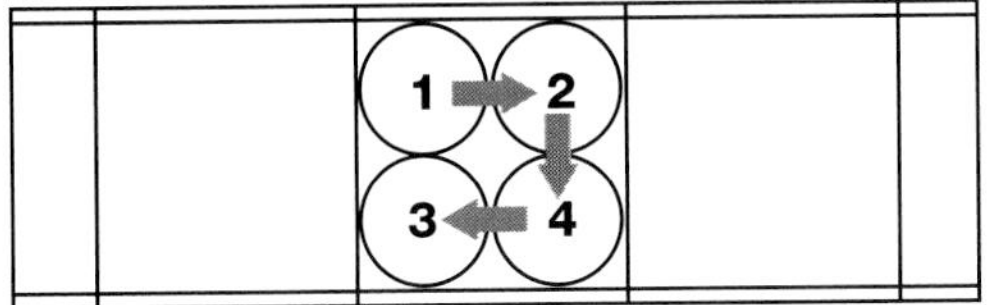

5. Count the live (unstained) cells and the dead (stained) cells in the four corners of a square using the pattern shown in the diagram above. In your data table, record the numbers of live cells and dead cells in the square.
6. Repeat step 5 until you have counted 6 squares on the slide. Complete Part B.
7. Find the total number of live cells in the 6 squares. Divide this total by 6 to find the average number of live cells per square. Record this number in your data table. Repeat this procedure for dead cells.
8. Estimate the population of live yeast cells in 1 mL (the amount in the test tube) by multiplying the average number of cells per square by 2,500. Record this number in your data table. Repeat this procedure for dead cells.
9. Repeat steps 1 through 8 each day for 4 more days.

Part B: Cleanup and Disposal

10. Dispose of solutions and broken glass in the designated waste containers. Do not pour chemicals down the drain or put lab materials in the trash unless your teacher tells you to do so.
11. Clean up your work area and all lab equipment. Return lab equipment to its proper place. Wash your hands thoroughly before you leave the lab and after you finish all work.

Analyze and Conclude

1. **Analyzing Methods** Why were several areas and squares counted and then averaged each day?
2. **Summarizing Results** Graph the changes in the numbers of live yeast cells and dead yeast cells over time. Plot the number of cells in 1 mL of yeast culture on the *y*-axis and the time (in hours) on the *x*-axis.
3. **Inferring Conclusions** What limiting factors probably caused the yeast population to decline?
4. **Further Inquiry** Write a new question about population growth that could be explored in another investigation.

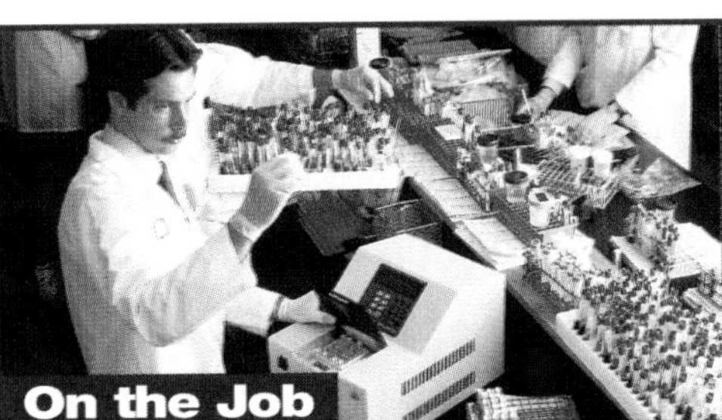

On the Job

Biologists must sometimes estimate the number of cells in a given volume. Do research in the library or media center or on the Internet to discover how a **hemocytometer** is used in medical tests. Summarize your findings in a written or oral report.

Lab Techniques

19 Chapter

Analyze and Conclude

Answers

1. An average was taken to allow for variation within the population.
2. Answers will vary. See the Sample Data Graph below for one example.
3. A lack of food and lack of space limit the yeast cells. They could also be poisoned by their own wastes.
4. Answers will vary. For example: Would the carrying capacity of the yeast's environment expand if the size of the environment increased?

On the Job

Answers

Biologists use hemocytometers to manually count the number of red blood cells in a given volume (usually 1 mm^3) of blood. A low red blood cell count can indicate a variety of medical conditions, including acute blood loss.

Resource Link

Assign **Basic Skills: Graphing** from the Holt BioSources Teaching Resources CD-ROM to allow capable and interested students to explore the issue of environmental preservation from one writer's perspective.
PORTFOLIO

birth rate—number of births in a population
death rate—number of deaths in a population
decline—decrease in population size

3. For example: How does the size of a population *with limited resources* change over time?

Procedure

Answers

5. Answers will vary.
6. Answers will vary.
7. Answers will vary.
8. Answers will vary.
9. Answers will vary.

Sample Data Graph

Estimate of Yeast Population Growth

Saving Biodiversity

Background

When it comes to saving biodiversity, both time and money are short, so conservationists are trying to focus their efforts on areas that are richest in biodiversity. Finding these areas is the problem, because biologists know surprisingly little about the Earth's inhabitants. Also, biological exploration cannot keep pace with habitat destruction—particularly in the species-rich tropics—because of a shortage of systematists and funds for systematic research.

While on a bird-watching expedition in South America, Murray Gell-Mann, a Nobel laureate in physics and a keen amateur ornithologist, and Ted Parker, a tropical bird expert at Louisiana State University, came up with a new way to focus conservation efforts. Both men agreed that traditional field research was gathering data too slowly, and they proposed conducting short, intensive surveys to quickly fill some of the gaps in the knowledge of biodiversity. From their conversation came the Rapid Assessment Program described in the feature. As head of the MacArthur Foundation's world environment and resources committee, Gell-Mann was in a position to help secure funding for RAP expeditions, which began in 1990 and are sponsored by Conservation International.

Teaching Strategies

- Point out that in the last 200 years, scientists have named only about 1.5 million species. About 15,000 new species are described and named each year.
- Remind students that extinction is a problem even in developed countries. Have students research a species that became extinct in the United States within the last century. Each student should give a short presentation to the class describing where the species lived and what caused it to become extinct.

Biologists race against time to save species threatened with extinction.

Saving Biodiversity

How many species of organisms live on Earth? That seems like a simple question, but no one knows the answer. The uncertainty over this basic question illustrates how little scientists actually know about biodiversity.

One thing is certain about biodiversity, though: it is disappearing fast. Biologists need good information about biodiversity in order to choose the actions that will save the most species. To focus conservation efforts, biologists agree that the first priority must be to take an inventory of global biodiversity, determining which species live where.

A New Approach

In the late 1980s, a group of biologists and conservationists decided a shortcut was needed. They launched the Rapid Assessment Program, or RAP, to speed the study of biodiversity. RAP sends small teams of experts to tropical habitats to quickly determine what kinds of organisms live there. The program focuses on "hot spots," threatened areas that are high in biodiversity and contain large numbers of unique species. The aim is to survey as many of these hot spots as possible and identify areas that need to be protected.

To conduct these surveys, Conservation International, the nonprofit organization that sponsors the program, brings together several of the world's leading tropical biologists. Each expedition also includes several scientists from the country where the team is working; one of the goals of RAP is to train biologists in tropical countries. The RAP team uses satellite photos, aerial reconnaissance, and discussions with scientists and government officials in tropical countries to choose the locations it will visit.

Instead of taking an exhaustive, time-consuming inventory of all species, the team focuses on a few well-known groups: mammals, birds, fishes, flowering plants, butterflies, and reptiles. These groups give a good indication of an area's total biodiversity. A habitat with many species of plants and birds, for example, also probably has many species of bacteria, insects, and other lesser-known groups.

Rare treasures
Loss of habitat threatens many tropical rain forest species.

- Explain that although most newly discovered species belong to poorly known groups like insects, eubacteria, and nematodes, new vertebrates—even new mammal species—turn up frequently. In 1994, for instance, scientists exploring the rain forests of New Guinea discovered a new species of kangaroo.
- Have students identify ways that the high standard of living in the United States and other developed countries affects biodiversity. Suggest that they consider how resource use, pollution, and agriculture might endanger species. Then ask them what they would give up or change to protect biodiversity.

Tropical rain forest
Tropical rain forests are home to many species.

The Rewards of RAP

RAP scientists describe their work as exhausting but exhilarating. "It's what we live for," says Tom Schulenberg, an ornithologist (a biologist who studies birds) and RAP team leader. "We're always scheming to get back to the field." While exploring new places, Schulenberg feels "an incredible sense of excitement, knowing no other biologist has been there and everything you see is being seen for the first time."

How RAP Helps Conserve Biodiversity

Though less exciting than exploring a rain forest, the next stage of RAP is just as important. The scientists return to the United States to analyze their data. They then present a report containing their recommendations to the host country. RAP scientists stress that their role is to provide scientifically sound advice, not to tell tropical countries how to manage their natural resources. They leave all decisions to the individual governments. Using the information from RAP reports, tropical countries can guide their land-use decisions to help preserve biodiversity. ■

Career

Biologist

RAP biologists

Profile

When biologists recognized that traditional field research was gathering data too slowly, the Rapid Assessment Program was born. Biologists help conduct short, intensive surveys to quickly fill some of the gaps in the knowledge of biodiversity.

Job Description

Biologists study all aspects of the biology of living things—anatomy, physiology, behavior, ecology, and evolutionary relationships. Their research may involve lab work, field studies, or a combination of both. Many biologists work for universities, museums, or government agencies.

Job Duties

Biologists who are members of a RAP team apply their knowledge of basic biology in parts of the world that are most important for saving species for future generations. Working in remote areas far from medical care has its dangers. Team members have been laid low by bubonic plague, malaria, and hepatitis. Besides disease, they must also watch out for poisonous snakes, biting insects and spiders, and falling trees.

Science/Math Career Preparation

- Biology
- Zoology
- Botany
- Evolutionary biology
- Mathematics
- Genetics
- Biochemistry

Analyzing STS Issues

Science and Society

1. **What is a RAP expedition like?** Read the article, "RAP: On the Fast Track in Ecuador's Tropical Forests," by Richard Conniff (*Smithsonian,* June 1991, pp. 36–49). How did ornithologist Ted Parker identify birds in the rain forest? What did the RAP team learn about the forests of western Ecuador?
2. **What are the benefits of preserving biodiversity?** About 25 percent of medicines are derived from chemicals made by plants. Research a medicine derived from a tropical plant. What plant was it isolated from? Where does the plant live? What disease or diseases is the medicine used to treat?

Technology: PCR

3. **Who should get the benefits of biodiversity?** The enzyme that copies DNA during the polymerase chain reaction (PCR) was isolated from an archaebacterium that lives in Yellowstone National Park. Although PCR generates more than $200 million in income each year, the federal government receives no royalties from the use of the enzyme. Research this issue and then write an essay supporting or opposing this statement: Companies that profit from PCR should be required to compensate the government for using a species discovered on federal land.

References

1. Conservation International. 800-429-5660; On-line. World Wide Web. Available http://www.conservation.org.
2. Abate, Tom. "Environmental Rapid-Assessment Programs Have Appeal and Critics." *Bioscience,* July/August 1992, 486–489.
3. Conniff, Richard. "RAP: On the Fast Track in Ecuador's Tropical Forests." *Smithsonian,* June 1991, 36–49.
4. Dobson, Andrew. *Conservation and Biodiversity.* New York: Scientific American Library, 1996.
5. Lipske, Michael. "Racing to Save the Hot Spots of Life." *National Wildlife,* April/May 1992, 40–49.
6. May, Robert. "How Many Species Are There?" *Scientific American,* October 1992, 42–48.
7. Pollock, Sarah. "Biological Swat Team Ranks for Diversity, Endemism." *Pacific Discovery,* Summer 1991, 6–7.
8. Reaka-Kudla, Marjorie, Don Wilson, and E. O. Wilson, eds. *Biodiversity II.* Washington, D.C.: Joseph Henry Press, 1997.
9. Roberts, Leslie. "Ranking the Rain Forests." *Science,* March 29, 1991, 1559–1560.
10. Stevens, William. "Deaths Set Back Plan to Assess Tropical Forests." *New York Times,* August 17, 1993, C1, C9.
11. Wilson, E. O. *The Diversity of Life.* Cambridge, Mass.: Belknap Press, 1992.

Analyzing STS Issues Answers

1. Parker identified birds by sound—he could distinguish nearly 4,000 species by their songs alone. The RAP team found that most of the forest had been cut, with only a few tiny patches remaining. You may want to tell students that two of the scientists discussed in this article—Ted Parker and Alwyn Gentry—were killed in a plane crash while on a RAP expedition in 1993.
2. Answers will vary. One possibility is the rosy periwinkle of Madagascar, which provides the anticancer drugs vincristine and vinblastine.
3. Answers will vary. Students' positions should be logical and supported by facts.

UNIT 5 Exploring Diversity

Chapters

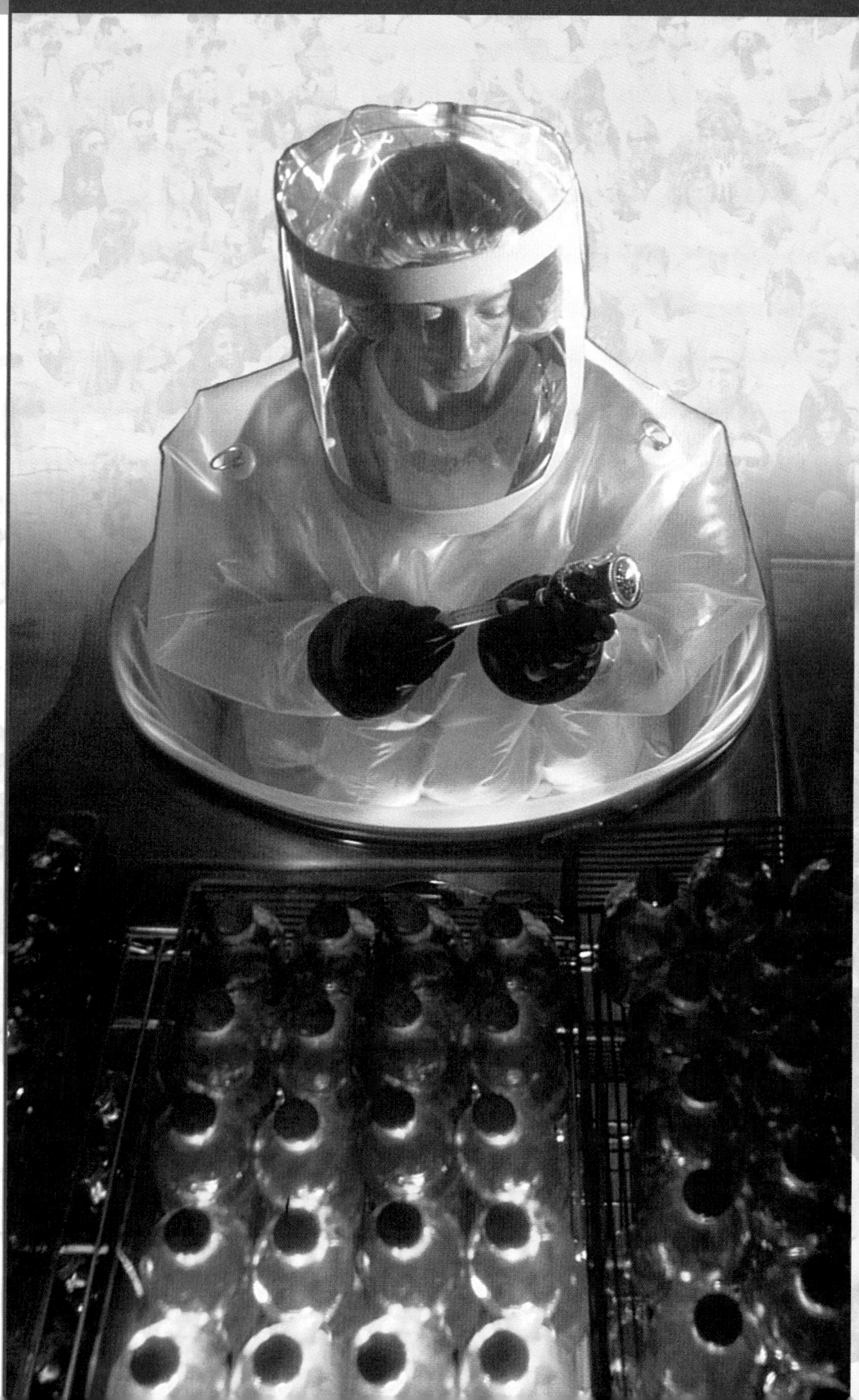

Surrounded by protective gear, a worker in a biotechnology clean room is isolated from contaminants in the surrounding environment.

in perspective

Microbes: Unseen Agents of Disease

Yesterday...

Surgeons without masks, gowns, and gloves? Although this early operation looks strange to us today, Dr. Joseph Lister was actually ahead of his time. In the mid-1850s, Dr. Lister became the first physician to treat patients with an antiseptic during surgery. Lister recognized that spraying an airborne mist of carbolic acid over a patient reduced the likelihood of infections. **See pp. 465–466 to learn how can bacteria can cause disease.**

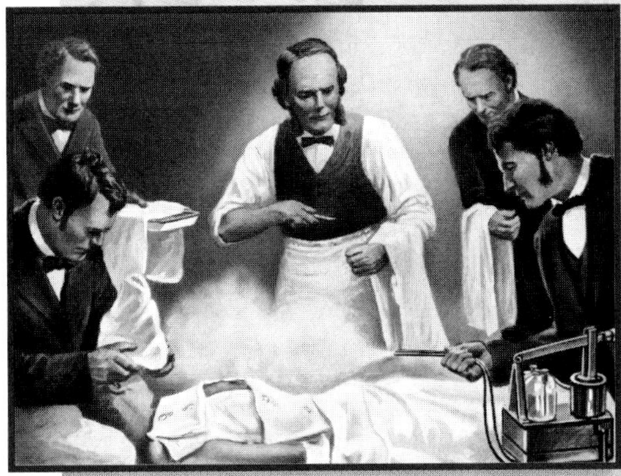

Dr. Joseph Lister and patient in 1867

Today...

Physicians fight bacteria with a far more sophisticated array of weapons than those used by Dr. Lister and his colleagues. Antibiotic drugs target infections throughout the body. **What is the Gram reaction?** **See page 463.**

Antibiotics

Tomorrow...

Some kinds of bacteria have become resistant to certain antibiotics. When an antibiotic fails to stop a bacterial infection, a physician will usually prescribe a different antibiotic—and hope that it works. As antibiotics are used with increasing frequency, strains of antibiotic-resistant bacteria (such as those that cause drug-resistant tuberculosis) continue to become more widespread. **Turn to page 467 to discover how antibiotics kill bacteria.**

Tuberculosis bacterium

internet**connect**

TOPIC: Bacteria
GO TO: www.scilinks.org
KEYWORD: HX429

Chapter 20

Introduction to the Kingdoms of Life

Planning Guide

	OBJECTIVES	MEETING INDIVIDUAL NEEDS	READING SKILLS AND STRATEGIES
BLOCK 1 45 min	**20-1 Simple Unicellular Organisms** • **Identify** the characteristics used to classify kingdoms. • **Differentiate** Eubacteria from Archaebacteria.	Learning Styles, p. 432 BASIC ADVANCED • pp. 432, 433, 435	ATE Active Reading, Technique: Brainstorming, p. 431 ■ Active Reading Guide, worksheet (20-1) ■ Directed Reading, worksheet (20-1)
BLOCK 2 45 min	**20-2 Advent of Multicellularity** • **Contrast** the terms *colony* and *aggregate.* • **List** the characteristics of protists. • **List** the characteristics of fungi.	Learning Styles, p. 436 BASIC ADVANCED • p. 439	■ Active Reading Guide, worksheet (20-2) ■ Directed Reading, worksheet (20-2)
BLOCKS 3 & 4 90 min	**20-3 Kingdoms of Plants and Animals** • **List** the levels of cellular organization that occur in plants and animals. • **Name** the characteristics of plants. • **Identify** the characteristics of animals. • **Differentiate** plants from animals.	Learning Styles, p. 440 BASIC ADVANCED • pp. 441, 442, 444	ATE Active Reading, Technique: Anticipation Guide, p. 441 ■ Active Reading Guide, worksheet (20-3) ■ Directed Reading, worksheet (20-3) ■ Supplemental Reading Guide, *The Andromeda Strain*

Review and Assessment Resources

BLOCKS 5 & 6 90 min

TEST PREP

- PE Study Zone, p. 445
- PE Performance Zone, pp. 446–447
- ■ Chapter 20 worksheets:
  - Vocabulary
  - Concept Mapping
  - Science Skills
  - Critical Thinking

 Audio CD Program

■ Test Prep Pretest

CHAPTER TESTING

- ■ Chapter Tests (blackline master)
- ▲ Test Generator
- ■ Assessment Item Listing

See the correlations table at the front of this book for a list of the ***National Science Education Standards*** addressed in this chapter.

Smithsonian Institution® Visit **www.si.edu/hrw** for additional on-line resources.

KEY

Biology Principles and Explorations

PE Pupil's Edition
ATE Teacher's Edition

Teaching Transparencies
Laboratory Program

Shows these resources within a customizable daily lesson plan:

■ Teaching Resources
▲ Test Generator

VISUAL STRATEGIES	CLASSWORK AND HOMEWORK	TECHNOLOGY AND INTERNET RESOURCES
ATE Fig. 20-3, p. 435 45A Six Kingdoms of Life 88A Comparing Eubacteria with Archaebacteria	Lab A16 Using Bacteria to Make Food **PE** Review, p. 435	Holt Biology Videodiscs Section 20-1 Audio CD Program Section 20-1 internet**connect** SCILINKS NSTA **TOPIC:** Domains of life
88 Two Kingdoms of Prokaryotes 105 Six Divisions of Algae 106 Classification of Fungi	**PE** Quick Lab How can you model true multicellularity?, p. 437 **PE** Review, p. 439	Holt Biology Videodiscs Section 20-2 Audio CD Program Section 20-2
ATE Fig. 20-14, p. 443	**PE** Lab Techniques Surveying Kingdom Diversity, pp. 448–449 **PE** Review, p. 444 **PE** Study Zone, p. 445	Holt Biology Videodiscs Section 20-3 Audio CD Program Section 20-3 internet**connect** SCILINKS NSTA **TOPICS:** • Classifying plants • Classifying animals

ALTERNATIVE ASSESSMENT

PE Portfolio Projects 24–27, p. 447
ATE Alternative Assessment, p. 445
ATE Portfolio Assessment, p. 445

Lab Activity Materials

Quick Lab
p. 437 15 ft lengths of rope (2), several objects in the classroom

Lab Techniques
p. 448 specimens from each of the six kingdoms; compound microscopes, hand lenses, or stereo microscopes

Lesson Warm-up
ATE pictures of internal structures of plant and animal cells, p. 440

Demonstrations
ATE photos of unusual habitats; micrographs of skin, teeth, gums, intestinal lining, p. 430
ATE a food product made with bacteria, p. 434
ATE specimen or photo of plant gall, p. 441

BIOSOURCES LAB PROGRAM
• Quick Labs: Lab A16, p. 31

Chapter 20

Chapter Theme

Evolution

Populations of organisms must be able to adapt to changing ecosystems to survive. As environmental conditions throughout the world have changed over time, a fantastic array of organisms has evolved to survive in nearly every environment found on Earth.

Establishing Prior Knowledge

Before beginning this chapter, make certain students can answer the questions in Study TIP Ready? on page 431. If they cannot, encourage them to reread the sections indicated.

Opening Demonstration

Collect photographs of places such as Antarctica, Old Faithful, the tundra, sea vents in the ocean floor, and the desert. Include micrographs of body structures such as skin, teeth, gums, and intestinal linings. Ask students what each environment has in common. After discussing possibilities, explain that they are all home to bacteria. Emphasize that bacteria have evolved to survive in nearly every conceivable environment.

CHAPTER 20 Introduction to the Kingdoms of Life

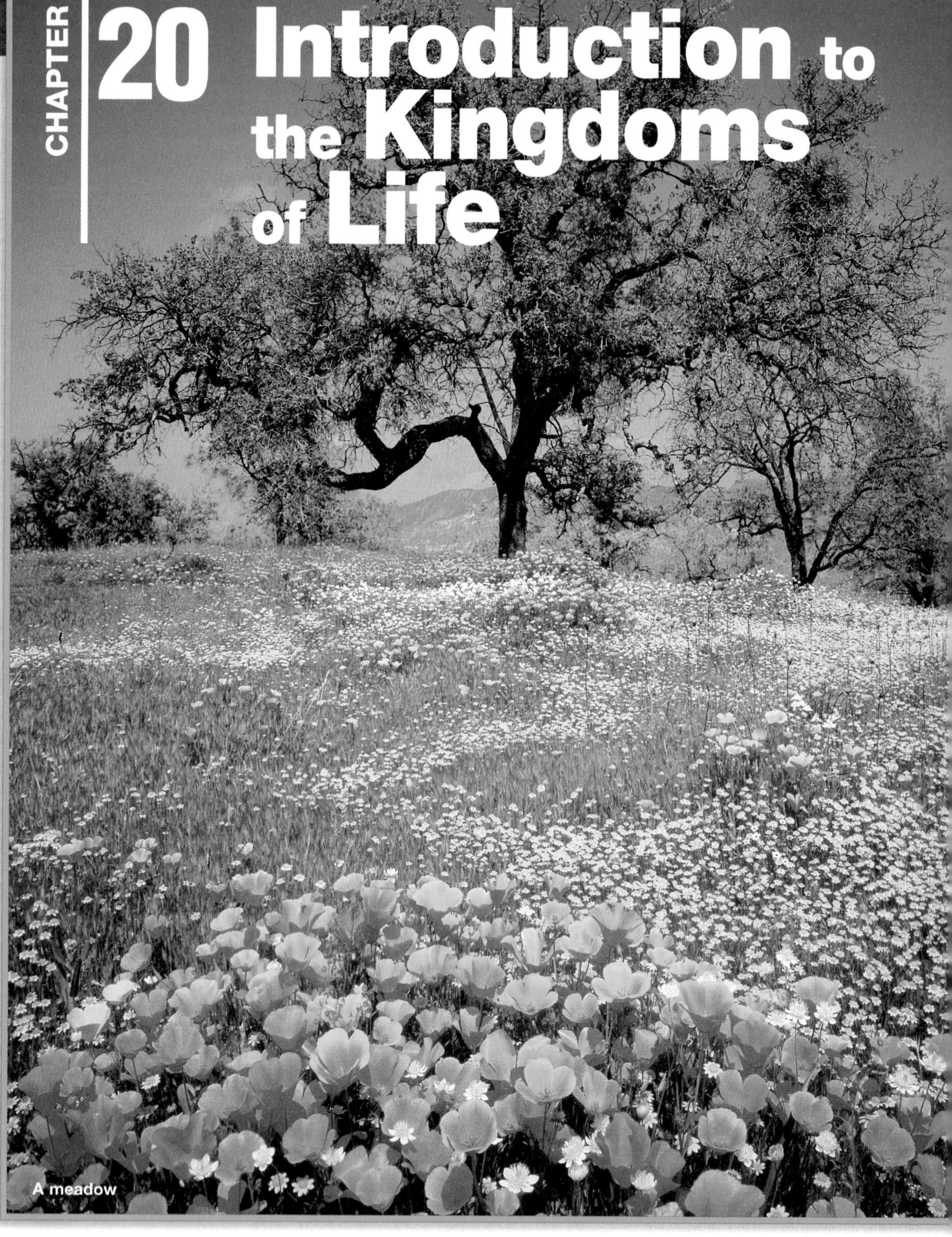

A meadow

What's Ahead — and Why

Chapter Outline

20-1 Simple Unicellular Organisms

- Organisms Are Grouped According to Their Similarities
- Kingdom Eubacteria Includes a Diverse Group of Prokaryotes
- Unusual Prokaryotes Are Found in Kingdom Archaebacteria

20-2 Advent of Multicellularity

- Multicellularity Has Many Forms
- Multicellularity Occurs in Protists
- Fungi Are Mostly Multicellular

20-3 Kingdoms of Plants and Animals

- Complex Multicellularity Defines Plants and Animals
- Autotrophs with Tissues Are in the Plant Kingdom
- The Animal Kingdom Contains Complex Heterotrophs

Students will learn

- the basic features of prokaryotes and how organisms are grouped into six kingdoms.
- how true multicellular organisms are made up of coordinated, communicating cells, and the basic features of fungi.
- how the cells of complex multicellular organisms are arranged to form tissues and basic features of plants and animals.

***Members of the kingdom Animalia,** as pictured above, have various characteristics in common, such as the level of cellular and body complexity, in the same way members of the plant kingdom also have common characteristics.*

Looking Ahead

20-1 Simple Unicellular Organisms
Prokaryotes are included in two kingdoms.

20-2 Advent of Multicellularity
Unicellular and multicellular eukaryotes are grouped in the same kingdoms.

20-3 Kingdoms of Plants and Animals
Tissues and organs occur in organisms of the plant and animal kingdoms.

Labs

- **Quick Lab**
  How can you model true multicellularity? (p. 437)
- **Lab Techniques**
  Surveying Kingdom Diversity (p. 448)

Features

- **Exploring Further**
  Domains of Life (p. 433)

Study TIP

Ready?

Check your understanding of these topics to help you prepare for what's ahead.

Can you . . .

- **calculate** the cell surface-to-volume ratio? (Section 3-2)
- **summarize** the characteristics of prokaryotes and eukaryotes? (Section 3-2)
- **differentiate** introns from exons? (Section 10-2)
- **identify** the terms *eubacteria* and *archaebacteria* (Section 12-2) and *kingdom*? (Section 15-2)
- **summarize** the system of classification of organisms? (Section 15-1)

If not, *review the sections indicated.*

internet**connect**

National Science Teachers Association *sci*LINKS Internet resources are located throughout this chapter.

Reading Strategies

Active Reading

Technique: Brainstorming

Point out that we often organize things—such as products in a store, books in a library, and subjects in a school curriculum—into fundamental classes or groups. Work with the class as a whole to develop a list of other things that are organized into classes or groups. After students have created a list, tell them that biologists have organized all life-forms into six major groups, or kingdoms.

Directed Reading

Assign **Directed Reading Worksheet Chapter 20** to help students interact with the text as they read the chapter.

Interactive Reading

Assign ***Biology: Principles and Explorations* Audio CD Program Chapter 20** to help reluctant readers achieve greater success in reading the chapter.

Reading Skills

Assign Reading Organizers and Reading Strategies Worksheets from the **One-Stop Planner CD-ROM with Test Generator** to help students learn and practice effective reading skills.

internet**connect**

Holt *Biology: Principles and Explorations*

On-line Resources: go.hrw.com

Visit the HRW Web site for a variety of resources related to this chapter. Just type in the keyword HXO Chapter 20.

Holt *Biology: Principles and Explorations*

On-line Resources: www.scilinks.org

The following *sci*LINKS Internet resources can be found in the student text for this chapter:

Page 433	Page 442	Page 443
TOPIC: Domains of life	**TOPIC:** Classifying plants	**TOPIC:** Classifying animals
KEYWORD: HX433	**KEYWORD:** HX442	**KEYWORD:** HX443

Chapter 20

1 Prepare

Scheduling

- **Block:** About 45 minutes will be needed to complete this section.
- **Traditional:** About one class period will be needed to complete this section.
- **Planning:** See the Planning Guide on pp. 429A and 429B for lecture, classwork, and assignment options for Section 20-1. Lesson plans for Section 20-1 can be found in a lesson cycle format in ***Biology: Principles and Explorations*** **One-Stop Planner CD-ROM with Test Generator.**

Resource Link

Assign **Active Reading Worksheet Section 20-1** to help students comprehend the key concepts and visuals in this lesson.

2 Teach

Lesson Warm-up

Ask students to name as many different living organisms as they can. Write their suggestions on the board or an overhead. After you have a list of at least 20 different organisms, have students organize them into broad categories. Compare their categories to the six kingdoms.

Teaching TIP — BASIC

Grouping by Complexity

Have students work with a partner, and give each pair a set of pictures that includes bacteria, protists, plants, fungi, and animals. Have students design a family tree using the pictures. Ask them to justify their answers. Give them additional pictures and ask them to place these on their trees as well. Have students transfer their family trees onto paper using words in place of pictures. Ask them to label each organism as unicellular or multicellular. PORTFOLIO

Section 20-1 Simple Unicellular Organisms

Objectives

- **Identify** the characteristics used to classify kingdoms. (pp. 432–433)
- **Differentiate** Eubacteria from Archaebacteria. (pp. 434–435)

Key Terms

None

Organisms Are Grouped According to Their Similarities

What would the sports page of your local newspaper look like if all the scores for all sports were printed on the page randomly? The confusion would cause many sports fans to give up looking for the results of their favorite team. Instead, the scores are usually arranged according to sport, age group, and whether they are professional or amateur. Basically, the game results are arranged according to their similarity to one another.

As you will recall from Chapter 15, biologists have organized living things into six large groups called kingdoms. Kingdoms are the largest taxonomic group of organisms, and they include several related phyla. Biologists group organisms in the different kingdoms based on their similarities. Most biologists today use the six kingdom system: Eubacteria, Archaebacteria, Protista, Fungi, Plantae, and Animalia, as illustrated in **Figure 20-1.** A few biologists still group Eubacteria and Archaebacteria together as kingdom Monera.

Biologists study life in many different ways. Depending on their field of study, biologists may live with gorillas, collect fossils, or listen to whales. They may identify bacteria, grow mushrooms and plants, or examine the structure of fruit flies. Still other biologists

Figure 20-1 Six kingdoms. Living organisms are divided into six kingdoms and are grouped according to their cell type, complexity, and method for obtaining nutrition.

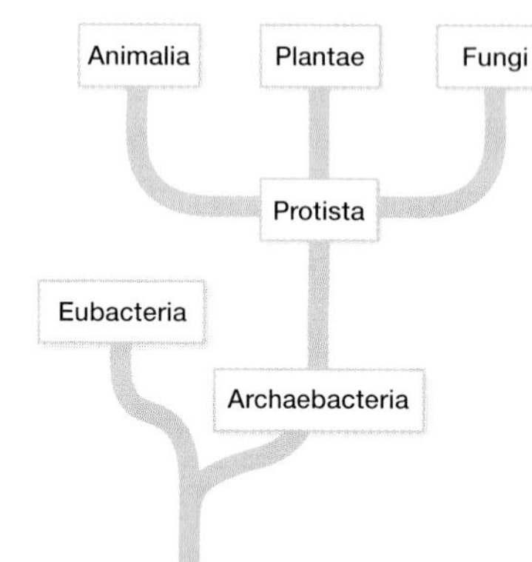

OVERVIEW Cell and Body Characteristics

1. **Cell type** Organisms are either prokaryotic or eukaryotic. Two of the kingdoms include prokaryotes, and the other four include eukaryotes.
2. **Cell construction** Cells are built differently. Some cells have cell walls made of different compounds, and some cells have no cell walls at all.
3. **Body type** Organisms may be unicellular or multicellular and may have tissues or organs. Only one kingdom includes organisms that have organs; organisms in the other kingdoms vary in body type.
4. **Nutrition** Organisms obtain their nutrition through photosynthesis or by heterotrophic means. Some kingdoms have organisms that use both methods; but organisms in other kingdoms use strictly one method or the other.

Integrating Different Learning Styles

Visual/Spatial	**PE**	Study Tip, p. 434; Performance Zone items 16, 17, 18, and 24, p. 447
	ATE	Teaching Tips, pp. 432, 433; Demonstration, p. 434
Interpersonal	**PE**	Performance Zone item 21, p. 447
	ATE	Closure, p. 435; Reteaching, p. 435
Verbal/Linguistic	**PE**	Real Life, p. 435; Performance Zone item 20, p. 447
	ATE	Teaching Tip, p. 433

may read the messages encoded in DNA or count how many times a hummingbird's wings beat each second. With all this diversity, it is easy to lose sight of the fact that all living things have much in common. But, there are a handful of key characteristics that we can examine that help us to understand the individual nature of each of the six kingdoms of life. The key characteristics of each of the six kingdoms are summarized in **Table 20-1.**

Table 20-1 Kingdom Characteristics

Characteristics	Kingdom					
	Eubacteria	Archaebacteria	Protista	Fungi	Plantae	Animalia
Cell type	Prokaryote	Prokaryote	Eukaryote	Eukaryote	Eukaryote	Eukaryote
Cell structure	Cell wall, peptidoglycan	Cell wall, no peptidoglycan	Mixed	Cell wall, chitin	Cell wall	No cell wall
Body type	Unicellular	Unicellular	Unicellular, multicellular	Unicellular, multicellular	Multicellular	Multicellular
Nutrition	Autotrophic and heterotrophic	Autotrophic and heterotrophic	Autotrophic and heterotrophic	Heterotrophic	Autotrophic	Heterotrophic
Example	*Bacillus subtilis*	*Methano-microbium mobile*	*Euglena gracilis*	*Penicillium notatum*	*Pinus radiata* (pine tree)	*Loxodonta africana* (elephant)

Exploring Further

Domains of Life

For many decades, scientists thought that there were two basic forms of life, eukaryotes and prokaryotes. In 1977, the American scientist Carl Woese and his colleagues proposed the idea that archaebacteria represented a third form of life, as distantly related to prokaryotes as they are to eukaryotes.

The case for the third form

Woese and his team first based their ideas on comparisons of ribosomal RNA sequences. They showed that archaebacteria are more closely related to eukaryotes than they are to eubacteria. Other scientists showed that archaebacteria have a type of lipid in their cell membrane that does not occur in either eubacteria or eukaryotes; also, their cell wall is composed of compounds not found in other cell walls.

In 1996, the first comparison was made between complete DNA sequences of an archaebacterium and a eubacterium. The differences confirmed the idea that archaebacteria are very different from eubacteria and may indeed be a third form of life. Recognizing this, biologists are increasingly adopting a classification that recognizes three superkingdoms, or domains.

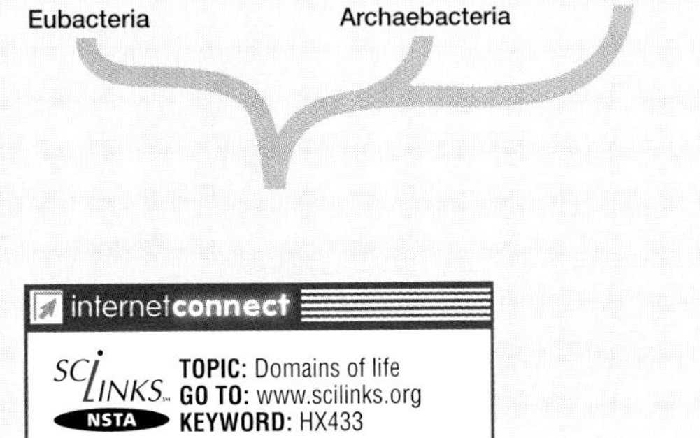

internet**connect**

SCILINKS NSTA
TOPIC: Domains of life
GO TO: www.scilinks.org
KEYWORD: HX433

did you know?

Not all bacteria are small.

The classification of the single-celled organism *Epulopiscium fishelsoni,* which lives in the intestines of surgeonfish, puzzled scientists. The organism looks like a bacterium on the inside, yet it is over a million times larger than the *E. coli* in our own intestines. Its massive size led scientists to believe it might be a protist. However, they now classify the organism as a bacterium. A comparison of the organism's genetic material with that of other organisms shows that the closest relative is a gram-positive bacterium. If *E. fishelsoni* is in fact a bacterium, science must redefine the size limits of bacteria.

Teaching TIP

Recognizing Relationships

Show students pictures of living organisms. Include organisms that are unusual in appearance, such as the slime mold *Plasmodium*. Ask students to determine in what kingdom they would place each organism.

Exploring Further

Domains of Life

Teaching Strategies

Tell students that organisms were once assumed to be closely related if they resembled each other physically. Explain that advances in molecular biology now allow scientists to determine the evolutionary relationships of organisms more accurately.

Discussion

- Why were fungi once considered part of the plant kingdom? *(Like plants, fungi are "rooted" and cannot move. They also have cell walls and produce fruiting bodies.)*
- If scientists organized prokaryotes incorrectly, should we trust the rest of today's scientific knowledge? *(Point out that science is a process, and that the knowledge it yields is always subject to challenge and possible revision.)*

Teaching TIP ADVANCED

Discoveries

Have each student prepare a report on a species of organism that was discovered within the last 10 years. Tell students to use on-line databases that may be available in their school or local library. Students' reports should include the common and scientific name of the organism, the kingdom to which it belongs, when and where it was discovered, and a description of its appearance, habitat, and interactions with other species.

PORTFOLIO

Demonstration

Bacteria and Our Food

Show students a food product that was made with the help of bacteria. Develop a list of other foods that are made with the use of bacteria, including yogurt, sour cream, buttermilk, olives, cheese, pickles, sauerkraut, vinegar, and sourdough bread. Ask students why we worry about bacteria in our food if we intentionally add them to certain foods. *(Some types of bacteria are harmless or even beneficial to our systems, while others can cause severe, life-threatening illnesses.)*

Career

Bacteriologist

Many bacteriologists are employed by health departments and medical institutions to protect the public from potential pathogens. Increasingly, however, bacteriologists are working for industry and agriculture. These scientists use bacteria as "microfactories" to produce medical and agricultural chemicals. A degree in biology, microbiology, or chemistry is required, with advanced studies in microbiology or immunology.

Study TIP

Organizing Information

Draw two overlapping circles. Label one circle "Eubacteria" and the other circle "Archaebacteria." In the circle labeled "Eubacteria," write down the characteristics of eubacteria. Do the same for the circle labeled "Archaebacteria." Finally, in the area where the circles overlap, write down the characteristics that the two kingdoms share.

Kingdom Eubacteria Includes a Diverse Group of Prokaryotes

Members of the kingdom Eubacteria (yoo bak TEHR ee uh) are prokaryotes that have the same kind of lipid in their cell membranes as do eukaryotes and that have a particular kind of compound in their cell wall. Most prokaryotes vary in size from 0.1µm to 15µm. Eubacteria are found in practically every environment on Earth, and they have an impact on humans every day. Eubacteria are similar in physical structure, with no internal compartments, and they obtain nutrients in many different ways. There are several key characteristics common among eubacteria.

Characteristics of eubacteria

Cell wall Eubacteria have strong exterior cell walls made of peptidoglycan *(pep tih doh GLY kan),* a molecule made of carbohydrate strands cross-linked by short peptide bridges.

Gene structure Unlike the genes of eukaryotes and archaebacteria, eubacterial genes have no introns. Instead, the entire gene is transcribed as a single mRNA transcript.

Gene translation apparatus The amino acid sequences of the ribosome proteins and RNA polymerases found in eubacteria are very different from those found in eukaryotes or in archaebacteria. These sequences are also used to establish the phylogenetic relationships between archaebacteria.

Kinds of eubacteria

Kingdom eubacteria includes species that normally live in and on your body. Some eubacteria cause disease. Other eubacteria are used by humans to process foods, such as the eubacteria shown in **Figure 20-2,** to control agricultural pests, to produce various chemicals, and to use in genetic engineering. Biologists group eubacteria based mostly on their shape, the nature of their cell wall, and their type of metabolism.

Some eubacteria obtain energy from inorganic compounds such as hydrogen sulfide, ammonia, and methane. Some are photosynthetic and are found in ocean and freshwater environments, where they are primary producers in these ecosystems. Other eubacteria are heterotrophs. Some heterotrophic eubacteria are capable of living in the absence of oxygen while others must have oxygen to live. Heterotrophic eubacteria are important decomposers in many ecosystems. They are responsible for the recycling of carbon, nitrogen, and phosphorus.

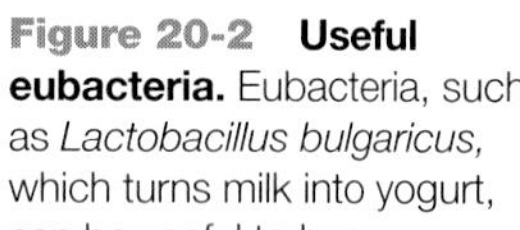

Figure 20-2 Useful eubacteria. Eubacteria, such as *Lactobacillus bulgaricus,* which turns milk into yogurt, can be useful to humans.

did you know?

Bacteria are used to mine minerals.

The first U.S. patent for using bacteria in mining was issued in 1958 to the Kennecott Copper Corporation. However, bacteria were first used to remove copper from ore by Roman miners over 2,000 years ago. The ancient Romans did not realize bacteria worked as a catalyst, but they recognized that the blue salts ($CuSO_4$ produced by the metabolism of bacteria) in the pools of some mines were copper-containing compounds.

Unusual Prokaryotes Are Found in Kingdom Archaebacteria

Archaebacteria are prokaryotes characterized by several unique biochemical characteristics. Although there are many kinds of archaebacteria, it is important to remember that all archaebacteria have certain features in common.

Characteristics of Archaebacteria

Archaebacteria are distinguished from eubacteria based on metabolic differences and key cell structural differences.

Cell wall and membrane The cell walls of archaebacteria do not contain peptidoglycan, as the cell walls of eubacteria do. Archaebacteria use distinctly different lipids than do bacteria or eukaryotes.

Gene structure and translation As with the genes of eukaryotes, the genes of archaebacteria are interrupted by introns, nontranslated segments scattered within the coding region. The ribosomal proteins of archaebacteria are very similar to those of eukaryotes and different from those of eubacteria.

Kinds of archaebacteria

There are at least three major groups of so-called "extremophile" archaebacteria, each living in very different hostile environments.

Methanogens obtain their energy by making methane gas (CH_4) from other organic compounds. These prokaryotes are found in environments such as swamps.

Thermophiles, as shown in **Figure 20-3,** live in very hot water, usually from 60°C to 80°C (140°F to 176°F), such as the water near deep-sea hydrothermal vents. Some thermophiles obtain their energy from elemental sulfur.

Halophiles live in very salty places, like the Great Salt Lake, in Utah. These bacteria live in water with a salinity of 15–20 percent.

Archaebacteria are also common in soil and even in seawater. Because many archaebacteria obtain their energy from inorganic sources, they are considered primary producers.

REAL LIFE

You have methanogens. Methanogens, a type of archaebacteria, can often be found in the digestive system of many animals, including humans and cows. Part of the digestive system has no oxygen, creating a nice habitat for methanogens, which use what you don't and turn it into methane.

Identifying *Find out what other kinds of animals are likely to have methanogens in their intestinal tracts.*

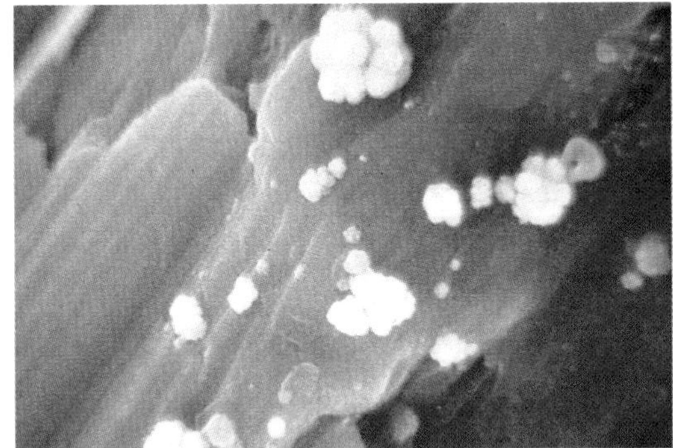

Figure 20-3
Thermophiles. This species of *Acidianus brierleyi* is a thermophilic bacterium and is found near volcanic vents.

Review *SECTION 20-1*

1. **Define** *kingdom.*
2. **List** the characteristics that distinguish the six kingdoms.
3. **Describe** how kingdom Eubacteria differs from kingdom Archaebacteria.

Critical Thinking

4. **SKILL Evaluating Conclusions** Justify the splitting of prokaryotes into two kingdoms.

Review *SECTION 20-1 ANSWERS*

1. Kingdoms are the broadest, most inclusive taxonomic groups of organims.
2. Kingdoms are distinguished by cell type (eukaryote vs. prokaryote); cell structure; body type (unicellular vs. multicellular); and nutrition (heterotrophic vs. autotrophic).
3. Eubacteria are prokaryotes with cell membrane lipids similar to those of eukaryotes. Eubacterial genes contain no introns, and the ribosomal proteins and RNA polymerase of eubacteria are very different from those found in archaebacteria and eukaryotes. Archaebacteria are prokaryotes with different membrane lipids than those of eukaryotes and a different cell wall composition than that of eubacteria. Archaebacteria also have genes with introns and ribosomal proteins that are similar to those of eukaryotes.
4. Although superficially similar, eubacteria and archaebacteria are very different at the molecular level.

Visual Strategy

Figure 20-3 — ADVANCED

Organisms living in harsh environments, such as the thermophiles shown in Figure 20-3, stretch our concept of life-supporting habitats. If life can exist in such places on Earth, ask students to consider the possibility of life on other planets.

REAL LIFE

Answer

Termites and other herbivores have methanogens in their intestinal tracts.

3 Assessment Options

Closure

Analyzing Information

Place students in cooperative groups. Give each group a list of 10 organisms. Next to each organism list a kingdom to which it may or may not belong. Have students determine if the kingdom associated with each organism is correct. Students must provide a reason for each decision. PORTFOLIO

Review

Assign Review—Section 20-1.

Reteaching — BASIC

Assign students to cooperative groups. Have them write down as many different forms of transportation as they can think of. Then ask them to place each mode of transportation into a transportation "kingdom." Have students compare kingdoms from each group. Ask them if some are more general, while others are more specific. Tell students that scientists face the same challenge when trying to organize the biological kingdoms. PORTFOLIO

Section 20-2 Advent of Multicellularity

1 Prepare

Scheduling

- **Block:** About 45 minutes will be needed to complete this section.
- **Traditional:** About one class period will be needed to complete this section.
- **Planning:** See the Planning Guide on pp. 429A and 429B for lecture, classwork, and assignment options for Section 20-2. Lesson plans for Section 20-2 can be found in a lesson cycle format in ***Biology: Principles and Explorations*** **One-Stop Planner CD-ROM with Test Generator.**

Resource Link

Assign **Active Reading Worksheet Section 20-2** to help students comprehend the key concepts and visuals in this lesson.

2 Teach

Lesson Warm-up

Ask students to list the advantages of being a multicellular organism and the advantages of being a unicellular organism. Discuss the advantages of each in class. *(Multicellular organisms are larger and more complex, and they reproduce sexually, giving them greater genetic diversity. Unicellular organisms are smaller and simpler, and they reproduce very rapidly.)*

Objectives

- **Contrast** the terms *colony* and *aggregate.* (p. 436)
- **List** the characteristics of protists. (p. 438)
- **List** the characteristics of fungi. (p. 439)

Key Terms

colonial organism
aggregation
multicellular organism
differentiation
protist
hypha
septum

Multicellularity Has Many Forms

More than half the biomass on Earth is composed of single-celled organisms, such as bacteria and some eukaryotes. For these organisms, unicellularity has been tremendously successful. However, many other organisms have found success not as individual cells but as members of a coordinated group of cells. Groups of cells that live together can have different levels of cooperation, as shown in **Figure 20-4.**

Colonies

Occasionally, the cell walls of bacteria adhere to one another. In fact, some bacteria, such as cyanobacteria, form filaments, sheets, or three-dimensional formations of cells. However, these formations cannot be considered truly multicellular because few cell activities are coordinated. Such bacteria may properly be considered colonial (living together). A **colonial organism** is a group of cells that are permanently associated but that do not communicate with one another. A colonial organism is shown in Figure 20-4.

Aggregates

An **aggregation** *(a gruh GAY shuhn)* is a temporary collection of cells that come together for a period of time and then separate. For example, a plasmodial slime mold, such as the one shown in Figure 20-4 (a member of the kingdom Protista) is a unicellular organism that spends most of its life moving about and feeding as a single-celled amoeba. When starved, however, these cells aggregate into a large group. This weblike mass produces spores, which are then dispersed to distant locations where there may be more food.

Figure 20-4
Multicellularity. *Volvox* is a colonial organism, and its wall is composed of hundreds or thousands of flagella-containing cells embedded in a jellylike layer. A plasmodial slime mold is an aggregate organism. Its cells form a large mass temporarily.

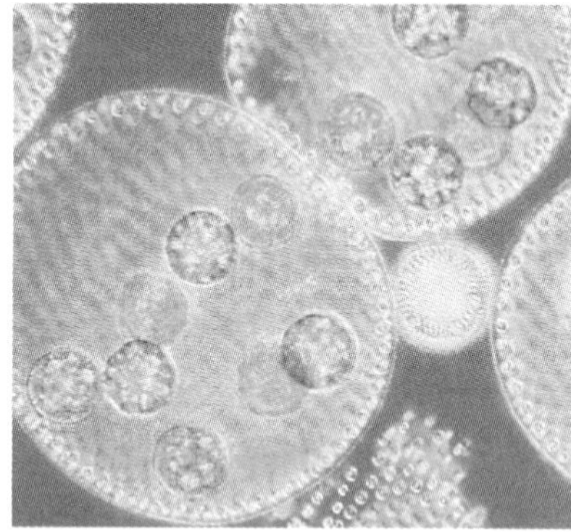

Volvox species

Plasmodial slime mold

Integrating Different Learning Styles

Visual/Spatial	**ATE**	Performance Zone item 1, p. 446
	PE	Teaching Tip, p. 438
Body/Kinesthetic	**PE**	Quick Lab, p. 437
Verbal/Linguistic	**ATE**	Study Tip, p. 438; Performance Zone item 2, p. 446
	PE	Reteaching, p. 439

True multicellularity

Figure 20-5 **Cell specialization.** This green algae has specialized cells that hold it to the bottom of ocean tidal pools. They have many other specialized cells.

All unicellular organisms are small. As the cell of a unicellular organism increases in size, surface-area-to-volume ratio decreases. That is, there is not enough surface area to properly maintain the cell volume. While single cells have difficulty maintaining a large size, multicellular organisms can be large.

A **multicellular organism** is an organism composed of many cells that are permanently associated with one another, such as the green algae shown in **Figure 20-5.** True multicellularity occurs when the activities of the individual cells are coordinated and the cells are in contact with one another. This occurs only in eukaryotes and is one of their major characteristics.

Multicellularity enables cells to specialize in different functions. With this division of labor, a multicellular organism can have cells that work specifically to protect the organism, other cells that help it move about, and still others involved in reproduction and feeding. Cell specialization begins during the development of a new organism. For example, during the development of a chicken in an egg, new cells are formed by cell division. These cells grow and undergo **differentiation,** the process by which cells become specialized in form and function.

How can you model true multicellularity?

In order to understand the advantage that true multicellular organisms have over colonial organisms, you will model multicellular and colonial life.

Materials

two 15 ft lengths of rope, several objects in the classroom

Procedure

1. Working as a class, divide into two groups. One group will model a colonial organism, and the other will model a true multicellular organism. Your teacher will loosely tie a rope around each group.
2. One student in each group will receive a set of instructions for collecting objects from around the classroom.
3. As each group carries out its instructions, students modeling the true multicellular organism may talk with one another, but students modeling the colonial organism must remain silent.

Analysis

1. **Identify** which group finished the assigned task first.
2. **Infer** why the first group to finish was able to accomplish its task so quickly.
3. **Choose** which type of organism is more advanced. Explain.
4. **Predict** how the more advanced organism could become more efficient.

Teaching TIP

Cell Specialization

Ask students to name different kinds of cells that make up their body. *(skin cells, muscle cells, bone cells, etc.)* Challenge students to explain how all these cells can be different if they all have the same genes. After discussing possibilities, tell them that cells begin to specialize very early in development, before birth. As cells develop in different parts of the body, various chemicals are produced that cause the cells to turn genes "on" or "off," which leads to differentiation.

How can you model true multicellularity?

Time 10–15 minutes

Process Skills Observing, interpreting results, applying information

Teaching Strategies

- Have the two groups collect different objects to avoid direct competition and to prevent the colony from copying the multicellular organism.
- Each group should collect the same number of objects and have equal access to those objects. For example, tape 10 purple circles around the room for one group and 10 red squares for the other.
- Make sure the colonial group does not use any communication, such as gestures or grunts.

Analysis Answers

1. The multicellular group should finish first.
2. Communication between "cells" allowed the multicellular group to move more efficiently and purposefully than the colony.
3. The multicellular organism is more advanced because its cells work in a coordinated manner.
4. Answers will vary. For example, if the organism had different kinds of tasks to perform, different groups of cells could become specialized to carry out specific tasks.

Chapter 20

Teaching TIP

Undersea Supermarket

Tell students that kelp is a marine alga that grows in coastal areas and may reach 100 m (330 ft) in length. Explain that some types are edible and are even eaten as candy. Others are harvested for algin, a gooey substance used in dairy products, car polish, glue, and some medical products.

CROSS-DISCIPLINARY CONNECTION

◆ History

Ireland has a long history of harvesting macroalgae (seaweed) along its western coast; the practice dates back to at least the twelfth century. Originally the seaweed was used as food, livestock feed, and fertilizer. Today the Irish Seaweed Industry Organization employs people in economically depressed areas to harvest and process seaweed for chemicals used in gels and gums.

Teaching TIP

Distinguishing Organisms

Set up microscope stations at which students can observe different structures of unicellular, colonial, aggregate, and multicellular algae. Have students make drawings of their observations and label the cell parts. PORTFOLIO

Study TIP

Compare and Contrast

To compare and contrast multicellular organisms with unicellular organisms, make a two-column list. In one column, write the ways in which the two are alike. In the other column, write the ways in which they are different.

Multicellularity Occurs in Protists

Of the six kingdoms of organisms, the kingdom Protista is the most diverse. Members of the kingdom Protista, **protists,** are defined on the basis of a single characteristic: they are eukaryotes that are not fungi, plants, or animals. Many are unicellular; in fact, all single-celled eukaryotes (except yeasts) are protists. Some protists, such as some kinds of algae, have cell specialization. Protists can vary widely in size, as shown in **Figure 20-6.**

Protists include heterotrophs and autotrophs. Some heterotrophic protists are detritivores. Other heterotrophic protists act as parasites. While protists often reproduce asexually, in times of environmental stress many can reproduce sexually.

Multicellular algae fill the evolutionary gap between unicellular protists and complex multicellular organisms (plants, animals, and fungi). The most ecologically important protists are probably the algae in the oceans. They help form the base of ocean food webs. Other heterotrophic protists are also part of the plankton and are primary consumers.

Figure 20-6 Protists range in size. Protists can be so small that a microscope is required to see them, or they can be as tall as 100 m.

Magnification: 1,500×

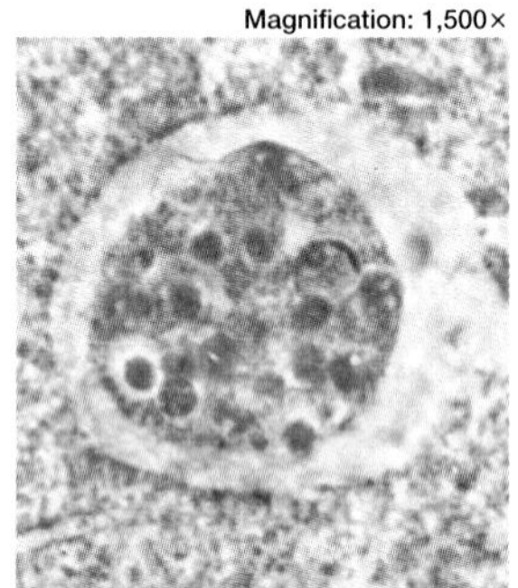

Entamoeba histolytica

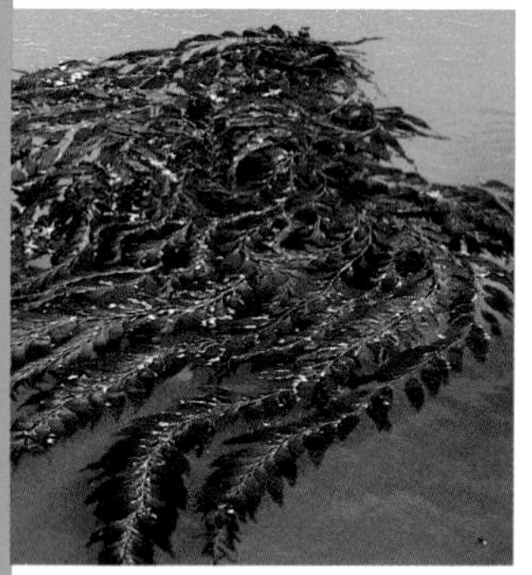

Macrocystis pyrifera

Kinds of Protists

Biologists recognize six general groups of protists. These groups are based on their physical or nutritional characteristics, as follows:

Protists that use pseudopodia Amoebas are protists that have flexible surfaces with no cell walls or flagella; they move by using extensions of cytoplasm called pseudopodia *(soo doh POH dee uh).* Forams, by contrast, have porous shells through which long, thin projections of cytoplasm can be extended.

Protists that use flagella Many protists, including autotrophs and heterotrophs, move by using flagella. The ciliates, which have large numbers of cilia, are so different from other protists that some biologists place them in a separate kingdom.

Protists with double shells Diatoms are photosynthetic protists with unique double shells made of silica, like boxes with lids. Diatoms are part of the plankton and may be found in fresh water or in marine environments.

Photosynthetic algae Algae are photosynthetic and are distinguished by the kinds of chlorophyll they contain. Many are multicellular and reproduce sexually. Algae may be found in marine and freshwater environments.

Funguslike protists Slime molds and water molds are often confused with fungi because they aggregate in times of stress to form spore-producing bodies. Slime molds are often found in fresh water, in damp soil, and on forest floors.

Spore-forming protists Sporozoans are nonmotile unicellular parasites that form spores. Responsible for many significant diseases, including malaria, they have complex life cycles.

Overcoming Misconceptions

Are Multicellular Organisms Superior?

Most people assume that bigger is better—that multicellular organisms have evolved to a form that is superior to unicellular prokaryotes and protists. Point out that while multicellular organisms may be more complex, they are not "better"—that is merely a value judgment.

Remind students that single-celled life-forms outnumber us, outweigh us, and are found in more diverse and extreme habitats. Many are motile, and some can live as either autotrophs or heterotrophs.

Fungi Are Mostly Multicellular

Figure 20-7 Fungus. The hyphae of fungi look like threads or filaments when they are viewed up close.

Fungi are composed of mostly multicellular organisms, and many express cell specialization. For example, many fungi reproduce sexually and produce sexual reproductive structures. Fungi exist mainly in the form of slender filaments barely visible with the naked eye called **hyphae** *(HY fee)* (singular, hypha). Hyphae are strings of connected fungus cells, as shown in **Figure 20-7.** In some species, the filaments weave tightly together to form reproductive structures such as mushrooms.

Fungal cells have an exterior wall made of chitin, the same tough material found in the hard shell of crabs, insects, and other arthropods. The walls dividing one cell from another are called **septa** *(SEHP tuh)* (singular, septum). The presence of septa is a key way in which fungi differ from all other multicellular organisms. Septa rarely form a complete barrier; this arrangement allows cytoplasm to freely flow from one cell to the next.

Fungi obtain nutrients by secreting digestive enzymes into their environment and then absorbing the digested organic molecules. In this way, fungi are able to break down the cellulose in wood. Fungi, together with bacteria, are the primary decomposers in the biosphere.

Kinds of fungi

There are three phyla of fungi, and they are distinguished by their type of reproductive structures.

Zygomycetes Zygomycetes *(zeye goh MEYE seets)* form structures for sexual reproduction called zygosporangia. Zygomycetes include species such as *Rhizopus stolonifer,* common bread mold.

Basidiomycetes Basidiomycetes *(buh sih dee oh MEYE seets)* include fungi that make mushrooms. Mushrooms are the sexual reproductive structures produced by basidiomycetes. Basidiomycetes almost always reproduce sexually.

Ascomycetes Ascomycetes *(as koh MEYE seets)* form sexual spores in special sac-like structures called asci. The sexual reproductive structure formed by ascomycetes often resembles a cuplike structure called an ascocarp.

biobit

Are all wild mushrooms dangerous?

Mushrooms, the fruiting body of a fungus, are eaten by many people throughout the world. However, every year in the United States up to 100 people die from eating poisonous mushrooms. Always consult an expert before trying to eat wild mushrooms.

Review SECTION 20-2

1. **Differentiate** an aggregate species from a species that forms a colony.
2. **Identify** the unique unifying character of protists.
3. **Describe** the characteristics of fungi.

Critical Thinking

4. **Evaluate** the argument that insects and fungi are closely related because both have chitin covering their bodies.

Teaching TIP

Saving Hawaii's Biodiversity

Tell students that Hawaii may have a new ally in the ongoing fight to control the island's non-native plant species: a fungus. Point out that researcher Eduardo Trujillo of the University of Hawaii found the fungus during his travels to the exotic plants' native habitat. There he searched for natural pathogens and predators of the unwanted weeds. After extensive testing to be sure the newly introduced fungus would not threaten any of Hawaii's native plant life, the fungus was successfully released into the infested areas.

3 Assessment Options

Closure

Inferring Function

Ask students to explain how multicellularity contributed to biodiversity. *(Cell specialization enabled different species to evolve many different structures and different ways of functioning.)*

Review

Assign Review—Section 20-2.

Reteaching BASIC

To help students compare unicellular and multicellular organisms, have them make two lists of tools. First have them list simple tools with minimal parts and no communication between parts. *(Examples include a wrench, a hammer, and a screwdriver.)* Then have them list tools that have multiple parts that communicate with each other. *(Examples include a personal computer, a pager, and a television.)* Ask students which tools are like unicellular organisms and which are like multicellular organisms.

Review SECTION 20-2 ANSWERS

1. An aggregate species consists of a group of cells that come together temporarily. A colonial species consists of a group of cells that are permanently associated but do not communicate with one another.
2. A protist is any eukaryote that is not a fungus, a plant, or an animal.
3. Fungi are multicellular (except yeast) eukaryotes that exist primarily as threadlike strands called hyphae. Their cell walls contain chitin. They are heterotrophic, digesting organic matter by secreting enzymes into their environment and then absorbing the nutrients.
4. The presence of chitin is a single, superficial trait and probably arose through convergent evolution.

Chapter 20

Section 20-3 Kingdoms of Plants and Animals

1 Prepare

Scheduling

- **Block:** About 90 minutes will be needed to complete this section.
- **Traditional:** About two class periods will be needed to complete this section.
- **Planning:** See the Planning Guide on pp. 429A and 429B for lecture, classwork, and assignment options for Section 20-3. Lesson plans for Section 20-3 can be found in a lesson cycle format in ***Biology: Principles and Explorations* One-Stop Planner CD-ROM with Test Generator.**

Resource Link

Assign **Active Reading Worksheet Section 20-3** to help students comprehend the key concepts and visuals in this lesson.

2 Teach

Lesson Warm-up

Show students pictures or overheads of the internal structures of plant and animal cells. Point out the chloroplasts and mitochondria. Remind students that many scientists think that these organelles were originally symbiotic prokaryotes living inside early eukaryotic cells (see theory of endosymbiosis in Chapter 12). Eventually these prokaryotes became permanent parts of the eukaryotic cell.

Making Biology Relevant

Technology

Scientists can now create human tissues in the laboratory. For example, skin grafts are created using living cells "seeded" onto growth media in a three-dimensional support structure. Burn victims and diabetes patients are some of the first to benefit. Scientists hope that blood vessels, liver, cartilage, bone, and nervous tissue replacements may be available in the near future.

Objectives

- **List** the levels of cellular organization that occur in plants and animals. (p. 440)
- **Name** the characteristics of plants. (p. 441)
- **Identify** the characteristics of animals. (p. 443)
- **Differentiate** plants from animals. (pp. 441, 443)

Key Terms

tissue
organ
organ system
vascular tissue
invertebrate
vertebrate

Complex Multicellularity Defines Plants and Animals

Plants and animals are the most easily recognizable life-forms. In fact, until the 1950s scientists classified all of life in only two kingdoms, plants and animals. While the number of kingdoms has since increased, we still have the plant and animal kingdoms. One of the most fundamental ways that we define the members of the plant and animal kingdoms is by the structure and complexity of their cells.

As you learned in Section 20-2, many fungi and protists have specialized cells. In contrast, *all* animals and plants have specialized cells. Most animals and plants have specialized cells organized into structures called tissues and organs. **Tissues** are defined as distinct types of cells with a common structure and function. For example muscle is a tissue composed of many muscle cells working together. Different tissues are organized into **organs,** which are specialized structures with specific functions. An example of an organ is the heart, which is composed of muscle, nerve, and other tissues that work together as a pump. The various organs that carry out major body functions make up **organ systems.** The circulatory system, which is composed of your heart, blood vessels, and the blood within them, is an example of an organ system. The relationships between tissues, organs, and organ systems are shown in **Figure 20-8.**

Figure 20-8 Complex multicellularity

Specialized cells form tissue that makes up an organ called the lung. The lungs and other organs constitute an organ system.

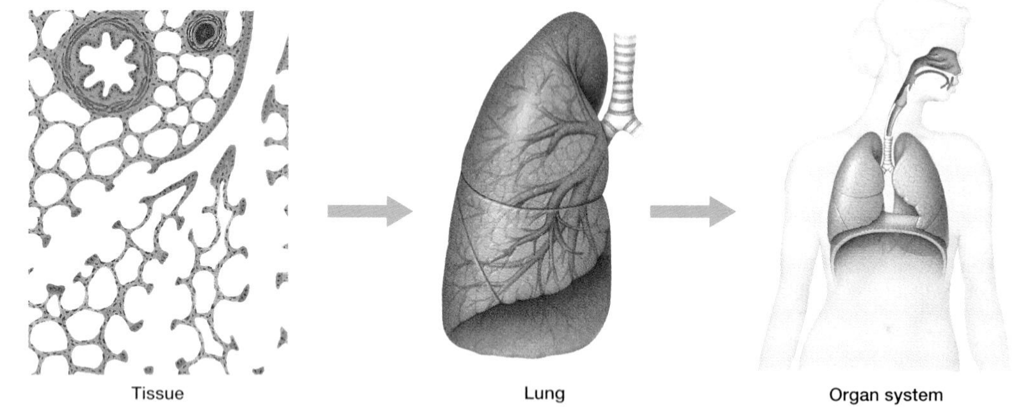

Integrating Different Learning Styles

Visual/Spatial	**PE**	Performance Zone items 1 and 13, p. 446
	ATE	Lesson Warm-up, p. 440; Demonstration, p. 441; Teaching Tips, pp. 441, 442, 443; Visual Strategy, p. 443; Biology in Your Community, p. 443; Reteaching, p. 444
Interpersonal	**ATE**	Multicultural Perspective, p. 442
Verbal/Linguistic	**PE**	Study Tip, p. 444; Performance Zone items 2, 19, 26, and 27, pp. 446, 447
	ATE	Active Reading, p. 441; Internet Connect, p. 442

Autotrophs with Tissues Are in the Plant Kingdom

Plants are complex multicellular autotrophs, which means they have specialized cells and tissues. Most plants have several different types of cells that are organized into many specialized tissues. For example, **vascular** *(VAS kyoo lur)* **tissue** is a group of specialized cells that transport water and dissolved nutrients. Plant cells are different from all other cells in that their cell walls are composed of cellulose, a complex carbohydrate.

Unlike many other organisms, plants cannot move from one place to another. A few groups have motile sperm, but most plants are rooted in the ground. Portable structures, such as spores and seeds, enable the dispersal of plants.

As autotrophs, plants are the primary producers in most terrestrial food webs. Thus, they provide the nutritional foundation for most terrestrial ecosystems. Plants also release oxygen gas to the atmosphere. They are also very important in the cycling of phosphorus, water, nitrogen, and carbon.

Plants, like the fungi, evolved on land and are the dominant organisms on the surface of Earth. They cover every part of the terrestrial landscape except the extreme polar regions and the highest mountaintops, as shown in **Figure 20-9.**

Plants provide food for humans and other animals. They are also a source of medicines, dyes, cloth, paper, and many other products. They vary in size from the 1 mm tall duckweed *(Wolffia microscopica)* to the giant sequoia redwood *(Sequoia gigantea),* which can grow to over 80 m (263 ft) tall, as shown in **Figure 20-10.**

Figure 20-9
Plant diversity

How many? There are about 265,000 known species of plants now in existence. Scientists are still discovering new species of plants.

Duckweed

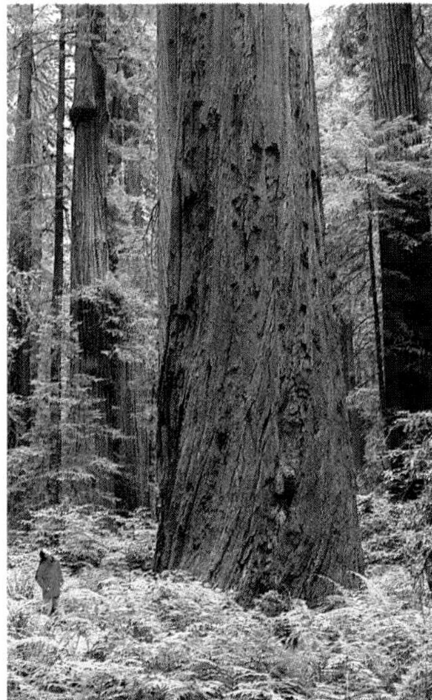

Redwood tree

Figure 20-10 Range of plant size. Plants can be as small as duckweed or as large as a redwood.

Section 20-3

Demonstration
Plant Cancer

Show students a specimen or photograph of a plant gall (one that was not induced by insects). Point out that the cells of complex, multicellular organisms must be highly regulated by internal mechanisms to prevent unwanted growth and changes in cell function. When cells of a particular tissue in an animal begin to grow out of control, the animal is said to have a tumor, or cancer. Tell students that such a growth in a plant is usually called a *gall.*

Active Reading
Technique: Anticipation Guide

Before students read further, have them create two lists of as many different kinds of plants and animals as they can think of. Have them write down broad categories rather than individual species. Ask them to compare their lists to the lists in the text. Have students then correct their lists.

Teaching TIP — BASIC
Organizing Cellular Complexity

Have students create a **Graphic Organizer** similar to the one at the bottom of page 441 to illustrate the increasing levels of complexity found in organisms. Ask students to include an example for each level of complexity.

PORTFOLIO

Graphic Organizer

*Use this graphic organizer with **Teaching Tip: Organizing Cellular Complexity** on page 441.*

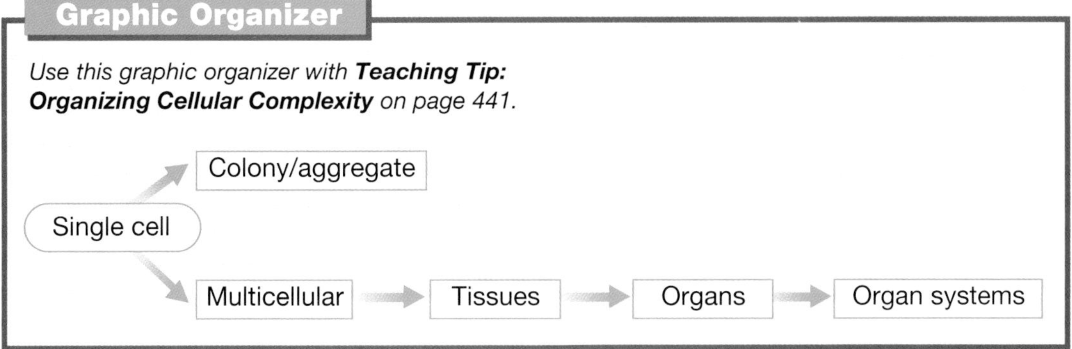

internetconnect

SCINKS. NSTA Have students research how botanists classify plants using the Web site in the Internet Connect box on page 442. Have students write a report summarizing their findings. PORTFOLIO

Teaching TIP BASIC

Gymnosperms

Tell students that *gymnosperm* means "naked seed," and that this group of plants produces seeds that are not enclosed in an ovary, or fruit, unlike angiosperms. Bring a pine cone to class and shake out the seeds of this gymnosperm structure for the class to examine.

Teaching TIP

Kinds of Plants

Have students make a table that compares the vascular and reproductive structures on the four different kinds of plants described on page 442. Students should include an example of each kind of plant.

Figure 20-11 Seedless plants. Mosses and ferns are two groups of plants that do not produce seeds.

Moss

Fern

Figure 20-12 Seed plants. Plants that make seeds are either nonflowering, such as the pine trees, or flowering, such as these bluebonnets.

Pine tree

Bluebonnet

Kinds of plants

There are four basic kinds of plants, as shown in **Figure 20-11** and **Figure 20-12.** They differ from one another according to their type of vascular and reproductive structures.

Nonvascular plants Plants without a well-developed system of vascular tissues are called nonvascular plants or bryophytes *(BREYE oh feyetz).* These plants are all relatively small. They lack the tissue to transport water and dissolved nutrients. They also lack true roots, stems, and leaves. Mosses, such as the one shown in Figure 20-11, are the most familiar example of nonvascular plants.

Plants with a well-developed system of vascular tissues are called vascular plants. Their larger, more-complex bodies are organized into roots, stems, and leaves. One group of vascular plants does not produce seeds. Most vascular plants produce seeds. Most plants are vascular plants.

Seedless vascular plants Ferns are the most common and familiar seedless vascular plants. They have roots, stems, and leaves, and their surfaces are coated with a waxy covering to prevent water loss. They reproduce with spores that are resistant to drying. Both haploid and diploid phases occupy significant parts of the life cycle.

Vascular plants that produce seeds are called seed plants. There are two general types.

Nonflowering seed plants Gymnosperms *(JIHM noh spuhrmz)* are vascular plants that reproduce by making seeds but do not produce flowers. Gymnosperms include plants that produce seeds in cones, such as pines and spruce. Seeds enable plants to scatter offspring and to survive long periods of harsh environmental conditions, such as drought and extreme temperatures.

Flowering seed plants Most plants that produce seeds also produce flowers. Flowering plants, such as roses, grasses, and oaks, produce seeds in fruits and are called angiosperms *(AN jee oh spuhrmz).* Fruits are structures that enable plants to disperse their seeds.

MULTICULTURAL PERSPECTIVE

Ethnobotany

Dr. Chuck Haines is a microbiologist who investigated the mutagenesis of eukaryotes at the University of Kansas. He is also an ethnobotanist. Combining his love of science with his fascination for the medicinal use of plants by indigenous people, Dr. Haines wrote his doctoral dissertation on the medicinal and microbiological properties of cinnamon. He teaches biology, ethnobotany, and microbiology at Haskell Indian Nations University in Lawrence, Kansas—the only school in the United States that brings together Native American students from over 160 tribes.

The Animal Kingdom Contains Complex Heterotrophs

Animals are multicellular heterotrophs. Their cells are mostly diploid, lack a cell wall, and are organized as tissues. In addition, their zygotes develop through several stages. These adaptations have enabled animals to be successful in different habitats. The specialized tissue called muscle enables animals to move on their own. The ability of animals to move more rapidly and in more complex ways than members of other kingdoms is their most interesting characteristic, as illustrated in **Figure 20-13.** A remarkable form of movement unique to animals is flying, an ability that is well developed among both insects and vertebrates. Movement enables animals to avoid predators and to look for food and mates.

Figure 20-13 A running animal. This impala is demonstrating the ability to avoid predators. Running is the result of several organ systems working well together.

Most animals reproduce sexually. In animals, cells formed in meiosis function directly as gametes. The haploid cells do not divide by mitosis first, as they do in plants and fungi, but rather fuse directly with one another to form the zygote. The zygote then gradually develops into an adult by going through several developmental stages.

Almost all animals (99 percent) are **invertebrates;** that is, they lack a backbone. Of the more than 1 million living species, only 42,500 have a backbone and are referred to as **vertebrates.** The animal kingdom includes about 35 phyla, most of which live in the sea. Far fewer phyla live in fresh water and fewer still live on land.

Kinds of animals

Animals are very diverse in form, as shown in **Figure 20-14.** They can range in size from 0.5 mm (0.02 in.) microscopic mites *(Demodex follicularum)* that live on your skin to enormous whales and giant squids. Blue whales can reach a length of 30 m (100 ft) and weigh up to 220 tons. The many kinds of animals can be roughly grouped into six categories.

Figure 20-14 Variety of animals. Organisms as small as the chigger mite share basic characteristics with organisms as large as the whale.

Demodex follicularum

Blue whale

Teaching TIP

Animal, Plant, or Protist?

Use a microprojector or a videomicroscope to show the students live *Euglena*. Have them determine whether they would place this organism in kingdom Plantae because it contains chlorophyll and is autotrophic, kingdom Animalia because it moves, or kingdom Protista. Point out that recent research at the molecular level is making such determinations easier.

Visual Strategy

Figure 20-14

Ask students to examine the mite shown in Figure 20-14. Point out that mites, although very small, are very common and may live on human skin. Tell students that chigger mites cause painful bites, but that other species of mites live harmlessly on the skin.

Biology in Your Community

A Zoological Expedition

Take the class on a field trip to a local zoo. Before the trip, explain that different animals have evolved different structures and tissues that help them exist in their natural habitat. While at the zoo, point out such structures on some of the animals that you see. After returning to class, have students write a report on the animals they saw and the unique adaptations that enable those animals to survive. PORTFOLIO

Teaching TIP

Ants and Termites

Tell students that arthropods—especially insects—make up most of the world's animal life. Then tell them some scientists estimate that ants account for 10 percent of the animal biomass in the world, and that termites account for another 10 percent.

3 Assessment Options

Closure

Analyzing Information

Have students retrieve the list of 10 organisms that they prepared during the **Closure** on page 435. Ask them to reevaluate the reasons they gave for placing the organisms in the selected kingdoms. Then have students list several key characteristics of each kingdom to see if their organisms have those characteristics.

Review

Assign Review—Section 20-3.

Reteaching — BASIC

Assign one of the six groups of animals discussed in the text to each student. Have each student create a colorful poster highlighting the key features of that group and a drawing of a representative species. Place posters around the room. PORTFOLIO

Study TIP

Reviewing Information

After reading this section, make a list of the questions you have about multicellularity. Exchange your list with a partner, and find answers to the questions.

Sponges and cnidarians *(nih DAYR ee uhnz)* Sponges, like the one shown in **Figure 20-15,** are the only animals that do not have tissue, but they do have specialized cells. Cnidarians are mostly marine animals; they include jellyfish, sea anemones, and corals.

Mollusks Mollusks have a saclike tissue called a coelom that encloses internal organs. They include snails, oysters, clams, octopuses, and squids. Mollusks are also the most diverse terrestrial animal, other than arthropods. Most mollusks have a hard external skeleton (shell).

Worms A variety of animals with cylinder-shaped bodies live in both aquatic and terrestrial habitats. These include worms with a specialized internal sac for organs and segmented worms, such as earthworms.

Arthropods By far the most diverse of all animals, arthropods have an external skeleton. They also have jointed appendages, such as antennae and jaws. These structures enable arthropods to sense their environment and obtain their food. Two-thirds of all named species of animals are arthropods, most of them insects. The high rate of reproduction of insects has contributed to their success.

Figure 20-15 Sponge. A sponge has specialized cells in its body that enable it to eat and reproduce sexually.

Echinoderms This group of invertebrates includes sea stars, sea urchins, and sand dollars. Many echinoderms *(ee KIH noh duhrmz)* are able to regenerate a lost limb. In fact some echinoderms deliberately lose a limb to a predator in order to escape that predator.

Vertebrates Vertebrates have an internal skeleton made of bone, a vertebral column (backbone) that surrounds and protects the spinal cord, and a head containing a skull and brain. Vertebrates include mammals, fish, birds, reptiles, and amphibians.

Ecological role

Animals fulfill various roles in an ecosystem. Some are detritivores, such as the microscopic mites that live on your body and eat dead skin cells. Others are primary consumers, such as buffalo that eat grass. Other animals, such as humans, bears, and lions, are secondary consumers that eat primary consumers. Finally, some animals, such as intestinal worms, act as parasites and may cause disease in other animals.

Review SECTION 20-3

1. **Identify** the type of cellular organization found only in plants and animals.
2. **Describe** the characteristics that best describe organisms in the plant kingdom.
3. **Relate** the characteristics of a bear with those of an earthworm.
4. **Compare** the characteristics of plants and animals.

Review SECTION 20-3 ANSWERS

1. Plants and animals have specialized cells that are organized into different tissues.
2. Plants are multicellular, photosynthetic autotrophs with specialized cells and tissues. They lack the ability to move from place to place.
3. The bear and earthworm both have specialized cells, tissues, and organs. The worm, however, lacks the internal skeleton, vertebral column, head, and brain of the bear.
4. Both plants and animals are multicellular eukaryotes that have specialized cells that form tissues. Plants are autotrophic, while animals are heterotrophic. Plants lack the ability of animals to move from place to place. Plant cells have a cell wall, but animal cells do not.

Study ZONE CHAPTER 20

Use the Key Concepts and Key Terms listed below to review the main ideas in this chapter. If you need to review the meaning of a term, turn to the page indicated.

Key Concepts

20-1 Simple Unicellular Organisms

- Members of the six kingdoms are grouped according to their cell type, cell complexity, body type, and nutrition.
- Eubacteria are heterotrophic and autotrophic prokaryotes that have peptidoglycan in their cell wall and no introns in their genes.
- Eubacteria are classified according to their nutrition, their cell shape, and the nature of their cell wall.
- Archaebacteria are prokaryotes that have unusual lipids in their cell membrane, have no peptidoglycan in their cell wall, and have introns in their genes.

20-2 Advent of Multicellularity

- A colonial organism is a group of cells that live together permanently but do not coordinate most cell activity. Aggregations are collections of cells that come together for a period of time.
- Only eukaryotes exhibit true multicellularity, which occurs when the activities of the individual cells are coordinated and the cells are in contact with one another.
- Protists include multicellular and unicellular eukaryotes and can be heterotrophs or autotrophs.
- Fungi are eukaryotic, principally multicellular heterotrophs that exist mainly as slender hyphae.

20-3 Kingdoms of Plants and Animals

- Specialized cells are organized into structures called tissues, organs, and organ systems. These cells have special functions and coordinate their activities with one another.
- Plants are photosynthetic eukaryotes with tissue. Their cells have cell walls.
- Plants are the primary producers in most terrestrial food webs. They release oxygen gas and aid in resource cycling.
- Animals are multicellular heterotrophs with cells that lack a cell wall, that are organized as tissues, and that are mostly diploid. They reproduce sexually, and their zygotes develop through several stages.

Key Terms

20-1

None

20-2

colonial organism (436)
aggregation (436)
multicellular organism (437)
differentiation (437)
protist (438)
hypha (439)
septum (439)

20-3

tissue (440)
organ (440)
organ system (440)
vascular tissue (441)
invertebrate (443)
vertebrate (443)

Study TIP Ready?

- *If you think you understand these Key Concepts and Key Terms, then you're ready for the Performance Zone.* 

Answer to Concept Map

The following is one possible answer to Performance Zone item 1 on page 446.

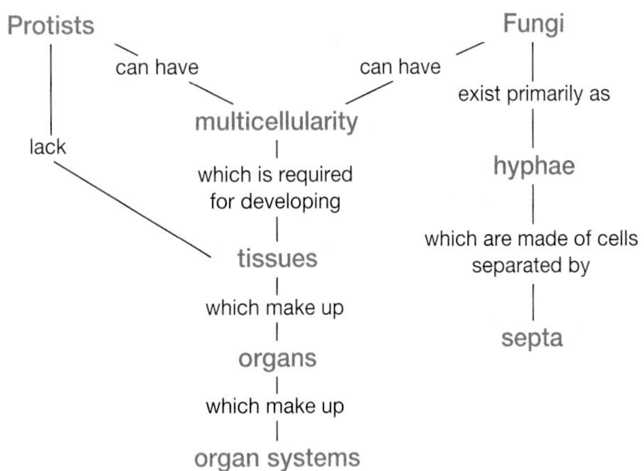

Study ZONE CHAPTER 20

Review and Practice

Assign the following worksheets for Chapter 20:

- **Student Review Guide**
- **Concept Mapping Worksheet**
- **Test Preparation Pretest**
- **Vocabulary Worksheet**
- **Critical Thinking Review Worksheet**
- **Science Skills Worksheet**

Assessment Options

The following assessment options are available for this chapter:

Resource Link

- Assign **Chapter Test 20** to assess students' understanding of the chapter.

Alternative Assessment

- Use the names of the six kingdoms as the categories for a game. Assign students to cooperative groups of four. Have each group develop four questions and answers for all six categories. Arrange competitions by having two groups use the questions and answers of a third group. Competitors choose a category and select an answer (already revealed) at random. They must then formulate a question to match the answer.

Portfolio Assessment

- *Portfolio Assessment* at the front of this book provides options for building and evaluating student portfolios. Students and teachers can select items from the areas listed for inclusion in a portfolio.
- See items labeled PORTFOLIO on pp. 432, 433, 435, 438, 441, 442, 443, and 444.

Performance ZONE
CHAPTER 20
REVIEW ANSWERS

Assignment Guide

Section	Review
20-1	3, 4, 5, 12, 16, 17, 18, 20, 21, 23, 24
20-2	1, 2, 6, 7, 8, 11, 14, 15
20-3	1, 2, 9, 10, 13, 19, 22, 25, 26, 27

Understanding and Applying Concepts

1. The answer to the concept map is found at the bottom of page 445.
2. **a.** A colony is a group of permanently associated cells that do not communicate with one another. An aggregation is a group of cells that come together only temporarily.

   b. A hypha is a threadlike fungal filament. A septum is a wall that separates fungal cells.

   c. An organ is a structure that is made of tissues and has specific functions. An organ system is a group of organs that carries out major body functions, such as circulation.

   d. Gymnosperms are seed-producing plants that do not flower. Angiosperms are plants that produce flowers and seeds.
3. b
4. a
5. c
6. d
7. a
8. d
9. a
10. b
11. Multicellular organisms evolved cell specialization, which enabled the development of tissues, organs, and organ systems.
12. The ribosomal RNA sequences, cell membrane lipids, and cell wall compounds of archaebacteria are different from those of other prokaryotes.

Performance ZONE CHAPTER 20 Review

Understanding and Applying Concepts

1. **Concept Mapping** Construct a concept map that shows the characteristics of protists and fungi. Try to include the following terms in your map: protists, fungi, multicellularity, tissues, organs, organ systems, hyphae, septa. Use additional terms in your map as necessary.
2. **Understanding Vocabulary** For each pair of terms, explain the difference in their meanings.
   a. colony/aggregation
   b. hypha/septum
   c. organ/organ system
   d. gymnosperm/angiosperm
3. Which of the following characteristics is not used to divide the kingdoms?
   a. cell type
   b. photosynthetic pigment
   c. body type
   d. nutrition
4. Eubacteria exhibit all of the following except
   a. introns.
   b. strong exterior walls made of peptidoglycan.
   c. a lack of internal compartments.
   d. flagella.
5. Archaebacteria differ from eubacteria because archaebacteria
   a. have cell walls made of peptidoglycan.
   b. have cell membranes made of phospholipids.
   c. contain genes interrupted by introns.
   d. cannot live in harsh environments.
6. Multicellularity occurs only in
   a. prokaryotes. c. aggregates.
   b. archaebacteria. d. eukaryotes.
7. An organism that has cells that are permanently associated but do not communicate with one another is called
   a. colonial. c. complex.
   b. aggregate. d. heterotrophic.
8. The cells of multicellular fungi are arranged in
   a. sheets. c. tissues.
   b. colonies. d. filaments.
9. Tissues occur only in
   a. animals and plants.
   b. animals.
   c. protists and plants.
   d. animals and protists.
10. Animals differ from plants in that animals
    a. have chlorophyll.
    b. have no cell walls.
    c. have eukaryotic cells.
    d. reproduce sexually.
11. **Theme Cellular Structure and Function** How did the advent of multicellularity enable organisms to grow in complexity?
12. **Exploring Further** What evidence led Carl Woese to propose that archaebacteria should be classified separately from other prokaryotes?
13. **Analyzing Information** What information can you find in the photograph below that indicates that the bird is a complex multicellular organism?

14. **Identifying Variables** What criteria are used to classify organisms as members of the kingdom Protista?
15. **Chapter Links** Infer how the theory of endosymbiosis could help explain how multicellular organisims evolved. (**Hint:** See Chapter 12, Section 12-2.)

13. Accept all reasonable answers, such as the relatively large size of the bird, the presence of complex structures (skin, feathers, beak, etc.), reproduction (chicks), and nutrition (eating insects).
14. Protists are eukaryotes that lack the distinguishing characteristics of other eukaryotes, such as tissues, organs, chitin, and septa. Therefore, they are not fungi, plants, or animals.
15. If cells incorporated chloroplast-like and mitochondrion-like bacteria into their cells, it would have been the beginning of cell cooperation. This cooperation could have given such cells a selective advantage and may have led to increasingly specialized and cooperative cells, which would have led eventually to increasing complexity and multicellularity.

Skills Assessment

Part 1: Interpreting Graphics

Study the phylogenetic tree below, and answer the following questions.

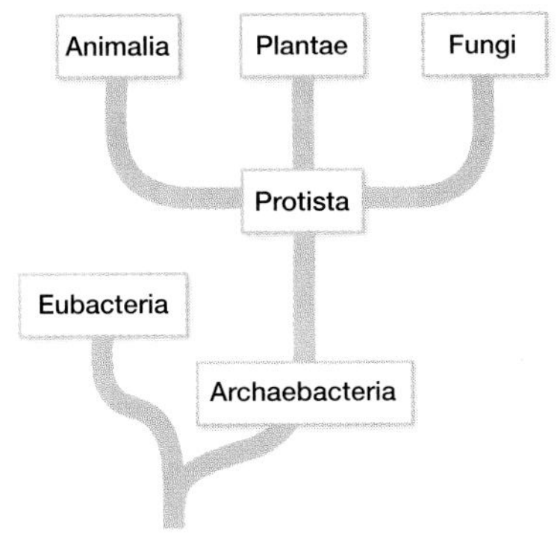

16. Explain why splitting the eubacteria and archaebacteria into two separate kingdoms is justified.
17. In terms of cell type and complexity, explain which two groups are most distantly related to the kingdom Animalia.
18. Does this tree reasonably separate prokaryotes from eukaryotes? Explain.

Part 2: Life/Work Skills

Use the media center or Internet resources to find out what kind of cell types and cell organization occurs in different kinds of organisms.

19. **Finding Information** What kind of animals are scientists using now to research basic questions about embryo development? What questions are scientists asking, and what information are they finding? Prepare a poster to summarize your findings.
20. **Selecting Technology** Find out what techniques scientists used to justify separating the eubacteria from the archaebacteria. Prepare a written report of your findings.
21. **Communicating** With a partner gather all of the arguments for a five-kingdom system and for a six-kingdom system. Organize a public debate in which you and your partner argue which system is more valid.

Critical Thinking

22. **Justifying Methods** Developmental biologists study embryonic development in fruit flies and echinoderms in order to understand embryonic development of more complex animals. How could studying the tissues and cells of simple animals contribute to the understanding of more complex animals?
23. **Evaluating Results** Paleontologists think that there were many different organisms that evolved in the history of Earth, but not one kind of cell nor level of complexity completely dominated the Earth. What mechanism could explain the diversity of life?

Portfolio Projects

24. **Organizing Information** Use cardboard and string to create a mobile that illustrates the hierarchy of classification. Remember that the broad end of the hierarchy is the kingdom, so make that section of the mobile the largest, while the smallest section should be the species.
25. **Comparing Structures** Visit the zoo, and list the scientific names of all the animals you see, or use your library to research 10 organisms. Record the scientific and common names of these organisms. For each organism, identify a trait that led taxonomists to classify the organism in its particular genus or family.
26. **Summarizing Information** Read "Lab-Grown Bladders Prove a Success in Dogs," by J. Travis, in *Science News*, February 18, 1999, vol. 155, p. 101. Explain why a temporary plastic shell was necessary in order to successfully produce a new bladder.
27. **Career Focus Zoologist** Research the field of zoology, and write a report on your findings. Your report should include a job description, training required, kinds of employers, growth prospects, and a starting salary.

Critical Thinking

22. Simple animals may have structures or processes that are more easily observed than they are in more complex animals. The mechanics of the processes in simple and complex animals are similar to some extent because of homology between cell structures and their functions.
23. Because there are so many different environments, the results of natural selection in those environments will vary. This variety of natural selection yielded the diversity of organisms.

Portfolio Projects

24. Accept all mobiles that correctly illustrate the hierarchical nature of the classification system.
25. Answers will vary depending on the organisms that are selected.
26. The cells needed a three-dimensional support structure as they grew to maintain proper shape and development of the bladder.
27. Answers will vary. Zoologists study animals, including their activities, how they interact with one another and their environment, and how different species are related to one another. Most zoology jobs require a college education. Certain zoology careers require additional training in graduate or professional schools. Employers include universities, zoos, and museums. Growth prospects are good. Starting salary will vary by region.

Skills Assessment

16. Eubacteria are as different from archaebacteria as animals, plants, and fungi are from one another.
17. Eubacteria and archaebacteria are most distantly related to the kingdom Animalia. Animals have multicellularity and a different cell structure, cell type, body type, and nutrient requirement.
18. No, archaebacteria, which are prokaryotes, are on the same branch of the tree that contains the four kingdoms of eukaryotes.
19. Answers will vary but should include a discussion of multicellularity, cell specialization, tissues, and organs.
20. Answers will vary but should include a discussion of the differences in molecular structure between the two kingdoms.
21. Answers will vary. Students arguing in favor of a five-kingdom system should base their argument on the similarities between eubacteria and archaebacteria. Students arguing for a six-kingdom system should focus on the differences between those two groups.

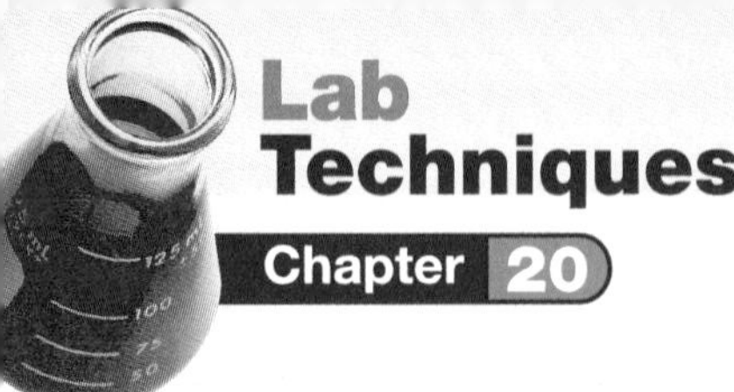

Lab Techniques

Chapter 20

Surveying Kingdom Diversity

Purpose

Students will examine representatives of six kingdoms of organisms.

Preparation Notes

Time Required: About 40 minutes

Materials

Materials for this lab can be ordered from WARD'S. See *Master Materials List* at the front of this book for catalog numbers.

Preparation Tips

- Provide students with preserved, mounted, or dried samples of archaebacteria, an *Anabaena* species, a protozoan, algae, fungi (mold and yeast), plant tissues, and animals.
- Set up the specimens at different stations and have students circulate among the stations.
- Have students make their data table and an answer sheet before class. This will make it easier for students to answer the questions, since not all students will start with Station 1.

Safety Precautions

- Caution students to use care when observing the preserved specimens. The preservative can leak if the jars are tilted.
- Tell students to wash their hands if they come into contact with the preservative.
- Have students wear safety goggles and a lab apron.
- Remind students about microscope procedures when they use oil-immersion lenses.
- Remind students to wash their hands before they leave the laboratory.

Procedural Tips

Students should work in groups of two to four and move clockwise from station to station.

Before You Begin

Answers

1. *kingdom*—the largest taxonomic group

Lab Techniques

Chapter 20 *Surveying Kingdom Diversity*

SKILLS

- Using a microscope
- Comparing

OBJECTIVES

- **Observe** representatives of each of the six kingdoms.
- **Compare** and **contrast** the organisms within a kingdom.
- **Analyze** the similarities and differences among the six kingdoms.

MATERIALS

- specimens from each of the six kingdoms
- compound microscopes
- hand lenses or stereomicroscopes

ChemSafety

CAUTION: Always wear safety goggles and a lab apron to protect your eyes and clothing.

CAUTION: Do not touch or taste any chemicals. Know the location of the emergency shower and eyewash station and how to use them. If you get a chemical on your skin or clothing, wash it off at the sink while calling to the teacher. Notify the teacher of a spill. Spills should be cleaned up promptly, according to your teacher's directions.

CAUTION: Glassware is fragile. Notify the teacher of broken glass or cuts. Do not clean up broken glass or spills with broken glass unless the teacher tells you to do so.

Before You Begin

Many biologists classify living things into six **kingdoms.** The organisms in a kingdom have fundamental characteristics in common. For example, the organisms of two kingdoms are made of **prokaryotic cells,** while the organisms in the other four kingdoms are made of **eukaryotic cells.** Some kingdoms contain only **unicellular** or **colonial** organisms, while others contain only **multicellular** organisms. In this lab, you will examine representatives of six kingdoms of organisms. You will see that each kingdom is distinct from the others.

1. Write a definition for each boldface term in the paragraph above and for each of the following terms: tissue, organ, organ system, autotroph, heterotroph.
2. Make a data table similar to the one below.
3. Based on the objectives for this lab, write a question you would like to explore about the kingdoms of organisms.

DATA TABLE

Kingdom name	Type of cells	Level of organization	Other characteristics	Examples

prokaryotic cell—a cell that has no organized nucleus or organelles

eukaryotic cell—a cell that has a nucleus and several different types of organelles

unicellular organism—an organism that consists of only one cell

colonial organism—a group of cells that are permanently associated but do not communicate with one another

multicellular organism—an organism made of many cells that are permanently associated with one another

tissue—a group of cells that work together to perform a certain function

organ—a specialized structure with a specific function

organ system—a group of organs that work together to perform a function

autotroph—an organism that obtains energy and makes organic compounds from inorganic materials

heterotroph—an organism that obtains energy and organic compounds by eating other organisms

2. Students' data tables should resemble the table on page 448.
3. Answers will vary. For example: How do organisms in the six kingdoms differ?

Procedure

Part A: Conducting a Survey

1. Put on safety goggles and a lab apron.
2. Visit the station for each kingdom listed below, and examine the specimens there. Answer the questions, and record observations in your data table.
3. **Archaebacteria** Examine the prepared slides.
   a. What does a microscope reveal about the structure of archaebacteria?
   b. How do these organisms get energy for life processes?
4. **Eubacteria** Examine the prepared slides.
   a. What does a microscope reveal about the structure of eubacteria?
   b. Would you consider *Anabaena* to be unicellular or multicellular? Explain.
   c. How does *Anabaena* appear to obtain energy for life processes? Explain.
5. **Protists** Examine the prepared slides.
   a. What does a microscope reveal about the structure of protozoans?
   b. How do protozoans appear to obtain energy for life processes? Explain.
   c. Are the algae unicellular or multicellular? Explain.
   d. How do algae differ from protozoans?
6. **Fungi** Examine the specimens.
   a. Are fungi unicellular or multicellular? Explain.
   b. What does a microscope reveal about the structure of fungi?
   c. How do the fungi appear to obtain energy for life processes? Explain.
7. **Plants** Examine the specimens.
   a. What is the most striking characteristic shared by these plants?
   b. What does a microscope reveal about the structure of plants?
8. **Animals** Examine the specimens.
   a. What is the most striking characteristic shared by these animals?
   b. What is the most striking difference among these animals?

Part B: Cleanup and Disposal

9. Dispose of broken glass and solutions in the designated waste containers. Do not pour chemicals down the drain or put lab materials in the trash unless your teacher tells you to do so.
10. Wash your hands thoroughly before you leave the lab and after you finish all work.

Analyze and Conclude

1. **Summarizing Data** What are the main differences observed among the six kingdoms?
2. **Recognizing Patterns** How does the size of bacterial cells compare with the cell size in the other kingdoms?
3. **Analyzing Methods** How did you determine the cell type for each kingdom?
4. **Inferring Conclusions** Which kingdom exhibits the most diversity?
5. **Further Inquiry** Write a new question about the kingdoms of life that could be explored with another investigation.

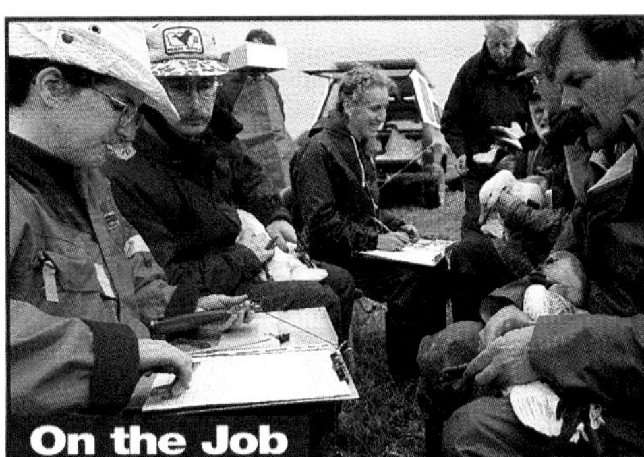

On the Job

A **survey** is a detailed study that is conducted through observation and analysis. Do research in the library or media center or on the Internet to learn about the surveys conducted by a famous biologist, such as Charles Darwin, Carolus Linnaeus, or John James Audubon. Summarize your findings in a written or oral report.

Procedure

Answers

3a. Archaebacteria are microscopic and prokaryotic.

3b. They can be either heterotrophic or autotrophic.

4a. Eubacteria are microscopic, prokaryotic, and many different shapes.

4b. *Anabaena* appears to be multicellular. It is made of chains of cells.

4c. *Anabaena* appears to be an autotroph that obtains energy through photosynthesis.

5a. Protozoans are unicellular and much larger than bacteria. They are eukaryotic, meaning that they have a nucleus and several types of organelles.

5b. Protozoans appear to be heterotrophic since they are not green and therefore do not contain chlorophyll.

5c. Some algae are unicellular, and some are multicellular.

5d. Algal cells have cell walls and pigments, such as chlorophyll, which enable them to be photosynthetic autotrophs.

6a. Some fungi are unicellular, but most are multicellular.

6b. Fungi are made of strands of undifferentiated cells. These strands are called hyphae.

6c. Fungi appear to be heterotrophic since they are not green and cannot photosynthesize.

7a. Answers will vary. For example, students may note that all of the plants are green.

7b. Plants are made of several types of specialized cells that are arranged in tissues and organs.

8a. Answers will vary. For example, students may note that all of the animals have an organized structure with specialized parts.

8b. Answers will vary. For example, students may describe differences in size, shape, or symmetry.

Analyze and Conclude

Answers

1. Answers will vary. Students should include differences in cell structure, body organization, cell specialization, and method of getting food.
2. Bacterial cells are much smaller than the cells of organisms in the other four kingdoms.
3. The cell type (prokaryotic or eukaryotic) is determined by microscopic inspection for a nucleus and organelles.
4. Protista is the most diverse, although that may not be evident from the specimens examined.
5. Answers will vary. For example: How does the method of reproduction differ among the kingdoms?

On the Job

Answers

Answers will vary. Darwin surveyed a wide variety of living organisms and fossils during his five-year voyage on HMS *Beagle*. Linnaeus classified and named more than 11,000 species of plants and animals. Audubon surveyed and painted American wildlife, especially birds.

Chapter 21 Viruses and Bacteria

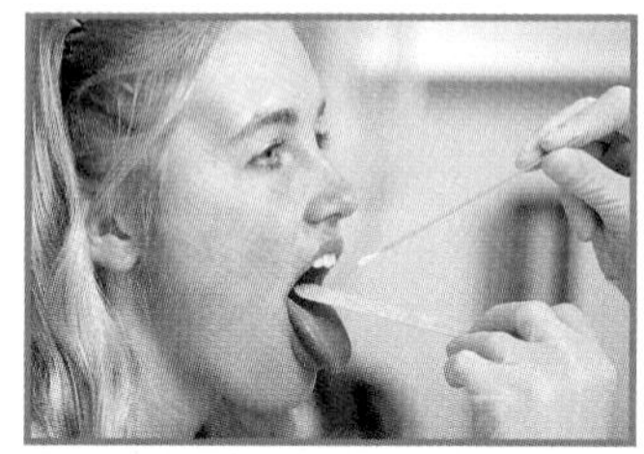

Planning Guide

	OBJECTIVES	MEETING INDIVIDUAL NEEDS	READING SKILLS AND STRATEGIES
BLOCKS 1 & 2 90 min	**21-1 Viruses** • Describe why a virus is not considered a living organism. • Summarize the discovery of the tobacco mosaic virus. • Describe the basic structure of a virus. • Infer how HIV causes illness.	Learning Styles, p. 452 BASIC ADVANCED • pp. 456, 459	ATE Active Reading, Technique: K-W-L, p. 451 ATE Active Reading, Technique: Paired Summarizing, p. 453 ■ Active Reading Guide, worksheet (21-1) ■ Directed Reading, worksheet (21-1) ■ Reading Strategies, K-W-L worksheet
BLOCKS 3 & 4 90 min	**21-2 Bacteria** • List seven differences between bacteria and eukaryotic cells. • Describe three different ways bacteria can obtain energy. • Describe the external and internal structure of *Escherichia coli.* • Distinguish two ways that bacteria cause disease. • Identify three ways that bacteria benefit humans.	Learning Styles, p. 460 BASIC ADVANCED • pp. 463, 465, 468	■ Active Reading Guide, worksheet (21-2) ■ Directed Reading, worksheet (21-2) ■ Supplemental Reading Guide, *The Andromeda Strain*

Review and Assessment Resources

BLOCKS 5 & 6 90 min

TEST PREP

- PE Study Zone, p. 469
- PE Performance Zone, pp. 470–471
- ■ Chapter 21 worksheets:
  - Vocabulary
  - Concept Mapping
  - Critical Thinking
- Audio CD Program
- ■ Test Prep Pretest

CHAPTER TESTING

- ■ Chapter Tests (blackline master)
- ▲ Test Generator
- ■ Assessment Item Listing

See the correlations table at the front of this book for a list of the ***National Science Education Standards*** addressed in this chapter.

Smithsonian Institution® Visit **www.si.edu/hrw** for additional on-line resources.

KEY

Biology Principles and Explorations

PE Pupil's Edition

PE Teacher's Edition

Teaching Transparencies

Laboratory Program

One-Stop Planner CD-ROM
Shows these resources within a customizable daily lesson plan:

■ Teaching Resources

▲ Test Generator

VISUAL STRATEGIES	CLASSWORK AND HOMEWORK	TECHNOLOGY AND INTERNET RESOURCES
ATE Fig. 21-2, p. 453 **ATE** Fig. 21-3, p. 454 93 Structures of Adenovirus and Bacteriophage 94 Structures of TMV and Influenza Virus 95 Tour of a Virus 96 Reproductive Cycle of HIV 97 Viral Diseases and Modes of Transmission	Lab C30 Screening for Resistance to Tobacco Mosaic Virus Lab C31 Using Aseptic Technique **PE** Review, p. 459	CNN Video Segment 16 Depression Virus Holt Biology Videodiscs Section 21-1 Audio CD Program Section 21-1 internetconnect SCILINKS NSTA TOPICS: • Viruses • Viral diseases
ATE Fig. 21-12, p. 466 89 Three Bacterial Cell Shapes 90 Gram Staining 91 Tour of a Bacterium 92 Bacterial Diseases and Modes of Transmission	**PE** Data Lab Identifying pathogenic bacteria, p. 467 **PE** Lab Techniques Staining and Observing Bacteria, pp. 472–473 Lab C32 Gram Staining of Bacteria Lab C33 Gram Staining of Bacteria—Treatment Options Lab D3 Genetic Transformation of Bacteria Lab D10 Ice-Nucleating Bacteria **PE** Review, p. 468 **PE** Study Zone, p. 469	Holt Biology Videodiscs Section 21-2 Audio CD Program Section 21-2 internetconnect SCILINKS NSTA TOPIC: Bacteria

ALTERNATIVE ASSESSMENT

PE Portfolio Projects 25–27, p. 471

ATE Alternative Assessment, p. 469

ATE Portfolio Assessment, p. 469

Lab Activity Materials

Data Lab
p. 467 pencil, paper

Lab Techniques
p. 472 wax pencil, microscope slides (3), safety goggles, lab apron, disposable gloves, rubbing alcohol, paper towels, culture tubes of bacteria (3—A, B, and C), test-tube rack, sterile cotton swabs, Bunsen burner with striker, microscope slide, forceps or wooden alligator-type clothespin, 150 mL beaker, methylene blue stain in dropper bottle, compound microscope

Demonstrations

ATE labels found on tomato plants, p. 452

ATE newspaper or magazine articles about infections, p. 465

ATE photos of an oil spill, a ski resort, and a growing crop, p. 468

BIOSOURCES LAB PROGRAM

- Laboratory Techniques and Experimental Design: Lab C30, p. 145
- Laboratory Techniques and Experimental Design: Lab C31, p. 149
- Laboratory Techniques and Experimental Design: Lab C32, p. 157
- Laboratory Techniques and Experimental Design: Lab C33, p. 163
- Biotechnology: Lab D3, p. 9
- Biotechnology: Lab D10, p. 51

Chapter 21

Chapter Themes

Cellular Structure and Function

Bacteria are successful in part because they have cellular structures that enable them to live in a wide variety of environments.

Establishing Prior Knowledge

Before beginning this chapter, make certain students can answer the questions in **Study TIP Ready?** on page 451. If they cannot, encourage them to reread the sections indicated.

Opening Question

Ask students which of the following diseases are caused by viruses and which by bacteria: measles, AIDS, tuberculosis, syphilis, influenza, chickenpox, botulism, bubonic plague, polio, mumps, Lyme disease, strep throat, and the common cold. *(Measles, AIDS, influenza, chickenpox, polio, mumps, and colds are caused by viruses; the other diseases are caused by bacteria.)* Explain that although antibiotics can cure many bacterial infections, they are not effective at curing viral infections.

CHAPTER 21 Viruses and Bacteria

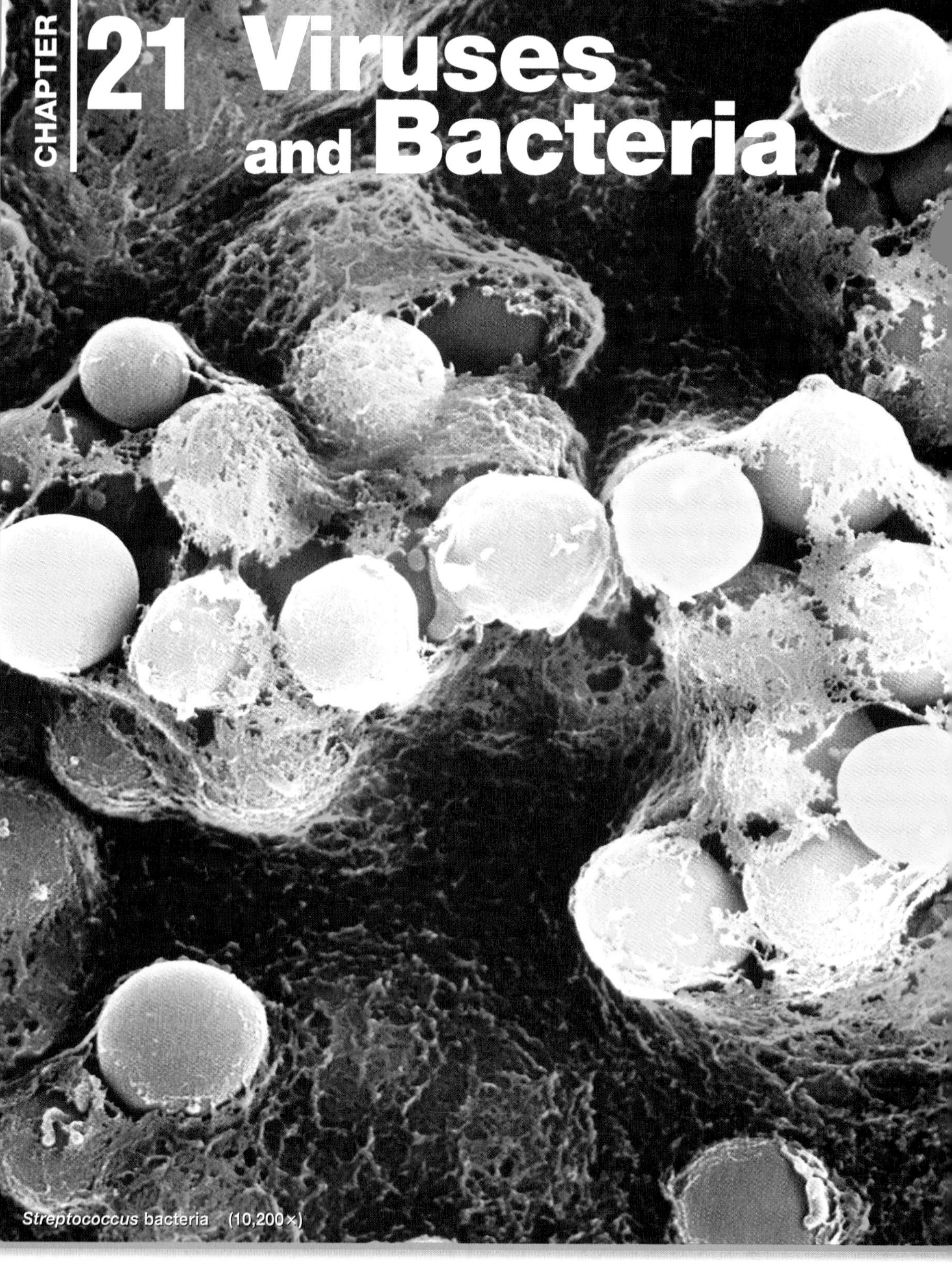

Streptococcus bacteria (10,200×)

What's Ahead — and Why
CHAPTER OUTLINE

21-1 Viruses
- Is a Virus a Living Organism?
- A Virus's Shape Is Determined by Its Parts
- Viruses Replicate Inside Living Cells
- Viruses Cause Many Diseases

21-2 Bacteria
- Bacteria Have a Simpler Structure than Eukaryotes
- Bacteria Are Grouped According to How They Obtain Energy
- Bacteria Cause Disease in Two Basic Ways
- Bacterial Disease Can Be Fought
- Bacteria Are Important

Students will learn

▷ about the structure and replication process of viruses.

▷ how the unique structure and diversity of bacteria make them an extremely successful group of organisms, and about their profound impact on humans and the rest of the ecosystem.

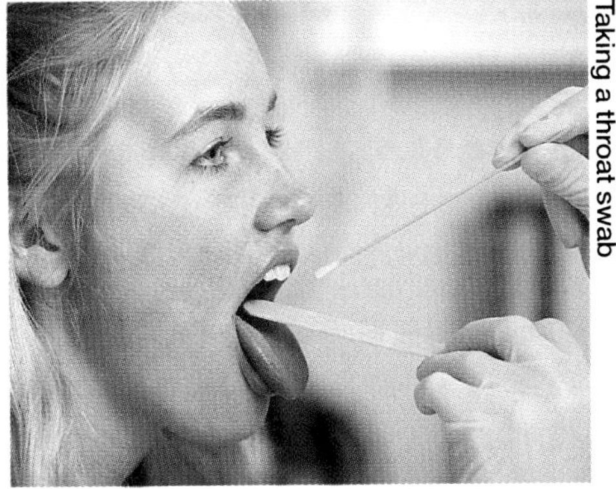
Taking a throat swab

Your sore throat *could be caused by a species of bacteria from the genus* Streptococcus, *shown at left. Bacteria and viruses are capable of causing illness in humans, but not all do. Many viruses are used in genetic engineering and many bacteria are used to make food, medicines, and other industrial products.*

Looking Ahead

Study TIP

Ready?

Check your understanding of these topics to help you prepare for what's ahead.

Can you . . .

- **list** the properties of life? (Section 1-1)
- **define** *prokaryotes*? (Section 3-2)
- **describe** a bacteriophage? (Section 9-1)
- **differentiate** DNA from RNA? (Section 10-1)

If not, *review the sections indicated.*

internet connect

National Science Teachers Association *sci*LINKS Internet resources are located throughout this chapter.

Chapter 21

Reading Strategies

Active Reading

Technique: K-W-L

Before students read this chapter, have them write short individual lists of all the things they already **K**now (or have heard) about viruses. When students have finished their lists, ask them to contribute their entries to a class list on the board or overhead. Then have students list things they **W**ant to know about viruses. Have students save their lists to determine what they have **L**earned in **Active Reading** on page 459.

Directed Reading

Assign **Directed Reading Worksheet Chapter 21** to help students interact with the text as they read the chapter.

Interactive Reading

Assign ***Biology: Principles and Explorations*** **Audio CD Program Chapter 21** to help reluctant readers achieve greater success in reading the chapter.

Reading Skills

Assign Reading Organizers and Reading Strategies Worksheets from the **One-Stop Planner CD-ROM with Test Generator** to help students learn and practice effective reading skills.

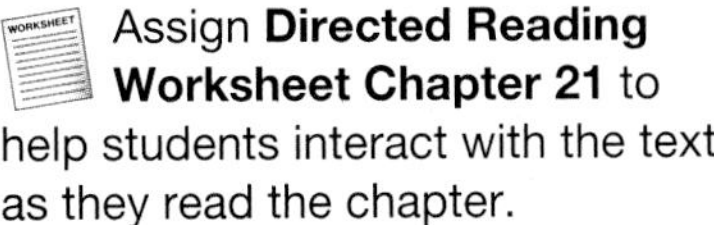

internet connect

Holt *Biology: Principles and Explorations*

On-line Resources: go.hrw.com

Visit the HRW Web site for a variety of resources related to this chapter. Just type in the keyword HXO Chapter 21.

Holt *Biology: Principles and Explorations*

On-line Resources: www.scilinks.org

The following *sci*LINKS Internet resources can be found in the student text for this chapter:

Page 452	Page 459	Page 461
TOPIC: Viruses	**TOPIC:** Viral diseases	**TOPIC:** Bacteria
KEYWORD: HX452	**KEYWORD:** HX459	**KEYWORD:** HX461

Chapter 21

Section 21-1 Viruses

1 Prepare

Scheduling

- **Block:** About 90 minutes will be needed to complete this section.
- **Traditional:** About two class periods will be needed to complete this section.
- **Planning:** See the Planning Guide on pp. 449A and 449B for lecture, classwork, and assignment options for Section 21-1. Lesson plans for Section 21-1 can be found in a lesson cycle format in ***Biology: Principles and Explorations* One-Stop Planner CD-ROM with Test Generator.**

Resource Link

Assign **Active Reading Worksheet Section 21-1** to help students comprehend the key concepts and visuals in this lesson.

2 Teach

Lesson Warm-up

Ask students to discuss what tomatoes, white potatoes, and garden peppers have in common. *(All three are from the nightshade family, Solanaceae, and they are vulnerable to the same diseases. Tobacco is another nightshade family member. A virus that plagues this family worldwide, the tobacco mosaic virus [TMV], was first recognized on tobacco.)*

Demonstration

TMV Resistance

Show students examples of labels found on tomato plants at a nursery or ads in nursery catalogs that state that the plants are resistant to TMV. Explain that nursery workers, farmers, and gardeners who handle tobacco products can carry the virus on their hands and can infect plants by touching them. An infected plant will develop a pattern of light and dark patches on its leaves—the mosaic-like symptoms of the disease.

Objectives

- **Describe** why a virus is not considered a living organism. (p. 452)
- **Summarize** the discovery of the tobacco mosaic virus. (p. 452)
- **Describe** the basic structure of a virus. (p. 453)
- **Summarize** the steps of viral replication. (pp. 454–455)
- **Infer** how HIV causes illness. (p. 456)

Key Terms

virus
capsid
envelope
glycoprotein
bacteriophage
pathogen
lytic cycle
provirus
lysogenic cycle
emerging virus
prion
viroid

internet**connect**

TOPIC: Viruses
GO TO: www.scilinks.org
KEYWORD: HX452

Is a Virus a Living Organism?

In Chapter 1 you learned about the properties of life. All living things are made of cells, are able to grow and reproduce, and are guided by information stored in their DNA. The smallest organisms that have these properties are bacteria. **Viruses** are segments of nucleic acids contained in a protein coat. They are smaller than bacteria and range in size from about 20 nm to 250 nm (0.02–0.25 μm) in diameter as shown in **Figure 21-1.** (One nanometer is equal to 0.001μm or 0.00000004 in.) Most viruses can be seen only with an electron microscope. Viruses replicate by infecting susceptible cells and using the cell to make more viruses. Because viruses do not have all the properties of life, biologists do not consider them to be living. Viruses do not grow, do not have homeostasis, and do not metabolize. However, they cause diseases in many organisms. For this reason, viruses have a major impact on the living world.

Magnification: 112,500×

Figure 21-1
Influenza virus

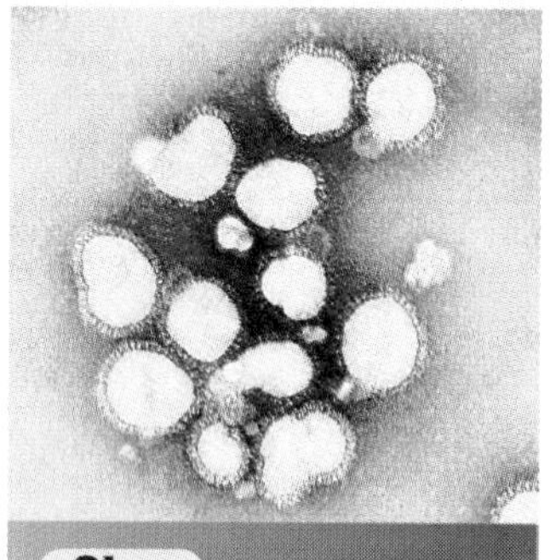

Size It would take about 100 influenza viruses stacked end-to-end to be as long as a bacterium.

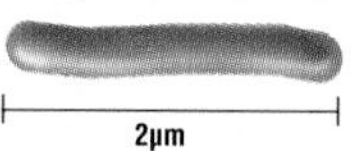

Discovery of viruses

Near the end of the nineteenth century scientists were trying to find the cause of tobacco mosaic disease, which stunts the growth of tobacco plants. They found that infected tobacco plant sap, from which bacteria had been filtered, still caused disease in the tobacco plants. Scientists concluded that the infectious agent must be smaller than a bacterium. They called the agent a *virus,* a Latin word meaning "poison."

For many years after this discovery, viruses were thought to be tiny cells. In 1935 biologist Wendell Stanley, of the Rockefeller Institute, purified tobacco mosaic virus (TMV). The purified virus was a crystal, which is a property of chemicals, and could infect healthy tobacco plants. Stanley concluded that TMV is a chemical rather than an organism.

Each TMV is made of RNA and protein. Scientists were able to separate the RNA from the protein and reassemble the virus so that it could infect plants.

Integrating Different Learning Styles

Logical/Mathematical	**PE**	Real Life, p. 456
	ATE	Teaching Tip, p. 454; Performance Zone item 23, p. 471
Visual/Spatial	**PE**	Performance Zone items 17, 18, 19, 20, and 25, p. 471
	ATE	Demonstration, p. 452; Teaching Tip, p. 453; Visual Strategy, p. 453; Health Watch, p. 455; Up Close, p. 457; Cross-Disciplinary Connection, p. 459
Interpersonal	**ATE**	Biology in Your Community, p. 457
Verbal/Linguistic	**PE**	Performance Zone items 3, 4, 5, 6, 14, 21, and 25, pp. 470–471
	ATE	Teaching Tip, p. 456; Up Close, p. 457; Active Reading, p. 459

A Virus's Shape Is Determined by Its Parts

The virus protein coat, or **capsid,** may contain either RNA or DNA, but not both. RNA viruses include the human immunodeficiency virus (HIV), which causes AIDS, influenza viruses, and rabies virus. DNA viruses include those viruses that cause warts, chickenpox, and mononucleosis. Many viruses, such as the influenza virus shown in **Figure 21-2,** have a membrane, or **envelope,** surrounding the capsid. The envelope helps the virus enter cells. It consists of proteins, lipids, and **glycoproteins** *(glie koh PROH teenz),* which are proteins with attached carbohydrate molecules that are derived from the host cell. Some viruses also contain specific enzymes.

Most viruses have one of two shapes. A helical virus, like the tobacco mosaic virus, is rodlike in appearance, with capsid proteins winding around the core in a spiral. A polyhedral virus has many sides and is roughly spherical, as shown in Figure 21-2. The capsid of most polyhedral viruses has 20 triangular faces and 12 corners. This shape is the most efficient one that can hold the viral genome. Figure 21-2 shows the polyhedral shape of an adenovirus, which can cause several different kinds of infections in humans.

Viruses that infect bacteria, called **bacteriophages,** have a complicated structure. A bacteriophage has a polyhedron capsid attached to a helical tail. A long DNA molecule is coiled within the polyhedron.

Figure 21-2 Viral structures

Viruses can have an envelope, be helical, or be polyhedral.

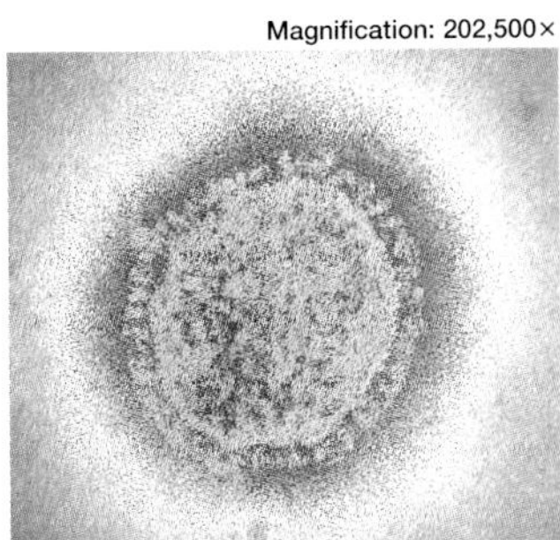

Influenza (enveloped)

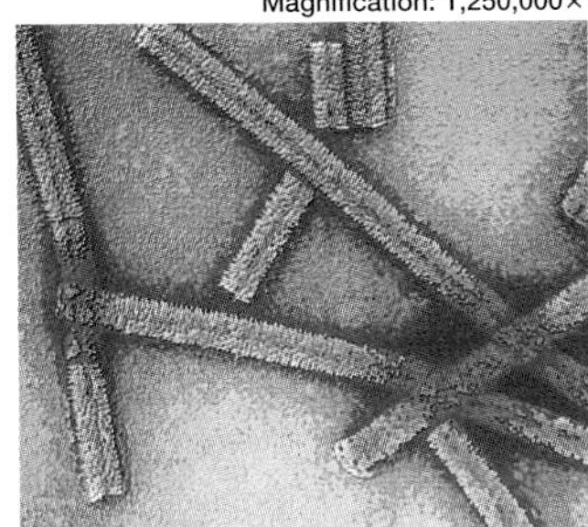

Tobacco mosaic virus (helical)

Magnification: 135,000×

Adenovirus (polyhedral)

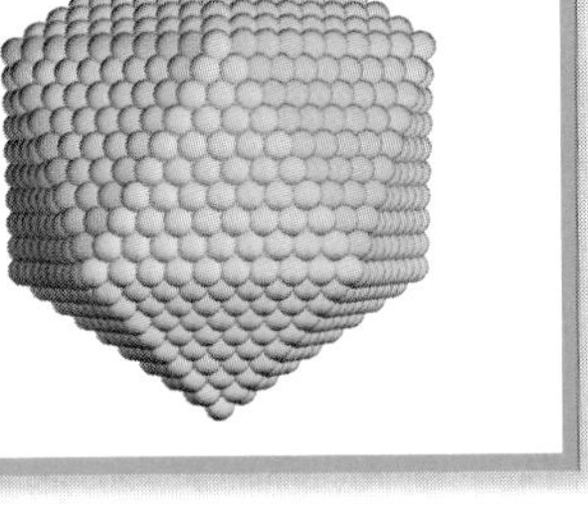

Study TIP

Compare and Contrast

To compare and contrast the properties of life as defined in Chapter 1 and the properties of viruses, make a two-column list. In one column, write the properties of life. In the other column, write the properties of life that viruses have.

Teaching TIP

Design a Virus

Have students create a fictitious virus. Students should produce a drawing or a model of their virus. They should also provide historical information about the origin of the virus, its components, the host it infects, how the virus is transmitted, and the effects of the virus on the host. Have students share their creations with the class. PORTFOLIO

Visual Strategy

Figure 21-2

Use this figure to point out two important parts of a virus: the capsid (proteins) and nucleic acid (RNA or DNA). Whereas most viruses contain DNA, some contain RNA. Ask students what constitutes the membranous envelope surrounding the capsid of the influenza virus. *(proteins, lipids, and glycoproteins from the host cell)* Emphasize that different viruses can have very different appearances; some are larger and more complex than others.

Active Reading

Technique: Paired Summarizing

Have pairs of students silently read the first part of the chapter. As they read, have them note passages they do not understand. After students finish, ask one member of each team to summarize what he or she understood, referring to the text as needed. The listener should add anything omitted and continue the discussion by pointing out what he or she did not understand. Partners should work together to clarify what was not clear and to formulate questions to present to the class about new concepts.

Visual Strategy

Figure 21-3

Point out that some of the capsids appear empty. Ask students to explain why. *(Those capsids that appear empty have already delivered their DNA to the bacterial cell. Those capsids that appear dark still contain their DNA.)*

Teaching TIP

Viral Reproduction

Tell students that viruses replicate rapidly. Then tell them that a single virus that infects a bacterial cell can produce about 100 new viruses in 20 minutes. Have students calculate how many viruses would exist 1 hour after a single virus infected one bacterium in a culture, assuming that there are sufficient cells to support continuous viral replication. *(1 million)*

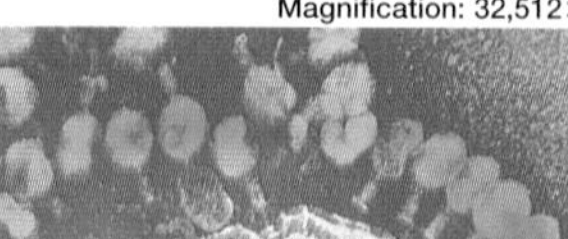
Magnification: 32,512×

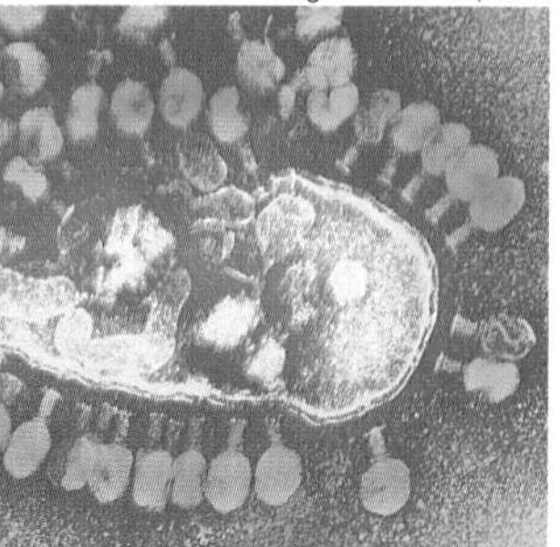

Figure 21-3 Bacteriophage infecting a bacterium. Bacteriophages first attach to a bacterial cell, push their DNA into it, and then produce more viruses.

Viruses Replicate Inside Living Cells

Viruses lack the enzymes necessary for metabolism and have no structures to make protein. Therefore, viruses must rely on living cells (host cells) for replication, as shown in **Figure 21-3.** Before a virus can replicate, it must first infect a living cell. A bacteriophage punches a hole in the bacterial cell wall and injects its DNA into the cell. A plant virus, like TMV, enters a plant cell through tiny rips in the cell wall at points of injury. An animal virus enters its host cell by endocytosis. Once inside a cell, a virus sets out on one of two different paths. Sometimes the virus will immediately begin to produce more viruses. This is called a lytic cycle. Other times the virus will enter the host DNA and cause no immediate harm. This type of cycle is called a lysogenic cycle.

Lytic cycle

Viruses cause damage when the viruses replicate inside the cells. The entry of the virus into the cell is not by itself harmful, but after the virus has replicated itself several hundred times and breaks out, then the cell is destroyed. Organ damage in an organism can become severe if enough tissue is damaged by the virus. Any agent that causes disease is called a **pathogen** *(PATH uh jehn)*.

The cycle of viral infection, replication, and cell destruction is called the **lytic cycle.** After the viral genes have entered the cell, they use the host cell to replicate viral genes and to make viral proteins, such as capsids. The proteins are then assembled with the replicated viral genes to form complete viruses. The host cell is broken open and releases newly made viruses. This cycle is shown in **Figure 21-4.**

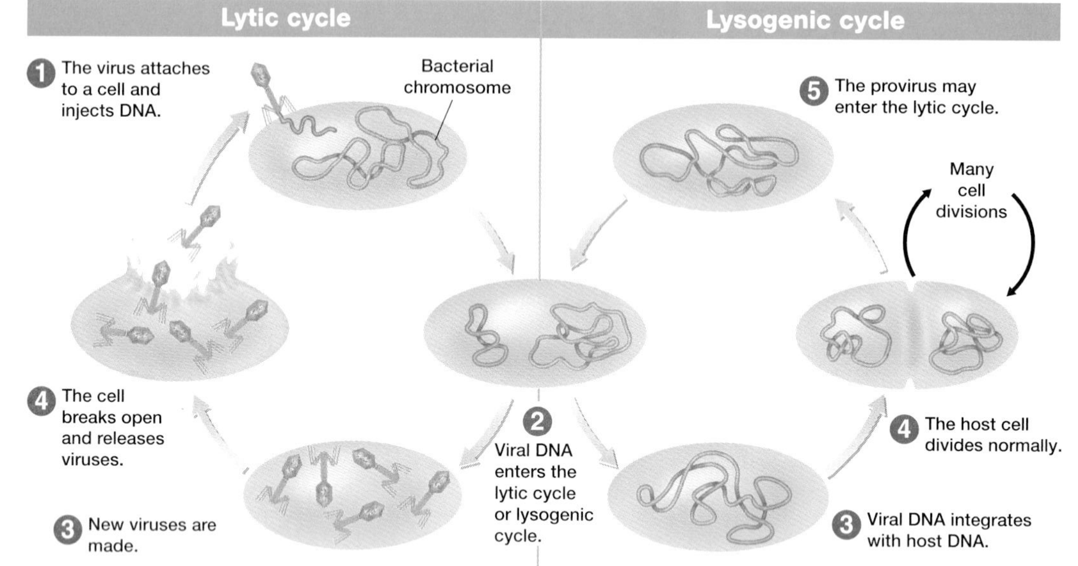
Lytic cycle
Lysogenic cycle
1 The virus attaches to a cell and injects DNA.
Bacterial chromosome
5 The provirus may enter the lytic cycle.
Many cell divisions
4 The cell breaks open and releases viruses.
2 Viral DNA enters the lytic cycle or lysogenic cycle.
4 The host cell divides normally.
3 New viruses are made.
3 Viral DNA integrates with host DNA.

Figure 21-4 Viral replication in bacteria. Bacterial viruses provide a model by which viruses replicate through the lytic cycle or lysogenic cycle.

MULTICULTURAL PERSPECTIVE

Accidental Spread of Viral Diseases

Isolated communities are at particular risk of epidemics when outsiders visit. Diseases spread by travelers can wipe out a small, isolated community. Spanish conqueror Hernán Cortés was aided in his conquest of the Aztecs by a smallpox epidemic that struck those Native Americans. Prior to the arrival of the Europeans, smallpox had been unknown in the New World. The disease killed millions of Native Americans. Today, the Yanomamo tribe of Brazil and Venezuela is dying rapidly because of the onslaught of malaria, influenza, measles, and chickenpox brought by miners in search of gold.

Lysogenic cycle

During an infection, some viruses stay inside the cells but do not make new viruses. Instead of producing virus particles, the viral gene is inserted into the host chromosome and is called a **provirus.** Whenever the cell divides, the provirus also divides, resulting in two infected host cells. In this cycle, called the **lysogenic** *(lie soh JEHN ihk)* **cycle,** the viral genome replicates without destroying the host cell. This cycle is shown in Figure 21-4. In some lysogenic viruses, a change in the environment can cause the provirus to begin the lytic cycle. This results in the destruction of the host cell.

Interrupting HIV Replication

In order for scientists to be able to treat HIV infection it is necessary to understand how HIV replicates. Understanding HIV replication enables scientists to develop drugs that will interrupt the HIV infection cycle.

Getting into the cell

Recall from Chapter 3 that every cell has a specific group of cell surface proteins that identifies it for other cells. HIV can attach to immune system cells because each HIV has a receptor protein on its surface that fits a cell-surface protein on the immune system cell. After attachment, the immune system cell takes in HIV.

Making more viruses

Once inside the host cell, HIV sheds its capsid. This releases the viral RNA and a viral enzyme. This enzyme, called reverse transcriptase, uses the viral RNA as a template to make a strand of complementary viral DNA. The immune system cell copies this strand of viral DNA to make double-stranded viral DNA. This double-stranded viral DNA may integrate into the host DNA and direct the host cell to produce many copies of HIV, as shown in the illustration. AIDS drugs such as AZT and 3TC interfere with the production of the complementary viral DNA strand.

When HIV multiplies within an infected immune system cell, its genes are translated into new viral proteins. These viral proteins include proteins used to make the viral envelope, capsid, and enzymes such as reverse transcriptase. New AIDS drugs such as protease inhibitors stop the processing of the viral proteins needed to assemble new HIV.

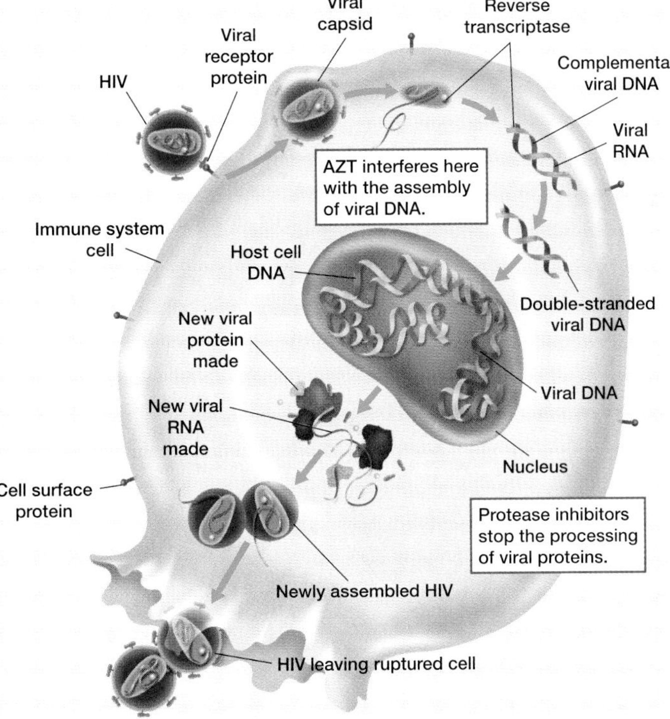

HEALTH WATCH

Interrupting HIV Replication

Teaching Strategies

- Use this figure to trace and review the steps involved in HIV replication within a white blood cell. After HIV builds an outer envelope that binds to a white blood cell's cell membrane, the virus is engulfed by the cell. Within the cell, reverse transcriptase generates DNA using the viral RNA template, and the genes are transcribed and translated into HIV proteins. The cell continues to manufacture HIV until it bursts, releasing many viruses.
- Azidothymidine (AZT) is an antiviral drug used to treat AIDS. AZT can suppress replication somewhat of HIV during reverse transcription. Although the drug initially appeared to be useful in slowing down the onset of AIDS in asymptomatic patients and in treating those mildly afflicted with AIDS-related illnesses, AZT by itself has not always been effective.
- Tell students that HIV proteins are transcribed as a very long protein. A virus-produced enzyme called protease cuts the protein into functional proteins, which are then assembled into a virus. Protease inhibitors prevent protease from processing viral proteins, thus preventing assembly of new viruses.

Discussion

- Why is it possible to cure some bacterial infections but not viral infections? *(Antibiotics can be used to kill bacteria. Because viruses are not alive, no antibiotics exist to kill them. Viral treatments focus on prevention.)*
- What becomes of the proteins made by the host cell under the direction of the viral genome? *(The proteins become the virus envelope, capsid, and replication proteins.)*

REAL LIFE

Answer

5,626,000 people would not know they have HIV.

Teaching TIP — ADVANCED

Future Headlines

Ask students to write headlines about HIV that they would expect to see in 10 years. Discuss the headlines and the predictions behind them. Then have students write short essays to explain their predictions about AIDS in the future. Post the headlines and essays around the room.

PORTFOLIO

CAREER

Epidemiologist

Epidemiologists are employed by hospitals, health departments, universities, and private consulting firms. The work is often a combination of biology, medicine, and detective-like problem solving. Epidemiologists study human health and disease, investigate and identify new cases, and work to prevent epidemics. A strong background in biology, especially microbiology, statistics, and evolution, is vital. Many epidemiologists also have advanced degrees in public health or medicine.

REAL LIFE

How many people have AIDS? The World Health Organization estimated that in 1998 5.8 million people were newly infected with HIV.

Calculating *If only three percent of newly infected people in 1998 know they have HIV, what is the number of people who don't know they have HIV?*

In animal cells, viruses can replicate slowly so that the host cell is not destroyed by the virus. For example, the virus that causes cold sores in humans hides deep in the nerves of the face. When the conditions in the body become favorable for the virus, such as when you are under stress, then the virus begins to cause tissue damage that you see as a cold sore or fever blister.

Viruses are host cell specific

Viruses are often restricted to certain kinds of cells. For example, TMV infects tobacco and related plants, but does not infect animals. Scientists hypothesize that this specificity may be due to the viruses' origin. Viruses may have originated when fragments of host genes escaped or were expelled from cells. The hypothesis that viruses originated from a variety of host cells may explain why there are so many different kinds of viruses. Biologists think there are at least as many kinds of viruses as there are kinds of organisms.

HIV can replicate in your cells

Acquired immunodeficiency syndrome (AIDS) is a disease in which an individual is unable to defend his or her own body against infections that do not normally occur in healthy individuals, as shown in **Figure 21-5.** This is why it is called an immunodeficiency. A syndrome is a disease defined by a set of many symptoms. This syndrome can be transmitted from one person to another, so it is called an acquired syndrome.

AIDS is caused by the human immunodeficiency virus (HIV). HIV-infected people do not develop AIDS symptoms until years after infection. As a result, an HIV-infected individual can feel healthy and still spread the virus to others.

HIV is not passed from an infected person to a healthy one through casual contact. It is transmitted in body fluids (such as semen or vaginal fluid) through sexual contact and in blood through the sharing of nonsterile needles. It is also transmitted to infants during pregnancy or through breast milk.

Figure 21-5 AIDS patient. People suffering from HIV infection have a weakened immune system and therefore become ill from unusual cancers and infectious organisms.

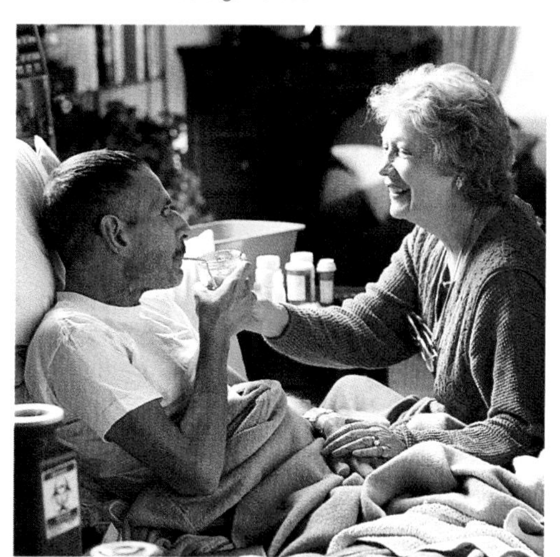

How HIV defeats the immune system

In healthy individuals, specialized cells attack and destroy any invading bacteria, virus-infected cell, or cancer cell. These cells make up our immune system. In AIDS patients, these cells are gradually infected and destroyed by HIV as more viruses are made. Without these "attack and destroy" cells, the body cannot defend itself against infection. After years of HIV infection, so many immune cells have been destroyed that AIDS patients usually die of infections that a healthy person would normally resist.

Up Close

AIDS Virus

- **Name:** Human immunodeficiency virus (HIV)
- **Size:** 125 nm
- **Habitat:** Interior of human white blood cells

Characteristics

Viral proteins Glycoproteins are embedded within the HIV envelope. They enable the virus to recognize the surface protein of the human white blood cell and to enter the cell.

Envelope The outer envelope of HIV is composed of a lipid bilayer derived from the membrane of the host cell. Beneath the envelope is a protein layer called a capsid.

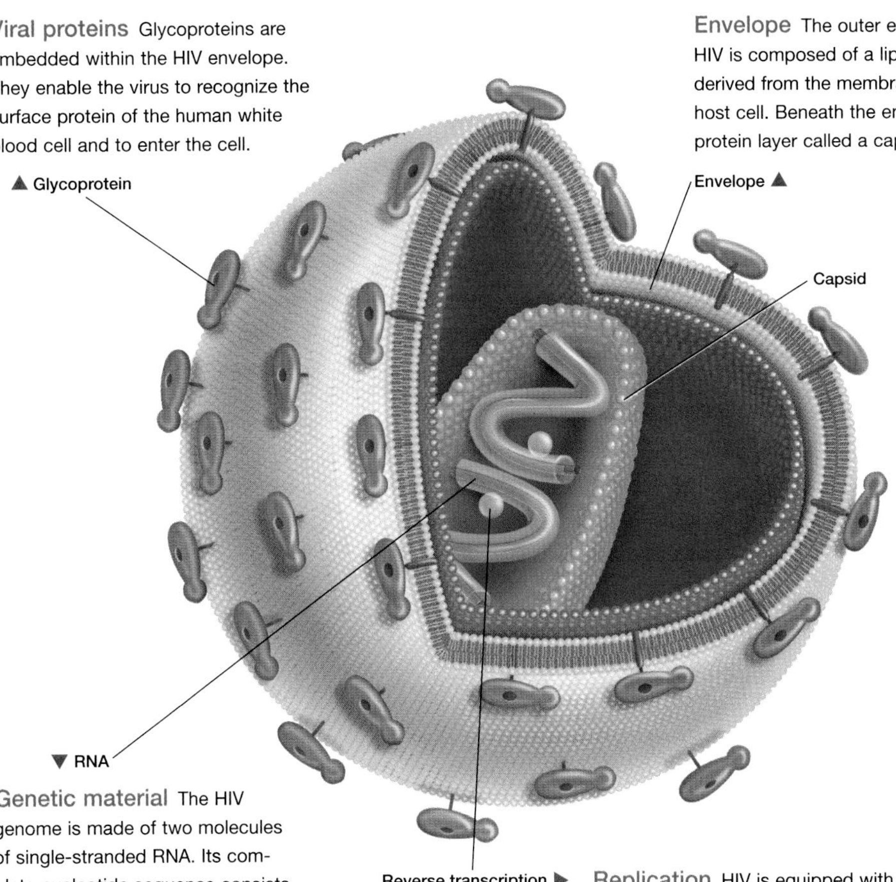

Genetic material The HIV genome is made of two molecules of single-stranded RNA. Its complete nucleotide sequence consists of approximately 9,000 nucleotides, which make up nine genes. Three of these genes are found in many different viruses.

Replication HIV is equipped with the enzyme reverse transcriptase. Once inside a cell, reverse transcriptase makes a DNA copy of the virus's RNA. Using the cell's gene translation machinery, this DNA then directs the production of thousands of new viruses. The new viruses are released when they bud out of the host cell or, occasionally, when the cell bursts.

Up Close

AIDS Virus

Teaching Strategies

- The first cases of AIDS were reported in the late 1970s. Have students obtain statistics on the spread of AIDS from its discovery to the present time. Ask students to present the information in brief reports, including a bar graph that compares the number of AIDS-related deaths over the past decades.
- Ask students to design a hypothetical drug to disable the infection cycle of HIV. Students' drugs can target any step within the cycle, such as preventing the virus from attaching to white blood cells, inhibiting endocytosis of the virus, or destroying reverse transcriptase. Ask students to make diagrams of the HIV infection cycle that show their drug at work. Remind students that the best way to fight AIDS is to prevent it. PORTFOLIO

Discussion

- Describe the HIV genome. *(The genome contains two single-stranded RNA molecules. The nucleotide sequence is composed of 9,000 nucleotides, which make up 9 genes.)*
- What is the role of reverse transcriptase in HIV reproduction? *(Reverse transcriptase makes a DNA copy of the virus's RNA. Without reverse transcriptase, the virus would be unable to use the cell's machinery to reproduce.)*

Biology in Your Community

Your County Health Department

Ask a county public health official to speak to your class about the kind of work the county public health department does to prevent viral diseases. Be sure to ask the speaker to discuss viral diseases that are of concern in your county (for example, mosquito-transmitted viral encephalitis, measles, polio, influenza, HIV). Check with your school's guidelines about the kinds of subject areas that are appropriate. Ask the speaker to bring samples of equipment that the county uses in its health work.

Chapter 21

Teaching TIP

Eradicating Smallpox

Tell students that smallpox is the first disease thought to have been completely eradicated across the Earth. Then tell them that through a highly effective worldwide vaccination strategy, the World Health Organization was able to wipe out the disease. The last known smallpox case outside of the lab was reported in 1977 in Somalia. Although officially declared eradicated in 1980, samples of the smallpox virus have been stored in Atlanta, Georgia, and Moscow, Russia. Controversy exists over whether to destroy the virus or maintain samples for future study.

Teaching TIP

Influenza

Explain to students that because influenza is an airborne virus, it is highly contagious. Then tell them that a common flu virus can be lethal to people who have respiratory ailments or compromised immune system as well as to elderly people and young children. Sometimes a new flu virus can be very lethal. For example, in 1918 influenza killed 20 million people—more people than were killed in combat during World War I.

Viruses Cause Many Diseases

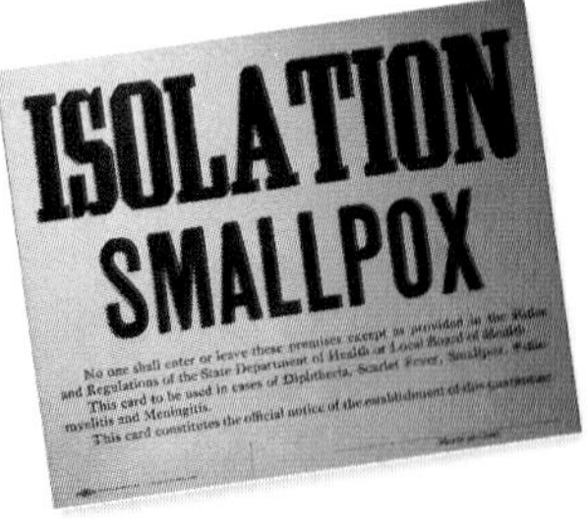

Diseases caused by viruses have been known and feared for thousands of years. Perhaps the most lethal virus in human history was the influenza virus. Commonly known as the flu, influenza is a viral disease of the lungs and throat characterized by chills, fever, and muscular aches. The virus infects cells of the upper respiratory tract. There the viruses replicate and spread to new cells. Some 22 million Americans and Europeans died of flu during an 18-month period in 1918–1919. **Table 21-1** lists a few viral diseases.

Recall from Chapter 6 that viruses (among other things) can also cause cancer. Cancer is a condition in which cells uncontrollably reproduce as a result of the inability of certain genes to control cell growth and division. Viruses associated with certain human cancers include hepatitis B (liver cancer), Epstein-Barr virus (Burkitt's lymphoma), and human papilloma virus (cervical cancer).

Table 21-1 Important Viral Diseases

Disease	Symptoms	How the disease is transmitted
Chickenpox	Blisters, rash, muscle soreness, fever	Inhalation
Influenza	Fever, chills, fatigue, cough, sore throat, muscle aches, weakness, headache	Inhalation
Rubella	Rash, swollen glands, fever, fatal to developing infant in pregnant woman	Inhalation
Mumps	Painful swelling in salivary glands	Inhalation
Smallpox	Blisters, lesions, fever, malaise, blindness, disfiguring scars; often fatal	Inhalation
Hepatitis A Hepatitis B	Fever, chills, nausea, swollen liver, yellow skin, painful joints, liver cancer	Contaminated blood, food, or water
Polio	Fever, headache, stiff neck, possible paralysis	Contaminated food or water
AIDS	Immune system failure; fatal	Sexual contact, contaminated blood, or contaminated needles
Cold	Sinus congestion, muscle aches, cough, fever	Inhalation, direct contact
Rabies	Mental depression, fever, restlessness, difficulty swallowing, paralysis, convulsions; fatal	Bite of infected animal

did you know?

Hantavirus is an emerging virus in North America.

On the Navajo reservation in the four corners region of Colorado, New Mexico, Utah, and Arizona, a viral infection previously unknown to Americans killed 16 people within 6 months in 1993. This pulmonary syndrome, which is spread by deer mice, is caused by a hantavirus, name after the Hantaan River in Korea, where U.S. military researchers first encountered it. The American hantavirus variant attacks the lungs primarily, unlike Asian and European strains, which cause hemorrhagic fever and kidney disease. Health workers still do not know whether the American hantavirus variant is really new or whether it has simply gone unnoticed until recently.

Emerging viruses

Viruses that evolve in geographically isolated areas and are pathogenic to humans are called **emerging viruses.** These new pathogens are dangerous to public health. People become infected when they have contact with the normal hosts of these viruses.

In the United States, the Hantavirus is considered an emerging virus. First detected in the southwestern United States, the Hantavirus occurs in wild rodents but causes a lethal illness in humans. Other types of Hanta viruses occur worldwide. Roughly 40 to 50 percent of its human victims die. Other emerging viruses include the Ebola (Africa), Lassa (Africa), and Machupo (South America) viruses.

TOPIC: Viral diseases
GO TO: www.scilinks.org
KEYWORD: HX459

Prions and viroids

In 1982, the American scientist Stanley Pruisner, of Stanford University, described a new class of infectious particles. He called these infectious particles **prions** *(PREE awnz)*. Prions are particles that are composed of proteins and have no nucleic acid. Prions are infectious even though they contain no genes. They cause infection by influencing how proteins fold into their active shape. A disease-causing prion is folded into a shape that does not allow it to function. Such a misfolded prion will cause a normal prion to misfold simply by contacting the normal prion. In this way the misfolding spreads like a chain reaction.

Prions were first linked to a sheep disease called scrapie. Later, brain diseases such as "mad cow disease," shown in **Figure 21-6,** and Creutzfeldt-Jakob disease were also associated with prions. Anyone unfortunate enough to eat food containing the disease-causing prion might become infected. Pruisner was awarded the 1997 Nobel Prize in medicine for his work.

A **viroid** *(VEER oid)* is a single strand of RNA that has no capsid. They are important infectious disease agents in plants. Viroids have affected economically important plants such as cucumbers, potatoes, avocados, and oranges.

Figure 21-6 **Mad cow.** This cow is showing signs of mad cow disease. Prions have begun to destroy this cow's brain, thus making it difficult for it to walk.

Review SECTION 21-1

1. **Compare** the properties of viruses with the properties of organisms.
2. **Describe** Stanley's experiment with the tobacco mosaic virus.
3. **Name** the parts of a virus.
4. **List** the steps by which viruses replicate.
5. **Describe** how HIV causes AIDS.

Critical Thinking

6. **Evaluate** the argument that emerging viruses are new viruses.

Review SECTION 21-1 ANSWERS

1. Viruses do not grow, do not exhibit homeostasis, and cannot metabolize—all key characteristics of life. Like organisms, viruses replicate (although not without the help of a host cell) and are guided by information encoded in their DNA (or RNA).
2. Stanley purified the tobacco mosaic virus and found it was crystalline in structure. This crystal was then used to infect healthy tobacco plants.
3. Viruses consist of a capsid (a DNA- or RNA-containing protein) and may be surrounded by an envelope (proteins, lipids, and glycoproteins).
4. Viruses enter the host cell. This is followed by replication of the viral genome, which is then used to create viral proteins. The proteins are assembled into new viruses, which cause the cell to burst, releasing many new viruses.
5. HIV enters and destroys the cells of the body's immune system. Without these cells, the body is unable to fight off the many infectious agents that routinely enter the body.
6. Emerging viruses are new in the sense that such viruses have newly been recorded as causing diseases. But such viruses may have simply gone unnoticed or unrecorded.

Section 21-1

CROSS-DISCIPLINARY CONNECTION

◆ **Chemistry**

Disease-causing prions have the same amino acid sequence as the normal prion protein, but the disease-causing prions are folded in a way that changes their function and physical characteristics. The abnormal prion is resistant to proteases, is insoluble in water, and has a high degree of β-sheet folding. Normal prion protein is susceptible to proteases, is soluble in water, and has a mostly α-helical structure. Remind students that the physical characteristics of proteins can change when the proteins are manipulated, but their amino acid sequence stays the same. For example, egg white solidifies when cooked; milk casein curdles when vinegar is added.

3 Assessment Options

Closure

Technique: K-W-L

Tell students to return to the list that they created using the **Active Reading** on page 451. Have them check off the questions that they can now answer. Students should then make a list of what they have Learned. Conclude by asking:

- Which questions are still unanswered?
- What new questions do you have?

You may wish to have students research their unanswered questions.

Review

Assign Review—Section 21-1.

Reteaching BASIC

Have each student write a brief report on a viral disease. Providing a list of diseases may be helpful. The report should concentrate on a brief description of the disease, the virus that causes it, its incidence, and its treatment. 

Section 21-2 Bacteria

1 Prepare

Scheduling

- **Block:** About 90 minutes will be needed to complete this section.
- **Traditional:** About two class periods will be needed to complete this section.
- **Planning:** See the Planning Guide on pp. 449A and 449B for lecture, classwork, and assignment options for Section 21-2. Lesson plans for Section 21-2 can be found in a lesson cycle format in ***Biology: Principles and Explorations* One-Stop Planner CD-ROM with Test Generator.**

Resource Link

Assign **Active Reading Worksheet Section 21-2** to help students comprehend the key concepts and visuals in this lesson.

2 Teach

Lesson Warm-up

Ask students to name some places where they might find bacteria. Write their responses on the board or overhead for discussion. Tell students that they are all correct because bacteria live everywhere—in the human body and on everything we touch and eat. Emphasize that bacteria are an important part of the environment.

Objectives

- **List** seven differences between bacteria and eukaryotic cells. (p. 460)
- **Describe** three different ways bacteria can obtain energy. (pp. 462–463)
- **Describe** the external and internal structure of *Escherichia coli.* (p. 464)
- **Distinguish** two ways that bacteria cause disease. (pp. 465–466)
- **Identify** three ways that bacteria benefit humans. (p. 468)

Key Terms

pilus
bacillus
coccus
spirillum
capsule
antibiotic
endospore
conjugation
anaerobic
aerobic
toxin

Bacteria Have a Simpler Structure than Eukaryotes

Bacteria, which outnumber all eukaryotes combined, differ from eukaryotes in at least seven ways. You can study the characteristics of *Escherichia coli,* eubacteria that reside in your intestinal tract, in Up Close: *Escherichia coli,* on page 464.

1. **Internal compartmentalization.** Bacteria are prokaryotes. Unlike eukaryotes, prokaryotes lack a cell nucleus. Bacterial cells have no internal compartments or membrane systems.
2. **Cell size.** Most bacterial cells are about 1 µm in diameter; most eukaryotic cells are more than 10 times that size.
3. **Multicellularity.** All bacteria are single cells. Some bacteria may stick together or may form strands. However, these formations are not truly multicellular because the cytoplasm in the cells does not directly interconnect, as is the case with many multicellular eukaryotes. Also, the activities of the cells are not specialized.
4. **Chromosomes.** Bacterial chromosomes consist of a single circular piece of DNA. Eukaryotic chromosomes are linear pieces of DNA that are associated with proteins.
5. **Reproduction.** Bacteria reproduce by binary fission, a process in which one cell pinches into two cells. In eukaryotes, however, microtubules pull chromosomes to opposite poles of the cell during mitosis. Afterward, the cytoplasm of the eukaryotic cell divides in half, forming two cells.
6. **Flagella.** Bacterial flagella are simple structures composed of a single fiber of protein that spins like a corkscrew to move the cell. Eukaryotic flagella are more-complex structures made of microtubules that whip back and forth rather than spin. Some bacteria also have shorter, thicker outgrowths called **pili** *(PIHL ee),* shown in **Figure 21-7.** Pili enable bacteria to attach to surfaces or to other cells.
7. **Metabolic diversity.** Bacteria have many metabolic abilities that eukaryotes lack. For example, bacteria perform several different kinds of anaerobic and aerobic processes, while eukaryotes are mostly aerobic organisms.

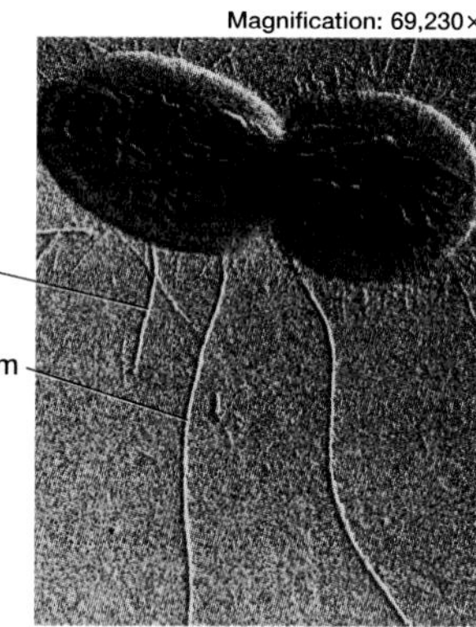

Figure 21-7 Flagella and pili. Bacteria have flagella that provide them with movement and pili that enable adherence to surfaces.

Integrating Different Learning Styles

Logical/Mathematical	**PE**	Data Lab, p.467; Performance Zone items 15 and 24, pp. 470–471
	ATE	Cross-Disciplinary Connection, p. 462; Did You Know?, p. 462; Teaching Tip, p. 465
Visual/Spatial	**PE**	Performance Zone item 1, p. 470
	ATE	Graphic Organizer, p. 463; Demonstrations, pp. 465, 468
Verbal/Linguistic	**PE**	Study Tip p. 461
	ATE	Multicultural Perspective, p. 461; Teaching Tips, pp. 462, 463; Up Close, p. 464

The structures of bacterial cells are unique

A bacterial cell is usually one of three basic shapes, as shown in **Figure 21-8:** **bacillus** *(buh SIHL uhs)*, a rod-shaped cell; **coccus** *(KAHK us)*, a round-shaped cell; or **spirillum** *(spy RIHL uhm)*, a spiral cell. A few kinds of bacteria aggregate into strands. Species that form filaments are indicated by the prefix *strepto-*, while species that form clusters are indicated by the prefix *staphylo-*.

Most bacteria have a cell wall that surrounds the cell membrane. Members of the kingdom Eubacteria have a cell wall made of peptidoglycan, a network of polysaccharide molecules linked together with chains of amino acids. Outside the cell wall and membrane, many bacteria have a gel-like layer called a **capsule.**

Cell walls Bacteria can have two types of cell walls, distinguished by a dye staining technique, called the Gram stain. One group is called Gram-negative, and the other Gram-positive.

Gram staining is important in medicine because the two groups of bacteria differ in their susceptibility to different antibiotics. **Antibiotics** are chemicals that interfere with life processes in bacteria. Thus, Gram-staining can help determine which antibiotic would be most useful in fighting an infection.

Endospores Some bacteria form thick-walled **endospores** *(EHN doh spohrz)* around their chromosomes and a small bit of cytoplasm when they are exposed to harsh conditions. These conditions can be the depletion of nutrients, a drought, or high temperatures. Endospores can survive environmental stress and may germinate years after they were formed and release new, active bacteria.

Pili Pili enable bacteria to adhere to the surface of sources of nutrition, such as your skin. Some kinds of pili enable bacteria to exchange genetic material through a process called conjugation. **Conjugation** *(kahn juh GAY shuhn)* is a process in which two organisms exchange genetic material. In prokaryotes, pili from one bacterium adhere to another bacterium, and genetic material is transferred from the bacterium that has pili to the one that doesn't have pili. Conjugation enables bacteria to spread antibiotic resistance genes from one genus to another.

Study TIP

Reviewing Information

Prepare flash cards for each of the **Key Terms** in this chapter. On each card, write the term on one side and its definition on the other side. Use the cards to review meanings of the Key Terms.

Figure 21-8 Bacterial shapes. Bacteria are usually one of three shapes.

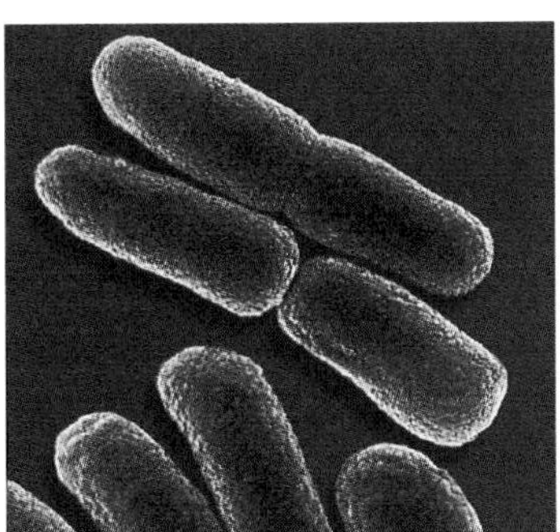

Bacillus (rod-shaped)
E. coli

Magnification: 117,300×

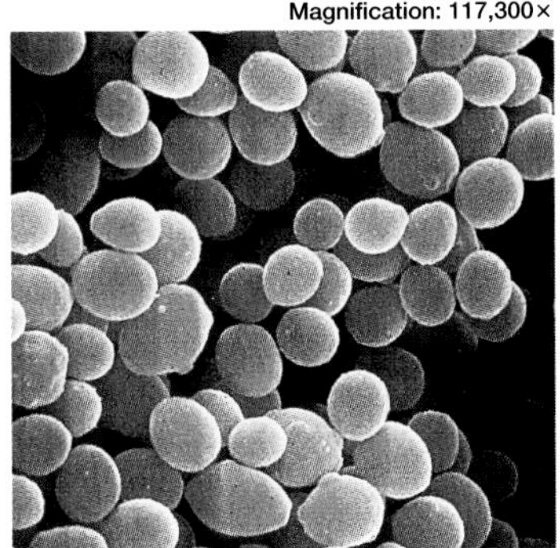

Coccus (round-shaped)
Micrococcus luteus

Magnification: 2295×

Spirillum (spiral-shaped)
Spirillum volutans

MULTICULTURAL PERSPECTIVE

Making Yogurt

The first yogurt was made by nomadic Middle Eastern tribes. The hot desert sun, aided by the gently rocking of camel transportation, provided the perfect environment for making yogurt. Yogurt is created from milk that has been curdled by two types of bacteria: *Lactobacillus bulgaricus* and *Streptococcus thermophilus.* Because yogurt is cultured, it is easier to digest than regular milk. In addition, the bacteria in yogurt can produce B vitamins in the intestines. Today many Americans prefer to eat yogurt that has been sweetened and flavored with fruit.

Teaching TIP

Movement and Locomotion

Explain that bacteria have several mechanisms of locomotion, including sliding over slimy surfaces, twisting through fluids, and propelling themselves with flagella. However, not all bacteria move. Bacteria that are capable of movement are called *motile* bacteria. Those that cannot move are called *nonmotile* bacteria.

internet connect

Have students research the structure of bacteria using the Web site in the Internet Connect box on this page. Have students write a report summarizing their findings. PORTFOLIO

Chapter 21

Teaching TIP

Bacteria in Cheese

Encourage students to research the roles of bacteria and heat in making cheese. The production of cheese involves bacteria that break down lactose in milk, producing lactic acid as a waste product. The acid causes the milk to separate into curds, solid components from which cheese is made, and whey, a liquid product that is removed. Cheeses require a heating process that can destroy the bacteria if the conditions are excessive, and different types of cheeses are made with different kinds of bacteria. For example, cheddar cheese is made with bacteria that require moderate temperatures, while Swiss cheese is made with bacteria that can tolerate higher temperatures.

CROSS-DISCIPLINARY CONNECTION

◆ Mathematics

Explain to students that in theory each colony on a Petri dish arose from a single bacterium and that bacteriologists use this understanding to count the number of bacteria in a sample. Tell the students that a 1 mL sample of bacterial solution was diluted 1,000 times in water and then a 1 mL sample was added to a Petri dish. Twenty-four hours later, 38 colonies were counted. Ask students to calculate how many bacteria were in the original sample. *($38 \times 1{,}000 = 38{,}000$ bacteria per mL)* Then ask students to calculate how many bacteria would have been in the sample if the dilution factor had been only 250 and only a 0.1 mL sample had been put on the Petri dish. *($38 \times 10 \times 250 = 95{,}000$ bacteria per mL)*

biobit

How big are the largest bacteria?

In 1999, scientists announced the discovery of the largest bacteria ever discovered. *Thiomargarita namibiensis* was found off the coast of Namibia. More than 100 times larger than the previously known largest bacterium, *T. namibiensis* is 0.5 mm wide.

Bacteria Are Grouped According to How They Obtain Energy

Over 4,000 species of bacteria have been named, and probably many more haven't yet been discovered. Bacteria occur in the widest possible range of habitats and play key ecological roles in nearly all of them. As you will recall from Chapter 20, bacteria thrive in hot springs, frigid arctic seas, and ground water. They are even found at high pressures in the deep sea and inside solid rock.

Bacteria can be classified in several different ways. Classifying bacteria by the different ways in which they obtain energy, for example, gives a good general sense of the great diversity among bacteria.

Bacteria can also be classified according to their phylogenetic relationships. By comparing the sequence of their ribosomal RNA, scientists have determined that there are at least 12 phyla of eubacteria and four phyla of archaebacteria.

Photosynthetic bacteria

Much of the world's photosynthesis is carried out by bacteria. Photosynthetic bacteria can be classified into four major groups based on the photosynthetic pigments they contain: purple nonsulfur bacteria, green sulfur bacteria, purple sulfur bacteria, and cyanobacteria. Green sulfur bacteria and purple sulfur bacteria grow in **anaerobic** (oxygen-free) environments. They cannot use water as a source of electrons for photosynthesis and instead use sulfur compounds, such as hydrogen sulfide, H_2S. Purple nonsulfur bacteria use organic compounds, such as acids and carbohydrates, as a source of electrons for photosynthesis.

Of particular importance are the cyanobacteria, which often clump together in large mats of filaments. Recall that cyanobacteria are thought to have made the Earth's oxygen atmosphere. Each filament is a chain of cells encased in a continuous jellylike capsule. Many cyanobacteria, such as species of *Anabaena*, shown in **Figure 21-9**, are capable of fixing nitrogen.

Figure 21-9
Photosynthetic bacteria. *Anabaena* is a photosynthetic cyanobacterium in which individual cells adhere in filaments. The two large orange-colored cells are encased in a structure where nitrogen fixation occurs.

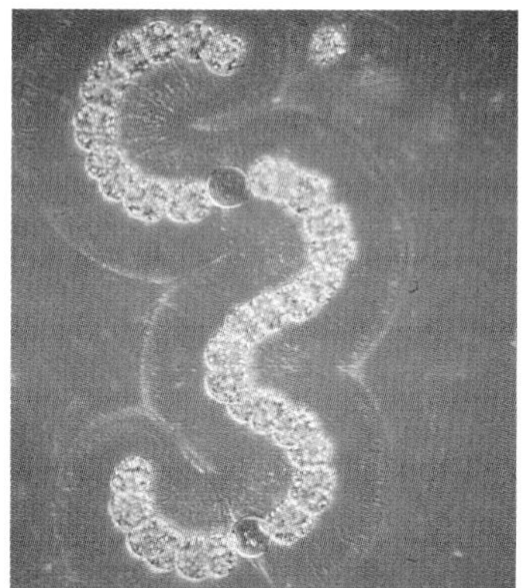

Chemoautotrophic bacteria

Bacteria called chemoautotrophs *(KEE moh awtoh trohfs)* obtain energy by removing electrons from inorganic molecules such as ammonia, NH_3, and hydrogen sulfide, H_2S, or from organic molecules such as methane, CH_4. In the presence of one of these hydrogen-rich chemicals, chemoautotrophic bacteria can manufacture all their own amino acids and proteins. Chemoautotrophic bacteria that live in the soil, such as *Nitrosomonas* and *Nitrobacter*, are of great importance to the environment and to agriculture. They have an important role in the nitrogen cycle called nitrification. Nitrification, as you will recall from Chapter 17, is the process in which bacteria oxidize ammonia into nitrate. Nitrate is the form of nitrogen most commonly used by plants.

did you know?

Certain kinds of bacteria are oil eaters.

An oil-eating bacterium that is used to clean up oil spills was the first organism protected under a United States patent in 1980. Ananda Chakrabarty engineered the bacterium.

Heterotrophic bacteria

Most bacteria are heterotrophs, feeding on organic material formed by other organisms. Together with fungi, heterotrophic bacteria are the principal decomposers of the living world; they break down the bodies of dead organisms and make the nutrients available to other organisms. Most of the odors associated with soil come from substances produced by heterotrophic bacteria. Many are **aerobic,** that is, they live in the presence of oxygen. Other bacteria can live with or without oxygen.

Other activities of heterotrophic bacteria may be helpful or harmful to humans. For example, more than half of our antibiotics are produced by several species of *Streptomyces,* a filamentous bacterium found in soil. On the other hand, one species of *Staphylococcus* can secrete a poison into food. This poison causes nausea, diarrhea, and vomiting in people who eat the *Staphylococcus*-contaminated food.

Species of the symbiotic bacteria *Rhizobium* are by far the most important of all nitrogen-fixing organisms. *Rhizobium* species are heterotrophic bacteria that usually live within lumps on the roots of legumes (plants such as soybeans, beans, peas, peanuts, alfalfa, and clover), as shown in **Figure 21-10.** Farmers take advantage of *Rhizobium's* nitrogen-fixing abilities when they "rotate" their crops every few years and grow legumes, which replenish the soil with nitrogen-containing compounds.

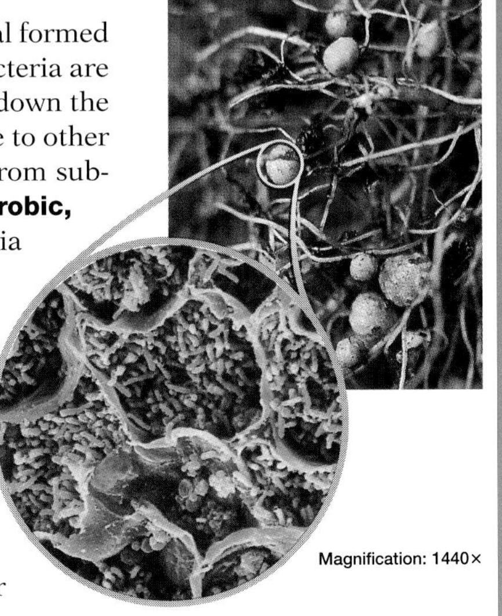

Figure 21-10 Nitrogen fixing bacteria. The bacteria inside these lumps on these soybean roots contain a species of nitrogen-fixing bacteria in the genus *Rhizobium.*

Exploring Further

The Gram Reaction

When you become sick with a bacterial infection, one of the first things a physician will want to know about the bacteria causing the disease is their Gram-reaction. Because Gram-positive and Gram-negative bacteria differ in their susceptibility to antibiotics, it is important to know their Gram-reaction.

The difference between Gram-negative and Gram-positive bacteria is that Gram-positive bacteria have a thicker peptidoglycan layer than do the Gram-negative bacteria.

How it's done

The bacteria are first stained with a violet dye, and then an iodine solution is applied to help hold the dye in the cells. The cells are then washed with alcohol. The alcohol wash removes the violet dye from those bacteria that cannot retain the dye. The bacteria are then stained with a light pink dye. Gram-positive bacteria are able to resist the destaining with alcohol and therefore retain the violet dye. Gram-negative bacteria are unable to retain the dye and are stained pink. The differences are shown in the figures above.

As a result of the Gram stain, a physician can choose the kind of antibiotic that is most effective against the kind of bacteria that are infecting you.

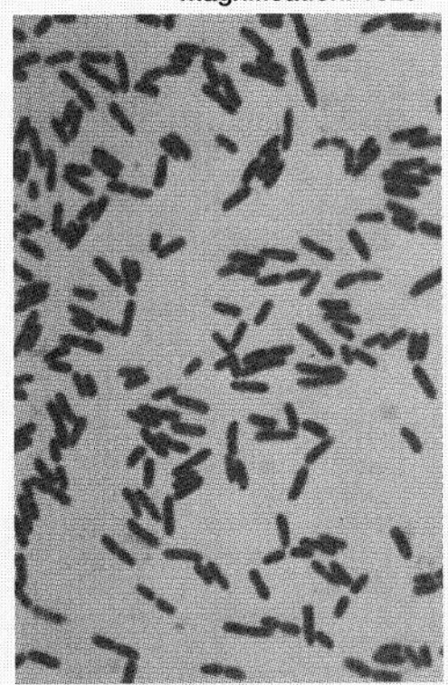

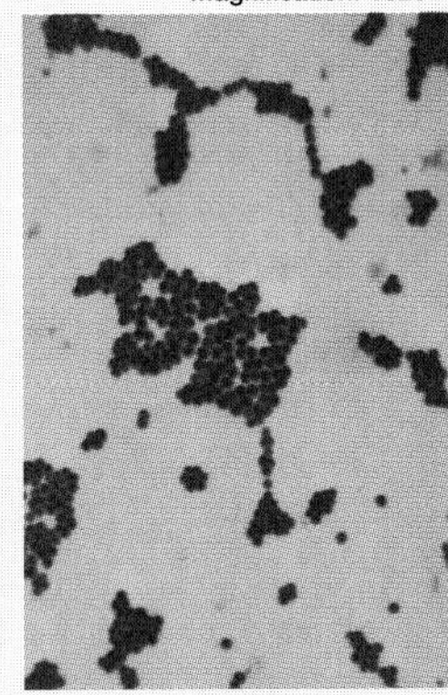

Graphic Organizer

Use this graphic organizer with the ***Teaching Tip: Classifying Bacteria*** on page 463.

Type of bacteria	Energy source	Examples
Photosynthetic	sunlight	cyanobacteria, green sulfur bacteria, purple sulfur bacteria
Chemoautotrophic	electrons in inorganic molecules	*Nitrosomonas, Nitrobacter*
Heterotrophic	organic material from other organisms	Streptomyces, Rhizobium

Teaching TIP — BASIC

Classifying Bacteria

Have students draw a table like the **Graphic Organizer** at the bottom of page 463 to classify bacteria by the way they obtain energy. Tables should divide bacteria into three groups: photosynthetic, chemoautotrophic, and heterotrophic. PORTFOLIO

Teaching TIP

Nitrogen-Fixing Bacteria

Ask students what would happen if all nitrogen-fixing bacteria were eliminated. *(The amount of available nitrogen would fall, plant growth would be reduced, and the amount of organic compounds available at higher trophic levels would decline.)*

Exploring Further

The Gram Reaction

Teaching Strategies

- Use the figure to point out the differences in the cell walls of gram-negative and gram-positive bacteria. Note that the thicker peptidoglycan wall of gram-positive bacteria remains intact and prevents the alcohol rinse from removing the dye.
- Because of their outer membranes, gram-negative bacteria tend to resist entry of the antibiotic penicillin, but permit entry of tetracycline. Some penicillins have been chemically altered to permit their passage through the cell wall.

Discussion

- How can the gram-staining technique accelerate the treatment of a patient? *(Doctors can better target their choice of antibiotic.)*
- How does the cell wall determine a bacterium's reaction to the gram stain? *(Because the cell wall regulates what leaves the cell, it determines whether the stain will leave the cell.)*

Up Close

Escherichia coli

Teaching Strategies
Tell students that in 1993, four children died from infection with a strain of *E. coli* O157:H7 found in undercooked hamburger meat, and more than 600 cases of food poisoning were reported in northwestern states. The symptoms of *E. coli* O157:H7 food poisoning include bloody diarrhea and kidney failure. Severe cases can cause permanent damage to organs or even death. The Centers for Disease Control and Prevention estimates that approximately 20,000 individuals suffer symptoms of *E. coli* O157:H7 poisoning each year from undercooked food. Ask students how they can avoid food poisoning. *(Possible answers include cooking meat thoroughly, choosing clean restaurants, and making sure that foods are served and stored at their proper temperatures.)*

Discussion
- Are *E. coli* gram-positive or gram-negative bacteria? *(gram-negative)*
- How do *E. coli* reproduce? *(binary fission)* How fast can they divide? *(as often as every 20 minutes)*
- Describe the genetic material of an *E. coli* cell. *(It has one circular DNA molecule containing about 5,000 genes.)*
- What is the function of pili? *(Pili serve to attach* E. coli *to surfaces and to join bacterial cells in conjugation.)*

Up Close

Escherichia coli

- **Scientific name:** *Escherichia coli*
- **Size:** Up to 1 μm
- **Habitat:** Inhabits the intestines of many mammals
- **Mode of nutrition:** Heterotrophic

Characteristics

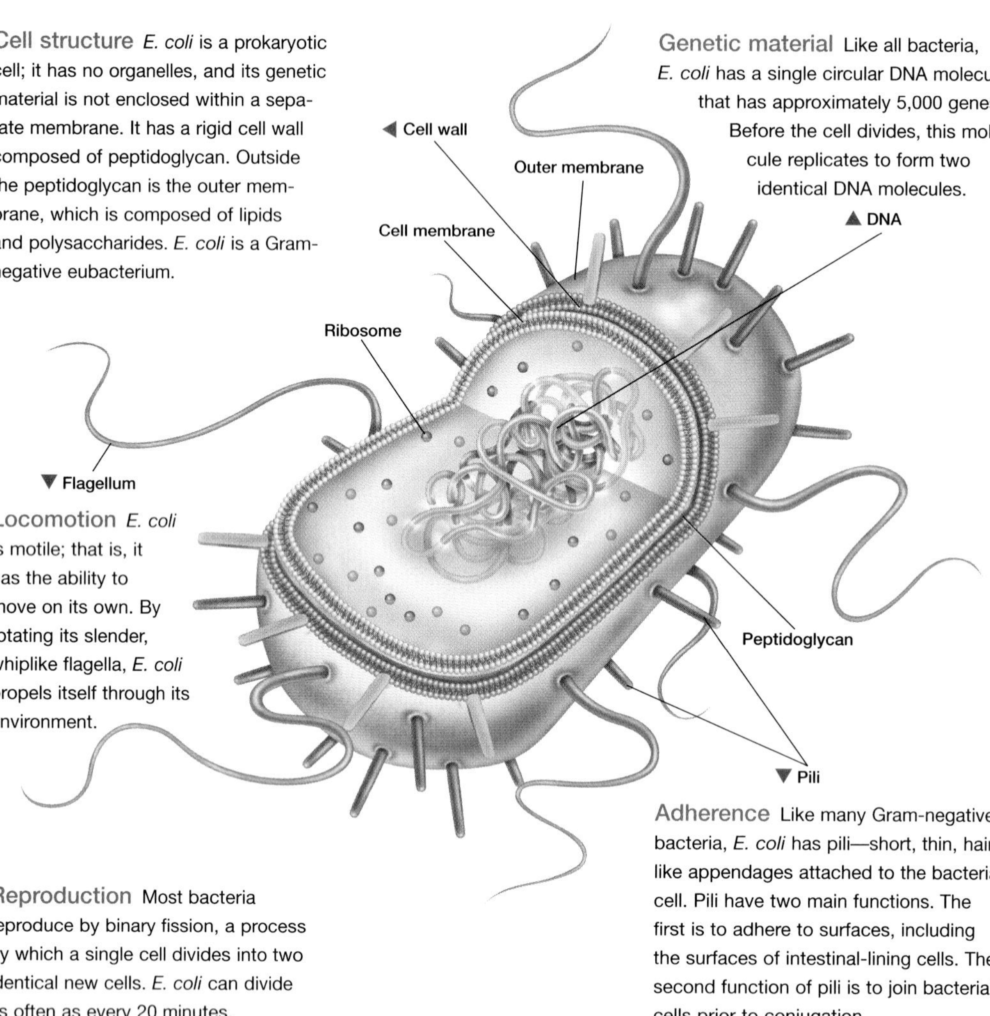

Cell structure *E. coli* is a prokaryotic cell; it has no organelles, and its genetic material is not enclosed within a separate membrane. It has a rigid cell wall composed of peptidoglycan. Outside the peptidoglycan is the outer membrane, which is composed of lipids and polysaccharides. *E. coli* is a Gram-negative eubacterium.

Genetic material Like all bacteria, *E. coli* has a single circular DNA molecule that has approximately 5,000 genes. Before the cell divides, this molecule replicates to form two identical DNA molecules.

Locomotion *E. coli* is motile; that is, it has the ability to move on its own. By rotating its slender, whiplike flagella, *E. coli* propels itself through its environment.

Reproduction Most bacteria reproduce by binary fission, a process by which a single cell divides into two identical new cells. *E. coli* can divide as often as every 20 minutes.

Adherence Like many Gram-negative bacteria, *E. coli* has pili—short, thin, hair-like appendages attached to the bacterial cell. Pili have two main functions. The first is to adhere to surfaces, including the surfaces of intestinal-lining cells. The second function of pili is to join bacterial cells prior to conjugation.

Bacteria Cause Disease in Two Basic Ways

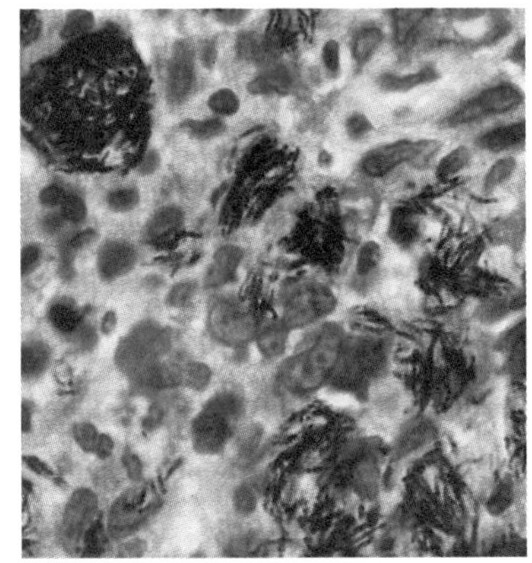

Figure 21-11 Tuberculosis in lung. The red-stained bacteria are *Mycobacterium tuberculosis,* which cause tuberculosis. This disease was once called consumption because its victims lost so much weight it appeared as if they were being consumed.

In order to understand infectious diseases, think of your body as a treasure chest full of resources. Your body has protein, minerals, fats, carbohydrates, and vitamins. You may want to keep and use these resources, but so do many other organisms, including the bacteria on and in your body. Bacteria have evolved various means of obtaining these resources from you. In some cases, the competition for the resources in your body can result in your becoming ill.

Bacteria can metabolize their host

Heterotrophic bacteria obtain nutrients by secreting enzymes that break down complex organic structures in their environment and then absorbing them. If that environment is your throat or lungs, this can cause serious problems.

For example, tuberculosis, a disease of the lungs, is caused by *Mycobacterium tuberculosis,* shown in **Figure 21-11.** Tuberculosis was once one of the most common causes of death. In most cases, infection occurs when tiny droplets of moisture that contain the bacteria are inhaled. Some bacteria settle in the lungs, where they grow using human tissue as their nutrients. The bacteria may also spread to other parts of the body. Symptoms may include coughing up sputum and blood, chest pain, fever, fatigue, weight loss, and loss of appetite. If left untreated, death may occur as quickly as within 18 months but more commonly within 5 years. Other important bacterial diseases are described in **Table 21-2.**

Table 21-2 Important Bacterial Diseases

Disease	Symptoms	Bacterium	How the disease is transmitted or caused
Bubonic plague	Fever, bleeding lymph nodes that form swellings called buboes; often fatal	*Yersinia pestis*	Bite of an infected flea
Cholera	Severe diarrhea and vomiting; fatal if not treated	*Vibrio cholerae*	Drinking contaminated water
Tooth cavities	Destruction of minerals in tooth	*Streptococcus mutans*	Dense collections of bacteria in mouth
Dysentery	Fever, bloody diarrhea, vomiting	*Shigella dysenteriae*	Contaminated food or water
Lyme disease	Rash, pain, swelling in joints	*Borrelia burgdorferi*	Bite of an infected tick
Typhoid fever	Headache, fever, diarrhea, rash	*Salmonella typhi*	Drinking or eating contaminated water or food

did you know?

Tuberculosis is back.

For decades the tuberculosis (TB) death rate in the United States was in decline. In the 1940s health workers believed TB would be eradicated because the disease seemed to be under control. By 1985 the number of reported cases had plunged to only 22,201—one-fourth of the number of cases reported in 1953. However, the number of cases began to rise in 1986. Although TB is treatable and preventable, once it infects a person's lungs, a cure takes about 6 months. Most people afflicted with the disease stop taking their medication before they are completely cured. Even those willing to endure the long treatment may not be cured because some TB bacteria have evolved resistance to the drugs used for treatment.

Teaching TIP

Disease Prevention

Ask students how to prevent bacterial and viral infections. Lead students to consider ways that pathogens can be spread such as through air, water, blood, food, pets, rodents, and so on. *(Students may suggest that diseases can be prevented by sanitation, good hygiene, lifestyle changes, vaccinations, healthy immune systems, and proper storage and preparation of food.)*

CROSS-DISCIPLINARY CONNECTION

◆ History

The plague that spread through Europe in the fourteenth century killed more than 25 million people. Introduced by rats from the Black Sea region, it spread through trade routes to Europe. The disease is caused by the bacterium *Yersinia pestis* and is transmitted by fleas. It causes enlarged lymph nodes, called buboes—hence, the name bubonic plague. Fever, pain, and necrosis, or blackening of the skin, can also occur. This last symptom lead to the disease's other name: Black Death.

Demonstration

Talking About Diseases BASIC

Ask students to bring newspaper and magazine articles about specific bacterial and viral infections to class. Make a list on the board or overhead of the diseases discussed in each article, and have students indicate whether each pathogen is a bacterium or virus. Have students discuss the means of contracting each disease, the people commonly affected, and potential treatments.

Visual Strategy

Figure 21-12

Point out to students that the zone of clearance around the bacteria represents the destruction of red blood cells by *Streptococcus.* Ask students if they think *Streptococcus* species are the only bacteria that lyse red blood cells. *(Hemolysins are also produced by* Staphylococcus *and* Clostridium *species.)*

Making Biology Relevant

Food

The Centers for Disease Control and Prevention recommends that food be kept hot at temperatures above 60°C (140°F) or cold at temperatures below 10°C (50°F) to control the growth of *Staphylococcus aureus* and prevent Staphylococcal intoxication. Other measures recommended by the CDC include frequent handwashing; cleaning fingernails; and not permitting individuals with boils, abscesses, and other infected open wounds to handle food.

Making Biology Relevant

Health

Some bacterial toxins can be used for medical purposes. Botulinum toxin is used to treat dystonia, an abnormal muscle rigidity that causes painful muscle spasms. The toxin inhibits the release of acetylcholine from the neurons in the neuromuscular junction, allowing the muscles to relax. A clot-breaking enzyme produced by *Streptococcus* is used to help dissolve clots that block coronary arteries during a heart attack.

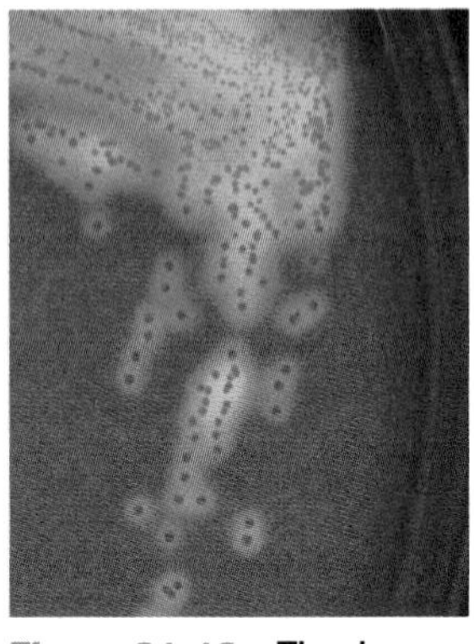

Figure 21-12 The devastation of bacterial toxins. This species of *Streptococcus* secretes a toxin that destroys red blood cells. The agar contains red blood cells and clearly shows a zone around the bacteria where the toxin has destroyed the red blood cells.

Not all bacteria are lethal. For example, some bacteria cause everyday health problems, such as acne. Acne occurs in about 85 percent of teenagers. Bacteria, such as *Propionibacterium acnes,* normally grow in an oil gland of the skin. They metabolize a certain kind of oil produced by those glands. During puberty the oil glands increase the amount of oil produced, and the bacterial population on the skin increases greatly. The bacteria plug the pores where the oil normally flows, and this causes the oil to accumulate in the skin, forming pimples and blackheads.

Bacterial toxins

The second way bacteria cause disease is by secreting chemical compounds into their environment. These chemicals, called **toxins,** are poisonous to eukaryotic cells, as shown in **Figure 21-12.** Toxins can be secreted into the body of an infected person or into a food in which bacteria are growing. For example, *Corynebacterium diphtheriae,* which is Gram-positive and causes diphtheria, grows in the throat, but the toxins attack the heart, nerves, liver, and kidneys. Other bacteria, such as Gram-negative bacteria, produce a type of toxin called endotoxin, which causes fever, body aches, and chills. Typical Gram-negative bacteria, such as *E. coli, Klebsiella pneumoniae, Pseudomonas aeruginosa,* and *Salmonella typhimurium*, produce endotoxins.

When bacteria grow in food and produce toxins, the toxins can cause illness in humans who eat those contaminated foods. This kind of illness is called an intoxication. For example, *Staphylococcus aureus* causes the most common type of food poisoning. The symptoms include nausea, vomiting, and diarrhea. This type of poisoning is painful but is seldom fatal.

Another type of intoxication that is fatal occurs when food is not canned properly, as shown in **Figure 21-13.** Sometimes canned food is not heated enough to kill endospore-forming bacteria, such as *Clostridium botulinum*. The bacteria can then grow and produce a deadly toxin that affects the nervous system. A person who eats food that contains this toxin then becomes ill with a disease called botulism, whose symptoms include double vision and paralysis. People with botulism may die because they are unable to breathe.

Some bacteria are responsible for other diseases reported in the news, such as *E. coli* O157:H7, the cause of several outbreaks of food poisoning in the United States. *E. coli* normally lives in our intestines. However, if it acquires DNA that codes for the toxin through conjugation, it can produce the toxin. *E. coli* poisoning is associated with raw or improperly cooked ground beef.

Study TIP

Reviewing Information

Work with a partner to review how bacteria cause disease, covered on pp. 465–466. Review the two ways that bacteria cause disease by quizzing each other on the ways that bacteria obtain nutrition.

Figure 21-13 Anaerobic growth. Endospore-forming bacteria can grow in the anaerobic atmosphere of a sealed can. Their metabolism produces such a large quantity of gas that the can bulges from the pressure.

Bacterial Disease Can Be Fought

Most bacteria can be killed by boiling water or various chemicals. Using hot, soapy water to prevent contamination of our food utensils and food supply is one way of preventing disease. Many commercial antibacterial products can also be used to prevent bacterial contamination in the kitchen and in industrial food factories.

Antibiotics

In 1928, the British bacteriologist Alexander Fleming noticed a fungus of the genus *Penicillium* growing on a culture of *S. aureus*. He saw that bacteria did not grow near the fungus. He concluded that the fungus was secreting a substance that killed the bacteria, as shown in **Figure 21-14.** Fleming isolated the substance and named it penicillin. In the early 1940s, scientists found that penicillin was effective in treating many bacterial diseases, such as pneumonia.

Different antibiotics interfere with different cellular processes. Because these processes do not occur in viruses, antibiotics are not effective against them. Other antibiotics, such as tetracycline and ampicillin, have been discovered in nature or imitated chemically.

In recent years, some bacteria have become resistant to antibiotics. Susceptible bacteria are eliminated from the population, and resistant bacteria survive and reproduce, thus passing on their resistance traits.

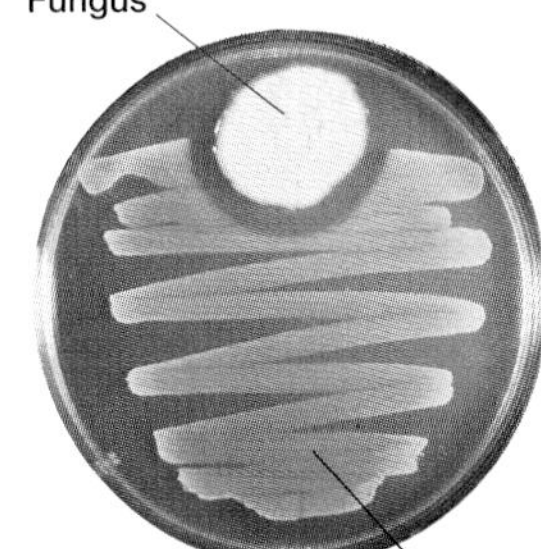

Figure 21-14 Antibiotics are naturally produced. Alexander Fleming saw a plate of agar very similar to this one. Notice how the bacteria do not grow next to the fungus.

DATA LAB

Identifying pathogenic bacteria

Background

Bacteria are typically identified with tests that determine the type of physiology they express, their Gram reaction, and their shape. These characteristics are used by lab technicians to identify bacteria isolated from patients. Use the chart below and information in the text to identify this commonly isolated pathogen and answer the questions.

Analysis

1. **Identify** the name of the unknown organism.
2. **Infer** whether the unknown sample came from a food sample suspected of causing illness.
3. **SKILL Predicting Outcomes** If the unknown sample came from a food sample that the patient had eaten but first thoroughly cooked, would antibiotics help the patient? Explain.

	TEST		
Organism	Gram reaction	A	B
E. coli	–	+	+
Salmonella enteritidis	–	+	–
Klebsiella pneumoniae	–	–	+
S. aureus	+	–	+
Streptococcus pyogenes	+	–	–
Unknown	+	–	+

Overcoming Misconceptions

"Bad" Bacteria

Students tend to think of bacteria as "bad" organisms because they are often described as sources of disease. However, of the thousands of kinds of bacteria, only a few are harmful. Bacteria are nature's recyclers, and scientists are finding ways to use bacteria to produce desirable materials and degrade wastes. Biotechnologists insert genes into bacterial cells that allow them to make plastics, pharmaceuticals, pesticides, and foods. Bacteria can even be used to mine metals, such as copper and gold, and to clean up industrial wastes.

Teaching TIP

"Rational" Drug Design

One strategy that researchers are using to control resistant bacterial strains is called "rational" drug design. By looking at the genes and enzymes involved in drug resistance, scientists are trying to create drugs that bind to the active sites of the enzymes and render them inactive. Ask students how binding the active sites of enzymes in bacterial cells can control resistant strains. *(Bacterial growth can be stopped by short-circuiting essential chemical reactions.)* Point out that using fewer antibiotics, avoiding antibiotics for viral disease treatment, and limiting their use in livestock may help prevent resistant bacterial strains.

Identifying pathogenic bacteria

Time 5–10 minutes

Process Skills Interpreting results, applying information, predicting outcomes

Teaching Strategies

Remind students that bacterial diseases may be the result of the metabolization of the host's cells by the bacteria or by toxins secreted by the bacteria.

Analysis Answers

1. The unknown matches the profile of *Staphylococcus aureus.*
2. *S. aureus* is a well-known agent of food poisoning, so it seems likely that the sample came from the food.
3. Antibiotics may not help the patient if the bacteria had already been killed by cooking. The patient may be reacting to toxins secreted by the bacteria.

Chapter 21

Demonstration
Bacteria Are Useful

Show students pictures of an oil spill and a field with a growing crop. Ask students what useful roles bacteria might play in these pictures. Explain that bacteria are being "domesticated" by biotechnologists to perform a variety of useful tasks, including cleaning up spilled petroleum products and making medicines. In agriculture, bacteria can take the place of some fertilizers and pesticides and can be used to prevent frost damage.

3 Assessment Options

Closure
Breakthroughs in Science

Organize students into small groups to discuss the technological use of bacteria. Have students propose future uses of bacteria and ask them to imagine the potential impact of their breakthroughs. Encourage students to consider potentially harmful, beneficial, and benign effects of their ideas on quality of life, economy, health, and the environment. Then have students produce a diagram illustrating the topic of their discussion to share with the class. Diagrams might feature the technological development in the center with its effects on society represented by spokes radiating outward.
PORTFOLIO

Review
Assign Review—Section 21-2.

Reteaching — BASIC
Have students research the use of bacteria in environmental restoration. How have they been used to clean up oil spills and other industrial accidents? What strains are used? How were these strains discovered or developed? What are the advantages to such treatments over alternatives? What are the disadvantages?

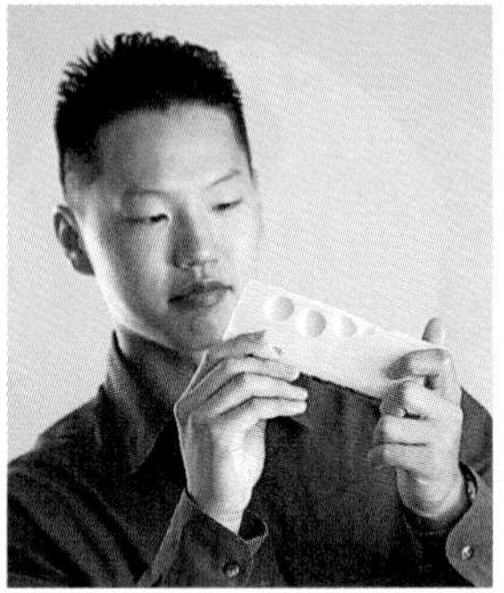
Figure 21-15 Swiss cheese. In making Swiss cheese, bacteria grow in the cheese and produce gas. As the cheese hardens these pockets of gas remain, giving the cheese its characteristic holes.

Bacteria Are Important

Despite the misery that some bacteria cause humans in the form of disease and food spoilage, much of what bacteria do is extremely important to our health and economic well-being.

Food and chemical production

Many of the foods that we eat are processed by specific kinds of bacteria. For example, many fermented foods are produced with the assistance of bacteria, as shown in **Figure 21-15.** These foods include pickles, buttermilk, cheese, sauerkraut, olives, vinegar, sourdough bread, and even some kinds of sausages.

Humans are able to use different bacteria to produce different kinds of chemicals for industrial uses, as shown in **Figure 21-16.** For example, different kinds of *Clostridium* species can make either acetone or butanol. These chemicals can be used to produce a large variety of other useful chemicals.

Genetic engineering companies use genetically engineered bacteria to produce their many products, such as drugs for medicine and complex chemicals for research.

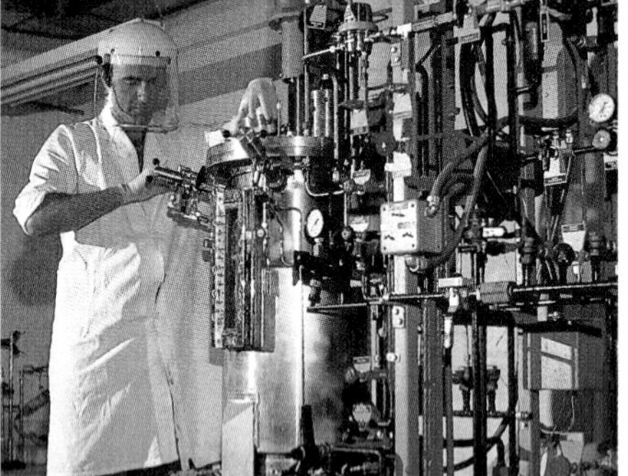
Figure 21-16 Industrial fermenter. Bacteria can be used to produce useful chemicals such as in this fermenter.

Mining and environmental uses of bacteria

Mining companies can use bacteria to concentrate desired elements from low-grade ore. Low-grade ore has a low percentage of the desired mineral, but it also has sulfur compounds. Chemoautotrophic bacteria can convert the sulfur into a soluble compound, leaving the desired mineral behind. The sulfur compound can be washed away with water, leaving only the desired mineral. This technique can be used to harvest copper or uranium.

Bacteria metabolize different organic chemicals and are therefore used to help clean up environmental disasters such as petroleum and chemical spills. Powders containing petroleum-metabolizing bacteria are used to help clean oil spills.

Review SECTION 21-2

1. **Construct** a table that lists the seven ways bacteria differ from eukaryotic cells.
2. **List** the structures found in *E. coli.*
3. **Identify** the relationship between photosynthesis, heterotrophic metabolism, and chemoautotrophic metabolism.
4. **Describe** the relationship between metabolism, toxins, bacteria, and disease.
5. **List** three ways bacteria are helpful.

Critical Thinking

6. **SKILL Defending a Theory** How does the growth of antibiotic resistance in bacteria support the theory of evolution by natural selection?

Review SECTION 21-2 ANSWERS

1. The table should include the seven differences listed on page 460.
2. *E. coli* has a cell wall, pili, flagella, ribosomes, and circular DNA.
3. Bacteria may obtain energy through photosynthesis, as heterotrophic consumers, or as chemoautotrophs.
4. Bacteria may cause disease by metabolizing the host's cells or by secreting toxins.
5. Bacteria are used in the processing of many foods, the creation of medical and industrial chemicals, the fixation of nitrogen, mining, and environmental restoration.
6. Natural selection favors those organisms best suited to their environment. Bacteria that are naturally resistant to antibiotics will be selected over nonresistant strains, eventually outnumbering the less well-adapted strains.

Study ZONE CHAPTER 21

Use the Key Concepts and Key Terms listed below to review the main ideas in this chapter. If you need to review the meaning of a term, turn to the page indicated.

Key Concepts

21-1 Viruses

- Viruses consist of segments of a nucleic acid contained in a protein coat, and some have an envelope.
- Viruses do not have all of the characteristics of life and are therefore not considered to be alive.
- Viruses replicate inside living cells. They enter a cell by injecting their genetic material into the cell, by slipping through tears in the plant cell wall, or by binding to molecules on the cell surface and triggering endocytosis.
- Viruses replicate through a lytic cycle or a lysogenic cycle.
- HIV replicates inside immune system cells, eventually destroying them, leaving the host without adequate defense against disease.
- Emerging viruses are geographically isolated viruses that cause disease in humans.
- Viroids are infectious RNA molecules that cause disease in plants and prions are infectious proteins that cause disease in certain animals.

21-2 Bacteria

- Bacteria differ from eukaryotes in their cellular organization, cell structures, and metabolic diversity.
- Bacteria can be classified into two groups according to their cell wall structure. Gram staining can be used to distinguish these two groups.
- Bacteria can transfer genes to one another by conjugation.
- Bacteria are grouped according to their ribosomal RNA sequences and the way they obtain energy.
- Bacteria cause disease by metabolizing nutrients in their host or by releasing toxins, which damage their host.
- Bacterial disease can usually be fought with soap, chemicals, and antibiotics.
- Bacteria are used to make foods, antibiotics, and chemicals; to fix nitrogen; to clean the environment; and to cycle important chemicals in the environment.

Key Terms

21-1

virus (452)
capsid (453)
envelope (453)
glycoprotein (453)
bacteriophage (453)
pathogen (454)
lytic cycle (454)
provirus (455)
lysogenic cycle (455)
emerging virus (459)
prion (459)
viroid (459)

21-2

pilus (460)
bacillus (461)
coccus (461)
spirillum (461)
capsule (461)
antibiotic (461)
endospore (461)
conjugation (461)
anaerobic (462)
aerobic (463)
toxin (466)

Study TIP Ready?

If you think you understand these Key Concepts and Key Terms, then you're ready for the Performance Zone. 

Answer to Concept Map

The following is one possible answer to Performance Zone item 1 on page 470.

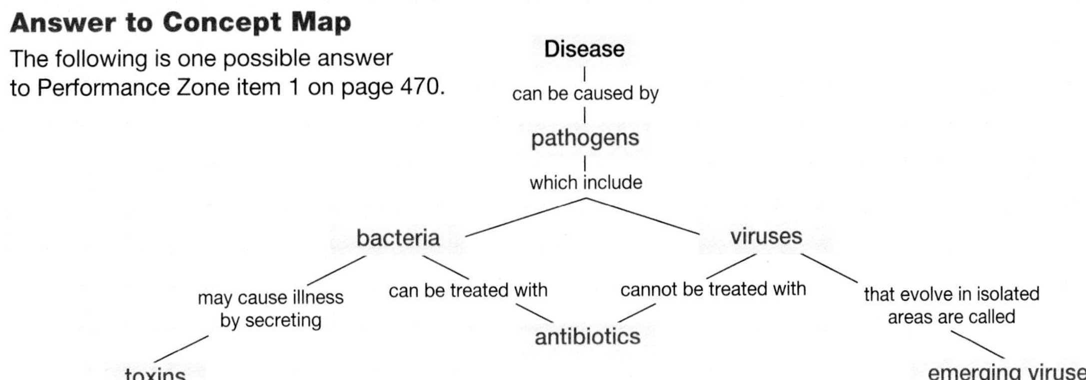

Study ZONE CHAPTER 21

Review and Practice

Assign the following worksheets for Chapter 21:

- **Student Review Guide**
- **Concept Mapping Worksheet**
- **Test Preparation Pretest**
- **Vocabulary Worksheet**
- **Critical Thinking Review Worksheet**

Assessment Options

The following assessment options are available for this chapter:

Resource Link

- Assign **Chapter Test 21** to assess students' understanding of the chapter.

Alternative Assessment

Have students choose two diseases, one caused by a bacterium and one caused by a virus. Students can use information from the textbook, do outside research, or create fictitious diseases. Ask students to draw plausible structures for the pathogens and to describe the method of infection, prevention, and treatment. Students' drawings must depict the characteristics of bacteria and viruses described in this chapter. PORTFOLIO

Portfolio Assessment

- *Portfolio Assessment* at the front of this book provides options for building and evaluating student portfolios. Students and teachers can select items from the areas listed for inclusion in a portfolio.
- See items labeled PORTFOLIO on pp. 453, 456, 457, 461, 463, 468, and 469.

Performance ZONE

CHAPTER 21

REVIEW ANSWERS

Assignment Guide

Section	Review
21-1	3, 4, 5, 6, 13, 14, 16, 17, 18, 19, 20, 21, 23, 25, 27
21-2	1, 2, 7, 8, 9, 10, 11, 12, 15 22, 24, 26

Understanding and Applying Concepts

1. The answer to the concept map is found at the bottom of page 469.
2. **a.** A capsid is the protein sheath that contains the viral nucleic acid. Many viruses also have an envelope, which surrounds the capsid and aids in cell penetration.

   b. During the lytic cycle, viral proteins are made and new viruses constructed and released when the host cell bursts. In the lysogenic cycle viruses are not made—the viral genome is inserted into the host genome and the viral genome multiplies when the host cell divides.

   c. Prions are nucleic acid-free pathogens consisting of protein particles. Viroids are protein-free pathogens consisting of a single strand of RNA.

   d. Bacillus is a rod-shaped cell, coccus is a spherically-shaped cell.
3. d
4. a
5. c
6. c
7. c
8. d
9. d
10. a
11. d
12. B
13. The body's immune system acts as a selective force, eliminating some of the virus while those viruses that evolve features that enable them to escape from being destroyed continue to multiply.

Performance ZONE

CHAPTER 21 Review

Understanding and Applying Concepts

1. **Concept Mapping** Make a concept map describing the relationships of bacteria and viruses to diseases. Try to include the following terms in your map: bacteria, viruses, pathogen, emerging viruses, antibiotics, toxin. Include additional concepts in your map as needed.
2. **Understanding Vocabulary** For each pair of terms below, explain the differences in their meanings.
   a. capsid/envelope
   b. lytic cycle/lysogenic cycle
   c. prion/viroid
   d. bacillus/coccus
3. Viruses do not
   a. grow.
   b. have homeostasis.
   c. metabolize.
   d. All of the above.
4. What evidence led Stanley to conclude that TMV is not a living organism?
   a. The extract of TMV crystallized.
   b. TMV is made of RNA and protein.
   c. TMV reproduces only in cells.
   d. The virus poisons tobacco plants.
5. The basic components of all viruses are a nucleic acid and a(n)
   a. endospore. c. protein coat.
   b. glycoprotein. d. icosahedron.
6. HIV infects and destroys
   a. skin cells c. immune cells
   b. red blood cells d. bacterial cells
7. Unlike eukaryotes, bacteria
   a. have flagella.
   b. are smaller than viruses.
   c. perform aerobic and anaerobic metabolism.
   d. have a nucleus.
8. *E. coli* can move by use of its
   a. pili. c. peptidoglycan.
   b. nucleus. d. flagella.
9. Bacteria that do not require sunlight and obtain energy by removing electrons from hydrogen-rich chemicals are called
   a. heterotrophs.
   b. photosynthetic bacteria.
   c. cyanobacteria.
   d. chemoautotrophs.
10. Bacterial disease occurs when bacteria
    a. secrete toxins. c. are heterotrophic.
    b. conjugate. d. live on humans.
11. Bacteria are beneficial because they make
    a. vinegar, sausages, and disease.
    b. drugs, minerals, and peptidoglycan.
    c. cytoplasm, antibiotics, and ribosomes.
    d. copper, acetone, and buttermilk.
12. Identify the pilus in the photo below.

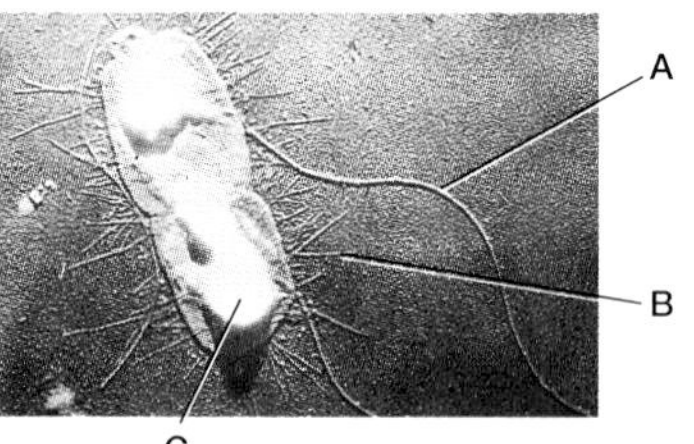

13. **Theme Evolution** If cold viruses invade your body, your body's immune system may destroy most but not all of these viruses. How does your body's immune system affect the evolution of the cold viruses?
14. **Health Watch** AZT (azidothymidine) is one of the drugs used to treat HIV infection. How does AZT interfere with HIV replication?
15. **Exploring Further** If you were making a Gram stain and ran out of the pink dye and substituted a green dye, what color would you expect a Gram-positive bacterium to be? Gram-negative bacterium?
16. **Chapter Links** How does the increase of resistance to antiviral drugs in HIV relate to the theory of evolution by natural selection? (**Hint:** See Chapter 13, Section 13-1.)

---

14. AZT interferes with the production of a complementary strand of DNA from the viral RNA.
15. Gram-positive bacteria would retain the purple dye; gram-negative bacteria would be stained green.
16. Certain mutations enable HIV to be resistant to antiviral drugs. As the virus is exposed to these drugs, those HIV that are resistant will survive and continue to multiply, becoming more common, while those HIV not resistant will be selected against and become less common.

Skills Assessment

Part 1: Interpreting Graphics

Study the illustration of virus replication shown below. Then answer the following questions:

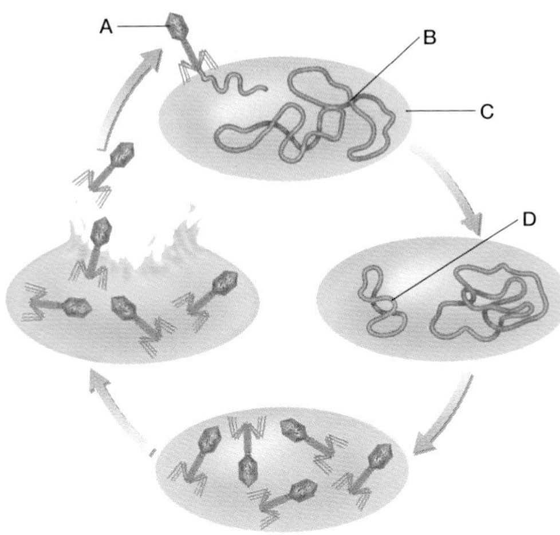

17. Identify the structures labeled *A–C*.
18. What is the name of this cycle?
19. Is this infection lethal to the host cell? Explain your answer.
20. Would you expect the released viruses to be capable of infecting similar host cells? Explain your answer.

Part 2: Life/Work Skills

Use the media center or Internet resources to learn how scientists study pathogenic viruses and bacteria.

21. **Finding Information** Research and write a report on a preventable viral disease, such as polio or smallpox. In your report, discuss the process scientists followed in identifying the cause of the disease, isolating the virus, formulating a vaccine, and testing the vaccine.
22. **Selecting Technology** Find out what techniques scientists are now using to study the biology of bacteria. How will these studies improve the health of humans and animals?

Critical Thinking

23. **Evaluating Results** In the 1520s, the Spanish explorer Cortés and his armies introduced smallpox to the Americas. The death rate among the native American people ranged between 50 to 90 percent compared with a death rate of about 10 percent among people in Europe. What accounts for the difference in death rates?
24. **Justifying Conclusions** Microbiologists criticize physicians for prescribing antibiotics when their patients have a viral infection. Explain why the microbiologists oppose use of antibiotics in patients with viral infections and why they say antibiotics do not help the patient.

Portfolio Projects

25. **Summarizing Information** AIDS, the disease caused by HIV, is a major health concern worldwide. Locate statistics on AIDS cases for as many countries as possible. Then draw a world map on poster-size paper, and devise a color legend for the map that shows the number of AIDS cases in countries for which you have data. Color the map to match the legend, and give it a title. Next write a set of questions that can be answered using the map. Display the map and the questions where they can be seen by other students. Different students can take responsibility for gathering the AIDS data, drawing and coloring the map, creating the legend, and writing the questions.
26. **Evaluating Viewpoints** Read "The Cholera Lesson" (*Discover*, February 1999, pp. 70–76). Was Rita Colwell justified in drawing her conclusions about the bacterium that causes cholera?
27. **Career Focus Virologist** Research the field of virology, and write a report on your findings. Your report should include a job description, training required, kinds of employers, growth prospects, and starting salary.

Skills Assessment

17. A. virus; B. bacterial chromosome; C. bacterium; D. viral DNA
18. the lytic cycle
19. Yes, the cell will burst once enough viruses have been replicated.
20. Yes, the released viruses should be genetically identical to the original invading virus.
21. Answers will vary depending on the viral disease chosen.
22. Answers will vary. Techniques include PCR, cycle sequencing, PFGE, subtractive hybridization, oligoribonucleotide technology, or automated DNA hybridization.

Critical Thinking

23. The Native Americans had never been exposed to such pathogens, and resistant individuals had therefore not yet become more common than those who were not resistant in that population.
24. Such use of antibiotics is ineffective because a virus does not carry out any of the life processes that the antibiotics interrupt. Such misuse and overuse of antibiotics encourages the evolution of resistant strains of bacteria.

Portfolio Projects

25. Answers will vary depending on the data available. Of HIV-infected individuals worldwide, most live in Africa and the Middle East, followed by Asia and Oceania, Latin America and the Caribbean, North America, and Europe.
26. Yes, Colwell was justified because she was able to repeatedly demonstrate that *Vibrio cholerea* exists in the environment as a permanent resident. *V. cholerea* also appears to be linked to the population movement and size of zooplankton.
27. Virologists work with viruses and tissue cultures that are the hosts for viruses. Most virologists have a medical degree or a graduate degree in microbiology or virology. Employers may include pharmaceutical companies, university research labs, government research labs, public health labs, and hospitals. Growth potential is good. Starting salary will vary by region.

Lab Techniques

Chapter 21

Staining and Observing Bacteria

Purpose

The purpose of this lab is to identify and compare and contrast different types of bacteria.

Preparation Notes

Time Required: 45 minutes

Materials

Materials for this lab can be ordered from WARD'S. See *Master Materials List* at the front of this book for catalog numbers.

Preparation Tips

- There should be sufficient wax pencils (or permanent marking pens) to supply three or four students with one pencil.
- You may purchase live cultures or lyophilized cultures from a biological supply house such as WARD'S. Cultures should be *Micrococcus luteus* (cocci), *Spirillum volutans* (spirillum), or *Bacillus megaterium* (bacillus).
- Standard isopropyl alcohol is sufficient for aseptic-technique washing down of the tables.
- Provide students with sufficient sterile cotton swabs that are individually or paired wrapped.
- Buy premixed nutrient broth from a biological supply house such as Ward's or prepare culture media from dehydrated concentrate and sterilize in an autoclave at 120°C at 15 psi for 20 minutes.
- About one day before the lab, grow the bacteria in nutrient broth cultures at 37°C. Grow a sufficient amount of culture broth to provide groups of three or four students with about 5 mL of each culture. Transfer about 5 mL of culture to each tube with a sterilized pipet. Do not allow students to access your stock cultures.
- If students' microscopes have an oil immersion lens you may provide them with immersion oil in order to view the bacteria at a higher magnification.

Lab Techniques

Chapter 21 **Staining and Observing Bacteria**

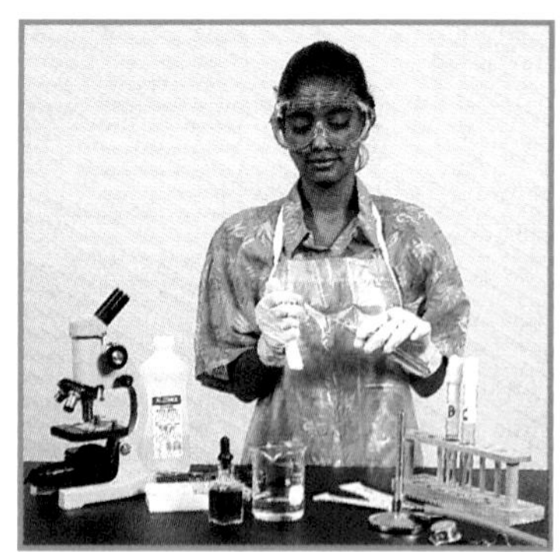

SKILLS

- Using aseptic techniques
- Using a microscope

OBJECTIVES

- **Prepare** and stain wet mounts of bacteria.
- **Identify** different types of bacteria by their shape.

MATERIALS

- wax pencil
- 3 microscope slides
- safety goggles
- lab apron
- disposable gloves
- rubbing alcohol
- paper towels
- 3 culture tubes of bacteria (A, B, and C)
- test-tube rack
- sterile cotton swabs
- Bunsen burner with striker
- microscope slide
- forceps or wooden alligator-type clothespin
- 150 mL beaker
- methylene blue stain in dropper bottle
- compound microscope

ChemSafety

CAUTION: Always wear safety goggles and a lab apron to protect your eyes and clothing.

CAUTION: Do not touch or taste any chemicals. Know the location of the emergency shower and eyewash station and how to use them. If you get a chemical on your skin or clothing, wash it off at the sink while calling to the teacher. Notify the teacher of a spill. Spills should be cleaned up promptly, according to your teacher's directions.

CAUTION: Glassware is fragile. Notify the teacher of broken glass or cuts. Do not clean up broken glass or spills with broken glass unless the teacher tells you to do so.

Before You Begin

Like all **prokaryotes,** bacteria are unicellular organisms that sometimes form filaments or loose clusters of cells. They are prepared for viewing by making a **smear,** a slide on which cells have been spread and dried. Treating the cells with a **stain** makes them more visible under magnification. In this lab, you will stain, identify, and compare and contrast different types of bacteria.

1. Write a definition for each boldface term in the paragraph above and for each of the following terms: strepto, staphylo, coccus, bacillus, spirillum.
2. Based on the objectives for this lab, write a question you would like to explore about different kinds of bacteria.

Procedure

Part A: Observing Live Bacteria

1.  Put on safety goggles, a lab apron, and disposable gloves.
2. Work with a partner. Use a wax pencil to label three microscope slides *A, B,* and *C.*
3. Use rubbing alcohol and paper towels to clean the surface of your lab table and gloves. Allow the table to air dry. **CAUTION: Alcohol is flammable. Do not use alcohol near an open flame.**
4. Light a Bunsen burner with a striker. **CAUTION: Keep combustibles away from flames. Do not light a Bunsen burner when others in the room are using alcohol.**

Safety Precautions

- Make sure students wear safety goggles, lab aprons, and disposable gloves before obtaining their bacterial cultures.
- In step 7, remind students to use methylene blue carefully and to keep it away from their faces, skin, and clothing.
- Be sure that students wipe their tables down with the rubbing alcohol before and after the lab.
- Make sure that students dispose of their materials properly and wash their hands before leaving the lab.

Procedural Tips

- Each lab group should be given a set of bacterial cultures. If you allow groups to obtain samples directly from stock cultures, the stocks are likely to become contaminated.
- Use broth cultures, which will enable students to see chains of cocci.

5. Beginning with culture A, make a smear of three different bacteria (A, B, and C) as follows. Remove the cap from a culture tube. *Note: Do not place the cap on the table.* Pass the opening of the tube through the flame of a Bunsen burner. Insert a sterile cotton swab into the tube, and lightly touch the tip of the swab to the bacteria in the culture. Pass the opening of the tube through the flame again, and replace the cap. Transfer a small amount of bacteria to the appropriately labeled microscope slide by rubbing the swab on the slide. Dispose of the swab in a proper container. Repeat for cultures B and C.
6. Allow your smears to air dry. Using microscope slide forceps, pick up each slide one at a time and pass it over the flame several times. Let each slide cool.
7. Using microscope slide forceps, place one of your slides across the mouth of a 150 mL beaker half-filled with water. Place 2–3 drops of methylene blue stain on the dried bacteria. *Note: Do not allow the stain to spill into the beaker.* **CAUTION: Methylene blue will stain your skin and clothing.** Let the stain stay on the slide for 2 minutes. Then dip the slide into the water in the beaker several times to rinse it. Blot the slide dry with a paper towel. *Note: Do not rub the slide.*
8. Repeat step 7 for your other two slides.
9. Allow each slide to air dry, and then observe them with a microscope. Make a sketch of a few cells on each slide. Compare your sketches to Figure 21-8, and identify the type of bacteria on each slide.

Part B: Cleanup and Disposal

10. Dispose of slides, used swabs, solutions, and broken glass in the designated waste containers. Do not pour chemicals down the drain or put lab materials in the trash unless your teacher tells you to do so.
11. Clean up your work area and all lab equipment. *Clean the surface of your lab table with rubbing alcohol.* Return lab equipment to its proper place. Wash your hands thoroughly before you leave the lab and after you finish all work.

Analyze and Conclude

1. **Summarizing Results** Describe the shape and grouping of the cells of each type of bacteria you observed.
2. **Analyzing Methods** Why should the test tube caps from the culture tubes (in Part B) not be placed on the table?
3. **Evaluating Viewpoints** Evaluate the following advice: Always use caution when handling bacteria, even if the bacteria is known to be harmless.
4. **Drawing Conclusions** How did you classify the bacteria in cultures A, B, and C, as a coccus, a bacillus, or a spirillum?
5. **Further Inquiry** Write a new question about bacteria that could be explored with another investigation.

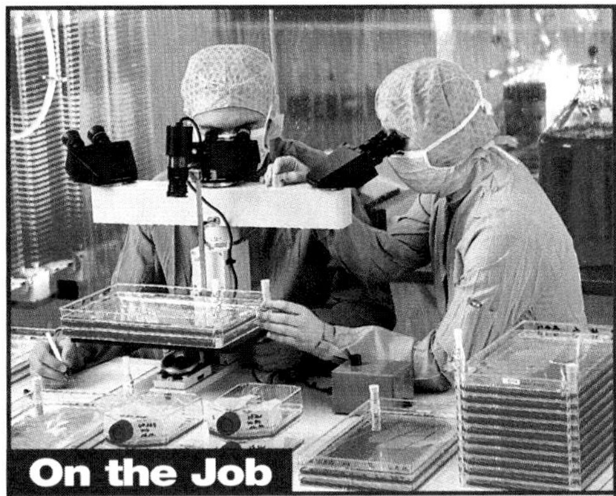

On the Job

Microbiologists are scientists who study organisms too small to be seen by the naked eye. Do research in the library or media center or on the Internet to discover how microbiologists better our lives. Summarize your findings in a written or oral report.

Disposal

- All cultures should be autoclaved before disposal.
- Have students place their used swabs in a centrally located container. The swabs may then be incinerated or autoclaved before disposal.

Before You Begin

Answers

1. *prokaryote*—organism that has no nucleus or membrane-enclosed structures
   smear—microscope slide that has bacterial cells smeared and dried on it
   stain—a substance that makes cells more visible under magnification
   strepto—prefix used to describe bacteria that form chains of cells
   staphylo—prefix used to describe bacteria that form clusters of cells
   coccus—spherical bacterium
   bacillus—cylindrical- or rod-shaped bacterium
   spirillum—corkscrew-shaped bacterium
2. Answers may vary. For example: What physical differences can we observe in different types of bacteria when seen under a compound light microscope?

Lab Techniques

21 Chapter

Procedure

Sample Data

Coccus bacteria

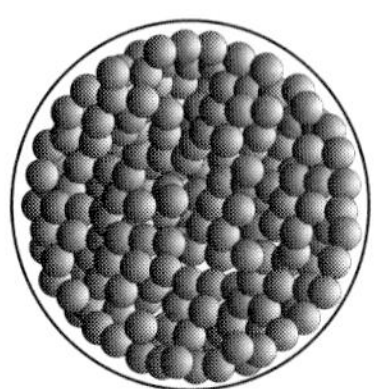

Bacillus bacteria

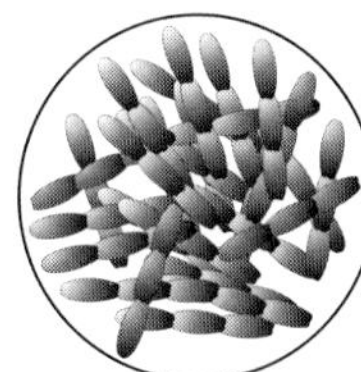

Spirillum bacteria

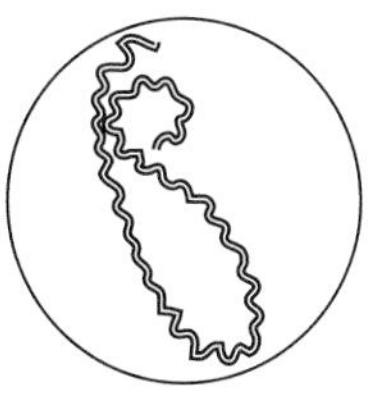

Analyze and Conclude

Answers

1. All three types of bacterial cells were observed. Coccus bacteria may form pairs, groups of four, chains, or clusters. Bacillus bacteria usually appear as single cells, although sometimes they are arranged in pairs or in chains. Spirillum bacteria almost always appear singly, and they differ in length and in the number and size of their spirals.
2. Culture tube caps should not be placed on lab tables because bacteria on the tables could contaminate the bacterial cultures and bacteria on the cap could contaminate the cultures.
3. Answers will vary. Students should mention that mistakes in identification, contamination, or mutations may result in the presence of harmful bacteria.
4. Answers will vary according to the type of bacteria placed into the marked tubes.
5. Answers will vary. For example: Are antiseptics equally effective against the three types of bacteria?

On the Job

Answer

Answers will vary. Microbiologists work closely with doctors and government health agencies to reduce the incidence of diseases caused by microorganisms.

Disease Detectives

Background

From the beginning of the twentieth century until the 1990s, deaths from infectious diseases declined in the United States. Death rates have been climbing recently for a number of reasons—AIDS, the evolution of drug-resistant strains of several pathogens, and the appearance of new, "emerging" diseases like hantavirus pulmonary syndrome. Hantavirus is typical of emerging diseases in many ways. It is not a new disease (caused by a pathogen that has never infected humans) but rather a newly discovered one. It came to light only after an environmental disruption (record precipitation the winter before) triggered an unusually high number of cases.

The swift response to the HPS outbreak was a triumph for the CDC, especially the Special Pathogens Branch, which deals with emerging diseases. Several factors made the quick discovery of the new hantavirus possible. First, the CDC committed more than 100 people to pursuing the virus—some of them worked 16 hours a day, 7 days a week, for months. Second, PCR has made the analysis of genetic data much easier and faster. Third, through years of work on hantaviruses in other parts of the world, scientists had accumulated a large database to draw from.

Teaching Strategies

- This story focuses on one geographical area. Have students locate the Four Corners area on a map. Explain how the area got its name—it is the only point where the borders of four states meet. Have them locate the Navajo Indian Reservation on the map. Explain that many of the first cases of HPS were found in Navajos.
- Ask students to describe the structure of a virus. Then have them explain how viruses reproduce. Ask students how viral reproduction differs from bacterial reproduction. Point out that most scientists do not consider viruses to be alive. Have students turn to the definition of *life* given on page 6 of Chapter 1, and then have them identify the criteria for life that viruses do not meet.
- Have students look in a field guide for mammals and find the map showing the range of the deer mouse. Have them locate the range of the cotton rat, which also carries hantavirus.

SCIENCE • TECHNOLOGY • SOCIETY

A thorough knowledge of biology—coupled with insight and hard work—helped scientists track down a killer.

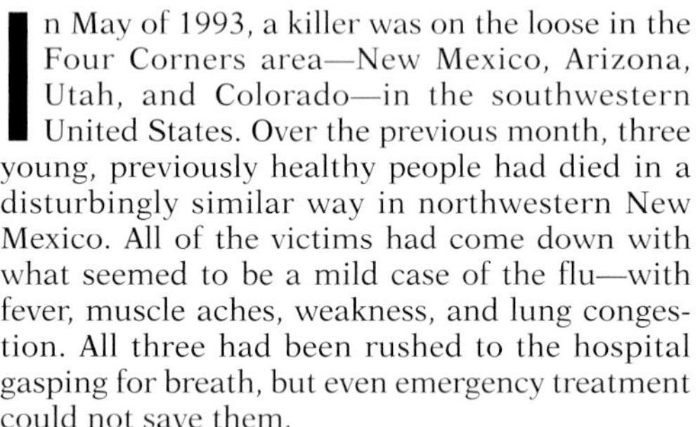

Disease Detectives

In May of 1993, a killer was on the loose in the Four Corners area—New Mexico, Arizona, Utah, and Colorado—in the southwestern United States. Over the previous month, three young, previously healthy people had died in a disturbingly similar way in northwestern New Mexico. All of the victims had come down with what seemed to be a mild case of the flu—with fever, muscle aches, weakness, and lung congestion. All three had been rushed to the hospital gasping for breath, but even emergency treatment could not save them.

Autopsies revealed that all three victims had drowned. Their lungs had filled with plasma—the liquid part of the blood—that had leaked from the tiny blood vessels within the lungs. What could cause such severe damage? Laboratory tests ruled out diseases with similar symptoms, including flu, bubonic plague, and anthrax.

As doctors and researchers began to investigate further, additional deaths came to light. Fearing that these cases might signal the beginning of a disease outbreak, officials decided to ask for help from the federal Centers for Disease Control and Prevention (CDC) in Atlanta, Georgia.

Catching a Killer Virus

The CDC sent a team of disease investigators to the Four Corners area while researchers in Atlanta tested blood and tissue samples from the victims. One of the tests revealed that the victims had been exposed to a kind of virus known as a hantavirus.

This result was surprising. All known hantaviruses from North America were harmless. Furthermore, the hantaviruses that did cause disease in Asia and Europe attacked the kidneys, not the lungs. But further testing confirmed the result. The killer was a hantavirus.

What was the virus like? Scientists used one of biology's most powerful tools to find out—the polymerase chain reaction, or PCR. PCR is a method of making many copies of a

Source of the virus
Deer mice, which thrive on the seeds of pinyon pine, were found to carry hantavirus, a dangerous pathogen.

Four Corners area
Researchers were surprised to find a deadly hantavirus here.

particular DNA strand. The copying process requires a primer, a short piece of DNA only a few nucleotides long that becomes the beginning of the new strand. To investigate the characteristics of the new hantavirus, CDC researchers made several primers that shared nucleotide sequences with known hantaviruses. By adding different primers to tissue samples from the victims, the scientists could "fish out" pieces of the virus's genes and then determine their nucleotide sequence.

Determining the Cause of the Outbreak

Within a month of being called in, CDC scientists had caught the killer and sequenced much of its genetic material. They named the disease hantavirus pulmonary syndrome, or HPS.

But scientists still needed to learn how HPS was transmitted and why the outbreak occurred when it did. Elsewhere, researchers knew that hantaviruses are carried by rodents—rats, mice, and their relatives. To find out whether rodents also carried the new hantavirus, researchers trapped thousands of rodents in and around the homes of the victims. About 30 percent of the deer mice they caught tested positive for the virus. These small mice—about 15 cm long—live throughout most of the United States.

Why had the disease not appeared before in the Four Corners area? An unusually wet winter the year before led to a bumper crop of pinyon pine seeds, the deer mouse's favorite food. In turn, the deer mouse population boomed. The larger population probably meant that mice entered homes more often and people were more likely to contact them outdoors. ■

Career

Infectious Disease Investigator

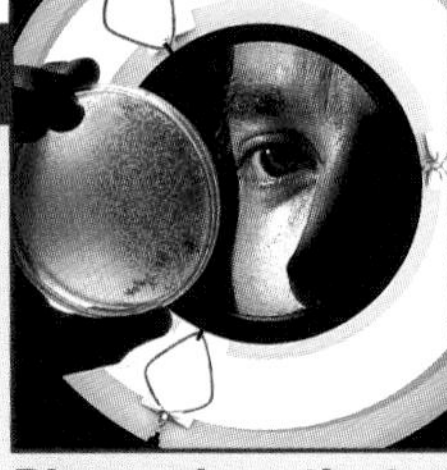

Disease investigator

Profile

Disease investigators never know what kind of phone calls they will get. Their work is often like detective work—trying to piece things together as they seek to find out who has been infected, where, and who else might have been exposed. The field of study that investigates the causes and controls of disease outbreaks is called epidemiology.

Job Description

Disease investigators are involved in preventing disease outbreaks and in stopping outbreaks that do occur from spreading. If there is an outbreak, they have to quickly identify the disease, determine how people are being infected, and take action to prevent more people from getting sick. Most disease investigators work for local, state, or federal government agencies.

Science/Math Career Preparation

Microbiology	Mathematics
Zoology	Genetics
Botany	Biochemistry

Analyzing STS Issues

Science and Society

1. **How did the HPS outbreak affect the people of the Four Corners area?** Read the article "Death at the Corners," by Denise Grady (*Discover,* December 1993, pp. 83–91). What evidence suggests that HPS may have been present in the Four Corners area for centuries?

2. **What does the CDC do?** Researchers from the CDC played a key role in solving the hantavirus mystery. Use library resources or an on-line database to find out more about the CDC. What is the CDC's mission? What kinds of people work there? Where does the CDC get its funding?

Technology: PCR

3. **What are some other newly discovered diseases?** HPS is just one of many emerging, or newly discovered, diseases. Use library resources or an on-line database to research three other emerging diseases that have come to light in the last 30 years. Find out where each disease was discovered, its cause, and its symptoms. How is PCR used in the study of emerging diseases?

References

1. Centers for Disease Control and Prevention. *Hantavirus Web Site.* On-line. World Wide Web. Available http://www.cdc.gov/ncidod/diseases/hanta/hps/noframes/intro.htm.
2. National Center for Infectious Diseases. *Emerging Infectious Diseases.* On-line. World Wide Web. Available http://www.cdc.gov/ncidod/EID/eid.htm.
3. Garrett, Laurie. *The Coming Plague.* New York: Penguin, 1995.
4. Le Guenno, Bernard. "Emerging Viruses." *Scientific American,* October 1995, 56–64.
5. Marshall, Eliot. "Hantavirus Outbreak Yields to PCR." *Science,* November 5,1993, 832–836.
6. Peters, C. J. and Mark Olshaker. *Virus Hunter.* New York: Anchor Books, 1997.
7. Rhodes, Richard. *Deadly Feasts.* New York: Simon and Schuster, 1997.

Analyzing STS Issues Answers

1. Biologists think that the similarity between the Four Corners hantavirus and two other viruses (a Scandanavian virus called Puumala and another American hantavirus strain known as Prospect Hill) suggests that the Four Corners hantavirus is not a recent Asian import. The fact that the hantavirus does not kill the deer mice who are its host also suggests that the hantavirus is a long-time resident of the Southwest.

2. The CDC is an agency of the Health and Human Services Administration of the federal government. The mission of the CDC is to reduce death and injury from all causes. The CDC employs epidemiologists, statisticians, pathologists, virologists, bacteriologists, public health experts, occupational injury experts, public relations specialists, and others. The CDC receives its funding from federal taxes.

3. Answers will vary. Possible answers include Lyme disease, Legionnaires' disease, Ebola fever, Lassa fever, and AIDS. PCR can be used to produce large quantities of DNA from pathogenic viruses and bacteria of unknown identity. The DNA is then used to identify the pathogenic viruses and bacteria.

Chapter 22

Protists

Planning Guide

OBJECTIVES	MEETING INDIVIDUAL NEEDS	READING SKILLS AND STRATEGIES
22-1 Characteristics of Protists (BLOCK 1 45 min) • **List** the characteristics of protists. • **Identify** the unifying features of protists. • **Distinguish** asexual and sexual reproduction of *Chlamydomonas.* • **Differentiate** two ways multicellular protists reproduce sexually.	**Learning Styles,** p. 478 BASIC ADVANCED • p. 481	**ATE** **Active Reading, Technique: L.I.N.K.,** p. 477 ■ **Active Reading Guide,** worksheet (22-1) ■ **Directed Reading,** worksheet (22-1) ■ **Reading Strategies,** Recognizing Cycles
22-2 Protist Diversity (BLOCKS 2 & 3 90 min) • **Identify** how amoebas and forams move. • **Describe** the structure of diatoms. • **Differentiate** three different kinds of flagellates. • **Summarize** the general characteristics of *Paramecium.*	**Learning Styles,** p. 482 BASIC ADVANCED • pp. 482, 484, 488, 489	■ **Active Reading Guide,** worksheet (22-2) ■ **Directed Reading,** worksheet (22-2)
22-3 Protists and Health (BLOCKS 4 & 5 90 min) • **Identify** two ways that protists affect human health. • **Name** three human diseases, other than malaria, caused by protists. • **Summarize** how malaria is transmitted. • **Evaluate** the methods used to control malaria.	**Learning Styles,** p. 490 BASIC ADVANCED • p. 492	**ATE** **Active Reading, Technique: Anticipation Guide,** p. 490 ■ **Active Reading Guide,** worksheet (22-3) ■ **Directed Reading,** worksheet (22-3) ■ **Supplemental Reading Guide,** *The Andromeda Strain*

Review and Assessment Resources

BLOCKS 6 & 7 90 min

TEST PREP

- **PE** **Study Zone,** p. 493
- **PE** **Performance Zone,** pp. 494–495
- ■ **Chapter 22 worksheets:**
  - Vocabulary
  - Concept Mapping
  - Critical Thinking
- Audio CD Program
- ■ Test Prep Pretest

CHAPTER TESTING

- ■ **Chapter Tests** (blackline master)
- ▲ **Test Generator**
- ■ **Assessment Item Listing**

See the correlations table at the front of this book for a list of the ***National Science Education Standards*** addressed in this chapter.

Smithsonian Institution® Visit **www.si.edu/hrw** for additional on-line resources.

Biology Principles and Explorations

One-Stop Planner CD-ROM
Shows these resources within a customizable daily lesson plan:

KEY

- **PE** Pupil's Edition
- **ATE** Teacher's Edition
- Teaching Transparencies
- Laboratory Program
- ■ Teaching Resources
- ▲ Test Generator

VISUAL STRATEGIES	CLASSWORK AND HOMEWORK	TECHNOLOGY AND INTERNET RESOURCES
ATE **Fig. 22-2,** p. 480 98 Protist Reproduction 103 Reproduction of *Chlamydomonas* 104 Reproduction of *Ulva* 89A Members of the Kingdom Protista 90A Phyla of Protists	**Lab A17** Observing Protists **PE** **Review,** p. 481	Holt Biology Videodiscs Section 22-1 Audio CD Program Section 22-1 internet**connect** SCILINKS NSTA **TOPIC:** Reproduction in protists
ATE **Fig. 22-4,** p. 482 **ATE** **Fig. 22-7,** p. 485 **ATE** **Fig. 22-9,** p. 488 99 Structure of *Euglena* 100 Tour of a Protist *(Paramecium)*	**PE** **Quick Lab** What are some characteristics of diatoms?, p. 483 **PE** **Data Lab** Interpreting competition among protists, p. 486 **Lab B13** Cryobiology **PE** **Review,** p. 489	Holt Biology Videodiscs Section 22-2 Audio CD Program Section 22-2 internet**connect** SCILINKS NSTA **TOPIC:** Diatoms
101 Life Cycle of *Plasmodium* 102 Diseases Caused by Protists	**PE** **Experimental Design** How Do Protozoans Respond to Light?, pp. 496–497 **Lab B14** Protists—A Comparison ■ **Occupational Applications:** Nurse Practitioner **PE** **Review,** p. 492 **PE** **Study Zone,** p. 493	CNN Video Segment 17 Killer Algae Holt Biology Videodiscs Section 22-3 Audio CD Program Section 22-3 internet**connect** SCILINKS NSTA **TOPICS:** • Malaria vaccine • Protozoans

ALTERNATIVE ASSESSMENT

- **PE** **Portfolio Projects 26–28,** p. 495
- **ATE** **Alternative Assessment,** p. 493
- **ATE** **Portfolio Assessment,** p. 493

Lab Activity Materials

Quick Lab
p. 483 pipet, water, microscope slide, toothpick, diatomaceous earth, coverslip, compound microscope

Data Lab
p. 486 pencil, paper

Experimental Design
p. 496 Detain™ (protist-slowing agent), microscope slides, plastic pipets with bulbs, mixed culture of protozoans, toothpicks, coverslips, compound microscope, protozoan references, black construction paper, scissors, paper punch, white paper, sunlit window sill or lamp, forceps

Lesson Warm-up
ATE culture of live amoebas, p. 482

Demonstrations
- **ATE** slides or pictures of protists, p. 476
- **ATE** folders containing pictures of protists, p. 478
- **ATE** water from a fish tank or pond, several containers, fertilizer, p. 479
- **ATE** live specimens of algae, p. 484

BIOSOURCES LAB PROGRAM
- Quick Labs: Lab A17, p. 33
- Inquiry Skills Development: Lab B13, p. 57
- Inquiry Skills Development: Lab B14, p. 63

Chapter 22

Chapter Theme

Reproduction

The amazingly diverse protists exhibit a range of sexual and asexual reproductive strategies not seen in the other kingdoms. Some reproductive strategies involve complex life cycles and multiple hosts, including humans, and are the cause of several important human diseases.

Establishing Prior Knowledge

Before beginning this chapter, make certain students can answer the questions in Study TIP Ready? on page 477. If they cannot, encourage them to reread the sections indicated.

Opening Demonstration

Show students prepared slides or pictures of various protists. Have students derive some general rules for classifying protists such as being microscopic, unicellular, eukaryotic, etc. Write students' suggestions on cards, and post them where students can view them during their study of protists. Remind them that protists do not have much in common with each other, but they are classified in one kingdom because they lack the specialized features found in members of other kingdoms.

CHAPTER 22 Protists

Pond water protists

What's Ahead — and Why
Chapter Outline

22-1 Characteristics of Protists
- Protists Are a Varied and Ancient Group of Organisms
- Protists Reproduce Sexually and Asexually

22-2 Protist Diversity
- Protists Can Move by Using Cytoplasmic Extensions
- Diatoms Have Double Shells
- Many Algae Are Multicellular
- Some Protists Move with Flagella
- Ciliates Are Complex Protists That Use Cilia to Swim
- Protistan Molds Are Not Fungi
- Some Protists Form Resistant Stages

22-3 Protists and Health
- How Do Protists Affect Humans?
- Malaria Is Caused by Several Species of *Plasmodium*

Students will learn

- 22-1: the distinguishing characteristics of protists and how protists reproduce.
- 22-2: how the diverse groups of protists can be distinguished by locomotion, reproduction, metabolism, multicellularity or unicellularity, and other unique features.
- 22-3: how protists affect the lives of humans as a vital component of the world's ecosystems and by causing disease.

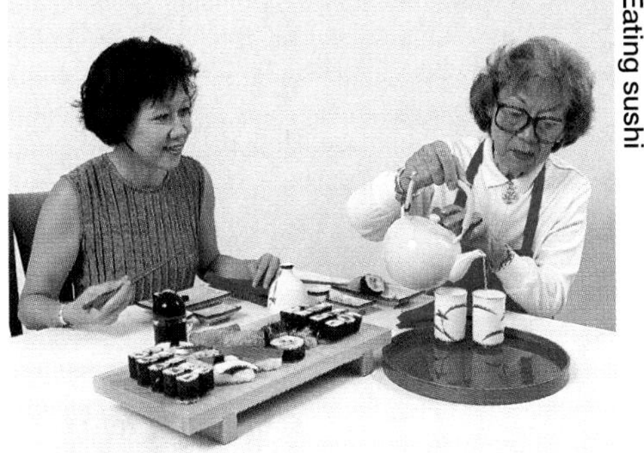
Eating sushi

The kingdom Protista contains *the largest variety of organisms among the six kingdoms. As you can see in the photo at left, microscopic protists include unicellular and multicellular organisms. Above, red algae is used to make the wrap for sushi. Protists obtain energy and reproduce in a variety of different ways.*

Study TIP

Ready?

Check your understanding of these topics to help you prepare for what's ahead.

Can you . . .

- **label** the structures of a eukaryotic cell? (Section 3-3)
- **describe** the different kinds of photosynthetic pigments that make photosynthesis possible? (Section 5-2)
- **summarize** the importance of sexual reproduction? (Section 7-2)
- **define** the term *plankton*? (Section 18-3)
- **summarize** the advantage of multicellularity? (Section 20-2)

If not, *review the sections indicated.*

Looking Ahead

internetconnect

National Science Teachers Association *sci*LINKS Internet resources are located throughout this chapter.

Chapter 22

Reading Strategies

Active Reading

Technique: L.I.N.K.

Write the word *Protist* on the board, and have students **L**ist on their own paper all the words, phrases, and ideas that they associate with the word. After several minutes, ask the students to share one or two of their ideas as you record them. Facilitate a discussion in which students **I**nquire of each other or the teacher to clarify each of the ideas listed. After all discussion is completed, have students write **N**otes on paper about what they remember. Allow about one minute. Have them look over their notes to see what they **K**now about protists based on experience and discussion.

Directed Reading

Assign **Directed Reading Worksheet Chapter 22** to help students interact with the text as they read the chapter.

Interactive Reading

Assign ***Biology: Principles and Explorations*** **Audio CD Program Chapter 22** to help reluctant readers achieve greater success in reading the chapter.

Reading Skills

Assign Reading Organizers and Reading Strategies Worksheets from the **One-Stop Planner CD-ROM with Test Generator** to help students learn and practice effective reading skills.

internetconnect

Holt *Biology: Principles and Explorations*

On-line Resources: go.hrw.com

Visit the HRW Web site for a variety of resources related to this chapter. Just type in the keyword HX0 Chapter 22.

Holt *Biology: Principles and Explorations*

On-line Resources: www.scilinks.org

The following *sci*LINKS Internet resources can be found in the student text for this chapter:

Page 479
TOPIC: Protista
KEYWORD: HX479

Page 480
TOPIC: Reproduction in protists
KEYWORD: HX480

Page 492
TOPIC: Malaria vaccine
KEYWORD: HX492

Page 497
TOPIC: Protozoans
KEYWORD: HX497

Chapter 22

1 Prepare

Scheduling

- **Block:** About 45 minutes will be needed to complete this section.
- **Traditional:** About one class period will be needed to complete this section.
- **Planning:** See the Planning Guide on pp. 475A and 475B for lecture, classwork, and assignment options for Section 22-1. Lesson plans for Section 22-1 can be found in a lesson cycle format in ***Biology: Principles and Explorations*** **One-Stop Planner CD-ROM with Test Generator.**

Resource Link

Assign **Active Reading Worksheet Section 22-1** to help students comprehend the key concepts and visuals in this lesson.

2 Teach

Lesson Warm-up

Ask students how an organism can ensure the survival of its genes in the next generation. Encourage students to propose strategies that could help organisms perpetuate their genes. How might protists ensure that their genes are passed on? *(Suggestions might include improved ways of storing and protecting genetic information or alternative forms of reproducing under different environmental conditions.)* Tell them that evolutionary pressures have led to the diversity of reproductive strategies in protists.

Demonstration

Classifying Protists into Groups

Provide small groups of students with folders that contain pictures of protists. (Try to include some representatives of each group shown in Table 22-1 on page 479.) Ask students to organize the protists into subgroups.

Section 22-1 Characteristics of Protists

Objectives

- **List** the characteristics of protists. (p. 478)
- **List** three environments where protists can be found. (p. 478)
- **Identify** the unifying features of protists. (p. 479)
- **Distinguish** asexual and sexual reproduction of *Chlamydomonas*. (p. 480)
- **Differentiate** two ways multicellular protists reproduce sexually. (p. 481)

Key Terms

protozoan
alga
zygospore
alternation of generations
sporangium

Protists Are a Varied and Ancient Group of Organisms

The most diverse of all organisms, protists are mostly unicellular, microscopic organisms, such as the green scum you might find growing on a rock in a pond. But a few protists are complex and multicellular, such as the massive house-high seaweed called kelp, which is found in the oceans.

Characteristics

The kingdom Protista *(proh TEES tuh)* consists of an unusually diverse assortment of eukaryotes that exhibit a broad array of characteristics, as summarized in **Figure 22-1.** For example, some protists are photosynthetic (like plants), some ingest food (like animals), and some absorb their food (like fungi).

Some protists have flagella or cilia, which they can use for locomotion or to collect nutrients. Other protists use other means of locomotion. Protists are found almost everywhere there is water. Many live in lakes and oceans, floating as plankton or anchored to rocks. They are also common inhabitants of damp soil and sand, and they thrive in moist environments, such as leaf litter. Some species are parasites.

Many protists have mechanisms for monitoring and responding to stimuli in their environment. For example, some protists have eyespots, small organelles containing light-sensitive pigments that detect changes in the quality and intensity of light.

The first eukaryotes

As you will recall from Chapter 12, protists—the first eukaryotes—are thought to have evolved about 1.5 billion years ago through the process of endosymbiosis. The kingdom Protista contains the lifeforms that gave rise to the three kingdoms of multicellular organisms—fungi, plants, and animals.

Protist Characteristics

- Sexual reproduction
- Multicellularity
- Mitosis and meiosis
- Complex flagella and cilia

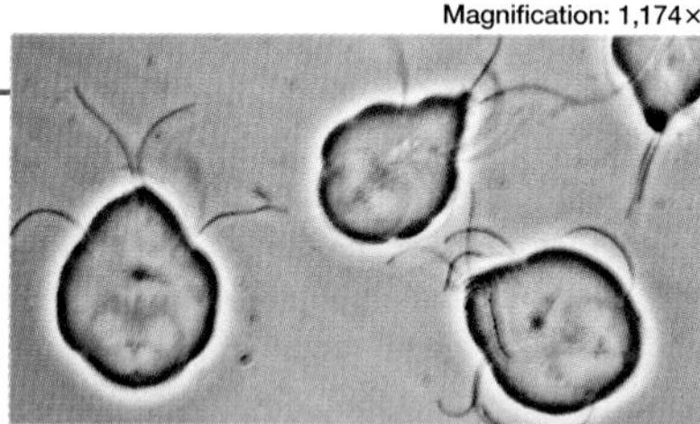

Figure 22-1 Protist characteristics. These characteristics, which are found in many eukaryotes, first evolved in protists.

Integrating Different Learning Styles

Visual/Spatial	**ATE**	Demonstrations, pp. 478 and 479; Visual Strategy, p. 480;
Interpersonal	**PE**	Study Tip, p. 481
Verbal/Linguistic	**PE**	Study Tip, p. 479; Performance Zone item 15, p. 494

Two important eukaryotic features that evolved among the protists are sexual reproduction and multicellularity. Many protists reproduce only asexually, by mitosis; some use meiosis and sexual reproduction only in times of environmental stress; and others reproduce sexually most of the time.

Multicellularity involves a significant amount of coordination among specialized cells. This trait evolved independently in different groups of protists at different times. Early during the evolution of protists, complex flagella and cilia also appeared.

Study TIP

Word Origins

The word *protist* is from the Greek *protistos,* meaning "first." Knowing this makes it easier to remember that protists were the first eukaryotes to evolve.

What unites protists?

The kingdom Protista contains all eukaryotes that cannot be classified as animals, plants, or fungi. Protists lack the specialized features that characterize the three other multicellular kingdoms. For example, unlike plants and animals, protists do not reproduce by forming embryos. Protists do not develop complex multicellular reproductive structures. The major phyla of protists are very different from one another and, with a few exceptions, are only distantly related. The major phyla of protists are shown in **Table 22-1.**

Historically, scientists have referred to heterotrophic protists as **protozoa** and to photosynthetic protists as **algae.** These commonly used terms are not formal classification categories used by scientists today. Although it is useful to refer to protists in these terms, it does not reflect their evolutionary relationships.

Table 22-1 Fourteen Phyla of Protists

Phylum	Distinguishing features	Mode of nutrition
Rhizopoda (amoebas) Foraminifera (forams)	Move using pseudopodia	Heterotrophic
Bacillariophyta (diatoms)	Have double shells made of silica	Photosynthetic
Chlorophyta (green algae) Rhodophyta (red algae) Phaeophyta (brown algae)	Photosynthetic protists; can be multicellular	Photosynthetic
Dinoflagellata Zoomastigina Euglenophyta	Move using flagella	Photosynthetic Heterotrophic Most are heterotrophic; some are photosynthetic
Ciliophora (ciliates)	Move using cilia	Heterotrophic
Acrasiomycota (cellular slime molds) Myxomycota (plasmodial slime molds) Oomycota	Funguslike protists	Heterotrophic
Apicomplexa (sporozoans)	Form resistant spores	Heterotrophic

did you know?

The first person to describe microorganisms was a cloth merchant.

Leeuwenhoek, a Dutch cloth merchant, is sometimes described as an amateur scientist. He is credited with the first descriptions of microorganisms such as bacteria and protists. Leeuwenhoek developed some of the first powerful microscopes (270×) to inspect cloth. He called the tiny organisms he viewed through his microscopes *animalcules* because he assumed that they were tiny animals.

Teaching TIP

Eyespots

Ask students to define the function of an eyespot. *(Eyespots detect the quality and intensity of light.)* Why might it be important for a protist to monitor the characteristics of light? Lead students to predict that some protists are photosynthetic, and these protists use information about light to maintain a homeostatic metabolism. This discussion will be continued in Section 22-2.

Teaching TIP

Pond Scum

When large numbers of green algae grow in a pond, they are often visible as pond scum. Pollution of fresh water can lead to a proliferation of green algae. When a small body of water is polluted with nitrates and phosphates from sewage, detergents, or runoff from fertilized soil, green algae reproduce rapidly. When the algae die, bacterial decomposers proliferate, resulting in a depletion of oxygen in the water. Ask students what is the overall effect of explosive algae growth on other organisms living in the pond? *(As decomposers consume the algae, oxygen-dependent organisms, such as fish, may die.)*

Demonstration

Creating an Algal Bloom

Obtain water from a fish tank or a nearby pond. Fill several containers with the water, and set them in the sun. Add small amounts of nitrogen- and/or phosphorous-based fertilizer to some of the containers and have students compare the resulting algae growth in each of the containers. Use tap water as a control. Ask students how this experiment simulates real-life conditions? *(The beakers represent bodies of water, and the added fertilizer represents agricultural runoff, lawn care products, and detergent-loaded wastewater.)* Note that some of the observed algae growth may actually be photosynthetic prokaryotes.

Visual Strategy

Figure 22-2

Use this figure to trace sexual and asexual reproduction in *Chlamydomonas.* Which form of reproduction does it use most frequently? *(asexual reproduction)* Which does it use under environmental stress? *(sexual reproduction)*

Teaching TIP

Reproduction in Multicellular Protists

Have students draw a **Graphic Organizer** similar to the one at the bottom of page 480 to summarize two modes of reproduction found among green algae. PORTFOLIO

Resource Link

Assign **Reading Strategies: Recognizing Cycles** from the Holt BioSources Teaching Resources CD-ROM. PORTFOLIO

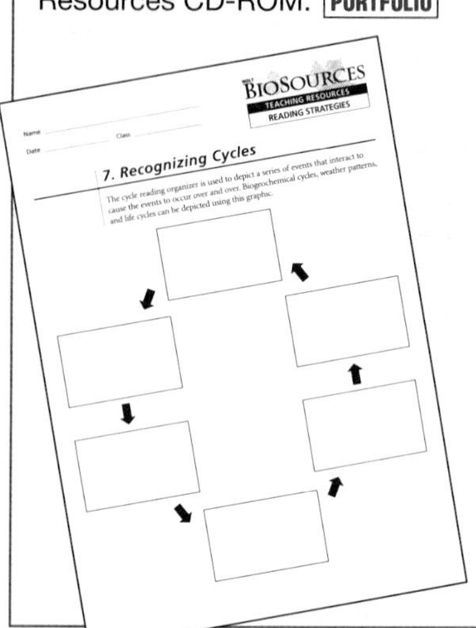

biobit

Do algae have eyes?

The light-sensitive pigment in the human eye, rhodopsin, also functions to detect light in *Chlamydomonas* species. A rhodopsin-containing organelle called an eyespot steers *Chlamydomonas* toward the light it needs for photosynthesis.

internetconnect

SciLinks NSTA

TOPIC: Reproduction in protists
GO TO: www.scilinks.org
KEYWORD: HX480

Protists Reproduce Sexually and Asexually

Reproduction in the unicellular green alga *Chlamydomonas* is typical of unicellular protists. *Chlamydomonas* is unusual in that it forms complex colonies with distinctly different cell types. The degree of functional specialization is not as great, however, as in the multicellular protists. *Chlamydomonas* species reproduce sexually and asexually.

As a mature organism, the single-celled protist is haploid. When it reproduces asexually, *Chlamydomonas* first absorbs its tail and divides by mitosis, producing two to eight haploid cells called zoospores, which remain within the wall of the parent cell. Mature zoospores break out of the parent cell and grow to become mature haploid cells.

Sexual reproduction among unicellular protists

During environmental stress, such as a shortage of nutrients, *Chlamydomonas* species reproduce sexually. The haploid cell divides first by mitosis to produce haploid gametes. After they are released, a pair of gametes from different *Chlamydomonas* individuals fuse to form a pair. This pair of gametes then shed their cell walls and fuse into a diploid zygote with a thick protective wall called a **zygospore** *(ZEYE goh spohr).*

A zygospore can withstand unfavorable environmental conditions for long periods of time. When environmental conditions become favorable again, meiosis within the zygospore produces haploid cells, which break out of the zygospore wall. These haploid cells grow into mature cells, completing the sexual life cycle, as shown in **Figure 22-2.**

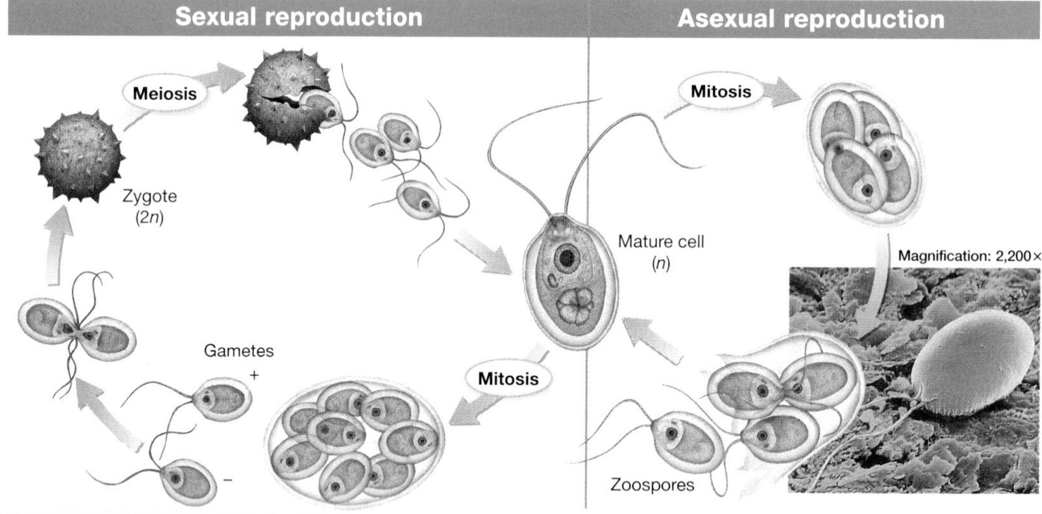

Figure 22-2 *Chlamydomonas* reproduction. The unicellular green algae in the genus *Chlamydomonas* reproduce sexually and asexually.

Graphic Organizer

Use this graphic organizer with ***Teaching Tip: Reproduction in Multicellular Protists*** *on page 480.*

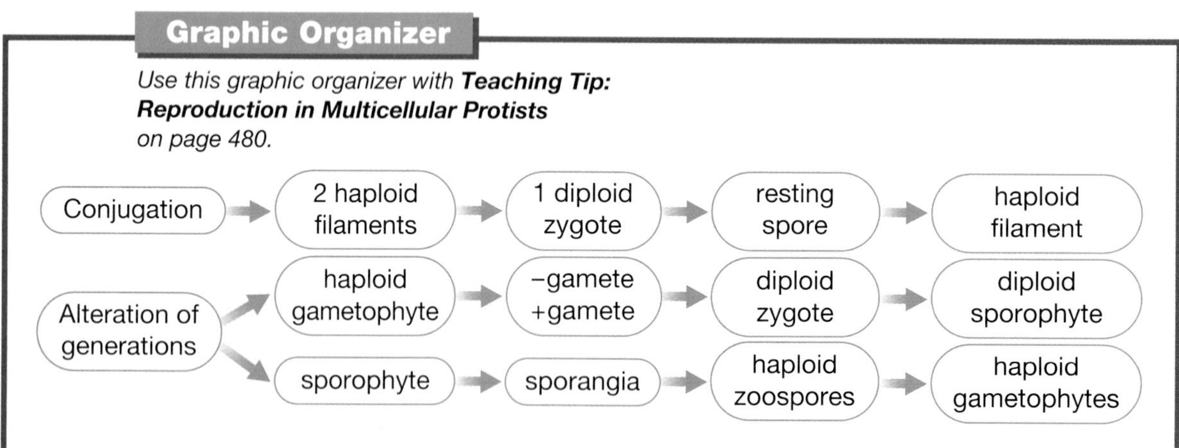

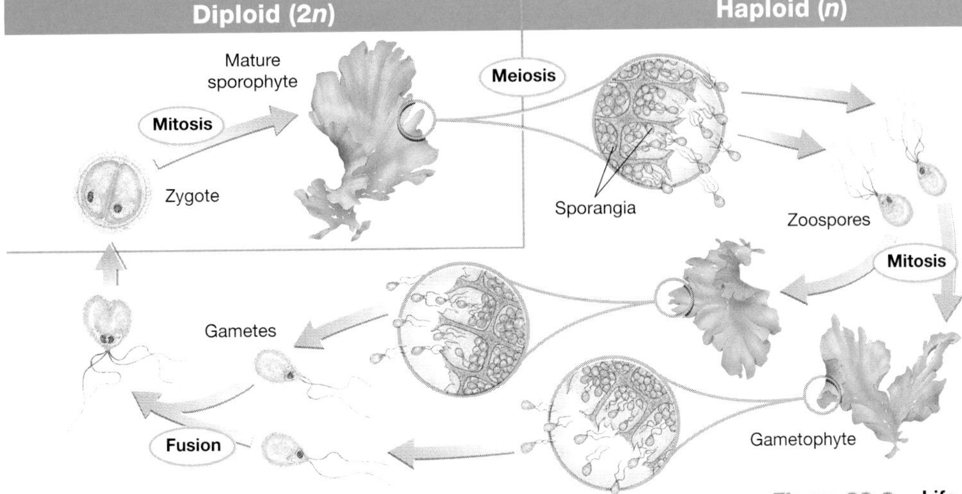

Figure 22-3 Life cycle of *Ulva.* The multicellular green alga *Ulva,* or sea lettuce, has a life cycle in which haploid and diploid individuals alternate.

Sexual reproduction among multicellular protists

Sexual reproduction among multicellular protists occurs in many different ways:

Alternation of generations *Ulva* is a common genus of marine green alga. **Figure 22-3** shows that the reproductive cycle of *Ulva,* called **alternation of generations,** is characterized by two distinct multicellular phases: a diploid, spore-producing phase, called the sporophyte generation, and a haploid, gamete-producing phase, called the gametophyte generation. The adult sporophyte alga has reproductive cells called **sporangia** *(spoh RAN jee uh),* which produce haploid spores by meiosis. The spores grow into multicellular haploid gametophytes. The mature gametophyte produces haploid gametes that fuse and complete the life cycle by dividing through mitosis to form a new diploid sporophyte.

Conjugation *Spirogyra,* a filamentous green alga species, reproduces sexually by conjugation. Conjugation is the temporary union of two protists to exchange nuclear material. The process begins when two filaments align side by side. Part of the cell walls between adjacent algae form a bridge between the cells. The nucleus of one cell then passes through the tube into the adjacent cell. The two nuclei eventually form a resting spore, which produces a new haploid filament.

Study TIP

Reading Effectively

To better understand the process of alternation of generations, read this page three times, then retell the information to a partner. Then switch roles with your partner and compare your understanding.

Review SECTION 22-1

1. **List** three characteristics of protists.
2. **Describe** the kinds of environments in which protists thrive.
3. **Summarize** why protists are not classified with the other three eukaryotic kingdoms.
4. **Summarize** the asexual life cycle of *Chlamydomonas* species.
5. **SKILL Relating Concepts** Describe two ways sexual reproduction can occur in multicellular protists.

Review SECTION 22-1 ANSWERS

1. Protists may reproduce sexually or asexually; be unicellular or multicellular; divide by mitosis and meiosis; be heterotrophic, autotrophic, or both; have complex flagella and cilia; and may have light-sensing organelles.
2. Answers could include moist environments such as bodies of water, damp soil or sand, rotting leaves, in the bloodstream, and the tissues of plants and animals.
3. Protists lack the specialized characteristics of plants, animals, and fungi.
4. The mature, single-celled *Chlamydomonas* is haploid and reproduces asexually by dividing by mitosis. This leads to the formation of a zoospore, which eventually breaks open to release new haploid organisms.
5. Multicellular algae may reproduce by undergoing an alternation of generations, characterized by a gamete-producing generation (gametophyte) that gives rise to a spore-producing generation (sporophyte), which then leads to a new gametophyte generation, and so on. Another method of sexual reproduction is conjugation—the merging of the nuclear material of two individuals.

CAREER

Limnologist

Limnologists are freshwater ecologists employed by universities and research and consulting firms, as well as local, state, and federal governments. Limnologists work with others to study and improve water quality and the health of aquatic ecosystems. Protists and other microorganisms are often at the center of their research. For example, Dr. Ruth Patrick, a limnologist, has invented the "diatometer"—a device used to identify diatoms, which are a type of algae. Encourage students to learn more about the work of limnologists.

3 Assessment Options

Closure

Summarizing Protists

Have students write a paragraph summarizing how the groups of protists are similar and how they are different. PORTFOLIO

Review

Assign Review—Section 22-1.

Reteaching — BASIC

Have students work individually or in small groups to make a concept map of the following terms: zoospores, haploid, diploid, unicellular, multicellular, zygospore, conjugation, gametes, alternation of generations, *Chlamydomonas, Spirogyra, Ulva,* zygote, and sporangia. PORTFOLIO

Section 22-2 Protist Diversity

1 Prepare

Scheduling

- **Block:** About 90 minutes will be needed to complete this section.
- **Traditional:** About two class periods will be needed to complete this section.
- **Planning:** See the Planning Guide on pp. 475A and 475B for lecture, classwork, and assignment options for Section 22-2. Lesson plans for Section 22-2 can be found in a lesson cycle format in ***Biology: Principles and Explorations*** **One-Stop Planner CD-ROM with Test Generator.**

Resource Link

Assign **Active Reading Worksheet Section 22-2** to help students comprehend the key concepts and visuals in this lesson.

2 Teach

Lesson Warm-up

Obtain a culture of live amoebas. Using a microprojector or videomicroscope, project an image of an amoeba for students to view. Discuss the general characteristics of the amoeba, pointing out the nucleus and means of locomotion (pseudopodia). If time permits, repeat the procedure with *Paramecium* and *Euglena* specimens.

Visual Strategy

Figure 22-4 — BASIC

Have students use this figure to create a step-by-step drawing or comic depicting the movement of the pseudopodia and the flow of cytoplasm as it engulfs its prey. Ask students to add a brief paragraph to explain their drawings.

PORTFOLIO

Objectives

- **Identify** how amoebas and forams move. (p. 482)
- **Describe** the structure of diatoms. (p. 483)
- **Contrast** three kinds of algae. (p. 484)
- **Differentiate** three different kinds of flagellates. (p. 485)
- **Summarize** the general characteristics of a *Paramecium.* (pp. 486–487)

Key Terms

amoeba
pseudopodium
diatom
zoomastigote
euglenoid
cilium
plasmodium

Protists Can Move by Using Cytoplasmic Extensions

One of the most easily recognized groups of protists consists of amoebas *(uh MEE buhs)* and forams. Amoebas and forams are unicellular heterotrophs that have a unique form of locomotion.

Amoebas

Amoebas, members of the phylum Rhizopoda *(reye zoh POH duh),* are protists that move by using flexible, cytoplasmic extensions. These extensions are called **pseudopodia** *(soo doh POH dee uh),* from the Greek words *pseudo,* meaning "false," and *podium,* meaning "foot." Because an amoeba has no cell walls or flagella, it is extremely flexible. A pseudopodium bulges from the cell surface, stretches outward, and anchors itself to a nearby surface. The cytoplasm from the rest of the amoeba then flows into the pseudopodium. Pseudopodia can surround and engulf food particles, as shown in **Figure 22-4.** Amoebas live in both fresh water and salt water and are especially abundant in soil. Meiosis and sexual reproduction do not occur in amoebas. They reproduce by fission, dividing into two new cells. The majority of amoebas are free-living, but some species are parasites, such as *Entamoeba histolytica,* the protist that causes amebic dysentery in humans. These organisms are transmitted in contaminated food or water.

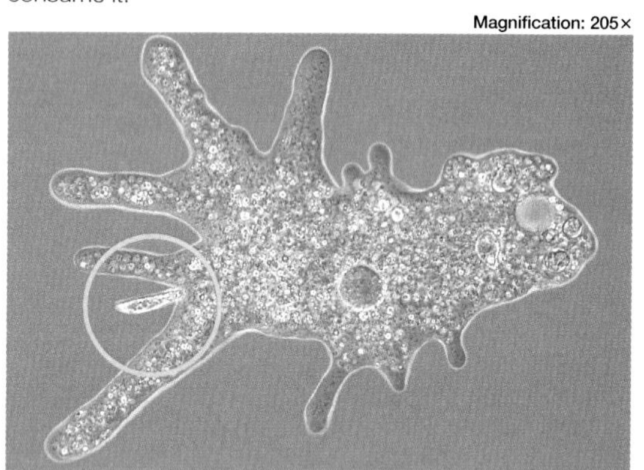

Figure 22-4
Pseudopodia. This amoeba is using its pseudopodia to engulf a smaller cell and consume it.

Foraminifera

Members of the phylum Foraminifera *(foh ram ih NIHF ur uh),* or forams, are marine protists that live in sand or attach themselves to other organisms or rocks. Forams are characterized by their porous shells, called tests. Tests usually have many chambers arranged in a spiral shape. They resemble a tiny snail and consist of organic material that contains grains of calcium carbonate. Long, thin projections of cytoplasm extend through the pores in the tests to aid in swimming and in catching prey. Some forams also obtain nourishment from algae that live symbiotically under their tests.

The tests of dead forams have accumulated on ocean floors over millions of years. Their calcium carbonate shells helped form limestone deposits and are important components of many land formations.

Integrating Different Learning Styles

Logical/Mathematical	**PE**	Data Lab, p. 486; Performance Zone item 25, p. 495
Visual/Spatial	**ATE**	Lesson Warm-up, p. 482; Visual Strategy, p. 482; Demonstration, p. 484
	PE	Quick Lab, p. 483; Up Close, p. 487; Performance Zone item 25, p. 495
Interpersonal	**ATE**	Teaching Tip, p. 486
Verbal/Linguistic	**ATE**	Teaching Tip, p. 483
	PE	Performance Zone items 22, 23, and 26, p. 495

Diatoms Have Double Shells

Diatoms *(DEYE uh tahmz),* members of the phylum Bacillariophyta, are photosynthetic, unicellular protists with unique double shells. The shells are made of silica and often have unique markings. Their shells are like small boxes with lids, one half fitting inside the other. Abundant in oceans and lakes, diatoms are important producers in the food chain. Diatoms can have one of two types of symmetry: radial (like a wheel) or bilateral (two-sided). The empty shells of diatoms form thick deposits that are mined commercially as diatomaceous earth. Diatomaceous earth is used as an abrasive or to add the sparkling quality to paint used on roads. Diatomaceous earth is also sold as a natural control for pests, such as slugs, and some insects, such as fleas. The sharp edges of the diatom shells cut into the body of the organism, leading to its death.

Diatoms secrete chemicals through holes in their shells, enabling them to move by gliding. Individuals are diploid and usually reproduce asexually. The two halves of the shell separate, and each half regenerates another matching half. As a consequence of this model of reproduction, diatoms tend to get smaller and smaller with each generation. When an individual gets too small because of repeated division, it slips out of its shell, grows to full size, and regenerates a new shell. Sexual reproduction in diatoms is rare.

Word Origins

The word *amoeba* is from the Greek *amoibe,* meaning "change." Knowing this makes it easier to remember that amoebas are able to change their shape.

What are some characteristics of diatoms?

Try this activity to find out why diatomaceous earth is used to make abrasives, fine filters, and reflective paints.

Materials

pipet, water, microscope slide, toothpick, diatomaceous earth, coverslip, compound microscope

Magnification: 240×

Procedure

1. Using a pipet, place a drop of water in the center of a clean microscope slide.
2. Use a toothpick to scoop up a small amount of diatomaceous earth and mix it with the water drop. Add a coverslip.
3. Observe your wet mount under both low and high power of a compound microscope.
4. Draw some of the diatoms you see.
5. Observe the wetmount under low power as your partner shines a flashlight (at a 45° angle) on the slide. Turn off the microscope's light source so that only the flashlight is lighting the slide. Record your observations.

Analysis

1. **Label** your drawings as radial or bilateral. Find out the meanings of these terms and how they apply to your diatom drawings.
2. **Select** some characteristics you observed that are useful in classifying particular species of diatoms.
3. **Interpret** what you observed when the flashlight was shone on the slide.

did you know?

Algae might be useful in space.

Green algae have been tested in the space program as a means of eliminating some of the cargo needed for long missions. The algae would feed shrimp, a food source for the astronauts, and would also recycle carbon dioxide and release oxygen. Since they regenerate continuously, the algae would add less weight and volume to the spacecraft than more traditional food and air sources.

What are some characteristics of diatoms?

Time 15–20 minutes

Process Skills Observing, interpreting observations

Teaching Strategies

- Have students experiment with different light angles using the flashlight.
- If a polarized light source is available (try using a polarizing filter from a camera), compare the appearance of the diatoms with the nonpolarized light source.

Analysis Answers

1. In radial symmetry, the two halves of the object are symmetrical no matter how the object is divided, provided the division passes through the center of the object. In bilateral symmetry, there are a limited number of planes—often just one—that can divide the object into two equal halves.
2. Students may use shape, symmetry, and texture as useful characteristics.
3. Changing the angle of the light will change the observed reflectance and alter the appearance of surface features.

Teaching TIP

Diatom Deposits

Diatoms make up a major portion of sediments on the ocean floor. One of the thickest deposits measured is about 1 km deep and is located off the coast of Lompoc, California. Diatom deposits are mined in Nevada, Washington, Kansas, and Oregon. Ask students to use library resources to find out what specific products contain diatoms. *(Fossil diatoms or diatomaceous earth are used in cleaning and abrasive powders, paints, soil additives, flea powders, and filtration processes.)*

Many Algae Are Multicellular

Algae are protists that are strict phototrophs. Some are unicellular, others multicellular. Algae are distinguished by the type of photosynthetic pigment they contain and by their cell or body shape.

Green algae

Most green algae (phylum Chlorophyta) are freshwater unicellular organisms, but some are large, multicellular marine organisms, as shown in **Figure 22-5.** They also exist as a major part of microscopic marine plankton, inhabit damp soil, or even thrive within the cells of other organisms as symbionts. Green algae contain the same pigments found in the chloroplasts of plants. Most green algae have sexual and asexual reproductive stages.

Red algae

Red algae (phylum Rhodophyta) are multicellular organisms found in warm ocean waters. Their red pigments are efficient at absorbing the light that penetrates deep waters. Some red algae, as shown in Figure 22-5, have calcium carbonate in their cell walls. Others are used to make agar and carrageenan *(kayr uh GEE nuhn).* Red algae have a complex life cycle, usually involving alternation of generations.

Brown algae

Brown algae (phylum Phaeophyta) are multicellular and are found mostly in marine environments. The larger brown algae known as kelp grow along coasts and provide food and shelter for many different kinds of organisms. They are among the largest organisms on Earth. Brown algae reproduce by alternation of generations.

REAL LIFE

People eat and use algae. Many red algae are used in making food and food additives. Agar is used to prevent baked goods from drying, to grow bacteria in Petri dishes, and to make gelatin capsules. Carrageenan is used in paints, dairy products, and cosmetics to help stabilize their different ingredients.

Finding Information
Read the labels on packages and cans at home, to find out if they contain any agar or carrageenan.

Figure 22-5 Three phyla of algae. *Ulva* is a species of green algae composed of a sheet of cells that is only two cells thick. Coralline algae is a red algae that contributes to the great coral reefs. *Macrocystis* is a brown algae and some species can grow to a length of more than 60 m.

Teaching TIP BASIC

Have You Had Your Algae Today?
Have students keep a tally of the algae-derived products that they eat or use in one week. Encourage them to read labels. Some key terms that identify algae extracts are *lecithin, carageenan, algin, emulsifier,* and *thickening agent.* Some of the many products that contain algae include pudding, jelly, jelly beans, ice cream, chocolate milk, marshmallow, processed cheese, and salad dressing. Have students write an essay about the algae in their life. PORTFOLIO

REAL LIFE

Answer
Answers will vary and may include ready-mix cakes, instant puddings, pie fillings, and canned pet foods.

Demonstration

Observing Live Algae
Collect live specimens of various algae or obtain preserved specimens from a biological supply house, or display pictures of algae from scientific reference books. Have students examine the algae and compare their characteristics.

CROSS-DISCIPLINARY CONNECTION

◆ Geography
The Sargasso Sea is named for the floating masses of *Sargassum natans*. This brown algae reproduces by fragmenting into small pieces, and these fragments cover millions of square miles between the Azores and the Bahamas. Nautical lore tells of ship graveyards trapped in the thick mat of algae. In truth, the area was avoided because it lies within the wind-sparse "horse" latitudes. Have students locate this area on a map or globe.

Overcoming Misconceptions

Hidden Green Pigments
Both brown and red algae contain chlorophyll in addition to the pigments that make them look brown and red. A misconception is that they lack chlorophyll; however, the green pigment is simply hidden by the other colors.

Some Protists Move with Flagella

Flagellates are protists that move using flagella. The three major phyla of flagellates are the dinoflagellates, the zoomastigotes, and the euglenoids.

Dinoflagellates

Dinoflagellates *(deye noh FLAJ uh layts),* members of the phylum Dinoflagellata, are unicellular phototrophs, most of which have two flagella. A few kinds of dinoflagellates are found in fresh water, but most are marine and make up part of the plankton. Most dinoflagellates have a protective coat made of cellulose that is often encrusted with silica, giving them unusual shapes, as shown in **Figure 22-6.** Their flagella beat in two grooves—one encircling the body like a belt, the other perpendicular to it. As a result, dinoflagellates spin through the water like a top. A few dinoflagellates produce powerful toxins. The poisonous "red tides" that occur frequently in coastal areas are often associated with population explosions of dinoflagellates. Dinoflagellates reproduce asexually by mitosis.

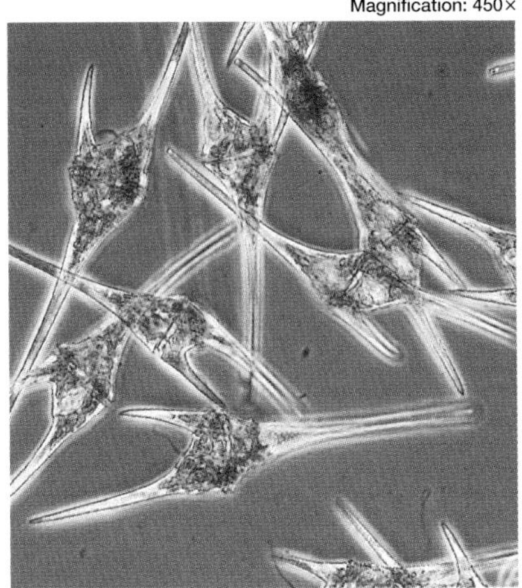

Figure 22-6
Dinoflagellate. This dinoflagellate, *Gonyaulax tamarensis,* is characterized by a pair of flagella set in perpendicular grooves.

Zoomastigotes

Zoomastigotes *(zoh oh MAS tih gohts),* members of the phylum Zoomastigina, are unicellular heterotrophs that have at least one flagellum, and some species have thousands. While most reproduce only asexually, some are known to produce gametes and reproduce sexually. Some zoomastigotes, such as *Trichonympha,* live symbiotically in the guts of termites, where they provide the enzymes that digest wood. Others, such as the trypanosomes, cause diseases in humans and domestic animals, such as African sleeping sickness.

Euglenoids

Euglenoids *(yoo GLEE noyds),* members of the phylum Euglenophyta, are freshwater protists with two flagella. This group clearly shows the difficulty of classifying protists as animals or plants. About one-third of the 1,000 known species of euglenoids have chloroplasts and are photosynthetic; other species lack chloroplasts, ingest their food, and are heterotrophic. Euglenoids are clearly related to zoomastigotes, and many taxonomists merge the two phyla together. A member of *Euglena,* shown in **Figure 22-7,** has a protein scaffold called a pellicle *(PEHL ih kuhl)* inside the cell membrane. Since the pellicle is flexible, the euglenoid can change shape. A light-sensitive organ called the eyespot helps orient the movements of these organisms toward light. Reproduction in this phylum occurs by mitosis.

Figure 22-7 ***Euglena.*** *Euglena* is a versatile protist. It contains chloroplasts and is photosynthetic, but it is also heterotrophic and can live without light.

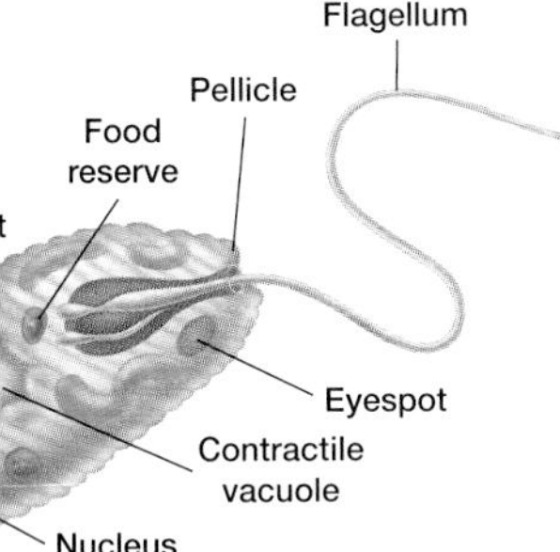

Teaching TIP

Red Tides

The toxic substance produced by the dinoflagellate responsible for "red tides" can kill fish and invertebrates that consume the algae. Red tides can also be a human health risk. Ask students how red tides might result in human poisoning. *(People who ingest mussels that feed on the toxic dinoflagellates can become ill and may even die.)*

Teaching TIP

Glowing Protists

The dinoflagellate *Noctiluca* glows because of a chemical reaction that produces light. The waters that contain these organisms create luminescent, or "sparkling," waves. Encourage students to find out more about luminescence and *Noctiluca.*

Visual Strategy

Figure 22-7

Use this figure to point out the parts of *Euglena*. Make sure students understand that the pellicle is inside the cell membrane. Ask students how protists in the Euglenophyta phylum represent a classification challenge. *(Some are photosynthetic, some are heterotrophic, and some are both.)*

did you know?

Dinoflagellates help form coral reefs.

Dinoflagellates are essential in the formation of coral reefs, coral islands, and atolls. By inhabiting the living tissues of reef corals, mollusks, and other marine organisms, dinoflagellates are able to use the intense sunlight that shallow reef waters receive. In turn, dinoflagellates contribute to the accumulation of reefs by speeding up the formation of calcium carbonate in coral skeletons.

Teaching TIP

Test the Eyespot

Have pairs of students design an experiment to test a euglenoid's response to light. Ask students to predict whether euglenoids will move toward or away from light. Guide students to conclude that euglenoids will move toward a light source because they use light to make food. PORTFOLIO

Interpreting competition among protists

Time 10–15 minutes

Process Skills Analyzing data, applying information, evaluating information

Teaching Strategies

- Based on the graph, ask students to predict what might happen to *P. caudatum* population after another few days. *(It may die off completely.)*
- Ask students to hypothesize conditions that might allow *P. caudatum* to thrive even in the presence of *P. aurelia*. *(In the presence of a predator or under certain environmental conditions,* P. caudatum *may be able to compete more effectively with* P. aurelia.*)*

Analysis Answers

1. *P. aurelia*
2. The variables were (1) growing the paramecia alone and (2) growing the two species in the presence of one another.
3. *P. aurelia* was the superior competitor, obtaining the necessary nutrients at the expense of *P. caudatum.*
4. Accept all well-supported answers. Students should realize that the results are valid only under these controlled conditions. Adding variables to simulate their natural habitat may yield results more closely resembling the natural world.

Ciliates Are Complex Protists That Use Cilia to Swim

Magnification: 148×

Figure 22-8 Ciliates. The cilia that surround this ciliate enable it to move and feed.

Ciliates are the most complex and unusual of the protists. They are so different that many biologists argue they should be placed in an entirely different kingdom.

All members of the phylum Ciliophora *(sihlee AWF oh ruh)* have large numbers of **cilia,** tightly packed rows of short flagella used for movement, as shown in **Figure 22-8.** Ciliates are complex unicellular heterotrophs. The body wall of ciliates is a tough but flexible outer pellicle that enables the organism to squeeze through or move around many obstacles. The pellicle consists of an outer membrane with numerous fluid-filled cavities beneath it. Ciliates, such as *Paramecium,* shown in *Up Close: Paramecium,* on page 487, form vacuoles for ingesting nutrients and regulating their water balance.

In addition to their characteristic cilia, most ciliates have two types of nuclei within their cells: micronuclei and larger macronuclei. The micronuclei contain normal chromosomes that divide by mitosis. Macronuclei contain small pieces of DNA derived from micronuclei.

Reproduction in ciliates is usually by mitosis, with the body splitting in half. Cells divide asexually for about 700 generations and then die if sexual reproduction has not occurred. Most ciliates can engage in sexual reproduction through conjugation, in which two cells unite and exchange genetic material.

Interpreting competition among protists

Background

Protists, like all organisms, must compete with one another for nutrients and therefore have evolved various strategies to compete. To examine the effects of competition between two species of *Paramecium,* equal numbers of the paramecia were grown together and separately. Study the graph at right, and answer the following questions.

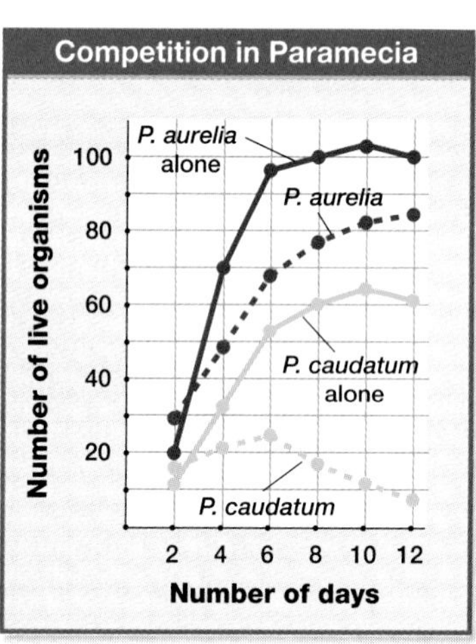

Analysis

1. **Identify** the *Paramecium* that grew best alone.
2. **Identify** the variables in this experiment.
3. **SKILL Analyzing Data** How would you explain the difference in the growth curves in the group that had both organisms?
4. **SKILL Predicting Outcomes** In a natural setting there would be more than two organisms present. Predict the effect that the presence of other organisms would have on the growth of the two species of *Paramecium.*

did you know?

Diatomaceous earth can stabilize dynamite.

Diatomaceous earth has been used in the manufacture of dynamite. When nitroglycerin is absorbed by diatomaceous earth, it is stable and safe to handle.

Paramecium

- **Scientific name:** *Paramecium caudatum*
- **Size:** Microscopic; up to 1 mm long
- **Habitat:** Lives in freshwater streams and ponds
- **Diet:** Bacteria, small protists, organic debris

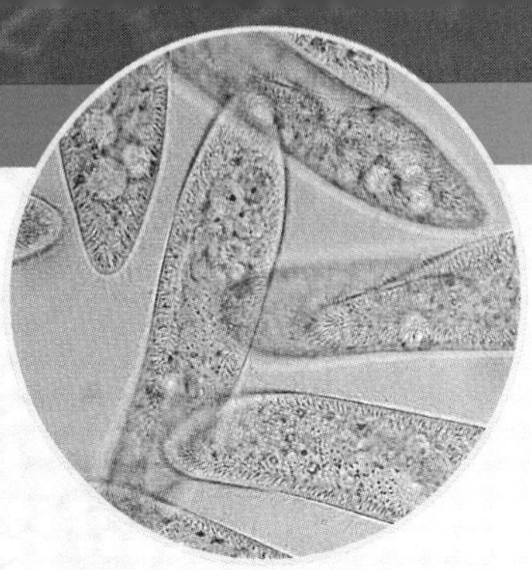

Magnification: 1,200×

Characteristics

Nuclei Members of the genus *Paramecium* have two nuclei. The macronucleus contains fragmented chromosomes used in routine cellular functions, and it divides by pinching in two. The micronucleus contains the cell's chromosomes and divides by mitosis.

▲ Macronucleus

Surface *P. caudatum*, a ciliate, is covered with thousands of cilia arranged in rows along the cell. Cilia beat in waves that move diagonally across the cell, causing the protist to spin through the water. *P. caudatum* is surrounded by a rigid protein covering called a pellicle.

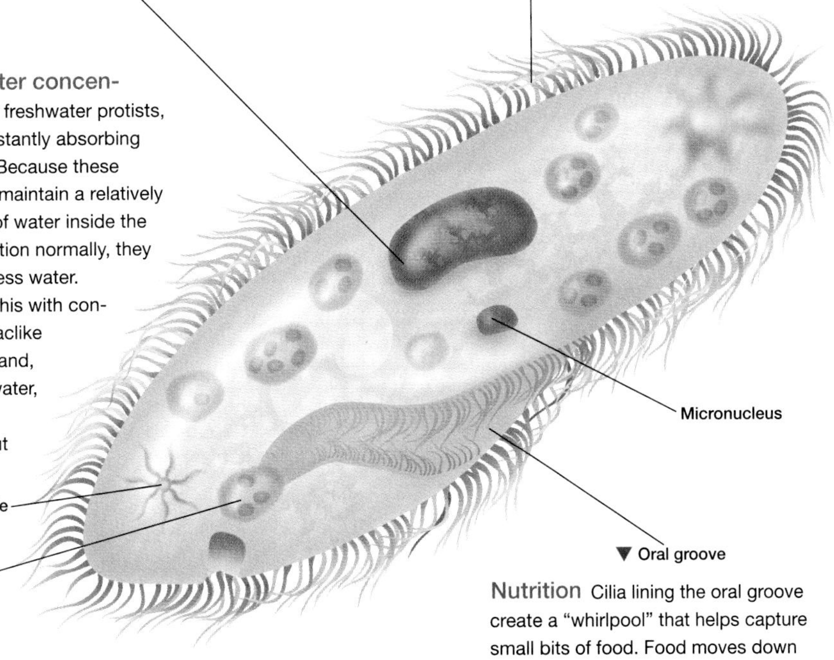

▲ Cilia

Maintaining water concentration Like other freshwater protists, *P. caudatum* is constantly absorbing water by osmosis. Because these organisms need to maintain a relatively low concentration of water inside the cell in order to function normally, they must get rid of excess water. *P. caudatum* does this with contractile vacuoles, saclike organelles that expand, collecting excess water, and then contract, squeezing water out of the cell.

▲ Contractile vacuole

Food vacuole

Micronucleus

▼ Oral groove

Nutrition Cilia lining the oral groove create a "whirlpool" that helps capture small bits of food. Food moves down the funnel-shaped groove and is engulfed in a food vacuole by endocytosis. As the food vacuole moves throughout the cell, it combines with digestive enzymes. Undigested food is released from the cell by exocytosis.

Genetic variation *P. caudatum* generally reproduces asexually by binary fission. Genes are shuffled during a sexual process called conjugation. During conjugation, two individuals exchange haploid micronuclei, which fuse to other micronuclei, forming a diploid nucleus that contains nuclei from both individuals.

Up Close

Paramecium

Teaching Strategies

- Ask students to draw a flowchart to trace the movement of food through a paramecium. *(Charts should begin with food entering the paramecium via the oral groove. At the cell mouth, food is engulfed by a food vacuole containing digestive enzymes. Undigested food is excreted by exocytosis.)* PORTFOLIO
- Ask students to compare the methods used by amoebas and paramecia for capturing food. Remind students that amoebas use their pseudopodia to surround and engulf food by endocytosis. Paramecia use cilia to create a whirlpool to pull food into the oral groove, where the food is engulfed by a food vacuole.

Discussion

- How would a paramecium living in a clean environment differ from one in a polluted environment? *(Paramecia in polluted water may be less efficient at maintaining water balance, acquiring nutrients, and moving due to the pollutants and subsequent damage to cilia.)*
- Describe an ideal environment for a paramecium. *(Answers will vary but should include a food source and a freshwater environment that will allow the organism to maintain homeostasis.)*

Chapter 22

Visual Strategy

Figure 22-9

Ask students what would cause *D. discoideum* to change its form, as observed in the figure. *(environmental stress such as food or water deprivation)*

Teaching TIP — BASIC

Comparing Protistan Molds

Ask students to make a table that summarizes the four protistan molds discussed in this section. Allow students to choose their own headings so that they can organize the information in their own terms. Possible headings include Phylum, Habitat, Appearance, and Type of mold.

PORTFOLIO

Teaching TIP

Grow Your Own Slime Mold

Let students grow their own slime molds in Petri dishes filled with dampened oatmeal. If a source of protistan molds is not readily available, check with a local university.

Figure 22-9 Cellular slime mold. The individual amoebas of *Dictyostelium discoideum* aggregate, move in a mass called a slug, and form a stalked structure that contains spores.

Protistan Molds Are Not Fungi

Protistan molds are heterotrophs with some mobility. They were once thought to be related to fungi because of a similar appearance and life cycle. Protistan molds have cell walls made of carbohydrates, like those found in other protists, whereas fungi do not. Also, protists carry out normal mitosis, whereas mitosis in fungi is unusual, as you will learn in Chapter 23.

Cellular slime molds

Cellular slime molds, members of the phylum Acrasiomycota *(uh KRAZ ee oh meye koh tuh),* resemble amoebas but have distinct features. The individual organisms behave as separate amoebas, moving through the soil and ingesting bacteria. During environmental stress, the individual amoebas gather together and move toward a fixed center. There they form multicellular colonies called slugs. Each slug develops a base, a stalk, and a swollen tip that develops spores, as shown in **Figure 22-9.** Each of these spores, when released, becomes a new amoeba, which then begins to feed and repeat the life cycle.

Plasmodial slime molds

Plasmodial slime molds, members of the phylum Myxomycota *(MIHKS oh meye koh tuh),* are a group of organisms that stream along as a **plasmodium,** a mass of cytoplasm that looks like an oozing slime. As they move, they engulf bacteria and other organic material, as shown in **Figure 22-10.** A plasmodial slime mold contains many nuclei, but they are not separated by cell walls. If the plasmodium begins to dry out or starve, it divides into many small mounds. Each mound produces a stalk tipped with a capsule in which haploid spores develop. The spores are highly resistant to hostile environmental conditions. When conditions are favorable, the spores germinate and become haploid cells that are either amoeboid or flagellated. These haploid cells are able to fuse into diploid zygotes, which undergo mitosis and form a new plasmodium.

Figure 22-10 Plasmodial slime mold. A plasmodial slime mold is a mass of cytoplasm containing many nuclei.

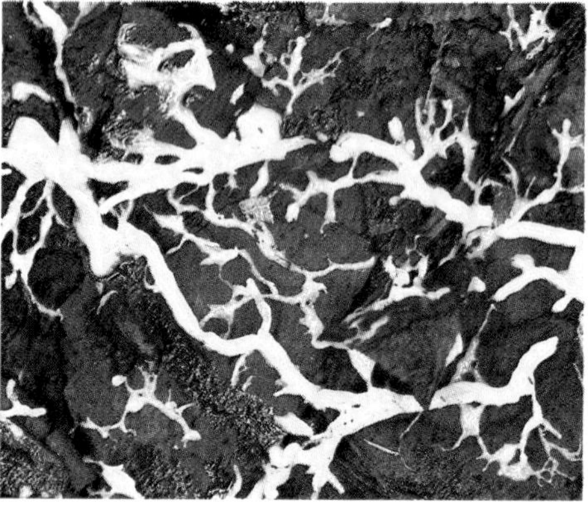

Other molds

Oomycetes *(oh oh MY seets),* members of the phylum Oomycota, are the water molds, white rusts, and downy mildews that often grow on dead algae and dead animals in fresh water. All members of the group either are parasites or feed on dead organic matter. Oomycetes are unusual in that their spores have two flagella: one pointed forward, the other pointed backward. Many oomycetes are plant pathogens, including *Phytophthora infestans,* which causes late blight in potatoes. This protist was responsible for the Irish potato famine of 1845–1847, during which about 400,000 people starved to death.

Some Protists Form Resistant Stages

Parasitic protists that form spores during their reproduction cycle are called sporozoans. They are members of the phylum Apicomplexa and are nonmotile, unicellular parasites. All sporozoans are parasitic and cause many serious diseases. Malaria, a sporozoan disease, kills more people than any other infectious disease. Sporozoans infect animals and are transmitted from host to host. There are about 4,500 named species in the phylum Apicomplexa.

Sporozoans have complex life cycles that involve both asexual and sexual reproduction. Sexual reproduction involves the fertilization of a large female gamete by a small, flagellated male gamete. The resulting zygote forms a thick-walled structure that makes the zygote resistant to drought and other unfavorable environmental conditions. A few sporozoans are listed in **Table 22-2.**

Many sporozoans are transmitted from one host to another by blood-feeding insects, such as mosquitoes, black flies, and midges. Other sporozoans are transmitted in the feces of an infected animal, such as the one shown in **Figure 22-11.** The parasites infect a new individual when that individual eats or drinks food or water that has been contaminated with infected feces. Some sporozoans are even transmitted through predation, such as when a cat becomes infected by eating an infected mouse.

Figure 22-11 ***Cryptosporidium.*** These infective stages of *Cryptosporidium* (stained purple) are found in the feces of infected animals. In the 1980s and 1990s, water supplies in several cities in the United States became infected with these sporozoans.

Table 22-2 Some Sporozoans and Their Hosts

Sporozoans	Disease	Hosts
Plasmodium species	Malaria	Humans and other vertebrates
Toxoplasma species	Toxoplasmosis	Humans and cats
Babesia species	Cattle tick fever	Cattle, mice, humans, deer, dogs
Cryptosporidium species	Cryptosporidiosis (diarrhea)	Cattle, humans, birds, deer, dogs, cats

Review SECTION 22-2

1. **Describe** a pseudopodium.
2. **Describe** the shells of diatoms.
3. **Construct** a table that compares three kinds of algae.
4. **Contrast** *Euglena* with two other phyla of flagellates.
5. **Describe** five general characteristics of *Paramecium.*

Critical Thinking

6. **SKILL Evaluating Viewpoints** A classmate makes the statement, "*Euglena* is a protozoan, not an alga." Is this an accurate statement? Why or why not?

Review SECTION 22-2 ANSWERS

1. A pseudopodium, or "false foot," is an extension of the amoeba's cell membrane that allows it to move and engulf prey.
2. Diatom shells are three-dimensional structures excreted around the bodies of certain algae and consisting primarily of silica.
3. Tables will vary but should reflect the descriptions of green, red, and brown algae on page 484.
4. Answers may vary, but should reflect the descriptions of flagellates on page 485.
5. Paramecia are unicellular, have cilia, contain micronuclei and macronuclei, reproduce asexually by mitosis, reproduce sexually by conjugation, are covered by a pellicle, and form vacuoles.
6. Yes, eugleniods cannot be classified as algae because they are not *strictly* autotrophic.

Section 22-2

Making Biology Relevant

Food

A new, more virulent form of the potato blight that was responsible for the Irish potato famine of 1845–1847 threatens current potato crops. The new blight, a type of water mold, is resistant to metalaxyl, the fungicide traditionally used against potato blight in the United States. The disease threatens a major global food source, as potatoes are grown and consumed around the world. Currently scientists are working with wild potatoes to produce new disease-resistant varieties.

3 Assessment Options

Closure

Name That Protist

Have students create a game in which they describe characteristics of a particular protist discussed in Section 22-2, and their partner guesses the organism's name.

Review

Assign Review—Section 22-2.

Reteaching BASIC

Tell students to pretend that they have been invited to give a presentation at a club of protist enthusiasts. Each student should focus on a particular protist discussed in Section 22-2. Presentations should include a drawing of the protist and a description of its characteristics, including reasons for classifying the organism in the kingdom Protista. Encourage student creativity, allowing them to name their "pet" and to make up anecdotes to describe their protist's special characteristics. Have students share their creations and post their work around the room. PORTFOLIO

Section 22-3 Protists and Health

1 Prepare

Scheduling

- **Block:** About 90 minutes will be needed to complete this section.
- **Traditional:** About two class periods will be needed to complete this section.
- **Planning:** See the Planning Guide on pp. 475A and 475B for lecture, classwork, and assignment options for Section 22-3. Lesson plans for Section 22-3 can be found in a lesson cycle format in ***Biology: Principles and Explorations* One-Stop Planner CD-ROM with Test Generator.**

Resource Link

Assign **Active Reading Worksheet Section 22-3** to help students comprehend the key concepts and visuals in this lesson.

2 Teach

Lesson Warm-up

Ask students to list some characteristics of an ideal habitat for a microbe. *(Students may suggest a moderate or warm climate in which the microbe can remain active throughout the year or poor sanitary conditions, which allow the microbe to spread.)* Lead students to conclude that populations facing such conditions suffer the most from microbial diseases.

Active Reading

Technique: Reading Effectively

Before reading this section, ask students to suggest how they would protect themselves from diseases caused by protists. After reading, have students amend their answers as needed. *(Precautions might include getting the appropriate vaccinations or medications, avoiding infested areas, watching for possible food or water contamination, and protecting themselves from the insects or other animals that can carry the pathogens.)*

Objectives

- **Identify** two ways that protists affect human health. (p. 490)
- **Name** three human diseases, other than malaria, caused by protists. (p. 490)
- **Summarize** how malaria is transmitted. (p. 491)
- **Evaluate** the methods used to control malaria. (p. 492)

Key Terms

sporozoite
merozoite

How Do Protists Affect Humans?

One of the greatest effects protists have on humans is that they cause disease. This effect can be measured in pain, death, and the medical costs of preventing and treating diseases. Some diseases caused by protists are listed in **Table 22-3.**

Protists also affect humans through the diseases they cause in livestock. The added cost of treating diseased livestock is passed on to consumers in the form of higher meat prices. This increased cost of living hinders progress in the developing world.

Beneficial protists

There are many commensal protists that live in the digestive tracts of humans and in the digestive tracts of the animals that humans eat. Cattle could not digest the cellulose in the hay and grass they eat without the aid of commensal protists in their digestive tract.

Protists, which make up much of the plankton in the ocean, help to support food chains. Protists are also the single largest group of photosynthesizers on the planet. Since we all breathe oxygen, we all benefit from the gas that protists produce. Many protists are also detritivores, so they help recycle important chemicals, such as nitrogen, carbon, and phosphorus, in the environment.

Table 22-3 Diseases Caused by Protists

Disease	Symptoms	Protist	How the disease is transmitted
Amebic dysentery	Bloody diarrhea, vomiting, extremely strong stomach cramps, fever	*Entamoeba histolytica*	Contaminated food or water
Giardiasis	Cramps, nausea, diarrhea, vomiting	*Giardia lamblia*	Contaminated food or water
African sleeping sickness	Fever, weakness, lethargy	*Trypanosoma gambiense, Trypanosoma rhodesiense*	Bite from infected tsetse fly
Malaria	Fever, chills, sweats	*Plasmodium* species	Bite from infected mosquito
Toxoplasmosis	Primary danger is fetal infection; can cause convulsions, brain damage, blindness, and death in fetuses	*Toxoplasma gondii*	Contact with infected cats or improperly cooked meat

Integrating Different Learning Styles

Logical/Mathematical	**PE**	Performance Zone items 18. 19, 20, and 21, p. 495
Visual/Spatial	**PE**	Performance Zone item 1, p. 494
Interpersonal	**ATE**	Closure, p. 492; Reteaching, p. 492
Verbal/Linguistic	**PE**	Health Watch, p. 492; Performance Zone items 27 and 28, p. 495

Malaria Is Caused by Several Species of *Plasmodium*

Malaria is one of the most deadly diseases in humans. Over 100 million people have malaria at any one time, and up to 3 million, mostly children, die from it every year.

The symptoms include severe chills, fever, sweating, confusion, and great thirst. Victims die of anemia, kidney failure, or brain damage unless the disease is treated.

Malaria life cycle

Malaria is caused by several species of *Plasmodium* and is spread by the bite of certain mosquitoes. There are three stages in the *Plasmodium* life cycle, as shown in **Figure 22-13.** When an infected mosquito bites a human to obtain blood, it injects saliva that contains a chemical that prevents the blood from clotting. If the mosquito is infected with *Plasmodium,* it will also inject about 1,000 protists with its saliva. This infective stage of *Plasmodium* is called the **sporozoite.** Sporozoites infect the liver, where they rapidly divide and produce millions of cells of the second stage of the life cycle, called the **merozoite.** Merozoites infect red blood cells and divide rapidly. In about 48 hours, the blood cells rupture, releasing more merozoites and toxic substances. This begins a cycle of fever and chills that characterizes malaria. The cycle repeats every 48–72 hours (depending on which species is causing the infection) as more blood cells are infected and destroyed.

In the third stage, some of the merozoites in the blood develop into gametes. After these gametes are eaten by a mosquito, they form a zygote. Eventually, many infectious sporozoites are formed in the zygote and migrate to the salivary glands of the mosquito. Unlike humans, the mosquito cannot regurgitate food or mix saliva and food in its mouth; therefore, the malaria parasite must mature in the mosquito before it can infect another human.

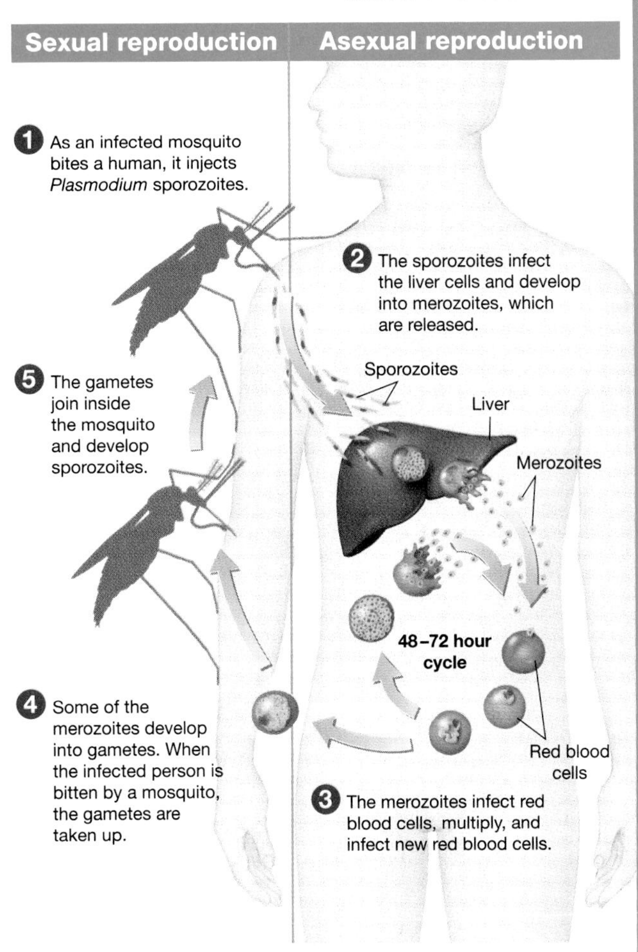

Sexual reproduction
Asexual reproduction
1 As an infected mosquito bites a human, it injects Plasmodium sporozoites.
2 The sporozoites infect the liver cells and develop into merozoites, which are released.
Sporozoites
Liver
Merozoites
5 The gametes join inside the mosquito and develop sporozoites.
48–72 hour cycle
4 Some of the merozoites develop into gametes. When the infected person is bitten by a mosquito, the gametes are taken up.
Red blood cells
3 The merozoites infect red blood cells, multiply, and infect new red blood cells.

Figure 22-13 Life cycle of *Plasmodium.* *Plasmodium* has a complex life cycle that involves a mosquito, human blood and liver cells.

Teaching TIP

Food Chains and DDT

Remind students that DDT use has been banned in the United States because it accumulates in the tissues of organisms as it passes through the food chain. The chemical is still in use in some malaria-infested countries to control mosquito populations. Review how some organisms, such as the bald eagle, were affected before and after the banning of DDT. *(Before the ban, the shells of the eggs were so thin that many broke, and the unhatched chicks died. The eagle is now making a comeback.)*

Teaching TIP

"New" Malaria Strain

Plasmodium falciparum is 95 percent more likely to cause death than other strains of malaria. Because *P. falciparum* has been observed in human populations for only the last 10,000 years, it is considered a "new" strain of malaria.

Teaching TIP

Global Warming and Malaria

Until efforts to control malaria began to restrict its geographical boundaries, it was found in areas where summer temperatures exceeded 16°C (61°F). Global warming has extended the areas of the world where conditions are favorable for the spread of malaria.

MULTICULTURAL PERSPECTIVE

Controlling Malaria in Africa

Ten percent of the world's population lives in tropical Africa, where 90 percent of the world's malarial infections occur. Three species of African mosquitoes carry and spread the pathogen to humans while feeding on their blood. In tropical Africa one person can receive more than 300 bites by infected mosquitoes per year. International researchers hope to introduce a pathogen-inhibiting gene into the genomes of these mosquitoes. Releasing transgenic mosquitoes to spread the new gene throughout the *Plasmodium*-carrying mosquito populations could spare other continents, such as South America, from future malaria epidemics.

Making a Malaria Vaccine

Teaching Strategies

- Ask students why it was important for scientists to develop more than one protein to use in a potential vaccine. (Plasmodium *goes through different stages in the human body with different surface proteins. Scientists want the body to recognize and attack any of the different stages.)*
- Remind students that heterozygous individuals are also resistant to malaria. Review the relationship between sickle cell anemia and malaria. *(The malaria-causing protists become trapped inside blood cells that become sickled. Once the cells are altered, the protists die.)*

Discussion

Is there any chance that the vaccines discussed will cause malaria in the patient? *(No, the vaccine consists only of surface proteins and not disease-causing organisms.)*

3 Assessment Options

Closure

Identifying Solutions

Tell students that they are to advise a town council on what to do about a malaria outbreak. Have students work in groups to come up with a proposal for the town council. Then have students present their proposal to the class. PORTFOLIO

Review

Assign Review—Section 22-3.

Reteaching — BASIC

Have students work in cooperative groups to research the life cycles of several disease-causing protists. Reports should include the method of infection, the symptoms, and where the disease is found. PORTFOLIO

Treating and preventing malaria

In the middle of the seventeenth century, quinine *(KWEYE nihn)*, a chemical derived from the bark of the cinchona tree (*Cinchona officialis,* found in South America) was discovered to be a remedy for malaria. Derivatives of quinine, such as chloroquine and primaquine, are now used to treat malaria.

Malaria can also be controlled by reducing mosquito populations. This is done by spraying insecticides, reducing mosquito breeding places, and introducing animals that will eat mosquito larvae, such as mosquito fish.

HEALTH WATCH: Making a Malaria Vaccine

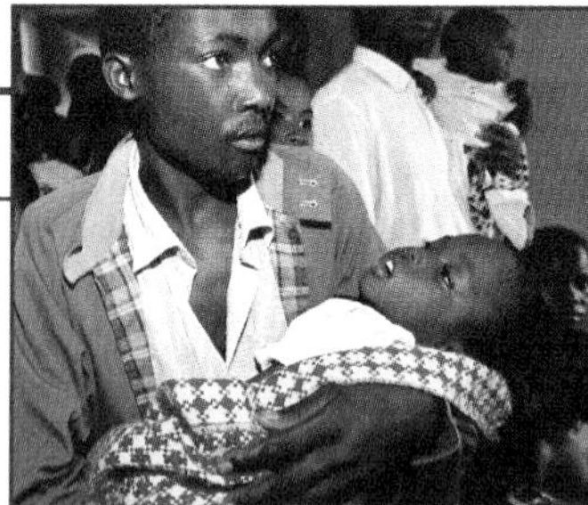

With their bodies racked by fever as high as 40°C (104°F) and chills so intense that their teeth chatter, up to 3 million people needlessly die of malaria every year. The worldwide epidemic of malaria is becoming more dangerous as malaria parasites become resistant to the standard drug treatments. To control the epidemic, scientists are trying to make a malaria vaccine.

Making the malaria vaccine work

For a vaccine to work against a parasite, it must stimulate the body's immune cells to recognize and attack specific molecules on the surface of the parasite. But *Plasmodium* goes through several stages inside the human body, and each stage can have many different surface molecules. Therefore, a vaccine may cause the immune system to attack only one stage. If a few parasites survive, they could produce millions of new parasites. A second problem is that *Plasmodium* parasites spend much of their time inside human liver and red blood cells, where the immune system cannot find them.

Current approaches

Scientists have identified surface proteins of *Plasmodium* sporozoites and merozoites and can make large amounts of these proteins. They hope that an injection of these proteins into human volunteers will give the volunteers' immune cells a chance to recognize the proteins and prepare for infection. If these people are later infected by a mosquito, their cells may be able to attack *Plasmodium* before the parasite can enter their liver cells.

World-wide efforts to develop an effective malaria vaccine have produced many failures and a few partial successes. Scientists are still working on ways to make the immune system's response to the vaccines stronger and longer lasting.

Review SECTION 22-3

1. **Summarize** two different ways protists affect human health.
2. **Describe** three human diseases caused by protists.
3. **Describe** how malaria is transmitted.
4. **Identify** the techniques used to control malaria.

Critical Thinking

5. **SKILL Evaluating an Argument** Some scientists argue that the relationship between some parasitic protists and their host is analogous to the relationship between cellular organelles and the cell as described in the theory of endosymbiosis. Is this argument valid? Explain.

Review SECTION 22-3 ANSWERS

1. Some protists cause disease, but others are beneficial inhabitants of the human gut, and many others contribute oxygen to the atmosphere and recycle carbon and nutrients in the environment.
2. Malaria is caused by *Plasmodium* sporozoans and causes severe fever and can lead to anemia, kidney failure, and brain damage. See Table 22-3 on p. 490 for a list and description of other diseases caused by protists.
3. Malaria is transmitted when a mosquito extracts *Plasmodium* gametes from an infected individual. The gametes mature into sporozoites in the mosquito, which can then be injected into a new individual, resulting in a new infection.
4. Primarily reducing mosquito populations using insecticides, eliminating breeding areas, and introducing larvae-eating fish to mosquito-laden waters controls malaria.
5. Accept all well-supported answers. Students should recognize that in endosymbiosis, the host cell and cellular organelle both benefit from the relationship—an example of mutualism. In contrast, the relationship between protist and host cell is parasitic.

Study ZONE CHAPTER 22

Use the Key Concepts and Key Terms listed below to review the main ideas in this chapter. If you need to review the meaning of a term, turn to the page indicated.

Key Concepts

22-1 Characteristics of Protists

- The kingdom Protista contains the most diverse groups of eukaryotic organisms of any kingdom.
- Protists live in moist environments and can be either free-living or parasitic.
- Some protists are able to reproduce sexually in times of stressful environmental conditions.

22-2 Protist Diversity

- Amoebas and forams are protists that move using cytoplasmic extensions called pseudopodia.
- Diatoms are unicellular protists with glasslike double shells.
- Algae are strictly photosynthetic protists that can be multicellular or unicellular. Algae are classified according to the type of photosynthetic pigment they contain.
- Flagellates move with the use of flagella.
- Ciliates are unicellular protists that use cilia to move.
- Protistan molds resemble fungi, but they are considered protists.
- Cellular slime molds normally live as individual cells and aggregate for sexual reproduction.
- Plasmodial slime molds live as colonial organisms and form sexual reproductive structures that form and spread spores.
- Oomycetes are molds that live as saprophytes or parasites.
- All sporozoans are parasitic and have complex life cycles.

22-3 Protists and Health

- Protists negatively affect human health by causing diseases in humans and their food sources.
- Protists positively affect human health through their participation in food webs, through commensal relationships with humans and their food sources, and by recycling vital resources.
- Protists cause diseases such as dysentery, giardiasis, malaria, and toxoplasmosis.
- Malaria is the most serious protist disease that affects humans. Drugs and mosquito control can be used to control malaria.

Key Terms

22-1

protozoan (479)
alga (479)
zygospore (480)
alternation of generations (481)
sporangium (481)

22-2

amoeba (482)
pseudopodium (482)
diatom (483)
zoomastigote (485)
euglenoid (485)
cilium (486)
plasmodium (488)

22-3

sporozoite (491)
merozoite (491)

Study TIP Ready?

If you think you understand these Key Concepts and Key Terms, then you're ready for the Performance Zone.

Answer to Concept Map

The following is one possible answer to the Performance Zone item 1 on page 494.

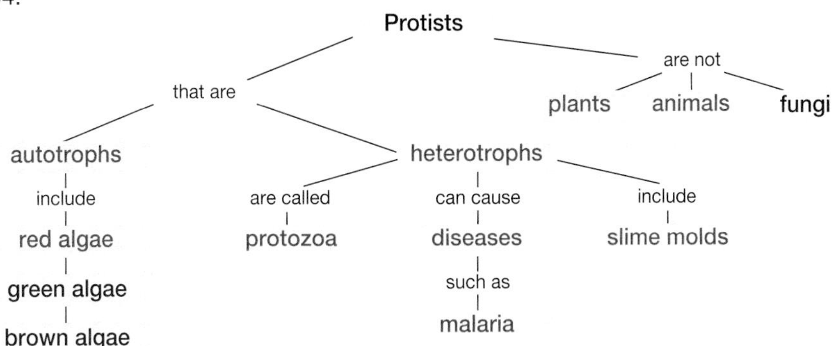

Study ZONE CHAPTER 22

Review and Practice

Assign the following worksheets for Chapter 22:

- **Student Review Guide**
- **Concept Mapping Worksheet**
- **Test Preparation Pretest**
- **Vocabulary Worksheet**
- **Critical Thinking Review Worksheet**

Assessment Options

The following assessment options are available for this chapter:

Resource Link

Assign **Chapter Test 22** to assess students' understanding of the chapter.

Alternative Assessment

- Ask students to write a short essay exploring the rise of drug-resistant protistan pathogens. Have students consider the phrase "survival of the fittest." Does evolution occur faster in microbes than in other organisms? *(Yes, they multiply more quickly and thus have more opportunity for mutation and recombination. By killing the nonresistant microbes with drugs, more resources are available for the resistant strains to multiply.)* PORTFOLIO

Portfolio Assessment

- *Portfolio Assessment* at the front of this book provides options for building and evaluating student portfolios. Students and teachers can select items from the areas listed for inclusion in a portfolio.
- See items labeled PORTFOLIO on pp. 480, 481, 482, 484, 486, 487, 488, 489, 492, and 493.

Performance ZONE
CHAPTER 22
REVIEW ANSWERS

Assignment Guide

Section	Review
22-1	3, 4, 5, 15
22-2	6, 7, 8, 9, 10, 11, 22, 23, 25, 26
22-3	1, 2, 12, 13, 14, 16, 17, 18, 19, 20, 21, 24, 27, 28

Understanding and Applying Concepts

1. The answer to the concept map is found at the bottom of page 493.
2. **a.** A protist is any member of the kingdom Protista; a protozoan is a general term for a heterotrophic protist.

   b. Zoomastigotes are unicellular heterotrophs and dinoflagellates are unicellular phototrophs.

   c. Cilia are short, tightly packed flagella used for locomotion; plasmodium is a mass of streaming cytoplasm found in plasmodial slime molds. *Plasmodium* is also the name of a genus of protists responsible for malaria.

   d. A *sporozoite* is the infectious stage of *Plasmodium* that is injected by mosquitoes; *merozoite* is the stage of *Plasmodium* that infects red blood cells, resulting in fever and chills.
3. c
4. c
5. c
6. c
7. a
8. b
9. d
10. c
11. a
12. a
13. b
14. d
15. The zygospore is a thick-walled zygote that forms after the union of two gametes. It can withstand harsh conditions and releases new haploid cells when favorable conditions return.
16. Creating a vaccine for malaria is difficult because the protists that cause the disease undergo several stages with different surface proteins. The protists also "hide" inside the human liver and blood cells.
17. The geographical range of the disease would increase.

Skills Assessment

18. malaria
19. Test the patient's blood for the presence of *Plasmodium*.
20. The disease could be transmitted to others via mosquitoes.
21. No, because the sporozoites need to mature in the salivary glands of a mosquito before they can infect someone.

Performance ZONE CHAPTER 22 Review

Understanding and Applying Concepts

1. **Concept Mapping** Make a concept map describing the protists. Try to include the following terms in your map: red algae, malaria, protozoa, slime mold, autotrophs, plants, diseases, heterotrophs, animals. Include additional terms as needed.
2. **Understanding Vocabulary** For each pair of terms listed below, explain the differences in their meanings.
   a. protist/protozoa
   b. zoomastigote/dinoflagellate
   c. cilia/plasmodium
   d. sporozoite/merozoite
3. Two important eukaryotic features that evolved among protists are
   a. photosynthesis and silica shells.
   b. forams and pseudopodia.
   c. sexual reproduction and multicellularity.
   d. spores and microtubules.
4. Which habitat is least likely to harbor any species of Protista?
   a. ocean waters
   b. the human liver
   c. a desert
   d. leaf litter
5. Which pair shows a correct match between a protist and its manner of reproduction?
   a. *Spirogyra:* sporangia
   b. *Ulva:* conjugation
   c. *Chlamydomonas:* gametes
   d. *Spirogyra:* alternation of generations
6. Pseudopodia are used by species of
   a. *Paramecium.*
   b. *Euglena.*
   c. *Entamoeba.*
   d. *Macrocystis.*
7. Photosynthetic protists with boxlike double shells are
   a. diatoms.
   b. plankton.
   c. zoomastigotes.
   d. euglenoids.
8. Red algae are different from green and brown algae because red algae
   a. are multicellular.
   b. have a red pigment.
   c. inhabit marine environments.
   d. display alternation of generations.
9. A photosynthetic single-celled protist that moves using flagella would likely be classified as a member of what phylum?
   a. Apicomplexa
   b. Oomycota
   c. Bacillariophyta
   d. Euglenophyta
10. Euglenoids and zoomastigotes both have ______, but only euglenoids are ______.
    a. flagella, unicellular
    b. multicellularity, photosynthetic
    c. flagella, photosynthetic
    d. a pellicle, pathogenic
11. The general characteristics of *Paramecia* do not include a
    a. flagellum.
    b. cilia.
    c. micronucleus.
    d. pellicle.
12. How is amebic dysentery spread?
    a. by drinking contaminated water
    b. from the bite of a mosquito
    c. from the bite of the tsetse fly
    d. by eating overcooked meat
13. Malaria is transmitted by
    a. tsetse flies.
    b. mosquitoes.
    c. contaminated water.
    d. *Trypanosoma gambiense.*
14. Malaria can be controlled by
    a. reducing mosquito breeding.
    b. using anti-malarial drugs.
    c. preventing mosquito bites.
    d. All of the above
15. **Theme Cellular Structure and Function** Describe the structure of the zygospore and its function in the life cycle of *Chlamydomonas.*
16. **Health Watch** Why has a successful malaria vaccine been so difficult to make?
17. **Chapter Links** If malaria-transmitting mosquitoes prefer a warm climate, infer what might happen to the incidence of malaria in the world if the greenhouse effect is not reversed. (**Hint:** See Chapter 19, Section 19-1.)

Skills Assessment

Part 1: Interpreting Graphs

Study the graph below showing the temperature variation in a sick person who recently returned from Africa. Then answer the following questions:

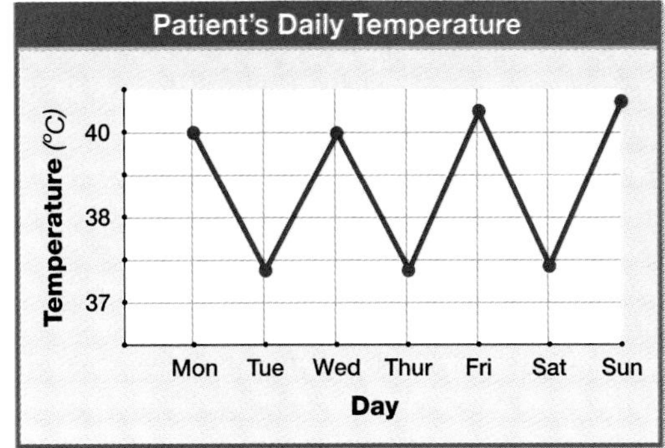

18. Given the travel history of the patient, what protist disease might you suspect?

19. How would you confirm your diagnosis?

20. If the patient checked into the hospital in the summertime, what kind of public health concerns would you have as a physician?

21. If while drawing blood from the patient, some of the blood splattered on your bare arm, could you catch this disease?

Part 2: Life/Work Skills

Use the media center or Internet resources to learn more about how protistan molds cause disease in plants.

22. Finding Information Find information about the history of research into plant diseases caused by protistan molds. What measures were taken by earlier scientists to control the diseases, and are those measures still used today?

23. Selecting Technology Prepare a written report about how scientists are able to exploit what they find out about the life cycles of molds in order to create new strategies of disease prevention.

Critical Thinking

24. Evaluating Methods Scientists were able to demonstrate the life cycle of malaria by feeding infected blood to mosquitoes and then letting the mosquitoes feed on uninfected human volunteers. What possible errors are there in using this method?

25. Justifying Conclusions A scientist has found two different euglenas, shown below, and concluded that specimen A is strictly heterotrophic, but specimen B is not. Justify her conclusions.

Magnification: 206×

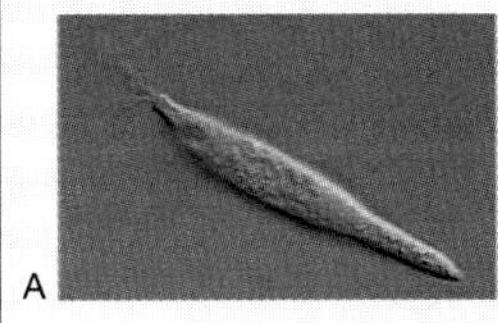
A

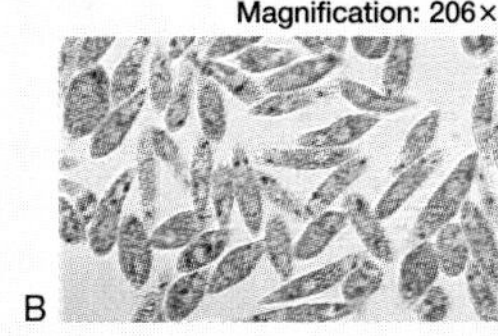
B

Portfolio Projects

26. Analyzing Data The Irish potato famine of 1845–1847 resulted in the deaths of many people and also the emigration of Irish people to other countries including the United States. Research this time period. Construct a graph that shows the number of Irish people who came to the United States between the years 1835 and 1857. Using the graph, write a paragraph that describes the impact of the Irish potato famine on the immigration of Irish people to the United States.

27. Interpreting Information Read "Collaring a Killer," by Robert H. Mazurek and Jeff Bohlander (*Popular Science,* March, 1999, vol. 254 pp. 50–53). Prepare a report that identifies what a DNA vaccine is, the problems with conventional vaccines, and the advantage of a DNA vaccine. How can breeding genetically engineered mosquitoes help fight malaria?

28. Career Focus Epidemiologist Research the field of epidemiology, and write a report on your findings. Your report should include a job description, training required, kinds of employers, growth prospects, and starting salary.

22. Answers will vary. An example is the discovery of potato blight, which is caused by the fungus *Phyophthora infestans.* Once researchers determined the life cycle of the mold and that the spores require a moist environment, they were able to recommend that the farmers keep the field around the infected plants dry. Today farmers use fungicides and are trying to genetically modify the potato to make it more resistant to blight.

23. Answers will vary. Once scientists know what is required for the organism to reproduce, they can try to inhibit the requirements and thus prevent reproduction of the organism.

Critical Thinking

24. Perhaps most seriously, humans were infected with the disease as a part of the experiment. There may also be doubt as to the role of the mosquito in the process—were the mosquitoes already infected? Could something else have transmitted the disease?

25. Her conclusions may be correct, as both heterotrophic and autotrophic euglenas exist. No chlorophyll is visible in specimen A, so it is most likely heterophobic.

Portfolio Projects

26. Answers will vary. The Irish potato famine led to a significant increase in the number of Irish immigrants entering the United States.

27. Answers will vary. DNA vaccines are based on the stimulation of an immune response through the introduction of foreign genes and the resulting production of foreign proteins. Because conventional vaccines use disabled or killed infectious organisms (or proteins specific to them) to generate an immune response, there is always a small risk that the person can get the disease. The advantages of DNA vaccines are that there is no risk of getting the disease and DNA vaccines stimulate production of cytotoxic T (killer) cells, which directly attack infected cellular targets. Also, by inserting different genes from the pathogen into the plasmid, researchers can easily produce many variations of the same vaccine. Genetically engineering mosquitoes that are able to kill malaria parasites will help in the fight against malaria.

28. Answers will vary. Epidemiologists study the distributions, causes, and controls of diseases and other health problems. An advanced degree is required, and jobs are often with universities, schools of public health, or government agencies. Growth prospects are good. Starting salary will vary by region.

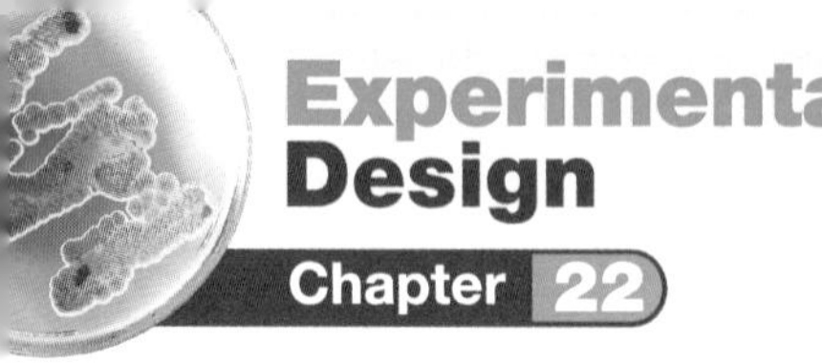

How Do Protozoans Respond to Light?

Purpose

Students will observe protozoans under a microscope and compare their structure, methods of locomotion and feeding, and behaviors.

Preparation Notes

Time Required: 45 to 75 minutes

Materials

Materials for this lab can be ordered from WARD'S. See *Master Materials List* at the front of this book for catalog numbers.

Preparation Tips

- This investigation may take more than one class period. To save time, you may wish to make the "sun shades" in advance.
- Before students begin the lab, ask the following question: How do you think protozoans will react to light? *(Students should consider that some protozoans benefit from light and others do not; this might affect their reactions.)*

Safety Precautions

Discuss all safety symbols and caution statements with students.

Procedural Tips

- You may want to have students repeat steps 1 and 2 using drops of liquid from the middle and top of the culture. Record any differences in the types and distribution of the protozoans found at different depths of the culture.
- In step 6, students should find that euglenoids congregate in the center area that is exposed to light.

Disposal

Protists and DETAIN™ may be washed down the drain.

Before You Begin

Answers

1. *protozoans*—heterotrophic protists

   kingdom Protista—a diverse group of eukaryotic organisms that cannot be classified as animals, plants, or fungi

   eukaryotes—organisms that have complex cells with the nucleus enclosed by a membrane

   producer—an organism that makes its own food from energy and carbon atoms in its environment

   consumer—an organism that must obtain energy to build its molecules by consuming other organisms

   decomposer—an organism that causes decay

   cilia—tightly packed rows of short hairlike structures used for movement

   flagellum—a whiplike structure used for movement

   pseudopod—a flexible cytoplasmic extension used for movement
2. Tables should be similar to the one on page 496.
3. Answers will vary. For example: How do *Euglena* and *Volvox* respond to light?

Procedure

Sample Procedure

1. Prepare five wet mounts of *Euglena* and five wet mounts of *Volvox,* all without any protist-slowing agent. View under the low power of

---

Experimental Design

Chapter 22 How Do Protozoans Respond to Light?

SKILLS

- Using scientific methods
- Using a microscope

OBJECTIVES

- **Identify** several different types of protozoans.
- **Compare** the structures, methods of locomotion and feeding, and behaviors of several different protozoans.
- **Relate** a protozoan's response to light to its method of feeding.

MATERIALS

- Detain™ (protist-slowing agent)
- microscope slides
- plastic pipets with bulbs
- mixed culture of protozoans
- toothpicks
- coverslips
- compound microscope
- protozoan references
- black construction paper
- scissors
- paper punch
- white paper
- sunlit window sill or lamp
- forceps

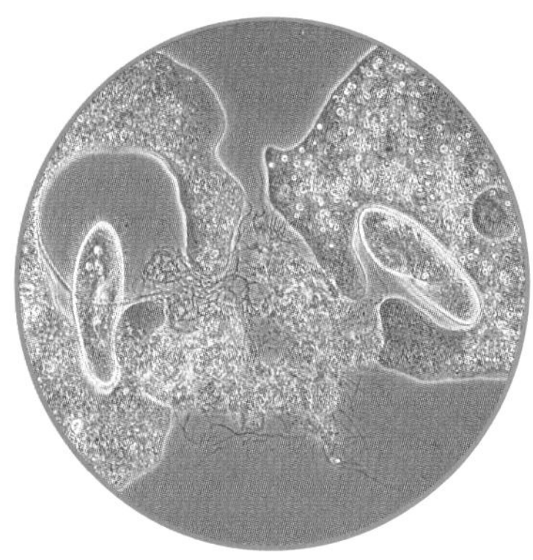

Before You Begin

Protozoans belong to the kingdom **Protista,** which is a diverse group of **eukaryotes** that cannot be classified as animals, plants, or fungi. All protozoans are unicellular. Among the protozoans, there are **producers, consumers,** and **decomposers.** In this lab, you will observe live protozoans and compare their structures, methods of locomotion and feeding, and behaviors.

1. Write a definition for each boldface term in the paragraph above and for each of the following terms: cilia, flagellum, pseudopod.
2. Make a data table similar to the one below.
3. Based on the objectives for this lab, write a question you would like to explore about protozoans.

Procedure

Part A: Make Observations

1. Place a drop of Detain on a microscope slide. Add a drop of liquid from the bottom of a mixed culture of protozoans. Mix the drops with a toothpick. Add a coverslip. View the slide under low power of a microscope. Switch to high power.
2. Use references to identify the protozoans. Record data for each type of protozoan you find in your data table.
3. Repeat step 1 *without* using the protist-slowing agent.
4. Punch a hole in a 40 × 20 mm piece of black construction paper that has a slight curl, as shown in the photo on page 497.

DATA TABLE

Protozoan	Color	Method of locomotion	Method of feeding	Other observations

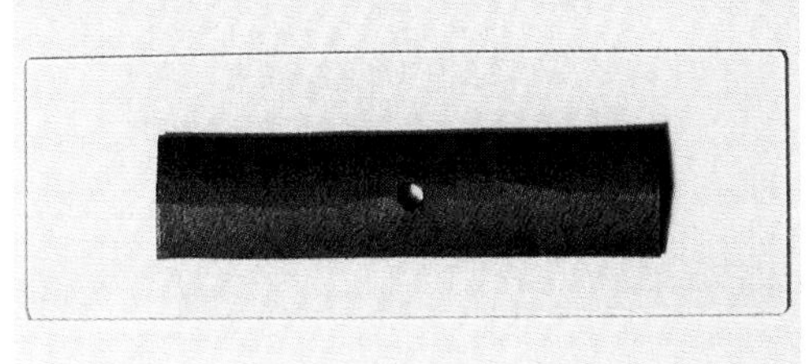

5. Place a wet mount of protozoans on a piece of white paper. Then put the paper and slide on a sunlit window sill or under a table lamp. Position the sun shade on top of the slide so that the hole is in the center of the coverslip.
6. To examine a slide, first view the area in the center of the hole under low power. *Note: Do not disturb the sun shade. Do not switch to high power.* Then have a partner carefully remove the sun shade with forceps while you observe the slide.

PART B: Design an Experiment

7. Work with members of your lab group to explore one of the questions written for step 3 of **Before You Begin.** To explore the question, design an experiment that uses the materials listed for this lab.

> **You Choose**
>
> **As you design your experiment, decide the following:**
> a. what question you will explore
> b. what hypothesis you will test
> c. how long you will expose protozoans to light
> d. how many times you will repeat your experiment
> e. what your control will be
> f. what data you will record and how you will make your data table

8. Write a procedure for your experiment. Make a list of all the safety precautions you will take. Have your teacher approve your procedure and safety precautions before you begin the experiment.
9. Set up and carry out your experiment.

PART C: Cleanup and Disposal

10. Dispose of lab materials and broken glass in the designated waste containers. Put protozoans in the designated containers. Do not put lab materials in the trash unless your teacher tells you to do so.
11.  Clean up your work area and all lab equipment. Return lab equipment to its proper place. Wash your hands thoroughly before you leave the lab and after you finish all work.

Analyze and Conclude

1. **Summarizing Results** Describe the different types of locomotion you observed in protozoans, and give examples of each.
2. **Analyzing Results** Identify which protozoans were affected by light, and describe how they were affected.
3. **Inferring Conclusions** What is the relationship between a protozoan's response to light and its method of feeding?
4. **Further Inquiry** Write a new question about protozoans that could be explored with another investigation.

Do You Know?

Do research in the library or media center to answer these questions:

1. What livestock diseases caused by parasitic protozoans are most commonly found in the United States?
2. How do backpackers avoid getting diseases caused by protozoans that are transmitted in water?

Use the following Internet resources to explore your own questions about protozoans.

internetconnect

SCILINKS NSTA
TOPIC: Protozoans
GO TO: www.scilinks.org
KEYWORD: HX497

Analyze and Conclude

Answers

1. Protists move by the movement of cilia, flagella, or pseudopodia. For example, *Amoeba* uses pseudopodia, *Paramecium* uses cilia, *Euglena* uses a flagellum, and *Volvox* uses flagella.
2. *Euglena* congregated within the area exposed to light. Once the shade was removed, they scattered. Euglenoids have an eyespot used to detect light needed for photosynthesis.
3. Protozoans that depend on light for food were attracted to the light.
4. Answers will vary. For example: Do samples from the middle and upper areas of the culture jar have different protist populations from those taken at the bottom?

Do You Know?

Answers

1. Answers will vary. Students should report on livestock diseases such as blackhead, coccidiosis, and giardia.
2. Answers will vary. Students should report on safe water treatments that are practical for backpackers such as portable water-filtering systems.

a compound microscope, and record the observations.

2. Place the wet mounts on top of a piece of white paper on a sunlit windowsill or under a table lamp. Position the "sun shade" so that the hole is in the center of the coverslip for each slide.
3. After 10 minutes, place each slide on the stage of a compound microscope and view the slide using the low-power objective. Record your observations.
4. Have one person remove the "sun shade" on each slide (using forceps) while another person continues to view the protists. Record your observations.

Sample Data Table

PROTOZOAN	COLOR	METHOD OF LOCOMOTION	METHOD OF FEEDING	OTHER OBSERVATIONS
Amoeba	Whitish	Pseudopod	Consumer	Changes shape
Blepharisma	Pink	Cilia	Consumer	
P. caudatum	Clear	Cilia	Consumer	
Euglena	Green	Flagellum	Producer	Has "eye-spot"
Stentor	Bluish	Cilia	Consumer	Trumpet-shaped
Volvox	Green	Flagellum	Producer	Colonial

Chapter 23

Fungi

Planning Guide

OBJECTIVES	MEETING INDIVIDUAL NEEDS	READING SKILLS AND STRATEGIES
23-1 Characteristics of Fungi (BLOCK 1 45 min) • List the characteristics of the kingdom Fungi. • Describe the structure of a typical fungus body. • Identify how fungi obtain nutrients. • Relate the way fungi obtain nutrients to their role in ecosystems. • Distinguish the ways that fungi reproduce.	Learning Styles, p. 500 BASIC ADVANCED • pp. 500, 502	ATE Active Reading, Technique: Anticipation Guide, p. 499 ■ Active Reading Guide, worksheet (23-1) ■ Directed Reading, worksheet (23-1)
23-2 Fungal Diversity (BLOCKS 2 & 3 90 min) • Describe the characteristics used to classify fungi. • List two commercial uses for fungi. • Describe three phyla of fungi. • Distinguish between the life cycles of zygomycetes, ascomycetes, and basidiomycetes. • Describe the mushroom *Amanita muscaria.*	Learning Styles, p. 503 BASIC ADVANCED • p. 506	ATE Active Reading, Technique: Reading Organizer, p. 503 ■ Active Reading Guide, worksheet (23-2) ■ Directed Reading, worksheet (23-2)
23-3 Fungal Partnerships (BLOCKS 4 & 5 90 min) • Distinguish two symbiotic relationships that involve fungi. • Summarize the ecological importance of mycorrhizae. • Describe lichens.	Learning Styles, p. 508 BASIC ADVANCED • pp. 509, 510	■ Active Reading Guide, worksheet (23-3) ■ Directed Reading, worksheet (23-3) ■ Supplemental Reading Guide, *The Andromeda Strain*

Review and Assessment Resources

BLOCKS 6 & 7 90 min

TEST PREP

- **PE** Study Zone, p. 511
- **PE** Performance Zone, pp. 512–513
- ■ Chapter 23 worksheets:
  - Vocabulary
  - Concept Mapping
  - Critical Thinking
- Audio CD Program
- ■ Test Prep Pretest

CHAPTER TESTING

- ■ Chapter Tests (blackline master)
- ▲ Test Generator
- ■ Assessment Item Listing

See the correlations table at the front of this book for a list of the ***National Science Education Standards*** addressed in this chapter.

Smithsonian Institution® Visit **www.si.edu/hrw** for additional on-line resources.

KEY

Biology Principles and Explorations

PE Pupil's Edition

ATE Teacher's Edition

HOLT BIOSOURCES

 Teaching Transparencies

 Laboratory Program

One-Stop Planner CD-ROM

Shows these resources within a customizable daily lesson plan:

■ Teaching Resources

▲ Test Generator

VISUAL STRATEGIES	CLASSWORK AND HOMEWORK	TECHNOLOGY AND INTERNET RESOURCES
ATE Fig. 23-2, p. 501 107 Structure of a Mycelium	**PE** Quick Lab What are some characteristics of bread mold?, p. 502 **PE** Review, p. 502	Holt Biology Videodiscs Section 23-1 Audio CD Program Section 23-1
ATE Fig. 23-6, p. 504 108 Life Cycle of a Zygomycete 109 Life Cycle of an Ascomycete 110 Life Cycle of a Basidiomycete	Lab C34 Limiting Fungal Growth **PE** Review, p. 506	CNN Video Segment 18 Lethal Mushrooms Holt Biology Videodiscs Section 23-2 Audio CD Program Section 23-2 internet**connect** SCILINKS NSTA **TOPIC:** Characteristics of fungi
ATE Fig. 23-10, p. 508 111 Up Close: Mushroom	**PE** Data Lab Analyzing the effect of mycorrhizae, p. 509 **PE** Experimental Design How Can Mold Growth Be Inhibited?, pp. 514–515 **PE** Review, p. 510 **PE** Study Zone, p. 511	Holt Biology Videodiscs Section 23-3 Audio CD Program Section 23-3 internet**connect** SCILINKS NSTA **TOPICS:** • Symbiosis of fungi • Lichens • Uses of fungi

ALTERNATIVE ASSESSMENT

PE Portfolio Projects 24–26, p. 513

ATE Alternative Assessment, p. 511

ATE Portfolio Assessment, p. 511

Lab Activity Materials

Quick Lab
p. 502 moldy bread, hand lens, microscope slide, coverslip, microscope, dropping pipet, paper towel, 50 mL beaker, toothpick

Data Lab
p. 509 pencil, paper

Experimental Design
p. 514 safety goggles, lab apron, disposable gloves, rubbing alcohol, paper towels, samples of molds, stereomicroscope, toothpicks, wax pencil, masking tape, sterile Petri dishes with plain nutrient agar, sterile Petri dishes with nutrient agar that contains propionic acid

Lesson Warm-up
ATE photos or live specimens of colorful lichens, p. 508

Demonstrations
ATE paper bag, mushroom, p. 498
ATE several scanning electron micrographs of zygospores, p. 504
ATE vinegar, ammonia, litmus paper, p. 509
ATE photos or live specimens of crustose, foliose, and fruticose lichens, p. 509

BIOSOURCES LAB PROGRAM
- Laboratory Techniques and Experimental Design: Lab C34, p. 167

Chapter 23

Chapter Themes

Cellular Structure and Function

Fungi have structures unique to their kingdom that facilitate their survival. Fungi have filamentous hyphae that spread and penetrate organic matter to obtain nutrients. Once the food source is exhausted or the environment changes, they develop new structures—reproductive bodies—to disperse their spores.

Interdependence

Fungi are important decomposers in the environment and help provide vital minerals and other basic components needed for the survival of other organisms.

Establishing Prior Knowledge

Before beginning this chapter, make certain students can answer the questions in Study TIP **Ready?** on page 499. If they cannot, encourage them to reread the sections indicated.

Opening Demonstration

Prior to class put a mushroom in a paper bag. Have students determine the contents of the bag by asking "yes" or "no" questions. (This strategy may also help you determine student misconceptions regarding fungi.) Once the class has identified the contents of the bag, analyze students' responses and point out any misconceptions.

CHAPTER 23 Fungi

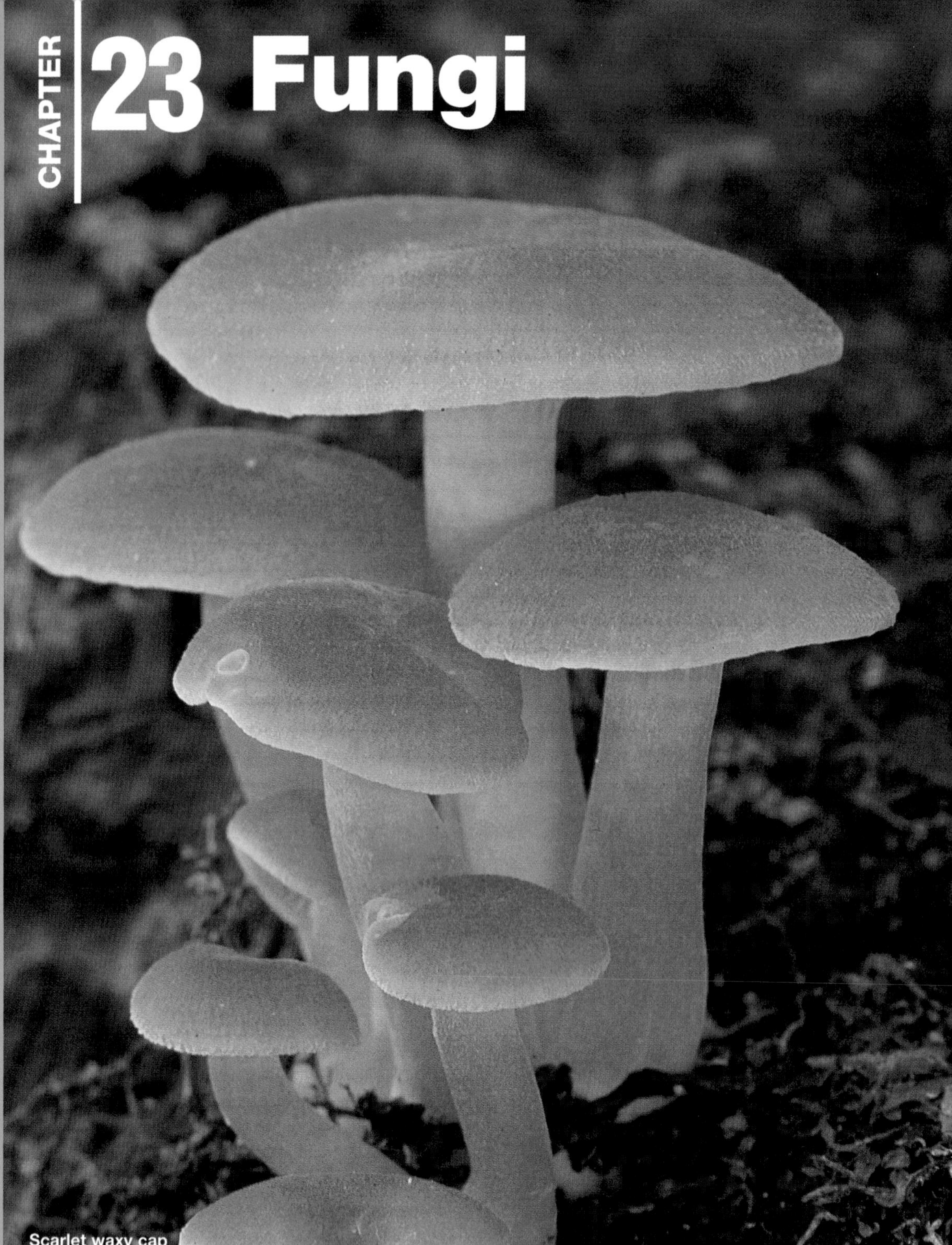

Scarlet waxy cap

What's Ahead — and Why
CHAPTER OUTLINE

23-1 Characteristics of Fungi
- Fungi Are Classified in Their Own Kingdom
- Fungi Are Well Suited for Absorbing Nutrients
- Fungi Reproduce Sexually and Asexually

23-2 Fungal Diversity
- Fungi Are Classified by Their Sexual Reproductive Structures
- Zygomycetes Form Thick-Walled Reproductive Structures
- Ascomycetes Form Sacs of Spores
- Basidiomycetes Produce Sexual Spores in Mushrooms

23-3 Fungal Partnerships
- Fungi Form Symbiotic Relationships with Plant Roots
- Fungi Form Symbiotic Relationships with Algae

Students will learn

- how the kingdom Fungi is characterized, how fungi obtain nutrients, and how fungi reproduce.
- how the three groups of fungi differ primarily in their means of sexual reproduction.
- how fungi play a key role in the environment as partners with plants and algae.

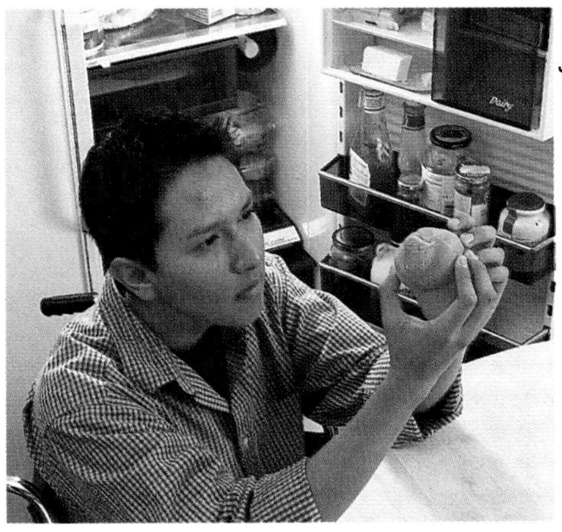

The mushrooms on the left

are in the same kingdom as the mold you might find growing on a peach or bread. The kingdom Fungi consists of varied organisms, including the yeast that is used to make bread.

Looking Ahead

23-1 Characteristics of Fungi
Fungi are unique and deserve their own kingdom.

23-2 Fungal Diversity
Reproductive structures make fungi different from one another.

23-3 Fungal Partnerships
Together with other organisms, fungi significantly impact the environment.

Labs

- **Quick Lab**
  What are some characteristics of bread mold? (p. 502)
- **Data Lab**
  Analyzing the effect of mycorrhizae (p. 509)
- **Experimental Design**
  How Can Mold Growth Be Inhibited? (p. 514)

Features

- **Up Close**
  Mushroom (p. 507)
- **Exploring Further**
  Lichens as Environmental Watchdogs (p. 510)

Study TIP

Ready?

Check your understanding of these topics to help you prepare for what's ahead.

Can you . . .

- **distinguish** meiosis from mitosis? (Section 7-1)
- **summarize** the importance of mycorrhizae? (Section 12-3)
- **define** the ecological role of lichen? (Section 17-1)
- **describe** the meaning of symbiosis? (Section 18-1)
- **relate** the importance of fungi to bacterial infection? (Section 21-2)

If not, *review the sections indicated.*

internetconnect

National Science Teachers Association *sci*LINKS Internet resources are located throughout this chapter.

Reading Strategies

Active Reading

Technique: Anticipation Guide

Before students begin this chapter, write the following statements on the board or overhead:

1. Fungi are closely related to plants.
2. Fungi are economically valuable.
3. Many fungi are beneficial to other organisms.

Give students a minute or two to consider the statements and decide whether they agree or disagree with each. Ask volunteers to share their opinions with the class. Have students read the chapter to see if their opinions are confirmed or changed. Rediscuss the statements. *(Fungi are not closely related to plants. Fungi are economically valuable as sources of food, baking, brewing, medicine, research, and environmental cleanup. Many fungi are beneficial to other organisms as recyclers of nutrients and materials and as mutualistic symbionts. Some species are also parasitic.)*

Directed Reading

Assign **Directed Reading Worksheet Chapter 23** to help students interact with the text as they read the chapter.

Interactive Reading

Assign ***Biology: Principles and Explorations* Audio CD Program Chapter 23** to help reluctant readers achieve greater success in reading the chapter.

Reading Skills

Assign Reading Organizers and Reading Strategies Worksheets from the **One-Stop Planner CD-ROM with Test Generator** to help students learn and practice effective reading skills.

internetconnect

Holt *Biology: Principles and Explorations*

On-line Resources: go.hrw.com

Visit the HRW Web site for a variety of resources related to this chapter. Just type in the keyword HX0 Chapter 23.

Holt *Biology: Principles and Explorations*

On-line Resources: www.scilinks.org

The following *sci*LINKS Internet resources can be found in the student text for this chapter:

Page 503
TOPIC: Characteristics of fungi
KEYWORD: HX503

Page 508
TOPIC: Symbiosis of fungi
KEYWORD: HX508

Page 510
TOPIC: Lichens
KEYWORD: HX510

Page 515
TOPIC: Uses of fungi
KEYWORD: HX515

Chapter 23

1 Prepare

Scheduling

- **Block:** About 45 minutes will be needed to complete this section.
- **Traditional:** About one class period will be needed to complete this section.
- **Planning:** See the Planning Guide on pp. 497A and 497B for lecture, classwork, and assignment options for Section 23-1. Lesson plans for Section 23-1 can be found in a lesson cycle format in ***Biology: Principles and Explorations* One-Stop Planner CD-ROM with Test Generator.**

Resource Link

Assign **Active Reading Worksheet Section 23-1** to help students comprehend the key concepts and visuals in this lesson.

2 Teach

Lesson Warm-up

Ask students to describe places where they have seen fungi growing and what these sites have in common. Ask students to describe the ideal environment in which fungi would thrive. List their responses on the board, and point out that fungi will survive almost anywhere with sufficient moisture and a carbon-rich food source.

Teaching TIP — BASIC

Comparing Fungi and Plants

Have groups of students write the following words on index cards or strips of paper (one word per card): *chitin, cellulose, cell wall, roots, hyphae, spore, seed, autotroph, heterotroph.* Have students sort the terms into a "plant" group and a "fungus" group. Ask students to discuss the differences and similarities between both groups. *(Plants and fungi have cell walls, and some plants make spores.)*

Section 23-1 Characteristics of Fungi

Objectives

- **List** the characteristics of the kingdom Fungi. (p. 500)
- **Describe** the structure of a typical fungus body. (p. 501)
- **Identify** how fungi obtain nutrients. (p. 501)
- **Relate** the way fungi obtain nutrients to their role in ecosystems. (p. 501)
- **Distinguish** the ways that fungi reproduce. (p. 502)

Key Terms

chitin
hypha
mycelium

Study TIP

Organizing Information

The numbered list on this page gives four important characteristics of fungi. Use this information to start a concept map that summarizes the characteristics of fungi.

Fungi Are Classified in Their Own Kingdom

Some of the most unusual organisms that exist today are members of the kingdom Fungi. Mushrooms and molds are common fungi that grow so rapidly they sometimes appear overnight.

The first fungi were probably unicellular eukaryotes. Fungi are at least 400 million years old. Traditionally, biologists grouped fungi with plants because fungi are immobile, have a cell wall, and appear "rooted" in the soil, as the mushrooms do in **Figure 23-1.** However, the unique features of fungi indicate that they are not closely related to any other group of organisms.

1. **Fungi are heterotrophic.** The stalk and cap of the mushroom are not green like the stem and leaves of a plant. Plants appear green because they contain chlorophyll, the pigment that aids in photosynthesis. Fungi do not contain chlorophyll. Because they are heterotrophs, they obtain energy by breaking down organic molecules that they absorb from their environment.
2. **Fungi have filamentous bodies.** Plants are made of many cell and tissue types, but fungi are made of long, slender filaments. These filaments weave tightly together to form the fungus body and reproductive structures, such as the mushroom.
3. **Fungal cells contain chitin.** The cells of the mushroom, like the cells of all fungi, have walls made of **chitin** *(KEYE tihn),* the tough polysaccharide found in the hard outer covering of insects. Plant cells have walls made of cellulose, a different polysaccharide.
4. **Fungi exhibit nuclear mitosis.** Mitosis in fungi is different from mitosis in plants and in most other eukaryotes. In most eukaryotes, the nuclear envelope disintegrates in prophase and re-forms in telophase. In dividing mushroom cells, however, the nuclear envelope remains intact from prophase to anaphase. As a result, spindle fibers form within the nucleus. The spindle fibers then drag chromosomes to opposite poles of the nucleus, rather than to opposite poles of the cell. Mitosis is complete when the nuclear envelope pinches in two.

Figure 23-1 Mushrooms. These mushrooms are actually the reproductive structures of a large network of filaments that makes up the body of a fungus.

Integrating Different Learning Styles

Visual/Spatial	**PE**	Quick Lab, p. 502
	ATE	Visual Strategy, p. 501; Teaching Tip, p. 501; Closure, p. 502
Interpersonal	**ATE**	Teaching Tips, pp. 500, 501
Verbal/Linguistic	**PE**	Performance Zone items 18, 20, 25, and 26, p. 513
	ATE	Lesson Warm-up, p. 500; Reteaching, p. 502

Fungi Are Well Suited for Absorbing Nutrients

In **Figure 23-2,** the fungus *Penicillium* is shown growing on an orange. The green and white fuzz you recognize as mold is actually the reproductive structures of the fungus. The body of the fungus lies within the tissues of the orange. All fungi except yeasts have bodies composed of slender filaments called **hyphae** *(HY fee).* When hyphae grow, they branch and form a tangled mass called a **mycelium** *(my SEE lee uhm),* shown in Figure 23-2. A mycelium can be made of many meters of individual hyphae. This body organization provides a high surface-area-to-volume ratio, which makes a fungus well suited for absorbing nutrients from its environment.

Each hypha is a long string of cells divided by partial walls. Some species have no walls. Cytoplasm flows freely throughout the hypha, as shown in Figure 23-2.

Figure 23-2 ***Penicillium*** **mold**

This orange is covered with a fungus from the genus *Penicillium.*

Mold

The green-and-white fuzz growing on the orange's surface is the fungus's reproductive structures.

Mycelium

Hyphae
Secrete enzymes to aid in absorbing food
Grow quickly by passing nutrients through incomplete septa

Reproductive structures

Throughout the rest of the orange, the fungus grows as a mycelium.

Fungal hyphae may be divided into cells by walls.

Magnification: 1,510×

How fungi absorb nutrients

All fungi obtain nutrients by secreting digestive enzymes that break down organic matter in their environment. Fungi then absorb the decomposed molecules. Many fungi decompose nonliving organic matter, such as leaves, branches, dead animals, and waste. So, fungi are resource recyclers. Other fungi are parasites that absorb nutrients from living hosts, such as the fungus that causes ringworm.

Because fungi compete for nutrients from our own bodies, they sometimes cause infectious diseases, such as athlete's foot and yeast infections. Other fungi, such as *Histoplasma capsulatum,* have evolved ways of invading our body's organs and causing life-threatening infections, as shown in **Figure 23-3.**

Fungi often compete with humans for other nutrients. Bread, fruit, vegetables, and meat are as nutritious to a fungus as is a fallen log. Fungi are also known to attack nonfood materials, such as paper, cardboard, cloth, paint, and leather.

Their ability to grow under a wide range of conditions makes them commercially valuable. Unicellular fungi called yeasts are useful in baking, brewing, and wine-making. Other fungi provide the flavor and aroma of certain cheeses. Many kinds of antibiotics, such as penicillin, are produced by fungi.

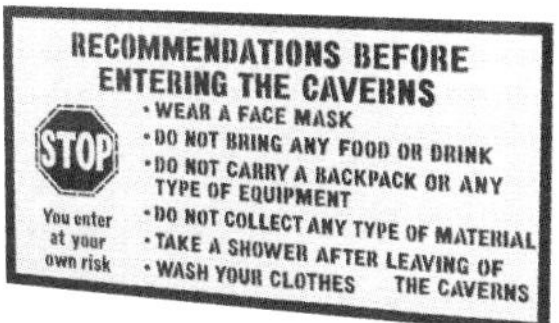

RECOMMENDATIONS BEFORE ENTERING THE CAVERNS
• WEAR A FACE MASK
• DO NOT BRING ANY FOOD OR DRINK
• DO NOT CARRY A BACKPACK OR ANY TYPE OF EQUIPMENT
• DO NOT COLLECT ANY TYPE OF MATERIAL
• TAKE A SHOWER AFTER LEAVING OF THE CAVERNS
• WASH YOUR CLOTHES
STOP
You enter at your own risk

Figure 23-3 Dangerous fungi. *Histoplasma capsulatum* grows in the feces of bats and birds and infects humans when the dried spores are inhaled. Symptoms of infection include fever, cough, chills, chest pain, and muscle aches.

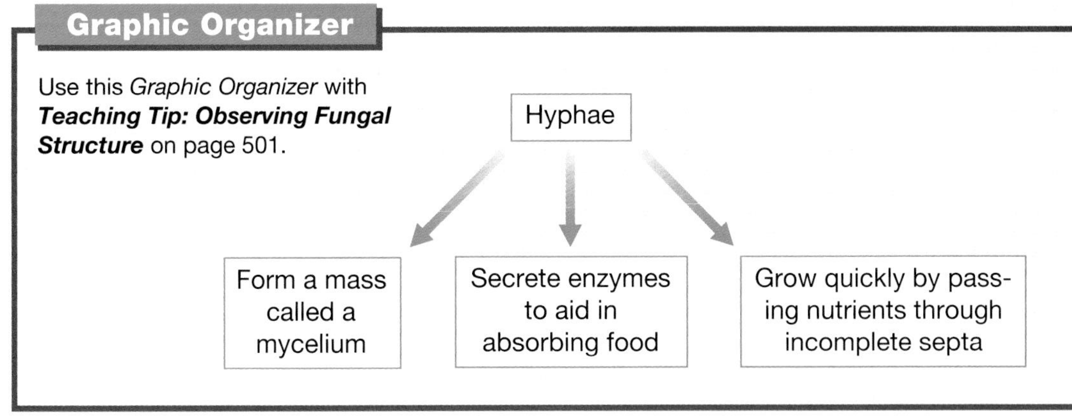

Visual Strategy

Figure 23-2

Bring an orange covered with mold to class. Cut the orange in half and compare it to the orange in the figure. Use this figure and the orange to illustrate the body, or mycelium, of the fungus and the reproductive structures. Have students locate each in the figure and in the sample you brought.

Teaching TIP

Observing Fungal Structure

Give pairs of students a piece of moldy fruit. Have them cut the fruit and examine the structures with a hand lens. Provide slides, coverslips, microscopes, and thin tissues slices. Have students make wet mounts and observe the cell structure of the hyphae. Then ask students to make a **Graphic Organizer** similar to the one at the bottom of page 501 to describe the structure and function of hyphae. PORTFOLIO

Making Biology Relevant

Technology

Fungi are used commercially to clean pollutants, including pesticides and chemicals produced in industrial waste. *Penicillium* is used to clean contaminants, such as selenium produced by industry from polluted waters. Growing "clean-up" fungi, such as white rot fungi, commercially is cost-effective because they can be grown on newspaper, corncobs, or wood chips. In addition, they grow at a rapid rate, and they are readily available. Fungi offer great potential as environmental "cleansers" because they break down more than one pollutant at a time.

What are some characteristics of bread mold?

QUICK LAB

Time 20 minutes

Process Skills Observing, interpreting results, applying information

Teaching Strategy
Have an alternative activity available using prepared slides for those students who might be allergic to mold.

Analysis Answers

1. Such a height allows greater dispersal of spores.
2. The hyphae, mycelium, and reproductive structures should be labeled.
3. The hyphae strands make up the mycelium. The mycelium secretes digestive enzymes and absorbs nutrients. The reproductive structures produce spores for dispersal.

3 Assessment Options

Closure

Design a Bumper Sticker
Have groups of students design bumper stickers that promote fungi as tools that can be used to clean up environmental waste. PORTFOLIO

Review

Assign Review—Section 23-1.

Reteaching — BASIC

Group students into teams of four. Tell students they are to complete a passage you read from the text. Read that passage, but do not complete the passage. The first group of students to complete the passage correctly wins that round. Play several rounds of this game until you have read all of the passages that pertain to the objectives of this section. Declare the group that finishes the most passages as the winner.

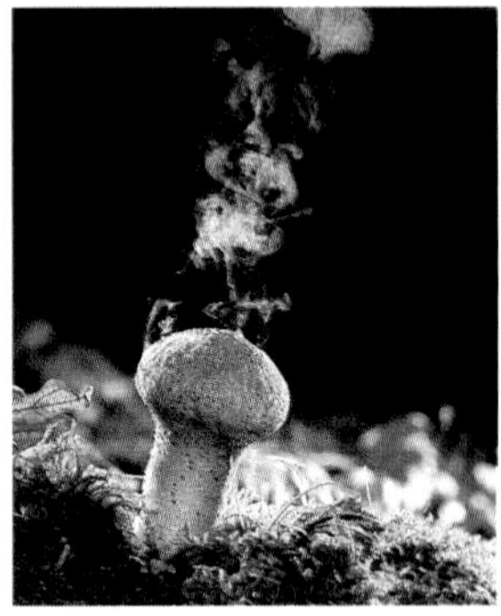

Figure 23-4 **Puffball.** *Lycoperdon perlatum,* a puffball, releases hundreds of thousands of spores through a small opening.

Fungi Reproduce Sexually and Asexually

Fungi reproduce by releasing spores formed sexually or asexually in reproductive structures at the tips of hyphae. Reproductive structures grow high above the food source. This adaptation allows air currents to carry the spores to a new habitat. As you can see in **Figure 23-4,** fungal spores are so small and light that they remain suspended in the air for long periods of time; the wind can carry them great distances.

Fungal spores are haploid. Most spores are formed by mitosis during asexual reproduction. In sexual reproduction, hyphae from two mating types fuse. The fused hyphae form a sexual reproductive structure. On this structure, the fungus forms spores through fusion of the two genetically different nuclei.

QUICK LAB

What are some characteristics of bread mold?

Like the individual threads that make up a ball of cotton, individual threads of cells make up the body of a fungus.

Materials

moldy bread, hand lens, microscope slide, coverslip, microscope, dropping pipet, paper towel, 50 mL beaker, toothpick

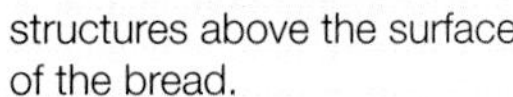

Procedure

1. Examine a piece of moldy bread with a hand lens.
2. Draw what you see at this magnification in your lab notebook. Be sure to use at least half a page.
3. Prepare a wet-mount slide of a small portion of the bread mold.
4. Examine the slide under low power.
5. Draw what you see in your lab notebook.

Analysis

1. **Explain** the advantage of the fungus keeping the elevated structures above the surface of the bread.
2. **Label** the drawings you made, using Figure 23-2 as a guide.
3. **Relate** the structures you drew to their functions described in the text.

Review SECTION 23-1

1. **Distinguish** the characteristics of fungi from those of plants.
2. **Compare** the structure of hyphae with the structure of mycelium.
3. **Summarize** the way fungi obtain nutrients.
4. **Describe** the role fungi play in the environment.
5. **Summarize** the different ways that fungi reproduce.

Critical Thinking

6. **SKILL Evaluating Conclusions** Two of your fellow students insist that yeasts should be classified as protists because they are eukaryotic unicellular organisms. Evaluate their claim.

Review SECTION 23-1 ANSWERS

1. Fungi are heterotrophic, have filamentous bodies, have cell walls containing chitin, and exhibit nuclear mitosis. Plants are autotrophic, have bodies made of various cell types and tissues, have cell walls containing cellulose, and do not exhibit nuclear mitosis.
2. The hyphae are thin filaments made of individual cells, which grow to form a tangled mass of hyphae called a mycelium.
3. The hyphae of the fungi secrete digestive enzymes that break down the organic material in which they are growing. The nutrients are then absorbed through the cell walls of the hyphae.
4. Fungi are decomposers, breaking down organic matter in the environment.
5. Fungi reproduce asexually when they produce spores, which develop into new individuals. Fungi reproduce sexually when hyphae from compatible mating types fuse, forming a sexual reproductive structure and spores.
6. Although yeasts are unicellular eukaryotes, they have chitin in their cell walls, as do all fungi.

Fungal Diversity

Section 23-2

Section 23-2

Fungi Are Classified by Their Sexual Reproductive Structures

You can see how diverse fungi are if you examine the types of reproductive structures that they have. Based on the types of structures produced during sexual reproduction, fungi can be classified in three phyla. Table 23-1 below lists some of these characteristics.

Asexual reproduction

A fourth group, the deuteromycetes, is composed of fungi in which no sexual stage has been seen. Traditionally, this group has been called a phylum, but through the use of molecular techniques, scientists have reclassified most of these asexually reproducing organisms into the phylum Ascomycota.

There are about 17,000 species without a sexual stage. Many of these fungi are economically important. For example, some species of *Penicillium* produce the antibiotic penicillin. Other species produce the unique flavors of some cheeses. Species of *Aspergillus* are used for fermenting soy sauce and producing citric acid. Most of the fungi that cause skin diseases, such as athlete's foot and ringworm, are also deuteromycetes.

Table 23-1 Three Sexually Reproducing Phyla of Fungi

Phylum	Distinctive characteristics	Examples
Zygomycota	Sexual spores are formed in zygosporangia; hyphae have no walls	black bread molds
Ascomycota	Sexual spores are formed in asci; hyphae are divided by walls	morels, truffles, yeasts, cup fungi
Basidiomycota	Sexual spores are formed in basidia; hyphae are divided by walls	mushrooms, puffballs, rusts, smuts

Objectives

- **Describe** the characteristics used to classify fungi. (p. 503)
- **List** two commercial uses for fungi. (p. 503)
- **Describe** three phyla of fungi. (pp. 504–506)
- **Distinguish** between the life cycles of zygomycetes, ascomycetes, and basidiomycetes. (pp. 504–506)
- **Describe** the mushroom *Amanita muscaria.* (p. 507)

Key Terms

zygosporangium
stolon
rhizoid
ascus
yeast
budding
basidium

internet connect

SCI LINKS. NSTA
TOPIC: Characteristics of fungi
GO TO: www.scilinks.org
KEYWORD: HX503

1 Prepare

Scheduling

- **Block:** About 90 minutes will be needed to complete this section.
- **Traditional:** About two class periods will be needed to complete this section.
- **Planning:** See the Planning Guide on pp. 497A and 497B for lecture, classwork, and assignment options for Section 23-2. Lesson plans for Section 23-2 can be found in a lesson cycle format in ***Biology: Principles and Explorations*** **One-Stop Planner CD-ROM with Test Generator.**

Resource Link

- Assign **Active Reading Worksheet Section 23-2** to help students comprehend the key concepts and visuals in this lesson.

2 Teach

Lesson Warm-up

Ask students to make a list of characteristics that could be used to identify and classify an organism. Once completed, check to see if anyone's list includes "reproductive structures." Introduce the lesson by explaining how fungal phyla are classified according to their methods of sexual reproduction.

Active Reading

Technique: Reading Organizer

Have students make a table on a piece of paper. Down the left side write these headings: *Zygomycota, Ascomycota,* and *Basidiomycota.* Across the top write these headings: *Method of Reproduction, Structures, Habitat,* and *Example.* As students read the section, they should complete the table.
PORTFOLIO

Integrating Different Learning Styles

Logical/Mathematical	**PE**	Performance Zone item 20, p. 513
	ATE	Cross-Disciplinary Connection, p. 504
Visual/Spatial	**PE**	Performance Zone items 1 and 13, p. 512
	ATE	Demonstration, p. 504; Visual Strategy, p. 504; Teaching Tip, p. 506; Reteaching, p. 506
Interpersonal	**ATE**	Closure, p. 506
Verbal/Linguistic	**PE**	Performance Zone items 11 and 24, pp. 512–513
	ATE	Lesson Warm-up, p. 503; Active Reading, p. 503

Chapter 23

Demonstration

Zygomycota zygospores

Display a series of scanning electron micrographs showing zygospore formation in *Rhizopus*. Ask students why zygospores are advantageous for fungi. *(The protective zygospore can survive until environmental conditions are favorable for germination.)*

CROSS-DISCIPLINARY CONNECTION

◆ **Physics**

Tell students that the zygomycete *Pilobolus* grows in animal dung, and produces a reproductive structure about 5 to 10 mm high. The clear structure below the spore acts as a lens, focusing the sunlight on a photoreceptor inside the fungus. Explain that the reproductive structure bends towards the light and eventually ejects the spore to a distance of up to 2 m at a velocity near 50 km/h. Ask students to calculate the time it would take a spore to travel 2 m, based on the velocity of 50 km/h. *($v = \frac{d}{t}$ where v is the velocity—50 km/h, d is the distance—2 m, and t is the time. The time it takes to travel 2 m is 0.14 s.)*

Visual Strategy

Figure 23-6

Use this figure to help students understand the complex life cycle of a zygomycete. Have students trace the life cycle of the mold, and ask them for an explanation at each illustrated stage.

Making Biology Relevant

Food

Ask students if they have ever used soy sauce. Tell them that soy sauce is made from soy beans with the help of a zygomycete fungus, *Aspergillus oryzae.* Other zygomycetes are used in producing foods such as miso (a soybean paste used in Japanese cooking) and tofu (a cheeselike soybean product used in different types of Asian cooking).

Zygomycetes Form Thick-Walled Reproductive Structures

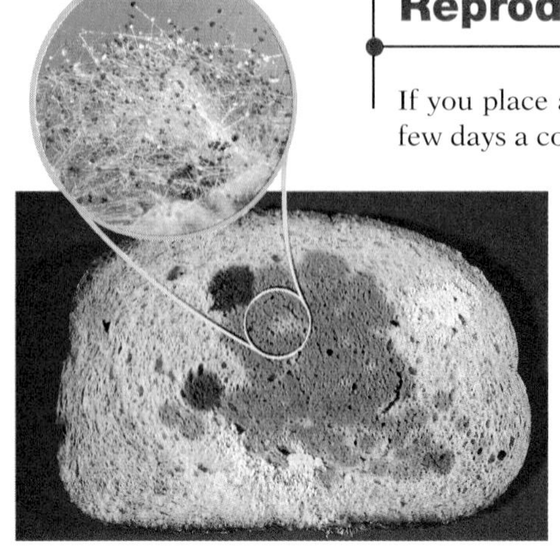
Magnification: 10×

Figure 23-5 Bread mold. *Rhizopus stolonifer* is often found growing on bread.

If you place an uncovered loaf of bread near a windowsill, after a few days a cottony mold will cover its surface. This common black bread mold, shown in **Figure 23-5,** is *Rhizopus stolonifer,* a member of the phylum Zygomycota *(zeye goh my COHT uh).* Members of the phylum Zygomycota are named for the thick-walled sexual structures called **zygosporangia** *(zeye goh spohr AN jee uh)* that characterize these members.

Species of *Rhizopus* and other zygomycetes usually live in the soil and feed on decaying plant and animal matter. The mycelia that grow along the surface of the bread are called **stolons** *(STOH lahnz).* The hyphae that anchor the fungus in the bread are called **rhizoids** *(REYE zoydz).* The hyphae of zygomycetes usually do not have walls.

Asexual reproduction in zygomycetes is much more common than sexual reproduction. During asexual reproduction, haploid spores are produced in the tips of specialized hyphae. It is the spores in fungi that cause fungal allergies in people. When they mature, these spores are shed and carried by the wind to new locations, where they germinate and start new mycelia. Reproduction in *Rhizopus* is shown in **Figure 23-6.**

Figure 23-6 Life cycle of zygomycetes. Zygomycetes may reproduce sexually or asexually.

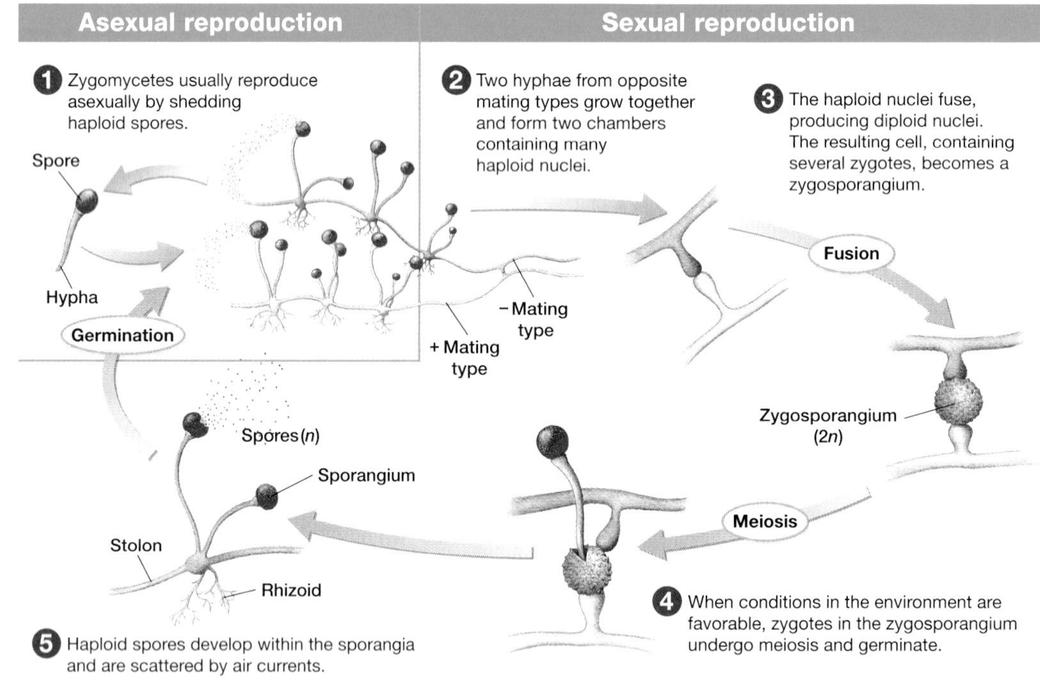
Asexual reproduction
Sexual reproduction
1 Zygomycetes usually reproduce asexually by shedding haploid spores.
Spore
Hypha
Germination
2 Two hyphae from opposite mating types grow together and form two chambers containing many haploid nuclei.
– Mating type
+ Mating type
3 The haploid nuclei fuse, producing diploid nuclei. The resulting cell, containing several zygotes, becomes a zygosporangium.
Fusion
Zygosporangium (2n)
Meiosis
4 When conditions in the environment are favorable, zygotes in the zygosporangium undergo meiosis and germinate.
Spores (n)
Sporangium
Stolon
Rhizoid
5 Haploid spores develop within the sporangia and are scattered by air currents.

MULTICULTURAL PERSPECTIVE

Bread-Mold Sandwich

An Indonesian food called tempeh is made of soybeans that are boiled, skinned, and inoculated with various species of *Rhizopus* (bread mold). The tempeh is then fried, roasted, or diced in preparation for a meal. Today tempeh can be found in health food and natural food stores, as well as in some large supermarkets.

Ascomycetes Form Sacs of Spores

The chestnut tree, *Castanea dentata,* was once a common tree in the eastern United States. Around 1890, a disease called chestnut blight wiped out virtually all the chestnut trees within a few years. Chestnut blight is caused by *Endothia parasitica,* a member of the phylum Ascomycota *(AS koh meye koh tuh)*. Other ascomycetes include flavorful morels and truffles prized by gourmet chefs.

The ascomycetes are named for their characteristic sexual reproductive structure, the microscopic **ascus** *(AS kuhs),* a saclike structure in which haploid spores are formed. Asci usually form within the interwoven hyphae of a cup-shaped fruiting body. Reproduction in a typical ascomycete is shown in **Figure 23-7.**

Ascomycetes usually reproduce asexually. Asexual spores form at the tips of the hyphae. The spores are not contained in any sac or structure. When the spores are released, air currents carry them to other places, where they may germinate and form new mycelia.

Yeast is the common name given to unicellular ascomycetes. There are about 350 species of yeasts. *Saccharomyces cerevisiae,* or baker's yeast, has been used for thousands of years to make bread and alcoholic beverages, such as beer. Other yeasts, such as *Candida albicans,* cause human disease. *C. albicans* causes thrush, a disease in which milk-white lesions form on the mouth, lips, and throat.

Most yeasts reproduce asexually by fission or budding. In **budding,** a small cell forms from a large cell and pinches itself off from the large cell.

biobit

What is that powder on your plants?

Many species of ascomycetes cause diseases, collectively known as powdery mildew, in different plants. Powdery mildew infects grapes, wheat, apples, and other plants.

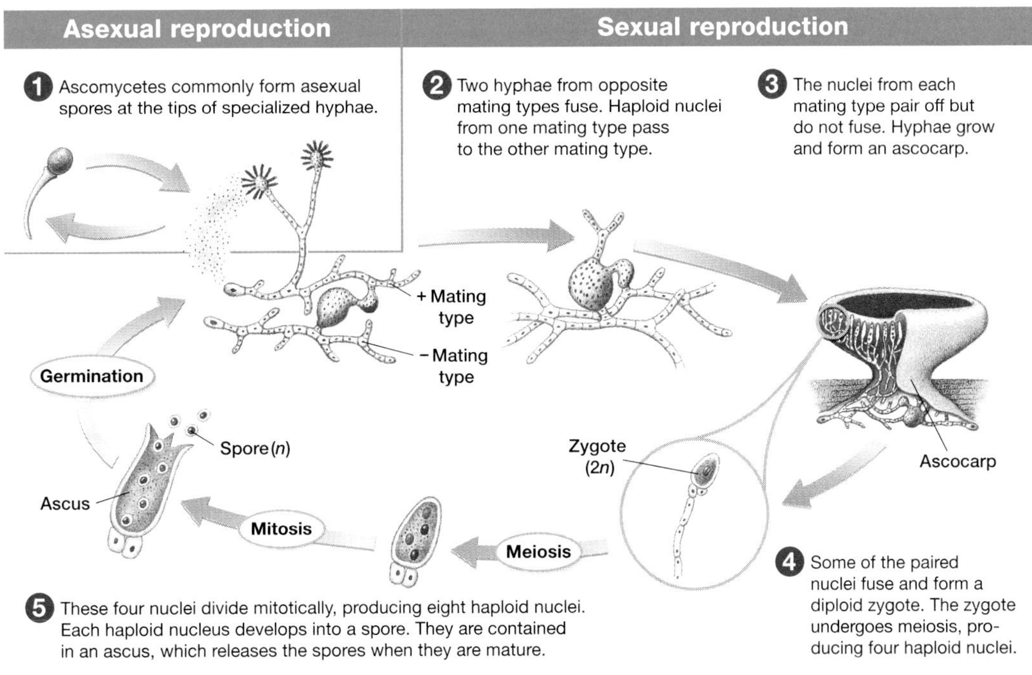

Asexual reproduction
Sexual reproduction
1 Ascomycetes commonly form asexual spores at the tips of specialized hyphae.
2 Two hyphae from opposite mating types fuse. Haploid nuclei from one mating type pass to the other mating type.
3 The nuclei from each mating type pair off but do not fuse. Hyphae grow and form an ascocarp.
+ Mating type
– Mating type
Germination
Spore (n)
Ascus
Mitosis
Meiosis
Zygote (2n)
Ascocarp
4 Some of the paired nuclei fuse and form a diploid zygote. The zygote undergoes meiosis, producing four haploid nuclei.
5 These four nuclei divide mitotically, producing eight haploid nuclei. Each haploid nucleus develops into a spore. They are contained in an ascus, which releases the spores when they are mature.

Figure 23-7 Life cycle of ascomycetes. Ascomycetes can reproduce sexually or asexually.

CROSS-DISCIPLINARY CONNECTION

◆ **History**

Ergotism, a disease caused when grain infected with the ergot fungus (an ascomycete) is harvested and milled, was common in Europe in the Middle Ages, when the disease was known as St. Anthony's Fire. Ergot contains several alkaloids that cause behaviors, such as seizures and hysteria. In the past, some victims of ergotism were accused of witchcraft.

Resource Link

Assign **Reading Strategies: Compare/Contrast Matrix** from the Holt BioSources Teaching Resources CD-ROM. Ask students to compare and contrast the three sexually reproducing phyla of fungi. PORTFOLIO

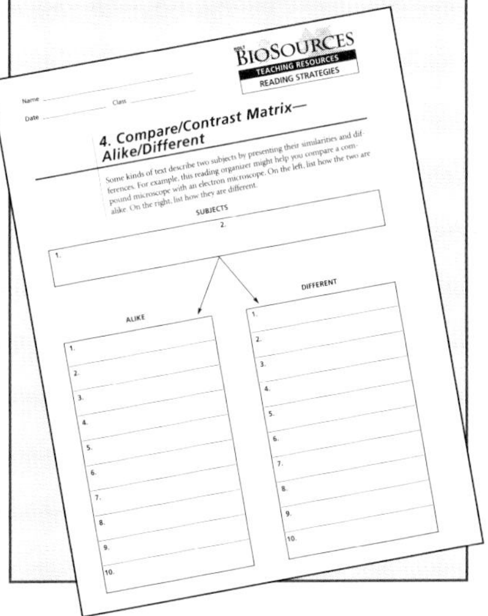

Biology in Your Community

Fungal Industries

Invite a speaker from a company that uses fungi in their production process (for example, a representative from a bakery, pharmaceutical company, winery, food-processing plant, or a cheese-making plant). Ask your guest to discuss the importance of fungi in his or her industry and how the fungi are cultivated for industrial use.

Chapter 23

Career

Plant Pathologist

Fungi are vitally important to the health of crops and forests as beneficial partners and as potential pathogens. Plant pathologists often wrestle with fungal diseases, including rusts, which have complex life cycles involving multiple hosts. Some researchers are investigating the use of fungi to control weeds.

Teaching TIP

Spore Prints

Have students place mushroom caps on paper, with the gill side down. Cover the mushrooms to keep them from drying and to keep the air still. After 1 or 2 days they will have prints made from the falling spores. Spore prints are used by mycologists to help identify and classify basidiomycetes.

3 Assessment Options

Closure

Comparing Structures

Have students use 3 × 5 in. cards to make flashcards. Have students label one side of the card with the name of one structure associated with one of the phyla of fungi. Label the other side of the card with the name of the phylum associated with that structure. Have students use their cards to quiz one another.

Review

Assign Review—Section 23-2.

Reteaching — BASIC

Collect or have students collect fungal fruiting bodies (mushrooms) growing nearby (or use prepared specimens). Caution students not to eat any of the samples. Using field guides and their knowledge of fungi, have students identify which phylum each sample represents.

Figure 23-8 Rust on wheat. Rust is a basidiomycete that attacks cereal crops, making them unfit for humans to eat.

Basidiomycetes Produce Sexual Spores in Mushrooms

The kind of fungi with which you are probably most familiar—mushrooms—are members of the phylum Basidiomycota *(buh SIHD ee oh meye koh tuh)*. Other basidiomycetes include toadstools, puffballs, jelly fungi, and shelf fungi. The **basidium** *(buh SIHD ee uhm)* is the club-shaped sexual reproductive structure for which the basidiomycetes are named. Spores are produced on this structure. You can see these spores in the *Up Close: Mushroom* feature on page 507. Asexual reproduction is rare among the basidiomycetes, except in some rusts and smuts. These two important groups of plant pathogens affect many crop plants, as shown in **Figure 23-8.** Sexual reproduction of a typical basidiomycete is illustrated in **Figure 23-9.** Many mushrooms are harmless, but many are also deadly, such as the *Amanita muscaria* (fly agaric), shown in the *Up Close* feature. Other *Amanita* species have names, such as death angel and destroying angel, that reflect the danger of their toxins for humans.

Figure 23-9 Life cycle of basidiomycetes

Basidiomycetes usually reproduce sexually by means of a fruiting body, also called a mushroom.

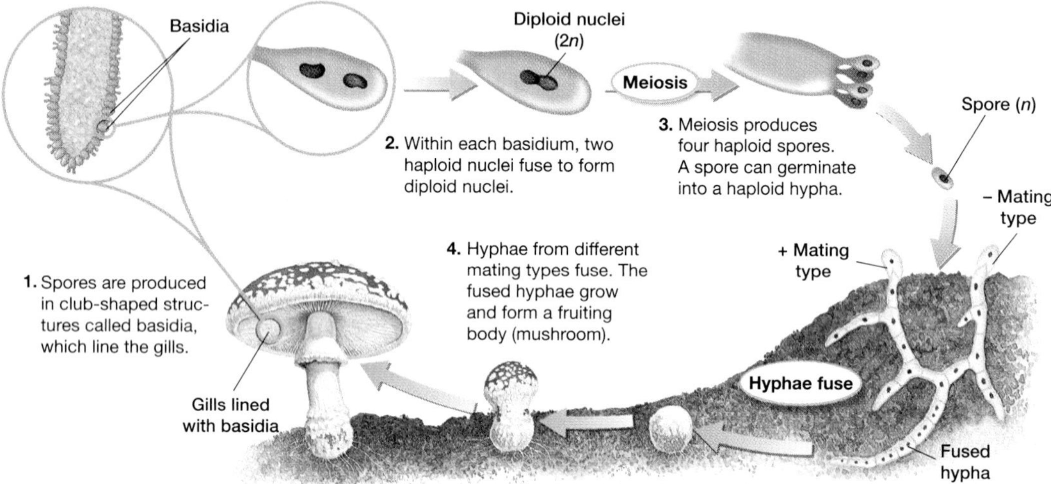

Basidia
Diploid nuclei (2n)
Meiosis
Spore (n)
2. Within each basidium, two haploid nuclei fuse to form diploid nuclei.
3. Meiosis produces four haploid spores. A spore can germinate into a haploid hypha.
– Mating type
4. Hyphae from different mating types fuse. The fused hyphae grow and form a fruiting body (mushroom).
+ Mating type
1. Spores are produced in club-shaped structures called basidia, which line the gills.
Hyphae fuse
Gills lined with basidia
Fused hypha

Review SECTION 23-2

1. **Summarize** how fungi are classified.
2. **Name** two commercial uses of fungi.
3. **Describe** the distinctive characteristics of the three phyla of fungi.
4. **Outline** the life cycles of the three phyla of fungi.
5. **Describe** the structure of the mushroom.
6. **Describe** how yeasts reproduce.

Critical Thinking

7. **SKILL Justifying Conclusions** The fruiting body of some ascomycetes, such as truffles, is found below ground and gives off a delicious scent. Scientists reason that this scent is related to a strategy for spreading the spores. How would a scent be related to spreading spores?

Review SECTION 23-2 ANSWERS

1. Fungi are classified into phyla based on their reproductive structures.
2. penicillin, cheese, soy sauce, citric acid
3. See Table 23-1 on p. 503.
4. See Figures 23-6, 23-7, and 23-8 on pp. 504, 505, and 506, respectively.
5. A mushroom consists of a stalk (stem) and flattened cap composed of interwoven hyphae. The underside of the cap is lined with gills, which house the spore-producing basidia.
6. Yeasts reproduce asexually by budding or fission.
7. If the spores could be spread by an animal, the scent would alert the animal to the presence of the underground fruiting body.

Up Close

Mushroom

- **Scientific name:** *Amanita muscaria*
- **Size:** 10–15 cm
- **Habitat:** Moist organic soils
- **Nutrition:** Heterotrophic

Characteristics

Cell structure *A. muscaria* and other fungi have cell walls made of chitin, a complex polysaccharide also found in the external skeleton of insects. In some fungi, hyphae are not divided into separate cells but have many nuclei in the same cytoplasm. In other fungi, hyphae are divided into cells by perforated walls called septa.

Reproduction Under proper conditions, underground hyphae grow upward and weave together to produce a mushroom, which is the reproductive structure of fungi such as *A. muscaria.* A mushroom has a flattened cap attached to a stem called a stalk. The underside of the mushroom cap is lined with rows of gills. Thousands of club-shaped reproductive cells called basidia form on the gills. Through fusion and meiosis, each basidium produces spores that are released and form new hyphae.

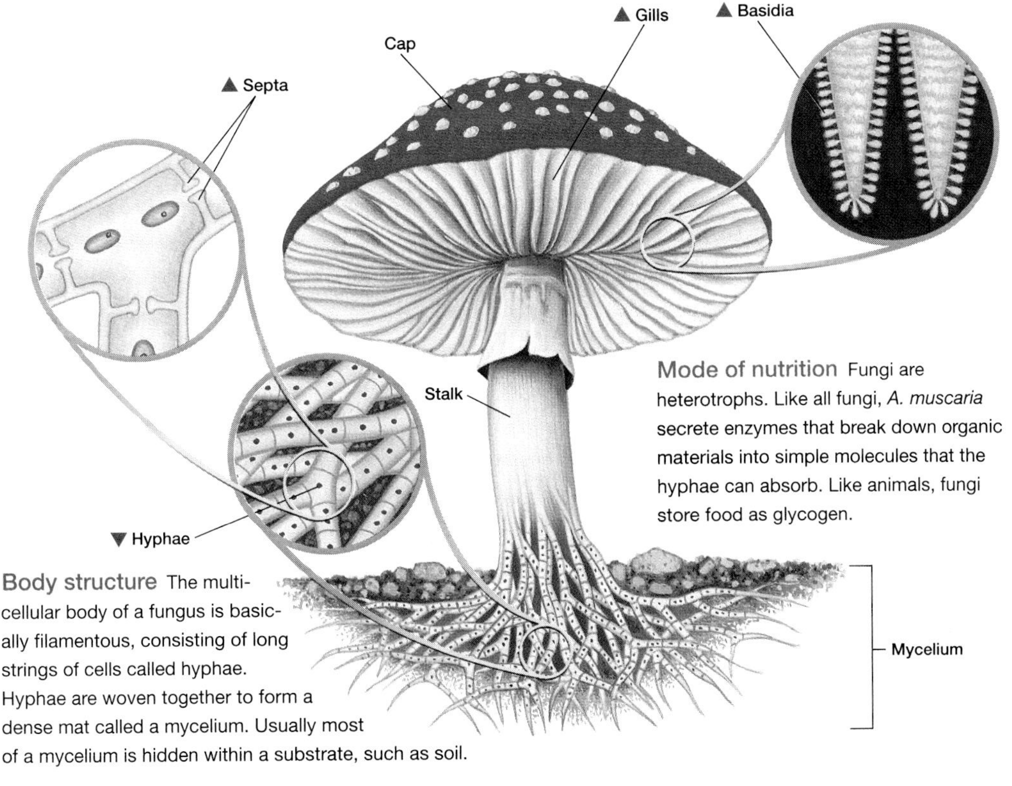

Mode of nutrition Fungi are heterotrophs. Like all fungi, *A. muscaria* secrete enzymes that break down organic materials into simple molecules that the hyphae can absorb. Like animals, fungi store food as glycogen.

Body structure The multicellular body of a fungus is basically filamentous, consisting of long strings of cells called hyphae. Hyphae are woven together to form a dense mat called a mycelium. Usually most of a mycelium is hidden within a substrate, such as soil.

Up Close

Mushroom

Teaching Strategies

- Bring a field guide of mushrooms and other fungi to class and have students look through it to gain an appreciation for fungal diversity. Have each student select a particular mushroom or fungus and draw a large, colored picture to display in the class. Each illustration should include the scientific name of the fungus and the group to which it belongs. PORTFOLIO
- Have students write and illustrate a brief report about mushrooms. Illustrations might include a spore print, a labeled drawing, or a recipe. PORTFOLIO

Discussion

- Why is it incorrect to classify a mushroom as a plant? *(Fungi are classified in their own kingdom because they are heterotrophic, have filamentous bodies, have cells that contain chitin, and exhibit nuclear mitosis. Plants lack these characteristics.)*
- How are septae different from the cell walls of plants? *(Septae have gaps that allow materials, even nuclei, to flow from one "cell" to another.)*
- Although gills and hyphae serve different purposes, what common characteristic do these structures share? *(Both gills and hyphae are structures with a large surface area, which allows the cells to carry out their role more efficiently.)*

Overcoming Misconceptions

Eating Mushrooms

Contrary to popular belief, poisonous mushrooms cannot be discerned by a quick test such as smearing liquid from a mushroom on white paper and looking for a color change or looking for a color change in a broken stem. Only a trained mycophagist should attempt to identify edible mushrooms.

Chapter 23

Section 23-3 Fungal Partnerships

1 Prepare

Scheduling

- **Block:** About 90 minutes will be needed to complete this section.
- **Traditional:** About two class periods will be needed to complete this section.
- **Planning:** See the Planning Guide on pp. 497A and 497B for lecture, classwork, and assignment options for Section 23-3. Lesson plans for Section 23-3 can be found in a lesson cycle format in ***Biology: Principles and Explorations*** **One-Stop Planner CD-ROM with Test Generator.**

Resource Link

Assign **Active Reading Worksheet Section 23-3** to help students comprehend the key concepts and visuals in this lesson.

2 Teach

Lesson Warm-up

Show students photos of colorful lichens or bring in live lichens growing on a rock. Ask students what kind of organism they think a lichen is and in which kingdom it is classified. *(A lichen is not an organism but a partnership of organisms; it is classified in kingdom Fungi.)*

Visual Strategy

Figure 23-10

Use this diagram to explain how mycorrhizae greatly increase the absorbing capacity of plant roots. Whether the fungus enters the root or wraps around the outside of the root, the results are the same: the fungus gains carbohydrates and the plant gains needed minerals.

Objectives

- **Distinguish** two symbiotic relationships that involve fungi. (pp. 508–509)
- **Summarize** the ecological importance of mycorrhizae. (p. 508)
- **Describe** lichens. (p. 509)

Key Terms

mycorrhiza
lichen

internetconnect
SciLinks NSTA
TOPIC: Symbiosis of fungi
GO TO: www.scilinks.org
KEYWORD: HX508

Fungi Form Symbiotic Relationships with Plant Roots

Fungi are involved in many kinds of symbiotic associations with algae and plants. These relationships play important roles in ecology. Recall from Chapter 18 that mutualism is a type of symbiosis in which each partner benefits. These symbiotic associations usually involve a sharing of abilities between a heterotroph (a fungus) and a photosynthesizer (a plant or an alga). The fungus provides minerals and other nutrients that it absorbs from the environment, and the photosynthesizer provides the ability to use sunlight to power the building of organic molecules.

Mycorrhizae

A **mycorrhiza** *(MEYE koh REYE zuh)* is a type of mutualistic relationship formed between fungi and vascular plant roots. The hyphae help transfer phosphorus and other minerals from the soil to the roots of the plant, while the plant supplies carbohydrates to the fungus.

In the mycorrhizae of most species of plants, the hyphae penetrate the outer cells of the root. The fungus is usually a zygomycete. In **Figure 23-10,** you can see the hyphae that grow in the roots. Fossils show that the rootlike structures of the earliest plants often had mycorrhizae, which may have played an important role in the invasion of land by plants. Scientists think that when plants invaded the land the soil of that time completely lacked organic matter. However, plants with mycorrhizae can grow successfully in infertile soil. Some vascular plants survive today by continuing this partnership as mycorrhizae.

In many plants, the mycorrhizae do not physically penetrate the plant root but instead wrap around it. These nonpenetrating mycorrhizae represent relationships in which a particular species of plant has become associated with a particular fungus, usually a basidiomycete. These kinds of mycorrhizae are important because they involve many commercially significant trees that grow in temperate regions, such as pines, oaks, beeches, and willows.

Some mycorrhizae produce economically important edible mushrooms, and the mycorrhizae of some ascomycete species produce an edible fruiting body called a truffle.

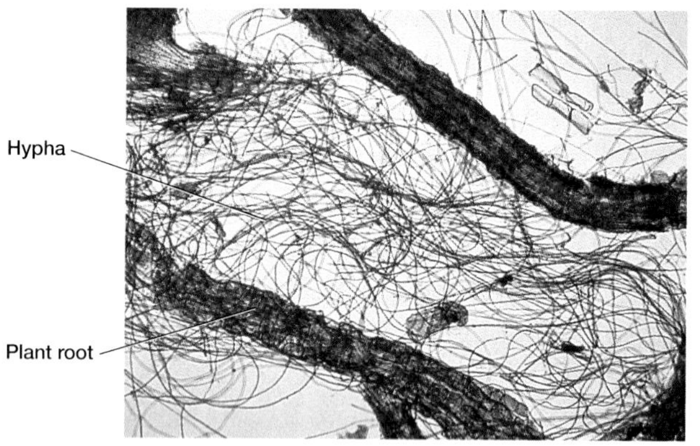

Figure 23-10
Mycorrhizae. The hyphae of the fungus in mycorrhizae sometimes appear as a tangled mass around the root of the plant. Both organisms benefit from this arrangement.

Integrating Different Learning Styles

Logical/Mathematical	**PE**	Data Lab, p. 509; Performance Zone items 15, 16, and 17, p. 513; Experimental Design, p. 514
	ATE	Data Lab, p. 509
Visual/Spatial	**ATE**	Lesson Warm-up, p. 508; Visual Strategy, p. 508; Demonstration, p. 509
Verbal/Linguistic	**PE**	Performance Zone item 12, p. 512
	ATE	Closure, p. 510; Reteaching, p. 510

Fungi Form Symbiotic Relationships with Algae

A **lichen** *(LY kuhn)* is a symbiosis between a fungus and a photosynthetic partner such as a green alga, a cyanobacterium, or both. The photosynthetic partner provides carbohydrates. It is protected from the environment by the fungal partner, which provides it with mineral nutrients. In most lichens, the fungi is an ascomycete. When you look at a lichen, such as the ones in **Figure 23-11,** you are seeing the fungus. The photosynthetic partner is hidden between the layers of hyphae. Sunlight penetrates the layers of hyphae and enables the photosynthetic partner to carry out photosynthesis.

The tough construction of the fungus combined with the photosynthetic abilities of the alga, or cyanobacterium, have enabled lichens to colonize harsh habitats. Lichens have been found in arid desert regions and in the Arctic; they grow on bare soil, on tree trunks, and on sunbaked rocks. Recall that during succession, lichens are often the first colonists. They break down rocks and prepare the environment for other organisms. Lichens are a key component of primary succession because they are able to carry out nitrogen fixation and introduce useful forms of nitrogen into their environment.

Figure 23-11 Lichens. The algae, shown as green cells in the micrograph, are the photosynthetic partners of the fungus growing in this British soldier lichen.

DATA LAB

Analyzing the effect of mycorrhizae

Background

Two groups of plants were planted in similar soils under similar conditions, but one group was grown in sterilized soil and the other group was grown in nonsterilized soil. After 18 weeks of growth, a photograph was taken of the plants. Examine the photographs, and answer the following questions:

Analysis

1. **Compare** the growth of the two groups. Which grew faster?
2. **Explain** why one group grew better than the other group.
3. **Infer** what the cause of slower growth in the smaller plants was.
4. **Recommend** a course of action to restore growth in the stunted plants.

did you know?

Lichens absorb radioactive fallout.

Lichens in Finland, Norway, and Sweden absorbed radioactive fallout from the cloud that carried the debris after Ukraine's 1986 Chernobyl nuclear power disaster. Reindeer, a major source of food for many people living in the area, consumed the radioactive lichen. Over 70,000 reindeer had to be destroyed because of the excessive radiation they absorbed from eating the contaminated lichens.

Demonstration

A Natural Indicator

Orchil is a commercial dye made from lichens that is used in making litmus paper. Using vinegar and ammonia, show students how litmus paper is used to determine whether a substance is an acid or a base. Ask them why the lichens are good indicators of pollution in the environment. *(Lichens absorb water from rain and atmospheric humidity and are sensitive to any pollutants dissolved in the water.)*

Demonstration

Describing Lichens BASIC

Provide students with pictures or live specimens of the three common forms of lichen: crustose, foliose, and fruticose. Ask students to describe the appearance of each lichen and to compare it to familiar objects. *(crustose—crusty; foliose—leafy; fruticose—upright, stalky)*

DATA LAB

Analyzing the effect of mycorrhizae

Time 5–10 minutes

Process Skills Interpreting results, applying information, predicting outcomes

Teaching Strategies

- In addition to increasing the uptake of minerals, mycorrhizae also increase water absorption and protect tiny rootlets from infection by soil pathogens.
- Despite their fertility, some prairie soils have trouble supporting pines and other trees because they lack the necessary mycorrhizal fungi.

Analysis Answers

1. Group B
2. Sterilizing the soil kills all of the mycorrhizal fungi, which provide mineral nutrients to plants.
3. Without the symbiotic fungi, they did not have as many nutrients available to them.
4. Inoculate the soil of the stunted plants with mycorrhizal fungi.

Exploring Further

Lichens as Environmental Watchdogs

Teaching Strategies
Inform students that organisms used to measure the health of the environment are called *indicator species.* Lichen serve as indicators for air quality; aquatic insects, such as mayfly larvae, are used by scientists to measure water quality. Encourage students to find out more about indicator species.

Discussion
Why don't the scientists simply test the rain directly for the presence of pollution? *(Testing the rain directly would give a picture of the pollution that day in one area, but studying the lichen can provide an idea of how conditions have changed over time or over a wide area, and may help identify sources of pollution.)*

3 Assessment Options

Closure

Early Partners

Have students prepare a two-column table comparing lichens with mycorrhizae. Their table should include type of photosynthesizers and ecological role. PORTFOLIO

Review

Assign Review—Section 23-3.

Reteaching — BASIC

Have students design a graphic organizer with three tiers. On the top tier place the word *Fungi.* The second tier should contain the following terms: *medicine, agriculture, industry, ecology,* and *economics.* The third tier should be developed by the students and should contain examples to illustrate the terms in the second tier. PORTFOLIO

Lichens are able to survive drought and freezing by becoming dormant. When moisture and warmth return, lichens resume normal activities. In harsh environments, lichens may grow slowly. Some lichens that grow in the mountains appear to be thousands of years old and cover an area no larger than a fist. These lichens are among the oldest living organisms on Earth. Although lichens are known to survive extremes of temperatures, they are susceptible to chemical changes in their environment and so have become a living indicator of the amount of pollution in the environment in which they live.

Exploring Further

Lichens as Environmental Watchdogs

Since the 1950s scientists have found that most lichens require clean air to thrive. In the Los Angeles Basin, for example, rising smog levels have been linked to the disappearance of lichens. In the Pacific Northwest, lichens are most abundant in old-growth forests with good air quality. For these reasons, scientists have been using lichens to monitor air pollution.

Why are lichens good indicators of air pollution?

Lichens have no roots, so the nutrients they take up must come from the air. Rain, fog, and dew wet the surface of a lichen. When they are wet, lichens absorb nutrients and any pollutants that are in the air.

Lichens can live for centuries, making them well suited for studies of air-pollution changes that occur over a long time. Many lichen species also have large geographical ranges. Thus, a single species can indicate air quality at different distances from a source of pollution, such as a factory or power plant.

How are lichens used?

To monitor air quality with lichens, scientists often map the distribution of lichens in an area. They count the number of lichen species, note how often each species occurs, and measure the total area covered by each species. Mapping studies done over many years can reveal long-term changes in lichen survival.

Pollutants from the stove pipe keep lichens from growing on part of this roof.

Scientists can obtain more detailed data on the effects of air pollution by determining the concentration of metals and other pollutants in lichen samples. They also assess the health of a lichen by measuring its chlorophyll content and its rate of photosynthesis.

To test air quality in places where lichens do not exist, scientists sometimes transplant healthy lichens from areas where they occur naturally. They then analyze the transplanted lichens for pollutants and look for any changes in the health of the lichens that may be caused by the move.

internet**connect**
SCI LINKS NSTA
TOPIC: Lichens
GO TO: www.scilinks.org
KEYWORD: HX510

Review SECTION 23-3

1. **Describe** two types of symbioses that involve fungi.
2. **Explain** how mycorrhizae benefit plants.
3. **Identify** the organisms found in lichens.
4. **Relate** the ability of lichens to become dormant to the great age of some lichens.

Review SECTION 23-3 ANSWERS

1. Fungal symbioses include mycorrhizae and lichens.
2. Mycorrhizal fungi provide additional mineral resources to the plant.
3. A lichen is a mutualistic relationship involving a fungus and an alga or a cyanobacterium.
4. By becoming dormant during periods of environmental stress, lichens survive conditions that would kill off other organisms, allowing them to survive for centuries.

Study ZONE

CHAPTER 23

Use the Key Concepts and Key Terms listed below to review the main ideas in this chapter. If you need to review the meaning of a term, turn to the page indicated.

Key Concepts

23-1 Characteristics of Fungi

- Fungi are eukaryotic heterotrophs. Their bodies are made up of slender woven filaments. Fungal cells contain chitin and go through nuclear mitosis.
- Fungi obtain nutrients by secreting digestive enzymes and absorbing the decomposed nutrients from their environment.
- Fungi decompose dead organic matter; they are an important resource recycler.
- Most fungi reproduce by releasing spores that are produced asexually and sexually.

23-2 Fungal Diversity

- Fungi are classified by their sexual reproductive structures.
- Fungi in which sexual reproduction has not been observed are referred to as deuteromycetes.
- Fungi in the phylum Zygomycota produce spores in thick-walled sexual structures called zygosporangia.
- Fungi in the phylum Ascomycota produce spores in a saclike structure called an ascus.
- Yeasts are unicellular ascomycetes that reproduce by budding.
- Fungi in the phylum Basidiomycota produce spores in a club-shaped structure called a basidium.

23-3 Fungal Partnerships

- Fungi can be involved in two types of symbioses, mycorrhizae or lichens.
- Mycorrhizae are symbiotic associations in which a fungus transfers minerals to a plant's roots, which in turn supply carbohydrates to the fungus.
- The fungal partner in a lichen protects the photosynthetic partner and provides the lichen with minerals. The photosynthetic partner provides the fungus with carbohydrates.

Key Terms

23-1

chitin (500)
hypha (501)
mycelium (501)

23-2

zygosporangium (504)
stolon (504)
rhizoid (504)
ascus (505)
yeast (505)
budding (505)
basidium (506)

23-3

mycorrhiza (508)
lichen (509)

Study TIP Ready?

If you think you understand these Key Concepts and Key Terms, then you're ready for the Performance Zone.

Study ZONE

CHAPTER 23

Review and Practice

Assign the following worksheets for Chapter 23:

- **Student Review Guide**
- **Concept Mapping Worksheet**
- **Test Preparation Pretest**
- **Vocabulary Worksheet**
- **Critical Thinking Review Worksheet**

Assessment Options

The following assessment options are available for this chapter:

Resource Link

Assign **Chapter Test 23** to assess students' understanding of the chapter.

Alternative Assessment

- Have students consider the following question: Which group of organisms has had the greatest impact on human life—bacteria, protists, or fungi? Have them defend their position in a one-page report. PORTFOLIO

Portfolio Assessment

- *Portfolio Assessment* at the front of this book provides options for building and evaluating student portfolios. Students and teachers can select items from the areas listed for inclusion in a portfolio.
- See items labeled PORTFOLIO on pp. 501, 502, 503, 507, and 510.

Answer to Concept Map

The following is one possible answer to Performance Zone item 1 on page 512.

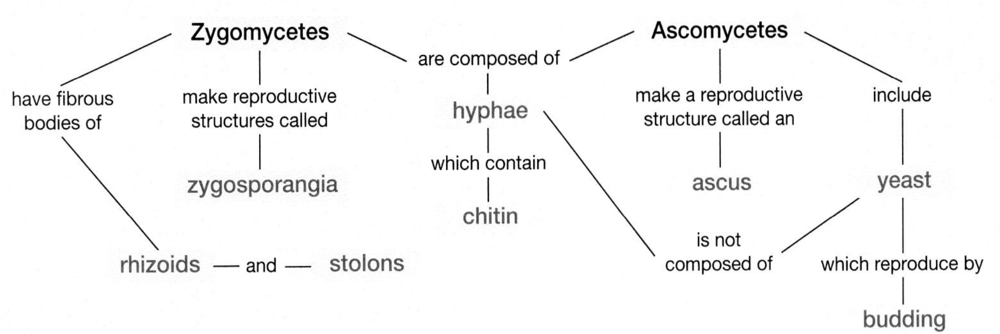

Performance ZONE
CHAPTER 23
REVIEW ANSWERS

Assignment Guide

Section	Review
23-1	3, 4, 5, 18, 19, 20, 25, 26
23-2	1, 2, 6, 7, 11, 13, 14, 21, 24
23-3	8, 9, 10, 12, 15, 16, 17, 22, 23

Understanding and Applying Concepts

1. The answer to the concept map is found at the bottom of page 511.
2. **a.** Hyphae are thin fungal filaments, which form a tangled mass called a mycelium.

   b. Stolons are the mycelia that grow on the surface of the fungus; rhizoids are mycelia that anchor the zygomycete in the substrate.

   c. Budding is a method of asexual reproduction in yeasts; a basidium is the characteristic sexual reproductive structure of the basidiomycetes.
3. d
4. d
5. a
6. d
7. c
8. c
9. d
10. c
11. The zygosporangium is a thick-walled structure that can remain dormant, enabling the fungus to survive until environmental conditions are more suitable for meiosis and germination.
12. By mapping the location, population density, and composition of lichens, scientists can monitor air quality. Extremely sensitive to pollutants, lichens cannot survive in areas of high pollution.
13. ***A.*** Members of phylum Zygomycota reproduce asexually by producing spores on the tips of specialized hyphae. They reproduce sexually when mating types fuse and produce a zygosporangium, which gives rise to a spore-producing hypha.

    B. Members of the phylum Ascomycota reproduce asexually by producing spores on the tips of specialized hyphae. They reproduce sexually when different mating types fuse their hyphae and produce a cup-shaped fruiting body. Spores form within an ascus.

    C. Members of the phylum Basidiomycota rarely reproduce asexually. For sexual reproduction, two mating types fuse their mycelia and form a fruiting body (mushroom). Spores are formed within structures called basidia.
14. Because of the pressures of natural selection, similar organisms in similar environments may develop similar structures in a phenomenon known as convergent evolution.

Performance ZONE
CHAPTER 23 Review

Understanding and Applying Concepts

1. **Concept Mapping** Construct a concept map that describes the structure and reproductive methods of different fungi. Use the following terms in your map: chitin, hyphae, zygosporangia, stolon, rhizoid, ascus, yeast, budding. Use additional terms as necessary.

2. **Understanding Vocabulary** For each pair of terms, explain the difference in their meanings.
   a. hyphae/mycelium
   b. stolon/rhizoid
   c. budding/basidium

3. Fungi differ from plants in that fungi
   a. are multicellular.
   b. are immobile.
   c. have cell walls.
   d. are heterotrophic.

4. The cell walls of all fungi are made of
   a. rhizoids.
   b. cellulose.
   c. asci.
   d. chitin.

5. Which of the following characteristics is shared by all fungi and helps them obtain their nutrients?
   a. external digestion
   b. phagocytosis
   c. feed on nonliving matter
   d. catching prey

6. Deuteromycetes are more difficult to classify than other fungi because
   a. they develop from zygosporangia.
   b. they are parasitic.
   c. they undergo meiosis.
   d. they do not reproduce sexually.

7. The common edible mushroom is classified in the phylum
   a. Zygomycota.
   b. Ascomycota.
   c. Basidiomycota.
   d. Deuteromycota.

8. Mycorrhizae are symbiotic relationships between fungi and
   a. algae.
   b. lichens.
   c. roots.
   d. chloroplasts.

9. Some fungal associations no larger than a fist appear to be thousands of years old. These have been found
   a. in temperate forests.
   b. on well-irrigated alluvial plains.
   c. in fields of corn.
   d. in harsh environments, high in the mountains.

10. One might expect that plants without mycorrhizae are
    a. more likely to get fungal diseases.
    b. less successful in the transfer of minerals from the soil to the roots.
    c. best suited to poor soil conditions.
    d. primitive and might soon become extinct.

11. **Theme Cellular Structure and Function** Describe the structure of the zygosporangium of *Rhizopus stolonifer.* How does the zygosporangium function to ensure survival of the species?

12. **Exploring Further** Explain how scientists use lichens to monitor air quality and why lichens were chosen by scientists to monitor air quality.

13. **Identify** the major group of fungi represented by each fungus below, and describe how each reproduces.

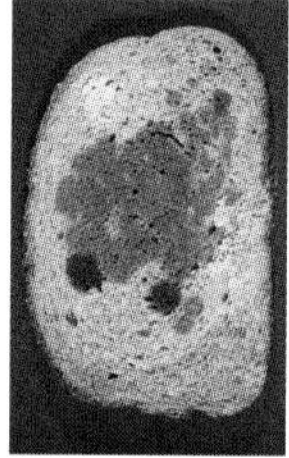
A

B

C

14. **Chapter Links** The morel is not a basidiomycete but is commonly called a mushroom. How do scientists explain this phenomenon in which distantly related organisms develop similar structures? (**Hint:** See Chapter 13, Section 13-2.)

Skills Assessment

Part 1: Interpreting Graphs

Study the graph below showing the amount of the truffle harvest and existing forest over the same period of time. Then answer the following questions:

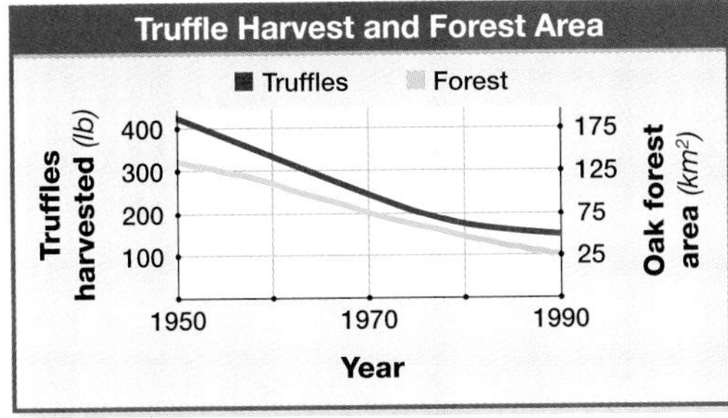

15. What is the relationship between the truffle harvest and forest clearing?

16. What would you advise landowners in the area to do if they would like to maintain the truffle harvest at its current level?

17. What would you infer is the relationship between truffles and trees?

Part 2: Life/Work Skills

Use the media center or Internet resources to learn about the economic impact of fungi on you.

18. **Finding Information** Prepare a written report about the importance of fungi as food or in the preparation or manufacture of food and how much money the food industries make. You may want to investigate mushrooms eaten as food, fungi used to produce certain types of cheeses and soy sauce, fungi used in baking bread, and fungi used to produce wine.

19. **Selecting Technology** Find out what techniques pharmaceutical companies are using to discover new antibiotics. Discover what kind of equipment manufacturers use to make large batches of antibiotics.

20. **Communicating** Prepare an oral report to summarize your findings. Create a display that demonstrates the amount of money spent by various industries on fungi-related materials and calculate that cost per person in the United States.

Critical Thinking

21. **Evaluating Results** Some fungi were discovered reproducing asexually and were given a scientific name. Other researchers found the same fungus reproducing sexually and gave it another scientific name. For many years scientists did not know they were both the same organism. Can a single organism have two names? Explain.

22. **Forming Reasoned Opinions** Given the kind of metabolism that lichens have, in what sort of environment do you think lichens evolved?

23. **Choosing Procedures** When purchasing plants to be planted, is it better to buy bare root plants or plants that are already established in a pot of soil? Explain your reasoning from an ecological point of view.

Portfolio Projects

24. **Summarizing Information** Downy mildew caused by the fungus *Plasmopara viticola* devastated French grape crops during the nineteenth century. Pierre Millardet invented the Bordeaux mixture, which cures this infection. Find out what chemicals are in the Bordeaux mixture and how it works. Also, learn how Millardet developed the mixture. Write a paper that details what you learn, and share your results with the class in an oral report.

25. **Analyzing Methods** Read "Alexander Fleming" (*Time,* March 29, 1999, pp. 117–119). Describe the role of luck in the discovery of penicillin. What was the reason given for Fleming not pursuing the isolation of penicillin?

26. **Career Focus Pharmaceutical Researcher** Research the field of antibiotic pharmaceutical research and write a report on your findings. Your report should include a job description, training required, kinds of employers, growth prospects, and starting salary.

Skills Assessment

15. Clearing the forest reduces the truffle harvest.
16. Stop clearing the forest.
17. Truffles are in some way dependent on trees.
18. Answers will vary. Students' reports might also address the historical significance of fungus-related foods, including the early, large-scale consumption of these foods in Europe and Asia, and their rise in popularity in the United States.
19. Answers will vary. Techniques may include testing natural and chemically modified microbial substances against harmful microbes or cancer cells. Equipment may include flasks and huge fermentation vats (for culturing the antibiotic-producing microbes).
20. Answers will vary. Students' displays should include data on both fungus-related foods and drugs.

Critical Thinking

21. A single organism can have several common names, but scientists have agreed upon a single scientific name for each described species to facilitate communication and understanding.
22. Answers will vary, but students should address the mutualistic relationship as evidence that some environments may have made such a relationship advantageous.
23. Plants already established in soil are better because they are more likely to have mycorrhizae than would bare-rooted plants.

Portfolio Projects

24. It is a mixture of lime and copper sulfate. The very insoluble copper sulfate is dissolved by the acid produced by germinating fungal spores. The dissolved copper kills the spores, but the remaining undissolved copper does not harm the grape plant. Millardet sprayed the grapes with copper sulfate and lime to prevent people from stealing and eating them, and he noticed that these grapes did not have downy mildew.
25. Fortunately, Fleming had decided not to store the bacteria culture in a warm incubator, but left it out where it was accidentally inoculated with the fungal spore. (Conditions were unusually cool, which also favored growth of the fungus.) Fleming did not pursue the isolation of the antibiotic because he lacked the chemical expertise and conviction that drugs could ever cure serious illness.
26. Pharmaceutical researchers isolate or prepare drugs to treat a variety of diseases. Many pharmaceutical researchers study fungi hoping to isolate chemicals that can be used to treat diseases or medical conditions. Training may include a graduate degree in chemistry or microbiology. Growth potential is good. Starting salary will vary by region.

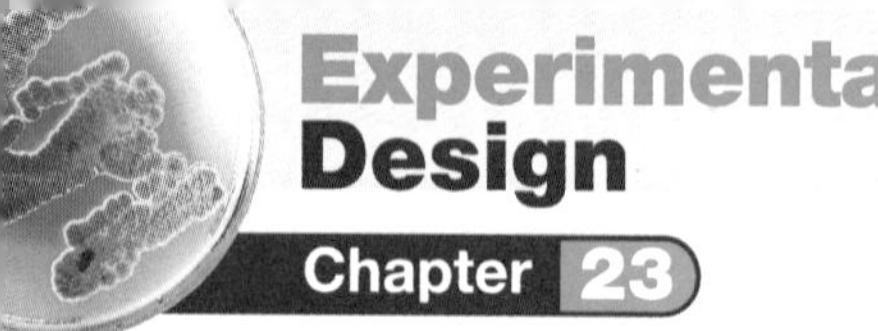

Experimental Design

Chapter 23

How Can Mold Growth Be Inhibited?

Purpose

Students will design and conduct a scientific experiment to study the effects of various environmental factors on the growth of fungi.

Preparation Notes

Time Required: one 50-minute lab period; 20 minutes 7 days later

Materials

Materials for this lab can be ordered from WARD'S. See *Master Materials List* at the front of this book for catalog numbers.

Preparation Tips

- Fungal samples can be grown on food items (e.g., bread without preservatives). Place a piece of damp filter paper under the food item in a bowl. Store the bowl at room temperature for a week. Keep the filter paper moist. To prevent contamination when putting the cultures out in the lab, cover each bowl with plastic wrap.
- You may want to homogenize food samples in a blender ahead of time to make it easier for students to obtain a small sample for their Petri dishes.
- Prepare media from Sabouraud dextrose agar and a small bottle of 0.3% propionic acid mold inhibitor. Melt the agar in a microwave or in a boiling water bath, and divide the contents evenly between two flasks. Add 0.75 mL of the propionic acid to one labeled flask. Sterilize both flasks and Petri dishes at 15 psi for 20 minutes (or buy sterile, disposable dishes). Label the Petri dishes. Pour the contents of each flask into the dishes. Make sure you flame the lip of the flask before you pour the media. Refrigerate dishes until lab time.

Safety Precautions

- Have students wear safety goggles, disposable gloves, and a lab apron during this investigation. They should wash their hands thoroughly after handling food samples or any fungal samples. After fungal samples have been added, be sure that all Petri dishes are sealed with tape and are not opened again. All fungal samples should be destroyed by incineration or in an autoclave.
- Have students dispose of the toothpicks and Petri dishes in a container designated for biohazardous materials only. Treat all growth in the Petri dishes as pathogenic.
- Remind students not to touch the inside of a Petri dish or allow it be exposed to the air.

Procedural Tips

- Have students prepare their data collection tables ahead of time.
- To prevent contamination, students should raise the lids of the Petri dishes as little as possible when transferring a fungal sample.
- Label three plates with each category: *Moist with acid, Moist without acid, Acid,* and *No acid.*

Disposal

After experiments are completed, autoclave the cultures. Petri plates may be safely thrown away after autoclaving.

Experimental Design

Chapter 23 *How Can Mold Growth Be Inhibited?*

SKILLS

- Using scientific methods
- Using aseptic techniques

OBJECTIVES

- **Observe** the effect of environmental conditions on mold growth.
- **Determine** the effect of propionic acid on mold growth.

MATERIALS

- safety goggles
- lab apron
- disposable gloves
- rubbing alcohol
- paper towels
- sterile Petri dishes with plain nutrient agar
- sterile Petri dishes with nutrient agar that contains propionic acid
- samples of molds
- stereomicroscope
- toothpicks
- wax pencil
- masking tape

ChemSafety

CAUTION: Always wear safety goggles and a lab apron to protect your eyes and clothing.

CAUTION: Do not touch or taste any chemicals. Know the location of the emergency shower and eyewash station and how to use them. If you get a chemical on your skin or clothing, wash it off at the sink while calling to the teacher. Notify the teacher of a spill. Spills should be cleaned up promptly, according to your teacher's directions.

CAUTION: Glassware is fragile. Notify the teacher of broken glass or cuts. Do not clean up broken glass or spills with broken glass unless the teacher tells you to do so.

Before You Begin

Molds are **fungi,** which obtain nutrients by releasing enzymes that break down living or nonliving matter then absorbing their molecules. **Spores** produced by molds are easily scattered by air currents. Thus, molds are often found growing on foods. In this lab, you will determine how environmental factors (light, moisture, and temperature) and the presence of **propionic acid** (a chemical added to many foods) affect mold growth.

1. Write a definition for each boldface term in the paragraph above.
2. Based on the objectives for this lab, write a question you would like to explore about mold growth.

Procedure

Part A: Transferring Mold to Agar

1.  Put on safety goggles, a lab apron, and disposable gloves.
2. Use rubbing alcohol and paper towels to clean the surface of your lab table and gloves. **CAUTION: Alcohol is flammable. Do not use alcohol near a flame.**
3. Look at a sample of mold with a stereomicroscope. Select an area of dense growth, and use a clean toothpick to scoop up a small amount of the mold.
4. Slowly lift the lid of a Petri dish containing sterile nutrient agar. *Note: Lift the lid just enough to allow a toothpick to reach the agar. Hold the lid over the agar at all times.* Gently touch the mold sample

to four different places on the nutrient agar. Dispose of the toothpick in a proper container.

5. Replace the lid, and seal the dish with masking tape. Label the dish with your name and the food source of the mold.

6. Incubate Petri dishes for one week. Look at any molds that grow with a stereomicroscope. **CAUTION: Do not open a sealed Petri dish.**

PART B: Design an Experiment

7. Work with the members of your lab group to explore one of the questions written for step 2 of **Before You Begin.** To explore the question, design an experiment that uses the materials listed for this lab.

You Choose

As you design your experiment, decide the following:

a. what question you will explore
b. what hypothesis you will test
c. which mold sample you will use
d. what conditions you will examine
e. where and how you will store your Petri dishes to maintain the appropriate conditions
f. what your control will be
g. how you will prevent contamination of samples
h. what data you will record in your data table

8. Write a procedure for your experiment. Make a list of all the safety precautions you will take. Have your teacher approve your procedure and safety precautions before you begin the experiment.

9. Set up and carry out your experiment.

PART C: Cleanup and Disposal

10. Dispose of Petri dishes, used toothpicks, and broken glass in the designated waste containers. Do not pour chemicals down the drain or put lab materials in the trash unless your teacher tells you to do so.

11. Clean up your work area and all lab equipment. *Clean the surface of your lab table with rubbing alcohol.* Return lab equipment to its proper place. Wash your hands thoroughly before you leave the lab and after you finish all work.

Analyze and Conclude

1. **Analyzing Results** What effect does propionic acid have on fungal growth?
2. **Analyzing Methods** What steps were taken in your experiment to avoid contamination of the Petri dishes?
3. **Drawing Conclusions** Which environmental conditions favor the growth of fungi, and which conditions inhibit it?
4. **Inferring Conclusions** Why do you think propionic acid is added to foods?
5. **Relating Conclusions** Based on your conclusions, how would you store a nonsterile food product that does not contain a preservative if you wanted to prevent it from becoming moldy?
6. **Further Inquiry** Write a new question about molds that could be explored with another investigation.

Do You Know?

Do research in the library or media center to answer these questions:

1. How are fungi used to clean pollutants such as pesticides and chemicals produced in industrial waste?
2. How are fungi used in medicine?

Use the following Internet resources to explore your own questions about fungi and food preservatives.

internet**connect**

TOPIC: Uses of fungi
GO TO: www.scilinks.org
KEYWORD: HX515

Before You Begin

Answers

1. *mold*—a fungus that grows as a mass of tangled filaments

   fungi—a group of heterotrophic, eukaryotic organisms belonging to the kingdom Fungi with structural and metabolic properties different from those of plants and animals

   spores—reproductive cells produced by molds and carried by wind or air currents

   propionic acid—a fungal growth inhibitor that is added to many foods
2. For example: What effect do different environmental conditions and propionic acid have on mold growth?

Procedure

Sample Procedure

1. Label three plates with each category: *Moist with acid, Moist without acid, Acid,* and *No acid.*

Sample Data Table

	MOIST WITH ACID	MOIST WITHOUT ACID	ACID	NO ACID
Desktop	Good growth	Good growth	Sparse growth	Sparse growth
Incubator	Light growth	Light growth	Sparse growth	Growth, large colonies
Refrigerator	No growth	No growth	No growth	No growth

2. Each type of plate is incubated either on a desktop (with light), in an incubator (without light), or in a refrigerator (cold).
3. Check and record the amount of growth on day 7.

Analyze and Conclude

Answers

1. Propionic acid inhibits fungal growth.
2. The lids of the Petri plates were raised only slightly, and new toothpicks were used for each transfer.
3. Answers will vary. Warmth and moisture favor fungal growth, while cold and dryness inhibit it. Light and darkness have little effect on fungal growth.
4. Propionic acid helps keep foods edible by retarding fungal growth.
5. Answers may vary. Students should recognize that foods that do not contain preservatives should be stored in a cool, dry place.
6. Answers will vary. For example: On what kinds of food do fungi grow best?

Do You Know?

Answers

1. Answers will vary. For example: *Penicillium* is used to clean contaminants, such as selenium produced by industry from polluted waters.
2. Answers will vary. For example: Secretions of fungi are used as antibiotics (e.g., penicillin). Yeasts are used as models for identifying human genes that influence cancer, Down syndrome, and other genetic disorders.

UNIT 6 Exploring Plants

Chapters

More than 20 billion bushels of corn are harvested worldwide every year.

in perspective

Corn: America's Crop

Used for food by Native Americans beginning 10,000 years ago, corn was first cultivated in what is today called Mexico in 6000 to 5000 B.C. The Aztec Indians, the Maya, and the Inca all grew corn and featured it prominently in their traditions and ceremonies. **Where is the Corn Belt? See page 536.**

Incas harvesting the corn crop

Cooked, steamed, roasted, ground into cornmeal, or popped, corn has long been a mainstay of the American diet. Corn is used to make cornstarch, corn syrup, breakfast cereals, cornmeal, salad dressing, corn oil, margarine, coloring agents, stabilizing agents, and countless other products. **To find out what corn and wheat have in common, see pp. 594–595.**

Agricultural scientist

Tomorrow...

Scientists continue to develop new uses for corn and its products. For example, ethanol, or ethyl alcohol distilled from corn, can be used as a lead-free octane booster and as a replacement for regular leaded gasoline. **Discover on page 577 how corn is able to conduct photosynthesis efficiently in intense heat.**

Automobile that runs on ethanol made from corn

internetconnect

TOPIC: Corn
GO TO: www.scilinks.org
KEYWORD: HX517

Chapter 24 Introduction to Plants

Planning Guide

OBJECTIVES	MEETING INDIVIDUAL NEEDS	READING SKILLS AND STRATEGIES
24-1 Adaptations of Plants (BLOCKS 1 & 2 90 min) • **Summarize** how plants are adapted to living on land. • **Distinguish** nonvascular plants from vascular plants. • **Relate** the success of plants on land to seeds and flowers. • **Describe** the basic structure of a vascular plant sporophyte.	Learning Styles, p. 520 BASIC ADVANCED • pp. 522, 524, 525	ATE Active Reading, Technique: Reading Organizer, p. 519 ■ Active Reading Guide, worksheet (24-1) ■ Directed Reading, worksheet (24-1)
24-2 Kinds of Plants (BLOCK 3 45 min) • **Describe** the key features of the four major groups of plants. • **Classify** plants into one of the 12 phyla of living plants.	Learning Styles, p. 526 BASIC ADVANCED • pp. 530, 531, 532, 533	■ Active Reading Guide, worksheet (24-2) ■ Directed Reading, worksheet (24-2)
24-3 Plants in Our Lives (BLOCKS 4 & 5 90 min) • **Identify** foods that come from plants and their dietary importance. • **Describe** several ways that wood is used. • **Explain** how plants are used to treat human ailments. • **Identify** plants that are used to make paper and cloth.	Learning Styles, p. 534 BASIC ADVANCED • pp. 538, 539, 540	ATE Active Reading, Technique: Anticipation Guide, p. 535 ATE Active Reading, Technique: Reading Organizer, p. 539 ■ Active Reading Guide, worksheet (24-3) ■ Directed Reading, worksheet (24-3) ■ Supplemental Reading Guide, *A Feeling for the Organism: The Life and Work of Barbara McClintock*

Review and Assessment Resources

BLOCKS 6 & 7 90 min

TEST PREP

- PE Study Zone, p. 541
- PE Performance Zone, pp. 542–543
- ■ Chapter 24 worksheets:
  - Vocabulary
  - Science Skills
  - Concept Mapping
  - Critical Thinking
- Audio CD Program
- ■ Test Prep Pretest

CHAPTER TESTING

- ■ Chapter Tests (blackline master)
- ▲ Test Generator
- ■ Assessment Item Listing

See the correlations table at the front of this book for a list of the ***National Science Education Standards*** addressed in this chapter.

Smithsonian Institution® Visit **www.si.edu/hrw** for additional on-line resources.

KEY

Biology Principles and Explorations

PE Pupil's Edition
ATE Teacher's Edition

 Teaching Transparencies
Laboratory Program

One-Stop Planner CD-ROM
Shows these resources within a customizable daily lesson plan:

■ Teaching Resources
▲ Test Generator

VISUAL STRATEGIES	CLASSWORK AND HOMEWORK	TECHNOLOGY AND INTERNET RESOURCES
ATE Fig. 24-2, p. 521 **ATE** Fig. 24-7, p. 524 **ATE** Fig. 24-8, p. 525 121 Structure of a Vascular Plant 113A Phyla of Living Plants 114A Phyla of Living Plants, cont.	**PE** Quick Lab Under what conditions are stomata open?, p. 521 Lab A18 Comparing Plant Adaptations ■ Occupational Applications: Botanist **PE** Review, p. 525	Holt Biology Videodiscs Section 24-1 Audio CD Program Section 24-1 internetconnect — SCILINKS NSTA — TOPICS: • Plant adaptations • Vascular plants • Structure and function of seeds
ATE Fig. 24-11, p. 528 115A Life Cycles of Plants 133 Life Cycle of a Conifer	**PE** Data Lab Analyzing the effect of climate on plants, p. 530 **PE** Review, p. 533	CNN Video Segment 19 Magnolia DNA Holt Biology Videodiscs Section 24-2 Audio CD Program Section 24-2 internetconnect — SCILINKS NSTA — TOPICS: • Nonvascular plants • Seedless vascular plants • Angiosperms
112 Meal Supplying a Complete Protein 113 Major Crop-Producing Regions of the World 114 Early Cultivated Plants 112A Medicines Originally Derived from Plants	**PE** Quick Lab When is a vegetable a fruit?, p. 535 **PE** Lab Techniques Surveying Plant Diversity, pp. 544–545 Lab C37 Growing Plants in the Laboratory **PE** Review, p. 540 **PE** Study Zone, p. 541	Holt Biology Videodiscs Section 24-3 Audio CD Program Section 24-3 internetconnect — SCILINKS NSTA — TOPICS: • Vegetarian diets • Medicines from plants

ALTERNATIVE ASSESSMENT

PE Portfolio Projects 27–29, p. 543
ATE Alternative Assessment, p. 541
ATE Portfolio Assessment, p. 541
■ Portfolio Projects: Plant Research Project, "Design Your Own Project"

Lab Activity Materials

Quick Labs
p. 521 clear nail polish, plant kept in light, plant kept in darkness, 4–5 cm strips of clear tape (2), microscope slides (2), compound microscope
p. 535 apple, banana, green bean, potato, squash, tomato, plastic knife

Data Lab
p. 530 pencil, paper

Lab Techniques
p. 544 live or preserved specimens of mosses, ferns, conifers, and flowering plants; stereomicroscope or hand lens; microscope; slides of fern gametophytes

Lesson Warm-up
ATE live specimens of a variety of plants, p. 526

Demonstrations
ATE pictures or live specimens of several plants from different phyla, p. 518
ATE samples of several kinds of ferns, p. 529
ATE samples of several kinds of cones and flowers, p. 532
ATE biodegradable packing pellets, water, iodine, p. 538
ATE swatches of fabric made of naturally colored cotton

BIOSOURCES LAB PROGRAM
- Quick Labs: Lab A18, p. 35
- Laboratory Techniques and Experimental Design: Lab C37, p. 181

Chapter Theme

Evolution

The development of structures that enabled plants to survive on land more efficiently can be traced by studying the great diversity of plants living today and the fossil record. The earliest plants to appear in the fossil record are also the simplest in structure, and the flowering plants, the most recent to appear in the fossil record, are also the most complex and most highly specialized plants.

Establishing Prior Knowledge

Before beginning this chapter, make certain students can answer the questions in **Study TIP Ready?** on page 519. If they cannot, encourage them to reread the sections indicated.

Opening Demonstration

Show students pictures or live specimens of several plants from different phyla. Ask students to point out the similarities and differences between the groups of plants. On the board or overhead projector, develop a chart organizing the characteristics that students point out.

CHAPTER 24 Introduction to Plants

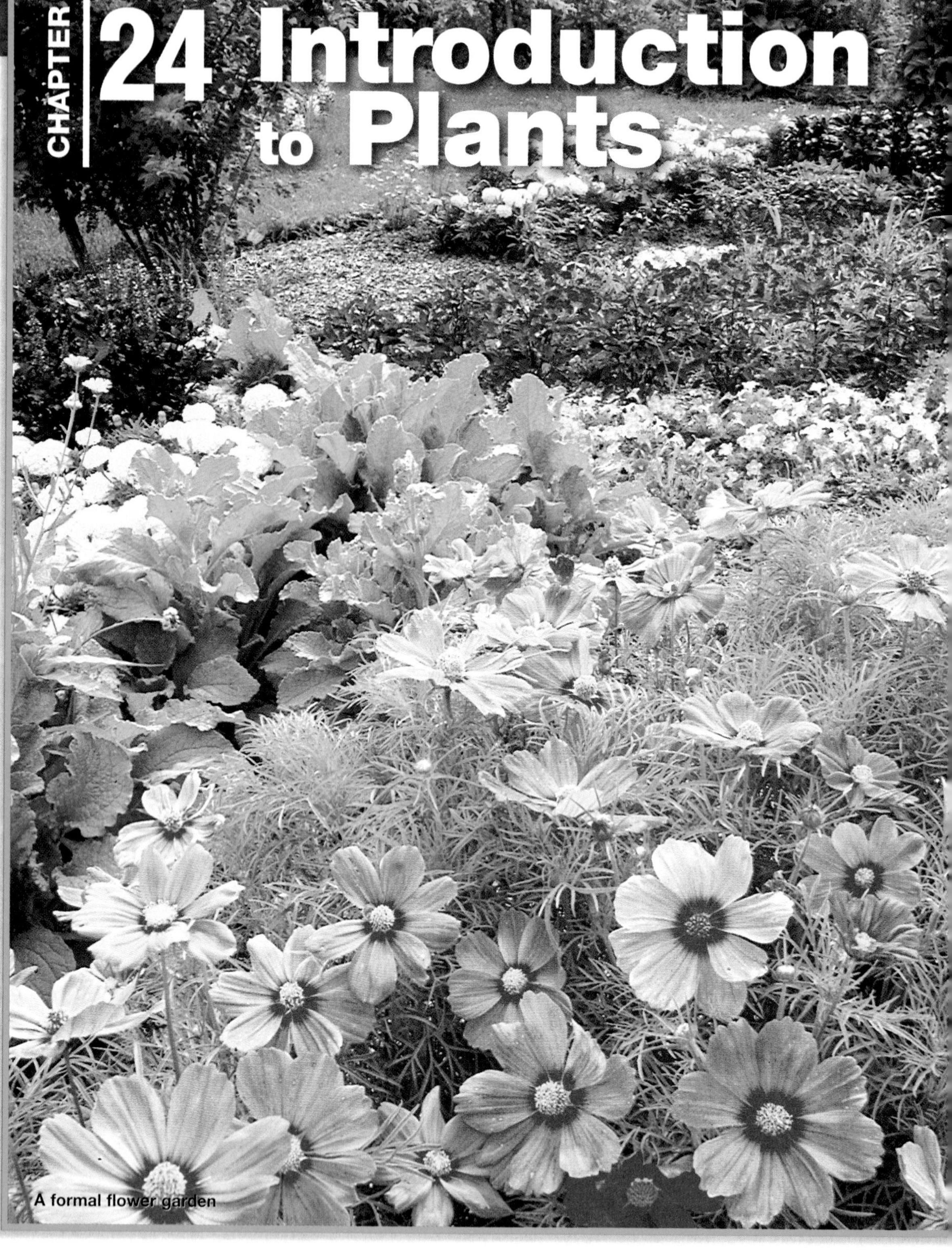

A formal flower garden

What's Ahead — and Why

CHAPTER OUTLINE

24-1 Adaptations of Plants
- Plants Became Established on Land
- Vascular Tissue, Seeds, and Flowers Made Plants Successful
- A Change Occurred in Plant Life Cycles

24-2 Kinds of Plants
- Nonvascular Plants Lack a Vascular System
- Seedless Vascular Plants Do Not Make Seeds
- Gymnosperms Are Seed Plants that Produce Cones
- Angiosperms Are Seed Plants that Produce Flowers

24-3 Plants in Our Lives
- Plants Provide Food for Animals
- Cereals Are the Most Important Source of Food
- Plants Have Many Nonfood Uses

Students will learn

- 24-1: how the adaptations of land plants enabled them to become established and thrive on land.
- 24-2: the characteristics of the major groups of land plants.
- 24-3: the many ways that plants are important to people.

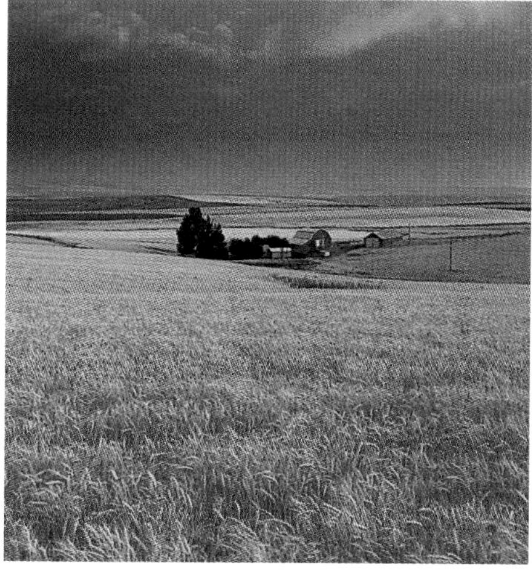
Wheat farm

We rely on plants for survival.

Plants provide us with the food and oxygen that make life possible. With their adaptations to land, plants also enrich our lives with the beauty of their vibrant colors and sweet scents and by providing materials for making strong buildings, fine papers, musical instruments, and comfortable clothes.

Study TIP

Ready?

Check your understanding of these topics to help you prepare for what's ahead.

Can you...

- **describe** the process of mitosis? (Section 6-3)
- **describe** the process of meiosis? (Section 7-1)
- **identify** life cycles with a gametophyte and a sporophyte? (Section 7-2)
- **describe** the role of mycorrhizae? (Sections 12-3 and 23-3)
- **list** the characteristics of the kingdom Plantae? (Section 20-3)

If not, *review the section indicated.*

Looking Ahead

24-1 Adaptations of Plants
How are plants adapted to living on land?

24-2 Kinds of Plants
There are many different kinds of plants.

24-3 Plants in Our Lives
In what ways do humans use plants?

Labs

- **Quick Labs**
  Under what conditions are stomata open? (p. 521)
  When is a vegetable a fruit? (p. 535)
- **Data Lab**
  Analyzing the effect of climate on plants (p. 530)
- **Lab Techniques**
  Surveying Plant Diversity (p. 544)

Features

- **Health Watch**
  Vegetarian Diets (p. 537)

internet connect

National Science Teachers Association *sci*LINKS Internet resources are located throughout this chapter.

Reading Strategies

Active Reading

Technique: Reading Organizer

Before students begin this chapter, explain that it traces the evolution of plants and provides an overview of the types of plants and how plants are used by humans. Have students survey each section before they read it and identify any subtitles, headings, and captions that signal the topic of discussion. As students read, ask them to locate other words in the body of the text that signal the sequential pattern.

Directed Reading

Assign **Directed Reading Worksheet Chapter 24** to help students interact with the text as they read the chapter.

Interactive Reading

Assign ***Biology: Principles and Explorations* Audio CD Program Chapter 24** to help reluctant readers achieve greater success in reading the chapter.

Reading Skills

Assign Reading Organizers and Reading Strategies Worksheets from the **One-Stop Planner CD-ROM with Test Generator** to help students learn and practice effective reading skills.

internet connect

Holt *Biology: Principles and Explorations*

On-line Resources: go.hrw.com

Visit the HRW Web site for a variety of resources related to this chapter. Just type in the keyword HX0 Chapter 24.

Holt *Biology: Principles and Explorations*

On-line Resources: www.scilinks.org

The following *sci*LINKS Internet resources can be found in the student text for this chapter:

Page 521
TOPIC: Plant adaptations
KEYWORD: HX521

Page 522
TOPIC: Vascular plants
KEYWORD: HX522

Page 523
TOPIC: Structure and function of seeds
KEYWORD: HX523

Page 527
TOPIC: Nonvascular plants
KEYWORD: HX527

Page 528
TOPIC: Seedless vascular plants
KEYWORD: HX528

Page 532
TOPIC: Angiosperms
KEYWORD: HX532

Page 537
TOPIC: Vegetarian diets
KEYWORD: HX537

Page 539
TOPIC: Medicines from plants
KEYWORD: HX539

Chapter 24

1 Prepare

Scheduling

- **Block:** About 90 minutes will be needed to complete this section.
- **Traditional:** About two class periods will be needed to complete this section.
- **Planning:** See the Planning Guide on pp. 517A and 517B for lecture, classwork, and assignment options for Section 24-1. Lesson plans for Section 24-1 can be found in a lesson cycle format in ***Biology: Principles and Explorations*** **One-Stop Planner CD-ROM with Test Generator.**

Resource Link

Assign **Active Reading Worksheet Section 24-1** to help students comprehend the key concepts and visuals in this lesson.

2 Teach

Lesson Warm-up

Ask students to name three things that are important for the survival of an organism on land. *(Students should mention that survival requires a means of getting water and nutrients, a means of reproduction, and a means of maintaining homeostasis.)*

Section 24-1 Adaptations of Plants

Objectives

- **Summarize** how plants are adapted to living on land. (pp. 520–521)
- **Distinguish** nonvascular plants from vascular plants. (pp. 522–524)
- **Relate** the success of plants on land to seeds and flowers. (pp. 522–523)
- **Describe** the basic structure of a vascular plant sporophyte. (p. 525)

Key Terms

cuticle
stoma
guard cell
vascular system
nonvascular plant
vascular plant
seed
embryo
seed plant
flower
phloem
xylem
shoot
root
meristem

Study TIP

Reading Effectively

Read the list of Key Terms on this page. As you read, when you find a Key Term, such as *cuticle,* be certain you can define the Key Term before you continue reading.

Plants Became Established on Land

Plants are the dominant group of organisms on land, based on weight. The kingdom Plantae is a very diverse group. Individuals range from less than 2 mm across to more than 100 m tall. Most plants are photosynthetic; they produce organic materials from inorganic materials by photosynthesis. A few plant species, like the one shown in **Figure 24-1,** live as parasites. Many parasitic plants cannot photosynthesize.

Figure 24-1 Rafflesia

Size The flowers of *Rafflesia keithii,* of Malaysia, measure almost 1 m across and weigh up to 11 kg (24 lb). The plant, which has no stems or leaves, is parasitic on the roots of grape vines.

Plants probably evolved from multicellular aquatic green algae that could not survive on land. Multicellularity enabled plants to develop features that helped them live more successfully on land. Before plants could thrive on land, they had to be able to do three things: absorb nutrients from their surroundings, prevent their bodies from drying out, and reproduce without water to transmit sperm.

Absorbing nutrients

Aquatic algae and plants take nutrients from the water around them. On land, most plants take nutrients from the soil with their roots. Although the first plants had no roots, fossils show that fungi lived on or within the underground parts of many early plants. So botanists think that fungi may have helped early land plants to get nutrients from Earth's rocky surface. Symbiotic relationships between fungi and the roots of plants are called mycorrhizae. Today, about 80 percent of all plant species form mycorrhizae.

Preventing water loss

The first plants lived at the edges of bodies of water, where drying out was not a problem. A watertight covering, which reduces water loss, made it possible for plants to live in drier habitats. This covering, called the **cuticle,** is a waxy layer that covers the nonwoody aboveground parts of most plants. But like the wax on a shiny car, the cuticle does not let oxygen or carbon dioxide pass through it.

Integrating Different Learning Styles

Logical/Mathematical	**PE**	Review items 3–5, p. 525; Performance Zone items 13, 16, and 24, pp. 542–543
Visual/Spatial	**PE**	Quick Lab, p. 521; Performance Zone items 13 and 27, pp. 542–543
	ATE	Visual Strategy, p. 521; Teaching Tips, p. 522, 523; Internet Connect, p. 523; Visual Strategies, p. 524, 525
Interpersonal	**ATE**	Reteaching, p. 525
Verbal/Linguistic	**PE**	Study Tip, p. 520; Review items 1 and 2, p. 525
	ATE	Lesson Warm-up, p. 520; Cross-Disciplinary Connection, p. 521; Teaching Tip, p. 522; Internet Connect, p. 522; Teaching Tips, p. 523, 524; Closure, p. 525

Figure 24-2 Stomata and guard cells

The surface of a leaf has numerous stomata, each of which is surrounded by a pair of guard cells.

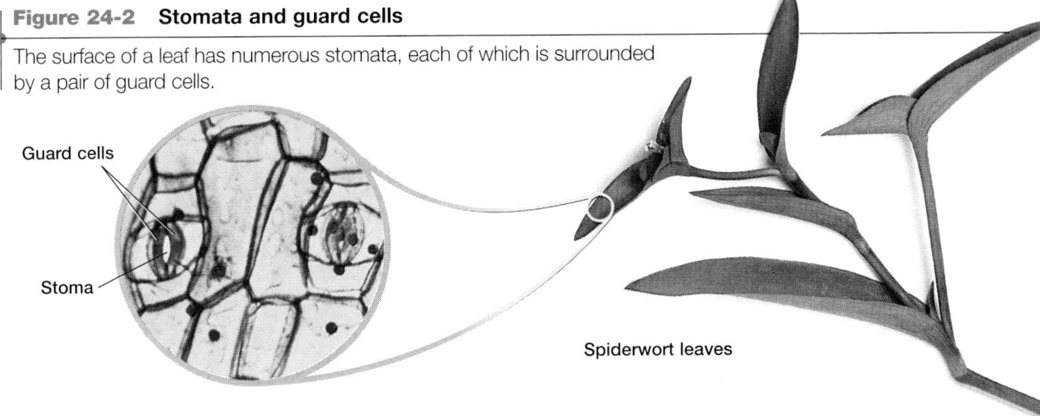

Pores called **stomata** *(STOH muh tuh)* permit plants to exchange oxygen and carbon dioxide. Stomata, which extend through the cuticle and the outer layer of cells, are found on at least some parts of most plants. A pair of specialized cells called **guard cells** border each stoma, as seen in **Figure 24-2.** Stomata open and close as the guard cells change shape.

internetconnect

SCiLINKS. NSTA

TOPIC: Plant adaptations
GO TO: www.scilinks.org
KEYWORD: HX521

Reproducing on land

Aquatic algae reproduce sexually when sperm swim through the water and fertilize eggs. The sperm of most plants, however, must be able to move without water. In most plants, sperm are enclosed in a structure that keeps them from drying out. The structures that contain sperm make up pollen. Pollen permits the sperm of most plants to be carried by wind or animals rather than by water.

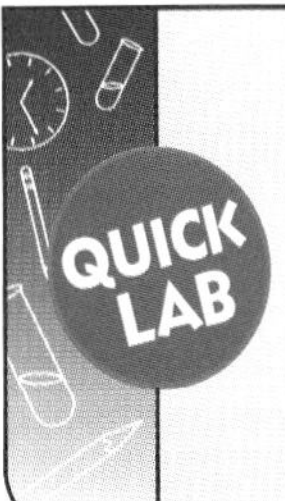

QUICK LAB

Under what conditions are stomata open?

You can use nail polish to see that a leaf has many stomata.

Materials

clear nail polish, plant kept in light, plant kept in darkness, two 4–5 cm strips of clear tape, 2 microscope slides, compound microscope

Procedure

1. Paint a thin layer of clear nail polish on a 1 × 1 cm area of a leaf on a plant kept in light. Do the same using a plant kept in darkness. Let the nail polish dry for 5 minutes.
2. Place a 4–5 cm strip of clear tape over the nail polish on each leaf. Press the tape to the nail polish to make sure the polish sticks to the tape.
3.  Carefully pull the tape off each leaf. Stick each piece of tape to a microscope slide. Label it appropriately.
4. View each slide with a microscope, first under low power and then under high power.
5. Draw and label what you see on each slide.

Analysis

1. **Describe** any differences in the stomata of the two plants.
2. **SKILL Drawing Conclusions** Which plant will lose water more quickly? Explain.

did you know?

Pine trees often have sunken stomata.

This adaptation helps them conserve water, an ability that is important for plants that grow in dry or cold climates. Many pine trees grow where soil is frozen and winds are dry.

Visual Strategy

Figure 24-2

Point out the guard cells and the stoma in the surface view and cross section of a leaf. Be sure students understand that the guard cells form the opening that is called a stoma. Ask students to point out where most of the stomata appear. *(lower surface of the leaf)* Tell students that in many species, the majority of the stomata are on the lower surface of a leaf.

CROSS-DISCIPLINARY CONNECTION

◆ Language Arts

Have students use a dictionary to find the origin of the term *stoma.* *(*Stoma *comes from the Greek word meaning "mouth.")*

QUICK LAB

Under what conditions are stomata open?

Time 20 minutes

Process Skills Analyzing results, making comparisons, drawing conclusions

Teaching Strategies

- Instruct students to put the nail polish on the lower leaf surface because there will likely be more stomata there than on the upper surface.
- Tell students that the nail polish is used to make an impression of the leaf surface that they can observe using a microscope. Tell them that the polish may also remove some of the cells on the leaf's surface.

Analysis Answers

1. The students should note that the plant kept in the light has more open stomata than the plant kept in the dark.
2. The plant kept in the light will lose water more quickly because it has more open stomata than the plant kept in the dark.

Teaching TIP ADVANCED

Tall Tendencies

Ask students to brainstorm and suggest characteristics that would enable plants to grow tall. *(a vascular system to carry water and nutrients; rigid cell walls to provide support; thickened stems)* Ask students to speculate on the advantages a plant might have because it is tall. *(Plants need sunlight for energy and to produce their organic molecules. Being tall positions leaves above competing plants, enabling the tall plants to obtain more light.)*

internet connect

Have students research vascular plants using the Web site in the Internet Connect box on this page. Have students write a report summarizing their findings. PORTFOLIO

Teaching TIP

Seed Dispersal

Bring a variety of fruits and cones containing seeds to class. Include some large, colorful fruits (apples, pears, grapes); some small fruits with wings (elm, maple) or umbrella-like appendages (dandelion); some cones with scales (pine, spruce, fir); and, if possible, a coconut. Ask students to examine the fruits and cones and explain whether they would likely be dispersed by the wind. *(small fruits with wings or other appendages, conifer seeds)* Ask students which fruits would likely be dispersed by animals. *(large, colorful fruits)* Ask students how they think the coconut fruit would be dispersed. *(It has an air pocket inside and is dispersed by floating on water.)*

Vascular Tissue, Seeds, and Flowers Made Plants Successful

biobit

How important is vascular tissue to the size of a plant?

Most nonvascular plants are no more that a few centimeters tall; some species may grow up to half a meter in length. On the other hand, many trees, the tallest vascular plants, grow to more than 100 m tall.

As plants adapted to land, they developed many features that helped in their success. For example, there was no basic difference in structure between the aboveground and underground parts of the earliest plants. Later plants evolved with roots, stems, and leaves. One of the most important changes in plants was the development of tissues that move water and other materials through the plant body.

Advantages of conducting tissue

The first plants were small. Materials were transported within their bodies by osmosis and diffusion. Today, most plants have strands of specialized cells that transport materials. These specialized cells are connected end to end like the sections in a pipeline, as shown in **Figure 24-3.** Some strands carry water and mineral nutrients from the roots to the leaves. Other strands carry organic nutrients from the leaves to wherever they are needed.

As you learned in Chapter 20, specialized cells that transport water and other materials within a plant are found in vascular tissues. The word *vascular* comes from the Latin word *vasculum,* meaning "vessel" or "duct." Vascular tissue permitted larger and more-complex plants to evolve. The larger, more-complex plants have a **vascular system,** a system of well-developed vascular tissues that distribute materials more efficiently. Today, three groups of living plants lack a vascular system. These relatively small plants that have no vascular system are called **nonvascular plants.** Plants that have a vascular system are called **vascular plants.**

internet connect

SCILINKS NSTA

TOPIC: Vascular plants
GO TO: www.scilinks.org
KEYWORD: HX522

Advantages of seeds

After vascular tissue, the seed was the next important adaptation to appear in plants. A **seed** is a structure that contains the embryo of a plant. An **embryo** is an early stage in the development of plants and animals. Most plants living today are **seed plants**—vascular plants that produce seeds. The first seed plants appeared about

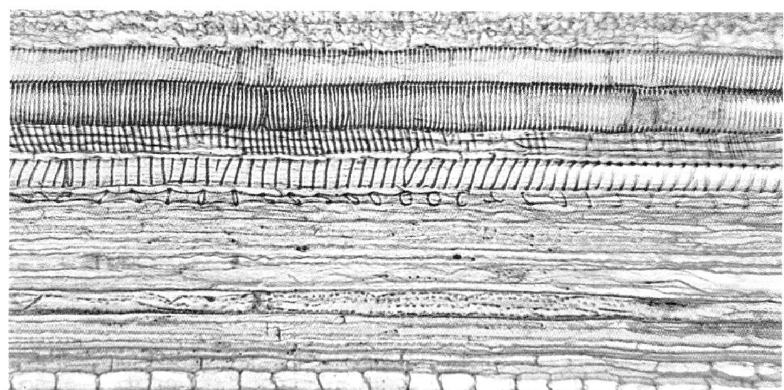

Figure 24-3 Vascular tissue. Thick-walled, tubular cells like these carry water from the tips of roots to the tips of leaves. Stacked end to end, these cells form tiny pipes called vessels.

did you know?

Seedless vascular plants formed coal.

The most common plants on Earth about 300 million years ago were seedless vascular plants. Many of these were very large trees, including a number of different tree ferns. When these plants died, their bodies became buried and only partly decomposed. The remains of their tissues formed underground coal deposits. People have dug up many of these deposits and burned the coal as a fossil fuel. The period of time in Earth's history when these plants were most abundant is called the Carboniferous period.

Figure 24-4 **Structure and function of seeds**

The structure of a seed helps it to perform its functions.

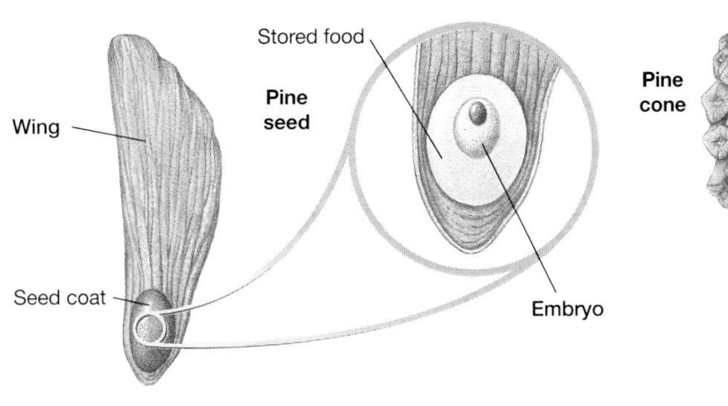

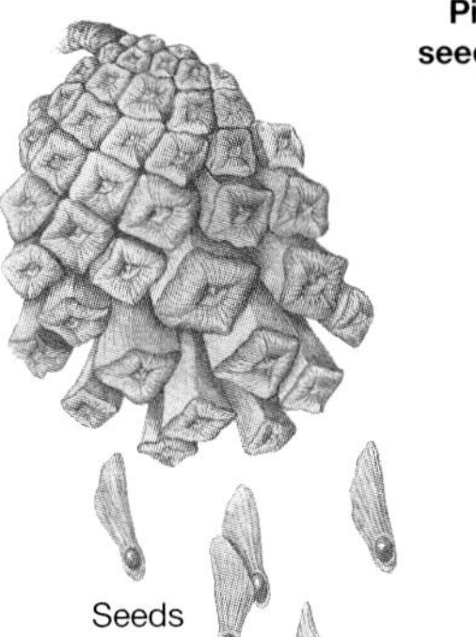

The seed coat of a pine seed covers and protects the embyro.

The stored food supply will nourish the embryo as it starts to grow.

A wing helps pine seeds disperse.

Pine seeds may not germinate for several years.

380 million years ago. Seeds offer a plant's offspring several survival advantages, which are summarized in **Figure 24-4.**

1. **Protection.** Seeds are surrounded by a protective cover called the seed coat. The seed coat protects the embryo from drying out and from mechanical injury and disease.
2. **Nourishment.** Most kinds of seeds have a supply of organic nutrients stored in them. These nutrients are a ready source of nourishment for a plant embryo as it starts to grow.
3. **Plant dispersal.** Seeds disperse (spread) the offspring of seed plants. Many seeds have appendages that help wind, water, or animals carry them away from their parent plant. Dispersal prevents competition for water, nutrients, light, and living space between parents and offspring.
4. **Delayed growth.** The embryo in a seed is in a state of suspended animation. Most seeds will not sprout until conditions are favorable, such as when moisture is present and the weather is warm. Thus, seeds make it possible for plant embryos to survive through unfavorable periods such as droughts or cold winters.

internet**connect**

TOPIC: Structure and function of seeds
GO TO: www.scilinks.org
KEYWORD: HX523

Advantage of flowers

The last important adaptation to appear as plants evolved was the **flower,** a reproductive structure that produces pollen and seeds. Flowers made plant reproduction more efficient. The pollen of the first seed plants was carried by wind. Large amounts of pollen are needed to ensure cross-pollination by wind—an inefficient system. Most plants living today are flowering plants—seed plants that produce flowers. The first flowering plants appeared more than 130 million years ago. Many flowers attract animals, such as insects, bats, and birds. As **Figure 24-5** shows, tiny pollen grains stick to animals, which carry pollen directly from one flower to another. Flowering plants that are pollinated by animals produce less pollen, and cross-pollination can occur between individuals that live far apart.

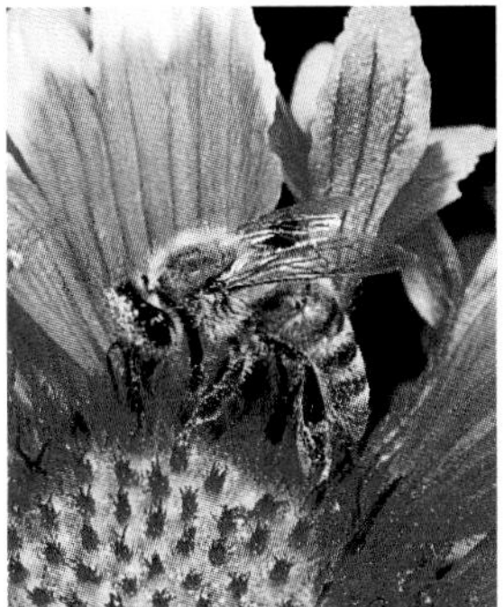

Figure 24-5 **Pollen grains.** This honeybee is covered with pollen grains containing the sperm of the plant it has just visited. The bee will transfer large quantities of the pollen to the next flower it visits.

did you know?

Orchids have evolved highly elaborate ways of attracting pollinators.

One orchid species has a petal that resembles a female wasp. Male wasps pick up some pollen while trying to mate with the modified petal. In another species, the flower holds a small pool of water. When a bee tries to collect nectar, it falls into the water and can escape only through a narrow passageway that forces the bee to brush against the pollen-bearing anthers.

Teaching TIP

Durability of Seeds

Tell students that seeds are adapted for survival in a wide variety of climates. Emphasize that seeds can survive in a dormant state if conditions are unfavorable for growth. Have students propose an explanation about why seedlings appear after a forest has been destroyed by fire. *(Seeds withstand and often benefit from the high temperature. Fire removes the accumulated leaf litter that often covers seed and in some cases opens fruits and cones to release their seeds.)*

internet**connect**

SCILINKS NSTA Have students research the structure or function of seeds using the Web site in the Internet Connect box on this page. Have students prepare posters illustrating their findings. PORTFOLIO

Teaching TIP

Observing Pollen

Have students make wet-mount slides of pollen from pine trees or observe prepared slides of pine pollen. Then have them draw and label the pollen grains and estimate their size based on the field of view. PORTFOLIO

Teaching TIP ADVANCED

Life Cycle Review

Ask students the following questions: What is the chromosome number of a gametophyte? *(n)* What is the chromosome number of a sporophyte? *(2n)* What is the name of the first sporophyte cell? *(zygote)* What process produces a zygote? *(fertilization)* What is the name of the first gametophyte cell? *(spore)* What process produces a spore? *(meiosis)*

Visual Strategy

Figure 24-7

Have students compare the relative sizes of sporophytes and gametophytes in the moss and the flowering plant. Students should notice that the moss sporophyte is smaller than the gametophyte, while the flowering-plant gametophyte is considerably smaller than the flowering-plant sporophyte.

A Change Occurred in Plant Life Cycles

In many algae, the zygote is the only diploid ($2n$) cell; it undergoes meiosis right after fertilization. So the bodies of these algae consist of haploid cells. However, meiosis was delayed in the ancestors of plants. The zygote divided by mitosis and grew into a multicelled sporophyte that was diploid and produced haploid (n) spores by meiosis. The spores grew into multicelled gametophytes that were haploid and produced gametes by mitosis. As a result, plants have a life cycle in which a haploid plant that makes gametes (a gametophyte) alternates with a diploid plant that makes spores (a sporophyte). As you learned in Chapter 7, a life cycle in which a gametophyte alternates with a sporophyte is called alternation of generations. The basic plant life cycle is shown in **Figure 24-6.**

Figure 24-6 Alternation of generations. In the life cycle of a plant, a haploid gametophyte generation alternates with a diploid sporophyte generation.

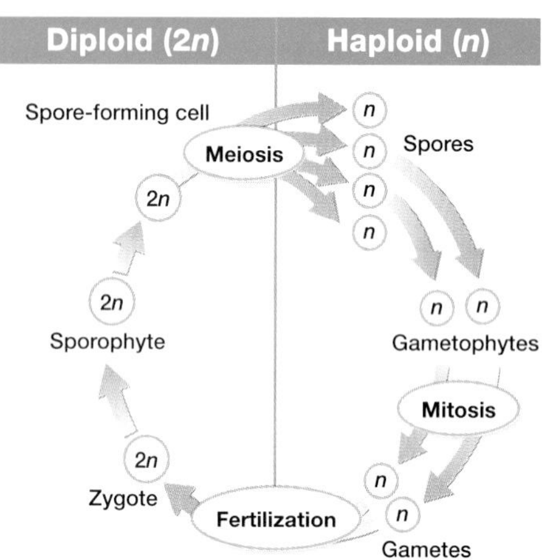

Unlike the green algae with alternation of generations, plants have gametophytes and sporophytes that look very different. In addition, the relative sizes of gametophytes and sporophytes changed as plants evolved, as **Figure 24-7** shows. In nonvascular plants, such as mosses, the gametophyte generation is dominant (most noticeable). In vascular plants, such as the flowering plants, the sporophyte generation is dominant. Like the presence of a vascular system, the relative sizes of gametophytes and sporophytes is a fundamental difference between the nonvascular plants and the vascular plants.

Figure 24-7 Nonvascular plants versus vascular plants

The relative size of the gametophyte and sporophyte is a fundamental difference between nonvascular plants and vascular plants.

Sphagnum moss

A sporophyte grows atop a gametophyte.

The gametophytes form inside a flower.

Tomato sporophytes

did you know?

Allergies to pollen are actually reactions to proteins.

The coats of the pollen grains contain proteins. Different pollens cause allergies in different people because each type of pollen has a different type of protein in the coats of its pollen grains.

The vascular-plant sporophyte

As the size of plant sporophytes increased, so did the complexity of their structure. An increase in size enables cell specialization and, therefore, the development of complex tissues and specialized structures. The following features characterize the sporophytes of most vascular plants.

Vascular system Larger bodies require an efficient vascular system for transporting materials internally. The sporophytes of vascular plants have a vascular system with two types of vascular tissue. Each type of vascular tissue contains strands of long, tubelike cells that are lined up end to end like sections of pipe. These strands of cells transport water and nutrients within a plant's body. Relatively soft-walled cells transport organic nutrients in a kind of tissue called **phloem** *(FLOH uhm)*. Hard-walled cells transport water and mineral nutrients in a kind of tissue called **xylem** *(ZEYE luhm)*. The walls of the water-conducting cells in xylem are thickened, which helps support the plant body. This makes it possible for vascular plants to grow to great heights.

Distinctive body form Nearly all plants have a body that consists of a vertical shaft from which specialized structures branch, like the plant shown in **Figure 24-8.** The part of a plant's body that grows mostly upward is called the **shoot.** In most plants, the part of the body that grows downward is called the **root.** Zones of actively dividing plant cells, called **meristems** *(MEHR uh stehmz)*, produce plant growth. The vertical body form results as new cells are made at the tips of the plant body. As vascular plants became better adapted to life on land, most developed the familiar plant structures—roots, stems, and leaves—which are complex structures made of several different types of specialized tissues.

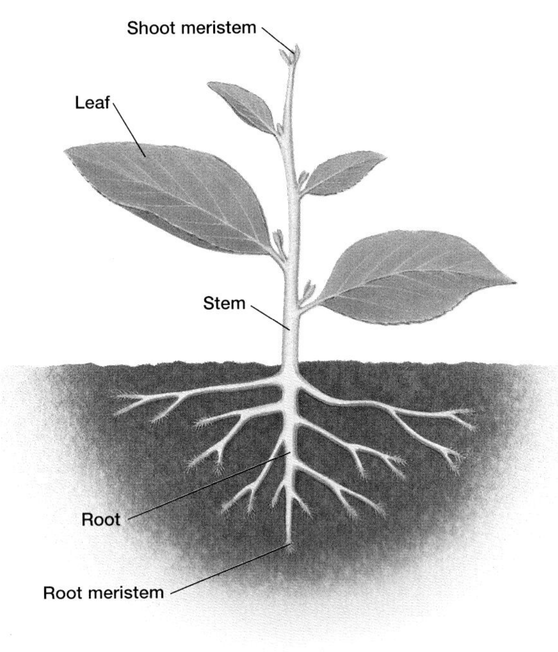

Figure 24-8 Vascular plant sporophyte. The sporophytes of the vast majority of vascular plants have an aboveground shoot with stems and leaves and an underground root. Growth occurs in regions called meristems.

Review SECTION 24-1

1. **Summarize** how plants are adapted to living successfully on land.
2. **Describe** two basic differences between nonvascular plants and vascular plants.
3. **SKILL Relating Concepts** How have seeds and flowers made plants more successful on land?
4. **SKILL Summarizing Information** What is the basic structure of the body of a vascular plant sporophyte?

Critical Thinking

5. **SKILL Evaluating Conclusions** Why do you think vascular plants are more successful as land plants than are nonvascular plants?

Review SECTION 24-1 ANSWERS

1. A waxy cuticle, vascular tissue, drought-resistant reproductive structures, and seed development are adaptations that help plants to live on land.
2. Vascular plants have a vascular system that transports water, minerals, and sugars throughout the plant; nonvascular plants do not. Nonvascular plants have a dominant gametophyte; vascular plants have a dominant sporophyte.
3. Seeds and flowers have enabled plants to reproduce and disperse offspring without the aid of water. Flowers have enabled plants to produce offspring from fewer numbers of gametes because fertilization is often linked to animals.
4. A vascular plant sporophyte is diploid and has roots, one or more stems, and leaves. All of these structures contain vascular tissues.
5. Vascular plants are more successful on land because many important parts of their life cycles are less dependent on water than are parts of the life cycles of nonvascular plants.

Visual Strategy

Figure 24-8

Point out that the aboveground portion of a plant, consisting of stems and leaves, is called the shoot. Roots are the underground, water-absorbing portion of a plant's body. Ask students if they think it is possible for a part of the shoot to grow underground. *(Yes,* Rhynia *is an example of a plant that has underground stems.)* Ask students to speculate how biologists distinguish underground stems from roots. *(Their structures are different.)*

3 Assessment Options

Closure

Relating Concepts

Ask students to think about some of the fundamental differences between plants and humans. Humans are land-living organisms that have to be able to move around to get food and meet their other needs for survival. Our skeletal system gives us a rigid frame so that with the help of muscles, we can move about quite efficiently. Plants do not have to move around to get food because they produce their own organic molecules during photosynthesis. However, they do need to be able to stay upright and capture light from the sun. Plants get structural support mainly from the thick-walled cells in their vascular tissue.

Review

Assign Review—Section 24-1.

Reteaching — BASIC

Have students work in pairs to review the key terms for this section. Encourage each pair of students to take turns asking and answering questions about the definitions of the terms.

Chapter 24

1 Prepare

Scheduling

- **Block:** About 45 minutes will be needed to complete this section.
- **Traditional:** About one class period will be needed to complete this section.
- **Planning:** See the Planning Guide on pp. 517A and 517B for lecture, classwork, and assignment options for Section 24-2. Lesson plans for Section 24-2 can be found in a lesson cycle format in ***Biology: Principles and Explorations*** **One-Stop Planner CD-ROM with Test Generator.**

Resource Link

Assign **Active Reading Worksheet Section 24-2** to help students comprehend the key concepts and visuals in this lesson.

2 Teach

Lesson Warm-up

Bring in living specimens or parts of specimens of a variety of plants. The collection should include as many of the following as possible: moss; liverwort; fern; horsetail; conifer; cycad; ginkgo; flowering monocot (corn, iris, a grass), and flowering dicot (rose, tomato, petunia, chrysanthemum). Ask students to describe the characteristics of each plant and note any major differences they observe. *(size; presence or absence of spores, flowers, or seeds; leaf shape and vein pattern; stiffness of stems)*

Section 24-2 Kinds of Plants

Objectives

- **Describe** the key features of the four major groups of plants. (pp. 526–533)
- **Classify** plants into one of the 12 phyla of living plants. (pp. 527–533)

Key Terms

rhizoid
rhizome
frond
cone
gymnosperm
angiosperm
fruit
endosperm
monocot
dicot

Nonvascular Plants Lack a Vascular System

The brilliant green carpet of mosses you see in **Figure 24-9** is made up of thousands of individual plants. Living carpets of mosses are often found near streams, coastlines, and other moist places. But these tough little plants also live in some surprising places, such as cracks in city sidewalks and rocky mountaintops—any place where a little moisture can collect. The moisture makes it possible for these nonvascular plants to survive.

Figure 24-9
A carpet of mosses

How many? Mosses grow in tightly packed mats that may contain dozens of plants per square inch.

Nonvascular plants do not have a vascular system for transporting water and other nutrients within their bodies. This means that all nonvascular plants lack true roots, stems, and leaves, although most have structures that resemble them. True roots, stems, and leaves are complex structures that contain vascular tissues.

Key features of nonvascular plants

In addition to the lack of true roots, stems, and leaves, nonvascular plants share several other features. These features are key adaptations that have enabled them to survive on land.

Small size All nonvascular plants are small and relatively simple. Water and other nutrients are transported within their bodies mostly by osmosis and diffusion, which move materials short distances. This greatly limits the size of a nonvascular plant's body.

Larger gametophyte The gametophytes of nonvascular plants are larger and more noticeable than the sporophytes. Hairlike projections called **rhizoids** anchor the gametophytes to the surfaces on which they grow. The smaller, usually nongreen sporophytes grow on the gametophytes and depend on them for nutrients.

Require water for sexual reproduction Nonvascular plants must be covered by a film of water in order for fertilization to occur. Eggs and sperm form in separate structures, which are often on separate plants. The gametophytes grow in mats of tightly packed individuals. When these mats are covered by a film of water, the sperm can easily swim to neighboring individuals and fertilize their eggs.

Study TIP

Organizing Information
Make a table to organize information about the kinds of plants. Across the top, write the headings *Phylum, Major Group, Characteristics,* and *Examples.* Add information to the table as you read Section 24-2.

Integrating Different Learning Styles

Logical/Mathematical	**PE**	Study Tip, p. 526; Data Lab, p. 530; Review items 2–4, p. 533; Performance Zone items 17–20, p. 543
	ATE	Cross-Disciplinary Connection, p. 527; Teaching Tips, p. 530, 532
Visual/Spatial	**PE**	Real Life, p. 532; Performance Zone items 1 and 27, pp. 542–543
	ATE	Lesson Warm-up, p. 526; Teaching Tip, p. 527; Visual Strategy, p. 528; Demonstration, p. 529; Demonstration, p. 532
Verbal/Linguistic	**PE**	Biobit, p. 527; Performance Zone item 28, p. 543
	ATE	Internet Connect, p. 528; Biology in Your Community, p. 528; Career and Resource Link, p. 531; Closure and Reteaching, p. 533

Kinds of nonvascular plants

The nonvascular plants include the mosses and the two simplest groups of plants—liverworts and hornworts. Examples of these plants are shown in **Figure 24-10.**

Mosses The mosses (phylum Bryophyta) are the most familiar nonvascular plants. The "leafy" green plants that you recognize as mosses are gametophytes. Moss sporophytes, which are not green, grow from the tip of a gametophyte. Each sporophyte consists of a bare stalk topped by a spore capsule. Most mosses have a cuticle, stomata, and some simple conducting cells. The walls of the water-conducting cells in mosses are not thickened, as they are in a vascular plant. Mosses never get very large because their water-conducting cells carry water only short distances.

Liverworts Like the mosses, liverworts (phylum Hepatophyta) grow in mats of many individuals. Liverworts have no conducting cells, no cuticle, and no stomata. Their gametophytes are green. In some species, such as the common liverwort shown in Figure 24-10, the gametophytes of liverworts are flattened and have lobes. Structures that resemble stems and leaves make up the gametophytes of most liverworts, like those of the mosses. The sporophytes of liverworts are very small and consist of a short stalk topped by a spore capsule.

Hornworts The hornworts (phylum Anthocerophyta) are a small group of nonvascular plants that, like the liverworts, completely lack conducting cells. Hornworts have both stomata and a cuticle. The gametophyte of a hornwort is green and flattened. Green hornlike sporophytes grow upward from the gametophytes.

biobit

How did liverworts get their name?

The name *liverwort* dates back to the Middle Ages, when it was thought that plants resembling certain body parts might contain substances that could cure diseases of those body parts. The word *wort* meant "herb" in Old English.

internetconnect

TOPIC: Nonvascular plants
GO TO: www.scilinks.org
KEYWORD: HX527

Figure 24-10 **Nonvascular plants**

There are three phyla of nonvascular plants.

Polytrichum, a moss (Phylum Bryophyta)

Marchantia, a liverwort (Phylum Hepatophyta)

Anthoceros, a hornwort (Phylum Anthocerophyta)

Overcoming Misconceptions

Mistaken Identity

A variety of plants and plantlike organisms are mistakenly called mosses. Examples include Irish moss, Spanish moss, and reindeer moss. Show photos of these examples, and reinforce the concept that common names can be very misleading. *(Irish moss is a red algae, Spanish moss is a flowering vascular plant, and reindeer moss is a lichen.)*

Teaching TIP

Mosses

Have students observe different types of mosses using a stereomicroscope. They should find both gametophyte and sporophyte stages. Point out that the structures resembling leaves on the moss gametophyte are not considered to be true leaves. Tell students that a leaf is a complex structure that consists of several specialized tissues. Ask them whether they think the sporophyte and the gametophyte are photosynthetic. *(The sporophyte is not photosynthetic; the gametophyte is photosynthetic.)* Tell students that the sporophyte is dependent upon the gametophyte for nutrients for its survival.

Cross-Disciplinary Connection

◆ Mathematics

Have students determine the mass of a small sample of dry sphagnum moss, which is available at garden centers. Have them place this sample in a large beaker of water for 10–15 minutes. Tell students that 1 kg of dry moss will take up approximately 25 kg of water. Have students make a prediction about the mass of the moss after soaking in water. Remove the wet moss from the beaker, and have students determine its mass. Have students compare their findings with their predictions. *(The mass of the wet moss should be about 25 times greater than the mass of the dry moss.)*

Cross-Disciplinary Connection

◆ Health

Tell students that moss has been used as an antiseptic. In emergency situations during World War I, moss was used to treat soldiers' wounds. The absorbency and natural acidity of moss make it an antiseptic.

internetconnect

SCILINKS NSTA

Have students research the importance of seedless vascular plants using the Web site in the Internet Connect box on this page. Have students write a report summarizing their findings.

PORTFOLIO

Visual Strategy

Figure 24-11

Ask students how Cooksonia resembles plants that are alive today. *(The upright growth habit of these plants and the location of reproductive structures at the tips of stems resemble modern plants.)* Ask students how they differ from the plants they are most familiar with. *(These primitive plants have no leaves, roots, flowers, seeds, or fruits.)*

Seedless Vascular Plants Do Not Make Seeds

internetconnect

SCILINKS NSTA

TOPIC: Seedless vascular plants
GO TO: www.scilinks.org
KEYWORD: HX528

Vascular plants that do not produce seeds are called seedless vascular plants. An early seedless vascular plant, *Cooksonia,* is shown in **Figure 24-11.** The sporophytes of these ancient plants had branched, leafless stems that were only a few centimeters long. Spore-forming sporangia were located at the tips of the stems. *Rhynia,* another early seedless vascular plant, also had horizontal underground stems, or **rhizomes.**

Figure 24-11 An early vascular plant

Cooksonia is the oldest known vascular plant.

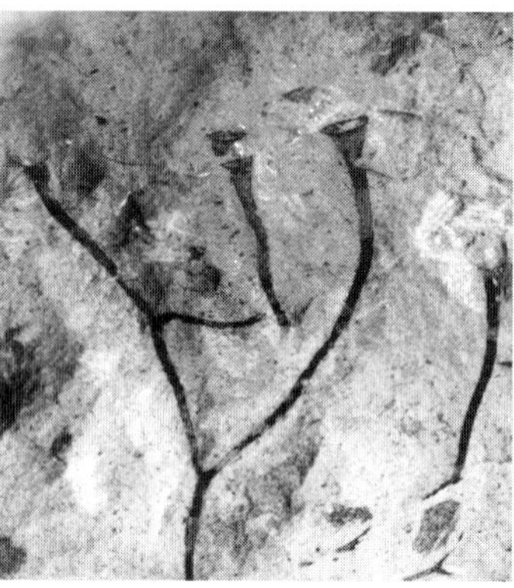

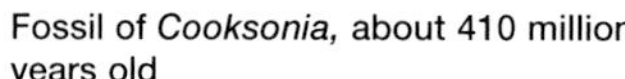

Fossil of *Cooksonia,* about 410 million years old

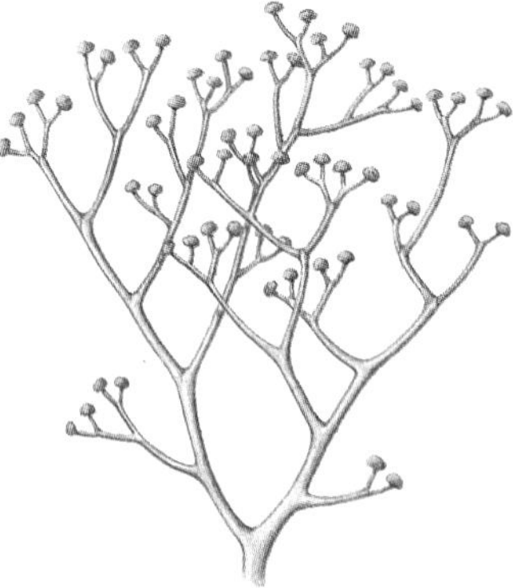

Artist's impression of how *Cooksonia* might have looked

Key features of seedless vascular plants

Seedless vascular plants are much larger and more complex than the nonvascular plants. Other key features enabled them to spread and adapt to drier habitats on land.

Vascular system Seedless vascular plants have a vascular system with both xylem and phloem. The water-conducting cells in the xylem are reinforced with lignin, a major part of wood. Because of their vascular system, seedless vascular plants grow much larger than nonvascular plants and also develop true roots, stems, and leaves.

Larger sporophyte The sporophytes of the seedless vascular plants are larger than the gametophytes. Their larger size makes it easier for the wind to carry away spores, which makes dispersal more efficient. The much smaller gametophytes of most seedless vascular plants develop on or below the surface of soil. As in the nonvascular plants, water is needed for fertilization. When there is enough water on or in the soil, the sperm swim to eggs and fertilize them.

Drought-resistant spores The spores of the seedless vascular plants have thickened walls that are resistant to drying. Such spores make it possible for a plant to live in drier habitats.

biobit

Are there any ways that people use plant spores?

The spores of a common club moss, *Lycopodium,* form a powder that has several uses. Herbalists use the spores to make a powder for treating skin disorders. Pharmaceutical companies use the spores to coat pills so they do not stick together. The spores are also used to make photographic flash powder.

Biology in Your Community

Local Habitats

Take students on a field trip to a park or other local area that has a variety of habitats. Ask the students to identify the habitats in which nonvascular plants would most likely be found. *(moist habitats)* Ask the students to identify the habitats in which vascular plants would most likely be found. *(any habitat)* Ask students to try to identify one nonvascular plant and one vascular plant. Have them explain what characteristics they used to identify a plant's category. *(nonvascular: moist habitat, small size, no seeds or flowers; vascular: any habitat, wide range of sizes, leaves with veins indicating vascular tissue)*

Kinds of seedless vascular plants

The seedless vascular plants include ferns and three other groups of plants known as fern allies—whisk ferns, club mosses, and horsetails. Like the ferns, the fern allies usually grow in moist places.

Ferns The ferns (phylum Pterophyta) are the most common and most familiar seedless vascular plants. Ferns grow throughout the world, but they are most abundant in the tropics. The plants you recognize as ferns are sporophytes. Most fern sporophytes have a rhizome that is anchored by roots and leaves called **fronds.** The coiled young leaves of a fern, shown in **Figure 24-12,** are called fiddleheads. Spores are produced in sporangia that grow in clumps on the lower side of fronds. The gametophytes of ferns are flattened, heart-shaped green plants that are usually less than 1 cm (0.5 in.) across.

Figure 24-12 **A fern.** This sword fern sporophyte has many fronds and fiddleheads. The inset shows a gametophyte at twice its actual size.

Club mosses Unlike true mosses, the club mosses (phylum Lycophyta), have roots, stems, and leaves. Their leafy green stems branch from an underground rhizome. Spores develop in sporangia that form on specialized leaves. In some species, such as the one seen in **Figure 24-13,** clusters of nongreen spore-bearing leaves form a structure called a **cone.**

Horsetails The horsetails (phylum Sphenophyta) also have roots, stems, and leaves. The vertical stems of horsetails, which grow from a rhizome, are hollow and have joints. Whorls of scalelike leaves grow at the joints. Spores form in cones located at the tips of stems.

Whisk ferns The whisk ferns (phylum Psilotophyta) probably most closely resemble the earliest vascular plants. Whisk ferns have highly branched stems and no leaves or roots. They produce spores in sporangia that form at the tips of short branches.

Figure 24-13 **Fern allies**

In addition to ferns, there are three other living phyla of seedless vascular plants that are known as fern allies.

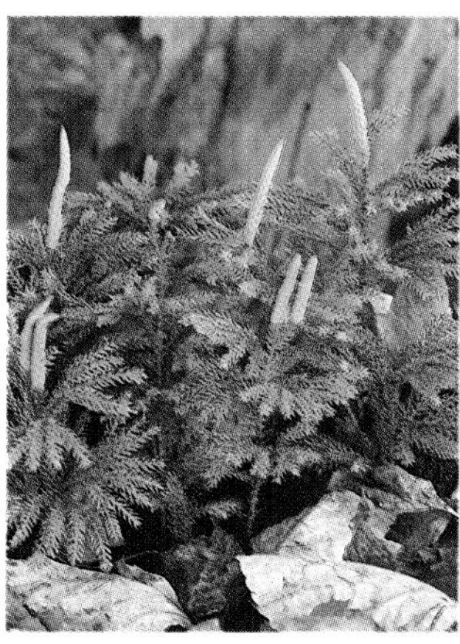

Lycopodium, a club moss (Phylum Lycophyta)

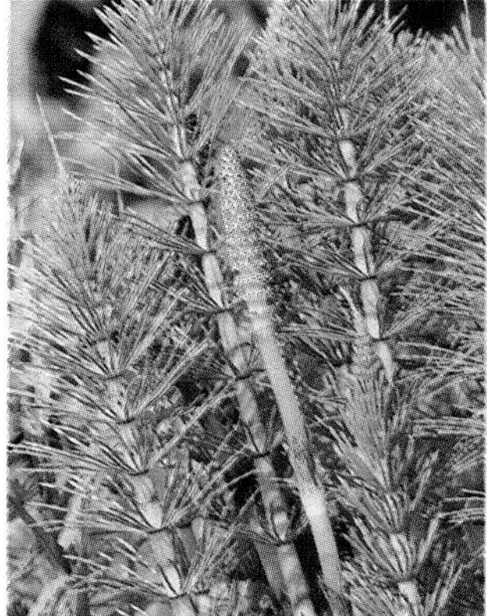

Equisetum, a horsetail (Phylum Sphenophyta)

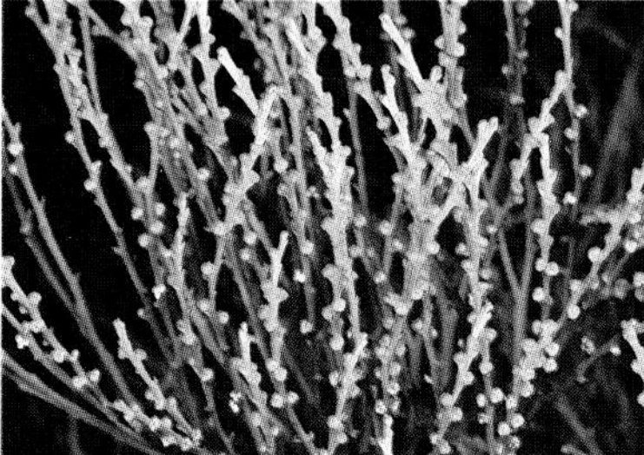

Psilotum, a whisk fern (Phylum Psilotophyta)

did you know?

Introduced animals and plants can replace native species.

Pigs were introduced on the island of Hawaii by settlers as a source of food, but these animals wiped out native vegetation in the area now called Hawaii Volcanoes National Park. Additionally, the settlers brough their own plants, which eventually took over areas of island plants. The people of Hawaii are now trying to reclaim their native plants. Tree ferns are one species that is now re-establishing in the area.

Demonstration

Ferns

Bring examples of several different kinds of ferns to class, and have students examine them. Also bring some pictures of tree ferns, which are native to many tropical regions. Have students note the shapes of the leaves and any sori present on the lower surfaces of the leaves. Point out that there are more than 12,000 species of ferns on Earth today. Among these species we find a tremendous diversity of leaf form (from simple, rounded leaves to lacy fronds with hundreds of leaflets); plant shape (mosslike to tree form); and plant size (about 2.5 cm to about 20 m tall).

Teaching TIP

Growing a Fern

Show students the clusters of sporangia (sori) on the lower surface of a fern frond. Let them remove a sorus and crush it on white paper using the eraser end of a pencil. Next, have them sprinkle some spores on a moist peat pellet and place the peat pellet in a sealed jar. The first structures they will observe will be the little green heart-shaped gametophytes.

Teaching TIP ADVANCED

Identifying Christmas Trees

This activity works well during the winter holiday season when many different conifers are available, but it can also be done at other times of the year. Have students identify several types of conifers using the following key.

1a Needles not in fascicles ...... Go to 2
1b Needles in fascicles .... Go to 3
2a Needles square .... Spruce
2b Needles not square ...... Fir
3a Needles in groups of 3 ...... Yellow pine
3b Needles not in groups of 3 ..... Go to 4
4a Needles in groups of 5 ...... White pine
4b Needles in groups of 2 ..... Black pine

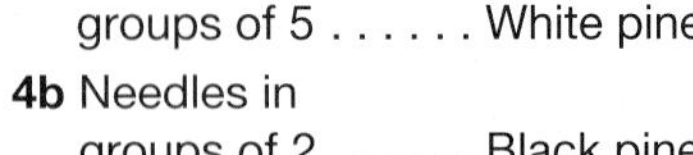

Analyzing the effect of climate on plants

Time 20 minutes

Process Skills Analyzing data, interpreting graphs, drawing conclusions

Teaching Strategies

- Have students examine the labels and units on the axes of the graph. Make sure that students understand that the sequence of letters along the *x*-axis represents the 12 months of the year.
- Point out that the bar graph represents precipitation data and that the line graph represents temperature data.

Analysis Answers

1. Anchorage has a cold, dry climate.
2. Other parts of the taiga should have similar climates because they are at similar latitudes.
3. Yes, the fact that the taiga is not found in other regions with different types of climates suggests that climate is a limiting factor in the growth of the conifers of the taiga.

Figure 24-14 Juniper pollen. This juniper, a type of gymnosperm, releases clouds of pollen in late fall or early winter.

Gymnosperms Are Seed Plants that Produce Cones

Gymnosperms *(JIHM noh spurmz)* are seed plants whose seeds do not develop within a sealed container (a fruit). The word *gymnosperm* comes from the Greek words *gymnos,* meaning "naked," and *sperma,* meaning "seed."

Key features of gymnosperms

Gymnosperms are among the most successful groups of plants. The following key features have made them successful on land.

Seeds All gymnosperms produce seeds. Seeds protect plant embryos, provide them with nutrients, and permit them to survive long periods of unfavorable conditions. In some plants, seeds also disperse new plants far from their parents.

Greatly reduced gametophytes All seed plants produce very tiny gametophytes of two types—male and female. The gametophytes form within the tissues of the sporophytes. Grains of pollen are male gametophytes. Female gametophytes form within structures that become seeds. In all but one species of gymnosperm, male and female gametophytes develop in male and female cones, respectively.

Wind pollination The sperm of gymnosperms do not swim through water to reach and fertilize eggs. Instead, the sperm are carried to the structures that contain eggs by pollen, which can drift on the wind, as seen in Figure **24-14.** Wind pollination makes sexual reproduction possible even when conditions are very dry.

Analyzing the effect of climate on plants

Background

The map at right shows the taiga of North America. The taiga is a vast forest of conifers, a type of gymnosperm. The graph shows average annual temperature and precipitation data for Anchorage, Alaska, which is located at the western edge of the taiga. Use the map and graph to answer the following questions.

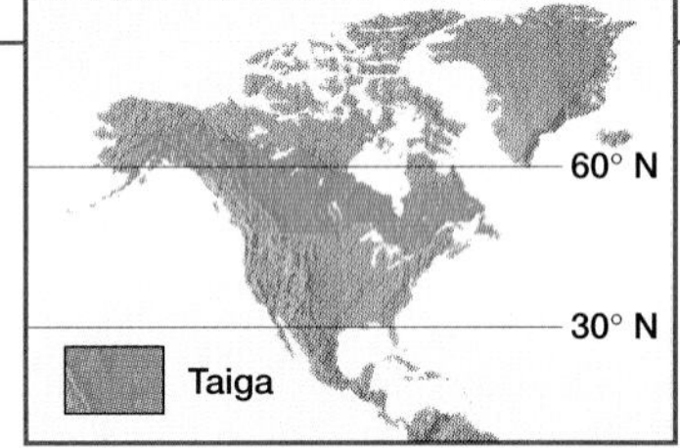

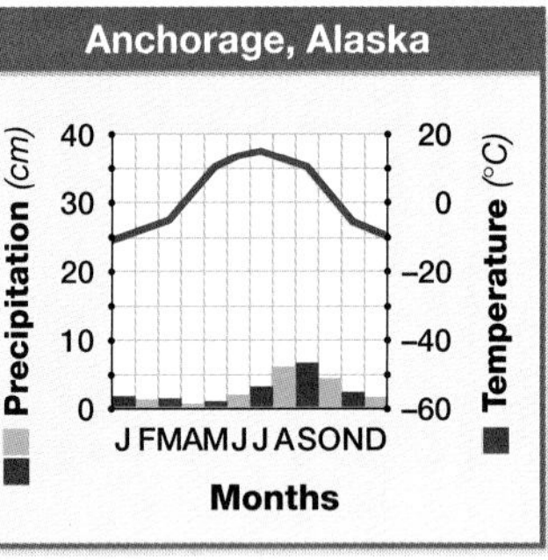

Analysis

1. **Describe** the climate of Anchorage, Alaska.
2. **SKILL Predicting Patterns** What type of climate is found in other parts of the taiga?
3. **SKILL Drawing Conclusions** Does climate appear to be an important factor in where the conifers of the taiga grow? Explain.

MULTICULTURAL PERSPECTIVE

Native Americans and Gymnosperms

Native Americans used gymnosperms in several ways. Using pine needles, they made a tea that was rich in vitamin C and helped them prevent scurvy. They also chewed resin from spruce trees as a type of chewing gum.

Kinds of gymnosperms

Four groups of living seed plants are referred to as gymnosperms—conifers, cycads, ginkgo, and gnetophytes. Examples of each of these four groups are shown in **Figure 24-15** and **Figure 24-16.**

Conifers The conifers (phylum Coniferophyta) are the most familiar, and most successful, gymnosperms. Conifers have leaves that are either needle-like or reduced to tiny scales, as Figure 24-15 shows. These leaves are an adaptation for limiting water loss. Some of the tallest living plants, the redwoods of coastal California and Oregon, are conifers. The oldest trees in the world are bristlecone pines, another species of conifer that grows in the Rocky Mountains. Some bristlecone pines are about 5,000 years old. Vast forests of conifers grow in cool, dry regions of the world.

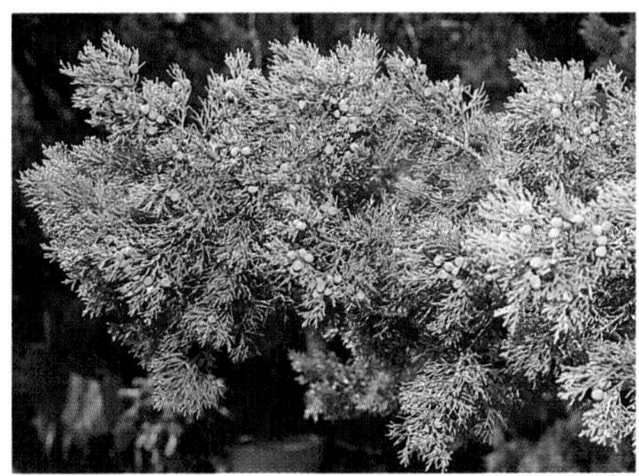

Figure 24-15 Juniper leaves and cones. The tiny scalelike leaves of junipers are an adaptation that limits water loss. The blue, berrylike structures are the female cones of this juniper.

Cycads The cycads (phylum Cycadophyta) have short stems and palmlike leaves. Cones that produce pollen and those that produce seeds develop on different plants. Cycads are widespread throughout the tropics.

Ginkgo The only living species of ginkgo (phylum Ginkgophyta), or maidenhair tree, has fan-shaped leaves that resemble the leaves of the maidenhair fern. The male and female gametophytes of ginkgo develop on separate trees. Ginkgo seeds do not develop within a cone.

Gnetophytes The gnetophytes (phylum Gnetophyta) are a diverse group of trees, shrubs, and vines that produce pollen and seeds in cones that resemble flowers. One type of gnetophyte, *Ephedra,* is common in the western United States.

Figure 24-16 Other gymnosperms

In addition to conifers, there are three other living phyla of gymnosperms.

Encephalartos, a cycad (Phylum Cycadophyta)

Leaves and seeds of *Ginkgo* (Phylum Ginkgophyta)

Ephedra (Mormon tea), a gnetophyte (Phylum Gnetophyta)

did you know?

Some gymnosperms are considered living fossils.

The ginkgo, or maidenhair tree, is considered a "living fossil" because it was once thought to be extinct and because specimens alive today very closely resemble fossil specimens. The ginkgo is the only broadleaf gymnosperm still surviving on Earth today. In 1994, another gymnosperm thought to be extinct since the Cretaceous period, 100 million years ago, was discovered in a rain forest in the Wollemi National Park near Sydney, Australia. A small population of Wollemi Pines, a prehistoric type of conifer belonging to a family that was thought to have become extinct along with the dinosaurs, was found by a park ranger. The once widespread distribution of this ancient family included the Northern Hemisphere.

CAREER — ADVANCED

Botanist

A botanist studies plants and their environments. Most botanists specialize in a particular area. For example, plant physiologists study the structural parts of plants and how they function; plant breeders study the ways in which plants reproduce and how their traits are inherited; horticulturists work with crops such as flowers, fruits, and vegetables; agronomists work with crops such as grains and dry beans; and foresters study all aspects of tree growth. Have students research a field of botany that interests them and prepare a written report on career opportunities in that field. PORTFOLIO

Resource Link

- Assign **Occupational Applications: Botanist** from the Holt BioSources Teaching Resources CD-ROM.

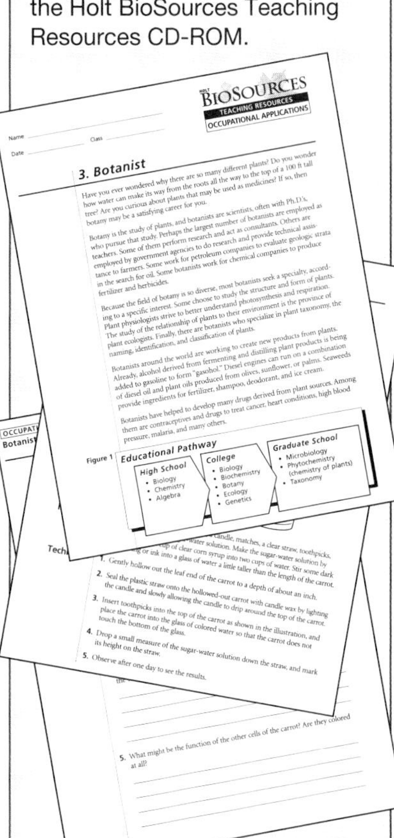

HOLT BIOSOURCES TEACHING RESOURCES
OCCUPATIONAL APPLICATIONS

Name ______ Class ______
Date ______

3. Botanist

Have you ever wondered why there are so many different plants? Do you wonder how water can make its way from the roots all the way to the top of a 100 ft tall tree? Are you curious about plants that may be used as medicines? If so, then botany may be a satisfying career for you.

Botany is the study of plants, and botanists are scientists, often with Ph.D.'s, who pursue that study. Perhaps the largest number of botanists are employed as teachers. Some of them perform research and act as consultants. Others are employed by government agencies to do research and provide technical assistance to farmers. Some work for petroleum companies to evaluate geologic strata in the search for oil. Some botanists work for chemical companies to produce fertilizer and herbicides.

Because the field of botany is so diverse, most botanists seek a specialty, according to a specific interest. Some choose to study the structure and form of plants. Plant physiologists strive to better understand photosynthesis and respiration. The study of the relationship of plants to their environment is the province of plant ecologists. Finally, there are botanists who specialize in plant taxonomy, the naming, identification, and classification of plants.

Botanists around the world are working to create new products from plants. Already, alcohol derived from fermenting and distilling plant products is being added to gasoline to form "gasohol." Diesel engines can run on a combination of diesel oil and plant oils produced from olives, sunflower, or palms. Seaweeds provide ingredients for fertilizer, shampoo, deodorant, and ice cream.

Botanists have helped to develop many drugs derived from plant sources. Among them are contraceptives and drugs to treat cancer, heart conditions, high blood pressure, malaria, and many others.

Figure 1 **Educational Pathway**

High School	College	Graduate School
• Biology • Chemistry • Algebra	• Biology • Biochemistry • Botany • Ecology • Genetics	• Microbiology • Phytochemistry (chemistry of plants) • Taxonomy

OCCUPAT...
Botanist

Tech...

...candle, matches, a clear straw, toothpicks, ...water solution. Make the sugar-water solution by ...up of clear corn syrup into two cups of water. Stir some dark ...g or ink into a glass of water a little taller than the length of the carrot.

1. Gently hollow out the leaf end of the carrot to a depth of about an inch.
2. Seal the plastic straw onto the hollowed-out carrot with candle wax by lighting the candle and slowly allowing the candle to drip around the top of the carrot.
3. Insert toothpicks into the top of the carrot as shown in the illustration, and place the carrot into the glass of colored water so that the carrot does not touch the bottom of the glass.
4. Drop a small measure of the sugar-water solution down the straw, and mark its height on the straw.
5. Observe after one day to see the results.

5. What might be the function of the other cells of the carrot? Are they colored at all?

Chapter 24

REAL LIFE

Answer

Answers will vary widely, depending on the region where you live. Plants that have flowers are angiosperms, including all broadleaf trees and shrubs. Conifers, mosses, and ferns are commonly seen plants that are not angiosperms.

Demonstration

Comparing Cones and Flowers BASIC

Bring several examples of cones (pine, spruce, fir, yew) and flowers (iris, daylily, gladiolus) to class, and have students examine and compare each of these types of reproductive structures. *(Both produce seeds; flowers have petals, stamens, and pistils, and cones do not.)*

Making Biology Relevant

Health

Scientists have found that chemicals from plants prevent or delay the growth of tumors in laboratory animals. Some of these new hopes for cancer prevention include members of the Brassicaceae family, such as broccoli, cabbage, cauliflower, and Brussels sprouts, which contain sulfur compounds found to prevent colon and lung cancer. Other plants that show promise in preventing cancer include garlic, citrus fruits, tea, and soybeans.

Teaching TIP

Charting Information

Have students make a Graphic Organizer similar to the one at the bottom of this page to compare and contrast the characteristics of plants in the major groups of plants studied in this section. Ask students to illustrate their graphic organizer with a drawing of a plant in each of the major groups. PORTFOLIO

REAL LIFE

Nearly all of the plants you see and depend on daily are angiosperms. For example, virtually all of your food comes directly or indirectly from angiosperms. About 90 percent of all living plants—more than 250,000 species—are angiosperms.

Finding Information *Use a field guide to identify the plants that grow in the area where you live. Make a list of those that are angiosperms and a list of those that are not.*

Angiosperms Are Seed Plants that Produce Flowers

Most seed plants are flowering plants, or **angiosperms** *(AN jee oh spurmz),* which produce seeds that develop while they are enclosed within a specialized structure, as seen in **Figure 24-17.** The word *angiosperm* comes from the Greek words *angeion,* meaning "case," and *sperma,* meaning "seed."

Key features of angiosperms

Angiosperms are the most recent group of plants to evolve. The following key features made them the most successful group of plants.

Flowers The male and female gametophytes of angiosperms develop within flowers, which promote pollination and fertilization more efficiently than do cones. Some flowers, such as roses, are brightly colored or have strong scents, which attracts insects and other animals that carry pollen and increases the likelihood of cross-pollination. Other flowers, such as garden peas, are adapted for self-pollination, which often occurs before the flowers open. The flowers of many angiosperms, such as oaks and grasses, have small greenish flowers that are adapted for wind pollination. The female reproductive part of a flower also provides a pathway that enables sperm to reach and fertilize eggs without swimming through water.

Fruits The structures in which the seeds of angiosperms develop are called **fruits.** Although fruits provide some protection for developing seeds, their primary function is to promote seed dispersal. The angiosperms produce many different types of fruits, which develop from parts of flowers. Many fruits are eaten by animals, which help disperse the seeds within. Others have structures that help them float on wind or water. Some even forcefully eject their seeds, flinging them away from the parent plant.

Endosperm The seeds of angiosperms have a supply of stored food called **endosperm** at some time during their development. In many angiosperms, the endosperm is absorbed by the embryo before the seeds mature.

Figure 24-17 Seeds in a fruit. These melons, which contain seeds, are the fruits of an angiosperm.

Graphic Organizer

*Use this graphic organizer with **Teaching Tip: Charting Information** on this page.*

Phylum	Vascular system	Roots	Seeds	Flowers	Dominant gametophyte	Dominant sporophyte
Liverworts					✔	
Mosses					✔	
Ferns	✔	✔				✔
Conifers	✔	✔	✔			✔
Flowering plants	✔	✔	✔	✔		✔

Kinds of angiosperms

The angiosperms are a very diverse group that contains most species of living plants. Botanists divide the angiosperms into two subgroups—monocots and dicots. **Figure 24-18** shows an example of each group. The **monocots** (class Monocotyledonae) are flowering plants that produce seeds with one seed leaf (cotyledon). Most monocots also produce flowers with flower parts that are in multiples of three and have long, narrow leaves with parallel veins. The **dicots** (class Dicotyledonae) are flowering plants that produce seeds with two seed leaves (cotyledons). Most dicots also produce flowers with flower parts in multiples of two, four, or five and have leaves with branching, or netted, veins. **Table 24-1** contains the names and examples of some of the most familiar families of angiosperms.

Figure 24-18 A monocot and a dicot

Monocots and dicots differ in several ways.

Daylilies are monocots.

Roses are dicots.

Table 24-1 Familiar Families of Angiosperms

Subgroup	Family	Examples
Monocots	Iridaceae (iris)	Irises, gladiolus, crocus, blue-eyed grass
	Liliaceae (lily)	Daylilies, tulips, asparagus, aloe vera
	Poaceae (grass)	Wheat, corn, rice, lawn grasses
Dicots	Asteraceae (composite)	Daisies, sunflowers, lettuce, ragweed
	Brassicaceae (mustard)	Broccoli, cauliflower, turnips, cabbage
	Fabaceae (legume)	Beans, clovers, peas, peanuts, soybeans
	Rosaceae (rose)	Roses, apples, peaches, pears, plums
	Solanaceae (nightshade)	Potatoes, tomatoes, peppers, petunias

Review SECTION 24-2

1. **List** three key features of each of the four major groups of plants.
2. **Classify** each of the following plants into one of the phyla of living plants: pine trees, carnations, sphagnum moss, and wood fern.
3. **SKILL Recognizing Patterns** How are spores and pollen grains adapted for their functions?

Critical Thinking

4. **SKILL Evaluating Conclusions** Why are angiosperms the most successful group of plants?

Review SECTION 24-2 ANSWERS

1. Nonvascular plants are small, lack a vascular system, lack roots and true leaves, have a larger gametophyte generation, and release spores. Seedless vascular plants are generally larger; have a vascular system, roots, leaves, and a larger sporophyte; and release spores. Gymnosperms are generally large; have a vascular system, roots, leaves, and a larger sporophyte; and produce seeds. Angiosperms can be small to large; have a vascular system, roots, leaves, and a larger sporophyte; and produce flowers, fruits, and seeds.
2. Pine trees are gymnosperms; carnations are angiosperms; sphagnum moss is a nonvascular plant; and wood fern is a seedless vascular plant.
3. Spores and pollen grains have outer coverings that are resistant to harsh environmental conditions. They are also often transported by wind because they are very small.
4. Angiosperms are efficient reproducers. They also produce fruits that provide a mechanism for seed dispersal by wind, water, or animals. Their seeds contain food reserves for the new plant's growth.

Teaching TIP

Classifying Plants as Monocots or Dicots

Provide examples and pictures of a variety of plants. Let students determine whether the examples are monocots or dicots. Include examples of plants that "stray" from the rules, such as a pothos ivy. Students will think the net-veined leaves make this plant a dicot; however, it is a monocot. Leaf sheaths on the stem are a strong indication of a monocot. Use this opportunity to show students that exceptions to the rules exist in the leaves of monocots and dicots. This is one reason why botanists use flowers to identify plants.

3 Assessment Options

Closure

Recognizing Relationships

Examples of all of the plant groups discussed in this chapter could be found in a forest ecosystem. Ask students to describe a role that each type of plant might play in an ecosystem. Then have them suggest plants that would play each role. For some groups, you may need to supply some examples of plants and explain what they contribute to everyday life in a forest.

Review

Assign Review—Section 24-2.

Reteaching BASIC

Have students write down the major characteristics of each major group of plants discussed in this section. For each major group, have students provide one example.

1 Prepare

Scheduling

- **Block:** About 90 minutes will be needed to complete this section.
- **Traditional:** About two class periods will be needed to complete this section.
- **Planning:** See the Planning Guide on pp. 517A and 517B for lecture, classwork, and assignment options for Section 24-3. Lesson plans for Section 24-3 can be found in a lesson cycle format in ***Biology: Principles and Explorations*** **One-Stop Planner CD-ROM with Test Generator.**

Resource Link

Assign **Active Reading Worksheet Section 24-3** to help students comprehend the key concepts and visuals in this lesson.

2 Teach

Lesson Warm-up

Ask students to make a list of the different types of plant parts that can be eaten as food. *(Students should list leaves, roots, stems, fruits, seeds, flowers, buds, bulbs, tubers, and shoots.)* Ask students to list any other products that plants provide for people. *(Examples are oxygen, wood, fibers for clothing, medicines, and chemicals.)*

Making Biology Relevant

Food

Leafy vegetables are an important source of nutrition. Most greens may be eaten cooked or raw. Many varieties are rich in beta carotene, vitamin C, folic acid, calcium, iron, and other substances that help keep us healthy. In general, darker leaves are more nutritious than lighter-colored leaves. Have students check references on nutrition (such as the USDA's *Composition of Foods*) for vitamin, mineral, and fiber content of a variety of leafy green vegetables.

Section 24-3 Plants in Our Lives

Objectives

- **Identify** foods that come from plants and their dietary importance. (pp. 534–537)
- **Describe** several ways that wood is used. (p. 538)
- **Explain** how plants are used to treat human ailments. (p. 539)
- **Identify** plants that are used to make paper and cloth. (p. 540)

Key Terms

vegetative part
cereal
grain

Plants Provide Food for Animals

Humans depend on plants in many ways. For one thing, plants store the extra nutrients they make or absorb in their bodies. Thus, plant parts contain organic nutrients (carbohydrates, fats, and proteins) and minerals (calcium, magnesium, and iron). All types of plant parts—roots, stems, leaves, flowers, fruits, and seeds—are eaten as food.

Fruits and vegetables

For marketing purposes, each of the foods that comes from a plant is identified as an agricultural commodity. Each type of food is classified by a term—such as *fruit* or *vegetable*—that is registered in Washington, D.C. But, these terms have different meanings in botany. For example, botanically, a fruit is the part of a plant that contains seeds, and a **vegetative part** is any nonreproductive part of a plant. The foods that you think of as fruits—such as apples, bananas, and melons—are also fruits in the botanical sense. Vegetables, on the other hand, may be any botanical part of a plant, as you can see in **Figure 24-19.** Fruits and vegetables provide dietary fiber and are important sources of essential vitamins and minerals.

Figure 24-19 Plant parts eaten as food

The vegetables you eat come from different parts of plants.

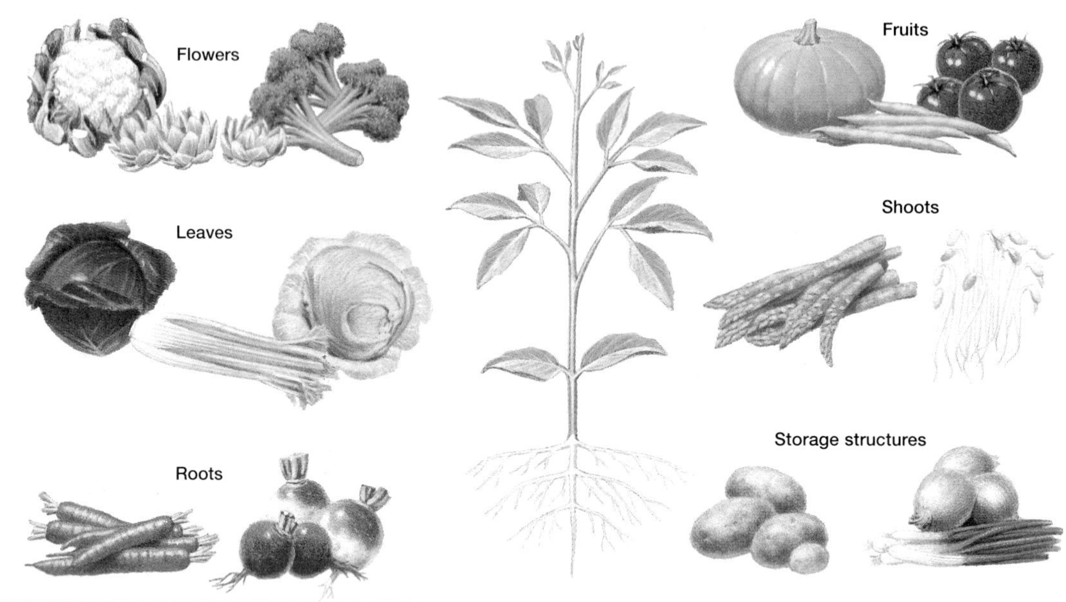

Integrating Different Learning Styles

Logical/Mathematical	**PE**	Review items 5 and 6, p. 540; Performance Zone items 21, 23, 25, and 26, p. 543
	ATE	Cross-Disciplinary Connection, p. 536
Visual/Spatial	**PE**	Quick Lab, p. 535; Performance Zone item 26, p. 543
	ATE	Demonstration, p. 538; Teaching Tips, p. 538; Demonstration, p. 540
Interpersonal	**PE**	Real Life, p. 539
	ATE	Reteaching, p. 540
Verbal/Linguistic	**PE**	Study Tip, p. 540; Review items 1–4, p. 540; Performance Zone items 15, 22, 26, and 29, pp. 542–543
	ATE	Active Reading, p. 535; Active Reading, p. 539; Closure, p. 540

Root crops

Potatoes are an important food staple in many parts of the world. Rich in calories and easy to grow, potatoes are an ideal crop for a small farm. Potatoes are classified as a root crop because they grow underground. But, potatoes are actually tubers, modified underground stems that store starch. Yams, an essential food crop in many tropical parts of the world, are also tubers. Sweet potatoes, carrots, radishes, turnips, beets, and cassava *(kuh SAH vuh)* are also important root crops. These vegetables are enlarged roots that store starch. Cassava, seen in **Figure 24-20,** is the staple food of more than 500 million people. Also known as manioc *(MAN ee awk),* cassava supplies more than one-third of the calories eaten in Africa.

Figure 24-20 Cassava. Cassava develops thick starch-filled roots up to 120 cm (4 ft) long. The roots are eaten like potatoes. Tapioca is made from cassava.

Legumes

Many members of the pea family, which are called legumes, produce protein-rich seeds in long pods. For example, about 45 percent of a soybean, the most important legume grown for food, is protein. Soybeans are often cooked and pressed into cakes called tofu *(TOH foo).* Peas, peanuts, and the many different types of beans are the seeds of legumes. Alfalfa, which is fed to livestock, is another important legume. Like many legumes, alfalfa has nitrogen-fixing bacteria, which add nitrogen compounds to the soil, in its roots. Therefore, alfalfa is also grown to enrich the soil.

QUICK LAB

When is a vegetable a fruit?

You can find out if a vegetable is a fruit by cutting it open and examining its internal structure.

Materials

apple, banana, green bean, potato, squash, tomato, plastic knife

Procedure

1. Look at several familiar fruits and vegetables. Classify each one as a commodity—either a fruit or a vegetable.
2.  **CAUTION: Sharp objects can cause injury. Handle knives carefully.** Use a plastic knife to cut open each fruit and vegetable.
3. Look at the fruits and vegetables again. Classify each by its botanical function—either a fruit or a vegetative part.

Analysis

1. **Compare** the commodity and botanical classifications you gave each fruit and vegetable.
2. **SKILL Analyzing Data** Which fruits and vegetables did you classify differently?
3. **SKILL Analyzing Results** Defend the classifications you made for item 2.
4. **SKILL Drawing Conclusions** Based on your data, when is a vegetable a fruit?

Multicultural Perspective

Manioc

The staple crop of Brazil and many other tropical and subtropical countries is a perennial shrub called manioc, or cassava. It is asexually propagated from its tuberous root. The starchy roots contain compounds that form cyanide, which must be processed out before the root is eaten. The roots are usually ground into flour and used for cooking. Processing methods are complex, labor intensive, and pose a risk of cyanide poisoning to processors. In Brazil, toasted manioc meal, or "farofa," is served at most meals. Manioc is also used to make tapioca and beer in some cultures. There are bitter and sweet varieties of manioc with varying levels of toxicity. Worldwide, manioc feeds more than 500 million people.

Active Reading

Technique: Anticipation Guide

Before students begin this section, write the following statements on the board or overhead projector:

1. Wood is the most valuable resource that people get from plants.
2. Most of the food that people eat comes from fruits.
3. Plants produce many chemicals that are useful as medicines and none that are poisonous.

Ask students to decide whether they agree or disagree with each statement. Ask volunteers to share the reasons for their opinions. Have students read the chapter to see if their opinions are confirmed or changed.

QUICK LAB

When is a vegetable a fruit?

Time 20 minutes

Process Skills Analyzing data, analyzing results, drawing conclusions

Teaching Strategies

Point out that some true fruits might not have obvious seeds. Commercially produced bananas are sterile triploids, their seeds are not viable, and the seeds are difficult to see.

Analysis Answers

1. Apples and bananas are true fruits that most people would identify as fruits. Green beans, squashes, and tomatoes are true fruits that most people would identify as vegetables. Potatoes are vegetative structures that most people would identify as vegetables.
2. After considering their botanical functions, students should reclassify green beans, squashes, and tomatoes as true fruits.
3. True fruits contain seeds, and these vegetables did contain seeds.
4. A vegetable is really a fruit if it is produced from a flower, contains parts of the flower (ovary), and contains seeds.

Cereals Are the Most Important Source of Food

Most of the foods that people eat come directly or indirectly from the fruits of **cereals,** which are grasses that are grown as food for humans and livestock. Cereal grasses produce large numbers of a type of edible, dry fruit called a **grain.** A grain contains a single seed with a large supply of endosperm. Each grain develops from a flower. The flowers of cereal grasses form in tightly packed clusters of many individual flowers. A grain is covered by a dry, papery husk called the bran, which includes the wall of the ovary and the seed coat. Cereal grains are rich in carbohydrates and also contain protein, vitamins, and dietary fiber. More than 70 percent of the world's cultivated farmland is used for growing cereal grains. In fact, more than half of the calories that humans consume come from just three species of cereal grasses: wheat, corn, and rice.

Figure 24-21 Wheat. Modern bread wheat is a hybrid of three wild species. The ripenend heads of bread wheat turn golden brown.

Wheat

For more than one-third of the world's population, wheat, seen in **Figure 24-21,** is the primary source of food. The endosperm of wheat grains, which is high in carbohydrates, is commonly ground into white flour and used to make breads and pasta. Vitamin-rich wheat germ consists of the embryos of wheat grains. Whole-wheat flour consists of the endosperm plus the germ and bran layers. Wheat grains are not always ground into flour. In the Middle East, wheat grains are often boiled or soaked, dried, and then pounded until they crack. The cracked grains, called bulgur *(BUL guhr),* are used in dishes such as tabbouleh *(tuh BOO lee)* and pilaf *(pih LAHF).* Most wheat is grown in temperate regions that have fertile soil and moderate rainfall. One of the world's best wheat-growing areas is the Great Plains region of the United States and Canada—a temperate grassland biome.

Figure 24-22 Corn

How many? Each of these ears of corn developed from a flower spike that consisted of more than 500 flowers.

Corn

Corn, seen in **Figure 24-22,** is the most widely cultivated crop in the United States. American colonists of the 1600s and 1700s first learned how to grow corn from Native Americans. In the southeastern United States, corn was more widely grown than wheat, which does not grow as well in hot climates. Thus, foods that are made from corn—corn bread, corn pone, hominy, and grits—are a traditional component of the southeastern American diet. Corn is also one of the world's chief foods for farm animals. About 70 percent of the corn crop harvested in the United States is consumed by livestock. Other uses for corn include the production of corn syrup, margarine, corn oil, cornstarch, and fuel-grade ethanol. Most of the corn grown in the United States today comes from a region known as the Corn Belt, which includes Iowa, Nebraska, Minnesota, Illinois, and Indiana.

Teaching TIP

The World Food Supply

Tell students that approximately 80,000 plant species produce edible parts, but only 50 of these are cultivated as major food sources. Yet out of these, only seven species—rice, wheat, corn, barley, sorghum, potatoes, and cassava—are the source of 75 percent of the world's food.

CROSS-DISCIPLINARY CONNECTION

◆ Mathematics

Give students the following statistics on the worldwide production of wheat, rice, and corn. Of the 560 million metric tons of wheat produced, 20 percent is used to feed livestock, and almost 80 percent is used to feed humans. Of the 520 million metric tons of rice produced, almost 100 percent goes toward feeding humans. Finally, 65 percent of the 530 million metric tons of corn produced is used to feed livestock, and most of the remaining yield is used to feed humans. Have students calculate the amount of each crop (in metric tons) eaten by humans, and then have them make bar graphs that represent the information. (Students can shade the total yield for each crop and the estimated percentage of each crop used for human consumption in different colors.) Have students write a paragraph analyzing the information on their graphs and describing how the crops rank according to their relative importance as a food source for humans. *(Rice is the most important food source, followed by wheat and then by corn.)* PORTFOLIO

did you know?

In 1944 new strains of wheat helped end food shortages in Mexico.

Norman Borlaug, who helped launch the Green Revolution, was awarded the Nobel prize in 1970 for developing new strains of wheat in Mexico. His research on wheat began in 1944. Within 20 years, Mexico had enough wheat not only to feed its own people but also to export to other countries. India and Pakistan experienced similar successes. Borlaug's high-yielding strains were produced from crosses with a dwarf variety.

Rice

Figure 24-23 **Rice.** Rice plants are grown in standing water. This rice paddy is in Sabah, Malaysia.

For more than half of the people in the world, rice is the main part of every meal. Although it is low in protein, rice is an excellent source of energy-rich carbohydrates. While brown rice still has its vitamin-rich bran layers, white rice has been processed to remove the bran layers. This processing helps to prevent spoilage in stored rice. In societies where people eat mainly rice, vitamin-rich sauces such as soy sauce are often added to white rice to make meals more nutritious. The white rice you buy at a grocery store is enriched with added vitamins. Rice is often added to processed foods such as breakfast cereal, soup, baby food, and flour to increase their energy content. Broken rice grains are used to make mash, an ingredient used to make beer. As you can see in **Figure 24-23,** rice is grown in standing water. In the United States, rice is grown in central California, in the Southeast, and along the Gulf Coast.

HEALTH WATCH: Vegetarian Diets

Many people are vegetarian; they eat only foods from plants. Like any diet, vegetarian diets must satisfy the body's nutritional needs to be healthy. Two important considerations in eating a healthy vegetarian diet are the essential amino acids and vitamins B-12 and D.

Getting the essential amino acids

The essential amino acids are amino acids that the human body cannot make. Most plant proteins contain all of the essential amino acids, but in different relative amounts. For example, the proteins in cereals are low in lysine and high in methionine; the opposite is true for the proteins in beans. Traditionally, cereals and beans are eaten together to obtain enough of the essential amino acids. However, vegetarians can get the essential amino acids they need even if they do not eat a variety of plant foods, as long as they eat enough protein. For example, if you eat enough rice to satisfy your daily protein requirement, you will get more than twice your daily requirement for lysine. The recommended daily requirement for protein is 44 g for a 55 kg (121 lb) woman and 56 g for a 70 kg (154 lb) man. Legumes, grains, nuts, broccoli, and potatoes are good sources of protein.

Vitamins B-12 and D

A vegetarian diet can provide enough of all but two vitamins, vitamin B-12 and vitamin D. To get enough vitamin B-12, vegetarians can eat fortified foods, add eggs or dairy products to their diet, or take vitamin B-12 supplements. Vitamin D is made in the skin when it is exposed to sunlight. Vegetarians may need to take vitamin D supplements if their exposure to sunlight is limited.

MULTICULTURAL PERSPECTIVE

Rice and Culture

Rice plays an important part in the customs and economies of many cultures where it is a major food source. For example, Japanese place a shiminawa, a rice-straw rope, over an entrance to ward off evil. The largest rope in the world, which hangs over the entrance to the Izumo Shinto Shrine, is a shiminawa that weighs 6 metric tons. Rice is the symbol of the goddess of wealth and prosperity and is often worshiped as a goddess by the Hindu. The act of pouring rice over someone's head signifies prosperity and happiness. Students may be familiar with a custom that seems similar—throwing rice at a wedding.

HEALTH WATCH

Vegetarian Diets

Teaching Strategies

- Bring in pictures of vegetarian meals that provide good quality and quantity of essential amino acids and are good sources of vitamins B-12 and D. Some examples are bean and cheese tacos on wheat or corn tortillas and tofu and rice with green vegetables.
- Point out that in many parts of the world, people consume most of their calories in combinations of foods that include one or more grains and one or more legumes. (South America: corn and beans; Asia: rice and soybeans)

Discussion

- Which essential amino acids are commonly abundant in meat products but low in cereal grains and legumes? *(lysine, which is low in cereal grains; methionine, which is low in legumes)*
- Why is a varied diet that includes animal products, such as milk, eggs, and cheese, especially important for vegetarians? *(A varied diet will be more likely than a monotonous diet to provide all the essential amino acids and vitamins that a vegetarian needs.)*

Resource Link

- Assign **Portfolio Projects: Plant Focus Worksheet "How Important Is Plant Life?"** from the Holt BioSources Teaching Resources CD-ROM. PORTFOLIO

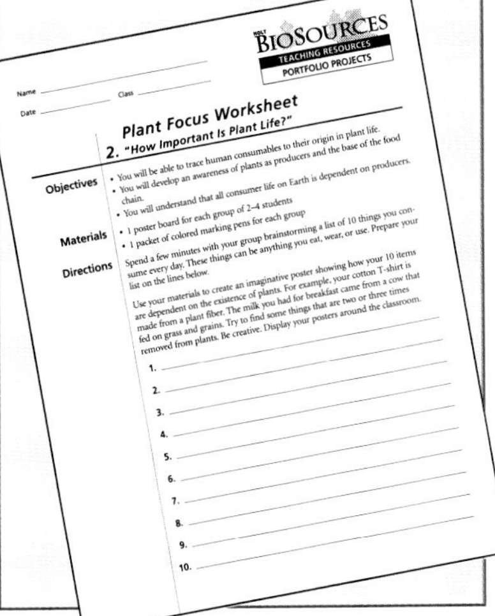

HOLT BIOSOURCES
TEACHING RESOURCES
PORTFOLIO PROJECTS

Name ______ Class ______
Date ______

Plant Focus Worksheet
2. "How Important Is Plant Life?"

Objectives
- You will be able to trace human consumables to their origin in plant life.
- You will develop an awareness of plants as producers and the base of the food chain.
- You will understand that all consumer life on Earth is dependent on producers.

Materials
- 1 poster board for each group of 2–4 students
- 1 packet of colored marking pens for each group

Directions
Spend a few minutes with your group brainstorming a list of 10 things you consume every day. These things can be anything you eat, wear, or use. Prepare your list on the lines below.

Use your materials to create an imaginative poster showing how your 10 items are dependent on the existence of plants. For example, your cotton T-shirt is made from a plant fiber. The milk you had for breakfast came from a cow that fed on grass and grains. Try to find some things that are two or three times removed from plants. Be creative. Display your posters around the classroom.

1. ______
2. ______
3. ______
4. ______
5. ______
6. ______
7. ______
8. ______
9. ______
10. ______

Chapter 24

Demonstration

Biodegradable Packing Pellets

Bring some biodegradable packing pellets, a beaker of water, and iodine in a dropper bottle to class. To show students that biodegradable products can replace petroleum-based products that cannot be recycled, drop a few of the packing pellets into a beaker of water. The pellets will dissolve. Then add a few drops of iodine to the beaker, and note the color change. Ask students to identify the composition of the pellets based on the color change that occurs. *(Purple indicates the presence of starch.)* Tell students that these pellets, which are replacing the plastic foam pellets derived from petroleum-based products, are made of cornstarch, which is derived from corn.

Teaching TIP BASIC

Observing Paper Types

Have students make wet-mount slides of various types of paper and then observe the slides under a microscope. Staining the paper with methylene blue will make the fibers more visible. Have students draw what they see, and then ask them to relate the fibers to the characteristics of the papers. PORTFOLIO

Teaching TIP

Products That Come from Trees

Have students make a Graphic Organizer similar to the one at the bottom of this page to illustrate the importance of trees. Ask them to use the vocabulary words applicable for this section together with any other words that they wish to include. PORTFOLIO

Plants Have Many Nonfood Uses

Plants are used by people for many purposes other than food. For example, rubber was first made from latex, the milky white sap of tropical trees of the genus *Hevea.* Latex is extracted from rubber trees by the method seen in **Figure 24-24.** Guayule *(gwah YOO lee),* a member of the sunflower family that is native to the southwestern United States, is another source of natural rubber. Most of today's rubber, however, is synthesized from petroleum, a nonrenewable resource. The most important nonfood products obtained from plants are wood and fibers.

Figure 24-24 **Latex.** The milky sap of certain plants is called latex. This man is collecting latex from a rubber tree in Java, Indonesia.

Wood

After food, wood is the single most valuable resource obtained from plants. Thousands of products, such as those shown in **Figure 24-25,** are made from wood. The wood from trees that have been cut down and sawed into boards is called lumber. Nearly 75 percent of the lumber cut in the United States is used for building construction. The rest is used to make products that contain wood, or it is ground and moistened to make wood pulp. Wood pulp is made into paper, rayon, and many other products. Finally, for more than a quarter of the world's people, wood is still the main source of fuel for heating and cooking.

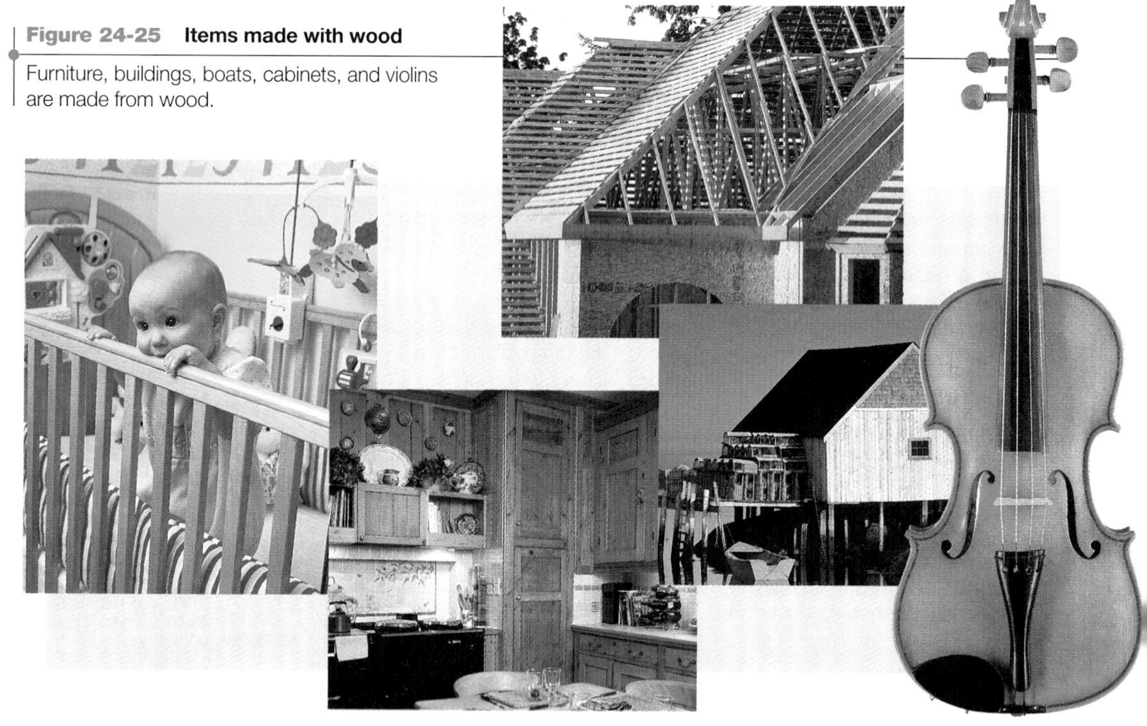

Figure 24-25 **Items made with wood**

Furniture, buildings, boats, cabinets, and violins are made from wood.

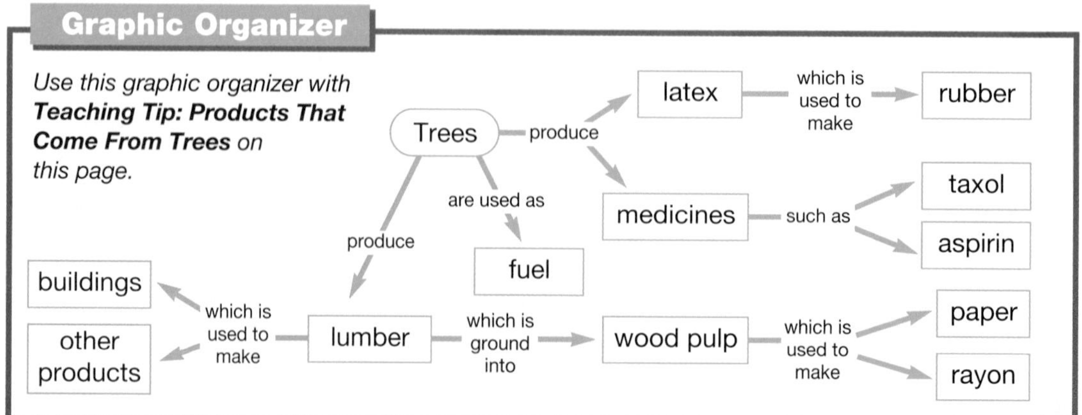

Figure 24-26 Sources of medicines

These two common garden plants are the sources of important medicines.

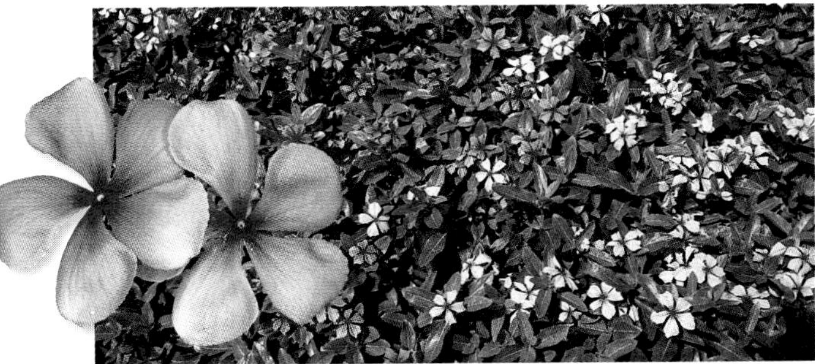

Rosy periwinkle, the original source of two cancer-fighting drugs

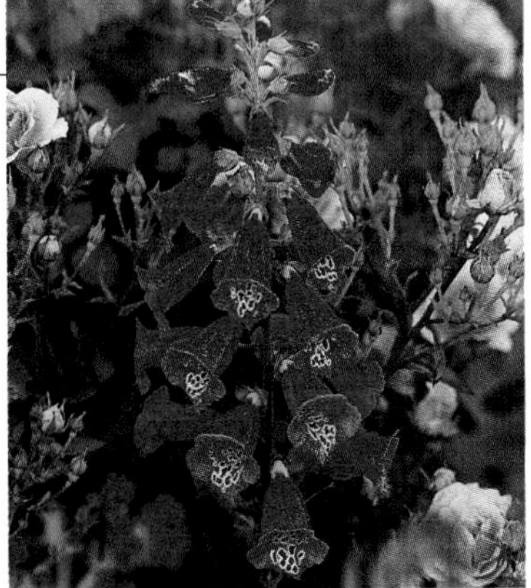

Foxglove, the source of a drug used to treat cardiac disorders

Medicines

People have always used substances obtained from plants to treat a variety of ailments. By studying the plants traditionally used to treat human ailments, researchers have developed many "modern" medicines. For example, solutions made by soaking the bark of willow trees, *Salix,* were a traditional cure for aches and pains. The pain-relieving chemical found in willows is called salicin *(SAL uh sihn).* Acetylsalicylic *(uh SEET l sal uh SIHL ihk)* acid, a derivative of salicin, was first sold in 1899 under the name aspirin. Today, aspirin is the most widely used drug in the world.

Two familiar garden plants, seen in **Figure 24-26,** are important sources of life-saving medicines. The extremely poisonous leaves of the foxglove, *Digitalis purpurea,* yield digitalis *(dihj ih TAL ihs),* a drug that is used to stabilize irregular heartbeats and to treat cardiac disorders. The rosy periwinkle, *Catharanthus roseus,* is the original source of two cancer-treatment drugs—vinblastine *(vihn BLAS teen)* and vincristine *(vihn KRIHS teen).* Vinblastine is used to treat Hodgkin's disease, a type of cancer that affects the lymph nodes. Vincristine is used to treat childhood leukemia. **Table 24-2** contains other examples of medicines originally derived from plants.

REAL LIFE

Many people are allergic to the pollen of wind-pollinated plants. Juniper pollen, oak pollen, and rag-weed pollen are among the most notorious allergens, substances that cause allergy symptoms.

Finding Information *Ask an allergist or your doctor what to do to avoid having an allergic reaction to pollen. Compile a list of tips for preventing pollen allergies.*

Table 24-2 Some Medicines Originally Derived from Plants

Name	Source	Action
Caffeine	Tea leaves	Acts as a stimulant
Codeine	Poppy fruits	Relieves pain
Cortisone	Yam tubers	Relieves symptoms of allergies
Ephedrine	Ephedra stems	Acts as a decongestant
Taxol	Yew tree bark	Reduces the size of cancerous tumors

internet**connect**

TOPIC: Medicines from plants
GO TO: www.scilinks.org
KEYWORD: HX539

did you know?

Biotech plants may help fight cancer.

Genetic engineering offers hope of introducing the anticancer proteins found in foods, such as broccoli, into other fruits and vegetables. People would then have a wider variety of plant foods to chose from to obtain these proteins. Those who do not like to eat broccoli and related plants could get these proteins from other sources.

Section 24-3

Active Reading

Technique: Reading Organizer

Have students keep a list of the uses of plants as they read this section. When they have finished reading, have them make a table of the items in their list.

REAL LIFE

Answer

Answers will vary. Some tips for avoiding allergic reactions to pollen include keeping windows closed, using an air filter, staying inside during pollen season, conducting outdoor activities when plants are wet with dew or rain, and wearing a pollen mask.

internet**connect**

Have students research medicines that come from plants using the Web site in the Internet Connect box on this page. Have students write a report summarizing their findings. PORTFOLIO

Making Biology Relevant

Medicine ADVANCED

The potato plant (*Solanum tuberosum*) is a member of the nightshade (Solanaceae) family. Many plants in this family produce alkaloids that are poisonous to people and animals. Some examples are tobacco, which produces the alkaloid nicotine; jimson weed and datura, which produce the alkaloids atropine and scopolamine; and belladonna, which produces atropine. Atropine and scopolamine are used medically to treat muscle spasms and related disorders. The belladonna ("beautiful woman") plant got its name because people once made extracts from this plant and placed drops of it in women's eyes, resulting in pupil dilation. Potato plants are normally not poisonous. However, if potatoes are left in storage for too long and they begin to sprout and turn green, these parts will contain alkaloids that could be harmful if ingested in large enough quantities.

Chapter 24

Demonstration

Naturally Colored Cotton

Show students swatches of fabrics made from naturally colored cotton. Tell students that biotechnology has produced cotton plants with fibers that are different colors, including green, brown, blue, and red. As naturally colored cotton fabrics are washed, the colors do not fade as they do with dyed fabrics; instead, they become darker. Cotton from plants with green and brown fibers was used in clothing and other supplies for the troops in Operation Desert Storm.

3 Assessment Options

Closure

Relating Concepts

Review this section by asking students the following questions: What are the most important cereal grains? *(wheat, rice, and corn)* What are some other plants used as food sources? *(lentils, soybeans, sugar cane, celery)* What are some of the important products of the timber industry? *(fuel, lumber, paper, turpentine)* What are some medicines derived from plants? *(aspirin, codeine, digitalis, vinblastine)*

Review

Assign Review—Section 24-3.

Reteaching — BASIC

Assign students to cooperative groups to prepare a lunch menu that consists only of plants but that guarantees a fully nutritional meal. Provide a library cart of reserved books selected for your class.

PORTFOLIO

Figure 24-27 Cotton

Cotton is the plant fiber that is most widely used to make cloth.

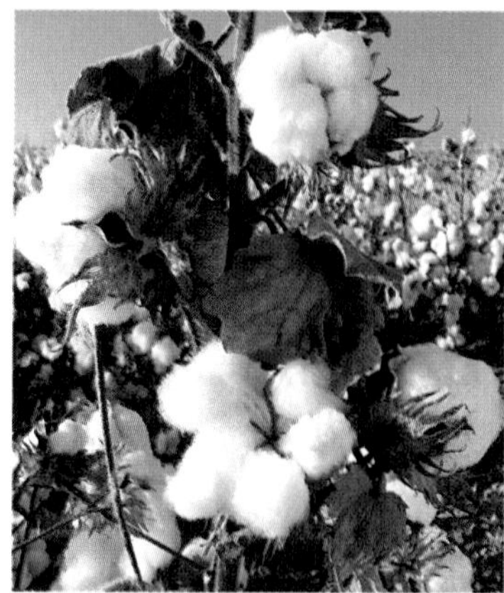

Cotton bolls that have split open, revealing cotton fibers

Indian woman spinning cotton into thread that will be woven into cloth

Study TIP

Reviewing Information

Prepare flashcards for each of the Key Terms in this chapter. On each card, write the term on one side and its definition on the other side. Use the cards to review meanings of the Key Terms.

Fibers

If you were to look at this sheet of paper very closely through a magnifying glass, you would see that it is made of many interlocking fibers. These fibers are strands of cellulose, which is a component of the cell walls of plants. In plants, fibers help provide support for the plant body. The strength and flexibility of plant fibers make them ideal materials for making paper, cloth, and rope. Most of the fibers used to make paper come from wood. Paper-making fibers are also obtained from many other plants, including cotton, flax, rice, bamboo, and papyrus *(puh PEYE ruhs)*.

For centuries, people have made clothing with cloth made of cotton, the world's most important plant fiber. As **Figure 24-27** shows, white fibers fill up the inside of a cotton boll *(bohl)*, the fruit of the cotton plant. Cotton thread is spun from the fine white fibers, which grow on cotton seeds. The stems of flax yield softer, more durable fibers that are used to make linen. More than 30 percent of the world's clothing is now made of synthetic fibers, but natural plant fibers are still prized for their durability and comfort. Sturdy fibers of hemp and sisal *(SEYE suhl)* plants are used to make rope.

Review SECTION 24-3

1. **Name** the three most important cereal grains.
2. **Describe** several ways that wood is used.
3. **List** five medicines that are derived from plants, and state how each is used.
4. **Name** three types of plants that provide fiber used in clothing.

Critical Thinking

5. **SKILL Forming Reasoned Opinions** How is research on crops that can be used to make products we now make from petroleum important?
6. **SKILL Evaluating Viewpoints** Justify the viewpoint that wood is the most important nonfood plant product.

Review SECTION 24-3 ANSWERS

1. wheat, rice, and corn
2. Wood is used for fuel and for the production of lumber, paper, and rayon.
3. Aspirin is used as a pain reliever; digitalis is used to treat heart disorders; vinblastine is used to treat cancer; codeine is used as a pain reliever; cortisone is used to treat allergies; ephedrine is used as a decongestant; and taxol is used to treat cancer.
4. cotton, flax, and trees
5. Answers will vary. This research could provide alternatives to the use of nonrenewable resources that would be renewable.
6. Answers will vary, but students should mention a number of the valuable products obtained from wood described in this section.

Study ZONE

CHAPTER 24

Use the Key Concepts and Key Terms listed below to review the main ideas in this chapter. If you need to review the meaning of a term, turn to the page indicated.

Key Concepts

24-1 Adaptations of Plants

- To survive on land, plants must absorb mineral nutrients, prevent their bodies from drying out, and reproduce without water to transmit male gametes.
- Vascular plants have a system of well-developed tissues that transport water within a plant. The nonvascular plants lack a vascular system.
- Seeds protect and nourish a plant's embryo, disperse the offspring, and delay the growth of the embryo until conditions are favorable. Flowers make reproduction more efficient by promoting pollination.
- The sporophytes of vascular plants have a vascular system. Their bodies consist of an aboveground shoot and an underground root.

24-2 Kinds of Plants

- Nonvascular plants are small and lack vascular tissue. Mosses, liverworts, and hornworts are nonvascular plants.
- Seedless vascular plants produce spores with thickened walls that prevent them from drying out. Ferns, club mosses, horsetails, and whisk ferns are seedless vascular plants.
- Gymnosperms are seed plants that produce cones. Conifers, cycads, ginkgoes, and gnetophytes are gymnosperms.
- Angiosperms are seed plants that produce flowers and fruits. The angiosperms are classified as either monocots or dicots.

24-3 Plants In Our Lives

- All types of plant parts—roots, stems, leaves, flowers, fruits, and seeds—provide food for humans. Rice, corn, and wheat are cereal grasses and are our most important sources of food.
- Wood is a source of wood pulp used for making paper, lumber used for building materials, and fuel.
- Many important medicines are currently made from plants or were originally derived from plants.
- Plant fibers are used to make paper, cloth, and rope. The most important sources of plant fibers are wood and cotton.

Key Terms

24-1

cuticle (520)
stoma (521)
guard cell (521)
vascular system (522)
nonvascular plant (522)
vascular plant (522)
seed (522)
embryo (522)
seed plant (522)
flower (523)
phloem (525)
xylem (525)
shoot (525)
root (525)
meristem (525)

24-2

rhizoid (526)
rhizome (528)
frond (529)
cone (529)
gymnosperm (530)
angiosperm (532)
fruit (532)
endosperm (532)
monocot (533)
dicot (533)

24-3

vegetative part (534)
cereal (536)
grain (536)

Study TIP Ready?

- *If you think you understand these Key Concepts and Key Terms, then you're ready for the Performance Zone.* 

Study ZONE

CHAPTER 24

Review and Practice

Assign the following worksheets for Chapter 24:

- **Student Review Guide**
- **Concept Mapping Worksheet**
- **Test Preparation Pretest**
- **Vocabulary Worksheet**
- **Critical Thinking Review Worksheet**
- **Science Skills Worksheet**

Assessment Options

The following assessment options are available for this chapter:

Resource Link

- Assign **Chapter Test 24** to assess students' understanding of the chapter.

Alternative Assessment

Have students use the information in Chapter 24 to design a procedure for a question-and-answer game. Students may choose to model their game after a television game show or a trivia-type board game. Encourage students to write questions that vary in difficulty.

Portfolio Assessment

- *Portfolio Assessment* at the front of this book provides options for building and evaluating student portfolios. Students and teachers can select items from the areas listed for inclusion in a portfolio.
- See items labeled PORTFOLIO on pp. 522, 523, 528, 531, 532, 536, 538, 539, and 540.

Answer to Concept Map

The following is one possible answer to Performance Zone item 1 on page 542.

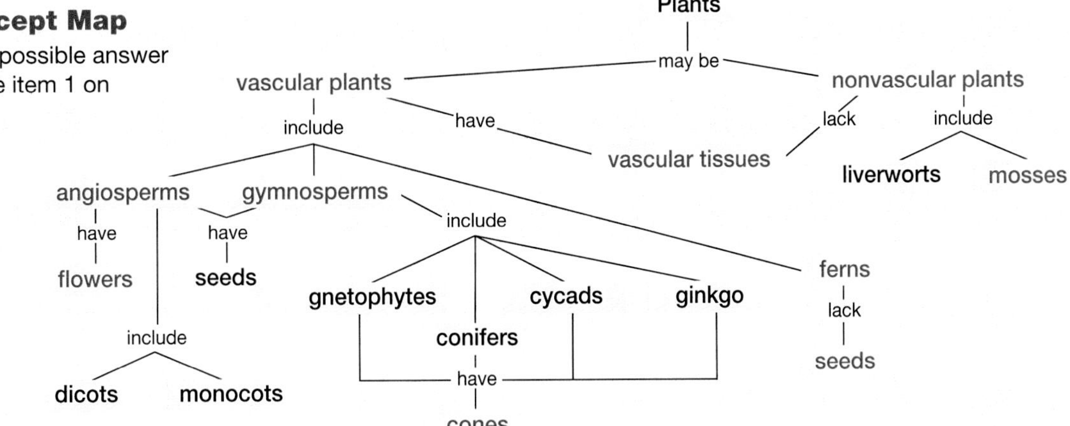

Performance ZONE
CHAPTER 24
REVIEW ANSWERS

Assignment Guide

SECTION	REVIEW
24-1	1, 2, 3, 4, 5, 6, 13, 14, 16, 24
24-2	1, 2, 7, 8, 9, 17, 18, 19, 20, 27, 28
24-3	10, 11, 12, 15, 21, 22, 23, 25, 26, 29

Understanding and Applying Concepts

1. The answer to the concept map is found on page 541.
2. **a.** Xylem is a water-transporting tissue; phloem is a tissue that transports sugar and other organic molecules.
   b. A shoot is the aboveground part of a plant; a root is the belowground part of a plant.
   c. A rhizoid is a rootlike structure that anchors a plant; a rhizome is an underground stem.
   d. A seed contains an embryo, a seed coat, and stored food; a fruit contains part of the flower (the ovary) and seeds.
3. b
4. a
5. a
6. c
7. b
8. d
9. a
10. c
11. b
12. d
13. The seeds would likely be dispersed by becoming attached to the fur of a mammal and being transported as the mammal travels around.
14. The sporophytes are large and have many leaves that can intercept light from the sun. A vascular system enables the sporophytes to grow very large by distributing water and nutrients efficiently. Their reproductive cells (eggs and sperm) are produced near the tops of the sporophyte, where they can readily be dispersed by wind or animals.

Performance ZONE
CHAPTER 24 Review

Understanding and Applying Concepts

1. **Concept Mapping** Make a concept map that shows how plants are classified. Include the following terms in your map: vascular plants, nonvascular plants, ferns, angiosperms, gymnosperms, mosses, cones, vascular tissue, seeds, flowers. Include additional terms in your map as needed.

2. **Understanding Vocabulary** For each pair of terms, explain the differences in their meanings.
   a. xylem, phloem
   b. shoot, root
   c. rhizoid, rhizome
   d. seed, fruit

3. What structure made it possible for plants to prevent water loss on land?
   a. mycorrhizae
   b. cuticle
   c. pollen
   d. seed

4. Plant life cycles include a diploid individual that is called a(n)
   a. sporophyte.
   b. gametophyte.
   c. zygospore.
   d. epiphyte.

5. Seeds helped plants adapt to life on land by
   a. providing nourishment for embryos.
   b. protecting embryos from air pollution.
   c. sprouting during unfavorable weather.
   d. limiting the dispersal of plant offspring.

6. Which of the following is *not* a characteristic of vascular plants?
   a. xylem and phloem
   b. stems and leaves
   c. a dominant gametophyte
   d. a diploid sporophyte

7. Which of the following are *not* seedless vascular plants?
   a. horsetails
   b. gymnosperms
   c. ferns
   d. club mosses

8. Unlike angiosperms, gymnosperms
   a. are pollinated by wind.
   b. do not have seeds.
   c. have a diploid sporophyte generation.
   d. do not bear fruit.

9. Which of the following are *not* dicots?
   a. grass family
   b. rose family
   c. mint family
   d. legume family

10. The most important sources of food are
    a. legumes.
    b. root crops.
    c. cereal grains.
    d. vegetables.

11. Drugs derived from the rosy periwinkle are used in the treatment of
    a. heart disease.
    b. leukemia.
    c. allergies.
    d. headaches.

12. Which of the following is *not* a source of fibers for both paper and cloth?
    a. flax
    b. cotton
    c. wood
    d. sisal

13. Look at the cocklebur in the photograph below. It is a fruit that contains the seeds of a cocklebur plant. Suggest how this plant's seeds might be dispersed.

14. **Theme Evolution** Most plants have a vascular system and a sporophyte that is much larger that the gametophyte. How have these features contributed to the success of plants as they evolved on land?

15. **Health Watch** A friend is concerned that your vegetarian diet is not healthy. Make a list of the measures you would take to ensure that your diet will provide you with all the nutrients you need.

16. **Chapter Links** How does meiosis result in the production of haploid spores? (**Hint:** See Chapter 7, Section 7-1.)

15. Answers may vary. Vegetarian diets should contain both grains and legumes, as well as sources of vitamins B-12 and D such as dairy products, fortified foods, and vitamin supplements.

16. Meiosis involves two divisions of the nucleus, which reduces the number of chromosomes in a cell and produces cells that are haploid; in plants, the cells that result directly from meiotic cell division are called spores.

Skills Assessment

Part 1: Interpreting Tables

Use the table below to answer the questions that follow:

Some Families of Angiosperms		
Subgroup	**Family**	**Examples**
Monocots	Amaryllis	Garlic, narcissus, onion
	Bromeliad	Pineapple, Spanish moss
	Palm	Coconut, date, palmetto
Dicots	Carrot	Carrot, celery, parsley
	Mallow	Cotton, hibiscus, okra
	Mint	Rosemary, sage, thyme

17. Which families in this table are dicots?
18. Which families in this table are monocots?
19. Which family in this table contains a plant that is an important source of fibers?
20. Identify the family to which each of the following plants belongs.
    a. okra
    b. coconuts
    c. onion
    d. celery

Part 2: Life/Work Skills

Use the media center or Internet sources to find out how and why measures are being taken to reduce the consumption of paper.

21. **Allocating Resources** Record every paper product you encounter for three days. Include items such as notebook paper, paper towels, newspaper, and product packaging. Identify each paper product you list as recyclable or nonrecyclable. Suggest ways to eliminate or reduce your use of nonrecyclable paper.
22. **Communicating** Prepare an oral report to summarize your suggestions. Create a visual that suggests a reason for reducing the consumption of nonrecylable paper.
23. **Selecting Technology** You are asked to compare the length and thickness of the fibers in several types of paper. Make a list of the laboratory equipment you would need for this task, and explain why you would need each piece.

Critical Thinking

24. **Justifying Conclusions** Many woody plants grow very tall. Botanists consider this to be a reproductive advantage. Why might it a reproductive advantage for a plant to be tall?
25. **Forming Reasoned Opinions** Why is the loss of tropical rain forests and other types of forests of concern to medical science?
26. **Justifying Conclusions** Suppose that a friend asks you why corn, which he or she considers to be a vegetable, is listed as a cereal crop in the encyclopedia. To answer this question, write a paragraph that explains why corn is a cereal crop agriculturally and why it is also a fruit botanically.

Portfolio Projects

27. **Cooperative Group Project** Investigate the plants in an area near your school. Then draw a map of the area on poster-sized paper. Mark the location of the different types of plants (liverworts, mosses, ferns, angiosperms, and gymnosperms) that you find on the map. To identify some plants, you may need to use an identification key. Include a map legend that lists the groups of plants found and the symbol used to identify the plants on the map. Display your map in your classroom or school.
28. **Interpreting Information** Read "On the Trail of the Lonesome Pine," by Wilson da Silva, (*New Scientist,* December 6, 1997, pp. 36–39). What about the Wollemi pines surprises botanists the most? What has DNA fingerprinting revealed about the pines? How old is this species of pine thought to be?
29. **Career Focus Ethnobotany** Use the media center or the Internet to research the field of ethnobotany. Write a report on your findings. Your report should include a job description, the training required, names of employers, growth prospects, and an average starting salary.

Skills Assessment

17. carrot, mallow, mint
18. amaryllis, bromeliad, palm
19. mallow
20. a. mallow; b. palm; c. amaryllis; d. carrot
21. Answers will vary, but students might include the suggestion to use reusable cloth towels instead of paper towels, buy products made with recycled paper, buy products with minimal paper packaging, and reuse paper as scratch paper.
22. Reports will vary, but students should include several helpful suggestions about ways to reduce the use of nonrecyclable paper. Visuals will vary, but students should represent effects of deforestation.
23. Answers will vary, but students should include microscope slides and a microscope.

Critical Thinking

24. The reproductive cells (eggs and sperm) of trees are produced near the tops of the sporophyte, where they have exposure to wind. This enhances wind pollination. The seeds can also be protected from animals until they are mature and then be dispersed more readily by wind.
25. Many scientists think that there are many as yet undiscovered plants in forests and that some of these plants have possible medicinal value.
26. Corn is a cereal crop because the part of the plant that is eaten consists of grains. All cereal grains are actually fruits, because they contain a seed that formed inside an ovary of a flower.

Portfolio Projects

27. Answers will vary but should reflect the students' knowledge of plant groups. Students' maps should mark the location of plants from each of the four major plant groups.
28. Botanists were most surprised that the Wollemi pines were not discovered sooner. DNA fingerprinting showed that there was no variation among the individuals tested. The Wollemi pines are members of a family that has existed since the Triassic period. The pollen most closely resembles fossilized pollen that is 94 million years old.
29. Ethnobotanists are scientists who study the uses of plants by different cultures. They also look for plants with medicinal value. Ethnobotanists usually work for universities or for private companies, such as pharmaceutical companies, that make or market products derived from plants. A bachelor's degree, master's degree, or Ph.D. in botany or horticulture is helpful if you want to become an ethnobotanist. Starting salary will vary by region.

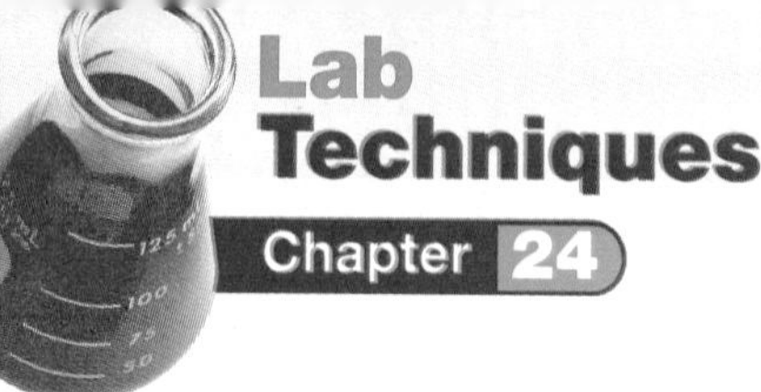

Surveying Plant Diversity

Purpose

Students will observe examples of the four major groups of plants.

Preparation Notes

Time Required: 45 minutes

Materials

Materials for this lab can be ordered from WARD'S. See *Master Materials List* at the front of this book for catalog numbers.

Preparation Tips

- Review the four major plant groups before doing this lab.
- Set up at least two stations for each group of plants.
- Provide a stereomicroscope or hand lens for stations with mosses and ferns.
- Provide a compound microscope for stations with prepared slides.

Safety Precautions

Discuss all safety symbols and caution statement with students.

Procedural Tip

Direct students to spread out among the stations and begin wherever there is space. They should then continue to the next consecutively numbered station.

Lab Techniques

Chapter 24 Surveying Plant Diversity

SKILLS

- Observing
- Comparing

OBJECTIVES

- **Identify** similarities and differences among four phyla of living plants.
- **Relate** structural adaptations of plants to their success on land.

MATERIALS

- live or preserved specimens of mosses, ferns, conifers, and flowering plants
- stereomicroscope or hand lens
- compound microscope
- prepared slides of fern gametophytes

Before You Begin

Most plants are complex photosynthetic organisms that live on land. The ancestors of plants lived in water. As plants evolved on land, however, they developed adaptations that made it possible for them to be successful in dry conditions. All plant life cycles are characterized by **alternation of generations,** in which a haploid **gametophyte** stage alternates with a diploid **sporophyte** stage. Distinct differences in the relative sizes and structures of gametophytes and sporophytes are seen among the 12 phyla of living plants. In this lab, you will examine representatives of the four most familiar plant phyla.

1. Write a definition for each boldface term in the paragraph above and for the following terms: sporangium, spore, frond, cone, flower, fruit.
2. Make a data table similar to the one below.
3. Based on the objectives for this lab, write a question you would like to explore about the characteristics of plants.

Procedure

PART A: Conducting a Survey

1. Visit the station for each of the plants listed below, and examine the specimens there. Answer the questions, and record observations in your data table.
2. **Mosses** Examine a clump of moss with a stereomicroscope or hand lens. Make a sketch of what you see.
3. **Mosses** Examine a moss gametophyte with a sporophyte attached to it. Draw what you see, and label the parts you recognize. Label each part as haploid or diploid.
   a. Which stage of a moss has rootlike structures?
   b. Where are the spores of a moss produced?

DATA TABLE

Phylum name	Dominant generation	Major characteristics	Examples

Before You Begin

Answers

1. *alternation of generations*—the alternating of a haploid gametophyte with a diploid sporophyte in the life cycle

   gametophyte—the haploid gamete-producing stage of a plant's life cycle

   sporophyte—the diploid spore-producing stage of a plant's life cycle

   sporangium—structure in which spores are produced

   spore—haploid asexual reproductive structure

   frond—leaf of a fern

   cone—structure of the sporophyte in many plants in which the gametophytes develop

   flower—structure of the angiosperm sporophyte in which the gametophytes develop

   fruit—structure in which the seeds of angiosperms develop
2. Answers will vary. For example: What characteristics are shared by all plants?

Procedure

Answers

2. Drawings will vary depending on the species.
3. Drawings will vary depending on the species.
   a. gametophyte
   b. spore capsule
4. Drawings will vary depending on the species.
   a. Water travels through the cells of the vascular system. Veins are visible in the leaves.
   b. spores
5. Drawings will vary but should resemble the plant shown in the insert in Figure 24-12 on page 529.

4. **Ferns** Examine the sporophyte of a fern, and look for evidence of reproductive structures on the underside of the fronds. Draw what you see. Label a leaf (frond), stem, root, and reproductive structure.
   a. How does water travel through a fern? List observations supporting your answer.
   b. What kind of reproductive cells are produced by fern fronds?
5. **Ferns** Examine a slide of a fern gametophyte with a compound microscope. Draw what you see, and label any structures you recognize.
6. **Conifers** Draw a part of a branch of one of the conifers at this station. Label a leaf, stem, and cone (if present).
   a. Is a branch of a pine tree part of a gametophyte or part of a sporophyte?
   b. In what part of a conifer would you look to find its reproductive structures?
7. **Conifers** Examine a prepared slide of pine pollen. Draw a few of the grains.
   a. What reproductive structure is found within a pollen grain?
   b. How does the structure of pine pollen aid in its dispersal by wind?
8. **Angiosperms** Draw one of the representative angiosperms at this station. Label a leaf, stem, root, and flower (if present). Indicate the sporophyte and location of gametophytes.
   a. Where do angiosperms produce sperm and eggs?
   b. How do the seeds of angiosperms differ from those of gymnosperms?
9. **Angiosperms** Examine several fruits. Draw and label the parts of one fruit.

PART B: Cleanup and Disposal

10. Dispose of broken glass in the designated waste containers. Do not put lab materials in the trash unless your teacher tells you to do so.
11. Wash your hands thoroughly before you leave the lab and after you finish all work.

Analyze and Conclude

1. **Analyzing Information** How are bryophytes different from the other major groups of plants?
2. **Recognizing Patterns** How do the gametophytes of gymnosperms and angiosperms differ from the gametophytes of bryophytes and ferns?
3. **Drawing Conclusions** What structures are present in both gymnosperms and angiosperms but absent in both bryophytes and ferns?
4. **Evaluating Hypotheses** Dispersal is the main function of fruits in angiosperms. Defend or refute this hypothesis. List observations you made during this lab to support your position.
5. **Inferring Conclusions** Based on their characteristics, which phylum of plants appears to be the most successful? Justify your conclusion.
6. **Further Inquiry** Write a new question about plant diversity that could be explored with another investigation.

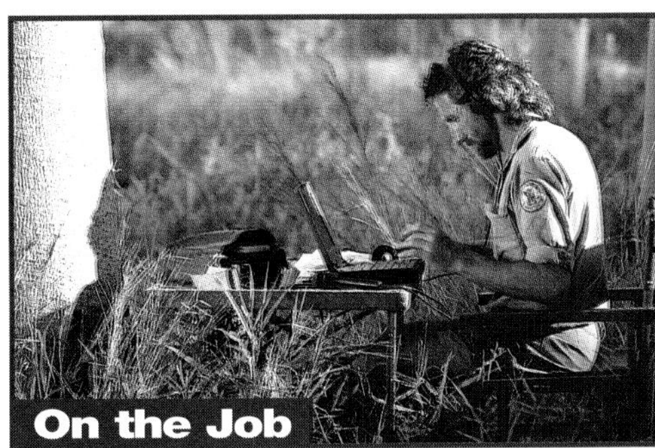

On the Job

Drawing accurate diagrams of organisms is an important part of a plant taxonomist's research. Do research in the library or media center or on the Internet to discover the names of some famous plant taxonomists and scientific artists and to learn about some of the techniques used in their work. Summarize your findings in a written or oral report.

Analyze and Conclude

Answers

1. The gametophyte is the most noticeable stage of the bryophyte life cycle; the sporophyte is the most noticeable stage in other groups of plants.
2. Gymnosperm and angiosperm gametophytes develop within tissues of the sporophyte; bryophyte and fern gametophytes are separate individuals.
3. pollen and seeds
4. Answers will vary. Students should recognize that fruits often have appendages or shapes that promote seed dispersal by wind, water, animals, or gravity and that many have fleshy parts that are edible and encourage seed dispersal by animals.
5. Answers will vary. Students should recognize that angiosperms are the most successful phylum of plants. The presence of flowers to attract pollinators and fruits to disperse seeds could be offered as reasons for the success of angiosperms. The great diversity among the angiosperms could be offered as evidence of their success.
6. Answers will vary.

On the Job

Reports will vary depending on which person the student selects.

6. Drawings will vary depending on the species.
   a. sporophyte
   b. cones
7. Drawings will vary depending on the species.
   a. sperm
   b. Appendages help the pollen grains to float on the wind.
8. Drawings will vary depending on the species.
   a. in flowers
   b. Angiosperm seeds are produced in fruits; gymnosperm seeds are produced in cones.
9. Drawings will vary depending on the fruit. Students should label seeds.

Sample Data Table

Phylum name	Dominant generation	Major characteristics	Examples
Bryophyta	gametophyte	nonvascular	moss
Pterophyta	sporophyte	vascular, seedless, sori	fern
Coniferophyta	sporophyte	vascular, seeds, cones	pine tree
Anthophyta	sporophyte	vascular, seeds, flowers	rose

Chapter 25

Plant Reproduction

Planning Guide

	OBJECTIVES	MEETING INDIVIDUAL NEEDS	READING SKILLS AND STRATEGIES
BLOCK 1 45 min	**25-1 Sexual Reproduction in Seedless Plants** • Summarize the life cycle of a moss. • Summarize the life cycle of a fern. • Compare and contrast the life cycle of a moss with the life cycle of a fern.	Learning Styles, p. 548 BASIC ADVANCED • pp. 549, 551	ATE Active Reading, Technique: Reading Effectively, p. 547 ■ Active Reading Guide, worksheet (25-1) ■ Directed Reading, worksheet (25-1)
BLOCKS 2 & 3 90 min	**25-2 Sexual Reproduction in Seed Plants** • Distinguish the male and female gametophytes of seed plants. • Describe the function of each part of a seed. • Summarize the life cycle of a conifer. • Summarize the life cycle of an angiosperm.	Learning Styles, p. 552 BASIC ADVANCED • pp. 554, 555, 558	ATE Active Reading, Technique: Reading Organizer, p. 553 ■ Active Reading Guide, worksheet (25-2) ■ Directed Reading, worksheet (25-2) ■ Reading Strategies, Recognizing Cycles
BLOCKS 4 & 5 90 min	**25-3 Asexual Reproduction** • Describe several types of vegetative reproduction in plants. • Distinguish sexual reproduction in kalanchoës from asexual reproduction in kalanchoës. • Recommend several ways to propagate plants.	Learning Styles, p. 559 BASIC ADVANCED • pp. 559, 562	■ Active Reading Guide, worksheet (25-3) ■ Directed Reading, worksheet (25-3) ■ Supplemental Reading Guide, *A Feeling for the Organism: The Life and Work of Barbara McClintock*

Review and Assessment Resources

BLOCKS 6 & 7 90 min

TEST PREP

- PE Study Zone, p. 563
- PE Performance Zone, pp. 564–565
- ■ Chapter 25 worksheets:
  - Vocabulary
  - Concept Mapping
  - Critical Thinking

 Audio CD Program

- ■ Test Prep Pretest

CHAPTER TESTING

- ■ Chapter Tests (blackline master)
- ▲ Test Generator
- ■ Assessment Item Listing

See the correlations table at the front of this book for a list of the ***National Science Education Standards*** addressed in this chapter.

Smithsonian Institution® Visit **www.si.edu/hrw** for additional on-line resources.

KEY

Biology Principles and Explorations

PE Pupil's Edition

ATE Teacher's Edition

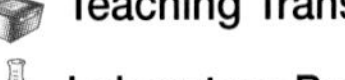

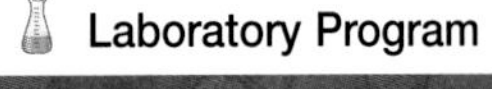

HOLT BIOSOURCES

Teaching Transparencies

Laboratory Program

One-Stop Planner CD-ROM
Shows these resources within a customizable daily lesson plan:

■ Teaching Resources

▲ Test Generator

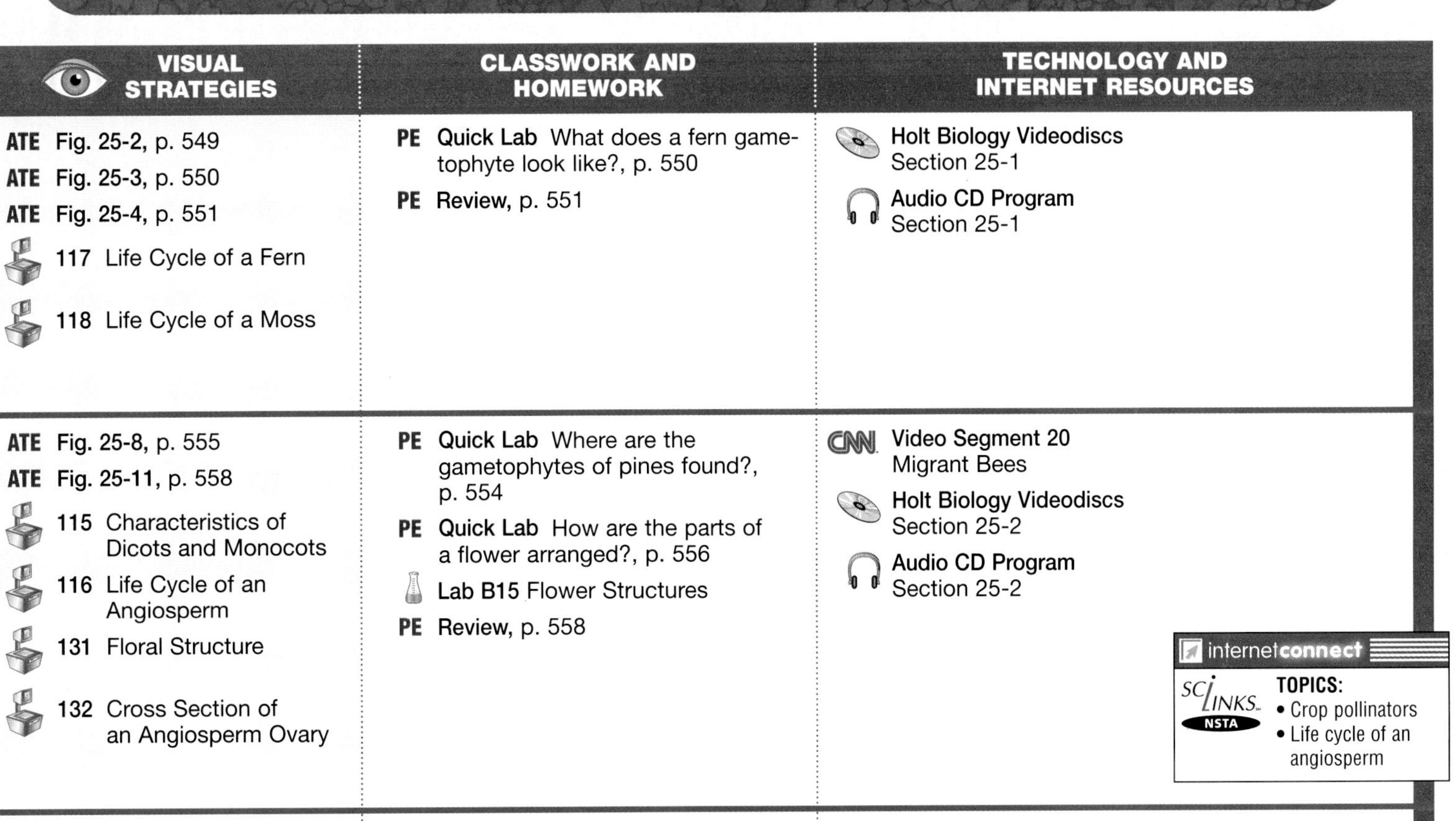

VISUAL STRATEGIES	CLASSWORK AND HOMEWORK	TECHNOLOGY AND INTERNET RESOURCES
ATE Fig. 25-2, p. 549 **ATE Fig. 25-3,** p. 550 **ATE Fig. 25-4,** p. 551 117 Life Cycle of a Fern 118 Life Cycle of a Moss	**PE Quick Lab** What does a fern gametophyte look like?, p. 550 **PE Review,** p. 551	Holt Biology Videodiscs Section 25-1 Audio CD Program Section 25-1
ATE Fig. 25-8, p. 555 **ATE Fig. 25-11,** p. 558 115 Characteristics of Dicots and Monocots 116 Life Cycle of an Angiosperm 131 Floral Structure 132 Cross Section of an Angiosperm Ovary	**PE Quick Lab** Where are the gametophytes of pines found?, p. 554 **PE Quick Lab** How are the parts of a flower arranged?, p. 556 Lab B15 Flower Structures **PE Review,** p. 558	CNN Video Segment 20 Migrant Bees Holt Biology Videodiscs Section 25-2 Audio CD Program Section 25-2 internet**connect** SCILINKS NSTA TOPICS: • Crop pollinators • Life cycle of an angiosperm
	PE Experimental Design How Do Nutrients Affect Vegetative Reproduction?, pp. 566–567 **PE Review,** p. 562 **PE Study Zone,** p. 563	Holt Biology Videodiscs Section 25-3 Audio CD Program Section 25-3 internet**connect** SCILINKS NSTA TOPIC: Duckweed

ALTERNATIVE ASSESSMENT

PE Portfolio Projects 28–30, p. 565

ATE Alternative Assessment, p. 563

ATE Portfolio Assessment, p. 563

Lab Activity Materials

Quick Labs

p. 550 prepared slide of a fern gametophyte with archegonia and antheridia, compound microscope

p. 554 prepared slides of the following: male pine cone, female pine cone, pine ovule, hand lens, compound microscope

p. 556 gloves, monocot flower, dicot flower, paper, tape

Experimental Design

p. 566 safety goggles, lab apron, duckweed culture, Petri dishes (5), stereomicroscope or hand lens, glass-marking pens, beakers, pond water, Knop's solution, 0.1% fertilizer solution, distilled water

Demonstrations

ATE plants with sexual and/or asexual reproductive structures, p. 546

ATE variety of seeds (gymnosperm, dicot, monocot), p. 554

ATE examples of bulbs, corms, runners, and tubers, p. 559

BIOSOURCES LAB PROGRAM

- Inquiry Skills Development: Lab B15, p. 69

Chapter Theme

Reproduction

Reproduction is essential to the survival of a species. Most plants can reproduce both sexually and asexually. Sexual reproduction mixes genes from the parents and promotes variation, which tends to help plants adapt to changes in their environment. Asexual reproduction produces many identical offspring with less energy expenditure, but it is most suitable when the environment is stable.

Establishing Prior Knowledge

Before beginning this chapter, make certain students can answer the questions in **Study TIP Ready?** on page 547. If they cannot, encourage them to reread the sections indicated.

Opening Demonstration

Bring some plants with sexual and asexual reproductive structures to class and display them. To show the sexual reproductive structures of seed plants, display cones from a conifer (pine, juniper, spruce, fir) and flowers from a flowering plant (rose, hibiscus, iris). To show asexual reproductive structures, display bulbs or corms (onion, tulip, gladiolus); rhizomes (iris, fern); plantlets on runners (strawberry, spider plant); and plantlets on leaves (kalanchoë). Ask students to identify the form of reproduction each structure represents.

CHAPTER 25 Plant Reproduction

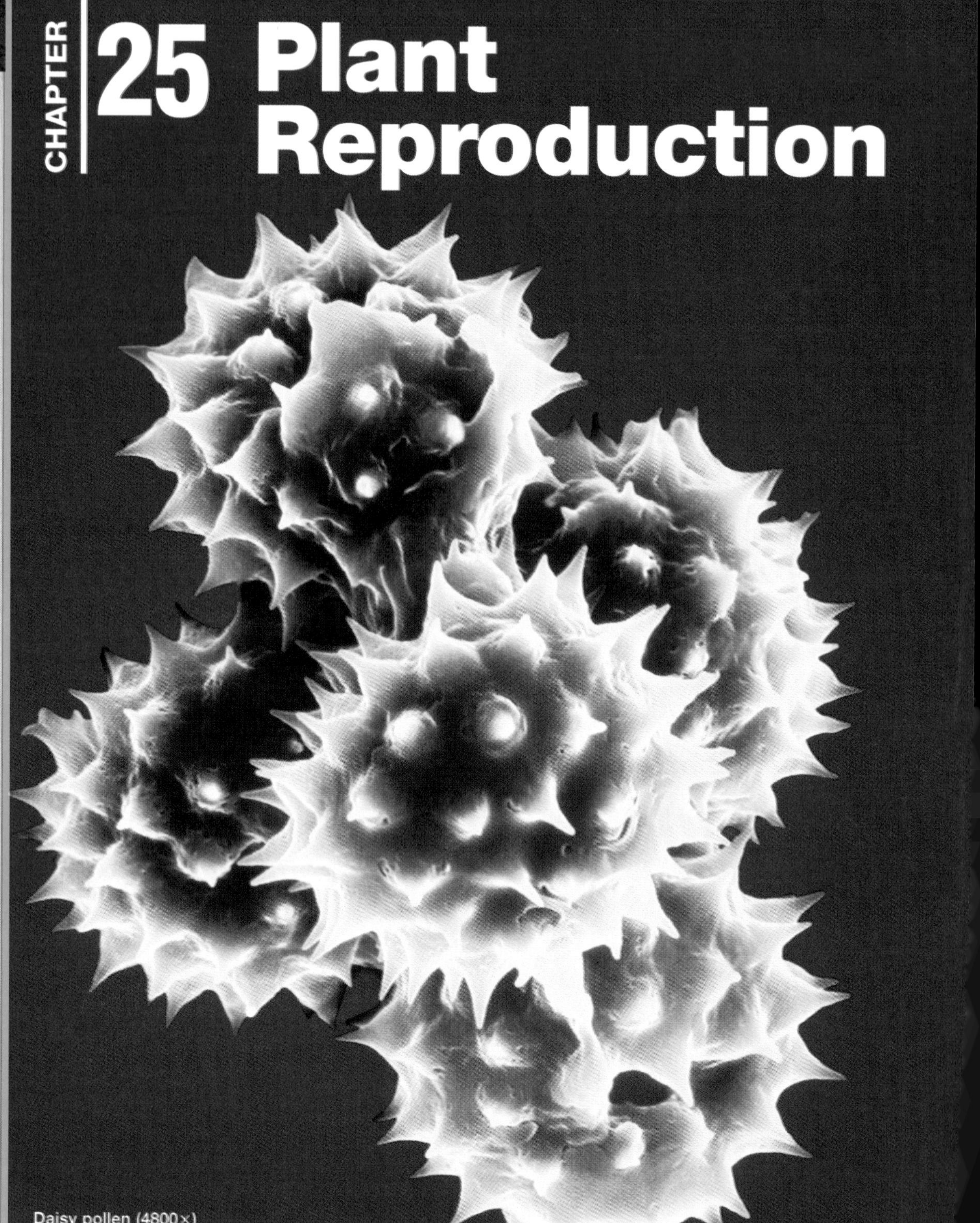

Daisy pollen (4800×)

What's Ahead — and Why
CHAPTER OUTLINE

25-1 Sexual Reproduction in Seedless Plants
- Nonvascular Plants Have Larger Gametophytes
- Seedless Vascular Plants Have Larger Sporophytes

25-2 Sexual Reproduction in Seed Plants
- Seed Plants Have Tiny Gametophytes
- Seeds Contain Plant Embryos
- Gymnosperms Produce Reproductive Structures in Cones
- Angiosperms Produce Reproductive Structures in Flowers

25-3 Asexual Reproduction
- Many Plants Reproduce Asexually
- People Grow Many Plants from Vegetative Parts

Students will learn

▷ about the sexual reproductive structures and life cycles of seedless plants.

▷ about the sexual reproductive structures and life cycles of seed plants.

▷ about how many plants reproduce asexually and how people propagate many plants vegetatively.

Chapter 25

Honeybees pollinate many of our crop plants. *In doing so, they collect the nectar they turn into honey. Beekeepers raise bees to produce honey and work with farmers to increase crop yields. Most plants require pollination for fruits to develop. Therefore, honeybees are essential for producing many of the foods we eat.*

Looking Ahead

Labs

Features

Study TIP

Ready?

Check your understanding of these topics to help you prepare for what's ahead.

Can you . . .

- **describe** the process of mitosis? (Section 6-3)
- **describe** the process of meiosis? (Section 7-1)
- **distinguish** sexual reproduction from asexual reproduction? (Section 7-2)
- **differentiate** between plant gametophytes and sporophytes? (Section 24-1)
- **define** the terms *cone* and *fruit*? (Section 24-2)

If not, *review the section indicated.*

internetconnect

National Science Teachers Association *sci*LINKS Internet resources are located throughout this chapter.

Reading Strategies

Active Reading

Technique: Reading Effectively

Before students begin to read this chapter, have them write down all of the key words for each of the three sections of the chapter. Ask them to write a definition next to each word that they have heard of. As the students read the chapter, have them write definitions next to the words that they did not previously know and modify as needed any definitions of words they had known.

Directed Reading

Assign **Directed Reading Worksheet Chapter 25** to help students interact with the text as they read the chapter.

Interactive Reading

Assign ***Biology: Principles and Explorations* Audio CD Program Chapter 25** to help reluctant readers achieve greater success in reading the chapter.

Reading Skills

Assign Reading Organizers and Reading Strategies Worksheets from the **One-Stop Planner CD-ROM with Test Generator** to help students learn and practice effective reading skills.

internetconnect

Holt *Biology: Principles and Explorations*

On-line Resources: go.hrw.com

Visit the HRW Web site for a variety of resources related to this chapter. Just type in the keyword HX0 Chapter 25.

Holt *Biology: Principles and Explorations*

On-line Resources: www.scilinks.org

The following *sci*LINKS Internet resources can be found in the student text for this chapter:

Page 557
TOPIC: Crop pollination
KEYWORD: HX557

Page 558
TOPIC: Life cycle of an angiosperm
KEYWORD: HX558

Page 567
TOPIC: Duckweed
KEYWORD: HX567

Chapter 25

1 Prepare

Scheduling

- **Block:** About 45 minutes will be needed to complete this section.
- **Traditional:** About one class period will be needed to complete this section.
- **Planning:** See the Planning Guide on pp. 545A and 545B for lecture, classwork, and assignment options for Section 25-1. Lesson plans for Section 25-1 can be found in a lesson cycle format in ***Biology: Principles and Explorations*** **One-Stop Planner CD-ROM with Test Generator.**

Resource Link

Assign **Active Reading Worksheet Section 25-1** to help students comprehend the key concepts and visuals in this lesson.

2 Teach

Lesson Warm-up

Tell students that the sperm of nonvascular plants must swim to an egg cell for fertilization to occur. Ask students to describe environmental conditions that would enable nonvascular plants to reproduce sexually. *(Nonvascular plants are able to reproduce asexually whenever moisture is abundant. Shady areas near streams are ideal habitats. In drier habitats, nonvascular plants can reproduce sexually after a rainfall or a heavy dew.)*

Section 25-1 Sexual Reproduction in Seedless Plants

Objectives

- **Summarize** the life cycle of a moss. (p. 549)
- **Summarize** the life cycle of a fern. (p. 551)
- **Compare** and **Contrast** the life cycle of a moss with the life cycle of a fern. (pp. 548–551)

Key Terms

archegonium
antheridium
sorus

Nonvascular Plants Have Larger Gametophytes

The carpet of green you often see near streams and in moist, shady places is usually made up of mosses or liverworts. As you learned in Chapter 24, these small, relatively simple plants are nonvascular plants; they do not have a vascular system for distributing water and nutrients. Mosses and liverworts do not usually thrive outside moist places because they must be covered by a film of water to reproduce sexually.

Like all plants, nonvascular plants have a life cycle called alternation of generations. In this type of life cycle, a gamete-producing stage, or gametophyte, alternates with a spore-producing stage, or sporophyte. Gametophytes produce gametes (eggs and sperm) in separate multicellular structures. The structure that produces eggs is called an **archegonium** *(ark uh GOHN ee uhm)*. The structure that produces sperm is called an **antheridium** *(an thuhr IHD ee uhm)*. Sporophytes produce spores in a sporangium. The gametophytes of nonvascular plants are larger and more noticeable than the sporophytes. This difference in size is very pronounced in the liverworts, as you can see in **Figure 25-1.**

Figure 25-1 Reproductive structures of a liverwort

The gametophytes of *Marchantia*, a common liverwort, produce male and female gametes on separate stalks.

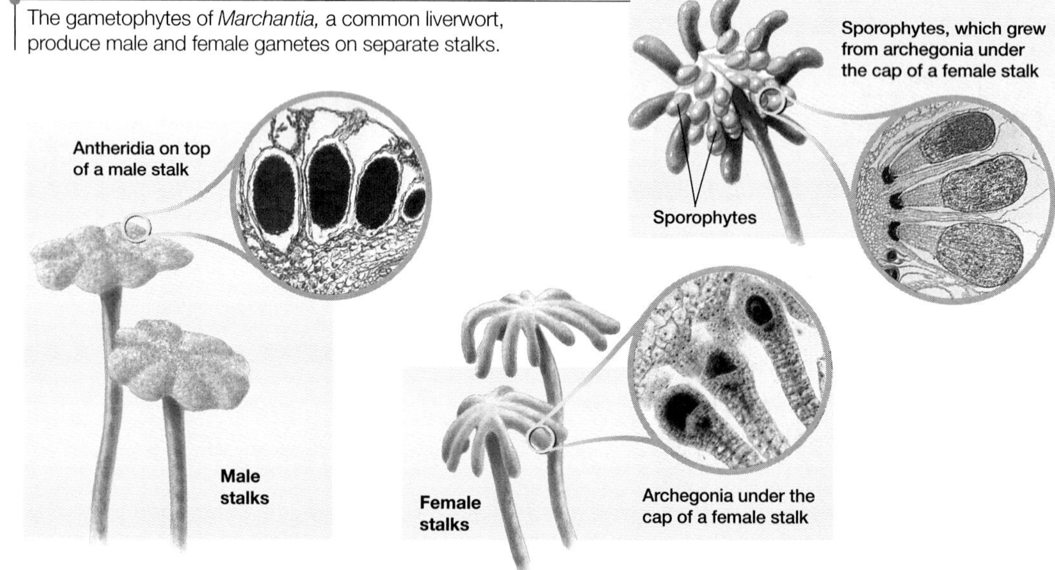

Integrating Different Learning Styles

Logical/Mathematical	**PE**	Review items 3 and 4, p. 551; Performance Zone item 25, p. 550
	ATE	Visual Strategy, p. 550
Visual/Spatial	**PE**	Quick Lab, p. 550
	ATE	Opening Demonstration, p. 546; Visual Strategy, p. 549; Visual Strategy, p. 551
Interpersonal	**ATE**	Lesson Warm-up, p. 548; Teaching Tip, p. 549
Verbal/Linguistic	**PE**	Study Tip, p. 549; Review items 1 and 2, p. 551
	ATE	Reading Strategies, p. 547; Resource Link, p. 548; Closure, p. 551; Reteaching, p. 551

Seeds Contain Plant Embryos

As you learned in Chapter 24, seeds contain the embryos of seed plants. A plant embryo is a new sporophyte. A seed forms from an ovule after the egg within it has been fertilized. The outer cell layers of an ovule harden to form the **seed coat** as a seed matures. The tough seed coat protects the embryo in a seed from mechanical injury and from a harsh environment. The seed coat also prevents the embryo from immediately growing into a young plant by keeping out water and oxygen. Deprived of water and oxygen, the embryo stops growing and cannot resume its growth until water and oxygen can pass through the seed coat. Often, a seed must be exposed to cold temperatures, or the seed coat must be damaged, before the seed can take in water and oxygen. Thus, seeds enable the embryos of seed plants to survive conditions that are unfavorable for plant growth for long periods of time.

Seeds also contain nutritious tissue that provides nutrients to plant embryos. In gymnosperms, this nutritious tissue is part of the female gametophyte. The seeds of angiosperms, however, develop a nutritious tissue called endosperm. Endosperm originates at the same time an egg is fertilized. In some angiosperms, such as corn and wheat, endosperm is still present in mature seeds. In other angiosperms, such as beans and peas, the nutrients in the endosperm have already been transferred to the embryo by the time a seed is mature.

Leaflike structures called **cotyledons** *(kah tuh LEE duhnz),* or seed leaves, are a part of a plant embryo. Cotyledons function in the transfer of nutrients to the embryo. The embryos of gymnosperms have two or more cotyledons. For example, pine embryos have eight cotyledons. In the flowering plants, the embryos of monocots have one cotyledon, and the embryos of dicots have two cotyledons. The structure of three types of seeds is shown in **Figure 25-6.**

Study TIP

Interpreting Graphics

After reading the chapter, trace or make a sketch of Figure 25-6 without the labels. On separate pieces of paper, write down the labels. Without referring to your book, match the labels with the correct parts of your sketch.

biobit

Do the embryos in seeds always develop from fertilized eggs?

In many plants, such as citrus, embryos in seeds usually do not develop from a fertilized egg. Instead, they develop from other cells in the ovule. Plants that grow from such embryos are identical to their female parent.

Figure 25-6 Seed structure

Seeds have many similarities and differences in structure.

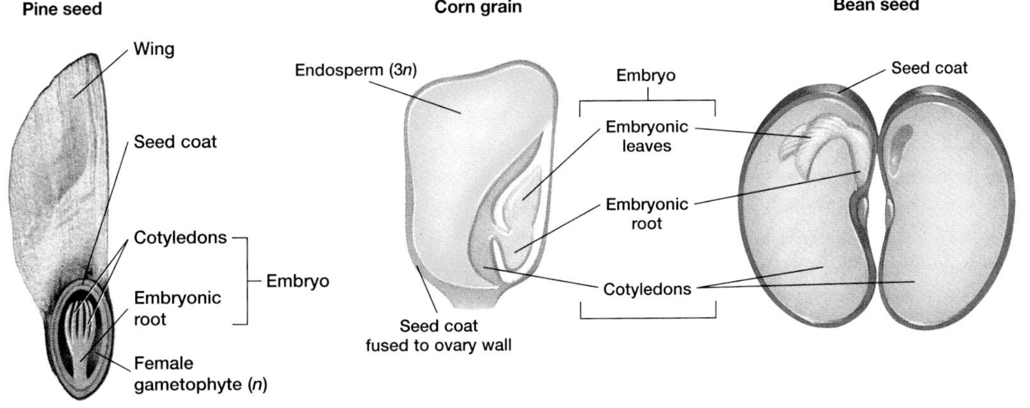

did you know?

A sunflower seed is not just a seed.

A sunflower seed is a dry fruit (called an achene) with a seed inside. The edible part is the actual seed, which consists mainly of cotyledons. A corn kernel is also not just a seed but a dry fruit (called a caryopsis) with a seed inside. The outer covering of a corn kernel is not easily removed, as is the outer covering of a sunflower seed. The edible part of the corn fruit is the whole kernel.

Active Reading

Technique: Reading Organizer

As students read about the plant life cycles in this chapter, have them complete a reading organizer, such as the one in the **Resource Link** below, for each cycle.

Resource Link

Assign **Reading Strategies: Recognizing Cycles Worksheet** from the Holt BioSources Teaching Resources CD-ROM. PORTFOLIO

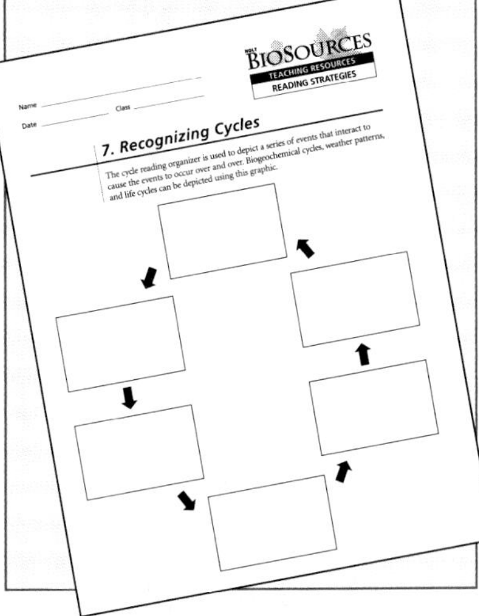
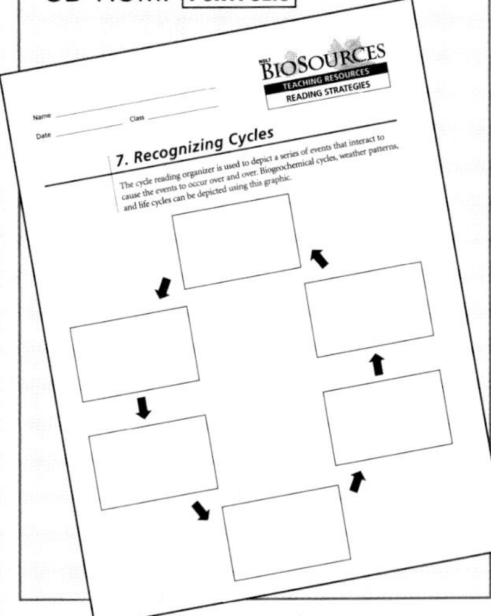

Demonstration

Dissecting Seeds — BASIC

Bring a variety of seeds to class, including at least one type of gymnosperm seed, one type of dicot seed, and one type of monocot seed. Make a longitudinal cut through at least one seed of each type. Have students examine the whole and opened seeds using their unaided eyes and a dissecting microscope. Ask students to compare the characteristics of the three different types of seeds.

Making Biology Relevant

Technology

Many products are made from gymnosperms. Wood products are the largest use of gymnosperms and include lumber, particle board, plywood, timbers, and veneers. Fiber products made from gymnosperm wood include hardboard, ceiling tile, paper, and cardboard. Gymnosperm wood is also used as a fuel. Chemical products made from gymnosperms include plastics, dyes, glues, rosin, linoleum, turpentine, lacquers, and varnishes. In your classroom, you can find paper, pencils, lumber, and linoleum that came from gymnosperms.

CAREER

Plant Propagator

Plant propagators work for wholesale and retail nurseries where they grow plants for sale. A plant propagator grows plants from seeds and by a variety of vegetative methods. Have students work in teams to use library references or on-line databases to find out about plant propagation in the nursery and floral industries. Have students write a report summarizing their findings. PORTFOLIO

Where are the gametophytes of pines found?

Time 20 minutes

Process Skills Observing, making comparisons, drawing conclusions

Teaching Strategies

- Refer to Figure 25-7 and point out that the male cones and female cones of conifers look different.
- Ask students to look for similarities between the two types of cones when they are examined under a microscope.

Analysis Answers

1. The structure is similar because both have whorls of scales, but male cones contain pollen and female cones contain eggs and eventually seeds.
2. The growth rate of pollen tubes in pines is extremely slow.

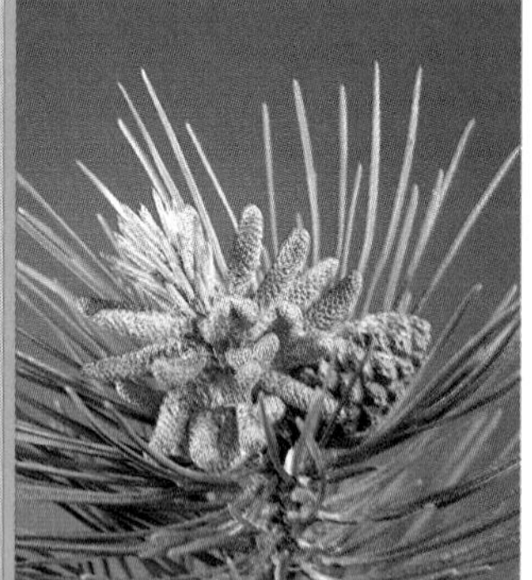

Figure 25-7 Male and female pine cones. This branch of an Austrian pine has an immature seed cone and many pollen cones.

Gymnosperms Produce Reproductive Structures in Cones

Seed plants are the most successful of all plants. The success of the seed plants is due in part to the specialized structures in which seeds develop. In angiosperms, the ovules (immature seeds) are completely enclosed by sporophyte tissue at the time of pollination. In gymnosperms, the ovules are not completely enclosed by sporophyte tissue until after pollination.

The gametophytes of gymnosperms develop in cones, which consist of whorls (circles) of modified leaves called scales. Gymnosperms produce two types of cones. Male cones, or pollen cones, produce pollen grains within sacs that develop on the surface of their scales. Female cones, or seed cones, produce ovules on the surface of their scales. Many gymnosperms produce both male and female cones on the same plant, as shown in **Figure 25-7.** In some gymnosperms, male and female cones form on separate plants. Pollen cones produce huge quantities of pollen grains that are carried by wind to female cones. At the time of pollination, the scales of a female cone are open, exposing the ovules. When a pollen grain lands near an ovule, a slender pollen tube grows out of the pollen grain and into the ovule. The sperm moves through the pollen tube and enters the ovule. Thus, the pollen tube delivers a sperm to the egg inside the ovule. Seed cones close up after pollination and remain closed until the seeds within them are mature. This process can take up to two years.

QUICK LAB

Where are the gametophytes of pines found?

You can observe the gametophytes of a pine with a microscope.

Materials

prepared slides of the following: male pine cone, female pine cone, pine ovule; hand lens; compound microscope

Immature female pine cone

Procedure

1. Examine prepared slides of male and female pine cones first with a hand lens and then under the low power of a microscope.
2. Make a sketch of each type of pine cone, and label the structures that you recognize.
3. Examine a prepared slide of a pine ovule under the low power of a compound microscope. Compare what you see with the photo above.
4. Draw a pine ovule, and label the following structures: scale, ovule, egg, pollen tube (if visible).

Analysis

1. **Compare** and **Contrast** the structure and contents of male and female pine cones.
2. **SKILL Applying Information** It takes 15 months for a pine pollen tube to grow through the wall of a pine ovule. How would you describe the rate of pollen-tube growth in pines?

Historical Note

George Washington Carver

George Washington Carver (1864–1943) was a horticulturist whose agricultural research revolutionized the economy of the southern United States by encouraging farmers to grow more than just cotton. He urged farmers to grow sweet potatoes, peanuts, pecans, and soybeans. In 1896, the same year he received his master's degree at Iowa State College of Agriculture and Mechanic Arts, Carver was asked to head the Department of Agriculture at Tuskegee Institute, in Alabama. There he helped to develop more than 300 industrial products that could be made from peanuts, 100 that could be made from sweet potatoes, and 75 that could be made from pecans.

Life cycle of a conifer

Most gymnosperms are conifers, a group that includes pines. You can trace the stages in the life cycle of a pine in **Figure 25-8.** As in all plants, a diploid zygote results from sexual reproduction in pines. The zygote develops into an embryo, which then becomes dormant (inactive). The embryo and the surrounding tissues form a seed. Seed cones open when their seeds are mature, and the seeds fall out. A pine seed has a wing that causes it to spin like the blades of a helicopter. Thus, pine seeds often travel some distance from their parent tree. When conditions are favorable for growth, the seeds grow into new sporophytes. An adult pine tree produces both male and female cones. Spores form by meiosis, which occurs inside immature cones. The spores grow into gametophytes, which produce eggs and sperm by mitosis. After pollination, a pollen tube begins to grow from each pollen grain toward the eggs inside an ovule. The pollen tube provides a way for the sperm to reach the eggs without having to swim through water. Fertilization occurs as a sperm fuses with an egg, forming a zygote that will grow into a new sporophyte.

Study TIP

Reviewing Information

Work with a partner to review the plant life cycles on pages 549, 551, 555, and 558. Review the steps of each life cycle by quizzing each other on the events of each step.

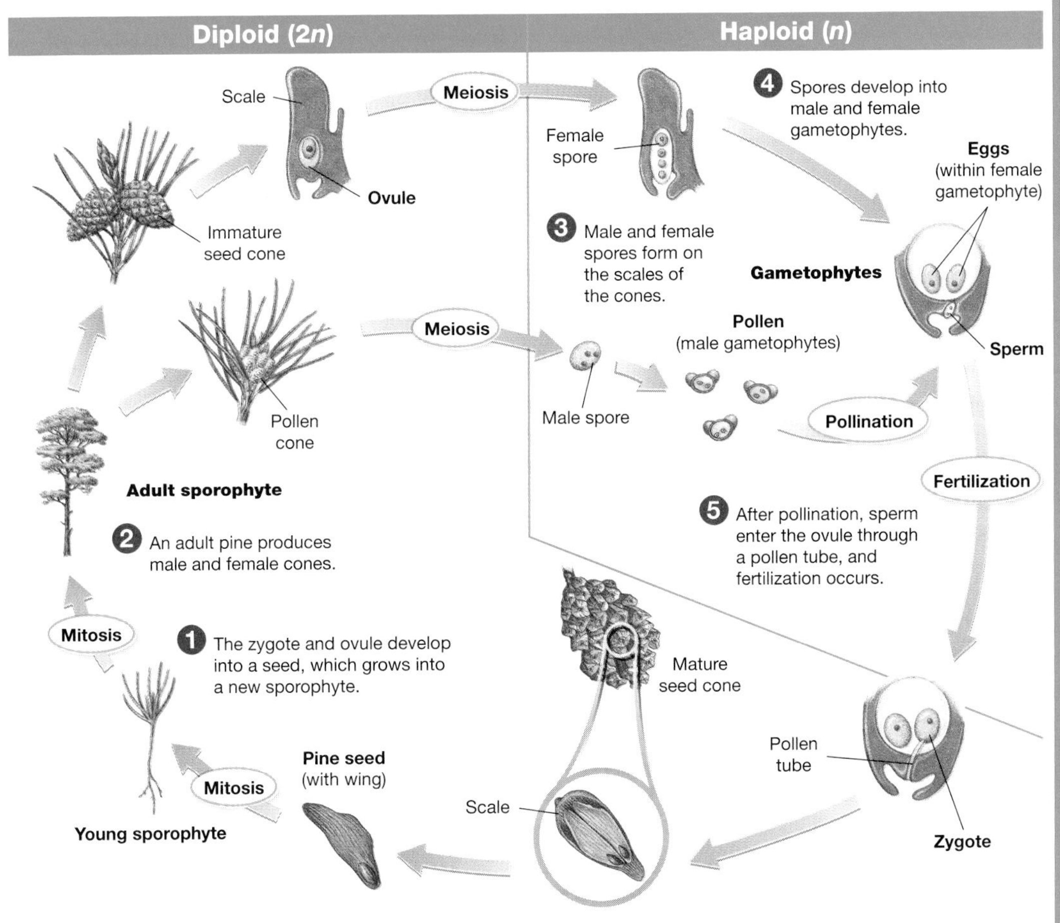

Figure 25-8 Conifer life cycle. In conifers, a very large sporophyte that produces cones alternates with tiny gametophytes that form on the scales of cones.

Visual Strategy

Figure 25-8

Walk students through the gymnosperm life cycle, starting with the zygote. Point out that, as in animals and other multicellular eukaryotic organisms, repeated mitotic cell divisions of the zygote result in a multicellular diploid organism. In plants, the multicellular diploid organism is called the sporophyte. Reproductive cells in the mature sporophyte undergo meiosis. During this process, the chromosome number is reduced from diploid to haploid. Point out that there are two kinds of spores produced—male spores (microspores) and female spores (megaspores)—and that these spores grow into two different types of gametophytes—pollen grains, or male gametophytes (microgametophytes), that produce sperm cells and female gametophytes (megagametophytes) that produce egg cells. Both kinds of gametophytes are haploid. Finally, point out that when fertilization (the uniting of an egg and a sperm) occurs and a new zygote is formed, the diploid chromosome number is restored, and a new sporophyte generation begins.

Teaching TIP — ADVANCED

Diversity of Pollen Grains

Provide students with a variety of flowers with pollen or give them pollen alone (for example, pollen from a pine tree). Have students make wet-mount slides of pollen samples to observe the intricately detailed coats of pollen grains. Have students draw what they observe. You could also provide them with pictures (for example, from a college biology textbook) showing electron micrographs of some pollen grains. PORTFOLIO

Multicultural Perspective

What is a "jumping bean"?

Jumping beans, which are grown in Central and South America, are the seeds of plants in the spurge family. A jumping bean contains a moth larva. The jumping movements occur when the moth larva inside changes position with a jerk. The jumping bean is also known as the "bronco bean" or "leaper."

CROSS-DISCIPLINARY CONNECTION

◆ **History**

The fossil remains of food plants found among the remains of the people of a past civilization tell us more about the people than just what they ate. The storage of wheat has been traced back to 10,000 B.C. People who stored crops were not nomadic hunter-gatherers; instead, they had permanent homes and grew crops.

How are the parts of a flower arranged?

Time 20 minutes

Process Skills Observing, making comparisons, drawing conclusions

Teaching Strategies

- Provide a variety of monocot and dicot flowers for the class, and have each student examine one of each type.
- Ask students to think about how the flowers they examine are pollinated. *(Flowers that are open and have stamens and stigmas exposed are likely cross-pollinated. Flowers that are closed and have stamens and stigmas that are not exposed are likely self-pollinated.)*

Analysis Answers

1. Answers will vary. Students should comment on color, size, and numbers of sepals and petals.
2. Answers will vary. Students should suggest that large, brightly colored petals attract an animal pollinator (such as an insect) while small, inconspicuous petals are wind-pollinated.
3. Answers will vary. Students should note the following: monocots have flower parts in multiples of three; dicots have flower parts in multiples of two, four, or five.

Angiosperms Produce Reproductive Structures in Flowers

In angiosperms, gametophytes develop within flowers. The basic structure of a flower is shown in **Figure 25-9.** Flower parts are arranged in four concentric whorls. The outermost whorl consists of one or more **sepals** *(SEE puhlz),* which protect a flower from damage while it is a bud. The second whorl consists of one or more **petals,** which attract pollinators. The third whorl consists of one or more **stamens** *(STAY muhnz),* which produce pollen. Each stamen is made of a threadlike filament that is topped by a pollen-producing sac called an **anther.** The fourth and innermost whorl of a flower consists of one or more **pistils,** which produce ovules. Ovules develop in a pistil's swollen lower portion, which is called the **ovary.** Usually, a stalk, called the style, rises from the ovary. Pollen lands on and sticks to the stigma—the swollen, sticky tip of the style.

Flowers may or may not have all four of the basic flower parts. A flower that has all four parts is called a complete flower. Flowers that lack any one of the four types of parts are called incomplete flowers. If a flower has both stamens and pistils, it is called a perfect flower. Flowers that lack either stamens or pistils are called imperfect flowers.

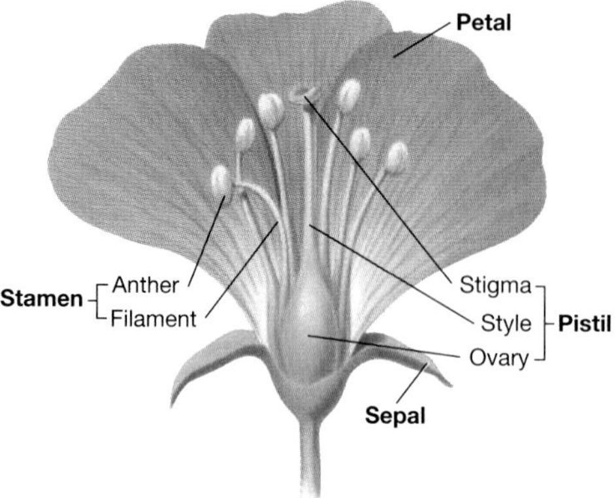

Figure 25-9 Basic flower structure. The four basic parts of a flower—sepals, petals, stamens, and pistils—are arranged in concentric whorls.

QUICK LAB

How are the parts of a flower arranged?

You can see how the parts of flowers are arranged by dissecting flowers.

Materials

gloves, monocot flower, dicot flower, paper, tape

Procedure

1. Put on gloves. Examine a monocot flower and a dicot flower. Locate the sepals, petals, stamens, and pistil of each flower.
2. Separate the parts of each flower, and tape them to a piece of paper. Label each set of parts.
3. Count the number of petals, sepals, and stamens in each flower. Record this information below each flower.

Analysis

1. **Compare** and **Contrast** the appearance of the sepals and petals of each flower.
2. **SKILL Forming a Hypothesis** For each flower, suggest a function for the petals based on their appearance.
3. **SKILL Justifying Conclusions** Explain why each flower is from either a monocot or a dicot.

did you know?

A sunflower is not a single flower.

Like the flowers of all members of the family that includes daisies and dandelions, a sunflower is a composite of many flowers. Examination with a magnifying lens of a yellow "petal" will reveal a pistil. This "petal" is actually a single flower. Sunflowers are called composites because their "flowers" are an aggregate of many flowers.

Flowers and their pollinators

Many flowers have brightly colored petals, sugary nectar, strong odors, and shapes that attract animal pollinators such as insects, birds, and bats. Flowers are a source of food for pollinators. For example, bees eat nectar and collect pollen, which is a rich source of protein they feed to their larvae. A bee gets coated with pollen as it visits a flower and then carries that pollen to other flowers.

Bees locate flowers by scent first and then by color and shape. Bee-pollinated flowers are usually blue or yellow and often have markings that show the location of nectar. Moths, which feed at night, visit heavily scented white flowers, which are easy to find in dim light. Flies pollinate flowers that smell like rotten meat.

Many flowers are not pollinated by insects. Red flowers, for instance, are typically pollinated by hummingbirds. Some large white flowers that open at night are pollinated by a nighttime visitor—bats, as seen in **Figure 25-10.** Many flowers, such as those of grasses and oaks, are pollinated by wind. Wind-pollinated flowers are usually small and lack bright colors, strong odors, and nectar.

Figure 25-10 Bat pollination. This lesser long-nosed bat pollinates an organ pipe cactus as it feeds on the pollen of the plant's flowers. Bats are important pollinators of cactuses and tropical fruit plants.

FOOD WATCH Bees Help Feed the World

Bees pollinate more species of plants than any other animal. Today there are more than 20,000 species of bees around the world. The most familiar species is the European honeybee, *Apis mellifera,* which was imported to the United States in the 1600s. Beekeepers raise European honeybees to collect the honey they produce. The bees also benefit farmers by pollinating crop plants. Wild and beekeeper-raised honeybees pollinate more than 90 kinds of cultivated crops, which are worth $10 billion a year.

Decline in honeybees

In the 1990s, the population of beekeeper-raised honeybees decreased by 25 percent in the United States. Many bees were killed by diseases or parasites. Pesticides used by farmers and home gardeners kill bees as well as pests. Honeybees continue to lose food sources as native plant habitat is replaced with buildings and parking lots.

Blue orchard bees

Farmers are turning to other species of bees to pollinate their crops. One such species is the blue orchard bee, *Osmia lignaria,* which is native to much of western North America. Blue orchard bees are more efficient pollinators than honeybees. Unlike honeybees, they will visit flowers under cloudy skies, in a light drizzle, and in colder weather. Also, blue orchard bees tend to move from tree to tree, while honeybees tend to stay in one tree. Therefore, blue orchard bees are better pollinators of trees that must be cross-pollinated, such as fruit trees. As a result, about 600 blue orchard bees can pollinate a 1 hectare (2.5 acre) apple orchard that it would take as many as 50,000 European honeybees to pollinate.

internet connect

SCILINKS NSTA

TOPIC: Crop pollination
GO TO: www.scilinks.com
KEYWORD: HX557

FOOD WATCH Bees Help Feed the World

Teaching Strategies

- Point out that many fruit trees have "self-incompatibility genes," which prevent pollen from one plant from fertilizing eggs on the same plant. Growers must plant about one row of a different variety of a fruit tree for every nine rows of a desirable variety to have a source of pollen for cross-pollination and fruit production. Pollinators are required to spread pollen to the desirable fruit trees.
- Have students research crop pollination using the Web site in the Internet Connect box on this page and write a report summarizing their findings. PORTFOLIO

Discussion

- In what ways are bees important in food production? *(They produce honey, and they pollinate the flowers of many food crops.)*
- Why have European honeybee populations declined in the United States in the last decade? *(Many European honeybees have been killed by diseases, parasites, and pesticides.)*
- Why are blue orchard bees considered to be more desirable for pollinating food crops than European honeybees? *(Blue orchard bees are active in weather conditions that cause European honeybees to be inactive; fewer of them are required to pollinate the same area; and they tend to move from tree to tree, which helps promote cross-pollination.)*

Graphic Organizer

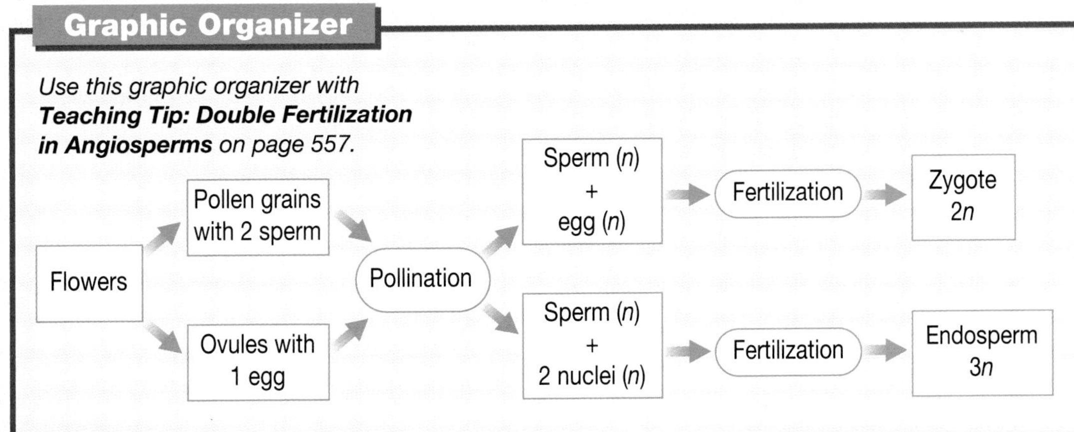

Teaching TIP

Double Fertilization in Angiosperms

Have students make a Graphic Organizer similar to the one on page 557 to illustrate the process of double fertilization in angiosperms. Ask students to use the applicable vocabulary words for this section, plus any other words that they wish to include. PORTFOLIO

internet**connect**

SCiLINKS. NSTA Have students research the life cycle of an angiosperm using the Web site in the Internet Connect box on this page. Have students write a report summarizing their findings. PORTFOLIO

Visual Strategy

Figure 25-11

Walk students through the angiosperm life cycle, starting with the zygote. Point out the similarities between this life cycle and the gymnosperm life cycle on page 555. *(Meiotic cell division in reproductive structures produces haploid male and female spores that grow into haploid male and female gametophytes. A pollen tube delivers sperm to the female gametophyte. Fertilization results in a diploid zygote that develops into the embryo of a seed and eventually into an adult sporophyte.)* Also point out the differences. *(flowers and double fertilization)*

3 Assessment Options

Closure

Classifying Structures

Provide students with a variety of cones and fruits. Try to include pine, spruce, and fir cones. Try to include peaches; tomatoes, grapes, or blueberries; apples; beans; nuts; strawberries; pineapple or figs; and elm or maple fruits. Provide a botany textbook that students can use to classify the fruits.

Review

Assign Review—Section 25-2.

Reteaching — BASIC

Have students construct a crossword puzzle using the terms presented in this section and the definitions they wrote as directed in **Teaching Tip: Reproductive Structures** on page 552. PORTFOLIO

internet**connect**

SCiLINKS. NSTA
TOPIC: Life cycle of an angiosperm
GO TO: www.scilinks.org
KEYWORD: HX558

Life cycle of an angiosperm

Figure 25-11 summarizes the life cycle of an angiosperm. Following fertilization, the zygote and the tissues of the ovule develop into a seed, which grows into a new sporophyte. The adult sporophytes of angiosperms produce spores by meiosis within the pistils and stamens of their flowers. These spores grow into gametophytes. The female gametophytes grow inside the ovules, which develop within the ovary of a pistil. The male gametophytes, or pollen grains, are produced in the anther of a stamen. A pollen grain contains two sperm cells. One sperm fuses with the egg, as in all sexually reproducing organisms, forming the zygote. The other sperm fuses with the haploid nuclei of two other cells produced by meiosis. The fusing of three haploid (n) cells forms a triploid ($3n$) cell that develops into endosperm. The term **double fertilization** is used to describe the process by which two sperm fuse with cells of a female gametophyte to produce both a zygote and endosperm.

Figure 25-11 Angiosperm life cycle. In angiosperms, a large sporophyte that produces flowers alternates with tiny gametophytes that form within the male and female structures of flowers.

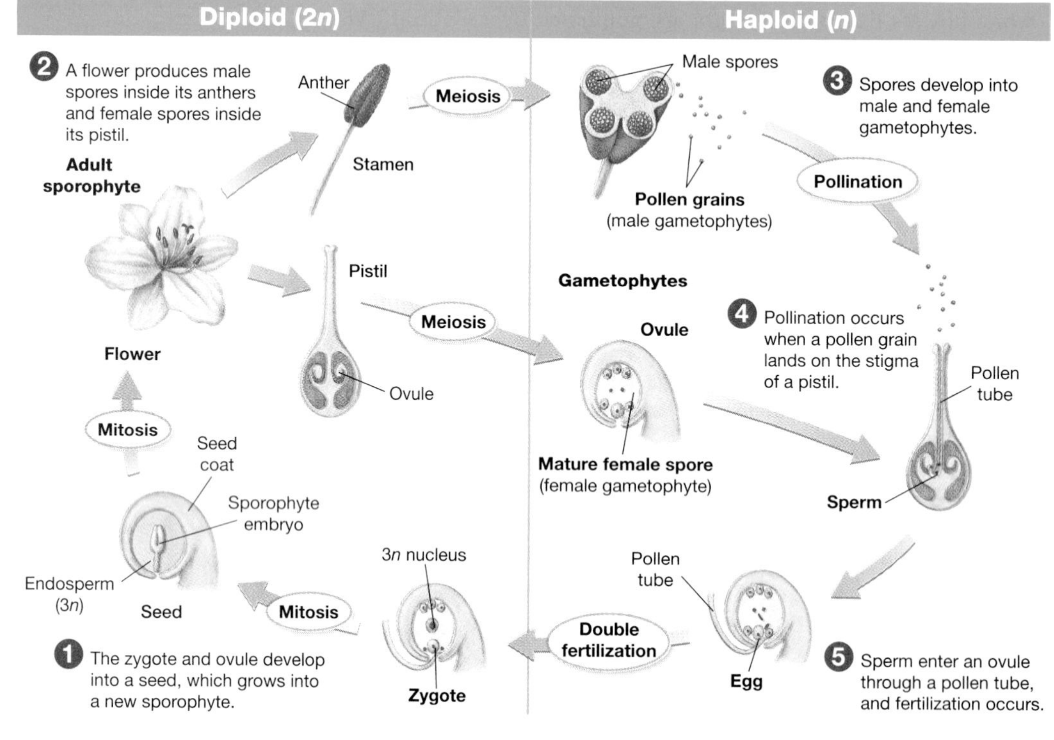

Review SECTION 25-2

1. **Distinguish** pollen grains from ovules.
2. **Describe** the function of each part of a seed.
3. **Summarize** the life cycle of a conifer.
4. **SKILL Relating Concepts** How is each part of a flower suited to its function?
5. **SKILL Summarizing Information** What are the main events in the life cycle of an angiosperm?

Review SECTION 25-2 ANSWERS

1. A pollen grain is a male gametophyte that contains cells that form sperm cells and a pollen tube. An ovule is a sporophyte structure that contains a female gametophyte, in which an egg cell forms.
2. The embryo is the young sporophyte, the food reserves nourish the young sporophyte when the seed germinates, and the seed coat protects the embryo from harsh environmental conditions.
3. First a zygote grows into a sporophyte by mitotic cell division. Next the sporophyte forms male and female spores in cones following meiotic cell division. Then the spores grow into male and female gametophytes by mitotic cell division. After the gametophytes form eggs and sperm by mitotic cell division, eggs are fertilized by sperm and a new zygote forms. The zygote and surrounding tissue develop into a seed.
4. Answers should summarize the discussion on page 556.
5. Answers should summarize the steps of Figure 25-11.

Asexual Reproduction

Section 25-3

Many Plants Reproduce Asexually

Most plants are able to reproduce asexually. The new individuals that result from asexual reproduction are genetically the same as the parent plant. Plants reproduce asexually in a variety of ways that involve vegetative (nonreproductive) parts, such as stems, roots, and leaves. The reproduction of plants from vegetative parts is called **vegetative reproduction.** Many of the structures by which plants reproduce vegetatively are modified stems, such as runners, bulbs, corms, rhizomes, and tubers. Table 25-1 describes these structures.

Vegetative reproduction is faster than sexual reproduction in most plants. A single plant can spread rapidly in a habitat that is ideal for its growth by reproducing vegetatively. Therefore, a mass of hundreds or even thousands of individuals, such as a stand of grasses or ferns, may have come from one individual. To learn about one unique method of vegetative reproduction in one plant, look at *Up Close: Kalanchoë,* on pp. 560–561.

Table 25-1 Stems Modified for Vegetative Reproduction

Name	Description	Examples
Runner	Horizontal, aboveground stem	Airplane plant, Bermuda grass
Bulb	Very short, stem with thick, fleshy leaves; only in monocots	Onion, daffodil, tulip
Corm	Very short, thickened, underground stem with thin, scaly leaves	Gladiolus, crocus
Rhizome	Horizontal underground stem	Iris, fern, sugar cane
Tuber	Swollen, fleshy, underground stem	Potato, caladium

Objectives

- **Describe** several types of vegetative reproduction in plants. (p. 559)
- **Distinguish** sexual reproduction in kalanchoës from asexual reproduction in kalanchoës. (p. 560)
- **Recommend** several ways to propagate plants. (p. 562)

Key Terms

vegetative reproduction
plant propagation
tissue culture

1 Prepare

Scheduling

- **Block:** About 90 minutes will be needed to complete this section.
- **Traditional:** About two class periods will be needed to complete this section.
- **Planning:** See the Planning Guide on pp. 545A and 545B for lecture, classwork, and assignment options for Section 25-3. Lesson plans for Section 25-3 can be found in a lesson cycle format in ***Biology: Principles and Explorations* One-Stop Planner CD-ROM with Test Generator.**

Resource Link

- Assign **Active Reading Worksheet Section 25-3** to help students comprehend the key concepts and visuals in this lesson.

2 Teach

Lesson Warm-up

Ask students to think of examples of how flowering plants reproduce without making seeds. Help them by asking about gardening, hiking, or other nature experiences they may have had. They may have seen clumps of plants of the same type, runners on lawn grasses or strawberries, or clusters of bulbs. Ask students why all the plants in a lawn might look the same. *(They are all clones from one plant.)*

Demonstration

Modified Stems and Vegetative Reproduction

Show students examples of bulbs (lily, onion, tulip); corms (crocus, gladiolus); rhizomes (iris); runners (airplane plant, strawberry); and tubers (potato). Point out that all of these are modified stems and therefore have buds that can grow into new plants. Demonstrate vegetative propagation by rooting an airplane plant or a potato. Be sure to include a bud in the section you root.

Integrating Different Learning Styles

Logical/Mathematical	**PE**	Review items 2, 3, and 4, p. 562; Performance Zone item 24, p. 565
Visual/Spatial	**PE**	Performance Zone item 1, p. 564; Up Close, pp. 560–561
	ATE	Demonstration, p. 559; Up Close, p. 560
Interpersonal	**PE**	Performance Zone item 28, p. 565
	ATE	Lesson Warm-up, p. 559; Up Close, p. 561; Biology in Your Community, p. 561
Verbal/Linguistic	**PE**	Review item 1, p. 562; Performance Zone item 23, p. 565
	ATE	Closure, p. 562; Reteaching, p. 562

Up Close

Kalanchoë

Teaching Strategies

- Bring specimens of *Kalanchoë daigremontiana* to class. Let students propagate the plant by planting stem cuttings, leaf cuttings, or plantlets.
- Point out to students that the formation of plantlets along the leaf margin that occurs in kalanchoës is a very unusual form of asexual reproduction that occurs in only a few species. Tell students that these plants are often grown commercially because of their unusual appearance and method of asexual reproduction. Also tell students that because kalanchoës are easy to grow, they often become serious greenhouse weeds. Tell students that this plant is often referred to as the "maternity plant."
- For interest, take students to a greenhouse where succulents are grown, or have a horticulturist talk to students about the characteristics and propagation of kalanchoës and other succulents.
- CAM photosynthesis may be difficult for students to comprehend. Help students understand CAM photosynthesis by having them illustrate the process.

Up Close

Kalanchoë

- **Scientific name:** *Kalanchoë daigremontiana*
- **Size:** Grows from 30 cm (1 ft) to 1 m (3 ft) tall
- **Range:** Native to southwestern Madagascar; cultivated worldwide
- **Habitat:** Semiarid tropical grassland with moist summers and well-drained, fertile soil
- **Importance:** Kalanchoës *(kal an KOH eez)* are grown as indoor potted plants and as outdoor perennials in warm climates.

External Structures

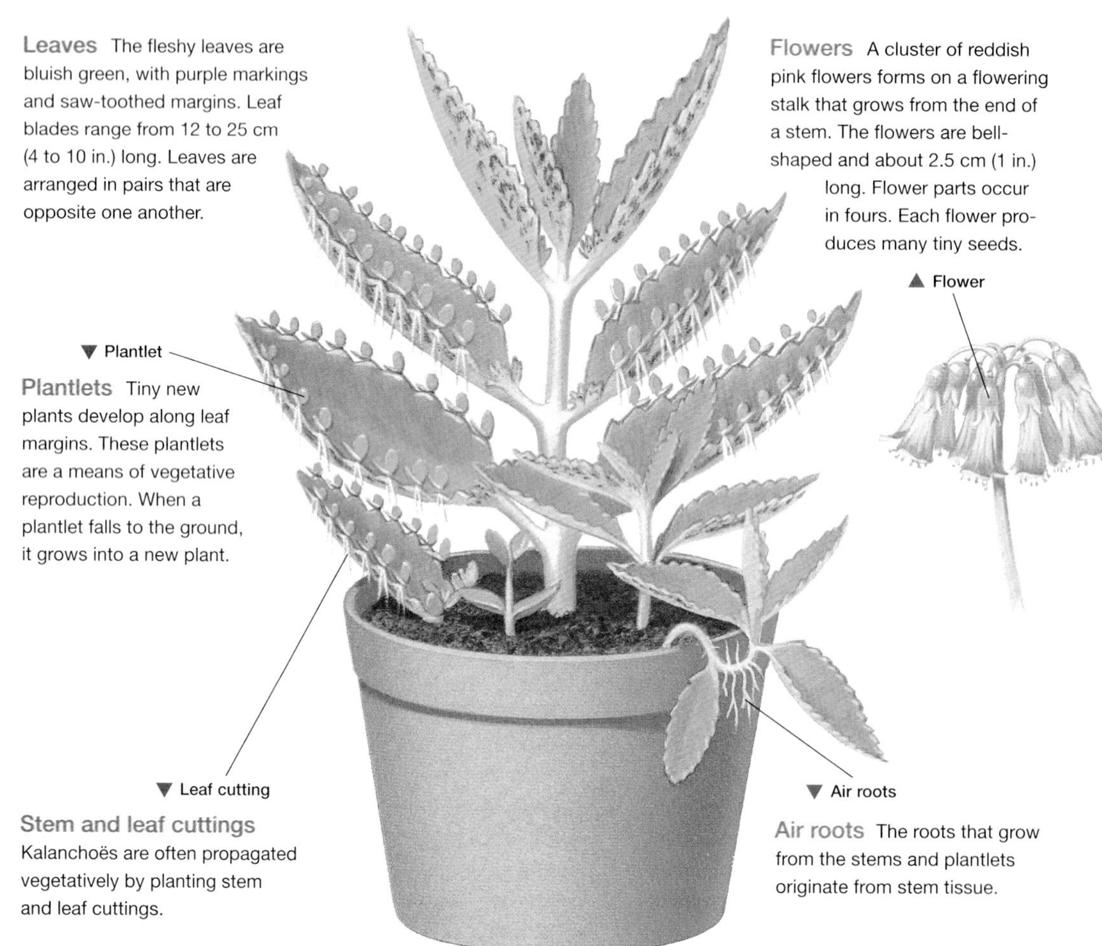

Leaves The fleshy leaves are bluish green, with purple markings and saw-toothed margins. Leaf blades range from 12 to 25 cm (4 to 10 in.) long. Leaves are arranged in pairs that are opposite one another.

Flowers A cluster of reddish pink flowers forms on a flowering stalk that grows from the end of a stem. The flowers are bell-shaped and about 2.5 cm (1 in.) long. Flower parts occur in fours. Each flower produces many tiny seeds.

Plantlets Tiny new plants develop along leaf margins. These plantlets are a means of vegetative reproduction. When a plantlet falls to the ground, it grows into a new plant.

Stem and leaf cuttings Kalanchoës are often propagated vegetatively by planting stem and leaf cuttings.

Air roots The roots that grow from the stems and plantlets originate from stem tissue.

did you know?

Many commercial fruit trees are composed of two different plants.

Fruit trees, such as those that produce apples and citrus fruits, are composed of one variety (genotype) of a species grafted onto another variety. The root stock is often that of a very hardy variety that can withstand harsh environmental conditions and is resistant to diseases common for that crop. The scion *(SEYE on),* or the part that is grafted onto the root stock, is a variety that has desirable characteristics (such as flavor, color, early fruit set), but it is not well adapted to the environmental conditions under which it is grown.

Internal Structures

Leaf structure Kalanchoës are succulents, which means they have fleshy leaves and stems that store water. A kalanchoë leaf shows how some succulents are adapted for conserving water. A thick cuticle covers the leaf, and the epidermis (outer layer of cells) consists of several layers of cells. Relatively few, very small stomata dot the leaf surfaces.

Large central vacuole The cells inside a leaf, called the mesophyll cells, have a large central vacuole that can hold a great deal of water.

Mesophyll cells

CAM photosynthesis Kalanchoës belong to the Crassulaceae family, a group of succulent plants that are adapted to hot climates. Photosynthesis in kalanchoës involves a process called crassulacean acid metabolism (CAM). The stomata of CAM plants open only at night, unlike those of other plants. At night, the plants fix carbon dioxide by using it to make malic acid. The malic acid is stored in the large central vacuoles of the mesophyll cells. In daytime, the stomata remain closed, which prevents water loss. Carbon dioxide is released from malic acid during the day and used by the Calvin cycle to make a sugar.

Mesophyll cell

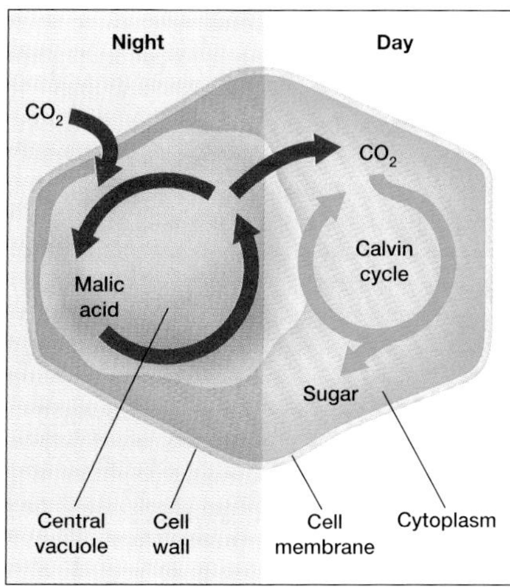

Up Close

Kalanchoë

Discussion

- Ask students to describe sexual and asexual reproduction in kalanchoës. *(Kalanchoës reproduce sexually by forming flowers and seeds following pollination and fertilization. They reproduce asexually by forming plantlets containing roots, stems, and leaves on the margins of the mother plant's leaves. They can also produce new plants from pieces of stems and leaves.)*
- Ask students how the genetic composition of a plantlet compares with that of the mother plant. *(They are identical because the plantlets are clones of the mother plant.)*
- Ask students to name some adaptations found in kalanchoës that enable them to survive in dry environments. *(Kalanchoës have thick, fleshy leaves and stems that store water; the leaves have a thick cuticle, several layers of epidermal cells, and few stomata; the plants photosynthesize using the CAM process, which enables them to have their stomata open at night and capture energy from the sun during the day.)*

Biology in Your Community

Bioprospecting

Like *Kalanchoë daigremontiana,* many other horticulturally important plants are natives of Madagascar, which is an island off the southeast coast of Africa. One such plant is the rosy periwinkle, *Catharanthus roseus,* which is a popular bedding plant in the United States and the original source of two cancer treatment drugs. Many such economically important plants have been collected in their native lands, which are often developing countries with very low incomes and standards of living.

Ask students to look for specimens of *K. daigremontiana* and rosy periwinkle at a local nursery or garden center. Have them research the origin of other ornamental plants. Then have students research the practice of bioprospecting and the concept of *intellectual property rights* and then debate whether people should have the right to use valuable plants for economic gain without compensation to the people of the countries in which the plants are collected.

Chapter 25

Teaching TIP

Asexual Versus Sexual Reproduction

Point out to students that asexual reproduction enables plants to produce many offspring by using less energy than is required for sexual reproduction. However, because offspring produced asexually are genetically identical to their parent, they are all equally susceptible to any environmental condition or disease that the parent was susceptible to. Remind students that sexual reproduction results in genetic variation among the offspring. This is particularly advantageous when the environment is very changeable or when new disease-causing agents enter the environment.

3 Assessment Options

Closure

Comparing and Contrasting

Ask students to identify the characteristic that distinguishes asexual reproduction from sexual reproduction. *(In asexual reproduction, new plants are produced from vegetative parts such as roots, stems, or leaves. In sexual reproduction, new plants are produced following meiosis and fertilization in reproductive structures.)* Then ask students how the offspring produced by each type of reproduction differ. *(Offspring produced by asexual reproduction are genetically identical to each other and to their parent; offspring produced by sexual reproduction are genetically different from each other and from their parents.)*

Review

Assign Review—Section 25-3.

Reteaching BASIC

Ask students to describe characteristics of plants that would indicate that they have been asexually propagated. *(uniformity of appearance, as in flower color, fruit color and size, plant size, etc.)*

People Grow Many Plants from Vegetative Parts

Figure 25-12 Budding pears. A bud from a desirable variety of pears is attached to a stem of another pear species. The bud grows into a branch that produces the desirable variety of pears.

People grow plants for many purposes, such as for food, to beautify homes, or to sell. Most field crops, such as cereal grains, vegetables, and cotton, are grown from seed. Many other plants are grown from vegetative parts. Growing new plants from seed or from vegetative parts is called **plant propagation.**

Plants are often propagated with structures the plants produce for vegetative reproduction. Bulbs and corms divide as they grow, forming many pieces that can each grow into a new plant. Rhizomes, roots, and tubers can be cut or broken into pieces with one or more buds that can grow into new shoots. But people also grow plants from vegetative parts that are not specialized for vegetative reproduction. For example, pieces of plants, such as the stems of ivys and the leaves of African violets, are cut from the parent plant. The cuttings are then used to grow new plants. Figure 25-12 shows a method of propagating trees called budding. In a technique called **tissue culture,** pieces of plant tissue are placed on a sterile medium and used to grow new plants. Table 25-2 summarizes some of the methods of vegetative plant propagation that are widely used to grow plants.

Table 25-2 Methods of Vegetative Plant Propagation

Method	Description	Examples
Budding and grafting	Small stems from one plant are attached to larger stems or roots of another plant.	Grape vines, hybrid roses, fruit and nut trees
Taking cuttings	Leaves or pieces of stems or roots are cut from one plant and used to grow new individuals.	African violets, ornamental trees and shrubs, figs
Tissue culture	Pieces of tissue from one plant are placed on a sterile medium and used to grow new individuals.	Orchids, potatoes, many houseplants

Review SECTION 25-3

1. **Describe** four types of vegetative reproduction in plants, and give an example of each.
2. **Classify** methods of reproduction in kalanchoës as sexual or asexual.
3. **Recommend** five ways to propagate plants.

Critical Thinking

4. **SKILL Justifying Conclusions** Why would someone choose to propagate a particular plant for commercial purposes by using vegetative structures instead of seed?

Review SECTION 25-3 ANSWERS

1. bulbs (daffodil, onion, tulip); corms (crocus, gladiolus); rhizomes (fern, iris, sugar cane); tubers (potato, caladium); rooting of aerial stems (airplane plant, kalanchoë); formation of plantlets on leaves (kalanchoë)
2. Formation of seeds is sexual reproduction. Formation of plantlets on the leaf margins or growth of new plants from pieces of stems or leaves is asexual reproduction.
3. planting seeds; planting vegetative reproductive structures such as bulbs, corms, tubers, rhizomes, or plantlets; budding or grafting; taking stem or leaf cuttings; tissue culture
4. Someone might choose to propagate a plant commercially using vegetative structures to produce a new crop rapidly and/or to produce a crop of identical plants.

Study ZONE

CHAPTER 25

Use the Key Concepts and Key Terms listed below to review the main ideas in this chapter. If you need to review the meaning of a term, turn to the page indicated.

Key Concepts

25-1 Sexual Reproduction in Seedless Plants

- In mosses, the "leafy" green gametophytes are larger than the sporophytes, which consist of a bare stalk and a spore capsule.
- In the life cycle of a fern, the sporophytes are much larger than the gametophytes. The thin, green, heart-shaped gametophytes produce both sperm and eggs.
- Nonvascular plants and seedless vascular plants need water for fertilization because sperm must swim to eggs.

25-2 Sexual Reproduction in Seed Plants

- The tiny gametophytes of seed plants develop from spores that remain within sporophyte tissues. Male gametophytes develop into pollen grains. Female gametophytes develop inside ovules.
- A seed contains an embryo, which is a new sporophyte, and a supply of nutrients for the embryo. The cotyledons of an embryo help transfer nutrients to the embryo. A seed coat covers and protects a seed.
- In gymnosperms, male and female gametophytes develop in separate cones on the sporophytes. After fertilization, ovules develop into seeds, which grow into new sporophytes.
- Flowers have four types of parts—petals, sepals, stamens, and pistils. Petals attract pollinators. Sepals protect buds and may also attract pollinators. Pollen forms in the anthers of stamens. Seeds develop in the ovary of a pistil.
- In angiosperms, male and female gametophytes develop in the flowers of the sporophytes. After fertilization, ovules develop into seeds, which grow into new sporophytes.

25-3 Asexual Reproduction

- Vegetative reproduction is the growth of new plants from non-reproductive plant parts, such as stems, roots, and leaves.
- Kalanchoës are succulents that are often grown as potted plants and readily reproduce either vegetatively or by seeds.
- People often grow plants from their vegetative structures. This is called vegetative propagation.

Key Terms

25-1

archegonium (548)
antheridium (548)
sorus (550)

25-2

pollen grain (552)
ovule (552)
pollination (552)
pollen tube (552)
seed coat (553)
cotyledon (553)
sepal (556)
petal (556)
stamen (556)
anther (556)
pistil (556)
ovary (556)
double fertilization (558)

25-3

vegetative reproduction (559)
plant propagation (562)
tissue culture (562)

Study TIP Ready?

If you think you understand these Key Concepts and Key Terms, then you're ready for the Performance Zone.

Study ZONE

CHAPTER 25

Review and Practice

Assign the following worksheets for Chapter 25:

- **Student Review Guide**
- **Concept Mapping Worksheet**
- **Test Preparation Pretest**
- **Vocabulary Worksheet**
- **Critical Thinking Review Worksheet**

Assessment Options

The following assessment options are available for this chapter:

Resource Link

- Assign **Chapter Test 25** to assess students' understanding of the chapter.

Alternative Assessment

Assign students to four cooperative groups. Have each group draw and label a life cycle diagram of a different type of plant (moss, fern, gymnosperm, angiosperm). Collect the diagrams, and then redistribute them to the groups for evaluation, making sure that none of the groups gets its own diagram back. Discuss each life cycle diagram, making corrections as necessary. PORTFOLIO

Portfolio Assessment

- *Portfolio Assessment* at the front of this book provides options for building and evaluating student portfolios. Students and teachers can select items from the areas listed for inclusion in a portfolio.
- See items labeled PORTFOLIO on pp. 553, 554, 555, 557, and 558.

Answer to Concept Map

The following is one possible answer to Performance Zone item 1 on page 564.

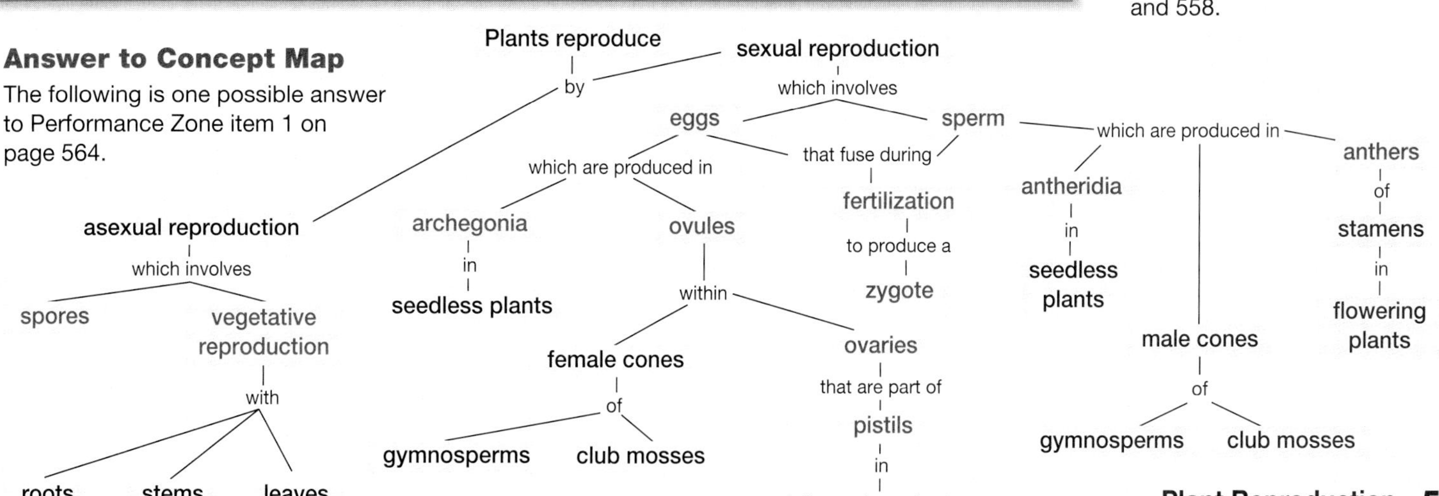

Performance ZONE
CHAPTER 25
REVIEW ANSWERS

Assignment Guide

Section	Review
25-1	3, 4, 5, 25
25-2	2, 6, 7, 8, 9, 13, 14, 15, 16, 17, 18, 19, 20, 21, 22, 26, 27, 29, 30
25-3	1, 10, 11, 12, 23, 24, 28

Understanding and Applying Concepts

1. The answer to the concept map is found at the bottom of page 563.
2. **a.** Pollen grains are mature male gametophytes that consist of only two or three cells and produce sperm cells. An ovule is a multicellular structure that develops in a female cone or in the ovary of a flower and in which a female gametophyte that contains an egg cell develops.

   b. A sepal encloses and protects a flower bud, while a petal usually attracts animal pollinators.

   c. A cotyledon is a structure of a plant embryo in all seeds, and endosperm is a 3*n* stored food supply that is not part of the embryo and is found only in angiosperm seeds.

   d. Reproduction is the process by which organisms produce offspring. Propagation is the process of producing plants (or other organisms) from seed or from vegetative structures.
3. c
4. a
5. c
6. d
7. b
8. a
9. d
10. d
11. b
12. c

Performance ZONE CHAPTER 25 Review

Understanding and Applying Concepts

1. **Concept Mapping** Make a concept map that explains how plants reproduce. Include the following terms in your map: archegonium, antheridium, egg, sperm, ovule, zygote, stamen, anther, pistil, ovary, fertilization, spore, vegetative reproduction. Include additional terms in your map as needed.
2. **Understanding Vocabulary** For each pair of terms, explain the difference in their meanings.
   a. pollen grain, ovule
   b. sepal, petal
   c. cotyledon, endosperm
   d. reproduction, propagation
3. Mosses and liverworts thrive in a moist environment because they need _______ for reproduction.
   a. bees c. water
   b. birds d. wind
4. A structure that produces eggs in mosses and ferns is called a(n)
   a. archegonium. c. ovule.
   b. sporangium. d. antheridium.
5. The life cycle of a moss differs from the life cycle of a fern in that
   a. the gametophyte is absent in ferns.
   b. the sporophyte is absent in mosses.
   c. moss spores do not form on leaves.
   d. the gametophytes of mosses are green.
6. In seed plants, the _______ transfers sperm from a pollen grain directly to an egg in an ovule.
   a. pollinator c. endosperm
   b. seed coat d. pollen tube
7. In conifers, the sporophyte produces spores and gametophytes in
   a. flowers. c. sori.
   b. cones. d. sporangia.
8. Which part of a flower produces eggs?
   a. pistil c. stamen
   b. petal d. sepal
9. In angiosperms, the zygote and the first cell of the endosperm form by
   a. mitosis. c. pollination.
   b. meiosis. d. double fertilization.
10. Vegetative reproduction has *not* occurred when a new plant grows from a
    a. leaf. c. stem.
    b. root. d. seed.
11. Which of the following structures do kalanchoës produce for vegetative reproduction?
    a. seeds c. flowers
    b. plantlets d. bulbs
12. Which of the following structures is *not* used to propagate dicots vegetatively?
    a. tubers c. bulbs
    b. rhizomes d. stem cuttings
13. Look at the flower in the photograph below. It is the flower of the unicorn plant. How do you think this flower is probably pollinated? Justify your answer.

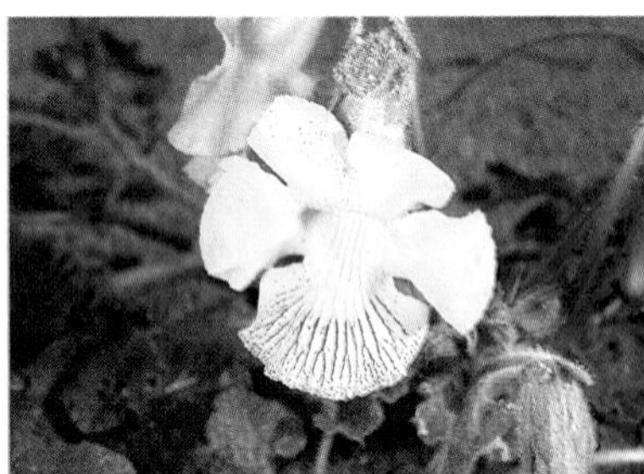

14. **Theme Reproduction** How does each of the four whorls of parts that make up the basic structure of a flower function in reproduction?
15. **Food Watch** What is being done to prevent the decline in honeybees from affecting American food sources?
16. **Chapter Links** What is the function of the fruits in which seeds mature? (**Hint:** See Chapter 24, Section 24-1.)

---

13. Students should recognize that this flower is probably bee pollinated because it is yellow, has a landing platform, and has pollen guides showing the location of nectar.
14. Sepals enclose and protect the flower bud, and, in some flowers, they resemble petals and help attract pollinators. Petals usually attract pollinators. Stamens produce sperm cells. Pistils produce egg cells.
15. To prevent the decline in honeybees from affecting American food sources, growers are turning to other pollinators such as blue orchard bees, which are more efficient pollinators than honeybees.
16. The dispersal of seeds is the main function of fruits.

Performance ZONE
CHAPTER 25

Skills Assessment

Part 1: Interpreting Graphics

Use the diagram below to answer the questions that follow:

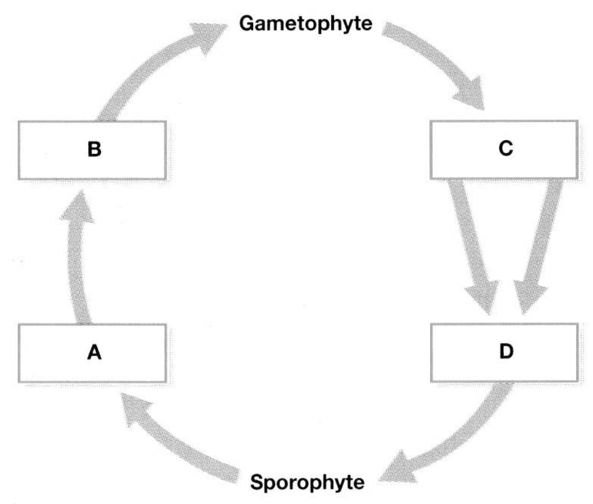

17. What process is shown in *A*?
18. What structures are produced at *B*?
19. What structures are produced at *C*?
20. What process takes place at *D*?
21. Which parts of the life cycle are haploid?
22. Which parts of the life cycle are diploid?

Part 2: Life/Work Skills

Use the media center or Internet sources to find out how the plants commonly sold in your local garden centers and plant nurseries are propagated.

23. **Finding Information** Write a report summarizing the most common method used to propagate each of the plants you researched. Explain why each plant is usually propagated by this method instead of another method.

24. **Selecting Technology** You are asked to grow a large number of identical potted plants for a florist. The plants can be grown from either seeds or cuttings. Which method of plant propagation would you use? Justify your choice.

Critical Thinking

25. **Evaluating Conclusions** All nonvascular plants require a film of water for sperm to swim through and fertilize eggs. Therefore, many people conclude that nonvascular plants are not able to survive in very dry climates, such as deserts. Is this a valid conclusion? Justify your answer.

26. **Justifying Conclusions** A classmate has found a plant whose flowers lack petals and have many stamens. Your classmate tells you that the plant is wind-pollinated. Justify this conclusion.

27. **Distinguishing Revelant Information** What evidence supports the conclusion that blue orchard bees are more efficient pollinators than European honeybees?

Portfolio Projects

28. **Working Cooperatively** Go with two or three of your classmates to visit a wholesale plant grower. Orchid growers would be an excellent choice if one is in your area. Find out how tissue culture is used to propagate different kinds of plants. Try to find out information on two methods of vegetative propagation that are not described in this chapter. Prepare an illustrated report of your findings to share with the class.

29. **Interpreting Information** Read the article "A Garden of Mutants," by Carol Fletcher (*Discover,* August 1995, pp. 49–53). Why do botanists study genetics using the *Arabidopsis* plant? What is the MADS-box? What causes the flower mutations in *Arabidopsis*?

30. **Career Focus Plant Breeding** Use the media center or Internet to find out about the field of plant breeding. Write a report on your findings. Your report should include a job description, the training required, names of employers, growth prospects, and an average starting salary.

Skills Assessment

17. meiosis
18. spores
19. gametes
20. fertilization
21. spores, gametophytes, and gametes
22. sporophyte
23. Answers will vary.
24. Answers will vary, but students should recommend a type of vegetative propagation, such as cuttings or tissue culture, since these methods produce plants that are genetically identical to each other and to their parent plant.

Critical Thinking

25. Answers will vary. Students should offer evidence to support their conclusion. Some students may recognize that this conclusion is not valid because nonvascular plants are found in dry climates. These plants survive in shady spots under trees and rocks, where moisture accumulates occasionally, and near springs, creeks, and rivers.
26. The primary role of petals is to attract animal pollinators. Therefore, the absence of petals is a strong indication that the plant is wind-pollinated.
27. Blue orchard bees visit flowers when skies are cloudy, during a light drizzle, and during cold weather. European honeybees do not. Also, blue orchard bees tend to move from tree to tree, while European honeybees tend to stay in one tree.

Portfolio Projects

28. Answers will vary.
29. *Arabidopsis* is small, easy to grow, has a short life cycle, produces thousands of seeds per plant, and has the smallest known genome of any plant. The MADS-box is a region of DNA in the *Arabidopsis thaliana* genome that codes for a protein that binds to another region of DNA that codes for a gene involved in flower development and turns on the gene. Each mutation was caused by the loss of a single class of gene.
30. Plant breeders are scientists who work to improve plants that have commercial value. Plant breeders make crosses between plants that have desirable characteristics and select plants with the most desirable combination of traits. Some plant breeders use genetic engineering in their work. Plant breeders work for universities, federal agencies, and private industries. To become a plant breeder, you must obtain a master's degree or a Ph.D. in agronomy, horticulture, or plant breeding. Starting salary will vary by region.

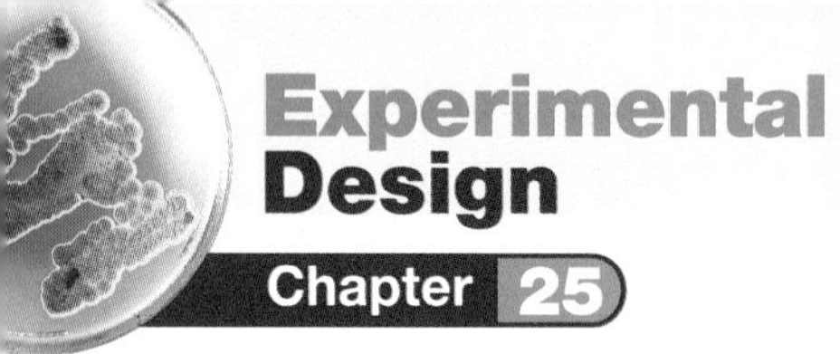

Experimental Design

Chapter 25

How Do Nutrients Affect Vegetative Reproduction?

Purpose
Students will design and conduct a scientific experiment to study the effects of mineral nutrients on vegetative reproduction in duckweed.

Preparation Notes
Time Required: Day 1, 45 minutes; then 10 minutes each day for 2 weeks

Materials
Materials for this lab can be ordered from WARD'S. See *Master Materials List* at the front of this book for catalog numbers.

Preparation Tips
- Obtain duckweed from a pond or a quiet stream. You can also purchase duckweed from a garden center or from a biological supply company.
- To make a 0.1 percent fertilizer solution, place 1 g of a commercial fertilizer (with an analysis of at least 23-19-17) in a large container, and add enough distilled water to make 1 L of solution. Stir until the fertilizer is dissolved.
- Place enough materials for a class on a supply table.
- Set up labeled containers for the disposal of solutions, broken glass, and duckweed.
- Before students begin the lab, conduct a class discussion in which you ask the following questions:
  1. What is vegetative reproduction? *(the process of making new individuals from nonreproductive parts of an organism)*
  2. How do mineral nutrients affect the growth of vegetative structures in plants? *(They provide additional elements needed for making organic macromolecules.)*

Safety Precautions
- Be sure that students have read and understand all of the safety rules for working in the lab.
- Remind students to read the **ChemSafety** section before beginning the lab. Discuss all safety symbols and caution statements with students.
- Make sure students wear safety goggles and a lab apron.
- Make sure students wash their hands thoroughly before leaving the laboratory.

Procedural Tips
- Require students to present a written procedure for their experiment and a list of all safety precautions before allowing them to gather materials for Part C of the lab.
- Students should see abundant vegetative reproduction in the 2-week period.

Disposal
- Dilute nutrient solutions with water, and pour them down the sink.
- Wrap duckweed in newspaper, and place into a trash can.

Experimental Design

Chapter 25 How Do Nutrients Affect Vegetative Reproduction?

SKILLS
- Using scientific processes
- Observing
- Graphing and analyzing data

OBJECTIVES
- **Identify** the structures of duckweed.
- **Compare** vegetative reproduction of duckweed in different nutrient solutions.

MATERIALS
- safety goggles
- lab apron
- duckweed culture
- 5 Petri dishes
- stereomicroscope or hand lens
- glass-marking pen
- beakers
- pond water
- Knop's solution
- 0.1% fertilizer solution
- distilled water

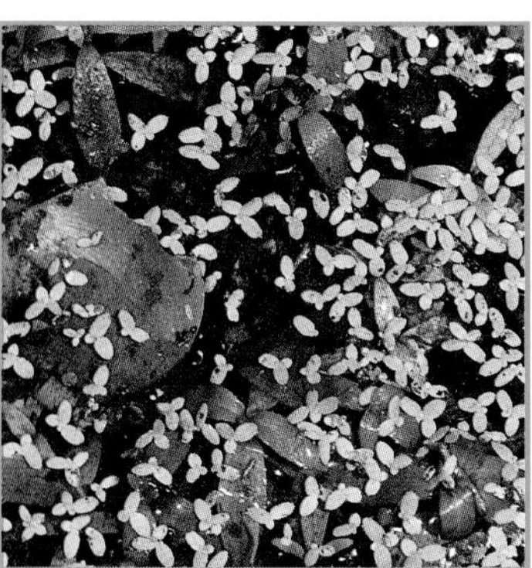

ChemSafety

CAUTION: Always wear safety goggles and a lab apron to protect your eyes and clothing.

CAUTION: Do not touch or taste any chemicals. Know the location of the emergency shower and eyewash station and how to use them. If you get a chemical on your skin or clothing, wash it off at the sink while calling to the teacher. Notify the teacher of a spill. Spills should be cleaned up promptly, according to your teacher's directions.

CAUTION: Glassware is fragile. Notify the teacher of broken glass or cuts. Do not clean up broken glass or spills with broken glass unless the teacher tells you to do so.

Before You Begin
Duckweed is a common aquatic plant. Like many flowering plants, duckweed reproduces readily by **vegetative reproduction,** which is a type of **asexual reproduction.** As individual plants grow, they divide into smaller individuals. Several individuals may remain joined together, forming a mat. All plants require certain **mineral nutrients,** such as nitrogen, phosphorus, and potassium, for the growth of vegetative parts. In this lab, you will investigate the effect of nutrients on the vegetative reproduction of duckweed.

1. Write a definition for each boldface term in the paragraph above.
2. Based on the objectives for this lab, write a question you would like to explore about vegetative reproduction in duckweed.

Procedure
Part A: Make Observations
1. Place a duckweed plant in a Petri dish. Then place a few drops of water on the plant.
2. Observe the duckweed plant with a stereomicroscope or a hand lens. Sketch what you see. Label the structures that you recognize.

Part B: Design an Experiment

3. Work with members of your lab group to explore one of the questions written for step 2 of **Before You Begin.** To explore the question, design an experiment that uses the materials listed for this lab.

You Choose

As you design your experiment, decide the following:

a. what question you will explore
b. what your hypothesis will be
c. what solutions to test
d. how much of each solution to use
e. how many individuals to use for each test
f. what your control will be
g. how you will judge which solution is the best
h. what data to record in your data table

4. Write a procedure for your experiment. Make a list of all the safety precautions you will take. Have your teacher approve your procedure and safety precautions before you begin the experiment.

PART C: Conduct Your Experiment

5. Put on safety goggles and a lab apron.
6. Set up your experiment. **CAUTION: Nutrient solutions are mild eye irritants. Avoid contact with your skin and eyes.** Complete step 8.
7. Conduct your experiment and collect data for two weeks.

PART D: Cleanup and Disposal

8. Dispose of solutions, broken glass, and duckweed in the designated waste containers. Do not pour chemicals down the drain or put lab materials in the trash unless your teacher tells you to do so.
9. Clean up your work area and all lab equipment. Return lab equipment to its proper place. Wash your hands thoroughly before you leave the lab and after you finish all work.

Analyze and Conclude

1. **Summarizing Results** Compare the appearance of plants growing in each nutrient solution with that of the plants in distilled water. Explain your observations.
2. **Analyzing Data** In which Petri dish did the greatest amount of growth (increase in numbers) take place?
3. **Analyzing Results** In which Petri dish did the least amount of growth take place?
4. **Evaluating Hypotheses** Did the results you observed agree with your hypothesis? If not, how are they different?
5. **Recognizing Patterns** As the number of new duckweed plants in a particular group increased, what happened to the group of plants?
6. **Graphing Data** Make a graph of your data. Label the *y*-axis "Number of plants," and the *x*-axis "Days." Use a different color to represent each solution you tested.
7. **Drawing Conclusions** What factors regulate the rate of vegetative reproduction in duckweed?
8. **Evaluating Methods** Why are the new duckweed plants produced by vegetative reproduction genetically the same as the parent plant?
9. **Further Inquiry** Write a new question about vegetative reproduction in duckweed that could be explored with another investigation.

Do You Know?

Do research in the library or media center to answer these questions:

1. What do the flowers of duckweeds look like?
2. What is the smallest species of flowering plants, and how small are they?

Use the following Internet resources to explore your own questions about duckweed.

internetconnect

SCILINKS NSTA
TOPIC: Duckweed
GO TO: www.scilinks.org
KEYWORD: HX567

Analyze and Conclude

Answers

1. Plants grown in Knop's solution should be the largest and darkest green. Plants grown in pond water should also look healthy and green. Plants grown in dilute fertilizer solution or distilled water should be smaller and lighter green.
2. The most amount of growth should occur in the dish with Knop's solution.
3. The least amount of growth should occur in distilled water.
4. Answers will vary. Students should state how their results differ from their hypothesis.
5. The groups will divide as they get larger.
6. Answers will vary.
7. Students should recognize that more mineral nutrients will produce more growth.
8. Vegetative reproduction involves mitotic cell division, and mitosis results in the production of genetically identical cells.
9. Answers will vary.

Do You Know?

Answers

1. The flowers of duckweeds consist of only one or two stamens and a pistil, which are located in a pocket on the plant.
2. Two species of the genus *Wolfia* are less than 1 mm across.

Sample Data Table

Day	Dist. Water	Pond Water	0.1% Fert.	Knop's Sol.
0	6	6	6	6
1	6	6	6	6
2	6	6	6	6
3	6	7	6	7
4	6	7	6	7
5	6	7	7	7
6	6	9	8	9
7	6	9	8	10
8	6	10	8	11
9	6	10	8	12
10	6	11	8	13
11	6	12	8	14
12	6	13	8	15
13	6	13	8	17
14	6	13	8	18

Before You Begin

Answers

1. *vegetative reproduction*—a type of asexual reproduction that results in the growth of new individuals from nonreproductive parts, such as leaves, stems (rhizomes, tubers, bulbs, corms), and roots

   asexual reproduction—reproduction that occurs without the union of gametes

   mineral nutrients—elements and inorganic compounds that are needed for growth and development
2. For example: Which nutrients are needed for vegetative reproduction in duckweed?

Procedure

Sample Procedure

1. Label four Petri dishes to indicate one of each of these test solutions: distilled water, pond water, Knop's solution, 0.1% fertilizer.
2. Fill each Petri dish three-quarters full with the solution indicated on the dish.
3. Place six duckweed plants in each Petri dish. Cover the dishes, and place them in a well lighted area.
4. Observe the dishes every day for 2 weeks. Record data on number of plants in a dish, number of plants in each cluster, and general appearance of plants.

Chapter 26 Plant Structure and Function

Planning Guide

	OBJECTIVES	MEETING INDIVIDUAL NEEDS	READING SKILLS AND STRATEGIES
26-1 The Vascular Plant Body (BLOCKS 1, 2, & 3 135 min)	• **Identify** the three kinds of tissues in a vascular plant's body, and state the function of each. • **Compare** the structures of different types of roots, stems, and leaves. • **Relate** the structures of roots, stems, and leaves to their functions.	Learning Styles, p. 570 BASIC ADVANCED • pp. 571, 572, 577	**ATE** Active Reading, Technique: Reading Organizer, p. 569 **ATE** Active Reading, Technique: Reading Effectively, p. 571 ■ Active Reading Guide, worksheet (26-1) ■ Directed Reading, worksheet (26-1)
26-2 Transport in Plants (BLOCKS 4 & 5 90 min)	• **Relate** transpiration to the movement of water up a plant. • **Describe** how guard cells regulate the rate of transpiration. • **Recognize** several distinguishing features of sugar maple trees. • **Describe** the process of translocation in a plant.	Learning Styles, p. 578 BASIC ADVANCED • pp. 578, 579, 582	■ Active Reading Guide, worksheet (26-2) ■ Directed Reading, worksheet (26-2) ■ Supplemental Reading Guide, *A Feeling for the Organism: The Life and Work of Barbara McClintock*

Review and Assessment Resources

BLOCKS 6 & 7 90 min

TEST PREP

- **PE** Study Zone, p. 583
- **PE** Performance Zone, pp. 584–585
- ■ Chapter 26 worksheets:
  - Vocabulary
  - Concept Mapping
  - Critical Thinking
- Audio CD Program
- ■ Test Prep Pretest

CHAPTER TESTING

- ■ Chapter Tests (blackline master)
- ▲ Test Generator
- ■ Assessment Item Listing

See the correlations table at the front of this book for a list of the ***National Science Education Standards*** addressed in this chapter.

Smithsonian Institution® Visit **www.si.edu/hrw** for additional on-line resources.

KEY

Biology Principles and Explorations

- **PE** Pupil's Edition
- **ATE** Teacher's Edition

- Teaching Transparencies
- Laboratory Program

One-Stop Planner CD-ROM
Shows these resources within a customizable daily lesson plan:

- ■ Teaching Resources
- ▲ Test Generator

VISUAL STRATEGIES	CLASSWORK AND HOMEWORK	TECHNOLOGY AND INTERNET RESOURCES
ATE Fig. 26-4, p. 572 **ATE** Fig. 26-9, p. 576 119 Structure of a Leaf 123 Structure of Stems 124 Structure of Roots 125 Structure of Xylem 126 Structure of Phloem 116A Modified Leaves 117A Modified Stems	**PE** Quick Lab How does the internal structure of roots and stems differ?, p. 575 **PE** Quick Lab What structures are found inside a leaf?, p. 576 **PE** Lab Techniques Separating Plant Pigments, pp. 586–587 Lab A19 Inferring Function from Structure Lab B16 Fruits and Seeds Lab C35 Staining and Mounting Stem Cross Sections ■ Occupational Applications: Forestry Technician **PE** Review, p. 577	Holt Biology Videodiscs Section 26-1 Audio CD Program Section 26-1 internetconnect SCILINKS NSTA TOPICS: • Vascular plants • Vascular tissue
127 Pressure-Flow Model of Translocation	**PE** Data Lab Inferring the rate of transpiration, p. 579 Lab A20 Relating Root Structure to Function Lab B17 Transpiration and Stem Structure Lab C36 Using Paper Chromatography to Separate Pigments **PE** Review, p. 582 **PE** Study Zone, p. 583	CNN Video Segment 21 Maple Syrup Holt Biology Videodiscs Section 26-2 Audio CD Program Section 26-2 internetconnect SCILINKS NSTA TOPIC: Transpiration

ALTERNATIVE ASSESSMENT

- **PE** Portfolio Projects 26–28, p. 585
- **ATE** Alternative Assessment, p. 583
- **ATE** Portfolio Assessment, p. 583
- ■ Portfolio Projects: Plant Focus Worksheet

Lab Activity Materials

Quick Labs
p. 575 compound microscope; slide of a cross section of a dicot root, a monocot root, a dicot stem, and a monocot stem
p. 576 prepared slide of leaf cross section, compound microscope

Data Lab
p. 579 pencil, paper

Lab Techniques
p. 586 safety goggles, lab apron, chromatography paper, scissors, metric ruler, pencil, capillary tube, simulated plant pigments extract, 10 mL graduated cylinder, 5 mL chromatography solvent, chromatography chamber

Demonstrations
ATE 3 very different plants, p. 568

ATE microscope, slides with coverslips (3), red onion, potato, celery, p. 571
ATE wax paper, water, liquid detergent, beakers (2), p. 571
ATE variety of modified plant organs, p. 573

BIOSOURCES LAB PROGRAM
- Quick Labs: Lab A19, p. 37
- Quick Labs: Lab A20, p. 39
- Inquiry Skills Development: Lab B16, p. 73
- Inquiry Skills Development: Lab B17, p. 77
- Lab Techniques and Experimental Design: Lab C35, p. 171
- Lab Techniques and Experimental Design: Lab C36, p. 177

Chapter Theme

Cellular Structure and Function

Plants are composed of structures (leaves, stems, and roots) that are modified to meet the needs of the plant. Specialized cells in leaves are adapted for capturing sunlight and carrying out photosynthesis. Stems contain cells that are adapted for holding leaves upward towards light and for transporting water, mineral nutrients, and sugars throughout the plant. Roots contain tissues that are adapted for anchoring the plant, providing storage, and absorbing water and mineral nutrients from the soil.

Establishing Prior Knowledge

Before beginning this chapter, make certain students can answer the questions in **Study TIP Ready?** on page 569. If they cannot, encourage them to reread the sections indicated.

Opening Demonstration

Show students three very different plants, such as a cactus, a tomato, and a Venus' flytrap. Have the class compose a list of things that the three plants have in common. *(All plants have the same needs—water, carbon dioxide, light energy, and mineral nutrients for making organic molecules.)* Ask students why these three plants look so different. *(They have specialized tissues to accommodate their needs.)*

CHAPTER 26 Plant Structure and Function

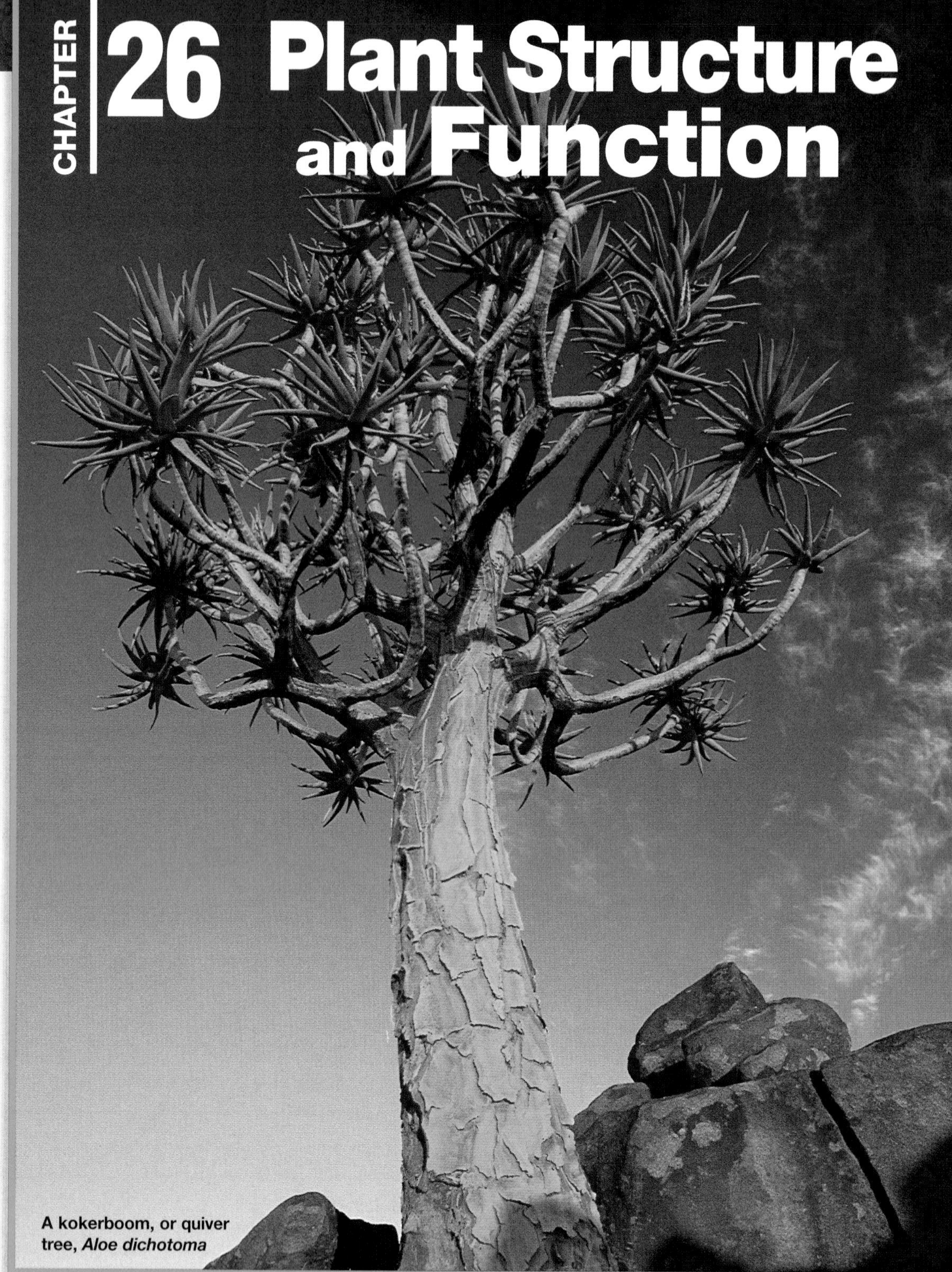

A kokerboom, or quiver tree, *Aloe dichotoma*

What's Ahead — and Why
CHAPTER OUTLINE

26-1 The Vascular Plant Body
- Vascular Plants Contain Three Kinds of Tissues
- There Are Two Types of Vascular Tissue
- Roots Absorb and Store Nutrients
- Stems Connect Roots to Leaves
- Leaves Carry Out Photosynthesis

26-2 Transport in Plants
- Water Is Pulled up Through a Plant
- Organic Compounds Are Pushed Through a Plant

Students will learn

- about the three types of tissues in plants and about the arrangement and functions these tissues have in roots, stems, and leaves.
- how water and mineral nutrients are transported from roots to leaves and how sugars are transported from where they are made or stored to where they are needed.

Form facilitates function. *The structure of the quiver tree enables it to survive in the Namibian desert, where the Hottentots use the tree's bark and branches to make arrow quivers. Vascular tissues make the branches strong and supple. And, like highway systems, vascular tissues also distribute materials, such as water and sugars.*

Looking Ahead

Study TIP

Ready?

Check your understanding of these topics to help you prepare for what's ahead.

Can you . . .

- **explain** what causes the adhesion and cohesion of water? (Section 2-2)
- **describe** the organization of a plant cell? (Section 3-3)
- **summarize** the steps in photosynthesis? (Section 5-2)
- **state** the relationship between stomata and guard cells? (Section 24-1)
- **define** the terms *cuticle, stoma, xylem, phloem,* and *meristem?* (Section 24-1)

If not, *review the sections indicated.*

internet**connect**

National Science Teachers Association *sci*LINKS Internet resources are located throughout this chapter.

Reading Strategies

Active Reading

Technique: Reading Organizer

Before students read this chapter, have them survey the subtitles, headings, captions, and words in boldface type. Ask students to identify the purpose of this chapter. *(The purpose of this chapter is to describe the three major parts of the plant body—roots, stems, and leaves—and to describe their functions.)* As students read this chapter, have them work individually or in small groups to create a descriptive reading organizer.

Directed Reading

Assign **Directed Reading Worksheet Chapter 26** to help students interact with the text as they read the chapter.

Interactive Reading

Assign ***Biology: Principles and Explorations*** **Audio CD Program Chapter 26** to help reluctant readers achieve greater success in reading the chapter.

Reading Skills

Assign Reading Organizers and Reading Strategies Worksheets from the **One-Stop Planner CD-ROM with Test Generator** to help students learn and practice effective reading skills.

internet**connect**

Holt *Biology: Principles and Explorations*

On-line Resources: go.hrw.com

Visit the HRW Web site for a variety of resources related to this chapter. Just type in the keyword HX0 Chapter 26.

Holt *Biology: Principles and Explorations*

On-line Resources: www.scilinks.org

The following *sci*LINKS Internet resources can be found in the student text for this chapter:

Page	TOPIC	KEYWORD
Page 571	Vascular plants	HX570
Page 572	Vascular tissue	HX572
Page 579	Transpiration	HX579

Chapter 26

Section 26-1 The Vascular Plant Body

1 Prepare

Scheduling

- **Block:** About 135 minutes will be needed to complete this section.
- **Traditional:** About three class periods will be needed to complete this section.
- **Planning:** See the Planning Guide on pp. 567A and 567B for lecture, classwork, and assignment options for Section 26-1. Lesson plans for Section 26-1 can be found in a lesson cycle format in ***Biology: Principles and Explorations*** **One-Stop Planner CD-ROM with Test Generator.**

Resource Link

Assign **Active Reading Worksheet Section 26-1** to help students comprehend the key concepts and visuals in this lesson.

2 Teach

Lesson Warm-up

Read or paraphrase the chapter theme, and ask students to describe the kinds of structural characteristics they would expect to find in roots, stems, and leaves that enable these organs to carry out their functions. *(roots—a large surface area, tubes through which water, minerals, and sugars can move; stems—rigid internal structures, tubes through which water, minerals, and sugars can move; leaves—cells specialized for photosynthesis, adapted for taking in carbon dioxide)*

internet connect

Have students research the structure of vascular plants using the Web site in the Internet Connect box on page 571. Have students write a report summarizing their findings.

PORTFOLIO

Objectives

- **Identify** the three kinds of tissues in a vascular plant's body, and state the function of each. (pp. 570–572)
- **Compare** the structures of different types of roots, stems, and leaves. (pp. 573–577)
- **Relate** the structures of roots, stems, and leaves to their functions. (pp. 573–577)

Key Terms

dermal tissue
ground tissue
epidermis
cork
vessel
sieve tube
cortex
root hair
root cap
herbaceous plant
vascular bundle
pith
heartwood
sapwood
petiole
mesophyll

Study TIP

Reading Effectively

Use the Objectives and the red subheadings at the top of the pages to help you identify the main ideas presented in Section 26-1.

Vascular Plants Contain Three Kinds of Tissues

Like your body, a plant's body is made of tissues that form organs. In vascular plants, there are three types of tissues—dermal tissue, ground tissue, and vascular tissue. As you learned in Chapter 24, vascular tissue forms strands that conduct water, minerals, and organic compounds throughout a vascular plant. **Dermal tissue** forms the protective outer layer of a plant. **Ground tissue** makes up much of the inside of the nonwoody parts of a plant, including roots, stems, and leaves. **Figure 26-1** shows how these three tissues are arranged in a nonwoody dicot. The tissues are organized a little differently in other types of vascular plants.

Each type of tissue contains one or more kinds of cells that are specialized to perform particular functions. Look again at the diagram of a plant cell on page 65, in Chapter 3. As you read farther in this chapter, you will learn that many specialized plant cells lack some of the organelles shown in the diagram. In fact, some plant cells cannot perform their functions until they have lost most of their organelles.

Figure 26-1 **Plant tissues**

The leaves, stems, and roots of a vascular plant contain all three kinds of plant tissues.

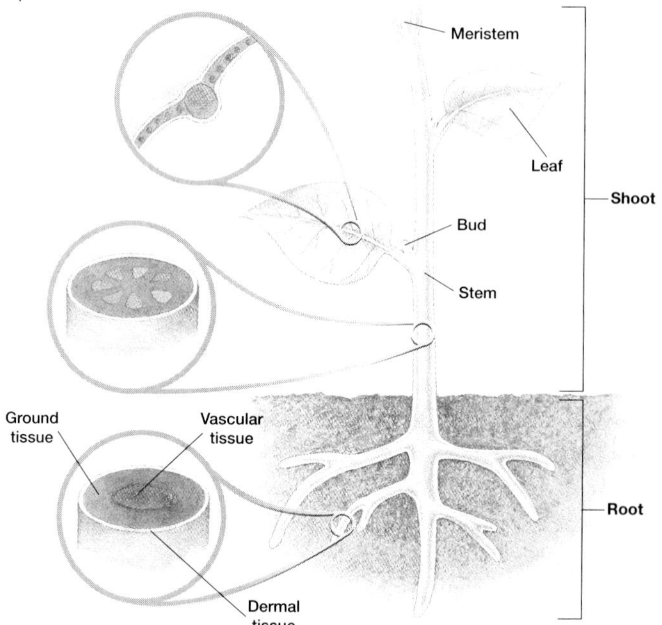

Integrating Different Learning Styles

Logical/Mathematical	**PE**	Review item 3, p. 577; Lab Techniques, pp. 586–587
	ATE	Teaching Tip, p. 572; Reteaching, p. 577
Visual/Spatial	**PE**	Real Life, p. 574; Quick Labs, pp. 575, 576; Performance Zone items 1 and 26, pp. 584–585
	ATE	Demonstration, p. 571; Graphic Organizer, p. 571; Visual Strategy, p. 572; Demonstration, p. 573; Teaching Tip, p. 575
Interpersonal	**ATE**	Active Reading, p. 569; Demonstration, p. 571
Verbal/Linguistic	**PE**	Study Tip, p. 570; Review items 1 and 2, p. 577; Performance Zone items 15, 21, and 22, pp. 584–585
	ATE	Active Reading, p. 569; Lesson Warm-up, p. 570; Closure, p. 577

Figure 26-2 Dermal tissue. Epidermis is one type of dermal tissue. Hairlike outgrowths of the epidermis, like the ones on this African violet leaf, help reduce water loss from the leaves of some plants.

Dermal tissue

Dermal tissue covers the outside of a plant's body. In the nonwoody parts of a plant, dermal tissue forms a "skin" called the **epidermis.** The word *epidermis* comes from the Greek words *epi,* meaning "upon," and *derma,* meaning "skin." The epidermis of most plants is made up of a single layer of flat cells. A waxy cuticle, which prevents water loss, coats the epidermis of the stems and leaves. Often, the cells of the epidermis have hairlike extensions or other structures, as Figure 26-2 shows. Extensions of the epidermal cells on leaves and stems often help to slow water loss. Extensions of the epidermal cells on root tips help increase water absorption. The dermal tissue on woody stems and roots consists of several layers of dead cells that are referred to as **cork.** Cork cells contain a waterproof chemical and are not covered by a waxy cuticle. In addition to protection, dermal tissue also functions in gas exchange and in the absorption of mineral nutrients.

TOPIC: Vascular plants
GO TO: www.scilinks.org
KEYWORD: HX570

Ground tissue

Ground tissue makes up much of the inside of most plants. Most ground tissue consists of thin-walled cells that remain alive and keep their nucleus after they mature. In addition, ground tissue contains some thick-walled cells. Look for these different types of cells in the ground tissue shown in Figure 26-3. Ground tissue has different functions, depending on where it is located in a plant. The ground tissue in leaves, which is packed with chloroplasts, is specialized for photosynthesis. The ground tissue in stems and roots functions mainly in the storage of water, sugar, and starch. Throughout the body of a plant, ground tissue also surrounds and supports the third kind of plant tissue—vascular tissue.

Figure 26-3 Ground tissue. A variety of cells make up the ground tissue visible in this cross section of a wheat stem. Thin-walled cells make up most of the ground tissue. Thick-walled cells strengthen the stem.

Magnification: 122×

Graphic Organizer

Use this graphic organizer with ***Teaching Tip: Summary of Plant Cells and Tissues*** *on page 572.*

Plant tissue	Cell types
Dermal tissue	Epidermal cells, guard cells, cork cells
Ground tissue	Mesophyll cells, cortex cells, pith cells
Vascular tissue	Vessel cells, tracheids, sieve-tube cells, companion cells

Active Reading

Technique: Reading Effectively— BASIC

Have students write down each of the key terms for this section on separate note cards. Have them also make a note card with the words *vascular tissue* on it. As the students read the section, have them write a brief definition of each term on the back of its card. Have students use their cards in the **Reteaching** exercise at the end of the section.

Demonstration

Wetting a Waxy Surface

Provide each group of students with a piece of wax paper, water, liquid detergent, and two beakers. Have students place two or three drops of water on a piece of wax paper and draw what they see. Ask students what happens when the wax paper is tilted. *(The drops of water roll off the paper.)* Then have students place two or three drops of soapy water on the wax paper and describe what they see. *(The drops of soapy water spread out over the surface of the wax paper.)* Tell students that the detergent acts as a "wetting agent" by interfering with the attraction water molecules have for each other. Wetting agents are mixed with chemicals that protect plants from pests and diseases before the chemicals are applied so that the chemicals will cover and stick to the leaves.

Demonstration

Three Types of Tissue

Provide each group of students with a microscope, three microscope slides, three coverslips, a red onion, a potato, and a bunch of celery. Have students prepare wet-mount slides of the following: red-onion skin (dermal tissue), potato (ground tissue), and celery (dermal, ground, and vascular tissue). Have students draw what they observe in the field of view under the microscope and label the types of tissues they find in each example. Remind them that the field of view should be drawn as a circle. PORTFOLIO

Visual Strategy

Figure 26-4

Have students examine the pictures and diagrams of xylem and phloem and the cells that compose them. Ask what similarities they notice between the conducting cells of the xylem and those of the phloem. *(The cells are elongated and connected end to end; they have openings in their cell walls at the ends of the cells and also some in the sides.)* Ask what these similarities indicate about the similarities in function of xylem and phloem. *(The two types of tissues are adapted for transporting substances over long distances, as well as for allowing lateral transport.)* Ask what differences they notice between the cells of the xylem and those of the phloem. *(There are two types of conducting cells in xylem but only one type in phloem. The sieve-tube cells have companion cells, but the xylem cells do not.)*

Teaching TIP — BASIC

Interconnected Vascular Tissues

Point out to the students that the conducting cells of the xylem and those of the phloem in a plant are all interconnected. The strands of vascular tissue in the roots extend down into every root branch, and the strands in the stems extend up into every branch and leaf. Tell students to consider the plumbing system in a home. It transports fresh water throughout the house and removes waste water, and all of its pipes are interconnected.

Teaching TIP

Summary of Plant Cells and Tissues

Have students construct a table summarizing the types of plant tissues and the cells they contain. In the table, students should classify each type of cell encountered as part of the dermal tissue, ground tissue, or vascular tissue. A completed table is found in the Graphic Organizer on page 571.

PORTFOLIO

There Are Two Types of Vascular Tissue

internet connect

SCiLINKS NSTA

TOPIC: Vascular tissue
GO TO: www.scilinks.org
KEYWORD: HX572

As you learned in Chapter 24, plants have two kinds of vascular tissue—xylem and phloem. Both xylem and phloem contain strands of cells that are stacked end to end and act like tiny pipes, as you can see in **Figure 26-4.** These pipes act as a plumbing system, carrying fluids and dissolved substances throughout a plant's body.

Xylem

Xylem has thick-walled cells that conduct water and mineral nutrients from a plant's roots through its stems to its leaves. The conducting cells in xylem must lose their cell membrane, nucleus, and cytoplasm before they can conduct water. At maturity, all that is left of these cells is their cell walls. One type of xylem cell found in all vascular plants is called a tracheid *(TRAY kee ihd)*. Tracheids are narrow, elongated, and tapered at each end. Water flows from one tracheid to the next through pits, which are thin areas in the cell walls. Gnetophytes and flowering plants also have a second type of xylem cell, which makes up conducting strands called **vessels.** The vessel cells are wider than tracheids and have large holes, called perforations, in their ends. Perforations allow water to flow more quickly between vessel cells.

Figure 26-4 Xylem and phloem

Both xylem and phloem contain strands of tubular conducting cells that are stacked end to end like sections of pipe.

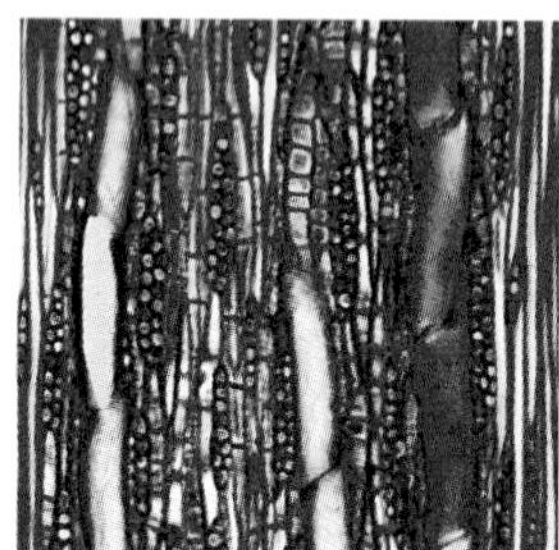

The conducting cells of xylem are tracheids and vessel cells. Tracheids are slender and have tapered ends. Vessel cells are larger and form vessels.

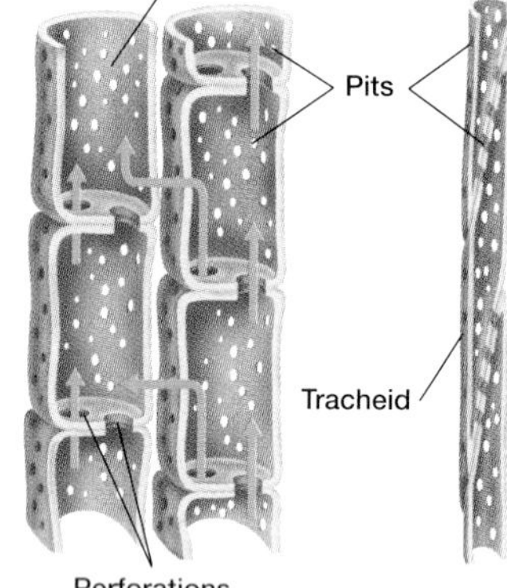

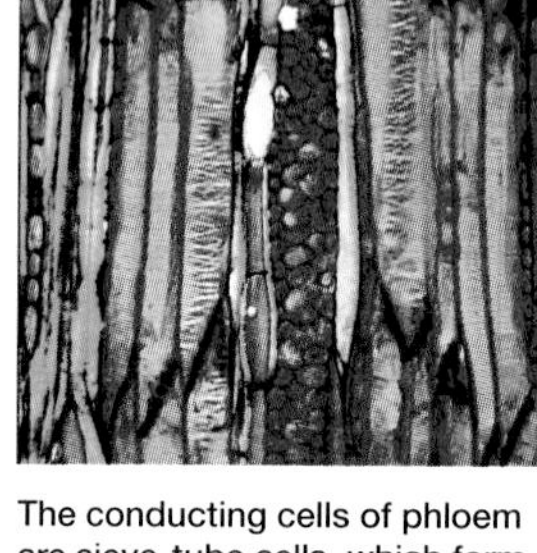

The conducting cells of phloem are sieve-tube cells, which form sieve tubes. Companion cells lie next to the sieve-tube cells in phloem.

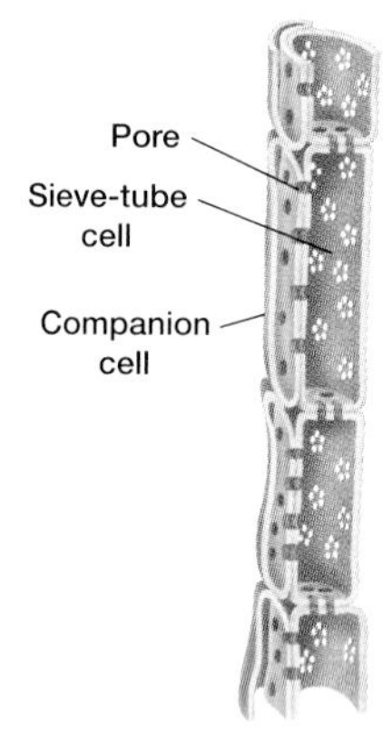

Phloem

Phloem contains cells that conduct sugars and other nutrients throughout a plant's body. The conducting cells of phloem have a cell wall, a cell membrane, and cytoplasm. These cells either lack organelles or have modified organelles. The conducting strands in phloem are called **sieve tubes.** Pores in the walls between neighboring sieve-tube cells connect the cytoplasms and allow substances to pass freely from cell to cell. Beside the sieve tubes are rows of companion cells, which contain organelles. Companion cells carry out cellular respiration, protein synthesis, and other metabolic functions for the sieve-tube cells.

did you know?

Aphids led scientists to the sugar-transporting tissue.

Scientists who were studying sugar transport in plants noticed that many aphids that feed on plants release a substance called "honeydew" from their anuses. Other insects commonly lick the honeydew off of the aphids. The scientists analyzed the honeydew and found that it was high in sugars. They concluded that when aphids feed on plants, they must insert their mouthparts, called *stylets,* directly into the sugar-transporting cells. The scientists then conducted some studies in which they quickly cut the aphids off from their stylets while they were feeding and then made cross-sections of the plant tissues. In these sections, they found that the tips of the aphids' stylets ended in sieve-tube cells. This led the scientists to conclude that phloem is the sugar-transporting tissue in plants.

Roots Absorb and Store Nutrients

Most plants are anchored to the spot where they grow by roots, which also absorb water and mineral nutrients. In many plants, roots also function in the storage of organic nutrients, such as sugar and starch. Many dicots, such as carrots and radishes, have a large central root from which much smaller roots branch. This type of root system is called a taproot system. In contrast, most monocots, such as grasses, have a highly branched, fibrous root system, shown in **Figure 26-5.** Some plants have roots that grow from aboveground stems or leaves. Such roots are called adventitious *(ad ven TIH shuhs)* roots. The prop roots of corn and the aerial roots of orchids are examples of adventitious roots.

As **Figure 26-6** shows, a root has a central core of vascular tissue that is surrounded by ground tissue. The ground tissue surrounding the vascular tissue is called the **cortex.** Roots are covered by dermal tissue. An epidermis covers the end of a root, or root tip. The epidermal cells just behind a root tip often produce **root hairs,** which are slender projections of the cell membrane. Root hairs greatly increase the surface area of a root and its ability to absorb water and mineral nutrients. A mass of cells called the **root cap** covers and protects the actively growing root tip. A layer of cork replaces the epidermis in the older sections of a root. Many plants have roots that become woody as they get older. Layers of xylem replace the ground tissue in woody roots.

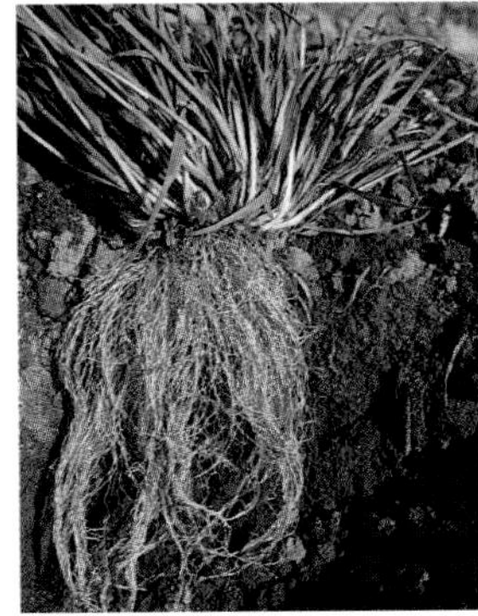

Figure 26-5 Fibrous root system. A fibrous roots system is made up of many roots that are about the same size.

Figure 26-6 Root structure. Roots have characteristic external and internal structures.

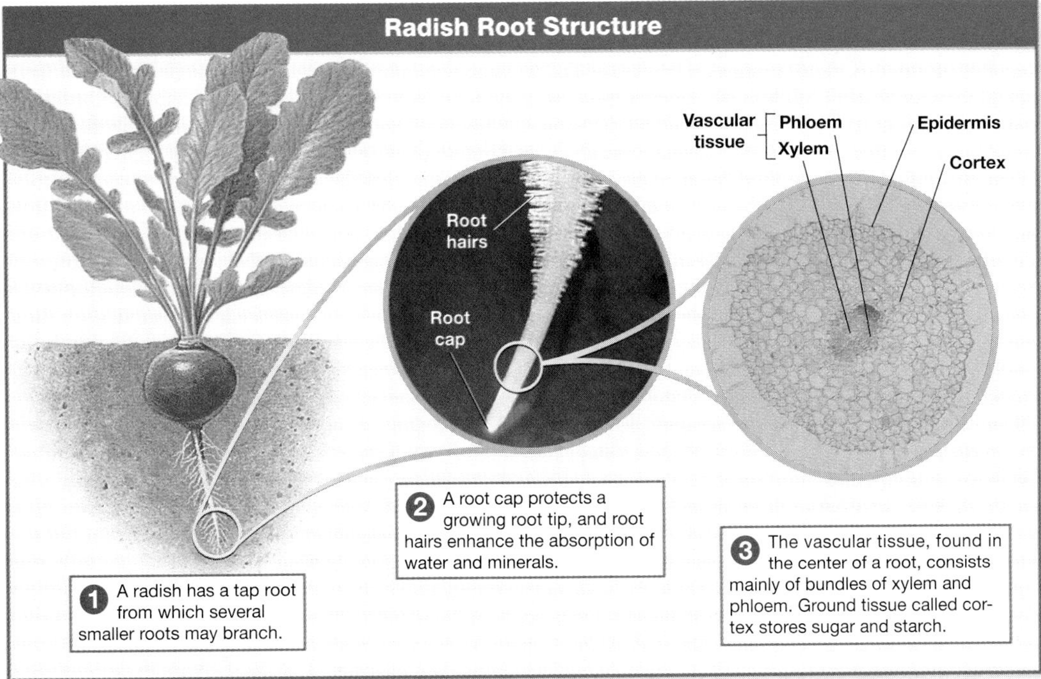

Radish Root Structure
Vascular tissue
Phloem
Xylem
Epidermis
Cortex
Root hairs
Root cap
1 A radish has a tap root from which several smaller roots may branch.
2 A root cap protects a growing root tip, and root hairs enhance the absorption of water and minerals.
3 The vascular tissue, found in the center of a root, consists mainly of bundles of xylem and phloem. Ground tissue called cortex stores sugar and starch.

Overcoming Misconceptions

Roots Require Oxygen for Survival

It is a common misconception that plants do not need oxygen because they produce it in photosynthesis. However, almost all plant cells have mitochondria, in which they break down carbon-containing molecules in the presence of oxygen and release energy for cellular functions. In leaves and green stems, oxygen is produced in photosynthesis. However, in roots, no oxygen is produced because there are no photosynthetic cells. Therefore, roots must obtain oxygen from the soil. This is why houseplants that are overwatered and crops that are flooded for extended periods of time die. This is also why gardeners try to keep their soil aerated, using organic matter, such as mulch and compost, and earthworms that dig holes in the soil. Trees such as mangroves and bald cypresses that live in swamps have roots that emerge above the water (pneumatophores) that obtain oxygen from the air.

Demonstration

Modified Plant Parts

Bring a variety of modified plant organs or pictures of these plant organs to class for the students to observe. Examples of roots may include fleshy taproots adapted for sugar and starch storage (carrot, beet, and sweet potato) and aerial roots that extract water from the air (orchids and bromeliads). Examples of stems may include bulbs (onion and lily), corms (gladiolus and crocus), rhizomes (iris), and tubers (Irish potato), all of which are adapted for sugar and starch storage; runners (spider plant and strawberry) and stolons (most lawn grasses), which grow new plants at nodes; stems adapted for water storage (cactus); stems adapted for defense with thorns (citrus and acacia); and stems with tendrils that are adapted for climbing (grape). Examples of leaves may include large leaves adapted for moist, understory environments (philodendron); needles adapted for cold, windy environments (pine); leaves adapted for sugar and starch storage (cabbage); leaves with fleshy petioles adapted for sugar, starch, and/or water storage (celery and rhubarb); bracts (flowering dogwood and poinsettia), which are modified leaves adjacent to flowers and are sometimes showy; and leaves adapted for asexual reproduction (kalanchoë). Ask the students to compare the functions of these modified organs with their usual functions.

Career

Farmer

Farmers grow crops that provide food for humans and livestock, fibers for clothing, and fuels for transportation. Have students work in teams to use library references or on-line databases to find information about the role of farmers in crop production. Ask students to research modern farming methods, and have them report their findings to the class. PORTFOLIO

REAL LIFE

Answers

Answers will vary. Many different types of trees may be cut to produce either straight-grained lumber or cross-grained lumber. Straight-grained lumber is used in making boards and planks for construction, decking, handles, and baseball bats. Plywood, veneers for furniture and paneling, musical instruments, and art objects are often made from cross-grained lumber.

Resource Link

Assign **Occupational Applications: Forestry Technician** from the Holt BioSources Teaching Resources CD-ROM. PORTFOLIO

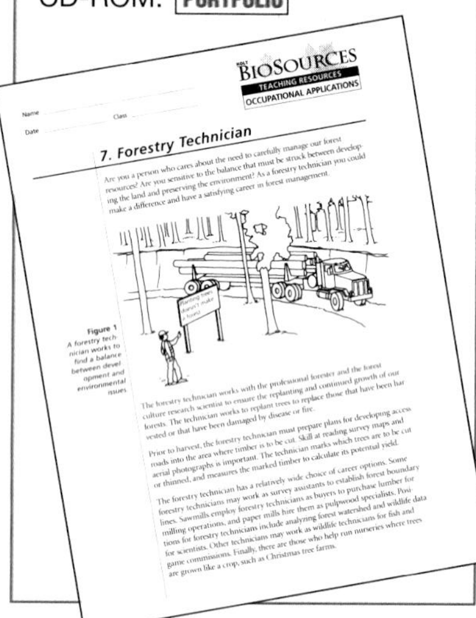
BioSources
TEACHING RESOURCES
OCCUPATIONAL APPLICATIONS
7. Forestry Technician

REAL LIFE

Xylem forms wood's characteristic markings, or grain. Wood is usually cut parallel to the axis of a tree trunk, producing straight-grained lumber. In some trees, however, xylem vessels form interesting patterns.

Recognizing Patterns
Examine several wooden objects. Decide whether they have straight grain or another type of grain.

Stems Connect Roots to Leaves

The shoots of most plants consist of stems and leaves. Stems support the leaves and house the vascular tissue, which transports substances between the roots and the leaves. Many plants have stems that are specialized for other functions. For example, the stems of cactuses store water. Potatoes are stems that are specialized for nutrient storage and for asexual reproduction.

Leaves are attached to a stem at points called nodes. The space between two nodes is called an internode. Buds that can grow into new branches are also located at the nodes on a stem. Look for these structures in **Figures 26-7** and **26-8.** Other features of a stem depend on whether the stem is woody or nonwoody.

Nonwoody stems

A plant with stems that are flexible and usually green is called a **herbaceous** *(huhr BAY shuhs)* **plant.** Herbaceous plants include violets, clovers, and grasses. As Figure 26-7 shows, the stems of herbaceous plants contain bundles of xylem and phloem called **vascular bundles.** The vascular bundles are surrounded by ground tissue. In monocot stems, the vascular bundles are scattered in the ground tissue. In dicot stems, however, the vascular bundles are arranged in a ring. The ground tissue outside the ring of vascular bundles is called the cortex. The ground tissue inside the ring is called the **pith.** Herbaceous stems are covered by an epidermis. Stomata in the epidermis enable the stems to exchange gases with the outside air.

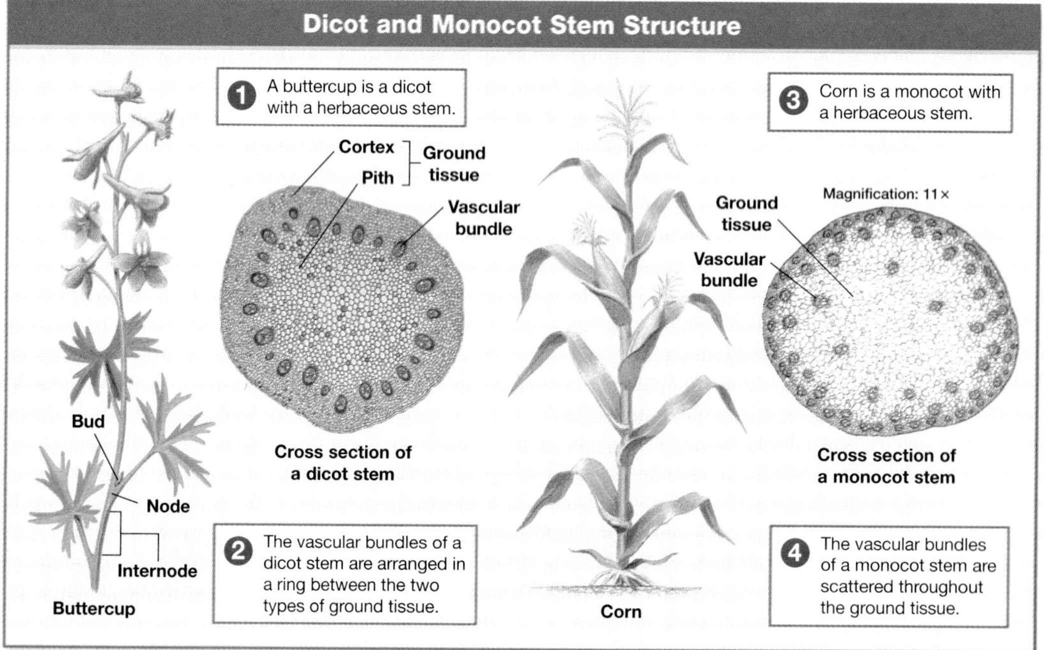
Dicot and Monocot Stem Structure
1 A buttercup is a dicot with a herbaceous stem.
Cortex
Pith
Ground tissue
Vascular bundle
Bud
Node
Internode
Buttercup
Cross section of a dicot stem
2 The vascular bundles of a dicot stem are arranged in a ring between the two types of ground tissue.
3 Corn is a monocot with a herbaceous stem.
Magnification: 11×
Ground tissue
Vascular bundle
Corn
Cross section of a monocot stem
4 The vascular bundles of a monocot stem are scattered throughout the ground tissue.

Figure 26-7 Herbaceous stems. Herbaceous stems are typically soft and green. The xylem and phloem are found in vascular bundles, which are arranged differently in dicots and monocots.

did you know?

Ants live in the spines of acacia shrubs.

Acacias are shrubs or trees of the genus *Acacia* in the pea family (Fabaceae). Some species, such as the "bullhorn" acacia of Mexico, have very sharp thornlike spines on their stems. These spines are hollow and are often inhabited by ants. The ants ward off potential insect predators from the acacia trees, while the trees provide a home for the ants.

Woody stems

Trees and shrubs, such as pines, oaks, roses, and hollies, have woody stems. As Figure 26-8 shows, woody stems are stiff and nongreen. Buds, which produce new growth, are found at the tips and at the nodes of woody stems. They exchange gases through pores in their bark. A young woody stem has a central core of pith and a ring of vascular bundles, which fuse into solid cylinders as the stem matures. Layers of xylem form the innermost cylinder and are the major component of wood. A cylinder of phloem lies outside the cylinder of xylem. Woody stems are covered by cork, which protects them from physical damage and helps prevent water loss. Together, the layers of cork and phloem make up the bark of a woody stem. A mature woody stem contains many layers of wood and is covered by a thick layer of bark. The wood in the center of a mature stem or tree trunk is called **heartwood.** The xylem in heartwood, which can no longer conduct water, provides support. **Sapwood,** which lies outside the heartwood, contains vessel cells that can conduct water.

Figure 26-8 Woody stems

Woody stems are typically stiff and nongreen.

Bud, Node, Internode, Pore

Bark, Cork, Phloem, Xylem, Pith

Young woody stem

Bark, Sapwood, Heartwood

Mature woody stem

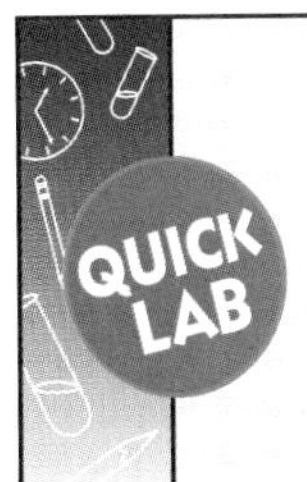

How does the internal structure of roots and stems differ?

You can use a microscope to see differences in the internal structure of roots and stems.

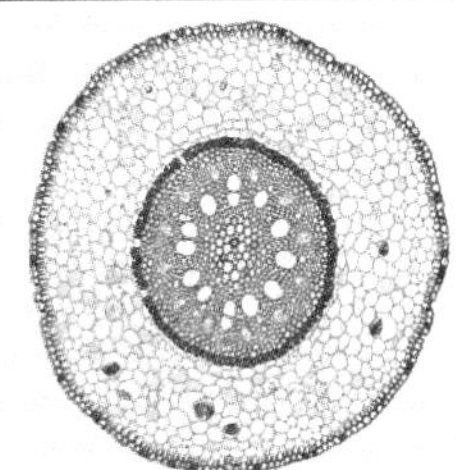

Cross section of a corn root

Materials

compound microscope, prepared slide of a cross section of the following: dicot root, monocot root, dicot stem, monocot stem

Procedure

1. View cross sections of dicot and monocot roots with a compound microscope. For each, draw and label what you see under low power. Then look at the vascular tissue in each root under high power. Draw what you see in each root, and label the xylem and phloem.
2. View cross sections of dicot and monocot stems with a compound microscope. For each, draw and label what you see under low power. Then look at a vascular bundle in each stem under high power. Draw each vascular bundle, and label the xylem and phloem.

Analysis

1. **Compare** and **Contrast** the location of xylem and phloem in roots and stems.
2. **Compare** and **Contrast** the arrangement and structure of the vascular bundles in monocot and dicot stems.

Overcoming Misconceptions

Roses Do Not Have Thorns

The structures that protect the stems of many members of the rose family are not really thorns. Botanically, thorns are modified stems that contain xylem. The "thorns" found on a rose are actually prickles, which are derived only from dermal tissue. How can you tell them apart without using a microscope? Prickles are easily removed; thorns are not.

Teaching TIP

Annual Rings in Woody Plants

Obtain a picture of a cross section of a tree trunk in which the rings are not regular. Ask students to relate what the rings might indicate about the tree's history. *(Closely spaced rings usually indicate slow growth, possibly due to drought; rings that are closer together on one side than on the other indicate growth against an obstacle; and a dark slash indicates exposure to fire.)* Have students count the rings to estimate the age of the tree. Explain that scientists use an increment borer to extract a thin cylinder-shaped sample from a living tree to obtain this information.

How does the internal structure of roots and stems differ?

Time 20 minutes

Process Skills Observing, comparing structures

Teaching Strategies

- Ask students to name some differences in the functions of roots and stems. *(Roots anchor plants and absorb water and mineral nutrients. Stems connect leaves to roots and conduct materials between leaves and roots.)*
- Ask students to think about the differences between monocots and dicots as they observe the different kinds of stems.

Analysis Answers

1. Xylem and phloem in the roots are located in a cylinder in the center of the root. Xylem and phloem in the stem are located in discrete vascular bundles in the cortex.
2. The vascular bundles in monocot stems are scattered throughout the cortex; the vascular bundles of dicot stems are arranged in a ring near the outside of the stem.

Visual Strategy

Figure 26-9

Have students describe the differences between simple and compound leaves. Point out that one way to tell a leaf from a leaflet is to look for the bud in the axil of the leaf (the angle between a leaf and the stalk or stem to which it is attached). Show students examples of simple and compound leaves from the local area. Ask students if there are buds in the axils of simple leaves. *(yes)* Ask students if there are buds in the axils of leaflets. *(no)* Also point out that compound leaves help plants maintain homeostasis in arid environments by reducing the leaf surface area and water loss.

What structures are found inside a leaf?

QUICK LAB

Time 20 minutes

Process Skills Observing, analyzing data, making conclusions

Teaching Strategies

- Tell students that the stomata may not be easy to see because the guard cells are small in cross section and the stomata may be very small.
- Ask students to note which leaf surface has the most stomata. *(lower)*

Analysis Answers

1. A stoma is a small, slitlike opening between two guard cells that are attached to each other at their ends. This opening allows carbon dioxide to enter the leaf, where it is used to make sugars in the process of photosynthesis.
2. The veins are distributed throughout the mesophyll. They contain both xylem and phloem tissue.
3. The palisade cells are located where the most light will hit the leaf. The spongy cells form a loosely packed layer (or layers) of irregularly shaped cells. The large open areas around these cells allow carbon dioxide to move freely from stomata on the lower leaf surface up to the palisade layer.

biobit

Which plants have the smallest leaves? Which have the largest?

The leaves of duckweed, an aquatic plant, may be less than 1 mm (0.04 in.) in length. Some palm trees, in contrast, have leaves that are over 20 m (66 ft) long.

Leaves Carry Out Photosynthesis

Leaves are the primary photosynthetic organs of plants. Most leaves have a flattened portion, called the blade, that is often attached to a stem by a stalk called the **petiole** *(PEHT ee ohl)*. A leaf blade may be divided into two or more sections called leaflets, as shown in **Figure 26-9.** Leaves with an undivided blade are called simple leaves. Leaves with two or more leaflets are called compound leaves. Leaflets reduce the surface area of a leaf blade. Many plants have highly modified leaves that are specialized for particular purposes. For example, the spines of a cactus and the tendrils of a garden pea are modified leaves. Cactus spines are specialized for protection and water conservation, while garden-pea tendrils are specialized for climbing.

Figure 26-9 **Simple and compound leaves**

Most leaves consist of a flattened blade and a petiole that attaches to a stem.

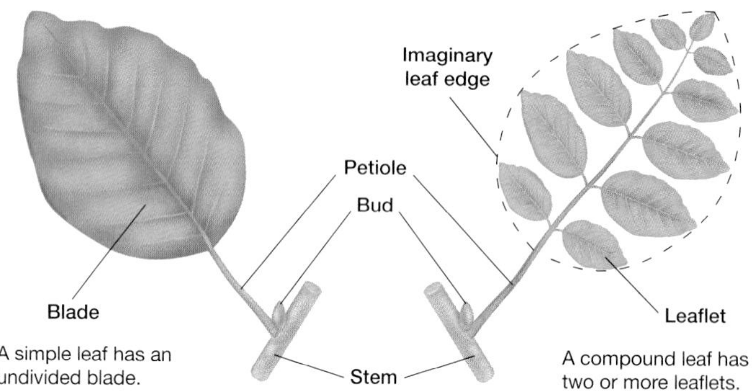
Imaginary leaf edge
Petiole
Bud
Blade
Leaflet
Stem
A simple leaf has an undivided blade.
A compound leaf has two or more leaflets.

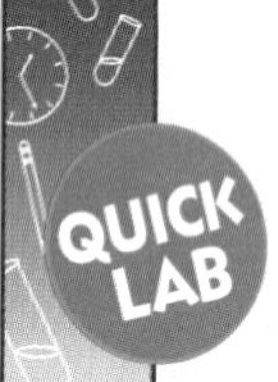

What structures are found inside a leaf?

QUICK LAB

With a microscope, you can see how a leaf is put together.

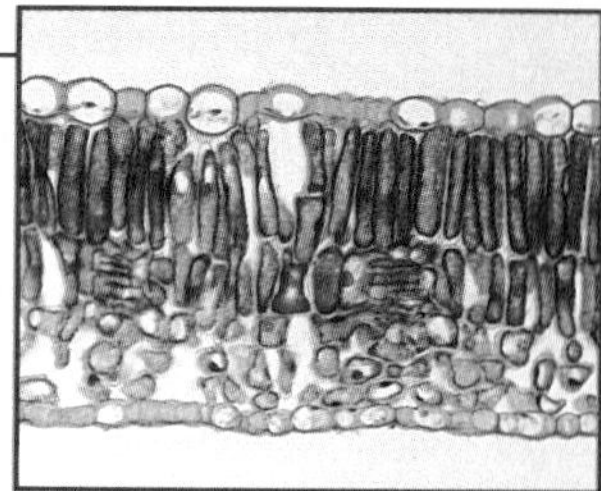

Cross section of a lilac leaf (530×)

Materials

prepared slide of a leaf cross section, compound microscope

Procedure

1. View a cross section of a leaf under low power of a compound microscope. Then switch to high power.
2. Use Figure 26-10, on page 577, to identify the following structures: stoma, guard cells, upper and lower epidermis, palisade layer, spongy layer, and vein.

Analysis

1. **Describe** a stoma, and relate the function of a stoma to your description.
2. **Describe** the location and contents of the veins.
3. **SKILL Relating Concepts** How do the location and structure of the palisade and spongy layers help a leaf perform photosynthesis?

MULTICULTURAL PERSPECTIVE

Cattails

Cattails, which grow in standing water, are a type of grass. Native Americans once ate the young shoots of cattails, which taste very much like the asparagus we eat today. The leaves of cattails were also used to make baskets.

A leaf is a mass of ground tissue and vascular tissue covered by epidermis, as **Figure 26-10** shows. A cuticle coats the upper and lower epidermis. The vascular tissue, both xylem and phloem, is found in the veins of a leaf. Veins are extensions of vascular bundles that run from the tips of roots to the edges of leaves. In leaves, the ground tissue is called **mesophyll** *(MEHS oh fihl)*. The word *mesophyll* comes from the Greek words *mesos,* meaning "middle," and *phyllon,* meaning "leaf." Mesophyll cells are packed with chloroplasts, where photosynthesis occurs. The chlorophyll in chloroplasts makes leaves look green.

Figure 26-10 Structure of a leaf

The internal structure of a leaf is highly organized.

Cuticle
Xylem
Phloem
Vein
Guard cell
Stoma
Upper epidermis
Mesophyll
Palisade layer
Spongy layer
Lower epidermis

Most plants have leaves with two layers of mesophyll. One or more rows of closely packed, columnar cells make up the palisade layer, which lies just beneath the upper epidermis. A layer of loosely packed, spherical cells, called the spongy layer, lies between the palisade layer and the lower epidermis. The spongy layer has many air spaces through which gases can travel. Stomata, the tiny holes in the epidermis, connect the air spaces to the outside air.

Exploring Further

C_3 Versus C_4 Leaves

Internal leaf structure varies among plants, depending on how they carry out photosynthesis. In photosynthesis, carbon dioxide from the air is added to organic molecules. This process, called carbon fixation, occurs during the Calvin cycle. Because the product of carbon fixation by the Calvin cycle is a three-carbon compound, plants that use *only* the Calvin cycle to fix carbon are called C_3 plants. The leaf structure shown in Figure 26-10 is typical of a C_3 plant. More than 90 percent of all plants are C_3 plants.

Plants such as corn and sugar cane use another chemical process to fix carbon before the Calvin cycle. The first products of this process are four-carbon compounds, so these plants are called C_4 plants. The leaves of C_4 plants have a different internal structure, as shown in the photo of a corn leaf above. C_4 plants are plentiful in the tropics because they conduct photosynthesis efficiently in high heat and intense light.

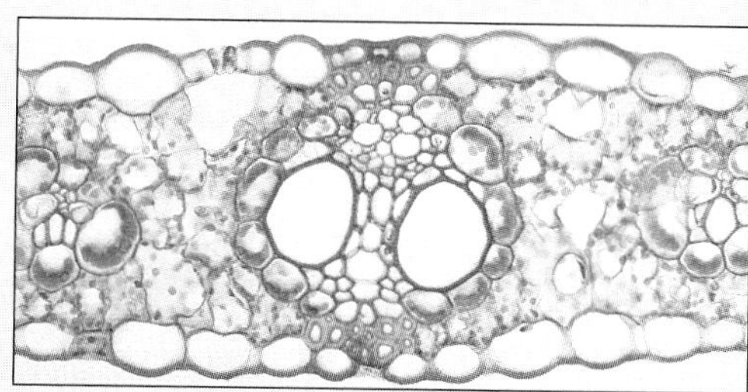

Cross section of a corn leaf (584×)

Review SECTION 26-1

1. **Name** the three kinds of tissue that make up a plant, and list two functions for each.
2. **Summarize** the differences in the structure of the stems of monocots and dicots.

Critical Thinking

3. **SKILL Forming Hypotheses** How do the structures of roots, stems, and leaves help them to carry out their functions?

Review SECTION 26-1 ANSWERS

1. Dermal tissue protects the plant and absorbs water and minerals. Ground tissue stores water, sugar, and starch and conducts photosynthesis. Vascular tissue transports water, mineral nutrients, and sugars throughout the plant.
2. The primary difference is that the vascular bundles of a dicot stem are arranged in a ring, and the vascular bundles of a monocot stem are scattered throughout the ground tissue.
3. Roots have lots of branches and root hairs, which help to anchor the plant and create a large surface area for absorption of water and mineral nutrients. Stems have vascular bundles containing cells with rigid walls that keep the stem upright. These cells transport water, mineral nutrients, and sugars. Leaves are thin and flat, which creates a large surface area for gathering light, and they have stomatal openings, which enable carbon dioxide to enter the leaves.

Section

Exploring Further

C_3 Versus C_4 Leaves

Teaching Strategies

Have students examine the photo of the cross section of a corn leaf and compare its structure with the structure of the leaf shown in Figure 26-10.

Discussion

How does the structure of the corn leaf shown differ from the structure of a typical C_3 plant leaf shown in Figure 26-10? *(The corn leaf has a ring of cells surrounding the veins, and it does not have palisade and spongy layers in the mesophyll.)*

3 Assessment Options

Closure

Relating Functions

Review this section by asking students the following question: What are the important functions of roots, stems, and leaves? *(Roots anchor a plant, store food and water, and absorb water and mineral nutrients from the soil. Stems hold leaves upright; store food and water; and transport water, mineral nutrients, and organic compounds throughout the plant. Leaves conduct photosynthesis.)*

Review

Assign Review—Section 26-1.

Reteaching — BASIC

Have students take out the note cards they prepared in the **Active Reading** on page 571. Ask them to arrange the cards in five sets: terms relating to roots; terms relating to stems; terms relating to leaves; terms relating to roots and stems; and terms relating to roots, stems, and leaves. Have students review Section 26-1 to check the arrangement of their cards.

1 Prepare

Scheduling

- **Block:** About 90 minutes will be needed to complete this section.
- **Traditional:** About two class periods will be needed to complete this section.
- **Planning**: See the Planning Guide on pp. 567A and 567B for lecture, classwork, and assignment options for Section 26-2. Lesson plans for Section 26-2 can be found in a lesson cycle format in ***Biology: Principles and Explorations* One-Stop Planner CD-ROM with Test Generator.**

Resource Link

Assign **Active Reading Worksheet Section 26-2** to help students comprehend the key concepts and visuals in this lesson.

2 Teach

Lesson Warm-up

Ask students how phloem and xylem are similar. *(The cells of both types of vascular tissue are elongated, tubular, and connected end-to-end.)* Ask how the two cell types differ. *(Xylem cells have thick walls and lack cellular contents, and phloem cells have thin walls and contain some cellular contents.)*

Teaching TIP 

Water Is Sometimes Pushed up a Plant

Tell students that sometimes in the early morning, droplets of water appear on the edges of the leaves of some plants. Show photos of leaves with these droplets, which look somewhat like dew except that they form at the tips of the veins. These droplets of water are forced out of the plant by root pressure, which develops as water moves into roots by osmosis. This phenomenon, called guttation, occurs only when the air is very moist and the stomata are closed.

Section 26-2 Transport in Plants

Objectives

- **Relate** transpiration to the movement of water up a plant. (p. 578)
- **Describe** how guard cells regulate the rate of transpiration. (p. 579)
- **Recognize** several distinguishing features of sugar maple trees. (pp. 580–581)
- **Describe** the process of translocation in a plant. (p. 582)

Key Terms

transpiration
source
sink
translocation

Water Is Pulled up Through a Plant

You know that water and mineral nutrients move up from a plant's roots to its leaves through xylem. However, some trees have leaves that are more than 100 m (330 ft) above the ground. How do plants manage to get water so high? Simply put, water is pulled up through a plant as it evaporates from the plant's leaves, as **Figure 26-11** shows.

Step 1 Recall that the surfaces of leaves are covered with many tiny pores, the stomata. When the stomata are open, water vapor diffuses out of a leaf. This loss of water vapor from a plant is called **transpiration.** Air moving across the surfaces of leaves carries away the water vapor. In most plants, more than 90 percent of the water taken in by the roots is ultimately lost through transpiration.

Step 2 The xylem contains a column of water that extends from the leaves to the roots. The cohesion of water molecules causes water molecules that have been lost by a plant to pull on the water molecules still in the xylem. This pull extends through the water in the xylem and draws water upward in the same way liquid is drawn up a drinking straw. As long as the column of water in the xylem does not break, water will keep moving upward as transpiration occurs.

Step 3 Roots take in water from the soil by osmosis. This water enters the xylem and replaces the water lost through transpiration.

Figure 26-11

BIO graphic — Water Movement in Plants

Transpiration drives the movement of water through a plant.

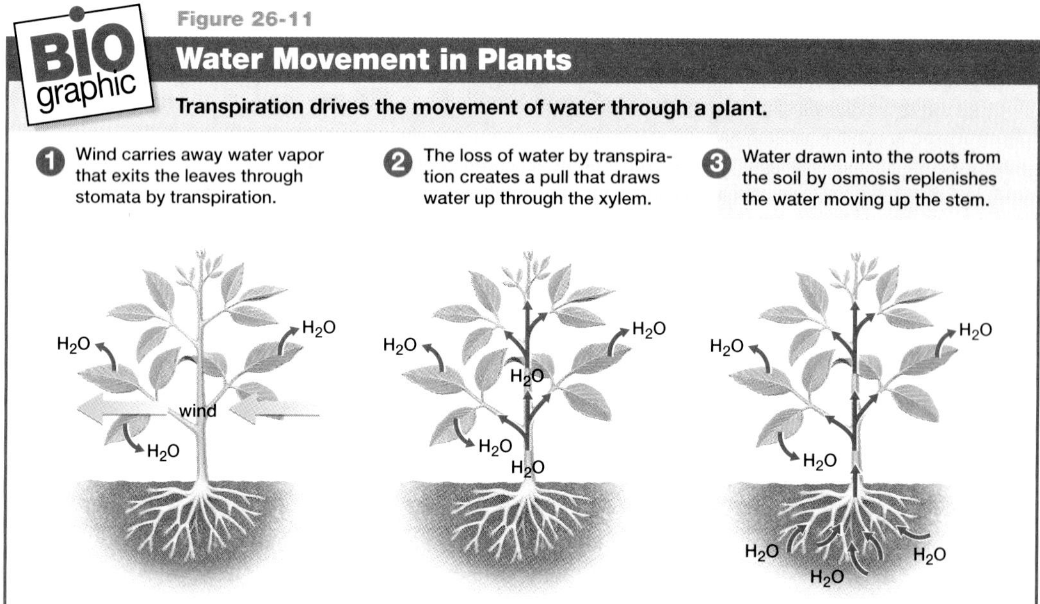

1 Wind carries away water vapor that exits the leaves through stomata by transpiration.
2 The loss of water by transpiration creates a pull that draws water up through the xylem.
3 Water drawn into the roots from the soil by osmosis replenishes the water moving up the stem.
H_2O
wind

Integrating Different Learning Styles

Logical/Mathematical	**PE**	Data Lab, p. 579; Review items 1 and 5, p. 582; Performance Zone items 17, 18, 19, and 20, p. 585
	ATE	Cross-Disciplinary Connection, p. 581
Visual/Spatial	**PE**	Up Close, pp. 580–581; Biographic, p. 582
	ATE	Graphic Organizer, p. 581; Teaching Tip, p. 582; Performance Zone item 1, p. 584
Interpersonal	**PE**	Study Tip, p. 582
Verbal/Linguistic	**PE**	Review items 2, 3, and 4, p. 582; Performance Zone item 27, p. 585
	ATE	Lesson Warm-up, p. 578; Teaching Tip, p. 580; Closure, p. 582; Reteaching, p. 582

Figure 26-12 Control of stomatal opening

Changes in the shape of guard cells cause stomata to open or close.

Guard cell
Stoma
H_2O
Chloroplasts
Epidermal cell
Nucleus

1. A stoma opens as the guard cells take in water, lengthen, and bow apart.
2. A stoma closes as the guard cells lose water, shorten, and come together.

Guard cells and transpiration

A stoma is surrounded by a pair of guard cells that are shaped like two cupped hands. Changes in water pressure within the guard cells cause the stoma to open or close, as shown in **Figure 26-12.** When the guard cells take in water, they swell. However, extra cellulose strands in their cell walls permit the cells to increase in length but not in diameter. As a result, guard cells that take in water bend away from each other, opening the stoma and allowing transpiration to proceed. When water leaves the guard cells, they shorten and move closer to each other, closing the stoma and stopping transpiration. Thus, the loss of water from guard cells for any reason causes stomata to close, stopping further water loss. This is an example of homeostasis in action.

internet connect

TOPIC: Transpiration
GO TO: www.scilinks.org
KEYWORD: HX579

Inferring the rate of transpiration

Background

The graph at right shows the rate of water movement in a plant during high humidity and during low humidity. The rate of water movement is assumed to indicate the rate of transpiration. Use the graph to answer the questions that follow.

Transpiration Rates

A B
Distance water moved (mm)
0 10 20 30 40
5 10 15
Time (in minutes)

Analysis

1. **Determine** how far water had moved after 10 minutes under the condition represented by curve A.
2. **SKILL Analyzing Results** After 15 minutes, how much farther had water moved under condition A than under condition B?
3. **SKILL Recognizing Relationships** Which curve indicates a lower transpiration rate? Explain your reasoning.
4. **SKILL Drawing Conclusions** Which curve shows the transpiration rate during low humidity? Justify your answer.

did you know?

Some leaves have hairs called trichomes.

A "fuzzy" or "woolly" leaf is covered with hairs called trichomes, which are appendages of certain epidermal cells. Some trichomes are stiff and serve a protective function, like horns. Other trichomes branch, sometimes elaborately, and aid in the establishment of a buffer zone between the surrounding air and the leaf surface. Such buffer zones protect leaves from temperature extremes and from drying conditions. By trapping water vapor around and near the surfaces of leaves, the hairs on leaves also slow down the rate of transpiration.

Teaching TIP BASIC

Benefits of Transpirational Water Loss

Point out to students that, as noted in the text, plants have many adaptations for reducing transpirational water loss. Some examples are sunken stomata and reduced numbers of stomata on plants that live in dry and cold winter climates (pine trees) or hot climates (cactuses), hairs on leaves that retain water, and closure of stomata during hot and dry conditions. Tell students that, in addition to needing to have stomata open to enable carbon dioxide to enter leaves for photosynthesis, plants need to have stomata open to transport mineral nutrients to all of the cells in the plant.

Inferring the rate of transpiration

Time 20 minutes

Process Skills Interpreting graphics, analyzing data, making conclusions

Teaching Strategies

Point out the following components of any properly prepared graph such as this one: descriptive title is neither too long nor too short; both axes labeled with an appropriate variable label and units; units on both axes cover the entire range of the data or go slightly outside the range, usually including zero; data points and/or lines are clearly indicated and differentiated.

Analysis Answers

1. 30 mm
2. about 17 mm farther
3. Curve B indicates a lower transpiration rate because it shows water moving a shorter distance than curve A shows. This correlates with a lower transpiration rate.
4. Curve A shows the transpiration rate during low humidity because it shows water moving a greater distance up a plant than curve B shows. This correlates with a higher transpiration rate, which would be expected when the humidity is low.

Up Close

Sugar Maple

Teaching Strategies

- Have students do library research on the sugar maple. Then have each student write an interesting fact about sugar maples on a piece of paper cut in the shape of a maple leaf. Draw the outline of a maple tree on a large sheet of butcher paper, and have students attach their leaves to the branches of the tree.
- Have students bring leaves and samaras from local maple trees and compare them with the leaves and samaras shown in this feature. Remind students to seek permission to collect leaves and samaras from someone else's trees.
- Display samples of maple wood and photographs of various types of maple trees at different times of the year.
- Provide examples of foods that contain maple flavoring or maple sugar.

Making Biology Relevant

Fall Color

Ask students why leaves are green. *(They contain chlorophyll, which reflects green light.)* Point out that there are normally also other pigments in leaves, such as carotenoids, which reflect yellow and orange light; xanthophylls, which reflect orange light; and anthocyanins, which reflect red light. Usually we do not see the colors of these other pigments because chlorophyll is more abundant and the green it reflects masks the other colors. In the fall, chlorophyll is broken down and the colors of the other pigments can be seen.

Up Close

Sugar Maple

- **Scientific name:** *Acer saccharum*
- **Size:** 12 to 37 m (40 to 120 ft) tall, canopy up to 14 m (45 ft) wide, trunk up to 1 m (3 ft) in diameter
- **Range:** Northeastern United States and adjacent regions of southeastern Canada
- **Habitat:** Northern temperate forests
- **Importance:** The wood of sugar maples is used to make furniture, musical instruments, and flooring. Their sap is made into maple syrup and maple sugar.

External Structures

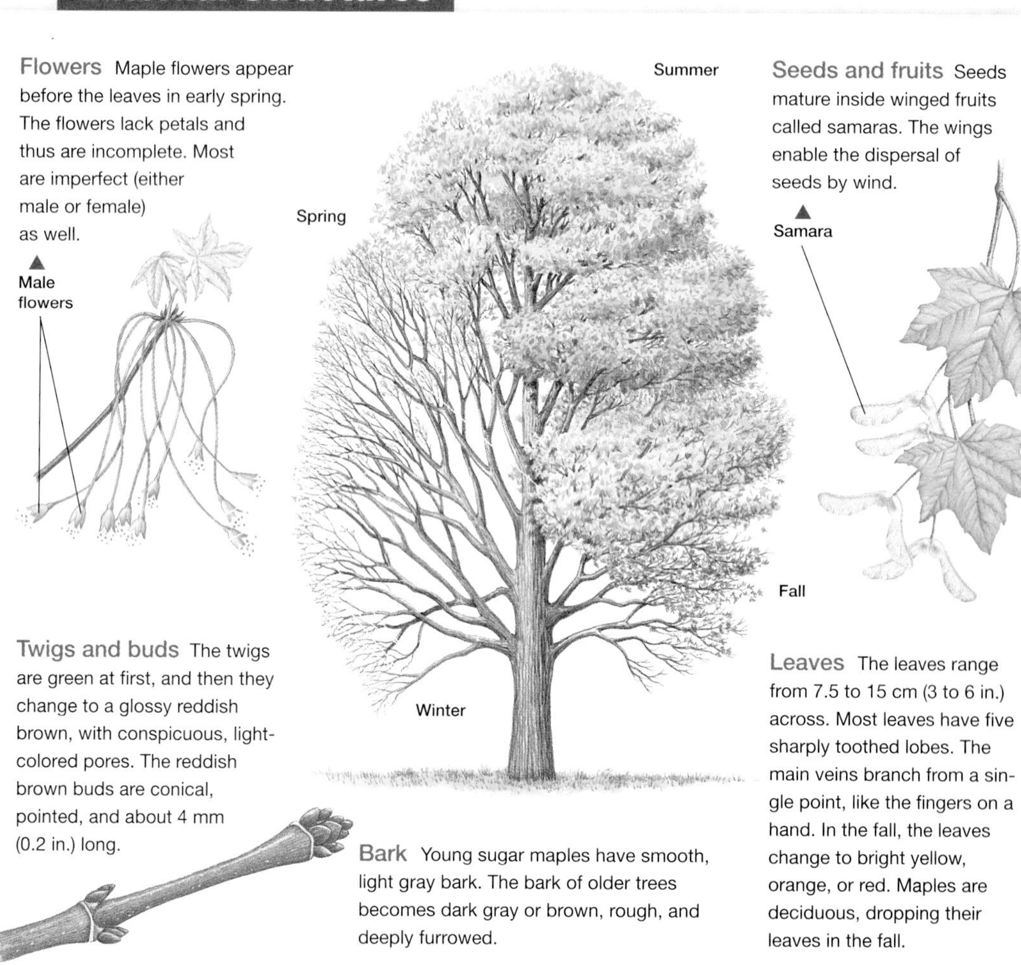

Flowers Maple flowers appear before the leaves in early spring. The flowers lack petals and thus are incomplete. Most are imperfect (either male or female) as well.

Seeds and fruits Seeds mature inside winged fruits called samaras. The wings enable the dispersal of seeds by wind.

Twigs and buds The twigs are green at first, and then they change to a glossy reddish brown, with conspicuous, light-colored pores. The reddish brown buds are conical, pointed, and about 4 mm (0.2 in.) long.

Bark Young sugar maples have smooth, light gray bark. The bark of older trees becomes dark gray or brown, rough, and deeply furrowed.

Leaves The leaves range from 7.5 to 15 cm (3 to 6 in.) across. Most leaves have five sharply toothed lobes. The main veins branch from a single point, like the fingers on a hand. In the fall, the leaves change to bright yellow, orange, or red. Maples are deciduous, dropping their leaves in the fall.

did you know?

Fungus can kill a tree by blocking the tree's xylem.

Dutch elm disease is a fungal disease that has killed thousands of American Elm (*Ulmus americana*) trees in this country, particularly in eastern areas. The fungal spores are transported from tree to tree by an elm bark beetle that burrows down in the bark and eats it. The fungus can also be transported from tree to tree through root grafts that form when trees of the same type are planted close together and their roots grow together. The fungus grows into the xylem. The tree isolates the area where the fungus is living by growing new cells around it. This cuts off water flow through the affected xylem cells. The first visible symptom of Dutch elm disease in a tree is called "flagging," which occurs when all the leaves on a branch high in the tree die. If enough of the xylem of a tree becomes affected, the tree will die.

Internal Structures

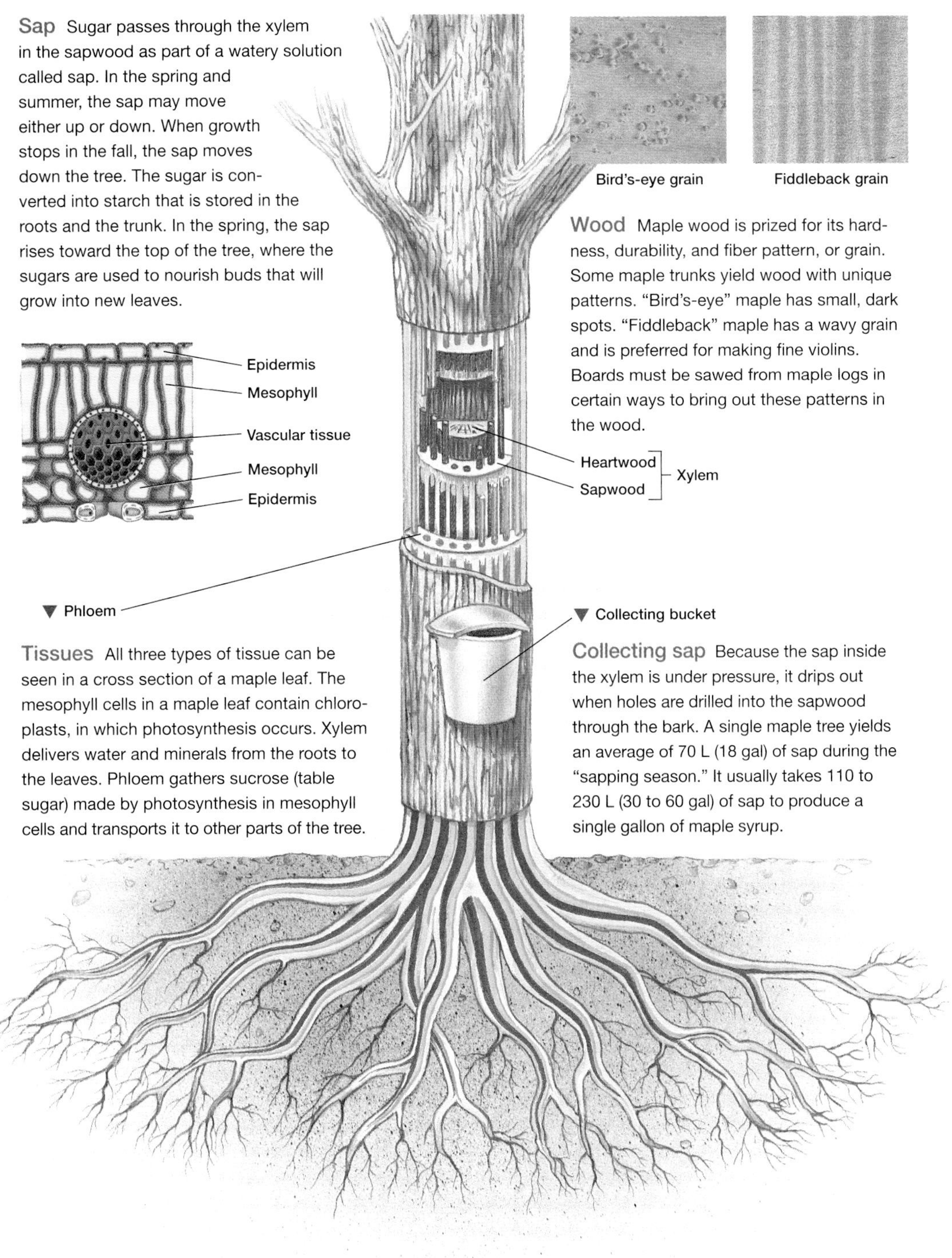

Sap Sugar passes through the xylem in the sapwood as part of a watery solution called sap. In the spring and summer, the sap may move either up or down. When growth stops in the fall, the sap moves down the tree. The sugar is converted into starch that is stored in the roots and the trunk. In the spring, the sap rises toward the top of the tree, where the sugars are used to nourish buds that will grow into new leaves.
Bird's-eye grain
Fiddleback grain
Wood Maple wood is prized for its hardness, durability, and fiber pattern, or grain. Some maple trunks yield wood with unique patterns. "Bird's-eye" maple has small, dark spots. "Fiddleback" maple has a wavy grain and is preferred for making fine violins. Boards must be sawed from maple logs in certain ways to bring out these patterns in the wood.
Epidermis
Mesophyll
Vascular tissue
Mesophyll
Epidermis
Heartwood
Sapwood
Xylem
Phloem
Collecting bucket
Tissues All three types of tissue can be seen in a cross section of a maple leaf. The mesophyll cells in a maple leaf contain chloroplasts, in which photosynthesis occurs. Xylem delivers water and minerals from the roots to the leaves. Phloem gathers sucrose (table sugar) made by photosynthesis in mesophyll cells and transports it to other parts of the tree.
Collecting sap Because the sap inside the xylem is under pressure, it drips out when holes are drilled into the sapwood through the bark. A single maple tree yields an average of 70 L (18 gal) of sap during the "sapping season." It usually takes 110 to 230 L (30 to 60 gal) of sap to produce a single gallon of maple syrup.

Graphic Organizer

Use this graphic organizer with ***Teaching Tip: Transpiration and Translocation*** *on page 582.*

- Leaves lose water by transpiration. →
- Leaves produce sugars in photosynthesis. →
- Sugars are transported downward in the stem. → Sugars are transported into flowers and fruits.
- Sugars are transported downward in the stem. ↓ Sugars are transported downward into roots for use or storage. ↖ Sugars are transported upward to leaf buds in the spring. → Leaves lose water by transpiration.
- Water enters the xylem of roots and moves upward into the xylem of the stem. ↑ Leaves lose water by transpiration.

Up Close

Sugar Maple

Discussion

- What is the economic importance of sugar maple trees in the United States? *(They provide raw materials for making important products such as maple syrup, maple sugar, furniture, musical instruments, and wood flooring and furniture.)*
- How might a sugar maple tree's production of maple sap be affected if it were growing in an environment where the temperature stayed the same year round, such as the tropics? *(In an environment with constant temperature, there might not be as much sugar converted to starch for storage. As a result, there would not be a large amount of sap rising at any one time, and collecting the sap would be difficult.)*
- Why is maple a more desirable wood for making furniture than pine? *(Maple is a hardwood, which produces more durable wood than a softwood such as pine.)*

Cross-Disciplinary Connection

◆ History

In 1932 the infant son of Charles Lindbergh, the first person to fly solo across the Atlantic Ocean, was kidnapped and killed in Hopewell, New Jersey. The most damaging piece of evidence that led to the conviction of the suspect was a ladder used in the kidnapping. A wood technologist established by microscopic evidence that four types of wood were used in making the ladder. A detailed analysis of the grain and wood from one of the rungs matched the floorboards from the suspect's attic.

Teaching TIP — BASIC

Transpiration and Translocation

Have students prepare a **Graphic Organizer** similar to the one at the bottom of page 581 that summarizes the processes of transpiration and translocation. PORTFOLIO

3 Assessment Options

Closure

Transpiration and Translocation

Have students compare and contrast transpiration and translocation by studying Figures 26-11 and 26-13. Point out that water moves upward through a plant from the roots to the leaves, and exits the leaves in the process of transpiration. The upward movement of water is not powered by energy from the plant's cells because the xylem cells lack cellular organelles. Instead, water is pulled upward by adjacent water molecules as other water molecules are lost from the leaves. The source of energy for this is the sun, which dries the air around the plant. Also point out that sugars usually move downward through a plant from the leaves to the roots, flowers, and fruits where it is used or stored. The movement of sugars within the phloem is a passive process that depends on water pressure that is built up in the sieve-tube cells. The source of energy for translocation is the water pressure that builds up because sugars are actively transported into the sieve-tube cells, where they accumulate. This causes water to move in by osmosis.

Review

Assign Review—Section 26-2.

Reteaching — BASIC

Have students write down the names of all of the cells involved in transpiration and translocation and then identify the role that each cell type plays in these processes.

Study TIP

Reviewing Information

Work with a partner to review the pressure-flow model of translocation, covered on page 582. Review the steps of model by quizzing each other on the events of each step.

Organic Compounds Are Pushed Through a Plant

Organic compounds move throughout a plant within the phloem. Botanists use the term **source** to refer to a part of a plant that provides organic compounds for other parts of the plant. For example, a leaf is a source because it makes starch during photosynthesis. A root that stores sugar is also a source. Botanists use the term **sink** to refer to a part of a plant that organic compounds are delivered to. Actively growing parts, such as root tips and developing fruits, are examples of sinks. The movement of organic compounds within a plant from a source to a sink is called **translocation.**

The movement of organic compounds in a plant is more complex than the movement of water for three reasons. First, water flows freely through empty xylem cells, but organic compounds must pass through the cytoplasm of living phloem cells. Second, water only moves up in xylem, while organic compounds move in all directions in phloem. Third, water can diffuse through cell membranes but organic compounds cannot. The German botanist Ernst Münch proposed a model of translocation in 1924. Münch's model, which is often called the pressure-flow model, is shown in Figure 26-13.

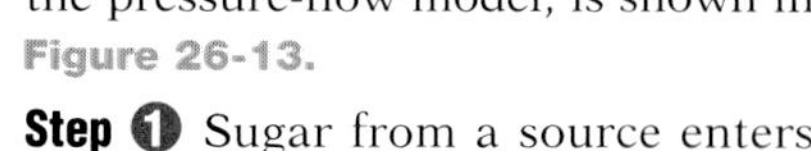

Step 1 Sugar from a source enters phloem cells by active transport.

Step 2 When the sugar concentration in the phloem increases, water enters the sieve tubes in phloem from xylem vessels by osmosis.

Step 3 Pressure builds up inside the sieve-tube cells and pushes sugar through the sieve tubes. The sugar moves at a rate as high as 100 cm/h.

Step 4 Sugar moves from phloem cells into a sink by active transport.

BIO graphic

Figure 26-13

The Pressure-Flow Model

Translocation is described by this model.

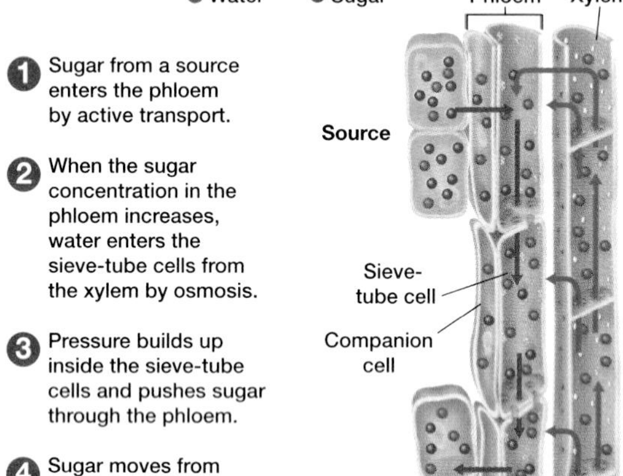

Water
Sugar
Phloem
Xylem
1 Sugar from a source enters the phloem by active transport.
2 When the sugar concentration in the phloem increases, water enters the sieve-tube cells from the xylem by osmosis.
3 Pressure builds up inside the sieve-tube cells and pushes sugar through the phloem.
4 Sugar moves from the phloem into the sink by active transport.
Source
Sieve-tube cell
Companion cell
Sink

Review SECTION 26-2

1. **Relate** the process of transpiration to the movement of water through a plant.
2. **Describe** how guard cells regulate transpiration.
3. **List** the features that make sugar maples economically important.
4. **Describe** how translocation occurs in plants.

Critical Thinking

5. **SKILL Predicting Outcomes** When the soil is dry and it is very hot, how can a plant reduce its water loss? Explain your answer.

Review SECTION 26-2 ANSWERS

1. Transpiration is the loss of water from leaves. Water moves through a leaf to replace the lost water molecules. This draws water into the leaf from the stem, up the stem, from the roots into the stem, and from the soil into the roots.
2. When guard cells swell with water, they form an opening, or stoma, through which water can exit a leaf in the process of transpiration. When guard cells lose water and shrink, they close the stoma.
3. Maple trees provide raw materials for making many important products such as maple syrup, maple sugar, furniture, musical instruments, and wood flooring and furniture.
4. Sugar from a source is actively transported into a sieve-tube cell. Water then enters the sieve-tube cell by osmosis, and pressure builds up in the sieve-tube cell and pushes the solution through the sieve tube. Sugar is then actively transported into a sink cell.
5. A plant can reduce its water loss by closing most of its stomata. This decreases water loss from transpiration and conserves water in the plant.

Study ZONE

CHAPTER 26

Use the Key Concepts and Key Terms listed below to review the main ideas in this chapter. If you need to review the meaning of a term, turn to the page indicated.

Key Concepts

26-1 The Vascular Plant Body

- A vascular plant's body contains three kinds of tissue—dermal tissue, ground tissue, and vascular tissue.
- Dermal tissue covers a plant. A thin layer of epidermis covers nonwoody parts. Several layers of cork cover woody parts.
- Ground tissue is specialized for photosynthesis in leaves and for storage and support in stems and roots.
- Vascular tissue conducts water, minerals, and organic compounds throughout the plant.
- Xylem contains vessels, which are made up of cells that conduct water only after they lose their cytoplasm. Water flows between cells through pits and perforations in their cell walls.
- Phloem contains sieve tubes, which are made up of cells that are still living. Substances pass between the cells through pores.
- Roots have a central core of vascular tissue that is surrounded by ground tissue and epidermal tissue. Root hairs on root tips increase the surface area available for absorption.
- Nonwoody stems contain bundles of xylem and phloem embedded in ground tissue. Woody stems have an inner core of xylem surrounded by a cylinder of phloem.
- Leaves are a mass of ground tissue and vascular tissue covered by epidermis. The ground tissue cells conduct photosynthesis. Gases are exchanged through the stomata in the epidermis.

26-2 Transport in Plants

- Transpiration, the loss of water from a plant's leaves, creates a pull that draws water up through xylem from roots to leaves.
- Guard cells control water loss by closing a plant's stomata when water is scarce. Thus, they also regulate the rate of transpiration.
- The sugar maple is a commercially valuable tree. Its hard wood is used to make furniture, musical instruments, and other products. Its sap is made into maple syrup and maple sugar.
- Organic compounds are pushed through the phloem from a source to a sink in a process called translocation.

Key Terms

26-1

dermal tissue (570)
ground tissue (570)
epidermis (571)
cork (571)
vessel (572)
sieve tube (572)
cortex (573)
root hair (573)
root cap (573)
herbaceous plant (574)
vascular bundle (574)
pith (574)
heartwood (575)
sapwood (575)
petiole (576)
mesophyll (577)

26-2

transpiration (578)
source (582)
sink (582)
translocation (582)

Study TIP Ready?

- *If you think you understand these Key Concepts and Key Terms, then you're ready for the Performance Zone.*

Study ZONE

CHAPTER 26

Review and Practice

Assign the following worksheets for Chapter 26:

- **Student Review Guide**
- **Concept Mapping Worksheet**
- **Test Preparation Pretest**
- **Vocabulary Worksheet**
- **Critical Thinking Review Worksheet**

Assessment Options

The following assessment options are available for this chapter:

Resource Link

- Assign **Chapter Test 26** to assess students' understanding of the chapter.

Alternative Assessment

Have students make a drawing of a plant, including its roots, stems, and leaves. (The drawings do not have to look very realistic, but they should clearly distinguish the structures shown.) Have students indicate the tissues and types of cells present in the roots, stems, and leaves. Have them also place labeled arrows indicating the movement of water and sugars in the processes of transpiration and translocation. PORTFOLIO

Portfolio Assessment

- *Portfolio Assessment* at the front of this book provides options for building and evaluating student portfolios. Students and teachers can select items from the areas listed for inclusion in a portfolio.
- See items labeled PORTFOLIO on pp. 570, 571, 572, 573, 574, 582, and 583.

Answer to Concept Map

The following is one possible answer to Performance Zone item 1 on page 584.

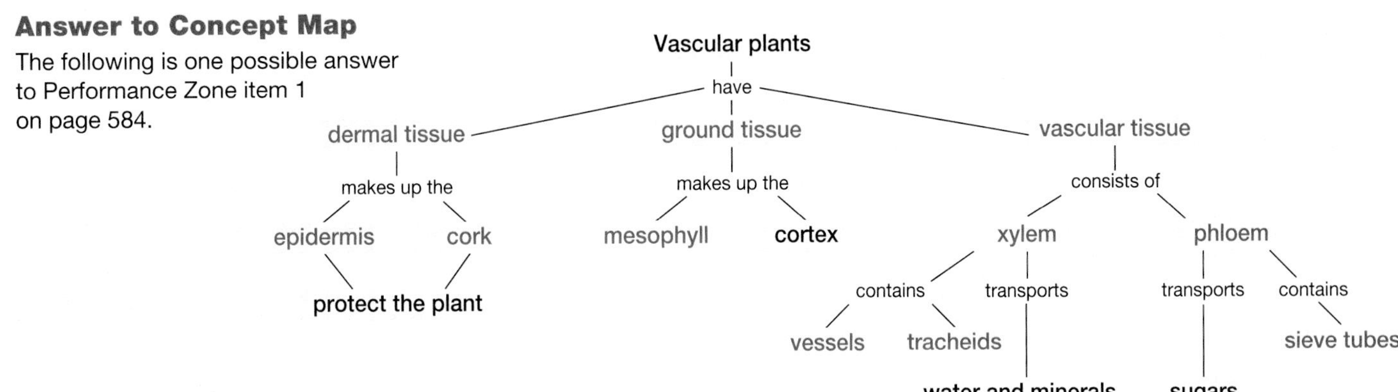

Performance ZONE
CHAPTER 26 REVIEW ANSWERS

Assignment Guide

Section	Review
26-1	2, 3, 4, 5, 6, 7, 8, 14, 15, 21, 22, 23, 24, 26, 28
26-2	1, 9, 10, 11, 12, 13, 16, 17, 18, 19, 20, 25, 27

Understanding and Applying Concepts

1. The answer to the concept map is found at the bottom of page 583.
2. **a.** Epidermis covers herbaceous stems, root tips, and young woody stems. Cork covers woody stems and roots.
   b. Cortex is the type of ground tissue that is outside of the vascular bundles in a stem or root. The pith is the type of ground tissue that is found at the center of a stem.
   c. Heartwood is in the center of a mature stem or tree trunk. Sapwood lies just outside the heartwood and contains xylem that can still conduct water.
   d. The loss to water vapor from a plant is called transpiration. The movement of organic compounds within a plant is called translocation.
   e. A source is a part of a plant where sugar and starch are made or stored. A sink is a part of a plant to which sugars are transported for use by the plant.
3. d
4. c
5. c
6. b
7. d
8. a
9. b
10. a
11. c
12. d
13. Wilting helps a plant prevent further water loss and maintain in its body the amount of water that is necessary for its survival. Thus, wilting helps maintain homeostasis.

Performance ZONE CHAPTER 26 Review

Understanding and Applying Concepts

1. **Concept Mapping** Make a concept map that describes the organization of the vascular plant body. Include the following terms in your map: cork, dermal tissue, epidermis, ground tissue, mesophyll, phloem, sieve tubes, tracheids, vascular tissue, vessels, xylem. Include additional terms as needed.

2. **Understanding Vocabulary** For each pair of terms, explain the difference in their meanings.
   a. epidermis/cork
   b. cortex/pith
   c. heartwood/sapwood
   d. transpiration/translocation
   e. source/sink

3. Which of the following is *not* a function of ground tissue?
   a. storage c. photosynthesis
   b. support d. protection

4. The tissue that transports water and carbohydrates through a plant is
   a. dermal tissue. c. vascular tissue.
   b. ground tissue. d. spongy tissue.

5. Roots that grow from aboveground stems or leaves are called
   a. fibrous roots. c. adventitious roots.
   b. taproots. d. petioles.

6. Xylem cells that no longer conduct water are found in the
   a. sapwood. c. pith.
   b. heartwood. d. cortex.

7. Which of the following phrases describes the structure of a monocot stem?
   a. contains a ring of vascular bundles that surrounds a core of ground tissue
   b. contains a core of vascular tissue that is surrounded by a ring of ground tissue
   c. contains several layers of xylem that are surrounded by a ring of phloem
   d. contains vascular bundles that are scattered throughout the ground tissue

8. The leaf in the photograph below is probably adapted for
   a. climbing. c. nutrient storage.
   b. water storage. d. conserving water.

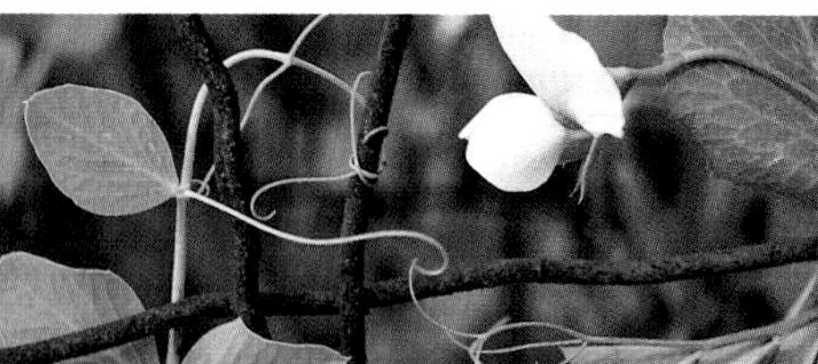

9. The movement of water through a plant is driven by the loss of water vapor from
   a. root hairs. c. buds.
   b. leaves. d. nodes.

10. The column of water in a plant's xylem can remain unbroken because of the
    a. cohesion of water molecules.
    b. repulsion between water molecules.
    c. strong walls in the xylem.
    d. stiff fibers in the bark.

11. Guard cells swell and become longer when
    a. carbon dioxide moves out of the cells.
    b. water moves out of the cells.
    c. water moves into the cells.
    d. oxygen moves into the cells.

12. Organic compounds move through phloem
    a. by diffusion from a sink to a source.
    b. by diffusion from the leaves to the roots.
    c. by active transport within a sieve tube.
    d. because of the pressure created by the movement of water into the sieve tubes.

13. **Theme Homeostasis** When a plant wilts, its stomata close. How does wilting help a plant maintain homeostasis?

14. **Exploring Further** How are C_3 leaves different from C_4 leaves?

15. **Up Close** What is the range and habitat of the sugar maple?

16. **Chapter Links** How is osmosis involved in translocation in a vascular plant? (**Hint:** See Chapter 4, Section 4-1.)

14. C_3 leaves have distinct palisade and spongy mesophyll layers; C_4 leaves lack distinct palisade and spongy mesophyll layers and have a layer of cells surrounding their vascular bundles.
15. The sugar maple's range is the northeastern United States and adjacent regions of southeastern Canada. Its habitat is northern temperate forests.
16. Water enters into sieve tubes by osmosis when the cells become loaded with sugars. The water pressure that builds up as a result of the influx of water causes the sap in the sieve tube to move into an adjacent sieve tube. Movement of sugars from sieve tube to sieve tube in this manner results in translocation.

Skills Assessment

Part 1: Interpreting Graphs

Use the graph below to answer the questions that follow:

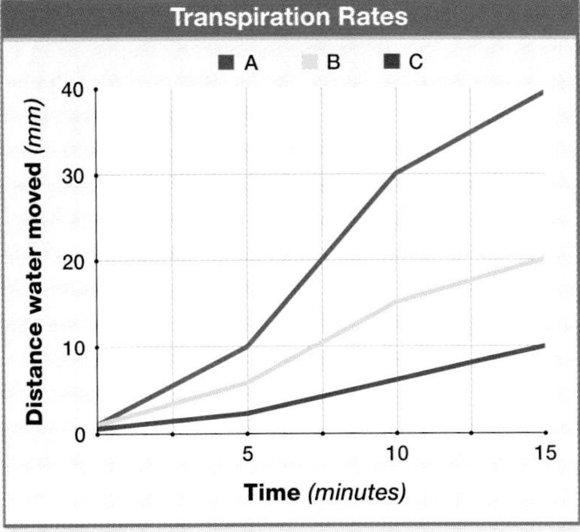

17. How far did water move during the first 10 minutes under the condition represented by curve *A*?
18. After 15 minutes, how much farther did water move under condition *A* than under condition *B*?
19. Which curve indicates a lower transpiration rate? Explain your reasoning.
20. Which curve shows the transpiration rate during low humidity? Justify your answer.

Part 2: Life/Work Skills

Use the media center or Internet resources to learn about three types of plants: the Venus' flytrap, the pitcher plant, and the sundew.

21. **Finding Information** Write a report summarizing the process by which each type of plant traps insects. Explain why trapping insects is important to the plants.
22. **Communicating** Prepare an oral presentation in which you describe the trapping process in each type of plant. Illustrate your presentation with drawings that show the steps in the process for each plant.

Critical Thinking

23. **Evaluating Results** Some herbicides (weed killers) contain a chemical that breaks down waxy substances. Explain why such a chemical might be useful in a herbicide.
24. **Forming Reasoned Opinions** Suggest whether plants that have a taproot system or plants that have a fibrous root system would be more likely to prevent soil erosion on a steep hillside. Explain your reasoning.
25. **Analyzing Information** Trace the path of a water molecule through a vascular plant, from the water molecule's entry into a root hair to its exit between two guard cells. Identify and describe the function of each structure the water molecule would encounter on its journey.

Portfolio Projects

26. **Organizing Information** Construct clay models of cross sections through a C_3 leaf and a C_4 leaf. Include the epidermis, mesophyll, veins, and guard cells in the model of each leaf. Using the models, explain to your class each leaf's structure and the function of its parts.
27. **Interpreting Information** Read "Botanical Cleanup Crews," by Tina Adler (*Science News*, July 20, 1996, pp. 42–43). What types of pollutants are plants being used to clean up? What types of plants have been used to clean up pollution? How does the structure and function of the roots of these plants help them to clean up pollution?
28. **Career Focus Forester** Foresters are involved in the management of forest trees for the commercial production of timber, wildlife habitat, and recreation. Research a forester's responsibilities, and write a report on your findings. Your report should include a job description, training required, kinds of employers, growth prospects, and starting salary.

Skills Assessment

17. 30 mm
18. 20 mm
19. Curve C. It shows water moving a shorter distance up the stem. This correlates with a lower transpiration rate.
20. Curve A. When the humidity is low, the transpiration rate will be high because a lot of water will be lost from the plant.
21. Students should find that the trapping and digestion of insects provides a source of nitrogen for plants adapted to living in low nitrogen habitats.
22. Presentations will vary. Visuals should show the trapping process in each plant.

Critical Thinking

23. Such a chemical would be useful because it would allow a herbicide to reach unprotected epidermal cells and enter leaves. This might make it easier for a herbicide to kill a plant.
24. Plants with a fibrous root system would be more likely to prevent erosion on a steep hillside because their roots would cover a larger area of the top surface of the soil, which is lost first in erosion.
25. The molecule would cross the epidermis of the root hair, enter the ground tissue, which functions in storage, and then enter the xylem in the center of the root. It would enter a vessel or a tracheid and then be pulled upward through the stem and into the leaves. In the leaves, the water molecule would be pulled out of the xylem and into the spongy mesophyll. The water molecule would then evaporate and diffuse out of the leaf through a stoma, the opening between two guard cells.

Portfolio Projects

26. Student models will vary but should resemble the leaves shown in Figure 26-10 and **Exploring Further** on page 577. The C_3 leaf should contain both palisade and spongy mesophyll. Both layers carry out Calvin cycle photosynthesis, and the spongy layer allows movement of carbon dioxide throughout the inside of the leaf. The C_4 leaf should contain only generalized mesophyll, which attaches carbon dioxide molecules to a 4-carbon molecule and then gives off carbon dioxide to a bundle sheath cell, which carries out Calvin cycle photosynthesis. The C_4 leaf should contain a ring of bundle sheath cells surrounding the vascular bundles.
27. Plants have been used to clean up heavy metals, such as mercury, and radioactive waste. Several different species, as well as some genetically engineered plants, have been used to clean up pollution. The structure of the plants' roots enables them to absorb the pollutants and store them in the root tissue.
28. Reports will vary. Foresters work in erosion control, timber production, habitat preservation, and recreation. Foresters have a college education and may work for the public park systems or private companies, such as timber producers. Starting salary will vary by region.

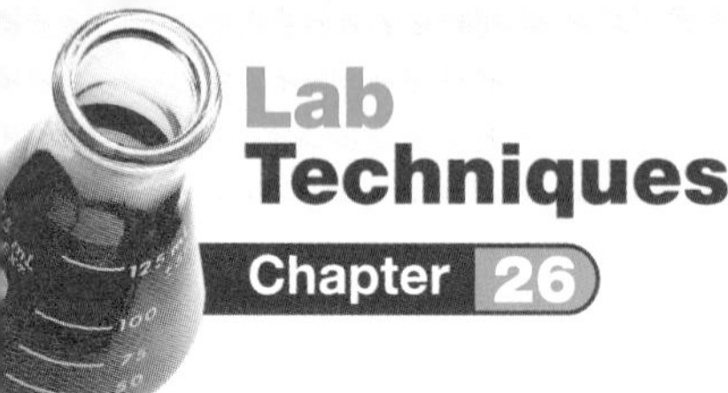

Lab Techniques

Chapter 26

Separating Plant Pigments

Purpose

Students will use paper chromatography to separate the individual pigments in a mixture of simulated plant pigments.

Preparation Notes

Time Required: 30 minutes

Materials

Materials for this lab can be ordered from WARD'S. See *Master Materials List* at the front of this book for catalog numbers.

Preparation Tips

- For every 30 students, order one WARD'S "Chromatography of Simulated Plant Pigments" kit. WARD'S refill kit contains all the consumable items.
- Set up labeled containers for the disposal of solutions, broken glass, and chromatograms.

Safety Precautions

- Be sure that students have read and understand all of the safety rules for working in the lab.
- Remind students to read the **ChemSafety** section before beginning the lab. Discuss all safety symbols and caution statements with students.
- Make sure students wear safety goggles and a lab apron.
- Caution students to use care when working with the stock solutions so that they do not spill.
- Caution students to take care when using the capillary tubes because they are breakable.
- Make sure students wash their hands before leaving the lab.

Procedural Tips

- Review the procedure for creating a chromatogram, on which pigments show up as colored streaks. Students' chromatograms will be almost identical to those of real plant pigments.
- Have students practice using a capillary tube to ensure getting a minute drop of extract on the chromatography paper.

Lab Techniques

Chapter 26 Separating Plant Pigments

SKILLS

- Performing paper chromatography
- Calculating

OBJECTIVES

- **Separate** the pigments that give a leaf its color.
- **Calculate** the R_f value for each pigment.
- **Describe** how paper chromatography can be used to study plant pigments.

MATERIALS

- safety goggles
- lab apron
- strip of chromatography paper
- scissors
- metric ruler
- pencil
- capillary tube
- drop of simulated plant pigments extract
- 10 mL graduated cylinder
- 5 mL of chromatography solvent
- chromatography chamber

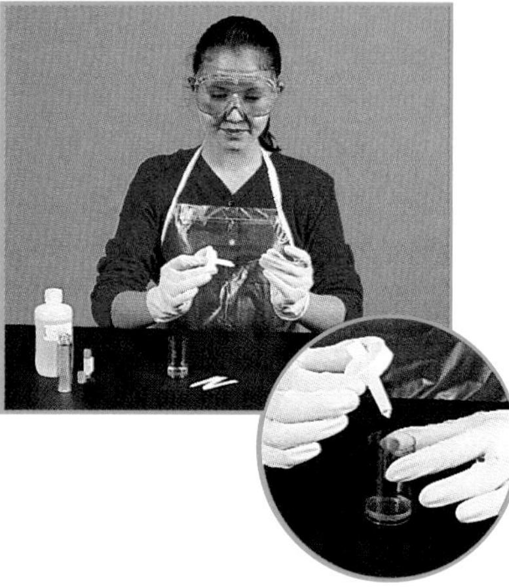

ChemSafety

CAUTION: Always wear safety goggles and a lab apron to protect your eyes and clothing.

CAUTION: Do not touch or taste any chemicals. Know the location of the emergency shower and eyewash station and how to use them. If you get a chemical on your skin or clothing, wash it off at the sink while calling to the teacher. Notify the teacher of a spill. Spills should be cleaned up promptly, according to your teacher's directions.

CAUTION: Glassware is fragile. Notify the teacher of broken glass or cuts. Do not clean up broken glass or spills with broken glass unless the teacher tells you to do so.

Before You Begin

Pigments produce colors by reflecting some colors of light and absorbing or transmitting others. Pigments can be removed from plant tissues using **solvents,** chemicals that dissolve other chemicals. The pigments can then be separated from the solvent and from each other by using **paper chromatography.** The word *chromatography* comes from the Greek words *chromat,* which means "color," and *graphon,* which means "to write." The R_f is the ratio of the distance that a pigment moved relative to the distance that a solvent moved. Since the R_f for a compound is constant, scientists can use it to identify compounds. In this lab, you will learn how to use paper chromatography to separate a mixture of pigments.

1. Write a definition for each boldface term in the previous paragraph and for each of the following terms: chlorophyll a, chlorophyll b, carotene, xanthophyll.
2. Make a data table similar to the one below.

DATA TABLE

Band no.	Color	Pigment	Migration (in mm)	R_f value
1 (top)				
2				
3				
4				
Solvent				

- Emphasize the importance of using a pencil when marking a chromatogram.
- Emphasize the importance of placing only the tip of the chromatography paper in the solvent. To prevent contamination of the solvent, the pigment should never make direct contact with the solvent.
- Tell students to mark the distances traveled by the pigments with a pencil as soon as the chromatogram is removed from the solvent. The simulated plant pigments fade over time.
- Allow students to use calculators to calculate R_f values.

Disposal

- Put on PPE (personal protective equipment). Dilute no more than 250 mL of waste solutions of simulated plant pigments with 20 times as much tap water. Test the mixture with pH or litmus paper, and neutralize, if necessary, by adding small amounts of a 1 M acid or base. Place a beaker with the neutralized solution into a sink that drains to a sanitary sewer, and run water into the beaker for 10 minutes.
- Wrap used chromatography paper and broken glass in newspaper, and place them in the trash.

Lab Techniques

26 Chapter

3. Based on the objectives for this lab, write a question you would like to explore about plant pigments or paper chromatography.

Procedure

PART A: Making a Chromatogram

1. Put on safety goggles and a lab apron. Use scissors to cut the bottom end of a strip of chromatography paper to a tapered end. **CAUTION: Sharp or pointed objects may cause injury. Handle scissors carefully.**
2. Draw a faint pencil line 1 cm above the pointed end of the paper strip, as shown in the photo on page 586. Use a capillary tube to apply a tiny drop of the simulated plant pigments extract on the center of the line.
3. Pour 5 mL of chromatography solvent into a chromatography chamber. Pull the chromatography paper through the opening of the cap, and adjust the length of the strip so that a small portion of the tip end is immersed in the solvent. DO NOT immerse the pigment in the solvent.
4. Place the cap over the chromatography chamber. Carefully bend the end of the strip of chromatography paper over the cap, as shown in the photograph on page 586. Be sure that the strip does not touch the walls of the chamber.
5. Remove the strip from the chromatography chamber when the solvent nears the top of the chamber (within 5–7 minutes).
6. With a pencil, mark the position of the uppermost end of the solvent and the farthest distance each pigment moved. Measure the distance that the solvent and each pigment moved. Record your observations and measurements in your data table. Tape or glue your chromatogram to your lab report. Label the pigment colors.
7. Use the formula below to calculate and record the R_f for each pigment.

$$R_f = \frac{\textit{Distance substance (pigment) traveled}}{\textit{Distance solvent traveled}}$$

PART B: Cleanup and Disposal

8. Dispose of chromatography paper, solutions, and broken glass in the designated waste containers. Do not pour chemicals down the drain or put lab materials in the trash unless your teacher tells you to do so.
9. Clean up your work area and all lab equipment. Return lab equipment to its proper place. Wash your hands thoroughly before you leave the lab and after you finish all work.

Analyze and Conclude

1. **Summarizing Results** Describe what happened to the simulated plant pigments during the lab.
2. **Analyzing Data** How do your R_f values compare with those of your classmates?
3. **Inferring Conclusions** What is a chromatogram?
4. **Further Inquiry** Write a new question about plant pigments that could be explored with another investigation.

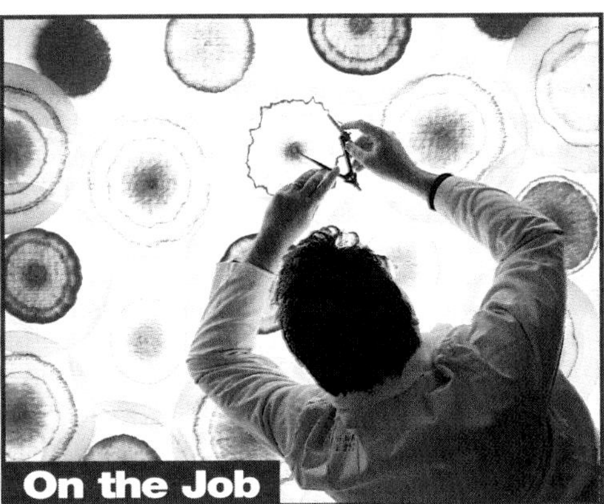

On the Job

Paper chromatography is used to separate a variety of chemicals from living tissue extracts. Do research in the library or media center or on the Internet to discover how paper chromatography is used in research laboratories. Summarize your findings in a written or oral report.

Analyze and Conclude

Answers

1. The greenish mixture separated into four different colors: dark green, light green, yellow, and orange. The pigments moved different distances along the strip of chromatography paper. The orange moved the farthest distance, and the dark green moved the shortest distance.
2. Answers will vary. Students should conclude that the R_f values are the same for each pigment separated.
3. A chromatogram is a strip of chromatography paper that has pigments separated from a mixture along its length.
4. Answers will vary. For example: Can pigments in other substances such as fruit juices and ink be separated with the same solvent?

On the Job

Answer

Paper chromatography is used to separate mixtures of organic compounds in many biochemical and plant physiology laboratories.

Before You Begin

Answers

1. *pigment*—molecule that absorbs some colors of light and reflect others

   solvent—chemical that can dissolve other chemicals

   paper chromatography—technique of separating a mixture of chemicals by making them pass through a strip of paper in which the chemicals move at different speeds

   chlorophyll a—green pigment found in plants

   chlorophyll b—green pigment found in plants

   carotene—orange pigment found in plants

   xanthophyll—yellow pigment found in plants
3. Answers will vary. For example: How many different pigments are found in the mixture?

Sample Data Table

Band No.	Color	Pigment	Migration distance (in mm)	R_f value
1 (top)	orange	carotene	37	0.92
2	yellow	xanthophyll	29	0.72
3	light green	chlorophyll a	17	0.42
4	dark green	chlorophyll b	8	0.20
Solvent	—	—	40	—

Chapter 27

Plant Growth and Development

Planning Guide

OBJECTIVES	MEETING INDIVIDUAL NEEDS	READING SKILLS AND STRATEGIES
27-1 How Plants Grow and Develop (BLOCKS 1, 2, & 3 135 min) • Compare seed germination in beans and corn. • Contrast annuals, biennials, and perennials. • Explain how primary and secondary growth are produced. • Describe several traits of bread wheat. • Contrast development in plants and animals.	Learning Styles, p. 590 BASIC ADVANCED • pp. 590, 591, 593, 596	ATE Active Reading, Technique: K-W-L, p. 589 ■ Active Reading Guide, worksheet (27-1) ■ Directed Reading, worksheet (27-1) ■ Reading Strategies, K-W-L worksheet
27-2 Regulating Plant Growth and Development (BLOCKS 4 & 5 90 min) • Identify the major nutrients plants need to grow. • Describe how plant hormones control plant growth. • Relate environmental factors to plant growth.	Learning Styles, p. 597 BASIC ADVANCED • pp. 600, 602	ATE Active Reading, Technique: Reading Organizer, p. 600 ■ Active Reading Guide, worksheet (27-2) ■ Directed Reading, worksheet (27-2) ■ Supplemental Reading Guide, *A Feeling for the Organism: The Life and Work of Barbara McClintock*

Review and Assessment Resources

BLOCKS 6 & 7 90 min

TEST PREP

PE Study Zone, p. 603

PE Performance Zone, pp. 604–605

■ Chapter 27 worksheets:
- Vocabulary
- Concept Mapping
- Critical Thinking

 Audio CD Program

■ Test Prep Pretest

CHAPTER TESTING

■ Chapter Tests (blackline master)

▲ Test Generator

■ Assessment Item Listing

See the correlations table at the front of this book for a list of the ***National Science Education Standards*** addressed in this chapter.

Smithsonian Institution® Visit **www.si.edu/hrw** for additional on-line resources.

KEY

Biology Principles and Explorations

PE Pupil's Edition
ATE Teacher's Edition

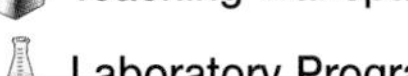

Teaching Transparencies
Laboratory Program

One-Stop Planner CD-ROM
Shows these resources within a customizable daily lesson plan:

■ Teaching Resources
▲ Test Generator

VISUAL STRATEGIES	CLASSWORK AND HOMEWORK	TECHNOLOGY AND INTERNET RESOURCES
ATE **Fig. 27-1,** p. 591 **ATE** **Fig. 27-4,** p. 593 128 Events in Germination 129 Secondary Growth 120A Stages of Plant Differentiation	**PE** **Lab Techniques** Comparing Bean and Corn Seedlings, pp. 606–607 **Lab B18** Mineral Deficiencies in Plants ■ **Portfolio Projects:** Plant Focus Worksheet "Roots and Shoots: Orientation of Growth" **PE** **Review,** p. 596	Holt Biology Videodiscs Section 27-1 Audio CD Program Section 27-1 internetconnect — SCiLINKS NSTA — **TOPICS:** • Primary growth in plants • Secondary growth in plants • Plant life spans • Genetic engineering
ATE **Fig. 27-6,** p. 598 **ATE** **Fig. 27-8,** p. 601 130 Photoperiodism in Plants 119A Went's Experiment	**PE** **Quick Lab** How does ethylene affect a plant?, p. 599 **PE** **Data Lab** Interpreting annual rings, p. 601 **PE** **Math Lab** Analyzing the effect of cold on seed germination, p. 602 **Lab B19** Gravitropism and Phototropism in Plants **PE** **Review,** p. 602 **PE** **Study Zone,** p. 603	CNN Video Segment 22 Plastic Farming Holt Biology Videodiscs Section 27-2 Audio CD Program Section 27-2 internetconnect — SCiLINKS NSTA — **TOPIC:** Plant hormones

ALTERNATIVE ASSESSMENT

PE **Portfolio Projects** 27–29, p. 605
ATE **Alternative Assessment,** p. 603
ATE **Portfolio Assessment,** p. 603

Lab Activity Materials

Quick Lab
p. 599 4 L glass jars with lids (2), plants in 5 cm pots (2), ripe apple

Data Lab
p. 601 pencil, paper

Math Lab
p. 602 pencil, paper

Lab Techniques
pp. 606–607 bean seeds soaked overnight (6), stereomicroscope, corn kernels soaked overnight (6), scalpel, paper towels, rubber bands (2), 150 mL beakers (2), glass-marking pen, metric ruler

Demonstrations
ATE slide of a woody stem, p. 593

BIOSOURCES LAB PROGRAM
- Inquiry Skills Development: Lab B18, p. 83
- Inquiry Skills Development: Lab B19, p. 89

Chapter 27

Chapter Theme

Homeostasis

As plants grow, they develop the specialized cells, tissues, and structures of their bodies in a coordinated manner. Plant growth and development require a balance of substances obtained from the environment and a balance of growth-regulating hormones produced by the plant.

Establishing Prior Knowledge

Before beginning this chapter, make certain students can answer the questions in **Study TIP Ready?** on page 589. If they cannot, encourage them to reread the sections indicated.

Opening Question

Tell students that if plants are grown in a shady location, they typically produce larger leaves than do plants grown in a sunny location. Why might this phenomenon occur? *(Each leaf of a plant grown in shade would have less light hitting it per unit area than would each leaf of a plant grown in full sunlight. By growing larger leaves in the shade, plants can capture more sunlight and compete better with other plants in that location.)*

CHAPTER 27 Plant Growth and Development

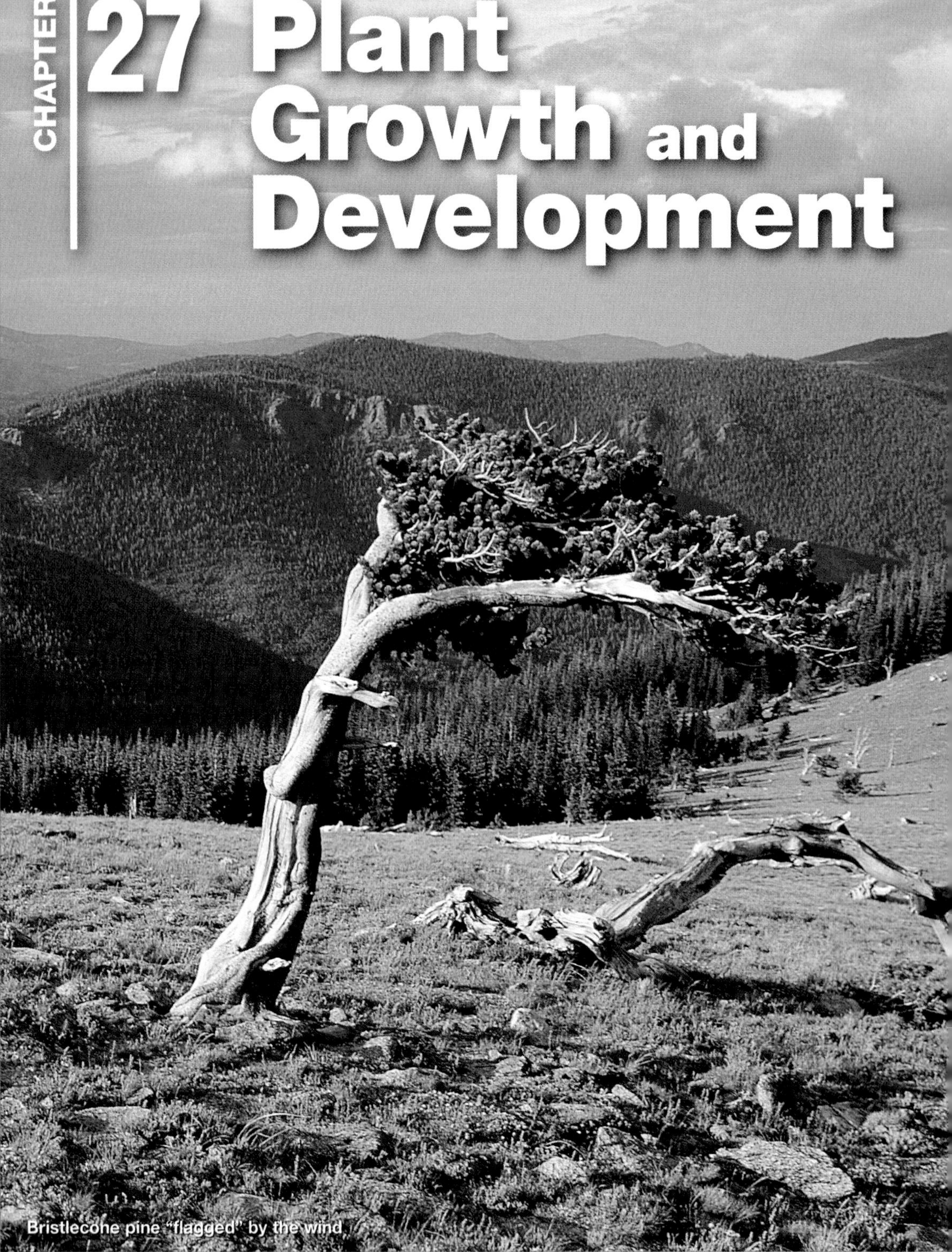

Bristlecone pine "flagged" by the wind

What's Ahead — and Why

CHAPTER OUTLINE

Students will learn

- how a whole, mature plant grows from the tiny embryo in a seed.
- how internal and external factors determine the timing and patterns of plant growth, development, and reproduction.

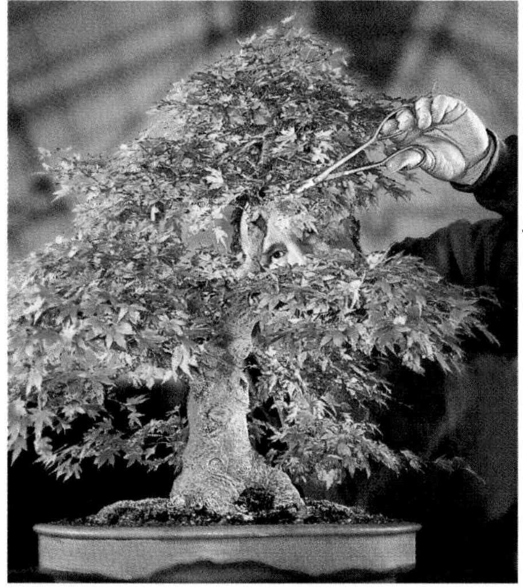

Bonsai mountain maple tree

Art imitates nature. *By practicing the ancient oriental art of bonsai, people produce miniature plants with artful shapes. Nature also "practices" the art of bonsai. Exposed to extreme cold and deprived of water, trees that grow near treeline in the mountains are stunted. Like flags, they "stream" away from the prevailing wind.*

Study TIP

Ready?

Check your understanding of these topics to help you prepare for what's ahead.

Can you . . .

- **identify** the requirements for photosynthesis and cellular respiration? (Sections 5-2 and 5-3)
- **define** *meristem, vascular tissue, xylem,* and *phloem*? (Section 24-1)
- **identify** the parts of a seed? (Section 24-1)
- **describe** the structure of stems and roots? (Section 26-1)

If not, *review the sections indicated.*

Looking Ahead

27-1 How Plants Grow and Develop

How is a seed transformed into a mature plant?

27-2 Regulating Plant Growth and Development

Plants adjust their form and activities in response to signals from their environment.

Labs

- **Quick Lab**
  How does ethylene affect a plant? (p. 599)
- **Data Lab**
  Interpreting annual rings (p. 601)
- **Math Lab**
  Analyzing the effect of cold on seed germination (p. 602)
- **Lab Techniques**
  Comparing Bean and Corn Seedlings (p. 606)

Features

- **Up Close**
  Bread Wheat (p. 594)
- **Tech Watch**
  Plant Tissue Culture (p. 596)
- **Exploring Further**
  Plant Hormones and Agriculture (p. 599)

internet connect

National Science Teachers Association *sci*LINKS Internet resources are located throughout this chapter.

Chapter 27

Reading Strategies

Active Reading

Technique: K-W-L

Before students read this chapter, have them write short individual lists of all the things they already **K**now (or think they know) about plant growth and development. Ask them to contribute their entries to a group list on the board or overhead projector. Then have students list things they **W**ant to know about plant growth and development. Have students save their lists for use later in the **Closure** exercise on page 596.

Directed Reading

Assign **Directed Reading Worksheet Chapter 27** to help students interact with the text as they read the chapter.

Interactive Reading

Assign ***Biology: Principles and Explorations* Audio CD Program Chapter 27** to help reluctant readers achieve greater success in reading the chapter.

Reading Skills

Assign Reading Organizers and Reading Strategies Worksheets from the **One-Stop Planner CD-ROM with Test Generator** to help students learn and practice effective reading skills.

internet connect

Holt *Biology: Principles and Explorations*

On-line Resources: go.hrw.com

Visit the HRW Web site for a variety of resources related to this chapter. Just type in the keyword HX0 Chapter 27.

Holt *Biology: Principles and Explorations*

On-line Resources: www.scilinks.org

The following *sci*LINKS Internet resources can be found in the student text for this chapter:

Page 592
TOPIC: Primary growth in plants
KEYWORD: HX592

Page 593
TOPIC: Secondary growth in plants
KEYWORD: HX593

Page 596
TOPIC: Plant life spans
KEYWORD: HX591

Page 596
TOPIC: Genetic engineering
KEYWORD: HX596

Page 599
TOPIC: Plant hormones
KEYWORD: HX599

Chapter 27

1 Prepare

Scheduling

- **Block:** About 135 minutes will be needed to complete this section.
- **Traditional:** About three class periods will be needed to complete this section.
- **Planning:** See the Planning Guide on pp. 587A and 587B for lecture, classwork, and assignment options for Section 27-1. Lesson plans for Section 27-1 can be found in a lesson cycle format in ***Biology: Principles and Explorations*** **One-Stop Planner CD-ROM with Test Generator.**

Resource Link

Assign **Active Reading Worksheet Section 27-1** to help students comprehend the key concepts and visuals in this lesson.

2 Teach

Lesson Warm-up

Ask students to describe the structure of a seed. *(Seeds consist of an embryo, a stored food supply, and a protective outer covering.)*

Visual Strategy

Figure 27-1 — BASIC

Have students compare the structures of the germinating corn seed and bean seed. Ask students how monocot and dicot seedlings differ in the way they emerge from the soil. *(Dicot seedlings have a stem that hooks as it emerges to protect the delicate tip, whereas monocots grow straight out of the soil. Monocot seedlings have a protective sheath that guards the tip.)* Ask students what other differences they notice. *(The cotyledon and seed coat of the corn seed remain underground, while the cotyledons of the bean emerge from the soil.)* Tell students that the seed coats of beans sometimes emerge with the cotyledons and that the cotyledons of some other dicot seeds also remain underground.

Section 27-1 How Plants Grow and Develop

Objectives

- **Compare** seed germination in beans and corn. (p. 590)
- **Contrast** annuals, biennials, and perennials. (p. 591)
- **Explain** how primary and secondary growth are produced. (pp. 592–593)
- **Describe** several traits of bread wheat. (pp. 594–595)
- **Contrast** development in plants and animals. (p. 596)

Key Terms

germination
perennial
annual
biennial
primary growth
secondary growth
apical meristem
cork cambium
vascular cambium
annual ring

Growth Resumes As a Seed Sprouts

A seed contains a plant embryo that is in a state of suspended animation. Some embryos can remain in suspended animation inside a seed for thousands of years. Seeds sprout with a burst of growth in response to certain changes in the environment. These changes, such as rising temperature and increasing soil moisture, usually signal the start of favorable growing conditions.

Many seeds must be exposed to cold or to light before they can sprout. The seed coats of other seeds must be broken before they can sprout. Exposure to fire, passing through the digestive system of an animal, and falling on rocks are several natural ways that seed coats are damaged. A seed cannot sprout until water and oxygen penetrate the seed coat. When water enters a seed, the tissues in the seed swell, and the seed coat breaks. If enough water and oxygen are available after the seed coat breaks, the young plant, or seedling, begins to grow.

Germination

A plant embryo resumes its growth in a process called **germination.** The first sign of germination is the emergence of the embryo's root. What happens next varies somewhat from one type of plant to another, as you can see in **Figure 27-1.** The young shoots of some plants, such as beans, form a hook. The hook protects the tip of the

Figure 27-1 Seed germination

Beans and corn show two characteristic patterns of seed germination.

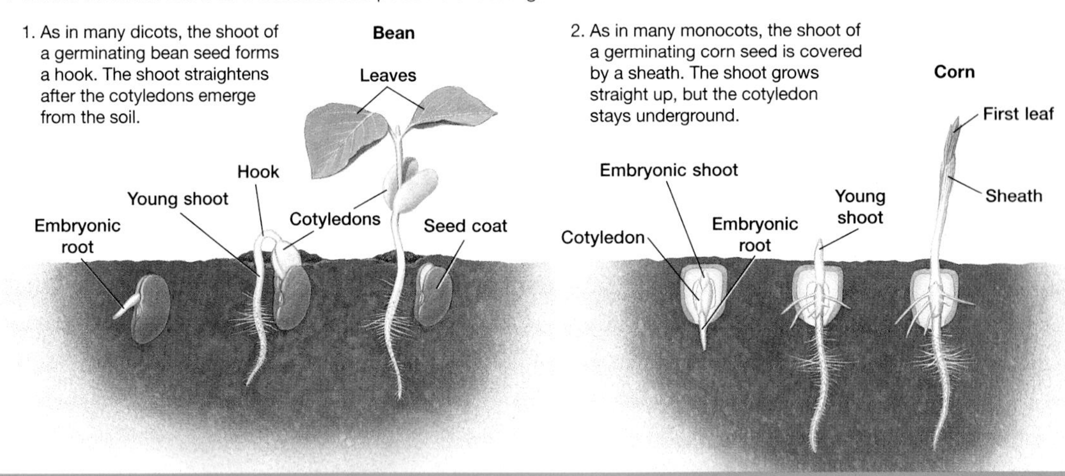

1. As in many dicots, the shoot of a germinating bean seed forms a hook. The shoot straightens after the cotyledons emerge from the soil.
Bean
Leaves
Hook
Young shoot
Embryonic root
Cotyledons
Seed coat
2. As in many monocots, the shoot of a germinating corn seed is covered by a sheath. The shoot grows straight up, but the cotyledon stays underground.
Corn
First leaf
Embryonic shoot
Sheath
Young shoot
Embryonic root
Cotyledon

Integrating Different Learning Styles

Logical/Mathematical	PE	Study Tip, p. 592; Review items 1, 3, 4, 5, and 6, p. 596; Performance Zone item 23, p. 605
Visual/Spatial	PE	Up Close, p. 594; Performance Zone items 1, 18, 19, and 20, pp. 604–605
	ATE	Teaching Tip, p. 591; Teaching Tip, p. 592; Demonstration, p. 593; Visual Strategy, p. 593; Teaching Tip, p. 593; Up Close Teaching Tip, p. 594
Interpersonal	ATE	Lesson Warm-up, p. 590; Teaching Tip, p. 591; Biology in Your Community, p. 591; Up Close Discussion, p. 595; Reteaching, p. 596
Intrapersonal	ATE	Making Biology Relevant, p. 594
Verbal/Linguistic	PE	Review item 2, p. 596
	ATE	Reading Strategies, p. 589; Resource Link, p. 590; Closure, p. 596

shoot from injury as it grows through the soil. The young shoots of other plants, such as corn, have a protective sheath around their shoots. In some plants, such as beans and marigolds, the cotyledons emerge from the soil and unfold aboveground. In other plants, such as corn and peas, the cotyledons remain underground. After the shoot of a seedling emerges, its roots and shoots continue to grow throughout its life span.

Plant life spans

As you learned in Chapter 24, the oldest known trees are bristlecone pines that are estimated to be about 5,000 years old. In contrast, some plants live for only a few weeks. Depending on how long it lives, a plant can be classified as one of three basic types: perennial, annual, or biennial.

Perennials Many herbaceous plants and all woody plants are perennials. A **perennial** is a plant that lives for several years. Most perennials reproduce many times during their life span. Others, like the herbaceous perennial shown in **Figure 27-2,** reproduce only once before they die. Chrysanthemums, daffodils, and irises are familiar herbaceous perennials. These plants store nutrients for the next season's growth in fleshy roots or underground stems. The aboveground shoots of herbaceous perennials often die after each season of growth. Trees, shrubs, and many vines are woody perennials. Some woody perennials drop their leaves each year. Plants that drop all of their leaves each year, such as elms, maples, and grapevines, are known as deciduous *(dee SIHJ oo uhs)* plants. Those that drop a few leaves at a time throughout the year, such as firs, pines, junipers, and honeysuckles, are called evergreens.

Annuals Sunflowers, beans, corn, and many weeds are annuals. An **annual** is a plant that completes its life cycle (grows, flowers, and produces fruits and seeds) and then dies within one growing season. Virtually all annuals are herbaceous plants. Most annuals grow rapidly when conditions are favorable. Individual plants can become quite large if they get enough water and nutrients.

Biennials Carrots, parsley, and onions are biennials. A **biennial** is a flowering plant that takes two growing seasons to complete its life cycle. During the first growing season, biennials produce roots and shoots. The shoots consist of a short stem and a rosette (circular cluster) of leaves. The roots store nutrients. In the second growing season, a biennial plant uses the stored nutrients to produce a flowering stalk. The plant dies after flowering and producing fruits and seeds.

biobit

How long do seeds remain viable?

Some seeds can germinate after waiting for thousands of years. Arctic lupine seeds, for example, were able to germinate after being buried in sheets of ice for 10,000 years. Egyptian wheat and Indian lotus plants have also been grown successfully from seeds that were several thousand years old.

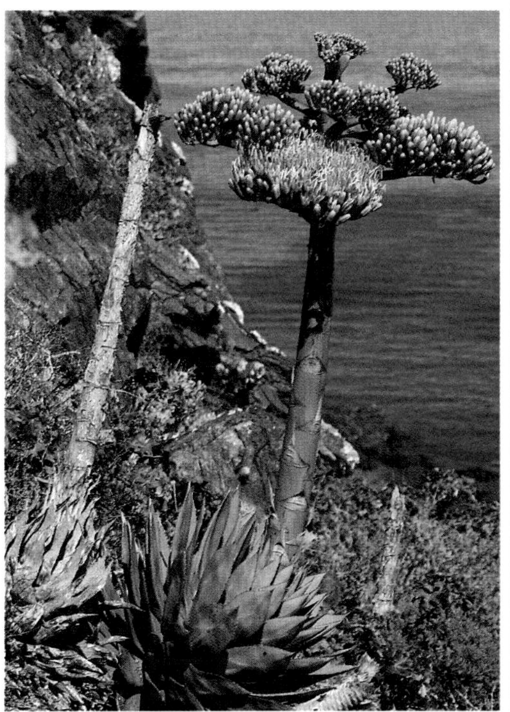

Figure 27-2 A herbaceous perennial. Century plants live for many years but reproduce only once. Like the dried-up plant on the left, this flowering century plant will die when its seeds are mature.

Resource Link

Assign **Portfolio Projects: Plant Focus Worksheet "Roots and Shoots: Orientation of Growth"** from the Holt BioSources Teaching Resources CD-ROM. PORTFOLIO

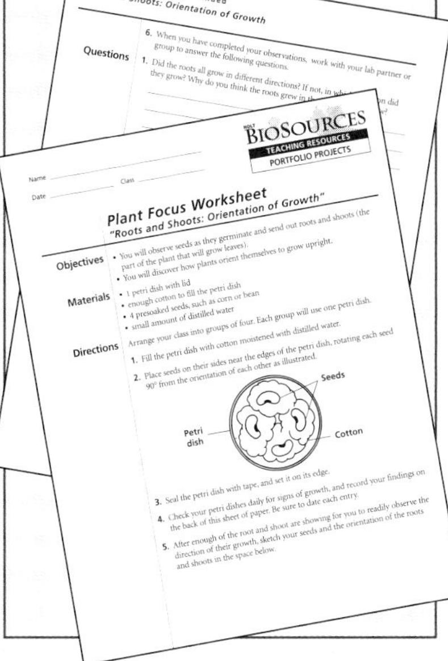

PORTFOLIO PROJECTS continued
Roots and Shoots: Orientation of Growth

Questions
6. When you have completed your observations, work with your lab partner or group to answer the following questions.
1. Did the roots all grow in different directions? If not, in wh... they grow? Why do you think the roots grew in th...

HOLT BIOSOURCES TEACHING RESOURCES PORTFOLIO PROJECTS

Name ____ Class ____
Date ____

Plant Focus Worksheet
"Roots and Shoots: Orientation of Growth"

Objectives
- You will observe seeds as they germinate and send out roots and shoots (the part of the plant that will grow leaves).
- You will discover how plants orient themselves to grow upright.

Materials
- 1 petri dish with lid
- enough cotton to fill the petri dish
- 4 presoaked seeds, such as corn or bean
- small amount of distilled water

Directions
Arrange your class into groups of four. Each group will use one petri dish.
1. Fill the petri dish with cotton moistened with distilled water.
2. Place seeds on their sides near the edges of the petri dish, rotating each seed 90° from the orientation of each other as illustrated.

Seeds
Petri dish
Cotton

3. Seal the petri dish with tape, and set it on its edge.
4. Check your petri dishes daily for signs of growth, and record your findings on the back of this sheet of paper. Be sure to date each entry.
5. After enough of the root and shoot are showing for you to readily observe the direction of their growth, sketch your seeds and the orientation of the roots and shoots in the space below.

Teaching TIP

Design a Flower Bed

Have students work in groups of four to design a flower garden that will provide color in every season (or in three seasons, if you live in an area with a harsh winter climate). Let students use nursery catalogs to find examples of annuals, biennials, and perennials, as well as to determine where each plant will grow and when it will bloom. Assign each group a set of conditions under which its garden must grow (e.g., size, distribution of sun and shade, soil type). Finally, tell them to consider aesthetic factors (e.g., color combinations, taller plants in back, shorter plants in front). Have each student design a graphic that shows the group's garden from the top during one of the four seasons and identifies their plants. PORTFOLIO

Biology in Your Community

Plants in Your Community

Take students on a walk around the school grounds or a park, and have them keep a log of the plants they see, using general terms such as flower, tree, grass, or shrub. Have students write a paragraph that answers the following questions based on what they observe. What types of plants did you see? Did you see more of one type of plant than another? If so, what might account for the difference? *(More annuals tend to be present in spring and summer; only woody perennials may be present in winter.)* PORTFOLIO

Chapter 27

Teaching TIP

Apical Meristems Up Close

Let students observe prepared slides of the apical meristems of the shoots and roots of several plants. Have them draw what they see and label the parts identified in Figure 27-3. Have students write a brief paragraph describing how the cells in the apical meristems differ from those in the surrounding tissues. *(The cells of the apical meristems are small and unspecialized. The surrounding cells tend to be larger and longer and have thicker walls, and some have specialized structures such as chloroplasts or root hairs.)* PORTFOLIO

internet connect

Have students research the growth of plants using the Web site in the Internet Connect box on this page. Have students write a report summarizing their findings. PORTFOLIO

Teaching TIP

Removal of the Apical Meristem

Ask students what they think would happen if the apical meristem of a stem or root were removed. *(Removing or damaging the apical meristems of a plant would stop or limit the vertical growth of the stem or root.)* This frequently occurs in nature, as tips of stems are broken off by wind or animals, killed by disease, or eaten by insects or other animals. Lateral buds that develop at the nodes of a stem often begin to grow after the apical meristem is removed. If the plant is a tree that grows with one main trunk, one of these lateral buds will take over as the main trunk, and the tree will take on its normal form again in a few years. If the plant is a shrub or herbaceous, the growth of many lateral buds will make the plant bushier and it will not grow as tall. Gardeners often remove apical meristems of shrubs and herbaceous plants to encourage fuller, more compact growth.

Study TIP

Compare and Contrast

To compare and contrast *primary growth* and *secondary growth,* make a two-column list. In one column write the ways in which the two are alike. In the other column, write the ways in which they are different.

internet connect

SciLINKS NSTA

TOPIC: Primary growth in plants
GO TO: www.scilinks.org
KEYWORD: HX592

Plant Growth Is Produced by Cell Division in Meristems

Plants grow by producing new cells in regions of active cell division called meristems. Almost all plants grow in length by adding new cells at the tips of their stems and roots. Growth that increases the length or height of a plant is called **primary growth.** Many plants also become wider as they grow taller. Growth that increases the width of stems and roots is called **secondary growth.** After new cells are formed by cell division, they grow and undergo differentiation. Recall from Chapter 20 that differentiation is the process by which cells become specialized in form and function.

Primary growth

Apical *(AP ih kuhl)* **meristems,** which are located at the tips of stems and roots, produce primary growth through cell division. As shown in **Figure 27-3,** apical meristems are regions of small, undifferentiated cells. To better understand how primary growth occurs in most plants, imagine a stack of dishes. As you add more dishes to the top, the stack grows taller but not wider. Similarly, the cells in the apical meristems of most plants add more cells to the tips of a plant's body. New cells are added through cell division. The cells then lengthen. Thus, primary growth makes a plant's stems and roots get longer without becoming wider. To learn about primary growth in a monocot, look at *Up Close: Bread Wheat* on pp. 594–595.

The tissues that result from primary growth are called primary tissues. The new cells produced by apical meristems differentiate into the primary dermal, ground, and vascular tissues of roots, stems, and leaves. Some of the cells produced by the root apical meristem also become part of the root cap. These cells replace cells that are worn away as the root pushes through the soil.

Figure 27-3 Apical meristems

Both shoot tips and root tips contain apical meristems, where cell division occurs.

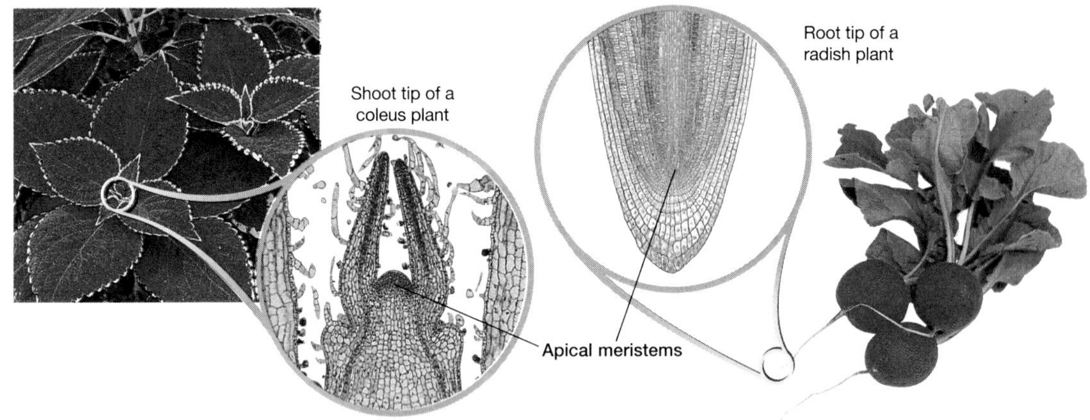

MULTICULTURAL PERSPECTIVE

The Asian Art of Bonsai

Bonsai is the cultivation of miniature specimens of plants that would normally be much larger. The effect is obtained by frequent, selective pruning of both the shoot and the roots. Bonsai plants are typically grown in small containers. The plants will usually form cones or fruits and flowers of normal size.

The word *bonsai* comes from the Japanese characters *bon,* meaning "tray," and *sai,* meaning "plant." Bonsai was practiced by the Chinese in approximately the third century B.C., but it was the Japanese Buddhist monks who perfected the art.

Secondary growth

Secondary growth occurs in parts of many herbaceous plants, such as in carrot roots. However, it is most dramatic in woody plants. Secondary growth is produced by cell division in two meristems, which form thin cylinders near the outside of woody stems and roots. One meristem, called the **cork cambium** *(KAM bee uhm),* lies within the bark and produces cork cells. The other meristem, called the **vascular cambium,** lies just under the bark and produces vascular tissues. The tissues that result from secondary growth are called secondary tissues. Figure 27-4 shows how woody stems develop.

Step 1 A young woody stem has a ring of vascular bundles between the cortex and the pith. Each vascular bundle contains primary xylem and primary phloem.

Step 2 Vascular cambium develops between the primary xylem and the primary phloem in each vascular bundle. Secondary phloem is produced toward the outside of the stem. Secondary xylem is produced toward the inside of the stem. The cork cambium forms when the epidermis is stretched and broken as the stem grows in diameter.

Step 3 Eventually, the vascular bundles merge into solid cylinders. The cork, cork cambium, and secondary phloem make up the bark. The vascular cambium and secondary xylem lie inside the bark. Thick layers of secondary xylem, or wood, often form rings. Since one new ring is usually formed each year, the rings are called **annual rings.**

Do all trees produce annual rings?

Annual rings form only in trees that grow in regions with distinct cold and warm seasons. Abrupt changes in cell diameter produce the rings. Vessel cells grow larger in diameter in the moist springs. Smaller cells form in dry summers.

TOPIC: Secondary growth in plants
GO TO: www.scilinks.org
KEYWORD: HX593

Figure 27-4

BIOgraphic: Development of a Woody Stem

The wood in a woody stem results from secondary growth.

1. Initially, the stem is covered by epidermis and contains cortex, pith, and a ring of vascular bundles with primary xylem and phloem.
2. Next, a vascular cambium forms between the xylem and phloem in each vascular bundle. Cork cambium forms as the epidermis is stretched and broken.
3. In a mature woody stem, no cortex or primary phloem remains. The vascular cambium has formed a solid cylinder that adds new layers of secondary xylem and phloem each year.

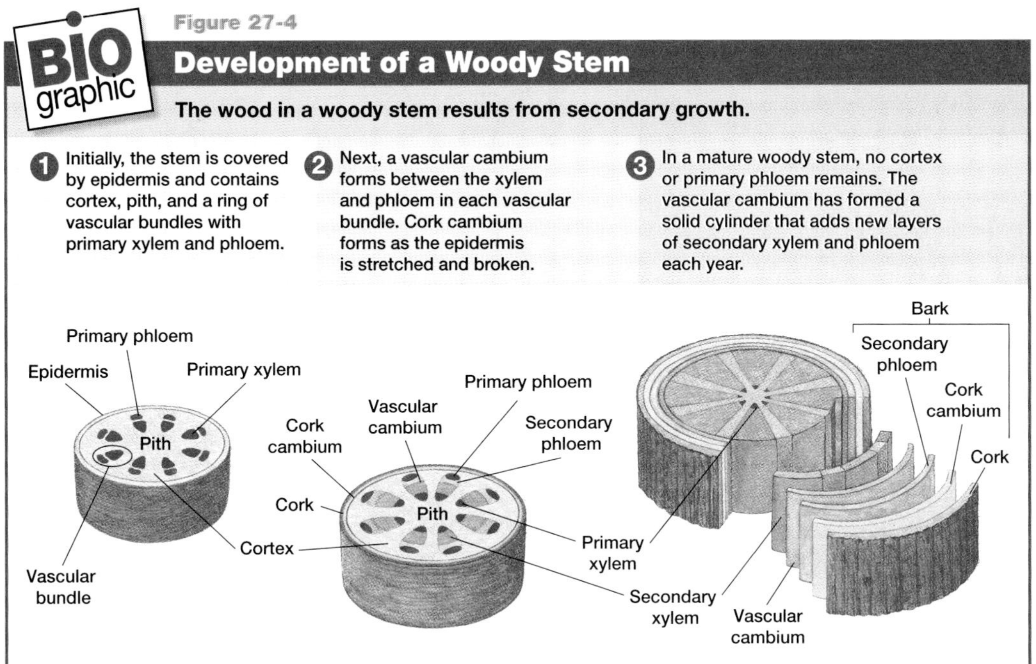

Demonstration

Finding Vascular Cambium

Provide students with prepared slides of a cross section of a woody stem. Have them examine the slides under a compound microscope. Point out that vascular cambium is only one layer of cells. It is found just under the bark of a woody stem and can be located in microscope slides of woody stem cross sections by finding the phloem cells and the large, thickened xylem cells. Also point out that both xylem and phloem develop from the vascular cambium cells, so they develop in rows on either side of this "meristematic" tissue. Have students locate the layer between the two types of cells in a woody stem cross section. This is the row of vascular cambium cells. Have students describe how xylem and phloem originate from vascular cambium. *(through cell division and cell differentiation)*

Visual Strategy

Figure 27-4 ADVANCED

Walk students through the diagrams showing the following stages:

1. A new portion of a woody stem forms from primary growth. The vascular bundles contain only primary xylem and phloem.
2. A vascular cambium forms between the primary xylem and phloem in each vascular bundle. The bundles grow in diameter as the vascular cambium produces secondary xylem and phloem. A cork cambium forms and replaces the epidermis with cork.
3. The vascular bundles fuse into solid cylinders. Secondary xylem becomes wood. Secondary phloem and cork form the bark.

Graphic Organizer

Use this graphic organizer with ***Teaching Tip: Summarizing Plant Growth*** *on this page.*

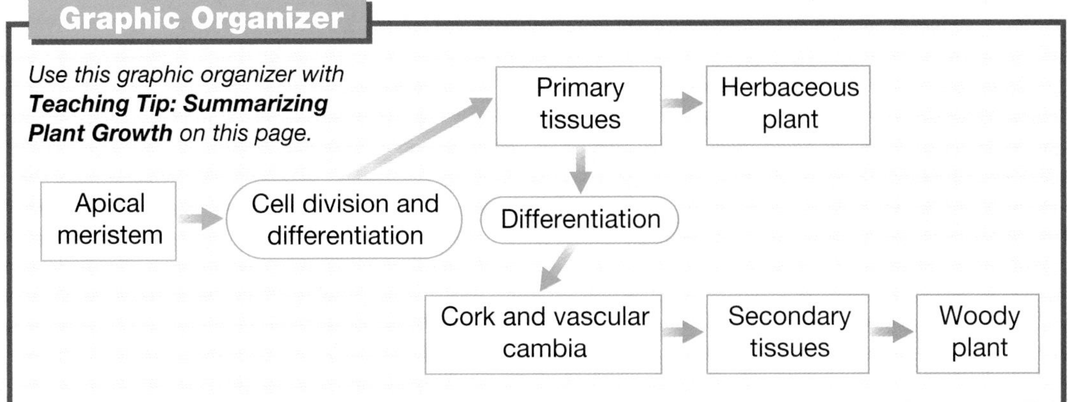

Teaching TIP — BASIC

Summarizing Plant Growth

Have students prepare a sequential diagram that summarizes how herbaceous plants grow by primary growth and how woody plants grow by both primary and secondary growth. An example of a completed diagram appears in the Graphic Organizer on this page. PORTFOLIO

Up Close

Bread Wheat

Teaching Strategies

- Show students specimens of living or preserved wheat plants.
- Let students dissect grains, or kernels, of wheat and locate the endosperm, bran, and germ of the kernels. Explain that the endosperm serves as a food supply for the embryo when it begins to grow. Remind students that in Chapter 24, they learned that white wheat flour contains only endosperm and that whole wheat flour contains both endosperm and bran.
- For interest, have students make bread so that they notice the elasticity of the dough, which is cause by gluten.

Making Biology Relevant

Health

In recent years, a great deal of research has been conducted to investigate the use of municipal sewage as a fertilizer. Such wastes must receive treatments that remove or kill organisms that might pose a health hazard to humans. One serious concern about these materials is the tendency they have to contain excesses of heavy metals such as lead and cadmium, especially if certain types of industries contribute to the waste. These metals can be harmful to humans even at low levels. Municipal wastes are not recommended for use in the fertilization of food crops for this reason. Ask students how they feel about the use of such fertilizers in their communities.

Up Close

Bread Wheat

- **Scientific name:** *Triticum aestivum*
- **Size:** 0.3 to 0.8 m (1 to 2.5 ft) tall
- **Range:** Agricultural regions worldwide
- **Habitat:** Cultivated fields in temperate and subtropical grasslands
- **Importance:** Wheat is the principal staple food in temperate regions of the world. The grains of *Triticum aestivum* are usually ground into flour that is used to make bread.

External Structures

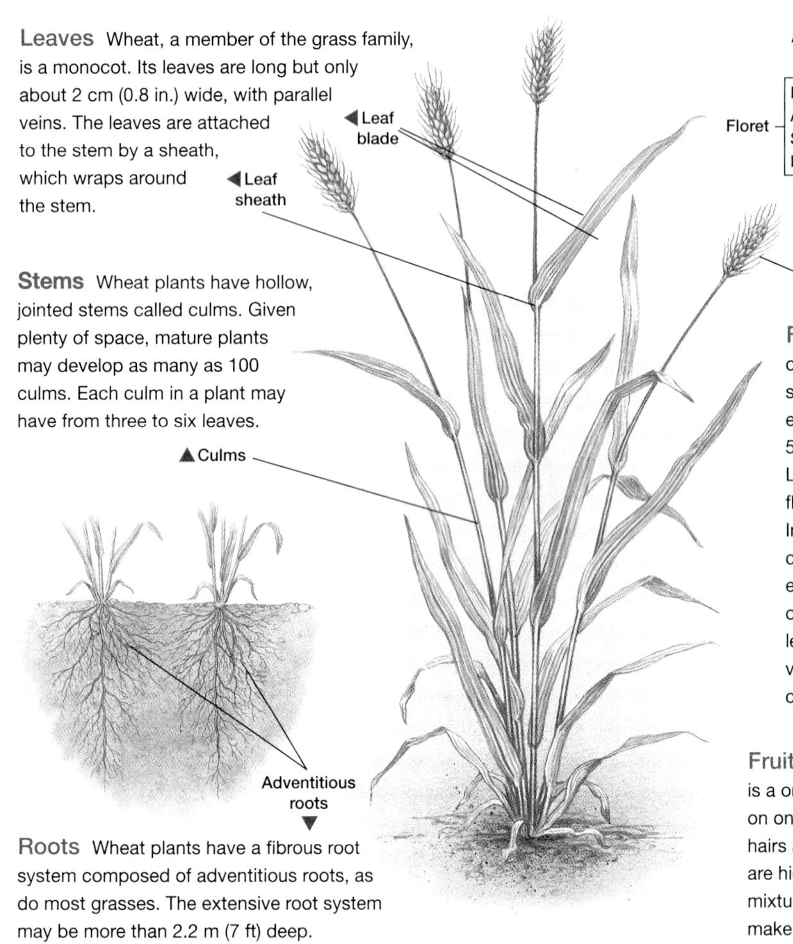

Leaves Wheat, a member of the grass family, is a monocot. Its leaves are long but only about 2 cm (0.8 in.) wide, with parallel veins. The leaves are attached to the stem by a sheath, which wraps around the stem.

Stems Wheat plants have hollow, jointed stems called culms. Given plenty of space, mature plants may develop as many as 100 culms. Each culm in a plant may have from three to six leaves.

Roots Wheat plants have a fibrous root system composed of adventitious roots, as do most grasses. The extensive root system may be more than 2.2 m (7 ft) deep.

Flowers The flowers, which occur in dense clusters called spikes, develop at the top of each culm. Spikes range from 5 to 13 cm (2 to 5 in.) in length. Like all grass flowers, wheat flowers lack petals and sepals. Instead, two modified leaves called the palea and the lemma enclose the stamens and pistil of each tiny flower, or floret. The lemmas of some bread-wheat varieties have a long bristle called an awn.

Fruit A kernel, or grain, of wheat is a one-seeded fruit with a crease on one side and a brush of tiny hairs at one end. The grains are high in gluten, a sticky mixture of proteins that make dough elastic.

did you know?

More than 200 varieties of wheat are grown in the United States.

Hard red wheats, the varieties grown for making bread flour, belong to the species *Triticum aestivum.* Durum wheats, or macaroni wheats, which are grown for making pasta because they contain less gluten, belong to the species *Triticum dicoccum.* Wheat may be planted in either the fall or the spring. The term *winter wheat* refers to a crop that is planted in the fall, overwinters as seedlings, and is harvested in late spring or summer. The term *spring wheat* refers to a crop that is planted in the spring for harvest in the fall.

Internal Structures

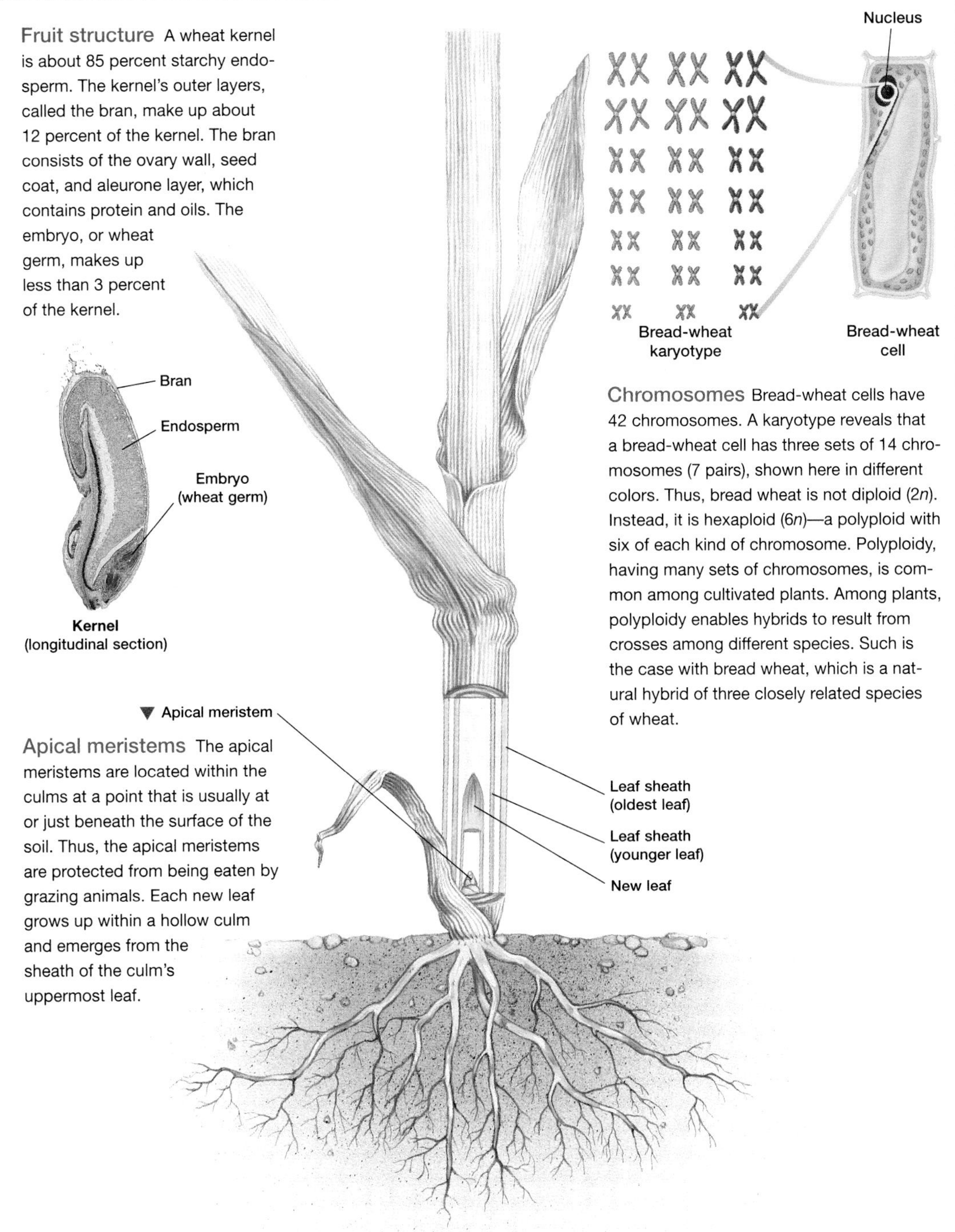

Fruit structure A wheat kernel is about 85 percent starchy endosperm. The kernel's outer layers, called the bran, make up about 12 percent of the kernel. The bran consists of the ovary wall, seed coat, and aleurone layer, which contains protein and oils. The embryo, or wheat germ, makes up less than 3 percent of the kernel.

Chromosomes Bread-wheat cells have 42 chromosomes. A karyotype reveals that a bread-wheat cell has three sets of 14 chromosomes (7 pairs), shown here in different colors. Thus, bread wheat is not diploid ($2n$). Instead, it is hexaploid ($6n$)—a polyploid with six of each kind of chromosome. Polyploidy, having many sets of chromosomes, is common among cultivated plants. Among plants, polyploidy enables hybrids to result from crosses among different species. Such is the case with bread wheat, which is a natural hybrid of three closely related species of wheat.

Apical meristems The apical meristems are located within the culms at a point that is usually at or just beneath the surface of the soil. Thus, the apical meristems are protected from being eaten by grazing animals. Each new leaf grows up within a hollow culm and emerges from the sheath of the culm's uppermost leaf.

Up Close

Bread Wheat

Discussion

- The chemical 2, 4-D is used to kill dicots. Will it kill wheat plants? Why or why not? *(No, wheat is a monocot, which can be determined by observing some typical characteristics of the plant such as parallel leaf veins, fibrous roots, and seeds with a single cotyledon.)*
- How did our ancestors help produce the bread wheat varieties that exist today? *(From wild hybrids, they selected plants with desirable characteristics such as higher yields, larger seeds, and more gluten.)*
- How would a human karyotype differ from the wheat karyotype? *(A human karyotype normally has 23 pairs of chromosomes, while a wheat karyotype has seven sets of six similar chromosomes, in which there are three homologous pairs. Also, wheat plants do not have sex chromosomes.)*
- Is polyploidy beneficial to all living organisms? Why or why not? *(No, polyploidy may be beneficial in plants because it causes plants to be larger, have showier flowers, and be more vigorous. However, it can also be detrimental in nature if the plant is sterile, which often occurs in triploids. In most animals, including humans, polyploidy is usually detrimental.)*

Chapter 27

TECH WATCH

Plant Tissue Culture

Teaching Strategies

Point out that tissue culture requires sterile conditions and that the new plants are grown in chambers in which the environmental conditions can be carefully controlled. Because of these requirements, tissue culture is an expensive process and is usually used only for high-dollar crops.

Discussion

What enables cells from part of a plant to grow into a whole plant? *(Cells of some parts of plants can form a mass of undifferentiated cells, and then these cells can differentiate into all the specialized types of cells found in a plant.)*

3 Assessment Options

Closure

Technique: K-W-L

Tell students to return to the lists of things they Want to know about plant growth and development that they created using the **K-W-L** technique on page 589. Have them place check marks next to the questions that they are now able to answer. Students should finish by making a list of what they have Learned. Conclude by asking students the following questions:

- Which questions are still unanswered?
- What new questions do you have?

You may wish to assign projects that require students to research their unanswered questions.

Review

Assign Review—Section 27-1.

Reteaching — BASIC

Have students work in cooperative groups to summarize the growth of a plant, either woody or nonwoody, from a seed to a mature adult.

Plant Development Is Continuous and Reversible

internetconnect

SCILINKS NSTA
TOPIC: Plant life spans
GO TO: www.scilinks.org
KEYWORD: HX591

Genes guide the development of both plants and animals, but their patterns of development are very different. As an animal develops, sets of genes that control development are inactivated and may not be used again. Most animals stop developing when they become adults. Plants, in contrast, continuously make new cells in meristems. These cells differentiate and replace or add to existing tissues. Thus, a plant continues to develop throughout its life.

Many cells in a mature plant can activate all of their genes. Such cells can divide and form masses of undifferentiated cells. In a sense, they can reverse their development. These cells can undergo differentiation and develop into a mature plant. A technique called tissue culture is used to grow new plants from tissue that can reverse its development. A tissue culture is prepared by placing tissue on a sterile nutrient medium. Masses of undifferentiated cells that form grow into plants that are genetically identical to the parent plant.

TECH WATCH: Plant Tissue Culture

Magnification: 810×

Tissue culture is used to propagate orchids, houseplants, and fruit plants. Thousands of cultures can be made from a single plant. Tissue culture can also be used to produce plants with new characteristics.

Protoplast fusion

Protoplast fusion has been used to produce hybrid petunias, potatoes, and carrots. A protoplast is a plant cell that has had its cell walls removed by enzymes. Certain chemicals or an electrical shock can cause two protoplasts to fuse, as the photo at right shows. If the protoplasts came from plants of different species, a hybrid cell results. The hybrid is then placed in a tissue culture and grown into an adult plant.

Genetic engineering

Tissue culture is also an essential part of producing genetically engineered plants. First, foreign genes are inserted into a plant's cells. The genetically altered cells are then grown into adult plants in tissue culture.

internetconnect

SCILINKS NSTA
TOPIC: Genetic engineering
GO TO: www.scilinks.org
KEYWORD: HX596

Review SECTION 27-1

1. **Compare** and **Contrast** the germination of bean seeds and corn seeds.
2. **Summarize** the basic differences between annuals, biennials, and perennials.
3. **Explain** how primary growth and then secondary growth produce a woody stem.
4. **SKILL Making Predictions** How might removing the bark from a tree affect the tree? Explain.
5. **Compare** the location of the apical meristem in bread-wheat stems with that of dicot stems.
6. **SKILL Analyzing Methods** How does plant development differ from animal development?

Review SECTION 27-1 ANSWERS

1. Corn seedlings grow straight up, and the cotyledon stays in the soil. Bean seedlings form a hook as they grow, and the two cotyledons emerge.
2. Annuals live for one growing season, reproduce once, and then die. Biennials live for two growing seasons, reproduce once during the second season, and then die. Perennials live for more than two years and may reproduce many times.
3. Primary growth lengthens the stem and forms new primary dermal, ground, and vascular tissues. Secondary growth increases the stem's diameter by producing layers of cork and secondary xylem and phloem.
4. Removing the bark from a tree might kill it because the phloem is part of the bark. If the phloem is damaged, sugars cannot get from the leaves to the roots, and the roots would die.
5. The apical meristem of a wheat plant is near the base of the plant. The apical meristem of a dicot stem is just behind the tip of the shoot.
6. Plant development is continuous and "reversible," while animal development is not.

Regulating Plant Growth and Development

Section 27-2

Plants Need Nutrients to Grow

Like all multicellular organisms, plants grow by adding new cells through cell division. Just like you, plants must have a steady supply of the raw materials they use to build new cells in order to grow. Plants need just two raw materials—carbon dioxide and water—to make all the carbohydrates in their bodies. As you learned in Chapter 5, these are the materials needed for photosynthesis.

Like animals, plants also need oxygen for cellular respiration. Although the green parts of a plant produce oxygen during photosynthesis, most of the oxygen used by leaves and stems comes from the air. Roots, which usually do not carry out photosynthesis, get oxygen from the air spaces between soil particles. If the soil around a plant's roots becomes compacted or saturated with water, it may not provide enough oxygen for the roots, and the plant could die.

However, carbon dioxide, water, and oxygen do not satisfy all of a plant's needs for raw materials. Plants also require small amounts of at least 13 **mineral nutrients,** which are elements absorbed mainly as inorganic ions. **Table 27-1** lists the six mineral nutrients needed in the greatest amounts for healthy plant growth and describes the importance of each nutrient. As **Figure 27-5** shows, commercial fertilizers may contain most of these mineral nutrients.

Objectives

- **Identify** the major nutrients plants need to grow. (p. 597)
- **Describe** how plant hormones control plant growth. (pp. 598–599)
- **Relate** environmental factors to plant growth. (pp. 600–602)

Key Terms

mineral nutrient
auxin
hormone
apical dominance
tropism
photoperiodism
dormancy

Table 27-1 Major Mineral Nutrients Required by Plants

Nutrient	Importance
Nitrogen	Part of proteins, nucleic acids, chlorophylls, ATP, and coenzymes; promotes growth of green parts
Phosphorus	Part of ATP, ADP, nucleic acids, phospholipids of cell membranes, and some coenzymes
Potassium	Needed for active transport, enzyme activation, osmotic balance, and stomatal opening
Calcium	Part of cell walls; needed for enzyme activity and membrane function
Magnesium	Part of chlorophyll; needed for photosynthesis and activation of enzymes
Sulfur	Part of some proteins and coenzyme A; needed for cellular respiration

Figure 27-5 Plant fertilizer. The three numbers on this bag indicate the amount of nitrogen, phosphorus, and potassium in the fertilizer. Most fertilizers also contain several other mineral nutrients.

1 Prepare

Scheduling

- **Block:** About 90 minutes will be needed to complete this section.
- **Traditional:** About two class periods will be needed to complete this section.
- **Planning:** See the Planning Guide on pp. 587A and 587B for lecture, classwork, and assignment options for Section 27-2. Lesson plans for Section 27-2 can be found in a lesson cycle format in ***Biology: Principles and Explorations*** **One-Stop Planner CD-ROM with Test Generator.**

Resource Link

Assign **Active Reading Worksheet Section 27-2** to help students comprehend the key concepts and visuals in this lesson.

2 Teach

Lesson Warm-up

Ask students to identify the substances plants need for survival and write these on the board. *(Initially, they will probably identify light, water, and carbon dioxide.)* Prompt them to identify other needs by asking why people fertilize their yards. Ask students how they think plants might regulate their growth and development without having a nervous system as animals do. *(They produce hormones.)*

Teaching TIP

Plant Growth Without Soil

Tell students that plants do not require soil in order to grow. Usually, soil provides minerals that plants need for growth. Using a practice known as hydroponics, plants are grown with their roots immersed in water through which oxygen is bubbled and to which the minerals needed for plant growth are added.

Integrating Different Learning Styles

Logical/Mathematical	**PE**	Quick Lab, p. 599; Math Lab, p. 602; Review items 2 and 4, p. 602; Performance Zone items 24, 25, and 26, p. 605
	ATE	Cross-Disciplinary Connection, p. 598
Visual/Spatial	**PE**	Data Lab, p. 601; Performance Zone item 11, p. 604; Lab Techniques, pp. 606–607
Interpersonal	**ATE**	Lesson Warm-up, p. 597; Reteaching, p. 602
Verbal/Linguistic	**PE**	Study Tip, p. 600; Review items 1 and 3, p. 602; Performance Zone items 2, 13, 14, 15, 16, 17, 21, 22, 27, 28, and 29, pp. 604–605
	ATE	Resource Link, p. 597; Teaching Tip, p. 597; Active Reading, p. 600

Chapter 27

REAL LIFE

Answer

The chemicals found in Agent Orange include 2,4-D and 2,4,5-T, which are synthetic auxins, and the dioxin 2,3,7,8-TCDD, which is a contaminant of 2,4,5-T. Dioxin has been shown to cause chloracne, which is a condition characterized by skin lesions on the head and upper body. Evidence suggests that there may be a link between exposure to dioxin and prostate cancer, multiple myeloma, and respiratory cancer in humans.

Visual Strategy

Figure 27-6

Walk students through this figure, which summarizes the steps of Fritz Went's classic experiment on the effects of auxin. Help students to see that when the agar is purple, it contains auxin. Ask students to state the significance of step 4. *(It is a control designed to show that agar does not cause the effect.)*

CROSS-DISCIPLINARY CONNECTION

◆ **Chemistry**

A fertilizer is considered "complete" if it contains nitrogen, phosphorus, and potassium. The "analysis" of a fertilizer is the percent of nitrogen in the form of elemental nitrogen, or N; the percent of phosphorus in the form of P_2O_5; and the percent of potassium in the form of K_2O, in that order. This is the traditional way of expressing the fertilizer's composition, even though most fertilizers do not actually contain P_2O_5 and K_2O. The amount of P_2O_5 can be converted to the amount of elemental phosphorus (P) by multiplying the percent by 0.437. The amount of K_2O can be converted to the amount of elemental potassium (K) by multiplying by 0.83.

REAL LIFE

The herbicide Agent Orange contains two synthetic auxins. This herbicide was used to defoliate plants in the jungles of Vietnam during the Vietnam War. One ingredient of Agent Orange is still commonly used as a herbicide.

Finding Information *Find out the names of the chemicals in Agent Orange, and identify the health problems it has been linked to.*

Hormones Control the Pattern of Plant Growth

For centuries, people have known that plants bend strongly toward a light source as their shoots elongate. In the 1920s, the Dutch biologist Frits Went showed that a chemical produced in the shoot tip causes this bending response. Went named the growth-promoting chemical that causes stems to bend **auxin** *(AWK sihn)*. The steps in Went's experiment are summarized in **Figure 27-6.**

Step 1 Went removed the tip of an oat shoot and placed the tip on an agar block. Auxin diffused from the tip into the block.

Step 2 Went then transferred the agar block to the cut end of a shoot, which caused the shoot to grow.

Step 3 When Went placed an agar block with auxin on either side of cut shoots, the shoots grew in the opposite direction.

Step 4 As a control, Went placed an agar block without auxin on the cut end of other shoots. These shoots did not grow.

How auxin works

Auxin is one of many plant hormones. The word *hormone* comes from the Greek word *horman,* meaning "to set in motion." A **hormone** is a chemical that is produced in one part of an organism and transported to another part, where it causes a response. Auxin

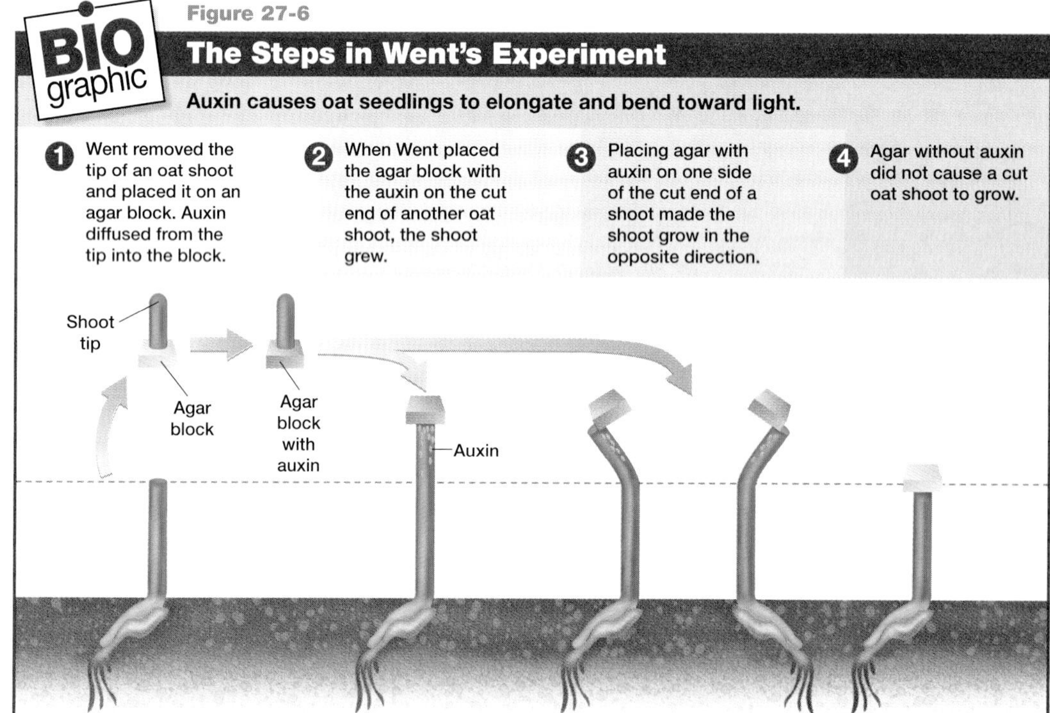

BIO graphic
1 Went removed the tip of an oat shoot and placed it on an agar block. Auxin diffused from the tip into the block.
2 When Went placed the agar block with the auxin on the cut end of another oat shoot, the shoot grew.
3 Placing agar with auxin on one side of the cut end of a shoot made the shoot grow in the opposite direction.
4 Agar without auxin did not cause a cut oat shoot to grow.
Shoot tip
Agar block
Agar block with auxin
Auxin

Figure 27-6

The Steps in Went's Experiment

Auxin causes oat seedlings to elongate and bend toward light.

did you know?

Charles Darwin did many experiments with plants.

Darwin and his son Francis Darwin did the earliest experiments on phototropism in the mid-1800s. They discovered that the tip of a grass seedling perceives light coming from one direction and that the growth response occurs lower, away from the tip, and causes the seedling to bend toward the light source.

causes plant cell walls to become more flexible, which allows the cells to elongate as they grow. Auxin accumulates on the dark side of a stem. As a result, the cells on the dark side of a stem elongate more than the cells on the light side. The difference in elongation causes the stem to grow toward the light. Auxin also inhibits the growth of the buds along a stem. This inhibition is called **apical dominance.** Cutting off the tip of a stem removes the source of auxin and enables the other buds to grow. That is why pruning the stems of a plant makes the plant become bushier.

Other plant hormones

More than a century ago, citrus farmers discovered that they could cause citrus fruits to ripen by storing them in a room heated by a kerosene stove. The ripening was caused by ethylene, which is a gaseous organic compound produced when kerosene is incompletely burned. Most plant tissues produce ethylene. Today, ethylene is used to promote the ripening of tomatoes, grapes, and other fruits that are harvested before they ripen. Ethylene also loosens the fruit of cherries, blackberries, and blueberries, making it easier to harvest these crops mechanically. You can read about more plant hormones in *Exploring Further: Plant Hormones and Agriculture* on this page.

Exploring Further

Plant Hormones and Agriculture

Many plant hormones are used in agriculture. Two examples are gibberellins and cytokinins.

Gibberellins

Gibberellins *(jihb uhr EHL ihnz)* are produced in developing shoots and seeds. They stimulate stem elongation, fruit development, and seed germination. Gibberellins are used to enlarge Thompson seedless grapes, shown in the photo at right. Other seedless fruits treated with gibberellins include apples, cucumbers, mandarin oranges, and peaches.

Cytokinins

Cytokinins *(seye toh KEYE nihnz)*, which are produced in root tips, stimulate cell division and slow the aging of some plant organs. Cytokinins are sprayed on cut flowers to keep them fresh and on fruits and vegetables to extend their shelf life. Cytokinins are added to tissue-culture media to cause undifferentiated cells to form shoots.

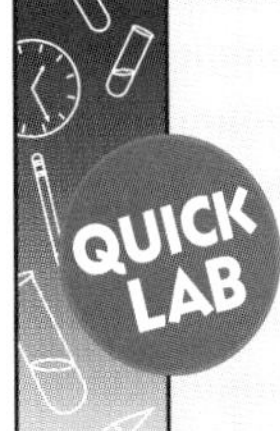

How does ethylene affect a plant?

You can use a ripe apple to see one of the effects of ethylene on plants.

Materials

4 L glass jars with lids (2), 2 plants in 5 cm pots, small ripe apple

Procedure

1. Place a plant inside one of the jars. Tightly secure the lid.
2. Place the other plant and the apple inside the other jar. Tightly secure the lid.
3. Observe both jars for several days. Record what you see.

Analysis

1. **Describe** any changes in the plant in each jar.

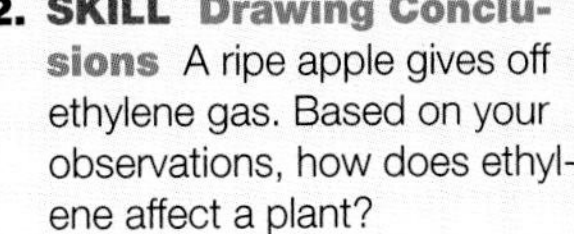

2. **SKILL Drawing Conclusions** A ripe apple gives off ethylene gas. Based on your observations, how does ethylene affect a plant?

did you know?

A rotten apple does spoil the bunch.

The expression "One rotten apple in the barrel spoils the whole bunch" is commonly used to refer to people who exhibit behaviors that are not socially acceptable. This expression comes from the observation by food growers and handlers that a rotten apple in a barrel of apples will cause the others to rot also. The explanation for this phenomenon is that ripe fruit gives off ethylene, which promotes the ripening of other nearby fruits.

Plant Hormones and Agriculture

Teaching Strategies

Tell students that seedless grapes, like many other plants that produce seedless fruits, are triploid (3*n*). Triploid plants are always sterile—they cannot produce seeds.

Discussion

- What natural source of gibberellin is being replaced by spraying Thompson seedless grapes with gibberellin sprays? *(seeds)*
- Why do the undifferentiated cells in tissue cultures need cytokinin to grow shoots? *(The masses of undifferentiated cells that develop in tissue cultures do not have roots, which are the natural source of cytokinins.)*

How does ethylene affect a plant?

Time 5–10 minutes to set up; 5 minutes per day over the next 2 weeks to collect data

Process Skills Observing, making comparisons, making conclusions

Teaching Strategies

Tell students that ethylene affects plants in many ways. For example, when some plants are exposed to ethylene gas, they do not elongate normally. Ethylene also promotes the production of male flowers in plants such as squash and cucumbers. Ethylene is used by citrus-fruit growers to ripen fruits that are picked while they are still green.

Analysis Answers

1. Answers will vary. Plants in jars containing an apple should begin to lose their leaves after a few days.
2. Answers will vary. From the results of this experiment, students should conclude that ethylene promotes plant aging and leaf drop.

Active Reading

Technique: Reading Organizer

Before students read this section, have them make a list of the following terms: auxin, tropism, phototropism, gravitropism, and thigmotropism. As they learn about the terms, have them write a brief definition of each term. Have them save their list of terms and definitions for use later in the **Reteaching** exercise on page 602.

Teaching TIP — ADVANCED

Tissue Culture

Tell students that the growth of whole plants from pieces of plants in tissue culture requires the addition of plant hormones in order for them to differentiate properly. And the relative amounts of the hormones affect the process of differentiation. Typically, auxins and cytokinins are first added to the culture medium to promote the formation of a mass of undifferentiated cells. Next a high level of auxins and a low level of cytokinins are added to the culture medium. This promotes root formation. Then a low level of auxins and a high level of cytokinins are added to the culture medium. This promotes shoot formation.

CAREER

Horticulturist

Horticulturists study fruits, vegetables, and ornamental plants. Horticulturists are interested in producing better fruit, vegetable, and ornamental crops with less pesticides, fertilizers, and water. Have students work in teams to use library references or on-line databases to find information about the role of horticulturists in food production. Have students investigate current research projects in horticulture, and have them report their findings to their classmates.

PORTFOLIO

Study TIP

Reading Effectively

The word *thigmotropism* comes from the Greek words *thigma,* meaning "touch," and *tropos,* meaning "turn."

Plant Growth Adjusts to Environmental Influences

Because most plants are anchored in one spot, they cannot move from an unfavorable environment to a more favorable one as animals do. Instead, plants respond to their environment by adjusting the rate and pattern of their growth. For example, a plant that receives plenty of water and mineral nutrients may grow much faster and larger than it would if it received very little water and mineral nutrients. Also, a plant grown in full sun may grow much faster and larger than it would if it were grown in the shade or indoors. So the availability of light and nutrients affect the rate of plant growth. Many of a plant's responses to environmental stimuli, however, are triggered by the hormones that regulate plant growth.

Tropisms

A **tropism** *(TROH piz uhm)* is a response in which a plant grows either toward or away from a stimulus. Auxin is responsible for producing tropisms. **Figure 27-7** shows examples of three common types of tropisms. Phototropisms are responses to light. Responses to gravity are called gravitropisms. A thigmotropism is a response to touch. If a plant grows toward a stimulus, the response is called a positive tropism. If a plant grows away from the direction of the stimulus, the response is called a negative tropism. Thus, a shoot that grows up out of the ground shows both positive phototropism (growing toward the light) and negative gravitropism (growing away from the pull of gravity).

Figure 27-7 Tropisms

Tropisms are growth responses that occur either toward or away from a stimulus.

The bending of an amaryllis toward the light is a positive phototropism.

The upward growth of shoots is a negative gravitropism; the downward growth of roots is a positive gravitropism.

The coiling of grapevine tendrils around a wire is a thigmotropism.

did you know?

Many woody plants exhibit bud dormancy.

Dormant flower buds must be exposed to a certain number of days of chilling temperatures (usually less than 5°C) in order for bud dormancy to be broken. In the Northern Hemisphere, some southern regions do not consistently receive chilling winter temperatures that meet the dormancy requirements of many plants. This is the reason for the southern limit of some fruit crops such as apples and peaches. The minimum number of hours of chilling temperatures varies not only among species, but also among varieties within a species. In fact, apples and peaches are bred so that the number of hours of cold that are required to break bud dormancy are reduced in order to extend these crops into warm locations.

Interpreting annual rings

Background

The annual rings of a woody stem provide important clues to annual variations in the rainfall an area receives over time. Thick rings form in years with heavy rainfall. Relatively thin rings form in dry years. Use the photo at right to answer the following questions.

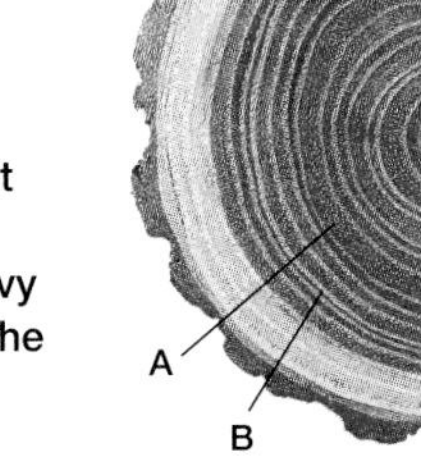

Analysis

1. **SKILL Interpreting Data** What do the annual rings indicate about the climate where this plant grew?
2. **SKILL Drawing Conclusions** Which ring, A or B, indicates a year when this plant received more rainfall?
3. **SKILL Making Predictions** How will the annual rings of a nearby tree of the same age and species compare with those of this tree?

Photoperiodism

Certain plants bloom in the spring and others bloom in the summer or fall. Some plants bloom as soon as they reach a mature size. In many plants, seasonal patterns of flowering and other aspects of growth and development are caused by changes in the length of days and nights. The response of a plant to the length of days and nights is called **photoperiodism.**

Most plants can be categorized as one of three types in reference to photoperiodism. A plant that responds when days become shorter than a certain number of hours is said to be a short-day plant. A plant that responds when days become longer than a certain number of hours is called a long-day plant. Plants whose growth and development are not affected by day length are known as day-neutral plants. However, it is really the length of the nights rather than the length of the days that controls photoperiodism, as Figure 27-8 shows. Knowledge of photoperiodism is very important to the nursery and floral industries. The length of days and nights is controlled artificially in greenhouses where plants such as poinsettias and chrysanthemums are grown. Thus, commercial growers force the plants to produce flowers at times of the year when they ordinarily would not. This makes it possible for poinsettias to be available for Christmas and chrysanthemums to be available year round.

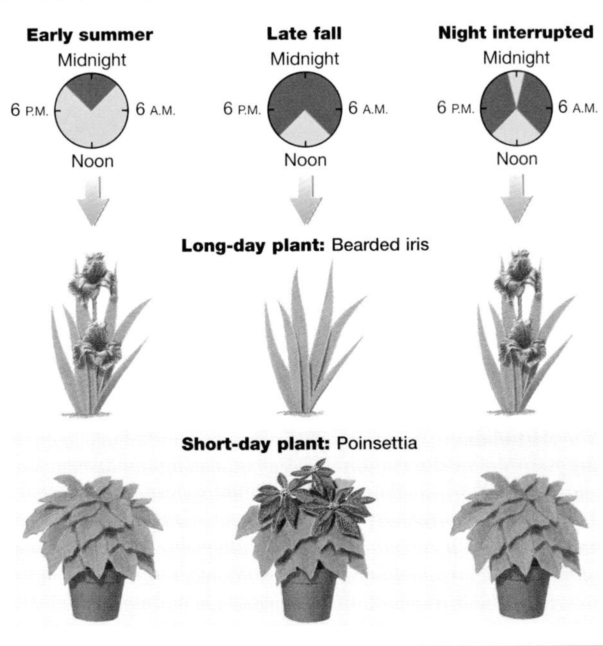

Figure 27-8 Flowering and photoperiodism. Long-day plants flower when nights are short. Short-day plants flower when nights are long. If a flash of light interrupts a long night, long-day plants flower and short-day plants do not.

Overcoming Misconceptions

Plant "Food"

Fertilizers are often referred to as plant "food." However, the term *food* is usually reserved for organic substances that provide a source of energy and organic molecules for living things when they are broken down. Because minerals do not contain usable energy and are not broken down in plants, they should not be referred to as food. Plants do not eat food because they produce their own organic molecules in photosynthesis.

Interpreting annual rings

Time 15–20 minutes

Process Skills Making observations, analyzing data, making conclusions

Teaching Strategies
Point out to students that it is the number of rings in a tree trunk that indicates its age, not the diameter of the trunk. A number of environmental factors can influence the width of an annual ring.

Analysis Answers

1. Answers will vary. The rings indicate that the tree grew in a temperate climate and that the amount of rainfall varied from year to year.
2. A
3. They should be very similar.

Visual Strategy

Figure 27-8

Walk students through this diagram by first following what happens to long-day (short-night) and short-day (long-night) plants when the days are long and the nights are short. Then follow what happens to long-day (short-night) and short-day (long-night) plants when the days are short and the nights are long. Finally, follow what happens to long-day (short-night) and short-day (long-night) plants when they are grown under long nights that are interrupted by a flash of light in the middle of the night. Tell students that poinsettias are short-day (long-night) plants that are desirable ornamentals for the Christmas holidays. Ask students if these plants would normally produce flowers in late December. *(yes)* Ask students what would happen if lights were turned on during the night in a greenhouse where poinsettias are grown? *(The plants would not bloom.)*

Analyzing the effect of cold on seed germination

MATH LAB

Time 20 minutes

Process Skills Interpreting graphs, analyzing data, making conclusions

Teaching Strategies

Tell students that another plant hormone, abscisic acid, promotes seed dormancy. Seeds with true dormancy cannot germinate until the abscisic acid is either washed away or broken down.

Analysis Answers

1. Cold temperatures promote the germination of apple seeds.
2. 45 days ÷ 7 days/week = 6.43 weeks
3. 30 percent
4. 100 percent

3 Assessment Options

Closure

Summarizing

Summarize the roles of the plant hormones discussed in this section. *(Auxins cause cell elongation, inhibit the growth of lateral buds, and produce tropisms. Gibberellins cause stem elongation, fruit development, and seed germination. Cytokinins stimulate cell division and slow aging. Ethylene causes fruit ripening and fruit and leaf drop.)* These hormones interact within plant tissues to regulate plant growth and development.

Review

Assign Review—Section 27-2.

Reteaching — BASIC

Have students return to the lists of terms and definitions they made for the **Active Reading** on page 600. Have them work in cooperative groups of three or four to create a crossword puzzle using the definitions as clues and the terms as words in the puzzle. PORTFOLIO

Figure 27-9 Bud dormancy. Thick scales cover the dormant buds on this twig from an apple tree.

Responses to temperature

Temperature affects growth and development in many plants. For example, most tomato plants will not produce fruit if the nighttime temperatures are too high. Many plants that flower in early spring will not produce flowers until they have been exposed to cold temperatures for a certain number of hours. Cold temperatures also trigger the temporary periods of inactivity that many perennial plants enter in the fall. These plants remain inactive through the winter, even during warm periods that often occur during winter. Most deciduous woody plants, for example, drop their leaves in the fall. Thick, protective scales develop around their buds, as **Figure 27-9** shows. After a period of low temperatures, the buds begin growing into new leaves or sections of woody stem.

Dormancy is the condition in which a plant or a seed remains inactive, even when conditions are suitable for growth. Chemicals that cause dormancy break down in response to cold temperatures. Thus, many plants and seeds remain dormant until they have been exposed to low temperatures for several weeks. Periods of dormancy are observed in many plants that live where winters are cold. Dormancy helps plants to survive by keeping buds from growing and seeds from germinating during warm spells before winter has ended.

MATH LAB: Analyzing the effect of cold on seed germination

Background

In some plants, a period of low temperatures is needed to break seed dormancy. The graph at right shows how being stored at a low temperature (4°C) affected the ability of apple seeds to germinate. Use the graph to answer the following questions.

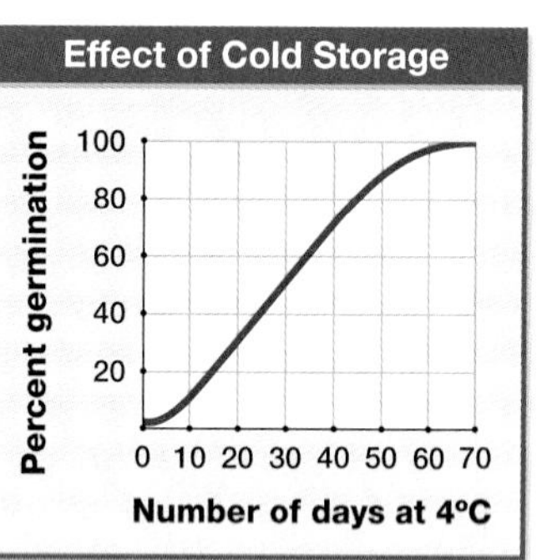

Analysis

1. **Summarize** the overall effect of cold temperatures on the germination of apple seeds.
2. **Calculate** the number of weeks that apple seeds must be stored at 4°C for at least 80 percent of the seeds to germinate.
3. **SKILL Interpreting Graphs** What percentage of apple seeds germinate after storage at 4°C for 20 days?
4. **SKILL Predicting Patterns** What percentage of apple seeds will germinate after being stored at 4°C for 80 days?

Review SECTION 27-2

1. **List** the six mineral nutrients that plants require in the greatest amounts.
2. **Explain** how auxin causes a stem to grow toward a light source.
3. **Describe** an example of a negative gravitropism.

Critical Thinking

4. **SKILL Predicting Outcomes** Why is it an advantage for plant growth and development to be regulated by environmental stimuli?

Review SECTION 27-2 ANSWERS

1. nitrogen, phosphorus, potassium, calcium, magnesium, sulfur
2. Auxin diffuses to the side of a stem away from a light source. There, it stimulates cells to grow longer than they do on the side of the stem facing the light source. This causes the stem to bend toward the light source.
3. A shoot growing upward from the soil after a seed germinates is an example of a negative gravitropism.
4. It is an advantage for plant growth to be regulated by environmental stimuli because plants are unable to move to a new environment when conditions become unfavorable.

Study ZONE CHAPTER 27

Use the Key Concepts and Key Terms listed below to review the main ideas in this chapter. If you need to review the meaning of a term, turn to the page indicated.

Key Concepts

27-1 How Plants Grow and Develop

- Germination is the resumption of growth by the embryo in a seed. Water and oxygen must penetrate the seed coat before germination can occur.
- Annuals complete their life cycle in one growing season. Biennials complete their life cycle in two growing seasons. Perennials live several years and may reproduce many times.
- Apical meristems located at the tips of shoots and roots produce primary growth. The tissues that result from primary growth are known as primary tissues.
- Secondary growth increases a plant's width. In woody stems, secondary growth is produced by the cork cambium and vascular cambium, two meristems near the outside of the stem.
- Bread wheat is a cereal grass with long leaves, hollow stems, and clusters of tiny flowers. The fruits, or grains, are usually ground into flour that is used to make bread.
- Plants develop throughout their lives. Plant development is reversible. Many mature plant cells can divide to form masses of undifferentiated cells, which can develop into a new plant.

27-2 Regulating Plant Growth and Development

- Plants need 13 mineral nutrients for growth. They also need carbon dioxide and water for photosynthesis and oxygen for cellular respiration.
- Hormones regulate plant growth and development. Auxin is a hormone that causes shoots to elongate and inhibits the growth of lateral buds.
- Plants modify their growth in response to the direction of light, gravity, and touch. Such growth responses are called tropisms.
- Seeds and many mature plants survive periods of unfavorable environmental conditions by becoming dormant.
- In many plants, seasonal patterns of flowering and other aspects of growth and development are caused by changes in the length of days and nights.

Key Terms

27-1

germination (590)
perennial (591)
annual (591)
biennial (591)
primary growth (592)
secondary growth (592)
apical meristem (592)
cork cambium (593)
vascular cambium (593)
annual ring (593)

27-2

mineral nutrient (597)
auxin (598)
hormone (598)
apical dominance (599)
tropism (600)
photoperiodism (601)
dormancy (602)

Study TIP Ready?

If you think you understand these Key Concepts and Key Terms, then you're ready for the Performance Zone. 

Study ZONE CHAPTER 27

Review and Practice

Assign the following worksheets for Chapter 27:

- **Student Review Guide**
- **Concept Mapping Worksheet**
- **Test Preparation Pretest**
- **Vocabulary Worksheet**
- **Critical Thinking Review Worksheet**

Assessment Options

The following assessment options are available for this chapter:

Resource Link

Assign **Chapter Test 27** to assess students' understanding of the chapter.

Alternative Assessment

Have students use the information in Chapter 27 to prepare a multimedia presentation on plant growth and development that they could give to a middle-school science class or to the general public. PORTFOLIO

Portfolio Assessment

- *Portfolio Assessment* at the front of this book provides options for building and evaluating student portfolios. Students and teachers can select items from the areas listed for inclusion in a portfolio.
- See items labeled PORTFOLIO on pp. 591, 592, 593, 600, and 602.

Answer to Concept Map

The following is one possible answer to Performance Zone item 1 on page 604.

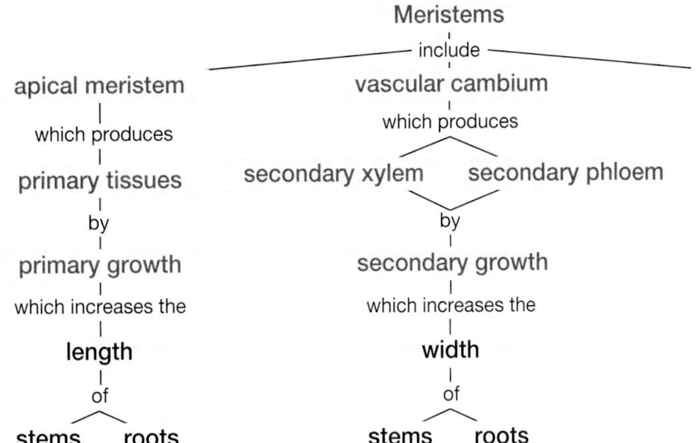

Performance ZONE
CHAPTER 27
REVIEW ANSWERS

Assignment Guide

SECTION	REVIEW
27-1	1, 2, 3, 4, 5, 6, 7, 13, 14, 15, 18, 19, 20, 23, 28, 29
27-2	2, 8, 9, 10, 11, 12, 16, 17, 18, 19, 20, 21, 22, 24, 26, 27, 29

Understanding and Applying Concepts

1. Answer to the concept map is found on page 603.
2. **a.** Germination is the term for the resumption of growth by the embryo in a seed.
   b. A perennial is a plant that lives for more than two years and may reproduce many times.
   c. Apical dominance is the inhibition of the growth of side buds that is caused by auxin produced in the tip of a shoot.
   d. Dormancy is the condition in which seeds or buds do not grow even when environmental conditions are favorable for growth to occur.
3. a
4. c
5. c
6. d
7. a
8. d
9. b
10. b
11. d
12. b
13. The energy for a carrot plant's second year of growth comes from compounds that are stored in the root.
14. Wheat cells have 3 distinct sets of 14 chromosomes that occur in 7 pairs. It actually has six of each kind of chromosome.
15. Hybrid cells, made from the fusion of two cells from two different types of plants, are grown into new, hybrid plants in tissue culture.
16. Cytokinins stimulate cell division and slow aging. They are used in agriculture to prolong freshness of cut flowers, fruits, and vegetables and to stimulate shoot growth in tissue cultures. Gibberellins stimulate stem elongation, fruit development, and seed germination. They are used in agriculture to induce fruit formation without seeds and to enlarge seedless fruits.
17. Phosphorus is a component of ATP, ADP, phospholipids, DNA, RNA, and some coenzymes.

Performance ZONE

CHAPTER 27 **Review**

Understanding and Applying Concepts

1. **Concept Mapping** Construct a concept map that describes growth in vascular plants. Include the following terms in your map: meristems, apical meristem, primary growth, primary tissues, cork cambium, secondary growth, cork, secondary phloem, secondary xylem, vascular cambium. Include additional terms as necessary.
2. **Understanding Vocabulary** Write a sentence that shows your understanding of each of the following terms.
   a. germination
   b. perennial
   c. apical dominance
   d. dormancy
3. The first sign of germination of a bean seed is the emergence of the embryo's
   a. root. c. sheathed shoot.
   b. hooked shoot. d. cotyledons.
4. Plants that live several years or more are known as
   a. annuals. c. perennials.
   b. biennials. d. terminals.
5. Providing cells for growth at the tips of a plant is the main function of the
   a. cork cambium. c. apical meristems.
   b. root cap. d. lateral meristems.
6. Cell division that increases the diameter of a woody stem occurs in the
   a. pith. c. apical meristems.
   b. primary tissues. d. vascular cambium.
7. Plant and animal development differ in that plant development
   a. is continuous and reversible.
   b. stops after a plant reaches maturity.
   c. is not affected by environment.
   d. is controlled by genes that cannot be reactivated.
8. Which of the following is *not* a raw material needed for plant growth?
   a. carbon dioxide c. oxygen
   b. water d. vitamins
9. Major mineral nutrients required by plants include all of the following except
   a. nitrogen. c. phosphorus.
   b. lead. d. potassium.
10. Auxin causes cells to
    a. have less flexible cell walls.
    b. elongate more as they grow.
    c. bend toward light.
    d. develop lateral buds.
11. Which of the following caused the growth patterns of the mums shown below?
    a. apical meristem c. apical dominance
    b. auxin d. all of the above

12. The response of a plant to the length of days and nights is
    a. gravitropism. c. phototropism.
    b. photoperiodism. d. thigmotropism.
13. **Theme Metabolism** Carrots are biennials. What role does the root of a carrot plant play in the plant's second year of growth?
14. **Up Close** How does the chromosome number of bread wheat differ from that of diploid organisms?
15. **Tech Watch** How is tissue culture used to produce hybrid varieties of plants?
16. **Exploring Further** How do cytokinins and gibberellins affect plant growth, and how are these hormones used in agriculture?
17. **Chapter Links** Why is phosphorus a major mineral nutrient required by plants? (**Hint:** See Chapter 2, Section 2-3.)

Skills Assessment

Part 1: Interpreting Graphics

Study the diagram of a cross section of a tree trunk shown below. Then answer the following questions:

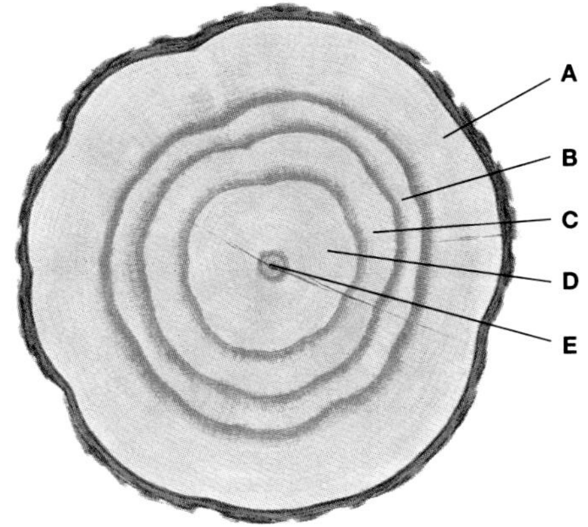

18. What was the approximate age of this tree when it was cut down?
19. At which labeleled point(s) would you find tissue(s) formed by primary growth?
20. Write the series of three letters that represents a period of time in which the climate became gradually drier.

Part 2: Life/Work Skills

Use the media center or Internet resources to learn how plants such as poinsettias, chrysanthemums, and Easter lilies are raised at commercial nurseries.

21. **Finding Information** Learn how commercial growers produce large numbers of plants that flower uniformly and at times of the year when they would not ordinarily flower in nature.
22. **Communicating** Summarize your findings in a written report that can be shared with your class.
23. **Selecting Technology** Suppose that you have discovered a chrysanthemum plant with flowers of a color you have never seen before. What method would you choose to produce thousands of plants just like this plant for sale to a florist?

Critical Thinking

24. **Analyzing Methods** Some seeds are soaked in acid before they are packaged and sold to farmers and home gardeners. What is the purpose of this treatment?
25. **Forming Reasoned Opinions** Why is it possible for people to grow new plants from pieces of leaf, stem, or root tissue in which the cells have already undergone differentiation?
26. **Evaluating Results** A student placed a green banana in each of 10 plastic bags. The student also placed a ripe pear in five of the bags and then sealed all of the bags. The bananas in the bags without pears took longer to ripen than the bananas in the bags with pears. Evaluate these experimental results.

Portfolio Projects

27. **Interpreting Information** Read "Blossoming talent," by Stephen Day (*New Scientist,* July 12, 1997, pp. 38–40). What are some advantages of being able to make plants flower "on cue?" What does the *LEAFY* gene do? How can "genetic flower arranging" be used to change a plant's shape?
28. **Organizing Information** Research the practice and philosophical basis of bonsai, the Asian art of growing miniature plants. Find out when and where bonsai originated and how bonsai plants are kept small. Relate your findings in a report that also explains how an understanding of plant growth and development is important to success in bonsai.
29. **Career Focus Agronomist** Agronomists study soil management and other aspects of crop production. Research an agronomist's responsibilities, and write a report on your findings. Your report should include a job description, training required, kinds of employers, growth prospects, and starting salary.

Skills Assessment

18. The tree was probably 5 years old.
19. E
20. D, C, B
21. Answers will vary. Growers use hormone treatments and pruning techniques to control the plants' environment to produce plants that flower uniformly when the grower wants them to.
22. Answers will vary.
23. tissue culture

Critical Thinking

24. Answers will vary. Students should realize that acid will probably damage the seed coat. Water can then penetrate the seed, and the seed will sprout. Seeds treated with acid germinate more quickly and more uniformly than do untreated seeds.
25. The cells in these plant parts are able to reverse their development. These cells can produce cells that are undifferentiated, which can then undergo differentiation to form all of the tissues necessary to grow an entire new plant.
26. Answers will vary. Students should suggest that the bananas in the bags with ripe pears ripened faster than those in bags without pears; ripe pears give off ethylene, which promotes fruit ripening.

Portfolio Projects

27. There is commercial value to being able to make plants flower at times of the year when they ordinarily would not flower in nature. The *LEAFY* gene is involved in the transformation of a plant from its vegetative phase to its reproductive phase. With genetic engineering, plants can be made to flower when they are a certain size, which changes their shape.
28. In the practice of bonsai, plants are shaped by selective pruning and training and are kept small by pruning their roots and shoots, potting them in small containers, and watering and fertilizing them sparingly. The bonsai gardener must have knowledge of the growth habits and nutritional needs of a plant to grow healthy miniature plants successfully.
29. Agronomists are plant scientists who traditionally study cereal crops, such as corn and wheat, fiber crops, such as cotton, and soil science. Many agronomists work in the field, in government and university research laboratories, and in college classrooms. Others work for private industry, state agricultural extension services, or the federal government. Positions are available for individuals with bachelor's degrees, master's degrees, and Ph.D's in agronomy or soil science. Starting salary will vary by region.

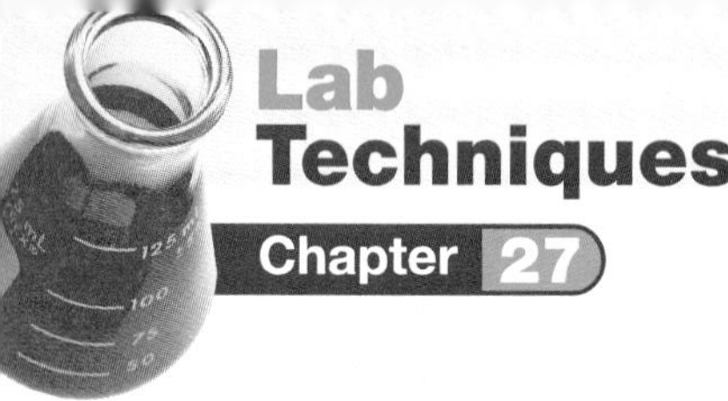

Lab Techniques

Chapter 27

Comparing Bean and Corn Seedlings

Purpose

Students will germinate beans and corn kernels and then compare the structures of bean seedlings and corn seedlings.

Preparation Notes

Time Required: Three 30-minute lab periods

Materials

Materials for this lab can be ordered from WARD'S. See *Master Materials List* at the front of this book for catalog numbers.

Preparation Tips

- Bean seeds and corn kernels may be purchased at a nursery, hardware store, or feed store, or they may be ordered from a biological supply company.
- Soak bean seeds and corn kernels prior to this investigation.
- Set up labeled containers for the disposal of solutions, broken glass, and seedlings.

Safety Precautions

- Be sure that students have read and understand all of the safety rules for working in the lab.
- Caution students to use scalpels carefully to avoid injuring themselves and others.
- Make sure students wash their hands before leaving the lab.

Procedural Tips

- This investigation must be done in three 30-minute periods over 5 days.
- The seed coat of the bean should be easy to remove if the seeds have soaked long enough.

Lab Techniques

Chapter 27

Comparing Bean and Corn Seedlings

SKILLS

- Comparing
- Drawing
- Relating

OBJECTIVES

- **Observe** the structures of bean seeds and corn kernels.
- **Compare** and **Contrast** the development of bean embryos and corn embryos as they grow into seedlings.

MATERIALS

- 6 bean seeds soaked overnight
- stereomicroscope
- 6 corn kernels soaked overnight
- scalpel
- paper towels
- 2 rubber bands
- 150 mL beakers (2)
- glass-marking pen
- metric ruler

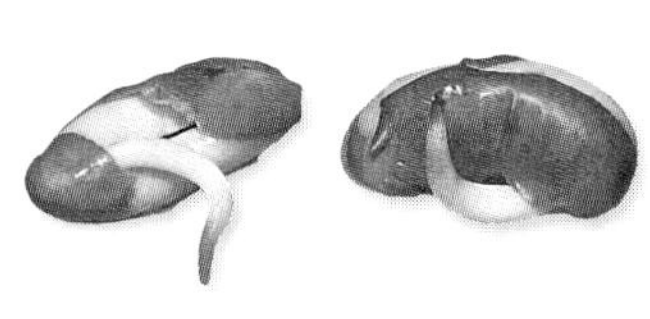

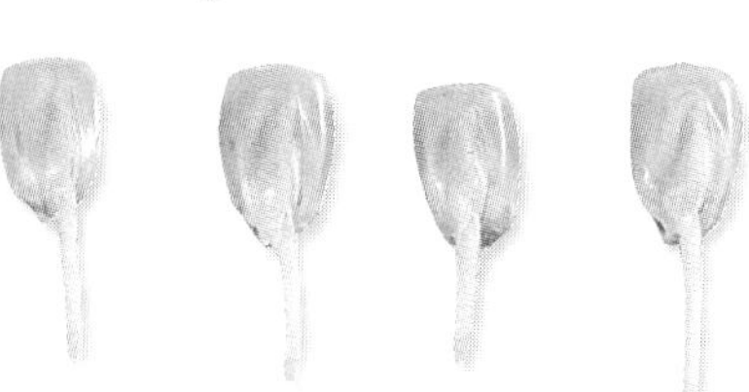

Before You Begin

A **seed** contains a dormant plant **embryo.** A plant embryo consists of one or more **cotyledons,** an embryonic **shoot,** and an embryonic **root.** Seeds also contain a supply of nutrients. In **monocots,** the nutrients are contained in the **endosperm.** In **dicots,** the nutrients are transferred to the cotyledons as seeds mature. A seed **germinates** when the embryo begins to grow and breaks through the protective **seed coat.** The embryo then develops into a young plant, or **seedling.** In this lab, you will examine bean seeds and corn kernels and then germinate them to observe the development of their seedlings.

1. Write a definition for each boldface term in the paragraph above.
2. Based on the objectives for this lab, write a question you would like to explore about seedling development.

Procedure

PART A: Observing Seed Structure

1. Remove the seed coat of a bean seed, and separate the two fleshy halves of the seed.
2. Locate the embryo on one of the halves of the seed. Examine the bean embryo with a stereomicroscope. Draw the embryo, and label the parts you can identify.
3. Examine a corn kernel, and find a small light-colored oval area. **CAUTION: Sharp or pointed objects may cause injury. Handle scalpels carefully.** Use a scalpel to cut the kernel in half along the length of this area.
4. Locate the corn embryo, and examine it with a stereomicroscope. Draw the embryo, and label the parts you can identify.

PART B: Observing Seedling Development

5. Fold a paper towel in half as shown in the photo at the top of the next page. Set five corn kernels on the paper towel. Roll up the paper towel, and put a rubber band around the roll. Stand the roll in a beaker with 1 cm of water in the bottom. Add water to the beaker as needed to keep the paper towels wet, but do not allow the corn kernels to be covered by water.
6. Repeat step 5 with five bean seeds.

Before You Begin

Answers

1. *seed*—reproductive structure of a plant that consists of an embryo, a stored food supply, and a seed coat

   embryo—an early stage in the development of an organism

   cotyledons—structures in a seed that are part of the embryo and function in its nourishment

   shoot—portion of a plant that consists of stems and leaves and that usually grows aboveground

   root—portion of a plant that anchors the plant, absorbs water and mineral nutrients, and usually grows belowground

   monocot—plant characterized by seeds with one cotyledon

   endosperm—triploid ($3n$) nutritive tissue that develops from the union of a sperm with two nuclei in an ovule

   dicot—plant characterized by seeds with two cotyledons

   germinate—to begin growth, as in a seed

   seed coat—protective outer covering of a seed

   seedling—a young plant that has developed from a germinating seed
2. Answers will vary. For example: What part of a seedling emerges first?

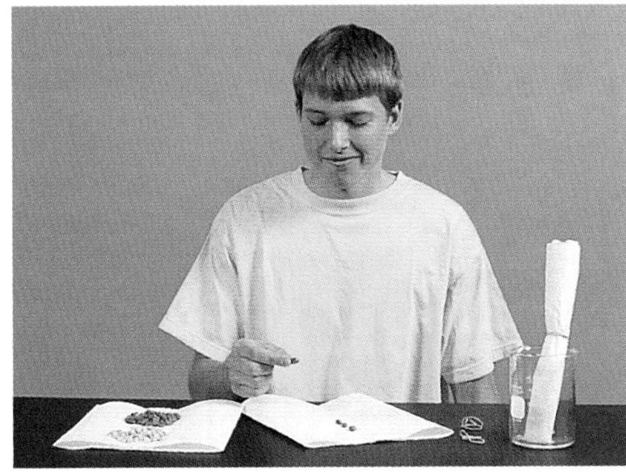

7. After three days, unroll the paper towels and examine the corn and bean seedlings. Use a glass-marking pen to mark the roots and shoots of the developing seedlings. Starting at the seed, make a mark every 0.5 cm along the root of each seedling. And again starting at the seed, make a mark every 0.5 cm along the stem of each seedling.
8. Draw a corn seedling and a bean seedling in your lab report. Label the parts of each seedling. Also show the marks you made on each seedling, and indicate the distance between the marks.
9. Using a fresh paper towel, roll up the seeds, place the rolls in the beakers, and add fresh water to the beakers.
10. After two more days reexamine the seedlings. Measure the distance between the marks. Repeat step 8.

PART C: Cleanup and Disposal

11. Dispose of seeds, broken glass, and paper towels in the designated waste containers. Do not put lab materials in the trash unless your teacher tells you to do so.
12. Clean up your work area and all lab equipment. Return lab equipment to its proper place. Wash your hands thoroughly before you leave the lab and after you finish all work.

Analyze and Conclude

1. **Relating Concepts** Corn and beans are often cited as representative examples of monocots and dicots, respectively. Relate the seed structure of each to the terms *monocotyledon* and *dicotyledon.*
2. **Summarizing Results** What parts of a plant embryo were observed in all seedlings on the third day?
3. **Drawing Conclusions** In which part or parts of bean seedlings and corn seedlings do the seedlings grow in length? Explain.
4. **Forming Hypotheses** How are the tender young shoots of bean seedlings and corn seedlings protected as the seedlings grow through the soil?
5. **Evaluating Viewpoints** Defend the following statement: There are both similarities and differences in seed structure and seedling development in beans and corn.
6. **Further Inquiry** Write a new question about seedling development that could be explored with another investigation.

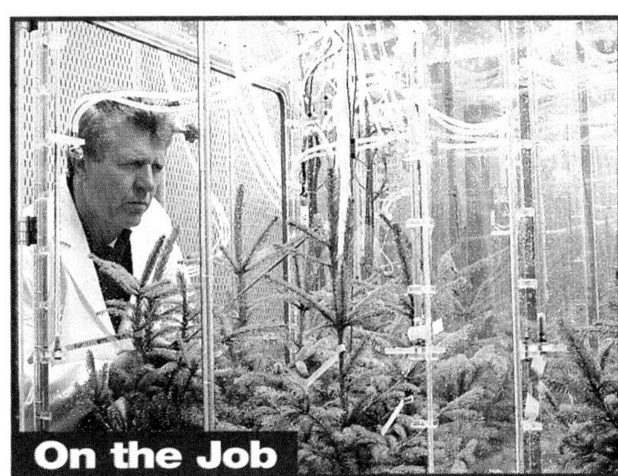

On the Job

Plant physiology is the study of the processes that occur in plants. Do research in the library or media center or on the Internet to discover where plant physiologists work and what types of research are currently being conducted in the field of plant physiology. Summarize your findings in a written or oral report.

Analyze and Conclude

Answers

1. Corn seeds have only one cotyledon and thus are monocotyledons; bean seeds have two cotyledons and thus are dicotyledons.
2. Embryonic leaves and roots are observed in all seedlings on the third day.
3. Bean seedlings grow at the tips of their roots and stems. The distance between the marks has changed at the tips of the stems and the tips of the roots. Corn seedlings grow in length at the tips of their roots. They do not have a visible stem. The distance between the marks changed only on the roots.
4. Bean shoots have a hook in their embryonic stems. The hook pushes through the soil before the cotyledons, which enclose the embryonic shoot. Corn shoots are surrounded and protected by a sheath as they push through the soil.
5. In both beans and corn, the embryo consists of an embryonic shoot, an embryonic root, and cotyledons. As bean embryos and corn embryos develop, their shoots grow up, become green, and form leaves, while their roots grow down and do not become green. Bean seeds have two cotyledons; corn kernels have only one. Corn seeds have endosperm; mature bean seeds do not. The shoots of beans hook as they germinate; a corn shoot grows straight up.
6. Answers will vary.

Procedure

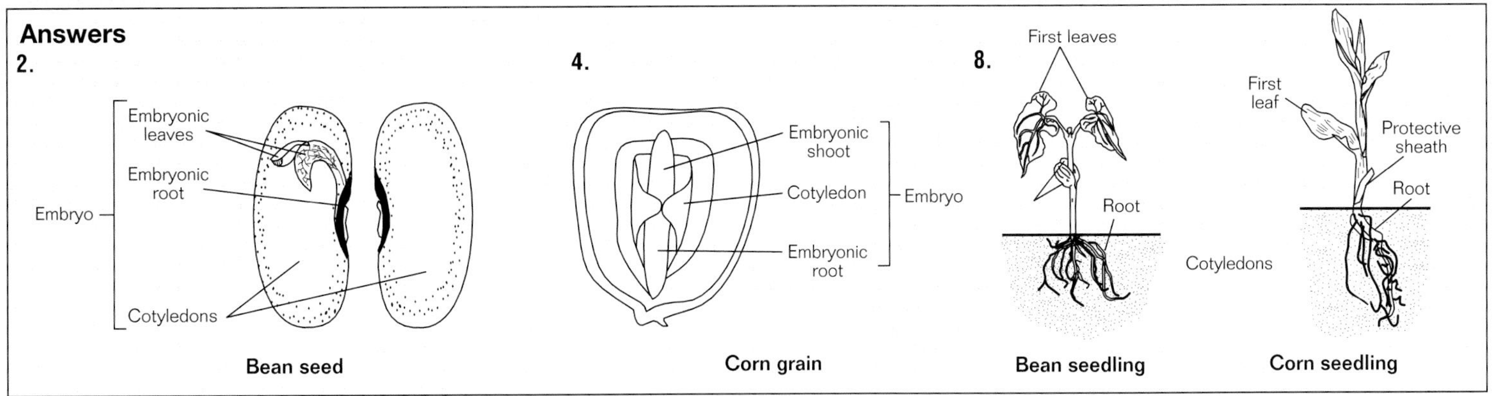

UNIT 7 Exploring Invertebrates

Chapters

Beekeeper examining a honeycomb frame full of honey

in perspective

Honeybees

Yesterday...

Humans have raised bees for their honey and their beeswax for thousands of years. American colonists brought bees from England and early pioneers took them along on the journey west. **What are the characteristic features of insects? See page 692.**

French beekeepers in the early 1700s

Today...

Honeybees are valued today not only as a source of honey but also for their role as pollinators of crops such as almonds, cherries, apples, and pears. Migratory beekeepers carry bees from state to state, renting their bees to farmers in time to ensure a successful harvest. **Discover on page 697 how a beehive is a highly organized social system.**

Honeybee

Tomorrow...

In the United States, bee-dependent crops are worth $10 billion a year. Scientists continue to study different varieties of bees in an effort to find those that are the most efficient pollinators.

Honeybee research

internet**connect**

TOPIC: Honeybees
GO TO: www.scilinks.org
KEYWORD: HX609

Chapter 28

Introduction to Animals

Planning Guide

OBJECTIVES	MEETING INDIVIDUAL NEEDS	READING SKILLS AND STRATEGIES
28-1 Animals—Features and Body Plans (BLOCKS 1 & 2 90 min) • **Identify** the features that animals have in common. • **Distinguish** radial symmetry from bilateral symmetry. • **Summarize** the importance of a body cavity. • **Identify** how scientists determine evolutionary relationships among animals.	Learning Styles, p. 612 BASIC ADVANCED • pp. 613, 619, 620, 621	**ATE** Active Reading, Technique: Reading Organizer, p. 611 **ATE** Active Reading, Technique: Reading Organizer, p. 614 ■ Active Reading Guide, worksheet (28-1) ■ Directed Reading, worksheet (28-1)
28-2 Animal Body Systems (BLOCKS 3 & 4 90 min) • **Summarize** the functions of the digestive, respiratory, circulatory, nervous, skeletal, and excretory systems. • **Compare** a gastrovascular cavity with a one-way digestive system. • **Differentiate** open from closed circulatory systems. • **Distinguish** asexual from sexual reproduction.	Learning Styles, p. 622 BASIC ADVANCED • pp. 623, 624, 625, 628	■ Active Reading Guide, worksheet (28-2) ■ Directed Reading, worksheet (28-2) ■ Supplemental Reading Guide, *Journey to the Ants: A Story of Scientific Exploration*

Review and Assessment Resources

BLOCKS 5 & 6 90 min

TEST PREP

PE Study Zone, p. 629

PE Performance Zone, pp. 630–631

■ Chapter 28 worksheets:
- Vocabulary
- Concept Mapping
- Science Skills
- Critical Thinking

 Audio CD Program

■ Test Prep Pretest

CHAPTER TESTING

■ Chapter Tests (blackline master)

▲ Test Generator

■ Assessment Item Listing

See the correlations table at the front of this book for a list of the ***National Science Education Standards*** addressed in this chapter.

Smithsonian Institution® Visit **www.si.edu/hrw** for additional on-line resources.

KEY

Biology Principles and Explorations

PE Pupil's Edition

ATE Teacher's Edition

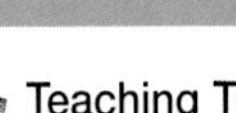

 Teaching Transparencies

Laboratory Program

One-Stop Planner CD-ROM
Shows these resources within a customizable daily lesson plan:

■ Teaching Resources

▲ Test Generator

VISUAL STRATEGIES	CLASSWORK AND HOMEWORK	TECHNOLOGY AND INTERNET RESOURCES
ATE **Fig. 28-1,** p. 614 **ATE** **Fig. 28-10,** p. 618 **ATE** **Fig. 28-11,** p. 619 134 Possible Evolutionary Relationships of Major Animal Groups 135 The Animal Body: An Evolutionary Journey 137A Evolutionary Tree of Vertebrates	**PE** **Quick Lab** How is symmetry recognized?, p. 617 **PE** **Data Lab** Exploring the animal kingdom, p. 621 **Lab A21** Recognizing Patterns of Symmetry **PE** **Review,** p. 621	CNN Video Segment 23 Mexico Sea Turtle Holt Biology Videodiscs Section 28-1 Audio CD Program Section 28-1 internet**connect** SCILINKS NSTA **TOPICS:** • Origin of tissues • Body symmetry • Body cavity • Animal groups
ATE **Fig. 28-16,** p. 625 **ATE** **Fig. 28-18,** p. 627	**PE** **Math Lab** Calculating filtration rate in the human kidney, p. 626 **PE** **Lab Techniques** Surveying Invertebrate Diversity, pp. 632–633 **Lab B20** Life in a Pine Cone **PE** **Review,** p. 628 **PE** **Study Zone,** p. 629	Holt Biology Videodiscs Section 28-2 Audio CD Program Section 28-2 internet**connect** SCILINKS NSTA **TOPICS:** • Body systems • Animal reproduction

ALTERNATIVE ASSESSMENT

PE **Portfolio Projects 26–29,** p. 631

ATE **Alternative Assessment,** p. 629

ATE **Portfolio Assessment,** p. 629

Lab Activity Materials

Quick Lab
p. 617 envelope containing the letters of the alphabet

Data Lab
p. 621 pencil, paper

Math Lab
p. 626 pencil, paper

Lab Techniques
p. 632 safety goggles; lab apron; preserved or living specimens of invertebrates; prepared slides of sponges, hydras, planarians, and nematodes; compound microscopes; hand lenses or stereomicroscopes; probes

Lesson Warm-up

ATE photo of *Daphnia* and a large, familiar mammal, p. 612

Demonstrations

ATE potted plant, natural sponge, p. 610

ATE common classroom objects, p. 616

ATE cup, tubing, water, p. 623

BIOSOURCES LAB PROGRAM
- Quick Labs: Lab A21, p. 41
- Inquiry Skills Development: Lab B20, p. 93

Chapter 28

Chapter Themes

Evolution

The diversity of animals today is the result of the evolution of animals in response to many different environments over millions of years.

Homeostasis

The balance between water and salts in animal bodies depends on a complex control system. The excretory system, which is introduced in this chapter, plays a vital part in regulating this balance.

Establishing Prior Knowledge

Before beginning this chapter, make certain students can answer the questions in **Study TIP Ready?** on page 611. If they cannot, encourage them to reread the sections indicated.

Opening Demonstration

Place a potted plant and a natural sponge on your desk. (If you have an aquarium with some freshwater sponges, use them instead.) Ask students what the plant and the sponge have in common. *(Both organisms are stationary; the plant is alive and the sponge was once alive.)* Ask students how the two are different. *(The plant can make its own food. Some students will say that the sponge is an animal.)* Tell students that the sponge is one animal that they will study.

CHAPTER 28 Introduction to Animals

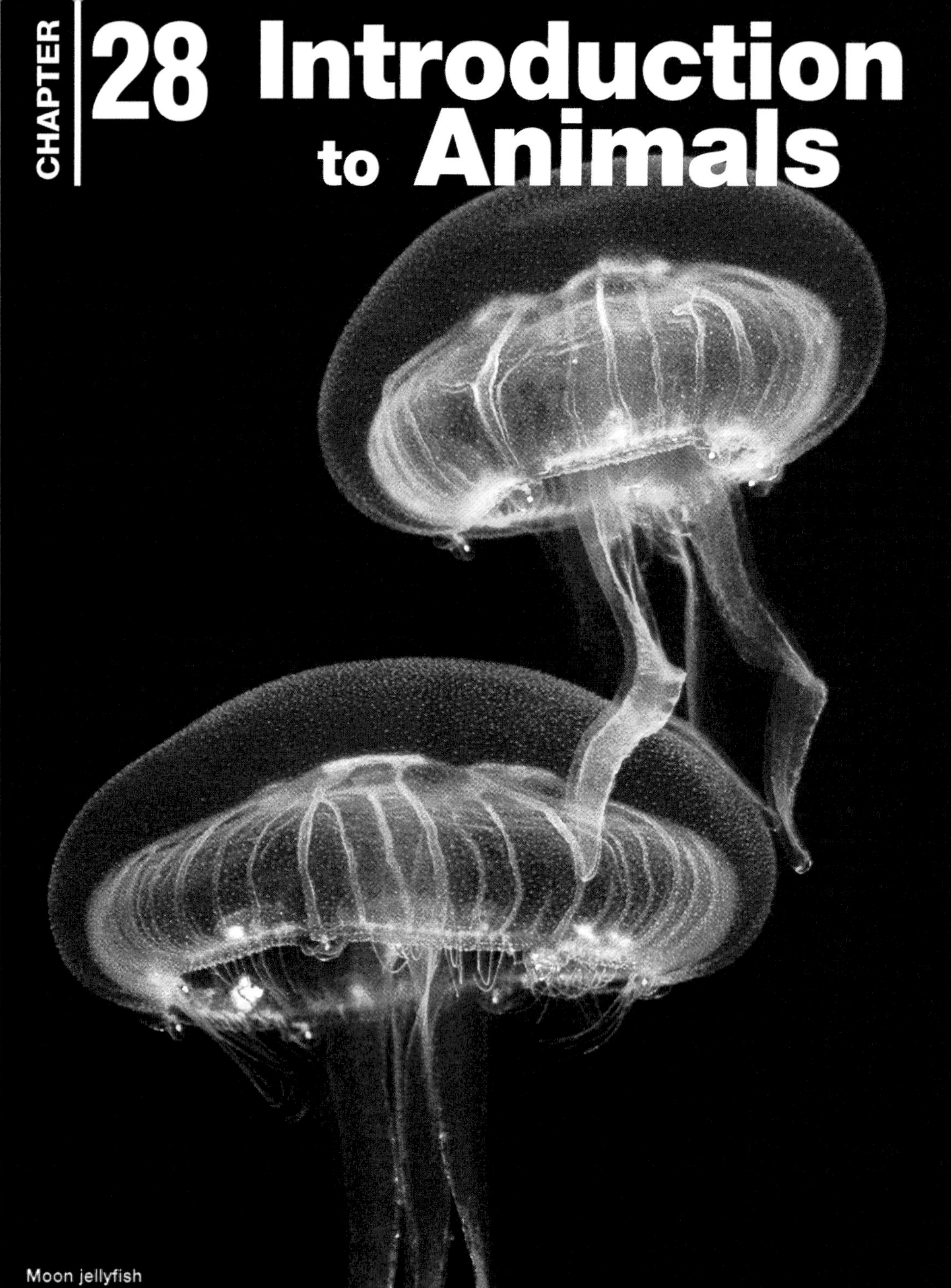

Moon jellyfish

What's Ahead — and Why
CHAPTER OUTLINE

28-1 Animals—Features and Body Plans
- Animals Share Several General Features
- Most Animals Show Body Symmetry
- Many Animals Have an Internal Body Cavity
- "Advanced" Animals Show Body Segmentation
- There Are Many Kinds of Animals

28-2 Animal Body Systems
- Tissues and Organs Perform Specific Tasks
- Animals Have Different Reproductive Strategies

Students will learn
- the features shared by all animals and how the basic animal body plan can vary.
- how animal bodies carry out different functions.

Beached moon jellyfish

Watch your step!

While often found traveling in large swarms in the open seas, the moon jellyfish is most commonly seen washed up on beaches. But be careful. Although moon jellyfish deliver a sting far less toxic than that of some of their relatives, even a dead one can cause an itchy rash that lasts for hours.

Study TIP

Ready?

Check your understanding of these topics to help you prepare for what's ahead.

Can you . . .

- **define** the term *metabolism*? (Section 1-2)
- **describe** the process of diffusion? (Section 4-1)
- **summarize** how gametes are formed? (Section 6-1)
- **state** the relationship of a phylum to a kingdom? (Sections 15-1, 15-2)
- **define** the terms *tissue* and *organ*? (Section 20-3)

If not, *review the sections indicated.*

Looking Ahead

internetconnect

National Science Teachers Association *sci*LINKS Internet resources are located throughout this chapter.

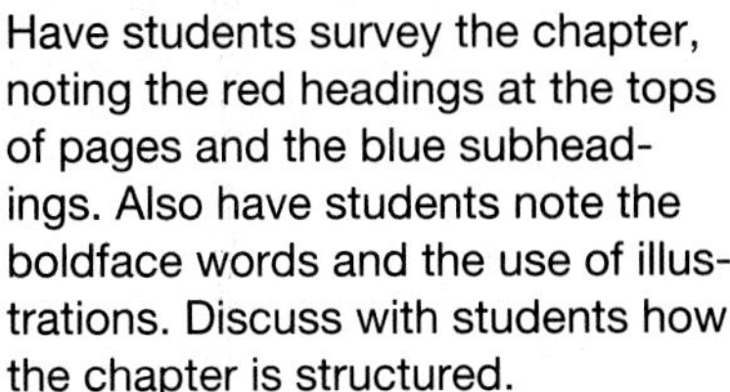

Reading Strategies

Active Reading

Technique: Reading Organizer

Have students survey the chapter, noting the red headings at the tops of pages and the blue subheadings. Also have students note the boldface words and the use of illustrations. Discuss with students how the chapter is structured.

Directed Reading

Assign **Directed Reading Worksheet Chapter 28** to help students interact with the text as they read the chapter.

Interactive Reading

Assign ***Biology: Principles and Explorations*** **Audio CD Program Chapter 28** to help reluctant readers achieve greater success in reading the chapter.

Reading Skills

Assign Reading Organizers and Reading Strategies Worksheets from the **One-Stop Planner CD-ROM with Test Generator** to help students learn and practice effective reading skills.

internetconnect

Holt *Biology: Principles and Explorations*

On-line Resources: go.hrw.com

Visit the HRW Web site for a variety of resources related to this chapter. Just type in the keyword HX0 Chapter 28.

Holt *Biology: Principles and Explorations*

On-line Resources: www.scilinks.org

The following *sci*LINKS Internet resources can be found in the student text for this chapter:

Page 615
TOPIC: Origin of tissues
KEYWORD: HX615

Page 616
TOPIC: Body symmetry
KEYWORD: HX616

Page 618
TOPIC: Body cavity
KEYWORD: HX618

Page 620
TOPIC: Animal groups
KEYWORD: HX620

Page 622
TOPIC: Body systems
KEYWORD: HX622

Page 627
TOPIC: Animal reproduction
KEYWORD: HX627

Chapter 28

1 Prepare

Scheduling

- **Block:** About 90 minutes will be needed to complete this section.
- **Traditional:** About two class periods will be needed to complete this section.
- **Planning:** See the Planning Guide on pp. 609A and 609B for lecture, classwork, and assignment options for Section 28-1. Lesson plans for Section 28-1 can be found in a lesson cycle format in ***Biology: Principles and Explorations*** **One-Stop Planner CD-ROM with Test Generator.**

Resource Link

Assign **Active Reading Worksheet Section 28-1** to help students comprehend the key concepts and visuals in this lesson.

2 Teach

Lesson Warm-up

Have students compare a projected image of a *Daphnia* with that of any large, familiar mammal. (If possible, use a flex camera to project an image of a live *Daphnia* on a projection microscope.) Have students list the similarities and differences between the two animals. *(Similarities are that both move and neither can make its own food. Differences include size, shape, body covering, complexity, and methods of reproduction.)*

Section 28-1 Animals—Features and Body Plans

Objectives

- **Identify** the features that animals have in common. (pp. 612–615)
- **Distinguish** radial symmetry from bilateral symmetry. (pp. 616–617)
- **Summarize** the importance of a body cavity. (p. 618)
- **Identify** how scientists determine evolutionary relationships among animals. (pp. 620–621)

Key Terms

blastula
ectoderm
endoderm
mesoderm
body plan
asymmetrical
radial symmetry
bilateral symmetry
cephalization
coelom
acoelomate
pseudocoelomate
coelomate
phylogenetic tree

Animals Share Several General Features

Humans have long marveled at and depended on animals. When Linnaeus first classified animals in the mid-1700s, he counted 4,236 kinds. Since then, many new animal species have been identified, and the count is now over a million. We share the planet with a fantastic array of animal forms, and our lives intersect in many ways. For example, if it were not for insects, such as the honeybee, many of our crops would not be pollinated. The prices of insect-pollinated fruits and vegetables would increase dramatically. And items such as apples and cucumbers might not be available at your local supermarket. Learning something about animal life will enable you to make informed decisions about issues that affect humans and other animals.

Think about a snail, a lizard, a hawk, an elephant—all are quite different. But the features these four animals share are just as important as their differences in size, shape, and behavior. Like all animals, they exhibit the following features—heterotrophy, mobility, multicellularity, diploidy, and sexual reproduction. They also exhibit the absence of a cell wall, blastula formation, and cells organized into tissues. These features are a legacy inherited from their common ancestor (most likely a unicellular protist with a flagellum).

Heterotrophy

Animals are heterotrophs—that is, they cannot make their own food. Some animals that live in the ocean remain in one place and consume tiny particles of food that they filter from passing sea water. But most animals, such as those shown in **Figure 28-1,** move from place to place searching for food. Once food is located, it is eaten and then digested in a cavity inside the animal's body.

Figure 28-1 Heterotrophy. Unlike plants, animals must seek out food sources in their environment.

Integrating Different Learning Styles

Logical/Mathematical	**ATE**	Cross-Disciplinary Connection, p. 615
Visual/Spatial	**PE**	Performance Zone items 1, 17, 18, 21, 23, and 27, pp. 630–631; Quick Lab, p. 617; Real Life, p. 617; Lab Techniques, pp. 632–633
	ATE	Lesson Warm-up, p. 612; Visual Strategies, pp. 614, 618, 619; Demonstration, p. 616; Teaching Tips, pp. 616, 620
Interpersonal	**ATE**	Teaching Tip, p. 613; Closure, p. 621; Reteaching, p. 621
Intrapersonal	**ATE**	Biology in Your Community, p. 618
Verbal/Linguistic	**PE**	Study Tips, pp. 614, 618; Data Lab, p. 621; Performance Zone items 22, 26, and 29, p. 631
	ATE	Teaching Tips, pp. 613, 620; Active Reading, p. 614; Career, p. 618

Mobility

Animals are unique among living things in being able to perform rapid, complex movements. They move by means of muscle cells, specialized cells that are able to contract with considerable force. Animals can swim, crawl, walk, run, and even fly. In fact, flight has evolved four times among the animals, in insects, pterosaurs (extinct reptiles from the time of the dinosaurs), birds, and bats. The tiger shown in **Figure 28-2** moves through its environment searching for food. The size of the tiger's home range depends on the availability of prey species.

biobit

What is the largest animal?

The blue whale is probably the largest animal ever to live on Earth. Blue whales reach lengths of up to 30 m (100 ft). Amazingly, they attain that size by eating only microscopic plankton.

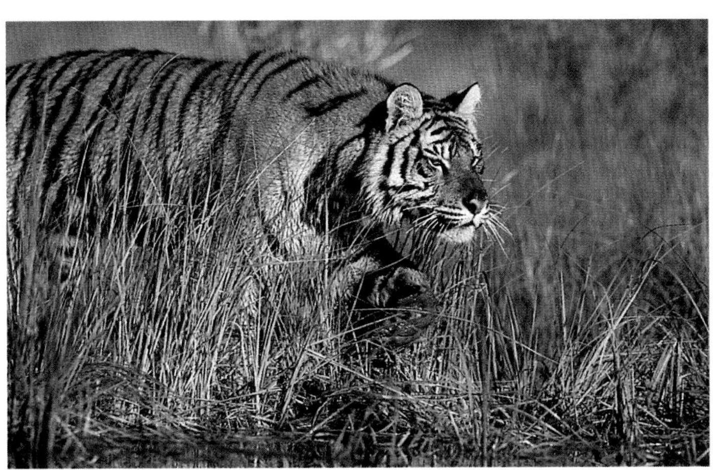

Figure 28-2 Mobility. A tiger finds food by roaming over its territory, which usually ranges from 10 to 30 mi^2.

Multicellularity

All animals are multicellular. Some are too small to be seen clearly with the naked eye, like *Daphnia,* shown in **Figure 28-3.** Others, such as some enormous whales, are larger than a city bus. In spite of their differences in body size, there is little difference in the size of most of the cells that make up these animals. The cells on the skin of your hand are roughly the same size as the cells in the heart of a whale or in the wing muscle of a hummingbird.

Figure 28-3 Life in a drop of water. *Daphnia,* commonly called water fleas, belong to a group of animals known as crustaceans.

Diploidy

With few exceptions, animals are diploid *(DIP loyd),* meaning adults have two copies of each chromosome, one inherited from their father and one from their mother. Only their gametes (egg and sperm) are haploid. In contrast, many plants have four or more copies of each chromosome, while fungi often have only one. A great advantage of diploidy is that it permits an animal to exchange genes between the two copies of a set of chromosomes, creating new combinations of genes.

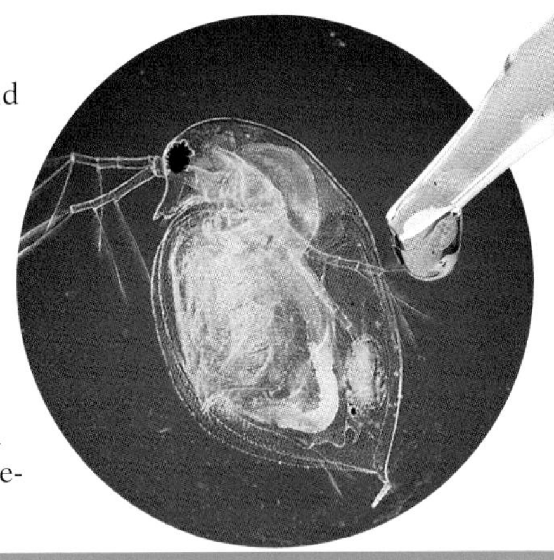

did you know?

Some fish are triploid.

Through gene technology, an egg can be made triploid. The eggs of some commercially raised salmon are exposed to a stimulus, such as a sudden temperature change, that interferes with the eggs' chromosomes as they separate during cell division. The result is triploid salmon that do not mature sexually. Since these salmon do not invest any energy in reproduction, they feed longer and grow larger, which increases the yield when the salmon are harvested.

Teaching TIP

Connecting the Abstract to the Concrete

Terms and phrases such as *heterotrophy, multicellularity, diploidy, absence of a cell wall,* and *reproduce sexually* are used to describe features of animals on pp. 612–614. Help students connect these terms and phrases to their own concrete experiences by asking them to choose any animal they are familiar with. Have them write the name of the animal and then present evidence that their animal is heterotrophic, multicellular, lacks cell walls, and reproduces sexually. Have a few students explain to the class how they know that their animals have the features common to all animals. 

Teaching TIP ADVANCED

Mobility

Have students work in groups of two or three to gather information on the speed that different animals can travel. Have them graph, illustrate, and display their data. Tell students to be certain to include both aquatic and terrestrial animals in their research. PORTFOLIO

Chapter 28

Visual Strategy

Table 28-1

Have students study Table 28-1. Then name a body structure and ask students which embryonic tissue layer gave rise to the structure. *(Structures of ectodermal origin include skin, hair, fingernails, brain, spinal cord, and nerves. Structures of mesodermal origin include the heart, arteries, veins, muscles, kidneys, bladder, uterus, testes, and thyroid gland. Structures of endodermal origin include the lungs, stomach, small intestine, gallbladder, pancreas, esophagus, and colon.)*

Active Reading

Technique: Reading Organizer

Have students start a list of words that can be explained by their etymologies. Ask them to split each word into its parts and to define each part. Tell them that knowing the meanings of word roots can help them understand other words. Have students read the **Study Tip** on page 614. As students read the chapter, they should add word roots to their list. PORTFOLIO

Magnification: 620×

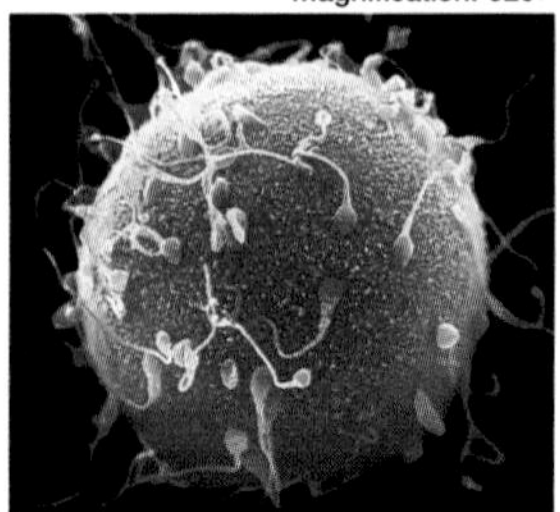

Figure 28-4 Egg and sperm. This human egg is surrounded by sperm, only one of which will fertilize the egg.

Sexual reproduction

Almost all animals reproduce sexually by producing gametes, as do many plants, fungi, and protists. As shown in **Figure 28-4,** the females' egg cells are much larger than the males' sperm cells. Unlike the egg cells, the sperm cells of animals have a flagella and are highly mobile.

Absence of a cell wall

Among the cells of multicellular organisms, only animal cells lack rigid cell walls. The absence of a cell wall has allowed animals mobility that other multicellular organisms do not have. You may not realize this, but there are cells moving about in your body all the time. Cells called macrophages, for example, act as mobile garbage collectors, crawling over tissues and removing debris. A cell with a rigid cell wall could never perform such a task.

Blastula formation

In all animals except sponges, the zygote (fertilized egg cell) undergoes cell divisions that form a hollow ball of cells called a **blastula** *(BLAS tyoo luh),* shown in **Figure 28-5.** Cells within the blastula eventually develop into three distinct layers of cells—**ectoderm, endoderm,** and **mesoderm.** These layers are called the primary tissue layers because they give rise to all of the tissues and organs of the adult body. (A few simple invertebrates, such as *Hydra* and their kin, develop only two tissue layers, endoderm and ectoderm.) **Table 28-1** lists the three primary tissue layers and summarizes the body tissues and organs to which they give rise. Note that the table includes some organs, such as the urinary bladder, found only in vertebrates. The organs of vertebrates are complex structures containing cells that arise from more than one primary tissue layer. For example, the digestive system is formed primarily from endoderm and mesoderm.

Figure 28-5 Blastula. It takes a scanning electron microscope (SEM) to see the individual cells of a blastula.

Magnification: 30×

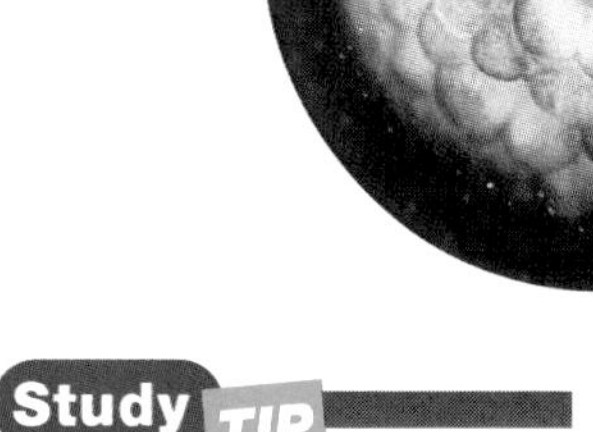

Study TIP

Word Origins

The root word *-derm* means "skin." Placing prefixes in front of *-derm* creates words that name the three primary tissue layers and indicate their origin. For example, *ecto-* means "outside," *endo-* means "within," and *meso-* means "middle."

Table 28-1 Origin of Animal Tissues and Organs

Primary tissue layer	Gives rise to
Ectoderm	Outer layer of skin; nervous system; sense organs, such as eyes
Endoderm	Lining of digestive tract; respiratory system; urinary bladder; digestive organs; liver; many glands
Mesoderm	Most of the skeleton; muscles; circulatory system; reproductive organs; excretory organs

Tissues

The cells of all animals except sponges are organized into structural and functional units called tissues. Recall that tissues are groups of cells with a common structure that work together to perform a specific function. For example, the cells of adipose *(AD uh pohs)* tissue, shown in **Figure 28-6**, are specialized for storing fat. The cells of muscle tissue are specialized to contract, producing movement. The cells of nerve tissue are specialized to conduct signals.

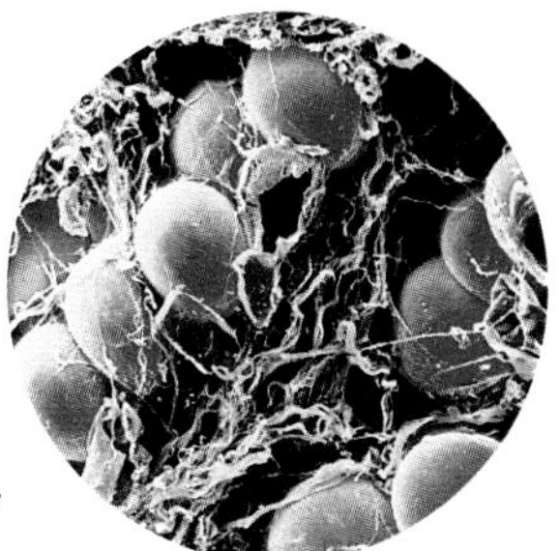

Figure 28-6 Adipose tissue. These adipose cells are specialized for storing fat, and each one contains a fat droplet. Together the cells make up adipose tissue.

Exploring Further

From Zygote to Gastrula

Just as you did, this sea urchin began life as a zygote but quickly developed into an embryo made up of layers of tissue—endoderm, ectoderm, and mesoderm. A variation of this pattern of development is found in all animals except sponges and is evidence that animal life arose from a common ancestor. The illustrations at right show the sea urchin's pattern of development in its first hours of life.

Cell numbers increase

During a process called cleavage, the cells of the zygote divide, doubling the number of cells with each division. However, the mass of the developing embryo does not increase, and the cells formed are progressively smaller. After about 3 hours, the zygote has become a solid ball of cells. Cell division continues until a blastula is formed.

Cells change locations

During a second process, called gastrulation, the blastula begins to collapse inward. At the same time, its cells move about, changing their loction within the blastula. The cells begin to vary in size and form three primary tissue (germ) layers. Gastrulation is complete, and the zygote is now an embryo.

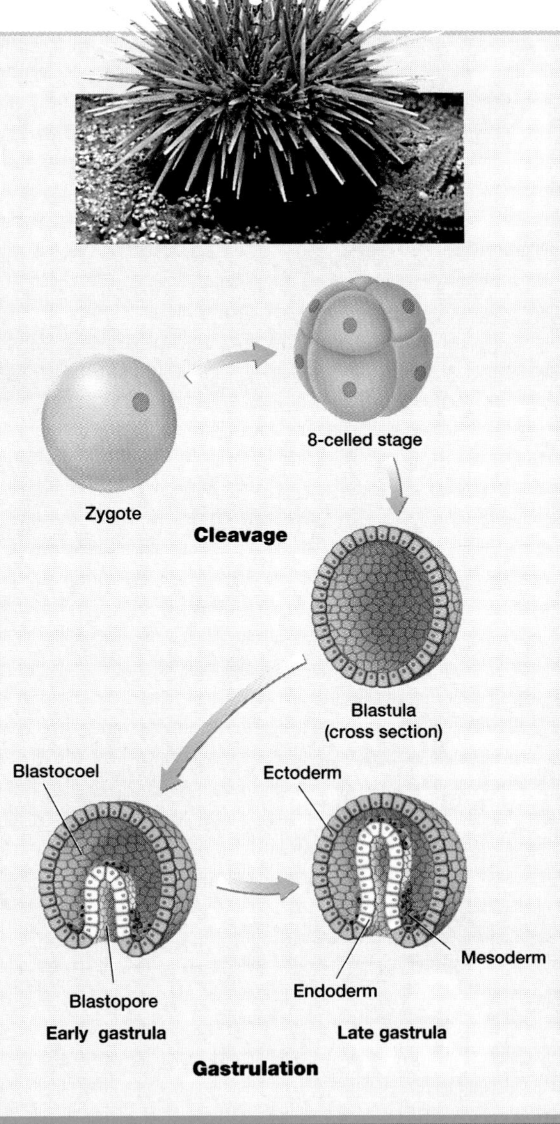

TOPIC: Origin of tissues
GO TO: www.scilinks.org
KEYWORD: HX615

did you know?

Twins form in one of two ways.

Monozygotic (identical) twins result when the embryo splits in two early in cleavage, giving rise to two genetically identical embryos.

Dizygotic twins result when ovulation produces two eggs that are fertilized by different sperm. Because dizygotic twins develop from different sperm and eggs, they can be as genetically dissimilar as any two siblings.

Exploring Further

From Zygote to Gastrula

Teaching Strategies

- Tell students that although all animals except sponges develop a blastula, the way the blastula develops can vary. The number of cells in a blastula and the length of time it takes for the blastula and the gastrula to form depend on the species.
- Tell students that not all blastulas look like that of a sea urchin. Show students photos or drawings of a frog's blastula. Ask them why these two blastulas might have different shapes. *(A frog's blastula has much more yolk.)*

Discussion

- Have students look at the five drawings. What do the colors represent on the bottom two drawings? *(blue, ectoderm; yellow, endoderm; red, mesoderm)*
- What is a zygote, and how many cells does it have? *(a fertilized egg; one cell)*
- What happens during cleavage? *(The cells divide. Each division doubles the number of cells in the embryo.)*
- What happens during gastrulation? *(The cells of the blastula begin to move about and vary in size. As they move, they form three distinct layers—ectoderm, endoderm, and mesoderm.)*

Cross-Disciplinary Connection

◆ Mathematics

Have students calculate how many cells will be present in a developing zygote after 3, 5, and 10 divisions. *(8, 32, 1024)* Have students write a general formula for calculating the number of cells after *n* divisions. *(number of cells = 2^n)* Ask students what assumption must be made for this formula to be valid. *(Every cell must divide during each cell division.)*

Demonstration

Symmetry of Common Objects

Use some common classroom objects to illustrate different types of symmetry. For instance, a crumpled piece of paper is asymmetrical, a roll of tape is radially symmetrical, and a stapler is bilaterally symmetrical. Point out that the symmetry does not have to be perfect in every detail to be considered a type of symmetry. For example, a sea anemone is radially symmetrical even though all its tentacles might not be exactly the same size. Then give students a series of food items to classify as asymmetrical *(scrambled eggs, mashed potatoes, popcorn),* radially symmetrical *(pizza, doughnut, bagel),* or bilaterally symmetrical *(taco, hot dog, loaf of bread).*

Teaching TIP

Following the Directions

After introducing the terms *anterior, posterior, ventral,* and *dorsal* to your students, reinforce this anatomical terminology by having students sketch the outline of a worm. Then direct them to draw different structures on their worm. For example, you might ask them to shade the dorsal surface black; draw rings around the posterior end; draw spots on the ventral surface; and draw eyes on the anterior end. Finish by drawing the completed worm on the board. PORTFOLIO

internetconnect

Have students research symmetry using the Web site in the Internet Connect box on page 616. Have students write a report summarizing their findings. Students should include illustrations or models with their report. PORTFOLIO

Most Animals Show Body Symmetry

All animals have their own particular **body plan,** a term used to describe an animal's shape, symmetry, and internal organization. An animal's body plan results from a pattern of development programmed into the animal's genes by natural selection. Sponges, such as the one shown in **Figure 28-7,** have the simplest body plan of all animals. Sponges are **asymmetrical,** or irregular in shape, and sometimes their shape depends on where they are growing. The body plans of virtually all other animals show a definite body shape and symmetry.

Figure 28-7 Asymmetry. Animals that grow in an irregular pattern, such as this sponge, show asymmetry.

Radial symmetry

The first animals to evolve in the ancient oceans had radial symmetry. Animals with **radial symmetry** have body parts arranged around a central axis, somewhat like the spokes around a bicycle wheel. A plane passing through the central axis divides the organism into roughly equal halves, as shown in **Figure 28-8.** Today's radially symmetrical animals are aquatic. Most move slowly or drift in ocean currents.

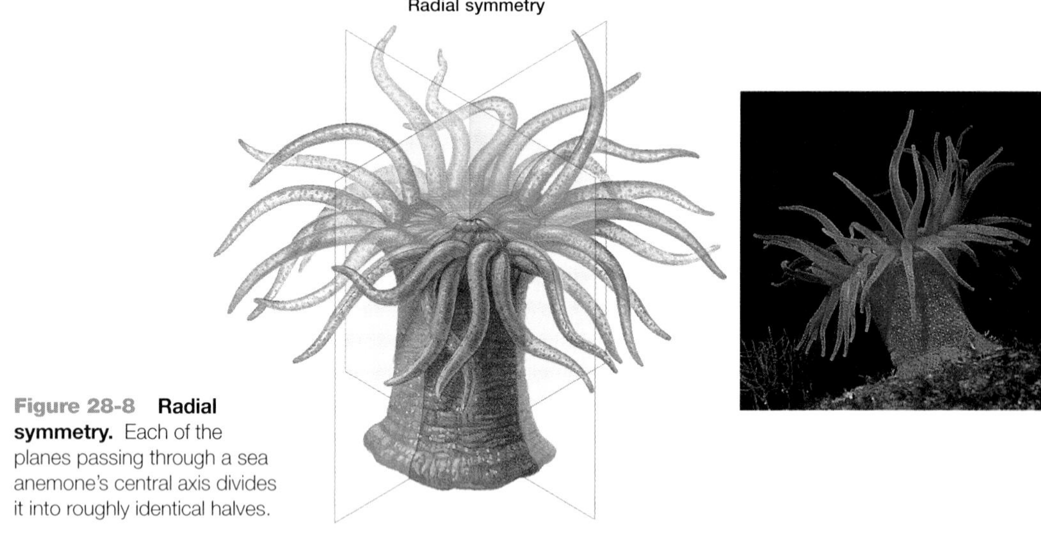

Figure 28-8 Radial symmetry. Each of the planes passing through a sea anemone's central axis divides it into roughly identical halves.

internetconnect

SCILINKS NSTA

TOPIC: Body symmetry
GO TO: www.scilinks.org
KEYWORD: HX616

Bilateral symmetry

The bodies of all other animals show **bilateral symmetry,** a body design in which there are distinct right and left halves. A plane passing through the animal's midline divides the animal into mirror image halves, as shown in **Figure 28-9.** There is a dorsal (top) and a ventral (bottom) surface plus an anterior (front) end and a posterior (back) end.

Bilateral symmetry was a major evolutionary change in animals because it enabled different parts of the body to become specialized in different ways. For example, most bilaterally symmetrical animals have evolved an anterior concentration of sensory structures and nerves, a process called **cephalization** *(SEF uhl lih ZAY shuhn)*. Animals with cephalic ends, or heads, are often active and mobile. With sensory organs concentrated in the front, such animals can more easily sense food and danger.

REAL LIFE

On the outside, you look symmetrical. However, if you could place a mirror down the midline of your interior, you would see that your insides are asymmetrical. For example, your spleen is on the left side of your body, while your liver is on the right.

Finding Information
Examine a detailed drawing or model of the interior of the human body to find additional examples of internal asymmetry.

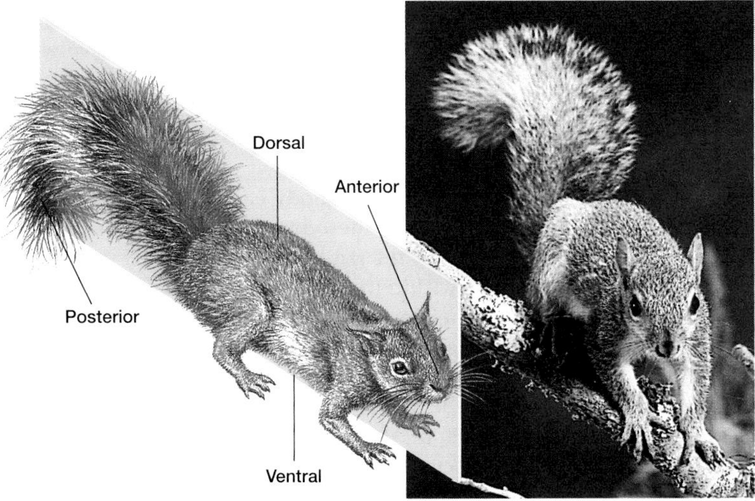

Figure 28-9 Bilateral symmetry. The body parts of this gray squirrel are arranged so that it has a right and a left half that are mirror images of each other.

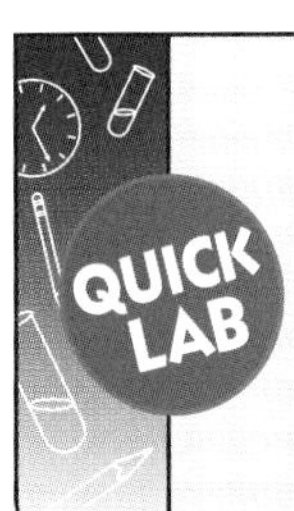

How is symmetry recognized?

You can use the letters of the alphabet to better understand the nature of symmetry.

Materials

envelope containing letters of the alphabet

Procedure

1. Spread the letters on the table in front of you so you can see all of them.
2. Sort the letters into groups based on their symmetry, using the terms *asymmetry, radial symmetry,* and *bilateral symmetry.* For example, the letters *A* and *T* show bilateral symmetry. The letter *J* is asymmetrical.

Analysis

1. **Propose** a definition for each kind of symmetry you found in the letters.
2. **List** any letters you found difficult to classify, explaining why it was difficult to classify these letters.
3. **Identify** the letters that show the same kind of symmetry as sponges.
4. **Identify** two or three animals that you might be familiar with that have the same kind of symmetry as the letter *M*.
5. **SKILL Evaluating Methods** What are some strengths and weaknesses of using symmetry as a way of classifying or describing organisms?

REAL LIFE

Answer

The heart is slightly left of center; the right lung has three lobes and the left lung has two; the stomach is on the left; and the appendix is on the right.

QUICK LAB

How is symmetry recognized?

Time 10–15 minutes

Process Skills Comparing, categorizing, applying information

Teaching Strategies
Tell students that they may identify some letters that do not fit neatly into a category. They should categorize these letters as they think best.

Analysis Answers

1. Definitions will vary but should be similar to those on page 616.
2. Answers will vary. Students may have difficulty with letters such as *O.* Some may say it shows radial symmetry, while others may classify it as having bilateral symmetry.
3. Answers will vary, but students may include *F, G, J, K, L, N, P, Q, R,* and *Z.*
4. Answers will vary; accept any bilaterally symmetrical animal.
5. Answers will vary. Students may think that considering symmetry is helpful when the type of symmetry is clear-cut but less helpful when there is a question of what type of symmetry is displayed.

MULTICULTURAL PERSPECTIVE

Totem Poles

Many Native American tribes have family totems, which are symbols for their family, clan, or tribe. A specific animal is associated with the totem, and some groups consider the animal totem to be the ancestor of the family. Often there is a prohibition against killing or eating that particular animal. Some tribes that live along the northwest coast of North America carve totem poles—tall log poles with depictions of animals carved into them. In addition to human figures, animals seen on totem poles include minks, orcas, frogs, seals, hawks, eagles, ravens, and small marine animals.

Visual Strategy
Figure 28-10

Have students compare the cross section of the acoelomate flatworm to the cross section of the pseudocoelomate roundworm. Point out the solid-body construction of the flatworm versus the hollow-body construction of the roundworm. Then have students compare the body plans of the flatworm and roundworm to that of the coelomate earthworm. Draw students' attention to the placement of the three primary tissues in each figure. Tell students that while pseudocoelomates have muscles lining their body wall, coelomates have an advantage because they have an additional layer of muscles surrounding their gut. Ask students to name some human structures that contain a muscle layer derived from mesoderm. *(blood vessels, digestive tract, and uterus)*

Career
Aquatic Biologist

An aquatic biologist is a scientist who specializes in the biology of organisms that live in water, including oceans, rivers, ponds, and lakes. Many aquatic biologists study the complex ecosystems of these environments. They investigate the physical and chemical features of the water in addition to the biological factors. For example, they might gauge the effects of pollutants on the organisms that live in a lake. Have students do research to find out more about the duties and daily activities of someone in this field. PORTFOLIO

Word Origins

The term *coelomate* is from the Greek *koilia,* meaning "body cavity." Placing prefixes in front of this word changes its meaning. For example, *a-* means "without" and *pseudo-* means "false."

Many Animals Have an Internal Body Cavity

Bilaterally symmetrical animals have one of three basic kinds of internal body plans, each illustrated in **Figure 28-10.** The body plan may include a body cavity, or **coelom** *(SEE luhm),* a fluid-filled space found between the body wall and the digestive tract (gut). This space is lined with cells that come from mesoderm. Animals with no body cavity are called **acoelomates** *(ay SEEL oh mayts).* The space between an acoelomate's body wall and gut is completely filled with tissues. Other animals, called **pseudocoelomates** *(SOO doh seel oh mayts),* have a body cavity located between the mesoderm and endoderm. Their body cavity is called a pseudocoelom (false coelom).

Coelomates have a true coelom, a body cavity located entirely within the mesoderm. Because the mesodermal layer lines the body wall and wraps around the gut, the gut and other internal organs of coelomates are suspended within the coelom.

A true coelom provides an internal space where mesoderm and endoderm can be in contact with each other during embryonic development. This aided the evolution of complex organs made of more than one type of tissue. Since these internal organs are suspended in a fluid-filled coelom, they are protected from the movement of surrounding muscles. Thus, an animal can move about without damaging the organs or interfering with their function.

Figure 28-10 Three body plans

Bilaterally symmetrical animals have one of three basic body plans.

Skin
Tissue-filled region
Gut
Skin
Body cavity (pseudocoelom)
Muscle tissue
Gut
Skin
Body cavity (coelom)
Muscle tissue
Gut

Acoelomate body plan **Pseudocoelomate body plan** **Coelomate body plan**

Endoderm Ectoderm Mesoderm

Biology in Your Community

Urban Animals

A variety of animals have adapted to life with humans. These animals are attracted to food, water, nesting sites, and the protection from predators they find within the human community. In the United States, examples of animals that have successfully adapted to urban or suburban life include gulls, peregrine falcons, owls, doves, geese, coyotes, rabbits, raccoons, squirrels, mice, chipmunks, opossums, turtles, lizards, snakes, cockroaches, ants, and wasps. Work with students to create a list of animals that are adapted to life in your community.

"Advanced" Organisms Show Body Segmentation

Just as tunnels can be constructed from a series of identical premade parts, segmented animals are "assembled" from a series of repeating, similar units called segments. Segmentation underlies the organization of all "advanced" animals and is easy to observe in some animals, such as earthworms. Crustaceans (lobsters and their kin), spiders, and insects also show some degree of body segmentation, but it may be difficult to observe during their adult stage. For example, segments are not apparent in the adult moth in **Figure 28-11,** but they are clearly visible in its immature caterpillar stage.

In vertebrates, segments are not visible externally, but there is evidence of segmentation in a vertebrate embryo. Vertebrate muscles develop from repeated blocks of tissue called somites *(SOH meyetz),* and the vertebral column (backbone) consists of a stack of very similar vertebrae.

In segmented worms, each segment can move independently, permitting great flexibility and mobility. Therefore, a segmented worm's long body can move in ways that are often quite complex. For example, consider an earthworm as it crawls along a flat surface. It lengthens some parts of its body while shortening others. Many scientists think segmentation evolved as an adaptation that permitted more-efficient burrowing.

In highly segmented animals, such as earthworms, each segment repeats many of the organs in the adjacent segment. As a result, an injured animal can still perform vital life functions. Segments are not totally independent of each other, however. Materials pass from one segment to another through a circulatory system that connects them. Nerves also connect each segment to a brain that coordinates the body's movements. Segmentation also offers evolutionary flexibility. A small change in an existing segment can produce a new type of segment with a different function. For example, segments can be modified for feeding or for reproduction.

biobit

How long is the human digestive tract?

From beginning to end, the adult human digestive system is a little over 8 m (26 ft) long. Unwound, it would measure more than four times your height.

Figure 28-11
Segmentation. This caterpillar shows clearly defined segments. In its adult form—a cecropia moth—the segments are fused and are more difficult to observe.

Section 28-1

Vertebrate Segmentation

Ask students to point out examples of segmentation in the external structure of the human body. Students will find this difficult, so tell them that although vertebrates are considered segmented animals, vertebrate segmentation is strictly internal. The backbone and rib cage are composed of repeating units, and the muscles are made up of multiple blocks of tissue. Segmentation is also evident in the development of the brain.

Visual Strategy

Figure 28-11 — BASIC

Direct students' attention to Figure 28-11. Have students count the number of segments they can see in the caterpillar. Tell students that most of the segments are fused in the adult form. Then have students try to count the number of fused segments on the moth's abdomen. Ask: What clue is present that can help you distinguish the segments? *(color pattern)*

Teaching TIP

Evolutionary Milestones

Draw a **Graphic Organizer** on the board similar to the one shown at the bottom of page 620, but do not include the names of the phyla. Scramble the order of the phyla and list them separately. Beginning with *Multicellularity,* have a student come to the board and write the correct phylum next to this evolutionary milestone. Then ask the student to explain what the description means. When you get to *Deuterostomes* and *Notochord,* tell students that they will learn these terms in Chapter 32.Explain that these evolutionary milestones will be indicated by an **Evolutionary Milestones** box in Chapters 29–32.

Teaching TIP ADVANCED

Other Phyla

Have students find out about one phylum that is not shown on the phylogenetic tree in Figure 28-12. *(Phyla that are not represented in Figure 28-12 include Ctenophora, comb jellies; Nemertea, ribbon worms; Bryozoa, bryozoans; Brachiopoda, brachiopods; and Phoronida, phoronids.)* Ask them to write a report about the phylum and tell where it would be located if it were shown on this tree.

PORTFOLIO

There Are Many Kinds of Animals

Kingdom Animalia contains about 35 major divisions called phyla (singular, phylum), depending on how certain organisms are classified. The animals in the various phyla show an extraordinary range of body forms, internal body systems, and behaviors. The members of no other kingdom occupy as many habitats in as many different ways as do the animals.

To visually represent the relationships among various groups of animals, scientists often use a type of branching diagram called a phylogenetic tree. A **phylogenetic tree** shows how animals are related through evolution. The phylogenetic tree in **Figure 28-12** shows the evolutionary relationships of nine of the major animal phyla. Scientists determine these relationships by comparing several different types of evidence. Clues to animal relationships can be found in the fossil record and by comparing the anatomy and physiology of living animals. Clues are also found by comparing patterns

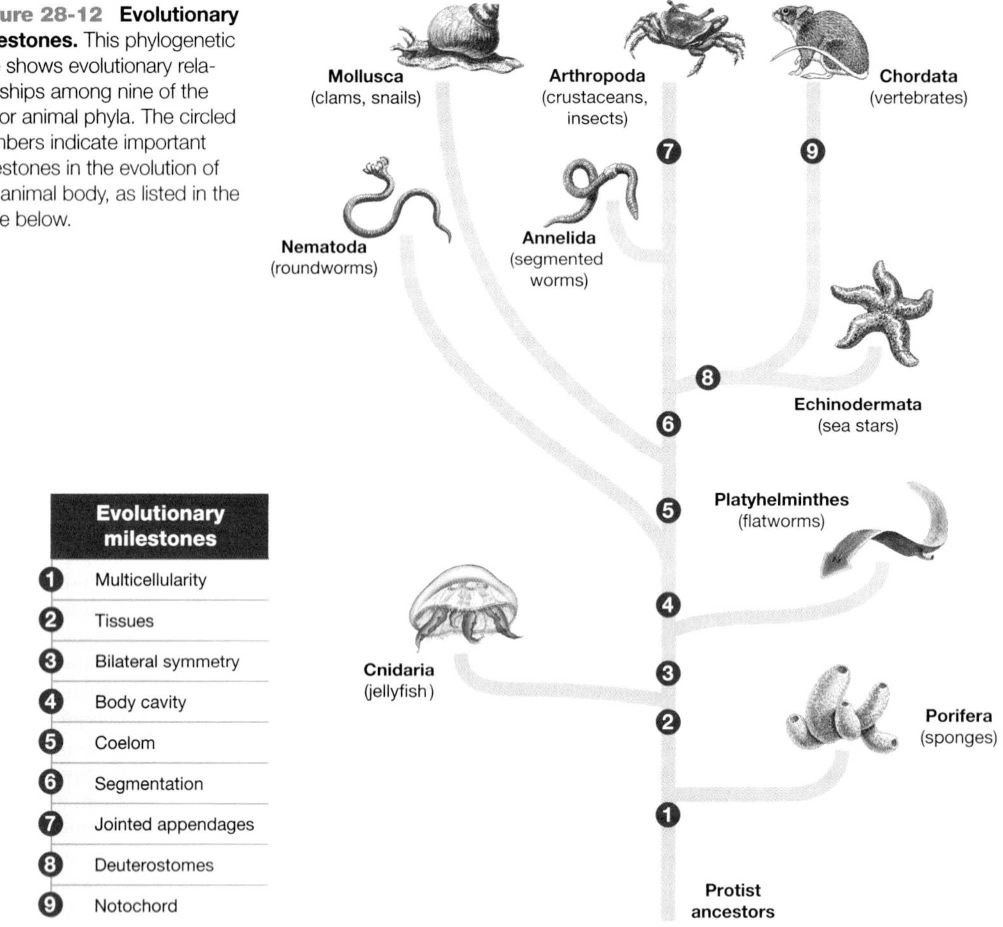

Figure 28-12 Evolutionary milestones. This phylogenetic tree shows evolutionary relationships among nine of the major animal phyla. The circled numbers indicate important milestones in the evolution of the animal body, as listed in the table below.

Graphic Organizer

*Use this graphic organizer with **Teaching Tip: Evolutionary Milestones** on page 620.*

Phylum	Evolutionary milestone
Chordata	Notochord
Echinodermata	Deuterostomes
Arthropoda	Jointed appendages
Annelida	Segmentation
Mollusca	Coelom
Nematoda	Pseudocoelom
Platyhelminthes	Bilateral symmetry
Cnidaria	Tissues
Porifera	Multicellularity

of development in animal embryos. The most direct evidence of evolutionary relationships, however, comes from comparing the DNA in the genes of various animal species. For example, scientists long debated whether giant pandas are related more closely to bears or to raccoons. Comparing their DNA showed that pandas are more closely related to bears than to raccoons.

The animal kingdom is often divided into two groups, invertebrates (animals without a backbone) and vertebrates (animals with a backbone). As you study these two groups, you will discover how a series of key evolutionary innovations in the body plan of animals has led to today's animals. The first animals to evolve did not have body cavities. After body cavities appeared, segmentation evolved. Later, jointed appendages (such as legs) evolved. Backbones evolved even later. These innovations, shown in Figure 28-12 as numbered milestones, help scientists determine the evolutionary relationships among the different groups of animal. Watch for these evolutionary stages and milestones as you continue studying invertebrates in Chapters 29–32.

Exploring the animal kingdom

Background

You can find out more about the animal phyla by referring to the section "A Six-Kingdom System of Classification" located in the Appendix. Turn to this section and locate the information for kingdom Animalia. Follow the procedure below to evaluate this information.

Procedure

1. Read the introductory paragraph for kingdom Animalia. Then quickly skim over the information presented. Do not read it word for word, but observe how the information is divided into sections.
2. Choose one phylum, and read all of the information about it.
3. As you read, notice how color is used and what types of information are given, for example, number of species found and habitat.

Analysis

1. **List** at least three types of information you found for the phylum you read about.
2. **Analyze** how color is used to distinguish between the different entries on a page.
3. **Propose** a way that you might use this information when studying about a particular animal phylum.

Review SECTION 28-1

1. **Describe** each of the eight features animals have in common.
2. **Summarize** the difference between radial symmetry and bilateral symmetry.
3. **Compare** the body plans of acoelomates, pseudocoelomates, and coelomates.
4. **SKILL Relating Concepts** In 1994, Western scientists first observed the Vietnamese saola, a hoofed mammal. At first, the saola was thought to be related to sheep and goats or to antelope. Later it was proved to be related to wild cattle and buffalo. How do you think scientists identified the saola's closest relatives?

Review SECTION 28-1 ANSWERS

1. Student descriptions will vary but should mention heterotrophy, mobility, multicellularity, diploidy, sexual reproduction, absence of a cell wall, blastula formation, and tissues.
2. Animals with radial symmetry can be divided into equal parts by a line drawn anywhere, as long as the line crosses their midpoint. They have no right or left side. Animals with bilateral symmetry can be divided into identical right and left halves by a line that passes down their midline. They have dorsal and ventral sides and anterior and posterior ends.
3. Acoelomates have no body cavity. Pseudocoelomates have a body cavity located between the mesoderm and the endoderm. Coelomates have a body cavity that is completely surrounded by mesoderm.
4. Scientists compared the saola's DNA with that of animals they thought the saola was related to. The DNA evidence pointed to wild cattle and buffalo.

Exploring the animal kingdom

Time 10–15 minutes

Process Skills Observing, comparing, summarizing

Teaching Strategies

Before students begin, write the names of the six kingdoms on the board. Tell students that they will look only at the introduction and the information about kingdom Animalia.

Analysis Answers

1. Acceptable answers include number of organisms in the phylum, general characteristics of the phylum, classes that belong to the phylum, and examples of organisms in the phylum and orders in its classes.
2. The kingdoms are in bold red print; the phyla and subphyla are in bold blue print; the classes are in bold green print; and the orders are in bold black print.
3. Answers will vary. Students may answer that they could use the information to compare the number of species in different phyla or orders and to refresh their memory about the classes in a phylum.

3 Assessment Options

Closure

Animal Characteristics

Form teams of three or four students. Assign each team one of the characteristics of animals. Have each team write a brief description of the characteristic and develop an accompanying visual. Teams should then present their information to the class. PORTFOLIO

Review

Assign Review—Section 28-1.

Reteaching — BASIC

Have students work in pairs to make two drawings: one showing the three types of symmetry and one showing the three animal body plans. PORTFOLIO

1 Prepare

Scheduling

- **Block:** About 90 minutes will be needed to complete this section.
- **Traditional:** About two class periods will be needed to complete this section.
- **Planning:** See the Planning Guide on pp. 609A and 609B for lecture, classwork, and assignment options for Section 28-2. Lesson plans for Section 28-2 can be found in a lesson cycle format in ***Biology: Principles and Explorations*** **One-Stop Planner CD-ROM with Test Generator.**

Resource Link

Assign **Active Reading Worksheet Section 28-2** to help students comprehend the key concepts and visuals in this lesson.

2 Teach

Lesson Warm-up

Make a two-column table on the board. Label one column *Think about* and the other *Don't think about*. Tell students that people engage in many activities that they make conscious decisions about, but they also engage in activities that they make no decisions about. Place an entry in each column: *Think about—how my hair looks; Don't think about—digesting my breakfast.* Ask students to suggest other entries. Tell students that in this section they will learn more about activities they do not usually think about.

Section 28-2 Animal Body Systems

Objectives

- **Summarize** the functions of the digestive, respiratory, circulatory, nervous, skeletal, and excretory systems. (pp. 622–626)
- **Compare** a gastrovascular cavity with a one-way digestive system. (p. 622)
- **Differentiate** open from closed circulatory systems. (p. 624)
- **Distinguish** asexual from sexual reproduction. (pp. 627–628)

Key Terms

gastrovascular cavity
respiration
gill
open circulatory system
closed circulatory system
hydrostatic skeleton
exoskeleton
endoskeleton
hermaphrodite
external fertilization
internal fertilization

internetconnect
SCiLINKS
NSTA
TOPIC: Body systems
GO TO: www.scilinks.org
KEYWORD: HX622

Tissues and Organs Perform Specific Tasks

As you go about your day, you make decisions that involve thinking—what to wear to school, how to solve a problem, where to sit during lunch. But a lot that happens during your day does not require your thought. You digest your food, you balance yourself as you walk, and your heart beats. These and thousands of other functions are carried out by your body. But how does your body, or the body of any animal, carry out these tasks?

Simple animals like sponges carry out the many tasks of living with little specialization in the cells of their body. More-complex animals have evolved tissues and organs that are specialized to perform specific functions. Six important functions of these tissues and organs are digestion, respiration, circulation, conduction of nerve impulses, support, and excretion.

Digestion

Single-celled organisms and sponges digest their food within their body cells. This means that their food source cannot be larger than their individual cells. All other animals digest their food extracellularly (outside of their body cells) within a digestive cavity, as shown in **Figure 28-13.** Digestive enzymes that are released into the cavity

Figure 28-13 Extracellular digestion

A hydra has a gastrovascular cavity, while a roundworm has a digestive tract in which food travels in one direction only.

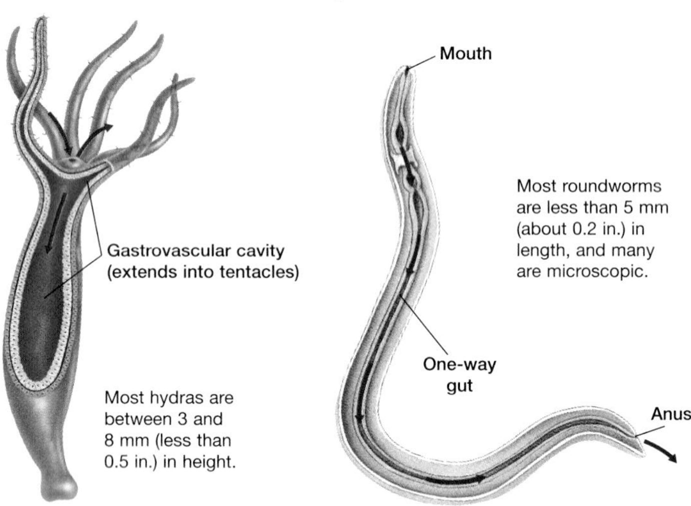
Gastrovascular cavity
(extends into tentacles)
Most hydras are between 3 and 8 mm (less than 0.5 in.) in height.
Mouth
Most roundworms are less than 5 mm (about 0.2 in.) in length, and many are microscopic.
One-way gut
Anus

Integrating Different Learning Styles

Logical/Mathematical	**PE**	Math Lab, p. 626
Visual/Spatial	**PE**	Performance Zone items 1, 19, and 20, pp. 630–631
	ATE	Demonstration, p. 623; Visual Strategies, pp. 625, 627
Interpersonal	**PE**	Study Tip, p. 624
	ATE	Reteaching, p. 628
Intrapersonal	**PE**	Real Life, p. 623
Verbal/Linguistic	**PE**	Performance Zone items 2 and 28, pp. 630–631
	ATE	Closure, p. 628

begin the breakdown of food, permitting the animal to prey on organisms larger than its body cells. Simple animals, such as the hydra and flatworms, have a **gastrovascular cavity,** a digestive cavity with only one opening. There can be no specialization within a gastrovascular cavity because every cell is exposed to all stages of food digestion.

Other animals have a digestive tract (gut) with two openings, a mouth and an anus. The anus is the opening through which undigested food leaves the body. In a digestive tract, food moves in one direction, from mouth to anus. Unlike a gastrovascular cavity, a one-way digestive system allows for specialization. For example, there may be a section of gut for food storage, a section for breaking down food into small pieces, and a section for the chemical digestion of food. Eventually the food is broken down into molecules small enough to pass through the lining of the gut and into the bloodstream.

Respiration

In simple animals, such as jellyfish, oxygen gas and carbon dioxide gas are exchanged directly with the environment by diffusion. The uptake of oxygen and the release of carbon dioxide, called **respiration,** can take place only across a moist surface, such as the damp skin of an earthworm. In larger, more complex animals, simple diffusion cannot provide for adequate gas exchange. Most large animals have specialized respiratory structures.

Some aquatic (and a few terrestrial) animals respire with **gills,** extremely thin projections of tissue that are rich in blood vessels, providing a large surface for gas exchange. **Figure 28-14** shows the gills of a mud puppy. Gills are not suitable for most terrestrial animals because gills do not function unless they are kept moist. Among terrestrial animals, a variety of respiratory organs, such as lungs, have evolved to meet this need.

Figure 28-14 **Gills.** The gills of this mud puppy are supported by its watery environment.

REAL LIFE

Lung tissue is delicate. Many products, such as spray paints and some glues, contain chemicals that, if inhaled, will damage or destroy lung tissue. When this occurs, the surface area available for gas exchange is reduced, causing oxygen intake to be insufficient for normal daily activities. To prevent lung damage when using chemical products, always follow the instructions on the warning label. If necessary, wear a protective mask.

Finding Information
Research diseases and conditions that cause lung damage, and find out what a person can do to prevent lung disease.

REAL LIFE

Answer

Lung diseases include asthma, bronchitis, emphysema, and lung cancer. Smoking and exposure to air pollution are major causes of lung damage.

Demonstration

Digestive Systems — BASIC

To demonstrate the difference between a gastrovascular cavity and a one-way gut, show students a cup and a length of tubing. Demonstrate how water can enter and leave the cup through only one opening. Swirl the water around a bit to show how it is exposed to all areas of the inner surface of the cup. Then fill the tube with water. Use your fingers to push the water through the tube, duplicating the action of muscles. Point out how the water moves in only one direction through the tube. Once the water passes through a section of the tube, it will not return to that section again.

Chapter 28

Teaching TIP — ADVANCED

Transport in Plants and Animals

Tell students to think about how nutrients are transported in plants. *(Mineral nutrients are transported in the xylem by water that rises from the roots to the leaves, drawn by evaporation from the stomata. Sugars are transported in the phloem.)* Have students contrast this with how nutrients are transported in animals. *(In most animals, nutrients enter the circulatory fluid from the gut. A heart pumps the fluid through the body, delivering nutrients to the body's cells.)*

Teaching TIP

Open versus Closed Circulatory Systems

Ask students to make a two-column list that will allow them to compare and contrast open and closed circulatory systems. In one column, ask them to write the ways in which the two systems are alike. In the other column, ask them to write the ways in which the two systems are different. PORTFOLIO

Study TIP

Reviewing Information

Work with a partner to review the body functions on pages 622–626. Review each function by asking each other questions that include the boldface terms. For example, ask, "What is the difference between a gatrovascular cavity and a one-way gut?"

Circulation

In simple animals, body cells are exposed to either the external environment or the gastrovascular cavity. None of the body cells are far away from sources of oxygen or nourishment. More-complex animals, however, have tissues that are several cell layers thick. Many of the cells in this tissue are not close enough to the surface of the cell layer to exchange materials directly with the environment. Oxygen and nutrients must be transported to these body cells by a circulatory system.

Two types of circulatory systems are shown in **Figure 28-15.** In an **open circulatory system,** a heart pumps fluid containing oxygen and nutrients through a series of vessels out into the body cavity. There the fluid washes across the body's tissues, supplying them with oxygen and nutrients. The fluid collects in open spaces in the animal's body and flows back to the heart. In a **closed circulatory system,** a heart pumps blood through a system of blood vessels. These blood vessels form a network that permits blood flow from the heart to all of the body's cells and back again. The blood remains in the vessels and does not come in direct contact with the body's tissues. Instead, materials pass into and out of the blood by diffusing through the walls of the blood vessels.

Figure 28-15 **Circulatory systems**

In an open circulatory system, fluid leaves the circulatory vessels, but in a closed circulatory system, blood remains in the blood vessels.

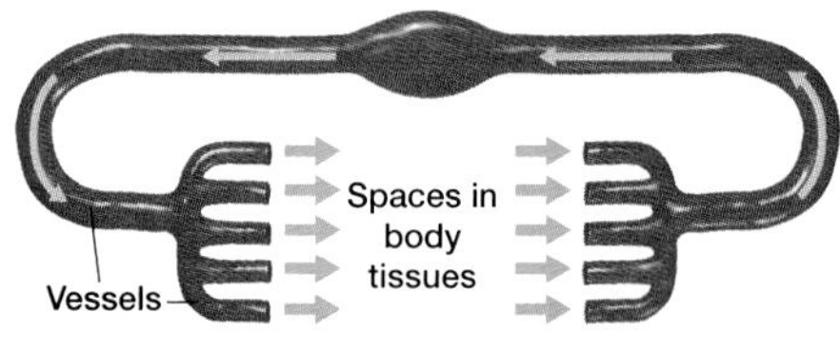

Open circulatory system

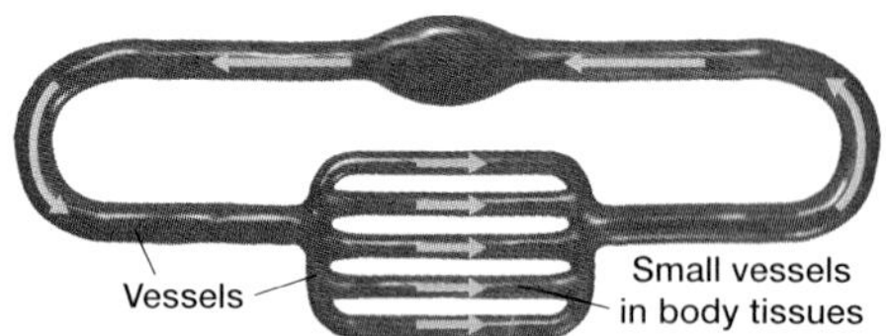

Closed circulatory system

Conduction of nerve impulses

Nerve cells (neurons) are specialized for carrying messages in the form of electrical impulses (conduction). These cells coordinate the activities in an animal's body, enabling the animal to sense and respond to its environment. Members of all of the major animal phyla except sponges have nerve cells. **Figure 28-16** shows the arrangement of nerve cells in three animals—a hydra, a flatworm, and a grasshopper. The simplest arrangements of nerves are found in animals like hydra and jellyfish. All of their nerve cells are similar and are linked to one another in a web called a nerve net. There is little coordination among the nerve cells in a nerve net.

As animals evolved and interacted with their environment in more-complex ways, they needed to process more information. To

Overcoming Misconceptions

Bird Brains

Students may have heard the expression "bird brain" used to indicate that someone is not smart. Tell them that birds actually have very complex brains and that many bird species are capable of learning complex tasks.

Explain that even though most animals may not appear smart to us, each species (except sponges) has evolved a nervous system that enables it to be successful in its environment.

Figure 28-16 **Nervous systems**

The hydra has a simple nerve net, while the flatworm and the grasshopper have more-complex nervous systems.

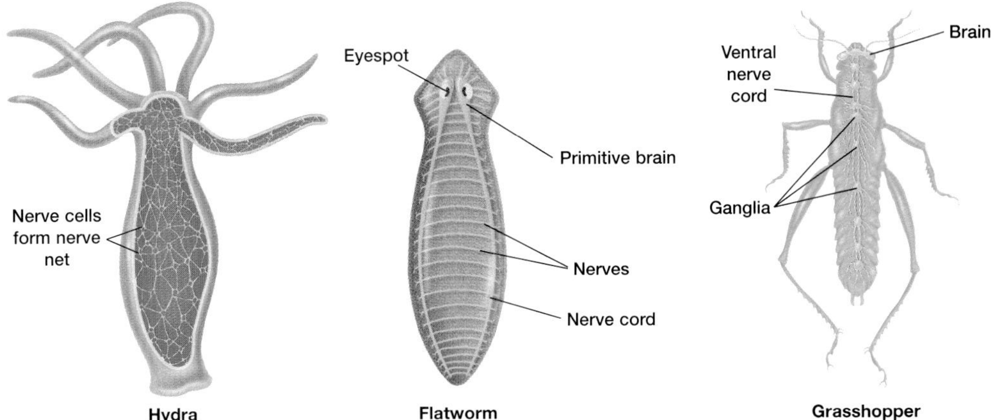

meet this need, clusters of neurons, called ganglia, developed. The ganglia at the anterior end of the animal body became larger and more complex, forming a brain-like structure, as seen in the flatworm. More-complex invertebrates, such as the grasshopper, evolved brains with sensory structures, such as eyes, associated with them.

Support

An animal's skeleton provides a framework that supports its body. It is also vital to animal movement. All animals move using the same force: the contraction (shortening) of muscle tissue against a framework such as the skeleton provides.

Hydrostatic skeleton Many soft-bodied invertebrates have a hydrostatic skeleton. A **hydrostatic skeleton** consists of water that is contained under pressure in a closed cavity, such as a gastrovascular cavity or a coelom. Imagine a balloon filled with water. The water presses against the balloon, supporting it. If pressure is applied to the balloon in any place, the water must shift, altering the shape of the balloon. The hydrostatic skeleton of a hydra, shown in **Figure 28-17,** is formed by its gastrovascular cavity. Other soft-bodied invertebrates, such as earthworms, have a fluid-filled coelom that serves as a hydrostatic skeleton. In both cases, muscle forces exerted against the hydrostatic skeleton aid movement.

Exoskeleton Other invertebrates, such as insects, clams, and crabs, have a type of skeleton known as an exoskeleton. An **exoskeleton** is a rigid external skeleton that encases the body of an animal. An exoskeleton supports movement in a different manner than a hydrostatic skeleton does. The muscles of animals with exoskeletons are attached to the inside of the skeleton, which provides a surface for them to pull against. Exoskeletons also protect an organism's soft internal parts.

Magnification: 70×

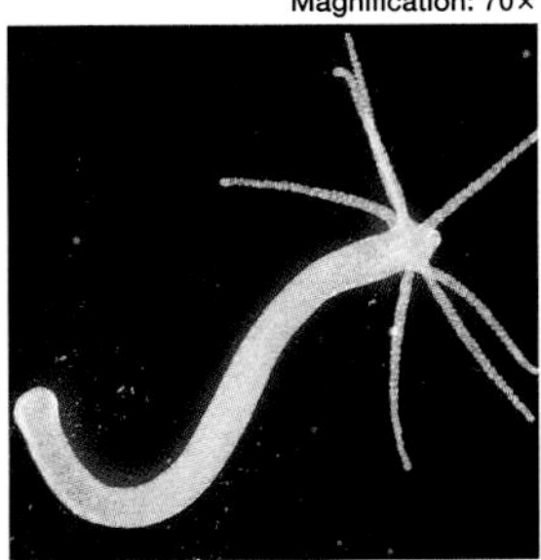

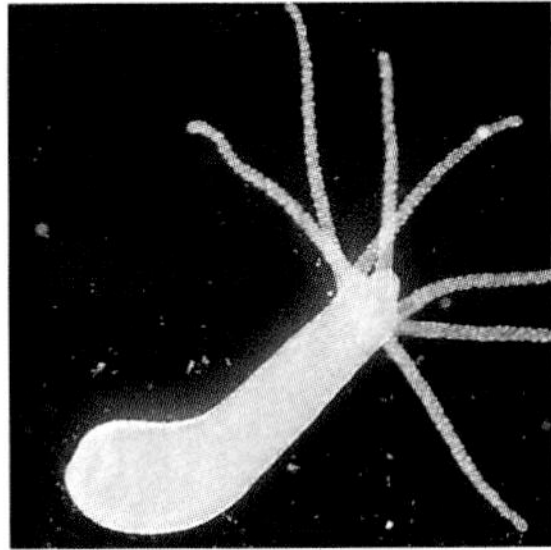

Figure 28-17 **Hydrostatic skeleton.** When the hydra closes the opening to its gastrovascular cavity and contracts muscles in its body wall, its body elongates. When water is released from the cavity and other muscles contract, the hydra's body shortens.

Visual Strategy

Figure 28-16

Have students look at the three nervous systems in Figure 28-16. Tell them that a hydra's nerve cells are found throughout its thin body wall. Each cell is linked to other nerve cells. In contrast to a flatworm, a hydra has no central region with a greater concentration of nerve cells. Point out that as animals became more complex, a more complex system of nerves was needed to coordinate activities and to handle sensory stimuli. A cluster of neurons at the anterior end (primitive brain) became increasingly complex until a true brain evolved.

Teaching TIP — ADVANCED

Hydrostatic Skeleton

Tell students that hydrostatic skeletons are found in some invertebrates with a coelom, such as an earthworm. Explain that the force of muscles against the fluid-filled coelom produces movement. Ask students why the coelom needs to be filled with fluid for this to happen. *(A gas can be compressed when force is applied to it but a fluid cannot.)*

Chapter 28

Calculating filtration rate in the human kidney

Time 15 minutes

Process Skills Calculating, applying information

Teaching Strategies

- Tell students that they will calculate how much fluid (filtrate) the kidneys filter from the blood each day. Remind them that most of the filtrate is reabsorbed by the kidneys; only a small percentage is excreted as urine.
- When students finish, have them look at the answer to analysis item 3. Tell them to visualize 180 1 L bottles of soda. As a class, calculate the percentage of fluid that is reabsorbed if a person excretes 2.0 L of urine a day. *(about 99%; 2.0 L ÷ 180 L = 1.1% filtrate excreted; 100% – about 1% = 99%)*

Analysis Answers

1. 7,500 mL per hour (125 mL/min × 60 min/h)
2. 180,000 mL per day (7,500 mL/h × 24 h/d)
3. 180 L per day (180,000 mL/d ÷ 1,000 mL/L)
4. If the kidneys could not return water to the body, a person would become dehydrated very quickly.

Making Biology Relevant

Health

Obtain a blank urinalysis form from a clinic, and have students study the list of substances that are measured by this test. Help students identify the nitrogen-containing wastes listed on the form. Point out the electrolytes (salts) that are measured by urinalysis. Explain that the kidneys help to keep the balance of these salts in our body fluids within a certain range. Explain the connection between salt balance and good health. For example, sodium and potassium are essential to the transmission of nerve impulses, and calcium is used in muscle contraction.

biobit

Is undigested food excreted?

The process of excretion should not be confused with the elimination of undigested food from the body. Food is eaten, or ingested. However, while undigested food travels through the digestive system, it never enters the body's cells. Thus, undigested food is *egested,* or passed out, rather than excreted.

Endoskeleton An **endoskeleton** is composed of a hard material, such as bone, embedded within an animal. The most familiar type of endoskeleton is that of humans and other vertebrates. As with exoskeletons, muscles attached to the endoskeleton alternately contract and relax, enabling movement to occur.

Excretion

The term *excretion* refers to the removal of wastes produced by cellular metabolism. These waste products leave the cell by crossing the cell membrane and are then removed from the body. Some waste products are highly toxic and will poison an organism if not removed. The most important of these toxic wastes is ammonia. As ammonia forms, it dissolves in body fluids, becoming more dilute and thus less toxic. Simple aquatic invertebrates and some fishes excrete ammonia into the water through their skin or gills by diffusion. This is effective, but results in the loss of a lot of water.

Other animals, especially terrestrial animals, need to minimize water loss. These animals have evolved a variety of ways to remove wastes from the body by converting ammonia to nontoxic chemicals, like urea. As the excretory system eliminates these wastes, water and other useful substances are returned to the body. Thus, eliminating metabolic wastes is linked to maintaining the body's water balance. For example, a mammal's kidneys filter fluid from the blood that flows through them. The kidneys then concentrate the metabolic wastes filtered from the fluid and excrete them as concentrated urine. Simultaneously, the kidneys regulate the water content of the body by making the urine more or less dilute as necessary. In Chapter 40 you will learn more about the kidney.

Calculating filtration rate in the human kidney

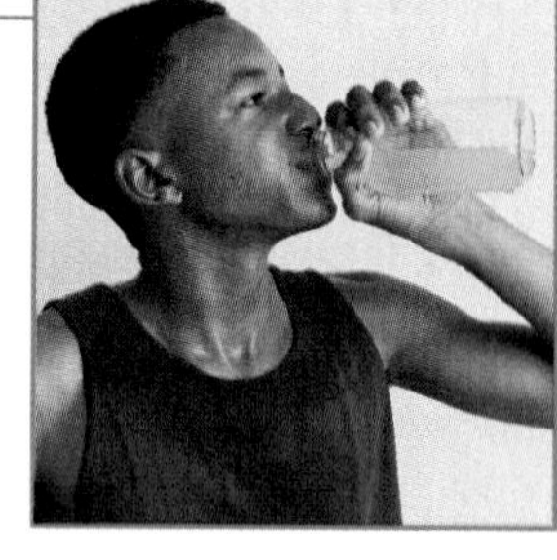

Background

The human kidney filters fluid from the blood at the rate of approximately 125 mL per minute. However, only a small percentage of this fluid is excreted as urine—adult humans normally excrete between 1.5 and 2.3 L of urine a day.

Analysis

1. **Calculate** how many milliliters of fluid the human kidneys filter each hour.
2. **Calculate** how many milliliters of fluid the kidneys filter each day.
3. **SKILL Analyzing Data** Convert your answer in item 2 from milliliters to liters. For help, see "Using SI Measurement" in the Appendix. To better visualize the quantity of fluid represented by your answer, think about the volume of fluid contained in a 1 L bottle of soda.
4. **SKILL Predicting Outcomes** What would happen if the kidneys could not return water to the body?

Animals Have Different Reproductive Strategies

While the reproductive system of an individual animal is not essential to its survival, reproduction is necessary if the species is to survive. There are two types of reproduction in animals, asexual and sexual.

Asexual reproduction

Reproduction that does not involve the fusion of two gametes is called asexual reproduction. A sponge, for example, can reproduce by fragmenting its body. Each fragment grows into a new sponge. Some species of sea anemone reproduce by pulling themselves in half, forming two new adult anemones, as shown in **Figure 28-18.**

An unusual method of asexual reproduction is parthenogenesis *(pahr thuh noh JEN uh sis),* in which a new individual develops from an unfertilized egg. Parthenogenesis is common among insects. In honeybees, for example, a queen bee mates only once and stores the sperm. Although she has mated, the queen bee has the ability to lay unfertilized eggs that develop by parthenogenesis into male bees, called drones. (Female bees develop by sexual reproduction, when the queen releases stored sperm to fertilize her eggs.) A few species of fishes, amphibians, and lizards reproduce by parthenogenesis. Animals that reproduce asexually usually reproduce sexually also.

biobit

Do any male animals become pregnant?

A female sea horse inserts her eggs into a pouch on the male's abdomen, where they are fertilized. The male incubates the eggs until they are fully developed and then gives birth to tiny sea horses.

Figure 28-18 Asexual reproduction. This pink-tipped surf anemone is in the process of pulling into two halves, each of which will be a new adult sea anemone.

Sexual reproduction

In sexual reproduction, a new individual is formed by the union of a male and a female gamete. Gametes are produced in the gonads, or sex organs. The testes *(TEHS teez)* produce the male gametes (sperm), and the ovaries produce the female gametes (eggs). Some species of animals, called **hermaphrodites** *(huhr MAF roh deyets),* have both testes and ovaries. Each individual functions as both a male and a female. But a hermaphrodite's sperm and eggs are

internetconnect

TOPIC: Animal reproduction
GO TO: www.scilinks.org
KEYWORD: HX627

Section 28-2

Teaching TIP

Egg Shells and Internal Fertilization

Ask students to hypothesize how a bird egg inside its hard shell can be fertilized. *(Answers may vary.)* Explain that internal fertilization occurs before the shell is formed. After fertilization, as the egg moves through the oviduct, albumin (egg white) is deposited around the egg. Next, a fibrous shell membrane covers the albumin, and finally a hard shell of calcium salts covers the egg. It takes about 24 hours after the egg is released into the oviduct, where it may or may not be fertilized, for the shell to be completed and the egg to be ready for laying.

Visual Strategy

Figure 28-18

Use this photograph to help explain asexual reproduction by splitting. Ask students why splitting into two halves is considered asexual reproduction. *(Only one parent is involved.)* Ask why it is beneficial for a sea anemone to reproduce asexually. *(If the parent is well adapted to its environment, so will both of the offspring.)*

internetconnect

SciLINKS NSTA Have students research animal reproduction using the Web site in the Internet Connect box on page 627. Have students write a report summarizing their findings. PORTFOLIO

did you know?

Some animals have ovaries and testes at different times during their lives.

Some fish, for example, all begin life as females. Only the largest individuals develop into males. If the male in a group dies, the largest surviving female in the group will then become a male. There are other animals that change from male to female during their lives. These sex changes are called sequential hermaphroditism.

Teaching TIP

Fertilization

Tell students that animals that reproduce by external fertilization usually produce a large number of eggs. For example, sea urchins and sea stars release millions of eggs into the ocean. In contrast, animals that reproduce by internal fertilization usually produce far fewer eggs. Ask students to list reasons for this relationship between type of fertilization and number of eggs produced. *(Externally fertilized eggs are vulnerable to predators, and only a small fraction of the offspring will survive. Internal development of the embryo and parental care ensure a higher survival rate among internally fertilized eggs, so fewer are needed.)*

3 Assessment Options

Closure

Body Systems

Write the terms *gastrovascular cavity, digestive tract, nerve net, hydrostatic skeleton, exoskeleton,* and *endoskeleton* on the board. Ask students to define each term and to give examples of animals that have these features. PORTFOLIO

Review

Assign Review—Section 28-2.

Reteaching — BASIC

Have students work in pairs to make a set of flash cards for the boldface words in the chapter. One side of each card should have a word, and the reverse side should have its definition. Ask students to test each other using their flash cards. PORTFOLIO

Figure 28-19
External fertilization

How many? Coral on the Great Barrier Reef, off the coast of Australia, release billions of gametes in a mass spawning that occurs once a year.

usually produced at different times, so self-fertilization does not occur. Among animals that rarely meet members of their own species, hermaphrodites can have a reproductive advantage. A hermaphrodite's chances of reproductive success are increased because it can either fertilize eggs or have its own eggs fertilized. Many simple invertebrates, including slugs and earthworms, and some fishes are hermaphrodites.

When sperm are released during sexual reproduction, they use their flagella to swim toward the egg. Most aquatic animals simply release the male and female gametes near one another in the water, where fertilization occurs. This method is called **external fertilization** because the egg is fertilized outside of the female's body. Often large numbers of gametes are released during external fertilization, but only a small percentage of the resulting fertilized eggs will survive to develop into adults. **Figure 28-19** shows the mass release of gametes by coral polyps on a coral reef.

External fertilization is not practical for animals that live on land because gametes dry out quickly when exposed to air. Therefore, most terrestrial animals reproduce sexually by means of internal fertilization. In **internal fertilization,** the union of the sperm and egg occurs within the female's body. The male places his semen *(SEE muhn),* a fluid containing sperm and fluid secretions, directly into the female's body. In this way, fertilization takes place in a moist environment, and the gametes are protected from drying out.

Figure 28-20 **Sea turtle.** This green sea turtle digs a nest in the sand, where she deposits her eggs.

Once fertilization occurs, developing eggs must be kept moist. This is not a problem for aquatic animals whose eggs are covered by a jellylike coat. The eggs of terrestrial animals need more protection. Their eggs have a shell that protects them from drying out and provides a degree of protection from physical damage. Some animals, such as the sea turtle shown in **Figure 28-20,** place their fertilized eggs in a safe place and leave them. Other animals remain with their eggs to protect them. In most mammals and a few other species, the eggs develop internally, and live young emerge from their mother's body.

Review SECTION 28-2

1. **Summarize** the functions of the six body systems discussed.
2. **Describe** how a gastrovascular cavity differs from a one-way gut.
3. **Compare** open and closed circulatory systems.
4. **Describe** how asexual reproduction differs from sexual reproduction.

Critical Thinking

5. **Justifying Conclusions** Which method of fertilization, external or internal, is more practical for most terrestrial animals? Justify your answer.

Review SECTION 28-2 ANSWERS

1. digestion—breaks down food for absorption; respiration—exchanges oxygen and carbon dioxide; circulation—transports nutrients and oxygen and removes wastes, including carbon dioxide; nervous—carries nerve impulses so an animal can sense and respond to its environment; support—provides a framework to support the body; excretion—removes cellular waste products
2. A gastrovascular cavity has one opening to the exterior environment; a one-way gut has two, a mouth and an anus.
3. Fluid leaves an open circulatory system and washes across the body cells. Fluid (blood) does not leave a closed circulatory system.
4. Asexual reproduction does not involve the fusion of two gametes. An organism produced by asexual reproduction is an exact genetic copy of its single parent. Sexual reproduction involves the union of a sperm and an egg.
5. For most terrestrial animals, internal fertilization is more practical. The sperm, egg, and zygote must remain moist to survive.

Study ZONE CHAPTER 28

Use the Key Concepts and Key Terms listed below to review the main ideas in this chapter. If you need to review the meaning of a term, turn to the page indicated.

Key Concepts

28-1 Animals—Features and Body Plans

- All animals share these general features: heterotrophy, mobility, multicellularity, sexual reproduction, diploidy, the absence of a cell wall, cells organized as tissues, and blastula formation.
- Animals with radial symmetry have body parts arranged around a central axis. Animals with bilateral symmetry have a distinct right and left half, and most display cephalization.
- Animals have one of three basic body plans: acoelomate, pseudocoelomate, and coelomate.
- Segmentation in body structure underlies the organization of all advanced animals.
- There are about 35 animal phyla, which contain an extraordinary range of body forms and body systems.
- Scientists classify animals using several different types of data.
- The animal kingdom is divided in two groups: vertebrates and invertebrates.

28-2 Animal Body Systems

- Body systems are specialized to carry out different tasks.
- Simple animals have a gastrovascular cavity with only one opening, while more-complex animals have a one-way gut.
- Simple animals exchange gases directly through their skin. More-complex aquatic animals use gills, while terrestrial animals use a variety of respiratory organs, such as lungs.
- In an open circulatory system, circulatory fluid leaves the vessels and enters the body cavity. In a closed circulatory system, blood remains in the vessels.
- While simple animals have little coordination among their nerve cells, complex animals have nerve cords and a brain with associated sensory structures.
- For most animals, eliminating wastes is linked to maintaining the correct water balance in their body.
- Asexual reproductive methods include fragmentation, splitting in two, and parthenogenesis. In sexual reproduction, male and female gametes combine to form a new individual.

Key Terms

28-1

blastula (614)
ectoderm (614)
endoderm (614)
mesoderm (614)
body plan (616)
asymmetrical (616)
radial symmetry (616)
bilateral symmetry (616)
cephalization (617)
coelom (618)
acoelomate (618)
pseudocoelomate (618)
coelomate (618)
phylogenetic tree (620)

28-2

gastrovascular cavity (623)
respiration (623)
gill (623)
open circulatory system (624)
closed circulatory system (624)
hydrostatic skeleton (625)
exoskeleton (625)
endoskeleton (626)
hermaphrodite (627)
external fertilization (628)
internal fertilization (628)

Study TIP Ready?

If you think you understand these Key Concepts and Key Terms, then you're ready for the Performance Zone.

Study ZONE CHAPTER 28

Review and Practice

Assign the following worksheets for Chapter 28:

- **Student Review Guide**
- **Concept Mapping Worksheet**
- **Test Preparation Pretest**
- **Vocabulary Worksheet**
- **Critical Thinking Review Worksheet**
- **Science Skills Worksheet**

Assessment Options

The following assessment options are available for this chapter:

Resource Link

Assign **Chapter Test 28** to assess students' understanding of the chapter.

Alternative Assessment

Assign students to cooperative work groups. Have each group write questions and answers for certain pages of the book. Collect the questions and answers and use them to quiz the groups. You can keep students in the same groups or form new ones. PORTFOLIO

Portfolio Assessment

- *Portfolio Assessment* at the front of this book provides options for building and evaluating student portfolios. Students and teachers can select items from the areas listed for inclusion in a portfolio.
- See items labeled PORTFOLIO on pp. 613, 614, 616, 618, 620, 621, 624, 627, 628, 629, and 633.

Answer to Concept Map

The following is one possible answer to Performance Zone item 1 on page 630.

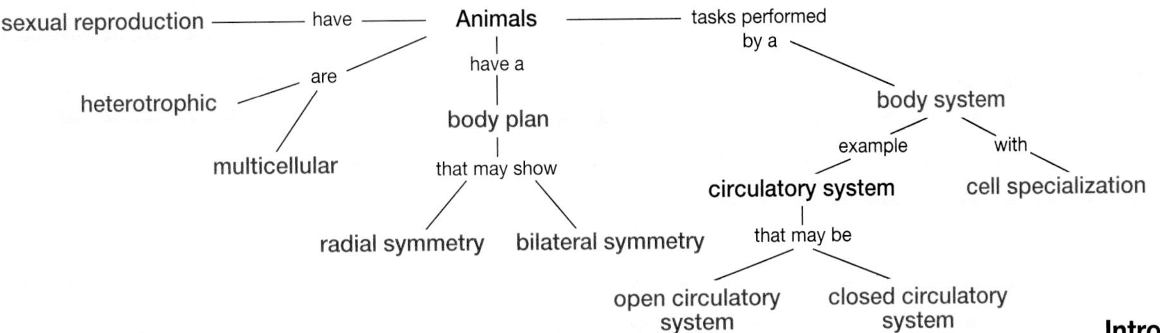

Performance ZONE
CHAPTER 28 REVIEW ANSWERS

Assignment Guide

SECTION	REVIEW
28-1	3, 4, 5, 6, 7, 8, 13, 15, 16, 17, 18, 21, 22, 23, 24, 25, 26, 27, 29
28-2	1, 2, 9, 10, 11, 12, 14, 19, 20, 28

Understanding and Applying Concepts

1. The answer to the concept map is found at the bottom of page 629.
2. **a.** A gastrovascular cavity has one opening to the exterior; a digestive tract has two.

   b. Fluid leaves an open circulatory system and washes across the body cells. Blood does not leave a closed circulatory system.

   c. An endoskeleton is inside an animal's body; an exoskeleton is outside.

   d. During internal fertilization, the male places sperm inside the female. During external fertilization, eggs and sperm are released into the environment, where their union occurs.
3. d
4. a
5. c
6. c
7. c
8. b
9. a
10. c
11. b
12. a
13. Segmentation is the division of the body into repeating, similar units. Segmentation offers evolutionary flexibility because a small change in one segment can produce a new type of segment.
14. Homeostasis is the maintenance of an inner stability. The excretory system helps maintain the balance of salts and water in an animal's body, keeping the concentration of salts stable.

Performance ZONE CHAPTER 28 Review

Understanding and Applying Concepts

1. **Concept Mapping** Construct a concept map that describes the characteristics of animals. Include the following terms in your concept map: heterotrophic, multicellular, cell specialization, body plan, radial symmetry, bilateral symmetry, body system, open circulatory system, closed circulatory system, sexual reproduction. Include additional terms as necessary.
2. **Understanding Vocabulary** For each pair of terms, explain the difference in their meanings.
   a. gastrovascular cavity, digestive tract (gut)
   b. open circulatory system, closed circulatory system
   c. endoskeleton, exoskeleton
   d. internal fertilization, external fertilization
3. Heterotrophic organisms
   a. remain in one place.
   b. make their own food.
   c. are capable of flight.
   d. obtain food from their environment.
4. Organisms that have two copies of each of their chromosomes are called
   a. diploid. c. gametes.
   b. haploid. d. germ cells.
5. Which of the following organisms do not have cells organized into tissues?
   a. hydras c. sponges
   b. sea urchins d. sponges and hydras
6. The developmental process that leads to the formation of tissue layers is
   a. asexual reproduction. c. gastrulation.
   b. evolution. d. fertilization.
7. An animal's body plan includes all of the following except
   a. internal organization. c. size.
   b. shape. d. symmetry.
8. The presence of a true body cavity (coelom) allows
   a. direct exchange of oxygen and carbon dioxide with the environment.
   b. specialization of the gut.
   c. cephalization.
   d. bilateral symmetry.
9. Gas exchange in many aquatic phyla takes place in the
   a. gills. c. kidneys.
   b. gut. d. lungs.
10. Which of the following does not have nerve cells?
    a. hydra c. sponge
    b. jellyfish d. flatworm
11. Skeletal systems provide all of the following except
    a. protection for an animal's soft parts.
    b. an attachment point for organ systems.
    c. a framework for supporting the body.
    d. a framework for muscles to pull against.
12. Which of the following is not true of kidneys?
    a. They produce dilute urine.
    b. They remove wastes from the bloodstream.
    c. They produce concentrated urine.
    d. They help balance the body's water content.
13. **Theme Evolution** What is segmentation? How does segmentation provide for evolutionary flexibility?
14. **Theme Homeostasis** Explain how the excretory system of an animal helps maintain homeostasis.
15. **Exploring Further** Summarize the process of gastrulation, relating it to the formation of tissue layers and the development of tissues and organs.
16. **Chapter Links** Explain how the terms *diploid* and *haploid* apply to what you have learned in this chapter. (**Hint:** See Chapter 6, Section 6-1.)

15. Gastrulation is the process in which the blastula folds inward and cells move or change in size. It leads to the formation of three primary tissue (germ) layers. Each layer gives rise to cells that become a part of specific tissues and organs.

16. All of the cells in an animal's body are diploid (have two sets of chromosomes) except for the sex cells (gametes), which are haploid (have one set).

Skills Assessment

Part 1: Interpreting Graphics

Observe the figure of the animal shown below. Then answer the following questions:

17. What kind of symmetry does the animal display?
18. Does the animal display cephalization?
19. How many openings does its digestive system have?
20. What is the function of the arrows in the illustration?
21. Compare the figure to the "Late gastrula" stage of the sea urchin on page 615. What primary tissue layer do you think yellow cells in the figure above arose from? Support your answer.

Part 2: Life/Work Skills

Use the media center or Internet resources to learn what role public aquariums play in educating the public about aquatic life.

22. **Allocating Resources** You are in charge of a city-owned aquarium, and a person has given your facility a $1,000,000 gift. The donor wants the money to be used to educate the public about the need to protect freshwater and marine species. Explain how you would spend the donation, and justify your proposed expenditures.
23. **Communicating** Develop a poster presentation designed to convince the donor and the board of directors of the merits of your plan. Present your proposed expenditures to your class.

Critical Thinking

24. **Forming Reasoned Opinions** Radially symmetric animals, such as hydra, are not found on land. However, bilaterally symmetrical animals live on land and in water. Propose a hypothesis to explain why radially symmetrical animals are best suited to aquatic life.
25. **Evaluating Hypotheses** Defend the position that cephalization gives terrestrial animals an advantage as they seek food.

Portfolio Projects

26. **Interpreting Information** Read "The Deep-Sea Floor Rivals the Rain Forests in Diversity of Life," by Cheryl Lyn Dybas (*Smithsonian*, Jan. 1996, pp. 96–106). Write a report that explains what "marine snow" is and how it supports the organisms that live on the deep-sea floor. Include examples in your report.
27. **Forming a Model** Use three different colors of modeling clay or any other material to construct models that show the formation of the primary tissue layers—endoderm, ectoderm, and mesoderm—during gastrulation. Alternately, make models that show the three major body plans—acoelomate, pseudocoelomate, and coelomate. Present your models to your class, and summarize the features shown on them.
28. **Summarizing Information** Do research to find out more about how male bees develop by parthenogenesis. How does their development differ from that of female worker bees? Summarize your findings in a written report.
29. **Career Focus Zookeeper** Zookeepers are responsible for the day-to-day care of animals in their charge. Research a zookeeper's responsibilities, and write a report on your findings. Your report should include a job description, training required, kinds of employers, growth prospects, and starting salary.

Portfolio Projects

26. Marine snow is all of the living and dead aquatic organisms that fall to the bottom of the ocean. So much of this material falls that it is referred to as snow. Marine snow supports large populations of organisms that live deep on the ocean floor.
27. Students' models should clearly show the primary tissue layers in their correct locations.
28. Students' reports should include information about differences in size, stingers, and the ability to feed themselves between drones (male bees) and worker bees.
29. Zookeepers feed animals, clean their cages, and are responsible for bathing and exercising some animals. Often a zookeeper will specialize in caring for certain types of animals, such as big cats or hoofed mammals. The zookeeper understands the behavior of the animals and is usually the first to notice any sign of sickness. Training includes a four-year college degree with training in animal science, zoology, marine biology, conservation biology, wildlife management, or animal behavior preferred. Employers include zoos and aquariums. Growth prospects are fair. Starting salary will vary by region.

Skills Assessment

17. radial
18. no
19. one
20. They show that food enters and undigested material exits through the same opening.
21. They arose from the endoderm. The endoderm gives rise to tissues that line the digestive system.
22. Answers will vary. Accept any answers that show a thoughtful consideration of the use of funds.
23. Posters and presentations will vary. They should support the proposed use of funds that the speaker presents.

Critical Thinking

24. Radial symmetry is beneficial for sessile animals or animals that float freely in water because food may come from any direction. Bilateral symmetry is better suited to terrestrial animals that must forage or hunt for food.
25. Cephalization is the concentration of sensory organs in the front end of an animal. Terrestrial animals tend to move through their environment head first. Having more sensory organs in the front end permits them to seek food and sense that they may be approaching danger.

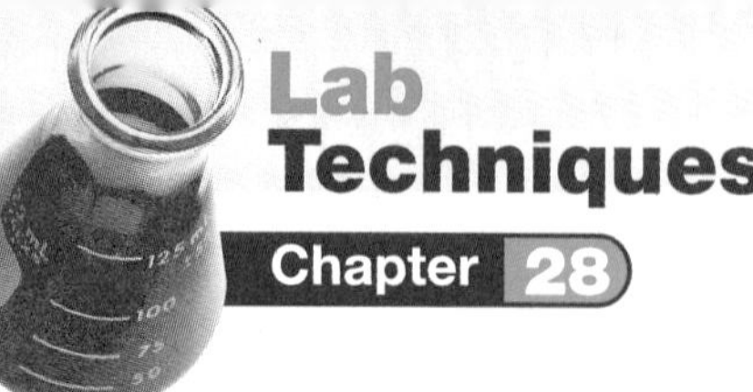

Lab Techniques

Chapter 28

Surveying Invertebrate Diversity

Purpose

Students will examine representatives of eight phyla of animals.

Preparation Notes

Time Required: Two 45-minute periods

Materials

Materials for this lab can be ordered from WARD'S. See *Master Materials List* at the front of this book for catalog numbers.

Preparation Tips

- Have students read pp. 632–633 before they come to class.
- Have materials on the lab tables when students arrive.

Safety Precautions

- Caution students to use extreme care when observing the preserved specimens. The preservative can leak if the jars are tilted. Since some animals may be preserved in glass jars, students should be careful not to break the jars. Tell students to wash their hands if they come into contact with the preservative. Also remind students to wash their hands before leaving the laboratory.

Procedural Tips

- Students should work in groups of two to four and move clockwise from station to station.
- Before class, have students make their data table and an answer sheet for the questions at each station. This will make it easier for students to answer the questions, because not all students will start with Station 1.

Before You Begin

Answers

1. *invertebrates*—animals without backbones

   body plan—the shape, symmetry, and internal organization of the body

   radial symmetry—symmetry in all directions about a central point

   bilateral symmetry—two-sided symmetry

   dorsal—the top side

   ventral—the bottom side

   anterior—the front end

   posterior—the rear end

   cephalization—the anterior concentration of sensory structures and nerves

   segmentation—the division of the body into repeating, similar units

2. An acoelomate has no body cavity. A pseudocoelomate has a body cavity between the mesoderm and the endoderm. A coelomate has a body cavity in the mesoderm.

4. Students' questions will vary. For example: What characteristics are shared by all invertebrates?

Lab Techniques

Chapter 28 *Surveying Invertebrate Diversity*

SKILLS

- Observing
- Comparing

OBJECTIVES

- **Observe** the similarities and differences among groups of invertebrates.
- **Relate** the structural adaptations of invertebrates to their evolution.

MATERIALS

- safety goggles
- lab apron
- preserved or living specimens of invertebrates
- prepared slides of sponges, hydras, planarians, and nematodes
- compound microscopes
- hand lenses or stereomicroscopes
- probes

ChemSafety

 CAUTION: Always wear safety goggles and a lab apron to protect your eyes and clothing.

CAUTION: Do not touch or taste any chemicals. Know the location of the emergency shower and eyewash station and how to use them. If you get a chemical on your skin or clothing, wash it off at the sink while calling to the teacher. Notify the teacher of a spill. Spills should be cleaned up promptly, according to your teacher's directions.

CAUTION: Glassware is fragile. Notify the teacher of broken glass or cuts. Do not clean up broken glass or spills with broken glass unless the teacher tells you to do so.

Before You Begin

Invertebrates include all animals except those with backbones. Every phylum of the kingdom Animalia except the phylum Chordata consists only of invertebrates. In this lab, you will examine representatives of eight phyla of animals. You will see many similarities and differences in **body plan**—shape, symmetry, and internal organization.

1. Write a definition for each boldface term in the paragraph above and for the following terms: radial symmetry, bilateral symmetry, dorsal, ventral, anterior, posterior, cephalization, segmentation.
2. Describe the three basic body plans found in animals.
3. Make a data table similar to the one on the right to record observations about invertebrates.
4. Based on the objectives for this lab, write a question you would like to explore about the characteristics of invertebrates.

DATA TABLE

Phylum	Symmetry	Body plan	Other	Examples

Procedure

PART A: Conducting a Survey

1.  Put on safety goggles and a lab apron.
2. Visit each invertebrate station, and examine the specimens there. Answer the questions, and record observations in your data table.

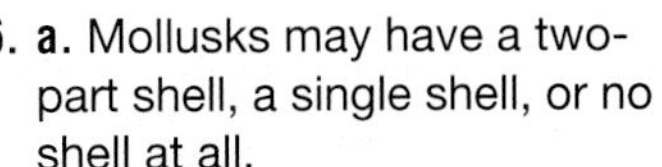

3. **Sponges** Examine each specimen.
   a. Describe the shape of a sponge.
   b. What do you think is the role of the many holes, or pores, in a sponge?
   c. Examine a prepared slide of a sponge with a compound microscope. What do you notice about the organization of the cells in sponges?
4. **Cnidarians** Examine each specimen.
   a. Divide the cnidarian specimens into two groups. What feature did you use to make your division?
   b. How many body openings does a cnidarian have?
   c. Examine a prepared slide of a hydra. What do you notice about the organization of the cells in cnidarians?
5. **Flatworms and Roundworms** Examine each specimen.
   a. How does a flatworm differ from a roundworm in external appearance?
   b. Do any of the worms appear to be segmented? Explain.
   c. Examine prepared slides of planarians and nematodes. How many body openings does each have?
6. **Mollusks** Examine each specimen.
   a. In what ways do the mollusks differ in external appearance?
   b. Which group of mollusks has the most noticeable "feet"?
7. **Annelids** Examine each specimen.
   a. How are an earthworm and a leech similar? How are they different?
   b. Describe any differences you see in the segments of the annelid worm.
8. **Arthropods** Examine each specimen.
   a. What characteristic do you observe in all arthropod appendages?
   b. How does the number of walking legs differ among these arthropods?
9. **Echinoderms** Examine each specimen.
   a. The word *echinoderm* means "spiny skin." Why is this name appropriate?
   b. What does an echinoderm's ventral surface look like?

PART B: Cleanup and Disposal

10. Dispose of broken glass and solutions in the designated waste containers. Do not pour chemicals down the drain or put lab materials in the trash unless your teacher tells you to do so.
11. Wash your hands thoroughly before you leave the lab and after you finish all work.

Analyze and Conclude

1. **Summarizing Data** Which animal phyla show cephalization, and which do not?
2. **Recognizing Patterns** What type of symmetry is found with cephalization?
3. **Recognizing Patterns** What characteristics do annelids and arthropods share?
4. **Analyzing Methods** Were you able to identify the type of body plan found in all of the specimens? Explain.
5. **Further Inquiry** Write a new question about invertebrates that could be explored with another investigation.

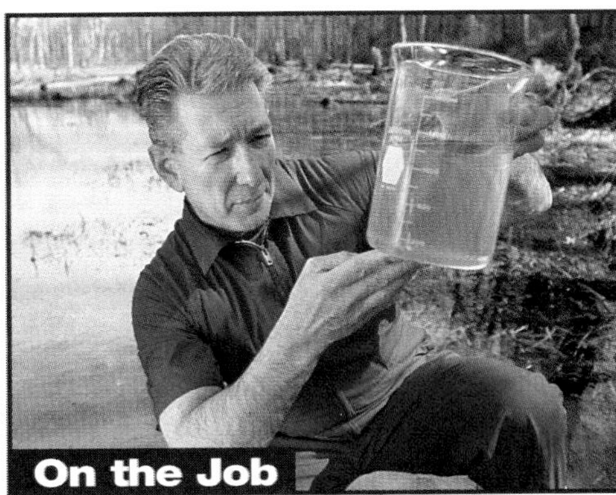

On the Job

Parasitology is the study of parasites. Do research in the library or media center or on the Internet to learn how parasitologists help to protect you and your pets from diseases caused by parasites. Summarize your findings in a written or oral report.

Procedure

Answers

3. a. Sponges have an irregular shape; they are asymmetrical.
   b. The pores allow water, which carries in food particles and oxygen and carries out wastes, to flow through the body of a sponge.
   c. The cells in a sponge are of several types that are loosely arranged in several layers but are not organized into tissues.
4. a. Body shape. There are polyps and medusas.
   b. one
   c. The cells of the hydra are of several types and are packed into two tissue layers.
5. a. Flatworms are flat and vary in length. Roundworms have long, tubular bodies.
   b. If students view planarians and flukes, they will respond "no." Students may respond "yes" if they view a tapeworm because its body looks somewhat segmented. Explain that only coelomates show true segmentation. Tapeworms do have a series of similar sections, but this is different from true segmentation.
   c. Planarians have one; nematodes have two.
6. a. Mollusks may have a two-part shell, a single shell, or no shell at all.
   b. Cephalopods—the feet are modified as several long tentacles.
7. a. Both have bilateral symmetry and a body divided into many segments. Earthworms are tubular and much longer than they are wide; leeches are flattened and not nearly as long.
   b. Segments at the anterior and posterior ends are smaller than those in the middle.
8. a. They are jointed.
   b. Millipedes have two pairs of legs per body segment, centipedes have one pair of legs per body segment, insects have a total of three pairs, spiders have four, and decapod crustaceans have five.
9. a. Most echinoderms are covered by numerous spines that grow out of the skin.
   b. The ventral surface of an echinoderm has several grooves in which there are numerous tube feet.

Analyze and Conclude

Answers

1. All but the sponges, cnidarians, and echinoderms exhibit cephalization.
2. bilateral symmetry
3. segmentation, bilateral symmetry, and cephalization
4. No, body plans cannot be determined by examining an animal's exterior.
5. Answers will vary. For example: How does an arthropod's outer covering differ from a worm's?

On the Job

Answer

Answers will vary. Parasitologists study the life cycles of parasites and search for medicines to treat diseases caused by parasites.

Chapter 29

Simple Invertebrates

Planning Guide

	OBJECTIVES	MEETING INDIVIDUAL NEEDS	READING SKILLS AND STRATEGIES
BLOCK 1 45 min	**29-1 Sponges** • **Summarize** the general features of sponges. • **Describe** how sponge cells receive nutrients. • **Describe** how a sponge's body is structurally supported. • **Distinguish** between sexual and asexual reproduction in sponges.	Learning Styles, p. 636 BASIC ADVANCED • pp. 637, 639	**ATE** Active Reading, Technique: K-W-L, p. 635 **ATE** Active Reading, Technique: Paired Summarizing, p. 638 ■ Active Reading Guide, worksheet (29-1) ■ Directed Reading, worksheet (29-1) ■ Reading Strategies, K-W-L worksheet
BLOCKS 2 & 3 90 min	**29-2 Cnidarians** • **Describe** the two cnidarian body forms. • **Summarize** how cnidocytes function. • **Summarize** the life cycle of *Obelia.* • **Compare** three classes of cnidarians. • **Compare** asexual and sexual reproduction in cnidarians.	Learning Styles, p. 640 BASIC ADVANCED • pp. 642, 644, 645, 646	■ Active Reading Guide, worksheet (29-2) ■ Directed Reading, worksheet (29-2) ■ Reading Strategies, Compare/Contrast Matrix
BLOCKS 4 & 5 90 min	**29-3 Flatworms and Roundworms** • **Compare** the three classes of flatworms. • **Summarize** the life cycle of a blood fluke. • **Describe** the body plan of a roundworm. • **Summarize** the life cycle of the roundworm *Ascaris.*	Learning Styles, p. 647 BASIC ADVANCED • pp. 651, 652	■ Active Reading Guide, worksheet (29-3) ■ Directed Reading, worksheet (29-3) ■ Supplemental Reading Guide, *Journey to the Ants: A Story of Scientific Exploration*

Review and Assessment Resources

BLOCKS 6 & 7 90 min

TEST PREP

- **PE** Study Zone, p. 653
- **PE** Performance Zone, pp. 654–655
- ■ Chapter 29 worksheets:
  - Vocabulary
  - Concept Mapping
  - Critical Thinking
- Audio CD Program
- ■ Test Prep Pretest

CHAPTER TESTING

- ■ Chapter Tests (blackline master)
- ▲ Test Generator
- ■ Assessment Item Listing

See the correlations table at the front of this book for a list of the ***National Science Education Standards*** addressed in this chapter.

Smithsonian Institution® Visit **www.si.edu/hrw** for additional on-line resources.

KEY

Biology Principles and Explorations

PE Pupil's Edition

ATE Teacher's Edition

HOLT BIOSOURCES

Teaching Transparencies

Laboratory Program

One-Stop Planner CD-ROM

Shows these resources within a customizable daily lesson plan:

■ Teaching Resources

▲ Test Generator

VISUAL STRATEGIES	CLASSWORK AND HOMEWORK	TECHNOLOGY AND INTERNET RESOURCES
ATE Fig. 29-2, p. 637 144 Structure of a Sponge 145 Sexual Reproduction of Sponges	**PE** Review, p. 639	Holt Biology Videodiscs Section 29-1 Audio CD Program Section 29-1
ATE Fig. 29-8, p. 642 **ATE** Fig. 29-10, p. 643 **ATE** Fig. 29-11, p. 644 147 Sexual Reproduction of *Obelia*  153 Exploration of a Cnidarian 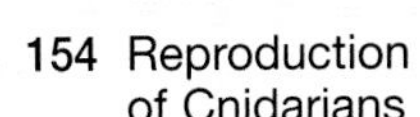 154 Reproduction of Cnidarians 155 Development of a Cnidarian Embryo	**PE** Math Lab Using the microscope to estimate size, p. 641 ■ Basic Skills: Microscope Magnification **PE** Review, p. 646	CNN Video Segment 24 Year of the Reef Holt Biology Videodiscs Section 29-2 Audio CD Program Section 29-2 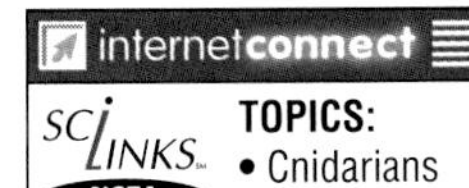 internetconnect SCILINKS NSTA TOPICS: • Cnidarians • Coral reefs
ATE Fig. 29-16, p. 650  149 Life Cycle of *Schistosoma*  150 Life Cycle of Beef Tapeworm  156 Development of Flatworm Embryo  157 Exploration of a Flatworm	**PE** Quick Lab How does a planarian locate food sources?, p. 649 **PE** Data Lab Identifying parasites, p. 652 **PE** Experimental Design How Do Hydras Respond to Stimuli?, pp. 656–657 Lab B21 Flatworm Behavior **PE** Review, p. 652 **PE** Study Zone, p. 653	Holt Biology Videodiscs Section 29-3 Audio CD Program Section 29-3  internetconnect SCILINKS NSTA TOPICS: • Flukes • Roundworms • Hydra

ALTERNATIVE ASSESSMENT

PE Portfolio Projects 25–28, p. 655

ATE Alternative Assessment, p. 653

ATE Portfolio Assessment, p. 653

Lab Activity Materials

Math Lab
p. 641 metric ruler, compound microscope or a dissecting microscope, prepared slide of a medusa or polyp

Quick Lab
p. 649 eyedropper, *Planaria* culture, Petri dish, pond water, hand lens or dissecting microscope, forceps, and 3–7 cm raw liver

Data Lab
p. 652 pencil, paper

Experimental Design
pp. 656–657 silicone culture gum, microscope slide, medicine droppers (2), *Hydra* culture, *Daphnia* culture, concentrated beef broth, filter paper, forceps, stereomicroscope

Lesson Warm-up

ATE photos of several cnidarians, p. 640

Demonstrations

ATE several natural sponges, several synthetic sponges, p. 634

ATE mixture of pebbles, water, spaghetti strainer, p. 637

BIOSOURCES LAB PROGRAM

- Inquiry Skills Development: Lab B21, p. 97

Chapter Theme

Evolution

Although the simple invertebrates introduced in this chapter are not as complex as many other animals, they are more complex than their name implies. Simple invertebrates have evolved to occupy a broad array of lifestyles, and their body plans and behaviors reflect this evolution.

Establishing Prior Knowledge

Before beginning this chapter, make certain students can answer the questions in **Study TIP Ready?** on page 635. If they cannot, encourage them to reread the sections indicated.

Opening Demonstration

Have several natural sponge skeletons and several synthetic sponges available for students to examine. Emphasize that observation is an important part of scientific inquiry. Ask students to make a list of how the natural and synthetic sponges are alike and how they are different. *(For example, natural sponges are asymmetrical, while synthetic sponges are symmetrical.)* Tell students that although sponges and the other organisms they will learn about in this chapter may not fit many people's image of an animal, they are all members of kingdom Animalia.

CHAPTER 29 Simple Invertebrates

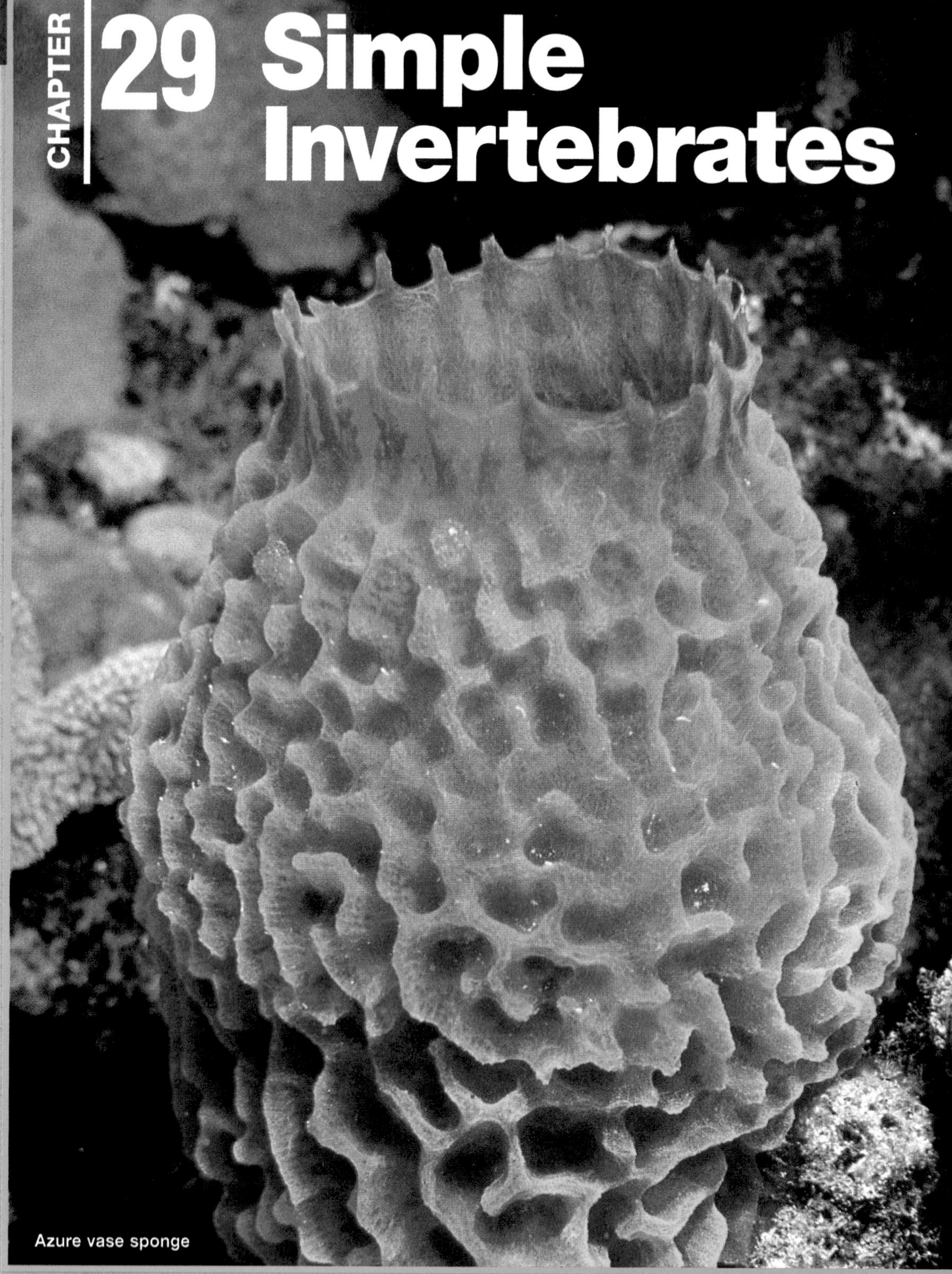

Azure vase sponge

What's Ahead — and Why
CHAPTER OUTLINE

Students will learn

- about the characteristics and evolutionary significance of sponges.
- about the similarities and differences between the three classes of cnidarians.
- about the characteristics of flatworms and roundworms.

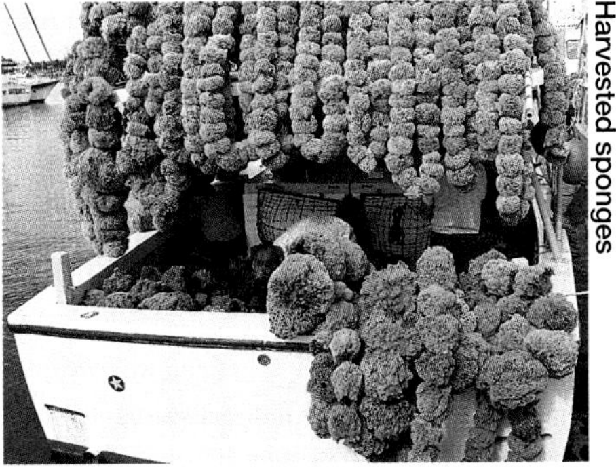
Harvested sponges

Since ancient times, *people have used sponges for a variety of functions. The ancient Greeks placed sponges in their helmets for padding, and the Romans used them for paintbrushes and mops. While people today often use manufactured, or synthetic, sponges, natural sponges are still the best choice for many tasks.*

Study TIP

Ready?

Check your understanding of these topics to help you prepare for what's ahead.

Can you...

- **describe** the process of osmosis? (Section 4-1)
- **identify** the organisms collectively called plankton? (Sections 18-3, 22-2)
- **define** the term *colonial organization*? (Section 20-1)
- **distinguish** between ectoderm, mesoderm, and endoderm? (Section 28-1)
- **distinguish** between acoelomates, pseudocoelomates, and coelomates? (Section 28-1)

If not, *review the sections indicated.*

Looking Ahead

internet connect

National Science Teachers Association *sci*LINKS Internet resources are located throughout this chapter.

Reading Strategies

Active Reading

Technique: K-W-L

Before students read the chapter, display a list of the following phyla: sponges, cnidarians, flatworms, and roundworms. Have students make a list of everything they already **K**now (or think they know) about the members of each phylum. Have them work together to form a class list. Then have students list things they **W**ant to know about the simple invertebrates in these phyla. Have students save their lists for use later in the **Closure** on page 652.

Directed Reading

Assign **Directed Reading Worksheet Chapter 29** to help students interact with the text as they read the chapter.

Interactive Reading

Assign ***Biology: Principles and Explorations*** **Audio CD Program Chapter 29** to help reluctant readers achieve greater success in reading the chapter.

Reading Skills

Assign Reading Organizers and Reading Strategies Worksheets from the **One-Stop Planner CD-ROM with Test Generator** to help students learn and practice effective reading skills.

internet connect

Holt *Biology: Principles and Explorations*

On-line Resources: go.hrw.com

Visit the HRW Web site for a variety of resources related to this chapter. Just type in the keyword HX0 Chapter 29.

Holt *Biology: Principles and Explorations*

On-line Resources: www.scilinks.org

The following *sci*LINKS Internet resources can be found in the student text for this chapter:

Page 640 **TOPIC:** Cnidarians **KEYWORD:** HX640

Page 646 **TOPIC:** Coral reefs **KEYWORD:** HX646

Page 650 **TOPIC:** Flukes **KEYWORD:** HX650

Page 652 **TOPIC:** Roundworms **KEYWORD:** HX652

Page 657 **TOPIC:** Hydra **KEYWORD:** HX657

1 Prepare

Scheduling

- **Block:** About 45 minutes will be needed to complete this section.
- **Traditional:** About one class period will be needed to complete this section.
- **Planning:** See the Planning Guide on pp. 633A and 633B for lecture, classwork, and assignment options for Section 29-1. Lesson plans for Section 29-1 can be found in a lesson cycle format in ***Biology: Principles and Explorations*** **One-Stop Planner CD-ROM with Test Generator.**

Resource Link

Assign **Active Reading Worksheet Section 29-1** to help students comprehend the key concepts and visuals in this lesson.

2 Teach

Lesson Warm-up

Challenge students to think about different structures for different lifestyles. Ask them how an animal that cannot move is able to feed and protect itself. Have them imagine that they are stuck on the floor and cannot ever move again. How will they get food? How will they protect themselves from large mobile predators? *(Answers may vary. Some students might say that they would just wait until food came close enough, and then they would reach out and grab them. For protection they might crawl up into a ball if a predator comes too close. This would serve to protect their head, which contains the brain and important sense organs.)*

Section 29-1 Sponges

Objectives

- **Summarize** the general features of sponges. (pp. 636–637)
- **Describe** how sponge cells receive nutrients. (p. 637)
- **Describe** how a sponge's body is structurally supported. (p. 638)
- **Distinguish** between sexual and asexual reproduction in sponges. (p. 639)

Key Terms

ostia
oscula
sessile
choanocyte
amoebocyte
spongin
spicule
gemmule

Sponges Are the Simplest Animals

Sponges are so unlike other animals that early naturalists classified them as plants. It wasn't until the mid-1800s that scientists using improved microscope technology began studying sponges closely. Scientists then realized that sponges are animals. The bodies of most sponges completely lack symmetry and consist of little more than masses of specialized cells embedded in a gel-like substance called mesoglea *(mehs oh GLEE uh)*. You could say that a sponge's body is somewhat like chopped fruit in gelatin. The chopped fruit represents the specialized cells, and the gelatin represents the mesoglea.

Sponge cells are not organized into tissues and organs. However, they do have a key property of all animal cells—cell recognition. The fact that sponge cells can recognize other sponge cells can be demonstrated with a simple lab experiment. A living sponge is passed through a fine silk mesh, causing the individual cells to separate. On the other side of the mesh, the individual sponge cells will recombine to form a new sponge.

Sponges have a body wall penetrated by tiny openings, or pores, called **ostia** *(AHS tee uh)*, through which water enters. The name of the phylum, Porifera, refers to this system of pores. Sponges also have larger openings, or **oscula,** through which water exits. You can see the many oscula of the sponge in **Figure 29-1.** Sponges are also **sessile** *(SEHS eyel)*. Early in their lives, sponges attach themselves firmly to the sea bottom or some other submerged surface, like a rock or coral reef. They remain there for life. Sponges can have a diameter as small as 1 cm (0.4 in.) or as large as 2 m (6.6 ft).

Figure 29-1
Sponge. Sponges pull water through their body wall through tiny openings (ostia). The water is expelled through larger openings (oscula), seen in the sponge at right.

Most sponges are bag-shaped and have a large internal cavity. One or more oscula (singular, osculum) are

Evolutionary Milestone

1 Multicellularity

The bodies of all animals, including sponges (phylum Porifera), are multicellular—made of many cells. Although the sponge is composed of several different cell types, these cells show only a small degree of coordination with each other.

Integrating Different Learning Styles

Logical/Mathematical	**PE**	Performance Zone items 12 and 23, pp. 654–655
	ATE	Cross-Disciplinary Connection, p. 637
Visual/Spatial	**PE**	Real Life, p. 638; Performance Zone item 27, p. 655
	ATE	Demonstration, p. 637; Visual Strategy, p. 637; Closure, p. 639
Interpersonal	**ATE**	Active Reading, p. 638
Verbal/Linguistic	**PE**	Study Tip, p. 639; Performance Zone item 25, p. 655

Figure 29-2 Sponge interior

Water enters the sponge through many small pores (ostia) in its body wall and exits through the osculum, an opening at the top of the sponge.

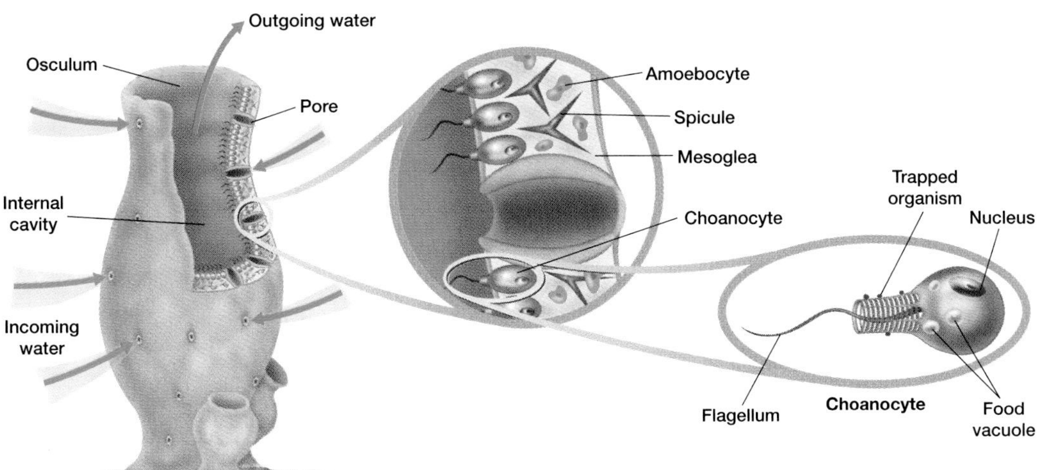

located in the top of the body wall, as shown in **Figure 29-2.** Lining the internal cavity of a sponge is a layer of flagellated cells called **choanocytes** *(koh AN oh seyets),* or collar cells. The flagella of these cells extend into the body cavity. As the flagella beat, water is drawn in through the pores in the body wall. The water is driven through the body cavity before it exits through the osculum.

As sea water passes through the sponge's body cavity, the collar cells function as sieves. These cells trap plankton and other tiny organisms in the small hairlike projections on the collar. The trapped organisms are then pulled into the interior of the collar cells, where they are digested intracellularly (within the cell). As sea water leaves the sponge, wastes are carried away in it.

How do the other sponge cells, such as those in the body wall, survive if the collar cells take in all of the food? The collar cells release nutrients into the mesoglea where other specialized cells, called amoebocytes *(uh MEE boh seyets),* pick up the nutrients. **Amoebocytes** are sponge cells that have irregular amoeba-like shapes. They move about the mesoglea, supplying the rest of the sponge's cells with nutrients and carrying away their wastes.

Protist ancestors

The choanocytes of sponges very closely resemble a kind of protist called a choanoflagellate, shown in **Figure 29-3.** Ancient choanoflagellates are thought by most scientists to be the ancestors of sponges. Other free-swimming colonial flagellates closely resemble sponge larvae, however, and some scientists believe these other flagellates are the true ancestors of sponges.

biobit

Do sponges move?

Time-lapse photographs of several species of sponges indicate that sponges do move, but not very fast—just a few millimeters a day. Some cells of the sponge form a protrusion and "crawl" along like amoebas, pulling the rest of the sponge with them.

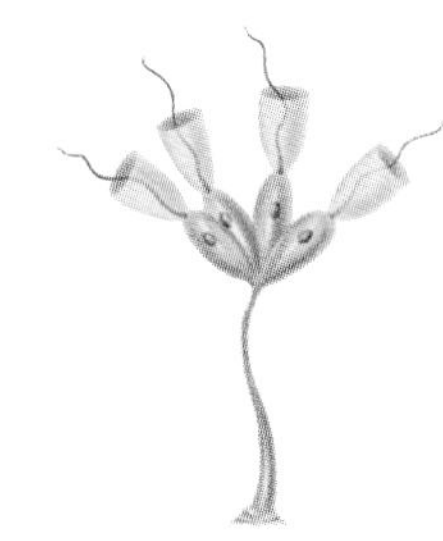

Figure 29-3
Choanoflagellate. Ancient choanoflagellates similar to the one shown above may be the ancestors of sponges.

did you know?

In 1994, scientists discovered a carnivorous sponge living in a marine cave near France.

The previously unknown species captures small crustaceans and other animals. It traps its prey using filaments covered with spikes of silica. Small animals become trapped in the filaments. Then certain of the sponge's cells migrate to the prey, envelope it, and slowly digest it.

Demonstration

Choanocytes in Your Kitchen — BASIC

Pour a mixture of pebbles and water through a spaghetti strainer, trapping the pebbles inside the strainer. Show the pebbles to your students, and tell them to imagine that the pebbles are food particles. Explain that the collars of the choanocytes that line the inside of the sponge trap food particles in water in a similar manner.

Visual Strategy

Figure 29-2

Work with students to help them understand the relationship of the two close-ups to the whole sponge. Point out that in the close-up on the left, the flagella of the collar cells are extending into the internal cavity of the sponge. The close-up on the right shows one collar cell.

CROSS-DISCIPLINARY CONNECTION

◆ Mathematics

Tell students that a sponge that is 10 cm tall and 1 cm in diameter pumps 22.5 L of water through its body every day. To give students an idea of how much water this is, have them make a life-sized clay model of the sponge and display it next to a 3 L soda bottle. Ask students to calculate the number of bottles it would take to hold the amount of water circulated by the sponge in one day. *(22.5 L ÷ 3 L per bottle = 7.5 bottles)* Then ask students to calculate the amount of water that the sponge circulates in a year. *(22.5 L per day × 365 days = 8,212.5 L per year)* How many bottles would it take to hold that amount of water? *(8,212.5 L ÷ 3 L per bottle = 2,737.5 bottles)* Have students display their calculations with their sponge models. PORTFOLIO

Chapter 29

REAL LIFE

Answer

Luffa sponges have a more symmetrical structure, exhibiting an almost radial symmetry, and are cylindrical. True sponges, however, are usually asymmetrical and globular.

Active Reading

Technique: Paired Summarizing

Have students read the information on sponge skeletons to themselves. Assign partners, and have one student summarize aloud what has been read while the other student listens. The listener should point out inaccuracies and add ideas that the other student left out. Have students alternate roles and repeat this process. Students should work together during a final clarification process.

Teaching TIP

Know Those Animals

On the chalkboard draw a chart similar to the Graphic Organizer on page 638. Leave the *Lifestyles* and the *Structures/functions* columns blank. Ask volunteers to come up and fill in the empty columns. List at least three major structures and their functions. Now extend your chart to create three new rows labeled *Cnidarians*, *Flatworms*, and *Roundworms*. Tell students that they will fill in these rows as they learn about the simple invertebrates in Sections 29-2 and 29-3.

REAL LIFE

What is a luffa sponge?

A luffa sponge really isn't a sponge at all but a gourd. When dried, the fibrous material found in the gourd forms a "skeleton" similar to that of some sponges, and it can be used for many of the same purposes.

Comparing Structures

Obtain a natural sponge and a luffa sponge, and compare the nature of their "skeletons."

Sponges Are a Diverse Phylum

As any snorkler can tell you, brilliantly colored sponges abound in warm, shallow sea waters. Other marine sponges live at great depths, and a few species even live in fresh water. Rather than being a simple baglike shape, the body wall of some sponges, such as the azure vase sponge on page 634, may contain hundreds of folds that are sometimes visible as fingerlike projections. These folds allow sponges to increase in size and surface area.

Sponge skeletons

To prevent the sponge from collapsing in on itself, the sponge body is supported by a skeleton. However, a skeleton does not have a fixed framework like a human skeleton does. Instead, the skeletons of most sponges are made of a resilient, flexible protein fiber called **spongin.** A few sponges have skeletons made of tiny needles of silica or calcium carbonate called **spicules.** Some sponges contain both spongin and spicules. These supporting structures are found throughout the mesoglea. Taxonomists group sponges into three classes based on the composition of their skeletons. Calcareous sponges have spicules composed of calcium carbonate. Glass sponges have spicules made of silica. Demosponges contain spongin. In some species the spongin is reinforced with spicules made of silica. The three classes of sponges are represented in Figure 29-4.

Figure 29-4 Three types of sponges

Sponges have skeletons made of spicules, spongin, or both.

Calcareous sponge

Magnification: 240×

Glass sponge

Magnification: 20×

Demosponge

Magnification: 15×

Graphic Organizer

*Use this graphic organizer with **Teaching Tip: Know Those Animals** on page 638.*

Animal groups	Lifestyles	Structures/functions
Sponges	Sessile filter feeders	Collar cells/filter food & intracellular wastes Spicules/support & protection Spongin/support

Sponges Reproduce Both Asexually and Sexually

A remarkable property of sponges is that they regenerate when they are cut into pieces. Each bit of sponge, however small, will grow into a complete new sponge. As you might suspect, sponges frequently reproduce by simply breaking off fragments, each of which develops into a new individual. Sponges also reproduce by budding. Another form of asexual reproduction occurs in some freshwater sponges. When living conditions become harsh (cold or very dry), some freshwater sponges ensure their survival by forming **gemmules** *(JEHM yools),* clusters of amoebocytes encased in protective coats. Sealed in with ample food, the cells survive even if the rest of the sponge dies. When conditions improve, the cells grow into a new sponge.

Sexual reproduction is also common among sponges. Most sponges are hermaphrodites, meaning they produce both eggs and sperm. Since eggs and sperm are produced at different times, self-fertilization is avoided. In most species of sponges, sperm cells from one sponge enter another sponge through its pores, as shown in **Figure 29-5.** Collar cells on the receiving sponge's interior pass the sperm into the mesoglea, where the egg cells reside, and fertilization occurs. The fertilized eggs develop into larvae and leave the sponge. After a brief free-swimming stage, the larvae attach themselves to an object and develop into new sponges.

Study TIP

Reviewing Information

Prepare flash cards for each of the **Key Terms** in this chapter. Use the cards to review the meaning of each term.

Figure 29-5 **Sexual reproduction in sponges**

In most species of sponges, sperm from one sponge fertilize eggs from another sponge.

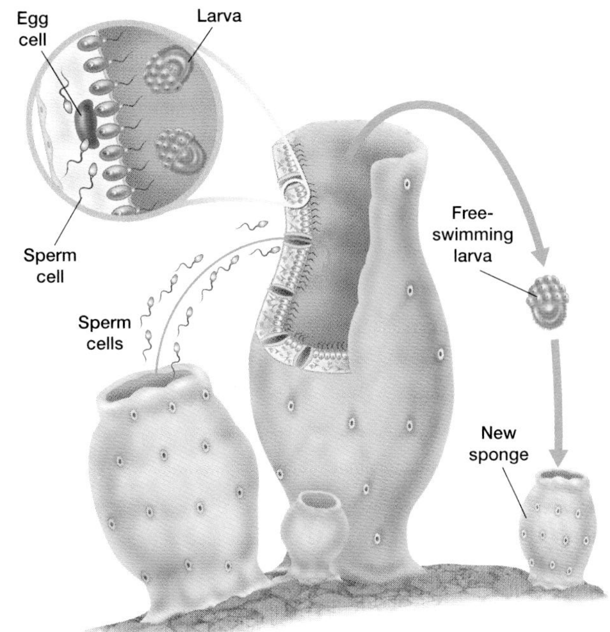

Review SECTION 29-1

1. **Draw** a simple sketch of a sponge body plan, and label all the parts you include.
2. **Summarize** how a sponge feeds and distributes nutrients.
3. **Describe** the three types of sponge skeletons.
4. **Compare** asexual and sexual reproduction in sponges.
5. **SKILL Forming Hypotheses** What advantage might there be to a free-swimming larval stage in sponges?

Critical Thinking

6. **SKILL Determining Factual Accuracy** Evaluate this statement: Sponges have two cell layers, mesoglea and collar cells.

Review SECTION 29-1 ANSWERS

1. Sketches will vary but should include the labels in Figure 29-2.
2. A sponge's collar cells draw water into the sponge's pores and trap small organisms. The cells digest the organisms and release nutrients into the mesoglea, where amoebocytes distribute the nutrients to the other cells.
3. Calcareous sponges have skeletons with calcium carbonate spicules. Glass sponges have silica spicules. Demosponges have skeletons made of spongin, a flexible protein fiber.
4. Sponges reproduce asexuality by budding. Some freshwater sponges reproduce by forming gemmules. Although sponges are hermaphrodites, they reproduce sexually by fertilizing other sponges. Sperm enters a sponge through the pores and passes into the mesoglea, where fertilization of the eggs occurs. The resulting larvae leave the sponge.
5. Free-swimming larvae can colonize new areas. If there were no free-swimming stage, all the sponges in one area could be wiped out by a local disaster.
6. This statement is false. The mesoglea is not a cell layer; it is a gel-like substance that has cells embedded in it.

Teaching TIP

Gemmule Survival Pods

Survival pods have been used for submarine and spacecraft, and they should be familiar to your students. Ask students what a properly equipped survival pod on a submarine should contain. *(water, food, oxygen, the means to get rid of waste, protection from the outside environment)* List suggestions on the board. Then take each item and inform students how a gemmule meets that need.

3 Assessment Options

Closure

Sponge Anatomy

Have students draw a simple diagram of a sponge and label all of its cells and structures. Tell students to be certain to include egg and sperm cells in their diagram. Students should add arrows to show the path water takes as it flows through the sponge. They should also indicate how sperm enter the sponge. PORTFOLIO

Review

Assign Review—Section 29-1.

Reteaching — BASIC

Show students a diagram of a sponge. Provide them with self-sticking labels bearing the names of sponge cells and structures. Have students place the labels on the diagram. Then ask students to describe the path of water as it flows through the sponge, the transport of nutrients within the sponge, and the process of sexual reproduction in sponges.

Chapter 29

1 Prepare

Scheduling

- **Block:** About 90 minutes will be needed to complete this section.
- **Traditional:** About two class periods will be needed to complete this section.
- **Planning:** See the Planning Guide on pp. 633A and 633B for lecture, classwork and assignment options for Section 29-2. Lesson plans for Section 29-2 can be found in a lesson cycle format in ***Biology: Principles and Explorations*** **One-Stop Planner CD-ROM with Test Generator.**

Resource **Link**

Assign **Active Reading Worksheet Section 29-2** to help students comprehend the key concepts and visuals in this lesson.

2 Teach

Lesson Warm-up

Show students photographs of jelly fish, hydra, sea anemones, and corals. Ask students about their own experiences with these cnidarians. Have they ever seen a jellyfish, a Portuguese man-of-war, a coral, or a sea anemone? Have they ever touched a cnidarian? What did it feel like? Where have they seen cnidarians—in fresh water or salt water? Take this opportunity to explain that although most cnidarians are marine, some, such as hydras, live in fresh water.

Section 29-2 Cnidarians

Objectives

- **Describe** the two cnidarian body forms. (p. 640)
- **Summarize** how cnidocytes function. (p. 641)
- **Summarize** the life cycle of *Obelia.* (p. 643)
- **Compare** three classes of cnidarians. (pp. 642, 644, 645)
- **Compare** asexual and sexual reproduction in cnidarians. (pp. 643–645)

Key Terms

medusa
polyp
cnidocyte
nematocyst
basal disk
planula

Cnidarians Have Two Body Forms

As the fragile bell of a jellyfish moves rhythmically through the water or the flowerlike sea anemone sways gently in the ocean currents, it's easy to be caught up in the mystery and beauty of these animals. But don't be deceived by their allure, for jellyfish and sea anemone are carnivores that can inflict a vicious sting. Along with hydras and corals, these animals belong to the phylum Cnidaria *(nih DAIR ee uh).* Cnidarians have two basic body forms, as shown in **Figure 29-6,** and both show radial symmetry. **Medusa** *(muh DOO suh)* forms are free-floating, jellylike, and often umbrella-shaped. **Polyp** *(PAHL ihp)* forms are tubelike and are usually attached to a rock or some other object. A fringe of tentacles surrounds the mouth, located at the free end of the body. Many cnidarians exist only as medusas, while others exist only as polyps. Still others alternate between these two phases during the course of their life cycle.

The cnidarian body has two layers of cells, as illustrated by the hydra in **Figure 29-7.** The outer layer is derived from ectoderm, and the inner layer is derived from endoderm. As in the sponge, there is a middle layer of mesoglea. But cnidarians differ from sponges in that cnidarians' cells are arranged into tissues.

Figure 29-6 Cnidarian body forms

The two body forms of cnidarians—medusa and polyp—consist of the same body parts arranged differently.

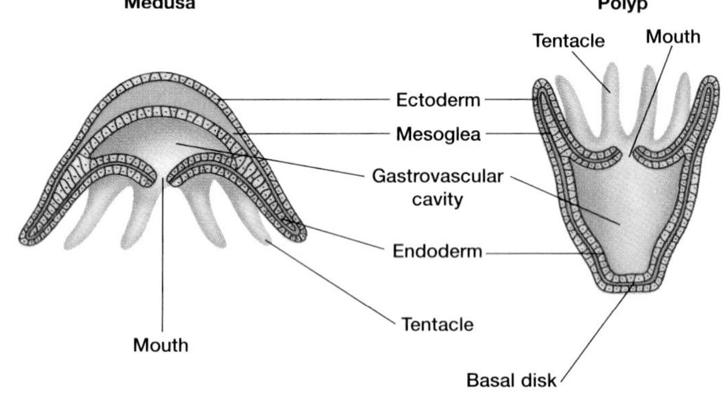

Evolutionary Milestone

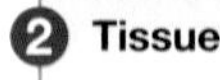
2 Tissues

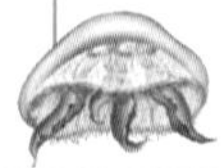

The cnidarian body plan is more complex than that of a sponge because it contains specialized tissues that carry out particular functions. However, the tissue is not organized into organs.

Integrating Different Learning Styles

Logical/Mathematical	**PE**	Math Lab, p. 641
	ATE	Resource Link, p. 641; Teaching Tip, p. 645
Visual/Spatial	**PE**	Performance Zone items 16, 17, 18, 19, 20, and 26, p. 655
	ATE	Lesson Warm-up, p. 640; Visual Strategies, pp. 642, 643, 644
Verbal/Linguistic	**PE**	Study Tips, pp. 642, 644, 645; Performance Zone item 28, p. 655
	ATE	Lesson Warm-up, p. 640; Teaching Tip, p. 644

Cnidocytes

Flexible fingerlike tentacles surround the opening to the gastrovascular cavity of cnidarians. Located on the tentacles are stinging cells called **cnidocytes** *(nih DOH seyets),* also shown in Figure 29-7. Cnidocytes are the distinguishing characteristic of cnidarians that gives them their phylum name, Cnidaria. Within each cnidocyte is a small barbed harpoon called a **nematocyst** *(NEHM uh toh sihst).* Nematocysts are used for defense and to spear prey. Some nematocysts contain deadly toxins, while others contain chemicals that stun but do not kill. When triggered, the nematocyst explodes forcefully and sinks into the cnidarian's prey. The captured prey is then pushed into the cnidarian's gastrovascular cavity by the tentacles.

Extracellular digestion

In cnidarians and all subsequent animal phyla, digestion begins extracellularly (outside the cell), in the gastrovascular cavity. Enzymes break food down into small fragments. Then cells lining the cavity engulf the fragments, and digestion is completed intracellularly. This allows cnidarians to feed on organisms larger than their own individual cells.

Figure 29-7 Cnidarian body plan

Like all cnidarians, this hydra is made up of tissues derived from endoderm and ectoderm.

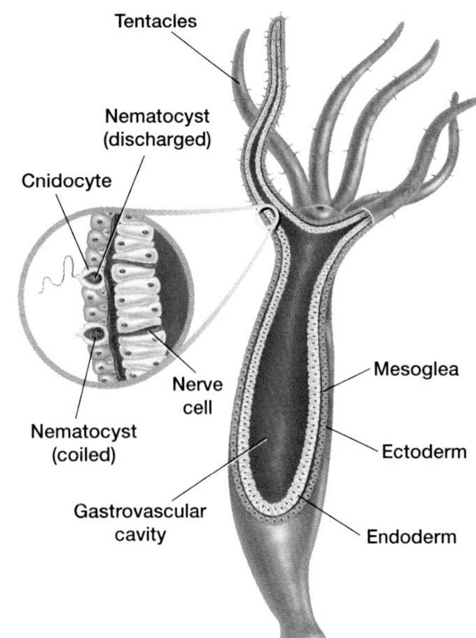

Resource Link

- Assign **Basic Skills Worksheet: Microscopic Magnification** from the Holt BioSources Teaching Resources CD-ROM.

Using the microscope to estimate size

MATH LAB

Time 20 minutes

Process Skills Calculating, applying information

Teaching Strategies

- Students can become confused when trying to measure the diameter of the field of view. Use a diagram on the board to show how to place the ruler to measure the field of view.
- Go over the math before starting the lab. Using a diagram on the board, determine the diameter of the field of view. Show mathematically that if an organism is two-fifths the size of the field of view, then its length can be calculated as $S = (2/5)D$. S is the size of the organism, and D is the diameter in millimeters of the field of view.

Analysis Answers

1. Answers will vary.
2. Answers will vary. Refer to Figure 29-6 on page 640 for reference.

Using the microscope to estimate size

MATH LAB

You can use the microscope to estimate the size of cnidarians that are too small to measure directly.

Materials

transparent millimeter ruler, compound microscope with low-power objective or a dissecting microscope, prepared slide of a medusa or polyp

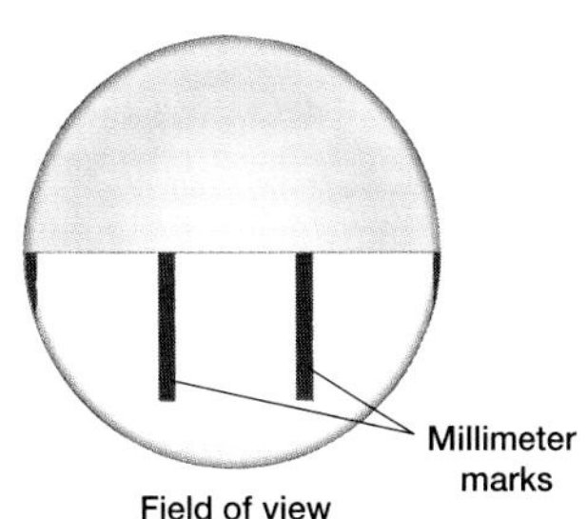

Procedure

1. Identify the millimeter marks along the edge of the ruler.
2. With the microscope on low power (4× or lower), place the ruler on the stage and focus on the millimeter marks.
3. Adjust the ruler so that one edge lies across the diameter of the field, as shown above. Then measure the diameter of the field of view in millimeters.
4. Remove the ruler, and place the prepared slide on the stage. Identify the tentacles, gastrovascular cavity, and mouth.
5. Estimate the length and width of your organism as a ratio of the width of the field of view. For example, the length of your organism may appear to cover about two-thirds of the field of view.

Analysis

1. **Calculate** the size of your organism in millimeters by multiplying the ratio you found in step 5 by the width of the field of view you found in step 3.
2. **Describe** the body plan of the organism you viewed using terms from step 4.

did you know?

Some sea slugs "steal" cnidocytes.

Some sea slugs (nudibranchs) feed on cnidarians and save the untriggered nematocysts within their own bodies. When other animals try to eat the sea slug, the stored-up nematocysts are triggered. As a result of this painful encounter, many predators learn to avoid sea slugs.

Visual Strategy

Figure 29-8 BASIC

Refer students to the illustration of the tumbling hydra in Figure 29-8. Tumbling may look like a difficult way to move, so compare it to the movement of a Slinky™ spring toy. Once a Slinky is started, it will continue down a stairway. A hydra is similar in that once it connects one end of its body to a solid surface, the rest of its body will flip over due to elasticity and momentum.

Teaching TIP

Hydrozoan Locomotion

Have students look at the figures on these two pages and reflect on the ways that hydrozoans move and on why even sessile cnidarians need to move from time to time.

Resource Link

Assign **Reading Strategies: Compare/Contrast Matrix** from the Holt BioSources Teaching Resources CD-ROM.

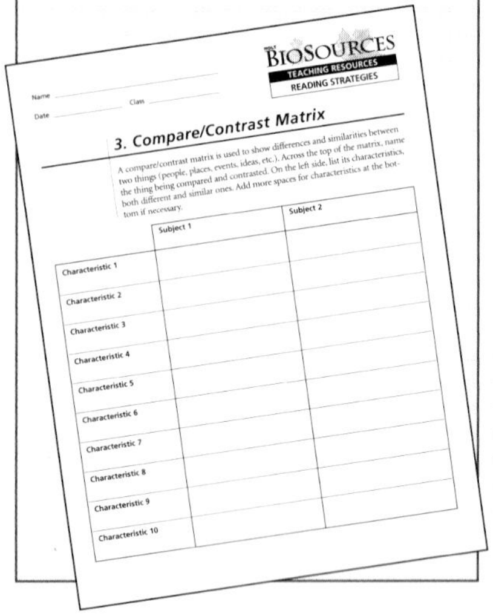

Hydrozoans Spend Most of Their Life as a Polyp

The most primitive cnidarians are members of class Hydrozoa. Most species of hydrozoans are colonial marine organisms whose life cycle includes both polyp and medusa stages. Freshwater hydrozoans are less common, but are familiar to many people because they are often studied in school laboratories.

Figure 29-8 **Freshwater hydra**

This tiny hydra is attached to the leaf of a small aquatic plant. One way a hydra can move is by tumbling.

Magnification: 34×

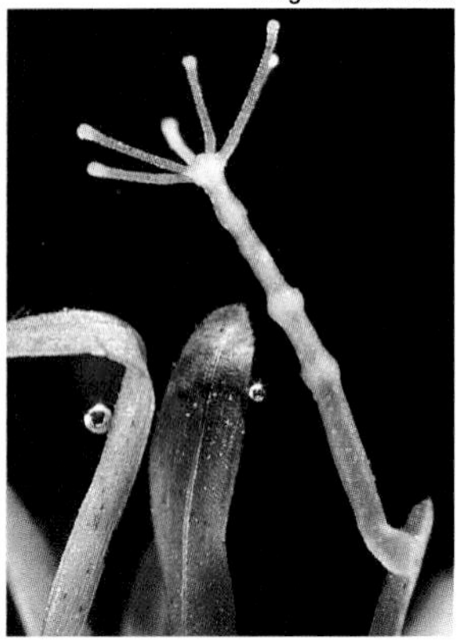

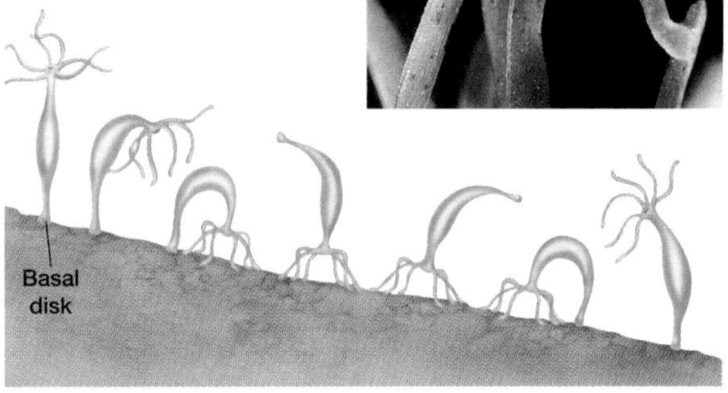

Basal disk

Freshwater hydrozoa

The abundant freshwater genus *Hydra* is unique among hydrozoans because it has no medusa stage and exists only as a solitary polyp. Hydras live in quiet ponds, lakes, and streams. They attach to rocks or water plants by means of a sticky secretion they produce by an area of their body called the **basal disk.** Hydras can glide around by decreasing the stickiness of the material secreted by their basal disk. Sometimes hydras move by tumbling, as shown in **Figure 29-8.** To tumble, the hydra bends its body over and touches the surface it is attached to with its tentacles. Then it pulls its basal disk free, flipping it over to the other side of its tentacles. The basal disk then reattaches, and the hydra returns to an upright position. Most hydras are brown or white, like the one in Figure 29-8. Others appear green because of the algae living beneath their outer cells.

Study TIP

Organizing Information

Make a table to organize the information about cnidarians. Across the top, write the headings *Hydra, Jellyfish, Sea anemone,* and *Coral.* Along the sides, write *Class, Body form, Reproduction, Feeding,* and *Habitat.* Add information to the table as you read this section.

Marine hydrozoa

Marine hydrozoans are typically far more complex than freshwater hydrozoans. Often many individuals live together, forming colonies. The cells of the colony lack the interdependence that characterizes the cells of multicellular organisms. However, they often exhibit considerable specialization. For example, the colonial Portuguese man-of-war (genus *Physalia*) incorporates both medusas and polyps. A gas-filled float (probably a highly modified polyp) allows *Physalia* to float on the surface of the water. Dangling below the float are tentacles that can reach 15 m (50 ft) long. These tentacles are used to stun and entangle prey. Their nematocysts are tipped with powerful neurotoxins (nerve poisons) dangerous and even fatal to humans. *Physalia,* shown in **Figure 29-9,** has other specialized polyps and medusas, each carrying out a different function, such as feeding or sexual reproduction.

Multicultural Perspective

A Mythological Hydra

The name *hydra* is a reference to the nine-headed serpent that was slain by Hercules, a figure in Greek mythology. According to Greek myth, the slaying of the Hydra was one of the twelve labors that Hercules was assigned to perform. Killing the Hydra was not an easy task for Hercules because when one head of the Hydra was cut off, two others would grow in its place.

Reproduction in hydrozoans

Most hydrozoans are colonial organisms whose polyps reproduce asexually by forming small buds on the body wall. The buds develop into polyps that eventually separate from the colony and begin living independently. Many hydrozoans are also capable of sexual reproduction. Some species of *Hydra* are hermaphrodites, but in most species the sexes are separate.

The genus *Obelia* is typical of many marine colonial hydrozoans. *Obelia* lives in colonies that form when one polyp asexually produces buds that do not separate from it. Eventually, there are numerous polyps attached to one stem, forming the colony. The *Obelia* colony shown in **Figure 29-10** is branched like deer antlers, with various polyps attached to the branched stalks. The reproductive polyps give rise asexually to male and female medusas. These medusas leave the polyps and grow to maturity in the ocean waters.

During sexual reproduction, the medusas release sperm or eggs into the water. The gametes fuse and produce zygotes that develop into free-swimming larvae called **planulae** *(PLAN yoo lee)*. The planulae eventually settle on the ocean bottom and develop into new polyps. Each polyp gives rise to a new colony by asexual budding, and the life cycle is repeated.

Figure 29-9 ***Physalia***

How many? As many as 1,000 individual medusas and polyps may compose a single Portuguese man-of-war colony. Their feeding tentacles can be as long as 15 m.

Figure 29-10 **Reproduction in *Obelia***

In *Obelia's* life cycle, the medusa stage (sexual) and the polyp stage (asexual) alternate.

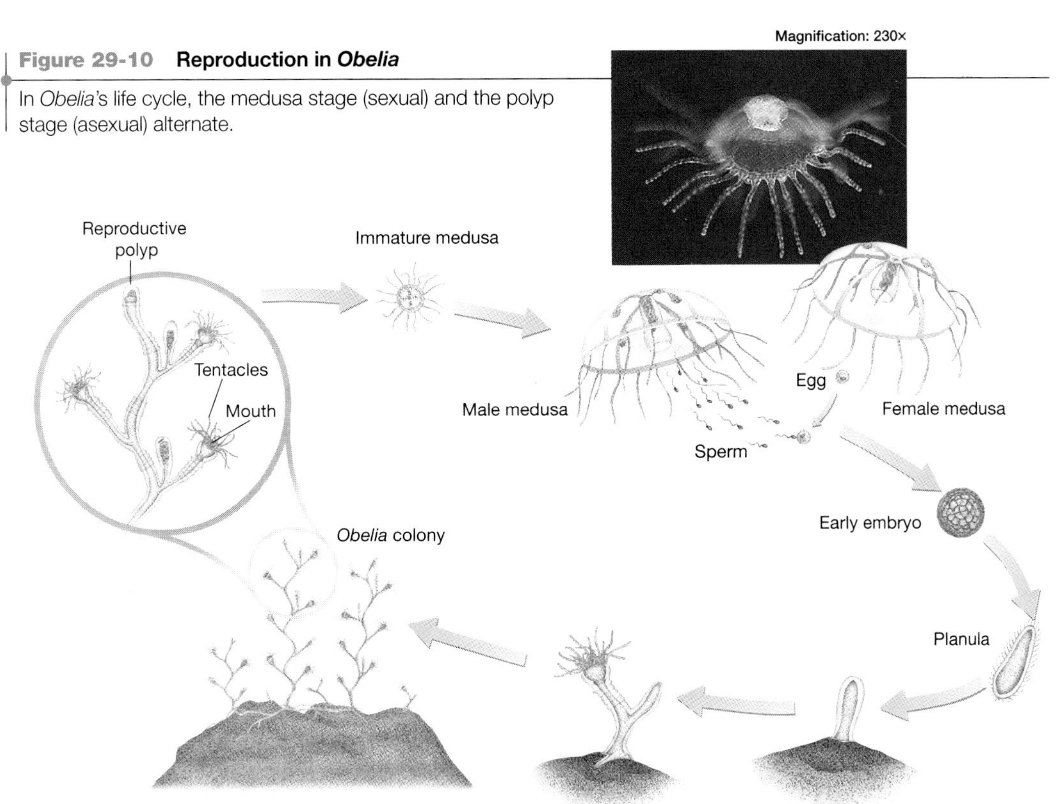

Section 29-2

Visual Strategy

Figure 29-10

Check for student understanding of the life cycle of *Obelia,* shown in Figure 29-10. Ask which stages are free-swimming and which are sessile. *(medusa, polyp)* Which stage reproduces asexually? *(polyp stage)* In which stage are there males and females that reproduce sexually? *(medusa stage)*

Teaching TIP

The Problems and the Pluses of Planulae

There are both negative and positive aspects to the existence of the free-swimming planulae of *Obelia*. Ask students about the dangers to free-swimming larvae in the ocean, and why it is an advantage for larvae to be able to swim far from their parents. *(Larvae may starve, be eaten, wash up on shore, or not find a suitable place to settle down. Motile larvae allow* Obelia *to spread to new places.)*

Overcoming Misconceptions

The Portuguese Man-of-war

Many people think that the Portuguese man-of-war *(Physalia)* is a jellyfish because it is shaped somewhat like one. However, the term *jellyfish* is used for scyphozoans, a class of cnidarians in which the medusa form dominates the life cycle. *Physalia* is actually a hydrozoan and exists as a colony of medusas and polyps.

Teaching TIP

Jellyfish Toxins

Jellyfish have toxins in their nematocysts that paralyze and kill fish. Ask students why it is important for a simple floating animal like a jellyfish to immobilize its prey. *(A jellyfish cannot stalk or chase prey, and it has no jaws or claws to capture prey. It must catch and immobilize any fish that accidentally swims into its tentacles.)*

Visual Strategy

Figure 29-11

Have students study Figure 29-11 to compare and contrast the medusa and polyp stages of a jellyfish. Ask students to name other animals that have different bodies for different stages of their life. *(hydrozoans, insects, frogs, etc.)* Then ask students to compare *Aurelia* medusae and polyps to those of *Obelia*. *(They are almost identical.)* Ask students to identify the big difference in the life cycles of hydrozoans and scyphozoans. *(Hydrozoans spend most of their life as a polyp; scyphozoans spend most of their live as a medusa.)*

Teaching TIP — ADVANCED

Symbiotic Relationships with Sea Anemones

Have students research the relationship between some species of sea anemones and certain types of fish, such as the damsel fish and the clownfish. Students should investigate how these fish are able to avoid discharging the anemone's nematocysts.

Study TIP

Word Origins

The name Scyphozoa is from the Greek *skyphos,* meaning "cup," and *zoia,* meaning "animal." The name refers to the fact that members of this class spend most of their lives as medusas, which have the shape of an inverted cup.

Jellyfish Spend Most of Their Life as a Medusa

Cnidarians belonging to the class Scyphozoa *(seye fuh ZOH uh)* are the organisms usually referred to as true jellyfish. Scyphozoans are active predators that ensnare and sting prey with their tentacles. The toxins contained within the nematocysts of some species are extremely potent. Scyphozoans range in size from as small as a thimble to as large as a queen-size mattress.

The jellyfish seen in the ocean are medusas, which reproduce sexually. However, most species of jellyfish also go through an inconspicuous polyp stage at some point in their life cycle. The stinging nettle, *Aurelia,* shown in **Figure 29-11,** is one of the most familiar jellyfishes. *Aurelia's* tiny polyps hang from rocky surfaces. Periodically the polyps release young medusas into the water. The *Aurelia* life cycle is similar to that of *Obelia,* pictured on page 643. The major difference is that *Aurelia* spends most of its life as a medusa, while *Obelia* spends most of its life as a polyp.

Polyps

Medusa

Figure 29-11 Marine jellyfish *Aurelia.* *Aurelia* polyps are about the size of hydras. The free-swimming medusas range from 10 cm (3.9 in.) to 25 cm (9.8 in.) in diameter.

Jellyfish relatives

Related to the jellyfish are members of the phylum Ctenophora *(TEHN uh fawr uh),* which includes the comb jellies. Comb jellies differ from true jellyfish in two major ways—they have only a medusa stage and they have no cnidocytes. Their tentacles are covered with a sticky substance that traps plankton, the comb jelly's main prey. Although a comb jelly is only about 2.5 cm (1 in.) in diameter, its tentacles can be 20 times as long.

Other relatives are the cubozoans, or box jellies. As their name implies, cubozoans have a cube-shaped medusa. Their polyp stage is inconspicuous, and in some species, it has never been observed. Most box jellies are only a few centimeters in height, although some are 25 cm (10 in.) tall. A tentacle or group of tentacles is found at each corner of the "box." Stings of some species, such as the sea wasp, can inflict severe pain and even death among humans. The sea wasp lives in the ocean along the tropical northern coast of Australia.

biobit

Are jellyfish edible?

In many Asian countries, some species of jellyfish are considered delicacies. Their taste has been compared with that of pickles. Although jellyfish have stinging cells, the salting process used to prepare them as food breaks down the toxins in these cells.

did you know?

Clownfish are not bothered by sea anemones.

An interesting example of symbiosis is a clownfish living in the tentacles of a sea anemone. Usually a sea anemone catches and eats fish, yet the clownfish seems to be immune to the barbs of the sea anemone's nematocysts. The clownfish gets protection by hiding in the tentacles, and the sea anemone gets cleaned as the clownfish eats undigested debris and parasites.

Anthozoans Have No Medusa Stage

The largest class of cnidarians is class Anthozoa. Anthozoans exist only as polyps. The most familiar anthozoans are the brightly colored sea anemones and corals. Other members of this class are known by such fanciful names as sea pansies, sea fans, and sea whips.

Anthozoans, such as the sea anemone shown in **Figure 29-12,** typically have a thick, stalklike body topped by a crown of tentacles that typically occur in groups of six. Nearly all of the shallow-water species contain symbiotic algae, such as dinoflagellates. The anthozoans provide a place for the dinoflagellates to live in exchange for some of the food that the dinoflagellates produce. The brilliant color of most anthozoans is actually that of dinoflagellates living within it.

Some anthozoans reproduce asexually by forming buds, but they also reproduce sexually by releasing eggs and sperm into the ocean, where fertilization occurs. The fertilized eggs develop into planulae that settle and develop into polyps.

Sea anemones

Sea anemones are a large group of soft-bodied polyps found in coastal areas all over the world. Many species are quite colorful, and most do not grow very large, only from 5 mm (0.2 in.) to 100 mm (4.0 in.) in diameter. Sea anemones feed on fish and other marine life that happen to swim within reach of their tentacles.

Sea anemones are highly muscular and relatively complex animals. When touched, most sea anemones retract their tentacles into their body cavity and contract into a tight ball. As you learned in Chapter 28 on page 627, sea anemones often reproduce asexually by slowly pulling themselves into two halves. This method of reproduction often results in large populations of genetically identical sea anemones.

Corals

Most coral polyps live in colonies called reefs, such as the one shown in **Figure 29-13.** Each polyp secretes a tough, stonelike outer skeleton of calcium carbonate that is cemented to the skeletons of

Study TIP

Word Origins

The class name Anthozoa is from the Greek *anthos,* meaning "flower," and *zoia,* meaning "animal."

Figure 29-12 Sea anemone. When threatened, the sea anemone quickly retracts its tentacles and compresses its body.

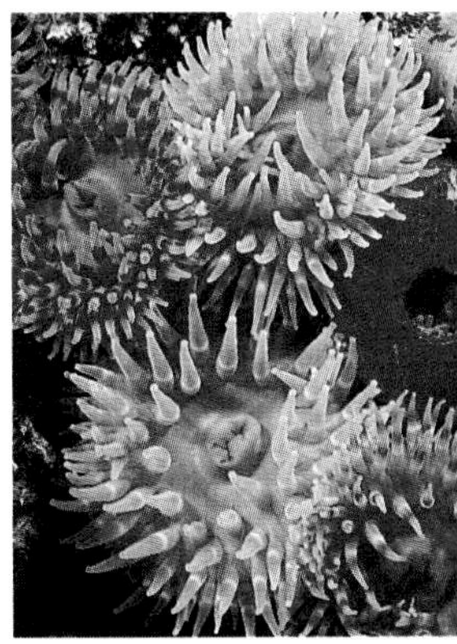

Figure 29-13 Coral. This coral reef is made up of hundreds of thousands of individual coral polyps. When the polyps feed (inset), they extend their tentacles from the protection of their stony skeleton.

Teaching TIP

Plant or Animal?

Some students find it difficult to believe that cnidarians, such as sea anemones and corals, are animals. Ask students to defend the classification of anthozoans in the animal kingdom. If necessary, have students return to pp. 612–615 in Chapter 28 and review how these anthozoans show the animal characteristics listed.

Visual Strategy

Figure 29-13 BASIC

Have students look closely at the insert of the coral polyp. Ask how the polyp resembles the polyp form of other cnidarians. *(It is cup-shaped and contains a ring of upward-facing tentacles surrounding the opening to its exterior.)*

Teaching TIP

Finding Coral Reefs

Obtain several world maps showing details of the oceans, and have students find some of the great reefs of the world. *(Florida Keys Reef, Australia's Great Barrier Reef, and the Eniwetok Atoll, in the Marshall Islands, for example)*

did you know?

Corals that form reefs live in a symbiotic relationship with single-celled algae.

The algae live in the coral polyp's tissues and produce food for the corals through photosynthesis. In turn, the polyps provide the algae with a home. The algae also help the corals secrete the limestone skeletons that, over many years, form the "backbone" of a coral reef.

Chapter 29

Exploring Further

Life on a Coral Reef

Teaching Strategies
Have students look at the photograph of the coral reef in this feature and identify as many animals as they can.

Discussion
Ask students why reefs might nurture such diverse assemblages of creatures. *(Answers will vary. Students should recognize that the physical structure of reefs provides a variety of places for finding food, shelter, and safety.)*

3 Assessment Options

Closure

Modeling Cnidarians
Using modeling clay, have students work in small teams to make models of a sponge, a typical polyp, and a typical medusa. Different colors of clay should be used for different structures. Then have each team present their models to the class.

Review

Assign Review—Section 29-2.

Reteaching — BASIC

Have students make a table that shows the similarities and differences between the three classes of cnidarians: *Hydrozoa*, *Scyphozoa*, and *Anthozoa*. *(Similarities include radial symmetry, aquatic lifestyle, and presence of cnidocytes. Differences include predominance of a polyp or medusa body form, and the maintenance of a solitary versus colonial lifestyle.)* PORTFOLIO

its neighbors. (Some corals called soft corals do not secrete hard exoskeletons.) Only the top layer of a coral reef contains living coral polyps. When coral polyps die, their skeletons remain and provide a foundation for new coral polyps. Over thousands of years, these formations build up into coral reefs where hundreds of thousands of polyps live together on top of old skeletons. Coral reefs are found primarily in tropical regions of the world, where the ocean water is warm and clear, an environment that is ideal for the corals and the dinoflagellates that live inside them.

Exploring Further

Life on a Coral Reef

The diversity of life on a coral reef is rivaled only by that of a tropical rain forest. Algae growing on the corals and microscopic invertebrates help cement the corals together. Tube worms, mollusks, and other organisms donate their hard skeletons to the reef. A variety of small animals, including sponges, flatworms, shrimps, and crabs, find protection from predators in crevices in the reef. Sea anemones, hydroids, and feather duster worms anchor themselves to the reef and snare food from the surrounding water. Predators such as sea stars, octopuses, eels, and sharks patrol the reef in search of prey. As many as 3,000 species of animals may live on a single reef.

The importance of coral reefs
Coral reefs directly benefit people by protecting coastlines from wave erosion and serving as resources for fisheries. Tourists who visit the reefs can be a significant source of income to nearby communities. Researchers are interested in reef-dwelling animals as sources of new medicines, including antibiotics and anticancer drugs. Today, coral reefs face a number of threats caused by human activity, including pollution of the waters surrounding them. Some nations have established marine sanctuaries around their most vulnerable coral reefs to ensure the survival of these natural treasures.

internet**connect**

TOPIC: Coral reefs
GO TO: www.scilinks.org
KEYWORD: HX646

Review SECTION 29-2

1. **Compare** the two body forms of cnidarians.
2. **Relate** cnidocytes and nematocysts to food gathering.
3. **Draw** and label the life cycle of *Obelia.*
4. **Summarize** the similarities and differences in the three classes of cnidarians described.
5. **Distinguish** between the two types of asexual reproduction found in cnidarians.
6. **SKILL Forming Hypotheses** Some cnidarians are unique in exhibiting polyp and medusa forms. How might their two body forms give them an advantage over species that live in the same environment but have only one body form?

Review SECTION 29-2 ANSWERS

1. The medusa form is free-floating, jellylike, and often umbrella shaped. The polyp form is tubelike and closed at one end.
2. Cnidocytes are stinging cells used in feeding or for defense. They contain a barbed harpoon called a nematocyst. When triggered, the nematocyst is shot into the cnidarian's prey.
3. Drawings will vary.
4. All cnidarians have stinging cells on their tentacles. Their body has two cell layers, a middle layer of mesoglea, and a gastrovascular cavity. Hydrozoans are primarily colonial organisms with both polyp and medusa forms. However, most of their life is spent as a polyp. Scyphozoans exist primarily as a medusa with only a brief polyp stage. Anthozoans have only a polyp stage and may be solitary or colonial.
5. A colonial cnidarian may reproduce by budding. Some, such as sea anemones, may reproduce asexually by slowly pulling themselves in half.
6. The two body forms exploit different environments and food sources, so these cnidarians have access to a broader array of resources while spreading their risks between two lifestyles.

Flatworms and Roundworms

Section 29-3

Flatworms Exhibit Bilateral Symmetry

When you think of a worm, you probably visualize a creature with a long, threadlike body, perhaps with a head end. It probably also shows bilateral symmetry, as do flatworms and roundworms. The flatworms are the largest group of acoelomate worms. Although the flatworm body plan is relatively simple, it is a great deal more complex than that of a sponge or cnidarian. Flatworms have a middle tissue layer, the mesoderm. And unlike sponges and cnidarians, the flatworm has tissues that are organized into organs.

The flatworm's body is flattened from top to bottom, like a piece of tape or ribbon, placing each cell in the animal's body very close to the exterior environment. This permits dissolved substances, such as oxygen and carbon dioxide, to pass efficiently through the flatworm's solid body by diffusion. In addition, portions of the flatworm's highly-branched gastrovascular cavity run close to practically all of its tissues. This gives each cell ready access to food molecules. As a result, flatworms have no need of a respiratory or circulatory system, and most have none.

Flatworms belong to phylum Platyhelminthes, which contains three major classes: Turbellaria, Cestoda, and Trematoda. They range in size from free-living forms less than 1 mm (0.04 in.) in length to parasitic intestinal tapeworms several meters long.

Turbellaria

Almost all members of class Turbellaria are free-living marine flatworms, such as the one shown in **Figure 29-14.** However, marine flatworms are rarely studied by students because they are difficult to raise in captivity. Instead, students usually study a freshwater turbellarian such as *Dugesia,* one of a group of flatworms commonly called planarians. *Dugesia* is shown in *Up Close: Planarian,* on page 648.

Figure 29-14 Marine flatworms. Most free-living flatworms are marine species that swim with graceful wave-like movements.

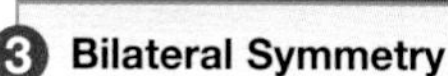

Evolutionary Milestone

3 Bilateral Symmetry

Flatworms were most likely the first animals to be bilaterally symmetrical, with left and right halves that mirror each other. Like all bilaterally symmetrical animals, flatworms have a distinct anterior (cephalic) end.

Objectives

- **Compare** the three classes of flatworms. (pp. 647, 649, 650)
- **Summarize** the life cycle of a blood fluke. (p. 650)
- **Describe** the body plan of a roundworm. (p. 651)
- **Summarize** the life cycle of the roundworm *Ascaris.* (pp. 651–652)

Key Terms

proglottid
fluke
tegument

Integrating Different Learning Styles

Logical/Mathematical	**PE**	Data Lab, p. 652
Visual/Spatial	**PE**	Up Close, p. 648; Performance Zone item 1, p. 654
	ATE	Teaching Tip, p. 651
Intrapersonal	**PE**	Study Tip, p. 650; Performance Zone item 21, p. 655
	ATE	Visual Strategy, p. 650
Verbal/Linguistic	**PE**	Quick Lab, p. 649; Performance Zone items 13 and 22, pp. 654–655
	ATE	Lesson Warm-up, p. 647; Teaching Tip, p. 648

Section 29-3

1 Prepare

Scheduling

- **Block:** About 90 minutes will be needed to complete this section.
- **Traditional:** About two class periods will be needed to complete this section.
- **Planning:** See the Planning Guide on pp. 633A and 633B for lecture, classwork and assignment options for Section 29-3. Lesson plans for Section 29-3 can be found in a lesson cycle format in ***Biology: Principles and Explorations*** **One-Stop Planner CD-ROM with Test Generator.**

Resource Link

Assign **Active Reading Worksheet Section 29-3** to help students comprehend the key concepts and visuals in the lesson.

2 Teach

Lesson Warm-up

The word *worm* can refer to a wide variety of organisms. Ask students to think about what a worm is, and ask them to name or describe the types of worms they can remember. Ask students where these worms live and what they eat. Have volunteers share their descriptions with the rest of the class.

Up Close

Planarian

Teaching Strategies
Explain to students that all organisms have mechanisms that help them maintain proper water balance. Tell them that water seeps into *Dugesia* like it seeps into a leaky boat. The excretory system expels water so that the animal does not swell up.

Discussion
Guide the discussion by posing the following questions:

- Instead of a circulatory system that delivers nutrients to tissues, what does a planarian have? *(Branches of the digestive tract reach the tissues directly.)*
- How do planarians reproduce asexually? *(They tear themselves in two, and each half regenerates to form a complete worm.)*
- How do planarians reproduce sexually? *(They are hermaphrodites that fertilize each other's eggs. Protective capsules surround groups of fertilized eggs, which hatch in 2–3 weeks.)*

Teaching TIP

Life of a Planarian
Have your students imagine a day in the life of a planarian and write a story about the planarian's adventures. Their story should include information about the planarian's nervous system, water balance, reproduction, feeding, digestion, and excretion. Caution them to avoid anthropomorphic descriptions of the planarian, though a little "creative license" should be allowed.
PORTFOLIO

Up Close

Planarian

- **Scientific name:** *Dugesia* sp.
- **Size:** Average length of 3–15 mm (0.1–0.6 in.)
- **Range:** Worldwide
- **Habitat:** Cool, clear, permanent lakes and streams
- **Diet:** Protozoans and dead and dying animals

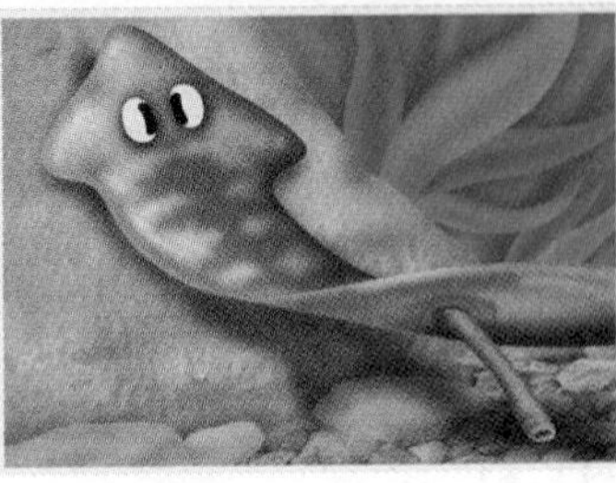
Dugesia feeding

Characteristics

Nervous System Sensory information gathered by the brain is sent to the muscles by two main nerve cords that are connected by cross branches. Light-sensitive structures called eyespots are connected to the brain by nerve cells. The eyespots are close to each other, giving *Dugesia* a cross-eyed appearance.

Feeding *Dugesia,* a free-living flatworm, must extend its muscular pharynx out of its centrally located mouth in order to feed.

Reproduction *Dugesia* reproduces asexually in the summer by attaching its posterior end to a stationary object and stretching into two parts, each of which will become a complete animal. Sexual reproduction also occurs. Individuals are hermaphrodites, and two individuals simultaneously transfer sperm to each other. Eggs of both individuals are fertilized and are released in clusters enclosed in a protective capsule. Several capsules are laid at a time, and the eggs inside hatch in 2 to 3 weeks.

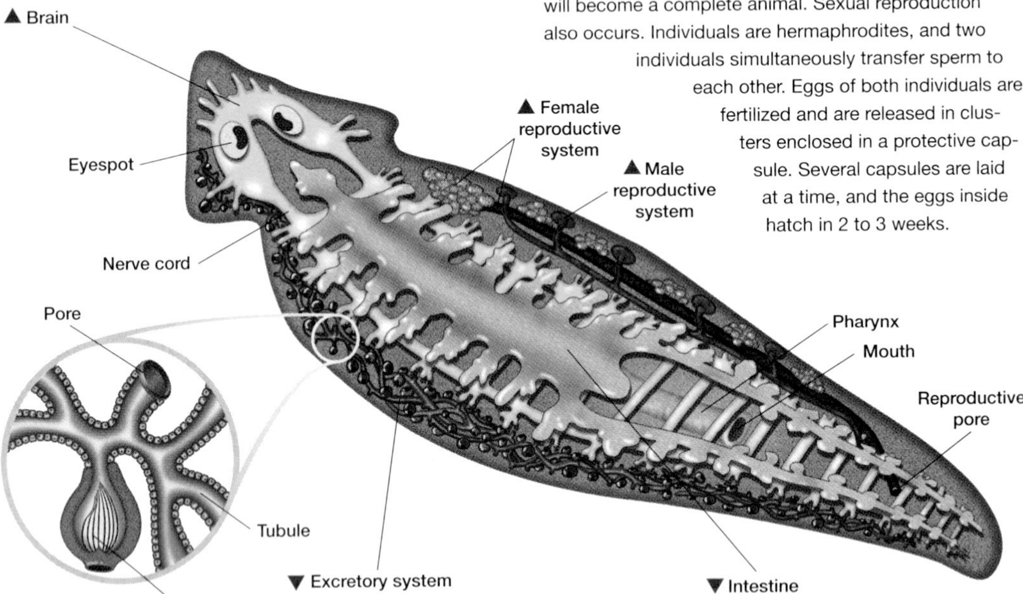

Water Balance Because *Dugesia*'s body cells contain more solutes than fresh water does, water continuously enters its body by osmosis. Excess water moves into a network of tiny tubules that run the length of *Dugesia*'s body. Side branches are lined with many flame cells, specialized cells with beating tufts of cilia that resemble a candle flame. The beating cilia draw water through pores to the outside of the worm's body.

Digestion The highly branched intestine enables nutrients to pass close to all of the flatworm's tissues. Nutrients are absorbed through the intestinal wall. Undigested food is expelled through the mouth.

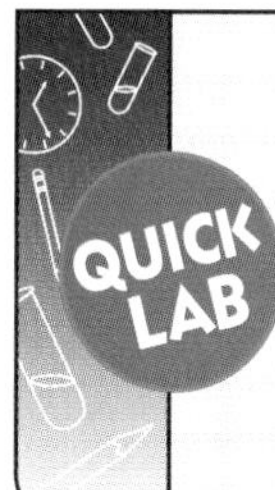

How does a planarian locate food sources?

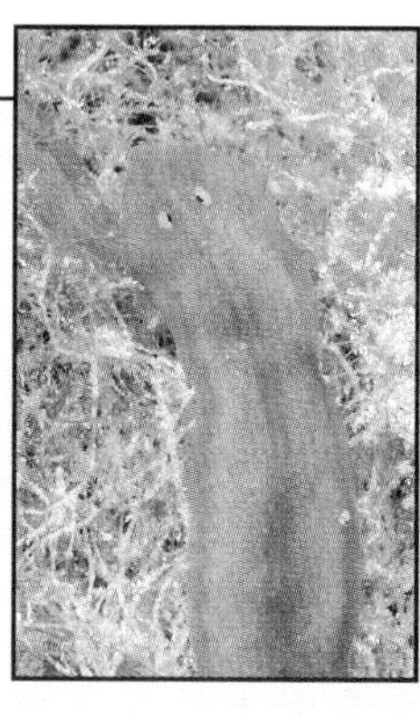

Most bilaterally symmetrical organisms have sense organs concentrated in one end of the animal. You can observe how this arrangement affects the way they explore their environment.

Materials

eyedropper, live culture of planaria, small culture dish with pond water, hand lens or dissecting microscope, forceps, and small piece of raw liver (3–7 cm)

Procedure

1. Using the tip of the eyedropper, place a planarian in the culture dish with pond water.
2. Using the hand lens or dissecting microscope, observe the planarian as it adjusts to its environment. Determine which end of the planarian contains sensory apparatus for exploring the environment.
3. Using forceps, place the liver in the pond water about 1 cm behind the planarian.
4. Observe the planarian's response. If the planarian approaches the liver, move the liver to a different position.
5. Continue observing the planarian for 5 minutes, moving the liver frequently.

Analysis

1. **Describe** the planarian's means of locomotion.
2. **Describe** how the planarian responded to the liver.
3. **Contrast** the feeding behavior of planarians with that of hydras, described earlier in this chapter.
4. **SKILL Evaluating an Argument** Evaluate this statement: Bilateral symmetry gives planaria an advantage when feeding because sensory organs are concentrated in one end. Support your opinion with the observations you made on planaria.

Cestoda

Class Cestoda is made up of a group of parasitic flatworms commonly called tapeworms. Tapeworms use their suckers and a few hooklike structures, shown in **Figure 29-15,** to permanently attach themselves to the inner wall of their host's intestines. Food is then absorbed from the host's intestine directly through the tapeworm's skin. Tapeworms grow by producing a string of rectangular body sections called **proglottids** *(proh GLAHT ihds)* immediately behind their head. (Each proglottid is a complete reproductive unit, a fact that makes it difficult to eliminate tapeworms once a person is infected.) These sections are added continually during the life of the tapeworm. The long, ribbonlike body of a tapeworm may grow up to 12 m (40 ft) long.

Figure 29-15 Tapeworm

A tapeworm's body consists of a head and a series of proglottids.

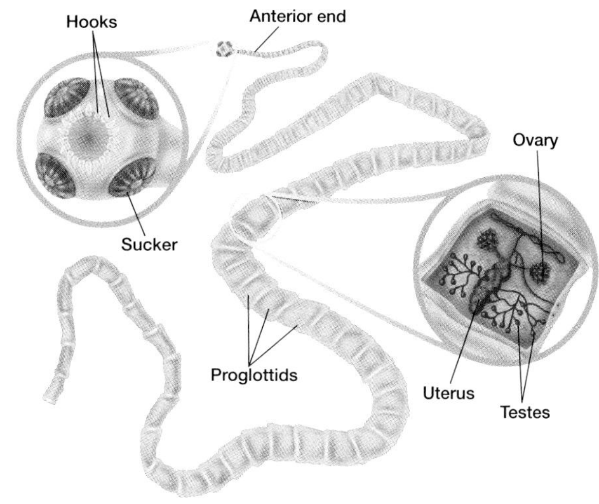

did you know?

Most adult tapeworms will not make a person sick.

However, they may lead to loss of appetite, abdominal discomfort, diarrhea, weakness, or nausea. Larval tapeworms, on the other hand, can be much more dangerous and can spread to a variety of different organs.

How does a planarian locate food sources?

Time 15–20 minutes

Process Skills Observing, comparing, evaluating conclusions

Teaching Strategies

- Before students begin, demonstrate how to place planarian in a culture quickly.
- Point out that the more gently the liver is placed in the dish, the quicker the planarian will approach it.
- Planarians are photonegative and should be kept in the dark as much as possible.
- Use opaque pans, and keep the water temperature close to 18°C.
- Prior to the lab, avoid feeding the planarians for a few days. This will increase their feeding response during the lab.

Analysis Answers

1. Planarians move by contracting and expanding their body as they grip the surface. They often turn their head from side to side as they move.
2. Planarians should turn their heads in the direction of the liver before moving toward it.
3. Hydras are relatively stationary feeders and use cnidocytes to sting and kill prey that is moving towards them. Planarians are bilaterally symmetrical and have sensory organs located at one end of their body. They are able to detect food sources and move toward them.
4. Answers will vary. Students should support their arguments with their observations.

Visual Strategy

Figure 29-16

Have students analyze Figure 29-16 from the perspective of a public health investigator. Divide the class into small groups of two or three students, and have them devise a plan to combat blood fluke infections.

internetconnect

SCILINKS. NSTA Have students research flukes using the Web site in the Internet Connect box on page 650. Have students write a report summarizing their findings.

PORTFOLIO

CAREER

Parasitologist

Parasitologists study the life cycles of parasites, animals, or plants that live in or on other organisms and take their nourishment from them. Some parasitologists specialize in studying animal parasites. Some medical doctors specialize in the diagnosis and treatment of people with parasitic infections. Other parasitologists specialize in studying plant parasites and their effects on agriculture. Group students into teams, and have them use library and on-line resources to research parasitology careers.

Study TIP

Reviewing Information

Learn the stages in the blood fluke life cycle by making a diagram similar to Figure 29-16. Quiz yourself by covering the text and explaining each stage.

Most tapeworm infections occur in vertebrates, and about a dozen different kinds of tapeworms commonly infect humans. One of the tapeworms that infects humans is the beef tapeworm, *Taenia saginata.* Beef tapeworm larvae live in the muscle tissue of infected cattle, where they form enclosed fluid-filled sacs called cysts. Humans become infected when they eat infected beef that has not been cooked to a temperature high enough to kill the larvae.

Trematoda

The largest flatworm class, Trematoda, consists of parasitic worms called **flukes.** Some flukes are endoparasites, or parasites that live *inside* their hosts. To avoid being digested by their host, endoparasites have a thick protective covering of cells called a **tegument.** Other flukes are ectoparasites, or parasites that live on the *outside* of their hosts.

Flukes have very simple bodies with few organs. Because they take their nourishment directly from their hosts, flukes do not have well-developed digestive systems. Flukes have one or more suckers that they use to attach themselves to their host. They use their muscular pharynx to suck in nourishment from the host's body fluids.

Most flukes have complex life cycles involving more than one host, one of which may be a human. Blood flukes of the genus *Schistosoma* are responsible for the disease schistosomiasis *(shihs tuh soh MEYE uh sihs),* a major public health problem in the tropics. Infection occurs when people use or wade in water contaminated with *Schistosoma* larvae. The larval parasites bore through a person's skin and make their way to blood vessels in the intestinal wall. Their presence causes blocked blood passages and results in bleeding of the intestinal wall and decay of the liver. As shown in **Figure 29-16,** the life cycle of blood flukes includes a particular species of snail as an intermediate host.

Figure 29-16 Blood fluke life cycle

In the life cycle of blood flukes, snails are intermediate hosts and humans are final hosts.

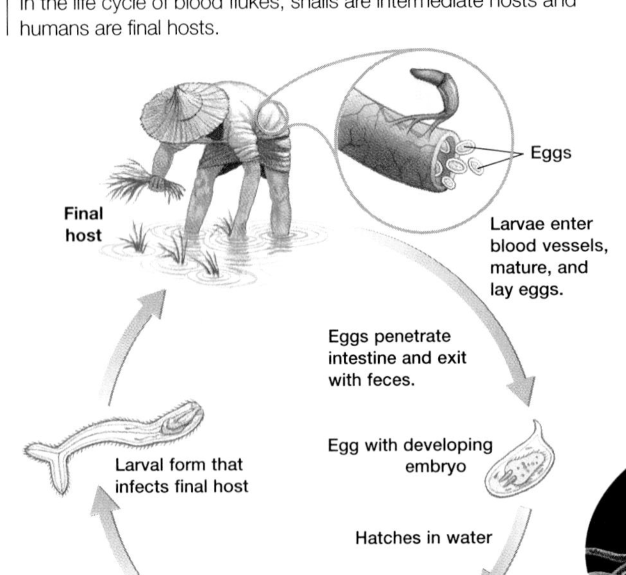

Adult male blood flukes are thick-bodied, while adult females are threadlike.

did you know?

Schistosomiasis, also called bilharziasis, is a major public health hazard, afflicting more than 200 million people worldwide.

Most cases occur in Africa, Asia, South America, and some Caribbean islands. An initial rash at the infection site is followed several weeks later by coughing, nausea, fever, and abdominal pain. The disease can also damage the liver, spleen, intestines, and nervous system.

Roundworms Have a Body Cavity

If you have a dog, you may be familiar with roundworms, some of which are canine parasites. Treatment for roundworms is a common reason for a trip to the vet, as shown in **Figure 29-17.** Roundworms (nematodes) are members of the phylum Nematoda and are characterized by the presence of a body cavity called a pseudocoelom. Movement of the fluid within the roundworm's pseudocoelom serves as a simple circulatory and gas exchange system. Oxygen and carbon dioxide move by diffusion into and out of the fluid. Nutrients from the digestive system also diffuse into the fluid and are distributed to the body cells.

Roundworms have long, cylindrical bodies and are the simplest animals to have a one-way digestive system. A flexible, thick layer of epidermis and cuticle form a protective cover and give the roundworm's body its shape. Beneath this cover, a layer of muscle extends along the length of the worm. These long muscles pull against the cuticle and the pseudocoelom (fluid-filled body cavity), whipping the worm's body from side to side. While some roundworms grow to be a foot or more in length, most are microscopic or only a few millimeters long. The vast majority of roundworms are free-living, active hunters.

Figure 29-17
Roundworms in pets. When a dog or cat has to be wormed, it is usually due to a roundworm infection caused by *Toxocara canis* or *Toxocara cati.*

Roundworm infections

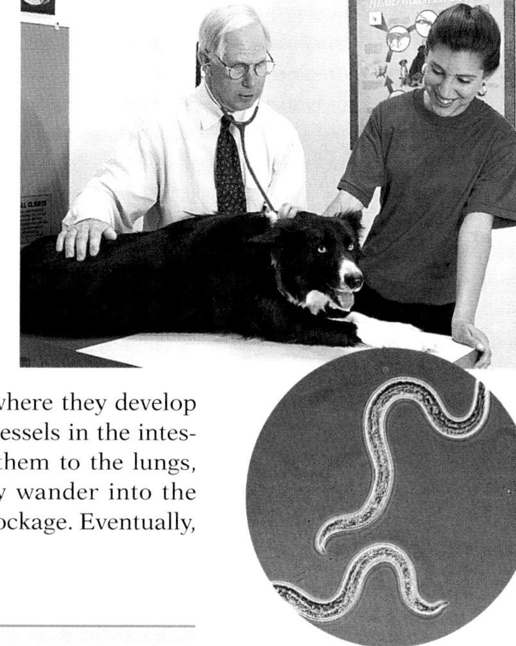

About 50 roundworm species are plant or animal parasites that cause considerable economic damage to crops and inflict terrible human suffering. Plant species may attack any part of the plant—leaves, stem, roots—depending on the species. They feed on the living plant cells, causing wilting and withering of the plant. At least 14 species of roundworms infect humans. Three sources of human infection are *Ascaris lumbricoides, Trichinella spiralis,* and members of the genus *Necator,* commonly called hookworms.

The eggs of *Ascaris* are carried through human waste to the soil, where they can live for years. If ingested, the eggs enter the intestine, where they develop into larvae. The larvae bore through the blood vessels in the intestine and enter the bloodstream, which carries them to the lungs, causing respiratory distress. Some larvae may wander into the ducts of the pancreas or gallbladder, causing a blockage. Eventually,

Magnification: 120×

Evolutionary Milestone

4 **Body Cavity**

Roundworms have a pseudocoelom, a body cavity that forms between the gut and the body wall. Like all pseudocoelomates, roundworms have a one-way gut in which food passes into the mouth and out of the anus.

Teaching TIP ADVANCED

Create a Life Cycle

Have students pick either *Ascaris* or *Necator*. Ask them to research their worm in a zoology textbook. Tell them to diagram a life cycle for their worm similar to the life cycles diagrammed earlier in the chapter. PORTFOLIO

Teaching TIP

Body Cavity

Have students read the **Evolutionary Milestone** on page 651. Remind them that unlike the other simple invertebrates they have learned about, roundworms have a pseudocoelom. Ask students why this is considered to be an evolutionary milestone. *(It permits the development of more-complex organs.)*

Making Biology Relevant

Food

Nematodes are organisms that gardeners have to deal with because they can destroy the health of a plant. When a gardener buys plants for a vegetable garden, they are sometimes marked with the letter *N.* The *N* stands for "nematode resistant." Because roundworms that infect plants are so common in some areas of the country, it is often necessary to breed plants that are resistant to them. But there are actually some free-living nematodes, called beneficial nematodes, that are good for plants. Beneficial nematodes kill many organisms in the soil that harm plants. Have interested students find out more about nematodes and plants.

Biology in Your Community

Trichinosis

Ask students if they have ever been told not to eat undercooked pork. Explain that the reason for this warning is that pork may contain the eggs of *Trichinella spiralis,* a roundworm that causes the disease trichinosis. If ingested by humans, the eggs will hatch and the larvae will develop in the host's gut. *Trichinella* infection can cause diarrhea, fever, muscle pain, and, in advanced cases, death.

Chapter 29

Identifying parasites

Time 10–15 minutes

Process Skills Interpreting graphs, analyzing data, inferring conclusions

Teaching Strategies
Have students review the information about *Schistosoma* and *Ascaris.*

Analysis Answers

1. Egg release in parasite A dropped after drug 2 was given. Egg release in parasite B dropped after drug 1 was given.
2. *Schistosoma:* blood vessels of the intestines, kidneys, and liver; *Ascaris:* intestines
3. Parasite A is *Schistosoma*, and parasite B is *Ascaris.*
4. Parasite B responded to drug 1, which targets the intestine, where *Ascaris* spends much of its life. Parasite A responded to drug 2, which targets blood vessels, where *Schistosoma* spends most if its life.
5. Dams create reservoirs, which provide habitat for snails. More snails can harbor more *Schistosoma* larvae, increasing infection rates.

3 Assessment Options

Closure

Technique: K-W-L

Tell students to return to the list of things they Want to know about simple invertebrates, which they created using the K-W-L active reading technique on page 635. Have students make a list of what they have Learned.

Review

Assign Review—Section 29-3.

Reteaching — BASIC

Tell students that flatworms lack circulatory and digestive systems. Flat bodies allow dissolved and digested substances to diffuse between their tissues, which all lie close to the water and to their branched intestine.

Figure 29-18
Roundworm *Ascaris.* These adult *Ascaris* are in the stomach of a brown pelican.

the larvae return to the intestine, where they mature and mate. Adult *Ascaris* may grow up to 0.3 m (1 ft) in length. **Figure 29-18** shows an *Ascaris* species whose final host is a bird.

Like *Ascaris, Trichinella* and *Necator* have complicated life cycles that can involve a human host. *Trichinella* infects pigs and causes a serious disease called trichinosis *(trihk ih NOH sihs)* in humans. Infection with *Trichinella* can be avoided by eating only pork that is fully cooked. Members of the genus *Necator* live mostly in the warm, moist soils of the tropics. Infection can occur when people step barefooted on soil containing hookworm larvae, which can enter the blood vessels if they penetrate the soles of the feet.

DATA LAB: Identifying parasites

Background

This graph shows how two drugs affect the release of eggs in a human infested with two parasites. Drug 1 works by killing adult parasites in the intestines. Drug 2 works by killing adult parasites in the blood vessels. Use the graph and your knowledge of parasitic infections to answer the analysis questions.

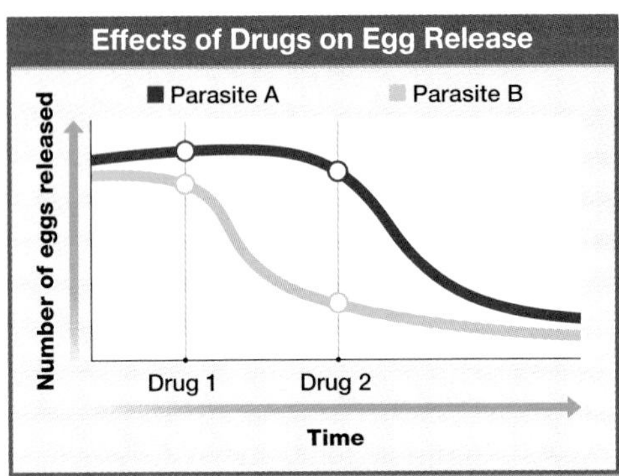

Analysis

1. **Describe** the response of the parasites to the two different drug treatments.
2. **Identify** the main human organs and tissues infected by the adult stages of *Schistosoma* and *Ascaris.* Use your textbook if necessary.
3. **Identify** which curve on the graph shows *Schistosoma* egg production and which shows *Ascaris* egg production.
4. **SKILL Justifying Conclusions** Explain why you made the identifications you did in item 3.
5. **SKILL Forming Hypotheses** *Schistosoma* spends part of its life cycle as a parasite of snails. Hypothesize a reason for an increase in the number of cases of schistosomiasis in villages near where hydroelectric dams have been built.

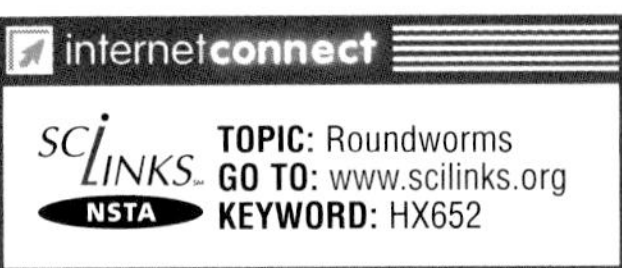

Review SECTION 29-3

1. **Compare** the internal and external anatomy of a planarian with that of a parasitic flatworm.
2. **Summarize** in words or with a diagram the life cycle of a blood fluke.
3. **Describe** a major innovation in body plan that first occurred in roundworms.

Critical Thinking

4. **SKILL Evaluating Conclusions** A student concludes that infection with *Schistosoma* is more difficult to prevent than is infection with *Trichinella.* Evaluate this conclusion.

Review SECTION 29-3 ANSWERS

1. Parasitic flatworms have suckers and hooks on their anterior end, while planarians do not. Planarians have eyespots, while parasitic flatworms do not. Planarians have a highly branched intestine, while parasitic flatworms do not have a digestive system.
2. Answers will vary but should include the basic steps shown on page 650.
3. Roundworms have a pseudocoelom.
4. Answers will vary. Many students may agree because *Trichinella* can be avoided simply by eating well-cooked pork or by not eating pork at all. *Schistosoma* is contracted by contact with infected water. It may be difficult for a person to avoid an infected water source, especially if it is the only water source avaliable.

Study ZONE CHAPTER 29

Use the Key Concepts and Key Terms listed below to review the main ideas in this chapter. If you need to review the meaning of a term, turn to the page indicated.

Key Concepts

29-1 Sponges

- Sponges lack symmetry and tissues.
- Sponges are sessile filter feeders that draw sea water through pores into an internal cavity, trapping tiny aquatic organisms.
- The sponge's supportive skeleton is composed of soft spongin fibers, hard spicules, or a combination of both.
- Sponges that reproduce sexually are usually hermaphrodites. Sponges also reproduce asexually.

29-2 Cnidarians

- Cnidarians are radially symmetrical, with bodies made up of tissue. Their body form may be a medusa or a polyp.
- Cnidocytes are stinging cells found in the tentacles of cnidarians. Harpoon-like nematocysts are located within the cnidocytes.
- Most hydrozoans are colonial organisms that reproduce asexually, though many forms can also reproduce sexually.
- Jellyfish are active predators, and some have extremely potent toxins within their nematocysts.
- Jellyfish spend most of their lives as medusas and usually reproduce sexually.
- Sea anemones and corals have thick, stalklike polyp bodies. Their life cycle includes no medusa form.

29-3 Flatworms and Roundworms

- Flatworms have flattened bodies that lack a body cavity. Most flatworms, such as planarians and marine flatworms, are free-living, but others, such as flukes and tapeworms, are parasites.
- Tapeworms are intestinal parasites that absorb food directly through their skin.
- Flukes are endoparasitic flatworms. They have a protective covering called a tegument that keeps them from being digested by their host.
- Roundworms have a pseudocoelom and a one-way gut. Most are free-living, but some are animal parasites.

Key Terms

29-1

ostia (636)
oscula (636)
sessile (636)
choanocyte (637)
amoebocyte (637)
spongin (638)
spicule (638)
gemmule (639)

29-2

medusa (640)
polyp (640)
cnidocyte (641)
nematocyst (641)
basal disk (642)
planula (643)

29-3

proglottid (649)
fluke (650)
tegument (650)

Study TIP Ready?

If you think you understand these Key Concepts and Key Terms, then you're ready for the Performance Zone.

Study ZONE CHAPTER 29

Review and Practice

Assign the following worksheets for Chapter 29:

- **Student Review Guide**
- **Concept Mapping Worksheet**
- **Test Preparation Pretest**
- **Vocabulary Worksheet**
- **Critical Thinking Review Worksheet**

Assessment Options

The following assessment options are available for this chapter:

Resource Link

- Assign **Chapter Test 29** to assess students' understanding of the chapter.

Alternative Assessment

- Assign students to cooperative groups. Have the groups develop a series of questions and answers for each section. Collect the questions but keep each group's questions together. Then have teams quiz each other with any set of questions except for the ones they made up. Students are free to challenge any answers they think are incorrect or incomplete.

Portfolio Assessment

- *Portfolio Assessment* at the front of this book provides options for building and evaluating student portfolios. Students and teachers can select items from the areas listed for inclusion in a portfolio.
- See items labeled PORTFOLIO on pp. 637, 639, 646, 648, 650, and 651.

Answer to Concept Map

The following is one possible answer to Performance Zone item 1 on page 654.

Simple invertebrates
include
sponges — roundworms — cnidarians — flatworms
sponges are sessile; sponges skeletons of spongin; sponges cells called choanocytes
cnidarians maybe polyp, medusa; cnidarians have cnidocytes
flatworms some have proglottids; flatworms covered by a tegument; flatworms including flukes

Performance ZONE
CHAPTER 29 REVIEW ANSWERS

Assignment Guide

Section	Review
29-1	3, 4, 5, 12, 23, 25
29-2	2, 6, 7, 8, 9, 14, 16, 17, 18, 19, 20, 26, 28
29-3	1, 10, 11, 13, 15, 21, 22, 24, 27

Understanding and Applying Concepts

1. The answer to the concept map is found at the bottom of page 653.
2. a. A medusa is the free-floating, often umbrella-shaped, form of a cnidarian. A polyp is the tube-shaped form of a cnidarian that lives attached to a rock or some other object.

   b. A gemmule is a cluster of sponge amoebocytes and their food that is encased in a protective coat. Gemmules can survive conditions that kill sponges, and they grow into new sponges when conditions improve. A planulae is the free-swimming larval form of a hydrozoan formed when two gametes fuse.

   c. Cnidocytes are the stinging cells found in cnidarians. Nematocysts are the small, barbed structures within a cnidocyte that spring out when triggered.
3. b
4. b
5. a
6. c
7. d
8. a
9. d
10. b
11. b
12. Answers will vary. Reasons why no animal groups evolved from sponges might include that they are unlike other animals and have no mouth or digestive tract, the structure of the sponge body is built around a water canal system, and the outer body layer is poorly developed.

Performance ZONE CHAPTER 29 Review

Understanding and Applying Concepts

1. **Concept Mapping** Make a concept map that shows the major characteristics of sponges, cnidarians, flatworms, and roundworms. Include the following terms in your map: sessile, choanocyte, spongin, medusa, polyp, cnidocyte, fluke, tegument, proglottid. Include additional concepts in your map as needed.
2. **Understanding Vocabulary** For each pair of terms, explain the difference in their meaning.
   a. medusa, polyp
   b. gemmule, planula
   c. cnidocyte, nematocyst
3. Which of the following is not a characteristic of sponges?
   a. body wall penetrated by many pores
   b. cells organized into tissues
   c. collar cells that trap food particles
   d. amoebocytes that transport food
4. A sponge's protein skeleton is composed of
   a. spicules.
   b. spongin.
   c. mesoglea.
   d. amoebocytes.
5. What prevents self-fertilization among sponges?
   a. Gametes are released at different times.
   b. Few male sponges exist.
   c. Sponges are hermaphrodites.
   d. Encounters between members of the same species are rare.
6. A Portuguese man-of-war and a hydra are similar in that both
   a. are colonial.
   b. contain medusas and polyps.
   c. are hydrozoans.
   d. produce planulae.
7. Cnidocytes are used
   a. by polyps only.
   b. by medusas only.
   c. in reproduction.
   d. for defense and in feeding.
8. Which sequence reflects the life cycle of *Obelia*?
   a. polyp ⟶ medusa ⟶ planula
   b. medusa ⟶ polyp ⟶ planula
   c. planula ⟶ medusa ⟶ polyp
   d. polyp ⟶ planula ⟶ medusa
9. Which is an anthozoan?
   a. hydra
   b. jellyfish
   c. Portuguese man-of-war
   d. sea anemone
10. The covering that protects endoparasites from the actions of digestive enzymes is called the
    a. osculum.
    b. tegument.
    c. proglottid.
    d. basal disk.
11. Identify the function of the structure shown below.
    a. respiration
    b. water removal
    c. feeding
    d. digestion

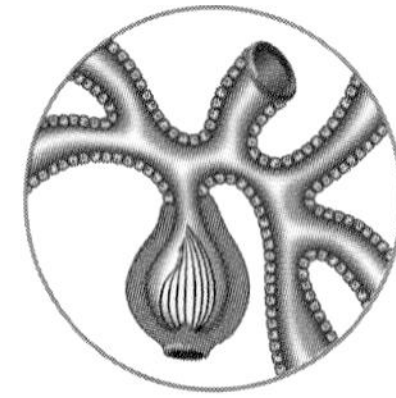

12. **Theme Evolution** Porifera has been called a dead-end phylum. List some possible reasons why no animal group evolved from the sponges.
13. **Theme Metabolism** How are parasitic flukes able to live when they no longer possess a well-developed digestive system?
14. **Exploring Further** What kinds of life-forms are supported by coral reefs, and why is it important for people to make sure that the reefs survive and develop?
15. **Chapter Links** Classify all of the organisms covered in this chapter as either acoelomate, pseudocoelomate, or coelomate. (**Hint:** See Chapter 28, Section 28-1.)

---

13. Parasitic flukes take their nourishment from their hosts by absorbing nutrients through their skin from the host's body fluids.
14. Life-forms found on and around a coral reef include algae and microscopic invertebrates, tube worms, mollusks, sponges, flatworms, shrimp, crabs, sea anemone, hydroids, feather duster worms, sea stars, octopuses, eels, sharks, and other fishes. Coral reefs protect coastlines from coastal erosion and serve as resources for fisheries. Tourists who visit them bring money into the local economy. Reef-dwelling animals are also sources of new medications.
15. acoleomates: hydrozoans, cnidarians, anthozoans, flatworms; pseudocoelomates: roundworms; coelomates: none of the animals are coelomates

Skills Assessment

Part 1: Interpreting Graphics

Study the illustration of a hydra below, and answer the following questions:

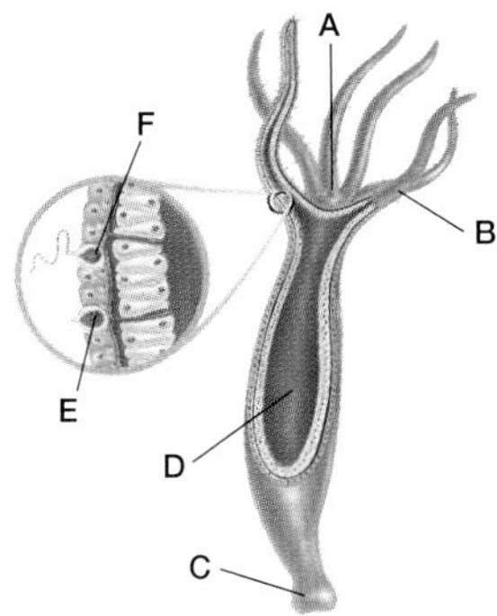

16. Identify the structures labeled *A–F* on the illustration above.
17. Identify the function of the structures labeled *E* and *F*, and explain the difference in their appearance.
18. How does the structure labeled *D* set the hydra apart from sponges?
19. What might be the relationship between the structures labeled *E* and *F* to the cells colored purple?
20. Identify the layers of the hydra colored blue, red, and yellow.

Part 2: Life/Work Skills

Use the media center or Internet resources to find more information about infections with tapeworms or roundworms.

21. **Working Cooperatively** Working with a partner, research at least three flatworms or roundworms that are parasites of humans. Find out how infection occurs, how it is treated, and if there are any programs aimed at decreasing infection rates.
22. **Communicating** Prepare an oral report to summarize your findings. Create a brochure that advises people of how they can avoid infection.

Critical Thinking

23. **Inferring Conclusions** Individuals of a single species of sponge may vary in appearance. Factors that affect sponge shape include differences in substrate, availability of space, and the velocity and temperature of water currents. How might these factors make the classification of sponges confusing?
24. **Forming Reasoned Opinions** When locating sunken ships, some treasure seekers use dynamite to blast away portions of the ocean floor. Consider what you know about some of the invertebrates that live in the ocean. Then give your opinion of why ocean blasting should or should not be used as a way to locate sunken ships.

Portfolio Projects

25. **Interpreting Information** Read "The Incredible Sponge," by Henry Genthe (*Smithsonian*, Aug. 1998, pp. 50–58). Prepare an oral report that discusses some of the symbiotic relationships mentioned in the article. Record your report on audiotape, and place a copy in your portfolio.
26. **Forming a Model** In groups of three, research how one of the three different types of coral reefs—fringing, barrier, and atoll—is formed. Then build a model of your reef type or make a map showing where such reefs are located. Set up an exhibit of your work, and use a tape recording to create a self-guided "tour."
27. **Identifying Structures** Make an anatomical drawing of the interior of a sponge, cnidarian, flatworm, or roundworm. Identify the species, and label at least 10 structures. Distribute copies of your drawing to your classmates.
28. **Career Focus Marine Biologist** Research the field of marine biology, and write a report on your findings. Your report should include a job description, training required, kinds of employers, growth prospects, and starting salary.

Skills Assessment

16. A. mouth; B. tentacle; C. basal disk; D. gastrovascular cavity; E. coiled nematocyst; F. nematocyst discharged
17. E and F are nematocysts within a cnidocyte cell. They can deliver a sometimes deadly sting, and their function is in feeding and defense. The nematocysts (E and F) differ in appearance because one (F) has been discharged, while the other (E) is still coiled within the cnidocyte.
18. Sponges do not have a gastrovascular cavity and must digest their food intracellularly. Hydrozoans begin digestion extracellularly, which means that they can eat food larger than their individual cells.
19. The purple cells are nerve cells. They trigger the firing of the cnidocytes.
20. blue: cells from ectoderm; red: mesoglea; yellow: cells from endoderm
21. Answers will vary depending on what parasitic worms the students research.
22. Reports and brochures will vary.

Critical Thinking

23. A biologist attempting to classify sponges would need to be very familiar with the environment in which they live. The variables that determine the shape of a sponge would need to be considered. If they were not, two similar-looking sponges might be mistakenly identified as the same species, or vice versa.
24. Answers will vary. Students might note that blasting disturbs the ocean floor, where many invertebrates live. Also, many invertebrates are filter feeders. The debris thrown up by the blast fills the water with particles, which can interfere with filter feeding.

Portfolio Projects

25. Reports will vary. Students might note that many small organisms find shelter within a sponge and have become adapted to the toxins many sponges produce. These organisms include small fish, shrimp, worms, and brittle stars. Sponges also live on some organisms, such as the sponge crab.
26. Models and exhibits will vary.
27. Drawings will vary.
28. Answers will vary. Marine biologists study organisms that live in salt water. They study the life forms found in various marine habitats and how the organisms interact with each other and with their physical environment. Marine biologists have a bachelor's degree and often a master's or Ph. D. Biologists with backgrounds in subfields as far-ranging as ecology and embryology can specialize on marine organisms and communities. Marine biologists work for federal and state government agencies, universities, saltwater fish and algae farming businesses, and nonprofit organizations. The growth potential for this field is fair to good. Starting salary will vary by region.

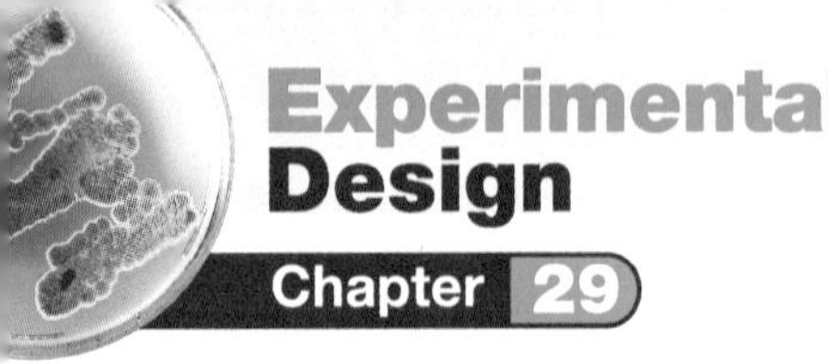

How Do Hydras Respond to Stimuli?

Purpose
Students will observe the feeding behavior of hydras to determine how they find and capture prey.

Preparation Notes
Time Required: 45 minutes

Materials
Materials for this lab can be ordered from WARD'S. See *Master Materials List* at the front of this book for catalog numbers.

Preparation Tips
- Demonstrate how to pick up hydras and *Daphnia* with a medicine dropper. If the tip of the dropper is too small, the dropper can be inverted.
- Instead of *Daphnia*, you may also use brine shrimp or a thread soaked in anything that has glutathione (e.g., liver) to elicit capture behavior in the hydra. The hydra will wave its tentacles in search of the "prey."

Safety Precautions
Caution students to handle glassware carefully so as to avoid injuring themselves or others.

Procedural Tips
- Part A will take about 30 minutes to complete. Part B will take about 15 minutes.
- Discuss some of the basic principles of sensory cells, neurons, and effector cells, relating the information to the way hydras capture prey.
- To observe whether a hydra responds to a chemical stimulus (nutrient), the students can hold a pennant-shaped piece of filter paper with forceps and move the long tip of the pennant near, but not touching, the hydra's tentacles (as a control). After observing and recording the hydra's response to the filter paper, they can dip the same piece of filter paper in beef broth and repeat the procedure.
- To observe how a hydra responds to touch, students can use the long tip of a clean pennant-shaped piece of filter paper and touch the hydra's tentacles, mouth, disk, and stalk.
- To observe feeding behavior, students can use a medicine dropper to transfer live *Daphnia* to the well with the hydra on the microscope slide.
- You may wish to have students return the fed hydra to a container marked "fed hydra" to ensure appropriate responses for other students conducting the lab.

Disposal
Solutions can be poured down the drain, and solids can be thrown in the trash.

Before You Begin
Answers
1. *cidarian*—organisms of the phylum Cnidaria that have a hollow gut with a single opening and flexible, fingerlike tentacles

   hydra—common freshwater cnidarian with a body stalk and several tentacles surrounding the mouth

Experimental Design

Chapter 29 How Do Hydras Respond to Stimuli?

SKILLS
- Using scientific processes
- Observing

OBJECTIVES
- **Observe** a hydra finding and capturing prey.
- **Determine** how a hydra responds to stimuli.

MATERIALS
- silicone culture gum
- microscope slide
- 2 medicine droppers
- *Hydra* culture
- *Daphnia* culture
- concentrated beef broth
- filter paper cut into pennant shapes
- forceps
- stereomicroscope

Magnification: 260×

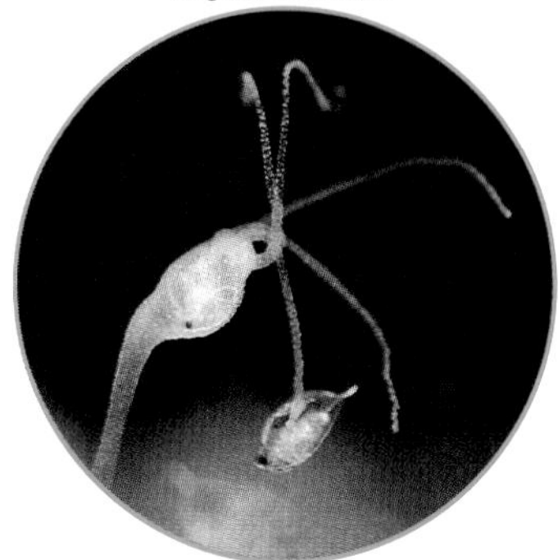

Hydra feeding on *Daphnia*

Before You Begin
Cnidarians are carnivorous animals. A common cnidarian is ***Hydra,*** a freshwater organism that feeds on smaller freshwater animals, such as water fleas *(Daphnia).* Hydras find food by responding to stimuli, such as chemicals and touch. The way an animal responds to stimuli is called **behavior.** The tentacles of a cnidarian are armed with **nematocysts,** shown below, which are used in defense and in capturing prey. When a hydra receives stimuli from potential prey, its nematocysts spring out and harpoon or entangle the prey. In this lab, you will observe the feeding behavior of hydras to determine how they find and capture prey.

1. Write a definition for each boldface term in the paragraph above.
2. Based on the objectives for this lab, write a question you would like to explore about the feeding behavior of hydras.

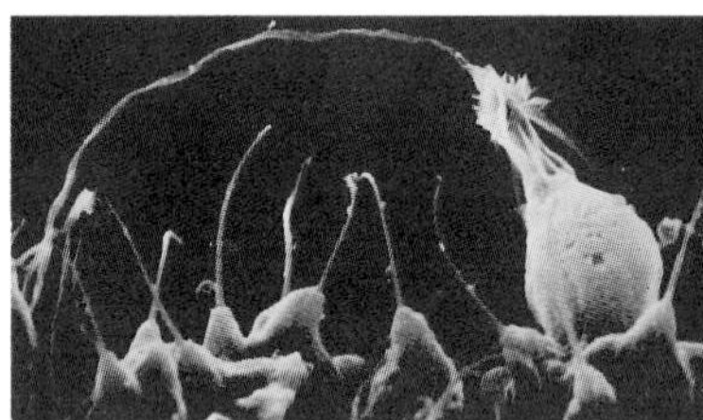

Nematocyst discharged

Procedure
Part A: Make Observations
1. To make an experimental pond for observing hydras, squeeze out a long piece of silicone culture gum. Arrange it to form a circular well on a microscope slide, as shown in the photograph below. **CAUTION: Glassware is fragile. Notify the teacher promptly of any broken glass or cuts.**

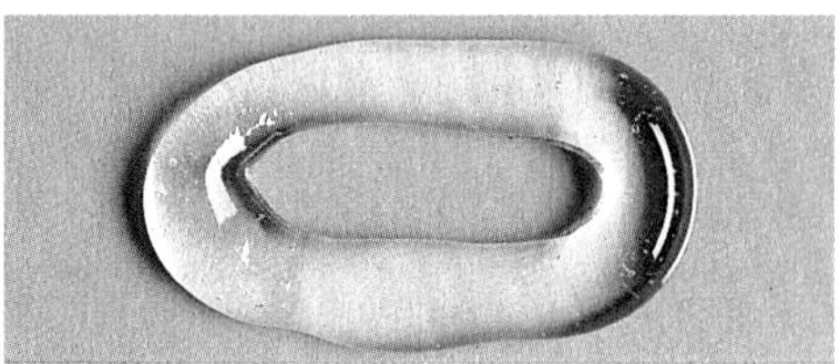

2. With a medicine dropper, gently transfer a hydra from its culture dish to the well on the slide, making sure the water covers the animal. **CAUTION: Handle hydras gently to avoid injuring them.** Allow the hydra to settle, then examine it under the high power of a stereomicroscope. Draw a hydra and label the body stalk, mouth, and tentacles.

Part B: Design an Experiment

3. Work with the members of your lab group to explore one of the questions written for step 2 of **Before You Begin.** To explore the question, design an experiment that uses the materials listed for this lab.
4. Write a procedure for your experiment. Make a list of all the safety precautions you will take. Have your teacher approve your procedure and safety precautions before you begin the experiment.

> **You Choose**
>
> **As you design your experiment, decide the following:**
>
> a. what question you will explore
> b. what hypothesis you will test
> c. how to observe a hydra's feeding behavior
> d. how to test a hydra's response to a stimulus, such as a chemical or a touch
> e. what your test groups and controls will be
> f. what to record in your data table

Part C: Conduct Your Experiment

5. Set up and carry out your experiment. **CAUTION: Handle hydras gently to avoid injuring them.**
6. Allow hydras to settle before exposing them to a test condition. If your hydra does not respond after a few minutes, obtain another hydra from the culture dish. Repeat your procedure.

Part D: Cleanup and Disposal

7. Dispose of lab materials and broken glass in the designated waste containers. Put hydras and daphnias in the designated containers. Do not put lab materials in the trash unless your teacher tells you to do so.
8. Clean up your work area and all lab equipment. Return lab equipment to its proper place. Wash your hands thoroughly before you leave the lab and after you finish all work.

Analyze and Conclude

1. **Analyzing Results** Describe a hydra's response to chemicals (beef broth).
2. **Analyzing Results** Describe a hydra's response to touch.
3. **Drawing Conclusions** How does a hydra detect its prey?
4. **Justifying Conclusions** Give evidence to support your conclusion about how hydras detect prey.
5. **Inferring Conclusions** Based on your observations, how do you think a hydra behaves when it detects a threat in its natural habitat?
6. **Inferring Conclusions** What happens to food that has not been digested by a hydra?
7. **Inferring Conclusions** How is a hydra adapted to a sedentary lifestyle?
8. **Further Inquiry** Write a new question about the behavior of hydras that could be explored with another investigation.

Do You Know?

Do research in the library or media center to answer these questions:

1. What different kinds of food does a hydra eat?
2. How is the feeding method of a hydra different from that of a sponge?

Use the following Internet resources to explore your own questions about sponges and cnidarians.

internetconnect

TOPIC: Hydra
GO TO: www.scilinks.org
KEYWORD: HX657

behavior—how an animal responds to a stimulus.

nematocysts—cells found on a cnidarian that are used in defense and in capturing prey

2. Answers will vary. For example: How does a hydra respond to a *Daphnia* that swims close to it?

Procedure

Sample Procedure

Use a medicine dropper to transfer live *Daphnia* to the well with the hydra on the microscope slide. Observe the hydra carefully under the stereomicroscope. Record observations in the data table. If the hydra does not respond, repeat the experiment with another hydra.

Sample Data Table

Observations of Hydra	
Response to filter paper	No response
Response to beef broth	Moves toward filter paper
Response to touch	Coils up
Feeding behavior	Nematocysts spear the *Daphnia*. Tentacles pull it to the hydra's mouth.

Analyze and Conclude

Answers

1. The hydra should show a feeding response to the beef broth, which includes expansion of its mouth, movement of its tentacles, elongation of its body, and the release of nematocysts.
2. The hydra contracts its body in response to touch.
3. Students should conclude that a hydra responds to chemicals in the water to detect the presence of prey.
4. The hydra exhibited predatory behavior in response to the beef broth, which contained proteins suspended in water. Yet the hydra displayed nonpredatory behavior when touched.
5. Based on the hydra's response to touch, students should conclude that a hydra contracts its body when threatened in its natural habitat.
6. Undigested food is released from the hydra's mouth.
7. Tentacles and nematocysts enable a hydra to capture prey that drift past it in slow-moving water. Its ability to contract when touched protects the hydra from predators.
8. Answers will vary. For example: How does a hydra respond to light? What kinds of food does a hydra prefer?

Do You Know?

Answers

1. Hydra will capture anything that is of the appropriate size, but the most common foods include small crustaceans, worms, and insect larvae.
2. Both sponges and hydras are sessile. However, sponges obtain food passively by filter feeding. They have no specialized structures for capturing food. Hydras are active feeders that have tentacles and nematocysts for capturing food.

Chapter 30

Mollusks and Annelids

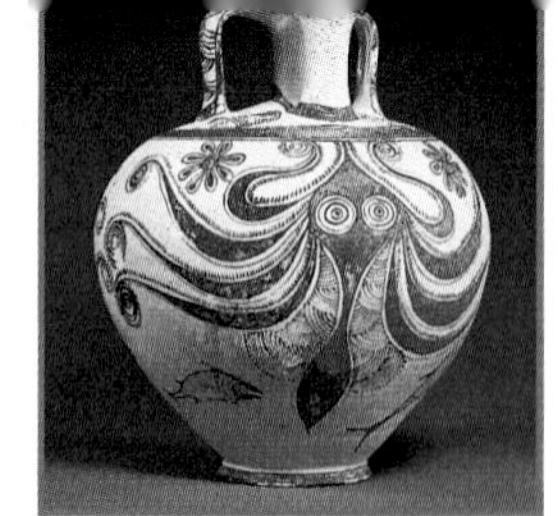

Planning Guide

	OBJECTIVES	MEETING INDIVIDUAL NEEDS	READING SKILLS AND STRATEGIES
BLOCKS 1 & 2 90 min	**30-1 Mollusks** • **Summarize** the evolutionary relationship between mollusks and annelids. • **Describe** the key characteristics of mollusks. • **Describe** excretion, circulation, respiration, and reproduction in mollusks. • **Compare** the body plans and feeding adaptations of gastropods, bivalves, and cephalopods.	Learning Styles, p. 660 BASIC ADVANCED • pp. 667, 668	**ATE** Active Reading, Technique: Brainstorming, p. 659 **ATE** Active Reading, Technique: Paired Summarizing, p. 661 ■ Active Reading Guide, worksheet (30-1) ■ Directed Reading, worksheet (30-1) ■ Reading Strategies, Compare/Contrast Matrix
BLOCKS 3 & 4 90 min	**30-2 Annelids** • **Identify** the major change in body plan that distinguishes annelids from mollusks. • **Describe** the basic annelid body plan. • **Describe** the annelid digestive system. • **Compare** the three classes of annelids.	Learning Styles, p. 669 BASIC ADVANCED • pp. 670, 674	■ Active Reading Guide, worksheet (30-2) ■ Directed Reading, worksheet (30-2) ■ Supplemental Reading Guide, *Journey to the Ants: A Story of Scientific Exploration*

Review and Assessment Resources

BLOCKS 5 & 6 90 min

TEST PREP

PE Study Zone, p. 675

PE Performance Zone, pp. 676–677

■ Chapter 30 worksheets:
- Vocabulary
- Concept Mapping
- Critical Thinking

 Audio CD Program

■ Test Prep Pretest

CHAPTER TESTING

■ Chapter Tests (blackline master)

▲ Test Generator

■ Assessment Item Listing

See the correlations table at the front of this book for a list of the ***National Science Education Standards*** addressed in this chapter.

Smithsonian Institution® Visit **www.si.edu/hrw** for additional on-line resources.

KEY

Biology Principles and Explorations

- **PE** Pupil's Edition
- **ATE** Teacher's Edition

HOLT BIOSOURCES

- Teaching Transparencies
- Laboratory Program

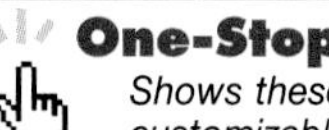

One-Stop Planner CD-ROM

Shows these resources within a customizable daily lesson plan:

- ■ Teaching Resources
- ▲ Test Generator

VISUAL STRATEGIES

CLASSWORK AND HOMEWORK

TECHNOLOGY AND INTERNET RESOURCES

Visual Strategies

ATE Fig. 30-3, p. 662

141 Anatomy of a Clam

142 Open Circulatory System in a Bivalve

159 Exploration of a Mollusk

Classwork and Homework

PE **Quick Lab** How can you model an open circulatory system?, p. 663

PE **Data Lab** Modifications of the mollusk body plan, p. 668

Lab B24 Snails

PE **Review,** p. 668

Technology and Internet Resources

Holt Biology Videodiscs Section 30-1

Audio CD Program Section 30-1

Visual Strategies

ATE Fig. 30-11, p. 671

152 Closed Circulatory System in an Earthworm

160 Exploration of an Annelid

136A Structure of a Marine Worm

Classwork and Homework

PE **Quick Lab** How can you model a closed circulatory system?, p. 670

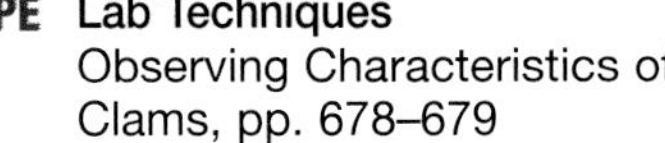

PE **Lab Techniques** Observing Characteristics of Clams, pp. 678–679

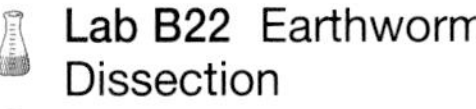

Lab B22 Earthworm Dissection

Lab B23 Live Earthworms

PE **Review,** p. 674

PE **Study Zone,** p. 675

Technology and Internet Resources

Holt Biology Videodiscs Section 30-2

Audio CD Program Section 30-2

ALTERNATIVE ASSESSMENT

PE **Portfolio Projects 26–29,** p. 677

ATE **Alternative Assessment,** p. 675

ATE **Portfolio Assessment,** p. 675

Lab Activity Materials

Quick Lab

p. 663 15 cm (about 6 in.) piece of surgical tubing, 15 cm (about 6 in.) and 7.5 cm (about 3 in.) pieces of clear plastic tubing, shallow pan filled with water, eyedropper, food coloring

p. 670 30 cm (about 12 in.) piece of clear plastic tubing, 15 cm (about 6 in.) piece of surgical tubing, shallow pan filled with water, eyedropper, food coloring

Data Lab

p. 668 pencil, paper

Lab Techniques

p. 678 safety goggles, lab apron, live clam, small beaker or dish, eyedropper, food coloring, glass stirring rod, clam shell, Petri dish, scalpel, stereomicroscope, 0.1 M HCl

Lesson Warm-up

ATE earthworms, p. 669

Demonstrations

ATE several living land snails, p. 664

ATE dissected fresh or preserved mollusk, p. 664

ATE fossil mollusks, p. 665

ATE suction cups (2), p. 667

ATE coffee filters (2), beakers (2), rubber bands (2), water, food coloring, p. 671

BIOSOURCES LAB PROGRAM

- Inquiry Skills Development: Lab B22, p. 101
- Inquiry Skills Development: Lab B23, p. 107
- Inquiry Skills Development: Lab B24, p. 111

Chapter Themes

Cellular Structure and Function

The development of a coelom in animals is a major evolutionary adaptation. The structure of the coelom led to the development of a closed circulatory system, provided a fluid environment within which organs could be suspended, and facilitated muscle-driven body movement.

Evolution

Mollusks and annelids are diverse groups of invertebrates that share common anatomical features. The shared pattern of development of trochophore larvae is evidence of their common ancestry.

Establishing Prior Knowledge

Before beginning this chapter, make certain students can answer the questions in **Study TIP Ready?** on page 659. If they cannot, encourage them to reread the sections indicated.

Opening Demonstration

Invite students to bring to class any shell collections they might have. Display the specimens with mollusks or mollusk shells. *(Examples include clams, mussels, scallops, oysters, octopuses, squids, slugs and snails.)* Ask students to look for similarities among the mollusk shells and the other shells.

CHAPTER 30 Mollusks and Annelids

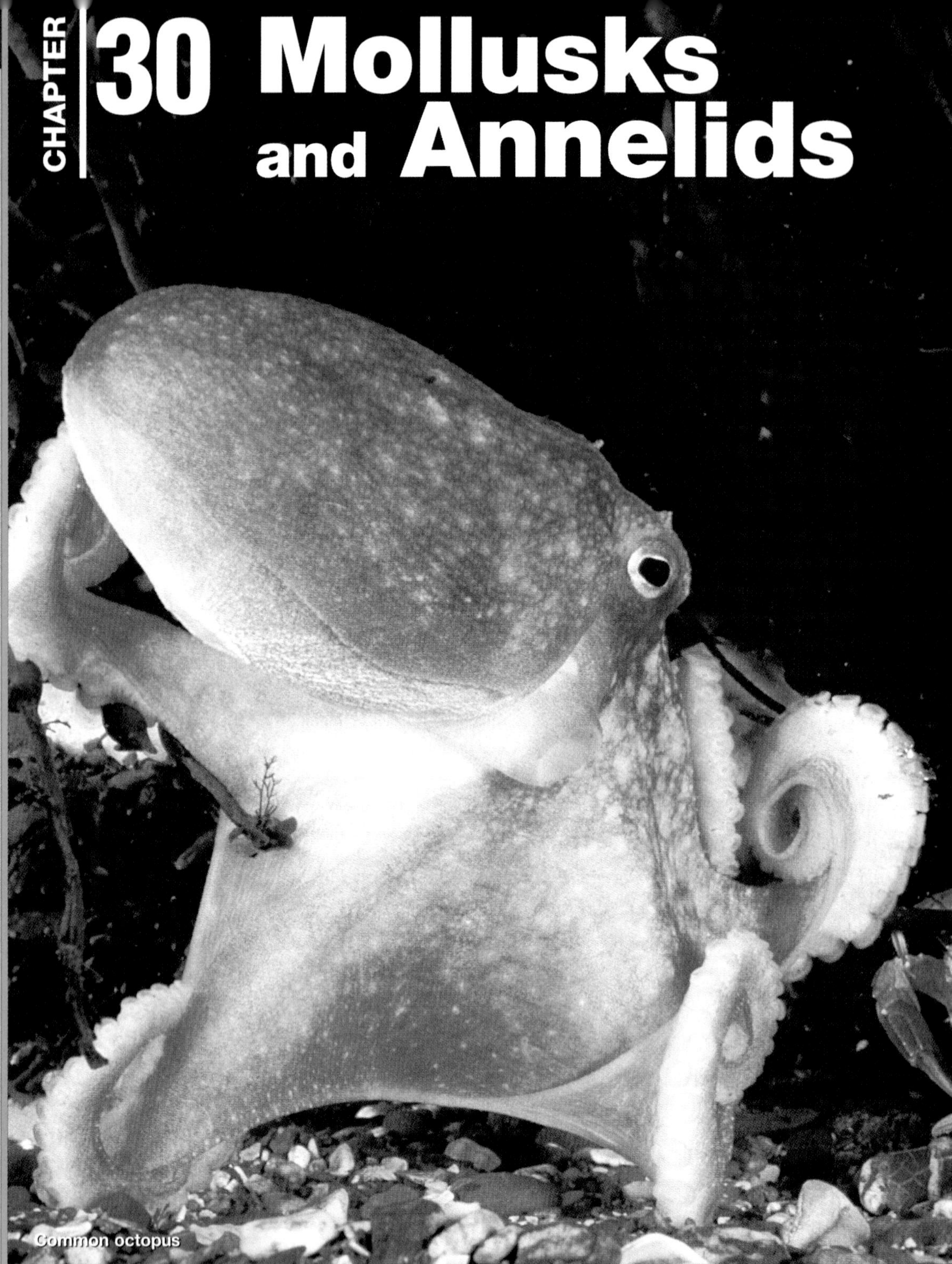

Common octopus

What's Ahead — and Why

CHAPTER OUTLINE

Students will learn

- about the similarities and differences between three major classes of mollusks—the bivalves, the gastropods, and the cephalopods.
- about the physical characteristics and various adaptations of worms.

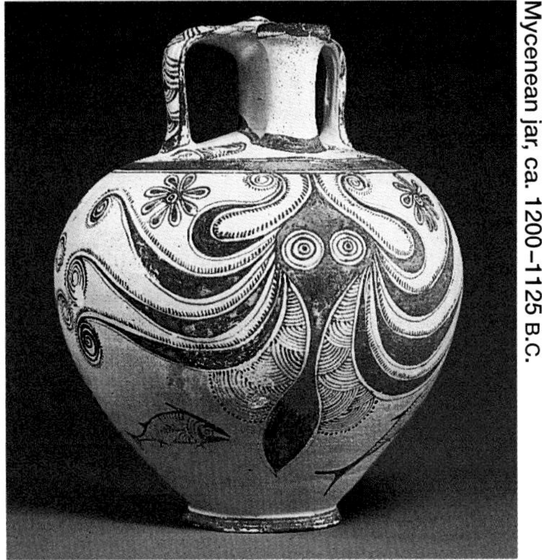

Mycenean jar, ca. 1200–1125 B.C.

Like their relatives *the segmented worms, mollusks have played a part in human life for many centuries. The ancient Mycenaeans often used the octopus as a motif on their urns. While worms never became the subject of great art, Darwin thought it unlikely that any creature played as important a role in the history of the world as did "these lowly organized creatures."*

Looking Ahead

30-1 Mollusks
There's more to this group than just snails.

30-2 Annelids
Find out why these "lowly" creatures are so important.

Labs

- **Quick Labs**
  How can you model an open circulatory system? (p. 663)
  How can you model a closed circulatory system? (p. 670)
- **Data Lab**
  Modifications of the mollusk body plan (p. 668)
- **Lab Techniques**
  Observing Characteristics of Clams (pp. 678–679)

Features

- **Food Watch**
  Eating Mollusks Safely (p. 666)
- **Up Close**
  Earthworm (p. 673)
- **Health Watch**
  Leeches Make a Comeback (p. 674)

Study TIP

Ready?

Check your understanding of these topics to help you prepare for what's ahead.

Can you . . .

- **describe** the process of diffusion? (Section 4-1)
- **define** the term *plankton*? (Section 18-3)
- **define** the terms *gill* and *hydrostatic skeleton*? (Section 28-2)
- **distinguish** between an open and a closed circulatory system? (Section 28-2)

If not, *review the sections indicated.*

internetconnect

National Science Teachers Association *sci*LINKS Internet resources are located throughout this chapter.

Reading Strategies

Active Reading

Technique: Brainstorming

Ask students to think about mollusks and annelids compared to humans. What body structures and systems do we have in common? Work with the class to create a list on the board or overhead of obvious similarities. After reading the chapter, add any items that may have been overlooked. *(respiratory system, reproductive system, circulatory system, nervous system, digestive system, excretory system, presence of a coelom)*

Directed Reading

Assign **Directed Reading Worksheet Chapter 30** to help students interact with the text as they read the chapter.

Interactive Reading

Assign ***Biology: Principles and Explorations* Audio CD Program Chapter 30** to help reluctant readers achieve greater success in reading the chapter.

Reading Skills

Assign Reading Organizers and Reading Strategies Worksheets from the **One-Stop Planner CD-ROM with Test Generator** to help students learn and practice effective reading skills.

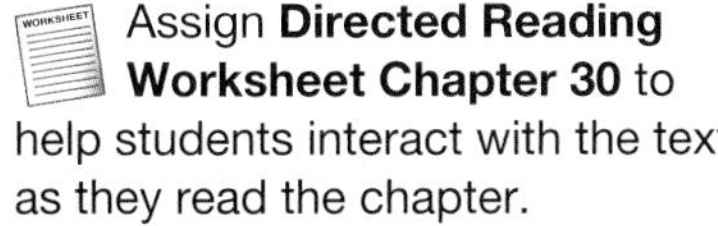

internetconnect

Holt *Biology: Principles and Explorations*

On-line Resources: go.hrw.com

Visit the HRW Web site for a variety of resources related to this chapter. Just type in the keyword HX0 Chapter 30.

Holt *Biology: Principles and Explorations*

On-line Resources: www.scilinks.org

The following *sci*LINKS Internet resources can be found in the student text for this chapter:

Page 661
TOPIC: Mollusks
KEYWORD: HX661

Page 666
TOPIC: Mollusk consumption
KEYWORD: HX666

Page 670
TOPIC: Annelids
KEYWORD: HX670

Page 672
TOPIC: Earthworms
KEYWORD: HX672

Page 674
TOPIC: Leeches
KEYWORD: HX674

Chapter 30

1 Prepare

Scheduling

- **Block:** About 90 minutes will be needed to complete this section.
- **Traditional:** About two class periods will be needed to complete this section.
- **Planning:** See the Planning Guide on pp. 657A and 657B for lecture, classwork, and assignment options for Section 30-1. Lesson plans for Section 30-1 can be found in a lesson cycle format in ***Biology: Principles and Explorations*** **One-Stop Planner CD-ROM with Test Generator.**

Resource Link

Assign **Active Reading Worksheet Section 30-1** to help students comprehend the key concepts and visuals in this lesson.

2 Teach

Lesson Warm-up

Ask students why some mollusks secrete a shell and other mollusks do not. Guide their thinking by asking how mollusks without shells might protect themselves. *(Students should recall that squids and octopuses spray ink and can propel themselves away from danger. Slugs secrete noxious substances that discourage predators.)* Lead students to conclude that the anatomical features of mollusks are adaptations for different lifestyles, or niches.

Section 30-1 Mollusks

Objectives

- **Summarize** the evolutionary relationship between mollusks and annelids. (p. 660)
- **Describe** the key characteristics of mollusks. (p. 661)
- **Describe** excretion, circulation, respiration, and reproduction in mollusks. (pp. 662–663)
- **Compare** the body plans and feeding adaptations of gastropods, bivalves, and cephalopods. (pp. 664–668)

Key Terms

trochophore
visceral mass
mantle
foot
radula
nephridium
adductor muscle
siphon

Mollusks Have a True Coelom

While most of the simple invertebrates you read about in Chapter 29 may be unfamiliar to you, chances are good that you have seen many mollusks and annelids. Snails, slugs, oysters, clams, scallops, octopuses, and squids are all mollusks. If you have seen an earthworm, then you know what an annelid is. While a snail may not seem to have much in common with an earthworm, these two very different-looking animals are related.

Mollusks and annelids were probably the first major groups of organisms to develop a true coelom. (Recall that one of the advantages of having a true body cavity is that the gut and other internal organs are suspended from the body wall and cushioned by the fluid within the coelom.) Another feature shared by mollusks and annelids is a larval stage called a **trochophore** *(TRAHK oh fawr),* which develops from the fertilized egg. In some species, the trochophore, shown in **Figure 30-1,** is free-swimming and propels itself through the water by movement of cilia on its surface. The presence of a trochophore larva in mollusks and annelids suggests that they share a common ancestor.

Members of the phylum Mollusca make up the second largest animal phyla, exceeded only by phylum Arthropoda (which you will read about in Chapter 31). Mollusks are abundant in almost all marine, freshwater, and terrestrial habitats. There are more species of terrestrial mollusks than there are of terrestrial vertebrates. These mollusks often go unnoticed because people are not accustomed to looking for them. Seven classes of mollusks make up the phylum Mollusca. The three major classes are Gastropoda (snails and slugs), Bivalvia (clams, oysters, and scallops), and Cephalopoda (octopuses and squids).

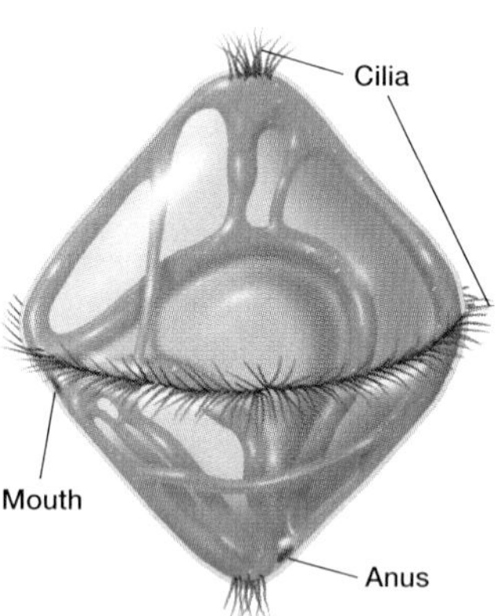

Figure 30-1 Trochophore larva. The microscopic trochophore larva has a belt of cilia that circles its body. The beating of the cilia propels the trochophore through the water.

Evolutionary Milestone

5 Coelom

A true coelom develops entirely within the mesoderm. This permits physical contact between the mesoderm and endoderm during the development of the embryo. Interaction of these two tissue layers leads to the development of complex organs.

Integrating Different Learning Styles

Logical/Mathematical	PE	Performance Zone items 24 and 25, p. 677
Visual/Spatial	PE	Quick Lab, p. 663; Data Lab, p. 668; Performance Zone item 19, p. 677
	ATE	Visual Strategy, p. 662; Demonstrations, pp. 664, 665
Interpersonal	PE	Study Tip, p. 662
	ATE	Active Reading, p. 661
Verbal/Linguistic	PE	Study Tip, p. 661; Performance Zone items 22, 26, and 28, p. 677
	ATE	Cross Disciplinary Connection, p. 662; Closure, p. 668

Mollusks Share Several Key Characteristics

Despite their varied appearance, the members of the different groups of mollusks share a number of key characteristics.

1. **Body cavity.** The body cavity in mollusks is a true coelom, although in most species it is reduced to a small area immediately surrounding the heart.
2. **Symmetry.** Most mollusks exhibit bilateral symmetry.
3. **Three-part body plan.** The body of every mollusk has three distinct parts: the visceral mass, the mantle, and the muscular foot, as shown in **Figure 30-2.** The **visceral** *(VIS uhr uhl)* **mass** is a central section that contains the mollusk's organs. Wrapped around the visceral mass like a cape is a heavy fold of tissue called the **mantle,** which forms the outer layer of the body. Finally, every mollusk has a muscular region called a **foot,** which is used primarily for locomotion.
4. **Organ systems.** Mollusks have organ systems for excretion, circulation, respiration, digestion, and reproduction.
5. **Shell.** Many mollusks have either one or two shells that serve as an exoskeleton, protecting their soft body. The shell is composed of protein that is strengthened by calcium carbonate, an extremely hard mineral.
6. **Radula.** All mollusks except bivalves have a **radula** *(RAJ uh luh)*, a rasping tongue-like organ located in their mouth. The radula, shown in Figure 30-2, has thousands of pointed, backward-curving teeth arranged in rows. When a mollusk feeds, it pushes its radula out of its mouth, and the teeth scrape fragments of food off rocks or plant matter. Mollusks that are predators use their radula for attacking their prey.

Study TIP

Reading Effectively

Before reading this chapter, write the objectives for each section on a sheet of paper. Rewrite each Objective as a question, and answer these questions as you read the section.

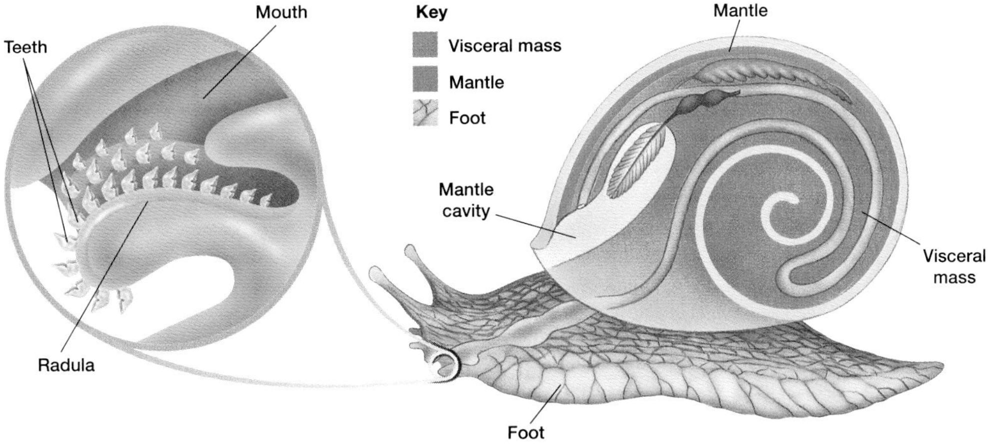

Figure 30-2 Three-part body plan. All mollusks have a three-part body plan that includes a visceral mass, a mantle, and a foot. Most mollusks also have a radula.

Making Biology Relevant

Food

Pass around menus from several seafood restaurants and seafood cookbooks from the library. Ask students to identify all the mollusks they can find. *(Students should be able to identify oysters, clams, and scallops. Depending on the menu or cookbook, there may also be snails [escargot], squids [calamari], octopuses, abalone, and mussels.)*

Active Reading

Technique: Paired Summarizing

Encourage students to pause after each paragraph they read to summarize the main points to a partner. You might model the technique by having a student read a paragraph out loud and summarize it. Then have another student read the next paragraph and summarize it.

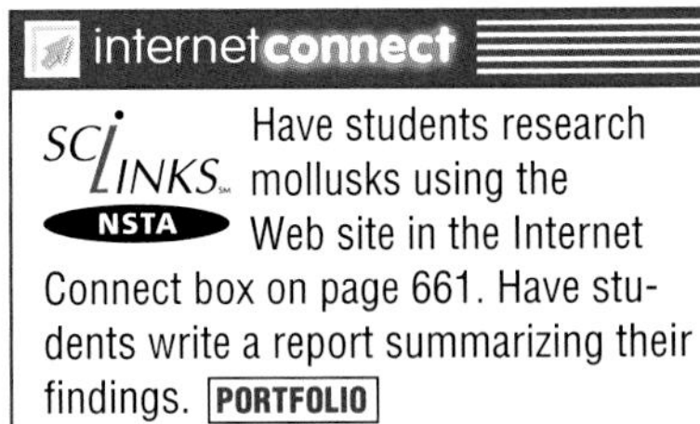

Have students research mollusks using the Web site in the Internet Connect box on page 661. Have students write a report summarizing their findings. PORTFOLIO

did you know?

Not all gastropod shells twist the same way.

Most gastropods are protected by single, spiraled shells into which the animals can retreat when threatened. Some shells spiral to the left and some spiral to the right, depending on the species.

Teaching TIP — BASIC

Excretory Strategies

Relate the excretion of wastes in nephridia to cleaning up a table after a meal. Nephridia get rid of wastes as well as substances that the mollusk needs; thus, some substances must be reabsorbed before the wastes are discharged. The process is similar to throwing away everything on the table after the meal, and then going through the trash to recover the plates, silverware, and leftovers that you want to keep.

CROSS-DISCIPLINARY CONNECTION

◆ Language Arts

Nephros is the Greek word for "kidney," the human excretory organ, and forms the root for many words in science and medicine. For example, nephritis is inflammation of the kidneys, a nephrectomy is the removal of a kidney, a nephron is the structure in a kidney that filters blood, and nephrosis is the degeneration of the kidneys. Ask students to think of a likely word for a kidney doctor and the field of medicine dedicated to the study of kidneys. *(nephrologist; nephrology)*

Visual Strategy

Figure 30-3

Use Figure 30-3 to point out that mollusks have advanced organ systems. Ask students to categorize each of the organs as part of one of the following systems: digestive, circulatory, respiratory, or excretory. *(digestive system—mouth, radula, intestine, stomach; circulatory system—heart; respiratory system—gill; excretory system—nephridium, coelom)*

Figure 30-3 Mollusk body plan

Although mollusks vary greatly in body form, they all have complex organ systems.

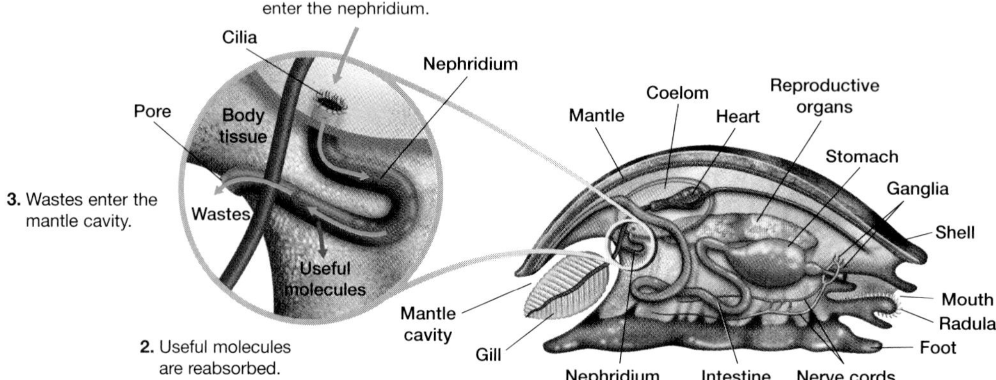

Study TIP

Reading Effectively

To better understand these mollusk organ systems, read this information three times and retell the information to a partner. Then switch roles with your partner and compare your understanding.

biobit

How fast is a "snail's pace"?

Many snails move at a speed of less than 8 cm (3 in.) per minute. This means that if a snail moved forward without stopping for 1 hour, it would travel only 4.8 m (almost 16 ft).

Organ systems perform tasks

Mollusks are the only coelomates without segmented bodies. Like the roundworms, mollusks have a one-way digestive system, but mollusks are more complex than roundworms. As a group, mollusks are quite diverse, and no one mollusk can represent the phylum as a whole. **Figure 30-3** shows the basic mollusk body with organs that are characteristic of the phylum.

Excretion Mollusks are one of the earliest evolutionary lines to have developed an efficient excretory system. Mollusks use their coelom as a collecting place for waste-laden body fluids. The beating of cilia pulls the coelomic fluid into tiny tubular structures called **nephridia** *(nee FRIHD ee uh)*, also shown in Figure 30-3. The nephridia recover useful molecules (sugars, salts, and water) from the coelomic fluid. The recovered molecules are reabsorbed into the mollusk's body tissues. The remaining fluid waste leaves the mollusk's body through a pore that opens into the mantle cavity. Nephridia are found in all coelomate animals except arthropods and chordates.

Circulation The digestive tube of mollusks and other coelomates is surrounded by mesoderm, which acts as a barrier to the diffusion of nutrients into the cells of the body. Thus, the circulatory system evolved. Recall that in a circulatory system, blood carries nutrients and oxygen to tissues and removes waste and carbon dioxide. Most mollusks have a three-chambered heart and an open circulatory system. Octopuses and squids are exceptions because they each have a closed circulatory system.

Respiration Respiration among mollusks is carried out in a variety of ways. Most mollusks respire with gills, which are located in the mantle cavity. Mollusk gills are very efficient. The gills extract 50 percent or more of the dissolved oxygen from the water that passes

Overcoming Misconceptions

The Shell and Mantle Difffer

Students often confuse the mantle with the shell or assume that the mantle composes part of the shell. Be sure students understand the distinction between the two: the mantle, which is composed of tissue, secretes the shell, which is composed of protein and calcium carbonate.

over them. In freshwater snails, the beating of ciliated gills on the inner surface of the mantle causes a continuous stream of water to pass over the gills. Most terrestrial snails have no gills. Instead, the thin membrane that lines their empty mantle cavity is modified to act like a primitive lung. This membrane must be kept moist for oxygen to diffuse across it. Therefore, terrestrial snails, shown in **Figure 30-4,** are most active at night or after it rains when air has a high moisture content. During dry weather, a terrestrial snail pulls back into its shell and plugs the opening with a wad of mucus to keep water in. Sea snails also lack gills, and gas exchange takes place directly through their skin.

Figure 30-4 Terrestrial snails. Terrestrial snails are most active when the air is moist.

Reproduction Most species of mollusks have distinct male and female individuals, although some snails and slugs are hermaphrodites. Certain species of oysters and sea slugs are able to change from one sex to the other and back again. Many marine mollusks are dispersed from place to place as their trochophore larvae drift in the ocean currents. Octopuses, squids, freshwater snails, and some freshwater mussels have no free-swimming larvae. In these mollusks, the larval stage occurs within the egg, and a juvenile-stage mollusk hatches from the egg.

How can you model an open circulatory system?

You can model an open circulatory system using simple items to represent the heart, blood vessels, blood, and body tissues of a living organism.

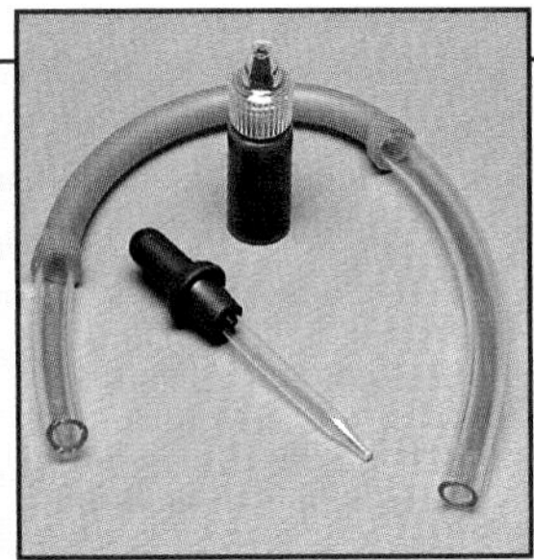

Materials

surgical tubing, 15 cm (about 6 in.) piece; clear plastic tubing, 15 cm (about 6 in.) and 7.5 cm (about 3 in.) pieces; shallow pan filled with water; eyedropper; food coloring

Procedure

1. Connect the surgical tubing to the two pieces of clear plastic tubing, as shown above.
2. Place the tubing into the tray filled with water. Allow the tubing to fill with water and rest on the bottom.
3. With the tubing still submerged, use an eyedropper to place two drops of food coloring into the short piece of clear plastic tubing.
4. With your thumb and index finger, squeeze along the piece of surgical tubing to pump the food coloring through the system.
5. As you continue to pump, observe the movement of food coloring.

Analysis

1. **Describe** what happened when you squeezed along the tubing.
2. **Identify** the structures represented by the pan of water, the surgical tubing, and the clear plastic tubing.
3. **SKILL Evaluating Results** Evaluate your model's efficiency at pumping blood through the system.
4. **SKILL Analyzing Methods** How does this model differ from a real circulatory system?
5. **SKILL Analyzing Methods** How could you modify the model to make it more accurate?

Teaching TIP

Oxygen Delivery Systems

Point out that it takes both respiratory and circulatory systems to deliver oxygen to the tissues of mollusks. If either system stops functioning, the mollusk will die. Ask students to consider the various ways a mollusk could suffocate. *(oxygen deficient water, nonfunctioning cilia, nonfunctioning heart, loss of blood, dried out gills or lungs)*

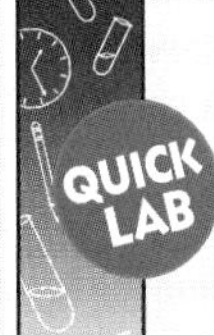

How can you model an open circulatory system?

Time 20 minutes

Process Skills Applying information, modeling, observing, evaluating

Teaching Strategies

- Make sure that students remove any air from the tubing once it is submerged and that the tubing stays submerged to prevent any air bubbles.
- Use a dark dye. If the dye diffuses too quickly in the water, try using an oil-based dye or small, lightweight colored beads.

Analysis Answers

1. The dye moves away from the pumping action but is also drawn back into the open tubing.
2. The pan of water represents the body and tissues of the mollusk, the surgical tubing represents the heart, and the clear plastic tubing represents the blood vessels.
3. Answers will vary.
4. In a real system, there would be many more vessels carrying the blood out to the tissues and back toward the heart. The oxygenated blood would also be absorbed by the various tissues, and the used blood would be reoxygenated by the gills.
5. Answers will vary but may include adding models for individual organs, adding more "blood," and adding more blood vessels.

Chapter 30

Demonstration

Snail Behavior

Allow students to explore snail behavior. Obtain several land snails from gardens, wooded areas, or a biological supply house. Have students devise careful ways to test the snails' response to touch, light, moisture, and gravity. When not in use, keep the snails in a cool, moist terrarium with pieces of lettuce.

Resource Link

Assign **Reading Strategies: Compare/Contrast Matrix** from the Holt BioSources Teaching Resources CD-ROM. PORTFOLIO

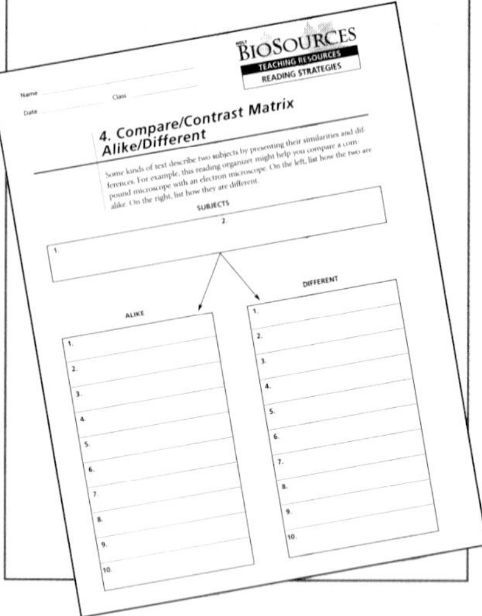

HOLT BIOSOURCES TEACHING RESOURCES READING STRATEGIES

Name Class

Date

4. Compare/Contrast Matrix
Alike/Different

Some kinds of text describe two subjects by presenting their similarities and differences. For example, this reading organizer might help you compare a compound microscope with an electron microscope. On the left, list how the two are alike. On the right, list how they are different.

SUBJECTS

1.

2.

ALIKE

1. 2. 3. 4. 5. 6. 7. 8. 9. 10.

DIFFERENT

1. 2. 3. 4. 5. 6. 7. 8. 9. 10.

Demonstration

Mollusk Anatomy — BASIC

Show students a dissected fresh or preserved mollusk such as a clam or oyster. Ask them to sketch the mollusk and label the following parts: shell, mantle, coelom, foot, gut, heart, and gills. PORTFOLIO

Study TIP

Organizing Information

Make a table to organize information about mollusks. Across the top, write the headings *Gastropods, Bivalves,* and *Cephalopods.* Along the side, write *Foot, Shell, Feeding, Reproduction.* Add information (and perhaps more terms along the side) to the table as you read.

Mollusks Show Many Variations on a Body Plan

The basic mollusk three-part body plan has been modified in each class of mollusks. As you read about the different classes of mollusks, you will see how the mollusk shell and foot have been adapted for many different living conditions.

Gastropods

Gastropods—snails and slugs—are primarily a marine group that has successfully invaded freshwater and terrestrial habitats. They range in size from microscopic forms to the sea hare *Aplysia,* which reaches 1 m (almost 40 in.) in length. Most gastropods have a single shell. During the evolution of slugs and nudibranchs *(NOO dih branks),* the shell was lost completely. **Figure 30-5** shows three terrestrial tree snails and a nudibranch (sea slug). The foot of gastropods is adapted for locomotion. Terrestrial species secrete mucus from the base of their foot, forming a slimy path that they can glide along. Most gastropods have a pair of tentacles on their head with eyes often located at the tips.

Gastropods display extremely varied feeding habits. Many are herbivores that scrape algae off rocks using their radula. Some terrestrial snails can be serious garden and agricultural pests by using their radula to saw off leaves. Sea slugs and many other gastropods are active predators. Whelks and oyster drills, for example, use their radula to bore holes in the shells of other mollusks. Then they suck out the tissue of their prey. In cone shells, shown in **Figure 30-6,** the radula has been modified into a kind of poison-tipped harpoon that is shot into prey. The poison paralyzes the prey, which is then swallowed whole.

Throughout human history, snails have been a source of food for humans. Land snails belonging to the genus *Helix* are raised on snail farms and are consumed in great quantities. While freshwater snails are rarely eaten, a few marine species, such as conchs *(KAHNGKS),* are considered delicacies.

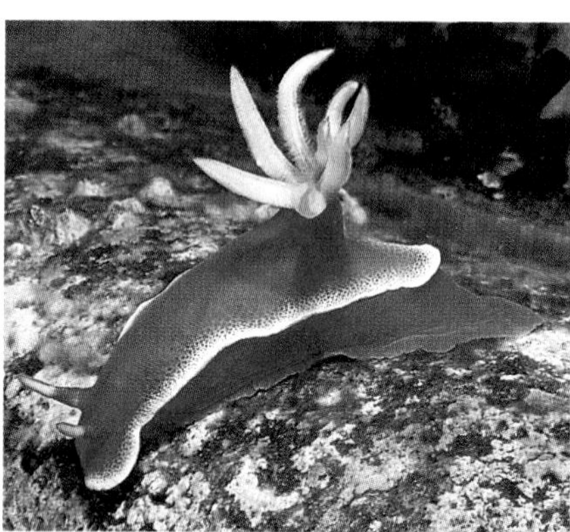

Figure 30-5 Snails and slugs. These Florida tree snails live where the air is moist enough to keep them from drying out. Sea slugs are often brilliantly colored, and most are under 15 cm (6 in.) in length.

did you know?

The beauty, diversity, and hardiness of mollusk shells accounts for their wide use as decorations.

Even the inside linings of shells are used. Shells of freshwater mussels are used to make pearl buttons, and mother-of-pearl from oysters is used to decorate ornamental boxes, jewelry, and musical instruments.

Figure 30-6 Cone shell. This cone shell searches the ocean bottom for prey. Once located, the prey is secured by the cone shell's radula and swallowed whole.

Bivalves

Most bivalves are marine, but some also live in fresh water. Many species of freshwater mussels are found throughout the rivers and lakes of North America and are important links in aquatic food chains. Throughout history, oysters and mussels have been important sources of food for humans. All bivalves have a two-part hinged shell. The valves, or shells, of a bivalve are secreted by the mantle. Two thick muscles, the **adductor muscles,** connect the valves. When these muscles are contracted, they cause the valves to close tightly. While most species of mollusks are sessile, some can move from place to place quite fast if necessary. For example, a swimming scallop opens and closes its valves rapidly, pushing itself along with the jets of water it expels when its valves snap shut.

Bivalves are unique among the mollusks because they do not have a distinct head region or a radula. A nerve ganglion above their foot serves as a simple brain. Most bivalves have some type of simple sense organs. For example, some bivalves have sensory cells located along the edge of their mantle that respond to light and touch.

Most bivalves are either male or female, but a few species are hermaphroditic. Bivalves reproduce sexually by releasing sperm and eggs into the water, where fertilization occurs. The fertilized eggs develop into free-swimming trochophore larvae. The larvae of some freshwater mussels are brooded in a special pouch within the mollusk's gills. The larvae are then released into the water, and they complete their larval stage as parasites on fish. This is a very unusual life cycle for a mollusk.

Most bivalves are filter feeders. Many, such as the clam illustrated in **Figure 30-7,** use their muscular foot to dig down into the sand. Once there, the cilia on their gills draw in sea water through hollow tubes called **siphons** *(SEYE fuhns)*. The water moves down one siphon tube, over the gills, and out the other siphon tube. The gills are used for feeding as well as respiration. A sticky mucus covers the gills, and as water moves over the gills, small marine organisms and organic material become trapped in the mucus. The cilia then direct the food-laden mucus to the bivalve's mouth.

Study TIP

Word Origins

The class name *Bivalvia* comes from the Latin *bi,* meaning "two," and *valva,* meaning "part of a door."

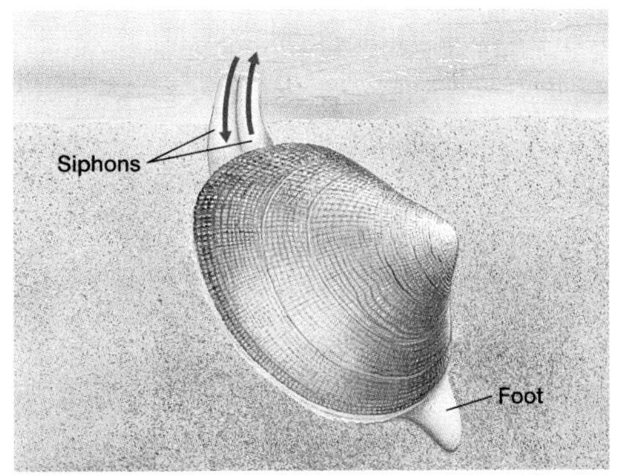

Figure 30-7 Clam in sediment. Many bivalves, like this clam, burrow into sand or mud and feed by drawing sea water in one siphon and expelling it out the other.

Demonstration
Fossil Mollusks

Obtain several fossil mollusks, and set them out for students to view. Fossil mollusks are some of the most common and inexpensive fossils you can obtain, and they are easily identifiable as bivalves, gastropods, or shelled cephalopods. Have students match modern varieties of mollusks to their fossils.

Cross-Disciplinary Connection

◆ Geography

Tell students about an increasingly infamous bivalve: the zebra mussel, *Dreissena polymorpha.* The zebra mussel is a small, freshwater bivalve native to Europe. At some point, zebra mussels hitchhiked across the Atlantic Ocean in the ballast water of ships. Although they were not here in detectable numbers before 1985, they have infested waters in the Great Lakes, where they have no natural predators. Although small, the mussels attach themselves to any hard surface, including ship hulls, docks, and even other mussels.

Multicultural Perspective

Cultural Uses of Shells

Mollusk shells have served many different functions in various cultures, especially in making jewelry and crafts. In addition, shells have been used for several practical purposes. Phoenicians and Romans used Murex sea snails to make a purple dye for coloring fabric. Native North Americans carved large clam shells into wampum beads and used other shells as money. Filipinos cut thin Placuna oyster shells to fit into wooden frames as windowpanes.

Eating Mollusks Safely

Teaching Strategies

- Point out that there are risks associated with eating any animal product raw, but the often polluted environment of many mollusks makes them especially risky.
- Some coastal areas are so polluted that bivalves have completely died off.

Discussion

- Ask for a show of hands of students who eat mollusks from time to time. Ask how many eat them raw. How will this new information affect their choices in the future about eating mollusks? *(Answers will vary.)*
- Ask students what can be done to clean up the coastal areas where bivalves are found? *(eliminate oceans as a place to dump waste, control agricultural runoff, increase coastal wetland acreage to improve water quality and provide more habitat for mollusks)*

internet**connect**

Have students research mollusk consumption using the Web site in the Internet Connect box on page 666. Have students write a report summarizing their findings. PORTFOLIO

Figure 30-8 Pearl oyster. The best pearls come from oysters belonging to the genus *Pinctada.*

Like clams, oysters and scallops use their gills to filter food from the water. Oysters are permanently attached to rocks in the open water, where they feed. Scallops swim, and water passes over their gills as they move.

Many species of bivalves, such as the oyster in **Figure 30-8,** produce pearls. Pearls form when a tiny foreign object, such as a grain of sand, becomes lodged between the mollusk's mantle and shell. Bivalves respond to these irritants by coating them with thin sheets of nacre *(NAY kuhr),* also called mother-of-pearl. Nacre is the same hard, shiny substance that composes the inner shell surface. Successive layers of nacre are added until the foreign body is completely enclosed in the newly formed pearl.

While many bivalves form pearls, only a few species produce the beautifully colored nacre essential for gem-quality pearls. In fine pearls, the nacre contains tiny, overlapping mineral crystals. These crystals act like prisms, breaking up any light that falls on them into rainbows of color. This is what gives these pearls their iridescence. The mineral crystals found in the nacre of ordinary pearls are large and do not reflect light beautifully.

FOOD WATCH

Eating Mollusks Safely

Since prehistoric times, mollusks have served as a human food resource. Today the annual worldwide harvest for bivalve mollusks alone amounts to an astounding 3 million metric tons (6,615,000,000 lb). But severe illness and even death have been associated with mollusk consumption. What causes these illnesses, and are mollusks really safe to eat?

Bivalves—oysters, mussels, clams, and scallops—are filter feeders. Any contaminant in a bivalve's environment circulates through and accumulates in its body. One source of contamination is water that is polluted with sewage, an ideal breeding ground for some bacteria and viruses. Because of this, harvesting is prohibited in such areas. The primary danger is from eating infected bivalves raw. Contrary to popular opinion, hot sauce will not kill dangerous organisms that infect them. But it is now evident that brief cooking, such as lightly steaming clams, does not destroy all pathogens either.

Follow the guidelines

An estimated 20 million Americans eat oysters, and the yearly consumption of mollusks is increasing. To reduce the risk of illness from eating mollusks, the Food and Drug Administration (FDA) has developed guidelines for preparing them. If you observe these guidelines, the health risk from eating mollusks is very low. Avoid eating raw bivalves. Make your purchases from reputable sources to assure that they were not illegally harvested. Test for freshness by trying to move the shells sideways. If the shells move, the mollusk is not fresh and should be discarded. Follow FDA recommendations for proper cooking times and temperatures. By following these guidelines, you can safely enjoy this delicious and healthful food choice.

internet**connect**

TOPIC: Mollusk consumption
GO TO: www.scilinks.org
KEYWORD: HX666

did you know?

Some bivalves have evolved to survive the crushing pressure of waves.

Tell students that mussels and other bivalves that often cling to exposed rocks must be able to withstand daily battering by waves. They survive the poundings by having shells that are very hard and are shaped so as to deflect onrushing water.

Cephalopods

Squids, octopuses, cuttlefish, and nautiluses are all cephalopods. Most of their body is made up of a large head attached to tentacles (a foot divided into numerous parts), as shown in **Figure 30-9.** The tentacles are equipped with either suction cups or hooks for seizing prey. Squids have 10 tentacles, while octopuses have eight. The nautilus has 80–90 tentacles, although they are not nearly as long as those of the other cephalopods. Although cephalopods evolved from shelled ancestors, most modern cephalopods lack an external shell. The nautilus is the only living cephalopod species that still has an outer shell. Squids as well as the cuttlefish have a small internal shell. Cuttlefish "bones" are often attached to bird cages to provide calcium for canaries and other pet birds.

Cephalopods are the most intelligent of all invertebrates. They have a complex nervous system that includes a well-developed brain. Cephalopods are capable of exhibiting complex behaviors, as demonstrated by their agility when catching prey. Octopuses can easily be trained to distinguish between classes of objects, such as between a square and a cross, and they are the only invertebrates with this ability.

The structure of a cephalopod eye is similar in many ways to that of a vertebrate eye, and some species have color vision. The eyes of a squid can be very large. A giant squid that washed up on a beach in New Zealand in 1933 had eyes that were 40 cm (almost 15 in.) across. At over 20 m (65 ft) in length, the giant squid is the largest of all invertebrates and has the largest eyes known in any animal, invertebrate or vertebrate.

Like most aquatic mollusks, cephalopods draw water into their mantle cavity and expel it through a siphon. Squids and octopuses, however, have modified this system into a means of jet propulsion.

Study TIP

Word Origins

The name *cephalopod* comes from the Greek *kephalicos*, meaning "head," and *pous*, meaning "foot."

Figure 30-9
Cephalopods. Like all cephalopods, the squid is an active predator. Cuttlefish are agile swimmers that hunt at night, seeking small fishes and crustaceans. The nautilus swims with its coiled shell positioned over its head.

Teaching TIP

A Good Defense

Ask students to list some ways that mollusks can escape from predators. *(Snails can close their shells with a plate, squids can speed away on a jet of water, sea slugs can excrete toxic substances, etc.)*

Teaching TIP

Major Characteristics of Mollusks

Ask students to develop a table similar to the **Graphic Organizer** at the bottom of page 667 to compare the major characteristics of bivalves, gastropods, and cephalopods. PORTFOLIO

Demonstration

Suction Cups — BASIC

Have students test the suction of dry and wet suction cups. Ask them whether suction cups work better when dry or wet. *(They work better when wet.)* Cephalopods have suction cups on their tentacles that have tremendous gripping power. Even sperm whales can receive suction disk scars during fights with giant squids. The suction disks hold so tight that they can damage the flesh.

Graphic Organizer

*Use this graphic organizer with **Teaching Tip: Major Characteristics of Mollusks** on page 667.*

Characteristics	Bivalves	Gastropods	Cephalopods
Have head, foot, and visceral mass	Yes; head undeveloped	Yes; torsion of visceral mass	Yes; foot divided into tentacles
Have true coelom	Yes	Yes	Yes
Have shells	Two shells	Most have one shell	No, except chambered nautilus
Use locomotion	Most do not	Yes	Yes

Modifications of the mollusk body plan

Time 5–10 minutes

Process Skills Observing, applying information, predicting

Teaching Strategies
If possible, display examples of shells from a cephalopod, gastropod, and bivalve.

Analysis Answers

1. A. cephalopod; B. gastropod; C. bivalve
2. The shells of the different classes have evolved to support different ways of life.
3. In cephalopods, the foot is modified into tentacles and aids in movement and in capturing food. The flat-bottomed foot of gastropods secretes a substance that allows the animal to glide over surfaces. The foot of a clam allows the bivalve to dig itself into the sand.
4. Slugs would be more abundant in areas with acidic soils because the environment cannot support many shell-building organisms.

3 Assessment Options

Closure

Mollusks and Humans
Have students choose a mollusk that affects humans. Then have them report on how the mollusk has been affected by humans and how humans have been affected by the mollusk.

Review

Assign Review—Section 30-1.

Reteaching — Basic

Ask students to list the different functions of the mantle with examples of mollusks that exhibit such functions. *(primitive lung—terrestrial snails; shell formation—bivalves; jet propulsion—squids and octopuses)*

When threatened, they quickly close their mantle cavity, causing water to shoot forcefully out of the siphon. Squids and octopuses can also release a dark fluid that clouds the water and conceals the direction of their escape. The ink of the cuttlefish contains a reddish brown pigment called sepia. For centuries this ink was used by artists as a pigment and is found on the canvases of many famous paintings.

All cephalopods are active marine predators. They feed on fish, mollusks, crustaceans, and worms. Once the prey has been snared by the tentacles, it is pulled to the mouth, where it is torn apart by strong, beaklike jaws. The cephalopod's radula then pulls the pieces into the mouth.

Modifications of the mollusk body plan

Background
Mollusks share many common characteristics, yet there is great variety among the classes. The drawings on the right show how the shell (brown) and foot (green) vary in three classes of mollusks. Use the drawings to answer the analysis questions.

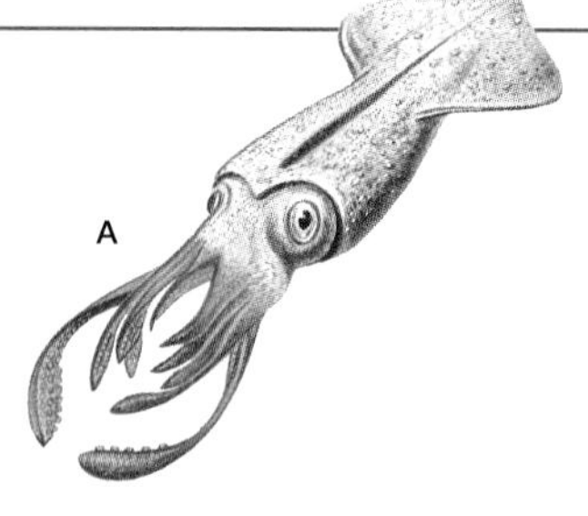

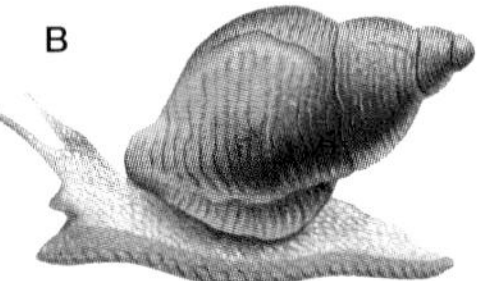

Analysis

1. **Determine** the class of mollusk *A*, mollusk *B*, and mollusk *C*.
2. **SKILL Comparing Structures** Compare the shell modifications. Why might a shell suited to one mollusk be inappropriate for another?
3. **SKILL Identifying Functions** For each class shown, explain how the foot is useful for the animal's environment or kind of movement.
4. **SKILL Predicting Outcomes** Terrestrial snails and slugs are nearly identical except that slugs do not have a shell. Acidic forest soils are often poor in minerals, including calcium. Alkaline or neutral soils are rich in minerals. In which kind of soil would you be more likely to find a slug? Explain your answer.

Review SECTION 30-1

1. **Identify** two characteristics that mollusks and annelids have in common.
2. **Summarize** six characteristics common to most groups of mollusks.
3. **Describe** how a nephridium functions in waste removal.
4. **Compare** the distinguishing features of each of the three major classes of mollusks.
5. **SKILL Forming Hypotheses** A chemical pollutant accidentally spills into a bay. One of the effects of this chemical is that it paralyzes cilia. The next day almost all of the oysters in the bay are dead. Develop a hypothesis that explains why the oysters died.

Review SECTION 30-1 ANSWERS

1. Mollusks and annelids both exhibit true coeloms, and most members of both groups develop from larvae called trochophores.
2. Most mollusks share the following key characteristics: a coelom, bilateral symmetry, a three-part body plan (visceral mass, mantle, and foot), organ systems, a shell, and a radula.
3. A nephridium funnels coelomic fluid, which includes wastes, into the mantle cavity, where it is excreted to the outside environment.
4. Most gastropods have a single shell, a single large foot, and, on their head, tentacles tipped with eyes. Bivalves have a two-part hinged shell and lack a head and radula. Most bivalves possess one or more siphons used for filter feeding. Cephalopods have a large head attached to numerous tentacles and large eyes.
5. As bivalves, oysters breathe by extracting oxygen from water flowing over gills. The beating of cilia moves water over the gills. Therefore, the oysters suffocated when the chemical paralyzed their cilia.

Annelids

Section 30-2

Annelids Were the First Segmented Animals

You have probably heard the expression "a can of worms," which calls up an image of a lot of wiggly, wriggly creatures. An earthworm may come to mind, but there are many different species of segmented worms. Worms might not look like much, but this second group of coelomates belong to an ancient group, phylum Annelida. Annelid fossils can be found in rock that is 530 million years old. Scientists think that annelids evolved in the sea, where two-thirds of today's annelid species live. Most other annelid species are terrestrial earthworms. Annelids range in size from less than 1 mm (0.04 in.) long to more than 3 m (10 ft) long.

Annelids, such as the earthworm and fireworm shown in **Figure 30-10,** are easily recognized by their segments, which are visible as a series of ringlike structures along the length of their body. Each segment contains digestive, excretory, circulatory, and locomotor (movement) organs. Some of the segments are modified for specific functions, such as reproduction, feeding, or sensation. A well-developed **cerebral ganglion,** or primitive brain, is located in one anterior segment. The brain is connected to a nerve cord that runs along the underside of the worm's body.

Objectives

- **Identify** the major change in body plan that distinguishes annelids from mollusks. (p. 669)
- **Describe** the basic annelid body plan. (p. 670)
- **Describe** the annelid digestive system. (p. 672)
- **Compare** the three classes of annelids. (pp. 671–674)

Key Terms

cerebral ganglion
septa
seta
parapodium

Earthworm

Fireworm

Figure 30-10 Annelids. The ringlike segments of this earthworm and marine fireworm identify them as annelids.

Evolutionary Milestone

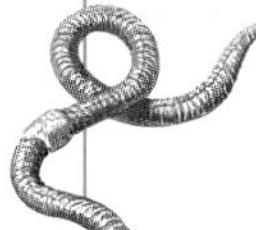

6 Segmentation

Annelids were the first organisms to evolve a body plan based on repeated body segments. Segmentation underlies the body organization of all coelomate animals except mollusks.

1 Prepare

Scheduling

- **Block:** About 90 minutes will be needed to complete this section.
- **Traditional:** About two class periods will be needed to complete this section.
- **Planning:** See the Planning Guide on pp. 657A and 657B for lecture, classwork, and assignment options for Section 30-2. Lesson plans for Section 30-2 can be found in a lesson cycle format in ***Biology: Principles and Explorations* One-Stop Planner CD-ROM with Test Generator.**

Resource Link

- Assign **Active Reading Worksheet Section 30-2** to help students comprehend the key concepts and visuals in this lesson.

2 Teach

Lesson Warm-up

Obtain earthworms from garden soil or a bait shop, and have students observe their behavior. Ask students to describe how they can tell which end of the earthworm is the head. If students say that the head is the leading end, remind them that squid sometimes move tail first. Direct their attention to the earthworm's mouth. Remind students to keep their worms moist. Ask them why dry worms die. *(They suffocate.)*

Integrating Different Learning Styles

Logical/Mathematical	**PE**	Performance Zone item 14, p. 676
Visual/Spatial	**PE**	Quick Lab, p. 670; Performance Zone items 1, 20, and 21, pp. 676–677
	ATE	Lesson Warm-up, p. 669
Verbal/Linguistic	**PE**	Performance Zone items 27 and 29, p. 677
	ATE	Teaching Tip, p. 672; Reteaching, p. 674

Teaching TIP

The Ultimate Worm

Ask students to list the anatomical features that make earthworms more advanced than flatworms or roundworms. Help students recall that earthworms have a segmented body, a true coelom, a highly specialized gut, a closed circulatory system, and external bristles.

QUICK LAB: How can you model a closed circulatory system?

Time 20 minutes

Process Skills Observing, modeling, interpreting results, applying information, evaluating

Teaching Strategies

- Have students conduct **Quick Lab: How can you model an open circulatory system?** on page 663 before conducting this lab.
- Ask students how oxygen and nutrients get to the tissues that need them if the blood never touches the body tissues directly. *(The oxygen and nutrients pass through the vessel walls.)*
- Use a dark dye. If the dye diffuses too quickly in the water, try using an oil-based dye or small, lightweight colored beads.

Analysis Answers

1. The "blood" moved through the blood vessels.
2. pan of water—body and tissues; surgical tubing—heart; clear plastic tubing—blood vessels
3. Answers will vary.
4. Answers will vary but may include adding model tissues and increasing the number of model blood vessels.
5. Students should note that the closed system exerted a greater pressure on the fluid than the open system.

Study TIP

Word Origins

The name *annelid* comes from the Latin word *annellus,* meaning "ring."

Internal body walls, called **septa,** separate the segments of most annelids. Nutrients and other materials pass between the segments through the circulatory system. Sensory information is delivered by a nerve cord that connects nerve centers in the segments to the brain.

Characteristics of annelids

In addition to segmentation, annelids share a number of other characteristics.

1. **Coelom.** The fluid-filled coelom is large and is located entirely within the mesoderm.
2. **Organ systems.** The organ systems of annelids show a high degree of specialization and include a closed circulatory system, excretory structures called nephridia, and a highly modified gut. The gut has evolved different regions that perform different functions in digestion.
3. **Bristles.** Most annelids have external bristles called **setae** *(SEET ee).* The paired setae located on each segment increase traction as the annelid crawls along. Some annelids, like the fireworm shown in Figure 30-10, also have fleshy appendages called **parapodia** *(par uh POH dee uh).*

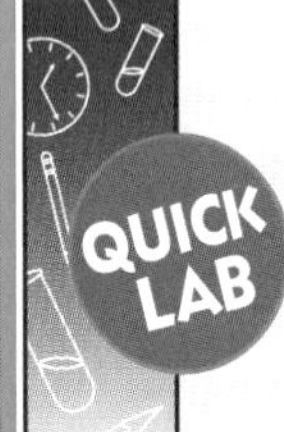

QUICK LAB: How can you model a closed circulatory system?

You can model a closed circulatory system using simple items to represent the heart, blood vessels, blood, and body tissues of a living organism.

Materials

clear plastic tubing, 30 cm (about 12 in.) piece; surgical tubing, 15 cm (about 6 in.) piece; shallow pan filled with water; eyedropper, food coloring

Procedure

1. Connect one end of the clear tubing to the surgical tubing. Submerge in the pan.
2. Use the eyedropper to insert several drops of food coloring into the surgical tubing. Close the tubing by attaching the two free ends together, as shown above.
3. With your thumb and index finger, squeeze along the piece of surgical tubing to pump the food coloring through the system.
4. Observe the food coloring as it moves through the tubing.

Analysis

1. **Describe** what happened when you pumped the food coloring through the system.
2. **Identify** what structures the pan of water, the surgical tubing, and the clear plastic tubing represent.
3. **Evaluate** your model's efficiency at pumping blood through the system.
4. **SKILL Analyzing Methods** How could you modify the model to make it more accurate?
5. **SKILL Comparing Structures** If you did the Quick Lab on page 663 that modeled an open circulatory system, recall what happened in that system. Which model do you think exerted a greater pressure on the fluid in the tube?

Overcoming Misconceptions

Grubs, Maggots, and Caterpillars

Many people mistake grubs, maggots, and caterpillars for worms. These animals are arthropods, not annelids, and are the larval form of insects (beetles, flies, and moths/butterflies, respectively). Have students list the annelid characteristics that are shared by these unrelated insects. *(Answers may vary, but arguments could be made for each of the four characteristics listed on page 670.)*

Annelids Are Grouped According to Their External Features

Annelids differ in the number of setae (bristles) they have on each segment, and not all annelids have parapodia. These two external characteristics are used to classify annelids.

Marine worms

Marine segmented worms are members of class Polychaeta *(PAHL ih keet uh)*, the largest group of annelids. Polychaetes live in virtually all ocean habitats. They are often beautiful, showing unusual forms and iridescent colors. A distinctive characteristic of polychaetes is the pair of fleshy, paddle-like parapodia that occur on most of their segments. The parapodia, which usually have setae, are used to swim, burrow, or crawl. Parapodia also greatly increase the surface area of the polychaete's body, making gas exchange between the animal and the water more efficient.

Many polychaetes are burrowing species, but others live in protective tubes formed by the hardened secretions of glands located on their segments. Grains of sand or other foreign material may be cemented into the tube. Such tubeworms, like the feather duster shown in **Figure 30-11,** live with only their head stuck out of the tube. Featherlike head structures trap food particles from the water that passes over them. Other species of polychaetes feed by pumping water through their body. Free-swimming polychaetes, such as *Nereis* shown in **Figure 30-12,** are predators that use their strong jaws to feed on small animals.

Figure 30-11 Feather duster. Feather dusters filter-feed by trapping food particles in their featherlike head structures.

Study TIP

Word Origins

The term *polychaete* comes from the Greek *poly,* meaning "many," and *chaite,* meaning "hair."

Figure 30-12 ***Nereis***

Nereis, a polychaete worm, grasps its prey in its jaws, which open when it thrusts out its pharynx.

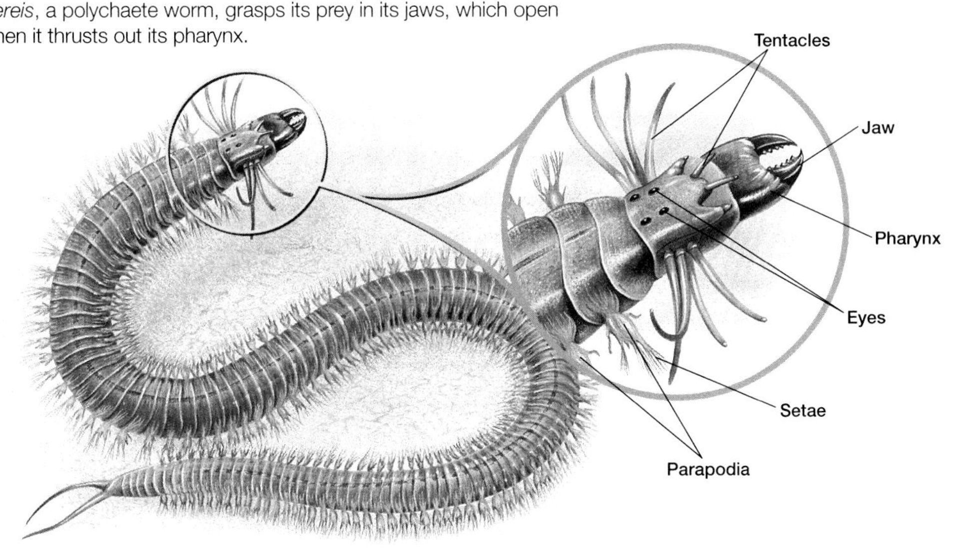

did you know?

Polychaetes are some of the most important animals in marine benthic (bottom-dwelling) communities.

Though few people notice polychaetes or even know of their existence, these worms exert a strong influence in marine benthic food webs. Some of the 10,000 species so far described eat animals, others eat plants, and many provide food for commercially important fish.

Visual Strategy

Figure 30-11

Ask a student volunteer to examine Figure 30-11, and explain how feather duster worms acquire food. *(Bristle-like appendages filter small organisms and organic matter out of water that flows over them.)* Have the class name an advantage and a disadvantage of this feeding strategy. *(Advantage: Allows for a sessile lifestyle; Disadvantage: Potential food must pass close by to be captured.)*

Demonstration

Surface Area and Gas Exchange

Perform the following demonstration to show students how a polychaete's textured parapodia provide a large area for respiratory exchange. Place a coffee filter over the mouth of each of two identical beakers. Make one filter flat, and form the other filter into an involution extending into the filter (creating a small, crumpled "cup"). Secure both filters with rubber bands. Then slowly pour colored water over the filters. Students should observe that liquid flows faster through the filter with a larger surface area exposed.

CAREER

Biological Oceanographer

Oceanography is a diverse field with branches in the biological, chemical, physical, geological, and engineering fields. A graduate degree is usually required, along with an undergraduate degree in biology, chemistry, geology, or engineering. Biological oceanographers study the interrelationships of marine life, energy transfer, and the impact of human activities on marine communities. Encourage students to learn more about this field at the library or on the Internet.

Teaching TIP

Comparing Mollusks and Annelids

Have students work in small groups to make a table similar to the one in the **Graphic Organizer** on the bottom of page 673 that compares mollusks and annelids. Students' tables should compare body cavity, body plan, organ systems, and larvae. Ask students to write a paragraph explaining the differences. PORTFOLIO

biobit

Why does "the early bird catch the worm"?

Because earthworms respire through their skin, they must stay moist. As a result, earthworms come to the surface of the soil only at night, when the air is more moist. The "early bird" catches worms that have not yet burrowed in for the day.

Earthworms

Earthworms and some related freshwater worms are members of the class Oligochaeta *(AHL ih goh keet uh)*. Oligochaetes have no parapodia and only a few setae on each segment. Earthworms lack the distinctive head region of polychaetes and have no eyes. They do, however, have light-sensitive and touch-sensitive organs located at each end of their body, and they have other sensory cells that detect moisture.

Earthworms, such as the ones in **Figure 30-13,** are highly specialized scavengers that literally eat their way through the soil, consuming their own weight in soil every day. As they tunnel, earthworms take in organic matter and other materials using their muscular pharynx. The ingested soil moves through their one-way gut, down the esophagus and into a storage chamber called the crop. From here, the soil moves to an area called the gizzard. The grinding action of the gizzard crushes the soil particles together, breaking them down. The crushed material moves to the intestine, which extends to the posterior end of the earthworm's body. Digestible food particles are absorbed into the intestinal wall, and the remaining material passes out through the anus in a form called castings. The tunneling activity of earthworms allows air to penetrate the soil, and their castings fertilize it. Rich, organic soil may contain thousands of earthworms per acre.

Hydrostatic skeleton

The fluid within the coelom of each body segment creates a hydrostatic skeleton that supports the segment. Each segment contains muscles that pull against this hydrostatic skeleton. Circular muscles wrap around the segment, while longitudinal muscles span its length. As shown in the *Up Close: Earthworm,* on page 673, when the circular muscles contract, the segment elongates. When the longitudinal muscles contract, the segment bunches up, increasing in diameter. An earthworm crawls by alternately contracting the two sets of muscles in its segments. The brain coordinates the muscular activity of each body segment, thus controlling movement.

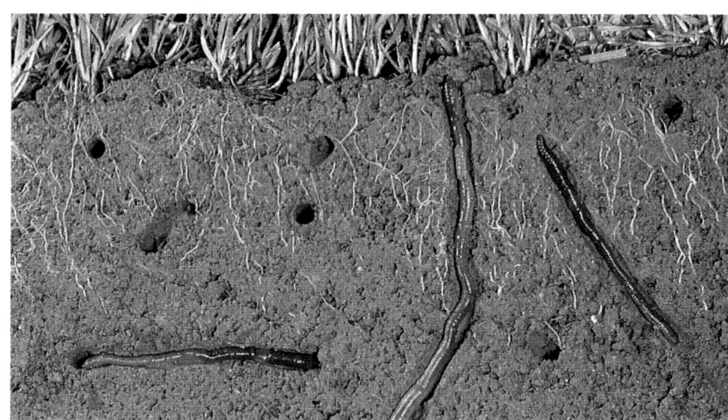

Figure 30-13
Earthworms burrowing. Earthworms come to the surface only at night or during heavy rains. During dry or cold weather, they burrow deep into the soil and become inactive.

did you know?

Charles Darwin was fascinated with earthworms.

Charles Darwin (1809–1882), the British naturalist famous for his theory of evolution by natural selection, was fascinated with the ways that earthworms aerate and enrich soil. In his last botanical book, *The Formation of Vegetable Mould Through the Action of Worms,* Darwin noted the service of earthworms in recycling organic matter in soil. He also observed that an earthworm could ingest its weight in soil each day and that in one year, the earthworms inhabiting one hectare could digest 22 to 40 metric tons of soil. The work was published only 6 months prior to Darwin's death and was called a pioneering study in quantifying ecology.

Earthworm

- **Scientific name:** *Lumbricus terrestris*
- **Size:** Grows up to 30 cm (12 in.) long
- **Range:** Europe; eastern and northwestern North America
- **Habitat:** Damp soil
- **Diet:** Organic matter contained in soil

Movement of an Earthworm

1. Rear anchor
2. Elongation
3. Front anchor
4. Pull

Characteristics

Respiration Oxygen and carbon dioxide diffuse through the earthworm's skin. This exchange can take place only if the worm's skin is kept moist.

Movement As shown in the diagram, 1 first the earthworm anchors several of its rear segments by sinking their setae into the ground. 2 The worm then contracts the circular muscles in front of the anchored segments. This causes the anterior segments to elongate. 3 Then the setae in front of the stretched region are anchored and the rear setae are released. 4 The circular muscles relax and the longitudinal muscles contract, pulling the rear segments forward.

Digestion Earthworms "eat" soil, which is ground up in a thick, muscular gizzard, breaking up the soil's organic matter. Food molecules pass across the walls of the intestine and are absorbed into the bloodstream.

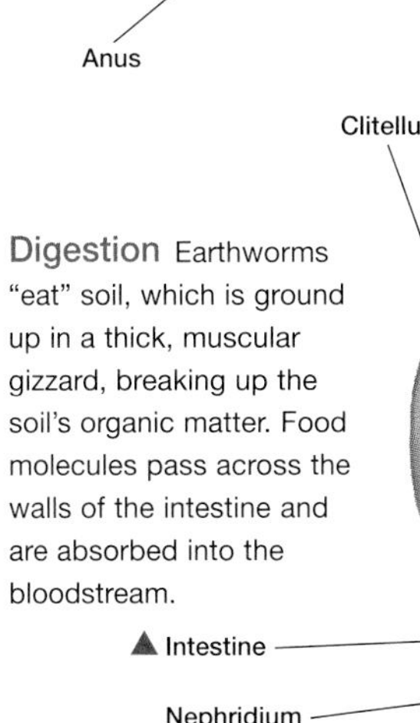

Reproduction Earthworms are hermaphrodites, each individual containing both sexes. Mating occurs when two earthworms join ventrally head to tail, exchanging sperm. During egg laying, the clitellum (a thickened, glandular ring of cells) of each worm secretes a mucous cocoon that encloses the fertilized eggs. Young worms emerge from the cocoon several weeks later.

Brain The brain coordinates the muscular activity of each body segment. It also processes sensory information from light-sensitive and touch-sensitive organs located at each end of the body.

Up Close

Earthworm

Teaching Strategies
Earthworms, which have no teeth, eat their way through soil and grind it up for nutrients. Ask students what structure earthworms use to grind up soil. *(gizzard)* Ask them which other toothless animals use gizzards to grind food. *(birds)* Finally, ask why food needs to be ground up. *(Smaller particles are digested more easily.)*

Discussion

- How might an earthworm try to escape if its anterior end were being pulled out of the ground by a bird? *(The earthworm would dig its posterior setae into the dirt and pull its anterior section back into the ground.)*
- How do earthworms loosen soil and help recycle soil nutrients? *(Earthworms make holes in the soil by moving through it, digesting soil matter into simpler components.)*
- What is the role of the clitellum? *(It secretes a protective layer of chitin for the fertilization and incubation of eggs.)*

Graphic Organizer

*Use this graphic organizer with **Teaching Tip: Comparing Mollusks and Annelids** on page 672.*

	Body cavity	Body plan	Organ systems	Larvae
Mollusks	True coelom	Muscular foot, head, visceral mass	Circulatory (usually open system), respiratory, digestive, excretory	Trochophore
Annelids	True coelom	Segmented with bristles	Circulatory (closed), respiratory, digestive, excretory	Trochophore

HEALTH WATCH

Leeches Make a Comeback

Teaching Strategies
Have students research past and present uses of leeches in medicine and write a brief (one-page) paper that describes their findings.

Discussion
Ask if any students have ever had a leech attach while swimming or wading in a pond or stream. Did it hurt as it attached? *(Usually there is no sensation.)* Did you bleed after the leech was removed? *(If the leech was fully attached and feeding, students should answer yes because leeches use anticoagulants to facilitate feeding. The bleeding stops soon after the leech is removed.)*

3 Assessment Options

Closure

Making an Annelid
Have students combine their new understanding of annelids with their artistic skills and draw a "new" species of annelid. Tell them to label their annelids and to make sure that the creatures include both the fundamental annelid traits as well as any reasonable adaptations to their designated environment. PORTFOLIO

Review

Assign Review—Section 30-2.

Reteaching — BASIC

Have students pick an annelid in this chapter and pretend that the animal is a pet. Ask them to provide a picture of the pet and prepare a report about it. In their report, students should identify the kind of habitat necessary to keep the pet, describe how and what the animal is fed, and provide any necessary warnings for handling.
PORTFOLIO

Figure 30-14 Leech. Well-developed, powerful muscles enable the leech to carry out the complex body movements it uses to crawl along.

Leeches

When people hear the word *leech,* they usually associate it with bloodsucking, and for a good reason. A leech, shown in **Figure 30-14,** has suckers at both ends of its body. Most species are predators or scavengers, but some have evolved into parasites that suck blood from mammals and other vertebrates. A few species are parasites on crustaceans.

Leeches are the only members of class Hirudinea *(hihr yoo DIHN ee uh).* Leeches lack both setae and parapodia. The body of a leech is flattened, and unlike other annelids, its segments are not separated internally. Most species of leeches are small, 2.5–5.0 cm (1–2 in.) long, although one tropical species grows up to 30.5 cm (1 ft) long.

HEALTH WATCH

Leeches Make a Comeback

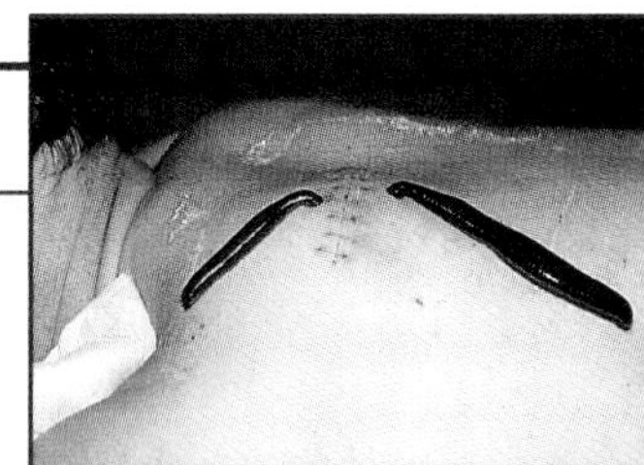

For many centuries, it was commonly believed that an excess amount of blood was the cause of a wide range of illnesses, from a fever or headache to severe heart disease. A standard treatment for these conditions was bloodletting with leeches. Physicians applied leeches to the patient's body, allowing the leeches to suck out the patient's "bad blood."

Use in microsurgery
Although doctors no longer believe in "bad-blood," leeches are making a comeback in the field of health care. One use of leeches is during surgery to reattach severed limbs, fingers, or toes. In this type of operation, called microsurgery, the surgeon uses tiny instruments and a microscope to reconnect tendons, blood vessels, and nerves. It is not possible to reconnect the smallest of the blood vessels, so circulation in the reattached part is usually poor. Often tissues in the region die, and the reattached part cannot heal. By applying leeches to suck out the accumulated blood, tissues remain healthy until new blood vessels grow and circulation is restored to normal. As a result, the success rate of surgery for reattachments has increased.

Other applications
Leeches possess other useful qualities. Their saliva contains chemicals that are anticoagulants—substances that prevent blood from clotting—and enzymes that can break up blood clots. It is not necessary to apply leeches to a patient to take advantage of these chemicals. Today these substances are produced through genetic engineering and have proven useful in the treatment of some heart patients.

Review SECTION 30-2

1. **Summarize** how you can tell if a wormlike organism is an annelid worm.
2. **Relate** an annelid's septa to its overall body plan.
3. **Describe** the major features of an earthworm's digestive system.
4. **Compare** the external appearance of marine annelids, earthworms, and leeches.

Critical Thinking

5. **SKILL Evaluating Conclusions** An earthworm is born with no moisture-secreting cells in its skin. A student concludes that the earthworm will not survive as long as an earthworm that has these cells. Evaluate this conclusion.

Review SECTION 30-2 ANSWERS

1. Annelid worms have segments, and most have external bristles.
2. The septa are internal body walls that separate the segments of most annelids. Blood vessels and a nerve cord connect the segments to one another and to the annelid's brain.
3. The earthworm's digestive system is a highly modified gut consisting of the following regions: mouth, pharynx, esophagus, crop, gizzard, intestine, and anus.
4. Marine annelids are often very colorful and have a head region with eyes; their segments usually have a pair of fleshy extensions called parapodia. Earthworms lack parapodia and have no head region. Usually there are several bristly setae on each segment. Leeches lack setae and parapodia, and their bodies are generally flat.
5. The student is probably right because the worm will likely dry out and suffocate without the ability to stay moist.

Study ZONE

CHAPTER 30

Use the Key Concepts and Key Terms listed below to review the main ideas in this chapter. If you need to review the meaning of a term, turn to the page indicated.

Key Concepts

30-1 Mollusks

- Many mollusks and annelids have a larval form called a trochophore.
- The mollusk body has three distinct parts: a visceral mass, a mantle, and foot.
- All mollusks except bivalves have a rasping tonguelike radula.
- Mollusks have a true coelom and well-developed organs.
- Most mollusks respire with gills but some respire with a primitive lung.
- Nephridia enable mollusks to recover the useful substances from their bodily wastes.
- Gastropods (snails and slugs) live in oceans, in fresh water, and on land.
- Bivalves (clams, oysters, and their kin) are aquatic and have hard shells called valves that protect their soft bodies.
- Gastropods and bivalves have an open circulatory system.
- Cephalopods (octopuses, squids, and their kin) have a well-developed head region, many tentacles, and a closed circulatory system. Most cephalopods have no external shell.

30-2 Annelids

- Annelids are coelomate worms that have segmented bodies and complex nervous systems.
- Annelids are classified according to the presence or absence of setae and parapodia.
- Annelids respire through their skin, and they have a closed circulatory system.
- Earthworms burrow through the soil, ingesting it as they crawl.
- Marine polychaetes have parapodia and setae. Some are active predators and others are filter feeders.
- Leeches lack parapodia and setae, and their segments are not separated internally. They may be aquatic or terrestrial, and some are parasites.

Key Terms

30-1

trochophore (660)
visceral mass (661)
mantle (661)
foot (661)
radula (661)
nephridium (662)
adductor muscle (665)
siphon (665)

30-2

cerebral ganglion (669)
septa (670)
seta (670)
parapodium (670)

Study TIP Ready?

- *If you think you understand these Key Concepts and Key Terms, then you're ready for the Performance Zone.*

Study ZONE

CHAPTER 30

Review and Practice

Assign the following worksheets for Chapter 30:

- **Student Review Guide**
- **Concept Mapping Worksheet**
- **Test Preparation Pretest**
- **Vocabulary Worksheet**
- **Critical Thinking Review Worksheet**

Assessment Options

The following assessment options are available for this chapter:

Resource Link

- Assign **Chapter Test 30** to assess students' understanding of the chapter.

Alternative Assessment

Have students design a new species of mollusk or annelid. Students might draw and describe their animal or create a model. They should give their creation a scientific name based on its features and describe its habitat, reproduction habits, and feeding habits. Designs must include the key characteristics of a mollusk or annelid. PORTFOLIO

Portfolio Assessment

- *Portfolio Assessment* at the front of this book provides options for building and evaluating student portfolios. Students and teachers can select items from areas listed for inclusion in a portfolio.
- See items labeled PORTFOLIO on pp. 661, 664, 666, 667, 672, 674, and 675.

Answer to Concept Map

The following is one possible answer to Performance Zone item 1 on page 676.

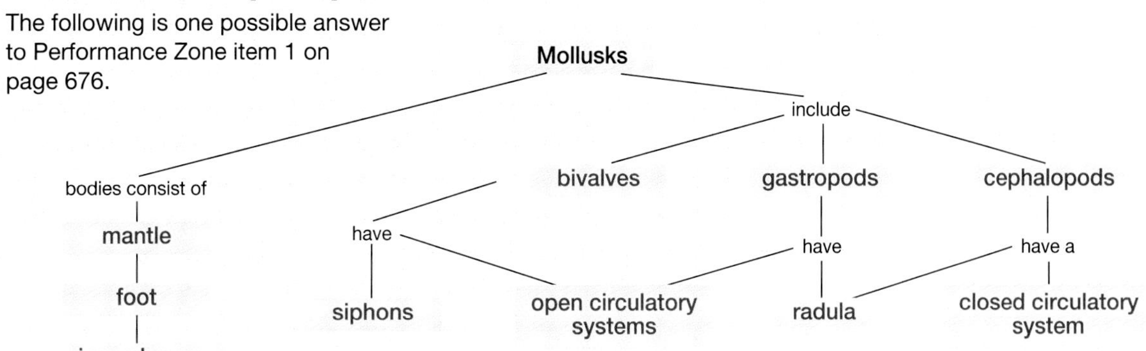

Performance ZONE CHAPTER 30 REVIEW ANSWERS

Assignment Guide

Section	Review
30-1	1, 5, 6, 7, 8, 9, 16,18, 19, 22, 23, 24, 25, 26, 28
30-2	2, 3, 4, 10, 11, 12, 13, 14,15, 17, 20, 21, 27, 29

Understanding and Applying Concepts

1. The answer to the concept map is found at the bottom of page 675.
2. **a.** Setae are bristlelike appendages found on the segments of earthworms. Parapodia are fleshy appendages found on some annelids.

   b. The mantle is the outer layer of the mollusk, surrounding the visceral mass. The visceral mass is the central, organ-containing section of a mollusk.
3. a
4. d
5. a
6. b
7. a
8. c
9. a
10. d
11. b
12. d
13. b
14. The food of aquatic annelids is not as coarse and does not require as much grinding for digestion as the food of terrestrial annelids.
15. Earthworms aerate the soil by tunneling and fertilize the soil with their castings. Earthworms' activity helps soil support crops and other beneficial plants, which humans use for food, fiber, and medicines.
16. The bivalves may harbor dangerous bacteria if they are not fresh or were harvested in contaminated waters.
17. Leeches remove blood from damaged areas, which helps keep the tissues healthy until new blood vessels can form.

Performance ZONE CHAPTER 30 Review

Understanding and Applying Concepts

1. **Concept Mapping** Make a concept map that shows the characteristics and diversity of the mollusks. Include the following words in your map: foot, visceral mass, mantle, radula, siphons, gastropod, bivalve, cephalopod, open circulatory system, closed circulatory system.

2. **Understanding Vocabulary** For each pair of terms, explain the difference in their meanings.
   a. setae, parapodia
   b. mantle, visceral mass

3. Both mollusks and annelids
   a. are coelomates.
   b. have at least a remnant of a shell.
   c. have no larval form.
   d. have a visceral mass.

4. The nephridia of annelids and mollusks function in
   a. respiration. c. digestion.
   b. circulation. d. excretion.

5. All of the following are a part of the mollusk's three-part body plan except a
   a. radula. c. visceral mass.
   b. mantle. d. foot.

6. Terrestrial snails respire
   a. with gills.
   b. with a primitive lung.
   c. through their skin.
   d. through their siphon.

7. Which of the following is not true of bivalves?
   a. They have a distinctive head region.
   b. They have sense organs.
   c. They have an open circulatory system.
   d. Most are filter feeders.

8. Cephalopods have all of the following characteristics except
   a. bilateral symmetry.
   b. a three-part body plan.
   c. an open circulatory system.
   d. a true coelom.

9. Which mollusk has an adductor muscle?
   a. clam c. squid
   b. snail d. nudibranch

10. Annelids are divided into three classes. This classification is based on the number of setae and the presence or absence of
    a. segments. c. a gizzard.
    b. hearts. d. parapodia.

11. Annelids are most easily recognizable by their
    a. cephalization. c. nephridia.
    b. segmentation. d. body cavity.

12. Blood in the circulatory system of an annelid
    a. flows into its body cavity.
    b. delivers carbon dioxide to its tissues.
    c. passes through gills.
    d. stays within its circulatory system.

13. Earthworm movement requires all of the following except
    a. circular muscles.
    b. secretion of mucus.
    c. muscle contractions.
    d. traction provided by setae.

14. **Theme Evolution** The gizzards of annelids that have returned to an aquatic environment are smaller and less muscular than those of terrestrial annelids, such as earthworms. How do the differences between the gizzards represent adaptations in aquatic and terrestrial feeding?

15. **Theme Interdependence** How does an earthworm affect the soil? How do you think humans benefit from these activities?

16. **Food Watch** What are some possible risks associated with purchasing bivalves from an unknown vendor?

17. **Health Watch** How can leeches be beneficial following microsurgery?

18. **Chapter Links** Explain the significance of the evolution of a coelom in mollusks. (**Hint:** See Chapter 28, Section 28-1.)

---

18. The coelom allowed for the evolution of complex organs composed of more than one tissue type. A coelom also allows the internal organs to be suspended from the body wall and cushioned by fluid. This allows mollusks to move around without damaging their organs or interfering with their function.

Skills Assessment

19. A. foot; B. mouth; C. mantle cavity; D. gill; E. mantle; F. visceral mass; G. intestine
20. The annelid belongs to the class Polychaeta.
21. The presence of parapodia body segments indicates that the annelid is a segmented marine worm, or polychaete.
22. Answers will vary. Students should list a variety of mollusks that are used for human consumption. They should also describe areas where each mollusk can be found.
23. Brochures will vary but should include a map showing habitats of the mollusks listed.

Skills Assessment

Part 1: Interpreting Graphics

Study the diagram of the mollusk below, and answer the following question:

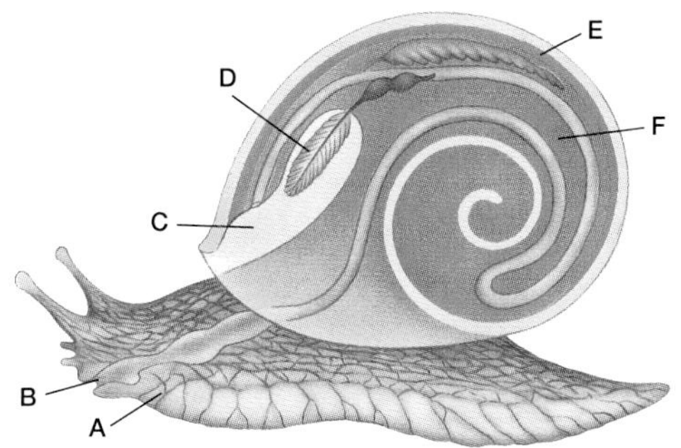

19. Identify the structures labeled *A–F*.

Study the photo below, and answer the following questions:

20. What class does this annelid belong to?

21. What characteristics of annelids did you identify that support your choice?

Part 2: Life/Work Skills

Use the library or Internet resources to compile a list of mollusks used by humans as food sources.

22. **Finding Information** For each mollusk listed, tell where it is harvested and which part of the mollusk is eaten.

23. **Communicating** Prepare a brochure to summarize your findings. Include a visual that informs the reader of the geographic location of the food source.

Critical Thinking

24. **Evaluating Results** A particular bivalve mollusk called a shipworm can do extensive damage to ocean pier pilings (supports) by burrowing into them with its radula. As a project, a student decides to determine if a new paint can reduce shipworm damage more than other paints do. What variables must the student control in order for the test results to be considered valid?

25. **Justifying Conclusions** Classification of organisms has traditionally been based on physical similarities. Given the physical differences of gastropods, bivalves, and cephalopods, how did they come to be grouped into the phylum Mollusca?

Portfolio Projects

26. **Summarizing Information** Throughout history, people around the world have used mollusk shells for adornment and other purposes. Research some specific ways humans have used shells, and prepare an oral or visual report of your findings.

27. **Interpreting Information** Read "Life Without Light," by Ian R. MacDonald and Charles Fisher (*National Geographic,* Oct. 1996, pp. 86–97). What do scientists hope to learn from this hidden food web?

28. **Summarizing Information** Chitons belong to a small class of mollusks that still have many characteristics of their mollusk ancestors. Use the library or Internet sources to find out more about chitons and why they are considered to be more primitive than other mollusks. Make a written or taped summary of your findings.

29. **Career Focus Worm Farmer** Research the field of growing segmented worms for use in research, as fishing bait, and for soil improvement. Your report should include a job description, training required, kinds of employers, growth prospects, and starting salary.

Portfolio Projects

26. Answers may vary but could include trade currency, tools, jewelry, and dishes.

27. Scientists hope to understand the importance of this deep-water community to the overall health of the Gulf of Mexico and even to the whole planet. For example, scientists have found that bacteria living within the mussels remove methane, a greenhouse gas, from the water.

28. Chitons have flat bodies and are covered by plates, traits that many scientists believe were found in ancestral mollusks. Chitons also have less sophisticated sensory organs than other mollusks and a ladderlike nervous system similar to the more primitive flatworms.

29. Answers will vary. Worm farmers have varying degrees of education, from high school through graduate school. They all share a working knowledge of growing, breeding, and harvesting worms on a large scale. Worm farmers usually own their own businesses or work as part of a larger agricultural products business. They may also be employed by waste management agencies. The growth potential for this field is fair. Starting salary will vary by region.

Critical Thinking

24. Every variable other than type of paint must be identical, including the shape and size of the boat, the type of wood, the thickness of paint, and the method of application.

25. Gastropods, bivalves, and cephalopods are grouped together as mollusks because they share a common evolutionary ancestor and because they share several key characteristics, including the presence of a coelom, a three-part body plan, and a trocophore larvae stage.

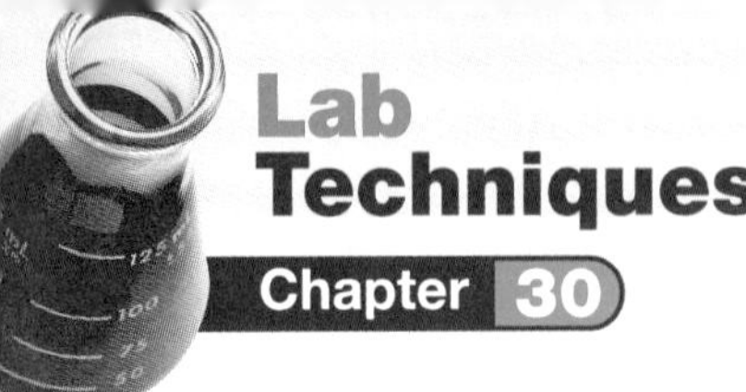

Lab Techniques

Chapter 30

Observing Characteristics of Clams

Purpose

Students will examine a live clam and a clam shell.

Preparation Notes

Time Required: 45 minutes

Materials

Materials for this lab can be ordered from WARD'S. See *Master Materials List* at the front of this book for catalog numbers.

Preparation Tips

- Be sure students wear safety goggles and lab aprons during the lab. Keep the clams in a freshwater aquarium with green algae and a layer of sand until lab time.
- Review the mollusk body plan before the lab.

Safety Precautions

- Make sure students wear safety goggles, lab aprons, and disposable gloves throughout the investigation.
- Before students begin step 8, remind then to keep hydrochloric acid (HCl) away from skin and clothing. If acid is spilled, thoroughly flush all affected areas with running water and mop up the spill with cloths designed for chemical spill cleanup. Avoid having students handle acid spills.

Procedural Tips

- Before students begin step 7, tell them to use extreme care when chipping clam shells with their scalpel. Have students use a scalpel that has a permanent blade.

Lab Techniques

Chapter 30

Observing Characteristics of Clams

SKILLS

- Observing
- Testing for the presence of a chemical

OBJECTIVES

- **Observe** the behavior of a live clam.
- **Examine** the structure and composition of a clam shell.

MATERIALS

- safety goggles
- lab apron
- live clam
- small beaker or dish
- eyedropper
- food coloring
- glass stirring rod
- clam shell
- Petri dish
- scalpel
- stereomicroscope
- 0.1 M HCl

ChemSafety

 CAUTION: Always wear safety goggles and a lab apron to protect your eyes and clothing.

CAUTION: Do not touch or taste any chemicals. Know the location of the emergency shower and eyewash station and how to use them. If you get a chemical on your skin or clothing, wash it off at the sink while calling to the teacher. Notify the teacher of a spill. Spills should be cleaned up promptly, according to your teacher's directions.

CAUTION: Glassware is fragile. Notify the teacher of broken glass or cuts. Do not clean up broken glass or spills with broken glass unless the teacher tells you to do so.

Before You Begin

Clams are **mollusks** that have a two-part shell. The body of a clam consists of a visceral mass and a muscular **foot.** There is no definite head. Two tubes, an **incurrent siphon** and an **excurrent siphon,** extend from the body on the side opposite the foot. Like all mollusks, clams use **calcium carbonate** to make their shells. A membrane called the **mantle** lines the shell and forms successive rings of shell as a clam grows. The **umbo** is the oldest part of a clam shell. In this lab, you will examine live clams and clam shells.

1. Write a definition for each boldface term in the paragraph.
2. Based on your objectives, write a question you would like to explore about clams.

Procedure

Part A: Observe a Live Clam

1.   Put on safety goggles and a lab apron.
2. Place a live clam in a small beaker or shallow dish of water. Using an eyedropper, apply two drops of food coloring near the clam, as shown above.
3. Observe and record what happens to the food coloring.
4. Using a stirring rod, touch the clam's mantle. **CAUTION: Touch the clam gently to avoid injuring it.**
5. Observe and record the clam's response to touch.

Disposal

- Broken glass and pieces of clam shell can be thrown in the trash.
- Unused 0.1 M HCl solution can be neutralized by carefully adding 0.1 M NaOH until the pH is between 6 and 8. The resulting solution can then be poured down the drain.

PART B: Observe a Clam Shell

6. Examine the concentric growth rings on the shell. Locate the knob-shaped umbo on the shell. Count and record the number of growth rings on the clam shell.

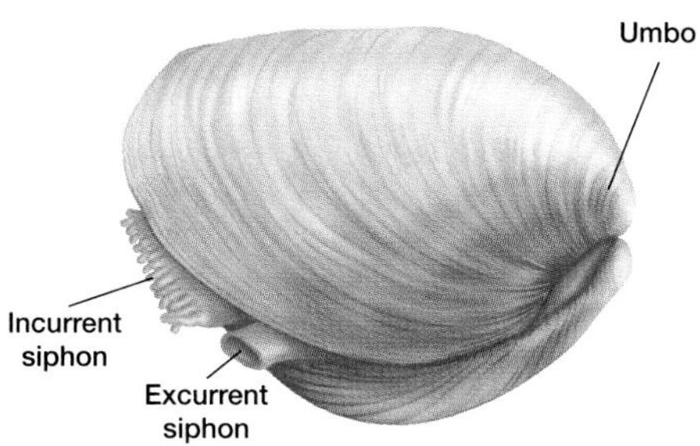

7. Place the clam shell in a Petri dish. Use a scalpel to chip away part of the shell to expose its three layers. **CAUTION: Sharp or pointed objects may cause injury. Handle scalpels carefully.** View the shell's layers with a stereomicroscope. The outermost layer protects the clam from acids in the water. The innermost layer is mother-of-pearl, the material that forms pearls.
8. The middle layer of the shell contains crystals of calcium carbonate. To test for the presence of this compound, place one drop of 0.1 M HCl on the middle layer of the shell. **CAUTION: Hydrochloric acid is corrosive. Avoid contact with skin, eyes, and clothing. Avoid breathing vapors.** If calcium carbonate is present, bubbles of carbon dioxide will form in the drop. Record your observations.

PART C: Cleanup and Disposal

9. Dispose of solutions, broken glass, and pieces of clam shell in the waste containers designated by your teacher. Do not pour chemicals down the drain or put lab materials in the trash unless your teacher tells you to do so.
10. Clean up your work area and all lab equipment. Return live clams to the stock container. Return lab equipment to its proper place. Wash your hands thoroughly before you leave the lab and after you finish all work.

Analyze and Conclude

1. **Analyzing Results** Find the incurrent and excurrent siphons of the clam in the illustration on this page. Using this information, explain your observations in step 2.
2. **Drawing Conclusions** What is the purpose of a clam's shell?
3. **Making Predictions** Based on your observations, how do you think clams respond when they are touched or threatened in their natural habitat?
4. **Forming a Hypothesis** What does a clam take in from water that passes through its body?
5. **Inferring Relationships** Water that enters a clam's incurrent siphon passes over the clam's gills. How does this help the clam respire?
6. **Further Inquiry** Write a new question about clams that could be explored with another investigation.

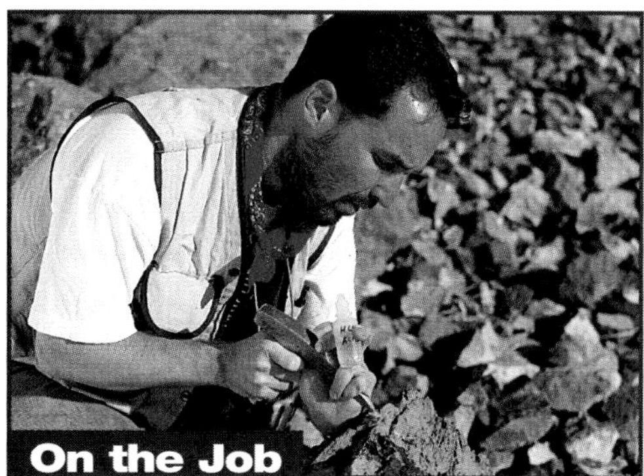

On the Job

Geologists use hydrochloric acid to test rocks for the presence of calcium carbonate. Do research in the library or media center or on the Internet to discover the role that mollusks play in the formation of rocks. Summarize your findings in a written or oral report.

Procedure

Answers

5. Answers will vary. Students will probably observe that the clam draws in its siphons and foot and closes its shell tightly when touched.
6. Answers will vary. The number of growth rings will depend on the clam's age.
8. Students should observe bubble formation, indicating the presence of calcium carbonate in the middle layer of the shell.

Analyze and Conclude

Answers

1. The food coloring was drawn into the clam through the incurrent siphon and released through the excurrent siphon.
2. The shell protects the clam from injury and predators.
3. Clams draw in their siphons and foot and close their shells tightly when threatened or touched.
4. Food and oxygen are taken in from the water.
5. A mollusk's gills can extract 50 percent or more of the dissolved oxygen from water that passes over them.
6. The clam's foot would be extended on the side of the shell across from the incurrent and excurrent siphons.
7. Answers will vary. Students should discuss the role that mollusks play in the formation of sedimentary rocks such as limestone.

Before You Begin

Answers

1. *mollusks*—animals with a soft body that secrete a shell containing calcium carbonate
   foot—part of the mollusk body that extends from the shell and aids in locomotion
   incurrent siphon—tube that takes water into a bivalve's shell
   excurrent siphon—tube that expels water from a bivalve's shell
   calcium carbonate—mineral found in a mollusk's shell
   mantle—tissue that lines a mollusk's shell and secretes the shell
   umbo—the oldest part of a bivalve's shell
2. Answers will vary. For example: How does a clam respond to touch?

On the Job

Answers

Bivalve shells are hard and fossilize well. Also, they live in aquatic environments where they are readily deposited with other sediments to form sedimentary rock, which contains almost all known fossils.

Chapter 31 Arthropods

Planning Guide

OBJECTIVES	MEETING INDIVIDUAL NEEDS	READING SKILLS AND STRATEGIES
31-1 Features of Arthropods (BLOCK 1 45 min) • **Summarize** the evolutionary relationship of arthropods and annelids. • **Identify** the three subphyla of arthropods. • **Describe** the characteristics of arthropods. • **Describe** how growth occurs in arthropods.	Learning Styles, p. 682 BASIC ADVANCED • pp. 683, 684, 687	ATE Active Reading, Technique: Reading Effectively, p. 681 ATE Active Reading, Technique: Reading Organizer, p. 683 ■ Active Reading Guide, worksheet (31-1) ■ Directed Reading, worksheet (31-1)
31-2 Spiders and Other Arachnids (BLOCK 2 45 min) • **Summarize** the characteristics of arachnids. • **Identify** the internal and external characteristics of brown recluse spiders. • **Compare** spiders, ticks, and mites. • **Describe** the health threats posed by some arachnids.	Learning Styles, p. 688 BASIC ADVANCED • p. 690	■ Active Reading Guide, worksheet (31-2) ■ Directed Reading, worksheet (31-2)
31-3 Insects and Their Relatives (BLOCKS 3 & 4 90 min) • **Describe** the characteristics of insects. • **Compare** complete and incomplete metamorphosis. • **Identify** the external and internal structures of the Eastern lubber grasshopper. • **Compare** millipedes and centipedes with insects.	Learning Styles, p. 691 BASIC ADVANCED • p. 697	■ Active Reading Guide, worksheet (31-3) ■ Directed Reading, worksheet (31-3)
31-4 Crustaceans (BLOCKS 5 & 6 90 min) • **Summarize** how crustaceans and insects are similar and dissimilar. • **Describe** the body plan of decapods.	Learning Styles, p. 698 BASIC ADVANCED • p. 700	■ Active Reading Guide, worksheet (31-4) ■ Directed Reading, worksheet (31-4) ■ Supplemental Reading Guide, *Journey to the Ants: A Story of Scientific Exploration*

Review and Assessment Resources

BLOCKS 7 & 8 90 min

TEST PREP

- PE Study Zone, p. 701
- PE Performance Zone, pp. 702–703
- ■ Chapter 31 worksheets:
  - Vocabulary
  - Concept Mapping
  - Critical Thinking
- Audio CD Program
- ■ Test Prep Pretest

CHAPTER TESTING

- ■ Chapter Tests (blackline master)
- ▲ Test Generator
- ■ Assessment Item Listing

See the correlations table at the front of this book for a list of the ***National Science Education Standards*** addressed in this chapter.

Smithsonian Institution® Visit **www.si.edu/hrw** for additional on-line resources.

KEY

Biology Principles and Explorations

PE Pupil's Edition
ATE Teacher's Edition

HOLT BIOSOURCES

 Teaching Transparencies
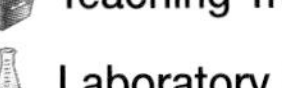
Laboratory Program

One-Stop Planner CD-ROM
Shows these resources within a customizable daily lesson plan:

■ Teaching Resources
▲ Test Generator

VISUAL STRATEGIES	CLASSWORK AND HOMEWORK	TECHNOLOGY AND INTERNET RESOURCES
ATE **Fig. 31-3**, p. 684 161 Exploration of an Arthropod 134A Phylogenetic Tree of Arthropods	**PE** **Quick Lab** What advantage do jointed appendages provide?, p. 685 **PE** **Review**, p. 687	Holt Biology Videodiscs Section 31-1 Audio CD Program Section 31-1 internet**connect** SCI LINKS NSTA **TOPICS:** • Arthropods • Head lice • Flea control
	PE **Review**, p. 690	Holt Biology Videodiscs Section 31-2 Audio CD Program Section 31-2 internet**connect** SCI LINKS NSTA **TOPIC:** Mites and disease
ATE **Fig. 31-12**, p. 692 136 Major Orders of Insects 137 Life Cycle of a Monarch Butterfly 139 Tracheal System of a Beetle 140 Insect Diversity	**PE** **Data Lab** Analyzing the effects of pesticide use, p. 696 **Lab A24** Observing Insect Behavior **Lab C39** Response in the Fruit Fly **PE** **Review**, p. 697	CNN Video Segment 25 Asthma Roaches Holt Biology Videodiscs Section 31-3 Audio CD Program Section 31-3
	PE **Data Lab** Relating molting to mortality rates, p. 700 **PE** **Experimental Design** How Do Pill Bugs Respond to Stimuli?, pp. 704–705 **Lab B25** Crayfish Dissection **PE** **Review**, p. 700 **PE** **Study Zone**, p. 701	Holt Biology Videodiscs Section 31-4 Audio CD Program Section 31-4 internet**connect** SCI LINKS NSTA **TOPICS:** • Crustaceans • Biological pest control

ALTERNATIVE ASSESSMENT

PE Portfolio Projects 26–30, p. 703
ATE **Alternative Assessment**, p. 701
ATE **Portfolio Assessment**, p. 701

Lab Activity Materials

Quick Lab
p. 685 meterstick, paper, and pencil

Data Labs
pp. 696, 700 pencil, paper

Experimental Design
pp. 704–705 adult pill bugs (4), Petri dishes (2), stereomicroscope or hand lens, blunt probe, fabrics with different textures, scissors, transparent tape, clock or watch with second hand

Demonstrations
ATE photos of several arthropods, p. 680
ATE specimens of insects, p. 691
ATE live crayfish, p. 699

BIOSOURCES LAB PROGRAM
- Quick Labs: Lab A24, p. 47
- Inquiry Skills Development: Lab B25, p. 115
- Lab Techniques and Experimental Design: Lab C39, p. 189

Chapter 31

Chapter Theme

Evolution

Arthropods are a diverse group of animals that have survived millions of years of change on Earth. Adaptations of arthropods have enabled crustaceans, arachnids, and insects to live in many different environments and in greater numbers than members of any other animal phylum.

Establishing Prior Knowledge

Before beginning this chapter, make certain students can answer the questions in **Study TIP Ready?** on page 681. If they cannot, encourage them to reread the sections indicated.

Opening Demonstration

Show students pictures of different arthropods, such as scorpions, crabs, centipedes, spiders, grasshoppers, and flies. Ask students to describe each animal. Have them make a list of differences among the animals. *(The animals differ in size, color, and method of movement.)* Then have students list the characteristics the animals have in common. *(They all have jointed legs and an exoskeleton.)*

CHAPTER 31 Arthropods

European swallowtail

What's Ahead — and Why
CHAPTER OUTLINE

Students will learn

- 31-1: the basic body plan of all arthropods.
- 31-2: the form and function of arachnids and how they fit into their niche in the environment.
- 31-3: the characteristics of insects and how they affect humans.
- 31-4: the benefits of crustaceans in the environment and their body plan.

Winged ant (24×)

Between 23 and 30 million years ago, *this ant became trapped in a drop of sticky tree sap. Today we know this fossilized sap as amber. Fossil remains of insects tell us much about their history. Some insects have survived almost unchanged for millions of years.*

Study TIP

Ready?

Check your understanding of these topics to help you prepare for what's ahead.

Can you . . .

- **relate** the term *vector* to the transmission of disease? (Section 11-1)
- **summarize** the evolutionary movement of arthropods from the ocean onto land? (Section 12-2)
- **define** the term *chitin*? (Section 23-1)
- **describe** the circulation of blood in an open circulatory system? (Section 28-2)
- **define** the term *sessile*? (Section 29-1)

If not, *review the sections indicated.*

Looking Ahead

internet connect

National Science Teachers Association *sci*LINKS Internet resources are located throughout this chapter.

Reading Strategies

Active Reading

Technique: Reading Effectively

As students read through this chapter, have them make an outline of the material presented. You may want to review outlining methods, such as using Roman numerals, letters, and Arabic numerals. Review the outlines to be sure students are including the main ideas.

Directed Reading

Assign **Directed Reading Worksheet Chapter 31** to help students interact with the text as they read the chapter.

Interactive Reading

Assign ***Biology: Principles and Explorations*** **Audio CD Program Chapter 31** to help reluctant readers achieve greater success in reading the chapter.

Reading Skills

Assign Reading Organizers and Reading Strategies Worksheets from the **One-Stop Planner CD-ROM with Test Generator** to help students learn and practice effective reading skills.

internet connect

Holt *Biology: Principles and Explorations*

On-line Resources: go.hrw.com

Visit the HRW Web site for a variety of resources related to this chapter. Just type in the keyword HX0 Chapter 31.

Holt *Biology: Principles and Explorations*

On-line Resources: www.scilinks.org

The following *sci*LINKS Internet resources can be found in the student text for this chapter:

Page	TOPIC	KEYWORD
Page 684	Arthropods	HX684
Page 687	Flea control	HX687
Page 698	Crustaceans	HX698
Page 687	Head lice	HX686
Page 690	Mites and disease	HX690
Page 705	Biological pest control	HX705

Chapter 31

1 Prepare

Scheduling

- **Block:** About 45 minutes will be needed to complete this section.
- **Traditional:** About one class period will be needed to complete this section.
- **Planning:** See the Planning Guide on pp. 679A and 679B for lecture, classwork, and assignment options for Section 31-1. Lesson plans for Section 31-1 can be found in a lesson cycle format in ***Biology: Principles and Explorations*** **One-Stop Planner CD-ROM with Test Generator.**

Resource Link

Assign **Active Reading Worksheet Section 31-1** to help students comprehend the key concepts and visuals in this lesson.

2 Teach

Lesson Warm-up

Have a discussion on the advantages *(protection, muscle attachment, prevention of water loss)* and disadvantages *(weight, molting)* of having an exoskeleton. Bring in the exoskeleton of an arthropod, such as a horseshoe crab, for students to examine.

Teaching TIP

Jointed Appendages

Have students read the **Evolutionary Milestone** on page 682. Remind them that the development of jointed appendages was a major evolutionary change. Have students point out the joints in the appendages shown in Figure 31-1.

Section 31-1 Features of Arthropods

Objectives

- **Summarize** the evolutionary relationship of arthropods and annelids. (p. 682)
- **Identify** the three subphyla of arthropods. (p. 683)
- **Describe** the characteristics of arthropods. (p. 684)
- **Describe** how growth occurs in arthropods. (p. 686)

Key Terms

appendage
thorax
cephalothorax
compound eye
molting
trachea
spiracle
Malpighian tubule

Arthropods Were the First Animals with Jointed Appendages

Whether you are looking at a scorpion or a leaf-footed bug, as shown in **Figure 31-1,** when you see an arthropod, you will probably notice its appendages. **Appendages** are structures that extend from the arthropod's body wall. Unlike the parapodia and setae of annelids, arthropod appendages have joints that bend. Their phylum name, Arthropoda, literally means "joint foot." A variety of jointed appendages are found in arthropods, including legs for walking, antennae for sensing the environment, and mouthparts for sucking, ripping, and chewing food.

Arthropods almost certainly share a common ancestor with the annelid worms. Like annelids, arthropods have a coelom and a segmented body. Arthropod fossils, some as much as 600 million years old, are among the oldest, best-preserved fossils of multicellular animals. Among the most numerous of the early arthropods were the now-extinct trilobites you read about in Chapter 12. Like modern arthropods, trilobites had segmented bodies and jointed appendages, and they were the first animals with eyes capable of forming images.

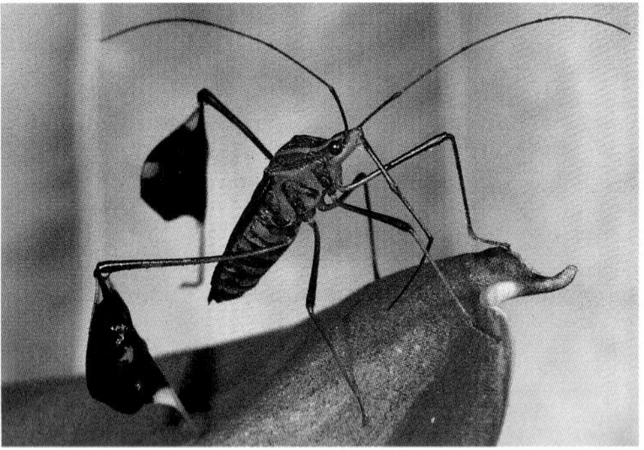
Leaf-footed bug

Scorpion

Figure 31-1 Arthropods. This leaf-footed bug and hairy desert scorpion belong to phylum Arthropoda.

Evolutionary Milestone

7 Jointed appendages

Arthropods were the first animals to have jointed appendages. Joints are an advantage because they permit more powerful movement, aiding locomotion. They also became specialized in many different ways, helping to create the vast diversity seen among the arthropods.

Integrating Different Learning Styles

Logical/Mathematical	**PE**	Study Tip, p. 683
Visual/Spatial	**PE**	Performance Zone items 1 and 18, pp. 702–703
	ATE	Lesson Warm-up, p. 682; Teaching Tip, p. 682; Visual Strategy, p. 684
Body/Kinesthetic	**PE**	Quick Lab, p. 685
Interpersonal	**PE**	Performance Zone items 22 and 23, p. 703
	ATE	Reteaching, p. 687
Verbal/Linguistic	**PE**	Performance Zone item 2, p. 702
	ATE	Active Reading, p. 683; Internet Connect, p. 684; Closure, p. 687

Arthropods Are the Most Diverse of All the Animal Phyla

If a prize were given for sheer numbers, it would go to the arthropods. The total number of arthropod species exceeds that of all other kinds of animals combined. Approximately 900,000 species of arthropods have been recorded, and probably at least as many remain to be classified. There are more species of beetles alone than there are of vertebrates. Scientists estimate that 10^{18} arthropods are alive at any one moment! The great majority of arthropods are small, about 1 mm (0.04 in.) in length, and the very smallest are parasitic mites only 80 µm (0.003 in.) long. The largest arthropods are gigantic crabs 3.6 m (12 ft) across, found in the sea near Japan.

Living arthropods are traditionally divided into two groups, arthropods with jaws and arthropods with fangs or pincers. As shown in **Figure 31-2,** arthropods with jaws belong to subphylum Uniramia *(yoo nuh RAY mee uh)*. Arthropods with fangs or pincers belong to either subphylum Chelicerata *(chuh LIS uh rahd uh)* or to subphylum Crustacea *(kruhs TAY shuh)*. Each of these three subphyla represents a distinct evolutionary line.

Study TIP

Reading Effectively

Read the number 10^{18} as "ten to the eighteenth." To get an idea of how large a quantity this number represents, write the number 1 followed by 18 zeros.

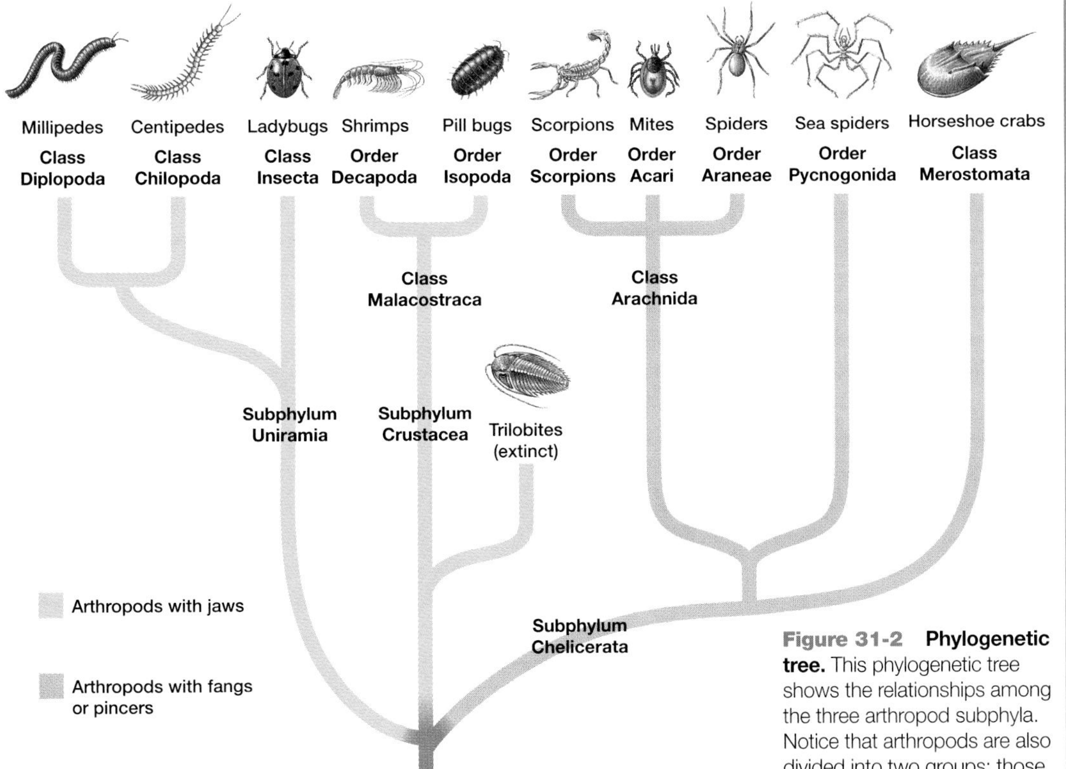

Figure 31-2 Phylogenetic tree. This phylogenetic tree shows the relationships among the three arthropod subphyla. Notice that arthropods are also divided into two groups: those with jaws and those with fangs or pincers.

Active Reading

Technique: Reading Organizer

As students read this section, have them put the various classes and their descriptions on index cards. The class name should be placed on one side and the description on the other. In addition, students could make cards with the class name on one side and examples on the other. Have students use the cards as flash cards to review the section. PORTFOLIO

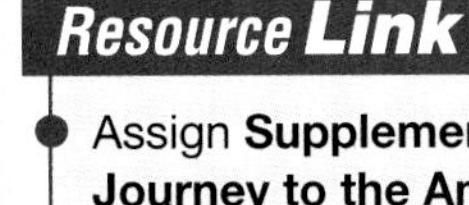

Resource Link

Assign **Supplemental Reading: Journey to the Ants:** from the Holt BioSources Teaching Resources CD-ROM. PORTFOLIO

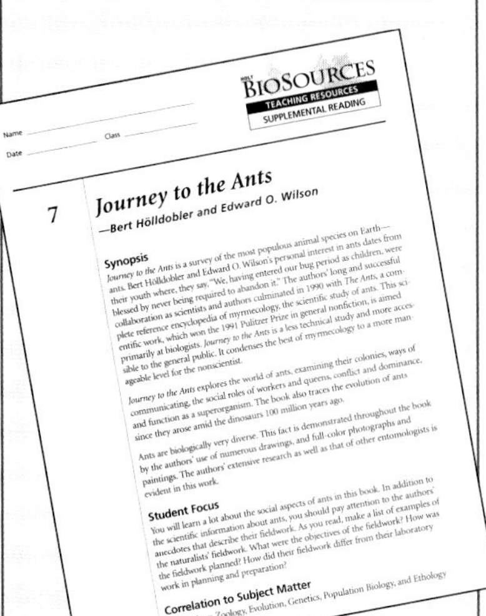

BIOSOURCES TEACHING RESOURCES SUPPLEMENTAL READING

Name ______ Class ______
Date ______

7 ***Journey to the Ants***

—Bert Hölldobler and Edward O. Wilson

Synopsis

Journey to the Ants is a survey of the most populous animal species on Earth—ants. Bert Hölldobler and Edward O. Wilson's personal interest in ants dates from their youth where, they say, "We, having entered our bug period as children, were blessed by never being required to abandon it." The authors' long and successful collaboration as scientists and authors culminated in 1990 with *The Ants*, a complete reference encyclopedia of myrmecology, the scientific study of ants. This scientific work, which won the 1991 Pulitzer Prize in general nonfiction, is aimed primarily at biologists. *Journey to the Ants* is a less technical study and more accessible to the general public. It condenses the best of myrmecology to a more manageable level for the nonscientist.

Journey to the Ants explores the world of ants, examining their colonies, ways of communicating, the social roles of workers and queens, conflict and dominance, and function as a superorganism. The book also traces the evolution of ants since they arose amid the dinosaurs 100 million years ago.

Ants are biologically very diverse. This fact is demonstrated throughout the book by the authors' use of numerous drawings, and full-color photographs and paintings. The authors' extensive research as well as that of other entomologists is evident in this work.

Student Focus

You will learn a lot about the social aspects of ants in this book. In addition to the scientific information about ants, you should pay attention to the authors' anecdotes that describe their fieldwork. As you read, make a list of examples of the naturalists' fieldwork. What were the objectives of the fieldwork? How was the fieldwork planned? How did their fieldwork differ from their laboratory work in planning and preparation?

Correlation to Subject Matter

Entomology, Zoology, Evolution, Genetics, Population Biology, and Ethology

Teaching TIP BASIC

Vocabulary

Ask students to define *arthritis*. *(inflammation of the joints)*. Then ask what kind of doctor a podiatrist is. *(foot)* Remind students that the word *arthropod* means "jointed leg."

Chapter 31

Teaching TIP BASIC

Arthropod Segmentation

Have students look through this chapter and find examples of trends in arthropod body plans. Ask: Which are more segmented, larvae or adults? *(larvae)* Which types of arthropods show more segmentation? *(Answers will vary. Millipedes and centipedes are highly segmented.)*

Visual Strategy

Figure 31-3

Have students discuss which characteristics from the list can be seen in the two arthropods pictured. *(Several arthropod characteristics should be evident; however, students will not be able to see internal structures in this illustration.)*

internet**connect**

SCILINKS NSTA Have students research arthropods using the Web site in the Internet Connect box on page 684. Have students write a report summarizing their findings. PORTFOLIO

internet**connect**

SCILINKS NSTA

TOPIC: Arthropods
GO TO: www.scilinks.org
KEYWORD: HX684

Arthropods Share a Similar Body Plan

While arthropods may be quite different in appearance, they share a number of internal and external features, which are summarized in **Figure 31-3.** There is great variation in appearance among arthropod species, and not every species will have every feature listed. However, these features are characteristic of the phylum as a whole.

Characteristics of Arthropods

- Jointed appendages
- Segmentation
- Distinct head, often with compound eyes
- Exoskeleton
- Tracheae and spiracles (respiratory structures)
- Open circulatory system
- Malpighian tubules (excretory structures)
- Wings

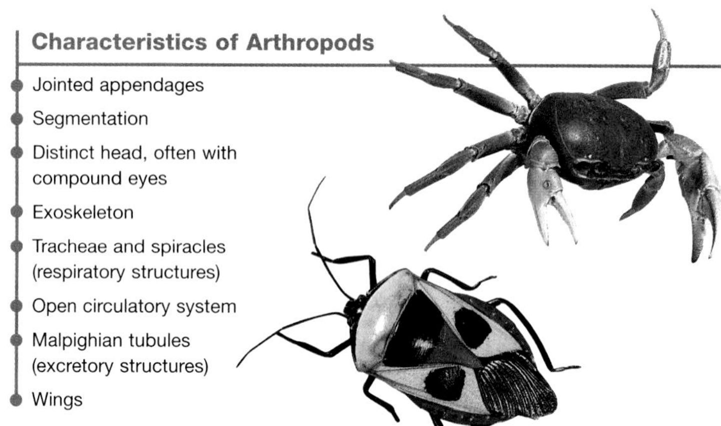

Figure 31-3 Arthropod characteristics. These eight characteristics are typical of arthropods, although not all show each characteristic.

Segmentation

In arthropods, individual body segments often exist only during the larval stage. For example, when you look at a butterfly larva (a caterpillar), you can easily see that it has many segments. However, if you look closely at an adult butterfly, you will see only three body regions. In most adult arthropods the many body segments fuse together during development to form three distinct regions—the head, **thorax** (mid-body region), and abdomen. In some arthropods, such as the crab in Figure 31-3, the head fuses with the thorax to form a body region called the **cephalothorax.**

Figure 31-4 Compound eyes. The compound eye of this house fly is made of thousands of individual units.

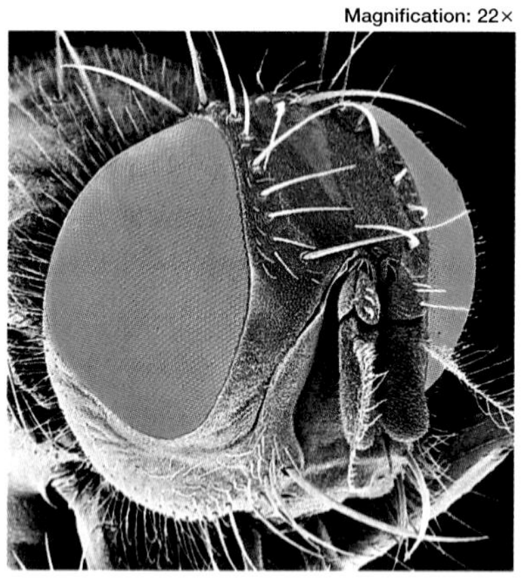

Compound eyes

Many arthropods have compound eyes, shown in **Figure 31-4.** A **compound eye** is an eye made of thousands of individual visual units, each with its own lens and retina. The brain receives input from each of the individual units, and then composes a detailed image of an object. While the image formed is not as clear as yours, motion is seen much more quickly, which is why it is so difficult to sneak up on a fly. Some arthropods also have simple, single-lens eyes that do not form images, but simply distinguish light from dark.

Most insects have both compound and simple eyes. In dragonflies and locusts, these simple eyes function as horizon detectors. The ability to see the horizon helps the insect stabilize its position during flight.

did you know?

The individual units in a compound eye do not all see the same image.

To study the lenses of ommatidia, photographers isolated the compound eyes of insects and used them to take photographs. The photographs taken through the lenses of ommatidia did not produce thousands of identical images as one might expect. Instead, the photographers found the visual effect to be similar to that of a mosaic.

Exoskeleton

The outer layer of the arthropod body has evolved into a rigid exoskeleton (often called a shell) made primarily of chitin. The exoskeleton is thinner and flexible where the joints of the appendages are located. Muscles attached to the interior surfaces of the exoskeleton close to the joints pull against the exoskeleton, causing the joint to bend. As shown in **Figure 31-5**, many arthropods can use their jointed appendages to perform complex movements. While chitin is tough, it is brittle and breaks easily. As arthropods increase in size, their exoskeletons must become thicker to withstand the pull of larger muscles without breaking. However, an increase in thickness of the exoskeleton adds weight, restricting the size arthropods can reach.

The exoskeleton of the different arthropod groups varies greatly in thickness. Crustaceans, for example, have a thick, relatively inflexible exoskeleton. If you have ever attempted to swat a large insect, you know that its exoskeleton can be difficult to crush. In comparison, the exoskeleton of other insects and some arachnids is fairly soft and flexible. Regardless of the nature of an arthropod's exoskeleton, it provides protection from injury and helps to prevent water loss.

Figure 31-5 Jointed appendages. The joints in the legs of this praying mantis permit it to perform many complex movements, such as manipulating prey.

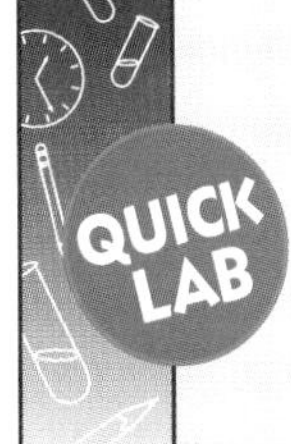

What advantage do jointed appendages provide?

To understand the importance of jointed appendages, test your range of movement without and with bending your joints.

Materials

meterstick, paper, and pencil

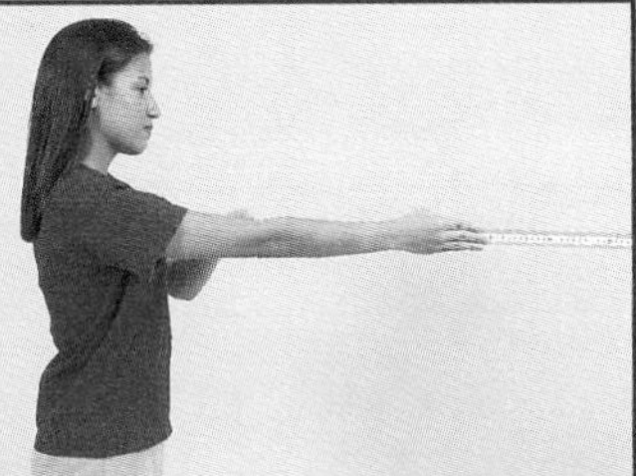

Procedure

1. Work in pairs, and assign one person to be the test subject and one person to record the data.
2. The test subject extends one arm straight out in front of the body. The subject then places a meterstick along the inside of the arm, as shown in the illustration. Do not bend the elbow.
3. The recorder measures and records the distance along the meterstick that the test subject can reach with extended (not bent) fingers.
4. The test subject now tries to increase the range of movement by bending the fingers only. The recorder measures and records the closest and farthest distance along the meterstick that can be reached.
5. The test subject now tries to increase the range of motion by bending the elbow. The recorder measures and records the closest and farthest distance along the meterstick that can be reached.

Analysis

1. **Describe** how eating breakfast might be different if you did not have joints on your fingers and at your elbows.
2. **Predict** the advantages an animal with jointed appendages has over an animal without jointed appendages when capturing and consuming food.
3. **Predict** the advantages for an arthropod that has sense organs (eyes and odor detectors) on the ends of jointed appendages.

MULTICULTURAL PERSPECTIVE

Crickets

To the Japanese, crickets are considered pets and a delightful source of musical and artistic pleasure. Favorite pet crickets are often given elaborate cages and beautiful porcelain water dishes. Japanese cricket owners even "tickle" their crickets with delicate hand-carved brushes to encourage them to sing. To the British, a cricket on the hearth is a sign of good luck.

Teaching TIP

Armor

Relate the weight of an exoskeleton to that of the body armor worn by the knights of the Middle Ages. The knights had to be assisted onto their horses, and the weight of the armor was very tiring. Even today, law enforcement officers sometimes dislike wearing bulletproof vests because they are too heavy and hot to wear.

What advantage do jointed appendages provide?

Time 10 minutes

Process Skills Measuring, predicting, drawing conclusions

Teaching Strategies

Bring in a crab exoskeleton to demonstrate the range of motion in one jointed appendage.

Analysis Answers

1. A person's arms would be useless for bringing food to one's own mouth.
2. An animal with jointed appendages would have a much greater range of motion, which would be better for manipulating food.
3. Answers will vary. One advantage is that the animal could move its sense organs by moving particular appendages instead of its whole body.

Chapter 31

Teaching TIP

Molting

Encourage students to think about some of the problems inherent in molting. Emerging from an exoskeleton can be hazardous to an arthropod. For example, a spider may lose a leg while molting. For about two hours after emerging from an outgrown exoskeleton, an arthropod without a hard exoskeleton for protection is vulnerable to predators and the environment. Ask students how arthropods can protect themselves during molting. *(Arthropods usually hide during molting until their new exoskeletons have hardened.)*

Teaching TIP

Arthropod Respiratory Organs

Insects have hollow branching tubes called tracheae that carry air into the body's tissues. Spiders have leaflike plates called book lungs that have a large surface area to exchange oxygen and carbon dioxide. Lobsters have gills that also have a large surface area for gas exchange.

REAL LIFE

Answer

Treatments include removing lice eggs and frequent shampooing. The spread of lice can usually be prevented if people avoid sharing combs, brushes, hats, towels, and other items that may transfer lice or eggs from one person to another.

Figure 31-6 **Molting.** This green cicada is emerging from its old exoskeleton, which it leaves behind as a colorless ghost of itself.

Molting

A tough exoskeleton protects an arthropod from predators and helps prevent water loss. But an exoskeleton cannot grow larger, so an arthropod cannot simply grow bigger, as many other animals do. Imagine blowing up a balloon inside a soft drink can—after a certain point, the balloon cannot get any bigger. Arthropods have the same problem, so they need to shed and discard their exoskeletons periodically, in a process called **molting,** or ecdysis *(EHK duh sihs).* Molting is triggered by the release of certain hormones. Just before molting, a new exoskeleton forms beneath the old one. When the new exoskeleton is fully formed, the old one breaks open. The arthropod emerges in its new, still-soft exoskeleton, as shown in **Figure 31-6.** The new exoskeleton hardens within a few hours or a few days, depending on the species.

Respiration

The majority of terrestrial arthropods respire through a network of fine tubes called **tracheae** *(TRAY kee ee),* as shown in **Figure 31-7.** Air enters the arthropod's body through structures called **spiracles** and passes into the tracheae, delivering oxygen throughout the body. Valves that control the flow of air through the spiracles and prevent water loss were a key adaptation for the first arthropods that invaded land more than 400 million years ago.

Figure 31-7 **Tracheal system of a beetle**

A complex series of hollow tubes called tracheae run through the bodies of most terrestrial arthropods.

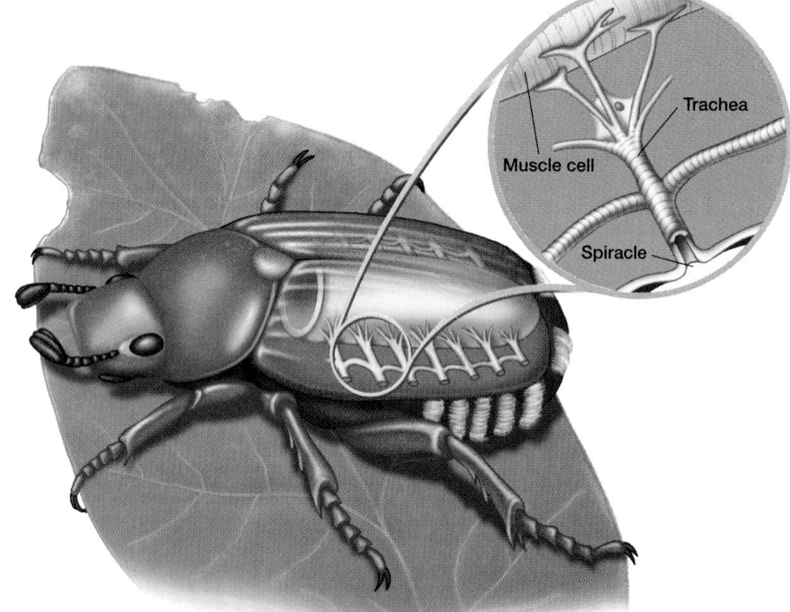

REAL LIFE

Head lice often infect entire schools. And they are notoriously difficult to control. Having clean hair will not prevent a head lice infection, and anti-lice shampoos often fail to kill immature head lice.

Finding Information

Research the latest remedies for treating head lice, and find out how to avoid infection.

Overcoming Misconceptions

Albino Roaches

People often see "albino" insects such as roaches. These white insects are not really albinos. They just shed their exoskeleton recently, and their full coloration has not returned.

Excretion

Terrestrial arthropods have a unique excretory system that efficiently conserves water and eliminates metabolic wastes. This system is composed of excretory units called Malpighian *(mal PIHG ee uhn)* tubules. **Malpighian tubules** are slender, fingerlike extensions from the arthropod's gut that are bathed by the blood that surrounds them. Water and small dissolved particles in the blood move through the tubules and into the arthropod's gut. As this fluid moves through the gut, most of the water, valuable ions, and metabolites from the fluid are reabsorbed into the arthropod's body tissues by osmosis. Metabolic wastes remain in the gut and eventually leave the body through the anus. You can see the Malpighian tubules on the grasshopper in *Up Close: Grasshopper,* on pages 694–695.

internet**connect**
SCILINKS NSTA
TOPIC: Head lice
GO TO: www.scilinks.org
KEYWORD: HX686

31-1

TECH WATCH

High-Tech Flea Control

Teaching Strategies

Ask students to describe common flea control methods. *(powders, sprays, oral medication, etc.)*

Discussion

- Why do pets scratch when they have fleas? *(Flea saliva, injected during feeding, often causes an allergic reaction and itching.)*
- How does this new method work? *(Lufenuron enters the flea when it feeds on the blood of a treated dog or cat; the drug prevents the flea's eggs from developing further.)*
- What are some possible drawbacks of using lufenuron? *(Answers will vary. Lufenuron does not kill adult or immature fleas or eggs deposited prior to treatment.)*
- Would you use lufenuron for your pet? Explain your answer. *(Answers will vary. Some students may want to try anything that will get rid of fleas. Others may be reluctant to introduce a pesticide into their pet's blood.)*

TECH WATCH

High-Tech Flea Control

As long as humans have had dogs and cats as pets, they have had to deal with fleas—those tiny but annoying insect pests that live in their pets' fur. Adult fleas feed by biting their host and sucking up a small amount of blood. As a flea feeds, it injects saliva into the wound. The saliva can cause an allergic reaction and intense itching, which is why animals scratch so much when they have fleas.

The fight against fleas

People have tried all kinds of ways to rid their pets of fleas, but until recently, none were completely successful. A new flea-control method that interrupts the flea life cycle with a drug called lufenuron is now available. Once a month, pet owners give their pet a tablet that contains lufenuron, and the drug enters the animal's bloodstream. When the female flea feeds on the blood, she consumes the lufenuron, which is then deposited into her eggs. Lufenuron prevents the eggs from developing any further. However, lufenuron will not kill adult or immature fleas. Nor will it affect eggs deposited prior to treatment. As a result, it may take 1–2 months before all adult fleas are gone.

internet**connect**
SCILINKS NSTA
TOPIC: Flea control
GO TO: www.scilinks.org
KEYWORD: HX687

Review SECTION 31-1

1. **State** the evolutionary relationship of the arthropods and the annelids.
2. **Describe** the three externally visible characteristics common to all arthropods.
3. **Summarize** how compound eyes function.
4. **Describe** how arthropods grow.
5. **SKILL Relating Concepts** Draw a concept map that shows the three subphyla of arthropods, and include two examples for each subphylum.
6. **SKILL Forming Hypotheses** Many people like to eat soft-shelled crabs. Using what you have learned, explain what soft-shelled crabs are and suggest a reason why they are available only at certain times of the year.

3 Assessment Options

Closure

Arthropod Trivia

Have students write review questions and answers for a trivia-type game. Pair each student with a partner, and have the students in each pair take turns asking their partner the questions. PORTFOLIO

Review

Assign Review—Section 31-1.

Reteaching — BASIC

Let students use the index cards they made for the **Active Reading** on page 683 to quiz each other.

Review SECTION 31-1 ANSWERS

1. Annelids and arthropods almost certainly share a common ancestor.
2. Jointed legs, an exoskeleton, and segmentation are three characteristics common to all arthropods.
3. Compound eyes receive images from each lens and send them to the brain, where one image is formed.
4. Arthropods cannot grow without shedding their exoskeleton. A new exoskeleton forms, and the old one splits and is molted. The new exoskeleton hardens, and the cycle begins again.
5. Concept maps will vary but should include the subphyla Uniramia, Crustacea, and Chelicerata, and two examples in each subphylum.
6. Soft-shelled crabs are crabs that recently molted. They are available only during crab-molting season.

Chapter 31

1 Prepare

Scheduling

- **Block:** About 45 minutes will be needed to complete this section.
- **Traditional:** About one class period will be needed to complete this section.
- **Planning:** See the Planning Guide on pp. 679A and 679B for lecture, classwork, and assignment options for Section 31-2. Lesson plans for Section 31-2 can be found in a lesson cycle format in ***Biology: Principles and Explorations*** **One-Stop Planner CD-ROM with Test Generator.**

Resource Link

Assign **Active Reading Worksheet Section 31-2** to help students comprehend the key concepts and visuals in this lesson.

2 Teach

Lesson Warm-up

Have students sketch spiders that they see around their homes. Ask them to display their sketches and to try to identify their spiders using a spider identification guide or other resource.

Making Biology Relevant

Health

Have students search the library or an on-line database for information about poisonous spiders around the world, such as the Australian funnel-web spider, whose bite can be lethal if left untreated. Have students make a pamphlet describing the spider and its habitat, the symptoms caused by the spider's bite, and the measures a person should take if bitten. PORTFOLIO

Section 31-2 Spiders and Other Arachnids

Objectives

- **Summarize** the characteristics of arachnids. (p. 688)
- **Identify** the internal and external characteristics of brown recluse spiders. (p. 689)
- **Compare** spiders, ticks, and mites. (p. 690)
- **Describe** the health threats posed by some arachnids. (p. 690)

Key Terms

chelicera
pedipalp
spinneret

Arachnids Have Fangs or Pincers

Perhaps no other group of animals is more disliked and feared by humans than are the arachnids—spiders, scorpions, ticks, mites, and daddy longlegs. While it's true that some spiders and scorpions are highly venomous, in general these creatures do more good than harm. For example, many spiders are major predators of insect pests, and gardeners usually welcome them. Arachnids *(uh RAK nihdz)* form the largest class in subphylum Chelicerata. Two minor classes, marine horseshoe crabs and sea spiders, also belong to this subphylum. The members of subphylum Chelicerata have mouthparts called **chelicerae** *(kuh LIS uh ray)* that are modified into fangs, as shown in **Figure 31-8,** or into pincers.

The arachnid body is made up of a cephalothorax and an abdomen. There are no antennae, and the first pair of appendages are chelicerae. The second pair of appendages are **pedipalps,** which are modified to catch and handle prey. (The pedipalps are sometimes specialized for sensory or even reproductive functions.) Following the pedipalps are four pairs of appendages called walking legs.

All arachnids except mites are carnivores, and most are terrestrial. Since arachnids do not have jaws, they are able to consume only liquid food. To do so, the arachnid first injects its prey with powerful enzymes that cause the prey's tissues to liquefy. Then the arachnid sucks the liquid food into its stomach.

Figure 31-8 Chelicerae. The baboon spider's pointed black chelicerae (fangs) and its two pair of pedipalps are clearly seen in this close-up of its head region.

Spiders

The chelicerae of spiders are modified into fangs. Poison glands located in the spider's anterior end secrete a toxin through these fangs. The toxin kills or paralyzes the prey. The spider then injects enzymes into the prey that digest its tissues and the spider sucks up the liquefied food. (Only two species of spiders living in the United States, the black widow and brown recluse, are dangerous to humans.) Spiders are important insect predators in virtually every terrestrial ecosystem. Not all spiders build beautiful webs as the orb-builders do, but most secrete sticky strands of silk from specially modified appendages called **spinnerets** located at the end of the abdomen. Tubes located on some spinnerets do not produce silk. Instead, they excrete a sticky substance that the spider can use to make some silk strands adhesive.

Integrating Different Learning Styles

Logical/Mathematical	**PE**	Review item 4, p. 690
Visual/Spatial	**PE**	Performance Zone items 1, 19 and 26, pp. 702–703
	ATE	Lesson Warm-up, p. 688
Verbal/Linguistic	**PE**	Study Tip, p. 690; Performance Zone item 2, p. 702
	ATE	Making Biology Relevant, p. 688; Cross-Disciplinary Connection, p. 690; Reteaching, p. 690

Up Close

Brown Recluse Spider

- **Scientific name:** *Loxosceles reclusa*
- **Size:** Length of females, up to 10 mm (0.5 in.); males are smaller
- **Range:** South-central United States, from central Texas to Alabama, north to southern Ohio
- **Habitat:** Dark, dry sheltered sites outdoors or indoors
- **Diet:** Small insects

Characteristics

Cephalothorax Six eyes, in pairs, form a semicircle around the front of the cephalothorax. Two chelicerae and two pedipalps are located next to the mouth. Four pairs of walking legs attach to the cephalothorax, which is marked on top with a distinctive violin shape.

Abdomen The abdomen contains most of the spider's organs. Spinnerets located here are used to spin small, irregular webs.

Reproduction During mating, the male uses its pedipalps modified into sperm storage organs to insert sperm into the female. The female lays an average of 20–50 eggs inside a silk cocoon that she spins and hangs in her web.

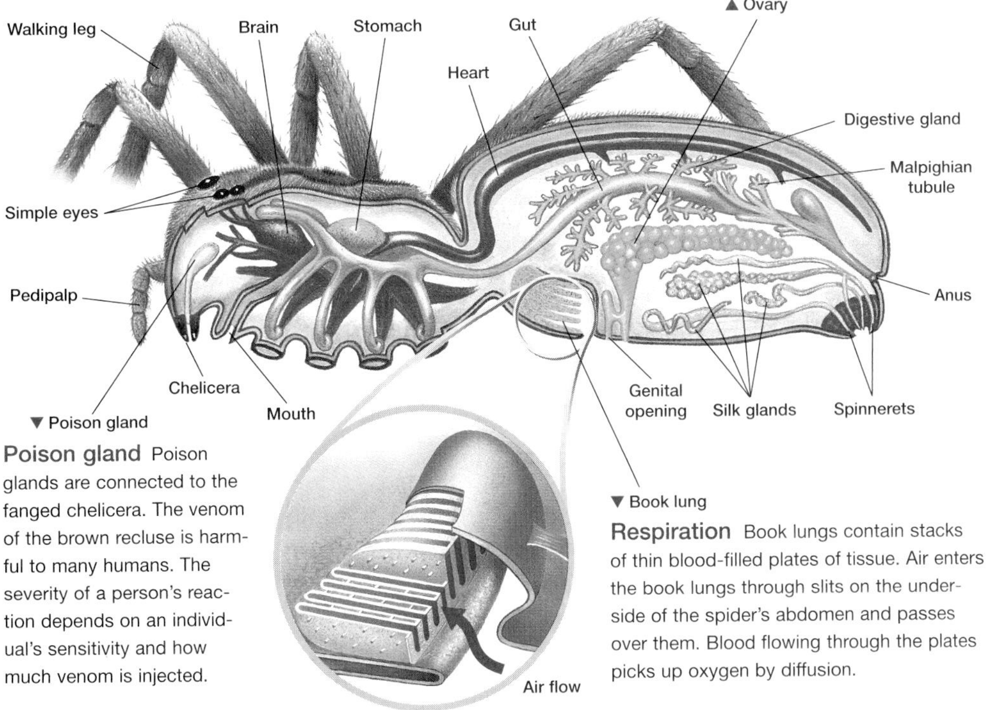

▼ Poison gland

Poison gland Poison glands are connected to the fanged chelicera. The venom of the brown recluse is harmful to many humans. The severity of a person's reaction depends on an individual's sensitivity and how much venom is injected.

▼ Book lung

Respiration Book lungs contain stacks of thin blood-filled plates of tissue. Air enters the book lungs through slits on the underside of the spider's abdomen and passes over them. Blood flowing through the plates picks up oxygen by diffusion.

did you know?

Spiders use silk in many ways.

All spiders have silk glands and organs called spinnerets that spin the silk. Some spiders spin beautiful webs that are used to catch prey. Others may use webs to protect their young. Spiders do not discard their old webs but eat them to recycle the protein in them. Spiders also use silk to wrap up prey for later use and to protect their eggs. Spiders that live in burrows line the burrow with silk. Some spiders actually spin sticky balls of silk and throw them at prey.

Up Close

Brown Recluse Spider

Teaching Strategies

- Ask students what the selective advantage of venom production might be in bees, ants, scorpions, spiders, and centipedes. *(Bees and ants use venom for defense. Scorpions, spiders, and centipedes use venom to capture prey.)*
- Explain that many spiders spin at least two varieties of silk threads. Dry threads are used to form the web's radial framework; sticky threads are arranged in a spiral around the radial threads. Ask students what advantage a spider gains by spinning both dry and sticky threads. *(Spiders avoid becoming trapped in their own webs by walking on the dry radial threads.)* Have students find information about types of silk that other spiders spin.
- Tell students that the most distinguishing feature of the brown recluse is a violin-shaped marking on the top of its body (the thorax). The neck of the violin points to the rear of its body (the abdomen).
- Have students look up the word *recluse* in a dictionary. Explain that brown recluses are active at night. During the day, they hide in corners or other dark areas, such as old boxes that are seldom opened. If you live in an area where brown recluses live, caution students to wear gloves when they clean out an area that has been left undisturbed for some time.

Discussion

- Look at the art showing the interior of the spider. What structures are present that are characteristic of arthropods in general? *(jointed appendages, segmentation, Malpighian tubules)*
- What characteristics that are found in arachnids are present? *(pedipalps, chelicerae)*

Chapter 31

CROSS-DISCIPLINARY CONNECTION

◆ **Language Arts**

Arachnids are named for Arachne, a weaver in Greek mythology. Arachne boastfully challenged the goddess Athena to a weaving contest. Arachne produced a beautiful tapestry. Athena was so enraged by Arachne's skill that she changed Arachne into a spider. Arachne's skill can still be found in the beautiful webs of spiders.

Teaching TIP

Arachnid-Borne Diseases

Ticks can carry a variety of diseases, such as Lyme disease, Rocky Mountain spotted fever, and ehrlichiosis. It is important to remember that these diseases are caused not by the ticks but by bacteria that the ticks carry. Tell students that whenever they come in from outdoors, it is a good idea to check themselves for ticks. Researchers believe that ehrlichiosis occurs only after the tick is on a person for 24 hours. "Early detection, early cure," is the motto for other tick-borne diseases as well.

3 Assessment Options

Closure

Arachnophobia

Have students suggest reasons why arachnids are feared and disliked. Ask them to counter each negative aspect with at least one positive thing arachnids do.

Review

Assign Review—Section 31-2.

Reteaching — BASIC

Have students create crossword puzzles using the vocabulary terms in this section. PORTFOLIO

Study TIP

Reviewing Information

In your own words, describe the three groups of arachnids discussed, and tell how they are alike and how they are different.

Scorpions and Mites Are Arachnids

Two other familiar groups of arachnids are scorpions and mites. Like spiders, they have chelicerae and pedipalps, but these structures are modified differently.

Scorpions

Scorpions have long, slender, segmented abdomens that end in a venomous stinger used to stun their prey. The stinger-tipped abdomen is usually folded forward over the rest of the scorpion's body, a trait that makes scorpions instantly recognizable. The pedipalps of scorpions have evolved into large, grasping pincers, which are used not for defense but for seizing food and during sexual reproduction.

Magnification: 140×

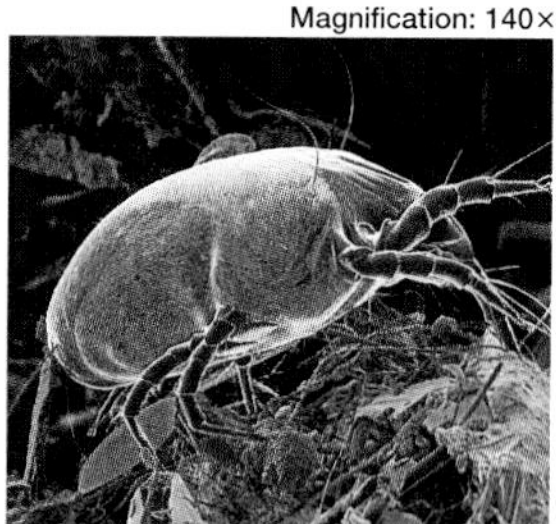

Figure 31-9 **Mite.** The house dust mite is a major cause of allergies in humans.

Mites

Mites are by far the largest group of arachnids. Some mites, including chiggers and ticks, are well known to humans because of their irritating bites. They are easily recognizable because their head, thorax, and abdomen are fused into a single, unsegmented body. Most adult mites, such as the one shown in **Figure 31-9,** are quite small, typically less than 1 mm (0.04 in.) long, but ticks grow larger. Many aquatic mites are herbivores, while terrestrial mites are usually predators.

Most mites are not harmful, but some are severe plant and animal pests. While feeding, plant mites may pass viral and fungal infections to the plant. Blood-sucking ticks attach themselves to a host, often a human. Because the bite of some ticks can transmit disease, ticks can be very dangerous. For example, Lyme disease is spread by bites from infected deer ticks, like the one shown in **Figure 31-10.**

Figure 31-10 **Adult deer tick**

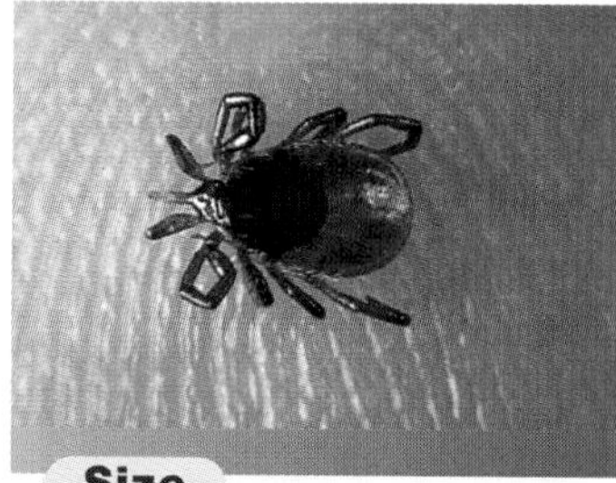

Size Immature deer ticks are aggressive feeders, but the bite of infected adults can also spread Lyme disease.

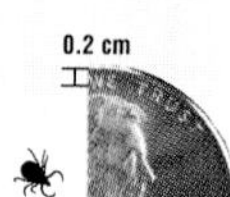

internet**connect**

SCI LINKS. NSTA

TOPIC: Mites and disease
GO TO: www.scilinks.org
KEYWORD: HX690

Review SECTION 31-2

1. **Compare** the body plan of spiders, scorpions, and mites, including differences in appendages and body segments.
2. **Summarize** how book lungs function in spiders.
3. **SKILL Summarizing Information** Explain why a tick bite is more a cause for concern than the bite of most spiders.

Critical Thinking

4. **SKILL Evaluating Conclusions** Someone brings you an organism that he has identified as an arachnid, probably a scorpion. The animal has grasping pincers, a segmented body, and two antennae. Give a reason why the identification is or is not correct.

Review SECTION 31-2 ANSWERS

1. Spider—fangs modified from chelicerae; pedipalps; two segments (cephalothorax and abdomen). Scorpion—stinger; pedipalps modified into pincers; long, slender, segmented body. mite—chelicerae and pedipalps; head, thorax, and abdomen fused into one
2. Blood flows through thin plates of tissue. Oxygen diffuses from the air into the blood.
3. Only two spiders in the United States are poisonous to humans. In contrast, many ticks transmit diseases, such as Lyme disease and Rocky Mountain spotted fever.
4. The identification is not correct. Scorpions have no antennae.

Insects and Their Relatives

Section 31-3

Insects Are the Most Numerous Animals

Anyone who has ever been on a picnic in a wooded area does not have to be told that insects are numerous. Ants, mosquitoes, gnats, flies, bees, crickets—they all want to join in while the cicadas sing in the background. These animals all belong to the arthropod subphylum Uniramia, an enormous group of mostly terrestrial arthropods that have chewing mouthparts called **mandibles** (jaws). Uniramians consist of three classes: Insecta (insects), Diplopoda (millipedes), and Chilopoda (centipedes).

The insects are by far the largest group of organisms on Earth, with more than 700,000 named species. Most scientists agree that there may be several million insect species in existence, with most of the undiscovered species living in the tropics. As shown in **Figure 31-11**, more than 50 percent of all named animal species are insects. More than 90 percent of these species belong to one of the four orders shown in **Table 31-1**. For other orders of insects, see "A Six-Kingdom System of Classification" in the Appendix.

Objectives

- **Describe** the characteristics of insects. (p.692)
- **Compare** complete and incomplete metamorphosis. (p. 693)
- **Identify** the external and internal structures of the Eastern Lubber grasshopper. (pp. 694–695)
- **Compare** millipedes and centipedes with insects. (p. 697)

Key Terms

mandible
metamorphosis
chrysalis
pupa
nymph
caste

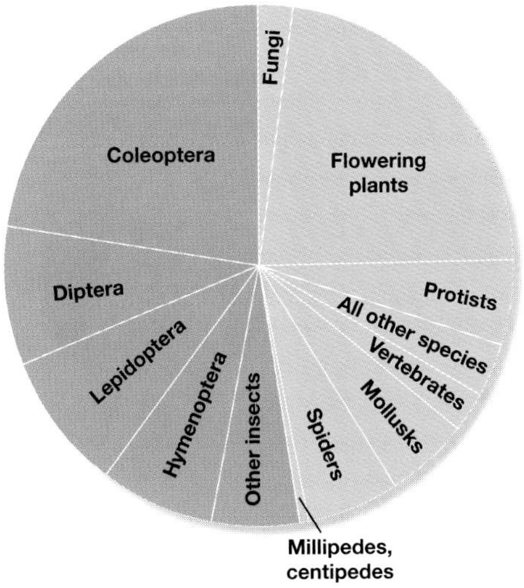

Figure 31-11 Species of insects. The predominance of insects, especially beetles, in the living world is demonstrated by the blue section of this pie chart.

Table 31-1 Four Orders of Insects

Order	Examples	Number of species
Coleoptera "shield winged"	Beetles, weevils	 350,000
Diptera "two winged"	Flies, mosquitoes	 120,000
Lepidoptera "scale winged"	Butterflies, moths	 120,000
Hymenoptera "membrane winged"	Ants, wasps, bees	 100,000

Integrating Different Learning Styles

Logical/Mathematical	PE	Data Lab, p. 696
Visual/Spatial	PE	Study Tip, p. 692; Performance Zone items 1, 19, 20 and 26, pp. 702–703
	ATE	Demonstration, p. 691; Visual Strategy, p. 692; Teaching Tip, p. 693
Interpersonal	ATE	Closure, p. 697; Reteaching, p. 697
Verbal/Linguistic	PE	Study Tip, p. 697; Performance Zone items 2, 27, and 30, pp. 702–703
	ATE	Career, p. 697

Section 31-3

1 Prepare

Scheduling

- **Block:** About 90 minutes will be needed to complete this section.
- **Traditional:** About two class periods will be needed to complete this section.
- **Planning:** See the Planning Guide on pp. 679A and 679B for lecture, classwork, and assignment options for Section 31-3. Lesson plans for Section 31-3 can be found in a lesson cycle format in ***Biology: Principles and Explorations*** **One-Stop Planner CD-ROM with Test Generator.**

Resource Link

Assign **Active Reading Worksheet Section 31-3** to help students comprehend the key concepts and visuals in this lesson.

2 Teach

Lesson Warm-up

Share these interesting facts with students. There are more arthropod species than all other animal species combined. Eighty-five percent of all animals are arthropods. Arthropods are our main competitors for food. If left unchecked (by nature), they could take over the world. There are 200 million insects for every human on earth. Insects destroy 10–15 percent of the world's food supply each year. Insects have been on Earth for approximately 300 million years. Bush crickets have ears on their knees.

Demonstration

Insect Collection

Bring examples of live or preserved insects to class for display, or bring detailed photographs of insects from a nature magazine. Allow students to compare the animals. Ask them to point out examples of the arthropod characteristics. *(jointed appendages, segmentation, exoskeleton, wings, spiracles, etc.)*

Chapter 31

CROSS-DISCIPLINARY CONNECTION

◆History

Dr. May Roberta Berenbaum, head of the entomology department at the University of Illinois, was terrified of insects as a child. While at Yale University, she enrolled in a course on terrestrial arthropods and was determined to learn about the insects that she had feared while growing up. Dr. Berenbaum founded several programs to promote learning about insects, including an annual Insect Fear Film Festival; a radio commentary called *Those Amazing Insects*; and two books, *Ninety-Nine Gnats, Nits and Nibblers* and *Ninety-Nine More Maggots, Mites and Munchers*. Dr. Berenbaum has won several awards for her research. She teaches some of the most popular courses at the University of Illinois and was recently elected to the National Academy of Sciences.

Visual Strategy

Figure 31-12

Use this figure to point out how the mouthparts of different insects are adapted for different functions. Grasshoppers cut and chew their food, mosquitoes puncture skin to get blood, and flies soak up fluids with their mouthparts.

Study TIP

Organizing Information

The numbered list on this page lists three important characteristics of insects. Use this information to start a concept map that summarizes insect characteristics. For more information about concept mapping, see pages 1030–1032 in the Appendix.

Insect body plan

Insects are primarily a terrestrial group, and aquatic insects probably had terrestrial ancestors. Although the great majority of insects are small (some are only a few centimeters in length), others are much larger. The African Goliath beetle, for example, exceeds 10 cm (4 in.) in length. Generally, the larger insects live in tropical areas. Despite great variation in their size, all insects share the same general body plan, made up of three body sections.

1. **Head.** Located on an insect's head are mandibles, specialized mouthparts, and one pair of antennae. The mandibles and mouthparts of different insect species are adapted for eating particular kinds of foods, as shown in **Figure 31-12.** In addition, an insect's head usually bears a relatively large pair of compound eyes and a pair of antennae. Like the mouthparts, antennae vary greatly in size and shape.
2. **Thorax.** The thorax is composed of three fused segments. Attached to the thorax are three pairs of jointed walking legs. Some insects, such as fleas, lice, and silverfish, lack wings, but all other adult insects have one or two pairs attached to the thorax.
3. **Abdomen.** The abdomen is composed of 9 to 11 segments. In adult insects, there are no wings or legs attached to the abdomen.

You can learn more about one particular insect, the grasshopper, in *Up Close: Eastern Lubber Grasshopper* on pp. 694–695.

Figure 31-12 **Insect mouthparts**

The mouthparts of the different insect species are adapted for different functions.

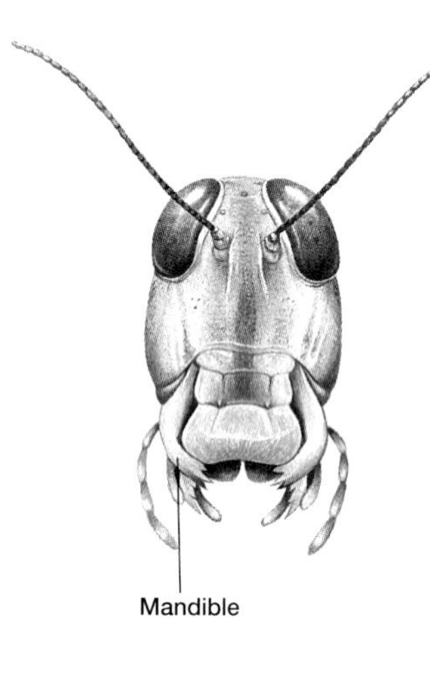

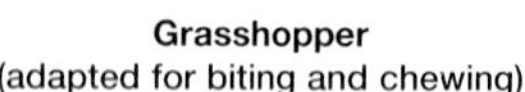

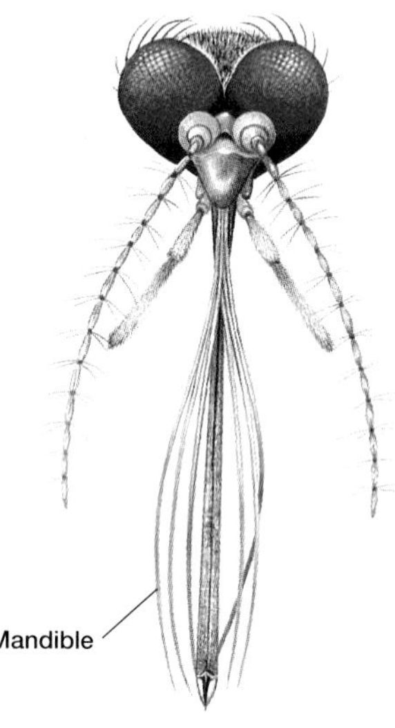

Mosquito
(adapted for piercing and sucking)

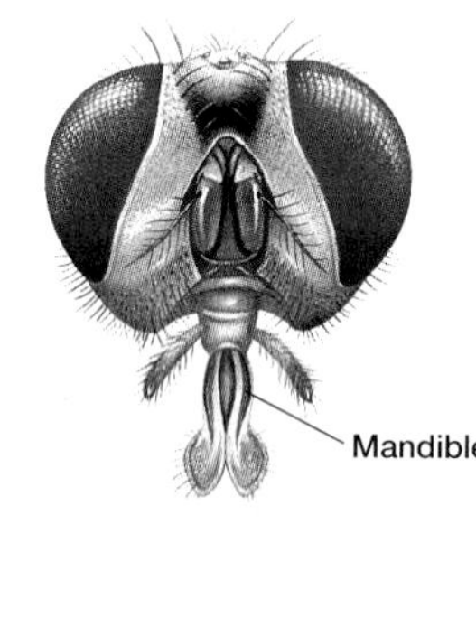

Fly
(adapted for sponging and lapping)

MULTICULTURAL PERSPECTIVE

Jumping Fleas

Dr. Miriam Louis A. Rothschild is a world expert on fleas. Using high-speed photography that records 10,000 frames per second, Dr. Rothschild concluded that a flea's ability to leap is equivalent to a human jumping over a tower 30,000 times before exhaustion sets in. She also found that a jumping rabbit flea has 230 times more acceleration than a spacecraft reentering the Earth's atmosphere after a trip to the moon. Although her family did not allow higher education for females, this self-taught zoologist, marine biologist, entomologist, and parasitologist received an honorary doctor of science degree and professorship from London University, as well as an honorary fellowship from St. Hugh's College at Oxford.

Insect life cycle

The life cycles of most insects are complex, and often several molts are required before the adult stage is reached. During the last molt, the young insect undergoes a dramatic physical change called **metamorphosis.**

Complete metamorphosis Almost all insect species undergo "complete" metamorphosis, as shown in **Figure 31-13.** In complete metamorphosis, the wingless, wormlike larva encloses itself within a protective capsule called a **chrysalis** *(KRIHS uh lihs).* Here, it passes through a **pupa** stage, in which it changes into an adult.

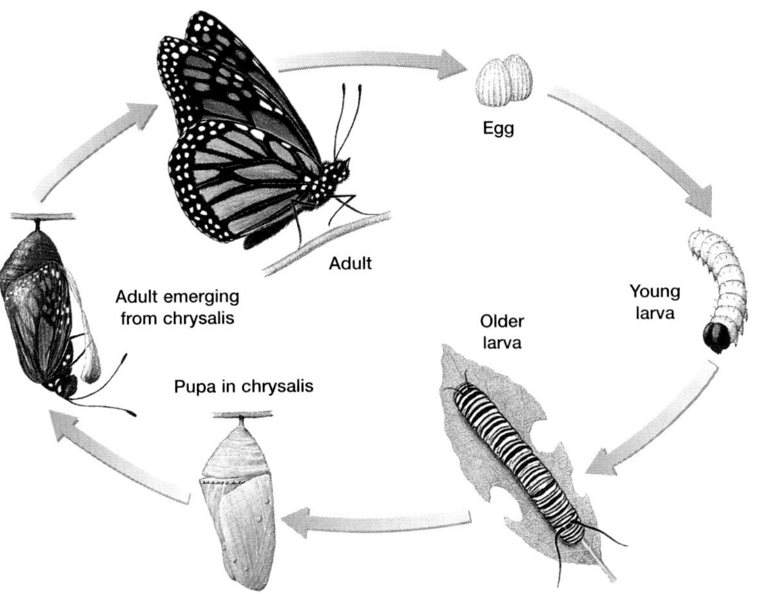

Figure 31-13 Complete metamorphosis. A complete metamorphosis includes a pupa stage passed inside a chrysalis.

Although a complete metamorphosis might seem unnecessarily complex, the larvae exploit different habitats and food sources than adults do. For example, the larvae of nectar-drinking butterflies are caterpillars that eat leaves! This ecological separation of young from adults eliminates competition, thus increasing the chance of survival for each phase of the life cycle.

Incomplete metamorphosis A smaller number of species develop into adults in a much less dramatic incomplete metamorphosis, as shown in **Figure 31-14.** In these species, the egg hatches into a juvenile, or **nymph** *(NIHMF),* that looks like a small, wingless adult. After several molts, the nymph develops into an adult.

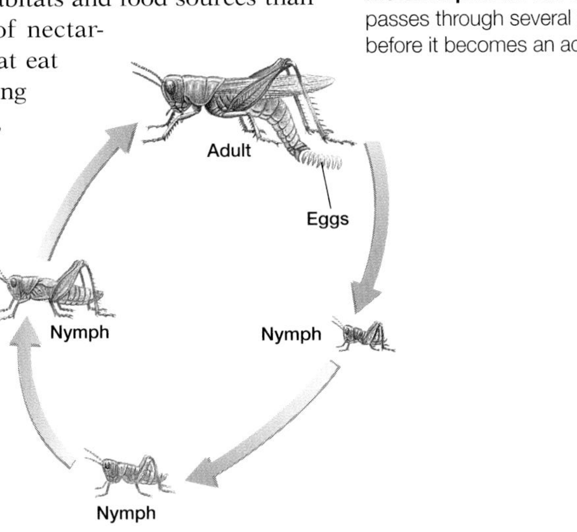

Figure 31-14 Incomplete metamorphosis. The nymph passes through several molts before it becomes an adult.

Teaching TIP

Silk

Encourage students to research the history of the discovery of silk. The Chinese were the first to produce silk fabrics, and for about 3,000 years they were the only people who knew how to cultivate silkworms. Chinese legends colorfully tell of the discovery of silk by Emperor Huangdi's wife in about 2700 B.C. Have students create a poster or diorama of the history of silk. PORTFOLIO

Teaching TIP

Metamorphosis

Ask students to compare complete and incomplete metamorphosis in a **Graphic Organizer** similar to the one at the bottom of page 693. Students may use either generalizations or specific insects in their comparisons. PORTFOLIO

Graphic Organizer

*Use this graphic organizer with **Teaching Tip: Metamorphosis** on page 693.*

Complete metamorphosis	Egg →	Larva →	Pupa →	Adult
Incomplete metamorphosis	Egg →		Nymph →	Adult

Chapter 31

Up Close

Eastern Lubber Grasshopper

Teaching Strategies

- Have students look up grasshoppers in field guides that are used for insect identification. Ask them to compare the external structures of the eastern lubber grasshopper with those of other grasshoppers. *(The eastern lubber grasshopper is a short-horned grasshopper. Long-horned grasshoppers, such as katydids and Mormon crickets, have long, threadlike antennae.)*
- Ask students to choose one of the four organ systems discussed on page 695 to illustrate in a cartoon or flowchart. Cartoons and flowcharts should show all of the organs involved in the system.
- Ask students to list the parts of the digestive system that help to physically break down food as well as the parts that chemically break down food. *(Physical digestion is performed by the labrum, labium, mandible, maxilla, and gizzard. Chemical digestion is performed by midgut and coelom.)*

Up Close

Eastern Lubber Grasshopper

- **Scientific name:** *Romalea microptera*
- **Size:** 5 cm (2 in.) to 6.5 cm (2.6 in.) in length
- **Range:** Eastern United States
- **Habitat:** Fields and meadows
- **Diet:** Grasses and other leafy vegetation

Grasshoppers feed on many types of vegetation.

External Structures

Thorax The thorax is composed of three fused segments, each with a pair of legs attached to it. The front two pairs are walking legs. The rear pair are much larger jumping legs. During mating season males "sing" to potential mates by rubbing a row of pegs on the inside of a jumping leg against ridges on a forewing.

Head Two antennae contain sense organs for both touch and smell. On each side of the head are a pair of very large compound eyes, each containing hundreds of six-sided lenses. Located high on the forehead are three light-detecting ocelli.

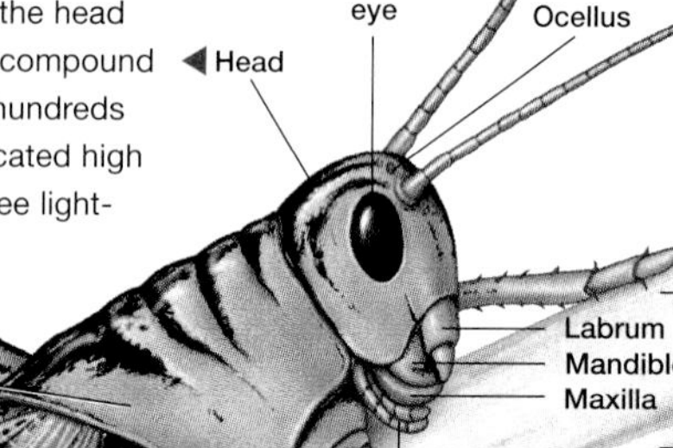

Wings Grasshoppers have a pair of leathery forewings that protect the more delicate flying wings.

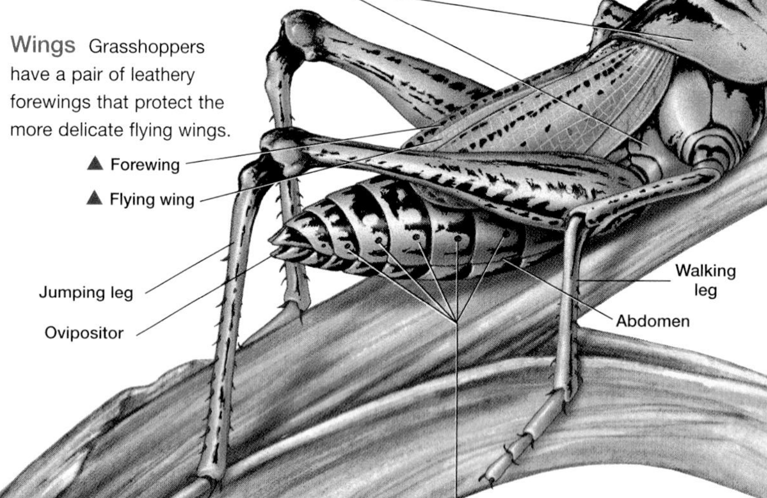

Mouthparts The mouthparts are composed of four kinds of appendages. The stiff upper labrum and lower labium (lips) hold a leaf or blade of grass in place while the mandibles (jaws), assisted by maxillas (graspers), tear off pieces of the plant.

Spiracles Tiny holes called spiracles admit air to an extensive branching system of internal tubes called tracheae that deliver oxygen throughout the body.

Internal Structures

Reproductive system The reproductive organs (testes and ovaries) are located in the abdomen. During the summer mating season, the female collects the male's sperm in a storage pouch called a seminal receptacle. Later, the female digs a hole using two pairs of pointed organs called ovipositors. As she releases the eggs into the hole, they are fertilized by the stored sperm. The eggs are dormant over the winter and hatch the following spring.

Circulatory system A long blood vessel with a series of muscular "hearts" runs along the grasshopper's back. Blood is pumped out of the open system and bathes the body tissues directly before returning to the heart.

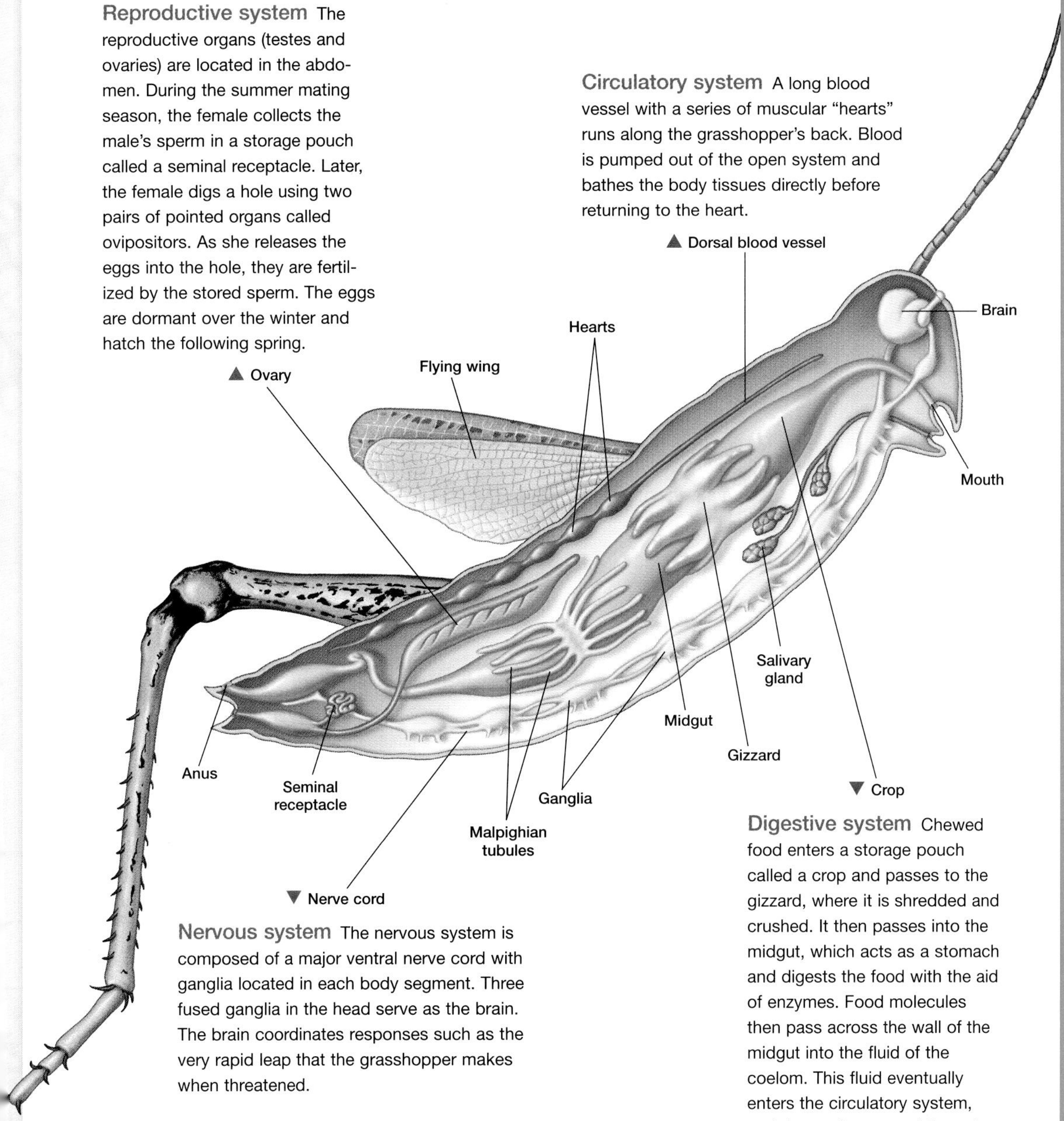

Nervous system The nervous system is composed of a major ventral nerve cord with ganglia located in each body segment. Three fused ganglia in the head serve as the brain. The brain coordinates responses such as the very rapid leap that the grasshopper makes when threatened.

Digestive system Chewed food enters a storage pouch called a crop and passes to the gizzard, where it is shredded and crushed. It then passes into the midgut, which acts as a stomach and digests the food with the aid of enzymes. Food molecules then pass across the wall of the midgut into the fluid of the coelom. This fluid eventually enters the circulatory system, and the nutrients are delivered to body tissues.

Up Close

Eastern Lubber Grasshopper

Discussion

- How do grasshopper eyes make it almost impossible to sneak up on this insect? *(Grasshoppers have compound eyes on each side of their head and three light-detecting ocelli on their forehead.)*
- How can you tell that this grasshopper is a female from its external structures? *(This grasshopper has ovipositors, which are used to deposit fertilized eggs.)*
- What structures do grasshoppers use to move slowly? How do they make quick escapes? *(Grasshoppers use their walking legs to move slowly. When they need to escape quickly, they use their jumping legs or wings.)*
- What are the three main sections of a grasshopper's body, and what is a major body function that occurs in each section? *(The grasshopper has a head, a thorax, and an abdomen. Touching, smelling, seeing, and chewing occur in the head; walking and jumping occur in the thorax; and respiration, reproduction, and excretion occur in the abdomen.)*
- What would happen if every grasshopper were to freeze during a cold winter? *(The species would continue to survive because grasshopper eggs stay dormant in winter and hatch in the spring.)*

did you know?

Silk is an agricultural product.

Silk is made today with cultivated silkworms, usually caterpillars of the species *Bombyx mori*. The female lays 300 to 500 eggs on special paper provided by silk farmers. When silkworm larvae are ready to begin metamorphosis, they spin a cocoon made of one continuous silk thread, in which they will develop into a pupa and from which they will emerge as a moth. Silk farmers do not allow most pupas to become moths because a moth will shatter the silk cocoon into many pieces as it leaves the cocoon. Instead, farmers kill most pupae and process the silk cocoons. First the cocoon is unwound. Then the threads are strengthened by several processes. Finally, the thread is dyed and woven into fabric.

Chapter 31

Analyzing the effects of pesticide use

Time 10 minutes

Process Skills Reading graphs, interpreting information

Teaching Strategies
- Ask students to name all the pesticides they can. Discuss the problems associated with their use.
- Have students contact a local garden center or on-line garden supply companies to find out how to purchase live beneficial insects.

Analysis Answers
1. Populations were stable during the first 4 years. Both populations cycled between high and low values that were the same each year.
2. As the population of the beneficial species increases, pest numbers decrease, and vice versa.
3. The number of beneficial species and insect pests decreased. The pest population eventually recovered. The beneficial species population did not recover.
4. Before the application of pesticide, the pest population reached an annual maximum of about 6,000 per hectare. After the pesticide application, the pest population rose to 10,000 per hectare.
5. Answers will vary. As the population of the beneficial species declines due to pesticide use, the pest population will increase.

Figure 31-15 Insect flight. This stop-action photograph of a green lacewing shows how the insect's wings move during flight.

Flight

Insects were the first animals to evolve wings. For more than 100 million years, until flying reptiles appeared, insects were the only flying organisms. Flight, shown in **Figure 31-15,** was a great evolutionary innovation, permitting insects to reach previously inaccessible food sources and to escape quickly from danger.

An insect's wings develop from saclike outgrowths of the body wall of the thorax. The wings of adult insects are composed entirely of chitin, strengthened by a network of tubes called veins (which carry air, not blood). In most insects, the power stroke of the wing during flight is downward, and it is produced by strong flight muscles. When at rest, most insects fold their wings over their abdomen, but a few insects are unable to do this. Dragonflies, for example, keep their wings outstretched when they rest beside a pond. Most insects have two pairs of wings, but a few species, such as flies, have lost one pair during the course of evolution. A few groups of insects, such as fleas and lice, are wingless.

In most insects only one pair of wings is functional for flight. In some species, the second pair of wings is modified for another purpose. For example, in grasshoppers and beetles, the forewings have evolved into protective wing covers. In flies, the hindwings are modified into knoblike structures that help control stability during flight.

Analyzing the effects of pesticide use

Background

In nature, insect pests are usually kept in balance by the presence of beneficial insects that prey on them. The use of some pesticides can upset this balance, as shown in the graph below. Examine the graph, and answer the analysis questions.

Analysis

1. **Identify** the years during which the two insect populations appear to maintain stability in relation to each other. Justify your answer with data from the graph.
2. **Describe** the relationship between the two insect species before year 5.
3. **Describe** the changes in the two populations after the use of a pesticide.
4. **Compare** the annual changes in population size of the pest species before and after the use of a pesticide.
5. **SKILL Developing Hypotheses** Propose a hypothesis that might explain the dramatic changes that occur in the insect populations after the use of pesticides.

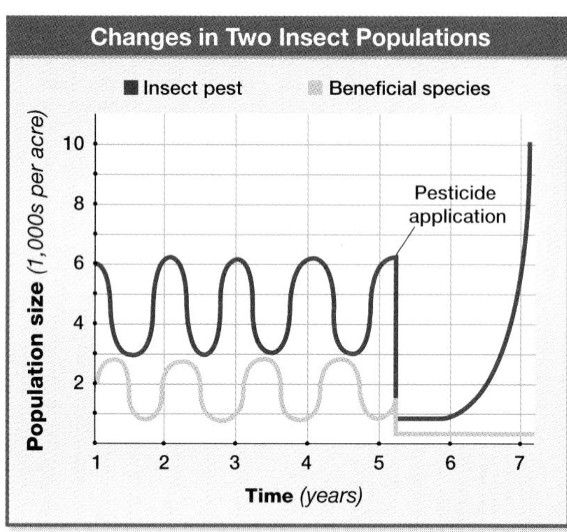

Biology in Your Community

Pest Control

Invite a pest-control authority to class to discuss the problems and risks associated with termites and other common household pests, such as ants and cockroaches. Ask the visitor to explain when such insects are considered a hazard. Have students ask questions about new methods used to control pests and about how to avoid infestations.

Social insects

Figure 31-16 **Termite colony.** Most of the members of this termite colony are sterile (unable to reproduce). The queen, with her distended abdomen, is the egg-laying machine of the colony.

Two orders of insects, Hymenoptera (ants, bees, and wasps) and Isoptera (termites), have evolved elaborate social systems. These insects often live in highly organized societies of genetically related individuals. Within these insect societies, there is a marked division of labor, with different kinds of individuals performing specific functions. The role played by an individual in a colony is called its **caste.** Caste is determined by a combination of heredity; diet, especially as a larva; hormones; and pheromones, chemical substances used for communication. In the termite colony shown in **Figure 31-16,** for example, small, active members called workers gather the food, raise the young, and excavate tunnels. Other, larger termites, called soldiers, defend the colony with their immense jaws. Both workers and soldiers are sterile. Reproduction is a function of only the queen and king.

Insect relatives

Centipedes and millipedes, shown in **Figure 31-17,** have similar body designs. Each has a head region followed by numerous segments that are all similar. Each segment bears one or two pairs of legs. Centipedes have one pair of legs per segment and can have up to 173 segments. Modern millipedes have from 11 to 100 or more body segments. As each segment evolved from two segments of their ancestors, most millipede segments have two pairs of legs. While centipedes are carnivores, most millipedes are herbivores.

Study TIP

Word Origins

The prefix *milli-* is from the Latin *mille,* meaning "thousand." The name *centipede* is from the Latin *centum,* meaning "hundred," and *pedis,* meaning "foot."

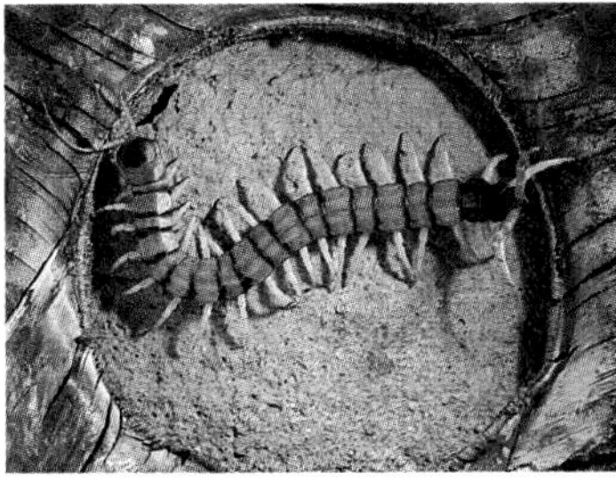

Centipede

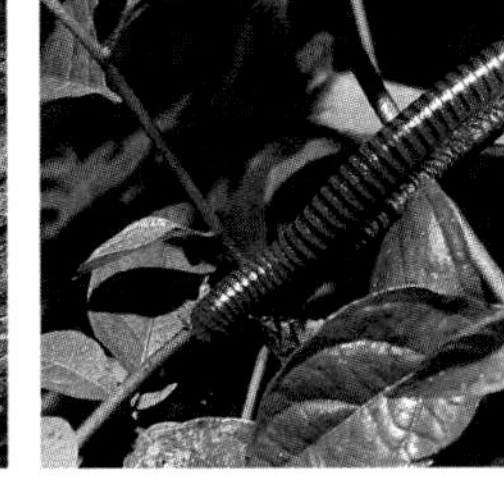

Millipede

Figure 31-17 **Centipedes and millipedes.** Centipedes are carnivorous predators, while millipedes are harmless herbivores that feed on decayed vegetation.

Review SECTION 31-3

1. **Relate** the Eastern Lubber grasshopper's body plan to that of a typical insect.
2. **Compare** the life cycles of grasshoppers and butterflies.
3. **Identify** the distinguishing characteristics of insects, millipedes, and centipedes.
4. **SKILL Forming Hypotheses** Based on the information given in Table 31-1, what characteristic is key to determining an insect's classification? Support your answer.

Review SECTION 31-3 ANSWERS

1. The eastern lubber grasshopper has a body plan like that of a typical insect. It has a head, a thorax, and an abdomen; three pairs of jointed legs; antennae; and an exoskeleton.
2. Grasshoppers go through several nymphal stages during incomplete metamorphosis. Butterflies go through complete metamorphosis, with egg, larval, pupal, and adult stages.
3. Insects have three body sections, three pairs of legs, and antennae; millipedes have many body segments with two pairs of legs per segment, and are herbivores; centipedes have many body segments with one pair of legs per segment, and are carnivores.
4. Wings are the key characteristic in classification. They are described for each class.

Teaching TIP — ADVANCED

Bees in Agriculture

Tell students that the widespread use of pesticide sprays has resulted in the loss of beehives. Have students discuss the effects the loss of these hives might have on consumers. *(Because bees are needed to pollinate many crops, reduced agricultural production and higher prices would result.)* Tell students that some farmers have to rent mobile beehives to pollinate their crops.

CAREER

Entomologist

Have students research the career of an entomologist and prepare a short report that describes the job, degrees and training needed, starting salary, employers, and growth outlook. PORTFOLIO

3 Assessment Options

Closure

Insect Orders

Divide the class into four groups and assign each group one of the insect orders in Table 31-1. Have the groups research their insect order and prepare an illustrated presentation for the class.
PORTFOLIO

Review

Assign Review—Section 31-3.

Reteaching — BASIC

Play an insect identification game with students to help them distinguish characteristics of the different orders of insects. Divide the class into four groups, and have each group choose an insect order from Table 31-1 on page 691. Then have each group ask questions of the other groups about insects in their order. The questions should be phrased so that they can be answered "yes" or "no."

Chapter 31

1 Prepare

Scheduling

- **Block:** About 90 minutes will be needed to complete this section.
- **Traditional:** About two class periods will be needed to complete this section.
- **Planning:** See the Planning Guide on pp. 679A and 679B for lecture, classwork, and assignment options for Section 31-4. Lesson plans for Section 31-4 can be found in a lesson cycle format in ***Biology: Principles and Explorations*** **One-Stop Planner CD-ROM with Test Generator.**

Resource Link

Assign **Active Reading Worksheet Section 31-4** to help students comprehend the key concepts and visuals in this lesson.

2 Teach

Lesson Warm-up

Invite students to name as many crustaceans as they can. If they have difficulty, suggest that they think about crustaceans they may have seen in seafood restaurants and aquaria.

Section 31-4 Crustaceans

Objectives

- **Summarize** how crustaceans and insects are similar and dissimilar. (p. 698)
- **Describe** the body plan of decapods. (p. 699)

Key Terms

nauplius
krill

Crustaceans Are Primarily Aquatic

Just as insect species dominate on land, crustaceans abound in the world's oceans. Their great numbers have earned them the nickname "the insects of the sea." Many are microscopic creatures that drift as plankton in the ocean currents. While primarily marine, members of subphylum Crustacea are also found in fresh water and in a few terrestrial habitats. Crustaceans include crabs, lobsters, crayfish, shrimps, barnacles, water fleas *(Daphnia)*, and pill bugs.

Almost all crustaceans have a distinctive larval form called a **nauplius** *(NAW plee uhs)*. The nauplius, shown in **Figure 31-18,** has three pairs of branched appendages. Like insects, the nauplius undergoes a series of molts before it takes on its adult form.

Adult crustaceans also have mandibles, as insects do, but they differ from insects in a number of important respects, as summarized in **Table 31-2.**

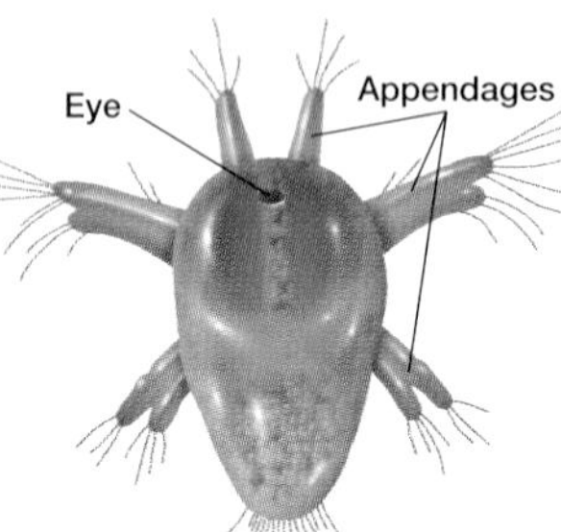

Figure 31-18 Nauplius. A microscopic, free-swimming nauplius larva is a developmental stage of almost all crustaceans.

Terrestrial crustaceans

Only a few crustacean groups have successfully invaded terrestrial habitats. The most widespread group of terrestrial crustaceans is composed of the pill bugs and sow bugs. They live among leafy ground litter found in gardens and woods. Pill bugs and sow bugs belong to a group called isopods and are the only crustaceans that are truly terrestrial. Another group, the sand fleas, includes several thousand species typically found along beaches. In addition, a few species of land crabs live in damp areas. Land crabs are only partly adapted to terrestrial living and are active primarily at night, when the air is more moist. Their life cycle is tied to the ocean, where the larvae live until maturity.

Table 31-2 Comparison of Crustaceans and Insects

Characteristic	Crustaceans	Insects
Nature of appendages	Most are branched at the end	Unbranched at the end
Antennae	Two pairs	One pair
Chewing appendages	Usually three	One
Location of appendages	Cephalothorax and abdomen	Head and abdomen
Respiration	Gills	Tracheal system

internetconnect
SCILINKS NSTA
TOPIC: Crustaceans
GO TO: www.scilinks.org
KEYWORD: HX698

Integrating Different Learning Styles

Logical/Mathematical	**PE**	Data Lab, p. 700
Visual/Spatial	**PE**	Performance Zone items 1 and 19, pp. 702–703; Experimental Design, pp. 704–705
	ATE	Demonstration, p. 699; Reteaching, p. 700
Interpersonal	**ATE**	Closure, p. 700
Verbal/Linguistic	**PE**	Study Tip, p. 699

Crustaceans Are a Major Food Source

Crustaceans are a major food source for humans and some animals. The members of some orders of crustaceans are quite small. Common are fairy shrimps, water fleas, minute ostracods, and tiny copepods *(KOH puh pahds)*. Copepods are among the most abundant multicellular organisms on Earth and are a key food source in the marine food chain. Another small marine crustacean, *Euphausia superba,* swarms in huge groups and is known by its common name, **krill.** Krill, shown in **Figure 31-19,** are the chief food for many marine species.

Decapods

Large marine crustaceans such as shrimps, lobsters, and crabs, along with the freshwater crayfish shown in **Figure 31-20,** have five pairs of legs and are often referred to as decapods. Almost one quarter of all crustaceans are decapods. The head and thorax of decapods are fused into a single cephalothorax, which is covered on top by a protective shield called a carapace.

In crayfish and lobsters, the anterior pair of legs are modified into large pincers called chelipeds *(KEE luh pehdz).* Appendages called swimmerets are attached to the underside of the abdomen and are used in swimming and in reproduction. Flattened, paddlelike appendages called uropods are at the end of the abdomen. Many decapods have a telson, or tail spine. Decapods can propel themselves through the water by forcefully flexing their abdomen.

Figure 31-19 Krill

Size Found in icy antarctic waters, small crustaceans called krill are the favorite food of many marine animals. As many as 60,000 krill can be found in 1m^3 of sea water.

Figure 31-20 Crayfish

Like all decapods, the crayfish has five pairs of legs.

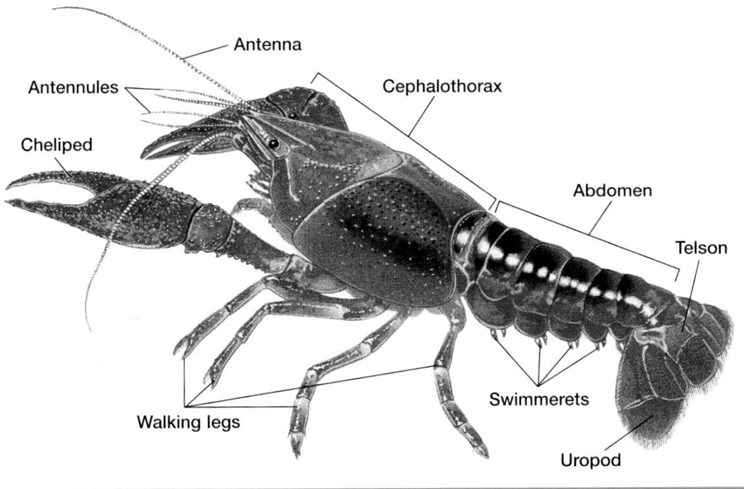

Study TIP

Word Origins

The term *decapod* is from the Greek *deka,* meaning "ten," and *pous,* meaning "foot."

Section 31-4

Demonstration

Crayfish

Bring in a few crayfish in a bucket for students to observe. Have students predict how the crayfish will move if a stick is suddenly placed in front of them. Students may be surprised when the crayfish shoot off backwards. Have students observe the crayfish's movements for several minutes.

Making Biology Relevant

Health

Some crabs have been found to carry human parasites. In Asia, freshwater crabs can carry lung flukes, which can then be transmitted to humans who eat the crabs. In Africa, onchocerciasis, an eye disease that is transmitted by flies, is also carried on the backs of river crabs, where the fly larvae live. However, most decapods are not harmful to people when they are cooked properly.

Teaching TIP

Blue Lobster

Homarus americanus, the blue lobster of the Atlantic Ocean, can grow to a length of 1 m and a weight of 18 kg. It has a cephalothorax, pincers, and several specialized appendages.

Overcoming Misconceptions

Pill Bugs

Although their name implies otherwise, pill bugs or potato bugs are actually terrestrial crustaceans, not insects. Pill bugs belong to the order Isopoda (the isopods). Whereas insects have three pairs of legs and breathe with a system of tracheae, pill bugs have seven pairs of legs and breathe with gills.

Relating molting to mortality rates

Time 15–20 minutes

Process Skills Reading graphs, applying information

Teaching Strategies
Bring in a menu that lists soft-shelled crabs. Have students infer what soft-shelled crabs are and when they are most readily available in restaurants.

Analysis Answers
1. Most crabs die in April, May, and June.
2. Most crabs molt between April and July.
3. During their molting season, many crabs fall prey to predators and disease.
4. When the crabs molt, more will die due to disease and predation than during the rest of the year.
5. Answers will vary. Crabs are fragile after molting and are more susceptible to disease.

3 Assessment Options

Closure

Importance of Crustaceans
Have students work in groups to develop ideas about the ecological and economic importance of crustaceans. Write their ideas on the board. Then have them design a concept map based on this information. PORTFOLIO

Review

Assign Review—Section 31-4.

Reteaching — BASIC

Show students photographs of a lobster and of a water flea, and ask them to identify the similarities and differences between the two animals.

Some Crustaceans Are Sessile

Barnacles are a group of crustaceans that are sessile as adults but have a free-swimming larval form. When mature, these larvae attach themselves to a rock, post, or some other submerged object, where they remain for life. Hard plates protect the barnacle's body and anchor it solidly. Two pairs of plates can open and close. When feeding, barnacles extend their jointed feeding appendages (legs) through the open plates. They then use their feathery legs to stir food from the water into their mouth. Although most crustaceans have separate sexes, barnacles are hermaphrodites. However, they do not usually fertilize their own eggs.

DATA LAB

Relating molting to mortality rates

Background
During the soft-shell stage that follows molting, many crustaceans die of disease or are eaten by predators. The bar graph below shows the percent mortality for crabs over a 9-month period. Study the data, and answer the analysis questions.

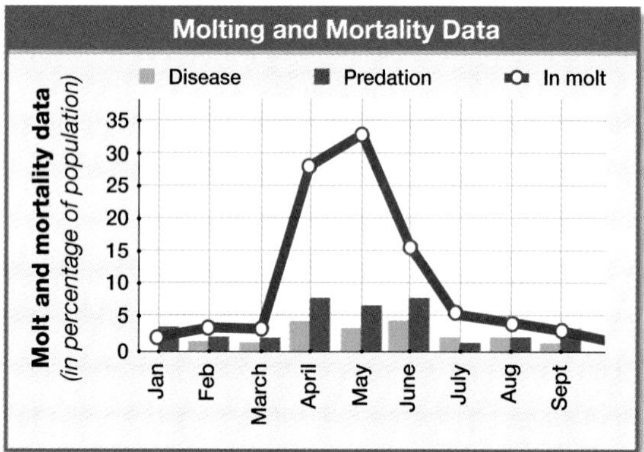

Analysis
1. **Summarize** what the data in the graph tell you about crab mortality.
2. **Summarize** what the data in the graph tell you about molting in crabs.
3. **Describe** the relationship between the mortality rates and molting periods of crabs.
4. **SKILL Developing Hypotheses** Propose a hypothesis that explains the relationship between the percent of crabs molting and mortality rates.
5. **SKILL Making Predictions** Most states have laws that require crab fishers to return molting crabs to the water. How might the length of time a molting crab is exposed to air or how roughly a crab is handled affect whether the crab survives being caught and released?

Review SECTION 31-4

1. **Compare** the body plan of a crustacean with that of an insect.
2. **Describe** the external body plan of a decapod.
3. **SKILL Making Predictions** Because they are crustaceans, pill bugs respire with gills. How might this affect the distribution of pill bugs in an ecosystem?

Review SECTION 31-4 ANSWERS

1. Answers should summarize the information given in Table 31-2 on page 698.
2. Decapods have antennae, pincers, a cephalothorax with walking legs, and a segmented abdomen.
3. You would most likely find pill bugs only in moist environments.

Study ZONE

CHAPTER 31

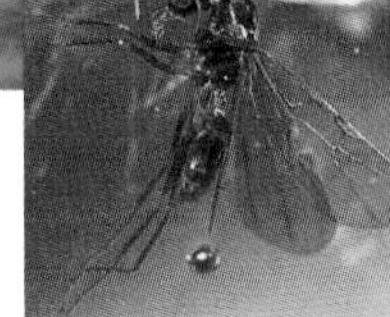

Use the Key Concepts and Key Terms listed below to review the main ideas in this chapter. If you need to review the meaning of a term, turn to the page indicated.

Key Concepts

31-1 Features of Arthropods

- All arthropods have a coelom, a segmented body, and jointed appendages that are modified to perform different functions.
- Arthropods have an exoskeleton made of chitin, which they discard periodically in a process called molting.
- Arthropods are grouped into three subphyla: Chelicerata, Uniramia, and Crustacea.

31-2 Spiders and Other Arachnids

- Members of subphylum Chelicerata have mouthparts, called chelicera, that are modified into fangs or pincers.
- Spiders have a head and a cephalothorax, no antennae, six or eight pairs of simple eyes, a pair of fangs and pedipalps, and four pairs of walking legs.
- Scorpions have pedipalps that have evolved into large, grasping pincers. A stinger is located at the end of their abdomen.
- Mites have body parts that are fused to form an unsegmented body. Many mites transmit diseases.

31-3 Insects and Their Relatives

- Insects make up more than half of all named animal species.
- All insects have a body plan with three body sections (head, thorax, abdomen), three pairs of legs (all attached to the thorax), and one pair of antennae.
- The life cycles of insects are complex and involve a process of physical change called metamorphosis, during which larvae change into the adult insects.

31-4 Crustaceans

- Crustaceans have a distinctive larval form called a nauplius.
- Copepods and krill, tiny marine crustaceans, are the chief food of many marine species.
- Most crustaceans have branched appendages, two pairs of antennae, three chewing appendages, walking legs attached to the thorax, and gills. Like insects, crustaceans have jaws called mandibles.

Key Terms

31-1
appendage (682)
thorax (684)
cephalothorax (684)
compound eye (684)
molting (686)
trachea (686)
spiracle (686)
Malpighian tubule (687)

31-2
chelicera (688)
pedipalp (688)
spinneret (688)

31-3
mandible (691)
metamorphosis (693)
chrysalis (693)
pupa (693)
nymph (693)
caste (697)

31-4
nauplius (698)
krill (699)

Study TIP Ready?

If you think you understand these Key Concepts and Key Terms, then you are ready for the Performance Zone.

Study ZONE

CHAPTER 31

Review and Practice

Assign the following worksheets for Chapter 31:

- **Student Review Guide**
- **Concept Mapping Worksheet**
- **Test Preparation Pretest**
- **Vocabulary Worksheet**
- **Critical Thinking Review Worksheet**

Assessment Options

The following assessment options are available for this chapter:

Resource Link

- Assign **Chapter Test 31** to assess students' understanding of the chapter.

Alternative Assessment

Have students make a concept map that shows the similarities and differences among the following arthropods: arachnids, insects, millipedes and centipedes, and crustaceans. PORTFOLIO

Portfolio Assessment

- *Portfolio Assessment* at the front of this book provides options for building and evaluating student portfolios. Students and teachers can select items from the areas listed for inclusion in a portfolio.
- See items labeled PORTFOLIO on pp. 683, 684, 687, 688, 690, 693, 697, 700, and 701.

Answer to Concept Map

The following is one possible answer to Performance Zone item 1 on page 702.

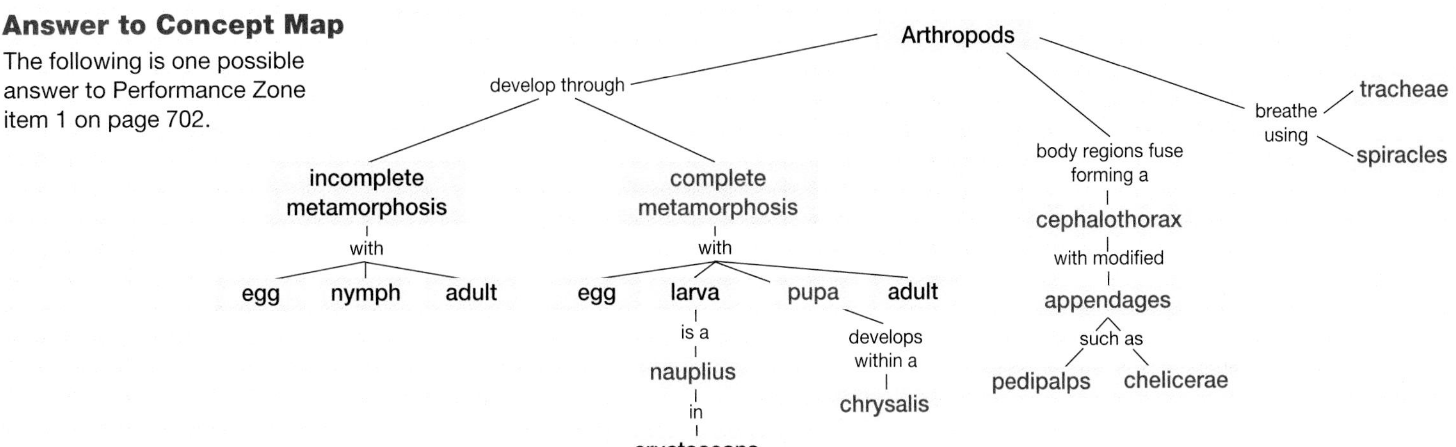

Performance ZONE
CHAPTER 31 REVIEW ANSWERS

Assignment Guide

Section	Review
31-1	1, 2, 3, 4, 5, 6, 16, 18, 22, 23
31-2	1, 2, 7, 8, 19, 24, 26
31-3	1, 2, 9, 10, 11, 12, 17, 19, 20, 21, 25, 26, 27, 28, 29, 30
31-4	1, 13, 14, 15, 19

Understanding and Applying Concepts

1. The answer to the concept map is found at the bottom of page 701.
2. a. The cephalothorax is the head-chest region of an arthropod; the exoskeleton is the hard outer covering of an arthropod.

   b. Tracheae are breathing tubes; spiracles are breathing holes in the exoskeleton.

   c. Pedipalps catch and handle prey; spinnerets make silk.

   d. Chelicerae are mouthparts of arachnids; mandibles are mouthparts of uniramians and crustaceans.
3. c
4. c
5. d
6. a
7. b
8. c
9. c
10. a
11. b
12. d
13. b
14. a
15. If we eat any seafood, it may have been ultimately dependent on copepods, since they are a key food source in the marine food chain.
16. It stops the development of eggs, so new generations are not produced.
17. Similarities include shared organs, such as the crop and gizzard. Differences are that worms do not chew their food and have no midgut.

Performance ZONE CHAPTER 31 Review

Understanding and Applying Concepts

1. **Concept Mapping** Construct a concept map that outlines the three major groups of arthropods and that gives the characteristics for each group. Try to include the following terms in your concept map: appendages, cephalothorax, tracheae, spiracles, chelicerae, pedipalps, complete metamorphosis, chrysalis, pupa, nauplius.

2. **Understanding Vocabulary** For each pair of terms, explain the differences in their meanings.
   a. cephalothorax/exoskeleton
   b. tracheae/spiracles
   c. pedipalps/spinnerets
   d. chelicera/mandible

3. What evidence suggests that arthropods are closely related to annelids?
   a. Arthropods and annelids have gills.
   b. Both groups have marine species.
   c. Segmentation is present in both groups.
   d. Arthropods have vestigial parapodia.

4. Which is a feature of the arthropod body plan?
   a. a hydrostatic support system
   b. pharyngeal slits
   c. an exoskeleton
   d. a nonsegmented body

5. Arthropods molt because
   a. their body grows faster than their shell.
   b. of damage to their exoskeleton.
   c. their exoskeleton cracks and lets in water.
   d. their hard exoskeleton cannot grow larger.

6. The chief organ of excretion in insects is the
   a. Malpighian tubule.
   b. pedipalp.
   c. nephridium.
   d. spiracle.

7. In spiders, the chelicerae are modified
   a. spinnerets.
   b. fangs.
   c. legs.
   d. antennae.

8. Spinnerets are located on the spider's
   a. thorax.
   b. cephalothorax.
   c. abdomen.
   d. pedipalps.

9. In adult insects
   a. the abdomen has wings.
   b. there are two pairs of antennae.
   c. the legs are attached to the thorax.
   d. the first appendages are chelicerae.

10. Which of the following sequences shows a complete metamorphosis?
    a. egg → larva → pupa → adult
    b. egg → larva → adult
    c. egg → young juvenile → older juvenile
    d. egg → pupa → winged juvenile → adult

11. Insects respire through structures called
    a. book lungs.
    b. spiracles.
    c. gills.
    d. ganglia.

12. Millipedes and centipedes differ in that millipedes
    a. are terrestrial and segmented.
    b. have one pair of legs on each segment.
    c. have poisonous fangs.
    d. are herbivores.

13. Which of the following is *not* a crustacean?
    a. lobster
    b. scorpion
    c. copepod
    d. water flea

14. Copepods are said to be the most important animals on Earth because they are
    a. a critical link in the marine food chain.
    b. found in both the ocean and fresh water.
    c. accomplished predators.
    d. easier to collect and study than other arthropods.

15. **Theme Interdependence** In what ways might humans be dependent on copepods?

16. **Tech Watch** What is lufenuron, and how does it affect flea reproduction?

17. **Chapter Links** Describe how the digestive system of a grasshopper is both similar to and different from that of an earthworm. (**Hint:** See Chapter 30, Section 30-2.)

Skills Assessment

18. A—Crustacea, B—Chelicerata, C—Uniramia
19. Crustaceans have two pairs of antennae. Chelicerates have mouthparts called chelicerae, which may be modified into pincers. Uniramians have one pair of antennae.
20. A—compound eye, B—antenna, C—ocellus, D—mandible, E—maxilla
21. cutting and chewing plant matter
22. Answers will vary but might include the following: malaria is carried by the *Anopheles* mosquito and causes chronic illness and eventual death; Lyme disease is carried by a deer tick and causes rash and damage to the nervous system; dust mites cause allergic reactions.
23. Answers will vary, but brochures should be well organized and present the group's findings accurately.

Skills Assessment

Part 1: Interpreting Graphics

Use the illustrations of arthropods below to answer the following questions:

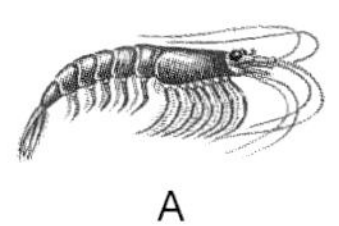

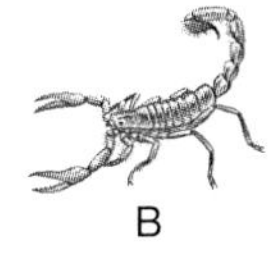

18. Identify the subphylum of arthropod *A, B,* and *C.*

19. What characteristic seen on each illustration is typical of its subphylum?

Use the illustration of the grasshopper's head below to answer the following questions:

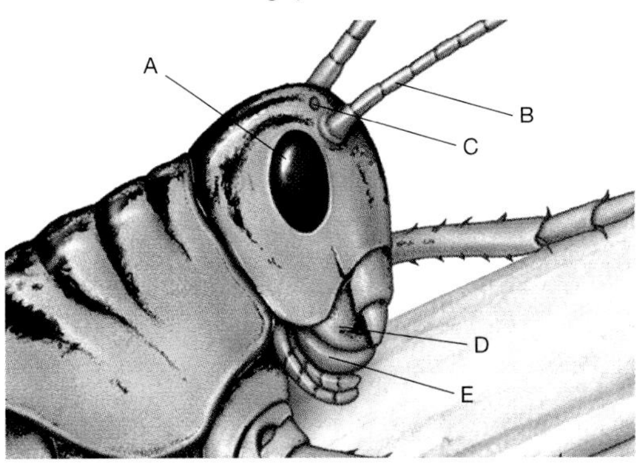

20. Name the structures labeled *A–E.*

21. The grasshopper's mouthparts are specialized for what functions?

Part 2: Life/Work Skills

Use the media center or Internet sources to learn more about diseases transmitted by arthropods.

22. Being a Team Member Working as a group, research diseases or allergic reactions caused by arthropods. Each group member could be responsible for a particular topic.

23. Communicating Develop a brochure that presents your findings about diseases transmitted by arthropods. Have each group member be responsible for a step, such as copyediting or design.

Critical Thinking

24. Recognizing Logical Inconsistencies A neighbor commented about the increased number of insects around her house. At the same time, she was removing every spider web and killing every spider she saw. How might the neighbor's actions affect the number of insects around her house?

25. Inferring Conclusions An unknown arthropod is found. It has three body segments, one pair of antennae, and three pairs of jointed legs. What kind of arthropod is it? Explain how you arrived at this conclusion.

Portfolio Projects

26. Forming a Model Using papier mâché or some other material, make a model of a grasshopper or a spider. Present your model to the class, and explain the structures shown on it.

27. Analyzing Information Ancient Egyptians worshipped the sun and held sacred the scarab beetle as a symbol of immortality. Research scarab beetles, and find out their relationship to eternal life. Prepare a written report of your findings.

28. Summarizing Information Read "The Seat of Insect Learning?" by Fred C. Dyer (*Natural History,* Sept. 1997, pp. 58–59). What are "mushroom bodies," and what role might they play in learning? Cite one experimental study that supports this position.

29. Summarizing Information Read "Planet of the Beetles," by Douglas H. Chadwick (*National Geographic,* Mar. 1998, pp. 100–119). What percentage of the world's identified animal species are beetles? What results of the beetle's evolution have aided its survival and proliferation?

30. Career Focus Agricultural Insect Inspector Research the field of inspecting field crops for the presence of harmful insects, and write a report on your findings. Your report should include a job description, training required, kinds of employers, growth prospects, and starting salary.

Critical Thinking

24. Spiders eat insects. When spiders are removed, insects have fewer predators. Killing the spiders probably led to an increase in the insect population.

25. It is an insect because it has three common insect characteristics.

Portfolio Projects

26. Models will vary but should show the external structures illustrated in the drawings on pp. 689 or 694.

27. Reports will vary. Scarab beetles are associated with the sun god Ra and eternal renewal of life.

28. Mushroom bodies are cup-shaped structures at the top of an insect's brain. They allow the insect to remember certain stimuli, such as odors. Several experiments support this position. For example, DeBelle's group destroyed fruit fly mushroom bodies during larval development, and those flies did poorly on odor-association tests. Strausfeld studied the cockroach's spatial learning of hot and cool zones. Menzel and Hammer taught honeybees to stick out their tongues. Farbach and Robinson studied changes in the mushroom bodies as honeybees aged.

29. Twenty-five percent of the world's identified animal species are beetles. Answers will vary but may include size (both tiny and huge); adaptations for walking upside down; the ability to "jet ski" on a substance secreted from the abdomen; the capacity to live in extreme temperatures (heat and cold) and environments (beaver fur, falcon nests); camouflage (looking) like bird droppings; use of dung as food; and laying of eggs in dung. Beetles also have a hard casing over their wings which provides protection and helps retain moisture.

30. Answers will vary. Agricultural insect inspectors work for crop producers to identify harmful and helpful insects. A college degree in entomology may be required. Employers include agricultural stations, universities, and government agencies. Growth prospects are good. Starting salary will vary by region.

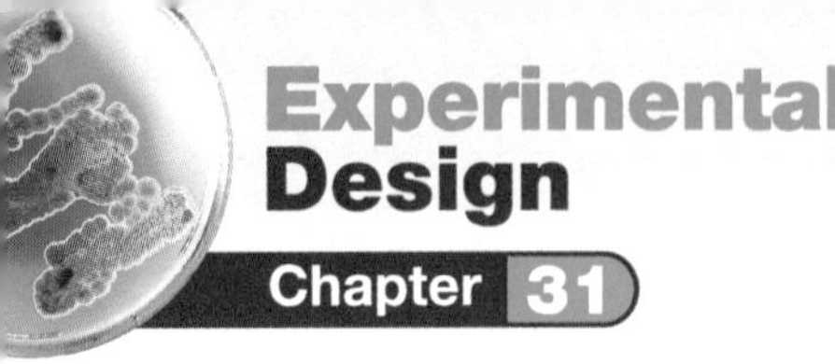

Experimental Design

Chapter 31

How Do Pill Bugs Respond to Stimuli?

Purpose

Students will look for arthropod characteristics in pill bugs and observe the behavior of pill bugs on surfaces with different textures.

Preparation Notes

Time Required: 40–50 minutes

Materials

Materials for this lab can be ordered from WARD'S. See *Master Materials List* at the front of this book for catalog numbers.

Preparation Tips

- Encourage students to read the lab and complete the **Before You Begin** section before they come to class on the day of the lab.
- Remind students that although pill bugs have an exoskeleton, they are delicate creatures and should be probed very gently.
- Before students begin the lab, conduct a class discussion in which you ask the following questions: What are the characteristics of a crustacean? What is the ideal habitat for land-dwelling isopods?

Safety Precautions

Discuss all safety symbols and caution statements with students. Remind students to be careful when using scissors.

Procedural Tips

- Distribute the pill bugs to students in small paper cups.
- Students should perform their experiment using at least three pill bugs.

Before You Begin

Answers

1. *pill bugs*—small crustaceans in the order Isopoda that live in moist terrestrial environments
   crustaceans—arthropods with mandibles, two pairs of antennae, and branched appendages.
   stimulus—any action or agent that causes or changes an activity in an organism
2. Answers will vary. For example: Do pill bugs prefer rough or smooth surfaces?

Procedure

Answers

1. Arthropod characteristics of pill bugs include an exoskeleton, jointed appendages, and a segmented body.
2. When students touch a pill bug with a blunt probe, the pill bug should curl into a ball.
4. One way of conducting the experiment would be as follows. Trace the outline of the bottom of the Petri dish on one of the fabrics. Cut out the circle from the fabric and fold it in half; then cut along the fold to produce two half-circles. Repeat this procedure for all fabrics used. Tape together two half-circles, each of a different fabric, and place them on the bottom of a Petri dish, tape side down. To study behavior, place a pill bug in the center of the circle and record the time it spends on each fabric and the path it follows. Repeat for each fabric used. To extend the investigation, students can tape together the four fabrics in all possible combinations, which would result in six different kinds of two-fabric

---

Experimental Design

Chapter 31 How Do Pill Bugs Respond to Stimuli?

SKILLS

- Using scientific methods
- Observing

OBJECTIVES

- **Identify** arthropod characteristics in a pill bug.
- **Observe** the behavior of pill bugs on surfaces with different textures.
- **Infer** the adaptive advantages of pill bug behaviors.

MATERIALS

- 4 adult pill bugs
- 2 Petri dishes
- stereomicroscope or hand lens
- blunt probe
- fabrics with different textures
- scissors
- transparent tape
- clock or watch with second hand

Before You Begin

Pill bugs live in moist terrestrial environments, such as under rocks and logs. Like other **crustaceans,** pill bugs respire with gills and have hard outer shells and jointed appendages. They respond to a **stimulus,** such as light, moisture, or touch, by moving toward or away from the stimulus or by curling into a ball. In this lab, you will look for arthropod characteristics in pill bugs and observe the behavior of pill bugs on surfaces with different textures.

1. Write a definition for each boldface term in the paragraph above.
2. Based on the objectives for this lab, write a question you would like to explore about pill bug characteristics and behavior.

Procedure

PART A: Make Observations

1. Place a pill bug in a Petri dish, and observe it with a stereomicroscope or hand lens. Observe it from a dorsal viewpoint as well as from the side. List the characteristics that tell you the pill bug is an arthropod.
2. Touch the pill bug with a blunt probe. **CAUTION: Touch pill bugs gently to avoid injuring them.** Record your observations.

PART B: Design an Experiment

3. Work with the members of your lab group to explore one of the questions written for step 2 of **Before You Begin.** To explore the question, design an experiment that uses the materials listed for this lab.

You Choose

As you design your experiment, decide the following:

a. what question you will explore
b. what hypothesis you will test
c. which four different fabrics to use
d. how many times to test each fabric
e. the length of each test
f. what your control will be
g. what data to record in your data table

4. Write a procedure for your experiment. Make a list of all the safety precautions you will take. Have your teacher approve your procedure and safety precautions before you begin the experiment.

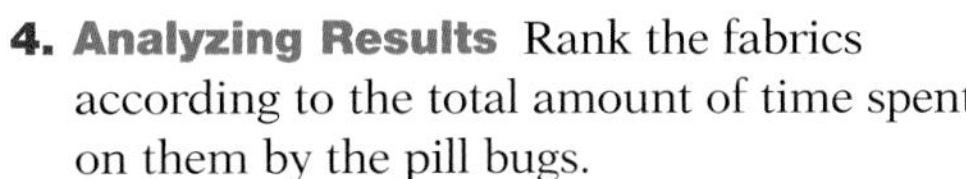

5. Set up and carry out your experiment. **CAUTION: Sharp or pointed objects can cause injury. Handle scissors carefully.**

PART C: Cleanup and Disposal

6. Dispose of filter paper and broken glass in the designated waste containers. Put pill bugs in the designated container. Do not put lab materials in the trash unless your teacher tells you to do so.
7. Clean up your work area and all lab equipment. Return lab equipment to its proper place. Wash your hands thoroughly before you leave the lab and after you finish all work.

Analyze and Conclude

1. **Analyzing Methods** Why did you test several pill bugs in this investigation instead of just one pill bug?
2. **Analyzing Results** Did all of your pill bugs show a similar pattern of movement? Explain.
3. **Graphing Results** Make a graph of your data. Plot the average time spent on the material on the *y*-axis and the type of material on the *x*-axis.
4. **Analyzing Results** Rank the fabrics according to the total amount of time spent on them by the pill bugs.
5. **Drawing Conclusions** Which fabric texture do pill bugs prefer?
6. **Inferring Conclusions** How is a pill bug's response to disturbances an advantage?
7. **Inferring Conclusions** How is being able to detect surface texture a good adaptation for pill bugs in their natural habitat?
8. **Further Inquiry** Write a new question about pill bugs that could be explored with another investigation.

Do You Know?

Do research in the library or media center to answer these questions:

1. What are the advantages and disadvantages of biological pest control when compared with the use of chemical pesticides?
2. What role do crustaceans play in the lives of humans?

Use the following Internet resources to explore your own questions about arthropods.

internetconnect

SCILINKS NSTA
TOPIC: Biological pest control
GO TO: www.scilinks.org
KEYWORD: HX705

circles. Then have students calculate the total time pill bugs spend on each type of fabric.

Analyze and Conclude

Answers

1. Several pill bugs were tested to improve the reliability of the data.
2. Answers may vary. The pill bugs should show a similar pattern of movement.
3. Graphs will vary but should accurately reflect the data. Students should make bar graphs, with each type of material represented by a bar.

Sample Graph

Time (minutes): 1, 2, 3, 4, 5
Fabric A — Fabric B

4. Answers will depend on the types of fabrics used. The highest ranking should go to the fabric on which the pill bugs spent the most time.
5. Answers will depend on the types of fabrics used. Pill bugs should spend more time on rough-textured fabrics, such as wool, than on smooth-textured fabrics, such as silk.
6. The pill bug's response to disturbances allows it to protect its gills and underside by rolling into a ball.
7. Being able to distinguish between rough textures and smooth textures may help pill bugs find soft substrates and avoid crawling over surfaces that might expose them to predators.
8. Answers will vary. For example: Do pill bugs prefer moist or dry environments? Do pill bugs prefer light or dark environments?

Do You Know?

Answers

1. Answers will vary. Insecticides have higher research, development and application costs; cause more environmental pollution; upset the natural balance of organisms; and require annual applications. Biological control has lower research, development and application costs; does not cause environmental pollution; and does not upset the natural balance of organisms. Both types of control are restricted in their general applicability.
2. Answers will vary. Crustaceans are a source of food and can also be crushed and used as fertilizer. Crustaceans can be destructive, damaging rice paddies and ship hulls. Some are used as bait for fishing.

Insect-Resistant Crops

Background

People have been fighting pests with chemicals since at least 1000 B.C. But synthetic chemical pesticides did not come into general use until after World War II, when DDT and related petroleum-based insecticides became widely available. At first, the new pesticides were so effective that some scientists predicted that insect pests could be wiped out. But the first DDT-resistant insects were discovered in 1946 in Sweden, and resistance quickly spread to other countries and species.

As resistance became more severe, scientists began to notice another problem with pesticides: they were killing large number of other organisms—birds, fish, helpful insects, spiders, and others. In 1962, Rachel Carson's book *Silent Spring* brought this problem to the attention of the public and helped start the environmental movement. In 1972, the EPA banned DDT for all but a few uses, and in 1976, dieldrin and aldrin—two widely used pesticides—were also banned. Recognition of the destructive effects of chemical pesticides stimulated the search for more environmentally friendly alternatives. *Bt* toxin is one example of such an alternative.

Teaching Strategies

- The approval of plants engineered to express *Bt* toxin has stirred up passions among scientists and non-scientists alike. Have students analyze newspaper and magazine articles expressing different views on this subject. Have students summarize each article's position and then list the arguments each article cites to support its position. Ask students whether each article resorts to emotional appeal or rhetoric instead of logic. Ask students to identify which article they find most convincing and to explain their choice.

SCIENCE • TECHNOLOGY • SOCIETY

Should the genes for insect resistance be transferred into crops?

Insect-Resistant Crops

Each year, Americans use about 600 million tons of pesticides to protect their homes, crops, forests, livestock, and pets from insects and other pests. A drawback of pesticides is that they remain effective only for a short time because pests evolve resistance to the effects of pesticides. Pest control has become a race between the scientists who are trying to invent new pesticides and the ever-evolving pests. And because of pests' adaptability—some insects have evolved resistance to a new insecticide within a single growing season—pests seem to have the edge.

Pesticides from Bacteria

One pesticide with many advantages is the *Bt* toxin, which is derived from the common soil bacterium *Bacillus thuringiensis*. This bacterium produces a toxin that binds to a receptor protein in the digestive system of some insects, killing the insect. But only a few groups of insects are susceptible, including beetles, butterflies, flies, and ants. *Bt* toxin does not harm most beneficial insects, spiders and other arthropod predators of insects, or vertebrates, including humans.

Bt toxin has two other things going for it. First, different strains of the bacterium make different toxins that are effective against different pests. This gives farmers a choice of weapons to use, depending on which insects are attacking their crops. It also reduces the likelihood that resistance will evolve, since the odds that an insect will be resistant to more than one toxin are small. Second, *Bt* toxin breaks down into harmless byproducts within 2 or 3 days of being sprayed on a field. As a result, pests are exposed

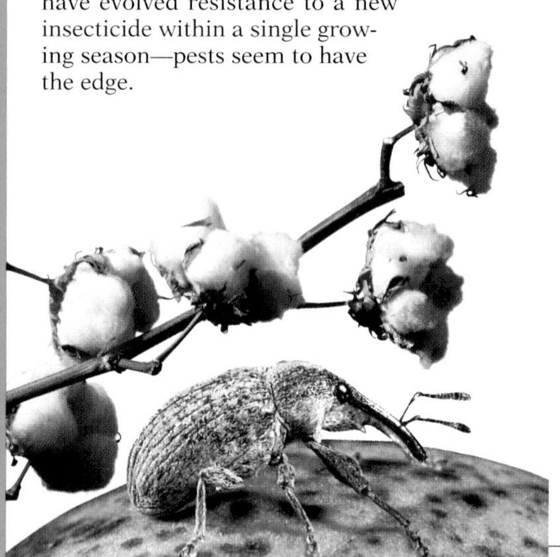

Boll weevil
Boll weevils cause extensive damage to United States cotton crops each year.

- Point out that the *Bt* toxin is one example of the importance of biodiversity. Before this valuable toxin was discovered, *Bacillus thuringiensis* was, as one writer put it, "just a bacterium found in ordinary dirt."

Harvesting cotton
Seed pods (called bolls) contain fibers that are used to make fabric.

to less of the toxin, and the probability of their evolving resistance is reduced.

Genetic Engineering of Insect Resistance

Through genetic engineering, the genes for *Bt* toxins have been inserted into several crop plants, including corn, cotton, and potatoes. These plants now produce their own insecticides. The Environmental Protection Agency, the federal agency that regulates pesticides, approved the sale of crops genetically engineered to produce *Bt* toxin in 1995. Since then, the area of farmland planted with the crops has grown rapidly.

Not everyone agrees with the EPA's decision that the crops are safe. Critics note that the question is not whether resistance will become widespread but how fast that will happen. They point out that the genetically engineered plants produce the toxin continuously, which increases the evolutionary pressure on the pests to adapt.

Supporters of the decision note that the farmers who buy the seeds are required to plant part of their land with nonengineered crops. This land will serve as a "refuge" on which susceptible insects can live. If any *Bt*-resistant insects appear, their resistance traits will be swamped when they interbreed with the nonresistant individuals from the refuge. Finally, because some pests attack both corn and cotton, the EPA prohibited the planting of genetically engineered corn and cotton in the same state. This should reduce the pests' exposure to *Bt* toxin, supporters say. ■

Career

Agricultural Scientist

Soil scientist

Profile

Agricultural scientists study farm crops and animals, seeking ways to increase crop yield and quality, control pests and weeds more effectively, and conserve soil and water.

Job Description

Agricultural scientists apply their knowledge of biology to agricultural issues, such as improving the nutritional value of crops. Crop scientists study the breeding and management of crops and sometimes use genetic engineering to develop crops resistant to pests and drought.

Science/Math Career Preparation

Botany	Biochemistry
Zoology	Technical writing
Microbiology	Mathematics
Chemistry	Economics

Analyzing STS Issues

Science and Society

1. **What are alternatives to synthetic chemical pesticides?** Use library resources or an on-line database to research integrated pest management. How does it differ from traditional methods of controlling pests? What are some of its shortcomings?
2. **Should genetically engineered crops be labeled?** Current regulations do not require labeling of genetically engineered crops that are sold in stores. Many consumer advocates contend that all such crops should be labeled so that consumers know what they are eating. Research both sides of this issue, and then write a logical, persuasive essay arguing either for or against labeling.
3. **What is biological control?** Biological control is an alternative to chemical pesticides. Using library resources or an on-line database, research the topic of biological control. What is biological control? Describe one example of biological control.

Technology: Genetic Engineering

4. **How do scientists genetically engineer plants?** Read the article "Transgenic Crops," by Charles Gasser and Robert Fraley (*Scientific American*, June 1992, pp. 62–69). What are the two main methods for inserting genes into plant cells? Why did scientists want to add genes for herbicide resistance to plants?

References

1. Beard, Jonathan. "Supermoths Spell Trouble for Natural Pesticide." *New Scientist,* March 8, 1997, 5.
2. Beardsley, Tim. "Picking on Cotton." *Scientific American,* October 1996, 45–46.
3. Bilger, Burkhard. "Wheat-Germ Warfare." *The Sciences,* January/February 1997, 8.
4. Holmes, Bob. "Caterpillar's Revenge." *New Scientist,* December 6, 1997, 7.
5. Macilwain, Colin. "Bollworms Chew Hole in Gene-Engineered Cotton." *Nature,* July 25, 1996, 289.
6. National Research Council. *Ecologically Based Pest Management.* Washington, D.C.: National Academy Press, 1996.
7. Perkins, S. "Transgenic Crops Provoke Petition." *Science News,* September 27, 1997, 199.
8. Snow, Allison and Pedro Moran Palma. "Commercialization of Transgenic Plants: Potential Ecological Risks." *Bioscience,* February 1997, 86–96.
9. Thayer, Ann. "Betting the Transgenic Farm." *Chemical and Engineering News,* April 28, 1997, 15–19.
10. Youssef, Ibrahim. "Genetic Soybeans Alarm Europeans." *New York Times,* November 7, 1996, D1, D24

Analyzing STS Issues Answers

1. Integrated Pest Management uses pesticides only as a last resort. It emphasizes other, more environmentally friendly pest-control techniques such as natural predators or diseases of the pests, crop rotation to prevent growth of large pest populations, and planting when insect populations are low. It has been criticized for relying too heavily on pesticides.
2. Answers will vary. Students' essays should be logical.
3. Biological control involves using the natural enemies or diseases of the pest to manage its population. Examples include Australian ladybird beetles used to control scale insects and parasitic wasps used to control caterpillars.
4. Scientists use two main techniques for inserting genes into plants: infecting the plant with a genetically engineered bacterium that carries the target gene or firing tiny metal pellets covered with DNA into the cell. Scientists wanted to put herbicide-resistant genes into plants in order to reduce the amount of herbicide used to control weeds.

Chapter 32

Echinoderms and Invertebrate Chordates

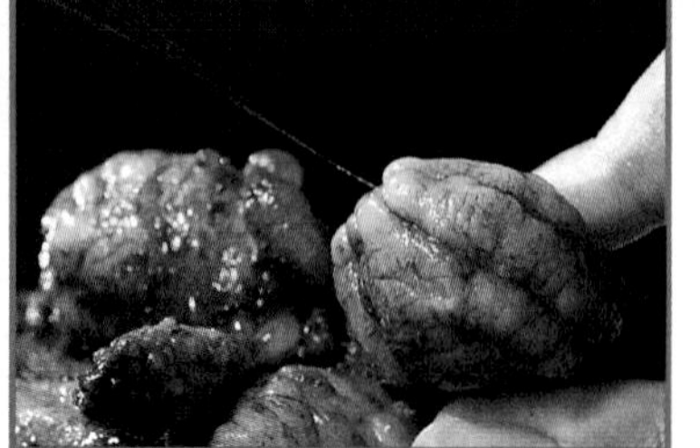

Planning Guide

OBJECTIVES	MEETING INDIVIDUAL NEEDS	READING SKILLS AND STRATEGIES
32-1 Echinoderms (BLOCK 1 45 min) • **Compare** the developmental pattern found in protostomes with that found in deuterostomes. • **Describe** the major characteristics of echinoderms. • **Summarize** how the sea star's water vascular system functions.	Learning Styles, p. 710 BASIC ADVANCED • pp. 716, 717	**ATE** Active Reading, Technique: K-W-L, p. 709 **ATE** Active Reading, Technique: Paired Summarizing, p. 715 ■ Active Reading Guide, worksheet (32-1) ■ Directed Reading, worksheet (32-1) ■ Reading Strategies, K-W-L worksheet ■ Reading Strategies, Reading Response Logs—Making Connections
32-2 Invertebrate Chordates (BLOCKS 2 & 3 90 min) • **Describe** the characteristics of chordates. • **Define** the term *invertebrate chordate*. • **Compare** tunicates and lancelets.	Learning Styles, p. 718 BASIC ADVANCED • p. 720	■ Active Reading Guide, worksheet (32-2) ■ Directed Reading, worksheet (32-2) ■ Supplemental Reading Guide, *Journey to the Ants: A Story of Scientific Exploration*

Review and Assessment Resources

BLOCKS 4 & 5 90 min

TEST PREP

PE Study Zone, p. 721

PE Performance Zone, pp. 722–723

■ Chapter 32 worksheets:
• Vocabulary • Concept Mapping
• Critical Thinking

 Audio CD Program

■ Test Prep Pretest

CHAPTER TESTING

■ Chapter Tests (blackline master)

▲ Test Generator

■ Assessment Item Listing

See the correlations table at the front of this book for a list of the ***National Science Education Standards*** addressed in this chapter.

Smithsonian Institution® Visit **www.si.edu/hrw** for additional on-line resources.

KEY

Biology Principles and Explorations

PE Pupil's Edition
ATE Teacher's Edition

Teaching Transparencies
Laboratory Program

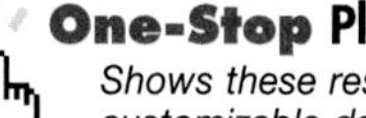

One-Stop Planner CD-ROM
Shows these resources within a customizable daily lesson plan:

■ Teaching Resources
▲ Test Generator

VISUAL STRATEGIES	CLASSWORK AND HOMEWORK	TECHNOLOGY AND INTERNET RESOURCES
162 Water Vascular System of the Sea Star	**PE** Data Lab Determining how predators affect prey, p. 713 **PE** Review, p. 717	Holt Biology Videodiscs Section 32-1 Audio CD Program Section 32-1 internet**connect** SCLINKS NSTA **TOPICS:** • Echinoderms • Bioassay
ATE Fig. 32-10, p. 718 **ATE** Fig. 32-11, p. 719 163 Exploration of a Lancelet	**PE** Quick Lab How do the notochord and nerve cord compare in structure?, p. 720 **PE** Lab Techniques Comparing Echinoderm and Chordate Development, pp. 724–725 **PE** Review, p. 720 **PE** Study Zone, p. 721	Holt Biology Videodiscs Section 32-2 Audio CD Program Section 32-2 internet**connect** SCLINKS NSTA **TOPIC:** Invertebrate chordates

ALTERNATIVE ASSESSMENT

PE Portfolio Projects 23–26, p. 723
ATE Alternative Assessment, p. 721
ATE Portfolio Assessment, p. 721

Lab Activity Materials

Data Lab
p. 713 pencil, paper

Quick Lab
p. 720 compound microscope, prepared slide of the cross section of an adult lancelet

Lab Techniques
p. 724 prepared slides of sea star development, compound microscope

Lesson Warm-up
ATE several preserved specimens of echinoderms, p. 710
ATE photos of embryological stages of several vertebrates, p. 718

Demonstrations
ATE phylogenetic tree of animal phyla, p. 708
ATE echinoderm specimens, p. 711
ATE eyedropper, p. 713

Chapter 32

Chapter Theme

Evolution

The anatomic diversity seen in animals today is the result of major innovations in body design that have evolved over millions of years. Echinoderms are among the oldest animal phyla. Fossil and embryological evidence shows that echinoderms are relatively close to chordates in evolutionary development.

Establishing Prior Knowledge

Before beginning this chapter, make certain students can answer the questions in **Study TIP Ready?** on page 709. If they cannot, encourage them to reread the sections indicated.

Opening Demonstration

Display a phylogenetic tree of the major animal phyla. Review the invertebrate animal phyla that have been discussed so far (Cnidaria, Platyhelminthes, Nematoda, Mollusca, Annelida, and Arthropoda). Point out Echinodermata and Chordata on the phylogenetic tree. Explain that both are deuterostomes. Ask students to name some animals from each phylum. *(sea star, sea urchin, sea cucumber, lancelet, tunicate, etc.)*

CHAPTER 32 Echinoderms and Invertebrate Chordates

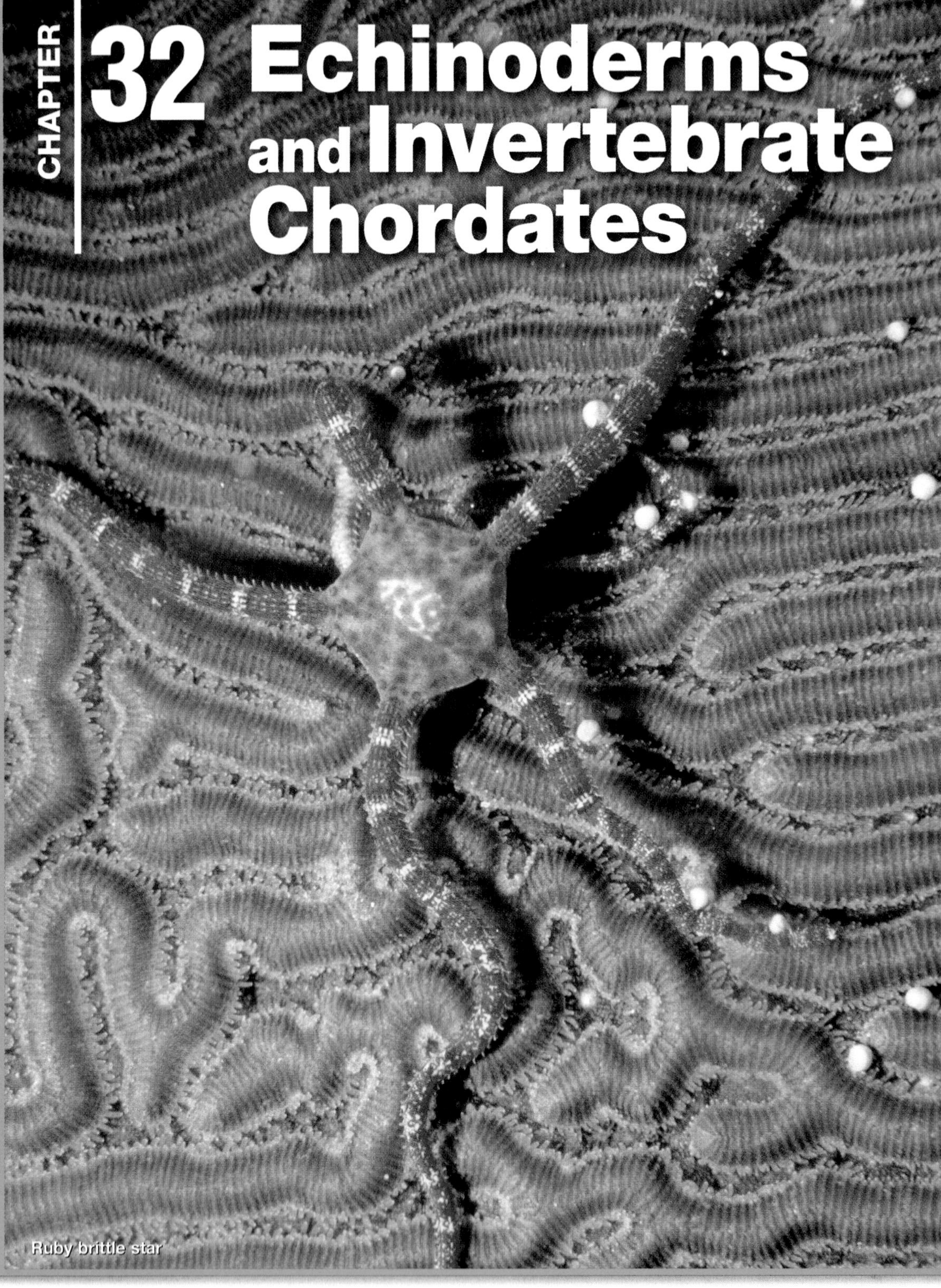

Ruby brittle star

What's Ahead — and Why
CHAPTER OUTLINE

32-1 Echinoderms

- Animals Develop in One of Two Ways
- Today's Echinoderms Share Four Major Characteristics
- Echinoderms Are a Diverse Group

32-2 Invertebrate Chordates

- Chordates Have a Completely Internal Skeleton
- Only a Few Chordate Species Are Invertebrates

Students will learn

- about how echinoderms develop, what characteristics all echinoderms display, and how the echinoderm classes differ.
- about the characteristics common to all chordates and about the two invertebrate chordate phyla.

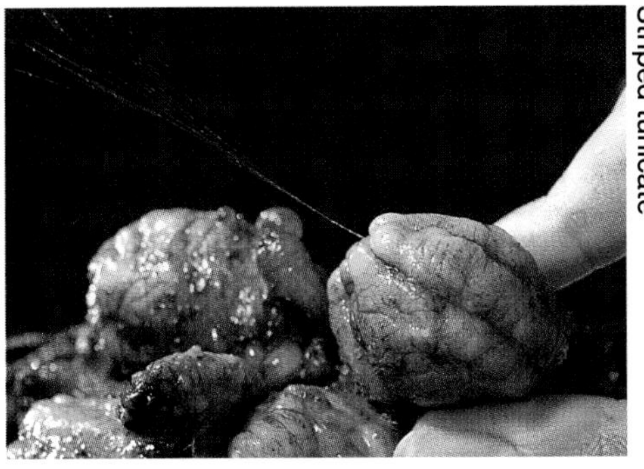

Striped tunicate

When exposed at low tide,

tunicates living in intertidal zones may suddenly eject a large squirt of water from their siphons—a trait that gives them their common name, sea squirt. To find out what sea squirts have in common with brittle stars, turn the page.

Looking Ahead

32-1 Echinoderms
How do they move? Take a look at these feet.

32-2 Invertebrate Chordates
They don't look like much, but they share a phylum with the mammals.

Labs

- **Data Lab**
  Determining how predators affect prey (p. 713)
- **Quick Lab**
  How do the notochord and nerve cord compare in structure? (p. 720)
- **Lab Techniques**
  Comparing Echinoderm and Chordate Development (pp. 724–725)

Features

- **Up Close**
  Sea Star (p. 714)
- **Health Watch**
  Monitoring Water Quality (p. 717)

Study TIP

Ready?

Check your understanding of these topics to help you prepare for what's ahead.

Can you . . .

- **distinguish** invertebrates from vertebrates? (Section 20-3)
- **define** the terms *gastrula* and *blastula*? (Section 28-1)
- **distinguish** between radial symmetry and bilateral symmetry? (Section 28-1)

If not, *review the sections indicated.*

internet**connect**

National Science Teachers Association *sci*LINKS Internet resources are located throughout this chapter.

Reading Strategies

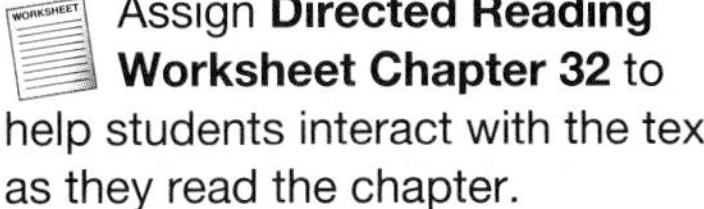

Active Reading

Technique: K-W-L

Before beginning this chapter, divide the class into two groups. Have students in one group create a list of all the things they **K**now about echinoderms; the students in the second group should create a list of all the things they **K**now about chordates. When the groups finish their lists, have them write their entries on the board. Then have students make individual lists of things they **W**ant to know about echinoderms and chordates. Have students save their lists for use later in the **Closure** on page 720.

Directed Reading

Assign **Directed Reading Worksheet Chapter 32** to help students interact with the text as they read the chapter.

Interactive Reading

Assign ***Biology: Principles and Explorations*** **Audio CD Program Chapter 32** to help reluctant readers achieve greater success in reading the chapter.

Reading Skills

Assign Reading Organizers and Reading Strategies Worksheets from the **One-Stop Planner CD-ROM with Test Generator** to help students learn and practice effective reading skills.

internet**connect**

Holt *Biology: Principles and Explorations*

On-line Resources: go.hrw.com

Visit the HRW Web site for a variety of resources related to this chapter. Just type in the keyword HX0 Chapter 32.

Holt *Biology: Principles and Explorations*

On-line Resources: www.scilinks.org

The following *sci*LINKS Internet resources can be found in the student text for this chapter:

Page 713
TOPIC: Echinoderms
KEYWORD: HX713

Page 717
TOPIC: Bioassay
KEYWORD: HX717

Page 719
TOPIC: Invertebrate chordates
KEYWORD: HX719

Chapter 32

1 Prepare

Scheduling

- **Block:** About 45 minutes will be needed to complete this section.
- **Traditional:** About one class period will be needed to complete this section.
- **Planning:** See the Planning Guide on pp. 707A and 707B for lecture, classwork, and assignment options for Section 32-1. Lesson plans for Section 32-1 can be found in a lesson cycle format in ***Biology: Principles and Explorations*** **One-Stop Planner CD-ROM with Test Generator.**

Resource Link

- Assign **Active Reading Worksheet Section 32-1** to help students comprehend the key concepts and visuals in this lesson.

2 Teach

Lesson Warm-up

Bring in several preserved or dried specimens of echinoderms, such as sea cucumbers, sea stars, and sand dollars. Allow students to inspect them and come up with as many common characteristics as they can. List those characteristics on the board or overhead, and refer to the list as you teach the specific characteristics.

Section 32-1 Echinoderms

Objectives

- **Compare** the developmental pattern found in protostomes with that found in deuterostomes. (p. 710)
- **Describe** the major characteristics of echinoderms. (pp. 712–713)
- **Summarize** how the sea star's water vascular system functions. (pp. 713–714)

Key Terms

blastopore
protostome
deuterostome
ossicle
water-vascular system
skin gill

Animals Develop in One of Two Ways

If you have been to a saltwater aquarium, you're sure to have seen echinoderms, spiny invertebrates that live on the ocean bottom. How could echinoderms like the brittle star shown on page 708 be related to an animal such as a sea squirt? They appear so different! And what relationship could either of them have to chordates, which are primarily vertebrates? The answer lies in their early development. As you learned in Chapter 28, during its development an embryo goes through a gastrula stage. As shown in **Figure 32-1,** a gastrula has an opening to the outside called the **blastopore.** In acoelomate animals, the mouth develops from or near the blastopore. This pattern of development also occurs in some coelomate animals, such as annelids, mollusks, and arthropods. Animals with mouths that develop from or near the blastopore are called **protostomes.**

Some animals follow a different pattern of development. In phylums Echinodermata and Chordata, the anus—not the mouth—develops from or near the blastopore. (The mouth forms later, on another part of the embryo.) Animals with this pattern of development are called **deuterostomes,** also shown in Figure 32-1. If you know the origin of these two terms, it's easy to remember the differences between the two developmental patterns. The term *protostome* is from the Greek *protos,* meaning "first," and *stoma,* meaning "mouth." The prefix *deutero-* is from the Greek *deuteros,* meaning "second." In deuterostomes, the anus develops first and the mouth develops second.

Figure 32-1 Embryonic development

The development of an animal embryo follows one of two different patterns.

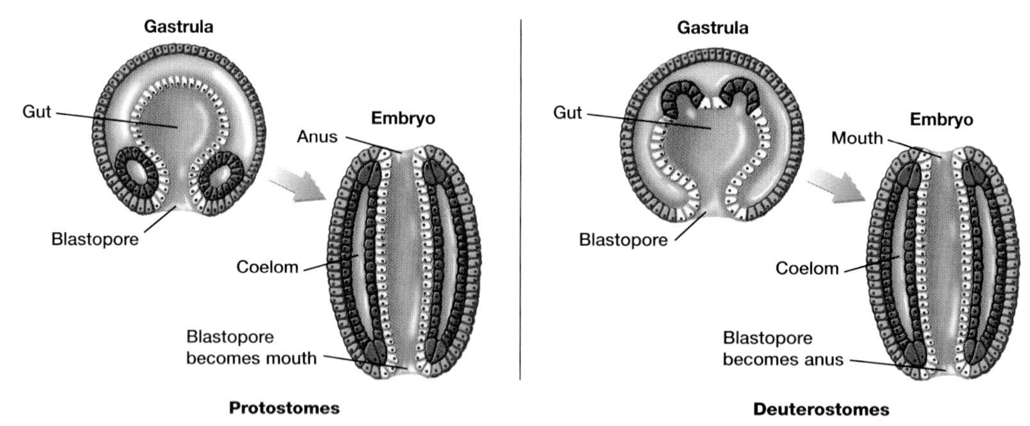

Integrating Different Learning Styles

Logical/Mathematical	**PE**	Data Lab, p. 713; Performance Zone items 21 and 22, p. 723
Visual/Spatial	**PE**	Performance Zone item 24, p. 723
	ATE	Demonstrations, pp. 711, 713; Teaching Tip, p. 711
Interpersonal	**ATE**	Active Reading, p. 715
Verbal/Linguistic	**PE**	Closure, p. 717; Performance Zone items 19, 20, 25, and 26, p. 723

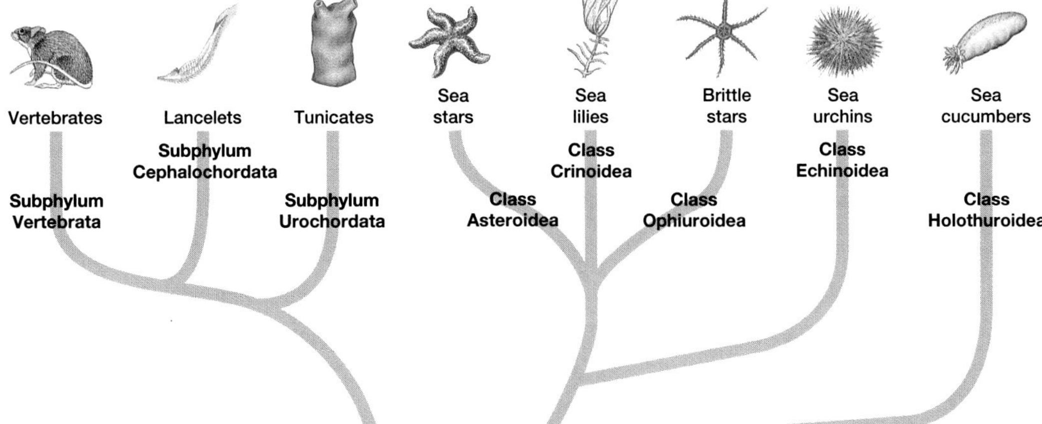

Figure 32-2 Evolution of chordates and echinoderms. This phylogenetic tree shows the relationship of the major chordate and echinoderm groups.

The first deuterostomes were marine echinoderms that evolved more than 650 million years ago. They were also the first animals to develop an endoskeleton. Today, most people are familiar with echinoderms known as "starfish," which are not really fish and are more properly called sea stars. In addition to sea stars, many other animals commonly seen along the sea shore—sea urchins, sand dollars, and sea cucumbers—are echinoderms. All are marine, and all are radially symmetrical as adults.

Chordates, which you will read about in Section 2, as well as a few other small phyla, are also deuterostomes. (Humans and all other vertebrates are chordates.) Like the echinoderms, chordates have an internal skeleton. This developmental similarity unites these seemingly dissimilar animal phyla. It also leads scientists to believe that chordates and echinoderms were derived from a common ancestor, as shown in the phylogenetic tree in **Figure 32-2.** The ancestral deuterostome is not known, but the fossil record, represented by the sea lily in **Figure 32-3,** indicates that echinoderms were abundant in the ancient seas.

Figure 32-3 Fossil sea lily. Sea lilies like the one seen in this fossil were plentiful in the ancient oceans.

Evolutionary Milestone

8 Deuterostomes

Echinoderms are coelomates that have a deuterostome pattern of embryo development. The same pattern of development occurs in the chordates.

Section 32-1

Demonstration

Observing Echinoderms

Display any echinoderms you or your students may have collected such as sea stars or sand dollars. Emphasize the five-part radial symmetry of the specimens. Point out that during drying, the skin of sand dollars disintegrates, and only the hard internal skeleton remains. Ask students how echinoderms differ from arthropods. *(Echinoderms have endoskeletons, no body segments, and five-part symmetry; arthropods have exoskeletons, body segments, and bilateral symmetry.)*

Teaching TIP

Symmetry

All adult echinoderms display pentaradial symmetry. *Penta-* means "five." Therefore, echinoderms have five or multiples of five arms. Have groups of students create displays that define and model various forms of symmetry. Encourage them to compare objects made by humans with those found in nature. They can use pictures and clay models to make the comparison.

Teaching TIP

Evolutionary Milestone

Have students read the **Evolutionary Milestone** on page 711. Remind them that unlike annelids, mollusks, and arthropods, echinoderms have a deuterostome pattern of development. Ask students why this is considered to be an evolutionary milestone. *(It shows an evolutionary connection to chordates.)*

Overcoming Misconceptions

Octopus Symmetry

Since animals with radial symmetry do not have centralized nervous systems, students may wonder how octopuses can be intelligent. Remind students that cephalopods are bilaterally symmetrical even though their tentacles may give them the appearance of being radially symmetrical.

Chapter 32

Making Biology Relevant

Medicine

The importance of chemical signals in the development of deuterostomes has significant medical implications for the regeneration of lost body parts. Scientists are working to discover factors that cause cells to change and grow so that someday it might be possible to regrow a severed spinal cord or an amputated limb.

Cross-Disciplinary Connection

◆ History

Clam farmers hate sea stars because sea stars devour clams. In the past, the farmers cut the sea stars into several pieces and flung them back into the sea. They did not know that sea stars could regenerate from a severed arm, as long as one piece of the central disc was present. The clam farmers were actually making the problem much worse! Today, clam farmers spread lime on the clam beds. Lime kills sea stars but leaves clams unharmed.

Teaching TIP

Five-Part System

The water vascular system of a typical sea star has five main parts. Water flows through this system to circulate oxygen and assist movement. Have students construct a **Graphic Organizer** similar to the one on page 712 to illustrate the path of water through a sea star. PORTFOLIO

biobit

Do sand dollars resemble the sun?

To the Chumash Indians of southern California, the lines radiating from a sand dollar's mouth resemble the sun's rays. Sand dollars, once plentiful in the Pacific Ocean, became the Chumash's symbol for the "newborn" sun of the winter solstice.

Study TIP

Word Origins

The term *echinoderm* is from the Greek *echinos,* meaning "spiny," and *derma,* meaning "skin."

Today's Echinoderms Share Four Major Characteristics

Many of the most familiar animals seen along the seashore—sea stars, sea urchins, sand dollars—are echinoderms. Echinoderms are also common in the deep ocean. While all echinoderms are marine, the different classes of echinoderms vary considerably in the details of their body design. Despite their apparent differences, all echinoderms share four fundamental characteristics.

1. **Endoskeleton.** Echinoderms have a calcium-rich endoskeleton composed of individual plates called **ossicles.** When ossicles first form in young echinoderms, they are enclosed in living tissue, so they are a true endoskeleton. Even though the ossicles of adult echinoderms appear to be external, they are covered by a thin layer of skin (although sometimes the skin is worn away). In adult sea stars and in many other echinoderms, a large number of these plates are fused together. The fused plates function much like an arthropod exoskeleton. They provide sites for muscle attachment and shell-like protection. In most echinoderms, the plates of the endoskeleton bear spines that project upward through their skin.
2. **Five-part radial symmetry.** All echinoderms are bilaterally symmetrical as larvae. During their development into adults, the larvae's body plan becomes radially symmetrical. Most adult echinoderms, such as the one shown on the left in **Figure 32-4,** have a five-part body plan with arms that radiate from a central point. However, the number of arms can vary. Echinoderms have no head or brain. Instead, the nervous system consists of a central ring of nerves with branches extending into each of the arms. Although echinoderms are capable of complex response patterns, each arm acts more or less independently. Many species , such as sea stars, can regenerate a new arm if a portion of an arm is lost. In some species of sea stars, a complete animal can regenerate from an arm connected to a portion of the central disk. However, a complete sea star cannot regenerate from an arm alone.

Figure 32-4 Five-part body plan. The echinoderm five-part body plan is easily seen in this colorful African species. Other sea stars, such as the sunstar, have more than five arms.

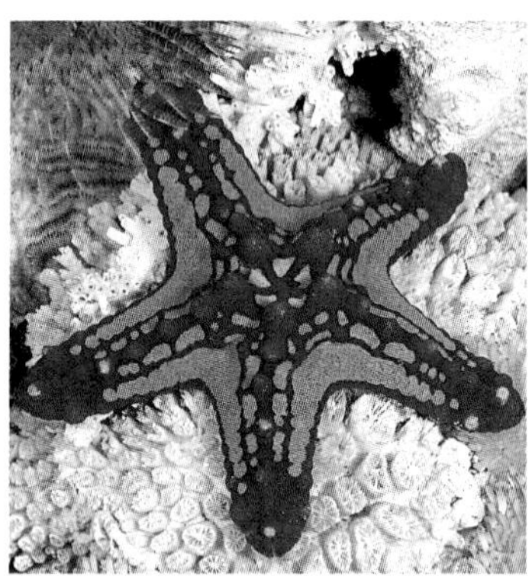

Graphic Organizer

Use this graphic organizer with ***Teaching Tip: Five-Part System*** *on page 712.*

Madeporite or Sieve Plate → Stone canal → Ring canal → Radial canal → Ampulla of Tube Feet

3. **Water-vascular system.** Echinoderms have a water-filled system of interconnected canals and thousands of tiny hollow tube feet called a **water-vascular system.** In some echinoderms, such as the sea star, the tube feet extend outward through openings in the ossicles. In some species, each tube foot has a sucker at its tip. Many echinoderms use their tube feet to crawl across the sea floor. The water-vascular system also functions in feeding and gas exchange. A sea star can use the hundreds of tube feet on its arms to pull the valves of a bivalve open. Some gas exchange and waste excretion takes place through the thin walls of the tube feet.

4. **Coelomic circulation and respiration.** The echinoderm body cavity functions as a simple circulatory and respiratory system. Particles, including respiratory gases, move freely throughout the large, fluid-filled coelom. Many echinoderms have skin gills that aid respiration and waste removal. **Skin gills,** shown in **Figure 32-5,** are small, fingerlike projections that grow between the echinoderm's spines. These projections create an increased surface area through which respiratory gases can be exchanged. Skin gills also function as excretory structures, and wastes that accumulate in them are released into the surrounding water.

You can learn more about the structure of one particular echinoderm, the sea star, in *Up Close: Sea Star,* on page 714.

Figure 32-5 Skin gills. An echinoderm's skin gills function as both respiratory and excretory organs.

internet**connect**

TOPIC: Echinoderms
GO TO: www.scilinks.org
KEYWORD: HX713

Determining how predators affect prey

Background

Sea stars can be very effective predators, and they frequently eat mollusks. The chart at right shows the relative number of two species of mollusks before and after the introduction of a predatory sea star. Study the chart, and answer the Analysis questions.

Sea Star Predation of Mollusks

Analysis

1. **Compare** the relative sizes of the two mollusk populations before the introduction of the sea star.
2. **Identify** the preferred prey of the sea star, and use the data presented in the graph to support your answer.
3. **SKILL Analyzing Data** When the sea star began preying on the nonpreferred species, the preferred species had dropped to what percent of its original population?
4. **SKILL Inferring Relationships** What factors might cause the sea star to begin consuming a nonpreferred species, even when its preferred prey is still present?
5. **SKILL Predicting Outcomes** Predict the relative abundance of the two species of mollusks if the sea star remains in the area indefinitely.

Determining how predators affect prey

Time 15 minutes

Process Skills Reading and interpreting graphs, analyzing data, predicting results

Teaching Strategies
Show students a picture of a sea star eating a mollusk.

Analysis Answers
1. Both had approximately the same percent of their original populations.
2. Species A; species A's population declined first and dropped more dramatically than species B's population.
3. about 50 percent
4. Accept any reasonable answer. The most obvious factor is relative abundance—nonpreferred prey will be easier to find as preferred prey becomes scarce.
5. There will probably be about 10 percent of the original population of species A and 25 percent of the original population of species B.

Demonstration

Tube Feet — BASIC

You can use an ordinary dropper to represent tube feet. Show students that when the ampulla is filling with water, suction is created at the open end. When the ampulla is expelling water, the suction is gone, and water leaves the open end. This may be easier for students to see if you add a small amount of glitter to the water.

did you know?

The legs of sea stars are delicate, but they can be very powerful when working together.

The tube feet of sea stars appear weak, and each foot can easily be pulled off a surface. However, when hundreds of tube feet pull together, a sea star can exert tremendous force—enough to pull apart the valves of a clam. People who shell clams have to use a knife to cut the muscles that hold the valves together.

Chapter 32

Up Close

Sea Star

Teaching Strategies
Ask students to explain the evolutionary link between sea stars and chordates. Point out that although the relationship between adult echinoderms and chordates may not be apparent, the embryos of both groups have certain developmental stages in common. Echinoderms not only share our deuterostome pattern of embryonic development, but also have bilateral symmetry as larvae and endoskeletons as adults.

Discussion
Guide the discussion by posing the following questions.

1. What makes a tube foot extend? *(contraction of the water-filled ampulla)*
2. What makes it withdraw? *(The ampulla relaxes and refills with water.)*
3. Why can't sea stars live in terrestrial environments? *(They wouldn't be able to move about because they need water to operate their water vascular system.)*
4. Why doesn't a sea star have a head? *(A sea star has no brain, so it has no need for a head to enclose one.)*

Up Close

Sea Star

- **Scientific name:** *Asterias vulgaris*
- **Size:** Typically from 15 to 30 cm (6 to 12 in.) in diameter
- **Range:** East coast of North America
- **Habitat:** Intertidal; often on hard, rocky surfaces
- **Diet:** Slow-moving or sessile species, including mollusks, crustaceans, polychaetes, and corals

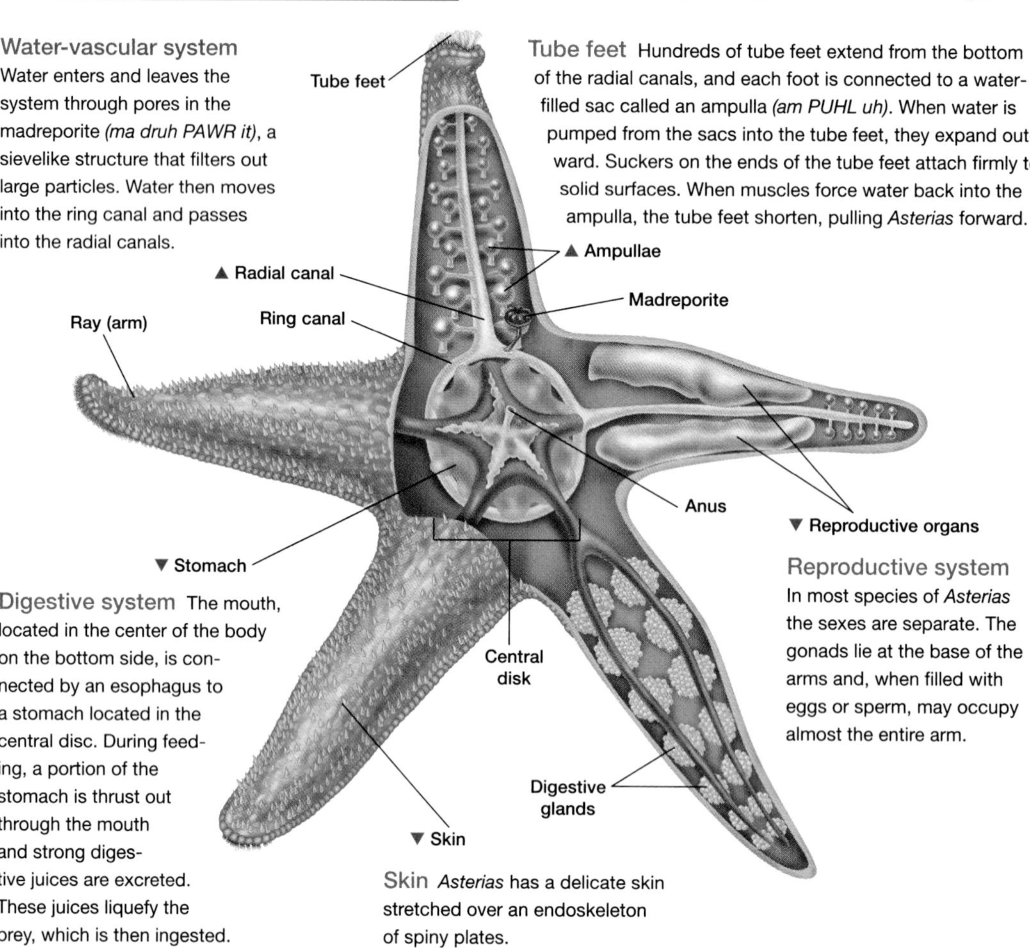

Characteristics
Water-vascular system Water enters and leaves the system through pores in the madreporite (ma druh PAWR it), a sievelike structure that filters out large particles. Water then moves into the ring canal and passes into the radial canals.
Tube feet Hundreds of tube feet extend from the bottom of the radial canals, and each foot is connected to a water-filled sac called an ampulla (am PUHL uh). When water is pumped from the sacs into the tube feet, they expand outward. Suckers on the ends of the tube feet attach firmly to solid surfaces. When muscles force water back into the ampulla, the tube feet shorten, pulling Asterias forward.
Tube feet
▲ Ampullae
▲ Radial canal
Madreporite
Ray (arm)
Ring canal
Anus
▼ Reproductive organs
▼ Stomach
Reproductive system In most species of Asterias the sexes are separate. The gonads lie at the base of the arms and, when filled with eggs or sperm, may occupy almost the entire arm.
Digestive system The mouth, located in the center of the body on the bottom side, is connected by an esophagus to a stomach located in the central disc. During feeding, a portion of the stomach is thrust out through the mouth and strong digestive juices are excreted. These juices liquefy the prey, which is then ingested.
Central disk
Digestive glands
▼ Skin
Skin Asterias has a delicate skin stretched over an endoskeleton of spiny plates.

did you know?

Sea stars can reproduce sexually and asexually.

In addition to asexual reproduction (any arm can regenerate a new body as long as part of the central disc is present), starfish normally undergo sexual reproduction. A female sea star can deposit as many as 2.5 million eggs at one time. The eggs are deposited in the water, where they are fertilized by the sperm of a male sea star.

Echinoderms Are a Diverse Group

Echinoderms are one of the most numerous of all marine phyla. In the past, they were even more plentiful than they are now. There are more than 20 extinct classes of echinoderms and an additional seven classes of living members. As you saw on the phylogenetic tree on page 711, the living classes include sea stars, sea lilies, brittle stars, sea urchins, and sea cucumbers. The recently discovered sea daisy does not appear on the phylogenetic tree because its relationship to the other echinoderms is not fully understood.

Figure 32-6 **Sea star.** This sea star is using its tube feet to pry open the shell of a clam. Then it will feed on the clam's soft tissues.

Sea stars

Sea stars are the echinoderms most familiar to people. Almost all species of sea stars are carnivores, and they are among the most important predators in many marine ecosystems. For example, the crown-of-thorns sea star eats coral polyps. In 1 hour, a single crown-of-thorns can graze along 20 m (65 ft) of a reef. Over time, this sea star can destroy an entire coral reef ecosystem. Other sea stars prey on bivalve mollusks, whose shells they pull open with their powerful tube feet, as shown in **Figure 32-6.**

The ossicles of many species of sea stars produce pincerlike structures called pedicellaria *(ped uh suh LAH ree uh).* Pedicellaria contain their own muscles and nerves, and they snap at anything that touches them. This action prevents small organisms from attaching themselves to the surface of the sea star.

Brittle stars

The sea star's relatives, the brittle stars and sea baskets, make up the largest class of echinoderms. Brittle stars have slender branched arms that they move in pairs from side to side to row along the ocean floor. Their arms break off easily, a fact that gives brittle stars their name. Brittle stars and sea baskets live primarily on the ocean bottom, and they usually hide under rocks or within crevices in coral reefs. Although a few species are predators, most brittle stars are filter feeders or feed on food in the ocean sediment.

Figure 32-7 **Feather star.** The feathery arms of these feather stars are adapted for filter feeding.

Sea lilies and feather stars

The sea lilies and feather stars are the most ancient and primitive living echinoderms. They differ from all other living echinoderms because their mouth is located on their upper, rather than lower, surface. Sea lilies are sessile and are attached to the ocean floor by a stalk that can be 1 m (3.3 ft) long. Feather stars, shown in **Figure 32-7,** use hooklike projections to attach themselves directly to the ocean bottom or a coral reef. They sometimes crawl or swim for short distances.

Biology in Your Community

Marine Zoology

Invite a marine zoologist or aquarium store employee to discuss echinoderms. If possible, have your guest bring in live specimens or videotapes to show how the echinoderms move and feed.

Section 32-1

Assign **Reading Strategies: Reading Response Logs—Making Connections** from the Holt BioSources Teaching Resources CD-ROM. **PORTFOLIO**

HOLT BIOSOURCES
TEACHING RESOURCES
READING STRATEGIES

Name ______ Class ______
Date ______

2. Reading Response Logs—Making Connections

Topic ______

What is the most interesting or exciting thing about this topic?

What are your attitudes or feelings (positive or negative) toward this topic?

What experiences can you share with others that may help them understand your views?

Active Reading

Technique: Paired Summarizing

Encourage students to pause after each paragraph they read to summarize the main points to a partner. You might model the technique by having a student read a paragraph out loud and summarize it. Then have another student read the next paragraph and summarize it.

Teaching TIP — ADVANCED

Moving Materials Into and Out of Cells

Tell students that the exchange of gases occurs on the surface of the sea star because particles tend to move from an area of greater concentration to an area of lower concentration. The uptake of oxygen molecules by skin gills occurs because ocean water contains a greater concentration of oxygen molecules than the skin gills contain. Therefore, oxygen enters the skin gills.

Teaching TIP — BASIC

Beach Treasures

After students read page 716, ask them if they would be more likely to find dead sea urchins or dead sand dollars washed up on a sandy beach. *(sand dollars because they live along sandy shores)* Find out if any students have ever found a sand dollar on the beach. Have them describe or display what they found.

Sand dollar

Sea urchin

Figure 32-8 Sea urchin and sand dollar. Sea urchins usually live on rocky ocean bottoms, while sand dollars live on sandy ocean bottoms.

Study TIP

Organizing Information

Make a six-column table to organize information about the classes of echinoderms. Across the top, write the class names. Under each class, write information about the class.

Sea urchins and sand dollars

The sea urchins and sand dollars, shown in **Figure 32-8,** lack distinct arms but have the basic five-part body plan seen in other echinoderms. Both sea urchins and sand dollars have a hard, somewhat flattened endoskeleton of fused plates covered with spines protruding from it. The spines provide protection and, in some species of sea urchins, contain a venom that causes a severe burning sensation. In some other species of sea urchin, a specialized type of pedicellaria contains a toxin used to paralyze prey. Sea urchins are found on the ocean bottoms while sand dollars live in sandy areas along the sea coast.

Sea cucumbers

Sea cucumbers are soft-bodied, sluglike animals without arms. They differ from other echinoderms in that their ossicles are small and are not fused together. Because of this, the sea cucumber's long, cylindrical body is soft. Often the body has a tough, leathery exterior. The sexes of most sea cucumbers are separate, but some species are hermaphrodites.

Sea cucumbers feed by trapping tiny organisms present in the sea water. Their mouth, located at one end of the body, is surrounded by several dozen tube feet that have been modified into tentacles. The tentacles are covered with a sticky mucus that entraps plankton. Periodically, the sea cucumber draws its tentacles into its mouth and cleans off the plankton and mucus. The tentacles are then coated with a fresh supply of mucus. When threatened, a sea cumber has an unusual means of defending itself. As shown in **Figure 32-9,** the sea cucumber can release a number of sticky threads from its anus to entrap its attacker.

Figure 32-9 Sea cucumber. When threatened, a sea cucumber releases sticky threads that entrap its attacker.

MULTICULTURAL PERSPECTIVE

California's Sea Urchin Fishery

A growing demand for traditional Japanese cuisine in the United States has created a growing market for California sea urchins. Although the sea urchins were once considered pests by kelp harvesters and sports divers, they are now commercially harvested. Most of California's sea urchin catch is exported to Japan.

Sea daisies

In 1986, a new class of echinoderm was discovered: strange disk-shaped little animals called sea daisies. Less than 1 cm (0.39 in.) in diameter, these creatures were first found in deep waters off the coast of New Zealand. Only two species are known. Sea daisies have five-part radial symmetry but no arms. Their tube feet are located around the edges of the disk rather than along the radial lines, like they are in other echinoderms.

HEALTH WATCH: Monitoring Water Quality

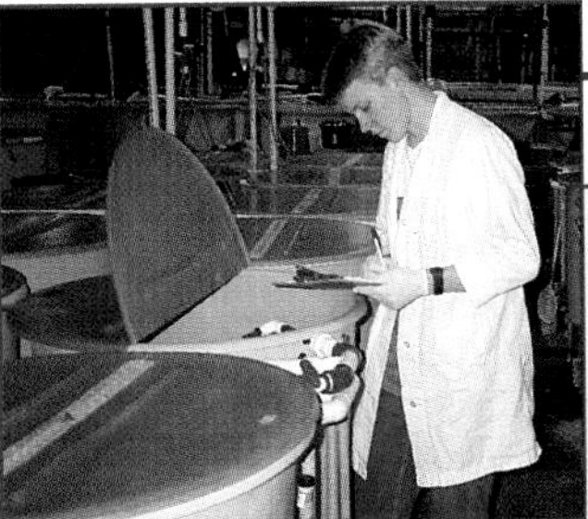
Recording water temperature in a bioassay tank

If you were swimming or fishing in coastal waters, you would not be able to detect the presence of any toxic chemicals in the water, sediments, or fish. To help protect humans and marine organisms, scientists have developed several tests to monitor marine environments for potential health hazards. Since sea urchin sperm and eggs are very sensitive to many pollutants, they are used in one of these tests, known as the sea urchin fertilization bioassay. (A bioassay is the use of a living organism or cell culture to test for the presence of a substance.)

Using sea urchins

Samples of ocean water, sediment, and industrial wastes that are discharged into the ocean are collected regularly from different sites. Then they are tested under controlled conditions in a lab. In this bioassay, sea urchin sperm and eggs are mixed together with the collected samples. After a short waiting period, scientists compare the fertilization success rate in the collected water samples with the fertilization success rate found in control water samples. If the test samples show a lower fertilization rate, scientists conclude that toxic contaminants are present.

Taking action

What happens when the test indicates the presence of contaminants? More specific tests may be run to determine exactly what contaminants are present. If the toxicity can be traced to runoff from a factory or sewage treatment plant, the plant may be forced to clean its waste before discharging it. Sediments may have to be removed or decontaminated. In the future, it may be possible to clean up some pollutants by using plants that have the ability to remove toxic chemicals from the water they are growing in. The use of this process, known as phytoremediation, in marine environments is an exciting new area of research.

internetconnect

SCILINKS NSTA
TOPIC: Bioassay
GO TO: www.scilinks.org
KEYWORD: HX717

Review SECTION 32-1

1. **Summarize** why echinoderms are considered to be more closely related to tunicates, lancelets, and vertebrates than to other animals.
2. **Summarize** the four major echinoderm characteristics.
3. **Describe** how the sea stars use their water-vascular system to move along the sea floor.

Critical Thinking

4. **SKILL Forming Reasoned Opinions** A scientist collects several specimens of an unidentified animal. After conducting an in-depth study of the mysterious species, the scientist observes that they have tube feet, an endoskeleton, and a protostome pattern of embryonic development. Why is the classification of these organisms difficult?

Review SECTION 32-1 ANSWERS

1. Echnioderms and chordates share a common pattern of embryonic development (both are deuterostomes), and both have an internal skeleton. These similarities indicate that the groups share a common ancestor.
2. All echinoderms have: an endoskeleton composed of ossicles; five-part radial symmetry; a water-filled vascular system; and coelomic circulation and respiration.
3. Water enters the sieve plate and travels into the ring canal and radial canals to the ampulla of the tube feet. When water enters the tube feet, they expand, and suckers attach to the surface. When water reenters the ampulla, the tube feet shorten and pull the sea star forward.
4. Although the organisms have characteristics of adult echinoderms, their embryonic development does not follow that of echinoderms. They do not exhibit hallmarks that quickly identify them as members of a larger group.

HEALTH WATCH
Monitoring Water Quality

Teaching Strategies

- Discuss the importance of water quality.
- Use articles from newspapers and magazines to point out recent problems and solutions.
- If possible, visit a water treatment plant or have someone from such a facility come to school to answer students' questions.

Discussion

- What is a bioassay? *(using living organisms or cell cultures to test for the presence of a substance)*
- Do you think this particular bioassay is a good practice? *(Answers will vary. Sample answers: Yes, human and vertebrate health is more important than sea urchin fertilization. No, there are parts of the world where sea urchins are scarce; it's not right to exploit them.)*
- How can scientists tell if water samples are contaminated? *(Fewer fertilizations occur.)*
- What should happen to companies found deliberately polluting our waterways? *(Answers will vary, but might include fining the companies and making them clean the waterways.)*

3 Assessment Options

Closure

Echinoderm Trivia

Have students write review questions for a trivia-type game about the groups of echinoderms.

Review

Assign Review—Section 32-1.

Reteaching — BASIC

Make an illustrated chart of echinoderm classes and the characteristics of each class.

Chapter 32

1 Prepare

Scheduling

- **Block:** About 90 minutes will be needed to complete this section.
- **Traditional:** About two class periods will be needed to complete this section.
- **Planning:** See the Planning Guide on pp. 707A and 707B for lecture, classwork, and assignment options for Section 32-2. Lesson plans for Section 32-2 can be found in a lesson cycle format in ***Biology: Principles and Explorations*** **One-Stop Planner CD-ROM with Test Generator.**

Resource Link

Assign **Active Reading Worksheet Section 32-2** to help students comprehend the key concepts and visuals in this lesson.

2 Teach

Lesson Warm-up

Show students pictures of the embryonic stages of several vertebrates, such as humans, sheep, fish, and chickens. Have them compare the animals and point out which structures they have in common *(gill slits, notochord, nerve cord, tail)*. Tell students that it is possible for an animal to be a chordate but not a vertebrate.

Visual Strategy

Figure 32-10

Point out the segmented muscles in Figure 32-10. Tell students that these muscle blocks are called *myomeres* and that they are actually modified body segments present in chordate muscle tissues. The "flaky" texture of many fish filets is an example of these stacked muscle layers.

Section 32-2 Invertebrate Chordates

Objectives

- **Describe** the characteristics of chordates. (p. 718)
- **Define** the term *invertebrate chordate.* (p. 719)
- **Compare** tunicates and lancelets. (pp. 719–720)

Key Terms

chordate
notochord
pharyngeal slit
invertebrate chordate

Chordates Have a Completely Internal Skeleton

The second major group of deuterostomes are the **chordates.** Chordates have a very different kind of endoskeleton, one that is completely internal. During the development of the chordate embryo, a stiff rod called the **notochord** develops along the back of the embryo. Using muscles attached to this rod, early chordates could swing their backs from side to side, enabling them to swim through the water. The evolution of an internal skeleton was an important step that led to the evolution of vertebrates. The endoskeleton, which muscles attach to, made it possible for animals to grow large and to move quickly.

Other chordate characteristics

Chordates also share three other characteristics. They have a single, hollow, dorsal nerve cord with nerves attached to it that travel to different parts of the body. Chordates also have a series of **pharyngeal slits** (openings) that develop in the wall of the pharynx, the muscular tube that connects the mouth to the digestive tract and windpipe. Another chordate characteristic is a postanal tail, which is a tail that extends beyond the anus. All chordates have all four of these characteristics at some time in their life, even if it is only briefly as embryos. **Figure 32-10** shows these chordate characteristics as seen in the body of an adult lancelet.

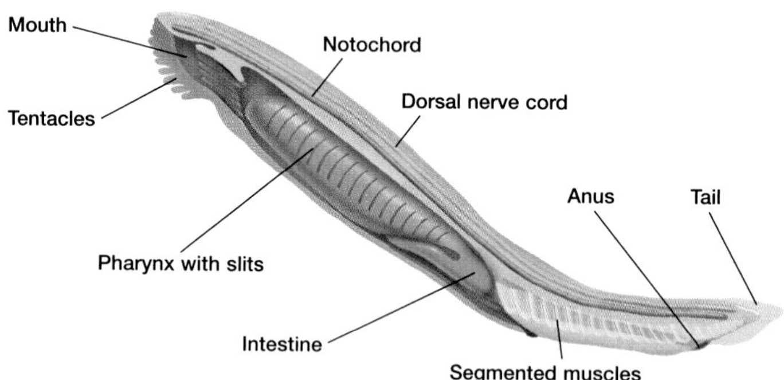

Figure 32-10 Lancelet interior. Adult lancelets possess all of the characteristics of chordates.

Evolutionary Milestone

9 Notochord

Tunicates, lancelets, and all the vertebrates belong to phylum Chordata (chordates). Chordates are coelomate animals that have a flexible, dorsal rod called a **notochord.** In vertebrate chordates, the notochord is replaced during embryonic development by a vertebral column (backbone).

Integrating Different Learning Styles

Logical/Mathematical	**PE**	Review item 4, p. 720
Visual/Spatial	**PE**	Quick Lab, p. 720; Performance Zone items 1, 16, and 18, p. 723
	ATE	Lesson Warm-up, p. 718; Visual Strategy, p. 719
Verbal/Linguistic	**PE**	Performance Zone item 23, p. 723
	ATE	Closure, p. 720

Only a Few Chordate Species Are Invertebrates

Phylum Chordata is divided into three subphyla. The vast majority of chordate species belong to subphylum Vertebrata, which you will study in the next unit. Two other subphyla, Urochordata (the tunicates) and Cephalochordata (the lancelets), contain a small number of species. Because members of these two subphyla are chordates that do not have backbones, they are called **invertebrate chordates.**

Tunicates

Only the free-swimming tunicate larvae have a nerve cord, notochord, and postanal tail. These features are lost during the larvae's transformation into adulthood. However, adult tunicates, shown in **Figure 32-11,** retain their pharyngeal slits. Most adult tunicates are sessile, filter-feeding marine animals. A tough sac, called a tunic, develops around the adult's body and gives tunicates their name. Cilia beating within the tunicate cause water to enter the incurrent siphon. The water circulates through the tunicate's body, passes though its pharyngeal slits, and leaves the body through the excurrent siphon. As water passes through the slits in the pharynx, food is filtered from it and passed into the stomach. Undigested food passes to the anus, which empties into the excurrent siphon.

All tunicates are hermaphrodites, and some are also able to reproduce asexually by budding. While some tunicates are solitary, budding can result in colonies of identical tunicates.

Study TIP

Organizing Information

Make a table to organize information about invertebrate chordates. Across the top, write the headings *Adult characteristics* and *Larval characteristics.* Along the sides, write *Lancelet* and *Tunicate.* Add information to the table as you read Section 32-2.

internetconnect

SCILINKS. NSTA

TOPIC: Invertebrate chordates
GO TO: www.scilinks.org
KEYWORD: HX719

Figure 32-11 **Adult tunicate**

Pharyngeal slits are the only chordate characteristic retained by adult tunicates.

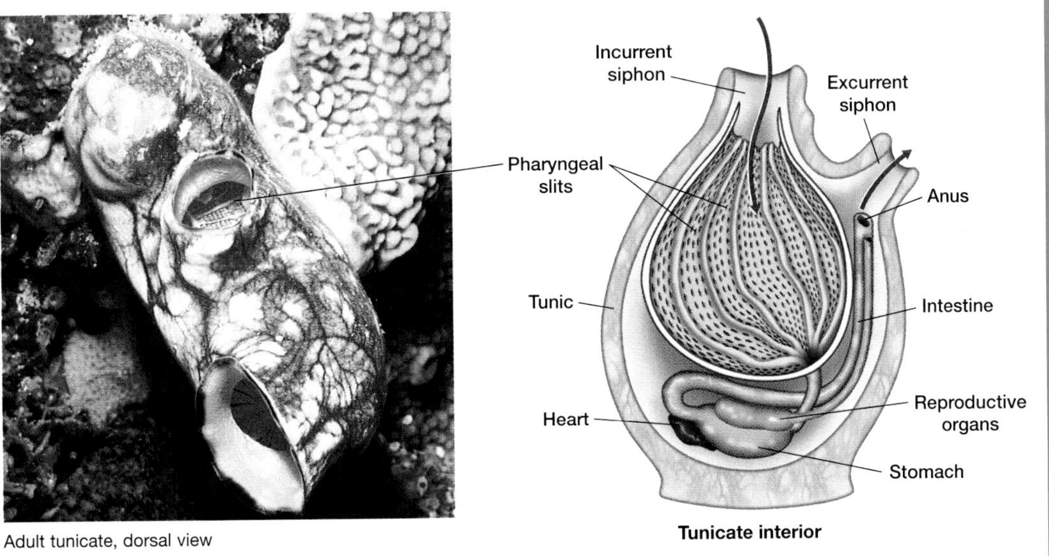

Adult tunicate, dorsal view

Tunicate interior

did you know?

Lancelets have transparent skin.

Unlike the skin of other vertebrates, the skin of a lancelet is transparent. This feature allows us to see the rows of muscles along the side of the notochord.

CAREER

Marine Biologist

Have interested students research the career of a marine biologist and write a report that includes the type of degree needed, classes to take in high school and college, summer experiences, helpful volunteer experiences, working environment, starting salary, and growth potential. PORTFOLIO

Teaching TIP

Other Deuterostome Phyla

In addition to echinoderms and chordates, there are two other deuterostome phyla. Both groups live in the sea. Arrow worms belong to the phylum Chaetognatha. Their size ranges from 2.5 to 10 cm in length, and they have straight bodies that resemble darts or torpedoes. Acorn worms, phylum Hemichordata, are characterized by their long proboscis. These bottom dwellers grow much longer than arrow worms, ranging from 20 mm to 2.5 m in length. Have students use library or Internet resources to find additional information and photographs of these marine deuterostomes.

Visual Strategy

Figure 32-11

Point out the arrows that indicate the direction of water flow in the tunicate interior. Have students trace the water's path with their finger.

Chapter 32

How do the notochord and nerve cord compare in structure?

QUICK LAB

Time 20 minutes

Process Skills Describing and identifying structures, forming hypotheses

Teaching Strategies
Explain the differences between cross sections and longitudinal sections. Ask students why we should use a cross section for this lab.

Analysis Answers

1. The nerve cord is located dorsally and has a very small diameter. The notochord, which is wider, lies just below the nerve cord.
2. bilateral symmetry
3. Like all chordates, lancelets have bilateral symmetry. Adult echinoderms have pentaradial symmetry.
4. Advantages include protection for the nerve cord, points for muscle attachment, and flexibility.

3 Assessment Options

Closure

Technique: K-W-L
Tell students to read over the lists that they created using the **Active Reading** on page 709. Have students list what they have Learned. Conclude by asking students which questions are still unanswered, and what are some new questions they may have.

Review

Assign Review—Section 32-2.

Reteaching —BASIC

Make a chart comparing the characteristics of lancelets and tunicates.

Figure 32-12 Lancelet. Lancelets spend most of their time with their tail buried in sediment.

Lancelets

Lancelets, shown in **Figure 32-12,** receive their name from their bladelike shape. Some biologists have argued that lancelets are degenerate fish. However, the discovery of lancelet fossils in rocks over 550 million years old—far older than any fish—argues that they are not. While lancelets may resemble fish, they are scaleless chordates only a few centimeters long. If you look closely at the lancelet's translucent body, you can see that its V-shaped bundles of muscles are arranged in a series of repeating segments. Lancelets are found worldwide in shallow ocean water. They spend most of their time with their tail end buried in mud or sand and their mouths protruding. The beating of cilia that line the front end of their digestive tract draws water through the mouth and pharynx and out the pharyngeal slits. Lancelets feed on microscopic protists that they filter out of the water. Unlike tunicates, the sexes are separate in lancelets.

QUICK LAB

How do the notochord and nerve cord compare in structure?

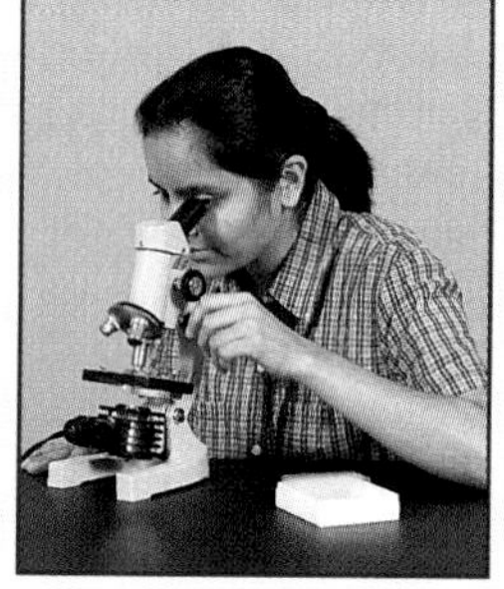

The notochord and hollow nerve cord are two important characteristics of all chordates. While both are located on an animal's dorsal side, they differ in size, structure, and location. You can compare the two when viewing a cross section of an adult lancelet.

Materials

compound microscope, prepared slide of the cross section of an adult lancelet

Procedure

1. Place a prepared slide of a cross section of an adult lancelet under the microscope.
2. Locate the dorsal side of the specimen, and turn the slide so the dorsal side is on top.
3. Locate the notochord and hollow nerve cord. If visible, locate the intestine.
4. Sketch the specimen, and label the dorsal and ventral sides, the notochord, the nerve cord, and the intestines.

Analysis

1. **Describe** the structure, location, and size of the nerve cord and the notochord.
2. **Identify** the kind of symmetry observed in the adult lancelet.
3. **Compare** the lancelet's symmetry with the symmetry of adult echinoderms.
4. **SKILL Forming Hypotheses** In vertebrate chordates, the notochord becomes a backbone that encases the nerve cord. Why might this arrangement be an advantage?

Review SECTION 32-2

1. **Describe** the characteristics common to all chordates.
2. **Summarize** why tunicates and lancelets are classified as invertebrate chordates.
3. **Compare** the chordate characteristics found in adult tunicates with those found in adult lancelets.
4. **SKILL Inferring Relationships** Explain why the larval form of an organism can be valuable in determining relationships among species.

Review SECTION 32-2 ANSWERS

1. At some point during their development, all chordates have a notochord, a hollow, dorsal nerve cord, pharyngeal slits, and a postanal tail.
2. Tunicates and lancelets possess all four basic chordate characteristics, but they lack a backbone.
3. Adult tunicates retain their pharyngeal slits. The nerve cord, notochord, and tail are lost after the larval stage. Adult lancelets have a notochord, a dorsal nerve cord, a postanal tail, and pharyngeal slits.
4. The larval form may be the only time certain characteristics are present in that organism.

Study ZONE

CHAPTER 32

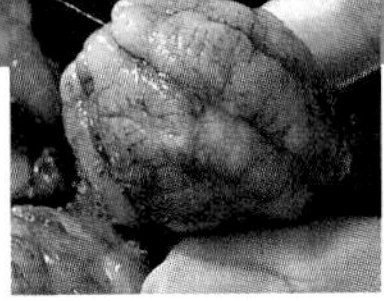

Use the Key Concepts and Key Terms listed below to review the main ideas in this chapter. If you need to review the meaning of a term, turn to the page indicated.

Key Concepts

32-1 Echinoderms

- During embryonic development, a protostome animal's mouth develops from the blastopore. In a deuterostome animal, the anus forms from the blastopore and the mouth forms later from a different opening.
- Because echinoderms and chordates are both deuterostomes, scientists believe that both groups were derived from a common ancestor.
- Echinoderms lack a head or brain. Their nervous system consists of a central ring of nerves with branches extending into each of the five parts of its body plan.
- Echinoderms share four characteristics: an endoskeleton composed of ossicles; five-part radial symmetry; a water-vascular system; and coelomic circulation and respiration.
- In many echinoderm species, respiration and waste removal are performed by skin gills.
- Echinoderms are a diverse group consisting of seven classes: sea stars, brittle stars, sea lilies, feather stars, sea urchins and sand dollars, sea cucumbers, and sea daisies.

32-2 Invertebrate Chordates

- At some point in their life, all chordates have a notochord, a dorsal nerve chord, pharyngeal slits, and a postanal tail.
- Phylum Chordata includes invertebrate and vertebrate chordates.
- Invertebrate chordates do not have a backbone (vertebral column). Two invertebrate subphyla are Urochordata (tunicates) and Cephalochordata (lancelets).
- Tunicates and lancelets are invertebrate chordates.
- Tunicate larvae have a nerve cord, notochord, pharyngeal slits, and a postanal tail. As adults, they lose all of these characteristics except the pharyngeal slits.
- Lancelets retain their notochord, dorsal nerve cord, pharyngeal slits, and postanal tail into adulthood.

Key Terms

32-1

blastopore (710)
protostome (710)
deuterostome (710)
ossicle (712)
water-vascular system (713)
skin gill (713)

32-2

chordate (718)
notochord (718)
pharyngeal slit (718)
invertebrate chordate (719)

Study TIP Ready?

- *If you think you understand these Key Concepts and Key Terms, then you are ready for the Performance Zone.*

Study ZONE

CHAPTER 32

Review and Practice

Assign the following worksheets for Chapter 32:

- **Student Review Guide**
- **Concept Mapping Worksheet**
- **Test Preparation Pretest**
- **Vocabulary Worksheet**
- **Critical Thinking Review Worksheet**

Assessment Options

The following assessment options are available for this chapter:

Resource Link

- Assign **Chapter Test 32** to assess students' understanding of the chapter.

Alternative Assessment

Have students write questions to be used in the exam for this chapter. They should write several types of questions: multiple choice, true/false, fill in, essay, etc. Be sure they provide the proper answer as well. Use the best questions in the test for this unit. PORTFOLIO

Portfolio Assessment

- *Portfolio Assessment* at the front of this book provides options for building and evaluating student portfolios. Students and teachers can select items from the areas listed for inclusion in a portfolio.
- See items labeled PORTFOLIO on pp. 712, 715, 719, and 721.

Answer to Concept Map

The following is one possible answer to Performance Zone item 1 on page 722.

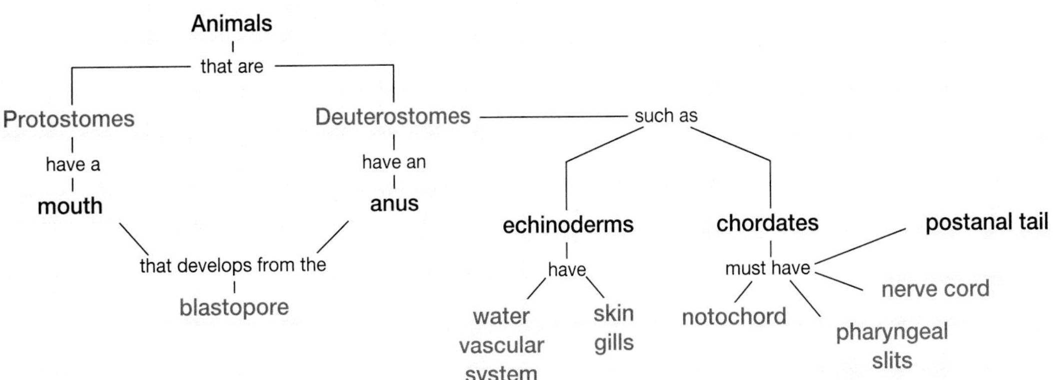

Performance ZONE

CHAPTER 32

REVIEW ANSWERS

Assignment Guide

SECTION	REVIEW
32-1	4, 5, 6, 7, 8, 9, 10, 14, 15, 19, 20, 21, 22, 24, 25, 26
32-2	1, 2, 3, 11, 12, 13, 16, 17, 18, 23

Understanding and Applying Concepts

1. The answer to the concept map is found at the bottom of page 721.
2. **a.** Protostomes are animals with a mouth that develops from or near the blastopore. Deuterostomes are animals with an anus that develops from or near the blastopore.

   b. Ossicles are hard, calcium-rich plates that make up the endoskeleton of echinoderms. Skin gills are soft, fingerlike projections that grow between an echinoderm's spines. Skin gills aid with respiration and waste removal.

   c. The nerve cord is dorsal and has a very small diameter. The notochord is dorsal, ventral to the nerve cord, and larger in diameter than the nerve cord.
3. b
4. d
5. c
6. d
7. a
8. c
9. d
10. c
11. b
12. b
13. The notochord represented the beginnings of an internal skeleton. An internal skeleton provided points for muscle attachment and allowed vertebrates to grow large and move quickly.
14. Scientists can run specific tests to determine what contaminants are present. If the toxicity can be traced to a specific source, the source may be forced to treat its waste. Contaminated sediments may be removed or decontaminated. In the future, plants may be used to remove toxic chemicals from contaminated water.
15. In most crustaceans, gas exchange occurs through gills. Insects use a system of tracheae and spiracles for gas exchange. In echinoderms, gas exchange occurs as dissolved oxygen moves through the water-filled coelomic cavity. Many echinoderms also have skin gills that provide extra surface area for respiratory exchange.

Skills Assessment

16. A. incurrent siphon; B. pharyngeal slits; C. tunic; D. excurrent siphon; E. anus; F. intestine; G. reproductive organs
17. Pharyngeal slits are important because they are the only chordate characteristic retained by adult tunicates.
18. The arrows show the direction of water flow throughout the tunicate.
19. Answers will vary. Steps might include raising sea urchins in captivity and reintroducing them into the wild, putting limits on the number of eggs taken, and establishing marine sanctuaries for sea urchins.

Performance ZONE

CHAPTER 32 Review

Understanding and Applying Concepts

1. **Concept Mapping** Construct a concept map that outlines the development and body plan of echinoderms and invertebrate chordates. Include the following terms in your map: blastopore, protostome, deuterostome, water-vascular system, skin gills, notochord, nerve cord, pharyngeal slits. Use additional terms in your map as needed.
2. **Understanding Vocabulary** For each pair of terms, explain the differences in their meanings.
   a. protostome, deuterostome
   b. ossicle, skin gills
   c. notochord, nerve cord
3. Which of the following patterns of development is characteristic of echinoderms and lancelets?
   a. protostome
   b. deuterostome
   c. acoelomate
   d. pseudocoelomate
4. In protostomes, the area labeled *A* in the diagram below becomes the
   a. anus.
   b. gastrula.
   c. coelom.
   d. mouth.

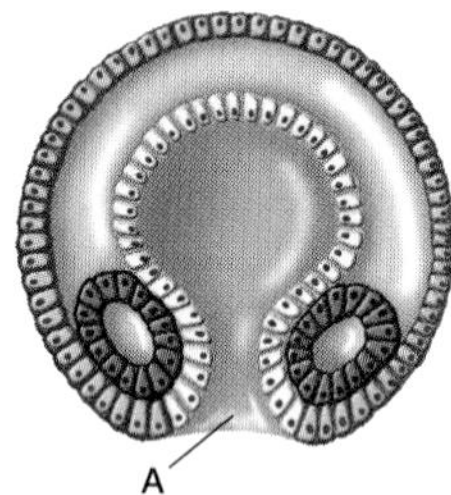

5. The phylum characterized by radial symmetry and a water-vascular system is
   a. Chordata.
   b. Arthropoda.
   c. Echinodermata.
   d. Cnidaria.
6. The presence of which of the following characteristics is typical among echinoderms?
   a. nauplius larvae
   b. a notochord
   c. an exoskeleton
   d. a water-vascular system
7. Which of the following pairs is a correct match between an adult echinoderm and a characteristic?
   a. sea cucumber: leathery epidermis
   b. sand dollar: five arms
   c. sea lily: free-swimming
   d. sea star: sessile
8. Which echinoderm group includes species that specialize in hunting bivalves?
   a. sea cucumbers
   b. sand dollars
   c. sea stars
   d. sea lilies
9. The ossicles of *Asterias* produce pincerlike structures called
   a. madreporites.
   b. tube feet.
   c. ampullae.
   d. pedicellaria.
10. Sea star movements are coordinated by the
    a. radula.
    b. mantle.
    c. nerve ring.
    d. tube foot.
11. Animals with a notochord, a dorsal nerve cord, and pharyngeal slits are members of the phylum
    a. Echinodermata.
    b. Chordata.
    c. Annelida.
    d. Cnidaria.
12. Which of the following do adult chordates and adult echinoderms have in common?
    a. a nonsegmented body
    b. an internal skeleton
    c. a water-vascular system
    d. bilateral symmetry
13. **Theme Evolution** In what way was the notochord an important development in the evolution of vertebrates?
14. **Health Watch** Once preliminary tests determine that coastal waters are polluted, what kinds of actions might be taken to reduce the pollution?
15. **Chapter Links** Gas exchange is essential to life. Compare how gas exchange occurs in most crustaceans, insects, and echinoderms. (**Hint:** See Chapter 31, Sections 31-1 and 31-4.)

Skills Assessment

Part 1: Interpreting Graphics

Study the illustration of a tunicate interior below. Then answer the following questions:

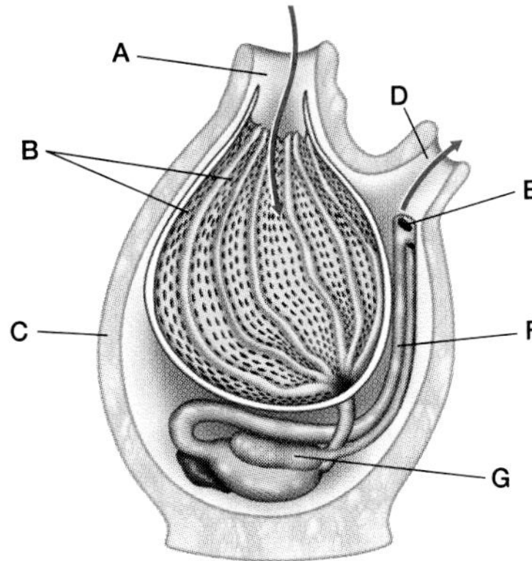

16. Identify the structures labeled *A–G* on the diagram.
17. Summarize why the structure labeled *B* is important in the classification of tunicates.
18. Describe the function of the arrows in the diagram.

Part 2: Life/Work Skills

Use the media center or Internet resources to learn more about two echinoderms, the sea urchins and sea stars.

19. **Finding Information** Sea urchin eggs, or roe, are highly prized for sushi, a Japanese dish that usually features raw seafood. In trying to supply this delicacy, divers have nearly wiped out sea urchin populations in some areas of California. Prepare a report that tells what steps could be taken to reestablish sea urchin populations.
20. **Communicating** Research the crown-of-thorns sea star, or *Acanthaster planci*, which is abundant on the Great Barrier Reef. Find out why this echinoderm poses such a threat to its environment, and prepare a report on your findings. Include steps being taken by the Australian government to deal with the problem.

Critical Thinking

21. **Evaluating Conclusions** Sea cucumbers and sea lilies are relatively sessile animals. Their larvae, however, are capable of swimming. Evaluate the claim that swimming larvae provide an advantage for these echinoderms.
22. **Forming Reasoned Opinions** Scientists have found many echinoderm fossils from the Cambrian period, but they have found few fossils of other species that must have lived then. What might explain the large number of fossilized echinoderms?

Portfolio Projects

23. **Summarizing Information** While none of the invertebrate chordates are of economic importance, in some parts of the world, both lancelets and tunicates are used as food sources for animals and humans. Use media center or Internet resources to find out more about the consumption of these invertebrate chordates. Present your findings as a written report or on audiotape.
24. **Identifying Functions** Use library or Internet resources to investigate how tube feet vary in the different groups of echinoderms. Make a series of drawings illustrating the differences.
25. **Interpreting Information** Read "Flowers of the Coral Seas," by Fred Bavendam (*National Geographic,* Dec. 1996, pp. 118–131). Describe how feather stars move. When and why do they move around? Describe two commensular relationships feather stars have with other animals. Does a feather star have to contend with predators or parasites?
26. **Career Focus Underwater Photographer** Research the field of underwater photography, and write a report on your findings. Your report should include a job description, training required, kinds of employers, growth prospects, and starting salary.

Portfolio Projects

23. Answers will vary. Students' findings may include that stingrays prey on lancelets.
24. Answers will vary. Students may note that the tube feet of many sea stars that live on soft surfaces lack suckers. Instead the feet are pointed, which enables them to dig into sand. Many brittle stars have worm-like feet that bring food particles to the animal's mouth. Sea cucumbers have tube feet that may or may not have suckers.
25. Feather stars swim or crawl slowly along the coral reef. Nocturnal ones venture out to feed and often move to the top of corals to catch food. Commensal relationships include those with *Periclimenes* shrimp, snapping shrimp, and galatheid crabs. Feather stars must contend with predators, such as the *Annulobacis* snail.
26. Answers will vary. Underwater photographers take still or moving pictures of underwater scenes. They usually focus on some aspect of marine or freshwater natural history. Underwater photographers have hands-on training in photography and oceanography. Some have a college degree in art with a concentration in photography. Underwater photographers are often self-employed, although some have permanent positions with museums, aquariums, and organizations such as the National Geographic Society. The growth potential for this field is limited. Starting salary will vary by region.

20. *A. planci* is dangerous to the Great Barrier Reef because it is a voracious eater. *A. planci* often eats in large groups, secreting a chemical that attracts more starfish to the area to feed. The typical *A. planci* can eat 5 to 6 m^2 of living coral each year. Marine biologists in Australia are taking steps to reduce the problems posed by the crown-of-thorns sea star. They have tried injecting starfish with poisons and asking amateur divers to remove them by hand.

Critical Thinking

21. Answers will vary. Swimming enables these echinoderm larvae to find underexploited living spaces and food sources.
22. Answers will vary. Students might note that the hard endoskeletons of echinoderms are easily fossilized. Echinoderms also live on the sea floor, where sediment piles up to form sedimentary rock, where most fossils are found.

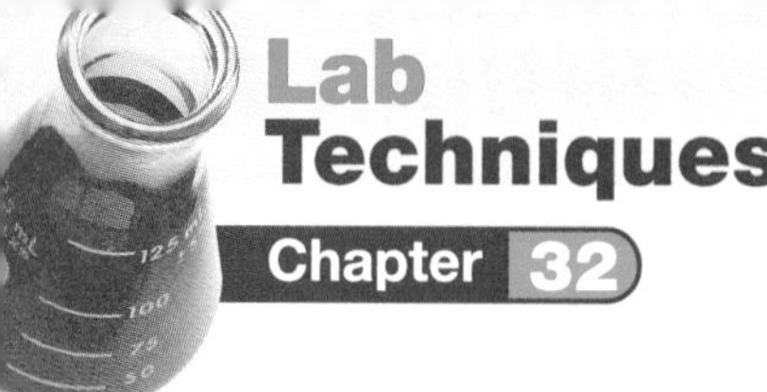

Lab Techniques

Chapter 32

Comparing Echinoderm and Chordate Development

Purpose

Students will examine prepared slides of developing sea stars and compare them with drawings of early stages of human development.

Preparation Notes

Time Required: 45 minutes

Materials

Materials for this lab can be ordered from WARD'S. See *Master Materials List* at the front of this book for catalog numbers.

Preparation Tips

- Obtain prepared slides of the following stages of sea star development: unfertilized egg, zygote, 2-cell stage, 4-cell stage, 8-cell stage, 16-cell stage, 32-cell stage, 64-cell stage, blastula, early gastrula, and young sea star larva.
- Each lab group will need a set of slides and a compound microscope for viewing the slides. If there are not enough sets of slides for each group to have its own set, set up a station for each developmental stage and have lab groups rotate from station to station.
- Before beginning this investigation, review the characteristics of echinoderms and chordates. Discuss the fact that humans are chordates. Remind students that the sequence of early embryonic development is only the very beginning of development. In humans, the process from zygote to blastocyst takes about seven days.

Lab Techniques

Chapter 32

Comparing Echinoderm and Chordate Development

SKILLS
- Observing
- Drawing

OBJECTIVES
- **Observe** prepared slides of sea star development.
- **Compare** the early stages of echinoderm development with the early stages of chordate development.
- **Describe** the changes that occur during embryo development.

MATERIALS
- prepared slides of sea star development
- compound microscope
- paper and pencil

Before You Begin

Both **echinoderms** and **chordates** have **deuterostome** embryo development. Sea stars, the most familiar echinoderms, begin life as a single cell—a fertilized egg, or **zygote.** After a series of cell divisions called **cleavage,** an **embryo** develops from the zygote. The early embryo is a hollow sphere of cells called a **blastula.** The space in the center of a blastula is called the **blastocoel.** The embryo is called a **gastrula** when cells begin to specialize and become part of one of the three embryonic tissue layers. As **cell differentiation** continues, a sea star embryo develops into a **larva.** In this lab, you will look at prepared slides of sea star (echinoderm) development and compare it with the early stages of human (chordate) development, shown in the illustration at right.

1. Write a definition for each boldface term in the paragraph above.
2. Based on the objectives for this lab, write a question you would like to explore about sea star development.

Procedure

PART A: Examine Sea Star Slides

1. Obtain a set of prepared slides showing sea stars at different stages of development. Choose slides labeled unfertilized egg, zygote, 2-cell stage, 4-cell stage, 8-cell stage, 16-cell stage, 32-cell stage, 64-cell stage, blastula, early gastrula, middle gastrula, late gastrula, and young larva.
2. Examine each slide with a compound microscope in the order in which they are listed in step 1. For each slide, focus on one good example of the developmental stage listed on the label under low power. Then switch to high power.
3. Draw a diagram of each developmental

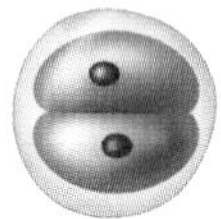

2-cell stage

4-cell stage

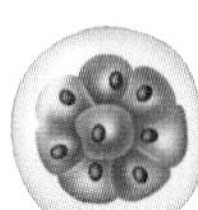

8-cell stage

64-cell stage

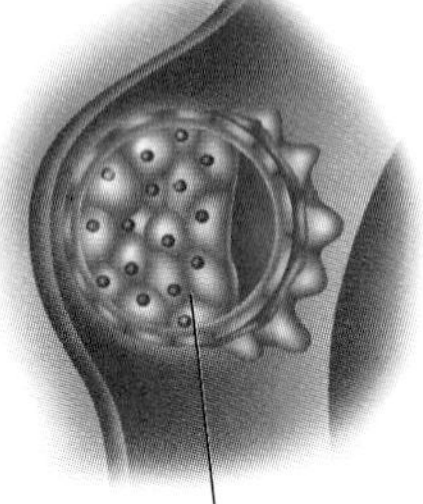

Blastocyst within uterus

Early stages of human development

Safety Precautions

Discuss all safety symbols and caution statements with students.

Procedural Tips

Have students compare the stages of sea star development to photos of mammalian stages, if available.

Disposal

Make sure students carefully dispose of any slides in the trash. Remind students to clean their work areas when finished with the lab.

Before You Begin

Answers

1. *echinoderms*—spiny-skinned animals with tube feet and radially symmetrical adults

   chordates—group of animals that have a notochord at some time during development.

   deuterostome—animal that has an anus that develops from or near the blastopore.

   zygote—a fertilized egg

   cleavage—a series of cell divisions that occurs during early embryo development

stage you observe. Draw what you see as you examine each slide. Label each diagram with the name of the stage it represents and the magnification.

4. Compare your diagrams with the diagrams of early human embryonic development shown on the previous page.

PART B: Cleanup and Disposal

5. Dispose of any broken glass in the waste container.
6. Clean up your work area and all lab equipment. Return lab equipment to its proper place. Wash your hands thoroughly before you leave the lab and after you finish all your work.

Analyze and Conclude

1. **Comparing Structures** Compare the size of a sea star zygote with that of a sea star blastula.
2. **Interpreting Graphics** How does cell size change as embryonic development proceeds?
3. **Recognizing Patterns** Compare the chromosome number of a fertilized sea star egg with that of one cell in each of the following: 2-cell stage, blastula, gastrula, and adult sea star.
4. **Drawing Conclusions** At what stage of development do the cells in a sea star embryo begin to look different?
5. **Drawing Conclusions** At what stage does a developing sea star embryo become bigger than the zygote?
6. **Forming Hypotheses** From your observations of changes in cellular organization, why do you think the blastocoel is important during embryonic development?
7. **Recognizing Patterns** How does the symmetry of sea star embryos and larvae compare with the symmetry of adult sea stars? Explain.
8. **Relating Concepts** Describe one way the cleavage of echinoderm zygotes and human zygotes is identical.
9. **Justifying Conclusions** Echinoderms and chordates are thought to share a common ancestor. Describe how your observations in this lab can be used to justify this conclusion.
10. **Further Inquiry** Write one or more questions about echinoderm development that could be explored with a new investigation.

On the Job

Embryology is the study of the embryonic development of animals. Do research in the library or media center or on the Internet to discover how embryology is used to determine the relationships among the animal phyla. Summarize your findings in a written or oral report.

Lab Techniques

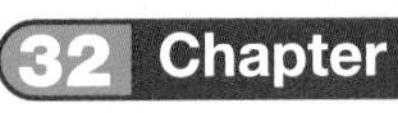

Analyze and Conclude

Answers

1. The zygote and the blastula are the same size.
2. From the first cleavage until the gastrula stage, the cells get progressively smaller. After that, some cells begin to grow larger.
3. The chromosome number is the same in each cell during embryonic development. In adult organisms, however, all cells have the same number of chromosomes except for eggs and sperm, which are haploid.
4. When the embryo reaches the gastrula stage, some cells look different from others.
5. By the time the embryo has developed into a gastrula, it is bigger than the zygote.
6. A blastocoel enables invagination (folding) to occur during gastrulation, when the three primary tissue types form.
7. Both sea-star embryos and larvae exhibit bilateral symmetry. The adult exhibits radial symmetry.
8. Cleavage in both echinoderms and humans results in equal-sized cells.
9. This lab illustrates that echinoderms and chordates exhibit similar stages of cell division and differentiation, indicating that they diverged from a common ancestor.
10. Answers will vary. Sample answer: How is the embryonic development of arthropods, such as insects, similar to that of sea stars?

embryo—name for a developing organism in the early stages of development

blastula—a hollow sphere of cells that comprises the early embryo of a sea star

blastocoel—the space in the center of a blastula

gastrula—name for the embryo after cell differentiation has begun

cell differentiation—process by which cells become different shapes and become specialized for different functions

larva—an advanced stage of sea star development

2. Answers will vary. For example: At what stage of echinoderm development does cell differentiation begin?

Procedure

Answers

3. Answers will vary. Compare the students' answers with the stages discussed in Before You Begin and with the illustrations in Figure 32-1, on page 710.

On the Job

Answers

Answers will vary. Students should note that embryologists compare animals during early stages of development to look for similarities. Developmental similarities often indicate a shared evolutionary past.

UNIT 8 Exploring Vertebrates

Chapters

Only a few wild tigers remain in the world.

in perspective

The Tiger

Yesterday...

Native to Asia, tigers once lived throughout much of the southern half of the continent. Tigers were killed in large numbers as hunters killed them for their skins and humans cleared the forests in which they lived. **What characteristics of tigers are representative of other mammals? See page 810.**

Prince of Wales, Edward VII, on a hunting expedition in India

Today...

Tigers everywhere are at critically low numbers. Poaching, the illegal killing of tigers for their bones, skin, and other body parts, continues in many countries. In India, Nepal, and Russia, protected areas have been set aside to help preserve tiger habitat. **Discover how a mammal's teeth provide clues about its diet on page 812.**

Tiger skin confiscated by the police in Delhi, India

Tomorrow...

The long-term survival of tigers may depend on breeding programs in zoos and nature reserves. In some areas, more rigorous enforcement of anti-poaching laws is helping to protect wild populations. **To which order of mammals do tigers belong? Find out on page 823.**

Bengal tiger cubs

TOPIC: Tigers
GO TO: www.scilinks.org
KEYWORD: HX727

Chapter 33

Introduction to Vertebrates

Planning Guide

OBJECTIVES	MEETING INDIVIDUAL NEEDS	READING SKILLS AND STRATEGIES
33-1 Vertebrates Spread from the Sea to the Land (BLOCKS 1 & 2 90 min) • Identify the key characteristics of vertebrates. • Describe two adaptations that evolved in early fishes. • Identify the relationship of fishes to amphibians. • Summarize the key adaptations of amphibians for life on land.	Learning Styles, p. 730 BASIC ADVANCED • pp. 731, 736, 738	**ATE** Active Reading, Technique: Reading Organizer, p. 729 ■ Active Reading Guide, worksheet (33-1) ■ Directed Reading, worksheet (33-1) ■ Reading Strategies, Series-of-Events Chain
33-2 Vertebrates Adapt to Terrestrial Living (BLOCKS 3 & 4 90 min) • Summarize why dinosaurs became the dominant land vertebrate. • Contrast ectotherms with endotherms. • Summarize why mammals replaced dinosaurs. • Identify the dinosaurlike and the birdlike features of *Archaeopteryx*.	Learning Styles, p. 739 BASIC ADVANCED • pp. 744, 748	**ATE** Active Reading, Technique: Paired Summarizing, p. 744 **ATE** Active Reading, Technique: Reading Effectively, p. 747 ■ Active Reading Guide, worksheet (33-2) ■ Directed Reading, worksheet (33-2) ■ Reading Strategies, Spider Map ■ Supplemental Reading Guide, *Through a Window: My Thirty Years with the Chimpanzees of Gombe*

Review and Assessment Resources

BLOCKS 5 & 6 90 min

TEST PREP

PE Study Zone, p. 749

PE Performance Zone, pp. 750–751

■ Chapter 33 worksheets:
- Vocabulary
- Concept Mapping
- Science Skills
- Critical Thinking

 Audio CD Program

■ Test Prep Pretest

CHAPTER TESTING

■ Chapter Tests (blackline master)

▲ Test Generator

■ Assessment Item Listing

See the correlations table at the front of this book for a list of the ***National Science Education Standards*** addressed in this chapter.

Smithsonian Institution® Visit **www.si.edu/hrw** for additional on-line resources.

KEY

Biology Principles and Explorations

PE Pupil's Edition

ATE Teacher's Edition

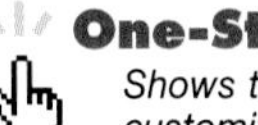

HOLT BIOSOURCES

 Teaching Transparencies

 Laboratory Program

One-Stop Planner CD-ROM
Shows these resources within a customizable daily lesson plan:

■ Teaching Resources

▲ Test Generator

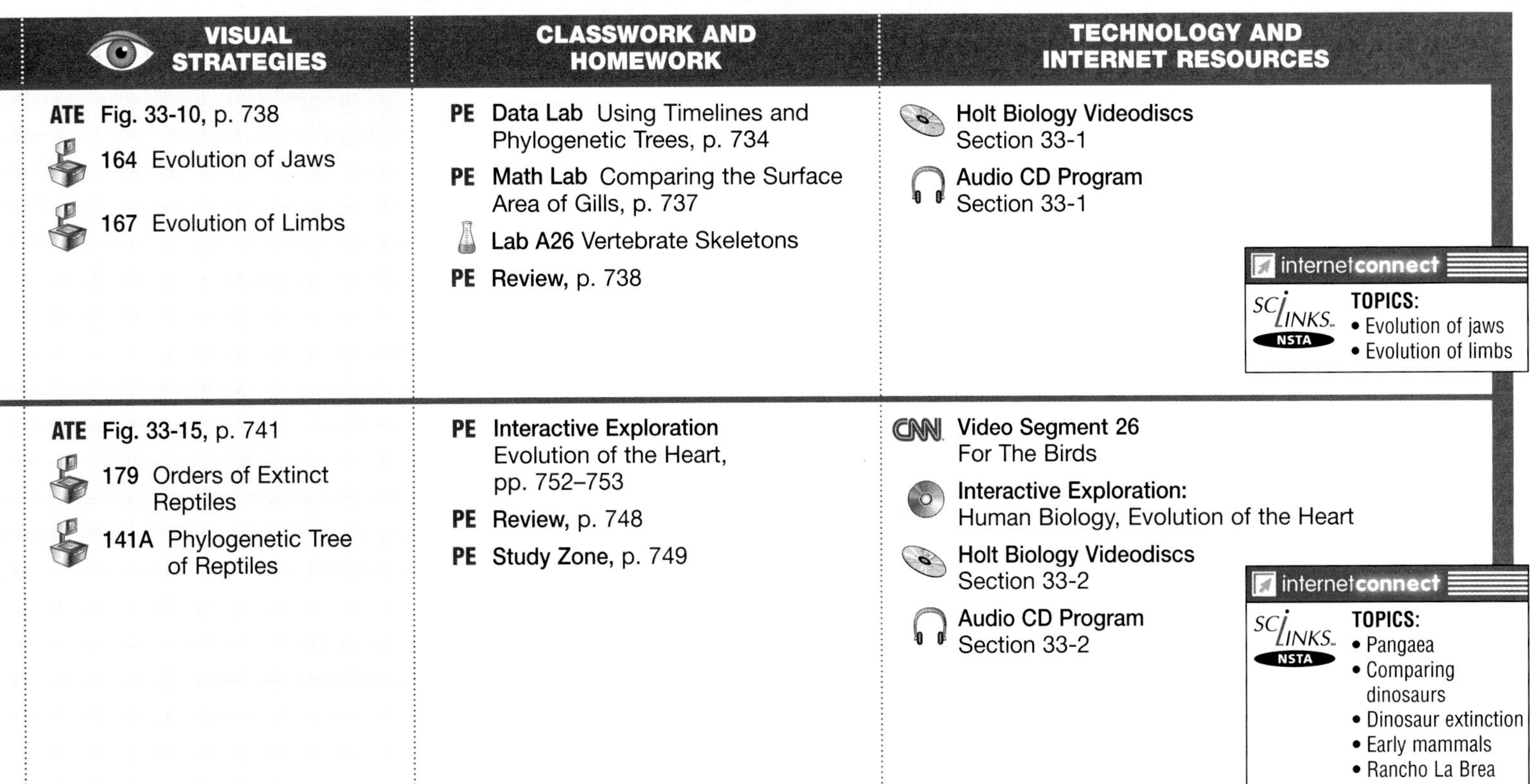

VISUAL STRATEGIES	CLASSWORK AND HOMEWORK	TECHNOLOGY AND INTERNET RESOURCES
ATE Fig. 33-10, p. 738 164 Evolution of Jaws 167 Evolution of Limbs	**PE** Data Lab Using Timelines and Phylogenetic Trees, p. 734 **PE** Math Lab Comparing the Surface Area of Gills, p. 737 Lab A26 Vertebrate Skeletons **PE** Review, p. 738	Holt Biology Videodiscs Section 33-1 Audio CD Program Section 33-1 internetconnect SciLinks NSTA TOPICS: • Evolution of jaws • Evolution of limbs
ATE Fig. 33-15, p. 741 179 Orders of Extinct Reptiles 141A Phylogenetic Tree of Reptiles	**PE** Interactive Exploration Evolution of the Heart, pp. 752–753 **PE** Review, p. 748 **PE** Study Zone, p. 749	CNN Video Segment 26 For The Birds Interactive Exploration: Human Biology, Evolution of the Heart Holt Biology Videodiscs Section 33-2 Audio CD Program Section 33-2 internetconnect SciLinks NSTA TOPICS: • Pangaea • Comparing dinosaurs • Dinosaur extinction • Early mammals • Rancho La Brea • Birds • Vertebrate evolution

ALTERNATIVE ASSESSMENT

PE Portfolio Projects 22–26, p. 751

ATE Alternative Assessment, p. 749

ATE Portfolio Assessment, p. 749

Lab Activity Materials

Data Lab
p. 734 pencil, paper

Math Lab
p. 737 cellophane wrap, scissors, toothpick, ruler, container of water

 **Interactive Exploration**
p. 752 CD-ROM, computer

Demonstrations

ATE photos (7), one each of a lamprey, shark, bony fish, amphibian, reptile, bird, mammal, p. 732

ATE 1 L bottles (2), water, p. 735

BIOSOURCES LAB PROGRAM
- Quick Labs: Lab A26, p. 51

Chapter 33

Chapter Theme

Evolution

Environmental changes over millions of years have resulted in changes in vertebrate structure and function through the process of natural selection. These changes have enabled vertebrates to exploit new habitats and adopt new ways of life. This chapter summarizes many of these changes and their possible origins.

Establishing Prior Knowledge

Before beginning this chapter, make certain students can answer the questions in **Study TIP Ready?** on page 729. If they cannot, encourage them to reread the sections indicated.

Opening Questions

Ask students how the internal, bony skeleton of vertebrates is an advantage over the external skeleton of many invertebrates. *(An internal skeleton provides for more mobility by producing a versatile system of levers and pulleys; calcium-containing bone is harder and stronger than invertebrate exoskeletons.)* Ask students why a turtle is considered a vertebrate even though it has a shell. *(Turtles have a backbone, which is fused to the shell.)*

CHAPTER 33 Introduction to Vertebrates

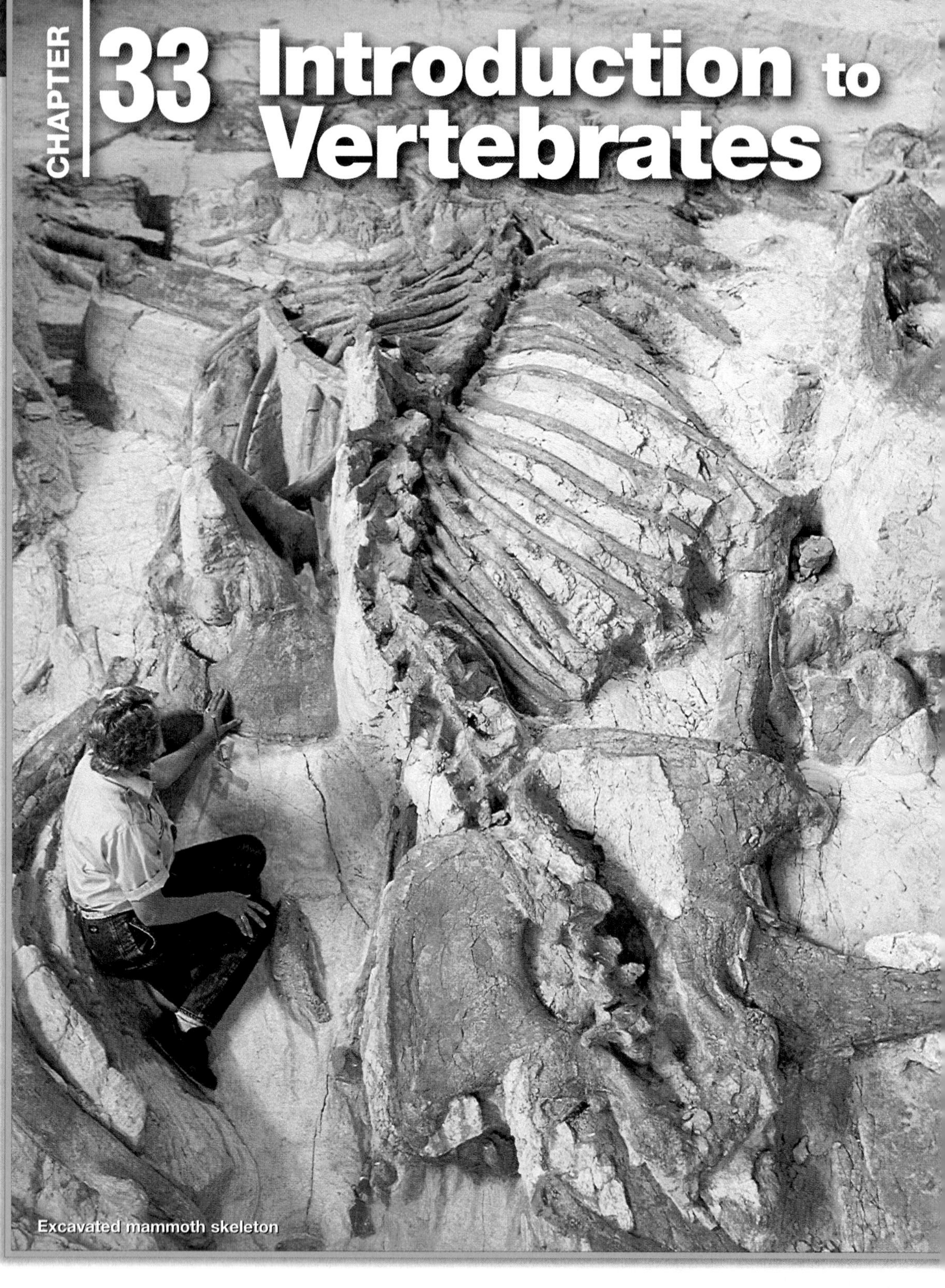

Excavated mammoth skeleton

What's Ahead — and Why

CHAPTER OUTLINE

33-1 Vertebrates Spread from the Sea to the Land

- Vertebrates Are Chordates with a Backbone
- The First Vertebrates Were Jawless Fishes
- Fishes Evolved Jaws and Paired Fins
- Today, Fish Are the Largest Group of Vertebrates
- The First Vertebrates to Venture onto Land Were Amphibians
- Today, There Are Three Groups of Amphibians

33-2 Vertebrates Adapt to Terrestrial Living

- Reptiles Became the First Fully Terrestrial Vertebrates
- Today, Only Four Groups of Reptiles Remain
- Mammals Replaced the Extinct Dinosaurs
- Today, There Are Three Groups of Mammals
- Birds May Have Evolved from Dinosaurs
- Today, Birds Are the Largest Group of Terrestrial Vertebrates

Students will learn

- the characteristics of vertebrates.
- the key adaptations that permit amphibians to survive in their environments.
- why dinosaurs became the dominant land vertebrate and possible reasons for their extinction.
- possible reasons for the rise of mammals and the relationship between reptiles and birds.

Brachiosaurus (The Field Museum)

Bones tell a tale. Although *this* Brachiosaurus *and the mammoth on the facing page lived at different times, their bones give us a window into the history of vertebrates.* Brachiosaurus *lived during the Jurassic period, long before humans inhabited the Earth. The mammoth skeleton is relatively young, only about 26,000 years old.*

Study TIP

Ready?

Check your understanding of these topics to help you prepare for what's ahead.

Can you . . .

- **summarize** how mass extinctions have affected life on Earth? (Section 12-2)
- **define** the terms *vertebrate* and *continental drift*? (Section 12-3)
- **distinguish** between a carnivore and an herbivore? (Section 17-2)
- **define** the terms *bilateral symmetry, cephalization, coelom,* and *phylogenetic tree*? (Section 28-1)

If not, *review the sections indicated.*

Looking Ahead

33-1 Vertebrates Spread from the Sea to the Land
What vertebrate first ventured out of its watery home?

33-2 Vertebrates Adapt to Terrestrial Living
What does an animal need to live on land?

Labs

- **Data Lab**
  Using timelines and phylogenetic trees (p. 734)
- **Math Lab**
  Comparing the surface area of gills (p. 737)
- **Interactive Exploration**
  Evolution of the Heart (p. 752)

Features

- **Exploring Further**
  La Brea Tar Pits (p. 745)

internet connect

National Science Teachers Association *sci*LINKS Internet resources are located throughout this chapter.

Reading Strategies

Active Reading

Technique: Reading Organizer

As students read pp. 732–733, have them complete a series-of-events reading organizer to describe the adaptations in body structure for agnathans, ostracoderms, acanthodians, and placoderms. Organizers will vary but should summarize the key features of each group.

Directed Reading

Assign **Directed Reading Worksheet Chapter 33** to help students interact with the text as they read the chapter.

Interactive Reading

Assign ***Biology: Principles and Explorations* Audio CD Program Chapter 33** to help reluctant readers achieve greater success in reading the chapter.

Reading Skills

Assign Reading Organizers and Reading Strategies Worksheets from the **One-Stop Planner CD-ROM with Test Generator** to help students learn and practice effective reading skills.

internet connect

Holt *Biology: Principles and Explorations*

On-line Resources: go.hrw.com

Visit the HRW Web site for a variety of resources related to this chapter. Just type in the keyword HX0 Chapter 33.

Holt *Biology: Principles and Explorations*

On-line Resources: www.scilinks.org

The following *sci*LINKS Internet resources can be found in the student text for this chapter:

Page 733
TOPIC: Evolution of jaws
KEYWORD: HX733

Page 736
TOPIC: Evolution of limbs
KEYWORD: HX736

Page 740
TOPIC: Pangaea
KEYWORD: HX740

Page 741
TOPIC: Comparing dinosaurs
KEYWORD: HX741

Page 742
TOPIC: Dinosaur extinction
KEYWORD: HX742

Page 744
TOPIC: Early mammals
KEYWORD: HX744

Page 745
TOPIC: Rancho La Brea
KEYWORD: HX745

Page 748
TOPIC: Birds
KEYWORD: HX748

Page 753
TOPIC: Vertebrate evolution
KEYWORD: HX753

Chapter 33

1 Prepare

Scheduling

- **Block:** About 90 minutes will be needed to complete this section.
- **Traditional:** About two class periods will be needed to complete this section.
- **Planning:** See the Planning Guide on pp. 727A and 727B for lecture, classwork, and assignment options for Section 33-1. Lesson plans for Section 33-1 can be found in a lesson cycle format in ***Biology: Principles and Explorations* One-Stop Planner CD-ROM with Test Generator.**

Resource Link

Assign **Active Reading Worksheet Section 33-1** to help students comprehend the key concepts and visuals in this lesson.

2 Teach

Lesson Warm-up

Instruct each student to make a list of 10 animals. Give no further instructions. Next have students put an asterisk by each animal that has a backbone. *(A high percentage of the animals listed will likely be vertebrates.)* Point out that while invertebrates, such as insects, mollusks, and echinoderms, are animals, many people use the term "animal" to refer only to animals with backbones—vertebrates.

PORTFOLIO

Section 33-1 Vertebrates Spread from the Sea to the Land

Objectives

- **Identify** the key characteristics of vertebrates. (pp. 730–731)
- **Describe** two adaptations that evolved in early fishes. (pp. 732–733)
- **Identify** the relationship of fishes to amphibians. (pp. 734–736)
- **Summarize** the key adaptations of amphibians for life on land. (p. 737)

Key Terms

vertebra
agnathan
acanthodian
cartilage

Vertebrates Are Chordates with a Backbone

If you go to a zoo, many of the animals you see—the lions, elephants, snakes, turtles, and birds—are vertebrates, just as people are. Even though these vertebrates appear different, they are very much alike internally, indicating that in the distant past they had a common ancestor. Their present-day differences reflect their different evolutionary paths, which is the topic of this chapter.

What are the internal similarities shared by vertebrates? First, vertebrates are chordates with a backbone. They take their name from the individual segments, called **vertebrae** *(VUR tuh bray),* that make up the backbone. (The singular form of vertebrae is vertebra.) **Figure 33-1** shows a typical vertebrate and its endoskeleton. In most vertebrates, the backbone completely replaces the notochord found in invertebrate chordates.

The backbone provides support for and protects a dorsal nerve cord. It also provides a site for muscle attachment. These functions paved the way for the development of an internal skeleton that

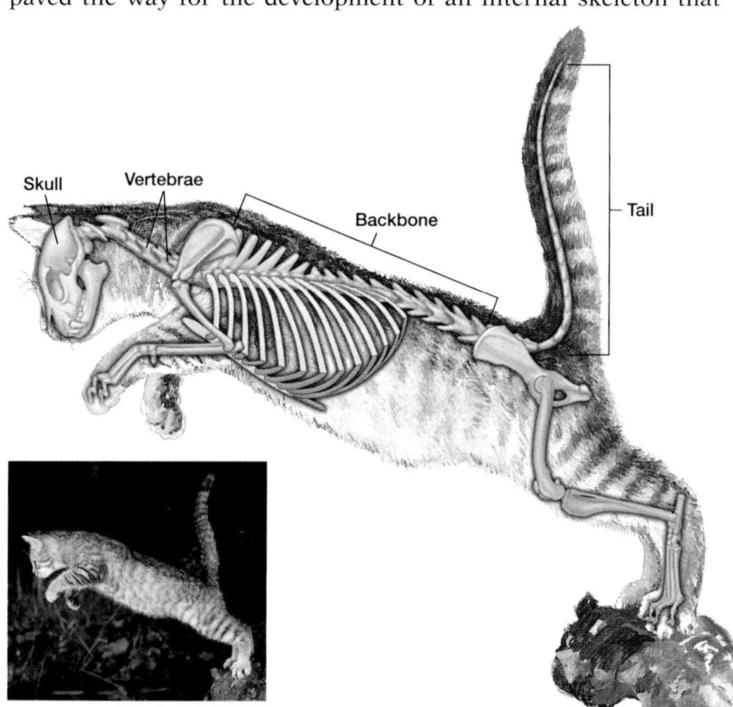

Figure 33-1 Vertebrate. This cat's skeleton includes a backbone made up of vertebrae and a skull.

Integrating Different Learning Styles

Logical/Mathematical	**PE**	Math Lab, p. 737; Interactive Exploration, pp. 752–753
Visual/Spatial	**PE**	Data Lab, p. 734; Study Tip, p. 736; Performance Zone items 1, 23, and 24, pp. 750–751
	ATE	Teaching Tip, p. 732; Demonstrations, pp. 732, 735; Biology in Your Community, p. 736
Body/Kinesthetic	**ATE**	Teaching Tips, pp. 731, 733; Visual Strategy, p. 736
Interpersonal	**ATE**	Teaching Tip, p. 736; Visual Strategy, p. 738
Intrapersonal	**PE**	Real Life, p. 735
Verbal/Linguistic	**PE**	Study Tips, pp. 730, 733; Performance Zone items 2, 26, and 27, pp. 750–751
	ATE	Teaching Tip, p. 732; Reteaching, p. 738

allowed vertebrates to grow larger than their invertebrate ancestors. In addition to a backbone, most vertebrates have a bony skull that encases and protects their brain.

Vertebrates share a number of other characteristics, including segmentation, bilateral symmetry, and two pairs of jointed appendages, such as limbs or fins. They exhibit cephalization and have complex brains and sense organs. All vertebrates have a true coelom and a closed circulatory system with a chambered heart.

The tissues of vertebrates are organized into organs. In turn, the organs are organized into organ systems. Vertebrate organ systems are far more complex than the organ systems found in higher invertebrates. For example, recall the grasshopper's digestive system, seen in the *Up Close* on page 695. Its digestive tract (gut) is divided into several specialized regions. However, the only associated structures are salivary glands. In contrast, the fish's digestive system consists of a gut plus two organs, the liver and the pancreas. These complex organs secrete enzymes that aid digestion. **Figure 33-2** shows the major organ systems of a typical vertebrate.

biobit

How many vertebrae do giraffes have in their neck?

Nearly all vertebrates, including giraffes, have seven neck vertebrae. At almost 19 feet, the giraffe is the tallest of all vertebrates, and its seven cervical (neck) vertebrae are greatly elongated.

Figure 33-2 Major organ systems. A vertebrate's organ systems perform a variety of functions.

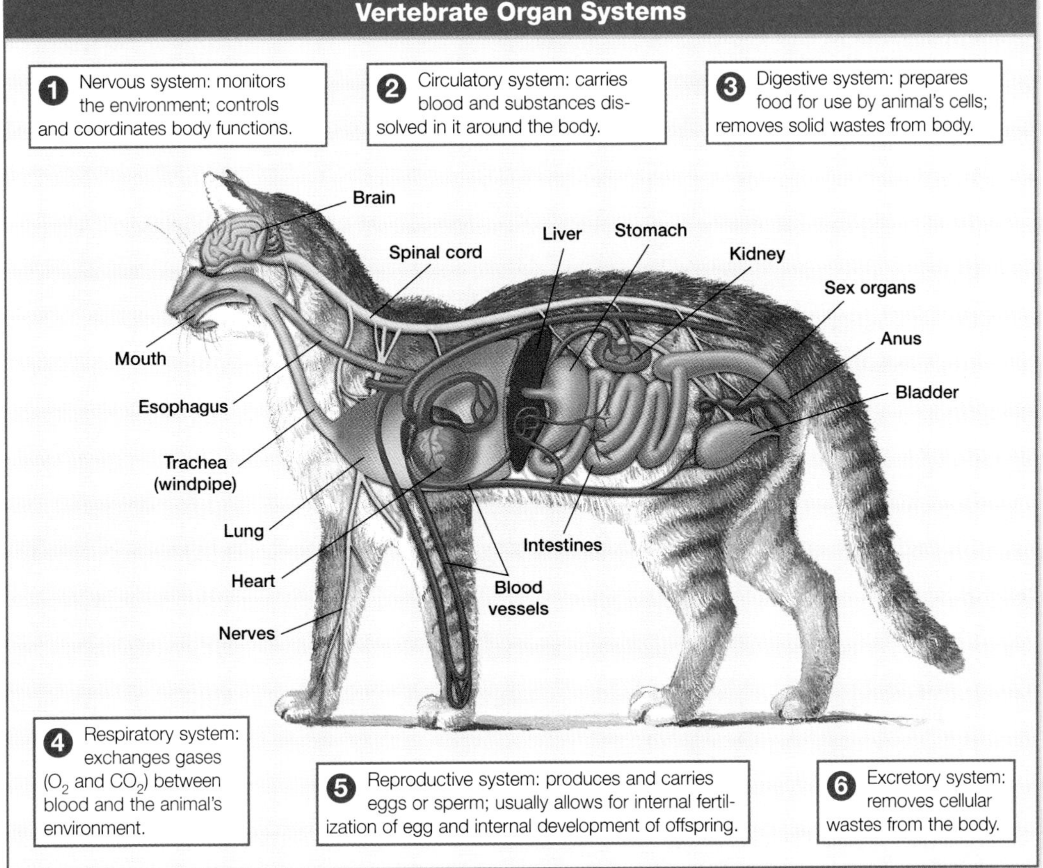
Vertebrate Organ Systems
1 Nervous system: monitors the environment; controls and coordinates body functions.
2 Circulatory system: carries blood and substances dissolved in it around the body.
3 Digestive system: prepares food for use by animal's cells; removes solid wastes from body.
Brain
Spinal cord
Liver
Stomach
Kidney
Sex organs
Anus
Bladder
Mouth
Esophagus
Trachea (windpipe)
Lung
Heart
Nerves
Intestines
Blood vessels
4 Respiratory system: exchanges gases (O_2 and CO_2) between blood and the animal's environment.
5 Reproductive system: produces and carries eggs or sperm; usually allows for internal fertilization of egg and internal development of offspring.
6 Excretory system: removes cellular wastes from the body.

Section 33-1

Teaching TIP — BASIC

Distinguishing Vertebrate, Vertebrae, and Vertebra

Write the words *vertebrate, vertebrae,* and *vertebra* on the board or overhead projector. Have students feel the large bump on the back of their neck. Point out that this bump is part of a single **vertebra,** that the entire column is made up of many **vertebrae,** and that animals that have this column of bones are called **vertebrates.**

CAREER — ADVANCED

Paleontologist

Paleontologists study Earth's rocky crust to identify ancient life forms. Using their observations, these scientists infer the relationships between extinct and living organisms. They also use evidence they find to make inferences about the ecology of previous periods in Earth's history. Paleontologists use a variety of methods to analyze the remains they examine. Have students work in teams to use library references or on-line data bases to find information about the different methods paleontologists use. For additional information, students can write to the Paleontological Society, U.S. Geological Survey, National Center MS970, Reston, VA 22092.

PORTFOLIO

Demonstration

Classes of Vertebrates

With all textbooks closed, display *in random order* seven large pictures—one each of a lamprey, a shark, a bony fish, an amphibian, a reptile, a bird, and a mammal. In game-show style, have a volunteer come forward and arrange the pictures in the order in which each class of vertebrate evolved. If the student is incorrect, have the student return to his or her seat. Tell the next volunteer how many pictures are out of order, but not which ones. Continue until the correct order is reached, at which point you can ring a bell or tap a spoon on a beaker (with great fanfare) to signify success.

Teaching TIP

Observing Fish

Give students an opportunity to observe fish, either on video or in a classroom aquarium. Their observations should include different types of fins and how each is used, different types of mouths for different diets, and different body shapes and movements of fish through water.

Teaching TIP — ADVANCED

Learning the Time Periods

Learning the sequence of the periods in Earth's history is difficult. A shortcut is to have students make up a creative mnemonic device. Divide the class into small groups, explain what a mnemonic device is, and have each group invent one for the Earth's time periods. Do not give any hints. Write each group's results on the board and discuss the merits of each one. *(Example: **Q**ueen **T**racy's **C**rown **J**ewels **T**aught **P**aleontologists **C**arbon-containing **D**iamonds **S**tand **O**ld **C**atastrophes.)* Then, if students learn that the Jurassic period began about 200 million years ago (mya), the Devonian about 400 mya, and the Cambrian about 600 mya, they can get a sense of how the other time periods fit in.

The First Vertebrates Were Jawless Fishes

The first chordates evolved about 550 million years ago. At that time, many different groups of organisms appeared in the shallow seas that covered a large portion of Earth's continents. According to the fossil record, the first vertebrates appeared about 50 million years later. These were fishes similar to the one shown in **Figure 33-3.** Unlike most of the fishes you are familiar with, the earliest fishes, called **agnathans,** had neither jaws nor paired fins. But they did have a backbone, which provided a central axis for muscle attachment. As their muscles pulled against the backbone, the agnathans propelled themselves along the ocean bottom.

Within another 50 million years, jawless fishes had diversified into a great variety of species. The major group was the ostracoderms *(AHS truh kah durms),* which had primitive fins and massive plates of bony tissue on their body. Jawless fishes dominated the oceans for about 100 million years, until they were replaced by new kinds of fishes that were hunters. **Figure 33-4** shows the relationship of the jawless fishes to the fishes that evolved from them.

Figure 33-3 **Ancient fish.** This fossil fish is typical of the early jawless fishes.

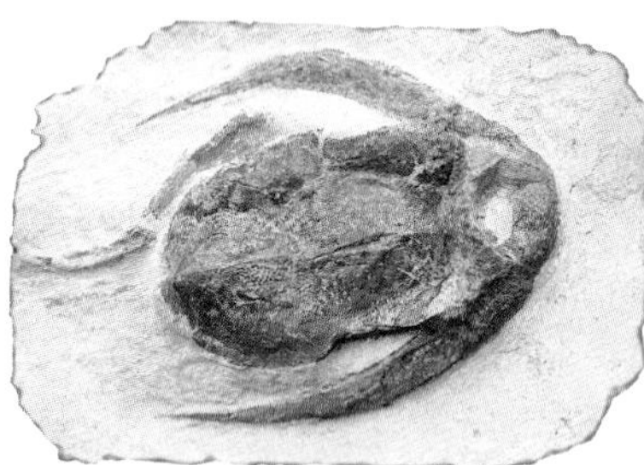

Figure 33-4 **Evolution of fishes**

This simplified phylogenetic tree shows the relationship of the early fishes to the fishes and amphibians that evolved from them.

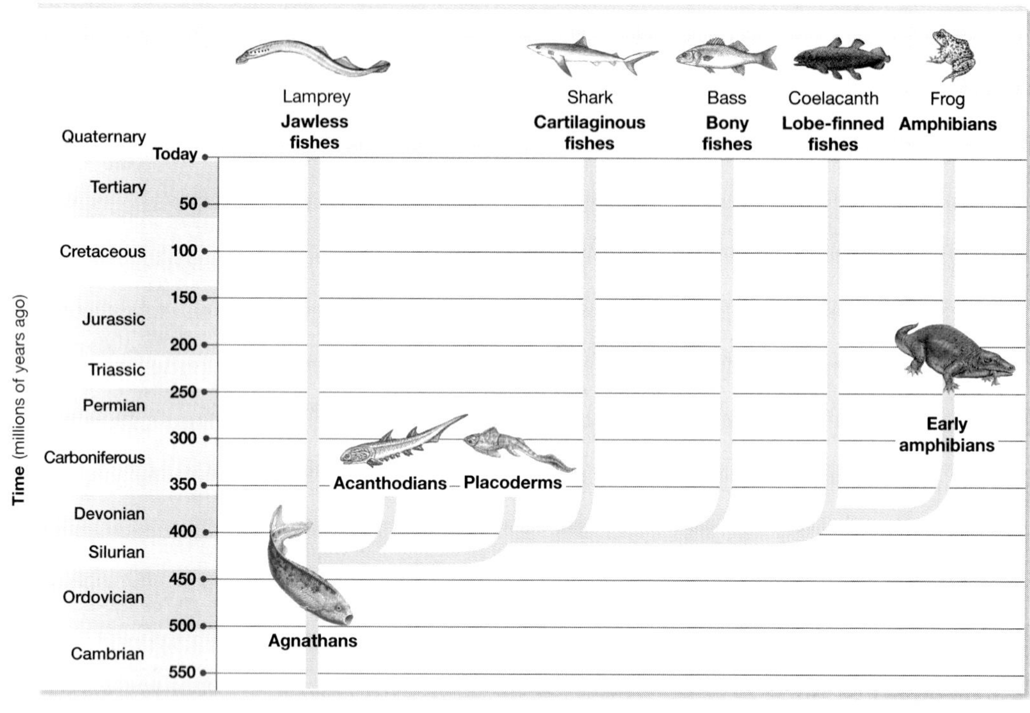

Biology in Your Community

Fisheries Biologist

Invite a local fisheries biologist, aquarium director, or pet shop owner to visit your class. Ask them to discuss the value, care, and problems with local native fish or aquarium fish. If possible, have the guest speaker bring some live fish.

Fishes Evolved Jaws and Paired Fins

To survive as predators in the water, fishes must have adaptations that meet two important challenges. The first challenge is that of pursuing prey through the water. Fishes responded to this challenge by becoming streamlined and flattened sideways, which made it easier for them to move through the water. They also evolved paired fins supported by spines, which provided finer control of movement.

The second challenge is that of grasping prey once it is within reach. About 430 million years ago, the **acanthodians** *(uh KAN thoh dee uhns),* or spiny fishes, appeared. Acanthodians had strong jaws with jagged bony edges that served as teeth, enabling them to hold onto prey. The development of jaws in fishes was a key evolutionary innovation. As shown in **Figure 33-5,** jaws are thought to have developed from gill arch supports made of **cartilage,** *(KAHRT lihj)* a lightweight, strong, flexible tissue.

The spiny fishes had internal skeletons of cartilage, although some fossils indicate that their skeletons also contained some bone. Their scales also contained small plates of bone. The presence of bone in the spiny fishes foreshadows the much larger role that bone would play in their descendants.

About 20 million years after the acanthodians appeared, the placoderms evolved. Placoderms *(PLAK uh durms)* were jawed fishes with massive heads armored with bony plates. By the end of the Devonian period, almost all of the early fishes, including the placoderms, had disappeared. After dominating the seas for almost 50 million years, they were replaced by swifter swimmers—the sharks and bony fishes.

Study TIP

Reading Effectively

As you read about the history of fishes and amphibians, use Figure 33-4 as a guide to the relationships between these organisms. For example, notice how the placoderms are related to the early amphibians.

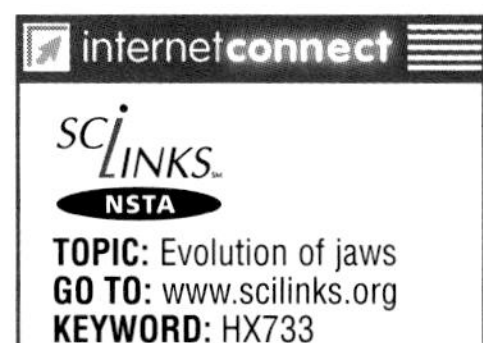

Figure 33-5 Evolution of jaws

The development of jaws was a key evolutionary innovation.

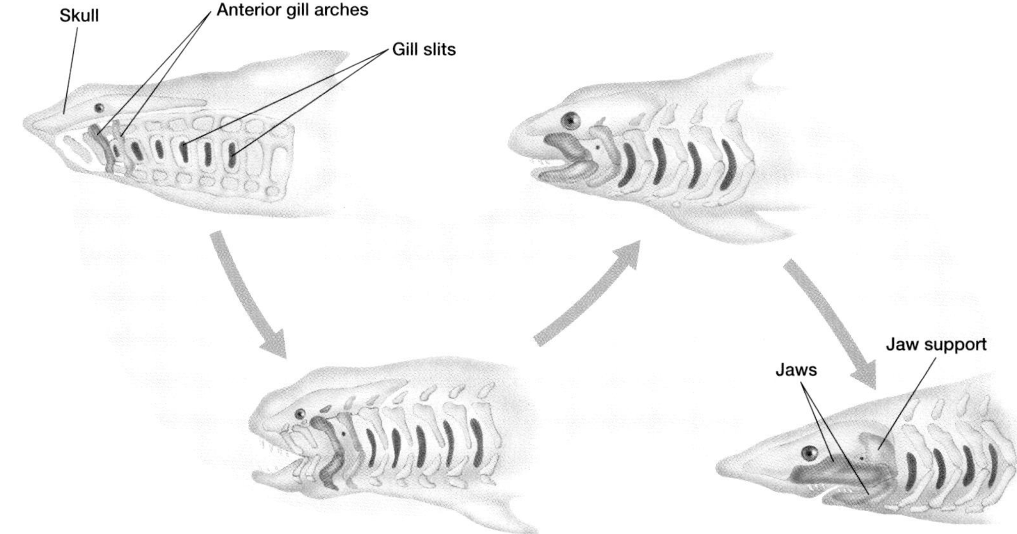

Teaching TIP

Fish Success

Ask students what they think contributes to the success of fishes. Write down their ideas. Ask: Why might a pond be full of perch, yet have no salmon, trout, sharks, or tropical fish? *(Answers should include adaptations such as size, ability to live in fresh water, ability to withstand low temperatures, and ability to get oxygen from oxygen-poor water.)* Have students apply the same thinking process to a new situation. Ask: Why might there be a large number of sharks in an area of the ocean, but no perch, salmon, trout, or tropical fish? *(Answers should include availability of food and adaptations that give sharks better access to food sources.)*

Teaching TIP

Sharks Are Not Boneheads

Remind students about the characteristics of cartilage and bone by having them find examples of each kind of material on their own head. Ask: Why is it advantageous that the end of the nose is cartilage instead of bone? *(It would break easily if it were made of bone.)* Why is the skull made of bone? *(The brain is delicate and needs a hard, protective covering.)* Why does a cartilaginous skeleton give sharks a maneuverability advantage? *(It is more flexible and lighter than a bony skeleton.)*

Using timelines and phylogenetic trees

Time 10–20 minutes

Process Skills Analyzing, interpreting

Teaching Strategies

Have students work in pairs, with one student analyzing Figure 33-4 on page 732 and the other analyzing the history timeline. After answering questions 1–3 using one of the graphics, students should double-check their answers against the other graphic. They should then compare their answers.

Analysis Answers

1. When the jawed fishes appeared, plants, arthropods, and fungi were invading the land.
2. The Ordovician period began about 500 million years ago and ended about 440 million years ago. The Silurian period began about 440 million years ago and ended about 410 million years ago.
3. Of the fishes living during the Devonian period, only the jawless fishes, cartilaginous fishes, bony fishes, and lobe-finned fishes have descendants living today.
4. Answers will vary. Some students may say a timeline would be more useful, because it would show when various musical groups appeared on the music scene. Others may say a tree would be more useful because it would show when certain related events occurred, such as when the lead singer of a band went solo, but the rest of the band continued on its own.

Sharks and bony fishes

By the end of the Devonian period, almost all of the early fishes had disappeared. At about the same time (400 million years ago), sharks and bony fishes appeared. Scientists hypothesize that the main reason sharks and bony fishes replaced earlier fishes is that they have a superior design for swimming. Most sharks and bony fishes have streamlined bodies that are well adapted for rapid movement through the water. Both have an assortment of movable fins that greatly aid their swimming. However, their evolutionary pathways differ. Sharks, such as the extinct *Hybodus,* shown in **Figure 33-6,** have a skeleton made of cartilage strengthened by the mineral calcium carbonate (the material oyster shells are made of). The calcium carbonate is deposited in the outer layers of the cartilage. A thin layer of bone covers this reinforced cartilage. The result is a very light but strong skeleton. Bony fishes, on the other hand, evolved a skeleton made of bone, which is heavier and less flexible than cartilage. But the bony fishes easily compensate for their increased weight. As you will see in Chapter 34, bony fishes possess a swim bladder. This gas-filled sac buoys them up in the water, just as an air-filled balloon buoys up a swimmer.

Figure 33-6 ***Hybodus.*** Early sharks, such as *Hybodus,* were among the first vertebrates to have jaws.

Using timelines and phylogenetic trees

Background

The timelines on pages 258–268 of Chapter 12 and inside the back cover of your textbook show the history of life on Earth, while the phylogenetic tree in Figure 33-4 shows how organisms are related to one another. Use these two graphic representations to answer the analysis questions. Tell which graphic you used to answer each question.

Analysis

1. **SKILL Interpreting Graphics** When the jawed fishes appeared, what was occurring on land?
2. **SKILL Interpreting Graphics** What are the approximate beginning and ending dates of the Ordovician period and the Silurian period?
3. **SKILL Interpreting Graphics** Of the fishes living during the Devonian period, which groups have descendants living today?
4. **SKILL Analyzing Methods** Imagine that you are giving a presentation on the history of a particular type of music, such as rock and roll. Which format, a timeline or a tree, would better suit your needs? Explain your answer.

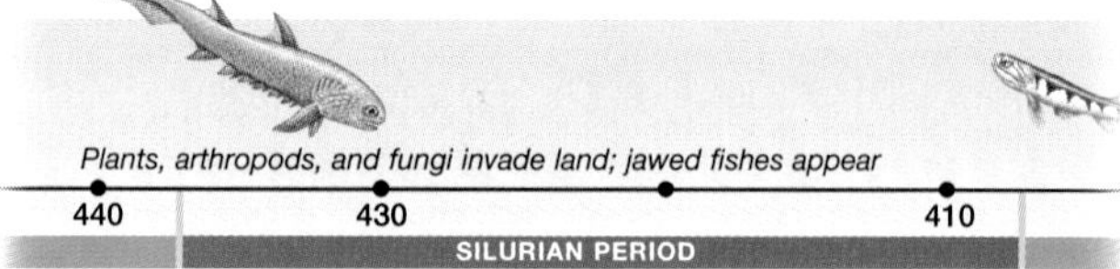

did you know?

Many health-food stores currently sell shark-related products.

These products supposedly give relief from various physical ailments. For example, shark cartilage capsules are advertised to ease the pain of arthritis, and shark-liver oil is advertised to help strengthen the immune and lymphatic systems.

Today, Fish Are the Largest Group of Vertebrates

Today there are more fishes than any other group of vertebrates, both in terms of numbers of individuals and numbers of species. If you consider that water covers three-fourths of Earth's surface and that fishes live in virtually every aquatic habitat, this fact is not surprising. Today's fishes belong to one of three major groups: the agnathans, the cartilaginous fishes *(KART'l aj uh nuhs),* and the bony fishes. The agnathans (hagfishes and the lampreys) are descendants of the early jawless fishes.

The first cartilaginous fishes (sharks) and the bony fishes evolved at about the same time, 400 million years ago. These two groups of fishes most likely evolved from the same early, jawless fishes that gave rise to the acanthodians and the placoderms. The shark's relatives, the skates and rays, evolved later.

Bony fishes make up about 95 percent of today's fish species. Because bony fishes have adapted to so many different environments, they vary greatly in size, color, and shape. **Figure 33-7** shows three examples of bony fishes.

REAL LIFE

Why eat fish? Several bony fishes, including sardines and salmon, produce a fatty acid called omega-3. Consuming this fatty acid appears to protect people against heart disease, arthritis, and some cancers.

Analyzing Information
Find an inexpensive way to add omega-3 fatty acids to your diet.

Figure 33-7 **Bony fishes**

Bony fishes have adapted to a wide variety of habitats, producing great variation in their appearance.

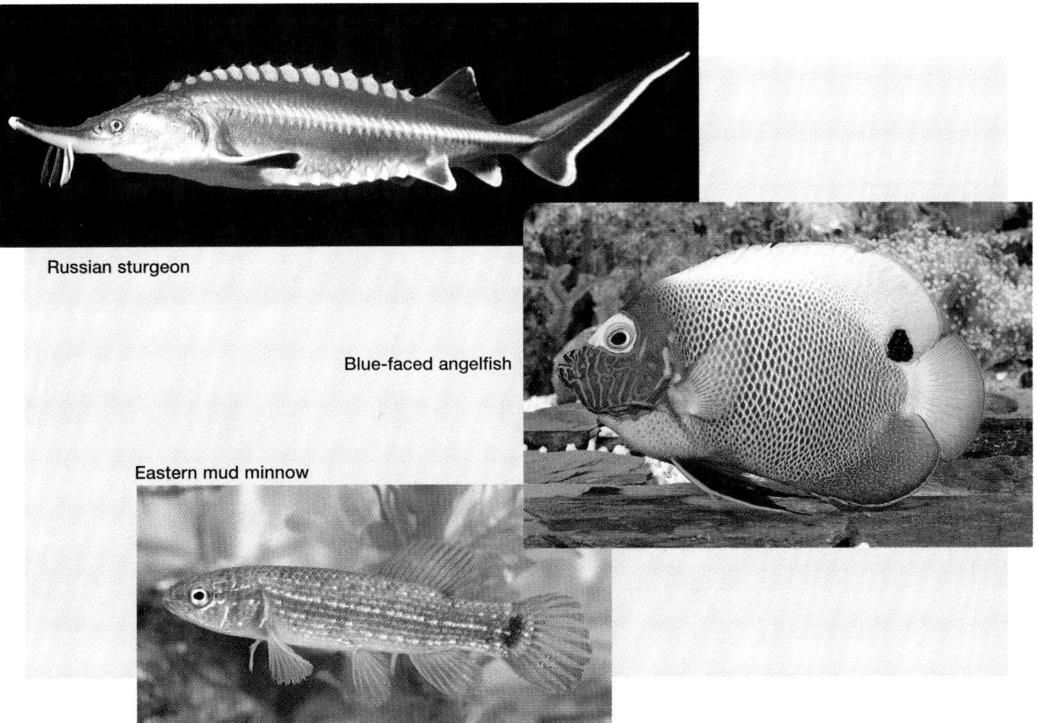

Making Biology Relevant

Health

The American Heart Association recommends eating fish regularly, because it is a good source of protein and does not contain the high levels of saturated fat often found in meat products. However, until further research is done, the AHA does *not* recommend the general use of fish-oil supplements, such as those that are claimed to lower blood-cholesterol levels.

Demonstration

Breathing Lessons

Fill a 1 L bottle with water and set it beside another 1 L bottle full of air. Ask: Which bottle is heavier? Which bottle contains more oxygen? *(The bottle of water is heavier. The bottle of air contains more oxygen.)* Explain that air contains about 21 percent oxygen, whereas water contains less than 1 percent. Stagnant, warm water contains even less oxygen. If fishes obtain oxygen only from the water that passes through their gills, they can suffocate in stagnant water. If a goldfish is kept in a fish bowl without an air pump, it will come to the surface to gulp air because the warm, still water in the bowl is so oxygen-poor.

REAL LIFE

Answer

Eat more vegetable oils which also contain omega-3 fatty acid.

Overcoming Misconceptions

Most Sharks Are Docile

Many people are afraid of sharks, believing all sharks to be man-eating killers. This belief was heightened for many after the release of the 1975 movie *Jaws,* which depicted an out-of-control great white shark terrorizing a small coastal community. In reality, most sharks are docile creatures, and many—such as the huge basking shark, which can be up to 14 m (45 ft) long—feed on plankton.

Chapter 33

Teaching TIP

Ocean Emigration

Split your class into small groups of two or three students, and have them discuss their answers to the following questions: What conditions might have caused vertebrates to leave the ocean and invade the land? What was unfavorable about an aquatic environment or favorable about a terrestrial one? Then have everyone in the class list the advantages of a terrestrial environment. *(new plant and animal food sources, less competition, fewer predators, and more room)*

Visual Strategy

Figure 33-8 ADVANCED

Have students examine the colored bones in the drawing of the lobe-finned fish in Figure 33-8. Tell them the name of each bone: femur—red; tibia—blue; fibula—purple; humerus—yellow; ulna—orange; radius—green. Have students find the corresponding bones in the drawing of the early amphibian. Then ask them to locate these bones on their own bodies. *(femur—thigh; tibia and fibula—lower leg; humerus—upper arm; ulna and radius—forearm.)*

Study TIP

Reading Effectively

Knowing the sequence of the periods in Earth's history will help you place events in time. Use the phylogenetic tree in Figure 33-4, on page 732, as a reference as you read.

The First Vertebrates to Venture onto Land Were Amphibians

The first group of vertebrates to live on land were the amphibians, which appeared about 370 million years ago. Amphibians probably evolved from a group of bony fishes called lobe-finned fishes. As shown in **Figure 33-8,** the pattern of bones in an amphibian's limbs bears a strong resemblance to that of a lobe-finned fish. While several species of lobe-finned fishes exist today, the species that amphibians evolved from are extinct.

The "Age of Amphibians"

Although amphibians first appeared in the Devonian period, they increased greatly in numbers during the Carboniferous period. During this time, which begins what biologists call the "age of amphibians," the number of amphibian families increased from 14 to about 34. By the late Carboniferous period, much of what was to become North America was covered by low-lying tropical swamplands. The amphibians thrived in this moist environment, sharing it with early reptiles.

In the Permian period that followed, amphibians reached their greatest diversity, increasing to 40 families. In the early Permian period, a remarkable change occurred among amphibians—many of them began to leave the marshes for dry uplands. By the middle Permian, 60 percent of all amphibian species were living in dry environments. *Eryops,* illustrated in **Figure 33-9,** was typical of amphibians of this period and was well adapted for life on land.

Figure 33-8 From fin to limb

As shown by the colors below, the pattern of bones in an early amphibian's limbs bears a distinct resemblance to that of the fin bones of lobe-finned fish.

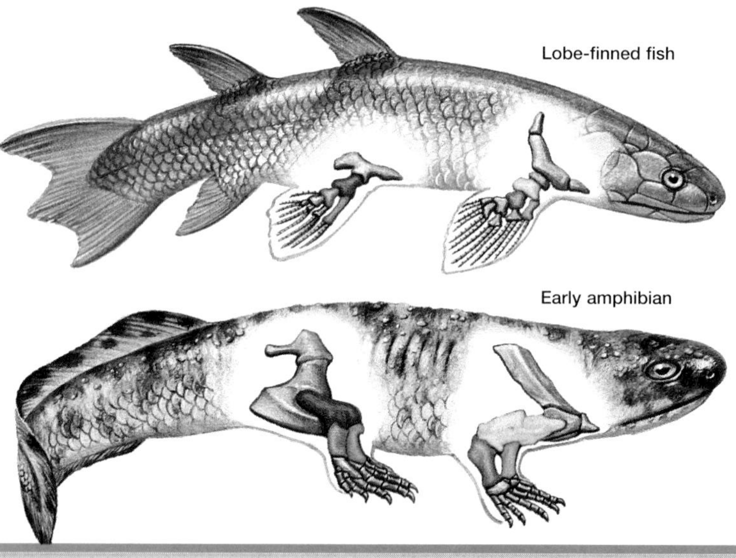

Biology in Your Community

Using a Field Guide

From a bookstore or library, obtain a field guide to amphibians of your region. Have students go through the range maps for all of the amphibians and find which ones live in your area. Make a list on the board of the students' findings.

Innovations for life on land

Life on land is quite different from life in the water. Thus, a number of major innovations were necessary for a species to successfully invade land.

1. **Legs.** Legs evolved to support body weight as well as to allow movement from place to place.
2. **Lungs.** The delicate structure of a fish's gills depends on the buoyancy of water for support. Thus, lungs evolved for gas exchange on land.
3. **Heart.** Walking around on land requires a greater expenditure of energy. Thus, a higher metabolism, which in turn uses greater amounts of oxygen, is necessary. The structure of the vertebrate heart evolved so that oxygen could be delivered to the body more efficiently.

Amphibians evolved such innovations, but they did not fully adapt to life on land. For example, amphibian eggs are not watertight, and modern amphibians must seek out water or damp areas in which to reproduce.

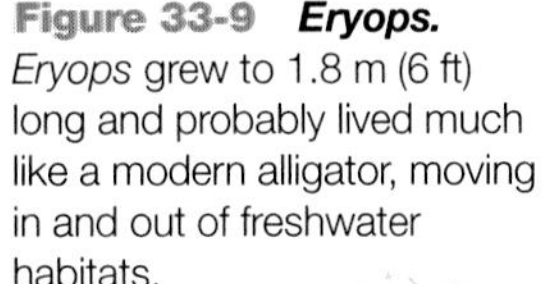

Figure 33-9 ***Eryops.*** *Eryops* grew to 1.8 m (6 ft) long and probably lived much like a modern alligator, moving in and out of freshwater habitats.

MATH LAB: *Comparing the surface area of gills*

Background

Air contains more oxygen than water, so why do fish die when removed from water? To understand what happens to gills when they are removed from water, follow the procedure below.

Materials

cellophane wrap, scissors, toothpick, ruler, container of water

6 cm

7 vertical cuts

Procedure

1. Cut a 8 x 6 cm piece of cellophane wrap. Use the piece of wrap and a toothpick to make a model of fish gills, as shown in the drawing.
2. Measure the approximate length and width of the model. Calculate the area of the model by using the formula $a = l \times w$.
3. Submerge the model in the water and allow it to float to the top. Notice any change in shape of the model.
4. Grasp the model by the toothpick and pull it gently through the water. Then remove the model and place it on the table.
5. Without rearranging the model in any way, measure its approximate length and width. Calculate the area of the model again.

Analysis

1. **Summarize** any difference you observed between the gill model in the water and the wet model out of the water.
2. **Compare** the area obtained in step 2 with that obtained in step 5. If the areas are different, identify which was larger.
3. **SKILL Analyzing Data** Consider what you know about the requirements for gas exchange across a gill's surface. Do the data you obtained suggest a reason why fishes cannot live out of water? Explain your answer.
4. **SKILL Comparing Structures** Your model is two-dimensional. To calculate the surface area of an actual gill, you would need to know another measurement. What is that measurement, and why is it important?

Graphic Organizer

*Use this graphic organizer with **Teaching Tip: Classification** on page 737.*

Name of Group	Characteristic	Example
Agnathans	(No jaws)	(Lamprey)
Cartilaginous fishes	(Cartilage skeleton)	(Shark)
(Bony fishes)	Bony skeleton	(Trout)
(Amphibians)	Soft, moist skin	(Frog)

MATH LAB: *Comparing the surface area of gills*

Time 10–15 minutes

Process Skills Comparing, calculating, interpreting results, applying information

Teaching Strategies

Before students begin, ask them how they would measure the classroom floor to order new floor covering. *(Multiply the length by the width.)* Ask students what units they would use in placing their order. *(square feet or square yards)* Emphasize that surfaces are always measured in units squared, and that their answers always need to indicate this. Ask students to think about how they would measure the room to fill it with water, but do not elicit answers.

Analysis Answers

1. The gill model in the water will appear larger and flatter.
2. The area calculated in step 2 will be larger than the area calculated in step 5.
3. The data suggest that when a fish is out of water, the surface area of its gills decreases. Because the surface area decreases, there is insufficient gas exchange, and the fish will die unless returned to the water.
4. One would need to know the thickness of the gill to determine the surface area of an actual gill. This is important because this third dimension increases the surface area of the gill.

Teaching TIP

Classification

Copy the **Graphic Organizer** at the bottom of page 737 onto the board, but do not copy the words in parentheses. Ask a volunteer to fill in one blank and to give an example of an animal that would fit in the same row. Continue until all blanks are filled in. Characteristics and examples may vary, as there are many correct responses.

Chapter 33

Visual Strategy

Figure 33-10

Have your class examine the photos of amphibians in Figure 33-10. Assign students to groups and ask them to think about why all four animals are classified as amphibians. *(Answers will vary depending on students' prior knowledge of amphibian characteristics.)* Then ask them which two amphibians are most closely related and why. *(The bullfrog and the toad—both have elongated hind legs and no tail.)*

Teaching TIP

Aquatic Amphibians

Point out that although many amphibians are both aquatic and terrestrial, some, such as the newt, are mostly aquatic. Other amphibians, such as toads, are mostly terrestrial, returning to water only to reproduce.

3 Assessment Options

Closure

Fishes and Amphibians

Have students make a two-column chart with the headings *Fishes* and *Amphibians.* For each column, have students enter information about the history of that group, starting with the earliest data. For example, the first entry under *Fishes* might be: *Jawless fishes appeared about 500 mya.* PORTFOLIO

Review

Assign Review—Section 33-1.

Reteaching — BASIC

Have students review the six major concept headings in Section 33-1. Then ask them to write a question for each heading that can be answered with the heading itself. *(For example: What are vertebrates? Vertebrates are chordates with a backbone.)* PORTFOLIO

Today, There Are Three Groups of Amphibians

The middle Permian period marked the peak of amphibian success. By the end of this period a new kind of vertebrate, a reptile called a therapsid, had become common and began to replace the amphibians. By the end of the Triassic period that followed, there were only 15 families of amphibians, including the first species of frogs. And by the Jurassic period, only two groups of amphibians remained—frogs and salamanders. The age of amphibians was over.

All of today's amphibians are descendants of the two kinds of amphibians that survived into the Jurassic period. They are found in aquatic and moist habitats throughout the temperate and tropical regions of the world. Frogs and toads make up the largest, and probably the most familiar, group of modern amphibians. Salamanders and newts are far fewer in number, while the caecilians account for less than 1 percent of today's amphibian species. **Figure 33-10** shows a representative sample of today's amphibians.

Figure 33-10 Amphibians

Today's amphibians are all descendants of one of the three amphibian groups that survived into the Jurassic period.

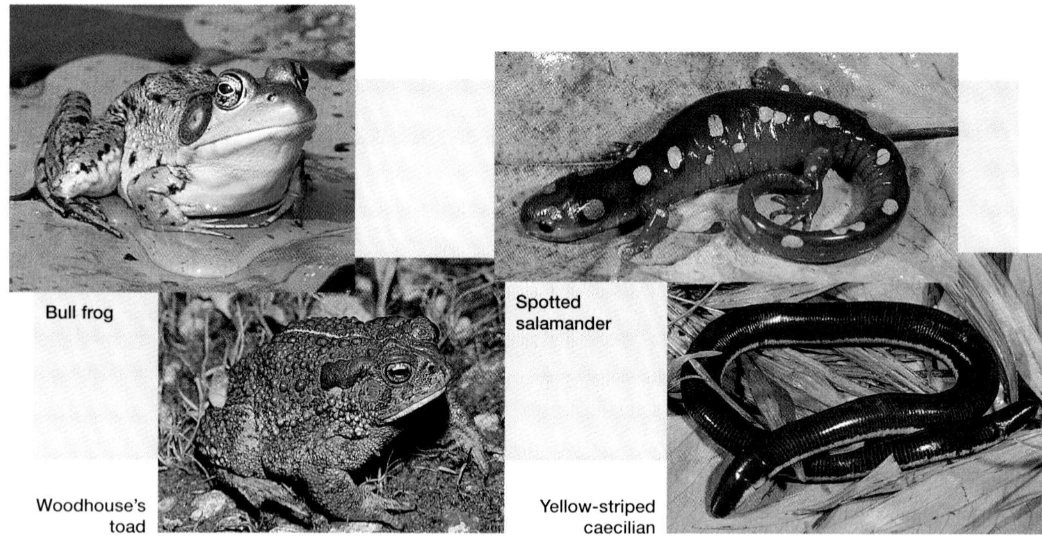

Review SECTION 33-1

1. **Describe** two characteristics that distinguish vertebrates from invertebrates.
2. **Discuss** two adaptations that enabled early fishes to dominate the oceans.
3. **Relate** the structure of the limbs of a lobe-finned fish to the evolution of amphibians.

Critical Thinking

4. **SKILL Evaluate** this statement: Amphibians are not fully adapted for life on land.

Review SECTION 33-1 ANSWERS

1. Vertebrates have a backbone made of individual vertebrae and a dorsal nerve cord. (Also accept a bony skull that encases and protects the brain.)
2. Early fishes had strong, biting jaws with primitive teeth. Streamlined bodies with paired fins enabled early fishes to move more efficiently when catching prey and escaping predators.
3. The pattern of bones in an early amphibian's limbs bears a distinct resemblance to that of the fin bones of a lobe-finned fish.
4. The statement is true. All amphibians are still dependent on water for reproduction. Because their eggs are not waterproof, amphibians must rely on an external water source to keep the eggs moist. If an egg dries out, the embryo within it will die.

Vertebrates Adapt to Terrestrial Living

Section 33-2

Section 33-2

Reptiles Became the First Fully Terrestrial Vertebrates

To understand how dangerous water loss can be, all you need to do is stay outside on a hot day. It won't be long before you are sweating and thinking about a cool drink. If you don't increase your fluid intake and cool off, you may start to feel dizzy and nauseous. Fluid loss is a problem for all **terrestrial** animals, animals with the ability to live on land. The adaptations that permitted amphibians to live on land were further developed in reptiles. However, the two most significant adaptations were the evolution of both a skin and an egg that were almost watertight.

When reptiles first evolved, about 320 million years ago, Earth was entering a long, dry period. Early reptiles were better suited to these conditions than the amphibians were, and they quickly diversified. Within 50 million years, reptiles had replaced amphibians as the dominant terrestrial vertebrates. **Figure 33-11** shows the evolutionary relationships of terrestrial vertebrates.

Objectives

- **Summarize** why dinosaurs became the dominant land vertebrate. (pp. 739–740)
- **Contrast** ectotherms with endotherms. (p. 742)
- **Summarize** why mammals replaced dinosaurs. (p. 744)
- **Identify** the dinosaurlike and the birdlike features of *Archaeopteryx*. (p. 747)

Key Terms

terrestrial
thecodont
Pangaea
ectothermic
endothermic
therapsid

Figure 33-11 **Evolution of terrestrial vertebrates**

This simplified phylogenetic tree shows the relationship of early reptiles to the diversity of vertebrates that evolved from them.

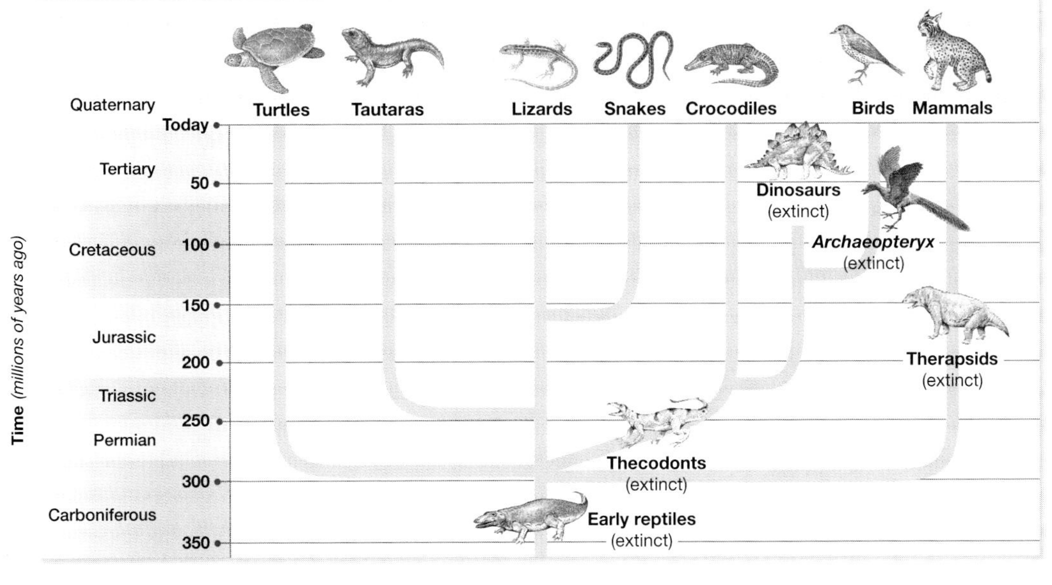

1 Prepare

Scheduling

- **Block:** About 90 minutes will be needed to complete this section.
- **Traditional:** About two class periods will be needed to complete this section.
- **Planning:** See the Planning Guide on pp. 727A and 727B for lecture, classwork, and assignment options for Section 33-2. Lesson plans for Section 33-2 can be found in a lesson cycle format in ***Biology: Principles and Explorations* One-Stop Planner CD-ROM with Test Generator.**

Resource Link

Assign **Active Reading Worksheet Section 33-2** to help students comprehend the key concepts and visuals in this lesson.

2 Teach

Lesson Warm-up

Ask students if any of them have pet reptiles. Then have your class read the **Real Life** on page 742. Ask the students who have pet reptiles to describe the pros and cons of owning their particular reptile. Ask these students what advice they would have for other students who might be interested in owning that type of reptile.

Integrating Different Learning Styles

Logical/Mathematical	**PE**	Interactive Exploration, pp. 752–753
Visual/Spatial	**PE**	Study Tip, p. 743; Performance Zone items 1, 16, 17, 18, and 19, pp. 750–751
	ATE	Teaching Tips, pp. 740, 741, 743, 747; Visual Strategy, p. 741
Interpersonal	**PE**	Real Life, p. 742; Performance Zone item 20, p. 751
Intrapersonal	**ATE**	Teaching Tip, p. 744
Verbal/Linguistic	**PE**	Performance Zone items 2 and 21, pp. 750–751
	ATE	Teaching Tips, pp. 740, 746; Active Readings, pp. 744, 747

Teaching TIP

Letter of Recommendation

Have your class pretend that your school is preparing a desert exhibit and is deciding what reptiles or amphibians to place in the exhibit. Ask each student to pick an animal to recommend for the exhibit and to describe how well the animal is adapted for a warm, dry environment. Have your class share some of the characteristics they believe would make an animal a good candidate for the exhibit. *(Even though many students will describe different reptiles, they should recommend similar traits, such as watertight eggs, watertight skin, feeding behaviors, and improved locomotion. Some students may recommend amphibians, such as desert toads. If so, they must tell how the amphibian is adapted for life in a desert.)*

Dinosaur Video Comparison

Obtain a video clip from an old movie that shows dinosaurs. Also get a clip from a recent videotape on dinosaurs. Ask students to compare the two depictions of dinosaurs. *(For example, the dinosaurs in each version will move quite differently.)* Point out that these differences are the result of new scientific discoveries based on fossil evidence and technological advances, such as 3-D computer analyses.

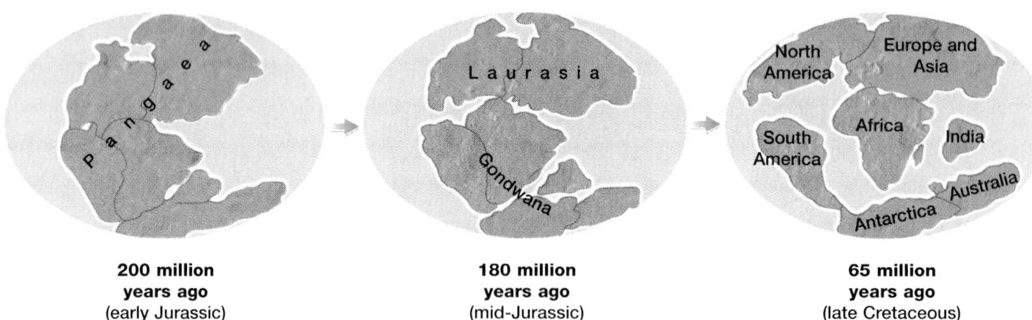

200 million years ago (early Jurassic) | **180 million years ago** (mid-Jurassic) | **65 million years ago** (late Cretaceous)

Figure 33-12 Breakup of Pangaea. The supercontinent, Pangaea, began to break up during the Jurassic period. By the late Cretaceous period today's continents began to be recognizable.

Dinosaurs dominate the land

For roughly 150 million years, dinosaurs dominated life on land. They evolved from the **thecodonts,** an extinct group of crocodile-like reptiles. During their long history, dinosaurs changed a great deal, in part because the world they inhabited changed. Thus, one cannot study dinosaurs as if they were one particular kind of animal. Rather, you have to look at dinosaurs more as a story, a long parade of change and adaptation—animals that lived at different times and were adapted to very different worlds.

One factor that affected dinosaur evolution was the movement of the continents, shown in **Figure 33-12,** which radically altered the Earth's climate. When the dinosaurs first appeared, all of the continents were joined in a single supercontinent called **Pangaea** (*pan GEE uh*). There were few mountain ranges over this enormous stretch of land, and the interior was dry. Coastal climates were much the same all over the world—quite warm, with a dry season followed by a very wet rainy season. As Pangaea broke apart, the climates of the various land masses varied. Some species of dinosaurs could not adapt and became extinct, while new kinds flourished.

Triassic dinosaurs

The oldest known dinosaur fossils are in rocks from the early Triassic period, some 235 million years old. The first known dinosaur, *Eoraptor,* illustrated in **Figure 33-13,** was a 30 cm (1 ft) long bipedal (two-footed) carnivore. By the end of the Triassic period some 22 million years later, small, carnivorous dinosaurs were very common and had largely replaced the thecodonts.

There are at least three reasons why dinosaurs were so successful.

1. **Leg structure.** Legs positioned directly under the body provided good support for the dinosaur's body weight, enabling dinosaurs to be faster and more agile runners than the thecodonts. They were better able to catch prey and escape from enemies.
2. **Drought resistance.** Dinosaurs were well adapted to dry conditions found in Pangaea in the late Triassic period.

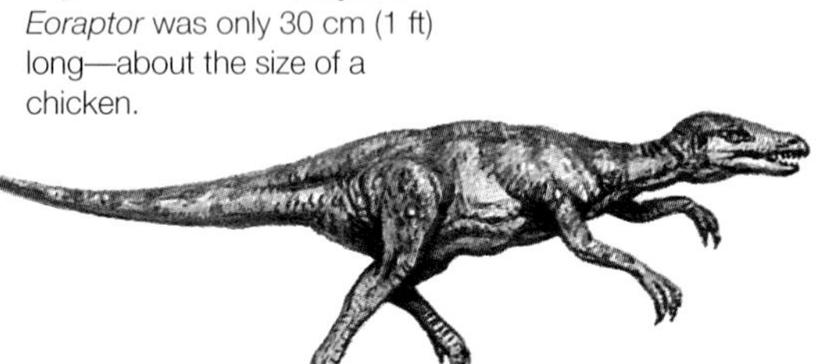

Figure 33-13 ***Eoraptor.*** *Eoraptor* was only 30 cm (1 ft) long—about the size of a chicken.

Biology in Your Community

Herpetologist

Invite a local herpetologist or reptile enthusiast to your class to show and discuss both exotic and local reptiles. The day before your guest visits, have each student make a list of five questions they would like to ask about reptiles. Ask your guest to bring turtles, lizards, and nonvenomous snakes, if possible. Encourage students to handle docile snakes to see that they are not "slimy." Remind students that reptiles can carry diseases, such as salmonellosis. Instruct all students who have handled the visitor's reptiles to wash their hands with soap and water.

3. **Extinction of other animal species.** At the end of the Triassic period a large meteorite landed in northeastern Canada. (Its impact site, the Manicuoagan Crater, is still visible today.) This event might have been responsible for the great loss of diversity that occurred at the end of the Triassic period. Thecodonts and many other species became extinct, but the dinosaurs survived.

Figure 33-14 Theropod. Like most raptors, the theropod *Velociraptor* was about 1.8 m (6 ft) long, about the size of a large human.

The Jurassic Period—Age of Dinosaurs

The Jurassic period is considered the golden age of dinosaurs because of the variety and abundance of dinosaurs that lived during this time. This included the largest land animals of all time, the sauropods *(SAWR oh pawdz),* which had enormous barrel-shaped bodies, heavy column-like legs, and very long necks and tails. Sauropods were the dominant herbivores of the Jurassic period.

By the late Jurassic period, a new type of dinosaur had evolved. These dinosaurs, the carnivorous theropods (*THEHR uh pawdz*), stood on two powerful legs and had short arms. Their large heads were equipped with sharp teeth, and each foot had sickle-shaped claws used for ripping open prey. This anatomy was well suited for rapid running and quick, slashing attacks. Theropods preyed on the large herbivorous dinosaurs and were the dominant terrestrial predators until the end of the Cretaceous period. **Figure 33-14** shows a representative theropod, and **Figure 33-15** shows a representative sauropod.

Figure 33-15 Sauropod

Size The sauropod *Brachiosaurus* stood 12.5 m (41 ft) tall and reached up to 23 m (75 ft) in length. It weighed 81,000 kg (89 tons), more than 14 African elephants or 1,500 students.

internetconnect

SciLINKS. NSTA

TOPIC: Comparing dinosaurs
GO TO: www.scilinks.org
KEYWORD: HX741

Teaching TIP

Dinosaur Adaptations

Obtain a set of high-quality plastic dinosaurs, such as those available through some museum shops. Divide the class into groups of three or four students, and give each group two or three dinosaur models, along with the names of the dinosaurs. Each group should appoint a recorder. The recorder should write the name of each dinosaur model at the top of a column on a sheet of paper. Each group should examine each of its models and determine how the dinosaur it represents was adapted for living. *(For example:* Brachiosaurus *—long neck for reaching tall vegetation;* Velociraptor*—sharp claws for slashing prey)* After 10–15 minutes, have each group present their models and speculate about each dinosaur's adaptations. Allow time for other students to ask questions.

Visual Strategy

Figure 33-15

Have students compare the *Brachiosaurus* shown in Figure 33-15 with the reconstructed *Brachiosaurus* skeleton shown in the photo on page 729. Tell them that the skeleton is the world's largest mounted dinosaur skeleton and is on display at the Field Museum in Chicago. Explain that it was excavated between 1899 and 1901 in Colorado by Elmer Riggs, a curator at the Field Museum.

Teaching TIP

Thermoregulation

Tell students that because reptiles are ectothermic they must avoid temperature extremes. Ask students how reptiles cope with hot climates and cold winters. *(Some reptiles that live in hot climates remain in the shade or come out only at night; some have a higher heat tolerance than mammals; and some that live in regions that have harsh winters hibernate.)*

CROSS-DISCIPLINARY CONNECTION

◆ Earth Science

Scientists have proposed that a nuclear war would result in disastrous climate changes. The explosion of a substantial number of nuclear bombs would cause billions of tons of dust and ash to enter the atmosphere. The dust and ash particles would act like a shield, blocking out so much of the sun's rays that darkness and below-freezing temperatures could last a year or longer. This scenario is called "nuclear winter." Discuss with students the similarities between nuclear winter and the hypothesis that suggests that the Cretaceous extinction was caused by the impact of a gigantic meteorite or asteroid.

REAL LIFE

Answer

Prospective pet owners should be aware of a reptile's needs for food, water, heat, shelter, and space.

REAL LIFE

Reptiles make fascinating pets! But there are drawbacks to their increasing popularity. Some reptiles are dangerous and difficult to care for, and many carry harmful bacteria in their intestinal tracts. Often reptiles offered for sale have been removed from their natural environment, causing a decline in their native populations.

Forming Reasoned Opinions *Evaluate the pros and cons of owning a reptile. What advice would you give to someone who wants a pet reptile?*

internetconnect

SciLINKS NSTA

TOPIC: Dinosaur extinction
GO TO: www.scilinks.org
KEYWORD: HX742

Were dinosaurs warmblooded?

Ectothermic animals, such as today's living reptiles, have a metabolism too slow to produce enough heat to warm their bodies. Such animals must absorb heat from their environment, and their body temperature changes as the temperature of their environment changes. Ectotherms are sometimes referred to as coldblooded. Other vertebrates, such as mammals and birds, are **endothermic;** that is, they maintain a high, constant body temperature by producing heat internally through a faster metabolism. Endotherms are sometimes called warmblooded.

Were dinosaurs ectotherms or endotherms? For a long time it was assumed that dinosaurs were ectotherms, as modern reptiles are, but this view is now the subject of much debate. New evidence indicates that at least some dinosaurs were warmblooded. The microscopic structure of the bones of some dinosaurs resembles that of modern endotherms, both in growth rate and growth pattern. The chemical analysis of selected dinosaur bones indicates that the bones of some Cretaceous dinosaurs probably formed under endothermic conditions. But there is no conclusive proof, and a lively debate will probably continue for some time.

The Cretaceous extinction

Toward the end of the Cretaceous period, sea levels began to fall and the climate began to cool. Many kinds of dinosaurs began to decrease in numbers. Then suddenly, 65 million years ago, nearly all dinosaurs abruptly disappear from the fossil record. No one knows for certain why this happened. Most scientists now agree that the major contributing cause was the impact on Earth of a gigantic meteorite or asteroid 8–16 km (5–10 mi) in diameter. There is physical evidence that a large object hit Earth at the end of the Cretaceous period. A 200-mile-wide crater off the coast of Mexico's Yucatán peninsula is very likely the site where a meteorite or asteroid collided with Earth approximately 65 million years ago. In addition, a thin line of sediment marks the end of the Cretaceous period in rocks throughout the world. This sediment is rich in iridium, a mineral rare in Earth's crust but common in meteorites.

Scientists believe that such an impact would have thrown massive amounts of material into the atmosphere, blocking out all sunlight for a considerable period of time and creating a prolonged worldwide cold period. The endothermic birds and mammals, insulated with feathers and fur, survived. The smaller ectothermic reptiles and amphibians also survived because they could lower their activity levels. But the dinosaurs did not survive.

Scientists cannot be certain why the dinosaurs did not survive this period of intense cold. Disease or sudden competition from mammals might have led to their extinction. However, if most Cretaceous dinosaurs were warmblooded, as scientists now think they were, the dinosaurs would not have been able to survive the cold. They lacked the insulation of birds and mammals and could not lower their activity level as the reptiles and amphibians could.

Overcoming Misconceptions

Dinosaurs Were Successful Land Vertebrates

Dinosaurs have been portrayed as large, stupid, sluggish creatures that were doomed to extinction. Have students evaluate popular images of dinosaurs versus what scientists now think to be true about them. Remind students that during the time when dinosaurs and mammals coexisted, dinosaurs were able to diversify and dominate the Earth, whereas mammals remained small and mainly nocturnal. It was not until the extinction of dinosaurs that mammals began to flourish. Also remind students that dinosaurs dominated the Earth for about 150 million years. Point out that extinction has been the rule rather than the exception throughout history.

Today, Only Four Groups of Reptiles Remain

Of the 16 orders of reptiles known to have existed, only four remain today. Representatives of these orders are shown in **Figure 33-16.** The turtles have the most ancient origins and have changed very little in structure since before the time of the dinosaurs. The vast majority of living reptiles belong to the second group to evolve—snakes and lizards. Tuataras belong to the third group of surviving reptiles to evolve.

The fourth line of living reptiles, the crocodiles and their relatives, appeared on the evolutionary scene much later than the first three groups. Like dinosaurs, crocodiles are descendants of the thecodonts and have changed very little in more than 200 million years. In some ways, such as the structure of their heart, crocodiles resemble birds far more than they resemble other living reptiles. And crocodiles are the only living reptiles that care for their young. What does this mean in terms of their relationships to other vertebrate species? Today, many biologists think that birds are direct descendants of the dinosaurs. If this is true, then crocodiles and birds are far more closely related to dinosaurs, and thus to each other, than they are to other living reptiles. This close evolutionary relationship would explain their similarities.

Study TIP

Reading Effectively

Locate the thecodonts in Figure 33-11, on page 739. Notice how the lines leading to the crocodiles and birds stem from the thecodonts. This suggests that today's crocodiles and birds come from a different lineage than the other reptile groups.

Figure 33-16 Reptiles

Of the four living orders of reptiles, the crocodiles and alligators are the youngest, and they differ in several ways from other reptiles.

Teaching TIP

Reptile Zoogeography

Tell students that no alligators or crocodiles are native to the Pacific Northwest or northeastern United States. Ask them why this is true. *(The climate in the winter is too cold for these large ectotherms to survive in these locations.)* Ask students how they think the number of reptile species found in the United States compares with the number found in Canada, and why. *(The United States has many more reptile species than Canada because of the difference in climates.)*

Teaching TIP

Crocodilian Parental Care

Show students a photograph of a female alligator or crocodile near her nest. Tell students that these reptiles lay 20 to 90 eggs and then bury them in sand, mud, or plant debris. Explain that female alligators and the females of some species of crocodiles remain near the nest to protect the eggs while the embryos inside are developing. After the young hatch, the mother may help them to escape from the nest and may even carry them in her mouth to water. Some mother alligators protect their young until they are over a year old.

MULTICULTURAL PERSPECTIVE

The Indian King Cobra

One of the deadliest reptiles, the Indian king cobra is also one of the most fascinating to the people of India. In a tradition that goes back at least 2,000 years, Indians compare areas of energy located throughout the human body, collectively termed kundalini ("coiled one"), to a cobra. They believe that this energy, like its cobra counterpart, must be treated with great respect. A very popular yoga exercise position is called The Cobra.

Active Reading

Technique: Paired Summarizing

Have students read pp. 744–746 silently. Then assign partners and have one student summarize orally what they have read while the other student listens. The listener should point out inaccuracies and add ideas that were left out. Have students alternate roles and repeat this process for each of the three pages. Students should work together during a final clarification process, referring to the text as needed.

Teaching TIP BASIC

Mammalian Insulation

Ask students which coat they would choose for a day of cold-weather activity: a lizard-skin coat or a wool coat. Have them relate their answer to the idea that the impact of a giant meteorite or asteroid contributed to the extinction of dinosaurs and the rise of mammals.

internetconnect

SciLINKS NSTA

TOPIC: Early mammals
GO TO: www.scilinks.org
KEYWORD: HX744

Mammals Replaced the Extinct Dinosaurs

The first mammals appeared about 220 million years ago, just as the dinosaurs were evolving from thecodonts. It is most likely that mammals were descendants of the **therapsids,** an extinct order of reptiles that were probably endotherms. Early mammals, such as *Eozostrodon,* illustrated in **Figure 33-17,** were small (about the size of mice), insect-eating tree dwellers that were active at night.

For 155 million years, while the dinosaurs flourished, mammals were a minor group that changed little. Only five orders of mammals arose in that time, and their fossils are scarce. In the Cretaceous extinction, most animal species larger than a small dog disappeared. The smaller reptiles, including lizards, turtles, crocodiles, and snakes, survived. Mammals and birds also survived.

Just as after earlier mass extinctions, the stage was again set for a great evolutionary play. But the world's climate was no longer dry, as it was after the Triassic extinction. The reptiles' advantages in dry climates that served them well then were no longer so important. Mammals and birds became the dominant vertebrates on land. In the Tertiary period, mammals rapidly diversified, taking over many of the ecological roles once dominated by dinosaurs. Mammals reached their greatest diversity in the late Tertiary period, about 15 million years ago. At that time, tropical conditions existed over much of the world. During the last 15 million years, world climates have changed, and the area covered by tropical habitats has decreased. As a result, the total number of mammalian species has declined.

As you learned in Chapter 14, early modern humans appeared about 34,000 years ago. These humans lived by hunting, and their lives depended on the success of the

Figure 33-17 ***Eozostrodon.*** Only 12 cm (5 in.) in total length, *Eozostrodon* is typical of the small, early nocturnal mammals.

Figure 33-18 Cave painting. This thundering herd of horses was painted on the wall of a cave in Lascaux, in the Dordogne region of France.

hunt—the hunt of other mammals. In caves throughout southern France, northern Spain, and Africa, prehistoric people painted magnificent pictures of the mammals that shared their habitat. **Figure 33-18** shows one of the hundreds of paintings, its colors still vivid. Humans arrived in North America from Asia about 12,000 years ago. Within a thousand years, almost all of the rich array of camels, mastodons, and other large North American animals were extinct. Most scientists think that these extinctions were the direct result of humans hunting.

Today, there are over 4,000 species of mammals, and all very large land animals are mammals. However, the largest land mammals reached their peak during the last ice age (about 2 million to 10,000 years ago). At that time, many species of enormous mammals such as mammoths, giant camels and ground sloths, and the Irish elk, illustrated in **Figure 33-19,** roamed Earth. The Irish elk is somewhat misnamed as it was actually a large species of deer that lived throughout Europe, northern Africa, and northern Asia.

Figure 33-19 Irish elk

Size

The shoulders of Irish elk were over 2.1 m (7 ft) high and the span of the antlers could be up to 3.65 m (12 ft).

Exploring Further

La Brea Tar Pits

Teaching Strategies
Have the class read this feature aloud. If possible, show pictures or slides of saber-toothed cats, giant ground sloths, mammoths, and the dire wolf. Point out that not all of the animals found at La Brea were vertebrates.

Discussion

- Why might large animals, which must have survived many perils to reach adulthood, have stepped into a "tar pit." *(Answers will vary. Some animals may have been chased by predators; others may have tried to eat animals already stuck and become stuck themselves; and some may have gone into the pit at night, when it was difficult for them to see.)*
- Why were so many complete or nearly complete skeletons found in the tar pits? *(Scavengers would have been unable to feed on the trapped animals and scatter their bones. Also, the sticky asphalt would have helped keep the bones together.)*
- How did the increasing dryness of the region affect the survival of the large mammals? *(Many of the plants that the large herbivorous mammals fed on did not survive the arid conditions. The predators of the large herbivores, in turn, lost their food supply.)*

Exploring Further

La Brea Tar Pits

One of the world's richest fossil deposits, Rancho La Brea (the tar ranch), is located within the city of Los Angeles. This area is criss-crossed with cracks and fissures that extend deep into Earth's surface, reaching down to a natural reserve of crude oil. Some of the oil seeps upward through the cracks, finally reaching the surface where it creates shallow pools. As the petroleum within the crude oil evaporates, a pool of soft, sticky asphalt, mistakenly called a "tar pit," remains. Today, there are only a few active tar pits, but at one time there were more than 100.

Stuck in the "tar"
When unwary animals stepped into the gummy pools, they became stuck and eventually buried. At La Brea, the remains of more than 140 species of plants and more than 420 species of animals (both invertebrates and vertebrates) have been found. The vertebrate species include saber-toothed cats, horses, bison, giant camels, giant ground sloths, mammoths, and dire wolf. Their fossilized remains record the variety of life that lived in the Los Angeles basin about 36,000 to 11,000 years ago.

The deposits found at La Brea are a valuable resource for paleontologists and geologists

seeking to understand the ecology and climate of this region at the end of the last ice age. Their studies show that the region became progressively drier and the vegetation changed dramatically. Animals that could not adapt to these changes—including all of the giant mammals—gradually disappeared. Those that did adapt, such as coyotes and skunks, can still be found in the area.

Topic: Rancho La Brea
Go To: www.scilinks.org
Keyword: HX745

Teaching TIP

Mammal Comparisons

Have students make a chart comparing monotremes, marsupials, and placental mammals. The chart headings should be *Type of Mammal, Examples, Where Embryo Develops,* and *Food Source for the Young. (The answers will vary according to type of mammal, except for* Food Source for the Young, *which is milk in every case.)* PORTFOLIO

Teaching TIP

Milk Is More Than Food

Mammalian milk is a rich source of food, but it contains other substances that affect the life of young mammals. Milk has factors important to the immunity of young mammals. Bring empty cartons of milk to class, and have students read the list of nutrients on the carton labels. Finish by asking what kinds of foreign substances may be present in milk and how they might get there. *(Answers include pollutants from the air, such as pesticides, and from water, such as fertilizers; and substances that the female may have ingested, such as medications or hormones. Vitamin D is also added during the processing of milk.)*

Today, There Are Three Groups of Mammals

The first group of mammals to evolve laid eggs, as did their reptilian ancestors. The only surviving direct descendants of these early mammals are two species of echidna (*ee KIHD nuh*) and the duck-bill platypus, shown in **Figure 33-20.** These egg-laying mammals are in a group of their own called the monotremes.

Mammals other than monotremes show one of two patterns of development. Both of these groups give birth to live young, but their embryos develop in different ways.

1. **Marsupials.** Marsupial (*mahr SOO pee uhl*) mammals include kangaroos, also shown in Figure 33-17, opossums, koalas, wombats, and even some mice. Their young are born at a very immature stage and complete their development in their mother's pouch (called a marsupium).
2. **Placentals.** Placental mammals include dogs, cats, horses, sheep, gorillas, humans, and most of the other mammals you are familiar with. Placental mammals, such as the goat seen in Figure 33-17, develop within their mother's body and are nourished by an organ called the placenta. In Chapter 36, you will learn more about the mammals.

Figure 33-20 Mammals. There are three types of mammals—monotremes, marsupials, and placentals—each represented below.

did you know?

The platypus was thought to be a hoax.

When the first platypus skin was sent back to London from Australia in 1799, zoologists thought that this creature with a ducklike bill and beaverlike tail must surely be a hoax. After careful investigation, the zoologists were amazed to discover that they were examining an extraordinary animal, one like no other on Earth.

Birds May Have Evolved from Dinosaurs

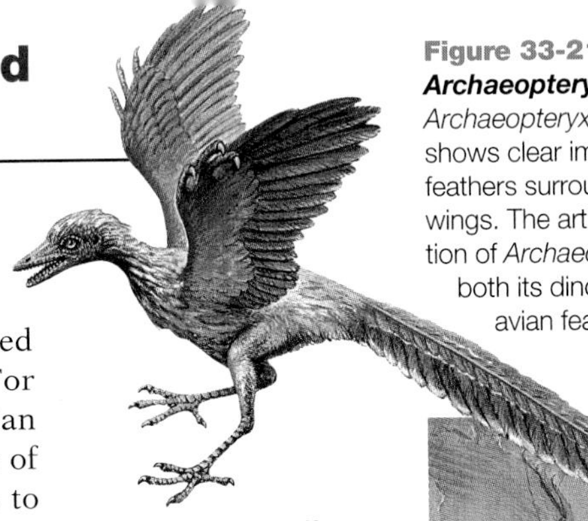

The earliest known bird is *Archaeopteryx* (meaning "ancient wing"), illustrated in **Figure 33-21.** *Archaeopteryx* was about the size of a crow and shared many features with small theropods. For example, it had teeth and a long reptilian tail. These are features of dinosaurs, not of birds. *Archaeopteryx* had no breastbone to anchor its flight muscles, as do modern birds. And unlike the hollow bones of modern birds, *Archaeopteryx*'s bones were solid. It also had the forelimbs of a dinosaur. Because of these dinosaurlike features, several *Archaeopteryx* fossils were originally classified as dinosaurs.

Once impressions of feathers were discovered on some *Archaeopteryx* fossils, it was clear that their classification as dinosaurs needed to be changed. The presence of feathers on its wings and tail classifies *Archaeopteryx* as avian (birdlike). *Archaeopteryx* also had other avian features, notably a fused collar bone, the wishbone, which dinosaurs did not have. Today, almost all biologists agree that *Archaeopteryx* is very closely related to the small dinosaur *Compsognathus.* Some biologists go so far as to classify *Archaeopteryx* and other birds as "feathered dinosaurs." Recent fossil discoveries in China, seen in **Figure 33-22,** support this opinion. However, most biologists continue to classify birds in a separate class, Aves, because of their distinct features.

The fossil record now reveals a diverse collection of toothed birds with the hollow bones and breastbones necessary for flight. By the early Cretaceous period, only 15 million years after *Archaeopteryx* lived, a variety of birds with many of the features of modern birds had evolved.

Figure 33-21
Archaeopteryx. The *Archaeopteryx* fossil below shows clear impressions of feathers surrounding the wings. The artist's interpretation of *Archaeopteryx* shows both its dinosaurian and avian features.

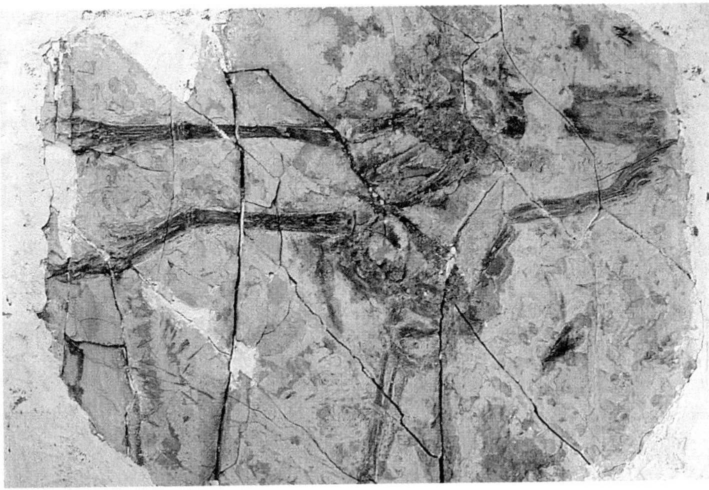

Figure 33-22 A feathered dinosaur? This dinosaur fossil found in China in 1996, shows traces of what appear to be feathers along its back.

Active Reading

Technique: Reading Effectively

Have the class scan pp. 747–748. Ask them to spot large words that they cannot pronounce, such as *Archaeopteryx*, Cretaceous, and *Compsognathus*. Make a list of these words on the board or overhead projector and pronounce them correctly for the class. Have the entire class repeat each word after you say it.

Teaching TIP

Fossilized Birds

Ask students for possible reasons why relatively few bird fossils have been found. *(The thin, hollow bones break easily and must have near-perfect conditions to fossilize.)* Show a skeleton of a bird and point out the fused pelvic bones and fused collarbones (wishbone, or furcula). Have students note the long neck and very short tail on the skeleton. Ask students to compare the bird tail with the tail of *Archaeopteryx.* Break or cut a chicken leg bone to show that the bone is hollow.

Teaching TIP

Dinochicken?

Read the article "New Limb on the Avian Family Tree" in the September 1993 issue of *Natural History* magazine. Summarize its findings for your students and show them the picture of the *Mononykus.* Point out that whenever a seemingly out-of-place fossil is discovered, scientists must collaborate and try to determine how the new fossil is related to ones already discovered. Explain that this determination can often be a very difficult task.

3 Assessment Options

Closure

Evolution of Birds and Mammals

Have students look at Figure 33-11 on page 739. Have them trace the line that leads from early reptiles to today's birds and mammals. Ask students to explain in their own words the events that led from the early reptiles to modern birds and mammals.

Review

Assign Review—Section 33-2.

Reteaching — BASIC

Tell students to make two columns on a sheet of paper. Have them label the first column *Reptiles* and the second column *Birds*. Then ask them to review Section 33-2 and to list as many key reptilian and avian facts or characteristics as possible under the appropriate headings. Tell students to skip a line between each fact and to write on only one side of the paper. Students should then compare their lists with those of a partner and add to their lists any important characteristics they missed. Ask students to cut out each of the facts, leaving the column headings intact. Then have them turn over and shuffle the pieces of paper. Each student should turn his or her partner's facts face-up, one-by-one, and place the fact under the appropriate column heading. When finished, partners should check each other's work. If necessary, students should use their textbook to check for accuracy. PORTFOLIO

Today, Birds Are the Largest Group of Terrestrial Vertebrates

internetconnect
SciLINKS NSTA
TOPIC: Birds
GO TO: www.scilinks.org
KEYWORD: HX748

There are more species of birds than of any other terrestrial vertebrate. All but a few of the modern orders of birds are thought to have arisen after the Cretaceous extinction. Since the impressions of feathers are rarely fossilized and since modern birds have hollow delicate bones, the fossil record of birds is incomplete. Relationships among the families of modern birds are mostly inferred from studies of the degree of DNA similarity among living birds. These studies suggest that the ostrich and its relatives belong to the oldest group of living birds. Ducks, geese, and waterfowl arose next and were followed by a diverse group of woodpeckers, parrots, swifts, and owls. Next came the songbirds, which include 60 percent of today's bird species. The most recent birds to appear are the more specialized orders, such as the birds of prey, flamingos, and penguins. Figure 33-23 shows representative modern birds.

Figure 33-23 Modern birds

Although birds vary greatly in appearance, they all have feathers, a key defining characteristic.

Review SECTION 33-2

1. **Describe** two key adaptations that allowed reptiles to live and reproduce completely on land.
2. **Summarize** the factors that contributed to the dinosaurs' success on land.
3. **Compare** ectothermic and endothermic animals.
4. **Relate** changes in the world's climate to the rise of mammals.

Critical Thinking

5. **SKILL Forming Reasoned Opinions** Evaluate the change in classification of *Archaeopteryx* from dinosaur to bird.

Review SECTION 33-2 ANSWERS

1. The watertight egg permitted eggs to be laid on land without the embryos drying out. Watertight skin prevented the loss of body fluids.
2. Legs positioned directly under the body permitted dinosaurs to be faster and more agile; dinosaurs were well adapted to living in the dry conditions found in Pangaea; and many other animal species became extinct by the end of the Triassic period.
3. Ectotherms must obtain heat from the environment because their metabolism is too slow to generate sufficient internal heat. Endotherms have a rapid metabolism, which produces enough heat to keep their bodies warm.
4. At the end of the Cretaceous period, the world began to cool. Mammals, with their high metabolic rate and insulating fur, flourished while large ectotherms such as dinosaurs did not.
5. Answers will vary but should indicate that this change was due to the discovery of new fossils that showed feathers—an indisputably avian characteristic.

Study ZONE CHAPTER 33

Use the Key Concepts and Key Terms listed below to review the main ideas in this chapter. If you need to review the meaning of a term, turn to the page indicated.

Key Concepts

33-1 Vertebrates Spread from the Sea to the Land

- Vertebrates are chordates with an endoskeleton, a major portion of which is a backbone.
- Vertebrates share the following characteristics: bilateral symmetry, cephalization, the presence of a coelom, a closed circulatory system, and a chambered heart. Their organs form organ systems that perform a variety of body functions.
- The first vertebrates evolved about 500 million years ago. They were the agnathans, fishes without jaws or paired fins.
- Spiny fishes (acanthodians) appeared 430 million years ago. They had paired fins, streamlined bodies, and jaws with teeth.
- Cartilaginous sharks and bony fishes evolved about 400 million years ago. Bony fishes account for more than 90 percent of today's fishes.
- Amphibians, which appeared about 370 million years ago, were the first group of vertebrates to live on land.

33-2 Vertebrates Adapt to Terrestrial Living

- Reptiles evolved about 320 million years ago, adapting completely to land with the evolution of watertight skin and a watertight egg.
- Dinosaurs dominated the land for 150 million years. Several factors contributed to their success.
- Ectothermic animals absorb heat from their environment, while endothermic animals produce heat internally.
- There are four orders of living reptiles: turtles, snakes and lizards, tuataras, and crocodilians.
- The first mammals appeared 220 million years ago. After the dinosaurs became extinct, mammals flourished, taking over many ecological roles once dominated by dinosaurs.
- The first mammals to evolve were the egg-laying monotremes. Placental mammals and marsupial mammals, which give birth to live young, evolved later.
- Modern birds may be the descendants of a group of small theropods, a type of dinosaur.

Key Terms

33-1

vertebra (730)
agnathan (732)
acanthodian (733)
cartilage (733)

33-2

terrestrial (739)
thecodont (740)
Pangaea (740)
ectothermic (742)
endothermic (742)
therapsid (744)

Study TIP Ready?

- *If you think you understand these Key Concepts and Key Terms, then you're ready for the Performance Zone.* 

Answer to Concept Map

The following is one possible answer to Performance Zone item 1 on page 750.

Vertebrates include: amphibians (evolved from bony fishes); fishes (can be bony fishes, cartilaginous fishes, jawless fishes; jawless fishes evolved but became extinct: spiny fishes, armored fishes); other reptiles (evolved from early reptiles); crocodiles (evolved from thecodonts; dinosaurs evolved from thecodonts; thecodonts evolved from early reptiles); birds (earliest known is Archaeopteryx, evolved from dinosaurs); mammals (evolved from therapsids; therapsids evolved from early reptiles).

Study ZONE CHAPTER 33

Review and Practice

Assign the following worksheets for Chapter 33:

- **Student Review Guide**
- **Concept Mapping Worksheet**
- **Test Preparation Pretest**
- **Vocabulary Worksheet**
- **Critical Thinking Review Worksheet**
- **Science Skills Worksheet**

Assessment Options

The following assessment options are available for this chapter:

Resource Link

- Assign **Chapter Test 33** to assess students' understanding of the chapter.

Alternative Assessment

- Assign each student to make up 12 questions, one from each of the red concept heads in Chapter 33. Have them provide two true/false, two short-answer, four multiple-choice, and four essay questions with answers. Collect the questions and use the best ones to conduct an oral quiz, reading them as students write the answers. Number each question you use so you can go back and have students check their answers. Discuss correct responses to items students have questions about. PORTFOLIO

Portfolio Assessment

- *Portfolio Assessment* at the front of this book provides options for building and evaluating student portfolios. Students and teachers can select items from the areas listed for inclusion in a portfolio.
- See items labeled PORTFOLIO on pp. 730, 731, 738, 746, 748 and 749.

Performance ZONE

CHAPTER 33 REVIEW ANSWERS

Assignment Guide

Section	Review
33-1	1, 2, 3, 4, 5, 6 ,12, 23, 24, 26, 27
33-2	1, 2, 7, 8, 9, 10, 11, 12, 13, 14, 15, 16, 17, 18, 19, 20, 21, 22, 25

Understanding and Applying Concepts

1. The answer to the concept map is found at the bottom of page 749.
2. **a.** A vertebrate is an animal with a backbone; vertebrae are the bones that make up the backbone.
   b. Endothermic animals produce enough body heat through their metabolism to keep their bodies warm. Ectotherms do not produce enough heat and therefore must get heat from the environment to carry on normal functions.
   c. Therapsids are an extinct order of reptiles from which mammals are believed to have descended; thecodonts are an extinct group of reptiles from which the dinosaurs are believed to have descended.
3. a
4. b
5. c
6. c
7. b
8. d
9. d
10. a
11. d
12. b
13. Crocodiles and birds have similar hearts, and both care for their young.
14. The tar pits contain exceptionally high quality fossils that date from the end of the last Ice Age. These fossils can be used to determine how the climate of the region changed and how this change affected the existing plant and animal life.
15. A key limiting factor on land is water. With the waterproof egg, reptiles were no longer tied to water for reproduction. Plants, with the development of vascular systems, were able to move into drier areas and get their water from the land.

Performance ZONE

Understanding and Applying Concepts

1. **Concept Mapping** Construct a concept map that shows the relationships among the various living and nonliving vertebrates. Include the following terms in your concept map: jawless fishes, spiny fishes, armored fishes, cartilaginous fishes, bony fishes, amphibians, thecodonts, dinosaurs, reptiles, therapsids, mammals, *Archaeopteryx*, birds. Include additional terms as necessary.
2. **Understanding Vocabulary** For each pair of terms, explain the differences in their meanings.
   a. vertebrate, vertebrae
   b. endothermic, ectothermic
   c. therapsids, thecodont
3. Which of the following is not considered a key characteristic of vertebrates?
   a. jaws c. vertebrae
   b. backbone d. skull
4. The jaws of spiny fishes evolved from
   a. plates of bone. c. muscles.
   b. gill arches. d. ray fins.
5. To survive as predators, fish had to adapt to which of the following two challenges?
   a. low oxygen conditions, shallow seas
   b. avoiding predators, grasping prey
   c. pursuing prey in water, grasping prey
   d. filtering invertebrates from the water, avoiding predators
6. Which of the following is not an amphibian adaptation for life on land?
   a. legs c. watertight skin
   b. lungs d. more-efficient heart
7. The first fully terrestrial vertebrates were the
   a. frogs. c. dinosaurs.
   b. reptiles. d. mammals.
8. As Pangaea broke apart,
   a. dinosaurs became more alike.
   b. mammals became dominant.
   c. dinosaur species flourished.
   d. world climates changed.
9. Mammals are thought to have survived the climatic changes that caused the extinction of dinosaurs because the bodies of mammals
   a. have little insulation.
   b. are ectothermic.
   c. maintain a constant temperature.
   d. have insulation.
10. After the dinosaurs became extinct, mammalian species flourished because
    a. many new habitats became available to them.
    b. the climate became warmer and dryer.
    c. they have four-chambered hearts.
    d. young mammals can learn necessary skills.
11. What evidence convinced scientists that *Archaeopteryx* should be classified as a bird rather than a small dinosaur?
    a. teeth
    b. hollow bones
    c. hard-shelled eggs
    d. feathers
12. Which sequence reflects the order in which the major groups of vertebrates are thought to have evolved?
    a. bony fishes → reptiles → amphibians
    b. bony fishes → amphibians → reptiles
    c. amphibians → mammals → reptiles
    d. birds → mammals → reptiles
13. **Theme Evolution** What evidence is there that crocodiles and birds are likely to be more closely related to each other than to other living reptiles?
14. **Exploring Further** Why are the La Brea Tar Pits an important site for scientists studying Ice Age animals and plants?
15. **Chapter Links** In what ways are the adaptations of reptiles to land similar to the adaptations of plants to land? (**Hint:** See Chapter 24, Section 24-1.)

Skills Assessment

Part 1: Interpreting Graphics

Use the maps below to answer the following questions:

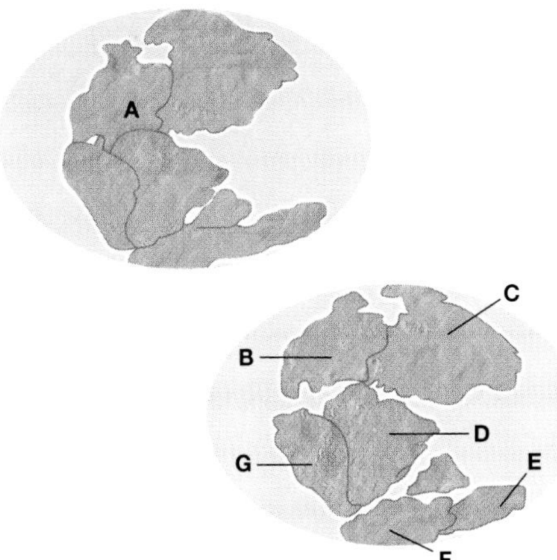

16. Identify the land mass labeled *A*.
17. What effect did the breakup of land mass *A* have on the evolution of the dinosaurs?
18. Identify what present-day continents were at one time located at the positions labeled *B–G*.
19. For each of the maps above, give the approximate date and time period it represents.

Part 2: Life/Work Skills

Use the media center or Internet resources to learn more about the types of dinosaurs that once lived on Earth.

20. **Being a Team Member** Work with a small group of students to develop a visual presentation showing the various types of dinosaurs and their relationship to one another. Make a rough outline of the information your team will present and decide who will be responsible for which tasks.
21. **Communicating** Prepare a written guide that will lead a viewer through your presentation. Have a taped version of the guide available for viewers who prefer an oral format.

Critical Thinking

22. **Evaluating Hypotheses** Fossil evidence collected in the 1980s in Arctic Alaska suggests that some dinosaurs were year-round residents of areas that reached freezing temperatures and were in total darkness during the winter months. Does this evidence of "Arctic dinosaurs" support or contradict the hypothesis that the extinction of dinosaurs was due to a period of intense cold produced by debris in the Earth's atmosphere? Explain your answer.

Portfolio Projects

23. **Recognizing Relationships** Vertebrates share several characteristics. Develop an illustrated guide that informs readers about these characteristics.
24. **Forming a Model** Make a series of models or create a mural that shows either the evolution of jaws or the evolution of limbs. Create a written guide or an audiotape that guides the viewer through your work.
25. **Interpreting Conclusions** Read "Dinosaur Fossils, in Fine Feather, Show Link to Birds" by Ann Gibbons (*Science*, June 26, 1998, vol. 281, p. 2051). Many paleontologists now think that feathers seem to have evolved for what purpose? What arguments do skeptics use to classify the new fossils as belonging to flightless descendants of birds that could fly?
26. **Finding Information** Research how amphibians are portrayed in literature. For example, in Shakespeare's play *Macbeth*, the witches call for "toe of frog" as part of their mysterious concoctions. Share your findings in an oral report to your class.
27. **Career Focus Exhibit Designers** Research the field of design that focuses on designing exhibits and displays for museums, such as institutes of paleontology. Write a report on your findings. Your report should include a job description, training required, kinds of employers, growth prospects, and starting salary.

Skills Assessment

16. Pangaea
17. With the breakup of Pangaea, the climates of the various land masses began to vary. Some species of dinosaurs could not adapt and became extinct. Other species flourished.
18. B—North America, C—Europe and Asia, D—Africa, E—Australia, F—Antarctica, G—South America
19. upper map—200 million years ago, early Jurassic; lower map—180 million years ago, mid-Jurassic
20. Presentations will vary. Students should describe the size and likely diet of each dinosaur as well as the approximate time during which it lived.
21. Guides should be organized in a logical way, such as by type of dinosaur or geological period.

Critical Thinking

22. This evidence contradicts the hypothesis that the extinction of dinosaurs was due to a period of intense cold. Since the large body mass of dinosaurs was difficult to heat up, and they had no insulating fur or feathers, a long period of cold weather would cause significant problems for dinosaurs. New hypotheses would have to be developed to account for these "Arctic dinosaurs."

Portfolio Projects

23. Guides will vary but should include the following information: backbone made of individual vertebrae (a spinal column), dorsal nerve cord protected by the backbone, bony skull, paired appendages, bilateral symmetry, cephalization with complex brain and sense organs, and complex organ systems.
24. Models and guides will vary but should show information similar to that presented in Figure 33-5 or Figure 33-8.
25. Evidence suggests that feathers evolved for insulation, display, or some purpose other than flight. Skeptics say that the new fossils show shortened tails and an ossified sternum, making them more birdlike and advanced than *Archaeopteryx*, not more primitive.
26. Answers will vary. Students should find literature portrayals of both major groups of amphibians.
27. Exhibit designers design, create, and lay out the exhibits and displays for museums. Technical knowledge of anatomy, behavior, and past environments are required, in addition to artistic skills. Employers include art galleries, museums, amusement parks, and universities. Many artists are hired on a commission basis to complete a particular project. Growth prospects will vary by region but generally are the best in urban areas. Starting salary will vary by region.

Evolution of the Heart

Purpose

Students will examine how the structure of the vertebrate heart changed as vertebrates evolved.

Preparation Notes

Time required: 45 minutes

Preparation Tips

- Depending on how many computers are available, set up work stations that enable students to work individually or in teams of two to four.
- Encourage students to read the lab and complete the **Before You Begin** section before they come to class on the day of the lab.
- Before students begin the lab, discuss the following:
  1. Tell students that in animals with a "one-cycle pump," blood flows from the heart to the gills, then to the body, and then back to the heart. Draw a simple diagram of this cycle on the board.
  2. Explain that the term *oxygenated blood* refers to blood that is rich in oxygen because it has acquired oxygen from the capillaries in the gills or lungs. The term *deoxygenated blood* refers to blood that contains little oxygen because it has given its oxygen to cells in the body.
  3. Explain that the heart's muscular ventricle is its main pumping chamber.
  4. Ask: Why would increasing the efficiency of the heart be advantageous to an animal? *(More oxygen can be transported to the tissues for a more efficient metabolic processing of nutrients.)*

Procedural Tips

- After students click on the navigation key, they will see a list of Key Concepts. To view the focus questions and related concepts, students should click on the forward arrow on the bottom left of the screen.

Interactive Exploration

Chapter 33 *Evolution of the Heart*

CD-ROM Lab

SKILL

- Computer modeling

OBJECTIVES

- **Observe** the action of the hearts of five different vertebrates.
- **Compare** the structure, action, and efficiency of five different vertebrate hearts.

MATERIALS

- computer with CD-ROM drive
- CD-ROM *Interactive Explorations in Biology: Human Biology*

Before You Begin

In many animals, a heart and a system of veins and arteries work together to distribute oxygen to cells and to carry away cellular wastes. The vertebrate heart is a muscular organ with two or more chambers. In this Interactive Exploration, you will examine how the structure of the vertebrate heart changed as vertebrates evolved. You will also compare the action and efficiency of the hearts of several groups of vertebrates.

1. Select *Evolution of the Heart* from the menu. Click on the key near the top of the navigation palette. Read the focus questions and review the following concepts: the circulatory system, the heart, the pulmonary vein, mixing within the heart, the heart's septum, and completing the septum.
2. Click *File* at the top left of the screen, and select *Interactive Exploration Help.* Listen to the instructions. Click the *Exploration* button at the bottom right of the screen to begin the exploration. You will see a diagram like the one below.

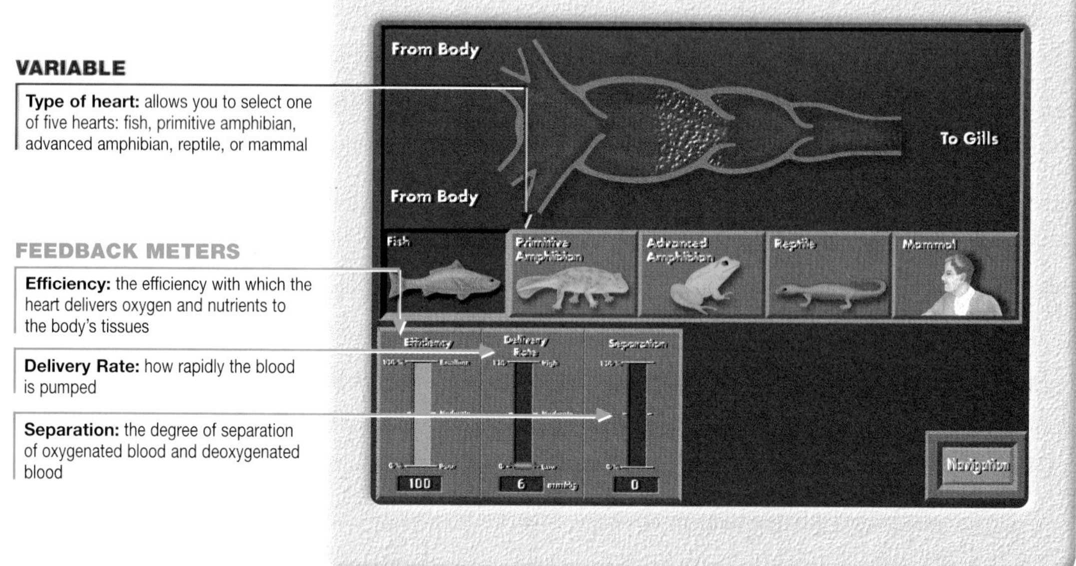

- Tell students to use the pull-down menu under *File* on the upper left of the screen to reach *Interactive Explorations*.

Procedure

Answers

1. The blood flows from the heart to the gills, then to the body, and then back to the heart.
2. The efficiency of the fish heart is 100 percent. The delivery rate is 6 mm Hg. The separation is 0 percent. The separation is zero because the fish heart pumps only one type of blood—oxygen-poor blood. The lack of separation does not matter because there is only one circulatory pathway. Because there is no mixing of oxygenated and deoxygenated blood in the fish heart, the efficiency of oxygen delivery is 100 percent.
3. No, the shape is basically the same. Blood now flows first to the lungs, then to the heart, then to the body, and then back to the heart.
4. The efficiency is 30 percent. The delivery rate is 40 mm Hg. The separation is 0 percent. The separation did not change because no septum is present.

3. Make a data table similar to the one below.

DATA TABLE

Type of animal	Efficiency (%)	Delivery rate (mm Hg)	Separation (%)

Procedure

PART A: Fish Heart

1. Examine the fish heart at the top of the screen. Notice the route that blood travels through the fish heart. How does the blood complete the circuit?
2. Look at the feedback meters at the bottom left of the screen. Record the numbers for the efficiency, delivery rate, and separation of this heart in your data table. Why is the separation so low? Why does the lack of separation not matter in a single-cycle heart? Why is the efficiency rate so high?

PART B: Primitive Amphibian Heart

3. Click the *Primitive Amphibian* button. Has the shape of the heart changed significantly? How has the route of blood flow through the body changed?
4. Record the numbers in the feedback meters in your data table. Did the amount of separation change? Explain.

PART C: Advanced Amphibian Heart

5. Click the *Advanced Amphibian* button. What significant change in the advanced amphibian's heart makes it different from a primitive amphibian's heart?
6. Record the numbers in the feedback meters. What do the changes in the numbers indicate about an advanced amphibian's heart in comparison with a primitive amphibian's heart?

PART D: Reptile Heart

7. Click the *Reptile* button. What changes took place in the diagram of the heart?
8. Record the numbers in the feedback meters. What do the changes in the numbers indicate about a reptile's heart in comparison with an advanced amphibian's heart?

PART E: Mammal Heart

9. Click the *Mammal* button. What is the major difference that occurs in the structure of the mammalian heart?
10. Record the numbers in the feedback meters. What do the changes in the numbers indicate about a mammal's heart in comparison with a reptile's heart?

Analyze and Conclude

1. **Analyzing Data** Compare the readings for the mammalian heart with the other readings. Are any of these readings the same? If so, which are the same? How do the other readings compare?
2. **Drawing Conclusions** What advantage does the fish heart have over the reptile and amphibian hearts? Explain.
3. **Inferring Conclusions** What advantages do the reptile and amphibian hearts have over a fish heart?
4. **Evaluating Conclusions** Evaluate the following conclusion: The mammalian heart is the most advanced of all the vertebrate hearts.

Use the following Internet resource to explore your own questions about the evolution of vertebrates.

Analyze and Conclude

Answers

1. The efficiency of the fish heart and the mammal heart is the same. The other readings show a steady increase in delivery rate and in separation.
2. The separation for the fish heart is higher; the fish heart delivers only oxygen-rich blood to the body's tissues.
3. Both the reptilian and amphibian hearts deliver blood to the tissues at a higher rate than fish hearts do.
4. This statement is true. The mammalian heart achieves complete separation of oxygen-rich and oxygen-poor blood. It pumps blood to the body cells at a higher rate than the other vertebrate hearts do.
5. Answers will vary. For example: How did the structure and function of the respiratory system change as vertebrates evolved?

5. The advanced amphibian heart has an incomplete septum that partially divides the ventricle.
6. The efficiency is 60 percent. The delivery rate is 55 mm Hg. The separation is 40 percent. The efficiency doubled, the delivery rate increased, and the separation of blood in the ventricle increased. These numbers indicate that the advanced amphibian's heart delivers more oxygen more rapidly to the body cells.
7. The septum has increased in size.
8. The efficiency is 75 percent. The delivery rate is 60 mm Hg. The separation is 60 percent. The reptilian heart is able to deliver blood to the tissues at a higher rate. The efficiency of the reptile's heart is higher.
9. A septum completely divides the ventricle into a right and a left ventricle.
10. The efficiency is 100 percent. The delivery rate is 120 mm Hg. The separation is 100 percent. The mammal's heart delivers more oxygen more rapidly to the body cells, and the separation of oxygen-rich and oxygen-poor blood is higher than in the reptile's heart.

Chapter 34

Fishes and Amphibians

Planning Guide

OBJECTIVES	MEETING INDIVIDUAL NEEDS	READING SKILLS AND STRATEGIES
34-1 The Fish Body (BLOCKS 1 & 2 90 min) • Describe the characteristics common to modern fishes. • Summarize how fish obtain oxygen. • Summarize how blood circulates through a fish. • Contrast how marine and freshwater fish balance their salt and water content. • Describe two methods of reproduction in fishes.	Learning Styles, p. 756 BASIC ADVANCED • p. 760	ATE Active Reading, Technique: K-W-L, p. 755 ■ Active Reading Guide, worksheet (34-1) ■ Directed Reading, worksheet (34-1) ■ Reading Strategies, K-W-L worksheet
34-2 Today's Fishes (BLOCKS 3 & 4 90 min) • Distinguish between the three general categories of modern fishes. • Describe the major external and internal characteristics of the yellow perch. • Summarize the unique features of bony fishes.	Learning Styles, p. 761 BASIC ADVANCED • pp. 763, 767	■ Active Reading Guide, worksheet (34-2) ■ Directed Reading, worksheet (34-2)
34-3 Amphibians (BLOCKS 5 & 6 90 min) • Summarize the characteristics common to all modern amphibians. • Compare the three orders of living amphibians. • Describe the major external and internal characteristics of the leopard frog.	Learning Styles, p. 768 BASIC ADVANCED • pp. 769, 771, 774	ATE Active Reading, Technique: Reading Effectively, p. 774 ■ Active Reading Guide, worksheet (34-3) ■ Directed Reading, worksheet (34-3) ■ Reading Strategies, Spider Map ■ Supplemental Reading Guide, *Through a Window: My Thirty Years with the Chimpanzees of Gombe*

Review and Assessment Resources

BLOCKS 7 & 8 90 min

TEST PREP

- **PE** Study Zone, p. 775
- **PE** Performance Zone, pp. 767–777
- ■ Chapter 34 worksheets:
  - Vocabulary
  - Concept Mapping
  - Critical Thinking
- Audio CD Program
- ■ Test Prep Pretest

CHAPTER TESTING

- ■ Chapter Tests (blackline master)
- ▲ Test Generator
- ■ Assessment Item Listing

See the correlations table at the front of this book for a list of the ***National Science Education Standards*** addressed in this chapter.

Smithsonian Institution® Visit **www.si.edu/hrw** for additional on-line resources.

KEY

Biology Principles and Explorations

PE Pupil's Edition

ATE Teacher's Edition

Teaching Transparencies

Laboratory Program

Shows these resources within a customizable daily lesson plan:

■ Teaching Resources

▲ Test Generator

VISUAL STRATEGIES	CLASSWORK AND HOMEWORK	TECHNOLOGY AND INTERNET RESOURCES
ATE Fig. 34-2, p. 757 165 Internal Anatomy of a Bony Fish 166 Structure of a Gill 168 Major Groups of Fishes	**PE Data Lab** Analyzing a graph, p. 759 **Lab B27** Schooling Behavior in Fishes **PE Review,** p. 760	Holt Biology Videodiscs Section 34-1 Audio CD Program Section 34-1 internetconnect SCILINKS NSTA TOPICS: • Fish heart • Kidneys
ATE Fig. 34-6, p. 762 138A Swim Bladder in Bony Fish 139A Lateral Line in Bony Fish 140A Operculum in Fish	**PE Quick Lab** How can you model the action of a swim bladder?, p. 766 **Lab B26** Perch Dissection **PE Review,** p. 767	CNN Video Segment 27 Fish Farming Holt Biology Videodiscs Section 34-2 Audio CD Program Section 34-2 internetconnect SCILINKS NSTA TOPICS: • Sharks • Bony fish
ATE Fig. 34-14, p. 769 **ATE Fig. 34-17,** p. 771 169 Life Cycle of a Frog 170 Orders of Amphibians 171 Comparison of Heart Structure 172 Fish and Amphibian Circulation	**PE Lab Techniques** Observing a Live Frog, pp. 778–779 **Lab A25** Observing a Frog **Lab B28** Frog Dissection **PE Review,** p. 774 **PE Study Zone,** p. 775	Holt Biology Videodiscs Section 34-3 Audio CD Program Section 34-3 internetconnect SCILINKS NSTA TOPIC: • Amphibian circulation

ALTERNATIVE ASSESSMENT

PE Portfolio Projects 26–29, p. 777

ATE Alternative Assessment, p. 775

ATE Portfolio Assessment, p. 775

Lab Activity Materials

Data Lab
p. 759 pencil, paper

Quick Lab
p. 766 100 mL beaker, or small glass; cold, clear carbonated soft drink; very dry raisins (2)

Lab Techniques
p. 778 live frog in a terrarium, live insects (crickets or mealworms), 600 mL beaker, aquarium half-filled with dechlorinated water

Lesson Warm-up
ATE pictures of lamprey, eel, hagfish, hammerhead shark, dog fish shark, ray, salmon, trout, bass, p. 761

Demonstrations

ATE several sizes and types of wrenches, p. 754

ATE aquarium siphon with bulb, p. 758

ATE 600 mL beakers (2), water, table salt, potato

ATE glass catfish, p. 766

ATE graduated cylinder, corn syrup, food coloring, water, p. 770

BIOSOURCES LAB PROGRAM
- Quick Labs: Lab A25, p. 49
- Inquiry Skills Development: Lab B26, p. 123
- Inquiry Skills Development: Lab B27, p. 127
- Inquiry Skills Development: Lab B28, p. 131

Chapter Theme

Evolution

Modifications of their skeletons and body systems gave fish an advantage in water and amphibians an advantage on land.

Establishing Prior Knowledge

Before beginning this chapter, make certain students can answer the questions in **Study TIP Ready?** on page 755. If they cannot, encourage them to reread the sections indicated.

Opening Demonstration

Bring several sizes and types of pliers to class. Have students pick up different objects using the tools. Ask students to decide which tools exert the most force and which are best for picking up small objects. Relate the function of the tools to jaw function. Ask students to name the vertebrate group in which jaws first evolved. *(fishes)*

CHAPTER 34 Fishes and Amphibians

Leopard frog

What's Ahead — and Why

CHAPTER OUTLINE

34-1 The Fish Body

- Today's Fishes Share Key Characteristics
- Fishes Respire with Gills
- Fishes Have a Simple Heart and Single-Loop Circulation
- Kidneys Help Balance Water and Salt
- Most Fishes Fertilize Their Eggs Externally

34-2 Today's Fishes

- Hagfishes and Lampreys Are Jawless
- Sharks, Skates, and Rays Have Flexible Skeletons
- Bony Fishes Have Three Structural Adaptations

34-3 Amphibians

- Today's Amphibians Share Five Key Characteristics
- Frogs and Toads Are Adapted for Jumping
- Salamanders and Caecilians Have Elongated Bodies

Students will learn

- 34-1: how fish are adapted for life in water.
- 34-2: how three groups of modern-day fishes differ from each other.
- 34-3: how amphibians differ from fish, and what adaptations amphibians have for life in and out of water.

Diver with shark

Swimming with sharks! *Like most sharks, this Hawaiian white-tipped reef shark is somewhat timid and would rather flee than attack. And like all other sharks, it is classified as a fish. But why is the shark in a chapter with frogs and other amphibians? In many ways an amphibian can be thought of as a "fish out of water." To find out why, turn the page.*

Study TIP

Ready?

Check your understanding of these topics to help you prepare for what's ahead.

Can you . . .

- **describe** the process of osmosis? (Section 4-1)
- **define** the term *gills*? (Section 28-2)
- **state** the function of the excretory system? (Section 28-2)
- **distinguish** between external and internal fertilization? (Section 28-2)
- **define** the term *metamorphosis*? (Section 31-3)

If not, *review the sections indicated.*

Looking Ahead

Labs

- **Data Lab**
  Analyzing ion excretion in fish (p. 759)
- **Quick Lab**
  How can you model the action of a swim bladder? (p. 766)
- **Lab Techniques**
  Observing a Live Frog (p. 778)

Features

- **Up Close**
  Yellow Perch (pp. 764–765)
- **Up Close**
  Leopard Frog (pp. 772–773)

internet connect

National Science Teachers Association *sci*LINKS Internet resources are located throughout this chapter.

Reading Strategies

Active Reading

Technique: K-W-L

Before students read the chapter, have them make short individual lists of all the things they already **K**now about fish. When students finish their lists, ask them to contribute their entries to a class list on the board or overhead. Then have students list things they **W**ant to know about fish. Have students save their lists for later use in the **Closure** on page 760.

Directed Reading

Assign **Directed Reading Worksheet Chapter 34** to help students interact with the text as they read the chapter.

Interactive Reading

Assign ***Biology: Principles and Explorations* Audio CD Program Chapter 34** to help reluctant readers achieve greater success in reading the chapter.

Reading Skills

Assign Reading Organizers and Reading Strategies Worksheets from the **One-Stop Planner CD-ROM with Test Generator** to help students learn and practice effective reading skills.

internet connect

Holt *Biology: Principles and Explorations*

On-line Resources: go.hrw.com

Visit the HRW Web site for a variety of resources related to this chapter. Just type in the keyword HX0 Chapter 34.

Holt *Biology: Principles and Explorations*

On-line Resources: www.scilinks.org

The following *sci*LINKS Internet resources can be found in the student text for this chapter:

Page 759
TOPIC: Kidneys
KEYWORD: HX759

Page 762
TOPIC: Sharks
KEYWORD: HX762

Page 770
TOPIC: Circulation in amphibians
KEYWORD: HX770

Chapter 34

Section 34-1 The Fish Body

1 Prepare

Scheduling

- **Block:** About 90 minutes will be needed to complete this section.
- **Traditional:** About two class periods will be needed to complete this section.
- **Planning:** See the Planning Guide on pp. 753A and 753B for lecture, classwork, and assignment options for Section 34-1. Lesson plans for Section 34-1 can be found in a lesson cycle format in ***Biology: Principles and Explorations*** **One-Stop Planner CD-ROM with Test Generator.**

Resource Link

Assign **Active Reading Worksheet Section 34-1** to help students comprehend the key concepts and visuals in this lesson.

2 Teach

Lesson Warm-up

Write the following words on the board: starfish, sea horse, sea cow, lamprey, eel, and stingray. Ask students to identify which organisms they think are fish. Ask them what characteristics they used to determine if an organism is a fish. Do not tell students the answers. Rather, have them do research at home (using an encyclopedia or the Internet) to gather information on the animals listed above. During the next class period, compile a summary of the information gathered by students and lead them in a discussion of the characteristics of fish.

Objectives

- **Describe** the characteristics common to modern fishes. (p. 756)
- **Summarize** how fish obtain oxygen. (p. 757)
- **Summarize** how blood circulates through a fish. (p. 758)
- **Contrast** how marine and freshwater fish balance their salt and water content. (p. 759)
- **Describe** two methods of reproduction in fishes. (p. 760)

Key Terms

gill filament
gill slit
countercurrent flow
nephron

Today's Fishes Share Key Characteristics

What makes a goldfish instantly recognizable as a fish? You might name characteristics such as its fins, gills, scales, and typical fish shape as traits that contribute to the goldfish's "fishiness." But some fishes don't look quite so fishy. This is because the term "fish" refers to any member of one of three general categories of vertebrates: Agnatha (jawless fishes), Chondrichthyes (cartilaginous fishes), or Osteichthyes (bony fishes). The great diversity of fishes found today reflects the various adaptations fish have evolved to live in the oceans and fresh waters around the world. Fishes vary in size from whale sharks that can be longer than a large moving van to tiny gobies no larger than your fingernail. Despite the variation seen among fishes, shown in **Figure 34-1,** all share certain key characteristics that enable them to live in aquatic environments.

1. **Gills.** Fishes normally obtain oxygen from the oxygen gas dissolved in the water around them. They do this by pumping a great deal of water through their mouths and over their gills.
2. **Single-loop blood circulation.** Blood is pumped from the heart to the capillaries in the gills. From the gills, blood passes to the rest of the body and then returns to the heart. (Lungfishes, which have double-loop circulation, are an exception.)
3. **Vertebral column (backbone).** All fishes have an internal skeleton made of either cartilage or bone, with a vertebral column surrounding the spinal cord. The brain is fully encased within a protective covering called the skull or cranium.

To learn about one common fish, read *Up Close: Yellow Perch* on pages 764–765.

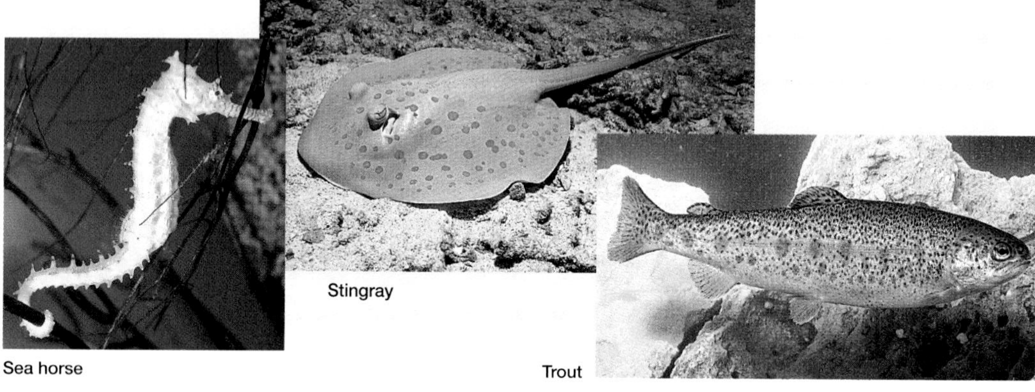

Figure 34-1 Fish diversity. While these three fish appear quite different externally, they share a number of characteristics.

Integrating Different Learning Styles

Logical/Mathematical	**PE**	Data Lab, p. 759
	ATE	Cross-Disciplinary Connection, p. 760
Visual/Spatial	**PE**	Study Tip, p. 758; Performance Zone item 26, p. 777
	ATE	Visual Strategy, p. 757; Demonstration, p. 758; Teaching Tip, p. 759
Interpersonal	**ATE**	Multicultural Perspectives, pp. 757, 758
Verbal/Linguistic	**PE**	Study Tips, pp. 757, 759; Performance Zone items 2 and 29, pp. 776–777
	ATE	Lesson Warm-up, p. 756; Closure. p. 760; Reteaching, p. 760

Fishes Respire with Gills

One challenge faced by all animals is the need to get enough oxygen for cellular respiration. Sponges, cnidarians, many flatworms and roundworms, and some annelids obtain oxygen by diffusion through the body surface. Other marine invertebrates, such as mollusks, arthropods, and echinoderms have gills, which are specialized respiratory organs. Fishes also respire with gills.

If you look closely at the face of a swimming fish, you will notice that as it swims, the fish continuously opens and closes its mouth, as if it were trying to eat the water. What looks like eating is actually breathing. The major respiratory organ of a fish is the gill, shown in **Figure 34-2.** Gills are made up of rows of **gill filaments**—fingerlike projections by which gases enter and leave the blood. The gill filaments hang like curtains between a fish's mouth and cheeks. At the rear of the cheek cavity is an opening called a **gill slit.** When a fish "swallows," water is forced over the gills and out through the gill slits.

This swallowing procedure is the core of a great change in gill design shown by fishes—one-way water flow through the gills combined with a specific arrangement of gill tissue that permits countercurrent water flow. In **countercurrent flow,** also shown in Figure 34-2, the water passing across the gills and the blood circulating in the capillary networks through the gills flow in opposite directions. Countercurrent flow ensures that oxygen diffuses into the blood over the entire length of the capillaries in the gills. Due to this arrangement, the gills of bony fishes are the most efficient respiratory organs known. Fish are able to extract up to 85 percent of the oxygen in the water passing over the gills.

Study TIP

Reviewing Information

Look closely at the portion of Figure 34-2 that shows blood flow and water flow. Identify what the arrows represent. Now, in your own words, explain what the term *countercurrent* means.

Figure 34-2 Fish gill structure. Countercurrent flow increases the gill's efficiency.

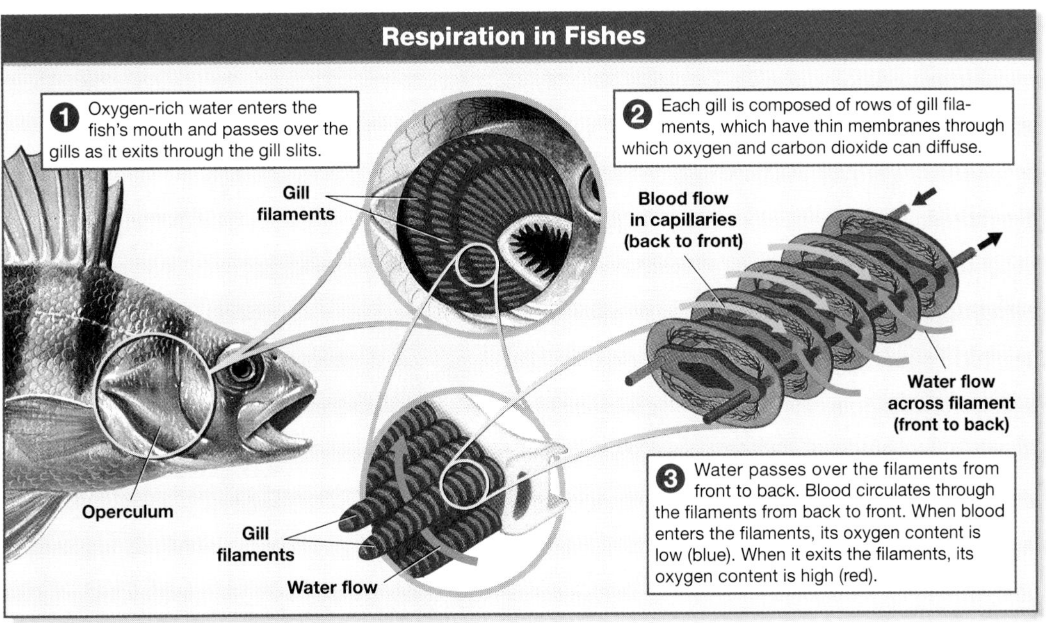

Respiration in Fishes
1 Oxygen-rich water enters the fish's mouth and passes over the gills as it exits through the gill slits.
2 Each gill is composed of rows of gill filaments, which have thin membranes through which oxygen and carbon dioxide can diffuse.
Gill filaments
Blood flow in capillaries (back to front)
Water flow across filament (front to back)
3 Water passes over the filaments from front to back. Blood circulates through the filaments from back to front. When blood enters the filaments, its oxygen content is low (blue). When it exits the filaments, its oxygen content is high (red).
Operculum
Gill filaments
Water flow

Section 34-1

CROSS-DISCIPLINARY CONNECTION

◆ **Chemistry**

Invite a chemistry teacher to your class to discuss the amazing chemical properties of water. Topics could include cohesion, surface tension, heat of vaporization, heat of fusion, polar covalent and ionic bonds, and the relationship between the temperature and density of water.

Visual Strategy

Figure 34-2

Have students locate the portion of the figure that shows countercurrent flow. Point out that water flows over the fish's gill from front to back. At the same time, the blood in the capillaries in the gills flows from back to front. Explain that this arrangement permits gas exchange to occur all along the capillaries in the gills. If the water and blood flowed in the same direction, the oxygen content of the two equilibrate at about 50 percent saturation, far lower than in countercurrent flow.

MULTICULTURAL PERSPECTIVE

Native Americans and Salmon

In the Pacific Northwest, salmon are an important part of many Native American religious ceremonies. Government treaties allow some Native American tribes, such as the Yakimas, to harvest set amounts of salmon. The mighty Columbia River, which forms the border between the states of Washington and Oregon, is often the site of Native American floats, which are attached to salmon nets.

Demonstration

Ancestral Vertebrate "Hearts"

To demonstrate the simple tubular "heart" of ancestral chordates, use an aquarium siphon (available at pet stores) that has a bulb attached to it. Insert a small piece of rubber or glass tubing into the siphon to keep the valve open. Place the end of the siphon in a jar of colored water and fill the rest of the tube with water. Place the other end of the siphon in another jar containing an equal volume of clear water. Squeeze the bulb, representing the primitive "heart," to show how blood is inefficiently moved in both directions. Water will flow out in both directions, and when the bulb is released, water will be sucked in from both ends. The level of each jar will go down very slightly when the bulb is squeezed and rise an equal amount when it is released. You can also demonstrate the value of valves by showing the efficiency of the pump when the valve is operating properly. With the valve enabled, water will move from one jar to the other.

Career

Aquaculturist

Aquaculture, or fish farming, is one of the fastest-growing sectors of agriculture. Both freshwater and saltwater species can be raised commercially. Operating a fish farm requires a knowledge of how to keep the fish's environment clean, proper feeding techniques, and how to control fish infections, such as those caused by bacteria and parasites.

Study TIP

Reviewing Information

Make a simple sketch of a fish heart. Add arrows to show the direction of the blood flow through the heart. Label the chambers the blood flows through.

Fishes Have a Simple Heart and Single-Loop Circulation

Chordates that were ancestral to the vertebrates had simple tubular "hearts." This heart was little more than a specialized zone of one artery that had more muscle tissue than the other arteries did. When a tube heart contracts, blood is pushed in both directions, and circulation is not very efficient.

For an organism with gills, such as a fish, a simple tube heart is not an adequate pump. The tiny capillaries in the fish's gills create resistance to the flow of blood, so a stronger pump is needed to overcome this resistance. In fishes, the tube pump of early chordates has been replaced with a simple chamber-pump heart, shown in **Figure 34-3.** The chamber-pump heart can be thought of as a tube with four chambers in a row.

1. **Sinus venosus** *(SEYE nuhs vuh NOH suhs).* This collection chamber acts to reduce the resistance of blood flow into the heart.
2. **Atrium.** Blood from the sinus venosus fills this chamber, which is large and has thin, muscular walls.
3. **Ventricle.** The third chamber is a thick-walled pump with enough muscle to contract strongly, forcing blood to flow through the gills and eventually the rest of the body.
4. **Conus arteriosus** *(KOH nuhs ahr TIHR ee oh suhs).* This chamber is a second pump that smoothes the pulsations and adds still more force.

The fish heart represents one of the great evolutionary changes found in vertebrates—a heart that pumps fully oxygenated blood through a single circulatory loop to the body's tissues.

Figure 34-3 Fish chamber-pump heart. These four steps show how blood flows through the heart of a fish.

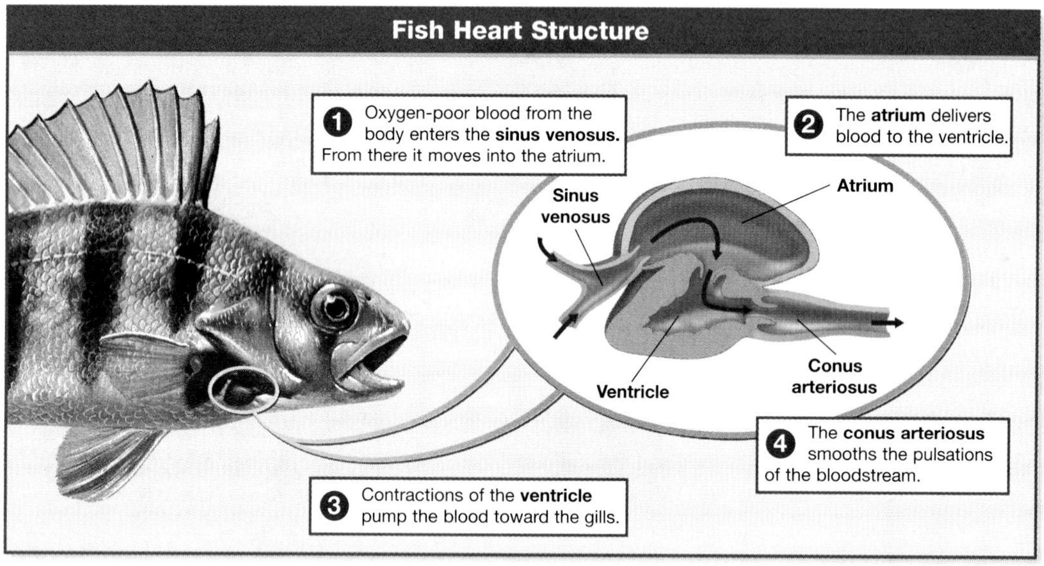

Fish Heart Structure
1 Oxygen-poor blood from the body enters the sinus venosus. From there it moves into the atrium.
2 The atrium delivers blood to the ventricle.
Sinus venosus
Atrium
Ventricle
Conus arteriosus
4 The conus arteriosus smooths the pulsations of the bloodstream.
3 Contractions of the ventricle pump the blood toward the gills.

Multicultural Perspective

Fish in Asia—Art and Food

Asian people have celebrated the elegant beauty of carp and goldfish for many centuries. Elaborately decorated fish bowls and kites with carp designs can be found throughout China, Japan, and Korea, as can serene water gardens with fish swimming among the willows and water lilies. Even a fish dinner becomes a celebration of beauty, as Asian chefs strive to create works of edible art.

Kidneys Help Balance Water and Salt

Vertebrates have a body that is about two-thirds water, and most will die if the amount of water in their body falls much lower than this. So minimizing dehydration (water loss) has been a key evolutionary challenge facing all vertebrates. Even some fishes must cope with the problem of water loss. If this seems strange to you, remember that the process of osmosis causes a net movement of water through membranes toward regions of higher ion concentration.

The ion (salt) concentration of sea water is three times that of the tissues of a marine bony fish. As a result, marine fishes lose water to the environment through osmosis. To make up for the water loss in their body, marine fishes drink a lot of sea water and actively pump excess ions out of their body. Freshwater fishes have the opposite problem. Because their bodies contain more ions than the surrounding water, they tend to take in water through osmosis. The additional water dilutes their body salts, so freshwater fishes regain salts by actively taking them in from their environment.

Although the gills play a role in maintaining a fish's salt and water balance, another key element is a pair of kidneys. Kidneys are organs made up of thousands of nephrons. **Nephrons** are tubelike units that regulate the body's salt and water balance and remove metabolic wastes from the blood. Excess water and bodily wastes leave the kidneys in the form of a fluid called urine. The concentration of the urine depends upon the environment in which an animal lives. Marine fishes excrete small amounts of urine while freshwater fishes excrete large amounts of dilute urine.

Organizing Information

To better understand how fish balance water and salt, make a two-column chart with the headings *Marine fish* and *Freshwater fish*. Under each column head, list facts about how that fish balances salt and water.

Analyzing ion excretion in fish

Background

A few species of fish, such as adult salmon, are able to move between saltwater and freshwater environments. The graph at right shows the excreted ion concentration of a fish as it travels from one body of water to another. Examine the graph, and answer the analysis questions.

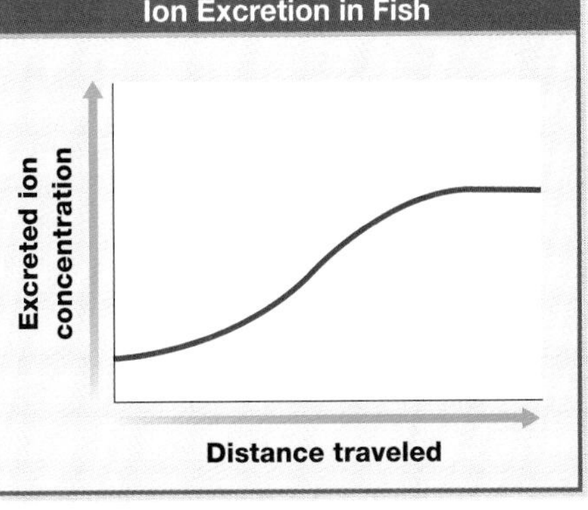

Analysis

1. **Determine** if the fish is losing or gaining ions as it travels.
2. **Infer** if the fish is traveling from fresh to salt water *or* from salt to fresh water.
3. **Summarize** the reasoning you used to answer item 2.

Teaching TIP

Osmosis Review

The day before this lesson, fill two beakers with distilled water. Add one tablespoon of salt to one beaker and mix well. Cut two equal lengths of potato ($7 \times 0.5 \times 0.5$ cm works well, but the size is not critical). Put one potato piece in each beaker, and let it sit overnight. The next day, tell students that each potato piece represents a freshwater fish. Tell them one fish has been placed in fresh water, and one has been placed in the ocean. Remove the potato pieces, and let students observe the rigidity and limpness of each. Ask students which "fish" was in fresh water and which was in the ocean. Ask them how they arrived at their conclusions. Lead a discussion into how and why fishes would have difficulty living in water environments other than their own. *(They would need some mechanism to regulate their water and salt concentrations.)* This could lead to a discussion about the kidneys.

Analyzing ion excretion in fish

Time 10 minutes

Process Skills Analyzing, inferring

Teaching Strategies

After finishing the three analysis questions, have students work in groups to speculate on what mechanisms allow fish such as salmon to move successfully between fresh water and salt water. *(Generally, they should come up with ideas about osmosis through the kidneys and gills.)*

Analysis Answers

1. losing ions
2. from fresh water to salt water
3. The graph shows that the farther the fish travels, the greater its excreted ion concentration. A fish's tissues have a lower concentration of ions than is found in salt water, and they lose water through osmosis. One would expect the fish to excrete more ions as the water becomes saltier.

CROSS-DISCIPLINARY CONNECTION

◆ **Mathematics**

Salmon return to spawn in the stream where they were hatched. During spawning, females lay from 2,000 to 10,000 eggs. The young salmon hatch and make their way downstream to the sea; most perish during this journey. A Canadian study reported that fewer than 1 percent of hatchlings return to spawn. Have students calculate the range of how many newly hatched salmon from each female survive to spawn. *(2,000 × 0.01 = 20, and 10,000 × 0.01 = 100, so approximately 20 to 100 hatchlings from each female survive to spawn.)*

3 Assessment Options

Closure

Technique: K-W-L

Tell students to return to their lists of things they want to know about fish, which they created using the **Active Reading** on page 755. Have them place check marks next to the questions that they are now able to answer. Students should finish by making a list of what they have Learned. Ask:

- Which questions are still unanswered?
- What new questions do you have?

Assign projects that require students to research their unanswered questions. PORTFOLIO

Review

Assign Review—Section 34-1.

Reteaching — BASIC

Have students make two columns on a sheet of paper. Have them label one column *External characteristics* and the other column *Internal characteristics.* Have them review the chapter and list the major internal and external characteristics of fishes. PORTFOLIO

biobit

How many eggs do female fish lay during spawning season?

The amount actually varies from species to species. Some fishes, such as seahorses, lay as few as a dozen or as many as several hundred eggs. These eggs are well protected and most develop into young seahorses. A female cod, on the other hand, may lay up to 9 million eggs in a season, most of which are eaten by predators.

Most Fishes Fertilize Their Eggs Externally

The sexes are separate in most fishes, and generally fertilization takes place externally. In a process called spawning, male and female gametes are released near one another in the water, as shown in **Figure 34-4.** A yolk sac within the egg contains nutrients the developing embryo will need for growth. The yolk sac remains attached to the hatching fish but is quickly used up. Then, the growing fish must seek its own food. More likely than not it will become food for some larger organism. To ensure that some individuals survive to maturity, many species of fishes release large numbers of eggs that are fertilized in a single spawning season.

The eggs of sharks, skates, and rays are fertilized inside the female's body. During mating, the male uses two organs called claspers to insert sperm into the female. In most species, the eggs develop inside the female and the young are born live. A few species of sharks lay eggs.

Figure 34-4 Fish spawning. These salmon spawn in shallow river waters. Because such a small percentage of hatchlings live until adulthood, thousands of eggs are released in a single mating.

Review SECTION 34-1

1. **Discuss** the key characteristics found in all fishes.
2. **Describe** how countercurrent flow aids a fish in obtaining oxygen.
3. **Summarize** why the fish heart and circulatory system are considered important evolutionary changes.
4. **Contrast** reproduction in sharks with that of most other fishes.
5. **SKILL Forming a Hypothesis** A student removes Fish A from a saltwater aquarium and Fish B from a freshwater aquarium. By mistake, the student returns each fish to the wrong aquarium. The next day, both fish are dead. Form a hypothesis that explains why the fish died.

Review SECTION 34-1 ANSWERS

1. Gills allow fish to extract oxygen from the water. Their single-loop blood circulation allows blood to be pumped through the gills and back to the heart. The vertebral column of bone surrounds and protects the spinal cord.
2. Countercurrent flow ensures that oxygen diffuses into the blood over the entire length of the gill capillaries, increasing the amount of oxygen that is taken up in the gills.
3. The fish heart pumps blood through a single-loop circulatory pathway. The blood goes to the gills and then to the body tissues. As a result, fully oxygenated blood is delivered to the tissues.
4. Fertilization in sharks is internal; most other fishes have external fertilization.
5. Fish A was unable to prevent the inflow of water by osmosis when placed in fresh water. Fish B was unable to prevent the outward flow of water by osmosis when placed in salt water. In each case, the resulting imbalance in the fish's salt concentration caused the fish's death.

Today's Fishes

Section 34-2

Hagfishes and Lampreys Are Jawless

Perhaps the most unusual fishes found today are the surviving jawless fishes, the lampreys and hagfish. These primitive creatures have changed little over the past 330 million years. Little is known about the hagfish, which is not surprising when you consider where it lives—1,700 m (about 1 mile) down on the ocean floor. Lampreys are better understood and are found in both salt and fresh water. Interestingly, all species of lamprey must return to fresh water to reproduce, suggesting that their ancestors arose in fresh water.

Lampreys and hagfishes have scaleless, eel-like bodies with multiple gill slits and unpaired fins. Their skeletons are made of cartilage, a strong fibrous connective tissue, and both retain their notochord into adulthood. Hagfishes, like the one shown in **Figure 34-5,** are scavengers of dead and dying fishes on the ocean bottom. Because of this, they are sometimes called the vultures of the sea. When threatened, hagfish produce huge quantities of slime from their roughly 200 slime glands. Recently, biologists have discovered that hagfishes are far more numerous than once thought and play a vital role in the ecology of the oceans.

Most lampreys, such as the one shown in Figure 34-5, are parasitic on other living fishes. Its mouth has a suction-cup-like structure that the lamprey uses to attach itself to its host. After attachment, the lamprey gouges out a wound with its rough tongue, feeding on blood and bits of flesh from the wound.

Figure 34-5 Hagfish and lamprey

These two modern jawless fishes have changed little over the past 330 million years.

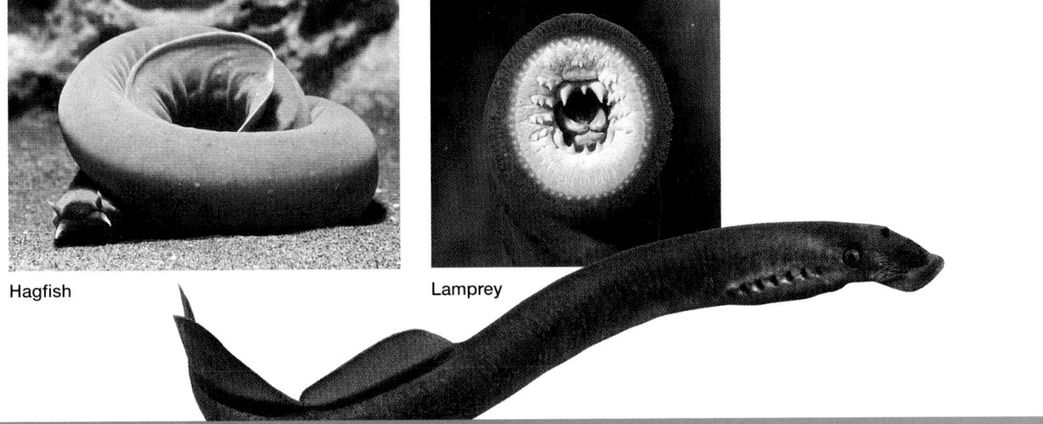

Hagfish

Lamprey

Objectives

- **Distinguish** between the three general categories of modern fishes. (pp. 761–763)
- **Describe** the major external and internal characteristics of the yellow perch. (pp. 764–765)
- **Summarize** the unique features of bony fishes. (pp. 763–767)

Key Terms

lateral line
operculum
swim bladder
teleost

REAL LIFE

Have you ever touched a hagfish? You may have touched a dead one and not known it. Products made of "eel skin," such as wallets, are actually made from the tanned skin of hagfishes.

Finding Information *Find out why the overharvesting of hagfishes for their skin is a concern to some scientists.*

Section 34-2

1 Prepare

Scheduling

- **Block:** About 90 minutes will be needed to complete this section.
- **Traditional:** About two class periods will be needed to complete this section.
- **Planning:** See the planning guide on pp. 753A and 753B for lecture, classwork, and assignment options for Section 34-2. Lesson plans for Section 34-2 can be found in a lesson cycle format in ***Biology: Principles and Explorations*** **One-Stop Planner CD-ROM with Test Generator.**

Resource Link

Assign **Active Reading Worksheet Section 34-2** to help students comprehend the key concepts and visuals in this lesson.

2 Teach

Lesson Warm-up

Provide students with pictures of the following fishes: lamprey, eel, hagfish, hammerhead shark, dogfish shark, stingray, salmon, trout, and bass. Display the pictures at the front of the room, and explain that they are classified into three distinct groups. Have students try to determine which fish belong in the same groups and why. List the characteristics students mention as similar or dissimilar for each fish. *(Students may note jaws or lack of jaws, paired or unpaired fins, scales or no scales, gill slits or no gill slits, etc.)* Discuss how structural characteristics are often important in classifying organisms, and how the fishes are good examples of this.

REAL LIFE

Answer

It may be causing the population of hagfish to decline.

Integrating Different Learning Styles

Visual/Spatial	**PE**	Study Tip, p. 763; Quick Lab, p. 766; Performance Zone items 1 and 24, pp. 776–777
	ATE	Lesson Warm-up, p. 761; Teaching Tip, p. 762; Visual Strategy, p. 762; Demonstration, p. 763
Interpersonal	**ATE**	Closure, p. 767
Intrapersonal	**ATE**	Making Biology Relevant, p. 763; Did You Know?, p. 766
Verbal/Linguistic	**PE**	Real Life, p. 761; Performance Zone items 2, 27 and 28, pp. 776–777
	ATE	Reteaching, p. 767

Teaching TIP

Natural Selection of Body Shape

To look for similarities in body shape, have students diagram the profile of several living and nonliving things that move quickly through water. Examples you might provide include a shark, a tuna, a bottlenose dolphin, a seal, a submarine, and a torpedo. All have the basic shape shown below. Ask students why this shape occurs in things that need to move rapidly through water. *(Direct students to think about water resistance and the need to direct movement through water using fins.)* PORTFOLIO

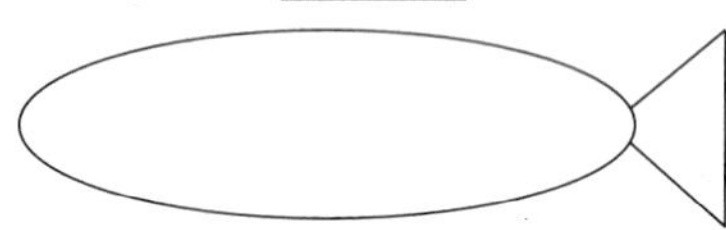

Visual Strategy

Figure 34-6

Point out how the shark's teeth are arranged in rows on its jaw, with the newer teeth behind the old ones. Explain that a shark's teeth and scales are very similar in structure and that the teeth probably evolved from scales. Both have an inner pulp cavity that is covered by a layer of dentine and then by enamel. The shark's teeth are not embedded in its jaw as human teeth are. Rather, they sit on top of the jaw and therefore break off easily.

Sharks, Skates, and Rays Have Flexible Skeletons

Sharks along with skates and rays are cartilaginous *(KAHRT'l aj uh nuhs)* fishes. Their skeletons are made of cartilage strengthened by the mineral calcium carbonate (the material oyster shells are made of). Calcium carbonate is deposited in the outer layers of cartilage and forms a thin layer that reinforces the cartilage. The result is a very light but strong skeleton.

The shark's light, streamlined body allows it to move quickly through the water in search of prey. Its skin contains cone-shaped placoid scales, which give the skin a rough texture. As you can see in **Figure 34-6,** the shark's scales and teeth are quite similar in structure. This is because the teeth are actually modified scales. The teeth are arranged in 6 to 10 rows along the shark's jaw. The teeth in front are pointed and sharp to do the work of biting and cutting. Behind the front teeth, rows of immature teeth are growing. When a functional tooth breaks or is worn down, a replacement tooth moves forward. One shark may use more than 20,000 teeth during its lifetime. This system of tooth replacement guarantees that the teeth being used are always new and sharp.

Two smaller groups of cartilaginous fishes, the skates and rays, have flattened bodies that are well adapted to life on the sea floor. Rays are usually less than 1 m (3.3 ft) long, while skates are typically smaller. However, the giant manta ray may be up to 7 m (23 ft) wide. Most species of skates and rays have flattened teeth that are used to crush their prey, mainly small fishes and invertebrates.

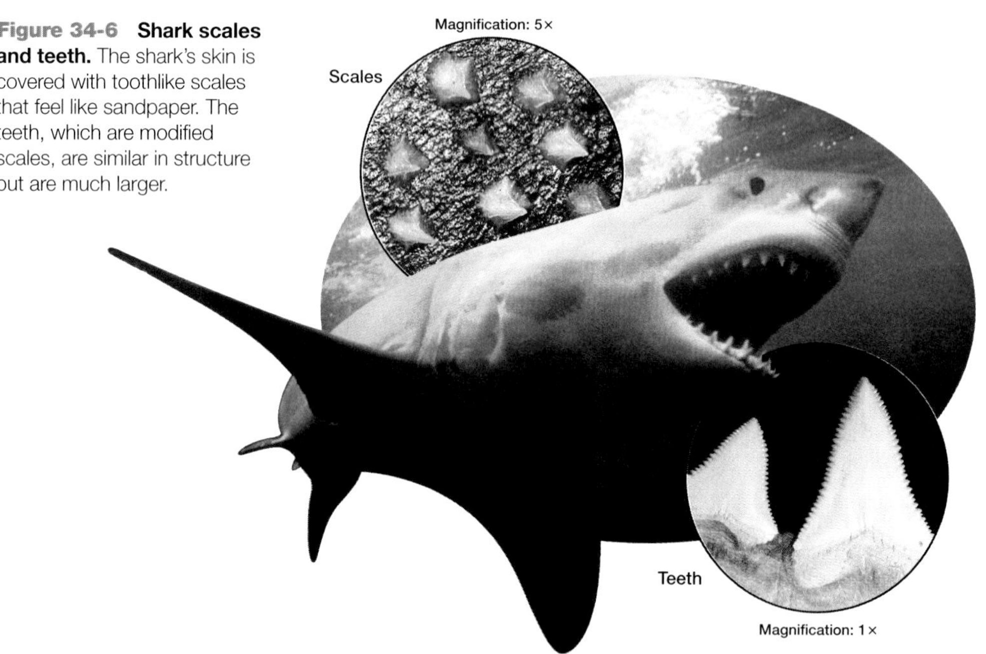

Figure 34-6 Shark scales and teeth. The shark's skin is covered with toothlike scales that feel like sandpaper. The teeth, which are modified scales, are similar in structure but are much larger.

Overcoming Misconceptions

Fear of Sharks

Shark attacks are often highly publicized. This has created a fear of shark attacks that is often greater than the actual probability of an attack. Bees, wasps, and snakes are responsible for far more fatalities each year. In the United States, the annual risk of death from being struck by lightning is 30 times greater than that from a shark attack. An attack by a shark is also less common than such beach-related injuries as spinal damage, dehydration, jellyfish and stingray stings, and sunburn.

Bony Fishes Have Three Structural Adaptations

Jawless and cartilaginous fishes do not approach the numbers or diversity seen in the bony fishes, the most numerous of all the fishes. In addition to their strong, internal skeleton made completely of bone, bony fishes have a series of unique structural adaptations that contribute greatly to their success.

1. **Lateral line system.** Although found to a limited degree in sharks, only bony fishes have a fully developed lateral line system. The **lateral line,** shown in **Figure 34-7,** is a specialized sensory system that extends along each side of a bony fish's body. As moving water presses against the lateral line, nerve impulses from ciliated sensory cells within it permit the fish to perceive its position and rate of movement. For example, a trout moving upstream to spawn uses its lateral line system to obtain the sensory information it needs to orient its head upstream.

   The lateral line system also enables a fish to detect a motionless object by the movement of water reflected off that object. In a very real sense, this is equivalent to hearing. The way that a fish detects an object with its lateral line and the way that you hear music with your inner ear are quite similar. They share the same basic mechanism—sensory cells with cilia detect vibrations and send this information to the brain.

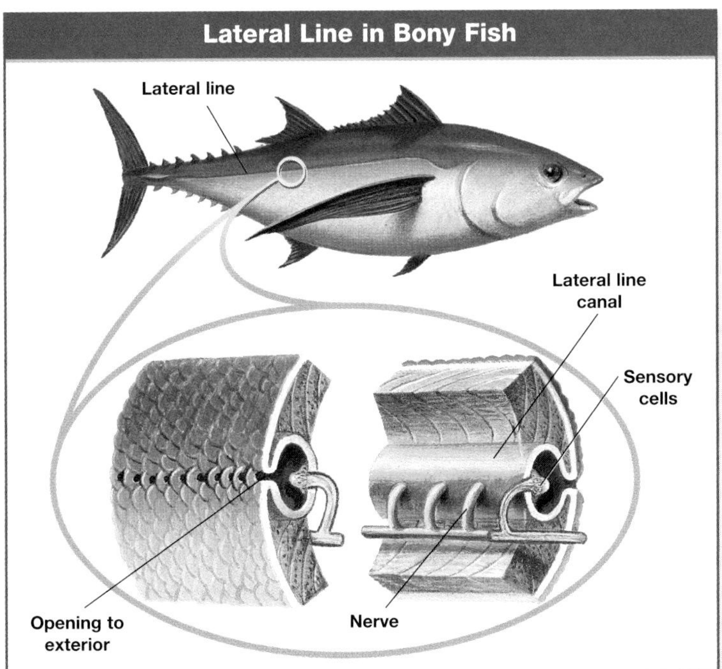

Figure 34-7 **Lateral line.** The lateral line contains sensory cells that help a fish perceive its position in the water.

Study TIP

Organizing Information

Use the information on pp. 763–767 to start a concept map that summarizes the characteristics of bony fish. Include information from *Up Close: Yellow Perch* on your concept map.

Demonstration

Internal Organs

Obtain a transparent tropical fish called a glass catfish, commonly available in pet stores. This fish gives a real-life view of the internal organs of a bony fish. Have students view the glass catfish and see what internal organs they can identify. Have them study the drawings on page 765 for help in identifying the organs.

Teaching TIP

Train Your Fish

Fish have the ability to detect pressure changes, such as sound waves, through their lateral line system, so it is possible to train a fish to come to the surface in response to a sound. Tell students who have fish to tap the side of their aquarium very gently and in the same manner each time they feed the fish. After a couple of weeks, the fish will come to the top to feed whenever the aquarium is tapped in this manner.

Making Biology Relevant

Health — ADVANCED

It is a good idea to include some fish in one's diet. Have students go to local grocery stores and record the different types of fish offered. Remind students to look for canned and frozen fish as well as fresh fish. Some students may want to survey local restaurants to determine what kind of fish they offer. Have students research these fish to determine where they were caught or raised; their habitats; their nutrient, vitamin, and mineral contents; and any other interesting characteristics of the fish, such as characteristic behaviors. PORTFOLIO

did you know?

Salmon return to their hatching stream.

Pacific salmon find their way to the stream where they hatched primarily by the sense of smell. Vegetation and minerals in the water give each stream a unique "flavor." Young salmon imprint on this flavor and can remember it 4 or 5 years later when they return to spawn. All Pacific salmon die shortly after spawning. Interestingly, when young salmon (smote stage) migrate to the sea, they go downstream *backward.*

Up Close

Yellow Perch

Teaching Strategies

- Have students relate the perch's external structures to its lifestyle and habitat. Have students pay special attention to the specialized senses and swimming adaptations of this successful fish.
- Then ask students to classify each labeled internal structure of the perch according to the following body systems: nervous system, circulatory system, reproductive system, digestive system, respiratory system, excretory (urinary) system, muscular system, and skeletal system.

Discussion

Guide a discussion about perch by posing the following questions:

1. How does the yellow perch control roll, pitch, and yaw when it swims? *(The dorsal fins prevent roll, the caudal fin controls pitch, and the anal fin controls yaw.)*

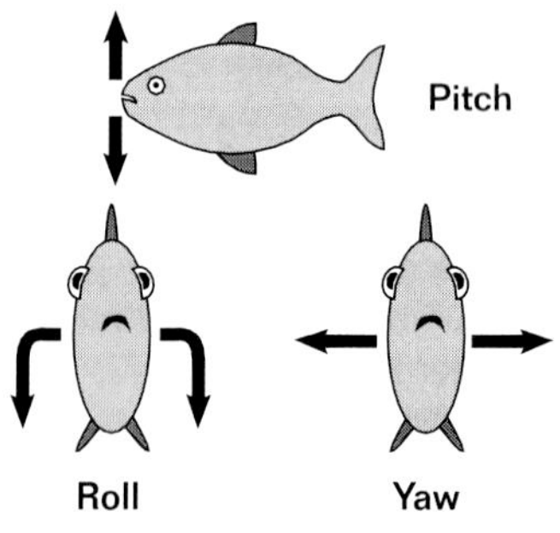

2. How would a perch find food if it were kept in the dark? *(It could smell the food and sense the food's movement with its lateral line.)*
3. Why are the mouth and opercula needed for a perch to respire? *(Movements of the opercula draw water in through the open mouth and force the water over the gills, where oxygen is removed from the water.)*

Up Close

Yellow Perch

- **Scientific name:** *Perca flavescens*
- **Size:** Grows to about 0.30 m (1 ft) long and up to 2.3 kg (5 lb)
- **Range:** Found in lakes and rivers from the Great Lakes to the Atlantic coast and as far south as South Carolina
- **Habitat:** Lives concealed among vegetation or submerged tree roots
- **Diet:** Feeds on insect larvae, crustaceans, and other fishes

External Structures

Lateral line The lateral line is a sense organ that detects vibrational disturbances in water caused by currents or pressure waves. The perch uses this sensory information to direct its movement as it swims and to detect objects in its environment, including predators and prey.

Opercula Movements of the opercula draw water into the perch's mouth. The water then moves over the gills, where oxygen and carbon dioxide are exchanged. Then the water is forced out through the opercular opening.

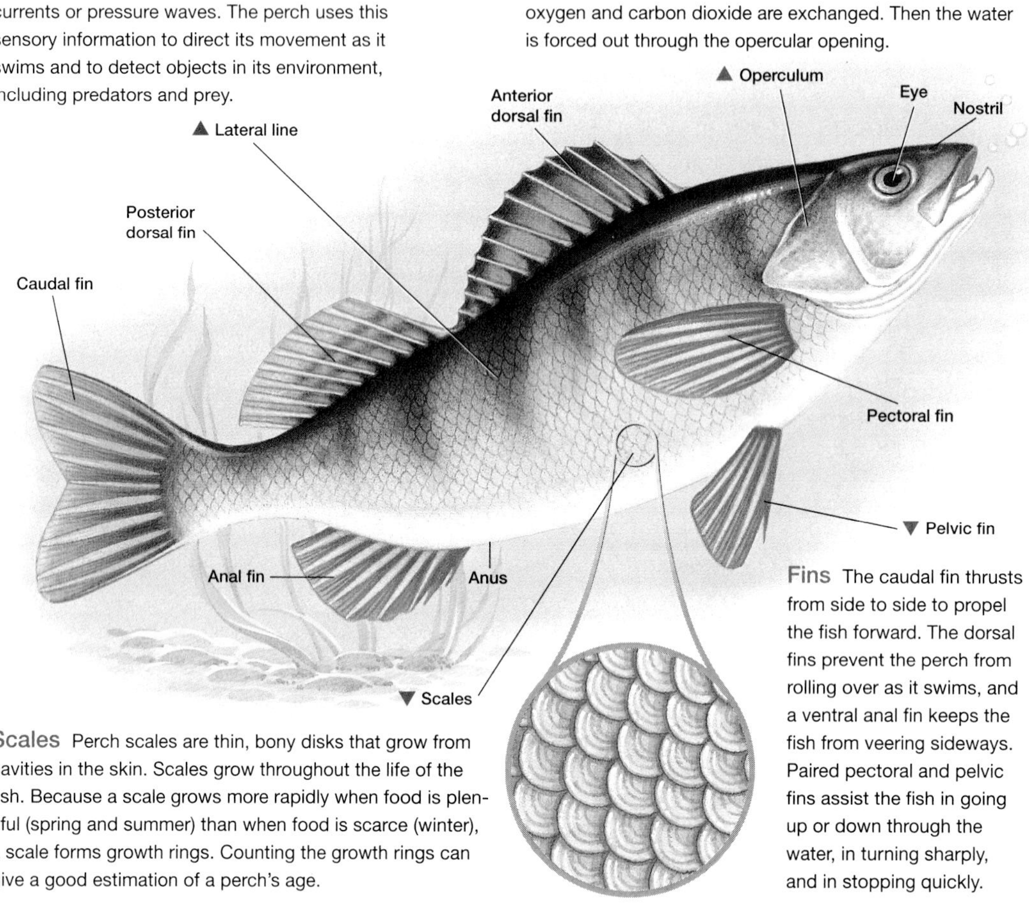

Scales Perch scales are thin, bony disks that grow from cavities in the skin. Scales grow throughout the life of the fish. Because a scale grows more rapidly when food is plentiful (spring and summer) than when food is scarce (winter), a scale forms growth rings. Counting the growth rings can give a good estimation of a perch's age.

Fins The caudal fin thrusts from side to side to propel the fish forward. The dorsal fins prevent the perch from rolling over as it swims, and a ventral anal fin keeps the fish from veering sideways. Paired pectoral and pelvic fins assist the fish in going up or down through the water, in turning sharply, and in stopping quickly.

Internal Structures

Reproductive organs Yellow perch produce gametes during their breeding season in the spring. The testes of males produce enormous numbers of sperm cells, and the ovaries of females become swollen with egg cells. The female lays strings of eggs that are wound among weeds and twigs in the water. Eggs are fertilized externally. In warm water, the young hatch within days and grow quickly. In cold water, the development of the eggs may take much longer.

Brain The optic lobe receives sensory information from the eyes, and the olfactory bulbs receive information concerning smell from chemical-sensing cells. The cerebrum processes this and other sensory information. The cerebellum coordinates muscle activity, and the medulla oblongata controls the function of many internal organs.

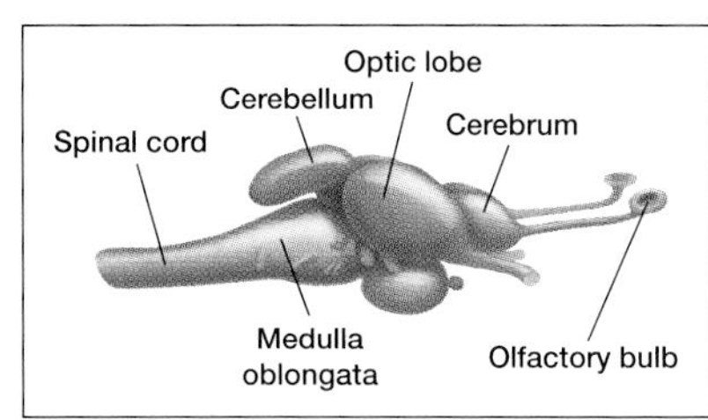

▲ Brain

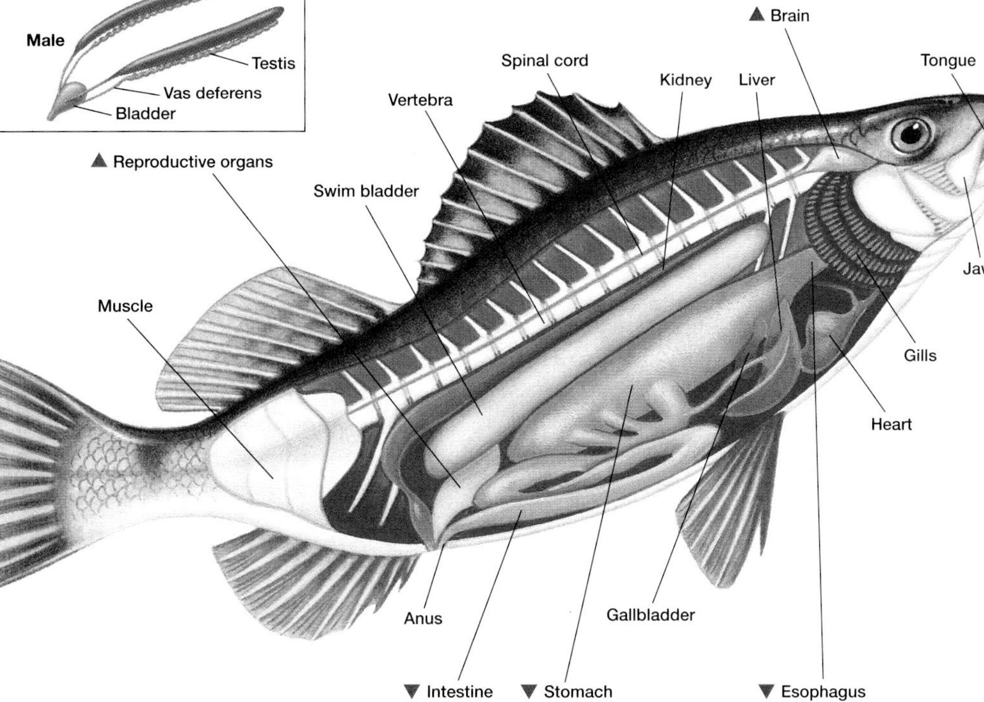

▲ Reproductive organs

▼ Intestine ▼ Stomach ▼ Esophagus

Digestive system The digestive system models a basic arrangement of structures found in all vertebrates. Food enters the mouth and passes from the esophagus to the stomach. The liver secretes bile, and the pancreas secretes enzymes into a short intestine. The bile and enzymes help break down the food. Absorption of digested food occurs through the inner lining of the intestine. Undigested food exits through the anus.

Up Close

Yellow Perch

4. What would happen to a yellow perch that was missing each body system? *(nervous system—it could not move or sense its environment; circulatory system—its cells could not get oxygen and nutrients or get rid of wastes; reproductive system—it could not produce offspring; respiratory system—it could not obtain oxygen or eliminate carbon dioxide; digestive system—it could not break down food; excretory (urinary) system—it could not get rid of nitrogen wastes; muscular system—it could not move; skeletal system—its body would collapse)*
5. What would happen if the swim bladder of a perch was suddenly unable to absorb more gas or get rid of gas? *(The perch would be unable to achieve neutral buoyancy at different levels in the water. It would have to use more energy to move up or down in the water.)*
7. What is the advantage of the one-way digestive system of the perch over a gastrovascular cavity? *(The one-way digestive system allows food to be processed continually as it moves from the mouth to the anus.)*
8. Why is the number of sperm cells produced by the perch so much greater than the number of egg cells? *(Because of external fertilization, most sperm cells will not reach the egg cells. A great number is needed to ensure that each egg is fertilized.)*

Teaching TIP

Uses for Opercula

Tell students that fish use their opercula in many ways besides respiration: for protection—when closed, the opercula provide protection for the delicate gills; for communication—Siamese fighting fish flare their opercula to exhibit aggression; and for water conservation—mudskippers close their opercula to hold water around their gills while out of water.

How can you model the action of a swim bladder?

QUICK LAB

Time 20 minutes

Process Skills Observing, analyzing, inferring, applying knowledge

Teaching Strategies

- Have all students drop their raisins into the soft drink at the same time.
- For a more dramatic effect, have students use a graduated cylinder instead of a beaker.

Analysis Answers

1. The raisins sank to the bottom. As bubbles collected on them, the raisins rose to the top. When they reached the top, they sank again.
2. The bubbles clung to the raisins and imparted buoyancy, causing the raisins to rise. When the raisins reached the top, the bubbles burst, and the raisins sank again.
3. The bubbles formed on the outside of the raisin. A swim bladder is internal. Also, the amount of gas in the swim bladder can be increased or decreased, unlike the bubbles on the raisins.
4. It would take a great deal of energy to stay in one position under water with two lungs full of air. A swim bladder offers a fish tremendous savings of energy because the fish controls the amount of gas in the swim bladder. This makes more energy available for daily activities, such as catching prey or escaping predators.

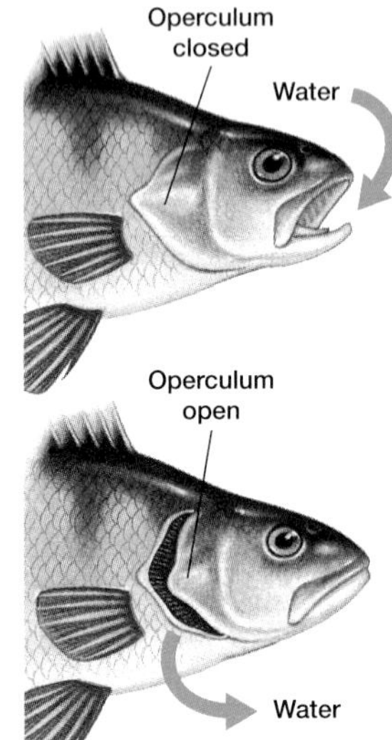

Figure 34-8 Operculum. When a fish's mouth opens, water enters and the operculum closes over the gills. When the mouth closes, the operculum opens and water moves across the gills and out of the fish.

2. **Gill cover.** Most bony fishes have a hard plate, an **operculum,** that covers the gills on each side of the head. Movements of certain muscles and of the opercula, shown in **Figure 34-8,** permit a bony fish to draw water over the gills, enabling the fish to take in oxygen. By using this mechanism, most bony fishes can move water over their gills while remaining stationary in the water. A bony fish doesn't have to swim forward with its mouth open to move water over its gills. This ability to respire without swimming enables a bony fish to conserve energy that can be spent chasing after prey and escaping from predators.
3. **Swim bladder.** The density of the fish body is slightly greater than sea water. So how do bony fish keep from sinking? Bony fishes contain a special gas sac called a **swim bladder.** By adjusting the gas content of the swim bladder, bony fishes can regulate their buoyancy. The effect is that as the bladder fills, the fish rises, and as it empties, the fish sinks. The swim bladder of early bony fishes was connected to their throat, and they could gulp air to fill it. The swim bladder of modern bony fishes, shown on page 765, does not have a direct passage to the mouth. Instead, gas is released from or absorbed into their bloodstream to fill or empty the bladder. This permits the fish to keep or change its depth in the water.

There are two groups of bony fishes, the ray-finned fishes and the lobe-finned fishes. The yellow perch described in *Up Close: Yellow Perch* on pp. 764–765 is a common type of ray-finned fish.

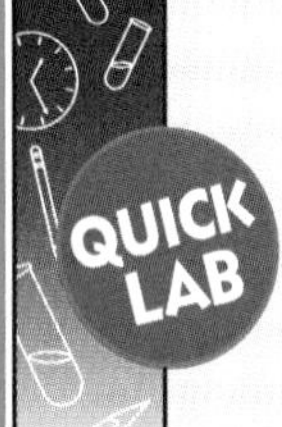

How can you model the action of a swim bladder?

QUICK LAB

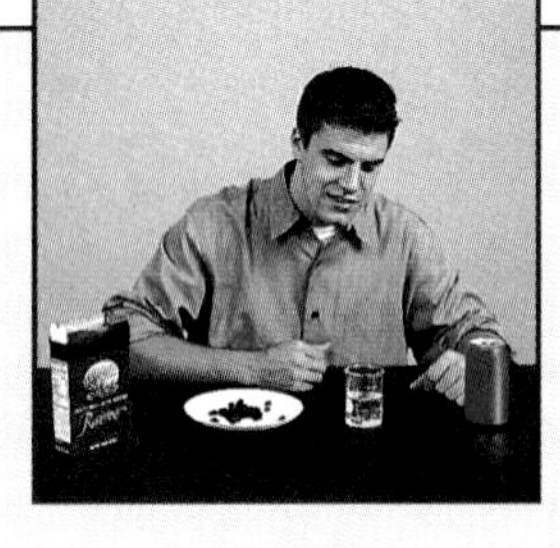

Most fish use a swim bladder to regulate their depth in water. As gas enters the swim bladder, the fish rises in the water. As gas is expelled, the fish sinks to a lower depth.

Materials

100 mL beaker or small glass; cold, clear carbonated soft drink; 2 very dry raisins

Procedure

1. Fill a 100 mL beaker with a cold, carbonated soft drink.
2. Drop two raisins into the beaker, and observe what happens over the next 5 minutes.

Analysis

1. **Describe** what happened after you dropped the raisins into the soft drink.
2. **SKILL Forming Hypotheses** Develop a hypothesis to explain your observations.
3. **SKILL Analyzing Results** How does the lifting of the raisins differ from the use of a swim bladder to control buoyancy?

Critical Thinking

4. **SKILL Forming Reasoned Opinions** Think about the energy you would have to expend to keep yourself in one position under water. What advantage might a swim bladder provide to a fish?

did you know?

Swim bladders can be damaged.

Swim bladders on some fish have been known to leak or rupture, releasing the gas into the body cavity. In some instances, the fish floats to the top of the water, causing it to be easy prey for hungry predators. Only by frenzied swimming can the fish manage to stay under water, and it will soon tire and return to float on the top. Ask students to imagine trying to dive under water while holding a large balloon.

Ray-finned bony fishes

Ray-finned bony fishes, such as the ones shown in **Figure 34-9,** comprise the vast majority of living fishes. Their fins are supported by bony structures called rays. **Teleosts** *(TEL ee ahsts),* such as the yellow perch you saw in *Up Close,* are the most advanced of the ray-finned bony fishes. Teleosts have highly mobile fins, very thin scales, and completely symmetrical tails. About 95 percent of all living fish species are teleosts.

Figure 34-9 **Pacific bluefin tuna**

Size Most of the Pacific bluefins caught range from about 7–14 kg (15–30 lb). At sexual maturity they weigh about 136 kg (300 lb), although some grow larger.

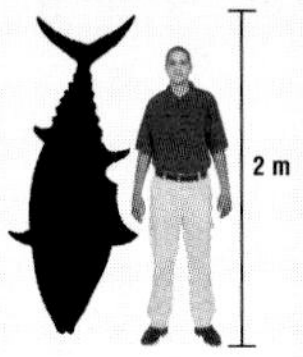

2 m

Lobe-finned bony fishes

Only seven species of lobe-finned fishes survive today. One species is the coelacanth *(SEE luh kanth),* shown in **Figure 34-10,** and the other six species are all lungfishes. The lobe-finned fishes have paired fins that are structurally very different from the fins of ray-finned fishes. In many lobe-finned fishes, each fin consists of a long, fleshy, muscular lobe that is supported by a central core of bones. These bones are connected by joints, like the joints between the bones in your hand. Bony rays are found only at the tips of each lobed fin. Muscles within each lobe can move the bony rays independently of each other.

Scientists have debated whether the direct ancestor of amphibians was a coelacanth or a lungfish. However, recent evidence has led biologists to believe that it was neither. The ancestor of the amphibians most likely was a third type of lobe-finned fish that is now extinct.

Figure 34-10 **Coelacanth.** Coelacanths were thought to have been extinct for millions of years, but one was caught off the coast of Africa in 1938. Individual coelacanths can reach up to nearly 3 m (9.8 ft) in length.

Review *SECTION 34-2*

1. **Compare** the three categories of modern fishes.
2. **Summarize** the role of the operculum in fish respiration.
3. **Summarize** how the swim bladder can be viewed as an energy-saving mechanism.
4. **Describe** the digestive process in a yellow perch.
5. **Relate** a yellow perch's lateral line system to the human ear.

Critical Thinking

6. **SKILL Evaluating Conclusions** An unidentified species of fish is caught. It has rough skin, several rows of teeth, and no opercula. Based on these characteristics, a student concludes that the fish has a swim bladder. Explain why you agree or disagree with this conclusion.

Review *SECTION 34-2 ANSWERS*

1. The agnathans are jawless and scaleless and have unpaired fins. The cartilaginous fishes have cartilaginous skeletons, streamlined bodies, and placoid scales. The bony fishes have bony skeletons, opercula, swim bladders, and a well-developed lateral line system.
2. The operculum pumps water through the mouth and over the gills for oxygen extraction.
3. The swim bladder permits a fish to maintain a particular depth in the water by adjusting the amount of gas in the bladder rather than by using muscular force.
4. Food enters the mouth and moves through the esophagus to the stomach. The liver and pancreas secrete substances into the intestine, where those substances help break down food. Absorption of digested food occurs in the intestine. Undigested food exits the body through the anus.
5. Sensory cells in both the lateral line and the human ear sense vibrations, which are converted to nerve impulses that are sent to the brain.
6. disagree; These are characteristics of cartilaginous fishes, which have no swim bladder.

Section 34-2

3 Assessment Options

Closure

Fish Characteristics

Have students work in small groups to make a table that compares each type of fish they have studied in this section. Across the top of their table they should write *Jawless Fish, Cartilaginous Fish,* and *Bony Fish.* The titles of the rows down the side of the table will vary from group to group. Have groups fill in the table with information they have learned. Each group can then present its table to the class. Have individual students copy his or her group's table and place it in his or her portfolio. PORTFOLIO

Review

Assign Review—Section 34-2.

Reteaching — BASIC

Have students review and summarize the differences between cartilaginous and bony fishes by preparing a two-column table. Across the top of the table, students should write *Sharks/rays* and *Bony fishes.* Down the left side, they should write the following questions: *Composition of skeleton, Skin covering, Has gills, Has an operculum, Has swim bladder, Has paired fins, Has lateral line system,* and *Type of fertilization* Have students complete the table without consulting their texts. PORTFOLIO

Chapter 34

1 Prepare

Scheduling

- **Block:** About 90 minutes will be needed to complete this section.
- **Traditional:** About two class periods will be needed to complete this section.
- **Planning:** See the Planning Guide on pp. 753A and 753B for lecture, classwork, and assignment options for Section 34-3. Lesson plans for Section 34-3 can be found in a lesson cycle format in ***Biology: Principles and Explorations*** **One-Stop Planner CD-ROM with Test Generator.**

Resource Link

Assign **Active Reading Worksheet Section 34-3** to help students comprehend the key concepts and visuals in this lesson.

2 Teach

Lesson Warm-up

Explain that a caecilian is a legless, wormlike animal. Ask students to close their books, and write the words *Apoda*, *Anura*, and *Urodela* at the top of three columns on the board. Tell students that these order names come from the Greek language. To the side write *a*—"without," *oura*—"tail," *pous*—"foot," and *delos*—"visible." Ask students to place the following amphibians into the proper columns by thinking of each animal's characteristics and what the three column headings mean: toad, salamander, bullfrog, caecilian, newt, and tree frog. *(Apoda—caecilian; Anura—toad, bullfrog, and tree frog; Urodela—salamander and newt)*

Section 34-3 Amphibians

Objectives

- **Summarize** the characteristics common to all modern amphibians. (pp. 768–770)
- **Compare** the three orders of living amphibians. (pp. 771–774)
- **Describe** the major external and internal characteristics of the leopard frog. (pp. 772–773)

Key Terms

lung
pulmonary vein
septum

Today's Amphibians Share Five Key Characteristics

Next time you see a frog, consider that it is a surviving member of an ancient amphibian group, the first vertebrates to walk on land. The croaking and peeping of frogs, such as the one shown in **Figure 34-11,** makes it difficult not to notice them, but their smaller, quieter relatives live nearby, hidden in damp habitats. Class Amphibia contains three orders of living amphibians: order Anura (frogs and toads), order Urodela (salamanders and newts), and order Apoda (caecilians). Most of these amphibians share five key characteristics.

1. **Legs.** The evolution of legs was an important adaptation for living on land. Frogs, toads, salamanders, and newts have four legs. Caecilians lost their legs during the evolutionary course of adapting to a burrowing existence.
2. **Lungs.** Although larval amphibians have gills, most adult amphibians breathe with a pair of lungs. Lungless salamanders are an exception.
3. **Double-loop circulation.** Two large veins called pulmonary veins return oxygen-rich blood from the lungs to the heart. Oxygen-rich blood is then pumped to the tissues at a much higher pressure than in the fish heart.
4. **Partially divided heart.** The atrium of the amphibian heart is divided into left and right sides, but the ventricle is not. A mixture of oxygen-rich and oxygen-poor blood is delivered to the amphibian's body tissues.
5. **Cutaneous respiration.** Most amphibians supplement their oxygen intake by respiring directly through their moist skin. Cutaneous respiration ("skin breathing") limits the maximum body size of amphibians because it is efficient only when there is a high ratio of skin surface area to body volume.

Figure 34-11 **Spring peeper.** In some areas, the call of the spring peeper is one of the first signs of spring.

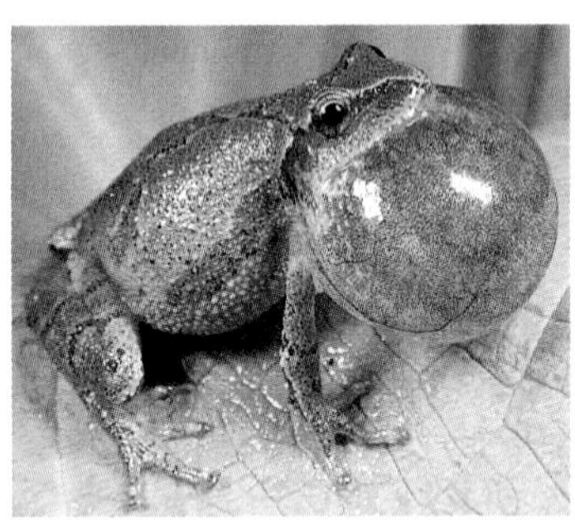

Lungs

Although air contains about 20 times as much oxygen as sea water does, gills cannot function as respiratory organs when out of water. Thus, one of the major challenges that faced the first land vertebrates was that of obtaining oxygen from air. The evolutionary solution to this need is the lung.

A **lung** is an internal, baglike respiratory organ that allows oxygen and carbon dioxide exchange between the air and the bloodstream. The amount of oxygen a lung can absorb depends on the size of its internal surface area. The greater the

Integrating Different Learning Styles

Visual/Spatial	**PE**	Performance Zone items 17, 18, 19, 20, 21, p. 777; Lab Techniques, pp. 778–779
	ATE	Teaching Tips, pp. 769, 770, 771; Visual Strategies, pp. 769, 770, 771; Demonstration, p. 770
Interpersonal	**ATE**	Active Reading, p. 774; Closure, p. 774
Verbal/Linguistic	**PE**	Study Tip, p. 769; Performance Zone items 2, 22, 23, and 25, pp. 776–777
	ATE	Lesson Warm-up, p. 768; Reteaching, p. 774

surface area, the greater the amount of oxygen that can be absorbed. In amphibians, the lungs are hardly more than sacs with folds on their inner membrane that increase their surface area, as shown in **Figure 34-12.** With each breath, fresh oxygen-rich air is drawn into the lungs. There it mixes with a small volume of air that has already given up its oxygen. Because the surface of the lung is exposed to a mixture of fresh air *and* air with a high carbon dioxide content, the respiratory efficiency of lungs is much less than that of gills. But there is much more oxygen in air than there is in water, so lungs do not have to be as efficient as gills. Many amphibians also obtain oxygen through their thin, moist skin.

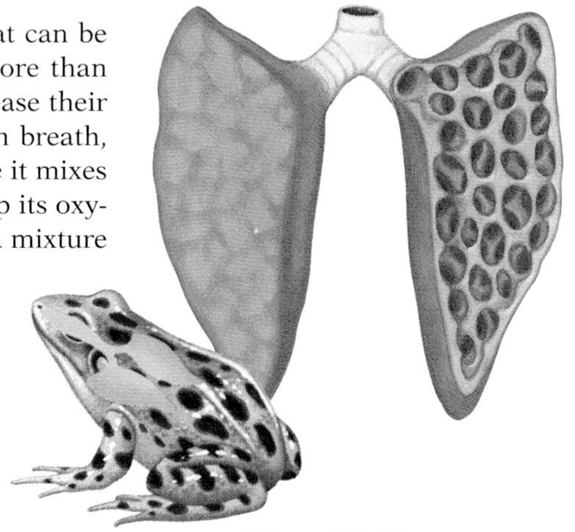

Figure 34-12
Amphibian lungs. The lungs of an amphibian are sacs with a folded internal membrane that provides a large surface for gas exchange.

Double-loop circulation

As amphibians evolved and became active on land, they needed more oxygen delivered to their muscles. So in amphibians (as well as in lungfish), the circulatory system underwent a change that created a second circulatory loop. **Figure 34-13** compares the single-loop circulation of fish with the double-loop circulation of amphibians. Notice that amphibians have a blood vessel not found in fish, the pulmonary vein. A pair of **pulmonary veins** carries oxygen-rich blood from the amphibian's lungs to its heart. The heart pumps the oxygen-rich blood out to the rest of the body. The advantage of this new arrangement is that oxygen-rich blood can be pumped to the amphibian's tissues at much higher pressures than it can in fishes. (Recall that in fish, blood is pumped through the gills before reaching the body organs. As a result, much of the force of the heartbeat is lost.)

Figure 34-13 **Circulatory loops**

Circulation in fishes is single-loop, while amphibians have a second loop that goes from the heart to the lungs and back.

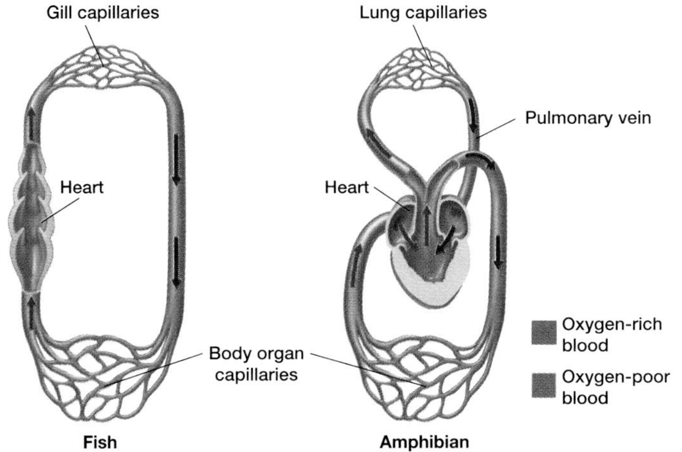

Study TIP

Compare and Contrast

To compare and contrast single-loop and double-loop circulation, make a two-column list. In one column, write the ways in which the two types of circulation are alike. In the other column, write the ways in which they are different.

Teaching TIP

Organizing Information

Have students create a Venn diagram similar to the one shown in the **Graphic Organizer** at the bottom of page 769. As students read Section 34-3, have them list, in the appropriate regions of the diagram, the characteristics that are exclusive to fish, exclusive to amphibians, and common to both groups. After the diagram is complete, have each student use the information in the diagram to write a paragraph that either supports or rejects the placement of fish and amphibians into two separate groups. PORTFOLIO

Teaching TIP — BASIC

Gas Exchange

Tell students that although most amphibians have lungs, some species of salamanders exchange oxygen and carbon dioxide through gills, even as adults. Ask students where they would expect to find these amphibians and why. *(in an aquatic environment; because gills collapse in air)* Tell students that many species of salamanders have neither lungs nor gills and respire entirely through their skin.

Visual Strategy

Figure 34-13

Use this figure to compare the circulatory pathways of fish and amphibians. Ask: Which system is a single loop and which is a double loop? *(The fish system is a single loop, and the amphibian system is a double loop.)* What is the function of the second loop in the amphibian system? *(The second loop creates a circulatory pathway to and from the lungs.)* What advantage does the double-loop system provide? *(The amphibian's heart pumps oxygen-rich blood to the amphibian's tissues at much higher pressures than can happen in the fish's single-loop system.)*

Graphic Organizer

Use this graphic organizer in ***Teaching Tip: Organizing Information*** *on page 769.*

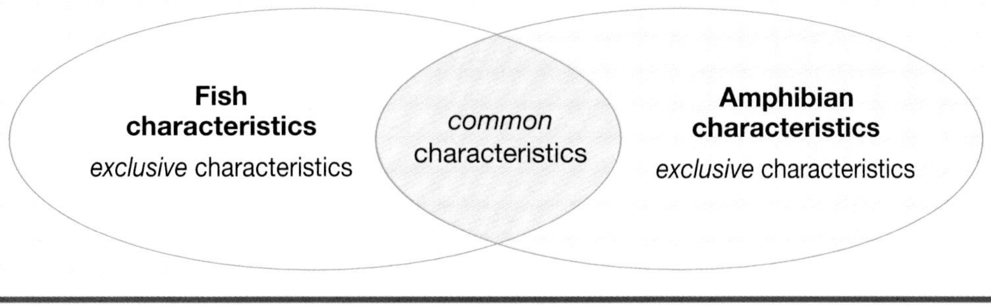

Fish characteristics
exclusive characteristics

common characteristics

Amphibian characteristics
exclusive characteristics

Teaching TIP

Comparing Heart Anatomy

Have students compare the heart of the amphibian in Figure 34-14 with that of the fish in Figure 34-3 on page 758. Ask them to list the similarities and differences between the two. Then have each student compare his or her list with that of a partner.

Demonstration

Separation of Blood

To illustrate that oxygen-rich and oxygen-poor blood tend to stay separate in the amphibian's ventricle, show how two liquids with different densities tend to separate. Fill a graduated cylinder about half full with corn syrup. Ask students what they think will happen if you add colored water to the cylinder. *(The clear corn syrup will remain at the bottom.)* Then fill the rest of the cylinder with colored water. In a second cylinder, reverse the order in which you add the liquids. Ask students to predict what will happen when you add the corn syrup to the colored water. *(The syrup will sink to the bottom.)* Tell students that the difference between the densities of oxygen-rich and oxygen-poor blood is not this large, but it is sufficient to keep them separated to a large degree.

Visual Strategy

Figure 34-14

Have students closely examine Figure 34-14. Tell them that the heart shown in the frog is drawn in *dorsal* view, with the blue sinus venosus on top. Explain that the heart shown in the enlargement is drawn in ventral view. The sinus venosus is not visible in the enlargement, but the three blue veins that carry blood to the sinus venosus are visible.

Heart

Not only did the path of circulation in amphibians change, but several important changes occurred in the heart. As you read about these changes, use **Figure 34-14** to trace the flow of blood through the amphibian heart.

The sinus venosus continues to deliver oxygen-poor blood from the body to the right side of the heart, as shown in step 1. In addition, oxygen-rich blood from the lungs enters the left side of the heart directly, as shown in step 2. The sinus venosus is on the dorsal side of the amphibian heart. In a ventral view, you can see only where blood from the sinus venosus enters the right atrium. (You can see the sinus venosus in the heart of the frog on the left side of Figure 34-14.)

A dividing wall known as the **septum** separates the amphibian atrium into right and left halves. You cannot see the septum between the right and left atria in Figure 34-14 as they are beneath the conus arteriosus. The septum prevents the mixing of oxygen-rich blood and oxygen-poor blood as each enters the heart. As shown in step 3, both types of blood empty into a single ventricle, where some mixing of oxygen-rich and oxygen-poor blood occurs. However, due to the anatomy of the ventricle, the two streams of blood remain somewhat separate, as shown in step 4. Oxygen-rich blood tends to stay on the side that exits toward the body. Oxygen-poor blood tends to stay on the side that exits toward the lungs.

A number of amphibians have a spiral valve that divides the conus arteriosus. The spiral valve helps to keep the two streams of blood separate as they leave the heart. Even so, some oxygen-poor blood is delivered to the body's tissues. Recall, however, that amphibians also obtain oxygen through their skin. This additional oxygen partly offsets the limitations of their circulatory system.

internetconnect

SciLINKS NSTA
TOPIC: Circulation in amphibians
GO TO: www.scilinks.org
KEYWORD: HX770

Figure 34-14 Amphibian heart. These four steps show how blood flows through the heart of an amphibian.

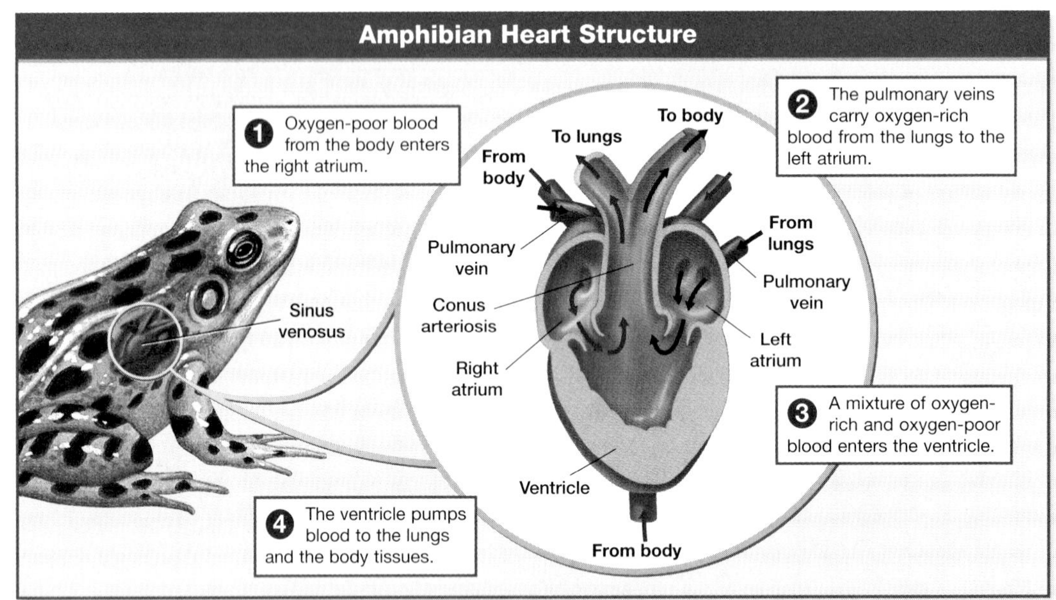
Amphibian Heart Structure
1 Oxygen-poor blood from the body enters the right atrium.
2 The pulmonary veins carry oxygen-rich blood from the lungs to the left atrium.
3 A mixture of oxygen-rich and oxygen-poor blood enters the ventricle.
4 The ventricle pumps blood to the lungs and the body tissues.
To body
To lungs
From body
From lungs
Pulmonary vein
Pulmonary vein
Conus arteriosis
Sinus venosus
Right atrium
Left atrium
Ventricle
From body

Frogs and Toads Are Adapted for Jumping

The order Anura is made up of frogs and toads that live in environments ranging from deserts to rain forests, valleys to mountains, and ponds to puddles. Adult anurans are carnivorous, eating a wide variety of insects. Some species have a sticky tongue that they extend with lightning speed to catch their prey. The anuran body, particularly its skeleton, is adapted for jumping, and its long muscular legs provide the power. Toads, such as the one shown in **Figure 34-15,** are very similar to frogs, but toads have squat bodies and shorter legs. Their skin is not smooth like that of a frog's but is covered with bumps. To learn about an anuran called the leopard frog, see *Up Close: Leopard Frog* on pp. 772–773.

Figure 34-15 **Toad.** Toads like this common Asian toad have dry, bumpy skin and short legs.

Reproduction in frogs

Like most living amphibians, frogs depend on the presence of water to complete their life cycle. The female releases her eggs into the water and a male's sperm fertilize them externally. After a few days, the fertilized eggs hatch into swimming, fishlike larval forms called tadpoles. Tadpoles breathe with gills and feed mostly on algae. After a period of growth, the body of the tadpole changes into that of an adult frog. The rate at which tadpoles develop depends on the species and the availability of food. This process of dramatic physical change, called metamorphosis, is shown in **Figure 34-16.**

Figure 34-16 **Frog life cycle**

The transition of a larval frog (tadpole) to an adult involves a complex series of external and internal body changes.

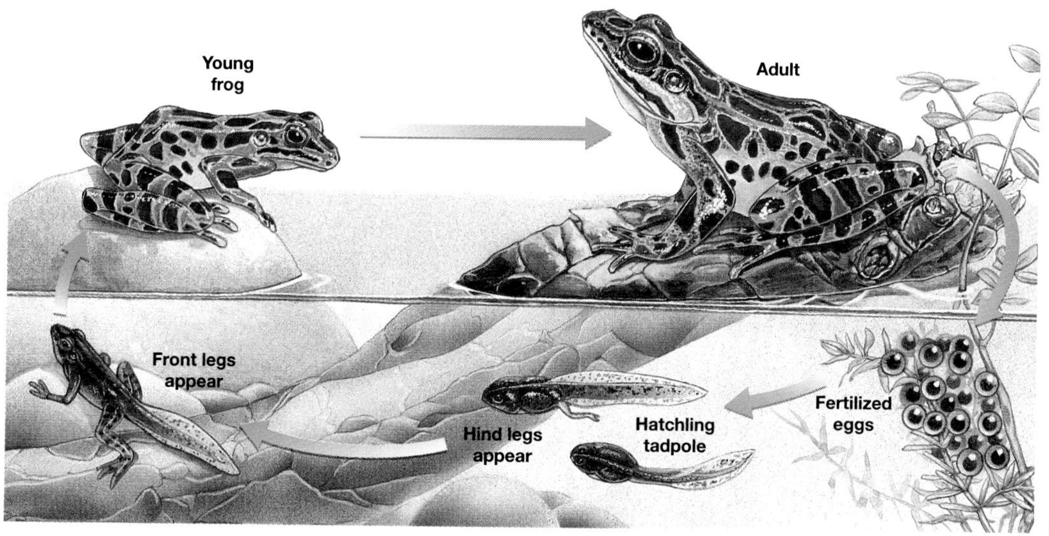

Teaching TIP — ADVANCED

Tadpole Metamorphosis

Purchase some "hindlimb bud-stage" tadpoles from a biological supply company. Keep the tadpoles in a small, aerated aquarium, and have students make written or drawn observations every few days. Students will note that the tail gets shorter, the body gets squarer, and the hindlimb buds develop into hind legs. Some tadpoles also may begin to develop front legs, but the mortality rate at this stage is high. Point out that while these dramatic external changes are taking place, important internal changes are occurring as well. In most species, the tadpole's digestive system changes to accommodate the switch from a herbivorous to a carnivorous diet. The respiratory system changes from gills to lung-and-skin respiration in the adult frog. If neither plant nor animal food is available to the metamorphosing tadpole at the precise time, it can starve. Likewise, if the tadpole does not have a place to rest its head above water, it can drown as its respiratory system changes. PORTFOLIO

Visual Strategy

Figure 34-16

Have students examine Figure 34-16. Have them note the external changes that are taking place in the frog. Then ask students to hypothesize about what internal changes are also occurring. Have them think about the different lifestyles of a frog and a tadpole. *(digestive system—herbivore to carnivore; respiratory system—gills to lungs)*

Overcoming Misconceptions

Do toads cause warts?

Some students may have the false idea that they can get warts from toads. Explain that this is not true. Warts are caused by viruses that are not carried on toads. The rough, "warty" appearance of the toad's skin is an adaptation for living successfully in dry habitats and is not due to skin problems.

Up Close

Leopard Frog

Teaching Strategies

- Have pairs of students make two columns on a sheet of paper. Ask them to label one *Water* and the other *Land*. As students read about the leopard frog, have them put the characteristics of the frog into one or both of the columns, depending on the nature of each characteristic. For example, *Tympanic membrane* would go in both columns, while *Webbed toes* would go under *Water*. Lead a discussion in which conflicting results are analyzed.
- Have students compare the organ systems of the frog with those of the yellow perch. Have them describe the similarities and differences.

Discussion

Guide discussion by posing the following questions:

1. How do you think the leopard frog got its name? *(its spotted appearance)*
2. How is the ear of the frog similar to the lateral line of the yellow perch? *(Both structures contain ciliated sensory cells.)*
3. Why do frogs produce so many eggs and sperm? *(Their eggs and larvae are eaten by many predators.)*
4. How do leopard frogs breathe? *(with lungs and through their skin)*
5. Why is a fish more agile in the water than a frog? *(A more-developed cerebellum in the fish enables it to have better muscle coordination. Also, a fish's streamlined body and fins are modifications for movement through water.)*
6. Why will a fish suffocate if it is taken out of water? *(Its gills will collapse, dry out, and no longer absorb oxygen.)*
7. How does a frog change its depth in water? *(It swims using its webbed feet.)*

Up Close

Leopard Frog

- **Scientific name:** *Rana pipiens*
- **Size:** Body length (legs excluded) of 5–9 cm (2–3.5 in.)
- **Range:** From northern Canada to southern New Mexico and from eastern California to the Atlantic coast
- **Habitat:** Lives in the short grass of meadows and around ponds
- **Diet:** Feeds on crickets, mosquitoes, and other insects

External Structures

Tympanic membrane When sound causes the tympanic membrane (eardrum) to vibrate, a tiny bone transmits the vibrations to the inner ear. There, ciliated sensory cells (similar to those found in the lateral line of a fish) detect the vibrations and help the frog maintain balance. Leopard frogs hear well in both water and air.

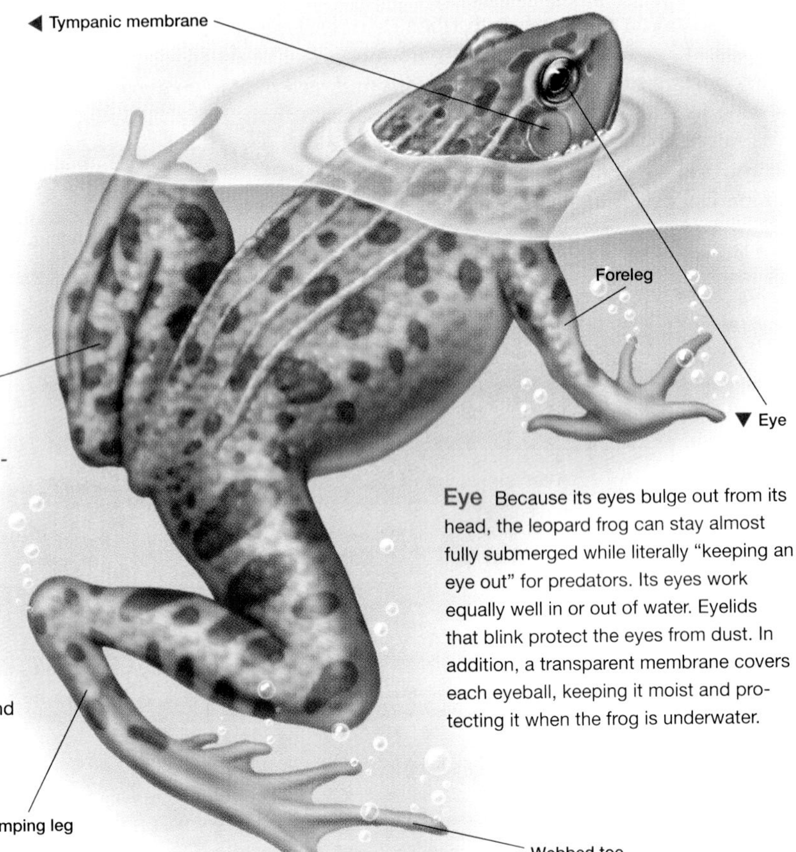

Skin Mucous glands embedded within the skin supply a lubricant that keeps the skin moist, a necessity for respiration. Unlike many frogs and toads, the leopard frog's skin glands do not secrete poisonous or foul-tasting substances. Instead, it must rely on its protective coloration and speed to evade predators.

Eye Because its eyes bulge out from its head, the leopard frog can stay almost fully submerged while literally "keeping an eye out" for predators. Its eyes work equally well in or out of water. Eyelids that blink protect the eyes from dust. In addition, a transparent membrane covers each eyeball, keeping it moist and protecting it when the frog is underwater.

did you know?

A frog's ears are external.

Most frogs and toads have a tympanic membrane on the surface of both sides of their head. The tympanic membranes of a human are called eardrums, and each is protected inside the ear canal. Ask: Why is it better for a frog or toad to have external tympanic membranes? Have students think about what happens to their ears when they go swimming.

Internal Structures

Brain The frog's brain differs from the fish's brain in that its components are more complex. For example, the larger, more complex cerebrum of a frog is able to process a wider assortment of sensory information than the cerebrum of a fish can.

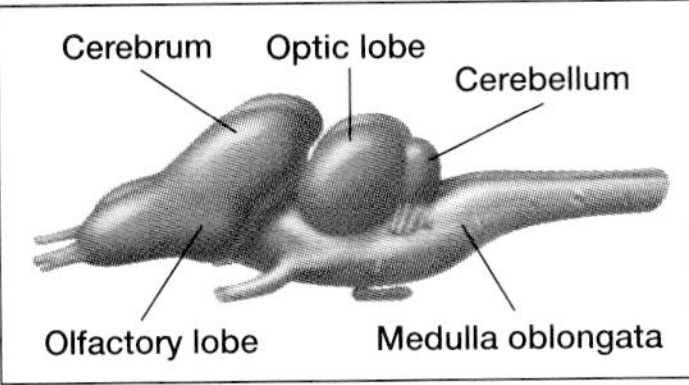

Skeleton The skeletal system of the leopard frog (and all other modern frogs) has only nine vertebrae and no ribs. The rest of the vertebrae are fused into a single bone (urostyle). When a frog is sitting, the urostyle points upward, giving the frog its characteristic humped back. When jumping, muscles extend the long hind legs, producing a powerful thrust forward.

Tongue and jaw The tongue flicks out with great speed, curls around the prey, then flicks back into the mouth. Two large teeth project inward from the roof of the mouth to impale struggling prey. The upper jaw is lined with small, sharp teeth that prevent the prey from escaping. Food is not chewed but swallowed whole.

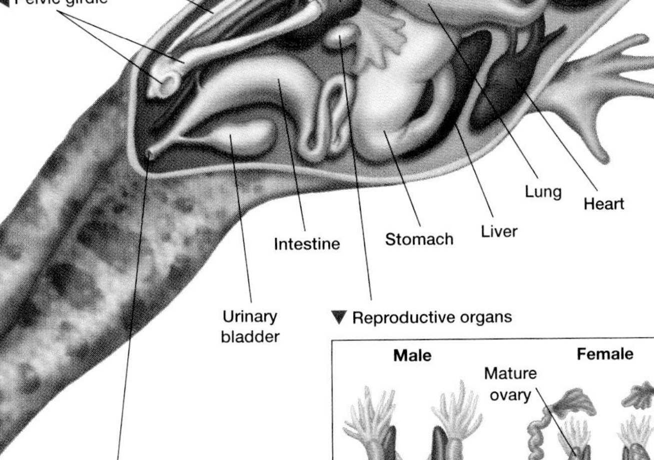

Cloaca Undigested food passes into the cloaca, a chamber that opens to the outside of the body. Urine from the kidneys travels to the bladder and then passes into the cloaca, as do gametes from the reproductive organs. All of these materials exit the body through the cloacal opening.

Reproductive organs Prior to breeding, the reproductive organs of male and female leopard frogs produce enormous numbers of gametes. The female releases a large cluster of eggs into the water. The male then discharges his sperm over them, fertilizing them externally.

Up Close

Leopard Frog

8. Why is it vital for male frogs to call to females? *(Frogs are camouflaged and may not find each other by sight alone. Each species of frog has its own mating call, which prevents females from mating with the wrong species of male.)*
9. What is the advantage of a frog of having its eyes on the top of its head? *(In water, the frog can keep most of its body hidden from predators while still being able to watch for them or for prey.)*
10. How can a frog stay underwater for so long with such small lungs? *(It can also absorb oxygen through its skin.)*

Chapter 34

Active Reading

Technique: Reading Effectively

After students read silently about salamanders and caecilians, ask volunteers to summarize the information without looking at the text. Other students should listen without interrupting. However, they should be prepared to point out any inaccuracies in the summary and add any ideas that were left out. Students may refer to the text during this clarification process.

3 Assessment Options

Closure — ADVANCED

A Frog's Life

Have students work in groups to prepare a poster showing the frog's life cycle. The poster should include drawings of each of the four stages of a frog's life. Next to each drawing, students should add a paragraph that explains what is illustrated by the drawing and what they have learned about it. For example, next to the drawing of frog eggs, the paragraph could explain how frog eggs are fertilized and where they are laid.

Review

Assign Review—Section 34-3.

Reteaching — BASIC

Have students make a list of characteristics that distinguish amphibians from fish. *(Answers might include that amphibians have legs and lungs, can live on land, exchange gases through their skin, and have a double-loop circulatory system.)* PORTFOLIO

Salamanders and Caecilians Have Elongated Bodies

Salamanders have elongated bodies, long tails, and smooth, moist skin. There are about 369 species of salamanders, all belonging to order Urodela. They typically range from 10 cm to 0.3 m (4 in. to 1 ft) in length. However, giant Asiatic salamanders of the genus *Andrias* grow as long as 1.5 m (5 ft) and weigh up to 41 kg (90 lb). Because salamanders need to keep their skin moist, they are unable to remain away from water for long periods, although some salamander species manage to live in dry areas by remaining inactive during the day.

Figure 34-17
Salamander. This four-toed salamander has deposited its eggs in a damp, mossy environment.

Salamanders lay their eggs in water or in moist places, as shown in **Figure 34-17.** Fertilization is usually external. A few species of salamanders practice a type of internal fertilization in which the female picks up a sperm packet that has been deposited by the male and places it in her cloaca. Unlike frog and toad larvae, salamander larvae do not undergo a dramatic metamorphosis. The young that hatch from salamander eggs are carnivorous and resemble small versions of the adults, except that the young have gills. A few species of salamanders, such as the North American mudpuppy and the Texas spring salamander, never lose their larval characteristics, retaining their external gills as adults.

Figure 34-18 Caecilian. Like most caecilians, this one from Colombia, South America, burrows beneath the soil and is rarely seen.

Caecilians

Caecilians (order Apoda) are a highly specialized group of tropical, burrowing amphibians with small, bony scales embedded in their skin. They live on small invertebrates found in soil. These legless, wormlike animals, shown in **Figure 34-18,** grow to about 0.3 m (1 ft) long, although some species can be up to 1.2 m (4 ft) long. During breeding, the male deposits sperm directly into the female. Depending on the species, the female may bear live young or lay eggs that develop externally. Caecilians are rarely seen, and scientists do not know a lot about their behavior.

Review SECTION 34-3

1. **Summarize** how amphibians take in oxygen.
2. **Contrast** the single-loop circulation of fish with the double-loop circulation of amphibians.
3. **Compare** the external characteristics of each order of amphibians.
4. **Compare** reproduction and development in frogs and salamanders.
5. **Relate** the tongue of the leopard frog to its feeding habits.
6. **State** why it is difficult to "sneak up" on a frog.

Critical Thinking

7. **SKILL Justifying Conclusions** Would you expect the digestive system of a tadpole to function like the digestive system of an adult frog? Justify your answer.

Review SECTION 34-3 ANSWERS

1. through gills, lungs, or skin
2. In single-loop circulation, blood passes through gills and then travels through the body. In double-loop circulation, blood is pumped to the lungs, where it receives oxygen, and then to the heart, which pumps it to the rest of the body. Blood flows faster and more efficiently in a double-loop system.
3. Apoda—small bony scales embedded in the skin, legless; Urodela—elongated bodies with tails; Anura—elongated hind legs and no tail
4. Except for a few species of salamanders, both have external fertilization. Tadpoles undergo a radical metamorphosis to become frogs; larval salamanders resemble adults with gills.
5. The tongue can be flicked out at great speed to capture prey.
6. Bulging eyes give a wide field of vision. Large tympanic membranes readily detect sounds.
7. No, most tadpoles are herbivores, whereas adult frogs are carnivores.

Study ZONE

CHAPTER 34

Use the Key Concepts and Key Terms listed below to review the main ideas in this chapter. If you need to review the meaning of a term, turn to the page indicated.

Key Concepts

34-1 The Fish Body

- All fishes have gills and a backbone, and they circulate oxygen-rich blood from their gills directly to body tissues.
- Countercurrent flow maximizes the amount of oxygen that can be extracted from water.
- The four-chambered fish heart collects oxygen-poor blood from the body and pumps it through the gills where it receives oxygen. Oxygen-rich blood then circulates to the rest of the body.
- A fish relies on its gills and a pair of kidneys to regulate its salt and water balance.
- Most fishes fertilize their eggs externally as males and females release their gametes near one another in the water.

34-2 Today's Fishes

- Hagfishes and lampreys are the only surviving jawless fishes.
- Sharks have light, highly streamlined bodies well suited for rapid swimming, making them swift and efficient predators.
- Bony fishes are the most diverse and abundant group of fishes.
- Bony fishes have an internal skeleton made completely of bone, a swim bladder, a lateral line sensory system, and a set of gill covers called the opercula.

34-3 Amphibians

- Amphibians have legs, breathe with lungs and through their skin, and have two circulatory loops.
- Most amphibians supplement their oxygen through cutaneous respiration—respiration through their moist skin.
- An amphibian lung is basically an air sac with a large surface area for gas exchange.
- The amphibian heart pumps oxygen-poor blood to the lungs and receives oxygen-rich blood from the lungs. The oxygen-rich blood is then pumped to the body.
- Salamanders are semiaquatic predators with tails, and caecilians are legless amphibians specialized for burrowing.

Key Terms

34-1

gill filament (757)
gill slit (757)
countercurrent flow (757)
nephron (759)

34-2

lateral line (763)
operculum (766)
swim bladder (766)
teleost (767)

34-3

lung (768)
pulmonary vein (769)
septum (770)

Study TIP Ready?

- *If you think you understand these Key Concepts and Key Terms, then you're ready for the Performance Zone.*

Study ZONE

CHAPTER 34

Review and Practice

Assign the following worksheets for Chapter 34:

- **Student Review Guide**
- **Concept Mapping Worksheet**
- **Test Preparation Pretest**
- **Vocabulary Worksheet**
- **Critical Thinking Review Worksheet**

Assessment Options

The following assessment options are available for this chapter:

Resource Link

- Assign **Chapter Test 34** to assess students' understanding of the chapter.

Alternative Assessment

- Have students make a timeline of the evolution of vertebrates from the first jawless fishes through amphibians. Next to each of the organisms listed on the timeline, students should describe the new characteristics that arose in the organism through natural selection. PORTFOLIO

Portfolio Assessment

- *Portfolio Assessment* at the front of this book provides options for building and evaluating student portfolios. Students and teachers can select items from the areas listed for inclusion in a portfolio.
- See items labeled PORTFOLIO on pp. 760, 762, 763, 767, 769, 771, 774, and 775.

Answer to Concept Map

The following is one possible answer to Performance Zone item 1 on page 776.

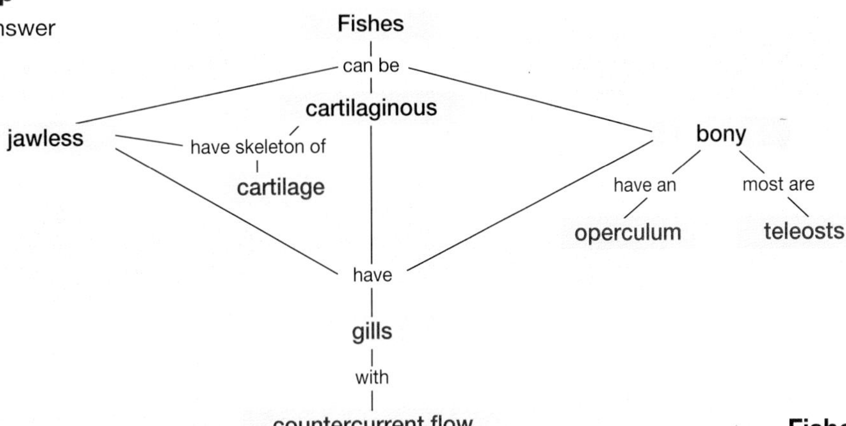

Performance ZONE

CHAPTER 34

REVIEW ANSWERS

Assignment Guide

SECTION	REVIEW
34-1	3,15, 26, 29
34-2	1, 4, 5, 6, 7, 8, 9, 24, 27, 28
34-3	2, 10, 11, 12, 13, 14, 16, 17, 18, 19, 20, 21, 22, 23, 25

Understanding and Applying Concepts

1. The answer to the concept map is found at the bottom of page 775.
2. **a.** A gill filament is the part of a gill in which capillaries for gas exchange are found. A lung is the entire organ used by many terrestrial animals for gas exchange.

   b. The swim bladder is an air-filled sac in bony fish that helps the fish maintain buoyancy in water. The lateral line is a structure in fish that contains groups of sensory cells that detect pressure and vibrations in the water.

   c. Pulmonary veins return blood to the heart from the lungs. The septum is a dividing wall in the heart.
3. d
4. d
5. b
6. d
7. a
8. c
9. c
10. d
11. b
12. a
13. a
14. a

Performance ZONE CHAPTER 34 Review

Understanding and Applying Concepts

1. **Concept Mapping** Construct a concept map describing the characteristics of jawless, cartilaginous, and bony fishes. Include the following terms in your concept map: gills, countercurrent flow, cartilage, operculum, teleosts. Include additional terms as necessary.

2. **Understanding Vocabulary** For each pair of terms, explain the difference in their meanings.
   a. gill filament, lung
   b. swim bladder, lateral line
   c. pulmonary veins, septum

3. Which of the following is not a key characteristic of fishes?
   a. vertebral column
   b. gills
   c. single-loop circulation
   d. tympanic membrane

4. Hagfish and lampreys are
   a. bony fishes.
   b. scavengers.
   c. scaled, finless fishes.
   d. primitive fishes.

5. Which of the following shark characteristics is not an adaptation for predation?
   a. streamlined body
   b. internal fertilization of eggs
   c. sharp, replaceable teeth
   d. lightweight skeleton

6. Yellow perch and sharks share all of the following characteristics except
   a. gills.
   b. an internal skeleton.
   c. a single-loop circulatory system.
   d. a swim bladder.

7. A shark's skeleton is
   a. made of cartilage.
   b. made of bone.
   c. very dense.
   d. quite rigid.

8. A bony fish's lateral line functions
   a. in respiration.
   b. as an aid in circulation.
   c. as a sensory system.
   d. as an excretory structure.

9. Bony fishes use their swim bladder
   a. to maintain salt balance.
   b. to sense their position in the water.
   c. to regulate their buoyancy.
   d. to move water over their gills.

10. Most adult amphibians respire
    a. through their skin.
    b. through their skin and gills.
    c. through their lungs.
    d. through their skin and lungs.

11. Which of the following is not a characteristic of amphibians?
    a. lungs
    b. heart with two ventricles
    c. cutaneous respiration
    d. double-loop circulation

12. Which of the following characteristics of leopard frogs is not an adaptation for avoiding predators?
    a. fast, flicking tongue
    b. skeleton adapted for jumping
    c. protective skin glands
    d. position of the eyes

13. How do tadpoles differ from frogs?
    a. Tadpoles have gills; frogs do not.
    b. Tadpoles are carnivorous; frogs are herbivorous.
    c. Frogs show body symmetry; tadpoles do not.
    d. Frogs live in water; tadpoles live in damp vegetation.

14. Caecilians are different from other amphibians in that they
    a. are legless.
    b. can breathe through their skin.
    c. have a partially divided heart.
    d. are aquatic.

15. **Theme Homeostasis** Explain how marine and freshwater fish differ in the way they maintain their salt and water balance.

16. **Chapter Links** Compare amphibian and insect metamorphosis. (**Hint:** See Chapter 31, Section 31-3.)

15. Marine fish lose water to their environment through osmosis, so they drink a lot of sea water. Because sea water contains a high concentration of salts, marine fish must actively pump excess salt out of their body. Freshwater fish actively take in salts from their environment.

16. Metamorphosis in amphibians takes place in gradual stages—from egg to tadpole to frog. In many insects, after the egg and larval stages, a pupa is formed. After a period of time, the adult form emerges. The adult looks quite different from the larva.

Skills Assessment

Part 1: Interpreting Graphics

Examine the two circulatory pathways labeled "A" and "B." Then answer the following questions:

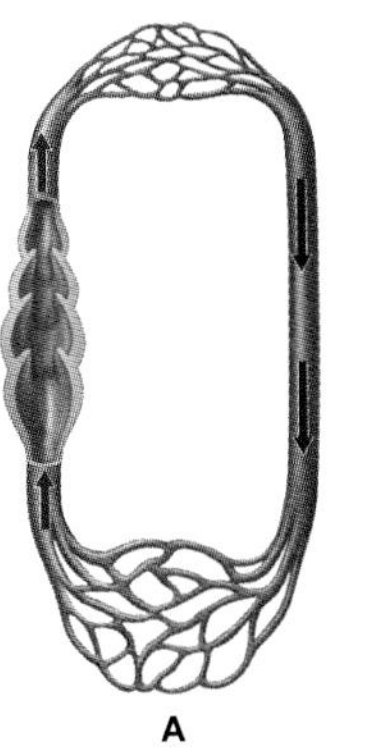

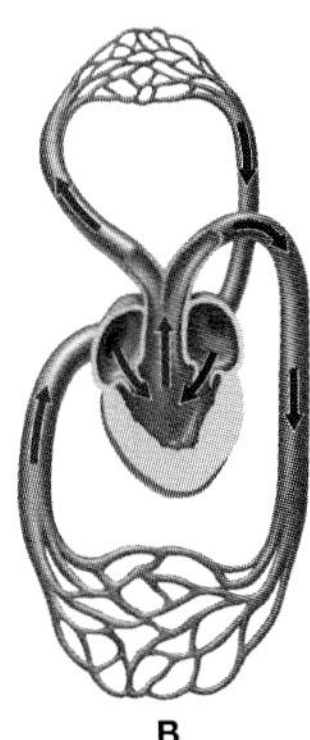

17. Identify the circulatory pathways *A* and *B* as either fish or amphibian. Defend your choice.
18. How does the heart in *A* differ from the heart in *B*?
19. In both *A* and *B*, what do the colors red and blue represent?
20. Describe the sequence of blood flow through each circulatory pathway, beginning and ending with the heart.
21. Compare the efficiency of the two circulatory pathways.

Part 2: Life/Work Skills

Use the media center or Internet resources to learn more about species of amphibians.

22. **Finding Information** Find out what amphibians live in your area of the country. Create a reference table that includes their scientific and common names and information about size, habitat, diet, and population size. Include illustrations in your table.
23. **Communicating** Prepare an oral or video presentation that explains how to use the reference table you created. Make available as a classroom reference a copy of the table.

Critical Thinking

24. **Recognizing Verifiable Facts** A newspaper article reports that some carp in a local pond are approximately 50 years old. While many readers think this is impossible, a representative from the State Department of Fish and Wildlife states that the claim can be verified. Can this claim be verified? If so, how?
25. **Distinguishing Relevant Information** A student is writing a paper on the evolution of the heart and has prepared the following list of vocabulary words: *sinus venosus, pulmonary veins, septum, osmotic balance, atrium, operculum, conus arteriosus, cutaneous respiration*. Review the list and identify any terms that do not relate to the topic, explaining why the terms are not needed.

Portfolio Projects

26. **Forming a Model** Make a model of the gill of a bony fish that shows how water passes over it. Then write a description of why countercurrent flow increases respiratory efficiency.
27. **Summarizing Information** Scientists are investigating an unexplained aspect of shark biology: sharks rarely get cancer. Use the media center or Internet resources to research this phenomenon. Write a report that includes information on the use of shark-cartilage dietary supplements.
28. **Interpreting Information** Read "Secrets of the Slime Hag," by Frederic H. Martini (*Scientific American*, Oct. 1998, vol. 279, pp. 70–75). What unusual method do slime hags use to repel attacks? Why is it risky to let the harvest of slime hags go unregulated?
29. **Career Focus Ichthyologist** Research the field of ichthyology (the study of fishes), and find out what types of career options are available to an ichthyologist. Your report should include job descriptions, training required, kinds of employers, growth prospects, and starting salaries.

Skills Assessment

17. A—fish; B—amphibian; Fishes have a single-loop circulation, which is shown in A. Amphibians have a double-loop circulation, which is shown in B.
18. A has only one atrium; B has two.
19. red—oxygen-rich blood; blue—oxygen-poor blood
20. A—heart, gills, body organs, heart; B—heart, lungs, pulmonary vein, heart, body organs, heart
21. The fish heart slowly delivers oxygen-rich blood to the body tissues. The amphibian heart delivers mostly oxygen-rich blood to the tissues at a higher rate.
22. Tables will vary. The table should list frogs and toads separately from salamanders and newts.
23. Presentations will vary. They should explain all features of the reference table.

Critical Thinking

24. Yes, one could determine the age of the carp by counting the number of growth rings on one of its scales.
25. Osmotic balance refers to the concentration of salts in the body fluids. The operculum is the cover over the gills. Cutaneous respiration is the process of taking in oxygen through the skin.

Portfolio Projects

26. Models should resemble the drawings in Figure 34-2. Countercurrent flow ensures that oxygen diffuses into the blood over the entire length of the capillaries in the gills.
27. Reports will vary. Some arthritis patients take shark-cartilage dietary supplements in the hope that the supplements will increase concentrations of cartilage components in their painful joints.
28. The hag oozes large quantities of slime that expand in sea water and coat the attacker. Slime hags are part of a complex, ocean-bottom ecosystem. Their role in this ecosystem is not well understood, so it is risky to permit overharvesting.
29. Ichthyologists study aspects of fish anatomy, physiology, and ecology. They are employed by federal and state wildlife agencies in park systems, by fish hatcheries, and by zoos and wildlife parks. Ichthyologists may also be hired to conduct environmental impact studies when commercial development plans might affect fish habitats. Training requirements vary according to position, but most positions require at least a master's degree. Growth prospects are fair. Starting salary will vary by region.

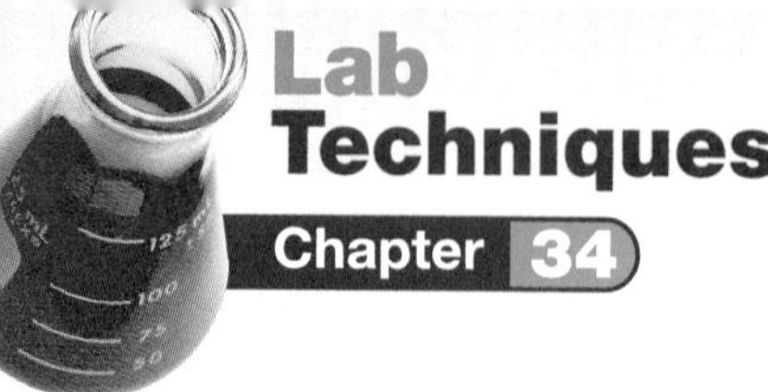

Lab Techniques

Chapter 34

Observing a Live Frog

Purpose
Students will examine a live frog in both a terrestrial environment and an aquatic environment.

Preparation Notes
Time Required: 45 minutes

Materials
Materials for this lab can be ordered from WARD'S. See *Master Materials List* at the front of this book for catalog numbers.

Preparation Tips
Provide each group with a frog in a small terrarium.

Safety Precautions
Have students follow the guidelines from the National Association of Biology Teachers for handling vertebrate animals.

Procedural Tips
Have groups take turns releasing their frog into the freshwater aquarium. After students complete their observations, remove each frog from the aquarium yourself to avoid injury to the frogs.

Disposal
- If you use frogs that are not native to your area, do not release them into the local environment when the investigation is completed.
- Indigenous frogs may be released into an appropriate habitat. Students may take indigenous frogs home to keep as pets if they have an appropriate habitat for them.

Lab Techniques

Chapter 34 *Observing a Live Frog*

SKILLS
- Observing
- Relating

OBJECTIVES
- **Examine** the external features of a frog.
- **Observe** the behavior of a frog.
- **Explain** how a frog is adapted to life on land and in water.

MATERIALS
- live frog in a terrarium
- live insects (crickets or mealworms)
- 600 mL beaker
- aquarium half-filled with dechlorinated water

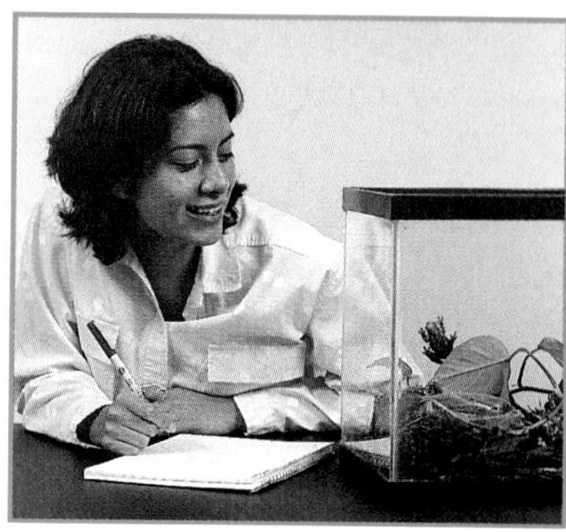

Before You Begin
Frogs, which are **amphibians,** are adapted for living on land and in water. For example, a frog's eyes have an extra eyelid called the **nictitating membrane.** This eyelid protects the eye when the frog is underwater and keeps the eye moist when the frog is on land. The smooth skin of a frog acts as a respiratory organ, exchanging oxygen and carbon dioxide with the air or water. The limbs of a frog enable it to move both on land and in water. In this lab, you will examine a live frog in both a terrestrial environment and an aquatic environment.

1. Write a definition for each boldface term in the paragraph above and for the following term: tympanic membrane.
2. Make a data table similar to the one below.

DATA TABLE

Behavior/structure	Observations
Breathing	
Eyes	
Legs	
Response to food	
Response to noise	
Skin	
Swimming behavior	

3. Based on the objectives for this lab, write a question you would like to explore about frogs.

Procedure
PART A: Observing a Frog

1. Observe a live frog in a terrarium. Closely examine the external features of the frog. Make a drawing of the frog. Label the eyes, nostrils, tympanic membranes, front legs, and hind legs.
2. Watch the frog's movements as it breathes air into and out of its lungs. Record your observations.
3. Look closely at the frog's eyes, and note their location. Examine the upper and lower eyelids as well as a third transparent eyelid called a nictitating membrane. Describe how the eyelids move.

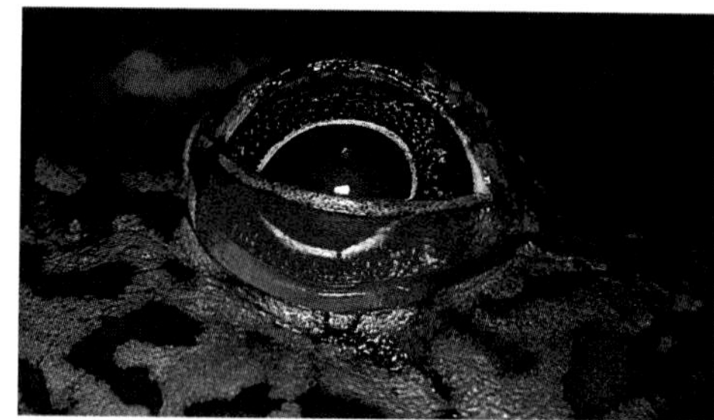

Before You Begin
Answers

1. *amphibians*—animals that are adapted for living both on land and in water, usually referring to vertebrates in the class Amphibia
   nictitating membrane—the extra eyelid of frogs' eyes
   tympanic membrane—membrane located on a frog's head that receives vibrations and transmits them to the inner ear
2. Students' data tables should resemble the one on page 778.
3. Answers will vary. For example: How are frogs adapted for living on land and in water?

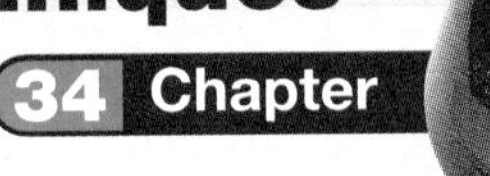

Lab Techniques

34 Chapter

4. Study the frog's legs, and note the difference between the front and hind legs.
5. Place a live insect, such as a cricket or a mealworm, into the terrarium. Observe how the frog reacts.
6. Tap the side of the terrarium farthest from the frog, and observe the frog's response.
7. Place a 600 mL beaker in the terrarium. **CAUTION: Handle live frogs gently. Frogs are slippery! Do not allow a frog to injure itself by jumping from a lab table to the floor.** Carefully pick up the frog, and examine its skin. How does it feel? Now place the frog in the beaker. Cover the beaker with your hand, and carry it to a freshwater aquarium. Tilt the beaker, and gently lower it into the water until the frog swims out.
8. Watch the frog float and swim. Notice how the frog uses its legs to swim. Also notice the position of the frog's head. As the frog swims, bend down to view the underside of the frog. Then look down on the frog from above. Compare the color on the dorsal and ventral sides of the frog.

PART B: Cleanup and Disposal

9. Dispose of broken glass in the designated waste containers. Put live animals in the designated containers. Do not pour chemicals down the drain or put lab materials in the trash unless your teacher tells you to do so.
10. Clean up your work area and all lab equipment. Return lab equipment to its proper place. Wash your hands thoroughly before you leave the lab and after you finish all work.

Analyze and Conclude

1. **Summarizing Information** How does a frog use its hind legs for moving on land and in water?
2. **Recognizing Relationships** How does the position of a frog's eyes benefit the frog while it is swimming?
3. **Analyzing Data** What features of an adult frog provide evidence that it has an aquatic life and a terrestrial life?
4. **Analyzing Methods** Were you able to determine in this lab how a frog hears? Explain.
5. **Inferring Conclusions** What can you infer about a frog's field of vision from the position of its eyes?
6. **Forming Hypotheses** How is the coloration on the dorsal and ventral sides of a frog an adaptive advantage?
7. **Further Inquiry** Write a new question about frogs that could be explored with another investigation.

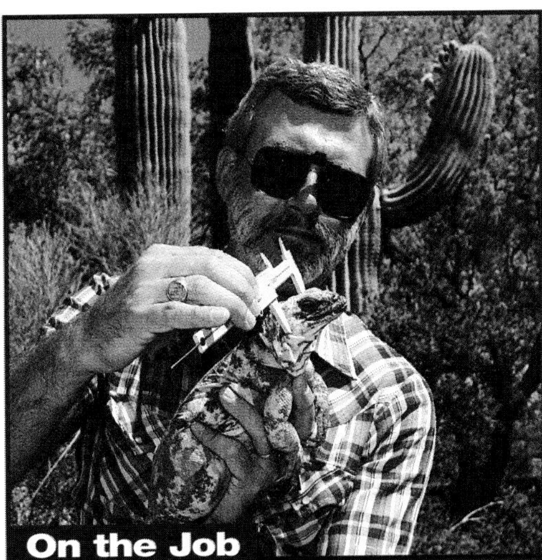

On the Job

Herpetology is the study of reptiles and amphibians. Do research in the library or media center or on the Internet to discover how herpetologists are working with the Declining Amphibian Task Force (FROGLOG) to solve the mystery of the worldwide decline in amphibian populations. Summarize your findings in a written or oral report.

Procedure

Answers

1. Students' drawing should resemble the illustrations on pp. 772–773.
2. The floor of the frog's mouth moves up and down as it breathes.
3. The eyes are on top of the head. The upper and lower eyelid are fixed. The third eyelid moves down to cover the eye.
4. The front legs are short. The hind legs are long and heavily muscled; they fold alongside the frog's body.
5. The frog may strike at the insect or worm, catching it with its tongue.
6. Answers will vary. Tapping might startle the frog.
7. The frog probably has smooth, moist skin.
8. The frog uses its hind legs to move through the water. The frog's head stays partially above water. The ventral side of the frog is light in color; the dorsal side is dark.

Analyze and Conclude

Answers

1. A frog uses its long hind legs for jumping on land and for swimming in water.
2. Having eyes on the top of its head enables a frog to see above the waterline while the rest of the frog is submerged.
3. aquatic life—webbed feet, eyes on top of head, respires through skin; terrestrial life—jumping legs, breathes with lungs
4. Answers will vary. Students should recognize that although they observed the tympanic membrane, they cannot determine how a frog hears by observing its actions.
5. A frog can see in almost all directions.
6. As viewed from above, the dark dorsal skin blends in with the color of the water, making the frog less visible to predators from above. Viewed from below, the light ventral skin blends in with the color of the sky, making the frog less visible to aquatic predators.
7. Answers will vary. For example: How do two frogs interact on land and in water?

On the Job

Answers

Herpetologists study the anatomy, physiology, and ecology of reptiles and amphibians. They are employed by federal and state wildlife agencies in park systems, wilderness areas, and animal refuges; by zoos; and by wildlife parks. Some may be professors at universities where they teach and conduct research. Training requirements vary according to position, but most positions require at least a master's degree. Starting salary will vary by region.

Chapter **35**

Reptiles and Birds

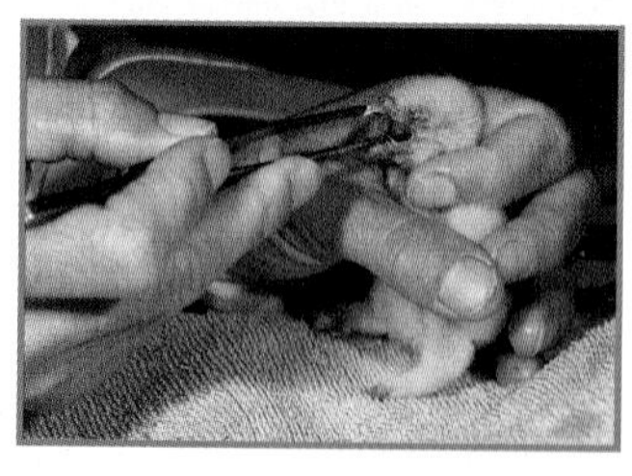

Planning Guide

	OBJECTIVES	MEETING INDIVIDUAL NEEDS	READING SKILLS AND STRATEGIES
BLOCKS 1 & 2 90 min	**35-1 The Reptilian Body** • **Describe** the key characteristics of reptiles. • **Relate** a reptile's endothermic metabolism to its activity level. • **Summarize** the adaptations that enable reptiles to live on land.	Learning Styles, p. 782 BASIC ADVANCED • pp. 786, 787	**ATE** Active Reading, Technique: Discussion, p. 781 ■ Active Reading Guide, worksheet (35-1) ■ Directed Reading, worksheet (35-1)
BLOCKS 3 & 4 90 min	**35-2 Today's Reptiles** • **Compare** the four living orders of reptiles. • **Describe** the timber rattlesnake's adaptations for locating and capturing prey. • **Compare** the parental care of crocodilians with that of other reptiles.	Learning Styles, p. 788 BASIC ADVANCED • pp. 788, 792, 793	■ Active Reading Guide, worksheet (35-2) ■ Directed Reading, worksheet (35-2)
BLOCKS 5 & 6 90 min	**35-3 Characteristics and Diversity of Birds** • **Describe** the key characteristics of birds. • **Describe** how a bird's feathers and bone structure aid flight. • **Summarize** how a bird's lungs and heart are adapted for high efficiency. • **Relate** the structure of a bird's feet and beak to its habits and diet.	Learning Styles, p. 794 BASIC ADVANCED • pp. 795, 802	**ATE** Active Reading, Technique: Paired Summarizing, p. 795 ■ Active Reading Guide, worksheet (35-3) ■ Directed Reading, worksheet (35-3) ■ Supplemental Reading Guide, *Through a Window: My Thirty Years with the Chimpanzees of Gombe*

Review and Assessment Resources

BLOCKS 7 & 8 90 min

TEST PREP

PE Study Zone, p. 803

PE Performance Zone, pp. 804–805

■ Chapter 35 worksheets:
- Vocabulary
- Concept Mapping
- Critical Thinking

 Audio CD Program

■ Test Prep Pretest

CHAPTER TESTING

■ Chapter Tests (blackline master)

▲ Test Generator

■ Assessment Item Listing

■ Performance-Based Assessment Activity: The Challenge of Water Retention

See the correlations table at the front of this book for a list of the ***National Science Education Standards*** addressed in this chapter.

Smithsonian Institution® Visit **www.si.edu/hrw** for additional on-line resources.

KEY

Biology Principles and Explorations

PE Pupil's Edition
ATE Teacher's Edition

HOLT BIOSOURCES

Teaching Transparencies

Laboratory Program

One-Stop Planner CD-ROM
Shows these resources within a customizable daily lesson plan:

■ Teaching Resources
▲ Test Generator

VISUAL STRATEGIES	CLASSWORK AND HOMEWORK	TECHNOLOGY AND INTERNET RESOURCES
 **ATE** Fig. 35-2, p. 783  173 Evolutionary Relationships of Reptiles and Descendants	**PE** **Data Lab** Identifying ectotherms, p. 783 **PE** **Quick Lab** How does watertight skin benefit reptiles?, p. 784 **PE** **Review,** p. 787	 Holt Biology Videodiscs Section 35-1 Audio CD Program Section 35-1  internetconnect SCLINKS NSTA TOPICS: • Characteristics of reptiles • Amniotic egg
ATE Fig. 35-10, p. 792 174 Orders of Living Reptiles	Lab A22 Comparing Animal Eggs ■ Occupational Applications: Emergency Medical Technician **PE** **Review,** p. 793	 Video Segment 28 What's Slithering in Guam  Holt Biology Videodiscs Section 35-2  Audio CD Program Section 35-2
 **ATE** Table 35-1, p. 801 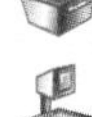 175 Major Orders of Birds  176 Major Orders of Birds, cont.  182 Avian Lung Structure 143A Structure of a Feather	**PE** **Math Lab** Calculating average bone density, p. 802 **PE** **Experimental Design** What Causes Anole Lizards to Change Color?, pp. 806–807 Lab C40 Conducting a Bird Survey **PE** **Review,** p. 802 **PE** **Study Zone,** p. 803	Holt Biology Videodiscs Section 35-3  Audio CD Program Section 35-3  internetconnect SCLINKS NSTA TOPICS: • Characteristics of birds • Flightless birds • Adaptations of reptiles

ALTERNATIVE ASSESSMENT

PE **Portfolio Projects 25–30,** p. 805
ATE **Alternative Assessment,** p. 803
ATE **Portfolio Assessment,** p. 803

Lab Activity Materials

Data Lab
p. 783 pencil, paper

Quick Lab
p. 784 forceps, grapes (2), balance, Petri dish

Math Lab
p. 802 pencil, paper

Experimental Design
p. 806 glass-marking pencil; large, clear jars with wide mouths and lids with air holes (2); live anoles (2); 6 shades each of brown and green construction paper, ranging from light to dark (2 swatches of each shade)

Lesson Warm-up
ATE photo of a skink, p. 788
ATE feather, p. 794

Demonstrations
ATE live, nonpoisonous snake, p. 780
ATE cooked, clean chicken bones, p. 795
ATE pictures of several birds, p. 797

BIOSOURCES LAB PROGRAM
- Quick Labs: Lab A22, p. 43
- Lab Techniques and Experimental Design: Lab C40, p. 193

Chapter Theme

Evolution

Students may have the misconception that the class Reptilia is unsuccessful because of the extinction of the dinosaurs. In fact, reptiles are successful and diverse, inhabiting all of the continents except Antarctica and numbering about 7,000 species. In diversity among terrestrial vertebrates, reptiles are second only to birds.

Establishing Prior Knowledge

Before beginning this chapter, make certain students can answer the questions in Study TIP **Ready?** on page 781. If they cannot, encourage them to reread the sections indicated.

Opening Demonstration

Have students make a list of some of their preconceptions about snakes. For instance, many students believe that a snake's skin is slimy, like that of a frog. Another misconception is that the tongue of a snake is poisonous. Show the class a living, nonpoisonous snake, such as a gopher snake or garter snake, to dispel these misconceptions. Allow students to examine and touch its skin, observe the flicking of its tongue, and watch its movements. Discuss with students how misconceptions contribute to the fear of snakes. (Be sensitive to the fact that many people fear snakes.) PORTFOLIO

CHAPTER 35 Reptiles and Birds

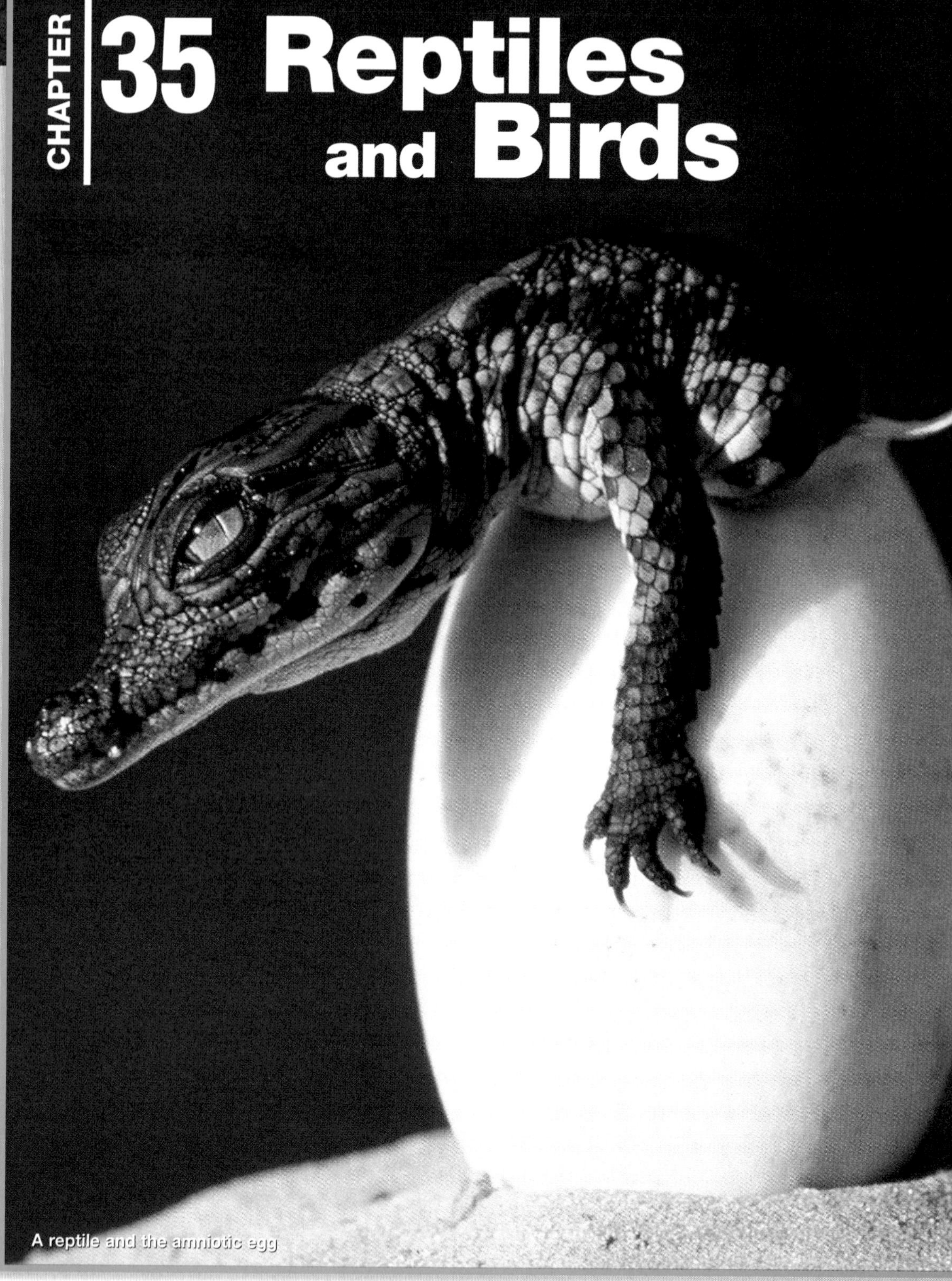

A reptile and the amniotic egg

What's Ahead — and Why
CHAPTER OUTLINE

35-1 The Reptilian Body
- Reptiles Share Several Key Characteristics
- A Key Adaptation to Terrestrial Life Is Water Retention
- Reptiles Need More Oxygen Than Amphibians
- Reptiles Have Internal Fertilization

35-2 Today's Reptiles
- Lizards and Snakes Have a Unique Jaw Design
- Other Orders of Reptiles Are Less Diverse

35-3 Characteristics and Diversity of Birds
- Birds Share Several Key Characteristics
- Birds Are Adapted for Different Ways of Life

Students will learn
- the key characteristics of reptiles.
- that reptiles have adaptations for life on land.
- how the members of each of the four orders of reptiles are adapted to their particular lifestyle.
- the key characteristics of birds and their adaptations for flight.

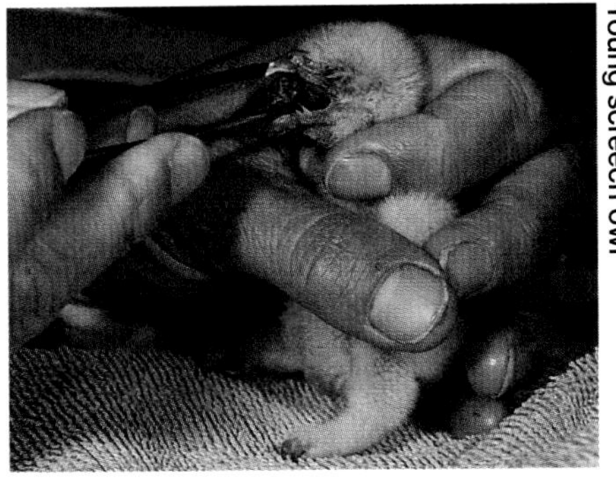
Young screech owl

At birth, reptiles look like their parents. *And most are on their own after hatching. Baby birds, on the other hand, appear quite different from their parents and many require a lot of care. This orphaned owl will need to be fed every 20 minutes, from sunrise to sunset, if it is to survive.*

Looking Ahead

35-1 The Reptilian Body
They're more than coldblooded creatures.

35-2 Today's Reptiles
Some are elusive, but others let you know they're around.

35-3 Characteristics and Diversity of Birds
What does it mean to be free as a bird?

Labs

- **Data Lab**
  Identifying ectotherms (p. 783)
- **Quick Lab**
  How does watertight skin benefit reptiles? (p. 784)
- **Math Lab**
  Calculating average bone density (p. 802)
- **Experimental Design**
  What Causes Anole Lizards to Change Color? (p. 806)

Features

- **Exploring Further**
  The Amniotic Egg (p. 785)
- **Up Close**
  Timber Rattlesnake (pp. 790–791)
  Bald Eagle (pp. 798–799)

internet**connect**

National Science Teachers Association *sci*LINKS Internet resources are located through-out this chapter.

Study TIP

Ready?

Check your understanding of these topics to help you prepare for what's ahead.

Can you . . .

- **describe** the process of molting in arthropods? (Section 31-1)
- **distinguish** between ectotherms and endotherms? (Section 33-2)
- **summarize** the evolutionary relationships between reptiles and birds? (Section 33-2)
- **relate** countercurrent flow to the efficiency of the fish gill? (Section 34-1)

If not, *review the sections indicated.*

Reading Strategies

Active Reading

Technique: Discussion

Before students begin reading the chapter, have a volunteer explain how the vertebrates they learned about in Chapter 34 were tied to water. Tell students that in this chapter they will learn about the first vertebrates to fully adapt to life on land—the reptiles—and about their relatives, the birds and mammals.

Directed Reading

Assign **Directed Reading Worksheet Chapter 35** to help students interact with the text as they read the chapter.

Interactive Reading

Assign ***Biology: Principles and Explorations* Audio CD Program Chapter 35** to help reluctant readers achieve greater success in reading the chapter.

Reading Skills

Assign Reading Organizers and Reading Strategies Worksheets from the **One-Stop Planner CD-ROM with Test Generator** to help students learn and practice effective reading skills.

internet**connect**

Holt *Biology: Principles and Explorations*

On-line Resources: go.hrw.com

Visit the HRW Web site for a variety of resources related to this chapter. Just type in the keyword HX0 Chapter 35.

Holt *Biology: Principles and Explorations*

On-line Resources: www.scilinks.org

The following *sci*LINKS Internet resources can be found in the student text for this chapter:

Page 782
TOPIC: Characteristics of reptiles
KEYWORD: HX782

Page 785
TOPIC: Amniotic egg
KEYWORD: HX785

Page 794
TOPIC: Characteristics of birds
KEYWORD: HX794

Page 802
TOPIC: Flightless birds
KEYWORD: HX802

Page 807
TOPIC: Adaptations of reptiles
KEYWORD: HX807

Chapter 35

1 Prepare

Scheduling

- **Block:** About 45 minutes will be needed to complete this section.
- **Traditional:** About one class period will be needed to complete this section.
- **Planning:** See the Planning Guide on pp. 779A and 779B for lecture, classwork, and assignment options for Section 35-1. Lesson plans for Section 35-1 can be found in a lesson cycle format in ***Biology: Principles and Explorations*** **One-Stop Planner CD-ROM with Test Generator.**

Resource Link

Assign **Active Reading Worksheet Section 35-1** to help students comprehend the key concepts and visuals in this lesson.

2 Teach

Lesson Warm-up

Have each student draw a picture of his or her favorite (or least favorite!) reptile on a sheet of paper. A simple sketch will do. Have volunteers share their drawings with the class and tell why they chose that particular reptile. Lead a discussion of how reptiles are viewed in popular culture (such as in films and books), and let students tell their favorite reptile stories. PORTFOLIO

Teaching TIP

Reptilian Characteristics

Tell students that although most reptiles have a ventricle that is partly divided by a septum, crocodiles are an exception. Crocodiles have a completely divided ventricle.

Section 35-1 The Reptilian Body

Objectives

- **Describe** the key characteristics of reptiles. (p. 782)
- **Relate** a reptile's ectothermic metabolism to its activity level. (p. 783)
- **Summarize** the adaptations that enable reptiles to live on land. (pp. 784–787)

Key Terms

amniotic sac
alveoli
oviparous
ovoviviparous

internetconnect

TOPIC: Characteristics of reptiles
GO TO: www.scilinks.org
KEYWORD: HX782

Reptiles Share Several Key Characteristics

Ugh! A coldblooded reptile! This is many people's reaction when they see a harmless little snake slither across a yard or field. But snakes and their reptile relatives are important members of most ecosystems, and they kill large numbers of insect pests and small rodents. It's true that some reptiles—venomous snakes and crocodilians—are dangerous, but most live quietly and go about their business, preferring to avoid humans.

Members of class Reptilia live throughout the world in a wide variety of habitats, except in the coldest regions, where it is impossible for ectotherms to survive. Reptiles share certain fundamental characteristics, features they retain from the time when reptiles replaced amphibians as the dominant terrestrial vertebrates. **Figure 35-1** summarizes these key features.

Reptiles have strong, bony skeletons, and most have two pairs of limbs, although snakes are legless. The legs of reptiles are positioned more vertically under their body than are the limbs of amphibians. Thus their legs support the weight of their body, enabling reptiles to move on land more easily than amphibians do. Unlike amphibians, reptiles have toes with claws, which are used for climbing and digging. Claws also enable reptiles to get a good grip on the ground, and many can run quickly for short distances.

The nervous system of a reptile is very similar to that of an amphibian. Like their dinosaur ancestors, modern reptiles have a brain that is small in relation to their body. For example, an alligator about 2.5 m (8 ft) long has a brain that is about the size of a walnut. Despite this small brain size, reptiles are capable of complex behaviors, including elaborate courtship, also shown in Figure 35-1.

Figure 35-1
Characteristics of living reptiles. This male anole is extending his dewlap, a display used during courtship and as an aggressive posture when defending territory.

Key Features of Reptiles

- Strong, bony skeletons and toes with claws
- Ectothermic metabolism
- Dry, scaly skin, almost watertight
- Amniotic eggs, almost watertight
- Respiration through well-developed lungs
- Ventricle of heart partly divided by a septum
- Internal fertilization

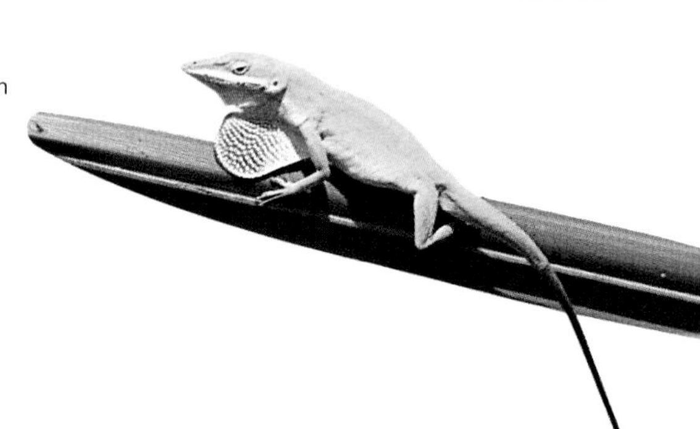

Integrating Different Learning Styles

Logical/Mathematical	**PE**	Data Lab, p. 783; Quick Lab, p. 784; Performance Zone items 17 and 23, pp. 804–805
	ATE	Teaching Tip, p. 786
Visual/Spatial	**PE**	Performance Zone item 1, p. 804
	ATE	Lesson Warm-up, p. 782; Visual Strategy, p. 783; Teaching Tip, p. 784
Interpersonal	**ATE**	Biology in Your Community, p. 786; Reteaching, p. 787
Verbal/Linguistic	**PE**	Performance Zone item 2, p. 804; Study Tips, pp. 784, 787
	ATE	Teaching Tip, p. 786; Cross-Disciplinary Connection, p. 786; Closure, p. 787

Ectothermic metabolism

Reptiles' ectothermic metabolism is too slow to generate enough heat to warm their bodies, so they must absorb heat from their surroundings. As a result, a reptile's body temperature is largely determined by the temperature of its environment. Many reptiles regulate their temperature behaviorally, by basking in the sun to warm up or seeking shade to cool down. **Figure 35-2** shows that a lizard can maintain a relatively constant body temperature throughout the day by moving between sunlight and shade. At very cold temperatures, most reptiles become sluggish and unable to function. Intolerance of cold generally limits their geographical range and, in temperate climates, forces them to hibernate through the winter.

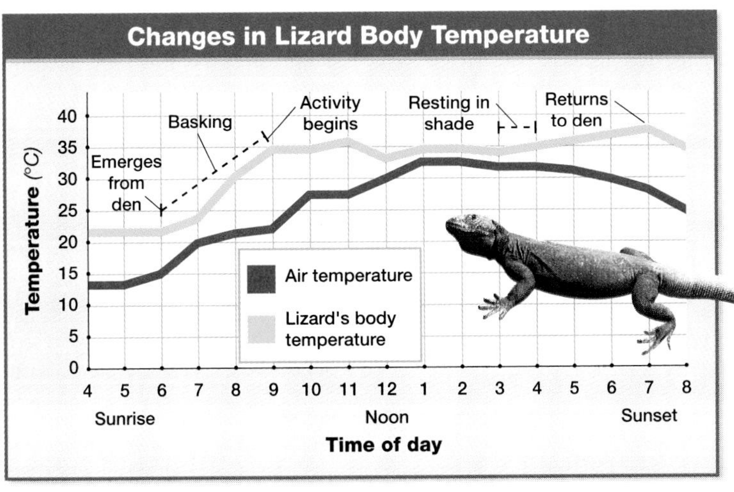

Figure 35-2 Body temperature in lizards. The body temperature data were transmitted by a temperature-sensitive instrument implanted in the lizard's abdomen.

Identifying ectotherms

Background

The body temperature of all animals changes during the course of a day. How it changes can help you identify an animal as an ectotherm or an endotherm.

Body Temperatures of Two Animals

Species A ■ Species B ■ Air

Temperature (°C): 10, 20, 30, 40

Time of day: 12 A.M., 6 A.M., 12 P.M., 6 P.M.

Analysis

1. **Analyze** the data and determine which animal species, A or B, is most likely an ectotherm. Explain your reasoning.
2. **Identify** the time of day the animal you identified as an ectotherm reaches its lowest body temperature.
3. **Identify** the time of day the animal you identified as an ectotherm reaches its highest body temperature.
4. **Propose** a reason why the ectotherm's body temperature is highest at this time.
5. **Predict** what the endotherm's graph line would look like if it were extended to show between 6 P.M. and midnight.

Overcoming Misconceptions

Ectothermic does not mean "coldblooded."

Emphasize that the terms *ectothermic* and *coldblooded* are not synonyms, although they are often mistakenly used that way. Many reptiles maintain body temperatures as high as or higher than human body temperature. For example, the desert iguana of western North America maintains a daytime body temperature of 42°C (108°F), which would be fatal for a human.

Visual Strategy

Figure 35-2

Have students draw a graph that depicts the changes in human body temperature for a 24-hour period. *(The graph should show that the human body remains at about 37°C for the entire day.)* Then ask them to compare their graphs to the one for the lizard's body temperature in Figure 35-2. Ask: Why are the graphs different? *(The lizard depends on external sources for heat, so its body temperature tracks the temperature of the environment. Humans produce their own heat, so body temperature is not dependent on environmental temperature changes.)* PORTFOLIO

Identifying ectotherms

Time 15 minutes

Process Skills Analyzing data, interpreting graphs, predicting patterns

Teaching Strategies

Remind students that *ecto-* refers to "outside," *endo-* refers to "within," and *therm* refers to "heat." Hence, *ectotherm* means "outside heat" and *endotherm* means "heat within."

Analysis Answers

1. Species A (yellow curve) probably is an ectotherm. Its temperature increases during the day, when the air temperature increases.
2. around 12 A.M.
3. around 6 P.M.
4. It has been exposed to sunlight-warmed air since early in the morning.
5. It would decline steadily from its value at 6 P.M. to its value at 12 A.M.

Teaching TIP

Regulating Body Temperature

Show students a photograph of a lizard basking in the sun. Point out the behaviors that enable the lizard to control its body temperature. The lizard may flatten its body against the surface it rests on, exposing more of its body surface to the sun. When some lizards are cold, their skin darkens, allowing them to absorb more solar radiation. Some desert lizards will actually stand on two diagonally opposed legs while the other two are held above the ground to cool. They then switch legs to cool their other two feet and may repeat this process several times. Point out that the easiest way for lizards to cool themselves is simply to go into a burrow or into the shade.

QUICK LAB

How does watertight skin benefit reptiles?

Time 20 minutes

Process Skills Calculating, analyzing data, inferring conclusions

Teaching Strategies

- Be sure to have students wipe the pan of the scale clean between weighings.
- Place an incandescent lamp on each table to simulate a desert environment.

Analysis Answers

1. The mass of the skinless grape should decrease. The mass of the intact grape should stay the same.
2. The skin prevents water from evaporating from an intact grape, so the grape's weight does not change. Without its skin, a grape loses water and becomes lighter.
3. The skinless grape represents an amphibian's skin. The intact grape represents a reptile's skin.
4. Watertight skin, as in the intact grape, prevents water loss. This allows an animal to be free from living in water. Most amphibians, like the skinless grape, would dry up if out of water for an extended period of time.

Study TIP

Reviewing Information

Reread the bulleted list of the characteristics of reptiles on page 782. Then write them down on a separate piece of paper, leaving room to write notes about each characteristic. As you read, summarize how the information relates to a particular characteristic.

A Key Adaptation to Terrestrial Life Is Water Retention

Amphibians such as frogs cannot be considered fully terrestrial because they lose too much water through their skin. Amphibians must stay moist to avoid dehydration, and their method of reproduction requires water. Reptiles have evolved adaptations that free them from the water requirements of amphibians.

Watertight skin

Terrestrial animals face a serious problem of water loss as water evaporates through their skin. Modern reptiles have evolved a skin made of light, flexible scales. These scales overlap and form a protective, almost watertight skin that minimizes water loss, as shown in **Figure 35-3.**

Figure 35-3 Reptilian scales. The scales of a reptile's skin form a tight seal that retains moisture within the reptile's body.

QUICK LAB

How does watertight skin benefit reptiles?

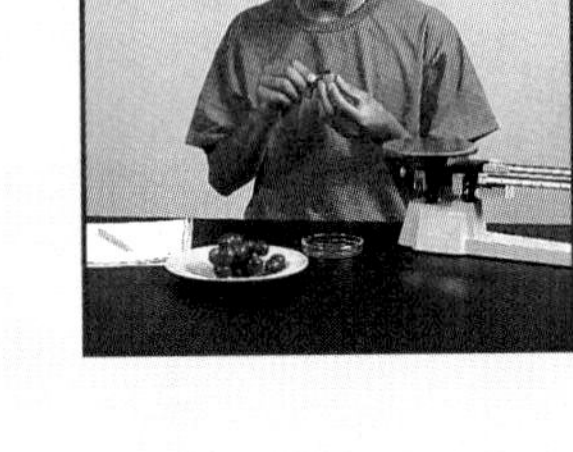

The scales of reptiles make their skin almost watertight. This is one of the reptiles' adaptations to terrestrial life. You can use grapes to model and compare water loss in different types of skin.

Materials

forceps, 2 grapes, balance, Petri dish

Procedure

1. Find the mass of one grape, and record it in a data table. Then place the grape in an open Petri dish.
2. Using forceps, remove the skin from the second grape. Find and record the mass of the peeled grape. Then place it in the same Petri dish, but do not let the two grapes touch.
3. Wait 15 minutes, and then find and record the mass of each grape again.

Analysis

1. **Calculate** any difference between the original and final masses of each grape.
2. **Propose** an explanation for any changes in mass you observed.
3. **Determine** which grape represents an amphibian's skin and which represents a reptile's skin.
4. **Describe** how a watertight skin is an adaptation to terrestrial life. Include information you have learned in this lab in your explanation.

did you know?

Sometimes solving a problem for one species results in harm to another.

In the western United States, the tansy ragwort plant is poisonous to cattle. The cinnabar moth, whose larvae eat the tansy ragwort, was introduced into the region to control this plant. However, the larvae accumulate toxins which kill the northern alligator lizards that prey on the larvae. Populations of these lizards may be at risk in certain areas of the west.

Watertight eggs

For a reptile living on dry land, reproduction presents another serious water-loss problem. Without a watery environment, both sperm and eggs will dry out. And a reptile's fertilized eggs need a moist environment in which to develop. As you will see on page 787, the first problem is overcome by internal fertilization.

The nature of the reptile's amniotic *(am nee AHT ic)* egg solves the problem of reproduction on dry land. An **amniotic egg** contains both a water supply and a food supply and is key to the reptile's success as a terrestrial animal. Because the egg's tough shell makes it essentially watertight, it does not dry out, even in very dry habitats. Most reptiles, all birds, and three species of mammals reproduce by means of amniotic eggs. (Other mammals produce amniotic eggs, but the embryo develops within the female's uterus rather than within a shell. You will learn about the development of these eggs in Chapter 44.) The formation of amniotic eggs with shells suggests that these three groups of animals evolved from a common ancestor.

Exploring Further

The Amniotic Egg

A reptile's egg is usually leathery compared with the hard shell of a bird, but internally the two are very much alike. The almost watertight shell is porous enough to allow oxygen to pass into, and carbon dioxide to pass out of, the egg. The shell and the albumen (egg white) lying beneath the shell protect and cushion the developing embryo. The albumen is also a protein and water source for the embryo.

Within the egg, four specialized membranes—the amnion, the yolk sac, the allantois, and the chorion—play important roles in maintaining a stable environment in which the embryo can develop.

The **amnion** *(AM nee awn)* encloses the embryo within a watery environment. In a sense, this membrane creates a little pond that substitutes for the water in which amphibians lay their eggs. This watery enclosure also protects the embryo by cushioning it.

The **yolk sac** contains the yolk, the developing embryo's food supply. The embryo absorbs nourishment from the yolk through blood vessels connecting its gut and the yolk sac.

The **allantois** *(uh LAHN toh is)* is a membrane-covered cavity that stores waste products from the embryo. It also serves as the embryo's organ for gas exchange. Blood vessels in its walls carry oxygen to and carbon dioxide from the embryo.

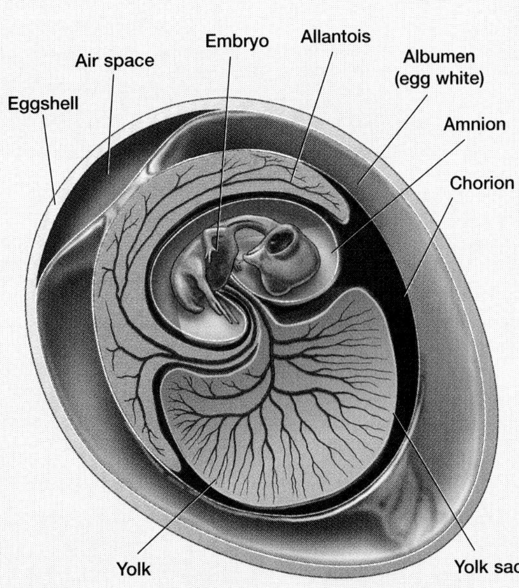

Surrounding the amnion, yolk sac, and allantois is a membrane called the **chorion** *(KAWR ee awn)*. The chorion allows oxygen to enter the egg and carbon dioxide to leave.

internetconnect

SciLinks NSTA

TOPIC: Amniotic egg
GO TO: www.scilinks.org
KEYWORD: HX785

did you know?

Some sea turtles undertake long migrations before laying their eggs.

Green turtles migrate more than 2,000 km (1,250 mi) from the waters off Brazil to the beaches of Ascension Island in the central Atlantic Ocean. A female green turtle will lay her eggs on the same beach where she hatched. Scientists do not know, however, how she finds her birthplace.

Exploring Further

The Amniotic Egg

Teaching Strategies

Give each lab group a hard-boiled egg. Have students crack it open and gently peel away the shell. Ask them to identify the chorion, yolk, and albumen in the egg. Have them note the location of the chorion and ask them how this corresponds to its function. *(It surrounds the amnion, yolk sac, and allantois and allows oxygen and carbon dioxide to pass between them and the environment.)*

Discussion

- What is the advantage of having an egg with a "pond" in it rather than just laying eggs in a pond? *(Eggs can be laid where water is not available.)*
- How is the amniotic egg similar to the high-tech fabrics used in outdoor clothing? *(It allows gases such as carbon dioxide and oxygen to pass through but prevents water from passing through.)*
- What is the advantage of having a storage sac for waste products (the allantois)? *(Without it, waste products could build up in the blood and become toxic to the embryo.)*
- Why does oxygen *enter* and carbon dioxide *leave* the egg? *(As the embryo gives off carbon dioxide, its concentration inside the egg increases, which causes it to diffuse out of the egg. Oxygen is more concentrated outside the egg, so it diffuses inward.)*

Teaching TIP BASIC

Surface Area

Asks students which melts faster—cubes of ice or a block of ice of the same weight? *(cubes of ice)* Lead students into a discussion of the importance of surface area in living things. Relate the cubes of ice to alveoli in the lungs.

Teaching TIP

Method of Development

Review the terms *oviparous, ovoviviparous,* and *viviparous.* Remind students that oviparous reptiles lay eggs. All turtles, tuataras, and crocodilians are oviparous. In ovoviviparous reptiles, the eggs are retained within the mother's body and the young are born live. There may be some transfer of materials, such as water and oxygen, from mother to offspring—often through a placenta. However, the yolk within the egg provides all of the nutrition needed by the developing embryo. In viviparous reptiles, a placenta furnishes not only water and oxygen but also nutrition to the embryo.

CROSS-DISCIPLINARY CONNECTION

◆ **Language Arts**

Snakes figure prominently in many literary works, such as *The Bible,* and in stories such as the Greek myth of Medusa. Even children's stories like *Aladdin* and *The Jungle Book* include snakes. Snakes are often portrayed as evil, coldblooded animals. Have students research this phenomenon in literature and write an essay on whether the reputation of the snake is deserved.

PORTFOLIO

Reptiles Need More Oxygen Than Amphibians

Because most reptiles are far more active than amphibians, they have greater metabolic requirements for oxygen. Their bodies meet this demand in several ways.

Lungs

The reptile's scaly skin does not permit gas exchange, so reptiles cannot use their skin for additional respiratory surface, as many amphibians can. Instead, most reptiles have a pair of lungs with small grape-shaped chambers called **alveoli** *(al VEE uh leye)* located on the inner surface of the lungs. The alveoli, shown in **Figure 35-4,** greatly increase the respiratory surface area of a reptile's lungs. In addition, reptiles have strong muscles attached to their rib cage. The action of these muscles helps to move air into and out of the lungs, increasing their efficiency.

Figure 35-4 Reptilian lungs. The lungs of reptiles contain numerous small chambers called alveoli.

Heart

Recall that the ventricle of the amphibian heart is not divided by a septum. Oxygen-poor blood and oxygen-rich blood mix in the amphibian's ventricle. In most reptiles, however, the septum extends into the ventricle, partly dividing it into right and left chambers (halves), as shown in **Figure 35-5.** As a result, there is much better, but still incomplete, separation of oxygen-rich and oxygen-poor blood. Still, oxygen is delivered to the body cells more efficiently than in amphibians. Unlike most reptiles, crocodilians (crocodiles and alligators) have a heart with a completely divided ventricle that forms two pumping chambers. This arrangement fully separates the lung circulation from the body circulation.

Figure 35-5 Reptilian heart. In most reptiles, the ventricle of the heart is partly divided by a septum.

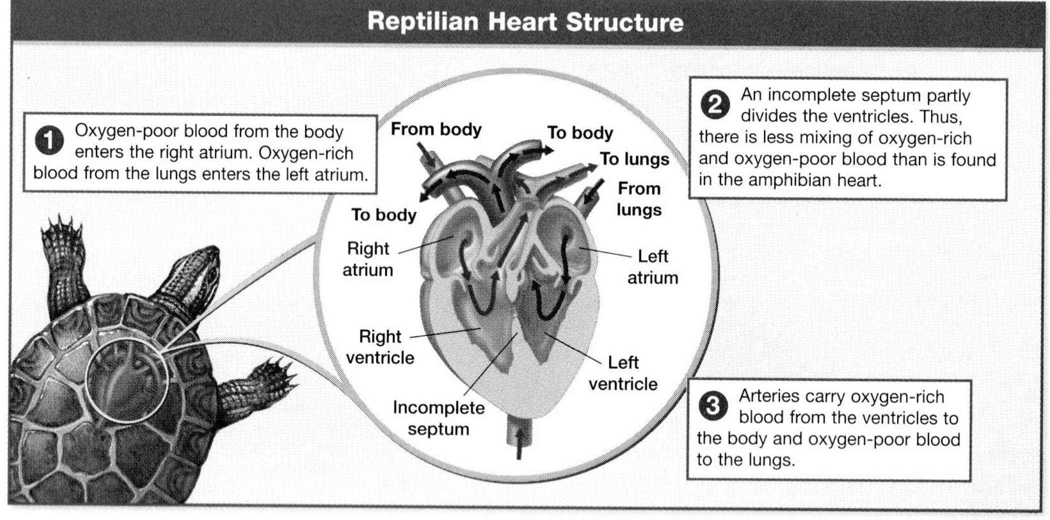

Reptilian Heart Structure
1 Oxygen-poor blood from the body enters the right atrium. Oxygen-rich blood from the lungs enters the left atrium.
From body
To body
To lungs
From lungs
To body
Right atrium
Left atrium
Right ventricle
Left ventricle
Incomplete septum
2 An incomplete septum partly divides the ventricles. Thus, there is less mixing of oxygen-rich and oxygen-poor blood than is found in the amphibian heart.
3 Arteries carry oxygen-rich blood from the ventricles to the body and oxygen-poor blood to the lungs.

Biology in Your Community

Snakes in Your Area

Use a field guide to determine what kinds of snakes are found in your area. Present students with a list of these snakes and their scientific names. Have groups research these snakes and report to the class on the snakes' habitat, range, reproduction, behavior, feeding habits, and any other relevant information.

PORTFOLIO

Reptiles Have Internal Fertilization

Unlike the eggs of most amphibians, reptilian eggs are fertilized within the female, a process called internal fertilization. The male reptile introduces his semen directly into the female's body. The semen contains sperm and fluid secretions. Internal fertilization protects the gametes from drying out, even though the adult animals are fully terrestrial.

Many reptiles are **oviparous** *(oh VIHP urh uhs)*, meaning the young hatch from eggs, as shown in **Figure 35-6.** The female deposits the eggs in a suitable place in the environment, and heat from the environment incubates them. In most cases, the eggs are not protected by the parents. Most snakes and lizards, all turtles and tortoises, and all crocodilians are oviparous. All birds and three species of mammals are also oviparous.

But eggs placed in the environment are vulnerable to predators. To reduce this risk, some species of snakes and lizards are **ovoviviparous,** which means the female retains the eggs within her body until shortly before hatching, or the eggs may actually hatch within the female's body. Although the embryos receive water and oxygen from the female, their nourishment comes from the yolk sac. The hog nose snake, also shown in Figure 35-6 gives birth to live young that are fully formed at birth and able to fend for themselves.

Study TIP

Word Origins

The term *ovoviviparous* comes from three different Latin words: *ovum*, meaning "egg," *vivus*, meaning "alive," and *parere*, meaning "to bring forth or bear."

Figure 35-6 Reptilian reproduction. Sea turtles hatch from eggs buried on sandy beaches. The eggs of some species of snakes are incubated within the female's body, and the young are born live.

Sea turtles

Hog nose snakes

Review SECTION 35-1

1. **Identify** seven characteristics of reptiles.
2. **Describe** how the ectothermic nature of reptiles influences their physical activity and feeding habits.
3. **Summarize** the skin and egg adaptations that allow reptiles to live on land.
4. **Describe** how reptiles meet their need for more oxygen than amphibians require.
5. **Compare** the way in which reptiles and amphibians fertilize their eggs.

Critical Thinking

6. **SKILL Forming Reasoned Opinions** Data show that an animal's temperature changes over the course of a day. A student asserts that this proves the animal is an ectotherm. What must the student consider before making such a claim?

Review SECTION 35-1 ANSWERS

1. strong, bony skeletons; ectothermic metabolism; watertight skin; amniotic eggs; well-developed lungs; partly or completely divided ventricle; internal fertilization
2. Because they are ectotherms, reptiles must thermoregulate behaviorally. They can be physically active and hunt for food only when their body temperature is within a certain critical range.
3. The watertight skin and amniotic eggs of reptiles minimize water loss on land.
4. In contrast to amphibians, reptiles have well-developed lungs with alveoli, as well as strong rib muscles.
5. Most amphibians have external fertilization. Reptiles have internal fertilization.
6. The student would need to compare the animal's temperature changes to changes in the environmental temperature. Even an endotherm has natural temperature fluctuations, but they are not linked to the environmental temperature.

3 Assessment Options

Closure

Life on Land

Have teams of students write brief descriptions of what they think are the five most important things they have learned about reptiles. Have teams present their descriptions to the class, with each team member participating in the presentation. PORTFOLIO

Review

Assign Review—Section 35-1.

Reteaching — BASIC

Have students pair up and list each of the seven key features of reptiles, which are given in Figure 35-1, on individual index cards. On the back of each card, have them write a one-word clue to identify the characteristic on the front of the card. Ask students to shuffle the cards and turn them so only the clue words are showing. Have one student choose a card at random and read the clue word. His or her partner should respond with the key characteristic. If the student answers correctly, his or her partner should set the card aside. Students should continue the game until all of the cards are set aside. Students should then switch roles and repeat the process. PORTFOLIO

Chapter 35

1 Prepare

Scheduling

- **Block:** About 45 minutes will be needed to complete this section.
- **Traditional:** About one class period will be needed to complete this section.
- **Planning:** See the Planning Guide on pp. 779A and 779B for lecture, classwork, and assignment options for Section 35-2. Lesson plans for Section 35-2 can be found in a lesson cycle format in ***Biology: Principles and Explorations* One-Stop Planner CD-ROM with Test Generator.**

Resource Link

Assign **Active Reading Worksheet Section 35-2** to help students comprehend the key concepts and visuals in this lesson.

2 Teach

Lesson Warm-up

Obtain a picture of a skink, and cover its body so that only its head shows. Ask students if the animal is a lizard or a snake. Then reveal the rest of the animal, pointing out how similar some lizards are to snakes. Ask students how to tell a lizard from a snake. *(Most lizards have legs and external ears, and many have movable eyelids. Snakes lack these features.)* Then ask how one would know if an animal is a snake or a legless lizard. Have students answer this question as they complete this section.

Section 35-2 Today's Reptiles

Objectives

- **Compare** the four living orders of reptiles. (pp. 788–793)
- **Describe** the timber rattlesnake's adaptations for locating and capturing prey. (pp. 790–791)
- **Compare** the parental care of crocodilians with that of other reptiles. (p. 793)

Key Terms

carapace
plastron

Lizards and Snakes Have a Unique Jaw Design

You've probably walked by a snake or lizard not even knowing it was there. Most are quiet, and their coloration conceals them from view. Even if you visited the jungles of South America, you might not notice the anaconda unless it moved. What's an anaconda? It's the world's largest snake, frequently reaching 5 m (about 16 ft) in length, with the longest recorded length being 10 m (33 ft). Very large anacondas have been known to prey on jaguars. After such a meal, the anaconda may not eat again for up to a year.

Snakes and lizards belong to order Squamata. A distinguishing characteristic of this order is a lower jaw that is only loosely connected to the skull. This allows the mouth to open wide enough to accommodate large prey and explains how an anaconda can swallow a jaguar. This ability is a contributing factor to the success of snakes and most lizards as predators.

Lizards

Common lizards include iguanas, chameleons, geckos, anoles, and horned lizards (often mistakenly called "horny toads"). A few species of lizards are herbivores, but most are carnivores. Most lizards are small, measuring less than 30 cm (1 ft) in length, but lizards that belong to the monitor family can become quite large. The Komodo dragon of Indonesia, shown in **Figure 35-7,** is the largest monitor lizard. It can be up to 3 m (10 ft) in length and weigh up to 250 kg (550 lb). The tail of some species of lizards, such as the gecko, also shown in Figure 35-7, breaks off easily when seized by a predator, allowing the lizard to escape. Lizards can regenerate a new tail, but the tail does not have any vertebrae in it.

Figure 35-7 Lizards. Geckos are the smallest reptiles, rarely exceeding 24 cm (10 in.) in length. The Komodo dragon is the world's largest lizard.

Integrating Different Learning Styles

Visual/Spatial	**PE**	Performance Zone items 18, 19, 20, 25, and 29, p. 805
	ATE	Lesson Warm-up, p. 788; Visual Strategy, p. 792
Body/Kinesthetic	**ATE**	Up Close, p. 790
Verbal/Linguistic	**PE**	Performance Zone item 27, p. 805
	ATE	Closure, p. 793; Reteaching, p. 793

Snakes

Snakes probably evolved from lizards during the Cretaceous period. The close relationship between lizards and snakes is reflected in their many similarities. In fact, it is often difficult to distinguish the legless species of lizards from snakes. Snakes lack movable eyelids and external ears, as do several species of lizards. Also, both snakes and lizards molt periodically, shedding their outer layers of skin.

The skeleton of snakes is unique. Most snakes have no trace of a pectoral girdle (the supporting bones for the bones of the forelimbs), which is found even in legless lizards. The snake's jaw is very flexible because it has five points of movement (your jaw has only one movement point). One of these points is the chin, where the halves of the lower jaw are connected by an elastic ligament. This attachment permits the lower jaw to spread apart when a large meal is being swallowed. The African egg-eating snake, shown in **Figure 35-8,** is swallowing a bird's egg, a process that can take an hour or more.

While many snakes simply seize their prey and swallow it whole, other snakes kill their prey by suffocation or by injecting lethal venom. Constrictors wrap their body around their prey, gradually squeezing tighter and tighter until the prey suffocates. Many very large snakes, such as anacondas, boas, and pythons, are constrictors. Many smaller species, such as king snakes, also use constriction to kill their prey. Constrictors swallow their prey whole, even if the prey is very large. Like all snakes, the constrictor's teeth are not suited for cutting and chewing.

While most species of snakes are not venomous, several kinds are dangerous to humans. In venomous snakes, modified salivary glands produce a venom (poison) that is injected into the victim through grooved or hollow teeth. Of the 13 families of snakes, only four are venomous: (1) cobras, kraits, and coral snakes; (2) sea snakes; (3) adders and vipers; and (4) rattlesnakes, water moccasins, and copperheads. The African boomslang and twig snakes produce venom but do not inject it. Instead, they bite their prey with fangs located at the back of their mouth. Grooved teeth direct the venom into their victim's wound. You can read more about the biology of snakes in *Up Close: Timber Rattlesnake,* on pp. 790–791.

REAL LIFE

Need a lizard? Instead of calling an exterminator to rid their homes of unwanted pests, a few adventuresome homeowners are using a type of lizard called a Tokay gecko. This gecko preys voraciously on mice and insects and is very efficient at ridding a home of these pests.

Finding Information
Investigate the pros and cons of using geckos for pest control. Who should try this method and who should avoid it?

Figure 35-8 **Snake feeding.** Snakes have flexible jaws that allow them to swallow prey much larger than their head.

did you know?

Some snakes have triple-jointed jaws.

The tiny snakes called threadsnakes feed on ants and other social insects. When a threadsnake burrows into an ant nest, parts of its lower jaw rotate like a pair of swinging doors. It is thought that this action helps the snake to eat quickly and escape injury or death from ant stings.

REAL LIFE

Answer

People who have other pets that might harass a gecko, such as a cat or a dog, should not try to use geckos for pest control.

Teaching TIP

Dangerous Snakes

Ask students what venomous snake is the most dangerous in the world. Record their responses. *(examples: king cobra, sea snake, coral snake, and Australian tiger snake)* Ask why a particular species is considered more dangerous than others. Tell them that many different venomous snakes are touted as the "most dangerous snake alive," or the "most poisonous snake on Earth." Point out that several factors determine how "dangerous" a snake is. Among these factors are the toxicity of its venom, the type of venom (hemolytic or neurotoxic), and the amount of venom a bite victim receives. In India, there are about 900,000 snakebites each year, resulting in about 9,000 deaths. Many of these bites are delivered by the aggressive king cobra, which is frequently encountered by India's large human population.

Making Biology Relevant

Health

The American Red Cross recommends the following first-aid treatment for snakebite:

- Wash the bite with soap and water.
- Immobilize the area and keep it lower than the heart.
- Get medical help.

Medical professionals are nearly unanimous in their views of what *not* to do for snakebite:

- No ice or any other type of cooling on the bite
- No tourniquets
- No electric shock
- No incisions in the wound

Up Close

Timber Rattlesnake

Teaching Strategies

- Help students understand the rattlesnake's ability to sense its prey in the dark by using a warm object such as a heating pad. Place the heating pad on a table and allow it to warm the surface. Remove the heating pad and have students move their hands above the table top without touching the table itself. They should be able to feel the heat radiating from the warmed spot. Have students determine how close their hands must be to detect the heat. Inform them that a rattlesnake can locate warm prey from a distance of 1 m (39 in.). These snakes, however, sense heat with their pit organs, not through their skin.
- On a sheet of paper, have students list each major heading in the **Up Close** feature. After they read each section, have them hypothesize how that particular feature of the timber rattlesnake is an adaptation for survival. Then have students exchange papers with a partner and read their partner's paper, noting how their partner may have identified a different survival value for some of the features.

Up Close

Timber Rattlesnake

- **Scientific name:** *Crotalus horridus*
- **Size:** Typically 90–150 cm (36–60 in.); maximum 189 cm (74 in.)
- **Range:** Eastern and central United States, from northern New York to northern Florida and central Texas
- **Habitat:** Prefers thick brush, dense woodland, or swamp
- **Diet:** Primarily small mammals

External Structures

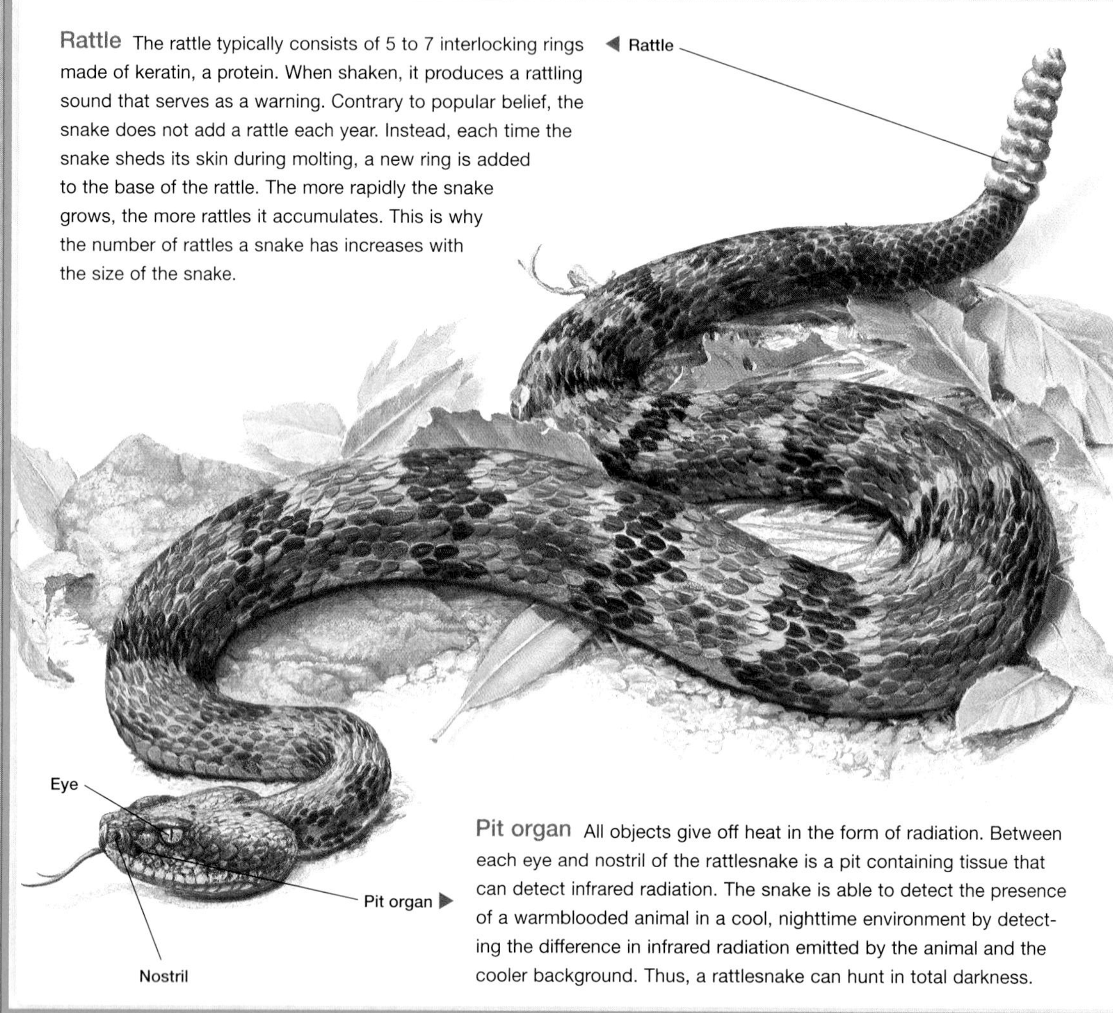

Rattle The rattle typically consists of 5 to 7 interlocking rings made of keratin, a protein. When shaken, it produces a rattling sound that serves as a warning. Contrary to popular belief, the snake does not add a rattle each year. Instead, each time the snake sheds its skin during molting, a new ring is added to the base of the rattle. The more rapidly the snake grows, the more rattles it accumulates. This is why the number of rattles a snake has increases with the size of the snake.

Pit organ All objects give off heat in the form of radiation. Between each eye and nostril of the rattlesnake is a pit containing tissue that can detect infrared radiation. The snake is able to detect the presence of a warmblooded animal in a cool, nighttime environment by detecting the difference in infrared radiation emitted by the animal and the cooler background. Thus, a rattlesnake can hunt in total darkness.

Internal Structures

Venom The timber rattlesnake has hollow upper front teeth, or fangs. When the rattlesnake strikes, these hinged fangs swing forward from the roof of the mouth and inject venom deep into the prey. The venom contains hemotoxins, proteins that attack the circulatory system, destroying red blood cells and causing internal hemorrhaging. Modified salivary glands in the upper jaw produce the venom.

Jacobson's organs

Fang

◀ Venom gland

Jacobson's organs Flicking its forked tongue into the air, the rattlesnake takes in chemical samples from the environment. These chemicals are transferred to two depressions in the roof of the mouth called Jacobson's organs, where the odor of the chemicals can be detected. The snake uses the odors to follow the scent trail of prey.

▲ Tongue

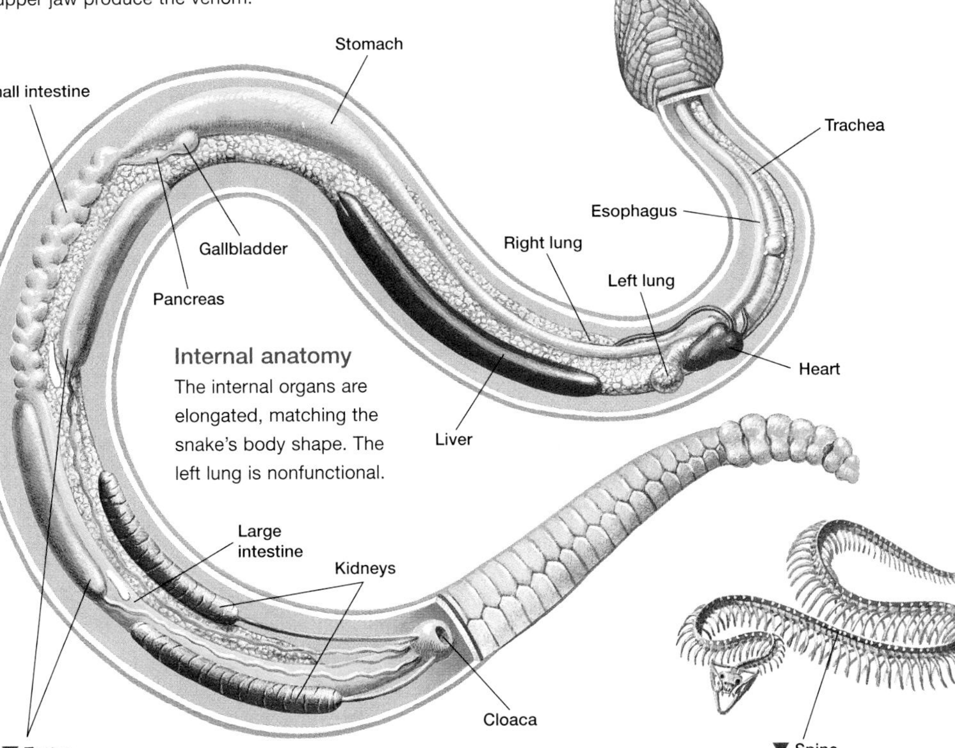

Internal anatomy The internal organs are elongated, matching the snake's body shape. The left lung is nonfunctional.

▼ Testes

▼ Spine

Reproduction This male rattlesnake produces sperm in his testes. Female timber rattlesnakes are ovoviviparous. A female carries her fertilized eggs in her body while they develop. Each egg has a thin membrane through which water and oxygen pass from the mother to the embryo. All nourishment is provided by the egg's yolk. After the eggs hatch in the mother's body, the live young are ejected and must fend for themselves.

Spine The rattlesnake's spine is made up of several hundred vertebrae, each with its own pair of attached ribs. It provides the framework for thousands of muscles that manipulate not only the skeleton but also the snake's skin, causing the overlapping scales to extend and contract.

Up Close

Timber Rattlesnake

Discussion

- Explain why it would be easier for a timber rattlesnake to kill a mouse than a lizard at night. *(The mouse is an endotherm and its body heat can be detected by the pit viper's heat-sensitive pits. The lizard is an ectotherm and would not necessarily be warmer than its surroundings.)*
- What is the advantage to a rattlesnake of announcing its presence to other animals by rattling? *(Answers may vary but may include that by rattling the snake keeps large mammals, such as cattle, from stepping on it.)*
- How is an ovoviviparous snake, such as the timber rattlesnake, different from a viviparous snake in the way developing embryos are sustained? *(Ovoviviparous mothers supply no nutrients to the developing young, while viviparous mothers do.)*

did you know?

Snakebites can be deadly.

About 99 percent of venomous snakebites in the United States are from pit vipers. About 8,000 snakebites occur in the United States each year, and 9–15 victims die. Every state except Maine, Alaska, and Hawaii is home to at least one poisonous snake species.

Visual Strategy

Figure 35-10 BASIC

Point out that a turtle's shoulders lie within its rib cage. Students should recognize that their own shoulders are outside of their ribs. Tell students that a turtle is attached to its shell and cannot crawl out of it, as cartoon turtles often do.

Teaching TIP

Endangered Crocodilians

Although crocodilians are fierce predators and some species occasionally eat humans, many of the 25 species of crocodilians are endangered or threatened. Overhunting of crocodilians for their hides, which are used to make leather goods, is the chief cause of the decline in their numbers.

Teaching TIP

Coral Snakes

The first casualty of the American Civil War was not caused by a bullet or saber but by the bite of the Eastern coral snake, according to Professor Janis Roze, the author of *Coral Snakes of the Americas.* The offending snake is preserved in a museum in Augusta, Georgia.

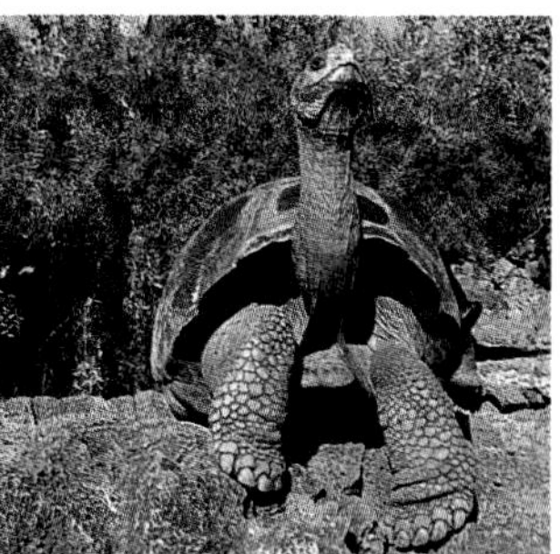

Figure 35-9 Turtle and tortoise. Like other sea turtles, this green sea turtle spends virtually its entire life in the sea. The Galápagos tortoise spends its life on land.

Other Orders of Reptiles Are Less Diverse

The remaining orders of living vertebrates contain far fewer species than the order Squamata does. There are about 250 species of turtles (which generally live in water) and tortoises (which live on land), all classified in the order Chelonia. The order Crocodilia is composed of 25 species of large, aquatic reptiles. The order Rhynchocephalia *(RING koh see FAY le a)* contains only two species of tuataras.

Turtles and tortoises

Turtles and tortoises, shown in **Figure 35-9,** differ from other reptiles in that their bodies are encased within a hard bony protective shell. Many of them can pull their head and legs into the shell for effective protection from predators. While most tortoises have a dome-shaped shell, water-dwelling turtles have a streamlined, disk-shaped shell that permits rapid maneuvering in water. Turtles and tortoises lack teeth but have jaws covered by sharp plates, forming powerful beaks. Many are herbivores but some, such as the snapping turtle, are aggressive carnivores.

Today's turtles and tortoises differ little from the earliest known turtle fossils, which date to more than 200 million years ago. This evolutionary stability may reflect the beneficial aspects of their basic shell-covered body design. The shell is made of fused plates of bone covered with horny shields or of tough leathery skin. In either case, the shell consists of two basic parts. The **carapace** is the dorsal (top) part of the shell, and the **plastron** is the ventral (bottom) portion. The vertebrae and ribs of most species are fused to the inside of the carapace, as shown in **Figure 35-10.** The shell provides the support for all muscle attachments.

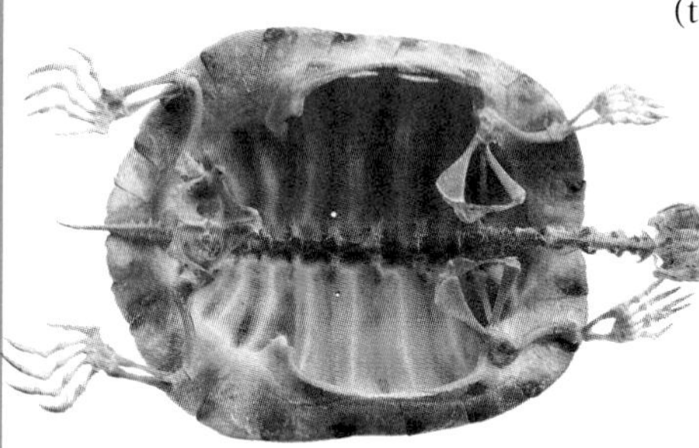

Figure 35-10 Turtle interior. In this ventral view, a turtle's plastron has been removed to show the relationship of the vertebral column, ribs, pelvis, and pectoral girdle to the carapace.

Crocodiles and alligators

Of all the living reptiles, the crocodilians are most closely related to the dinosaurs. In addition to crocodiles and alligators, shown in **Figure 35-11,** the order Crocodilia includes the alligator-like caimans and the long snouted gavial. Crocodilians are aggressive carnivores. Some are quite large. American alligators can reach 5.5 m (18 ft) in length. Nile crocodiles are enormous, reaching up to 6 m (20 ft) in length and weighing up to 750 kg (1,650 lb). Crocodilians generally capture prey by stealth, often floating just beneath the water's surface near the shore. If an animal comes to the water to drink, the crocodilian explodes out of the water and seizes its prey. The crocodilian then hauls the prey back into the water to be drowned and eaten. The bodies of crocodilians are well adapted for this form of hunting. Their eyes are high on the sides of the head, and their nostrils are on

did you know?

Alligators are unusual among reptiles in being able to make definite vocalizations.

The male alligator bellows loudly during mating season. Vocal sacs on each side of his throat inflate when he calls. When alligators hatch from their buried eggs, they make a sound that is almost like a bark. The sound signals their mother that it is time to open the nest. Hatchlings also have a distress call they use to alert their mother if they feel threatened.

American alligator

Australian crocodiles

top of the snout. This enables them to see and breathe while lying nearly submerged in the water. Crocodilians have a very strong neck and an enormous mouth studded with sharp teeth. A valve in the back of the mouth prevents water from entering the air passages when crocodilians feed underwater.

Unlike other living reptiles, crocodilians care for their young after hatching. For instance, a female American alligator builds a nest for her eggs from rotting vegetation. After the eggs hatch, the mother may tear open the nest to free the hatchlings. The young alligators will remain under her protection for up to a year.

Figure 35-11 **Crocodilians.** In general, the snouts of alligators are shorter and broader than those of crocodiles.

Tuataras

The two living species of tuataras are members of the genus *Sphenodon* and are native to New Zealand. *Sphenodon punctatus,* the more common species, is shown in **Figure 35-12.** Tuataras are lizardlike reptiles up to 70 cm (2 ft) long. Unlike most reptiles, tuataras are most active at low temperatures. They burrow or bask in the sun during the day and feed on insects, worms, and small animals at night. Tuataras are sometimes called living fossils because they have survived almost unchanged for 150 million years. Since the arrival of humans in New Zealand about 1,000 years ago, the tuatara's range has diminished, and their numbers are declining.

Figure 35-12 **Tuatara.** Tuataras survive on only a few small islands in New Zealand.

Review SECTION 35-2

1. **Describe** the characteristics shared by lizards and snakes.
2. **Describe** the function of two different organs that help snakes locate their prey.
3. **Summarize** the ways turtles and tortoises differ from other reptiles.
4. **Compare** the parental care shown by alligators with that shown by most other reptiles.
5. **SKILL Recognizing Relationships** How does the position of a crocodile's nostrils and eyes relate to its method of hunting?

Critical Thinking

6. **SKILL Forming a Hypothesis** When a piece of a lizard's tail breaks off, the separated portion may wiggle about forcefully. How might this adaptation be an advantage for the lizard?

Review SECTION 35-2 ANSWERS

1. Snakes and lizards both have scaly skin and molt periodically. Their lower jaw is loosely connected to the skull.
2. Jacobson's organs, located in the roof of the mouth, detect microscopic airborne particles, which the brain interprets as scent. Pit vipers use their pit organs to detect heat.
3. Their bodies are encased in a protective shell, and they lack teeth.
4. Alligators and other crocodilians are the only reptiles to care for their young. The female American alligator builds a nest for her young and cares for them for up to a year after they hatch.
5. The crocodile is able to keep most of its body submerged and hidden while still being able to breathe and look for prey.
6. Predators are distracted by the wiggling tail part, often attacking it while the lizard escapes.

3 Assessment Options

Closure

Know Your Reptiles

On individual sheets of paper, have students write a series of statements about each group of reptiles without stating the name of the group. Draw a six-column table on the board with the following headings: *Snakes, Lizards, Turtles, Tortoises, Crocodilians,* and *Tuataras.* Have students take turns placing statements in the correct column. PORTFOLIO

Review

Assign Review—Section 35-2.

Reteaching — BASIC

Write the following words or phrases on the board: *lizards and snakes, turtles and tortoises, crocodiles and alligators,* and *tuataras.* Have students copy them, leaving space for additional information. Have students write distinguishing traits of each group in the appropriate spaces without using their textbooks. *(lizards and snakes—lower jaw has loose connection to skull, no pectoral girdle in snakes, molt periodically, may lack eyelids; turtles and tortoises—hard shell, lack teeth, jaws have a powerful beak; crocodiles and alligators—long snout, eyes high on head, nostrils on top of head, large mouth with sharp teeth; tuataras—lizardlike, active at night)* When the students have finished their work, have them share their results with a partner, recording any characteristics they omitted but their partner identified. Then have students use their books to identify and record any characteristics that do not appear on their lists. PORTFOLIO

Chapter 35

1 Prepare

Scheduling

- **Block:** About 90 minutes will be needed to complete this section.
- **Traditional:** About two class periods will be needed to complete this section.
- **Planning:** See the Planning Guide on pp. 779A and 779B for lecture, classwork, and assignment options for Section 35-3. Lesson plans for Section 35-3 can be found in a lesson cycle format in ***Biology: Principles and Explorations*** **One-Stop Planner CD-ROM with Test Generator.**

Resource Link

- Assign **Active Reading Worksheet Section 35-3** to help students comprehend the key concepts and visuals in this lesson.

2 Teach

Lesson Warm-up

Bring a feather to class and ask students what kind of animal it is from. When students answer "a bird," ask them to think of any bird that does *not* have feathers. *(There are none.)* Ask them if they can think of any living animals that have feathers but are *not* considered birds. *(no)* Emphasize that having feathers is the key classification characteristic for the class Aves.

Section 35-3 Characteristics and Diversity of Birds

Objectives

- **Summarize** the key characteristics of birds. (p. 794)
- **Describe** how a bird's feathers and bone structure aid flight. (pp. 794–795)
- **Summarize** how a bird's lungs and heart are adapted for high efficiency. (pp. 796–797)
- **Relate** the structure of a bird's feet and beak to its habits and diet. (pp. 800–801)

Key Terms

contour feather
down feather
preen gland

internet**connect**

SCILINKS. NSTA

TOPIC: Characteristics of birds
GO TO: www.scilinks.org
KEYWORD: HX794

Birds Share Several Key Characteristics

Why do people use the expression "free as a bird"? Most likely it comes from a bird's ability to fly seemingly wherever it wishes. Through human history, the gift of flight has been celebrated in stories, poetry, and songs. But there is more to birds than flight; in fact, some species of birds can't fly.

The birds you see today are the modern members of class Aves. Unlike their reptilian relatives, birds lack teeth and have a tail that is greatly reduced in length. But they do retain some reptilian characteristics. For instance, birds lay amniotic eggs that are very similar to those of reptiles, and the feet and legs of birds are covered with scales. Other characteristics unique to birds distinguish them from all other animals. The most obvious is the presence of feathers and the modification of the forelimbs into wings. **Figure 35-13** lists some distinguishing features of birds. To find out more information on the anatomy and habits of one bird, see *Up Close: Bald Eagle,* on pp. 798–799.

Feathers

Feathers are modified reptilian scales that develop from tiny pits, called follicles, in the skin. Like snakes and lizards, birds molt and replace their feathers. However, few birds shed all of their feathers at one time.

Feathers serve two major functions. They provide lift for flight, and they conserve body heat. **Contour feathers** cover the bird's body and give adult birds their shape. Specialized contour feathers, called flight feathers, are found on a bird's wings and tail. **Down feathers** cover the body of young birds and are found beneath the

Figure 35-13
Characteristics of birds. Like most birds, this tern is well adapted to flight.

Characteristics of Birds

- Forelimbs modified into wings
- Body covered with feathers
- Lightweight bones
- Endothermic metabolism
- Super-efficient respiratory system
- Heart with completely divided ventricle

Integrating Different Learning Styles

Logical/Mathematical	**PE**	Math Lab, p. 802
Visual/Spatial	**PE**	Performance Zone items 1, 14, 22, and 26, pp. 804–805
	ATE	Lesson Warm-up, p. 794; Demonstrations, pp. 797, 801; Up Close, p. 798
Interpersonal	**ATE**	Visual Strategy, p. 801; Biology in Your Community, p. 801; Closure, p. 802; Reteaching, p. 802
Verbal/Linguistic	**PE**	Study Tip, p. 796; Performance Zone items 2, 21, 28, and 30, pp. 804–805
	ATE	Active Reading, p. 795; Career, p. 796; Teaching Tip, p. 800

Figure 35-14 **Structure of a feather**

The structure of a contour feather helps create a smooth, aerodynamic surface, aiding flight.

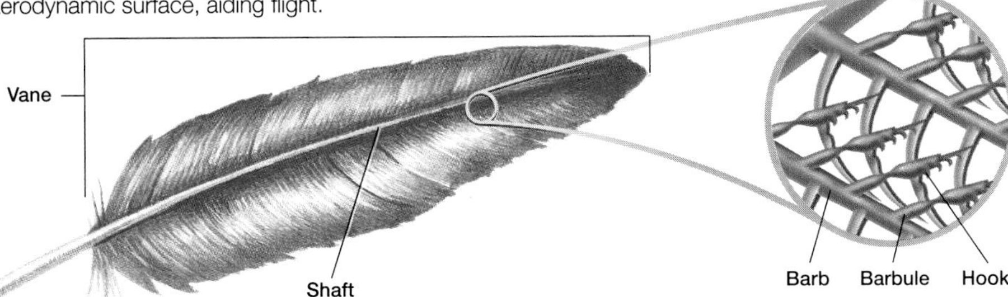

contour feathers of adults. Their soft, fluffy structure provides good insulation for the bird. As shown in **Figure 35-14,** a contour feather has many branches called barbs. Each barb has many projections, called barbules, that are equipped with microscopic hooks. These hooks link the barbs to one another, giving the feather a continuous surface and a sturdy but flexible shape. With use, the connections become undone. When you see a bird pulling its feathers through its beak, it is relinking these connections. This process is called preening. Preening also serves another function. Most birds have a special gland called a **preen gland** which secretes oil. When a bird preens, it spreads the oil over its feathers, cleaning and waterproofing them.

Feathers are important for other reasons too. Their coloration may be protective (as camouflage) or may be important in the selection of a mate. For example, the feathers of some birds allow them to blend in with their surroundings. In other species, the males will develop special plumage during the breeding season.

Strong, lightweight skeleton

If you have ever picked up a bird, such as a parakeet, you may have been surprised at how light it was compared to a mammal of a similar size. This is because the bones of birds are thin and hollow. Many of the bones are fused, making a bird's skeleton more rigid than a reptile's. The fused sections form a sturdy frame that anchors muscles during flight. The power for flight (or for swimming underwater in the case of some birds, like penguins) comes from large breast muscles that can make up 30 percent of a bird's total body weight. These muscles stretch from the wing to the breastbone. The breastbone is greatly enlarged and bears a prominent keel for muscle attachment, as illustrated in **Figure 35-15.** Muscles also attach to the fused collarbones (wishbone). No other living vertebrates have a keeled breastbone or a fused collarbone.

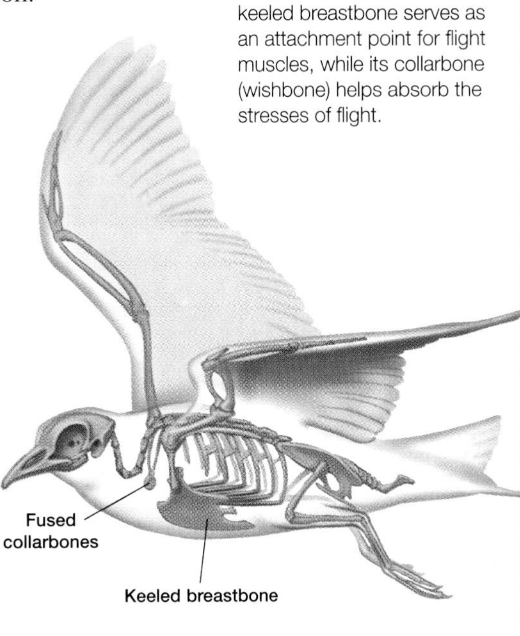

Figure 35-15 **Avian skeleton.** A bird's large, keeled breastbone serves as an attachment point for flight muscles, while its collarbone (wishbone) helps absorb the stresses of flight.

Overcoming Misconceptions

Most birds cannot smell well.

Children who find bird nests are often admonished by parents not to touch them. The children are warned, "the mother won't return to the nest if you touch it." In reality, most birds (exceptions are flightless birds, ducks, and vultures) have a very poor sense of smell and probably would not know if their nest had been touched by a human. Ask students to identify senses that would help birds compensate for a poor sense of smell. *(Birds have keen vision and hearing.)*

Active Reading

Technique: Paired Summarizing

After students have read silently about birds, have pairs of students summarize what they have read without looking at the textbook. One student should listen without interrupting but should be prepared to point out any inaccuracies in the summary and to add any ideas that were left out. Students may refer to the textbook during this clarification process.

Teaching TIP — ADVANCED

Preening

Birds often sit and preen their feathers. Students have probably observed this behavior. Have students hypothesize about the function of preening and the reason that birds spend so much time doing it. After a period of discussion, point out that when birds preen, they smooth out their feathers, making them more aerodynamic. Also note that birds spread oil from a gland near the base of their tail to waterproof the feathers and make them more resistant to breakage, much the way hair conditioner works.

Demonstration

Bird Bones

Bring cooked, cleaned chicken bones to class for students to examine. Cut a few bones in half so students can see the the hollow structure. If you have a wishbone, allow students to pull it gently to note its flexibility. The cartilage keel on the breast of a chicken is also interesting to examine because it provides an attachment surface for flight muscles.

Teaching TIP

Heat Source

The high temperature of birds is a by-product of their rapid metabolism. You may wish to use the analogy of a car engine for metabolism. If the engine is running, heat is produced. The faster the engine is running, the more heat is produced. Ectothermic reptiles are like cars whose engines run very slowly all the time, producing little heat. In contrast, birds (and mammals) keep their engines running at high speed all of the time, producing much heat.

CAREER

Veterinarian

When most people think of a veterinarian, they think of a "dog or cat doctor." Many students will know that veterinarians also treat livestock. However, most people probably do not know that veterinarians also treat birds and reptiles. Who, then, treats a sick parakeet or a lethargic lizard? Have students investigate the training required to become a veterinarian, and have them find out where vet schools are located in your region of the country. PORTFOLIO

Endothermic metabolism

Birds are endotherms; that is, they generate enough heat through metabolism to maintain a high body temperature. Birds maintain body temperatures ranging from 40°C to 42°C (104°F to 108°F), which is significantly higher than the body temperature of most mammals. For comparison, your body temperature is 37°C (98°F). These high temperatures are due to a high rate of metabolism, which contributes to the increased energy requirements of flight.

Study TIP

Interpreting Graphics

After studying Figure 35-18, use your own words to summarize how one breath of air circulates through a bird's lungs. Remember to include both the inhalation and the exhalation cycle.

Completely divided ventricle

As in crocodilians, the ventricle of birds is completely divided by a septum, as shown in **Figure 35-16.** Oxygen-rich and oxygen-poor blood are kept separate, meaning that oxygen is delivered to the body cells more efficiently. The sinus venosus, which you saw in the fish heart, is not a separate chamber of the heart in birds (or mammals). However, a small amount of tissue from it remains and plays an important role. This tissue is found in the wall of the right atrium and is the point of origin of the heartbeat. You may know it as the heart's pacemaker.

Highly efficient lungs

When birds, such as the geese shown in **Figure 35-17,** fly, they use a considerable amount of energy. Since birds often fly for long periods of time, their cellular demand for energy exceeds that of even the most active mammals. How do birds get the energy they need?

Reptiles meet their increased need for oxygen with lungs that have a larger surface area than the lungs of amphibians do. But there is a limit to how much the efficiency of a lung can be improved by just increasing its surface area. Another way to increase the

Figure 35-16 Avian heart. The avian heart has a complete septum.

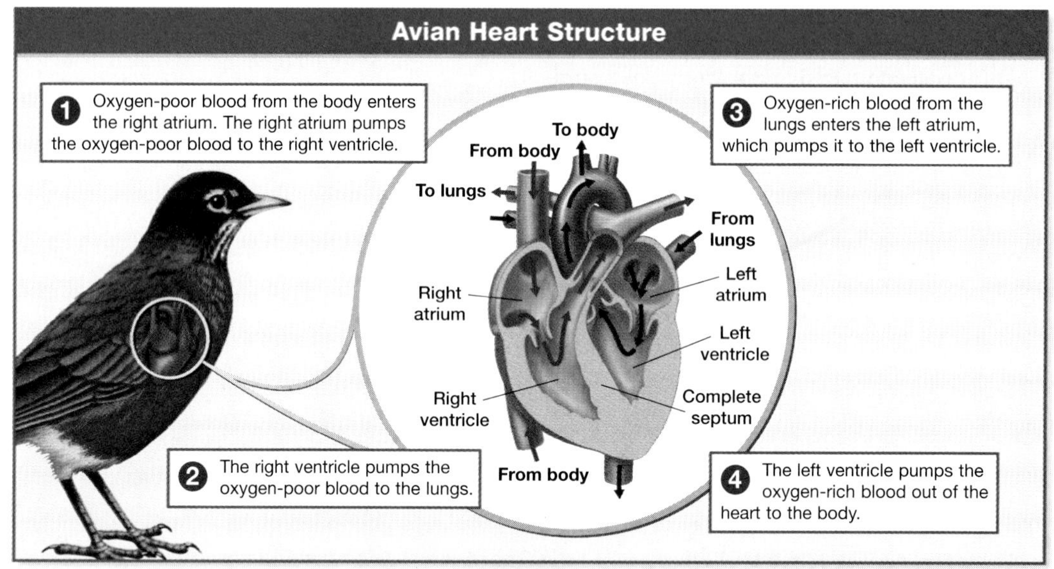

Avian Heart Structure
1 Oxygen-poor blood from the body enters the right atrium. The right atrium pumps the oxygen-poor blood to the right ventricle.
2 The right ventricle pumps the oxygen-poor blood to the lungs.
3 Oxygen-rich blood from the lungs enters the left atrium, which pumps it to the left ventricle.
4 The left ventricle pumps the oxygen-rich blood out of the heart to the body.
To body
From body
To lungs
From lungs
Right atrium
Left atrium
Left ventricle
Right ventricle
Complete septum
From body

MULTICULTURAL PERSPECTIVE

Eagle Feathers

The only people in North America who can legally own an eagle feather are Native Americans. The eagle is so highly valued by tribes throughout the United States that its feathers must be earned through personal sacrifice, and then they may be used only in special ceremonies. For instance, if a Winnebago pow-wow dancer accidentally drops an eagle feather during a performance, the dance is stopped until the feather is purified by an elder and then reclaimed by the dancer, who is not allowed to dance again for a year.

efficiency of a lung is to modify it so that air passes over its respiratory surface in one direction only, just as water flows over a fish's gills in one direction. This is what happened in birds. One-way air flow was made possible by the evolution of a series of air sacs connected to the bird's lungs, as shown in **Figure 35-18.** There is no gas exchange in the air sacs. They simply act as holding tanks.

There are two important advantages to one-way air flow. First, the lungs are exposed only to almost fully oxygenated air, increasing the amount of oxygen transported to the body cells. Second, the flow of blood in the lungs runs in a different direction than the flow of air does. While the flow of air and blood are not completely opposite (countercurrent), as in fish gills, the difference in direction does increase oxygen absorption.

These three characteristics—endothermic metabolism, a completely divided ventricle, and highly efficient lungs—provide the energy a bird needs for takeoff and for sustained flight. They enable a hummingbird to flap its wings rapidly (20–80 beats per second) as it hovers by a flower. They also permit migrating birds to fly thousands of kilometers without stopping. One group of shorebirds called the lesser yellowlegs fly across the open ocean from Massachusetts to Martinique in the West Indies. Incredibly, some yellowlegs can cover this distance of 3,220 km (about 2,000 mi) in less than 6 days. Note, however, that many birds, such as gulls and vultures, remain aloft for long periods of time using little energy. These birds take advantage of upward air movements that lift them.

Figure 35-17 Flight. These barnacle geese exert an enormous amount of energy during take off and flight.

Figure 35-18 Avian respiration. A single breath of air stays in a bird's respiratory system for two cycles of inhalation and exhalation.

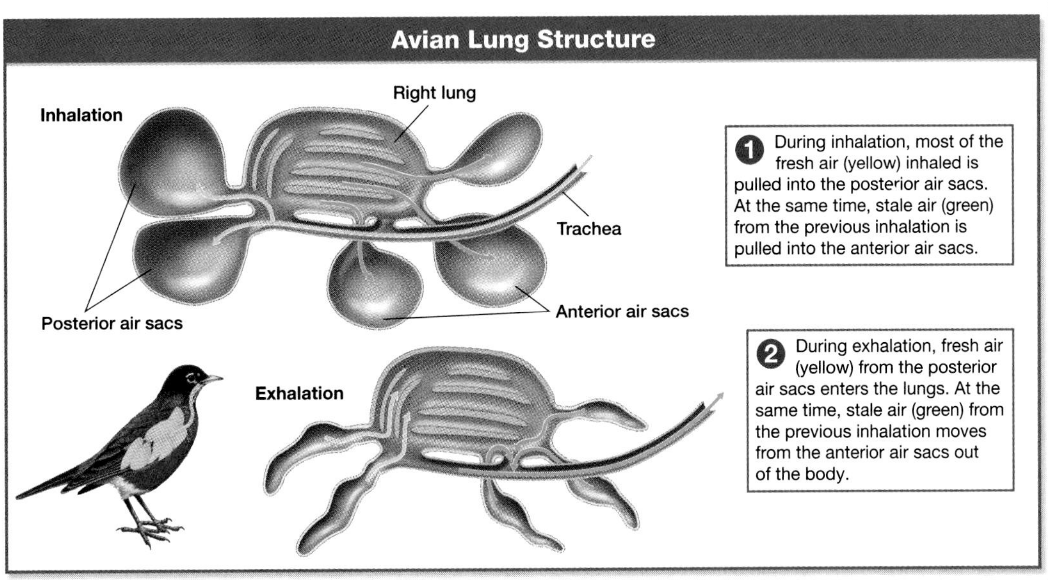

Avian Lung Structure
Inhalation
Right lung
Trachea
Posterior air sacs
Anterior air sacs
1 During inhalation, most of the fresh air (yellow) inhaled is pulled into the posterior air sacs. At the same time, stale air (green) from the previous inhalation is pulled into the anterior air sacs.
Exhalation
2 During exhalation, fresh air (yellow) from the posterior air sacs enters the lungs. At the same time, stale air (green) from the previous inhalation moves from the anterior air sacs out of the body.

Teaching TIP

White or Dark Meat?

The dark color of some chicken meat is partly due to the presence of myoglobin, a hemoglobin-related molecule that helps provide oxygen to muscles that must contract vigorously and repeatedly. Ask students why a duck has dark breast meat but a chicken has white breast meat. *(The duck flies long distances, but the chicken does not.)* Ask what chickens usually use for locomotion and what color that meat is. *(They usually use their legs to walk, and the legs contain dark meat.)*

Teaching TIP

Origin of Birds

Ask students to identify the living reptiles that are most closely related to birds. *(crocodilians)* Have them name two characteristics that crocodilians and birds share. *(a heart with completely divided ventricle, parental care of young, amniotic eggs)* Have students research recent fossil discoveries that have changed palentologists' views on the origin of birds.

did you know?

The brown-headed cowbird is a brood parasite.

A female brown-headed cowbird will fly to a nest full of eggs, roll one or two eggs out, and lay the same number as the number she has displaced. When the owner of the nest returns, it will incubate all the eggs, including the cowbird's. The cowbird eggs often have a shorter incubation time and hatch before the eggs of the incubating species. Therefore, the cowbird nestlings are fed first and get a head start in their development over the other chicks in the nest. Cowbird nestlings may even eject the nestlings of the incubating species, thereby eliminating competition for food.

Up Close

Bald Eagle

Teaching Strategies

- The largest bald eagle nests in North America were found in Vermillion, Ohio—3.7 m (12 ft) deep, 2.6 m (8.5 ft) in diameter, and weighing 1,800 kg (2 tons)—and in Florida—6.1 m (20 ft) deep and 2.9 m (9.5 ft) across. An eagle nest is called an aerie.
- After World War II, the numbers of bald eagles in the lower 48 states fell drastically because of the widespread use of the pesticide DDT. Because DDT breaks down very slowly, it accumulates in the food chain and causes the eggshells of bald eagles and many other birds to become thin and fragile. Affected eggs break when a parent attempts to incubate them. In 1972, the use of DDT was banned in the United States. By 1978, fewer than 500 breeding pairs of bald eagles remained in the United States outside Alaska. A slow recovery of the eagles began, and by the summer of 1999, six states had removed bald eagles from the endangered species list.
- Show students a picture of a large but immature bald eagle and ask them what kind of bird it is. Point out that the bald eagle does not get its white head and tail until its fourth or fifth year of life.
- Eagles have much better vision than humans. To simulate how much better, tape a worksheet on the wall and have students stand about 3 m away and try to read it. Then have them move 1 m away and read the worksheet. Tell them that an eagle could see the paper at 3 m as well as humans can at 1 m.

Up Close

Bald Eagle

- **Scientific name:** *Haliaeetus leucocephalus*
- **Size:** Wingspan is typically over 2 m (6.5 ft), and body mass often exceeds 7 kg (15 lb)
- **Range:** Nearly all of North America, from Florida to northern Alaska
- **Habitat:** Forested areas near water that have tall trees for perching and nesting
- **Diet:** Fish, small mammals, birds, carrion

External Structures

Feathers The body of the bald eagle is covered with feathers everywhere except the feet, which are bare. Both sexes develop the characteristic white head and neck at maturity.

Eyes Vision is the most important sense the bald eagle has. Its keen eyesight allows it to see prey at great distances. The bald eagle's visual acuity is 3–4 times higher than ours.

Grasping feet The bald eagle has large feet and talons—the hind claw may be 5 cm (2 in.) long. The talons are used to snatch fish from the water while the eagle is flying. When the muscles of the legs contract, the tendons in the lower legs are pulled, and the talons lock together around the fish.

Beak The beak is massive, with an elongated, sharp, downward-curving tip. Because they have no teeth, bald eagles do not chew their food. Instead, they use their beak to tear their prey into portions that they swallow whole.

Internal Structures

Brain In ratio of brain size to body size, birds rank second among vertebrates, behind only mammals. The large cerebellum receives and integrates information from the muscles, eyes, and inner ears. This makes possible the precise control of movement and balance necessary for flight. The optic lobe is large because it processes input from the eagle's most important sense organs—the eyes. The cerebrum performs many functions, including evaluation of sensory information, control of behavior, and learning.

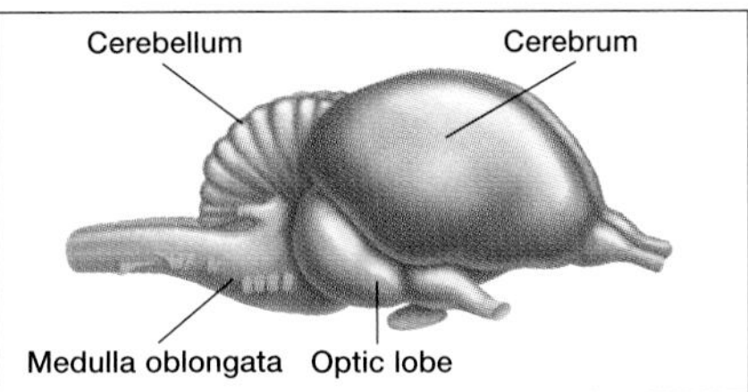

▲ Brain

Excretory system The excretory system is efficient and lightweight. It does not store waste liquids in a bladder. Instead, the bald eagle (and other birds) converts its nitrogenous wastes to uric acid, which is concentrated into a harmless white paste. The uric acid travels to the cloaca and is eliminated.

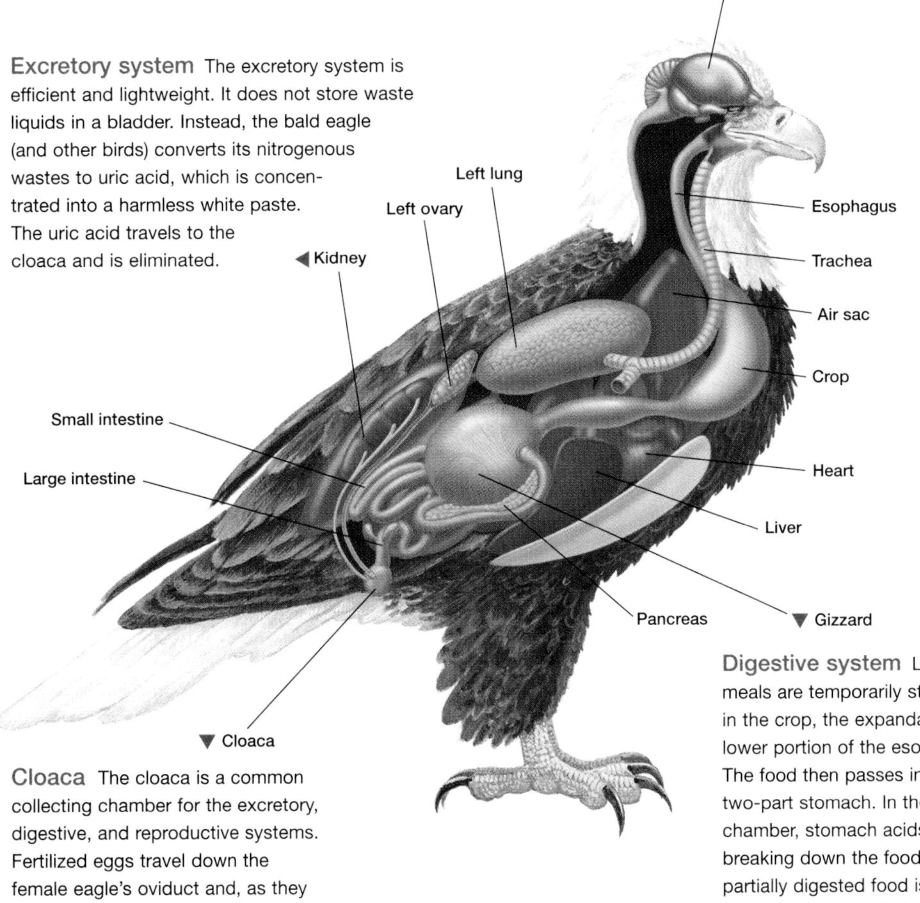

Cloaca The cloaca is a common collecting chamber for the excretory, digestive, and reproductive systems. Fertilized eggs travel down the female eagle's oviduct and, as they pass through it, egg white, membranes, and the shell are added. The completed egg then passes into the cloaca and out of the female's body.

Digestive system Large meals are temporarily stored in the crop, the expandable lower portion of the esophagus. The food then passes into a two-part stomach. In the first chamber, stomach acids begin breaking down the food. The partially digested food is then passed to a second chamber, the gizzard, where it is ground and crushed. Undigested food is eliminated through the cloaca.

Up Close

Bald Eagle

Discussion

- Identify three differences between the internal structures of the timber rattlesnake and the bald eagle. *(The rattlesnake has one functional lung, lacks air sacs, has neither a crop nor a gizzard, has venom-producing glands, and does not have hollow bones.)*
- Would the heat-sensitive pits of the timber rattlesnake be an effective way of sensing prey for a bald eagle? *(No, the rattlesnake's pits are sensitive only to nearby heat sources, and the bald eagle spots its prey from great distances.)*
- Why might it be a disadvantage for an eagle to have a urinary bladder? *(Urine contains a greater amount of water, which is heavy. The extra weight would be a disadvantage in flight.)*
- Why might teeth be a disadvantage to a bird? *(Teeth are heavy.)*

Teaching TIP

Reptiles and Birds

Have students make a three-column chart like the one in the **Graphic Organizer** shown at the bottom of page 800. They should use the information on birds in Section 35-3 and the material on reptiles in Sections 35-1 and 35-2. Students should use the following characteristics for both reptiles and birds: type of body covering, type of heart, mode of reproduction, type of metabolism, and parental care of offspring. PORTFOLIO

Teaching TIP

Introduced Pests

House sparrows and European starlings are two common birds that students probably are familiar with. Inform students that these two species were introduced into the United States from Europe. Both species have become serious pests. Starlings outcompete native songbirds for food and nesting sites. In fact, they are thought to be at least partly responsible for the decline in the numbers of bluebirds.

Figure 35-19 Penguin. The penguin's wings are adapted for swimming rather than flying.

Birds Are Adapted for Different Ways of Life

While there is great diversity among the 28 orders of birds, 60 percent of all bird species belong to order Passeriformes. These birds, also know as the songbirds, number approximately 5,276 species and are, by far, the largest group of terrestrial vertebrates. Birds are adapted for different ways of life, and you can tell a great deal about the habits and diet of birds by examining their beaks (bills), legs, and feet. Carnivorous birds such as hawks have curved talons for seizing prey and a sharp beak for tearing apart their meal. The beaks of ducks are flat for shoveling through water or mud, and their webbed feet enable them to swim. Finches are seed eaters and their short, thick beak is adapted for crushing seeds while their curved toes enable them to cling to branches. Other birds, such the penguins shown in **Figure 35-19,** are flightless, and their wings and feet are modified for swimming.

During their evolutionary history, the beaks, legs, and feet of birds have been adapted to the particular environment the birds live in, as shown in **Table 35-1.** Some birds are more highly specialized than others, and many birds are highly flexible in their eating habits. The song sparrow, for example, has a strong bill that it uses in winter to crack hard seed. In summer, the sparrow's diet changes and it uses its bill to catch soft-bodied insects.

Table 35-1 Avian Adaptations

Type of bird	Beak adaptations	Foot adaptations
Songbirds (e.g., cardinal, robin)	**Seed-cracking:** Short, thick, strong beak **Insect-catching:** Long, slender beak for probing	**Perching:** Toes can cling to branches; one toe points backward
Hummingbirds	**Probing:** Thin, slightly curved beak for inserting into flowers to sip nectar	**Hovering:** Legs so small the bird cannot walk on the ground; tiny feet

Graphic Organizer

*Use this graphic organizer with **Teaching Tip: Reptiles and Birds** on page 800.*

Characteristic	Reptiles	Birds
Body covering	Scaly skin	Feathers
Type of heart	Partially or completely divided ventricle	Completely divided ventricle
Reproduction	Oviparous or ovoviviparous	Oviparous
Metabolism	Ectothermic	Endothermic
Parental care	Little or none	Until offspring can fly

Type of bird	Beak adaptations	Foot adaptations
Woodpeckers	**Drilling:** Strong, chisel-like beak	**Grasping:** Feet with two toes pointing forward and two pointing backward
Parrots	**Cracking, tearing:** Short, stout hooked beak used to crack seeds and nuts and to tear vegetation	**Climbing/grasping:** Strong toes, two facing forward, two backward; adapted for perching, climbing, and holding food
Birds of prey	**Tearing:** Curved, pointed beak for pulling apart prey	**Grasping:** Powerful, curved talons for seizing and gripping prey
Ducks	**Sieving:** Long, flattened, rounded bill	**Swimming:** Three toes linked by webs for improved swimming
Long-legged waders	**Fishing:** Long, slender, spear-shaped beak for fishing	**Wading:** Long legs; toes spread out over large surface to support bird on soft surface

Visual Strategy

Table 35-1

Have students pair up. Instruct the students to cover the two columns on the right side of Table 35-1 with a sheet of paper. Have one student uncover the *Beak adaptations* column, choose one, and read the description. The other student should try to identify the type of bird whose beak is described. This process should continue until the second student has correctly identified each bird. The students should then switch roles and repeat the process using the *Foot adaptations* column.

Demonstration

Bird Adaptations

Show students pictures of several different types of birds. Be sure to include some with distinctly different beaks and feet. A typical assortment would be: woodpecker, heron, pelican, grosbeak, chickadee, goose, and hummingbird. Ask students to examine the different types of beaks and feet and to hypothesize about each bird's habitat and diet.

Biology in Your Community

Birding

Birding (formerly called birdwatching) has been described as the second-favorite outdoor activity in the United States, after gardening. Invite a member of a local or nearby chapter of the Audubon Society to visit your classroom and discuss tips on how to attract and identify backyard birds. Make a list of local birds and show slides of each bird to your class. Slides can be purchased from many sources, including the Cornell University Laboratory of Ornithology.

Calculating average bone density

Time 15 minutes

Process Skills Analyzing data, calculating, relating

Teaching Strategies
After students read the explanatory paragraph, have them glance at the data and predict which animal has the higher average bone density.

Analysis Answers
1. Animal 1—1.3 g/cm^3; Animal 2—2.0 g/cm^3
2. A certain amount of variation is normal in biological systems.
1. Animal 1

3 Assessment Options

Closure

Avian Adaptations
Divide the class into small groups, and have each group choose one bird adaptation. Groups should write a brief summary of the adaptation. If time permits, they should include a drawing with their summary. Make a copy of each of the summaries for each student. Have students bind the summaries together to make their own booklet. PORTFOLIO

Review

Assign Review—Section 35-3.

Reteaching — BASIC

Divide the class into small groups, and assign each group a particular feeding strategy and habitat where birds are found. Have each group use a field guide to identify five species of birds that share the assigned feeding strategy and habitat. Students should then use drawings or written descriptions to demonstrate the similarities in beaks, feet, and behavior among birds with similar lifestyles. Have each group share its findings with the entire class. PORTFOLIO

TOPIC: Flightless birds
GO TO: www.scilinks.org
KEYWORD: HX802

Other adaptations

The chart on the preceding pages does not cover all of the groups of birds, each of which is adapted to its particular living conditions. For example, shorebirds, such as gulls and terns, have streamlined bodies that are adapted for flying over the water in search of fish. Owls have large eyes that are directed forward. Their excellent low-light vision enables them to survive as nocturnal hunters. For a list of the orders of birds, see "A Six-Kingdom System of Classification" in the Appendix.

MATH LAB: Calculating average bone density

Background

Bird bones are hollow, which makes them less dense than the bones of other vertebrates and provides an advantage for flight. Density is the ratio of the mass of an object to its volume. Several teams of students determined the density of several samples from two different animals. You can use their data to practice calculating average bone density.

DATA TABLE

Bone type	Team 1	Team 2	Team 3	Team 4
Animal 1	1.6 g/cm^3	1.0 g/cm^3	1.2 g/cm^3	1.4 g/cm^3
Animal 2	2.3 g/cm^3	1.8 g/cm^3	1.8 g/cm^3	2.1 g/cm^3

1. **Add the densities of one bone type.** For example, if three bone samples have densities of 3.0, 3.1, and 2.9 g/cm^3, their sum would be 9 g/cm^3.
2. **Divide the sum of the densities by the number of samples.**

$$\text{Average density} = \frac{\text{sum of the densities}}{\text{number of samples}} = \frac{9 \text{ g/cm}^3}{3} = 3.0 \text{ g/cm}^3$$

Analysis

1. **Calculate** the average bone density for each of the two animals in the data table. Express your answer in grams per cubic centimeter.
2. **SKILL Evaluating Methods** Why is it important to analyze several samples and obtain the average of your data?
3. **SKILL Drawing Conclusions** Based on your answer to item 1, which of the two animals is most likely to be a bird, Animal 1 or Animal 2?

Review SECTION 35-3

1. **Identify** the adaptations birds have for flight.
2. **Summarize** how birds obtain the energy necessary for flight.
3. **Relate** the bald eagle's methods of hunting and feeding to its external body plan.

Critical Thinking

4. **SKILL Evaluating Conclusions** A student examines a bird that has delicate, perching feet with long, slender toes. Its beak is small but slightly long and pointed. The student concludes that the bird is a seed-eating songbird. Evaluate this conclusion using facts to support your reasoning.

Review SECTION 35-3 ANSWERS

1. forelimbs modified as wings; feathers; streamlined shape; lightweight bones, endothermic metabolism; highly efficient lungs; heart with completely divided ventricle
2. Birds have an endothermic body and a rapid metabolism, which provides the energy necessary for flight.
3. The bald eagle has strong feet with sharp talons for grabbing prey. The massive, hooked beak allows the eagle to tear the prey into pieces it can eat.
4. The small, pointed beak indicates that it is *not* a seed eater because seed eaters have thick beaks for crushing seeds. The foot structure, however, is consistent with the songbird anatomy.

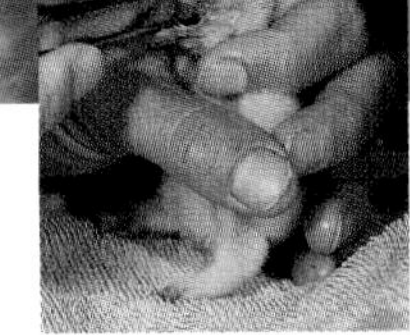

Study ZONE CHAPTER 35

Use the Key Concepts and Key Terms listed below to review the main ideas in this chapter. If you need to review the meaning of a term, turn to the page indicated.

Key Concepts

35-1 The Reptilian Body

- Reptiles have strong, bony skeletons and well-developed lungs.
- Reptiles are ectothermic.
- Reptiles have watertight skin and watertight eggs, both of which enable them to be terrestrial animals.
- Reptiles have paired lungs that have a greater surface area for gas exchange than the lungs of amphibians.
- Reptiles have a double-loop circulatory system with incomplete separation of oxygen-rich and oxygen-poor blood. Most reptiles have a ventricle that is partly divided into a right and left half.
- Reptilian fertilization is internal.

35-2 Today's Reptiles

- Snakes and lizards (order Squamata) share many characteristics, such as periodic molting, but snakes have no legs.
- The shells of turtles and tortoises (order Chelonia) are made of fused plates of bone covered with horny or leathery skin.
- Crocodilians (order Crocodilia), unlike other reptiles, care for their young after hatching. They also have a completely divided ventricle.
- There are only two species of tuataras (order Rhynchocephalia).

35-3 Characteristics and Diversity of Birds

- Birds are endotherms. Their high body temperature helps them meet the large energy requirements for flight.
- A bird's down feathers provide insulation. Its contour feathers give the bird its shape and aid flight.
- The bones of birds are thin and hollow, and many of them are fused; both are adaptations for flight.
- One-way airflow during respiration provides the large amounts of oxygen birds need for flight.
- The ventricle of the bird heart is completely divided by a septum.

Key Terms

35-1

amniotic egg (785)
alveoli (786)
oviparous (787)
ovoviviparous (787)

35-2

carapace (792)
plastron (792)

35-3

contour feather (794)
down feather (794)
preen gland (795)

Study TIP Ready?

- *If you think you understand these Key Concepts and Key Terms, then you're ready for the Performance Zone.* 

Study ZONE CHAPTER 35

Review and Practice

Assign the following worksheets for Chapter 35:

- **Student Review Guide**
- **Concept Mapping Worksheet**
- **Test Preparation Pretest**
- **Vocabulary Worksheet**
- **Critical Thinking Review Worksheet**

Assessment Options

The following assessment options are available for this chapter:

Resource Link

- Assign **Chapter Test 35** to assess students' understanding of the chapter.

Alternative Assessment

There are 23 subheadings in Chapter 35. Have each student make a subheading into a question, and then accurately answer it. For example, "Reptiles Share Several Key Characteristics," would become "What key characteristics do reptiles share?" PORTFOLIO

Portfolio Assessment

- *Portfolio Assessment* at the front of this book provides options for building and evaluating student portfolios. Students and teachers can select items from the areas listed for inclusion in a portfolio.
- See items labeled PORTFOLIO on pp. 780, 782, 783, 786, 787, 793, 796, 800, 802, and 803.

Answer to Concept Map

The following is one possible answer to Performance Zone item 1 on page 804.

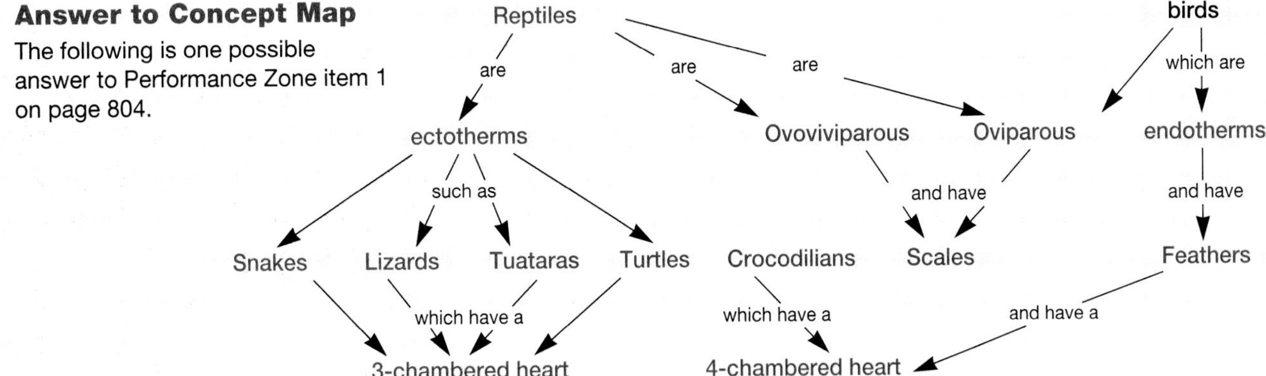

Performance ZONE
CHAPTER 35
REVIEW ANSWERS

Assignment Guide

SECTION	REVIEW
35-1	1, 2, 3, 4, 5, 16, 17, 23
35-2	6, 7, 8, 9, 10, 11, 18, 19, 20, 25, 27, 29
35-3	12, 13, 14, 15, 21, 22, 24, 26, 28, 30

Understanding and Applying Concepts

1. The answer to the concept map is found at the bottom of page 803.
2. **a.** Oviparous animals lay eggs. In ovoviviparous animals, embryos develop inside eggs within the mother's body, and the eggs hatch inside the mother's body.
b. Contour feathers are large and help make a bird aerodynamically sound. Down feathers are small and are found under contour feathers; they function only to insulate.
3. b
4. b
5. a
6. b
7. a
8. c
9. a
10. c
11. d
12. c
13. d
14. c
15. Reptilian metabolism is much slower than avian metabolism. Reptiles are ectotherms, while birds are endotherms.
16. The amnion holds water inside; the yolk sac stores food; the allantois stores wastes; and the chorion allows oxygen to enter the egg and carbon dioxide to exit.
17. Snakes, unlike frogs, have waterproof skin, efficient lungs, and an amniotic egg.

Performance ZONE CHAPTER 35 Review

Understanding and Applying Concepts

1. **Concept Mapping** Construct a concept map that describes the characteristics of both reptiles and birds. Include the following terms in your map: ectotherm, endotherm, oviparous, ovoviviparous, scales, feathers, three-chambered heart, four-chambered heart. Use additional terms in your map as needed.

2. **Understanding Vocabulary** For each pair of terms, explain the differences in their meanings.
a. oviparous, ovoviviparous
b. contour feather, down feather

3. Which is *not* an adaptation of reptiles for life on land?
a. watertight skin
b. external fertilization
c. amniotic egg
d. kidneys

4. The heart of most reptiles has
a. no septum.
b. a partly divided ventricle.
c. a fully divided ventricle.
d. two pumping chambers.

5. In reptiles, fertilization
a. is internal.
b. is external.
c. always occurs in water.
d. does not occur.

6. All of the following reptiles belong to the order Crocodilia, except
a. alligators.
b. tuataras.
c. crocodiles.
d. gavial.

7. The two basic parts of a turtle's shell are the
a. carapace and plastron.
b. septum and amnion.
c. chorion and allantois.
d. keratin and columella.

8. Snakes are different from lizards because snakes do not have
a. external ears.
b. movable eyelids.
c. a pectoral girdle.
d. a flexible jaw.

9. The Jacobson's organ is most like our sense of
a. smell.
b. hearing.
c. sight.
d. touch.

10. Compared with other reptiles, crocodiles
a. are harmless and tame.
b. have weak jaw muscles.
c. provide care for their young.
d. can tolerate drier conditions.

11. Which group of living reptiles is most closely related to birds?
a. snakes
b. turtles
c. rhynchocephalians
d. crocodiles

12. What structure has taken over the function of teeth in birds?
a. the beak
b. the crop
c. the gizzard
d. the claws

13. The feathers of most birds are well adapted for
a. swimming and repelling water.
b. expelling heat and feeding.
c. flying and conducting heat.
d. flying and insulating.

14. The illustration below is most likely of a bird adapted for
a. wading.
b. grasping.
c. perching.
d. swimming.

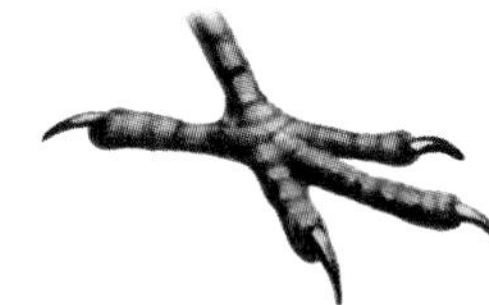

15. **Theme Metabolism** How do the metabolisms of reptiles and birds differ?

16. **Exploring Further** Name the four membranes contained in an amniotic egg, and describe how they make the egg an independent life-support system.

17. **Chapter Links** In what ways is a snake better suited for life on land than a frog? (**Hint:** See Chapter 34, Section 34-3.)

Skills Assessment

18. tuataras and crocodilians
19. lizards
20. Turtles, lizards, and snakes can tolerate both the heat of the desert day and the low temperatures of the desert night.
21. Answers will vary. Students should acknowledge the sources of their information.
22. Answers will vary. Guides should include an illustration of each bird.

Critical Thinking

23. Because the young are carried inside the body, they receive some heat as a result of the mother's muscular movements and basking behavior.
24. Because it mixes oxygen-rich and oxygen-poor blood, a three-chambered heart would not distribute enough oxygen to the hummingbird's body to sustain flight.

Skills Assessment

Part 1: Interpreting Graphs

Study the graph and answer the following questions:

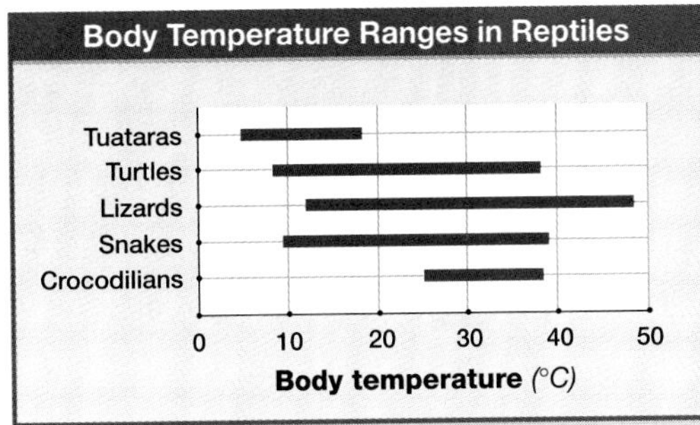

18. Which two groups show a much narrower range of temperatures?

19. Which group shows the largest range in body temperature?

20. Which groups could probably best tolerate the temperature extremes found in deserts? Explain your answer.

Part 2: Life/Work Skills

Use the media center or Internet resources to find out more about reptiles and birds that are common to your area of the country.

21. **Being a Team Member** Work with two or three of your classmates to find out what kinds of birds are common in your area. Select at least six birds to explore in depth and research the following: what habitat the bird occupies, what it eats, how its beak and feet are adapted for its lifestyle, and where it winters.

22. **Communicating** Present the information you gathered in an illustrated guide. Plan who will be responsible for which aspects of the guide. For example, each student could be responsible for all the information about two birds. Or each student could take on certain tasks, such as researching all the foot adaptations and habitat information. Make copies of your guide available for interested students.

Critical Thinking

23. **Forming Hypotheses** Many viviparous snakes and lizards live in cold climates. Why might viviparity be advantageous in such environments?

24. **Predicting Outcomes** How might having a three-chambered heart, like that of most reptiles, affect a hummingbird in flight?

Portfolio Projects

25. **Organizing Information** Create a habitat in a terrarium for a small lizard, and observe the lizard's behavior. Make a drawing of the environment you created. Keep a journal of your observations.

26. **Applying Information** Design and build a bird feeder. Place the feeder where it can be easily observed. Keep a journal of your observations to share with your class.

27. **Analyzing Information** Read "Snake Charmer," by D. Bruce Means (*National Wildlife,* Feb./Mar. 1999, pp. 36–41). How do eastern diamondback rattlers differ from their western cousins? What adaptive advantages might this difference provide? Evaluate this snake's role in its ecosystem.

28. **Interpreting Conclusions** Read "Dinosaurs Take Wing," by Jennifer Ackerman (*National Geographic*, July 1998, pp. 74–99). What evidence links dinosaurs to birds? What arguments are offered by those with opposing viewpoints? In your opinion, is it important for scientists to determine if birds evolved from dinosaurs?

29. **Summarizing Information** Use library references or an on-line database to find out about the different ways that snakes move. Prepare an illustrated written report to share your findings with your class.

30. **Career Focus Ornithologist** Research the field of ornithology (the study of birds), and write a report on your findings. Your report should include a job description, training required, kinds of employers, growth prospects, and starting salary.

Portfolio Projects

25. Students should consult an authoritative source about how to construct a terrarium. Suggest that they pattern their journals after standard animal behavior-recording protocols, which identify all possible behaviors and use standard and consistent terminology. Students should also record variables such as light, heat, and noise, as these affect animal behavior.

26. Students should research which type of bird feeder attracts which birds. Students should also follow the procedures for observing animals described in item 25.

27. The eastern diamondback is larger than its western cousin, and its venom is more potent. It is also more placid than the western diamondback. The tendency to be still and not rattle helps the snake remain camouflaged. This may be an advantage in open pine forests where there are not many places to hide. Answers will vary as to the role of the eastern diamondback in its ecosystem. Answers should include that it preys on small mammals and helps keep their populations in check.

28. Newly discovered fossils show that dinosaurs and birds share more than a hundred features, including a wishbone, similar skulls, and three forward-pointing toes. Some dinosaur fossils show tiny filaments along the neck and long fibers along the back and tail. These may be protofeathers, from which flight feathers evolved. There is strong evidence that the dinosaurs that may be the ancestors of birds lived in family groups and raised their young. Critics say that the fossil record is too incomplete to support the conclusion that birds evolved from dinosaurs. Answers will vary as to the importance of this issue.

29. Student reports should illustrate types of movement with examples from nature. Different species have different movements, and an experienced snake tracker usually can identify a snake by the path it leaves in soft soil or sand.

30. Answers will vary. Ornithologists study all aspects of bird biology, including the anatomy, physiology, behavior, and ecology of birds. The Cornell Laboratory of Ornithology and the National Audubon Society can provide extensive information about the field of ornithology. An ornithologist would complete a graduate program in ornithology after college. Professional ornithologists usually are engaged in university-based research. Growth prospects will vary with training. Starting salary will vary by region.

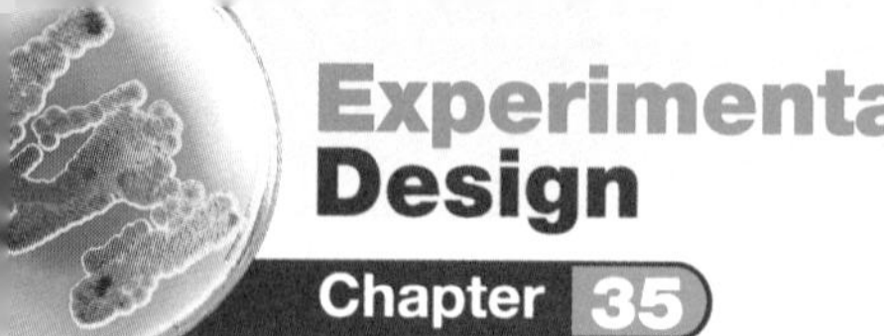

Experimental Design

Chapter 35

What Causes Anole Lizards to Change Color?

Purpose

Students will observe the ability of anoles to change color when placed on different background colors. Students will also determine how this ability might be an advantage to anoles.

Preparation Notes

Time Required: Two 45-minute lab periods

Materials

Materials for this lab can be ordered from WARD'S. See *Master Materials List* at the front of this book for catalog numbers.

Preparation Tips

- At least one hour before the lab exercise begins, place half the anoles in a terrarium lined with black construction paper. These anoles should turn brown. Place the remaining anoles in a terrarium lined with white construction paper. These anoles should turn green.
- Each terrarium should be illuminated by a fluorescent light and should contain a shallow dish of water.

Safety Precautions

Remind students to handle anoles with care. These lizards are easily frightened and can run quickly. If an anole is picked up by its tail, the tail may break off.

Procedural Tips

- Because anoles move quickly, you may want to place them in the jars for students in advance. Cover jars with lids that have air holes. Keep all jars in the same environment.
- Color change may be subtle and may be affected by adrenaline, if the lizard is frightened, or by feeding patterns. Very cool temperatures can produce a change to brown.

Experimental Design

Chapter 35

What Causes Anole Lizards to Change Color?

SKILLS

- Using scientific methods
- Observing

OBJECTIVES

- **Observe** live anoles.
- **Relate** the color of an anole to the color of its surroundings.

MATERIALS

- glass-marking pencil
- 2 large, clear jars with wide mouths and lids with air holes
- 2 live anoles
- 6 shades each of brown and green construction paper, ranging from light to dark (2 swatches of each shade)

Before You Begin

Lizards are a group of **reptiles.** There are 250–300 species of anoles, lizards in the genus *Anolis.* Like chameleons, anoles can change color, ranging from brown to green. Anoles live in shrubs, grasses, and trees. Light level, temperature, and other factors, such as whether the animal is frightened or whether it has eaten recently, can all affect the color of an anole. When anoles are frightened, they usually turn dark gray or brown and are unlikely to respond to other **stimuli.** Anoles generally change color within a few minutes. In this lab, you will observe the ability of anoles to change color when placed on different background colors. You will also determine how this ability might be an advantage to anoles.

1. Write a definition for each boldface term in the paragraph above.
2. Make a data table similar to the one below.
3. Based on the objectives for this lab, write a question you would like to explore about the color-changing behavior of anoles.

DATA TABLE

Anole	Color 1		Color 2	
	Change	Time	Change	Time
1				
2				

Procedure

PART A: Make Observations

1. Observe live anole lizards in a terrarium. Make a list of characteristics that you can use to determine that anoles are reptiles.
2. Work with a partner to place anoles to be studied in separate glass jars. **CAUTION: Handle anoles gently, and follow instructions carefully. Anoles run fast and are easily frightened. Plan your actions before you start.** By working efficiently, you can keep your anole from becoming overly frightened. Carefully pick up one anole by grasping it firmly but gently around the shoulders. Do *not* pick up anoles by their tail. Place the anole in a glass jar. Quickly and carefully place a lid with air holes on the jar.
3. When anoles become overly frightened, they remain dark in color. While you are designing your experiment, do not disturb your anoles, and let them recover from your handling.

PART B: Design an Experiment

4. Work with members of your lab group to explore one of the questions written for step 3 of **Before You Begin.** To explore the question, design an experiment that uses the materials listed for this lab.

Disposal

- Clean terrariums and jars after you have removed the anoles.
- Tell students that anoles that are no longer needed in the classroom make good pets at home. You may want to show students how to set up a vivarium for their anoles.

Before You Begin

Answers

1. *reptiles*—ectothermic vertebrates with scaly, watertight skin, lungs, and a heart with a partially or completely divided ventricle

   stimuli—environmental factors that influence the behavior of an organism
2. Student data tables should resemble the example on page 806.
3. For example: How quickly do anoles change color after they move to a new background color?

You Choose

As you design your experiment, decide the following:

a. what question you will explore
b. what hypothesis you will test
c. how many anoles you will need
d. what background colors you will use
e. how many times you will test each background with an anole
f. how long you will observe each test and how you will keep track of time
g. what your control will be
h. what data to record in your data table

5. Write a procedure for your experiment. Make a list of all the safety precautions you will take. Have your teacher approve your procedure and safety precautions before you begin the experiment.
6. Set up and carry out your experiment.

Part C: Cleanup and Disposal

7. Dispose of construction paper and broken glass in the designated waste containers. Put anoles in the designated container. Do not put lab materials in the trash unless your teacher tells you to do so.
8. Clean up your work area and all lab equipment. Return lab equipment to its proper place. Wash your hands thoroughly before you leave the lab.

Analyze and Conclude

1. **Summarizing Results** Briefly state how the variable you tested influenced the color-changing behavior of anoles.
2. **Evaluating Results** Did any unplanned variables influence your data? (For example, was there a loud noise, or was a jar suddenly moved?)
3. **Analyzing Methods** How could your experiment be modified to improve the certainty of your results?
4. **Analyzing Data** Were there any inconsistencies in your data? (For example, two anoles reacted in different ways.) If so, offer an explanation for them.
5. **Drawing Conclusions** After considering your data, make a statement about color-changing behavior in anoles.
6. **Further Inquiry** Write a new question about anoles that could be explored with another investigation.

Do You Know?

Do research in the library or media center to answer these questions:

1. What other behaviors are characteristics of anoles?
2. How is a chameleon different from an anole?

Use the following Internet resources to explore your own questions about lizards that change color.

internetconnect

SCILINKS NSTA
TOPIC: Adaptations of reptiles
GO TO: www.scilinks.org
KEYWORD: HX807

Procedure

Answers

1. Characteristics include scaly skin and toes with claws.
3. Answers will vary. For example: Place each jar containing an anole on top of a sheet of paper that is a different color than the anole. Note whether each anole changes color and, if it does, the color it becomes and the time required for the color change.
5. Answers will vary depending on the question explored. Check that safety precautions are listed.

Sample Data Table

	Green paper		Brown paper	
Anole	**Change**	**Time**	**Change**	**Time**
Brown at start	green	3 s	stayed brown	
Green at start	stayed green		brown	3 s

Analyze and Conclude

Answers

1. Students' answers should clearly describe how each test was conducted and should state their results in terms of color changes in the anoles. See Sample Data Table at bottom of page 807.
2. Anoles may react to stimuli other than the independent variable selected. Students may list several uncontrolled variables, such as fear or temperature, that could have affected their results.
3. Answers will depend on the groups' experimental designs. Students should suggest improving their methods by eliminating or controlling as many uncontrolled variables as possible.
4. Answers will vary. Students should include a possible explanation for any inconsistency observed.
5. Answers will vary. For example: Anoles change to a color that most closely matches the color of their environment.
6. Answers will vary. For example: Would anoles react differently if there were two in each jar?

Do You Know?

Answers

1. Male anoles extend their dewlap as a display during courtship and when defending their territory.
2. Chameleons have opposable toes and a strong, curled tail, two features that help them grasp tree branches. Chameleons also have large eyes that can move independently of each other.

Chapter **36**

Mammals

Planning Guide

	OBJECTIVES	MEETING INDIVIDUAL NEEDS	READING SKILLS AND STRATEGIES
BLOCK 1 45 min	**36-1 The Mammalian Body** • **Describe** three functions of hair. • **Relate** a mammal's teeth to its diet. • **Summarize** how mammals maintain a high body temperature. • **Relate** the characteristics of mammals to one mammal, the grizzly bear. • **Describe** parental care in mammals.	**Learning Styles**, p. 810 BASIC ADVANCED • p. 816	**ATE** **Active Reading, Technique: K-W-L**, p. 809 ■ **Active Reading Guide,** worksheet (36-1) ■ **Directed Reading,** worksheet (36-1) ■ **Reading Strategies,** K-W-L worksheet
BLOCKS 2 & 3 90 min	**36-2 Today's Mammals** • **Compare** reproduction in monotremes, marsupials, and placental mammals. • **Relate** the limited distribution of monotremes and marsupials to the breakup of Pangaea. • **Recognize** how mammals are adapted to many different environments.	**Learning Styles**, p. 817 BASIC ADVANCED • pp. 821, 824, 825, 826	**ATE** **Active Reading, Technique: Paired Reading**, p. 820 ■ **Active Reading Guide,** worksheet (36-2) ■ **Directed Reading,** worksheet (36-2) ■ **Supplemental Reading Guide,** *Through a Window: My Thirty Years with the Chimpanzees of Gombe*

Review and Assessment Resources

BLOCKS 4 & 5 90 min

TEST PREP

PE **Study Zone**, p. 827

PE **Performance Zone**, pp. 828–829

■ **Chapter 36 worksheets:**
- Vocabulary
- Concept Mapping
- Critical Thinking

 **Audio CD Program**

■ **Test Prep Pretest**

CHAPTER TESTING

■ **Chapter Tests** (blackline master)

▲ **Test Generator**

■ **Assessment Item Listing**

See the correlations table at the front of this book for a list of the ***National Science Education Standards*** addressed in this chapter.

Smithsonian Institution® Visit **www.si.edu/hrw** for additional on-line resources.

KEY

Biology Principles and Explorations

PE Pupil's Edition

ATE Teacher's Edition

 Teaching Transparencies

 Laboratory Program

One-Stop Planner CD-ROM
Shows these resources within a customizable daily lesson plan:

■ Teaching Resources

▲ Test Generator

VISUAL STRATEGIES	CLASSWORK AND HOMEWORK	TECHNOLOGY AND INTERNET RESOURCES
ATE Fig. 36-3, p. 812 181 Mammalian Lung Structure 144A Comparison of Coyote and Beaver Skulls	**PE** Quick Lab Is mammalian hair a good insulator?, p. 811 **PE** Review, p. 816	Holt Biology Videodiscs Section 36-1 Audio CD Program Section 36-1 internetconnect — SCiLINKS NSTA — TOPIC: Mammals
ATE Fig. 36-21, p. 825 177 Major Orders of Mammals 178 Major Orders of Mammals, cont.	**PE** Lab Techniques Exploring Mammalian Characteristics, pp. 830–831 Lab B29 Fetal Pig Dissection ■ Occupational Applications: Wildlife Biologist **PE** Review, p. 826 **PE** Study Zone, p. 827	Holt Biology Videodiscs Section 36-2 Audio CD Program Section 36-2 internetconnect — SCiLINKS NSTA — TOPICS: • Dairy products • Placental mammals • Manatee

ALTERNATIVE ASSESSMENT

PE Portfolio Projects 24–28, p. 829

ATE Alternative Assessment, p. 827

ATE Portfolio Assessment, p. 827

Lab Activity Materials

Quick Lab
p. 811 MBL or CBL system with appropriate software, temperature probe, wool sock (1), beaker of ice, graph paper (optional), cotton sock (1)

Data Lab
p. 821 pencil, paper

Lab Techniques
p. 830 hand lens or stereomicroscope, prepared slide of mammalian skin, compound microscope, mirror, specimens or pictures of vertebrate skulls (some mammalian, some nonmammalian)

Demonstrations

ATE selection of different horns and antlers, p. 821

ATE selection of animal hooves, p. 824

BIOSOURCES LAB PROGRAM
- Inquiry Skills Development: Lab B29, p. 137

Chapter Theme

Evolution

Mammals are very diverse anatomically and occupy a variety of habitats. But for most of their history, mammals were small and not very diverse. The diversification of mammals occurred only after the extinction of the dinosaurs.

Establishing Prior Knowledge

Before beginning this chapter, make certain students can answer the questions in **Study TIP Ready?** on page 809. If they cannot, encourage them to reread the sections indicated.

Opening Question

Ask students to make two lists. On one list they should write down the names of as many large terrestrial herbivores as they can think of. On the other list they should write down the names of as many large terrestrial carnivores as possible. Have students determine what percentage of the animals on their lists are mammals. Point out that today these two ecological roles are dominated by mammals. During the age of dinosaurs, however, reptiles dominated these roles.

CHAPTER 36 Mammals

Mountain lion

What's Ahead — and Why
CHAPTER OUTLINE

36-1 The Mammalian Body
- Mammals Share Five Key Characteristics
- Mammals Have Diverse, Specialized Teeth
- Mammals Maintain a High Body Temperature
- Mammals Nurse and Care for Their Young

36-2 Today's Mammals
- Mammals Have the Most Diverse Body Form of All Vertebrates
- Mammals Reproduce in One of Three Ways
- Most of Today's Mammals Are Placental

Students will learn

▷ the characteristics of mammals.

▷ about the adaptations that make mammals successful in a wide range of habitats.

Duckbill platypus

Using its soft beak as a scanner,

the platypus swims along underwater, locating worms and other small creatures by tracking the faint electric signals emitted by their beating hearts. How is this odd-looking, egg-laying creature related to the mountain lion on the facing page? Strange as it may seem, they're both mammals.

Looking Ahead

Study TIP

Ready?

Check your understanding of these topics to help you prepare for what's ahead.

Can you . . .

- **define** the term *metabolism*? (Section 1-2)
- **compare** the diets of carnivores and herbivores? (Section 17-2)
- **summarize** the difference between endothermic and ectothermic animals? (Section 33-2)
- **describe** the breakup of Pangaea? (Section 33-2)

If not, *review the sections indicated.*

internetconnect

National Science Teachers Association *sci*LINKS Internet resources are located throughout this chapter.

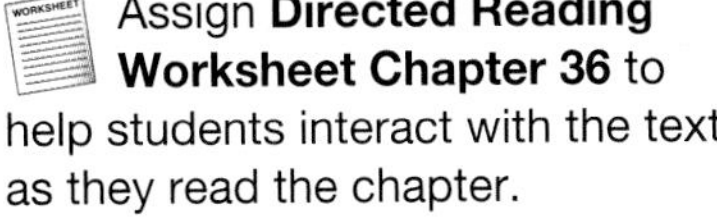

Reading Strategies

Active Reading

Technique: K-W-L

Before students read this chapter, have them write short individual lists of all the things they already **K**now (or think they know) about mammals. Ask them to contribute their entries to a group list on the board or overhead. Then have students list things they **W**ant to know about mammals. Have students save their lists for use later in the **Closure** on page 826.

Directed Reading

Assign **Directed Reading Worksheet Chapter 36** to help students interact with the text as they read the chapter.

Interactive Reading

Assign ***Biology: Principles and Explorations* Audio CD Program Chapter 36** to help reluctant readers achieve greater success in reading the chapter.

Reading Skills

Assign Reading Organizers and Reading Strategies Worksheets from the **One-Stop Planner CD-ROM with Test Generator** to help students learn and practice effective reading skills.

internetconnect

Holt *Biology: Principles and Explorations*

On-line Resources: go.hrw.org
Visit the HRW Web site for a variety of resources related to this chapter. Just type in the keyword HX0 Chapter 36.

Holt *Biology: Principles and Explorations*

On-line Resources: www.scilinks.org
The following *sci*LINKS Internet resources can be found in the student text for this chapter:

Page 812
TOPIC: Mammals
KEYWORD: HX812

Page 819
TOPIC: Dairy products
KEYWORD: HX819

Page 820
TOPIC: Placental mammals
KEYWORD: HX820

Page 825
TOPIC: Manatee
KEYWORD: HX825

Chapter 36

1 Prepare

Scheduling

- **Block:** About 45 minutes will be needed to complete this section.
- **Traditional:** About one class period will be needed to complete this section.
- **Planning:** See the Planning Guide on pp. 807A and 807B for lecture, classwork, and assignment options for Section 36-1. Lesson plans for Section 36-1 can be found in a lesson cycle format in ***Biology: Principles and Explorations*** **One-Stop Planner CD-ROM with Test Generator.**

***Resource* Link**

Assign **Active Reading Worksheet Section 36-1** to help students to comprehend the key concepts and visuals in this lesson.

2 Teach

Lesson Warm-up

Ask how many students have pets. Then find out how many of those pets are mammals. Ask students why they think most people prefer mammals to fish, amphibians, reptiles, or birds. For fun, have students share their favorite pet tricks or stories.

Section 36-1 The Mammalian Body

Objectives

- **Describe** three functions of hair. (pp. 810–811)
- **Relate** a mammal's teeth to its diet. (p. 812)
- **Summarize** how mammals maintain a high body temperature. (p. 813)
- **Relate** the characteristics of mammals to one mammal, the grizzly bear. (pp. 814–815)
- **Describe** parental care in mammals. (p. 816)

Key Terms

hair
mammary gland
weaning

Mammals Share Five Key Characteristics

If you were to look out over an African landscape, you would notice the big mammals—lions, elephants, antelopes, and zebras. And your eye would not as readily pick out the many birds, snakes, lizards, and frogs that also live there. The fact that almost all of today's large, land-dwelling vertebrates are mammals makes them much more easily noticed. While most mammals are terrestrial, some, like whales, swim in the sea, and others—the bats—fly in the air. In spite of their differences, mammals share key characteristics that are summarized in **Figure 36-1.**

Mammals are well adapted for terrestrial living and are able to retain water more efficiently than reptiles. Why is this? The mammalian kidney has an exceptional ability to concentrate waste products in a small volume of fluid called urine. In Chapter 40 you will find out how the mammalian kidney functions.

Hair

Of all animal species only the mammals have hair. Even whales and dolphins, which appear to be hairless, have a few sensitive bristles on their snout. A **hair** is a filament composed mainly of dead cells filled with the protein keratin. The evolutionary origin of hair is unknown, but it is probably not derived from reptilian scales.

The primary function of hair is insulation. Mammals, such as the polar bear shown in Figure 36-1, typically maintain body temperatures higher than the temperature of their surroundings. As a result, they tend to lose body heat. To reduce the loss of body heat to the environment, most mammals are covered with a dense coat of hair that holds heat in. Like other mammals that live in cold environments, the polar bear has a layer of fat under its skin that provides additional

Figure 36-1 Characteristics of mammals. Despite their vast external differences, mammals share a number of distinctive features.

Characteristics of Mammals

- Some hair
- Diverse and specialized teeth
- Endothermic metabolism
- Mammary glands that produce milk
- Unborn young usually nourished by a placenta

Integrating Different Learning Styles

Logical/Mathematical	**PE**	Quick Lab, p. 811
	ATE	Did You Know?, p. 812; Teaching Tip, p. 813
Visual/Spatial	**ATE**	Visual Strategy, p. 812; Performance Zone items 1, 18–20, pp. 828–829
Interpersonal	**PE**	Study Tip, p. 816
	ATE	Closure, p. 816
Intrapersonal	**ATE**	Did You Know?, p. 811; Making Biology Relevant, p. 812
Verbal/Linguistic	**PE**	Study Tip, p. 816; Performance Zone item 2, p. 828
	ATE	Teaching Tip, p. 811

insulation from the cold. Humans have a sparse covering of hair and a limited amount of body fat; in most climates we need clothes to provide adequate insulation.

Hair has functions other than insulation. The coloration and pattern of a mammal's coat often enable it to blend in with its surroundings. A small brown mouse is almost invisible against the dark forest floor; and the orange and black stripes of the Bengal tiger, shown in **Figure 36-2,** conceal it in the tall, orange-brown grass in which it hunts. Some animals, like the arctic fox you saw on page 177, show a seasonal change in the color of their coat that provides protective coloration year round. The color of a mammal's coat may also be a clear signal; the black and white fur of a skunk, for instance, warns would-be predators to stay away.

In some animals, special hairs serve a sensory function. The whiskers of cats and dogs are stiff hairs that are very sensitive to the touch. Mammals that are active at night or that live underground often rely on their whiskers for information about the environment. Other special hairs can be used as a defensive weapon. For example, when threatened, porcupines defend themselves by raising their sharp, barbed quills, also shown in Figure 36-2.

Figure 36-2 Functions of hair

This tiger's stripes and reddish fur blend in with the surrounding grasses. Porcupines often use their quills for defense.

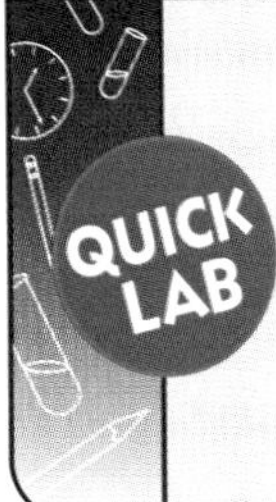

Is mammalian hair a good insulator?

When you are getting dressed on a cold day, why are you are more likely to choose a wool sweater than a cotton one? In this lab you will compare the insulating abilities of the animal fiber wool and the vegetable fiber cotton.

Materials

MBL or CBL system with appropriate software, temperature probe, 1 wool sock, beaker of ice, graph paper (optional), 1 cotton sock

Procedure

1. Set up an MBL/CBL system to collect data from a temperature probe at 6-second intervals for 100 data points.
2. Find and record the room temperature.
3. Insert the end of the probe into one thickness of wool sock. Then place the sock-covered probe into a beaker of ice. Collect temperature data for 10 minutes.
4. View the graph of your data. If possible, print out the graph. Otherwise, plot the graph on graph paper.
5. Repeat steps 1–4 using a cotton sock and fresh ice.

Analysis

1. **Analyze** your data and determine which sock was the better insulator.
2. **Summarize** why these results are of importance to mammals.

did you know?

The hair on your head is actually dead cells.

Sprays, conditioners, mousses, dyes, gels, and all kinds of special hair treatments cost Americans billions of dollars each year. Yet the very hair on which we spend so much money to "bring to life" is actually nothing more than dead cells. As hair cells are produced in the scalp, they die and are pushed outward. These long strands of dead cells are combed, brushed, dried, sprayed, and even "mayonnaised" by some people in the quest for the perfect "do."

Teaching TIP

Functions of Hair

Before students read pages 810–811, ask them to think of as many functions of hair as they can. Make a list on the board. Then have students read this page and add to the list as they identify or think of other functions of hair.

QUICK LAB

Is mammalian hair a good insulator?

Time 20 minutes per MBL/CBL system

Process Skills Observing, organizing data, analyzing data, inferring

Teaching Strategies

- Be sure students let the ice sit in the water for a sufficient time so the temperature can stabilize.
- Remind students to keep the probe in the tip of the sock for both trials. The probe should be at the same depth in the beaker for both trials.

Analysis Answers

1. Answers will vary. If done correctly, the wool sock should be the better insulator.
2. Mammals' lives often depend on the conservation of their body warmth by hair or fur.

Visual Strategy

Figure 36-3

Point out that the coyote's teeth function like a pair of scissors: two sharp surfaces slide across one another as the coyote closes its mouth. Students should see that the coyote's teeth are unsuited for grinding plant material. The deer's molars are suited for this task because they function like a grinding mill: two rough surfaces come together and slide across one another when the deer closes its mouth. Ask students to identify what distinctive type of tooth is not found in deer and why it is not important to them. *(Deer do not have canine teeth, which are used for puncturing and ripping apart prey. Because deer are herbivores, they do not need canines.)*

Making Biology Relevant

Health

Adult humans usually have 12 molars—six in each side of their mouth. The back molars, the so-called wisdom teeth, are the last to erupt, typically emerging between the ages of 17 and 25. Not everyone has four wisdom teeth; some people have two or none. If the wisdom teeth grow in at an angle, they can force the other teeth out of alignment and may have to be removed. Ask how many students have had their wisdom teeth removed, and have them share their experiences if they wish.

biobit

Do all mammals have teeth?

No. Baleen whales, for example, have long, bristle-covered plates called baleen that hang from the roof of their mouth. The baleen are used to strain plankton from the water.

Mammals Have Diverse, Specialized Teeth

Unlike some other vertebrates whose teeth are constantly being lost and replaced, mammals usually have only two sets of teeth. The first set, commonly called baby teeth or milk teeth, is replaced by permanent teeth, which are not replaced if lost or damaged. Animals use their teeth in a variety of ways—to secure and chew food, for protection or as a threat signal, and to perform tasks, as when a beaver cuts down trees to make a dam.

In most mammals, four types of teeth can be recognized: incisors, canines, premolars, and molars. Each type of tooth performs a different function in eating. Incisors, the front teeth, are for biting and cutting. Behind them are canines used for stabbing and holding. Lining the jaw are the premolars and molars. As a mammal chews, its upper and lower molars fit together, crushing and grinding the food more thoroughly than a reptile's teeth can. The resulting smaller bits of food can be quickly digested, which permits a mammal to eat enough food to fuel its endothermic metabolism.

A mammal's teeth are specialized to fit the food it eats, and it is usually possible to determine a mammal's diet by examining its teeth. **Figure 36-3** shows the differences between the teeth of a coyote (a carnivore) and those of a deer (a herbivore). The coyote has long canine teeth that are suited for grasping prey, and its sharp molars can cut off pieces of flesh. In contrast, the deer's canines are small, and it uses its incisors to nip off selected pieces of plant material. The deer's premolars and molars are flat and covered with ridges that form a surface on which plant material can be ground.

Figure 36-3 Specialized teeth

A mammal's teeth provide clues about its diet.

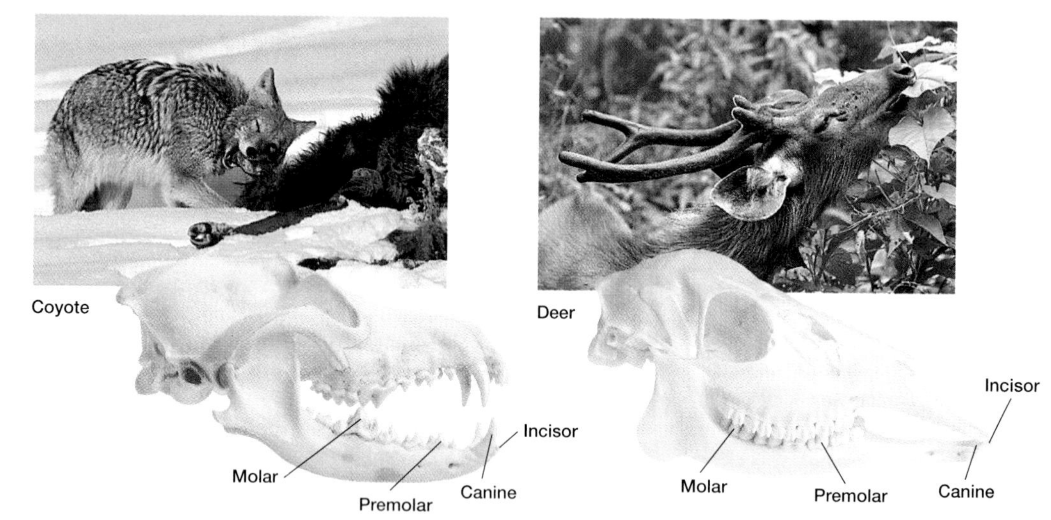

did you know?

Zoologists often identify mammal skulls by their teeth.

The kind and number of teeth make up what is called a dental formula. Many orders of mammals have characteristic dental formulas. An example of a dental formula is 2/2, 1/1, 2/2, 2/3. The numbers refer to the teeth in *half* of a jaw. The first number of each ratio refers to the teeth in the upper half of the jaw. The second number of each ratio refers to the teeth in the lower half of the jaw. The sequence refers to the following order: incisors, canines, premolars, and molars. Ask students how many teeth the animal with the above dental formula would have. *(30)*

Mammals Maintain a High Body Temperature

Like birds, mammals are endothermic, generating heat internally through the rapid metabolism of food. Because a mammal's body temperature remains relatively constant regardless of the temperature of its surroundings, mammals can be active at any time of day or night. They also can live in very cold climates, where most ectothermic amphibians and reptiles cannot. In addition, endothermic metabolism also permits mammals to sustain activities that require high levels of energy, such as running or flying long distances.

To maintain the high metabolic rate required by an endotherm, a mammal must eat about 10 times as much food as an ectotherm of similar size. Metabolizing this food requires a considerable amount of oxygen. Thus, mammals, like birds, have evolved respiratory and circulatory systems that are very efficient at acquiring and distributing oxygen.

Respiratory system

Mammalian lungs, shown in **Figure 36-4,** have a large internal surface area to aid the exchange of O_2 and CO_2, so they are much more efficient at obtaining oxygen from the air than reptilian and amphibian lungs are. Respiration in mammals is aided by the diaphragm, a sheet of muscle that separates the chest cavity from the abdominal cavity. When the diaphragm contracts, the chest cavity enlarges, drawing air into the lungs.

Some mammals are far more active than others, but the more active ones do not necessarily have proportionally larger lungs. Instead, in active mammals the individual alveoli, also shown in Figure 36-4, are smaller and more numerous, further increasing the surface area for diffusion.

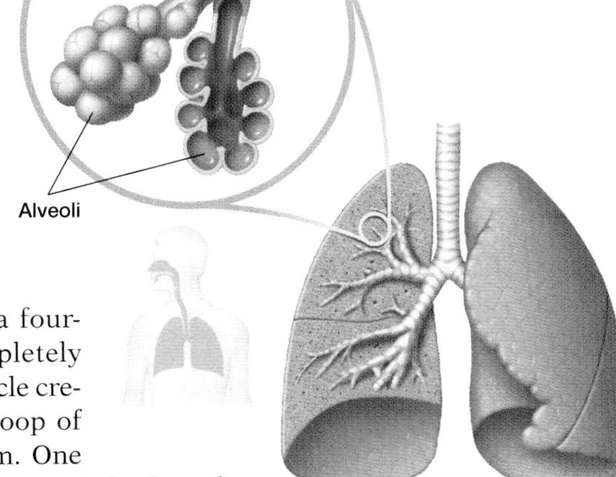

Figure 36-4 Mammalian lungs. The lungs of mammals contain clusters of alveoli that dramatically increase the lung's surface area.

Heart and circulatory system

Like crocodiles and birds, mammals have a four-chambered heart with a septum that completely divides the ventricle. The division of the ventricle creates two pumping chambers, one for each loop of the mammal's double-loop circulatory system. One chamber pumps oxygen-rich blood to the body, while the other pumps oxygen-poor blood to the lungs. Since the two do not mix, only oxygen-rich blood is delivered to the tissues, a condition vital for meeting the oxygen needs of endothermic animals.

In Chapter 39 you will learn more about the mammalian respiratory and circulatory systems. You can read more about the anatomy of one mammal in *Up Close: Grizzly Bear* on pages 814–815.

REAL LIFE

Some wild mammals love city life! Rats and mice have adapted most successfully. It is estimated that most large cities contain as many rodents as people. But when they decide to move in to our living space, they cause trouble. In addition to being destructive, some harbor diseases.

Finding Information
Find out what people can do to discourage wild mammals from invading their dwellings.

did you know?

Contrary to popular belief, bears do not undergo true hibernation.

In true hibernation, an animal's body temperature drops to nearly that of the environment. Bears of the northern forests den for several months but do not hibernate in the true sense. Their heart rate may decrease from 40–70 beats per minute to about 8–12 beats per minute. However, their body temperature only drops 3–7°C (5–9°F). Other mammals, such as ground squirrels, undergo true hibernation. During hibernation, a ground squirrel's respiratory rate decreases from 200 breaths per minute to 4 or 5 per minute, and its heart rate decreases from 150 to 5 beats per minute.

REAL LIFE

Answer

Invasion of a home by wild mammals can be discouraged by removing food and water sources and by modifying exterior openings such as dryer and attic vents.

Teaching TIP

Metabolic Rate and Body Size

Explain that the rate at which mammals metabolize food depends on their body size, and that body size determines surface-area-to-volume ratio. Small mammals, such as shrews, have a high surface-area-to-volume ratio. As a result, shrews lose heat so rapidly that they use 80 to 90 percent of their energy just trying to stay warm. Large mammals, such as elephants, have trouble dissipating excess body heat because their surface-area-to-volume ratio is low. Ask students which animal, the shrew or the elephant, they would expect to eat more with respect to body mass. *(shrew)* Because the shrew loses so much heat to the environment, it must replace that heat by consuming more fuel. A shrew needs so much energy that it must eat an amount of food equal to its own body mass each day.

Teaching TIP

Hibernation

Ask students how some small mammals that have a high metabolic rate, like squirrels, can survive cold winters when there is little food available. *(They hibernate.)* Ask students to speculate on how hibernation helps them survive. *(During hibernation the animal's body temperature drops almost to the temperature of the environment. As the body temperature decreases, the metabolism slows and therefore the animal's need for food is decreased.)*

Chapter 36

Up Close

Grizzly Bear

Teaching Strategies
Tell students that the grizzly bear is classified as a threatened species in the contiguous United States, a vulnerable species in Canada, and an at-risk species in the Canadian province of Alberta. Several factors are involved in the decline of the grizzly bear population. Human-caused mortality is the major factor. Bear deaths may be caused by hunting or through loss and fragmentation of habitats. Where habitats are fragmented, bear populations frequently occur at low densities. This is an important factor because grizzly bears have very few offspring during their lifetime, so their populations cannot recover quickly.

Resource Link

- Assign **Occupational Applications Worksheet: Wildlife Biologist** from the Holt BioSources Teaching Resources CD-ROM. **PORTFOLIO**

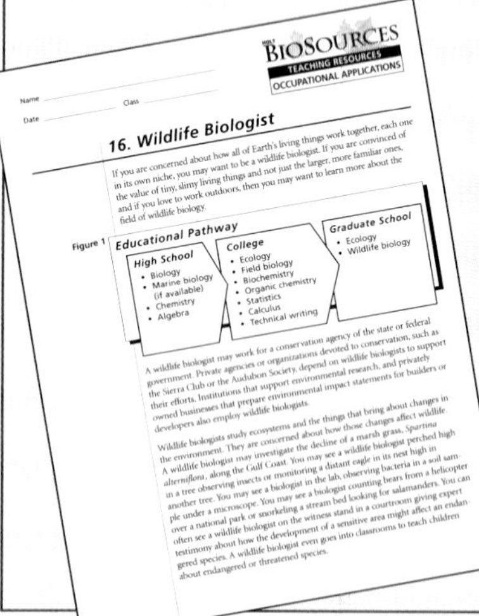

HOLT BIOSOURCES TEACHING RESOURCES
OCCUPATIONAL APPLICATIONS

Name ______ Class ______
Date ______

16. Wildlife Biologist

If you are concerned about how all of Earth's living things work together, each one in its own niche, you may want to be a wildlife biologist. If you are convinced of the value of tiny, slimy living things and not just the larger, more familiar ones, and if you love to work outdoors, then you may want to learn more about the field of wildlife biology.

Figure 1 **Educational Pathway**

- **High School**
  - Biology
  - Marine biology (if available)
  - Chemistry
  - Algebra
- **College**
  - Ecology
  - Field biology
  - Biochemistry
  - Organic chemistry
  - Statistics
  - Calculus
  - Technical writing
- **Graduate School**
  - Ecology
  - Wildlife biology

A wildlife biologist may work for a conservation agency of the state or federal government. Private agencies or organizations devoted to conservation, such as the Sierra Club or the Audubon Society, depend on wildlife biologists to support their efforts. Institutions that support environmental research, and privately owned businesses that prepare environmental impact statements for builders or developers also employ wildlife biologists.

Wildlife biologists study ecosystems and the things that bring about changes in the environment. They are concerned about how those changes affect wildlife. A wildlife biologist may investigate the decline of a marsh grass, *Spartina alterniflora*, along the Gulf Coast. You may see a wildlife biologist perched high in a tree observing insects or monitoring a distant eagle in its nest high in another tree. You may see a biologist in the lab, observing bacteria in a soil sample under a microscope. You may see a biologist counting bears from a helicopter over a national park or snorkeling a stream bed looking for salamanders. You can often see a wildlife biologist on the witness stand in a courtroom giving expert testimony about how the development of a sensitive area might affect an endangered species. A wildlife biologist even goes into classrooms to teach children about endangered or threatened species.

Up Close

Grizzly Bear

- **Scientific name:** *Ursus arctos horribilis*
- **Size:** Males average 160 kg (350 lb) and can reach 120 cm (4 ft) at the shoulder; females are smaller
- **Range:** Alaska and western Canada, with small populations in Washington, Idaho, Montana, and Wyoming
- **Habitat:** Tundra and mountainous forests and meadows
- **Diet:** Omnivorous; primary diet is vegetation; hunts insects, small mammals, fish; eats carrion

Female with cub

External Structures

Fur Thick fur—ranging from yellowish-brown to black—covers the body. The name "grizzly" comes from silver-tipped hairs that are often sprinkled over the bear's head and back.

Senses Grizzlies have good hearing but relatively poor eyesight. They rely primarily on their excellent sense of smell to follow an odor trail or catch the scent of distant food.

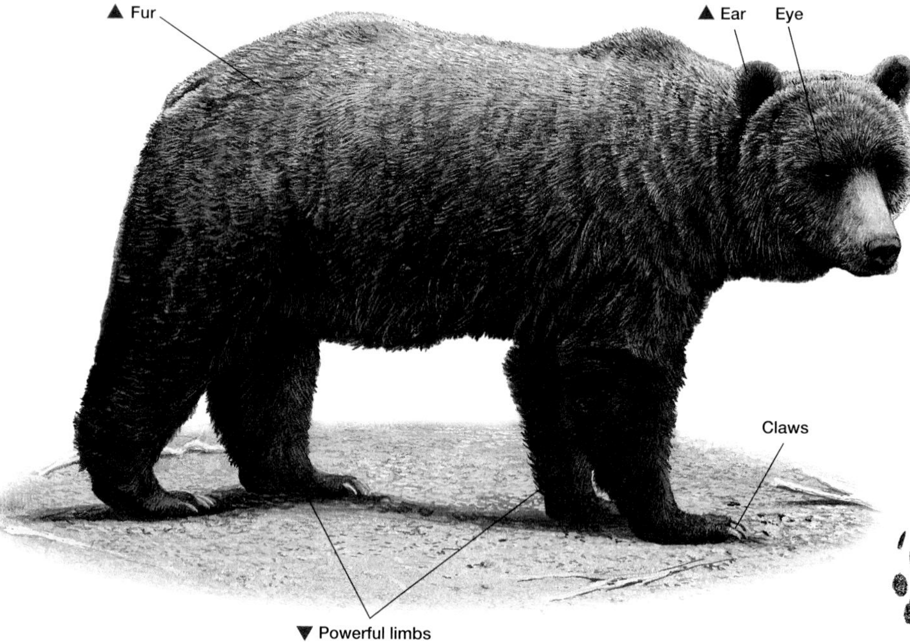

Strong limbs Grizzlies are extremely strong and have great endurance. The hump atop the bear's back is a knot of strong muscles that power the forelimbs. The paws are tipped with curved claws up to 10 cm (4 in.) in length. One swat of the forepaw can kill an adult moose or elk. Unlike a cat's, the bear's claws do not retract, and they are not adapted for climbing.

Forepaw

Hind paw

Internal Structures

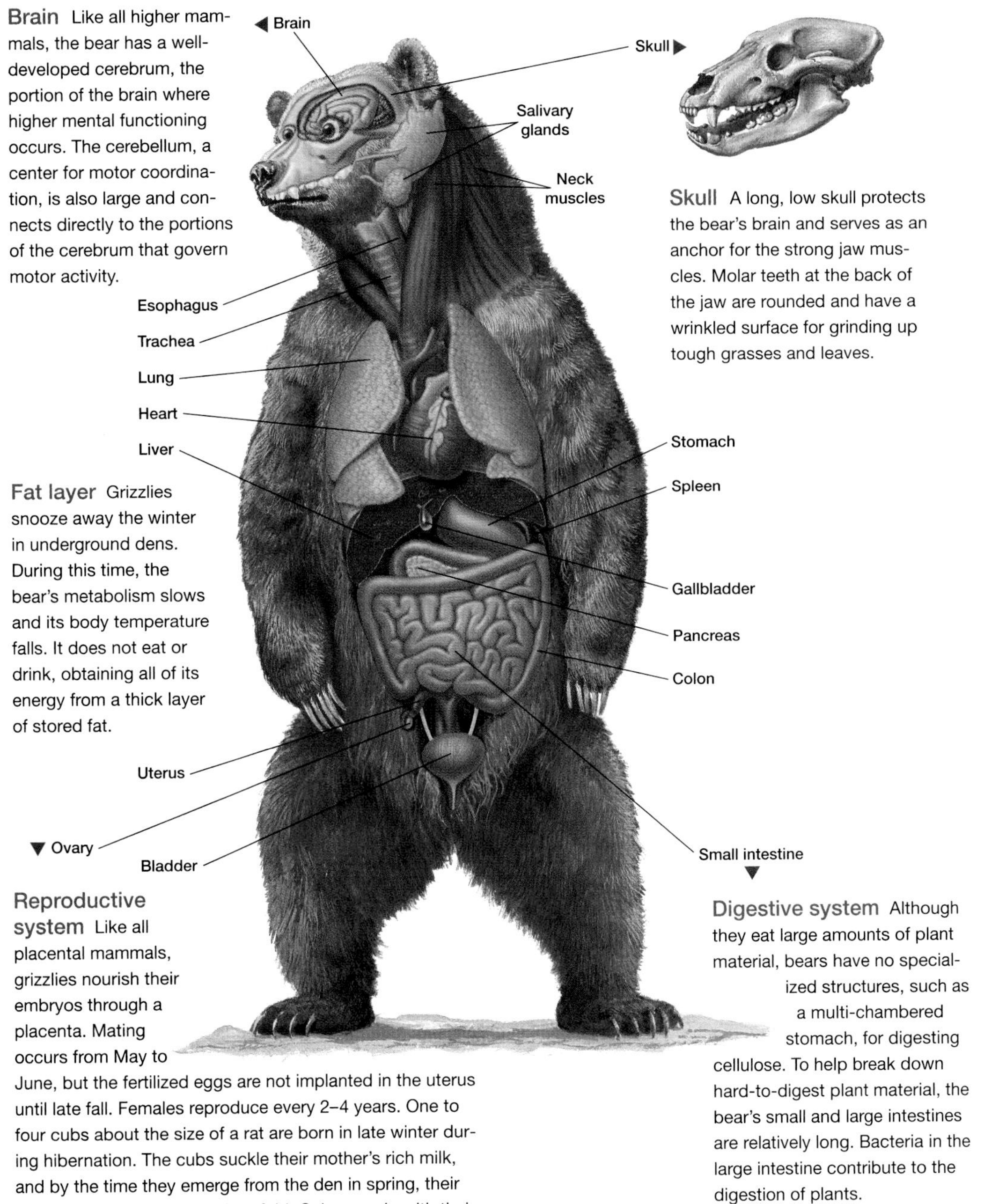

Brain Like all higher mammals, the bear has a well-developed cerebrum, the portion of the brain where higher mental functioning occurs. The cerebellum, a center for motor coordination, is also large and connects directly to the portions of the cerebrum that govern motor activity.

Skull A long, low skull protects the bear's brain and serves as an anchor for the strong jaw muscles. Molar teeth at the back of the jaw are rounded and have a wrinkled surface for grinding up tough grasses and leaves.

Fat layer Grizzlies snooze away the winter in underground dens. During this time, the bear's metabolism slows and its body temperature falls. It does not eat or drink, obtaining all of its energy from a thick layer of stored fat.

Reproductive system Like all placental mammals, grizzlies nourish their embryos through a placenta. Mating occurs from May to June, but the fertilized eggs are not implanted in the uterus until late fall. Females reproduce every 2–4 years. One to four cubs about the size of a rat are born in late winter during hibernation. The cubs suckle their mother's rich milk, and by the time they emerge from the den in spring, their weight may have increased 20-fold. Cubs remain with their mother until their second spring, and usually longer.

Digestive system Although they eat large amounts of plant material, bears have no specialized structures, such as a multi-chambered stomach, for digesting cellulose. To help break down hard-to-digest plant material, the bear's small and large intestines are relatively long. Bacteria in the large intestine contribute to the digestion of plants.

Up Close

Grizzly Bear

Discussion

- Identify several differences between the internal organs of the grizzly bear and the bald eagle, shown on page 799. *(The bald eagle has a crop for storing food, a two-part stomach, air sacs connected to the lungs, and a cloaca.)*
- How is the period of dormant inactivity during the winter advantageous to the grizzly bear? *(By lowering its metabolic rate, the bear can survive the winter when food is scarce.)*
- Most experts agree that if a person were to come upon a grizzly bear unexpectedly in the wild, one of the worst things to do is to turn and run. Why? *(Grizzly bears can easily outrun the swiftest runner. Furthermore, running stimulates the bear to treat a person as prey, and an attack is much more likely.)*
- Fertilized eggs do not implant in the female's uterus until late fall. What is the adaptive value of this timing? *(It ensures that the young will be born in springtime, when there will be ample food for the mother so she can produce adequate milk.)*

Overcoming Misconceptions

Digestive Systems

Students may think that the digestive systems of all vertebrate species are roughly the same. In birds, the crop, an enlarged segment of the esophagus, can store food temporarily. The bear has no such feature. In the gizzard of birds, partially digested food from the stomach is mechanically digested. The digestive system of the bear, however, lacks these specializations and is very similar to other mammalian, including human, digestive systems.

3 Assessment Options

Closure

Tell students to suppose that they are writing a guide to mammalian characteristics for a natural history museum. Have students work in small groups to develop a simple illustrated guide. PORTFOLIO

Review

Assign Review—Section 36-1.

Reteaching — BASIC

Write the following characteristics on the board: *Hair, Endothermy, Mammary glands, Specialized teeth,* and *Placental nourishment.* Ask students to explain the adaptive value of each characteristic. Elicit specific examples.

Study TIP

Reviewing Information

Work with a partner to review the five characteristics of mammals. For each characteristic, both partners should write a question. Exchange questions, and find the answers to your partner's questions.

Mammals Nurse and Care for Their Young

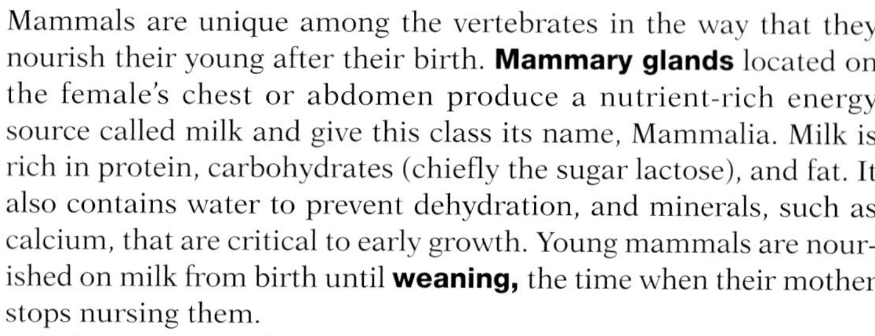

Mammals are unique among the vertebrates in the way that they nourish their young after their birth. **Mammary glands** located on the female's chest or abdomen produce a nutrient-rich energy source called milk and give this class its name, Mammalia. Milk is rich in protein, carbohydrates (chiefly the sugar lactose), and fat. It also contains water to prevent dehydration, and minerals, such as calcium, that are critical to early growth. Young mammals are nourished on milk from birth until **weaning,** the time when their mother stops nursing them.

Unlike other vertebrates, young mammals are dependent on their mother for a relatively long period, receiving milk and other food, protection, and shelter from her. For most animals, once the young can fend for themselves, the mother leaves them. The participation of the father in raising the young varies from species to species. Young mammals, such as the sea otter shown in **Figure 36-5,** often learn necessary survival skills during the time they are dependent on their mother. For primates, such as the chimpanzees in Figure 36-5, this early learning is especially important, and primates have the longest period of parental dependency of all mammals.

Figure 36-5 Parental care. This young otter will spend 6 to 8 months with its mother. In contrast, this baby chimp will remain close to its mother for several years.

Review SECTION 36-1

1. **Describe** three functions of hair.
2. **Compare** the functions of the different types of mammalian teeth.
3. **Relate** the mammal's heart and respiratory systems to its endothermic metabolism.
4. **Summarize** the ways in which mammals provide parental care.
5. **Summarize** how the grizzly bear exhibits characteristics typical of mammals.

Critical Thinking

6. **SKILL Justifying Conclusions** You and your lab partner examine a mammalian skull and jaw that contains only incisor teeth. Your partner concludes that you do not have enough information to identify the specimen as a herbivore or carnivore. Evaluate this conclusion.

Review SECTION 36-1 ANSWERS

1. insulation, protection, and sensory function
2. incisors: biting and cutting; canines: stabbing and holding; premolars/molars: crushing and grinding
3. The four-chambered heart keeps oxygen-rich and oxygen-poor blood separate. A diaphragm aids respiration, and numerous alveoli in the lungs provide a large surface for gas exchange. These features supply oxygen to the mammal's tissues at a high rate to meet its energy demands.
4. All female mammals provide nourishment (milk). The adults provide protection and teach survival skills. The degree to which this occurs depends on the species.
5. The grizzly's fur insulates its endothermic body, and its teeth are specialized for omnivorous feeding. The female produces milk and nourishes her developing young through a placenta.
6. Answers will vary. Students may conclude that because incisors are important teeth in herbivores, the jaw is that of a herbivore. Or they may conclude that the missing teeth are canines and that the skull is that of a carnivore.

Today's Mammals

Section 36-2

Mammals Have the Most Diverse Body Form of All Vertebrates

Cave paintings, some perhaps 40,000 years old, show that human life has long been intertwined with the lives of animals. The vast majority of the animals shown in ancient cave paintings are mammals. While some of the mammals represented are extinct, the ancestors of many (horses, hyenas, rhinoceroses, panthers, bison, and lions) still live. They are a part of the astounding array of mammals that share our planet with us. In their diversity of size, anatomy, and habitat, mammals surpass all other vertebrate groups. Mammals range in size from tiny shrews that weigh about 1.5 g (less than 0.1 oz) to gigantic blue whales that can weigh up to 136,000 kg (150 tons).

Consider also the differences between a bat and a whale, compared in **Figure 36-6.** These two mammals are adapted to live in very different environments. Bats fly and are active primarily at night, while whales are permanently aquatic. However, both groups face a similar challenge—how to navigate in an environment where visibility is often limited. Bats and some whales have evolved a similar solution to this problem: they use echolocation which works something like the sonar of a ship. In echolocation, animals emit high-frequency sounds. As the sound waves travel, they strike objects in the environment and a portion of each wave is reflected back to the animal. The brain interprets the reflected sound wave, or echo, revealing the object's size and location.

Objectives

- **Compare** reproduction in monotremes, marsupials, and placental mammals. (pp. 818–821)
- **Relate** the limited distribution of monotremes and marsupials to the breakup of Pangaea. (pp. 818–819)
- **Recognize** how mammals are adapted to many different environments. (p. 817, pp. 822–826)

Key Terms

placenta
gestation period
ungulate
cud

Comparison of Bats and Whales

Bats

- Forelimbs modified into wings; covered with leathery skin
- Body covered with hair
- Active at night; use echolocation to navigate

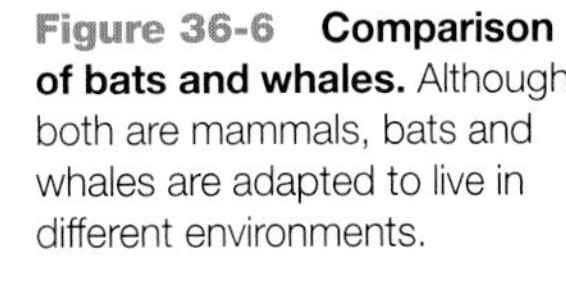

Whales

- Forelimbs flattened and paddle-shaped; no hind limbs
- Nearly hairless, streamlined body
- Communicate with sound; some use echolocation

Figure 36-6 Comparison of bats and whales. Although both are mammals, bats and whales are adapted to live in different environments.

Section 36-2

1 Prepare

Scheduling

- **Block:** About 90 minutes will be needed to complete this section.
- **Traditional:** About two class periods will be needed to complete this section.
- **Planning:** See the Planning Guide on pp. 807A and 807B for lecture, classwork, and assignment options for Section 36-2. Lesson plans for Section 36-2 can be found in a lesson cycle format in ***Biology: Principles and Explorations*** **One-Stop Planner CD-ROM with Test Generator.**

Resource Link

- Assign **Active Reading Worksheet Section 36-2** to help students comprehend the key concepts and visuals in this lesson.

2 Teach

Lesson Warm-up

As a springboard for a discussion on mammalian diversity write the following headings on the board: *Terrestrial, Marine, Carnivorous, Herbivorous,* together with any other heading you wish to include. Have students suggest mammals for each category. Discuss the differences in the mammals listed. If a kangaroo is not listed, ask students if they know how it is different from the other mammals listed. Tell students that in this section they will learn about how the orders of mammals differ.

Integrating Different Learning Styles

Logical/Mathematical	**PE**	Data Lab, p. 821
	ATE	Cross-Disciplinary Connection, p. 823
Visual/Spatial	**ATE**	Teaching Tip, p. 820; Demonstration, p. 824
Body/Kinesthetic	**ATE**	Demonstration, p. 821
Interpersonal	**ATE**	Lesson Warm-up, p. 817; Active Reading, p. 820; Career, p. 823; Performance Zone items 21, 22, p. 829
Intrapersonal	**ATE**	Food Watch, p. 819
Verbal/Linguistic	**PE**	Study Tips, pp. 818, 823; Food Watch, p. 819
	ATE	Teaching Tip, p. 818

Teaching TIP

Platypus

When the first specimens of the platypus arrived in Europe in the 1790s, the scientists who examined them suspected a hoax. The platypus seemed to be composed of the body parts of two different animals. Like mammals, it has hair, a single jawbone, and three middle-ear bones. But it also has a duck's bill. When scientists could find no evidence of stitching or tampering, they concluded that the platypus, however odd, is indeed a mammal. Because the reproductive tract of the platypus is similar to a bird's, scientists suspected that the platypus is oviparous. This fact was not established conclusively until almost a hundred years later in 1883, when platypus eggs were discovered. Have students construct the table suggested in the **Study Tip** on page 818 in their text and complete the first column. PORTFOLIO

Teaching TIP

More Oddities of Monotreme Reproduction

Female monotremes have no nipples, so the young do not suckle. The milk oozes out through pores in the skin, and the young lap it up. Male platypuses have a poisonous spur on their hind legs, which they use in combat with other males over control of breeding sites.

Mammals Reproduce in One of Three Ways

All mammals reproduce by internal fertilization: The male releases sperm into the female's reproductive tract where one or more eggs are fertilized. But there is some variation among mammals in how and where the fertilized egg develops. Present-day mammals are divided into three groups—monotremes, marsupials, and placental mammals—based on their pattern of development.

Monotremes

The monotremes, the most primitive of the mammals, are represented by only one order, Monotremata. All monotremes live in Australia and New Guinea where they originally evolved. This limited distribution is the result of the isolation of Australia and New Guinea from the other continents by continental drift.

The three living monotreme species—the duckbill platypus and two species of echidnas *(ee KIHD nuhz)*, or spiny anteaters—show a curious mix of characteristics. They have shoulders and forelimbs that are quite reptilian in appearance. Among living mammals, only monotremes reproduce by laying eggs. Their eggs, like reptile eggs, have leathery shells. Like birds, the female monotreme incubates her eggs with her body heat, and at hatching, the newborns are only partially developed. Also like birds, monotremes have a cloaca, a common passageway for the digestive, reproductive, and urinary systems. A cloaca is found in reptiles and birds, but monotremes are the only mammals that have one. Unlike other mammals, adult monotremes do not have true teeth. These characteristics lead biologists to think that monotremes are more closely related to the early mammals than the other living mammal species are.

Figure 36-7 Platypus. Although these platypus young hatched from eggs, they drink milk produced by their mother's mammary glands.

Despite these unusual features monotremes have other mammalian features; they have hair, and the females produce milk. Unlike other mammals, the female does not have nipples and young monotremes do not nurse. Instead, the young lap up milk that oozes from glands located on their mother's belly.

The platypus, shown in **Figure 36-7,** inhabits lakes and streams in eastern and southern Australia. Its broad, flat tail and webbed front feet make it an excellent swimmer. The platypus uses its flat bill to probe for shellfish, worms, and snails.

Echidnas, shown in **Figure 36-8,** are found in parts of Australia and in New Guinea. They are terrestrial and have very strong, sharp claws and an elongated beaklike snout which they use for burrowing and for digging out insects and other invertebrates.

Figure 36-8 Echidna. Echidnas are covered with sharp, barbless quills. They use their powerful claws and long beaklike snout to dig for their main food, termites and ants.

Multicultural Perspective

Makah Whale Hunt

In 1855 the Makah Indian Tribe signed a treaty with the United States in which the Makahs were promised the right to engage in whaling, a tribal tradition for more than 2,000 years. Whales are central to the Makah culture. They are represented in their songs, dances, designs, and basketry. The conduct of a whale hunt requires rituals and ceremonies that are deeply spiritual. In the 1920s, due to the scarcity of gray whales, whaling was halted. In 1994 the gray whale was taken off the endangered species list, and in 1996, the gray whale population was estimated at 22,263. Amid much controversy, protest, and constant media watch, the Makahs resumed the hunt for whales in the spring of 1999. Offshore from Washington State, they were successful in harvesting a gray whale. Its meat was distributed to tribal members.

Marsupials

Order Marsupialia includes not just the well-known kangaroos, but wombats, wallaroos, koalas, and the opossums shown in **Figure 36-9.** In marsupial mammals the young are born only days or weeks after fertilization—tiny and incompletely developed except for their front limbs. Without any parental help, the newborns must crawl to their mother's nipples, usually located in a pouch (the marsupium) on her abdomen. There the newborns attach themselves to a nipple and continue their growth and development for several months. When the young marsupials are able to function on their own, they emerge from the pouch, although they continue to return there to nurse.

Today, almost all marsupial species are found in the Australian region—Australia, New Guinea, and a few nearby islands—where the vast majority of mammalian species are marsupials. This limited distribution is the result of the breakup of Pangaea. About 70 million years ago, the Australian region became separated from the continents of Antarctica and South America. As placental mammals had not yet reached the Australian region, the marsupials there developed in isolation.

Figure 36-9 Virginia opossum. At birth these four young opossums looked like the pink newborn opossum, which is smaller than your thumb.

FOOD WATCH: Selecting Dairy Products

Unlike all other mammals, humans continue to drink milk (and eat products made from it) far into adulthood. Today, most of our dairy products are made from cow's milk, although goat milk products are becoming increasingly popular. In other parts of the world, milk from other mammals such as water buffalo, ewes (sheep), mares, and even reindeer is used as a food source.

Next time you are in the dairy section of a supermarket, take a look at what is available. Just deciding which milk to purchase—whole, skimmed, 2 percent, lactose reduced, flavored—may take some time. And then there is the array of yogurts, kefirs, cottage cheeses, hard cheeses, and string cheeses.

Read the label

Although dairy products are nutritious and an excellent source of calcium, some are high in fat. This does not mean that you should never eat them, but you can make informed decisions about how much to eat. Begin by locating the information on the label that tells you the product's per-serving calorie and fat content.

Once you've made a purchasing decision, be certain to check the expiration date on the packaging. Consume perishable dairy products while they are still fresh. Dairy products require refrigeration, so do your other shopping first, and pick up perishable items last.

internet**connect**

SCILINKS NSTA
TOPIC: Dairy products
GO TO: www.scilinks.org
KEYWORD: HX819

Section 36-2

Teaching TIP

Natural Selection at Birth

Marsupial newborns are so underdeveloped that their hind limbs have not yet formed. They must creep to their mother's pouch using their forelimbs, guided by their sense of smell. Each newborn attaches to a nipple and stays attached to it while it completes its development. But a female marsupial may give birth to more babies than she has nipples, and any newborn that does not get a nipple will not survive. Ask students to describe the features of newborn marsupials that would enhance their chances of survival. *(Their sense of smell must be keen enough to find the pouch, they must have strong forelimbs to crawl to the pouch, and they must be able to crawl quickly. In addition, the order in which the young are born may affect their survival.)*

FOOD WATCH

Selecting Dairy Products

Teaching Strategies

- Use the information about dairy products to stimulate a discussion about including these products in the diet.
- Ask if any students have tasted milk from mammals other than cows.

Discussion

- What is two percent milk, and why might it be a better choice than whole milk? *(It is milk that contains two percent fat; it has the same calcium content as whole milk but with a lower fat content.)*
- Some students may be lactose-intolerant and willing to share their experiences. Ask why it is important to find out from a health professional if you are lactose-intolerant before altering your intake of dairy products. *(A person experiencing digestive problems could have a different condition that needs treatment. Also, a person could unnecessarily reduce calcium intake and jeopardize the health of his or her bones.)*

Chapter 36

Teaching TIP

Mammalian Evolution

Have students create a Graphic Organizer that features a cladogram similar to the one at the bottom of page 820. Use the graphic organizer to show the relationships among monotremes, marsupials, and the placental mammals. Then have students locate the positions on the cladogram where the following characters should be written in: hair, milk production, live birth, and a placenta.

PORTFOLIO

Active Reading

Technique: Paired Reading

Have students work in pairs to complete the table they created after reading the **Study Tip** on page 818. Then create a class table as pairs of students suggest entries for the columns.

Placental mammals

The young of placental mammals develop within the female's uterus, where they are nourished by nutrients from her blood. An organ called the **placenta** allows the diffusion of nutrients and oxygen from the mother's blood, across placental membranes, and into the blood of the developing fetus. Waste materials from the fetus diffuse in the opposite direction and are eliminated by the mother's excretory system.

> internet**connect**
>
> SCI LINKS. NSTA
>
> **TOPIC:** Placental mammals
> **GO TO:** www.scilinks.org
> **KEYWORD:** HX820

The period of time between fertilization and birth is called the **gestation** *(jeh STAY shuhn)* **period.** Placental mammals have a far longer gestation period than marsupial mammals do, and their young are more completely developed at birth. Some placental mammals, like the foal in **Figure 36-10,** can stand and walk within a few hours of birth. Others, like the rabbits, also seen in Figure 36-10, are born blind, deaf, and helpless.

Placental mammals vary greatly in size, shape, diet, and habits. They live in a variety of habitats, from hot, moist rainforests to the frigid tundra. Although mammals share many similarities, each species is adapted for a particular way of living. For example, although most mammals walk on four legs, some are adapted for running, leaping, swimming, or flying.

Alone among vertebrates, some placental mammals have hooves, horns, or antlers. These structures are made primarily of keratin. Keratin is a versatile protein that is a component of many mammalian structures. Hooves, such as the zebra hoof shown in **Figure 36-11,** are specialized pads that cover the toes of many running animals. The horns of sheep, cattle, and antelopes are composed of a core of bone surrounded by a sheath of keratin. This bony core is firmly attached to the skull, and the horn is never shed. (The horn of a rhinoceroses is composed not of bone but of hairlike fibers of keratin that form a hard structure.)

Other placental mammals, such as deer and elk, grow and shed a set of antlers each year. Antlers, grown only by the male, are made

Figure 36-10 Young mammals

Newborn foals are on their feet nursing within a couple of hours, while young rabbits do not even open their eyes until they are about 10 days old.

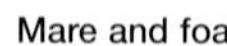

Mare and foal

Baby rabbits

Graphic Organizer

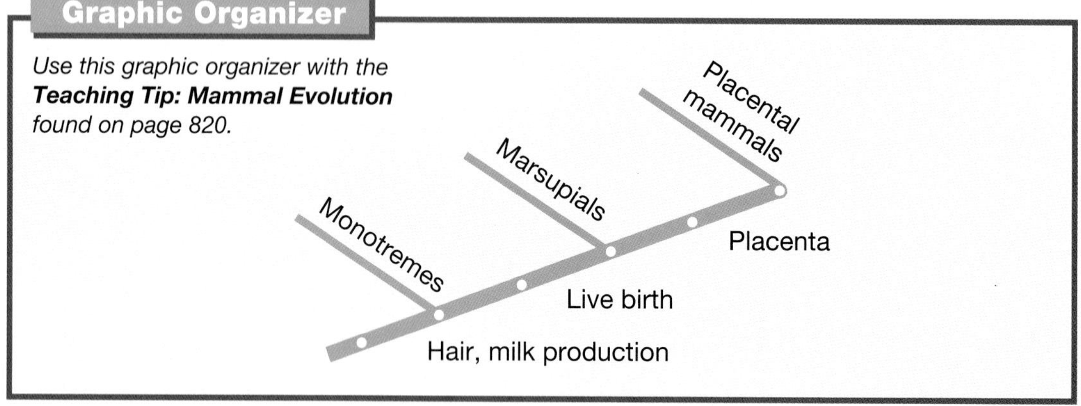

of bone but are not covered by a keratin sheath. While they are growing, the antlers are covered by a thin layer of soft skin called velvet. When the antlers are fully grown, the velvet dries up and comes off. The male uses his antlers during breeding season to attract females and to combat other males. After the breeding season is over, the antlers are shed, and a new pair grows the next year. Since the male grows a larger pair of antlers each year, antler size gives an indication of a male deer's age.

Domestic animals Domestic animals are animals that have been kept and bred by people for special purposes. Most domestic animals are placental mammals whose association with humans dates back at least 2,000 years. These animals include dogs, cats, cattle, horses, asses, mules, rabbits, sheep, goats, pigs, camels, llamas, and alpacas.

The different breeds of domestic mammals have been developed through selective breeding. For example, some breeds of goats produce more milk than others. The milk is used to produce a variety of dairy products. Other goats, such as angora goats, are bred for their fine hair, which is spun into yarn. Some domestic mammals are hybrids of two different species. Mules, for example, are the offspring of a female horse (mare) and a male ass. Like most hybrids, mules are sterile. The mule's strength, endurance, and surefootedness make it valuable as a pack animal and for chores such as plowing.

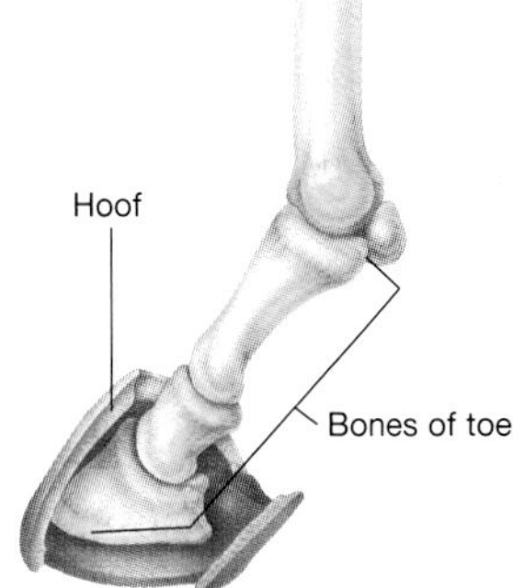

Figure 36-11 **Hoof.** A zebra's foot is modified so that it walks on a single toe that is covered with a hoof made of keratin.

Comparing gestation periods

Background

If you have ever raised gerbils or hamsters, you know that they can produce several litters of young each year. One reason is that they have a very short gestation period compared to other mammals. Use the table below to find out more about gestation periods.

Analysis

1. **SKILL Recognizing Patterns** State a generalization about the relationship of the length of the gestation period to the number of offspring per pregnancy.
2. **SKILL Forming Hypotheses** Propose a hypothesis to explain this relationship.
3. **Recommend** a way that the information in the table could be rearranged to show this relationship more clearly.

Gestation Periods in Mammals

Mammal	Gestation period	Offspring per pregnancy
Vampire bat	210 days	1
Gerbil *	19–21 days	4–7
Quarter horse	332–342 days	1
Monkey (black spider)	226–232 days	1
Grey squirrel	44 days usually	3
Rabbit *	about 31 days	3–6
Sperm whale	420–430 days	1
Arctic wolf	63 days	4–5

*More than two litters per year

Demonstration

Horns and Antlers — BASIC

Allow students to handle both horns and antlers. Students should be able to feel that antlers are made of bone, while horns are not. (State wildlife agencies are possible sources of both antlers and horns. Point out that the keratin in horns is the same as is found in fingernails and hair, and is nonliving tissue. Bone is a living tissue complete with an internal blood supply, (although examining antlers in the classroom will not reveal this).

Comparing gestation periods

Time 10–15 minutes

Process Skills Recognizing relationships, forming a hypothesis, organizing information

Teaching Strategies

- Have each student work independently. After students have completed the data lab, they should compare their answers with those of another student.
- Students should pay special attention to their answer to item 3, as each student may have rearranged the table differently. As a class, discuss the advantages and disadvantages of each new arrangement.

Analysis Answers

1. In general, the longer the gestation period is, the fewer offspring are born.
2. Student hypotheses will vary.
3. The table could be rearranged to show the gestation period or number of offspring per pregnancy in decreasing or increasing numbers.

Teaching TIP

Placental Marsupials?

A few marsupials have a placenta, but aren't considered placental mammals. These "placental" marsupials are the bandicoots (order Peramelina), insectivorous inhabitants of New Guinea and Australia. The size of bandicoots varies from that of a rat to that of a small dog. The bandicoot's placenta is not as efficient as that of a placental mammal, nor is it as intimately associated with the mother's tissues. In bandicoots, the fertilized egg adheres to the lining of the uterus. Blood vessels in the lining of the uterus exchange materials with blood vessels in the egg. By contrast, the fertilized eggs of a placental mammal bury themselves in the lining of the uterus. Fingerlike outgrowths penetrate into the uterine lining and greatly increase the surface area through which nutrients can be absorbed. The bandicoot's placenta cannot provide all the embryo's nutritional requirements, and supplementary nutrition is furnished by the secretions of the uterus, as in other marsupials. The gestation period in bandicoots is short, as in other marsupials.

Teaching TIP

Blind as a Bat

In the late 1700s, the Italian scientist Lazzaro Spallanzani found that blinded bats could still fly and capture prey. However, deafened bats could not avoid objects in their path, and they could not capture prey. He concluded that bats depend on sound to navigate.

Most of Today's Mammals Are Placental

There are 19 orders of placental mammals that include more than 90 percent of all mammal species. Terrestrial placental mammals inhabit all continents except Antarctica, and aquatic placental mammals inhabit all oceans. The following 12 orders contain the vast majority of mammal species.

Figure 36-12 Mouse

Order 1 Order Rodentia

Over 40 percent of all placental mammals are gnawing mammals called rodents. Rodents are distinguished from most other mammals by their teeth, which are specialized for gnawing. All rodents have two pairs of large, curving incisor teeth that grow continuously. As the rodent gnaws, the back side of the incisors wears away faster than the front, creating a sharp chisel edge on the teeth. The success of this group of herbivores can be attributed to their intelligence, small size, and rapid rate of reproduction. Most rodents, such as the mouse shown in **Figure 36-12,** are small, but a few are larger. For example, beavers can weigh between 18 and 43 kg (40–95 lb).

Figure 36-13 Bat

Order 2 Order Chiroptera

This order is composed of bats, the only mammals capable of true flight. The bat's front limbs are modified into wings. The thumb, which is not attached to the wing, has a curved claw that can be used for clinging or grasping. Bats generally live in groups and are active only at night. During the day, bats hang upside down in caves or some other protected place, as shown in **Figure 36-13.** Most bats are carnivorous, using echolocation to find insects, which they catch while in flight. Other bats eat fruit or nectar from night-blooming flowers.

Figure 36-14 Shrew

Order 3 Order Insectivora

Insectivores are the mammals most similar to the ancestors of the placental mammals. As their name implies, these small mammals eat mainly insects, but their diet may include fruit, small birds, and snakes, as well as other insectivores. Insectivores have an enormous appetite, and some, such as the shrew shown in **Figure 36-14,** must eat more than two times their body weight daily to fuel their extremely high metabolic rate.

Overcoming Misconceptions

The Truth About Bats

Apart from snakes, bats have been the subject of more misconceptions, superstitions, and old wives' tales than any other group of animals. For example, some people think that all bats suck blood (many bats are fruit eaters). Have students research these misconceptions and write a report summarizing their findings. PORTFOLIO

Order 4 Order Carnivora

Some of the best-known animals are the flesh-eating hunters called carnivores. Scientists generally divide this order into two subgroups: the cat family and the dog family, represented by the fox shown in **Figure 36-15.** Carnivores are strong, extremely intelligent, and have keen senses of smell, vision, and hearing—characteristics that have enabled them to become successful hunters. Their long canine teeth are specialized for capturing prey and tearing flesh. Some members of this order are no longer carnivorous: raccoons and bears are omnivores, while pandas are herbivores.

Figure 36-15 Fox

Order 5 Order Pinnipedia

This order of marine carnivores includes seals and sea lions that feed at sea but return to land to mate, rear their young, and rest. All four limbs are modified as flippers for swimming, and their bodies are streamlined for rapid movement through the water. An insulating layer of blubber protects them from the cold ocean waters. Most species live in large colonies called rookeries which are headed by a large male, such as the walrus shown in **Figure 36-16.**

Figure 36-16 Walrus

Order 6 Order Primate

Members of this order include apes, orangutans, chimpanzees, baboons, monkeys, and lemurs, as well as humans. Most, like the monkey in **Figure 36-17,** are tree dwellers, and many of their characteristic features are adaptations for living in trees. Flexible, grasping hands, and feet with big toes aid them in climbing, and most primates have a tail that is used for balance. Their excellent depth perception is critical for movement in the trees. Primates are extremely curious and their ability to learn is exceptional.

Figure 36-17 Monkey

Study TIP

Recognizing Differences

To understand the differences between the orders of placental mammals, make a flash card for each order. On one side write the order name. On the other side write an adaptation that distinguishes the order.

MULTICULTURAL PERSPECTIVE

Folk Medicine Threatens the Tiger's Survival

Tigers are disappearing at a rapid rate because humans are encroaching on their habitat and overhunting them. Poachers kill tigers for their pelts, which can be worth as much as $15,000. But tigers are also slaughtered to feed the large market for tiger-based folk medicines, which are used mainly in Asia. Folk remedies made from the bones, eyes, and other parts of the tiger purportedly lengthen life, instill vigor, and cure diseases and impotence. There is no medical evidence, however, that these medications are effective.

CAREER

Wildlife Biologist

Many colleges offer degrees in wildlife management or similar fields. Have students work in teams to use library references or on-line databases to find information about careers working with mammals. Invite a wildlife biologist from a national or state wildlife service to visit your class and discuss training, responsibilities, and rewards of a career in this field. PORTFOLIO

CROSS-DISCIPLINARY CONNECTION

◆ Mathematics

Tell students that the average life span of a mouse is 2 years and that each mouse needs 1 m^2 of grass-covered land to survive. Suppose that every 3 months each female mouse gives birth to a litter of eight, half of which are female.

A pair of adult mice—a male and a pregnant female—arrives, floating on a log, at a grassy island. The island is the size of a football field (91 m x 49 m). The day they arrive, the female gives birth. There are no predators on the island, and the mice and their descendants survive until the island can no longer provide enough space and food. Have students calculate the following:

1. How many mice will there be after 3 months? after 6 months? after 9 months? after 1 year? *(after 3 months, 42; after 6 months, 210; after 9 months, 1,050; after 1 year, 5,250)*
2. When can the island no longer support the entire population of mice? *(The island has an area of 4,459 m^2 and can no longer support the mice before the end of the first year.)* PORTFOLIO

Demonstration

Comparing Structures — BASIC

Allow students to observe real hooves. Pet stores often sell goat or cow hooves for dogs to chew. Provide students with a hand lens to examine the hooves. Remind students that hooves are composed of keratin, and have them compare the animal hooves with their own fingernails.

Teaching TIP

Comparing Whales with Pinnipeds

Whales and pinnipeds (seals, sea lions, and walruses) are adapted for life in the water. Have students construct a table that lists similarities and differences between these aquatic mammals. *(similarities: they have modified flipper-forelimbs, they feed in the water, they breathe air, they have streamlined bodies; differences: whales lack hind limbs—pinnipeds have paddlelike hind limbs; whales cannot leave the water—pinnipeds mate and sleep on land)*

Figure 36-18 Sheep

Order 7 Order Artiodactyla

Mammals belonging to this order and the following order, Perissodactyla, are classified as **ungulates** *(UNG gyoo lits),* mammals with hoofs. Ungulates walk, not on their entire foot as most vertebrates do, but with their weight supported by their hoof-covered toes. Most ungulates are herbivores that live together in herds. At birth the young are well developed and can move along with the herd within a day or two. An ungulate's main means of defense are the security of the herd and its ability to run very fast when danger approaches.

Artiodactyls have an *even* number of toes within their hooves. Many of these mammals have a stomach with a storage chamber called a rumen. Microbes in the rumen break down the cellulose in the plant material artiodactyls eat. Mammals with a rumen regurgitate partly digested food, called **cud,** rechew it, and swallow it again for further digestion. This order includes pigs, hippopotamuses, camels, deer, antelope, cattle, goats, and giraffes as well as sheep, such as the one shown in **Figure 36-18.**

Figure 36-19 Rhinoceros

Order 8 Order Perissodactyla

Ungulates with an *odd* number of toes within their hooves are classified as perissodactyls. This order includes horses, zebras, tapirs, and rhinoceroses, such as the two shown in **Figure 36-19.**

Perissodactyl animals do not chew their cud. Instead of a rumen, they have a cecum, a pouch branching from their large intestine. The cecum contains microbes that digest the cellulose in their diet. Perissodactyls are far fewer in number than artiodactyls are.

Order 9 Order Cetacea

Cetaceans are divided into two groups: the predatory toothed whales, dolphins, and porpoises, and the filter-feeding baleen whales. Whales such as the orca shown in **Figure 36-20** are probably descendants of land mammals that returned to the sea about 50 million years ago and have adapted to a fully aquatic life. Their streamlined bodies have front limbs modified into flippers, no hind limbs, and a broad, flat tail for swimming. A nostril called a blowhole is located on top of their head. Cetaceans are very intelligent animals that communicate with each other by making sounds that we hear as a series of clicks.

Figure 36-20 Orca

did you know?

The American wild horse is a recent arrival.

Herds of wild horses are part of the lore of the American West. These horses, however, are all descendants of horses brought over from Europe by early explorers and colonists, after Columbus's arrival in 1492. Ironically, horses probably first evolved in North America about 4 million years ago. They migrated to Asia and Europe, but became extinct in North America about 12,000 years ago.

Order 10 Order Lagomorpha

This order is composed of rabbits and hares. Like rodents, they have one pair of long, continually growing incisors, but they also have an additional pair of peg-like incisors that grow just behind the front pair. Rabbits and hares have long hind legs and are specialized for hopping, as shown in **Figure 36-21.** Hares do not construct nests. Their young are born with fur and their eyes are open. Rabbits build nests that the female lines with fur. The young are born furless and their eyes are closed.

Figure 36-21 Rabbit

Order 11 Order Sirenia

These somewhat barrel-shaped marine animals include the dugongs and manatees, shown in **Figure 36-23.** Like whales, they have front limbs modified as flippers and no hind limbs. A flattened tail is used for propulsion through the tropical oceans, estuaries, and rivers where sirenians live grazing on aquatic plants. Despite their appearance and habitat, sirenians are closely related to elephants and are often called sea cows.

Figure 36-22 Manatee

Order 12 Order Proboscidea

The two living species of this order, the African elephant and the Indian elephant shown in **Figure 36-23,** are the largest land animals alive today. Their long, boneless trunk is really an elongated nose and upper lip and is used for a variety of tasks. In some elephants, the upper incisor teeth are modified into long ivory tusks. Elephants live in herds made up of a dominant male, a number of females, and young of varying ages.

Figure 36-23 Indian elephant

Visual Strategy

Figure 36-21

Have students examine the picture of the jackrabbit running in Figure 36-21. Ask students to point out two specialized adaptations that help the jackrabbit survive. *(large hind legs for swift running, large ears for acute hearing, coloration that provides camouflage)* Ask students if they can think of a function of the large ears that is not associated with hearing. *(They act as radiators to dissipate heat and cool the rabbit.)*

Teaching TIP — ADVANCED

Orders of Mammals

Assign students to cooperative groups of two or three. Assign each group one of the mammalian orders listed in Section 36-2. Have each group become expert about its order by researching a pertinent area of interest. For example, groups could research migratory habits, if any; general physical and behavioral characteristics; feeding habits; mode of living; and significance of the order to humans or to the biosphere. Group members should compile their findings in a report and present it to the class.

PORTFOLIO

did you know?

Armadillos are expanding their range.

The nine-banded armadillo *(Dasypus novemcinctus)* is one of the few mammalian species whose range is expanding. In the late 1800s, armadillos lived only as far north as central Texas. Now they inhabit most of Texas and all of Louisiana. Armadillos also have spread to parts of Kansas, Oklahoma, Arkansas, Mississippi, New Mexico, Georgia, Florida, and Alabama.

Chapter 36

3 Assessment Options

Closure

Technique: K-W-L

Tell students to return to the lists of things they **W**ant to know about mammals, which they created using the **Active Reading** exercise on page 809. Have them place check marks next to the questions that they are now able to answer. Students should finish by making a list of what they have **L**earned. Ask students the following:

- Which questions are still unanswered?
- What new questions do you have?

You may wish to assign projects that require students to research their unanswered questions.
PORTFOLIO

Review

Assign Review—Section 36-2.

Reteaching — BASIC

Show students pictures of different mammals, including monotremes and marsupials. Include pictures of as many kinds of mammals as possible, preferably in their native habitats. Also include some non-mammals—birds, fishes, or reptiles—to test students' understanding. Ask students to identify the mammals and then classify them according to the orders described in Section 36-2.

The seven remaining orders of placental mammals are summarized in **Table 36-1.** These orders contain few species—some quite interesting. The hyrax, for example, looks as if it could be kin to rabbits or rodents. But traits such as its hooflike nails lead biologists to think the hyrax is most closely related to elephants or odd-toed ungulates.

Table 36-1 Orders of Placental Mammals

Order	Description
Edentata anteaters, armadillos	Toothless or poorly developed teeth without enamel; found only in Western Hemisphere
Macroscelidea elephant shrews	Ground-dwelling insect eaters; long, flexible snout; hop about somewhat like small kangaroos
Scandentia tree shrews	Omnivorous; small, squirrel-like mammals; long snout, sharp teeth; live mainly on the ground, despite their name
Pholidota pangolins (spiny anteaters)	Body covered with overlapping scales; no teeth; very long tongue for capturing ants
Hyracoidea hyraxes	Rabbitlike bodies; short ears; four hoofed toes on front feet; three hoofed toes on back feet
Dermoptera flying lemurs	Squirrel-like; glide on a sheet of skin stretching between their forelegs and hind legs
Tubulidentata aardvarks	Nocturnal; piglike body; big ears; long snout used to feed on ants and termites

Review SECTION 36-2

1. **Compare** the reproductive patterns of monotremes, marsupials, and placental mammals.
2. **Relate** the location of modern-day marsupials to the breakup of Pangaea.
3. **Describe** how artiodactyls and perissodactyls are adapted for digesting plant material.
4. **Summarize** the ways in which aquatic mammals are adapted to life in the water.

Review SECTION 36-2 ANSWERS

1. Monotremes lay eggs; marsupials are born underdeveloped and attach themselves to a nipple in the mother's pouch, where development is completed; and placental mammals develop in the mother's uterus, getting nourishment through the placenta.
2. About 70 million years ago, the Australian region of Pangaea, which contained many marsupials but lacked placental mammals, became separated from the rest of Pangaea as it broke apart.
3. Many artiodactyls are cud chewers. Perissodactyls, in contrast, do not chew their cud but have a special pouch, the cecum, that contains microbes that digest the cellulose in their food.
4. Aquatic mammals such as cetaceans and sirenians have streamlined bodies, front limbs modified as flippers, and flat tails for swimming. In pinnipeds, all four limbs are modified for swimming. All aquatic mammals have a layer of blubber to insulate them.

Study ZONE CHAPTER 36

Use the Key Concepts and Key Terms listed below to review the main ideas in this chapter. If you need to review the meaning of a term, turn to the page indicated.

Key Concepts

36-1 The Mammalian Body

- Mammals are the only animals with hair. The primary function of hair is to insulate an animal's body, though it can also provide camouflage or a clear signal through coloration, serve a sensory function, or be a defensive weapon.
- Mammals usually have two sets of teeth in their lifetime, the second of which are not replaced, even if lost or damaged.
- The four types of mammalian teeth are highly specialized: incisors are for biting and cutting; canines are for stabbing and holding; premolars and molars crush and grind the food.
- Mammals are endothermic, generating heat internally through the rapid metabolizing of food. This is made possible by the highly efficient respiratory and circulatory systems.
- Mammals nurse their young with milk from the mammary glands of the female.

36-2 Today's Mammals

- In terms of anatomy and habitat, mammals are the most diverse of all vertebrate groups.
- All mammals reproduce by internal fertilization.
- Monotremes have hair and the females produce milk, but they lay eggs.
- Marsupials give birth to incompletely developed young that complete their development in the mother's pouch.
- Placental mammals nourish their unborn young in the uterus through the placenta.
- The different mammalian species have evolved a variety of adaptations that permit them to live in a wide range of habitats.
- Many aquatic mammals have a layer of blubber that insulates them from the cold. Their forelimbs are modified into flippers; they have no hind limbs; and a flattened tail aids in swimming.
- Ungulates have digestive systems modified for digesting cellulose. Even-toed ungulates chew their cud.

Key Terms

36-1

hair (810)
mammary gland (816)
weaning (816)

36-2

placenta (820)
gestation period (820)
ungulate (824)
cud (824)

Study TIP Ready?

- *If you think you understand these Key Concepts and Key Terms, then you're ready for the Performance Zone.*

Study ZONE CHAPTER 36

Review and Practice

Assign the following worksheets for Chapter 36:

- **Student Review Guide**
- **Concept Mapping Worksheet**
- **Test Preparation Pretest**
- **Vocabulary Worksheet**
- **Critical Thinking Review Worksheet**

Assessment Options

The following assessment options are available for this chapter:

Resource Link

- Assign **Chapter Test 36** to assess students' understanding of the chapter.

Alternative Assessment

Assign students to cooperative groups. Have each group develop a set of questions and answers for specific pages of the chapter. Have each group pass its questions (but not answers) to another group for it to answer. Continue in this manner until every group has answered every set of questions. PORTFOLIO

Portfolio Assessment

- *Portfolio Assessment* at the front of this book provides options for building and evaluating student portfolios. Students and teachers can select items from the areas listed for inclusion in a portfolio.
- See items labeled PORTFOLIO on pp. 814, 816, 818, 820, 822, 823, 825, 826, and 827.

Answer to Concept Map

The following is one possible answer to Performance Zone item 1 on page 828.

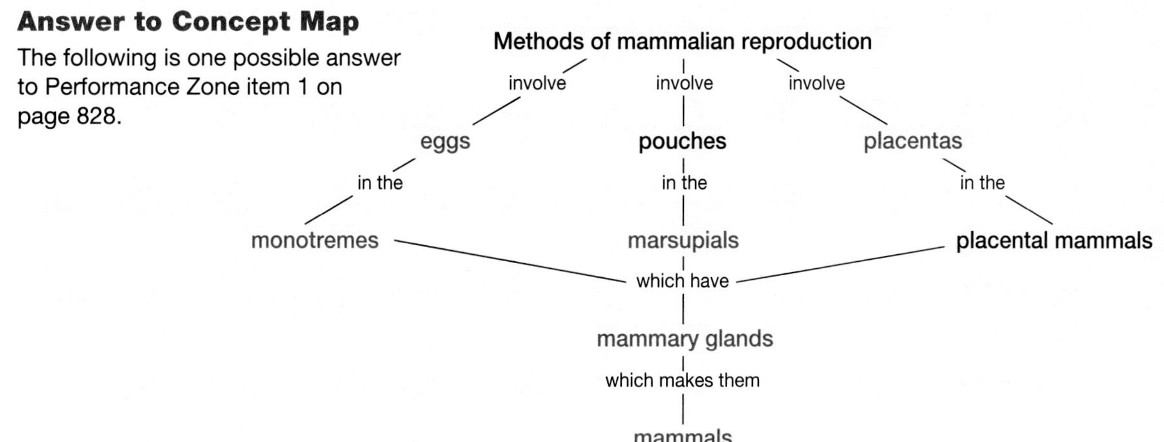

Performance ZONE
CHAPTER 36 REVIEW ANSWERS

Assignment Guide

SECTION	REVIEW
36-1	1, 2, 3, 4, 5, 6, 7, 8, 13, 14, 15, 17, 18, 19, 20, 23
36-2	9, 10, 11, 12, 16, 21, 22, 24, 25, 26, 27, 28

Understanding and Applying Concepts

1. The answer to the concept map is found at the bottom of page 827.
2. a. Hair is an external insulator; blubber is an internal insulator.

   b. The gestation period is the time from fertilization until birth; weaning occurs much later—when the mother ceases nursing the young.
3. b
4. b
5. d
6. d
7. c
8. c
9. a
10. a
11. a
12. d
13. c
14. The high metabolic rate produces much heat, and hair helps prevent heat loss.
15. A mammal requires large amounts of food to fuel its metabolism, and it must also take in large amounts of oxygen to "burn" (metabolize) the food.
16. Inform them that some dairy products are high in fat. Tell them to check the product label and to be sure to keep dairy products refrigerated to avoid spoilage.
17. Both hair and down feathers are light and trap air for insulation.

Performance ZONE CHAPTER 36 Review

Understanding and Applying Concepts

1. **Concept Mapping** Make a concept map that describes the different methods of mammalian reproduction. Include the following terms in your map: monotreme, marsupial, mammal, mammary gland, egg, placenta. Use additional terms as needed.
2. **Understanding Vocabulary** For each pair of terms, explain the differences in their meanings.
   a. hair, blubber
   b. weaning, gestation period
3. The primary function served by mammalian hair is
   a. camouflage. c. defense.
   b. insulation. d. sensory.
4. Which of the following is not a type of mammal tooth?
   a. canine c. molar
   b. baleen d. incisor
5. Endothermic metabolism permits mammals to do all of the following except
   a. live in very cold climates.
   b. generate heat internally.
   c. run or fly for long distances.
   d. adjust their body temperature.
6. The diaphragm allows mammals to
   a. carry the young inside the uterus.
   b. have a divided ventricle.
   c. provide nourishment for their young.
   d. breathe efficiently.
7. Which of the following groups have a four-chambered heart?
   a. reptiles, birds, and mammals
   b. reptiles and mammals
   c. crocodiles, birds, and mammals
   d. crocodiles, amphibians, and birds
8. The care given by mammals to their young
   a. begins after the young are weaned.
   b. is similar to that given by reptiles.
   c. involves nursing and teaching survival skills.
   d. ends soon after the young are born.
9. Monotremes differ from marsupials in that monotremes
   a. lay eggs.
   b. are viviparous.
   c. nourish unborn via the placenta.
   d. do not have mammary glands.
10. Which of these mammals is a marsupial?
    a. kangaroo c. lion
    b. duckbill platypus d. echidna
11. Artiodactyls differ from perissodactyls in that artiodactyls
    a. chew their cud.
    b. do not have a rumen.
    c. seldom live in herds.
    d. have an odd number of toes.
12. Which of the following is not a mammalian adaptation for aquatic living?
    a. limbs modified as flippers
    b. layer of blubber
    c. streamlined body
    d. keen sense of smell
13. Which of the following grizzly bear features is not typical of mammals in general?
    a. thick hair
    b. specialized teeth
    c. powerful claws
    d. placental reproduction
14. **Theme Homeostasis** How do hair and a high rate of metabolism help a mammal maintain homeostasis?
15. **Theme Metabolism** A mammal must eat about ten times as much food as an ectotherm of similar size. What role does the respiratory system play in this need?
16. **Food Watch** What advice would you give to someone who wanted to include more dairy products in their diet?
17. **Chapter Links** In what way are a mammal's hair and a bird's down feathers alike? How are they different? (**Hint:** See Chapter 35, Section 35-3.)

Skills Assessment

18. A. incisor; B. canine; C. premolar; D. molar
19. incisors—cutting and biting; canines—stabbing and holding; premolars and molars—grinding and chewing
20. It is a herbivore. The canines are very small, and the premolars and molars are broad and flat.
21. Answers will vary. Among animals that were domesticated early in human history are horses, cattle, sheep, pigs, and dogs.
22. Answers will vary. Students' presentations should be balanced and should convey information clearly.

Skills Assessment

Part 1: Interpreting Graphics

The illustration below shows a skull that is typical for a certain group of mammals. Study the illustration, and answer the following questions:

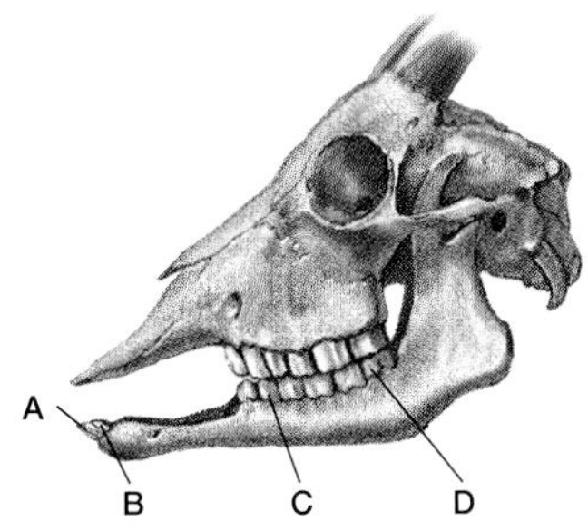

18. Identify the types of teeth labeled *A, B, C,* and *D.*

19. Describe the function of each of the teeth you identified.

20. From what you observe, identify the animal as a carnivore or herbivore and justify your answer.

Part 2: Life/Work Skills

Use the media center or Internet resources to research the history of domestic animals.

21. Being a Team Member Work in a group of students to find out what mammals were among the first to be domesticated. Assign each team member one of the animals identified. Each team should research when and where their particular mammal was domesticated and how the animal was probably used.

22. Communicating As a team, decide how you will present your information to the class. Possibilities include oral reports, posters or other visual representations, and multimedia presentations. Once you have decided how to present your information, assign different tasks to each team member. For example, some people may develop the visual component of the presentation while others write scripts. Still others may practice giving the presentation.

Critical Thinking

23. Evaluating Conclusions Some mammal species must care for their young for many years before they reach maturity and can survive on their own. Longer periods of parental care offer the young food, protection, and time to learn necessary survival skills. Would it be correct to conclude that all vertebrate young would benefit from this type of parental care? Explain why or why not.

Portfolio Projects

24. Summarizing Information Research the importance of bison to Native Americans who lived on the Great Plains. Write a report that summarizes what you have learned.

25. Interpreting Information Read "Polar Bears: Stalkers of the High Arctic," by John L. Eliot (*National Geographic,* Jan. 1998, pp. 52–71). What do polar bears prey on and how do they acquire this prey? How are bears adapted to withstand and thrive in the arctic climate? What dangers do humans pose to the bears' habitat? Do all polar bears hibernate? Explain.

26. Recognizing Patterns Find out more about how the theory of continental drift can explain the pattern of distribution of mammals on Earth. Write a report on your findings. Include a description of how the breakup of Pangaea led to the predominance of placental mammals on all the continents except Australia.

27. Comparing Functions Research some unusual functions of mammalian hair, such as the porcupine's quills. Present your findings in a set of annotated drawings.

28. Career Focus Veterinary Technician Use library and Internet resources to research the educational background necessary to become a veterinary technician. Write a report that includes a job description, training required, kinds of employers, growth prospects, and starting salary.

Performance ZONE

CHAPTER 36

Critical Thinking

23. Answers will vary. Students may conclude that vertebrates that mature quickly do not need long periods of parental care. Some students may point out that the process of natural selection shapes each species' behavior in a way that is optimal for survival and reproduction.

Portfolio Projects

24. Answers will vary, but should include information on how Plains tribes used bison for food, clothing, and shelter.

25. The polar bear's primary food is seals, which it hunts by stalking and ambush. The bear's keen sense of smell enables it to locate seals, and bears must hunt from ice platforms. Humans threaten bears by overhunting them and by polluting Arctic waters. The bear's dense fur protects it, and its long snout warms and moistens the frigid air before it reaches the bear's lungs. Only pregnant females hibernate.

26. Answers will vary, but should include a more detailed account than is found in the chapter.

27. Answers will vary, but may include information about the formation of rhinoceros horns from hair, color changes according to seasons or at some other time, raised hair that serves as a warning sign, or the sensory function of body hair and whiskers.

28. Veterinary technicians provide support and assistance to veterinarians. A two- or three-year college diploma as a veterinary technician is required. Veterinary technicians work in a variety of settings, including veterinary clinics and hospitals, zoos, and wildlife parks. Many veterinary technicians work only with small domestic animals, but others may work with larger animals such as cattle and horses, or even with wild animals such as gazelles or big cats. Growth prospects are good. Starting salary will vary by region.

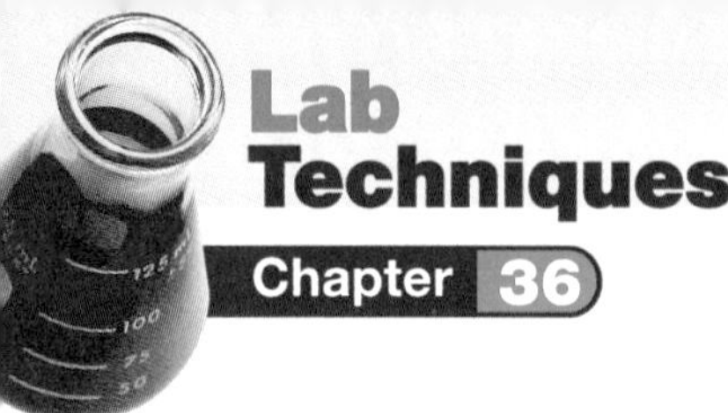

Lab Techniques

Chapter 36

Exploring Mammalian Characteristics

Purpose

Students will examine distinguishing characteristics of mammals.

Preparation Notes

Time required: 45 minutes

Materials

Materials for this lab can be ordered from WARD'S. See *Master Materials List* at the front of this book for catalog numbers.

Preparation Tips

Provide specimens, photos, or diagrams of a variety of mammalian and nonmammalian skulls.

Safety Precautions

- Advise students to handle specimens and laboratory equipment with care.
- Advise students to wash their hands before and after counting the teeth in their mouth. Tell students to examine only their own teeth.
- Warn students not to touch their faces or eyes after handling samples.
- Make certain that students thoroughly wash their hands after this lab.

Procedural Tips

Point out features of skulls that vary among animals, such as the position of the foramen magnum, the opening where the brain stem exits the skull. In bipedal animals, it is at the bottom of the skull, but in quadrupeds, it is at the back of the skull.

Disposal

Make certain to retrieve all samples from students. Store samples safely.

Lab Techniques

Chapter 36

Exploring Mammalian Characteristics

SKILLS

- Observing
- Drawing
- Inferring

OBJECTIVES

- **Examine** distinguishing characteristics of mammals.
- **Infer** the functions of mammalian structures.

MATERIALS

- hand lens or stereomicroscope
- prepared slide of mammalian skin
- compound microscope
- mirror
- specimens or pictures of vertebrate skulls (some mammalian, some nonmammalian)

Magnification: 28×

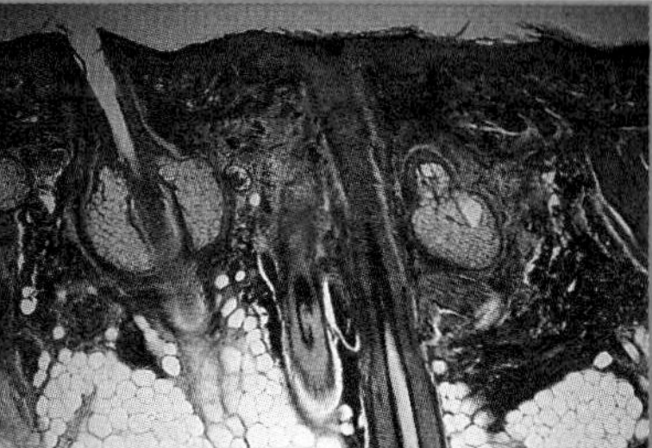

Mammalian skin (human scalp)

Before You Begin

Mammals are vertebrates with **hair, mammary glands,** a single lower jawbone, and specialized teeth. Other characteristics of mammals include **endothermy** and a four-chambered heart. Mammals also have **oil (sebaceous) glands** in their skin, and most have **sweat glands.** In this lab, you will examine some of the characteristics of mammals that distinguish them from other vertebrates.

1. Write a definition for each boldface term in the paragraph above.
2. Make a data table similar to the one below.
3. Based on the objectives for this lab, write a question you would like to explore about the characteristics of mammals.

DATA TABLE

Mammal	Incisors	Canines	Premolars	Molars
Human				

Procedure

Part A: Examine Mammalian Skin

1. Use a hand lens to look at several areas of your skin, including areas that appear to be hairless. Record your observations.
2. Look at a prepared slide of mammalian skin under low power of a compound microscope. Notice the glands in the skin. Look for the oil (sebaceous) glands and the sweat glands. Draw and label an example of each type of gland.

Part B: Examine Mammalian Teeth and Skulls

3. Wash your hands thoroughly with soap and water. Use a mirror to look in your mouth. Identify the four kinds of mammalian teeth you see.
4. Count how many of each you have on one side of your lower jaw. Multiply the number of each kind of tooth by four, and record these numbers in the appropriate columns of your data table. Wash your hands again before continuing.
5. Look at the skulls of several mammals. Identify the kinds of teeth in each skull. For each skull, find the number of each kind of tooth as you did in step 4.
6. Look at the skulls of several nonmammalian vertebrates, and compare nonmammalian teeth to mammalian teeth.
7. Compare the jaws of mammalian skulls to those of nonmammalian vertebrates. As you look at each skull, notice the structure of

Before You Begin

Answers

1. *mammal*—vertebrate that has hair, mammary glands, and specialized teeth

   hair—strands of protein that grow from the skin of mammals

   mammary glands—milk-producing glands of female mammals

   endothermy—the quality of maintaining a constant internal body temperature

   oil (sebaceous) glands—oil-producing glands located in the hair follicles of mammalian skin

   sweat glands—water-secreting glands in the skin of some mammals
3. Answers will vary. For example: How do the teeth of humans compare to those of other mammals?

the lower jawbone and how the upper jawbone and the lower jawbone connect.

Part C: Cleanup and Disposal

8. Dispose of broken glass in the waste container designated by your teacher.
9. Clean up your work area and all lab equipment. Return lab equipment to its proper place. Wash your hands thoroughly before you leave the lab and after you finish all work.

Analyze and Conclude

1. **Summarizing Information** List the characteristics that distinguish mammals from other vertebrates.
2. **Interpreting Graphics** Compare the amount of hair on humans to that on the mammals shown in the photographs below.
3. **Inferring Relationships** What, if any, role might hair or fur play in enabling mammals to be endotherms?
4. **Forming Hypotheses** Besides the role of hair you identified in item 3 above, what other roles do you think hair might play in mammals?
5. **Recognizing Patterns** Where are the oil (sebaceous) glands located in the skin of mammals?
6. **Forming Hypotheses** Do you think the mammals in the photos below have more sweat glands or fewer sweat glands than humans? Explain.
7. **Comparing Structures** How is the mammalian jaw different from nonmammalian jaws?
8. **Inferring Conclusions** Based on the shape of your teeth, would you classify humans as carnivores (meat eaters), herbivores (plant eaters), or omnivores (meat and plant eaters)? Explain.
9. **Evaluating Conclusions** Justify the following conclusion: The kinds and shapes of a mammal's teeth can be used to determine its diet.
10. **Further Inquiry** Write a new question about the characteristics of mammals that could be explored with a new investigation.

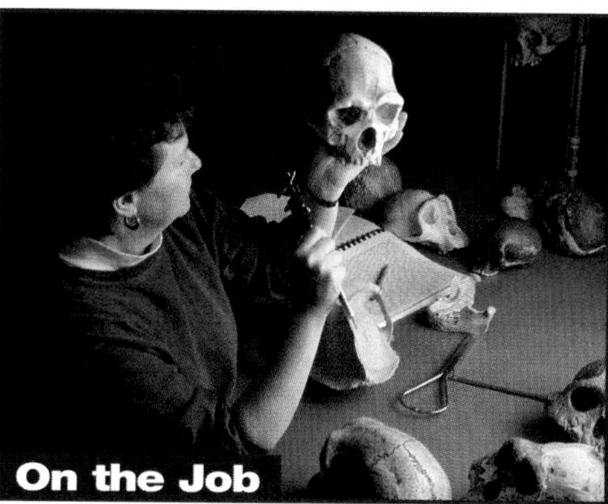

On the Job

Comparative anatomy is the study of the anatomical similarities and differences between organisms. Do research in the library or media center or on the Internet to discover how comparative anatomy is used to determine the relationships among the animal phyla. Summarize your findings in a written or oral report.

2. Humans have far less hair than skunks and more hair than dolphins.
3. Hair serves as insulation from extreme external temperatures and helps maintain a constant internal temperature.
4. Answers will vary. Hair enables a mammal to gather sensory information from its surroundings. It also helps to protect a mammal's skin from water loss, sunburn, and injury.
5. inside the hair follicles
6. Answers will vary. Students should recognize that skunks, as well as dolphins, are not likely to have as many sweat glands as humans. Humans have very little hair for insulation. Sweat glands help to cool the body. Skunks have sufficient hair for efficient insulation. Dolphins live in water and have no need for sweat glands.
7. Mammals have a single jawbone that connects to the skull at a joint.
8. Answers may vary. Students should recognize that humans are omnivores. Humans have sharp incisors and canines for tearing meat and flat molars and premolars for grinding plant material.
9. Answers will vary. Carnivores have long, sharp canines for grabbing prey and tearing meat. Herbivores have flattened teeth for grinding and crushing plant material.
10. Answers will vary. For example: How is a mammal's muzzle shape related to its diet?

Procedure

Answers

1. Students should note that they can see pores and hairs (except on the fingers, palms, toes, and soles of the feet).
2. Students' drawings should show oil and sweat glands.
3. incisors, canines, premolars, and molars
4. A typical answer is two incisors, one canine, two premolars, and three molars.
5. Answers will vary. The number and types of teeth will vary according to the animal.
6. Answers will vary. Nonmammalian teeth will differ significantly from mammalian teeth.

Analyze and Conclude

Answers

1. hair, single jawbone, specialized teeth, and mammary glands

On the Job

Answers

Students' reports should highlight the importance of the comparative anatomy of adult animals and of the embryological development of animals to our understanding of evolutionary relationships among animal phyla.

Chapter 37 Animal Behavior

Planning Guide

OBJECTIVES	MEETING INDIVIDUAL NEEDS	READING SKILLS AND STRATEGIES
37-1 Evolution of Behavior (BLOCK 1 45 min) • Distinguish between "how" and "why" questions about behavior. • Describe how natural selection shapes behavior. • Compare innate and learned behaviors. • Summarize how behavior is influenced by both heredity and learning.	Learning Styles, p. 834 BASIC ADVANCED • pp. 836, 837, 839	ATE Active Reading, Technique: Brainstorming, p. 833 ■ Active Reading Guide, worksheet (37-1) ■ Directed Reading, worksheet (37-1) ■ Reading Strategies, LINK worksheet
37-2 Types of Behavior (BLOCKS 2 & 3 90 min) • Discuss six types of animal behavior. • Discuss how animals use signals. • Summarize how sexual selection can influence evolution.	Learning Styles, p. 840 BASIC ADVANCED • p. 846	ATE Active Reading, Technique: Reading Organizer, p. 841 ■ Active Reading Guide, worksheet (37-2) ■ Directed Reading, worksheet (37-2) ■ Supplemental Reading Guide, *Through a Window: My Thirty Years with the Chimpanzees of Gombe*

Review and Assessment Resources

BLOCKS 4 & 5 90 min

TEST PREP

- **PE** Study Zone, p. 847
- **PE** Performance Zone, pp. 848–849
- ■ Chapter 37 worksheets:
  - Vocabulary
  - Concept Mapping
  - Critical Thinking

 Audio CD Program

■ Test Prep Pretest

CHAPTER TESTING

- ■ Chapter Tests (blackline master)
- ▲ Test Generator
- ■ Assessment Item Listing

See the correlations table at the front of this book for a list of the ***National Science Education Standards*** addressed in this chapter.

Smithsonian Institution® Visit **www.si.edu/hrw** for additional on-line resources.

KEY

Biology Principles and Explorations

PE Pupil's Edition

ATE Teacher's Edition

 Teaching Transparencies

 Laboratory Program

 **One-Stop Planner CD-ROM** *Shows these resources within a customizable daily lesson plan:*

■ Teaching Resources

▲ Test Generator

VISUAL STRATEGIES	CLASSWORK AND HOMEWORK	TECHNOLOGY AND INTERNET RESOURCES
ATE Fig. 37-4, p. 836	**PE** Quick Lab How can you recognize learned behavior?, p. 838 ■ Occupational Applications: Veterinary Technician **PE** Review, p. 839	Video Segment 29 Animal Communication Holt Biology Videodiscs Section 37-1 Audio CD Program Section 37-1 internetconnect SCILINKS NSTA TOPICS: • Animal behavior • Conditioning • Imprinting
	PE Quick Lab When do crickets exhibit territorial behavior?, p. 844 **PE** Lab Techniques Studying Nonverbal Communication, pp. 850–851 Lab C20 Observing Animal Behavior **PE** Review, p. 846 **PE** Study Zone, p. 847	Holt Biology Videodiscs Section 37-2 Audio CD Program Section 37-2 internetconnect SCILINKS NSTA TOPICS: • Animal communication • Tracking animals by satellite • Selective mating

ALTERNATIVE ASSESSMENT

PE Portfolio Projects 23–25, p. 849

ATE Alternative Assessment, p. 847

ATE Portfolio Assessment, p. 847

Lab Activity Materials

Quick Lab
p. 838 small wads of paper toweling (one moist and one dry), T-maze made of two 6 cm (about 2.25 in.) pieces of 0.5 in. clear vinyl tubing, sow bug, blunt object for prodding sow bug

Quick Lab
p. 844 male crickets (5), each marked with a different color; unmarked female crickets (5); covered aquarium; slice of apple and of potato; small plastic jar; 5 cm (about 2 in.) square of cardboard

Lab Techniques
p. 850 stopwatch or clock with a second hand, paper, pencil

Chapter 37

Chapter Theme

Evolution

Scientists have learned that some behavior is influenced by genes and therefore can be acted upon by natural selection. Looking at different kinds of behavior among animals reinforces the idea that some aspects of behavior better adapt particular animals to their way of life. However, behavior is not determined by genes alone. In this chapter, students examine how genes and learning interact to shape behavior.

Establishing Prior Knowledge

Before beginning this chapter, make certain students can answer the questions in **Study TIP Ready?** on page 833. If they cannot, encourage them to reread the sections indicated.

Opening Question

Ask several students who own pets to describe one behavior of their pet. *(example: a dog "shaking hands")* Then ask if each particular behavior is learned or not. *(The example behavior is learned.)* Ask students what criteria they used to determine whether or not a behavior is learned. *(Answers will vary.)* If all students describe learned behaviors, ask them to think of some behaviors that do not have to be learned. *(example: eating)*

CHAPTER 37 Animal Behavior

Great egret and chicks

What's Ahead — and Why

CHAPTER OUTLINE

37-1 Evolution of Behavior

- A Behavior Is an Action Performed in Response to a Stimulus
- Natural Selection Shapes Behavior
- Many Behaviors Are Genetically Based
- Learning Plays a Large Role in Behavior
- Many Behaviors Have Both Genetic and Learned Aspects

37-2 Types of Behavior

- Animal Behaviors Fall into Several Broad Categories
- Animals Often Communicate in Complex Ways
- Choosing a Mate Involves the Interplay of Many Behaviors

Students will learn

▷ what influences behavior in animals.

▷ how innate and learned behaviors differ.

▷ how animal behaviors are categorized, how animals communicate, and how they choose a mate.

Hyacinth macaw

It may look like a trick, *but training an animal takes advantage of its natural tendency to engage in certain types of behavior. In the wild, macaws and other birds use their beak to hold on to and move objects, such as food. This hyacinth macaw has learned to use its beak to match colored shapes, an unusual ability among animals.*

Looking Ahead

Study TIP

Ready?

Check your understanding of these topics to help you prepare for what's ahead.

Can you . . .

- **describe** the relationship of genes to inherited traits? (Sections 6-1, 8-2)
- **relate** natural selection to adaptation? (Sections 13-1, 13-3)

If not, *review the sections indicated.*

internetconnect

National Science Teachers Association *sci*LINKS Internet resources are located throughout this chapter.

Reading Strategies

Active Reading

Technique: Brainstorming

Before beginning this chapter, ask each student to write down five specific examples of animal behavior (these may include human behaviors). Then have the students read the first page of the chapter. They should formulate one "how" and one "why" question for each of the behaviors they have written down. Have them brainstorm in groups to come up with possible answers to their questions. Each group should share its questions, answers, and reasons for its answers with the rest of the class. Have the class discuss some of the more interesting ideas.

Directed Reading

Assign **Directed Reading Worksheet Chapter 37** to help students interact with the text as they read the chapter.

Interactive Reading

Assign ***Biology: Principles and Explorations* Audio CD Program Chapter 37** to help reluctant readers achieve greater success in reading the chapter.

Reading Skills

Assign Reading Organizers and Reading Strategies Worksheets from the **One-Stop Planner CD-ROM with Test Generator** to help students learn and practice effective reading skills.

internetconnect

Holt *Biology: Principles and Explorations*

On-line Resources: go.hrw.com

Visit the HRW Web site for a variety of resources related to this chapter. Just type in the keyword HX0 Chapter 37.

Holt *Biology: Principles and Explorations*

On-line Resources: www.scilinks.org

The following *sci*LINKS Internet resources can be found in the student text for this chapter:

Page 835
TOPIC: Animal behavior
KEYWORD: HX835

Page 837
TOPIC: Conditioning
KEYWORD: HX837

Page 839
TOPIC: Imprinting
KEYWORD: HX839

Page 842
TOPIC: Animal communication
KEYWORD: HX843

Page 842
TOPIC: Tracking animals by satellite
KEYWORD: HX842

Page 845
TOPIC: Selective mating
KEYWORD: HX845

Chapter 37

1 Prepare

Scheduling

- **Block:** About 45 minutes will be needed to complete this section.
- **Traditional:** About one class period will be needed to complete this section.
- **Planning:** See the Planning Guide on pp. 831A and 831B for lecture, classwork, and assignment options for Section 37-1. Lesson plans for Section 37-1 can be found in a lesson cycle format in ***Biology: Principles and Explorations*** **One-Stop Planner CD-ROM with Test Generator.**

Resource Link

Assign **Active Reading Worksheet Section 37-1** to help students comprehend the key concepts and visuals in this lesson.

2 Teach

Lesson Warm-up

Identify students in your class who have pets. Ask them to describe at least one thing they have taught their pet to do. Then have them describe in detail the procedure they used to teach their pet. Ask them why they used one particular method over another. For example, did students house-train their dog by using a punishment (such as a swat with a newspaper) after an "accident," or did they use a reward (such as a dog treat or praise) if their dog urinated while outdoors? Ask students which method (positive or negative) they believe is more effective, and why.

Section 37-1 Evolution of Behavior

Objectives

- **Distinguish** between "how" and "why" questions about behavior. (pp. 834–835)
- **Describe** how natural selection shapes behavior. (p. 835)
- **Compare** innate and learned behaviors. (pp. 836–838)
- **Summarize** how behavior is influenced by both heredity and learning. (p. 839)

Key Terms

behavior
innate behavior
fixed action pattern behavior
learning
conditioning
reasoning
imprinting

A Behavior Is an Action Performed in Response to a Stimulus

A squirrel buries a nut; a hungry baby cries. A frog jumps into a pond to avoid a predator, and a person applies the brakes when approaching a red traffic light. These are all examples of animal behavior. A **behavior** is an action or series of actions performed by an animal in response to a stimulus. The stimulus might be something in the environment, such as a sound, a smell, a color, or another individual. The stimulus can also be related to the internal state of the animal, such as being hungry or cold. For example, when under threat, the lizard shown in **Figure 37-1** flares out the folds of skin around its head, giving it a frightful appearance which tells potential enemies to stay away.

Scientists studying behavior investigate two kinds of questions—"how" questions and "why" questions. "How" questions are about how a behavior is triggered, controlled, and performed. For instance, consider the squirrel, also shown in Figure 37-1. "How" questions about squirrel behavior might include "How does a squirrel select which nuts to bury?", "How does it choose where to bury the nut?", and "How does it remember where the nut is?"

However, answering "how" questions provides only a partial understanding of a behavior. Scientists also try to answer the "why" question, such as "Why do squirrels bury nuts?" "Why" questions concern the reasons a behavior exists and are really questions about the evolution of behavior. To truly understand why a behavior exists, it is necessary to understand why the behavior evolved in the first place and why the behavior continues today.

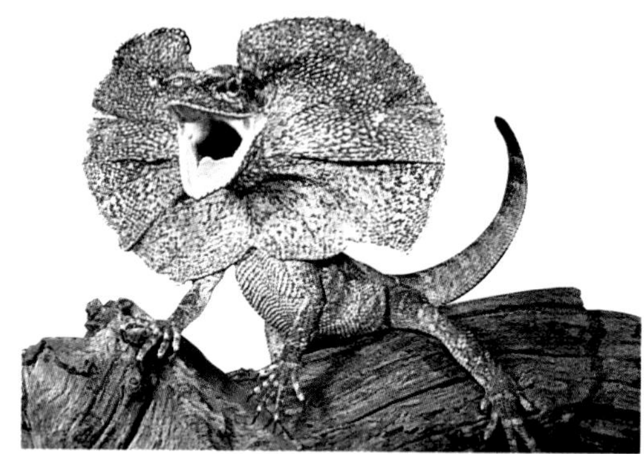

Figure 37-1 Animal behavior. This Australian frilled lizard and gray squirrel are engaging in innate behavior typical of their species.

Integrating Different Learning Styles

Logical/Mathematical	**ATE**	Teaching Tip, p. 836
Visual/Spatial	**PE**	Quick Lab, p. 838; Performance Zone items 1 and 17, pp. 848, 849
	ATE	Visual Strategy, p. 836
Body/Kinesthetic	**ATE**	Demonstration, p. 838
Interpersonal	**PE**	Performance Zone items 19 and 20, p. 849
	ATE	Biology in Your Community, p. 835; Teaching Tip, p. 837
Intrapersonal	**ATE**	Lesson Warm-up, p. 834; Teaching Tip, p. 837
Verbal/Linguistic	**PE**	Study Tip, p. 837; Performance Zone items 2 and 24, pp. 848–849
	ATE	Closure, p. 839; Reteaching, p. 839

Natural Selection Shapes Behavior

Recall that natural selection is a process by which populations change in response to their environment. Natural selection favors traits that improve the likelihood that an individual will survive and reproduce. Over time, traits that provide a survival advantage become more common. Traits that do not, become less common and may disappear.

An understanding of natural selection can help answer a "why" question. A good example of this is seen in East Africa lions, which live in small groups called prides. Each pride contains several adult females, several youngsters (called cubs), and one or more adult males. The adult males father all the cubs and defend the pride against other males. But a male or group of males usually can control a pride for only a couple of years. Then they are forced out by younger males who take over the pride. When this happens, the new males often kill all the young cubs in the pride, as shown in **Figure 37-2.** In contrast, male lions are usually quite tolerant of their own offspring, also shown in Figure 37-2.

Figure 37-2 Behavior in male lions. After taking over a pride, a male lion often kills the young cubs in the pride. However, they are usually tolerant of their own cubs.

To understand this behavior, you need to understand how the new males would benefit from it. The new males will be in the pride for only a few years before they also are forced out, so they have a very short time in which to reproduce. But female lions with cubs will not breed until their cubs are grown, which may take more than two years. If a female's cubs die, however, she will mate again almost immediately.

Why do the new males kill the cubs? One hypothesis suggests that the new males will father more cubs as a result of this behavior. Note, however, that this does not suggest that male lions are aware that they are killing the offspring of other males, or that they understand how they will benefit from this behavior.

Individuals benefit

You may have heard it said that a trait or behavior "ensures the survival of the species." This once popular belief is now considered false, and most scientists now agree that natural selection favors traits that contribute to the survival and reproduction of individuals, not species. The actions of male lions support this idea. Cub-killing increases the already high death rate among cubs and actually reduces the likelihood that the species will survive. Because natural selection favors traits that benefit individuals, the male lions usually will behave in ways that are favorable for them, not for the pride as a whole.

internetconnect

TOPIC: Animal behavior
GO TO: www.scilinks.org
KEYWORD: HX835

Teaching TIP

Recognizing Relationships

Inform students that scientists seek to answer "how" and "why" questions in all areas of biology. For example, geneticists have learned much about how meiosis occurs—how the chromosomes line up and separate and how cytokinesis occurs. But they are also interested in why meiosis occurs and what benefits it brings to the organism. Ask students to come up with some "how" and "why" questions about subjects they studied in previous chapters. Then ask students to answer their questions.

Teaching TIP

Point of Emphasis

Emphasize the last paragraph of the first section on page 835. Students often believe that an animal understands why it behaves in a certain way. Emphasize that male lions do not kill cubs because they know it might help them father more cubs of their own, even though they may actually father more cubs. Add that dogs don't eat because they know they will die if they don't. They eat because they are hungry—a genetically influenced sensation that keeps them alive, allowing them to reproduce and pass on their genes.

Biology in Your Community

Animal Trainer

Invite a dog-obedience trainer to visit your class and to discuss the most effective methods he or she uses to train dogs. Ask the trainer to share his or her favorite stories of incidents that have occurred during training. Beforehand, have each student formulate five questions to ask the trainer. Be sure to ask for the trainer's opinion on negative reinforcement versus positive reinforcement. If school rules allow, have the trainer bring a dog and demonstrate training skills in the classroom.

Teaching TIP — ADVANCED

Design an Experiment

Ask students how biologists know that the orb spider does not *learn* to build its web. After some discussion, ask each student to design an experiment that would demonstrate that web building by the orb spider is an innate behavior. *(The experimental design should include some form of isolation of the eggs so that the young spiders have no contact with the parent. The design of the isolated spiders' webs should be compared against that of the parents' web.)* PORTFOLIO

Teaching TIP

Innate Behaviors

Ask students to name several behaviors that are innate in animals. *(for example, suckling, eating, and drinking)* Ask them how each of these helps the animal to survive. Next ask them to come up with a hypothetical behavior that would be a disadvantage to an animal. *(for example, a bird refusing to fly when it is capable of doing so)* Ask: What would most likely happen to this bird? Would its genes be passed on to the next generation? Point out that this example illustrates how natural selection acts on behavior.

Visual Strategy

Figure 37-4

Have students examine the hybrid lovebird in the figure. Ask: What nesting behaviors does it exhibit that are similar to those of the Fischer's lovebird? *(It carries nesting materials in its beak.)* What nesting behaviors does it exhibit that are like those of the peach-faced lovebird? *(It places nesting materials in its feathers.)* What problem does it encounter in nest building? *(It fails to let go of the nesting material, so the material does not stay tucked into its feathers.)*

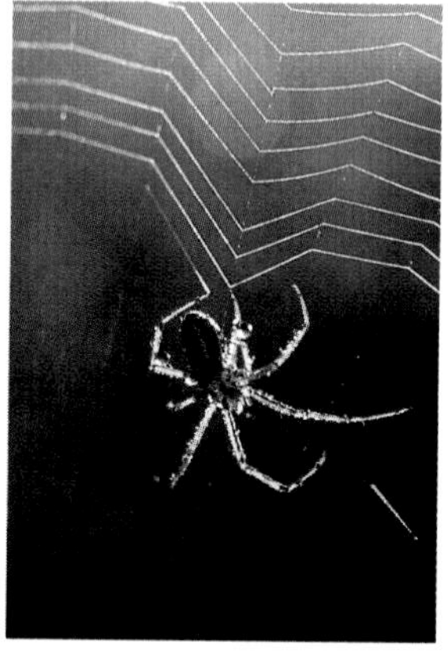

Figure 37-3 Fixed action pattern behavior. Like all web-building spiders, the orb spider is genetically programmed to build her web in the same way each time.

Many Behaviors Are Genetically Based

From years of observation and experimentation, biologists have learned that many kinds of animal behaviors are influenced by genes. Genetically-programmed behavior is often called **innate behavior**, or instinct. The orb spider, shown in **Figure 37-3,** builds her web in exactly the same way every time. There is little or no variation in what she does, and her female offspring will build their webs in exactly the same manner without being taught. This type of innate behavior is called **fixed action pattern behavior** because the action always occurs in the same way.

Establishing the genetic basis

Another innate behavior is nest building in most birds, including two closely-related species of African lovebirds. These small parrots construct their nests from materials that they collect and carry back to the nest site. One species, Fischer's lovebird, carries a single long strip of nesting material in its beak. A second, closely related species, the peach-faced lovebird, carries several short strips of nesting material tucked into the feathers near its tail.

Evidence that these behaviors have a genetic basis comes from studies in which the two types of lovebirds were interbred. As shown in **Figure 37-4,** the resulting hybrid birds showed nesting behaviors that resembled that of both parents. They chose medium length strips of nesting material and tried to place the strips in their feathers. But the hybrid birds were rarely successful because they did not let go of the strips after placing them in the feathers near their tail. Eventually, some of the hybrid birds learned to carry the nesting material in their beak.

Figure 37-4 Nest building behavior

The hybrid offspring of two lovebird species show nest-building behavior similar to both parents.

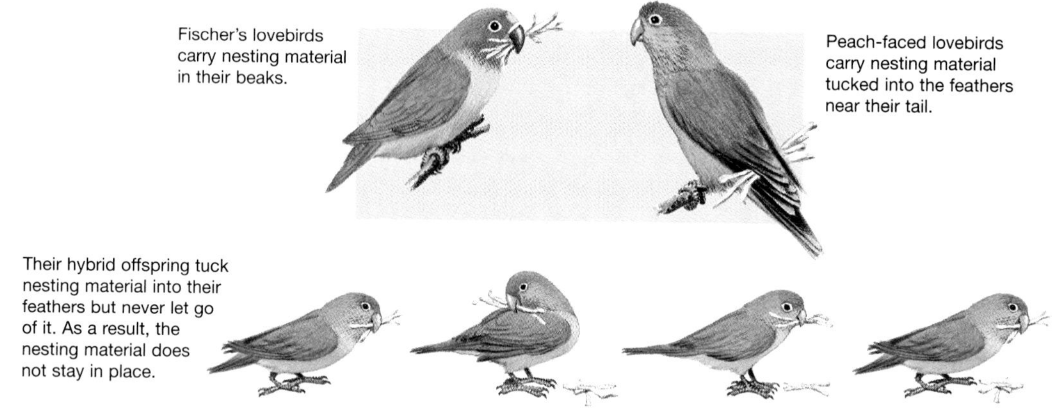
Fischer's lovebirds carry nesting material in their beaks.
Peach-faced lovebirds carry nesting material tucked into the feathers near their tail.
Their hybrid offspring tuck nesting material into their feathers but never let go of it. As a result, the nesting material does not stay in place.

MULTICULTURAL PERSPECTIVE

Australian Aborigines

To the early European colonists of Australia, the absence of Aborigine technology indicated that they were a very primitive people. The Aborigines could produce only a limited variety of tools because they did not know how to mine, smelt, or work metal. They had to rely on hunting and gathering because they did not practice agriculture. Most groups did not wear clothes, even in winter. The largest "structures" they created were piles of discarded shells. However, as the Europeans soon learned, the Aborigines compensated for their technological "deficiencies" with a deep and detailed knowledge of their environment, which allowed them to survive in areas where European explorers and settlers perished. In particular, the Aborigines were experts on the behavior of animals, and they used this knowledge to capture game and to find water in dry areas.

Learning Plays a Large Role in Behavior

Behaviors are influenced by genes, but to what degree can genetic behavior be modified by experience? The development of behaviors through experience is called **learning.** In many animals, learning is very important in determining the final nature of innate behavior. One simple kind of learning is habituation. In habituation, an animal learns to ignore a frequent, harmless stimulus. For example, birds may at first stay away from a garden that has a new scarecrow. But if the position of the scarecrow is not changed on a regular basis, the birds learn to ignore it and go into the garden unafraid.

Classical conditioning

A more complex type of learning is **conditioning,** learning by association. One of the most famous studies of conditioning is Russian psychologist Ivan Pavlov's work with dogs, carried out in the late 1890s and early 1900s. When Pavlov presented meat powder (a stimulus) to a hungry dog, the dog salivated, an innate response to food. At the same time the dog received the meat powder, Pavlov also presented the dog with a second, unrelated stimulus—a ringing bell. After repeated trials, the dog learned to associate the ringing bell with the meat powder and would salivate in response to the bell alone. The dog became conditioned to associate the ringing of the bell with a reward (meat powder). This type of conditioning, in which an animal comes to associate an unrelated response with a stimulus, is called classical conditioning.

Trial-and-error learning

Animals learn by trial-and-error that performing a certain action will result in a reward or a punishment. For example, a dog may learn to avoid a particular cat after being scratched on the nose once or twice. When trial-and-error learning occurs under highly controlled conditions, it is called operant conditioning. Operant conditioning is demonstrated in another famous set of experiments conducted by the American psychologist B. F. Skinner. Skinner studied learning in rats by placing them in a "Skinner box," illustrated in **Figure 37-5.** Once inside, the rat would explore the box. Occasionally, it would accidentally press a lever, and a pellet of food would appear. At first, the rat would ignore the lever and continue to move about, but it soon learned to press the lever to obtain food. This sort of trial-and-error learning is of major importance to most vertebrates, and it influences many behaviors essential to survival, such as searching for food.

Figure 37-5 Skinner box. When placed in a Skinner box, this rat learned by trial-and-error to push a lever to receive a reward of food.

Study TIP

Recognizing Differences

To understand the difference in classical and operant conditioning, think about how the behavior is learned. In classical conditioning, the stimulus has no relationship to an activity. In operant conditioning, there is a direct relationship.

TOPIC: Conditioning
GO TO: www.scilinks.org
KEYWORD: HX837

Teaching TIP ADVANCED

Learning in Pets

Divide the class into two-person teams, and have each team design an experiment to test the ability of a pet to learn a new behavior. Suggest some possibilities, such as teaching a dog a new behavior or having a mouse perform some sort of activity to obtain food. Encourage students to test different conditions on their own time to determine if one method is more successful. For example, a dog that performs a certain behavioral response can be rewarded with food. In a second conditioning experiment, the same dog can be rewarded with petting when it exhibits the desired response. Students can then correlate the learning time with the type of reward. If students plan to carry out their experiments, be sure they treat all animals in a humane manner. Guidelines for working with vertebrates are available from the National Association of Biology Teachers. Have students write a report describing their procedure and the results of their experiments. PORTFOLIO

Teaching TIP BASIC

Pavlov's Experiments

Pavlov's dogs initially would salivate only in response to the meat-powder stimulus. Salivation is an unconditioned response, meaning it occurs automatically and does not have to be learned. Remind students that they show a similar behavior when they sense food. Ask students to speculate about what the function of this response might be. *(Salivation prepares the mouth to receive food by lubricating the mouth and increasing the concentration of a digestive enzyme found in saliva.)*

did you know?

Four scientists who studied behavior have received Nobel prizes.

The first was the Russian scientist Ivan Pavlov, who is famous for his experiments on learning in dogs. Pavlov received his award in 1904. In 1973, Karl von Frisch and Konrad Lorenz, both of Austria, and Niko Tinbergen, from the Netherlands, were honored for their pioneering studies of behavior. All four scientists received Nobel prizes in the category of physiology or medicine. There is no Nobel prize for behavior.

Demonstration

Innate Behavior—Reflexes

Ask a volunteer to come to the front of the room. As the student stands in front of you, facing the class, quickly and without warning clap your hands in front of the student's face. The student should blink. Ask students if this is a learned response. *(no)* Ask students how this behavior helps the student. *(It prevents damage to the eyes, which, for many animals, are essential to survival.)*

How can you recognize learned behavior?

QUICK LAB

Time 10 minutes a day for 3 days

Process Skills Observing, collecting data, analyzing data, evaluating

Teaching Strategies
Have students work in small groups. Before they begin, have each group formulate a hypothesis about what the sow bugs will do.

Analysis Answers

1. Answers will vary depending on sow bug behavior.
2. Answers will vary depending on sow bug behavior.
3. Answers will vary, but make sure that students' answers are supported by evidence.
4. If the sow bugs show a preference for the wet paper towel and if no other factors influence the direction they turn, the bugs should exhibit random turning behavior if dry paper toweling is used.

Reasoning

The ability to analyze a problem and think of a possible solution is called **reasoning.** Reasoning involves using past experiences to develop an insight into how to solve a new problem. The dog shown in **Figure 37-6** is unable to reason and cannot think out a solution to the situation it is in. Humans and some other primates show the ability to reason. For example, in one experiment a chimpanzee placed in a room with some boxes and a banana hung high overhead appeared to think out the situation. Although the chimp had not been in a similar situation, it stacked up the boxes in order to reach the banana, a behavior that required reasoning. There has been much research and debate over the ability of some other animals, such as dolphins, some parrots, and even octopuses, to reason. So far, there is no clear cut evidence that they can, although they have remarkable abilities to learn often complex tasks. Certainly animals can learn complex tasks. Some birds living in cities have learned to remove the foil covering from bottles of nonhomogenized milk in order to reach the cream beneath. Japanese macaque monkeys have learned to float grain on water to separate it from sand.

Figure 37-6 An inability to reason. Although it seems obvious to humans, this dog is unable to figure out how to reach the food.

How can you recognize learned behavior?

QUICK LAB

Recall that sow bugs must keep moist to survive. Follow the procedure below to see if sow bugs will learn to seek moisture.

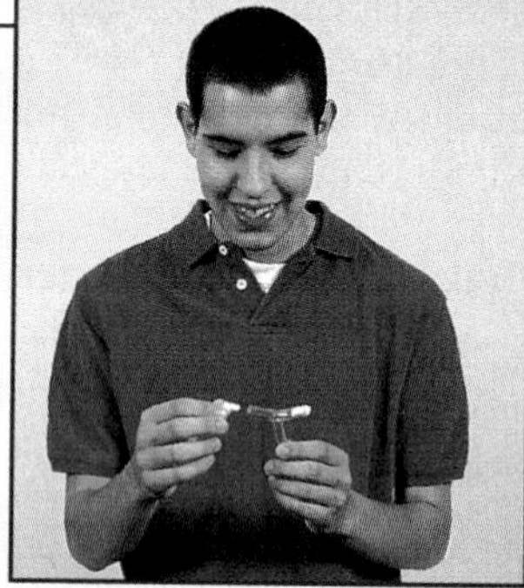

Materials

small wads of paper toweling (one moist and one dry), T-maze made of two 6 cm (about 2.25 in.) pieces of 0.5 in. clear vinyl tubing, sow bug, blunt probe for prodding sow bug

Procedure

1. Place the moist paper wad in the open end of the left side of the T, and place the dry paper wad in the right side.
2. Place the sow bug at the bottom of the T. If it does not start to crawl, gently prod it to move.
3. Observe what the sow bug does when it reaches the T section. Retrieve the sow bug and perform as many trials as time allows.
4. Keep a record of the results of each trial.
5. Using the same sow bug, repeat this procedure for three days.

Analysis

1. **Summarize** your sow bug's behavior, in writing or on a graph.
2. **Describe** any trend in behavior that you observed.
3. **Determine** if your sow bug modified its behavior through learning, using evidence to support your answer.
4. **Evaluate** the value of performing a final trial in which the T-maze contains only dry paper toweling.

did you know?

The capacity for learning seems to be somewhat related to life span and to brain size.

In small animals that have a short life span and therefore little time for learning (such as some insects), most behaviors appear to take the form of fixed action patterns. In large animals with a large brain and a long life span, such as humans, many behaviors depend on experience.

Many Behaviors Have Both Genetic and Learned Aspects

Do genes determine behavior, or do animals learn how to behave from experience? Over the last century, this topic has been debated. Some scientists argued that most behaviors are genetically programmed because different individuals in the same species act in the same ways. Other scientists claimed that behaviors are shaped by an animal's experiences. Most biologists who study animal behavior have come to think that animal behavior, particularly the complex behavior of vertebrates, has both genetic and learned components.

internetconnect
SCILINKS. NSTA
TOPIC: Imprinting
GO TO: www.scilinks.org
KEYWORD: HX839

Imprinting

Learning that can occur only during a specific period early in the life of an animal and cannot be changed once learned is called **imprinting.** Imprinting is easily observed in young geese and ducks who have no innate recognition of their mother. Instead, these birds are genetically programmed to follow the first moving object they see during a short period immediately following hatching. There is great survival value to this behavior, as the young must follow their mother as she leads them to water, helps them find food, and keeps them out of danger. However, the young will follow any object they see during this period just as they would their mother—including toy wagons, boxes, and balloons. Once the young birds imprint on an object, they prefer to follow it, even when given the opportunity to follow a member of their own species.

Figure 37-7 Imprinting. These goslings imprinted on Konrad Lorenz and followed him around just as if he were their mother.

Konrad Lorenz, a Nobel Prize-winning pioneer in the study of animal behavior, observed imprinting when he raised a group of newly hatched goslings (young geese) by hand and found that they imprinted on him. **Figure 37-7** shows Konrad Lorenz leading his "family" of goslings. The goslings ability to imprint on an object during a sensitive period is not a learned behavior; it is programmed into their genes. However, the process of imprinting is a form of learned behavior. Thus learning determines the final shape of a genetically based behavior.

Review SECTION 37-1

1. **Describe** the difference between "how" and "why" questions in regard to animal behavior.
2. **Summarize** how cub killing by male lions supports the hypothesis that natural selection shapes behavior.
3. **Distinguish** between and give an example of innate and learned behavior.
4. **Analyze** the behaviors involved in imprinting.

Critical Thinking

5. **SKILL Forming Reasoned Opinions** A friend is teaching his dog a new trick in which it is rewarded each time the trick is performed correctly. The friend says his method is called classical conditioning. Evaluate your friend's use of this term.

Review SECTION 37-1 ANSWERS

1. "How" questions are about how a behavior is triggered, controlled, and performed. "Why" questions concern the adaptive value of a behavior.
2. By killing cubs, the male lion increases his chances of fathering his own cubs, thereby increasing the likelihood that his "cub-killing" genes will be passed on.
3. Innate behavior, such as eating, requires no experience. Learned behavior, such as a dog salivating in response to a ringing bell, is modified by experience.
4. In imprinting, an animal is genetically programmed to learn something during a critical period early in its life. What the animal imprints *on* is a learned behavior.
5. The friend's usage is incorrect. In classical conditioning, the stimulus has no relationship to the activity. This is a case of trial and error learning.

Teaching TIP — ADVANCED

Behavior Flash Cards

Ask the students to prepare flash cards for each of the **Key Terms** on page 834. Have the students write the term on one side and its definition on the other side. Students should then work in pairs and use the cards to review the key terms.

3 Assessment Options

Closure

Observing Animals

Have students observe the behavior of a pet or other animal and write a report describing their observations. In their reports, students should describe the behaviors of their subject and the context for each behavior. Students should also explain the possible adaptive value of each behavior they observe. PORTFOLIO

Review

Assign Review—Section 37-1.

Reteaching — BASIC

Have students select any behavior that is described in Section 37-1. Ask them to write a list of "why" and "how" questions that a scientist could ask about that particular behavior. *(For example: Why does a gosling follow any object shortly after hatching? How does this behavior change, if at all, over time?)* PORTFOLIO

Section 37-2 Types of Behavior

1 Prepare

Scheduling

- **Block:** About 90 minutes will be needed to complete this section.
- **Traditional:** About two class periods will be needed to complete this section.
- **Planning:** See the Planning Guide on pp. 831A and 831B for lecture, classwork, and assignment options for Section 37-2. Lesson plans for Section 37-2 can be found in a lesson cycle format in ***Biology: Principles and Explorations*** **One-Stop Planner CD-ROM with Test Generator.**

Resource Link

Assign **Active Reading Worksheet Section 37-2** to help students comprehend the key concepts and visuals in this lesson.

2 Teach

Lesson Warm-up

Have the class come up with 10 different examples of animal behavior. Write each example on the board. Then write the six categories of behavior from Figure 37-9 on the board. Have students place the 10 examples into the 6 categories. Allow healthy disagreement among students to lead to a discussion of the characteristics of each category.

Objectives

- **Discuss** six types of animal behavior. (pp. 840–842)
- **Discuss** how animals use signals. (pp. 843–844)
- **Summarize** how sexual selection can influence evolution. (pp. 845–846)

Key Terms

sexual selection

Study TIP

Reading Effectively

Before reading this section, write the Objectives on a sheet of paper. Rewrite each Objective as a question, and answer these questions as you read the section.

Animal Behaviors Fall into Several Broad Categories

As you sit on a park bench, a pigeon approaches, and you soon realize that it expects food. You toss out a bit of your sandwich and the pigeon eats it, immediately looking up for more. This urban pigeon has been conditioned to seek food from people. Has its human supplier been conditioned too?

Behavior is an animal's most immediate way of dealing with its environment. Because the environment is complex and can change rapidly, most animals have many different kinds of behavior, each suited to a particular situation. For instance, a squirrel may perform one kind of behavior when it finds a nut on the ground—it digs a hole. It performs a completely different behavior when a snake approaches—it runs for shelter—because digging a hole would not help it escape from the snake.

Like the squirrel, the musk oxen of the Arctic display many different types of behavior, some of them cooperative. When predators, usually wolves, appear, the adult musk oxen form a defensive circle around their young, as shown in **Figure 37-8.** The tight circle and the danger of injury from the adult's horns and hooves usually prevent a successful attack. While musk oxen can run, running would not protect the herd in the way that their group defense does.

Biologists have classified the behaviors animals perform into several broad categories. **Figure 37-9** shows some of these categories and gives an example of each.

Figure 37-8 Musk oxen. When threatened, the adult musk oxen form a defensive circle around their young.

Integrating Different Learning Styles

Logical/Mathematical	**PE**	Quick Lab, p. 844; Lab Techniques, pp. 850–851
Visual/Spatial	**PE**	Performance Zone items 1 and 17, pp. 848–849
	ATE	Teaching Tip, p. 841; Cross-Disciplinary Connection, p. 843
Interpersonal	**PE**	Real Life, p. 844
Intrapersonal	**ATE**	Making Biology Relevant, p. 845
Verbal/Linguistic	**PE**	Study Tip, p. 840; Performance Zone items 23 and 25, p. 849
	ATE	Active Reading, p. 841; Tech Watch, p. 842; Career, p. 843; Multicultural Perspective, p. 843; Teaching Tip, p. 845; Reteaching, p. 846

Foraging behavior

While many of the behaviors illustrated in Figure 37-9 may seem different from one another, they all tend to favor survival and reproductive success. To gain a better idea of the nature of animal behavior, let's examine one of these behaviors—foraging behavior—in detail.

Animals can be divided into two broad groups based on the range of food items each group consumes. Specialists feed primarily or exclusively on one kind of food. Some species of ants, for example, eat only spider eggs. Generalists, in contrast, consume many different kinds of food. For example, some insects eat the leaves of a wide variety of plants. Generalists are typically less efficient than specialists at feeding on any one type of food. However, generalists have the

Figure 37-9 Animal behavior. Although their methods will differ, all animals will engage in at least some of the behaviors shown below.

Parental care

Ensure survival of young

This robin is feeding an insect to its offspring.

Courtship behavior

Attract a mate

During its breeding season, the male stickleback fish develops a bright coloring and builds an elaborate nest to attract a female.

Defensive behavior

Protection from predators

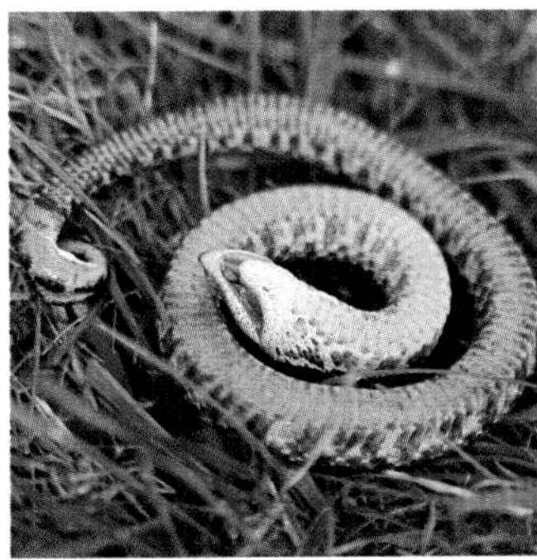

When threatened, a hognose snake flips onto its back and plays dead.

Foraging behavior

Locate, obtain, and consume food

A raccoon searches along streams and ponds for fish, frogs, crayfish, and small rodents. It also hunts for insects and fruit in woodlands.

Migratory behavior

Move to a more suitable environment as seasons change

Monarch butterflies migrate thousands of kilometers, from the United States to central Mexico.

Territorial behavior

Protect a resource for exclusive use

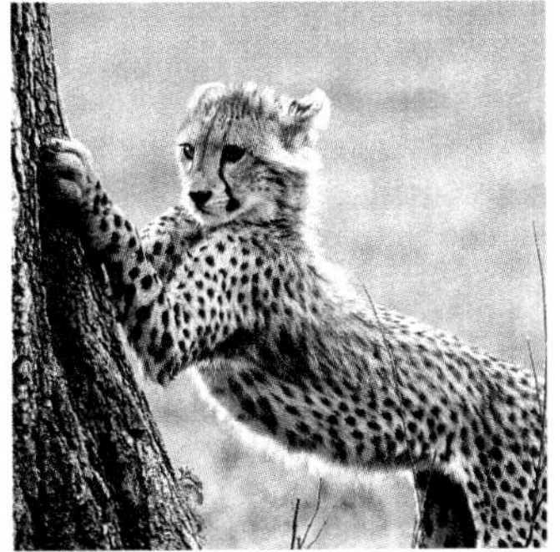

Like many wild cats, this young cheetah claws on trees, leaving a scent that marks its territory.

Active Reading

Technique: Reading Organizer

Have students make a chart with the following headings: *Parental Care, Courtship Behavior, Defensive Behavior, Foraging Behavior, Migratory Behavior,* and *Territorial Behavior*. As students read Section 37-2, they should assign each behavior mentioned in the text to one of the categories in their chart. Inform students that some behaviors will not fit into any of the categories, so it will not be possible to classify all the behaviors described in Section 37-2. PORTFOLIO

Teaching TIP

Animal Behavior

Copy the Graphic Organizer at the bottom of page 841 onto the board or an overhead, but do not copy the entries in parentheses. Have volunteers come up one at a time and fill in the blanks. Students' answers may vary from the ones given. PORTFOLIO

Teaching TIP

Migratory Behavior

Migration may bring a bird to a much more favorable environment where resources, such as food and nesting sites, are available, where predators are rare, or where the climate is moderate. Costs of migration include the energy required to fly long distances and the possibility of dying during a long migration.

Graphic Organizer

*Use this graphic organizer with **Teaching Tip: Animal Behavior** on page 841.*

Type of behavior	Function of behavior	Example of behavior
(Parental care)	Ensuring survival of young	(Bird feeds worm to nestling.)
(Courtship behavior)	(Attracting a mate)	Male frogs croak to attract a mate.
(Defensive behavior)	Protection from predators	(Hognose snake plays dead.)
(Foraging behavior)	(Locating, obtaining, and consuming food)	A raccoon searches along streams and ponds for rodents.
(Migratory behavior)	(Moving to a more suitable environment)	Birds fly south for the winter.
Territorial behavior	(Protecting a resource for exclusive use)	(An aquarium fish forces all other fish to one end of the tank.)

Tracking Animal Movements by Satellite

Teaching Strategies
Ask students to research this topic using Internet and library resources. Each student should identify an animal that has been tracked by electronic means (not necessarily by satellite—some animals are followed by airplane or other means). Have each student write a brief report that tells why the animal is tracked, what data are collected, and how this information is used to help the animal. PORTFOLIO

Discussion
- What is the purpose of the leather hood the bird is wearing? *(Leather hoods are often placed on large birds when they are handled. The hood keeps the bird calm and does not harm it. This protects both the bird and the handler from injury.)*
- What are the advantages of tracking electronically versus getting data by trapping animals? *(Specific information on the animals' travels is supplied continuously by electronic tracking; trapping only pinpoints one location the animal has traveled to. If a student mentions that trapping may injure the animal, remind them that most animals have to be trapped for electronic equipment to be fastened to them.)*
- Why would electronic tracking be especially helpful in tracking marine mammals, such as whales? *(This method provides information about a whale's specific location and also about the depths at which it travels. Also, whales are often not visible while they are traveling.)*

internet**connect**
SciLinks NSTA
TOPIC: Animal communication
GO TO: www.scilinks.org
KEYWORD: HX843

advantage of being able to collect more than one kind of food. Which approach to foraging is better? When one kind of food source is plentiful, specialists forage more successfully. But when food sources are diverse and no particular one of them is more common, generalists find more to eat.

For predators, food typically comes in a variety of sizes. Larger food items contain more energy. But larger items are harder to capture, and they are usually less abundant.

Foraging thus involves a trade-off between a food's energy content and its availability. Animals tend to feed on prey that maximize their energy intake per unit of foraging time. This approach is called optimal foraging. Natural selection has favored the evolution of foraging behaviors of this sort. Sometimes, however, animals will consume foods that are low sources of energy. Often this is because that food supplies an important nutrient. Alternately, the location of the food source allows the consumer to avoid the danger of capture by some other predator.

TECH WATCH

Tracking Animal Movements by Satellite

The migrations of whales, birds, butterflies, bats, and other animals are among the most fascinating examples of animal behavior. Some animals migrate thousands of kilometers between the same two places every year. To track these movements, biologists have turned to satellites for help.

How satellite tracking works

As shown on the right, a small transmitter containing electronic circuitry, a battery, and an antenna is attached to an animal. For birds or other small animals, these parts must be kept as light as possible. Satellites orbiting about 850 km (530 mi) above Earth pick up the radio signals produced by the transmitter and relay the signals to a central computer on the ground. The computer analyzes the information and calculates the animal's location. By connecting to this computer on the Internet, biologists all over the world can get information on the movements of the animals they are interested in tracking.

Why use satellites?

Satellite tracking allows biologists to track animals that would be impossible to follow on foot or in a vehicle. For example, biologists used satellite tracking to trace the winter migration of Swainson's hawks. Over the past few decades, the number of these hawks has been declining in some areas of North America. Satellites showed that the hawks spend the winter in a specific region in central Argentina. By traveling to that region, researchers discovered that thousands of the hawks were being killed accidentally as a result of the use of pesticides. The Argentinian government and the pesticide manufacturers are now working together to protect the hawks.

Swainson's hawk

Horned lizard

internet**connect**
SciLinks NSTA
TOPIC: Tracking animals by satellite
GO TO: www.scilinks.org
KEYWORD: HX842

did you know?

Many animals travel great distances during annual migrations.

The bobolink, a small bird, travels 22,500 km (14,000 mi) each year between its nesting sites in North America and its wintering range in Argentina. The caribou of Canada and Alaska make mass migrations of 160–1,100 km (100–700 mi) every other year. One of the longest mammal migrations is made by the fur seal. From wintering grounds off Southern California, the pregnant females journey as far as 2,800 km (1,700 mi) across open ocean to reach the Pribilof Islands, off the coast of Alaska. The young are born a few hours or days after the females reach the islands. After 3 months of nursing, the cows (females) and juveniles leave for the long migration southward. The bulls (males) remain in the Gulf of Alaska during the winter.

Animals Often Communicate in Complex Ways

You approach an unfamiliar dog and it begins to bark. You know that if you go closer, the dog might bite you, so you stop and talk to it. The dog continues to bark but not so aggressively, and it begins to wag its tail. You and the dog have each responded to a signal given by the other.

A signal can be a sound, posture, movement, color, scent, or a facial expression. These signals are sent and received through all of the senses familiar to us—sight, hearing, smell, touch, taste. Animals use signals to influence the behavior of other animals. Because they face a variety of social situations in which communication is needed, animals usually have several different signals, each suited to a different situation, as shown in **Figure 37-10.**

biobit

Do fishes make noise?

Researchers are finding out that they do—it's just that we can't hear them. By using equipment that picks up frequencies humans can't normally hear, it is possible to listen in. What will you hear? Well, one type of fish called a grouper lets out a creaky bark when it spots a predator.

Figure 37-10 Animal communication. A dog's play bow means "I'm available to play," while a snarl is a signal of aggression and sometimes fear.

Natural selection has shaped animal signals so that they reach the intended receiver efficiently and stimulate a response. To be transmitted efficiently, a signal must be able to travel through the environment from sender to receiver. A signal must also be recognizable to the receiver, or it won't have any effect on behavior. Consider the loud mating call emitted by a male frog such as the tungara frog. The call carries a long distance, reaching even far-off females. At night, when tungara frogs are active, a loud call is the best way to communicate. Visual signals, such as colors and movements, would be visible from only a short distance away and would not be nearly as effective at attracting a mate.

Figure 37-11 Primate communication. This howler monkey is signaling to its troop.

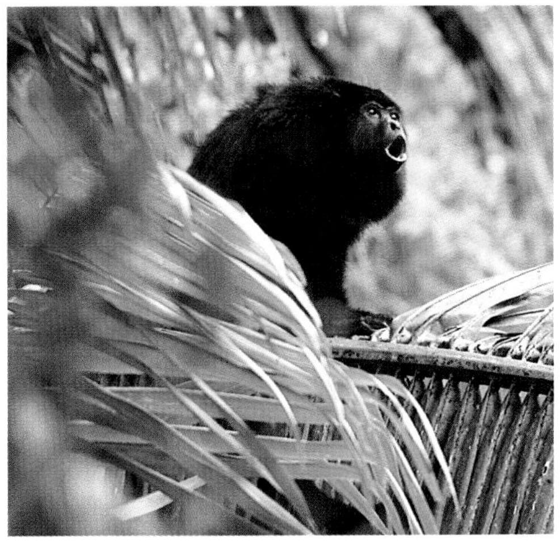

Primate communication

Among animals, vocal communication is most developed in the primates. Many primates have a "vocabulary" of sounds that allows individuals to communicate the identity of specific predators, such as eagles, leopards, and snakes, as shown in **Figure 37-11.** Chimpanzees and gorillas can learn to recognize and use a large number of

CROSS-DISCIPLINARY CONNECTION

◆ Geography

Teachers and students can track animal movements throughout the world on the Internet using WhaleNet. This educational Web site provides updated satellite data on blue whales, harbor porpoises, and loggerhead turtles, as well as archived satellite data on right whales, sperm whales, elephant seals, gray seals, harbor seals, and hooded seals. Another Web site is maintained by the Sea Turtle Survival League. Both sites show the dates and tracking locations of animals on a world or regional map, which is updated periodically. Have students make maps and plot the movements of individual animals. Keep the maps on the wall for other students to observe. Some of the transmitter batteries function for only 2–3 months, so it is a good idea for your students to choose several animals to track.

PORTFOLIO

CAREER

Animal Behaviorist

Animal behaviorists are scientists who study how animals behave. These scientists consider all the possible influences on animal behavior, such as heredity, physiology, environment, food, weather, and the behavior of other animals. Invite an animal behaviorist to the class to speak. Ask the speaker to include the following in his or her talk: recommended high school courses, college degree needed, working conditions, employment opportunities, and salary ranges. Have each student prepare several questions to ask the speaker. Have students write a short paper describing what they have learned.

PORTFOLIO

MULTICULTURAL PERSPECTIVE

Instrument Communication

Many ethnic groups use musical instruments to communicate with other members of their culture. In the United States, many Native American tribes use the drum as an important communication tool within the tribe. There has been a resurgence of drumming in the West, and every year dozens of powwows are held in which drumming competition is a featured event. Have students research native instruments of other cultures and how the sounds produced by these instruments are used to communicate. Examples of instruments include the djembé, sabar, didgeridoo, flute, rattle, and thumb piano. Students should write a short report to share with the class.

PORTFOLIO

Chapter 37

Teaching TIP

Territory and Range

Point out that in biology the term *territory* has a different meaning than it does in geography. An animal's territory is not just the area where it is found, but an area that the animal defends against intruders. An animal's range is the area the animal travels in its normal activities, such as searching for food or gathering nesting materials.

QUICK LAB
When do crickets exhibit territorial behavior?

Time 20 minutes

Process Skills Observing, collecting data, organizing data, analyzing

Teaching Strategies

- Tell students to remain quiet and as still as possible so the crickets are not affected by any noise or movements.
- Have students place the objects around the crickets so the crickets are about the same distance from each object.

Analysis Answers

1. Answers will vary depending on the crickets' behavior.
2. Answers will vary depending on the crickets' behavior.
3. Answers will vary. Students should not attribute a sense of purpose to the crickets' actions.
4. Answers will vary. Hypotheses should describe the adaptive value of each behavior.

REAL LIFE

Answer

Answers will vary depending on the foreign language of the interviewee.

REAL LIFE

Just say /r/. If you are having trouble mastering the throaty sound of a French /r/, just remember that English is just as troublesome to French students. A French student struggles with the English /th/ sound, and will probably pronounce *the* as *zee.*

Finding Information *Interview a native speaker of a foreign language. Ask them to identify three sounds in their language that are difficult for Americans to master.*

symbols to communicate abstract concepts. They cannot talk, however, because they are physically unable to produce the sounds of speech. Some researchers believe that chimpanzees can combine symbols they have learned in meaningful ways. But chimpanzees cannot rearrange symbols to form a new sentence with a different meaning. That requires a very complex brain structure, a complexity found only in humans.

In humans, language develops at a very early age. Infants begin to learn language by trial-and-error during the "babbling baby" phase of childhood, about six months of age. At first, infants all over the world all babble the same consonant sounds, including sounds they have never heard. Soon, however, the infants pick out the sounds used by the people around them and repeat only those sounds. The other sounds begin to drop away and are forgotten. Children quickly and effortlessly learn a vocabulary of thousands of words, a feat no chimpanzee can do. This ability to learn language rapidly seems to be genetically programmed. But language is not the only form of human communication. Evidence suggests that odor and nonverbal signals (body language) may also be important.

QUICK LAB
When do crickets exhibit territorial behavior?

The male cricket chirps to attract females and warn other males to stay away from his territory. You can study chirping behaviors by observing crickets in an aquarium.

Materials

5 male crickets, each marked with a different color; 5 unmarked female crickets; covered aquarium; slice of apple and of potato; small plastic jar; 5 cm (about 2 in.) square of cardboard

DATA TABLE

Cricket Behavior

Cricket	Apple	Potato	Jar	Tent	Female
Blue					
Yellow					
Red					
Green					
White					

Procedure

1. Place the crickets and food in an aquarium. Make two shelters by turning the plastic jar on its side and by folding the cardboard in half to form a tent-like structure.
2. Make a chart like the one above to record the behavior of the male crickets.
3. Observe the crickets for 10 minutes. Among the males, look for territorial (aggressive) behaviors—chirping, stroking others with antennae, pushing others away, etc.
4. For each observation of aggressive behavior, record the color of the aggressive male and where the behavior occurred—for example, next to the shelter.
5. For each cricket, tally the number of times aggressive behavior was observed. Make a list that ranks each cricket, placing the cricket with the highest tally on top.
6. Then tally the numbers for where the behaviors occurred. Rank the locations.

Analysis

1. **SKILL Analyzing Data** Was any cricket (or crickets) more aggressive than the others? Give evidence to support your answer.
2. **Describe** the circumstances in which most aggressive behavior occured.
3. **Propose** a reason to explain your answer to item 2.
4. **SKILL Forming Hypotheses** For each aggressive behavior you observed, form a hypothesis that explains its function.

did you know?

Gorillas can learn sign language.

Koko, a female lowland gorilla born in 1971, and Michael, a male lowland gorilla born in 1973, use sign language and understand spoken English. Their linguistic development has been studied extensively since they were born. Koko has a vocabulary of over 500 signs, and she understands approximately 2,000 words of spoken English. Koko initiates most conversations with her human companions and typically constructs statements averaging three to six words. She has a tested IQ of between 70 and 95 on a human scale, where 100 is considered average.

Choosing a Mate Involves the Interplay of Many Behaviors

When ready to mate, animals produce signals to communicate with potential partners. Each species usually produces a unique courtship signal to ensure that individuals do not mate with individuals of another species. For example, the flash patterns of fireflies are coded for species identity. A female firefly recognizes males of her own species by the number of flashes in his pattern, and she will ignore any male with a different flash pattern. The chemical produced by a female silk moth attracts only males of her own species. Many species of insects, amphibians, and birds produce unique sounds or songs to attract mates. A white-crowned sparrow will respond to the song of another white-crowned sparrow, but totally ignore the song of an English sparrow.

During the breeding season, animals make several important "decisions" concerning mating and parenting. These include how many mates to have and how much time and energy to devote to rearing offspring. These decisions are all aspects of an animal's reproductive strategy, a set of behaviors that have evolved to maximize reproductive success.

TOPIC: Selective mating
GO TO: www.scilinks.org
KEYWORD: HX845

Mate choice

Males and females usually differ in their reproductive strategies. In many animals, females do not mate with the first male they encounter. Instead, the female seems to evaluate the male before she decides whether to mate. This behavior, called mate choice, has been observed in many invertebrate and vertebrate species. Female tungara frogs, for instance, have been observed "shopping around" among calling males. A female will sit near a male and listen to his call for several minutes, then move on to another male and listen to his call. She may evaluate several males before choosing one with which to mate.

What characteristics do animals use in choosing a mate? When Charles Darwin considered this question more than a century ago, he made an important discovery about evolution. Darwin noticed that males often had extreme characteristics that they used in their courtship displays. Take, for example, widowbirds, shown in **Figure 37-12.** During the breeding season, the male widowbird grows an extremely long tail, up to five times longer than the female's. How have these differences between the sexes evolved? The long tail of the male widowbird cannot be essential for survival, since the female bird survives quite well without them.

Figure 37-12
Widowbirds. During breeding season, the tail of the male widowbird sometimes grows to more than three times the length of his body. At other times of the year, it is similar in length to that of the female.

Section 37-2

Making Biology Relevant

Health

In the United States, the greatest health problem that affects behavior is depression. It is known as "the common cold of mental health problems." Students should be aware that it is not unusual for a person to be saddened by divorce, the death of a loved one, or other major losses. But when the sadness does not go away, it becomes what authorities call "clinical depression." More than 15 million Americans suffer clinical depression at any given moment. The dark cloud of depression has an even darker lining—suicide. Suicide is now a leading cause of death among teens and young adults. Fortunately, depression is highly treatable when diagnosed accurately and attacked with some combination of antidepressant medication and the many nondrug therapies. John McIntyre, M.D., a past president of the American Psychiatric Association, says, "There is virtually no one who cannot be helped."

Teaching TIP

Ethology

The study of behavior is known as *ethology*, from the Greek word *ethos*, meaning "character" or "custom."

Overcoming Misconceptions

Anthropomorphism

Attributing a human trait to a nonhuman is known as anthropomorphism. People often attribute human emotions to animals. For example, someone might say that a mother bird is sad because one of her chicks fell out of the nest. While the female bird may engage in behavior that shows it understands and is concerned about the situation, the bird probably does not feel sadness as humans do.

Chapter 37

3 Assessment Options

Closure

Types of Behavior

Have students review the types of behavior shown in Figure 37-9. Then have them locate every example of animal behavior mentioned in the chapter. If a behavior fits in one of the categories listed in the table, have them place it there as an example. *(For example, Defensive behavior—Musk oxen form a defensive circle around their young.)* If students identify a behavior that does not fit into a category, have them make a new category for it. Students should justify the need for any category they create. PORTFOLIO

Review

Assign Review—Section 37-2.

Reteaching — BASIC

Have students go through Section 37-2 and turn each subhead into a question on a sheet of paper. Then ask them to write an answer to each question. Ask volunteers to share their questions and answers with the class. PORTFOLIO

Figure 37-13 Gorillas

Size On average, male gorillas are 50 percent larger than females. Full grown males can weigh as much as 200 kg (440 lb).

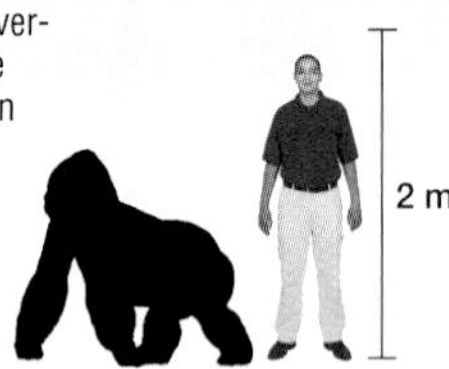

Darwin recognized that extreme traits, such as the male widowbird's tail, could have evolved if they helped males attract or acquire mates. He proposed the mechanism of sexual selection to account for them. **Sexual selection** is an evolutionary mechanism in which traits that increase the ability of individuals to attract or acquire mates appear with increased frequency. Even traits that have a negative effect on survival can evolve in this way, provided that their benefits to reproduction are high enough.

Typically females are the sex that select mates based on the physical traits of the mates, as their parental investment is usually far greater. One explanation for this is that eggs are much larger than sperm. (In humans, they are about 195,000 times larger!) In mammals, females are responsible for gestation and milk production, which are costly reproductive functions. Males may show mate choice as well if their parent involvement is high. For example, this happens in crickets where 30 percent of the male's body weight (the sperm packet he deposits) contributes nutrition for the female and helps her develop eggs.

What kinds of traits are advantageous for acquiring mates? Because male animals usually compete among themselves for the chance to mate with a female, sexual selection has often favored traits that make males more intimidating or better at combat. For example, in many animal species, such as the gorillas shown in **Figure 37-13,** the male is much larger than the female, and the largest male will have the most opportunities to mate. Other examples of extreme traits in males are antlers in deer and moose, horns in bighorn sheep, and manes in lions. The extreme traits found in some male animals, such as size and tusks in walrus, often permit males to size each other up. Males that are not physically matched rarely engage in serious fights. Thus the trait actually reduces conflict among males rather than increasing it.

Competition among males can also take subtle forms. In some species, a male can gain a reproductive advantage over other males by interfering with their reproduction. For instance, in some species of worms, butterflies, and snakes, the male seals the female's reproductive tract after mating so that no other males can mate with her.

Review SECTION 37-2

1. **Describe** the function of six different animal behaviors.
2. **Summarize** in words and with examples the ways in which animals use signals.
3. **Discuss** how the mechanism of sexual selection can account for the extreme traits found in the males of some species.

Critical Thinking

4. **SKILL Evaluating Conclusions** A student concludes that humans who feed ducks at a pond are just as conditioned to this behavior as are the ducks that come to feed. Evaluate the validity of the student's conclusion.

Review SECTION 37-2 ANSWERS

1. parental care—ensures the survival of offspring; courtship—attracts mate; defensive—predator protection; foraging—food location; migratory—movement to better environment; territorial—protects a resource
2. for example: sound—dog barking indicating alarm; posture—person with hands on hips indicating irritation; movement—threatening display indicating impending attack; color—bright plumage indicating quality of genes; scent—dog marking territory; facial expression—frowning indicating displeasure
3. The most intimidating male (for example, the deer with the largest antlers) has a better chance of becoming the dominant male. He will mate with more females and pass on more of his genes than subordinate males will.
4. It is a valid conclusion. The ducks have become conditioned to approach (a response) when a human nears the pond (a stimulus); humans are conditioned to feed the ducks (a response) when ducks swim over to them (a stimulus).

Study ZONE CHAPTER 37

Use the Key Concepts and Key Terms listed below to review the main ideas in this chapter. If you need to review the meaning of a term, turn to the page indicated.

Key Concepts

37-1 Evolution of Behavior

- A behavior is an action or a series of actions performed in response to a stimulus.
- "How" questions are about how a behavior is triggered, and "why" questions concern the reason the behavior exists.
- Natural selection favors behavioral traits that increase the likelihood of an individual's survival and reproduction.
- Genetically programmed behaviors are called innate behaviors, or instinct, and there is little or no variation in how they are performed.
- Learning is the modification of behavior by experience. Learning may occur by association to an unrelated stimulus (classical conditioning) or by trial-and-error (one type is operant conditioning).
- Reasoning is the ability to think out the possible solution to a problem.
- Many behaviors, especially complex behaviors, have both genetic and learned aspects.
- Learning determines the final shape of many genetically based behaviors, such as imprinting.

37-2 Types of Behavior

- Animal behaviors fall into several broad categories, which include: parental care, courtship behavior, defensive behavior, foraging behavior, migratory behavior, and territorial behavior.
- Most animals use signals, often vocal or visual, to communicate to one another.
- Primates are unique among animals in using symbols to communicate.
- The human ability to learn language rapidly during childhood seems to be genetically programmed.
- According to the mechanism of sexual selection, traits that increase the ability of an individual to attract a mate appear with increasing frequency, driving evolution.

Key Terms

37-1

behavior (834)
innate behavior (836)
fixed action pattern behavior (836)
learning (837)
conditioning (837)
reasoning (838)
imprinting (839)

37-2

sexual selection (846)

Study TIP Ready?

If you think you understand these Key Concepts and Key Terms, then you're ready for the Performance Zone.

Study ZONE CHAPTER 37

Review and Practice

Assign the following worksheets for Chapter 37:

- **Student Review Guide**
- **Concept Mapping Worksheet**
- **Test Preparation Pretest**
- **Vocabulary Worksheet**
- **Critical Thinking Review Worksheet**

Assessment Options

The following assessment options are available for this chapter:

Resource Link

Assign **Chapter Test 37** to assess the students' understanding of the chapter.

Alternative Assessment

Insect societies, like those of humans, exhibit complex behavioral patterns. Have students use library and on-line references to obtain information about a behavioral pattern exhibited by a particular social insect. Have them prepare a brief written report and share their findings with the class. PORTFOLIO

Portfolio Assessment

- *Portfolio Assessment* at the front of this book provides options for building and evaluating student portfolios. Students and teachers can select items from the areas listed for inclusion in a portfolio.
- See items labeled PORTFOLIO on pp. 836, 837, 839, 841, 842, 843, 846, and 847.

Answer to Concept Map

The following is one possible answer to Performance Zone item 1 on page 848.

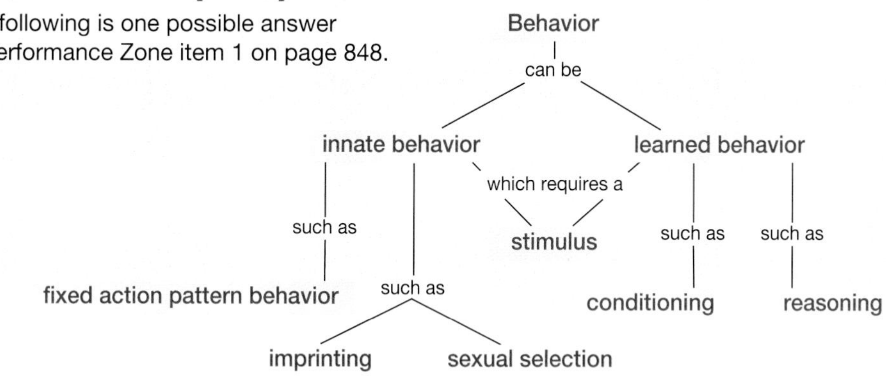

Performance ZONE
CHAPTER 37
REVIEW ANSWERS

Assignment Guide

SECTION	REVIEW
37-1	1, 2, 3, 4, 5, 6, 7, 8, 9, 16, 17, 18, 19, 20, 21, 24
37-2	1, 10, 11, 12, 13, 14, 15, 22, 23, 25

Understanding and Applying Concepts

1. The answer to the concept map is found at the bottom of page 847.
2. **a.** "How" questions are about how a behavior is triggered, controlled, and performed. "Why" questions concern the adaptive value of a behavior.
   b. Innate behavior is genetically programmed, whereas learned behavior comes from experience.
   c. In classical conditioning, two unrelated stimuli are paired, and the response to one becomes associated with the other. In operant conditioning, an apparently random response is linked with a stimulus if the response is rewarded or punished repeatedly.

3. c	**9.** c
4. c	**10.** a
5. c	**11.** d
6. b	**12.** d
7. c	**13.** c
8. a	

14. In some bird species, females prefer males that are more brightly colored or extravagantly plumed. Since the female chooses her mate, males with these traits are more likely to be chosen. As a result, these traits appear with greater frequency in the male population.
15. Satellite tracking saves the expense and time researchers must spend to follow an animal that has a radio collar. The radio collar also has a limited range, and the animal can travel to areas that are inaccessible to people.

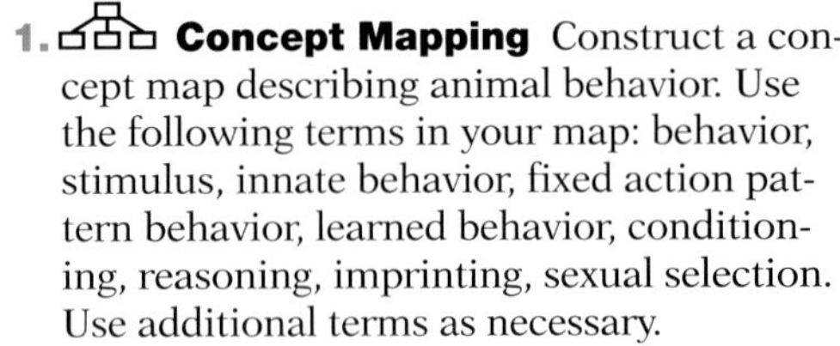

Performance ZONE — CHAPTER 37 Review

Understanding and Applying Concepts

1. **Concept Mapping** Construct a concept map describing animal behavior. Use the following terms in your map: behavior, stimulus, innate behavior, fixed action pattern behavior, learned behavior, conditioning, reasoning, imprinting, sexual selection. Use additional terms as necessary.

2. **Understanding Vocabulary** For each pair of terms, explain the differences in their meanings.
   a. "how" questions, "why" questions
   b. innate behavior, learned behavior
   c. classical conditioning, operant conditioning

3. A male lion who takes over a pride may kill all the young cubs in order to ensure the survival and reproduction of
   a. the females. c. himself.
   b. the pride. d. his siblings.

4. The orb spider builds her web in exactly the same way every time. This is an example of
   a. abnormal behavior.
   b. learned behavior.
   c. fixed action pattern behavior.
   d. random behavior.

5. The nest building behavior of Fischer's lovebirds is
   a. learned.
   b. a result of operant conditioning.
   c. innate.
   d. gradually learned.

6. Learning by association is called
   a. reasoning. c. imprinting.
   b. conditioning. d. assuming.

7. Which of the following best represents classical conditioning?
   a. the feeding behavior shown by rats in Skinner boxes
   b. a male tungara frog calling to a female
   c. a dog salivating at the sound of a bell
   d. a primate giving a warning signal to members of its troop

8. The ability to analyze a problem and think of a possible solution is called
   a. reasoning. c. imprinting.
   b. conditioning. d. assuming.

9. Learning that can only occur during a specific period early in an animal's life is called
   a. reasoning. c. imprinting.
   b. conditioning. d. assuming.

10. Parental care is performed in order to
    a. ensure survival of young.
    b. protect parents from predators.
    c. protect a resource for exclusive use.
    d. locate, obtain, and consume food.

11. The purpose of foraging behavior is to
    a. ensure survival of young.
    b. protect individuals from predators.
    c. protect a resource for exclusive use.
    d. locate, obtain, and consume food.

12. Which of the following behaviors is least likely to be associated with food resources?
    a. territorial c. foraging
    b. migratory d. defensive

13. Which of the following *best* demonstrates the mechanism of sexual selection?
    a. a female tungara frog listening carefully to the call of a potential mate
    b. the flash pattern found in fireflies
    c. the long tail of a male widowbird
    d. the killing of cubs by male lions

14. **Theme Evolution** In many bird species, the male is far more brightly colored and extravagantly plumed than the female. Explain this in terms of sexual selection and evolution.

15. **Tech Watch** What advantages does satellite tracking of animals provide over other methods, such as radio collars?

16. **Chapter Links** Summarize the mechanism of natural selection, and explain why behaviors are just as important to survival and reproduction as physical features are. (**Hint:** See Chapter 13, Section 13-1.)

---

16. In natural selection, traits that help an animal survive to reproductive age tend to be passed on to the next generation. Deleterious traits will be selected against. Like physical traits, many behaviors are important for survival and reproduction. For example, it is just as important to survival for an animal to locate, obtain, and consume food (foraging behavior) as it is for it to be able to run swiftly from predators. Animals that are poor at foraging may not live to reproduce.

Skills Assessment

17. The dog in photo A is in play-bow position, a signal form of communication. Signals are intended to reach the receiver efficiently and stimulate a response. In photo B, the goslings have imprinted on the person. Imprinting is an innate behavior. What the goslings imprint on, however, is learned. In photo C, the parrot is exhibiting a learned behavior. The trainer most likely used conditioning to teach the parrot.

Skills Assessment

Part 1: Interpreting Graphics

The photos below show four different types of behavior. Study the photos, and answer the following questions:

A

B

C

D

17. Identify and discuss the origins of the behavior shown in photos *A, B,* and *C* above.

18. How does photo *D* above support the assertion that natural selection favors individuals, not species?

Part 2: Life/Work Skills

Use the media center or Internet resources to learn more about imprinting.

19. Finding Information Imprinting does not necessarily involve the recognition of a mother. Young salmon, for example, imprint on the smell of the stream in which they hatch. Work with a small group to find examples of imprinting, and develop a class presentation on the subject.

20. Communicating Before you start, have a brainstorming session to decide how you will begin your work. Will everyone be involved in all phases of your project? Or will you delegate individual jobs such as research, writing, and creating visual aids?

Critical Thinking Skills

21. Evaluating an Argument "A child's behavior closely resembles the behavior of its parents. Therefore, most human behaviors are genetically controlled." Explain the logical flaw in this argument.

22. Forming Reasoned Opinions A scientist is studying sexual selection in widowbirds. The scientist captures several male widowbirds, cuts off part of their long tail, and then releases the birds. Data collected later shows that the males with the cut tails mate with half as many females as the males with the longer tails. The scientist concludes that female widowbirds prefer males with longer tails. What is an alternate explanation for this observation? How can the scientist alter the experiment to determine which explanation is correct?

Portfolio Projects

23. Interpreting Information Read "The Importance of Getting Clean," by Doug Stewart (*International Wildlife,* March/April 1998, pp. 12–21). What is COBS? What functions, besides cleanliness, are served by an animal grooming itself? List some other examples the article mentions of grooming signals in animals. How might grooming behavior be related to speech?

24. Summarizing Information At times, some species of whales seem to deliberately swim into waters that are shallow. Often they end up stranded on the beach, where they cannot survive. Using the Internet or other resources, search for information you can use to answer the "how" and "why" questions about this phenomenon. Present your findings in a report to your class.

25. Career Focus Veterinary Technician Research the field of veterinary technology, and write a report on your findings. Your report should include a job description, training required, kinds of employers, growth prospects, and salary ranges.

18. By killing male cubs that are not his own, the male lion decreases the chance that the species will survive, since he is eliminating viable members of the species. However, his actions make the females more likely to mate with him, so he increases the chances that his genes will be passed on to the next generation.

19. Answers will vary. Another example of imprinting is that shown by goats and sheep. The young imprint on their mother during the first few hours after birth. This type of imprinting is also based on smell.

20. Answers will vary. All students in the group should make a significant contribution to the presentation.

Critical Thinking

21. The flaw in this argument is that the variable of learning ("nurture") has not been eliminated and therefore cannot be excluded as a factor in the child's behavior.

22. Capture and handling may have affected the behavior of the widowbirds. The scientist should include a control group of captured and manipulated males whose tails are not cut off or are cut off and then reattached.

Portfolio Projects

23. *COBS* stands for "care of the body surface." Grooming is also used for signaling, courtship, coalition building, and appeasement. Examples of grooming signals will vary. One example is seen in mountain rams: the defeated ram will groom the victor. Just as humans use language to maintain friendship networks, animals such as apes often use grooming to maintain friendships.

24. Answers will vary. There are several ideas that may explain the "how" of this phenomenon. For example, the whales may follow prey into shallow waters. One possible "why" answer is that whales that do this (without becoming stranded) are more likely to find enough food to survive and reproduce.

25. Answers will vary. Veterinary technicians are knowledgeable about animal life processes, and they assist in the care and handling of animal patients. They play an important role in diagnostic, medical, and surgical procedures. The career requires at least two years of college leading to an Associate of Applied Science degree. Many veterinary technicians work in private veterinary practices, but others work in zoos or other wildlife facilities. The demand for veterinary technicians outside private practices is growing rapidly. Starting salary will vary by region.

Lab Techniques

Chapter 37

Studying Nonverbal Communication

Purpose

Students will observe and analyze how stance changes during conversations between pairs of people who are standing.

Preparation Notes

Time Required 45 minutes of classroom time; 30 minutes of time outside of class

Preparation Tips

- Form groups of two or three students. It is important that members of each group have time together outside of class to conduct their observations.
- Each observation will require the presence of at least two members of the group.
- Students will need a stopwatch or a clock with a second hand, paper, and a pencil.

Procedural Tips

- Caution students to observe people in a safe setting and in an unobtrusive way that cannot be construed as offensive.
- Behavior changes quickly. Suggest that at least one student of the pair or group observe the behavior while another student records observations.

Before You Begin

Answers

1. *posture*—body position
   stance—position of the body while standing
   equal stance—stance in which the body weight is supported equally by both legs
   unequal stance—stance in which more of the body weight is supported by one leg than by the other
2. Data tables should be similar to the sample on page 850.
3. Answers will vary. For example: What effects do conflicting postures have on an observer?

Lab Techniques

Chapter 37

Studying Nonverbal Communication

SKILLS

- Observing
- Analyzing
- Graphing

OBJECTIVES

- **Recognize** that posture is a type of nonverbal communication.
- **Observe** how human posture changes during a conversation.
- **Determine** the relationship of gender to the postural changes that occur during a conversation.

MATERIALS

- stopwatch or clock with a second hand
- paper
- pencil

Before You Begin

People communicate nonverbally with their **posture,** or body position. The position of the body while standing is called the **stance.** In an **equal stance,** the body weight is supported equally by both legs. In an **unequal stance,** more weight is supported by one leg than by the other. In this lab, you will observe and analyze how stance changes during conversations between pairs of people who are standing.

1. Write a definition for each boldface term in the paragraph above.
2. Make a data table similar to the one below. The sample data entered in row one shows how to enter data. Do not copy it.

DATA TABLE

	Gender		15-s intervals		
Pairs	Involved	Observed	15 s	30 s	45 s
1	F, M	M	U, W	E	E
2					
3					

3. Based on the objectives for this lab, write a question you would like to explore about nonverbal communication.

Procedure

PART A: Observing Behavior

1. Work in a group of two or three to observe conversations between pairs of people. Each conversation must last between 45 seconds and 5 minutes. One person in your group should be the timekeeper and the other group members should record data. **Note:** Be sure that your subjects are unaware of being observed.
2. Observe at least three conversations. Record the genders of the two participants in each conversation and the gender of the one person whose posture you observe. **Note:** Be sure that the timekeeper accurately clocks the passage of each 15-second interval.
3. For each 15-second interval, record all of the changes in stance by the person you are observing. For example, note every time your subject shifts from an equal stance to an unequal stance, or vice versa. To record the stance simply, you may write *E* to identify an equal stance or *U* to identify an unequal stance.

Procedure

Answers

Observations in steps 2–6 will vary. Check that students have correctly entered all observations in their data table.

7. a. Answers will vary. Students may find that the most common stance is with equal weight on each leg. Check that the bar graphs reflect the combined class data.
7. b. Answers will vary. Students may find that the number of weight shifts increased toward the end intervals. Check that the bar graphs reflect the combined class data.
8. Answers will vary. Check that students have correctly analyzed the combined class data by gender. The bar graphs should reflect the analysis by gender.
9. Answers will vary. Check that students have correctly compiled the data. The bar graphs should reflect the compiled data.

4. If the subject assumes an unequal stance, also record the number of weight shifts from one foot to the other. Indicate a weight shift simply by writing *W*.
5. When a conversation ends, write down whether the pair departed together or separately. To record this, write *T* to indicate departing together or *S* to indicate departing separately.
6. After you have completed each observation, tally the total number of weight shifts within each 15-second block. **IMPORTANT!** Retain data only for conversations that last at least 45 seconds. If a conversation ends before you have collected data for 45 seconds, observe another conversation.
7. After all observations have been completed, combine the data from all of the groups in your class. Analyze the data, without regard to gender.
   a. Determine the most common stance during the first 15 seconds of a conversation, the middle intervals, and the last 15 seconds. Make a bar graph to summarize the class data.
   b. Find the average number of weight shifts for the beginning, middle, and end intervals. Make a bar graph to summarize the class data.
8. Repeat step 7, but analyze the data according to gender this time.
9. Compile the data and make bar graphs for each of the following: males talking with a male, males talking with a female, females talking with a male, and females talking with a female. Compare these graphs with the ones you made for Step 7.

Part B: Cleanup and Disposal

10. Clean up your work area. Return lab equipment to its proper place. Wash your hands thoroughly before you leave the lab and after you finish all work.

Analyze and Conclude

1. **Analyzing Results** Which is the most common stance that was used during a conversation?
2. **Recognizing Relationships** Which behavior most often signals that a conversation is about to end: stance change or weight shift?
3. **Drawing Conclusions** Do males and females differ in their departure signals or are their departure signals the same? Justify your conclusion.
4. **Forming Hypotheses** What do you think might be an adaptive significance of a departure signal?
5. **Forming Reasoned Opinions** What other observed behaviors were forms of nonverbal communication? Justify your answer.
6. **Further Inquiry** Write a new question about animal behavior that could be explored with a new investigation.

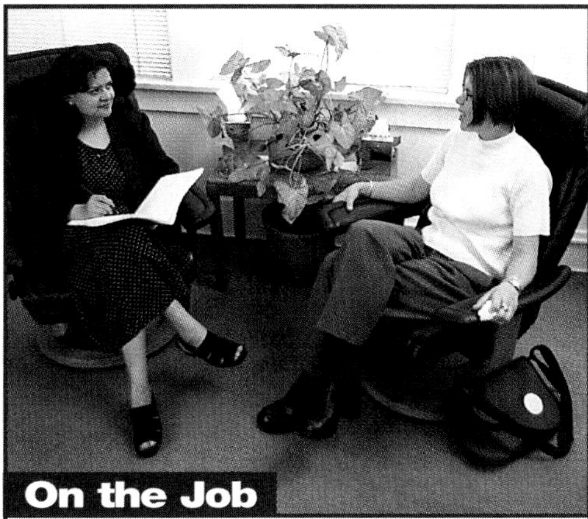

On the Job

Psychology is the study of human and animal behavior. Do research in the library or media center or on the Internet to discover how psychology is used to treat behavioral disorders in humans or in pets. Summarize your findings in a written or oral report.

Lab Techniques

37 Chapter

Analyze and Conclude

Answers

1. Answers will vary. Students will most likely find that the most common stance is with equal weight on each leg.
2. Answers will vary. Students should support their answers by citing their data.
3. While researchers have not found significant differences between the sexes in these signals, your students may see some trends due to the small sample size. This could lead to an interesting discussion about statistics.
4. Answers will vary, but students may suggest that nonverbal cues may make the departure expected and less abrupt. This may ensure that the person leaving does not insult the other person.
5. Answers will vary, but students may mention facial expressions and head, hand, arm, or eye movements.
6. Answers will vary. For example: What postural changes occur when two dogs greet each other?

On the Job

Answers

Answers will vary. Behavior modification techniques commonly used by psychologists include systematic desensitization, in which an individual is trained to relax in the presence of fear-producing stimuli; aversion therapy, in which an aversive stimulus is paired with an undesirable behavior; and biofeedback, in which individuals are trained to monitor their own physiological processes in cases in which a behavioral disorder may have a physiological basis.

UNIT 9 Exploring Human Biology

Chapters

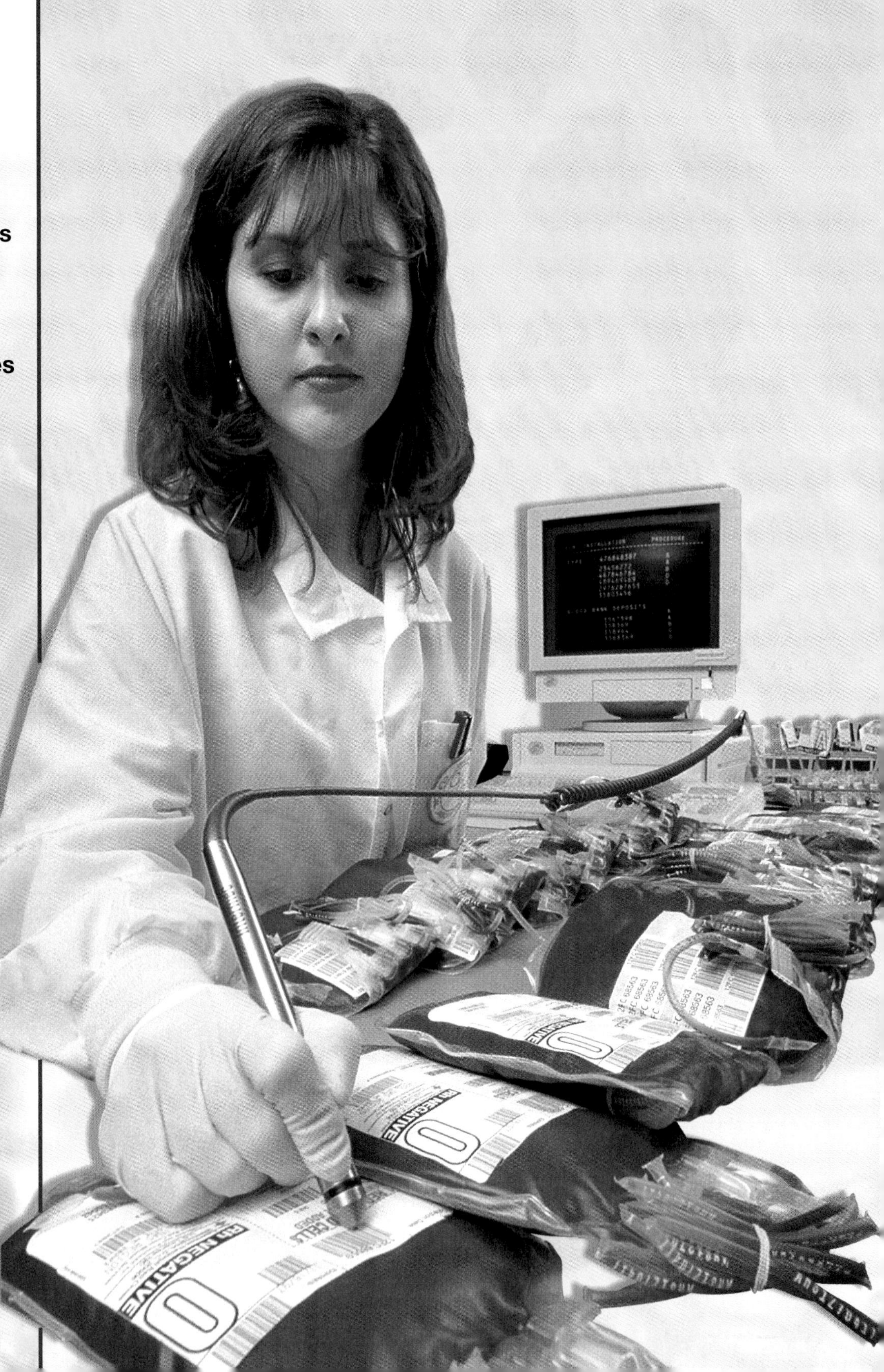

Donated blood saves thousands of lives every year.

in perspective

Circulation

Colleagues of William Harvey, a 17th century English anatomist, considered his conclusion that blood flows one way in a continuous system of vessels to be a strange idea. **See page 884 to learn how valves prevent blood from flowing backwards.** In the 1940s, Dr. Charles Drew, the first African-American to earn an M.D. at Columbia University, developed commercial procedures for the safe transfusion of plasma, the liquid component of blood.

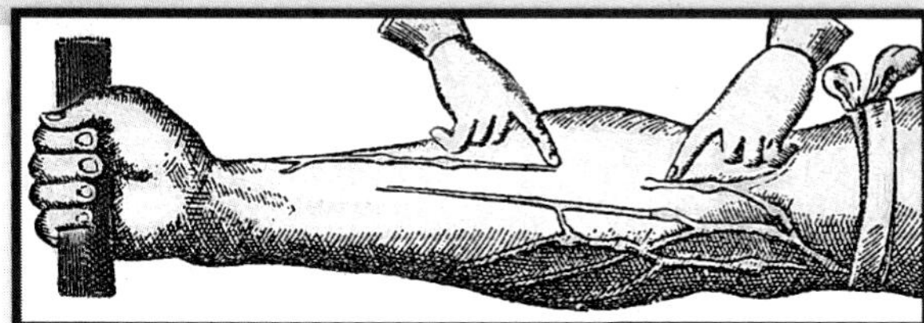

William Harvey's theory of blood circulation

Dr. Charles Drew

Scientists know that lifestyle choices—for example, what we eat and how much we exercise—affect the health of our circulatory system. Eating a healthful diet and exercising regularly will reduce the likelihood of hypertension (commonly called high blood pressure) and lower the risk of heart attack. **What are the dangers of hypertension? See page 893.**

Moderate exercise

Researchers are currently exploring artificial substitutes for red blood cells. **What are the components of blood? Find out on pp. 886–889.**

Artificial blood research

internet**connect**

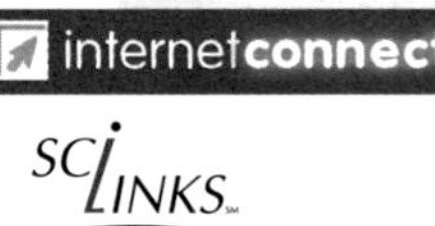

TOPIC: Blood
GO TO: www.scilinks.org
KEYWORD: HX853

Chapter 38

Introduction to Body Structure

Planning Guide

	OBJECTIVES	MEETING INDIVIDUAL NEEDS	READING SKILLS AND STRATEGIES
BLOCK 1 45 min	**38-1 Body Organization** • Identify four levels of structural organization within the human body. • Analyze the four kinds of body tissues. • List the body's major organ systems. • Evaluate the importance of endothermy in maintaining homeostasis.	Learning Styles, p. 856 BASIC ADVANCED • p. 859	**ATE** Active Reading, Technique: K-W-L, p. 855 ■ Active Reading Guide, worksheet (38-1) ■ Directed Reading, worksheet (38-1) ■ Reading Strategies, K-W-L worksheet
BLOCKS 2 & 3 90 min	**38-2 Skeletal System** • Distinguish between the axial skeleton and the appendicular skeleton. • Analyze the structure of bone. • Summarize the process of bone development. • List two ways to prevent osteoporosis. • Identify the three main classes of joints.	Learning Styles, p. 860 BASIC ADVANCED • pp. 861, 863, 864, 865	**ATE** Active Reading, Technique: Paired Reading, p. 862 ■ Active Reading Guide, worksheet (38-2) ■ Directed Reading, worksheet (38-2)
BLOCKS 4 & 5 90 min	**38-3 Muscular System** • Describe the action of muscle pairs in moving the body. • Relate the structure of a skeletal muscle to the muscle's ability to contract. • Describe how energy is supplied to muscles for contraction.	Learning Styles, p. 866 BASIC ADVANCED • pp. 867, 868, 870	■ Active Reading Guide, worksheet (38-3) ■ Directed Reading, worksheet (38-3)
BLOCKS 6 & 7 90 min	**38-4 Skin, Hair, and Nails** • Analyze the structure and function of the epidermis. • Describe how the dermis helps the body maintain homeostasis. • Summarize how hair and nails are formed. • Identify various types of skin disorders.	Learning Styles, p. 871 BASIC ADVANCED • p. 874	■ Active Reading Guide, worksheet (38-4) ■ Directed Reading, worksheet (38-4) ■ Supplemental Reading Guide, *Microbe Hunters*

Review and Assessment Resources

BLOCKS 8 & 9 90 min

TEST PREP

- **PE** Study Zone, p. 875
- **PE** Performance Zone, pp. 876–877
- ■ Chapter 38 worksheets:
  - Vocabulary
  - Concept Mapping
  - Science Skills
  - Critical Thinking
- Audio CD Program
- ■ Test Prep Pretest

CHAPTER TESTING

- ■ Chapter Tests (blackline master)
- ▲ Test Generator
- ■ Assessment Item Listing

See the correlations table at the front of this book for a list of the ***National Science Education Standards*** addressed in this chapter.

Smithsonian Institution® Visit **www.si.edu/hrw** for additional on-line resources.

KEY

Biology Principles and Explorations

PE Pupil's Edition
ATE Teacher's Edition

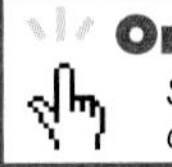

HOLT BIOSOURCES

Teaching Transparencies
Laboratory Program

One-Stop Planner CD-ROM
Shows these resources within a customizable daily lesson plan:

■ Teaching Resources
▲ Test Generator

VISUAL STRATEGIES	CLASSWORK AND HOMEWORK	TECHNOLOGY AND INTERNET RESOURCES
187 Inside the Human Coelom 193 Three Types of Muscle Tissue	**PE** **Quick Lab** Under what conditions can muscle activity continue?, p. 857 **PE** **Review,** p. 859	Holt Biology Videodiscs Section 38-1 Audio CD Program Section 38-1 internetconnect SCILINKS NSTA TOPIC: Organ systems
184 Structure of the Human Knee 185 Human Skeleton 186 Structure of Bone Tissue 198 Types of Joints	**Lab A27** Comparing Skeletal Joints ■ **Basic Skills:** Mass and Density ■ **Occupational Applications:** Physical Therapist **PE** **Review,** p. 865	Holt Biology Videodiscs Section 38-2 Audio CD Program Section 38-2 internetconnect SCILINKS NSTA TOPIC: Joint disorders
ATE **Fig. 38-11,** p. 867 **ATE** **Fig. 38-12,** p. 868 **ATE** **Fig. 38-13,** p. 869 190 Structure of Skeletal Muscle 191 Three Types of Skeletal Tissue 192 Contraction of a Muscle	**Lab C41** Evaluating Muscle Exhaustion **PE** **Review,** p. 870	Holt Biology Videodiscs Section 38-3 Audio CD Program Section 38-3 internetconnect SCILINKS NSTA TOPICS: • Muscle structure • Muscle contraction
188 Cross Section of Skin	**PE** **Interactive Exploration** Lifespan and Lifestyle pp. 878–879 **PE** **Review,** p. 874 **PE** **Study Zone,** p. 875	CNN Video Segment 30 Tanning Effects **Interactive Exploration:** Human Biology, Life Span and Lifestyle, pp. 878–879 Holt Biology Videodiscs Section 38-4 Audio CD Program Section 38-4 internetconnect SCILINKS NSTA TOPICS: • Forensic analysis • Skin cancer • Life span

ALTERNATIVE ASSESSMENT

PE **Portfolio Projects** 24–27, p. 877
ATE **Alternative Assessment,** p. 875
ATE **Portfolio Assessment,** p. 875

Lab Activity Materials

Quick Lab
p. 857 clock or watch with a second hand

Interactive Exploration
p. 878 CD-ROM, computer

Demonstrations
ATE cooked chicken bones (2), vinegar, p. 863
ATE plastic wrap, blocks of wood (2), tape, p. 865

BIOSOURCES LAB PROGRAM
- Quick Labs: Lab A27, p. 53
- Lab Techniques and Experimental Design: Lab C41, p. 197

Chapter 38

Chapter Theme

Interdependence

The human body is composed of over 100 trillion cells. These cells do not function independently. Rather, they require interactions with other cells. The organization of the body allows for an efficient interdependence of all cells. This interdependence extends to the tissues, organs, and organ systems, which together make up the human body.

Establishing Prior Knowledge

Before beginning this chapter, make certain students can answer the questions in **Study TIP Ready?** on page 855. If they cannot, encourage them to reread the sections indicated.

Demonstration

Have a volunteer perform a simple task, such as opening a jar, tuning a radio, or eating a cracker. Ask what systems are being used to accomplish each task. *(nervous system, motor system)*

CHAPTER 38 Introduction to Body Structure

Hurdler

What's Ahead — and Why

CHAPTER OUTLINE

38-1 Body Organization
- The Body Has Four Levels of Structural Organization
- Tissues Are Grouped into Organs

38-2 Skeletal System
- The Skeleton Has Two Main Parts
- Bones Grow Through Adolescence
- Joints Fasten Bones Together

38-3 Muscular System
- Muscles Move the Body
- The Sarcomere Is the Functional Unit of Muscle Contraction

38-4 Skin, Hair, and Nails
- Skin Has Two Distinct Layers
- Skin Disorders Often Result from Damage to the Epidermis

Students will learn

- about the internal organization of the body from cells to organ systems.
- how bones grow and work with each other to move the body.
- about the mechanics of muscle movement.
- about the structure of the skin and common skin disorders.

Resistance exercise

Energy... Control... Coordination...

Doing pull-ups requires coordination, strength, and endurance to overcome the resistance provided by the body's weight. An activity as complicated as jumping hurdles while running full speed requires not only a tremendous amount of energy but also the coordination of many body systems.

Study TIP

Ready?

Check your understanding of these topics to help you prepare for what's ahead.

Can you...

- **describe** the role of ATP in metabolism? (Section 5-1)
- **summarize** the two stages of cellular respiration? (Section 5-3)
- **distinguish** between tissues, organs, and organ systems? (Section 20-3)

If not, *review the sections indicated.*

Looking Ahead

internet connect

National Science Teachers Association *sci*LINKS Internet resources are located throughout this chapter.

Reading Strategies

Active Reading

Technique: K-W-L

Before students begin this chapter, divide the class into three groups. Have one group list all the things they Know about bones; have the second group list all the things they Know about muscles; have the third group list all the things they Know about skin. Ask them to write their lists on the board or the overhead. Then have students list things they Want to know about bones, muscles, and skin. Have them save their lists for use in **Closure** on page 874.

Directed Reading

Assign **Directed Reading Worksheet Chapter 38** to help students interact with the text as they read the chapter.

Interactive Reading

Assign ***Biology: Principles and Explorations* Audio CD Program Chapter 38** to help reluctant readers achieve greater success in reading the chapter.

Reading Skills

Assign Reading Organizers and Reading Strategies Worksheets from the **One-Stop Planner CD-ROM with Test Generator** to help students learn and practice effective reading skills.

internet connect

Holt *Biology: Principles and Explorations*

On-line Resources: go.hrw.com

Visit the HRW Web site for a variety of resources related to this chapter. Just type in the keyword HX0 Chapter 38.

Holt *Biology: Principles and Explorations*

On-line Resources: www.scilinks.org

The following *sci*LINKS Internet resources can be found in the student text for this chapter:

Page 858
TOPIC: Organ systems
KEYWORD: HX858

Page 865
TOPIC: Joint disorders
KEYWORD: HX865

Page 867
TOPIC: Muscle structure
KEYWORD: HX867

Page 869
TOPIC: Muscle contraction
KEYWORD: HX869

Page 872
TOPIC: Forensic analysis
KEYWORD: HX872

Page 874
TOPIC: Skin cancer
KEYWORD: HX874

Page 879
TOPIC: Life span
KEYWORD: HX879

1 Prepare

Scheduling

- **Block:** About 45 minutes will be needed to complete this section.
- **Traditional:** About one class period will be needed to complete this section.
- **Planning:** See the Planning Guide on pp. 853A and 853B for lecture, classwork, and assignment options for Section 38-1. Lesson plans for Section 38-1 can be found in a lesson cycle format in ***Biology: Principles and Explorations*** **One-Stop Planner CD-ROM with Test Generator.**

Resource Link

Assign **Active Reading Worksheet Section 38-1** to help students comprehend the key concepts and visuals in this lesson.

2 Teach

Lesson Warm-up

Ask students if they have ever had a black eye. Ask them if they know what caused it. Have them read the information in the **BioBit** on page 857. Remind students that blood is a tissue.

Section 38-1 Body Organization

Objectives

- **Identify** four levels of structural organization within the human body. (p. 856)
- **Analyze** the four kinds of body tissues. (pp. 856–857)
- **List** the body's major organ systems. (p. 858)
- **Evaluate** the importance of endothermy in maintaining homeostasis. (p. 859)

Key Terms

epithelial tissue
nervous tissue
connective tissue
muscle tissue
body cavity

The Body Has Four Levels of Structural Organization

How do your body's cells work together? The human body contains more than 100 trillion cells and more than 100 kinds of cells. The body is structurally organized into four levels: cells, tissues, organs, and organ systems. Recall that a tissue is a group of similar cells that work together to perform a common function. The cell types of the body are grouped by function into four basic kinds of tissues: epithelial, nervous, connective, and muscle tissues. These tissues, shown in **Figure 38-1,** are the building blocks of the human body.

Four kinds of body tissues

Epithelial tissue There are many different kinds of epithelial *(ehp ih THEE lee uhl)* tissue. **Epithelial tissue** lines most body surfaces, and it protects other tissues from dehydration and physical damage. An epithelial layer is usually no more than a few cells thick. These cells are typically flat and thin, and they contain only a small amount of cytoplasm. Epithelial tissue is constantly being replaced as cells die.

Figure 38-1 Body tissues

Cells of the body are grouped into different kinds of tissues.

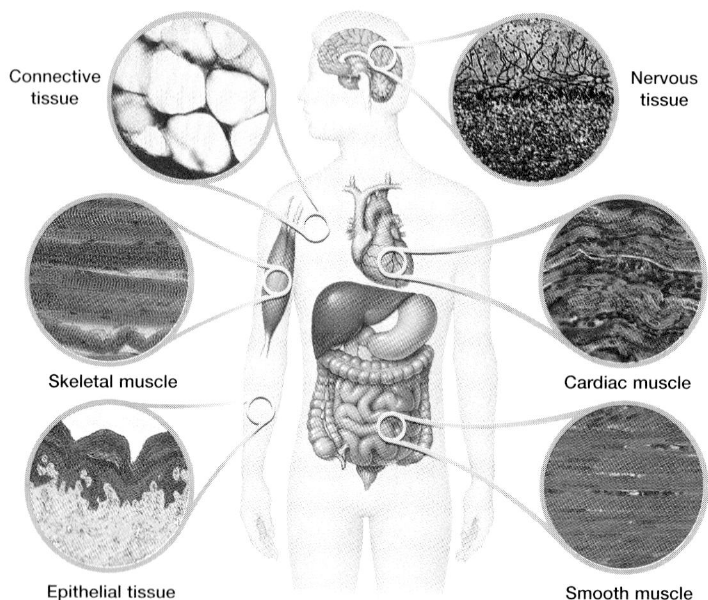

Integrating Different Learning Styles

Logical/Mathematical	**PE**	Quick Lab, p. 857
Visual/Spatial	**PE**	Performance Zone item 1, p. 876
	ATE	Internet Connect, p. 858; Teaching Tip, p. 858
Body/Kinesthetic	**PE**	Quick Lab, p. 857
	ATE	Demonstration, p. 854
Interpersonal	**PE**	Quick Lab, p. 857
Intrapersonal	**ATE**	Lesson Warm-up, p. 856
Verbal/Linguistic	**PE**	Multicultural Perspective, p. 857; Performance Zone items 2 and 26, pp. 876–877
	ATE	Teaching Tip, p. 859; Closure, p. 859

Nervous tissue The nervous system is made of **nervous tissue.** Nervous tissue consists of nerve cells and their supporting cells. Nerve cells carry information throughout the body. You will learn more about the nervous system in Chapter 42.

Connective tissue Various types of **connective tissue** support, protect, and insulate the body. Connective tissue includes fat, cartilage, bone, tendons, and blood. Some connective tissue cells, such as those in bone, are densely packed. Others, such as those found in blood, are farther apart from each other.

Muscle tissue Three kinds of **muscle tissue** enable the movement of body structures by muscle contraction. The three kinds of muscle tissues are skeletal muscle, smooth muscle, and cardiac muscle.

1. **Skeletal muscle.** Skeletal muscle is called voluntary muscle because you can consciously control its contractions. Skeletal muscles move bones in the trunk and limbs.
2. **Smooth muscle.** Smooth muscle is called involuntary muscle because you cannot consciously control its slow, long-lasting contractions. Some smooth muscles, such as those lining the walls of blood vessels, contract only when stimulated by signal molecules. Other smooth muscles contract spontaneously.
3. **Cardiac muscle.** Cardiac muscle is found in the heart. The powerful, rhythmic contractions of cardiac muscle pump blood to all body tissues. Cardiac muscle consists of interconnected cells. Groups of neighboring cells contract all at once, stimulating adjacent groups of cells to contract.

What is a black eye?

A black eye is a bruise around the eye that may also be swollen. The dark color of the bruise is caused by blood oozing from tiny blood vessels that were broken when the eye was injured.

Under what conditions can muscle activity continue?

You can observe how different conditions, such as the resistance exercise shown at right, affect muscle activity.

Materials

clock or watch with a second hand

Procedure

1. Working in groups of four, assign two subjects and two observers. Form two teams of two—Team A and Team B—each consisting of a subject and an observer.
2. If you are a subject, place your feet hip-distance apart. If you are on Team A, do activity (a). If you are on Team B, do activity (b).

   a. Rise repeatedly onto your toes as quickly as possible for 2 minutes. Rest as needed.

   b. Rise repeatedly onto your toes rhythmically and slowly for 2 minutes. Rest as needed.
3. If you are an observer, count the number of toe rises and the number of rest periods completed by your partner. Record your data in a table.

Analysis

1. **Compare** the numbers of toe rises and rest periods of subject A with those of subject B.
2. **Evaluate** the difference in the number of toe rises completed by subjects A and B.
3. **SKILL Applying Concepts** Which energy supply pathway, aerobic or anaerobic, enables muscles to remain active for long periods of time?

Multicultural Perspective

Qi and the Body

In some Asian countries, many people believe that the body has an energy field, called *qi* (chee), which moves through the body through channels called meridians. The function of qi is thought to be similar to the functions of the nervous and circulatory systems—moving information and energy within the body. Acupuncture, an ancient treatment for pain and many other ailments, is based on the concept of qi. Many scientists now believe that acupuncture is effective for pain because it activates the body's endogenous opiates.

Making Biology Relevant

Sports

There are two types of skeletal muscle fibers, fast-twitch and slow-twitch. Fast-twitch muscle fibers work quickly and powerfully for performing tasks such as weight-lifting. Slow-twitch muscle fibers contract slowly, use less energy, and are used for longer activities, such as distance running. Both types of fibers exist in each of the body's skeletal muscles in varying proportions depending on the muscle type.

Under what conditions can muscle activity continue?

Time 10–15 minutes

Process Skills Recognizing relationships, inferring conclusions

Teaching Strategies

If students have difficulty answering the SKILL question, use leading questions such as:

- What does *aerobic* mean? *(with oxygen)*
- What does *anaerobic* mean? *(without oxygen)*
- Which team did aerobic exercise? *(team B)*
- Which team did anaerobic exercise? *(team A)*

Explain to students that upon prolonged contraction, fast-twitch muscle fibers quickly fatigue and become almost completely nonfunctional. This is in contrast to slow-twitch muscle fibers, which can function for long periods of time.

Analysis Answers

1. Subject A probably will be able to rise on the toes fewer times and will require more rest periods.
2. Subject A will be able to rise fewer times because he or she tires quickly.
3. aerobic; Oxygen is needed for efficient energy production.

Chapter 38

internetconnect

SCILINKS NSTA

Ask students to choose an organ system from Table 38-1 on page 858 and research it using the Web site in the Internet Connect box on page 858. Have students draw pictures of the individual organs that make up the system. PORTFOLIO

Teaching TIP

Recognizing Relationships

Ask students to think of body parts that have been transplanted in humans. Have them identify whether the part is a tissue or an organ. *(Their list may include organs such as the heart, kidney, liver, lung, skin, and pancreas, and tissues such as blood, bone marrow, cornea, and bone.)* Explain that the screening of cell and tissue type before a transplant is vital to avoid transplant rejection, the destruction of transplanted tissue by the recipient's immune system. Students will better understand this concept when they study the immune system. Explain that screening also checks for infectious agents, such as viruses and bacteria.

Teaching TIP

Organizational Hierarchy

Have students make a Graphic Organizer similar to the one at the bottom of page 858 to demonstrate that simpler structures in the body are components of more complex structures. Ask students to use the terms *cell, tissue, organ, organ system,* and *body.* PORTFOLIO

Tissues Are Grouped into Organs

Body organs are made of combinations of two or more types of tissues. Tissues work together to accomplish a specific function. The heart, for example, contains cardiac muscle tissue and connective tissue, and the heart is stimulated by nervous tissue. Each organ belongs to at least one organ system, which is a group of organs that work together to carry out major activities or processes. The different organs in an organ system interact to perform a certain function, such as digestion. The digestive system is composed of the stomach, intestines, liver, gallbladder, and pancreas. Some organs function in more than one organ system. The pancreas, for example, functions in both the digestive system and the endocrine system. Table 38-1 lists the body's major organ systems.

Table 38-1 Major Organ Systems of the Body

System	Major structures	Functions
Circulatory	Heart, blood vessels, blood, lymph nodes and vessels, lymph	Transports nutrients and wastes
Digestive	Mouth, esophagus, stomach, liver, pancreas, small and large intestines	Extracts and absorbs nutrients from food; removes wastes; maintains water and chemical balances
Endocrine	Hypothalamus, pituitary, endocrine glands	Regulates body temperature, metabolism, development, and reproduction; maintains homeostasis; regulates other organ systems
Excretory	Kidneys, urinary bladder, ureters, urethra	Removes wastes from blood; regulates concentration of body fluids
Immune	White blood cells, lymph nodes and vessels, skin	Defends against pathogens and disease
Integumentary	Skin, nails, hair	Protects against injury, infection, and fluid loss; helps regulate body temperature
Muscular	Skeletal, smooth, and cardiac muscle tissues	Moves limbs and trunk; moves substances through body; provides structure and support
Nervous	Brain, spinal cord, sense organs	Regulates behavior; maintains homeostasis; regulates other organ systems; controls sensory and motor functions
Reproductive	Testes and penis (in males); ovaries and uterus (in females)	Produces gametes and offspring
Respiratory	Lungs, nose, mouth, trachea	Moves air into and out of lungs; controls gas exchange between blood and lungs
Skeletal	Bones and joints	Protects and supports the body and organs; interacts with skeletal muscles

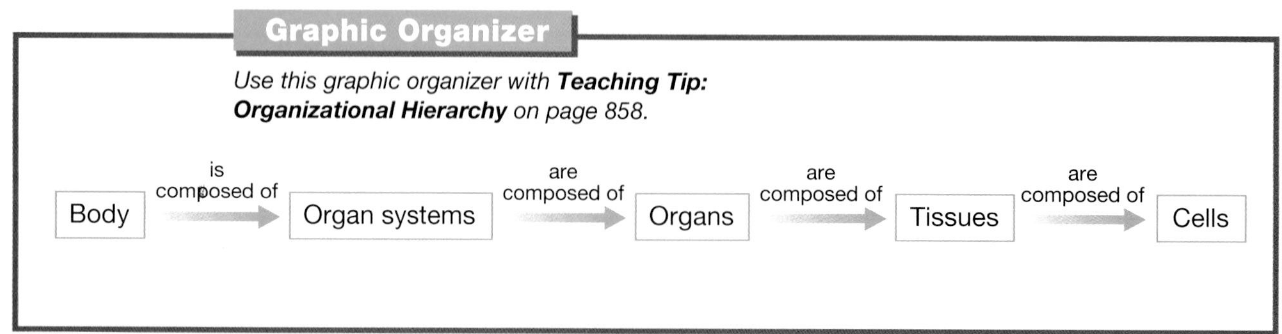

Body cavities

The body contains four large fluid-filled spaces, or **body cavities,** that house and protect the major internal organs. Within the body cavities, shown in **Figure 38-2,** organs are suspended in fluid that supports their weight and prevents them from being deformed by body movements. These organs are also protected by bones and muscles. For example, your heart and lungs are protected by the rib cage and the sternum inside the thoracic *(thoh RAS ik)* cavity. Your brain, encased within the cranial *(KRAY nee uhl)* cavity, is protected by the skull. Your digestive and reproductive organs, located in the abdominal cavity, are protected by the pelvis and abdominal muscles. In addition, your spinal cord is protected by the many vertebrae that make up the spinal cavity.

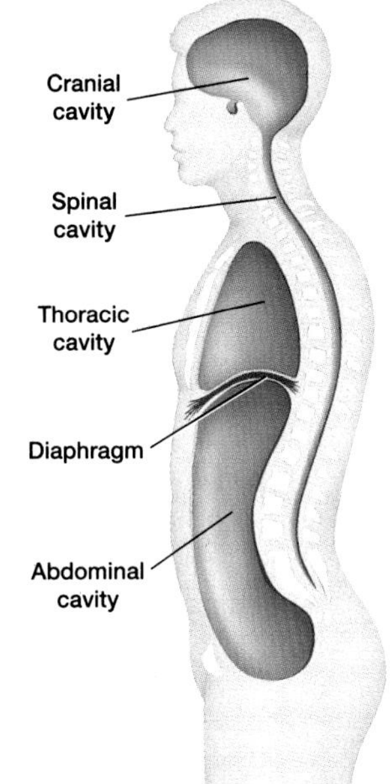

Figure 38-2 Body cavities. Many organs and organ systems are encased in protective body cavities.

Endothermy enables homeostasis

Like all mammals, humans are endothermic. Humans maintain a fairly constant internal temperature of about 38°C (99°F). Your body uses a great deal of energy to maintain a stable internal condition. For example, a large percentage of the energy you consume in food is devoted to maintaining your body temperature. You would not survive very long if your temperature fell much below the normal range. Very high temperatures, such as occur with fever, are also dangerous because they can inactivate critical enzymes. Your body keeps its temperature constant by regulating the flow of blood through blood vessels just under the skin. To release heat to the air, your body increases blood flow to these vessels; to retain heat, it restricts blood flow. Why does your body invest so much energy to maintain a constant temperature? As an endotherm, you can remain active at external temperatures that would slow the activity of nonendothermic organisms. In addition, endothermy enables you to sustain strenuous activity, such as exercise, for a long time.

In order to maintain homeostasis, the body's organ systems must function smoothly together. The nervous system and the endocrine system regulate and monitor other organ systems to make sure stability is maintained throughout the body. In addition to temperature regulation, homeostasis involves adjusting metabolism, detecting and responding to environmental stimuli, and maintaining water and mineral balances. You will learn more about the nervous system and the endocrine system in Chapter 42 and Chapter 43.

Review SECTION 38-1

1. **Summarize** the four levels of structural organization in the body.
2. **List** the four different kinds of body tissues, and give an example of each kind.
3. **Describe** the relationship between organs and organ systems.
4. **SKILL Relating Concepts** How is endothermy advantageous to humans?

Critical Thinking

5. **SKILL Inferring Relationships** Why should fever be controlled during an illness?

Review SECTION 38-1 ANSWERS

1. cell—individual unit; tissue—group of similar cells that work together to perform a common function; organ—combination of two or more types of tissues that work together; organ system—group of organs that work together to perform a specific function
2. epithelial—lines body surfaces; nervous—nerve cells; connective—fat, cartilage, bone, tendon, blood; muscle—skeletal, smooth, cardiac
3. An organ system is composed of two or more organs that work together.
4. Endothermy allows humans to function in many environments and to engage in strenuous activities for long periods.
5. It is important to control a fever because high temperatures can inactivate essential enzymes.

Section 38-1

Teaching TIP — ADVANCED

Homeostasis

Emphasize that maintenance of homeostasis in a multicellular organism is a difficult and complex task. Communication must be maintained among organ systems, organs, tissues, and even individual cells. Multicellular organisms have elaborate signaling systems that are only now being discovered. Many diseases result from the disruption of homeostasis. Have students select an endocrine or metabolic disorder and write a short report about it. Diseases to suggest include diabetes mellitus, goiter, hypo- or hyperthyroidism, Cushing's syndrome, and porphyria. PORTFOLIO

3 Assessment Options

Closure

Interdependence of Organ Systems

Ask volunteers to give examples of ways the systems in the body interact and depend on each other. For instance, the cells of the immune system move through the body in the circulatory system. Similarly, the respiratory system captures oxygen, which is used by all the cells of the body.

Review

Assign Review—Section 38-1.

Reteaching — BASIC

Use a relationship that students are more familiar with to help them learn the levels of organization. For example, a house is part of a city, which is part of a county, which is part of a state, which is part of the United States.

Chapter 38

1 Prepare

Scheduling

- **Block:** About 90 minutes will be needed to complete this section.
- **Traditional:** About two class periods will be needed to complete this section.
- **Planning:** See the Planning Guide on pp. 853A and 853B for lecture, classwork, and assignment options for Selection 38-2. Lesson plans for Section 38-2 can be found in a lesson cycle format in ***Biology: Principles and Explorations*** **One-Stop Planner CD-ROM with Test Generator.**

Resource Link

Assign **Active Reading Worksheet Section 38-2** to help students comprehend the key concepts and visuals in this lesson.

2 Teach

Lesson Warm-up

Ask students what they already know about different types of joints. Tell them the names of the types of joints. Ask them to provide an example of each type and to describe the movement for which each type is adapted.

Section 38-2 Skeletal System

Objectives

- **Distinguish** between the axial skeleton and the appendicular skeleton. (p. 860)
- **Analyze** the structure of bone. (p. 861)
- **Summarize** the process of bone development. (p. 862)
- **List** two ways to prevent osteoporosis. (p. 863)
- **Identify** the three main classes of joints. (p. 864)

Key Terms

axial skeleton
appendicular skeleton
bone marrow
periosteum
Haversian canal
osteocyte
osteoporosis
joint
ligament

The Skeleton Has Two Main Parts

What keeps your body from collapsing like a limp balloon? An internal skeleton of bones shapes and supports your body. This skeleton provides protection for internal organs and, along with muscles, enables movement with a versatile system of levers and joints. Muscles pull against bones at joints, moving the limbs and the trunk. Your skeleton is made mostly of bone, a type of hard connective tissue that is constantly being formed and replaced. The skeleton contains 206 individual bones. Of these, 80 bones form the **axial skeleton,** which includes bones of the skull, spine, ribs, and sternum. The other 126 bones, including those of the arms, legs, pelvis, and shoulder, form the **appendicular** *(ap uhn DIHK yoo luhr)* **skeleton.** The skeleton is shown in **Figure 38-3.**

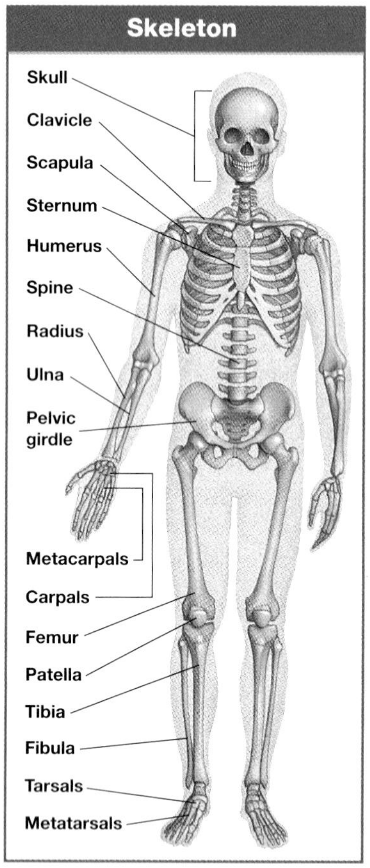

Skeleton

Skull
Clavicle
Scapula
Sternum
Humerus
Spine
Radius
Ulna
Pelvic girdle
Metacarpals
Carpals
Femur
Patella
Tibia
Fibula
Tarsals
Metatarsals

Figure 38-3 Skeleton. Bones of the appendicular skeleton "hang" from bones of the axial skeleton (purple).

Axial skeleton

The most complex part of the axial skeleton is the skull. Of the 29 bones in the skull, eight bones form the cranium, which encases the brain. The skull also contains 14 facial bones, six middle-ear bones, and a single bone that supports the base of the tongue. The skull is attached to the top of the spine, or backbone, which is a flexible, curving column of 26 vertebrae that supports the center of the body. Curving forward from the middle vertebrae are 12 pairs of ribs, which form a protective rib cage around the heart and lungs.

Appendicular skeleton

The appendicular skeleton forms the appendages or limbs—the shoulders, arms, hips, and legs. The arms and legs are attached to the axial skeleton at the shoulders and hips, respectively. The shoulder attachment, called the pectoral girdle, contains two large, flat shoulder blades, or scapulas, and two slender, curved collarbones, or clavicles. The

Integrating Different Learning Styles

Logical/Mathematical	**PE**	Performance Zone items 21 and 23, p. 877
	ATE	Did You Know?, p. 863
Visual/Spatial	**ATE**	Teaching Tip, p. 861; Demonstration, p. 863 Reteaching, p. 865
Body/Kinesthetic	**ATE**	Did You Know?, p. 861; Demonstration, p. 865
Interpersonal	**ATE**	Active Reading, p. 862
Verbal/Linguistic	**PE**	Study Tip, p. 861; Real Life, p. 862; Performance Zone item 2, p. 876
	ATE	Teaching Tip, p. 861; Active Reading, p. 862

clavicles connect the scapulas to the upper region of the sternum and hold the shoulders apart. This arrangement enables full rotation of the arms about the shoulder. The hip attachment, called the pelvic girdle, contains two large pelvic bones. The pelvic bones distribute the weight of the body evenly down the legs.

Structure of bone

As shown in **Figure 38-4,** bones are made of a hard outer covering of compact bone surrounding a porous inner core of spongy bone. Compact bone is a dense connective tissue that provides a great deal of support. Spongy bone is a loosely structured network of separated connective tissue. Some cavities in spongy bone are filled with a soft tissue called **bone marrow.** Red bone marrow begins the production of all blood cells and platelets. The hollow interior of long bones is filled with yellow bone marrow. Yellow bone marrow consists mostly of fat, which stores energy. Bones are surrounded and protected by a tough exterior membrane called the **periosteum** *(pair ee AHS tee uhm).* The periosteum contains many blood vessels that supply nutrients to bones.

Word Origins

The word *periosteum* is from the Greek *peri,* meaning "around," and *osteon,* meaning "bone."

Figure 38-4 **Structure of bone**

Many bones contain bone marrow, blood vessels, and both compact and spongy bone tissue.

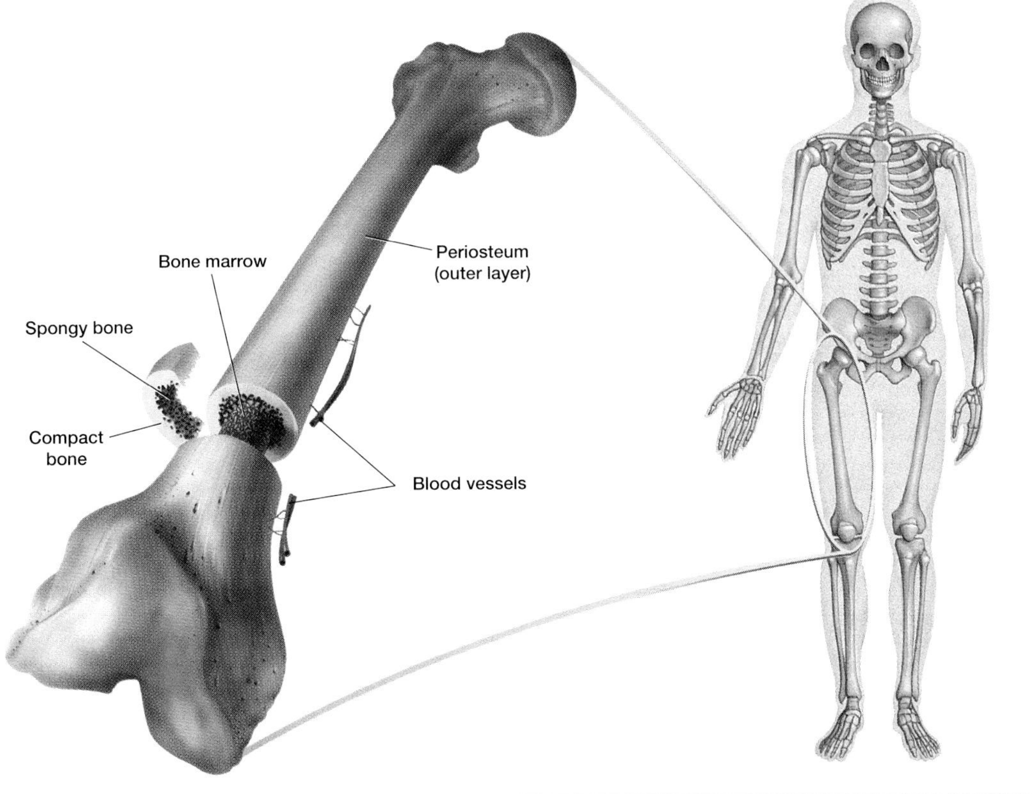

Teaching TIP

Axial and Appendicular Skeletons

Have students color a diagram of the human skeleton. Have them color the axial skeleton red and the appendicular skeleton green. Check their diagrams carefully. The ribs and sternum are part of the axial skeleton; the shoulder and pelvis are part of the appendicular skeleton. PORTFOLIO

Teaching TIP BASIC

Mnemonics for Bone Names

Tell students some simple ways to remember the names of bones. For example, the scapula is large and flat like a spatula; the ulna ends at the elbow (both first letters are vowels); the radius radiates around the ulna; the metacarpals are where fingers meet the carpals; the sternum looks like a tie (stern people wear ties); the fibula is the little bone in the lower leg (a little lie is a fib); tarsals are bones in your feet (you get tar on the soles of your feet)

did you know?

The hyoid bone is the only bone not articulated to any other bone.

The hyoid bone is the origin of the tongue muscles. You can find it by feeling the neck, lateral to the Adam's apple, directly beneath the carotid arteries. If you push gently on one side of the throat, you can feel the bone easily on the opposite side. This bone can be broken by choking, making it very painful to speak or swallow.

REAL LIFE

Answer

Answers will vary. Encourage them to use the media center to find nonfictional accounts of crime solving.

Teaching TIP

Bone Reabsorption

Tell students that bone is the body's storehouse for calcium. If calcium is needed elsewhere in the body (such as for nerve function), it may be released from the bones in a process called reabsorption. Conversely, if calcium is abundant in the body, it may be added to the bones. The amount of calcium in the body is a function of diet, and the recruitment of calcium in the body is monitored by the brain and the parathyroid glands.

Active Reading

Technique: Paired Reading

Pair each student with a partner. Have each student read silently about bone growth and osteoporosis. As students read, have them place a check mark on a sticky note next to passages they understand and a question mark on a sticky note next to passages they find confusing. After students finish reading, ask one student from each pair to summarize what he or she understood, referring to the text when needed. The second student should add anything omitted. Both students should then describe what they found confusing and should work together to understand all the passages. Have them prepare a list of questions to pose to the class about passages that are still unclear after their collaboration.

REAL LIFE

Years after an unsolved murder is committed, the victim's bones may hold clues to the crime. Forensic anthropologists have solved many cases by analyzing bones and other human remains found at crime scenes.

Finding Information *Read some accounts of crimes solved by forensic anthropologists to learn how these scientists analyze bones.*

Bones Grow Through Adolescence

In early development, the skeleton is made mostly of cartilage, a type of connective tissue that serves as a template for bone formation. During development, cartilage is gradually replaced by bone as minerals are deposited. Deposits of calcium and other minerals harden bones so they can withstand stress and provide support. In compact bone, new bone cells are added in layers around narrow, hollow channels called **Haversian canals.** Haversian canals extend down the length of a bone, and they contain blood vessels that enter the bone through the periosteum.

As shown in **Figure 38-5,** layers of new bone cells form several concentric rings around Haversian canals. These concentric rings form columns that enable the bone to withstand tremendous amounts of stress. Eventually, bone cells called **osteocytes** *(AHS tee oh seyets)* become embedded within the bone tissue. Osteocytes maintain the mineral content of bone. The blood vessels that run through each Haversian canal supply the osteocytes with nutrients needed to maintain bone cells. Bones continue to thicken and elongate through adolescence as bone cells replace cartilage. Bone elongation occurs at the ends of long bones. Cartilage degenerates as new bone cells are added, causing bones to lengthen.

Figure 38-5 Compact bone

In compact bone, concentric rings of bone surround Haversian canals.

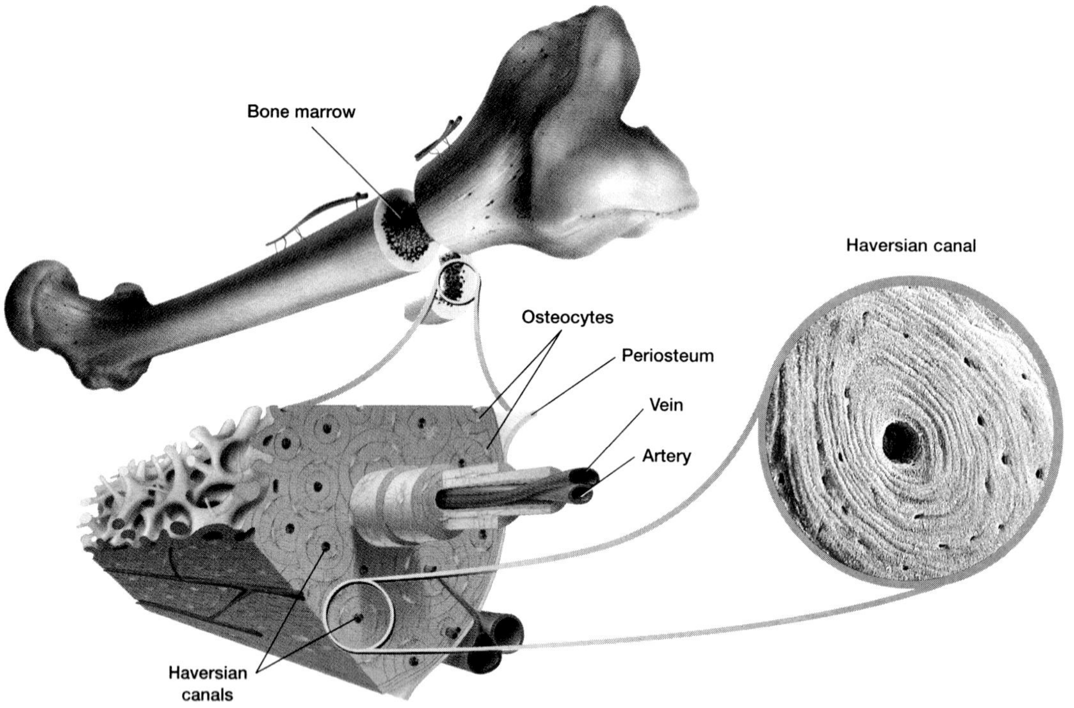

Overcoming Misconceptions

Living Bones

Students may think that bone is a rocklike substance with few of the traits of living tissue. Emphasize that the bony endoskeleton of vertebrates grows as the animal matures. This is unlike the hard exoskeleton of invertebrates, which is incapable of growth or change. The nonliving components of bone—crystalline material that surrounds the living bone cells—give bone its strength and hardness. Anyone who has broken a bone knows that bones contain nerves that convey pain and other information to the brain.

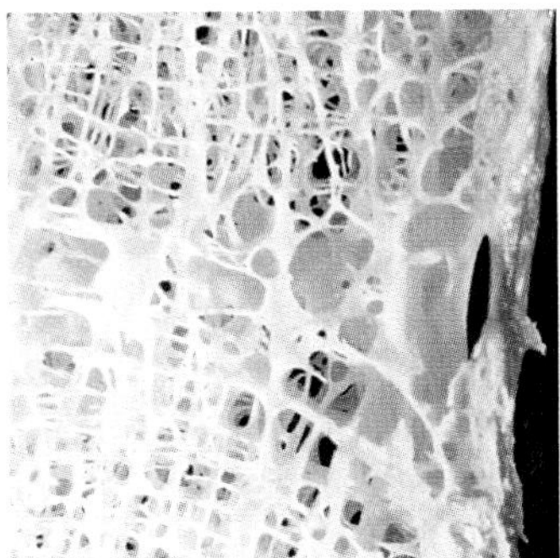

Normal bone density

Reduced bone density

Dowager's hump

Osteoporosis results from bone loss

In young adults, the density of bone usually remains constant. However, around the age of 35, bone replacement gradually becomes less efficient and some bone mass is lost. Severe bone loss, as shown in **Figure 38-6,** can lead to a condition called **osteoporosis** *(ahst ee oh puh ROH sihs),* which means "porous bone." Bones affected by osteoporosis become brittle and are easily fractured.

Although both women and men lose bone density as they age, more women than men are affected by osteoporosis. Because women's bones are usually smaller, women cannot afford to lose as much bone tissue as men. In addition, the production of sex hormones, which help maintain bone density, decreases after menopause. This decrease in hormone production has been linked to an increased rate of bone loss in women following menopause.

You can take action now to prevent future osteoporosis. Building strong bones now will make you less likely to be affected by osteoporosis later in life. Bone density can be increased with regular exercise and a mineral-rich diet that includes green leafy vegetables, whole grains, legumes, and dairy products, such as those shown in **Figure 38-7.** Regular exercise and physical activity throughout your life will also help maintain bone density.

Figure 38-6 Effects of osteoporosis. Compare the density of a normal bone with that of a bone weakened by osteoporosis. One sign of osteoporosis is the familiar "dowager's hump" of the back, caused by curvature of the spine.

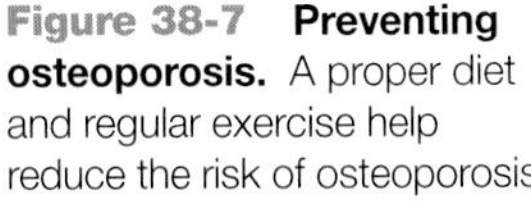

Figure 38-7 Preventing osteoporosis. A proper diet and regular exercise help reduce the risk of osteoporosis.

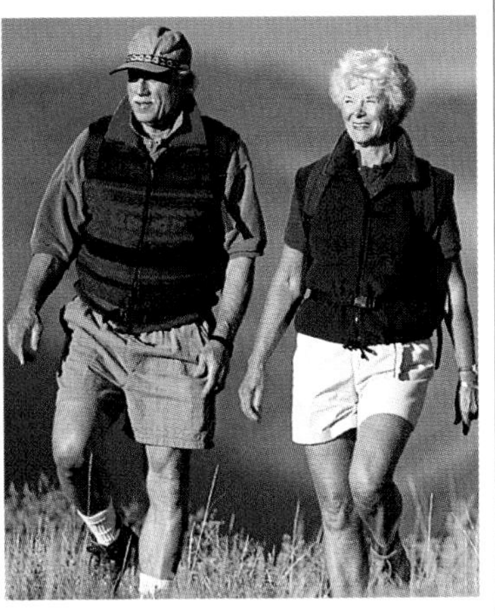

did you know?

Bones are stronger for their weight than iron.

A human long bone has the tensile strength of cast iron, but it weighs only one-third as much. A bone can resist 25,000 lb/in.^{2} (1,800 kg/cm^2) of compression and 15,000 lb/in.^{2} (1,100 kg/cm^2) of tension.

Section 38-2

Demonstration

Decalcification of Bone BASIC

Soak a cooked, clean chicken bone in vinegar overnight to remove much of the calcium from the bone. Compare the flexibility of the soaked bone with that of a cooked, clean bone that has not been soaked. Relate these ideas to what happens to bone due to osteoporosis.

Resource Link BASIC

- Assign **Basic Skills: Mass and Density** from the Holt BioSources Teaching Resources CD-ROM. PORTFOLIO

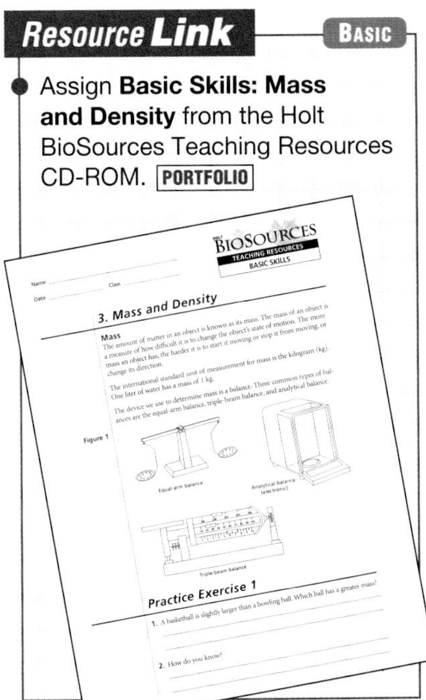

BioSources Teaching Resources Basic Skills

3. Mass and Density

Practice Exercise 1

Teaching TIP

The Spine

Point out that the small range of motion of the joints between each vertebra has a compounding effect. Although each joint allows only a small amount of movement between two individual bones, when added across all 26 vertebrae, the human spine is very flexible.

Teaching TIP

The Soft Spot

Many students who have younger siblings know that a baby has a soft spot on the top of its head. Explain that the soft spot is an area where sections of the skull have not grown together and fused. This area will become an immovable joint as the baby matures. Tell students that several layers of tough membranes protect the brain, but care must be taken not to injure the baby's vulnerable head.

Resource Link BASIC

Assign **Occupational Applications: Physical Therapist** from the Holt BioSources Teaching Resources CD-ROM. PORTFOLIO

Joints Fasten Bones Together

A junction between two or more bones is called a **joint.** Pads of cartilage cushion the ends of the bones of a joint, enabling the joint to withstand great pressure and stress. The bones of a joint are held together by strong bands of connective tissue called **ligaments.** Ligaments not only help stabilize joints but also prevent joints from moving too far in any one direction. For example, many sports-related injuries to ligaments are caused by an impact that causes a joint to overextend. Injury occurs because the impact exceeds the tension that the ligaments can withstand.

Three main types of joints

The skeletal system contains three main types of joints that enable varying degrees of movement: immovable joints, slightly movable joints, and freely movable joints. Examples of the three types of joints are shown in **Figure 38-8.**

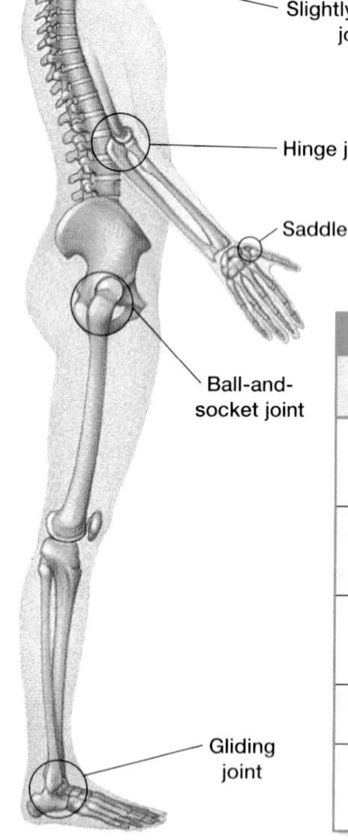

Figure 38-8 Types of joints. The body contains immovable, slightly movable, and freely movable joints.

Immovable joints Tight joints that permit little or no movement of the bones they join are called immovable joints. The cranial bones of the skull are joined by sutures, a type of immovable joint in which the bones are separated by only a thin layer of connective tissue.

Slightly movable joints Joints that permit limited movement of the bones they join are called slightly movable joints. For example, the vertebrae of the spine are joined by cartilaginous joints, which are a kind of slightly movable joint in which a bridge of cartilage connects the bones. Slightly movable joints are also located between bones of the rib cage.

Freely movable joints In joints that permit movement, the direction of bone movement is determined by the structure of the joint. Joints that permit the most move-

Table 38-2 Freely Movable Joints

Joint	Type of movement	Examples
Ball-and-socket joint	All types	Shoulders and hips
Pivot joint	Rotation	Top of spine (turning of head)
Hinge joint	Bending and straightening	Elbows, knuckles of fingers and toes
Gliding joint	Sliding motion	Wrists and ankles
Saddle joint	Rotation, bending, and straightening	Base of thumbs

did you know?

In 1963, a British orthopedic surgeon revolutionized arthritic hip therapy when he created an artificial hip joint.

He created a metal (vitalium) ball and plastic (polyethylene) socket. The plastic socket is anchored to the pelvis by a special cement. This invention has helped many people, including children who are born with a hip socket that is too shallow and bones that are not aligned properly as well as arthritis victims who suffer disabling pain in the hip joint.

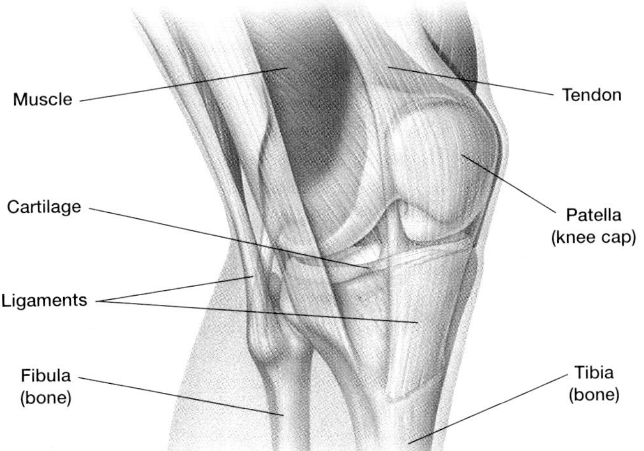

Figure 38-9 **Knee.** The knee is an example of a freely movable joint.

ment are called freely movable joints. Some kinds of freely movable joints are listed in **Table 38-2.** The structure of these joints also determines their range of movement. You are probably most familiar with freely movable joints, which include elbows, hips, shoulders, and the knee, shown in **Figure 38-9.**

Disorders of joints

Recall that ligaments hold the bones of a joint together. A lining of tissue that surrounds the joint secretes a lubricating fluid that reduces friction at the ends of the bones. When a disease afflicts the bones, connective tissue, or lubricating tissues in a freely movable joint, the joint's ability to move may be greatly impaired.

Rheumatoid arthritis is a painful inflammation of freely movable joints. This condition occurs when cells of the immune system attack the tissues around joints, severely damaging the joints. Symptoms of rheumatoid arthritis include stiffening and swelling of the joints. Osteoarthritis is a similar disorder that causes the degeneration of cartilage that covers the surfaces of bones. As the cartilage wears away, the joint becomes less stable and bones rub together, causing pain and discomfort.

internetconnect

SCILINKS. NSTA

TOPIC: Joint disorders
GO TO: www.scilinks.org
KEYWORD: HX865

Review SECTION 38-2

1. **Distinguish** between the axial skeleton and the appendicular skeleton.
2. **Differentiate** between compact bone and spongy bone.
3. **Describe** how bones elongate in development.
4. **List** the three main types of joints, and give an example of each type.
5. **SKILL Analyzing Information** What are two reasons that women are more likely than men to develop osteoporosis?

Critical Thinking

6. **SKILL Relating Concepts** The bones of a newborn baby are made mostly of cartilage. Why is that an advantage during childbirth?

Review SECTION 38-2 ANSWERS

1. The axial skeleton makes up the axis or center of the skeleton and is composed of the skull, spine, ribs, and sternum. The appendicular skeleton is composed of the remainder of the bones.
2. Compact bone is dense; spongy bone is loosely connected and porous.
3. Bone cells gradually replace cartilage at the ends of the bones.
4. immovable—cranium; slightly movable—vertebrae; freely movable—knee, elbow, hip, and shoulder
5. Women usually have less bone tissue to lose, and after menopause their production of sex hormones, which help maintain bone density, is reduced.
6. Cartilage is more flexible than bone, making passage through the birth canal easier.

Demonstration

Synovial Fluid — BASIC

Stretch two pieces of plastic wrap over two blocks of wood, and tape them in place. Allow students to demonstrate that the blocks do not slide over each other readily. Put a few drops of oil on one block and have students try again. They should slide easily. Ask students what body structure each component of the demonstration represents. *(The wood represents bone, the plastic wrap represents cartilage, and the oil represents synovial fluid, the lubricating fluid in joints.)*

3 Assessment Options

Closure

Justifying Conclusions

Have students list advantages and disadvantages of having an endoskeleton rather than an exoskeleton. *(An advantage would be that an endoskeleton is lighter and easier to move, and it does not need to be shed to allow the body to grow. A disadvantage would be that it does not provide full body protection.)* PORTFOLIO

Review

Assign Review—Section 38-2.

Reteaching — BASIC

Have students examine prepared slides of cartilage and bone. Ask them to explain how the structure of cartilage and bone is related to the function each performs in the body. *(Students' responses will vary but should describe the hardness of bone, the role of bone in support and protection, the flexibility of cartilage, and the role of cartilage as a precurser and model of bones before birth.)*

Section 38-3 Muscular System

1 Prepare

Scheduling

- **Block:** About 90 minutes will be needed to complete this section.
- **Traditional:** About two class periods will be needed to complete this section.
- **Planning:** See the Planning Guide on pp. 853A and 853B for lecture, classwork, and assignment options for Section 38-3. Lesson plans for Section 38-3 can be found in a lesson cycle format in ***Biology: Principles and Explorations*** **One-Stop Planner CD-ROM with Test Generator.**

Resource Link

Assign **Active Reading Worksheet Section 38-3** to help students comprehend the key concepts and visuals in this lesson.

2 Teach

Lesson Warm-up

Have a volunteer hold his or her right arm straight out, palm up, at shoulder level. Ask him or her to place their left thumb and index finger lightly on their right biceps, about two inches apart, and to raise their right palm to their shoulder. Have him or her explain what happens to the fingers. *(The fingers should move closer together, demonstrating that the muscle is actually shortening.)* Repeat the demonstration with the triceps, showing that it shortens as the arm is extended.

Objectives

- **Describe** the action of muscle pairs in moving the body. (p. 866)
- **Relate** the structure of a skeletal muscle to the muscle's ability to contract. (p. 867–869)
- **Describe** how energy is supplied to muscles for contraction. (p. 870)

Key Terms

tendon
flexor
extensor
actin
myosin
myofibril
sarcomere

Muscles Move the Body

Every time you move, you use your muscles. Walking and running, for example, require precisely timed and controlled contractions of many skeletal muscles. When you lift a heavy object, the total force produced by muscle contractions in your arm must overcome the weight of the object. Muscles in your jaw contract and enable you to chew food with your teeth. Even when you are idle, many skeletal muscles, including those in your back and neck, remain partially contracted to maintain balance and posture.

Movement of the skeleton

Muscles can move body parts because muscles are attached to bones of the skeleton. As shown in **Figure 38-10,** most skeletal muscles are attached to bones by strips of dense connective tissue called **tendons.** One attachment of the muscle, the origin, is a bone that remains stationary during a muscle contraction. The muscle pulls against the origin. The other attachment, the insertion, is the bone that moves when the muscle contracts. Movement occurs when a muscle contraction pulls the muscle's insertion toward its origin.

Skeletal muscles are generally attached to the skeleton in opposing pairs. One muscle in a pair pulls a bone in one direction, and the other muscle pulls the bone in the opposite direction. In the limbs, each opposing pair of muscles includes a flexor muscle and an extensor muscle, as shown in Figure 38-10. A **flexor** muscle causes a joint to bend. An **extensor** muscle causes a joint to straighten.

Figure 38-10 Muscle pair

Pairs of opposing muscles work together to move bones at joints.

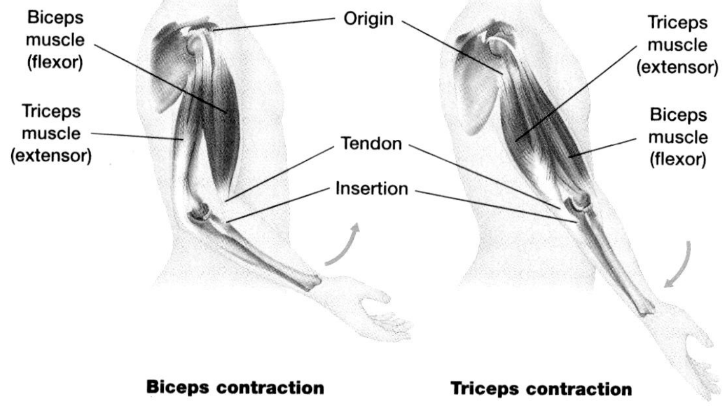

Integrating Different Learning Styles

Visual/Spatial	**PE**	Performance Zone items 17 and 18, p. 877
	ATE	Internet Connect, p. 869; Closure, p. 870; Reteaching, p. 870
Body/Kinesthetic	**PE**	Performance Zone item 24, p. 877
	ATE	Lesson Warm-up, p. 866; Teaching Tip, p. 867; Visual Strategies, pp. 868, 869; Teaching Tip, p. 868
Interpersonal	**ATE**	Visual Strategy, p. 869
Verbal/Linguistic	**PE**	Performance Zone items 2 and 24, pp. 876–877
	ATE	Cross-Disciplinary Connection, p. 867; Career, p. 870

Figure 38-11 **Skeletal muscle**

In skeletal muscle, muscle contraction occurs within sarcomeres of muscle fibers.

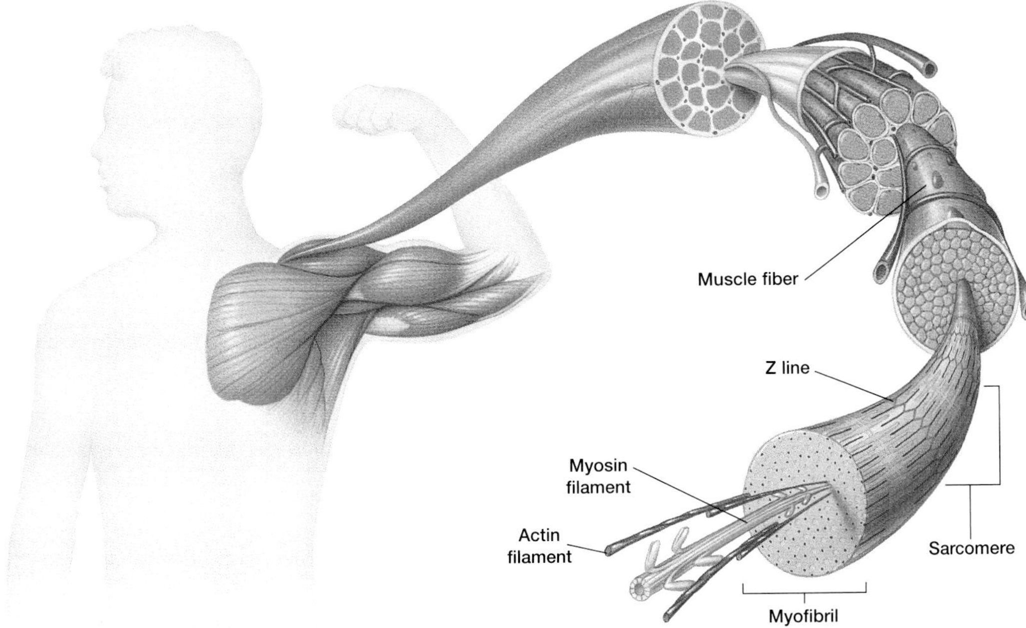

Muscle structure enables contraction

Muscles contain some connective tissue, which holds muscle cells together and provides elasticity. Muscle tissue also contains large amounts of protein filaments. These protein filaments, called **actin** and **myosin** *(MEYE oh sihn),* enable muscles to contract. Actin and myosin are usually found in the cytoskeleton of eukaryotic cells, but they are far more abundant in muscle cells. Other characteristics of muscle tissue include the ability to stretch or expand and the ability to respond to stimuli, such as signal molecules released by nerve cells.

Skeletal muscle tissue consists of many parallel elongated cells called muscle fibers. As shown in **Figure 38-11,** each muscle fiber is made of small cylindrical structures called **myofibrils** *(meye oh FEYE bruhlz).* Myofibrils have alternating light and dark bands that produce a characteristic striated, or striped, appearance when viewed under a microscope. In the center of each light band is a structure called a Z line, which anchors actin filaments. The area between two Z lines is called a **sarcomere** *(SAHR koh mihr).* Thus, a myofibril is a grouping of sarcomeres linked end to end. Each sarcomere contains overlapping thin and thick protein filaments that move and interact with each other. The thin filaments are actin, and the thick filaments are myosin. The filaments run parallel to one another along the length of the sarcomere. The dark bands that occur in the middle of the sarcomere are regions where the thick filaments and the thin filaments overlap.

internetconnect

TOPIC: Muscle structure
GO TO: www.scilinks.org
KEYWORD: HX867

CROSS-DISCIPLINARY CONNECTION

◆ Language Arts

The Achilles tendon is the thickest and strongest tendon in the human body. According to ancient Greek mythology, Achilles' mother dipped him in the magical River Styx to make him an invulnerable warrior. He was killed when an arrow pierced his one weak point, the place on his heel where his mother held on to him. The Achilles tendon is named for this mythical hero.

Teaching TIP — **ADVANCED**

Muscle Terminology

Refer to the exercise in the **Lesson Warm-up** on page 866. Have all students repeat this exercise in their seats. Ask them to name the flexor, extensor, origin, and insertion. Try the same demonstration for other muscle pairs, such as the gastrocnemius (calf muscle) and tibialis anterior (shin muscle).

Visual Strategy

Figure 38-11

Use this figure to identify the structure of a skeletal muscle. Point out the muscle fibers, myofibrils, sarcomeres, Z lines, actin, and myosin. Make sure that students understand the separate levels of organization shown in the figure. Help students understand that muscles comprise many bundles of muscle fibers, that a muscle fiber contains many myofibrils, and that a myofibril consists of actin and myosin. Remind students that muscle contractions begin at the lowest level—with actin and myosin.

Teaching TIP BASIC

Finger Movement

Have students curl their middle finger under and place their fingertips down on a table. Have them try to raise their fingers one by one. They should be able to lift all fingers except the ring finger. Explain that each finger has a tendon, and those tendons attach to a muscle in the arm. The tendons in the middle and ring finger are linked together in the hand, connecting their movement.

Visual Strategy

BIO graphic Figure 38-12

Work with students to help them understand the events of muscle contraction. Interlace the fingers of your hands to show how actin and myosin slide past each other in the sarcomeres. Have students contract a muscle (such as the gastrocnemius of the calf), and point out that they should feel the muscle become shorter and thicker. Explain that this is the result of the contraction of many sarcomeres within the muscle.

The Sarcomere Is the Functional Unit of Muscle Contraction

How does a muscle contract? Muscle contraction occurs in the sarcomeres of myofibrils. The overlapping arrangement of the thick and thin protein filaments in a sarcomere enables muscle contraction. The events of a muscle contraction are summarized in **Figure 38-12.**

Step 1 Before a muscle is stimulated, the sarcomere is relaxed. Myosin and actin filaments partially overlap one another.

Step 2 A muscle contraction usually begins when a muscle fiber is stimulated by signal molecules released by a nerve cell. This causes myosin and actin filaments to "slide" along one another so that they overlap even more. The sarcomere becomes shorter as the Z lines are pulled closer together.

Step 3 The sarcomere is fully contracted, and myosin and actin completely overlap one another. This shortening of sarcomeres occurs down the entire length of the muscle fiber.

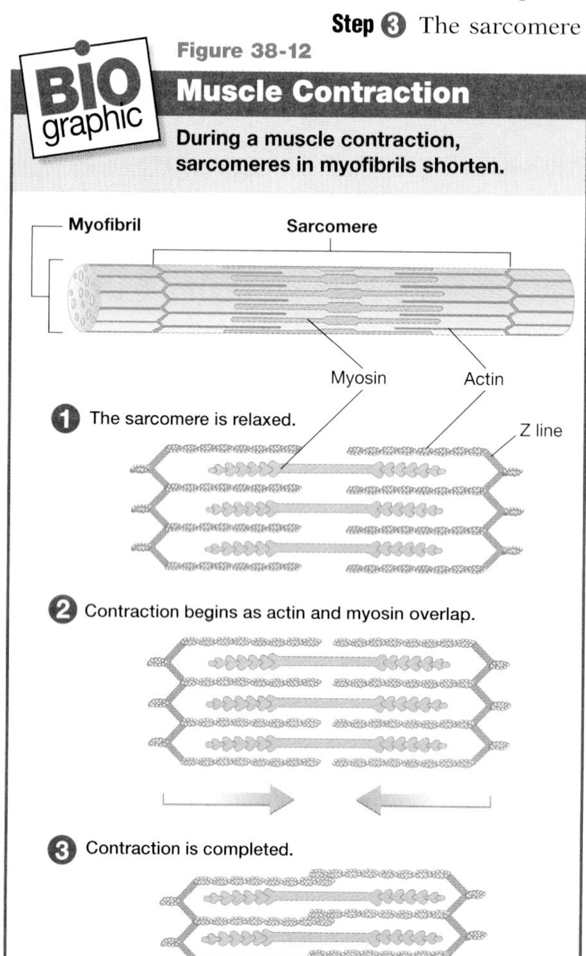
Myofibril
Sarcomere
Myosin
Actin
Z line
1 The sarcomere is relaxed.
2 Contraction begins as actin and myosin overlap.
3 Contraction is completed.

BIO graphic Figure 38-12 **Muscle Contraction**

During a muscle contraction, sarcomeres in myofibrils shorten.

What determines the force of contraction? A muscle exerts the greatest force when all of its fibers are contracted. The contraction of a muscle fiber is an all-or-nothing event. When a fiber is stimulated, its sarcomeres attempt to contract fully. The total amount of force a muscle exerts depends on how often muscle fibers are stimulated and how many muscle fibers contract.

How is the force of muscular contraction controlled? By contracting different numbers of fibers in a muscle at one time, the total force generated by contraction varies. For example, the total amount of force needed to lift a pencil is much less than the force needed to lift a brick. Thus, fewer muscle fibers in your arm contract when you lift a pencil than when you lift a brick.

The set of muscle fibers activated by a nerve cell is called a motor unit. Every time a nerve cell activates its motor unit, all the fibers in that unit contract. Muscles that require a finer degree of control, such as muscles that move the fingers, have only a few muscle fibers in each motor unit. Large muscles, such as muscles in the leg, have several hundred muscle fibers in each motor unit.

Overcoming Misconceptions

ATP and Muscle Contractions

Because movement requires energy, students might assume that ATP is consumed at the moment muscles contract. A key discovery made by the British physiologist and biophysicist Archibald V. Hill, however, helped biologists determine the sequence of events. Hill discovered that muscle cells use ATP after, not during, contraction. ATP does not cause contraction, rather, ATP prepares the muscle for the next contraction. In 1922 Hill received a Nobel Prize in medicine and physiology for his work on muscles. His work was continued by Hugh Huxley, who formulated a theory that explained the role of ATP in cocking myosin heads prior to the contraction of the sarcomere.

Interaction of myosin and actin

How do actin and myosin cause sarcomeres to shorten during a muscle contraction? Myosin filaments have long, finger-like projections with an enlarged "head" at one end. Actin filaments contain many sites to which myosin can bind during a muscle contraction. Stimulation of a muscle fiber leads to the exposure of binding sites on actin filaments. As shown in **Figure 38-13,** the myosin heads attach to sites on the actin filaments and then rotate, causing myosin to move relative to actin.

Step 1 Muscle contraction begins as a myosin head attaches to an exposed binding site on an actin filament.

Step 2 The myosin head rotates, causing the actin filament to "slide" against the myosin filament. This sliding causes the filaments to overlap one another.

Step 3 ATP is used as the myosin head detaches and snaps back into its original position. The myosin head reattaches to actin at a binding site farther along the actin filament. When the myosin heads cannot move farther, they release momentarily and reposition themselves to grab the actin and pull again. Thus, the myosin heads "walk" along actin filaments, essentially "stepping" at each available binding site. This grabbing and pulling action is repeated, causing the sarcomere to shorten as the Z lines are pulled closer together.

A lot of energy is needed to power a muscle contraction. ATP is used each time a myosin head moves from one binding site on an actin filament to another. Without ATP, myosin heads would remain attached to actin filaments, keeping the muscle contracted. ATP is also used to move calcium ions, Ca^{2+}, into muscle cells. Calcium ions are needed for myosin to bind to actin because the presence of calcium ions in a sarcomere exposes binding sites on actin filaments. Without calcium ions and ATP, a muscle could not contract.

internet connect
SCiLINKS
NSTA

TOPIC: Muscle contraction
GO TO: www.scilinks.org
KEYWORD: HX869

BIO graphic

Figure 38-13

Interaction of Muscle Protein Filaments

During a muscle contraction, myosin and actin move against one another.

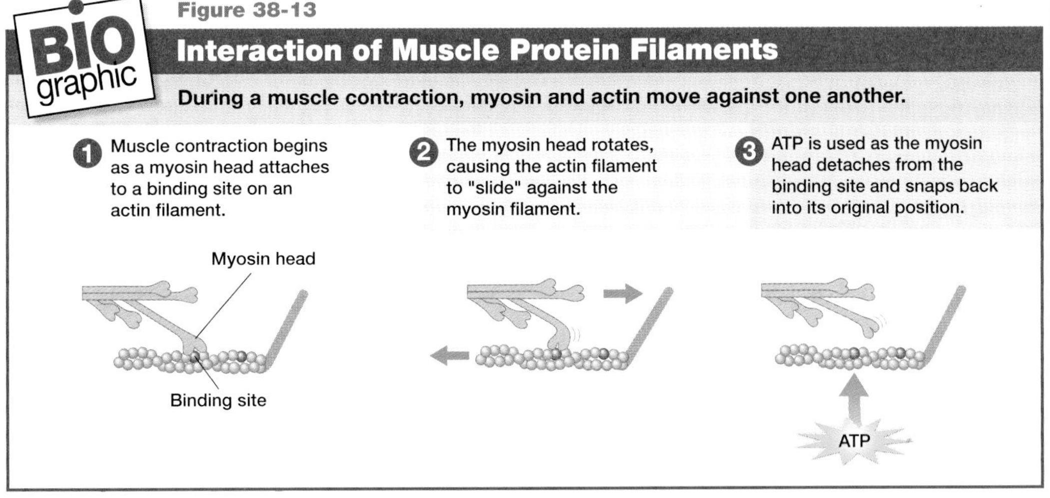
1 Muscle contraction begins as a myosin head attaches to a binding site on an actin filament.
2 The myosin head rotates, causing the actin filament to "slide" against the myosin filament.
3 ATP is used as the myosin head detaches from the binding site and snaps back into its original position.
Myosin head
Binding site
ATP

did you know?

Muscles become rigid for a brief period after an animal dies.

After death, ATP production stops. Without ATP, myosin heads attach to actin filaments but cannot detach. As a result, all the muscles are locked at one length, making the body rigid. This condition is known as rigor mortis, and it persists for up to 25 hours after death.

internet connect

SCiLINKS NSTA Ask students to research muscle contraction using the Web site in the Internet Connect box on page 869. Have them draw a diagram of a sarcomere before and during a contraction, and ask them to annotate their drawings with explanations of the steps of muscle contraction. PORTFOLIO

Visual Strategy

Figure 38-13

BIO graphic Recruit five volunteers. Have four of the students stand in a straight line, one behind the other, with their hands on their hips. Have the fifth student face the leader of the line and slowly pull himself or herself along the line by grasping the elbow of each member of the line in succession. Ask students to relate Figure 38-13 to what they are seeing in this demonstration. *(The fifth student represents myosin; the others are actin. The fifth student's hand is the myosin head; the other students' elbows are binding sites.)*

Chapter 38

CAREER

Physical Medicine

Physical medicine is a relatively new specialty of orthopedic medicine. Physiatrists specialize in orthopedics, physical therapy, and pain management. Have students contact a local health maintenance organization, hospital, or physician to find out what a physiatrist does. Students should then write a brief report explaining what they have learned. PORTFOLIO

3 Assessment Options

Closure

The Role of ATP

Ask students to draw a diagram to illustrate the role of ATP in muscle contraction. Their diagrams should show each step in the muscle contraction process and indicate where ATP is involved. PORTFOLIO

Review

Assign Review—Section 38-3.

Reteaching — BASIC

Have students create a flip book of 20 or more pages showing the mechanics of muscle contraction. The first and last pages should have the following items labeled: actin, myosin, myosin head, and sarcomere. PORTFOLIO

biobit

What is the best way to increase muscle size, strength, and endurance?

If you want to have large muscles and be strong enough to lift large loads, you should train with heavy weights. On the other hand, repeatedly lifting lighter weights builds muscle endurance.

Aerobic and anaerobic energy pathways

The intensity of muscle activity determines which energy pathway is used for muscle contractions. The ATP used to power contractions is usually supplied by aerobic respiration. During prolonged exercises, such as long-distance running, breathing may be more rapid than usual but not as intense. Thus, oxygen is consumed at a sustainable, steady rate, and aerobic respiration yields most of the ATP. However, in rapidly contracting muscles, the oxygen supply soon becomes inadequate, and anaerobic processes take over. During brief, intense activities, such as sprinting and weight lifting, breathing is rapid and intense, and the body quickly becomes tired. Thus, most of the ATP comes from glycolysis, an anaerobic process, as the oxygen available to muscle cells rapidly decreases.

When both anaerobic and aerobic energy pathways become insufficient for muscle contraction, muscles begin to use glycogen as an energy source. As glycogen is used up, the body must begin to use fat molecules as an energy source. When ATP consumption exceeds ATP production, muscle fatigue and soreness may result, leaving muscle fibers unable to recover from contraction.

Figure 38-14 Resistance exercise. Weight lifting and other resistance exercises primarily involve anaerobic energy pathways in muscles.

Exercise and fitness

Consistent aerobic exercise makes the heart pump more efficiently and thus increases the energy available to muscles as a result of improved blood circulation. More oxygen enters the body with each breath, increasing the oxygen supply to muscles. More ATP is available for muscle contractions, thereby reducing muscle fatigue. The increase in muscle efficiency results in greater endurance, or the ability to continue exercising.

Resistance exercises, such as weight lifting, shown in **Figure 38-14,** can increase muscle size and strength. Resistance exercises are mostly anaerobic, so they do not usually improve the delivery of oxygen to muscles. Muscle mass is increased by resistance training. The amount of tension and the rate of exercise are both important factors. However, the short-term demands of such strength training do not cause the circulatory changes that increase endurance.

Excessive exercise or failure to warm up properly can lead to muscle injury. Muscles can tear if they are stretched too far during strenuous exercise. If excessive stress causes tendons to become inflamed, a painful condition called tendinitis results.

Review SECTION 38-3

1. **Describe** how muscle pairs work together to move body parts.
2. **Summarize** the process that causes sarcomeres to shorten during muscle contraction.
3. **Compare** the roles of thick and thin filaments in muscle contraction.
4. **SKILL Analyzing Information** Which energy pathway is primarily involved with exercises that increase muscle size and strength?

Critical Thinking

5. **SKILL Applying Information** What causes muscle cramping after rigorous exercise or a repeated movement?

Review SECTION 38-3 ANSWERS

1. Muscles exist in opposing pairs. Flexors bend joints; extensors straighten them.
2. When contraction begins, actin and myosin overlap and pull together, shortening sarcomeres.
3. Thick (myosin) filaments have heads that attach to thin (actin) filaments. The heads rotate, causing actin to slide against myosin. This process repeats many times during a single muscle contraction.
4. the anaerobic pathway
5. When ATP consumption exceeds ATP production, muscle cramping results.

Skin, Hair, and Nails

Skin Has Two Distinct Layers

The skin, which makes up about 15 percent of your total body weight, is the largest organ of the body. Many specialized structures are found in the skin, which along with the hair and nails, forms the integumentary system. The skin protects the body from injury, provides the first line of defense against disease, helps regulate body temperature, and prevents the body from drying out through evaporation. As shown in **Figure 38-15,** the skin is made mostly of connective tissue and layers of epithelial tissue. The two primary layers of skin are the epidermis and the dermis.

Epidermis

The **epidermis** is the outermost layer of the skin. About as thick as this page, the epidermis is made of several layers of epithelial cells. The part of the epidermis you see when you look in a mirror is a thin layer of flattened, dead cells that contain keratin. **Keratin** is a protein that makes skin tough and waterproof. The cells of the epidermis are continuously damaged by the environment. They are scraped, ripped, worn away by friction, and dried out because of moisture loss. Your body deals with this damage not by repairing the cells, but by replacing them.

Objectives

- **Analyze** the structure and function of the epidermis. (pp. 871–872)
- **Describe** how the dermis helps the body maintain homeostasis. (pp. 872–873)
- **Summarize** how hair and nails are formed. (p. 873)
- **Identify** various types of skin disorders. (p. 874)

Key Terms

epidermis
keratin
melanin
dermis
hair follicle
subcutaneous tissue
sebum

Figure 38-15 **Structure of skin**

Skin has two distinct layers that contain many blood vessels, nerve cells, muscles, hairs, and glands.

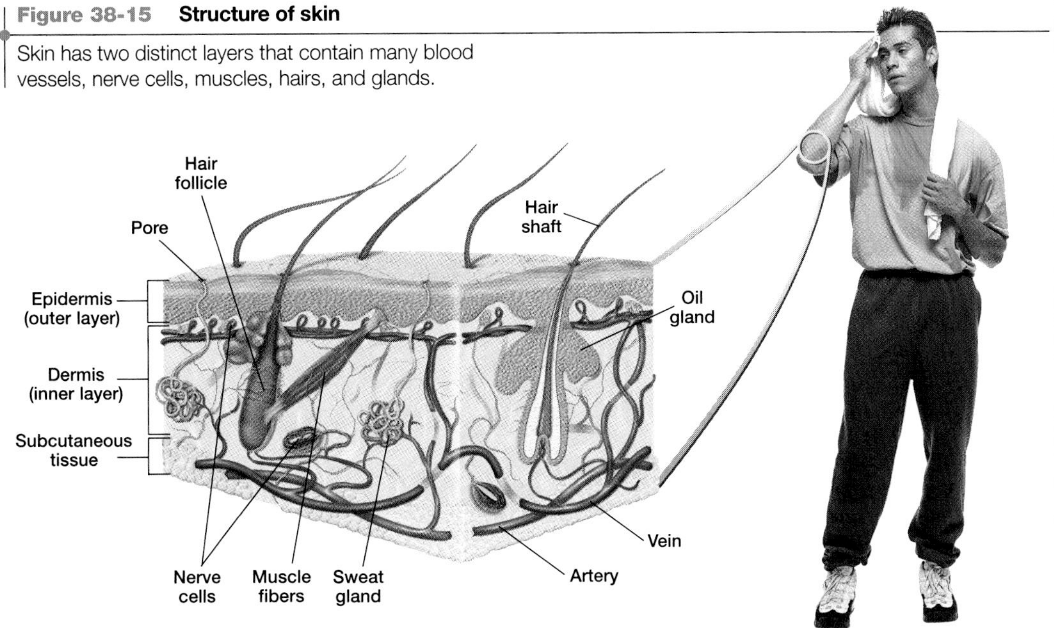

Integrating Different Learning Styles

Logical/Mathematical	**PE**	Performance Zone item 22, p. 871
Visual/Spatial	**ATE**	Teaching Tip, p. 873; Reteaching, p. 874
Intrapersonal	**ATE**	Lesson Warm-up, p. 871; Closure, p. 874
Verbal/Linguistic	**PE**	Performance Zone items 2, 25, and 27, pp. 876–877
	ATE	Internet Connects, pp. 872, 874

Section 38-4

1 Prepare

Scheduling

- **Block:** About 90 minutes will be needed to complete this section.
- **Traditional:** About two class periods will be needed to complete this section.
- **Planning:** See the Planning Guide on pp. 853A and 853B for lecture, classwork, and assignment options for Section 38-4. Lesson plans for Section 38-4 can be found in a lesson cycle format in ***Biology: Principles and Explorations*** **One-Stop Planner CD-ROM with Test Generator.**

Resource Link

- Assign **Active Reading Worksheet Section 38-4** to help students comprehend the key concepts and visuals in this lesson.

2 Teach

Lesson Warm-up

Ask students for a show of hands to the following questions: Who likes to go to the beach? Who tans? Who burns? Explain that a suntan is actually a defense mechanism of the skin. Melanin is a dark pigment that is present in varying degrees in different individuals. Additional melanin is produced by the skin when it is exposed to ultraviolet radiation. We call this additional melanin a tan; the darkened skin protects the dermis somewhat from damaging radiation. Emphasize the importance of using SPF 15 or higher sunscreen to prevent sunburn. Burns early in life increase the chances of skin cancer later.

REAL LIFE

Answer

Have students bring in different kinds of antiperspirants and deodorants. Have them list the active ingredients in the products. Have students find out why antiperspirants are regulated by the U.S. Food and Drug Administration, whereas deodorants are not. *(Deodorants do not alter any body functions.)*

Courtroom Science

Teaching Strategies

Have students collect newspaper, magazine, or Internet articles about the uses of forensic technologies in the courtroom.

Discussion

- Should these relatively new technologies be used in a courtroom? *(Answers will vary but may include that although they are scientifically valid, they have not been tested sufficiently.)*
- What are some of the drawbacks of using hair for drug testing? *(Answers will vary but may include that some individuals may cut their hair to avoid detection of their drug use.)*
- Should employers be able to do random drug testing on hair? *(Answers will vary but may include that it is an accurate indicator of drug use and provides a long-term record, but it may violate an individual's rights in certain cases.)* PORTFOLIO

internetconnect

Have students research forensic analysis using the Web site in the Internet Connect box on page 872. Have students write a report on their findings. PORTFOLIO

REAL LIFE

How can body odor be controlled? Antiperspirants and deodorants are both designed to combat body odor, but they work in different ways. While antiperspirants keep you from perspiring, many deodorants mask the odors produced by bacteria that live on sweat.

Finding Information

Research how antiperspirants and deodorants work.

The outermost cells of the skin are continuously shed and replaced by a layer of actively dividing cells at the base of the epidermis. As new skin cells form, they migrate upward and produce large amounts of keratin. After reaching the skin's surface, these cells are shed after about a month. The inner layer of the epidermis also contains cells that produce the pigment melanin. **Melanin** *(MEHL uh nihn)* ranges in color from yellow to reddish brown to black, and it helps determine skin color. People with more melanin tend to have darker skin, and people with less melanin usually have lighter skin. Melanin also absorbs ultraviolet (UV) radiation, protecting the skin from exposure to sunlight. Exposure to UV radiation increases the production of melanin. This is why some people become "tan" after lying outside. However, UV radiation has been shown to cause skin cancer, especially in people with light skin. Thus, excessive exposure to sunlight should be avoided, and sunscreen should be worn when outdoors.

Dermis

The **dermis** is the functional layer of skin that lies just beneath the epidermis. Connective tissue in the dermis makes the skin tough and elastic. The dermis contains many nerve cells, blood vessels, **hair follicles,** and specialized skin cells. Sensations of touch,

Courtroom Science

In the United States, medical evidence is often used in criminal trials. Hairs left at a crime scene can provide enough information to make or break a case. To the unaided eye, the few differences between hairs amount to no more than color, coarseness, and whether the hairs are curly or straight. Using a microscope, however, scientists can distinguish more than two dozen characteristics in a single hair. Experts compare hairs found at a crime scene with hairs of suspects. These comparisons can rule out a suspect when the hairs are different, but they usually are not enough to convict a suspect without other evidence.

DNA fingerprinting

Hair consists of cells that contain DNA. Thus, scientists can get additional evidence from hair through DNA fingerprinting. First used as forensic evidence in the 1980s, DNA fingerprinting is now fairly common in many kinds of criminal trials. By 1997, DNA fingerprinting had been used in more than 50,000 cases in the United States.

Drug testing

Hair can also provide evidence of drug use. Hair collects drugs that are delivered to hair follicles by blood. Hair provides a longer record of drug exposure than blood or urine does. Head hair grows an average of about 1 cm (0.4 in.) per month; a 10 cm (4 in.) length of hair gives evidence of a person's drug use during the last 10 months. Such evidence has been introduced in cases involving illegal drug use.

Forensic scientist

did you know?

An average hair grows about 2 mm per week.

Hair grows at different rates in different regions of the body. The growth rate also differs with age and sex. Hair follicles go through cycles that average 4 years of growth followed by several months of inactivity. With the beginning of a new active growth cycle, the old hair is pushed out (if it has not fallen out already). Out of more than 100,000 hairs, only about 50 hairs a day are lost. Several factors can increase hair loss. These include dietary factors, high fever, stress, drugs, radiation, and some hormonal factors related to pregnancy.

temperature, and pain originate in nerve cells within the dermis. The dermis also contains tiny muscles that are attached to hair follicles in your skin. When you get cold, these muscles contract and pull the hair shafts upright, helping to insulate the body. These muscles also cause goose bumps on the skin's surface.

Temperature regulation A network of blood vessels in the dermis provides nourishment to the living cells of the skin. These blood vessels also help regulate body temperature by either radiating heat into the air or conserving heat. If your body gets too hot, blood vessels just under the skin dilate so that blood flows near the skin's surface, releasing heat from the body. This is why people with light complexions turn slightly red during strenuous exercise, as shown in **Figure 38-16.** If your body gets too cold, the blood vessels constrict so heat loss is reduced.

Sweat glands in the dermis also help remove excess body heat. The evaporation of sweat from the skin's surface removes heat more efficiently than the dilation of blood vessels. Most sweat is about 99 percent water and 1 percent dissolved salts and acids. Certain sweat glands located in body areas with dense hair, such as the armpits, also secrete proteins and fatty acids. Because these substances provide a rich food source for bacteria, stale sweat often releases the offensive odor of bacterial waste products.

Figure 38-16 Cooling mechanism. Flushed (reddened) skin is a sign that the body is overheated and is trying to cool itself.

Subcutaneous tissue

Subcutaneous tissue, located beneath the skin just under the dermis, is a layer of connective tissue made mostly of fat. Subcutaneous tissue acts as a shock absorber, provides additional insulation to help conserve body heat, and stores energy. Subcutaneous tissue also anchors the skin to underlying organs. The thickness of subcutaneous tissue varies in different parts of the body. For example, the eyelids have very little, while the buttocks and thighs may have a lot. The pads of subcutaneous tissue in the soles of your feet may be more than one-fourth of an inch thick.

Hair and nails

Hair and nails are derived from the epidermis. Hair follicles produce individual hairs, which help protect and insulate the body. Shown in **Figure 38-17,** hair is made mostly of dead, keratin-filled cells. A shaft of hair grows up from the hair follicle and through the skin's surface. Each hair on your head grows for several years. Then the follicle enters a resting phase for several months, and the hair is eventually shed. Hair color is primarily determined by the presence of the pigment melanin. Blonde hair and red hair typically contain less melanin than brown hair and black hair.

Nails are produced by specialized epidermal cells located in the light, semicircular area at the base of each nail. These cells become filled with keratin as they are pushed outward by new cells. Nails protect the tips of the fingers and toes, and nails continue to grow throughout life.

Figure 38-17 Hair

How many? An average human head contains about 100,000 hairs.

Teaching TIP

Homeothermy

If a person's body temperature rises above 41°C (106°F), enzymes in the brain begin to denature. Have students describe the appearance of a raw egg white. *(clear, runny, slimy)* Then have them describe the appearance of a cooked egg white. *(white, firm)* Explain that the egg white changes when it is cooked because its proteins become misshapen, or denatured. Enzymes also are proteins, and at high temperatures they no longer function normally.

Teaching TIP

Skin Structures

Use rapid-fire questions to review the layers of the skin and in which layer important structures can be found. Some key items:

artery *(dermis)*

melanin *(epidermis)*

dead cells *(epidermis)*

fat *(subcutaneous tissue)*

middle layer *(dermis)*

Teaching TIP

Skin Layers

Have students make a **Graphic Organizer** similar to the one at the bottom of page 873 to demonstrate the relationship between the skin's two primary layers and the subcutaneous tissue. Below the name of each of the three layers, have students describe its components.

PORTFOLIO

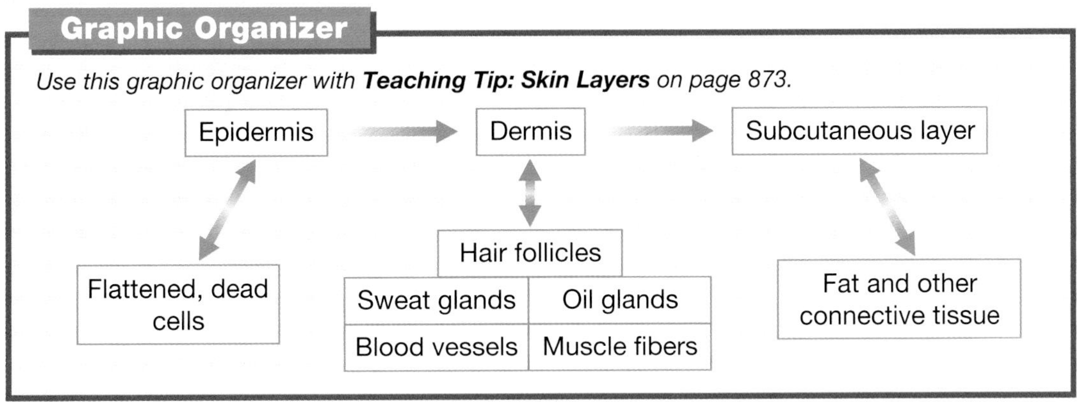

internet**connect**

SCI*LINKS* NSTA Ask students to research skin cancer using the Web site in the Internet Connect box on page 874. Have them write a paragraph on the risk factors and symptoms of two relatively common carcinomas: basal cell carcinoma and squamous cell carcinoma. Ask students to identify the primary risk factor for nearly all types of skin cancer. *(exposure of skin to the sun)* PORTFOLIO

3 Assessment Options

Closure

Technique: K-W-L

Tell students to return to their lists of things they Want to know, which they created using the K-W-L **Active Reading** on page 855. Have them place check marks next to questions that they are now able to answer. Students should finish by making a list of what they've Learned. Conclude by asking students:

- Which questions are still unanswered?
- What new questions do you have?

Review

Assign Review—Section 38-4.

Reteaching — BASIC

Have students make a three-dimensional model of the skin using construction paper, tissue paper, yarn, glue, or other art materials. Ask them to show and explain their models to the class when they finish. PORTFOLIO

Skin Disorders Often Result from Damage to the Epidermis

internet**connect**

SCI*LINKS* NSTA

TOPIC: Skin cancer
GO TO: www.scilinks.org
KEYWORD: HX874

The skin is the most exposed part of the body, and it is therefore continuously exposed to damaging factors such as insect bites, microorganisms, and ultraviolet (UV) radiation from the sun. Injuries such as scrapes and blisters are often minor and usually heal rapidly without permanent scarring. Burns, however, can be very serious and can result in permanent scarring or even death. Some skin disorders are the result of changes that occur within the body over time.

Acne

The most common skin problem for teenagers is acne *(AK nee),* a chronic inflammatory condition that involves the skin's oil-producing glands. Oil glands in the dermis release **sebum** *(SEE buhm),* an oily secretion that lubricates the skin. Sebum is released through ducts, or pores, into nearby hair follicles. These oil glands are especially active during adolescence. Acne is caused by excessive secretion of sebum, which blocks pores with oil, dirt, and bacteria. Makeup and other cosmetic products can contribute to clogging. As a result, the surrounding tissue becomes infected and inflamed, and the pores accumulate pus, producing pimples. Serious acne may need to be treated using antibiotics. Although acne cannot be prevented, it can usually be managed with proper skin care.

Figure 38-18 Skin cancer. In its early stages, a carcinoma may look like a wart. A malignant melanoma often looks like a mole that changes in size, shape, or color.

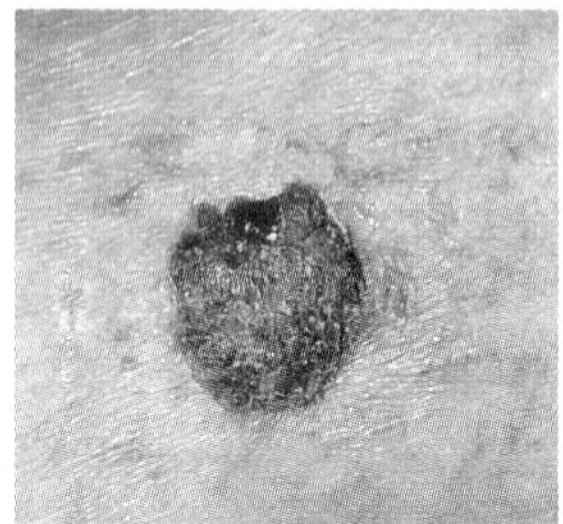

Carcinoma

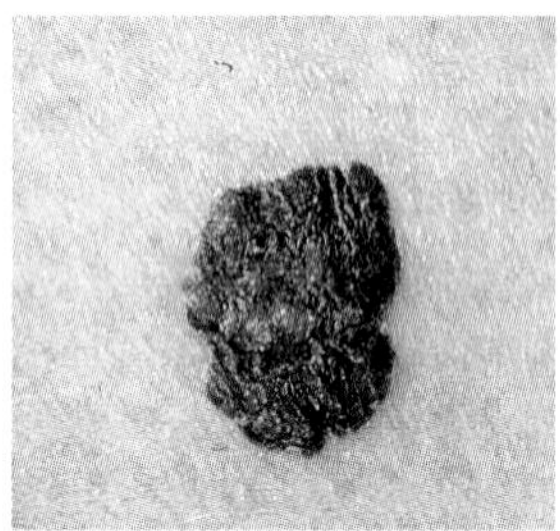

Malignant melanoma

Skin cancer

Skin cancer can result from genetic mutations caused by overexposure to UV radiation. The most common types of skin cancer are carcinomas *(kahr sih NOH mahz),* which originate in skin cells that do not produce pigments. If they are detected early, carcinomas can be treated. A small percentage of skin cancers are caused by mutations that occur in pigment-producing skin cells. These cancers, called malignant melanomas *(mehl uh NOH mahz),* grow very quickly and spread easily to other parts of the body. About 8 out of 10 skin cancer deaths are from malignant melanomas. A carcinoma and a malignant melanoma are shown in **Figure 38-18.** You can reduce the risk of skin cancer by avoiding overexposure to either natural or artificial UV radiation and by using protective sunscreens.

Review SECTION 38-4

1. **Describe** the structure of the epidermis.
2. **List** two ways that the dermis helps regulate body temperature.
3. **Identify** the most common cause of skin cancer and how it can be avoided.

4. **SKILL Identifying Variables** What factors that contribute to acne are controllable?

Critical Thinking

5. **SKILL Recognizing Relationships** Why is a third-degree burn, which destroys the epidermis and dermis of the skin, such a serious injury?

Review SECTION 38-4 ANSWERS

1. The epidermis is composed of a thin layer of flattened, keratin-containing dead cells that have arisen in the dermis below.
2. The dermis helps regulate body temperature by controlling the diameter of blood vessels. A large diameter vessel dissipates heat; a small diameter vessel conserves heat. Also, the dermis releases sweat to cool the body.
3. Overexposure to UV radiation is the most common cause of skin cancer. Using sunscreen and limiting sun exposure can help reduce the risk.
4. dirty skin and use of pore-clogging cosmetics
5. The epidermis and dermis perform several vital functions, including waterproofing the body, keeping pathogens out, and regulating body temperature. Loss of the dermis and epidermis removes the barrier between the interior tissues and the external environment.

Study ZONE CHAPTER 38

Use the Key Concepts and Key Terms listed below to review the main ideas in this chapter. If you need to review the meaning of a term, turn to the page indicated.

Key Concepts

38-1 Body Organization

- Cells are grouped into four types of body tissues: epithelial tissue, nervous tissue, connective tissue, and muscle tissue.
- Body organs contain a mixture of body tissues.
- Organs are grouped into organ systems in which organs interact to perform a certain function, such as digestion.
- Endothermy enables the body to maintain homeostasis at all times, regardless of the temperature outside the body.

38-2 Skeletal System

- The skeleton supports the body, provides protection for internal organs, and enables movement.
- The 206 bones of the skeleton are divided into the axial skeleton and the appendicular skeleton.
- Bones are made of hard compact bone surrounding porous spongy bone.
- Early in development, the skeleton is mostly cartilage. Bones harden as calcium and other mineral deposits build up.
- Bones thicken and elongate as development continues.
- Three kinds of joints fasten bones together: immovable joints, slightly movable joints, and freely movable joints.

38-3 Muscular System

- Muscles are attached to bones by tendons.
- Muscle pairs move parts of the body by pulling on bones.
- Sarcomeres shorten during muscle contraction.
- During muscle contraction, actin and myosin interact.
- Energy is required for muscles to contract.

38-4 Skin, Hair, and Nails

- The skin consists of two layers: the epidermis and the dermis.
- Subcutaneous tissue anchors skin to underlying organs.
- Hair and nails are derived from epidermis.
- Most skin disorders are caused by damage to the epidermis.

Study TIP Ready?

- *If you think you understand these Key Concepts and Key Terms, then you're ready for the Performance Zone.*

Key Terms

38-1

epithelial tissue (856)
nervous tissue (857)
connective tissue (857)
muscle tissue (857)
body cavity (859)

38-2

axial skeleton (860)
appendicular skeleton (860)
bone marrow (861)
periosteum (861)
Haversian canal (862)
osteocyte (862)
osteoporosis (863)
joint (864)
ligament (864)

38-3

tendon (866)
flexor (866)
extensor (866)
actin (867)
myosin (867)
myofibril (867)
sarcomere (867)

38-4

epidermis (871)
keratin (871)
melanin (872)
dermis (872)
hair follicle (872)
subcutaneous tissue (873)
sebum (874)

Study ZONE CHAPTER 38

Review and Practice

Assign the following worksheets for Chapter 38:

- **Student Review Guide**
- **Concept Mapping Worksheet**
- **Test Preparation Pretest**
- **Vocabulary Worksheet**
- **Critical Thinking Review Worksheet**
- **Science Skills Worksheet**

Assessment Options

The following assessment options are available for this chapter:

Resource Link

- Assign **Chapter Test 38** to assess students' understanding of the chapter.

Alternative Assessment

Have students write trivia questions for a review game. These questions can be based on the review questions found at the end of each section. PORTFOLIO

Portfolio Assessment

- *Portfolio Assessment* at the front of this book provides options for building and evaluating student portfolios. Students and teachers can select items from the areas listed for inclusion in a portfolio.
- See items labeled PORTFOLIO on pp. 858, 859, 861, 863, 864, 865, 869, 870, 872, 873, and 874.

Answer to Concept Map

The following is one possible answer to Performance Zone item 1 on page 876.

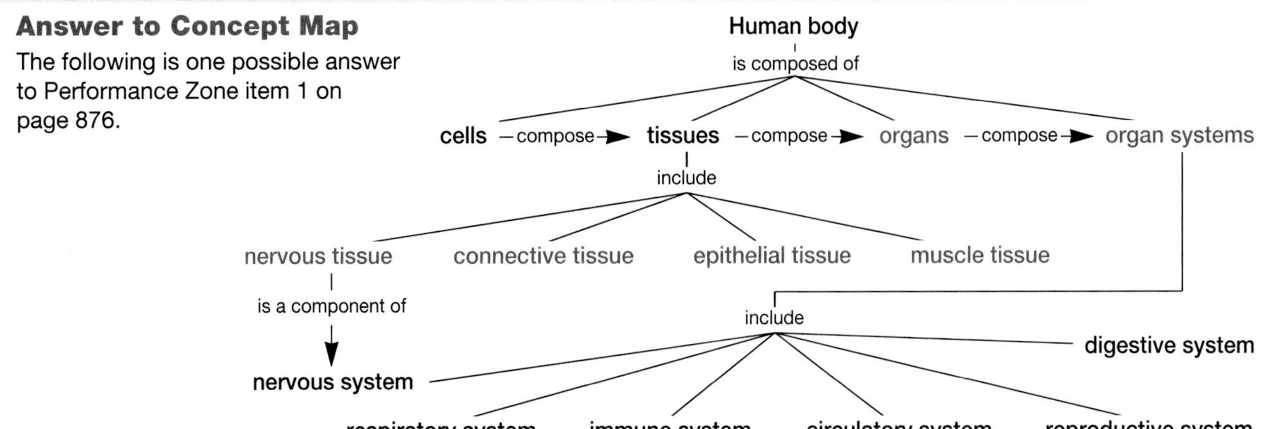

Performance ZONE
CHAPTER 38
REVIEW ANSWERS

Assignment Guide

SECTION	REVIEW
38-1	1, 2, 3, 26
38-2	4, 5, 6, 14, 21, 23
38-3	7, 8, 9, 16, 17, 18, 19, 20, 24
38-4	10, 11, 12, 13, 15, 22, 25, 27

Understanding and Applying Concepts

1. The answer to the concept map is found at the bottom of page 875.
2. **a.** A tissue is a group of cells working together; an organ is a group of tissues working together to perform a specific function.
   b. The epidermis is the outermost layer of the skin; the dermis is the layer of skin that lies just beneath the epidermis.
   c. The axial skeleton includes the bones of the skull, spine, ribs, and sternum; the appendicular skeleton includes the bones of the arms, legs, pelvis, and shoulders.
   d. A flexor is a muscle that causes a joint to bend; an extensor is a muscle that causes a joint to straighten.
3. c   8. a
4. d   9. c
5. b   10. b
6. c   11. a
7. c   12. d
13. increased secretion of oil, clogged pores, dirt, makeup, and bacteria
14. ball and socket—shoulder, pivoting; hinge—knee, bending and straightening
15. DNA fingerprinting and analysis of hair, blood, and urine samples. Answers will vary, but identification by DNA sampling is the most specific.
16. Muscle contraction requires ATP, which is produced when food is broken down during cellular respiration.

Performance ZONE CHAPTER 38 Review

Understanding and Applying Concepts

1. **Concept Mapping** Make a concept map that illustrates the body's four levels of structural organization. Try to include the following terms in your map: muscle tissue, connective tissue, epithelial tissue, nervous tissue, organ, organ system. Use more terms as needed.
2. **Understanding Vocabulary** For each pair of terms, explain the differences in their meanings.
   a. tissue, organ
   b. epidermis, dermis
   c. axial skeleton, appendicular skeleton
   d. flexor, extensor
3. The thoracic cavity contains
   a. the spinal cord.
   b. the brain.
   c. organs of the respiratory system.
   d. organs of the reproductive system.
4. Which of the following is a function of the skeletal system?
   a. support   c. movement
   b. protection   d. all of the above
5. The bones of the arms and legs form
   a. the axial skeleton.
   b. the appendicular skeleton.
   c. the pectoral girdle.
   d. vertebrae.
6. The outer membrane of bone is the
   a. marrow.   c. periosteum.
   b. spongy bone.   d. Haversian canal.
7. The bone at the end of a muscle attachment that does not move during muscle contraction is called the
   a. tendon.   c. origin.
   b. insertion.   d. flexor.
8. Shortening of sarcomeres causes
   a. muscles to contract.
   b. Z lines to move apart.
   c. muscles to relax.
   d. None of the above
9. Resistance exercises
   a. decrease muscle endurance.
   b. decrease muscle strength.
   c. increase muscle size.
   d. increase the number of muscle cells.
10. Which of the following is made mostly of connective tissue?
   a. epidermis   c. hair
   b. dermis   d. none of the above
11. The dermis helps regulate body temperature by producing
   a. sweat.   c. oil.
   b. acne.   d. sebum.
12. The risk of developing skin cancer is increased by
   a. eating oily foods.   c. exercising.
   b. using sunscreen.   d. sunbathing.
13. What factors contribute to acne, shown in the photograph below?

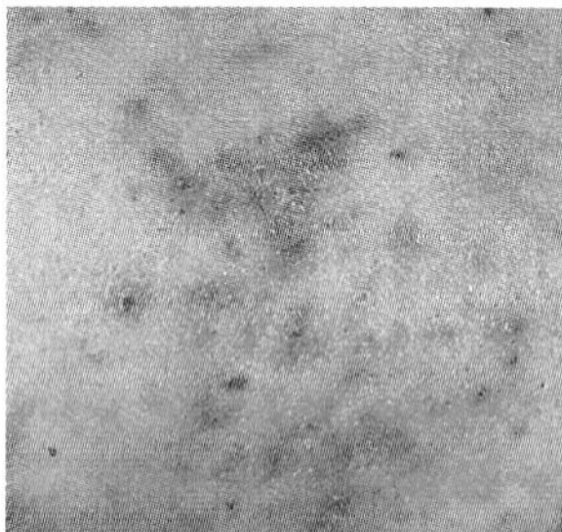

14. **Theme Structure and Function** Give an example of a ball-and-socket joint and a hinge joint. What type of movement is permitted by each of these joints?
15. **Tech Watch** What are some ways in which forensic science can help in identifying a criminal? Which method produces the most compelling evidence?
16. **Chapter Links** Relate the events of muscle contraction to cellular respiration. (**Hint:** See Chapter 5, Section 5-3.)

Skills Assessment

17. a sarcomere
18. It has contracted.
19. Answers will vary but should include that muscles marbled with fat are more tender and that the more a muscle is used, the tougher it tends to be.
20. Neuromuscular disorders are usually genetic in nature, although some muscular dystrophies may be autoimmune in origin. Symptoms of Lou Gehrig's disease (amyotrophic lateral sclerosis) include muscle tissue degeneration and paralysis, followed by death. In the muscular dystrophies, muscle tissue gradually is replaced by fat, leading to immobility. Muscle degeneration often affects the ability to breathe, leading to respiratory infections.

Skills Assessment

Part 1: Interpreting Graphics

Study the figure below, and then answer the following questions:

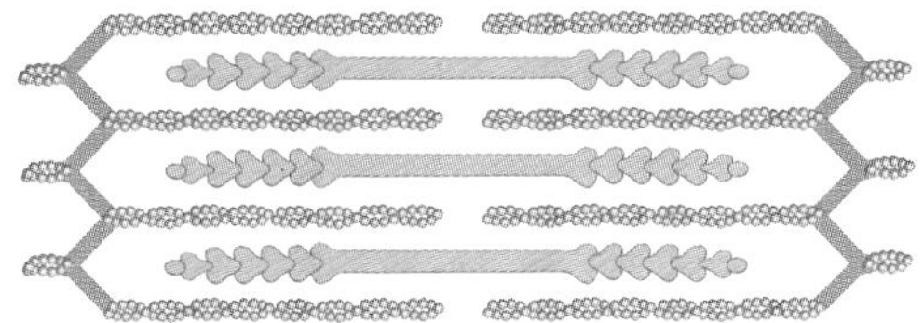

A

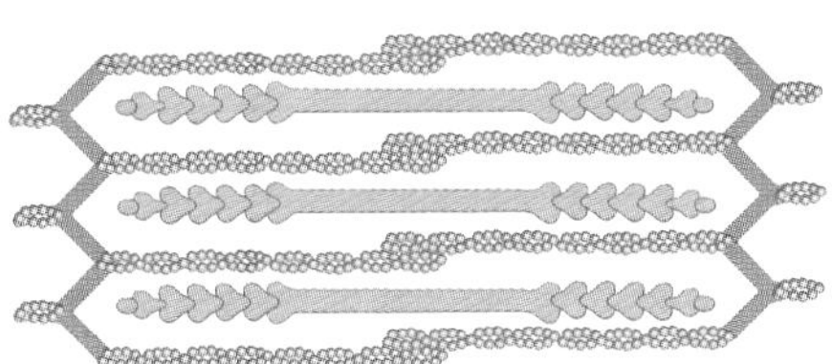

B

17. What structure is shown in part *A*?

18. What has happened to this structure in part *B*?

Part 2: Life/Work Skills

Use the media center or Internet resources to learn more about muscles.

19. Finding Information Visit a grocery store and go to the butcher section. Make a list of the different cuts of meat, such as beef and pork, that are available for human consumption. Find out the names of the different muscle groups from which these pieces are cut. Why are some cuts of meat more tender than other cuts? Why are some cuts of meat tough?

20. Communicating Find out about muscular disorders such as muscular dystrophy and Lou Gehrig's disease. What are the causes and symptoms of these disorders? Find out how these disorders are treated. Summarize your findings in an oral report.

Critical Thinking

21. Evaluating Conclusions Young thoroughbred horses that are raced too early in life have an increased risk of breaking the bones in their legs. Using this information, what can you infer about bone development in horses?

22. Predicting Outcomes Oil glands in the skin secrete a substance that helps kill bacteria. What might happen if you washed your skin too often?

23. Inferring Relationships Red blood cells are produced in red bone marrow. How are red blood cells transported to the rest of the body?

Portfolio Projects

24. Identifying Structures Visit a gym that has exercise machines. Test several of the machines, and determine which muscle groups are worked by each machine. Identify other activities that work the same muscle groups. Document the visit with photographs or a videotape. Make a list of each machine and the muscle groups worked. Present your findings in an oral report to your class.

25. Recognizing Patterns Research the incidence of skin cancer in different parts of the world. Identify similarities and differences in the types of skin cancer. Present your findings in a written or oral report.

26. Interpreting Information Read the article titled "How to Build a Body Part" (*Time,* March 1, 1999, pp. 54–55). How are researchers using existing cells to "grow" new organs for organ transplant? What is the benefit of being able to make new organs available for transplant?

27. Career Focus Cosmetologist Research the field of cosmetology, and write a report on your findings. Your report should include a job description, academic degrees and training required, and an average starting salary.

Critical Thinking

21. Bone development in horses is similar to that in humans. The bones grow longer and stronger with age.

22. Washing skin too often can deplete the natural bacteria-killing substance, resulting in an overgrowth of bacteria.

23. Red blood cells enter small blood vessels in bone and are transported to the rest of the body by the circulatory system.

Portfolio Projects

24. Answers will vary, but students should find machines that work the triceps, biceps, quadriceps, deltoid, and pectoralis muscle groups.

25. Answers will vary. Skin cancer is more common in fair-skinned people and in places with intense sunshine, such as Arizona and Hawaii. It is very common in Australia because many residents are fair-skinned people of European descent who have little natural melanin to protect them from the sun.

26. Researchers assemble frameworks out of very fine polymers and infuse them with cells. They add nutrients, which encourage the cells to develop. The demand for transplantable organs exceeds supply. Growing organs for transplantation could solve this problem. Moreover, if the organ is produced from cells from a person's own body, there is less chance for organ rejection.

27. Answers will vary. Cosmetologists perform many services, including caring for the hair, skin, and hands. They need to be able to give shampoos, rinses, and scalp treatments; style, set, cut, dye, tint, bleach, and apply permanent waves to hair; give facials and manicures; and massage hands and arms. Cosmetologists may study bacteriology, anatomy, hygiene, sanitation, salon management (including record keeping), and customer relations. Vocational or instructional schooling and a license are required. Growth prospects are very good. Starting salary will vary by region.

Life Span and Lifestyle

Purpose

Students will explore how exercise, diet, smoking, and drinking alcohol affect life span.

Preparation Notes

Time Required: 40–50 minutes

Preparation Tips
Students should work individually to have the opportunity to estimate their own life span.

Procedural Tips
Before students begin the lab, they should be able to answer the following questions:

- Why do physicians encourage their patients to eat a diet low in saturated fat? *(Diets high in saturated fat increase the likelihood of cardiovascular problems, such as stroke and heart disease.)*
- Which foods contain saturated fat? *(In general, animal products, such as meat, eggs, milk, and cheese, are high in saturated fat. Some plant products, such as coconut oil and palm oil, also contain saturated fat.)*
- What is cancer? *(Cancer is a disease characterized by uncontrolled cell division and cellular immortality caused by mutations in the genes that control cell reproduction.)*

Additional Background

A century ago, infectious diseases like pneumonia and influenza were the leading causes of death in the United States. Today, the two leading killers are cardiovascular disease (including heart attack and stroke) and cancer. The risk of dying from either of these two diseases depends, to a degree, on factors such as diet, amount of exercise, and number of cigarettes smoked.

Interactive Exploration

Chapter 38

Life Span and Lifestyle

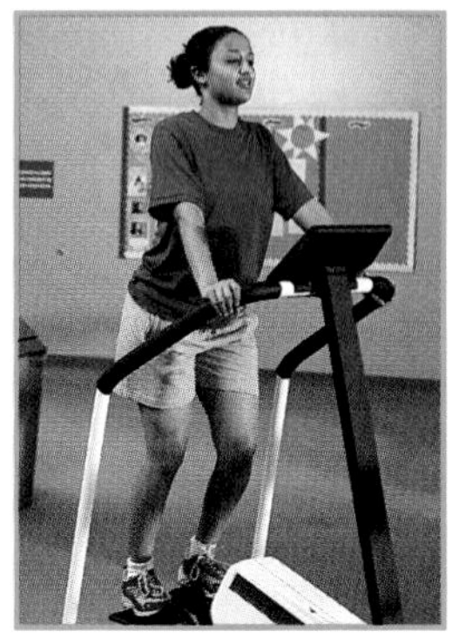

CD-ROM Lab

SKILL
- Computer modeling

OBJECTIVES
- **Compare** the average life spans of males and females.
- **Simulate** how lifestyle affects life expectancy.
- **Predict** your life expectancy based on your current lifestyle.

MATERIALS
- computer with CD-ROM drive
- CD-ROM *Interactive Explorations in Biology: Human Biology*

Before You Begin

How long will you live? Although genes play a role in longevity, your lifestyle is also very important. Some important factors include how much you exercise, what you eat, and whether you smoke or drink alcohol. In this Interactive Exploration, you will explore how these factors affect life span.

1. Estimate how much energy you use daily, the percentage of animal fat in your diet, and the energy content of your diet. Your teacher will explain how to do this.
2. Select *Life Span and Lifestyle* from the menu. Click on the key near the top of the navigation palette. Read the focus questions, and review the following concepts: life expectancy, smoking and health, diet and health, and exercise and health.
3. Click *File* at the top left of the screen, and select *Interactive Exploration Help.* Listen to the instructions. Click the *Exploration* button at the bottom right of the screen to begin the exploration. You will see a diagram like the one below.
4. Make a data table like the one on page 879.

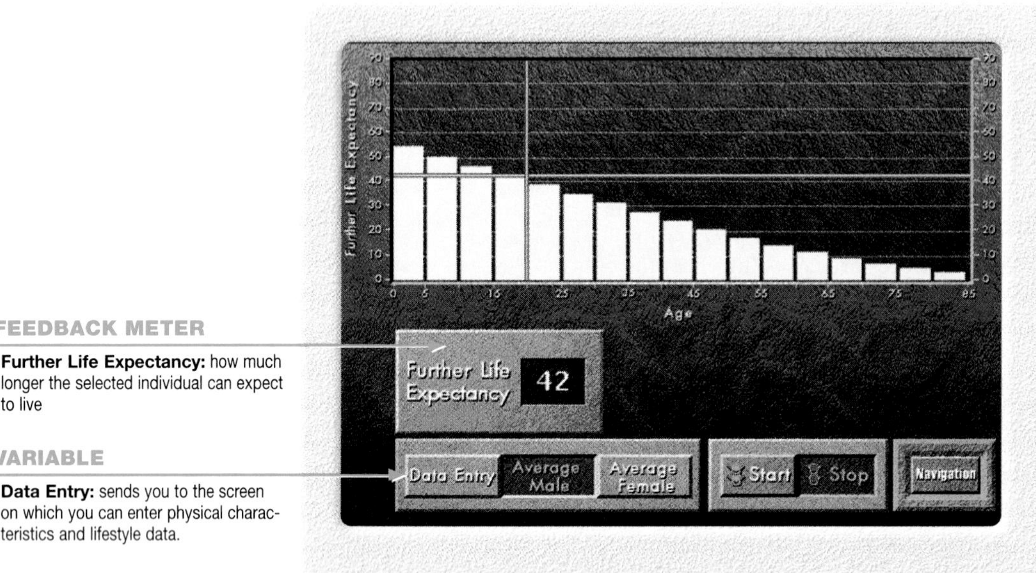

Before You Begin

Answers

Before beginning this exploration, students will need to calculate three values: the caloric content of their daily diet, the percentage of their caloric intake contributed by animal fat (saturated fat), and the average amount of energy they use each day. Tell students to choose an average day that is representative of their routine. On this day, students should make a detailed list of all the foods they eat, being careful to record all snacks and the number and sizes of portions. They should also record which activities they took part in and how long they spent at each activity. Students should use a nutrition book or one of the calorie-counting guides available in bookstores to determine the total number of calories in the food they ate. In the exploration, they should enter this value as the variable, Calories in diet. Have students use the same resource to calculate what percentage of their calories came from saturated fat. They will enter this value in the program as the variable, Animal fat in diet. Remind students that this value should be a percentage. Using their log of daily activities, students should determine

DATA TABLE

Age	Sex	Height (in.)	Weight (lb)	Exercise (Cal/day)	Animal fat in diet (%)	Smoking (cigarettes/day)	Calories in diet (kcal/day)	Further life expectancy

Procedure

PART A: Average Life Spans

1. The graph on the screen is called a life expectancy histogram. Each bar shows how many more years an individual of a certain age is expected to live. This graph is for an average male. The *Further Life Expectancy* box shows how many more years an average 20-year-old male is expected to live. Add this number to 20 to find the average life span of males.
2. Now click the *Average Female* button. Add the number for further life expectancy to 20 to find the average life span of females. How does the average life span of males compare with that of females?

PART B: Effect of Lifestyle

3. Click the *Data Entry* button. Slide the indicator for each scale to the following settings: height, 70 in.; weight, 130 lb; exercise, 760 Cal/day; animal fat in diet, 10 percent; smoking, 0 cigarettes/day; alcohol, 0 drinks/day; and diet, 2,480 kcal/day. Set the age to 15 years, and click the *Female* button under *Sex*. Record the value for each variable in your data table. Click the *Close* button and then the *Start* button. Record the value for further life expectancy in your data table.
4. Repeat step 3, but change the exercise scale to 1,520 Cal/day.
5. Click the *Data Entry* button. Enter the following data for an 18-year-old male: height, 70 in.; weight, 170 lb; exercise, 1,000 Cal/day; animal fat in diet, 20 percent; smoking, 0 cigarettes/day; alcohol, 0 drinks/day; and diet, 2,240 kcal/day. Record the values in your table. Click the *Close* button and then the *Start* button. Record the value for further life expectancy in your data table.
6. Repeat step 5, but change the smoking scale to 5 cigarettes/day.
7. Repeat step 5 again, but change the smoking scale to 20 cigarettes/day.
8. Click the *Data Entry* button. Enter the data for your age, sex, height, weight, and lifestyle, including the data from step 1 of **Before You Begin.** Click the *Close* button and then the *Start* button to determine your further life expectancy.

Analyze and Conclude

1. **Analyzing Data** How does smoking affect an 18-year-old male's life expectancy?
2. **Analyzing Data** How does exercise affect a 15-year-old female's life expectancy?
3. **Evaluating Conclusions** Assume that the 18-year-old male in step 5 smokes 10 cigarettes/day. He increases his exercise to 1,200 Cal/day, thinking that more exercise will counteract the effect of smoking and extend his life expectancy. Enter the data, and evaluate his decision.

Use the following Internet resources to explore your own questions about the relationship between lifestyle and life span.

Interactive Exploration

38 Chapter

Procedure

Answers

1. The average life expectancy of an average 20-year-old male is 72 years.
2. The average life expectancy of an average 20-year-old female is 79 years, 7 years longer than that for males.
3. The life expectancy is an additional 73 years, or 88 years total.
4. The second (and athletic) 15-year-old is expected to die 8 years later than the first, sedentary 15-year-old does.
5. The life expectancy for the smoking 15-year-old is an additional 51 years, or a total of 69 years.
6. Smoking five cigarettes per day reduces life expectancy by about 2 years.
7. Smoking 20 cigarettes per day reduces life expectancy by about 8 years.
8. Students' predicted life spans will vary. Among the ways they might increase their life span are exercising more, eating less, reducing the fat content of their diet, and not smoking.

Analyze and Conclude

Answers

1. For an 18-year-old male, smoking five cigarettes per day decreases life expectancy by about 2 years. Smoking 20 cigarettes per day reduces life expectancy by about 8 years.
2. Exercise increases a 15-year-old female's life expectancy by 8 years.
3. Although exercise does increase his life expectancy relative to no exercise, this gain is more than offset by the effects of smoking.

their daily expenditure of energy. The list at right shows estimates for the amount of energy that a person of average weight uses during various activities. Students can find more information in a health book or in various books on exercise.

The program presents two kinds of data: a single value for further life expectancy and a histogram showing the further life expectancy for each of several age classes. Students may wonder why the histogram shows values for ages above the life expectancy. Point out that the single life-expectancy value is an average, meaning that some people are expected to die before they reach this age, while others are expected to live longer.

Activity	Calories used/hour
Sleeping	130
Sitting	140
Reading	140
Classwork	160
Walking	400–480*
Cycling	360–780*
Football	860
Swimming	420*
Tennis	480–720*
Running	660–1,560*
Soccer	600

* Energy expenditure varies with effort.

Chapter 39 Circulatory and Respiratory Systems

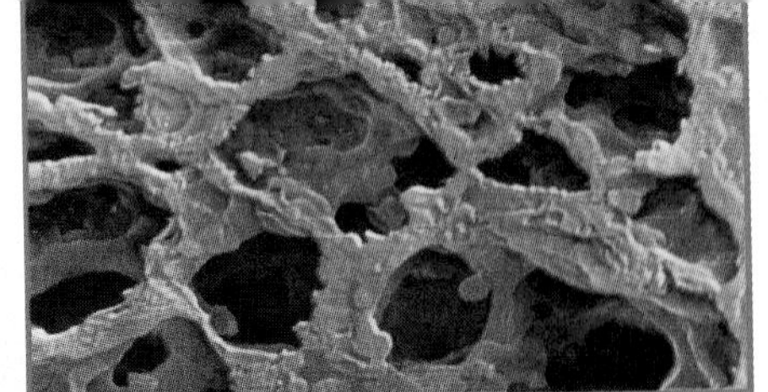
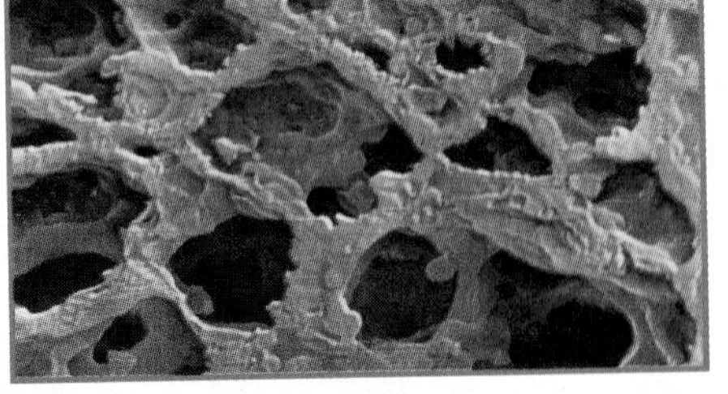

Planning Guide

	OBJECTIVES	MEETING INDIVIDUAL NEEDS	READING SKILLS AND STRATEGIES
BLOCKS 1 & 2 90 min	**39-1 The Circulatory System** • **List** five types of molecules that are transported by the circulatory system. • **Differentiate** between arteries, capillaries, and veins. • **Relate** the function of the lymphatic system to the functions of the circulatory and immune systems. • **Relate** each component of blood to its function. • **Summarize** how a person's blood type is determined.	Learning Styles, p. 882 BASIC ADVANCED • p. 889	**ATE** Active Reading, Technique: Reading Effectively, p. 881 ■ Active Reading Guide, worksheet (39-1) ■ Directed Reading, worksheet (39-1)
BLOCK 3 45 min	**39-2 The Heart** • **Differentiate** the pulmonary circulation loop from the systemic circulation loop. • **Summarize** the path that blood follows through the heart. • **Name** the cluster of heart cells that initiates contraction of the heart. • **Describe** three ways to monitor the health of the circulatory system. • **State** two vascular diseases, and identify factors that contribute to their development.	Learning Styles, p. 890 BASIC ADVANCED • p. 894	■ Active Reading Guide, worksheet (39-2) ■ Directed Reading, worksheet (39-2)
BLOCKS 4 & 5 90 min	**39-3 The Respiratory System** • **Summarize** the path that air follows when it enters the body through the nose or mouth. • **Describe** the role of the rib muscles and diaphragm in breathing. • **Describe** how the rate of breathing is regulated. • **Summarize** how oxygen and carbon dioxide are transported in the blood. • **Identify** three respiratory diseases.	Learning Styles, p. 895 BASIC ADVANCED • p. 900	**ATE** Active Reading, Technique: Reading Organizer, p. 898 ■ Active Reading Guide, worksheet (39-3) ■ Directed Reading, worksheet (39-3) ■ Supplemental Reading Guide, *Microbe Hunters*

Review and Assessment Resources

BLOCKS 6 & 7 90 min

TEST PREP

- **PE** Study Zone, p. 901
- **PE** Performance Zone, pp. 902–903
- ■ Chapter 39 worksheets:
  - Vocabulary
  - Concept Mapping
  - Critical Thinking
- Audio CD Program
- ■ Test Prep Pretest

CHAPTER TESTING

- ■ Chapter Tests (blackline master)
- ▲ Test Generator
- ■ Assessment Item Listing

See the correlations table at the front of this book for a list of the ***National Science Education Standards*** addressed in this chapter.

Smithsonian Institution® Visit **www.si.edu/hrw** for additional on-line resources.

KEY

Biology Principles and Explorations

PE Pupil's Edition
ATE Teacher's Edition

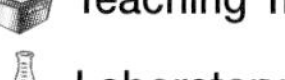

HOLT BIOSOURCES

Teaching Transparencies
Laboratory Program

One-Stop Planner CD-ROM
Shows these resources within a customizable daily lesson plan:

■ Teaching Resources
▲ Test Generator

VISUAL STRATEGIES	CLASSWORK AND HOMEWORK	TECHNOLOGY AND INTERNET RESOURCES
ATE **Fig. 39-6,** p.887 186A Lymphatic System 188A Clotting Cascade 189A Blood Types 213 Circulation Pathway in the Human Body	**PE** **Quick Lab** Can you map the valves in veins?, p. 885 Lab A32 Relating Structure to Function Lab C42 Blood Typing Lab C43 Blood Typing—Whodunit? ■ Occupational Applications: Blood-Bank Technologist **PE** Review, p. 889	Holt Biology Videodiscs Section 39-1 Audio CD Program Section 39-1 internetconnect SCILINKS NSTA TOPIC: White blood cells
ATE **Fig. 39-10,** p. 891 185A Electrocardiogram 210 Structure of the Human Heart 214 Circulatory Loops in the Human Body	■ **Problem-Solving:** Computing Rates and Heart Efficiency **PE** **Review,** p. 894	CNN Video Segment 31 High Blood Pressure Holt Biology Videodiscs Section 39-2 Audio CD Program Section 39-2
215 Respiratory System in the Human Body 217 Overview of Respiration	**PE** **Math Lab** Calculating the amount of air respired, p. 896 **PE** **Quick Lab** What is the role of bicarbonate in homeostasis?, p. 899 **PE** **Lab Techniques** Determining Lung Capacity, pp. 904–905 ■ Occupational Applications: Respiratory Therapist **PE** **Review,** p. 900 **PE** **Study Zone,** p. 901	Holt Biology Videodiscs Section 39-3 Audio CD Program Section 39-3

ALTERNATIVE ASSESSMENT

PE **Portfolio Projects** 27–30, p. 903
ATE **Alternative Assessment,** p. 901
ATE **Portfolio Assessment,** p. 901

Lab Activity Materials

Quick Lab
p. 885 felt-tip pen
p. 899 250 mL beakers (2), 250 mL distilled water, 2.8 g baking soda, glass stirring rod, strips of wide-range pH paper (4), drinking straws (2)

Math Lab
p. 896 pencil, paper

Lab Techniques
p. 904 spirometer, mouthpiece

Lesson Warm-up
ATE plastic syringes (3), 10 in. plastic tubing, p. 890

Demonstrations
ATE water, red food coloring, test tube, vegetable oil, p. 886
ATE balloon, glass tube, 1-holed rubber stopper, bell jar, rubber band, thin sheet of rubber, p. 897
ATE bromothymol blue, water, drinking straw, large beaker, p. 898

BIOSOURCES LAB PROGRAM
- Quick Labs: Lab A32, p. 63
- Lab Techniques and Experimental Design: Lab C42, p. 201
- Lab Techniques and Experimental Design: Lab C43, p. 207

Chapter 39

Chapter Theme

Structure and Function

A close relationship exists between the circulatory and respiratory systems. The circulatory system transports gases and other important solutes throughout the body. The respiratory system captures and releases gases and provides a moist surface across which gases are exchanged between the lungs and the blood.

Establishing Prior Knowledge

Before beginning this chapter, make certain students can answer the questions in Study TIP **Ready?** on page 881. If they cannot, encourage them to reread the sections indicated.

Opening Demonstration

Have one student take the resting pulse of a volunteer, and have another student determine the volunteer's breathing rate. Have the volunteer do jumping jacks or other aerobic exercises, and then remeasure and record his or her pulse and breathing rate. Ask students: How do the pulse and breathing rate vary before and after exercise? *(Both increase after exercise.)* Why would the pulse rate increase during exercise? *(The heart must pump faster to ensure that active muscles get enough food and oxygen.)* Why does the breathing rate increase during exercise? *(The body needs more oxygen and needs to get rid of more carbon dioxide.)*

CHAPTER 39 Circulatory and Respiratory Systems

Teen playing soccer

What's Ahead — and Why

CHAPTER OUTLINE

39-1 The Circulatory System
- Materials Are Transported and Heat Is Distributed in the Body
- Blood Circulates Through a Network of Vessels
- Blood Has Both Liquid and Solid Components

39-2 The Heart
- The Heart Is a Muscular Pump That Cycles Blood
- Blood Follows a Specific Path

39-3 The Respiratory System
- Gases Must Be Exchanged with the Environment
- Breathing Is Caused by Pressure Changes Within the Chest Cavity
- O_2 and CO_2 Are Transported in the Blood
- Respiratory Diseases Limit Lung Function

Students will learn

- about the structure and function of blood and vessels.
- about the structure of the heart and the pathways of blood through the body.
- how breathing works and about the process of gas exchange.

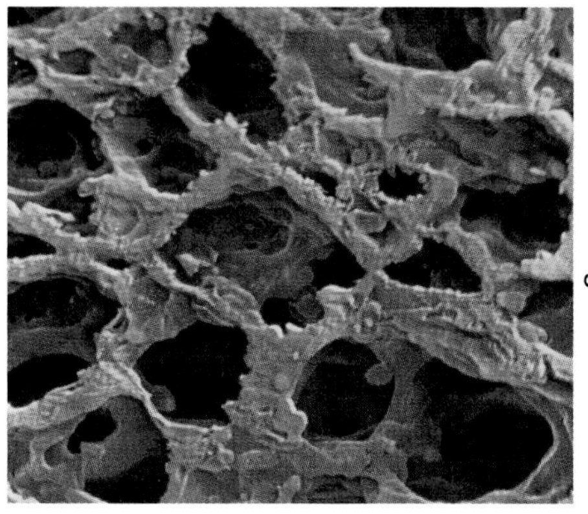

Section of human lung

Blood and air are needed by *soccer players, not only to play but also to stay alive. Human bodies have two elaborate systems that provide these essential components—the circulatory system and the respiratory system.*

Study TIP

Ready?

Check your understanding of these topics to help you prepare for what's ahead.

Can you . . .

- **define** *homeostasis*? (Section 1-1)
- **define** the terms *diffusion* and *osmosis*? (Section 4-1)
- **summarize** the role of oxygen in aerobic respiration? (Section 5-3)
- **describe** cardiac muscle, smooth muscle, epithelial tissue, and connective tissue? (Section 38-1)

If not, *review the sections indicated.*

Looking Ahead

Labs

Features

internetconnect

National Science Teachers Association *sci*LINKS Internet resources are located throughout this chapter.

Chapter 39

Reading Strategies

Active Reading

Technique: Reading Effectively

Have students read a paragraph and write one sentence to summarize the main point of that paragraph. Repeat for each section or chapter.

Directed Reading

Assign **Directed Reading Worksheet Chapter 39** to help students interact with the text as they read the chapter.

Interactive Reading

Assign ***Biology: Principles and Explorations* Audio CD Program Chapter 39** to help reluctant readers achieve greater success in reading the chapter.

Reading Skills

Assign Reading Organizers and Reading Strategies Worksheets from the **One-Stop Planner CD-ROM with Test Generator** to help students learn and practice effective reading skills.

internetconnect

Holt *Biology: Principles and Explorations*

On-line Resources: go.hrw.com

Visit the HRW Web site for a variety of resources related to this chapter. Just type in the keyword HX0 Chapter 39.

Holt *Biology: Principles and Explorations*

On-line Resources: www.scilinks.org

The following *sci*LINKS Internet resources can be found in the student text for this chapter:

Page 887
TOPIC: White blood cells
KEYWORD: HX887

Chapter 39

1 Prepare

Scheduling

- **Block:** About 90 minutes will be needed to complete this section.
- **Traditional**: About two class periods will be needed to complete this section.
- **Planning:** See the Planning Guide on pp. 879A and 879B for lecture, classwork, and assignment options for Section 39-1. Lesson plans for Section 39-1 can be found in a lesson cycle format in ***Biology: Principles and Explorations*** **One-Stop Planner CD-ROM with Test Generator.**

Resource Link

Assign **Active Reading Worksheet Section 39-1** to help students comprehend the key concepts and visuals in this lesson.

2 Teach

Lesson Warm-up

Ask students to name possible functions of the circulatory system, and write them on the board. *(The circulatory system carries oxygen, carbon dioxide, and nutrients; it regulates body temperature; and it is important to immune function.)* Add to the list as students read this chapter.

Section 39-1 The Circulatory System

Objectives

- **List** five types of molecules that are transported by the circulatory system. (p. 882)
- **Differentiate** between arteries, capillaries, and veins. (pp. 883–884)
- **Relate** the function of the lymphatic system to the functions of the circulatory and immune systems. (pp. 884–885)
- **Relate** each component of blood to its function. (pp. 886–888)
- **Summarize** how a person's blood type is determined. (pp. 888–889)

Key Terms

artery
capillary
vein
valve
lymphatic system
plasma
red blood cell
anemia
white blood cell
platelet
ABO blood group system
Rh factor

Materials Are Transported and Heat Is Distributed in the Body

Regardless of your activities—whether you are roller-blading, swimming, singing, reading, or just sleeping—your body must transport nutrients, hormones, and gases, and it must get rid of wastes. Two body systems play major roles in these functions. The circulatory system transports these materials to different parts of the body. The respiratory system exchanges gases with the environment—it takes in oxygen, O_2, and releases carbon dioxide, CO_2.

The human circulatory system, shown in **Figure 39-1,** functions like a network of highways. The circulatory system connects the muscles and organs of the body through an extensive system of blood vessels that transport blood, a mixture of specialized cells and fluid. The heart, a muscular pump, propels the blood through the blood vessels.

Different kinds of molecules move through the circulatory system:

1. Nutrients from digested food are transported to all cells in the body through the blood vessels.
2. The circulatory system also transports oxygen from the lungs, where the oxygen is taken in, to all cells.
3. Metabolic wastes, such as carbon dioxide, are transported in blood vessels to the organs and tissues that excrete them.
4. Hormones, substances which help coordinate the many activities of the body, are transported through the blood vessels.
5. The circulatory system also distributes heat more or less uniformly in order to maintain a relatively constant body temperature. In a warm environment, blood vessels in the skin dilate (relax) to allow more heat to move from the body into the environment. In a cold environment, blood vessels constrict to conserve heat by diverting blood to deeper tissues. This diversion of blood prevents heat from escaping to the outside environment.

Circulatory System

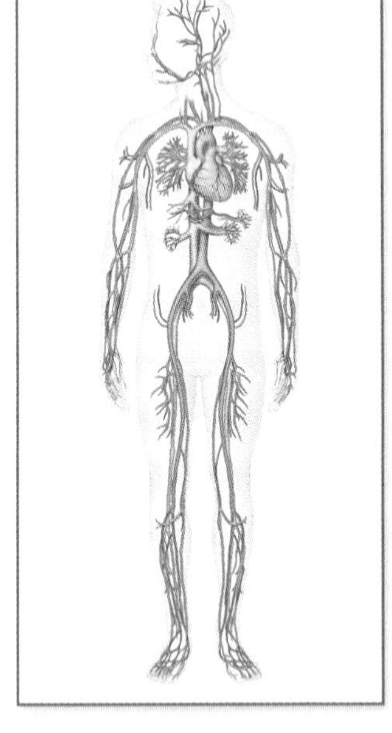

Figure 39-1 Blood vessels, blood, and a heart. The circulatory system transports materials throughout the body and distributes heat.

Integrating Different Learning Styles

Logical/Mathematical	**PE**	Performance Zone item 25; p. 903
	ATE	Cross-Disciplinary Connection, p. 886
Visual/Spatial	**PE**	Biobit, p. 884; Performance Zone items 1 and 3, p. 902
	ATE	Demonstration, p. 886; Visual Strategy, p. 887; Teaching Tip, p. 887
Body Kinesthetic	**PE**	Quick Lab, p. 885
Verbal/Linguistic	**PE**	Study Tip, p. 883; Performance Zone items 2 and 27, pp. 902–903
	ATE	Lesson Warm-up, p. 882; Resource Link, p. 887

Blood Circulates Through a Network of Vessels

Blood circulates through the body through a network of vessels. **Arteries** *(AHRT uh reez)*, shown in **Figure 39-2,** are blood vessels that carry blood away from the heart. The blood passes from the arteries into a network of smaller arteries called arterioles *(ahr TIHR ee ohls)*. Eventually, the blood is pushed through to the capillaries.

Capillaries are tiny blood vessels that allow the exchange of gases, nutrients, hormones, and other molecules in the blood. The molecules are exchanged with the cells of the body. From the capillaries, the blood flows into small vessels called venules *(VEHN yools)* before emptying into larger vessels called veins *(vaynz)*. **Veins** are blood vessels that carry the blood back to the heart.

Arteries

With each contraction, the heart forcefully ejects the blood into the arteries. To accommodate each forceful pulse of blood, an artery's wall expands and then returns to its original size. Elastic fibers in the walls of arteries allow arteries to expand.

The wall of an artery is made up of three layers of tissue, as shown in Figure 39-2. The innermost layer is a thin layer of epithelial tissue called the endothelium. The endothelium is made up of a single layer of cells. Surrounding the endothelium is a layer of smooth muscle tissue with elastic fibers. Finally, a protective layer of connective tissue with elastic fibers wraps around the smooth muscle tissue. Just as a balloon expands when you blow more air into it, the elastic artery expands when blood is pumped into it.

Study TIP

Reviewing Information

You can remember that arteries take blood away from the heart and veins take blood toward the heart by remembering the letter *a* at the beginning of the word *artery* and at the beginning of the word *away.*

Figure 39-2 Blood vessels

Blood vessels transport blood and allow for the exchange of substances.

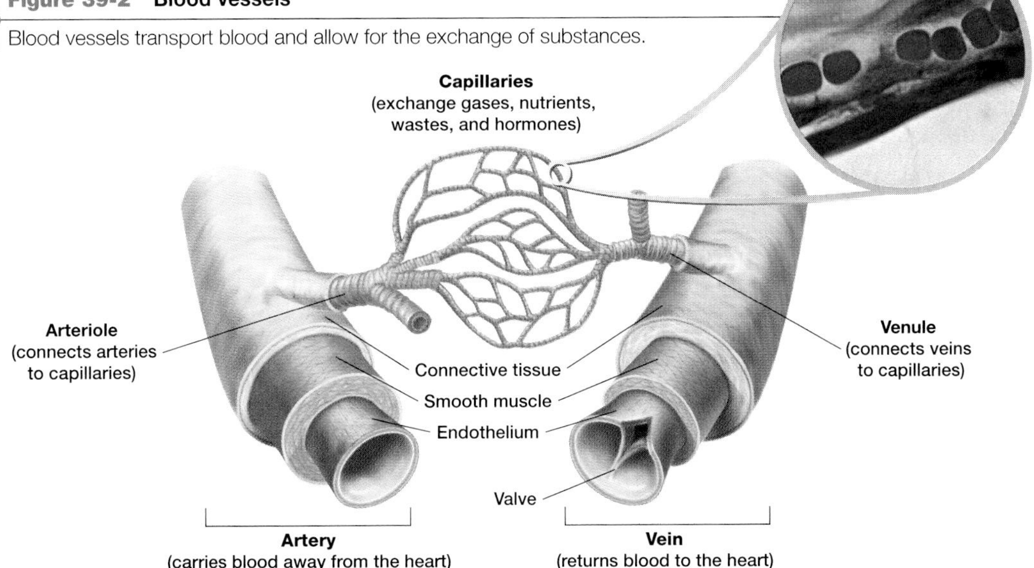

CROSS-DISCIPLINARY CONNECTION

◆ History

Prior to the mid-1500s, the movement of blood through the body was not understood. William Harvey, an English physician and anatomist, postulated the muscular nature of the heart and its function in pumping blood into the arteries. Harvey's mentor, Fabricus ab Aquapendente, an accomplished Italian surgeon and anatomist, discovered valves in veins. Building on this knowledge, Harvey used tourniquets to ascertain the direction of blood flow in the arteries and veins, and he demonstrated that venous blood can flow only toward the heart. In 1628, Harvey published his hypothesis describing the circulation of blood.

Teaching TIP

Solutions

Tell students that plasma, the liquid portion of blood, is a solution. Like all solutions, plasma contains solutes that are dissolved in a solvent. The solutes in plasma are metabolites, salts, ions, proteins, and wastes, and the solvent is water.

did you know?

Arteries are classified as either conducting arteries or distributing arteries.

While both kinds have substantial amounts of smooth muscle in their walls, conducting arteries are large and thick walled. They serve as low-resistance conduits that carry blood to the more numerous and smaller distributing arteries. Distributing arteries play a role in maintaining adequate blood pressure. For example, they bolster blood pressure by narrowing if blood pressure drops because of blood loss.

Teaching TIP

Veins

Show the class a picture of a person in a hospital bed. Ask why hospital workers try to get patients to walk, if only for a few minutes each day. *(Students should recognize that muscle movements squeeze the walls of veins, thus preventing blood from accumulating and clotting in parts of the body such as the legs. Other methods of enhancing circulation used in the hospital include special compression hoses and pumps that encourage blood flow through the lower extremities.)*

Teaching TIP

A Closed System

Remind students that the circulation of blood is a closed system, one in which the blood does not leave the heart and blood vessels. Other materials, however, such as nutrients, oxygen, and wastes, are transferred between the blood and other tissues by passing through the walls of the smallest blood vessels, the capillaries.

biobit

How long are all of your capillaries?

If all of the capillaries of your body were laid end to end, they would extend all the way across the United States! The network of capillaries in the body is several thousand miles long.

Capillaries

No cell in your body is more than about 100 µm (0.004 in.) away from a capillary. At any moment, about 5 percent of your blood is in capillaries. In capillaries, gases, nutrients, hormones and other molecules are transferred from the blood to the body's cells. Carbon dioxide and other wastes are transferred from the body's cells to the capillaries.

The extensive back-and-forth traffic in the capillaries is possible because of two key properties. Capillaries are only one cell thick, so gas and small food molecules easily pass through their thin walls. Capillaries are also very narrow, with an internal diameter of about 8 µm (0.0003 in.)—a diameter only slightly larger than the diameter of a red blood cell. Thus, blood cells passing through a capillary slide along the capillary's inner wall, as shown in the photo in Figure 39-2. This tight fit makes it easy for oxygen to diffuse from red blood cells through the capillary to body cells.

Veins

Because the pressure in the veins is not as great as that in the arteries, veins do not have to accommodate the pulsing pressure that arteries do. The walls of veins consist of a much thinner layer of smooth muscle and elastic fiber than the walls of arteries.

As shown in Figure 39-2, veins also differ from arteries in that they are larger in diameter. A large blood vessel offers less resistance to blood flow than a narrower one, so the blood can move more quickly through large veins. The largest veins in the human body are about 3 cm in diameter—about the same diameter as your thumb.

Most veins have one-way valves. A **valve** is a flap of tissue that ensures that the blood or fluid that passes through does not flow back. Valves in veins, such as the one shown in **Figure 39-3,** prevent the blood from flowing backward during its trip to the heart. When the skeletal muscles in your arms and legs contract, they squeeze against the veins, causing the valves to open and thus, allowing the blood to flow through. The pressure helps carry the blood—against gravity—toward the heart. When the skeletal muscles relax, the valves close, preventing the backflow of blood.

Figure 39-3 Valves in veins. Valves are most abundant in the veins of the arms and legs, where the upward flow of blood is opposed by gravity.

Magnification: 122×

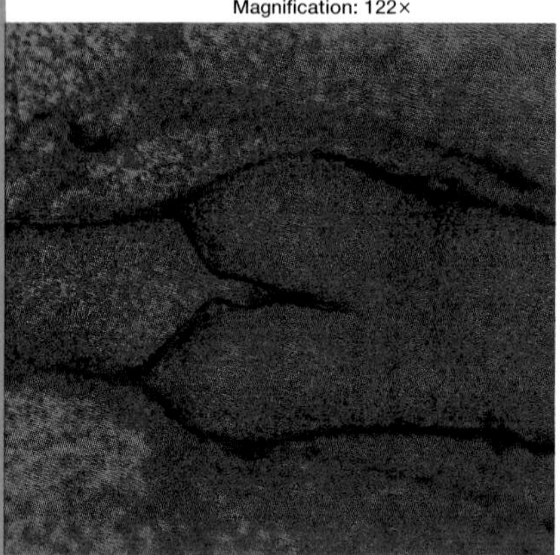

Sometimes the valves in the veins become weak and the veins become dilated (larger in diameter). Veins that are dilated because of weakened valves are called varicose veins. Dilated veins that occur in the anal area are called hemorrhoids.

Leaked fluids

Every time the heart pumps, it generates pressure, and fluids are forced out of the thin walls of the capillaries. Because the blood plasma is rich in proteins, most of the fluid returns to the capillaries. The fluid that does not return to the capillaries collects in spaces around the body's cells. The fluid that collects around the cells is eventually

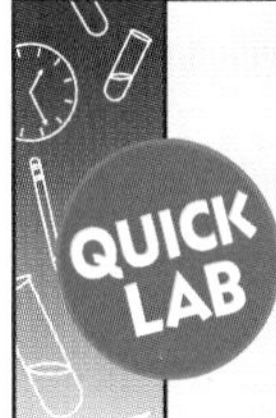

Can you map the valves in veins?

QUICK LAB

By applying pressure to your arm, you can locate the valves in the veins of your arm.

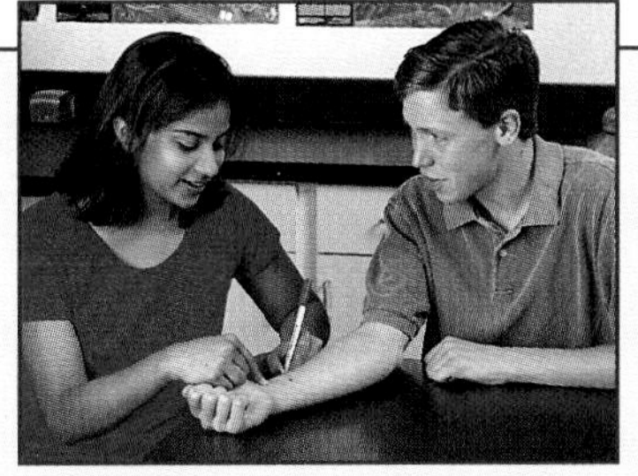

Materials

felt-tip pen

Procedure

1. Have a classmate make a fist and extend his or her arm, with the hand palm up and slightly below elbow level. Locate a prominent vein on the inside of the forearm. Using one finger, press down on the vein at a point near the wrist to block the blood flow.
2. Gently place a second finger along the vein about 5 cm from the first finger (toward the elbow). Release the second finger, but not the first. The vein should refill partway. Mark this point, which indicates the location of a valve, with a pen. You may have to try more than one vein to locate a valve.

Analysis

1. **Identify** the direction blood flows in the vein you chose.
2. **Propose** why the subject must make a fist and hold his or her arm slightly down.
3. **Infer** what effect standing in one place for long periods of time might have on the veins in the legs.

picked up by the lymphatic system and returned to the blood supply.

The **lymphatic system** is a system of the body that collects and recycles fluids leaked from the circulatory system and is involved in fighting infections. As shown in **Figure 39-4,** the lymphatic system is made up of a network of vessels called lymphatic vessels and tiny bean-shaped structures called lymph nodes. Lymph tissue is also located in various places throughout the body, including the thymus, tonsils, spleen, and bone marrow.

Lymphatic vessels eventually take the leaked fluid, called lymph, back to two major veins in the neck. Similar to veins, lymphatic vessels contain valves that prevent the backflow of the fluid. The fluid is pushed through the lymphatic vessels when the skeletal muscles in the arms and legs contract.

The lymphatic system also acts as a key element in the immune system. Immune cells in the lymph nodes and lymphatic organs help defend the body against cancerous cells, bacteria, viruses, and other infecting microbes. Lymph nodes, which are concentrated in the armpits, neck, and groin, sometimes get tender and swell when they are filled with dead bacteria and other cellular debris produced by immune cells. Health-care professionals are trained to detect certain types of infections by feeling for the lymph node swellings on the body.

Lymphatic System

Tonsils
Thymus
Lymph node
Spleen
Lymphatic vessel
Bone marrow

Figure 39-4 Lymphatic tissues. Lymphatic tissues are located throughout the body.

Can you map the valves in veins?

QUICK LAB

Time 10 minutes

Process Skills Inferring relationships, identifying functions

Teaching Strategies

- Instruct the students to allow their arms and hands to hang in a relaxed manner for 1–2 minutes before beginning the experiment. This demonstration will also work on the veins in the back of the hand.
- The effect will be more pronounced if the blood is squeezed out of the vein between the first and second finger. Have students do this by gently stroking the vein toward the heart.

Analysis Answers

1. toward the elbow, and so toward the heart
2. Blood will have to go against gravity, enlarging the vein slightly.
3. Blood will pool in the veins in the legs.

Teaching TIP

Stability and Homeostasis

Students should understand that swollen lymph nodes contain large numbers of white blood cells and immune cells that are actively engulfing bacteria or virus particles. For this reason, the nodes become inflamed and tender when the body is fighting infection.

did you know?

Our bodies have several different lymphoid organs.

They include the lymph nodes, the spleen, the thymus gland, and the tonsils. Specialized lymph vessels, called lacteals, in the intestines absorb fats from digested foods. They deliver their contents to other lymph vessels, which then conduct the fats to the bloodstream.

Chapter 39

REAL LIFE

Answer

Answers will vary. Drugs may include new cholesterol-lowering drugs (statins), new beta-blockers, and specifically designed estrogens (for older women).

Cross-Disciplinary Connection

◆ **Mathematics**

A human adult has about 5 L (1.25 gal) of blood. Plasma is more than 90 percent water. The body contains some 30 trillion red blood cells and about 60 billion white blood cells. Every second about 2 million new red blood cells are made in the bone marrow to replace those that die at the end of their 120-day life span. Have students calculate how many new red blood cells are made every hour. *(2,000,000 cells/s × 60 s/min × 60 min/h = 7,200,000,000 cells/h)*

Demonstration

Blood Components

Add red water to a test tube to make it about 40 percent full. Then carefully add vegetable oil until the test tube is full. Explain to students that the fluids in the test tube resemble a centrifuged blood sample. Centrifuging pulls the heavier components to the bottom of the tube. Lighter components stay at the top. The vegetable oil represents plasma, a yellow fluid. The red water represents red blood cells. Note that platelets and white blood cells, normally found between the two layers, are not represented in this model.

REAL LIFE

Several proteins from vampire bats stop blood from clotting. These proteins, including one named *draculin,* are being used to develop drugs to fight heart disease.

Finding Information
Investigate other drugs that are currently being developed to help fight heart disease.

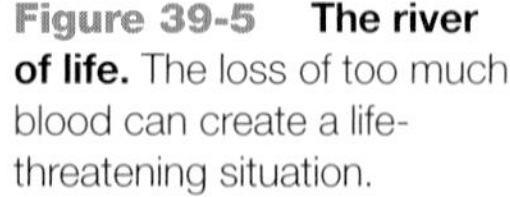

Blood Has Both Liquid and Solid Components

Blood has been called the river of life because it is responsible for transporting so many substances throughout the body. In life-threatening situations, a person's blood is carefully monitored, as shown in **Figure 39-5.** Typically, blood appears to us as a red, watery fluid. Blood is composed of water, but it also contains a variety of molecules dissolved or suspended in the water, and three kinds of cells.

Blood plasma

About 60 percent of the total volume of blood is **plasma,** the liquid portion of blood. Plasma is made of 90 percent water and 10 percent solutes. The solutes include metabolites, wastes, salts, and proteins.

Water Water in the plasma acts as a solvent. It carries other substances.

Metabolites and wastes Dissolved within the plasma are glucose and other food molecules. Vitamins, hormones, gases, and nitrogen-containing wastes are also found in plasma.

Salts (ions) Salts are dissolved in the plasma as ions. The chief plasma ions are sodium, chloride, and bicarbonate. The ions have many functions, including maintaining osmotic balance and regulating the pH of the blood and the permeability of cell membranes.

Proteins Plasma proteins, the most abundant solutes in plasma, play a role in maintaining the osmotic balance between the cytoplasm of cells and that of plasma. Water does not move by osmosis from the plasma to cells because the plasma is rich in dissolved proteins. In fact, the total amount of protein in cells and in plasma is the same, making cytoplasm and plasma essentially isotonic to each other.

Some plasma proteins help thicken the blood. The thickness of blood determines how easily it flows through blood vessels. Other plasma proteins serve as antibodies, defending the body from disease. Still other proteins, called clotting proteins or blood-clotting factors, play a major role in blood clotting. When blood is collected for clinical purposes, the blood-clotting factors are removed from the blood and stored for later use.

Figure 39-5 The river of life. The loss of too much blood can create a life-threatening situation.

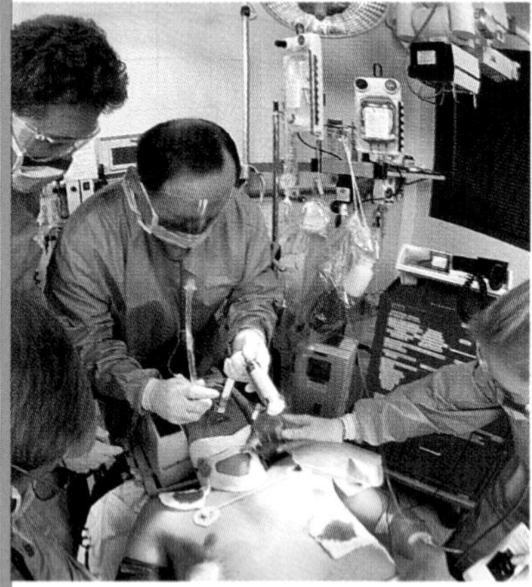

Blood cells and cell fragments

About 40 percent of the total volume of blood is cells and cell fragments that are suspended in the plasma. There are three principal types of cells in human blood: red blood cells, white blood cells, and platelets.

Red blood cells Most of the cells that make up blood are **red blood cells**—cells in the blood that carry oxygen. Each milliliter of your blood contains about 5 million red blood cells. Red blood cells are also called erythrocytes *(eh RIHTH roh seyets).*

Overcoming Misconceptions

Color of Blood

Students might think that blood is one of two colors, red or blue. Diagrams that portray bright red arteries and deep blue veins may enhance students' confusion. Oxygenated blood is, in fact, scarlet in color, while deoxygenated blood is maroon, closer to dark red than to blue. The blue color seen in the superficial veins of a person with fair skin is misleading: the skin and connective tissue that overlie the veins distort the true color of the blood within the veins.

Magnification: 5240×

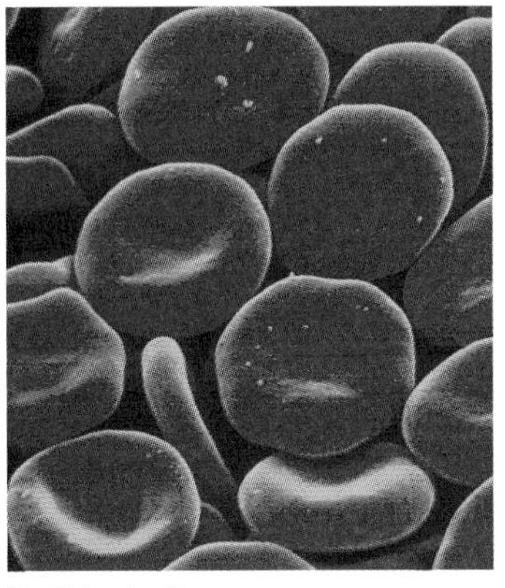
Red blood cells

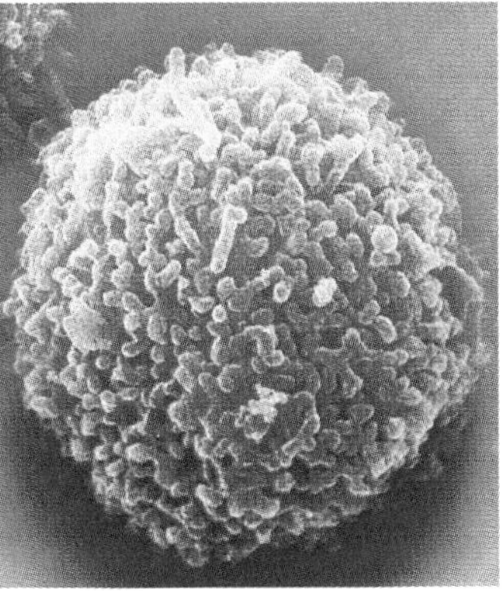
White blood cell

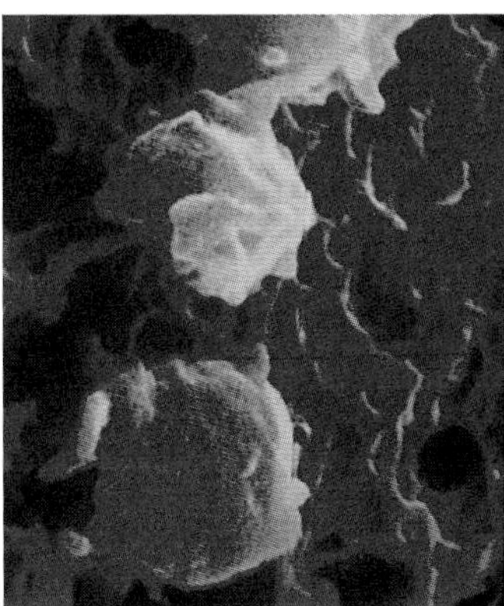
Platelets

Figure 39-6 Three kinds of blood cells. Red blood cells transport oxygen and some carbon dioxide. White blood cells help defend the body against disease. Platelets are involved in blood clotting.

Most of the interior of a red blood cell is packed with hemoglobin. Hemoglobin is an iron-containing protein that binds oxygen in the lungs and transports it to the tissues of the body. Red blood cells do not have nuclei and therefore cannot make proteins or repair themselves. The absence of a nucleus gives a red blood cell its biconcave shape, as shown in **Figure 39-6,** and a short life span (about 4 months). New red blood cells are produced constantly by stem cells, specialized cells in bone marrow.

An abnormality in the size, shape, color, or number of red blood cells results in anemia. **Anemia** *(uh NEE mee uh)* is a condition in which the oxygen-carrying ability of the blood is reduced. Anemia may result from blood loss or nutritional deficiencies.

White blood cells There are only 1 or 2 white blood cells, or leukocytes *(LOO koh seyets),* for every 1,000 red blood cells. **White blood cells** are cells whose primary job is to defend the body against disease. White blood cells, shown in Figure 39-6, are larger than red blood cells and contain nuclei.

There are many different kinds of white blood cells, each with a different immune function. For example, some white blood cells take in and then destroy bacteria and viruses. Other white blood cells produce antibodies, proteins that mark foreign substances for destruction by other cells of the immune system.

Platelets Certain large cells in bone marrow regularly pinch off bits of their cytoplasm. These cell fragments, called **platelets** *(PLAYT lihts),* are shown in Figure 39-6. Platelets play an important role in the clotting of blood. If a hole develops in a blood vessel wall, rapid action must be taken by the body, or blood will leak out of the system and death could occur.

When circulating platelets arrive at the site of a broken vessel, they assume an irregular shape, get larger, and release a substance that makes the platelets very sticky. They then attach to the protein fibers on the wall of the broken blood vessel and eventually form a sticky clump that plugs the hole.

Section 39-1

Visual Strategy

Figure 39-6

Have students relate the shapes and sizes of the cells and cell fragments to their functions. *(The shape of the red blood cells provides greater surface area for gas diffusion, and their small size enables them to squeeze through tiny capillaries. The shape of the white blood cells enables them to engulf other particles, and their size ensures that they can defend the body against a large number of invaders. The shape of the platelets allows them to stick together to form a small plug.)* Have students compare the pictures with prepared, sealed microscope slides of human blood.

Teaching TIP

Blood Composition

Blood has several different components, and learning them can be confusing. Have students construct a **Graphic Organizer,** such as the one shown at the bottom of page 887, that describes the structure and function of each of the blood components discussed in the text. Students should also show the composition of plasma, the fluid portion of the blood. PORTFOLIO

Resource Link

Assign **Occupational Applications: Blood-Bank Technologist** from the Holt BioSources Teaching Resources CD-ROM. PORTFOLIO

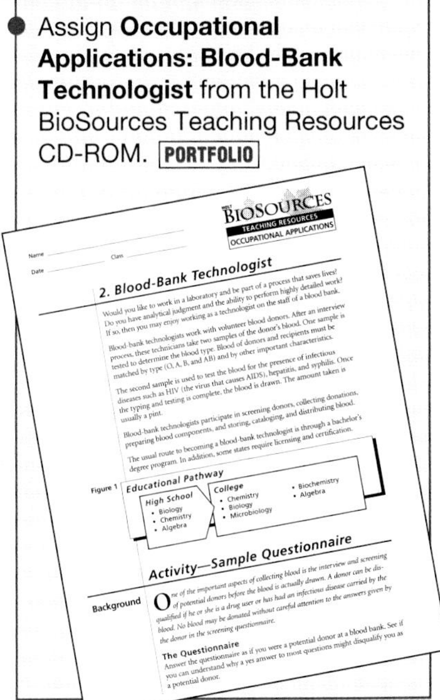
BIOSOURCES
TEACHING RESOURCES
OCCUPATIONAL APPLICATIONS

2. Blood-Bank Technologist

Figure 1 Educational Pathway
High School: Biology, Chemistry, Algebra
College: Chemistry, Biology, Microbiology, Biochemistry, Algebra

Activity—Sample Questionnaire

Background

The Questionnaire

Graphic Organizer

*Use this graphic organizer with **Teaching Tip: Blood Composition** on page 887.*

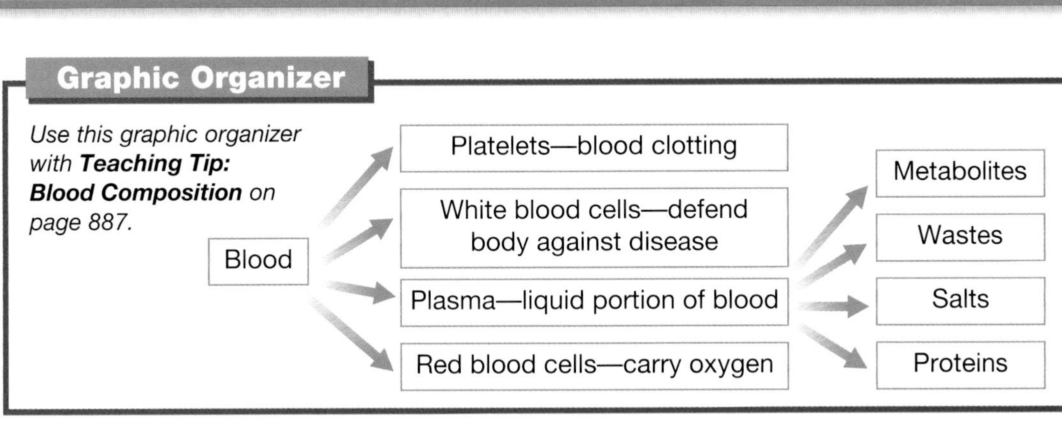

Teaching TIP

Blood Clotting

When people cut themselves shaving, they often put a piece of tissue over the cut. The fibers in the tissue act like platelets, beginning the clotting process by providing a framework for red blood cells to be caught.

Blood Doping

Teaching Strategies
Bring in articles about other performance-enhancing (but often prohibited) practices, such as the use of steroids and nutritional supplements for athletes. Explain how they help an athlete's performance but harm an athlete's health.

Discussion

1. Why do you think the plasma is reinjected immediately? *(to keep blood volume constant)*
2. Explain the physiological effects of blood doping and the reasons for these effects. *(increased endurance due to more oxygen in the blood; decreased heart rate because not as much blood flow is needed; decreased lactic acid in muscles due to increased oxygen, increased aerobic respiration, and decreased anaerobic respiration)*
3. Why would doped blood clot more easily? *(There are more solids in it.)*

Figure 39-7 Blood-clotting cascade

The release of enzymes from platelets at the site of a damaged blood vessel initiates a "clotting cascade."

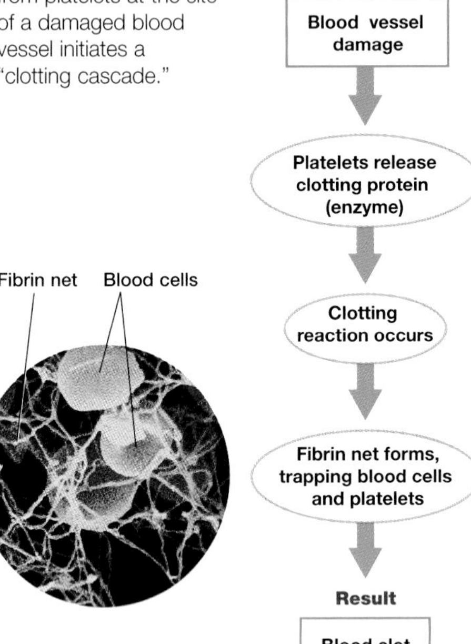

For more-severe wounds, such as an open cut, the platelets release a clotting enzyme that activates a series of chemical reactions. Eventually, a protein called fibrin is formed, as shown in **Figure 39-7.** The fibrin threads form a net, trapping blood cells and platelets. A mesh of fibrin and platelets develops into a mass, or clot, that plugs the blood vessel hole. A mutation in a gene for one of the blood-clotting proteins causes hemophilia, a blood clotting disorder.

Blood type

Occasionally, an injury or disorder is so serious that a person must receive blood or blood components from another person. The blood types of the recipient, the person receiving the blood, and that of the donor, the person giving the blood, must match. Blood type is genetically determined by the presence or absence of specific proteins found on the surface of red blood cells.

One protein group used to type blood is the **ABO blood group system.** Under this system, the primary blood types are A, B, AB, and O. The letters *A* and *B* refer to proteins on the surface of red blood cells that act as antigens, substances

Blood Doping

Most athletes try to gain an edge over their competitors through conditioning programs and longer hours in training. Some, however, turn to artificial methods, such as blood doping.

What is blood doping?
Blood doping is a practice used by some athletes to increase the amount of oxygen their blood can carry by elevating the number of red blood cells in their body. Several months before a competition, the athlete draws some of his or her blood and separates the plasma from the cells. The blood cells are frozen. The plasma is immediately injected into the athlete's bloodstream.

The body steps up its production of red blood cells. After 1 to 2 months, the entire procedure may be repeated. A few days before an athletic competition, the frozen blood cells are thawed and injected into the athlete, temporarily bringing the athlete's red blood cell count above normal.

Does blood doping work?
Athletes who use blood doping show an increase in endurance while running on a treadmill, a decrease in their heart rate during exercise, and a reduction in the level of lactic acid in their blood.

The risks of blood doping
The risks of blood doping far outweigh the benefits. The blood clots more easily, raising the risks of heart attack and stroke. Air bubbles are sometimes accidentally injected, and blood cells are sometimes contaminated while out of the body.

Athletes who inject blood cells taken from another person face the dangers of blood incompatibility and transmission of diseases, including AIDS and hepatitis. For these and other reasons, blood doping has been prohibited by many sports-governing bodies.

Biology in Your Community

Blood Donation

Invite a member of the American Red Cross or a local blood bank to come and talk about blood and plasma donation. Ask the speaker to discuss how blood is typed and screened for the presence of pathogens. The speaker should stress the importance of periodic blood donation to keep the blood supply plentiful and inform students of minimum age and weight requirements for blood donation.

Table 39-1 Blood Types

Blood type	Antigen on the red blood cell	Antibodies in plasma	Can receive blood from	Can donate blood to
A	A	B	O, A	A, AB
B	B	A	O, B	B, AB
AB	A, B	Neither A nor B	O, A, B, AB	AB
O	Neither A nor B	A, B	O	O, A, B, AB

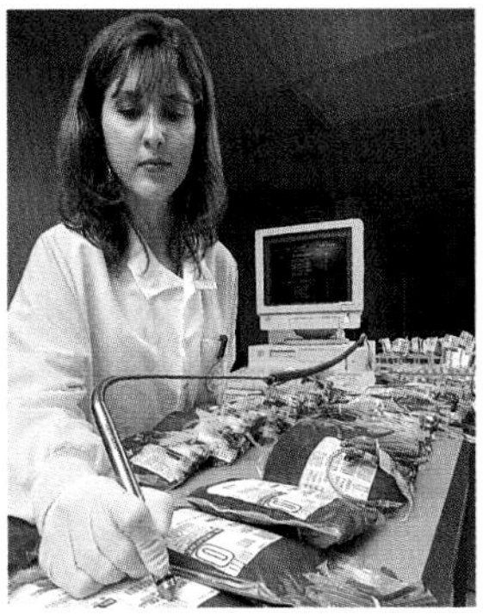

that can provoke an immune response. As summarized in **Table 39-1,** people with type A blood have the A antigen on their red blood cells. Individuals with type B blood have the B antigen. People with type AB blood have both the A and the B antigen, while those with type O blood have neither the A nor the B antigen. As shown in **Figure 39-8,** blood is typed before it is donated to a person that needs it.

Antibodies are defensive proteins made by the immune system. Individuals with type A blood produce antibodies against the B antigen, even if they have never been exposed to it. The red blood cells clump and can block blood flow. For this reason, blood transfusion recipients must receive blood that is compatible with their own.

Individuals with type AB blood are universal recipients (they can receive A, B, AB, or O blood) because they do not have anti-A or anti-B antibodies. Type O individuals are universal donors (they can donate blood to those with A, B, AB or O blood) because their blood cells do not carry A or B antigens and therefore do not react with either anti-A or anti-B antibodies.

Figure 39-8 Donated blood. Blood that is donated is carefully screened and typed.

Rh factor

Another important antigen on the surface of red blood cells is called **Rh factor,** which was originally identified in rhesus monkeys. People who have this protein are said to be Rh^+, and those who lack it are Rh^-. When an Rh^- mother gives birth to an Rh^+ infant, the Rh^- mother begins to make "anti-Rh" antibodies. The mother's antibodies may be passed to an Rh^+ fetus in a future pregnancy and cause the fetus's red blood cells to clump, which can lead to fetal death.

Review SECTION 39-1

1. **Name** the system that transports nutrients, oxygen, wastes, hormones, and heat.
2. **Compare** the structures and functions of arteries, capillaries, and veins.
3. **Describe** the role of the lymphatic system.
4. **Summarize** the functions of water, red blood cells, white blood cells, and platelets.
5. **Predict** the blood types that would be safe for a person with type A blood to receive during a transfusion.

Review SECTION 39-1 ANSWERS

1. the circulatory system
2. arteries—thick, elastic walls and thick smooth muscle layers that carry blood under high pressure away from heart; capillaries—very narrow, single cell layers that transport oxygen to individual cells and transport carbon dioxide away from cells; veins— thin walls and thin smooth muscle layers that carry blood under low pressure to heart
3. The lymphatic system collects and recycles fluids leaked from the circulatory system and is involved in fighting infections.
4. Water is the solvent for dissolved substances in the blood. Red blood cells carry oxygen to body tissues. White blood cells defend the body from pathogens. Platelets help keep the venous system intact by patching tears in blood vessels.
5. A and O are both safe.

Teaching TIP — BASIC

Antigens and Blood Type

An analogy for antigens on blood cells is decorations on a cake. The cake is a red blood cell. Frosting represents A antigens, sprinkles represent B antigens, and candles represent Rh antigens. For example, a person who has AB^- blood has a cake with frosting and sprinkles; a person who is O^+ would have a cake with only candles.

3 Assessment Options

Closure

Location, Location, Location

Have students explain the adaptive value of the location of arteries (generally found far below the body's surface), veins (located closer to the surface), and capillaries (distributed all over the body). *(Arteries carry blood under high pressure, so a tear or rupture is more life threatening than it is in a vein. Capillaries must come into contact with every cell in the body to perform their exchange function.)*

Review

Assign Review—Section 39-1.

Reteaching — BASIC

Have students describe abnormalities of a blood sample taken from someone with either leukemia (cancer of the white blood cells) or an iron deficiency. *(With leukemia, the number of white cells may be greater. An iron deficiency may cause the red cells to be lighter in color or fewer in number.)*

Section 39-2 The Heart

1 Prepare

Scheduling

- **Block:** About 45 minutes will be needed to complete this section.
- **Traditional**: About two class periods will be needed to complete this section.
- **Planning:** See the Planning Guide on pp. 879A and 879B for lecture, classwork, and assignment options for Section 39-2. Lesson plans for Section 39-2 can be found in a lesson cycle format in ***Biology: Principles and Explorations*** **One-Stop Planner CD-ROM with Test Generator.**

Resource Link

Assign **Active Reading Worksheet Section 39-2** to help students comprehend the key concepts and visuals in this lesson.

2 Teach

Lesson Warm-up

You will need plastic syringes and two pieces of hosing that fit on their ends. One hose should be about 8 cm (3 in.) long and the other about 15 cm (6 in.). To show students that when blood is far from the heart, it travels with less pressure, fill the syringe and force the water out. The stream should travel about 8–15 cm (3–6 in.) from the end of the hose. Repeat for capillaries (the syringe and short hose). Water should travel 15–30 cm (6–12 in.) from the end of the hose. Repeat once more for arteries (syringe with no hose). Water should travel 30 cm–3 m (1–10 ft). Be sure to apply equal pressure to the plunger each time so students appreciate the fact that it is distance from plunger tip to exit, not pressure on the plunger, that causes the change.

Objectives

- **Differentiate** the pulmonary circulation loop from the systemic circulation loop. (p. 890)
- **Summarize** the path that blood follows through the heart. (pp. 891–892)
- **Name** the cluster of heart cells that initiates contraction of the heart. (p. 892)
- **Describe** three ways to monitor the health of the circulatory system. (pp. 892–893)
- **State** two vascular diseases, and identify factors that contribute to their development. (p. 894)

Key Terms

atrium
ventricle
vena cava
aorta
coronary artery
sinoatrial node
blood pressure
pulse
heart attack
stroke

The Heart Is a Muscular Pump That Cycles Blood

Blood vessels allow for the movement of blood to all cells in the body. The pumping action of the heart, however, is needed to provide enough pressure to move the blood throughout the body. The heart is made up mostly of cardiac muscle tissue, which contracts to pump blood.

Two separate circulatory loops

As shown in **Figure 39-9,** the human heart has two separate circulatory loops. The right side of the heart is responsible for driving the pulmonary *(PUHL muh nehr ee)* circulation loop, which pumps oxygen-poor blood through the pulmonary arteries to the lungs. Gas exchange—the release of carbon dioxide and pick up of oxygen—occurs in the lungs. The oxygenated blood is then returned to the left side of the heart through pulmonary veins.

The left side of the heart is responsible for driving the systemic circulation loop, which pumps oxygen-rich blood through a network of arteries to the tissues of the body. Oxygen-poor blood is then returned to the right side of the heart through the veins.

Figure 39-9 Two systems of circulation

The pulmonary circuit transports blood between the heart and lungs; the systemic circuit transports blood between the heart and the rest of the body.

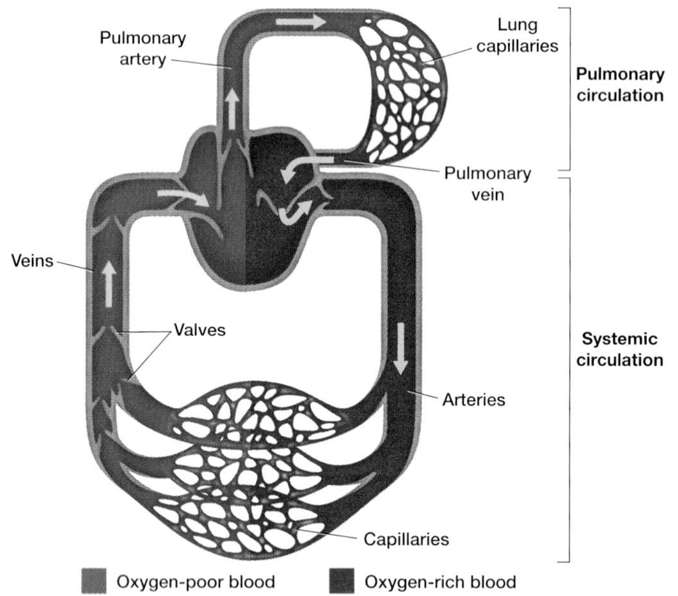

Integrating Different Learning Styles

Logical/Mathematical	**PE**	Performance Zone items 20, 21, 22, 23, 24, and 28, p. 903
	ATE	Resource Link, p. 891; Closure, p. 894
Visual/Spatial	**PE**	Study Tip, p. 891; Biobit, p. 891
	ATE	Lesson Warm-up, p. 890; Visual Strategy, p. 891
Intrapersonal	**PE**	Teaching Tip, p. 893
Verbal/Linguistic	**PE**	Study Tip, p. 893; Performance Zone items 2 and 29, pp. 902–903
	ATE	Reteaching, p. 894

Blood Follows a Specific Path

As shown in **Figure 39-10,** the heart has a wall that divides the right and left sides of the heart. At the top of the heart are the left and right atria *(AY tree uh)*. The **atria** (singular, *atrium*), are chambers that receive blood returning to the heart. Below the atria are the left and right **ventricles,** thick-walled chambers that pump blood away from the heart. A series of one-way valves in the heart prevent blood from moving backward. Figure 39-10 summarizes the path blood follows through the heart:

1. Two large veins called the inferior and superior **venae cavae** (singular, *vena cava*) collect all of the oxygen-poor blood from the body. The venae cavae empty blood directly into the right atrium of the heart.
2. The blood from the right atrium moves into the right ventricle.
3. As the right ventricle contracts, it sends the blood into the pulmonary arteries.
4. The pulmonary arteries carry the blood to the right and left lungs. At the capillaries of the lungs, oxygen is picked up and carbon dioxide is unloaded.
5. The freshly oxygenated blood returns from the lungs to the left side of the heart through the pulmonary veins, which empty the blood directly into the left atrium.
6. From the left atrium, the blood is pumped into the left ventricle.

Figure 39-10 **Blood flow through the heart**

The arrows trace the path of blood as it travels through the heart.

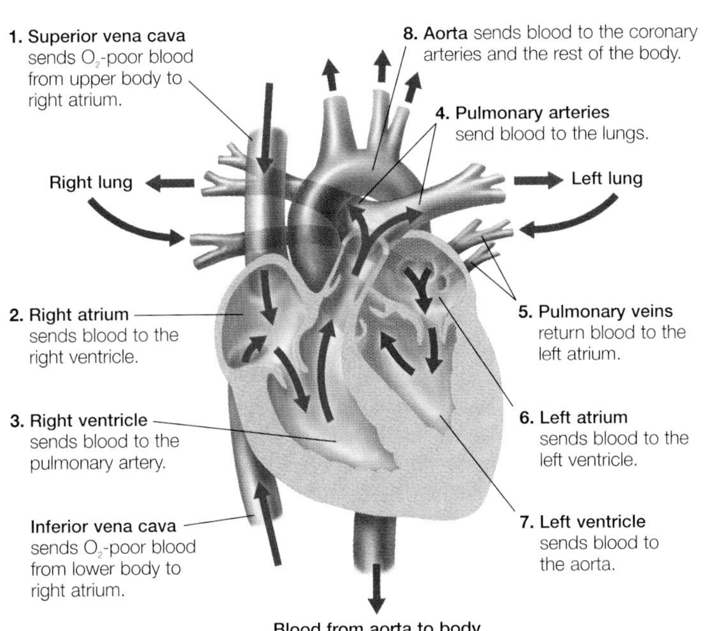

1. Superior vena cava sends O_2-poor blood from upper body to right atrium.
8. Aorta sends blood to the coronary arteries and the rest of the body.
4. Pulmonary arteries send blood to the lungs.
Right lung
Left lung
2. Right atrium sends blood to the right ventricle.
5. Pulmonary veins return blood to the left atrium.
3. Right ventricle sends blood to the pulmonary artery.
6. Left atrium sends blood to the left ventricle.
Inferior vena cava sends O_2-poor blood from lower body to right atrium.
7. Left ventricle sends blood to the aorta.
Blood from aorta to body

Study TIP

Interpreting Graphics

In human anatomy, the terms *left* and *right* always refer to the left and right from the perspective of the subject. This will help you understand why the terms *left* and *right* appear reversed in anatomical drawings, such as that of the heart in Figures 39-9 and 39-10.

biobit

Is the blood in veins really blue?

When oxygen is attached to hemoglobin, the blood is bright red; without oxygen, blood is dark red. The dark red blood in veins appears blue when it shows through the vein walls and skin.

did you know?

The heart cannot get food and oxygen from the blood in its own chambers.

This is because the heart muscle is too thick for diffusion to be an effective means of distribution. Instead, the heart must rely on the coronary arteries to sustain it. These two arteries lie in grooves that spiral around the heart. An obstruction in one of their branches often necessitates bypass surgery to restore proper blood flow to the heart. Restricted blood flow can result in angina pectoris, which is chest pain, or a myocardial infarction, which is a heart attack. Heart cells die during an infarction, which limits the heart's pumping efficiency.

Making Biology Relevant

Sports

When a person engages in strenuous activity, his or her heart rate climbs rapidly and so does the stroke volume—the amount of blood that passes through the heart with each beat. Although an unconditioned person's heart also beats faster during vigorous exercise, it does so with significantly less increase in stroke volume. Consequently, less blood is pumped out to the body, causing him or her to tire more quickly.

Resource Link

Assign **Problem Solving: Computing Rates and Heart Efficiency** from the Holt BioSources Teaching Resources CD-ROM. **PORTFOLIO**

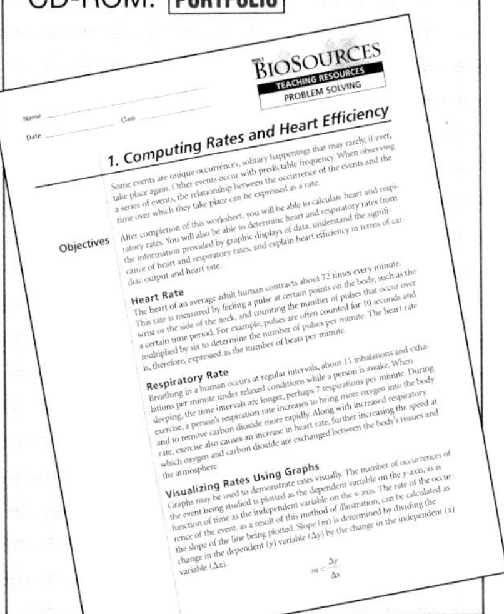

BioSources Teaching Resources Problem Solving

Name ______ Class ______ Date ______

1. Computing Rates and Heart Efficiency

Objectives

Heart Rate

Respiratory Rate

Visualizing Rates Using Graphs

Visual Strategy

Figure 39-10

Have students trace the path that blood follows through the heart. Tell students that although the heartbeat is described for convenience as occurring first on the right side and then on the left, the movement of blood occurs on both sides simultaneously. Have students compare the sizes of the left and right ventricle walls. Ask why the left ventricle wall is thicker than the right. *(Blood is pumped to the entire body from the left ventricle. Blood from the right ventricle goes only as far as the lungs.)*

Chapter 39

Teaching TIP

Heart Murmur

Invite the school nurse to demonstrate how a stethoscope can be used to hear the sounds of a heartbeat. Tell students that the "lubb" and "dubb" sounds that are heard are caused in part by the closing of the heart valves. A heart murmur is an abnormal sound that can be heard when listening to a person's heartbeat. Some murmurs are caused by "rough" blood flow through valves due to abnormal anatomy (which can be inherited) or increased volume of flow.

Making Biology Relevant

Health

Malfunctioning of the pacemaker (SA node) can result in a disorder known as fibrillation, a condition in which the heart contractions become irregular and rapid. In such cases, surgeons can implant an artificial pacemaker powered by batteries. This pacemaker operates by delivering electrical stimuli to the heart at regular intervals.

7. A… eft atrium to empty com-
r… e walls of the left ventricle
… ntraction is forceful.

8. … ry of the body, the **aorta**
…
T… orta are the **coronary**
(KC… …shly oxygenated blood
to … ch from the aorta and
carry oxygen-rich blood to all parts of the body.

After delivering oxygen to the cells of the body and picking up carbon dioxide, the cycle continues when blood returns to the heart through the inferior or superior venae cavae.

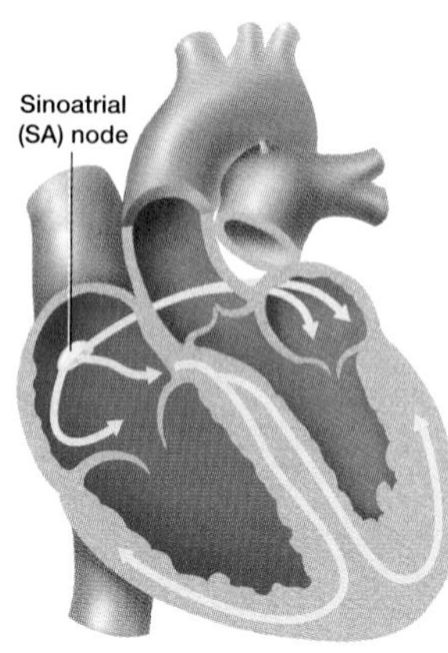

Figure 39-11 Electrical regulation of the heart. The SA node, or pacemaker, starts each heart contraction. The wave of contraction spreads across both atria and delays for an instant before it travels to the ventricles.

Initiating contraction

Contraction of the heart is initiated by a small cluster of cardiac muscle cells called the **sinoatrial** *(SEYE noh ay tree uhl)* **node,** which is embedded in the upper wall of the right atrium. The cells that make up the sinoatrial node (SA node, for short) act as the pacemaker of the heart, spontaneously starting contractions with a regular rhythm. Each contraction travels quickly in a wave that causes both atria to contract almost simultaneously, as shown in **Figure 39-11.**

The wave of contraction spreads from the atria to the ventricles, but almost one-tenth of a second passes before the ventricles start to contract. The delay permits the atria to finish emptying blood into the ventricles before the ventricles contract simultaneously. The wave of contraction is conducted rapidly over both ventricles by a network of fibers in the heart.

On average, heart contractions occur at a rate of about 72 times per minute. During sleep the rate decreases, and during exercise it increases. The SA node is controlled by two sets of nerves with antagonistic (opposite) signals and is influenced by many factors, including hormones, temperature, and exercise.

Monitoring the heart

Heart disease is one of the leading causes of death among humans. Health professionals have several different methods available for monitoring the health of the circulatory system.

Blood pressure Doctors routinely measure patients' blood pressure. **Blood pressure** is the force exerted by blood as it moves through blood vessels. Blood pressure readings provide information about the conditions of the arteries.

Blood pressure is measured with a device called a sphygmomanometer *(sfihg moh muh NAHM uht uhr)*, shown in **Figure 39-12.** The first number, the systolic pressure, tells how much pressure is exerted when the heart contracts and blood flows through the arteries. The second number, the diastolic pressure, tells how much pressure is exerted when the heart relaxes. Blood pressure is expressed in terms of millimeters of mercury (mm Hg) and is usually reported as the systolic pressure written over the diastolic pressure.

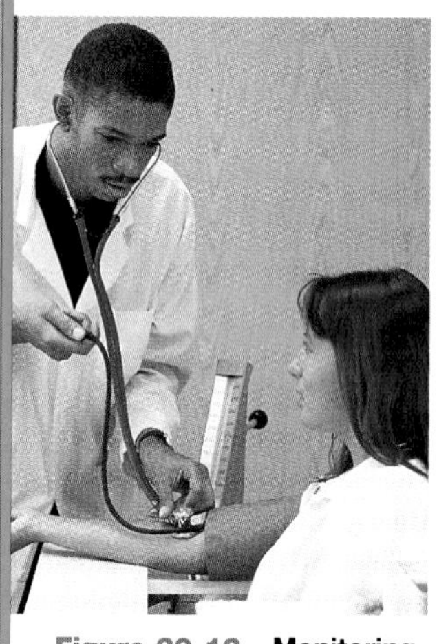

Figure 39-12 Monitoring blood pressure. Blood pressure is measured with a blood pressure cuff, a stethoscope, and a mercury column. Blood pressure readings provide information about the conditions of arteries.

did you know?

Cardiac muscle cells have an intrinsic, or self-initiated, beat.

Contractions of the heart muscle are not produced by stimulation from nerves. Nerves control only the rate of the heart's contractions. In addition, cardiac muscle cells influence each other. They will synchronize their rhythms in response to other cardiac muscle cells with which they have contact.

Normal blood pressure values are from 100 to 130 for systolic pressure and from 70 to 90 for diastolic pressure. An example of a normal reading would be written 120/80 mm Hg. These figures indicate the blood is pushing against the artery walls with a pressure of 120 mm Hg as the heart contracts and 80 mm Hg as the heart rests.

Many Americans suffer from a condition called high blood pressure, or hypertension. High blood pressure places a strain on the walls of the arteries and increases the chance that a vessel will burst. Left untreated, hypertension can lead to heart damage, brain damage, or kidney failure. Regular aerobic activity can help people to maintain a healthy blood pressure. Hypertension can be easily diagnosed and controlled by medicine, diet, and exercise.

Electrocardiograms (ECGs or EKGs) Another way to monitor the heart's function is to measure the tiny electrical impulses produced by the heart muscle when it contracts. Because the human body is composed mostly of water, it conducts electrical currents rather well. A small portion of the heart's electrical activity reaches the body surface. As shown in **Figure 39-13,** an instrument called an electrocardiograph uses special sensors to detect the electrical activity. A recording of these electrical impulses is called an electrocardiogram, abbreviated as ECG or EKG. In one normal heartbeat, three successive electrical-impulse waves are recorded, as shown in Figure 39-13.

Heart Rate It takes only a watch with a second hand to measure your pulse. Your **pulse** is a series of pressure waves within an artery caused by the contractions of the left ventricle. A person's pulse is an indicator of his or her heart rate—how fast or slow the heart is beating. Each time the blood surges from the aorta, the elastic walls in the blood vessels expand and stretch. This rhythmical expansion can be felt as a pulse in areas where the vessel nears the surface of the skin. The number of pulses counted per minute represents the number of heartbeats per minute. The most common sites for taking a pulse are at a radial artery, on the thumb side of each wrist, or at a carotid artery, on either side of the neck. The average pulse rate ranges from 70 to 90 beats per minute for adults.

Study TIP

Word Origins

The Greek word for heart is *kardia.* Because of the *k* in this word, *electrocardiogram* is sometimes abbreviated as EKG instead of ECG.

Figure 39-13 Monitoring heart contractions. The electrical changes with each heart contraction can be detected with an electrocardiograph. The characteristic up-and-down waves are analyzed to assess the health of the heart.

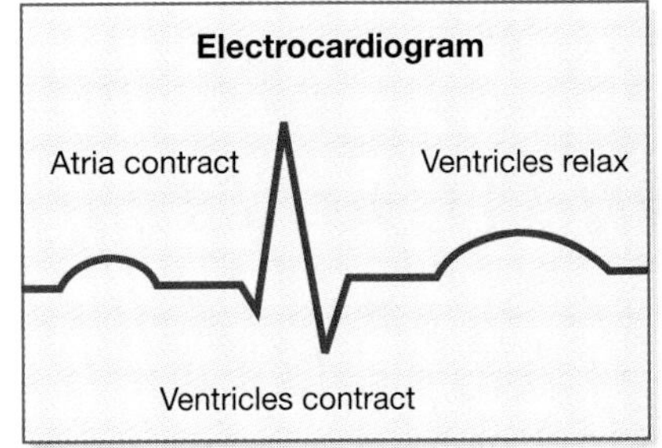

Teaching TIP

Sphygmomanometer

Have the school nurse demonstrate how a sphygmomanometer works. Let students try it as well. Remind students that blood pressure readings include two numbers, the systolic and diastolic pressures, represented as a ratio. The systolic reading, the top value, is the pressure on the artery walls exerted by contracting ventricles. The diastolic reading, the bottom value, is the pressure on the blood vessel walls during ventricle relaxation.

Teaching TIP

Pulse Points

Have students work in pairs to obtain pulse-rate readings. Some students feel faint when they press against the carotid artery to determine their pulse. It is better to have them feel the radial artery, brachial artery (inside the elbow), popliteal artery (back of the knee), or an artery on the ankle.

did you know?

Because it usually develops without other symptoms in its early stages, hypertension is often called the silent killer.

High blood pressure, or hypertension, is a serious health condition because it is the major cause of heart failure, kidney failure, and stroke. Ask students to use outside sources to research the topic of hypertension and prepare a report that explains the causes of this condition, how hypertension can be prevented or treated, and the rate of hypertension in the United States compared with that of other countries. PORTFOLIO

Chapter 39

Exploring Further

What Is a Heart Attack?

Teaching Strategies
Bring in a sheep heart or a picture of a human heart, and point out the coronary arteries to students.

Discussion
- Identify the ways arteries become blocked. *(blood clots, atherosclerosis, arteriosclerosis)*
- Why do some commercials for aspirin recommend taking an aspirin a day? *(Aspirin thins the blood.)*
- How can aspirin help prevent heart attacks? *(Thinner blood is less likely to clot, but it will not prevent plaques from coming loose or other causes for heart attacks.)*

3 Assessment Options

Closure

Blood Pressure
Review the **Lesson Warm-up** on page 890. Have students infer which blood vessel would have the highest blood pressure. *(the aorta, as blood is leaving the left ventricle)* Which blood vessel would have the lowest blood pressure? *(the vena cava, as blood is returning to the right atrium)* Have them justify their choices.

Review

Assign Review—Section 39-2.

Reteaching — BASIC

Have students design a crossword puzzle using key vocabulary words for the heart and circulation.
PORTFOLIO

Diseases of the blood vessels serving the heart and brain are leading causes of premature death and disability in the United States. When either the heart or the brain does not get enough blood, the organ slowly dies. A **heart attack** occurs when an area of the heart muscle dies and stops working. When an area of the brain dies the result is a **stroke.** Death or varying degrees of disability may result. Important factors that contribute to heart attacks and strokes are smoking tobacco, lack of physical activity, diets high in saturated fats, and unmanaged stress.

Exploring Further

What Is a Heart Attack?

You feel a sharp, crushing, squeezing pain in your chest. You have difficulty breathing. You break into a cold sweat. You feel nauseous. These symptoms occur to almost 2 million Americans each year when they experience a heart attack. Some people experience no warning signs.

Why do heart attacks occur?
Heart attacks usually happen when the arteries that deliver oxygen to the heart, the coronary arteries, become blocked. Heart cells begin to die after 4 to 6 hours without blood. When white blood cells in the body sense the low oxygen levels, they release toxic substances that damage the tissues of the body, including the heart. If a large part of the heart is affected, the victim can die immediately or within a few weeks.

How do arteries become blocked?
A blood clot formed somewhere else in the body can break loose and flow to the heart or to the brain, where it blocks the flow of blood. Blood flow is also blocked by the buildup of fatty deposits, including cholesterol, a condition called atherosclerosis *(ath uhr oh skluh ROH sihs)*.

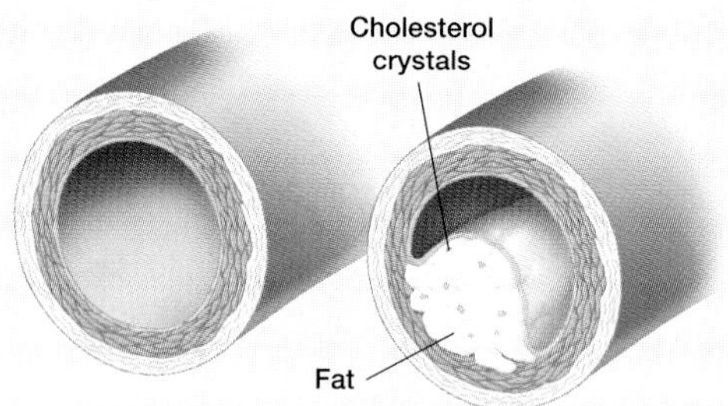

When calcium is deposited in the fatty buildup, the condition is called arteriosclerosis *(ahr tihr ee oh skluh ROH sihs)*, or hardening of the arteries. Hardened arteries cannot expand to handle the volume of blood that enters every time the heart contracts. Pressure builds up in the artery and feeds back to the heart, causing the heart to work harder.

Prevention
High blood pressure, high cholesterol levels, and cigarette smoking are all controllable risk factors in heart disease. Smoking less (or not at all), early diagnosis and treatment of high blood pressure, regular medical checkups, a healthy diet, and regular exercise can all help prevent a heart attack or decrease the severity of one.

Review SECTION 39-2

1. **Summarize** the path of blood through the body starting and ending with blood that has just returned from the lungs to the heart.
2. **List** the sequence of events that results in atrial and ventricular contraction.
3. **Identify** three ways that the condition of the circulatory system can be monitored.
4. **Differentiate** between a heart attack and a stroke.
5. **SKILL Inferring Conclusions** Identify the only arteries that carry oxygen-poor blood.

Critical Thinking

6. **SKILL Evaluating Conclusions** Is the blood that your heart pumps to your stomach part of the pulmonary circulatory loop? Explain.

Review SECTION 39-2 ANSWERS

1. left atrium, left ventricle, aorta, arteries, capillaries, veins, vena cava, right atrium, right ventricle, pulmonary arteries, lungs, and pulmonary veins
2. When the sinoatrial node is activated, it causes the atria to contract. After about a one tenth of a second pause, the ventricles contract.
3. EKG, blood pressure, and pulse can be used to monitor circulatory system condition.
4. A heart attack is a blocked blood vessel in the heart. A stroke is a blocked blood vessel in the brain.
5. pulmonary arteries
6. No, pulmonary circulation goes only to the lungs. Systemic circulation goes to the body, including the stomach.

The Respiratory System

Gases Must Be Exchanged with the Environment

A person can go without water for a few days and without food for more than a week. But if a person stops breathing for more than a few minutes, he or she will die. Breathing is the means by which your body obtains and releases gases. Oxygen is used by your cells to completely oxidize glucose and then make ATP, the main energy currency in your cells. Without oxygen, your body cannot obtain enough energy from food to survive. Excess carbon dioxide produced as a waste product of aerobic respiration is toxic to cells and must be removed.

The path that air follows

Breathing, however, is only one part of gas exchange. The gases must be transported by the circulatory system and then exchanged at the cells. All of the organs and tissues that function in this exchange of gases make up the respiratory system, as shown in **Figure 39-14.**

A breath of air enters the respiratory system through the nose or mouth. Air is made up of many gases. About 21 percent of air is oxygen gas. Hairs in your nose filter dust and particles out of the air. Tissues that line the nasal cavity moisten and warm the air.

Objectives

- **Summarize** the path that air follows when it enters the body through the nose or mouth. (pp. 895–896)
- **Describe** the role of the rib muscles and diaphragm in breathing. (p. 897)
- **Describe** how the rate of breathing is regulated. (p. 897)
- **Summarize** how oxygen and carbon dioxide are transported in the blood. (pp. 898–899)
- **Identify** three respiratory diseases. (p. 900)

Key Terms

pharynx
larynx
trachea
bronchus
alveolus
diaphragm

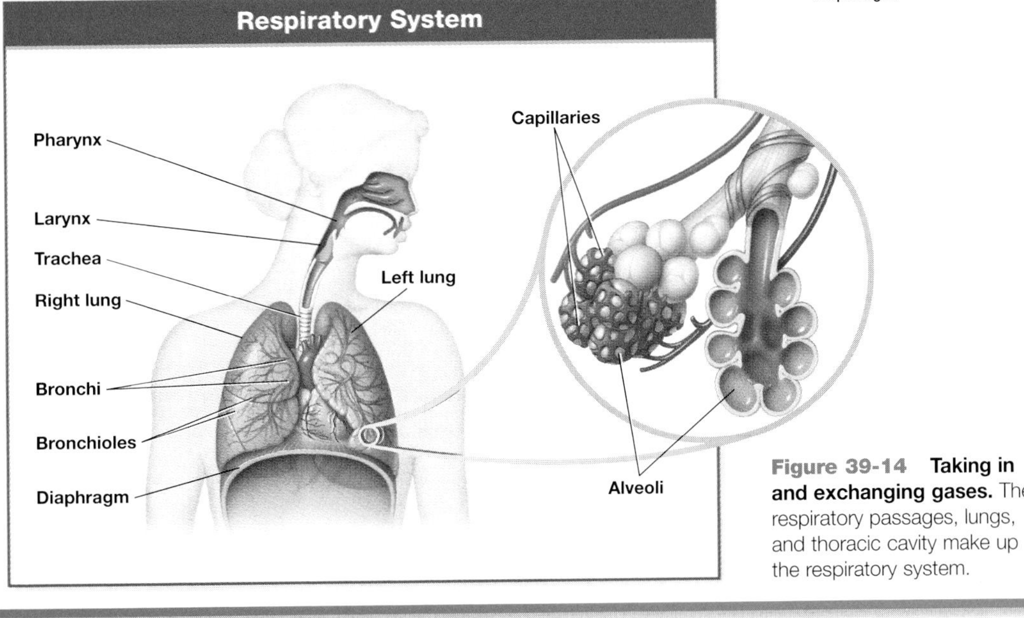

Figure 39-14 Taking in and exchanging gases. The respiratory passages, lungs, and thoracic cavity make up the respiratory system.

Integrating Different Learning Styles

Logical/Mathematical	**PE**	Math Lab, p. 896; Quick Lab, p. 899; Performance Zone item 26, p. 903
	ATE	Lesson Warm-up, p. 895
Visual/Spatial	**PE**	Study Tip, p. 896
	ATE	Demonstrations, pp. 897, 898; Reteaching, p. 900; Closure, p. 900
Body Kinesthetic	**ATE**	Demonstration, p. 898
Verbal/Linguistic	**PE**	Performance Zone item 30, p. 903
	ATE	Career, p. 897; Active Reading, p. 898; Resource Link, p. 900

1 Prepare

Scheduling

- **Block:** About 90 minutes will be needed to complete this section.
- **Traditional:** About two class periods will be needed to complete this section.
- **Planning:** See the Planning Guide on pp. 879A and 879B for lecture, classwork, and assignment options for Section 39-3. Lesson plans for Section 39-3 can be found in a lesson cycle format in ***Biology: Principles and Explorations*** **One-Stop Planner CD-ROM with Test Generator.**

Resource Link

- Assign **Active Reading Worksheet Section 39-3** to help students comprehend the key concepts and visuals in this lesson.

2 Teach

Lesson Warm-up

Ask students why a curve ball does not break as well at Coors Field (in Colorado) as at 3 Com Park (in San Francisco). *(In Colorado, the air is thinner, so there is less friction across the baseball's seams.)* Point out that this also affects respiration in humans. Gases are exchanged between the air in the alveoli and the blood in the capillaries entirely by diffusion. Have students postulate why a person breathes deeper and faster at higher altitudes, where there is decreased pressure. *(At high altitudes, hemoglobin's affinity for oxygen is lower. At 6,000 m [9,000 ft], the oxygen saturation of hemoglobin is only 67 percent, compared to 98 percent at sea level. One must become acclimatized over time to the low levels of oxygen.)* Athletes who train in high altitudes tend to perform better in lower altitudes than those who train in low altitudes.

Teaching TIP

Larynx

Point out to students where the vocal cords are found. Explain that sound is produced when air is forced through these cords. Also inform students that laryngitis is simply an inflammation of the vocal cords in which the cords lose their ability to vibrate. Hence, people "lose their voice." Another interesting note is that birds have a syrinx, or song box, while humans have a larynx, or voice box.

Teaching TIP

Mucus

A person normally swallows more than a pint of nasal mucus each day, and even more if he or she has an allergy or infection.

MATH LAB *Calculating the amount of air respired*

Time 10–15 minutes

Process Skills Analyzing data, calculating, inferring relationships

Teaching Strategies
Ensure that students understand that units are important in dimensional analysis.

Analysis Answers
1. 20 breaths/min × 0.5 L/breath = 10 L/min
2. 10 L/min × 60 min/hr = 600 L/hr
3. Answers will vary but may include that infants have smaller lungs, a faster heart rate, and a faster metabolism.

Study TIP

Reading Effectively

To better understand the branching of the bronchi, think of the bronchi dividing much as a tree limb divides into branches and the branches divide into twigs.

From the nose, air passes through a muscular tube in the upper throat called the **pharynx** *(FAIR ingks),* which serves as a passageway for air and food. The air continues on through a passageway for air, called the **larynx** *(LAIR ingks),* or voice box located in the neck. A flap of tissue, the epiglottis *(ehp uh GLAHT ihs),* covers the opening to the larynx when you swallow food and liquids, which prevents food and liquids from passing into your lungs.

From the larynx the air passes into the **trachea** *(TRAY kee uh),* a long, straight tube in the chest cavity. The trachea, or windpipe, divides into two smaller tubes, the **bronchi** *(BRAHNG keye),* which lead to the lungs. Within the lung, the bronchi (singular, *bronchus*) divide into smaller and smaller tubes called bronchioles *(BRAHNG kee ohls).* The smallest bronchioles end in clusters of air sacs called **alveoli** *(al VEE uh leye),* where gases are actually exchanged. As shown in Figure 39-14, each of the 300 million small alveoli is surrounded by capillaries. Alveoli increase the surface area of your lungs to as much as 42 times the surface area of your body.

The cells that line the bronchi and trachea secrete mucus that traps foreign particles in the air. The mucus is directed upward by cilia to the epiglottis, where the mucus is swallowed and digested. Microbes in the mucus are destroyed by acids and enzymes in the stomach.

Lungs

The lungs, which are among the largest organs in the body, are suspended in the chest cavity, bounded on the sides by the ribs and on the bottom by the diaphragm *(DEYE uh fram).* The **diaphragm** is a powerful muscle nestled in the bottom of the rib cage, and it aids in respiration. A double membrane surrounds both lungs. The outermost membrane is attached to the wall of the thoracic cavity, and the inner membrane is attached to the surface of the lungs. Between both membranes is a small space filled with fluid.

MATH LAB *Calculating the amount of air respired*

Background

Most adults take in about 0.5 L of air with each breath. The normal breathing rate is 18 to 20 breaths per minute.

Analysis

1. **Calculate** the volume of air in liters an adult breathes per minute if his or her breathing rate is 20 breaths per minute.
2. **Calculate** the volume of air in liters an adult breathes per hour if his or her breathing rate is 20 breaths per minute.
3. **SKILL Inferring Conclusions** The breathing rate of an infant is about 30 to 50 breaths per minute. Why might infants have higher respiratory rates than adults?

Overcoming Misconceptions

Passages

Emphasize to students that although food and air both pass through the pharynx, they do not share the same pathway to the rest of the body. Have students gently feel their throat for the cartilaginous trachea. The esophagus lies behind the trachea. The epiglottis covers the trachea during swallowing, ensuring that food makes it down the esophagus.

Breathing Is Caused by Pressure Changes Within the Chest Cavity

Air is drawn into and pushed out of the lungs by the mechanical process known as breathing. Breathing occurs because of air pressure differences between the lungs and the atmosphere, as shown in **Figure 39-15.** To draw air into the lungs, a process called inhalation, the rib muscles contract, drawing the rib cage up and out, and the diaphragm contracts, moving downward. The volume of the chest cavity increases, which reduces the air pressure within the cavity below the atmospheric pressure. Because air flows from a high pressure area to a low pressure area, air is drawn into the lungs.

Normal exhalation (breathing out) is a passive process. The rib cage and diaphragm muscles relax, which returns the rib cage and diaphragm to their resting position. The relaxation of these muscles decreases the volume in the chest cavity and increases the air pressure in the lungs. Because the air pressure is now higher in the lungs than in the atmosphere, air is forced out—from a high pressure area to a low pressure area.

Many factors influence breathing rate

You took your first breath within moments of being born. Since then, you have repeated the process more than 200 million times. What controls how fast or slow you breathe? Receptors in the brain and circulatory system continually monitor the levels of oxygen and carbon dioxide in the blood. The receptors enable the body to automatically regulate oxygen and carbon dioxide concentrations by sending signals to the brain. The brain responds by sending nerve signals to the diaphragm and rib muscles in order to speed or slow the rate of breathing.

It may surprise you to know that carbon dioxide levels have a greater effect on breathing than do oxygen levels. For example, if the concentration of carbon dioxide in your blood increases, such as during exercise, you breathe more deeply, ridding your body of excess carbon dioxide. When the carbon dioxide level drops, your breathing slows. Factors such as stress, pain, and fear also influence breathing rate.

The signals that travel from the breathing center of the brain are not subject to voluntary control. You cannot simply decide to stop breathing indefinitely. You can hold your breath for a while, but your respiratory control center will take over and force your body to breathe.

Figure 39-15 **Inhalation and exhalation**

The diaphragm and the muscles between the ribs are involved in the movement of the chest cavity during breathing.

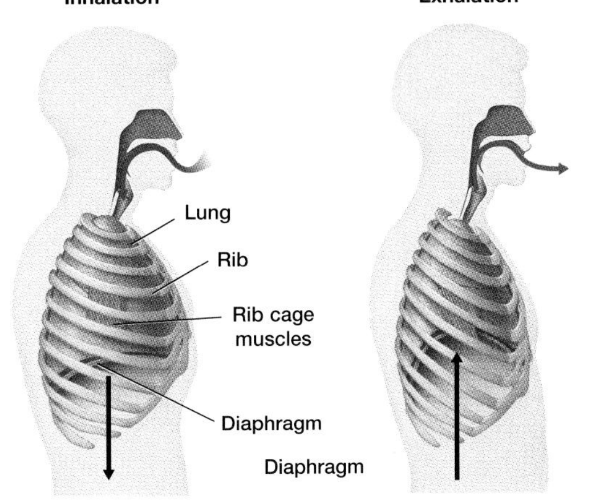

When the diaphragm contracts, it moves down and air rushes in.

When the diaphragm relaxes, it moves up and air is forced out.

biobit

What causes hiccups?

Hiccups are involuntary, sudden intakes of air that occur when the diaphragm contracts suddenly and strongly. The inhaled air causes the vocal cords in the larynx to open and then quickly close, producing a loud gasp—a hiccup. Overstretching or irritation of the stomach or other nearby organs, and certain diseases, can cause hiccups.

Demonstration

Breathing

Demonstrate how breathing works by making a model of the lungs and diaphragm. Attach a small, deflated balloon to one end of a short glass tube that is inserted through a one-holed rubber stopper. Use the stopper to plug the mouth of a bell jar, allowing the balloon to hang inside the jar. Use a rubber band to fasten a thin sheet of rubber over the mouth of the jar. Pull down on the rubber sheet, and have students observe what happens to the balloon. Have students relate their observations to inhalation. Release the rubber sheet. Have them relate their observations to exhalation. Ask students the drawbacks of this model. *(Chest muscles do not move; only one lung is shown.)*

Teaching TIP

Spirometer

People in hospitals are at risk for lung infections because they are inactive. If a patient is bedridden, he or she must breathe deeply to prevent pneumonia. Hospital staff members check a patient's tidal volume several times a day to ensure enough air is being expelled and inhaled.

CAREER

Anesthesiologist/ Anesthetist

Anesthesiologists and anesthetists administer anesthetic drugs and monitor heart and breathing rates during surgery. Have students use outside resources to research anesthesiology and write a short summary of job descriptions, education and training, and salary and employment outlook. PORTFOLIO

Chapter 39

Demonstration

Bromothymol Blue Indicator

Have a volunteer gently blow through a straw into a beaker that contains a weak solution of bromothymol blue (BTB) and water. The color will change slowly to green then yellow, indicating the presence of CO_2. Actually BTB is an indicator of acid, so bicarbonate ions and carbonic acid are causing the color change. (Make certain that the student understands that he or she is to blow rather than suck.)

Teaching TIP

Coronary Artery Disease

Have students present a short oral report on one of the following topics: atherosclerosis (hardening of the arteries), angina (chest pains), intermittent claudication (atherosclerosis of the leg arteries), transient ischemic attacks (ministrokes caused by reduced blood flow to the brain), or the effects of high blood pressure and high cholesterol on the heart.

PORTFOLIO

Active Reading

Technique: Reading Organizer

Encourage students to keep a vocabulary notebook, a stack of vocabulary cards, or some other way of recording and reviewing unfamiliar vocabulary. Each entry should include not only the correct spelling of the word but also a definition, preferably one in the student's own words. This strategy will be especially beneficial in a chapter like this one, where students encounter many new words.

O_2 and CO_2 Are Transported in the Blood

Breathing is the first step to getting oxygen to the millions of cells in your body. When oxygen molecules diffuse from the air into your alveoli, their journey has just begun, as shown in **Figure 39-16.** As oxygen passes into the plasma of the bloodstream, it is picked up by red blood cells that contain an oxygen-carrying protein called hemoglobin.

Oxygen transport

Each hemoglobin molecule contains four atoms of iron. The iron atoms in the hemoglobin give red blood cells their red color. The iron atoms bind reversibly with oxygen. Reversible binding means that at the appropriate time, the oxygen can be released elsewhere in the body and be taken up by the cells that need it.

Figure 39-16 summarizes the path of oxygen and carbon dioxide through the body:

1. Oxygen from the outside air reaches the lungs.
2. The oxygen diffuses from the alveoli to the capillaries. At the high oxygen levels that occur in the blood within the lungs, most hemoglobin molecules carry a full load of oxygen.
3. The oxygen-rich blood then travels to the heart. The heart pumps the blood to the tissues of the body.
4. Oxygen diffuses into the cells for use during aerobic respiration. In the tissues, oxygen levels are lower, causing the hemoglobin to release its oxygen.
5. In tissues, the presence of carbon dioxide produced by cellular respiration makes the blood more acidic and causes the hemoglobin molecules to assume a different shape, one that gives up oxygen more easily. The carbon dioxide diffuses from the cells to the blood.
6. Most of the carbon dioxide travels to the heart as bicarbonate (HCO_3^-) ions.
7. The heart pumps the blood to the lungs. In the lungs, carbon dioxide is released in its gaseous form to the alveoli.
8. The carbon dioxide is expelled.

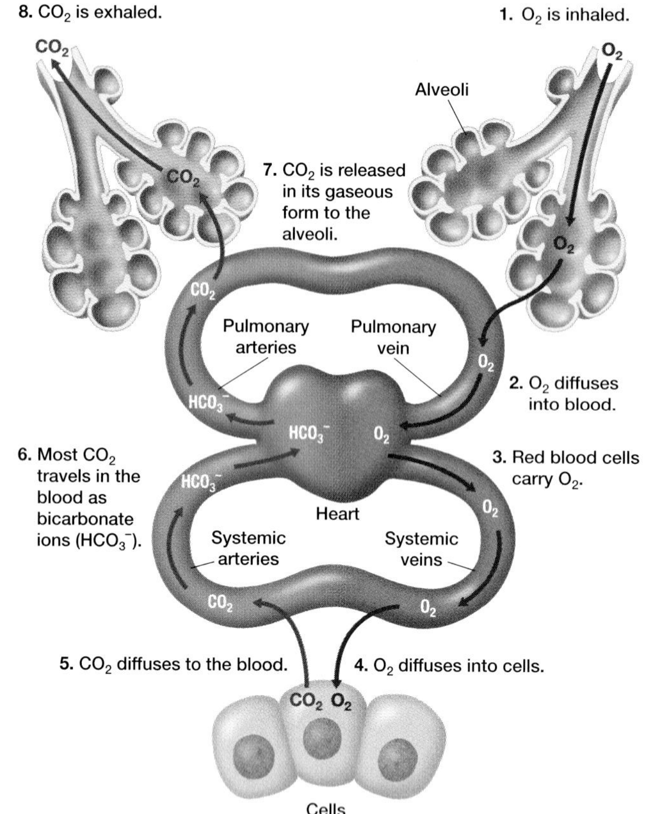
8. CO_2 is exhaled.
CO_2
1. O_2 is inhaled.
O_2
Alveoli
CO_2
7. CO_2 is released in its gaseous form to the alveoli.
O_2
CO_2
Pulmonary arteries
Pulmonary vein
O_2
2. O_2 diffuses into blood.
HCO_3^-
HCO_3^-
O_2
6. Most CO_2 travels in the blood as bicarbonate ions (HCO_3^-).
HCO_3^-
3. Red blood cells carry O_2.
Heart
O_2
Systemic arteries
Systemic veins
CO_2
O_2
5. CO_2 diffuses to the blood.
4. O_2 diffuses into cells.
CO_2 O_2
Cells

Figure 39-16 O_2 and CO_2 transport in the blood. Hemoglobin molecules inside red blood cells transport oxygen, while most carbon dioxide is transported as bicarbonate ions in the plasma.

Multicultural Perspective

Saving Lives

In 1914, Barrett A. Morgan, an African American, invented the world's first gas mask. Called by many names—breathing device, safety hood, gas inhalator—this clever invention won an award at the Second International Exposition of Safety and Sanitation in New York in 1914. Two years later it saved the lives of construction workers trapped in a smoke-filled tunnel beneath Lake Erie.

Carbon dioxide transport

At the same time that the red blood cells are unloading oxygen to tissues, they are also taking in carbon dioxide from the tissues. Carbon dioxide is carried by the blood in three forms:

1. About 7 percent of CO_2 is dissolved in the blood plasma.
2. About 23 percent of CO_2 is attached to hemoglobin molecules inside red blood cells.
3. The majority of CO_2, 70 percent, is carried in the blood as bicarbonate ions.

How is CO_2 carried as bicarbonate ions? In the presence of an enzyme, carbon dioxide combines with water to form carbonic acid, H_2CO_3, as shown in the equation below. The carbonic acid then breaks up to form a bicarbonate ion, HCO_3^-, and a hydrogen ion, H^+:

$$H_2O + CO_2 \rightleftarrows H_2CO_3 \rightleftarrows HCO_3^- + H^+$$

Thus, most of the CO_2 travels in the blood as bicarbonate ions. The hydrogen ions make the blood more acidic. When the blood reaches the lungs, the series of reactions is reversed:

$$HCO_3^- + H^+ \rightleftarrows H_2CO_3 \rightleftarrows H_2O + CO_2$$

A bicarbonate ion combines with a hydrogen ion to form carbonic acid, which in turn forms CO_2 and water. The CO_2 diffuses out of the capillaries into the alveoli and is exhaled into the atmosphere.

REAL LIFE

The lungs of human fetuses do not function until birth. A fetus's umbilical cord contains blood vessels that lead to and from the placenta, which contains fetal and maternal capillaries.

The O_2 in the mother's capillaries diffuses to the fetus's capillaries, and the CO_2 in the fetus's capillaries diffuses into the mother's capillaries.

Finding Information
Determine the signal that stimulates the baby to start breathing at birth.

Teaching TIP

First Breaths

In addition to using lungs for the first time, a newborn's heart and circulatory system change with the first breath. An opening, the foramen ovale, lies between the left and right atria. Another structure, the ductus arteriosus, connects the pulmonary artery to the aorta. Upon inspiration for the first time, the lungs are inflated, closing the opening and separating pulmonary circulation from the systemic circulation.

QUICK LAB: *What is the role of bicarbonate in homeostasis?*

Time 15–20 minutes

Process Skills Inferring, analyzing

Teaching Strategies
Carefully review the chemical equations on page 899.

Analysis Answers
1. A becomes acidic; B becomes basic.
2. $H_2O + CO_2 \rightleftarrows H_2CO_3$
3. sodium bicarbonate, $H_2CO_3^-$
4. It buffered the carbonic acid.
5. Most of the CO_2 in the blood travels as bicarbonate ions.

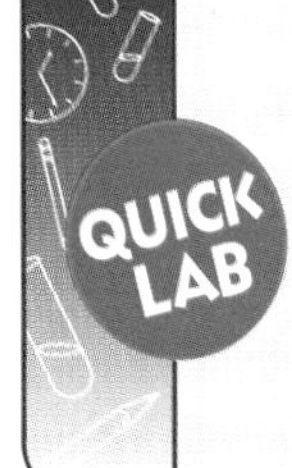

QUICK LAB: *What is the role of bicarbonate in homeostasis?*

You can use pH indicator paper, water, and baking soda to model the role of bicarbonate ions in maintaining blood pH levels in the presence of carbon dioxide.

Materials

two 250 mL beakers, 250 mL distilled water, 2.8 g baking soda, glass stirring rod, 4 strips of wide-range pH paper, 2 drinking straws

Procedure

1. Label one beaker "A" and another "B." Fill each beaker halfway with distilled water.
2. Add 1.4 g of baking soda to beaker B, and stir well.
3. Test and record the pH of the contents of each beaker.
4. *Gently* blow through a straw, into the solution in beaker A. Test and record the pH of the resulting solution.
5. Repeat step 4 for beaker B.

Analysis

1. **Describe** what happened to the pH in the two beakers during the experiment.
2. **Propose** the chemical reaction that might have caused a change in pH in beaker A.
3. **State** the chemical name for baking soda.
4. **Summarize** the effect the baking soda had on the pH of the solution in beaker B after blowing.
5. **Relate** what happened in beaker B to what occurs in the bloodstream.

did you know?

The respiratory control center is located in the medulla oblongata region of the brain.

The medulla oblongata is stimulated by changes in the pH of the cerebrospinal fluid that bathes the brain. When carbon dioxide levels in the blood rise, high levels of carbon dioxide diffuse into the cerebrospinal fluid, lowering its pH. The medulla oblongata responds by increasing respiration rate.

Resource Link

Assign **Occupational Applications: Respiratory Therapist** from the Holt BioSources Teaching Resources CD-ROM. PORTFOLIO

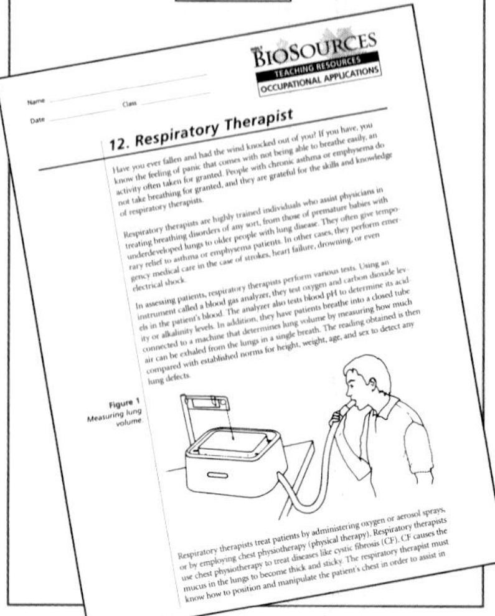

BioSources Teaching Resources
Occupational Applications

Name ____ Class ____
Date ____

12. Respiratory Therapist

Have you ever fallen and had the wind knocked out of you? If you have, you know the feeling of panic that comes with not being able to breathe easily, an activity often taken for granted. People with chronic asthma or emphysema do not take breathing for granted, and they are grateful for the skills and knowledge of respiratory therapists.

Respiratory therapists are highly trained individuals who assist physicians in treating breathing disorders of any sort, from those of premature babies with underdeveloped lungs to older people with lung disease. They often give temporary relief to asthma or emphysema patients. In other cases, they perform emergency medical care in the case of strokes, heart failure, drowning, or even electrical shock.

In assessing patients, respiratory therapists perform various tests. Using an instrument called a blood gas analyzer, they test oxygen and carbon dioxide levels in the patient's blood. The analyzer also tests blood pH to determine its acidity or alkalinity levels. In addition, they have patients breathe into a closed tube connected to a machine that determines lung volume by measuring how much air can be exhaled from the lungs in a single breath. The reading obtained is then compared with established norms for height, weight, age, and sex to detect any lung defects.

Figure 1 *Measuring lung volume.*

Respiratory therapists treat patients by administering oxygen or aerosol sprays, or by employing chest physiotherapy (physical therapy). Respiratory therapists use chest physiotherapy to treat diseases like cystic fibrosis (CF). CF causes the mucus in the lungs to become thick and sticky. The respiratory therapist must know how to position and manipulate the patient's chest in order to assist in

3 Assessment Options

Closure

Draw a simplified version of Figure 39-16 on the board. Have students close their books, and ask volunteers to draw arrows showing the directions of circulation and gas exchange. Have other volunteers annotate the diagram with brief statements the describe key events.

Review

Assign Review—Section 39-3.

Reteaching — BASIC

Have students draw pictures of lungs during inspiration and expiration. Students should use labeled arrows to show the direction of diaphragm and rib muscle movement and oxygen and carbon dioxide movement.

Respiratory Diseases Limit Lung Function

Respiratory diseases affect millions of Americans. A chronic pulmonary disease is one for which there is no cure. Afflicted patients must learn how to manage their disease.

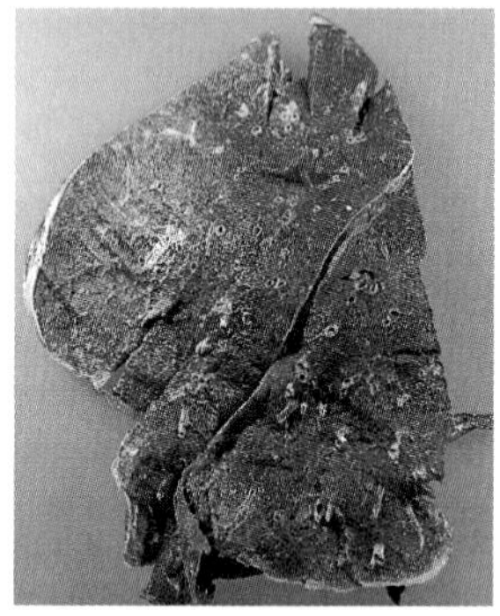

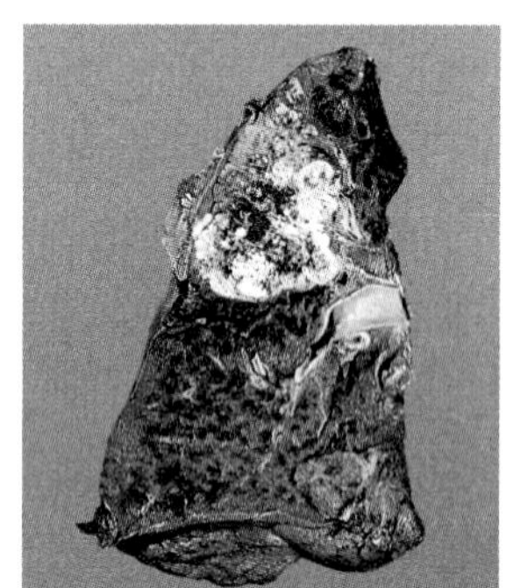

Figure 39-17 Healthy lungs and cancerous lungs. The top photo shows a healthy lung. The bottom photo shows a cancerous lung which contains cells that divide uncontrollably.

Asthma

Asthma *(AZ muh)* is a disease in which the bronchioles of the lungs constrict, or become narrow, because of their sensitivity to certain stimuli in the air. As the airway constricts, the air flow to the lungs is reduced. In severe allergic asthma attacks, the alveoli may swell enough to rupture. Stressful situations and strenuous exercise may trigger an asthma attack. Left untreated, asthma can be deadly. Fortunately, prescribed inhalant medicines may help to counteract an asthma attack by dilating (expanding) the bronchioles. People of all ages have asthma.

Emphysema

In emphysema *(ehm fuh SEE muh)*, another chronic pulmonary disease, the lungs' alveoli lose their elasticity, making it difficult to release air during exhalation. Because the body is forced to use more and more energy just to breathe, a person with emphysema suffers from constant fatigue and breathlessness. Severely affected individuals must breathe from tanks of pure oxygen in order to live. Emphysema often results from cigarette smoking, as does lung cancer. Both emphysema and lung cancer claim millions of lives annually.

Lung cancer

Lung cancer is one of the leading causes of death in the world today. As shown in **Figure 39-17**, cancer is a disease characterized by abnormal cell growth. In the United States alone, about 30 percent—180,000—of all cancer deaths each year are attributed to lung cancer. Once cancer is detected, the affected lung is usually removed surgically. Even with such drastic measures, fewer than 10 percent of lung cancer victims live more than 5 years after diagnosis.

Review SECTION 39-3

1. **Sequence** the path a breath of air follows through the respiratory system. (Begin with air that enters through the mouth.)
2. **State** the direction in which the diaphragm and rib muscles move before inhalation.
3. **Name** the main factor that regulates the rate of breathing.
4. **Evaluate** the role that bicarbonate plays in transporting carbon dioxide in the blood.

Critical Thinking

5. **SKILL Justifying Conclusions** Would a person with emphysema have trouble climbing stairs? Explain.

Review SECTION 39-3 ANSWERS

1. mouth, trachea, bronchi, bronchioles, and alveoli
2. The diaphragm moves down; the ribs move up and out.
3. Levels of oxygen and carbon dioxide in the blood regulate the rate of breathing. Increased blood carbon-dioxide concentration triggers the brain to send a message to the diaphragm and rib muscles to speed up the rate of breathing.
4. Seventy percent of the carbon dioxide in the blood is carried as bicarbonate ions.
5. Yes, any physical effort increases the oxygen needed and energy required to breathe.

Study ZONE

CHAPTER 39

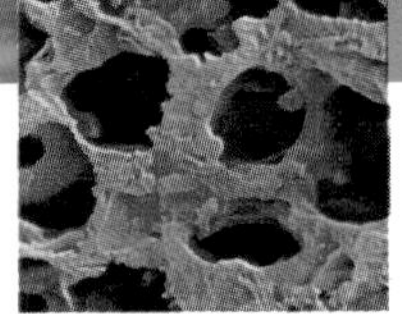

Use the Key Concepts and Key Terms listed below to review the main ideas in this chapter. If you need to review the meaning of a term, turn to the page indicated.

Key Concepts

39-1 The Circulatory System

- The human circulatory system is made up of blood vessels, blood, and a heart that together function to transport materials, remove wastes, and distribute heat.
- Arteries carry blood away from the heart. Materials are exchanged at the capillaries. Veins contain valves and carry blood back to the heart. Fluids not returned to the capillaries are picked up by lymphatic vessels.
- Blood consists of plasma (water, metabolites, wastes, salts, and proteins), red blood cells, white blood cells, and platelets.
- Blood types are defined by the presence or absence of proteins on the surface of red blood cells.

39-2 The Heart

- The right side of the heart receives oxygen-poor blood from the body and circulates it to the lungs. In the lungs, gases are exchanged. The left side of the heart receives oxygenated blood from the lungs and circulates it to the rest of the body.
- Atria receive blood entering the heart. Ventricles circulate blood away from the heart.
- Contraction of the heart is initiated by the sinoatrial node. The health of the heart can be monitored by measuring blood pressure, electrical impulses, and pulse rate.
- Blockages in blood vessels can lead to heart attacks or strokes.

39-3 The Respiratory System

- A series of tubes and bunched air sacs (alveoli) take in oxygen and remove carbon dioxide.
- Breathing is caused by pressure changes within the chest cavity.
- The concentration of carbon dioxide in the blood is the most critical factor affecting a person's breathing rate and depth.
- Oxygen is transported to tissues by combining with hemoglobin molecules inside red blood cells. Most carbon dioxide is transported to the lungs as bicarbonate ions.
- Asthma, emphysema, and lung cancer limit lung function.

Key Terms

39-1

artery (883)
capillary (883)
vein (883)
valve (884)
lymphatic system (885)
plasma (886)
red blood cell (886)
anemia (887)
white blood cell (887)
platelet (887)
ABO blood group system (888)
Rh factor (889)

39-2

atrium (891)
ventricle (891)
vena cava (891)
aorta (892)
coronary artery (892)
sinoatrial node (892)
blood pressure (892)
pulse (893)
heart attack (894)
stroke (894)

39-3

pharynx (896)
larynx (896)
trachea (896)
bronchus (896)
alveolus (896)
diaphragm (896)

Study TIP Ready?

- *If you think you understand these Key Concepts and Key Terms, then you're ready for the Performance Zone.* 

Study ZONE

CHAPTER 39

Review and Practice

Assign the following worksheets for Chapter 39:

- **Student Review Guide**
- **Concept Mapping Worksheet**
- **Test Preparation Pretest**
- **Vocabulary Worksheet**
- **Critical Thinking Review Worksheet**

Assessment Options

The following assessment options are available for this chapter:

Resource Link

- Assign **Chapter Test 39** to assess students' understanding of the chapter.

Alternative Assessment

Have students make a board game with questions about normal and abnormal function of the circulatory and respiratory systems. PORTFOLIO

Portfolio Assessment

- *Portfolio Assessment* at the front of this book provides options for building and evaluating student portfolios. Students and teachers can select items from the areas listed for inclusion in a portfolio.
- See items labeled PORTFOLIO on pp. 887, 891, 893, 897, 898, 900, 901, and 905.

Answer to Concept Map

The following is one possible answer to Performance Zone item 1 on page 902.

Circulatory system
is composed of
blood vessels
lymphatic system
heart
such as
arteries
capillaries
veins
largest is
aorta
largest are
venae cavae
has
atria
ventricles
regulates
systemic circulation
pulmonary circulation

Performance ZONE

CHAPTER 39

REVIEW ANSWERS

Assignment Guide	
SECTION	**REVIEW**
39-1	1, 2, 3, 4, 5, 6, 13, 16, 17, 19, 25, 27
39-2	7, 8, 9, 18, 20, 21, 22, 23, 24, 28, 29
39-3	10, 11, 12, 14, 15, 26, 30

Understanding and Applying Concepts

1. The answer to the concept map is found at the bottom of page 901.
2. **a.** Plasma is the liquid part of the blood; lymph is fluid that leaks out of blood vessels and into surrounding tissues.
   b. Blood pressure is the pressure exerted by blood as it moves through the blood vessels. Blood pressure is expressed in mm Hg. Pulse is a series of pressure waves within an artery caused by a contraction of the left ventricle. Pulse is an indicator of the number of times the heart beats in a minute.
   c. A heart attack occurs when an area of the heart muscle dies and stops working and is usually caused by a blocked blood vessel in the heart. A stroke is a a sudden attack of weakness or paralysis that occurs when an area of the brain dies after blood flow to the brain is interrupted.
3. b
4. b
5. d
6. c
7. a
8. b
9. b
10. c
11. b
12. c
13. plasma—transport of substances; white blood cells—fight infections; red blood cells—carry oxygen; platelets—clot the blood
14. Oxygen is carried by hemoglobin in red blood cells. Carbon dioxide is primarily carried as bicarbonate ions in the blood.

Performance ZONE

CHAPTER 39 Review

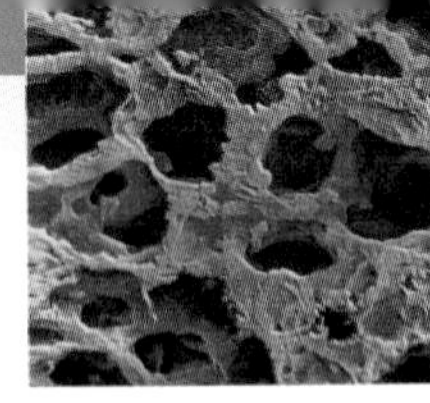

Understanding and Applying Concepts

1. **Concept Mapping** Make a concept map that outlines the path of blood through the body. Try to include the following terms in your map: artery, capillary, vein, lymphatic system, pulmonary circulation, systemic circulation, atrium, ventricle, aorta, venae cavae.
2. **Understanding Vocabulary** For each pair of terms, explain the differences in their meanings:
   a. plasma, lymph
   b. blood pressure, pulse
   c. heart attack, stroke
3. The layer labeled *A* is
   a. connective tissue.
   b. smooth muscle.
   c. endothelium.
   d. a valve.

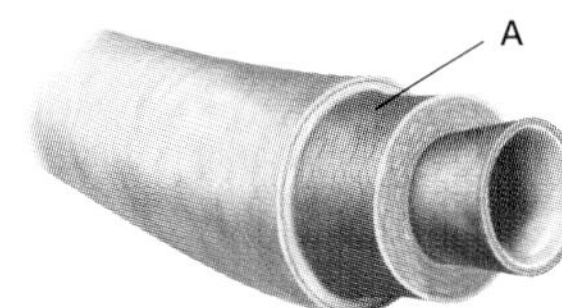

4. Lymphatic vessels
   a. transport blood.
   b. return fluid to the blood.
   c. produce antibodies.
   d. control blood clotting.
5. An abnormality involving the platelets will affect the process of
   a. breathing.
   b. locomotion.
   c. digestion.
   d. clotting.
6. What type of blood contains A antibodies (but not B antibodies) in the plasma and lacks Rh antigens?
   a. AB negative
   b. A positive
   c. B negative
   d. O positive
7. Blood in the pulmonary veins is
   a. oxygen-rich.
   b. iron-poor.
   c. oxygen-poor.
   d. calcium-rich.
8. The chamber of the heart that sends blood to the lungs is the
   a. left ventricle.
   b. right ventricle.
   c. left atrium.
   d. right atrium.
9. The pacemaker of the heart is the
   a. left ventricle.
   b. sinoatrial node.
   c. coronary sinus.
   d. inferior vena cava.
10. The diaphragm contracts and the pressure in the chest cavity decreases during
    a. bronchitis.
    b. exhalation.
    c. inhalation.
    d. asthma attacks.
11. Breathing rate will automatically increase when
    a. blood pH is high.
    b. the amount of carbon dioxide in the blood increases.
    c. blood acidity decreases.
    d. hemoglobin is unloaded.
12. A disease in which the alveoli lose their elasticity is called
    a. atherosclerosis.
    b. arteriosclerosis.
    c. emphysema.
    d. asthma.
13. **Summarizing Information** Identify the main components of blood, and describe the function of each.
14. **Analyzing Information** Describe how oxygen and carbon dioxide are transported in the blood.
15. **Identifying Structures** Name the key structures involved in respiration. How does each structure function during inhalation and exhalation?
16. **Theme Homeostasis** How is body temperature regulated by blood vessel constriction and dilation?
17. **Health Watch** For what types of athletic events would blood doping be most likely to increase an athlete's performance? Explain your reasoning.
18. **Exploring Further** One of the effects of aspirin is that it thins the blood. Why is aspirin prescribed for people who are at risk of having a heart attack?
19. **Chapter Links** What role do cell surface markers play in blood typing? (**Hint:** See Chapter 3, Section 3-2.)

15. trachea—windpipe; bronchi—two major air passages in lungs; bronchioles—small tubes in lungs; alveoli—air sacs in lungs. When the diaphragm and rib muscles contract, a vacuum is created in the chest cavity, which pulls air in (inhalation). Air moves through the trachea, bronchi, and bronchioles. Gas exchange occurs at the alveoli.
16. Constriction of blood vessels conserves heat. Dilation of blood vessels increases heat dissipation.
17. Answers will vary but may include any event that has a significant aerobic component, such as running, cycling, or swimming.
18. Thinner blood reduces chances of clot formation.
19. Cell surface markers called antigens exist on blood cells. Blood types are named for the antigens present.

Skills Assessment

Part 1: Interpreting Graphics

Salt intake and blood pressure data were collected from people representing more than 20 cultures around the world. The data are summarized in the graph below. Use the graph to answer the following questions:

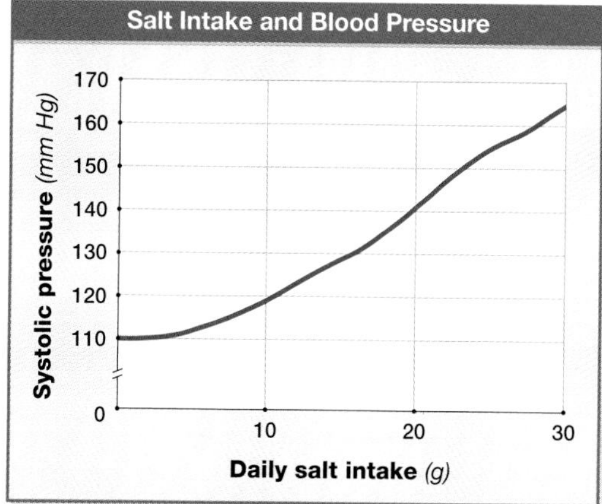

20. What is the relationship between salt intake and blood pressure?

21. If an American businessman averages a daily salt intake of 26 g, what would you predict his systolic pressure to be?

22. Knowing that high blood pressure can lead to a stroke or kidney failure, how might you suggest that this businessman reduce his salt intake?

Part 2: Life/Work Skills

Use the media center or Internet resources to learn about heart disease.

23. **Finding Information** Find out which foods are recommended as foods that prevent heart disease.

24. **Selecting Technology** Find out the medical options that are available for a person who has survived a heart attack. What are the benefits of each medical option? What are the disadvantages of each medical option?

Critical Thinking

25. **Forming Reasoned Opinions** The frequency of blood clots and heart attacks is much lower among the Inuit, the nomadic hunters of the North American Arctic, than among other North Americans and Europeans. This difference is credited to fish oils in the Inuit diet that cause blood platelets to be more slippery. How do you think the clotting ability of the Inuit's blood is affected by the slippery platelets?

26. **Evaluating Results** As altitude increases, the gases that make up the atmosphere become more scarce. When a runner who trained at sea level competes at a location 500 m above sea level, how will his performance at this altitude compare with his performance during training?

Portfolio Projects

27. **Multicultural Perspective** Dr. Charles Drew was an African-American physician. Write a report that discusses the life of Dr. Drew and his many accomplishments. How did his work with blood help save many lives?

28. **Calculating** Assume that the average heart beats 70 times per minute. How many times does the heart beat in a person who lives 75 years? Assume that the heart of an overweight person beats 10 additional times each minute. Explain why being overweight can put a strain on the heart.

29. **Interpreting Information** Read "The Puzzle of Hypertension in African-Americans," by Richard S. Cooper, Charles N. Rotimi, and Ryk Ward (*Scientific American,* Feb. 1999, vol. 280, pp. 57–63). Write a report that discusses the authors' hypotheses for why 35 percent of African-Americans suffer from chronic high blood pressure.

30. **Career Focus Respiratory Therapist** Research the field of respiratory therapy, and write a report on your findings. Your report should include a job description, training required, kinds of employers, growth prospects, and starting salary.

Critical Thinking

25. Clotting ability is most likely diminished because platelets will not stick as readily.
26. It will decrease because the runner will not get as much oxygen as obtained at sea level.

Portfolio Projects

27. Answers will vary. Dr. Drew found a way to keep blood plasma fresh. He also started the first blood bank.
28. 70 beats/1 min × 60 min/1 hr × 24 hr/1 day × 365 days/1 year × 75 years/lifetime = 2,759,400,000 beats in a lifetime. Being overweight causes the heart to beat 3,153,600,000 times in a lifetime—an extra 394,200,000 beats!
29. Answers will vary but should include genes, behavioral risk factors, being overweight, psychological and social stresses, inactivity, alcohol, and poor diet.
30. Respiratory therapists treat patients who have breathing difficulties. Respiratory therapists use respirators and oxygen tents. They develop treatment programs for long-term care, and they operate and maintain oxygen delivery equipment. Most attend 2 years of college with an emphasis on math and science courses. Respiratory therapists must be professionally certified. Many are employed by hospitals or physicians. The growth prospects of the field are excellent. Starting salary will vary by region.

Skills Assessment

20. They are positively correlated. Increased salt intake results in increased blood pressure.
21. about 155 mm Hg
22. Answers will vary. For example: Eat less fast food and reduce intake of food such as chips, fries, and popcorn.
23. Answers will vary but should include fish, fresh fruits, vegetables, and foods low in fat and saturated fat.
24. Answers will vary but should include cardiovascular drugs (anticoagulants, thrombolytics, antiplatelets, and beta-blockers); surgical procedures such as angioplasty, bypass surgery, and intracoronary stenting; and devices such as pacemakers and anti-tachycardia devices.

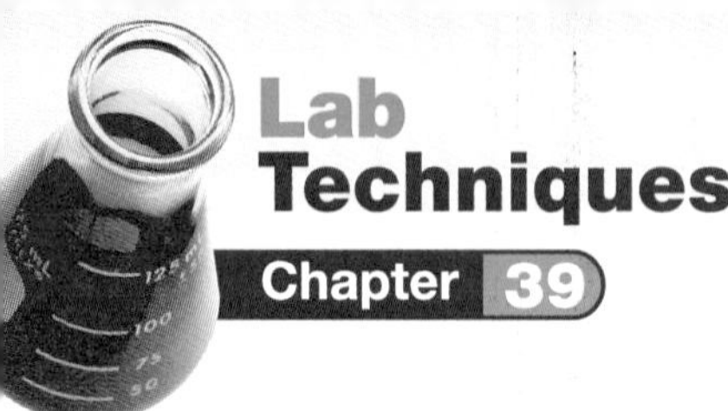

Lab Techniques

Chapter 39

Determining Lung Capacity

Purpose

Students will use a spirometer and data on lung capacity and residual volume of average young adults to estimate their lung capacity.

Preparation Notes

Time Required: 30–40 minutes

Materials

Materials for this lab can be ordered from WARD'S. *See Master Materials List* at the front of this book for catalog numbers.

Preparation Tips

- Place enough materials for a class on a supply table. Provide plenty of clean spirometer mouthpieces.
- Set up a labeled container for the disposal of spirometer mouthpieces.

Safety Precautions

- Discuss all safety symbols and caution statements with students.
- Direct students not to share spirometer mouthpieces.
- Make sure students discard used spirometer mouthpieces in the designated container.

Procedural Tips

- Review the use of the spirometer before students begin the lab.
- You may wish to have students record their data in a class data table on the board. They need not identify themselves by name, only by gender, noting whether they are an athlete or nonathlete, and a smoker or nonsmoker. Calculate means for males, females, smokers, athletes, nonathletes and nonsmokers.

Disposal

Have students dispose of their spirometer mouthpieces in a designated waste container.

Lab Techniques

Chapter 39 *Determining Lung Capacity*

SKILLS

- Measuring
- Organizing data
- Comparing

OBJECTIVES

- **Measure** your tidal volume, vital capacity, and expiratory reserve volume.
- **Determine** your inspiratory reserve capacity and lung capacity.
- **Predict** how exercise will affect tidal volume, vital capacity, and lung capacity.

MATERIALS

- spirometer
- spirometer mouthpiece

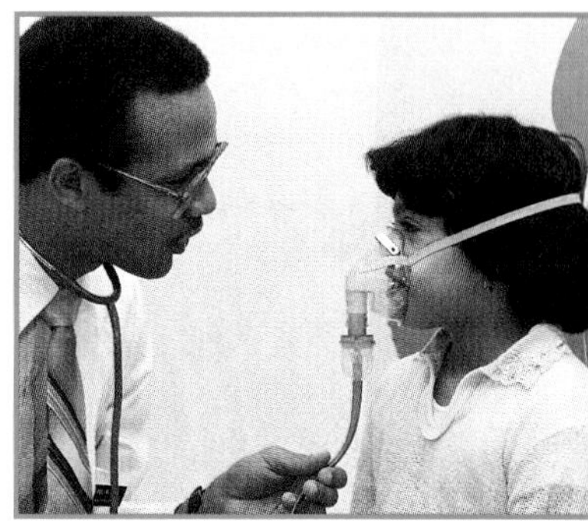

Before You Begin

Lung capacity is the total volume of air that the lungs can hold. The lung capacity of an individual is influenced by many factors, such as gender, age, strength of diaphragm and chest muscles, and disease.

During normal breathing, only a small percentage of your lung capacity is inhaled and exhaled. The amount of air inhaled or exhaled in a normal breath is called the **tidal volume.** An additional amount of air, called the **inspiratory reserve volume,** can be forcefully inhaled after a normal inhalation. The **expiratory reserve volume** is the amount of air that can be forcefully exhaled after a normal exhalation. **Vital capacity** is the maximum amount of air that can be inhaled or exhaled. Even after you have exhaled all the air you can, a significant amount of air called the **residual volume** still remains in your lungs.

In this lab, you will determine your lung capacity by using a **spirometer,** which is an instrument used to measure the volume of air exhaled from the lungs.

1. Write a definition for each boldface term in the paragraph above.
2. Make a data table similar to the one shown.

DATA TABLE	
Tidal volume	
Expiratory reserve volume	
Inspiratory reserve volume	
Vital capacity	
Estimated residual volume	
Estimated lung capacity	

3. Based on the objectives for this lab, write a question about breathing that you would like to explore.

Procedure

PART A: Measuring Volume

1. Place a clean mouthpiece in the end of a spirometer. **CAUTION: Many diseases are spread by body fluids, such as saliva. Do not share a spirometer mouthpiece with anyone.**
2. To measure your tidal volume, first inhale a normal breath. Then exhale a normal breath into the spirometer through the mouthpiece. Record the volume of air exhaled in your data table.
3. To measure your expiratory reserve volume, first inhale a normal breath and then exhale normally. Then forcefully exhale as much air as possible into the spirometer. Record this volume.

Before You Begin

Answers

1. *lung capacity*—the amount of air that can be inhaled and exhaled by the lungs

   tidal volume—the volume of air breathed into or out of the lungs in a normal breath

   inspiratory reserve volume—the volume of air that can be forcefully inhaled after a normal inhalation

   expiratory reserve volume—the volume of air that can be forcefully exhaled after a normal exhalation

   vital capacity—the maximum volume of air that can be inhaled or exhaled

   residual volume—the volume of air that remains in the lungs after you have exhaled all the air that you can

   spirometer—an instrument used to measure the volume of air exhaled from the lungs
3. Answers will vary. For example: What is my vital capacity and lung capacity?

4. To measure your vital capacity, first inhale as much air as you can, and then forcefully exhale as much air as you can into the spirometer. Record this volume.

PART B: Calculating Lung Capacity

The table below contains average values for residual volumes and lung capacities for young adults.

Residual Volumes and Lung Capacities

	Males	Females
Residual volume*	1,200 mL	900 mL
Lung capacity*	6,000 mL	4,500 mL

*Athletes can have volumes 30–40% greater than the average for their gender.

5. Inspiratory reserve volume (IRV) can be calculated by subtracting tidal volume (TV) and expiratory reserve volume (ERV) from vital capacity (VC). The formula for this calculation is as follows:

$$IRV = VC - TV - IRV$$

Use the data in your data table and the equation above to calculate your estimated inspiratory reserve volume.

6. Lung capacity (LC) can be calculated by adding residual volume (RV) to vital capacity (VC). The formula for this calculation is as follows:

$$LC = VC + RV$$

Use the data in your data table and the table above to calculate your estimated lung capacity.

PART C: Cleanup and Disposal

7. Dispose of your mouthpiece in the designated waste container.

8. Clean up your work area and all lab equipment. Return lab equipment to its proper place. Wash your hands thoroughly before you leave the lab and after you finish all work.

Analyze and Conclude

1. **Interpreting Data** How does your expiratory reserve volume compare with your inspiratory reserve volume?
2. **Interpreting Tables** How does the residual volume and lung capacity of an average young adult female compare with those of an average young adult male?
3. **Analyzing Data** How did your tidal volume compare with that of others?
4. **Recognizing Relationships** Why was the value you found for your lung capacity an estimated value?
5. **Analyzing Methods** Why didn't you measure inspiratory reserve volume directly?
6. **Inferring Conclusions** Why would males and athletes have greater vital capacities than females?
7. **Justifying Conclusions** Use data from your class to justify the conclusion that exercise increases lung capacity.
8. **Further Inquiry** Write a new question that could be explored with another investigation.

On the Job

Spirometry is the use of a spirometer to study respiratory function. Nurses and respiratory therapists use spirometers to evaluate patients with respiratory diseases. Do research in the media center or on the Internet to discover how spirometry is used to distinguish different respiratory diseases. Summarize your findings in a written or oral report.

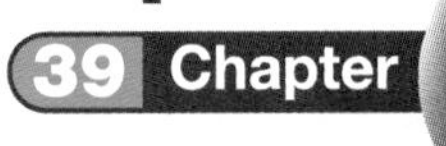

Lab Techniques

39 Chapter

average value, which is probably not the same as a student's actual value. Therefore, the lung capacity found, which includes the residual volume, is also an estimate.

5. Answers will vary. Students should state that there would be no way to judge when they had exhaled the amount of air inhaled after a normal inhalation because they must breathe out to take a reading.
6. Answers may vary. Vital capacity depends mostly on body size. Respiratory muscle strength, lung expansion capacity, and chest size are also important. Males are usually larger than females. Athletes have stronger respiratory muscles and greater lung expansion capacity.
7. Answers will vary. Students should see that athletes have a greater lung capacity than nonathletes.
8. Answers will vary. For example: How do asthma and emphysema reduce the efficiency of the respiratory system as measured by a spirometer?

On the job

Answers

Answers will vary. Respiratory diseases affect the lungs in different ways. For example, lung tumors and lung inflammations reduce the capacity of the lungs. Asthma and bronchitis reduce the rate of air flow. Spirometers can measure these and other types of changes in lung function.

Procedure

Sample Data Table

Tidal volume	500 mL
Expiratory reserve volume	100 mL
Inspiratory reserve volume	4,000 mL
Vital capacity	4,600 mL
Estimated residual volume	1,200 mL
Estimated lung capacity	6,000 mL

Analyze and Conclude

Answers

1. Answers will vary. The expiratory reserve volume should be less than half the inspiratory reserve volume.
2. The average value for females is less than the average value for males.
3. Answers will vary.
4. Answers will vary. The value for residual volume cannot be measured with a spirometer. The figure given for residual volume is an

Resource Link

- Assign **Problem Solving: Pressure Gradients and Calculating Respiratory Volume** from the Holt BioSources Teaching Resources CD-ROM. PORTFOLIO

Chapter 40

Digestive and Excretory Systems

Planning Guide

OBJECTIVES	MEETING INDIVIDUAL NEEDS	READING SKILLS AND STRATEGIES
40-1 Your Body's Need for Food (BLOCKS 1 & 2 90 min) • **Identify** five nutrients found in foods. • **Relate** the role of carbohydrates, proteins, lipids, vitamins, minerals, and water in maintaining a healthy body. • **Describe** each of the parts of the USDA food-guide pyramid. • **Name** one health disorder associated with high levels of saturated fats in the diet.	Learning Styles, p. 908 BASIC ADVANCED • pp. 910, 913	**ATE** Active Reading, Technique: Anticipation Guide, p. 907 ■ Active Reading Guide, worksheet (40-1) ■ Directed Reading, worksheet (40-1)
40-2 Digestion (BLOCKS 3 & 4 90 min) • **Relate** the four major functions of the digestive system to the processing of food. • **Summarize** the path of food through the digestive system and the major digestive processes that occur in the mouth, stomach, small intestine, and large intestine. • **Describe** how nutrients are absorbed from the digestive system into the bloodstream or lymphatic system. • **Identify** the role of the pancreas and liver in digestion.	Learning Styles, p. 914 BASIC ADVANCED • pp. 915, 919	**ATE** Active Reading, Technique: Reading Organizer, p. 918 ■ Active Reading Guide, worksheet (40-2) ■ Directed Reading, worksheet (40-2)
40-3 Excretion (BLOCKS 5 & 6 90 min) • **Identify** major wastes produced by humans and the organ or tissue where they are eliminated from the body. • **Relate** the role of nephrons to the filtering of blood in the kidneys. • **Summarize** how nephrons form urine. • **Describe** the path of urine through the human urinary system. • **Predict** how kidney damage might affect homeostasis and threaten life.	Learning Styles, p. 920 BASIC ADVANCED • p. 924	**ATE** Active Reading, Technique: Anticipation Guide, p. 924 ■ Active Reading Guide, worksheet (40-3) ■ Directed Reading, worksheet (40-3) ■ Supplemental Reading Guide, *Microbe Hunters*

Review and Assessment Resources

BLOCKS 7 & 8 90 min

TEST PREP

- **PE** Study Zone, p. 925
- **PE** Performance Zone, pp. 926–927
- ■ Chapter 40 worksheets:
  - • Vocabulary • Concept Mapping
  - • Critical Thinking
- Audio CD Program
- ■ Test Prep Pretest

CHAPTER TESTING

- ■ Chapter Tests (blackline master)
- ▲ Test Generator
- ■ Assessment Item Listing

See the correlations table at the front of this book for a list of the ***National Science Education Standards*** addressed in this chapter.

Smithsonian Institution® Visit **www.si.edu/hrw** for additional on-line resources.

KEY

Biology Principles and Explorations

PE Pupil's Edition

ATE Teacher's Edition

HOLT BIOSOURCES

 Teaching Transparencies

Laboratory Program

One-Stop Planner CD-ROM
Shows these resources within a customizable daily lesson plan:

■ Teaching Resources

▲ Test Generator

VISUAL STRATEGIES	CLASSWORK AND HOMEWORK	TECHNOLOGY AND INTERNET RESOURCES
ATE **Fig. 40-2,** p. 909 **ATE** **Table 40-1,** p. 912 222 USDA Food Pyramid 190A Nutrition Label 191A Vitamins 192A Trace Elements	**PE** **Data Lab** Analyzing the nutrition of your diet, p. 911 Lab A33 Reading Labels: Nutritional Information Lab C46 Identifying Food Nutrients ■ **Problem-Solving:** Using Food Labels to Calculate Percentage of Nutrients and Calories **PE** **Review,** p. 913	CNN Video Segment 32 Going Vegetarian Holt Biology Videodiscs Section 40-1 Audio CD Program Section 40-1
ATE **Fig. 40-9,** p. 917 216 Digestive System in the Human Body 219 Cross Section of the Small Intestine 193A Digesting Food Molecules	**PE** **Quick Lab** How does bile work?, p. 919 Lab C47 Identifying Food Nutrients—Food Labeling Lab E7 Diet and Weight Loss **PE** **Review,** p. 919	Holt Biology Videodiscs Section 40-2 Audio CD Program Section 40-2 internetconnect SCILINKS NSTA **TOPIC:** Poisons
ATE **Fig. 40-12,** p. 921 218 Human Kidney Structure 220 How a Kidney Machine Works 221 Excretory System in the Human Body	**PE** **Experimental Design** How Can You Improve Lactose Digestion?, pp. 928–929 Lab C48 Urinalysis Testing **PE** **Review,** p. 924 **PE** **Study Zone,** p. 925	Holt Biology Videodiscs Section 40-3 Audio CD Program Section 40-3 internetconnect SCILINKS NSTA **TOPICS:** • Kidney diseases • Kidney dialysis • Lactose intolerance

ALTERNATIVE ASSESSMENT

PE **Portfolio Projects** 27–30, p. 927

ATE **Alternative Assessment,** p. 925

ATE **Portfolio Assessment,** p. 925

Lab Activity Materials

Data Lab
p. 911 pencil, paper

Quick Lab
p. 919 250 mL beakers (2), water, cooking oil, dish detergent, stirring rod, graduated cylinder

Experimental Design
p. 928 milk-treatment product (liquid), toothpicks, depression slides, droppers, whole milk, glucose solution, glucose test strips

Lesson Warm-up

ATE clothesline or rope (8 m), clothespins (5), markers, p. 914

ATE clear plastic bag, tape, p. 920

Demonstrations

ATE 1 pair pantyhose, tennis ball, p. 915

ATE test tubes with stoppers (2), saltine crackers (4), water, soda straw, iodine solution, p. 915

ATE beakers (2), tap water, sugar cubes (2), stirring rods (2), p. 917

BIOSOURCES LAB PROGRAM

- Quick Labs: Lab A33, p. 65
- Lab Techniques and Experimental Design: Lab C46, p. 219
- Lab Techniques and Experimental Design: Lab C47, p. 223
- Lab Techniques and Experimental Design: Lab C48, p. 227
- Interactive Explorations in Biology Laboratory Manual: Lab E7, p. 37

Chapter 40

Chapter Theme

Homeostasis

When it functions normally, the human body remains in homeostasis—all of its systems are in balance, and they maintain their own dynamic equilibrium. The metabolic processes that each system performs require materials and energy ultimately obtained from food. Likewise, the same metabolic processes generate toxic waste products. Eventually, these materials are removed from the body by the process of excretion.

Establishing Prior Knowledge

Before beginning this chapter, make certain students can answer the questions in **Study TIP Ready?** on page 907. If they cannot, encourage them to reread the sections indicated.

Opening Questions

Have students make a list of their favorite foods. Write some of their answers on the board and then list the more obvious nutrient components of each food. Ask students to predict the following: How do these nutrients reach every individual cell in the body? What happens to the nutrients once they have been used by the mitochondria? How do waste products leave? Have students record their responses and revisit the list after reading the chapter.

CHAPTER 40 Digestive and Excretory Systems

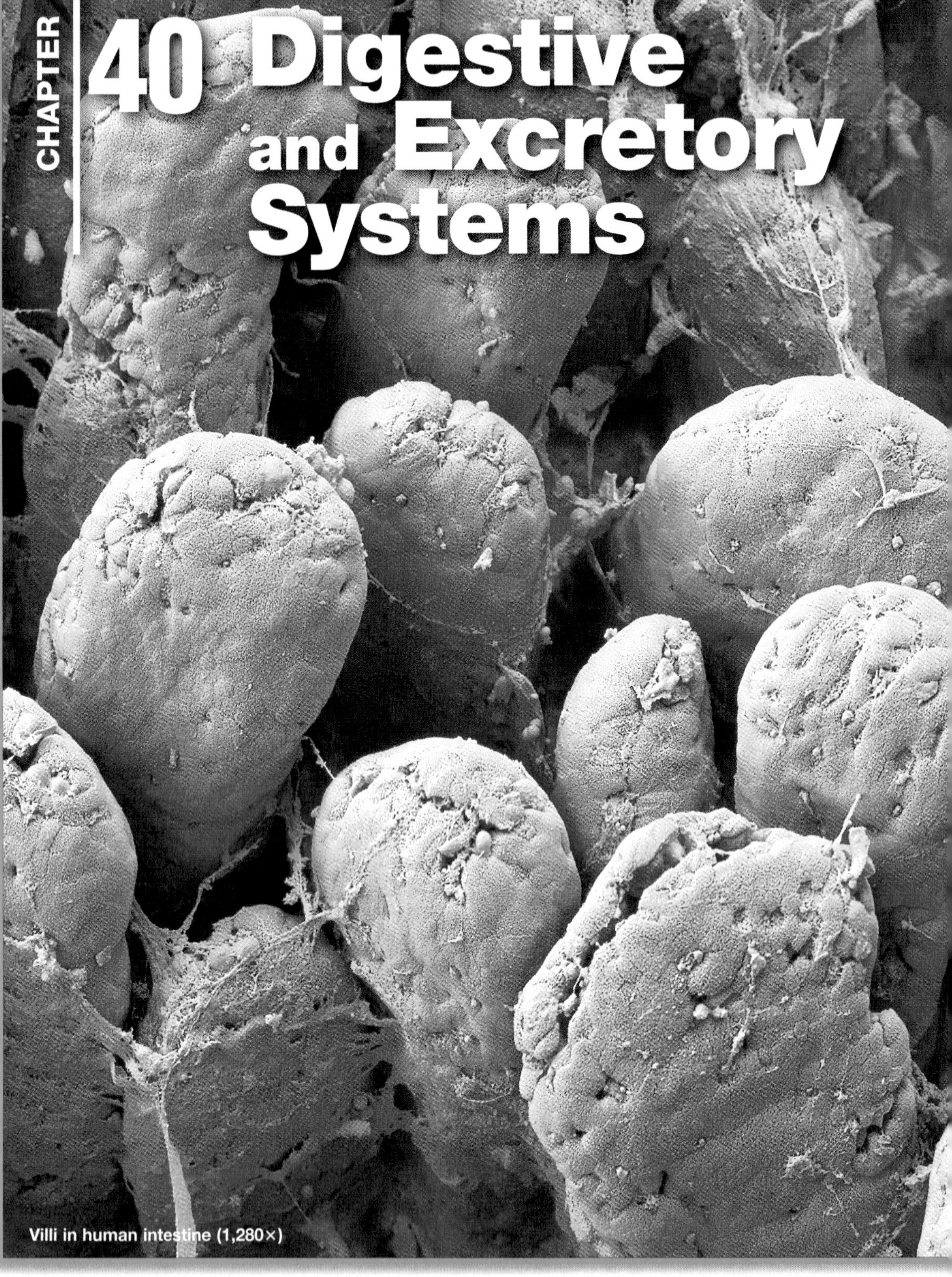

Villi in human intestine (1,280×)

What's Ahead — and Why
CHAPTER OUTLINE

Students will learn

- 40-1: what role each nutrient plays in maintaining a healthy body.
- 40-2: the functions of the digestive organs, and how food is broken down and absorbed.
- 40-3: how metabolic wastes are removed from the blood and excreted from the body.

Family eating

Appearance... Aroma... Taste...

When you sit down to a meal, these are the first things you notice about the food. However, it is also important that the food provide the proper nutrients—the substances needed for body growth and maintenance.

Study TIP

Ready?

Check your understanding of these topics to help you prepare for what's ahead.

Can you...

- **differentiate** between carbohydrates, fats, and proteins? (Section 2-3)
- **compare** saturated and unsaturated fatty acids? (Section 2-3)
- **describe** the role of enzymes in chemical reactions? (Section 2-4)
- **summarize** the function of cellular respiration? (Section 5-3)
- **discuss** the balance of water and salt in vertebrates? (Section 34-1)

If not, *review the sections indicated.*

Looking Ahead

40-1 Your Body's Need for Food
Why is what you eat so important?

40-2 Digestion
What happens to food as it moves through a hollow, muscular tube that is 8 m (26 ft) long?

40-3 Excretion
Why is your blood cleaned?

Labs

Features

internetconnect

National Science Teachers Association *sci*LINKS Internet resources are located throughout this chapter.

Chapter 40

Reading Strategies

Active Reading

Technique: Anticipation Guide

Before students begin this chapter, write the following statements on the board or overhead:

- Fifty percent of your body's fuel should come from carbohydrates.
- The majority of nutrients in food are released and absorbed in the stomach.
- Both the kidneys and the liver cleanse the blood of toxic substances.

Give students time to form opinions about each statement. Ask volunteers with differing views about each statement to share the reasons for their opinions with the class. Then have students read the chapter to see if their opinions are confirmed or changed in any way. Have students reevaluate their opinions in **Closure** on page 924.

Directed Reading

Assign **Directed Reading Worksheet Chapter 40** to help students interact with the text as they read the chapter.

Interactive Reading

Assign ***Biology: Principles and Explorations* Audio CD Program Chapter 40** to help reluctant readers achieve greater success in reading the chapter.

Reading Skills

Assign Reading Organizers and Reading Strategies Worksheets from the **One-Stop Planner CD-ROM with Test Generator** to help students learn and practice effective reading skills.

internetconnect

Holt *Biology: Principles and Explorations*

On-line Resources: go.hrw.com
Visit the HRW Web site for a variety of resources related to this chapter. Just type in the keyword HX0 Chapter 40.

Holt *Biology: Principles and Explorations*

On-line Resources: www.scilinks.org
The following *sci*LINKS Internet resources can be found in the student text for this chapter:

Page 919
TOPIC: Poisons
KEYWORD: HX919

Page 923
TOPIC: Kidney diseases
KEYWORD: HX923

Page 924
TOPIC: Kidney dialysis
KEYWORD: HX924

Page 929
TOPIC: Lactose intolerance
KEYWORD: HX929

Chapter 40

1 Prepare

Scheduling

- **Block:** About 90 minutes will be needed to complete this section.
- **Traditional:** About two class periods will be needed to complete this section.
- **Planning:** See the Planning Guide on pp. 905A and 905B for lecture, classwork, and assignment options for Section 40-1. Lesson plans for Section 40-1 can be found in a lesson cycle format in ***Biology: Principles and Explorations* One-Stop Planner CD-ROM with Test Generator.**

Resource Link

Assign **Active Reading Worksheet Section 40-1** to help students comprehend the key concepts and visuals in this lesson.

2 Teach

Lesson Warm-up

Tell students there is an old saying: "You are what you eat." Have students list the nutrients necessary for proper body functioning and food sources where each nutrient can be found. *(For example: Calcium is necessary for healthy teeth and bones; calcium can be found in milk.)* Have students revisit the list after reading Section 40-1.
PORTFOLIO

Section 40-1 Your Body's Need for Food

Objectives

- **Identify** five nutrients found in foods. (p. 908)
- **Relate** the role of carbohydrates, proteins, lipids, vitamins, minerals, and water in maintaining a healthy body. (pp. 909–913)
- **Describe** each of the parts of the USDA food guide pyramid. (p. 910)
- **Name** one health disorder associated with high levels of saturated fats in the diet. (p. 911)

Key Terms

nutrient
digestion
calorie
vitamin
mineral

Food Provides the Nutrients Required for Life

Your body uses energy to move, to grow, and even to lie still and sleep. The amount of energy you need depends on many factors, including your age, your sex, your rate of growth, and your level of physical activity. Different activities use different amounts of energy, as shown in **Figure 40-1.**

You obtain energy from the nutrients in the foods and beverages you consume. A **nutrient** is a substance needed by the body for energy, growth, repair, and maintenance. Nutrients in food and beverages include carbohydrates, lipids, proteins, vitamins, and minerals. Each nutrient plays a different role in keeping your body healthy. Water in food and beverages is also needed to maintain a healthy body.

The large molecules in food must be broken down in order to be absorbed into the blood and carried to cells throughout the body. The process of breaking down food into molecules the body can use is called **digestion.** Your cells then break the chemical bonds of the digested food molecules and use the energy that is released to make ATP during the process of cellular respiration.

The energy available in food is measured using a unit called a calorie. A **calorie** is the amount of heat energy required to raise the temperature of 1 g of water 1°C (1.8°F). The greater the number of calories in a quantity of food, the more energy the food contains. Since a calorie represents a very small amount of energy, nutritionists use a unit called the Calorie (with a capital C), which is equal to 1,000 calories. On food labels the word *calorie* represents Calories (1,000 calories).

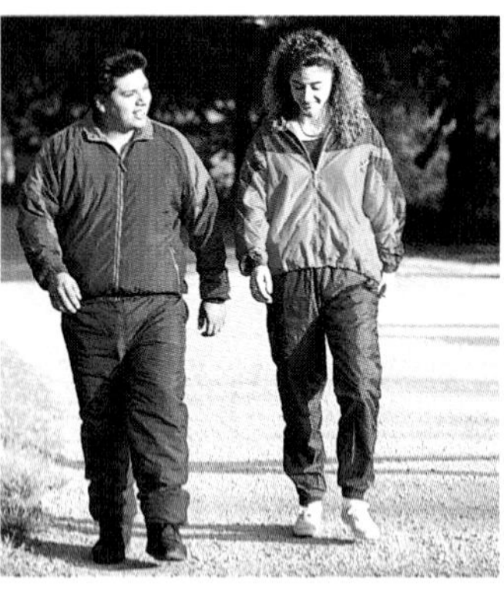

Figure 40-1 Energy required for common activities. Quiet activities require just a little more energy than what it takes to keep you alive. Strenuous activities require more energy.

Integrating Different Learning Styles

Logical/Mathematical	**PE**	Performance Zone items 25 and 26, p. 927
	ATE	Making Biology Relevant, p. 910; Resource Link, p. 911; Teaching Tip, p. 912
Visual/Spatial	**PE**	Performance Zone items 13, 18, 19, 20, and 21, pp. 926–927
	ATE	Visual Strategies, pp. 909 and 912; Teaching Tip, p. 910; Reteaching, p. 913
Interpersonal	**ATE**	Biology in Your Community, p. 910
Intrapersonal	**PE**	Data Lab, p. 911
Verbal/Linguistic	**PE**	Study Tip, p. 909; Performance Zone item 30, p. 927
	ATE	Alternative Assessment, p. 925

Some Nutrients Provide Energy and Building Materials

Each nutrient plays a different role in maintaining a healthy body. Carbohydrates, proteins, and lipids are involved in providing both energy and building materials to the cells.

Carbohydrates

Carbohydrates that exist as single sugar molecules are called monosaccharides or simple carbohydrates. Carbohydrates made of two or many sugar molecules linked together by chemical bonds are called complex carbohydrates. Complex carbohydrates must be digested (broken down) into simple sugars before cells can use their energy.

Many foods contain carbohydrates, as shown in **Figure 40-2.** Glucose, fructose, and other simple sugars are found in fruits, honey, and onions. Glucose, a simple sugar, is used by cells for energy, and it can be directly absorbed into your bloodstream. Table sugar contains sucrose, two simple sugars linked together. Starches are long chains of sugars found in cereal grains and in vegetables such as potatoes, beans, and corn. Cellulose is a major component of plant cell walls and is found in all foods that come from plants. Cellulose, which is a major part of fiber, does not provide energy since we do not have the enzymes to digest it. However, cellulose aids in human digestion by stimulating the walls of the digestive tract to secrete mucus, which helps pass food through the digestive tract.

If excess carbohydrates are consumed, they are stored as the carbohydrate glycogen in the liver and in some muscle tissue. Glycogen can later be broken back down into glucose when the body needs energy. The remainder of the excess glucose is converted to fat and stored in fatty tissue.

Study TIP

Organizing Information

Make a table to organize information about food nutrients and water. Across the top, write the headings *Carbohydrates, Proteins, Lipids, Vitamins, Minerals,* and *Water*. Along the left side, write *Functions, Food sources,* and *Additional comments.* Add information to the table as you read Section 40-1.

Figure 40-2 **Nutrients in food**

Although most foods contain a mix of nutrients, some foods are richer than others in a specific nutrient.

Carbohydrate-rich foods

(4 calories per gram)
Breads, pasta, grains, cereals, potatoes, fruits

Protein-rich foods

(4 calories per gram)
Fish, eggs, poultry, beef, pork, nuts, legumes, milk, cheese, tofu

Fat-rich foods

(9 calories per gram)
Milk, cheese, meats, butter, olives, avocados, fried foods, oils, chips

MULTICULTURAL PERSPECTIVE

Mediterranean Diets and Heart Disease

The French, Greeks, and other southern Europeans suffer from less heart disease than either their northern European neighbors or Americans. Researchers have studied various aspects of their life to determine if there is a major factor that contributes to the lower incidence of heart diseases. The Mediterranean diet is thought to play a role. The Mediterranean diet is rich in canola oil, fish, fruits, cereals, and beans. Even when researchers account for age, sex, alcohol intake, and lifestyle differences, eating a Mediterranean diet appears to reduce the probability of heart problems.

CROSS-DISCIPLINARY CONNECTION

◆ **History**

ATP is the energy currency of all living cells. Fritz Lipmann isolated ATP from muscle tissue in 1929. In research he began in 1941, he discovered that ATP and similar compounds shuttle energy around the cell, once the energy is captured from carbohydrates, fats, and proteins during cellular respiration. In 1953 Lipmann shared the Nobel Prize in physiology or medicine for his discoveries.

Visual Strategy

Figure 40-2

Tell students that the body takes longer to digest carbohydrates. A longer digestion time means your body uses more of the energy for cellular processes and stores less in the form of fat. Have students study Figure 40-2. Ask students why carbohydrate-rich foods are an especially good choice for weight-conscious people? *(Carbohydrate-rich foods take the body longer to digest, while releasing only 4 calories per gram of carbohydrate.)* Ask students to compare the three lists of foods and then decide why it is better to eat carbohydrate-rich foods than protein-rich foods. *(Protein-rich foods tend to contain hidden fat. See the foods which appear on both the fat and protein lists.)*

Chapter 40

Making Biology Relevant

Food ADVANCED

Have students make a list of protein food sources and note whether the food source is a source of "complete" or "incomplete" proteins. That is, does it contain all of the essential amino acids or only some of them? Then discuss the different types of vegetarian diets such as those that (1) omit only red meat, (2) omit all meat but fish, (3) omit all meat but include animal products such as milk and eggs, and (4) omit all animal products. Have students create a meal of each type of "vegetarian" diet and decide if it is difficult or easy to obtain a nutritionally sound diet in each category. PORTFOLIO

Teaching TIP

Food Preparation

Bring some French fries and a plain baked potato to class, or show students a picture of these foods. Have students examine the baked potato and the French fries carefully, and ask them to brainstorm differences between the two foods with regard to cooking, taste, and nutritional value. *(Unlike the baked potato, French fries are peeled, sliced, fried in oil or fat, and salted.)* Ask students how the body is affected by a diet high in oils, fats, and salt. *(Oils and fats increase caloric intake. Saturated fats may increase cholesterol levels. A diet high in salt can cause fluid to be retained in the body. Over a long period of time, high salt intake can contribute to hypertension.)*

Proteins

The digestive products of proteins—amino acids—are normally used by the body for making other protein molecules, such as enzymes and antibodies. When more protein is eaten than is needed by the cells, the amino acids are used for energy or converted to fat.

The body requires 20 different amino acids to function. A child's or teen's body can make 10 of the amino acids from other amino acids. The other 10, called essential amino acids, must be obtained directly from food. Most animal products, such as eggs, milk, fish, poultry, and beef, contain all of the essential amino acids. No single plant food contains all of the essential amino acids. But eating certain combinations of two or more plant products can supply all the essential amino acids. Eight amino acids are essential to adults.

The guidelines for healthy eating are summarized in the USDA (U.S. Department of Agriculture) food guide pyramid, as shown in **Figure 40-3.** The pyramid lists the daily number of servings needed from each food group to obtain a variety of nutrients in your diet.

Lipids

Lipids, organic compounds that are insoluble in water, are used to make steroid hormones and cell membranes and to store energy. Excess fats, lipids that store energy, are stored around tissues

Figure 40-3 The USDA food guide pyramid

Food groups placed at the bottom of the pyramid should be eaten in greater amounts than those placed at the top.

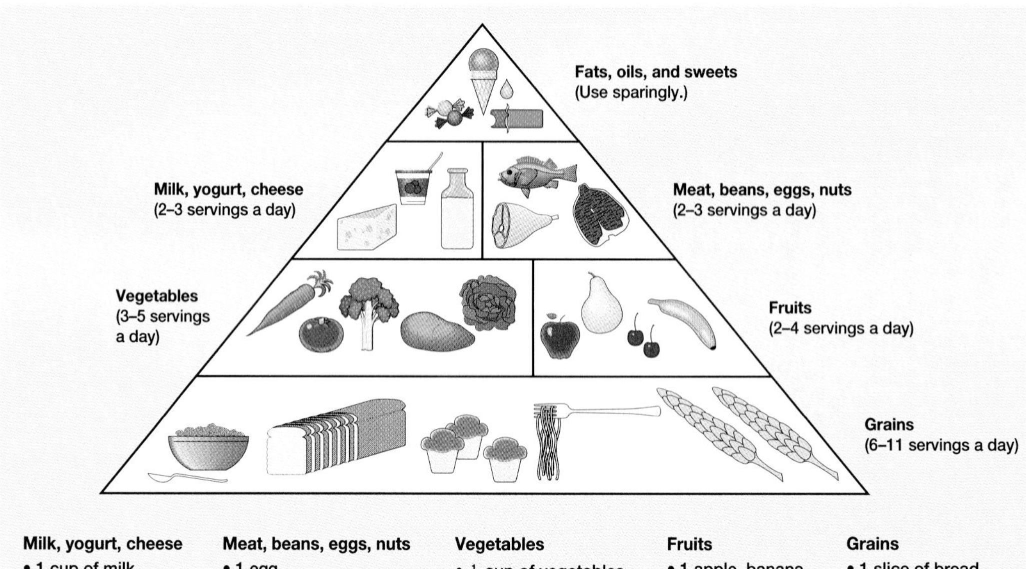

Milk, yogurt, cheese
- 1 cup of milk
- 1 cup of yogurt
- 1.5 oz of cheese
- 2 oz of processed cheese

Meat, beans, eggs, nuts
- 1 egg
- 2 tbsp of peanut butter
- 2–3 oz of meat, chicken, or fish
- $\frac{1}{2}$ cup of cooked beans

Vegetables
- $\frac{1}{2}$ cup of vegetables
- $\frac{3}{4}$ cup of vegetable juice
- 1 cup of raw, leafy vegetables

Fruits
- 1 apple, banana, or orange
- $\frac{1}{2}$ cup of fruit
- $\frac{3}{4}$ cup of fruit juice

Grains
- 1 slice of bread
- $\frac{1}{2}$ cup of rice or pasta
- $\frac{1}{2}$ cup of hot cereal
- 1 oz cold cereal
- 1 tortilla

Biology in Your Community

Athletes, Proteins, and Carbohydrates

A common misconception is that athletes need to eat more protein to improve their athletic performance and maintain greater muscle mass. Actually, a diet rich in complex carbohydrates is much more effective in sustaining intense muscle activity than a diet high in protein. Invite a marathon runner, cyclist, or some other person who participates in endurance events to speak to the class about the carbohydrate-rich foods he or she consumes for a race (carbohydrate loading). Ask the guest to explain why he or she eats carbohydrate-rich food not only prior to an event but also as part of a regular diet.

as padding and insulation. Fats also act as solvents for fat-soluble vitamins.

Although lipids are essential nutrients, too much fat in the diet is known to harm several body systems. For example, a diet high in saturated fats is linked to high blood cholesterol levels, which in turn may be connected to heart and circulatory diseases. It is recommended that no more than 30 percent of your day's calories come from fats and that most of these fats be unsaturated. Examples of foods containing lipids are shown in Figure 40-2.

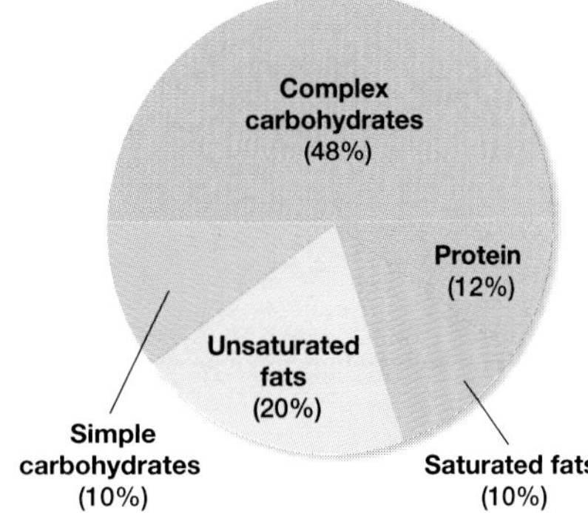

Figure 40-4 A balanced meal. The percentage of the day's total calories that should come from each nutrient are shown. At least half of your day's calories should come from foods high in complex carbohydrates.

Balancing nutrients and energy

Regardless of their source, the excess calories you eat will be stored as either glycogen or body fat, and you will gain weight. If you use more calories than you take in, additional energy will be obtained from your body's energy stores, and you will lose weight. Your diet and overall activity level determine in part whether you store excess calories as glycogen or as fat. **Figure 40-4** summarizes what percentage of the day's total calories should come from each nutrient.

Obesity is described as being more than 20 percent heavier than your ideal body weight. Obesity significantly increases an individual's risk of diabetes, heart disease, osteoarthritis, and many other disorders. Regular physical activity is important in maintaining energy balance.

DATA LAB: *Analyzing the nutrition of your diet*

Background

Make a data table similar to the one below. In the column labeled *Food,* list each food item that you ate yesterday. Use the food guide pyramid in Figure 40-3 to answer the questions below.

Analysis

1. **Determine** the food group and the equivalent number of servings for each food item you listed.
2. **Determine** the total number of servings in each food group.
3. **Identify** which food group(s) had too few daily servings.
4. **Identify** which food group(s) had too many daily servings.
5. **SKILL Applying Information** Suggest foods that can be added to or removed from your daily diet in order to make it a more balanced diet.

DATA TABLE

Food	Number of servings in each food group					
	Milk, yogurt, cheese	Meat, beans, eggs, nuts	Vegetables	Fruit	Grains	Fats, oils, sweets

MULTICULTURAL PERSPECTIVE

Food Guides Around the World

Countries around the world use different graphics to illustrate their dietary guidelines. The graphic used by each country represents the cultural norms and symbols for the country. Canada uses a four-banded rainbow, with each color representing one of its food groups and the width of each band representing the amount recommended. Israel uses a chalice with water at the top and largest section to emphasize its importance and grains at the wide base of the chalice. The Philippines use a six point star. The United Kingdom, Germany, and Norway use a wheel or dinner plate divided into sections for each food group. The size of the sections indicates its relative proportion to the total diet.

Section 40-1

Resource Link

- Assign **Problem Solving: Using Food Labels to Calculate Percentage of Nutrients and Calories** from the Holt BioSources Teaching Resources CD-ROM. **PORTFOLIO**

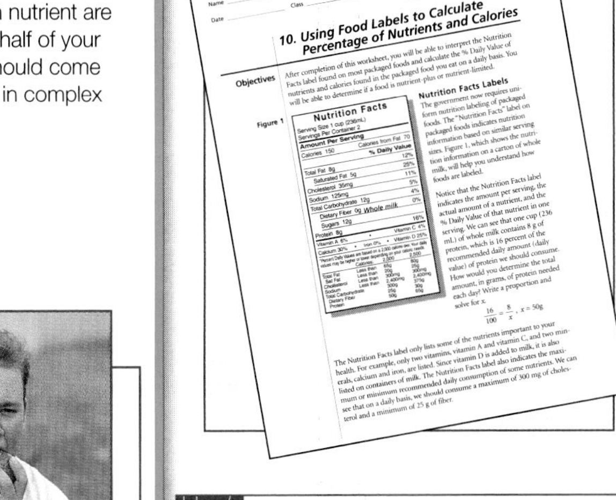
BIOSOURCES TEACHING RESOURCES PROBLEM SOLVING

10. Using Food Labels to Calculate Percentage of Nutrients and Calories

DATA LAB: *Analyzing the nutrition of your diet*

Time 20 minutes

Process Skills Calculating, applying information

Teaching Strategies

The day before assigning the lab, have students list the kind and amount of each food they eat for the next 24 hours.

Analysis Answers

1. Answers will vary. Use the information in Figure 40-3 on page 910 to verify answers.
2. Answers will vary. Check for correct addition in each column.
3. Answers will vary. Use the information in Figure 40-3 on page 910 (the daily number of servings needed from each food group) to check answers.
4. Answers will vary.
5. Answers will vary.

Chapter 40

Teaching TIP

Diagnosing Deficiencies

Tell students to imagine that they are physicians. Ask them to use information in Table 40-1 to answer the following questions: If a child suffers from frequent internal infections and has difficulty seeing in dim light, what problem would you suspect? *(The child likely is suffering from a vitamin A deficiency.)* What foods would you prescribe to someone who has sore gums and tends to bruise easily? *(fruit, especially citrus fruits)*

Visual Strategy

Table 40-1

Pass around the class a few boxes of cereals that are eaten cold. Have students examine the food labels and identify the vitamins that are listed on the labels. Have students use Table 40-1 to list the role of each vitamin and the effects of a deficiency in that vitamin. Have students compare the vitamin content when consuming the cereal with and without milk. PORTFOLIO

Teaching TIP — ADVANCED

Enzymes, Vitamins, and Minerals

Review the role of enzymes as biological catalysts. Point out that some enzymes cannot work unless they are first activated by nonprotein helpers. Cofactors are inorganic helpers such as the minerals zinc, iron, and copper. Coenzymes are organic helpers such as most vitamins. Remind the students that enzymes can be reused many times. Have students explain why vitamins and minerals that function as coenzymes and cofactors are needed in only small amounts. *(They can be reused many times.)*

Vitamins, Minerals, and Water Help Regulate Metabolism

Vitamins, minerals, and water are required in our diets. They do not provide energy, but they contribute to many different functions including regulating the reactions that release energy.

Vitamins Many different **vitamins,** organic substances that occur in many foods in small amounts, are necessary in trace amounts for the normal metabolic functioning of the body. They enhance the activity of enzymes.

Vitamins dissolve in either water or fat, as summarized in **Table 40-1.** Fat-soluble vitamins—vitamins A, D, E, and K—can be stored in body fat. Excess fat-soluble vitamins can be stored in the body,

Table 40-1 Vitamins

Vitamin	Food sources	Role	Effects of deficiency
Water-soluble			
Vitamin B_1 (thiamin)	Most vegetables, nuts, whole grains, organ meats	Assists in carbohydrate metabolism, helps nerves and heart to function properly	Digestive disturbances, impaired senses
Vitamin B_2 (riboflavin)	Fish, poultry, cheese, yeast, green vegetables	Needed for healthy skin and tissue repair, assists in carbohydrate metabolism	Blurred vision, cataracts, cracking of skin, lesions of intestinal lining
Vitamin B_3 (niacin)	Whole grains, fish, poultry, tomatoes, legumes, potatoes	Keeps skin healthy, assists in carbohydrate metabolism	Mental disorders, diarrhea, inflamed skin
Vitamin B_{12} (cobalamin)	Meat, poultry, green vegetables, milk, dairy products	Needed for formation of red blood cells	Reduced number of red blood cells
Vitamin C (ascorbic acid)	Citrus fruits, strawberries, potatoes	Needed for wound healing, healthy gums and teeth	Swollen and bleeding gums, loose teeth, slow-healing wounds
Fat-soluble			
Vitamin A (retinol)	Butter, eggs, liver, carrots, green leafy vegetables, sweet potatoes	Keeps eyes and skin healthy, needed for strong bones and teeth	Infections of urinary and digestive systems, night blindness
Vitamin D (cholecalciferol)	Salmon, tuna, fish liver oils, fortified milk, cheese	Assists in calcium uptake by the gut, needed for strong bones and teeth	Bone deformities in children, loss of muscle tone
Vitamin E (tocopherol)	Many foods, especially wheat and other vegetable oils, olives, whole grains	Protects cell membranes from damage by reactive oxygen compounds (free radicals)	Reduced number of red blood cells; nerve tissue damage in infants
Vitamin K (menadione)	Leafy green vegetables, liver, cauliflower	Needed for normal blood clotting	Bleeding caused by prolonged clotting time

Overcoming Misconceptions

Vitamin Overdoses

Some people have the idea that if a little is good, then a lot must be better. Unfortunately, this is not the case with certain vitamins. For example, overdoses of vitamin A can cause enlargement of the liver and spleen. Overdoses of vitamin D can cause kidney damage and calcification of soft tissues. Although vitamins A and D are both fat soluble, making them more difficult for the body to eliminate, even the water-soluble vitamins have some negative effects if overused. Massive doses of vitamin C can cause kidney stones and enhance blood coagulation. Megadoses of niacin, one of the B vitamins, can cause liver damage and gout.

but excessive amounts of vitamins A and D can be toxic. Excess water-soluble vitamins—vitamin C and the B vitamins—are excreted in urine. For this reason you must have regular sources of water-soluble vitamins to prevent deficiency diseases.

Minerals Different minerals are required to maintain a healthy body. **Minerals** are naturally occurring inorganic substances that are used to make certain body structures and substances, for normal nerve and muscle function, and to maintain osmotic balance. Some minerals are essential for enzyme function. Minerals are not produced by living organisms. Minerals must be replaced on a daily basis because they are soluble in water. Teeth and bones require the minerals calcium and phosphorus. Iron is required for transporting oxygen. Magnesium, calcium, sodium, potassium, and zinc help regulate function of the nerves and muscles.

Trace elements, such as those listed in Table 40-2, are minerals present in the body in small amounts. Humans usually obtain adequate amounts of the required trace elements directly from the plants they eat or indirectly from animals that have eaten plants.

Water You can survive only a few days without water, though you can live several weeks without food. Water is used by the body as a medium to transport gases, nutrients, and waste products. Water also plays a role in regulating body temperature. Two-thirds of the body's weight is water.

biobit

Does the sun help make vitamins?

Sunlight triggers a chemical reaction that starts the conversion of a molecule in our skin to vitamin D. The process is eventually completed in the kidneys.

Table 40-2 Trace Elements

Trace element	Food sources	Role
Iodine	Seafood, plants grown in iodine-rich soil, iodized table salt	Synthesis of thyroid hormones
Cobalt	Leafy vegetables, liver, kidney	Synthesis of vitamin B_{12}
Zinc	Meat, shellfish, dairy products	Synthesis of digestive enzymes, proper immune function
Molybdenum	Legumes, cereals, milk	Protein synthesis
Manganese	Whole grains, nuts, legumes	Hemoglobin synthesis, urea formation
Selenium	Meat, seafood, cereal grains	Preventing chromosome breakage

Review SECTION 40-1

1. **Predict** four nutrients that would be found in a serving of green beans.
2. **Compare** the function of carbohydrates and proteins in maintaining a healthy body.
3. **Describe** the type of information the USDA food guide pyramid provides.
4. **Evaluate** the roles vitamins, minerals, and water play in maintaining a healthy body.

Critical Thinking

5. **SKILL Applying Information** Your friend wants to feed her elderly grandmother more food in order to keep her healthy. Do you think this is a good idea? Explain.

Review SECTION 40-1 ANSWERS

1. carbohydrates, vitamins, minerals, protein
2. Both can be used as a source of energy, but the body usually uses amino acids from proteins as building blocks for other proteins.
3. The food guide recommends the number of servings of each different type of food group a person should eat daily to obtain a variety of nutrients in their diet.
4. Vitamins enhance enzyme activity. Minerals are necessary for making certain body structures and substances, for normal nerve and muscle functions, for maintaining osmotic balance, and for enzyme function. Water transports gases, nutrients, and waste and regulates body temperature.
5. Answers will vary. Students who disagree might argue that elderly people have a lower metabolic rate, and thus need fewer calories in a balanced diet. Students who agree might point out that elderly people tend to dislike eating, and may get sick if they become malnourished.

Teaching TIP

Calcium and Vitamin D

Ask students to identify the predominant mineral in milk. *(calcium)* Point out that vitamin D is added to milk and then ask students why. *(Accept any answers that tie vitamin D and calcium together.)* Tell students that vitamin D is needed for the absorption of calcium from the intestines. Explain that rickets, a disease that affects growing bones and is caused by a vitamin D deficiency, used to be common in the United States. In the 1930's the federal government mandated the enrichment of milk with vitamin D. Today rickets is virtually unknown in the United States.

3 Assessment Options

Closure

Fad Diets

Have students describe any fad diets they may have read about or even tried. Have them analyze these diets in terms of how they may cause problems by failing to provide certain nutritional requirements. Explain to students that 500 g (about a pound) of body fat contains about 3500 calories of energy. Given that there are seven days in a week, the best way to lose weight is to cut (or use up) an additional 500 calories per day.

Review

Assign Review—Section 40-1.

Reteaching — BASIC

Have students bring in a label from a cereal box. Prepare a table on the board or overhead, listing the nutrients found in each. Have students write the function of each nutrient. PORTFOLIO

Chapter 40

1 Prepare

Scheduling

- **Block:** About 90 minutes will be needed to complete this section.
- **Traditional:** About two class periods will be needed to complete this section.
- **Planning:** See the Planning Guide on pp. 905A and 905B for lecture, classwork, and assignment options for Section 40-2. Lesson plans for Section 40-2 can be found in a lesson cycle format in ***Biology: Principles and Explorations* One-Stop Planner CD-ROM with Test Generator.**

Resource Link

Assign **Active Reading Worksheet Section 40-2** to help students comprehend the key concepts and visuals in this lesson.

2 Teach

Lesson Warm-up

Hang an 8 m (26 ft) piece of clothesline or rope at the front of the room. Explain to students that it represents the approximate length of the digestive system. Use clothespins and markers to label the clothesline as follows:

Part	Length	Average time food spends here
Mouth	8 cm (3 in.)	5–30 s
Pharynx/ esophagus	25 cm (10 in.)	5–10 s
Stomach	15 cm (6 in.)	2–6 h
Small intestine	4–6 m (13–20 ft)	3–6 h
Large intestine	1.2 m (4 ft)	up to 2 d

Section 40-2 Digestion

Objectives

- **Relate** the four major functions of the digestive system to the processing of food. (p. 914)
- **Summarize** the path of food through the digestive system and the major digestive processes that occur in the mouth, stomach, small intestine, and large intestine. (pp. 914–918)
- **Describe** how nutrients are absorbed from the digestive system into the bloodstream or lymphatic system. (p. 917)
- **Identify** the role of the pancreas and liver in digestion. (pp. 917–919)

Key Terms

amylase
esophagus
peristaltic contraction
pepsin
ulcer
lipase
villus
colon
hepatitis

Food Must First Be Broken Down

Imagine you just ate your favorite meal. What happens to that food? Before your body can use the nutrients in the food you eat, the large food molecules must be broken down physically and chemically. The process of breaking down food into molecules the body can use is called digestion. The digestive system is the body system that is involved in the taking in and processing of food for use by your body cells. The major functions of the digestive system are taking in food, breaking the food down into molecules small enough for the body to absorb, taking up the small molecules, and getting rid of undigested molecules and waste.

As shown in **Figure 40-5,** the digestive system is made up of a long, winding tube, the digestive tract, that begins at the mouth and winds through the body to the anus. Food travels more than 8 m (26 ft) through your digestive system. The digestive tract includes the mouth, pharynx, esophagus, stomach, small intestine, large intestine, and rectum. Although the liver and pancreas *(PAN kree uhs)* are not part of the digestive tract, they are part of the

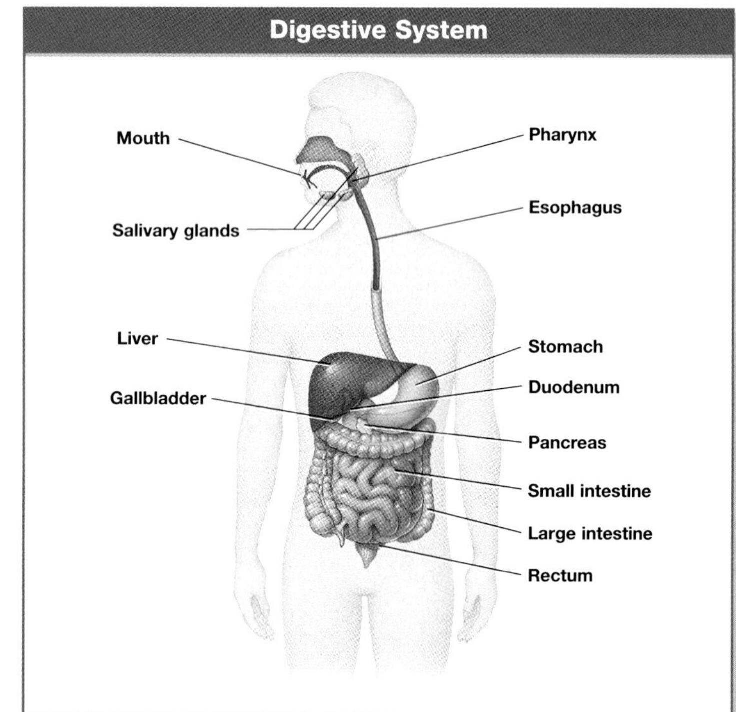

Figure 40-5 Processing food. The digestive system breaks down food into individual nutrient molecules that can be absorbed into the bloodstream.

Integrating Different Learning Styles

Visual/Spatial	**PE**	Quick Lab, p. 919; Performance Zone item 1, p. 926
	ATE	Lesson Warm-up, p. 914; Demonstrations, pp. 915, 917; Visual Strategy, p. 917; Teaching Tip, p. 918; Closure, p. 919; Reteaching, p. 919
Interpersonal	**PE**	Health Watch, p. 916; Performance Zone item 28, p. 927
	ATE	Career, p. 916
Verbal/Linguistic	**PE**	Performance Zone items 22, 23, and 29, p. 927
	ATE	Active Reading. p. 918;

digestive system because they deliver secretions into the digestive tract through ducts (tubes).

Starting digestion

The digestion of food begins as soon as the food is ingested. The teeth rip and chew food into shreds, and the tongue mixes the pieces with a watery solution called saliva. Taste buds on the tongue help the person taste the meal. Saliva is secreted into the mouth by three pairs of salivary glands, shown in Figure 40-5. Saliva moistens and lubricates the food so that it can be swallowed more easily.

Saliva also contains **amylases** *(AM uh lay sehs)*, enzymes that begin the breakdown of carbohydrates such as starch, into monosaccharides (single sugars). The mechanical action of chewing and the chemical action of amylase are both part of the digestion of carbohydrates in the mouth.

Notice in **Figure 40-6** that the structure of our teeth helps in the breakdown of food. The two front teeth, the incisors, cut food. The cuspids, or canines, shred food. The back teeth, the molars, crush and grind food.

After passing through the region in the back of the throat called the pharynx *(FAIR ihnks)*, the food triggers a swallowing reflex. The reflex moves the epiglottis (a flap of cartilage) over the opening of the trachea—the tube that leads to the lungs. This prevents food from entering the trachea and eventually the lungs. Instead, food enters the esophagus *(ih SAHF uh guhs)*.

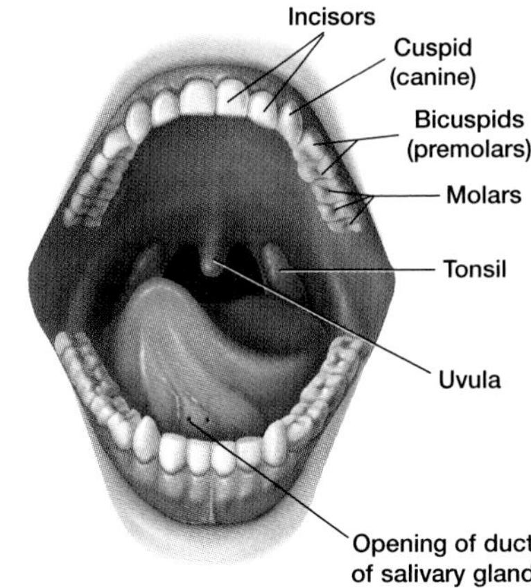

Figure 40-6 Teeth and digestion. Canine and incisor teeth, used for cutting and tearing food, are located toward the front of the jaws. The molars, located toward the back of the jaws, are used to grind food.

Getting food to the stomach

The **esophagus** is a long tube that connects the mouth to the stomach. No digestion takes place in the esophagus. Its role is to act as a kind of descending escalator, moving food down to the stomach.

The esophagus is about 25 cm (10 in.) long. The lower two-thirds of the esophagus is wrapped in sheets of smooth muscle. Food does not simply fall into the stomach; it is "pushed" down, as shown in **Figure 40-7.** Successive rhythmic waves of smooth muscle contraction in the esophagus, called **peristaltic** *(pehr uh STAHL tihk)* **contractions,** or peristalsis, move the food toward the stomach. Peristalsis can be thought of as waves moving through the muscle with the area where the wave is passing causing the muscle to narrow. It takes about 5 to 10 seconds for food to pass down the esophagus and into the stomach.

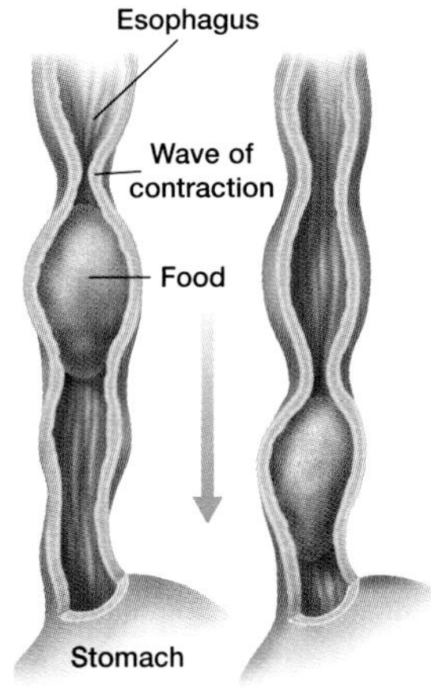

Figure 40-7 Peristalsis moves food. Food is pushed down the esophagus and toward the stomach by waves of smooth muscle contractions that occur in the wall of the esophagus.

Digestion continues in the stomach

Food exits the esophagus and enters the stomach through a muscular valve called a sphincter *(SFIHNGK tuhr)*. The sphincter prevents acid-soaked food in the stomach from making its way back into the esophagus. The stomach is a saclike organ located just beneath the diaphragm. Besides temporarily storing food, the stomach, shown

did you know?

Halitosis (bad breath) results when decomposing food particles accumulate in the mouth, allowing bacteria to flourish.

Saliva aids in washing away the food particles. Therefore, people who suffer from any disease that inhibits saliva secretion usually have problems with dental caries and bad breath, as well as difficulty talking, swallowing, and eating.

Demonstration

Salivation

Gather two test tubes with stoppers, four saltine crackers, water, a straw or stirring rod, and an iodine solution. In one test tube stir two crushed crackers with water, then add a few drops of iodine and invert to mix. The solution in the test tube will turn blue-black, indicating the presence of starch. Have a volunteer chew up the other two crackers *without swallowing*, and spit them into the second test tube. Add some water, stir, and then add a few drops of iodine, and invert to mix. The solution starts out blue-black, but as salivary amylase enzymes start to break the starch down into sugars, the color will fade to purple and eventually disappear.

Making Biology Relevant

Medicine

Heartburn is a condition caused by the reflux, or backward movement, of acidic stomach fluids into the esophagus or when the esophagus undergoes intense spasms. The esophagus, lacking the mucous coating of the stomach, becomes irritated by the acidic fluid, and a burning sensation in the chest results. Occasional indigestion may be caused by overeating, upset stomach, or stress, but it is usually not serious. Because the stomach acid can more easily flow into the esophagus if a person bends forward or lies down, heartburn often occurs when a person lies down less than two hours after eating a heavy meal. A physician should evaluate cases of persistent heartburn, since this could indicate a more serious condition.

Demonstration

Peristalsis BASIC

Cut the legs off of an old pair of pantyhose. Place a tennis ball in the toe and tie off the other end. You can show the process of peristalsis by rhythmically squeezing the tennis ball from one end to the other. Remind students that this happens automatically after swallowing, and several "swallows" (boli) can travel down the esophagus at one time.

Chapter 40

CAREER

Counselor

Counselors give clients moral support and an opportunity to discuss personal or work-related problems. Counselors may be professional social workers, psychologists, or community workers interested in helping people. Employers include treatment centers, hospital-based programs, or government agencies. Each state has its own requirements for counselors. In some states counselors must earn a certificate. Have students interview a counselor that works with persons with eating disorders such as anorexia or bulimia. Ask each student to compile a list of 10 questions and answers from the interview.

PORTFOLIO

Eating Disorders

Teaching Strategies

Eating disorders are a sensitive issue with teens. You may have a student with an eating disorder that has not yet been diagnosed. Be sure to address these issues openly and without judgment. Be prepared to direct students to a person or place where they can seek help.

Discussion

- What is the difference between anorexia and bulimia? *(Anorexics do not eat sufficient nutrients. Bulimics eat, but purge their digestive system through vomiting or the use of laxatives.)*
- Why does weight stay nearly constant for bulimics? *(Because food is actually ingested, some nutrients are absorbed. If laxatives are the method of purging, the food has the chance to travel throughout the whole digestive system, just at a more rapid pace.)*
- Should advertisers seek models with fuller figures to reduce the incidence of eating disorders? *(Answers will vary. For example: Yes, people might feel better about themselves with less cultural emphasis on being thin.)*

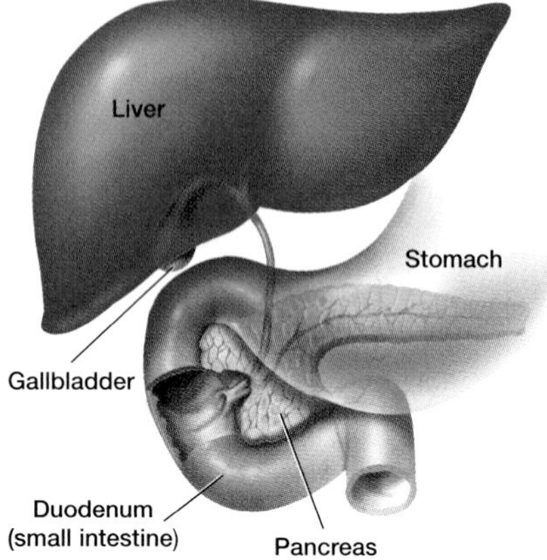

Figure 40-8 The stomach and accessory digestive organs. Many organs are involved in the complete breakdown of nutrients.

in **Figure 40-8,** also mechanically breaks down food and chemically unravels and breaks down proteins.

When food enters the stomach, gastric juice is secreted by the cells that line the inside of the stomach. Gastric juice is a combination of hydrochloric *(HEYE droh klawr ihk)* acid (HCl) and pepsin. The acid breaks the bonds in proteins and unfolds large protein chains into single protein strands. **Pepsin,** a digestive enzyme secreted by the stomach, cuts the single protein strands into smaller chains of amino acids. Pepsin is effective only in an acidic environment.

The stomach mixes its contents using peristaltic waves. Swallowed food can spend from 2 to 6 hours in the stomach. Your stomach secretes about 2 L (2.11 qt) of HCl every day, creating a solution about 3 million times more acidic than your bloodstream. The hormone gastrin regulates the synthesis of HCl, permitting it to be made only when the pH in the stomach is higher than about 1.5.

A coating of mucus protects the lining of the stomach from gastric acid. Excessive acid may result in an **ulcer,** a hole in the wall of the stomach or small intestine.

HEALTH WATCH

Eating Disorders

Millions of teens are on weight-reducing diets. Although maintaining a healthy body weight is important, *obsessive* dieting can lead to eating disorders. Eating disorders can disrupt lives and sometimes even result in death.

Starving oneself

Anorexia nervosa is an eating disorder in which people starve themselves. Anorexics have an irrational fear of becoming fat and see themselves as much heavier than they really are. They shun food and may exercise for hours every day in an attempt to lose weight.

Female anorexics may never mature sexually or may stop menstruating because they lose too much body fat, causing the level of sex hormones to fall. Their body temperature and pulse rate become abnormally low; their hair becomes thin and dry; and their resistance to infection drops.

Bingeing and purging

Patients with bulimia engage in frequent episodes of bingeing (eating a large amount of food in a short time). After a binge, they almost always purge by self-induced vomiting or by using laxatives. Therefore, their weight stays fairly constant. Unlike anorexic patients, bulimics often recognize that their behavior is abnormal and often feel guilty, depressed, and helpless.

The frequent purging removes salts from the body, which can eventually lead to muscle weakness, heart failure, and kidney disease. The vomiting gradually destroys tooth enamel and results in unhealthy gums.

Help for people with eating disorders

Little is known about what causes anorexia nervosa and bulimia. Many people believe that western society's emphasis on thinness in women is largely to blame. However, recent research suggests that genetic factors might also be involved. Both eating disorders can be managed with a combination of medical treatment, counseling, and family support.

did you know?

Different parts of the body have different normal pH levels.

The pH of blood is 7.35–7.45. If the pH drops below 7.35 (producing a condition called acidosis), disorientation and coma might occur. If the pH rises above 7.45 (producing a condition called alkalosis), nervousness and even convulsions might occur. Either condition can lead to death. The pH of stomach fluids is between 1 and 3, low enough to kill many bacteria and stop carbohydrate digestion, yet high enough for pepsin to break down proteins.

Most Digestion and Absorption Occurs in the Small Intestine

Food passes from the stomach into the small intestine when a sphincter between the two organs opens. The small intestine is a coiled tubular organ about 6 m (19.8 ft) long that is continuous with the stomach and that functions mainly in the digestion and absorption of nutrients. The word *small* refers to the small diameter of the small intestine as compared with the diameter of the large intestine—not to its length. Peristalsis mixes the food which remains in the small intestine for about 3 to 6 hours.

The first part of the small intestine, the duodenum *(doo oh DEE nuhm)*, receives secretions from the pancreas, liver, and gallbladder, as shown in Figure 40-8. Cells that line the small intestine and the pancreas secrete digestive enzymes involved in completing the digestion of carbohydrates into monosaccharides, proteins into amino acids, and lipids into fatty acids and glycerol.

Before fats can be digested by pancreatic enzymes called **lipases** *(LEYE pays uhs)*, the fats must first be treated with bile, a greenish fluid produced by the liver. Bile breaks up fat globules into tiny fat droplets, a process called emulsification *(ee MUHL suh fih kay shuhn)*. The gallbladder, a green muscular sac attached to the liver, stores bile until it is needed in the small intestine.

Most absorption (passage of nutrients to the blood or lymph) occurs in the small intestine. The lining of the small intestine is covered with fine fingerlike projections called **villi** (singular form, villus), shown in **Figure 40-9,** each of which is too small to see with the naked eye. In turn, the cells covering each villus have projections on their outer surface called microvilli. The villi and microvilli greatly increase the area available for absorption of nutrients. Sugars and amino acids enter capillaries in the villi and are carried in the blood to the liver for further metabolism. Fatty acids and glycerol enter lymphatic vessels in the villi and eventually enter the bloodstream.

biobit

What causes ulcers?

Ulcers were long thought to result from spicy food, stress, alcohol, or high levels of acid. It is now known that a bacterium, *Helicobacter pylori,* causes most ulcers. Treatment involves a combination of antibiotics and other drugs that kill the bacteria. Scientists are trying to develop a vaccine to prevent *H. pylori* infection.

Figure 40-9 Villi in the small intestine

Inside each villus are capillaries and lymphatic vessels where nutrients enter the bloodstream.

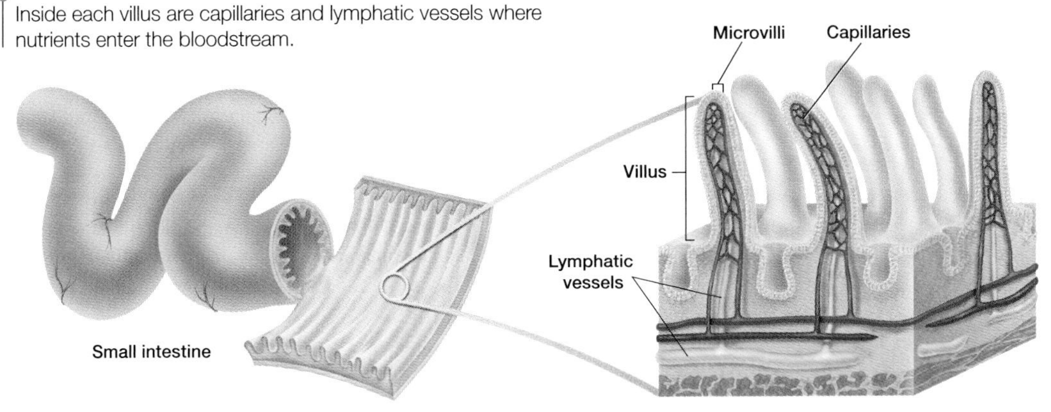

Section 40-2

Teaching *TIP*

Nutrient Digestion Times

Carbohydrates spend 2–3 hours in the stomach, proteins 3–4 hours, and fats 4–6 hours. The longer the food is in your stomach, the longer you tend to feel satisfied and not crave food.

Demonstration

Surface Area

You will need two beakers of tap water, a sugar cube, a crushed sugar cube, and two stirring rods. At the same time, drop the sugar cube and crushed sugar into two separate beakers. Stir rapidly and observe which sugar dissolves first. Ask students for an explanation. *(The crushed sugar cube has more surface area, so it dissolves faster.)* Relate this principle to the increase of surface area in the intestine brought about by the villi and microvilli.

Visual Strategy

Figure 40-9

Use two pieces of yarn to show how surface area increases with the number of villi. Make a circle with a short piece of yarn. Use a much longer piece of yarn to simulate villi inside of the shorter piece. When you compare the lengths of the yarn, it becomes clear that villi (and microvilli) drastically increase surface area.

Overcoming Misconceptions

Digestion and Absorption

Although some digestion occurs in the mouth and stomach, most chemical digestion occurs in the small intestine. All nutrient absorption into the bloodstream occurs in the small intestine, *not* the stomach. However, both alcohol and aspirin are absorbed from the stomach directly into the bloodstream, accounting for their fast-acting effects on the body.

Teaching TIP

Vitamin K and Intestinal Bacteria

Vitamin K, essential for blood coagulation, is produced by intestinal bacteria *(E. coli)*. Large doses of oral antibiotics can kill *E. coli*, reducing the blood's ability to coagulate, which may lead to nosebleeds or excessive bleeding.

Active Reading

Technique: Reading Organizer

Have students create a reading organizer that lists the organs of the digestive system, their functions, and associated enzymes. If students are confused about any of these, encourage them to use outside sources of information such as an encyclopedia. PORTFOLIO

Teaching TIP

Enzymes and Nutrients

Help students keep enzymes and the nutrients they break down straight by making a **Graphic Organizer** similar to the one at the bottom of page 918. Remind students that the suffix *-ase* means "to break apart."

Figure 40-10 X ray of the large intestine (colon)

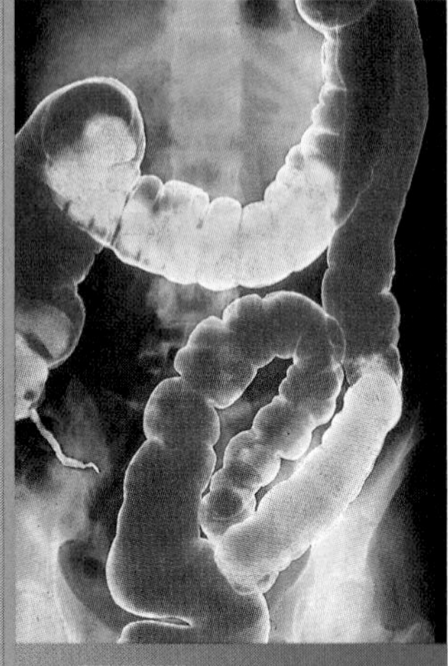

Size Stretched out, the large intestine is about 1 m (3.3 ft) long. The large intestine absorbs water and packs undigested material from the small intestine.

1 m

The Large Intestine Absorbs Water and Packs Together Solids

All components of food that are not for energy production (for example, cellulose) are considered wastes. The wastes move into the large intestine. The large intestine, or **colon** *(KOH luhn)*, shown in **Figure 40-10,** is much shorter than the small intestine. However, the diameter of the large intestine is about three times larger than the diameter of the small intestine. The colon is not coiled like the small intestine. Instead, it is made of three relatively straight segments. No digestion takes place in the colon.

The volume of material that flows through the digestive system each day is large—about 10 L of food, drink, and secretions enter the small intestine. But the amount of material that leaves the body as waste is small. This is because almost all of the fluids and solids (about 90–95 percent) are absorbed during their passage through the small intestine. Mostly mineral ions and water are absorbed through the wall of the large intestine.

Most of the colon's contents are dead cells, mucus, digestive secretions, bacteria, and yeast. A thriving colony of bacteria live in the human colon. These microbes synthesize many compounds that your body needs and cannot get easily from the food you eat, including vitamin K and several B vitamins. In addition, bacteria aid in transforming and compacting the undigested materials into the final waste product, feces.

The final segment of the large intestine is the rectum. Solids in the colon pass into the rectum as a result of peristalsis in the large intestine. From the rectum, the solid feces are eliminated from the body through the anus. Undigested material passes through the large intestine and is expelled through the anus in 12 to 24 hours.

Balancing water absorption in the intestine is important. Wastes rushed through the large intestine before the remaining water is absorbed result in diarrhea (watery feces). When food remains in the colon for long periods of time, causing much water to be absorbed, constipation (hard feces) results. The feces are difficult to pass.

The liver's role in digestion

The human liver, shown in Figure 40-8, is about the size of a football and weighs more than 1.4 kg (3.1 lb). It presses upward against the diaphragm and occupies the upper right side of the abdominal cavity.

The liver plays several roles in digestion. The liver secretes bile, which aids in the emulsification of fats. Bile also promotes the absorption of fatty acids and the fat-soluble vitamins A, D, E, and K. Bile pigments (the products of hemoglobin breakdown) give bile a yellowish green color. Jaundice, a condition in which the eyeballs, skin, and urine become abnormally yellow, is a result of increased amounts of bile pigments in the blood. Jaundice often occurs as a result of **hepatitis,** an inflammation of the liver.

Graphic Organizer

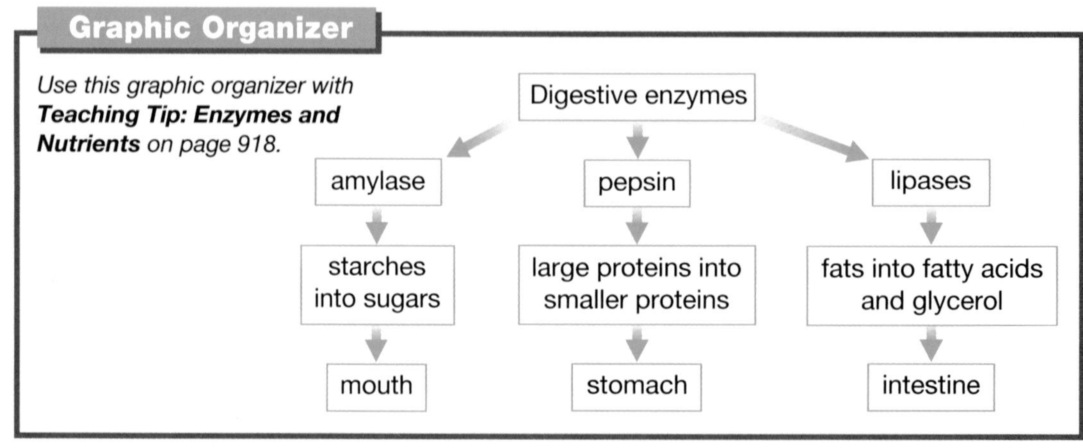

The liver's role in metabolism

Digested food molecules absorbed from the small intestine into the bloodstream are transported through a vein to the liver. The liver maintains proper levels of blood sugar. The liver converts extra sugar to glycogen for storage and retrieves and breaks down the glycogen when it is needed for energy. The liver also modifies amino acids for several uses, including to make nonessential amino acids and plasma proteins. Fat-soluble vitamins and the mineral iron are stored in the liver. The liver monitors the production of cholesterol and detoxifies poisons. If the liver is unable to change a substance's harmful form, it stores it. In this way, toxins, including heavy metals and pesticides, accumulate in the liver.

The liver can also be damaged by viral infections, chronic drug and alcohol use, and traumatic injury. As a result of any of these, healthy liver cells are destroyed and replaced by scar tissue. The scarring of the liver is called cirrhosis *(suh ROH sihs)*.

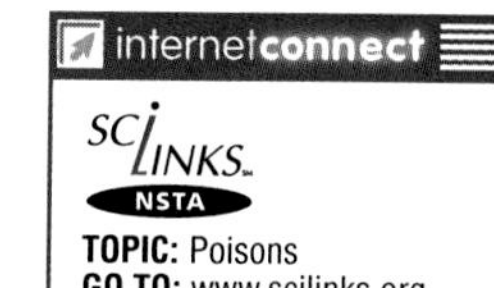

internet**connect**

SCLINKS. NSTA

TOPIC: Poisons
GO TO: www.scilinks.org
KEYWORD: HX919

QUICK LAB

How does bile work?

You can use a detergent and cooking oil to simulate the effect bile has on breaking up (emulsifying) fats as part of digestion.

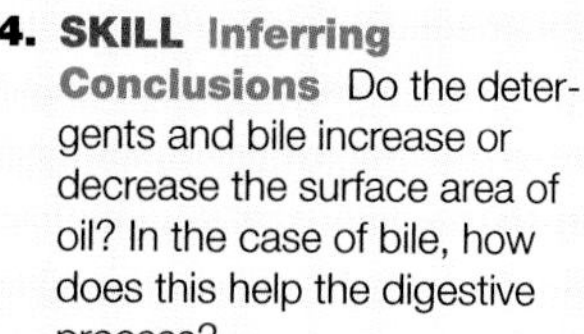

Materials

two 250 mL beakers, water, cooking oil, dish detergent, stirring rod, graduated cylinder

Procedure

1. Label one beaker *A* and one beaker *B*. Fill each beaker halfway with water.
2. Add 10 mL of the cooking oil to each beaker.
3. While stirring, slowly add 10 drops of dish detergent to beaker *B* only.

Analysis

1. **Describe** how oil reacts with the water.
2. **Describe** what happened to the oil when the dish detergent was added.
3. **Compare** the effect of dish detergent on oil with the effect of bile on fats.
4. **SKILL Inferring Conclusions** Do the detergents and bile increase or decrease the surface area of oil? In the case of bile, how does this help the digestive process?

Review SECTION 40-2

1. **Summarize** the path a piece of cheese pizza would follow through the digestive system.
2. **Relate** the role of the mouth, stomach, small intestine, and large intestine in the digestion of a piece of cheese pizza.
3. **Locate** the area of the digestive system where nutrients are absorbed into the bloodstream or lymphatic system.
4. **State** how the liver and pancreas are involved in digestion.

Critical Thinking

5. **SKILL Applying Information** A person has a small intestine that has villi but a reduced number of microvilli. Would you expect this person to be underweight or overweight? Explain.

Review SECTION 40-2 ANSWERS

1. mouth, pharynx, esophagus, stomach, small intestine, large intestine, rectum
2. mouth—mechanical and chemical digestion of carbohydrates in the pizza; stomach—mechanical digestion and chemical digestion of the proteins in the pizza; small intestine—chemical digestion of the carbohydrates, proteins, and lipids in the pizza, and absorption; large intestine—reabsorption of water and compaction of waste
3. small intestine
4. Both the pancreas and the liver provide digestive enzymes. The liver also provides bile to the small intestine, maintains blood sugar levels, modifies amino acids, stores vitamins and minerals, regulates cholesterol production, and detoxifies poisons.
5. The person would be underweight—more food will pass out as waste. The person will probably also be malnourished.

QUICK LAB

How does bile work?

Time 10–15 minutes

Process Skills Relating concepts, inferring conclusions, comparing

Teaching Strategies

- Explain to students that the oil and water in a vinegar-and-oil salad dressing separate because the water molecules hydrogen-bond to one another and exclude the oil.
- Tell the students that detergents are made of molecules with hydrophobic and hydrophilic ends. The hydrophobic ends surround the grease. The hydrophilic ends then help pull the grease toward the water, breaking up the grease.

Analysis Answers

1. The oil spreads away from the water and the oil droplets associate.
2. The oil was broken up.
3. Detergent breaks up oil like bile breaks up fat.
4. They both increase surface area. For the bile this speeds up the breakdown of foods.

3 Assessment Options

Closure

Digestive System

Refer to the clothesline from the **Lesson Warm-up** on page 914. Remove the labels, and have students identify the parts of the digestive system and what happens to food in each of those parts.

Review

Assign Review—Section 40-2.

Reteaching — BASIC

Have students make flashcards with a digestive organ named on one side and its function on the other. Have students arrange them in order from mouth to anus and identify each as part of the digestive tract or as accessory organs.

1 Prepare

Scheduling

- **Block:** About 90 minutes will be needed to complete this section.
- **Traditional:** About two class periods will be needed to complete this section.
- **Planning:** See the Planning Guide on pp. 905A and 905B for lecture, classwork, and assignment options for Section 40-3. Lesson plans for Section 40-3 can be found in a lesson cycle format in ***Biology: Principles and Explorations*** **One-Stop Planner CD-ROM with Test Generator.**

Resource Link

Assign **Active Reading Worksheet Section 40-3** to help students comprehend the key concepts and visuals in this lesson.

2 Teach

Lesson Warm-up

Have a volunteer put his or her hand in a clear plastic bag and tape it loosely to the wrist. Study the bag after 10 or 15 minutes. What do students observe? *(tiny droplets of moisture)* Explain that sweat is constantly evaporating from the skin. If time permits, have the student rub some of their own sweat onto individual slides and observe salt crystals. Red dye will make the crystals easier to see.

Section 40-3 Excretion

Objectives

- **Identify** major wastes produced by humans and the organ or tissue where they are eliminated from the body. (p. 920)
- **Relate** the role of nephrons to the filtering of blood in the kidneys. (pp. 921–922)
- **Summarize** how nephrons form urine. (p. 922)
- **Describe** the path of urine through the human urinary system. (p. 922)
- **Predict** how kidney damage might affect homeostasis and threaten life. (pp. 923–924)

Key Terms

excretion
urea
nephron
urine
ureter
urinary bladder
urethra

The Body Excretes Water and Metabolic Wastes

Cleaning up, though not always a pleasant chore, must be done to maintain a healthy living environment. In the same way, our bodies must get rid of wastes to maintain a healthy body. Food residues are eliminated from the body in the form of feces. Other wastes produced as a result of metabolic reactions that occur in the body must also be eliminated. For example, water and carbon dioxide are produced during cellular respiration. During the metabolism of proteins and nucleic acids, a toxic nitrogen-containing waste, ammonia, is formed.

The body must remove wastes. It must also maintain osmotic balance and pH by either excreting or conserving salts and water. **Excretion** is the process that rids the body of toxic chemicals, excess water, salts, and carbon dioxide and maintains osmotic and pH balance.

The organs involved in excretion are shown in **Figure 40-11.** Carbon dioxide (and some water vapor) is transported to your lungs by the circulatory system and excreted every time you exhale. Excess water is excreted through the skin in sweat and through the kidneys in urine. In the liver, ammonia is converted to a much less toxic nitrogen waste called **urea** *(yoo REE uh),* which is then carried by the bloodstream to the kidneys, where it is removed from the blood.

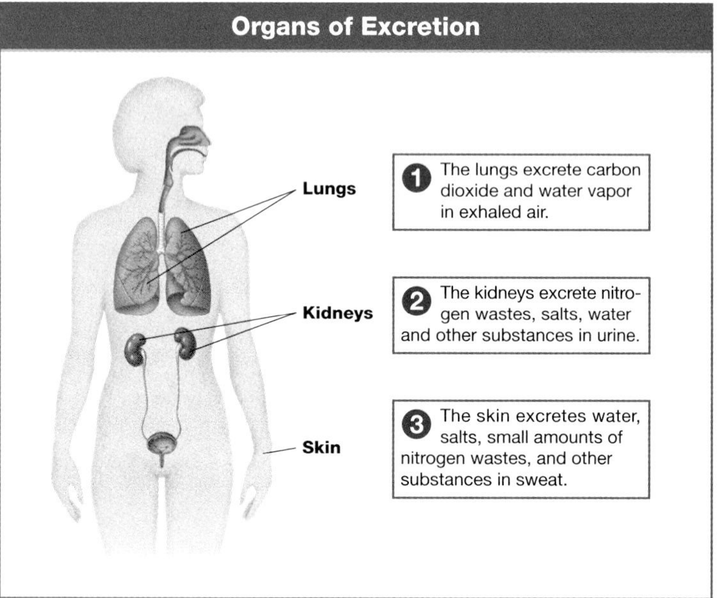

Organs of Excretion
Lungs
Kidneys
Skin
1 The lungs excrete carbon dioxide and water vapor in exhaled air.
2 The kidneys excrete nitrogen wastes, salts, water and other substances in urine.
3 The skin excretes water, salts, small amounts of nitrogen wastes, and other substances in sweat.

Figure 40-11 Organs of excretion. The lungs, the kidneys, and the skin all function as excretory organs. The main excretory products are carbon dioxide, water, and nitrogen wastes (urea).

Integrating Different Learning Styles

Logical/Mathematical	**PE**	Performance Zone item 24, p. 927
	ATE	Cross-Disciplinary Connection, p. 922
Visual/Spatial	**ATE**	Lesson Warm-up, p. 920; Visual Strategy, p. 921; Teaching Tip, p. 923
Interpersonal	**ATE**	Biology in Your Community, p. 922
Intrapersonal	**ATE**	Teaching Tip, p. 921; Overcoming Misconceptions, p. 921
Verbal/Linguistic	**PE**	Performance Zone item 27, p. 927
	ATE	Career, p. 923; Reteaching, p. 924

The Kidneys Clean the Blood

The kidneys are extremely important organs because of their role in regulating the amount of water and salts contained in blood plasma. The kidneys are a pair of bean-shaped, reddish brown organs located in the lower back. Each kidney is the size of a small fist. The body has to maintain a certain level of salts in the blood plasma and in the fluid surrounding cells or it can cause serious harm to the body's cells and organ systems.

Blood filters

Each kidney is a complex organ composed of roughly 1 million microscopic blood-filtering units called nephrons *(NEHF rahns)*, as shown in **Figure 40-12.** **Nephrons** are tiny tubes in the kidneys with cup-shaped capsules surrounding a tight ball of capillaries that filter wastes from the blood, retain useful molecules, and produce urine. Three different phases occur as the blood flows through a nephron: filtration, reabsorption and secretion, and urine formation.

Filtration Filtration begins at the cup-shaped capsule called Bowman's capsule. Within each Bowman's capsule an arteriole enters and splits into a fine network of capillaries called a glomerulus *(gloo MEHR yoo luhs)*. The glomerulus acts as a filtration device. The blood pressure inside the glomerulus forces a fluid composed of water, salt, glucose, amino acids, and urea into the hollow interior of the Bowman's capsule. This fluid is called filtrate. Blood cells, proteins, and other molecules too large to cross the membrane remain in the blood.

Study TIP

Word Origins

The word *glomerulus* is from the French *gloméruli*, meaning "a compact cluster." Knowing this makes it easier to remember that the glomerulus is a compact cluster of capillaries within a nephron.

Figure 40-12 Kidneys and nephrons

The kidneys filter out toxins, urea, water, and mineral salts from the blood as fluid passes through the microscopic filtering units called nephrons.

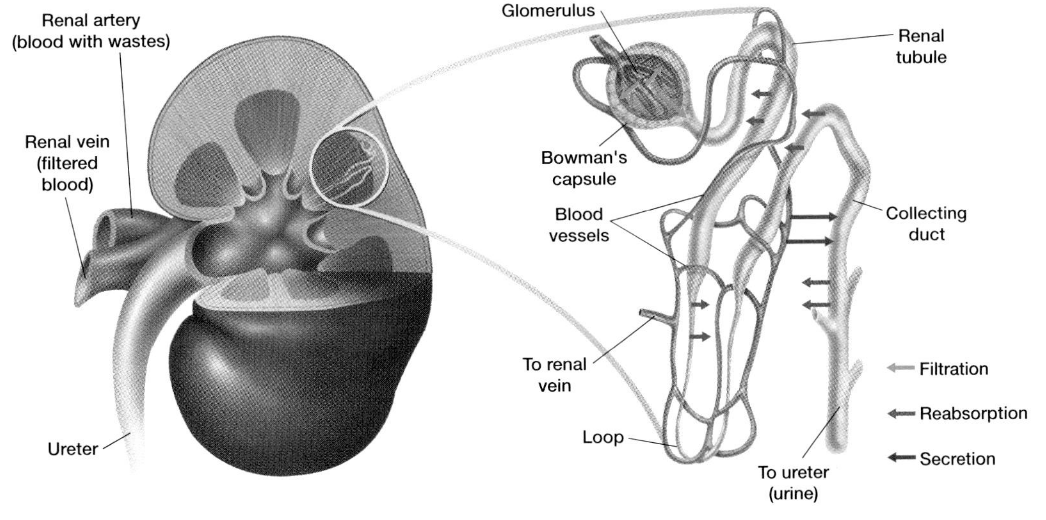

Overcoming Misconceptions

Excretion

Tell students that they are presently excreting! Excretion is more than just urine and fecal matter. In fact, fecal matter is considered an end product of digestion and not necessarily part of excretion, which focuses on metabolic wastes. Carbon dioxide and water are excreted with every breath you exhale.

Teaching TIP

Kidneys

Ask students to point to the location of their kidneys. They are located in the middle of the back, near the spinal cord, just under the last ribs.

Visual Strategy

Figure 40-12

Review how passive transport processes *(blood pressure)* account for the filtration of blood in the nephron. Have students locate where filtration occurs in the nephron shown in Figure 40-12. *(in Bowman's capsule)* Review how active transport processes account for the reabsorption of materials as blood passes through the nephron and both passive and active processes are involved in secretion. Have students locate where reabsorption and secretion occur in the nephron shown in Figure 40-12. *(the renal tubules)* Point out that because of active transport, the kidney has an extremely high energy requirement, even higher than that of the heart, when comparing their relative weights.

Making Biology Relevant

Sports

Exercise and sweating deplete the body of many necessary materials. Sports drinks are beverages that consist of water with about a 6 percent concentration of carbohydrates and electrolytes (salts, particularly sodium). Sports drinks are thought to improve performance in athletes that are exercising for more than 90 minutes because the carbohydrates maintain blood sugar (glucose) levels during exercise when muscle glycogen (stored glucose) levels start to run critically low. The electrolytes in the sports drink are supposed to allow the carbohydrates and water in the drink to be absorbed into the body faster.

Making Biology Relevant

Health

It is ironic that as food conscious as Americans are, they often pay little attention to the other side of the coin—their fluid intake. In fact, people can survive without food much longer than they can survive without water. On average, an adult needs 1.5–3 L of water per day, some of it being supplied by foods with high water content such as lettuce, watermelon, and citrus fruits. Any combination of foods and beverages that provides the needed amount of water is acceptable, but most nutritionists recommend 6–8 glasses of liquid a day over and above the amount contained in food.

REAL LIFE

Answers

Answers will vary. Researchers have found nanobacteria in kidney stones. It is thought that the bacteria cause kidney stones in genetically susceptible humans. The mineral shells that cause calcium deposits in the body form around the bacteria. Nanobacterial infections can be treated with antibiotics.

CROSS-DISCIPLINARY CONNECTION

◆ Mathematics

Tell students that the heart pumps about 5 L (2.6 gal) of blood per minute. Keeping in mind that one-fourth of this volume travels to the kidneys, have students calculate the volume of blood that is filtered each day. *(5 L/min × 60 min/hr × 24 h × 0.25 = 1,800 L per day)* Ask students to calculate how much of the blood is actually filtered if about 55–60 percent of the 5 L of blood is plasma, and only 18–20 percent of the plasma, water, and small solutes are actually filtered out of the blood and into the nephron capsule. *(1,800 L × 0.55 × 0.18 = 178.2 L are actually filtered per day.)* PORTFOLIO

REAL LIFE

Kidney stones are painful. Kidney stones form when mineral and urea salts crystallize and become insoluble. The salts form "stones" that block the passage of urine and result in extreme pain. Ultrasound waves can be used to crush kidney stones until they are small enough to pass with the urine.

Finding Information
Research the latest findings on the role of tiny bacteria called nanobacteria in the formation of kidney stones.

Reabsorption and secretion Reabsorption begins when the filtrate passes from the Bowman's capsules into the renal tubules—long, narrow tubes connected to Bowman's capsules. Renal tubules bend at their center, forming a loop. As the filtrate passes through the renal tubules, the tubules extract from the filtrate a variety of useful molecules, including glucose, ions, and some water. These substances reenter the bloodstream through capillaries that wrap around the tubule, which prevents these molecules from being eliminated from the body in the urine. When the filtrate reaches the last portions of the renal tubule, substances such as wastes and toxic materials pass from the blood into the filtrate in a process called secretion.

Urine formation The water, urea, and various salts that are left after the reabsorption and secretion processes together make up **urine.** The urine is formed when the renal tubule empties the fluid that remains after reabsorption and secretion into a larger tube called the collecting duct. The collecting duct removes much of the water from the filtrate that passes through it. As a result, human urine is very concentrated—in fact, it is four times more concentrated than blood plasma. Millions of collecting ducts in each kidney empty the urine into areas of the kidneys that lead to the **ureters** *(yoo REET uhrs)*—tubes that carry the urine from the kidney to the urinary bladder.

Urine elimination

The ureters, shown in **Figure 40-13,** have smooth muscle in their walls. The slow, rhythmic contractions of this muscle move the urine through the ureters. The ureters direct the urine into the **urinary bladder,** a hollow, muscular sac that stores urine. The urinary bladder gradually expands as it fills. The urinary bladder can hold up to 0.6 L (0.63 qt) of urine.

Figure 40-13 The organs of urinary excretion. Urine exits the kidneys by way of two ureters that empty into a storage organ called the urinary bladder. Urine exits the body through the urethra.

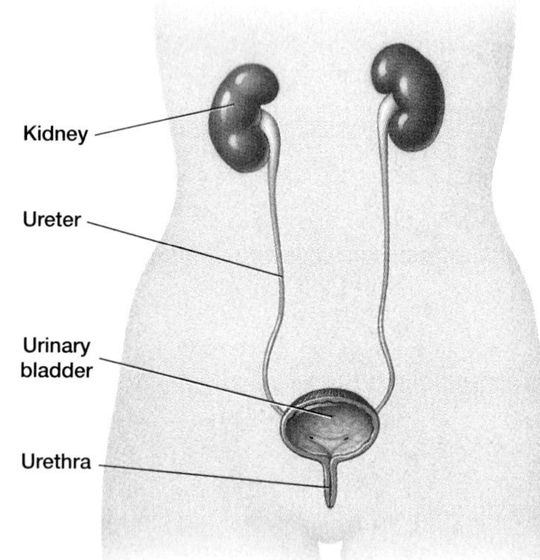

Muscular contractions of the bladder force urine out of the body. Urine leaves the bladder and exits the body through a tube called the **urethra** *(yoo REE thruh)*. A healthy adult eliminates from 1.5 L (1.6 qt) to 2.3 L (2.4 qt) of urine a day.

In females the urethra lies in front of the vagina and is only about 2.5 cm (1 in.) long. Such a short length makes it easy for bacteria and other pathogens to invade the female urinary system, which explains why females are more prone to urinary infections than males are. There is no connection between the urethra and the genital (reproductive) system in females.

In males the urethra passes through the penis. In males, both sperm and urine exit the body through the urethra because the tube that carries sperm from the testes eventually merges with the urethra.

Biology in Your Community

Water Treatment Plant

Ask the manager from a water treatment plant to come and discuss the processes the plant uses to remove substances from water. Have students write a short report that relates this information to how the kidney removes substances from the blood. PORTFOLIO

The elimination of urine from the body through the urethra is called urination. When the bladder fills with urine, stretch receptors in the bladder's wall send nerve impulses to the spinal cord. In response, the spinal cord returns impulses to the bladder and urethra simultaneously. These impulses cause contraction of the bladder's muscular walls and relaxation of the rings of muscle closing off the urethra. The bladder then empties its contents through the urethra. In older children and adults, the brain overrides this urination reflex, delaying the release of urine until a convenient time.

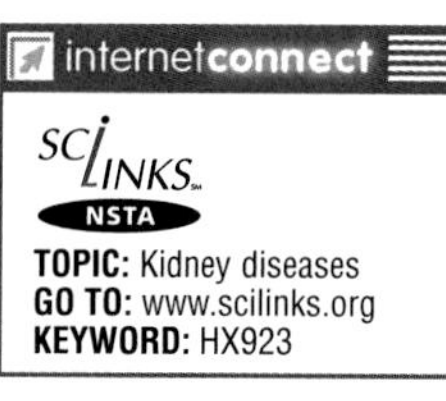

Damage to the kidneys

Because of the vital role played by the kidneys in maintaining homeostasis, diseases affecting these organs may eventually threaten life. If one kidney is lost in an accident or by disease, the other may enlarge and do the work of both. Nephrotic syndrome refers to a number of signs and symptoms that result from damage to the glomeruli, leading over time to kidney failure. The most common causes of kidney failure are infection, diabetes, high blood pressure, and damage to the kidneys by the body's own immune system.

Because of their function in excretion, kidneys often are exposed to hazardous chemicals that have entered the body through the lungs, skin, or gastrointestinal tract. Household substances that, in concentration, can damage kidneys include paint, varnishes, furniture, oils, glues, aerosol sprays, air fresheners, and lead. When kidneys fail, toxic wastes, such as urea, accumulate in the plasma, and blood-plasma ion levels increase to dangerous levels. If both kidneys fail, there are only two treatment options.

Kidney dialysis Kidney dialysis, also called hemodialysis *(HEE moh deye AL uh sihs)*, is a procedure for filtering the blood using a dialysis machine, as shown in **Figure 40-14.** A dialysis machine, just as the nephrons in the kidney, sorts small molecules in the blood, keeping some and discarding others. Dialysis machines are sometimes used when the kidneys are damaged but can either be repaired or will eventually be replaced by a kidney transplant.

Kidney transplants A more permanent solution to kidney failure is transplantation of a kidney from a healthy donor. A major problem with kidney transplants is common to all organ transplants—rejection of the transplanted organ by the recipient's immune system. Recall that the cells of your body have "self-markers," or antigens, on their surfaces that identify the cells to your immune system so it will not attack them. The combination of these antigens displayed on your body's cells is as unique as your fingerprints.

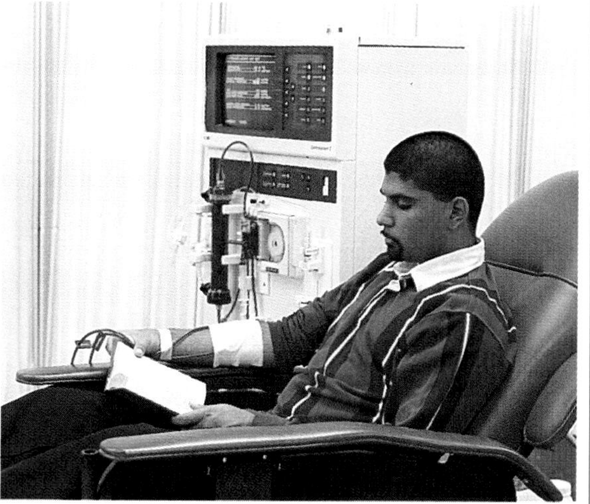

Figure 40-14
Hemodialysis. Hemodialysis has prolonged the lives of many people with damaged or diseased kidneys. The dialysis machine functions like a kidney in that it filters urea and excess ions from the blood.

did you know?

High blood pressure, drinking lots of fluids, and cold weather promote increased urine output.

High blood pressure increases the filtration pressure in the kidneys. Increased fluid consumption raises the blood pressure and decreases the output of antidiuretic hormone (ADH). Cold weather increases the metabolic rate and thus increases the rate of filtration. Low blood pressure, reduced intake of fluids, and hot weather decrease urine output for the converse reasons.

Teaching TIP

Bladder Muscle Weakness

Aging, childbirth, or disease often causes the weakening of muscles in the bladder. Although the conscious override for bladder control is intact, sometimes the muscles leak small amounts of urine during laughter, coughing, or other times.

Teaching TIP

Kidney Anatomy

Obtain a pig or sheep kidney from a local butcher. Cut the kidney in half lengthwise. Have students identify the obvious structures. Ask students if they can identify the smell. *(ammonia)* Ask students how kidneys are prepared before cooking. *(You must soak the kidney in salt water to release the ammonia before cooking it.)*

CAREER

Dialysis Technician

Dialysis technicians operate dialysis machines and monitor and record a patient's weight and vital signs before, during, and after dialysis. Have students research the career of a dialysis technician and prepare a short report. The report should include a job description, training required, employer, growth potential, starting salary, and job satisfaction. PORTFOLIO

Chapter 40

Exploring Further

Kidney Dialysis

Teaching Strategies
To demonstrate osmosis, bring in some dialysis tubing and show students how permeable it is. Pour a starch solution into the tubing, which is immersed in an iodine and water solution outside. The iodine will diffuse into the tubing, turning the starch solution blue-black.

Discussion
- Describe how a dialysis machine works. *(A catheter removes blood from an artery. Wastes and ions pass out of the blood into a fluid similar to plasma. Filtered blood is returned to the veins through a different catheter.)*
- Why are dialysis machines less effective than kidneys? *(Machines do not dialyze constantly as the body does. Levels of salt, protein and water cannot be closely regulated with a machine.)*
- If the tissue types matched, would you donate a kidney to someone who needed one? Does the recipient matter? Would you donate to a stranger? a distant relative? a close relative? *(Answers will vary.)*

3 Assessment Options

Closure

Technique: Anticipation Guide
Have students return to their statements in the **Active Reading** on page 907. Ask students if their opinions have been confirmed or changed in any way, now that they have read the chapter.

Review
Assign Review—Section 40-3.

Reteaching BASIC
Ask students to summarize the function of the urinary system in a paragraph or two, using as many science terms as possible in their summary. PORTFOLIO

Only identical twins have the same set of antigens. The more closely related two individuals are, the more likely they are to have common antigens. This is why tissue transplants are more likely to succeed if the donor and recipient are closely related. But even in close matches, there is some chance of transplant rejection. To reduce chances of rejection, the recipient is treated with drugs that suppress the activity of the immune system. However, when such drugs are given, there is an increased risk of infection.

Exploring Further

Kidney Dialysis

People whose kidneys are damaged cannot filter their blood. Kidney dialysis is one option for artificially filtering the blood. In kidney dialysis, tubes called catheters are surgically inserted into an artery and a vein, usually on a lower arm. The catheters are equipped with valves. Every few days the catheters are connected to a dialysis machine, as shown.

Blood is filtered
Blood passes from the patient's artery into the dialysis machine. Inside the machine, the blood travels through many hollow tubes, each of which is surrounded by a thin permeable membrane. Waste materials and ions that have accumulated in the person's blood diffuse through the membrane into a fluid that has the same makeup as normal blood plasma and is free of wastes. The filtered blood is then returned to the person's vein.

Dialysis is not a permanent solution to kidney failure. A single healthy kidney can meet all of the homeostatic needs of the body, but no dialysis machine can. Dialysis patients must carefully manage their salt, protein, and water intake because the dialysis machine cannot regulate these blood components as well as the kidney can.

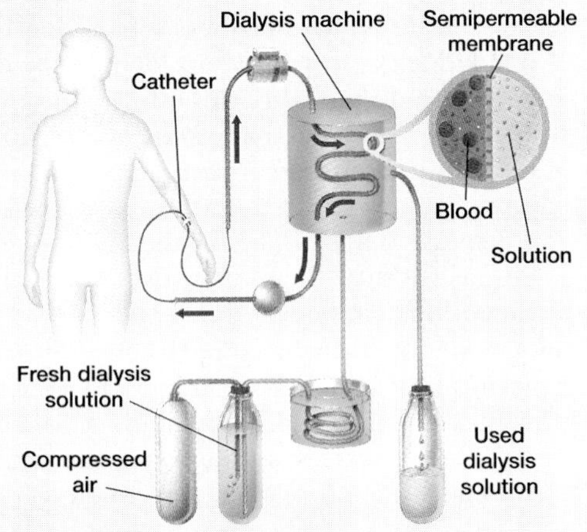

Review SECTION 40-3

1. **Identify** how the carbon dioxide in your body is produced and excreted.
2. **Relate** the following terms to the formation of urine: filtration, reabsorption, and secretion.
3. **Name** the liquid stored inside the collecting duct of a nephron.
4. **Summarize** how urine is stored and eliminated from the body.
5. **Identify** two effects of kidney damage.

Critical Thinking

6. **Skill Applying Information** A doctor has just informed a patient that his urine sample contains a high concentration of sugar. Explain why this may indicate damaged kidneys.

Review SECTION 40-3 ANSWERS

1. Carbon dioxide is produced during cellular respiration and excreted through the lungs.
2. filtration—kidneys filter blood, removing water, salt, glucose, amino acids, and urea; reabsorption—useful molecules and water are reabsorbed by the bloodstream; secretion—wastes and toxic materials pass from blood into filtrate
3. Urine
4. Urine is stored in the bladder. When stretch receptors in the bladder send nerve impulses to the spinal cord, the bladder muscles relax and urination occurs.
5. the accumulation of toxins such as urea and an increase of plasma ions
6. Sugar should be reabsorbed by the bloodstream in the kidney.

Study ZONE — CHAPTER 40

Use the Key Concepts and Key Terms listed below to review the main ideas in this chapter. If you need to review the meaning of a term, turn to the page indicated.

Key Concepts

40-1 Your Body's Need for Food

- Food and beverages provide the nutrients and water required by the body for growth, energy, repair, and maintenance.
- Carbohydrates and lipids provide most of the body's energy. Proteins are normally used for making other proteins.
- The USDA food guide pyramid graphically summarizes the daily recommended servings from each food group.
- Vitamins enhance the activity of enzymes and regulate the release of energy. Minerals are used to make certain body structures and substances, for normal nerve and muscle function, to maintain osmotic balance, and for enzyme function.
- Water acts as a lubricant, solvent, and coolant, and as a support medium for cells and tissues.

40-2 Digestion

- Teeth break down food into smaller pieces. Amylase begins the breakdown of starch to sugars. The stomach stores and mechanically breaks down food. Stomach acid and pepsin chemically break down proteins.
- Most chemical digestion occurs in the small intestine with the help of secretions from the pancreas, liver, and gallbladder.
- Usable compounds are absorbed into capillaries or lymphatic vessels in villi. Compounds not absorbed are eventually excreted as feces.
- The liver releases bile, helps to maintain blood sugar levels, and detoxifies poisons.

40-3 Excretion

- The skin, lungs, and kidneys are specialized to excrete wastes.
- Nephrons in the kidneys filter wastes from the blood. Most of the water, some of the salts, and all of the sugar and amino acids in the filtrate are reabsorbed into the bloodstream. The water, urea, and salts that remain in the nephron are eliminated as urine.
- Kidney dialysis and organ transplants are treatment options when both kidneys fail.

Key Terms

40-1

nutrient (908)
digestion (908)
calorie (908)
vitamin (912)
mineral (913)

40-2

amylase (915)
esophagus (915)
peristaltic contraction (915)
pepsin (916)
ulcer (916)
lipase (917)
villus (917)
colon (918)
hepatitis (918)

40-3

excretion (920)
urea (920)
nephron (921)
urine (922)
ureter (922)
urinary bladder (922)
urethra (922)

Study TIP Ready?

- *If you think you understand these Key Concepts and Key Terms, then you're ready for the Performance Zone.* 

Answer to Concept Map

The following is one possible answer to Performance Zone item 1 on page 926.

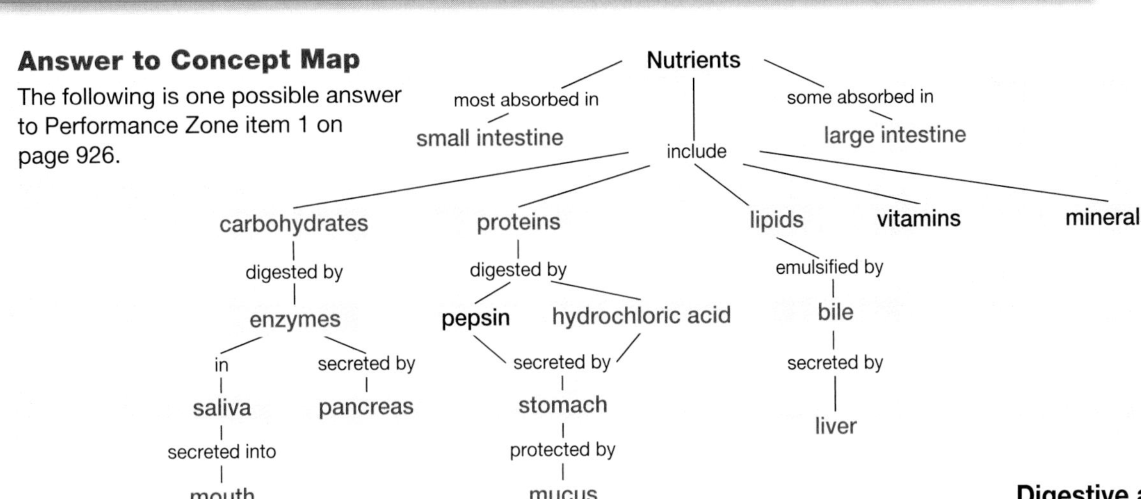

Study ZONE — CHAPTER 40

Review and Practice

Assign the following worksheets for Chapter 40:

- **Student Review Guide**
- **Concept Mapping Worksheet**
- **Test Preparation Pretest**
- **Vocabulary Worksheet**
- **Critical Thinking Review Worksheet**

Assessment Options

The following assessment options are available for this chapter:

Resource Link

- Assign **Chapter Test 40** to assess students' understanding of the chapter.

Alternative Assessment

Commercial diet programs are big business. Have students research commercial diet programs and report how each claims to bring about weight loss. Encourage students to look for information on how these programs might affect either the digestive or excretory systems. For example, a high-protein diet places added stress on the liver and kidneys, because the body must produce and then rid itself of more urea, secondary to higher nitrogen intake. PORTFOLIO

Portfolio Assessment

- *Portfolio Assessment* at the front of this book provides options for building and evaluating student portfolios. Students and teachers can select items from the areas listed for inclusion in a portfolio.
- See items labeled PORTFOLIO on pp. 908, 910, 911, 912, 913, 916, 918, 922, 923, 924, and 925.

Performance ZONE
CHAPTER 40 REVIEW ANSWERS

Assignment Guide

Section	Review
40-1	3, 4, 5, 6, 13, 15, 18, 19, 20, 21, 25, 26, 28, 30
40-2	1, 7, 8, 9, 22, 23, 29
40-3	2, 10, 11, 12, 14, 16, 17, 24, 27

Understanding and Applying Concepts

1. The answer to the concept map is found at the bottom of page 925.
2. **a.** absorption—the other terms refer to processes involved in the filtering of blood by nephrons.
   b. villi—the other terms refer to structures in the kidney
3. c
4. a
5. c
6. b
7. d
8. c
9. a
10. d
11. d
12. c
13. A. grains, 6–11 servings a day; B. fruits, 2–4 servings a day
14. Materials that are not too large move from high concentration to low concentration (blood to Bowman's capsule).
15. In anorexia the levels of sex hormones may fall, the body temperature and pulse rate become abnormally low, the hair becomes thin and dry, and resistance to infections drops. In bulimia the teeth, gums, and esophagus often are damaged, and muscle weakness, heart failure, and kidney disease can occur.
16. Blood moves through tubes composed of semipermeable membranous material, which are bathed in fluid with the same composition as clean plasma. Wastes diffuse out of the tubes and into the "plasma." The filtered blood circulates into the patient's body.
17. Hypertonic

Performance ZONE CHAPTER 40 Review

Understanding and Applying Concepts

1. **Concept Mapping** Make a concept map that shows how nutrients are digested by the body. Try to include the following words in your map: carbohydrates, proteins, lipids, enzymes, saliva, mouth, pancreas, hydrochloric acid, stomach, mucus, bile, liver, small intestine, large intestine.
2. **Understanding Vocabulary** For each set of terms below, choose the one that does not belong, and explain why it does not belong.
   a. absorption, filtration, secretion, reabsorption
   b. nephron, villi, glomerulus, renal tubule
3. The primary function of carbohydrates is to
   a. break down molecules.
   b. aid in digestion.
   c. supply the body with energy.
   d. regulate the flow of acid.
4. Food from the _______ food group should be eaten in the greatest abundance.
   a. grains
   b. fats, oils, and sweets
   c. vegetables
   d. milk, yogurt, cheese
5. Proteins are made up of
   a. fatty acids.
   b. glycerol.
   c. amino acids.
   d. monosaccharides.
6. The body needs vitamins because they
   a. supply energy.
   b. activate enzymes.
   c. function as enzymes.
   d. act as hormones.
7. During digestion, food passes through the mouth, esophagus, stomach, _______, and large intestine.
   a. pancreas
   b. lungs
   c. gallbladder
   d. small intestine
8. Which pair matches the enzymes with the food molecules they digest?
   a. lipases: starches
   b. amylases: fats
   c. lipases: fats
   d. pepsin: starches
9. Nutrients in the small intestine enter the bloodstream by passing through
   a. villi.
   b. glomeruli.
   c. gastric pits.
   d. nephrons.
10. Which substance is *not* a waste eliminated from the body through the kidneys?
   a. urea
   b. water
   c. salts
   d. oxygen
11. The _______ are involved in excretion.
   a. kidneys and stomach
   b. liver and pancreas
   c. pancreas and kidneys
   d. kidneys and lungs
12. During secretion in the kidney, substances move from
   a. the filtrate to the blood.
   b. the urethra to the bladder.
   c. the blood to the filtrate.
   d. the bladder to the urethra.
13. **Identifying Information** Write the food groups for letters *A* and *B*. Indicate the number of servings that should be eaten daily for *A* and *B*.

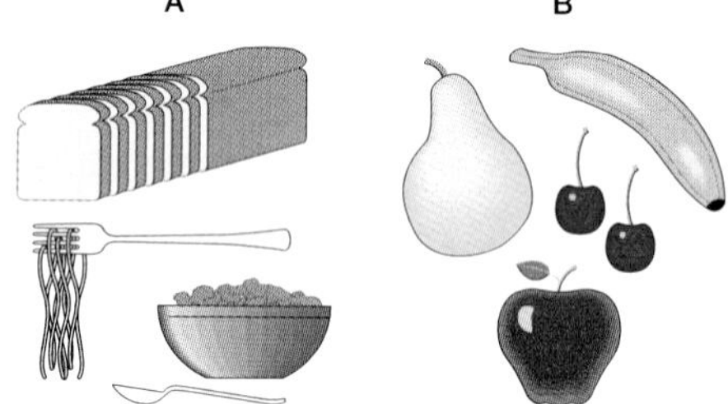

14. **Theme Metabolism** How does passive transport account for the filtration of blood in the nephron?
15. **Health Watch** Describe the changes caused by anorexia and bulimia.
16. **Exploring Further** Summarize how a kidney dialysis machine filters blood.
17. **Chapter Links** Is urine hypotonic, hypertonic, or isotonic relative to blood? (**Hint:** See Chapter 4, Section 4-1.)

Skills Assessment

18. 210
19. thiamin, riboflavin, niacin, iron
20. no
21. carbohydrates
22. Answers will vary. Report topics might include gastroesophageal reflux or acid reflux (the backward flow of stomach acid into the esophagus); ulcers (an area of the stomach or duodenum that has been destroyed by stomach acids); inflammatory bowel disease (inflammation of the intestinal lining, which may lead to bleeding, fever, diarrhea, cramping, and increased white blood cell count); and food intolerance (the ingestion of certain food substances causes symptoms such as nausea, gas, abdominal pain, and diarrhea).
23. Answers will vary. Students may start by contacting your local public health authority or mental health agency.

Skills Assessment

Part 1: Interpreting Graphics

"Nutrition Facts" is the title for the food labels required by law. The food labels help consumers make food choices based on the nutrients that are most important to health. The serving size is based on the amount of the food people usually eat. Use the illustration below, of a label from a box of pasta, to answer the following questions:

Nutrition Facts	**Amount/serving**	**% DV***	**Amount/serving**	**% DV***
	Total fat 1 g	1%	**Total carb.** 43 g	**14%**
Serv. size 2 oz (56 g/$\frac{1}{8}$ box)	Sat. fat 0 g	**0%**	Dietary fiber 2 g	**8%**
Servings per container 8	**Cholesterol** 0 mg	**0%**	Sugars 3 g	
Calories 210 **Fat Cal.** 10	**Sodium** 0 mg	**0%**	**Protein** 6 g	
*Percent Daily Values (DV) are based on a 2,000 Calorie diet.	Vitamin A 0% • Vitamin C 0% • Calcium 2% • Iron 10% Thiamin 30% • Riboflavin 10% • Niacin 15%			

18. How many calories are found in one serving of pasta?

19. The Percent Daily Values (%DV) tells you what percentage of the daily requirement for a nutrient you are getting. For vitamins and minerals, a %DV of about 10 percent indicates that a food is a good source of these essential nutrients. Which vitamins and minerals in pasta provide 10% or more of the %DV?

20. Foods that get more than one-third of their calories from fat should be avoided. Should pasta be avoided?

21. Which major nutrient, fats, carbohydrates, or proteins, is most abundant in one serving of pasta?

Part 2: Life/Work Skills

Use the media center or Internet sources to find out about common disorders of the digestive system.

22. Communicating Prepare an oral report using graphics to interpret and summarize your findings.

23. Finding Information Compile a list of agencies and programs in your community that work with teens who have eating disorders. Develop a fact sheet for each eating disorder and include them with your list.

Critical Thinking

24. Evaluating Conclusions The looped tubule in a nephron conserves water by reabsorbing water. Its length varies among mammal species. A friend believes the looped tubules of animals that live in watery environments would be shorter than the looped tubule found in humans. Do you agree or disagree? Explain.

25. Recommending Information Calcium needs an acidic environment for optimum absorption. What kinds of foods would you recommend be combined with calcium-rich foods in order to maximize absorption?

26. Determining the Validity of a Claim A friend believes a vegetarian diet would decrease his intake of saturated fat and cholesterol. Do you agree or disagree? Explain.

Portfolio Projects

27. Finding Information Write an article for your school or local newspaper that discusses diuretics (substances that increase urine excretion). Emphasize diuretics that most people have heard of, such as the caffeine in coffee and soft drinks.

28. Relating Concepts Look at books that describe the cultures and customs of other nations. Find information relating to ideas about body shape, ideal body weight, and eating habits for at least five different countries. Describe how these ideas differ from those in the United States.

29. Applying Information Read "Proteins that Produce Hunger..." (*Science News*, March 7, 1998, vol. 153, p. 159). Describe the experiments Yanagisawa and his team conducted to discover and then explore orexins—proteins that produce hunger.

30. Career Focus Dietitian Research the field of dietetics, and write a report on your findings. Your report should include a job description, training required, kinds of employers, growth prospects, and starting salary.

Portfolio Projects

27. Answers will vary. Students' lists might include caffeine, celery seed, dandelion, parsley, juniper, corn silk, couch grass, buchu leaves, uva ursi, and other herbs, as well as prescription drugs.

28. Answers will vary.

29. Yanagisawa's team made cells with many surface proteins (which transmit signals) and put them in brain tissue extracts. Then they monitored the resulting signals. Once they identified orexins, they injected them into mouse brains, which increased the animals' appetites. When animals were starved, the gene in the brain that codes for orexins' precursor became more active.

30. Dieticians provide information on nutrition, plan menus, and develop new foods. Training varies by region but usually requires a B.S. degree in dietetics, nutrition, or food science. Many also complete a clinical internship and a qualifying exam. Employers include hospitals, nursing homes, health professionals, food industries, and health agencies. Growth prospects are good. Starting salary will vary by region.

Critical Thinking

24. Answers will vary. Students who agree might argue that water is more readily available so that their need to conserve water is not as critical as in humans.

25. acidic foods such as tomatoes, oranges, and other citrus fruits

26. Answers will vary. Students who agree might point out that in most nonvegetarian diets, most fats and cholesterol come from animal products.

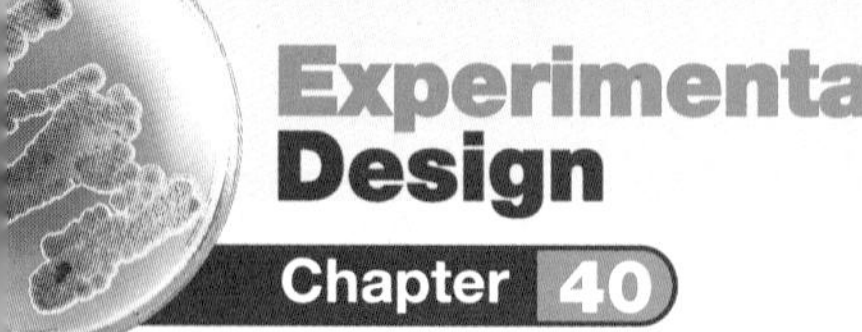

Chapter 40

How Can You Improve Lactose Digestion?

Purpose
Students will design an experiment that investigates a commercially available milk-treatment product designed to aid lactose digestion.

Preparation Notes
Time Required: two 45–50 minute lab periods

Materials
Materials for this lab can be ordered from WARD'S. See *Master Materials List* at the front of this book for catalog numbers.

Preparation Tips
- Glucose test strips and commercial milk-treatment products that contain yeast-derived lactase are available from most pharmacies. Buy several boxes of the glucose test strips so the lab groups each have a color guide. Avoid buying the yeast-derived lactase in solid form. Do not use milk-treatment product tablets. The tablets will give positive glucose results.
- Each lab group will need copies of the information sheet that accompanies the product.
- When you read the students' protocols for approval, direct students to use only drops of milk and milk-treatment product instead of full quarts of milk.
- Glucose test strips are more convenient than Benedict's test because they do not require a hot water bath and they give faster results.

Safety Precautions
Discuss all safety symbols and caution statements with students.

Procedural Tips
- Before students begin the lab ask how you could tell if a milk-treatment product digests lactose. *(The presence of glucose and/or galactose in the milk after addition of the treatment product would indicate that lactose has been broken down.)*

Experimental Design

Chapter 40

How Can You Improve Lactose Digestion?

SKILLS
- Using scientific methods
- Observing
- Comparing

OBJECTIVES
- **Describe** the relationship between enzymes and the digestion of food molecules.
- **Evaluate** the ability of a milk-treatment product to promote lactose digestion.
- **Infer** the presence of lactose in milk and foods that contain milk.

MATERIALS
- milk-treatment product (liquid)
- toothpicks
- depression slides
- droppers
- whole milk
- glucose solution
- glucose test strips

ChemSafety

CAUTION: Always wear safety goggles and a lab apron to protect your eyes and clothing.

CAUTION: Do not touch or taste any chemicals. Know the location of the emergency shower and eyewash station and how to use them. If you get a chemical on your skin or clothing, wash it off at the sink while calling to the teacher. Notify the teacher of a spill. Spills should be cleaned up promptly, according to your teacher's directions.

CAUTION: Glassware is fragile. Notify the teacher of broken glass or cuts. Do not clean up broken glass or spills with broken glass unless the teacher tells you to do so.

Before You Begin
People with a condition known as **lactose intolerance** often experience stomach and intestinal pain, vomiting, and diarrhea when they eat foods that contain milk. These symptoms result from an inability to digest lactose, a sugar found in milk. **Lactose** is a disaccharide made of one glucose unit and one galactose unit. Lactose molecules are broken down into glucose and galactose molecules during **digestion.** People who cannot digest lactose do not produce **lactase,** the digestive enzyme that aids the breakdown of lactose. In this lab, you will investigate a milk-treatment product that is designed to aid lactose digestion.

1. Write a definition for each boldface term in the paragraph above.
2. List at least 10 foods that contain milk.
3. Make a data table similar to the one below.
4. Based on the objectives for this lab, write a question you would like to explore about enzymes and digestion.

DATA TABLE

Solution	Result (+ or –)	Interpretation

Procedure
Part A: Design an Experiment
1. Read the information sheet that comes with the milk-treatment product. Discuss with your lab group what the product is and what it does. Write a summary of your discussion for your lab report.

- Some students will be unsure about what controls to use. They may not think about testing the milk-treatment product itself for sugar.
- Instruct students to test the glucose test strip for a positive glucose test in a known glucose solution.

Disposal
All milk products can be poured down the drain. Glucose test strips can be placed in the trash.

Before You Begin
Answers
1. *lactose intolerance*—an inability to digest the lactose in milk or milk-containing products due to the lack of the enzyme lactase

   lactose—a disaccharide found in milk and composed of a glucose and galactose

   digestion—the breakdown of food molecules into simpler molecules that can be taken up by cells

   lactase—the enzyme that catalyzes the breakdown of lactose
2. Answers will vary. Examples may include milk, cheese, butter, ice cream, and cottage cheese.
4. Answers will vary. For example: How effective is a milk-treatment product in breaking down lactose?

2. Work with the members of your lab group to explore one of the questions written for step 4 of **Before You Begin.** To explore the question, design an experiment that uses the materials listed for this lab.

> **You Choose**
>
> **As you design your experiment, decide the following:**
>
> a. what question you will explore
> b. what hypothesis you will test
> c. what your controls will be
> d. how much milk and milk-treatment product to use for each test
> e. how to determine whether lactose was broken down
> f. what data to record in your data table

3. Write the procedure for your group's experiment. Make a list of all the safety precautions you will take. Have your teacher approve your procedure and safety precautions before you begin the experiment.

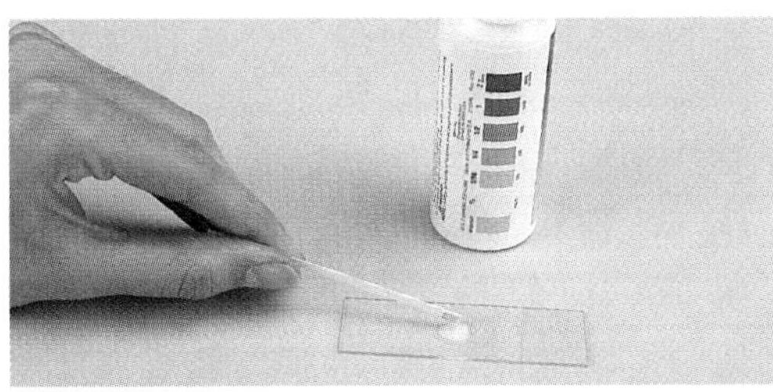

4.   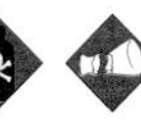 Set up your group's experiment, and collect data.

Part B: Cleanup and Disposal

5. Dispose of solutions, broken glass, and glucose test strips in the designated waste containers. Do not pour chemicals down the drain or put lab materials in the trash unless your teacher tells you to do so.

6. Clean up your work area and all lab equipment. Return lab equipment to its proper place. Wash your hands throughly before you leave the lab and after you finish all work.

Analyze and Conclude

1. **Summarizing Information** What are the milk-treatment product's ingredients?
2. **Recognizing Relationships** What is the relationship between lactose and lactase?
3. **Analyzing Methods** What role did the glucose solution play in your experiment?
4. **Drawing Conclusions** What does the milk-treatment product do to milk?
5. **Analyzing Conclusions** How do your results justify your conclusion?
6. **Evaluating Methods** Why should you test the milk-treatment product with glucose test strips?
7. **Analyzing Results** What do you infer from the results of this lab about treatments for other medical problems resulting from enzyme deficiencies?
8. **Forming Reasoned Opinions** As a person grows older, will he or she be more likely or less likely to develop lactose intolerance? Explain your answer.
9. **Predicting Patterns** Do you think lactose intolerance might be inherited? Explain your answer.
10. **Further Inquiry** Write a new question about enzymes and digestion that could be explored with another investigation.

Do You Know?

Do research in the library or media center to answer these questions:

1. What are some other food-treatment products that contain digestive enzymes?
2. Why does the improper breakdown of certain food molecules cause symptoms such as stomach pain, gas, and diarrhea?

internetconnect

TOPIC: Lactose intolerance
GO TO: www.scilinks.org
KEYWORD: HX929

Analyze and Conclude

Answers

1. yeast-derived lactase
2. Lactose is a disaccharide found in milk. Lactase is an enzyme that catalyzes the breakdown of lactose into glucose and galactose. Lactose is the substrate for the enzyme lactase.
3. The glucose solution is a positive control for the experiment.
4. It breaks down the lactose in the milk into glucose and galactose.
5. Answers may vary. Students indirectly tested for the breakdown of lactose by testing for the presence of glucose. Milk that was treated with the milk-treatment product tested positive for glucose, while milk that was not treated was negative for glucose.
6. to eliminate the milk-treatment product as a possible source of glucose
7. Other disorders caused by enzyme deficiencies might be treated with products that contain the missing enzymes.
8. Answers may vary. The risk of lactose intolerance increases with age because lactase is required most during infancy and early childhood when milk is a major part of the diet.
9. Answers may vary. As an enzyme, lactase is a protein and proteins are encoded in DNA, the molecule responsible for inheritance. Certain populations have a higher incidence of lactose intolerance than others.
10. Answers will vary. For example: How would blood glucose levels compare in people eating treated dairy products with those eating untreated dairy products?

Do You Know?

Answers

1. Answers will vary. Students could list products such as bean-treatment products.
2. Stomach pain results from a buildup of gas, which is produced by bacteria as a waste product when breaking down food.

Procedure

Sample Procedure

1. Use a glucose strip to test water treated with 3 drops of milk-treatment product (as a negative control).
2. Use a glucose strip to test an untreated glucose solution (as a positive control).
3. Use a glucose strip to test the milk-treatment product itself (to verify the product does not contain glucose).
4. Use a glucose strip to test untreated milk and other untreated foods.
5. Use a glucose strip to test milk and other foods treated with the milk-treatment product.

Sample Data Table

Solution	Result (+ or -)	Interpretation
Water (treated with milk-treatment product)	–	negative control
Glucose (untreated)	+	positive control for the presence of glucose
Milk-treatment product	–	milk-treatment product itself does not contain glucose
Milk (untreated)	–	milk does not contain glucose
Milk (treated with milk-treatment product)	+	the lactose in milk is broken down into glucose and galactose

Chapter 41 The Body's Defenses

Planning Guide

OBJECTIVES	MEETING INDIVIDUAL NEEDS	READING SKILLS AND STRATEGIES
41-1 Nonspecific Defenses (BLOCK 1 45 min) • Describe how skin and mucous membranes defend the body. • Compare the inflammatory response with the temperature response. • Identify proteins that kill or inhibit pathogens. • Analyze the roles of white blood cells in combating pathogens.	Learning Styles, p. 932 BASIC ADVANCED • p. 934	ATE Active Reading, Technique: Paired Reading, p. 931 ■ Active Reading Guide, worksheet (41-1) ■ Directed Reading, worksheet (41-1)
41-2 Immune Response (BLOCKS 2 & 3 90 min) • List four kinds of immune-system cells, and describe their functions. • Describe how white blood cells recognize pathogens. • Identify the role of helper T cells in the immune response. • Compare the role of T cells with that of B cells in the immune response.	Learning Styles, p. 935 BASIC ADVANCED • p. 937	■ Active Reading Guide, worksheet (41-2) ■ Directed Reading, worksheet (41-2)
41-3 Disease Transmission and Prevention (BLOCKS 4 & 5 90 min) • List five ways diseases can be transmitted to humans. • Summarize Koch's postulates for identifying pathogens. • Analyze how the body produces immunity to pathogens. • Describe how vaccines produce immunity to pathogens.	Learning Styles, p. 938 BASIC ADVANCED • p. 940	■ Active Reading Guide, worksheet (41-3) ■ Directed Reading, worksheet (41-3)
41-4 Disorders of the Immune System (BLOCKS 6, 7 & 8 135 min) • Describe several autoimmune diseases. • Summarize how HIV disables the immune system. • List five ways HIV is transmitted. • Identify causes of an allergic reaction.	Learning Styles, p. 941 BASIC ADVANCED • p. 944	■ Active Reading Guide, worksheet (41-4) ■ Directed Reading, worksheet (41-4) ■ Supplemental Reading Guide, *Microbe Hunters*

Review and Assessment Resources

BLOCKS 9 & 10 90 min

TEST PREP

- PE Study Zone, p. 945
- PE Performance Zone, pp. 946–947
- ■ Chapter 41 worksheets:
  - Vocabulary
  - Concept Mapping
  - Critical Thinking
- Audio CD Program
- ■ Test Prep Pretest

CHAPTER TESTING

- ■ Chapter Tests (blackline master)
- ▲ Test Generator
- ■ Assessment Item Listing

See the correlations table at the front of this book for a list of the ***National Science Education Standards*** addressed in this chapter.

Smithsonian Institution® Visit **www.si.edu/hrw** for additional on-line resources.

KEY

Biology Principles and Explorations

PE Pupil's Edition
ATE Teacher's Edition

HOLT BIOSOURCES

Teaching Transparencies
Laboratory Program

One-Stop Planner CD-ROM
Shows these resources within a customizable daily lesson plan:

■ Teaching Resources
▲ Test Generator

VISUAL STRATEGIES	CLASSWORK AND HOMEWORK	TECHNOLOGY AND INTERNET RESOURCES
235 Inflammatory Response	**PE** **Review,** p. 934	Holt Biology Videodiscs Section 41-1 Audio CD Program Section 41-1 internet connect — SCILINKS NSTA — **TOPIC:** Immune systems
ATE Fig. 41-6, p. 936 236 Events of Immune Response 202A Immune System Response Time	Lab B32 Antigen-Antibody Interaction **PE** **Review,** p. 937	Holt Biology Videodiscs Section 41-2 Audio CD Program Section 41-2
237 How a Killer T Cell Recognizes an Infected Cell	**PE** **Quick Lab** How can you observe antigen activity?, p. 940 Lab B33 Transmission of a Communicable Disease **PE** **Review,** p. 940	CNN Video Segment 33 Salmonella Outbreak Holt Biology Videodiscs Section 41-3 Audio CD Program Section 41-3
238 Known Routes of HIV Transmission 203A AIDS Cases Among Young Adults	**PE** **Data Lab** Studying the spread of AIDS, p. 943 **PE** **Lab Techniques** Simulating Disease Transmission pp. 948–949 **PE** **Review,** p. 944 **PE** **Study Zone,** p. 945	Holt Biology Videodiscs Section 41-4 Audio CD Program Section 41-4 internet connect — SCILINKS NSTA — **TOPICS:** • HIV transmission • Asthma

ALTERNATIVE ASSESSMENT

PE **Portfolio Projects** 23–26, p. 947
ATE **Alternative Assessment,** p. 945
ATE **Portfolio Assessment,** p. 945

Lab Activity Materials

Quick Lab
p. 931 goggles, gloves, apron, blood-typing trays (2), simulated blood (AB and O), simulated anti-A and anti-B blood-typing serums, toothpicks (4)

Data Lab
p. 943 pencil, paper

Lab Techniques
p. 948 goggles, apron, gloves, bottle of unknown solution, test tube, indophenol indicator

Demonstrations
ATE photo of David, the "boy in the bubble," p. 930
ATE Petri dishes with nutrient agar (5), tape, swabs, p. 933

BIOSOURCES LAB PROGRAM
- Inquiry Skills Development: Lab B32, p. 153
- Inquiry Skills Development: Lab B33, p. 157

Chapter 41

Chapter Theme

Cellular Structure and Function

Cells of the immune system continually monitor the body for disease-causing agents such as pathogens and cancerous cells. How do immune system cells distinguish invaders from normal body cells? Immune-system cells have protein receptors that recognize protein markers on normal body cells. The surface proteins of pathogens differ from body cells, however, allowing the immune system to recognize and attack them.

Establishing Prior Knowledge

Before beginning this chapter, make certain students can answer the questions in **Study TIP Ready?** on page 931. If they cannot, encourage them to reread the sections indicated.

Opening Demonstration

To illustrate our dependence on a functional immune system, show students a picture of David, the "boy in the bubble." David was born with severe combined immune deficiency (SCID), a disease characterized by the inability to produce B cells and T cells, which play key roles in the immune system. SCID patients must live in a protected environment because the slightest infection can lead to death. David died at the age of 12 after an unsuccessful bone marrow transplant.

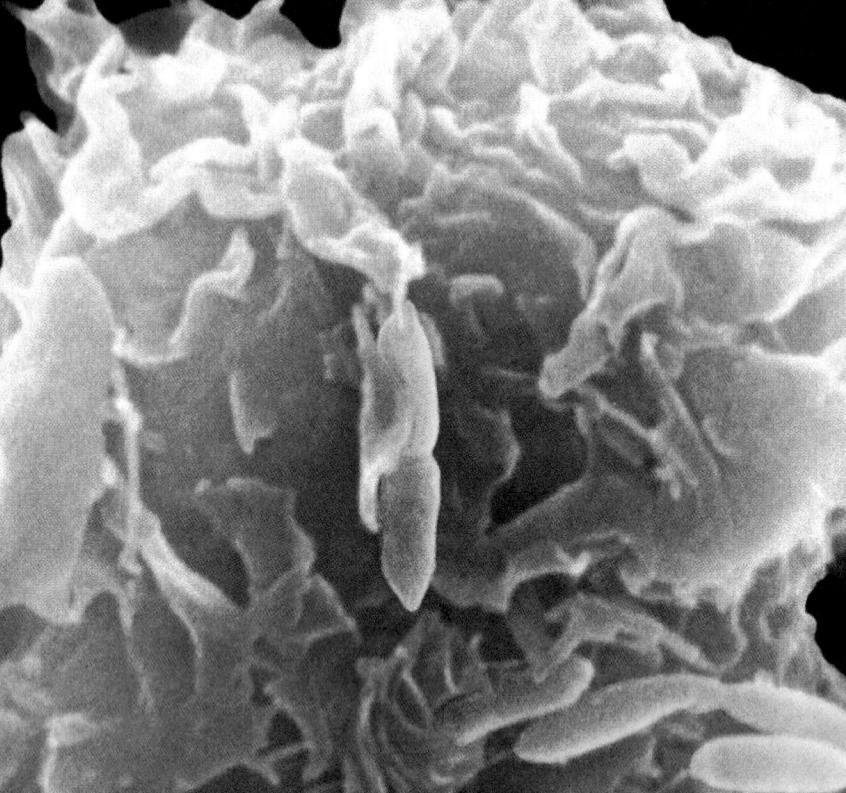

CHAPTER 41 The Body's Defenses

White blood cell attacking bacteria (6,480×)

What's Ahead — and Why

CHAPTER OUTLINE

41-1 Nonspecific Defenses
- Nonspecific Defenses Protect Against Infection

41-2 Immune Response
- The Third Line of Defenses Is Specific

41-3 Disease Transmission and Prevention
- Diseases Are Transmitted in Various Ways

41-4 Disorders of the Immune System
- The Immune System Sometimes Attacks Body Tissues
- HIV Infection Leads to Immune System Collapse
- Some Antigens Can Cause Allergic Reactions

Students will learn

- how the skin and mucous membranes reduce the chances of infection.
- how B and T cells locate and destroy pathogens.
- how diseases are detected and how vaccines work.
- routes of HIV transmission and why allergic reactions take place.

Disease prevention

Communicable diseases

like the common cold and the flu often cause people to miss work. In many Japanese cities, people have begun to wear masks when they contract a communicable disease. The masks help prevent the transmission of such diseases. Those who are infected can go out in public while minimizing the spread of disease.

Looking Ahead

Study TIP

Ready?

Check your understanding of these topics to help you prepare for what's ahead.

Can you . . .

- **identify** the role of receptor proteins in cellular communication? (Section 4-2)
- **recognize** the relationship between HIV and AIDS? (Section 21-1)
- **differentiate** between antibodies and antigens? (Section 39-1)

If not, *review the sections indicated.*

internet**connect**

*sci*LINKS NSTA

National Science Teachers Association *sci*LINKS Internet resources are located throughout this chapter.

Reading Strategies

Active Reading

Technique: Paired Reading

Pair each student with a partner. Have each student identify paragraphs or pages with topics or concepts they find difficult to understand. One student should read the paragraph or section aloud, and then the other student should repeat the information in his or her own words. Both students should agree on the context of the second student's response.

Directed Reading

Assign **Directed Reading Worksheet Chapter 41** to help students interact with the text as they read the chapter.

Interactive Reading

Assign ***Biology: Principles and Explorations*** **Audio CD Program Chapter 41** to help reluctant readers achieve greater success in reading the chapter.

Reading Skills

Assign Reading Organizers and Reading Strategies Worksheets from the **One-Stop Planner CD-ROM with Test Generator** to help students learn and practice effective reading skills.

internet**connect**

Holt *Biology: Principles and Explorations*

On-line Resources: go.hrw.com

Visit the HRW Web site for a variety of resources related to this chapter. Just type in the keyword HX0 Chapter 41.

Holt *Biology: Principles and Explorations*

On-line Resources: www.scilinks.org

The following *sci*LINKS Internet resources can be found in the student text for this chapter:

Page 933
TOPIC: Immune systems
KEYWORD: HX933

Page 943
TOPIC: HIV transmission
KEYWORD: HX943

Page 944
TOPIC: Asthma
KEYWORD: HX944

Chapter 41

1 Prepare

Scheduling

- **Block:** About 45 minutes will be needed to complete this section.
- **Traditional:** About one class period will be needed to complete this section.
- **Planning:** See the Planning Guide on pp. 929A and 929B for lecture, classwork, and assignment options for Section 41-1. Lesson plans for Section 41-1 can be found in a lesson cycle format in ***Biology: Principles and Explorations*** **One-Stop Planner CD-ROM with Test Generator.**

Resource Link

Assign **Active Reading Worksheet Section 41-1** to help students comprehend the key concepts and visuals in this lesson.

2 Teach

Lesson Warm-up

A recent public health information campaign has combined classical literature with messages about washing hands. Ask students why it is so important for people to wash their hands frequently throughout the day. *(to reduce the spread of pathogens)*

Section 41-1 Nonspecific Defenses

Objectives

- **Describe** how skin and mucous membranes defend the body. (p. 932)
- **Compare** the inflammatory response with the temperature response. (p. 933)
- **Identify** proteins that kill or inhibit pathogens. (p. 934)
- **Analyze** the roles of white blood cells in combating pathogens. (p. 934)

Key Terms

pathogen
mucous membrane
inflammatory response
histamine
complement system
interferon
neutrophil
macrophage
natural killer cell

Nonspecific Defenses Protect Against Infection

Some animals, including turtles, clams, and armadillos, defend themselves with their hard armor shells. However, even armor will not protect against the most dangerous enemies the human body faces—harmful bacteria, viruses, fungi, and protists. You survive because your body's immune system defends against these pathogens. A **pathogen** is a disease-causing agent. The immune system consists of cells and tissues found throughout the body. The body uses both nonspecific and specific defense mechanisms to prevent infection and to detect and destroy pathogens.

First line of nonspecific defenses

The body's surface defenses are nonspecific, meaning they do not target specific pathogens. Your skin is the first of your immune system's nonspecific defenses against pathogens. Skin acts as a nearly impenetrable barrier to invading pathogens, keeping them outside the body. This barrier is reinforced with chemical weapons. Oil and sweat make the skin's surface acidic, inhibiting the growth of many pathogens. Sweat also contains the enzyme lysozyme, which digests bacterial cell walls.

Internal surfaces of the body through which pathogens can pass are covered by mucous membranes. **Mucous** *(MYOO kuhs)* **membranes** are layers of epithelial tissue that produce a sticky, viscous fluid called mucus. Mucous membranes line the digestive system, nasal passages, lungs, respiratory passages, and the reproductive tract. Like the skin, mucous membranes serve as a barrier to pathogens and produce chemical defenses. Cells lining the bronchi and bronchioles in the respiratory tract secrete a layer of mucus that traps pathogens before they can reach the warm, moist lungs, which are an ideal breeding ground for microorganisms. Cilia on cells of the respiratory tract, shown in Figure 41-1, continually sweep mucus toward the opening of the esophagus. Mucus then can be swallowed, sending pathogens to the stomach, where they are digested by acids and enzymes.

Skin and mucous membranes work to prevent any pathogens from entering the body. Occasionally these defenses are penetrated. You take pathogens into your body when you breathe, because many microbes and microbial spores are suspended in the air. Other pathogens may be present in the food you eat. Pathogens can also enter through wounds or open sores. When invaders reach deeper tissue, a second line of nonspecific defenses takes over.

Figure 41-1 Cilia in the lungs

Size It would take about 20 cilia (green) lined up end to end to extend from the top of Lincoln's head to the edge of the penny.

Integrating Different Learning Styles

Visual/Spatial	**ATE**	Opening Demonstration, p. 930; Demonstration, p. 933; Reteaching, p. 934
Interpersonal	**ATE**	Reteaching, p. 934
Verbal/Linguistic	**ATE**	Internet Connect, p. 933; Reteaching, p. 934

Second line of nonspecific defenses

What happens when pathogens break through your body's first line of defense? When the body is invaded, four important nonspecific defenses take action: the inflammatory response; the temperature response; proteins that kill or inhibit pathogens; and white blood cells, which attack and kill pathogens.

internet**connect**
SCI LINKS NSTA
TOPIC: Immune systems
GO TO: www.scilinks.org
KEYWORD: HX933

Inflammatory response Injury or local infection, such as a cut or a scrape, causes an inflammatory response. An **inflammatory response** is a series of events that suppress infection and speed recovery. Imagine that a splinter has punctured your finger, creating an entrance for pathogens, as shown in **Figure 41-2.** Infected or injured cells in your finger release chemicals, including histamine. **Histamine** *(HIHST uh meen)* causes local blood vessels to dilate, increasing blood flow to the area. Increased blood flow brings white blood cells to the infection site, where they can attack pathogens. This also causes swelling and redness in the infected area. The whitish liquid, or pus, associated with some infections contains white blood cells, dead cells, and dead pathogens.

Temperature response When the body begins its fight against pathogens, body temperature increases several degrees above the normal value of about 37°C (99°F). This higher temperature is called a fever, and it is a common symptom of illness that shows the body is responding to an infection. Fever is helpful because many disease-causing bacteria do not grow well at high temperatures. Although fever may slow the growth of bacteria, very high fever is dangerous because extreme heat can destroy important cellular enzymes. Temperatures greater than 39°C (103°F) are considered dangerous, and those greater than 41°C (105°F) can be fatal.

Figure 41-2 Inflammatory response

When pathogens penetrate your body, an inflammatory response is triggered.

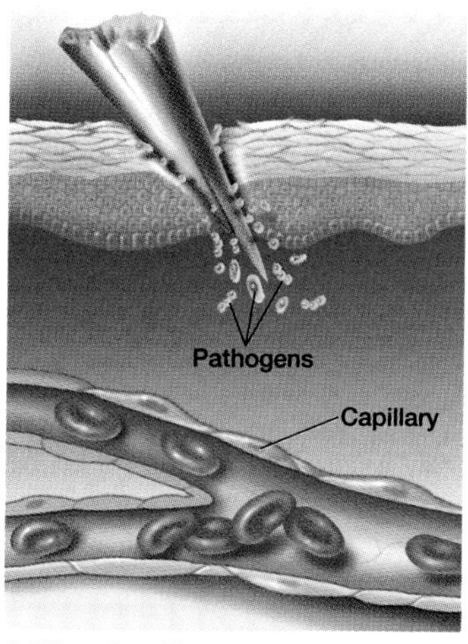

1. When the skin is punctured, pathogens enter the body.

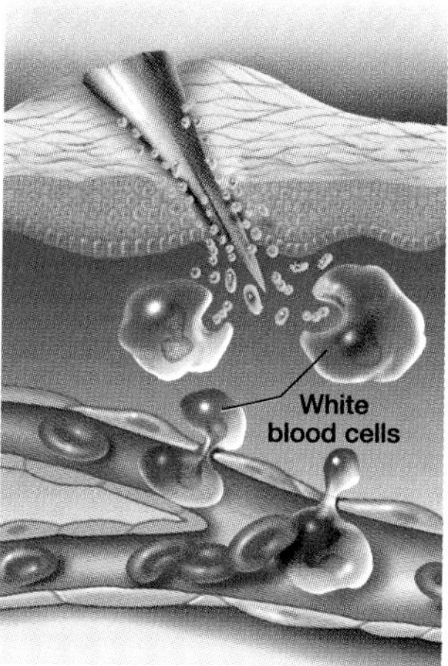

2. Blood flow to the area increases, causing swelling and redness.

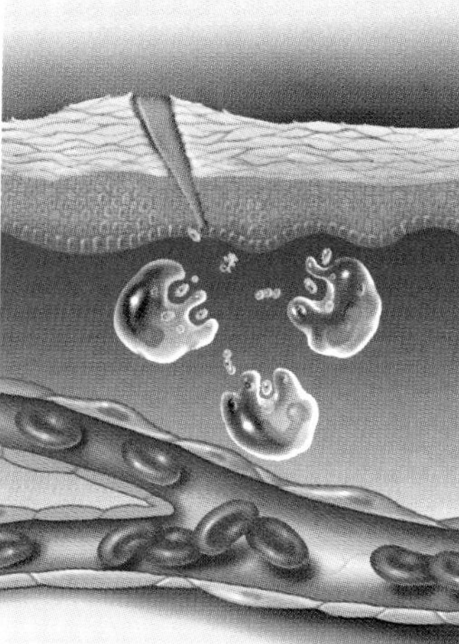

3. White blood cells attack and destroy the pathogens.

did you know?

Some of the body's protection from pathogens comes from symbiotic bacteria.

For example, in the vagina, beneficial bacteria inhibit the growth of harmful bacteria and fungi. Use of antibiotics to kill pathogens often kills the body's beneficial bacteria as well, which can lead to vaginal and urinary tract infections.

SCI LINKS NSTA Have students research the immune system using the Web site in the Internet Connect box on page 933. Have students write a report summarizing their findings. PORTFOLIO

Demonstration

Microorganisms Around Us

To demonstrate that microorganisms are ubiquitous, gather five sterile Petri dishes filled with sterile nutrient agar. Take samples from five different places in the classroom. Collect microorganisms from surfaces such as desktops and doorknobs by sweeping them with sterile swabs moistened with sterile distilled water. Seal each Petri dish with masking tape, label each Petri dish, and place all five upside down in a warm, dark place or in an incubator for several days. At the end of this period, show students the colonies of microorganisms that have grown on the agar. Point out that most of the microorganisms growing on the dishes probably are harmless but that some may be pathogens.

CAUTION: Do not allow students to handle Petri dishes—they might contain pathogens. Dispose of swabs and Petri dishes after sterilizing them in an autoclave or soaking them for 30 minutes in chlorine bleach.

Teaching TIP

Mosquito Bites

Mosquito bites produce an inflammatory response. Before drawing blood, the mosquito injects a small amount of saliva which contains a substance that keeps the victim's blood from clotting. This substance initiates the inflammatory response, making the bite site swell, turn red, and itch.

Chapter 41

Teaching TIP

Immune Cells

Lymphocytes, such as cytotoxic T cells and natural killer cells, are very similar in shape. Natural killer cells lyse (break up) virus-infected cells or tumor cells. Cytotoxic T cells are more restricted in their actions; they must recognize a specific antigen (a foreign molecule) before they act.

3 Assessment Options

Closure

Smoking Damage

Smoking not only introduces carcinogens into the lungs but also damages and destroys the cilia lining the respiratory tract. Without the continual sweeping function of the cilia, mucus accumulates in the respiratory passages and stimulates coughing—the "smoker's cough." Ask students why smokers are more likely to develop respiratory infections than are nonsmokers. *(In a person who does not have normally functioning cilia, pathogens can more readily reach the lungs.)*

Review

Assign Review—Section 41-1.

Reteaching — BASIC

Assign students to cooperative groups of three. Have each group write a comic sketch or draw cartoons to show how the first line of defense operates in protecting the body against pathogens. *(Sketches and drawings may depict the skin keeping out invaders, pathogens sticking to mucous membranes, phagocytes engulfing bacteria, and aspects of the inflammatory response.)* PORTFOLIO

Magnification: 2,280×

Figure 41-3 Macrophage. Cytoplasmic extensions of this macrophage (yellow) are capturing bacteria (blue).

Magnification: 14,250×

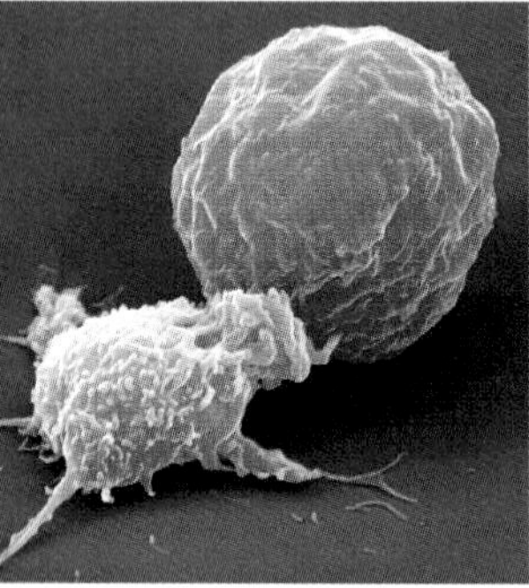

Figure 41-4 Natural killer cell. This natural killer cell (yellow) is attacking a cancer cell (pink).

Proteins Various proteins also provide nonspecific defenses. One defense mechanism, called the **complement system,** consists of about 20 different proteins. Complement proteins circulate in the blood and become active when they encounter certain pathogens. Then some of these proteins form a membrane attack complex (MAC), a ring-shaped structure that ruptures the cell membrane of pathogens. A MAC punches a hole in the cell membrane, causing the cell to leak and eventually die. Another nonspecific defense is **interferon** *(in tuhr FEER ahn),* a protein released by cells infected with viruses. Interferon causes nearby cells to produce an enzyme that prevents viruses from making proteins and RNA.

White blood cells The most important counterattacks in the second line of nonspecific defenses are carried out by three kinds of white blood cells: neutrophils, macrophages, and natural killer cells. These cells patrol the bloodstream, wait within the tissues for pathogens, and then attack the pathogens. Each kind of cell uses a different mechanism to kill pathogens.

1. **Neutrophils.** A **neutrophil** *(NOO truh fihl)* is a white blood cell that engulfs and destroys pathogens. The most abundant type of white blood cell, neutrophils sacrifice themselves to defend the body. Neutrophils engulf bacteria and then release chemicals that kill the bacteria—and themselves. Neutrophils can also squeeze between cells in the walls of capillaries to attack pathogens at the site of an infection.
2. **Macrophages.** White blood cells called **macrophages** *(MA kroh fay jez),* shown in **Figure 41-3,** ingest and kill pathogens they encounter. They also clear dead cells and other debris from the body. Most macrophages travel through the body in blood, lymph, and fluid between cells. Macrophages are concentrated in particular organs, especially the spleen and lungs.
3. **Natural killer cells.** A **natural killer cell** is a large white blood cell that attacks cells infected with pathogens. Natural killer cells destroy an infected cell by puncturing its cell membrane. Water then rushes into the infected cell, causing the cell to swell and burst. One of the body's best defenses against cancer, natural killer cells can detect and kill cancer cells before a tumor can develop, as shown in **Figure 41-4.** Natural killer cells also attack body cells infected with viruses.

Review SECTION 41-1

1. **Summarize** how skin and mucous membranes help protect the body against infection.
2. **Describe** how the inflammatory and temperature responses help defend against infection.
3. **Identify** the role of white blood cells in the second line of nonspecific defenses.
4. **SKILL Comparing Functions** How is the action of interferon different from the action of the complement system?

Critical Thinking

5. **SKILL Relating Concepts** Explain why taking a drug that reduces fever might slow rather than hasten your recovery from an infection.

Review SECTION 41-1 ANSWERS

1. Skin presents a physical barrier, and its surface acids and oils provide a hostile environment to pathogens. Mucous membranes also are physical barriers; mucus traps and destroys pathogens, and cilia sweep them out of the body.
2. The inflammatory response increases blood flow, which brings white blood cells to an injury site. The temperature defense raises tissue temperature above that at which most pathogens grow well.
3. White blood cells kill pathogens in different ways. Neutrophils engulf and destroy pathogens. Macrophages ingest and kill pathogens and rid the body of dead cells and other debris. Natural killer cells puncture the cell membrane of infected cells.
4. The complement system ruptures the cell membrane of pathogens. Interferon causes cells to manufacture enzymes that prevent viruses from making proteins and RNA.
5. The drug would short-circuit the temperature response, reducing body temperature to a range at which the pathogen could thrive.

Immune Response

Section 41-2

The Third Line of Defenses Is Specific

What happens when pathogens occasionally overwhelm your body's nonspecific defenses? Pathogens that have survived the first and second lines of nonspecific defenses still face a third line of specific defenses—the immune response. The immune response consists of an army of individual cells that rush throughout the body to combat specific invading pathogens. The immune response is not localized in the body, nor is it controlled by a single organ. It is more difficult to evade than the nonspecific defenses.

Cells involved in the immune response

White blood cells are produced in bone marrow and circulate in blood and lymph. Of the 100 trillion cells in your body, about 2 trillion are white blood cells. Four main kinds of white blood cells participate in the immune response: macrophages, cytotoxic T cells, B cells, and helper T cells. Each kind of cell has a different function. Macrophages consume pathogens and infected cells. **Cytotoxic** *(seye toh TAHKS ihk)* **T cells** attack and kill infected cells. **B cells** label invaders for later destruction by macrophages. **Helper T cells** activate both cytotoxic T cells and B cells. These four kinds of white blood cells interact to remove pathogens from the body.

Recognizing invaders

To understand how the third line of defenses works, imagine that you have just come down with the flu. You have inhaled influenza viruses, but they were not trapped by mucus in the respiratory tract. The viruses have begun to infect and kill your cells. At this point, macrophages begin to engulf and destroy the viruses.

An infected body cell will display antigens of an invader on its surface. An **antigen** *(AN tih jihn)* is a substance that triggers an immune response. Antigens typically include proteins and other components of viruses or pathogen cells present on the cell surface. White blood cells of the immune system are covered with receptor proteins that respond to infection by binding to specific antigens on the surfaces of the infecting microbes. These receptors recognize and bind to antigens that match their particular shape, as shown in **Figure 41-5.**

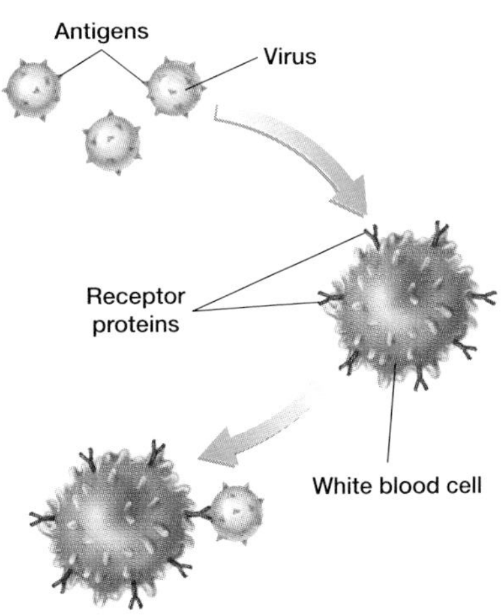

Figure 41-5 Antigens. White blood cells have receptor proteins that bind to specific antigens.

Objectives

- **List** four kinds of immune-system cells, and describe their functions. (p. 935)
- **Describe** how white blood cells recognize pathogens. (p. 935)
- **Identify** the role of helper T cells in the immune response. (pp. 936–937)
- **Compare** the role of T cells with that of B cells in the immune response. (pp. 936–937)

Key Terms

cytotoxic T cell
B cell
helper T cell
antigen
plasma cell
antibody

Section 41-2

1 Prepare

Scheduling

- **Block:** About 90 minutes will be needed to complete this section.
- **Traditional:** About two class periods will be needed to complete this section.
- **Planning:** See the Planning Guide on pp. 929A and 929B for lecture, classwork, and assignment options for Section 41-2. Lesson plans for Section 41-2 can be found in a lesson cycle format in ***Biology: Principles and Explorations*** **One-Stop Planner CD-ROM with Test Generator.**

Resource Link

Assign **Active Reading Worksheet Section 41-2** to help students comprehend the key concepts and visuals in this lesson.

2 Teach

Lesson Warm-up

Show the class an artist's rendering of viruses bursting out of an infected cell. Point out that viruses kill their host cell. Students should recognize that in viral infections many cells in the body are infected and destroyed, making the destructive effects of the virus widespread. Students should also understand that each infected cell produces many new virus particles, each of which can go on to infect other cells, amplifying the damaging effects of the virus.

Integrating Different Learning Styles

Visual/Spatial	**PE**	Performance Zone item 1, p. 946
	ATE	Lesson Warm-up, p. 935; Teaching Tip, p. 936; Visual Strategy, p. 936
Interpersonal	**ATE**	Reteaching, p. 937
Verbal/Linguistic	**ATE**	Closure, p. 937; Reteaching, p. 937

Chapter 41

Teaching TIP

B-Cell Production

When the influenza virus invades the body, it stimulates the division of antibody-producing B cells. The antibodies produced by these B cells match the shape of the proteins on the surface of the virus. This correspondence of shape occurs because only cells with antibodies matching the shape of the virus are activated. The body makes millions of types of B cells, and each type makes only one uniquely shaped type of antibody.

Teaching TIP

Immune Cells

Have students create a Graphic Organizer similar to the one at the bottom of page 936. Along the top of the table they should list the following categories: *Type of cell, Function,* and *Location in the body.* In the first column, they should list all the types of cells that play a role in the body's defenses: *macrophage, neutrophil, natural killer cell, helper T cell, cytotoxic T cell, B cell, plasma cell,* and *memory cell.* Then ask students to complete the information in the second and third columns. PORTFOLIO

Visual Strategy

Figure 41-6

BIO graphic Once you have discussed this figure, have students answer the following questions: What types of white blood cells are involved in the immune response? *(helper T, cytotoxic T, and B cells)* What type of cell is involved in a passive, humoral defense? *(B cell)* What type of cell is active and destroys pathogens? *(cytotoxic T cell)* What is the role of helper T cells? *(They regulate the T- and B-cell actions.)*

Study TIP

Reading Effectively

Antigens trigger an immune response. Remember that an antigen is an *anti*body-*gen*erating substance.

The immune response has two main parts

Two distinct processes work together in an immune response. One is the B cell response, a passive, humoral defense that aids the removal of pathogens from the body. The other is the T cell response, an active, cell-mediated defense that involves the destruction of pathogens by cytotoxic T cells. Both the T cell response and the B cell response are regulated by helper T cells. Both responses, which happen simultaneously, are summarized in **Figure 41-6.**

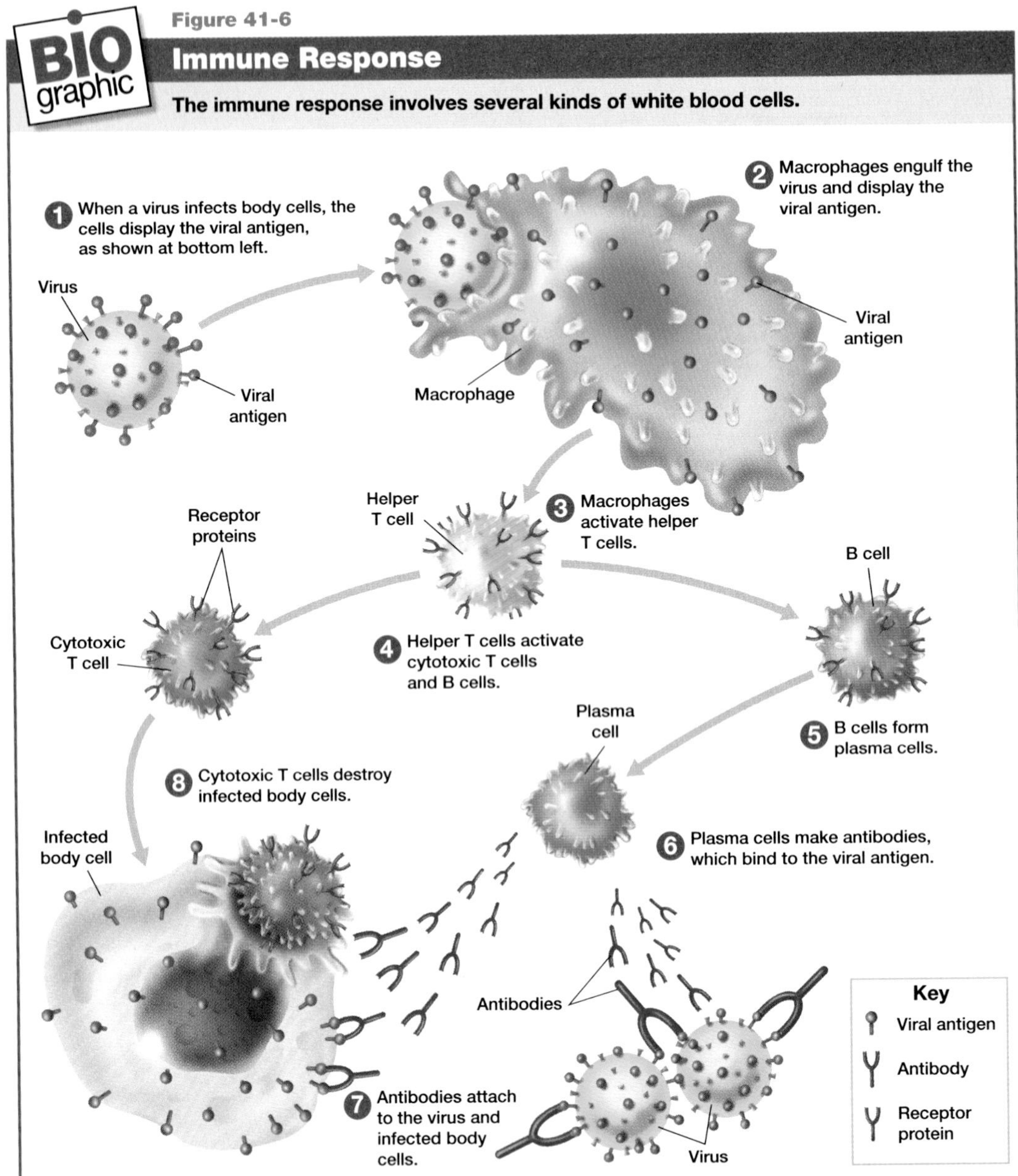
BIO graphic
Figure 41-6
Immune Response
The immune response involves several kinds of white blood cells.
1 When a virus infects body cells, the cells display the viral antigen, as shown at bottom left.
Virus
Viral antigen
2 Macrophages engulf the virus and display the viral antigen.
Viral antigen
Macrophage
Helper T cell
3 Macrophages activate helper T cells.
Receptor proteins
B cell
Cytotoxic T cell
4 Helper T cells activate cytotoxic T cells and B cells.
Plasma cell
5 B cells form plasma cells.
8 Cytotoxic T cells destroy infected body cells.
Infected body cell
6 Plasma cells make antibodies, which bind to the viral antigen.
Antibodies
7 Antibodies attach to the virus and infected body cells.
Virus
Key
Viral antigen
Antibody
Receptor protein

Graphic Organizer

Use this graphic organizer with **Teaching Tip: Immune Cells** on page 936.

Type of cell	Function	Location in the body
Macrophage	Ingests and kills pathogens	Spleen, lungs, blood, lymph, interstitial fluids
Neutrophil	Engulfs and destroys pathogens	Bloodstream, infection sites
Natural killer cell	Punctures infected cells	Infected cells
Helper T cell	Activates cytotoxic T and B cells	Bloodstream, infection sites
Cytotoxic T cell	Punctures infected cells	Infection sites
B cell	Labels invaders for destruction by macrophages	Infection sites
Plasma cell	Releases antibodies	Bloodstream
Memory cell	Protects against defeated pathogens	Bloodstream

Step 1 When a virus infects body cells, the infected cells display the viral antigen on their surface.

Step 2 Macrophages engulf the virus and display the viral antigen on their surface.

Step 3 Receptor proteins on helper T cells bind to the viral antigen displayed by the macrophages. The macrophages release a protein called interleukin-1 *(ihn tuhr LOO kihn).*

Step 4 Interleukin-1 activates helper T cells, but helper T cells do not attack pathogens directly. Instead, helper T cells activate cytotoxic T cells and B cells. Stimulation by interleukin-1 causes helper T cells to release interleukin-2. Interleukin-2 stimulates further division of helper T cells and cytotoxic T cells, amplifying the body's response to the infection.

Step 5 Interleukin-2 released by helper T cells also activates B cells. When activated by interleukin-2, B cells divide and develop into plasma cells. **Plasma cells** are cells that release special defensive proteins into the blood. These specialized proteins are called antibodies. An **antibody** is a Y-shaped molecule that is produced by plasma cells upon exposure to a specific antigen and that can bind to that antigen.

Step 6 Plasma cells divide repeatedly and make large amounts of antibodies. Plasma cells either release antibodies into the bloodstream or attach them directly to the virus. Antibodies bind to the viral antigen on the virus and on infected cells. Antibodies mark the virus and infected cells for destruction.

Step 7 When a plasma cell encounters a virus with an antigen that matches its antibodies, it binds to the virus. This causes other viruses to stick together, forming a clump that can be easily identified and destroyed by macrophages.

Step 8 With the help of antibodies and plasma cells, cytotoxic T cells destroy infected cells by puncturing the cell membrane of the infected cells. How do cytotoxic T cells recognize antigens? Your body makes millions of different T cells, each with receptor proteins that bind to a specific antigen. Receptor proteins on cytotoxic T cells bind to the viral antigen displayed by infected cells. For example, any of your body's cells that bear traces of an influenza virus will be destroyed by cytotoxic T cells with receptor proteins that bind to the antigen of that virus.

REAL LIFE

How often do you wash your hands? Hand washing is an effective way to prevent the spread of disease. A recent study found that only 68 percent of adults wash their hands after using public restrooms.

Inferring Relationships
Why is it crucial that food preparers wash their hands after using a restroom?

Review SECTION 41-2

1. **List** the different kinds of white blood cells involved in the immune response.
2. **Describe** how white blood cells recognize and bind to pathogens.
3. **Compare** the roles of B cells and T cells in the immune response.
4. **SKILL Recognizing Relationships** Explain the role of helper T cells in the immune response.

Critical Thinking

5. **SKILL Predicting Outcomes** How would an enzyme that destroys interleukins affect the immune response?

Review SECTION 41-2 ANSWERS

1. macrophages, cytotoxic T cells, B cells, helper T cells
2. White blood cells recognize invaders because infected body cells display antigens on their surface. White blood cells are covered with receptors that bind to these antigens.
3. B cells become plasma cells that release antibodies that bind to antigens to mark them for destruction by macrophages. T cells are stimulated by interleuken-1 to produce interleuken-2, which activates helper T cells and cytotoxic T cells that puncture cell membranes.
4. Helper T cells bind to viral antigens on macrophage surfaces, which stimulates the release of interleukin-1 by macrophages. Helper T cells also activate B cells and cytotoxic T cells.
5. Destroying interleukens would short-circuit the immune response.

Teaching TIP

Negative Feedback

Interleuken-2 released by helper T cells also stimulates the division of another type of T cell, the suppressor T cell. These cells divide slowly and become common a week or more after infection. They gradually shut down the immune response by inhibiting the production of B cells and cytotoxic T cells.

Making Biology Relevant

Health

Monoclonal antibodies are produced by cloning a single cell that has been obtained by fusing a B cell with a tumor cell. Known as a *hybridoma,* this new cell has properties of both parents—it divides forever (like a cancer cell), and it produces a specific antibody (like a B cell). The use of monoclonal antibodies is currently being studied as a way of focusing a massive attack on a particular pathogen and perhaps even killing cancer cells without harming normal cells.

3 Assessment Options

Closure

Immune Competence

Ask students to name all of the infectious diseases they can. Then ask why people get fewer infectious diseases as they age. *(The immune system builds up a repertoire of specific defenses over exposures to various pathogens.)*

Review

Assign Review—Section 41-2.

Reteaching — BASIC

Have students review the function of T cells in the immune response. Have them make index cards with words and drawings depicting T-cell function. Have students pair off. One student should shuffle the cards, and the other student should put them in the correct order.

Chapter 41

1 Prepare

Scheduling

- **Block:** About 90 minutes will be needed to complete this section.
- **Traditional:** About two class periods will be needed to complete this section.
- **Planning:** See the Planning Guide on pp. 929A and 929B for lecture, classwork, and assignment options for Section 41-3. Lesson plans for Section 41-3 can be found in a lesson cycle format in ***Biology: Principles and Explorations*** **One-Stop Planner CD-ROM with Test Generator.**

Resource Link

Assign **Active Reading Worksheet Section 41-3** to help students comprehend the key concepts and visuals in this lesson.

2 Teach

Lesson Warm-up

Instruct students to find and bring in a copy of their vaccination records if they are readily available. Ask them why it is important to maintain such records. Ask students if they know about other less common vaccinations such as those that protect against anthrax, typhoid fever, yellow fever, and so on.

Section 41-3

Disease Transmission and Prevention

Objectives

- **List** five ways diseases can be transmitted to humans. (p. 938)
- **Summarize** Koch's postulates for identifying pathogens. (p. 938)
- **Analyze** how the body produces immunity to pathogens. (p. 939)
- **Describe** how vaccines produce immunity to pathogens. (p. 939)

Key Terms

Koch's postulates
immunity
vaccination
vaccine
antigen shifting

Diseases Are Transmitted in Various Ways

In general, you can get infectious diseases in any of five different ways: through person-to-person contact, air, food, water, and animal bites. Diseases transferred from person to person are considered contagious, or communicable. For example, when a person sneezes, droplets of saliva and mucus carrying pathogens are expelled from the mouth and nose, as shown in **Figure 41-7.** If another person breathes these droplets, the pathogens can infect that person. People directly transmit some diseases by kissing, shaking hands, touching open wounds or sores, or having sexual contact. People can also transmit diseases indirectly through objects contaminated with pathogens, such as drinking glasses and toilets.

By minimizing exposure to pathogens, you can decrease your chances of becoming ill. For example, to prevent illnesses caused by bacteria found in potentially hazardous foods, such as meat, poultry, and eggs, these foods should always be cooked thoroughly. Utensils and other surfaces that foods touch should be sanitized.

Detecting disease

The German physician Robert Koch (1843–1910) established a procedure for diagnosing causes of infection. Koch determined that bacteria cause anthrax, a disease that afflicts cattle, sheep, and goats, and can be spread to humans. Anthrax is a serious disease threat whenever a case occurs. In an experiment, Koch isolated bacteria from a cow with anthrax and then infected a healthy cow with the bacteria. The healthy cow developed anthrax and had the same bacteria that the first cow had. In his research, Koch developed the following four-step procedure, known as **Koch's postulates,** as a guide for identifying specific pathogens. Biologists have used Koch's postulates to identify many pathogens.

Figure 41-7 Disease transmission. When a person sneezes, pathogens are expelled from the mouth and nose.

1. The pathogen must be found in an animal with the disease and not in a healthy animal.
2. The pathogen must be isolated from the sick animal and grown in a laboratory culture.
3. When the isolated pathogen is injected into the healthy animal, the animal must develop the disease.
4. The pathogen should be taken from the second animal and grown in a laboratory culture. The pathogen should be the same as the original pathogen.

Integrating Different Learning Styles

Logical/Mathematical	**PE**	Quick Lab, p. 940
Interpersonal	**ATE**	Reteaching, p. 940
Verbal/Linguistic	**ATE**	Lesson Warm-up, p. 938; Closure, p. 940; Reteaching, p. 940

Long-term protection

The immune response is very powerful, and it can be a long-lasting defense. After an immune response, some B cells and T cells become memory cells that continue to patrol your body's tissues. Some memory cells provide lifelong protection against previously encountered pathogens. If a pathogen ever appears again, memory cells activate antibody production against that pathogen. As shown in **Figure 41-8,** antibodies help macrophages destroy the pathogen before you become ill. You are not even aware of the battle going on in your body and are said to be "immune," or resistant, to the pathogen or to disease caused by that pathogen.

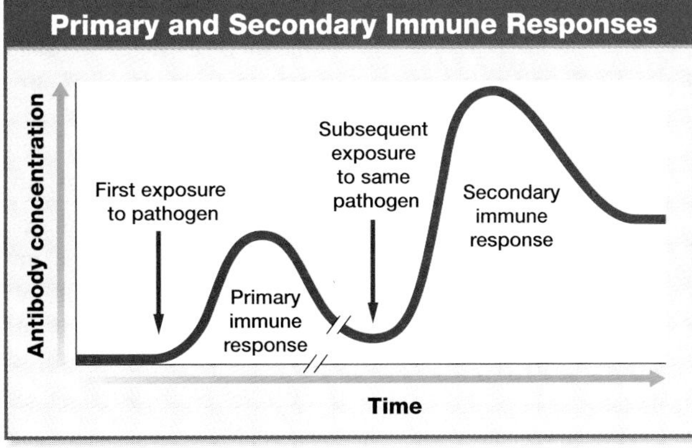

Figure 41-8 Immune responses. The first time you are exposed to a pathogen, your immune system responds normally. If you become exposed to the same pathogen again, antibody production increases quickly.

Resistance to disease

Resistance to a particular disease is called **immunity.** It has long been observed that individuals who recover from an infectious disease develop an immunity to that disease. This knowledge preceded the development of immunology, a branch of science that deals with antigens, antibodies, and immunity. Immunologists study the body's defenses and ways to help protect against disease.

In 1796, an English doctor named Edward Jenner, shown in **Figure 41-9,** performed an experiment that marks the beginning of immunology. Smallpox, which is caused by a virus, was a common and deadly disease then. Jenner observed that milkmaids who had contracted cowpox, a mild form of smallpox, rarely became infected with smallpox. Jenner hypothesized that cowpox produced protection against smallpox. To test his hypothesis, Jenner infected healthy people with cowpox. As Jenner had predicted, many of the people he infected never developed smallpox, even though they had been exposed to the virus. We now know that smallpox and cowpox are caused by two similar viruses. The cowpox infection caused an immune response that later prevented smallpox infection in Jenner's patients.

Vaccination Jenner's procedure of injecting the cowpox virus to produce resistance to smallpox is called vaccination. **Vaccination** *(vak sih NAY shuhn)* is a medical procedure used to produce immunity. You have probably been to the doctor for vaccination to guard against various diseases. Modern vaccination involves an injection, or "shot," of a vaccine under the skin using a hypodermic needle. A **vaccine** *(vak SEEN)* is a solution that contains a dead or modified pathogen that can no longer cause disease.

A vaccine triggers an immune response against the pathogen without symptoms of infection. For several days

Born in England — Discovers procedure of vaccination. — Dies

Figure 41-9 Edward Jenner

Section 41-3

Teaching TIP

Evolution

Changes in the surface proteins of a virus can prevent the immune system from recognizing the virus as one that it has encountered before. These small changes allow many viruses, such as cold and influenza viruses, to reinfect individuals many times with slightly different versions of the virus.

Teaching TIP

Vaccination

Emphasize that vaccination is effective because it exposes the body to the surface proteins of a pathogen that has been stripped of its disease-causing ability. AIDS and influenza are among the diseases that have defeated attempts to produce vaccines. The viruses that cause these diseases have rapidly evolving surface proteins. Ask students why such proteins make vaccination development difficult. *(The immune system only produces defenses against those surface proteins to which it was exposed by the vaccine. The immune system probably will not recognize the altered proteins of a mutant pathogen.)*

CROSS-DISCIPLINARY CONNECTION

◆History

Have students research the contributions of the English physician Edward Jenner to the development of vaccination. Students should write a report summarizing what they have learned.
PORTFOLIO

did you know?

AIDS is just one of many diseases that has appeared in the last few decades.

This table lists a few examples of new diseases and their place and time of appearance.

Disease	Cause	Origin and Time
Lassa fever	Virus	Western Africa, 1969
Lyme disease	Bacterium	Eastern U.S., early 1970s
Ebola fever	Virus	Central Africa, 1976
Hantavirus ARDS	Virus	Southwestern U.S., 1993

Chapter 41

How can you observe antigen activity?

QUICK LAB

Time 15–20 minutes

Process Skills Inferring relationships, predicting outcomes

Teaching Strategies
Review the difference between antigens and antibodies before beginning the lab.

Analysis Answers
1. AB
2. Antibodies are bound to antigens on the simulated blood cells.
3. Type A would clot only with anti-A serum.
   Type B would clot only with anti-B serum.

3 Assessment Options

Closure

Disease Transmission
Ask students if all infectious diseases have outward signs. *(no)* Ask if diseases with long incubation periods (the length of time between exposure and the start of symptoms) are more or less likely to be transmitted than are diseases with shorter incubation periods. Why? *(If the period of harboring an undiscovered infection is long, the infected person may expose many other people.)* Remind students why it is always necessary to take precautions against infection, even if there is no obvious source of disease in their environment.

Review

Assign Review—Section 41-3.

Reteaching — BASIC

Invite a local public health official or the school nurse to speak to the class about disease transmission and prevention. Have students contribute to a class list of questions to ask the speaker.

biobit

How deadly is influenza?

In 1918, an influenza epidemic killed more than 20 million people. To prevent such a disaster from happening again, scientists track the antigen shifting of influenza viruses and target the new viral antigens for vaccines.

after you are vaccinated, your immune system develops antibodies and memory cells against the pathogen. You develop an immunity to the disease. In 1977, smallpox became the first infectious disease to be eradicated from the public by vaccination. Vaccination has also reduced the incidence of many other diseases, including measles, polio, tetanus, and diphtheria.

Antigen shifting If memory cells provide immunity, why can you get the flu even if you have already been infected or vaccinated? Influenza viruses constantly mutate over time. The viruses produce new antigens that your immune system does not recognize. This way of deceiving the immune system is known as **antigen shifting.** With subsequent exposure to the virus, your body must make new antibodies.

How can you observe antigen activity?

QUICK LAB

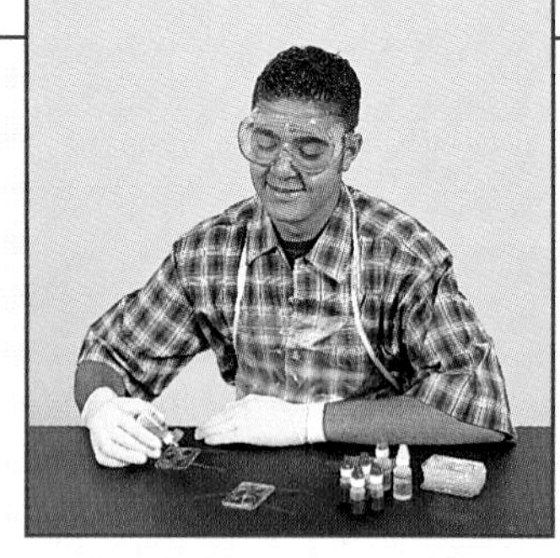

Using simulated blood, you can see what happens when antigens encounter their antibodies.

Materials

safety goggles, disposable gloves, lab apron, 2 blood-typing trays, simulated blood (types AB and O), simulated anti-A and anti-B blood-typing serums, 4 toothpicks

Procedure

1. Put on safety goggles, disposable gloves, and a lab apron.
2. Place 3–4 drops of type AB simulated blood into each well in a clean blood-typing tray. **CAUTION: Use only simulated blood provided by your teacher.**
3. Add 3–4 drops of anti-A blood-typing serum to one well. Stir the mixture for 30 seconds using a toothpick. Add 3–4 drops of anti-B blood-typing serum to the other well. Use a new toothpick to stir the mixture. Look for clumps separating from the mixtures.
4. Repeat steps 2 and 3 using simulated type O blood.
5. Dispose of your materials according to your teacher's directions. Clean up your work area and wash your hands.

Analysis

1. **Determine** which blood type, AB or O, contains antigens to the blood-typing serums.
2. **SKILL Evaluating Results** What does clumping of the blood mixtures indicate?
3. **SKILL Predicting Outcomes** What would happen if you did the same experiment using type A blood and type B blood?

Review SECTION 41-3

1. **List** two ways that diseases can be transmitted between people.
2. **Summarize** Koch's postulates for identifying specific pathogens.
3. **Describe** how vaccination produces immunity.
4. **SKILL Relating Concepts** Explain why you cannot get most diseases more than once.

Critical Thinking

5. **SKILL Applying Information** Most common "colds" are caused by viruses. Why can you get colds more than once?

Review SECTION 41-3 ANSWERS

1. Answers might include kissing, shaking hands, sharing eating utensils, engaging in sexual activity, and so on.
2. The pathogen is found in sick, but not in healthy, animals. The pathogen can be isolated and grown in the laboratory. Injection of the pathogen into a healthy animal causes the disease. The pathogen isolated from the second animal is the same as the one isolated from the first animal.
3. Vaccination induces the immune system to develop antibodies and memory cells for a specific pathogen.
4. Memory cells immediately recognize the pathogen, and the immune system mounts a response.
5. The common cold is caused by many variations of the cold virus.

Disorders of the Immune System

Section 41-4

The Immune System Sometimes Attacks Body Tissues

The ability of your immune system to distinguish cells and antigens of your body from foreign cells and antigens is crucial to the fight against pathogens. In some people, the immune system cannot distinguish between the body's antigens and foreign antigens, causing an autoimmune disease. In an **autoimmune disease,** the body launches an immune response against its own cells, attacking body cells as if they were pathogens. This inability of the immune system to distinguish between antigens of "self" and "nonself" may be caused by the inappropriate production of antibodies specific to the antigens of body cells.

Autoimmune diseases affect organs and tissues in various areas of the body. For example, multiple sclerosis *(sklih ROH sihs)* is an autoimmune disease that usually strikes people between the ages of 20 and 40. In people with multiple sclerosis, the immune system attacks and gradually destroys insulating material surrounding nerve cells in the brain, in the spinal cord, and in the nerves leading from the eyes to the brain. This impairs and may eventually stop the functioning of these nerve cells. Multiple sclerosis causes problems with vision, speech, and coordination. Table 41-1 lists and describes several autoimmune diseases.

Objectives

- **Describe** several autoimmune diseases. (p. 941)
- **Summarize** how HIV disables the immune system. (p. 942)
- **List** five ways HIV is transmitted. (p. 943)
- **Identify** causes of an allergic reaction. (p. 944)

Key Terms

autoimmune disease
AIDS
HIV
CD4
allergy

Table 41-1 Autoimmune Diseases

Disease	Areas affected	Symptoms
Graves' disease	Thyroid gland	Weakness, irritability, heat intolerance, increased sweating, weight loss, insomnia
Multiple sclerosis	Nervous system	Weakness, loss of coordination; problems with vision and speech
Rheumatoid arthritis	Joints	Severe pain, fatigue, disabling inflammation of joints
Systemic lupus erythematosus (SLE)	Connective tissue, joints, kidneys	Facial skin rash, painful joints, fever, fatigue, kidney problems, weight loss
Type I diabetes	Insulin-producing cells in pancreas	Increased blood glucose level, excessive urine production, problems with vision, weight loss, fatigue, irritability

Integrating Different Learning Styles

Logical/Mathematical	**PE**	Data Lab, p. 943
	ATE	Lesson Warm-up, p. 941
Interpersonal	**ATE**	Closure, p. 944
Verbal/Linguistic	**ATE**	Career, p. 942; Health Watch, p. 944; Internet Connect, p. 943

Section 41-4

1 Prepare

Scheduling

- **Block:** About 90 minutes will be needed to complete this section.
- **Traditional:** About two class periods will be needed to complete this section.
- **Planning:** See the Planning Guide on pp. 929A and 929B for lecture, classwork, and assignment options for Section 41-4. Lesson plans for Section 41-4 can be found in a lesson cycle format in ***Biology: Principles and Explorations*** **One-Stop Planner CD-ROM with Test Generator.**

Resource Link

Assign **Active Reading Worksheet Section 41-4** to help students comprehend the key concepts and visuals in this lesson.

2 Teach

Lesson Warm-up

Drugs such as cyclosporine and corticosteroids, which are immunosuppressants, are given to recipients of organ transplants. Have students hypothesize why this is necessary and what precautions must be taken by transplant recipients. *(Immunosuppressive therapy reduces the recipient's immune response to foreign antigens found in the transplanted organ. Precautions against infection are necessary because the patient's immune system activity is reduced.)* Have students explain why transplant patients are more likely than others to develop cancer. *(One of the functions of the immune system is to destroy cancer cells before they spread.)*

CAREER

Organ Transplant Coordinator

Organ transplant coordinators match donated organs to recipients. This matching includes determining who among potential recipients is a good tissue match and who among tissue-matched recipients is in the most dire need. Organ transplantation must be undertaken quickly—it is opportunistic in nature—and harvested organs often are flown from the donor to the recipient. Have students use outside resources to learn about the career of an organ transplant coordinator. They should prepare a short summary including a job description, education required, potential employers, and starting salary. PORTFOLIO

Teaching TIP

The Origin of HIV

The apparently sudden appearance of a virus as deadly as HIV inspires many questions. Where and when did HIV originate? How did it spread from its point of origin? Do any similar viruses infect other species?

Scientists are slowly piecing together the puzzle of HIV. They now know that HIV did not appear suddenly. Using retrospective studies of stored tissues, scientists have shown that the virus was present in the United States at least a decade before AIDS was described. A blood sample taken in what is now the Democratic Republic of the Congo in 1959 contains antibodies to HIV.

Most scientists think HIV arose in Africa. A similar virus, simian immunodeficiency virus (SIV), infects monkeys and apes in Africa. No HIV-like viruses have been found, however, in wild primates in Asia or the New World. SIV, therefore, might be an ancestor of HIV. HIV seems to be most closely related to a strain of SIV found in chimpanzees. No one is sure, however, how the virus first infected humans or where this occurred.

HIV Infection Leads to Immune System Collapse

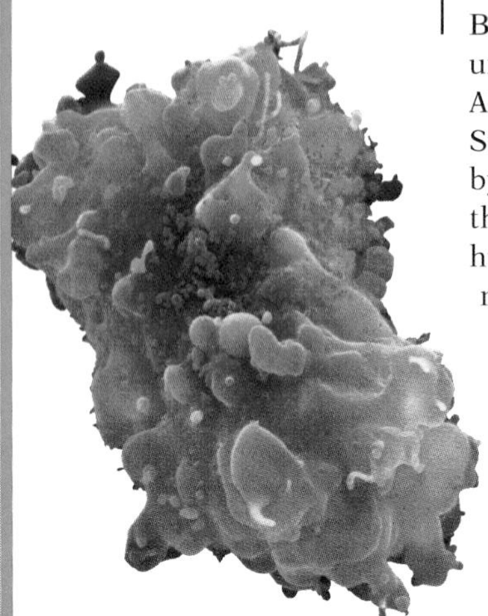

Figure 41-10 HIV. Small (purple) HIV particles surround a helper T cell (orange).

Before 1981, **AIDS,** or acquired immunodeficiency syndrome, was unknown. Since 1981, more than 385,000 Americans have died of AIDS, and the total number of AIDS cases reported in the United States has increased to more than 680,000. AIDS is a disease caused by **HIV,** or the human immunodeficiency virus. Many scientists think HIV evolved from a virus similar to one that infects nonhuman primates in Africa. A mutation enables HIV to recognize a receptor protein called **CD4** on some human cells. HIV, shown in **Figure 41-10,** enters white blood cells by binding to CD4. HIV usually invades helper T cells, which begin to produce HIV soon after infection. As helper T cells die, the immune system gradually weakens and becomes overwhelmed by pathogens that it would normally detect and destroy. The body becomes susceptible to other diseases, called opportunistic infections, that generally persist only in people with weakened immune systems.

Testing for HIV

Antibodies to HIV can be detected in blood. Someone whose blood contains antibodies to HIV is said to be HIV positive. When the number of helper T cells in a person infected with HIV is less than 200 cells/mL of blood, the person is diagnosed with AIDS, as shown in **Figure 41-11.** Many scientists think almost everyone infected with HIV who is not treated will eventually develop AIDS.

The time between HIV infection and the onset of AIDS can exceed 10 years, and this time period is increasing as new treatments for HIV infection are developed. A person with HIV may feel and appear healthy but can infect other people. In the United States, the number of deaths caused by AIDS has dropped from more than 45,000 in 1995 to about 35,000 in 1996, and to about 20,000 in 1997. This decrease does not reflect a decline in HIV infection, but rather more effective drug therapies, which postpone onset of the disease.

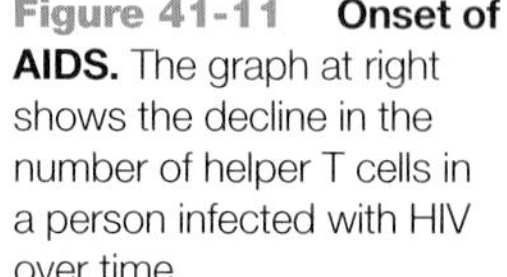

Figure 41-11 Onset of AIDS. The graph at right shows the decline in the number of helper T cells in a person infected with HIV over time.

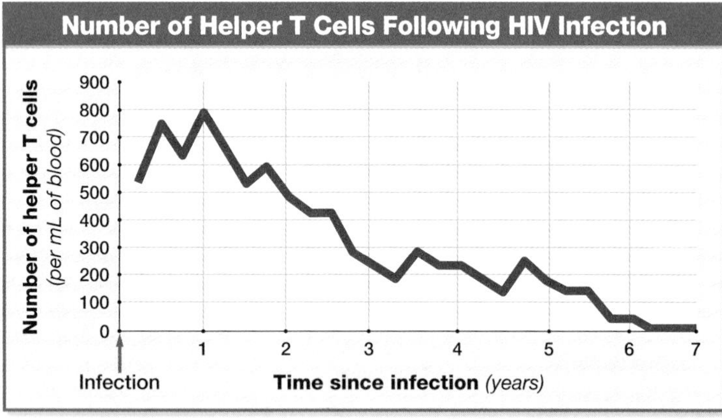

MULTICULTURAL PERSPECTIVE

Sir Frank Macfarlane Burnet

Sir Frank Macfarlane Burnet was an Australian scientist who discovered and described acquired immunological tolerance to transplanted tissues. He also developed techniques of virus culture in chick embryos, which have become standard laboratory practice. Sir Macfarlane Burnet was elected to the Royal Society in 1942, received the Order of Merit in 1958, and was awarded the Nobel prize for medicine in 1960. His book, *Natural History of Infectious Disease*, which was co-authored with David O. White, was published by Cambridge University Press in 1972.

Transmission of HIV

You can become infected with HIV if you receive HIV-infected white blood cells, which are present in many body fluids. The most common method of HIV transmission is through sexual contact. Because semen, vaginal fluid, and mucous membranes may contain HIV, both males and females can become infected with HIV during vaginal, anal, or oral intercourse. Use of a latex condom during sexual intercourse reduces but does not eliminate the risk of getting or spreading HIV.

HIV can be passed between drug users who share a hypodermic needle if HIV-infected blood remains in the needle or syringe. In the late 1970s and early 1980s, many people became infected with HIV after receiving transfusions of HIV-contaminated blood. This is very unlikely now because blood made available for transfusion is tested for HIV. In addition, pregnant or nursing mothers can pass HIV to their infants through blood and breast milk.

HIV is not transmitted through the air, on toilet seats, by kissing or handshaking, or by any other medium where HIV-infected white blood cells could not survive. Although HIV has been found in saliva, tears, and urine, these body fluids usually contain too few HIV particles to cause an infection. Mosquitoes and ticks do not transmit HIV because they do not carry HIV-infected white blood cells.

internet**connect**

SciLINKS NSTA

TOPIC: HIV transmission
GO TO: www.scilinks.org
KEYWORD: HX943

DATA LAB

Studying the spread of AIDS

Background

The World Health Organization estimates that nearly 12 million deaths worldwide have been caused by AIDS. The graph below shows the total AIDS cases reported in the United States since 1987. Use the graph to answer the following questions:

Analysis

1. **Describe** how the number of people with AIDS has changed since 1987.
2. **SKILL Inferring Relationships** Is the number of Americans infected with HIV most likely greater than or less than the number of people with AIDS? Explain why.
3. **SKILL Evaluating Data** The graph indicates that the number of new AIDS cases reported each year has decreased since 1995. Suggest a possible reason for this decline.

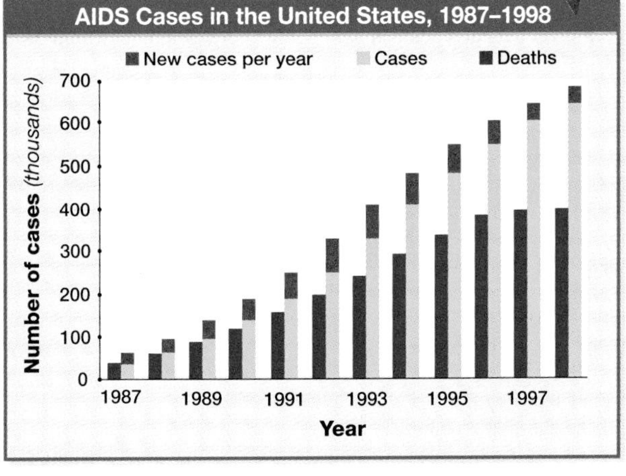

internet**connect**

SciLINKS NSTA Have students research HIV transmission using the Web site in the Internet Connect box on page 943. Have students write a report summarizing their findings. PORTFOLIO

DATA LAB

Studying the spread of AIDS

Time 10–15 minutes

Process Skills Interpreting graphs, analyzing data, inferring relationships

Teaching Strategies

Review with students how to read a bar graph.

Analysis Answers

1. AIDS cases have risen sharply, from less than 50,000 cases to more than 600,000 cases.
2. greater; Everyone who has AIDS is infected with HIV, but not everyone who is infected with HIV has developed AIDS.
3. New drugs postpone the onset of AIDS in HIV-infected individuals.

Overcoming Misconceptions

AIDS-Related Infections

Some students may think that HIV itself causes the death of AIDS patients. Actually, AIDS results in cancers and infections by opportunistic pathogens that are unusual in individuals with healthy immune systems. Infections and cancers associated with AIDS include Kaposi's sarcoma, *Pneumocystis carinii* pneumonia, candidiasis, and others. Kaposi's sarcoma is a cancer of the skin and mucous membranes that often spreads through other organs. *Pneumocystis carinii* pneumonia is the most common of these AIDS-related diseases. Candidiasis, commonly known as thrush, is a fungal infection producing a thick, white coating of the mouth, tongue, and other components of the digestive system. Other opportunistic pathogens include cytomegalovirus, which produces inflammation of the retina, colon, and adrenal glands; and tuberculosis and *Salmonella* infection, both caused by bacteria.

Making Biology Relevant

Health

Allergies can be life threatening, particularly those allergies to food, drugs, insect venom, or other chemicals that lead to anaphylactic shock, a severe, systemic reaction to an allergen. Anaphylactic shock is associated with constriction of the bronchioles and circulatory collapse, which can result in death.

Asthma

Teaching Strategies

Invite a person who has asthma to class to share his or her experiences with asthma.

Discussion

- What causes of asthma are cited? *(pollution, emotional stress, substances in the air)*
- Do you agree or disagree with the reasons given for high inner-city asthma rates? *(Answers will vary.)*

3 Assessment Options

Closure

Immune System Problems

Have students cooperate to list as many allergies and disorders of the immune system they can think of. Students should rank them on a scale of 1–10, from most critical to least critical. (Example: Hay fever might be ranked as an 8, whereas environmental allergies may be a 1 or 2.)

Review

Assign Review—Section 41-4.

Reteaching — BASIC

Have students write and act out a 1-minute commercial about allergy relief. They should address what causes allergies, what the symptoms are, and how allergy medication works. PORTFOLIO

biobit

What causes food allergies?

A food allergy often occurs when the immune system reacts to a harmless antigen in the food, triggering an immune response. Symptoms may include hives or difficulty breathing.

Some Antigens Can Cause Allergic Reactions

Many health problems are caused by overreactions of the immune system. One type of overreaction is allergic reaction. An **allergy** is the body's overreaction to a normally harmless antigen. Allergy-causing antigens include pollen, the feces of dust mites, fungal spores, and substances found in some foods and drugs. Most allergic reactions are merely uncomfortable. Cells exposed to allergy-causing antigens release histamine. Histamine causes swelling, redness, increased mucus production, runny nose, itchy eyes, and nasal congestion. These are common symptoms of an allergic reaction. Most allergy medicines contain antihistamines, which are drugs that prevent the action of histamine. Severe allergic reactions, such as asthma, can be life threatening if they are not treated immediately.

HEALTH WATCH

Asthma

If you have ever had difficulty breathing, you could have asthma. Asthma is an inflammation of the respiratory tract often caused by an allergic reaction to substances in the air. Asthma affects about 15 million Americans and causes more than 5,000 deaths each year. Inner-city residents get asthma three times as often as people who live outside cities. In some cities, the death rate from asthma is eight times the national average. Some scientists think increased asthma in inner-city residents is caused by a combination of pollution, emotional stress, and limited access to health care. One study suggests that cockroach excrement is causing asthma in many inner-city children.

Asthma attack

Three things happen during an asthma attack. First the respiratory passages become inflamed and swollen. Then mucus collects in the lungs, restricting airflow. Finally, muscles that surround the bronchial tubes tighten, restricting air flow and causing shortness of breath.

Treating asthma

Asthma sufferers can take medicines to relieve symptoms. Some medicines increase airflow by relaxing bronchial-tube muscles, but their effects wear off after a few hours. Other medicines provide long-lasting relief by preventing or reducing inflammation.

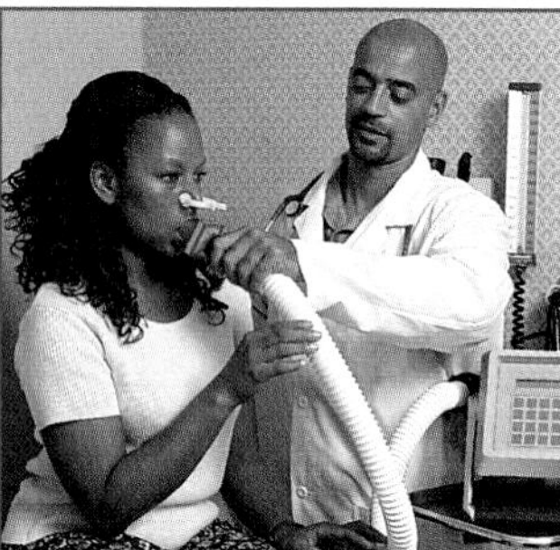

Measuring lung capacity

internet**connect**

SCILINKS NSTA
TOPIC: Asthma
GO TO: www.scilinks.org
KEYWORD: HX944

Review SECTION 41-4

1. **Describe** the cause of autoimmune diseases.
2. **List** two ways that HIV can be transmitted and two ways that it cannot.
3. **Distinguish** between HIV infection and AIDS.
4. **SKILL Summarizing Information** What causes an allergic reaction?

Critical Thinking

5. **SKILL Recognizing Relationships** Explain why it might take several weeks after exposure to HIV for a person's HIV antibody test to be positive.

Review SECTION 41-4 ANSWERS

1. Autoimmune diseases occur because the body lacks the ability to distinguish between self and nonself cells.
2. HIV can be transmitted by sexual contact, hypodermic needles, tainted blood, breast milk, and from a mother to a fetus. HIV cannot be transmitted by shaking hands, toilet seats, insect bites, or through the air.
3. HIV infection is diagnosed by the presence of antibodies in the blood. AIDS is diagnosed if there are less than 200 helper T cells/mL blood.
4. An allergic reaction is an overreaction of the immune system to a relatively harmless antigen.
5. The immune system might take several weeks to build up enough HIV antibodies to be detected.

Study ZONE

CHAPTER 41

Use the Key Concepts and Key Terms listed below to review the main ideas in this chapter. If you need to review the meaning of a term, turn to the page indicated.

Key Concepts

41-1 Nonspecific Defenses

- Skin and mucous membranes act as barriers to pathogens.
- The inflammatory response increases blood flow to an infected area, while the temperature response inhibits bacterial growth.
- Complement proteins form a membrane attack complex (MAC). Interferon stimulates cells and inhibits viruses.
- Neutrophils, macrophages, and natural killer cells use different methods to attack and destroy invading pathogens.

41-2 Immune Response

- Receptor proteins on white blood cells bind to specific antigens.
- The T cell response is an active, cell-mediated defense in which cytotoxic T cells destroy pathogens.
- The B cell response is a passive, humoral defense in which antibodies mark pathogens for destruction by white blood cells.

41-3 Disease Transmission and Prevention

- Diseases are transmitted to humans through person-to-person contact, air, food, water, and animal bites.
- Biologists use Koch's postulates to identify pathogens.
- Memory cells can produce immunity to pathogens.
- Vaccination produces immunity to pathogens.
- Antigen shifting deceives the immune system.

41-4 Disorders of the Immune System

- In an autoimmune disease, the immune system attacks body cells as if they were pathogens.
- HIV, the virus that causes AIDS, invades helper T cells, causing them to produce more HIV particles and eventually die.
- HIV is transmitted through HIV-infected white blood cells in body fluids, most commonly through sexual contact or by the sharing of a hypodermic needle with an infected person.
- An allergic reaction is an inappropriate response to normally harmless antigens.

Key Terms

41-1

pathogen (932)
mucous membrane (932)
inflammatory response (933)
histamine (933)
complement system (934)
interferon (934)
neutrophil (934)
macrophage (934)
natural killer cell (934)

41-2

cytotoxic T cell (935)
B cell (935)
helper T cell (935)
antigen (935)
plasma cell (937)
antibody (937)

41-3

Koch's postulates (938)
immunity (939)
vaccination (939)
vaccine (939)
antigen shifting (940)

41-4

autoimmune disease (941)
AIDS (942)
HIV (942)
CD4 (942)
allergy (944)

Study TIP Ready?

If you think you understand these Key Concepts and Key Terms, then you're ready for the Performance Zone.

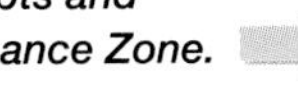

Study ZONE

CHAPTER 41

Review and Practice

Assign the following worksheets for Chapter 41:

- **Student Review Guide**
- **Concept Mapping Worksheet**
- **Test Preparation Pretest**
- **Vocabulary Worksheet**
- **Critical Thinking Review Worksheet**

Assessment Options

The following assessment options are available for this chapter:

Resource Link

Assign **Chapter Test 41** to assess students' understanding of the chapter.

Alternative Assessment

Have students work in cooperative groups. Ask each group to develop a question to send to a newspaper health advice columnist dealing with any topic covered in this chapter. Collect the questions and redistribute them to other groups. Each group should then write an appropriate response to the question. PORTFOLIO

Portfolio Assessment

- *Portfolio Assessment* at the front of this book provides options for building and evaluating student portfolios. Students and teachers can select items from the areas listed for inclusion in a portfolio.
- See items labeled PORTFOLIO on pp. 933, 934, 936, 939, 942, 944, and 945.

Answer to Concept Map

The following is one possible answer to Performance Zone item 1 on page 946.

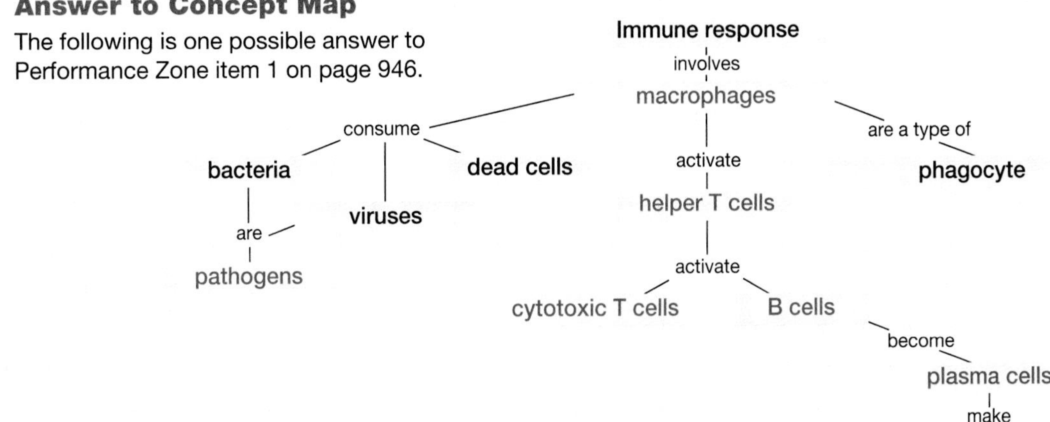

Performance ZONE

CHAPTER 41

REVIEW ANSWERS

Assignment Guide

Section	Review
41-1	2, 3, 4, 21
41-2	1, 5, 6, 15, 16, 17, 18
41-3	7, 8, 13, 23, 24, 26
41-4	9, 10, 11, 12, 14, 19, 20, 22, 25

Understanding and Applying Concepts

1. The answer to the concept map is found at the bottom of page 945.
2. **a.** Macrophages ingest and destroy pathogens and clear debris. Neutrophils engulf and poison pathogens.
   b. Helper T cells activate cytotoxic T cells, which seek and destroy pathogens.
   c. Immunity is the body's resistance to pathogens that have been previously encountered. A vaccine is a dose of dead or deactivated pathogen given to encourage immunity.
   d. An allergy is the body's overreaction to a normally harmless antigen. Histamine causes an allergic response that includes swelling, redness, increased mucus production, runny nose, itchy eyes, and nasal congestion.
3. d
4. b
5. c
6. c
7. b
8. a
9. b
10. b
11. a
12. about 49 months
13. More robust bacterial cells that are not killed by the antibiotic replace the weaker cells and multiply, producing cells that have greater and greater resistance to drugs.
14. The respiratory passages become inflamed and swollen, mucus collects in the lungs, and muscles around the bronchial tubes tighten, all leading to a reduced airflow.
15. The Golgi apparatus and endoplasmic reticulum are responsible for protein mobilization. Because plasma cells produce antibodies, high protein-synthesis activity would be expected.

Skills Assessment

16. antigens
17. to bind antigens
18. Antibodies bind specific antigens.
19. Answers will vary. Some important developments and treatments include a clinical trial of a potential HIV vaccine conducted in 15 U.S. medical centers and a new medication, which combines two proven medications, 3Tc and AZT, into one tablet and may help people with HIV live longer, healthier lives. Another medication, alitretinoin, treats lesions of Kaposi's sarcoma. AIDS awareness campaigns include the red ribbon campaign, the AIDS quilt, and World AIDS Day.
20. Food allergies often involve shellfish, nuts, peanuts, fruits, and food additives, such as dyes, thickeners, and preservatives. Food allergies can cause asthma attacks, hives, and anaphylactic shock. Food intolerances cause unpleasant symptoms but are not life threatening as are some food allergies.

Performance ZONE — CHAPTER 41 Review

Understanding and Applying Concepts

1. **Concept Mapping** Make a concept map that describes the immune response. Include the following terms in your map: pathogen, macrophage, helper T cell, cytotoxic T cell, B cell, plasma cell, antibody.

2. **Understanding Vocabulary** For each pair of terms, explain the differences in their meanings.
   a. macrophage, neutrophil
   b. helper T cell, cytotoxic T cell
   c. immunity, vaccine
   d. allergy, histamine

3. Nonspecific defenses include
   a. the T cell response.
   b. the B cell response.
   c. antibodies.
   d. the inflammatory response.

4. Mucous membranes
   a. activate helper T cells.
   b. secrete mucus, which traps pathogens.
   c. prevent blood clots.
   d. produce antibodies.

5. B cells and cytotoxic T cells are stimulated by interleukin-2, which is released by
   a. macrophages.
   b. neutrophils.
   c. helper T cells.
   d. natural killer cells.

6. Plasma cells
   a. are directly stimulated by interleukin-1.
   b. result from cytotoxic T cells.
   c. produce antibodies.
   d. engulf pathogens.

7. Robert Koch
   a. treated smallpox patients.
   b. established a four-step procedure for identifying pathogens.
   c. perfected vaccination.
   d. identified complement proteins.

8. Flu vaccinations are given each year because
   a. influenza viruses mutate often.
   b. influenza is caused by bacteria.
   c. very few memory cells are produced.
   d. macrophages cannot engulf flu viruses.

9. Rheumatoid arthritis is an example of
   a. an allergic reaction.
   b. an autoimmune disease.
   c. an AIDS-related infection.
   d. a bacterial infection.

10. HIV disables the immune system by
    a. blocking the action of macrophages.
    b. destroying helper T cells.
    c. activating production of B cells.
    d. All of the above

11. HIV can be transmitted by
    a. sexual contact.
    b. mosquito bites.
    c. shaking hands.
    d. vaccination only.

12. The graph below shows the decrease in the number of helper T cells in a person with AIDS. In this person, how many months after infection did the onset of AIDS occur?

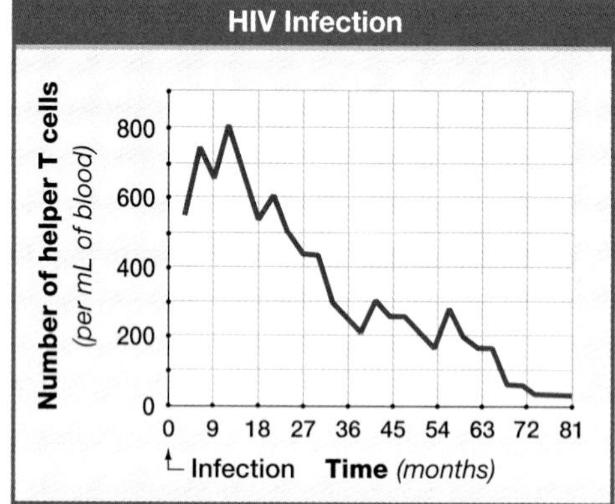

13. **Theme Evolution** Why have many disease-causing bacteria become resistant to antibiotics?

14. **Health Watch** What symptoms are usually associated with an asthma attack?

15. **Chapter Links** Plasma cells contain a large Golgi apparatus and large amounts of rough endoplasmic reticulum. How is the presence of these organelles related to the function of plasma cells? (**Hint:** See Chapter 3, Section 3-3.)

Skills Assessment

Part 1: Interpreting Graphics

Study the diagram below, and then answer the following questions:

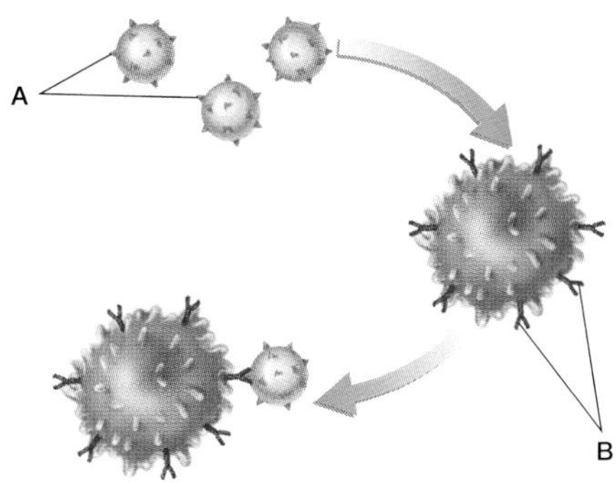

16. Identify the structures labeled *A*.

17. What is the function of the structures labeled *B*?

18. How do structures *A* and structures *B* interact with one another?

Part 2: Life/Work Skills

Use the media center or Internet resources to learn more about disease prevention and treatment.

19. **Finding Information** Scientific research into treatments and a possible cure for AIDS is an ongoing process. Find out about the latest research into HIV and prospective cures for AIDS. Research the most recent pharmaceutical developments and other treatments. Evaluate public awareness and education programs and campaigns. Also find out how much money is being spent in each of these areas. Present your findings in a written report.

20. **Finding Information** Find out what kinds of foods and food additives are known to cause allergic reactions in some people. List common symptoms associated with food allergies. Research the difference between food allergy and food intolerance. Present your findings in an oral report to your class.

Critical Thinking

21. **Inferring Relationships** People who are severely burned often die from infection. Given what you know about disease transmission, explain why this is common.

22. **Forming Reasoned Opinions** A government agency is reviewing the following two proposals for HIV research but can fund only one. Suppose you are asked to provide input on the decision. Which proposal would you recommend that the agency fund? You should consider not only the likely effectiveness of the treatment but also possible side effects. Explain how you made your choice. **Proposal 1:** Develop a drug that interferes with protein production. **Proposal 2:** Develop a substance that binds to CD4 receptors on helper T cells.

Portfolio Projects

23. **Summarizing Information** Use the *Physician's Desk Reference* at your local library to research five different antibiotics made by drug companies. Make a large chart or table on poster board listing the brand name of each drug, the generic name of the drug, uses of the drug, the name of the company that makes the drug, and the disease the drug controls. Present your chart to your class.

24. **Finding Information** Research the causes of various foodborne infections and intoxications. Identify the types of microorganisms and toxins involved, the types of foods involved, and symptoms of the illnesses.

25. **Interpreting Information** Read the article titled "New AIDS Vaccine Stimulates Hope" (*Science News*, January 16, 1999, p. 36). How are tumor cells used in this vaccine?

26. **Career Focus Immunologist** Research the field of immunology, and write a report on your findings. Your report should include a job description, education and training required, kinds of employers, growth prospects, and starting salary.

Critical Thinking

21. The barrier of the skin prevents many infections. Severe burns result in skin loss and subsequent infection.

22. The solution described in proposal 1 might interfere with all protein production in the body, and it would work only after infection by HIV. Proposal 2 is a better choice because if CD4 binding sites on helper T cells were occupied, the sites would not be available for the HIV virus to attach.

Portfolio Projects

23. Answers will vary.

24. Answers will vary. For example, botulism is caused by the *Clostridium botulinum* bacteria's toxins. Humans get the disease by ingesting the toxins from improperly canned foods. The toxin attacks the nervous system, causing symptoms such as blurred or double vision, dry mouth, paralysis, and breathing difficulties. Salmonellosis is caused by bacteria from the *Salmonella* genus. Humans are infected when they ingest the bacteria in undercooked meat, chicken, and eggs, and in unpasteurized milk, and other dairy products. It can also be passed on by fecal contamination and in contaminated drinking water. The symptoms include diarrhea, fever, abdominal pain, headache, and vomiting.

25. Tumor cells are used to make gp120 and gp41, proteins that attach HIV to cells to form antibodies.

26. Allergist/immunologists are physicians who diagnose and treat allergic diseases such as allergic rhinitis, asthma, gastrointestinal disorders, anaphylaxis, and contact dermatitis and immunologic issues, such as adverse drug reactions, autoimmune diseases, bone marrow and solid organ transplantation, and gene replacement therapy. Allergist/immunologists attend medical school after college. The American Board of Allergy and Immunology (ABAI) provides for certification for allergist/immunologists; the requirements are current certification as a member of either the American Board of Internal Medicine (ABIM) or the American Board of Pediatrics; completion of at least 2 years of full-time residency or fellowship in an allergy/immunology program; and completion of the ABAI certification examination. Allergist/immunologists work in private practice, in hospitals, and for pharmaceutical companies. The growth potential of the field is excellent. Starting salary will vary by region.

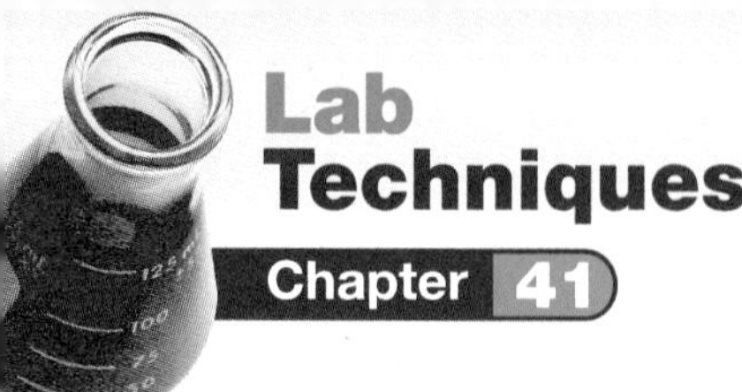

Lab Techniques

Chapter 41

Simulating Disease Transmission

Purpose

Students will simulate the transmission of a communicable disease and then try to identify the original infected person in the closed class population.

Preparation Notes

Time Required: About 30 minutes

Materials

Materials for this lab can be ordered from WARD'S. See *Master Materials List* at the front of this book for catalog numbers.

Preparation Tips

- Use stock dropper bottles of distilled water for half the students and dropper bottles containing 10 percent ascorbic acid for the rest of the students. Both solutions will be clear.
- Ascorbic acid and indophenol indicator solution can be obtained from a biological supply company such as WARD'S.
- A 10 percent solution of ascorbic acid can be prepared by dissolving 10 g of ascorbic acid in 50 mL of water and diluting it to 100 mL with water.
- Each student will need a dropper bottle of unknown solution, a large test tube, and indophenol indicator solution.

Safety Precautions

- Students should wear a lab apron, goggles, and disposable gloves throughout the investigation.
- Caution students not to allow solutions to touch their skin or clothing. If this happens, flush the area with water.
- After completing the investigation, students should rinse the test tubes well.

Lab Techniques

Chapter 41

Simulating Disease Transmission

SKILLS

- Modeling
- Organizing and analyzing data

OBJECTIVES

- **Simulate** the transmission of a disease.
- **Determine** the original carrier of the disease.

MATERIALS

- safety goggles
- lab apron
- disposable gloves
- dropper bottle of unknown solution
- large test tube
- indophenol indicator

ChemSafety

CAUTION: Always wear safety goggles and a lab apron to protect your eyes and clothing.

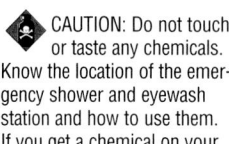

CAUTION: Do not touch or taste any chemicals. Know the location of the emergency shower and eyewash station and how to use them. If you get a chemical on your skin or clothing, wash it off at the sink while calling to the teacher. Notify the teacher of a spill. Spills should be cleaned up promptly, according to your teacher's directions.

CAUTION: Glassware is fragile. Notify the teacher of broken glass or cuts. Do not clean up broken glass or spills with broken glass unless the teacher tells you to do so.

Before You Begin

Communicable diseases are caused by **pathogens** and can be transmitted from one person to another. You can become infected by a pathogen in several ways, including by drinking contaminated water, eating contaminated foods, receiving contaminated blood, and inhaling infectious **aerosols** (droplets from coughs or sneezes). In this lab, you will simulate the transmission of a communicable disease. After the simulation, you will try to identify the original infected person in the closed class population.

1. Write a definition for each boldface term in the paragraph above.
2. Make data tables similar to the ones shown at right.
3. Based on the objectives for this lab, write a question you would like to explore about disease transmission.

DATA TABLE 1

Round number	Partner's name

DATA TABLE 2

Name of infected person	Names of infected person's partners		
	Round 1	Round 2	Round 3

Procedure

Part A: Simulate Disease Transmission

1.  Put on safety goggles, a lab apron, and gloves.
2. You will be given a dropper bottle of an unknown solution. When your teacher says to begin, transfer 3 dropperfuls of your solution to a clean test tube.

Procedural Tips

- Before students begin, remind them of safety practices, and emphasize that they should not allow solutions to touch their skin or clothes.
- If the procedure is followed precisely, the route of transmission can be traced easily.

3. Select a partner for Round 1. Record the name of this partner in Data Table 1.
4. Pour the contents of one of your test tubes into the other test tube. Then pour half the solution back into the first test tube. You and your partner now share any pathogens either of you might have.
5. On your teacher's signal, select a new partner for Round 2. Record this partner's name in Data Table 1. Repeat step 4.
6. On your teacher's signal, select another new partner for Round 3. Record this partner's name. Repeat step 4.
7. Add one dropperful of indophenol indicator to your test tube. "Infected" solutions will stay colorless or turn light pink. "Uninfected" solutions will turn blue. Record the results of your test.

PART B: Trace the Disease Source

8. If you are infected, write your name and the name of your partner in each round on the board or on an overhead projector. Mark your infected partners. Record all the data for your class in Data Table 2.
9. To trace the source of the infection, cross out the names of the uninfected partners in Round 1. There should be only two names left. One is the name of the original disease carrier. To find the original disease carrier, place a sample from his or her dropper bottle in a clean test tube, and test it with indophenol indicator.
10. To show the disease transmission route, make a diagram similar to the one below. Show the original disease carrier and the people each disease carrier infected.

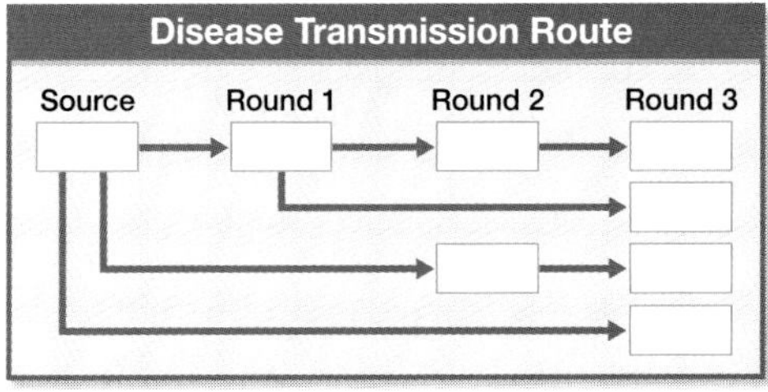

PART C: Cleanup and Disposal

11. Dispose of solutions and broken glass in the designated waste containers. Do not pour chemicals down the drain unless your teacher tells you to do so.
12. Clean up your work area and all lab equipment. Return lab equipment to its proper place. Wash your hands thoroughly before you leave the lab and after you finish all work.

Analyze and Conclude

1. **Interpreting Data** After Round 3, how many people were "infected"? Express this number as a percentage of your class.
2. **Relating Concepts** What do you think the clear fluids each student started with represent? Explain why.
3. **Drawing Conclusions** Can someone who does not show any symptoms of a disease transmit that disease? Explain.
4. **Further Inquiry** Write a new question about disease transmission that could be explored with another investigation.

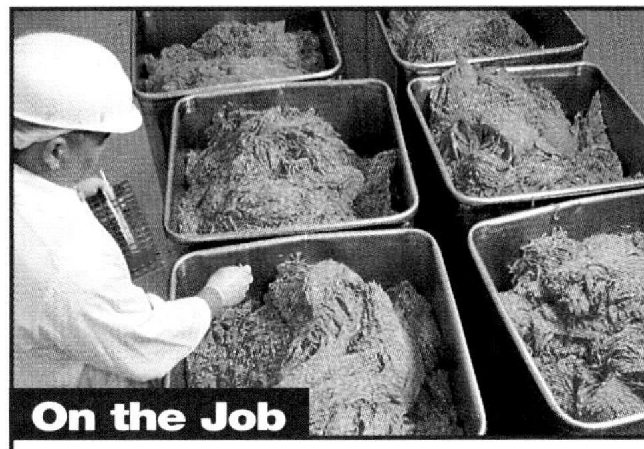

On the Job

Public health officials, such as food inspectors, study and work to prevent the spread of diseases in human populations. Use the media center or Internet resources to find out how public health officials trace the origin of communicable diseases. Summarize your findings in a written or oral report.

Lab Techniques

Procedure

Answers

2. One half of the students will have an acidic solution.
3. Students should record their partner's name in Data Table 1.
4. After they share solutions, most students will have acidic mixtures ("infected" + "noninfected" or "infected" + "infected" solutions). A few students will have nonacidic solutions ("noninfected" + "noninfected" solutions).
5. Students should record the name of their second partner in Data Table 1.
6. Each repetition of the infection activity will increase the number of "infected" individuals.
7. Most solutions will remain colorless or turn pink.
8. All students will need to announce whether or not they are infected for this step.
9. If the procedure is followed precisely, the route of transmission can be traced easily.
10. A student volunteer should construct a diagram of the route of infection on the board or overhead.

Analyze and Conclude

Answers

1. Answers will vary depending on class size.
2. The clear fluids represent pathogen-containing aerosols, or contaminated food, water, or blood.
3. yes
4. Answers will vary. For example: Would dividing the class into two or three quarantined groups before repeating the experiment affect disease transmission rate?

On the Job

Answers

Public health officials study information obtained from doctors and health clinics and conduct interviews to trace the origin of a communicable disease

Disposal

Dilute the volume of ascorbic acid with 20 times as much tap water. Test pH with litmus or other indicator; if necessary, adjust pH to neutrality by adding small amounts of 1 M acid, base, or other reagent as required. Place a beaker containing the diluted mixture in the sink, and run water to overflowing for 10 minutes, flushing to a known sanitary sewer.

Before You Begin

Answers

1. *communicable disease*—a disease that can be transmitted from one person to another

   pathogen—a disease-causing agent such as a bacterium or a virus

   aerosol—a mist of tiny droplets of a liquid
3. Answers will vary. For example: Can you determine whether a disease has been caused by the passing of pathogens from person to person or by environmental conditions?

Chapter 42

Nervous System

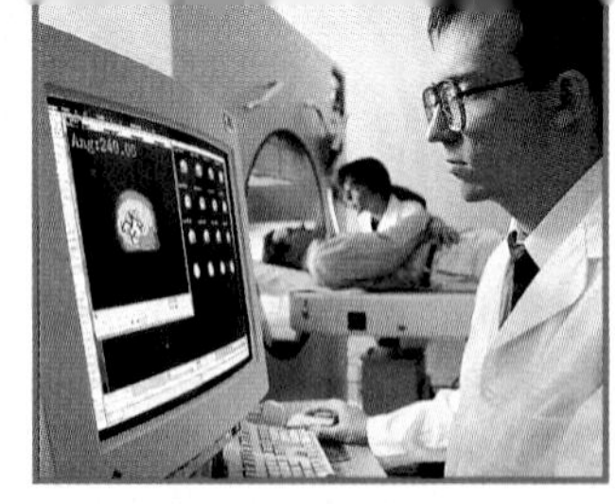

Planning Guide

OBJECTIVES	MEETING INDIVIDUAL NEEDS	READING SKILLS AND STRATEGIES
42-1 Neurons and Nerve Impulses (BLOCKS 1 & 2 90 min)		
• Analyze the structure and function of neurons. • Describe how the resting potential is established in a neuron. • Sequence the steps of a nerve impulse. • List the events that occur in synaptic transmission of a nerve impulse.	Learning Styles, p. 952 BASIC ADVANCED • pp. 953, 957	ATE Active Reading, Technique: Anticipation Guide, p. 951 ■ Active Reading Guide, worksheet (42-1) ■ Directed Reading, worksheet (42-1)
42-2 Structures of the Nervous System (BLOCKS 3 & 4 90 min)		
• Distinguish between the central nervous system and the peripheral nervous system. • Identify the major parts of the brain and their functions. • Describe the structure of the spinal cord. • Sequence the events of a spinal reflex. • Compare the somatic nervous system with the autonomic nervous system.	Learning Styles, p. 958 BASIC ADVANCED • pp. 959, 963	■ Active Reading Guide, worksheet (42-2) ■ Directed Reading, worksheet (42-2) ■ Reading Strategies, Spider Map worksheet
42-3 Sensory Systems (BLOCKS 5 & 6 90 min)		
• List five types of sensory receptors and the stimuli to which they respond. • Identify sites of sensory processing in the brain. • Analyze the structure of the eye and its role in the visual system. • Describe how the ear detects sound and helps maintain balance. • Compare the senses of taste and smell.	Learning Styles, p. 964 BASIC ADVANCED • p. 968	■ Active Reading Guide, worksheet (42-3) ■ Directed Reading, worksheet (42-3) ■ Supplemental Reading Guide, *Microbe Hunters*
42-4 Drugs and the Nervous System (BLOCKS 7 & 8 90 min)		
• Identify several types of psychoactive drugs, and describe their effects. • Describe how drug use can lead to drug addiction. • Summarize the effects of commonly abused drugs on the nervous system.	Learning Styles, p. 969 BASIC ADVANCED • pp. 971, 972, 974	ATE Active Reading, Technique: Anticipation Guide, p. 974 ■ Active Reading Guide, worksheet (42-4) ■ Directed Reading, worksheet (42-4)

Review and Assessment Resources

BLOCKS 9 & 10 90 min

TEST PREP

- **PE** Study Zone, p. 975
- **PE** Performance Zone, pp. 976–977
- ■ Chapter 42 worksheets:
  - Vocabulary
  - Concept Mapping
  - Critical Thinking
- Audio CD Program
- ■ Test Prep Pretest

CHAPTER TESTING

- ■ Chapter Tests (blackline master)
- ▲ Test Generator
- ■ Assessment Item Listing

See the correlations table at the front of this book for a list of the ***National Science Education Standards*** addressed in this chapter.

Smithsonian Institution® Visit **www.si.edu/hrw** for additional on-line resources.

KEY

Biology Principles and Explorations

PE Pupil's Edition

ATE Teacher's Edition

HOLT BIOSOURCES

 Teaching Transparencies

 Laboratory Program

One-Stop Planner CD-ROM
Shows these resources within a customizable daily lesson plan:

■ Teaching Resources

▲ Test Generator

VISUAL STRATEGIES	CLASSWORK AND HOMEWORK	TECHNOLOGY AND INTERNET RESOURCES
ATE Fig. 42-2, p. 953 **ATE** Fig. 42-3, p. 955 **ATE** Fig. 42-4, p. 956 **ATE** Fig. 42-5, p. 957 194 Structure of a Neuron 195 Physiology of a Nerve Impulse	**PE** Data Lab Analyzing changes during a nerve impulse, p. 954 Lab A30 Collecting Data Through a Survey **PE** Review, p. 957	Holt Biology Videodiscs Section 42-1 Audio CD Program Section 42-1 internetconnect SCILINKS NSTA TOPIC: Neurons
196 Peripheral Nervous System 197 Action of a Reflex 197A Divisions of the Nervous System	Lab A28 Bias and Experimentation Lab B30 Touch Receptors in the Skin **PE** Review, p. 963	CNN Video Segment 34 Stroke Brain Repair Holt Biology Videodiscs Section 42-2 Audio CD Program Section 42-2 internetconnect SCILINKS NSTA TOPICS: • Spinal cord • Autonomic nervous system
199 Structure of the Human Brain 200 Anatomy of the Ear 201 Structure and Function of Semicircular Canals 202 Anatomy of the Eye 203 How the Brain Senses Light	**PE** Quick Lab How can you demonstrate your blind spot?, p. 966 Lab B31 Exploring Vision **PE** Review, p. 968	Holt Biology Videodiscs Section 42-3 Audio CD Program Section 42-3 internetconnect SCILINKS NSTA TOPICS: • Sensory receptors • Chemical senses
ATE Fig. 42-15, p. 970 233 Drug-Affected Synapse 234 Physiology of Addiction	**PE** Interactive Exploration Drug Addiction, pp. 978–979 ■ Occupational Applications: Substance Abuse Counselor ■ Problem-Solving: Determining Percentage Composition **PE** Review, p. 974 **PE** Study Zone, p. 975	Holt Biology Videodiscs Section 42-4 Audio CD Program Section 42-4 internetconnect SCILINKS NSTA TOPICS: • Blood alcohol concentration • Drug addiction

ALTERNATIVE ASSESSMENT

PE Portfolio Projects 25–27, p. 977

ATE Alternative Assessment, p. 975

ATE Portfolio Assessment, p. 975

Lab Activity Materials

Data Lab
p. 954 pencil, paper

Quick Lab
p. 966 unlined 3 × 5 index card, pencil

Interactive Exploration
p. 978 CD-ROM, computer

Lesson Warm-up
ATE Animaniacs Sing-Along music video *The Senses,* p. 964

Demonstrations
ATE See pp. 959, 961, 966–969, and 973

BIOSOURCES LAB PROGRAM
- Quick Labs: Lab A28, p. 55
- Quick Labs: Lab A30, p. 59
- Inquiry Skills Development: Lab B30, p. 145
- Inquiry Skills Development: Lab B31, p. 149

Chapter Theme

Homeostasis

The nervous system helps control the human body by transmitting information from muscles, organs, and sensory receptors to the brain. In the brain, the information is integrated, and signals are sent to muscles and glands. This function is the crux of homeostasis. The actions of the nervous system are rapid but of a shorter duration than chemical messages such as hormones. Through the transmission of nerve impulses, an organism can respond quickly to internal or external stimuli.

Establishing Prior Knowledge

Before beginning this chapter, make certain students can answer the questions in Study TIP Ready? on page 951. If they cannot, encourage them to reread the sections indicated.

Opening Demonstration

Ask students to brainstorm about the functions of the nervous system. Write their responses on the board or overhead. Group the responses into the following categories: (1) monitoring internal and external environments; (2) coordinating bodily activities; and (3) intellectual function and memory.

CHAPTER 42 Nervous System

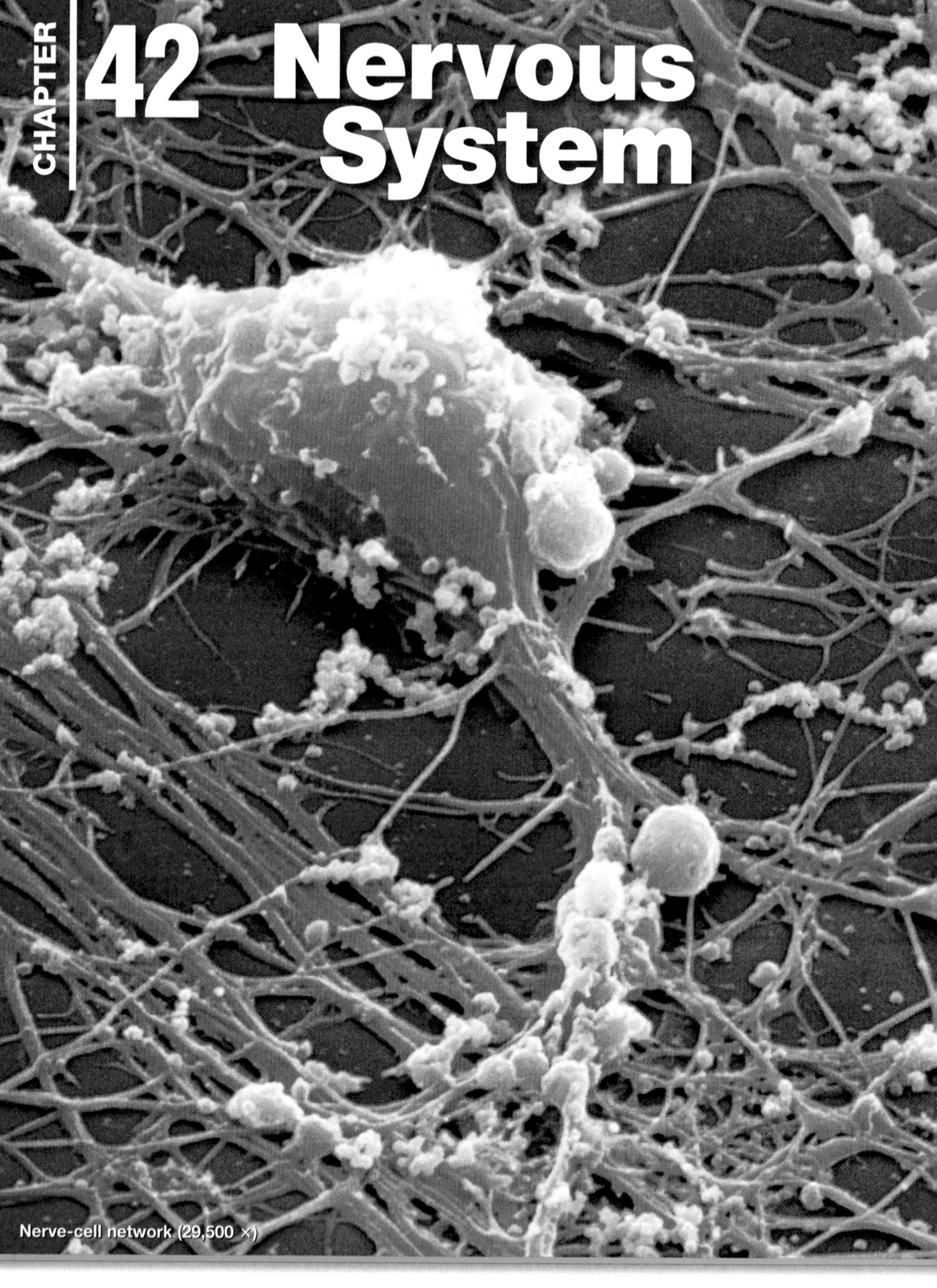

Nerve-cell network (29,500 ×)

What's Ahead — and Why

CHAPTER OUTLINE

Students will learn

- how nerve cells conduct electrochemical signals.
- about the parts and functions of the nervous system.
- how the sense organs detect sensory stimuli.
- how addictive drugs affect the nervous system.

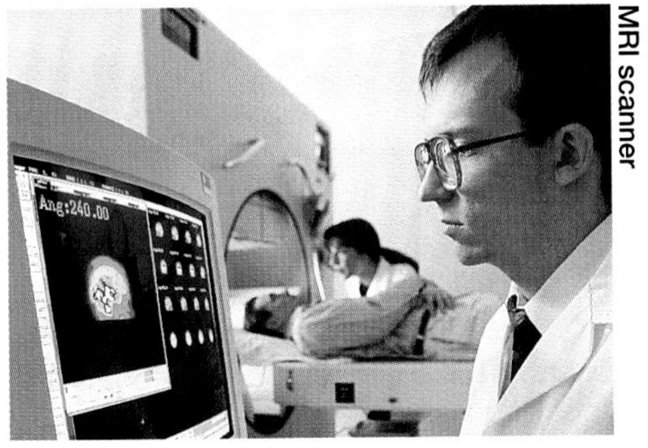

MRI scanner

Thought... emotion... sensation...

These are made possible by the nervous system. This patient is about to be scanned by a machine that produces magnetic resonance images (MRI). An MRI is a detailed picture of a body structure, such as the brain. The use of MRI scanners is helping scientists learn a great deal about the nervous system.

Study TIP

Ready?

Check your understanding of these topics to help you prepare for what's ahead.

Can you...

- **describe** the importance of ion channels in cell transport? (Section 4-1)
- **identify** the role of sodium-potassium pumps in cells? (Section 4-2)
- **distinguish** between endocytosis and exocytosis? (Section 4-2)
- **list** three functions of receptor proteins? (Section 4-2)

If not, *review the sections indicated.*

Looking Ahead

internetconnect

sciLINKS NSTA — National Science Teachers Association *sci*LINKS Internet resources are located throughout this chapter.

Reading Strategies

Active Reading

Technique: Anticipation Guide

Write the following statements on the board or overhead:

1. Addiction is a purely psychological response to drug use.
2. Reflexes occur before the brain is aware of danger.
3. Prolonged exposure to loud noise can cause permanent hearing loss.

Ask students whether they agree or disagree with each statement. Have them save their lists for use in the **Closure** on page 974.

Directed Reading

Assign **Directed Reading Worksheet Chapter 42** to help students interact with the text as they read the chapter.

Interactive Reading

Assign ***Biology: Principles and Explorations*** **Audio CD Program Chapter 42** to help reluctant readers achieve greater success in reading the chapter.

Reading Skills

Assign Reading Organizers and Reading Strategies Worksheets from the **One-Stop Planner CD-ROM with Test Generator** to help students learn and practice effective reading skills.

internetconnect

Holt *Biology: Principles and Explorations*

On-line Resources: go.hrw.com

Visit the HRW Web site for a variety of resources related to this chapter. Just type in the keyword HX0 Chapter 42.

Holt *Biology: Principles and Explorations*

On-line Resources: www.scilinks.org

The following *sci*LINKS Internet resources can be found in the student text for this chapter:

Page 952
TOPIC: Neurons
KEYWORD: HX952

Page 961
TOPIC: Spinal cord
KEYWORD: HX961

Page 963
TOPIC: Autonomic nervous system
KEYWORD: HX963

Page 964
TOPIC: Sensory receptors
KEYWORD: HX964

Page 968
TOPIC: Chemical senses
KEYWORD: HX968

Page 973
TOPIC: Blood alcohol concentration
KEYWORD: HX973

Page 979
TOPIC: Drug addiction
KEYWORD: HX979

Chapter 42

1 Prepare

Scheduling

- **Block:** About 90 minutes will be needed to complete this section.
- **Traditional:** About two class periods will be needed to complete this section.
- **Planning:** See the Planning Guide on pp. 949A and 949B for lecture, classwork, and assignment options for Section 42-1. Lesson plans for Section 42-1 can be found in a lesson cycle format in ***Biology: Principles and Explorations*** **One-Stop Planner CD-ROM with Test Generator.**

Resource Link

Assign **Active Reading Worksheet Section 42-1** to help students comprehend the key concepts and visuals in this lesson.

2 Teach

Lesson Warm-up

Ask students to recall the game "Grapevine," in which one person whispers a statement to the next person, who whispers to the next, and so on. Discuss the similarities and differences between this game and nerve impulses passing through the body. *(Similarities: a message travels along passing from one part [or person] to the next. Differences: the message usually is garbled in the game, but it is not in a nerve impulse.)*

internet**connect**

Have students research neurons using the Web site in the Internet Connect box on page 952. Have students write a report summarizing their findings. PORTFOLIO

Section 42-1 Neurons and Nerve Impulses

Objectives

- **Analyze** the structure and function of neurons. (pp. 952–953)
- **Describe** how the resting potential is established in a neuron. (p. 954)
- **Sequence** the steps of a nerve impulse. (p. 955)
- **List** the events that occur in synaptic transmission of a nerve impulse. (p. 956)

Key Terms

neuron
dendrite
axon
nerve
membrane potential
resting potential
action potential
synapse
neurotransmitter

internet**connect**

TOPIC: Neurons
GO TO: www.scilinks.org
KEYWORD: HX952

Neurons Conduct Electrical Signals

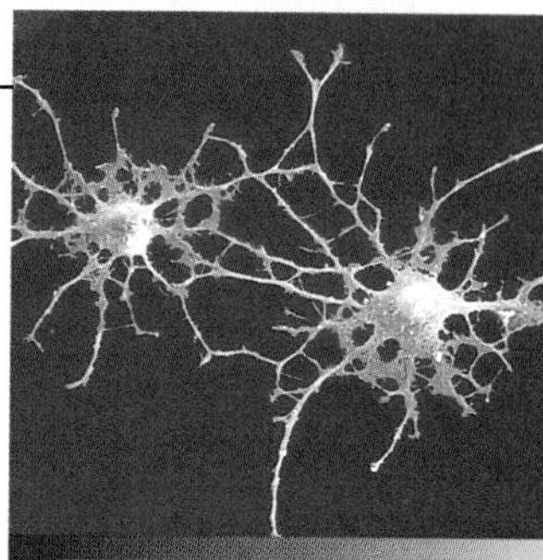

Figure 42-1 Neurons

How many? An average adult human brain contains about 100 billion neurons.

If your body used only chemical signals to send messages, your body's interaction with the environment would be slow. A quicker means of communication is needed, especially if your brain has an urgent message for the muscles in your legs, such as "Contract quickly, a speeding car is headed this way!" In addition to chemical signals, your nervous system uses electrical signals to send messages rapidly throughout your body.

The nervous system contains a complex network of nerve cells, or **neurons** *(NOO rahns)*. Neurons, such as the cells shown in **Figure 42-1,** are specialized cells that transmit information throughout the body. Neurons enable many important functions, such as movement, perception, thought, emotion, and learning.

Structure of neurons

A neuron's unique structure enables it to conduct electrical signals called nerve impulses. Neurons communicate by transmitting nerve impulses to body tissues and organs, including muscles, glands, and other neurons. Neurons vary greatly, but a typical neuron is similar to the one shown in **Figure 42-2. Dendrites** *(DEHN dreyets),* which extend from the cell body of the neuron, are the "antennae" of the neuron. Dendrites receive information from other cells. The cell body collects information from dendrites, relays this information to other parts of the neuron, and maintains the general functioning of the neuron. An **axon** is a long extension of the cytoplasm that conducts nerve impulses. The end of an axon is called an axon terminal. When the neuron transmits a nerve impulse to other cells, the nerve impulse leaves from axon terminals.

Nervous tissue consists mostly of neurons and their supporting cells. Bundles of neurons are called **nerves,** which contain the axons of many different neurons. The arrangement of axons in a nerve is similar to a telephone line with many different communication channels. Nerves appear as fine, white threads when viewed with the unaided eye.

Integrating Different Learning Styles

Logical/Mathematical	**PE**	Data Lab, p. 954
	ATE	Cross-Disciplinary Connection, p. 956
Visual/Spatial	**PE**	Study Tip, p. 955; Performance Zone items 16, 17, 18, and 19, p. 977
	ATE	Visual Strategies, pp. 953, 955, 956; Reteaching, p. 957
Verbal/Linguistic	**PE**	Study Tip, p. 954
	ATE	Internet Connect, p. 952; Closure, p. 957

Some neurons are insulated

Many neurons have a fatty outer layer called a myelin *(MEYE uh lihn)* sheath, as shown in Figure 42-2. Myelin insulates the axons of these neurons. Myelin is produced by supporting cells that surround the axon. The presence of myelin causes nerve impulses to move faster down the axon. The myelin sheath is interrupted at intervals, leaving gaps called nodes of Ranvier *(RAHN vee ay)*, where the axon membrane is exposed to the surrounding fluid. Conduction of nerve impulses is faster in myelinated axons because nerve impulses "jump" from node to node as they move down the length of the axon. Axons with a large diameter conduct nerve impulses faster than axons with a small diameter. Thus, a myelinated axon conducts nerve impulses much faster than an unmyelinated axon, assuming that both axons had the same diameter before the myelin sheath formed. Myelin is especially beneficial in neurons that must conduct nerve impulses very rapidly, such as those involved with quick movement.

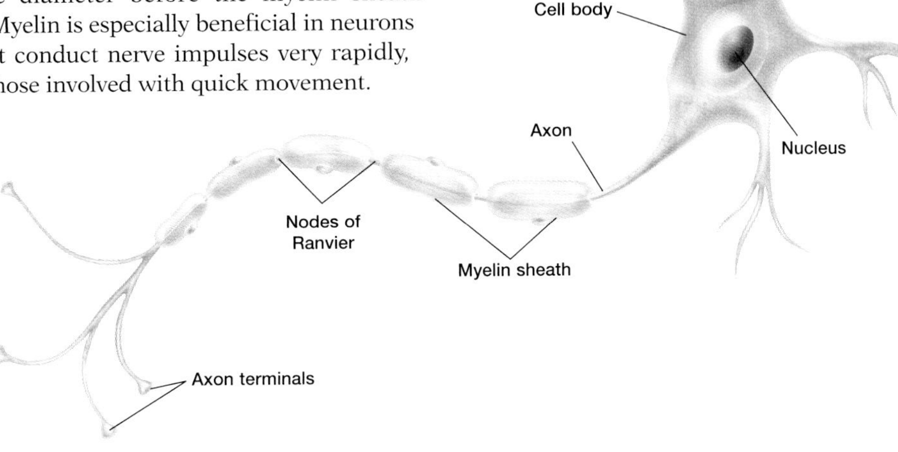

Figure 42-2 Myelinated neuron. A myelin sheath covers the axons of many neurons. Myelin increases the speed of nerve impulses.

Neuron function depends on electrical activity

Neurons have an electrical charge that is different from the electrical charge of the fluid that surrounds them. The difference in electrical charge across the cell membrane is called the **membrane potential.** The membrane potential, which is measured in units called volts (V), depends on the movement of ions into and out of the cell. This movement depends on the relative concentration of ions inside and outside the cell, the ability of the ions to pass through the cell membrane, and the electrical charge of the ions.

Ions diffuse across a neuron's cell membrane through special proteins called voltage-gated ion channels. These channel proteins are permeable only to specific ions. The channels are voltage-gated because whether they are open or closed depends on the membrane potential. Voltage-gated ion channels are embedded in the cell membrane along the entire length of the axon, so even a small change in the membrane potential can affect the permeability of the axon. Thus, the permeability of the cell membrane to certain ions depends on the membrane potential. Yet, the membrane potential depends on whether certain voltage-gated ion channels are open or closed.

biobit

What happens when myelin goes bad?

A disease called Guillain-Barré syndrome attacks the myelin sheaths on neurons that stimulate muscles and skin. The conduction of nerve impulses slows and may even stop completely, severely impairing responses to stimuli.

Teaching TIP — BASIC

Myelin Sheath

Tell students that the fatty myelin sheath around an axon acts like the insulating plastic around electrical wires. It prevents "short-circuiting" in the nervous system.

Visual Strategy

Figure 42-2

Have students observe prepared slides of neurons with simple morphology (such as stellate neurons). Remind students to look for cell bodies, axons, and dendrites. Have students draw and label what they see and compare their drawings with the neuron shown in Figure 42-2.

did you know?

Phenylketonurotics must not ingest phenylalanine.

Phenylketonuria, or PKU, is a disease caused by an enzyme deficiency that leads to a buildup of phenylpyruvic acid in the body, leading to the destruction of the myelin sheaths on neurons. If undetected, it can lead to severe mental retardation and even death. PKU is easy to detect, so all newborns born in the United States are tested for it. If found, a low phenylalanine diet is prescribed. Food products such as diet sodas carry a message to phenylketonurotics warning them that the product contains phenylalanine.

Teaching TIP

Ions

Remind students that electrons are negatively charged particles that move around the nucleus of an atom and that an ion is an atom or molecule that has either a positive charge (it has lost one or more electrons) or a negative charge (it has gained one or more electrons).

Teaching TIP

Resting Potential

The term *resting potential* is misleading because there is constant activity in a neuron even when the neuron is not conducting a nerve impulse. Ions are continuously being transported and are diffusing across the membrane.

Analyzing changes during a nerve impulse

Time 10 minutes

Process Skills Interpreting graphs, inferring relationships

Teaching Strategies
Remind students that a voltage-gated ion channel acts like the gate of a toll booth—if the correct change (voltage) is not present, the gate does not open.

Analysis Answers

1. about 1.5 milliseconds
2. open
3. open
4. an influx of positively charged ions (sodium)
5. an outflow of positively charged ions (potassium)

Study TIP

Reading Effectively

The membrane potential of a neuron is expressed in millivolts (mV). A millivolt is equal to one-thousandth of a volt (V). As shown in the graph in the Data Lab below, the membrane potential can be positive or negative. The resting potential of an average neuron is about –70 mV. During an action potential, the membrane potential of the neuron reaches about +40 mV.

Resting potential

When a neuron is not conducting a nerve impulse, the neuron is said to be at rest. The membrane potential of a neuron at rest is called the **resting potential.** In a typical neuron, the resting potential is negative, about –70 millivolts (mV). At the resting potential, the inside of the cell is negatively charged with respect to the outside of the cell. Why is the resting potential negative? Recall that sodium-potassium pumps actively transport sodium ions, Na^+, out of a cell and potassium ions, K^+, into the cell. This results in a greater concentration of sodium ions outside the cell than inside the cell, and a greater concentration of potassium ions inside the cell than outside the cell. In a neuron, voltage-gated sodium channels are closed at the resting potential. Thus, very few sodium ions can diffuse into the cell, despite their strong concentration gradient. Voltage-gated potassium channels are open at the resting potential. Potassium ions can therefore diffuse out of the cell down their concentration gradient, carrying their positive charge with them. Neurons also contain negatively charged proteins that are too large to exit the cell.

Action potential

When a neuron is conducting a nerve impulse, changes occur in the cell membrane of the neuron. A nerve impulse is also called an action potential. An **action potential** is a local reversal of polarity inside the neuron. An action potential moves down an axon like a flame burning down a fuse. The events of an action potential are summarized in **Figure 42-3.**

Analyzing changes during a nerve impulse

Background

The graph below illustrates changes that occur in the membrane potential of a neuron during an action potential. Use the graph to answer the following questions. Refer to Figure 42-3 as needed.

Analysis

1. **Determine** about how long an action potential lasts.
2. **State** whether voltage-gated sodium channels are open or closed at point *A*.
3. **State** whether voltage-gated potassium channels are open or closed at point *B*.
4. **SKILL Recognizing Relationships** What causes the increase in the membrane potential shown at point *A*?
5. **SKILL Recognizing Relationships** What causes the decrease in the membrane potential shown at point *B*?

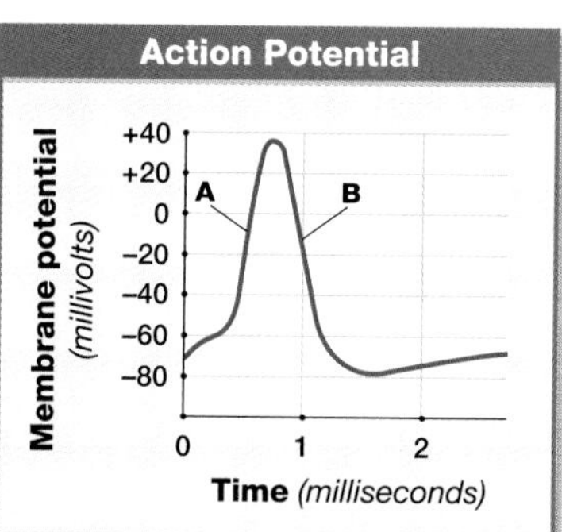

did you know?

The squid (Loligo sp.) *was one of the first organisms used to measure the resting potential of a neuron.*

Because of its large, long axons that innervate muscles responsible for expelling water from the body cavity, the squid makes an ideal specimen for studies in cellular neurophysiology. Some of these axons are as large as 1 mm in diameter, making them easy to work with.

Figure 42-3

Conduction of a Nerve Impulse

An action potential moves rapidly down an axon.

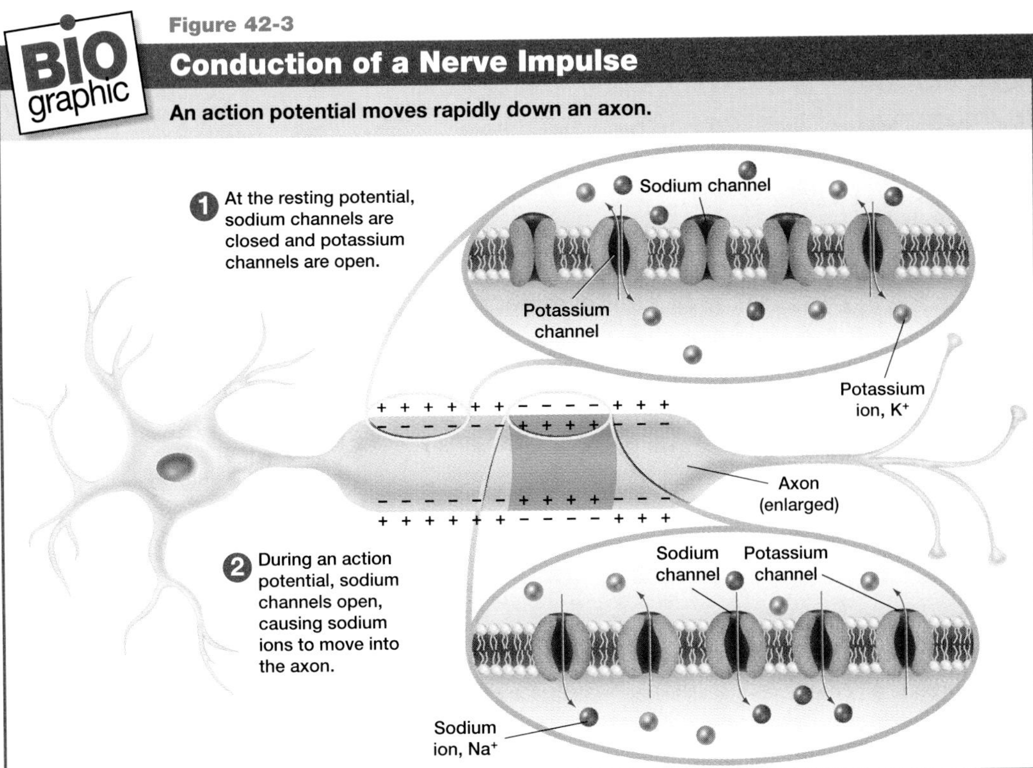

1 At the resting potential, sodium channels are closed and potassium channels are open.
Sodium channel
Potassium channel
Potassium ion, K+
+ + + + + + – – – – + + +
– – – – – – + + + + – – –
Axon (enlarged)
2 During an action potential, sodium channels open, causing sodium ions to move into the axon.
Sodium channel
Potassium channel
Sodium ion, Na+

Step 1 At the resting potential, the inside of the neuron is negatively charged with respect to the outside of the neuron. The neuron is ready to conduct an action potential.

Step 2 An action potential begins when a stimulus, such as a signal molecule, causes a positive change in the membrane potential near the axon. This change causes voltage-gated sodium channels to open, and sodium ions rapidly flow into the axon. For a brief moment, the membrane potential increases to about +40 mV as the inside of the axon becomes positively charged. This sudden local reversal of polarity begins a chain reaction that causes voltage-gated sodium channels to open down the entire length of the axon. As each sodium channel opens, sodium ions flow into the axon. The action potential conducts rapidly down the axon toward the axon terminals.

Voltage-gated sodium channels close immediately after the action potential has passed. Then voltage-gated potassium channels open, causing potassium ions to flow out of the axon. As a result, the membrane potential becomes negative again immediately after the action potential. The resting potential is fully restored as sodium-potassium pumps reestablish the original concentrations of sodium ions and potassium ions inside and outside the axon. The neuron cannot conduct another action potential until then.

Study TIP

Interpreting Graphics

As you look at Figure 42-3, notice that an action potential moves down the axon in one direction only—away from the cell body and toward the axon terminals.

Multicultural Perspective

Dr. Emmeline Edwards

Dr. Emmeline Edwards, an African-American scientist who grew up in Haiti, studies the role that neurochemicals play in coping with stress. Information gained from her research can be applied toward the development of drugs for treating human depression. Dr. Edwards works at the Department of Psychiatry at the State University of New York, in Stony Brook.

Making Biology Relevant

Medicine

The cause of multiple sclerosis is unknown, but many scientists think that it results from the actions of the immune system against the myelin sheaths of the central nervous system. The myelin sheaths in the brain and spinal cord become sclerotic, or hard, causing poor impulse conductivity. Sometimes the symptoms of multiple sclerosis disappear altogether, but they may return after periods of remission. Current therapies include corticosteroids and beta interferon.

Visual Strategy

Figure 42-3

This figure illustrates action potential conduction down an axon. The axon of this neuron has been enlarged to show that the membrane potential becomes positive during an action potential. Tell students that the reversal of polarity during an action potential is caused by the rapid influx of sodium ions through voltage-gated sodium channels. After an action potential has passed, voltage-gated potassium channels remain open, causing potassium ions to diffuse out of the cell. Thus, the membrane repolarizes shortly after the action potential.

Chapter 42

CROSS-DISCIPLINARY CONNECTION

◆ **Mathematics**

A single neuron may have synapses with as many as 1,000 other neurons. Have the class calculate how many neurons could be stimulated if each of these 1,000 neurons forms synapses with another 1,000 neurons. Students should readily see how quickly and widely nerve impulses can travel. In reality, some of these synapses are excitatory, while others are inhibitory. Each neuron "evaluates" the impulses it receives, averages the input, and responds accordingly. For example, if the overall total input is excitatory, the receiving neuron transmits an impulse to a neighboring cell.

Visual Strategy

Figure 42-4

Use this figure to illustrate how impulses are transmitted between neurons. Have students identify the dendrites and the cell body, which receive impulses. Then have students identify the axon terminals, which allow the signal to pass to the next neuron in a circuit. Tell students that neurotransmitter molecules are released at the axon terminals, and they then travel to adjacent dendrites and cell bodies. Thus, an impulse can travel in only one direction—away from the cell body of the neuron and toward its axon terminals.

biobit

What is Alzheimer's disease?

Alzheimer's disease is a progressive deterioration of the brain that affects about 3 million people—most of them elderly—in the United States. In people with Alzheimer's disease, neurotransmitter function is impaired, resulting in confusion and disorientation.

Neurons Use Neurotransmitters to Communicate

When a nerve impulse arrives at axon terminals, the impulse can be transmitted to other cells. A junction at which a neuron meets another cell is called a **synapse** *(SIHN aps),* shown in **Figure 42-4.** At synapses, neurons usually do not touch one another or the cells they stimulate. Between an axon terminal and a receiving cell is a tiny gap called a synaptic cleft. At a synapse, the transmitting neuron is called a presynaptic neuron, and the receiving cell is called a postsynaptic cell. When a nerve impulse arrives at an axon terminal of a presynaptic neuron, the impulse must cross the synaptic cleft to continue the signal to a postsynaptic cell.

In the transmission of a nerve impulse across a synapse, electrical energy is converted to chemical energy and then back to electrical energy. Most nerve impulses are transmitted across the synaptic cleft by signal molecules called **neurotransmitters.** Neurotransmitters are produced by neurons and are stored inside vesicles. There are many different neurotransmitters and several mechanisms of neurotransmitter action. For example, in human muscles the principal neurotransmitter is acetylcholine *(as ee tihl KOH leen),* while in the brain it is glutamate *(GLOO tuh mayt).*

Figure 42-4 Synapse. A synapse is a junction at which signals are transmitted between a neuron and another cell.

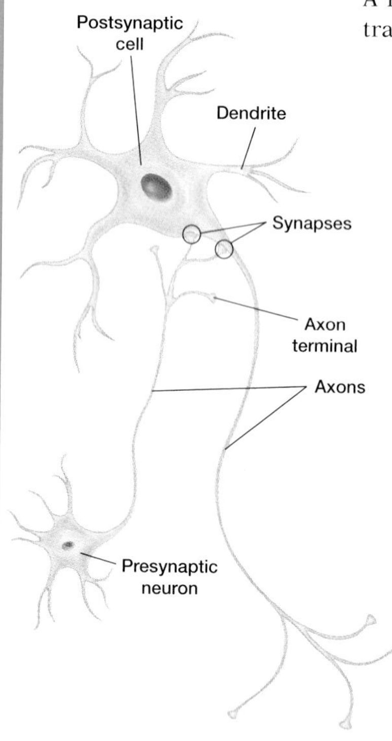

Release of neurotransmitters

A nerve impulse causes a presynaptic neuron to release a neurotransmitter into the synaptic cleft. When an action potential reaches an axon terminal of the presynaptic neuron, vesicles that contain neurotransmitters fuse with the cell membrane. This releases neurotransmitter molecules into the synaptic cleft by exocytosis. Neurotransmitter molecules diffuse across the synaptic cleft and interact with the postsynaptic cell. As shown in **Figure 42-5,** some neurotransmitters bind to receptor proteins on the postsynaptic cell. In some cells, ion channels open when a neurotransmitter binds to these receptor proteins. Such channels are called chemical-gated ion channels; whether these channels are open or closed depends on the binding of a chemical—in this case a neurotransmitter.

A neurotransmitter may either excite or inhibit the activity of a postsynaptic cell it binds to. For example, when the neurotransmitter opens chemical-gated ion channels, ions move across the cell membrane of the postsynaptic cell. This causes the membrane potential of the postsynaptic cell to change depending on the charge of the ions that move into or out of the cell. If positively charged ions enter a postsynaptic neuron, an action potential may be produced (excitation). On the other hand, if positively charged ions flow out of the postsynaptic neuron, or if negatively charged ions enter the neuron, an action potential may be suppressed (inhibition).

did you know?

Extremely small amounts of neurotransmitters underlie all of our moods, from ecstasy to depression.

The neurotransmitters norepinephrine and dopamine are involved in feelings of pleasure. If the levels of these neurotransmitters are too low in the brain, we might feel depressed. Endorphins and enkephalins are substances that block pain signals. Their production increases during physical stress, and runners who experience the increase refer to the sensation as a "runner's high." Another compound, substance P, is released in response to certain painful stimuli, such as ingestion of capsaicin, a chemical irritant in hot peppers. Our neurotransmitter levels react to emotional and physical stressors, such as our thoughts and environment. You can literally have an "attitude" that can affect your health.

Not all neurotransmitters bind to receptor proteins once they are released into the synaptic cleft, but unused neurotransmitters do not remain in the synaptic cleft indefinitely. Instead, most neurotransmitters are cleared from the synaptic cleft very shortly after they are released. Many presynaptic neurons reabsorb neurotransmitters and use them again. At other synapses, neurotransmitters are broken down in the synaptic cleft by enzymes or other chemicals. This happens, for example, at the synapses between motor neurons and muscle cells. The reuptake or breakdown of neurotransmitters ensures that postsynaptic cells do not become overstimulated.

Figure 42-5 Synaptic transmission

Neurotransmitters are released from a presynaptic neuron and diffuse across the synaptic cleft, stimulating a postsynaptic cell.

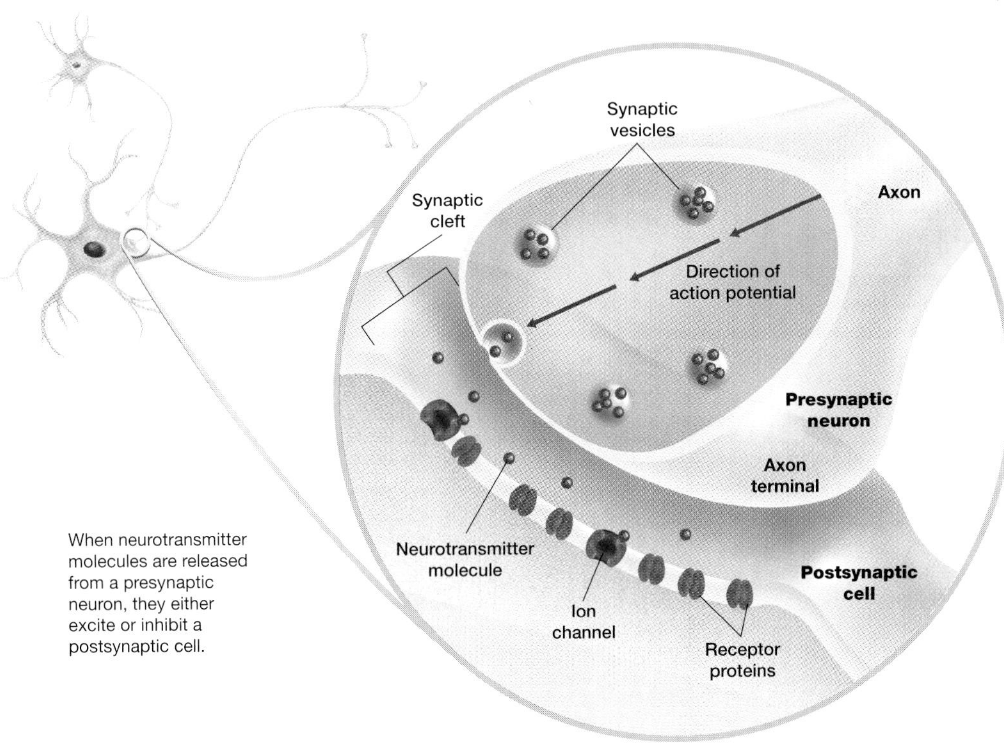

Review SECTION 42-1

1. **Describe** the structure of a typical neuron.
2. **Describe** how the movement of ions across the cell membrane determines the membrane potential.
3. **Summarize** the events involved in the synaptic transmission of a nerve impulse.
4. **SKILL Relating Concepts** What must happen to the cell membrane of a neuron for an action potential to occur?
5. **SKILL Inferring Relationships** How does the membrane potential affect the permeability of a neuron's cell membrane?

Review SECTION 42-1 ANSWERS

1. A bulky cell body is studded with dendrites, which are membrane-covered processes. One or more axons, which are long, membrane-covered processes extend from the cell body to adjacent cells.
2. When ions move across the cell membrane, the relative concentration of ions changes inside and outside the cell. This results in a change of charge. The difference in electrical charge across a membrane is the membrane potential.
3. At the synaptic cleft, the presynaptic neuron releases a neurotransmitter substance. The neurotransmitter diffuses across the synaptic cleft, and binds to the postsynaptic cell, which converts the signal back into an electrical signal.
4. The membrane potential must become sufficiently depolarized (less negative) so that voltage-gated sodium channels can open. Then sodium ions flow into the axon, producing an action potential.
5. The membrane potential determines whether voltage-gated ion channels are opened or closed.

Visual Strategy

Figure 42-5

This figure illustrates chemical synaptic transmission. Neurotransmitter molecules released from a presynaptic neuron bind to receptor proteins on a postsynaptic cell, opening ion channels in the postsynaptic cell. Emphasize to students that neurotransmitters can be either excitatory or inhibitory depending on the nature of the neuron they stimulate and the mechanism of their action.

3 Assessment Options

Closure

Summarizing an Action Potential

Have students describe the events involved in a neuron's propagation of an action potential, including depolarization, movement of the action potential down the axon, release of neurotransmitter substance from the axon terminal, diffusion across the synaptic cleft, and binding to the post-synaptic neuron's membrane.

Review

Assign Review—Section 42-1.

Reteaching — BASIC

Have students use clay or another moldable material to build a model of a neuron and synapse. Ask them to label the model, showing at what points the message is electrical and at what points the message is chemical. Have the students use arrows to show the direction of the impulse. PORTFOLIO

Chapter 42

1 Prepare

Scheduling

- **Block:** About 90 minutes will be needed to complete this section.
- **Traditional:** About two class periods will be needed to complete this section.
- **Planning:** See the Planning Guide on pp. 949A and 949B for lecture, classwork, and assignment options for Section 42-2. Lesson plans for Section 42-2 can be found in a lesson cycle format in ***Biology: Principles and Explorations* One-Stop Planner CD-ROM with Test Generator.**

Resource Link

Assign **Active Reading Worksheet Section 42-2** to help students comprehend the key concepts and visuals in this lesson.

2 Teach

Lesson Warm-up

Have a volunteer stand at the front of the room. While talking to the class, suddenly snap your fingers directly in front of the student's face. What reaction does he or she have? Reflexive blinking occurs even though the student may not think you will poke him or her in the eye. Try snapping again a few more times. Can the student overcome the reflex when not surprised?

Section 42-2

Structures of the Nervous System

Objectives

- **Distinguish** between the central nervous system and the peripheral nervous system. (p. 958)
- **Identify** the major parts of the brain and their functions. (pp. 958–959)
- **Describe** the structure of the spinal cord. (p. 960)
- **Sequence** the events of a spinal reflex. (p. 962)
- **Compare** the somatic nervous system with the autonomic nervous system. (pp. 962–963)

Key Terms

central nervous system
peripheral nervous system
sensory neuron
motor neuron
brain
cerebrum
cerebellum
brain stem
thalamus
hypothalamus
spinal cord
reflex
interneuron

The Central Nervous System Controls the Body

Neurons are the most important cells of the nervous system. Functions of the nervous system depend on the complex interaction between billions of neurons. Networks of neurons constantly gather, integrate, interpret, and respond to information about the body's internal condition and environmental conditions. How are neurons organized in the nervous system? As shown in **Figure 42-6,** there are two main divisions of the nervous system—the central nervous system, shown in orange, and the peripheral nervous system, shown in purple. The **central nervous system** (CNS) consists of the brain and the spinal cord. The CNS is the control center of the body. The CNS interprets and responds to information from the environment and from within the body. The **peripheral nervous system** (PNS) contains sensory neurons and motor neurons. **Sensory neurons** send information from sense organs, such as the skin, to the CNS. **Motor neurons** send commands from the CNS to muscles and other organs.

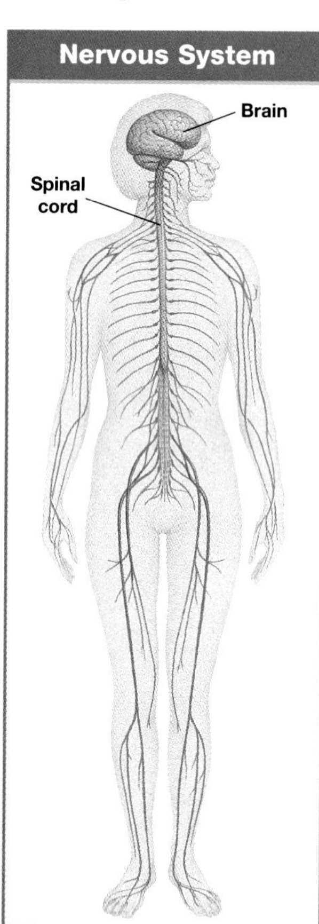

Figure 42-6 Nervous system. The central nervous system (orange) consists of the brain and the spinal cord. The peripheral nervous system (purple) branches throughout the body.

Brain

The **brain** is the body's main processing center. Encased entirely within the skull, the brain contains about 100 billion neurons. An average adult brain weighs about 1.5 kg (3 lb). Thoughts, feelings, emotions, behavior, perception, and memories are controlled by your brain. Your brain also enables you to learn and process information, such as the text in this book. Scientists have determined the location of various functions in the brain. The brain consists of three major parts, shown in **Figure 42-7**—the cerebrum, the cerebellum, and the brain stem.

Integrating Different Learning Styles

Logical/Mathematical	**PE**	Performance Zone item 24, p. 977
Visual/Spatial	**PE**	Study Tips, pp. 960, 962; Performance Zone item 1, p. 976
	ATE	Demonstration, pp. 959, 961; Health Watch, p. 961; Teaching Tip, p. 962
Body/Kinesthetic	**ATE**	Lesson Warm-up, p. 958; Demonstrations, pp. 961, 962; Closure, p. 963
Interpersonal	**ATE**	Demonstrations, pp. 961, 962; Closure, p. 963
Verbal/Linguistic	**PE**	Real Life, p. 959; Performance Zone item 26, p. 977
	ATE	Making Biology Relevant, p. 960; Internet Connect, p. 963

Cerebrum The **cerebrum** *(seh REE bruhm)* is the largest part of the brain. The capacity for learning, memory, perception, and intellectual function resides in the cerebrum. The cerebrum has a folded outer layer with many bumps and grooves. A long, deep groove down the center of the brain divides the cerebrum into right and left halves, or hemispheres. The cerebral hemispheres communicate through a connecting band of axons called the corpus callosum *(KOR puhs kuh LOH suhm)*. In general, the left cerebral hemisphere receives sensations from and controls movements of the right side of the body. The right cerebral hemisphere receives sensations from and controls movements of the left side of the body.

Most sensory and motor processing occurs in the cerebral cortex *(KOHR teks)*, the folded, thin (2–4 mm) outer layer of the cerebrum. The cerebral cortex contains about 10 percent of the brain's neurons. The folds on the outer surface of the cerebrum accommodate the large surface area of the cerebral cortex. The cerebral cortex is primarily involved with the functioning of sensory systems.

Cerebellum The **cerebellum** *(ser uh BEL uhm)*, which is located at the posterior base of the brain, regulates balance, posture, and movement. The cerebellum smooths and coordinates ongoing movements, such as walking, by timing the contraction of skeletal muscles. The cerebellum integrates and responds to information about body position from the cerebrum and the spinal cord to control balance and posture.

Brain stem At the base of the brain is the stalklike **brain stem.** The brain stem is a collection of structures leading down to the spinal cord and connecting the cerebral hemispheres with the cerebellum. The lower brain stem consists of the midbrain, the pons,

REAL LIFE

Each year, nearly 250 bicyclists die because of brain injuries. Wearing a bicycle helmet reduces the risk of head trauma by more than 70 percent.

Evaluating Viewpoints
Should there be helmet laws for bicyclists just as there are seatbelt laws for automobile drivers?

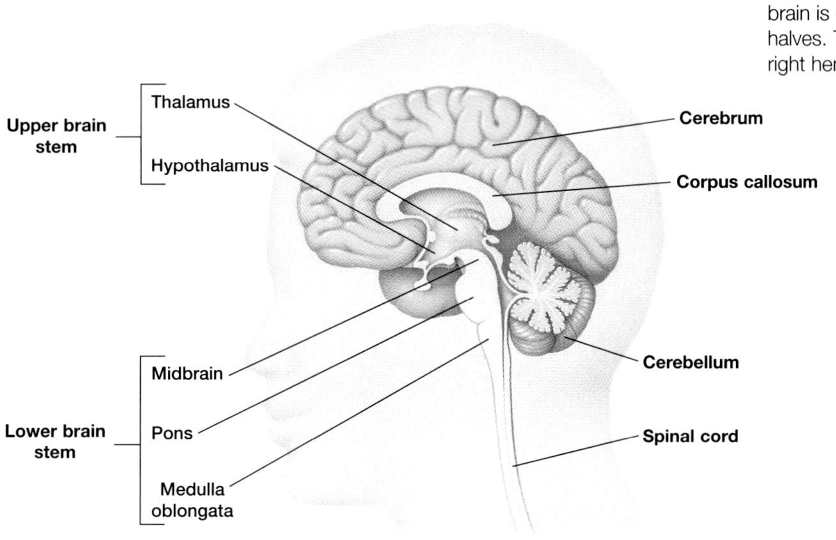

Right hemisphere of brain
Upper brain stem
Thalamus
Hypothalamus
Cerebrum
Corpus callosum
Lower brain stem
Midbrain
Pons
Medulla oblongata
Cerebellum
Spinal cord

Figure 42-7 **Brain.** The brain is divided into two halves. This view shows the right hemisphere of the brain.

REAL LIFE

Answer

Students' answers will vary. Proponents of helmet laws may cite the large cost, usually born by the public, of caring for head-injury victims.

Demonstration

Tricking the Brain

Bring in a simple maze or make one yourself. Have a student complete the maze. Here's the catch: Stand two textbooks on the table. Place the maze at the base of the books, and have the student hold a mirror behind the maze. Ask the student to put his or her chin on top of the textbooks. They can only look in the mirror to do the maze! Discuss why it was difficult to accomplish this task.

Teaching TIP — ADVANCED

Vocabulary

Tell students that the raised surfaces and grooves on the brain are called *gyri* (sing. *gyrus*) and *sulci* (sing. *sulcus*).

Teaching TIP

Meningitis

Remind students that the brain is encased in the skull. In addition, tough protective membranes known as meninges cover the brain. Inflammation of these meninges, caused by bacteria or viruses, is known as meningitis and can be fatal if not treated successfully.

Overcoming Misconceptions

Early Hominids

Dispel any misconception that modern *Homo sapiens* have the largest brain of any hominid in history. Remind students that Neanderthals had larger brains than those of some modern humans. Obviously, overall brain size is not an accurate indicator of intelligence. The brain of modern humans is much more convoluted, however, than that of the extinct Neanderthals, suggesting that the extent of convolutions (which is related to the size of the cortex) might be related to intelligence. The relative mass of the brain in relation to the mass or even the surface area of the body may be a better indicator of relative intelligence among different species. For example, whales and porpoises have brains larger than those of humans. When differences in body mass are factored out, however, humans have the largest brains in relation to the rest of the body.

Making Biology Relevant
Health

Medical scientists think that some psychosomatic illnesses may be mediated by the hypothalamus through its connections to the autonomic nervous system and the endocrine system. Many stress-related illnesses, such as ulcers, asthma, and high blood pressure, are clearly connected to hypothalamic functions.

Career
Medical Imaging Technician

Have students prepare a report on the job of a medical imaging technician. They should include descriptions of the different medical procedures these workers perform such as computerized axial tomography (CAT), magnetic resonance imaging (MRI), and positron emission tomography (PET). They should also include a description of training necessary and starting salary. PORTFOLIO

Making Biology Relevant
Health

When no electrical activity in the brain can be registered, a person may be declared "brain dead." In such people, the vital body processes controlled unconsciously by the brain stem in some cases can be sustained by a life-support system. Have students research the legal definition of death in different states and countries. Each student could be assigned one state, for example. PORTFOLIO

Study TIP

Interpreting Graphics

As you look at Figure 42-8, think of the spinal cord as a busy two-way highway with sensory traffic going north and motor traffic going south.

and the medulla oblongata *(mi DUHL uh ahb lahn GAHT uh)*. These structures relay information throughout the CNS and play an important role in homeostasis by regulating vital functions, such as heart rate, breathing rate, body temperature, and sleep.

The upper brain stem contains important relay centers that direct information to and from different parts of the brain. The **thalamus** *(THAL uh muhs)* is a critical site of sensory processing. Sensory information from all parts of the body converges on the thalamus, which relays the information to appropriate areas of the cerebral cortex. Below the thalamus, at the base of the brain, is the hypothalamus. The **hypothalamus,** along with the medulla oblongata, regulates many vital homeostatic functions, such as breathing and heart rate. The hypothalamus is responsible for feelings of hunger and thirst. It also regulates many functions of the endocrine system by controlling the secretion of many hormones.

The thalamus and hypothalamus are linked to some areas of the cerebral cortex by an extensive network of neurons called the limbic system. The limbic system includes structures of both the brain stem and the cerebrum. The limbic system has an important role in memory, learning, and various emotions, such as pleasure, anger, and lust.

Spinal cord

The **spinal cord,** shown in **Figure 42-8,** is a dense cable of nervous tissue that runs through the vertebral column. The spinal cord extends from the medulla oblongata through the vertebrae to a level just below the ribs. The spinal cord links the brain to the PNS. The brain receives information that travels upward through the spinal cord. Down through the spinal cord, the brain also sends commands that control the rest of the body. In addition to relaying messages, the spinal cord functions in reflexes. A **reflex** is a sudden, involuntary movement of muscles in response to a stimulus.

Figure 42-8 Spinal cord

Spinal nerves have a dorsal root and a ventral root that diverge near the spinal cord.

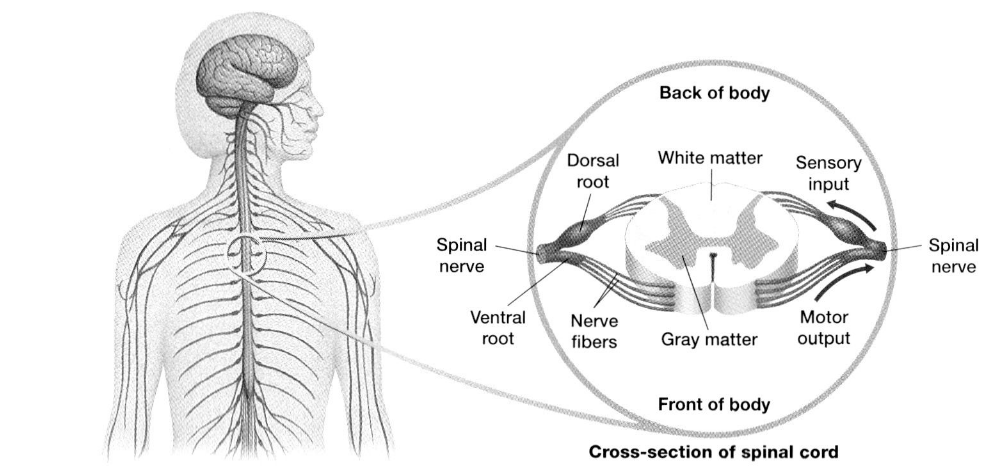

Overcoming Misconceptions
Brain Tumors

Because most neurons do not undergo mitosis in adult humans, students may think that a brain cannot develop a tumor. Brain tumors, however, are not formed from neurons. They arise from the supportive glial cells in the nervous system, which do undergo mitosis.

The spinal cord is linked to the PNS through 31 pairs of spinal nerves. The spinal nerves, which branch from the spinal cord, carry information to and from the CNS. Spinal nerves in the upper part of the spinal cord stimulate the arms and upper body, and spinal nerves in the lower part of the spinal cord stimulate the legs and lower body. Each spinal nerve has a dorsal root and a ventral root. Dorsal roots contain sensory neurons, which carry information from areas of sensory input to the CNS. Ventral roots contain motor neurons, which carry motor responses from the CNS to muscles, glands, and other organs. As shown in Figure 42-8, dorsal and ventral roots diverge from the spinal nerves near the spinal cord.

The spinal cord contains a core of gray matter covered by a sheath of white matter, as shown in Figure 42-8. Gray matter contains the cell bodies of neurons, whereas white matter contains the axons of neurons. Also within the spinal cord are **interneurons,** which link neurons to each other.

HEALTH WATCH

Spinal Cord Injury

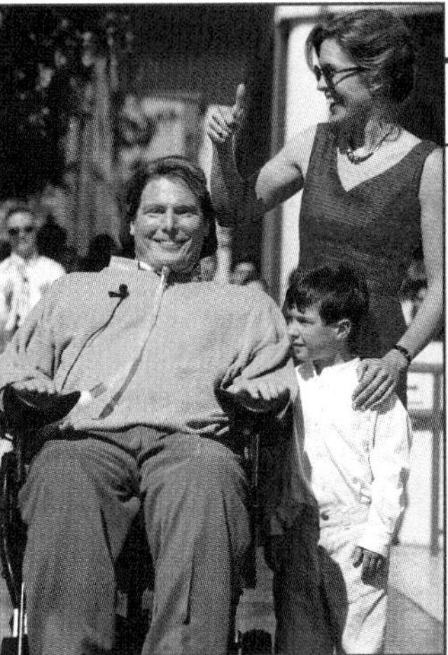

Unlike most other parts of the body, the spinal cord does not heal after an injury. Damaged neurons stop conducting nerve impulses at the site of injury, permanently paralyzing the legs or, in some cases, all four limbs.

Every year, spinal cord injuries—whether incurred in athletics or automobile accidents—leave nearly 15,000 Americans partially or totally paralyzed. In 1995, actor Christopher Reeve, shown in the photo at right, injured his spinal cord after falling headfirst from a horse. The fall broke vertebrae in Reeve's neck, paralyzing him from the neck down.

A treatment currently available for people with spinal cord injuries is an anti-inflammatory drug called methylprednisolone. If given within 8 hours after the spinal cord is injured, the drug can improve chances of recovery. Even with this drug, however, recovery is usually far from complete.

Stopping cell death

Cells continue to die near the site of a spinal cord injury for several weeks after injury occurs. Myelin-producing cells die, leaving neurons in the spinal cord unable to function. Some scientists think that stopping the death of these cells could help avoid paralysis. In experiments on rats, researchers have found that a cell-death inhibitor improves the rats' ability to use their hind legs after a spinal cord injury. Researchers are investigating other cell-death inhibitors that could be used on humans.

Bridging the gap

After the spinal cord is injured, damaged axons begin to regrow. However, their growth is inhibited by substances in the spinal cord. Peripheral nerves lack these substances, so the axons in these nerves can regrow quite well. To stimulate the growth of axons in the injured spinal cord, researchers have grafted pieces of peripheral nerves into the spinal cord. The nerve grafts provide tunnels for regrowing axons. Rats with such nerve grafts begin to show signs of recovery within 3 weeks. Within a year, they can support their own weight. Similar grafts have not yet been tried on humans.

Section 42-2

Demonstration

Reaction Time

Reaction time is difficult to measure without elaborate equipment, but students can get a general idea using this demonstration. Use a pair of student volunteers. Have one student hold a ruler vertically. Have the other student position his or her fingers 1 inch below the ruler, in a pinching mode. (This works best if the forearm rests on a table; that way the student will be less likely to move.) Have the first student drop the ruler without warning. The other student should try to catch it between the thumb and index finger. The shorter the segment allowed to fall through the fingers, the quicker the reaction time.

Spinal Cord Injury

Teaching Strategies

Bring in pictures of Christopher Reeve and other paraplegics and quadriplegics. If the information is available, talk about how they were injured. Were any of the injuries preventable?

Discussion

- Why does paralysis occur? *(Nerve impulses cannot cross the site of injury in the spinal cord.)*
- Is it possible for a person to have arms but not legs paralyzed as the result of a spinal cord injury? Explain your answer. *(No—everything below the spinal cord injury is affected. If the injury is above the spinal cord area that controls the arms, all four limbs are affected.)*
- Would you volunteer to be a test case for the methods described if you had a spinal cord injury? Why or why not? *(Answers will vary.)*

Teaching TIP

Central Nervous System

As students read about the central nervous system, have them construct a **Graphic Organizer** similar to the one at the bottom of page 962 that describes the main structures of the central nervous system. Have them use the following terms: autonomic nervous system, brain, central nervous system, nervous system, peripheral nervous system, spinal cord, and somatic nervous system. PORTFOLIO

Teaching TIP

Referred Pain

Referred pain is pain that originates in a location different from where it is felt. Referred pain is important in clinical diagnoses. For example, inadequate oxygenation of the heart muscle often results in pain being referred to the chest wall and shoulder. Inflammation of the appendix, in the lower right quadrant of the abdomen, is sometimes felt as pain in the navel area.

Demonstration

Visual Feedback and Touch

Have students work in pairs. One student should hold his or her arms out and crossed so that the palms meet, interlocking the fingers. Bending at the elbows, the student should move the hands down and in toward the body. Keep the elbows bent, and bring the hands close to the body, twisting the hand upward. The partner should point to a finger without touching it. The student should try to move that finger quickly. What happens? Is it easier if the partner touches that finger? *(Your eyes confirm body positions that are monitored by positional receptors, or proprioceptors. If the hands are in an unusual position, proprioception and the eyes send conflicting messages to the brain.)*

Study TIP

Interpreting Graphics

As you look at Figure 42-9, notice that in a spinal reflex, motor neurons stimulate muscles in the same region in which the stimulus that caused the reflex originated.

The Peripheral Nervous System Branches Throughout the Body

The peripheral nervous system connects the brain and the spinal cord to the rest of the body. In addition to the 31 pairs of spinal nerves, 12 pairs of cranial nerves connect the brain with areas in the head and neck. The PNS contains two principal divisions—the sensory division and the motor division. The sensory division directs the flow of sensory information to the central nervous system. You will learn more about sensory systems in Section 42-3. The motor division carries out motor responses to sensory information. The motor division of the PNS consists of two independent systems—the somatic nervous system and the autonomic nervous system.

Somatic nervous system

Peripheral motor neurons that stimulate skeletal muscles are under our conscious control. These neurons are part of the somatic nervous system. Some activity in the somatic nervous system, such as spinal reflexes, is involuntary. A spinal reflex is a self-protective motor response. Spinal reflexes are extremely rapid because they usually involve the spinal cord and the PNS but do not involve the brain. The knee-jerk reflex, shown in **Figure 42-9,** is an example of a spinal reflex. When you tap the ligament below your kneecap, your lower leg suddenly kicks forward. Tapping the ligament stimulates a sensory neuron, shown in red in Figure 42-9, in the attached quadriceps. The sensory neuron sends a nerve impulse to the spinal cord and excites a motor neuron, shown in green, which causes the

Figure 42-9 Knee-jerk reflex

When the ligament below the patella is tapped, the quadriceps contracts, the hamstrings relax, and the leg rapidly extends.

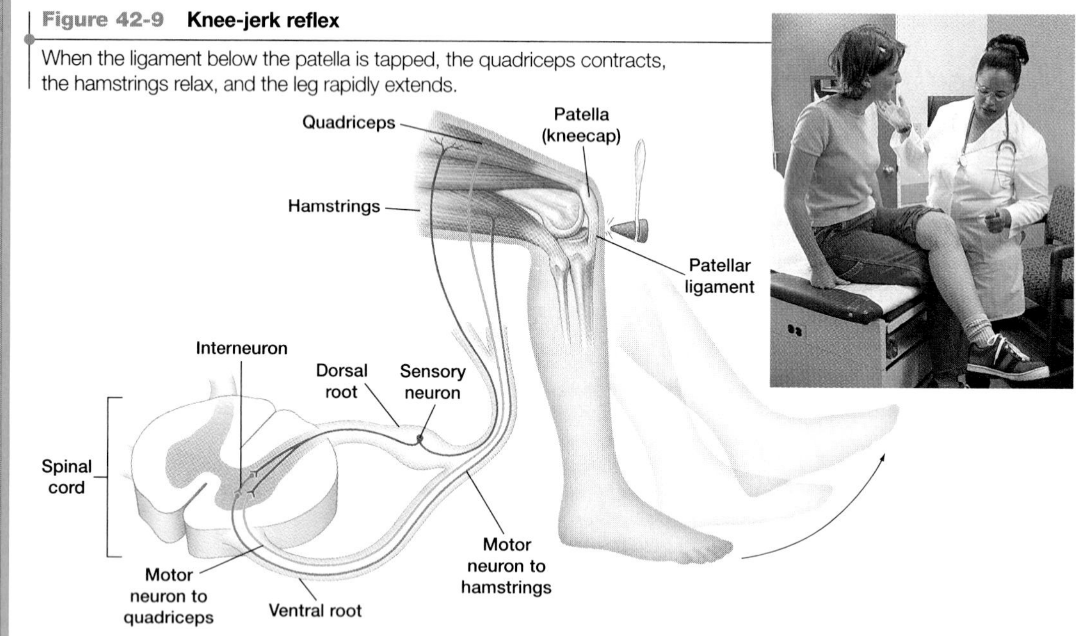

Graphic Organizer

*Use this graphic organizer with **Teaching Tip: Central Nervous System** on page 962.*

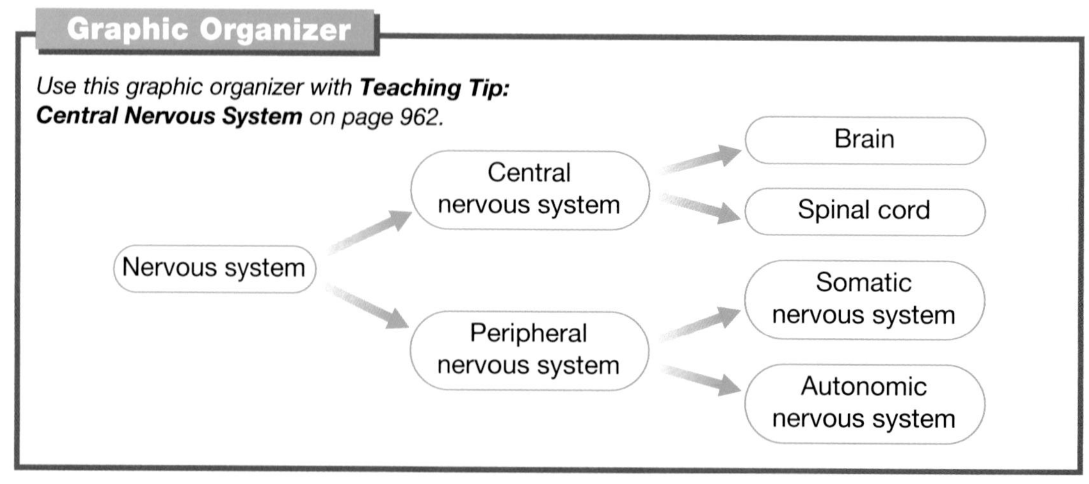

quadriceps to contract. This causes the leg to extend rapidly. The sensory neuron also stimulates an interneuron, shown in blue. The interneuron inhibits a motor neuron that would normally cause the hamstrings to contract. Thus, the hamstrings relax and do not oppose contraction of the quadriceps.

Autonomic nervous system

Peripheral motor neurons that regulate smooth muscles do not require our conscious control. These neurons are part of the autonomic nervous system, which regulates heart rate and blood flow by controlling contractions of cardiac muscle in the heart and contractions of smooth muscle lining the walls of blood vessels. It also controls muscles in the digestive, urinary, respiratory, and reproductive systems, as well as the secretions of many glands.

Two divisions of the autonomic nervous system—the parasympathetic division and the sympathetic division—maintain stability in the body by counterbalancing each other's effects. The parasympathetic division is most active under normal conditions. It keeps your body functioning even when you are not active. For example, you continue to breathe when you fall asleep. The sympathetic division dominates in times of physical or emotional stress. It controls the "fight-or-flight" response that you experience during a stressful situation, such as "nervousness" when taking a pop quiz. The sympathetic division increases blood pressure, heart rate, and breathing rate. It also directs blood flow toward your heart and skeletal muscles. Effects of the autonomic nervous system are summarized in Table 42-1.

internet connect

SCiLINKS NSTA

TOPIC: Autonomic nervous system
GO TO: www.scilinks.org
KEYWORD: HX963

Table 42-1 Physiological Effects of the Autonomic Nervous System

Organ	Effect of sympathetic stimulation	Effect of parasympathetic stimulation
Eyes	Pupils dilate	Pupils constrict
Heart	Heart rate increases	Heart rate decreases
Lungs	Bronchioles dilate	Bronchioles constrict
Intestines	Gastric secretions decrease	Gastric secretions increase
Blood vessels	Blood vessels dilate	Little or none

Review SECTION 42-2

1. **Name** the two main divisions of the nervous system, and state their general functions.
2. **Compare** the functions of the cerebellum and the brain stem.
3. **Distinguish** between dorsal roots and ventral roots of the spinal cord.
4. **SKILL Comparing Functions** Why is a spinal reflex more rapid than a voluntary movement?

Critical Thinking

5. **SKILL Forming a Hypothesis** A reflex is an unlearned, innate behavior. What is the evolutionary advantage of reflexes?

Review SECTION 42-2 ANSWERS

1. The central nervous system, consisting of the brain and spinal cord, is the control center of the body. The peripheral nervous system connects the brain and spinal cord to the rest of the body.
2. The cerebellum regulates balance and the timing of movement, while the brain stem regulates vital functions such as breathing, heart rate, body temperature, and sleep.
3. Dorsal roots contain sensory neurons; ventral roots contain motor neurons.
4. A spinal reflex travels only as far as the spinal cord, whereas a voluntary movement involves the brain and therefore takes longer than a spinal reflex.
5. The evolutionary advantage of a reflex behavior is that the response to a potentially harmful stimulus is very rapid.

internet connect

SCiLINKS NSTA Have students research the autonomic nervous system using the Web site in the Internet Connect box on page 963. Have students write a report summarizing their findings. PORTFOLIO

3 Assessment Options

Closure

Reflex Reactions

Have students take turns lightly tapping each other's leg slightly below the knee. If possible, have them use a rubber-headed hammer borrowed from the school nurse. Ask students to write a brief report summarizing their observations and the cause of the effect produced. Tell them that their report must include the following terms: reflex, axon, synapse, motor neuron, sensory neuron, and spinal cord. PORTFOLIO

Review

Assign Review—Section 42-2.

Reteaching — BASIC

On the board or overhead, list the major concepts presented in this section. Have the class review this section to identify examples that can be used to illustrate each of these concepts.

Chapter 42

1 Prepare

Scheduling

- **Block:** About 90 minutes will be needed to complete this section.
- **Traditional:** About two class periods will be needed to complete this section.
- **Planning:** See the Planning Guide on pp. 949A and 949B for lecture, classwork, and assignment options for Section 42-3. Lesson plans for Section 42-3 can be found in a lesson cycle format in ***Biology: Principles and Explorations*** **One-Stop Planner CD-ROM with Test Generator.**

Resource Link

Assign **Active Reading Worksheet Section 42-3** to help students comprehend the key concepts and visuals in this lesson.

2 Teach

Lesson Warm-up

Have students close their eyes and press lightly on their eyelids until they see an effect. Ask what they see. *(patterns of colors)* These patterns are called phosphenes. The nerves in the eyes send messages to the brain, which interprets these messages as colors. Point out to students that they are not really "seeing" anything. Then ask what this phenomenon tells them about the sense of sight. *(The eyes receive information that is sent via nerve impulses to the brain, which analyzes and interprets these.)*

Section 42-3 Sensory Systems

Objectives

- **List** five types of sensory receptors and the stimuli to which they respond. (p. 964)
- **Identify** sites of sensory processing in the brain. (p. 965)
- **Analyze** the structure of the eye and its role in the visual system. (p. 966)
- **Describe** how the ear detects sound and helps maintain balance. (p. 967)
- **Compare** the senses of taste and smell. (p. 968)

Key Terms

sensory receptor
retina
rod
cone
optic nerve
cochlea
semicircular canal

internetconnect

SCILINKS. NSTA
TOPIC: Sensory receptors
GO TO: www.scilinks.org
KEYWORD: HX964

Sensory Systems Enable Perception

The perception of everything you respond to in the environment, such as the horn of a passing car or cold rain on your face, is made possible by sensory systems. Sensory systems are essential to survival, and they enable us to experience both pleasurable and painful stimuli. Sensory systems help maintain homeostasis by constantly adjusting body conditions to respond to changes in the environment. This requires the integration of the peripheral nervous system and the central nervous system. The sensory division of the PNS collects information about sensory stimuli in and around the body. The sensory information moves through the spinal cord to the brain. The brain then processes this information and, if necessary, generates a motor response to the stimuli.

How does the nervous system detect sensory stimuli? Specialized neurons called **sensory receptors** detect sensory stimuli and then convert the stimuli to electrical signals, in the form of nerve impulses, that can be interpreted by the brain. Although sensory receptors are located throughout the body, they are most concentrated in the sense organs—the eyes, ears, nose, mouth, and skin. **Table 42-2** lists several types of sensory receptors.

Sensory receptors

Mechanoreceptors throughout the body respond to physical stimuli—such as pressure and tension—that cause distortion or bending of tissue. These stimuli alter the electrical activity of mechanoreceptors. Many mechanoreceptors are found in the skin, and they are concentrated in very sensitive areas, including the face, hands, fingertips, and neck. Pain receptors, which respond to potentially

Table 42-2 Types of Sensory Receptors

Receptor type	Stimulus	Location
Thermoreceptors	Temperature change	Skin, hypothalamus
Pain receptors	Tissue damage	All tissues and organs except the brain
Mechanoreceptors	Movement, pressure, tension	Skin, ears
Photoreceptors	Light	Eyes
Chemoreceptors	Chemical	Tongue, nose

Integrating Different Learning Styles

Visual/Spatial	**PE**	Quick Lab, p. 966
Body/Kinesthetic	**PE**	Quick Lab, p. 966
	ATE	Teaching Tip, p. 965; Did You Know?, p. 966; Demonstrations, pp. 966, 967, 968
Music/Rhythmic	**ATE**	Teaching Tip, p. 967
Interpersonal	**ATE**	Demonstrations, pp. 966, 967, 968
Intrapersonal	**ATE**	Demonstrations, pp. 966, 967, 968
Verbal/Linguistic	**PE**	Study Tip, p. 967; Performance Zone item 27, p. 977
	ATE	Closure, p. 968

Figure 42-10 Processing sites and lobes of the brain

Specific areas of the brain control different regions and functions of the body.

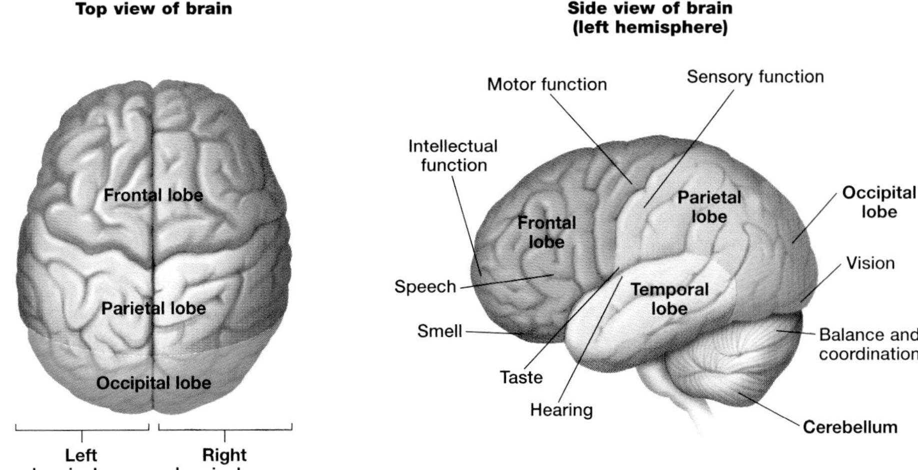

harmful stimuli—such as intense heat or cold and tissue damage—are responsible for painful sensations. Pain is an extremely important sensation because it informs you that something is wrong in your body. Many self-protective responses, such as reflexes, are initiated by pain receptors. Thermoreceptors, located in the skin and hypothalamus, detect changes in temperature. Thermoreceptors play an important role in homeostasis, helping to keep the body temperature within its normal range.

Sensory receptors are located throughout the body, and sensory input from these receptors enters the central nervous system in an organized fashion. Sensory stimuli that originate in the lower body enter the lower part of the spinal cord. Sensory stimuli that originate in the upper body enter the upper part of the spinal cord.

Processing of sensory information

Recall that the cerebral cortex contains a large percentage of the brain's neurons. A large percentage of neurons in the cerebral cortex are responsible for processing incoming sensory information from the sense organs. The thalamus relays information from the sense organs to certain regions of the cerebral cortex. Processing sites for sensory systems tend to be localized in specific regions of the cerebrum. As shown in **Figure 42-10,** deep grooves divide the cerebral hemispheres into four general areas, or lobes: the occipital *(ahk SIP ih tuhl)* lobe, the parietal *(puh REYE uh tuhl)* lobe, the temporal lobe, and the frontal lobe. Sensory neurons leading from each sense organ to the brain come together at a common region in the cerebral cortex. For example, most visual processing takes place in the occipital lobe, located near the back of the skull. Similarly, auditory processing is localized within the temporal lobe.

did you know?

On September 13, 1848, an accidental explosion on a Vermont railroad caused an extremely unusual injury.

A railway foreman named Phineas Gage was setting explosive charges with a tamping iron, a heavy iron rod. One of the charges exploded, sending the tamping iron through the left side of his skull. It entered under his left cheekbone, exited through the top of his head, and landed 25 yards away. Most of the left frontal lobe of his brain was destroyed, but it is said that Gage never even lost consciousness. After 7 months of recuperation, Gage was well enough to return to work. His physical and intellectual function was unchanged, but his personality was completely different. Once polite and patient, he was now profane, impatient, and rude, and was unable to resume his work as a foreman. His horrific injury, however, greatly influenced physicians and scientists who were attempting to determine how function is localized in the brain.

Teaching TIP

Two-Point Touch Discrimination

Tell students that there are touch receptors spread throughout the skin, but some areas have a larger concentration than others. For example, your lips and hands have large numbers of touch receptors situated close together. It is easy to distinguish two points at very close distances. On your back, however, there are areas where those touch receptors are several millimeters apart. Two points may feel like just one if you touch certain areas of the skin.

Making Biology Relevant

Health

Glaucoma is an eye disease in which increased pressure inside the eye causes damage to the retina and the optic nerve. The cause of the increased pressure is a buildup of the aqueous humor, the fluid interior of the eye. Glaucoma is the second most common cause of blindness.

internet**connect**

SCiLINKS NSTA Have students research sensory receptors using the Web site in the Internet Connect box on page 964. Have students write a report summarizing their findings. PORTFOLIO

Demonstration

Visual Field

You will need a cardboard tube from a roll of paper towels. Tell a student volunteer to keep both eyes open, and to place the tube over his or her left eye. Have the student hold his or her right hand next to the end of the tube, and move it slowly along the tube toward his or her face. Ask the student what he or she observes. *(It should appear as though the tube passes through the palm. The brain automatically combines the images seen with both eyes into one picture. Usually the eyes see very similar things. Even when they do not, however, the brain still combines the two images.)*

Demonstration

Colorblindness

Show the class a chart used to test people for colorblindness. Determine if any students are colorblind. Point out that colorblindness is caused by a chemical disorder in the cones of the eye. Complete colorblindness is extremely rare. A deficiency in the red or green cones is most common, affecting about 5 percent of the population.

How can you demonstrate your blind spot?

Time 10 minutes

Process Skills Inferring relationships, relating concepts

Teaching Strategies
Make sure students are closing their right eye and are looking at the O. It should disappear about 1 ft from the end of the nose. Try it with the X and the left eye as well.

Analysis Answers

1. rods, cones
2. There are no photoreceptors (rods and cones) at that point.
3. The X disappears when it is in the part of the visual field mediated by the part of the retina where there are no photoreceptors.

The Eyes Detect Light

Humans have extremely good eyesight; our eyes enable us to see in color and to distinguish fine details and movement. The structure of the eye is shown in **Figure 42-11.** Light enters the eye through the pupil. Light then passes through the lens, a thick, transparent disk that focuses light on the retina. The **retina** is a lining on the back inner surface of the eye that consists of photoreceptors and neurons. The retina contains two types of photoreceptors—rods and cones—which convert light energy to electrical signals that can be interpreted by the brain. **Rods** respond best to dim light. **Cones** respond best to bright light and enable color vision. Cones are also sensitive to edges and fine detail, so they produce sharp images. The retina also contains many other neurons that process visual information. The axons of some of these neurons make up the **optic nerve.** The optic nerve exits through the back of the eye and runs along the base of the brain to the thalamus. The thalamus then relays visual information to the occipital lobe of the cerebral cortex, where the information is processed.

Figure 42-11 Structure of eye. Light enters the eye through the pupil and is focused on the retina, which contains photoreceptors.

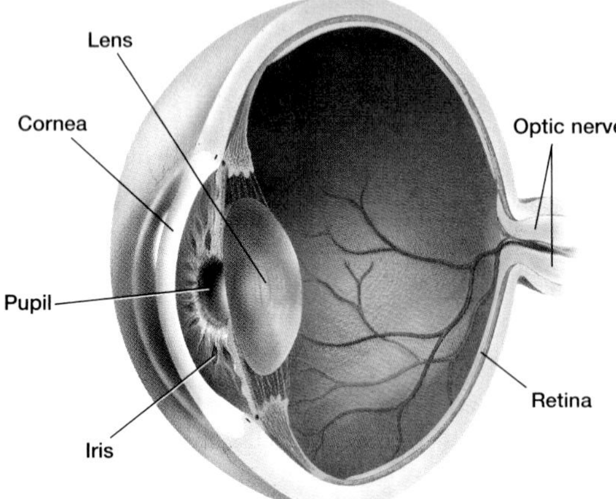

How can you demonstrate your blind spot?

The blind spot is the site where the optic nerve exits the back of the eye through the retina. There are no photoreceptors at the blind spot. Use the procedure below to demonstrate your blind spot.

Materials

unlined 3 × 5 index card, pencil

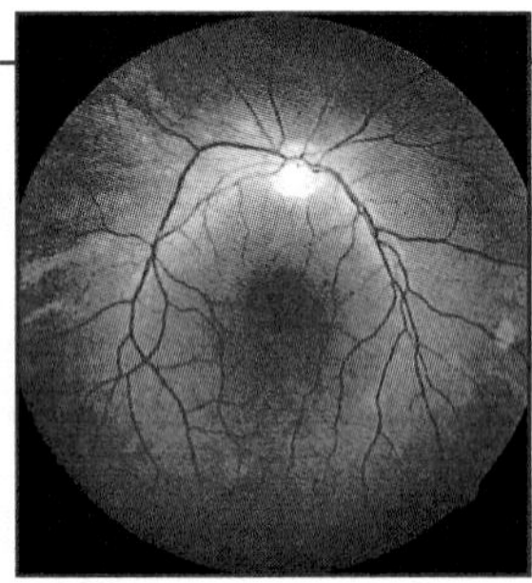

Retina

Procedure

1. On the index card, draw an X about 1 in. from the left side of the card. Draw an O about the same size 3 in. to the right of the X.
2. Hold your index card in front of you at arm's length. Close your right eye and stare at the circle with your left eye. Slowly move the card toward you while continuing to stare at the circle until the X "disappears" from view.

Analysis

1. **Name** the two kinds of photoreceptors found in the retina.
2. **Propose** why you cannot see images that fall on the site where the optic nerve exits the eye.
3. **SKILL Relating Concepts** What is the relationship between the structure of the retina and the disappearance of the X on the index card?

did you know?

Your eyes and brain sometimes change color information.

If you stare at a picture long enough and then look at a white wall, the image shapes are the same, but the colors are different. Try this with a piece of black construction paper with a large blue spot in the middle. When you then look at a white wall, you should see a red spot on a white background. The retina has three different types of cones. Each type detects one color of light: red, green, or blue. Our perception of color is a combination of the action of these three types of cones. Cones sometimes stop working after they have been exposed to their color for long periods of time. Other cones then take over, and we see a "ghost" color different from the original color.

The Ears Detect Sound and Help Maintain Equilibrium

How do your ears enable you to hear? Your ears convert the energy in sound waves to electrical signals that can be interpreted by your brain. **Figure 42-12** shows the structure of the ear. Sound waves enter the ear through the ear canal and strike the tympanic *(tim PAN ik)* membrane, or eardrum, causing the tympanic membrane to vibrate. Behind the eardrum, three small bones of the middle ear—the hammer, anvil, and stirrup—transfer the vibrations to a fluid-filled chamber within the inner ear. This chamber, called the **cochlea** *(KAHK lee uh),* is coiled like a snail's shell, and it contains mechanoreceptors called hair cells. Hair cells rest on a membrane that vibrates when waves enter the cochlea. Waves of different frequencies cause different parts of the membrane to vibrate and thus stimulate different hair cells. When hair cells are stimulated, they generate nerve impulses in the auditory nerve. The electrical signals travel to the brain stem through the auditory nerve. The thalamus then relays the information to the temporal lobe of the cerebral cortex, where the auditory information is processed.

Study TIP

Word Origins

The word *cochlea* is from the Greek *kochlias,* meaning "snail shell."

Keeping your balance

The ears not only enable you to hear but also help you maintain equilibrium. The **semicircular canals** are fluid-filled chambers in the inner ear that contain hair cells. Clusters of these hair cells respond to changes in head position with respect to gravity. When your head moves or rotates, the hair cells bend according to the magnitude and direction of the fluid's movement and send electrical signals to the brain. Signals generated by the hair cells enable the brain to determine the orientation and position of the head.

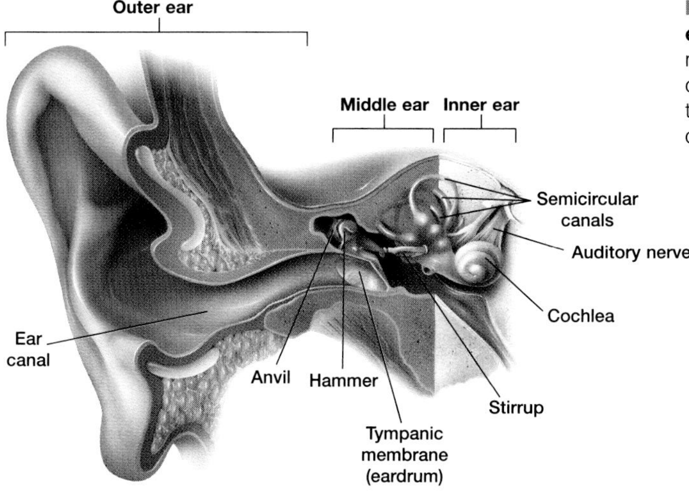

Figure 42-12 **Structure of ear.** Sound waves are transmitted to the inner ear and are detected by mechanoreceptors. The semicircular canals detect the position of the head.

Teaching TIP

Cochlea Shape

Show students a snail shell, and tell them that the human cochlea takes up very little space in the inner ear because it has a similar spiral shape.

Teaching TIP

Measuring Sound

Play recorded music at a volume high enough to be disconcerting to an average listener. Tell students that the loudness of sounds is measured in decibels, ranging from a sound of 0 decibels (dB), which is just audible, to a sound of 140 dB, which is painful. A jet taking off can produce a 150 dB sound. People who work on the tarmac at an airport wear protective earplugs and headsets to prevent hearing loss.

Demonstration

Sound Through Steel

Sound travels at 343 m/s (1,125 ft/s) through the air, approximately 1482 m/s (4,861 ft /s) through water, and 5960 m/s (19,549 ft/s) through steel. Sounds passing through media other than air often lose high and low tones, leading to a muffled sound. Have students try this activity: Tie a spoon in the middle of a 1 m (40 in.) piece of kite string. Wrap the ends of the string around your index fingers. Letting the spoon hang freely, rest the tips of your fingers in your ears. Have another student tap the spoon with another spoon. Compare this sound to the sound you hear when your fingers are removed from your ears.

Overcoming Misconceptions

The "Five" Senses

We've all heard of the five traditional senses: hearing, smell, taste, sight, and touch. But in fact, these are misleading categories. Touch actually consists of hot, cold, pressure, and pain, all of which are detected by distinctly different sensory receptors. Another sense, equilibrium or balance, is equally important.

The senses may be classified as general or special depending on the distribution of their sensory receptors. Special senses are those served by either large, complex sensory organs, such as the eyes and ears, or those with small clusters of receptors, such as the taste buds and olfactory (smell) organs.

Chapter 42

internet connect

SciLinks NSTA Have students research chemical senses using the Web site in the Internet Connect box on page 968. Have students write a report summarizing their findings. PORTFOLIO

Demonstration

Taste and Smell

You will need jelly beans for this activity. Have a volunteer close his or her eyes and pinch his or her nose closed. Give the student a jelly bean, instructing him or her to chew it slowly, keeping the nose closed. Ask the student to guess the flavor. Keeping the eyes closed, the student should try it again with the same flavor, but with the nose open. Can the student guess the flavor now? *(Much of what we call flavor in foods is actually the result of smell rather than taste. That is why when your nose is congested, food does not have much taste.)*

3 Assessment Options

Closure

Sensory Deprivation

Sensory deprivation chambers have been developed to isolate the occupant from as much sensory input as possible. Have students describe how a person might react in such a situation. *(Answers will vary, but may include disorientation, loss of the sense of passage of time, and hallucinations.)* Have them research how such chambers are costructed and how humans respond to them.

Review

Assign Review—Section 42-3.

Reteaching — BASIC

Assign students to cooperative groups of four. Have each group choose the sensory system they would least like to lose. Instruct each group to defend its choice.

Taste and Smell Are Related Chemical Senses

internet connect

SciLinks NSTA
TOPIC: Chemical senses
GO TO: www.scilinks.org
KEYWORD: HX968

Embedded within the surface of the tongue are 2,000–5,000 taste buds. A taste bud, shown in **Figure 42-13,** is a cluster of 50–100 taste cells. Taste cells are chemoreceptors that detect four basic chemical substances: sugars (sweet), acids (sour), alkaloids (bitter), and salts (salty). The tip of the tongue is most sensitive to sweet tastes, the sides to salty and sour tastes, and the back to bitter tastes. Each taste cell is generally sensitive to all four tastes but is most sensitive to only one of them. A taste bud is stimulated when food molecules dissolved in saliva bind to taste cells. Taste cells generate electrical signals that can be interpreted by the brain.

Chemoreceptors that detect odors, called olfactory *(ahl FAK tuh ree)* receptors, are located in the roof of the nasal passage. Chemicals in the air stimulate olfactory receptors, which generate electrical signals that can be interpreted by the brain. The sense of smell affects the enjoyment of food. When you have a bad cold and your nose is stuffed up, your food may seem to have little taste.

Figure 42-13 Structure of tongue

A taste bud is a cluster of taste cells surrounding a taste pore.

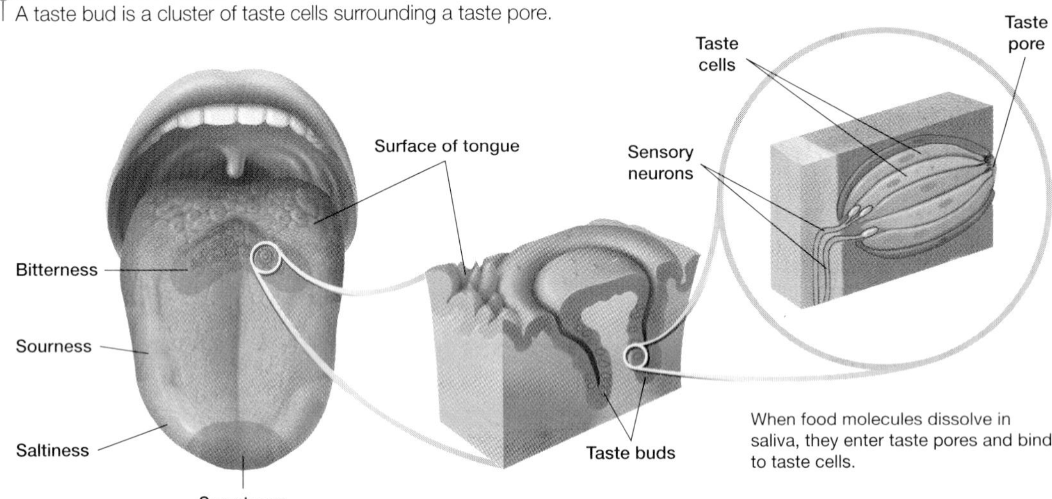

Review SECTION 42-3

1. **List** two different types of sensory receptors and the kinds of stimuli to which they respond.
2. **Sequence** the events that occur when light enters the eye.
3. **Describe** how sound waves are transmitted through the ear.
4. **SKILL Comparing Structures** Distinguish between taste cells and olfactory receptors.

Critical Thinking

5. **SKILL Recognizing Relationships** Suggest a possible defect of the retina that would cause colorblindness, a condition in which a person cannot distinguish between certain colors.

Review SECTION 43-3 ANSWERS

1. Answers will vary. For example: pain receptors respond to potentially harmful stimuli; thermoreceptors respond to temperature changes; and mechanoreceptors respond to mechanical stimuli.
2. Rods and cones convert light energy into electrical signals sent along the axons that constitute the optic nerve, which runs to the thalamus, and finally to the occipital lobe of the cerebral cortex where the information is processed.
3. Sound waves enter the ear and strike the tympanic membrane, making it vibrate. The vibrations pass through the hammer, anvil, and stirrup to the cochlea where mechanoreceptors send impulses via the auditory nerve to the thalamus, then to the temporal lobe of the cerebrum.
4. Chemicals in food are detected by one or more of four kinds of taste receptors on the tongue. Olfactory receptors bind chemicals in the air.
5. A malfunction of the cones could cause colorblindness.

Drugs and the Nervous System

Section 42-4

Psychoactive Drugs Affect the Central Nervous System

Many different kinds of drugs are available to the public. Advertisements tell you about pain relievers, antacids, cough syrups, and other medications that can help you feel better. Drugs can prevent, treat, or cure many different illnesses. However, drugs, whether legal or illegal, can also be misused or abused.

In the broadest sense, a drug is a chemical that alters body structures or biological functions. Drugs that alter the functioning of the central nervous system are known as **psychoactive drugs.** Alcohol, nicotine, cocaine, and heroin are examples of commonly abused psychoactive drugs. Psychoactive drugs also include many other substances, such as inhalants. Caffeine, found in coffee and soft drinks, is also a psychoactive drug, as is THC, the active ingredient in marijuana. In addition, many medications, such as those prescribed by doctors to treat mental disorders, contain psychoactive drugs. All psychoactive drugs can produce psychological dependence. Many also can produce physical dependence, or addiction. **Table 42-3** lists several classes of commonly abused psychoactive drugs.

Objectives

- **Identify** several types of psychoactive drugs, and describe their effects. (p. 969)
- **Describe** how drug use can lead to drug addiction. (pp. 970–971)
- **Summarize** the effects of commonly abused drugs on the nervous system. (pp. 972–974)

Key Terms

psychoactive drug
addiction
tolerance
withdrawal
stimulant
depressant

Table 42-3 Classes of Psychoactive Drugs

Drug	Examples	Psychoactive effects	Risks associated with use
Depressants	Barbiturates (sedatives), tranquilizers, alcohol	Decrease activity of the central nervous system	Drowsiness, depression, brain or nerve damage, coma, respiratory failure
Stimulants	Caffeine, cocaine, crack, nicotine, amphetamines	Increase activity of the central nervous system	Heart failure, high blood pressure, brain damage, aggressive behavior, paranoia
Inhalants	Nitrous oxide, ether, paint thinner, glue, cleaning fluid, aerosols	Disorientation, confusion, memory loss	Brain damage, kidney and liver damage, respiratory failure
Hallucinogens	LSD, PCP, peyote (mescaline), psilocybe mushroom (psilocybin)	Sensory distortion, anxiety, hallucinations, numbness	Depression, paranoia, aggressive behavior
THC	Marijuana, hashish	Short-term memory loss, impaired judgment	Lung damage, loss of motivation
Narcotics	Heroin, morphine, codeine, opium	Feeling of well-being, sedation, impaired sensory perception, impaired reflexes	Coma, respiratory failure

Section 42-4

1 Prepare

Scheduling

- **Block:** About 90 minutes will be needed to complete this section.
- **Traditional:** About two class periods will be needed to complete this section.
- **Planning:** See the Planning Guide on pp. 949A and 949B for lecture, classwork, and assignment options for Section 42-4. Lesson plans for Section 42-4 can be found in a lesson cycle format in ***Biology: Principles and Explorations*** **One-Stop Planner CD-ROM with Test Generator.**

Resource Link

Assign **Active Reading Worksheet Section 42-4** to help students comprehend the key concepts and visuals in this lesson.

2 Teach

Lesson Warm-up

Demonstrate the effect of sidestream smoke. Put several smoking cigarettes in an ashtray. **Caution: Burn cigarettes only under a vent hood.** Place a piece of glass approximately 5 cm (2 in.) above the cigarettes. After the cigarettes have burned completely, wipe the glass with a white cloth and show the cloth to the class.

Integrating Different Learning Styles

Logical/Mathematical	**PE**	Real Life, p. 971
	ATE	Resource Link, p. 970
Visual/Spatial	**PE**	Lesson Warm-up, p. 969; BioGraphic, pp. 970–971
	ATE	Visual Strategy, p. 970; Demonstration, p. 973; Reteaching, p. 974
Interpersonal	**ATE**	Teaching Tip, p. 972; Biology in Your Community, p. 973; Closure, p. 974
Intrapersonal	**ATE**	Closure, p. 974
Verbal/Linguistic	**PE**	Performance Zone items 20, 22, and 25, p. 977
	ATE	Resource Link, p. 971; Teaching Tips, p. 972; Closure, p. 974

Resource Link

Assign **Problem Solving: Determining Percentage Composition** from the Holt BioSources Teaching Resources CD-ROM. PORTFOLIO

Visual Strategy

BIO graphic **Figure 42-15**
Emphasize that addiction is physiological as well as psychological. Have students explain the relationship between the craving for an addictive drug and the changes in the number of receptor proteins on post-synaptic cells.

Drug Addiction Involves Changes in Neuron Function

Figure 42-14 Coca plant. Cocaine is derived from the coca plant, *Erythroxylon coca.*

Addiction is a physiological response caused by repeated use of a drug that alters the normal functioning of neurons and synapses. Once a neuron or synapse has been altered by a drug, it cannot function normally unless the drug is present. With repeated exposure to a drug, a person addicted to the drug develops tolerance to the drug. **Tolerance** is a characteristic of drug addiction in which increasing amounts of the drug are needed to achieve the desired sensation. The severity of drug addiction is evident in recovering addicts who experience withdrawal when they stop taking an addictive drug. **Withdrawal** is a set of emotional and physical symptoms caused by removal of the drug from the body. Symptoms of withdrawal may include vomiting, headache, depression, and seizures.

Cocaine as a model of drug addiction

Cocaine is a highly addictive stimulant found in the leaves of the coca plant, *Erythroxylon coca,* shown in Figure 42-14. A **stimulant** is a drug that generally increases the activity of the central nervous system. Despite being illegal, cocaine is still used by many people.

Recall that in synaptic transmission, neurotransmitters are released from a presynaptic neuron and bind to receptor proteins on

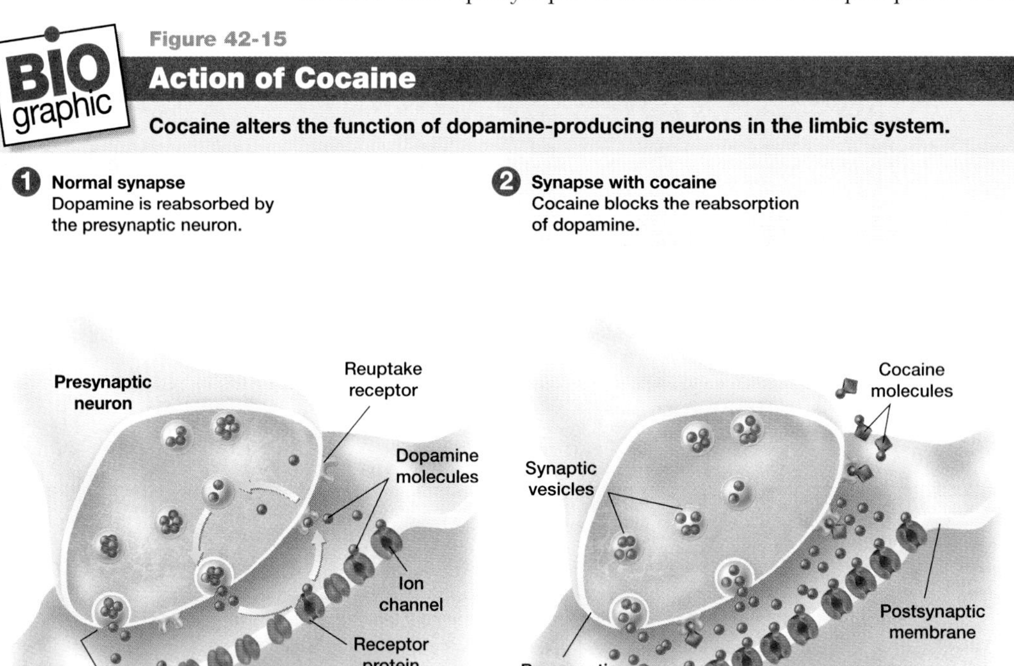

Figure 42-15 Action of Cocaine
Cocaine alters the function of dopamine-producing neurons in the limbic system.

did you know?

Street drugs often are not pure.

A drug with the street name of "croak" is produced by mixing methamphetamine, which stimulate the release of norepinephrine and dopamine, with crack cocaine, which prevents the reabsorption of dopamine. Extremely high heart rate and blood pressure can result when large amounts of dopamine enter and remain in synapses. Drug dealers mix deadly combinations of drugs to make more money, to keep the users "hooked," and in some cases to fool the users into thinking they are using a "pure" drug.

a postsynaptic cell. Some neurotransmitters are reabsorbed by presynaptic neurons after they have been released into the synaptic cleft. Cocaine is an example of a drug that causes addiction by interfering with a presynaptic neuron's ability to reabsorb, or reuptake, a neurotransmitter, in this case dopamine *(DOH pah meen)*. Cocaine targets dopamine-producing neurons in the limbic system, which plays an important role in the sensation of pleasure. The mechanism of cocaine action is summarized in **Figure 42-15.**

Step ❶ At a normal synapse, reuptake receptors move molecules of dopamine in the synaptic cleft back into the presynaptic neuron.

Step ❷ Cocaine blocks the reuptake of dopamine molecules by interfering with reuptake receptors in the synaptic cleft.

Step ❸ As a result, excess dopamine remains in the synaptic cleft, overstimulating the postsynaptic cell. Overstimulation produces an intense feeling of exhilaration and well-being. Because the postsynaptic cell has been overstimulated, the number of dopamine receptors on it decreases.

Step ❹ Cocaine not only increases the intensity of dopamine action on the postsynaptic cell but also causes the cell to change its structure. If cocaine is removed from the synaptic cleft, the number of dopamine molecules returns to normal, a level too low to adequately stimulate the postsynaptic cell. Addiction occurs because more cocaine must be taken to maintain adequate stimulation of the postsynaptic cell.

REAL LIFE

Caffeine is a stimulant found in many foods and beverages. Chocolate, coffee, tea, and some soft drinks often contain caffeine.

Calculating *Make a list of everything you eat and drink (including medicines) for 24 hours; then use a reference table to determine your total caffeine intake.*

❸ Overstimulated postsynaptic cell
The number of receptor proteins on the postsynaptic cell decreases.

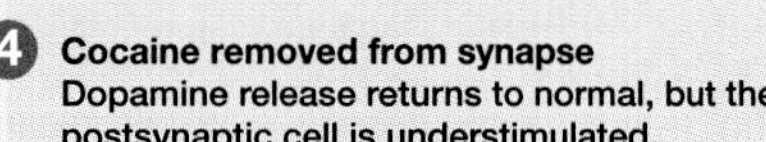
❹ Cocaine removed from synapse
Dopamine release returns to normal, but the postsynaptic cell is understimulated.

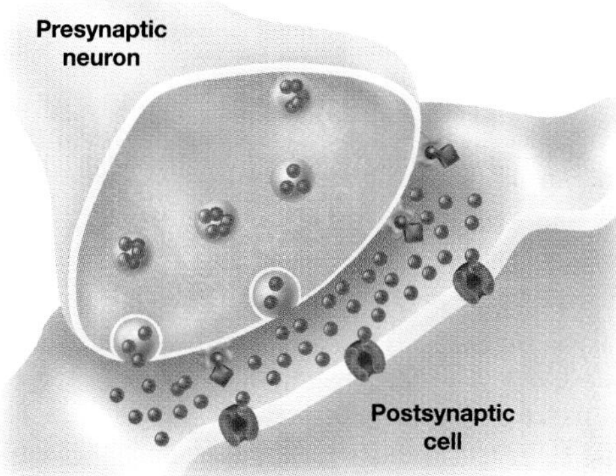

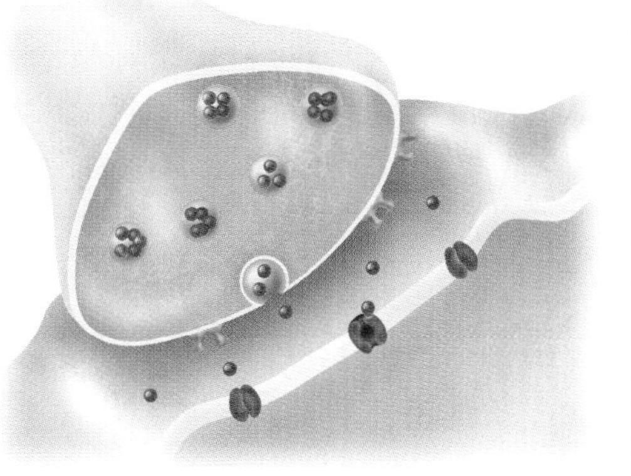

did you know?

Alcohol and caffeine are diuretics, which cause the excretion of excessive amounts of urine.

Alcohol inhibits the release of antidiuretic hormone, causing the kidneys to reabsorb water from the blood. The subsequent increase in the permeability of the kidney to water causes more water to be excreted in the urine. Caffeine circulates in the bloodstream to the kidneys where it inhibits the reabsorption of solutes and water. Thus, caffeine also increases urine production.

Section 42-4

Teaching TIP — **ADVANCED**

Drug Addiction

A number of potentially addictive substances, including opiates and cocaine, have now been linked to specific receptors in the brain. Have students research how the work of neurochemist Solomon H. Snyder led to a greater understanding of how potentially addictive substances affect the operation of receptors in the brain. Students should write a report, describing the evidence that led Snyder to infer the existence of endorphins, the naturally occurring opiate-like substances in brain tissue. **PORTFOLIO**

Resource Link

Assign **Occupational Applications: Forensic Toxicologist; Occupational Applications: Pharmacist;** and/or **Occupational Applications: Substance Abuse Counselor** from the Holt BioSources Teaching Resources CD-ROM. **PORTFOLIO**

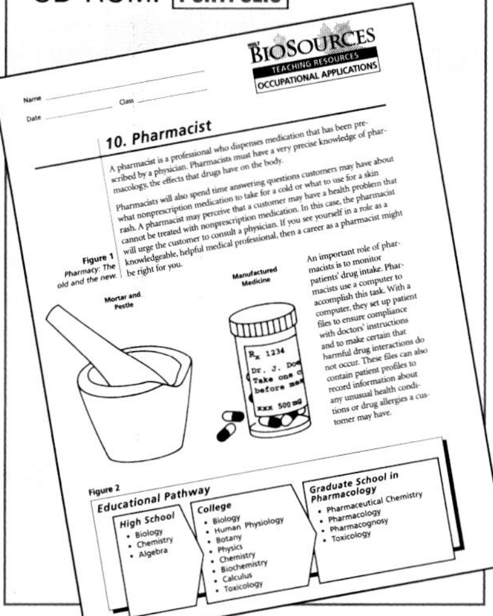
HOLT BIOSOURCES TEACHING RESOURCES
OCCUPATIONAL APPLICATIONS

Name ______ Class ______
Date ______

10. Pharmacist

A pharmacist is a professional who dispenses medication that has been prescribed by a physician. Pharmacists must have a very precise knowledge of pharmacology, the effects that drugs have on the body.

Pharmacists will also spend time answering questions customers may have about what nonprescription medication to take for a cold or what to use for a skin rash. A pharmacist may perceive that a customer may have a health problem that cannot be treated with nonprescription medication. In this case, the pharmacist will urge the customer to consult a physician. If you see yourself in a role as a knowledgeable, helpful medical professional, then a career as a pharmacist might be right for you.

Figure 1 *Pharmacy: The old and the new.*

Mortar and Pestle

Manufactured Medicine

An important role of pharmacists is to monitor patients' drug intake. Pharmacists use a computer to accomplish this task. With a computer, they set up patient files to ensure compliance with doctors' instructions and to make certain that harmful drug interactions do not occur. These files can also contain patient profiles to record information about any unusual health conditions or drug allergies a customer may have.

Figure 2 **Educational Pathway**

High School	College	Graduate School in Pharmacology
• Biology • Chemistry • Algebra	• Biology • Human Physiology • Botany • Physics • Chemistry • Biochemistry • Calculus • Toxicology	• Pharmaceutical Chemistry • Pharmacology • Pharmacognosy • Toxicology

Teaching TIP BASIC

Social Drugs and the Media

Have students bring in newspaper and magazine articles that deal with smoking cigarettes and drinking alcohol. Display the articles on a bulletin board, and encourage the class to read them as they study this section. Ask students to consider the image of social drug use that is portrayed in the media.

Teaching TIP

Classifying Drugs

Assign students to groups of four. Have each group make a table listing different kinds of drugs they can think of (both commonly abused drugs and medicinal drugs), their short-term effects, their long-term effects, and their desired effects. Have them categorize the drugs as medicinal or social, legal or illegal, and nonprescription or prescription. Their tables might include stimulants, depressants, narcotics, hallucinogens, inhalants, hormone replacements, analgesics, antibiotics, antihistamines, and appetite suppressants. Students should find that many drugs have more side effects than desired effects.

Figure 42-16
Tobacco plant

How many?

More than 60 million Americans smoke cigarettes, and about 3 million Americans under the age of 18 smoke about half a billion cigarettes each year.

Nicotine Is an Addictive Stimulant

About 50 million Americans smoke cigarettes despite convincing evidence that smoking causes mouth cancer, heart disease, lung cancer, and emphysema. So why do people continue to smoke? Many smokers say they would like to stop smoking but find the habit too difficult to overcome. They are addicted to nicotine, a chemical in cigarette smoke.

Effects of nicotine

Nicotine is the highly addictive stimulant found in the leaves of the tobacco plant, *Nicotiana tabacum,* shown in **Figure 42-16.** Nicotine is extremely toxic; a dose of only 60 mg is lethal in humans. Tobacco leaves are dried or crushed and are then smoked in cigarettes, cigars, and pipes. Tobacco is also chewed and snuffed.

Nicotine quickly enters the bloodstream and circulates through the body. In the brain, nicotine mimics the action of the neurotransmitter acetylcholine. The way that nicotine causes addiction has been studied by neurobiologists, who have found that nicotine binds to brain nerve cells at specific sites usually reserved for acetylcholine. These sites are the central controls of the brain—mechanisms the brain uses to adjust levels of many of its activities. Like twisting the dial on a central control, the binding of nicotine to these sites produces many changes. After a while, the smoker's body makes adjustments, and systems almost return to normal—as long as the smoker keeps smoking. Take away the nicotine, however, and all those adjustments throw everything out of balance all at once. The only way to keep things "normal" is to keep smoking. The smoker is addicted.

Effects of tobacco

Smokers get more than nicotine from cigarette smoke. Inhaled smoke contains hundreds of toxic and mutagenic chemicals that pass through the mouth, air passages, and lungs. These chemicals, also called tars, are produced by burning tobacco. Because tars and other chemicals in tobacco smoke are powerful mutagens, smoking causes lung cancer. Almost all cases of lung cancer, a major cause of death in the United States, are attributed to smoking.

In the United States, smoking-related illnesses cause more than 400,000 deaths each year. Smoking is associated with cancer of the mouth and larynx, and smoking may increase the risk of cancer of the pancreas and bladder. Smoking is also a major contributor to often-fatal respiratory disorders, such as emphysema. The tars in smoke irritate mucous membranes in the mouth, nose, and throat. They accumulate in the lungs and paralyze cilia that move debris from the lungs. Tars also blacken lung tissue and decrease breathing capacity. People who are exposed to secondhand smoke are at risk for the same diseases as people who smoke. Women who smoke during pregnancy are more likely to have miscarriages or to give birth to stillborn babies.

biobit

Is smokeless tobacco harmful?

The use of smokeless tobacco, such as chewing tobacco, causes cancers of the lips, mouth, and gums. When chewing tobacco is placed between the cheek and gum, nicotine and other chemicals are absorbed into the bloodstream.

did you know?

Nicotine is a powerful poison.

Nicotine is extremely toxic. In fact, a sulfate derivative of nicotine has been used as a commercial insecticide for years. Nicotine's effects in the lower dosages experienced by smokers are not immediately noticeable as toxic, however. Initially, nicotine elevates blood pressure by constricting blood vessels and increasing the heart rate. It stimulates the release of free fatty acids into the bloodstream. The added carbon monoxide in the blood causes these fatty acids to be deposited on the inside of the arteries and thus is a major contributor to cardiovascular disease. The combined effect of the strain of cardiovascular disease and the lack of oxygen contributes heavily to the damage seen in smokers' cardiovascular systems.

Alcohol Is an Addictive Depressant

Of all the psychoactive drugs, alcohol (ethanol) is one of the most widely used and abused. Alcohol, found in wine, beer, and liquor, is a depressant that produces a sense of well-being when taken in small amounts. A **depressant** is a drug that generally decreases the activity of the central nervous system. As more alcohol is consumed, however, reaction time increases, and coordination, judgment, and speech become impaired. This produces a state of intoxication known as being "drunk." Drunkenness results as the blood-alcohol concentration (BAC) increases. As shown in **Figure 42-17,** BAC can be measured by a breath test that detects the level of alcohol vapors in the breath. **Table 42-4** shows the effects of alcohol at various concentrations in the blood.

Figure 42-17 Breath test. Law enforcement officials use a device that detects the level of alcohol vapors in the breath to estimate the BAC of drunk-driving suspects.

Table 42-4 Effects of Blood-Alcohol Concentration

BAC*	Condition
0.02–0.04	Slight impairment and sedation
0.05–0.06	Slight impairment of coordination; increased reaction time
0.07–0.09	Slurred speech; blurred vision; intoxication
0.10–0.15	Severe intoxication; impaired coordination, vision, and balance
0.15–0.30	Dizziness; confusion; inability to walk; extremely severe intoxication
0.30–0.50	Unconsciousness
0.50–0.60	Coma or death

*in mg of alcohol per mL of blood

Alcohol is absorbed into the blood through the stomach and small intestine. Alcohol alters neurons throughout the nervous system, changing the shape of receptor proteins. Altered receptor proteins may become more or less sensitive to regular stimuli. Such widespread changes in receptor proteins have various effects on normal brain functioning.

Addiction to alcohol, or alcoholism, is the most prevalent drug-abuse problem in the United States. People who drink excessive amounts of alcohol over long periods of time develop serious health problems. For example, many alcoholics do not eat properly when drinking heavily. This can lead to malnutrition, abnormalities in the circulatory system, and inflammation of the stomach lining. In addition, the liver begins to use alcohol as an energy source. After several years of drinking, the liver accumulates fat deposits. If heavy drinking continues, a potentially fatal liver condition called cirrhosis *(sih ROH sis)* may develop. In a cirrhotic liver, cells are replaced with scar tissue, and liver functioning is impaired.

internet**connect**

SCLINKS NSTA

TOPIC: Blood alcohol concentration
GO TO: www.scilinks.org
KEYWORD: HX973

Section 42-4

Demonstration

Motor Function

Show the class a videotape, in slow motion if possible, of someone performing a complicated high dive or balance-beam routine. Have students relate this type of performance to the function of the cerebellum. Alcohol has a significant effect on the synapses in the cerebellum. Have students discuss alcohol's effects on balance and coordination.

Teaching TIP

Alcohol Content

Tell students that beer, wine, and liquor differ in their ethanol concentrations. However, a 12 oz can of beer, a 4 oz glass of wine, and a 1 oz shot of liquor all contain about the same amount of absolute ethanol. Tell students that "proof" reflects the alcohol concentration in a beverage. Alcohol content is determined by dividing the proof number in half—80 proof liquor has a 40 percent alcohol content.

internet**connect**

Have students research laws for blood alcohol concentration using the Web site in the Internet Connect box on page 973. Have students write a report summarizing their findings.

PORTFOLIO

Biology in Your Community

Alcohol and Automobiles

Invite a local law enforcement officer to discuss topics such as the differences between DUI and DWI, incidence of DUI and DWI, and the dangers of drinking and driving. Have the officer tell students what blood alcohol concentration is considered "legally drunk" in your state. Have students find this BAC level in Table 42-4 on this page.

Chapter 42

Teaching TIP

Effects of Drugs

Have students choose a class of frequently abused drugs (such as opiates, depressants, or stimulants) or a specific drug (such as cocaine, marijuana, or alcohol). Ask them to make a pamphlet about the effects of the drug(s), the dangers of using the drug(s), and where to seek help for drug addiction. PORTFOLIO

3 Assessment Options

Closure

Technique: Anticipation Guide

Ask volunteers to discuss how their opinions formed in the **Active Reading** exercise on page 951 changed or stayed the same as a result of their reading. Students should point to specific passages in the text that support their reasoning.

Review

Assign Review—Section 42-4.

Reteaching — BASIC

Have students design an informative poster to educate people about the dangers of smoking. PORTFOLIO

Some Drugs Mimic Neurotransmitters

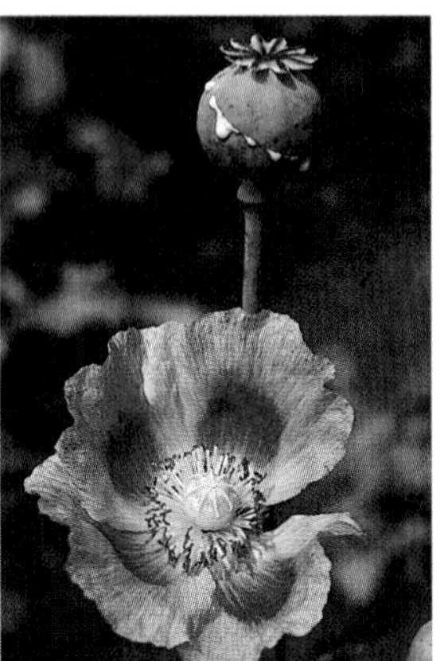

Figure 42-18 Opium poppy. Opium is a narcotic derived from the poppy plant, *Papaver somniferum.*

Narcotics are extremely addictive psychoactive drugs that relieve pain and induce sleep. Many of the most potent narcotics are derived from the poppy plant, *Papaver somniferum,* shown in **Figure 42-18.** The sap that oozes from the cut seed pod forms a thick, gummy substance called opium. Drugs derived from opium, called opiates or narcotics, include codeine *(KOH deen),* morphine, and heroin, a more potent form of morphine. Codeine is widely prescibed by doctors for pain relief. Morphine is one of the most effective pain-relieving drugs used today. Heroin addiction and abuse are among the most serious illegal-drug problems in society.

Stopping pain

Recall that pain receptors throughout the body detect painful stimuli. As uncomfortable as it may feel, pain plays a very important role in the body. Pain notifies you that body tissues have been injured or damaged. Imagine how your body would look and function today if you did not have the ability to sense pain. Pain begins as a signal at damaged nerve endings. Nerve impulses generated by pain receptors travel to the spinal cord toward the brain. After reaching the spinal cord, a pain signal is suppressed by a class of neurotransmitters called enkephalins *(ihn KEHF uh lihnz).* When enkephalins bind to neurons in the spinal cord, they prevent pain signals from reaching the brain.

Narcotics mimic the action of enkephalins by binding to the same receptor proteins in the spinal cord. These receptor proteins are called opiate receptors because scientists observed opiates binding to them before enkephalins were ever discovered. Narcotics also affect the limbic system, producing a feeling of well-being.

Figure 42-19 Hemp. Marijuana is produced from the hemp plant, *Cannabis.*

Marijuana

In addition to alcohol and tobacco, marijuana, though illegal, is a widely consumed drug. Marijuana comes from various species of the hemp plant, *Cannabis,* shown in **Figure 42-19.** Hashish also comes from the hemp plant. The active ingredient in marijuana and hashish is commonly known as THC. When marijuana is smoked, it may cause disorientation, impaired judgment, short-term memory loss, and general loss of motivation. Scientists continue to research the effects of THC on the nervous system.

Review SECTION 42-4

1. **Describe** how tolerance to a drug develops.
2. **Summarize** how cocaine produces addiction.
3. **Distinguish** between stimulants and depressants. Give an example of each.
4. **SKILL Recognizing Relationships** What do all psychoactive drugs have in common?
5. **SKILL Applying Information** Why is drug addiction considered a physiological condition?

Review SECTION 42-4 ANSWERS

1. Repeated exposure to a drug leads to tolerance, a condition in which increased amounts of the drug are needed to achieve the desired effect.
2. Cocaine produces addiction because it overstimulates postsynaptic neurons by blocking dopamine reuptake, ultimately decreasing the number of dopamine receptors on the postsynaptic cells. When cocaine is removed, insufficient dopamine is available to stimulate post-synaptic cells at a normal level.
3. Stimulants, such as nicotine, increase central nervous system activity. Depressants, such as alcohol, decrease central nervous system activity.
4. All psychoactive drugs alter the function of the central nervous system.
5. Drug addiction is considered physiological because addictive drugs alter the normal functioning of neurons and synapses.

Study ZONE

CHAPTER 42

Use the Key Concepts and Key Terms listed below to review the main ideas in this chapter. If you need to review the meaning of a term, turn to the page indicated.

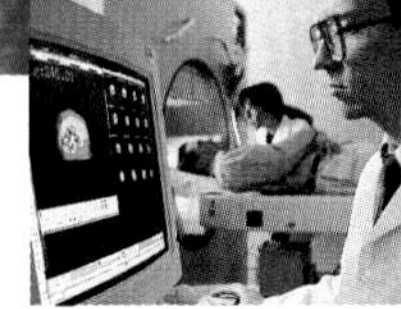

Key Concepts

42-1 Neurons and Nerve Impulses

- Neurons are specialized cells that rapidly transmit information as electrical signals throughout the body.
- At the resting potential, the inside of a neuron is negatively charged with respect to the outside of the neuron.
- An action potential moves rapidly down an axon.
- Synaptic transmission of nerve impulses involves the release of neurotransmitters at synapses.

42-2 Structures of the Nervous System

- The central nervous system consists of the brain and spinal cord.
- The brain contains three major parts: the cerebrum, the cerebellum, and the brain stem.
- The spinal cord links the brain to the peripheral nervous system, which branches throughout the body.

42-3 Sensory Systems

- Sensory receptors detect various sensory stimuli.
- Photoreceptors in the eyes convert light into electrical signals that are interpreted by the brain.
- The ear converts sound into electrical signals that are interpreted by the brain.
- The semicircular canals monitor the position of the head.
- Taste and smell are related chemical senses.

42-4 Drugs and the Nervous System

- Psychoactive drugs affect the central nervous system.
- Drug addiction involves physiological changes in neurons.
- Nicotine is an addictive stimulant found in tobacco products.
- Alcohol is an addictive depressant that widely affects the central nervous system.
- Narcotics are addictive drugs used to stop pain.
- THC in marijuana causes short-term memory loss and impaired judgment.

Key Terms

42-1

neuron (952)
dendrite (952)
axon (952)
nerve (952)
membrane potential (953)
resting potential (954)
action potential (954)
synapse (956)
neurotransmitter (956)

42-2

central nervous system (958)
peripheral nervous system (958)
sensory neuron (958)
motor neuron (958)
brain (958)
cerebrum (959)
cerebellum (959)
brain stem (959)
thalamus (960)
hypothalamus (960)
spinal cord (960)
reflex (960)
interneuron (961)

42-3

sensory receptor (964)
retina (966)
rod (966)
cone (966)
optic nerve (966)
cochlea (967)
semicircular canal (967)

42-4

psychoactive drug (969)
addiction (970)
tolerance (970)
withdrawal (970)
stimulant (970)
depressant (973)

Study TIP Ready?

If you think you understand these Key Concepts and Key Terms, then you're ready for the Performance Zone.

Study ZONE

CHAPTER 42

Review and Practice

Assign the following worksheets for Chapter 42:

- **Student Review Guide**
- **Concept Mapping Worksheet**
- **Test Preparation Pretest**
- **Vocabulary Worksheet**
- **Critical Thinking Review Worksheet**

Assessment Options

The following assessment options are available for this chapter:

Resource Link

Assign **Chapter Test 42** to assess students' understanding of the chapter.

Alternative Assessment

- Have students write questions and answers for a trivia-type review game. PORTFOLIO

Portfolio Assessment

- *Portfolio Assessment* at the front of this book provides options for building and evaluating student portfolios. Students and teachers can select items from the areas listed or inclusion in a portfolio.
- See items labeled PORTFOLIO on pp. 953, 957, 960, 962, 963, 964, 968, 970, 971, 973, and 974.

Answer to Concept Map

The following is one possible answer to Performance Zone item 1 on page 976.

Nervous system
has divisions called
central nervous system
is composed of
spinal cord
brain
peripheral nervous system
is composed of
neurons
made of
nerves
pass signals across
synapse
using chemicals called
neurotransmitters
such as
sensory neurons
motor neurons
send signals to
receive signals from
is connected to the CNS by
nerves
which travel to the

Performance ZONE

CHAPTER 42

REVIEW ANSWERS

Assignment Guide

SECTION	REVIEW
42-1	2, 3, 4, 5, 6, 13, 15, 16, 17, 18, 19
42-2	1, 7, 8, 9, 12, 14, 21, 24, 26
42-3	10, 23, 27
42-4	11, 20, 22, 25

Understanding and Applying Concepts

1. The answer to the concept map is found at the bottom of page 975.
2. **a.** The cerebellum smooths and coordinates ongoing movements and helps maintain balance. The cerebrum is involved with learning, memory, reasoning, and voluntary actions.

   b. A neuron is a nerve cell composed of dendrites, a cell body, and one or more axons. A nerve is a bundle of axons.

   c. Addiction is a physiological dependence on a drug. As tolerance builds, increased amounts of a drug are needed to achieve the desired effect.

   d. A stimulant increases central nervous system activity. A depressant decreases central nervous system activity.
3. c
4. c
5. a
6. d
7. b
8. b
9. b
10. d
11. c
12. The fish cerebrum is much smaller than a human cerebrum (relative to other parts of the brain). The sense of smell probably is very important to the fish.
13. Voltage-gate sodium channels, which close shortly after an action potential, cannot open again for a brief moment after the action potential. This prevents the nerve impulse from moving in the opposite direction.
14. Cell-death inhibitors are used for myelin-forming cells. Peripheral nerves are graphed to stimulate axon growth.
15. When binding to receptor proteins on the postsynaptic cell, some neurotransmitters cause ion channels to open, some stimulate the formation of second messengers in the cell, and others stimulate enzyme activity in the cell.

Skills Assessment

16. *axon terminal*—transmits nerve impulses
17. *myelin sheath*—insulates axon
18. *axon*—conducts nerve impulses
19. *dendrite*—receives nerve impulses
20. Answers will vary. For example: Some antidepressants boost serotonin levels by inhibiting its reuptake and are used to treat depression.
21. For example: *Disease:* Alzheimer's disease, *cause:* unknown—origin of plaque buildups in the brain may be neurochemical, environmental, infectious, or genetic; *symptoms:* impaired memory, loss of intellectual function, atrophy of brain tissue, degradation of language ability; *part of the nervous system:* brain; *available treatment:* none available—symptoms are treated with medication, physical therapy, and occupational therapy. *Disease:* Parkinson's disease; *cause:* progressive deterioration of the nerve cells in a part of the brain responsible for controlling muscle movement and producing dopamine; *symptoms:* muscle rigidity,

Performance ZONE CHAPTER 42 Review

Understanding and Applying Concepts

1. **Concept Mapping** Make a concept map that describes the structures and functions of the nervous system. Try to include the following terms: spinal cord, brain, neuron, nerve, synapse, neurotransmitter. Include additional terms as needed.

2. **Understanding Vocabulary** For each pair of terms, explain the differences in their meanings.
   a. cerebellum, cerebrum
   b. neuron, nerve
   c. addiction, tolerance
   d. stimulant, depressant

3. A myelin sheath on the axon of a neuron
   a. covers the axon completely.
   b. decreases the rate of impulse conduction.
   c. increases the rate of impulse conduction.
   d. has no effect on nerve impulse conduction.

4. When a neuron is at the resting potential,
   a. the inside is positively charged.
   b. the outside is negatively charged.
   c. the inside is negatively charged.
   d. None of the above

5. During an action potential,
   a. sodium ions diffuse into a neuron.
   b. sodium ions diffuse out of a neuron.
   c. potassium ions flow into a neuron.
   d. there is no movement of ions.

6. At a synapse, neurotransmitters
   a. diffuse across the synaptic cleft.
   b. bind to receptors on the postsynaptic cell.
   c. may excite or inhibit the postsynaptic cell.
   d. All of the above

7. Which of the following might result from damage to the cerebellum?
   a. insatiable thirst c. indigestion
   b. loss of balance d. loss of hearing

8. The thalamus
   a. is part of the peripheral nervous system.
   b. relays signals to the cerebral cortex.
   c. coordinates movement.
   d. regulates emotions.

9. In a spinal reflex, the signal travels
   a. immediately to the brain.
   b. to the spinal cord and out to a muscle.
   c. only through sensory neurons.
   d. only through motor neurons.

10. Photoreceptors are stimulated by
    a. heat. c. chemicals.
    b. pressure. d. light.

11. Drug addiction is considered a physiological condition because addictive drugs
    a. can be purchased illegally.
    b. must be injected.
    c. alter the functioning of neurons.
    d. are used in social settings.

12. The diagram below shows the brain of a fish. How is the cerebrum of the fish brain different from that of a human brain? What do the large olfactory bulbs of the fish brain indicate about the relative importance of the sense of smell to the fish?

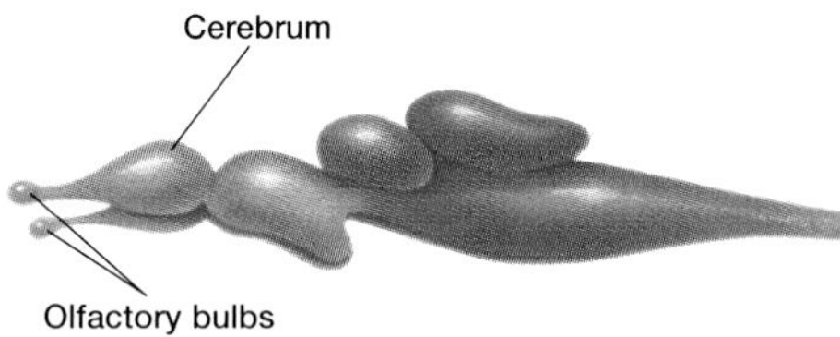

13. **Theme Cellular Structure and Function** Action potentials travel in only one direction along a neuron—toward the axon terminals and away from the cell body. What structures of the neuron ensure that this pattern is always followed?

14. **Health Watch** Although the spinal cord does not heal when it has been injured, scientists have developed ways to help prevent paralysis. Describe two of these ways.

15. **Chapter Links** List three ways that the binding of a neurotransmitter to a receptor protein on a postsynaptic cell could cause changes in the cell. (**Hint:** See Chapter 4, Section 4-2.)

Skills Assessment

Part 1: Interpreting Graphics

Use the diagram of the neuron below to answer the following questions:

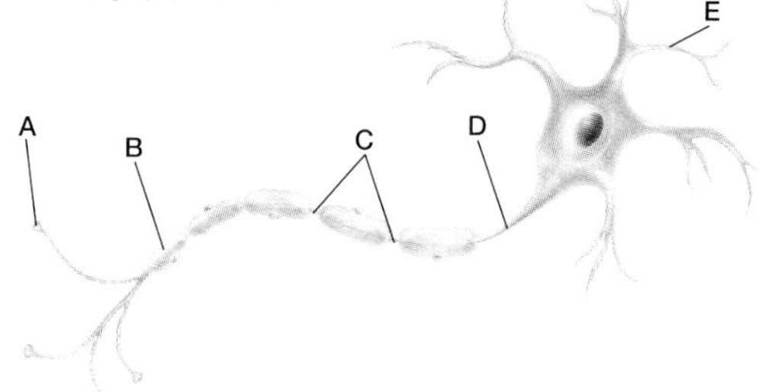

16. What part of the neuron is labeled *A*? What is its function?

17. What part of the neuron is labeled *B*? How does this structure affect the conduction of nerve impulses?

18. What part of the neuron is labeled *D*? What is its function?

19. What part of the neuron is labeled *E*? What is its function?

Part 2: Life/Work Skills

Use the media center or Internet resources to learn more about disorders of the nervous system.

20. Finding Information Research different types of prescription drugs used to treat various mental disorders. Find out how these drugs affect the nervous system. Summarize your findings in an oral report to your class.

21. Summarizing Information Research the causes and symptoms of various disorders of the nervous system caused by the degeneration of neurons in the brain. Some disorders include Alzheimer's disease, Parkinson's disease, and multiple sclerosis. Find out how these disorders affect the nervous system. List the types of drugs or other methods used to treat these disorders.

22. Finding Information Depression affects several million Americans. Find out about the symptoms, causes, and treatments of depression. Summarize your findings in a written report.

Critical Thinking

23. Inferring Relationships What is the function of the semicircular canals of the inner ear? People who suffer from vertigo feel dizzy and disoriented in certain situations. What is the relationship between vertigo and the semicircular canals?

24. Distinguishing Relevant Information Epilepsy affects about one of every 200 Americans. Neurons in the brain normally produce small bursts of action potentials in varying patterns. During an epileptic seizure, many neurons in the brain produce large bursts of action potentials simultaneously. This can cause the body to become rigid and to jerk or convulse. From what you know about the brain's control of muscles and posture, how might you explain these symptoms?

Portfolio Projects

25. Analyzing Information Research the way general anesthetics work. Then write a short report in which you discuss a specific anesthetic, the effect of the anesthetic on the brain, and the effect of body weight and the passage of time on the action of the anesthetic.

26. Finding Information Read the article titled "Adult Human Brain Adds New Cells" (*Science News,* October 31, 1998, p. 276). How are cancer patients being used to investigate the possible growth of new neurons in adult brains? What evidence suggests that the growth of neurons in animals continues into adulthood?

27. Career Focus Optometrist Optometrists are professionals who provide and adjust eyeglasses and contact lenses for people with vision problems. Research the field of optometry, and write a report on your findings. Find out the difference between optometry and ophthalmology. Your report should include a job description, training and education required, kinds of employers, growth prospects, and starting salary.

stiffness, tremors, shuffling gait, "masked" appearance of face, muscle atrophy, memory loss; *part of the nervous system:* brain, specifically basal ganglia; *available treatment:* no known cure—symptoms can be controlled to some extent using medication such as L-dopa, together with good nutrition, health, exercise, physical therapy, occupational therapy, and speech therapy; *Disease:* multiple sclerosis; *cause:* inflammation in the central nervous system which destroys myelin, leaving scar tissue that slows nerve impulse transmission. The root cause is unknown, but it may be environmental, geographical, genetic, or a disorder of the immune system; *symptoms:* weakness, paralysis of extremities, muscle atrophy and spasticity, vision loss, dizziness, decreased attention span; *part of the nervous system:* brain and spinal cord; *available treatment:* no known cure—medication, physical therapy, speech therapy, and occupational therapy and counseling are all helpful.

22. Symptoms of depression include downheartedness, sadness, dejection, poor communication, withdrawal, and indifference to surroundings. Some episodes of depression are triggered by the loss of friend or relative, a major disappointment in work, some prescription drugs, prolonged illness, alcohol or drug withdrawal, or hormones. Treatments include counseling, drugs, and exercise.

Critical Thinking

23. The semicircular canals detect acceleration and orientation of the body. Eyes detect movement, sending that message to the brain. If the hair cells in the semicircular canals do not, however, detect the same degree of movement, they send a conflicting signal to the brain. These conflicting signals can cause dizziness and vertigo.

24. Many muscles are being stimulated all at once.

Portfolio Projects

25. Answers will vary. For example, the drug methohexital induces general anesthesia. It depresses the central nervous system by lowering neural activity. Effective dose rises with body weight.

26. Patients who died of cancer donated their bodies so that the brain can be examined for the presence of new neurons. Researchers found 200 new, healthy neurons per cubic millimeter of the dentate gyrus.

27. Answers will vary. A doctor of optometry, also called an optometrist, diagnoses and treats disorders of the visual system. Optometrists examine the eyes and prescribe glasses or contact lenses for correctable problems with vision. An optometrist also diagnoses diseases of the visual system, such as glaucoma and diabetic retinopathy. A doctor of optometry is required to complete a four-year postgraduate degree program at an accredited school of optometry. Optometrists work in private practices, hospitals, or for public-health agencies. The growth potential for this field is good. Starting salary will vary by region.

Interactive Exploration
Chapter 42

Drug Addiction

Purpose

Students will use a computer simulation to study the effects of cocaine on the brain.

Preparation Notes

Time Required: 50–60 minutes

Preparation Tips

- Depending on how many computers are available, set up work stations that allow students to work individually or in teams of two to four.
- Encourage students to read the lab and construct the data table described in the **Before You Begin** section before coming to class on the day of the lab.
- Before students begin the lab, conduct a class discussion in which you ask the following questions:
  1. What is a synapse? *(A synapse is the gap and functional connection between two neurons.)*
  2. How is an impulse transmitted across a synapse? *(Neurotransmitter molecules diffuse across the synapse.)*

Procedural Tips

- Be sure that students are familiar with the instructions for loading and running an interactive investigation before they begin the lab.
- The animation shows a schematic cross-section of a synapse in the limbic system. When the investigation is running, dopamine molecules, represented by blue spheres, diffuse across the synapse and bind to receptors on the postsynaptic neuron (represented by gray, cuplike structures), starting a pleasure-inducing nerve impulse. Cocaine, represented by red spheres, can be introduced into the synapse at five different dosages and four frequencies of use.

Interactive Exploration

Chapter 42 *Drug Addiction*

CD-ROM Lab

SKILL

- Computer modeling

OBJECTIVES

- **Simulate** the effects of cocaine use on a synapse in the brain.
- **Observe** the effect of cocaine use on the number of receptors in a synapse.

MATERIALS

- computer with CD-ROM drive
- CD-ROM *Interactive Explorations in Biology: Human Biology*

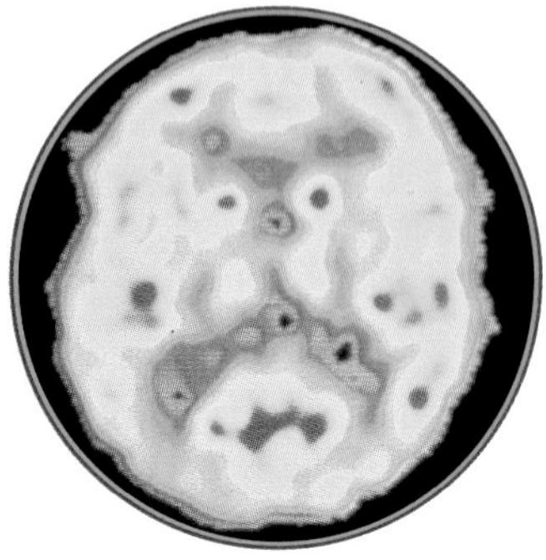
PET scan of brain

Before You Begin

Drug addiction results when a drug alters the function of neurons in the brain. Cocaine is an addictive drug that affects the limbic system, a group of brain structures that control the sensation of pleasure. Cocaine produces euphoria (an intense feeling of well-being) by interfering with the reabsorption of the neurotransmitter dopamine. In this Interactive Exploration, you will explore the consequences of introducing cocaine into a synapse within the limbic system.

1. Select *Drug Addiction* from the menu. Click on the key near the top of the navigation palette. Read the focus questions, and review these concepts: synapse, neurotransmitters, transporters, neuromodulators, desensitization, and addiction.
2. Click *File* at the top left of the screen, and select *Interactive Exploration Help*. Listen to the instructions. Click *Exploration* at the bottom right of the screen to begin the exploration. You will see a diagram like the one below.

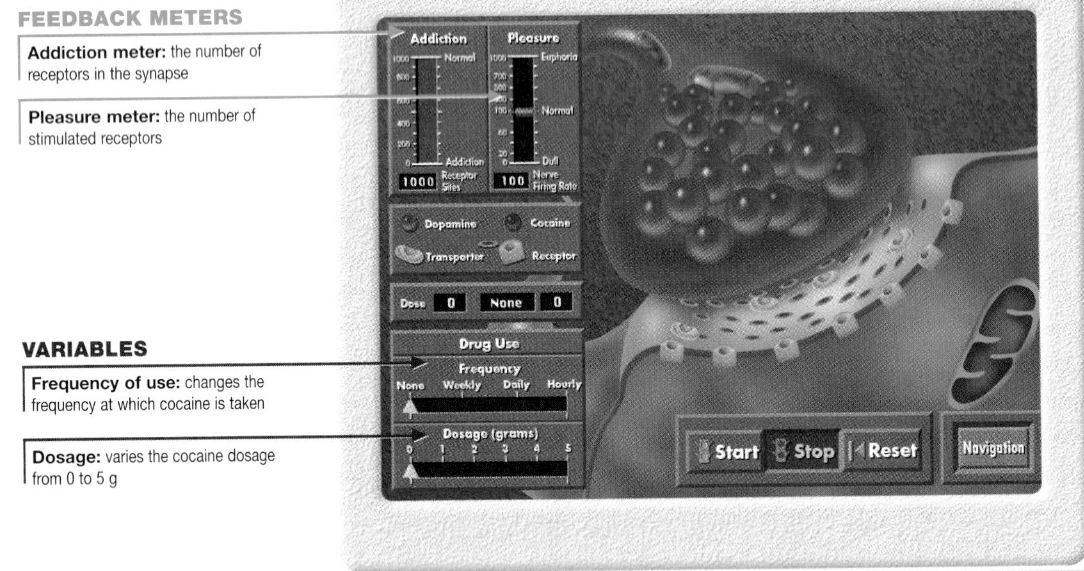

Additional Background

The limbic system is a part of the brain that is important in the sensation of pleasure. The neurotransmitter dopamine transmits signals between neurons in the limbic system. Dopamine is removed from synapses and returned for recycling by transporters. Cocaine binds to these transporter molecules and blocks reabsorption of dopamine. Thus, dopamine neurotransmitters are “trapped” in the synapses and continue to stimulate the postsynaptic neurons, stimulating the brain's pleasure centers repeatedly. This is why cocaine produces a sense of euphoria. The nervous system responds to long-term cocaine use by reducing the number of receptors on postsynaptic neurons. Eventually, the number of receptors falls so low that, even in the presence of cocaine, the synapses fire no more often than non-drugged synapses. There are only two ways to achieve a normal rate of firing in drug-altered synapses. One way is through withdrawal—waiting for post-synaptic neurons to make more receptors. The other way to reestablish a high rate of firing is to take more cocaine. Over time, however, more and more cocaine is required as the post-synaptic neurons continually reduce the number of receptors they have.

3. Make a data table similar to the one below.

DATA TABLE						
Amount of cocaine	Addiction meter			Pleasure meter		
	Start	End	Change	Start	End	Change
None						
1 g/wk						
2 g/wk						

Procedure

PART A: Normal Synapse

1. Look at the two meters in the upper left corner of the screen. Notice the number of receptor sites on the postsynaptic neuron, indicated by the *Addiction* meter, and the number for the nerve firing rate, indicated by the *Pleasure* meter. Record these starting numbers in your data table.
2. Click the *Start* button to begin the simulation. Observe what happens when the neurotransmitter dopamine is released into the synapse. Notice that each contact with a receptor on the other side of the synapse "fires" the postsynaptic neuron. This produces a sensation of pleasure. Also notice how the transporters recycle neurotransmitter molecules. This keeps dopamine from building up in the synapse.
3. Let the simulation run for 3–4 minutes. Click the *Stop* button to end the simulation. Record the numbers in the *Addiction* meter and the *Pleasure* meter in your data table. Also record any change that occurred in these numbers during the simulation.

PART B: Effect of Cocaine

4. Click the *Reset* button. Then slide the *Frequency* indicator to Weekly and the *Dosage* indicator to 1 gram.
5. Record the numbers in the *Addiction* meter and the *Pleasure* meter in your data table. Then click the *Start* button to begin the simulation. You will see several red spheres, representing cocaine, enter the synapse. What happens to the cocaine molecules when they enter the synapse?
6. The simulation, which should take about 5 minutes, will run until five doses have been taken. As the simulation runs, notice how the number of receptor sites and the nerve firing rate change. Record the highest and lowest numbers reached on each meter between each dose of cocaine. What happens to these numbers when a new dose of cocaine enters the synapse?
7. When the simulation ends, record the numbers in the *Addiction* meter and the *Pleasure* meter. Also record any change that occurred in these numbers.
8. Click the *Reset* button to clear the simulation. Now set the *Frequency* indicator to Weekly and the *Dosage* indicator to 2 grams. Repeat steps 5–7.

Analyze and Conclude

1. **Summarizing Results** How did the number of receptor sites in the synapse change with repeated doses of cocaine?
2. **Analyzing Results** What happens to the number of receptor sites if the weekly dose of cocaine is more than 2 grams?
3. **Analyzing Methods** How did the simulation in Part A serve as a control for the simulation in Part B?
4. **Drawing Conclusions** Is a weekly dose of 2 grams of cocaine likely to cause addiction? Justify your answer.
5. **Relating Concepts** Based on what happens to the red spheres in Part B, explain why cocaine causes pleasurable sensations.
6. **Forming Reasoned Opinions** Explain why people addicted to cocaine need higher and higher doses to feel pleasure.

Use the following Internet resources to explore your own questions about mechanisms of drug addiction.

Interactive Exploration

Chapter 42

Analyze and Conclude

Answers

1. The number of receptors was reduced.
2. The number of receptor sites decreases sharply.
3. The simulation in Part A illustrates activity in a normal synapse, which will serve as the baseline, or control, level.
4. There is no safe dose of cocaine. Even the lowest dose taken at the lowest frequency produces the condition necessary for addiction: loss of receptors in the synapse.
5. Cocaine blocks the removal of dopamine from the synapse. Because dopamine is not removed, it continues to stimulate the post-synaptic neuron and therefore increases pleasurable sensations.
6. The more cocaine is taken, the smaller the number of receptors becomes. Therefore, more cocaine is needed to restore normal sensation. Because tolerance increases, more cocaine must be taken to achieve the desired effect.

Procedure

Answers

1. See sample data table on page 979.
3. See sample data table on page 979.
5. The red spheres bind to the transporter molecules that remove dopamine from the synapse.
6. The number of receptors declines when each new dose of cocaine enters the synapse.
7. See sample data table on page 979.
8. Students will discover that the Pleasure Meter reaches a much higher value, and the number of receptors in the synapse falls to a lower value, by the end of the simulation.

Sample Data Table

DATA TABLE						
AMOUNT OF COCAINE	ADDICTION METER			PLEASURE METER		
	START	END	CHANGE	START	END	CHANGE
None						
1 g/wk	1, 000	750	250	100	75	25
2 g/wk	1, 000	500	500	100	50	50

Chapter **43**

Hormones and the Endocrine System

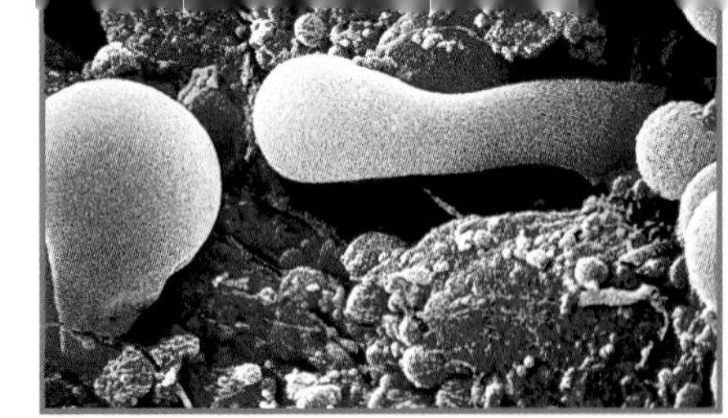

Planning Guide

BLOCK	OBJECTIVES	MEETING INDIVIDUAL NEEDS	READING SKILLS AND STRATEGIES
BLOCK 1 45 min	**43-1 Hormones** • **Identify** four major functions of hormones. • **Differentiate** between endocrine and exocrine glands. • **Compare** the endocrine and nervous systems. • **Distinguish** the functions of endorphins from the functions of prostaglandins.	Learning Styles, p. 982 BASIC ADVANCED • p. 984	**ATE** Active Reading, Technique: Reading Effectively, p. 981 **ATE** Active Reading, Technique: Paired Summarizing, p. 983 ■ Active Reading Guide, worksheet (43-1) ■ Directed Reading, worksheet (43-1)
BLOCK 2 45 min	**43-2 How Hormones Work** • **Relate** how hormones act only on specific cells. • **Summarize** how amino-acid-based hormones produce responses. • **Summarize** how steroid and thyroid hormones produce responses. • **Relate** how negative feedback is used to regulate hormone levels.	Learning Styles, p. 985 BASIC ADVANCED • p. 989	■ Active Reading Guide, worksheet (43-2) ■ Directed Reading, worksheet (43-2)
BLOCKS 3 & 4 90 min	**43-3 The Major Endocrine Glands** • **Evaluate** the roles of the hypothalamus and the pituitary gland in controlling other hormones. • **Summarize** the roles of the thyroid and parathyroid hormones. • **Compare** the roles of the hormones secreted in each area of the adrenal gland. • **Relate** how each of the two hormones secreted by the pancreas regulates blood sugar levels. • **Describe** the roles of reproductive hormones and of melatonin.	Learning Styles, p. 990 BASIC ADVANCED • pp. 991, 993, 996	■ Active Reading Guide, worksheet (43-3) ■ Directed Reading, worksheet (43-3) ■ Supplemental Reading Guide, *Microbe Hunters*

Review and Assessment Resources

BLOCKS 5 & 6 90 min

TEST PREP

- **PE** Study Zone, p. 997
- **PE** Performance Zone, pp. 998–999
- ■ Chapter 43 worksheets:
  - Vocabulary
  - Concept Mapping
  - Critical Thinking
- Audio CD Program
- ■ Test Prep Pretest

CHAPTER TESTING

- ■ Chapter Tests (blackline master)
- ▲ Test Generator
- ■ Assessment Item Listing

See the correlations table at the front of this book for a list of the ***National Science Education Standards*** addressed in this chapter.

Smithsonian Institution® Visit **www.si.edu/hrw** for additional on-line resources.

KEY

Biology Principles and Explorations

PE Pupil's Edition

ATE Teacher's Edition

BIOSOURCES

Teaching Transparencies

Laboratory Program

One-Stop Planner CD-ROM

Shows these resources within a customizable daily lesson plan:

■ Teaching Resources

▲ Test Generator

VISUAL STRATEGIES	CLASSWORK AND HOMEWORK	TECHNOLOGY AND INTERNET RESOURCES
ATE Fig. 43-2, p. 983 230 Nerves Versus Hormones	PE Review, p. 984	Holt Biology Videodiscs Section 43-1 Audio CD Program Section 43-1
ATE Fig. 43-4, p. 986 ATE Fig. 43-5, p. 987 229 How Hormones Work 231 Action of Thyroxine 199A How Parathyroid Hormone Affects Blood Calcium Levels	PE Quick Lab How can you model negative feedback?, p. 988 PE Review, p. 989	Holt Biology Videodiscs Section 43-2 Audio CD Program Section 43-2 internetconnect SCLINKS NSTA TOPIC: Anabolic steroids
194A Major Endocrine Glands and Hormones 195A Major Endocrine Glands and Hormones, cont.	PE Math Lab Analyzing blood-sugar regulation after eating different meals, p. 994 PE Experimental Design How Does Epinephrine Affect Heart Rate?, pp. 1000–1001 ■ Occupational Applications: Pharmacist PE Review, p. 996 PE Study Zone, p. 997	CNN Video Segment 35 Obesity Hormone Holt Biology Videodiscs Section 43-3 Audio CD Program Section 43-3 internetconnect SCLINKS NSTA TOPICS: • Hormones and body fat • Melatonin • Hormones

ALTERNATIVE ASSESSMENT

PE Portfolio Projects 25–28, p. 999

ATE Alternative Assessment, p. 997

ATE Portfolio Assessment, p. 997

Lab Activity Materials

Quick Lab

p. 988 water (250 mL), 500 mL beaker, thermometer, hot plate, ring stand with rings, wire mesh, heat-resistant gloves, apron, safety goggles, tongs, stirring rod, ice

Math Lab

p. 994 pencil, paper

Experimental Design

p. 1000 medicine droppers, *Daphnia, Daphnia* culture water, depression slides, petroleum jelly, coverslips, compound microscope, watch or clock with second hand, paper towels, 100 mL beaker, 10 mL graduated cylinder, epinephrine solutions (0.001%, 0.0001%, 0.00001%, 0.000001%)

Lesson Warm-up

ATE several locks with keys, tape, paper, p. 985

Demonstrations

ATE photos of persons with visible hormonal imbalances, p. 980

ATE small beakers (4), gelatin, cooking oil, vitamin E capsules, water, p. 987

ATE photos of persons with dwarfism, giantism, goiter, and Cushing's syndrome, p. 990

Chapter 43

Chapter Theme

Homeostasis

The endocrine system is essential in maintaining homeostasis in the body. Among other things, hormones regulate blood sugar, coordinate the body's responses to stress, control the amount of urine produced, and regulate the menstrual cycle in females. Disorders of the endocrine system can be life threatening.

Establishing Prior Knowledge

Before beginning this chapter, make certain students can answer the questions in **Study TIP Ready?** on page 981. If they cannot, encourage them to reread the sections indicated.

Opening Demonstration

Show the class photographs of people who obviously have a permanent or temporary hormonal imbalance (for example, someone who is extremely short or tall or someone who is covered with excessive body hair). Ask students if they know the type of molecule that is responsible *(hormone).* Tell students that hormones are very important to maintaining homeostasis. They are secreted in tiny concentrations in the blood or extracellular fluid. Either too much or too little of a hormone can have serious consequences.

CHAPTER 43 Hormones and the Endocrine System

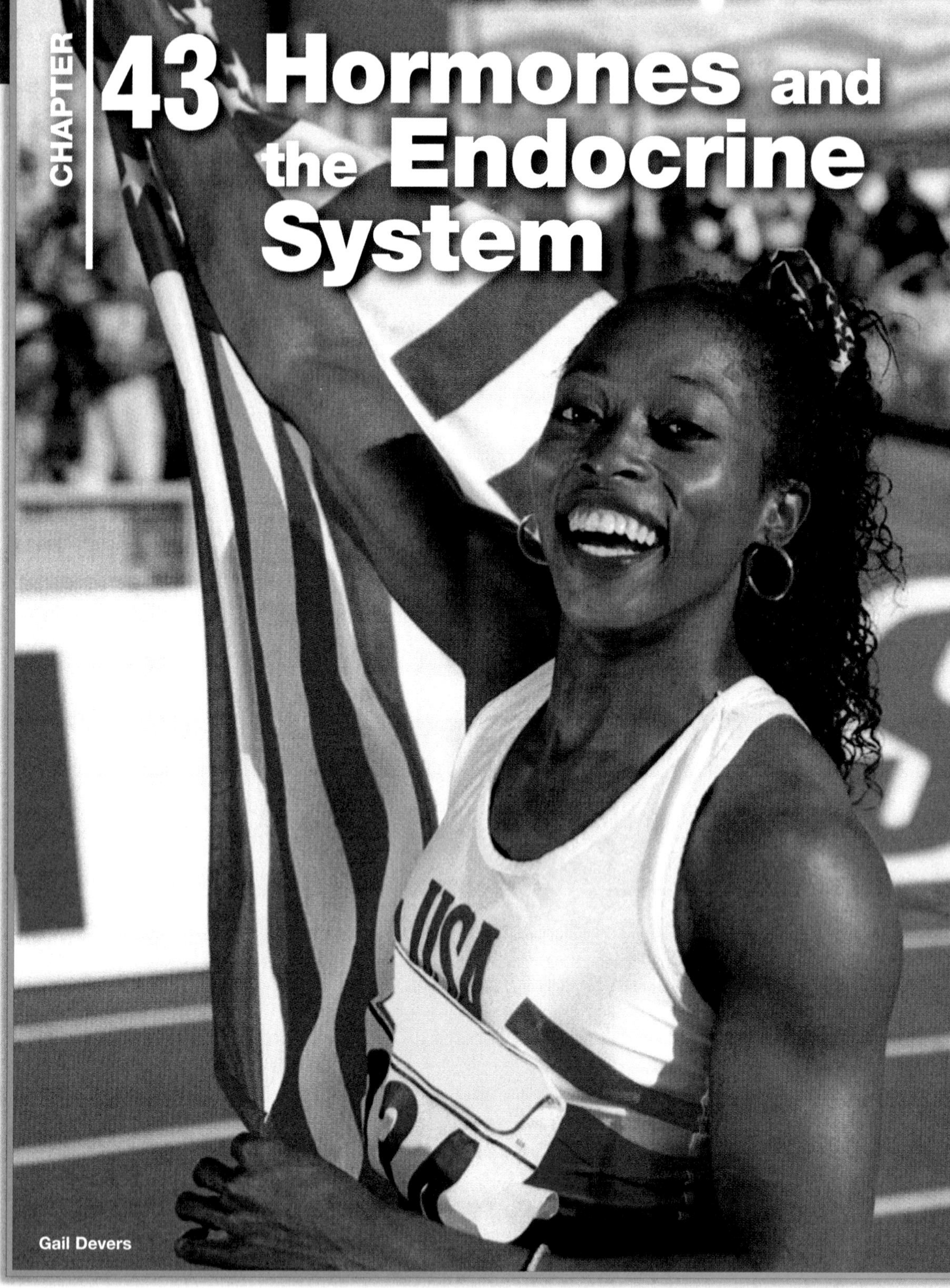

Gail Devers

What's Ahead — and Why
CHAPTER OUTLINE

43-1 Hormones
- The Body's Activities Are Coordinated
- Hormones Are Made in Certain Organs and Tissues

43-2 How Hormones Work
- Hormones Act on Specific Cells
- Hormones Bind to Receptors
- The Release of Hormones Must Be Regulated

43-3 The Major Endocrine Glands of the Body
- Two Endocrine Glands Are A Major Control Center
- The Thyroid Regulates Metabolism, Development, and Calcium Levels
- The Adrenals Respond to Stress
- The Pancreas and Other Organs Also Produce Hormones

Students will learn
- how hormones coordinate the body's activities.
- how a hormone binding to its receptor can cause a cell to change its activities.
- how the endocrine system is coordinated and what role different hormones play.

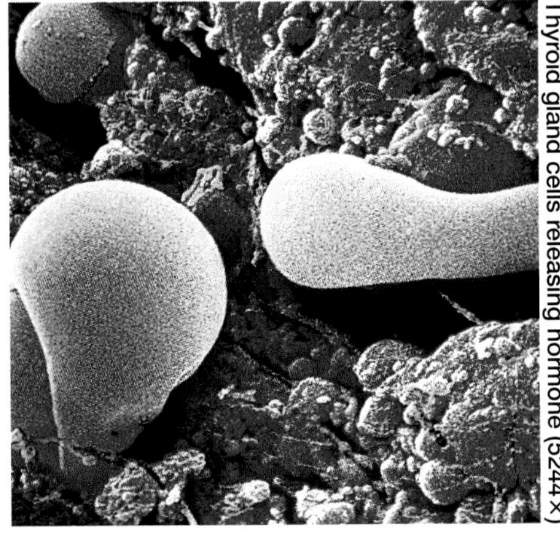
Thyroid gland cells releasing hormone (5244×)

Severe weight loss, shaking fits, and loss of vision in one eye *were symptoms experienced in 1988 by American sprinter Gail Devers. She was later diagnosed with Grave's disease, a condition in which excessive thyroid hormones are made by the thyroid gland. Devers went on to win the gold medal in the 100 m dash in two Summer Olympics.*

Study TIP

Ready?

Check your understanding of these topics to help you prepare for what's ahead.

Can you . . .

- **describe** the function and structure of glycogen? (Section 2-3)
- **describe** the action of enzymes? (Section 2-4)
- **summarize** the location and function of receptors? (Section 3-2)
- **summarize** the role of DNA and mRNA in protein synthesis? (Section 10-1)
- **describe** the role of the sympathetic nervous system? (Section 42-2)

If not, *review the sections indicated.*

Looking Ahead

43-1 Hormones
What kinds of substances help coordinate all of the activities your body carries out?

43-2 How Hormones Work
How do hormones cause cells to change what they are doing?

43-3 The Major Endocrine Glands of the Body
What types of body activities do hormones coordinate?

Labs

- **Quick Lab**
  How can you model negative feedback? (p. 988)
- **Math Lab**
  Analyzing blood-sugar regulation after eating different meals (p. 994)
- **Experimental Design**
  How Does Epinephrine Affect Heart Rate? (p. 1000)

Features

- **Health Watch**
  Anabolic Steroids Are Dangerous (p. 989)
  Hormones and Body Fat (p. 995)

internet connect

National Science Teachers Association *sci*LINKS Internet resources are located throughout this chapter.

Reading Strategies

Active Reading

Technique: Reading Effectively

As students read through this chapter, have them make an outline of the material presented. You may want to review outlining methods, such as using Roman numerals, letters, and numbers. Go over the outlines to be sure students are developing the main ideas.

Directed Reading

Assign **Directed Reading Worksheet Chapter 43** to help students interact with the text as they read the chapter.

Interactive Reading

Assign ***Biology: Principles and Explorations* Audio CD Program Chapter 43** to help reluctant readers achieve greater success in reading the chapter.

Reading Skills

Assign Reading Organizers and Reading Strategies Worksheets from the **One-Stop Planner CD-ROM with Test Generator** to help students learn and practice effective reading skills.

internet connect

Holt *Biology: Principles and Explorations*

On-line Resources: go.hrw.com
Visit the HRW Web site for a variety of resources related to this chapter. Just type in the keyword HX0 Chapter 43.

Holt *Biology: Principles and Explorations*

On-line Resources: www.scilinks.org
The following *sci*LINKS Internet resources can be found in the student text for this chapter:

Page 989
TOPIC: Anabolic steroids
KEYWORD: HX989

Page 995
TOPIC: Hormones and body fat
KEYWORD: HX995

Page 996
TOPIC: Melatonin
KEYWORD: HX996

Page 1001
TOPIC: Hormones
KEYWORD: HX1001

1 Prepare

Scheduling

- **Block:** About 45 minutes will be needed to complete this section.
- **Traditional:** About one class period will be needed to complete this section.
- **Planning:** See the Planning Guide on pp. 979A and 979B for lecture, classwork, and assignment options for Section 43-1. Lesson plans for Section 43-1 can be found in a lesson cycle format in ***Biology: Principles and Explorations*** **One-Stop Planner CD-ROM with Test Generator.**

Resource Link

Assign **Active Reading Worksheet Section 43-1** to help students comprehend the key concepts and visuals in this lesson.

2 Teach

Lesson Warm-up

At the beginning of a class, while students are settling down and paying little attention to you, make a loud, startling noise. Immediately afterward, ask students to make a list of their feelings in response to the noise. List some responses on the board or overhead. At intervals, ask if anyone still feels jittery. Tell students that the effects they felt were the result of a hormone called epinephrine.

Section 43-1 Hormones

Objectives

- **Identify** four major functions of hormones. (p. 982)
- **Differentiate** between endocrine and exocrine glands. (p. 983)
- **Compare** the endocrine and nervous systems. (pp. 983–984)
- **Distinguish** the functions of endorphins from the functions of prostaglandins. (p. 984)

Key Terms

hormone
endocrine gland

The Body's Activities Are Coordinated

Reacting to fear, growing taller, and developing male or female characteristics are all activities in the body that are partially regulated by hormones *(HAWR mohnz)*. **Hormones** are substances secreted (released) by cells that act to regulate the activity of other cells in the body.

As shown in **Figure 43-1,** hormones are involved in coordinating the activities that your tissues and organs carry out to keep your body functioning properly.

1. Hormones regulate many processes including growth, development, behavior, and reproduction.
2. Hormones coordinate the production, use, and storage of energy.
3. Hormones are involved in maintaining homeostasis (nutrition, metabolism, excretion, and water and salt balance).
4. Hormones react to stimuli from outside the body.

Hormones act as chemical messengers, carrying instructions to the cells to change their activities. For example, some hormones carry instructions to the cells of the heart to increase the rate at which the heart is beating. In the past, it was thought that hormones, once secreted from a cell, had to be transported through the bloodstream before they reached the cell they were to act on. Today, we know that some hormones act directly on adjacent cells without traveling through the blood.

Figure 43-1 Hormones and homeostasis. Combining activities, such as water balance and temperature regulation, requires coordination. Such coordination is maintained by hormones.

The instructions hormones carry are determined by both the hormone itself and the cell it affects. For example, a hormone may instruct the cell to make a specific protein or to activate a specific enzyme. The same hormone can also instruct a different cell to alter the permeability of its cell membrane or even to release another hormone. Hormones can instruct muscle cells to relax and nerve cells to fire.

Each hormone is very specific about which type of cells receive their instructions. Each hormone acts like a key that opens a lock on or inside the cell. A hormone will only act on cells with the right lock. The locks, as discussed later in this chapter, are receptors on or inside the cell.

Integrating Different Learning Styles

Visual/Spatial	**ATE**	Opening Demonstration, p. 980; Visual Strategy, p. 983; Reteaching, p. 984
Interpersonal	**ATE**	Biology in Your Community, p. 983
Intrapersonal	**ATE**	Lesson Warm-up, p. 982
Verbal/Linguistic	**PE**	Study Tip, p. 984; Performance Zone item 28, p. 999
	ATE	Active Reading, p. 983; Closure, p. 984

Hormones Are Made in Certain Organs and Tissues

A gland is an organ whose cells secrete materials into other regions of the body. **Endocrine** *(EN doh krihn)* **glands** are ductless organs that secrete hormones directly into either the bloodstream or the fluid around cells (extracellular fluid). In addition to the endocrine glands, several other organs contain cells that secrete hormones. These organs include the brain, stomach, small intestine, kidney, liver, and heart.

All of the endocrine glands and tissues collectively make up the endocrine system, as shown in **Figure 43-2.** The endocrine system coordinates all of the body's sources of hormones.

Some organs, such as the pancreas, are both endocrine and exocrine *(EHKS oh krihn)* glands. Exocrine glands deliver substances through ducts (tubelike structures). The ducts transport the chemicals to specific locations inside and outside the body. Sweat glands, mucous glands, salivary glands, and other digestive glands are examples of exocrine glands. The exocrine part of the pancreas produces digestive enzymes and delivers them to the small intestine through ducts. The endocrine part of the pancreas secretes two hormones into the bloodstream that regulate blood glucose levels.

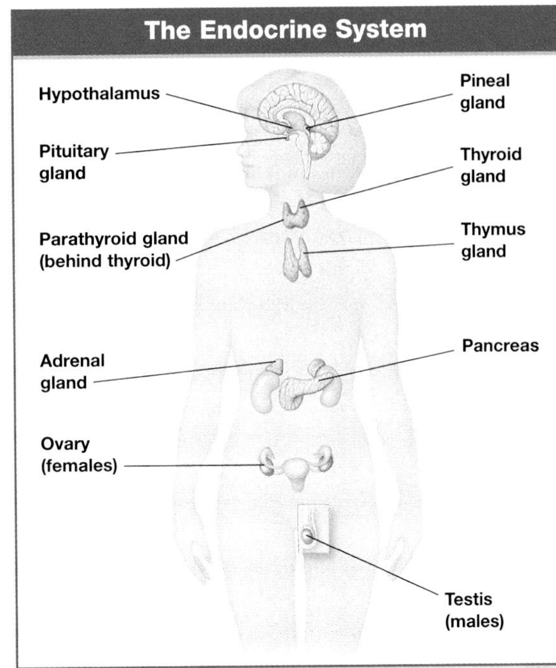

Figure 43-2 Coordinating the body's activities. Endocrine glands are located throughout the human body. As research continues, many of the body's organs have been found to contain cells that secrete hormones.

Hormones and neurotransmitters are chemical messengers

As you learned in Chapter 42, the nervous system is also involved in coordinating the body's activities. The endocrine system and nervous system interact in their overall job of coordinating the body's activities. However, with some exceptions, each system acts through different chemical messengers and in different ways.

The chemical messengers for the nervous system are known as neurotransmitters, while the chemical messengers for the endocrine system are called hormones. Some nerve cells, however, are capable of secreting hormones, and several chemicals serve as both hormones for the endocrine system and as neurotransmitters for the nervous system. For example, the neurotransmitter epinephrine is chemically identical to the hormone epinephrine. When secreted from a nerve cell, epinephrine conveys messages between neurons. When secreted from an endocrine cell in the adrenal gland, epinephrine acts as a "fight-or-flight" hormone.

Visual Strategy

Figure 43-2

Use this figure to show how the endocrine glands are widely distributed in the body. Point out to students the parts of the brain illustrated in this figure so that they recognize the close association between the endocrine and nervous systems.

Teaching TIP

Overlap Between Endocrine and Nervous Systems

Point out to students that although it is convenient to distinguish between the endocrine and nervous systems, the lines between these two systems are blurred. There is considerable overlap between them, which helps the body maintain homeostasis.

Active Reading

Technique: Paired Summarizing

Have students read silently the first page of Section 43-1. Then assign partners and have one student, the "reteller," summarize orally what has been read while the other student, the "listener," listens to the retelling. The listener should point out inaccuracies and add ideas that were left out. Have students alternate roles and repeat this process for each page of Section 43-1. Students should work together during a final clarification process, referring to the text as needed.

Biology in Your Community

Endocrinologist

Invite a medical professional who specializes in hormone therapy to discuss hormones, hormone disorders, and treatments available. Ask the guest to bring photographs of people with endocrine disorders and micrographs of endocrine tissues, if possible. Have the class prepare a list of questions to ask your guest speaker. PORTFOLIO

Chapter 43

REAL LIFE

Answers

Long-term use of drugs that inhibit prostaglandin production can lead to stomach ulcers or kidney damage. Drugs that inhibit prostaglandin production work by inhibiting the body's production of cyclooxygenase, an enzyme involved in prostaglandin synthesis.

Making Biology Relevant

Health

Although prostaglandins are known for increasing pain, they play a vital role in labor and delivery. Prostaglandins not only stimulate uterine muscle contractions, but also promote blood clotting and thus minimize hemorrhaging during and after delivery. Prostaglandin-inhibiting drugs, such as aspirin and ibuprofen, may delay labor, or they may inhibit blood clotting and promote excessive bleeding, which could result in severe blood loss or even death.

3 Assessment Options

Closure

Test Questions

Have students submit test questions (with answers) formulated from the information in this section. Choose several of the best and put them on the test for this section. PORTFOLIO

Review

Assign Review—Section 43-1.

Reteaching — BASIC

Have students label a diagram similar to that in Figure 43-2 with the names of the endocrine glands shown in Figure 43-2. PORTFOLIO

REAL LIFE

Prostaglandins can cause pain, fever, and inflammation. These symptoms, however, can be treated. Many nonprescription drugs, such as aspirin, ibuprofen, naproxen sodium and ketoprofen, work by inhibiting prostaglandin production.

Finding Information
What symptoms can occur with long-term use of these drugs?

Study TIP

Compare and Contrast

To compare and contrast the endocrine and neural systems, make a two-column list. In one column, write the ways in which the two systems are alike. In the other column, write the ways in which the two systems are different.

Another difference between the endocrine and nervous systems is that neurotransmitters are fast-acting and usually short-lived messengers. Hormones are usually slower-acting and longer-lived messengers. The effect of a hormone can last for days, weeks, or even years.

Finally, neurotransmitters are released from nerve cells directly to adjacent nerve cells. Endocrine cells can release hormones either into the bloodstream, where they travel to the cell they are to act on, or into the extracellular fluid to act on nearby cells.

Hormonelike substances

The human body has many substances that regulate cellular activities much as hormones do. Although these substances were not initially considered hormones because they were not secreted into the bloodstream, most scientists today classify them as hormones. These substances include a large number of chemical signals secreted by the brain and nerves called neuropeptides, as well as chemicals called prostaglandins *(prahs tuh GLAN dihnz),* which are secreted by most cells.

There are several different groups of neuropeptides. Enkephalins *(ehn KEHF uh lihnz),* which were discussed in Chapter 42, are a group of neuropeptides that inhibit pain messages traveling toward the brain. Endorphins *(ehn DOHR fihnz),* which are thought to regulate emotions, influence pain, and affect reproduction, are another important group of neuropeptide hormones. Unlike neurotransmitters, enkephalins and endorphins tend to affect many cells near the nerve cells that produce them.

Prostaglandins are modified fatty acids that have a variety of functions. They tend to accumulate in areas where tissue is disturbed or injured. There are dozens of different prostaglandins, and they produce a variety of effects. Some prostaglandins stimulate smooth muscle contractions that cause the constriction of blood vessels. The constricted blood vessels in turn affect blood pressure and body temperature. Other prostaglandins cause blood vessels to dilate, causing inflammation. A headache may result when blood vessels swell and their walls press against nerves in the brain. Aspirin relieves headaches and reduces fever and inflammation by inhibiting prostaglandin production.

Review SECTION 43-1

1. **Describe** four ways in which hormones coordinate the activities of the body.
2. **Name** the type of glands that secrete substances into the bloodstream or extracellular fluid.
3. **Differentiate** the actions and chemical messengers in the endocrine system from those in the nervous system.
4. **SKILL Recognizing Relationships** Compare endorphins to neurotransmitters.

Critical Thinking

5. **SKILL Validating Viewpoints** When asked to give an example of a hormone that is influenced by a drug, a friend offered the example of prostaglandins being inhibited by aspirin. Do you agree that prostanglandins are hormones? Explain.

Review SECTION 43-1 ANSWERS

1. Answers will vary but might include regulating growth, development, behavior, and reproduction; producing, using, and storing energy; maintaining homeostasis; and reacting to external stimuli.
2. endocrine glands
3. The endocrine system relies on hormones, which are relatively slow acting with longer-lived responses. The nervous system relies on neurotransmitters, which are faster acting with shorter-lived responses than hormones.
4. Unlike neurotransmitters, endorphins affect many cells near the nerve cells that produce them. Both endorphins and neurotransmitters are secreted by neurons.
5. Answers will vary. In arguing their points, students should cite the similarity of prostaglandins to hormones in that both are secreted and both act as chemical messengers.

How Hormones Work

Hormones Act on Specific Cells

After hormones are released from the cell in which they are made, they bind and act only on target cells. A **target cell** is a specific cell a hormone binds to and acts on (carries the message to). Imagine what would happen if hormones were not specific. All the cells in the body would respond to the hormone, resulting in uncoordinated activities, such as activation of many different enzymes.

A hormone recognizes a target cell by its "address." The address is a receptor located either on the surface of the target cell (cell membrane) or inside the cell (cytoplasm or nucleus). Recall that a receptor is a protein to which a molecule binds. A hormone's shape matches a particular receptor protein on its target cells much like a key fits into a lock, as shown in **Figure 43-3.** Thus, a hormone binds only to cells that have a particular receptor protein, ignoring all other cells.

Types of hormones

Most hormones are classified as either amino-acid-based or steroid hormones. **Amino-acid-based hormones** are hormones made of amino acids (either a modified amino acid or any number from 3 to 200 amino acids linked together as a protein). Most amino-acid-based hormones are water soluble. **Steroid hormones** are lipid hormones the body makes from cholesterol. Steroid hormones are fat soluble.

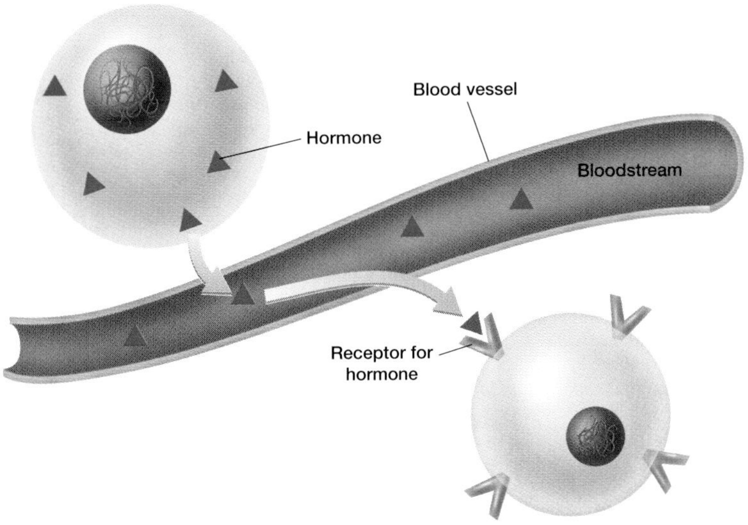

Blood vessel

Hormone

Bloodstream

Receptor for hormone

Figure 43-3 Hormones act on target cells. Hormones travel in the blood or in the fluid around cells to reach their target cells. The binding of the hormone with its receptor signals the cell to change its activity.

Objectives

- **Relate** how hormones act only on specific cells. (p. 985)
- **Summarize** how amino-acid-based hormones produce responses. (p. 986)
- **Summarize** how steroid and thyroid hormones produce responses. (p. 987)
- **Relate** how negative feedback is used to regulate hormone levels. (pp. 988–989)

Key Terms

target cell
amino-acid-based hormone
steroid hormone
second messenger
negative feedback

Section 43-2

1 Prepare

Scheduling

- **Block:** About 45 minutes will be needed to complete this section.
- **Traditional:** About one class period will be needed to complete this section.
- **Planning:** See the Planning Guide on pp. 979A and 979B for lecture, classwork, and assignment options for Section 43-2. Lesson plans for Section 43-2 can be found in a lesson cycle format in ***Biology: Principles and Explorations*** **One-Stop Planner CD-ROM with Test Generator.**

Resource Link

Assign **Active Reading Worksheet Section 43-2** to help students comprehend the key concepts and visuals in this lesson.

2 Teach

Lesson Warm-up

Obtain several different locks, each opened by a different key. On each lock, tape a slip of paper with a cellular activity (for example, activate an enzyme, constrict muscle tissue, increase blood flow, etc.). Each lock should have a different cellular activity. Tell students that the keys represent hormones. Give one of the keys to one student and ask the student to try to open all of the padlocks. The student will be able to open only one lock. Have a different student read out the cellular activity. Tell the students that in the same way, hormones can act only on specific cells, not on all of them. Only the cell with the right "lock-hole" will respond to the hormone. The hormone's message is then read by a second messenger and results in a change in activity.

Integrating Different Learning Styles

Logical/Mathematical	**PE**	Performance Zone item 17, p. 998
	ATE	Closure, p. 989
Visual/Spatial	**PE**	Quick Lab, p. 988
	ATE	Lesson Warm-up, p. 985; Visual Strategies, pp. 986, 987; Demonstration, p. 987
Verbal/Linguistic	**PE**	Performance Zone item 25, p. 999
	ATE	Teaching Tip, p. 986

Teaching TIP

Types of Hormones

Have students organize information about the two main classes of hormones in a **Graphic Organizer** similar to the one at the bottom of page 986.

PORTFOLIO

Teaching TIP

Receptor Sites

Remind students that the interaction between a hormone and its receptor protein is much like that between an enzyme and its substrate—the two must "fit together" to work.

Visual Strategy

BIO graphic **Figure 43-4**

Point out to students how the binding of the hormone to the receptor in **Step 1** results in enzyme activations (**Step 4**). The hormone is the "first messenger" because it is the first molecule to bring the message to the cell telling the cell what it must do. Point out that amino-acid-based hormones cannot pass through the cell membrane. Thus, in **Step 1** the hormone binds to a receptor located on the cell membrane, and not inside of the cell. When the hormone binds to the receptor, the receptor changes its shape. The shape change leads to a series of steps that eventually results in the conversion of ATP to cyclic AMP, as shown in **Step 2**. Cyclic AMP then acts as the "second messenger" by initiating a series of enzyme activations (or deactivations, depending on the hormone) as shown in **Step 3**. In the example in the figure, **Step 4** illustrates how cyclic AMP starts the activation of enzymes that catalyze the breakdown of glycogen to individual glucose molecules.

biobit

Who figured out how hormones work?

Alfred Gilman and Martin Rodbell were two of the first scientists to determine how one of the amino-acid-based hormones works. They received the Nobel Prize for medicine in 1994 for their discoveries.

Hormones Bind to Receptors

When a hormone binds to a specific receptor on a target cell, the hormone brings the target cell a message. What happens after the hormone binds, however, depends on the type of hormone.

Amino-acid-based hormones

Because amino-acid-based hormones are not fat soluble, most bind to cell membrane receptors, as shown in **Figure 43-4.**

Step 1 When an amino-acid-based hormone binds to a receptor protein, the shape of the receptor protein changes. For example, when glucagon, an amino-acid-based hormone made in the pancreas, binds to receptors on liver cells, the receptors change their shape.

Step 2 This change in shape eventually results in activation of a **second messenger,** a molecule that passes the message from the first messenger (the hormone) to the cell. For example, when glucagon binds to a receptor, an enzyme is activated that converts ATP to a second messenger called cyclic AMP (cAMP).

Step 3 The second messenger then activates or deactivates certain enzymes in a cascade fashion. That is, one enzyme activates another enzyme, which activates yet another, and so on.

Step 4 Eventually the activity of the target cell is changed by the final enzyme in the cascade—even though the hormone never enters the cell! In the case of glucagon, the cAMP molecules alter the activity of liver cells by activating a series of enzymes that break down glycogen into many individual glucose molecules.

Figure 43-4

BIO graphic **How Amino-Acid-Based Hormones Work**

Most amino-acid-based hormones, such as glucagon, bind to cell-membrane receptors, which activate second messengers that relay the hormone's message.

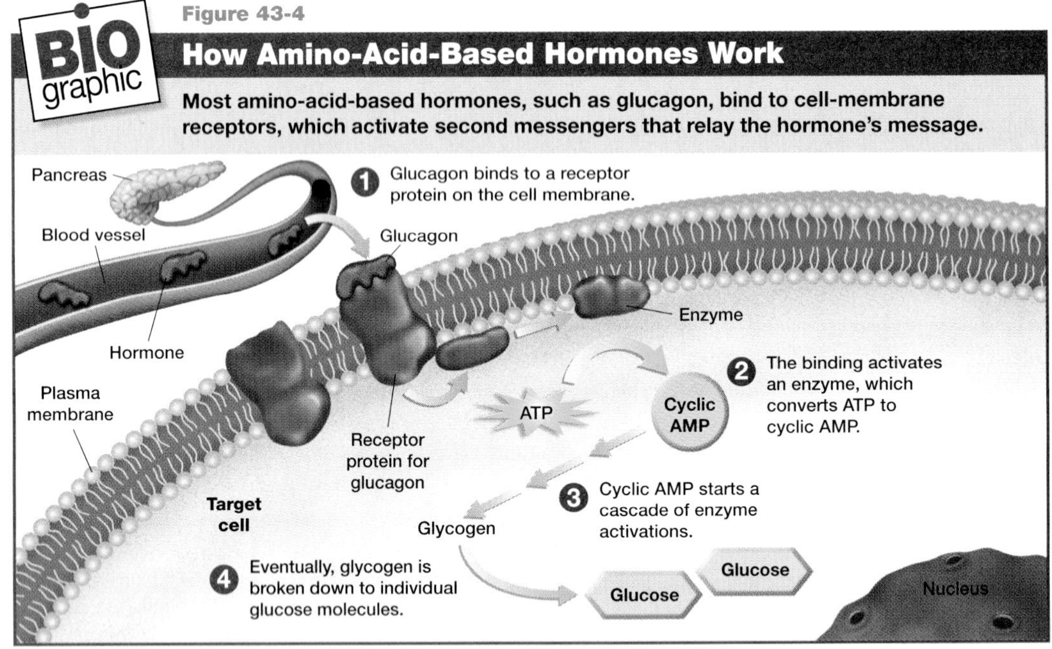
Pancreas
Blood vessel
Hormone
Plasma membrane
Target cell
1 Glucagon binds to a receptor protein on the cell membrane.
Glucagon
Receptor protein for glucagon
Enzyme
ATP
Cyclic AMP
2 The binding activates an enzyme, which converts ATP to cyclic AMP.
3 Cyclic AMP starts a cascade of enzyme activations.
Glycogen
4 Eventually, glycogen is broken down to individual glucose molecules.
Glucose
Glucose
Nucleus

Graphic Organizer

*Use this graphic organizer with **Teaching Tip: Types of Hormones** on page 986.*

Amino-acid-based hormones	Steroid and thyroid hormones
Water soluble	Fat soluble
Bind to receptors on the surface of the cell membrane	Bind to receptors in the cytoplasm or nucleus
Example: glucagon	Example: cortisol
Activate or deactivate enzymes	Activate or deactivate gene expression

Steroid and thyroid hormones

Because steroid and thyroid hormones are fat soluble, they readily pass through the cell membranes of their target cells. Steroid hormones bind to receptors located in a target cell's cytoplasm or its nucleus; thyroid hormones bind to receptors in a target cell's nucleus.

Cortisol is a steroid hormone made in the adrenal glands and released in response to stressful situations such as the one shown in **Figure 43-5.** How steroid hormones such as cortisol work is summarized in **Figure 43-6.**

Step 1 The hormone diffuses through the cell membrane and attaches to its receptor. The hormone and receptor form a hormone-receptor complex in the cytoplasm.

Step 2 The hormone-receptor complex enters the nucleus of the cell and binds to DNA.

Step 3 Depending on the hormone and the target cell, the binding either activates or inactivates a gene; that is, the gene either undergoes transcription and translation (protein synthesis), or transcription and translation is inhibited.

Step 4 Specific proteins, such as enzymes, are made. The enzymes alter the cell's activities. For example, cortisol stimulates the making of enzymes that lead to the breakdown of proteins and fats and their conversion to glucose. The glucose is used as energy so the cells can respond to the stress.

If the receptor for a steroid or thyroid hormone is located in the nucleus, the hormone enters the nucleus and binds to the receptor there. The hormone-receptor complex then binds to and affects the DNA in the same manner as it does with receptors in the cytoplasm.

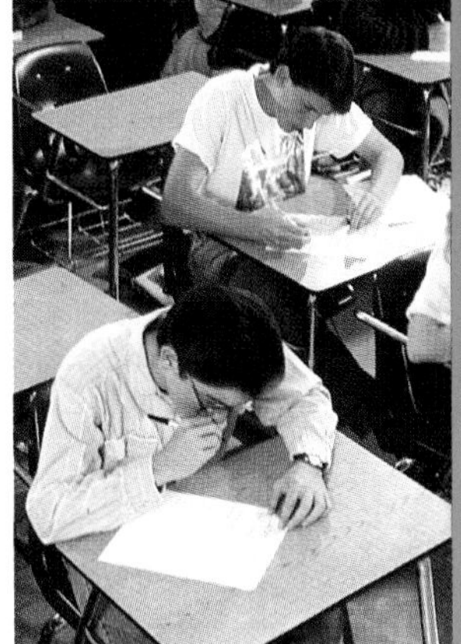

Figure 43-5 Steroid hormones and stress. Stressful situations, such as taking an exam, result in the release of the steroid hormone cortisol.

BIO graphic

Figure 43-6

How Steroid Hormones Work

The steroid hormone–receptor complex binds to DNA in the nucleus and activates or inactivates transcription and translation of a gene.

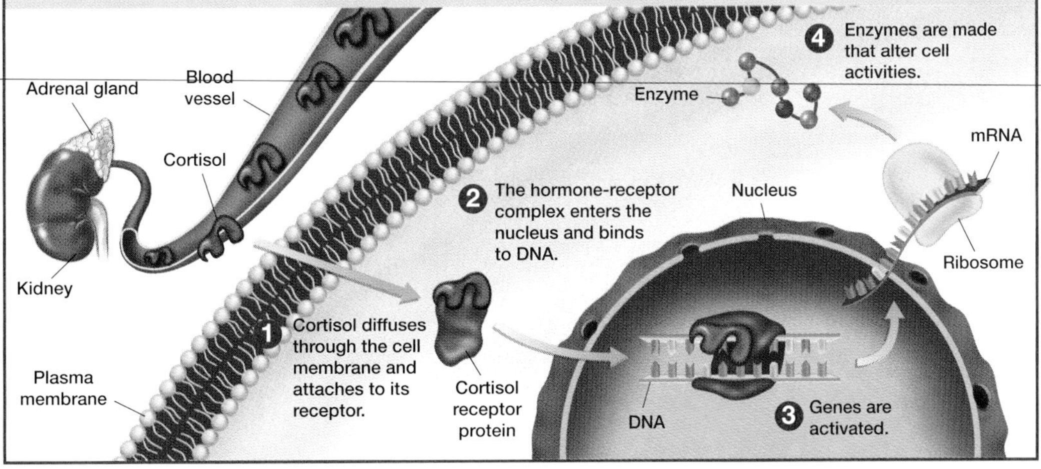

Adrenal gland
Blood vessel
Cortisol
Kidney
Plasma membrane
1 Cortisol diffuses through the cell membrane and attaches to its receptor.
Cortisol receptor protein
2 The hormone-receptor complex enters the nucleus and binds to DNA.
Nucleus
DNA
3 Genes are activated.
mRNA
Ribosome
Enzyme
4 Enzymes are made that alter cell activities.

did you know?

Some steroid hormones help transplant patients.

The class of steroid hormones called corticosteroids inhibits the immune system when the hormones are present at high levels. Patients who undergo a transplant operation might receive a corticosteroid, such as cortisol, to reduce the immune defenses of the body, allowing the patient to accept, rather than reject, the transplanted organ.

Demonstration

Solubility of Proteins and Lipids

Obtain 4 small glass beakers or flasks, water, plain gelatin (a protein), light-colored cooking oil, and vitamin E capsules (oil-filled). To demonstrate that lipids are fat soluble and proteins are not, place a small amount of gelatin in each of two flasks, one half filled with water and one half filled with cooking oil. Repeat the procedure with the vitamin capsules. Swirl the liquids in the flasks until each substance has dissolved in one of the liquids. Ask students which substance is fat soluble? *(the vitamin)* Which is not? *(the protein—it is water soluble)* Which type of substance can pass through a plasma membrane by diffusion? *(a fat-soluble substance)* Why? *(The plasma membrane has a double layer of phospholipids.)* Could the gelatin enter a cell? *(no)* Tell students that like gelatin, protein-based peptide hormones cannot enter cells, while fat-soluble steroid hormones can.

Visual Strategy

BIO graphic **Figure 43-6**

Have students compare Figures 43-4 and 43-6. Point out that because steroid hormones are fat soluble, they penetrate the cell membrane to bind to a receptor either in the cytoplasm (as shown in **Step 1**) or in the nucleus (not shown). Once the hormone binds to the receptor, the hormone-receptor complex enters the nucleus of the cell, as shown in **Step 2**. **Step 3** shows how the hormone-receptor complex binds to DNA and activates a gene. Depending on the hormone, a gene can also be deactivated. In the figure, the gene is activated, mRNA is transcribed, and a new protein is synthesized (**Step 4**). Point out to students that amino-acid-based hormones activate or deactivate enzymes or proteins, whereas steroid hormones activate or deactivate genes.

Chapter 43

Teaching TIP

Feedback Loops

Have students describe what would happen if part of the endocrine system were damaged and began to function as a positive feedback system, wherein a change in one direction promotes a continued change in the same direction. *(There would be a constant oversecretion of those hormones.)*

Teaching TIP

Homeostasis and Feedback

Homeostasis is maintained by feedback mechanisms. Feedback loops constantly monitor and adjust our endocrine levels for optimum bodily function. Often our hormones act in opposition to one another. Insulin and glucagon work in opposition to maintain a certain level of glucose in the blood. Other hormones work synergistically. Estrogens and progesterone work together to prepare the uterus for the implantation of an embryo.

How can you model negative feedback?

Time 20 minutes

Process Skills Relating information, analyzing, inferring

Teaching Strategies

- Review all safety symbols with students before beginning the lab.
- Ask students to explain in their own words the graph in Figure 43-7 on page 988.

Analysis Answers

1. Answers will vary.
2. They both represent negative feedback mechanisms.

The Release of Hormones Must Be Regulated

Since the human body makes more than 40 hormones, the body must regulate the release of the hormones. Nerve impulses alone can increase or decrease secretion of some hormones. For example, a baby nursing on a mother's breast stimulates the release of the hormone oxytocin, which in turn stimulates the release of milk from the mother's mammary glands.

In many cases the level of a hormone in the blood turns production of the hormone off and on through feedback mechanisms. Feedback mechanisms detect the amount of hormones in circulation or the amount of other chemicals produced because of hormone action. The endocrine system then adjusts the amount of hormones being made or released.

If high levels of a hormone stimulate the output of even *more* hormone, the regulation is called positive feedback. In humans, the release of most hormones is regulated through negative feedback as shown in **Figure 43-7.** In **negative feedback** a change in one direction

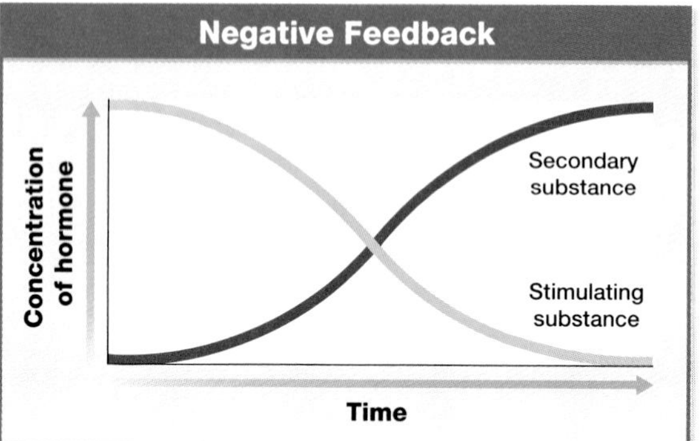

Figure 43-7 Negative feedback. In negative feedback, a secondary substance inhibits production of its initial stimulating substance.

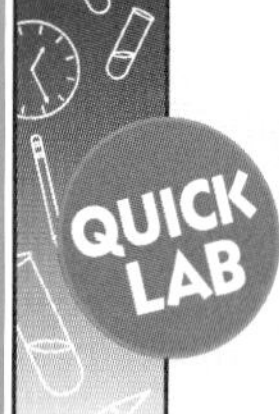

How can you model negative feedback?

You can try to maintain the temperature of a beaker of water to model how the human body maintains hormone levels.

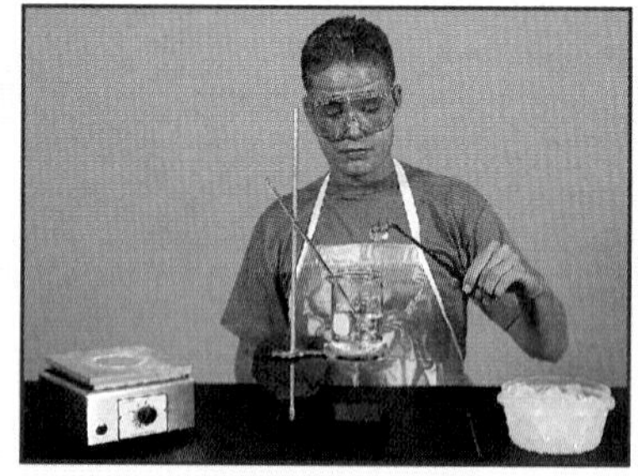

Materials

250 mL of water, 500 mL beaker, thermometer, hot plate, ring stand with ring, wire mesh, heat resistant glove, apron, goggles, tongs, stirring rod, ice

Procedure

1. Set up the ring stand, ring, and wire mesh.
2. Pour 250 mL of water into a 500 mL beaker. Determine and record the temperature of the water.
3.  After putting on your goggles, apron, and glove, heat the water for 1 minute. Record the temperature.
4. With your gloved hand, place the beaker on the wire mesh. Try to maintain the original temperature for 7 minutes by adding ice or heat. Record the amount of ice or heating time used each time.

Analysis

1. **Analyze** how your addition of ice or heat differed from that of other groups.
2. **SKILL Inferring Relationships** How does maintaining the temperature of the water compare to a hormonal negative feedback system?

did you know?

Researchers have found four heart hormones.

The first heart hormone was discovered in the early 1990s. This hormone, atrial natriuretic factor, results in lower blood volume and pressure. In 1999, three more heart hormones were discovered: vessel dilator, which removes sodium and lowers blood pressure; long acting natriuretic peptide, which removes salt from the blood; and kaliuretic peptide, which removes potassium from the blood. Researchers are currently testing whether one or more of these heart hormones can be used to treat congestive heart failure.

stimulates the control mechanism to counteract further change in the same direction. For example, high levels of a hormone inhibit the production of more hormone, whereas low levels of a hormone stimulate the production of more hormone. Negative feedback works like a person trying to maintain a certain speed in a car by using the gas pedal (accelerator). If the car goes too fast, the driver releases the gas pedal. If the car goes too slow, the driver presses on the gas pedal. The liver also plays a role in feedback by removing hormones from the blood and breaking them down.

Anabolic Steroids Are Dangerous

Many athletes use anabolic (protein-building) steroids and other hormone therapies to increase the size of their muscles and improve their performance. The unnatural use of steroids disrupts the feedback mechanisms that regulate hormone concentrations in the body.

Which steroid hormones are used?
The steroids used by athletes include synthetic hormones that mimic the male sex hormone testosterone. Many precursors to testosterone (such as androstenedione) are also used. Testosterone is secreted during puberty, when it stimulates many of the characteristics associated with being a man. For example, hair grows on the face, the underarms, and the pubic area; the voice deepens; and bigger muscles develop in the arms, legs, shoulders, and elsewhere.

Do steroids really improve an athlete's performance?
When athletes inject steroids, they are trying to stimulate the production of proteins in the muscle cells as a way of increasing muscle mass and strength. In large doses, steroids can promote increases in mass, strength, and endurance. But no studies to date have shown that steroids enhance performance, agility, skill, or cardiovascular capacity.

Many side effects are associated with the use of steroids.
There are many side effects that accompany steroid use. When steroids are used before the skeleton matures completely, they stop the bones from growing. The body never reaches adult height and may look distorted. Liver cancer and other liver disorders may also result from steroid use. Some males who use anabolic steroids develop enlarged breasts and shriveled testes. Females who use these chemicals may develop facial hair, deepening of the voice, and male-pattern baldness. Finally, the virus that causes AIDS can be transmitted if shared needles are used to inject the steroids. *The long-term risks to health are often greater than any benefits from the use of steroids.*

internet**connect**
SCI LINKS NSTA
TOPIC: Anabolic steroids
GO TO: www.scilinks.org
KEYWORD: HX989

Review SECTION 43-2

1. **Name** the structures found on or inside cells that allow hormones to recognize their target cells.
2. **Relate** how an amino-acid-based hormone changes a cell's activity.
3. **Relate** how a steroid or thyroid hormone changes a cell's activity.
4. **SKILL Analyzing Graphics** Use Figure 43-7 to describe how hormone levels are regulated by negative feedback.

Review SECTION 43-2 ANSWERS

1. receptors
2. When an amino-acid-based hormone binds to a receptor protein, that receptor protein changes shape, which then causes the activation of a second messenger. The second messenger activates or deactivates enzymes in a cascade fashion.
3. Steroid and thyroid hormones enter the cell and bind to receptors in the cell; the hormone-receptor complexes then bind DNA, initiating or inhibiting protein synthesis.
4. High hormone levels cause hormone production to be reduced. Low hormone levels cause production to increase.

HEALTH WATCH

Anabolic Steroids Are Dangerous

Teaching Strategies
Have students bring in articles from the newspaper or magazines about athletes who have been disqualified from competitions because of steroid use.

Discussion
- Why do athletes take steroids? *(to increase muscle mass, muscle strength, and endurance)*
- Why do you think anabolic steroids are banned from athletic events? *(Answers will vary. For example: they are dangerous and they give an athlete an unfair advantage.)*
- What side effects are associated with steroid use? *(liver cancer; enlarged breasts and shrunken testes in men; facial hair and deeper voices in women)*
- Should there be more drug testing of athletes? *(Answers will vary.)*

3 Assessment Options

Closure

Comparing Hormones
Ask students to make a two-column list. In one column, write the ways in which amino-acid-based hormones and steroid hormones are alike. In the other column, write the ways in which they are different. PORTFOLIO

Review

Assign Review—Section 43-2.

Reteaching — BASIC

Have students describe how protein hormones and neurotransmitters are similar in the ways they affect their target cells.

Chapter 43

1 Prepare

Scheduling

- **Block:** About 90 minutes will be needed to complete this section.
- **Traditional:** About two class periods will be needed to complete this section.
- **Planning:** See the Planning Guide on pp. 979A and 979B for lecture, classwork, and assignment options for Section 43-3. Lesson plans for Section 43-3 can be found in a lesson cycle format in ***Biology: Principles and Explorations* One-Stop Planner CD-ROM with Test Generator.**

Resource Link

Assign **Active Reading Worksheet Section 43-3** to help students comprehend the key concepts and visuals in this lesson.

2 Teach

Lesson Warm-up

Show the class pictures of people with hormone disorders (for example, dwarfs, giants, acromegaly, goiter, Cushing's syndrome, etc.). Point out which hormone is responsible for the symptoms, and whether the hormone is lacking or is overproduced. Explain to students that the first part of Section 43-3 describes how the endocrine system is coordinated to maintain normal hormone levels.

Section 43-3 The Major Endocrine Glands of the Body

Objectives

- **Evaluate** the roles of the hypothalamus and the pituitary gland in controlling other hormones. (pp. 990–991)
- **Summarize** the roles of the thyroid and parathyroid hormones. (p. 992)
- **Compare** the roles of the hormones secreted in each area of the adrenal gland. (p. 993)
- **Relate** how each of the two hormones secreted by the pancreas regulates blood sugar levels. (pp. 994–995)
- **Describe** the roles of reproductive hormones and of melatonin. (p. 996)

Key Terms

hypothalamus
pituitary gland
adrenal gland
epinephrine
norepinephrine
insulin
glucagon
diabetes mellitus

Two Endocrine Glands Are A Major Control Center

Feedback mechanisms fine-tune the levels of hormones in circulation, but two endocrine glands control the initial release of many hormones. The hypothalamus *(HEYE poh thal uh muhs)* and the pituitary *(pi TOO uh tehr ee)* gland, shown in **Figure 43-8,** together serve as a major control center for the rest of the endocrine system.

The hypothalamus

The **hypothalamus** is the area of the brain that coordinates the activities of the nervous and endocrine systems and controls many body functions, including body temperature, blood pressure, and emotions. The hypothalamus receives information about external and internal conditions from other brain regions. The hypothalamus responds to these signals from the nervous system as well as to blood concentrations of circulating hormones. The hypothalamus responds by issuing instructions—in the form of hormones—to the pituitary gland.

Figure 43-8 The hypothalamus and pituitary gland

Many hormones are released in a cascade starting with the release of hormones from the hypothalamus.

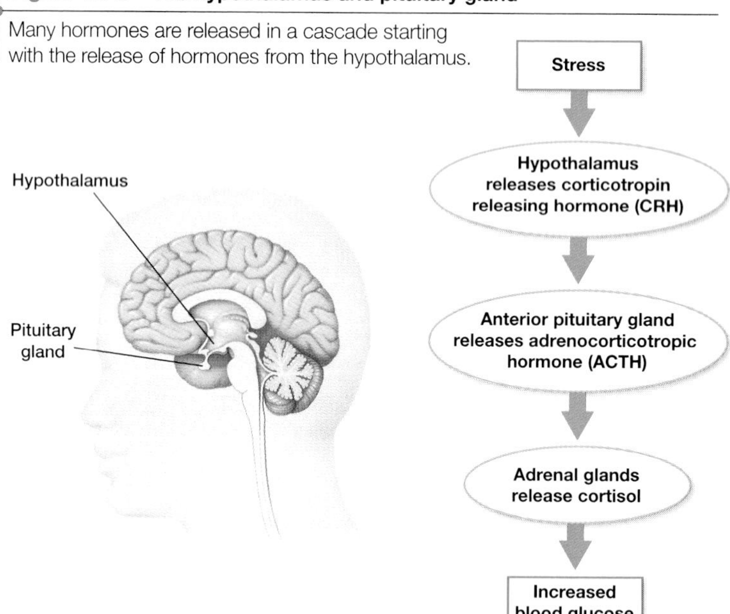
Hypothalamus
Pituitary gland
Stress
Hypothalamus releases corticotropin releasing hormone (CRH)
Anterior pituitary gland releases adrenocorticotropic hormone (ACTH)
Adrenal glands release cortisol
Increased blood glucose

Integrating Different Learning Styles

Logical/Mathematical	**PE**	Math Lab, p. 994; Performance Zone items 21, 22, 23, and 24, p. 999; Experimental Design, pp. 1000–1001
	ATE	Cross-Disciplinary Connection, p. 995
Visual/Spatial	**PE**	Performance Zone items 1, 13, 18, 19, and 20, pp. 998–999
	ATE	Lesson Warm-up, p. 990; Teaching Tip, p. 992
Interpersonal	**PE**	Performance Zone item 22, p. 999
	ATE	Alternative Assessment, p. 997
Verbal/Linguistic	**PE**	Study Tip, p. 995; Performance Zone items 26 and 27, p. 999
	ATE	Career, p. 992; Closure, p. 996; Reteaching, p. 996

The pituitary gland

As shown in Figure 43-8, the **pituitary gland** is an endocrine gland suspended from the hypothalamus by a short stalk. The pituitary gland secretes many hormones, including some that control endocrine glands elsewhere in the body.

The nerve cells in the hypothalamus make at least six hormones that are released into a special network of blood vessels between the hypothalamus and the pituitary gland. Some of these hormones are "releasing" hormones, which cause the front part of the pituitary gland, the anterior pituitary, to make and then release a corresponding pituitary hormone. "Inhibiting" hormones signal the anterior pituitary to stop secretion of one of its hormones.

Certain pituitary hormones travel to a distant endocrine gland and cause the gland to begin producing its particular hormone. One example of this cascade of events is shown in Figure 43-8. Other pituitary hormones act directly on their target cells, as summarized in **Table 43-1.**

The nerve cells of the hypothalamus also have axons that extend to the back part of the pituitary gland, the posterior pituitary. The nerve cells in the hypothalamus make two hormones that are stored in the posterior pituitary and released when needed: oxytocin *(ahks ih TOH sihn)* and antidiuretic hormone (ADH or vasopressin). Oxytocin triggers milk ejection during nursing and uterine contractions for childbirth. ADH causes the kidneys to form more-concentrated urine, thereby conserving water in the body.

biobit

What happens when excess growth hormone is secreted?

Excess amounts of growth hormone (GH) can result in a disorder called gigantism. The tallest known human, Robert Wadlow, who grew to a height of 8 ft 11 in., had a childhood pituitary tumor that secreted excess GH.

Table 43-1 Hormones Secreted by the Pituitary Gland

Hormone	Target tissue	Effects
Adrenocorticotropic hormone (ACTH)	Adrenal glands	Stimulates the release of cortisol and other steroid hormones from the adrenal cortex
Follicle-stimulating hormone (FSH)	Ovaries and testes	Regulates the development of male and female gametes
Luteinizing hormone (LH)	Ovaries and testes	Stimulates the release of an egg (ovulation) from an ovary; stimulates secretion of sex hormones from ovaries and testes
Prolactin	Mammary glands	Stimulates milk production in breasts
Growth hormone (GH)	All tissues	Stimulates protein synthesis and bone and muscle growth
Thyroid-stimulating hormones (TSH)	Thyroid gland	Stimulates synthesis and release of the thyroid hormones by the thyroid gland
Antidiuretic hormone (ADH)	Kidneys, blood vessels	Stimulates reabsorption of water from the kidney; constricts blood vessels
Oxytocin	Mammary glands, uterus	Stimulates uterine contractions and milk secretion

MULTICULTURAL PERSPECTIVE

Choh Hao Li—Endocrinologist

Choh Hao Li, a Chinese American endocrinologist, isolated and identified five hormones of the pituitary gland. He discovered that growth hormone consists of a chain of 256 amino acids. In 1970 he discovered a method for synthesizing the hormone, and he set the record for creating the largest synthesized protein molecule.

Teaching TIP — ADVANCED

Somatotropin

Tell students that somatotropin, or growth hormone (GH), promotes the growth and division of many body cells, although its major targets are bone and muscle cells. GH enables protein synthesis, cellular uptake of amino acids from the blood, and the mobilization of fats in the blood to areas of high energy demand. Point out the connection between the hypothalamus and pituitary gland by telling students that the release of GH is itself controlled by the hypothalamic hormones growth hormone releasing hormone (GHRH) and growth hormone inhibiting hormone (GHIH). As the names imply, the hypothalamic hormones are antagonistic hormones.

Making Biology Relevant

Medicine

The hormone oxytocin stimulates uterine contractions during labor. A woman having parturition difficulties may be given an injection of oxytocin to speed up delivery. After parturition, oxytocin helps the uterus to shrink and blood vessels to constrict, reducing the danger of hemorrhage.

Making Biology Relevant

Health

When the body is unable to secrete antidiuretic hormone (ADH), large volumes of dilute urine are produced. This condition is called diabetes insipidus. Dehydration can lead to illness or death, unless large quantities of water are consumed regularly.

Chapter 43

Making Biology Relevant

Health

List the symptoms of hypothyroidism on the board—fatigue, general aches and pains, slow movement, constipation, slow heart rate, dry hair and skin, puffy face, deeper voice, and numbness and tingling in the hands. Emphasize that we all feel this way sometimes, but that if these symptoms persist and the thyroid gland is enlarged, a doctor should be consulted. Hypothyroidism usually manifests itself during the teenage years, when hormonal change is most rapid. Hypothyroidism is relatively common and tends to run in families. Therefore, students should be made aware of this disorder.

Teaching TIP

Hyperthyroidism

Tell students that thyroid hormones play an important role in homeostasis. Thyroid hormones help maintain heart rate, digestion, blood pressure, muscle tone, and reproductive functions. Show the class a photograph of a person with pronounced bulging eyes (exopthalmos). Point out that hyperthyroidism causes fluid to accumulate in the fat behind the eyes, making them protrude. Other symptoms of hyperthyroidism include high body temperature, profuse sweating, weight loss, irritability, and high blood pressure. Ask students to relate each of these symptoms to the functions of thyroid hormones in maintaining homeostasis.

CAREER

Pharmacologist

Have students research pharmacology as a career. They should explain why a pharmacologist must understand the functioning of the endocrine system. Have them write a short report in which they also describe a pharmacologist's job, education and training, working environment, employment outlook, and starting salary. PORTFOLIO

The Thyroid Regulates Metabolism, Development, and Calcium Levels

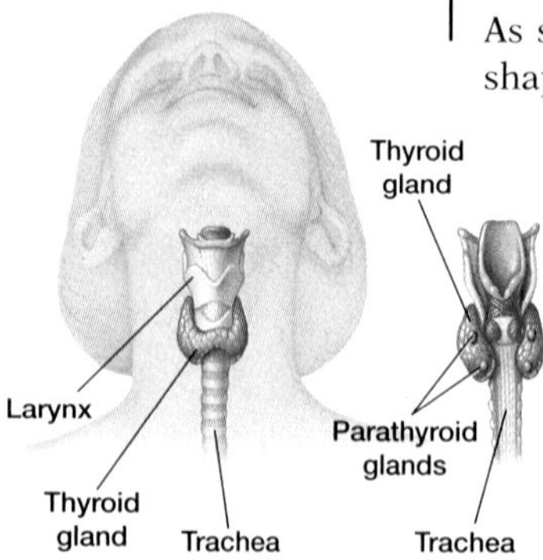

As shown in **Figure 43-9,** the thyroid gland is an endocrine gland shaped like a shield of armor, located just below the Adam's apple in the front of the neck. The name *thyroid* comes from the Greek word *thyros,* which means "shield."

Figure 43-9 The thyroid and parathyroid glands. The thyroid gland, located in the neck, is wrapped around the trachea. The parathyroid glands are located on the back of the thyroid gland.

Regulating metabolism and development

The thyroid gland makes and releases thyroid hormones. Thyroid hormones regulate the body's metabolic rate and promote normal growth of the brain, bones, and muscles during childhood. Thyroid hormones also affect reproductive functions and maintain mental alertness in adults.

Thyroid hormones are modified amino acids produced by the addition of iodide to the amino acid tyrosine. If iodide salts are lacking in the diet, the thyroid gland becomes greatly enlarged as a result of futile attempts to make more thyroid hormones. An enlarged thyroid gland, like the one shown in **Figure 43-10,** is called a goiter *(GOY tuhr).* Goiters resulting from iodide deficiency are now rare in the United States because of the addition of iodide to commercially available table salt.

The underproduction of thyroid hormones is known as hypothyroidism. In childhood hypothyroidism, an underproduction of thyroid hormones can cause permanently stunted growth, mental retardation, or both. In adults, hypothyroidism can cause a lack of energy, dry skin, and weight gain. Overproduction of thyroid hormones, or hyperthyroidism, can cause nervousness, sleep disorders, an irregular heart rate, and weight loss.

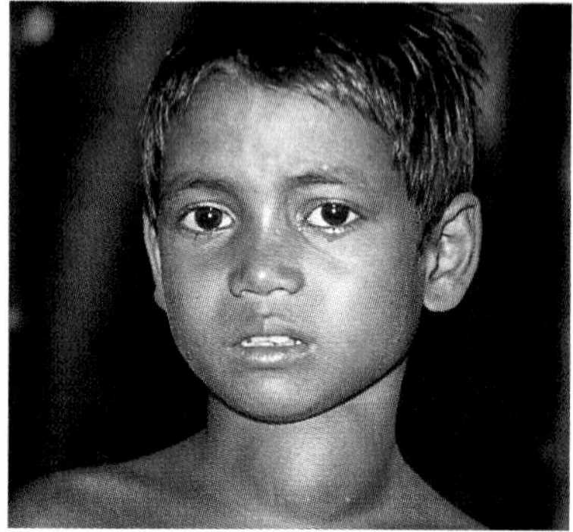

Figure 43-10 Goiter. Goiters result from a lack of iodide in the diet or improper functioning of the thyroid gland.

Regulating calcium levels

A high level of calcium in the blood stimulates the thyroid glands to produce a hormone called calcitonin. Calcitonin stimulates calcium to be deposited in bone tissue rapidly, lowering the blood-calcium level. Calcium is used for different purposes. For example, calcium ions are required for muscle contraction and for the release of certain substances from cells.

Parathyroid hormone (PTH) is a hormone that is produced by four parathyroid glands attached to the back part of the thyroid gland, as shown in Figure 43-9. PTH is made and released in response to a falling level of calcium in the blood. PTH stimulates bone cells to break down bone tissue and release calcium into the bloodstream. PTH also causes the kidneys to reabsorb calcium ions from urine, and it leads to activation of vitamin D, which is necessary for calcium absorption by the intestine.

did you know?

Some synthetic chemicals can disrupt hormones in humans and animals.

These chemicals are known as endocrine disrupters or hormone mimics. They are thought to cause reproductive abnormalities, cancer, birth defects, and immune problems in developing organisms including alligators, fish, and beluga whales. The disrupters include pesticides such as DDT, and manufacturing products or by-products such as PCBs (polychlorinated biphenyls) and dioxins. Many of the chemicals have structures similar to either estrogens, androgens, or thyroid hormones and so can bind to hormone receptors. Once the chemicals bind, they either disrupt or enhance the message normally sent by a hormone to the cell.

The Adrenals Respond to Stress

The body has two **adrenal glands,** which are endocrine organs located above each kidney. Each almond-size adrenal gland is actually two glands in one, as seen in **Figure 43-11:** the inner core, the adrenal medulla, and the outer shell, the adrenal cortex.

Immediate response to stress

The adrenal medulla acts as a warning system in times of stress by releasing the "fight-or-flight" hormones **epinephrine** *(ehp uh NEHF rihn)* and **norepinephrine** (formerly called adrenaline and noradrenaline, respectively). The effects of these hormones, which prepare the body for action in emergencies, are identical to the effects of the sympathetic nervous system in response to a stressful situation, but longer lasting. In stressful situations, such as the one in Figure 43-11, the fight-or-flight hormones increase heart rate, blood pressure, blood-sugar level, and blood flow to the heart and lungs.

Longer-term response to stress

The adrenal cortex makes several hormones, including cortisol and aldosterone. The adrenal cortex hormones provide a slower, more long-term response to stress than epinephrine and norepinephrine. Cortisol makes more energy available to the body. For example, cortisol causes the body to increase the level of blood glucose and to break down proteins for energy. High levels of cortisol, such as occurs when the body is under stress for long periods of time, suppresses the immune system. Artificial derivatives of this hormone, such as prednisone *(PREHD nih sohn),* are widely used as anti-inflammatory drugs.

Aldosterone *(al DAHS tuh rohn)* helps retrieve sodium ions from the fluids removed by the kidneys so that these ions are not lost in the urine. The overall effect of aldosterone to prolonged stress is that the volume of blood is increased, which raises blood pressure. In contrast, aldosterone stimulates the kidneys to secrete potassium ions into the urine. When the aldosterone level is too low, the potassium level in the blood may rise to a dangerous level.

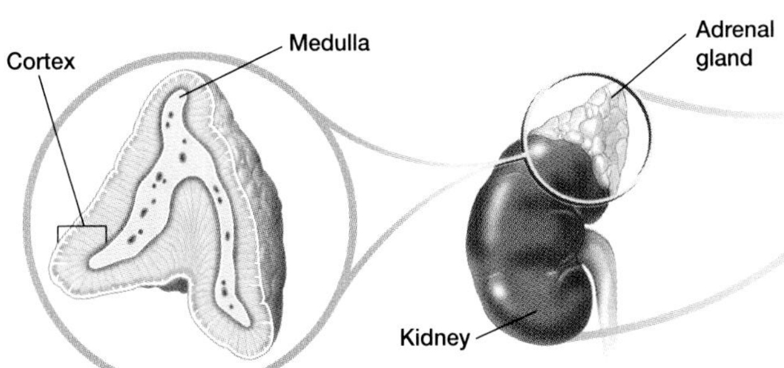

Figure 43-11 Fight-or-flight hormones at work. The fight-or-flight hormones are triggered during stressful situations. Cells in the adrenal medulla secrete these hormones.

biobit

Can women have beards?

The adrenal cortex secretes very small concentrations of male sex hormones. A tumor in the adrenal cortex of a woman or a deficiency of an enzyme involved in cortisol synthesis can result in increased production of male sex hormones. The appearance of male secondary sex characteristics, such as facial hair, is enhanced.

Overcoming Misconceptions

Cholesterol Is Not All Bad

The most important steroid is cholesterol, which has earned a bad reputation for its role in coronary artery diseases, but is absolutely essential for human survival. In fact, the body synthesizes cholesterol in the liver, regardless of dietary intake of cholesterol. Cholesterol is not just a building block for the steroid hormones; it also is found in cell membranes and is the precursor to vitamin D. Inform students that they take in cholesterol in a variety of foods, and the body makes cholesterol. The amount of cholesterol in one fast-food hamburger is much more than the body needs in a single day.

Making Biology Relevant

Health

Overstimulation of the adrenal medulla by stress can quickly deplete the body's energy reserve. High-level secretions of corticosteroids from the adrenal cortex can impair the immune system. Prolonged stress can lead to disease susceptibility, and many studies have demonstrated that prolonged stress can lead to learning and memory impairments.

Teaching TIP — ADVANCED

Adrenal Cortex Sex Hormones

Tell students that the adrenal cortex in both sexes produces sex hormones—the gonadocorticoids. Most of these hormones are androgens, but some female hormones (estrogen and progesterone) are also produced. Gonadocorticoid production is particularly high in the fetus and during puberty. Production falls rapidly in late puberty and is never again significant in comparison with the production of sex hormones by the gonads. The gonadocorticoids may function in maintaining the sex drive in adult women while supplying small amounts of estrogen after menopause.

CROSS-DISCIPLINARY CONNECTION

◆ **History**

Edward Kendall synthesized the first artificial corticosteroid in 1944. Prior to this time, a physician named Philip Showalter Hench had been studying rheumatoid arthritis. He observed that symptoms of rheumatoid arthritis often abate when the body is subjected to stresses such as pregnancy or jaundice. After further research, he suspected that corticosteroids were responsible. In 1949, Hench began using cortisone as a treatment for rheumatoid arthritis. In 1950, Kendall, Hench, and a third contributor to corticosteroid research, Tadeus Reichstein, received a Nobel prize for their work.

Teaching TIP

Endocrine and Exocrine

Remind students that the pancreas, like many organs, is also an exocrine gland. In fact, the approximately one million islets of Langerhan make up only 1–2 percent of the weight of the pancreas. The exocrine portion of the pancreas produces digestive enzymes and bicarbonate ions that are secreted into a duct, the pancreatic duct, that leads to the small intestine.

Analyzing blood-sugar regulation after eating different meals

Time 5–10 minutes

Process Skills Analyzing graphics, calculating, applying information

Teaching Strategies

Use this lab to discuss the importance of breakfast, often the meal many teens skip. Hypoglycemia often develops during the teenage years, when hormonal change is most rapid. Hypoglycemia is relatively common in a mild form, and students should be made aware of this disorder.

Analysis Answers

1. about 5–10 minutes
2. meal #1—the milk, waffles, and banana breakfast
3. More meals spread out throughout the day will help keep blood sugar levels constant. Limiting simple sugars decreases the amount of insulin produced, keeping blood glucose levels at an appropriate level.

The Pancreas and Other Organs Also Produce Hormones

In addition to those mentioned so far, several other organs and glands produce hormones. For example, the stomach, small intestine, thymus, kidney, liver, and heart all contain endocrine cells. Recall that the stomach and small intestine secrete hormones, such as gastrin, that regulate the release of acids and digestive enzymes.

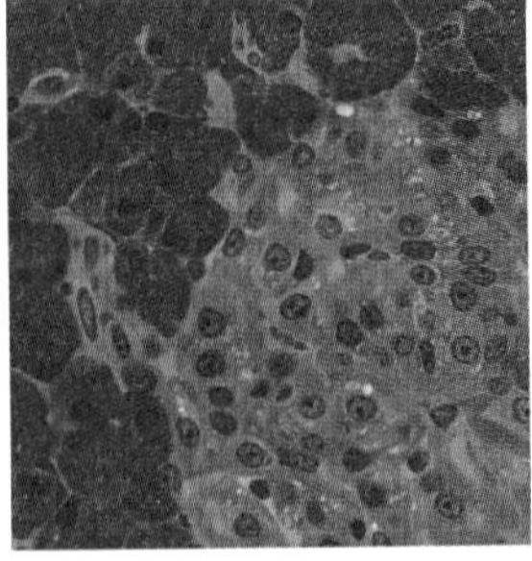

Figure 43-12 Islets of Langerhans. Islets of Langerhans are clusters of cells in the pancreas. The lighter stained cells produce glucagon. The darker stained cells produce insulin.

Regulating blood sugar levels

The pancreas contains clusters of specialized cells, called the islets *(EYE litz)* of Langerhans *(LAHNG uhr hahns)*, as shown in **Figure 43-12.** Two hormones made by the islets interact to control the level of glucose in the blood. **Insulin** is a hormone that lowers blood glucose levels by promoting the accumulation of glycogen in the liver. Insulin also stimulates glucose uptake by muscles and conversion to glycogen as an energy source. **Glucagon** has the opposite effect of insulin—it raises blood glucose levels. Glucagon causes liver cells to release glucose that was stored as glycogen.

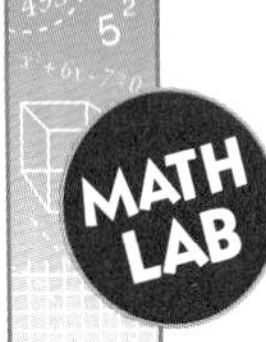

Analyzing blood-sugar regulation after eating different meals

Background

Eating simple sugars causes glucose to enter the bloodstream faster than eating complex carbohydrates or proteins. The rise in sugar levels triggers the production of insulin, which decreases blood sugar levels.

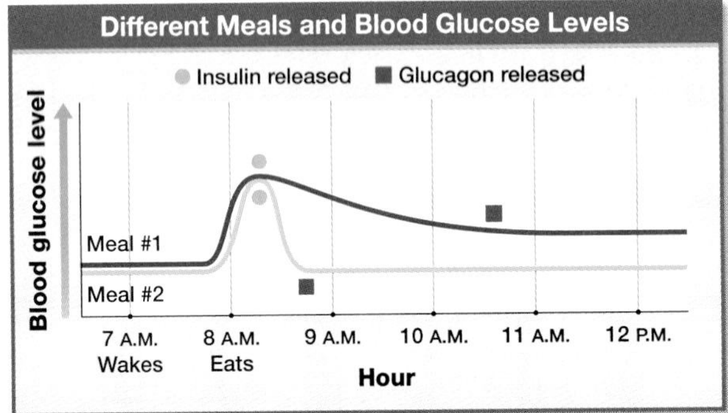

Analysis

1. **Calculate** for each meal the time in minutes for blood-sugar levels to begin to rise after eating.
2. **Determine** which meal has complex carbohydrates and proteins that allow glucose to be released into the bloodstream more slowly.
3. **SKILL Applying Information** Hypoglycemic people have low blood-sugar levels. They are often advised to eat six small meals a day containing little or no simple sugars. Why are these individuals given such advice?

Diabetes *(deye uh BEET eez)* **mellitus** *(meh LIHT uhs)* is a serious disorder in which cells are unable to obtain glucose from the blood, resulting in high blood-glucose levels. The kidneys excrete the excess glucose, and water follows, resulting in excessive volumes of urine and persistent thirst. The cells use the body's supply of fats and proteins for energy. The fat breakdown results in acidic products that accumulate in the blood, leading to low blood pH, coma, and, in extreme cases, death.

There are two kinds of diabetes mellitus. About 10 percent of affected individuals suffer from Type I diabetes and 90 percent suffer from Type II diabetes. Type I diabetes is a hereditary autoimmune disease. The immune system attacks the islets of Langerhans, causing low insulin levels. Also called insulin-dependent diabetes because it is treated with daily injections of insulin, Type I diabetes usually develops before age 20.

People with Type II diabetes (non-insulin-dependent) often have an abnormally low number of insulin receptors, while the level of insulin in their blood is often higher than normal. Type II diabetes often develops in people over 40 as a consequence of obesity and an inactive lifestyle. Type II diabetes is usually treated with diet and exercise and sometimes daily doses of medication other than insulin.

Study TIP

Organizing Information

Make a table to organize information about endocrine disorders. Across the top, write the headings *Endocrine disorder, Hormone involved,* and *Effects on body.* Along the sides, write each disorder discussed in Section 43-2 and then add information to the table.

CROSS-DISCIPLINARY CONNECTION

◆ Mathematics

Taking the correct dosage of insulin is vital to balancing blood glucose levels. In normal individuals, the blood glucose level ranges between 80 and 90 mg per 100 mL of blood. Present a case in which a diabetic determines that her glucose level is 220 mg per 100 mL of blood. She knows that one "unit" of insulin will reduce her glucose level by 3 mg per 100 mL of blood. Have students calculate how many units of insulin she should take to bring her glucose to a normal level. *(220 mg/100 mL − 90 mg/100 mL = 130 mg/100 mL over normal blood glucose levels; 130 mg/100 mL ÷ 3 mg/100 mL/unit = 43 units. She should take about 43 units.)* PORTFOLIO

Hormones and Body Fat

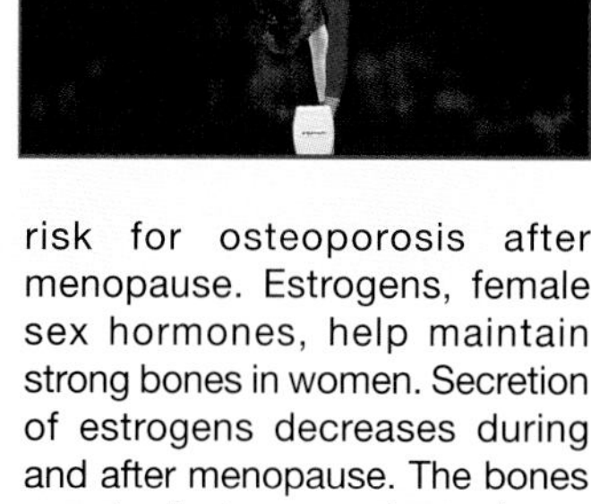

People with very little body fat, including many long-distance runners and gymnasts, often have disrupted reproductive systems. Very thin women may stop having menstrual periods, and very thin men may have lower testosterone levels.

During puberty, girls accumulate body fat before their first menstrual period. If they are very thin, their first period may be delayed by a year or more. Scientists are searching for hormones that tie a person's reproductive state to his or her body-fat content.

Hormone made by fat cells

In 1994, researchers discovered that fat cells secrete a hormone called leptin that helps control metabolism.

When injected into young female mice, leptin causes the mice to reach sexual maturity sooner. Mutant female mice that cannot make leptin do not produce eggs and thus are infertile. If the mutant mice are injected with leptin, they begin to produce eggs and can become pregnant.

The more body fat a person has, the more leptin that exists in the blood. Leptin is involved in regulating body weight. Scientists are unsure how leptin controls human reproduction. Cells in the ovaries and hypothalamus have leptin receptors.

Female hormones

Some women suffer severe bone loss (osteoporosis). Women with more body fat tend to have stronger bones and are at lower risk for osteoporosis after menopause. Estrogens, female sex hormones, help maintain strong bones in women. Secretion of estrogens decreases during and after menopause. The bones eventually become less dense and break more easily. After menopause, a woman's ovaries and adrenal cortex secrete small amounts of the "male" hormone testosterone, which is converted to estrogen by enzymes in fat cells.

TOPIC: Hormones and body fat
GO TO: www.scilinks.org
KEYWORD: HX995

Hormones and Body Fat

Teaching Strategies

- Discuss the eating disorders anorexia nervosa and bulimia.
- Tell students that researchers have discovered many other roles for leptin including triggering blood vessel growth.

Discussion

- Why do you think girls with eating disorders do not start menstruating, or stop menstruating? *(They lack adequate body fat.)*
- What function does the hormone leptin serve in mice? *(Female mice reach sexual maturity sooner.)*
- Why is body fat important in older women? *(Enzymes in body fat convert testosterone into estrogen, which may help prevent osteoporosis.)*

did you know?

The menstrual cycles of women living in the same dwelling tend to synchronize.

This effect is called the "dormitory effect." A study showed that the synchronization does not occur in a new roommate if her nasal passages are blocked. This suggests that pheromones are responsible for the dormitory effect. Pheromones are small chemical signals that function in communication between animals and act like hormones in influencing physiology and behavior. The study involving the blocked nasal passages was the first time researchers had shown the possible existence of a human pheromone.

Chapter 43

Teaching TIP

Sex Hormones

Tell students that the gonads produce and secrete three main types of sex hormones: androgens (the main hormone is testosterone), estrogens (the main hormone is estradiol), and progestins (such as progesterone). The three types of sex hormones are found in both males and females, but in different proportions. Females have a high ratio of estrogens to androgens and males have a high ratio of androgens to estrogens. Remind students that sex hormones are also secreted by the adrenal cortex, as discussed in **Teaching Tip** on page 993.

3 Assessment Options

Closure

Organizing Information

Have students make a chart with the following headings: *Hormone, Gland, Location in the Body,* and *Action of the Hormone.* Ask students to use the information in the chapter to fill in the chart for all of the hormones mentioned in this section. PORTFOLIO

Review

Assign Review—Section 43-3.

Reteaching — BASIC

Have students write riddles about the effects of a specific hormone. They can then trade riddles and try to solve them. For example, "No matter how hard I try, I cannot lose weight. I'm always tired and never have any energy. I went to the doctor, and she told me I have a very low metabolic rate. What hormone is my body underproducing?" *(thyroxine)* PORTFOLIO

Regulating reproduction

The ovaries and the testes, which also produce gametes, secrete hormones that regulate reproduction. The ovaries secrete estrogens and progesterone, and the testes produce testosterone. These hormones affect the formation of gametes and control sexual behavior and cycles. They also stimulate the development of secondary sex characteristics, such as breast size, hair growth, and muscle development.

internetconnect

SCILINKS. NSTA

TOPIC: Melatonin
GO TO: www.scilinks.org
KEYWORD: HX996

Regulating daily rhythms

The pineal *(PIHN ee uhl)* gland is a pea-sized gland located in the brain. The pineal gland secretes the hormone melatonin, which is a modified form of the amino acid tryptophan.

The function of the pineal gland in humans is not yet known. The hormone melatonin seems to be released by the human pineal gland as a response to darkness. Therefore, the pineal gland is thought to be involved in establishing daily biorhythms, such as the one seen in **Figure 43-13.** The pineal gland has been implicated in mood disorders such as winter depression, also called seasonal affective disorder (SAD) syndrome, and in a variety of aspects of sexual development.

Figure 43-13 **A daily biorhythm.** Sleep is an example of a daily biorhythm thought to be influenced by melatonin.

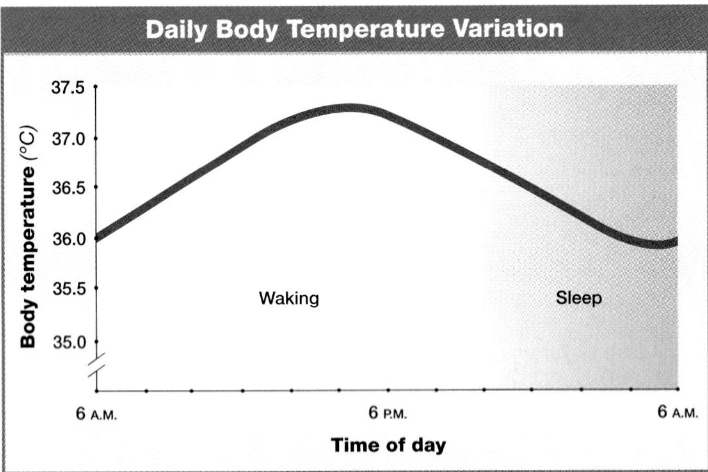

Review SECTION 43-3

1. **Infer** why the hypothalamus and pituitary gland are considered the major control center of the endocrine system.
2. **Evaluate** the consequences of an underproduction of thyroid hormones during childhood.
3. **Summarize** the hormonal control of calcium levels in the blood.
4. **Identify** the functions of reproductive hormones.
5. **SKILL Recognizing Relationships** Compare the effects of glucagon and insulin on blood sugar levels.
6. **SKILL Applying Information** A classmate states that hormones from the adrenal medulla, but not the adrenal cortex, are secreted in response to stress. Do you agree? Explain.

Review SECTION 43-3 ANSWERS

1. The hypothalamus and pituitary regulate other endocrine glands.
2. Hypothyroidism can lead to stunted growth and mental retardation.
3. Calcitonin stimulates the deposition of calcium in bone; parathyroid hormone stimulates calcium release from bones.
4. Reproductive hormones affect the formation of gametes and control sexual behavior cycles and secondary sex characteristics.
5. Glucagon increases blood sugar levels. Insulin decreases blood sugar levels, inducing storage of glucose in the form of glycogen in the liver.
6. Students should disagree. Hormones from both the adrenal medulla and adrenal cortex are secreted in response to stress. The adrenal cortex hormones, however, provide a slower, more long-term response.

Study ZONE CHAPTER 43

Use the Key Concepts and Key Terms listed below to review the main ideas in this chapter. If you need to review the meaning of a term, turn to the page indicated.

Key Concepts

43-1 Hormones

- Hormones are chemical messengers secreted by cells that act to regulate the activity of other cells. The ductless glands called endocrine glands make most of the body's hormones.
- Hormones are usually slower-acting but longer-lasting than nerve signals.
- Similar to hormones, endorphins, enkephalins, and prostaglandins act on nearby cells to regulate cellular activities.

43-2 How Hormones Work

- Amino-acid-based hormones bind to cell-membrane receptors, activating a second messenger. The second messenger then activates or deactivates enzymes in a cascade fashion.
- Steroid and thyroid hormones bind to receptors inside the cell. The hormone-receptor complex binds to DNA in the nucleus and turns genes either on or off.
- Most hormones are regulated by negative feedback.

43-3 The Major Endocrine Glands of the Body

- The hypothalamus and pituitary gland serve as the major control center for the release of many hormones.
- The thyroid hormones regulate metabolism and development. Calcitonin and parathyroid hormone regulate blood calcium levels.
- The inner medulla of the adrenal glands produces the fight-or-flight hormones. The outer cortex of the adrenal glands produces cortisol, aldosterone, and other steroid hormones.
- The pancreas secretes insulin and glucagon, which are involved in regulating blood-sugar levels.
- Abnormally low levels of insulin or a lack of insulin receptors makes an individual's cells unable to take up glucose from the blood and results in diabetes mellitus.
- Hormones in the ovaries and testes regulate reproductive functions.
- Melatonin is thought to regulate daily body rhythms.

Key Terms

43-1

hormone (982)
endocrine gland (983)

43-2

target cell (985)
amino-acid-based hormone (985)
steroid hormone (985)
second messenger (986)
negative feedback (988)

43-3

hypothalamus (990)
pituitary gland (991)
adrenal gland (993)
epinephrine (993)
norepinephrine (993)
insulin (994)
glucagon (994)
diabetes mellitus (995)

Study TIP Ready?

- *If you think you understand these Key Concepts and Key Terms, then you're ready for the Performance Zone.*

Study ZONE CHAPTER 43

Review and Practice

Assign the following worksheets for Chapter 43:

- **Student Review Guide**
- **Concept Mapping Worksheet**
- **Test Preparation Pretest**
- **Vocabulary Worksheet**
- **Critical Thinking Review Worksheet**

Assessment Options

The following assessment options are available for this chapter:

Resource Link

- Assign **Chapter Test 43** to assess students' understanding of the chapter.

Alternative Assessment

Have students work in cooperative groups. Each group should develop a question to send to a medical advice columnist dealing with any topic covered in this chapter. Collect the questions and redistribute them to other groups. Each group should then write an appropriate response to the question. PORTFOLIO

Portfolio Assessment

- *Portfolio Assessment* at the front of this book provides options for building and evaluating student portfolios. Students and teachers can select items from areas listed for inclusion in a portfolio.
- See items labeled PORTFOLIO on pp. 983, 984, 986, 989, 992, 995, 996, and 997.

Answer to Concept Map

The following is one possible answer to Performance Zone item 1 on page 998.

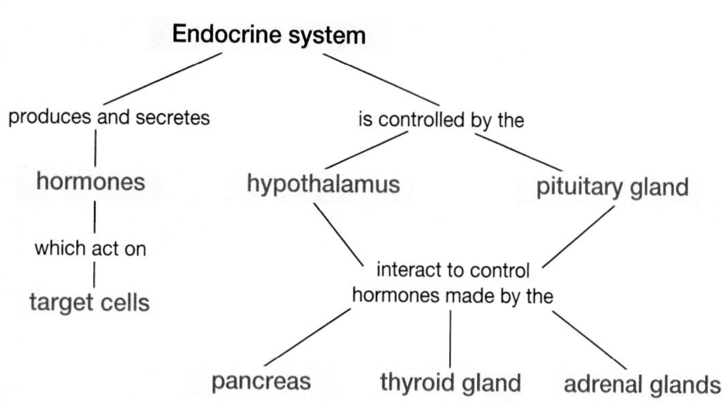

Performance ZONE
CHAPTER 43
REVIEW ANSWERS

Assignment Guide

Section	Review
43-1	2, 3, 4, 28
43-2	5, 6, 7, 14, 15, 17, 25
43-3	1, 8, 9, 10, 11, 12, 13, 16, 18, 19, 20, 21, 22, 23, 24, 26, 27

Understanding and Applying Concepts

1. The answer to the concept map is found at the bottom of page 997.
2. a. Endorphins are neuropeptides that have hormonelike properties; prostaglandins are modified fatty acids that have hormonelike properties
   b. A receptor is a protein located on the surface of a cell membrane or in the cytoplasm or nucleus to which a signal molecule, such as a hormone, binds. A second messenger is a molecule, such as cyclic AMP, that passes the "message" received from a hormone, which results in a change in the cell's activities.
   c. A hormone is a chemical messenger in the endocrine system. A neurotransmitter is a chemical messenger in the nervous system
3. c
4. a
5. b
6. c
7. a
8. d
9. d
10. d
11. b
12. c
13. The endocrine gland is the adrenal gland. Hormones made by the adrenal gland include epinephrine, norepinephrine, cortisol, and aldosterone.
14. The hormone must fit in the receptor protein in order to start the cascade of events that lead to a change in the cell's activity.

Performance ZONE
CHAPTER 43 Review

Understanding and Applying Concepts

1. **Concept Mapping** Make a concept map that describes the endocrine system. Try to include the following terms in your map: hypothalamus, pituitary gland, thyroid gland, hormones, adrenal glands, pancreas, target cell.

2. **Understanding Vocabulary** For each set of terms, summarize the differences in their meanings.
   a. endorphins, prostaglandins
   b. receptor, second messenger
   c. hormone, neurotransmitter

3. The chemical messengers of the endocrine system are
   a. neurons. c. hormones.
   b. blood cells. d. carbohydrates.

4. Exocrine glands
   a. release products through ducts.
   b. release products into the bloodstream.
   c. only function after puberty.
   d. All of the above

5. Amino-acid-based hormones may use cyclic AMP as a
   a. receptor. c. target cell.
   b. second messenger. d. coenzyme.

6. Steroid hormones
   a. bind to cell membrane receptors.
   b. bind to mRNA.
   c. eventually form hormone-receptor complexes that bind to DNA.
   d. never enter cells.

7. ________ leads to a decrease in hormone levels when hormone levels rise above normal.
   a. Negative feedback c. Positive feedback
   b. Neutral feedback d. None of the above

8. The ________ interact to control the secretion of other hormones.
   a. pancreas and thyroid gland
   b. hypothalamus and pineal gland
   c. adrenal gland and pancreas
   d. pituitary gland and hypothalamus

9. Thyroid hormones
   a. slow growth.
   b. inhibit insulin production.
   c. promote sperm production.
   d. control metabolic activities.

10. What adrenal cortex hormone acts, at high levels, to reduce inflammation?
    a. calcitonin c. prostaglandin
    b. aldosterone d. cortisol

11. Insulin leads to
    a. higher blood-sugar levels.
    b. lower blood-sugar levels.
    c. release of additional insulin.
    d. glycogen breakdown.

12. Which of the following endocrine glands secretes melatonin and is believed to be involved in establishing biorhythms?
    a. pituitary gland c. pineal gland
    b. thyroid gland d. adrenal gland

13. **Summarizing Information** Identify the endocrine gland labeled *A* and name two hormones it makes.

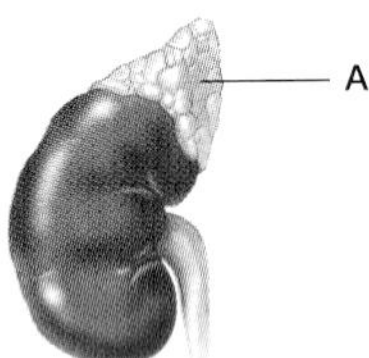

14. **Theme Cellular Structure and Function** Describe the importance of "fit" between a receptor protein and a hormone.

15. **Health Watch** Summarize the major side effects experienced by athletes that use anabolic steroids.

16. **Health Watch** If having more body fat lowers a woman's risk of developing osteoporosis, why don't doctors advise their female patients to gain weight?

17. **Chapter Links** What is the relationship between transcription factors and steroid and thyroid hormones? (**Hint:** See Chapter 10, Section 10-2.)

15. Anabolic steroids halt bone growth; contribute to liver disorders, including cancer; cause females to develop a deeper voice and facial hair; and cause males to develop enlarged breasts.
16. The health risks of being overweight, including stress on musculoskeletal and cardiovascular systems, outweigh the benefits.
17. Steroid and thyroid hormones act as transcription factors because they turn gene expression on or off when they bind to a gene.

Skills Assessment

Part 1: Interpreting Graphics

The graph below shows the blood-glucose levels measured over time of three experimental rats. At time T_1, two rats received injections of a saline solution plus one hormone. The control rat was injected with a saline solution only. Answer the following questions.

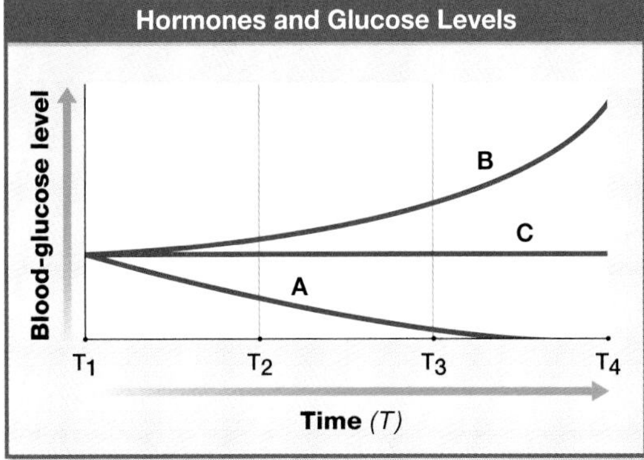

18. Which rat, *A, B,* or *C,* received insulin? Explain your answer.

19. Which rat, *A, B,* or *C,* received glucagon? Explain your answer.

20. Which rat, *A, B,* or *C,* was the control? Explain your answer.

Part 2: Life/Work Skills

Use the media center or Internet resources to learn about diabetes.

21. **Allocating Resources** Contact a local pharmacy for information about the average cost of insulin and the average daily dosage taken by an adult Type II diabetic. Using this information, calculate how much money an adult would spend in 1 year to maintain his or her blood-glucose levels.

22. **Finding Information** Research suggested diets for diabetics. Prepare a one-week menu that uses the types and amounts of foods recommended for diabetics. Share your findings with your class.

Critical Thinking

23. **Justifying Information** Before iodide was added to table salt, goiters were common among people living in inland regions but rare among people living in coastal areas. Why do you think this was so?

24. **Distinguishing Relevant Information** After a medical examination, a person is found to be unable to move glucose, stored as glycogen, from the liver into the blood. Further tests show that glucagon levels are normal, as is the structure of the hormone. Why do you think glucagon is unable to carry out its function in this case?

Portfolio Project

25. **Evaluating Information** Urine tests that detect steroid use by athletes are being advocated by many high school coaches. Interview several coaches to determine their attitudes toward steroid testing. Write an article for your school newspaper that discusses your findings and explains how steroids affect the body.

26. **Analyzing Information** The connection between melatonin and sleep patterns has kindled interest in this hormone. Check the library and on-line references for information about other uses that have been proposed for melatonin. Write a report to summarize your findings.

27. **Interpreting Information** Read "A New Breadth to Estrogen's Bisexuality," by R. Monastersky (*Science News,* Feb. 22, 1997, vol. 151, p. 116). Is estrogen only a female sex hormone? What physical deficiency did the mutant male mice have that prevented them from responding to estrogen?

28. **Career Focus Endocrinologist** Research the field of endocrinology, and write a report on your findings. Your report should include a job description, training required, kinds of employers, growth prospects, and starting salary.

Portfolio Projects

25. Answers will vary. Side effects include stunted growth, liver problems, enlarged breasts and shriveled testes in males, and facial hair and deepening voice in females.

26. Answers will vary. Melatonin may be used to treat insomnia and prevent jet lag.

27. No, estrogen appears to play a role in determining gender-specific behaviors in both males and females. The mutant mice lacked estrogen receptors on their cells.

28. Answers will vary. Endocrinologists are researchers that study hormones or physicians who care for patients with hormonal disorders. Endocrinologists who are researchers usually have post-graduate training. Endocrinologists who are physicians attend medical school after college, and must complete an internship, a residency in endocrinology, and become certified. Research endocrinologists are employed by universities, and research industries. Endocrinologists who are physicians are employed in private practice, university health centers, and by hospitals and clinics. The growth potential of endocrinology is excellent. Starting salary will vary by region.

Skills Assessment

18. A
19. B
20. C
21. Answers will vary. For example: 1 bottle = 10 mL @ 100 units/mL = $28.00; a daily dose might be about 66 units; 1,000 units/bottle × 1 dose/66 units = 15.16 d/bottle; 1 bottle/15.16 d × 365 d/year = 24 bottles/year; 24 bottles/year × $28.00/bottle = $672/year
22. Answers will vary, but the diet should be low in fat, low in simple sugars, low in salt, high in fiber, and high in complex-carbohydrate foods.

Critical Thinking

23. Marine animals, such as fish, have relatively high iodine levels. Many coastal peoples eat fish as a regular part of their diet.
24. The glucagon receptors on the cell membrane could be faulty.

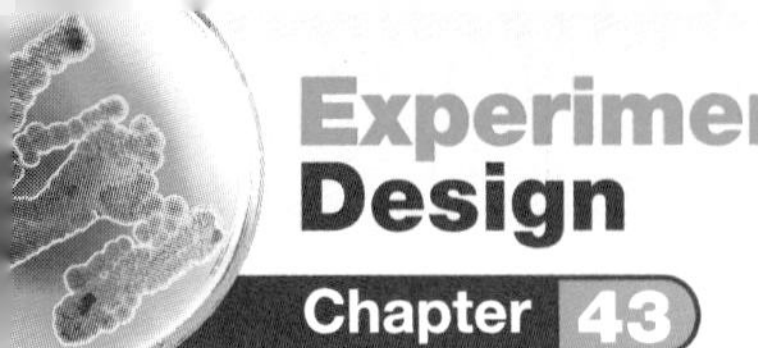

How Does Epinephrine Affect Heart Rate?

Purpose

Students will observe the effect of epinephrine on heart rate in the crustacean *Daphnia*.

Preparation Notes

Time Required: 45–60 minutes

Materials

Materials for this lab can be ordered from WARD'S. See *Master Materials List* at the front of this book for catalog numbers.

Preparation Tips

- When preparing solutions, mix each solution continuously for 5 min., reversing the direction of stirring every 10 to 20 seconds.
- To make a 0.0001 percent solution of epinephrine, dilute 10 mL of the 0.001 percent solution with distilled water to make 100 mL. To make a 0.00001 percent solution, dilute 1 mL of 0.001 percent solution with distilled water to make 100 mL. To make 0.000001 percent solution, dilute 0.1 mL of 0.001 percent solution with distilled water to make 100 mL.
- Prepare separate containers for the disposal of solutions and broken glass.

Safety Precautions

- Review all safety symbols with students.
- Tell students that epinephrine is toxic and is absorbed through intact skin. Students should not be allowed to clean up spills or broken glass. You should wear sturdy leather gloves over latex gloves when cleaning up broken glass and spills.
- Remind students that chemicals used in the laboratory should not be touched or tasted. Spills should be reported immediately.
- Remind students to report all injuries to you. Instruct students how to disinfect the affected surfaces following a spill.

Experimental Design

Chapter 43 How Does Epinephrine Affect Heart Rate?

SKILLS

- Using scientific methods
- Graphing
- Calculating

OBJECTIVES

- **Determine** the heart rate (HR) of *Daphnia*.
- **Observe** the effect of the hormone epinephrine on heart rate in *Daphnia*.
- **Determine** the threshold concentration for the action of epinephrine on *Daphnia*.

MATERIALS

- medicine droppers
- *Daphnia*
- *Daphnia* culture water
- depression slides
- petroleum jelly
- coverslips
- compound microscope
- watch or clock with second hand
- paper towels
- 100 mL beaker
- 10 mL graduated cylinders
- epinephrine solutions (0.001%, 0.0001%, 0.00001%, and 0.000001%)

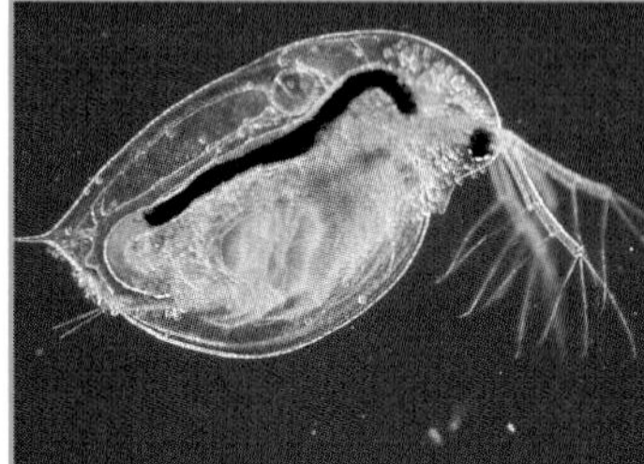

Daphnia

ChemSafety

CAUTION: Always wear safety goggles and a lab apron to protect your eyes and clothing.

CAUTION: Do not touch or taste any chemicals. Know the location of the emergency shower and eyewash station and how to use them. If you get a chemical on your skin or clothing, wash it off at the sink while calling to the teacher. Notify the teacher of a spill. Spills should be cleaned up promptly, according to your teacher's directions.

CAUTION: Glassware is fragile. Notify the teacher of broken glass or cuts. Do not clean up broken glass or spills with broken glass unless the teacher tells you to do so.

Before You Begin

Epinephrine is a hormone released in response to stress. It increases blood pressure, blood-sugar level, and **heart rate** (HR). The lowest concentration that stimulates a response is called the **threshold concentration.** In this lab, you will observe the effect of epinephrine on heart rate using the crustacean *Daphnia*. Epinephrine affects the heart rate of *Daphnia* and humans in similar ways.

1. Write a definition for each boldface term in the paragraph above.
2. Make a data table similar to the one above.
3. Based on the objectives for this lab, write a question you would like to explore about the action of hormones.

DATA TABLE

Solution	HR (beats/s) Trial 1 (A)	HR (beats/s) Trial 2 (B)	HR (beats/s) Trial 3 (C)	Average HR (beats/s) [(A+B+C) / 3]	Average HR (beats/min)

Procedure

Part A: Observing Heart Rate in *Daphnia*

1. Using a clean medicine dropper, transfer one *Daphnia* and fluid to the well of a clean depression slide. Place a small dab of petroleum jelly in the well. Add a coverslip, and observe the *Daphnia* with a compound microscope under low power.

Procedural Tips

Instruct students to get *Daphnia* from the classroom stock, and then return the *Daphnia* to a "recovery container."

Disposal

- After the lab is finished, place *Daphnia* into a 100 mL beaker with 50 mL of *Daphnia* culture water to allow the *Daphnia* to recover.
- Collect all wastes containing epinephrine. To the combined liquid, add 5 mL of full-strength chlorine bleach. Stir the mixture, and then heat to boiling. Boil gently for 5 minutes. Let cool and pour down the drain.

Before You Begin

Answers

1. *epinephrine*—amino-acid-based hormone released by the adrenal medulla in times of stress

   heart rate (HR)—the number of heart beats per unit of time; usually expressed as beats/sec

   threshold concentration—the lowest concentration of a compound that stimulates a response

2. Count the *Daphnia*'s heartbeats for 10 seconds. Divide this number by 10 to find the HR in beats/s. Record this number under Trial 1 in your data table. Turn off the microscope light, and wait 20 s. Repeat the count for Trials 2 and 3.
3. After calculating the average HR in beats/s, calculate the HR in beats/min by using the following formula: HR (in beats/min) = Average HR (in beats/s) × 60 s/min.

PART B: Design an Experiment

4. Work with the members of your lab group to explore one of the questions written for step 3 of **Before You Begin.** To explore the question, design an experiment that uses the materials listed for this lab.

You Choose

As you design your experiment, decide the following:

a. what question you will explore
b. what hypothesis you will test
c. how many *Daphnia* to use
d. what your controls will be
e. what concentrations of epinephrine to test
f. how many trials to perform
g. what data to record in your data table

5. Write a procedure for your experiment. Make a list of all the safety precautions you will take. Have your teacher approve your procedure and safety precautions before you begin the experiment.

PART C: Conduct Your Experiment

6.    Put on safety goggles, gloves, and a lab apron.
7. To add a solution to a prepared slide, first place a drop of the solution at the edge of the coverslip. Then place a piece of paper towel along the opposite edge to draw the solution under the coverslip. Wait 1 minute for the solution to take effect.
8. Set up your group's experiment, and collect data. **Caution: Epinephrine is toxic and is absorbed through the skin.**

PART D: Cleanup and Disposal

9. Dispose of solutions and broken glass in the designated waste containers. Place treated *Daphnia* in a "recovery container." Do not pour chemicals down the drain or put lab materials in the trash unless your teacher tells you to do so.
10.  Clean up your work area and all lab equipment. Return lab equipment to its proper place. Wash your hands thoroughly before you leave the lab and after you finish all work.

Analyze and Conclude

1. **Summarizing Results** Make a graph of your group's data. Plot "Epinephrine concentration (%)" on the *x*-axis. Plot "Average heart rate (beats/min)" on the *y*-axis.
2. **Analyzing Data** Which solutions affected the heart rate of *Daphnia*?
3. **Drawing Conclusions** What was the threshold concentration of epinephrine?
4. **Predicting Patterns** Based on the information you have and on your data, predict how epinephrine concentration would affect human heart rates.
5. **Further Inquiry** Write a new question about hormones that could be explored with another investigation.

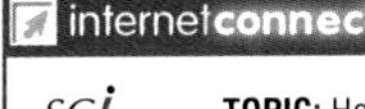

Do You Know?

Do research in the library or media center to answer these questions:

1. What is anaphylactic shock?
2. Why is epinephrine used to treat anaphylactic shock?

Use the following Internet resources to explore your own questions about hormones.

SCiLINKS NSTA
TOPIC: Hormones
GO TO: www.scilinks.org
KEYWORD: HX1001

Procedure

3. Answers will vary. For example: How does epinephrine affect heart rate in *Daphnia*?

Sample Data Table

SOLUTION	TRIAL 1	TRIAL 2	TRIAL 3	AVERAGE HR (BEATS/S)	AVERAGE HR (BEATS/MIN)
Control	24	24	25	24	146
0.000001%	24	25	25	25	148
0.00001%	25	26	25	25	152
0.0001%	31	32	33	32	192
0.001%	34	37	36	36	214

Sample Procedure

1. Put on safety goggles, disposable gloves, and lab aprons.
2. Using a clean medicine dropper, collect a fresh *Daphnia* with some fluid from the stock to serve as a control. Place it into the well of a depression slide with a small dab of petroleum jelly to slow the *Daphnia's* movement. Add a coverslip and observe under the low-power objective of a microscope.
3. Count the heartbeats for 10 seconds. Calculate and then record the number of beats/sec (Trial 1). Turn off the microscope light and wait 20 seconds. Repeat the procedure for trials 2 and 3, turning off the light and waiting 20 seconds between counts.
4. Place a drop of 0.0001 percent epinephrine solution on the depression slide as directed in step 7, on page 1001.
5. Wait 60 seconds for the solution to take effect, and then repeat step 3.
6. Place the *Daphnia* in a 100 mL beaker with culture water. Clean the slide and coverslip with soap and water. Rinse thoroughly. Prepare a slide with a fresh *Daphnia* as in step 2.
7. Repeat steps 4–6 for the remaining test solutions, using a clean dropper for each new solution.
8. Clean up the work area and wash hands.
9. Complete the calculations in the data table.

Analyze and Conclude

Answers

1. Answers will vary.
2. Answers may vary. The 0.001 percent and 0.0001 percent solutions should increase heart rate.
3. Answers may vary. Students should find that 0.0001 percent is the threshold concentration.
4. Answers will vary. Students may predict a similar pattern for humans.
5. Answers will vary. For example: How do other hormones affect heart rate in *Daphnia*?

Do You Know?

Answers

1. Anaphylactic shock is an acute systemic allergic reaction. Symptoms include agitation, flushing, heart palpitations, and difficulty breathing, followed by collapse of the cardiovascular system.
2. Epinephrine raises heart rate and blood pressure, and increases smooth muscle tone in the vascular system.

Controlling Diabetes

Background

Along with heart disease, stroke, and some kinds of cancer, diabetes is a so-called disease of affluence, or a lifestyle disease. These diseases, rather than infectious diseases, now cause the greatest mortality in industrialized nations like the United States. Heart disease, cancer, and diabetes share similar risk factors. The typical Western diet, with high levels of saturated fat and sugar but low amounts of fiber, contributes to the high rates of all these diseases, as does physical inactivity and obesity.

Diabetes provides a prime example of how an environmental change can alter the selective advantage of genes. Before Western diets and sedentary ways arrived, diabetes was almost unknown among American Indians, Australian Aborigines, Polynesians, and many other groups. Westernization has brought devastating epidemics of diabetes to all these groups. In all these groups, the high-calorie, low-effort Western diet has turned the genes for quick fat storage, which were once advantageous, into killers.

Teaching Strategies

- Remind students how the hormones glucagon and insulin act antagonistically to control blood sugar levels. Insulin, released by the β cells of the pancreas after a meal, stimulates cells to take in glucose and inhibits the breakdown of glycogen in the liver. Thus, insulin lowers blood glucose levels. Glucagon, released by pancreatic α cells, has the opposite effect—it raises blood glucose levels by stimulating the liver to hydrolyze glycogen. Working together, insulin and glucagon maintain blood glucose levels close to the body's set point of about 90 mg/100 mL.
- Inform students that diabetes is normally diagnosed through a fasting glucose test. The patient fasts overnight, and a blood sample is taken the next day and is analyzed for glucose. High levels of glucose—above 126 mg/100 mL—indicate diabetes. Ask students to explain why eating breakfast on the day of the test might give a false indication of diabetes. (Blood sugar levels surge after a meal, so high levels would probably be the result of the meal.)
- Have students bring a snack food to class. Have them look at the nutritional label on the package and list the food's fat, carbohydrate, and fiber content. Compile a table of these values on the board. Point out that eating a diet containing foods that are high in sugar and fat but low in fiber is one of the risk factors for diabetes. A diet rich in these foods provides so much sugar that the body is frequently faced with high blood glucose levels. Over time, excess glucose can overwhelm the pancreas's ability to secrete insulin or can make the body's cells resistant to insulin.

SCIENCE • TECHNOLOGY • SOCIETY

Healthier lifestyles can lower the risk of diabetes, while new treatments help reduce the damage caused by the disease.

Controlling Diabetes

It is the seventh leading cause of death in the United States, killing more than 180,000 people each year. But almost one-third of the people who have this disease—more than 5 million individuals—are unaware they are sick. During the 7 to 10 years that typically elapse between the onset of the illness and diagnosis, the heart, eyes, kidneys, and nerves in the hands and feet can suffer irreversible damage. In later stages, the disease can cause blindness, kidney failure, heart disease, stroke, and infections so severe that often the affected limb must be amputated.

The disease is diabetes mellitus. It affects nearly 16 million Americans and is responsible for nearly $100 billion a year in medical expenses, disability payments, and other costs. Although all racial and ethnic groups suffer from diabetes, it is much more common among blacks, American Indians, and Hispanics.

With a combination of drugs, diet, and exercise, many diabetics can prevent the development of complications and delay the progression of the disease—although no cure yet exists. Even better news is that people at risk for diabetes can take some simple steps to dramatically lower their likelihood of becoming sick. The bad news is that, despite these advances in treatment and prevention, the number of diabetics is rising fast in the United States and throughout the world.

What Is Diabetes?

Diabetes is a disease of the endocrine system in which the body loses the ability to regulate the amount of glucose in the blood. It stems from a defect in the body's production or use of insulin, a hormone secreted by the pancreas that stimulates cells to take in glucose. Blood glucose levels,

Smart choices
A healthful diet helps reduce the likelihood of diabetes.

Young diabetic
Diabetes affects both young and old.

normally tightly controlled, rise abnormally high, and the excess glucose is excreted in the urine.

There are two forms of diabetes. Type I diabetes usually begins suddenly when the person is a child or young adult. The immune system attacks the insulin-producing cells of the pancreas. What stimulates this attack remains a mystery, although some recent research suggests a virus may be responsible. People with Type I diabetes cannot make insulin and must take daily injections of the hormone to survive.

By contrast, Type II diabetes is rarely diagnosed before the age of 40 and takes years to develop. Over time, the pancreas either stops producing enough insulin or the body's cells become insensitive to insulin's effects, taking in less glucose. In both cases, chronically high blood glucose levels result. Between 90 and 95 percent of diabetics have this form of the disease, which can often be controlled by diet and exercise instead of insulin injections.

Type II diabetes results from a combination of genetic and environmental causes. Susceptibility to the disease runs in families—the odds of becoming sick are higher if a parent or sibling suffers from the disease. How you live and what you eat are just as important, since the disease is usually brought on by an environmental trigger. The environmental risk factors are a diet high in fat and sugar but low in fiber, lack of exercise, high blood pressure, and obesity (weighing more than 20 percent over your ideal body weight).

Career

Community Health Educator

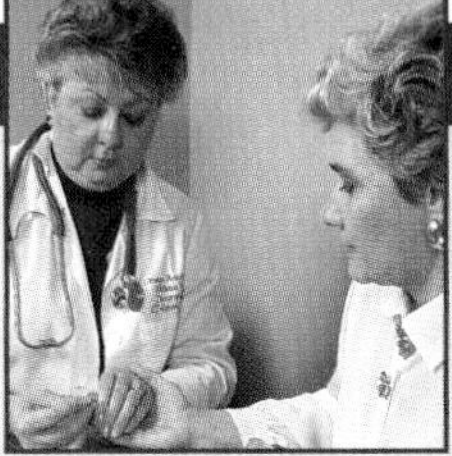
Diabetes educator

Profile

Community health educators counsel individuals and groups on health practices designed to prevent disease and promote good health.

Job Description

Persons who work in community health education may specialize in a particular disease (such as diabetes) and its effects and treatments. Many community health educators are employed by state and local governments, public health clinics, social service agencies, or resident care facilities. Employment is expected to grow rapidly as the population ages and public interest in health education continues to increase.

Science/Math Career Preparation

Biology	Biochemistry
Microbiology	Psychology
Chemistry	Sociology

Although it's more often diagnosed in people over 40, Type II diabetes results from damage that accumulates over years or decades. Eating a healthful diet, getting plenty of exercise, and avoiding obesity will reduce your likelihood of diabetes and can pay off in other ways, such as reducing your risk of heart disease, stroke, and some kinds of cancers. ■

Analyzing STS Issues

Science and Society

1. **Why hasn't natural selection eliminated the genes for diabetes?** Read the article "Sweet Death," by Jared Diamond (*Natural History,* February 1992, pp. 2–6). What benefit might the genes for diabetes have provided to our ancestors? How did this benefit become a disadvantage after the arrival of the Western diet?

2. **What are some other lifestyle diseases?** Diabetes is often called a lifestyle disease because it frequently results from how one chooses to live—diet, activity level, and so forth. Using library resources or an on-line database, research some other lifestyle diseases, such as certain kinds of cancer and heart disease. What are the risk factors for these diseases? How have the frequencies of these diseases changed over the last century?

Technology: Biomedical Devices

3. **How might new technology further improve diabetes treatment?** Over the last few decades, new methods for treating diabetes have come on the market. These include new ways of delivering insulin and better methods for monitoring blood glucose levels. Research a new technology for treating diabetes that is being developed or tested. Then write a short report describing what you learned.

References

1. American Association of Diabetes Educators. On-line. World Wide Web. Available http://www.diabetesnet.com/aade.html.
2. American Diabetes Association. On-line. World Wide Web. Available http://diabetes.org.
3. Centers for Disease Control and Prevention. *Diabetes Information.* On-line. World Wide Web. Available http://www.cdc.gov/diabetes.
4. Juvenile Diabetes Foundation International. On-line. World Wide Web. Available http://www.jdfcure.com.
5. National Institutes of Health. *National Institute of Diabetes and Digestive and Kidney Diseases.* On-line. World Wide Web. Available http://www.niddk.nih.gov.
6. Coghlan, Andy. "Acid Test Turns Out Sweet." *New Scientist,* April 12, 1997, 22.
7. Day, Michael. "Undercover Agents." *New Scientist,* August 9, 1997, 6.
8. Fackelman, K. A. "Panel Urges Widespread Testing For Diabetes." *Science News,* July 5, 1997, 4.
9. Murphy, Kate. "New Weapons to Beat Back Diabetes." *Business Week,* June 24, 1996, 163–164.
10. Sapolsky, Robert. "Requiem for an Overachiever." *The Sciences,* January/February 1997, 15–19.
11. Travis, J. "Surprising Pair of Diabetes Genes Debuts." *Science News,* December 7, 1996, 359.

Analyzing STS Issues Answers

1. The genes for diabetes were beneficial because they allowed the body to quickly store fat during times of plenty, allowing people to survive lean times. These genes were adaptive only when food was hard to get and famines were common. However, when food is easy to obtain and humans can consume unlimited amounts of food with almost no effort, quick fat storage leads to obesity and increases the risk of diabetes. After the arrival of the high-fat, high-sugar twentieth-century Western diet, the genes for diabetes became a disadvantage.

2. Heart disease and some kinds of cancer—including colon cancer, lung cancer, and skin cancer—are lifestyle diseases. The risk factors for heart disease and colon cancer are similar to those for diabetes—inactivity, obesity, and a high-fat diet. For skin cancer, one risk factor is the amount of time spent in the sun. For lung cancer, smoking is a risk factor. Lifestyle diseases have gone from being minor causes of death 100 years ago, when infectious diseases topped the list of causes of death, to being the biggest killers in America. Cancer is the leading killer, and heart disease is the second-leading killer.

3. Answers will vary. A recent alternative to injecting insulin with a hypodermic needle is an automatic pump that provides continual doses of insulin throughout the day, better mimicking the pancreas's release of the hormone. One form of new treatment that is currently being tested is an insulin pump that is implanted in the abdomen. Another treatment under development is a synthetic pancreas that could sense blood glucose levels and release insulin when needed.

Chapter 44

Reproduction and Development

Planning Guide

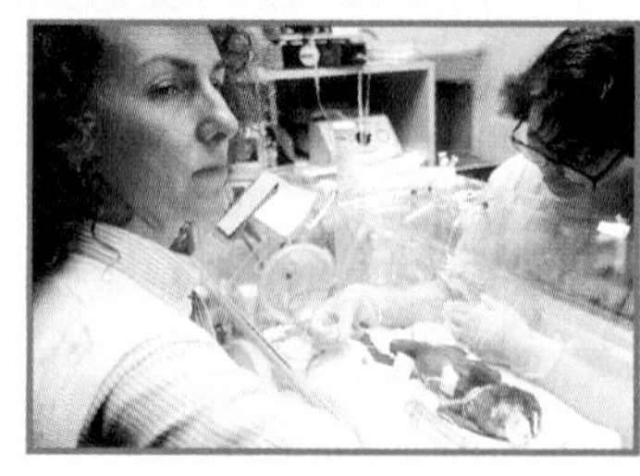

	OBJECTIVES	MEETING INDIVIDUAL NEEDS	READING SKILLS AND STRATEGIES
BLOCK 1 45 min	**44-1 Male Reproductive System** • Describe how sperm are produced. • Identify the major structures of the male reproductive system. • Relate the structure of a sperm cell to its functions. • Sequence the path taken by sperm as they leave the body.	**Learning Styles,** p. 1006 BASIC ADVANCED • pp. 1007, 1008	ATE **Active Reading, Technique: K-W-L,** p. 1005 ■ **Active Reading Guide,** worksheet (44-1) ■ **Directed Reading,** worksheet (44-1) ■ **Reading Strategies,** K-W-L worksheet
BLOCK 2 45 min	**44-2 Female Reproductive System** • Describe how eggs are produced. • Identify the major structures of the female reproductive system. • Analyze the events of the ovarian and menstrual cycles.	**Learning Styles,** p. 1009 BASIC ADVANCED • p. 1013	■ **Active Reading Guide,** worksheet (44-2) ■ **Directed Reading,** worksheet (44-2)
BLOCKS 3 & 4 90 min	**44-3 Development** • Sequence the events of fertilization, cleavage, and implantation. • Summarize the three trimesters of pregnancy. • Describe the effects of drug use on development.	**Learning Styles,** p. 1014 BASIC ADVANCED • pp. 1016, 1017	■ **Active Reading Guide,** worksheet (44-3) ■ **Directed Reading,** worksheet (44-3)
BLOCKS 5 & 6 90 min	**44-4 Sexually Transmitted Diseases** • Identify the causes and symptoms of several bacterial STDs. • Identify the causes and symptoms of some viral STDs. • Compare the treatment and cure rates of viral STDs with those of bacterial STDs.	**Learning Styles,** p. 1018 BASIC ADVANCED • p. 1020	■ **Active Reading Guide,** worksheet (44-4) ■ **Directed Reading,** worksheet (44-4) ■ **Supplemental Reading Guide,** *Microbe Hunters*

Review and Assessment Resources

BLOCKS 7 & 8 90 min

TEST PREP

- PE **Study Zone,** p. 1021
- PE **Performance Zone,** pp. 1022–1023
- ■ **Chapter 44 worksheets:**
  - Vocabulary
  - Concept Mapping
  - Critical Thinking
- Audio CD Program
- ■ Test Prep Pretest

CHAPTER TESTING

- ■ Chapter Tests (blackline master)
- ▲ Test Generator
- ■ Assessment Item Listing

See the correlations table at the front of this book for a list of the ***National Science Education Standards*** addressed in this chapter.

Smithsonian Institution® Visit **www.si.edu/hrw** for additional on-line resources.

KEY

Biology Principles and Explorations

PE Pupil's Edition

ATE Teacher's Edition

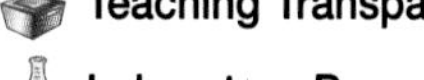

Teaching Transparencies

Laboratory Program

One-Stop Planner CD-ROM
Shows these resources within a customizable daily lesson plan:

■ Teaching Resources

▲ Test Generator

VISUAL STRATEGIES	CLASSWORK AND HOMEWORK	TECHNOLOGY AND INTERNET RESOURCES
ATE **Fig. 44-3,** p. 1007 224 Male Reproductive System 225 Male Hormones and Reproduction	**PE** **Review,** p. 1008	Holt Biology Videodiscs Section 44-1 Audio CD Program Section 44-1 internet**connect** SCILINKS NSTA **TOPIC:** Transportation of sperm
ATE **Fig. 44-6,** p. 1010 223 Female Reproductive System 226 Ovarian and Menstrual Cycles	**PE** **Data Lab** Analyzing hormone secretion in ovarian and menstrual cycles **PE** **Review,** p. 1013	Holt Biology Videodiscs Section 44-2 Audio CD Program Section 44-2 internet**connect** SCILINKS NSTA **TOPIC:** Ovarian cycle
227 Events Leading to Implantation 228 Structure of the Placenta 196A Events of Human Fetal Development	**Lab A34** Culturing Frog Embryos **Lab B34** Embryonic Development **PE** **Review,** p. 1017	Holt Biology Videodiscs Section 44-3 Audio CD Program Section 44-3 internet**connect** SCILINKS NSTA **TOPIC:** Ultrasound
	PE **Interactive Exploration** Meiosis: Down Syndrome, pp. 1024–1025 **PE** **Review,** p. 1020 **PE** **Study Zone,** p. 1021	Interactive Exploration: Cell Biology and Genetics, Meiosis: Down Syndrome Holt Biology Videodiscs Section 44-4 Audio CD Program Section 44-4

ALTERNATIVE ASSESSMENT

PE **Portfolio Projects 26–28,** p. 1023

ATE **Alternative Assessment,** p. 1021

ATE **Portfolio Assessment,** p. 1021

Lab Activity Materials

Data Lab
p. 1005 pencil, paper

Interactive Exploration
p. 1024 CD-ROM, computer

BIOSOURCES LAB PROGRAM
- Quick Labs: Lab A34, p. 67
- Inquiry Skills Development: Lab B34, p. 161

Chapter 44

Chapter Theme

Evolution

Development is the change of a single organism over time. Among animals, similarities in early stages of embryonic development suggest evolutionary relationships.

Establishing Prior Knowledge

Before beginning this chapter, make certain students can answer the questions in **Study TIP Ready?** on page 1005. If they cannot, encourage them to reread the sections indicated.

Opening Demonstration

Draw a small dot on the board. Ask students how the size of this dot compares with a newly fertilized egg. *(It is larger than a fertilized egg.)* Tell students that a fertilized egg is about 0.14 mm (0.005 in.) in diameter. Then show students a doll that is about the size of a newborn baby. Tell students that the average length of a newborn baby is approximately 51 cm (20 in.). Have them calculate a fetus's average rate of growth per month *(5.7 cm [2.2 in.])* and their own average rate of growth per month since birth *(for a 65-in.-tall 16-year old, about 1 cm [0.5 in.])*. How does the increase in size during prenatal development compare with that of postnatal growth? *(It is much greater during prenatal development.)*

CHAPTER 44 Reproduction and Development

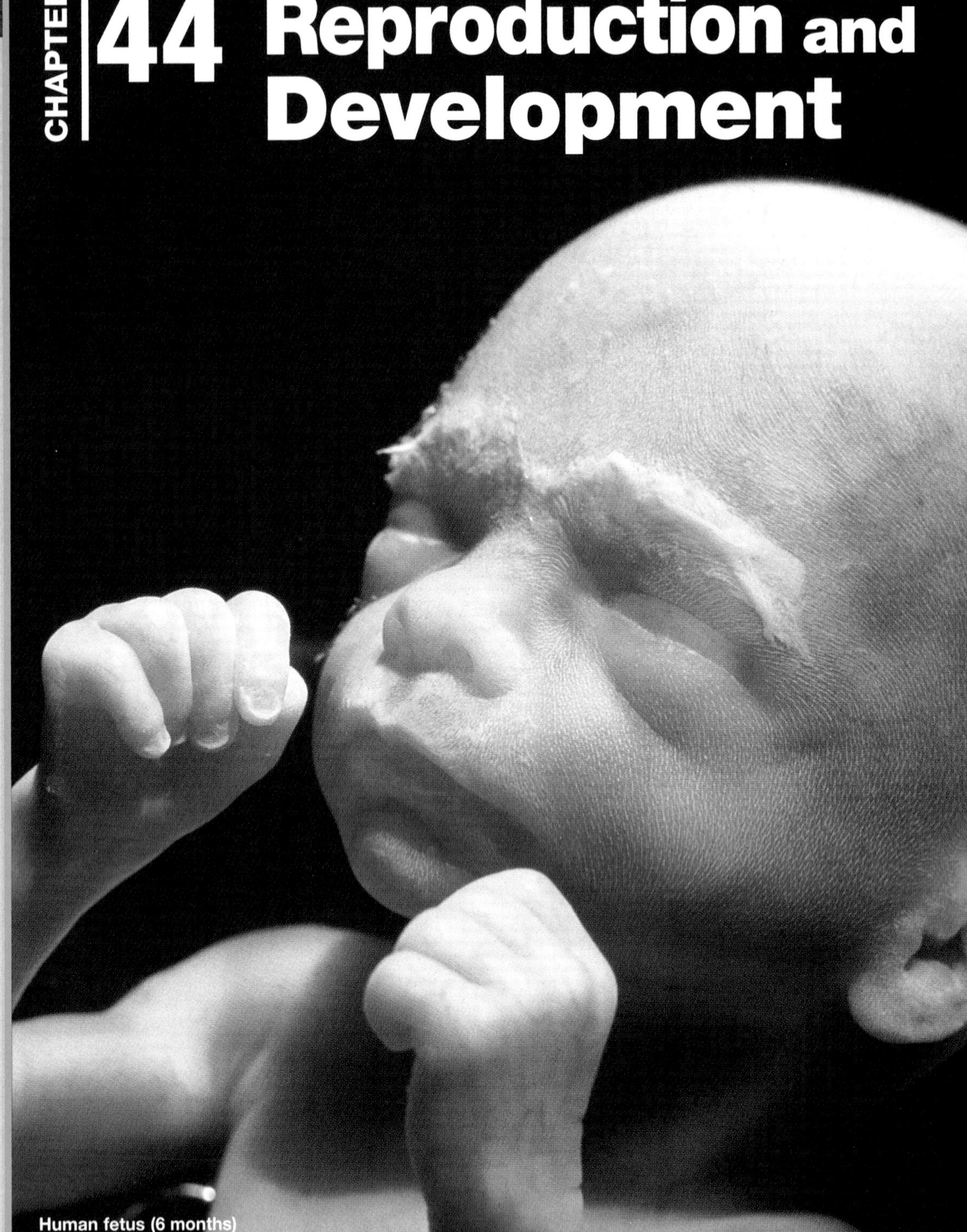

Human fetus (6 months)

What's Ahead — and Why

Chapter Outline

44-1 Male Reproductive System
- The Testes Produce Sperm Cells

44-2 Female Reproductive System
- The Ovaries Produce Egg Cells
- An Egg Cell Matures During the Ovarian Cycle

44-3 Development
- Fertilization Produces a Zygote
- Pregnancy Is Divided into Three Trimesters

44-4 Sexually Transmitted Diseases
- STDs Are Spread by Sexual Contact

Students will learn

- about the structure and function of the male reproductive system.
- about the structure and function of the female reproductive system.
- about the changes that take place within the fetus during pregnancy.
- about the symptoms, treatments, and prevention of STDs.

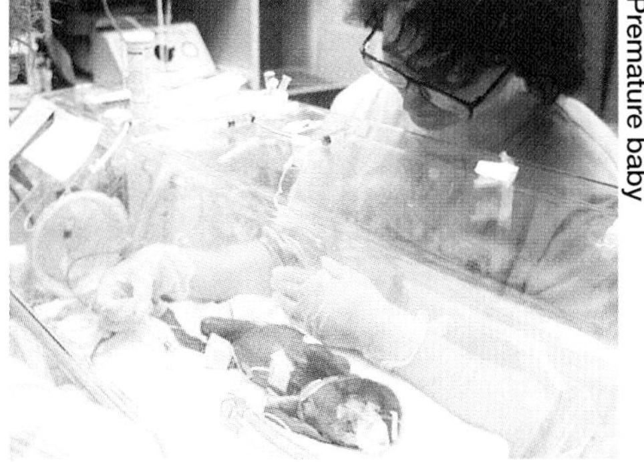

Premature baby

This baby was born prematurely.

The infant has been placed in an incubator to complete development. A normal pregnancy lasts about 9 months. Infants born before completing 7 months of development usually weigh less than 1 kg (2.2 lb) and have about a 50 percent chance of survival. Infants born after 7 months of development typically weigh more than 1 kg and have a greater chance of survival.

Looking Ahead

44-1 Male Reproductive System
How are sperm produced?

44-2 Female Reproductive System
How are eggs produced?

44-3 Development
What happens during pregnancy?

44-4 Sexually Transmitted Diseases
What causes sexually transmitted diseases?

Labs

Data Lab
Analyzing hormone secretion in ovarian and menstrual cycles (p. 1013)

Interactive Exploration
Meiosis: Down Syndrome (p. 1024)

Features

Tech Watch
Ultrasound Imaging (p. 1017)

Study TIP

Ready?

Check your understanding of these topics to help you prepare for what's ahead.

Can you . . .

- **sequence** the phases of meiosis? (Section 7-1)
- **list** the stages of animal development? (Section 28-1)
- **describe** the functions of sex hormones? (Section 43-3)
- **summarize** the role of the hypothalamus in the endocrine system? (Section 43-3)

If not, *review the sections indicated.*

internetconnect

SCiLINKS NSTA — National Science Teachers Association *sci*LINKS Internet resources are located throughout this chapter.

Chapter 44

Reading Strategies

Active Reading

Technique: K-W-L

Before students begin this chapter, have them write short lists of all the things they **K**now about reproduction and development. Write some of these ideas on the board or overhead. Then have students list things they **W**ant to know about reproduction and development. Have students save their lists for use later in the **Closure** on page 1020.

Directed Reading

Assign **Directed Reading Worksheet Chapter 44** to help students interact with the text as they read the chapter.

Interactive Reading

Assign ***Biology: Principles and Explorations* Audio CD Program Chapter 44** to help reluctant readers achieve greater success in reading the chapter.

Reading Skills

Assign Reading Organizers and Reading Strategies Worksheets from the **One-Stop Planner CD-ROM with Test Generator** to help students learn and practice effective reading skills.

internetconnect

Holt *Biology: Principles and Explorations*

On-line Resources: go.hrw.com
Visit the HRW Web site for a variety of resources related to this chapter. Just type in the keyword HX0 Chapter 44.

Holt *Biology: Principles and Explorations*

On-line Resources: www.scilinks.org
The following *sci*LINKS Internet resources can be found in the student text for this chapter:

Page 1008
TOPIC: Transportation of sperm
KEYWORD: HX1008

Page 1011
TOPIC: Ovarian cycle
KEYWORD: HX1011

Page 1017
TOPIC: Ultrasound
KEYWORD: HX1017

Chapter 44

1 Prepare

Scheduling

- **Block:** About 45 minutes will be needed to complete this section.
- **Traditional:** About one class period will be needed to complete this section.
- **Planning:** See the Planning Guide on pp. 1003A and 1003B for lecture, classwork, and assignment options for Section 44-1. Lesson plans for Section 44-1 can be found in a lesson cycle format in ***Biology: Principles and Explorations*** **One-Stop Planner CD-ROM with Test Generator.**

Resource Link

Assign **Active Reading Worksheet Section 44-1** to help students comprehend the key concepts and visuals in this lesson.

2 Teach

Lesson Warm-up

Have students name male secondary sex characteristics. *(deepening of the voice, increased facial hair, body hair, muscle mass, and so on)* List these on the board or overhead, and add to the list as you work through the chapter.

Section 44-1 Male Reproductive System

Objectives

- **Describe** how sperm are produced. (pp. 1006–1007)
- **Identify** the major structures of the male reproductive system. (p. 1007)
- **Relate** the structure of a sperm cell to its functions. (p. 1007)
- **Sequence** the path taken by sperm as they leave the body. (p. 1008)

Key Terms

testes
seminiferous tubules
epididymis
vas deferens
seminal vesicles
prostate gland
bulbourethral glands
semen
penis

The Testes Produce Sperm Cells

What are the roles of a male in sexual reproduction? Recall that sexual reproduction involves the formation of a (diploid) zygote from two (haploid) sex cells, or gametes, through fertilization. The roles of a male in sexual reproduction are to produce sperm cells—the male gametes—and to deliver the sperm cells to the female reproductive system to fertilize an egg cell—the female gamete.

Where are sperm produced? Two egg-shaped **testes** *(TEHS teez),* or testicles, are the gamete-producing organs of the male reproductive system. The testes are located in the scrotum *(SKROHT uhm),* an external skin sac. The testes first form inside the abdominal cavity then move down into the scrotum either before or shortly after birth. The normal body temperature of 37°C (98°F) is too high for sperm to complete development. In the scrotum the temperature is about 3°C lower than it is in the rest of the body, making it an ideal location for sperm production.

Production of sperm

The testes begin to produce sperm during the adolescent stage of development known as puberty *(PYOO buhr tee).* As shown in **Figure 44-1,** each testis contains hundreds of compartments packed with many tightly coiled tubules, called **seminiferous** *(sehm uh NIHF uhr uhs)* **tubules.** Sperm cells are produced through meiosis in the lining of the seminiferous tubules. Thus, sperm cells contain only 23 chromosomes (the haploid number) instead of the usual 46 chromosomes (the diploid number) found in other body cells. Two hormones released by the anterior pituitary regulate the functioning of the testes. Luteinizing hormone (LH) stimulates secretion of the sex hormone testosterone. Follicle-stimulating hormone (FSH), along with testosterone, stimulates sperm production in the seminiferous tubules. Cells located between the seminiferous tubules secrete testosterone.

Figure 44-1 Testes

The testes produce sperm cells.

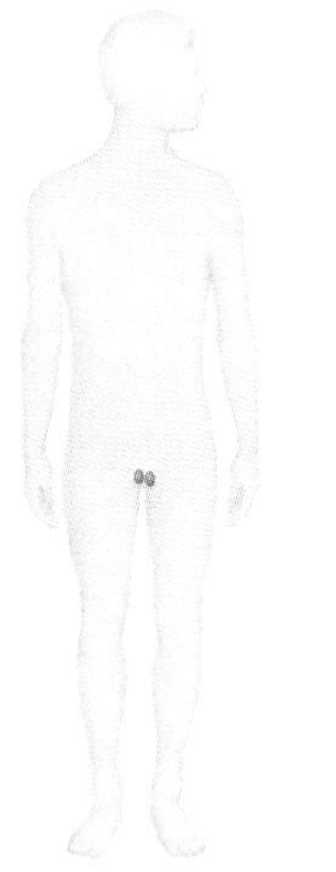

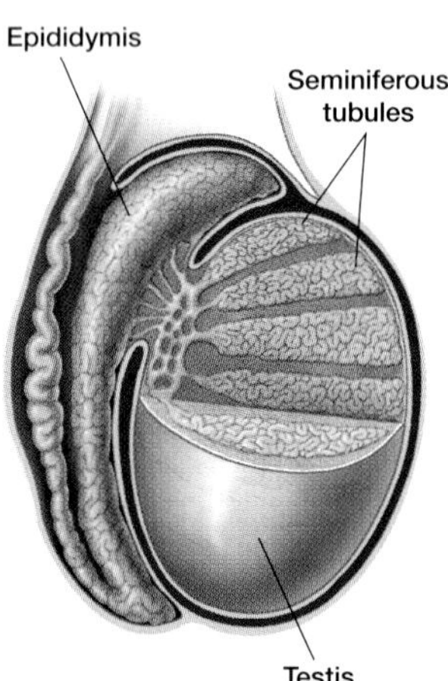

Integrating Different Learning Styles

Logical/Mathematical	**PE**	Review item 5, p. 1008
Visual/Spatial	**PE**	Performance Zone item 12, p. 1022
	ATE	Opening Demonstration, p. 1004; Teaching Tip, p. 1007; Visual Strategy, p. 1007; Reteaching, p. 1008
Verbal/Linguistic	**ATE**	Active Reading, p. 1005; Lesson Warm-up, p. 1006; Teaching Tip, p. 1007

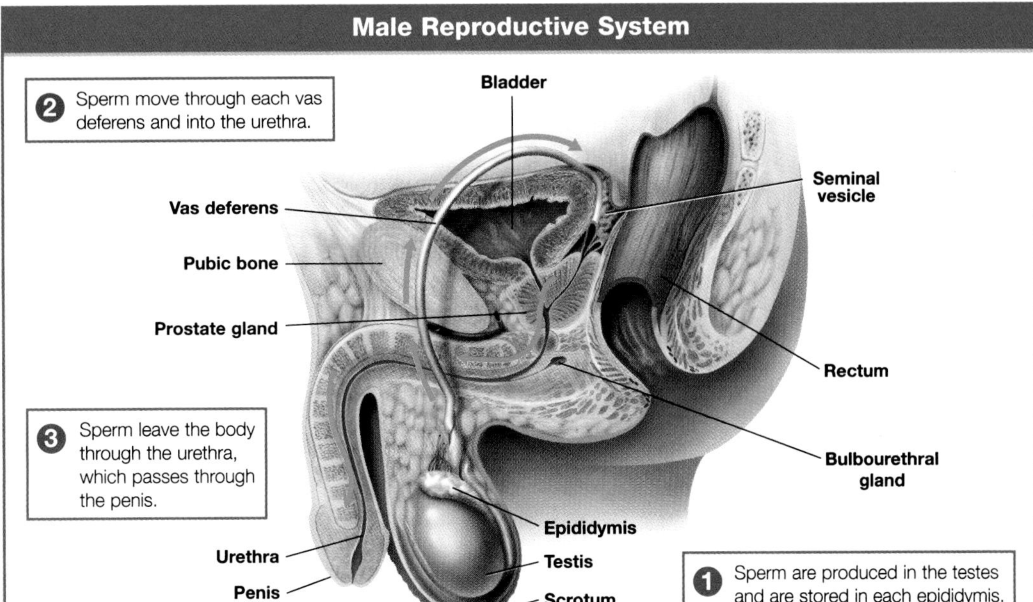
Male Reproductive System
2 Sperm move through each vas deferens and into the urethra.
Bladder
Seminal vesicle
Vas deferens
Pubic bone
Prostate gland
Rectum
3 Sperm leave the body through the urethra, which passes through the penis.
Bulbourethral gland
Epididymis
Urethra
Testis
Penis
Scrotum
1 Sperm are produced in the testes and are stored in each epididymis.

Figure 44-2 Male reproductive system. The arrows indicate the path taken by sperm cells from the testes as they exit the body.

Maturation and storage of sperm

A typical adult male produces several hundred million sperm cells each day. After being produced in the seminiferous tubules, the sperm mature and travel through a series of long tubes. Sperm then enter a long coiled tube called the **epididymis** *(ehp uh DIHD ih mihs),* shown in Figure 44-1. When the sperm leave the seminiferous tubules, they are not yet capable of swimming. Within each epididymis, the sperm mature and become capable of moving.

The epididymis is also the site where most of the sperm are stored. From the epididymis, some sperm move to another long tube, the **vas deferens** *(vas DEHF uh rehnz).* Sperm move through the vas deferens and into the urethra, as shown by the arrows in **Figure 44-2.** Sperm leave the body by passing through the urethra, the same duct through which urine exits the body.

Structure of mature sperm

As shown in **Figure 44-3,** a mature sperm cell consists of a head with very little cytoplasm, a midpiece, and a long tail. Enzymes at the tip of the head help the sperm cell penetrate an egg cell during fertilization. The midpiece contains many mitochondria that supply sperm with the energy needed to propel themselves through the female reproductive system. The tail of a sperm cell is a powerful flagellum that whips back and forth, enabling the sperm cell to move. ATP produced in the mitochondria power the whiplike movements of the tail. During fertilization, only the head of a sperm enters an egg, so a father's mitochondria are not passed to offspring.

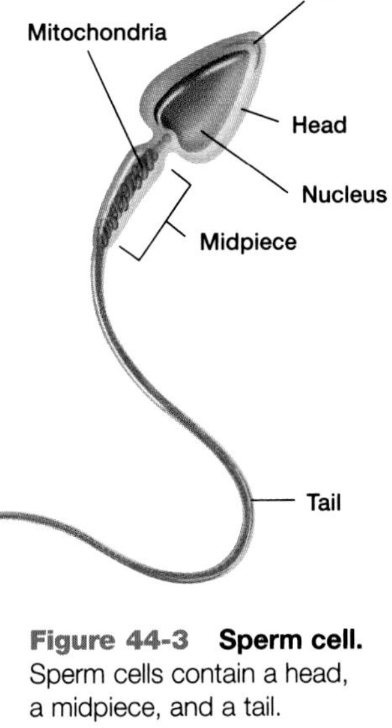
Enzymes
Mitochondria
Head
Nucleus
Midpiece
Tail

Figure 44-3 Sperm cell. Sperm cells contain a head, a midpiece, and a tail.

Graphic Organizer

Use this graphic organizer with ***Teaching Tip: Path of Sperm*** *on page 1007.*

Seminiferous tubules → Epididymus → Vas deferens → Urethra

Section 44-1

Teaching TIP

Androgens

Male hormones are called androgens. Androgens have a generalized anabolic effect; for example, they increase protein synthesis, especially in muscles, and they promote bone growth. Androgens are derived from cholesterol, as are all steroid hormones. The principal androgen is testosterone.

Teaching TIP

Vasectomy

A vasectomy is a minor operation performed on adult males as a method of birth control. The vas deferens are cut to short-circuit the path of sperm. A vasectomy does not have any significant effect on ejaculation because sperm account for only 1 percent of the total volume of semen. Because normal levels of male hormones continue to be produced, a vasectomy also does not affect secondary sex characteristics.

Teaching TIP — BASIC

Path of Sperm

Have students make a **Graphic Organizer** similar to the one at the bottom of page 1007 to trace the movement of sperm cells from the seminiferous tubules to the urethra. PORTFOLIO

Visual Strategy

Figure 44-3

Have students identify what part of the sperm shown in this figure must penetrate the egg. *(The head region, which contains DNA representing paternal genes, must be delivered into the egg.)* What parts of a sperm cell help it move toward a waiting egg and deliver its DNA? *(Sperm cells have a tail, which propels the sperm and its genes toward an awaiting egg, and mitochondria, which provide energy for swimming.)*

Chapter 44

Making Biology Relevant

Male Infertility

Many cases of male infertility are caused solely by insufficient sperm production. However, the abnormal composition of fluids in the semen may also be responsible for infertility. Insufficient fructose content, hormonal or pH imbalances, and prostaglandin concentrations that are too high or too low all can cause infertility. Other causes of infertility include malformations of the sperm and obstructions of the delivery system. Poor circulation in the testes can raise the temperature of the testes, leading to the release of immature sperm in reduced amounts. Mumps, radiation, or trauma can also cause damage to the testes.

3 Assessment Options

Closure

Body Chemistry

Have students explain why sperm can survive for only about 48 hours once they enter the female reproductive tract *(the acidic fluid constitutes a hostile environment),* yet survive much longer in the testes *(specialized cells and fluids nourish and protect them).*

Review

Assign Review—Section 44-1.

Reteaching — BASIC

Show the class an overhead transparency of the male reproductive system. Have students identify the structures and functions of the various parts. Review quickly, calling on students to name the structure or function rapidly.

---

internetconnect

SciLINKS NSTA

TOPIC: Transportation of sperm
GO TO: www.scilinks.org
KEYWORD: HX1008

Transportation of sperm

As sperm cells move into the urethra, they mix with fluids secreted by three exocrine glands: the seminal vesicles, the prostate gland, and the bulbourethral glands. These fluids nourish the sperm and aid their passage through the female reproductive system. The **seminal vesicles,** which lie between the bladder and the rectum, produce a fluid rich in sugars that sperm use for energy. The **prostate** *(PRAHS tayt)* **gland,** which is located just below the bladder, secretes an alkaline fluid that neutralizes the acids in the female reproductive system. Before semen leaves the body, the **bulbourethral** *(buhl boh yoo REE thruhl)* **glands** also secrete an alkaline fluid that neutralizes traces of acidic urine in the urethra. The mixture of these secretions with sperm is called **semen** *(SEE muhn).*

Delivery of sperm

The urethra passes through the **penis,** the male organ that deposits sperm in the female reproductive system during sexual intercourse. During sexual arousal, blood flow to the penis increases. As shown in **Figure 44-4,** the penis contains three cylinders of spongy tissue. Small spaces separate the cells of the spongy tissue. Blood collects within these spaces, causing the penis to become rigid and erect.

Sperm exit the penis through ejaculation *(ee jak yoo LAY shun),* the forceful expulsion of semen. During ejaculation, muscles around each vas deferens contract, moving sperm into the urethra. Muscles at the base of the penis force semen out of the urethra.

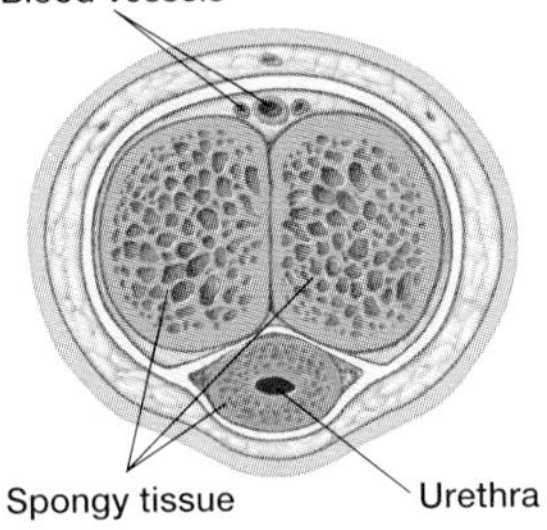

After the semen is deposited in the female reproductive system, sperm swim until they encounter an egg cell. If sperm are unable to reach an egg, fertilization does not occur. One way to prevent fertilization is to block the path of the sperm. Covering the penis with a thin rubber sheath called a condom helps prevent fertilization during sexual intercourse. Abstaining from sexual intercourse is the surest way to prevent fertilization.

About 3.5 mL of semen is expelled during ejaculation. This amount of semen normally contains 300–400 million sperm. However, few sperm ever reach an egg cell because most sperm die in the female reproductive system. Because only a small percentage of sperm survive after ejaculation occurs, fertilization usually requires a high sperm count. Males with fewer than 20 million sperm per mL of semen are generally considered sterile.

Blood vessels
Spongy tissue
Urethra

Figure 44-4 Cross section of penis. The penis contains three cylinders of spongy tissue. When the spaces in these cylinders fill with blood, the penis becomes erect.

Review SECTION 44-1

1. **Describe** the functions of the testes.
2. **Sequence** the path that mature sperm take from the testes to the outside of the body.
3. **Describe** the role of each part of a mature sperm cell.
4. **SKILL Recognizing Relationships** How do secretions by exocrine glands help the delivery of sperm to the female reproductive system?

Critical Thinking

5. **SKILL Inferring Relationships** If a male's left vas deferens is blocked, how is his sperm count affected? Explain your answer.

Review SECTION 44-1 ANSWERS

1. Testes produce gametes called sperm.
2. Mature sperm travel from the epididymis, through the vas deferens, and out the urethra.
3. The head contains genetic information and enzymes that penetrate the egg. The midpiece contains mitochondria, which supply energy. The tail is composed of a flagellum, which allows the sperm cell to move.
4. Exocrine secretions decrease acidity of the semen and provide nourishment and a fluid environment in which the sperm can swim.
5. The sperm count likely is reduced by half.

Female Reproductive System

Section 44-2

The Ovaries Produce Egg Cells

Each month, the female reproductive system prepares for a possible pregnancy by producing a mature egg cell—the female gamete. After sperm have been deposited and fertilization has occurred, the role of the male in reproduction is complete. If pregnancy occurs, the female reproductive system will nourish and protect the fertilized egg through nine months of development.

Production of eggs

Two egg-shaped ovaries, shown in **Figure 44-5,** are located within the abdominal cavity. The **ovaries** *(OH vuh reez)* are the gamete-producing organs of the female reproductive system. Females are born with all of the egg cells they will ever produce. At birth, the ovaries contain about 2 million immature egg cells that already have begun the first division of meiosis. Like sperm cells, egg cells contain 23 chromosomes (the haploid number) because eggs also are formed through meiosis.

After meiosis begins, egg cells become stalled in prophase of the first meiotic division. When a female reaches puberty, the increased production of sex hormones enables meiosis to resume. However,

Objectives

- **Describe** how eggs are produced. (p. 1009)
- **Identify** the major structures of the female reproductive system. (p. 1010)
- **Analyze** the events of the ovarian and menstrual cycles. (pp. 1011–1013)

Key Terms

ovary
ovum
fallopian tube
uterus
vagina
ovarian cycle
ovulation
follicle
corpus luteum
menstrual cycle
menstruation

Figure 44-5 Ovaries

The ovaries produce egg cells.

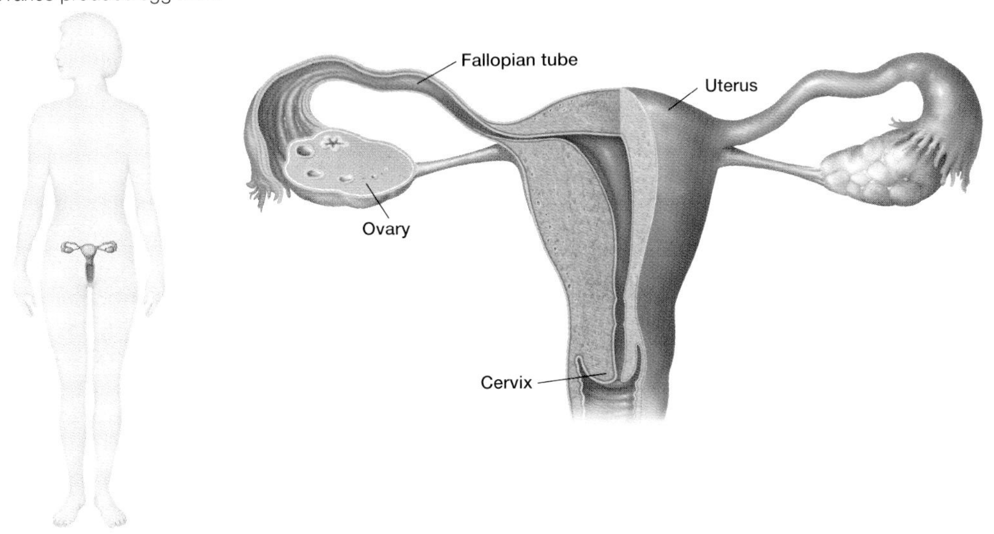

Section 44-2

1 Prepare

Scheduling

- **Block:** About 45 minutes will be needed to complete this section.
- **Traditional:** About one class period will be needed to complete this section.
- **Planning:** See the Planning Guide on pp. 1003A and 1003B for lecture, classwork, and assignment options for Section 44-2. Lesson plans for Section 44-2 can be found in a lesson cycle format in ***Biology: Principles and Explorations*** **One-Stop Planner CD-ROM with Test Generator.**

Resource Link

Assign **Active Reading Worksheet Section 44-2** to help students comprehend the key concepts and visuals in this lesson.

2 Teach

Lesson Warm-up

Have students name female secondary sex characteristics. *(breast development, increased pubic and armpit hair, widening of the hips, menstruation, and so on)* List them on the board or overhead, and add to the list as you work through the chapter.

Integrating Different Learning Styles

Logical/Mathematical	**PE**	Data Lab, p. 1013; Performance Zone items 16–18, p. 1023
	ATE	Cross-Disciplinary Connection, p. 1011
Visual/Spatial	**ATE**	Visual Strategy, p. 1010
Verbal/Linguistic	**PE**	Study Tip, p. 1012
	ATE	Lesson Warm-up, p. 1009; Cross-Disciplinary Connection, p. 1010

Visual Strategy

Figure 44-6

Have students locate the point at which sperm enter the female reproductive system illustrated in this picture. *(the vagina)* Have them trace the path the egg travels and locate where the sperm meets the egg. *(in the upper third of the fallopian tubes)*

Teaching TIP

Sperm and Egg Size

The difference in the sizes of sperm and eggs is a result of a fundamental difference between the processes of sperm formation (spermatogenesis) and egg formation (oogenesis). During spermatogenesis, equal divisions of the cytoplasm follow meiosis I and meiosis II, resulting in four equal-sized sperm cells. During oogenesis, unequal divisions of the cytoplasm follow meiosis I and meiosis II, resulting in one large egg and three tiny cells that degenerate. In humans, meiosis II occurs just as a sperm penetrates an ovum.

Cross-Disciplinary Connection

◆ **Language Arts**

The word *cervix* means "neck" in Latin. Have students explain why this is an appropriate term for the ring of strong muscles located in the lower part of the uterus. *(Answers will vary but could include the similarity of the cervix to the neck of a bottle.)* Ask students where in the body cervical vertebrae would be found. *(Cervical vertebrae are found in the part of the spine forming the neck.)*

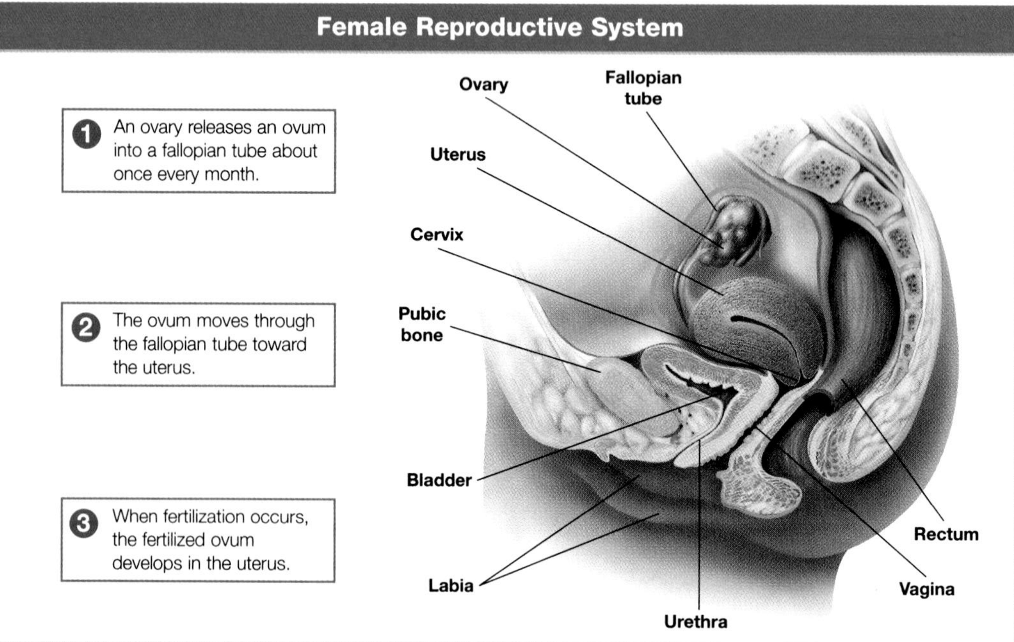
Female Reproductive System
1 An ovary releases an ovum into a fallopian tube about once every month.
2 The ovum moves through the fallopian tube toward the uterus.
3 When fertilization occurs, the fertilized ovum develops in the uterus.
Ovary
Fallopian tube
Uterus
Cervix
Pubic bone
Bladder
Labia
Urethra
Rectum
Vagina

Figure 44-6 Female reproductive system. The arrows indicate the path taken by an ovum from an ovary to the uterus.

normally only one immature egg cell completes its development each month. In the lifetime of a female, only 300–400 egg cells will mature. When an egg cell matures, it is called an **ovum** *(OH vuhm)*.

Structures of the female reproductive system

An ovum is released from an ovary about every 28 days. Cilia sweep the ovum into a fallopian tube. Each **fallopian** *(fuh LOH pee uhn)* **tube** is a passageway through which an ovum moves from an ovary toward the uterus. Smooth muscles lining the fallopian tubes contract rhythmically, moving the ovum down the tube and toward the uterus, as shown by the arrows in **Figure 44-6.** An ovum's journey through a fallopian tube usually takes three to four days to complete. If the ovum is not fertilized within 24–48 hours, it dies. An ovum, shown in **Figure 44-7,** is many times larger than a sperm cell and can be seen with the unaided eye.

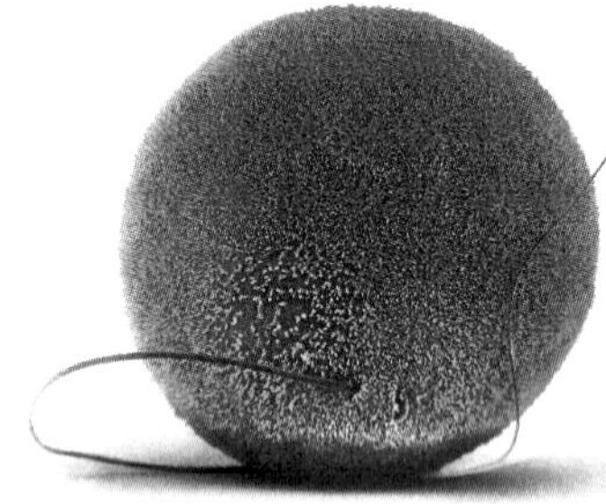

Figure 44-7 Ovum. Notice the great difference in size between the sperm and the ovum.

The **uterus** *(YOO tuh ruhs)* is a hollow, muscular organ about the size of a small fist. If fertilization occurs, development will take place in the uterus. During sexual intercourse, sperm are deposited inside the **vagina** *(vuh JEYE nuh)*, a muscular tube that leads from the outside of the female's body to the entrance to the uterus, called the cervix *(SUR vihks)*. A soft rubber cap called a diaphragm *(DEYE uh fram)* can be used to cover the cervix and help prevent fertilization by blocking the passage of sperm into the uterus. A diaphragm is more effective when used with a sperm-killing chemical, or spermicide. During childbirth, a baby passes through the cervix and leaves the mother's body through the vagina.

did you know?

Young children think a baby develops in a woman's stomach.

Older students are often confused about the exact location of the uterus as well. The uterus sits above the bladder, yet under the intestine. What happens to the mother's internal organs as the fetus grows? *(The organs become compressed.)* This compression of the abdominal and thoracic organs can become a problem, especially late in the pregnancy. The mother ultimately may experience fatigue, backache, indigestion, circulatory problems, reduced lung capacity, and changes in her gait. Most of these symptoms are reversed upon delivery of the baby.

An Egg Cell Matures During the Ovarian Cycle

The ovaries prepare and release an ovum in a series of events collectively called the **ovarian cycle.** The release of an ovum from an ovary is called **ovulation** *(ahv yoo LAY shuhn).* The ovum is then swept into the fallopian tube and begins to move toward the uterus, awaiting fertilization. Although the duration of the ovarian cycle varies from female to female, the cycle generally spans about 28 days.

internet**connect**

SCILINKS. NSTA

TOPIC: Ovarian cycle
GO TO: www.scilinks.org
KEYWORD: HX1011

Phases of the ovarian cycle

Follicular phase The ovarian cycle has two distinct phases: the follicular phase *(fuh LIK yoo luhr)* and the luteal phase. These phases are regulated by hormones released by the hypothalamus and the anterior pituitary. The events of the ovarian cycle are summarized in **Figure 44-8.** In an ovary, egg cells mature within follicles. A **follicle** *(FAHL i kuhl)* is a cluster of cells that surrounds an immature egg cell and provides the egg with nutrients. During the follicular phase of the ovarian cycle, hormones regulate the completion of an egg cell's maturation. The follicular phase, which marks the beginning of the ovarian cycle, begins when the anterior pituitary releases follicle-stimulating hormone (FSH) and luteinizing hormone (LH) into the bloodstream. Both FSH and LH cause the follicle to produce estrogen, a sex hormone that aids in the growth of the follicle.

Figure 44-8 Ovarian cycle

During the ovarian cycle, ovulation occurs about every 28 days.

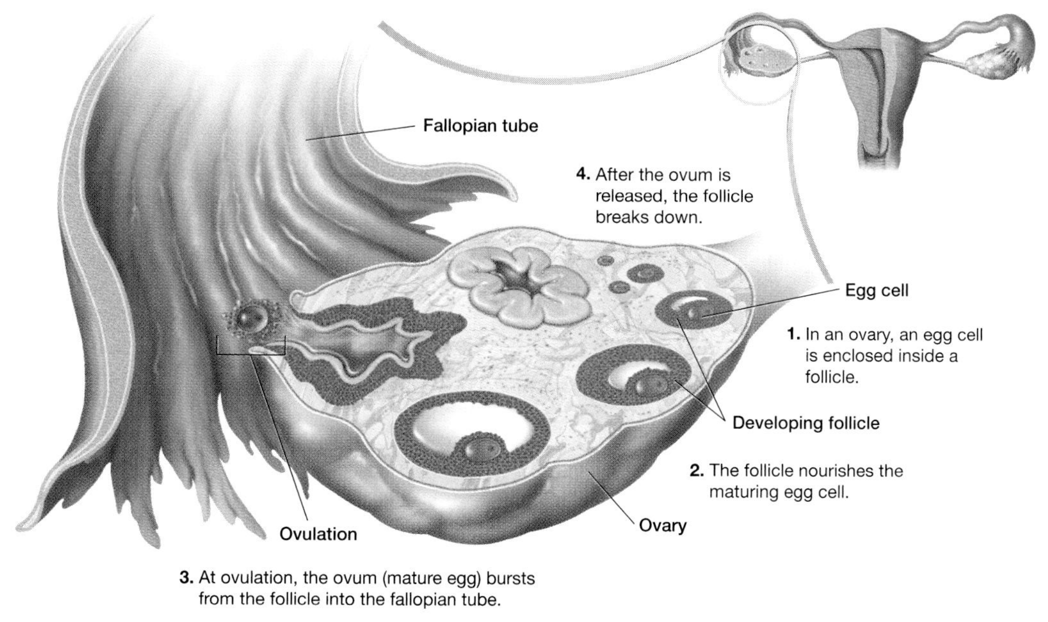

Fallopian tube
4. After the ovum is released, the follicle breaks down.
Egg cell
1. In an ovary, an egg cell is enclosed inside a follicle.
Developing follicle
2. The follicle nourishes the maturing egg cell.
Ovary
Ovulation
3. At ovulation, the ovum (mature egg) bursts from the follicle into the fallopian tube.

Section 44-2

CROSS-DISCIPLINARY CONNECTION

◆ Mathematics

At puberty, females have more than 400,000 immature egg cells in their ovaries. Have students calculate the percentage of these eggs that will be ovulated, assuming that 13 ovarian cycles occur per year and that the reproductive life span of a woman is approximately 40 years. *(40 years × 13 eggs/year = 520 eggs; 520 eggs/400,000 eggs = 0.13 %)*

Teaching TIP

PMS

Premenstrual syndrome (PMS) can cause drastic mood changes in some women just before menstruation. Hormonal changes due to the menstrual cycle have been blamed for the syndrome. Steroid hormones have been used with differing degrees of success to treat PMS.

Teaching TIP

Cervical Cancer

Cervical cancer is a relatively easily detected and treated problem. The cells in the cervix change in a characteristic manner early in the disease. For sexually mature women, a yearly Pap smear is advised. For a Pap smear, a sample of cervical cells is removed with a swab and then examined under a microscope. A yearly Pap smear can disclose the cellular changes that precede cervical cancer, as well as full-blown cases of cervical cancer.

Overcoming Misconceptions

Fertility and Youth

It is a common belief that a girl cannot become pregnant before she menstruates. However, ovulation precedes a menstrual period. Therefore, a young girl must ovulate before her first menstrual period (menarche). It is therefore possible, though unlikely, for a girl to become pregnant before menarche.

Chapter 44

Making Biology Relevant

Sports

In females, an exercise program that is too vigorous can delay menarche (the first menstrual flow), disrupt the normal menstrual cycle in adult women, and even cause amenorrhea—a complete cessation of the normal menstrual cycle. These problems, seen extensively in highly trained female athletes, seem to stem from low body fat content, which impairs the body's ability to produce estrogen. Abnormal function of the hypothalamus also may contribute to low hormone levels. Although these effects abate once the athlete assumes a more normal body composition, other problems may arise. For example, amenorrhea can cause otherwise healthy, young women to lose bone mass, as postmenopausal women do. In an effort to counteract this loss, sports physicians encourage female athletes to increase their daily intake of calcium to 1.5 g per day, which is about the amount of calcium in a quart of milk.

Study TIP

Word Origins

The term *corpus luteum* is from the Latin words *corpus,* meaning "body," and *luteum,* meaning "yellow."

Ovulation At first the small increase in the level of estrogen prevents further release of FSH and LH from the anterior pituitary. This is caused by a negative feedback mechanism. But as the follicle approaches maturity, it begins to secrete large amounts of estrogen. The anterior pituitary responds to this high level of estrogen by greatly increasing secretion of LH. This increase in LH secretion is caused by a positive feedback mechanism. This surge of LH causes the egg cell to complete the first meiotic division, and it causes the follicle and the ovary to rupture. When the follicle bursts, ovulation occurs, as shown in Figure 44-8.

Luteal phase The luteal *(LOOT ee uhl)* phase of the ovarian cycle follows the follicular phase, as shown in **Figure 44-9.** After ovulation occurs, LH causes the cells of the ruptured follicle to grow, forming a corpus luteum. A **corpus luteum** *(KOHR puhs LOOT ee uhm)* is a yellowish mass of follicular cells that functions like an endocrine gland. LH causes the corpus luteum to secrete estrogen and progesterone, another sex hormone. Estrogen and progesterone cause a negative feedback mechanism that inhibits the release of FSH and LH. This prevents the development of new follicles during the luteal phase.

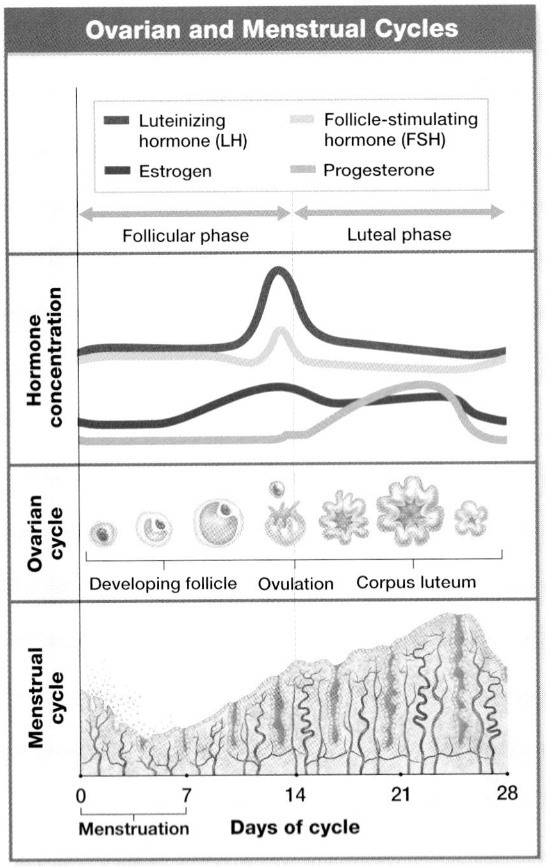

Figure 44-9 Ovarian and menstrual cycles. The ovarian cycle is regulated by hormones produced by the hypothalamus and the pituitary gland. The menstrual cycle is regulated by hormones produced by the follicle and the corpus luteum.

Preparation for pregnancy

Progesterone signals the body to prepare for fertilization. If fertilization occurs, the corpus luteum continues to produce progesterone for several weeks. If fertilization does not occur, production of progesterone slows and eventually stops, marking the end of the ovarian cycle. Prescription drugs containing relatively large doses of synthetic estrogen and progesterone-like hormones disrupt the ovarian cycle and prevent ovulation.

Menstrual cycle While changes occur in the ovaries during the ovarian cycle, changes also occur in the uterus, as shown in Figure 44-9. The series of changes that prepare the uterus for a possible pregnancy each month is called the **menstrual** *(MEN struhl)* **cycle.** The menstrual cycle lasts about 28 days.

The events of the menstrual cycle are influenced by the changing levels of estrogen and progesterone during the ovarian cycle. Prior to ovulation, increasing levels of estrogen cause the lining of the uterus to thicken. After ovulation, high levels of both estrogen and progesterone cause further thickening and maintain the uterine lining. If pregnancy does not occur, the levels of estrogen and progesterone decrease. This

did you know?

Pregnancy disrupts the normal homeostasis of the female body.

At least four major systems—the digestive, urinary, respiratory, and cardiovascular systems—must change significantly to accommodate pregnancy. Until a woman can adjust to the elevated progesterone and estrogen levels in her body, she may experience excessive salivation and nausea, which is called "morning sickness." In addition, pressure in her abdomen may cause stomach acid to wash backward into her esophagus, causing chronic heartburn. Total blood volume rises, and both blood pressure and pulse rate increase. The pregnant woman's heart pumps 20 to 40 percent more blood, depending on the stage of pregnancy. Pressure by the uterus on pelvic blood vessels may cause varicose veins to result from the restricted blood flow back from the legs.

decrease causes the lining of the uterus to shed, marking the end of the menstrual cycle. The end of the menstrual cycle coincides with the end of the luteal phase of the ovarian cycle.

Menstruation

When the lining of the uterus is shed, blood vessels break and bleeding results. A mixture of blood and discarded tissue then leaves the body through the vagina. This process, called **menstruation** *(men STRAY shuhn),* usually occurs about 14 days after ovulation. At the end of the ovarian and menstrual cycles, neither estrogen nor progesterone is being produced. In the absence of estrogen and progesterone, the pituitary again begins to produce FSH and LH, starting the cycles again.

Women eventually stop menstruation, usually between the ages of 45 and 55. After this event, called menopause, a woman no longer ovulates and thus moves out of the childbearing phase of her life. During menopause, many women experience symptoms, such as hot flashes, caused by a decrease in estrogen production. Estrogen, which can be taken to relieve symptoms of menopause, is a widely used prescription drug in the United States.

REAL LIFE

Many women have cramps during their menstrual periods.
The release of a prostaglandin stimulates contractions of muscles in the wall of the uterus, causing cramps in the uterus.

Finding Information
Research ways that menstrual cramps can be treated.

Analyzing hormone secretion in ovarian and menstrual cycles

Background

The ovarian and menstrual cycles are regulated by hormones secreted by the hypothalamus, the pituitary gland, and the ovaries. Feedback mechanisms play a major role in these cycles. Use Figure 44-9 to answer the following questions.

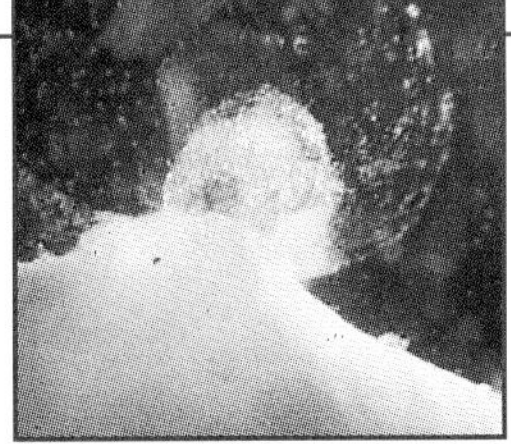

Ovulation

Analysis

1. **Identify** the hormones that are secreted in large amounts prior to ovulation.
2. **Describe** the effect of estrogen production on the secretion of luteinizing hormone.
3. **SKILL Analyzing Concepts** What type of feedback mechanism causes a decrease in the secretion of luteinizing hormone and follicle-stimulating hormone during the luteal phase?
4. **Skill Analyzing Concepts** What type of feedback mechanism causes the surge of luteinizing hormone secretion during the follicular phase?

Review *SECTION 44-2*

1. **Identify** three functions of the female reproductive system.
2. **Describe** the function of follicles in the production of egg cells.
3. **Compare** the roles of LH and FSH in regulating the ovarian cycle.
4. **SKILL Recognizing Relationships** What causes the lining of the uterus to thicken and then to be shed during the menstrual cycle?

Critical Thinking

5. **SKILL Relating Concepts** How could the maturation of an egg cell be blocked during the ovarian cycle?

Review *SECTION 44-2 ANSWERS*

1. Three important functions of the female reproduction system include producing gametes (eggs), providing an environment for internal fertilization, and nourishing and protecting the fetus.
2. The follicle encloses the egg and secretes estrogen, which encourages ripening of the follicle.
3. LH and FSH both induce the follicle to produce estrogen. Increased secretion of LH causes the follicle to divide and ovulation to occur. Continued secretion of LH stimulates the growth of the corpus luteum. Estrogen and progesterone secreted by the corpus luteum suppress further follicular maturation.
4. Rising estrogen and progesterone levels cause the uterine lining to thicken in preparation for pregnancy. When estrogen and progesterone levels drop, the lining is shed.
5. Reduced levels of FSH and/or LH could halt egg cell development.

REAL LIFE

Answer

Menstrual cramps often are treated with drugs that inhibit prostaglandin formation, such as aspirin and naproxin sodium.

Analyzing hormone secretion in ovarian and menstrual cycles

Time 10–15 minutes

Process Skills Interpreting graphs, analyzing data, inferring relationships

Teaching Strategies

Remind students to use the key to properly assign hormone names to the colored curves.

Analysis Answers

1. LH, estrogens, and to some extent, FSH spike just prior to ovulation.
2. A peak in estrogen production precedes a spike in LH production.
3. Negative feedback—as progesterone increases, LH and FSH decrease.
4. Positive feedback—as estrogen increases, LH also increases.

3 Assessment Options

Closure

Stress Tests

Ask the class why certain environmental factors, such as stress and prolonged, strenuous exercise, can affect the female reproductive cycle. *(These can influence hormone levels, thus affecting the female reproductive cycle.)*

Review

Assign Review—Section 44-2.

Reteaching — BASIC

Use Figure 44-9 to correlate hormone levels with uterine lining formation and menstruation.

Section 44-3 Development

1 Prepare

Scheduling

- **Block:** About 90 minutes will be needed to complete this section.
- **Traditional:** About two class periods will be needed to complete this section.
- **Planning:** See the Planning Guide on pp. 1003A and 1003B for lecture, classwork, and assignment options for Section 44-3. Lesson plans for Section 44-3 can be found in a lesson cycle format in ***Biology: Principles and Explorations*** **One-Stop Planner CD-ROM with Test Generator.**

Resource Link

Assign **Active Reading Worksheet Section 44-3** to help students comprehend the key concepts and visuals in this lesson.

2 Teach

Lesson Warm-up

Have students draw a 0.5 mm dot at the top of a sheet of paper. Tell them to label the mark *14 days.* Then have them draw a line segment about 1 mm in length below the dot and label it *18 days.* Tell them to continue down the sheet in a similar manner, drawing line segments of 4 mm, 8 mm, 19 mm, 38 mm, and 63 mm in length and labeling them *4 weeks, 6.5 weeks, 9 weeks, 11 weeks,* and *15 weeks,* respectively. Tell students that these are the actual lengths of a developing human through the 15th week of pregnancy. PORTFOLIO

Objectives

- **Sequence** the events of fertilization, cleavage, and implantation. (p. 1014)
- **Summarize** the three trimesters of pregnancy. (pp. 1015–1017)
- **Describe** the effects of drug use on development. (p. 1015)

Key Terms

cleavage
blastocyst
implantation
gestation
pregnancy
embryo
placenta
fetus

Fertilization Produces a Zygote

If sperm are present in the female reproductive system within a few days after ovulation, fertilization may occur. To fertilize an ovum, a sperm cell must swim to a fallopian tube, where fertilization usually occurs. During fertilization, a sperm cell penetrates an ovum by releasing the enzymes at the tip of its head. These enzymes break down the jellylike outer layers of the ovum. The head of the sperm enters the ovum, and the nuclei of the ovum and sperm fuse together. This produces a diploid cell called a zygote.

Cleavage and implantation

In the first week after fertilization, the zygote undergoes a series of internal divisions known as **cleavage,** as shown in **Figure 44-10.** Cleavage produces many smaller cells—first two cells, then four, then eight, and so on—within the zygote. Cleavage continues as the zygote moves through the fallopian tube toward the uterus. By the time it reaches the uterus, the zygote is a hollow ball of cells called a **blastocyst** *(BLAS toh sist)*. About six days after fertilization, the blastocyst burrows into the lining of the uterus in an event called **implantation.** There it will undergo development, eventually forming a living human.

Figure 44-10 Formation of zygote

Fertilization, cleavage, and implantation occur after ovulation.

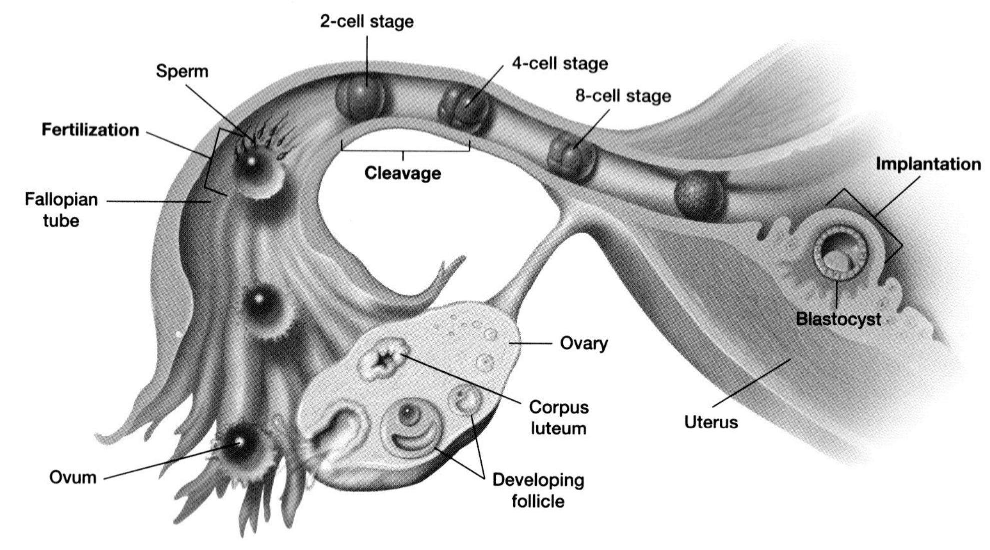

Integrating Different Learning Styles

Logical/Mathematical	**PE**	Tech Watch, p. 1017
	ATE	Multicultural Perspective, p. 1015
Visual/Spatial	**PE**	Performance Zone item 1, p. 1022
	ATE	Reteaching, p. 1017
Body Kinesthetic	**ATE**	Demonstration, p. 1016
Verbal/Linguistic	**ATE**	Career, p. 1016; Biology in Your Community, p. 1016

Pregnancy Is Divided into Three Trimesters

Development begins with a single diploid cell from which billions of other cells arise. The uterus provides protection and nourishment during development. Human development takes about 9 months—a period known as **gestation** *(jes TAY shuhn)*, or pregnancy. The 9 months of **pregnancy** are often divided into three trimesters, or 3-month periods. For the first 8 weeks of pregnancy, the developing human is called an **embryo** *(EHM bree oh)*.

First trimester

Supportive membranes The most crucial events of development occur very early in the first trimester. In the second week after fertilization—shortly after implantation—the embryo grows rapidly. Membranes that will protect and nourish it also develop. One of these membranes, the amnion *(AM nee ahn)*, encloses and protects the embryo. Another membrane, the chorion *(KOHR ee ahn)*, interacts with the uterus to form the placenta. The **placenta** *(plah SEHN tah)* is the structure through which the mother nourishes the embryo. As shown in **Figure 44-11,** the mother's blood normally never mixes with the blood of the embryo. Instead, nutrients in the mother's blood diffuse through the placenta and are carried to the embryo through blood vessels in the umbilical *(uhm BIL i kuhl)* cord.

The waste products of the embryo also pass through the placenta into the mother's blood. Most other substances, including drugs and pathogens, also diffuse through the placenta. Thus, if the mother ingests any harmful substances, the embryo is also affected. For example, alcohol use by pregnant women, especially during early

REAL LIFE

In the United States, miscarriages occur in about 20 percent of pregnancies. Sometimes miscarriages occur so early that the woman is unaware she was pregnant. The rate of miscarriage is high for teenagers, lowest (12 percent) by age 20, and high (41 percent) by age 42.

Finding Information *Find out why the miscarriage rate changes with age.*

Figure 44-11 Placenta

The developing human is nourished through the placenta.

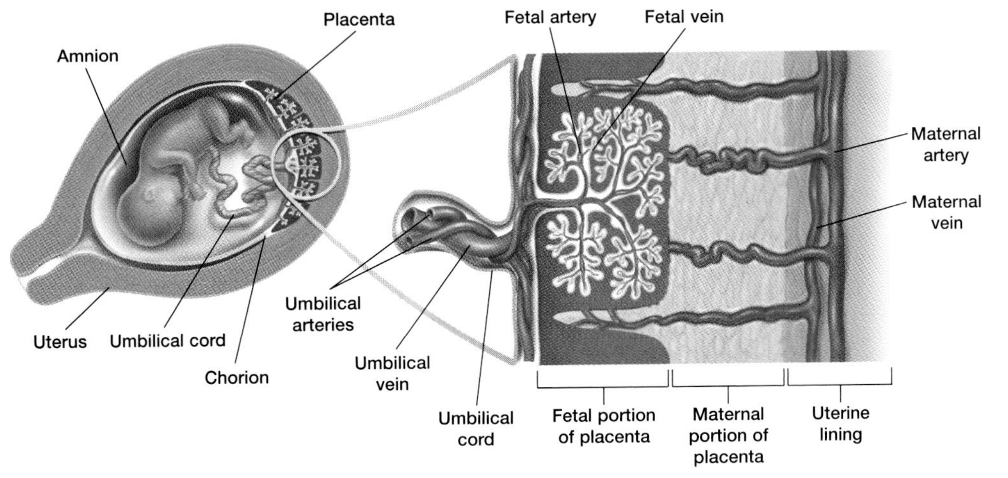

MULTICULTURAL PERSPECTIVE

Calculating Age

We count our age in terms of time that has passed since our birth. Some Asian cultures, however, take the approximate time spent in the uterus into account; therefore, a newborn is considered to be 1 year old.

REAL LIFE

Answer

Most miscarriages involve an abnormally developing fetus. In many of these cases, the fetus has genetic abnormalities. An older mother's eggs have been in place for many years and may have been exposed to mutagens, which damaged their chromosomes.

Teaching TIP

Tubal Ligation

Cutting or tying the fallopian tubes, an operation called tubal ligation, prevents fertilization by blocking sperms' path to the ovum.

Teaching TIP

In Vitro Fertilization

If a male has a low sperm count, or if a female does not ovulate normally or has a blocked fallopian tube, natural fertilization may be impossible. In such cases, the events that ordinarily happen in the fallopian tube can be caused in a culture dish in a process known as *in vitro* fertilization. In the process, hormones are given to the woman to ensure precise timing of ovulation. Within hours of ovulation, one or more eggs are surgically removed from the ovary and placed in a culture dish. Sperm taken from the male are then added. If the procedure is successful, one or more fertilized eggs are then implanted in the woman's uterus to complete development.

Making Biology Relevant

Medicine

Placenta previa is a condition in which the embryo is implanted near the cervix. This may cause hemorrhaging as the fetus grows. Abruptio placentae is a condition in which the placenta begins tearing away from the uterine wall, causing hemorrhaging. Both of these conditions can lead to miscarriage and even death of the mother.

Demonstration

Umbilical Cord

Have students put their fingers on their navel, and ask students what their navel is. *(Once it is separated from the mother, a baby's umbilical cord is clamped, tied off, and cut. What remains on the baby slowly shrivels and falls off, leaving a scar known as the navel.)*

Making Biology Relevant

Health

An insufficient supply of vitamin B6 can put fetuses at risk for nervous-system disorders. Mothers-to-be must make sure they are eating a balanced and healthy diet before conception so that their babies will be born healthy. Oral contraceptives deplete the body's store of vitamin B6, and a pregnant woman might require more than the recommended daily dose for adults. Pregnant women should follow their doctor's advice about all vitamin intake.

Teaching TIP — BASIC

Development

Emphasize to students that much more than growth is happening to the fetus from fertilization to birth. Development, including cell differentiation, is essential to produce the human form.

CAREER

Genetic Counselor

Genetic counselors help couples assess their risk of having a child with an inherited disorder. Genetic counselors review evidence of genetic disorders in the family backgrounds of prospective parents, and then they make predictions. Many genetic counselors are physicians with intensive training in genetics. Have students research the career of a genetic counselor. They should prepare a pamphlet that includes a job description, education and training, starting salary, and career outlook. PORTFOLIO

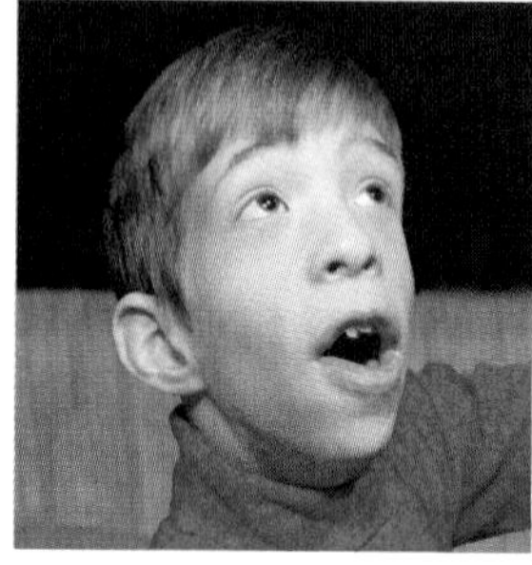

Figure 44-12 Fetal alcohol syndrome. This child has several signs of fetal alcohol syndrome, including an abnormally small head, deformed facial features, and widely spaced eyes.

pregnancy, is a leading cause of birth defects. Fetal alcohol syndrome is a birth defect in which a baby is born with a deformed face, as shown in **Figure 44-12,** and often severe mental and physical retardation. Women should abstain from alcohol and avoid all unnecessary drugs throughout pregnancy.

Development of embryo As the placenta forms, the inner cells of the blastocyst form the three primary tissue layers—endoderm, mesoderm, and ectoderm. By the end of the third week, blood vessels and the gut begin to develop, and the embryo is about 2 mm (0.08 in.) long. In the fourth week the arms and legs also begin to form, and the embryo more than doubles in length to about 5 mm (0.2 in.). By the end of the fourth week, all of the major organs begin to form, and the heart begins to beat.

During the second month, the final stage in embryonic development takes place. The arms and legs take shape. Within the body cavity, the major internal organs, including the liver and pancreas, are evident. By the end of the second month, the embryo is about 22 mm (0.9 in.) long and weighs about 1 g (0.002 lb.).

Development of fetus From the eighth week of pregnancy until childbirth, the developing human is called a **fetus** *(FEET uhs)*. By the end of the first trimester, the sex of the fetus has been established. A fetus has recognizable body features, and its organ systems have begun to form, as shown in **Figure 44-13.**

Figure 44-13 Development of fetus

A fetus has recognizable body characteristics.

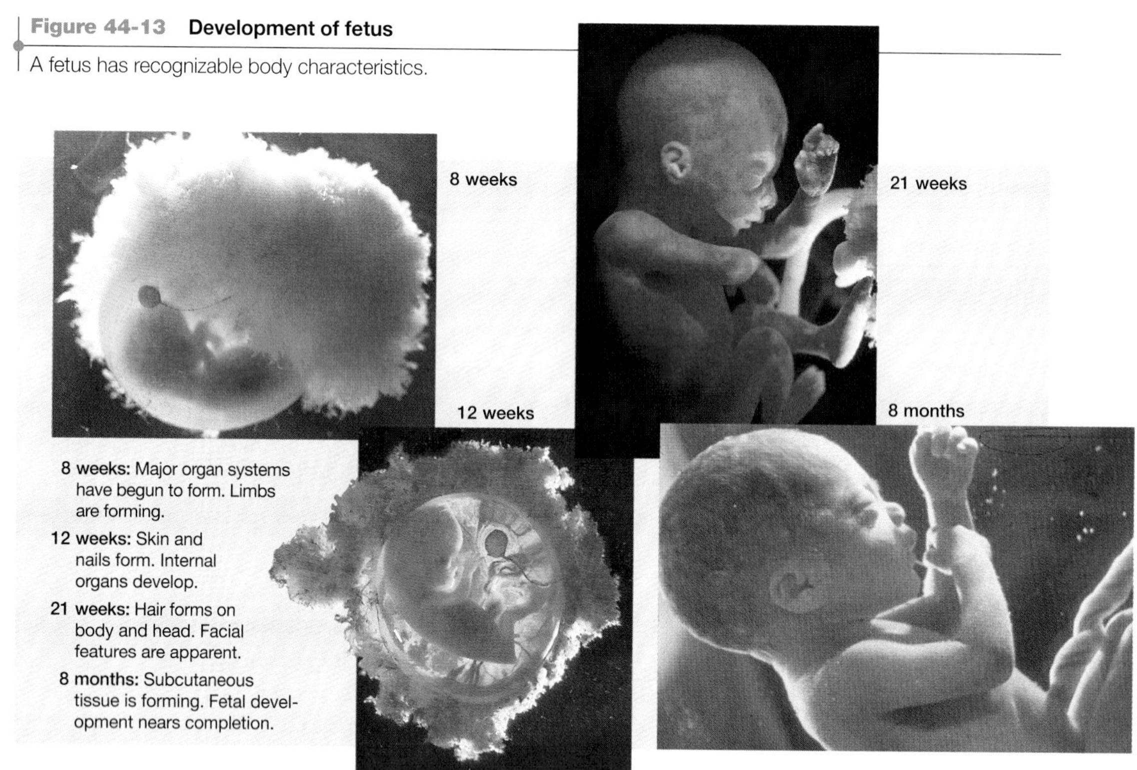

8 weeks
21 weeks
12 weeks
8 months
8 weeks: Major organ systems have begun to form. Limbs are forming.
12 weeks: Skin and nails form. Internal organs develop.
21 weeks: Hair forms on body and head. Facial features are apparent.
8 months: Subcutaneous tissue is forming. Fetal development nears completion.

Biology in Your Community

Substance Abuse During Pregnancy

Have a crisis pregnancy counselor, a health teacher, or the school nurse discuss the effects of substance abuse during pregnancy. Ask the speaker to emphasize that in many cases, doctors do not know how much of a drug is necessary to harm a fetus. Consequently, a pregnant woman is best advised to avoid all unnecessary drugs, including alcohol, nicotine, and caffeine. The speaker could also address proper nutrition, including vitamin intake, during pregnancy. Have students write a short report summarizing what they learned. PORTFOLIO

Second and third trimesters

During the second and third trimesters, the fetus grows rapidly as its organs become functional. By the end of the third trimester, the fetus is able to exist outside the mother's body. After about 9 months of development, the fetus leaves the mother's body in a process called labor, which usually lasts several hours. During labor, the walls of the uterus contract, expelling the fetus from the uterus, as shown in **Figure 44-14.** The fetus leaves the mother's body through the vagina. The placenta and the umbilical cord are expelled after the baby is born. After birth, development is far from complete. Physical growth and neurological development continue after birth.

Figure 44-14 Childbirth

During childbirth, the fetus exits the mother's body through the vagina.

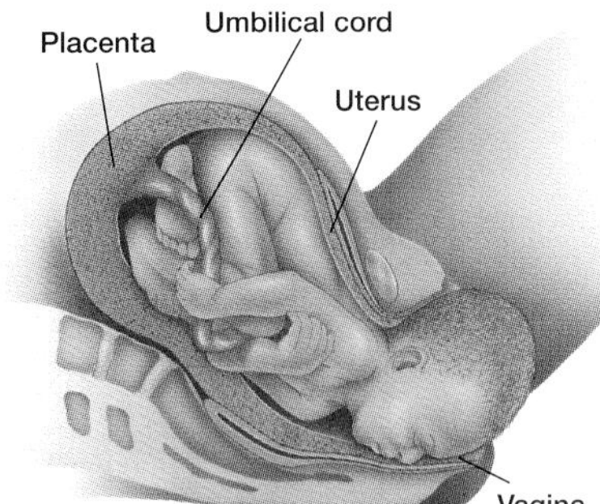

TECH WATCH: Ultrasound Imaging

Until the 1960s, X rays were the only way to study the inside of the body without performing surgery. Since then, several other methods have been developed for looking inside the body. One such method is ultrasound imaging. To make an ultrasound image, a physician holds a small probe against a patient's skin. The probe emits high-frequency sound waves, which produce echoes when they reflect off structures in the patient's body. The probe detects the echoes, which are converted to an image on a video screen.

Medical uses of ultrasound

Ultrasound imaging is often used on pregnant women. It can detect pregnancies as early as 4 weeks after fertilization. Ultrasound measurements of the size of the embryo or fetus accurately indicate its age, while other signs indicate its health. For example, heart motions can usually be seen by 7 weeks. In addition, many fetal abnormalities can be diagnosed using ultrasound.

Advantages of ultrasound

The biggest advantage of ultrasound imaging is its safety.

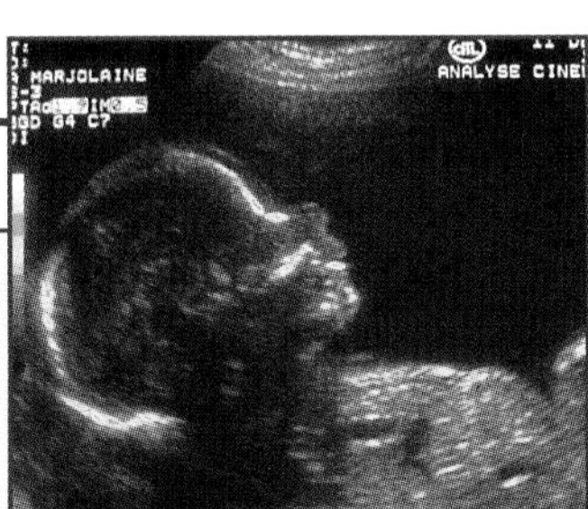

Ultrasound image of fetus

Unlike X rays, ultrasound does not involve ionizing radiation, which can cause mutations. Ultrasound imaging has been used for more than 35 years without any known harmful effects.

internet**connect**

SCILINKS NSTA — **TOPIC:** Ultrasound **GO TO:** www.scilinks.org **KEYWORD:** HX1017

Review SECTION 44-3

1. **Distinguish** between fertilization and cleavage.
2. **Summarize** the events in embryonic development that occur in the first month after fertilization.
3. **Describe** the function of the placenta.
4. **Describe** fetal alcohol syndrome.
5. **SKILL Relating Concepts** Why can certain drugs affect development when they are taken by a woman during pregnancy?

Critical Thinking

6. **SKILL Predicting Outcomes** What might happen if more than one egg were released from the ovaries prior to fertilization?

Review SECTION 44-3 ANSWERS

1. Fertilization is the joining of an egg and sperm to form a zygote. Cleavage is the beginning of cell division in the zygote.
2. In the first month after fertilization, the zygote implants into the uterine wall; the amnion, chorion, and placenta form; and differentiation begins.
3. In the placenta, the blood systems of the mother and fetus come in close association (but do not mix) for the passage of necessary oxygen and nutrients to the fetus and the removal of carbon dioxide and wastes from the fetus.
4. Fetal alcohol syndrome is caused by the consumption of alcohol by a pregnant woman. Her child is born with mental and physical deficits, including an abnormal head and facial features.
5. Many drugs can cross the placenta from mother to fetus.
6. a multiple birth

TECH WATCH: Ultrasound Imaging

Teaching Strategies

Obtain a sonograph on videotape. Use the videotape to stimulate discussion about the activity.

Discussion

- How is an ultrasound image made? *(A probe emits high frequency sound waves that produce echoes when they are reflected by structures in the patient's body. The probe records the echoes and produces an image on a screen.)*
- What is ultrasound used for during pregnancy? *(to detect pregnancy, determine the size and age of a fetus, check for fetal health and abnormalities)*
- Other than pregnancy, can you think of other medical uses for ultrasound? *(Answers will vary. Ultrasound also is used to examine internal organs such as the heart.)*

3 Assessment Options

Closure

Fetal Development

Obtain a copy of *LIFE* magazine's August 1990 cover article titled "How Life Begins" or a similar article that is well illustrated with photographs. Have students read the article, observe the pictures, and relate the information to this chapter.

Review

Assign Review—Section 44-3.

Reteaching — BASIC

Show the class the opening segment of the movie *Look Who's Talking*, and have them describe the sequence of events that occurs from conception to birth. Point out that the effects of the painkiller administered to the mother also affected the baby.

Chapter 44

1 Prepare

Scheduling

- **Block:** About 90 minutes will be needed to complete this section.
- **Traditional:** About two class periods will be needed to complete this section.
- **Planning:** See the Planning Guide on pp. 1003A and 1003B for lecture, classwork, and assignment options for Section 44-4. Lesson plans for Section 44-4 can be found in a lesson cycle format in ***Biology: Principles and Explorations* One-Stop Planner CD-ROM with Test Generator.**

Resource Link

Assign **Active Reading Worksheet Section 44-4** to help students comprehend the key concepts and visuals in this lesson.

2 Teach

Lesson Warm-up

Based on students' previous knowledge, generate a class list of STDs, including their symptoms and treatments. Amend the information as students read this section.

Section 44-4 Sexually Transmitted Diseases

Objectives

- **Identify** the causes and symptoms of several bacterial STDs. (pp. 1018–1019)
- **Identify** the causes and symptoms of some viral STDs. (p. 1020)
- **Compare** the treatment and cure rates of viral STDs with those of bacterial STDs. (pp. 1018–1020)

Key Terms

gonorrhea
syphilis
chlamydia
pelvic inflammatory disease
genital herpes

STDs Are Spread by Sexual Contact

Recall that disease-causing pathogens are transmitted in many ways. Pathogens present in body fluids, such as semen, can be passed from one person to another through sexual contact. Diseases spread by sexual contact are called sexually transmitted diseases, or STDs. Both viruses and bacteria can cause STDs. **Table 44-1** lists several types of STDs.

Bacterial STDs

STDs caused by bacteria are called bacterial STDs. Most bacterial STDs can be successfully treated and cured using antibiotics. Unfortunately, the early symptoms of most bacterial STDs are very mild and often are not detected. Early detection and treatment are necessary to prevent serious consequences that can result from infection. For example, untreated bacterial STDs can cause sterility in both men and women. Three major bacterial STDs are gonorrhea, syphilis, and chlamydia.

Gonorrhea *(gahn uh REE uh)* is a bacterial STD that causes painful urination and a discharge of pus from the penis in males. In females, gonorrhea sometimes causes a vaginal discharge but more

Table 44-1 Sexually Transmitted Diseases

Disease	Symptoms	Pathogen
AIDS	Immune-system failure; susceptibility to opportunistic infections	Human immunodeficiency virus (HIV)
Chlamydia	Painful urination and penile discharge in males; vaginal discharge and abdominal pain in females	*Chlamydia trachomatis* (bacterium)
Genital herpes	Painful blisters on genital region, thighs, or buttocks; flulike symptoms	Herpes simplex virus (HSV)
Genital warts	Warts on genital or anal region	Human papilloma virus (HPV)
Gonorrhea	Painful urination and penile discharge in males; vaginal discharge and abdominal pain in females	*Neisseria gonorrhoeae* (bacterium)
Hepatitis B	Flulike symptoms; yellowing of skin	Hepatitis B virus
Syphilis	Chancre on penis in males; chancre in vagina or on cervix in females; fever; rash	*Treponema pallidum* (bacterium)

Integrating Different Learning Styles

Logical/Mathematical	**PE**	Review item 5, p. 1020
Interpersonal	**ATE**	Active Reading, p. 1020
Verbal/Linguistic	**ATE**	Lesson Warm-up, p. 1018; Making Biology Relevant, p. 1019; Closure, p. 1020

often has no symptoms. In males, untreated gonorrhea can spread to the vas deferens, epididymis, or testes. In females, it can spread to the fallopian tubes and cause scarring that may lead to infertility. Some strains of gonorrhea are resistant to commonly used antibiotics, such as penicillin.

Syphilis *(SIHF uh lihs)* is a serious bacterial STD that usually begins with the appearance of a small, painless ulcer called a chancre *(SHAHN kuhr)* 2–3 weeks after infection. In males, the chancre usually appears on the penis. In females, the chancre may form inside the vagina or on the cervix. If syphilis is not treated, it may cause fever, swollen lymph glands, or a rash like the one shown in **Figure 44-15** a few weeks after infection. These symptoms disappear without treatment. Years later, however, syphilis may cause destructive lesions on the nervous system, blood vessels, bones, and skin. A pregnant woman infected with syphilis can also transmit the disease to the fetus. As a result, the fetus may be stillborn or suffer serious damage to organ systems.

Figure 44-15 **Syphilis.** A rash such as this one is a symptom of the second stage of syphilis. Even at this stage, syphilis can be cured by treatment with antibiotics.

Chlamydia *(kluh MIHD ee ah)* is the most common bacterial STD in the United States. The symptoms of chlamydia are similar to those of a mild case of gonorrhea: painful urination in males and vaginal discharge in females. Like gonorrhea, chlamydia often is not detected. Chlamydia, even more than gonorrhea, is likely to cause scar tissue in infected fallopian tubes, leading to infertility, or the inability to become pregnant.

Pelvic inflammatory disease One of the most common causes of infertility in women is **pelvic inflammatory disease,** or **PID.** PID is a severe inflammation of the uterus, ovaries, fallopian tubes, or abdominal cavity that results from a bacterial STD that has gone untreated. **Figure 44-16** shows the damage that PID can cause in the fallopian tubes. Most cases of PID are the result of gonorrhea or chlamydia infections.

Figure 44-16 **Pelvic inflammatory disease**

Most cases of PID result from gonorrhea or chlamydia infections.

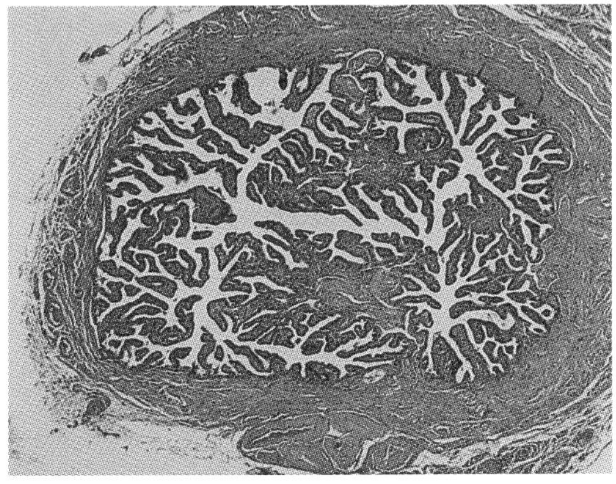

A normal fallopian tube has a highly folded lining and many spaces through which gametes can pass.

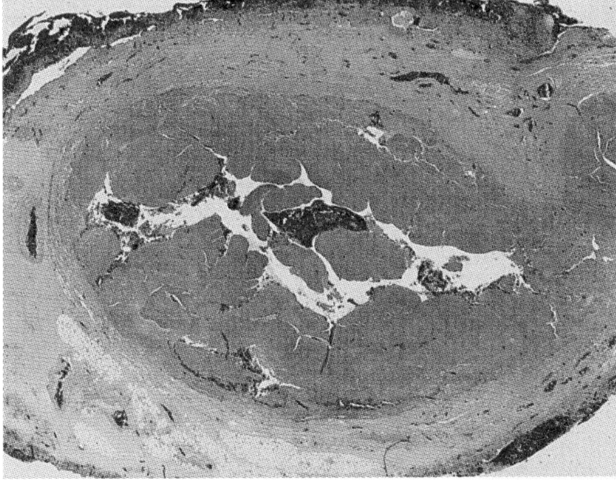

In a fallopian tube scarred by PID, many of these spaces have become blocked with tissue.

Multicultural Perspective

Dr. William Hinton

Dr. William Hinton, a pathologist and renowned authority on sexually transmitted diseases, developed the Davies-Hinton Test for the detection of syphilis. Hinton was the first African-American professor to teach at Harvard Medical School.

Teaching TIP

STDs and Risk Assessment

Venereal diseases, or STDs, are infections transferred from one person to another during sexual contact. Pathogens that cause STDs thrive in warm, moist environments. In most cases, if they are exposed to conditions outside the body for longer than several minutes, the pathogen will die or be destroyed. The chances of contracting a sexually transmitted disease increase with the number of sexual partners. Studies indicate that if a person contracts one STD, he or she is probably at risk for contracting others.

Making Biology Relevant

Health

Remind students that there are at least five forms of viral hepatitis, or inflammation of the liver. Hepatitis A generally is spread by fecal-oral contact. Hepatitis B is spread through needle punctures, tattooing, blood transfusions, and sexual contact. A vaccine for hepatitis B is available. Hepatitis C is similar to hepatitis B in transmission and can lead to chronic active hepatitis. These diseases can cause liver failure and death.

Teaching TIP

Herpes

The human herpes simplex virus (HSV) belongs to a group of viruses that includes the Epstein-Barr virus and others. These viruses are all extremely difficult to control. HSV-2, the most common cause of genital herpes, is transmitted via infectious secretions. Symptoms of genital herpes include painful, itching sores in or around the genitals. These sores usually appear between 2 and 20 days after exposure to the virus. They may last as long as 3 weeks. Other symptoms include a burning sensation during urination and a fever.

Teaching TIP

Working with AIDS Patients

Ask students how people who work with AIDS patients protect themselves from becoming infected with HIV. *(Health workers wear gloves and protective clothing and eye wear if necessary, and take precautions against contact with blood, saliva, and other body fluids.)*

3 Assessment Options

Closure

Technique: K-W-L

Tell students to return to the lists of things they **W**ant to know that they created using the **Active Reading** on page 1005. Have them place check marks next to questions that they are now able to answer. Students should finish by making a list of what they have **L**earned. Conclude by asking students the following questions:

- Which questions are still unanswered?
- What new questions do you have?

Review

Assign Review—Section 44-4.

Reteaching — BASIC

Have students develop a 1-minute public service announcement on STDs. Have the announcements judged by the school nurse, the health teacher, and a local physician, and see if a local radio station will broadcast the winning entry. PORTFOLIO

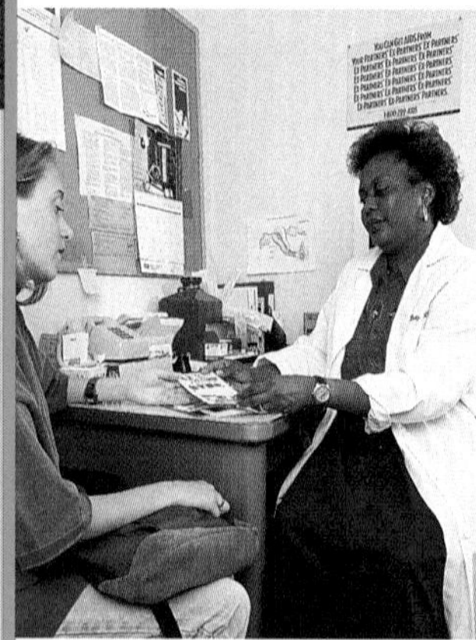

Figure 44-17 Counseling. Counseling is available to people with HIV.

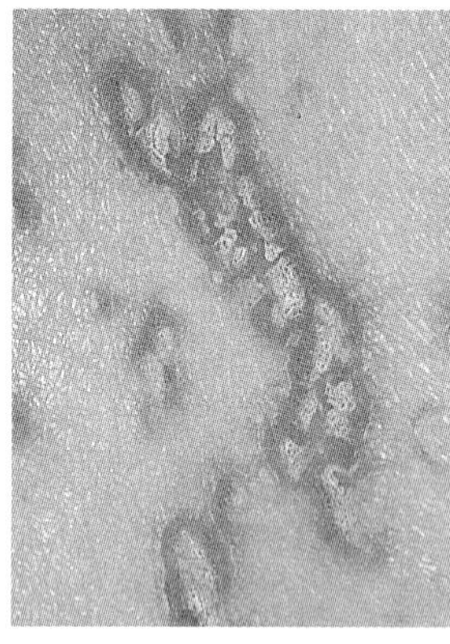

Figure 44-18 Genital herpes. Blisters caused by genital herpes may appear on or near the genitalia.

Viral STDs

STDs caused by viruses are called viral STDs. Because viruses are not affected by antibiotics, viral STDs cannot be treated and cured using antibiotics. AIDS and genital herpes are two common viral diseases that are transmitted through sexual contact. Other common viral STDs include genital warts and hepatitis B.

AIDS is a fatal disease caused by the human immunodeficiency virus (HIV). As you learned in Chapter 41, transmission through sexual contact is the most common way that people become exposed to HIV. HIV destroys the immune system of infected individuals by attacking white blood cells. People with AIDS generally die from opportunistic infections that persist only in people with weakened immune systems.

The number of HIV infections among teenagers and young adults has increased dramatically over the last decade. AIDS is now the leading killer of American men between the ages of 25 and 44. More than 385,000 people in the United States have already died from AIDS. The number of young adults in the United States with AIDS has increased drastically over the last 15 years. While the number of new AIDS cases reported has decreased each year since 1993 due to improved drug treatments, new HIV infections have not decreased and are most frequent among young adults. Researchers are trying to develop new treatments for AIDS. Counseling is available to people infected with HIV, as shown in **Figure 44-17.**

Genital herpes is caused by herpes simplex virus (HSV). Two types of HSV can cause genital herpes. About 80 percent of genital herpes infections are caused by HSV-2. The rest are caused by HSV-1, which more commonly causes cold sores, or fever blisters, around and inside the mouth.

Symptoms of genital herpes include periodic outbreaks of painful blisters in the genital region, as shown in **Figure 44-18,** and flulike aches and fever. Antiviral drugs can temporarily eliminate the blisters caused by genital herpes, but they cannot eliminate HSV from the body. Although genital herpes is not life threatening, it can have serious consequences. Women with genital herpes have a greater risk of developing cervical cancer. Like HIV, herpes simplex virus can be passed from mother to fetus during pregnancy or birth. Infants infected with HSV may suffer severe damage to their nervous system or even die as a result of the infection.

Review SECTION 44-4

1. **Name** three common STDs caused by bacteria. Why is early detection of these diseases important?
2. **Describe** how HIV weakens the immune system of an infected individual.
3. **List** three symptoms of genital herpes.
4. **SKILL Recognizing Differences** What is the main difference between the treatment of viral STDs and the treatment of bacterial STDs?
5. **SKILL Applying Information** How can you best protect yourself from contracting a sexually transmitted disease?

Review SECTION 44-4 ANSWERS

1. Gonorrhea, syphilis, and chlamydia are bacterial STDs. Left untreated, they can cause sterility and other serious, permanent damage.
2. HIV attacks white blood cells, eventually destroying the immune system.
3. Symptoms of genital herpes include painful genital blisters, flulike aches, and fever.
4. Bacterial STDs can be treated and usually cured with antibiotics. Drugs for viral STDs can provide relief from symptoms but cannot cure the disease.
5. Answers will vary. Students may suggest abstinence, use of condoms, and monogamy with another monogamous person.

Study ZONE

CHAPTER 44

Use the Key Concepts and Key Terms listed below to review the main ideas in this chapter. If you need to review the meaning of a term, turn to the page indicated.

Key Concepts

44-1 Male Reproductive System

- Sperm cells are produced by meiosis in the testes.
- Sperm mature and are stored in each epididymis.
- A mature sperm cell consists of a head, a midpiece, and a long, powerful tail.
- Sperm move through the epididymis and the vas deferens and exit the body through the urethra.

44-2 Female Reproductive System

- Egg cells are produced by meiosis in the ovaries.
- An egg cell matures in a 28-day ovarian cycle.
- The menstrual cycle prepares the uterus for pregnancy.
- The menstrual and ovarian cycles are regulated by hormones.

44-3 Development

- After fertilization, cleavage and implantation occur.
- The human gestation period is about 9 months.
- The most crucial events of development occur during the first trimester of pregnancy.
- The mother nourishes the fetus through the placenta.
- Primary tissue layers develop into organs and tissues.
- The fetus leaves the mother's body during labor.

44-4 Sexually Transmitted Diseases

- Many STDs are caused by bacteria and viruses.
- Syphilis is a severe bacterial STD that can have destructive effects on the nervous system, bones, and skin if untreated.
- Gonorrhea and chlamydia are common bacterial STDs that can scar the fallopian tubes and lead to infertility.
- AIDS is a viral STD in which HIV destroys immune-system cells, leaving the body vulnerable to opportunistic infections.
- Genital herpes is a viral STD that causes blistering.

Key Terms

44-1

testes (1006)
seminiferous tubules (1006)
epididymis (1007)
vas deferens (1007)
seminal vesicles (1008)
prostate gland (1008)
bulbourethral glands (1008)
semen (1008)
penis (1008)

44-2

ovary (1009)
ovum (1010)
fallopian tube (1010)
uterus (1010)
vagina (1010)
ovarian cycle (1011)
ovulation (1011)
follicle (1011)
corpus luteum (1012)
menstrual cycle (1012)
menstruation (1013)

44-3

cleavage (1014)
blastocyst (1014)
implantation (1014)
gestation (1015)
pregnancy (1015)
embryo (1015)
placenta (1015)
fetus (1016)

44-4

gonorrhea (1018)
syphilis (1019)
chlamydia (1019)
pelvic inflammatory disease (1019)
genital herpes (1020)

Study TIP Ready?

- *If you think you understand these Key Concepts and Key Terms, then you're ready for the Performance Zone.*

Study ZONE

CHAPTER 44

Review and Practice

Assign the following worksheets for Chapter 44:

- **Student Review Guide**
- **Concept Mapping Worksheet**
- **Test Preparation Pretest**
- **Vocabulary Worksheet**
- **Critical Thinking Review Worksheet**

Assessment Options

The following assessment options are available for this chapter:

Resource Link

- Assign **Chapter Test 44** to assess students' understanding of the chapter.

Alternative Assessment

Have students work in cooperative groups. Each group should develop a question to send to a newspaper or magazine health-advice columnist dealing with any topic covered in this chapter. Collect the questions and redistribute them to other groups. Each group should then write an appropriate response to the question. PORTFOLIO

Portfolio Assessment

- *Portfolio Assessment* at the front of this book provides options for building and evaluating student portfolios. Students and teachers can select items from areas listed for inclusion in a portfolio.
- See items labeled PORTFOLIO on pp. 1007, 1014, 1016, 1020, and 1021.

Answer to Concept Map

The following is one possible answer to Performance Zone item 1 on page 1022.

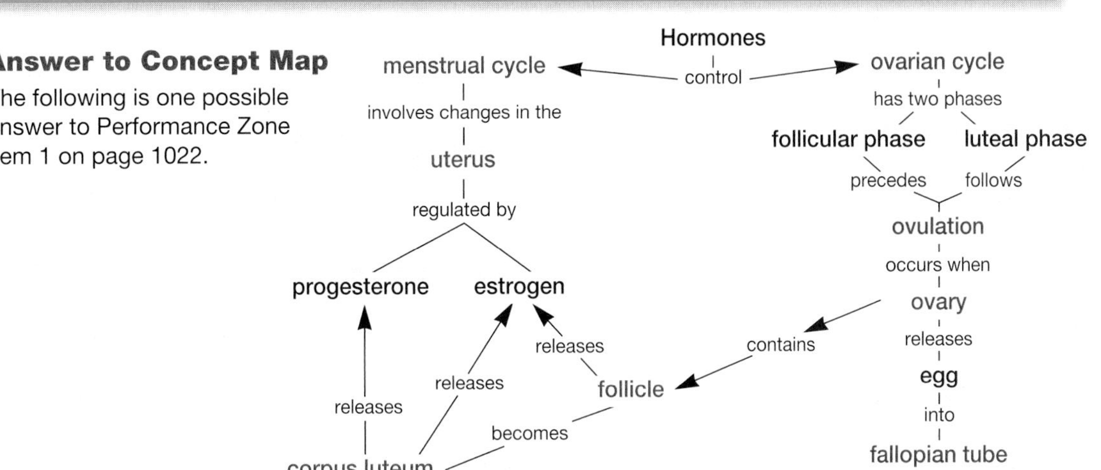

Performance ZONE
CHAPTER 44 REVIEW ANSWERS

Assignment Guide

Section	Review
44-1	2, 3, 12, 21, 22, 23, 26
44-2	1, 4, 5, 6, 7, 13, 16, 17, 18, 19
44-3	8, 9, 14, 15, 19, 20, 24, 25, 27
44-4	10, 11

Understanding and Applying Concepts

1. The answer to the concept map is found at the bottom of page 1021.
2. **a.** The ovary is the female sex organ that produces gametes. The testis is the male sex organ that produces gametes.
   b. The uterus is the hollow organ that protects and nourishes the fetus. The placenta connects the mother to the fetus.
   c. The epididymis provides storage for maturing sperm. The seminiferous tubules are the site of sperm production.
   d. Semen is composed of fluids and sperm. Sperm are gametes.
3. c
4. d
5. a
6. d
7. d
8. d
9. c
10. a
11. d
12. The head, *A*, contains genetic information and enzymes that penetrate the egg. The midpiece, *B*, contains mitochondria, which supply energy. The tail, *C*, is composed of a flagellum, which allows the sperm cell to move.
13. A small increase in estrogen has a negative feedback effect, reducing the continued release of FSH and LH. Later in the cycle, increasing estrogen causes a positive feedback effect, stimulating LH release.
14. Ultrasound can be used to detect pregnancy, estimate fetal age, and detect abnormalities in the developing baby. It is also used to examine internal organs.
15. Prolactin stimulates the secretion of milk from the mammary glands.

Performance ZONE CHAPTER 44 Review

Understanding and Applying Concepts

1. **Concept Mapping** Make a concept map that describes the ovarian and menstrual cycles. Include the following terms in your map: ovary, fallopian tube, uterus, ovarian cycle, follicle, ovulation, corpus luteum, menstrual cycle. Include additional terms in your map as needed.

2. **Understanding Vocabulary** For each pair of terms, explain the difference in their meanings.
   a. ovary, testis
   b. uterus, placenta
   c. epididymis, seminiferous tubules
   d. semen, sperm

3. The correct pathway of sperm is from the
   a. testes to vas deferens to epididymis.
   b. epididymis to urethra to vas deferens.
   c. testes to epididymis to vas deferens.
   d. urethra to vas deferens to testes.

4. Which of the following is not a function of the female reproductive system?
   a. production of gametes
   b. nourishment of the fetus
   c. maturation of eggs
   d. secretion of FSH

5. Fertilization normally takes place in the
   a. fallopian tubes.
   b. cervix.
   c. epididymis.
   d. vas deferens.

6. The maturation of an egg cell during the ovarian cycle is regulated by the
   a. hypothalamus.
   b. corpus luteum.
   c. fallopian tubes.
   d. follicles.

7. The follicular phase of the ovarian cycle
   a. occurs when LH levels drop to zero.
   b. starts when fertilization occurs.
   c. stops estrogen production.
   d. ends when ovulation occurs.

8. An embryo develops endoderm, mesoderm, and ectoderm during
   a. cleavage.
   b. implantation.
   c. fertilization.
   d. None of the above

9. Which of the following is *not* true for human development?
   a. Alcohol and drugs taken during pregnancy may harm the embryo or fetus.
   b. Crucial events occur during the first trimester of pregnancy.
   c. Drugs and alcohol taken during pregnancy cannot cause birth defects.
   d. Normal development may be affected by viral diseases.

10. Which of the following sexually transmitted diseases cannot be treated with antibiotics?
    a. genital herpes
    b. syphilis
    c. gonorrhea
    d. chlamydia

11. A symptom associated with the earliest stage of syphilis is
    a. painful urination.
    b. blisters in the genital area.
    c. fever blisters and cold sores.
    d. a painless chancre.

12. The diagram below shows a mature human sperm cell. Explain the roles of the structures labeled *A, B,* and *C* in the sperm's ability to fertilize an egg.

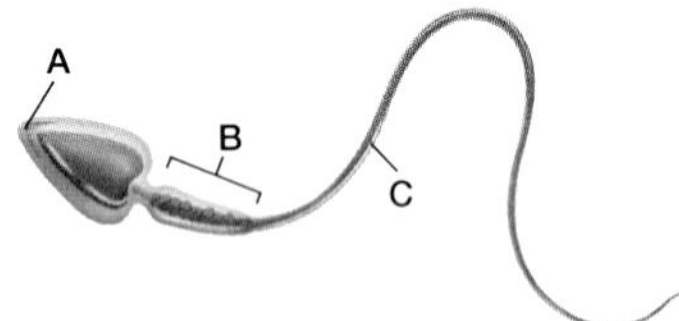

13. **Theme Homeostasis** Describe the role of feedback mechanisms in the maturation of an egg cell during the ovarian cycle.

14. **Tech Watch** Describe two different medical uses for ultrasound imaging. What can be learned about a fetus using this technology? Find out other ways ultrasound imaging is used in medicine.

15. **Chapter Links** Describe the function of prolactin, a hormone secreted by the pituitary gland during pregnancy. (**Hint:** See Chapter 43, Section 43-3.)

Skills Assessment

16. the follicular phase
17. the luteal phase
18. at day 14—the point where follicular phase changes to luteal phase
19. During an in vitro fertilization procedure, an egg is taken from the female and sperm from the male, and fertilization occurs externally (in vitro). The zygote is released into the woman's uterus 2–6 days later, so that implantation can proceed. The first IVF baby was born in England in July 1978.
20. Amniocentesis is performed using a hollow needle inserted into the amniotic sac through the woman's abdominal wall. Ultrasound is used to guide the needle into the correct position. A small volume of amniotic fluid is extracted and tested. Analysis of the fluid can detect chromosomal abnormalities such as Down syndrome. A check of alpha-fetoprotein (AFP) is used to detect neural tube defects such as anencephaly or spina bifida.

Skills Assessment

Part 1: Interpreting Graphics

Use the graph of the ovarian cycle below to answer the following questions:

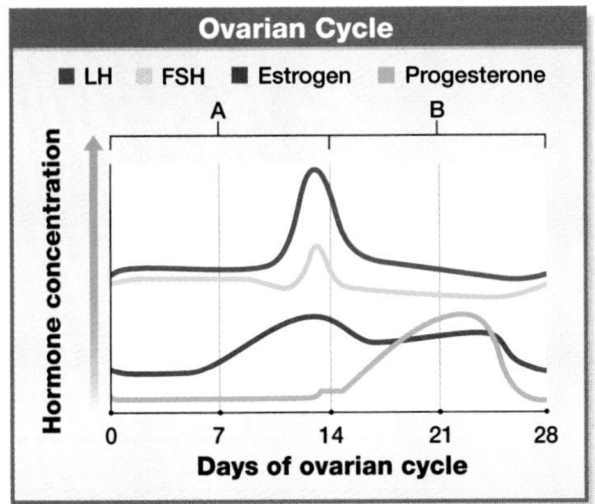

16. What phase of the ovarian cycle is occurring in the period labeled *A*?
17. What phase of the ovarian cycle is occurring in the period labeled *B*?
18. When during the ovarian cycle does ovulation usually occur?

Part 2: Life/Work Skills

Use the media center or Internet resources to find out more about various kinds of medical technology used before and during pregnancy.

19. **Finding Information** Research the process of in vitro fertilization. Find out how in vitro fertilization is performed. Who was the first person born as a result of this process? Summarize your findings in a written report.
20. **Summarizing Information** Research the procedure of amniocentesis. Find out how doctors use this procedure to study fetuses. Summarize your findings in an oral report to your class.
21. **Interpreting Information** Research some causes of and treatments for infertility. What are the benefits and shortcomings of each type of treatment? Summarize your findings in a written report.

Critical Thinking

22. **Making Inferences** A man interested in fathering children wants to know his sperm count. He finds out that he has a sperm count of fewer than 60 million sperm in a 3.5 mL sample of semen. If you were his physician, what would you tell him about the results of the test?
23. **Predicting Outcomes** What do you think would happen if more than one sperm were able to fertilize an egg?
24. **Inferring Relationships** Why should a pregnant woman eat many healthy foods?
25. **Evaluating Conclusions** In the 1960s, many women who took a tranquilizer called thalidomide early in pregnancy gave birth to babies with serious limb defects. Other women who took the drug later in pregnancy gave birth to normal babies. What does this tell you about the pattern of development?

Portfolio Projects

26. **Finding Information** Concerned about the rapid increase in the world's population, several countries are attempting to reach zero population growth (ZPG). These countries have encouraged women to have fewer children, or have instituted laws that limit the number of children that each woman may have. Check Internet resources to identify one such country and write a report summarizing what measures the country has taken to attain ZPG.
27. **Interpreting Information** Read the article titled "Boy? Girl? Up to You" (*Time*, September 21, 1998, pp. 82–83). How do the scientists mentioned in the article distinguish sperm that carry a Y chromosome from those that carry an X chromosome?
28. **Career Focus Obstetrician** Use the media center or Internet resources to find information on the career of an obstetrician. Your report should include a job description, the training and education required, and an average starting salary.

Portfolio Projects

26. Answers will vary. China has attempted to attain zero population growth through advocating late marriage (at age 20 for females, 22 for males) and late childbirth (24-year-old mothers). Contraceptive use is also encouraged. A one-child policy was established in 1979. There are financial benefits for those with only one child.
27. Sperm that carry Y chromosomes have slightly less DNA, so they take up less dye and shine less brightly.
28. An obstetrician is a medical doctor who cares for women during pregnancy and childbirth. Obstetricians attend college, followed by medical school, and they complete specialty training and a residency in obstetrics; they must also complete licensing requirements of the state in which they practice. Obstetricians are employed by hospitals, private practices, and clinics. The growth potential of this field is good. Starting salary will vary by region.

21. Many factors cause infertility, including hormone imbalance, decreased sperm count, ovarian or fallopian-tube scarring from STDs, ovulation dysfunction, and chronic disease. Some treatments are in vitro fertilization (which increases the chances of conceiving but is very expensive) and fertility drugs (which increase chances of conceiving but also increase chances for multiple births).

Critical Thinking

22. His low sperm count means that he has a very reduced chance of producing a pregnancy.
23. The cell would not survive.
24. The fetus must have essential nutrients to develop properly.
25. Critical periods of limb formation occur early in pregnancy.

Interactive Exploration
Chapter 44

Meiosis: Down Syndrome

Purpose

Students will use a computer simulation to determine whether parental age affects the probability of having a child with Down syndrome.

Preparation Notes

Time Required: Approximately 1 hour

Preparation Tips

If enough computers are available, students can work individually. Otherwise, they should work in teams of two to four.

Procedural Tips

- Remind students that the animation of normal and abnormal meiosis is greatly simplified to facilitate understanding. Only two pairs of chromosomes are shown.
- Students may find the computer's graph of parental age versus incidence of Down syndrome confusing. Tell students that the program plots two points for each combination of parents. Blue represents the incidence of Down syndrome for the father's age, and red represents the incidence for the mother's age. Inform students that the meter labeled "Incidence of Down Syndrome per 1,000 Live Births" that is located at the top left of the screen provides one value for the incidence of Down syndrome given the ages of both parents. Students should use this value when recording data.

Additional Background

Down syndrome is a disorder caused by the presence of an extra copy of chromosome number 21. The disorder is named for J. Langdon Down, the English doctor who described its characteristic features in 1866. Down syndrome involves a variety of abnormalities, including reduced growth, a heavy eyefold, short hands, intestinal deformities, and heart defects. Although the severity of impairment varies, most individuals with Down syndrome also have some degree of mental retardation. The life expectancy of children with Down syndrome is about 17 years.

In most cases, Down syndrome results from nondisjunction, the failure of chromosomes to separate during meiosis. The incidence of Down syndrome rises dramatically with the age of the mother. This age effect is thought to be due to the increasing amount of time the egg has remained in "suspended animation" within the mother's body. All of the eggs a woman will ever produce form in her ovaries before birth. These eggs begin meiosis, but they stop at prophase I and enter a state of stasis that lasts until puberty, after which several eggs resume meiosis each month. The older a woman is, therefore, the longer her eggs have remained in the quiescent state, and the more time they have had to accumulate genetic damage. The age of the father seems to have little effect on the incidence of Down syndrome, probably because sperm production does not begin until puberty and continues throughout the life of the male.

Interactive Exploration

Chapter 44 *Meiosis: Down Syndrome*

CD-ROM Lab

SKILL

- Computer modeling

OBJECTIVES

- **Compare** normal meiotic cell division with meiotic cell division that results in Down syndrome.
- **Relate** the age of two parents to the likelihood of having a child with Down syndrome.

MATERIALS

- computer with CD-ROM drive
- CD-ROM *Interactive Explorations in Biology: Cell Biology and Genetics*
- graph paper

Magnification: 14,040×

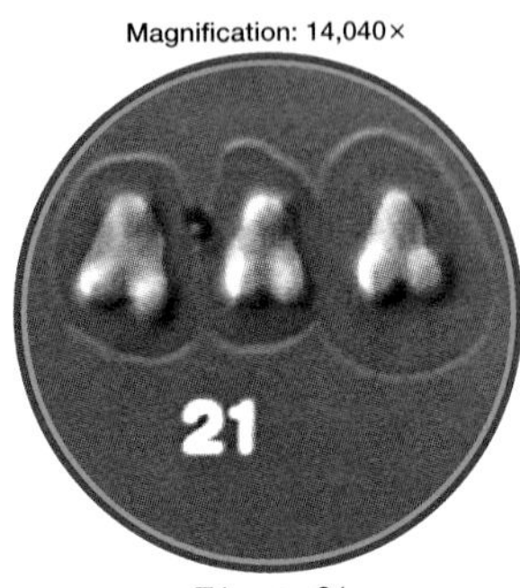

Trisomy 21

Before You Begin

About one in 1,000 children is born with Down syndrome. Down syndrome is caused by nondisjunction, the failure of chromosomes to separate normally during meiosis. Characteristics of Down syndrome include mental retardation, a short stature, stubby hands and feet, and a heavy eyefold. In this exploration, you will examine the events of meiotic cell division and discover how the age of each parent affects the incidence of Down syndrome.

1. Select *Meiosis: Down Syndrome* from the menu. You will see a diagram like the one below. Click the word *Navigation,* and then click the key on the navigation palette. Read the focus questions and review the following concepts: ploidy, aneuploidy, meiosis, and nondisjunction.

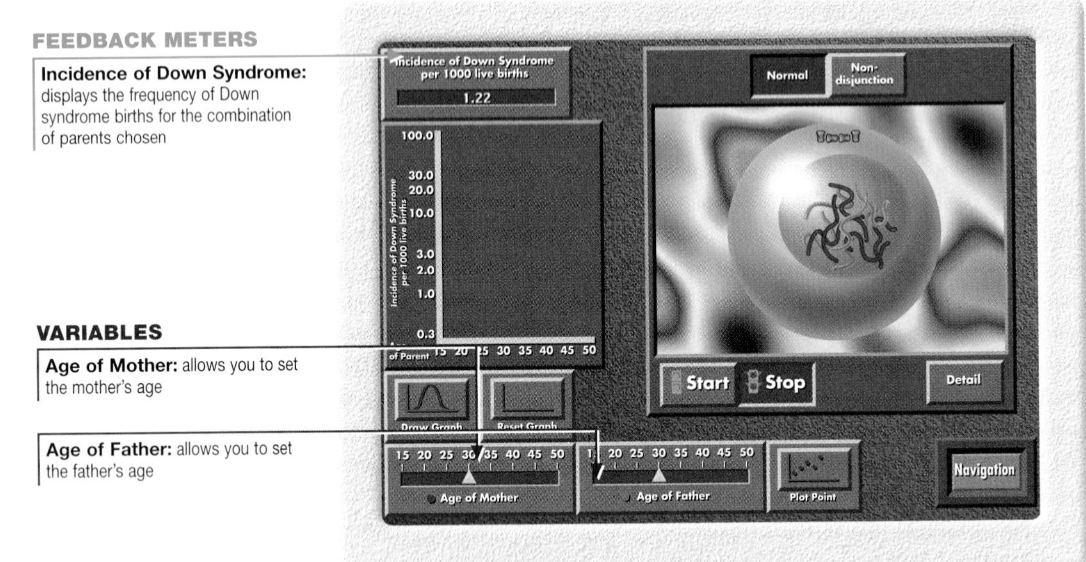

2. Click *Help* at the top left of the screen, and select *How To Use This Exploration.* Listen to the instructions. Click the magnifying glass on the navigation palette to begin the exploration.
3. Make a data table like the one below.

DATA TABLE

Father's age	Incidence of Down syndrome	Mother's age	Incidence of Down syndrome

Procedure

PART A: Normal Meiosis

1. Click the *Normal* button to see a simulation that shows what happens during normal meiotic cell division.
2. Click the *Detail* button. Then click the right-facing arrow to advance through the steps of normal meiotic cell division. As you view each step, read the caption and study the illustration.
3. Click the *Close* button and then the *Start* button to see this simulation in motion.

PART B: Nondisjunction

4. Click the *Nondisjunction* button to see a simulation that shows what happens when nondisjunction occurs during meiotic cell division.
5. Click the *Detail* button. Then click the right-facing arrow to advance through the steps of meiotic cell division that involves nondisjunction. As you view each step, read the caption and study the illustration. What is unusual about the gametes shown in the upper-right and lower-left corners of Frame 9?
6. Click the *Close* button and then the *Start* button to see this simulation in motion.

PART C: Effect of the Father's Age

7. Click and slide the left-hand indicator so that the age of the mother is 25.
8. Click and slide the other indicator so that the age of the father is 25.
9. Click the *Plot Point* button. Record in your data table the age of the father and the incidence of Down syndrome per 1,000 live births for these parents, shown by the meter in the upper-left corner of the screen.
10. Repeat steps 8–9 five times, increasing the age of the father by 5 years each time. Make a graph of the age of the father versus the incidence of Down syndrome.

PART D: Effect of the Mother's Age

11. Click the *Reset Graph* button.
12. Click and slide the left-hand indicator so that the age of the father is 25.
13. Click and slide the other indicator so that the age of the mother is 25.
14. Click the *Plot Point* button. Record in your data table the age of the mother and the incidence of Down syndrome per 1,000 live births for these parents.
15. Repeat steps 13–14 five times, increasing the age of the mother by 5 years each time. Make a graph of the mother's age versus the incidence of Down syndrome.

Analyze and Conclude

1. **Analyzing Methods** In step 10, why did you hold the mother's age constant while varying the father's age?
2. **Drawing Conclusions** Which has a greater effect on the likelihood of having a child with Down syndrome, the father's age or the mother's age?
3. **Making Predictions** In the United States, the age at which women have their first child has been increasing. How might this affect the incidence of Down syndrome?

10. Apparently the father's age has no effect on the incidence of Down syndrome because the incidence does not vary with the father's age. Students' graphs should reflect the data shown in the Sample Data Table below.
15. The likelihood of having a child with Down syndrome rises with the mother's age. Students' graphs should reflect the data shown in the table below.

Analyze and Conclude

Answers

1. The mother's age was held constant so that the influence of the father's age could be evaluated.
2. The mother's age has the greater effect.
3. This may lead to an increase in the number of children with Down syndrome.

Procedure

Answers

5. One pair of chromatids fails to separate during the first meiotic division. The gamete in the upper-right corner has an extra chromosome. The gamete in the lower-left corner is missing a chromosome.

Sample Data Table

Father's age	Incidence of Down syndrome	Mother's age	Incidence of Down syndrome
25	0.70	25	0.70
30	0.70	30	1.22
35	0.70	35	3.38
40	0.70	40	12.3
45	0.70	45	56.0
50	0.70	50	145.0

Reading and Study Skills

Becoming an Active Reader

What Is an Active Reader?

Active reading means thinking and interacting with the text before, during, and after reading it. You might be wondering, "Isn't everyone *thinking* while they are reading?" Not necessarily. Many students read without consciously directing their thinking. Active readers are different—they have developed habits of thinking to use *before, during,* and *after* they read. They use these habits to successfully reach their goal: to understand what they're reading about and remember it well enough to be able to pass the test.

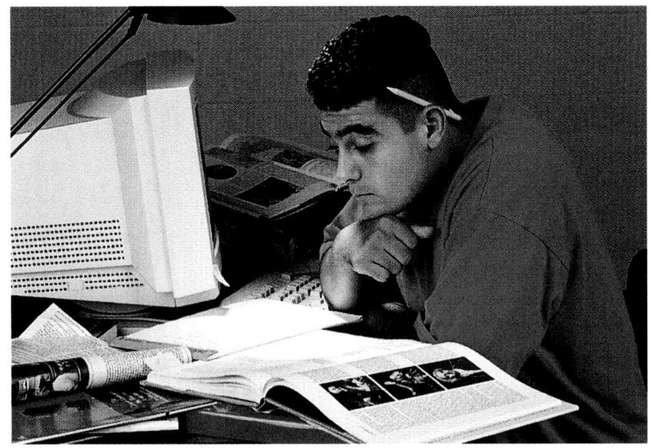

Before You Read

Active readers begin by getting familiar with the reading assignment. This is called previewing. The word *preview* means "to look before." Looking over the chapter before you read helps you predict what you will be learning. It can also help you recall what you already know about a subject—this can help you learn more material and remember it longer.

Scan the horizon

An active reader will turn to the first page of a chapter, ask questions, and use the information on the page to find answers. Here are some questions to ask.

1. **What's here?** If you're in an unfamiliar city, you might use a road map to find your way around. In this textbook, the section titled Looking Ahead is your map, preparing you for what's coming up. Take time to really look at the chapter title so you'll remember it. The title tells you the focus and the purpose of the chapter. Some readers skip over the title, not realizing its importance.
2. **What do I know?** Questions at the beginning of each chapter in **Study TIP Ready?** prompt you to remember vital information from previous chapters.
3. **What do I *need* to know?** Ask yourself questions, such as, "What do I expect to learn from the section I am about to read?"

Get to know the neighborhood

Now you're ready to begin the first section. Start by recognizing the elements that are there to help you:

- **Objectives** tell you what you should be able to do after reading the section. Each Objective includes the page number where the information is covered. Read these before you read the section.
- **Key Terms** are the lesson's key vocabulary words listed in order of appearance. They appear in **bold** black print within the lesson.

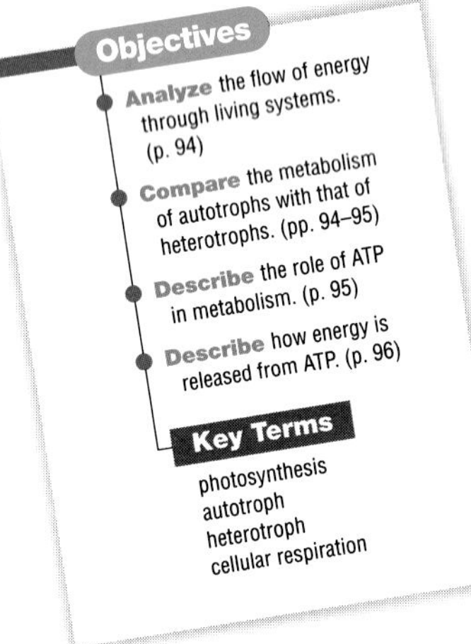

Find your way

As you look over the chapter, you will notice text headings in **bold** red print and text subheadings in **bold** blue print. Each red heading is a complete sentence and is a main idea for the lesson. Red headings are always located at the top of a page, making it easy for you to scan the chapter and find the main ideas. Blue subheadings are words or phrases. These subheadings identify important topics that are covered under each red heading. As you read, notice that bold blue print is used for references to nearby photos and illustrations (ex. **Figure 8-9**). Bold green print identifies tables (ex. **Table 8-1**).

As You Read

As they read, active readers remain self-aware. They ask themselves questions about how well they understand what they are reading. As you read the lesson, try asking yourself questions.

Monitor your understanding.
- What did I just learn from that paragraph?
- Do I need to read this paragraph again?
- What can I say about that illustration?
- Can I respond to the Objectives now?

Express what you know.
- What can I say about what I just learned?

Ask more questions.
- What is it that I want to know now?
- How does this affect my life?

Study Tips along the way will give you tips for improvement. Look for the following types of *Study Tips* as you read.

Reading Effectively strategies for finding meaning, understanding content, and identifying main ideas

Organizing Information tips on how to organize what you're learning

Comparing and Contrasting recognizing similarities and differences

Reviewing Information how to review—and remember—what you're reading

Word Origins tips for using word origins to learn important terms

Interpreting Graphics suggestions for understanding graphs and illustrations

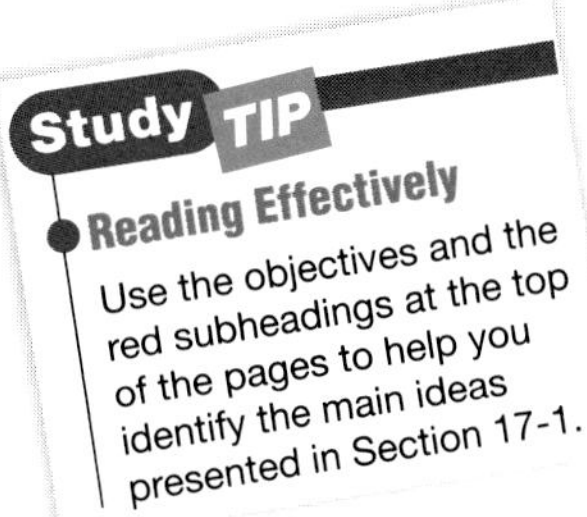

After You Read

When finished reading, active readers take a few moments to assess their overall level of understanding. Ask yourself:

- What can I say about each illustration or figure?
- Did I learn what I expected to learn?
- What do I know about each lesson title, Objective, heading, subheading, figure, or Key Term in the text?
- What can I tell others about what I know?

Check your comprehension

At the end of each section, *Review* questions will let you see how much you have learned. Each *Review* question measures how well you mastered an Objective at the beginning of the section. If you cannot answer a *Review* question, go back to the page indicated in the Objective and re-read the information on that page.

Are you ready for the test?

When you have finished reading the chapter, turn to the Study Zone page at the end of the chapter to review the main ideas (Key Concepts) and the Key Terms. If you've forgotten the meaning of a term, turn to the page number indicated. If you understand the Key Concepts and Key Terms, then you may answer the questions in the Performance Zone to see how well you've learned the material.

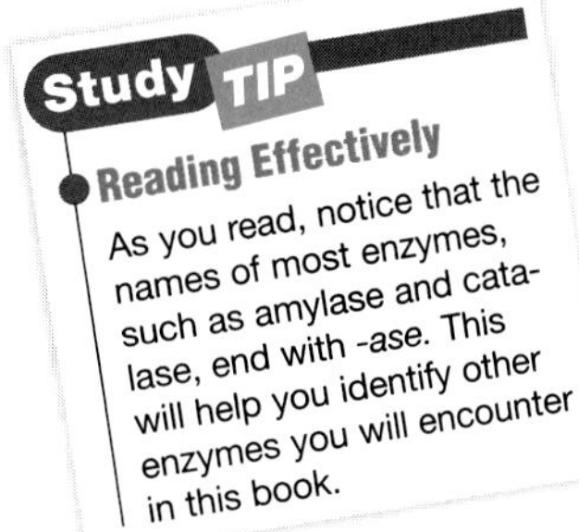

Study TIP
Reading Effectively
As you read, notice that the names of most enzymes, such as amylase and catalase, end with -ase. This will help you identify other enzymes you will encounter in this book.

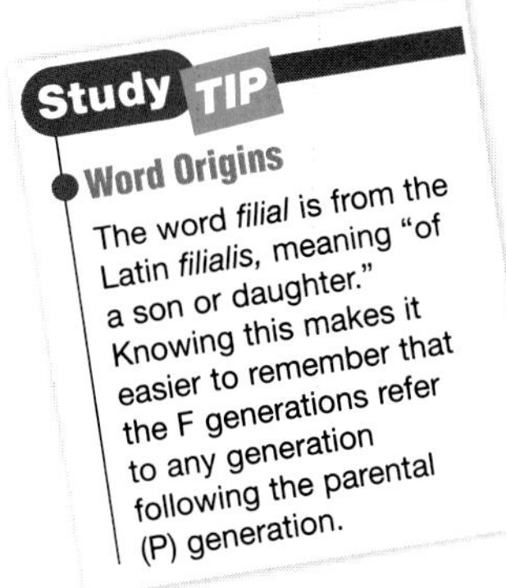

Study TIP
Word Origins
The word filial is from the Latin filialis, meaning "of a son or daughter." Knowing this makes it easier to remember that the F generations refer to any generation following the parental (P) generation.

Reading and Study Skills Concept Mapping

What's the Best Way to Study?

Although there are many ways to study, concept mapping may help you understand ideas by showing you their connections to other ideas. A concept map not only identifies the major concepts from a chapter or your class notes but also shows the relationships between the concepts, much as a road map illustrates how highways and other roads are linked to cities.

Identifying concepts

Suppose you have just finished reading a section of a chapter and you would like to make a concept map as a study aid. How do you begin? First you need to identify the concepts in that section. For instance, examine the following words:

- pool
- grass
- tree
- water
- sky
- playing
- biking
- raining
- thinking

All of these words are concepts, which usually form a picture in your mind. Now read the following series of words:

- the
- to
- has
- when
- was
- be
- with
- can

Are they concepts? No. They do not form a picture in your mind. They are linking words. Linking words play an important role in concept mapping; you use them to connect concepts in your map. See if you can identify the concepts and linking words in the concept map shown in **Figure A** on page 1029. As you can see, the concepts are listed in boxes, and the linking words and lines are used to connect the concepts. Notice that one linking word can be used to link several concepts. You can also use arrows to help link concepts more clearly.

Study TIP

Organizing Information

You can use concept maps as a study guide to help you organize and review information.

Organizing concepts

Some concepts are more general and include other concepts. These general concepts will be the main ideas in your map. Determine the main idea for the following list of concepts:

- Tokyo
- Mexico City
- Seoul
- New York
- Bombay

They are cities, of course. Now determine the main idea for the following list:

- car
- bus
- train
- bicycle
- truck

Each of these concepts is an example of a vehicle.

In a concept map, the main concept should go at the top. The more specific concepts and examples should go below. Capitalize the first letter of the main concept. Write the other concepts using all lowercase letters. For example, in the concept map shown in Figure A, the main concept is "Biology." The more specific concepts listed below it in the map, such as "taxonomy," describe biology. Study this concept map before you try one yourself. Then read the following paragraph and make a list of any important words you think should be defined.

What is life? What is the difference between living and nonliving things? If you were in a wilderness area, it would be easy for you to pick out the living and nonliving things. The animals and plants are the living organisms. Organisms are made of substances organized into living systems. The rocks, air, water, and soil you see are nonliving. They contribute substances to the living organisms.

Connecting ideas

Two general ideas prevail in the preceding paragraph: living and nonliving. You could make two separate concept maps, the main concept of one being "Living" and the other being "Nonliving,"

Figure A

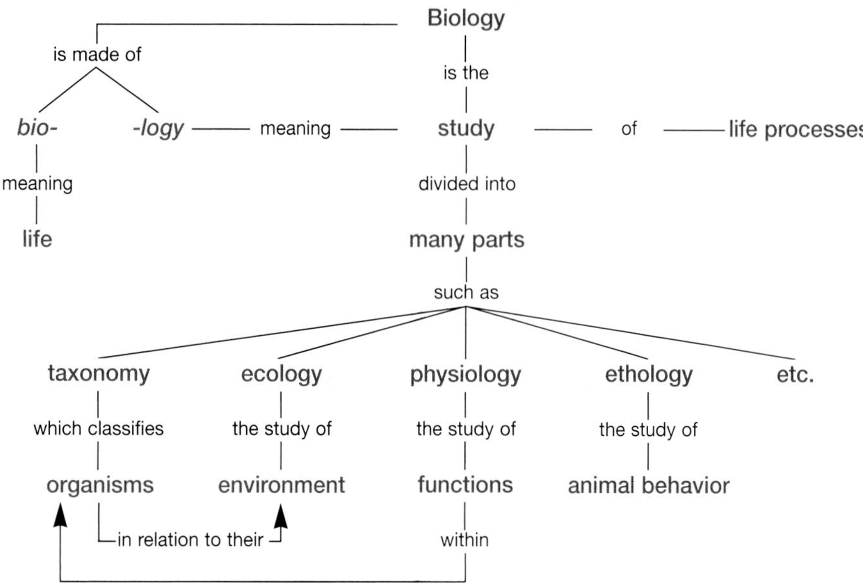

or you could make one map and use "Natural things" as the main concept. Natural things can be living or nonliving. Therefore, the concepts "living" and "nonliving" are parts of the main concept "Natural things" and should be placed below it in the map.

You have learned that living things are organisms and that they are made of substances that are organized into living systems. Living things are different from but related to nonliving things. So you should make the connection between living and nonliving things near the top of the concept map. The rest of the map should relate living and nonliving things. Nonliving things contribute substances that are used by living systems.

Listing concepts

The next step is to put the concepts in order from the most general to the most specific. You can do this by writing the concepts on pieces of paper. Some concepts will share the same rank and be equally specific. Remember that the examples are the most specific and will be at the bottom of the map.

Now begin to rearrange the concepts you have written on the pieces of paper. Start with the most general; get the main idea. Then, if there are two or more concepts that are equally specific, place them on the same level. For example, "Natural things" is the main idea. Then "living" and "nonliving" are placed under "Natural things" on the same row or level.

Now continue to lay out all the other concepts under the subconcepts in the first row or level until you have used them all. You can rearrange your pieces of paper any time, so keep pushing them around as if they were pieces of a jigsaw puzzle until you have arranged them the way you think they belong.

Linking concepts

Now make the connections between the concepts. Use lines to connect the concepts, and write linking words on the lines to show or tell why they are connected. Use linking words for all the lines connecting all the concepts. Glue or tape down your concept papers if you want to make the map permanent, or use a separate piece of paper to draw a sketch showing the way you have arranged the concepts.

Now you have the completed concept map shown in **Figure B** on page 1030. If you had a choice between reading the paragraph or looking at this map, you would probably agree that the map shows the concepts more clearly. This map gives you the main idea more quickly, and it is easier to understand all the ideas because their relationships to other ideas are shown.

Features of Good Concept Maps

Remember, practice is the key to good concept mapping. You will get better as you go along. Here are some things to remember:

- A concept map does not have to be symmetrical. It can have more concepts on one side than on the other.
- There are no "perfect" concept maps, only maps that come closer to the meanings of the concepts. As the mapmaker, you must make your map work for you.
- Do not put more than three words in a concept box.
- Do not have more than four concept boxes in a row without branching out.
- Connect every pair of concepts with linking words. Use as few linking words as possible.

If the relationships you have made between any two concepts are wrong, your teacher will help you sort out your misconception. Even if your relationships are absolutely correct, maps made by your classmates may be different. These maps could be equally correct, even though they may look nothing like yours. Everyone thinks a little bit differently, and as a result, other people may see different relationships between certain concepts.

As you practice making concept maps, your teacher will examine your linking words more closely. Because the linking words and the lines between concept boxes relate concepts, your linking words will tell you if you really understand how concepts are connected.

A concept map should always have the following characteristics:

- It is two dimensional—not just a list of concepts connected by lines.
- It shows concepts in order of importance.
- It contains many branches with no more than four concept boxes in a row and no more than three words in each concept box.
- It contains only concepts in the boxes and only linking words on the lines.

Evaluating your skills

For the first map you make on your own, think about something you know very well. Do you play a team sport or an individual sport? Do you have a hobby? Do you enjoy a particular kind of music? Whatever topic you choose, use it as the main concept for your concept map. This will be more fun and easier because you know this topic so well.

Figure B

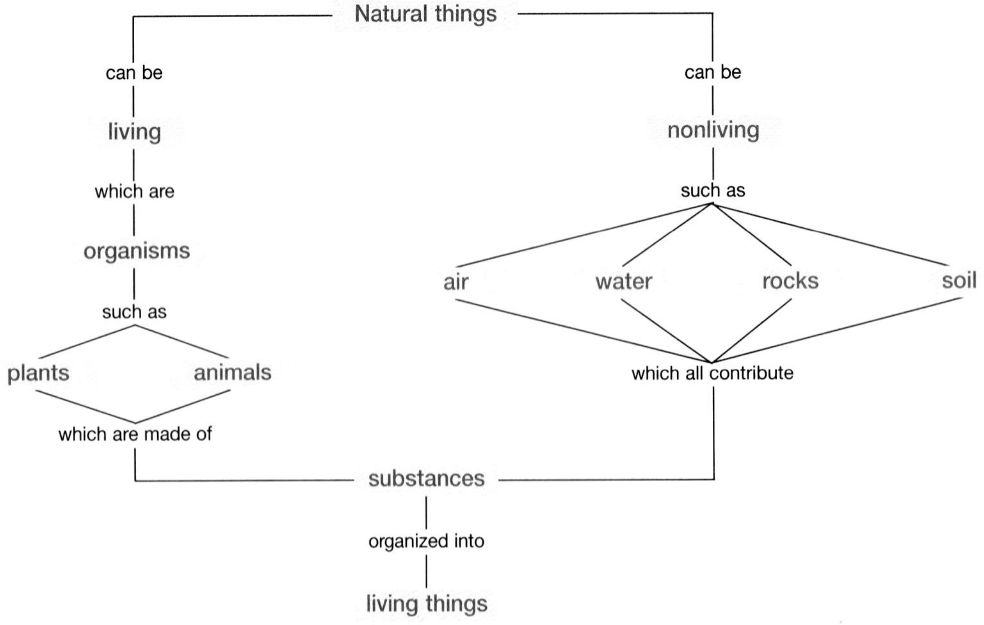

Reading and Study Skills Analyzing Word Parts

Determining the Meanings of Words

The challenge of understanding a new word can often be simplified by carefully examining the word. Many words can be divided into three parts: a prefix, root, and suffix. The **prefix** consists of one or more syllables placed in front of a **root.** The root is the main part of the word. The **suffix** consists of one or more syllables at the end of a root. Prefixes and suffixes modify or add to the meaning of the root. A knowledge of common prefixes, roots, and suffixes can give you clues to the meaning of unfamiliar words and can help make learning new words easier. For example, each of the word parts in **Table 1** can be combined with the root *derm* to form a word.

Table 2 lists word prefixes and suffixes commonly used in biology. Each word part is followed by its usual meaning, an example of a word in which it is used, and a definition of that word. Examine the definition and the example. Decide whether each word part in the first column is a prefix or suffix, depending on how the word part is used in the example.

Table 1 Word Parts

Prefix	Root	Suffix
hypo-	derm	*-ic*
pachy-	derm	
	derm	*-atology*
	derm	*-atologist*
	derm	*-atitis*

Use Table 1 to form five words using the root *derm.* Then use the list of word parts and their definitions to write what you think is each word's meaning. An example is shown below.

Example:	**Dermatologist**
derm (root):	skin
-logy (suffix):	the study of
-ist (suffix):	someone who practices or deals with something
dermatologist:	someone who studies or deals with skin

Table 2 Word Prefixes and Suffixes

Prefix or suffix	Definition	Example
a-	not, without	asymmetrical: not symmetrical
ab-	away, apart	abduct: move away from the middle
-able	able	viable: able to live
ad-	to, toward	adduct: move toward the middle
amphi-	both	amphibian: type of vertebrate that lives both on land and in water
ante-	before	anterior: front of an organism
anti-	against	antibiotic: substance, such as penicillin, capable of killing bacteria
arche-	ancient	*Archaeopteryx: a* fossilized bird
arthro-	joint	arthropod: jointed-limbed organism belonging to the phylum Arthropoda
auto-	self, same	autotrophic: able to make its own food
bi-	two	bivalve: mollusk with two shells
bio-	life	biology: the study of life
blast-	embryo	blastula: hollow ball stage in the development of an embryo
carcin-	cancer	carcinogenic: cancer-causing
cereb-	brain	cerebrum: part of the vertebrate brain
chloro-	green	chlorophyll: green pigment in plants needed for photosynthesis

Analyzing Word Parts *continued*

Table 2 Word Prefixes and Suffixes

Prefix or suffix	Definition	Example
chromo-	color	chromosome: structure found in eukaryotic cells that contains DNA
chondro-	cartilage	Chondrichthyes: cartilaginous fish
circ-	around	circulatory: system for moving fluids through the body
-cide	kill	insecticide: a substance that kills insects
co-, con-	with, together	conjoined twins: identical twins physically joined by a shared portion of anatomy at birth
-cycle	circle	pericycle: layer of plant cells
cyt-	cell	cytology: the study of cells
de-	remove	dehydration: removal of water
derm-	skin	dermatology: study of the skin
di-	two	diploid: full set of chromosomes
dia-	through	dialysis: separating molecules by passing them through a membrane
ecol-	dwelling, house	ecology: the study of living things and their enviroments
ecto-	outer, outside	ectoderm: outer germ layer of developing embryo
-ectomy	removal	appendectomy: removal of the appendix
endo-	inner, inside	endoplasm: cytoplasm within the cell membrane
epi-	upon, over	epiphyte: plant growing upon another plant
ex-, exo-	outside of	exobiology: the search for life elsewhere in the universe
gastro-	stomach	gastropod: type of mollusk
-gen	type	genotype: genes in an organism
-gram	write or record	climatogram: depicting the annual precipitation and temperature for an area
hemi-	half	hemisphere: half of a sphere
hetero-	different	heterozygous: different alleles inherited from parents
hist-	tissue	histology: the study of tissues
homeo-	the same	homeostasis: maintain a constant condition
hydro-	water	hydroponics: growing plants in water instead of soil
hyper-	above, over	hypertension: blood pressure higher than normal
hypo-	below, under	hypothalamus: part of the brain located below the thalamus
-ic	of or pertaining to	hypodermic: pertaining to under the skin
inter-	between, among	interbreed: breed within a family or strain
intra-	within	intracellular: inside a cell
iso-	equal	isogenic: having an identical set of genes
-ist	someone who practices or deals with something	biologist: someone who studies life
-logy	study of	biology: the study of life
macro-	large	macromolecule: large molecule, such as DNA or proteins

Table 2 Word Prefixes and Suffixes

Prefix or suffix	Definition	Example
mal-	bad	malnourishment: poor nutrition
mega-	large	megaspore: larger of two types of spores produced by some ferns and flowering plants
meso-	in the middle	mesoglea: jellylike material found between outer and inner layers of coelenterates
meta-	change	metamorphosis: change in form
micro-	small	microscopic: too small to be seen with unaided eye
mono-	one, single	monoploid: one set of alleles
morph-	form	morphology: study of the form of organisms
neo-	new	neonatal: newborn
nephr-	kidney	nephron: functional unit of the kidneys
neur-	neuron	neurotransmitter: chemical released by a neuron
oo-	egg	oogenesis: gamete formation in female diploid organisms
org-	living	organism: living thing
-oma	swelling	carcinoma: cancerous tumor
orth-	straight	orthodontics: the practice of straightening teeth
pachy-	thick	pachyderm: thick-skinned animal, such as an elephant
para-	near, on	parasite: organism that lives on and gets nutrients from another organism
path-	disease	pathogen: disease-causing agent
peri-	around	pericardium: membrane around the heart
photo-	light	phototropism: bending of plants toward light
phyto-	plants	phytoplankton: plankton that consists of plants
poly-	many	polypeptide: sequence of many amino acids joined together to form a protein
-pod	foot	pseudopod: false foot that projects from the main part of an amoeboid cell
pre-	before	prediction: a forecast of events before they take place
-scope	instrument used to see something	microscope: instrument used to see very small objects
semi-	partially	semipermeable: allowing some particles to move through
-some	body	chromosome: structure found in eukaryotic cells that contains DNA
sub-	under	substrate: molecule on which an enzyme acts
super-, supra-	above	superficial: on or near the surface of a tissue or organ
syn-	with	synapse: junction of a neuron with another cell
-tomy	to cut	appendectomy: operation in which the appendix is removed
trans-	across	transformation: the transfer of genetic material from one organism to another
ur-	referring to urine	urology: study of the urinary tract
visc-	organ	viscera: internal organs of the body

Laboratory Skills Using the Interactive Explorations

Imagine that you could watch what happens in a plant cell during photosynthesis or that you could observe how drug use affects the human nervous system. You cannot do these things in a conventional laboratory exercise, but you can in an Interactive Exploration. By using the *Interactive Explorations in Biology* CD-ROMs, you can explore biological processes that are important to both human biology and cell biology. The instructions on these two pages describe how to use the CD-ROMs.

Using the CD-ROM *Interactive Explorations in Biology: Human Biology*

1. Get the *CD-ROM Interactive Explorations in Biology: Human Biology* from your teacher. Load the disc into the CD-ROM drive as directed by your teacher.
2. An icon for the CD-ROM will appear on the screen. Click it, and then click the icon labeled "Start Explorations."
3. After a few moments, the title screen will appear. Click the button labeled "Menu" at the bottom right of the screen.
4. The computer will then display the menu screen, which lists all the Interactive Explorations on the disc. Select the Interactive Exploration you would like to run, and click its title.
5. After a few moments, the computer will display a screen containing a column of buttons along the right edge. This column is the Navigation Palette, and the buttons it contains allow you to access the various features found in each Interactive Exploration. To begin, click the top button of the column. This button displays a magnifying glass and is labeled "Interactive Investigation."
6. If you have questions about an Interactive Exploration, you can play the instructions for operating the Interactive Exploration by clicking the File menu at the top of the screen and selecting "Interactive Exploration Help." You may also refer to the User's Guide for additional information or help.

Using the CD-ROM *Interactive Explorations in Biology: Cell Biology and Genetics*

1. Get the CD-ROM *Interactive Explorations in Biology: Cell Biology and Genetics* from your teacher. Load the disc into the CD-ROM drive as directed by your teacher.
2. An icon for the CD-ROM will appear on the screen. Click it, and then click the icon labeled "Start Explorations."
3. After a few moments, the title screen will appear. Click the button labeled "Menu" at the bottom right of the screen.
4. The computer will then display the menu screen, which lists all the Interactive Explorations on the disc. Select the Interactive Exploration you would like to run, and click its title.
5. After a few moments, the computer will display the Interactive Exploration and play a short narration that describes the problem you will be studying. (To shut off this narration, select "Sound Off" from the Sound menu at the top right of the screen.) After this narration has played, the Interactive Exploration is ready to run.
6. If you have questions about an Interactive Exploration, you can play the instructions for operating the Interactive Exploration by clicking the Help menu at the top of the screen and selecting "How to Use This Exploration." Clicking the Help button on the Navigation Palette will call up instructions on how to use the features of the Navigation Palette. You may also refer to the User's Guide.

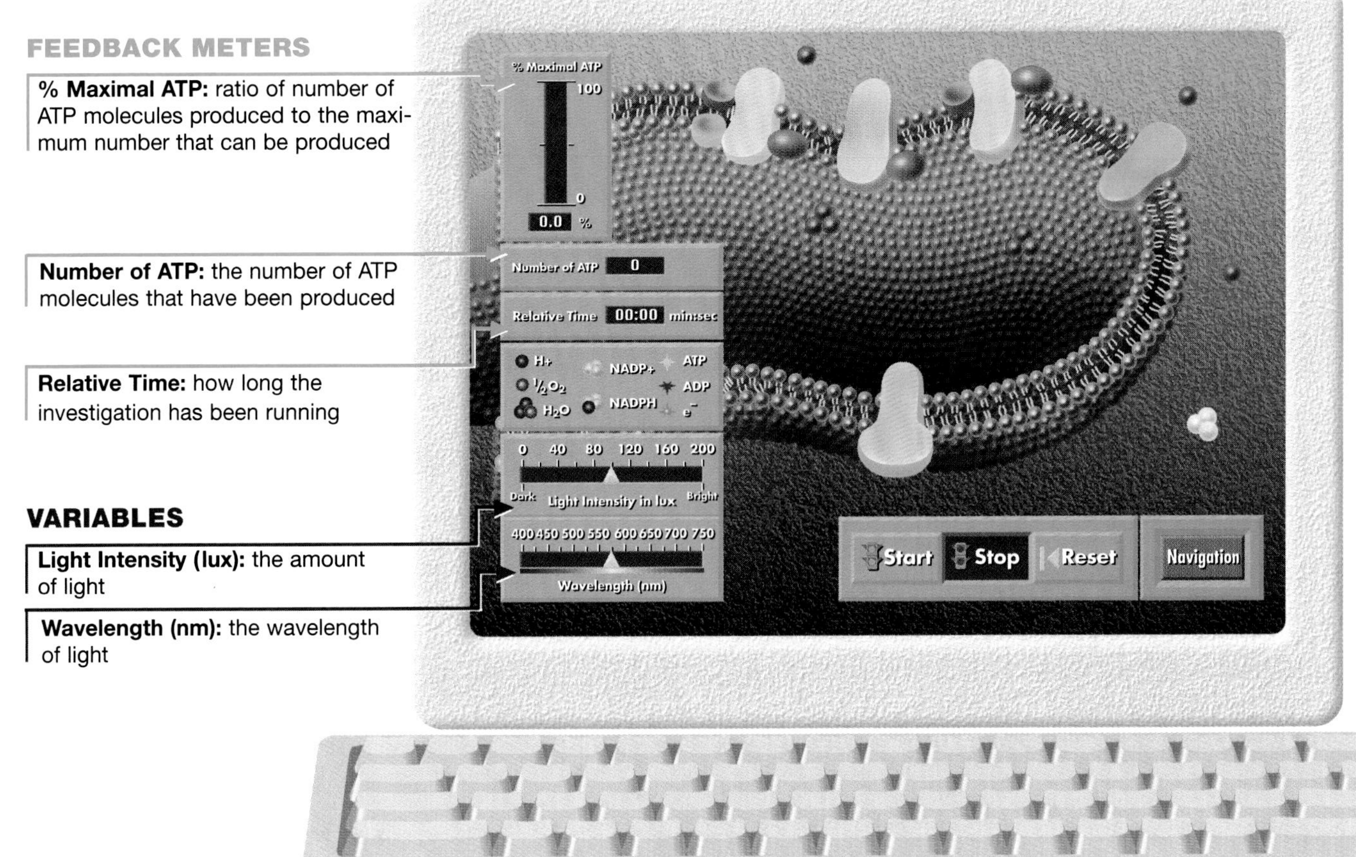

FEEDBACK METERS
% Maximal ATP: ratio of number of ATP molecules produced to the maximum number that can be produced
Number of ATP: the number of ATP molecules that have been produced
Relative Time: how long the investigation has been running
VARIABLES
Light Intensity (lux): the amount of light
Wavelength (nm): the wavelength of light
Start
Stop
Reset
Navigation

Safe Laboratory Practices

It is your responsibility to protect yourself and other students by conducting yourself in a safe manner while in the laboratory. You can avoid accidents in the laboratory by following directions, handling materials carefully, and taking your work seriously. Read the following general safety guidelines below before attempting to do work in the laboratory. Make sure you understand all safety guidelines before entering the laboratory. If necessary, ask your teacher for clarification of laboratory rules and procedures.

General Guidelines for Laboratory Safety

- **Only perform experiments specifically assigned by your teacher.** Do not attempt any laboratory procedure without your teacher's direction, and do not work alone in the laboratory.
- **Familiarize yourself with the investigation and all safety precautions before entering the lab.** Be aware of the potential hazards of the required materials and procedures. Before you begin, ask your teacher to explain any parts of an investigation that you do not understand.
- **Before beginning work, tie back long hair, roll up loose sleeves, and put on any required personal protective equipment as directed by your teacher.** Avoid or confine loose clothing that could knock things over, catch on fire, or absorb chemical solutions. Nylon and polyester fabrics burn and melt more readily than does cotton. Do not wear open-toed shoes, sandals, or canvas shoes in the laboratory.
- **Always wear a lab apron and safety goggles.** Wear this equipment even if you are not working on an experiment. Laboratories contain chemicals that can damage your clothing, skin, and eyes. If your safety goggles cloud up or are uncomfortable, ask your teacher for help. Lengthening the strap slightly, washing the goggles with soap and warm water, or using an anti-fog spray may help the problem.
- **No contact lenses are allowed in the lab.** Even if you are wearing safety goggles, chemicals could get between contact lenses and your eyes and cause irreparable eye damage. If your doctor requires that you wear contact lenses instead of glasses, then you should wear eye-cup safety goggles—similar to goggles worn for underwater swimming—in the lab. Ask your doctor or your teacher how to use eye-cup safety goggles to protect your eyes.
- **Know the location of all safety and emergency equipment used in the laboratory.** Ask your teacher where the nearest eyewash stations, safety blankets, safety shower, fire extinguisher, first-aid kit, and chemical spill kit are located.
- **Immediately report any accident, incident, or hazard—no matter how trivial—to your teacher.** Any incident involving bleeding, burns, fainting, chemical exposure, or ingestion should also be reported immediately to the school nurse or to a physician.
- **In case of fire, alert your teacher and leave the lab.** Standard fire-safety procedures should be followed.
- **Do not have or consume food or drink in the lab.** Do not store or eat food in the laboratory.

- **Do not fool around in the lab.** Take your lab work seriously and behave appropriately in the laboratory. Be aware of your classmates' safety as well as your own at all times.
- **Do not apply cosmetics in the lab.** Some hair-care products and nail polish are highly flammable.
- **Keep your work area neat and uncluttered.** Have only books and other materials that are needed to conduct the experiment in the laboratory.
- **Clean your work area at the conclusion of each lab period as directed by your teacher.** Broken glass, chemicals, and other laboratory waste products should be disposed of in separate special containers. Dispose of waste materials as directed by your teacher.
- **Wash your hands with soap and hot water after each lab period.** Wash your hands at the conclusion of each lab period and before leaving the laboratory to avoid contamination.

Key to Safety Symbols and Their Precautions

Before you begin working in the laboratory, familiarize yourself with the following safety symbols, which are used throughout this textbook, and guidelines that you should follow when you see these symbols.

Eye Safety

- **Wear approved safety goggles as directed.** Safety goggles should always be worn in the laboratory, especially when you are working with a chemical or solution, a heat source, or a mechanical device.
- **In case of eye contact, do the following:** Go to an eyewash station immediately and flush your eyes (including under the eyelids) with running water for at least 15 minutes. Hold your eyelids open with your thumb and fingers, and roll your eyeball around. While doing this, have another student notify your teacher.
- **Do not wear contact lenses in the lab.** Chemicals can be drawn up under a contact lens and into the eye. If you must wear contacts prescribed by a physician, tell your teacher. You must also wear approved eye-cup safety goggles to help protect your eyes.

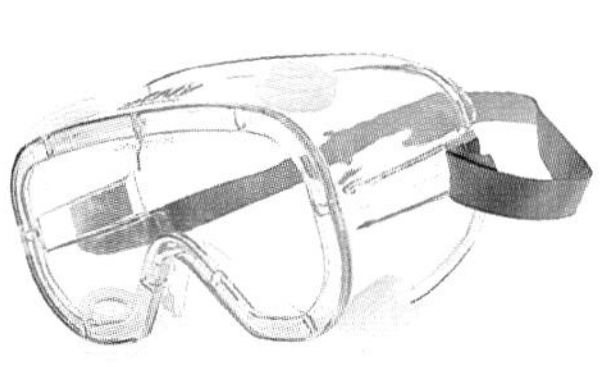

- **Do not look directly at the sun through any optical device or lens system, and do not reflect direct sunlight to illuminate a microscope.** Such actions concentrate light rays to an intensity that can severely burn your retinas, possibly causing blindness.

Hand Safety

- **Do not cut objects while holding them in your hand.** Dissect specimens in a dissecting tray.
- **Wear protective gloves when working with an open flame, chemicals, solutions, or wild or unknown plants.**

Safety with Gases

- **Do not inhale any gas or vapor unless directed to do so by your teacher.** Never inhale pure gases.
- **Handle materials that emit vapors or gases in a well-ventilated area.** This work should be done in an approved chemical fume hood.

Sharp-Object Safety

- **Use extreme care when handling all sharp and pointed instruments, such as scalpels, sharp probes, and knives.**
- **Do not use double-edged razor blades in the laboratory.**
- **Do not cut objects while holding them in your hand.** Cut objects on a suitable work surface. Always cut in a direction away from your body.

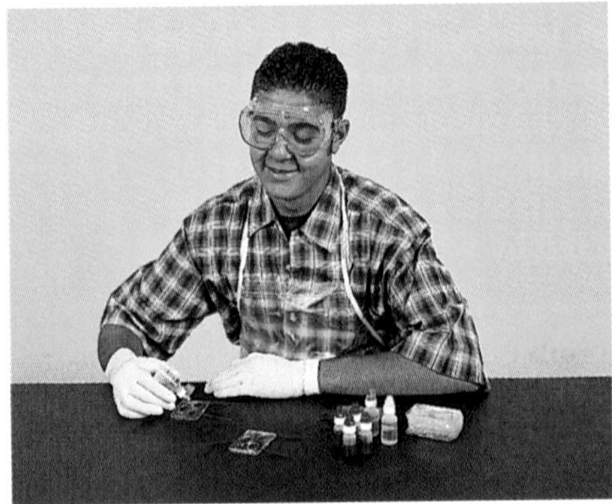

Animal Care and Safety

- **Do not approach or touch any wild animals.** When working outdoors, be aware of poisonous or dangerous animals in the area.
- **Always get your teacher's permission before bringing any animal (including pets) into the school building.**
- **Handle animals only as directed by your teacher.** Mishandling or abusing any animal will not be tolerated.

Heating Safety

- **Be aware of any source of flames, sparks, or heat (open flames, electric heating coils, hot plates, etc.) before working with flammable liquids or gases.**
- **When heating chemicals or solutions in a test tube, do not point the test tube toward anyone.**
- **Avoid using open flames.** If possible, work only with hot plates that have an "On-Off" switch and an indicator light. Do not leave hot plates unattended. Do not use alcohol lamps. Turn off hot plates and open flames when they are not being used.
- **Know the location of laboratory fire extinguishers and fire-safety blankets.**
- **Use tongs or appropriate insulated holders when handling heated objects.** Heated objects often do not appear to be hot. Do not pick up an object with your hand if it could be warm.
- **Keep flammable substances away from heat, flames, and other ignition sources.**

Hygienic Care

- **Keep your hands away from your face and mouth while working in the lab.**
- **Wash your hands thoroughly before leaving the laboratory.**
- **Remove contaminated clothing immediately.** If you spill caustic substances on your skin or clothing, use the safety shower or a faucet to rinse. Remove affected clothing while under the shower, and call to your teacher. (It may be temporarily embarrassing to remove clothing in front of your classmates, but failure to rinse a chemical off your skin could result in permanent damage.)
- **Launder contaminated clothing separately.**
- **Use the proper technique demonstrated by your teacher when handling bacteria or other microorganisms.** Do not open Petri dishes to observe or count bacterial colonies.
- **Return all stock and experimental cultures to your teacher for proper disposal.**

Glassware Safety

- **Inspect glassware before use; do not use chipped or cracked glassware.** Use heat-resistant glassware for heating materials or storing hot liquids.
- **Do not attempt to insert glass tubing into a rubber stopper without specific instruction from your teacher.**
- **Immediately notify your teacher if a piece of glassware breaks. Do not attempt to clean up broken glass.**

Proper Waste Disposal

- **Clean and sanitize all work surfaces and personal protective equipment after each lab period as directed by your teacher.**
- **Dispose of all sharp objects (such as broken glass) and other contaminated materials (biological or chemical) in special containers as directed by your teacher.**

Electrical Safety

- **Do not use equipment with frayed electrical cords or loose plugs.**
- **Fasten electrical cords to work surfaces using tape.** This will prevent tripping and will ensure that equipment cannot fall off the table.
- **Do not use electrical equipment near water or with wet hands or clothing.**
- **Hold the rubber cord when you plug in or unplug equipment.** Do not touch the metal prongs of the plug, and do not unplug equipment by pulling on the cord.

Clothing Protection

- **Wear an apron or laboratory coat at all times in the laboratory to prevent chemicals or chemical solutions from contacting skin or clothes.**

Plant Safety

- **Do not ingest any plant part used in the laboratory (especially commercially sold seeds).** Do not touch any sap or plant juice directly. Always wear gloves.
- **Wear disposable polyethylene gloves when handling any wild plant.**
- **Wash hands thoroughly after handling any plant or plant part (particularly seeds). Avoid touching your face and eyes.**
- **Do not inhale or expose yourself to the smoke of any burning plant.** Smoke contains irritants that can cause inflammation in the throat and lungs.
- **Do not pick wildflowers or other plants unless directed by your teacher.**

Chemical Safety

- **Always wear safety goggles, gloves, and a lab apron or coat when working with any chemical or chemical solution to protect your eyes and skin.**
- **Do not taste, touch, or smell any chemicals or bring them close to your eyes, unless specifically instructed by your teacher.** If you are directed by your teacher to note the odor of a substance, do so by waving the fumes toward you with your hand. Do not pipette any chemicals by mouth; use a suction bulb as directed by your teacher.
- **Know the location of the emergency lab shower and eyewash and how to use them.** If you get a chemical on your skin or clothing, wash it off at the sink while calling to your teacher.
- **Always handle chemicals or chemical solutions with care.** Check the labels on bottles, and observe safety procedures. Do not return unused chemicals or solutions to their original containers. Return unused reagent bottles or containers to your teacher.
- **Do not mix any chemicals unless specifically instructed by your teacher.** Two harmless chemicals can be poisonous if combined.
- **Do not pour water into a strong acid or base.** The mixture can produce heat and splatter.
- **Report any spill immediately to your teacher.** Spills should be cleaned up promptly as directed by your teacher.

ChemSafety

CAUTION: Always wear safety goggles and a lab apron to protect your eyes and clothing.

CAUTION: Do not touch or taste any chemicals. Know the location of the emergency shower and eyewash station and how to use them. If you get a chemical on your skin or clothing, wash it off at the sink while calling to the teacher. Notify the teacher of a spill. Spills should be cleaned up promptly, according to your teacher's directions.

CAUTION: Glassware is fragile. Notify the teacher of broken glass or cuts. Do not clean up broken glass or spills with broken glass unless the teacher tells you to do so.

Using a Compound Light Microscope

Parts of the Compound Light Microscope

- The **eyepiece** magnifies the image, usually 10×.
- The **low-power objective** further magnifies the image, up to 4×.
- The **high-power objectives** further magnify the image, from 10× to 43×.
- The **nosepiece** holds the objectives and can be turned to change from one objective to another.
- The **body tube** maintains the correct distance between the eyepiece and the objectives. This is usually about 25 cm (10 in.), the normal distance for reading and viewing objects with the unaided eye.
- The **coarse adjustment** moves the stage up and down in large increments to allow gross positioning and focusing of the objective lens.
- The **fine adjustment** moves the stage slightly to bring the image into sharp focus.
- The **stage** supports a slide that contains the viewed specimen.
- The **stage clips** secure the slide in position for viewing.
- The **diaphragm** (not labeled), located under the stage, controls the amount of light allowed to pass through the object being viewed.
- The **light source** provides light for viewing the image. It can be either a light reflected with a mirror or an incandescent light from a small lamp. Never use reflected direct sunlight as a light source.
- The **arm** supports the body tube.
- The **base** supports the microscope.

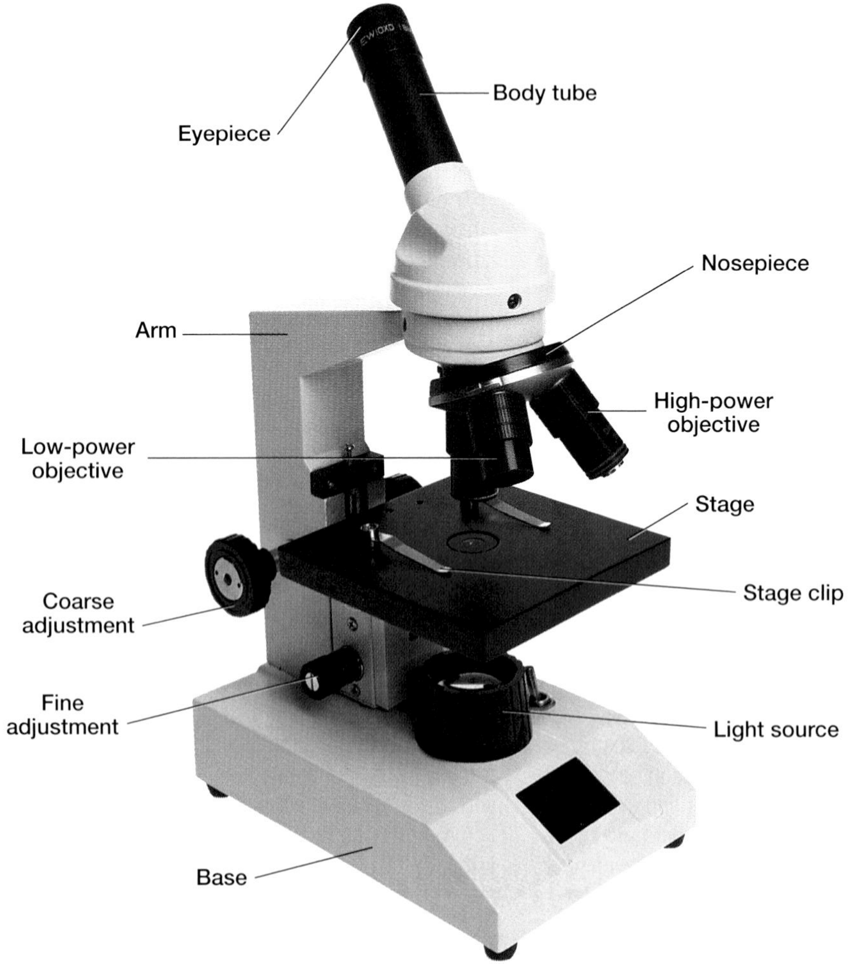

Proper Handling and Use of the Compound Light Microscope

1. Carry the microscope to your lab table using both hands, one supporting the base and the other holding the arm of the microscope. Hold the microscope close to your body.
2. Place the microscope on the lab table at least 5 cm (2 in.) from the edge of the table.
3. Check to see what type of light source the microscope has. If the microscope has a lamp, plug it in, making sure that the cord is out of the way. If the microscope has a mirror, adjust it to reflect light through the hole in the stage.

   CAUTION: If your microscope has a mirror, do not use direct sunlight as a light source. Using direct sunlight can damage your eyes.
4. Adjust the revolving nosepiece so that the low-power objective is aligned with the body tube.
5. Place a prepared slide over the hole in the stage, and secure the slide with the stage clips.
6. Look through the eyepiece, and move the diaphragm to adjust the amount of light that passes through the specimen.
7. Now look at the stage at eye level. Slowly turn the coarse adjustment to raise the stage until the objective almost touches the slide. Do not allow the objective to touch the slide.
8. While looking through the eyepiece, turn the coarse adjustment to lower the stage until the image is in focus. Never focus objectives downward. Use the fine adjustment to achieve a sharply focused image. Keep both eyes open while viewing a slide.
9. Make sure that the image is exactly in the center of your field of vision. Then switch to the high-power objective. Focus the image with the fine adjustment. Never use the coarse adjustment at high power.
10. When you are finished using the microscope, remove the slide. Clean the eyepiece and objectives with lens paper, and return the microscope to its storage area.

Procedure for Making a Wet Mount

1. Use lens paper to clean a glass slide and coverslip.
2. Place the specimen that you wish to observe in the center of the slide.
3. Using a medicine dropper, place one drop of water on the specimen.
4. Position the coverslip so that it is at the edge of the drop of water and at a 45° angle to the slide. Make sure that the water runs along the edge of the coverslip.
5. Lower the coverslip slowly to avoid trapping air bubbles.
6. If a stain or solution will be added to a wet mount, place a drop of the staining solution on the microscope slide along one side of the coverslip. Place a small piece of paper towel on the opposite side of the coverslip.
7. As the water evaporates from the slide, add another drop of water by placing the tip of the medicine dropper next to the edge of the coverslip, just as you would if adding stains or solutions to a wet mount. If you have added too much water, remove the excess by using the corner of a paper towel as a blotter. Do not lift the coverslip to add or remove water.

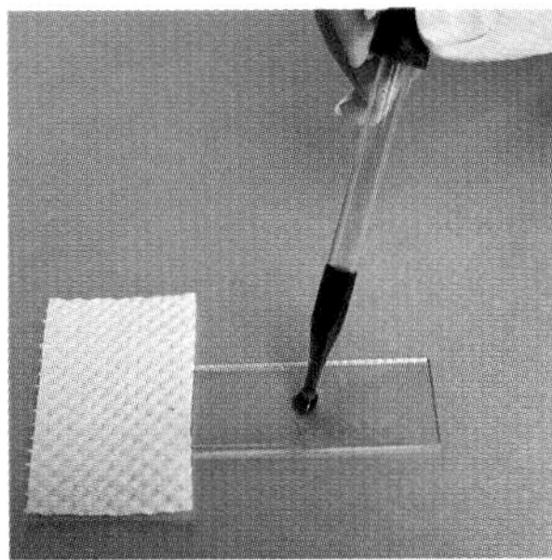

Laboratory Skills Determining Mass and Temperature

Reading a Balance

A single-pan balance, such as the one shown at right, has one pan and three or four beams. The scale of measure for each beam depends on the model of the balance. When an object is placed on the pan, the riders are moved along the beams until the mass on the beams equals the mass of the object in the pan.

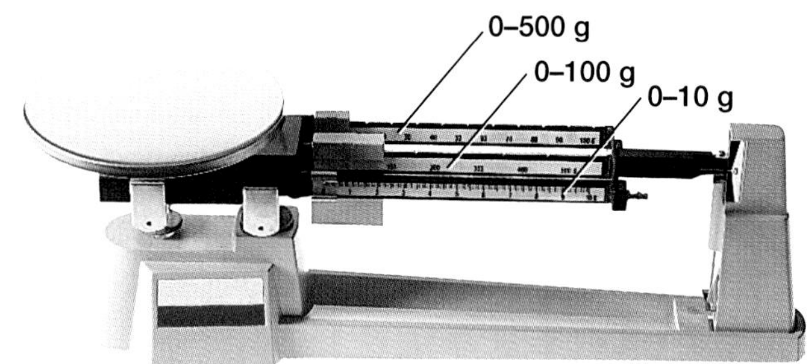

Measuring mass

When determining the mass of a chemical or powder, use weighing or filter paper. Determine the mass of the paper, and subtract that mass from the total mass. Use the following procedure for determining the mass of objects:

1. Make sure the balance is on a level surface and the pan is allowed to move freely. Position all the riders at zero. If the pointer does not come to rest in the middle of the scale, calibrate the balance using the adjustment knob (usually located under and to the left of the pan).

   CAUTION: Never place a hot object or chemical directly on a balance pan.

2. Place the object on the pan.
3. Move the largest rider along the beam to the right until it is at the last notch that does not move the pointer below the zero point in the middle of the scale.
4. Follow the same procedure with the next rider.
5. Move the smallest rider until the pointer rests at zero in the middle of the scale.
6. Add up the readings on all the beams to determine the mass of the object.

Practice Exercises

1. **Determine** the mass of each of the following items using a single-pan balance: Answers will vary.
   - **a.** an empty 250 mL beaker
   - **b.** 250 mL beaker filled with 100 mL of water
   - **c.** 250 mL beaker filled with 100 mL of vegetable oil
   - **d.** a house key
   - **e.** a small book
   - **f.** a paper clip or small safety pin
2. **Determine** the mass of each object represented by the balance readings shown.

a.

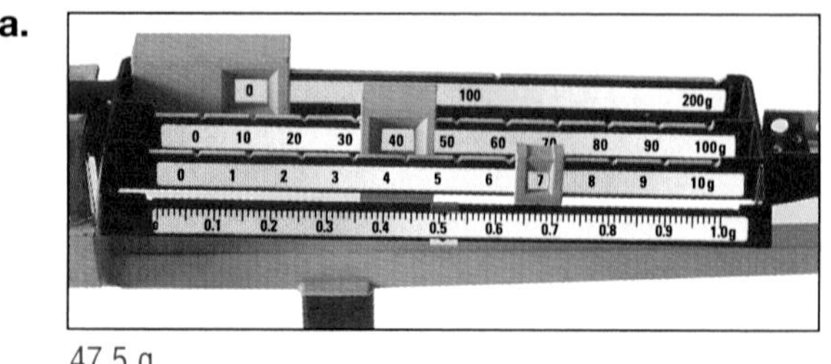

47.5 g

b.

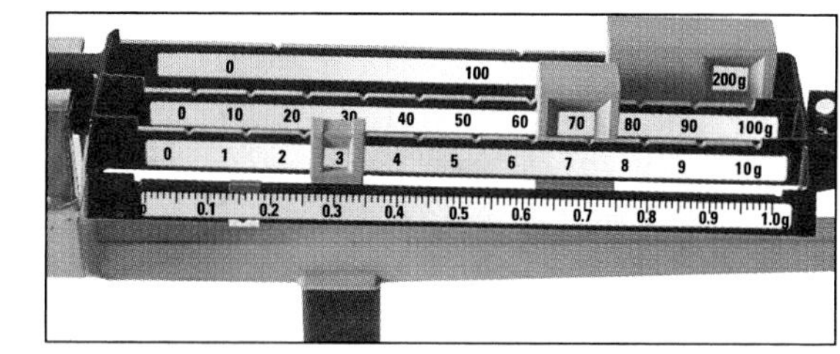

273.15 g

c.

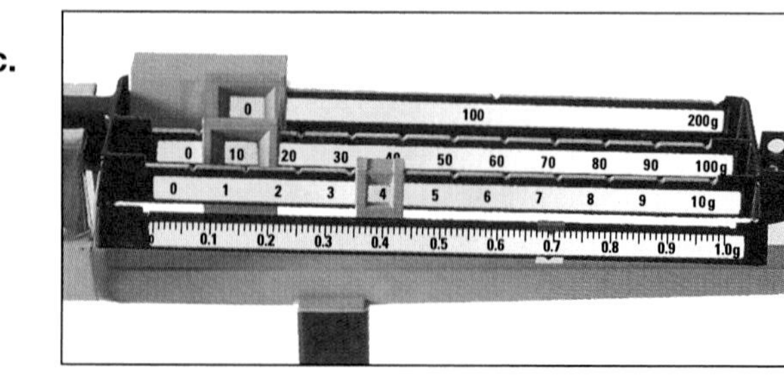

14.7 g

Reading a Thermometer

Most laboratory thermometers are the bulb-type shown below. The sensing bulb of the thermometer is filled with a colored liquid (alcohol or mercury) that expands when heated. When the liquid expands, it moves up the stem of the thermometer through the capillary tube. Thermometers usually measure temperature in degrees Celsius (°C).

Measuring temperature

Use the following procedure when measuring the temperature of a substance.

1. Carefully lower the bulb of the thermometer into the substance. The stem of the thermometer may rest against the side of the container, but the bulb should never rest on the bottom where heat is being applied. If the thermometer has an adjustable clip for the side of the container, the thermometer can be suspended in the liquid.

CAUTION: Do not hold a thermometer in your hand while measuring the temperature of a heated substance.

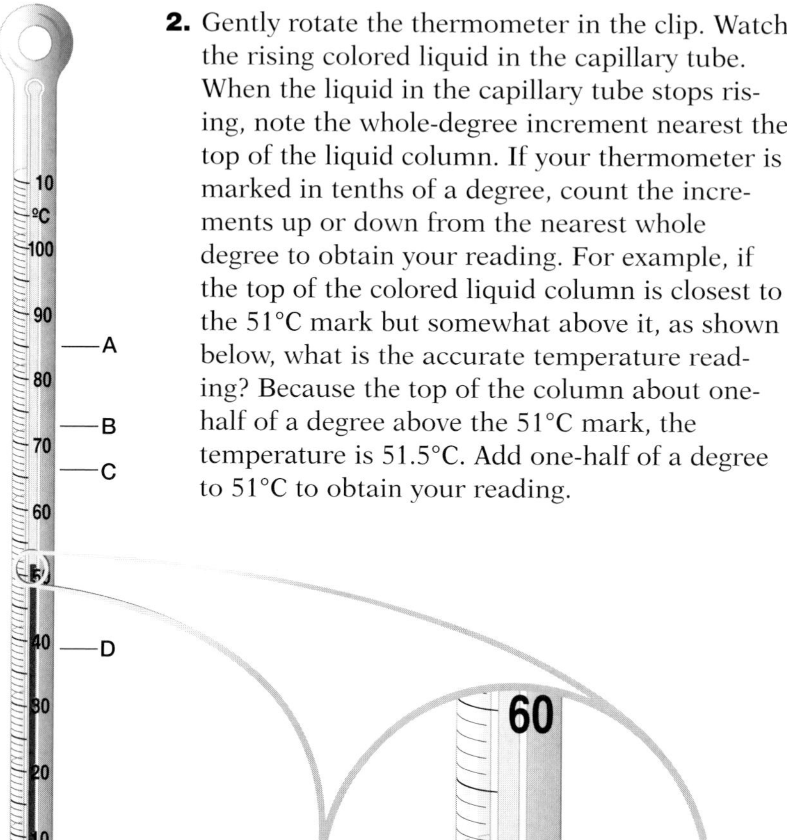

2. Gently rotate the thermometer in the clip. Watch the rising colored liquid in the capillary tube. When the liquid in the capillary tube stops rising, note the whole-degree increment nearest the top of the liquid column. If your thermometer is marked in tenths of a degree, count the increments up or down from the nearest whole degree to obtain your reading. For example, if the top of the colored liquid column is closest to the 51°C mark but somewhat above it, as shown below, what is the accurate temperature reading? Because the top of the column about one-half of a degree above the 51°C mark, the temperature is 51.5°C. Add one-half of a degree to 51°C to obtain your reading.

Practice Exercises

Use the thermometer shown above to answer the following questions:

1. Identify the scale used for this thermometer.

2. Determine whether this thermometer is marked only in whole degrees or in tenths of degrees.

3. Estimate the temperature reading on this thermometer.

1. Celsius; 2. whole degrees; 3. 51.5°C

4. SKILL Interpreting Variables What would be the temperature reading if the top of the column were resting at each of the following points?

a. *A* 85°C

b. *B* 73°C

c. *C* 66°C

d. *D* 39°C

e. *E* 8°C

f. *F* −12°C

Math and Problem-Solving Skills: SI Measurement

SI Units

Scientists throughout the world use the metric system. The official name of the metric system is the Système International d'Unités, or the International System of Measurements. It is usually referred to simply as SI. Most measurements in this book are expressed in metric units. You will always use metric units when you take measurements in the lab.

SI prefixes

SI is a decimal system; that is, all relationships between SI units are based on powers of 10. Most units have a prefix that indicates the relationship of that unit to a base unit. For example, the SI base unit for length is the meter. A meter equals 100 centimeters (cm), or 1,000 millimeters (mm). A meter also equals 0.001 kilometer (km). **Table 1** summarizes the prefixes used in SI units.

Table 1 SI Prefixes

Prefix	Symbol	Factor of base unit
giga-	G	1,000,000,000
mega-	M	1,000,000
kilo-	k	1,000
hecto-	h	100
deka-	da	10
deci-	d	0.1
centi-	c	0.01
milli-	m	0.001
micro-	μ	0.000001
nano-	n	0.000000001
pico-	p	0.000000000001

Conversion factors

Conversion between SI units requires a conversion factor. For example, to convert from meters to centimeters, you need to know the relationship between meters and centimeters.

$$1 \text{ cm} = 0.01 \text{ m} \quad \text{or} \quad 1 \text{ m} = 100 \text{ cm}$$

If you need to convert 15.5 centimeters to meters, you could do either of the following:

$$15.5 \text{ cm} \times \frac{1 \text{ m}}{100 \text{ cm}} = 0.155 \text{ m}$$

or

$$15.5 \text{ cm} \times \frac{0.01 \text{ m}}{1 \text{ cm}} = 0.155 \text{ m}$$

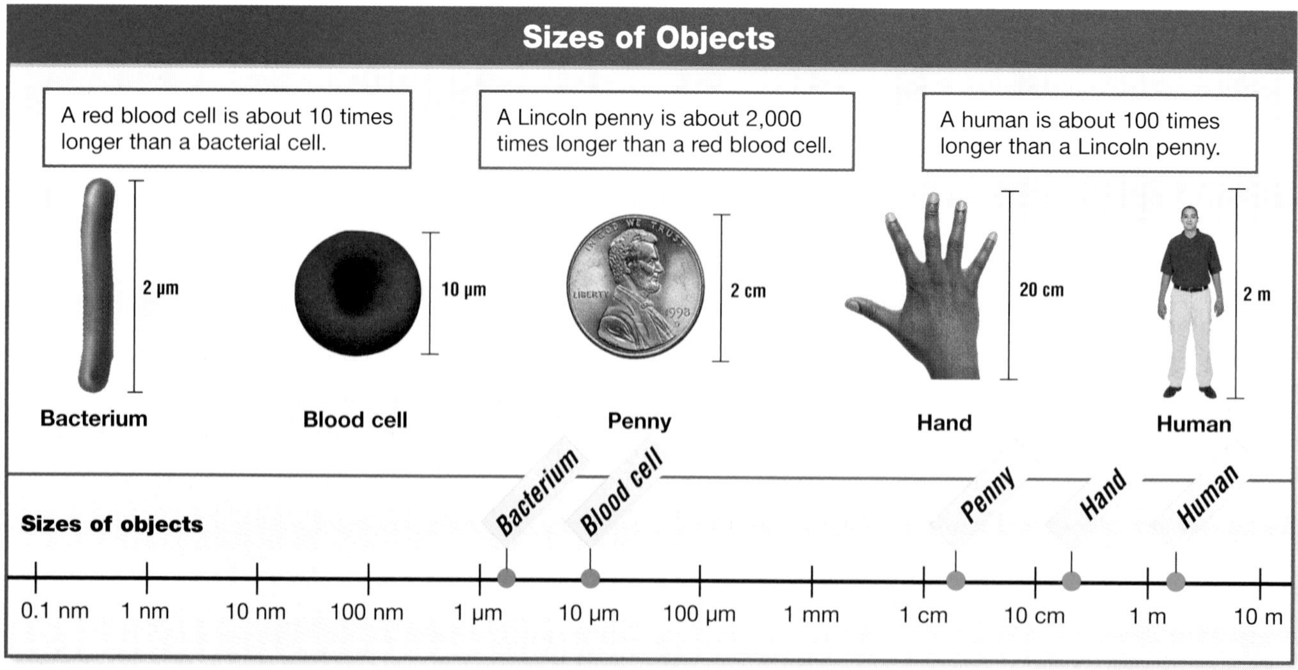

Base units

In this book, you will see three fundamental quantities represented by base units in SI: mass, length, and time. The base units of these quantities are the kilogram (kg), the meter (m), and the second (s). These quantities, their abbreviations, and their equivalent measurements are listed in **Table 2.**

Derived units

Other important quantities, such as area (m^2) and liquid volume (m^3), are expressed in derived units. A derived unit is a combination of one or more base units. Like base units, derived units can be expressed using SI prefixes. These quantities are listed in **Table 3.**

Table 2 Conversions for SI Base Units
Mass: unit = kilogram (kg)
1 kilogram (kg) = 1,000 g
1 gram (g) = 0.001 kg
1 milligram (mg) = 0.001 g
1 microgram (μg) = 0.000001 g
Length: unit = meter (m)
1 kilometer (km) = 1,000 m
1 meter (m) = 100 cm
1 centimeter (cm) = 0.01 m
1 millimeter (mm) = 0.001 m
1 micrometer (μm) = 0.000001 m
Time: unit = second (s)
1 minute (min) = 60 s
1 hour (h) = 3,600 s = 60 min
1 day (d) = 24 h

Table 3 Conversions for SI Derived Units
Area: unit = square meter (m^2)
1 square kilometer (km^2) = 100 ha
1 hectare (ha) = 10,000 m^2
1 square meter (m^2) = 10,000 cm^2
1 square centimeter (cm^2) = 100 mm^2
Liquid volume: unit = cubic meter (m^3)
1 cubic meter (m^3) = 1 kL
1 kiloliter (kL) = 1,000 L
1 liter (L) = 1,000 mL
1 milliliter (mL) = 0.001 L
1 cubic centimeter (cm^3) = 1 mL
Mass density: unit = kilograms per cubic meter (kg/m^3)
Temperature: unit = degrees Celsius (°C)
Velocity: unit = meters per second (m/s)

Temperature

In SI, the Celsius scale is used to express temperature measurements. In the Celsius scale, 0°C (32°F) is the freezing point of water, and 100°C (212°F) is the boiling point of water. You can use the temperature scale shown below to convert between the Celsius scale and the Fahrenheit scale, which is commonly used in the United States. You can also use the following equation to convert between degrees Celsius (T_C) and degrees Fahrenheit (T_F):

$$T_F = \frac{9}{5}T_C + 32$$

For example, to convert 0°C to degrees Fahrenheit, do the following:

$$T_F = \frac{9}{5}(0°C) + 32°F = 0 + 32°F = 32°F$$

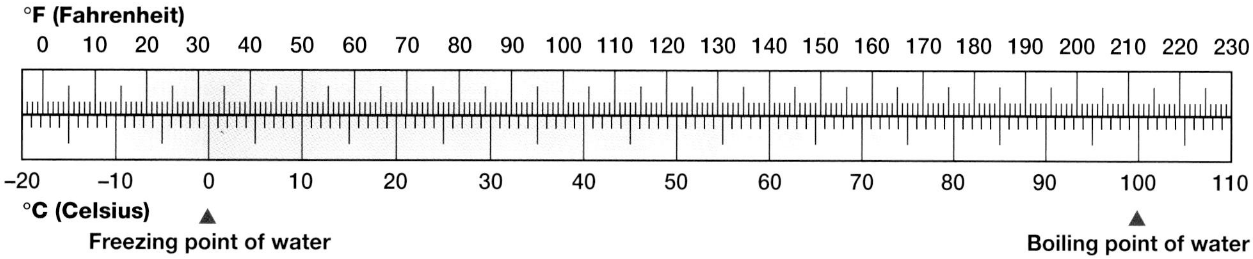

Line Graphs

Line graphs, such as the one shown at right, are most often used to compare or relate one or more sets of data that show continuous change. In the graph shown at right, both "Daily salt intake" and the "Systolic pressure" are the **variables,** or sets of data that are being compared.

In graphs, values are assigned to the independent variable. In this case, "Daily salt intake" is the **independent variable.** "Systolic pressure" is called the **dependent variable** because blood pressure is affected by salt intake, according to the graph. Each set of data—the independent and dependent variables—is called a data pair.

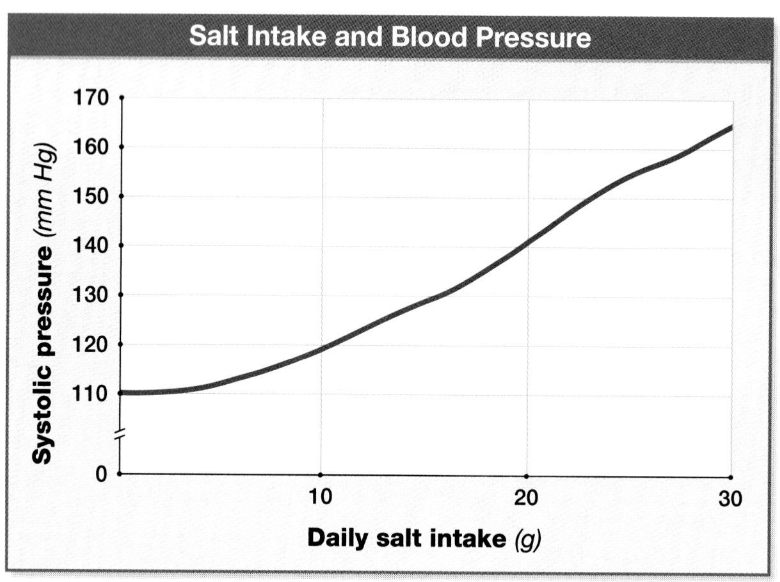

Another way to think about independent and dependent variables is to think about the amount of sleep you get. You know that how alert or tired you feel often depends on the number of hours of sleep that you had the night before. The amount of sleep is the independent variable; your alertness is the dependent variable. Studying biology, you will see many examples of dependent and independent variables represented in graphs.

When you are making a line graph using data pairs, first organize data pairs into a table. Then draw the horizontal and vertical axes of your graph. Be sure to label each axis, including units where appropriate. Refer to your data table to determine the scale and interval of each axis. Make sure that the scale and interval of each axis are consistent. Plot each data pair on the graph. Then connect the plotted data points to make a line, or curve. Finally, give the graph a title that clearly indicates the relationship between the data shown by the graph.

Characteristics of line graphs

Important characteristics of line graphs include the following:

- The independent variable is put on the horizontal (x) axis.
- The dependent variable is put on the vertical (y) axis.
- Both axes are labeled.
- An appropriate scale and interval are used on each axis. The same scale and interval must be used for the total length of the axis.
- Reasonable starting points are used for each axis.
- The data pairs are plotted as accurately as possible.
- The title of the graph accurately reflects the data presented.
- If more than one set of data is presented on a graph, a key must accompany the graph.
- The graph is easy to understand and interpret.

Bar Graphs

Sometimes it is not appropriate to use a line graph to represent data. A bar graph, such as the one shown at right, is appropriate for data that are not continuous. A bar graph is a good indicator of trends if the data are taken over a sufficiently long period of time. For example, studying color variations in moths requires that data be collected over a long period of time. Even after years of study, predictions can still be difficult to make with certainty. Notice that a bar graph is also useful in comparing multiple sets of data, such as those for the light and dark moths found in the woods near Birmingham and Dorset.

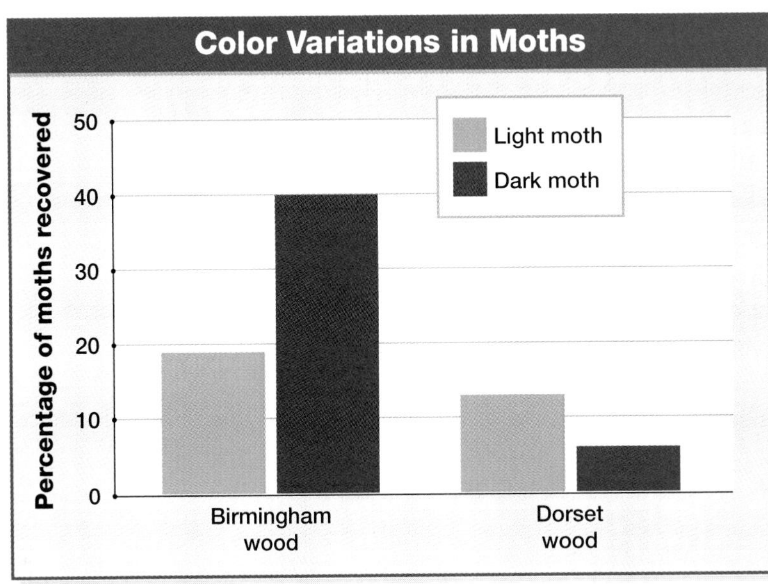

Characteristics of bar graphs

Important characteristics of bar graphs include the following:

- An appropriate scale is used on each axis.
- Reasonable starting points are used for each axis.
- A key accompanies the graph.
- The axes are labeled.
- Data are accurately plotted.
- The title of the graph accurately reflects the data presented.
- The graph is easy to understand and interpret.

Using Graphs to Make Predictions

Graphs show trends in data that may not be obvious from a data chart or table. Examine the graph of salt intake and blood pressure on page 1046. Do you notice any trends? Do you think you can conclude that salt intake affects blood pressure? The process of going beyond the data points in a graph to determine a relationship between the data is called **extrapolation.** The further we extrapolate, the less certain we can be of our predictions.

We can also use graphs to link information quickly between two sets of data pairs. For example, we can read the graph to determine that when the subject takes in 10 g of salt daily, the subject's systolic blood pressure is about 120 mm Hg. Linking information between two sets of data pairs in this way is called **interpolation.** Interpolation will help you identify relationships between sets of data in data pairs.

Study TIP

Interpreting Graphics

When reading graphs, identify the dependent and independent variables. Then see if you can determine a relationship between the two variables by interpolation.

A Six-Kingdom System of Classification

This is a classification system based on the six-kingdom system presented in Chapter 20. The information that follows does not include every group in each kingdom, and the placement of some groups is controversial. For example, biologists do not agree on how the kingdoms Eubacteria and Archaebacteria should be divided into phyla. Here these two kingdoms are divided into broad, generally recognized groups. There is also disagreement about the number of species in various groups. Unless stated otherwise, the numbers given represent approximate numbers of living, named species.

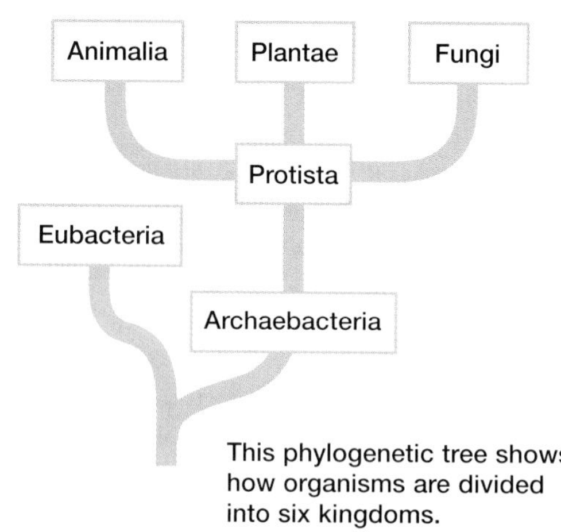

This phylogenetic tree shows how organisms are divided into six kingdoms.

Kingdom Eubacteria

More than 4,000 species
Typically unicellular; prokaryotic; without membrane-bound organelles; nutrition mainly heterotrophic (by absorption), but some are photosynthetic or chemosynthetic; reproduction usually by fission or budding.

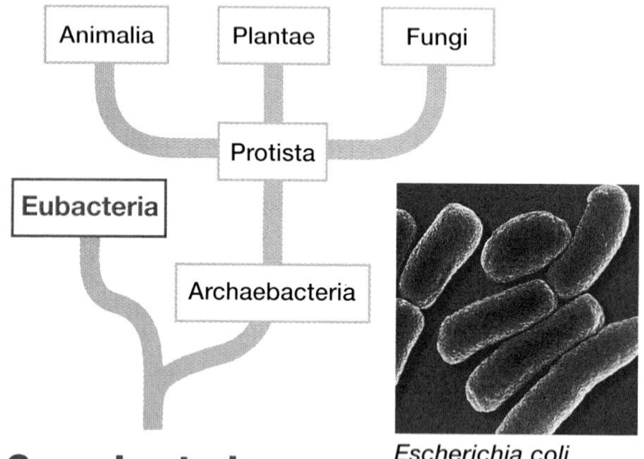

Escherichia coli

Cyanobacteria

Photosynthetic; surrounded by a pigmented covering; common on land and in the ocean; probably ancestors of chloroplasts in some protists.
Examples: *Anabaena, Oscillatoria, Spirulina*

Chemoautotrophs

Ancient bacteria that can grow without sunlight or other organisms; derive energy from reduced gases—ammonia (NH_3), methane (CH_4), hydrogen sulfide (H_2S); play critical roles in Earth's nitrogen cycles; includes nitrobacteria and sulfur bacteria.
Examples: *Nitrosomonas, Nitrobacter*

Enterobacteria

Typically rigid, rod-shaped, heterotrophic bacteria; can be aerobic or anaerobic; have flagella; responsible for many serious diseases of plants and humans.
Examples: *Escherichia coli, Salmonella typhimurium*

Pseudomonads

Straight or curved rods with flagella at one end; strict aerobes; common in soil; many are plant pathogens.
Example: *Pseudomonas aeruginosa*

Spirochaetes

Long, spiral cells; flagella originating at each end; responsible for several serious diseases.
Examples: *Treponema pallidum, Borrelia burgdorferi*

Actinomycetes

Filamentous bacteria that are often mistaken for fungi; spore-producing; sources of antibiotics including streptomycin, tetracycline, and chloramphenicol; cause diseases including dental plaque, leprosy, and tuberculosis.
Example: *Mycobacterium tuberculosis*

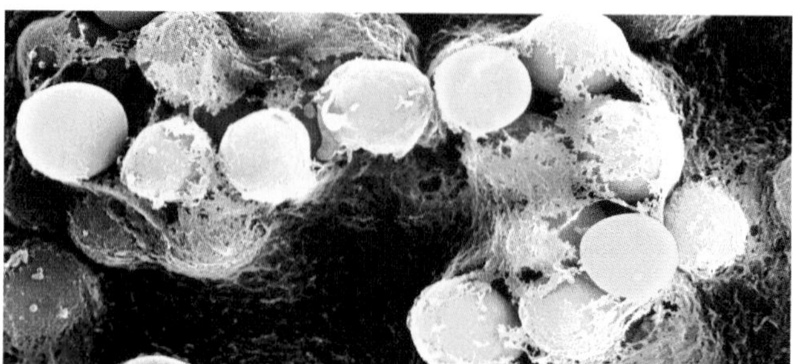

Streptococcus

Rickettsias
Parasitic bacteria found within the cells of vertebrates and arthropods; cause serious diseases.
Example: *Rickettsia rickettsii*

Gliding and budding bacteria
Rod-shaped cells; secrete slimy polysaccharides; often aggregate into gliding masses; live mainly in soil.
Example: *Myxobacteria*

Kingdom Archaebacteria

Fewer than 100 species
Includes anaerobic and aerobic bacteria adapted to extreme environments; prokaryotic; differ from eubacteria in structure of cell wall and cell membrane; similarities to eukaryotes suggest that archaebacteria are more closely related to eukaryotes than to eubacteria; asexual reproduction only; divided into three broad groups. There is evidence they also exist freely in oceans but have not yet been cultured in labs.

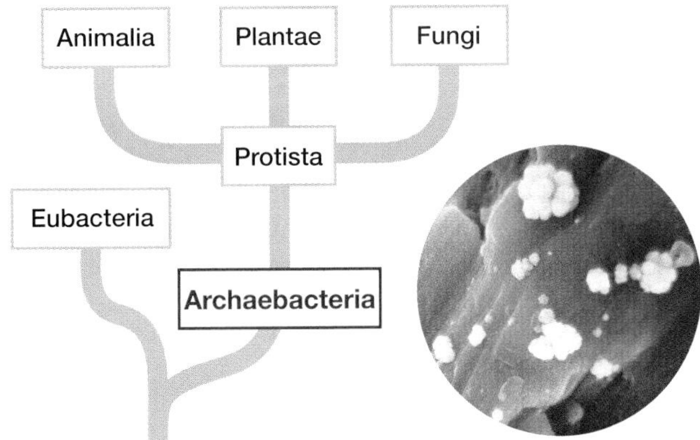

Acidianus brierleyi

Methanogens
Anaerobic methane producers; most species use carbon dioxide as a carbon source; inhabit soil, swamps, and the digestive tracts of animals, particularly grazing mammals such as cattle; produce nearly 2 trillion kilograms (2 billion tons) of methane gas annually.
Example: *Methanobrevibacter ruminatium*

Thermoacidophiles
Inhabit hot, acidic environments; can tolerate high temperatures; require sulfur; mostly anaerobic.
Example: *Sulfolobus solfataricus*

Extreme halophiles
Inhabit environments with very high salt content (salinity 15 to 20 percent), including the Dead Sea and the Great Salt Lake; many aerobic; gram-negative.
Example: *Halobacteroides holobius*

Kingdom Protista

About 43,000 species
Includes eukaryotes that are not plants, fungi, or animals; the most structurally diverse kingdom; both unicellular and multicellular; membrane-bound nucleus; nearly all have chromosomes, mitochondria, and internal compartments; many have chloroplasts; most have cell walls; reproduce sexually and asexually; aquatic or parasitic; many live in soil.

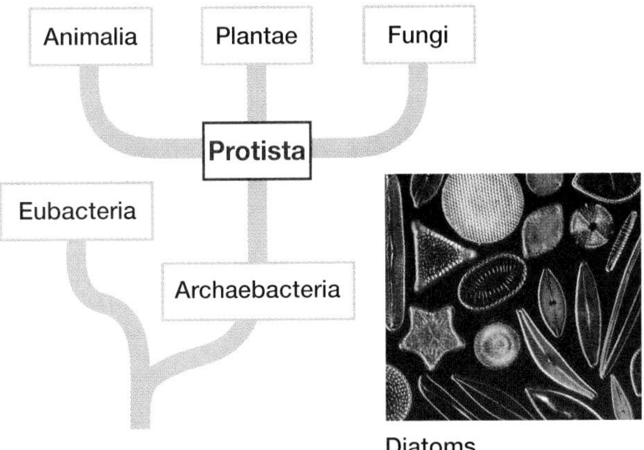

Diatoms

Phylum Rhizopoda
About 300 species
Unicellular and heterotrophic; amorphously shaped cells that move using cytoplasmic extensions called pseudopods; includes amoebas.

Phylum Foraminifera
About 300 species
Unicellular and heterotrophic; marine; have shells of organic material with pores through which many cytoplasmic threads project; includes forams.
Example: *Heterostegina depressa*

Phylum Bacillariophyta
More than 11,500 species
Unicellular and photosynthetic; secrete a unique shell made of opaline silica that resembles a box with a lid; chloroplasts resemble those of brown algae; contain chlorophylls *a* and *c* and fucoxanthin; includes diatoms.

Phylum Chlorophyta
About 7,000 species
Unicellular, colonial, and multicellular; photosynthetic; contain chlorophylls *a* and *b;* contain chloroplasts similar to those of plants; scientists think plants descended from this group; includes green algae.
Examples: *Chlamydomonas, Chorella, Oedogonium, Spirogyra, Ulva, Volvox*

Phylum Rhodophyta
About 4,000 species
Almost all are multicellular; photosynthetic; most are marine; contain chlorophyll *a* and phycobilins; chloroplasts probably evolved from symbiotic cyanobacteria; includes red algae.
Example: *Porphyra*

Phylum Phaeophyta
About 1,500 species
Multicellular and photosynthetic; nearly all are marine; contain chlorophylls *a* and *c* and fucoxanthin, which is the source of their brownish color; includes brown algae.
Examples: *Fucus, Laminaria, Postelsia, Sargassum*

Phylum Dinoflagellata
More than 2,100 species
Unicellular; heterotrophic and autotrophic species; mostly marine; body enclosed within two cellulose plates; contain chlorophylls *a* and *c* and carotenoids; includes dinoflagellates.
Examples: *Gonyaulax, Noctiluca*

Phylum Zoomastigina
About 3,000 species
Mostly unicellular; heterotrophic; all have at least one flagellum; includes zoomastigotes.
Examples: *Giardia, Leishmania, Trypanosoma*

Phylum Euglenophyta
About 1,000 species
Unicellular; both photosynthetic and heterotrophic species; asexual; most live in fresh water; chloroplasts are similar to those of green algae and are thought to have evolved from the same symbiotic bacteria; includes euglenoids.
Example: *Euglena*

Phylum Ciliophora
About 8,000 species
Very complex single cells; heterotrophic; have rows of cilia and two types of cell nuclei; includes ciliates.
Examples: *Didinium, Paramecium, Stentor, Vorticella*

Phylum Acrasiomycota
About 70 species
Heterotrophic; amoeba-shaped cells that aggregate into a moving mass called a slug when they are deprived of food; cells within the slug retain their membranes and do not fuse; a slug produces spores that form new amoebas elsewhere; includes cellular slime molds.
Example: *Dictyostelium*

Phylum Myxomycota
About 500 species
Heterotrophic; individuals stream along as a multinucleate mass of cytoplasm; can give rise to spores that start a new individual in a more favorable environment; includes plasmodial slime molds.
Example: *Physarum*

Phylum Oomycota
About 580 species
Heterotrophic; unicellular parasites or decomposers; cell walls composed of cellulose, not chitin as in fungi; includes water molds, white rusts, and downy mildews.
Example: *Phytophthora*

Phylum Apicomplexa
About 3,900 species
Unicellular; heterotrophic; nonmotile; spore-forming parasites of animals; have complex life cycles; asexual and sexual reproduction; includes sporozoans.
Examples: *Plasmodium, Toxoplasma*

Kingdom Fungi

About 77,000 species
Eukaryotic heterotrophs with nutrition by absorption; all but yeasts are multicellular; nearly all are terrestrial; body is typically composed of filaments (called hyphae) and is multinucleate, with incomplete divisions (called septae) between cells; cell walls made of chitin; about 17,000 species (known as deuteromycetes) without a sexual stage.

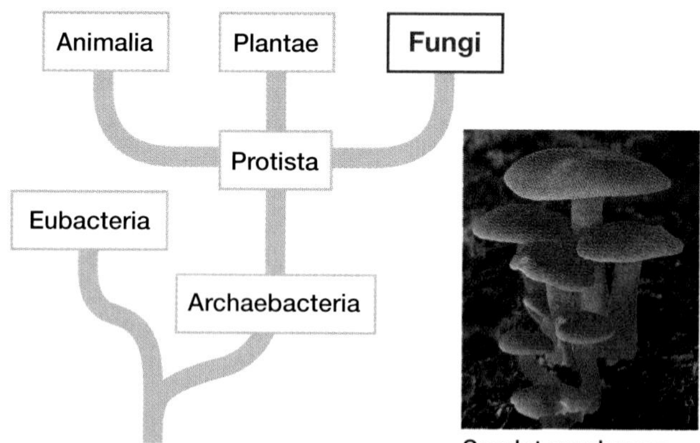

Scarlet mushroom

Phylum Zygomycota
About 665 species
Usually lack septae; fusion of hyphae leads to formation of zygote, which divides by meiosis when it germinates; terrestrial or parasitic; includes bread molds.
Examples: *Pilobolus, Rhizopus*

Phylum Ascomycota

About 30,000 species

Hyphae usually have perforated septae; fusion of hyphae leads to formation of densely interwoven mass that contains characteristic microscopic reproductive structures called asci (singular, ascus); many fungi formally grouped as Fungi Imperfecti are now grouped here; terrestrial, marine, and freshwater species; includes brewer's and baker's yeasts, molds, morels, and truffles.

Examples: *Neurospora, Saccharomyces*

Phylum Basidiomycota

About 16,000 species

Hyphae usually have incomplete septae; reproduction is typically sexual; fusion of hyphae leads to the formation of densely interwoven reproductive structure (mushroom) with characteristic microscopic structures called basidia (singular, basidium); includes mushrooms, toadstools, shelf fungi, rusts, and smuts.

Fungal Associations

About 20,000 species

Fungi form symbiotic associations with plants, green algae, and cyanobacteria.

Lichen

Lichens

About 15,000 species

Mutualistic relationships between fungi (almost always ascomycetes) and cyanobacteria, green algae, or both; the photosynthetic partners actually live among the hyphae of the fungus; the fungus derives energy from its photosynthetic partners.

Mycorrhizae

About 5,000 species

Mutualistic relationships between fungi and the roots of plants; 80 percent of all plants have mycorrhizae associated with their roots; the plant provides sugars to the fungi; in return, the fungi serve as accessory roots, greatly increasing the surface area available for the absorption of nutrients.

Kingdom Plantae

About 280,000 species

Multicellular; eukaryotic; mostly autotrophic; mostly terrestrial organisms containing tissues and organs; cell walls with cellulose; contain chlorophylls *a* and *b* in plastids; life cycle is alternation of generations.

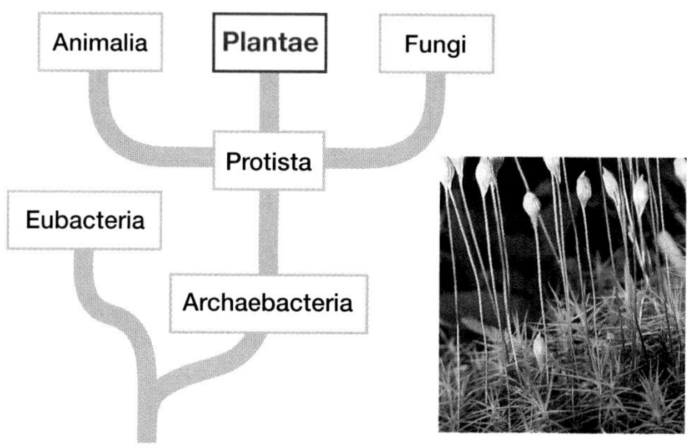

Moss

Phylum Bryophyta

About 10,000 species

Nonvascular plants; gametophytes are larger than sporophytes; sporophytes grow on gametophytes; have simple conducting tissue; lack roots, stems, and leaves; includes mosses.

Example: *Sphagnum*

Phylum Hepatophyta

About 6,000 species

Nonvascular plants; gametophytes are larger than sporophytes; sporophytes grow on gametophytes; lack stomata, roots, stems, and leaves; includes liverworts.

Example: *Marchantia*

Phylum Anthocerophyta

About 100 species

Nonvascular plants; gametophytes are larger than sporophytes; sporophytes grow on gametophytes; sporophytes have stomata; lack roots, stems, and leaves; includes hornworts.

Example: *Anthoceros*

Phylum Pterophyta

About 11,000 species

Seedless vascular plants; sporophytes are larger than gametophytes; sporophytes have roots, stems, and leaves that produce spores on their lower surfaces; gametophytes are small, flat, and independent; includes ferns.

Example: *Salvinia*

Phylum Lycophyta

About 1,000 species

Seedless vascular plants; sporophytes are larger than gametophytes; sporophytes produce spores in cones, resemble moss gametophytes, and have roots, stems, and leaves; gametophytes are small, flat, and independent; includes club mosses.

Examples: *Lycopodium, Selaginella*

Phylum Sphenophyta

15 species

Seedless vascular plants; sporophytes are larger than gametophytes; sporophytes produce spores in cones and have roots, leaves, and jointed stems; gametophytes are small, flat, and independent; includes horsetails.

Example: *Equisetum*

Phylum Psilotophyta

Several species

Seedless vascular plants; sporophytes are larger than gametophytes; sporophytes produce spores in sporangia at tips of stems and have roots and stems but no leaves; gametophytes are small, flat, and independent; includes whisk ferns.

Example: *Psilotum*

Phylum Coniferophyta

About 550 species

Gymnosperms, seed plants that produce naked seeds; sporophytes are mostly evergreen trees or shrubs with needlelike or scalelike leaves; male and female gametophytes are microscopic and develop from spores produced within cones on sporophytes; includes pines, spruces, firs, larches, and yews.

Examples: *Pinus, Taxus*

Phylum Cycadophyta

About 100 species

Gymnosperms, seed plants that produce naked seeds; sporophytes are evergreen trees and shrubs with palmlike leaves; male and female gametophytes are microscopic and develop from spores produced within cones on separate sporophytes; includes cycads.

Example: *Cycas*

Phylum Ginkgophyta

1 species

Gymnosperm, seed plant that produces naked seeds; sporophyte is a deciduous tree with fan-shaped leaves and fleshy seeds; male and female gametophytes are microscopic and develop from spores produced by separate sporophytes; includes *Ginkgo biloba.*

Phylum Gnetophyta

About 70 species

Gymnosperms, seed plants that produce naked seeds; sporophytes are shrubs or vines with some angiosperm characteristics; male and female gametophytes are microscopic and develop from spores produced within cones on sporophytes; includes gnetophytes.

Examples: *Ephedra, Welwitschia*

Phylum Anthophyta

About 250,000 species

Angiosperms, seed plants that produce seeds within a fruit; sporophytes are trees, shrubs, herbs, or vines that produce flowers; male and female gametophytes are microscopic and develop from spores produced within the reproductive structures of a flower; includes flowering plants.

Examples: *Aster, Prunus, Quercus, Zea*

Class Monocotyledones

About 70,000 species

Embryos have one cotyledon; flower parts in multiples of three; leaf veins parallel; vascular bundles scattered through stem tissue; includes grasses, sedges, lilies, irises, palms, and orchids.

Class Dicotyledones

About 180,000 species

Embryos have two cotyledons; flower parts in multiples of two, four, or five; leaves with netlike veins; vascular bundles in stems are arranged in rings; includes daisies, roses, maples, and elms.

Rose

Kingdom Animalia

More than 1 million species
Multicellular, eukaryotic, heterotrophic organisms; nutrition mainly by ingestion; most have specialized tissues, and many have complex organs and organ systems; no cell walls or chloroplasts; sexual reproduction predominates; both aquatic and terrestrial forms.

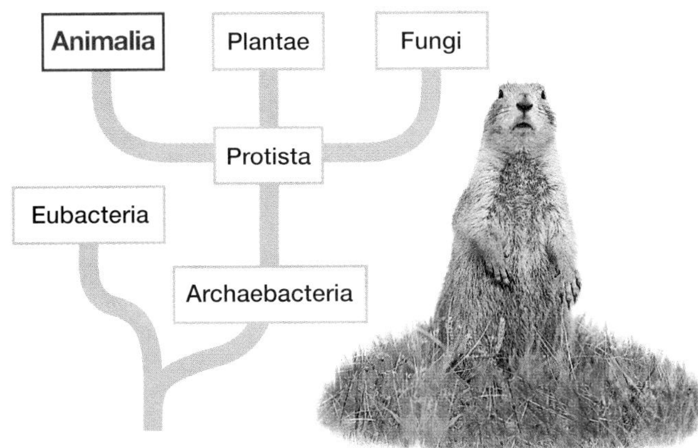

Prairie dog

Phylum Porifera

About 9,000 species
Asymmetrical; lack tissues and organs; body wall consists of two cell layers, penetrated by numerous pores; internal cavity is lined with unique food-filtering cells called choanocytes; sexual and asexual reproduction; mostly marine; includes sponges.

Phylum Cnidaria

About 10,000 species
Radially symmetrical; most have distinct tissues; baglike body of two cell layers; gelatinous; marine and freshwater species.

Class Hydrozoa

About 2,700 species
Most have both polyp and medusa stages in life cycle; includes hydras.
Examples: *Hydra, Obelia, Physalia*

Class Scyphozoa

About 200 species
Exclusively marine; medusa stage dominant; includes jellyfish.
Example: *Aurelia*

Class Anthozoa

About 6,200 species
Marine; solitary or colonial; medusa stage absent; includes sea anemones, corals, and sea fans.

Phylum Ctenophora

About 100 species
Radially symmetrical; transparent, gelatinous bodies resembling jellyfish; marine; includes comb jellies.

Phylum Platyhelminthes

About 20,000 species
Bilaterally symmetrical acoelomates; body flat and ribbonlike, without true segments; organs present; three germ layers; includes flatworms.

Class Turbellaria

More than 3,000 species
Mostly free-living aquatic or terrestrial forms; includes planarians.
Example: *Dugesia*

Class Cestoda

About 1,500 species
Specialized internal parasites; no digestive system; body sections called proglottids; hooked scolex for attaching to host; includes tapeworms.
Example: *Taenia saginata*

Class Trematoda

About 6,000 species
Internal parasites, with mouth at anterior end; often have complex life cycle with alternation of hosts; cause disease in humans and other animals; includes flukes.
Examples: *Schistosoma, Chlonorchis sinensis*

Phylum Nematoda

More than 12,000 species
Tiny, parasitic, unsegmented worms; long, slender body; pseudocoelomates; includes roundworms.
Examples: *Ascaris, Trichinella spiralis, Necator, Toxocara canis, Toxocara cati*

Phylum Mollusca

About 110,000 species
Soft-bodied animals with a true coelom; three-part body consisting of foot, visceral mass, and mantle; protostomes; most have a unique rasping tongue (radula); terrestrial, freshwater, and marine.

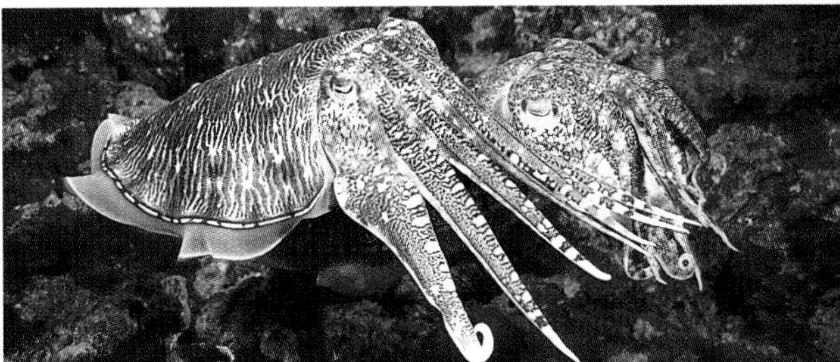

Cuttlefish

Class Polyplacophora
About 600 species
Elongated body and reduced head; similar to ancestral mollusk form; includes chitons.

Class Gastropoda
About 80,000 species
Visceral mass twisted during development; head, distinct eyes, and tentacles usually present; includes gastropods, such as snails, slugs, and whelks.

Class Bivalvia
10,000 species
Two shells connected by a hinge; no radula; large, wedge-shaped foot; includes bivalves, such as oysters, clams, and scallops.

Class Cephalopoda
More than 600 species
Foot modified into tentacles; includes cephlapods, such as squids, octopuses, nautilus, and cuttlefish.

Phylum Annelida
About 12,000 species
Serially segmented worms; bilaterally symmetrical; protostomes.

Class Polychaeta
About 8,000 species
Fleshy outgrowths called parapodia extend from segments; many bristles (setae); marine; includes feather dusters.
Example: *Nereis*

Class Oligochaeta
About 3,100 species
Head not well developed; no parapodia; few setae; terrestrial and freshwater forms; includes earthworms.

Class Hirudinea
About 600 species
Body flattened; no parapodia; usually suckers at both ends; many are external parasites; includes leeches.

Phylum Arthropoda
About 1 million species
Segmented bodies with paired, jointed appendages; bilaterally symmetrical; chitinous exoskeleton; protostomes; aerial, terrestrial, and aquatic forms.

Subphylum Chelicerata
Distinguished by absence of antennae and presence of chelicerae; all appendages unbranched; four pairs of walking legs; two body regions (cephalothorax and abdomen); predominantly terrestrial.

Class Arachnida
About 57,000 species
Terrestrial; use book lungs and tracheae for respiration; four pairs of legs; includes spiders, scorpions, ticks, and mites.

Class Merostomata
5 species
Cephalothorax covered by protective "shell"; sharp spike on tail; includes horseshoe crabs.

Class Pycnogonida
About 1,000 species
Small marine predators or parasites; usually four pairs of legs; includes sea spiders.

Subphylum Crustacea
About 35,000 species
Two pairs of antennae, mandibles, and appendages with two branches; predominantly aquatic.

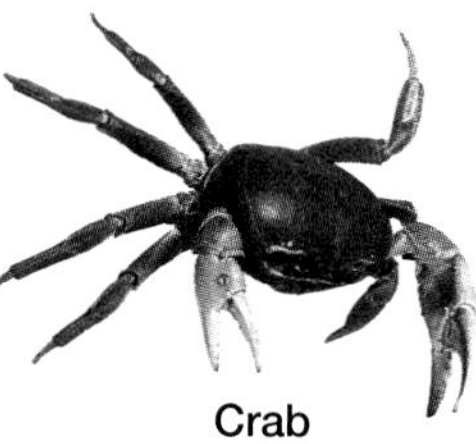
Crab

Class Malacostraca
About 20,000 species
Typically five pairs of legs; two pairs of antennae; most are aquatic; includes crayfish, lobsters, crabs, shrimp, sow bugs, and krill.

Subphylum Uniramia
Antennae, mandibles, and unbranched appendages.

Class Insecta
About 750,000 species
Head, thorax, and abdomen; three pairs of legs, all attached to thorax; usually two pairs of wings.

Order Coleoptera: includes beetles, ladybugs, and weevils.
Order Diptera: includes flies, mosquitoes, gnats, and midges.
Order Lepidoptera: includes butterflies and moths.
Order Hymenoptera: includes bees, ants, wasps, hornets, and ichneumon fly.

Beetle

Order Hemiptera: includes water striders, water boatmen, back swimmers, bedbugs, squash bugs, stink bugs, and assassin bugs.
Order Homoptera: includes cicadas, aphids, leaf hoppers, and scale insects.
Order Orthoptera: includes grasshoppers, cockroaches, walking sticks, praying mantises, and crickets.
Order Odonata: includes dragonflies and damsel flies.
Order Neuroptera: includes ant lions and lacewings.

Order Thysanura: includes silverfish, bristletails, and firebrats.
Order Anoplura: includes sucking lice.
Order Isoptera: includes termites.
Order Ephemeroptera: includes mayflies.
Order Siphonaptera: includes fleas.
Order Dermaptera: includes earwigs.

Class Chilopoda
About 2,500 species
Body flattened and consisting of 15–170 or more segments; one pair of legs attached to each segment; includes centipedes.

Class Diplopoda
About 10,000 species
Elongated body of 15–200 segments; two pairs of legs per segment; herbivorous; includes millipedes.

Phylum Echinodermata
About 6,000 species
Deuterostomes; adults radially symmetrical with five-part body plan; most forms have water vascular system with tube feet for locomotion; marine.

Class Asteroidea
About 1,500 species
Body usually with five arms and double rows of tube feet on each arm; mouth directed downward; includes sea stars.

Sea star

Class Ophiuroidea
About 2,000 species
Usually with five slender, delicate arms or rays; includes brittle stars and basket star.

Class Crinoidea
About 600 species
Mouth faces upward and is surrounded by many arms; includes sea lilies and feather stars.

Class Echinoidea
About 900 species
Body spherical, oval, or disk-shaped; arms lacking but five-part body plan still apparent; includes sea urchins and sand dollars.

Class Holothuroidea
About 1,500 species
Elongated, thickened body with tentacles around the mouth; includes sea cucumbers.

Phylum Chordata
About 42,500 species
Bilaterally symmetrical; deuterostomes; coelom present; have notochord, dorsal nerve cord, pharyngeal slits, and tail; aquatic and terrestrial.

Subphylum Urochordata
About 1,250 species
Saclike covering, or tunic, in adults; larvae are free-swimming and have nerve cord and notochord; marine; includes tunicates.

Subphylum Cephalochordata
23 species
Small and fishlike with a permanent notochord; filter feeders; includes lancelets.

Subphylum Vertebrata
About 40,000 species
Most of the notochord is replaced by a spinal column composed of vertebrae that protect the dorsal nerve cord; recognizable head containing a brain.

Jawless fishes
63 species
Freshwater or marine eel-like fishes without true jaws, scales, or paired fins; cartilaginous skeleton; includes lampreys and hagfish.

Cartilaginous fishes
About 850 species
Fishes with jaws and paired fins; gills present; no swim bladder; cartilaginous skeleton; includes sharks, rays, and skates.

Bony fishes
About 18,000 species
Freshwater and marine fishes with gills attached to gill arch; jaws and paired fins; bony skeleton; most have swim bladder; includes rayfins, such as tuna, sea horse, trout, perch, sturgeon, and angelfish; and lobefins, such as coelacanth and lungfish.

Class Amphibia
About 4,200 species
Freshwater or terrestrial; gills present at some stage; skin often slimy and lacking scales; eggs typically laid in water and fertilized externally.

Order Anura: includes frogs and toads.
Order Urodela: includes salamanders and newts.
Order Apoda: includes caecilians.

Class Reptilia
About 7,000 species
Terrestrial or semiaquatic vertebrates; breathe using lungs at all stages; body covered by scales; most species lay amniotic eggs covered with a protective shell; fertilization internal.

Order Squamata: includes lizards and snakes.
Order Chelonia: includes turtles and tortoises.

A Six-Kingdom System *continued*

Order Crocodilia: includes alligators, crocodiles, gavials, and caimans.
Order Rhynchocephalia: includes tuataras.

Class Aves

About 9,000 species
Body covered with feathers; forelimbs modified into wings; four-chambered heart; endothermic; lay shelled, amniotic eggs.

Order Passeriformes: includes robins, bluebirds, sparrows, warblers, and thrushes.
Order Apodiformes: includes swifts and hummingbirds.
Order Piciformes: includes woodpeckers, sapsuckers, flickers, and toucans.
Order Psittaciformes: includes parrots, parakeets, macaws, and cockatoos.
Order Charadriiformes: includes snipes, sandpipers, plovers, gulls, terns, auks, puffins, and ibises.

Tern

Order Columbiformes: includes pigeons and doves.
Order Falconiformes: includes hawks, falcons, eagles, kites, and vultures.
Order Gaviiformes: includes loons.
Order Gruiformes: includes cranes, coots, gallinules, and rails.
Order Anseriformes: includes ducks, geese, and swans.
Order Strigiformes: includes owls.
Order Ciconiiformes: includes herons, bitterns, egrets, storks, spoonbills, and ibises.
Order Pelecaniformes: includes pelicans, cormorants, and gannets.
Order Galliformes: includes pheasants, turkeys, quails, partridges, and grouse.
Order Procellariiformes: includes albatrosses and petrels.
Order Cuculiformes: includes cuckoos and roadrunners.
Order Caprimulgiformes: includes goatsuckers, whippoorwills, and nighthawks.
Order Coraciiformes: includes kingfishers.
Order Sphenisciformes: includes penguins.
Order Struthioniformes: includes ostriches.
Order Apterygiformes: includes kiwis.

Class Mammalia

About 4,400 species
Hair on at least part of body; young nourished with milk secreted by mammary glands; endothermic; breathe with lungs.

Order Rodentia: includes squirrels, woodchucks, mice, rats, muskrats, and beavers.
Order Chiroptera: includes bats.

Bat

Order Insectivora: includes moles and shrews.
Order Marsupialia: includes opossums, kangaroos, koalas, and wallabies.
Order Carnivora: includes bears, weasels, mink, otters, skunks, lions, tigers, and wolves.
Order Primates: includes monkeys, lemurs, gibbons, orangutans, gorillas, chimpanzees, and humans.
Order Artiodactyla: includes hippopotamuses, camels, llamas, deer, giraffes, cattle, sheep, and goats.
Order Cetacea: includes whales, porpoises, and dolphins.
Order Lagomorpha: includes rabbits, hares, and pikas.
Order Pinnipedia: includes seals, sea lions, and walruses.
Order Edentata: includes armadillos, sloths, and anteaters.
Order Macroscelidea: includes elephant shrews.
Order Perissodactyla: includes tapirs, rhinoceroses, horses, and zebras.
Order Scandentia: includes tree shrews.
Order Hyracoidea: includes hyraxes.
Order Pholidota: includes pangolins.
Order Sirenia: includes sea cows, dugongs, and manatees.
Order Monotremata: includes duckbill platypus and spiny anteaters.
Order Dermoptera: includes flying lemurs.
Order Proboscidea: includes elephants.
Order Tubulidentata: includes aardvark.

Polar bear

Glossary

Pronunciation Key

Sound	As In	Phonetic Respelling
ahy	bat	(BAT)
ay	face	(FAYS)
ah	lock argue	(LAHK) (AHR gyoo)
ow	out	(OWT)
ch	chapel	(CHAP uhl)
eh	test	(TEHST)
ai	rare	(RAIR)
ee	eat feet ski	(EET) (FEET) (SKEE)
ih	bit	(BIHT)
eye	idea	(eye DEE uh)
y	ripe	(RYP)

Sound	As In	Phonetic Respelling
ihng	going	(GOH ihng)
k	card kite	(KAHRD) (KEYET)
ng	anger	(ANG guhr)
oh	over	(OH vuhr)
aw	dog horn	(DAWG) (HAWRN)
oy	foil	(FOYL)
u	pull	(PUL)
oo	pool	(POOL)
s	cell sit	(SEHL) (SIHT)
sh	sheep	(SHEEP)

Sound	As In	Phonetic Respelling
th	that thin	(THAT) (THIHN)
uh	cut	(CUHT)
ur	fern	(FURN)
y	yes	(YEHS)
yoo	globule	(GLAHB yool)
yu	cure	(KYUR)
z	bags	(BAGZ)
zh	treasure	(TREHZH uhr)
uh	medal pencil onion	(MEHD uhl) (PEHN suhl) (UHN yuhn)
uhr	paper	(PAY puhr)

A

ABO blood group system a system used to classify human blood by proteins found on the surface of red blood cells (888)

abiotic factors physical aspects of a habitat (360)

acandothian early fish; first to exhibit jaws (733)

acid compound that increases the concentration of hydrogen ions in a solution (33)

acid rain precipitation with below-normal pH, often the result of industrial pollution and automobile exhaust (406)

acoelomate an animal that lacks a coelom, or body cavity (618)

actin thin protein filament found in muscles that functions in contraction (867)

action potential sudden reversal of polarity across a neuron membrane (954)

activation energy amount of energy required to start a chemical reaction (39)

active site site on an enzyme that attaches to a substrate (41)

active transport movement against a concentration gradient requiring the use of energy by the cell (81)

adaptation process of becoming adapted to an environment; an anatomical structure, physiological process, or behavioral trait that improves an organism's likelihood of survival and reproduction (280)

addiction physiological dependence on a drug (970)

adductor muscle thick muscle that attaches the two valves of a mollusk and causes the shell to open and close (665)

adhesion attraction between different substances (31)

adrenal gland endocrine gland located above each kidney (993)

aerobic term for processes that require oxygen (104, 463)

aggregation a temporary collection of cells that comes together for a period of time and then separates (436)

agnathan earliest fish; had backbone but no jaws or paired fins (732)

AIDS (acquired immunodeficiency syndrome) disease caused by infection by HIV that results in a depressed immune system (942)

alga historical term for photosynthetic protists—not accepted by all scientists (479)

allele an alternative form of a gene (164)

allergy a reaction by the body's immune system to a harmless antigen (944)

alternation of generations life cycle in which a haploid gamete-producing individual (gametophyte) alternates with a diploid spore-producing individual (sporophyte) (481)

alveolus microscopic air sac in the lung where oxygen and carbon dioxide are exchanged (786, 896)

amino acid organic molecule that is the building block of proteins (36)

amino-acid-based hormone water-soluble hormone made of amino acids (985)

amniotic egg egg containing a water and food supply, key to reproduction on land (785)

amoeba a protist that moves using flexible, cytoplasmic extensions (482)

amoebocyte amoeba-like cell that supplies nutrition and removes wastes from sponge body cells (637)

amylase enzyme that breaks down starches into sugars (915)

anaerobic term for processes that do not require oxygen (104, 462)

analogous characters similar features of organisms that evolve independently (327)

anemia condition in which the oxygen-carrying ability of red blood cells is reduced (887)

angiosperm seed plant that produces reproductive structures in flowers and seeds within a fruit (532)

annual plant that completes its life cycle, reproduces, and dies within one growing season (591)

annual ring layer of secondary xylem that forms each year in a woody stem (593)

anther sac at the tip of a stamen in which pollen grains form (556)

antheridium reproductive structure that produces sperm in seedless plants (548)

anthropoid day-active primates including monkeys, apes, and humans (304)

antibiotic chemical used as a drug to kill bacteria (461)

antibody defensive protein released by plasma cells during an immune response (937)

anticodon a three-nucleotide sequence on tRNA that recognizes a complementary codon on mRNA (210)

antigen substance that triggers an immune response (935)

antigen shifting production of new antigens by a virus as it mutates over time (940)

aorta main artery in the body; receives blood from the left ventricle (892)

apical dominance inhibition of lateral bud growth on the stem of a plant by auxin produced in the terminal bud (599)

apical meristem meristem in the tips of stems and roots (592)

appendage in arthropods, a structure that extends from the body wall (682)

appendicular skeleton bones of the arms and legs (860)

aquifer ground water trapped within porous rock (413)

archaebacteria one of the two kingdoms of prokaryotes; differentiated from eubacteria by various important chemical differences (258)

archegonium reproductive structure that produces eggs in seedless plants (548)

artery vessel that carries blood away from the heart to the body's organs (883)

arthropod member of the phylum Arthropoda, which includes invertebrate animals such as insects, crustaceans, and arachnids (266)

ascus sac that forms on the surface of an ascocarp in which haploid spores are formed (505)

asexual reproduction reproduction that involves only one parent and results in genetically identical offspring (148)

asymmetrical irregular in shape (616)

atom smallest unit of matter that cannot be broken down by chemical means (28)

ATP (adenosine triphosphate) organic molecule composed of a base, a sugar, and three phosphate groups that acts as the main energy currency of cells (37)

atrium a chamber that receives blood returning to the heart (891)

autoimmune disease disease in which the immune system cannot distinguish body cells from pathogens (941)

autosome a chromosome that is not directly involved in determining sex (122)

autotroph organism that obtains energy from sunlight or inorganic substances (94)

auxin type of plant hormone that stimulates cell elongation, causing plants to bend toward a stimulus, and inhibits lateral bud growth (598)

axial skeleton bones of the skull, spine, ribs, and sternum (860)

axon an elongated extension of a neuron that conducts nerve impulses (952)

B

bacillus rod-shaped bacterium (461)

bacteriophage virus that infects bacteria (190, 453)

basal disk a small area on hydra that secretes a sticky substance, which enables them to adhere to rocks or plants in the water (642)

base compound that reduces the concentration of hydrogen ions in a solution (33)

base-pairing rules the rule stating that in DNA adenine on one strand always pairs with a thymine on the opposite strand and guanine on one strand always pairs with a cytosine on the opposite strand (195)

basidium club-shaped sexual reproductive structure that forms on the gills of basidiomycetes (506)

B cell white blood cell that labels pathogens for destruction by macrophages (935)

behavior an action or series of actions performed by an animal in response to a stimulus (834)

biennial plant that completes its life cycle, reproduces, and dies within 2 growing seasons (591)

bilateral symmetry arrangement of body parts so there are distinct left and right halves that mirror each other (616)

binary fission form of asexual reproduction that produces identical offspring (119)

binomial nomenclature a system for giving each organism a two-word scientific name that consists of the genus followed by the species (320)

biodiversity the variety of species living within an ecosystem (361, 390)

biogeochemical cycle cycle in which a substance, such as carbon or water, enters into an environment's living reservoir where it remains for a certain period of time and is returned to a nonliving reservoir (370)

biological magnification process by which molecules passing upward through the trophic levels of a food chain become increasingly concentrated (411)

biological species group of actually or potentially interbreeding natural populations, reproductively isolated from other such groups (325)

biology science of life (6)

biomass the dry weight of tissue and other organic matter found in a specific ecosystem (369)

biome major type of terrestrial ecological community, such as a grassland and desert (392)

biotic factors living organisms in a habitat (360)

bipedal term used to denote the ability to walk erect on two feet (306)

blastocyst hollow-ball embryo that becomes implanted in the uterus (1014)

blastopore opening in the gastrula of protostomes that develops into a mouth (710)

blastula hollow ball of cells that gives rise to all the tissues and organs of an adult body (614)

blood pressure force exerted by blood as it moves through vessels (892)

body cavity fluid-filled spaces that house and protect major internal organs (859)

body plan an animal's shape, symmetry, and internal organization (616)

bone marrow soft tissue inside bones where red blood cells are produced (861)

brain body's main processing center; major organ of the nervous system (958)

brain stem the part of the brain that connects to the spinal cord; contains nerves controlling breathing, swallowing, digestive processes, heartbeat, and blood vessel diameter (959)

bronchus one of the two branches of the trachea that leads to the lungs (896)

budding asexual reproduction where a small cell forms from a large cell and either pinches itself off or remains attached to the large cell (505)

bulbourethral glands glands in male reproductive system that secrete a fluid that neutralizes traces of acidic urine in the urethra (1008)

C

calorie amount of energy needed to raise the temperature of 1 gram of water 1°C. The Calorie used to indicate the energy content of food is a kilocalorie. (908)

Calvin cycle common method of carbon dioxide fixation in photosynthesis (102)

cancer a disease characterized by abnormal cell growth (12, 126)

capillary tiny blood vessel that allows exchange between blood and cells in tissue (883)

capsid a protein sheath that surrounds the nucleic acid core in a virus (453)

capsule a gelatinous outer layer enclosing many bacterial cell walls (461)

carapace dorsal part of a turtle's shell; shieldlike plate covering the cephalothorax of decapods (792)

carbohydrate organic compound composed of carbon, hydrogen, and oxygen; used by living things as an energy source (34)

carbon dioxide fixation process by which carbon dioxide is incorporated into organic compounds (102)

carnivore flesh-eating organism (366)

carotenoids yellow and orange plant pigments that aid in photosynthesis (98)

carrier protein transport protein that carries a specific substance across a cell membrane (80)

carrying capacity population size that an environment can sustain (342)

cartilage strong, lightweight, flexible connective tissue (733)

caste role played by an individual insect in a colony (697)

CD4 receptor protein on human cells targeted by HIV (942)

cell smallest unit that can perform all the life processes (7)

cell cycle repeating five-phase sequence of eukaryotic cell growth and division (125)

cell membrane lipid bilayer with embedded proteins that encloses the cytoplasm; essential to the cell's cytoplasm; also called the plasma membrane (56)

cell theory principle that states all organisms are made of one or more cells that arise from other cells (55)

cell wall the structure that surrounds the cell membrane and provides support for the cell (57)

cellular respiration process by which living organisms harvest the energy in food molecules (95)

central nervous system controls the body; system composed of the brain and spinal cord (958)

central vacuole membrane-bound cavity in plant cells used for storage (66)

centromere region joining two chromatids (119)

cephalization in bilaterally symmetric animals, the development of a head end with a concentration of nerves and sensory structures (617)

cephalothorax the body region of some arthropods that consists of a head fused with a thorax (684)

cereal type of grass plant that produces edible fruits called grains and is grown as food for humans and livestock (536)

cerebellum region of the brain that controls coordination and balance (959)

cerebral ganglion the primitive brain contained in one anterior segment of an annelid (669)

cerebrum largest portion of the brain; center of memory, learning, emotion, and other highly complex functions (959)

chelicera paired mouthparts of arachnids and their relatives that are modified into fangs or pincers (688)

chitin tough carbohydrate found in many fungi and in the exoskeletons of all arthropods (500)

chlamydia bacterial STD marked by painful urination and vaginal discharge (1019)

chlorofluorocarbon any of a group of compounds that contain carbon, chlorine, and fluorine, often used as coolants, propellants, or foaming agents (407)

chlorophyll pigment responsible for trapping light energy in photosynthesis (98)

chloroplast organelle that uses light energy to make carbohydrates from carbon dioxide and water (65)

choanocyte unique flagellated cell with a collar-like structure, found in sponges (637)

chordate deuterostome with a completely internal endoskeleton, notochord, pharyngeal slits, and post anal tail (718)

chromatid one of a pair of strands of DNA that make up a chromosome during meiosis or mitosis (119)

chromosome structure made of DNA and associated proteins on which genes are located (119)

chrysalis protective capsule enclosing the transforming larva in insects (693)

cilium in cells, one hairlike structure in tightly packed rows that are used for movement (58, 486)

cladistics phylogenetic method in which relationships are inferred based on presence of derived characters (327)

cladogram diagram based on patterns of shared, derived traits that shows the evolutionary relationships among groups of organisms (328)

class taxonomic category containing orders with common characteristics (322)

cleavage in development, the rapid, mitotic division of the zygote (1014)

climate prevailing weather conditions in a given area (391)

clone organism produced by asexual reproduction that is genetically identical to its parent (148)

closed circulatory system system in which the blood does not leave the blood vessels and materials diffuse across the walls of the vessels (624)

cnidocyte stinging cell used by cnidarians to stun prey (641)

coccus spherical-shaped bacterium (461)

cochlea fluid-filled chamber of the inner ear that is involved in hearing (967)

codominance condition in which both alleles for a gene are expressed when present (176)

codon a three-nucleotide sequence in DNA or mRNA that encodes an amino acid or signifies a stop signal (209)

coelom fluid-filled body cavity that forms between the body wall and the digestive tract (618)

coelomate animal with a body cavity located entirely within the mesoderm (618)

coevolution reciprocal evolutionary adjustments between interacting members of an ecosystem (382)

cohesion attraction between substances of the same kind (31)

colon organ that compacts waste for excretion; also called the large intestine (918)

colonial organism collection of cells that are permanently associated but in which little or no integration of cell activities occurs (436)

commensalism ecological interaction in which one species benefits and the other species is neither harmed nor helped (384)

community the many different species that live together in a habitat (360)

competition ecological interaction between two or more species that use the same scarce resource such as food, light, and water (385)

competitive exclusion local extinction of one species due to competition (389)

complement system defensive proteins that circulate in the bloodstream (934)

complementary characteristic of nucleic acids in which the sequence of bases on one strand determines the sequence of bases on the other (195)

compound substance that is made of more than one kind of atom (29)

compound eye eye made of thousands of individual visual units (684)

concentration gradient a difference in the concentration of a substance across a space (74)

conditioning learning by association(837)

cone in plants, a cluster of nongreen spore-bearing leaves; in animals, photoreceptor of the retina of the eye that detects color (529, 966)

conjugation temporary union of two organisms to exchange nuclear material (461)

connective tissue tissue that supports, protects, and insulates the body (857)

consumer organism that must obtain energy to build its molecules by consuming other organisms; heterotroph (365)

continental drift common name for movement of the continents over geologic time (268)

contour feather feather that provides insulation and shape to adult birds (794)

control group a group in an experiment that does not receive experimental treatment (17)

convergent evolution process by which unrelated species become similar as they adapt to similar environments (327)

cork type of dermal tissue that replaces the epidermis of woody stems; part of the bark (571)

cork cambium meristem in the bark of a woody plant that produces cork cells (593)

coronary artery artery that branches from the aorta and carries oxygenated blood to the heart muscle (892)

corpus luteum structure that forms from the ruptured follicle in the ovary after ovulation and releases hormones (1012)

cortex outer layer of ground tissue in roots and stems of plants (573)

cotyledon leaflike structure of a plant embryo; functions in the transfer of stored nutrients to the embryo, in which food is stored (553)

countercurrent flow arrangement in fish respiratory system where water going over the gills and blood in the gill filaments flows in opposite directions (757)

crossing-over the exchange of reciprocal segments of DNA by homologous chromosomes at the beginning of meiosis; source of genetic recombination (142)

cud partly digested food that is regurgitated, rechewed, and reswallowed for further digestion by mammals with a rumen (824)

cuticle waxy, watertight, outer covering of nonwoody aboveground parts of a plant (520)

cyanobacteria group of photosynthetic eubacteria (258)

cystic fibrosis fatal genetic disorder in which excessive amounts of mucus are secreted, blocking intestinal and bronchial ducts and causing difficulty in breathing (12)

cytokinesis division of the cytoplasm to form two separate cells (125)

cytoplasm the interior of a cell (56)

cytoskeleton cytoplasmic network of protein filaments that plays an essential role in cell movement, shape, and division (56)

cytotoxic T cell white blood cell that attacks and kills infected cells (935)

D

decomposer organism that causes decay (367)

dendrite cytoplasmic extension of a neuron that receives stimuli (952)

density dependent factor limited resources whose rates of depletion depend on the density of the population using them (342)

density independent factor factors, such as climate, that affect the growth of populations. These factors are unaffected by the density of populations. (344)

deoxyribose five-carbon sugar that is a component of DNA nucleotides (192)

dependent variable variable that is measured in an experiment (17)

depressant drug that decreases the activity of the central nervous system (973)

derived trait unique characteristic of a particular group of organisms (328)

dermal tissue type of plant tissue that covers and protects the body of a plant (570)

dermis thick, functional layer of skin beneath the epidermis (872)

detritivore organism that obtains its energy by feeding on dead organisms or wastes (366)

deuterostome an animal whose anus forms from the blastopore (710)

diabetes mellitus serious disorder in which cells are unable to obtain glucose from the blood; caused by a deficiency of insulin or lack of response to insulin (995)

diaphragm in mammals, sheet of muscle at the bottom of the rib cage that aids in respiration (896)

diatom photosynthetic unicellular protist of the phylum Bacillariophyta (483)

dicot dicotyledon; angiosperm that produces seeds with two cotyledons (533)

differentiation process in which the cells of a multicellular individual become specialized during development (437)

diffusion random movement of particles from an area of high concentration to an area of lower concentration (75)

digestion the process of breaking food down into molecules the body can use (908)

diploid term used to indicate a cell containing two sets of chromosomes ($2n$), one set inherited from each parent (121)

directional selection natural selection that causes the frequency of a particular allele to move in one direction (352)

dispersion the pattern of distribution of organisms in a population (341)

diurnal term describing animals that are active during the day and sleep at night (304)

divergence accumulation of differences between groups; can lead to the formation of new species (293)

DNA (deoxyribonucleic acid) double-stranded, helical nucleic acid that stores hereditary information (37)

DNA fingerprint the pattern of bands that results when an individual's DNA fragments are separated by gel electrophoresis (235)

DNA helicase enzyme that breaks the hydrogen bonds between bases during DNA replication (196)

DNA polymerase enzyme that catalyzes the replication of DNA by adding complementary nucleotides (197)

DNA replication the process of making a copy of DNA (196)

dominant genetic trait that is expressed when its allele is homozygous or heterozygous (164)

dormancy condition in which a seed or plant remains inactive for a period of time, even when conditions are favorable for growth (602)

double fertilization process by which two sperm fuse with cells of the female gametophyte, producing both a diploid ($2n$) zygote and a triploid ($3n$) endosperm (558)

double helix spiral-staircase structure characteristic of the DNA molecule (192)

down feather feather that covers young birds and insulates adult birds (795)

E

ecological race population of a species that differs genetically because of adaptations to different living conditions (293)

ecology the study of the interactions of living organisms with one another and with their environment (9, 360)

ecosystem ecological system encompassing a community and all the physical aspects of its habitat (360)

ectoderm in animals, the outer layer of embryonic tissue from which the skin and nervous system develop (614)

ectothermic referring to an animal whose body temperature is determined by the temperature of the environment (742)

electron microscope microscope that focuses a beam of electrons on an object to form an enlarged image (51)

electron transport chain series of molecules through which electrons are passed to make ATP (100)

electrophoresis a technique that uses an electrical field within a gel to separate molecules by their size and charge (229)

element substance composed of a single type of atom (28)

embryo early stage in the development of plants and animal; term for a developing human during the first 8 weeks of pregnancy (522, 1015)

emerging virus a virus that evolves in a geographically isolated area and is pathogenic to humans (459)

endocrine gland ductless gland in the body that releases its product directly into the bloodstream or the fluid around cells (983)

endocytosis movement of a substance by a vesicle to the inside of a cell (83)

endoderm inner layer of embryonic tissue from which the digestive organs develop in animals (614)

endoplasmic reticulum (ER) cell membranes in the cytoplasm that transport substances made by the cell (62)

endoskeleton an internal skeleton composed of a hard material, such as bone (626)

endosperm nutritious triploid ($3n$) tissue that develops in the seeds of angiosperms (532)

endospore dormant bacterial cell enclosed by a tough coating that is highly resistant to environmental stress (461)

endosymbiosis theory that proposes that mitochondria are the descendants of symbiotic, aerobic eubacteria (259)

endothermic refers to an animal that generates its own body heat (742)

energy capacity for doing work; the ability to move or change matter (38)

energy pyramid diagram that depicts each trophic level in an ecosystem and how much energy is stored in each trophic level (368)

envelope outer layer covering the capsid of many kinds of viruses (453)

enzyme substance that speeds up a chemical reaction (40)

epidermis outermost layer of tissue, consisting of from one to several layers of dead cells (571, 871)

epididymis long, coiled tube on the surface of the testes where sperm mature (1007)

epinephrine amino-acid-based hormone released by the adrenal medulla in times of stress; formerly called adrenaline (993)

epithelial tissue thin, flat tissue layer that lines most body surfaces and protects other tissues from dehydration and physical damage (856)

equilibrium state in which the concentration of a substance is equal throughout a space (74)

esophagus tube that connects the mouth to the stomach (915)

eubacteria one of the two kingdoms of prokaryotes; characterized by presence of peptidoglycan in cell wall (258)

euglenoids members of the phylum Euglenophyta (485)

eukaryote organism whose cells have a nucleus enclosed by a membrane (58)

evolution change in the genetic makeup of a population or species over time (9)

evolutionary systematics method of constructing phylogenies that involves weighing characters by their presumed evolutionary significance (330)

excretion the process of eliminating wastes (920)

exocytosis movement of a substance by a vesicle to the outside of a cell (64)

exon sequence of nucleotides on a gene that gets transcribed and translated (216)

exoskeleton hard external covering of some invertebrates (625)

experiment a planned procedure to test a hypothesis (17)

exponential growth curve J-shaped curve showing the rapid increase in an exponentially growing population (342)

extensor skeletal muscle that causes a joint to straighten (866)

external fertilization union of egg and sperm occurring outside the body of either parent (628)

extinct term used to indicate species that have disappeared permanently (282)

F

F_1 generation the first offspring from a cross of two varieties in the parental (P) generation (162)

F_2 generation the offspring from crosses among individuals of the F_1 generation (162)

$FADH_2$ electron carrier produced during the Krebs cycle (107)

facilitated diffusion transport of substances through a cell membrane down a concentration gradient aided by carrier proteins (80)

fallopian tubes organs of the female reproductive system that lead from the ovaries to the uterus (1010)

family taxonomic category containing genera with similar properties (322)

fermentation anaerobic process of cellular respiration that recycles NAD^+ needed to continue glycolysis (108)

fertilization the process by which haploid gametes join to form a diploid zygote (151)

fetal alcohol syndrome (FAS) birth defect resulting from alcohol use by the mother during pregnancy (1016)

fetus developing human from 8 weeks until birth (1016)

fixed action pattern behavior an innate behavior that always occurs the same way (836)

flagellum threadlike structure that grows out of a cell and enables it to move (57)

flexor skeletal muscle that causes a joint to bend (866)

flower reproductive structure of angiosperms; produces pollen in the anthers of stamens and seeds in a fruit; the mature ovary of a pistil (523)

fluke parasitic flatworm of the class Trematoda (650)

follicle cluster of cells that surrounds an immature egg cell in an ovary (1011)

food chain linear pathway of energy transfer in an ecosystem (366)

food web a network of interconnected food chains in an ecosystem (367)

foot in a mollusk, muscular region used for locomotion (661)

fossil preserved or mineralized remains or traces of an organism that lived long ago (258)

frond long, highly divided leaf of a fern (529)

fruit structure that contains one or more seeds and develops from the ovary of the pistil of a flower (532)

fundamental niche the entire range of conditions an organism can tolerate (386)

G

gamete haploid cell that participates in fertilization by fusing with another haploid cell (118)

gametophyte haploid phase in the type of life cycle known as alternation of generations; haploid individual that produces gametes (152)

gastrovascular cavity a digestive cavity with only one opening; found in simple animals (623)

gemmules clusters of amoebocytes encased in a protective coat; produced by freshwater sponges in cold or dry weather (639)

gene section of a chromosome that codes for a protein or RNA molecule (8, 119)

gene cloning process in which many copies of a specific gene are made each time the host cell reproduces (227)

gene expression two-stage processing of information encoded in DNA to produce proteins (206)

gene flow movement of alleles into or out of a population due to the migration of individuals to or from the population (348)

gene therapy technique that places a healthy copy of a gene into the cells of a person whose copy of the gene is defective (234)

genetic code sequence of nucleotides that specifies the amino acid sequence of a protein (209)

genetic drift random change in allele frequency in a population (348)

genetic engineering process of isolating a gene from the DNA of one organism and transferring it to the DNA of another organism (226)

genetics study of heredity (160)

genital herpes sexually transmitted disease caused by herpes simplex virus (1020)

genotype the genetic makeup of an organism as indicated by its set of alleles (166)

genus taxonomic category containing similar species (321)

germination resumption of growth by the plant embryo in a seed (590)

gestation period in mammals, length of time between fertilization and birth (820)

gestation period of human development; pregnancy (1015)

gill extremely thin projection of tissue rich in blood vessels that is the site of gas exchange in aquatic and a few terrestrial animals (623)

gill filament fingerlike projection from a gill where respiratory gases enter and leave the blood (757)

gill slit opening at the rear of a fish's cheek cavity where water exits (757)

glucagon a peptide hormone produced by the pancreas that causes liver cells to release glucose stored in glycogen (994)

glycolysis biochemical pathway that breaks down glucose to pyruvate (105)

glycoprotein protein with carbohydrate molecules attached (453)

Golgi apparatus organelle that packages and distributes molecules produced by a eukaryotic cell (63)

gonorrhea sexually transmitted disease caused by bacteria that results in inflammation of the mucous membranes in the urinary and reproductive tracts (1018)

gradualism model of evolution in which gradual change over a long period of time leads to species formation (289)

grain edible dry fruit of a cereal grass (536)

greenhouse effect atmospheric warming resulting from heat trapped by gases such as carbon dioxide (408)

ground tissue type of plant tissue other than vascular tissue that makes up much of the inside of a plant (570)

ground water water found beneath Earth's surface (371)

guard cell one of a pair of specialized cells that border a stoma (521)

gymnosperm seed plant that produces seeds that do not develop within a fruit (530)

H

habitat place where a particular population of a species lives (360)

hair in mammals, filament of dead cells and keratin used for insulation (810)

hair follicle specialized dermal structure that produces hair (872)

half-life the period of time that it takes for one-half of a given amount of a radioisotope to change (252)

haploid having only one set of chromosomes (121)

Hardy-Weinberg principle principle stating that the frequency of alleles in a population does not change unless evolutionary forces act on the population (346)

Haversian canal hollow channel in bone through which blood vessels pass (862)

heart attack when an area of the heart muscle dies and stops working (894)

heartwood nonconducting wood in the center of a tree trunk (575)

helper T cell white blood cell that activates cytotoxic T cells and B cells in an immune response (935)

hepatitis inflammation of the liver (918)

herbaceous plant plant with a stem that is flexible and usually green (nonwoody) (574)

herbivore organism that eats only plants or algae (366)

heredity transmission of genetic traits from parent to offspring (8, 160)

hermaphrodite organism with both testes and ovaries (627)

heterotroph organism that must get energy from food sources (75)

heterozygous refers to an individual with two different alleles for a trait (165)

histamine chemical released by injured cells in an inflammatory response (933)

HIV human immunodeficiency virus; the infectious virus that causes AIDS (12, 942)

homeostasis the maintenance of stable internal conditions in spite of changes in the external environment (8)

hominid member of the family Hominidae of the order Primates; characterized by opposable thumbs, no tail, longer lower limbs, and bipedalism (306)

homologous chromosome a member of a chromosome pair, both of which are similar in shape, size, and the genes they carry (120)

homologous structures structures that share a common ancestry (287)

homozygous refers to an individual with two identical alleles for a trait (165)

hormone substance secreted by cells that acts to regulate the activity of other cells (598, 982)

Human Genome Project research effort to sequence and locate the entire collection of genes in human cells (236)

hydrostatic skeleton support structure that consists of water contained under pressure in a closed cavity (625)

hypertonic solution a solution that causes a cell to shrink because of osmosis (77)

hypha slender filament that is part of the body of a multicellular fungus (439, 501)

hypothalamus region of the brain located below the thalamus that coordinates the activities of the nervous and endocrine systems and that controls many body activities related to homeostasis (960, 990)

hypothesis a proposed explanation for an observation that can be tested by additional observations or experimentation (16)

hypotonic solution a solution that causes a cell to swell because of osmosis (77)

immunity resistance to a disease (939)

implantation burrowing of a blastocyst into lining of the uterus (1014)

imprinting learning that can occur only during a specific period early in an animal's life and cannot be changed once learned (839)

incomplete dominance condition in which a trait in an individual is intermediate between the phenotype of its two parents (175)

independent assortment random distribution of homologous chromosomes during meiosis (144)

independent variable factor that is varied in an experiment (17)

industrial melanism darkening of populations of organisms over time in response to industrial pollution (290)

inflammatory response series of events, initiated by an injury or local infection, that suppress infection and promote healing (933)

innate behavior genetically programmed behavior; instinct (836)

insulin amino-acid-based hormone produced by the islets of Langerhans in the pancreas that lowers blood glucose levels by promoting the accumulation of glycogen in the liver (994)

interferon protein released by cells that inhibits viruses (934)

internal fertilization fertilization that occurs within the body of the female parent (628)

interneuron a neuron that links neurons to each other (961)

interphase period between two mitotic or meiotic divisions of a eukaryotic cell during which the cell carries out routine functioning, copies its DNA, and prepares to divide (125)

intron segment of mRNA transcribed from eukaryotic DNA but removed before translation of mRNA into a protein (216)

invertebrate animal that lacks a backbone (443)

invertebrate chordate chordates without a backbone; tunicates and lancelets (719)

ion electrically charged atom or molecule (30)

ion channel transport protein in a cell membrane through which ions can pass (78)

isolation condition in which two populations of a species are separated so that they cannot interbreed (281)

isotonic solution a solution that causes no change in cell size (77)

joint junction of two or more bones (864)

karyotype array of the chromosomes found in an individual's cells arranged in order of size and shape (122)

keratin protein that makes skin tough and waterproof (871)
kingdom taxonomic category that contains phyla with similar characteristics (322)
Koch's postulates four-stage procedure used to identify specific pathogens (938)
Krebs cycle cyclic biochemical pathway of cellular respiration that uses carbon dioxide and produces ATP, NADH, and $FADH_2$ (106)
krill small marine crustacean that is the chief food source for many marine species (699)
***K*-strategist** species characterized by slow maturation, few young, slow population growth, and reproduction late in life (345)

***lac* operon** gene system with a promoter, an operator gene, and three structural genes that control lactose metabolism (214)
larynx voice box; structure at the upper end of the trachea containing the vocal cords (896)
lateral line specialized sensory system running the length of both sides of a fish's body (763)
law of independent assortment second law of heredity stating that pairs of genes separate independently of one another in meiosis (167)
law of segregation first law of heredity stating that pairs of alleles for a trait separate when gametes are formed (167)
learning the development of behaviors through experience (837)
lichen symbiotic association between a fungus and an alga or cyanobacterium (509)
life cycle entire lifespan of an organism; summary of all of the stages of an organism (150)
ligament band of connective tissue that holds together the bones of a joint (864)
light microscope microscope that uses a beam of light passing through one or more lenses to magnify an object (51)
limnetic zone area in a freshwater habitat away from the shore but still close to the surface (396)
lipase enzyme that breaks down fat molecules into fatty acids and glycerol (917)
lipid one of a family of nonpolar organic molecules that are not soluble in water; includes fats, phospholipids, and steroids (35)
lipid bilayer foundation of cell membranes; composed of two layers of phospholipids (59)
littoral zone shallow zone near the shore in a freshwater habitat (396)
logistic model model of population growth that assumes finite resource levels limit population growth (343)
lung internal, baglike respiratory organ of a vertebrate that enables gas exchange between the air and the blood (768)
lymphatic system system of the body that collects and recycles fluids leaked from the cardiovascular system and is involved in fighting infections (885)
lysogenic cycle cycle in which a viral genome replicates as a provirus without destroying the host cell (455)
lysosome organelle in a eukaryotic cell that contains digestive enzymes (63)
lytic cycle cycle of viral infection; results in replication of virus and cell destruction (454)

macrophage large white blood cell that engulfs pathogens (934)
magnification enlargement or enlarging of an image (51)
Malpighian tubule slender, fingerlike organ of excretion that opens into the gut of certain arthropods (687)
mammary gland gland located in the chest or abdomen of a female mammal that secretes milk to nourish her young (816)
mandible chewing mouthpart found in many arthropods (691)
mantle heavy fold of tissue that surrounds the visceral mass of mollusks (661)
mass extinction episode during which large numbers of species become extinct (263)
medusa free-swimming, jellylike, often umbrella-shaped body form of a cnidarian; jellyfish (640)
meiosis process during cell division in which the nucleus of a cell completes two successive divisions that produce four cells (gametes), each with a chromosome number that has been reduced by half (142)
melanin pigment that helps determine skin color (872)
membrane potential difference in electrical charge across a cell membrane (953)
menstrual cycle series of hormone-induced changes that prepare the uterine lining for a possible pregnancy each month (1012)
menstruation discharge of blood and discarded uterine tissue during the menstrual cycle (1013)
meristem region (or zone) of actively dividing, undifferentiated plant cells that are capable of developing into specialized plant tissues (525)

merozoite stage of the life cycle of the protist *Plasmodium* in which the protist divides rapidly in red blood cells and produces many more merozoites (491)

mesoderm middle layer of embryonic tissue in animals from which the skeleton and muscles develop (614)

mesophyll ground tissue of a leaf (577)

messenger RNA RNA copy of a gene used as a blueprint for the making of a protein during translation (209)

metabolism sum of all chemical processes occurring in an organism (7, 39)

metamorphosis dramatic physical change through which an immature organism passes as it grows to adulthood (693)

microsphere tiny, abiotically produced vesicle formed by short chains of amino acids (256)

mineral naturally occurring inorganic substance used to make certain body structures and substances, for normal nerve and muscle function, to maintain osmotic balance, and for enzyme function (913)

mineral nutrient element required by plants that is absorbed mainly as an inorganic ion (597)

mitochondrion organelle that produces much of the ATP made by a eukaryotic cell (64)

mitosis process during cell division in which the nucleus of a cell divides into two nuclei, each with the same number and kind of chromosomes (125)

molecule group of atoms held together by covalent bonds (29)

molting a periodic shedding of an arthropod's exoskeleton; also found in some reptiles (686)

monocot monocotyledon; angiosperm that produces seeds with a single cotyledon (533)

monohybrid cross cross involving one pair of contrasting traits (162)

monosaccharide simple sugar that is the base unit of carbohydrates (34)

motor neuron neuron that sends motor responses from the central nervous system to muscles, glands, and other organs (958)

mucous membrane layer of epithelial tissue covering internal surfaces of the body that secretes mucus and functions in nonspecific defenses (932)

multicellular organism an organism that consists of many cells that are permanently associated and integrated in their activities (437)

multiple alleles the existence of more than two alleles (versions of the gene) for a genetic trait (176)

muscle tissue body tissue that enables movement (857)

mutation change in the DNA of a gene or chromosome (8)

mutualism symbiotic association in which both partners benefit (265, 384)

mycelium mass of hyphae forming the body of a fungus (501)

mycorrhiza mutualistic association between a fungus and a plant's roots, in which the fungus absorbs water and nutrients for the plant and the plant supplies food to the fungus (265, 508)

myofibril cylindrical components of muscle, containing many myosin and actin filaments (867)

myosin thick protein filament in muscle that functions in contraction (867)

N

NADH electron carrier that stores energy used to make ATP (105)

NADPH electron carrier that provides high-energy electrons for photosynthesis (101)

natural killer cell immune system cell that attacks cells infected with pathogens (934)

natural selection process by which populations change in response to their environment as individuals better adapted to the environment leave more offspring than those individuals not suited to the environment (9, 279)

nauplius larval form of a crustacean (698)

negative feedback mechanism used in homeostasis to keep a monitored variable within a certain range. A change in one direction stimulates two control mechanisms to counteract further change in the same direction. (988)

nematocyst barbed harpoon within a cnidocyte of a cnidarian; used to spear prey (641)

nephridium tiny tubular organ of excretion that filters cellular wastes from the coelom of certain invertebrates (662)

nephron tubelike structure in the kidneys that filters wastes from the body and retains useful molecules; also regulates the body's salt and water balance (759, 921)

nerve bundle of neurons that appears as a fine white thread to the unaided eye (952)

nervous tissue nerve cells and their supporting cells (857)

neuron cell that transmits nerve impulses (952)

neurotransmitter signal molecule that transmits nerve impulses across a synapse (956)

neutrophil white blood cell that kills pathogens and itself through the release of chemicals (934)

niche functional role of a species in an ecosystem (385)

nitrogen fixation process of combining nitrogen gas with hydrogen to form ammonia (373)

nonrandom mating mating between individuals of the same phenotype or by those who live nearby (348)

nonvascular plant plant that does not have a vascular system (522)

norepinephrine amino-acid-based molecule released as a hormone by the adrenal medulla in times of stress and as a neurotransmitter in the nervous system; formerly called noradrenaline (993)

normal distribution bell-shaped curve that results when the values of a trait in a population are plotted against their frequency (351)

notochord flexible rod of tissue along the back of a chordate that aids in locomotion (718)

nucleic acid organic molecule made of nucleotides; stores hereditary information for cell function; DNA or RNA (37)

nucleotide subunit of nucleic acids consisting of a nitrogenous base, a sugar, and a phosphate group (37, 192)

nucleus organelle that houses the DNA of eukaryotic cells (58)

nutrient substance needed by the body for energy growth, repair, and maintenance (908)

nymph juvenile stage of some insects that is a smaller version of the adult (693)

O

observation the act of noting or perceiving objects or events using the senses (14)

omnivore animal that eats both plants and animals (366)

oogenesis process by which gametes are produced in female animals (146)

open circulatory system system in which blood leaves the blood vessels and bathes the body's tissues (624)

operator region of DNA that controls RNA polymerase's access to a set of genes with related functions (214)

operculum hard plate that covers the gills on each side of the head of bony fishes (766)

operon segment of DNA that controls gene regulation in a set of genes with related functions in prokaryotes (214)

opposable thumb the finger that stands out at an angle from the other fingers and can be bent for grasping (304)

optic nerve nerve that transmits signals from the eye to the brain (966)

order taxonomic category consisting of families with similar characteristics (322)

organ collection of tissues that work together to perform a particular function (440)

organelle structure in eukaryotic cells that has a specialized function (58)

organ system group of organs that function together to carry out a major activity of the body (440)

oscula large pores in a sponge's body wall through which water exits the body (636)

osmosis movement of water through a selectively permeable membrane from an area of high concentration to an area of lower concentration (76)

ossicle calcium-rich plates that compose the endoskeleton of echinoderms (712)

osteocyte cell that maintains the mineral content of bone (862)

osteoporosis reduction of bone mass that produces porous bone (863)

ostia small pores in a sponge's body wall through which water enters the body (636)

ovarian cycle series of hormone-induced changes in which the ovaries prepare and release a mature ovum each month (1011)

ovary reproductive structure that produces eggs; in flowering plants, the lower part of a pistil that produces eggs in ovules; in animals, the gamete-producing organ of the female reproductive system (556, 1009)

oviparous term that describes organisms that produce eggs that hatch outside the mother's body (787)

ovoviviparous term that describes organisms that produce eggs that are retained in the mother's body until hatching or just before hatching (787)

ovulation the release of an ovum from a follicle (1011)

ovule structure of a seed plant sporophyte in which a female spore forms and then develops into a female gametophyte that contains an egg; structure in the ovary of a pistil that develops into a seed (552)

ovum a mature egg cell (147, 1010)

P

paleontologist scientist who studies fossils (285)

Pangaea the single supercontinent of the Triassic period that included all presently known continents (740)

parapodium fleshy appendage of marine annelids (670)

parasitism type of predation in which the predator feeds on but usually does not kill a larger organism (382)

passive transport movement of a substance without the use of energy by the cell (74)

pathogen a disease-causing agent (454, 932)

pedigree family history that shows how a trait is inherited over several generations (173)

pedipalps in arachnids, a second pair of appendages that are modified to catch and handle prey (688)

pelvic inflammatory disease severe inflammation of the uterus, ovaries, fallopian tubes, or pelvic cavity usually caused by bacterial STDs (1019)

penis male reproductive organ that delivers sperm to the female reproductive tract (1008)

pepsin digestive enzyme secreted by the stomach that cuts single protein strands into smaller chains of amino acids (916)

perennial plant that lives for more than 2 years and may produce flowers, fruits, and seeds many times during its life (591)

periosteum tough exterior membrane of bones (861)

peripheral nervous system system of sensory and motor neuron that branch throughout the body (958)

peristaltic contraction contraction of the smooth muscle in the wall of the gut that moves food through the digestive system (915)

petal structure of a flower that is often colorful and may attract pollinators to the flower (556)

petiole slender stalk that attaches the blade of a leaf to a stem (576)

P generation (parental generation) the first two individuals that mate in a genetic cross (162)

pH measure of the acidity of a solution, representing the hydrogen ion concentration (16)

pharyngeal slits openings in the wall of the pharynx in chordates (718)

pharynx upper portion of the throat leading to the esophagus (896)

phenotype observable characteristics of an organism (166)

phloem type of vascular tissue in plants, that contains soft-walled conducting cells through which organic compounds are transported throughout the body of a plant (525)

phospholipid lipid molecule in all membranes of a cell (59)

photoperiodism response of plants to the length of days and nights (601)

photosynthesis process by which organisms use light energy to produce organic compounds (94)

phylogenetic tree branching diagram that shows how animals are related through evolution (620)

phylogeny evolutionary history of a species (327)

phylum a taxonomic category containing classes with similar characteristics (322)

pigment molecule that absorbs certain wavelengths of light (98)

pilus short, thick outgrowth of a bacterium that allows it to attach to another bacterium (460)

pioneer species small, fast-growing plants that are the first organisms to live in a new habitat (363)

pistil female reproductive part of a flower, consisting of an ovary, style, and stigma (556)

pith inner core of ground tissue in the stem or root of a plant (574)

pituitary gland endocrine gland at the base of the brain that stores and releases hormones produced by the hypothalamus and hormones that control endocrine glands elsewhere in the body (991)

placenta organ through which the mother nourishes an embryo or fetus (820, 1015)

plankton biological community drifting freely in the upper waters of the ocean and consisting mostly of microscopic organisms (398)

plant propagation practice of growing plants from seeds or from vegetative parts (562)

planula free-swimming larva that developed from cnidarian zygotes (643)

plasma noncellular portion of blood (886)

plasma cell cell that releases antibodies in response to a specific antigen (937)

plasmid a circular DNA molecule, usually found in bacteria, that can replicate independently from the main chromosome (227)

plasmodium protist of the phylum Myxomycota; characterized by multinucleate, amoeboid appearance (488)

plastron the bottom, or ventral, portion of a turtle's shell (792)

platelet un-nucleated cell fragment that aids in blood clotting (887)

point mutation mutation in which one or just a few nucleotides in a gene are changed (217)

pollen grain structure consisting of a few haploid cells surrounded by a thick protective wall that contains a male gametophyte of a seed plant (552)

pollen tube structure that grows from a pollen grain to an ovule, enabling a sperm to pass directly to an egg (552)

pollination transportation of pollen grains from a male reproductive structure of a flower to a female reproductive structure of a flower (552)

polygenic trait characteristic of an organism that is influenced by several genes (175, 351)

polyp cylindrical, pipe-shaped body form of a cnidarian, usually attached to a rock or other object (640)

population group of individuals that belong to the same species, live in the same area, and breed with others in the group (279, 340)

population density in a population, the number of individuals in a given area (341)

population model hypothetical population that attempts to exhibit the key characteristics of a real population (342)

population size total number of individuals in a population (341)

predation an ecological interaction in which one organism feeds on another (382)

prediction the expected outcome if a hypothesis is correct (16)

preen gland in birds, a special gland that secretes oil that birds spread over their feathers to clean and waterproof them (795)

pregnancy period of time in which embryonic and fetal development occur (1015)

primary growth growth that causes a plant to grow in length or height and occurs from the formation of new cells at the tips of stems and roots (592)

primary productivity the amount of organic material that the photosynthetic organisms of an ecosystem produce (365)

primary succession succession that occurs in a newly formed habitat that has never before sustained life (363)

primate order of mammals that includes prosimians, monkeys, apes, and humans and is characterized by limb structure, dentition, digital mobility, flat nails, and binocular vision (302)

prion infectious particles composed of protein with no nucleic acid (459)

probability the likelihood that a specific event will occur (171)

probe labeled RNA or single-stranded DNA molecule used to bind with and identify a specific gene in genetic engineering (229)

producer organism that makes its own food from energy and carbon atoms in its environment; autotroph (365)

profundal zone zone in a freshwater habitat that is below the limits of effective light penetration (396)

proglottid rectangular body section of tapeworms (649)

prokaryote single-celled organism without a nucleus; bacterial cell (57)

prosimian a group of primates that includes lorises, tarsiers, and lemurs (303)

prostate gland gland in males that secretes an alkaline fluid necessary to neutralize the acids produced by the female reproductive tract (1008)

protein organic compound made of amino acids (36)

protist a member of the kingdom Protista (261, 438)

protostome an animal whose mouth develops from or near the blastopore, the opening in the gastrula (710)

protozoan historic term for a heterotrophic protist (479)

provirus the viral DNA which inserts into a host cell's chromosome (455)

pseudocoelomate animal with a body cavity located between the endoderm and the mesoderm (618)

pseudopodium extension of cytoplasm of the amoeba that enables it to move (482)

psychoactive drug drug that alters the functioning of the central nervous system (969)

pulmonary vein vessel that carries oxygen-rich blood from the lungs to the heart (769)

pulse series of pressure waves within an artery caused by a contraction of the left ventricle; indicator of heart rate (893)

punctuated equilibrium model of evolution in which short periods of rapid change in species are separated by long periods of little or no change (289)

Punnett square diagram used by biologists to predict the probable outcome of a genetic cross (168)

pupa stage of complete metamorphosis in which an insect changes from larva to adult (693)

radial symmetry arrangement of body parts around a central axis (616)

radioisotope unstable isotope of an element (252)

radiometric dating dating of objects through the measurement of relative proportions of certain radioisotopes and the products of their radioactive decay (252)

radula rasping tonguelike organ of mollusks used in obtaining food (661)

realized niche the part of its fundamental niche that a species actually occupies (387)

reasoning the ability to analyze a problem and think of a possible solution (838)

receptor protein protein that binds to specific signal molecules, causing the cell to respond (84)

recessive genetic trait that is not expressed when the contrasting form of the trait is present (164)

recombinant DNA DNA made from two or more different organisms (226)

red blood cell cell in the blood that carries oxygen (886)

reflex sudden, rapid, and involuntary self-protective response to a stimulus (960)

replication fork a Y-shaped point that results when a double helix of DNA separates so that it can be copied (197)

repressor protein that binds to the operator in an operon to switch off transcription (214)

reproduction the process by which organisms make more of their own kind (7)

reproductive isolation prevention of mating between formerly interbreeding groups, or the inability of these groups to produce fertile offspring (294)

resolution in microscopes, a measure of the clarity of an image (51)

respiration simultaneous uptake of oxygen and release of carbon dioxide, by the lungs, gills, or through the skin (623)

resting potential membrane potential of a neuron that is not conducting a nerve impulse (954)

restriction enzyme bacterial enzyme that cuts DNA at a specific sequence of nucleotides (227)

retina light-sensitive area at the back of the eye that contains photoreceptors (966)

Rh factor protein antigen on the surface of red blood cells (889)

rhizoid hypha that acts as an anchor for a fungus; hairlike projection that anchors a nonvascular plant (504, 526)

rhizome horizontal underground stem (528)

ribonucleic acid See RNA

ribosomal RNA type of RNA molecule that plays a structural role in ribosomes (210)

ribosome organelle on which proteins are made (56)

RNA (ribonucleic acid) a type of nucleic acid involved in protein synthesis (37, 206)

RNA polymerase enzyme that adds and links complementary RNA nucleotides during transcription (207)

rod photoreceptor in the eye that responds to dim light (966)

root portion of a plant that grows mostly downward and below ground (525)

root cap protective layer covering the tip of a root (573)

root hair extension of an epidermal cell near the root tip that aids in the absorption of water (573)

***r*-strategist** species characterized by rapid growth, high fertility, short lifespan, and exponential population growth (344)

S

sapwood wood in the outer part of a tree trunk containing xylem that conducts water (575)

sarcomere the functional unit of muscle contraction (867)

scanning tunneling microscope microscope that measures differences in voltage of electrons leaving the surface of an object and creates a three-dimensional image of the object; can be used to study living organisms (54)

sebum oily secretion that lubricates the skin; released by oil glands in the dermis (874)

secondary compound defensive chemical compound produced by a plant (383)

secondary growth growth that causes a plant's stems or roots to increase in width (592)

secondary succession an episode of succession that occurs in areas where there has been previous growth, such as abandoned fields or forest clearings (363)

second messenger signal molecule produced in response to the binding of a chemical signal; alters the chemical activity within the cell (85, 986)

seed structure that consists of a plant embryo surrounded by a protective coat (522)

seed coat the protective, outer covering of a seed (553)

seed plant vascular plant that produces seeds (522)

semen the mixture of secretions and sperm produced by male reproductive organs (1008)

semicircular canal fluid-filled chamber in the inner ear that contains hair cells involved in maintaining equilibrium (967)

seminal vesicles glands of male reproductive system that produce fluids rich in sugars to nourish sperm cells (1008)

seminiferous tubules tightly coiled tubes within the testes in which sperm are produced (1006)

sensory neuron neuron that sends information from the sense organs to the central nervous system (958)

sensory receptor specialized neurons that detect sensory (964)

sepal structure of a flower that encloses and protects a flower bud (556)

septum in fungi, a wall-like division between cells within a hypha; wall that internally divides body segments in annelids; thick wall that divides the heart's atrium or ventricle into right and left halves (439, 670, 770)

sessile describes an organism that remains attached to a surface for its entire life (636)

setae external bristles of annelids (670)

sex chromosome one of a pair of chromosomes that are involved in determining the sex of an individual (122)

sex-linked trait a trait that is determined by a gene found on the X chromosome (173)

sexual reproduction reproduction in which gametes from opposite sexes or mating types unite to form a zygote (148)

sexual selection evolutionary mechanism by which traits that increase the ability of individuals to attract or acquire mates appear with increasing frequency in a population (845)

shoot portion of a plant that grows mostly upward and aboveground (525)

sieve tube conducting strand in phloem that consists of a series of sieve-tube cells stacked end to end (572)

sink part of a plant to which organic compounds are delivered (582)

sinoatrial node a small cluster of cardiac muscle cells in the upper wall of the right atrium that initiates and regulates contraction of the heart (892)

siphon hollow tube of bivalves used for sucking in and expelling sea water (665)

skin gills small, fingerlike projections that grow between the spines of echinoderms (713)

sodium-potassium pump carrier protein that transports sodium ions out of a cell and potassium ions into the cell (81)

solution homogeneous mixture of two or more substances (32)

sorus cluster of sporangia on a fern frond (550)

source part of a plant that provides organic compounds for other parts of the plant (582)

speciation process by which new species are formed (293)

species group of organisms that look alike and are capable of producing fertile offspring in nature (9)

sperm male gamete (146)

spermatogenesis process by which gametes are produced in male animals (146)

spicule needle of silica or calcium carbonate in the skeleton of some sponges (638)

spinal cord cable of nerve tissue extending from the base of the brain through the backbone, to level just below the ribs (960)

spindle structure composed of centrioles and individual hollow protein fibers that moves chromosomes apart during cell division (128)

spinnerets modified appendages in arthropods that secrete silk (688)

spiracle respiratory opening of certain arthropods that allows for the passage of air into the body (686)

spirillum spiral-shaped bacterium (461)

spongin flexible, structural protein fibers in the mesoglea of most sponges (638)

spontaneous origin origin of life through natural chemical and physical processes (253)

sporangium spore capsule in which haploid spores are produced by meiosis (481)

spore an asexual, resting, reproductive haploid cell (152)

sporophyte diploid phase in the type of life cycle known as alternation of generations; diploid individual that produces spores (152)

sporozoite one of three stages of *Plasmodium* that lives in mosquitoes and is injected into humans (491)

stabilizing selection type of natural selection in which the average form of the trait is favored and becomes more common (352)

stamen male reproductive part of a flower (556)

steroid hormone fat-soluble hormone derived from cholesterol (985)

stimulant drug that increases the activity of the central nervous system (970)

stolon stemlike structures of a mold (504)

stoma opening in a leaf or a stem of a plant that enables gas exchange to occur (521)

stroke a sudden attack of weakness or paralysis that occurs when an area of the brain dies after blood flow to the brain is interrupted (894)

subcutaneous tissue layer of fat-rich cells just below the dermis (873)

substrate the substance on which an enzyme acts during a chemical reaction (41)

succession the regular progression of species replacement that occurs after disturbance or the development of new habitat (363)

swim bladder the gas-filled sac of bony fishes used to regulate their buoyancy (766)

symbiosis ecological interaction in which two or more species live together in a close, long-term association (384)

synapse junction at which a neuron meets another cell (956)

syphilis sexually transmitted disease caused by the bacterium *Treponema pallidum* (1019)

target cell a specific cell a hormone binds to and acts on to produce a specific effect (985)

taxonomy science of naming and classifying organisms (320)

tegument thick protective covering of endoparasite cells that helps prevent their being digested by their host (650)

teleost group of bony fishes with highly mobile fins, thin scales, a swim bladder, and symmetrical tails; largest group of living fish (767)

tendon dense connective tissue that attaches muscles to bones (866)

terrestrial term that describes an organism with the ability to live on land (739)

test cross a genetic cross of an individual whose phenotype is dominant but whose genotype is unknown, with a homozygous recessive individual (170)

testes gamete-producing organs of the male reproductive system (1006)

thalamus part of the brain that directs incoming sensory information to the proper region of the cerebral cortex (960)

thecodont extinct crocodilelike reptile from which the dinosaurs evolved (740)

theory explanation based on a set of related hypotheses that have been tested and confirmed many times (19)

therapsid extinct order of reptiles thought to be endothermic that likely gave rise to mammals (744)

thorax mid-body region in arthropods (684)

thylakoid internal membrane-bound sac of a chloroplast (99)

tissue group of cells with a common structure and function (440)

tissue culture technique for growing pieces of living tissue in artificial media (562)

tolerance condition of drug addiction in which increasing amounts of the drug are needed to achieve the desired effect (970)

toxin chemical compound excreted into the environment by bacteria that are poisonous to eukaryotic cells (466)

trachea tube that carries air from the larynx to the lungs (696)

tracheae in certain arthropods, fine tubes that extend into the interior of the body; used for gas exchange (686)

transcription stage of protein synthesis in which the information in DNA for making a protein is transferred to an RNA molecule (206)

transfer RNA RNA molecule that temporarily carries a specific amino acid to a ribosome during translation (210)

transformation the transfer of genetic material from one organism to another; first observed by Griffith (189)

transgenic animal an animal that has had foreign DNA introduced into its cells (239)

translation stage of gene expression in which the information in mRNA is used to make a protein (206)

translocation movement of organic compounds within a plant from a source to a sink (582)

transpiration loss of water from a plant through its stomata (371, 578)

trochophore larval stage in many mollusks and annelids (660)

trophic level a group of organisms that have the same source of energy; a step in a food chain (365)

tropism growth response to a stimulus in which the direction of growth is determined by the direction from which the stimulus comes (600)

true-breeding displaying only one form of a particular trait in offspring (162)

U

ulcer hole in the lining of the stomach or small intestine (916)

ungulate mammal with hoofs (824)

uracil nitrogen-containing base of RNA, complementary to adenine when RNA base pairs with DNA (206)

urea principal nitrogenous waste of mammals and a chief component of urine (920)

ureter tube through which urine produced in the kidneys passes to the bladder (922)

urethra tube through which urine leaves the bladder and exits the body (922)

urinary bladder hollow, muscular sac that stores urine (922)

urine water and metabolic wastes left after filtering process of the kidneys; expelled from the body (922)

uterus hollow, muscular organ of the female reproductive system in which the embryo and fetus develop (1010)

V

vaccination the application of a vaccine to produce immunity (939)

vaccine substance prepared from killed or weakened pathogens and introduced into a body to produce immunity (188, 232, 939)

vagina muscular tube that leads from the uterus to the outside of the female body (1010)

valve flap of tissue that ensures that the blood or fluid that passes through does not flow back (884)

vascular bundle cluster of vascular tissues in a leaf or a herbaceous stem that contains both xylem and phloem (574)

vascular cambium meristem of woody plants that produces secondary vascular tissue (593)

vascular plant plant that has a vascular system (522)

vascular system system of vascular tissues that transport water and other materials in plants (522)

vascular tissue type of plant tissue that contains specialized cells that distribute water and other nutrients within the body of a plant (441)

vas deferens tube through which sperm move from the epididymis to the urethra (1007)

vector agent used to carry a DNA fragment into a cell (227)

vegetative part any nonreproductive part of a plant (534)

vegetative reproduction type of reproduction in which new plants grow from vegetative parts such as roots, stems, and leaves (559)

vein large vessel that carries blood toward the heart (883)

vena cava one of two large veins that collect all of the oxygen-poor blood from the body (891)

ventricle thick-walled heart chamber that pumps blood away from the heart (891)

vertebra individual segment of a backbone (730)

vertebrate animal with a backbone (267, 443)

vesicle membrane-enclosed sac in a cell's interior (62)

vessel conducting strand in xylem that consists of a series of vessel cells stacked end to end (572)

vestigial structure structure reduced in size and function; considered to be evidence of an organism's evolutionary past (287)

villus one of many fine, fingerlike projections that cover the lining of the small intestine, increasing its absorptive surface (917)

viroid in plants, infectious disease agent composed of a single strand of RNA with no capsid (459)

virulent referring to the deadliness of a disease-causing agent (188)

virus a strand of nucleic acid encased in a protein coat that can infect cells and replicate within them (452)

visceral mass central section of a mollusk's body that contains its organs (661)

vitamin organic substance that enhances the activity of enzymes and that occurs naturally in foods (912)

water vascular system water-filled system of interconnected canals and tube feet that aids echinoderms in movement (713)

weaning the time when a female mammal stops nursing its young (816)

white blood cell cell in blood whose primary job is to defend the body against disease (887)

withdrawal set of symptoms associated with the removal of an addictive drug from the body (970)

xylem type of vascular tissue in plants that contains hard-walled conducting cells that transport water and dissolved minerals upward from the roots to the leaves (525)

yeast common name given to unicellular ascomycetes (505)

zoomastigote unicellular, heterotrophic protist that is a member of the phylum Zoomastigina (485)

zygosporangium thick-walled sexual structure that characterizes members of the phylum Zygomycota (504)

zygospore diploid zygote that results from the pairing of gametes of opposite mating types (480)

zygote fertilized egg cell (121)

Index

Page numbers preceded by the letter t *refer to tables.*
Page numbers in **boldface** *type refer to figures.*

Index

Index

H

Index

J

Index

Credits

ILLUSTRATION

Abbreviation Code

AS=Art & Science, Inc.; AT=Alexander & Turner; BH=Barbara Hoopes-Ambler; CF=Chris Forsey; CK=Christy Krames; FP=Felipe Passalacqua; JB=John Bindon; JK=John Karapelou; JW=John White; KC=Kip Carter; KK=Keith Kasnot; MCA=Morgan-Cain & Associates; MW=Michael Woods; PG=Pond & Giles; RH=Robert Hynes; SD=Stephen Durke; SR=Steve Roberts; SW=Sarah Woodward; WAA=Wildlife Art Agency; WN=Will Nelson

iv PG; vi (t) SD; (b) MCA; vii (t) MCA; (b) RH; ix, xii MCA; xiii (t) WN; (b) JW, xv MCA; xxi Sam Collins; xxv (t, b) MCA; CHAPTER 1: 11, 16, 19 MCA; 22 SD; 23 MCA; CHAPTER 2: 28 Kristy Sprott; 29 (t) MCA; (b) Photri; 30 MCA; 33, 34 MCA; 35 MCA/Kristy Sprott; 36 Kristy Sprott; 37, 39, 40 MCA; 41 Robert Margulies/MCA; 42 MCA; CHAPTER 3: 51 (graphics) MCA; 55, 58, 59, 60, 61, 62, 63, 64, 65, 69 MCA; CHAPTER 4: SD/MCA; 76, 77, 78, 80, 82, 83, 84, 85, 88, 89 MCA; CHAPTER 5: 94 RH; 95, 96, 97, 98, 99, 100, 102, 104, 105, 106, 107, 108, 110, 113, 114 MCA; CHAPTER 6: 119 PG/MCA; 120 MCA; 123 CK; 124, 125, 126, 128, 130, 131, 132, 134, 135 MCA; CHAPTER 7: 142-143, 144, 145, 146, 149, 150, 151, 152, 154 MCA; CHAPTER 8: 161, 162 (spots) JW; 164; 168, 169, 170 (spots) JW; (graphics) MCA; 172, 173, 174, 176 MCA; 179 JK; 182-183, 184 MCA; CHAPTER 9: 189 (graphics) MCA; (spots) SR; 190, 192, 193, 195, 196, 198, 201 MCA; CHAPTER 10: 206, 207, 210-211, 212, 213, 215, 216, 217, 218, 220, 221 MCA; CHAPTER 11: 227 MCA; 228 (t) MCA; (b) SD; 229 MCA; 232 SD/David Fischer; 233, 236 MCA; 238-239 (graphics) MCA; (sheep) JW; 242, 243, 245 MCA; CHAPTER 12: 252, 254 MCA; 255 JW; 256 MCA; 258 (timeline) (l) SD; (r) PG; (graphics) MCA; 259 (timeline) SD; (graphics) MCA; 260 (t) MCA; (timeline) (l) SD; (r) Steven Kirk; (graphics) MCA; 261 (timeline) PG; (graphics) MCA; 262 (t) BH; (timeline) PG; (graphics) MCA; 263 (timeline) PG; (graphics) MCA; 264 (t) Rob Schuster; (timeline) PG, ; (graphics) MCA; 265 (timeline) PG; (graphics) MCA; 266 (t) CF; (timeline) PG; (graphics) MCA; 267 (timeline) (l,c) PG; (r) BH; (graphics) MCA; 268 (timeline) PG; (graphics) MCA; 271 MCA; CHAPTER 13: 277 (map) MCA/Mountain High Map resources; 278 (spots) MW; (insect eater, graphics) MCA; 279 MCA; 281 JW; 284 (tl, bl) BH; (tr, br) Rob Wood (Wood Ronsaville Harlin, Inc.); 285 (tl, bl) BH; (tr) Rob Wood (Wood Ronsaville Harlin, Inc.); (br) JB; 286 David Ashby; (graphics) MCA; 287 JK/MCA; 288 Molly Babich; 289 MW; 291, 292, 293, 294 MCA; 297 MW; (graphics) MCA; 298 MCA; 299 MCA; CHAPTER 14: 302 Steven Kirk; 305 David Ashby; (humans) Ralph Voltz; (graphics) MCA; 306 CK; (graphics) MCA; 308 PG; (graphics) MCA; 310 CF; 311 MCA/Mountain High Map resources; 314 CK; 315 PG; (graphics) MCA; 316 CK; 317 CK; CHAPTER 15: 322, 323 MCA; 328 PG; (graphics) MCA; 329 (t) PG; (b) Stansbury, Ronsaville, Wood, Inc.; 330 BH; (graphics) MCA; 333 MCA; CHAPTER 16: 341, 342, 343, 345, 346, 351, 352, 354, 355, 357 MCA; CHAPTER 17: 365 JW; 366, 367 David Beck; 368 Carlyn Iverson; (graphics) MCA; 369 Ralph Voltz; 371, 372, 373 RH; 376 MW; 377 MCA; CHAPTER 18: 383 JW; 385 MW; 386 (birds) MW; (trees) JW; (graphics) MCA; 387 MCA; 388 RH; 389 MCA; 391 JW; (graphics) MCA; 392 MCA; 396 RH; 400 JW; (graphics) MCA; 401 MCA; CHAPTER 19: 407 MCA; 408 Jared Schneidman Design; 409 MCA; 411 MCA; (animals) JB; 413 RH; 414, 415, 423, 425 MCA; CHAPTER 20: 432, 433 MCA; 440 JK; 447 MCA; CHAPTER 21: 453 (l,r) MCA; (c) George Kelvin; 454, 455, 457, 458, 464, 471 MCA; CHAPTER 22: 480, 481 Henry Hill/John Edwards & Associates; 485, 486, 487 MCA; 491 JK; 495 MCA; CHAPTER 23: 501 MCA; 503 PG; 504, 505 PG; (graphics) MCA; 506 MW; (zooms, graphics) MCA; 507 WN; 513 MCA; CHAPTER 24: 523 PG; 524 (t) MCA; (b) JW; 525 JW; 528 PG; 530 MCA; 533, 534 JW; CHAPTER 25: 548 JW; 549, 551, 552 PG; (graphics) MCA; 553 (l); (c,r) JW; 555 PG; (graphics) MCA; 556 JW; 558 PG; (graphics) MCA; 559, 560 JW; 561 (t,b) MCA; (c); 565 MCA; CHAPTER 26: 570 PG; (graphics) MCA; 572 MCA; 573, 574, 575 JW; 576 MCA; 577 MCA; (leaf) PG; 578 JW; (graphics) MCA; 579 (t) PG; (b) MCA; 580; PG; 581 (l) MCA; (c) PG; 582, 585 MCA; CHAPTER 27: 590 JW; 593 PG; (graphics) MCA; 594 PG; 595 (l) MCA; (r) PG; 598 MCA; 601 JW; (graphics) MCA; 602 MCA; 605 JW; CHAPTER 28: 615 MCA; 616, 617 BH; (graphics) MCA; 618 MCA; 620 (spots) BH; (mouse) MW; (jellyfish) SW; (graphics) MCA; 622 (l) MCA; (r) Molly Babich; 624 SD; 625 (l,r) MCA; (c) AT; 631 MCA; CHAPTER 29: 636 BH; 637, 639 MCA; 640 SW; 641, 642 MCA; 643 Henry Hill/John Edwards & Associates; 648, 649 MCA; 650 David DeGasperis/John Edwards & Associates; (human) Walter Stuart; 651 (spot) BH; 652, 654, 655 MCA; CHAPTER 30: 660 AT; 661 (zoom) MCA; (r) SW; 662 MCA; 665, 668 SR/WAA; 669 BH; 671 SR/WAA; 673 JB Woolsey Associates/Molly Babich; 677 SW; 679 JB Woolsey & Assoc.; CHAPTER 31: 682 BH; 683 (spots) SR/WAA; (graphics) MCA; 686 AS/Sam Collins; 689 SR/WAA; 691 (l) MCA; (spots) SR/WAA; 692 SR/WAA; 693 (t) Bridgette James/WAA; (b) SR/WAA; 694, 695 AS/Sam Collins; 696 MCA; 698 AT; 699 FP; 700 MCA; 703 (t) SR/WAA; (b) AS/Sam Collins; CHAPTER 32: 710 MCA; 711 (spots) Barbe Dekeyser/John Edwards & Associates; (lancelet, sea star) BH; (mouse) MW; (graphics) MCA; 713, 714 MCA; 718 (t) AT; (b) BH; 719, 722, 723, 724 MCA; CHAPTER 33: 730, 731 (exterior) JB; (interior) KC; 732 (spots) BH; CF, David Ashby, PG; (graphics) MCA; 733 Laurie O'Keefe; (graphics) MCA; (t) 734 BH; (b) PG; 736 BH; 737 (t) CF; (b) MCA; 739 (animals) PG; (archaeopteryx) BH; (dinosaurs) CF; (graphics) MCA; 740 (t) MCA; (b) BH; 741 BH; (spot) MCA; 744 Howard Friedman; 747 BH; 751 MCA; 752 MCA; CHAPTER 34: 757 (fish) BH; (zooms, spots) WN; (graphics) MCA; 758 (fish) BH; (hearts) KC; (graphics) MCA; 759 MCA; 763 BH; (graphics) MCA; 764 WN; 765 WN; (spots) KC; 766 WN; 769 (t) WN; (b) KC; 770 (frog) WN; (hearts) KC; (graphics) MCA; 771 FP; 772 AS/Sam Collins; 773 AS/Sam Collins; (spots) KC; 777 KC; CHAPTER 35: 783 MCA; 785 KC; 786 (t) WN; (b) (turtle) WN; (hearts) KC; (graphics) MCA; 790, 791 BH; 795 (tl) RH; (tr) MCA; (b) KC; 796 (bird) JW; (hearts) KC; (graphics) MCA; 797 (bird) JW; (lungs) KC; (graphics) MCA; 798 JB; 799 (external) JB; (internal, spot) KC; 800, 801, 804 JW; 805 MCA; CHAPTER 36: 813 JK; (graphics) MCA; 814 JB; 815 (external) JB; (internal) KC; (skull) BH; 821 John Daugherty/Lynn Prentice/LPI (Lynn Prentice Illus.); 826 Graham Allen; 829 BH; CHAPTER 37: 836 JW; 838 JB; 846 WN; CHAPTER 38: 856 JK; (graphics) MCA; 858 (t) KK/CK; (c) JK; (b) KK; 859 KK; (silhouette) JK; 860 JK; 861 (l) KK; (r) JK; 862 KK; 864, 865 JK; 866 KK; (silhouette) JK; 867 JK; 868, 869 MCA; 871, 877 MCA; CHAPTER 39: 882 MCA; 883 CK/MCA/Walter Stuart; 885 JK; 888, 890 MCA; 891 KC; 892 CK; 893 MCA; 894 JK; 895 JK; (graphics) MCA; 897, 898 JK; 902 CK/MCA; 903 MCA; CHAPTER 40: 910 Rolin Graphics, Inc.; 911 MCA; 914 JK; 915, 916 CK; 917 MCA; 920 JK; (graphics) MCA; 921 (l) CK; (r) MCA; 922 CK; 924 Yemi; (silhouette) JK; 926 Rolin Graphics, Inc.; 927 MCA; CHAPTER 41: 933 CK; 935, 936, 939, 942, 943, 946, 947 MCA; CHAPTER 42: 953, 954, 955, 956, 957 MCA; 958 JK; 959 MCA; 960 (l) JK; (r) CK; (graphics) MCA; 962 MCA; 965 CK; 966, 967 KK; 968 CK; 970-971 MCA; 976 Todd Buck; 977 MCA; 978 MCA; CHAPTER 43: 983 JK; 985, 986, 987 MCA; (pancreas, kidney) FP; 988 MCA; 990 (l) JK; (r) MCA; 992 JK; 993 (l) CK; (r) FP; 994, 996 MCA; 998 FP; 999 MCA; CHAPTER 44: 1006 (l) JK; (r) CK; 1007 (t) KK; (graphics) MCA; (b) MCA; 1008 CK; 1009 (l) JK; (r) CK; 1010 KK; (graphics) MCA; 1011 CK; 1012 MCA; 1014, 1015 CK; 1017 AT; 1022, 1023, 1024 MCA; APPENDIX 1035, 1043, 1044, 1045, 1046, 1047, 1048, 1049, 1050, 1051, 1053 MCA

PHOTOGRAPHY

Abbreviation Code

AA=Animals, Animals; BC=Bruce Coleman, Inc.; DRK=DRK Photo; ES=Earth Scenes; GH=Grant Heilman Photography; HRW/SD=Holt, Rinehart and Winston/Sam Dudgeon; MP=Minden Pictures; PA=Peter Arnold, Inc.; PR=Photo Researchers; PT=Phototake; RLM=Robert & Linda Mitchell; SP/FOCA= Sergio Purtell/Foca Co., NY, NY; TF=Tim Fuller; TSI=Tony Stone Images; VU=Visuals Unlimited

COVER AND TITLE PAGE Tony Stone Images; ii © Don Riepe/PA; iii © UNEP (Brunner)/PA; v © Richard Scheill/ES; vii (t) © Mark Gamba/The Stock Market; (b) © Arthur Tilley/FPG International; x © Jerry L. Ferrara/PR; xi (t) © Gerry Ellis/ENP Images, (b) © Seth Resnick; xiii © William H. Mullins/NASC/PR; vii (t) © Lefever/Grushow/GH, (b) © PhotoDisc; xv © Art Wolfe/TSI; xvi (t) © Andrew J. Martinez/PR, (b) © Bill Kamin/VU; xvii © Richard During/TSI; xix (t) © David Young-Wolff/PhotoEdit, (m) © David M. Phillips/PR, (b) © Tom McHugh/PR; xx (t) © Michael A. Keller/The Stock Market, (b) © TF, (b) © Telegraph Colour Library/FPG International; xxi (t) © Marty Cordano/DRK, (b) © Jim Cummins/FPG International; xxii © HRW/SD; xxiii © Hans Reinhard/BC; xxiv (t) Charlie Winters, (b) © HRW/SD; CHAPTER 1: 2 © PhotoDisc; 2-3 © Dr. Dennis Kunkel/PT; 3 (b) © Meckes/Ottawa/PR; 3 (c) © Wood Ransaville Harlin, Inc; 3 (t) © Corbis; 4 © TF; 5 © TF; 6 © UNEP (Brunner)/PA; 7 (b) © TF; 7 (c) © Zig Leszczynski/AA; 7 (t) © M. Abbey/VU; 8 (b) © TF; 8 (t) © Dan Guravich/NASC/PR; 9 (b) © Gerard Lacz/PA; 9 (t) © Runk/Shoenberger/GH; 10 HRW/Randal Alhadeff; 11 (b) © Runk/Schoenberger/GH; 11 (t) © Agricultural Research Service/USDA; 12 (t) © TF, (b) © Ray Pfortner/PA; 13 © Frans Lanting/MP; 14 (bc) © John Harte, (bl) © Zig Leszczynski/AA; 15 (bc) © John Harte, (br) © Jeff Smith/FOTOSMITH; 16 (bc) © RLM, (c) HRW/SD; 17 HRW/SD; 18 (bc) © Brian Payne/Black Star, (bl) Courtesy of Linda Rosa, (tl) © Dan Nedrelo; 19 © Will & Deni McIntyre/PR; 20 © Sinclair Stammers/SPL/PR; 24 HRW/SD; CHAPTER 2: 26 © Georg Gerster/PR; 27 © Mike Severns/TSI; 29 HRW photo; 30 (bc) © Dennis Kunkel/PT, (bl) © SP/FC, (cl) © Pal Hermansen/TSI; 31 © Marc Epstein/DRK; 33 (all) HRW/SD; 34 © SP/FC; 35 (all) HRW/SD; 36 © TF; 38 (bl) HRW photo, (br) © G.I. Bernard/AA; 44 (cr) © RLM; 46 HRW/SD; CHAPTER 3: 48 © Robert Brons/BPS/TSI; 49 © Ray Juno/TSM; 51 (bc) HRW photo, (bl) HRW/SD, (br) HRW/SD, (tr) © Mike Abbey/PR; 52 (bl) © SP/FC, (br) © Manfred Kage/PA; 53 (bl) © Sinclair Stammers/SPL/PR, (br) © Dr. Tony Brain/SPL/PR (tr) HRW/SD; 54 (tl) © Philippe Plailly/Eurelios/SPL/PR, (tr) © David M. Phillips/VU; 56 © Michael Abbey/PR; 57 (br) © Chris Bjornberg/PR, (cr) © John Cardmore/BPS/TSI; 58 © Michael Gabridge/VU; 61 © Don Fawcett/VU; 62 © R. Bolender-D. Fawcett/VU; 64 © Don Fawcett/VU; 65 © Jean Claude Revy/PT; 66 (cr) © Cell Robotics International, Inc., (tl) © Dr. Jeremy Burgess/SPL/PR; 68 (cr) Michael Gabridge/VU; 70 © E. R. Degginger; 71 © Barry Bomzer/TSI; CHAPTER 4: 72 © Dr. Dennis Kunkel/PT; 73 © Mitch Kezar/TSI; 75 (all) HRW/SD; 76 HRW/SD; 78 © Victor Scocozza/FPG International; 79 © Dr. David Scott/PT; 81 © Doug Wechsler; 84 © TF; 85 HRW/SD; 86 © Charles Gupton/TSI; 90 © WARD'S Natural Science; CHAPTER 5: 92 © Steve Gettle/ENP Images; 93 © David Young-Wolff/PhotoEdit; 99 © E. R. Degginger; 101 © C. Milkins/ES; 103 © Alan & Linda Detrick/PR; 107 © Dimaggio/Kalish/The Stock Market; 108 © John Colwell/GH; 109 © Mark Burnett/PR; 112 (br) © BPA/SS/PR; 114 © Bob Thomason/TSI; CHAPTER 6: 116 © Professors P.M. Motta & J. van Blerkom/SPL/PR; 117 © Tom McHugh/NASC/PR; 118 (bc) HRW/SD, (br) © Stephen J. Kraseman/DRK; 119 Institut Pasteur/CNRI/PT; 120 (bc) © David M. Phillips/VU, (bl) © PhotoDisc, (br) © PhotoDisc; 121 (bc) HRW/SD, (bl) © Evelyn Gallardo/PA, (c) © Dr. Dennis Kunkel/PT, (tr) © Gunther F. Bahr/AFIP/TSI; 122 (bl) © CNRI/SPL/PR, (br) © TF, (cl) © 1990 CMSP; 127 (c) © TF, (cr) © Biophoto Associates/SS/PR; 129 (cr) © Ariel Skelley/TSM, (tr) © Dr. P. Marazzi/SPL/PR; 130 (bc), (br) © John D. Cunningham/ VU; 131 (bl), (br) © John D. Cunningham/VU, (tr) © David M. Phillips/VU; 132 (cr) HRW/SD, (tl) © R. Calentine/VU; 136 (l) HRW/SD,6 (tr) © John D. Cunningham/VU; 137 HRW/SD; 138 © Roger Tully/TSI; 138-139 © Dee Breger/PR; 139 (b) © Larry Lefever/GH; 139 (c) © Francoise Sauze/Science Photo Library/PR, (t) © W & D Downey/Hulton-Getty/TSI; CHAPTER 7: 140 © Yorgos Nikas/TSI; 141 © Mark Gamba/ The Stock Market; 146 © Norbert Wu/PA; 147 © C. Edelmann/La Villette/SS/PR; 148 © BPA/SS/PR; 149 (br) HRW/SD, (tc), (tr) © PhotoDisc; 150 © Robert Calentine/VU; 151 (br) © L.J. Vitt, (cr), (tr) © PhotoDisc; 152 © Alan & Linda Detrick/PR; 155 (bc) © David M. Phillips/VU, (cc) © Jason Burns/Ace/PT, (cl) © Petit Format/Nestle/PR, (cr) © Yorgos Nikas/TSI, (tc), (tl), (tr) © PhotoDisc; 156 (l) HRW/SD, (tr) © Paul A. Zahl/PR.; **157** HRW/SD; **CHAPTER 8: 158** © Gary J. Benson; **159** © Wm. Munoz/PR; **160** (b) © Barry Runk/GH, (t) © Culver Pictures, Inc.; **163** (c) © Nigel Cattlin/Holt Studios International/PR; **165** (bc) © Barry Runk/GH, (bl) © Christian Grzimek/Okapia/PR, (br) © Christian Grzimek/Okapia/PR, (t) © Mark S. Skalny/VU; **166** (b) © Stephen Simpson/ FPG International, (t) © Jeff Noble/International Stock; **167** © Angelina Lax/PR; **170** © Runk/Schoenberger/GH; **171** © Image Shop/PT; **172** (b) © Lynn M. Stone/DRK, (cl) © PhotoDisc, (cr) © Lynn M. Stone/DRK, (t) HRW/Ken Lax; **173** (b) © G.R. 'Dick' Roberts, (t) © TSI; **174** © Ben Klaffke; **175** © Arthur Tilley/FPG International; **176** (b) © Yoav Levy/PT, (c) © Larry Lefever/GH, (t) © Larry Lefever/GH; **177** (bl) © Dan Guravich, (br) © Tom Walker/TSI, (c) © E.R. Degginger/ES, (t) © Barry L. Runk/GH; **178** (b) © Meckes/Ottawa/PR, (t) © PhotoDisc; **179** © Meckes/Ottawa /PR; **180** © Michael A. Keller/The Stock Market; **184** © Jurgen Berger, Max-Planck Institute/SPL/ PR; **CHAPTER 9: 186** © Tony Stone Imaging/TSI; **187** © Stefan Lawrence/International Stock; **188** © Alfred Pasieka/PA; **191** © Phil Degginger; **193** HRW/Victoria Smith; **194** (b) © SS/PR, (c) © SPL/PR, (t) © SS/PR; **197** © Dr. Gopal Murti/SPL/PR; **202**(tl) © SP/FC, (tr) HRW/SD; **203** HRW/SD; **CHAPTER 10: 204** © Runk/Schoenberger/GH; **205** © Tony Freeman/PhotoEdit; **207** © Minh Hoang Cong/PR; **208** (b) HRW/SD, (t) © O.L. Miller, B.R.Beatty, D.W. Fawcett/VU; **212** HRW/J. Langford; **214** (b) © Diego Goldberg/ Sygma, (t) HRW/Russell Dian; **216** © TF; **222** HRW/SD; **223** HRW/SD; **CHAPTER 11: 22**[illegible] © Ted Horowitz/The Stock Market; **225** © Peter Poulides/TSI; **226** (bc) © Biophoto Associates/PR, (bl) © James Gerholdt/PA, (br) © David M. Phillips/VU; **230** HRW/SD; **231** © G. Thomas Bishop/CMSP; **234** © Ted Thai/Time Magazine; **235** (b) Courtesy of Cellmark Diagnostics, Germantown, Md, (c) © E. R. Degginger; **237** © Agricultural Research Service, USDA; **239** (b) © Roslin Institute/PT, (c) ©Lou Hawthorne 199[illegible] Courtesy Bio-Arts and Research Corp.; **240** © Associated Press YONHAP; **244** © Jeff Smith/FOTOSMITH; **245** © Matt Meadows/PA; **246** (bl) © Adrian T. Sumner/TSI; **246** (t) © Jorge Ramierez/International Stock, (t) © Mark Joseph/TSI; **247** © Charles Thatcher/TSI; **248–249** © Gregory G. Dimijian/PR **248** (c), © RLM; **249** (b) © Phillip Hayson/Photo Researchers, INC, (t) © The Granger Collection; **249** (c) © Alejandro Balaguer/TSI; **CHAPTER 12: 250** © Steve Bloom-TCL/ Masterfile; **251** © William H. Mullins/NASC/PR; **253** HRW/SD; **254** Sojourner ®, Mars Rover ® and spacecraft design and images © 1996-97, California Institute of Technology. All rights reserved. Further reproduction prohibited; **257** © Sydney Fox/VU; **259** © Newcomb & Wergin/BPS; **261** (b) © Lewis Trusty/AA, (c) © A.M. Siegelman/VU, (t) © Mike Abbey/VU; **26**[illegible] Breck P. Kent/ES; **263** © Martin Wendler/PR; **265** (l) © Richard Thatcher, (r) R. Frazier Photolibrary, (r) © Dana Richter/VU; **266** © Runk/Schoenberger/GH; **267** © Eric Kamp/PT; **268** © Wolfgang Bayer/Bruce Coleman, Inc.; **272** HRW/SD; **273** © Rich Frishman/TSI; **CHAPTER 13: 274** © RLM; **275** © Michael J. Minardi/PA; **2**[illegible] Bridgeman/Art Resource; **280** © SS/PR; **281** (bc) © Tom & Pat Leeson/DRK, (br) Tom & Pat Leeson/DRK; **282** HRW/SD; **283** (bl) © RLM, (br) © E.R. Degginger/ES; [illegible] Gunter Ziesler/PA; **290** © N. Pecnik/VU; **291** © Breck B. Kent/AA; **292** © S. Frieco; **293** (b) © Gary Meszaros/VU, (c) © Zig Leszczynski/AA; **296** (b) © Andrew G. Wood/NASC/PR, (c) © Stephen G. Maka/DRK; **298** © TF; **299** © Brian Milne/ES; **CHAPTER** [illegible]**: 300** © TF; **301** © Scala/Art Resource, © Scala/Art Resource; **303** (b) HRW/SD, (t) Frans Lanting/MP; **304** (l) © Dani/Jeske/AA, (r) © Frans Lanting/MP; **307** (c) © TF, (cr) © David L. Brill \ Atlanta, artifact credit, Transvaal Museum, Pretoria, (tl) David L. Brill \ Atlanta, artifact credit, National Museum of Ethiopia, (tr) © David L. Brill \ Atlanta, artifact credit, National Museum of Tanzania, Dar es Salaam; **30**[illegible] © Gordon Gahan/National Geographic Society, (r) © David L. Brill \ Atlanta, artifact credit, National Museums of Kenya, Nairobi; **310** © TF; **312** © TF; **317** © Ken Garrett/NGS; **CHAPTER 15: 318** © Jack Dykinga; **319** © Doug Wechsler; **320** © [illegible] Ferrara/PR; **321** (bc), (bl) © SP/FC, (br) © TF, (c) © SP/FC, (cl) © Leonard Lee Rue III/NASC/PR, (cr) © Grant Heilman/GH; **323** (b) © Stephen Dalton/AA, (t) © Jack Clark/University of California; **324** HRW/SD; **325** (c) © Jeff Lepore/NASC/PR, © Zig Leszczynski/AA, (r) © Ron Kimball Photography; **326** (b) © Scala/Art Resource, © TF Flannery; **327** (bl) © Gregory Ochocki/ PR, (br) © RLM, (c) © Doug Wechsler/DRK, (t) © James Watt/AA; **334** (bc) © BIOS (Klein/Hubert)/PA, (bl) © RLM, (br) Barry Runk/GH, (tc) © BIOS (Klein/Hubert)/PA, (tl) © John Gilmore Runk/Schoenberger/GH; **335** HRW/SD; **336** © Art Wolfe/ TSI; **336-337** © Carr Clifton/MP; **337** (b) © Stephen J Krasemann/DRK; **337** (c) © Art Wolfe/ © Stouffer Prod./AA/ES; **CHAPTER 16: 338** © Frans Lanting/MP; **339** © TF; **340** (b) McHugh/Steinhart Aquarium/PR, (br) © Peter Unmack, (t) © Alan Schein/TSI, (c) © Manfred Danegger/PA, (l) © Manfred Mehlig/TSI, (r) © Jack Wilburn; **343** © Shin Yoshino/MP; **344** (l) © Tom McHugh/PR, (r) © Johnny Johnson/DRK, (t) © David Scharf/PA; **347** HRW/Michelle Bridwell; **348** (b) © R.H./GH, (t) © Stan Osolinski; **349** HRW/SD; **350** © UPI/Bettman; **351, 356,** HRW/SD, HRW/SD; **CHAPTER 17: 358** © Marc Chamberlain/TSI; **359** © Rosing, N. OSF; **360** © Merlin D. Tuttle/Bat Conservation International/PR; **361** (bl) © RLM, (br) Tuten/ES, (t) © Jim Strawser/GH; **362** (b) HRW/Randal Alhadeff, (tl) © Patti Murray, (tr) © Runk/Schoenberger/GH; **363** (b) © Richard Kolar/ES, (t), (c) © Tom Bean; **364** HRW/SD; **370** © Grant Heilman/GH; **372** © Breck P. Kent/AA/ © Steve Strickland/VU; **378** HRW/SD; **379** HRW/SD; **CHAPTER 18: 380** © Jack [illegible]; **381** © Superstock; **382** © Anthony Mercieca/PR; **383** (b) © Patti Murray/ Bill Beatty/VU; **384** (b) © RLM, (t) © Hans Pfletschinger/PA; **388** © Anne Wertheim/AA; **390** (b) © Runk/Schoenberger/GH, (t) © RLM; **391** © D. Cavagnaro/DRK; © W.H. Hodge/PA, (t) © Tom Bean/TSI; **394** (b) © Gerry Ellis/ENP Images, (c) Wechsler/ES, (t) © Art Wolfe/TSI; **395** (b) HRW/SD, (c) © Arthur C. Smith III © Leo Keeler/AA/ES; **397** (b) © Jeff Greenberg/PR, (t) © David B. Fleetham; **398** (b) © Norbert Wu/PA, (t) © Manfred Kage/PA; **402** (b) HRW/Victoria Smith; Manfred Kage/PA; **CHAPTER 19: 404** © Johnny Johnson/DRK; **405** © Kennan Ward/DRK; **406** © Phil Degginger; **407** © NOAA; **410** © John S. Lough/VU; **412** (b) © Hugh H. Iltis, (t) © Richard Schiell/ES; **414** © Chad Ehlers/TSI; **416** (l) © Lionel Delevingue/PT, (r) © E. R. Degginger; **417** (b) © Bios (Frebet)/PA, (t) © Tony Freeman/PhotoEdit; **418** (b) HRW/ Larry Kolvooro, (c) Courtesy of Chesapeake Bay Foundation, (t) © Superstock; **419** (b) © Randy Wells, (t) © Seth Resnick; **420** (b) © TF, (c) HRW/SD, (t) © Steve Smith/FPG International; **422** (b) © Pat & Tom Leeson/PR; **424** © LSF OSF/ES; **425** © Tom Tracey/ TSI; **426** (bc) © Enrico Ferorelli/Masterfile, (bl) © Tom Boyden, (br) © Stuart Westmorland/TSI; **426** (t) © VCG/FPG International; **427** © Michael P. Moore; **428** © Jeff Corwin/TSI; **428-429** © Ray Pfortner/PA; **429** (b) © NIBSC/Science Photo Library/PR, (c) © James King-Holmes/Science Photo Library/PR, (t) © Corbis; **CHAPTER 20: 430** © Larry Ulrich/DRK; **431** © Grant Heilman/GH; **434** (l) © Andrew Syred/SPL/ PR, (r) © TF; **435** © Corale L. Brierley/VU; **436** (l) © Rakosky/Custom Medical Stock Photo, (r) © Robert Calentine/VU; **437** (b) © James R. McCullagh/VU, (t) © N. Pecnik/ VU; **438** (b) © Gerald & Buff Corsi/VU,(t) © Despommier/PR; **439** © Ken Wagner/PT; **441** (bl) © Rod Planck Photography, (r) © Tom Bean/DRK, (t) © Robert Frerck/TSI; **442** (b) © Barbara Gerlach/DRK, (c) © Jim Steinberg/PR, (tl) © Wayne Lynch/DRK, (tr) © John Gerlach/DRK; **443** (bl) © David Scharf, (br) © Flip Nicklin/Minden, (t) © Patrick Ward-TCL/Masterfile; **444** © Fred Bavendam/PA; **446** (b) © Hal Harrison/GH; **448** (c) © E. R. Degginger, (l) © Gerry Ellis/ENP Images, (r) © PhotoDisc; **449** © Fred Bruemmer/DRK; **CHAPTER 21: 450** © S. Lowry/Univ. Ulster/TSI; **451** © Will and Deni McIntyre/PR; **452** © John Cardmore/BPS/TSI; **453** (c) © Dr. O Bradfute/PA, (l) © NIBSC/SPL/PR, (r) © E.O.S./Gelderblom/PR; **454** © Lee D. Simon/SS/PR; **456** © Bruce Ayres/TSI; **457** © Oliver Meckes/Gelderblom/PR; **458** © Runk/Schoenberger/GH; **459** © Germain Rey/Gamma Liaison; **460** © Fred Hossler/VU; **461** (c) © G. Shih - R. Kessel/ VU, (l) © Fran Heyl Associates, Color Enhancement by Pix Elation (r) © John D. Cunningham/VU; **462** © Rakosky/Custom Medical Stock Photo; **463** (bl) © Science VU/VU, (br) © George J. Wilder/VU, (c) © C.P. Vance/VU, (t) © E. Webber/VU; **464** © Fred Hossler/VU; **465** © Cecil H. Fox/SS/PR; **466** (b) © Laura Reyes/Fran Heyl Assoc., (t) © Michael Abbey/VU; **467** © Christine Case/ VU; **468** (b) © Richard T. Nowitz/PT, (t) © TF; **470** (b) © Fred Hossler/VU; **472** HRW/SD; **473** © Michael Rosenfeld/TSI; **474** © Maximilian Stock LTD/Science Photo Library/PR, (b) © Gilbert S. Grant/PR, (bl) © Judd Cooney/PT, (c) © Scott Camezine/ Science Source/PR, (t) © A. Weber/Index Stock; **CHAPTER 22: 476** © RLM; **477** © Gary A. Conner/PhotoEdit; **478** © George J. Wilder/VU; **479** (b) © Dr. William Fry/Fran Heyl Associates, (c) © M. Abbey/PR, (t) © Runk/Schoenberger/GH; **480** © Microfield Scientific Ltd./SPL/PR; **482** © Robert Brons/BPS/TSI; **483** © Kent Wood; **484** (c) © Gary R. Robinson/VU, (l) © Mark E. Gibson/VU, (r) © Darryl Torckler/TSI; **485** © Roland Birke/PA; **486** © Manfred Kage/PA; **487** © RLM; **488** (b) © Bill Beatty/ES, (c) © Cabisco/VU, (t) © Cabisco/VU; **489** (c) © Tetsu Yamazaki/International Stock, (t) © Centers for Disease Control; **492** © Reuters/Corinne Dufka/Archive Photos; **495** (l) Courtesy of Richard Triemer, Rutgers University, (r) © M. Abbey/VU; **496** © Robert Brons/BPS/TSI; **497** HRW/SD; **CHAPTER 23: 498** © Jeff Lepore/PR; **499** © TF; **500** © Michael Gadomski/ES; **501** (b) Courtesy of Marshall Lyon/CDC, (c) © Mike Eichelberger/VU, (t) © Noble Proctor/PR, (t) HRW/SD; **502** (b) © Tom Bean/DRK, (t) © Stephen Dalton/NASC/PR; **504** (b) © E. R. Degginger, (t) © Barry L. Runk/GH; **506** Runk/Schoenberger/GH; **507** © Marty Cordano/DRK; **508** © BPA/SS/PR; **509** (b) Runk/Schoenberger/GH, (l) © V. Ahmadjian/VU, (r) © Doug Wechsler/AA/ES; **510** © D. Cavagnaro/DRK; **512** (c) © Ray Coleman/PR, (l) © E. R. Degginger, (r) © Marty Cordano/DRK; **514** © Jeff Smith/FOTOSMITH; **516-517** © George Lepp/TSI; **516** © Corbis; **517** (b) © Howard Buffett/GH, (c) © Keith Weller/ Agricultural Research Service, USDA; **517** (t) © The Granger Collection; **CHAPTER 24: 518** © David Cavagnaro/PA; **519** © Tom Bean/DRK; **520** © Frans Lanting/MP; **521** (b) HRW/SD,(tl) © L. Mellichamp/VU, (tr) HRW/SD; **522** © RLM; **523** © RLM; **524** (l) © Michael P. Gadomski/PR, (r) © Jane Grushow/GH; **526** © Uniphoto; **527** (c) © Runk/ Schoenberger/GH, (l) © Ed Reschke/PA, (r) © Runk/Schoenberger/GH; **528** Courtesy of Hans Steur; **529** (bc) © Ed Reschke/PA, (bl) © Runk Schoenberger/GH, (br) © RLM, (tl) © Dwight Kuhn, (tr); © Michael Ederegger/DRK; **530** © RLM; **531** (bc) © Runk/ Schoenberger/GH; **531** (bl) © Patti Murray/ES, (br) © Gerald & Buff Corsi/VU, (t) HRW/SD; **532** © Guy Marche/FPG International; **533** (l) © Alan & Linda Detrick/PR, (r) © William H. Mullins/NASC/PR; **535** (b), (tl) © TF, (tr) © Peter Holden/VU; **536** (b) HRW/SD, (t) © Photodisc; **537** (b) © TF, (t) © James P. Rowan/DRK; **538** (bc) © Bob Thomas/TSI, (bl) © Robert Kusel/TSI, (br) © Howard Kingsnorth/TSI, (cl) © Nick Gunderson/TSI, (cr) © Tom Till/TSI, (t) © Paolo Koch/PR; **539** (c) © Patti Murray/ES, (l) © RLM, (r) © Larry Lefever/GH; **540** (l) © Inga Spence/VU, (r) © Ric Ergenbright/ CORBIS; **542** (b) © David M. Schleser, Nature's Images/PR; **544** HRW/SD, Bill Bachman/PR; **CHAPTER 25: 546** © S Lowry/Univ. Ulster/TSI; **547** © Annie Griffiths Belt/DRK; **548** (b) © E. R. Degginger, (c), (l), (r) © RLM, (t) © Ed Reschke/PA; **552** (l), (r) © RLM; **554** (b) © E.R. Degginger/ES, (t) © Runk/Schoenberger/GH; **556** HRW/SD; **557** (b) © Scott Bauer/ ARS/USDA, (t) © Merlin D. Tuttle, Bat Conservation International; **560** HRW/SD; **562** (l) © William D. Adams, (r) © William D. Adams; **564** (b) © RLM; **566** © RLM; **CHAPTER 26: 568** © Art Wolfe/PR; **569** © Ken Biggs/TSI; **571** (b) © Ed Reschke/PA, (l) © Lefever/ Grushow/GH, (r) © E. R. Degginger; **572** (l) © George J. Wilder/VU, (r) © George J. Wilder/VU; **573** (bl) © Cabisco/VU, (br) Herb Charles Ohlmeyer/Fran Hely Assocs., (t) © R. Calentine/VU; **574** (l), (r) Runk/Schoenberger/GH; **575** (b), (c) © Runk/ Scheonberger/GH, (t) © John D. Cunningham/VU; **576** © RLM; **577** © Dwight R. Kuhn/DRK; **580** © Walter H. Hodge/PA; **581** (l), (r) © Dominic Oldershaw; **584** © Scott Camazine/PR; **586** (b) HRW/SD, (t) HRW/SD; **587** © Geoff Tomkinson/Science Photo Library/PR; **CHAPTER 27: 588** © Arthur C. Smith/GH; **589** © Alan Thornton/TSI; **591** © D. Cavagnaro/VU; **592** (cl) © RLM, (cr) © R.F. Evert, (l) © Robert P. Comport/AA/ES, (r) © TF; **594** © Darrell Gulin/DRK; **595** © R.F. Evert; **596** © Dr. Jeremy Burgess/PR; **597** (b) © SP/FC, (t) © SP/FC; **599** (b) HRW/SD, (c), (t) © Abbott Laboratories; **600** (c) © R. Calentine/VU, (l) © Barry L. Runk/GH, (r) © Ed

Credits continued

Reshke/PA; **601** © Runk/ Schenberger/GH; **602** © Runk/Schoenberger/GH; **604** (b) © Walter Chandoha ; **606** (b) © Runk/Shoenberger/GH, (t) © Michael P. Gadomski/PR; **607** (b) © Michael Rosenfeld/ TSI, (t) HRW/SD; **608-609** © Gerry Ellis /MP; **608** © Maria Stenzel/National Geographic Image Collection; **609** (b) © Scott Camazine/PR, (c) © Christoph Burki/ TSI, (t) © TSI; **CHAPTER 28: 610** © Bill Kamin/VU; **611** © Richard Herrmann; **612** (l) © PhotoDisc, (r) © TF; **613** (bl) © Manfred Kage/PA, (br) © Jeff Smith/FOTOSMITH, (t) © Tim Fitzharris/MP; **614** (b) © Carolina Biological Supply/PT, (t) © David Phillips/The Population Concil/PR; **615** (b) Jeffrey L. Rotman/PA, (t) © Quest/SPL/PR; **616** (b) © Neil McDaniel/PR, (r) © Don Riepe/PA, (t) © Mary Beth Angelo/NASC/PR; **617** (b) © TF, (t) © W.J. Weber/VU; **619** (l) © E. R. Degginger; **621** © Stephen Dalton/AA; **623** (b) © E. R. Degginger, (t) © Mitch Kezar/TSI; **625** (b), (t) © Runk/Schoenberger/GH; **626** © SP/FC; **627** (b) © Brandon D. Cole, (t) © Dr. Paul Zahl/NASC/PR; **628** (b) © Frans Lanting/MP, (t) © Franklin J. Viola; **632** © Jeff Smith/FOTOSMITH; **633** © Agricultural Research Service, USDA; **CHAPTER 29: 634** © J.W. Mowbray/NASC/PR; **635** © Robert Rattner; **636** © Jeffrey L. Rotman/PA, (c) © Tom Bean; **638** (bc) © Manfred Kage/PA, (bl) © Cabisco/VU, (br) © RLM, (cc) © Neil G. McDaniel/NASC/PR, (cl) © David J. Wrobel/BPS, (cr) © Runk/Schoenberger/GH, (t) © TF; **642** © Runk/Schoenberger/GH; **643** (b) © RLM, (t) Runk/Schoenberger/GH; **644** (b) © Catherine Karnow, (cl) © David J. Wrobel, (cr) © Bryan Mullennix/TSI; **645** (bl) © Fred Bavendam, (br) © Herv Chaumeton/PR, (t) © Fred Bavendam/MP; **646** © Franklin J. Viola; **647** © Hal Beral/ VU; **649** © RLM; **650** © Sinclair Stammers/SS/PR; **651** (b) © Sinclair Stammers/ SPL/PR, (c) © TF, (t) HRW/SD; **652** © RLM; **656** (b) © G. Shih-R. Kessel/VU, (c) HRW/SD, (t) © Carolina Biological Supply/PT; **CHAPTER 30: 658** © John Seagrim-TCL/ Masterfile; **659** The Metropolitan Museum of Art, Purchase, Louisa Eldridge McBurney Gift, 1953. (53.11.6) Photograph © 1985 The Metropolitan Museum of Art; **663** (b) HRW/SD, (t) © Robert Maier/AA; **664** (l) © J.H. Robinson/AA, (r) © MC Chamberlain/ DRK; **665** (l) © A. Kerstitch/VU, (r) © A. Kerstitch/VU; **666** (b) © TF, (t) © George H. Harrison/GH; **667** (c) © Joyce & Frank Burek/AA, (l) © Runk/Schoenberger/GH, (r) © Douglas Faulkner/NASC/PR; **669** (l) © Ed Reschke/PA, (r) © Hal Beral/VU; **670** HRW/SD; **671** © John Anderson/AA; **672** (b) © J.P. Ferraro/Jacana/PR, (t) © Barbara Gerlach/DRK; **674** (b) © St. Bartholomew's Hospital/SPL/PR, (t) © RLM; **677** © Mary Beth Angelo/PR; **678** © Jeff Smith/FOTO-SMITH; **679** © TF; **CHAPTER 31: 680** © Mervyn Rees/TSI; **681** © Runk/Schoenberger/GH; **682** (l) © Raymond A. Mendez/AA, (r) © Joe McDonald/ AA; **684** (b) © Dr. Jeremy Burgess/SPL/PR, (b) © TF, (c) © Art Wolfe/TSI, (t) © E. R. Degginger; **685** (b) HRW/SD, (t) © Ed Reschke/PA; **686** (t) © RLM; **687** © TF; **688** © RLM, (l) © David Scharf/PA, (r) © Scott Camazine/NASC/PR; **689** © RLM; **690** (l) © David Shauf/PA, (r) © Scott Camazine/NASC/PR; **694** © R. Al Simpson/VU; **696** © Stephen Dalton/NASC/PR; **697** (bl) © Marty Cordano/DRK, (br) © Karl Weidmann/PR, (t) © RLM; **699** (t) © Flip Nicklin/MP; **704** © RLM; **705** HRW/SD; **706** (bl) © Henry Ausloos/AA/ES, (br) © Grant Heilman/GH, (t) © Jim Strawser/GH; **707** © Scott Bauer/Agricultural Research Service, USDA; **CHAPTER 31: 708** © Marilyn Kazmers/PA; **709** © Michael Todd Walker; **711** © Phil Degginger/Geological Enterprises; **712** (bl) © E. R. Degginger, (br) © Andrew J. Martinez/PR, (t) © TF; **713** © Joyce & Frank Burek/AA; **714** © Andrew J. Martinez/PR; **715** (b) © Fred Bavendam, (c) HRW/SD, (t) © Fred Winner/Jacana/PR; **716** (b) © Kelvin Aitken/PA, (tl) © D. Shale/ AA, (tr) © Bruce Watkins/AA; **717** HRW/David Thielk; **719** © Gregory G. Dimijian, M.D.; **720** (b) HRW/SD, (t) © Grant Heilman/GH; **724** © W. Gregory Brown/AA; **725** © SIU/PA; **726** © Tim Davis/TSI; **726-727** © Nigel J.H. Smith/AA/ES; **727** (b) © K. Schafer & M. Hill/TSI, (c) © Valmik Thapar /PA, (t) © Hulton Getty/TSI; **CHAPTER 33: 728** © Tom Bean/DRK; **729** © Vito Palmisano/TSI; **730** © Robert Maier/AA; **731** © Jim Tuten/AA; **732** © John Cancalosi/Natural Selection; **735** (b) © E.R. Degginger/AA, (cl) © Tom McHugh/PR, (cr) © M. Gibbs/AA, (t) © Stock Editions; **738** (bl) © William Leonard/ DRK, (br) © Zig Leszczynski/AA, (tl) © Carl R. Sams II/PA, (tr) © James P. Rowan/ DRK; **742** © TF; **743** (bl) © Joe McDonald/VU, (br) © Tom J. Ulrich/VU, (tl) © C.C. Lockwood/DRK, (tr) © M.H. Sharp/PR; **744** © Douglas Mazonowicz/Art Resource, NY; **745** (b) © Natural History Museum of Los Angeles County, (t) © E. R. Degginger; **746** (c) © Tom McHugh/PR, (l) © D. Cavagnaro/VU;, (r) © Tom McHugh/PR; **747** (b) © O. Louis Mazzatenta/NGS, (t) © S. Nielsen/DRK; **748** (bc) © David Welling/AA, (bl) © E. R. Degginger, (br) © Stephen J. Krasemann/DRK, (tc) © Rob & Ann Simpson/VU; **752** © TF; **CHAPTER 34: 754** © Dalton, S. OSF/AA/ES; **755** © Ed Robinson/Pacific Stock; **756** (c) © Fred Bavendam, (l) © Bill Kamin/VU, (r) © Kelvin Aitken/PA; **760** (l) © George Bernard/AA, (r) © Stuart Westmorland/TSI; **761** (b) © Hans Reinhard/Okapia/NASC/PR, (cl) © Tom McHugh/NASC/PR, (cr) © Tom McHugh/NASC/PR; **762** (c) © Kevin Aitken/ PA, inset (b) © Wayne & Karen Brown, inset (t) © Doug Perrine (J. Castro)/PA,(t) HRW/SD; **764** © John G. Shedd Aquarium; **766** HRW/SD; **767** (b) © Tom McHugh/PR, (t) © Richard Herrmann; **768** © R. Lindholm/VU; **771** © RLM; **772** © John Mitchell/PR; **774** (b) Juan M. Renjifo/AA, (t) © Zig Leszczynski/AA; **778** (b) © RLM, (t) HRW/SD; **779** © T.A. Wiewandt/DRK; **CHAPTER 35: 780** © Frans Lanting/MP; **781** © Joe McDonald/VU; **782** © Joe McDonald/AA/ES; **783** © G.C. Kelley/PR; **784** (b) HRW/SD, (tl) © Mark A. Chappell/AA, (tr) © Mark A. Chappell/AA; **787** (l) © Zig Leszczynski/AA, (r) © Michael Fogden/DRK; **788** (l) © Tom McHugh/PR, (r) © Roland Seitre/PA; **789** (c) © Michael Fogden/DRK, (l) © Michael Fogden/DRK, (r) © Michael Fogden/DRK; **790** © E. R. Degginger; **792** (b) © SP/FOCA, (c) © Tom Walker/TSI, (t) © DR & TL Schrichte/TSI; **793** (b) © Zig Leszczynski/AA, (c) © David Austen/TSI, (t) © Fred Whitehead/AA; **794** © Richard During/TSI; **797** © Keith Scholey/Masterfile; **798** © Alan & Sandy Carey/PR; **800** © Tim Davis/TSI; **806** © E. R. Degginger; **807** Jeff Smith/ FOTOSMITH; **CHAPTER 36: 808** © Darrell Gulin/TSI; **809** © Taronga Zoo, Sydney/Tom McHugh/PR; **810** © Dan Guravich/PR; **811** (b) HRW/SD, (c) © J. Sneesby/B Wilkins/TS, (t) © Stanley Breeden/DRK; **812** (bl) © SP/FC, (br) © SP/FC, (tl) © Carl R. Sams II/PA, (tr) © RLM; **814** © Kevin Schafer; **816** (l) © Jeff Foott, (r) © Tom McHugh/PR; **817** (b) © James D. Watt/Masterfile, (t) © Joe McDonald/AA/ES; **818** (b) © Tom McHugh/PR, (t) © Jean Phillippe Varin/NASC/PR ; **819** (b) © TF, (c) © John Cancalosi/PA, (t) © E.R. Degginger/AA; **820** (l) © Robert Maier/AA/ES, (r) © Fred Habegger/GH; **822** (b) © Sturgis McKeever/PR, (c) © Art Wolfe/TSI, (t) © Rod Planck/PR; **823** (b) © Norbert Wu/Masterfile, (c) © Leo Keeler/AA/ES, (t) © Jeremy Woodhouse/DRK; **824** (b) © Fred Felleman/TSI, (c) © M. Harvey/DRK, (t) © Stephen J. Kraseman/DRK; **825** (b) © John Downer/Masterfile, (c) © Douglas Faulkner/PR, (t) © Thomas Kitchen/Natural Selection; **830** © Ed Reschke/PA; **831** (b) © Flip Nicklin/MP, (t) © Corbis/Phil Schermeister, (tl) © Arthur C. Smith III/GH; **CHAPTER 37: 832** © J. Eastcott & Y. Momatiuk/Masterfile; **833** © T. McHugh/Parrot Jungle, Miami Fl./PR; **834** (l) © Klaus Uhlenhut/AA, (r) © Neil McIntyre/Masterfile; **835** (b) © Tim Davis/PR, (t) © Robin Brandt/Natural Selection; **836** © S.J. Kraseman/PR; **837** © Omikron/Science Source/ PR; **838** HRW/SD; **839** © Thomas McAvoy/Life Magazine; **840** © Jeff Lepore/PR; **841** (bc) © Patti Murray/AA, (bl) © Stephen J. Krasemann/PR, (br) © N. Rosing/AA/ES, (tc) © Dwight R. Kuhn, (tl) © Gary W. Griffen/AA, (tr) © William H. Beatty/VU; **842** (b) © Wayne Lynch/DRK, (t) © Brian Woodbridge; **843** (b) © Peter Weimann/AA/ES, (tl) © Francois Gohier/PR, (tr) © Charles Palek/AA/ES; **845** (l) © Hans Reinhard/BC, (r) © J.C. Carton/BC; **846** © Martin Harvey/TWC; **849** (bl) © T. McHugh/Parrot Jungle, Miami Fl./PR, (br) © Robin Brandt/Natural Selection, (tl) © Francois Gohier/PR, (tr) © Thomas McAvoy/Life Magazine; **850** © TF; **851** © TF; **852** © Lester L kowitz/TSM; **852-853** © G W Willis/BPS/TSI; **853** (b) © Richard Nowitz /PT,) © Jim Cummins/ FPG International, (tl), (tr) © The Granger Collection; **CHAPTER 38: 54** © Telegraph Colour Library/FPG International; **855** © TF; **856** (bl) © GW Willis/PS/TSI, (br) © Carolina Biological Supply/PT, (cl) © John D. Cunningham/VU, (cr© Eric Grave/SS/ PR, (tl) © Ed Reschke/PA, (tr) © John D. Cunningham/VU; 8 © TF; **862** © Andrew Syred/TSI; **863** (bl) HRW/Coronado Frank Wing, (br) © Caroli Wood/International Stock, (tl, tc) © Dr. M. Klein/PA, (tr) © Yoav Levy/PT; **870** © T **71** © SP/FOCA; **872** (b) © Telegraph Colour Library/FPG International, (t) HRW/SD**73** (b) © TF, (t) © TF; **874** (b) James Stevenson/SS/PR, (t) © Dr. P. Marazzi/SS/PR; (b) © Dr. P. Marazzi/ SPL/PR; **878** © TF; **CHAPTER 39: 880** © Uniphoto, Inc.; **881** © CI/SPL/PR; **883** © Ed Reschke/PA; **884** © John D. Cunningham/VU; **885** © TF; **886** (b Uniphoto, (t) © Merlin D. Tuttle/BCI; **887** (c) © Don Fawcett/PR, (l) © David M. Phi/VU, (r) © Robert Caughey/VU; **888** © Veronica Burmeister/VU; **889** © Lester Lefk z/TSM; **892** © TF; **893** © Jon Feingersh/Tom Stack & Associates; **896** © Blaitz/PR; **899** (b) HRW/SD, (t) © Petit Format/Nestle/SS/PR; **900** (b) © James StevenSPL/PR, (t) © A. Glauberman/SS/PR; **904** © Yoav Levy/PT; **905** © Geoff Tompkinson/PR; **CHAPTER 40: 906** © Meckes/Ottawa/PR; **907** © TF; **908** (l), (r) HRWchelle Bridwell; **909** (c) © Dominic Oldershaw, (l), (r) © Dominic Oldershaw; **911**/Michelle Bridwell; **912** (b), (c), (t) © SP/FC; **913** © Davies & Starr Inc./TSI, (t) HDiane Russell; **916** © TF; **918** © CNRI/SPL/PR; **919** HRW/Victoria Smith; **922** nk/Schoenberger/GH; **923** © TF; **928** © TF; **929** © Jeff Smith/FOTOSMITH; **CHAPTE930** © Dr. Dennis Kunkel/ PT; **931** © Mike Yamashita; **932** (t) © Prof. P. Motta, Cor Nottola/Univ. "La Sapienza", Rome/SS/PR; **934** (b) © Meckes/Ottawa/PR, (t r. Dennis Kunkel/PT; **937** HRW/SD; **938** © Mednet/PT; **939** © THE GRANGER CO TION, New York; **940** HRW/SD; **942** © Lennart Nilsson/Boehringer Ingelheim Int'l H; **943** © Paul Hardy/ The Stock Market; **944** © David M. Grossman/PR; **948** ©**49** © Mark Richards/ PhotoEdit; **CHAPTER 42: 950** © Oliver Meckes/Ottawa/PR; Roger Tully/TSI; **952** © Andrew Leonard/PR; **959** © TF; **961** AP/Wide World; 9TF; **966** © A.L. Blum/VU; **970** © Gregory G. Dimijian/PR; **971** HRW/Michelle ll; **972** © Jim Strawser/GH; **973** © Jeff Smith/FOTOSMITH; **974** (b) © E. R. Deg, (t) © Scott Camazine/PR; **978** © Dr. Monty Buchsbaum/PA; **CHAPTER 43: 980** © rs/Wolfgang Rattay/Archive Photos; **981** © Dee Breger/PR; **982** © David Young-WotoEdit; **984** HRW/SD; **987** © Bob Daemmrich/The Image Works; **988** HRW/Victmith; **989** © Terry Allen/ International Stock; **992** © John Paul Kay/PA; **993** r Skinner/PR; **994** (b), (c) © SP/FC, (t) © Carolina Biological Supply/PT; **995** © T © TF; **1000** © M.I. Walker/ PR; **1002** (b) © Stephen Simpson/FPG, (t) © St Bmew's Hospital/Science Photo Library/PR; **1003** © SIU/PA; **CHAPTER 44: 1004**nart Nilsson/Albert Bonniers Förlag; **1005** © Ed Kashi/PT; **1010** © David Mps/PR; **1013** © C. Edelmann/La Villette/SS/PR; **1016** (bl) © Lennart Nilsson/Abnniers Förlag, (br) © Petit Format/Nestle/Science Source/PR, (cl) © Lensson/Albert Bonniers Förlag, (cr) © Bromhall, D. OSF/AA/ES, (t) © David H. WRBIS; **1017** © Eurelios/PT; **1019** (bl) © Cabisco/VU, (br) © Frederick C. Skvar (t) © L. Moskowitz M.D./ Medichrome; **1020** (b) © SPL/Custom Medic Photo, (t) HRW/Michelle Bridwell; **1024** © CNRI/SPL/PR; **1026** HRW/n Steen ; **1034** (b), (t) © SP/FOCA; **1036** HRW/SD; **1037** (l), (r) © SP/FOCA; **1038**D; **1040** © SP/FOCA; **1041** (l), (r) © SP/FOCA; **1042** (bl), (br), (c), (t) © SP/FOC(bl) © Dr. Dennis Kunkel/PT; **1048** (l) © Fran Heyl Associates (r) © S. Lowry/Un/TSI; **1049** (l) © Corale L. Brierley/VU, (r) © Kent Wood; **1050** © Jeff L; **1051** (l) © Doug Wechsler/ AA/ES, (r) © Ed Reschke/PA; **1052** © Williamins/NASC/PR; **1053** (l) © Gerry Ellis/ENP Images, (r) Joyce and Frank Burek**4** (t) © E. R. Degginger, (b) © Art Wolfe/ TSI; **1055** © RLM; **1056** (c) © RichardTSI, (t) Joe McDonald/AA, (b) Dan Guravich/PR

Timeline of the History of Life on Earth

Age *(in millions of years ago)*

Earliest fossil bacteria

Origin of O_2 by photosynthesis

3,500

2,500

PRECAMBRIAN ERA

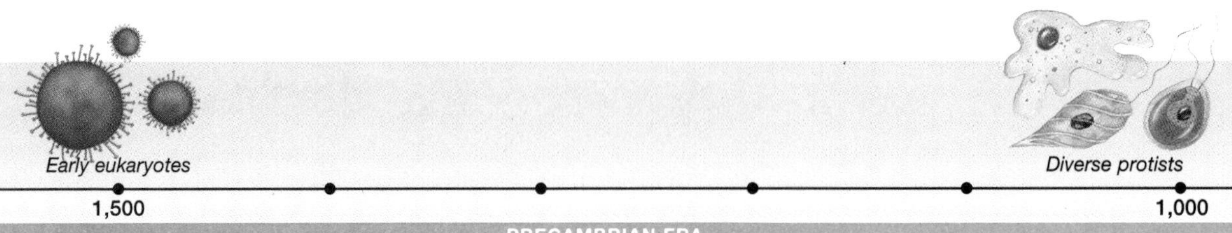

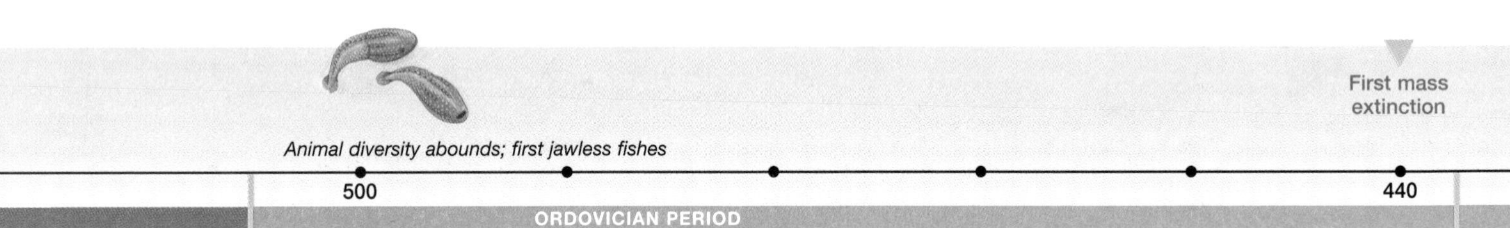

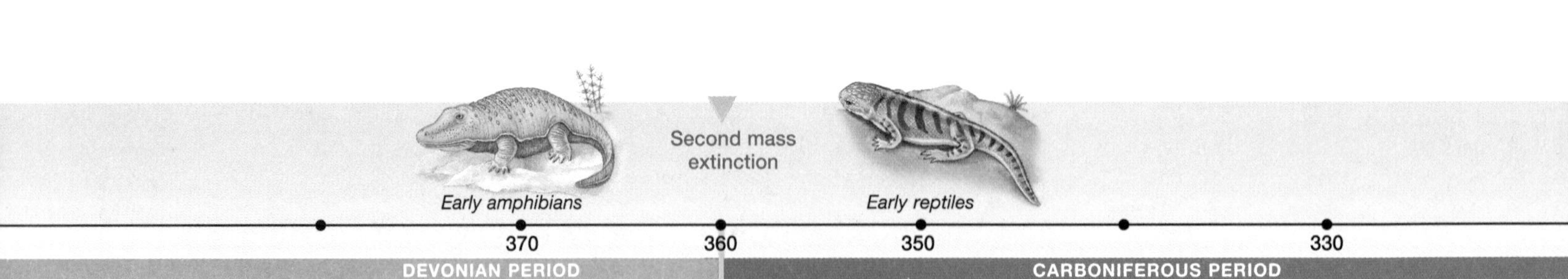

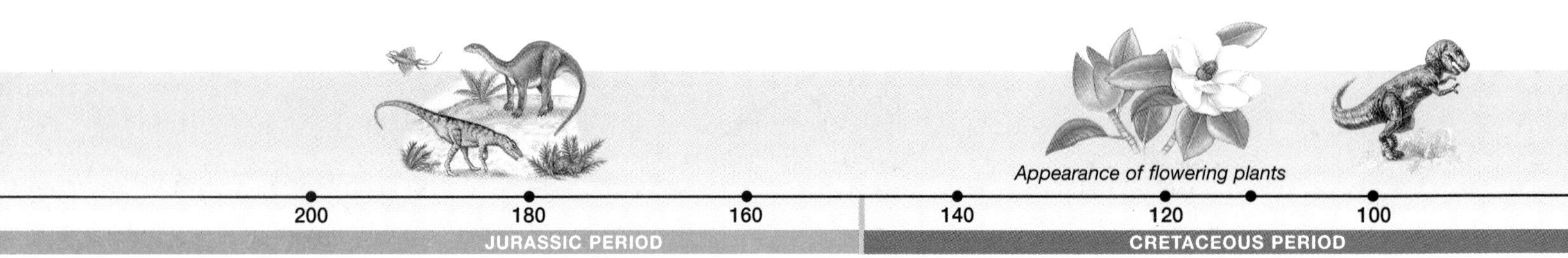